What we do today, determines how the world will look tomorrow.

Lower emissions, low fuel consumption, high driving comfort all require creativity. This is nothing special for us as suppliers to the automotive industry.

We develop bearings and precision components for engines, chassis and transmissions in close collaboration with manufacturers. From classic idler pulley units and strut bearing units to complete internal gearshift systems.

For us, being creative today means infinitely variable camshaft phasing units that have a positive influence on power, fuel consumption and emissions. Or switching valve lash adjustment elements – from variable valve strokes to cylinder deactivation.

For tomorrow's environmentally-friendly vehicle.

Schaeffler KG · 91072 Herzogenaurach (Germany) · www.ina.com

Let's boost the performance of you engine components

Dylyn® Plus diamond-like coatings

Through the unique combination of an extremely low coefficient of friction and a high surface hardness Dylyn® Plus coatings reduce frictional losses and extend the lifetime of components.

For different valve train parts Dylyn® Plus is a customized solution leading to lower fuel consumption, power increase, higher life time & reliability.

Dylyn® Plus significantly reduces friction losses and extends the lifetime of piston rings.

Low friction of Dylyn® Plus allows reduction of the complexity (eliminating bushings). High wear resistance of Dylyn® Plus permits use of lighter wrist pin.

Bekaert is always close to you:

Bekaert (Belgium) +32 9 338 59 10
Bekaert/Sorevi (France) + 33 450 31
Bekaert (NC, USA) +1 919 485 8900
or visit www.bekaert.com/bac

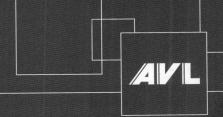

Richard van Basshuysen
Fred Schäfer

Modern Engine Technology from A to Z

Other SAE titles of interest:

Internal Combustion Engine Handbook—
Basics, Components, Systems, and Perspectives
By Richard van Basshuysen and Fred Schäfer
(Order No. R-345)

Introduction to Engine Valvetrains
By Yushu Wang
(Order No. R-339)

Vehicular Engine Design
By Kevin L. Hoag
(Order No. R-369)

For more information or to order a book, contact SAE at 400 Commonwealth Drive, Warrendale, PA 15096-0001; phone (724) 776-4970; fax (724) 776-0790; e-mail CustomerService@sae.org; Web site http://store.sae.org.

Richard van Basshuysen/Fred Schäfer

Modern Engine Technology

from A to Z

Warrendale, Pennsylvania, USA

For permission and licensing requests contact:

SAE Permissions
400 Commonwealth Drive
Warrendale, PA 15096-0001-USA
Email: permissions@sae.org
Tel: 724-772-4028
Fax: 724-772-4891

Library of Congress Cataloging-in-Publication Data

Lexikon Motorentechnik English.
   Modern engine technology : from A to Z / Richard van Basshuysen and
   Fred Schäfer, editors.
      p. cm.
   Includes bibliographical references.
   ISBN 978-0-7680-1705-2
      1. Internal combustion engines—Dictionaries.   2. Internal combustion
engines—Encyclopedias.   I. Van Basshuysen, Richard, 1932– .   II. Schäfer,
Fred, 1948– .   III. Title.

TJ785.L4913 2007
629.25--dc22   2006100615

SAE
400 Commonwealth Drive
Warrendale, PA 15096-0001 USA
Tel: 877-606-7323 (inside USA and Canada)
Tel: 724-776-4970 (outside USA)
Fax: 724-776-1615
Email: CustomerService@sae.org

Copyright © 2007 SAE International
ISBN 978-0-7680-1705-2
SAE Order No. R-373

Printed in China

Translated from the German language edition:
Lexicon Motorentechnik: Der Verbrennungsmotor von A-Z by Richard van Basshuysen and Fred Schäfer
Copyright © Friedr.Vieweg & Sohn Verlag/GWV Fachverlage GmbH, Wiesbaden, Germany, 2004

Foreword

Modern Engine Technology from A to Z is a reference work with about 4500 keywords covering all aspects of the combustion engine. It is derived from "Shell-Lexikon Verbrennungsmotor" (Shell Lexicon of the Combustion Engine), which appeared in the form of 90 supplements to issues of *ATZ* (*Automobiltechnische Zeitschrift*; Journal of Automotive Engineering) and *MTZ* (*Motortechnische Zeitschrift*; Journal of Engine Engineering) over a period of eight years, with the generous support of Deutsche Shell AG. The lexicon was thoroughly revised, updated, and expanded to reach its present form. The entire lexicon project was made possible by technical and material assistance provided by Siemens VDO Automotive AG.

No one author could possibly provide an in-depth view of so many special areas—nearly one hundred authors contributed to the exhaustive presentation of the subject matter. We would like to extend our thanks to everyone involved. The text for keywords that do not appear in the List of Subjects and Authors was written by the publishers.

The explosion of engineering insights and detailed knowledge engendered by more than one hundred years of development of the combustion engine reflects the wide variety of engineering demands met, the many structural components involved, and the intricacies of their interactions. The technological essentials of the combustion engine and its peripheral modules are covered in the approximately 1100 pages and 1700 illustrations of this volume. Using a detailed system of cross-references, readers are directed to the keyword entries of subentries by arrows (for example: Piston dimensions →Piston ~Dimensions). This positions each keyword within its relevant context, reflecting the reality that the keywords are by no means isolated terms, but rather derive much of their significance from their relevance within the whole picture.

In addition to accepting the challenge of presenting the current state of the art of engine development, the publishers took particular pains to present theory and practice in a balanced ratio. This was made possible mainly by contributions from so many top-grade experts from industry and science. With their help, a reference work has been created that we hope will be as useful in the contexts of theory and research as in everyday practice.

This book is intended, above all, for specialists in the automotive, engine, petroleum, and supporting industries, be they scientists or practitioners, and to serve as a study guide and reference for students. It also can provide useful advice and information to patent lawyers, automotive salespeople, official bodies, journalists, and interested laypersons.

Despite the efforts of everyone involved, it is unavoidable in a volume of this scope that keywords may have been left out entirely or may have been given insufficient coverage. We are therefore well aware that the book could be improved and would appreciate any ideas and information that serve this purpose.

Richard van Basshuysen
Fred Schäfer
Bad Wimpfen/Hamm, April 2006

About the Editors

Dr.-Ing. E.h. Richard van Basshuysen, VDI, was born in Bingen/Rhein, Germany, in 1932. Following an apprenticeship and then qualification as an automobile mechanic, he studied at the Technical University of Braunschweig/Wolfenbüttel from 1953 to 1955, graduating as a mechanical engineer. In 1982, he earned his university degree of engineering. From 1955 to 1965, he held a position in engineering sciences with Mineralölgesellschaft Aral AG/BP in Germany.

In 1965, van Basshuysen joined NSU AG, where he directed the experimental unit working on engine and transmission development, including development of the Wankel engine, and rose to the position of vice director of experimental automotive engineering. In this position, he contributed to the development of a number of vehicles, including the Prinz 4, NSU 1000 and 1200, RO 80 and K 70. In 1969, NSU was taken over by what is now Audi AG, at which he assumed the post of director of development for luxury vehicles and director of engine and transmission development; he also was elected by the managerial employees to a seat on the supervisory board of Audi AG. His most significant development was the first passenger-car diesel engine with detoxified emissions, direct injection, and turbocharging, which van Basshuysen established in the face of considerable opposition in his own company within the VW Group. This engine featured a 20% reduction in fuel consumption, compared with its prechamber engine predecessor, coupled with high power output and torque levels, which established its preeminence worldwide. In Europe, the market share for this type of engine has grown from about 12% in 1989 to more than 50% in 2007. Subsequent to his corporate career, van Basshuysen founded an engineering office in 1992, which he still manages. His efforts now are dedicated to publishing the internationally recognized scientific and engineering journals *ATZ* (*Automobiltechnische Zeitschrift*; Journal of Automotive Engineering), and *MTZ* (*Motortechnische Zeitschrift*; Journal of Engine Engineering), performing consulting work for international automakers, providing engineering services, and writing and publishing scientific and engineering books, most of which have been translated into English. All told, he is the author or contributing author of more than 50 publications. His editorials have appeared regularly in *ATZ* and *MTZ* for more than 15 years.

Prof. Dr.-Ing. Fred Schäfer was born in 1948 in Neuwied am Rhein, Germany. Following an apprenticeship in mechanical engineering, he continued studying in that field at the State School of Engineering at Koblenz. He went on to complete studies at the University of Kaiserslautern, Institute for Engines and Machines, earning the degree of Dipl.-Ing. His doctoral thesis was "Studies of the Reaction Kinetics of Hydrogen/Methanol Combustion in the Otto Engine."

Schäfer's career then took him to Audi AG in Neckarsulm, Germany, to fill the position of assistant head of development. During his ten years with the company, other positions held included manager of the experimental engine group, followed by his tenure as director of the engine design department.

In 1990, Schäfer was appointed professor for engines and machines at what was then the Technical University of Iserlohn, now part of the Technical University of Southern Westfalia in Iserlohn, where he directed the laboratory for combustion engines and flow machines. He is an active member of many university bodies, including the University Senate. In his current position as vice dean of theory and research, he is a member of the steering committee for the mechanical engineering faculty.

Schäfer is also active in the field of research and development of engine technology, both in a self-employed capacity and in cooperation with the company mpa-engineering.

With Richard van Basshuysen, he published the journal supplement "Shell-Lexikon Verbrennungsmotor" (Shell Lexicon of the Combustion Engine) from 1996 to 2003, which appeared in book form in 2004 under the title *Lexikon Motorentechnik* (Lexicon of Engine Technology).

Schäfer has been a member of the VDI (Association of German Engineers) and of the Society of Automotive Engineers for many years.

Information for Users

This lexicon contains keywords followed, where relevant, by subentries extending to as many as four sublevels, each sublevel indicated by a tilde (~). The keywords are in blue, semi-bold letters in a larger font size to make them easier to find. The first sublevel headings are also in blue, but in the text size.

Terms are arranged using the word-by-word system, where alphabetizing continues up to the end of the first word (hyphenated compounds count as one word). When the first word is the same as another entry, the second word is considered, and so on. Composite terms (e.g., "Variable valve control") are listed according to the alphabetic position of the adjective (first word). In some cases, a clarifying term follows the keyword, separated from it by a comma (e.g., "Cylinder charge, volumetric efficiency."

List of Subjects and Authors

Actuators	Dipl.-Ing. Schahier El-Garrahi
	Dipl.-Ing. Stefan Klöckner
	Dipl.-Ing. Jörg Ohlenburger
	Dipl.-Ing. Axel Tuschik
Air cycling valve	Dipl.-Ing. Kay Brodesser
	Dr.-Ing. Alfred Elsäßer
Air intake system	Matthias Alex
	Dr.-Ing. Olaf Weber
Balancing of masses	Dipl.-Ing. Klaus Hirschfelder
	Dipl.-Ing. Helmut Schönfeld
Bearings	Dr.-Ing. Eckhart Schopf
Calculations	Dr. Peter Nefischer
Camshaft	Dipl.-Ing. Norbert Gatzka
Carburetor	Dr. Dietrich Großmann
Catalytic converter	Dipl.-Ing. Michael Jäger
Chain drive	Dr. Peter Bauer
Charge movement	Dipl.-Ing. Götz Tippelmann
Combustion process, diesel/gasoline engine	Prof. Dr.-Ing. habil. Günter P. Merker
Commercial vehicle crankcase	Prof. Dr.-Ing. Stefan Zima
Commercial vehicle cylinders/liners	Prof. Dr.-Ing. Stefan Zima
Commercial vehicle cylinder head	Prof. Dr.-Ing. Stefan Zima
Connecting rod	Dipl.-Ing. Achim Voges
Coolant	Dipl.-Ing. Ladislaus Meszaros
Coolant pump	Peter Amm
	Franz Pawellek
Cooling circuit	Dipl.-Ing. Klaus Daniel
Control/gas transfer	Prof. Dr.-Ing. Stefan Zima

Crankcase	Dipl.-Ing. Günter Helsper Dipl.-Ing. Karl B. Langlois
Crankcase ventilation systems	Dr.-Ing. Stephan Wild
Crankshaft	Dipl.-Ing. Hans-Jürgen Narcis Dipl.-Ing. Jochen Schmidt
Cylinder head	Dr. Ralf Marquard Prof. Dr.-Ing. Fred Schäfer
Engine~Historic engines	Prof. Dr.-Ing. Stefan Zima
Engine~Racing engines	Dipl.-Kaufm. Roland Schedel
Engine acoustics	Prof. Dr. Hartmut Bathelt Dr.-Ing. Hans-Walter Wodtke
Engine bolts	Dipl.-Ing. Siegfried Jende Dipl.-Ing. Hermann Köhler
Engine concepts	Prof. Dr.-Ing. Stefan Zima
Engine damage	Prof. Dr.-Ing. Stefan Zima
Electronic/mechanical engine and transmission control	Dipl.-Ing. Gerwin Höreth Dipl.-Ing. Rainer Riecke Dipl.-Ing. Karl Smirra Dipl.-Ing. Christian Weinzierl
Exhaust gas analysis equipment	Dipl.-Ing. Mathias Goldhahn Prof. Dr.-Ing. Fred Schäfer
Filter	Dipl.-Ing. Markus Kolczyk Dipl.-Ing. Jochen Reyinger Dr.-Ing. Pius Trautmann Dr.-Ing. Olaf Weber
Fuel consumption	Prof. Dr.-Ing. Rudolf Flierl
Fuel, diesel engine	Dr. Herbert Krumm Wolfgang Lange Klaus Reders Dr. Andrea Schütze
Fuel, gasoline engine	Dr. Herbert Krumm Wolfgang Lange Klaus Reders Dr. Andrea Schütze
Heat accumulator	Dr. Olaf Schatz

Heat exchanger	Dr. Dragi Antonijevic
	Dr. Peter Diehl
Ignition system, diesel/gasoline engine	Dr. Thomas Heinze
Injection functions	Dipl.-Ing. Achim Koch
	Dipl.-Ing. Frank Lohrenz
Injection functions~Diesel engine	Dr. Wolfgang Oestreicher
Injection functions~Gasoline engine	Dr. Erwin Achleitner
	Dipl.-Ing. Jörg Neugärtner
Injection system~Diesel engine	Dipl.-Ing. Wolfgang Bloching
	Dipl.-Ing. Thomas Grossner
	Dipl.-Ing. Steffen Jung
	Prof. Dr.-Ing. Helmut Tschöke
Injection system~Gasoline engine	Dr. Erwin Achleitner
	Dipl.-Ing. Johannes Deichmann
	Dr. Franz Finzenhagen
	Dipl.-Ing. Bernhard Schmitt
	Dipl.-Ing. Ralph Schröder
Injection systems~Diesel engine	Dipl.-Ing. Wolfgang Bloching
Injection systems~Gasoline engine	Dr. Erwin Achleitner
	Dipl.-Ing. Bernhard Schmitt
Injection valves~Diesel engine	Dr. Erwin Achleitner
	Dipl.-Ing. Wolfgang Bloching
	Prof. Dr.-Ing. Helmut Tschöke
Injection valves~Gasoline engine	Dr. Erwin Achleitner
	Dipl.-Ing. Bernhard Schmitt
Intake system	Dr.-Ing. Alfred Elsäßer
Lubrication	Dr. Josef Affenzeller
Mixture formation	Univ.-Prof. Dr. techn. Dipl.-Ing. Hans Peter Lenz
Oil	Peter Busse
	Eric Froböse
	Dr. Helmut Leonhardt
	Jan Miller
	Hans Dieter Müller
	Volker Null
Oil circulation	Dipl.-Ing. Klaus Daniel
Particles (Particulates)	Dipl.-Ing. Andreas C. R. Mayer

Performance characteristic maps	Dipl.-Ing. Bernd Haake Dr. Peter Wolters
Piston	Dr. Manfred Röhrle
Piston ring	Dipl.-Ing. Frank Münchow Dr. Jochem Neuhäuser
Pollutant after-treatment	Dipl.-Ing. Frank Terres
Radiator	Dipl.-Ing. Jürgen Eitel Dr. Wolfgang Kramer
Sealing systems	Dipl.-Ing. (FH) Klaus Bendl Dipl.-Ing. (FH) Armin Diez Dipl.-Ing. (FH) Eberhard Griesinger Dipl.-Ing. (FH) Uwe Georg Klump Dipl.-Ing. (TU) Wilhelm Kullen
Sensors	Dr. Anton Grabmaier Dipl.-Ing. Manfred Glehr
Software	Dipl.-Ing. Stefan Weber
Spark plug	Dr. Thomas Heinze
Starter	Carsten Bohnenkamp Dirk Schulte Dipl.-Ing. Peter Skotzek
Torsional vibrations	Dipl.-Ing. Bernd Jörg
Two-stroke engine	Dr.-Ing. Uwe Meinig
Variable valve control	Dr. Peter Heuser

List of Authors

Achleitner, Erwin, Dr.

Siemens VDO Automotive, Regensburg
www.siemensvdo.de

Affenzeller, Josef, Dr.

AVL List GmbH, A-Graz
www.avl.com

Alex, Matthias

Mann + Hummel GmbH, Ludwigsburg
www.mann-hummel.com

Amm, Peter

Geräte- und Pumpenbau GmbH, Dr. Eugen
Schmidt, Merbelsrod
www.gpm-merbelsrod.de

Antonijevic, Dragi, Dr.

Visteon Technologie Zentrum, Kerpen
www.visteon.com

Bathelt, Hartmut, Prof. Dr.

Akustikzentrum GmbH, Lenting
www.akustikzentrum.de

Bauer, Peter, Dr.

IWIS ketten, Joh. Winkelhofer & Söhne
GmbH & Co KG, München
www.iwis.de

Bendl, Klaus, Dipl.-Ing. (FH)

Elring Klinger AG, Dettingen/Erms
www.elringklinger.de

Bloching, Wolfgang, Dipl.-Ing.

Siemens VDO Automotive, Regensburg
www.siemensvdo.de

Bohnenkamp, Carsten

Siemens VDO Automotive, Regensburg
www.siemensvdo.de

Brodesser, Kay, Dipl.-Ing.

MAHLE Filtersysteme GmbH, Stuttgart
www.mahle.com

Busse, Peter

Deutsche Shell GmbH, Shell Global Solutions
(Deutschland), PAE-Labor, Hamburg
www.deutsche-shell.de

Daniel, Klaus, Dipl.-Ing.

APL Automobil-Prüftechnik Landau GmbH, Landau
www.apl-landau.de

Deichmann, Johannes, Dipl.-Ing.

Siemens VDO Automotive, Regensburg
www.siemensvdo.de

Diehl, Peter, Dr.

Visteon Technologie Zentrum, Kerpen
www.visteon.com

Diez, Armin, Dipl.-Ing. (FH)

Elring Klinger AG, Dettingen/Erms
www.elringklinger.de

Eitel, Jürgen, Dipl.-Ing.
Behr GmbH & Co. KG, Stuttgart
www.behrgroup.com

El-Garrahi, Schahier, Dipl.-Ing.
Siemens VDO Automotive, Regensburg
www.siemensvdo.de

Elsäßer, Alfred, Dr.-Ing.
MAHLE Filtersysteme GmbH, Stuttgart
www.mahle.com

Finzenhagen, Franz, Dr.
Siemens VDO Automotive, Regensburg
www.siemensvdo.de

Flierl, Rudolf, Prof. Dr.-Ing.
Universität Kaiserslautern, Lehrstuhl
Verbrennungskraftmaschinen, Kaiserslautern
www.uni-kl.de

Froböse, Eric
Deutsche Shell GmbH, Shell Global Solutions
(Deutschland), PAE-Labor, Hamburg
www.deutsche-shell.de

Gatzka, Norbert, Dipl.-Ing.
retired from Dr. Schrick GmbH, Remscheid
www.drschrick.de

Glehr, Manfred, Dipl.-Ing.
Siemens VDO Automotive, Regensburg
www.siemensvdo.de

Goldhahn, Mathias, Dipl.-Ing.
Vaihingen-Enz

Grabmaier, Anton, Dr.
Siemens VDO Automotive, Regensburg
www.siemensvdo.de

Griesinger, Eberhard, Dipl.-Ing. (FH)
Elring Klinger AG, Dettingen/Erms
www.elringklinger.de

Großmann, Dietrich, Dr.
retired from Pierburg GmbH, Neuss
www.kolbenschmidt.de

Grossner, Thomas, Dipl.-Ing.
Siemens VDO Automotive, Regensburg
www.siemensvdo.de

Haake, Bernd, Dipl.-Ing.
FEV Motorentechnik GmbH, Aachen
www.fev.com

Heinze, Thomas, Dr.
AES Automotive Engineering Services, Finnentrop

Helsper, Günter, Dipl.-Ing.
F Porsche AG, Weissach
www.porsche.de

Heuser, Peter, Dr.
Meta Motoren- und Energie-Technik GmbH,
Herzogenrath
www.metagmbh.de

Hirschfelder, Klaus, Dipl.-Ing.
BMW AG, München
www.bmw.de

Höreth, Gerwin, Dipl.-Ing.

Siemens VDO Automotive, Regensburg
www.siemensvdo.de

Jäger, Michael, Dipl.-Ing.

Emitec GmbH, Lohmar
www.emitec.com

Jende, Siegfried, Dipl.-Ing.

TEXTRON Peiner Umformtechnik GmbH, Peine
www.textron.de

Jörg, Bernd, Dipl.-Ing.

Vibracoustic GmbH & Co. KG., Neuenburg
www.vibracoustic.de

Jung, Steffen, Dipl.-Ing.

Siemens VDO Automotive, Regensburg
www.siemensvdo.de

Klöckner, Stefan, Dipl.-Ing.

Siemens VDO Automotive, Regensburg
www.siemensvdo.de

Klump, Uwe Georg, Dipl.-Ing. (FH)

Elring Klinger AG, Dettingen/Erms
www.elringklinger.de

Koch, Achim, Dipl.-Ing.

Siemens VDO Automotive, Regensburg
www.siemensvdo.de

Köhler, Hermann, Dipl.-Ing.

KAMAX-Werke Rudolf Kellermann GmbH &
Co KG, Homberg/Ohm
www.kamax.de

Kolczyk, Markus, Dipl.-Ing.

Mann + Hummel GmbH, Ludwigsburg
www.mann-hummel.com

Kramer, Wolfgang, Dr.

Behr GmbH & Co. KG, Stuttgart
www.behrgroup.com

Krumm, Herbert, Dr.

Deutsche Shell GmbH, Shell Global Solutions
(Deutschland), PAE-Labor, Hamburg
www.deutsche-shell.de

Kullen, Wilhelm, Dipl.-Ing. (TU)

Elring Klinger AG, Dettingen/Erms
www.elringklinger.de

Lange, Wolfgang

Deutsche Shell GmbH, Shell Global Solutions
(Deutschland), PAE-Labor, Hamburg
www.deutsche-shell.de

Langlois, Karl B., Dipl.-Ing.

F Porsche AG, Weissach
www.porsche.de

Lenz, Hans Peter, Univ.-
Prof. Dr. techn. Dipl.-Ing.

Institut für Verbrennungskraftmaschinen und
Kraftfahrzeugbau der TU Wien, A-Wien
www.tuwien.ac.at

Leonhardt, Helmut, Dr.

Deutsche Shell GmbH, Shell Global Solutions
(Deutschland), PAE-Labor, Hamburg
www.deutsche-shell.de

Lohrenz, Frank, Dipl.-Ing.   Siemens VDO Automotive, Regensburg
www.siemensvdo.de

Marquard, Ralf, Dr.   MAN B&W Diesel AG, Augsburg
www.manbw.com

Mayer, Andreas C. R., Dipl.-Ing.   TTM Technik Thermische Maschinen,
CH-Niederrohrdorf

Meinig, Uwe, Dr.-Ing.   Hengst GmbH & Co.KG, Münster
www.hengst.de

Merker, Günter, P., Prof. Dr.-Ing. habil.   Institut für Technische Verbrennung,
Universität Hannover, Hannover
www.uni-hannover.de

Meszaros, Ladislaus, Dipl.-Ing.   BASF Aktiengesellschaft, Ludwigshafen
www.basf.de

Miller, Jan   Deutsche Shell GmbH, Shell Global Solutions
(Deutschland), PAE-Labor, Hamburg
www.deutsche-shell.de

Müller, Hans Dieter   Deutsche Shell GmbH, Shell Global Solutions
(Deutschland), PAE-Labor, Hamburg
www.deutsche-shell.de

Münchow, Frank, Dipl.-Ing.   Federal Mogul Burscheid GmbH
(formerly Goetze AG), Burscheid
www.federal-mogul.com

Narcis, Hans-Jürgen, Dipl.-Ing.   Maschinenfabrik Alfing Kessler GmbH, Aalen
www.alfing.de

Nefischer, Peter, Dr.   BMW Motoren GmbH, A-Steyr
www.bmw-werk-steyr.at

Neugärtner, Jörg, Dipl.-Ing.   Siemens VDO Automotive, Regensburg
www.siemensvdo.de

Neuhäuser, Jochem, Dr.   Federal Mogul Burscheid GmbH
(formerly Goetze AG), Burscheid
www.federal-mogul.com

Null, Volker   Deutsche Shell GmbH, Shell Global Solutions
(Deutschland), PAE-Labor, Hamburg
www.deutsche-shell.de

Oestreicher, Wolfgang, Dr.   Siemens VDO Automotive, Regensburg
www.siemensvdo.de

Ohlenburger, Jörg, Dipl.-Ing.   Siemens VDO Automotive, Regensburg
www.siemensvdo.de

Pawellek, Franz	Geräte- und Pumpenbau GmbH, Dr. Eugen Schmidt, Merbelsrod www.gpm-merbelsrod.de
Reders, Klaus	Deutsche Shell GmbH, Shell Global Solutions (Deutschland), PAE-Labor, Hamburg www.deutsche-shell.de
Reyinger, Jochen, Dipl.-Ing.	Mann + Hummel GmbH, Ludwigsburg www.mann-hummel.com
Riecke, Rainer, Dipl.-Ing.	Siemens VDO Automotive, Regensburg www.siemensvdo.de
Röhrle, Manfred, Dr.	RÖ-CONSULT, Ostfildern
Schäfer, Fred, Prof. Dr.-Ing.	FH Südwestfalen, Iserlohn www.fh-swf.de
Schatz, Olaf, Dr.	Schatz Thermo Engineering, Kröppen-Pirmasens
Schedel, Roland, Dipl.-Kaufm.	TEXT-COM, Taunusstein www.text-com.de
Schmidt, Jochen, Dipl.-Ing.	Maschinenfabrik Alfing KesslerGmbH, Aalen www.alfing.de
Schmitt, Bernhard, Dipl.-Ing.	Siemens VDO Automotive, Regensburg www.siemensvdo.de
Schönfeld, Helmut, Dipl.-Ing.	retired from Schenck RoTec GmbH, Darmstadt
Schopf, Eckhart, Dr.-Ing.	Federal-Mogul Wiesbaden GmbH & Co. KG, Wiesbaden www.federal-mogul.com
Schröder, Ralph, Dipl.-Ing.	Siemens VDO Automotive, Regensburg www.siemensvdo.de
Schulte, Dirk	Siemens VDO Automotive, Regensburg www.siemensvdo.de
Schütze, Andrea, Dr.	Deutsche Shell GmbH, Shell Global Solutions (Deutschland), PAE-Labor, Hamburg www.deutsche-shell.de
Skotzek, Peter, Dipl.-Ing.	Siemens VDO Automotive, Regensburg www.siemensvdo.de
Smirra, Karl, Dipl.-Ing.	Siemens VDO Automotive, Regensburg www.siemensvdo.de

Terres, Frank, Dipl.-Ing.   Tenneco Automotive Heinrich Gillet
                                   GmbH & Co KG, Edenkoben
                                   www.tenneco.com

Tippelmann, Götz, Dipl.-Ing.   Drallmesstechnik GmbH, Neuenstadt

Trautmann, Pius, Dr.-Ing.   Mann + Hummel GmbH, Ludwigsburg
                                   www.mann-hummel.com

Tschöke, Helmut, Prof. Dr.-Ing.   Otto-von-Guericke-Universität Magdeburg,
                                   Magdeburg
                                   www.uni-magdeburg.de/imko

Tuschik, Axel, Dipl.-Ing.   Siemens VDO Automotive, Regensburg
                                   www.siemensvdo.de

van Basshuysen, Richard, Dr.-Ing. E.h.   Publisher, Automobiltechnische Zeitschrift and
                                   Motortechnische Zeitschrift, Bad Wimpfen

Voges, Achim, Dipl.-Ing.   Mahle GmbH, Stuttgart
                                   www.mahle.com

Weber, Olaf, Dr.-Ing.   Mann + Hummel GmbH, Ludwigsburg
                                   www.mann-hummel.com

Weber, Stefan, Dipl.-Ing.   Siemens VDO Automotive, Regensburg
                                   www.siemensvdo.de

Weinzierl, Christian, Dipl.-Ing.   Siemens VDO Automotive, Regensburg
                                   www.siemensvdo.de

Wild, Stephan, Dr.-Ing.   Mann + Hummel GmbH, Ludwigsburg
                                   www.mann-hummel.com

Wodtke, Hans-Walter, Dr.-Ing.   AFT Atlas Fahrzeugtechnik GmbH, Werdohl
                                   www.aft-werdohl.de

Wolters, Peter, Dr.   FEV Motorentechnik GmbH, Aachen
                                   www.fev.com

Zima, Stefan, Prof. Dr.-Ing.   FH Gießen-Friedberg, FB Maschinenbau, Friedberg
                                   www.fh-giessen-friedberg.de

List of Abbreviations

ACEA	*Association des Constructeurs Européens d'Automobiles*; Association of European Automobile Manufacturers
API	American Petroleum Institute
ATIEL	*Association Technique de l'Industrie Européenne des Lubrifiants*; Association of the European Lubricant Industry
AVL	private company that develops powertrain systems for internal combustion engines, as well as instrumentation and test systems
ATZ	*Automobiltechnische Fachzeitschrift*; Journal of Automotive Engineering
BDC	bottom dead center (of piston travel)
bhp	measurement of engine power: BHP = torque (ft/lbs) x rpm/5252
BUWAL	*Bundesamt für Umwelt, Wald und Landschaft*; Swiss Agency for the Environment, Forests and Landscape
°CA	degrees crank angle
CARB	California Air Resources Board
CDI	capacitor discharge ignition, a high-voltage capacitor ignition system.
CEC	Coordinating European Council for the development of performance tests for lubricants and engine fuels
CFD	Computational flow dynamics
CCMC	*Comité des Constructeurs d'Automobiles du Marché Commun* (predecessor of ACEA)
DI	direct injection
DIN	*Deutsches Institut für Normung*; German Institute for Standardization
DEF	UK military specifications (Ministry of Defense)
ECE	Economic Commission for Europe
EEA	European Environment Agency
EEC	European Economic Community

EGR	exhaust gas recirculation
EU4, EU5	European Union standards 4 and 5
FID	flame ionization detector
FVV	*Forschungsvereinigung Verbrennungskraftmaschinen*; Research Association for Combustion Engines
IDI	indirect injection
IEC	International Electrotechnical Commission
ILSAC	International Lubricant Standardization and Approval Committee
ISO	a network of the national standards institutes of 157 countries, and the world's largest developer of standards
JASO	Japan Automobile Standards Organization
LEV	low-emissions vehicle
LEV2	CARB low-emissions vehicle standards
MIL	US military specification
MTZ	*Motortechnische Zeitschrift*; Journal of Engine Engineering
OBD	on-board diagnostics
OSEK	*Offene Systeme und deren Schnittstellen für die Elektronik im Kraftfahrzeug* [open systems and the correponding interfaces for automotive electronics]; German standards committee
SAE	Society of Automotive Engineers (US)
TDC	top dead center (of piston travel)
TÜV	*Technischer Überwachungsverein*; Technical Inspection Association (Germany), an independent product testing and standards organization
ULEV	ultra-low-emissions vehicle
VDI	*Verein Deutscher Ingenieure*; Association of German Engineers
ZEV	zero-emissions vehicle
ZVDI	*Zentralverband Deutscher Ingenieure*; Association of German Engineers

A classification →Engine acoustics ~Assessment curves

Abnormal combustion process →Glow ignition; →Knocking

Abrasive wear →Piston ring ~Wear

Absolute fuel consumption →Fuel consumption

Absolute muffler →Engine acoustics

Absorption →Engine acoustics ~Noise absorption

Absorption curve →Supercharging ~Engine air mass flow curve

Absorption-type muffler →Engine acoustics; →Exhaust system ~Muffler

Acceleration (*also*, →Crankshaft drive). A rapid load change of the engine (such as an increase from part load in the direction of full load) or an engine speed change, is called acceleration. The inertias of the masses in the engine that need to be accelerated work produce a resistance against the acceleration.

The acceleration of the engine depends very much on the mass to be accelerated. It comprises reciprocating and rotating masses. Low masses to be accelerated (e.g., in the flywheel, vibration dampers, crankshaft) result in shorter acceleration times, because the acceleration period declines in proportion to the mass (moments of inertia) to be accelerated, assuming that the same force (torque) is applied. Sports engines are normally designed in such a way that a small moment of inertia allows a fast acceleration of the engine, which means that a fast load and engine speed change is achieved.

Delayed acceleration is seen in engines with exhaust gas turbocharging, especially at lower engine speeds, because the turbocharger needs a certain time to make the appropriate boost pressure available (referred to as "turbocharger lag"). There is, therefore, always a difference between steady state status and the instantaneous nonsteady status.

The flywheel mass is an important parameter with respect to the "rotational engine ability" in the constructional design of an engine. A compromise is always required between a small rotational mass and, therefore, rotational engine ability and the degree of nonuniformity.

Acceleration device (*also*, →Acceleration enrichment; →Carburetor). The acceleration device is a system used to supply the engine with an additional volume of fuel during acceleration. This can be implemented,

for example, with additional injection impulses (injection system) or with an accelerator pump (carburetor system). Acceleration devices are required in particular for engines with external mixture formation and also in gasoline engines.

Acceleration enrichment (*also*, →Carburetor). In gasoline engines, a proportion of the fuel attaches itself to the relatively cold intake manifold walls in the form of a film. This was especially true in carbureted engines or engines with central fuel injection. Additionally, when the throttle is opened rapidly, the relatively heavy fuel, in contrast to the air, arrives later in the combustion chamber, resulting in a leaning of the mixture. This may cause misfires or reductions in engine torque. Acceleration enrichment is used to deliver an additional fuel supply into the intake manifold to compensate for the above effect. However, this normally results in an undesired increase in hydrocarbons and carbon monoxide. In single-point fuel injection systems, the mixture is normally tuned in such a way that the additional volume injected is calculated to be just sufficient to enable the acceleration to be achieved without delay or torque reduction. This has advantages with respect to fuel consumption and emissions, in particular for carbon monoxide and hydrocarbon emissions. Excessive mixture enrichment, as was used in older carburetor gasoline engines through an accelerator pump, is not required for injections directly upstream of the intake valve.

Modern injection systems activate the acceleration enrichment function if the increase in air volume flow, or the engine load, exceeds a certain threshold between power cycles. The amount of mixture enrichment is dependent on load, engine speed, engine temperature, and the speed of the load change. Additionally, an increase in the volume of fuel supplied can be achieved by prolongation of the normal injection impulse or through an intermediate injection. At the same time, the ignition angle can be reduced dynamically to prevent knocking under acceleration. An additional increase of the injected fuel volume for acceleration is generally not required for diesel engines. The addition of a starter volume with the help of injection adaptation to compensate for condensation and leakage losses and to increase the engine torque during the start-up phase only makes sense after the start.

A load pressure-dependent on the injection volume is required after the diesel engine is loaded to minimize the black smoke.

Accelerator pedal (*also*, →Actuator ~E-gas). The accelerator pedal is used to control the load on an engine. The position of the throttle valve actuator in a gasoline engine is changed by the accelerator pedal, whereas in a diesel engine the control rod of the injection pump is moved—or on common rail systems the injection valve is changed—via a signal, and this determines the mass of the diesel fuel to be injected. The connection of the accelerator pedal to the engine was mechanical in the past (control cable or linkage). However, electronic accelerator pedals are used increasingly in modern automobiles.

Accelerator pedal, electronic →Electronic open- and closed-loop control

Accelerator pedal module →Actuators ~E-gas

Accelerator pump →Acceleration enrichment; →Carburetor ~Constant pressure carburetor, ~Dashpot pump

Accessories →Engine accessories

Accessory noises →Engine acoustics

ACEA (Association des Constructeurs Européens d'Automobiles) →Oil ~Classification, specifications, and quality requirements

Acetone →Fuel, gasoline engine ~Alternative fuels

Acoustics →Air intake system ~Acoustics; →Engine acoustics

Activated carbon container →Activated carbon filter

Activated carbon filter (ACF). The volatile hydrocarbons evaporate in the tank of a vehicle with a gasoline engine depending on the pressure, temperature, composition, and vapor pressure of the fuel. They must be fed to the engine for combustion to prevent them from getting into the atmosphere. This is either done directly through a bypass pipe or after interim storage in an activated carbon container, where the hydrocarbons are absorbed.

The activated carbon filter is regenerated using the vacuum in the intake manifold to suck the hydrocarbons from the activated carbon container. To achieve this, part of the combustion air is sucked through the activated carbon filter. Precontrol is performed using an ACF pulse valve. The feedback volume is approximately 2% of the air intake amount. The charge status of the air with hydrocarbons from the ACF system is not known, but microcontrol through the lambda probe signal is performed to avoid over-enrichment. The feedback can be performed for naturally aspirated engines over the total load range from idle to full load.

Normal activated carbon container sizes range from 0.5 to 3 liters, depending on the size of the gasoline tank. The carbon filling in the activated carbon container is slightly preloaded by a spring load to minimize abrasion due to shock. The container material is sheet steel or plastic. The air is guided through a fleece

or the equivalent to prevent fouling of the carbon filling. Also, the activated carbon must be protected from water, because this reduces the active surface that is used for the attachment of hydrocarbons.

As a result, a larger activated carbon filter would be required to bind the volatile hydrocarbons during filling of the vehicle. These are fed back to the engine during its operation. This achieves a higher level of efficiency than the oscillation method (extraction of the volatile hydrocarbons during fuelling and recirculation into the tank system) used today. Disadvantageous is the larger volume (approximately 10 liters) of the container and the additional cost.

Activation energy →Chemical reaction

Active bearing assembly →Engine acoustics ~Engine/accessory mount

Active engine bearing →Engine acoustics ~Engine/accessory mount

Active engine speed sensor →Sensors ~Speed sensors

Active noise control →Air intake system ~Acoustics ~~Future systems

Active sensor →Sensors ~Speed sensors

Activity →Catalytic converter

Actuators (*also*, →Electronic open- and closed-loop control ~gasoline engine). Actuators are actuation devices, which are usually activated pneumatically or electrically.

~Actuators for exhaust gas recirculation (EGR). Actuators supply exhaust gas to the intake mixture. The combustion speed is lowered by the addition of exhaust gas and this reduces the peak combustion temperature, which in turn reduces the nitrogen oxide (NOx) emissions. The recirculation of the burned exhaust gas also reduces the oxygen mass in the combustion chamber, which results in a decrease in fuel consumption. Exhaust gas recirculation is divided into external and internal exhaust gas recirculation.

~~EGR bypass valve. Depending on the engine operating point, the recirculated exhaust gas will either be fed to the EGR radiator or around it. This bypass function is implemented by integrating an EGR bypass valve. The valve consists of a pneumatic and electric actuator and a valve seat. Butterfly valves are widely used.

~~EGR valve. The EGR valve controls the amount of exhaust gas that gets recirculated into the intake manifold during external exhaust gas recirculation. The control is based on engine speed and load. Exhaust gas recirculation valves are divided into pneumatic, electropneumatic, and electrical systems.

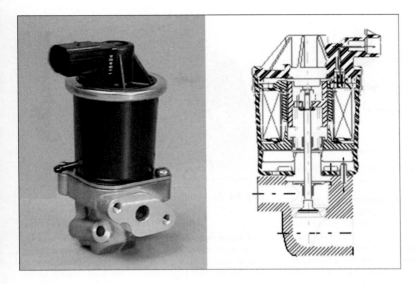

Fig. A1
Electric EGR valve

~~~**Electric EGR valves.** These systems permit the utilization of exhaust gas recirculation at higher operating points, depending on the intake vacuum.

Electric EGR valves (**Figs. A1 and A2**) consist of an electrical actuator (DC motor, solenoid coil, or stepper motor) and valve seat (disk valve, needle valve, butterfly valve, or rotary valve). Electric EGR valves have integrated sensors, which make monitoring of the valve position possible to satisfy present accuracy requirements. Very accurate control of the quantity of exhaust gas recirculated and reduced setting times is possible in comparison to pneumatic and electro-pneumatic valves. The requirement for highly accurate electrical EGR valves is justified by continuously lowered emission limits.

~~~**Electronic-pneumatic EGR valves.** The electronic-pneumatic valves are integrated into the system as pressure controllers or are mounted in front of the vacuum cell to control the quantity of exhaust gas recirculated independent of the engine operating point. The control of the quantity of exhaust gas recirculated is limited to the range determined by the force of the intake vacuum, which opens the valve against the retractor spring force and the applied pressures.

~~~**Pneumatic EGR valves.** This system consists of disk valves with pneumatic actuation by a vacuum cell. They are subjected to the intake manifold vacuum, which permits adjustment of the EGR valve independent of the engine operating point. The range of pneumatic EGR valves is limited by pneumatic delay valves or pressure limiting valves to avoid negative influences from unacceptable exhaust gas recirculation amounts. For the same reason, the exhaust gas recirculation is switched off in some systems for certain operating points by using electrical switchover valves.

~~**External exhaust gas recirculation.** A portion of the combusted exhaust gas is recirculated from the exhaust manifold through a pipe to the intake tract (**Fig. A3**). The amount of the exhaust gas recirculation is controlled depending on engine speed and load by an EGR valve.

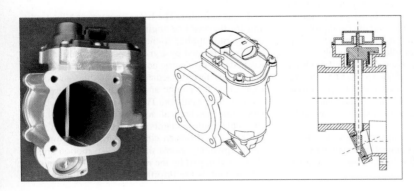

Fig. A2
Butterfly valve with electric motor

3

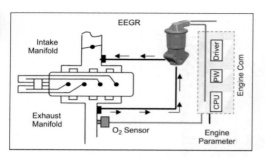

Fig. A3 Schematic diagram of exhaust gas recirculation

~~Internal exhaust gas recirculation. The proportion of exhaust residuals in the combustion chamber will be influenced by the overlap of intake and exhaust valves. Based on the particular design, only small quantitative control of the exhaust gas is possible. More effective exhaust gas recirculation can only be achieved by means of variable valve drives. A further reduction of the nitrogen oxide (NOx) emissions can be achieved by EGR cooling. Depending on the operating condition, the exhaust gas is recirculated through the exhaust gas cooler or around the exhaust gas cooler to the intake manifold. An EGR bypass valve performs the control function.

~Air control valve (ACV). This component was developed to be used in systems for diesel engines. The purpose of the actuator is to produce a defined vacuum in the intake manifold. It gets expanded to a complete EGR system for diesel engines by adding an exhaust gas recirculation valve.

The concept permits adaptation of the actuator for customization. This makes it possible to adapt the actuator to the existing installation space, the connection geometry, and the air duct diameter.

The emissions from combustion engines must be reduced over the next few years due to legal regulations. Exhaust gas recirculation is gaining importance, because it offers the possibility of bringing the emission of pollutants in line with the environmental obligations. It will be necessary to control the EGR rate for all operational conditions to meet the requirements of the EU4 pollution standard.

The ACV (**Fig. A4**) generates a defined vacuum in the intake air tract to promote exhaust gas recirculation. It controls the throttle in the air duct and provides defined pressure conditions in the intake manifold, which in turn permits a controlled EGR supply.

The ACV has a contactless sensor, which provides a feedback signal of the nozzle position to the ECU (engine control unit). The built-in DC motor actuates the nozzle via a two-stage transmission. The nozzle is opened fully by a retractor spring if no electrical power is applied.

~E-gas (electric motor–driven throttle valve actuator). For E-gas, no mechanical connection exists between gas pedals and actuators for engine load and engine speed changes.

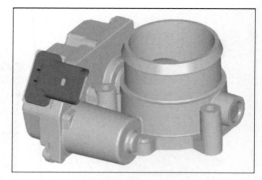

Fig. A4 Air control valve

~~Cruise control. This is a control system for the vehicle speed based on a speed signal through a separate cruise control actuator or the electrical throttle actuator. The control algorithm for the load control of the engine is integrated in the actuator through separate electronics or in the engine electronics.

~~~Cruise control actuator. The cruise control actuator moves the load actuation parts of the combustion engine independent of the gas pedal position during the speed-controlled operation. This is implemented by direct connection of the throttle actuator or by Bowden cables or linkages. Separate cruise control actuators are now often replaced by electrical throttle actuators.

~~Dashpot function. *See below,* ~~Load-reversal damping

~~Drive-by-wire. This is the mechanically decoupled actuation of the load control mechanisms of a gasoline engine by electrically powered actuators. It permits load control independent of the gas pedal position of the combustion engine for the implementation of additional functions such as speed control, load-reversal damping, idle control, antislip control, and support of gear changes for automatic and semi-automatic transmissions.

~~E-gas actuator. The E-gas operates as a load actuator for the gasoline engine. The throttle controls the air volume supplied to the engine. The throttle actuation is performed by a mechanical connection to the gas pedal or by electric actuators such as a throttle actuator. The throttle actuator controls the air volume supplied to the engine and, therefore, operates as the primary load actuator. Throttle actuators can be divided into mechanical and electrical actuators. Mechanical throttle actuators are connected directly to the gas pedal by Bowden cables or linkages. Nonlinear transformations against the gas pedal motion are implemented in part by the mechanical connection or in the air duct of the throttle. Integrated idle actuators

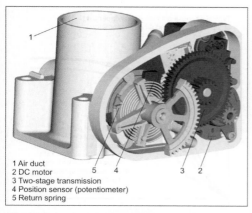

1 Air duct
2 DC motor
3 Two-stage transmission
4 Position sensor (potentiometer)
5 Return spring

Fig. A5 Drive-by-wire throttle actuator with DC motor

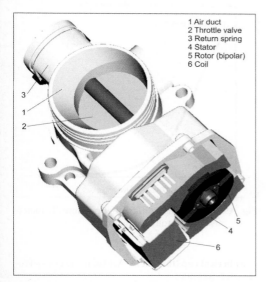

1 Air duct
2 Throttle valve
3 Return spring
4 Stator
5 Rotor (bipolar)
6 Coil

Fig. A6 Drive-by-wire throttle actuator with torque motor

make additional functions possible. Electrical throttle actuators operate without mechanical connection to the gas pedal and are actuated by different types of electrical actuators. DC motor actuators are used in most cases (**Figs. A5 and A6**).

~~**Electronic gas pedal.** *See above,* ~~Drive-by-wire

~~**Electronic stability control.** This is control of the vehicle stability around the vertical axis by sensing the yaw angle and then by electronic intervention on brakes and throttle actuator.

~~**FMVSS124.** This is the safety standard (US) for the retraction of a throttle actuator into a safe position

in case of a failure. Two independent energy sources are required for the retraction and the retraction times for different temperatures, which must not be over-ridden by the total gas pedal/control/throttle actuator system if one of the two energy sources fails.

~~**Gas pedal module.** This is a gas pedal with an integrated position sensor for the incorporation of the combustion engine load status determined by the driver. The position sensors are potentiometers or contactless sensors, such as Hall effect and inductive sensors (drive-by-wire).

~~**Idle control (gasoline engine).** Gasoline engine idle control is performed by controlled opening of a bypass valve to the idle actuator of the main air duct or by position control of the throttle for the electrical throttle actuator (**Fig. A7**).

~~~**Idle actuator (gasoline engine).** This is control equipment (valve) for the air volume during idle operations of the engine. Control of the air volume is achieved by opening and closing the bypass valve to the main air duct or by direct actuation of the throttle. Stepper motors (**Fig. A8**), or solenoid actuators, are primarily used as actuators.

~~~**Idle valve.** *See above,* ~~~Idle actuator (gasoline engine)

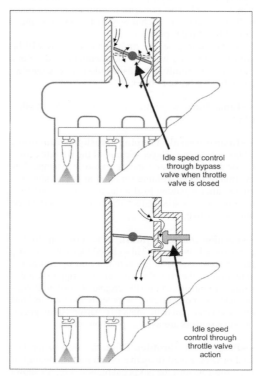

Idle speed control through bypass valve when throttle valve is closed

Idle speed control through throttle valve action

Fig. A7 Comparison of idle control principles

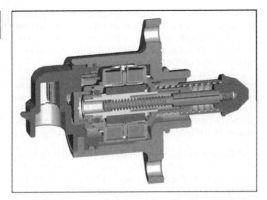

Fig. A8 Bypass idle actuator with stepper motor

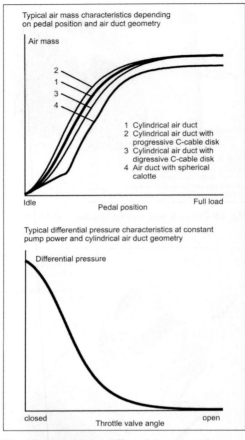

Fig. A9 Air mass and differential pressure characteristics of throttle actuators

~~**Limp-home air mass.** *See below,* ~~Progression
(throttle)

~~**Limp-home feature.** Continued operation of the
gasoline engine is possible in limited mode if the electrical actuator of the throttle actuator fails. This is
made possible by positioning the throttle into the limp-
home position. The throttle will not be closed completely when the electrical power fails if a throttle actuator is used. The remaining limp-home air mass
will be adjusted in such a way that a slow continua-
tion of the drive is possible.

At the same time, the limp-home air mass will be
adjusted in such a way that the vacuum that develops
in the intake manifold in this position is large enough
to supply the brake booster.

~~**Limp-home position.** *See below,* ~~Progression
(throttle)

~~**Load-reversal damping (dashpot function).**
This refers to the delayed closing of the throttle actuator or the idle actuator and, therefore, to damped reduction of the air volume supplied to the combustion
engine after a sudden load reduction by gas pedal retraction. The braking effect of the engine is delayed
but fully active.

~~**Position control of throttle.** The position control
of the throttle is executed in a closed loop by analyzing
the position sensor in the actuator and by controlled
actuation of the throttle actuator. The required algorithms are integrated into the engine controller. Potentiometers or contactless sensor systems, such as Hall
effect sensors, magnetic-inductive sensors, or magneto-
resistive sensors are used as position sensors.

~~**Progression (throttle).** This term refers to accelerated or decelerated (negative progression) opening of
the throttle against the gas pedal motion. **Fig. A9** shows
the air mass flow characteristics.

~~**Spherical cap (throttle)** →Actuators ~E-gas ~~Pro-
gression (throttle)

~**General purpose actuator (GPA).** The products of
the GPA family (**Fig. A10**) are powered by the vehicle
battery. This is in contrast to the pneumatic actuators,
which need a vacuum system. GPAs are used to move
or position an external system (e.g., a lever, linkage, or
rod) within a predefined range.

Application areas may include:

- Intake manifold length and/or volume adjustments
- Turbocharger with variable turbine geometry
- Wobble and swirl throttle systems
- Waste-gate valves
- Starter-generator
- Exhaust gas throttles
- Exhaust gas cooler
- Transmission

The following components are available for the system design.

A

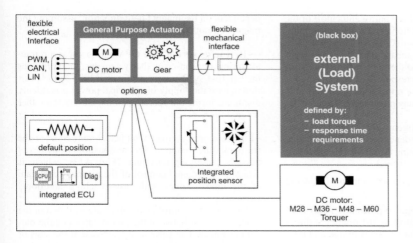

Fig. A10
Overview of possible
GPA components

~~**Electric motors.** The electric motors used are DC motors.

~~~**DC motor.** This is a classic parallel flow motor, which is normally used in conjunction with a transmission system. Actuators with worm-gear drive (self-locking) or with spur gear (not self-locking) can be designed depending on the transmission type.

Fast actuation speeds and high surplus torques can be achieved by using transmissions.

~~~**Torque motor.** The torque motor is a DC motor. Its coil is wound onto the stator and the magnets are attached to the rotor. This motor preferably is used for applications that require fast reaction times and small strokes. The torque motor features contactless function and, therefore, provides resistance to wear (no carbon brushes).

~~**Integrated electronic control unit.** Depending on the application, it may be necessary to equip the GPA with its own electrical circuit. This can be achieved by, for example, installing an H-bridge on the carrier substrate in the rotational direction of the electric motor and the microcontroller, which is responsible for the bearing control. Different carrier substrates can be used depending on environmental conditions and system design. The electrical circuit is implemented on a thick-film hybrid due to the temperature requirements, caused by close proximity to the combustion engine. Printed wiring boards are also used in some applications.

~~~**GPA subgroups.** Some product examples of possible subgroups are shown in **Fig. A11**.

The subgroups are divided as follows:

• *Basic GPA*. This is a GPA that travels only between two predefined end positions and can only be stopped

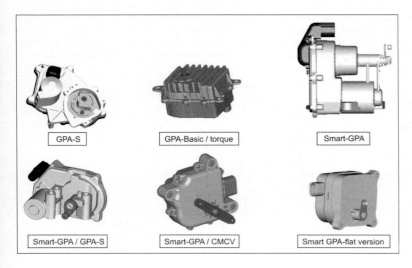

Fig. A11
Examples of general
purpose actuators

at these end positions. The control and cutoff of the GPA are handled completely by the external system.

- *GPA with position sensor* (GPA-S). This GPA has a position sensor, which means that the actual position can be detected and then fed back to the external system through corresponding pins and plugs. The external system can now position the movable system accurately into any position between the defined end positions.
- *Smart GPA*. This intelligent GPA offers the maximum flexibility and independence for the engine management system during development, because it possesses integrated electronics.

These electronics are responsible for the following functions:

- Position feedback (integrated position sensor)
- Bearing control by controlling the rotational direction of the DC motor (integrated H bridge and integrated microcontroller)
- Diagnostics and communication (integrated logic)
- Temperature monitoring

~~**Plug connector.** The electrical plug connector is the interface between the GPA and the external system. Customized pins are normally used in the plug connector for the supply of power for the vehicle electrical system and for communication with external systems. Communication with external systems can be realized by means of pulse width modulation (PWM), controller area network (CAN), or local interconnect network (LIN) bus.

~~**Position sensors.** The following sensors are, for example, used for accurate position acquisition:

- Potentiometer
- Hall effect element with magnet pole wheel
- Magneto-resistive sensor (MR sensor)

In contrast to the potentiometer, the Hall effect element with magnet pole wheel and the MR sensor are contactless. The primary advantages of the contactless sensors include a reduction of the probability of failure and an increase in total system reliability. The MR sensor is an absolute value sensor whose signal is only

dependent on the magnet field direction, assuming the magnet circuit is designed accordingly.

~~**Retractor spring.** The GPA may possess a retractor spring, depending on the system design. It provides the external system with a so-called limp-home position in case the supply of electrical power is suddenly disrupted during the operation of the GPA or if another severe failure occurs.

~Pulse supercharging. A higher compression of the fresh mixture required for combustion is achieved through pulse supercharging (**Fig. A12**), which is achieved by targeted control of the air mass upstream of the intake valves with fast-switching additional valves.

These freely controllable valves are installed in the intake manifold between the intake valve and the plenum chamber. The valve will be closed at the start of the intake cycle, when the piston moves downward. A vacuum develops in the combustion chamber. Shortly after opening the electrical valve and just before the reversal point of the piston, the air is suddenly released, whereupon it flows into the cylinder powered by the pressure difference. A pressure wave is generated simultaneously, and it moves with the speed of sound into the combustion chamber. The inlet valve closes before the overpressure wave escapes, increasing the cylinder charge. This charging effect is available without delay within a power cycle.

Each cylinder requires one valve (**Fig. A13**). These valves are controlled by an electronic system, which in turn requires signals from the crankshaft and camshaft sensors, which mirror the phases of the operating cycle. An engine control interface delivers, among other things, the information for timing of the start of pulse charging.

The volumetric efficiency, λ_a, and the indicated mean effective pressure, p_{mi}, for the lower speed range can be increased by up to 50%. Without pulse supercharging, the cylinder charge is particularly unsatisfactory for lower speeds at full load in dynamic terms. Exhaust gas turbochargers also deliver the desired pressure increase, but only after a delay time and at a minimum speed. Pulse supercharging reacts spontaneously in

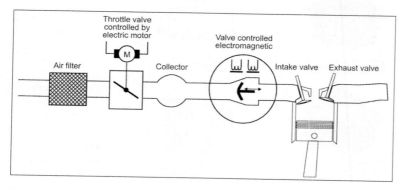

Fig. A12
Diagram for pulse supercharging

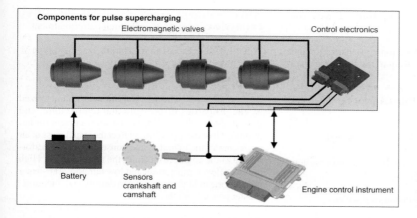

Components for pulse supercharging
Electromagnetic valves   Control electronics

Battery   Sensors   Engine control instrument
                 crankshaft and
                 camshaft

Fig. A13 System design for a four-cylinder engine

these situations and delivers the desired torque increase during the next power cycle. The increased torques are basically available without delay during the operation of the vehicle. A large torque increase is available for lower speed ranges in particular (**Fig. A14**).

Another side effect resulting from the optimized charge control during pulse supercharging is improved flushing of residual exhaust gas as a result of the larger quantity of fresh mixture between the intake valve and the pulse supercharger valve at the end of the exhaust phase.

Pulse supercharging, intake runner supercharging, and resonance tube supercharging are all part of the dynamic charging class. Pulse supercharging processes can be used for gasoline and diesel engines. Combinations with mechanical superchargers, such as exhaust gas turbochargers, are possible and result in further power increases, a widening of the speed range for high torque levels, and an improvement in vehicle response.

ACV (air control valve) →Actuators

Adaptation →Injection system, fuel~Diesel engine ~~Serial injection pump ~~~Add-on modular assemblies

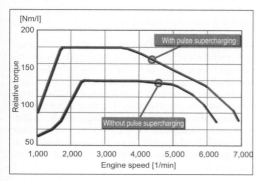

Fig. A14 Torque improvement in gasoline engine with pulse supercharging

Adaptation devices, diesel engine (*also*, →Injection system, fuel ~Diesel engine). Adaptation devices are, for example, systems for the control of inline injection pumps or distributor pumps for diesel engines. The primary adaptation devices are:

* *Control lever and control rod stops.* The control lever stop is used for limiting the idle and full-load fuel delivery or the overspeed delivery; it is also used for limiting the full-load and start fuel delivery.
* *Supercharging pressure dependent full-load stop (LDA).* The engine can burn a higher fuel mass when the supercharging pressure increases. The control rod will be influenced in such a way that the injection fuel quantity increases while the supercharging pressure increases.
* *Full-load stop dependent on atmospheric pressure.* The control rod will be moved to smaller fuel quantities by using a barometric cell to compensate for the lower air density at higher altitudes.
* *Surge damper.* It is used for the suppression of the jerking oscillation of the vehicle during acceleration or deceleration. The torsional vibrations at the flywheel are a measure of the jerking behavior. The injection fuel quantity is adjusted based on this value.
* *Start-of-injection sensor.* The start of injection has a major influence on the operating characteristics of the diesel engine. The combustion process and the resulting parameters such as emissions, noise behavior, and fuel consumption are especially influenced by the start of injection. This requires an accurate relationship between the start of injection and the position of the crankshaft.

Other adaptation devices are pneumatic idle-speed increase and shut-off device, electronic idle-speed control, control rod travel sensor, and temperature-dependent starting stop.

Literature: Kraftfahrtechnisches Taschenbuch, VDI Publishers, 1984. — Bosch technical training: Diesel-Einspritztechnik, Bosch, VDI Publishers, 1993.

Adaptive knock control system →Knock control; →Knocking

Added accessories →Electronic/mechanical engine and transmission control ~Requirements for mechanical and housing concepts

Additives, fuel →Fuel, diesel engine; →Fuel, gasoline engine

Additives, oil →Oil ~Body ~~Additives

Add-on modular assembly →Injection system, fuel ~Diesel engine

Adhesive abrasion →Piston ring ~Wear

Adiabatic combustion chamber. Attempts were made in the 1970s and 1980s to design and build so-called adiabatic combustion chambers for decreasing the heat losses through the walls, thus enhancing the cycle efficiency of the engine. One disadvantage of this approach was that a major part of the thermal energy additionally available for effective work ended up in the exhaust gas as an increase in enthalpy.

Adiabatic combustion chamber (heat insulated). Adiabatic combustion chambers have been analyzed in terms of reduction of heat losses and thermal efficiency increases. An improvement of the efficiency and, as a result, a reduction of fuel consumption were expected. However, the results actually revealed a large increase in the cylinder surface temperatures and the heat transfer coefficient between the gas and wall. An explanation for the resulting increase of the heat flow density is the increased temperature gradient near the walls. It is triggered by the flame burning near the wall and this results in higher turbulence near the wall.

Significant fuel consumption increases were seen with increasing engine loads. Improvements were only achieved in the lower part-load range and these were, however, clearly overwhelmed by the increases at higher loads. The knocking tendency also increases in gasoline engines.

Literature: G. Woschni, K. Kolesa, F. Bergbauer, K. Huber: Einfluss von Brennraumisolierungen auf den Kraftstoffverbrauch und die Wärmeströme bei Dieselmotoren, MTZ 49 (1988) 7/8.

Adiabatic expansion →Expansion ~Adiabatic

Afterburning (*also,* →Exhaust gas treatment system). Afterburning or subsequent reactions in the exhaust system decompose primarily unburned hydrocarbons; small quantities of carbon monoxide are also eliminated. The oxygen required for the reaction comes either from air fed in with the aid of a secondary air valve or from "residual oxygen" still present after the combustion process is completed.

Aftertreatment concepts using three-way catalytic converters →Pollutant aftertreatment

Aging test →Emission measurements ~Test type V

Air →Intake air

Air condition. The air condition is characterized by the properties of pressure and temperature, and thus at given volumes density and water content (or humidity). All of these properties affect the actual mass of air or oxygen in the cylinder, and consequently the performance of the engine, especially the power output. High pressure, low temperature, and low humidity (air-water content) mean high air or oxygen density and, therefore, greater power from the engine.

Air conditioning compressor noises →Engine acoustics

Air cooling →Cooling, engine

Air correction →Carburetor ~Air correction

Air cycling valve. The air cycling valve represents an innovative concept in optimizing the charge transfer processes in reciprocating piston engines. With an additional valve in the manifold directly before the cylinder head, which closes and reopens the intake cross section extremely quickly on every intake stroke (**Fig. A15**), considerable increases in torque can be achieved through pulse supercharging across the entire engine speed range, but especially at low engine speeds. Pressure waves are generated in the intake duct through the selected control of flow and oscillation processes using this air cycling valve, and these are used to obtain pulse turbocharging and numerous other additional effects. The benefit of this system is primarily that the air cycling valve can be installed in an additional interim flange so that the cylinder head and the intake and exhaust valve timing can remain unchanged.

~Air cycling valve functions. The air cycling valve makes possible a series of functions that can affect the operating characteristics of the engine.

~~Cylinder shut-off. One process option is the alternating shut-off of the intakes of individual cylinders in part-load operation, which is realized simply through a closed air cycling valve during the intake process. This process can also be used in conjunction with a conventional mechanical valve drive. The displacement of the working cylinder is used here at a higher load. In contrast to electromagnetic valve control with the complete closure of all valves, additional displacement work has to be generated by the engine as the exhaust valve remains open.

~~Exhaust gas recirculation. Analogous to recharging in turbocharged engines, exhaust gas recirculation

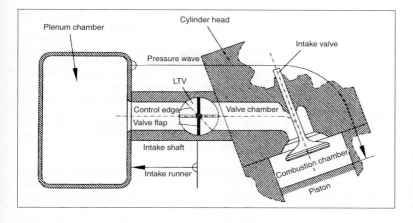

Fig. A15
Air cycling valve in the intake port (Source: Schatz Thermo Engineering)

can also be controlled selectively for each cylinder so that exhaust gas is first drawn in and fresh air can then be selected during the intake stroke (**Fig. A16**). Defined stratification of the charge in the combustion chamber is possible as a result, which offers further possible options with respect to intake flow speed and charge transfer.

~~Heat loading systems (cold-starting aid). The heat loading system enables an increase in the temperature of the air drawn into the combustion chamber, rather than an increase in the air mass, in order to improve the mixture formation in gasoline and diesel engines during a cold start. This process can be applied during the first few engine revolutions of a cold start. If it succeeds in raising the air temperature to an appropriately high level, in principle the glow plug in diesel engines can be omitted. In addition, the exhaust treatment system starts to operate considerably more quickly, which simplifies fulfillment of the 2004 European D4 exhaust standards for the diesel engine and also improves cabin heating in the vehicle. It is also possible to reduce the compression ratio in diesel engines, for fuel consumption reasons, without affecting their cold-start capability. The temperature increase results from a change in condition of the fresh air. First, the pressure in the cylinder falls considerably as a result of the piston movement when the air cycling valve is closed and the intake valve is open. The trapped mixture of residual gas and fresh air is expanded. After the air cycling valve opens, the cylinder fills in a very short period. The air is accelerated to very high speeds as a result of the large vacuum. Thermodynamically, this process corresponds to a compression of the intake air within the cylinder, which results in an increase in temperature. The size of the temperature difference achieved depends on the time of opening and period the air cycling valve is open and on the leakage at the closed valve (**Fig. A17**).

The large self-adjusting volumetric efficiency, the high speed of the induced flow and the increased end temperature at the end of compression allow multiple

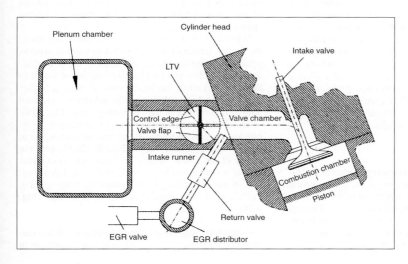

Fig. A16
Controlling the exhaust gas recirculation through the air control valve (Source: Schatz Thermo Engineering)

A

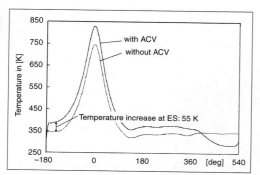

Fig. A17 Computation of the heat loading for a typical two-liter, four-cylinder engine (Source: Mahle Filtersysteme GmbH)

injections of fuel with improved combustion quality. This leads to higher exhaust temperatures during cold start, reduced combustion noise, faster warm-up and improved engine response. The requirement for this process is a very rapidly switching valve, which facilitates a sufficiently fast flow of the intake air into the cylinder and prevents reverse flow at the end of the intake phase.

~~Throttle-free load control. Throttle-free load control is an important step toward considerable reductions in part-load fuel consumption in gasoline engines through the minimization of the charge transfer loop.

The opening period of the air cycling valve is adjusted according to the air requirement of the engine. The opening of the air cycling valve and the intake valve can be phased so that both opening periods only slightly overlap—this results in the smallest air mass flow. The fast switching time of the valve then becomes less important for this process. In contrast, the leakage at the valve and the size of the volume between the air cycling valve and the intake valve become more important, as the air mass trapped there in the intake process is not available in the combustion chamber.

When using a valve that can be switched independently of the engine valve timings, various processes such as early intake closing or late intake opening can be achieved simply with a conventional valve mechanism in order to optimize the part-load operation of the engine. This places very high requirements on the precision of the valve switching in order to restrict the air mass drawn into the engine at idle.

~~Torque increase.

~~~Cold charging of turbocharged engines. Corresponding to cold charging, the air cycling valve can also be used to reduce the temperature of the air in the combustion chamber of turbocharged engines. Through the early-intake closing (*see above*, ~~Throttle-free load control), the highly compressed charge air, which also has been precooled by the heat exchanger, is

trapped in the combustion chamber, expanded by the piston movement, and thus greatly cooled further.

This temperature reduction causes a reduction of temperature and pressure in the combustion chamber at the end of the compression, which in gasoline and diesel engines reduces NO_x formation and the tendency of gasoline engines to knock. Thus a higher pressure ratio can be realized in the turbocharger with the resulting higher end of compression temperature, which in turn can be used to further increase the torque and power of the engine without further thermal loading of the combustion chamber.

~~~Dynamic charging (pulse charging). Pulse charging, known as dynamic charging, is achieved as a result of the fact that the opening and closing processes of the air cycling valve can be controlled selectively depending on the operating parameters of the engine relative to the movement of the intake valve and piston. The engine intake valve opens and the piston begins the induction stroke while the air cycling valve initially remains closed (**Fig. A18**). The air in the space between the air cycling valve and the intake valve is expanded into the combustion chamber. If a sufficiently low pressure is generated, the air cycling valve opens and fresh air flows in at high speed. As a superimposed mass transport effect, a rarefaction wave flows from the valve to the collector and is reflected as a compression wave into the manifold inlet of the combustion chamber. The inflowing air is decelerated again at the piston crown and is reflected anew toward the manifold, forming a reverse flow. The pressure increase as a result of the conversion of the kinetic energy into potential energy and the oscillations in the manifold work to increase the air mass either by closing the air cycling valve just before the start of the reverse flow or by the process being timed such that the engine intake valve closes when there is increased pressure in the combustion chamber. The inflowing air mass can be increased through the air cycling valve process, and the reverse flow, occurring toward the end of the intake process over a wide range of engine speeds, can be prevented.

When there is a valve with freely controllable drive, the double intake stroke is possible as another process option in the intake phase. As a result of the double activation of inflow and oscillation processes of the air in the manifold while the intake valve is open, a further increase in cylinder charge can be achieved, compared with a single stroke of the air cycling valve (**Fig. A19**).

The double stroke in the intake cycle leads to a significant increase in torque at low engine speeds, but at engine speeds above approximately 3000 rpm the charge effect greatly decreases. Therefore, a combination of a double stroke, a single stroke, and deactivation of the air cycling valve system by fully opening the valve at high engine speeds is beneficial (**Fig. A20**).

~~~Recharging supercharged engines. When mechanical superchargers or exhaust turbochargers are in

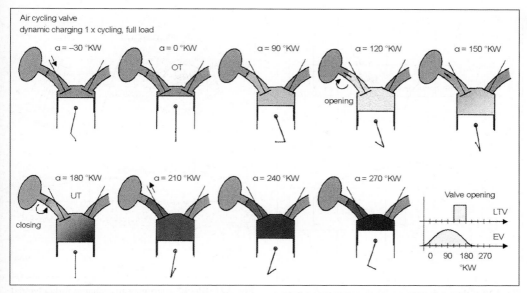

Fig. A18 Switching principle of the single air cycling valve stroke in the intake phase to increase torque (Source: Mahle Filtersysteme GmbH)

use, the air cycle valve offers further possibilities for increasing the engine output.

The response of the turbocharger is accelerated considerably through the increased mass flow, especially with exhaust gas turbochargers. Also, when the engine is started, an additional trapped mass of air is available in the combustion chamber, which results in higher temperatures at the end of compression and thus im-

proves cold start properties. As the air mass flow is already increased at low engine speeds as a result of the air cycling valve , the charge pressure can be reduced and the specific loading of the compressor is reduced as a result. Therefore, it is possible to use smaller turbochargers and there is no need for variable geometry turbochargers with their costly guide vane movement in the part-load range.

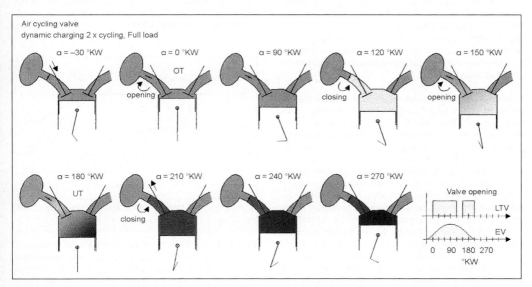

Fig. A19 Switching principle of the double air cycling valve stroke in the intake phase to increase torque (Source: Mahle Filtersysteme GmbH)

A

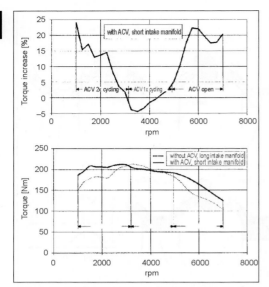

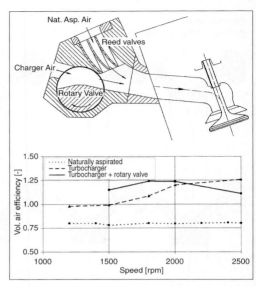

Fig. A20 Calculated torque with dynamic turbocharging through the air cycle valve for a typical two-liter, four-cylinder engine (Source: Mahle Filtersysteme GmbH)

Fig. A21 Block experiment to recharge turbocharged engines fitted with an air cycle valve designed as a rotating roll valve (Source: Schatz Thermo Engineering).

Another process option is sensible in energy terms for supercharged engines: uncompressed fresh air can be drawn into the combustion chamber through a non-return valve, and then a recharge of the system can take place with compressed air through the air cycling valve (**Fig. A21**). The advantage of this principle lies in the dramatically reduced flow through the supercharger at reduced pressure. This leads to a compact construction size and lower energy requirement for the supercharger, as only the air mass flow additionally required for the recharge has to be compressed. There is no induction work by the supercharger for the directly drawn air mass.

~Concept. In combustion engines, it is desirable to obtain as high a torque as possible and for this to remain constant from idle to high engine speeds. This situation, however, can only be achieved at a high cost as many techniques for torque optimization are only effective over a limited engine speed range because of pressure waves in the intake manifold with corresponding resonant frequencies at various engine speeds.

Naturally aspirated engines often operate with variable manifold lengths or with variable geometry in order to configure the torque characteristics as well as possible in the part-load range. Fully variable, mechanical, electrohydraulic or electromechanical valve drives represent other methods of increasing the torque. In these, the charge transfer can be optimized through displacement of the intake valve timing. High torque at low engine speeds leads to improved drive comfort, more agile engine response in transient operation, and

a noteworthy fuel-saving potential as a result of the possibility of using higher gear ratios. A method used primarily in diesel engines but also quite often in gasoline engines to increase the torque below the nominal engine speed is to force more air into the combustion chamber with exhaust turbochargers or mechanical superchargers. However, at rpms less than 1500–2000, these systems have a pronounced weakness in attaining high torques (**Fig. A22**).

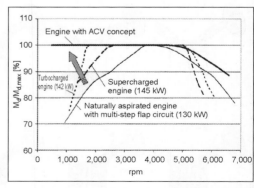

Fig. A22 Comparison of the torque characteristics of gasoline engines with switchover intake systems, exhaust turbochargers, and compressors and gasoline engines with air cycle valve concepts (Source: Mahle Filtersysteme GmbH)

In addition, there are restrictions in transient operation with respect to the dynamics of the engine, as the rotor of the turbocharger first has to be accelerated to provide the air mass flow corresponding to the required operating conditions.

With the air cycling valve , the nominal torque of the engine can theoretically be maintained down to the idle engine speed without length or resonance valve switching in the manifold. This concept can, specifically, move the required air mass in the event of a load step in the next cycle without response or delay times. With the air cycling valve, the length of the intake runners can remain very short so that there are considerable package benefits, compared with the usual resonance tube systems. The combination of an air cycling valve with an exhaust turbocharger therefore promises a considerable increase in the dynamics of the vehicle, especially for small-volume turbocharged engines, as a result of the increased mass flow immediately following the request of the driver for more power. Additionally, a reduction in fuel consumption and a benefit with respect to pollutant emissions can be expected.

~Design. The air cycling valve process was protected by patents in 1987, which generally describe two possible embodiments. Other possibilities such as valve systems or stop disks mechanically fixed to the camshaft are not suitable because of their lack of timing flexibility and the difficulty of air cycling valve activation at high engine speeds.

~~Controlled nonreturn valve. The simplest approach in terms of cost is the controlled nonreturn valve, where the energy for opening the valve is provided by the inducing piston, which produces a pressure difference between the collector of the intake system and the combustion chamber (Fig. A23).

A spring is tightened when the valve opens, which provides the energy to close the nonreturn valve. The arrangement can now be operated through the opening pressure difference and the closing spring force as the inflow reduces or backflow occurs. However, electromagnetic control of the nonreturn valve with magnets offers additional benefits so that the above effects, such as the creation of sufficient backpressure, can be used selectively through tuning at the particular engine operating point. Results of tests with a controlled nonreturn valve have been presented in [1] and show the considerable possibilities for increasing the charge in such a system.

Literature: Kreuter, Bey, and Wensing: Impulslader für Otto- und Dieselmotoren, 22nd Vienna International Motor Symposium 2001.

~~Valve with external drive. An expansion of the system performance, compared with that achieved with the controlled nonreturn valve, is obtained by fitting the air cycling valve valve with an external drive as indicated in Fig. A24. The control of the valve is therefore independent of the current pressure difference between collector and combustion chamber and the local gas speed at the nonreturn valve. More favorable switching times can be obtained as a result in order to exploit charge effects while throttle-free load control of the engine is possible also as a result of the time delay between the air cycling valve opening and inlet valve opening. The same valve control leads to the cold charging of turbocharged engines in turbocharged and charge-air intercooler engines.

When a valve with freely controllable drive is used, another option for increasing the torque lies in the double cycles for flowing in the intake phase (*see above,* ~Air cycling valve functions ~~Torque increase ~~~Dynamic charging) as described in [1].

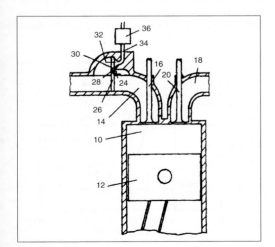

Fig. A23 Schematics of the principle controlled nonreturn valve (Source: Schatz; Patent DE 37 37 828 A1 [1987]).

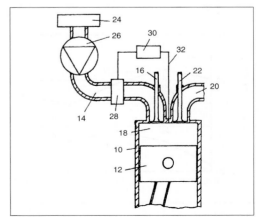

Fig. A24 Schematics of the principle air cycle valve (Source: Schatz; Patent DE 37 37 824 A1 [1987]).

A further increase in cylinder charge can be achieved as a result of the double activation of inflow and oscillation processes of the air in the manifold while the intake valve is open, compared with a single stroke of the air cycling valve .

Literature: [1] Schatz: Patent Specification DE 43 08 931 C2 (1993), http://www.depatisnet.de.

~Realization. A prototype with external drive has been built to demonstrate the ACV process, which is switchable using an electromagnetic control at any time during the intake phase [1]. The required switching speed of 2 ms was achieved by means of a spring-mass oscillator as direct drive for the butterfly valve in the intake port.

Literature: Elsäßer, Schilling, Schmidt, Brodesser, Schatz: Impulsaufladung und Laststeuerung von Hubkolbenmotoren durch ein Lufttaktventil, MTZ 62 (2001) 12.

~~Dynamic charging. The volumetric efficiency, λ_L, is considered as a reference for dynamic charging. The volumetric efficiency defines the ratio between the measured mass of air trapped in the cylinder and the theoretical mass of air evaluated from the displace-ment, V_H, and engine speed, n. As the volumetric efficiency with a stoichiometric air-fuel ratio, $\lambda = 1$, increases linearly, to a first approximation, with the torque on the operating engine, it provides a suitable parameter for judging the charge transfer processes. **Fig. A25** shows λ_L on a motored single-cylinder engine as a function of the engine speed. Because of the short manifold length, λ_L is quite small at low engine speeds. If the air cycling valve is opened twice in the intake phase at the same engine speed, a considerable increase in the volumetric efficiency can be seen at optimum timings, which increases greatly as the engine speed decreases. At $n = 1000$ rpm, the measured gain in volumetric efficiency is around 13%, compared with that achieved with the basic intake without valve switching.

The pressure diagram measured in the manifold of the motored engine as a function of the crank angle is shown in **Fig. A26** as an illustration of the charge transfer processes in the intake system.

The pressure measuring position upstream from the air cycling valve characterizes the pressure level of the manifold near the collector. The second measuring position is located in the volume between the air cycling valve and the intake valve. The lift curves on the intake and exhaust valves and the movement of the air cycle valve are also included in relation to the engine cycle.

The pressure diagram shown using an engine speed of $n = 1000$ rpm as an example clearly highlights the physical processes in the dynamic charging of the engine.

Because of the large valve overlap between the intake and exhaust valve, it is necessary initially to keep the air cycling valve closed when the intake valve is first opened in order to prevent exhaust gases from flowing into the intake manifold. The intake manifold is opened by the air cycling valve and the delayed intake flow process starts only after the exhaust valve has almost closed. The valve is closed again as part of the intake valve stroke. The pressure in the valve chamber between the air cycling valve and the intake

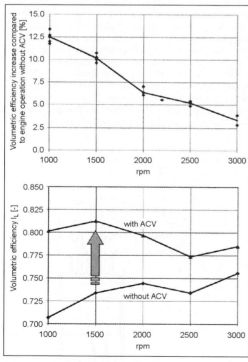

Fig. A25 Increase in volumetric efficiency with air cycle valve with double-stroke in the intake phase on a motored single-cylinder engine (Source: Mahle Filtersysteme GmbH)

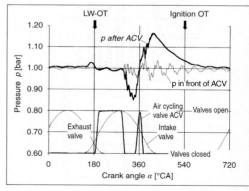

Fig. A26 Pressure development at double-stroke in the intake phase of a motored single-cylinder engine (Source: Mahle Filtersysteme GmbH)

valve quickly falls by 150 mbar as a result of the piston movement. Shortly before the charge transfer at bottom dead center, the valve is briefly opened a second time. The air flow shoots into the combustion chamber as a result of the pressure difference, where pressure increased by nearly 100 mbar compared with the manifold pressure as a result of reflections on the piston crown. This pressure peak in the combustion chamber is maintained for the dynamic charging and thus produces a significant increase in trapped density, which causes the additional mass in the combustion chamber. In **Fig. A26** the pressure decrease in the chamber between air cycling valve and intake valve is detectable after the intake valve closes, which is caused by leakage through the air cycling valve.

The increases in volumetric efficiency achieved in a motored engine were also confirmed in a functioning engine. The functioning engine was driven with and without the air cycling valve at a constant stoichiometric air-fuel ratio, $\lambda = 1$, and with a constant ignition timing in order to make possible a direct comparison of the effect of the air cycling valve on the engine. The results showed that the torque increased by a greater percentage than the volumetric efficiency at $\lambda = 1$. This indicates that the double-stroke of the air cycling valve with the resulting high air speeds intensifies the charge transfer in the combustion chamber and as a result has a positive effect on the mixture preparation.

~~**Heat loading.** Along with dynamic charging in a single-cylinder engine, the feasibility of heat loading was examined further using a single, short opening of the air cycling valve in the cycle. A characteristic result at $n = 1000$ rpm is represented in **Fig. A27**. The opening duration of the air cycling valve here is 30°CA, which corresponds to the shortest possible opening duration with a switching time of 2.25 ms. The 30°CA air cycling valve opening window was dis-

placed across the opening phase of the intake valve. The increase of temperature in the combustion chamber and the trapped mass of air in the combustion chamber were calculated for the respective cases from the measured pressure and temperature data.

When the intake process is complete, temperature differences of up to $\Delta T_{ES} = 45$ K were measured, compared with engine operation without air cycling valve, which is sufficient for air preheating in diesel engines when starting cold without a glow plug. The indicative measurements show that at these times the combustion chamber pressure behind the closed valve falls to 0.46 bar by the time the valve opens. As a result of the intensive intake flow due to this pressure difference, the mass of fresh air is raised because of the dynamic charging effect, despite the very short valve opening period, so that with heat loading there is no restriction with respect to the mass of air available for combustion. The total mass of gas in the combustion chamber increases by nearly 25% as a result of the residual gas proportion, which results in the temperature at the end of compression being raised further. Further temperature increases are possible if the sealing of the air cycling valve was ideal.

~**Requirements.** The technical realization of the air cycle valve places the highest requirements on mechanical components, drive design, and control electronics, because large changes in cross-sectional area, with extremely short switching times, have to be performed at defined times in the engine cycle. In particular, dynamic charging and heat loading require switching speeds for the opening and closing processes of approximately 2 ms. Furthermore, there are considerable requirements in terms of the reliability of the air cycling valve system when using it for throttle-free load control, when the air cycling valve is activated in every engine cycle.

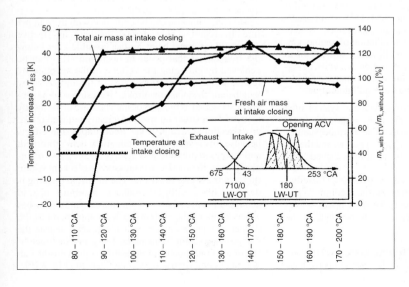

Fig. A27
Air temperature at the end of the intake process and the mass of air in the combustion chamber with heat loading (Source: Mahle Filtersysteme GmbH)

Air deficiency →Carburetor ~Icing

Air density →Intake air

Air distributor (*also*, →Intake system). The air distributor is a part of the manifold, from which the air is drawn after each firing order of the engine. The volume, shape, and feed from the individual manifolds into the air distributor, affect the engine response.

Air filter →Air intake system; →Filter ~Intake air filter

Air injection, combustion chamber →Injection system (components)

Air injection, exhaust system →Pollutant after-treatment

Air injection systems →Injection system (components); →Pol-lutant aftertreatment

Air intake system. Air intake systems for modern combustion engines currently fulfill a range of different functions in addition to air management and filtration. The requirements for air intake systems will increase with the increasing complexity of modern engines. Design trends include:

- The air management from the intake orifice to the cylinder head, which is designed, manufactured, and delivered ready for installation by the supplier, is treated as a system. A system concept for air management as a whole must be a requirement for the supplier, including the exhaust system with regard to charge transfer.
- There is an increasing modularization of the intake system. This modularization is beneficial because the air management is distributed across large parts of the engine and is suitable, because of its size, for different components to be attached. These components do not have to be part of the air management system. An example of this is the installation of engine controls in the air filter, which allows the flowing air to cool the electronics. Modularization demands, alongside an understanding of the system, increased competence in terms of production and integration.

Fig. A28 shows the air management system of a four-cylinder engine with the most important functions and some component parts. The thermodynamics of the air management are explained below, and the relationship to acoustics and filtration is described.

~Acoustics. Noise is caused by mechanical oscillations and produces waves in an elastic medium. When the piston is moved after the intake valve opens, a low-pressure wave is generated, which moves against the direction of flow.

~~Acoustic elements of tubing systems. Diverse acoustic principles can be applied to dampen intake

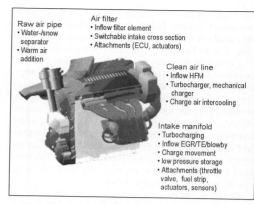

Fig. A28 Air management of combustion engine

noises (**Fig. A29**). The most important damper design is constructed using the principles of the so-called series resonator.

This includes a form of Helmholtz resonator in which a tube is connected to a damping chamber. In principle, such a resonator works as a spring/mass system does, in which the spring force is generated by the compressible air in the chamber and the oscillating mass is the volume of air in the tube. Depending on the dimensions of the components, a resonance frequency, f_0, can be calculated, in which such a resonator amplifies the applied noise. The frequency is calculated using the formula

$$f_0 = \frac{c}{2\pi}\sqrt{\frac{A_W}{l_{acoust}V}},$$

where A_W is the mean cross-sectional area of the resonator neck, l_{acoust} is the effective acoustic length of the neck, and V is the chamber volume. In contrast, frequencies above $\sqrt{2}f_0$ are dampened.

This phenomenon can be exploited with the damper filter. In order to achieve dampening that is as effective as possible, f_0 must be as low as possible—far below the frequencies that occur in operation. This can be achieved by enlarging the volume of the air filter, by reducing the cross-sectional area of the intake, or by lengthening the intake manifold. The casing volume cannot be increased as much as desired because of the mostly restricted construction space. A greatly reduced intake cross-sectional area also has undesirable side effects because this results in the intake flow being throttled, and increased pressure loss always means a loss of engine power. In practice, the pressure loss is kept within limits in the intake manifold. The manifold opening is designed like a diffuser, similar to a venturi tube, to achieve this. An extension of the intake manifold pushes at the limits of the system: such an approach has the danger of pipe resonance, which can again negate the damping at certain frequencies. Only an exact matching of the components in the entire system ensures an optimum compromise of cost and benefit (**Fig. A29**).

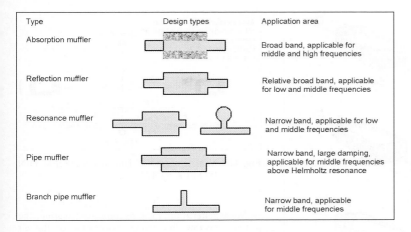

Fig. A29
Construction designs of
acoustic dampers with
their area of application

~~**Acoustic measurement and simulation tools.**
Many methods are available for designing an intake
system; in particular, simulation tools have become
important in the last few years, because information
about the acoustic behavior of the system can be
gained at a very early stage in the development of the
component. In addition to the finite element (FE)
method, one-dimensional computer programs based on
the transfer matrix or finite difference methods have become popular. The advantage of the latter is that thermodynamic values can be calculated alongside the
acoustic values. As soon as the first prototypes are available, the calculation results can be validated on simple
component test beds. A conclusive optimization with
near-production parts then takes place on an engine
acoustics test bed or in the vehicle.

Noise quality optimization, as well as minimization
of the noise level, is playing a more important role in
development. Noises are recorded using artificial
"heads" in order to allow them to be evaluated subjectively in sound comparisons.

The tools available are shown in **Fig. A30**.

~~**Future systems.** Ever more adaptive measures
are being used in intake systems alongside passive
measures. Switchover intake systems are being used
to increase the volumetric efficiency, and these
components can also be used in air management
systems to optimize acoustic behavior. For example,
a smaller intake cross-sectional area can be used in
a switchable intake manifold at lower engine speeds,
if the engine does not need its full volume flow, to
achieve low-frequency harmonization of the Helmholtz resonator. **Fig. A31** shows an example of such
a design.

~~~**Active noise control.** With the introduction of
electronics into the intake system, entirely different
systems became feasible, such as the use of antinoise
to eliminate noises. If the noise coming from the engine is countered by a wave with the same amplitude,
which is 180° antiphase, the two waves eliminate
each other. This principle is also called active noise
cancellation (antinoise) and is represented in **Fig.
A32**.

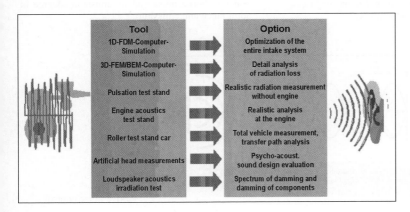

Fig. A30
Acoustic measurement
and simulation tools

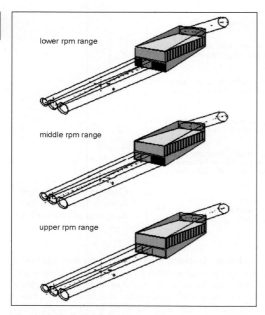

lower rpm range

middle rpm range

upper rpm range

Fig. A31 Switchable intake manifold

~~**Legislation.** The problem of noise emissions from a vehicle can be divided into two parts:

- Internal vehicle noise
- Passing noise

Whereas the reduction in internal noise goes hand-in-hand with the increased comfort requirements of the vehicle passengers, a statutorily fixed limit for accelerated passing according to DIN ISO 362 applies to the external noise.

The procedure for measuring this limit is represented in **Fig. A33**. Since October 1, 1996, automobiles must not exceed 74 dB(A) during accelerated passing.

The total acoustic level can be traced to the total of the individual small sources of noise. For passing noise, these are primarily engine noise, intake and exhaust noise, tire noise, and wind noise.

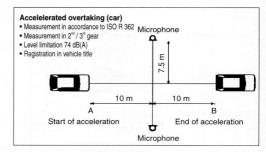

Accelelerated overtaking (car)
- Measurement in accordance to ISO R 362
- Measurement in 2nd / 3rd gear
- Level limitation 74 dB(A)
- Registration in vehicle title

Microphone

7.5 m

10 m | 10 m

A | B

Start of acceleration | End of acceleration

Microphone

Fig. A33 Vehicle noise, EU Directive 92/97/EU, DIN ISO EEA 84/424

~~**Noise generation.** In a reciprocating engine, the pistons generate fluctuating movements of the air through their upward and downward motion (air pulsations) and, as a consequence of this, sound is propagated though the air. The pistons, therefore, act as an air-pulsating noise source; also, as a result of disruptions to the air flow along the intake system, aerodynamic sources can contribute to the intake noise.

Noise is primarily emitted through the intake entry and directly reaches the surrounding area. A second part of the pulsation energy causes structure-borne oscillations within the elastic structure. These are then transferred from the external surfaces of the components to the surrounding air and through fixing points to the chassis. These connections are represented schematically in **Fig. A34**.

~~**Optimization measures.** The primary aim of measures for optimizing intake noise is consistent acoustic development, by which noise is reduced at the design stage. The methods for optimizing the noise are divided as follows:

- *Primary measures* influence the noise source. For noises generated by air movement, this means a reduction in transfer pressures. Noises generated by vibrating structures require a reduction in the generating powers and a change in the behavior of the structure-borne noise and its emission (degree of admittance and emission).

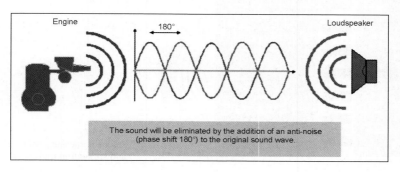

Engine

180°

Loudspeaker

The sound will be eliminated by the addition of an anti-noise (phase shift 180°) to the original sound wave.

Fig. A32
Active noise control

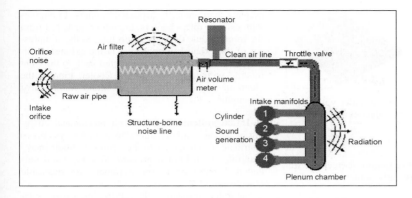

Fig. A34
Noise sources in an intake system

- *Secondary measures* reduce the noise generated by air movement after they are produced and reduce the noise emissions through sound dampers or acoustic encapsulation or both.

  The air-pulsating noise source of the intake system is the engine, and influencing this often conflicts with the thermodynamic objectives. Therefore, secondary measures such as damper filters and partial flow resonators are used to reduce the intake noise.

  The effect of an acoustic measure on the orifice noise and charge transfer is represented as an example in **Fig. A35**.

~~**Orifice noise.** Noise emitted from the ends of the intake or exhaust system. Orifice noise comprises pressure fluctuations, which are emitted as noise from the ends of the intake or exhaust system. This noise is not always perceived as pleasant, which is why noise limits have to be required for all vehicles.

~~**Structure-borne noise.** This is noise that is transmitted through a solid medium or on its surface. Waves generated by flexible surfaces are particularly important here as they cause very good air noise waves

~**Helmholtz resonator.** *See above,* ~Acoustics ~~Acoustic elements of tubing systems. *Also,* →Engine acoustics ~Helmholtz resonator

~**Interior noise.** *See above,* ~Acoustics ~~Legislation. *Also,* →Engine acoustics

~**Thermodynamic air management system.** The thermodynamics of the air management system depend on the type of combustion process (gasoline or diesel) and on the charging principle (**Fig. A36**). Raw air line and air filters are similar in all variants. Depending on the charging principle, the systems differ considerably downstream from the air filter, however.

Turbocharged engines have a large clean air area with rectifier and charge air intercooler, while the intake manifold is designed simply, as it merely serves to distribute the air to the cylinders. Naturally aspirated engines, in contrast, mostly have complex switchable intake manifolds to improve the cylinder charge.

~~**Air filter.** Air filters are generally the casing for the air filter element. Apart from the acoustic effect, the function of the air filter is air management for optimal air flow into the filter element. "Optimal" in this context means as consistent a flow as possible; the "normal" or "vertical" speed to the filter element must be homogeneous across the entire filter surface (**Fig. A37**). In the event of irregular flow, the pressure loss at the filter element increases and the efficiency of the engine deteriorates. In particular, the dirt bearing capability (dust retention capacity) of the filter material is used optimally with a homogeneous flow.

Three-dimensional flow simulation (computational fluid dynamics, or CFD) is used at a very early stage when designing the flow in air filters. Thus, it is

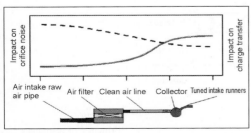

Fig. A35 Effect on orifice noise and charge transfer

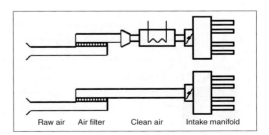

Fig. A36 Air management for naturally aspirated engines (*bottom*) and in turbocharged engines (*top*)

21

Fig. A37 Filter element inflow: irregular with high-pressure loss (*left*); approaching optimum (*right*)

possible to determine the optimum geometry with low experimental costs very early in the development of the filter.

In some cases the construction size can be reduced by 30% with the same pressure loss and dust retention capacity, and the flow can approach the ideal test bed value to within a few percent.

~~Clean air line. The flow of intake air mass through the flowmeter on the clean air side in new intake systems is simulated using CFD to ensure a consistent flow. Reliable functioning of the mass-flowmeter is required in all operating states and for the life of the vehicle if the stricter emission limits are to be met. Creeping degradation of the mass-flowmeter through the accumulation of oil drops from the crankcase or from exhaust gas recirculation (EGR) on the sensor can also be dramatically reduced using flow management based on CFD simulations.

On the clean-air side, downstream, the gas pulsations generated by the engine become more intensive. If the thermodynamics and acoustics are not considered in principle as a whole, they must be addressed in the clean air line at the latest because both disciplines impact on the air management. There are acoustic components (partial flow resonators, λ/4 tubes) in the clean-air area that also affect the charge transfer. Today simulation tools are used to calculate volumetric efficiency and orifice noise at a very early stage of the design phase. Modeling costs can be reduced considerably as a result because a computation model provides both results, acoustic and thermodynamic, simultaneously. Turbocharged engines have longer air management

systems than naturally aspirated engines. In engines with exhaust turbochargers, the intake air is fed from the front module through the air filter to the compressor of the exhaust turbocharger at the exhaust manifold. The compressed air is then fed back to the front module with its charge air intercooler. Finally, the clean air line leads back to the intake manifold of the engine.

~~Intake systems. Engines with mechanical charging or exhaust turbochargers need intake systems to distribute the charge air to the cylinders. Short intake manifolds with limited pressure loss and good balanced distribution to the cylinders are desirable objectives.

Naturally aspirated engines utilize the waves generated by the pistons to charge the combustion chamber. The process of resonance supercharging is shown in **Fig. A38**.

After the inlet valve opens, the downward movement of the piston generates a rarefaction (low pressure) wave, which moves from the combustion chamber against the flow direction along the tuned intake runner. The rarefaction wave is reflected at the collector as a result of the area change, and the pressure wave returning to the combustion chamber can be used to improve the cylinder charging if it reaches the combustion chamber before the inlet valve closes.

The optimum length of the pipe for constant speed of sound, a, is inversely proportional to the engine speed, n. Switchover intake systems, which can switch between short and long tuned intake runners, are increasingly being used in all classes of vehicle in order to achieve good cylinder charging. A typical torque characteristic for a switchover intake system with two lengths of tuned intake runner is shown in **Fig. A39**.

With the increasing complexity of intake systems, the rise in volumetric efficiency depends increasingly on the quality of the production process and the material. **Fig. A40** shows an example of how volumetric efficiency depends on seal quality and the associated leakage. A seal, which facilitates an increase in volumetric efficiency with a two-stage switchover intake system, can lead to a reduction in the volumetric efficiency with three-stage or multistage intake manifolds.

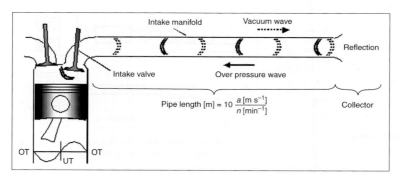

Fig. A38
Principle of resonance supercharging

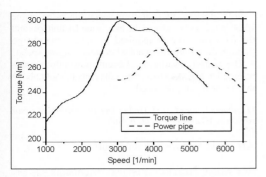

Fig. A39 Torque characteristic of a six-cylinder engine with switchover intake system

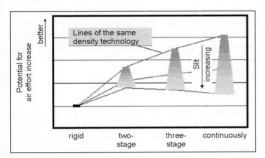

Fig. A40 Effect on volumetric efficiency of the seal technology in inlet manifold flaps

Apart from the leakage through the switch elements, a series of other factors also affects the charge transfer in the inlet manifold. **Fig. A41** provides an overview of the possible loss sources.

For the supplier of modern intake systems, this results in the requirement to define the entire intake system both thermodynamically and mechanically at an early stage of its development. This requires the incorporation and networking of all CAE (computer-aided engineering) tools at the start of development.

~~**Raw air line.** The raw air line, the area of the air intake system between the intake orifice and the air filter, performs the functions of warm air mixing and dirt removal as well as flow management. The mixing of warm air affects the operating characteristics of the engine, especially in the cold start phase. This function will become more important with stricter limits in exhaust legislation. Drying the filter element and melting snow are also tasks of the warm air mixing. Fuel consumption can be influenced positively through intelligent control of the temperature of the intake air. Warm air is added through a second induction point close to the exhaust system, which is activated through valves. The valves are activated through expansion elements or actuators such as a vacuum unit or electric motor.

A suitable raw air management system is capable of removing large particles (drops, snow, and dirt) with limited pressure loss through deflections. This removal reduces the dust retention in the air filter and protects the air filter element against dampness. Particle removal and pressure loss at the deflector are now predetermined by computer using the flow simulations (computational fluid dynamics, or CFD).

Literature: K. Müller, W. Mayer: Einfluss der Ventilgeometrie auf das Einströmverhalten in den Brennraum, 3rd Edition, Braunschweig/Wiesbaden, Vieweg Publishers, 1999.

Air limiter. Air limiters serve to throttle the intake air to limit the power output and engine speed. They are or were prescribed in motor sport, e.g., in Formula 3 engines. The air limiter restricts the volumetric efficiency above an engine speed limit, which depends on the capacity of the engine.

The flow conditions at the throttling point can be described with good approximation using the isentropic expansion for compressible flow of a perfect gas. The intake of fresh air from the environment into the plenum chamber of the combustion engine, also called the airbox, can be approximated as the flow out of an infinitely large container. The state variables of the environment do not change as a result of the induction by the engine.

The Saint Venant-Wantzel formula follows from the law of conservation of energy and the general gas

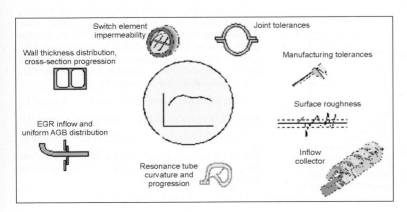

Fig. A41
Factors affecting the volumetric efficiency in switchover intake systems

A

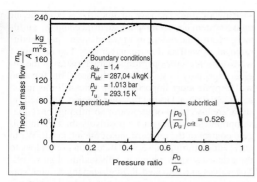

Fig. A42 Theoretical air mass flow through a throttling point, based on the throttle cross-sectional area

equation for the gas velocity in the narrowest cross section. Application of the continuity equation gives the theoretical, loss-free, air mass flow.

Fig. A42 shows the theoretical air mass flow based on the throttle cross section as a function of the pressure ratio. It shows that the mass flow of a compressible fluid through a throttle can increase only to a maximum value, which is achieved if speed of sound occurs in the narrowest cross section. The accompanying pressure ratio is called the critical pressure ratio. The mass flow remains constant even with further reductions of the pressure in the airbox.

The maximum possible air mass flow through a throttling point at the ambient pressure for the given general conditions, (e.g., density of the ambient air and outflow function) is therefore proportional to the area of the narrowest cross section.

It follows from this fundamental consideration that an ideal engine with a constant efficiency and delivery ratio in the unthrottled region up to a "critical" engine speed, at which the induced mass flow corresponds exactly to the maximum possible mass flow, would produce a constant torque. Above the "critical" engine speed, the engine is throttled by the air restrictor; and as the engine speed increases further, it produces an almost constant power output because of the falling volumetric efficiency (**Fig. A43**).

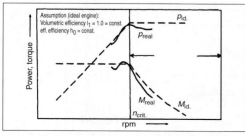

Fig. A43 Torque and power output of a perfect Formula 3 engine, compared with a produced racing engine

In a real engine, the overall efficiency falls as the engine speed increases because of the increasing friction loss and the increasing power required to induce the charge through the throttle above the "critical" engine speed.

The measured variation of engine power output represented in **Fig. A43** therefore falls considerably above the "critical" engine speed.

Literature: F. Indra, U.D. Grebe: Der Formel-3-Rennmotor von Opel, MTZ 54 (1993) 11, p. 567.

Air mass →Flowmeter intake air

Air mass characteristics →Actuators ~E-gas ~~Progression (throttle)

Air mass flow. The mass of air flowing through the engine is to a first approximation proportional to the engine power output and depends on the displacement of the engine. If the air mass flow increases, the power output increases at the same air-fuel ratio (turbocharged engine). The air mass flowing through the engine is calculated as

$$\dot{m}_L = i \cdot n \cdot (V_H \cdot \rho_E \cdot \lambda_l + m_{sp})$$

where $\dot{m}_L$ = air mass flow, i depends on two-stroke or four-stroke, V_H = displacement of the engine, ρ_E = density of the air at the engine intake, λ_l = volumetric efficiency, and m_{Sp} = scavenging air mass.

Air mass flow sensor →Sensors

Air mass flowmeter →Flowmeter intake air; →Sensors

Air movement →Mixture formation

Air preheating →Starting aid ~Intake air preheating

Air pressure →Intake air

Air ratio →Air-fuel ratio

Air requirement (stoichiometric). The stoichiometric air requirement is a dimensionless figure which denotes how much air (kg) is necessary for complete combustion of 1 kg of fuel; that is, all the fuel, typically hydrocarbons, is oxidized to form CO_2 and H_2O, and the sulfur content in the fuel is converted into SO_2.

Represented in chemical equations:

$$C + O_2 \rightarrow CO_2 \; 1 \text{ mol } C + 1 \text{ mol } O_2 \rightarrow 1 \text{ mol } CO_2$$

$$H_2 + 1/2\,O_2 \rightarrow H_2O \; 1 \text{ mol } H_2 + 1/2 \text{ mol } O_2 \rightarrow 1 \text{ mol } H_2O$$

$$S + O_2 \rightarrow SO_2 \; 1 \text{ mol } S + 1 \text{ mol } O_2 \rightarrow 1 \text{ mol } SO_2$$

The molar masses (in kg/kmol) for the substances are: $M_C = 12$; $M_O = 16$; $M_S = 32$; $M_H = 1$. This gives the following mass calculations:

$$12 \text{ kg } C + 32 \text{ kg } O_2 \rightarrow 44 \text{ kg } CO_2$$

$$1 \text{ kg C} + 32/12 \text{ kg O}_2 \rightarrow 44/12 \text{ kg CO}_2$$

$$2 \text{ kg H} + 32/2 \text{kg O}_2 \rightarrow 18 \text{ kg H}_2\text{O}$$

$$1 \text{ kg H} + 8 \text{ kg O}_2 \rightarrow 9 \text{ kg H}_2\text{O}$$

$$32 \text{ kg S} + 32 \text{ kg O}_2 \rightarrow 64 \text{ kg SO}_2$$

$$1 \text{ kg S} + 1 \text{ kg O}_2 \rightarrow 2 \text{ kg SO}_2$$

The stoichiometric air requirement also differs, depending on the fuel used (different composition).

The composition can be determined in mass particles from the chemical analysis of the elements in the fuel. The primary components of the fuel are carbon, hydrogen, sulfur, oxygen, water, ash, and nitrogen. The respective proportions related to 1 kg of fuel are

$$m_C + m_H + m_S + m_O + m_N + m_W + m_A = 1.$$

The dimension, for example, for m_C is kg C/kg fuel. Thus, the stoichiometric mass of oxygen required to burn this fuel is

$$m_{O,st} = 32/12 m_C + 8 m_H + m_S - m_O \text{ [kg O/kg fuel]},$$

with m_O taking into account the oxygen content in the fuel, e.g., alcohols.

As the mass percentage of oxygen in the air is 0.232, the result for the stoichiometric air mass is

$$m_{L,st} = m_{O,st} / 0.232 \text{ [kg air / kg fuel]}.$$

The stoichiometric air requirement influences the composition of the fuel mixture, partly depending on the mixture form (injection, spray) and partly on the fuel.

If the actual air mass is compared with the stoichiometric air mass and with the fuel mass, this results in the air-fuel ratio

$$\lambda = m_L / (m_{L,st} \, m_{Kr}).$$

During engine combustion there is more or less deviation from the stoichiometric air ratio depending on the conditions. A stoichiometric mixture is necessary in engines with mixture control and a catalytic converter. A substoichiometric (rich) mixture is used at full load to achieve high specific work and to avoid knocking. The smallest fuel consumption is achieved with an overstoichiometric (weak) mixture.

Literature: F. Pischinger: Verbrennungsmotoren, Lecture Reprint Volume 1.

Air resistance →Fuel consumption ~Variables ~~Road resistances

Air restrictor →Air limiter

Air rotation →Charge movement ~Swirl

Air shrouding →Injection system, fuel ~Gasoline engine ~~Direct injection systems ~~~Air-shrouded direct injection

Air sound →Engine acoustics ~Structure-borne sound

Air sound radiation →Engine acoustics ~Sound radiation

Air storage process →Combustion process, diesel engine ~Coal dust engine

Air swirl →Charge movement ~Swirl

Air temperature →Intake air

Air volume →Flowmeter intake air

Air volume flowmeter →Flowmeter intake air; →Sensors

Air-cooled crankcase →Crankcase ~Crankcase construction

Aircraft engine →Engine

Aircraft engine oils →Oil ~Classification, specifications, and quality requirements ~~Special engine oils

Air-distributed injection →Mixture formation ~Mixture formation, diesel engine

Airflow →Delivery degree

Air-fuel mixture. The air-fuel mixture is a significant parameter in the operation of the engine. In a λ-controlled gasoline engine with three-way catalytic converter, the ratio of air to fuel in the mixture moves in a relatively narrow range ($\lambda \approx 0$). Diesel engine mixtures range from $\lambda = 1.05$ to $\lambda \approx 8$, the upper limit dependent on the friction losses in the engine.

~**Air-fuel mixture formation** →Mixture formation

~**Calorific value** →Fuel, diesel engine; →Fuel, gasoline engine

~**Capability of achieving leaner mixtures.** The capability of achieving leaner mixtures depends on how far a mixture of air and fuel can be made leaner without putting reliable ignition at risk. A leaner mixture can be achieved by adding air or exhaust (exhaust gas recirculation). This is then called a "weak" or lean mixture.

It is important in a gasoline engine that there is always an ignitable mixture at the spark plug. The leaner the mixture, the lower the fuel consumption; however, the power output is also less. This can be seen in principle in **Fig. A44**. However, as the ignition limit is approached from the lean side, the specific fuel consumption increases again. The air-fuel ratio, at which this occurs, depends on the mixture preparation; that is, on the conditions existing at the time of ignition at the spark plug.

With specific measures to stabilize combustion, lean-burn engines can use lean mixtures of up to $\lambda =$

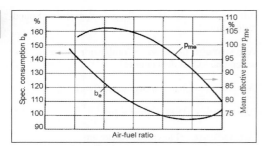

Fig. A44 Specific fuel consumption and mean pressure over the air-fuel ratio

1.8. Stratified charge engines or gasoline engines with direct injection achieve λ values up to 4.

~Enrichment (rich mixture). The specific fuel consumption and the power output increase with enrichment of the mixture—that is, the addition of fuel. As the maximum flame speed occurs around λ = 0.85, this range of enrichment is also the range of the maximum mean effective pressure (**Fig. A44**).

~Formation →Mixture formation

~Homogeneous mixture. A homogeneous mixture is an air-fuel mixture that, to a first approximation, has a locally constant composition of air and fuel. In conventional gasoline engines, the mixture is approximately homogeneous; this is generated through a formation of the mixture outside the combustion chamber. In diesel engines, there is an extremely heterogeneous mixture in the combustion chamber—that is, fuel is injected into the precompressed air. There is a spectrum of lambda (λ) from zero to infinity. Similar conditions exist in gasoline engines with direct injection.

Homogeneity and heterogeneity have extreme effects on combustion and, therefore, on the products of combustion—for example, exhaust composition, power output, noise, and fuel consumption.

~Ignition. The air and fuel must be mixed such that reliable ignition occurs at a given time. In a conventional gasoline engine, the spark causes the ignition, so an ignitable mixture has to be present in the region of the spark plug at least.

In diesel engines, the energy for igniting the fuel is provided through compression of the air, and the resulting compression temperature must be high enough for ignition to occur.

~Lean mixture →Capability of achieving leaner mixtures

~Misfire limit. In a gasoline engine, the misfire limit is the air-fuel ratio at which the mixture does not ignite reliably or at which, statistically, a certain number of combustion cycles do not result in acceptable ignition of the mixture. There is a misfire limit for lean, λ > 1, and for rich, λ < 1, mixtures. The misfire limit greatly depends on the preparation of the mixture and the movement of the mixture in the cylinder. When the misfire limit is reached, there are high emissions of unburned hydrocarbons, and reductions in efficiency also have to be taken into account. Frequent misfires lead to increased temperatures in the catalytic converter and can cause damage.

~Mixture enrichment →Mixture formation ~Mixture formation, gasoline engine ~~Mixture metering

~Nonheterogeneous mixture. *See above,* ~Homogeneous mixture

~Preparation →Mixture formation ~Mixture formation, gasoline engine ~~Mixture preparation

~Rich air-fuel mixture. A rich air-fuel mixture is one in which there is a deficiency of air. In a gasoline engine, the tuning is generally at λ = 1. The engine is only operated rich in order to achieve higher specific work—that is, with a rich air-fuel mixture (λ < 1).

In diesel engines, there is a difference between a chamber engine and an engine with direct injection. The former has a rich mixture on average in the pre- or swirl chamber and a lean mixture on average in the main chamber. In a diesel engine with direct injection, there is, on average, always a mixture in the greater-than-stoichiometric range over the combustion chamber—that is, a lean mixture.

~Rich cloud. In lean-tuned gasoline engines, which are desirable for their low fuel consumption, mixture stratification is necessary because there has to be a rich—that is, ignitable—mixture in the region of the spark plug (rich cloud). Gasoline engines with direct injection, which work according to this principle, are being developed worldwide for reasons of consumption. Fuel consumption savings of up to 20% can be achieved in the part-load range, as they can be quality controlled ideally.

~Rich mixture. A rich mixture is one in which there is less air present than theoretically required to achieve a complete reaction (complete conversion of the fuel into H_2O and CO_2). A rich mixture brings about the internal cooling of the engine, prevents knocking, and is typically used at full load. However, fuel consumption is increased, as are HC emissions.

~Stoichiometric mixture. In a stoichiometric mixture, the air-fuel ratio is precisely measured such that exactly the mass of air is present theoretically to oxidize all the fuel to form H_2O and CO_2. Stoichiometric mixtures are necessary when using a three-way catalytic converter, because all three pollutants—CO, HC, and NO_x—are only converted with a sufficiently high level of efficiency at this mixture composition.

~Wall film formation. Unsatisfactory preparation of the fuel in conjunction with deviations in the flow of the mixture in the manifold leads to a wall film, the thickness of which depends on the current wall and mixture temperatures. This film primarily occurs in the manifolds of older engines with mixture formation systems such as carburetor and central injection. The proportion of fuel at full load, which forms the film, can contribute up to half of the fuel quantity delivered. The intensity with which the fuel attaches to the wall depends primarily on the condition of the mixture upon entry into the manifold. Barriers to the flow, such as edges and throttle valves, have a significant effect on film formation. During cold start and warm running, formation of a wall film is also possible even with very good fuel preparation.

Literature: H.P. Lenz: Gemischbildung bei Ottomotoren, Springer Publishers, Vienna/New York, 1990.

Air-fuel mixture preparation →Mixture formation ~Mixture formation, gasoline engine ~~Mixture preparation

Air-fuel ratio. The complete combustion of 1 kg of fuel requires around 14.5 kg of air. At this stoichiometric ratio of air (crucial is the oxygen content therein) and fuel, the air ratio or air-fuel ratio is:

$$\lambda = \frac{\text{Added Air Quantity}}{\text{Theoretical Air Requirements}} = 1$$

Air-guided combustion process →Injection valves ~Gasoline engine ~~Direct injection ~~~Installation positions

Air-shrouded injection →Injection system, fuel ~Gasoline engine ~~Direct injection systems

Alcohol →Fuel, diesel engine ~Alternative fuels; →Fuel, gasoline engine

Alcohol content →Fuel, gasoline engine ~Requirements

Aldehydes →Exhaust gas analysis; →Fuel, diesel engine ~Alternative fuels ~~Vegetable oil

Alternating load damping (gasoline engine) →Actuators ~E-gas

Alternating torque →Balancing of masses ~Ideal balancing

Alternative combustion processes →Combustion process, diesel engine ~Homogeneous combustion, diesel engine

Alternative connecting rod materials →Connecting rod ~Material

Alternative engines →Engine ~Alternative engines

Alternative fuels →Fuel, diesel engine; →Fuel, gasoline engine

Alternative (HC) SCR function for NOx →Pollutant aftertreatment ~Pollutant aftertreatment lean concepts

Alternator (*also*, →Starter ~Starter-generators ~~Operating modes of the starter-generator). The alternator or generator is an electrical generator that powers the onboard network of the vehicle (Fig. A45). It is usually driven by the engine from the belt drive, as is the power steering pump. The task of the alternator is

• To provide sufficient electrical energy to all electrical consumers required for engine and vehicle operation, e.g., ignition, injection system, electronics, lights; and
• To charge the battery during travel.

The generator produces electrical energy by electromagnetic induction, in which an electrical conductor loop moves through the field lines of a magnetic field. In new vehicles, practically, only three-phase alternators are used. These generate a three-phase alternating current, which is rectified using power diodes. With the greater use of the integrated starter-generator, this function is assumed by the alternator.

Literature: Bosch (ed.): Autoelektrik, Autoelektronik am Ottomotor, 2nd edition, VDI Publishers, − 1994. Bosch (ed.): Kraftfahrtechnisches Taschenbuch, 21st edition, VDI Publishers, 1991.

Alternator and retarder →Starter ~Starter-generators ~~Operating modes of the starter-generator

Alternator balancing →Balancing of masses ~Balancing ~~Accessory equipment

Alternator noise →Engine acoustics ~Generator noises

Altitude compensation (carburetor). The physical properties of the atmosphere change with

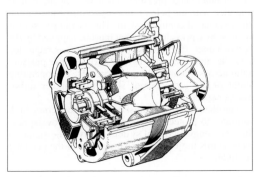

Fig. A45 Cross section of an alternator (Source: Bosch)

the geodetic altitude. The following equation is valid for the barometric pressure at altitude in an isothermal atmosphere:

$$p(h) = \rho_0 \cdot e^{(-g \cdot \rho_0 \cdot h)/(p_0)},$$

where $p(h)$ is the air pressure at altitude; h, p_0, and ρ_0 are the pressure and density at sea level; and g is the gravitational acceleration.

The variables of pressure, temperature, and humidity of the air have important influences on the performance and torque of an engine. The actual influence of the humidity is small (1% to 2% at 100% humidity). The following equation describes the relationship among pressure, temperature, volume, and the mass of air—it defines the density of the charge supplied to the engine,

$$p \cdot V = m \cdot R \cdot T,$$

where p is pressure, V is volume, m is mass, R is the gas constant, and T is the absolute temperature. A lower atmospheric pressure results in smaller mass of air induced if the same volume is induced and the temperature is kept constant.

The mass of fuel in gasoline and diesel engines that contributes to the appropriate air-fuel ratio must be adjusted to compensate for the conditions at altitude in those cases where the quantity of induced air is measured volumetrically and not related to mass. This is implemented by measuring the barometric pressure and using this to perform the fuel adjustment—the air-fuel ratio can be maintained at a constant level with this procedure.

The mass of fuel in a diesel engine must be limited depending on the geodetic altitude to ensure combustion with little soot. The impact of the temperature of the fuel on the mass of fuel supplied for electronic direct injection is included by measuring the temperature of the fuel.

The altitude compensation cannot prevent a reduction in engine performance, because the performance of the engine always depends on the mass of air or mixture supplied to the engine. This always will reduce with naturally aspirated engines.

The situation is different for turbocharged engines. A constant charge pressure (absolute pressure) can be supplied with the help of an exhaust turbocharger and the associated control system. The charge pressure is defined by a pressure ratio that must be supplied by the turbocharger. If a certain charge pressure is requested and the atmospheric pressure declines, the pressure ratio across the turbocharger increases. The operating point of the turbocharger shifts to a higher turbocharger speed.

The charge pressure must be limited for certain high altitudes to protect the turbocharger from overspeeding. The engine control system receives the signal to impose this limit from a measurement of the geodetic altitude.

Aluminum alloy →Bearings ~Materials

Aluminum crankcase →Crankcase ~Weight, crankcase ~~Reduction of specific weight of material

Aluminum cylinder head →Cylinder head

Aluminum floor plate →Electronic/mechanical engine and transmission control ~Requirements for mechanical and housing concepts

Aluminum oxide →Catalytic converter

American LeMans Series/Le Mans →Engine ~Racing engines

Analysis process →Exhaust gas monitoring check

Angle, valve →Valve; →Valve arrangement

Angle sensor →Sensors ~Position sensor

Annular gear pumps →Gear pump ~Internal gear pump

Antifoam additive →Coolant

Antifoaming agent →Oil ~Body ~~Additives

Antifreeze →Coolant

Antifriction device →Oil ~Body ~~Additives

Anti-icing →Fuel, gasoline engine ~Additives

Antiknock additive →Fuel, gasoline engine ~Octane number

Antiknock control →Knock control

Antioxidants →Oil ~Power during operation ~~Aging of oil ~~~Oxidation; →Fuel, diesel engine ~Additives; →Fuel, gasoline engine ~Additives

API →Oil ~Classification, specifications, and quality requirements

Application areas, computation →Calculation processes

Arbitration →Electronic/mechanical engine and transmission control ~Electronic components ~~CAN interface

Architecture, electronics/mechanicals →Electronic/mechanical engine and transmission control ~Software

Arcing voltage →Ignition system, gasoline engine

Aromatics →Fuel, diesel engine ~Composition; →Fuel, gasoline engine ~Aromatics contents

Articulated skirt piston →Piston ~Design

Ash content →Fuel, diesel engine ~Properties; →Oil ~Properties and characteristic values

Assembled flanged bearing →Bearings ~Design solutions

Assembly, piston ring →Piston ring

Assessment curves →Engine acoustics

Atmospheric pressure plasma process →Combustion process, gasoline engine ~Stratified charge process

Atmospheric pressure–dependent full-load stop →Injection system, fuel ~Diesel engine ~~Serial injection pump ~~~Add-on modular assemblies

Atomization →Mixture formation

Automatic choke →Carburetor ~Starting systems ~~Design types

Automatic drive belt jockey pulley →Toothed belt drive ~Belt tensioning systems ~~Automatic tension pulleys

Automatic lubrication →Two-stroke engine ~Lubrication

Automatic stop-start →Electronic open- and closed-loop control ~Electronic open- and closed-loop control, gasoline engine ~~Functions

Automatic tensioning device →Toothed belt drive ~Belt tensioning systems ~~Automatic tension pulleys

Autotermic piston →Piston ~Design

Auxiliary air jet →Carburetor

Auxiliary fan →Electronic/mechanical engine and transmission control ~Requirements for mechanical and housing concepts ~~Electronics box

Auxiliary nozzle →Carburetor ~Systems

Axial bearing →Bearings ~Design solutions; →Crankshaft ~Guide bearing

Axial bearing crankshaft →Crankshaft ~Guide bearing

Axial clearance →Piston ring ~Versions ~~Compression rings ~~~Keystone rings

Axial piston distributor pump →Injection system, fuel ~Diesel engines ~~Distributor pumps

B

B classification →Engine acoustics ~Assessment curves

Background emissions →Emission measurements ~Test type IV ~~Calculation of evaporative emissions

Balance shaft (*also*, →Balancing of masses). The moving masses of a crankshaft drive generate inertia forces, which result in couples, via the corresponding levers, for example, the cylinder gap. A difference is made between rotating (contribution of the crankshaft, contribution of the piston rod) and reciprocating masses (pistons, piston pins, piston rings, contribution of the piston rod). The rotating masses generate a centrifugal force, which can be balanced by corresponding counterweights offset from the crankshaft.

The forces generated by the reciprocating masses (e.g., divided into forces of the first and second order) are periodically varying forces depending on the crankshaft angle.

A complete balancing of all inertia forces created by moving masses is only possible if the common center of gravity of all moving parts is identical with the rotational center of gravity of the crankshaft for each crankshaft position.

The analysis of single-cylinder engines shows that the complete balancing of the reciprocating mass forces of the first and second order can be achieved by contra-rotating mass systems. For this to work, they must rotate either at crankshaft speed or double it.

The generated forces must have the value of the mass forces of the first and second order, but they must rotate in the opposite direction. They must be mounted symmetrically to the cylinder axis to ensure that the additional masses do not generate a torque. **Fig. B1** shows a schematic design of the balance shaft for single-cylinder engines. A complete balancing of the first order is possible by an increased counterweight at the crankshaft and a mass system with the same speed but rotating in the opposite direction. It must be en-

sured, however, that the effect of the counter-rotating mass force appears on the operational level of the reciprocating forces. Another alternative to implement a complete balancing of masses is the use of a Lancaster balancer. **Fig. B2** shows an example on how this balance is implemented.

This example uses, for instance, two piston rods and two balance shafts, which counter-rotate. These balance shafts, which are also called balance gears and which deliver a complete balancing of masses, are very demanding in cost, weight, and installation space.

Systems used in practice achieve only partial balance—for example, a balance shaft mounted sideways. This results in a balance of the mass forces of the first order on the operating level of the reciprocating forces, but it leads to additional changing torques around the length axis of the engine due to an offset from the cylinder level.

The reciprocating forces of the first and second order are balanced in engines with crankshafts with multi-offsets if the crankshaft scheme is symmetrical, which means the crankshaft web and crank arrangements are symmetrical. This is not the case for forces of the second order in a four-cylinder engine. These forces can be balanced partly or completely with balance shafts or differential gears. The balance shafts are in this case driven counter-rotating—for example, by gears with twice the crankshaft speed.

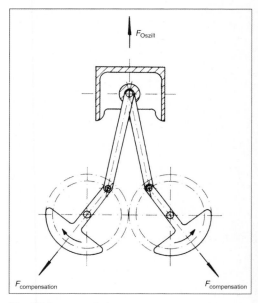

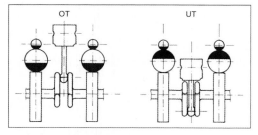

Fig. B1 Complete balancing of masses for a single-cylinder engine

Fig. B2 Lancaster balancer

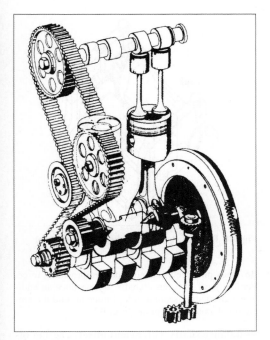

Fig. B3 Balance shaft for a twin-cylinder engine (Source: VW)

Fig. B3 shows an example for the design of a balance shaft installed at a twin-cylinder engine.

Despite the mass forces balancing, couples are generated for multi-offset crankshafts which are the result of cylinder center displacements. These torques are present for inline engines and for V-engines, and they can also be reduced or balanced by balance shafts. For example, a six-cylinder V60° engine with a firing interval of 120° for the balancing of the second order couples needs a balance shaft with counter masses offset by 180°, which rotates at twice the crankshaft frequency.

Literature: H. Maass, H. Klier: Kräfte, Momente und deren Ausgleich in der Verbrennungskraftmaschine, Volume 2, Vienna/ New York, Springer Publishers, 1981. — A. Urlaub: Verbrennungsmotoren, Springer Publishers, ISBN 3-540-58194-4, 1995. — P. Bauer, K. Bruchner: Entwicklung des Steuer- und Lanchestertriebes des DaimlerChrysler-Vierzylindermotors M271. MTZ (2003) 7–8. — R. Flierl, R. Jooßt: Ausgleichswellen für den BMW-Vierzylindermotor im neuen 316i und 318 i. MTZ 60 (1990) 5.

Balance weight →Flywheel class

Balance weights crankshaft →Crankshaft ~Balance weights

Balancing →Balancing of masses

Balancing accessory equipment →Balancing of masses ~Balancing

Balancing complete engine →Balancing of masses ~Balancing

Balancing of masses (*also*, →Engine acoustics). Balancing is used to compensate the forces and couples caused by the motion of the crankshaft drive system; this is done as close as possible to the location at which they occur.

Measures for balancing are required to:

- Compensate forces acting only inside the engine. Counterweights on the crankshaft have this effect, and they reduce the basic bearing forces, and, consequently, deformation of the engine block and the internal bending moments on the shaft.
- Reduce forces and couples that are effective externally. The so-called out-of-balance inertia force and out-of-balance couples cause the entire engine and transmission unit to vibrate and the inertia of the assembly must be taken up by the engine/transmission mounts.

As a result of the slider-crank relationship, the forces occur primarily in the vertical and horizontal engine planes. On multicylinder engines, these excitation forces can partly or completely cancel themselves out. However, they can also be additive, and this leads to a reduction of the comfort of the passengers. Additional compensating mechanisms can be attached to the engine to reduce the remaining inertia effects. However, such external compensating elements result in additional weight, frictional losses, and costs.

The accelerating motion of the connecting rod and piston has an influence on the torque curve of the crankshaft. These moments, which are centered round the longitudinal axis of the engine, are called alternating torques; they can only be influenced by external measures where the first harmonic of the rotary force and the compensation shafts present coincide in terms of order.

~**Alternating torque.** The torque at the crank pin results from superimposition of the tangential components of the gas and inertia forces. The tangential force periodically accelerates and decelerates the crankshaft. The sum of the tangential force components for all throws (multiplied by the crank radius) results in the alternating torque (**Fig. B4**). Normally, the magnitude of fluctuations in the alternating torque decreases as the number of cylinders increases. Alternating torques are absorbed by the engine block and transferred by the engine mounts to the body of the vehicle.

~**Balance shaft** (*also*, →Balance shaft; →Mass balancing mechanism). A single balance shaft can produce only one rotating force and/or one rotating couple. A reciprocating force and/or reciprocating couples can be produced using two shafts rotating in opposite directions (**Fig. B5**). This effect can be adjusted to the required engine order by using a transmission stage.

Frequently, two balance shafts rotating in opposite directions at twice the speed of the crankshaft are used

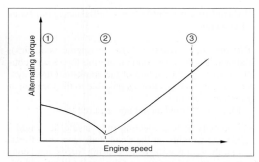

Fig. B4 Alternating torque in relation to speed (Source: Maas/Klier)

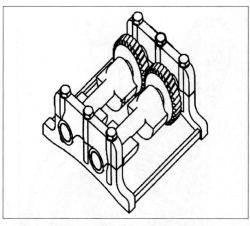

Fig. B6 Balance shaft unit without vertical offset

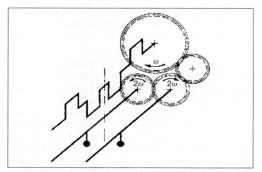

Fig. B5 Four-cylinder crankshaft with balance shafts rotating in opposite directions, driven by a gear stage (Source: Mass/Klier)

in large four-cylinder engines to reduce the second-order out-of-balance forces and, therefore, significantly reduce the vibration of the engine.

The center of gravity of the unbalanced masses should be located on the center plane of the engine. If the balance shafts are located at the same height (**Fig.**

B6), they produce a second-order reciprocating force. Vertically offset balance shafts will produce an additional alternating second-order torque around the longitudinal axis of the engine (**Fig. B7**).

~~**Balance shafts with height offset.** The out-of-balance force curve for a four-cylinder engine is dominated by the second-order effects. These cause periodic torsion on the engine around its longitudinal axis, which must be taken up by the engine/transmission mounts or the inertia of the unit. This is caused by the reciprocating force produced by the connecting rod and piston as well as the gas pressure. Two balance shafts with offset height (**Fig. B8**) rotating at twice the engine speed produce a second-order moment which counteracts the rotary force curve of the tangential force. Because the magnitude and direction of the rotary force curve depend on the actual load on the engine, balance shafts with offset height only act optimally at one operating point.

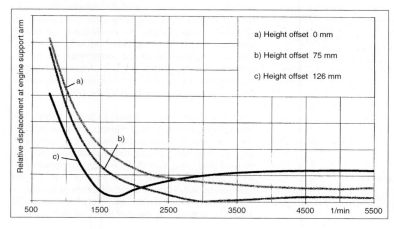

a) Height offset 0 mm

b) Height offset 75 mm

c) Height offset 126 mm

Fig. B7
Effect of balance shaft offset height on rotation of block around longitudinal axis

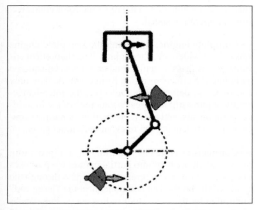

Fig. B8 Schematic representation of balance shafts with offset height

~Balance weight →Balancing of masses ~Counterweight

~Balancing. External mass balancing is primarily a design problem; the problem of production quality is of secondary importance. Balancing is an operation in which the mass distribution of a rotor is checked and, if necessary, corrected. This ensures that vibrations at the rotational frequency are kept within limits. Vibrations in opposite directions or at different frequencies such as those that can be caused by the reciprocating masses in engines with a small number of cylinders cannot be influenced by balancing. The limits of acceptable imbalance are defined by the level of comfort required in the vicinity of the engine.

Production-related errors in the distribution of mass can either be random or of a systematic type. **Fig. B9** shows the vector diagram of a typical balancing operation for a mass-production engine.

The scatter field indicates the imbalance present; its off-center position indicates a systematic error. The permissible imbalance occurs at the radius of a tolerance circle into which the rotors must be brought by "balancing."

On reciprocating piston engines, this consideration is not limited to the rotating part. The parts connected to the crankpins, some of which only reciprocate, must be included. The mass distribution multiplied by the crankshaft radius indicates an effect similar to imbalance. **Fig. B10** provides a summary of the parts mentioned explicitly below in terms of balancing technology. It is necessary to answer two basic questions:

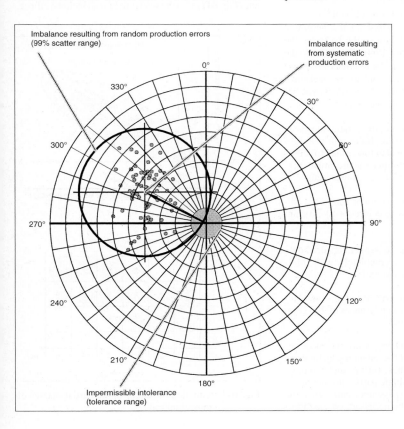

Fig. B9
Vector diagram for "balancing in mass production"

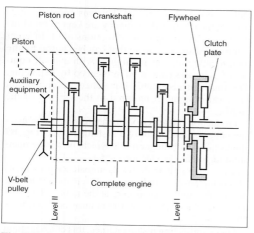

Fig. B10 Assemblies relevant for balancing technology

1. Where is mass correction possible?
   Removing material (drilling, milling) is preferable to adding material for reasons of operational reliability, rational production, and metering accuracy. This requires a sufficient quantity of "dead" material to be available on the shaft.
2. What amount of imbalance is permissible?
   Balancing tolerances are based on the permitted amount of vibration: this is also the basis for the general standard DIN ISO 1940. Considering the similarities, this means that the tolerances increase in proportion to the rotor mass but are inversely proportional to the "rotor" speed. Decisive is whether the imbalance excites resonances.

~~Accessory equipment. Here, the rotational speeds have a tendency to be increasing. Balancing is important not only for comfort, but also for service life. Frequently, the priority is to eliminate static imbalance. The following accessories should be considered:

~~~Alternator. High speed generator rotors are balanced today as single parts with tolerances of 10–20 gmm per plane (passenger cars).

~~~Cooling air fan. Fans in passenger cars driven by electric motors are balanced completely with tolerances of 20–30 gmm (priority static).

~~~Starter. High-speed electric armatures are balanced today as single parts with tolerances of 2–3 gmm (passenger cars, priority static).

~~~Turbochargers. Turbochargers are the most demanding element in terms of the balancing technology on internal combustion engines. Turbine and compressor wheels are balanced as single parts with tolerances of 0.2–0.5 gmm per plane (passenger cars). In some cases, corrections may still be necessary on the turbo-

charger as a whole due to the dynamic rotor characteristics at operating speed.

~~Complete engine. Balancing the complete engine was once a widely used technique. The theoretical advantage was that the actual total effect was considered including the effects resulting from eccentric assembly and dynamic deformation. However, the interchangeability of parts still requires "balancing" of the individual components, and today, production techniques concentrate on balancing the relevant individual parts.

~~Connecting rod. The rotating or reciprocating masses are divided up into partial masses for purposes of classification. The reference planes pass through the center of the bores. The components at the "large end" multiplied by the crankshaft radius act as imbalances on the crankshaft. Components at the "small end" must be added to the reciprocating masses. As with the out-of-balance masses, it is necessary to differentiate here between random and systemic errors. Systemic errors are permissible in engines with particular crank-pin arrangements (inline 4, inline 6, V-12; here, only the mass equality of all connecting rods in the engine is considered). Such systemic errors are not permissible in any other types of engine.

The random scatter of the two partial masses correlate (**Fig. B11**), and the scatter field is an inclined ellipse.

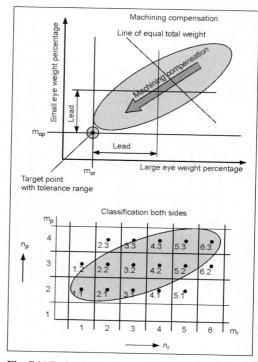

Fig. B11 Reduction of scatter in the partial masses of connecting rods

Correction of the masses by milling requires additional material to be available on the component. Classification leads very quickly to a large number of classes due to the two dimensions; however, the remaining scatter width is greater than the milling compensation. Because minimizing the second-order reciprocating forces has top priority on engines with dominating systemic errors, milling compensation is frequently accomplished here only on the large end.

The scatter of the masses for connecting rods depends on the production method (forging, casting, sintering) and decreases in the sequence specified. Today, sintered connecting rods generally do not require any correction; however, the quality of conventional milling compensation is not achieved.

~~**Crankshaft.** Balancing is accomplished in two planes for (normal) two-piece bearings. The selection of the bearing location is optimized in terms of static sag on long shafts. The speed is relatively low; the magnitude is approximately 1/10 of maximum operating speed. Material is removed by drilling radially into the counterweights and in special cases also by drilling out the crankpins. In the event of discrepancies between the effective direction of the imbalance and the direction of compensation possible, conversion for other web planes and/or subdivision of the force vectors into components is required. Optimization programs exist to minimize the removal of material.

On large crankshafts in which the counterweights are bolted on, coarse compensation of the imbalance is possible by replacing the counterweights.

The usual tolerances for balancing passenger car crankshafts are between 50 and 200 gmm per plane (in relation to outer main bearing or counterweights); for commercial vehicle crankshafts they are between 300 and 2000 gmm.

Crankshafts with asymmetrical arrangement of crankpins (e.g., inline 3, inline 5, V-6, V-8) must either have substitute masses to simulate the effect of the crank drives installed later or their resulting effect must be simulated on special balancing machines by mechanical or electrical means (**Figs. B10, B12**).

The latter widely used method has the advantage of rationality but the disadvantage of ignoring the actual position of the crankpin. Individual measurement of the geometry can eliminate this disadvantage.

~~**Flywheel.** Normally, balancing in one plane is sufficient because the flywheel is mounted on the long crankshaft, which provides stability in the other plane. It is necessary to take into consideration the fact that play and eccentricity of the seat on the crankshaft multiplied by the mass of the flywheel (and clutch) results in additional imbalance, which becomes apparent only after assembly. However, balancing of the complete crankshaft and flywheel assembly based on these considerations is no longer common practice today.

Measurement of the imbalance is accomplished on machines with vertical spindles, the upper end of

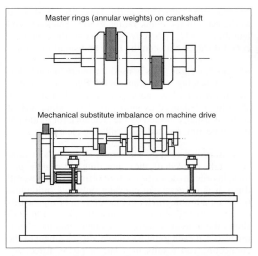

Master rings (annular weights) on crankshaft

Mechanical substitute imbalance on machine drive

Fig. B12 Consideration of connecting rod and piston masses

which hold a mount that clamps precisely into the bore of the flywheel. Compensation of the imbalance is accomplished by drilling parallel or normal to the axis. Here it is necessary to take into consideration restrictions on the strength of the remaining material. Regions with "prohibited" angles require that the compensation be split up into its components.

~~**Piston.** Great advances have been achieved recently in reducing the piston mass. The production-related scatter is approximately ±1%. Today, correction of the mass is no longer required: because of their highly accurate surfaces, piston pins and rings do not have any significant deviation in mass.

~~**Pulley.** Balancing tolerances are similar to those for flywheels. Due to the low mass of the pulley, the production-related imbalance and assembly-related offset imbalance are lower. Compensation of imbalance can be made by drilling, but when the pulley has the combined function of a rotary vibration damper, compensation on the outer part may have a different effect than compensation on the inner part.

Pulleys produced from sheet metal are usually balanced by welding or clamping on additional material.

~**Balancing of out-of-balanced couples.** If a resulting out-of-balanced couple remains in the engine, it can be reduced or eliminated by balancing. Because the balancing couple is a free vector, its position is not important as long as it is parallel to the couple responsible for the excitation.

~**Connecting rod motion diagram** (*also*, →Crankcase). This diagram defines the outer contour of the connecting rod along its path.

~Connecting rod/crank ratio. This ratio is defined as the ratio of the crank radius to the connecting rod length, $\lambda = r/l_{pl}$, whereby l_{pl} represents the center-to-center distance between the large and small rod ends.

~Counterweight. This is used for internally or externally balancing the crankshaft. External counterweights usually have additional material to allow the shaft to be balanced dynamically. The counterweight radius is usually limited by the clearance for motion of the connecting rod and the crankcase clearance.

~Crank drive system. The crank drive system consists of the crankshaft, the connecting rod, and the pistons. In order to describe the force effects in the overall drive, it is necessary to consider only one individual throw on a multicylinder engine and superimpose this on the overall system in relation to the phase. **Fig. B13** shows the individual force components and the resulting forces at each joint. As shown, displacement of the connecting rod results in a lateral piston force, F_N, which is taken up by the reaction force in the main bearing. The road force, F_S, results in forces in the radial and tangential direction. Radial components result in loads to the components only; they must be taken up by the main bearings or adjacent throws.

The tangential force, F_T, multiplied by the crank radius results in the instantaneous torque. The reciprocating mass force is periodic and has the first- and second-order components. The resulting tangential force supplies an alternating torque.

~Crank drive system with articulated connecting rod. If the cylinders in V- or radial engines are located

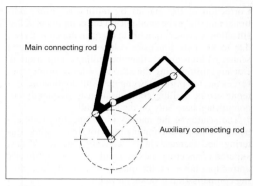

Fig. B14 Crank drive system with articulated connecting rod

in the same effective plane, it is necessary to use a forked connecting rod (the secondary rod connects with a fork placed over primary connecting rod) or an articulated connecting rod (**Fig. B14**). The articulated connecting rod is subject to kinematics different from those of the main connecting rod. Due to modern space requirements and bearing problems for the auxiliary connecting rod, this design is hardly used anymore.

~Crank sequence. The crank sequence is laid out according to the following three criteria:

• Minimum free mass effects to obtain maximum smoothness
• Firing order favorable for charging the cylinders
• Harmonic main bearing load and minimum excitation of torsional vibrations

~Crank set. A crank set of the first order can be assessed by looking on the front of the crankshaft and drawing the positions of the individual crank throws. If a force vector is then drawn for each throw, it is possible to determine directly whether the free mass forces of the corresponding order are present and how they can be compensated. **Fig. B15** shows that the first-order forces are balanced on all shafts. The out-of-balance forces for higher orders can be generated by multiplying the throw angle by the corresponding order. The second-order out-of-balance force diagram for a four-cylinder engine shows that the forces all act in the same direction. This results in the familiar comfort problems with four-cylinder engines.

~Firing order. The firing order determines the crankshaft geometry and, therefore, the out-of-balance inertia forces. With the same crankshaft geometry, it is necessary to check how different firing orders affect the out-of-balance forces and couples on the valve train.

~Flywheel. Rotational energy is stored in the flywheel, according to $E_{rot.} = 1/2 \cdot \dot{\varphi}^2 \cdot \Theta_{tot.}$. This energy is stored when starting up and compensates for the compression and combustion strokes in the near-idle speed range. The

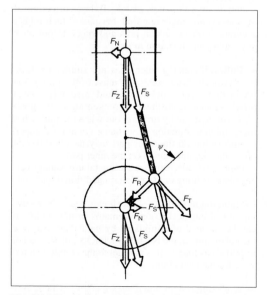

Fig. B13 Breakdown of forces on crank drive system (Source: Mass/Klier)

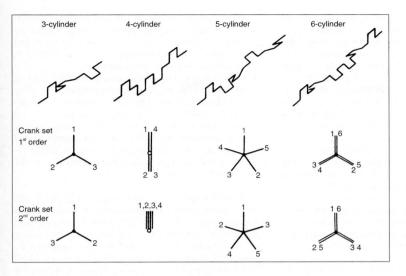

Fig. B15
First- and second-order slider-crank kinematic diagrams

torque curve at the crankshaft is determined by the reciprocating masses and gas force. The change in the angular velocity ($\dot{\varphi}$) of the crankshaft depends on the moment of inertia (Θ) of the entire engine and the applied torque (M) according to the equation $\ddot{\varphi} = M/\Theta_{tot.}$ The flywheel itself usually represents the greatest percentage of the moment of inertia in the crank drive (80–90%).

~**Gas force.** The cylinder pressure acts on the crankshaft through the piston surface. A torque is developed only when the crankpin is not located at top dead center—that is, when a tangential component is present at the crankshaft.

~**Gas pressure curve.** This has no influence on the out-of-balance effects. It affects only the alternating torque on the crankshaft. Because this torque is also taken up by the engine mounts, its effect is similar to that of out-of-balance excitation. However, because

there is no direct relationship to the speed of the engine—that is, it does not increase with engine speed—it acts primarily at low speeds.

~**Ideal balancing.** First- and second-order balance shafts rotating in opposite directions are required to completely balance out the first- and second-order reciprocating inertia forces on a single cylinder. **Fig. B16** shows such an arrangement. Extremely smooth operation of a single-cylinder engine can be achieved with this approach. However, the lower order (0.5, 1, 1.5, etc.) alternating torques are still present (from the gas forces) and their effect can only be compensated by a large flywheel. Such designs are used practically only in tests and research engines.

~**Ignition (firing) interval.** Modern engines have a constant firing interval. This provides the greatest uniformity of rotational speed for a given number of cylinders.

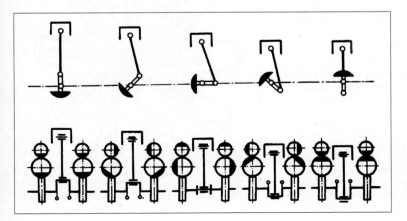

Fig. B16
Perfect first- and second-order balancing on a single-cylinder engine (Source: Maas/Klier)

37

~Inertia. A mass (m) always attempts to continue its current state of motion. This has positive as well as negative aspects. According to the equation $F = m\ddot{s}$, this means forces (F) occur only where there is an acceleration ($\ddot{s}$). An ideally resilient mounted engine block with rotating crank drive (without torque output) has an overall center of gravity that is at equilibrium. If this crank drive has unbalanced forces, they are transferred to the engine block and move it. The magnitude of the movement of the block is dependent on its mass in this case.

The moment of inertia of the flywheel is used to increase the radial uniformity of the engine drive. The moment of inertia (Θ) of a point mass (m) about its axis of rotation (r) can be calculated with $\Theta = mr^2$.

Because the moment of inertia increases with the square of the distance from the axis of rotation, an attempt is always made to design the diameter of the flywheel as large as possible to achieve the maximum moment of inertia with the lowest possible mass.

~Inertia forces. A force must be present to change the state of motion of a body. The reciprocating masses in a crank drive continuously change speed and ultimately, direction: this motion causes fluctuating forces which are transferred to the block structure through the main bearings. If the inertia forces of the individual throws are not compensated, the block is also accelerated.

Balancing of inertia forces: The composite center of gravity can be kept steady by the design of the crankshaft, the layout of the counterweights and, where necessary, attachment of external balancing mechanisms.

~~Horizontally opposed engine. Horizontally opposed engines are a special form of V-engine with a V-angle of 180°. They have one crankpin per connecting rod and are frequently designed as four- and six-cylinder engines. The slider-crank kinematic diagram shows that the free mass forces are balanced.

On four-cylinder engines, the minimum remaining effects are the second-order reciprocating moments due to the offset of the opposing cylinders; on six-cylinder engines, the first- and second-order forces are balanced.

~~Inline engine. The compact design of the four-cylinder engine and simple and low production costs have made this type of engine one of the primary propulsion sources for motor vehicles.

- Three-cylinder: The three-cylinder version is used in passenger cars with low displacements to achieve reduced levels of fuel consumption that result from the reduction in bearing and piston friction, as well as the more favorable volume of the individual cylinders. This arrangement generates first- and second-order out-of-balance couples. The cyclic variation of speed can be reduced by the flywheel moment of inertia.
- Four-cylinder: With four-cylinder engines, there are no out-of-balance couples; the first-order forces are also balanced. However, the second-order out-of-balance inertia forces for all the cylinders are in

phase and this results in considerable loss of comfort at higher engine speeds. Balance shafts are being increasingly used to compensate these and are capable of compensating the second-order out-of-balance inertia forces completely.

- Five-cylinder: These have usually resulted from an expansion of an engine in series production, by using a modular approach. First- and second-order out-of-balance couples remain uncompensated.
- Six-cylinder: On inline six-cylinder engines, the first- and second-order inertia forces are compensated completely without additional measures. The cyclical speed variation is also very small due to the number of cylinders.

~~Radial engine. The radial engine (**Fig. B17**) was developed to meet the requirements of aircraft manufacturers for a short multiple-cylinder engine with noncritical vibration characteristics and good cooling of all cylinders. Construction was accomplished by using articulated connecting rods—that is, all auxiliary connecting rods were joined to one main connecting rod. The auxiliary connecting rods have slightly different slider-crank dynamics from the main connecting rod. If this difference is neglected, radial engines have a complete first- and second-order compensation of the inertia forces.

~~Reciprocating mass force. This is caused by acceleration of the masses moving in a direction normal to the crankshaft axis.

~~Resultant inertia force. This is the total of all rotating and reciprocating inertia forces.

~~Rotating inertia force. This produces a rotating force or rotating moment.

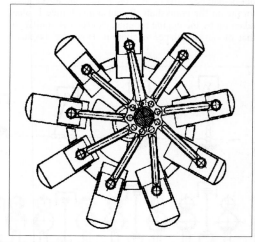

Fig. B17 Radial engine with main and auxiliary connecting rods (Source: Maas/Klier)

B

~~RV-6 engine. This design resulted from the desire to develop a multicylinder engine with extremely compact dimensions which ran very smoothly. In principle, this is a V-engine with extremely small V-angle (10–20°). The intersecting point for the cylinder axes is significantly below the crankshaft axis to maintain sufficient space between the cylinder barrels. This results in a double offset crank drive. The offset should be kept the smallest possible due to its negative effect on the lateral piston forces (**Fig. B18**).

As with the inline six-cylinder engine, there are no out-of-balance inertia forces; however, first- and second-order out-of-balance couples remain. The V-angle and offset affect these inertia effects. The smoothness is comparable to that of an inline six-cylinder engine due to the uniformity of the firing interval.

~~V-engine. Engines with more than six cylinders are arranged in a V shape more and more frequently because they can be constructed more compactly and have better torsional vibration characteristics. Common V-angles are 60° and 90°. Selection of the V-angle frequently depends on the family of engines on which the engine is based.

~~~60-degree V-engine. For each individual throw, the rotating forces can be compensated by the counterweight. The first-order out-of-balance forces result in superimposition of a vertical ellipse. The second-order forces result in a rotational force of constant value. This is the ideal configuration for a V-12 engine with 60-degree firing interval.

~~~90-degree V-engine. For each individual throw, the rotating forces can be compensated by the counterweight. The first-order reciprocating inertia forces from the two connecting rods result in a rotational force with constant value. Because the direction of rotation coincides with the crankshaft, these forces act as a rotating imbalance and can therefore be compensated with suitable counterweights. Superimposition of the second-order reciprocating inertia forces results in a reciprocating force of $\sqrt{2}F_{osc}$, II in the horizontal engine direction.

On engines such as the 90° V-8 engines, the second-order force is also compensated by the location of the throws (four throws on the crankshaft), thereby achieving almost perfect balance. This is the ideal configuration for a V-8 engine with a 90° firing interval.

~Internal bending moment. This is the result of uncompensated imbalance of an individual throw. No free force or moment acts outside the crankshaft; however, the shaft itself and the main bearings are subjected to high loads (**Fig. B19**). This can be reduced by internal balancing. The effect of reciprocating force from the connecting rod and piston cannot be compensated completely by balancing the crankshaft.

~Internal mass balancing (based on tests). Even when the external masses are perfectly balanced and the moving parts are produced to highest quality, the inertia forces on the crank drive result in bending moments in the crankshaft. Compensating crank drives as well as counterweights operate in offset planes. This results in bending deformation forces in the shaft; these are dependent on the engine speed. Even though they are taken up by the main bearings, the limited rigidity of the engine block transfers these reactions to the outside. The shaft cants in the bearings, and this is clearly shown by the displacement and inclination of the flywheel. This results in unbalanced forces, depending on the engine speed and, therefore, the inertia forces.

Fig. B20 shows test results for three different engine types with symmetrical crankshafts. The engine imbalance on the flywheel side is within close limits at low speeds thanks to the good production quality. The increase in imbalance at higher speeds is the result of the

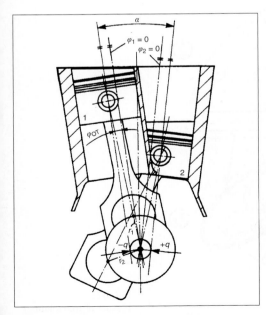

Fig. B18 RV arrangement (Source: *MTZ*)

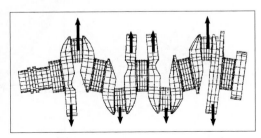

Fig. B19 Crankshaft end deformation resulting from centrifugal force as well as rotating and reciprocating mass forces

B

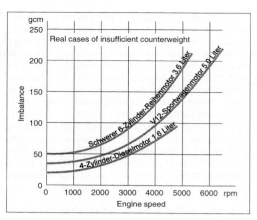

Fig. B20 Imbalance resulting from displacement of the flywheel and clutch

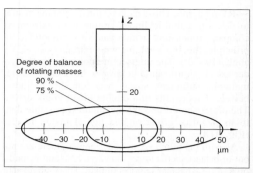

Fig. B21 Flexure of a six-cylinder inline engine at 5000 rpm measured amplitude of the first-order deflection in relation to variations in the degree of crankshaft balancing

design—that is, more or less avoidable internal bending moments. The dominant parameters here are the degree of compensation by the counterweights and the rigidity of the block. Good balancing in a six-cylinder engine can be negated by low rigidity. In the opposite case, the relatively poor balancing of a V-12 engine can be overshadowed by the higher rigidity of the block. Four-cylinder engines benefit from their short overall length.

Fig. B21 shows the primarily U-shaped flexure of an inline six-cylinder engine. The unbalanced reciprocating masses in the direction of the cylinder axes (z) result in high bending forces; however, these are absorbed by the high rigidity of the block in this direction. In the lateral axis, y, in which only relatively well bal-

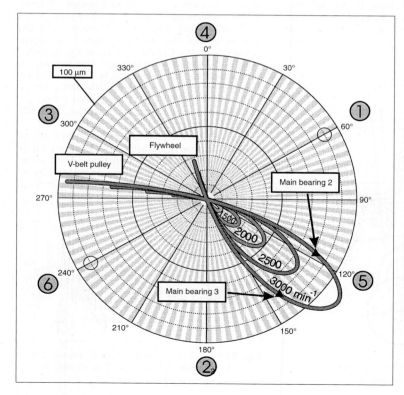

Fig. B22 Flexure of free rotating V-6 crankshaft

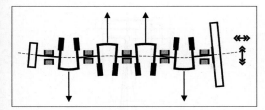

Fig. B23 Deformation of throw on four-cylinder crankshaft

anced rotating masses are effective, the lower rigidity of the block still leads to greater deformation. The anisotropy of the inline engine in regard to inertia forces and rigidities leads to the recommendation that the counterweights should be dimensioned so that they compensate the bending moments from the rotating inertia forces by 100% and the reciprocating inertia forces to approximately 10% of the forces. However, the rotating masses are usually balanced out only to 70–90% on inline engines actually in production, due to many other factors.

On asymmetrical crankshafts, an S-shaped flexure is superimposed on the primary U-shaped flexure in the lateral direction. **Fig. B22** shows an example (zone as projected view) of the typical intrinsic shape of a free-rotating V-6 crankshaft. The counterweights on such crankshafts are dimensioned primarily for external balancing. Internal bending moments are taken up relatively well by the stiff block on V-engines, and the slight outer deformation obscures the high bearing load.

At higher engine speeds, the primary shaft flexure is subject to deformation dominated by the individual crank throws (**Fig. B23**). This leads to edge loads in the bearings and longitudinal vibrations in the crankshaft.

Nearly all deformations resulting from errors in the internal balance are reflected by axial motion at the flywheel. **Fig. B24** shows the vectors for the first order tumbling motion at the flywheel on a four-cylinder engine in the speed range of 1000 to 7000 rpm. The reversal in direction at 3000 rpm is caused by the deformation of the throws dominating the motion at high speeds, as shown in **Fig. B23**.

~Lancaster harmonic balancer (first order). According to Lancaster, perfect first-order balancing can be obtained by using two shafts rotating in opposite directions with the connecting rods each connected to the same piston (**Fig. B25**). However, use of that design has remained insignificant. Today, the balancing mechanisms used in four-cylinder inline engines are called Lancaster harmonic balancers.

~Local curves. Periodic rotational forces are frequently illustrated in the form of a polar diagram— that is, a force vector is drawn corresponding to its effective direction and its magnitude (**Fig. B26**). Clarity is lost with a large number of force vectors, so only the ends of the vectors are connected, thereby obtaining the so-called local curve. The illustration shows the unbalanced force on one cylinder. The angles define the crank angle position at which the corresponding force occurs.

~Mass reduction. The reduction of the mass of the engine parts directly affects the size of the components. As a rule, a significant amount of weight can again be saved by adapting the balancing measures. This should also be attempted even on completely balanced crank drives, because the internal loads and, therefore, the vibration excitations also decrease.

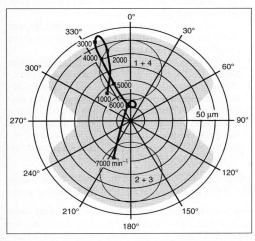

Fig. B24 Dynamic wobble of flywheel relative to engine block

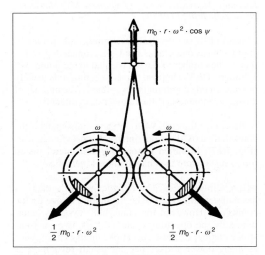

Fig. B25 Lancaster harmonic balancer, first-order with two connecting rods (Source: Mass/Klier)

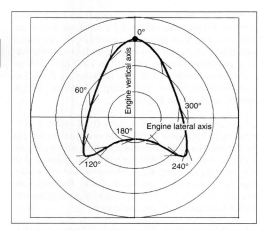

Fig. B26 Local curve of out-of-balance inertia force on single-cylinder engine

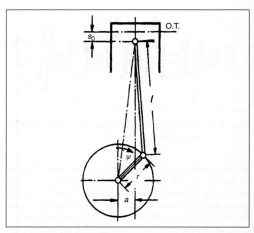

Fig. B27 Offset crank drive, offset between bore axis and crankshaft axis (offset a) (Source: Mass/Klier)

~**Number of cylinders.** Normally (on inline and V-engines), this is the dominating parameter for the out-of-balance inertia effects and the uniformity of rotational speed.

The number of cylinders alone does not provide any information on the quality of the balancing of the masses. It is important to consider the entire system. This becomes very clear by using a 90° V-8 engine as an example. In the normal version with four crank throws, the first- and second-order forces and couples are compensated. If a single plane crankshaft is installed in the same engine, high second-order inertia forces result in the horizontal direction. This design is used for racing engines, because it allows a uniform firing order on each bank of cylinders, providing advantages in cylinder charging.

Literature: Maas Klier/Springer Publishers: MTZ Motortechnische Zeitschrift 51 (1990).

~**Number of throws.** Inline engines and special V-engines have one crank throw per connecting rod—that is, the number of throws is equal to the number of cylinders. On V- engines, two connecting rods usually operate on one common throw—that is, the number of throws is equal to half the number of cylinders.

~**Offset crank drive.** With offset crankshafts, the crankshaft axis does not intersect the cylinder axes (**Fig. B27**). Modified piston rods and lateral piston forces result from the change in the kinematics.

~**Operating process.** The cylinder pressure diagram has a direct influence on the fluctuating torque on the crankshaft. However, the crankshaft layout is determined by the operating cycle—that is, two- or four-stroke engine. Because two-stroke engines have twice as many power strokes per crankshaft revolution as four-stroke engines with the same number of cylinders, they have a more uniform torque curve.

~**Order.** The frequency with which an event occurs in relation to one rotation of the crankshaft is known as the order. For example, with the slider-crank relationship, integral multiples of the crankshaft angle, φ, occur. The gas pressure curve for an individual cylinder has a period of 720° in relation to the 360° of the crankshaft, meaning that the gas pressure curve also has nonintegral orders, designated 0.5, 1.0, 1.5, and so on. Subdivision of the orders allows auxiliary measures to be easily established at the desired speed ratio. The balance shafts always have a constant gear ratio related to the crankshaft, usually first- or second-order. Balance shafts which reduce the second-order out-of-balance rotate at twice the speed of the crankshaft.

- *First order* compensating devices can be attached to the crankshaft when it is not necessary for them to rotate in the opposite direction.

  Example of reverse direction of rotation: Balance shafts for compensation of first-order out-of-balance couples on 90° V-6 engines.
- *Second order* compensating devices usually require higher gearing in the form of a chain or gear drive.
- *Higher orders* are not compensated, as a rule, because the higher orders normally have smaller amplitudes of excitation. At the same basic excitation amplitude, the effect decreases proportionally to the square of the order. Shafts with high-order compensation would have accordingly high speeds and this would lead to higher frictional losses.

~**Out-of-balance couples.** This results from inertia forces that are not compensated in the operating plane.

~~**Compensation.** On asymmetrical crankshafts such as on inline, three-cylinder, V-6, and V-8 engines, the location of the individual throws results in couples that can be compensated or reduced only by an asymmetrical arrangement of the counterweights.

B

~~Resulting out-of-balance couples. Depending on its development from first-, second-, or higher-order inertia forces, the resulting out-of-balance couple can have various orders. Considering the overall engine, all uncompensated force effects on the crank drive, valve train, and auxiliary equipment must be taken into consideration.

~Out-of-balance force effect. This is the resultant force of all force components in a dynamic system that is transmitted to the outside. It can occur in the form of forces and couples, and this depends on the engine design (inline or V-engine) and the selected crankshaft geometry. If the engine is bolted rigidly to a foundation in an ideal manner, this force is transferred to the foundation as a reaction force. If the engine is suspended in ideally resilient mounts, the effect of the out-of-balance forces is to make the entire system move. A familiar example is the second-order out-of-balance force in four-cylinder engines, resulting from the reciprocating masses causing the engine to vibrate in the direction of the cylinder centerlines.

~Piston acceleration. The second derivative of the slider-crank equation provides the piston acceleration. The following equation for piston acceleration shows only integral orders and indicates that the amplitude percentages of the higher orders decrease rapidly.

$$\ddot{x} = \frac{s_0}{r\omega^2} = A_1 \cdot cos\psi + A_2 \cdot cos2\psi \\ + A_4 \cdot cos \cdot 4\psi \\ + A_6\, cos6\psi$$

The coefficients correspond to the coefficients for the slider-crank relationship. The diagram shows the standard acceleration for a number of connecting rod ratios.

Fig. B28 shows that the ideal sinusoidal acceleration curve is approached with small connecting rod ratios. The magnitudes of higher order (second, fourth, etc.) forces decrease, but the crankcase height also increases. The connecting rod ratio results in higher piston accelerations at TDC and lower values at BDC; the force also increases in proportion to the accelerations.

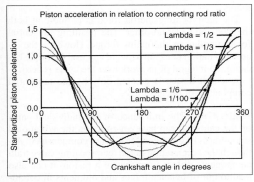

Piston acceleration in relation to connecting rod ratio

Lambda = 1/2
Lambda = 1/3
Lambda = 1/6
Lambda = 1/100

Standardized piston acceleration

Crankshaft angle in degrees

Fig. B28 Piston acceleration in relation to connecting rod ratio

~Reciprocating masses. All masses included in the slider-crank relationship are called reciprocating masses; these are normally the piston mass and the reciprocating proportion of the connecting rod mass.

~Reduced mass. The unbalanced mass and unbalance radius can be different for each crank web. This makes the overall calculation more difficult and the clarity suffers because information on the effect is reduced to only the product of the two values. This problem can be eliminated by introducing the so-called reduced mass for which the crank radius serves as a reference.

$$m_{red} = \frac{m_{web}\, r_{centerweb}}{r_{crankradius}}$$

~Residual imbalance. On engines in which complete balance is achieved only after installation of the piston rods and pistons (e,g., 90° V-8), it is necessary for the crankshaft to have a defined residual imbalance (reserve weight). Due to manufacturing tolerances, all crank drives have a residual imbalance in the installed state; however, this usually is in the range of only a few gcm.

~Rotating masses. Rotating masses are all masses which rotate around an axis with a constant out-of-balance radius. The rotating proportion of the connecting rod mass is also treated as such.

~Rotational nonuniformity. The crankshaft is continuously accelerated and decelerated due to the temporal variation in the excitation from the gas pressure and the effects of reciprocating masses of the connecting rod and piston. The resulting excitation force is taken up by the crankcase. The rotational nonuniformity, cyclic variation, is super-imposed on the crankcase rotation as rigid body motion. Therefore, at low speed the flywheel serves as an energy storage mechanism to ensure uniform smoothness of engine speed. On four-stroke engines, half the number of cylinders usually defines the primary order of speed fluctuations. The uniformity of the speed increases with an increasing number of cylinders, increasing engine speed and higher moment of inertia of the crankshaft drive.

~Scotch yoke engine. If the piston is not guided by a pivoting connecting rod as illustrated in **Fig. B29**, it is subject only to the sinusoidal motion of the crankpin.

~Slider-crank kinematics (*also*, →Crankshaft drive). The piston performs a reciprocating motion. **Fig. B30** shows the arrangement and dimensions. The piston travel depends on the crank angle, ψ, and connecting rod pivot angle, β. Because the connecting rod pivot angle is a function of the crank angle, it can be expressed in an equation containing only the angle, ψ, as a variable.

The following relationship exists between the crank angle and piston stroke for a simple crank drive mechanism.

$$s_0 = r \cdot cos\psi + l \cdot cos\beta = r + l$$

B

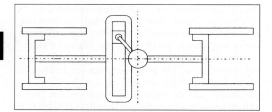

Fig. B29 Schematic representation of engine with Scotch yoke mechanism

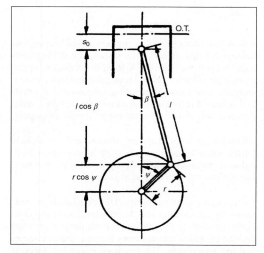

Fig. B30 Slider-crank mechanism (Source: Maas/Klier)

The connecting rod ratio is defined as $\lambda = r/l$. The following trigonometric relationship exists between the conrod pivot angle, β, the piston rod, and the crank angle, ψ:

$$\sin\beta = r/l \cdot \sin\psi = \lambda \cdot \sin\psi$$

and

$$\cos\beta = \sqrt{1 - \lambda^2 \cdot \sin^2\psi} \, ;$$

and the following relationship results for the piston travel, x:

$$x = \frac{s_o}{r} = 1 - \cos\psi + \frac{1}{\lambda} - \frac{1}{\lambda}\sqrt{1 - \lambda^2 \cdot \sin^2\psi} \, .$$

Series expansion results in the relationship between piston travel and the crank radius:

$$x = \frac{s_0}{r} = A_0 - A_1\cos\psi - \frac{A_2}{4}\cos2\psi$$
$$- \frac{A_4}{16}\cos4\psi - \frac{A_6}{36}\cos6\psi - \dots$$

where

$$A_0 = 1 + \frac{1}{4}\lambda + \frac{3}{64}\lambda^3 + \frac{5}{256}\lambda^5$$

$$A_1 = 1$$

$$A_2 = \lambda + \frac{1}{4}\lambda^3 + \frac{15}{128}\lambda^5 + \dots$$

$$A_4 = -\frac{1}{4}\lambda^3 - \frac{3}{16}\lambda^5$$

$$A_6 = \frac{9}{128}\lambda^5 + \dots$$

It is easy to see that the higher order-terms quickly lose significance.

~Substitute system. Frequently, a simplified arrangement of masses is used to give an easier description of the entire mechanical system. For example, a crank throw is divided up into the masses for the web, main journals, and connecting rod journals as well as the counterweights. The mass of the connecting rod is subdivided into the weight at the large and small ends. As an analogy to the primary directions of motion, it is usual to speak of reciprocating and rotating masses. The piston is, therefore, only a reciprocating mass. The masses and inertias of the individual elements as well as their acceleration are taken into consideration. The elliptical path of the center of gravity of the connecting rod is generally ignored because it has hardly any influence on the out-of-balance effects and only a marginal influence on the torque curve.

~Tangential force. The gas force curves for full load and idle are illustrated at the top in **Fig. B31**. The component of the pressure curve acting tangentially on the crankpin is shown in the center diagram for both load states. The lowest illustration shows the curve for the tangentially acting inertia force at low and high speeds, represented as a tangential inertia force for better comparability. The tangential component of force acting on the crankpin is always a superimposition of the two components. As can be seen, the tangential inertia force acts against the component of gas force.

The torque available at the crankshaft can be calculated by multiplying the tangential force by the crank radius.

The reciprocating inertia force is periodic at both the first- and second-order. The resulting tangential force supplies an alternating torque. However, the inertia excitation does not provide any usable torque because the integral of its work over the cycle is zero. A useful torque is created only by the tangential components of the gas force.

~Valve train. The imbalance of the individual cams on the camshaft can have an effect on the vibrational smoothness of the engine if it is not compensated. Although the acceleration of the reciprocating valve train components differs from that of a piston, it can also lead to out-of-balance forces and moments in combination with the valve train masses: balancing of the valve train is possible and practical.

Balancing of out-of-balance couples →Balancing of masses ~Balancing of out-of-balanced couples

Balancing the connecting rod →Balancing of masses ~Balancing

Balancing the pulley →Balancing of masses ~Balancing

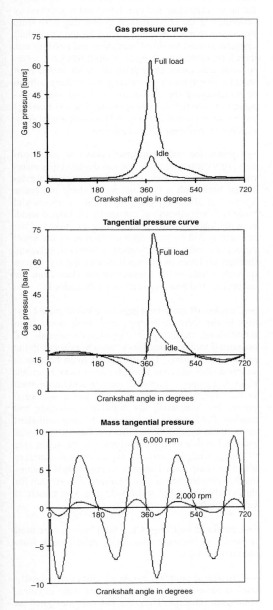

Fig. B31 Gas pressure and tangential force curve

BAP (barometric absolute pressure sensor) →Sensors ~Pressure sensors

Bare die →Electronic/mechanical engine and transmission control ~Electronic components

Basic gasoline →Fuel, gasoline engine

Basic liquids →Oil ~Safety and environmental aspects ~~Used oil

Basic oil →Oil ~Body

Battery. The battery is an electrical power supply (through storage, not generation) in the vehicle that is always available. Its tasks include the storage of electrical energy produced by the generator; to supply energy for the starter, the ignition system, and the electronic control of the engine; and to make electrical energy available to all of the vehicle's electrical system.

Many types of electrical storage have been considered but the only type accepted in vehicles today is the galvanic secondary cell in the form of a lead- (sulfuric) acid battery. The basic elements of this battery are the electrodes with a set of positive and negative plates per cell. It is based on plates (lead grids and active mass) with microporous insulation material between the plates of different polarity, and it is filled with an electrolyte, generally dilute sulfuric acid (H_2SO_4). The acid density is normally about 1.28 kg/L. A schematic description of the battery design can be seen in **Fig. B32**.

The nominal voltage for each cell of a lead battery is specified as 2V in accordance with the DIN standard. Therefore, six cells must be connected in series to achieve the required nominal voltage of 12V for starter batteries. The required voltage for trucks, which have a vehicle electrical system of 24V, is achieved by two 12V batteries that are connected in series. The voltage increases during charging and decreases when the battery is loaded. Lead oxide and lead are converted by the dilute sulfuric acid into lead sulfate when the battery gets discharged. The proportion of the sulfate ions and, therefore, the acid density in the electrolyte, then declines. The acid density can be used as a measure for the charge status.

The capacity of a battery is determined by the current volume in ampere hours (A·h) that can be delivered

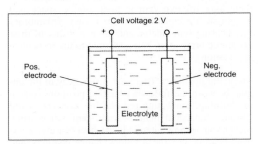

Fig. B32 Battery design

B

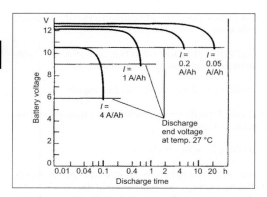

Fig. B33 Battery voltage curve (Source: Bosch)

under certain conditions. The nominal capacity is determined in accordance with EN 60095-1. It states that the electrical capacity in A·h can be used up with a predetermined discharge current in 20 hours until a discharge voltage of 10.5 V is achieved. The dischargeable current volume decreases as the discharge current increases and if the temperature decreases. **Fig. B33** shows the battery voltage curve depending on the discharge time for different discharge currents.

Literature: Bosch Fachwissen Kfz-Technik, Edition 2002.

Beading →Sealing systems ~Cylinder head gaskets

Bearing assembly →Engine acoustics

Bearing semi-shell →Bearings ~Design solutions

Bearing wear →Bearings ~Operational damage

Bearings. The various bearings in combustion engines serve to fix rotating or oscillating components to their location or to guide them on a prescribed path, whereby they have to absorb the impinging forces. In principle, both rolling bearings such as ball bearings or needle bearings and plain bearings are used. However, in practice, plain bearings are used in almost all applications. The main reasons for this are their good silencing and high-impact capacity, their low noise even at high speeds, their good limp-home qualities, their long life-expectancy, their simple space and weight-saving construction, their easy separability into bearing semi-shells, and the possibility of their economic production being tailored to the respective application.

~Bearing positions. There are differences in the engine between the following bearing positions: piston pin bushing, connecting rod bushing, connecting rod bearing, crankshaft main bearing, camshaft bearing, rocker arm bearing, intermediate wheel bearing, crosshead bearing, crosshead guides, flange bearings, and axial bearings (stop disks).

~~Camshaft bearings. Camshaft bearings are typically designed as bushings made from materials based on bronzes (copper alloys) or aluminum alloys or as bearing semi-shells made from materials based on bronzes.

~~Connecting rod bearings. Connecting rod bearings in the form of bearing semi-shells made from double-layer materials or triple-layer materials are linked to the connecting rod by press fit in the bore of the large connecting rod eye. Lubricant is supplied by crankpins in the crankshaft. A connecting bore from the adjacent main bearing journal to the crankshaft crankpin is used for this. Connecting rod bearings are, therefore, typically designed without lubricant holes. In some cases, however, there are holes in the shaft shell for supplying the connecting rod bushing with oil or for spraying the piston crown. Usual wall thicknesses of automobile engine connecting rod bearings are between 1.4 and 2.0 mm and for commercial vehicle engines between 1.8 and 3.0 mm.

~~Connecting rod bushings. Connecting rod bushings from copper alloys, mostly as double-layer materials and more rarely as single-layer materials, are designed both with a constant width along the entire length and also with changing widths ("trapezoidal bushing," "stage bushing"), whereby the largest width is in the load-supporting area. They are linked to the connecting rod by press fit into the bore of the small connecting rod eye. Lubrication is mostly by oil spray through oil holes in the small connecting rod eye or through a hole in the connecting rod shaft from the connecting rod bearings (pressure lubrication).

~~Crankshaft main bearings. Crankshaft main bearings in the form of bearing semi-shells made from double-layer materials or triple-layer materials are connected to the engine block through press fit in the main bearing bores. The crankshaft main bearing is supplied with lubricant through one or more lubricant holes in the upper crankshaft main bearing semi-shell in the engine block and typically has a rotational groove for lubricating the adjacent connecting rod bearing. The lower crankshaft main bearing semi-shell in the main bearing cover also can have a rotational groove, but that mostly is omitted to increase the size of the load-bearing surface. Typical wall thicknesses for crankshaft main bearings in automobile engines are 2–3 mm for thin-walled bearings, 3.0–3.5 mm for thick-walled bearings, and 2.5–3.5 mm in commercial vehicle engines.

~~Crosshead bearings. Crosshead bearings are used in the form of bearing semi-shells, mostly made from aluminum alloys.

~~Crosshead guides. Crosshead guides are mostly used in slow-rotating diesel engines (ship engines), stationary motors). They are ususally designed as slideways.

~~**Intermediate wheel bearings.** In control motors, these are designed as copper alloy or aluminum alloy bushings.

~~**Main bearings** →Bearings ~Bearing positions ~Crankshaft main bearing

~~**Piston pin bearings.** These are bearings for piston pins in the piston and the connecting rod (bushings).

~~**Piston pin bushing.** The piston pin bushing, also called simply "bushing," is used for high loads in the piston boss of high-performance diesel engines and is typically made of copper alloys.

~~**Rocker arm bearings.** They are bearings for rocker arms for controlling the engine's gas transfer. The design is mostly bushings made from copper alloys.

~**Computation.** The computation is the calculation of operating factors of the plain bearing tribological system, comprising the elements of bearing shell, lubricant, main journal, and bearing casing. The operating factors calculated for the critical operating conditions have to be compared with approximate operating values (thresholds calculated in tests, pragmatical values) and are checked for reliability to establish operating safety. The basis of the computation is the Reynolds differential equation, in which the flow of the lubricant in the lubricating joint between bearings and main journal, through which the motion equation is linked with the continuity condition (Navier-Stokes equation), is described. Improvements are being made to forecasting bearing function by expanding the computations from the simplified HD (hydrodynamic) method with absolutely rigidly assumed bearing surfaces to the EHD (elasto-hydrodynamic) method, taking into account operational mechanical elastic deformations of the bearing surfaces using the finite element method, or from the HD method to the TEHD (thermo-elasto-hydrodynamic) method, which allows for thermic operational deformations as well as mechanical deformations, and also by taking into account the effect of the lubricant viscosity within the lubricating joint as it changes temperature. The lengthy computation method, for which especially powerful computers are required, allows simulations of altered constructive designs of the engine components relevant to the bearings and provides useful results of the operating behavior, as a result of which the empirics can be restricted to a minimum.

~~**Bearing clearance.** Diameter difference between bearing bore and bearing journal.

~~**Displacement path.** The displacement path (precise journal displacement path) is the path on which the crankshaft crankpin in the connecting rod and main bearings moves as a result of the changing bearing loads and the reaction forces as a result of the hydrodynamics (**Fig. B34**). The distance of the displacement path from the bearing surface is a measure of the lubricant film thickness.

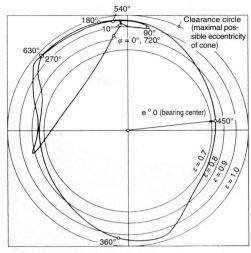

Fig. B34 Displacement path of the connecting rod journal in the connecting rod bearing

~~**Eccentricity.** Eccentricity is the center offset compared with the bearing bore axis resulting from operating forces.

~~**Lubricant film thickness.** This is the distance between bearing surface and journal surface during operation.

~~**Lubricating film pressure.** Lubricating film pressure is the locally and temporally changing oil pressure in the bearings lubricant gap that is a result of the pressure field created hydrodynamically through turning and radial movement of the journal (**Fig. B35**).

~~**Operational characteristics.** The most important operational characteristics for evaluating the operating safety relate to the specific load with the risk of fatigue, the lubricant film thickness with the risk of wear and tear, and the operating temperature of the bearings with the risk of overheating. Other operational characteristics are friction loss performance and the required lubricant throughput.

~~**Sommerfeld number.** The Sommerfeld number is a similarity factor in hydrodynamics, also known as load rating, which is defined as

$$So = \frac{p \cdot \Psi^2}{\eta \cdot \omega}$$

where p = specific load, Ψ = relative bearing clearance (bearing clearance divided by the internal bearings diameter), η = dynamic lubricant viscosity, and ω = journal speed.

~~**Specific load, p.** This is defined as bearing load force, F, divided by the projected bearing area (supporting width, B, multiplied by internal diameter, D):

B

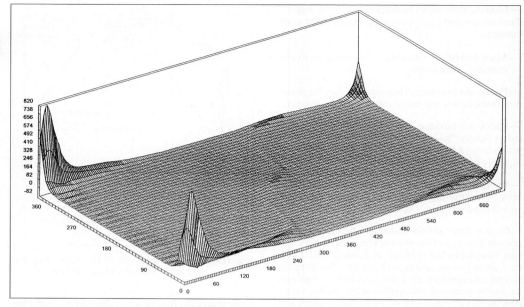

Fig. B35 Lubricant film pressure in a connecting rod bearing

$$p = \frac{F}{B \cdot D}$$

~Design solutions. Plain bearings in combustion engines are designed according to the bearing location as radial bearings, axial bearings, or flanged bearings. Overwhelmingly, thin-walled bearings as 180° bearing semi-shells (connecting rod bearings, crankshaft main bearing) or 360° bearings (casings) made from composite materials are used whereby a steel band laminated with different materials is typically used. When light metal casings are used, thick-walled bearings can be advantageous (crankshaft main bearings). Various lubrication elements are used.

~~**Axial bearings.** Plain bearings as axial bearings are mostly used in the form of stop disks or flanges of flanged bearings, which serve to absorb axial forces— for example, axial guide of the crankshaft or absorption of the clutch pressure (**Fig. B36**).

~~**Bearing semi-shell.** Bearing semi-shells (**Fig. B37**) are 180° plain bearings (made from a blank). In engine plain bearings, two bearing semi-shells always form the complete bearing (e.g., crankshaft main bearing, connecting rod bearing).

~~**Bushing.** Bushings (**Fig. B38**) are 360° plain bearings with (made from a blank, possibly with a linked pulse) or without pulse (pipe).

~~**Flanged bearings.** Flanged bearings (**Fig. B39**) are radial bearings as bearing semi-shells sometimes

with one flange, but more often with two, which can serve as axial bearings. They are, therefore, a combination of radial and axial bearings—for example, to guide the crankshaft or absorb clutch pressure.

~~~**Assembled flange bearing.** This is a flange bearing whose flange is assembled as stop disks (axial

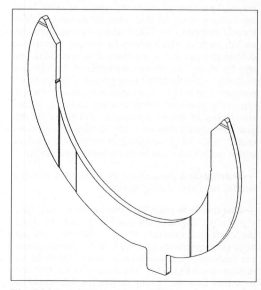

Fig. B36 Axial bearings (stop disk)

Wieland

Metall
ist unsere Welt

Quality

Performance

Solutions

Slide bearings and precision parts for the automotive and commercial vehicle industry.

Each Wieland bearing is especially designed for use in applications subjected to high mechanical load and thermal stress in modern engines, transmissions, axles and braking systems.

With an accredited production from pre-material casting until manufacturing of the finished machined part. All process steps are done according to latest QM standards and customer satisfaction.

Wieland-Werke AG
Slide Bearing Division
Graf-Arco-Str. 36
89079 Ulm, Germany
Fon: +49 (0) 7 31 9 44-0
Fax: +49 (0) 7 31 9 44-28 71
info@wieland.de
www.wieland.com

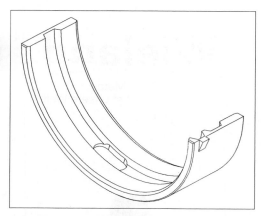

Fig. B37 Bearing semi-shells

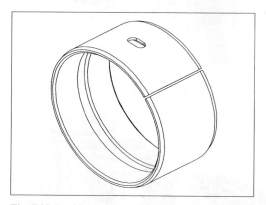

Fig. B38 Bushing

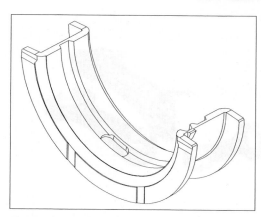

Fig. B39 Flanged bearings

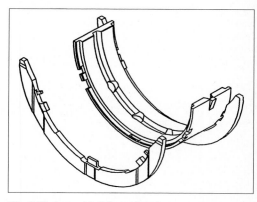

Fig. B40 Assembled flange bearings

bearings) and is assembled with the radial bearing (e.g., linked) (**Fig. B40**).

~~Lubricating elements. They serve to lubricate the bearings and can be in the form of a lubricating hole, groove, or pocket.

~~~Lubricating groove. This is an indentation in the bearing surface for distributing lubricant within the bearing surface and preferably runs in the rotational direction (**Fig. B37**).

~~~Lubricating hole. A lubricating hole is a hole in the plain bearing or journal through which the lubricant is fed to the bearing or forwarded to other lubrication points. Both round and straight (oval-shaped) holes are typical.

~~~Lubricating pocket. The lubricating pocket is an indentation in the bearing surface and acts as a lubricant reservoir.

~~Radial bearings. Radial bearings are used in the form of bushings or bearing semi-shells and serve to absorb radial (cross) forces.

~Friction loss conditions. In engine plain bearings, mixed friction or fluid friction can occur.

~~Fluid friction. This occurs with a perfect separation of the bearing surfaces from journal and bearings and is the desired optimum (low) friction condition without wear and tear.

~~Mixed friction. This occurs when the bearing surfaces of bearings and journal are not fully separated by lubricant during operation (lubricant film thickness too small), and the condition leads to wear.

~Function. Plain bearings in combustion engines work on the principle of hydrodynamics. Internal pressure developments occur in the lubricant as a result of turning and radial displacement of the main journal (or bearing casing) within the bearing clearance (in contrast to hydrostatic plain bearings where an external high-pressure pump generates the lubricant pressure), by which bearing surface and main journal surface are separated.

~~**Hydrodynamics** →Bearings ~Function

~Materials. The materials must be suitable for the operational loads placed on them when the engine is running. The crucial criterion, which makes a certain material a plain bearing material, is its compatibility with the sliding partner (e.g., crankshaft bearing surface) under the dominant operating conditions. The bearings must never scuff, as this mostly means the total failure of the bearing. Requirements of the plain bearing material are adaptability, break-in capability, single-bed capability, limp-home capability, wear resistance, scuff resistance, and fatigue resistance. Additionally, good heat conducting qualities and sufficient corrosion resistance and economical manufacturability are necessary. The different, partly contradictory requirements are best fulfilled by multilayer alloys with hard and soft components adapted to the respective application case.

For low loads, lead or tin alloys are used in whose soft matrix hard mixed crystals ("base crystals") with good compatibility, adaptability, and scuff resistance are embedded with stibium to improve the wear and fatigue resistance. For medium loads, aluminum alloys are frequently used, and for high loads copper alloys are used in whose relatively hard matrix soft components with good wear and fatigue resistance are embedded (tin in aluminum and lead in copper) to improve compatibility, adaptability, and scuff resistance.

As a result of increased environmental awareness and recent statutory requirements, lead is no longer supposed to be used as a bearing material. As lead has advantages both with respect of limp-home capabilities and scuff resistance, the challenge is to find a lead-free bearing material that satisfies these requirements.

~~**Aluminum alloys.** Aluminum alloys, mostly with tin and copper but also with lead (soon to be prohibited under law) and silicon or for improved loadability with manganese and nickel, have good adaptability, break-in capability, single-bed capability, limp-home capability, and corrosion resistance. Their wear and fatigue resistance is higher, because of their greater hardness compared to lead and tin alloys. However, they have a low scuff resistance. In automobile engines, they typically are used for crankshaft main bearings and as axial bearings (stop disks) and more rarely for connecting rod bearings. In commercial vehicle engines, aluminum alloys are used frequently for axial bearings (stop disks) and more rarely for crankshaft main bearings or connecting rod bearings.

~~**Bronzes.** Bronzes are alloys of copper with tin and/or lead and copper with aluminum. Lead as an alloy element will soon be prohibited in bronzes.

~~**Copper alloys.** Copper alloys, mostly with lead (lead will soon be banned as an alloy element) and tin, in some cases also with aluminum or zinc, have a moderate adaptability, break-in capability, single-bed capability, limp-home capability, corrosion resistance, and limited scuff resistance. Their wear resistance and fatigue resistance, however, are considerably greater because of their improved hardness compared to aluminum alloys. These advantages can only be fully exploited for applications with relatively low sliding speeds (e.g., in piston pin bushing or connecting rod bushing) because of the simultaneous drawbacks. With the sliding speed usual in crankshaft main bearings, their drawbacks have to be reduced with additionally fitted, very thin, more compatible sliding layers (triple-layer materials).

~~**Deep-groove ball bearing.** Deep-groove ball bearings are plain bearings made from a triple-layer material into whose second layer (copper or aluminum alloy) fine grooves are scored in the rotational direction, which are filled with a soft material (e.g., PbSnCu) so that the bearing surfaces next to each other have hard (second layer) and soft (groove filling) areas, which is supposed to produce a compromise between compatibility and wear resistance. This type of bearing is still used occasionally in medium-speed engines and large engines, but they are rarely used in commercial vehicle engines and not at all in automobile engines because of their insufficient life expectancy.

~~**Double-layer materials.** Double-layer materials can be created by cast cladding, sinter cladding, or roll cladding steel bands. With cast cladding, the bearing alloy (lead, tin, or copper alloy) is cast onto steel as a molten mass. A representative of cast clad double-layer materials used frequently in combustion engines is lead-tin bronze $CuPb_{10}Sn_{10}$ on steel (**Fig. B41**), which is mostly used for connecting rod bushing, piston pin bushing, rocker arm bushing, or camshaft bushing. Tin bronzes, aluminum bronzes, or brass are also used for these bushings now in order to avoid lead. Lead or tin

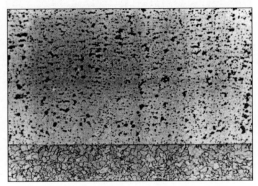

Fig. B41 Lead-tin-bronze $CuPb_{10}Sn_{10}$ on steel

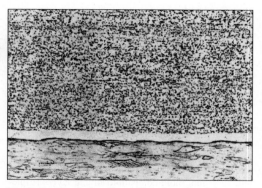

Fig. B42 Aluminum alloy $AlSn_{20}Cu$ on steel

alloys are only seldom used, and then mostly for camshaft bushing.

With sinter cladding, the alloy (copper alloys) is sintered onto the steel as sintered powder through heating. With roll cladding, the bearing alloy is rolled onto steel (through cast aluminum alloy rolled into bands). The most common roll clad double-layer material in combustion engines is the aluminum alloy $AlSn_{20}Cu$ on steel (**Fig. B42**), which is often used for crankshaft main bearings or stop disks and sometimes as connecting rod bearings.

~~Lead alloys. Lead alloys, mostly with stibium (**Fig. B43**) have an especially good adaptability, break-in capability, limp-home capability, and scuff resistance. However, because of their limited hardness, their wear and fatigue resistance are only moderate. They are used in some commercial vehicle engines for camshaft bearings. Their continued use is not permitted as a result of statutory requirements.

~~PVD process. PVD or physical vapor deposition is a laminating process by means of which the laminating material is physically (kinetic energy, impact energy) forced onto the material being laminated under vacuum conditions. The advantage compared with normal galvanic layers lies in the greater variety of materials that can be used (sputter bearings).

~~Qualities. The main qualities of bearing materials must be adaptability, single-bed capability, break-in capability, fatigue resistance, scuff resistance, corrosion resistance, limp-home capability, wear resistance, and compatibility.

~~~Adaptability. This describes the ability of the bearing material to compensate for initial geometric faults in the bearing surfaces through deformations or break in wear (break-in capability).

~~~Break-in capability. Break-in capability is the capability of the bearing material to reduce friction, wear, and heating during the break-in process (adaptability).

~~~Compatibility. This defines the ability of a bearing material to function safely with the sliding partner, which is determined, in particular, through good qualities with respect to adaptability, break-in capability, and wear resistance.

~~~Corrosion resistance. The ability of a material to be damaged as little as possible by aggressive media (e.g., in the engine oil, is called corrosion resistance).

~~~Fatigue resistance. This is the ability of a material to bear loads occurring as a result of dynamic forces without fatigue damage. The dynamic load in plain bearings in combustion engines is a pulsating compression load, which causes pull-push alternating tensions and push tensions in the bearing material.

~~~Limp-home capability. This is the ability of a bearing material to maintain bearing function even with insufficient lubrication, at least for a certain period.

~~~Scuff resistance. Scuff resistance is a material's resistance against scuffing. Bearing scuffing caused by formation of connections with the sliding partner as a result of adhesion or reaction, notably with insufficient lubrication, can lead to total damage.

~~~Single-bed capability. This is the capability of a bearing material to embed hard particles into the bearing surface, whereby wear and the risk of scuffing is reduced.

~~~Wear resistance. The ability of a material to keep levels of abrasion (wear) during operation as low as possible is called wear resistance.

~~Single-layer materials. Single-layer materials are rarely used in combustion engines for plain bearings. In some cases, they are used for connecting rod bushings or piston bushings and are made from single-layer copper alloys. In the overwhelming number of all

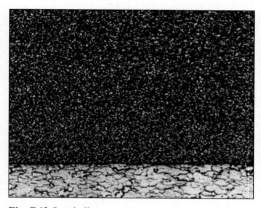

Fig. B43 Lead alloy

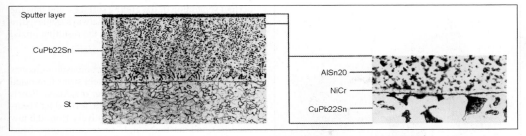

Fig. B44 Sputter bearings: copper alloy $CuPb_{22}Sn$ on steel with a PVD third layer of $AlSn_{20}$

cases, however, only multilayer composite materials are used because of their greater strength.

~~Sputter bearings. Sputter bearings are high-performance bearings, preferred in diesel engines with a bearing surface produced using the PVD process of, for instance, $AlSn_{20}$ (**Fig. B44**).

~~Tin alloys. Tin alloys, mostly with stibium and copper, have particularly good adaptability, break-in capability, single-bed capability, limp-home capability, scuff resistance, and corrosion resistance. However, because of their limited hardness, their wear and fatigue resistance are only moderate. They are not suitable for higher temperatures either, because of their lower melting point. Because of their drawbacks, they are rarely used in modern combustion engines, and then mostly as camshaft bearings.

~~Triple-layer materials. Triple-layer materials are used if double-layer materials are no longer suitable for the requirements. This is especially the case with hard bearing alloys for connecting rod bearings or crankshaft main bearings—for example, copper alloys— and more rarely with aluminum alloys, which are used because of their higher fatigue and wear resistance, but whose adaptability, break-in capability, limp-home capability, and scuff resistance are not sufficiently safe because of their high sliding speeds. To create triple-layer materials, a third bearing material layer is applied to double-layer materials, either galvanically or with the PVD (physical vapor deposition) method ("third layer").

The most frequent representative of the triple-layer materials with a galvanic layer found in combustion engines is the copper alloy $CuPb_{22}Sn$ on steel with a third layer of PbSnCu (**Fig. B45**). Newer triple-layer materials are made without lead—for example, $CuSn_8$ with a third layer of $SnCu_6$.

Bearings with PVD third layers are called "sputter bearings." The advantage of these compared with bearings with a galvanic third layer lies in the fact that the PVD layer has substantially greater fatigue resistance, wear resistance, and corrosion resistance while maintaining sufficient adaptability, break-in capability, limp-home capability, and scuff resistance. Sputter bearing semi-shells are typically only fitted as stan-

dard in the high load area of connecting rod bearings (shaft shells), although also occasionally as crankshaft main bearings (main bearing cover shells) in automobile and commercial vehicle diesel engines with high performance. The most frequent representative of the triple-layer materials with a sputter layer found in combustion engines is the copper alloy $CuPb_{22}Sn$ on steel with a third layer of $AlSn_{20}$ (**Fig B44**). Lead-free bronzes are also now being used here.

~Operational damage. Operating damage can occur as a result of wear and tear, fatigue, corrosion, overheating, cavitation, and scuffing if the corresponding qualities of the materials are not sufficient.

~~Cavitation. Lower pressure zones occur in the lubricant as a result of unfavorable lubricant flows or high-frequency oscillations in the plain bearings. Through accelerating, the lubricant drops near the low-pressure zones in the lubricant layer; fine, deep holes can occur in the bearing lubricant, holes which frequently are sprouted extensively through the lubricant after longer periods.

~~Corrosion. This is caused by the chemical breakdown of the lubricant layer—for example, aggressive media in the lubricant.

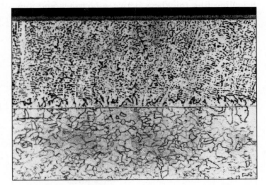

Fig. B45 Three-layer material: copper alloy $CuPb_{22}Sn$ on steel with a galvanized third layer of PbSnCu

B

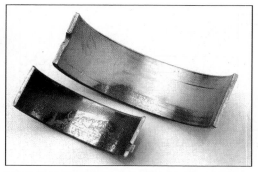

Fig. B46 Fatigue of the galvanic layer of triple-layer bearings

~~Fatigue. Fatigue occurs through cracking or crumbling of the lubricant layer if the material's fatigue resistance is exceeded (**Fig. B46**).

~~Overheating. High bearing temperatures lead to low-melting components of the lubricant material weakening, which can lead to total loss of the bearings.

~~Scuffing. Scuffing is adhesion between the bearing surface and main journal, frequently with the material tearing off from the surface until the journal jams in the bearing bore, which often leads to total damage.

~~Wear. Wear is an abrasive, erosive, or adhesive material degradation from the bearing surface (**Fig. B47**).

~Operational loads. Plain bearings in combustion engines are mechanically subject to the gas forces, inertial forces, control forces, and initial forces there. The changing hydrodynamic lubricating film pressure leads to an increasingly occurring compression on the bearing surfaces, resulting in push-pull alternating loads within the lubricant layer, which can lead to fatigue. The thermal load results from the friction heat and heat from the combustion chambers. As the lubricating film thickness between bearings and journals is not sufficient in all operating conditions for a

perfect separation of the gliding surfaces, and thus the desired pure fluid friction is not always achieved, a load results from wear because of the resulting mixed friction.

~Production. Plain bearings for combustion engines are typically manufactured from steel strips laminated with specific materials (double-layer materials). Panels are punched out of the bimetal band, formed into bearing semi-shells or casings (alternatively from the raw material), and machine cut. In some cases, other bearing material layers are created through galvanization or through the PVD (physical vapor deposition) method. In rarer cases, bearings also are produced from solid materials (without steel ridges).

Beat →Engine acoustics

Beauty cover →Sealing systems ~Modules

Beehive spring →Valve spring ~Design types

Belt →Toothed belt

Belt drive calculation →Calculation processes ~Application areas ~~Engine mechanics ~~~Belt drive

Belt drive rotary oscillations →Torsional vibrations ~Belt drive

Belt tension →Toothed belt drive ~Belt tensioning systems

Benchmark (Crank angle and TDC markers; *also*, →Sensors). The benchmark delivers a signal once for each rotation of the crankshaft and, therefore, provides the angle of the crankshaft. This is required to provide the engine control system with a clear correlation—for example, for the ignition or the injection impulse—with respect to a cylinder. An additional benchmark from the camshaft provides the correlation to the correct stroke of a four-stroke engine.

Benzene →Fuel, gasoline engine

Bifurcated front pipe →Exhaust system

Bimetal spring →Carburetor ~Design types, ~Starting systems;

Bimetal valve →Valve ~Gas transfer valves ~~Valve designs

Binary O_2 sensor probe →Sensors ~Lambda sensors

Binary O_2 sensor probe interface →Electronic/mechanical engine and transmission control ~Electronic components

Biogas →Fuel, diesel engine ~Biogas

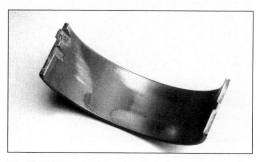

Fig. B47 Wear on the galvanic layer of a triple-layer bearing

Biological decomposition, oil →Oil ~Safety and environmental aspects ~~Used oil ~~~Biodegradability

Biological fuel →Fuel, diesel engine ~Alternative fuels

Biomass (*also*, →Fuel, gasoline engine ~Fuels containing ethanol). All materials of organic origin are part of the biomass. They originate in animals, in plants, or in their wastes or residual materials. The annual natural biomass production is very high; however, only a small part (about 10%) can be utilized. **Fig. B48** shows the distribution of biomass on the earth. It is generated by photosynthesis, which splits water in organic cells by using the radiation of the sun in the range of visible light by means of pigment molecules, in particular chlorophyll. Hydrogen is generated by this photolysis. The hydrogen, in combination with the carbon dioxide in the air, generates the biomass. Molecular oxygen is released during this process. The annual growth of the biomass generated this way is approximately 10% of the total.

The achievable energy potential from biomass is theoretically five to six times higher than the world energy demand. It is estimated that only 1% of the worldwide biomass production is used for energy purposes. This would represent approximately 10% of the primary energy consumption. Germany could produce approximately 3–12 million tCE/annum (tCE = tons coal equivalent) as primary energy through biomass (by using surplus areas). This is approximately equivalent to the potential biomass from residue and waste.

Important products for the production of biomass are, for example, oil seed rape, sugar cane, corn, and sugar beets. The efficiency rate is 3–6%. Synthetic gasoline, diesel fuel, methyl alcohol, ethyl alcohol, methane, and hydrogen can be produced from these raw materials.

Pure alcohols, mixed fuels (e.g., gasoline/alcohol and diesel/alcohol), and vegetable oil can be produced from biomass for the operation of gasoline and diesel

Raw Material	Ethanol Yield liter/ton	Biomass Yield ton/hectare	Specific liter/ hectare/year
Cane sugar	70	50	3500
Manioc	180	12	2150
Corn	370	6	2200
Sugar millet	83	34	3000

Fig. B49 Alcohol yield and hectare yields (Source: Heber et al.)

engines. The following biomasses are suitable in principle, if alcohol production by fermentation is taken into account:

- Biomass with sugar content, such as sugar cane, molasses, sugar beets, sugar millet
- Biomass with cellulose content, such as organic waste, wood, straw
- Biomass with starch content, such as potatoes, grain, manioc

The individual raw materials deliver different alcohol yields. **Fig. B49** shows, for example, the ethanol yield for different raw materials.

Literature: M. Kleemann, M. Meliß: Regenerative Energiequellen, Berlin/Heidelberg/Tokyo, Springer Publishers, 1988. — M. Meliß: Biomassepotenzial: Conference Energie aus Biomasse. — Erfahrungen mit verschiedenen technischen Lösungen und Zukunftsaussichten, TU Munich, Nov. 1985. Bine: Informationspaket Energie aus Biomasse, Karlsruhe 1981. — G. Heber et al.: Nutzungsmöglichkeiten alternativer Kraftstoffe in Entwicklungsländern, GTZ No. 153, 1983.

Bitter compound →Coolant

Bi-turbo →Supercharging

Area, millions of square meters		Net Primary Productivity billions of t/a	Calorific Value MJ/kg	Annual Energy Equivalent		Yield of Solar Radiation %
				kWh/square meter	billions of Mwh	
Forests	50	64	18	6.5	322	0.55
Forrest land	7	4	16.7	3.3	23	0.30
Brush	26	2.4	18.8	0.5	12	0.04
Grassland	24	15	17.6	2.9	70	0.30
Desert	24	—	16.7	—	—	—
Cultivated land	14	9	17.2	3.1	44	0.30
Freshwater	4	5	18	6.3	25	0.50
Land total	149	100	18*	3.4*	496	0.3*
Oceans	361	55	18.8	0.8	303	0.07
Earth	510	155	18.4*	1.6*	799	0.14*
* weighted average						

Fig. B48 Productivity distribution of biomass on the surface of the earth (Source: Lieth/Whittaker)

Black sludge →Oil ~Power during operation ~~Aging of oil

Black smoke. Black smoke is the visible part of the soot in engine exhaust gas that gives the exhaust gas its opacity. Black smoke occurs especially with diesel engines, when there is not sufficient air for complete combustion. In diesel engines with exhaust gas turbocharging, the turbocharger needs a certain amount of time to accelerate up to its steady state speed, and if the amount of fuel injected is not boost (charge) pressure limited, black smoke can be emitted during acceleration because of an insufficient supply of air.

Blank production →Piston

Blended fuels →Fuel, gasoline engine ~Methanol fuel

Blind hole nozzle →Injection valves ~Diesel engine ~~Hole-type nozzle

Block diagram →Electronic/mechanical engine and transmission control ~Electronic components

Block height →Crankcase ~Crankcase construction ~~Primary dimensions

Blowby (*also*, →Piston ~Blowby). Losses due to blowby are primarily caused by so-called leakage losses between the piston and cylinder liner. These result in a reduction of the effective usable work in the cylinder and, therefore, a reduction in efficiency. Modern pistons are normally equipped with two compression rings and one oil scraper ring to minimize blowby losses.

The piston rings ensure the necessary level of sealing through their geometrical design and contact pressure. The contact pressure must not be too high or the friction losses would increase by too much. Additionally, the cylinder liner must not be unduly deformed by the impact forces from the piston or by the temperatures. The blowby losses are increased by ring and liner wear. The blowby gases, with their water content and the unburned hydrocarbons, mix with the oil in the crankcase, and this results in lubrication oil dilution. Increased oil consumption can also occur due to the recirculation of the blowby gases in the intake tract (crankcase ventilation) of the engine. The potential negative impact on the catalytic converter involves poisoning due to oil additives.

Normal blowby volumes at full load and high engine speed are 30–50 liters per minute.

Blowby supply →Crankcase ventilation systems; →Intake system ~Intake manifold ~~Intake manifold design

Blowby system computation →Calculation processes ~Application areas ~~Fluid circulation

Blow-off valve →Supercharging; →Valve ~Exhaust gas control valves ~~Supercharging pressure control valve

Blue smoke. Blue smoke is engine exhaust gas that has a high content of partially burned or unburned engine oil.

Boiling characteristics →Fuel, diesel engine ~Properties ~~Latent heat of vaporization; →Fuel, gasoline engine ~Boiling characteristics ~Volatility

Bolts →Engine bolts

Bonding →Electronic/mechanical engine and transmission control ~Electronic components ~~Bare die

Boost function →Starter ~Starter-generators ~~Operating modes of the starter-generator

Bore →Bore-stroke ratio; →Displacement; →Crankcase ~Crankcase design ~~Cylinder

Bore diameter →Crankcase ~Crankcase design

Bore wear at top ring TDC →Cylinder running surface ~Wear

Bore-stroke ratio. The bore-stroke ratio is the quotient of the stroke and the bore diameter. The bore-stroke ratio and the cylinder volume have an impact on the exhaust emissions and the fuel consumption of an engine. **Fig. B50** shows how the relative hydrocarbon (HC) emissions in a gasoline engine depend on these parameters. The HC emissions and the part-load fuel consumption decline as the stroke of the engine

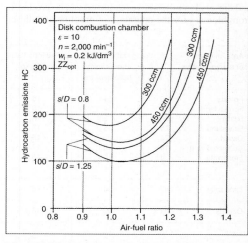

Fig. B50 Hydrocarbon emission as a function of air-fuel ratio, showing the effect of bore-stroke ratio (Source: B. Gand)

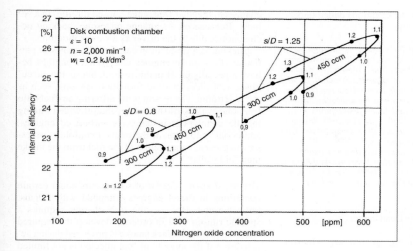

Fig. B51
Efficiency and emissions
of nitrogen oxides
(Source: B. Gand)

increases (with the bore remaining constant). This correlation cannot be freely applied, because, for automobiles use, only small variations can be allowed in
criteria such as inertia forces, combustion chamber design, existing production facilities, and so on.

The variation of surface-volume ratio can be influenced by varying the bore-stroke ratio. This ratio or
the combustion chamber surface is larger for short-
stroke engines than for long-stroke engines, which
means it is less favorable. Also important are the compression ratio, the change in gas temperatures at the
end of the combustion, and, therefore, the secondary
reaction of hydrocarbons.

Fig. B51 shows the relationship between concentrations of oxides of nitrogen and the different parameters
for a gasoline engine. The long-stroke engine generally offers a large potential for improvements in efficiency, especially when used in connection with four-
valve technology. The reason for this is that the efficiency increases when the compression increases, and
this can be higher for engines with a long stroke than
for engines with a short stroke. The explanation for
this effect is that the engine with a short stroke has an
unfavorable combustion chamber shape.

As expected, the engine with a long stroke has disadvantages with respect to concentrations of NO_x because the combustion temperature is higher due to the
compact combustion chamber. A larger displacement
increases the concentration of nitric oxides in the exhaust gas. This is in contrast to the characteristics for
unburned hydrocarbons.

Many analyses have shown that the optimum cylinder displacement with respect to the fuel consumption
in gasoline engines is 450–500 cm³.

Fig. B52 shows the relationship of the above-
mentioned variables to the indicated efficiency and the
air-fuel ratio of a gasoline engine. Engines with long
strokes also have advantages with respect to lean-burn
capabilities.

The higher indicated efficiency results in higher

torques at lower speeds, and this can be translated into
a lower vehicle fuel consumption because of a larger
transmission ratio.

The efficiency of diesel engines increases with increasing displacement, and the bore-stroke ratio should
be as large as possible for thermodynamic reasons.
Low-speed large-capacity four-stroke diesel engines
have bore-stroke ratios up to 1.5, while large-capacity
two-stroke engines have a ratio of more than three.
However, engines with long strokes have acoustic disadvantages, and their highest speed is below that of engines with shorter strokes.

~Oversquare. The bore-stroke ratio is called over-
square if it is less than one.

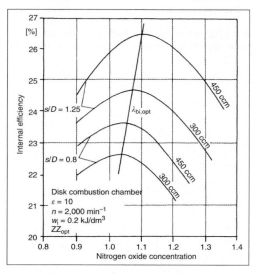

Fig. B52 Impact of S/D and displacement on the internal efficiency (Source: B. Gand)

B

~Undersquare. The bore-stroke ratio is called under-square if it is greater than one.

Literature: F. Schäfer, R. van Basshuysen: Schadstoffreduz-ierung und Kraftstoffverbrauch von Pkw-Verbrennungsmo-toren, Die Verbrennungskraftmaschine, Volume 7, Vienna/New York, Springer Publishers, 1993. — P. Kreuter, B. Gand, W. Bick: Beeinflussbarkeit des Teillastverhaltens von Ottomo-toren durch das Verdichtungsverhältnis bei unterschiedlichen Hub-Bohrungs-Verhältnissen, 2nd Aachen Colloquium on Ve-hicle and Engine Technology, 1989. — B. Gand: Einfluss des Hub-Bohrungs-Verhältnisses auf den Prozessverlauf des Ot-tomotors, Diss. RWTH Aachen University, 1986. — W. Bick: Einflüsse geometrischer Grunddaten auf den Arbeitsprozess des Ottomotors bei verschiedenen Hub-Bohrungs-Verhältnis-sen, Diss. RWTH Aachen University, 1990. — F. Pischinger: Gedanken über den Automobilmotor von morgen, Lecture VW-AG, July 1990.

Bottom dead center →Dead center

Boxer engine →Balancing of masses ~Inertia forces; →Engine concepts ~Reciprocating piston engines ~~Single shaft engines

Braided fiber filter →Particles (Particulates) ~Particulate filter system ~~Particulate filter ~~~Filter media

Brake adjustment →Emission measurements ~FTP-75

Brake thermal efficiency →Engine efficiency

Braking operation →Variable valve control ~Engine brake

Bronzes →Bearings ~Materials

Bucket tappet →Valve gear ~Actuation of valves, ~Gear components

Bucking on overrun. Bucking on overrun occurs when the change from part load and/or full load changes to the load-free range (closed throttle) in a ve-hicle. Because of the change of torque, bucking oc-curs, noticeable to the driver as a weak vibration.

Bucking is a problem in carbureted engines in par-ticular. Two-stroke engines and Wankel engines be-have especially badly in this respect; but so, too, do en-gines with older mixture formation systems, such as the K-Jetronic. The reason is the differing air-fuel ratio from one cycle to the next. Fuel shutoff with a trailing throttle counteracts bucking. In carbureted engines, additional ventilation of the intake manifold is neces-sary to prevent vaporization of the fuel from the intake manifold walls.

Burst of soot. A burst of soot occurs under certain conditions in diesel engines equipped with exhaust turbocharging. Because the turbocharger requires a specific run-up time in order to develop the required charging pressure, black smoke is produced during ac-celeration if the amount of fuel injected is not limited based on the charging pressure. The burst of soot also occurs if soot deposits in the exhaust system are re-leased during an acceleration process.

Bushing →Bearings ~Design solutions

Butane →Fuel, diesel engine ~Liquefied petroleum gas (LPG); →Fuel, gasoline engine ~Alternative fuels

Bypass catalytic converter →Catalytic converter

Bypass oil filter →Filter ~Lubricating oil filter; →Lubrication ~Filter

Bypass scavenging →Two-stroke engine ~Fresh mixture losses

Bypass valves →Lubrication ~Control and safety components, ~Lubricating systems

C

C classification →Engine acoustics ~Assessment curves

Cab heater →Heat accumulator ~Applications

Calculation →Calculation processes; →Engine acoustics ~Engine noise ~~Simulation

Calculation of the component connection →Sealing systems ~Development methods

Calculation processes. Calculation or simulation processes provide information for the design and characteristics of a component or a component group at an early stage in the development process. Using calculation processes reduces test requirements and avoids unnecessary loops in the development cycle, consequently reducing cost and development time.

~Application areas. The primary applications analyzed here are stability, engine mechanics, structural dynamics, thermodynamics, and numerical flow mechanics.

~~Engine mechanics.

~~~Belt drive. Belts are used to drive equipment such as steering pumps, generators, water pumps, air-conditioning compressors, and so forth. The forces in the belt and those that the belt applies to the equipment are important for the service life and the acoustics characteristics. For example, the number of redirections, the smallest redirection radii, the transferable tangential forces, and the slip are important for the service life of belts. The belts have V-belt profiles to increase the transferable tangential force. The cross-section height can be kept small by using poly-V belts.

Contact angle. The tangential force transferable by the belts depends heavily on the contact angle of the belt on the pulley. The contact angle does not have a significant impact on the service life of the belt, because there is only one bending load change of the belt when entering and exiting the pulley.

Poly-V belts. Wedge profiles are used to increase the transferable tangential force of belts. This increases the vertical force between belt and pulley and, therefore, the friction force. Poly-V belts have several small wedge profiles next to each other, resulting in a smaller belt cross-sectional area for the same transferable tangential force. Poly-V belts are used to ensure that the belt height does not get too large, because the bend changes would stress the belts.

Slip. Belt slip is always present, because the belt transfers forces through friction. The slip contributes sig-nificantly to the heating and wear of the belt and should not exceed certain limits.

~~~~Timing gear.

Dynamics. The moving masses of a timing gear and the elasticity of the components result in a system with vibration capacity. Multibody systems calculations are used to evaluate the dynamics. Results from the dynamics calculations are:

- Free inertia forces of the single timing gear
- Dynamic loads in timing gear
- Forces and pressures for all modeled components of the timing gear

Kinematics. The characteristics of the single timing gear with respect to valve lift, speed, and acceleration are determined by the kinematic calculations for the gear. This results in tuning of the valve spring characteristics to the cam lift curve based on the desired valve movement or the backward projection of a specified cam contour to achieve particular valve kinematics. Results of the valve kinematics calculation are:

- Valve acceleration, lift curve, speed, based on a specified cam contour
- Cam contour based on a specified/optimized valve lift curve
- Camshaft torques
- Geometrical opening cross-sectional areas
- Contact point speeds

Literature: R. van Basshuysen, F. Schäfer: Handbuch Verbrennungsmotor, 2nd Edition, Wiesbaden, Vieweg Publishers, 2002.

~~~Chain drive. Engine start-up is simulated to predict critical operating conditions of this component. This includes stimulations from the crankshaft, fuel pump drag torques, fuel pump torques, the inertias of the chain and gear box wheels, chain stress forces and chain dynamic stiffness, antirotation flank clearances of gears in the system, and much more. Resonance frequencies of the chain drive and chain oscillation can, for example, be calculated. Based on these calculations, it is possible to reduce the maximum forces in the chain links by means of specific targeted measures.

Chain forces. The maximum chain forces cannot exceed a limit that depends on the design during chain operation.

Chain oscillations. The chain drive system (chain wheels, chains, all powered rotating masses, chain tensioners, etc.) has resonance frequencies that depend heavily on not only the dynamic chain rigidity and the powered masses, but also on the tensioning force from the chain tensioner. The design of the chain drive should take into consideration that these frequencies do not match a dominant order of the chain resonance frequency, because this can lead to severe chain oscillation and, therefore, to high chain forces.

Resonance frequencies. It is the objective here, as it is in other dynamic systems, to keep the resonance frequency away from the stimulation spectrum, because the stimulation of a resonance frequency always leads to high system stress levels. The level of the stimulation must be kept as low as possible if it is not possible to move the resonance frequency completely from the stimulation spectrum. The stimulation for chain drives can be reduced by measures at the crankshaft (e.g., improved torsional vibration damper), and the resonance frequency of the chain drive system can be moved by changing the chain tensioning forces, inertia of rest at the control assembly, or using a different chain with higher dynamic chain rigidity.

Influences. The chain drive is primarily stimulated by the uniformity of rotational speed and the torsional vibrations of the crankshaft. For example, the torsional vibration damper, which is mounted directly on the crankshaft, has a strong impact on the torsional vibrations. The combustion forces from the individual cylinders are the main cause of torsional vibrations. This means that the main stimulation order is determined by the number of cylinders. The order for four-stroke engines is (number of cylinders)/2.

Based on improved emission values, diesel engines are operating at continually higher injection pressures, and this leads to higher fuel pump driving torques. The change from pressure control to volume control of the pump is also implemented, resulting in a reduction of power loss and, therefore, in higher peak torques in the pump. These torques must be transferred to the pump by the timing chain and can result in unfavorable increases in chain load.

The masses or the mass moments of inertia of components play a decisive role in determining the forces that occur in the system, as they do for all dynamic systems. This is the reason why these masses should be kept low. This not only has a positive effect in terms of chain forces in the engine, but also generally results in better dynamics.

~~~Crankshaft drive

Balancing of masses. Forces are present due to the acceleration and deceleration of components in the drive parts of the piston engine; these are called inertia forces. These inertia forces must first be determined at the individual cylinder with the associated powertrain parts (piston rod stem, piston). The overall inertia forces that occur through interaction of the individual cylinders must be calculated in multiple-cylinder engines with multithrow crankshafts.

All activities that completely or partly compensate the crankshaft drive inertia forces, and the resulting moments of inertia or inertia forces are termed "mass balance."

Rotating masses. The rotating masses of the engine include:

- The crankshaft and the counterweights
- Rotating piston rod parts
- Other masses rotating with the crankshaft: flywheel, damper, pulley screw

Rotary inertia force (centrifugal force), F_r (**Fig. C1**):

$$F_r = m \cdot r_s \cdot \left(\frac{d}{dt}\psi\right)^2$$

where m = rotating mass point, r_s = distance from the axis of rotation, ($\neq 0$), ψ = angle of rotation, t = time.

Torque, M:

$$M = \Theta \cdot \frac{d^2}{dt^2}\psi$$

where Θ = polar moment of inertia, ψ = angle of rotation, t = time.

Conclusion

- The centrifugal force only acts if the center of gravity is outside of the center of rotation.
- Torque only acts if the rotational movement is unsteady.

Reciprocating masses. The reciprocating masses include:

- The piston with
- Piston pin and rings and
- The upper part of the connecting rod (piston rod stem): m_{po}.

Balancing of masses. The rotary inertial forces are balanced by counterweights. The reciprocating mass forces can be balanced by counter-rotating balancer systems.

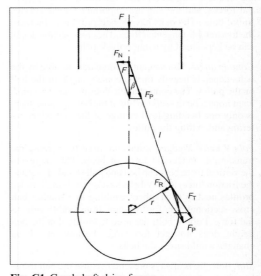

Fig. C1 Crankshaft drive forces

The reciprocating mass forces are balanced in practice by first-order balance shafts rotating at crankshaft speed and second-order balance shafts rotating at twice crankshaft speed.

Calculation of the balancing of the masses results in the following information:

- Size of the balancing degree
- Mass and position of the center of gravity of the counterweights
- Mass, position of the center of gravity, and position of the balance shaft(s) and transmission

Bearing forces. The determination of bearing stress is most important. This requires the knowledge of the changes in gas forces produced in the cylinder over time and the knowledge of the change in size and direction of the inertia forces. The stress characteristics are repeated periodically in each combustion cycle.

Piston pin bearing force. The piston force separates itself at the piston pin into the (piston rod) stem force and the guide-way force, which is used by the piston to support itself at the cylinder wall. The piston pin bearing transmits the gas force and the reciprocating mass force from piston, piston pin, piston ring, and the reciprocating portion of the piston rod.

Piston rod bearing forces (crankpin bearing). Forces of the (piston rod) stem load are transmitted here to the crankshaft. Also present is the rotary inertial force of the piston rod. High inertia forces are generated at high engine speeds, which means that the piston rod bearings are the most highly stressed bearings in the internal combustion engine.

Base bearing (main bearing) forces. The base bearings receive the gas forces and the reciprocating mass forces transmitted by the piston rod as well as the rotary inertial forces of the piston rod and the crankshaft. The crankshaft is in most cases supported by bearings between each cylinder (full support). One of the base bearings is designed as the thrust bearing (guide bearing), and it accepts axial forces in addition to radial forces.

Axial forces. These are caused by actuation of the clutch, axial thrust from helical gears on the crankshaft, manufacturing inaccuracies, engine tilt at certain driving conditions, and slanting of rollers in rolling bearings.

Results of the bearing force calculation are:

- Stress characteristics for a combustion cycle (for the total engine speed range)
- Maximum bearing loads
- Fourier analysis of the bearing forces (stimulation of the bearing pedestal)

Torsional vibrations. The torsional vibrations of the crankshaft can be calculated to an acceptable accuracy level with experience and by using a simple calculation model, the so-called torsional vibration chain.

The order of vibration which results in one vibration node between the crankshaft flanges is of interest for the piston engine (torsion second type).

The inertia impact of the vibrating engine masses is included in a conventional calculation model by increasing the mass moment of inertia of the rotating components (Frahm coefficient) and by an additional vibration stimulation that increases with the square of the engine speed (moment of inertia). Results of the torsional vibration analysis are:

- Resonance frequency forms of the crankshaft
- Maximum angle of twist amplitudes
- Fourier transformation of the twist vibration — maximum vibration amplitudes in the individual engine designs
- Tuning of the damping characteristics with measurement results
- Impact/design of the torsional vibration damper

Balanced state. Calculation of the mass distribution at the crankshaft axis is necessary to achieve the required degree of balance in the engine. The dimensioning of the counterweights is done in such a way that the individual imbalances at the crankshaft start and at the crankshaft end point in the direction of the counterweights. This permits a balancing process through removal of material (balance drilling).

The mass balance is achieved by precisely tuned counterweights for crankshafts with free moments of inertia (longitudinal tilt moments) such as for 3- and 5-cylinder in-line engines and V-6, V-8, and V-10 engines. Proportional piston rod and piston masses must, therefore, be replaced during balancing by so-called master weights.

Results of the balancing calculations are:

- Design of the counterweight and crank engine masses and center of gravity positions for achieving the defined (rough part) imbalance

Literature: J. Affenzeller, H. Gläser: Lagerung und Schmierung der Verbrennungskraftmaschine, Die Verbrennungskraftmaschine, Volume 8, Vienna/New York, Springer, 1996.

~~**Fluid circulation.** The piston blowby system, the coolant circuits, and the fuel and lubrication oil circulations are discussed for the fluid circuits.

~~~**Blowby system computation.** A one-dimensional calculation method is used for the calculation of the blowby system. The camshaft space, the oil return beams, and the total crankshaft space will be described as a hydraulic network. The blowby gas, which comes from the combustion chamber and flows over the piston rings into the crankcase, is superimposed on the reciprocating movement of the pistons, which can be related to the change in cylinder volume over time. The blowby volume, its variation with time, and the incoming air volume, which comes from the vacuum pump, are known from measurements and are described by volume flow sources with a corresponding development of volume flow over time. Straight and

bent pipe pieces with arbitrary cross sections and lengths, throttles, and general flow loss components are available for modeling of the remaining flow geometry. The main tasks are the calculation of the transient and time-averaged volume flows, flow speeds, and total pressure distributions in the system and the establishment of possible correction measures to guarantee return of oil from the camshaft into the crankcase at all times.

~~~**Coolant circuit.** Computational fluid dynamics (CFD) is used frequently to ensure sufficient engine cooling, which enables the avoidance of inadmissible component temperatures (**Fig. C2**). The nominal power point is analyzed for the determination of coolant flow speeds, pressure losses, heat transfer coefficients and other flow parameters. An important objective for multicylinder engines is an even temperature distribution over the different cylinders. This is ensured by the appropriate design of the flow cross sections for the changeover in the cylinder head gaskets. However, a particular challenge is the description of boiling processes in the system because these cause very high local heat transfer coefficients. Zero- and one-dimensional models are applicable for calculations that include the total cooling system. The main task is the calculation of the volume flow distribution in the individual flow branches and also of the pressure losses therein. In addition, calculation of the thermal characteristics (e.g., the warm-up phase) within the framework of thermal management can also be based on these models.

CFD calculations are also used for the calculation of air throughput through individual coolers or for the total engine space through-flow. Investigation of parameter variations, however, should be performed with one-dimensional methods because these are much more cost-effective.

The individual calculation processes and methods are becoming more tightly linked due to the increasing complexity of the systems. This approach is an attempt to combine the advantages of the respective methods without amplifying the disadvantages too much. An example of this is the linking of the finite element (FE) method with CFD calculations within the framework of the conjugate heat transfer-method. This method iteratively exchanges the results of the CFD calculation (heat transfer coefficients) and the results of the FE simulation (component temperatures).

~~~**Fuel systems.** Modern injection systems are divided into devices with low- and high-pressure circulation.

Low-pressure circulation. The low-pressure circulation of the fuel in the injection system of a combustion engine can be simulated by a one-dimensional calculation method. The method takes into account the tank, fuel pumps, fuel filter, and low pressure fuel lines. The objective of the calculation is to determine the volume flows and pressures in the system, with the purpose of

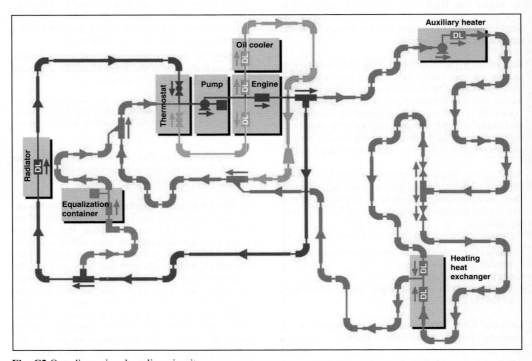

Fig. C2 One-dimensional cooling circuit

checking the functioning of the system, taking into account the tolerances. Possible design modifications can be undertaken at this stage to guarantee the required system characteristics—primarily, the pressure to be provided for the satisfactory operation of the high-pressure circulation at the intake and exit of the high-pressure pump, the fuel volume required for full-load operation, and the minimum volume required in the tank for return flow (which may be required for a potentially present ejector pump).

High-pressure circulation. The high-pressure circulation of the fuel in the injection system of a combustion engine is also simulated by one-dimensional calculation methods. The objective here is calculation of the transient variation of pressures and volume flows. An example of a high-pressure pump with time-synchronized piston components is the common rail system. The required presupply pressure upstream of the suction valve of the high-pressure pump is supplied by the low-pressure system. Much like all other high-pressure pipes, the high-pressure tank (rail) comprises pipe segments with the appropriate cross-sectional areas and lengths; knowledge of these permits the analysis of pressure wave propagation in the system.

Sufficient components are available in the model for manipulation of the rail pressure or the fuel volume supplied to the rail. The model includes measurement components for the acquisition of pressures and volume flows and control components, with process measurement values and calculated control variables. This means, for example, that the control of the metering unit, located in the high-pressure pump of the pressure valve mounted at the rail, and the injector are impacted.

The 1-D software provides models for such elements as controllable mechanical and electromechanical valves, pressure sources, volume flow sources, and mechanical elements such as springs, masses, dampers, end stops, travel limiters, and so on. The parameters required for the description of the injector nozzle flow characteristics, which also permit the consideration of cavitations processes, are usually determined by measurements.

~~~**Lubrication oil circuit.** Zero- and one-dimensional methods are primarily used for the calculation of lubrication circuits. The methods describe the lubrication circuit as a hydraulic network. The main objectives include the determination of the pressures at the individual lubrication points and the oil throughput of the individual components.

Semi-empirical submodels of the individual components based on measurements, for example, of bearings and piston cooling, are often used for this purpose. An objective of these calculations is the reduction of the power required to drive the oil pump while ensuring adequate lubrication.

~~**Stability**

~~~**Distortions.** Cylinder liners and bearing deformations are the main components evaluated for distortions.

Liner deformation. The deformations of the cylinder barrel for the cold static and thermal load cases are used for the analysis of the liner deformations (**Fig. C3**). The basis for the mathematical preparation of the FEM calculation results is the description of the deviation, ΔR, from the ideal circular shape with the help of a Fourier series.

$$\Delta R = A_0 + A_1 \cos \varphi + A_2 \cos 2\varphi + \ldots + A_i \cos i\varphi$$
$$+ B_1 \sin \varphi + B_2 \sin 2\varphi + \ldots + B_i \sin i\varphi$$

where A_i, B_i = Fourier coefficients and i = ordinal number.

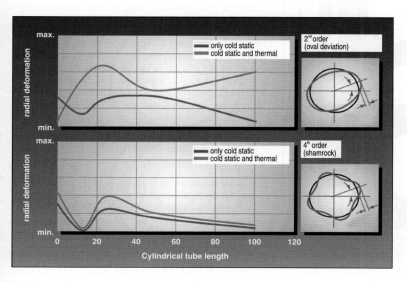

Fig. C3
Liner deformation cylinder crankcase

The deviation, ΔR, will result from the total model calculation with the help of local, axially arranged cylindrical coordination systems, and the Fourier coefficients A_i and B_i are then evaluated. The Fourier coefficients permit the description of the deviation of cylinder shape with few values and also permit statements that are important for the assessment of the piston and piston ring characteristics. The Fourier coefficients of the second and fourth order are especially significant for oil consumption or for the blowby values. The Fourier coefficients of the first and third order are small and, therefore, negligible. The Fourier coefficients of the zero order are normally of no importance for the functional characteristics of the piston rings. The calculated results of the cold static and liner thermal deformations are used as input data for the piston design (grinding patterns, piston diameters, shape fill capacity) and for the piston ring design (e.g., impact clearance calculations)

Bearing deformations. Based on the results of the FEM manufacturing calculations, the bearing pedestal deformations can be evaluated with the help of a cylindrical coordination system positioned on the axis of the main bearing (**Fig. C4**). The expansion or the constriction of the main bearings is decisive for their function, namely, the oil supply and the lubrication of the main bearing. Additionally, large deformations of the bearing pedestal result in unfavorable noise stimulations of the engine. The deformations of the bearing pedestal are normally evaluated at the operating temperature and at the maximum combustion load.

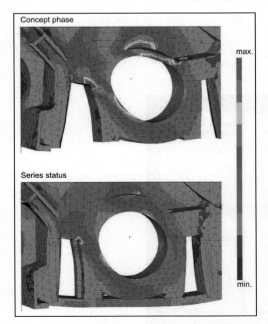

Concept phase

max.

Series status

min.

Fig. C4 Finite element (FE) results of a bearing deformation.

Bearing loads. The bearing forces and the bearing torques are determined by a dynamic simulation depending on the engine speed and the angle of the crankshaft. A multibody model of the engine transmission assembly is constructed with the help of multibody simulation software. The gas pressure, which gets delivered to the engine depending on crankshaft angle in the firing order, is the main load in this model.

Compression stress distribution. Based on the results of the FEM stability calculations, the bearing friction points and the compression stress distribution can be determined or evaluated with the help of nonlinear contact models at operational loads under the maximum bearing load.

~~~**Material behavior.** The characteristics of the material used significantly influence the calculation result. Depending on the production process, the local manufacturing characteristics are dependent on the structure distribution and the internal stress status.

Structure impact. The structural characteristics can vary to a great extent for different components (also within a component) due to locally different cooling conditions during the casting process (**Fig. C5**). Temperature and stress analyses can be performed with stability characteristics that depend on location by using applicable material models if the cooling temperature distribution is available from a casting simulation.

Literature: P. Nefischer, F. Steinparzer, H. Kratochwill, G. Steinwender: "Neue Ansätze bei der Lebensdauerberechnung von Zylinderköpfen" 12th Aachen Colloquium on Vehicle and Engine Technology, Aachen, 2003.

Internal stress. Simulations are usually performed for ideal, perfect components. However, real components often have significant imperfections. These imperfections include deviations and variations in the material characteristics (e.g., in a die-cast structure) and internal stresses as a result of the manufacturing process; these effects are superimposed on the assembly and operational loads and can have a decisive impact on durability. Internal stresses from heat treatments (e.g., quenching) can be determined by a transient temperature calculation with the relevant nonlinear (plastic) stress analysis, while internal casting stress can be calculated in a cast simulation.

Literature: R.J.A. Ehart, P. Nefischer: Fatigue Analysis of Engine Components Including the Heat Treatment Process, Fifth World Congress on Computational Mechanics, Vienna, Austria, 2002.

~~~**Service strength.** Fatigue includes material damage, fissuring, and crack growth due to repetitive fluctuating operational loads (**Fig. C6**). Three areas are defined depending on the number of load cycles, N, the load changes, LW, and the crack or break size:

$N <$ approx. 10,000 LW
classified as low cycle fatigue (LCF)

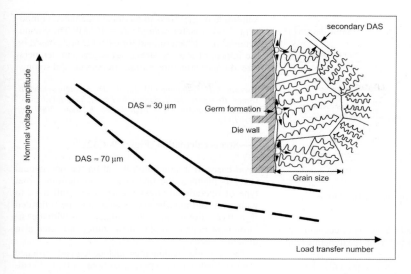

Fig. C5
DAS impact for a
specific material

$N >$ approx. 1^6–10^6 LW
classified as long life fatigue (LLF).

High cycle fatigue (HCF) exists between LCF and
LLF. Differentiations according to the load type are
also used, such as fatigue due to thermal load (TF,
thermal fatigue) and fatigue due to a combination of
thermal and mechanical load (TMF, thermal mechanical fatigue).

*Literature: E. Haibach: Betriebsfestigkeit, Verfahren und
Daten zur Bauteilberechnung, 2nd Edition, Berlin, Springer
Publishers, 2002. Material models. The calculation of the operational stability by using service life software requires a detailed description of the material characteristics.*

Wöhler characteristics curves, with information
about the test parameters, document the predicted service life depending on the amplitude of stress or elongation (**Fig. C7**). Together with the static stability
data, they determine the Haigh or Smith diagram. The
cyclically stabilized stress elongation curve describes
local plasticity in approximate terms. Ratios, such as
bend numbers, which relate specifically to individual
materials are also used for the influence of notches
(e.g., gradient impact according to Eichlseder).

*Literature: W. Eichlseder: Lebensdauervorhersage auf Basis von Finite Elemente Ergebnissen, in Mat.-wiss. u. Werkstofftech Wiley-VCH Publishers GmbH, 34 (2003) No. 9, pp.
843–849.*

Load history. The operational stress can be random or determined (Gassner test). Ideal cases are also called oscillating operational stress, such as, for example, sinusoi–dally-shaped stress variations with time. The operational
stability test refers primarily to this simple case of an

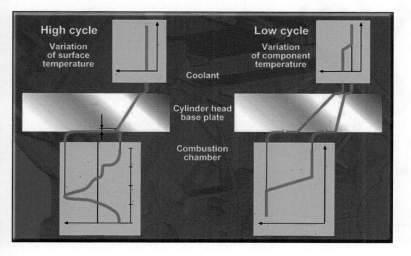

Fig. C6
Cylinder head
operational stress

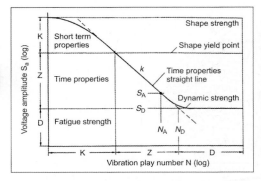

Fig. C7 Wöhler material characteristics curve

oscillating load (Wöhler test). A steady or changing average stress can be superimposed on the dynamic load.

Service life calculation. Engine components, depending on the material used and the manufacturing technology, must be dimensioned in such a way that they will withstand the specified, expected, or possible loads for the duration of the planned life (**Fig. C8**).

For engine development, the service life before testing is calculated with CAE tools based on finite element (FE) models. Experience shows that the calculation correlates well with the tests if the ratio of the local

amplitude of the stress to the local calculated component stability is, for example, $N = 10^6 \, LW$. The component stability is determined for each FE node based on the impact of average stress, stress gradient, temperature, die-cast structure, surface treatment, and so on.

Literature: H. Dannbauer, W. Eichlseder, G. Steinwender, B. Unger: Rechnerische Kerbmodelle–Anwendung auf nicht geschweißte Bauteile, DVM Report 127, Koblenz, 2000, pp. 121–133.

~~~Stress calculation (Figs. C9–C12)

Load cases. If a simulation based on the operational stress is planned, the first task is to determine what type of operational stress a component will have to withstand. Initially the component must be separated from its environment, and the interactions with the environment must be replaced by static, and often also dynamic, loads.

The particular loading must include logically grouped external loads and clamping conditions. A typical loading situation for an engine component is, for example, the case of the combustion load; here the combustion and inertia forces act at a previously calculated temperature distribution.

Temperature fields. The material characteristics such as the modulus of elasticity, E, are often significantly impacted by temperatures, and there can also be a high

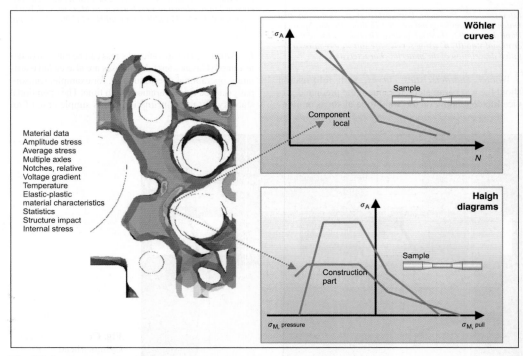

Fig. C8 Service strength calculation

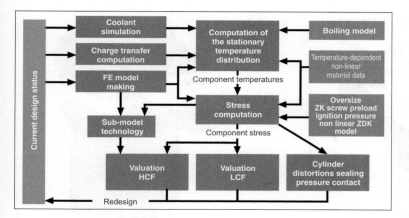

C

Fig. C9
Calculation of engine
components under
thermal load (cylinder
head, crankcase)

Fig. C10 FE model segment cylinder crankcase

internal stress due to the temperature field in the component (**Fig. C13**). This internal stress is the result of local temperature differences and, therefore, the locally varying thermal expansion in the component.

Temperature fields, like stress fields, are determined with the help of established FE programs and are often used as the basis for subsequent stress calculations. Temperatures must be included in the determination of the resulting stress, because material characteristics, such as the modulus of elasticity, E, depend significantly on them.

Literature: P. Nefischer, S. Blumenschein, A. Keber, B. Seli: Verkürzter Entwicklungsablauf beim neuen Achtzylinder-Dieselmotor von BMW, Part 1: Mechanik und Festigkeit, MTZ 60 (1999), No. 10.

Boiling point impact. Boiling of the coolant within the cooling ducts may occur if a component under thermal load is cooled by a liquid (**Figs. C14 and C15**). The initiation of the boiling process characteristically requires thermal energy to transform the coolant into a gaseous state; the onset of boiling becomes obvious through locally increasing heat removal by the coolant (the heat transfer coefficients increase locally). Ultimately a saturation effect occurs if the temperature continues to increase, which means that boiling com-

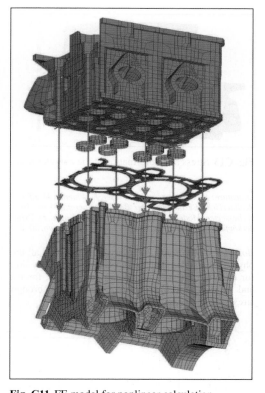

Fig. C11 FE model for nonlinear calculation

mences; the local heat transfer coefficients decrease at the same time. This is an unstable operating point. The initiation and location of boiling (nucleate boiling) depend on many system parameters (e.g., coolant, pressure in the cooling system, fluid flow speeds, medium and wall temperature, steady state heat transfer coefficient, etc.). The possibility of boiling must be considered if temperature fields need to be calculated accurately.

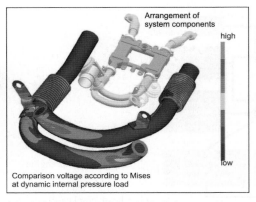

Fig. **C12** EGR system stress calculation

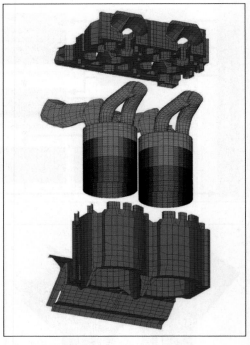

Fig. **C14** FE mesh for heat transfer

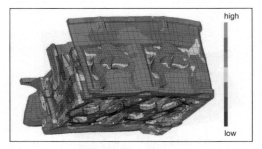

Fig. **C13** Stress and deformation at the cylinder head

Literature: P. Nefischer, A. Ennemoser, A. Wimmer, M. Pflügl: Rechnerische Bestimmung der Bauteiltemperaturen mit Hilfe verbesserter Modellierung des Wärmeüberganges in Zylinderköpfen, 23, International Vienna Engine Symposium, 2002.

Boundary conditions. Boundary conditions are all the physical variables that impact the system. They are, for example, the material data, temperature, forces, and torques that impact from the outside through stress, electric or magnetic fields, and so on.

Contacts. Contacts are significant nonlinear connections between component surfaces. Contact occurs at common contact surfaces where components touch each other. The degrees of freedom of a component are limited or restricted at that contact surface (contact pressure, contract friction, contact temperatures, etc.), so contacts are, therefore, an important basis for the simulation of interaction between components.

Contacts can be closed or opened depending on the particular loading of the system, and this in turn can change the system characteristics completely.

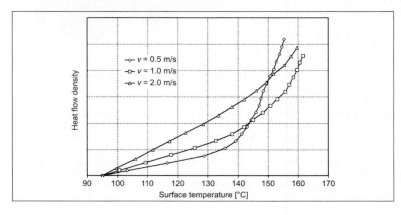

Fig. **C15**
Impact of nucleate boiling on the heat transfer

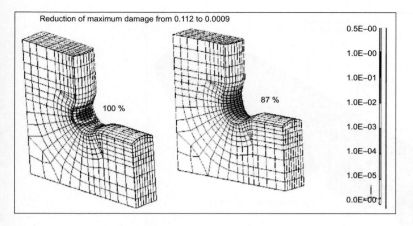

Reduction of maximum damage from 0.112 to 0.0009

100 % 87 %

0.5E−00
1.0E−00
1.0E−01
1.0E−02
1.0E−03
1.0E−04
1.0E−05
0.0E−00

Fig. C16
Numerical optimization
(shape optimization)

~~~**Structure optimization** (*also, see below*, ~~Numerical processes). Reliable calculation results are the basis for optimizations. The objective of the optimization is to achieve improvements in the design directly from the calculation results. The modified model is also evaluated afterward, which means that the optimum design is achieved through iterations (if possible by an automated sequence).

The optimization strategy typically comprises general mathematical algorithms (e.g., gradient-based processes) and techniques based on experience (adaptive growth).

Shape optimization. Shape optimization changes surfaces in such a way that stress concentrations are avoided (**Fig. C16**). An adaptive growth (but also shrinking) takes place depending on the applied stress, but the topology of the component (and of the FE model) remains unchanged. Restrictions such as prismatic shapes for stamped or sintered parts and, for example, roundings or certain radii of drilled holes can be included through link conditions.

Topology optimization. The best material distribution can be found with the help of topology optimization by specifying weight or rigidity to achieve a design adapted to power flow (**Fig. C17**). The lightly loaded material is removed in steps, based on the so-called installation space, the maximum possible component version. Exact knowledge of all possible load cases is of great importance because the optimized components react poorly to loads for which they are not designed, very much as an athlete performs badly in sports that have not been practiced.

Parameter optimization. It is advantageous to optimize the parameters by means of mathematical sensitivity analyses if the geometry of a component can be described with a few salient parameters (radii, wall thicknesses, angles, and so forth). Large component changes can occur when this method is used, however, and the method requires a parameter-based CAD model and an automatic transfer between the CAD and simulation software. Another example of parameter optimization is a variation of shape optimization that describes the shape changes by means of trial functions and their coefficients. However, the possible component design is in both cases predetermined and, therefore, limited by the choice of the parameters or the approach function.

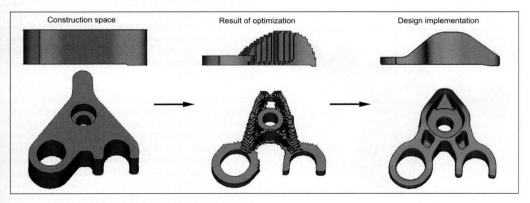

Construction space Result of optimization Design implementation

Fig. C17 Numerical optimization topology optimization

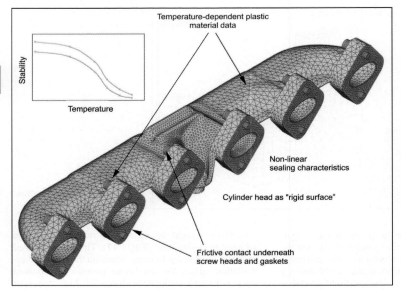

Temperature-dependent plastic material data

Non-linear sealing characteristics

Cylinder head as "rigid surface"

Frictive contact underneath screw heads and gaskets

Fig. C18
FE model of an exhaust gas manifold

Topography optimization. Favorable swage lines for the stiffening of sheet metal parts can, for example, be found with the help of topography optimization. The topology of the component is unchanged, but the wall thickness does not change in contrast to shape optimization.

~~~**Thermal-shock calculation.** When a component is subject to high temperature loads that change over time, the resulting stresses can have a significant impact on its service strength (**Figs. C18 and C19**). The mechanism producing the resulting component stress is caused by the interference of the free thermal component expansions. These interferences are produced by external boundary conditions and/or by fluctuating and regionally nonuniform heating of the component.

Thermal-shock calculations are thermally linked stress calculations. They are normally performed for load change numbers of about 10,000 LW (LCF range).

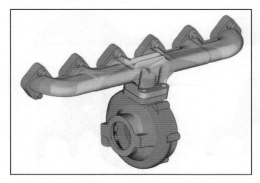

Fig. C19 T-map of an exhaust gas manifold

The exhaust manifold or the exhaust gas turbocharger are typical thermal-mechanically stressed components.

Literature: D. Niederhauser, F. Durstberger, P. Nefischer: Thermomechanische Analyse von abgasführenden Bauteilen aufgeladener Dieselmotoren, 8th Conference on Supercharging, Dresden, 2002.

~~**Structure dynamics**

~~~**Resonance frequency calculation.** The resonance frequency analysis is the first step in the dynamic analysis. By solving the intrinsic value problem it delivers the frequencies and vibration forms of the free system vibrations. These are the potential resonance frequencies for a dynamic stimulation, but the analysis does not provide information about the actual amplitudes. However, it offers the opportunity to move the resonance frequencies out of the stimulation range of the engine processes by implementing targeted rigidity changes. The modular processes of the response calculation are also based on a degree-of-freedom reduction through an upstream resonance frequency analysis. Therefore, the total shift is composed of the parts of the modular shift.

Vibrations of the engine transmission assembly. FE modules are established for the engine transmission assembly, including the auxiliary equipment and attached parts, to permit a comprehensive analysis of the total structure. The results of dynamics calculations with engine transmission assembly models provide information about the global characteristics (e.g., engine transmission deflections), but also about local vibrations of the equipment. The individual component vibrations are often linked, which means that the separation of a partial structure and its separate analysis will deliver incorrect results.

Vibrations of attached parts. The resonance frequencies of attached parts and their fasteners should be outside the range of critical engine stimulations, so that these are not functionally influenced or destroyed by vibration resonances. Strengthening (supercritical design) is typically used; however, a targeted decoupling from the stimulating structure can also avoid resonance (subcritical design).

Stimulation of forced vibrations. Forced vibrations are vibrations that are stimulated by external forces. This is in contrast to free vibrations, which are generated in vibration-sensitive systems that are released after they have been forced out of their stable condition. The vibration unit also reacts to periodic stimulation with a periodic motion sequence after the onset phase. The reaction of the system can either be determined in the time domain or in the frequency domain. The movement is calculated by direct, mostly numerical integration when considered in the time domain.

Calculation in the frequency domain is normally preferred for its faster and more direct calculation of complex structures in combination with a broad stimulation spectrum, such as for engine transmission assemblies. The resonance frequencies and the associated modes of vibration of the structure are determined by solving the intrinsic value problem. The stimulation forces are converted by a Fourier transformation into a sum of single harmonic stimulations. With the selection of suitable damping parameters, the resulting system of equations can be decoupled in such a way that only a given number of individual differential equations have to be solved. Additionally, the calculation effort can further be reduced by eliminating inconsequential vibration forms.

Acoustic stimulation. A typical application of forced vibration calculations can be found in the engine acoustics. The generated structure-borne noise is transferred to the air—that is, radiated into the air—due to stimulation of acoustic waves by a fixed body in contact with air (**Fig. C20**). Therefore, the evaluation of the engine structure in regard to acoustics includes the resulting speeds in the direction normal to the structure surface. Response calculations for the total speed range of the engine with a defined step size can be performed to identify critical engine speeds for a certain number of significant analysis points that are sensitive to increases in resonance. More accurate calculations can subsequently be performed with a higher number of evaluation points at the surface of the FE structure for engine speeds that are subject to increased vibration. The vibration amplitude normal to the surface is transferred with the level values normally used in acoustics and subsequently filtered (mostly A-classification), to make the values approximate the effect on the human ear.

~~~Rigid body dynamics; multibody system.

Multibody simulation of a multibody system permits the evaluation of systems comprising several components that perform large relative movements in relation to each other. Examples in engine design are the crankshaft drive and crankshaft, piston rods and pistons, and the timing gears with camshafts, transfer levers, and valves. The introduction of flexible bodies in the form of FE models permits inclusion of the impact of component deformation on the coupling forces and allows a detailed evaluation of parts regarding stress and service life.

Crankshaft. Commercial software tools are available for the dynamic crank angle simulation (**Fig. C21**). However, modeling is also possible with a general multibody system program. The crankshaft, the flywheel, the torsional vibration damper, and the main bearing blocks are available as elastic structures depending on the level of detail required. Pistons and piston rods are primarily described as rigid bodies with corresponding mass characteristics. The nonlinear, statically indeterminate bearing forces in the main bearings can be obtained with different coupling models. The cylinder pressure characteristics are used for stimulation, and the drive moment is supported by drivetrain rigidity.

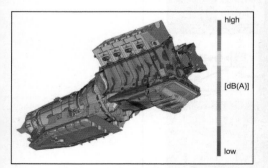

Fig. C20 Surface mobility (acoustics) engine transmission assembly

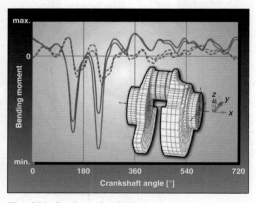

Fig. C21 Crank shaft calculation

C

The results of the dynamic crank angle simulation are the main bearing forces and moments, which are included as loads in the stability calculations of the crankcase. They are also used for the bearing design.

Several models are available for the plain bearing. The use of nonlinear springs is the simplest solution, but this does not include feedback for the pivotal rotation or shift in position and speed of the pivot center on the pressure distribution. They require the use of analytical lubricating film models. Differentiation is made between pure hydrodynamic (HD, surfaces are rigid) and hydrodynamic-elastohydrodynamic (EHD, elastic bearing points, reciprocal feedback between pressure distribution and bearing deformation) methods of modeling. Other developments also include cavitation, mixed friction (micro-contact of the slip partners) and temperature distribution inside the lubrication film (T-EHD). The impact of component geometry, such as the convexity of journals or the deviation of the bearing shells from a circular shape, can also be included in the simulation.

Multibody system simulation either delivers the transient shifts directly or through inverse transformation to the FE models of the elastic bodies. This allows, among other things, the analysis of pintle displacement paths, angled bearing positions, flywheel wobble, and torsional vibration damper movements.

The use of elastic bodies permits the inverse transformation to the time-based stress processes. These are the input values for the service life analysis of the crankshaft.

~~**Thermodynamics**

~~~**Charge exchange calculation.** The charge exchange simulation (CCS) is used for evaluating the gas flow in the engine and its air and exhaust gas carrying components (**Figs. C22 and C23**). Overall models of

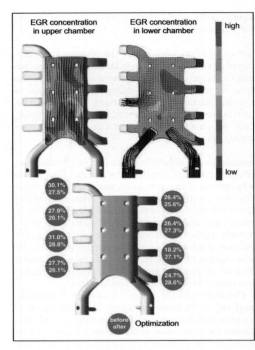

Fig. C23 EGR distribution in the plenum chamber

the engine are built from a number of predefined components (pipes, branches, plenum volumes, etc.), supplemented by the relevant geometrical and flow data. The models are used for the calculation and design of engine power output, volumetric efficiency, valve timing, pipe lengths, and pipe cross sections, as well as nonsteady operating characteristics; they also simplify,

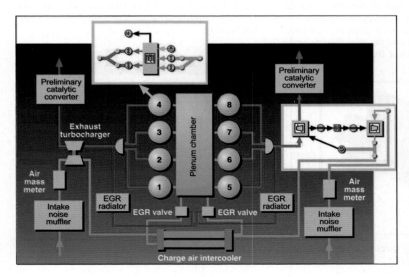

Fig. C22
Charge change model

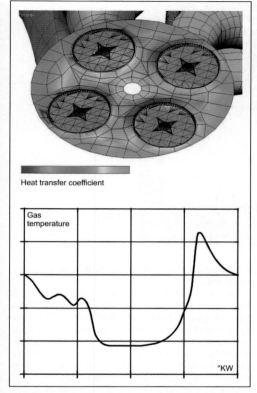

Heat transfer coefficient

Gas temperature

°KW

Fig. C24 Heat transfer on the gas side in the cylinder head

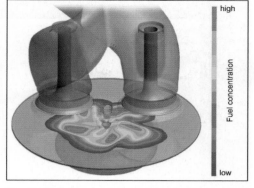

high

Fuel concentration

low

Fig. C26 Injection and combustion simulation

for example, the tuning of the turbocharger map or the intake and exhaust noise characteristics of the engine.

The initialization and tuning of the established model is performed by means of adaptation to existing measurement data and maps.

Simulations at different loads are often used for parameter studies or complex optimization tasks involving a large number of free parameters due to the relatively small modeling and calculation effort.

The CCS also delivers starting and boundary condi-

tions for 3-D flow simulations and can be linked through special interfaces directly to 3-D simulations.

~~~Internal cylinder flow. The current development of the combustion process is caught up in the conflict between the partly contradicting objectives of high power/torque, low fuel consumption, and minimum pollution development (**Figs. C24–C26**). Flow calculations play an increasingly important role in this development process. The simulation of internal flows in the cylinder can provide valuable information about the complex system characteristics. It has become standard to design the intake ports with the help of steady state three-dimensional flow calculation. Measurements of the flow through the port on the through-flow bench for fixed valve lifts are used to adjust the models. **Fig. C27** shows an example of an iterative process used to reduce the swirl level of a swirl channel and increase the flow coefficient.

Literature: P. Nefischer, P. Grafenberger, C. Hoelle, E. Kranawetter: Verkürzter Entwicklungsablauf durch Einsatz von CAE-Methoden beim neuen Achtzylinder-Dieselmotor von BMW, Part 2: Thermodynamik und Strömung, MTZ 60 (1999) No. 11.

The modeled region not only covers the intake ports with steady state flow in the channels (possibly with the upstream intake systems), but also, if required, the

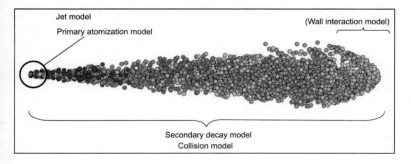

Jet model

Primary atomization model

(Wall interaction model)

Secondary decay model
Collision model

Fig. C25
Simulation of diesel injection

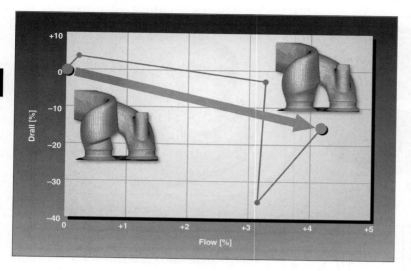

Fig. C27
Iterative swirl
optimization

combustion chamber and the exhaust ports if the gas exchange flow in the channels over time is of interest. Boundary conditions from the one-dimensional unsteady flow calculations can be superimposed at the boundaries of the calculation, and piston and valve movements are transferred through a movable calculation grid for these simulations; however, this increases the modeling and calculation effort significantly.

Modeling of fuel injection, evaporation, and combustion is even less validated than for gas flows, if the simulation of the flow process is considered to be state-of-the-art. The quality of the partial models used for the latter processes can still bear improvement. Adjustment of the simulation results at regular intervals by aligning the model parameters with accurate measurement data, which often require extensive measurement efforts, is usually necessary. The calculation of the emission of pollutants becomes increasingly important for internal cylinder flow simulation. This presents an even bigger challenge than the modeling of the injection and combustion processes due to the extraordinary complexity of the underlying processes.

~~~**Simulation of flow-through components.** The design of gas flow-optimized engine components (air passage, exhaust gas components, etc.) is an important part of engine development supported by simulation techniques. The main areas are the determination and minimization of flow losses to achieve optimum flow approach characteristics, analysis for the mixing of fluids, and determination of the heat transfer. The total pressure loss between the pipe intake and exit is determined in a steady-state three-dimensional flow simulation to minimize flow losses; reverse flow regions are identified and removed to a large extent by modification of the component geometry. The flow distribution at the exit of the pipe can be influenced by geometrical modifications or mountings (such as rectification grids) to achieve optimized approach flow for the component

(e.g., uniform flow distribution at the catalytic converter intake).

The mixing of the recirculated exhaust gas (EGR distribution) in the intake air can be evaluated with a transient three-dimensional calculation (with boundary conditions from a one-dimensional total engine simulation). Reasons for nonuniformity can be identified and a uniform recirculated exhaust gas distribution to the individual cylinders can be achieved by means of appropriate measures (**Fig. C23**).

Another application area is the analysis of dispersion flows. This means that, for example, water droplets in the intake air must be deposited quickly on the manifold wall to avoid an impact on the operation of sensitive components (e.g., air mass meters). The description of a droplet path in the existing steady flow field, taking into account the effect of gravity, can provide information about the movement of the drops. Determination of the heat transfer between a flowing fluid and the component wall is a standard application. It is used for the specifications of the thermal boundary conditions for downstream component temperature and stress calculations.

~**Numerical processes**

~~**Finite element (FE) method.** The basic process steps of the finite element method (**Figs. C28 and C29**), are the formulation of the task to be solved as a variation problem or as a relationship of the weighted residual; the discretization of this formulation with finite elements; and, finally, the effective solution of the resulting finite element equations. The continuous body will be viewed as a grouping of discrete elements linked to each other at the intersections of the element edges. The unknown starting parameters are equivalent to the unknown generalized shifts of the element intersections. The unit shift equation is used here as the starting function, which is also known as the shape function or

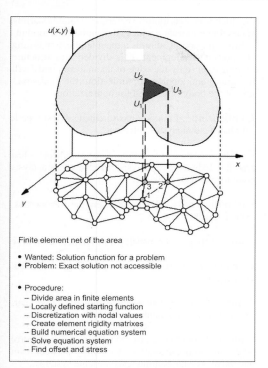

Finite element net of the area

- Wanted: Solution function for a problem
- Problem: Exact solution not accessible

- Procedure:
  - Divide area in finite elements
  - Locally defined starting function
  - Discretization with nodal values
  - Create element rigidity matrixes
  - Build numerical equation system
  - Solve equation system
  - Find offset and stress

Fig. C28 Finite element method, procedure

shift interpolation function. The finite element method is an approximation procedure because the starting functions generally represent mere approximations.

The application areas include linear and nonlinear calculations of solid body problems and structure mechanics, heat transfer and field problems, and flow mechanics problems.

Literature: Bathe: Finite-Elemente-Methoden, Berlin/Heidelberg/New York/London/Paris/Tokyo/Hong Kong, Springer Publishers, 1990.

~~Finite volume method. The finite volume method has now been accepted as the standard solution for simulation of fluid flows, having been viewed at first as a derivate of the finite difference method.

The finite volume method is divided into the following items as part of a flow simulation:

- Division of the calculation domain into partial areas (= finite volume)
- Balancing of the mass, impulse, and energy values in relation to the finite volume

This approach is chosen because the change in value of a parameter, Φ, within a finite volume (= cell of the calculation area) can be described as flow balance.

[Change of Φ inside the cell] = [convective flow of Φ across the cell limits] þ [diffusion of Φ across the cell limits] þ [source or depression flow of Φ inside the cell]

The resulting algebraic equation system is solved by a suitable iterative method.

~~Model construction. The objective of model construction is to describe a real system with a mathematical or computerized structure in such a way that a condition can be described and analyzed with acceptable accuracy. The choice of appropriate assumptions presents a challenge, because the system characteristics of interest must be determined accurately enough so that they do not become more complicated than necessary.

Possible differentiators are:

- *Starting basis*. Empirical models describe systems as a function of relevant parameters, without immediately formulating the physical laws in mathematical terms. Physical models start with known axi-

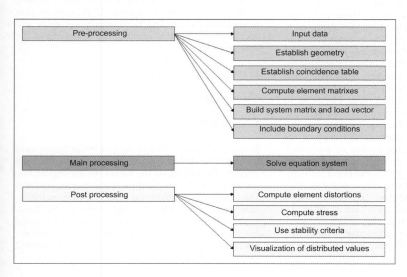

Fig. C29
Finite element method, work flow

C

oms or physical effects and are systematically expanded by balance equations.

- *Time-related and local solution.* Multidimensional models are the result if local dependence of the values is not included. They are based on the global inclusion of the physical effects. Single- or multidimensional models include the changes in the variables in the form of one or more local coordinates.

Steady-state (stationary) models deal with systems without including the time component. Transient models describe the time-dependent variation of the model conditions.

The models can also be divided with respect to predictability (stochastic or deterministic), parameterization, and other viewpoints.

~~**Multibody simulation.** Multibody simulation deals with the dynamics of rigid bodies connected to each other through linked degrees of freedom (joints). Modern multibody system tools permit not only the solution of the resulting movement equations but also the automated establishment of the models with graphic support. Flexible bodies are increasingly being used in multibody systems. The combination of large rigid body movements and the elastic deformation of the components involved is a specialty of flexible multibody simulation. The commercially available simulation tools generally use a modular approach. This provides an opportunity to expose the components directly to their stresses in the simulation in contrast to a calculation of cutting forces or shifts and superimposing these onto an FE structure, as is often done. The latter method can result in nonnegligible errors, especially for dynamically highly stressed components. That is why this method is especially used for components such as crankshafts and piston rods.

~~**Numerical optimization.** This optimization is viewed as the selection of the best option from a number of possible options. The characteristics to be optimized are formulated as target functions that need to be minimized or maximized. These target functions generally depend on several parameters freely variable in their permitted range and determined by restrictions of individual parameters or desired relationships between the parameters. A number of algorithms are available for numerical optimization. They can be divided into two classes: gradient-based and stochastic algorithms. The main requirements for optimization algorithms are robustness (finding of the optimum independent of the choice of parameter starting values), accuracy (the calculated optimum is as close as possible to the actual optimum), and efficiency (finding of the optimum with as small an effort as possible). Gradient-based algorithms provide a high accuracy and efficiency, but their success often depends on the selected starting value. Stochastic algorithms, on the other hand, are very robust; however, they are not as accurate and efficient. Advantage can be taken of the contrary behavior of both classes by using the stochastic algorithms to determine suitable starting values for optimization with a gradient-based algorithm.

Structure optimization is a main application during component development. It is divided into structure optimization (only the edge of an analyzed body will be varied) and topology optimization (the total structure of the body is released for optimization).

Calibration →Emission measurements ~Test type I; →Exhaust gas analysis equipment

Calibration, gas →Emission measurements ~Test type I; →Exhaust gas analysis; →Exhaust gas analysis equipment ~NO_x converter

Calorific value →Fuel, diesel engine ~Properties; →Fuel, gasoline engine

Cam contour →Camshaft ~Cam shape, ~Cam nose

Cam engines →Engine concepts ~Engines without crankshaft

Cam follower →Valve gear ~Indirect valve gear; →Valve gear components

Cam follower control →Variable valve control ~Systems with camshaft ~~Variability between cam and valve ~~~Discontinuously variable systems

Cam follower drive →Valve gear ~Indirect valve gear

Cam lift →Camshaft ~Exhaust camshaft

Cam lift curve →Camshaft; →Valve lift curve

Cam nose (*also*, →Camshaft). The section of the cam contour is termed as cam nose on which the base circle passes into the cam ramp. Depending on cam nose/quieting ramp design, corresponding speed and acceleration responses result at the beginning of the valve motion.

Cam profile →Camshaft ~Cam shape

Cam shape →Camshaft

Cam widths →Camshaft

Camshaft (*also*, →Valve gear). Camshafts are used to control the charge processes on internal combustion engines. Depending on the version of the engine, one common camshaft is used for the intake and exhaust valves or two camshafts are used, one each on the intake and exhaust sides. The latter design is predominant on engines with four or more valves. If an engine has five or six valves per cylinder, two camshafts are also used when three cams per cylinder can be located on one camshaft.

SIMPACK –
Technology Leading Engine Simulation

SIMPACK has been setting the standards for the simulation of high-end multibody systems for more than a decade.

Used extensively by the leading Formula One engine manufacturers for a number of years, SIMPACK has become their simulation tool of choice. The new integrated module - SIMPACK Engine - is a specialised tool for modelling both individual engine components as well as entire engine models.

Combining the accuracy, stability and speed of the SIMPACK solvers, SIMPACK has enabled the latest technologies in engine design to be efficiently analysed. SIMPACK Engine's additional functionality can be used to predict structural loads, life cycles, as well as the vibration and dynamic performance of valve trains, crank trains and timing systems (chain, belt, gear wheel).

Courtesy of DaimlerChrysler AG

 + +  =

SIMPACK Crank Train   SIMPACK Valve Train   SIMPACK Chain Drive   SIMPACK Engine

For further information visit www.simpack.com
INTEC GmbH, Argelsrieder Feld 13,
82234 Wessling, Germany, intec@simpack.de

C

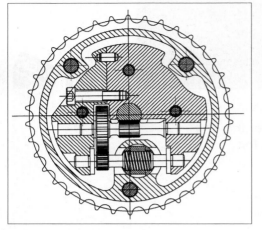

Fig. C30 Camshaft adjustment system with self-locking (Dr. Schrick)

~**Adjustment** (*also*, →Charge transfer; →Control units; →Control/gas transfer ~Four-stroke engine ~~Variable valve timing; →Valve gear; →Valve timing). Crankshaft adjustment means the change in the angle of the camshaft in relation to the crankshaft to achieve optimum timing for the various states of engine speed and load. A variety of different designs have been used. The best known include:

- Axial piston controller
- Vane cell controller
- Adjustment device for tensioning rail integrated into chain tensioner

In terms of function, these can be differentiated between the so-called black-white adjustment, with stop positions at "advanced" and "retarded," and continuous advance where it is also possible to move to intermediate positions and stop there. Examples of a few systems in production are shown in **Figs. C30–C32**.

~**Arrangement.** It is necessary to differentiate between three basic arrangements for camshafts. In the cylinder block in the vicinity of the crankshaft,

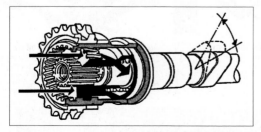

Fig. C31 Phase converter camshaft adjustment system with helical gears

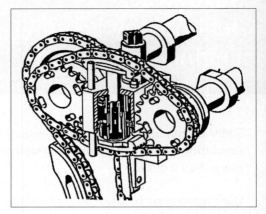

Fig. C32 Camshaft adjustment with chain system (hydraulic ring)

these are called in-block camshafts. This design provides:

- Cost advantages
- Low overall height of cylinder head
- Disadvantages for the engine dynamics

Fig. C33 shows the basic layout with an in-block camshaft. If the camshaft is located in the cylinder head, it is called an overhead camshaft. The primary characteristics of this design are:

- Advantages for engine dynamics
- Low moving masses
- Higher cylinder head
- Complicated camshaft drive

Figs. C34 and C35 show a cylinder head with overhead camshaft and the camshaft drive.

If the camshaft is located in the "V" between the cylinder banks on V-engines, it is called a central camshaft. The advantages of this arrangement are:

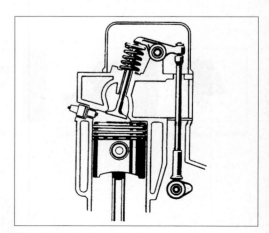

Fig. C33 In-block camshafts

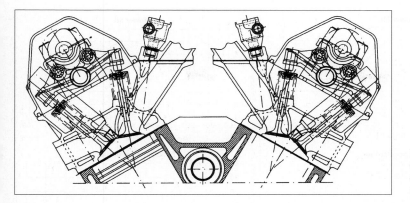

C

Fig. C34
Cylinder head with
overhead camshaft and
associated chain drive

- Compact engine
- Low-cost camshaft drive
- Low overall height of cylinder head

~**Base radius.** The base radius is the geometrical location at which the valve lift is zero. The base radius is marked with RG in **Fig. C36**.

~**Bearings.** Generally, camshaft bearings are designed as bushings made from materials based on bronzes (copper alloys) or aluminum alloys; or as bearing semi-shells made from copper alloys. Due to the larger space requirement and higher costs, rolling bearings are seldom used.

Bearings are dimensioned according to the drive forces and actuating forces for the valves; this is important for the friction loss (power required to overcome the friction resistances).

~**Cam contour.** *See below*, ~Cam shape

~**Cam lift curve** (*also*, →Valve lift curve). The cam lift curve is a plot of the cam lift over the cam rotation angle.

~**Cam nose.** The cam contour can be subdivided into the cam nose and the main cam. The acceleration is relatively low in the area of the cam nose so that changes in the valve play do not result in any major shock pulses within the scope of usual values.

~**Cam profile.** *See below*, ~Cam shape

~**Cam shape.** The cam shape (cam contour, cam profile) can be represented in different manners. Usual forms of representation are:

- Polar coordinates (points on cam contour) (**Fig. C37**)
- Tangents to cam contour (flat probe) (**Fig. C38**)
- Circles tangent to the cam contour (round probe) (**Fig. C39**)

The cam contour results from the desired valve motion, taking into consideration the geometry of the transfer element between the valve and cam. For example, a system with the associated lever geometry, shown in **Fig. C40**, provides a symmetrical valve lift diagram with an asymmetrical cam shape. **Fig. C41** shows the symmetrical valve lift curve and its second derivation, the specific valve deceleration.

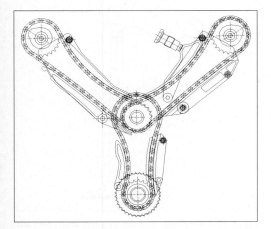

Fig. C35 Chain for camshaft drive

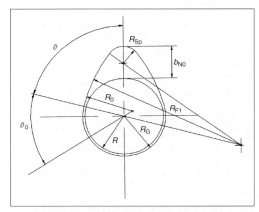

Fig. C36 Geometrical relationships on a cam

C

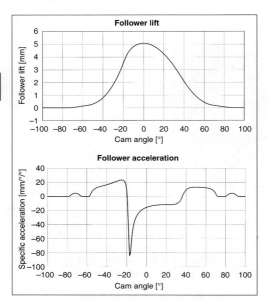

Fig. C37 Probe stroke and specific acceleration at probe radius R = 0 (polar coordinates)

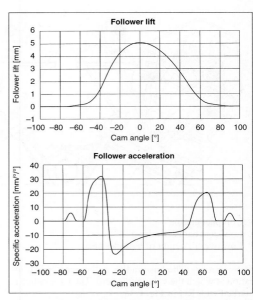

Fig. C39 Probe stroke and specific acceleration at probe radius R = 10 mm (radius probe)

~**Cam width.** The cam width is determined primarily by the surface pressure between the cam and tappet. Because these two elements slide against one another with high surface pressure, the material pairing is important.

~**Drive.** Camshafts are usually driven by toothed belts, chains, or spur gears. In rare cases, other types of transfer elements are used such as vertical shaft drives with bevel gears or push rod crank gears.

On engines with overhead camshafts, the preciseness of the angular association between the camshaft

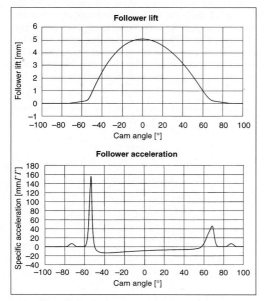

Fig. C38 Probe stroke and specific acceleration at probe radius R = ∞ (flat probe)

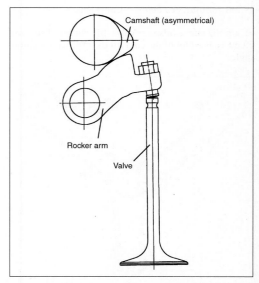

Fig. C40 Lever geometry for symmetrical valve lift and asymmetrical cam shape

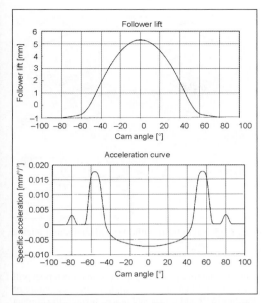

Fig. C41 Symmetrical valve lift curve and specific valve acceleration

and the crankshaft increases in the sequence listed above; however, the complexity and costs also increase.

On engines with in-block camshafts or central camshafts, the spur gear drive can be produced at lower cost due to the minimum distance between the camshaft and the crankshaft.

~**Exhaust camshaft.** This is the exhaust analog of the intake camshaft, and actuates the exhaust valves. The frequently lower cam lift compared to the intake camshaft results from the higher pressure difference in the cylinder at exhaust opening for exhausting the burned gas.

~**Fatigue strength (pitting).** *See below,* ~Wear

~**Hertzian stress.** The Hertzian stress is an important parameter for evaluating the load on cams and cam fol-

lowers. It is determined decisively by the cam shape and cam width.

~**Intake camshaft.** If two camshafts are used—for example, on multivalve engines—the intake camshaft actuates only the intake valves. The cam lift on the intake camshaft is usually the same as or greater than the lift on the exhaust camshaft to achieve maximum charge.

~**Low-jerk cams.** The jerk is the third derivative of the path with respect to time (i.e., d^3x/dt^3), it is the rate of change of acceleration. Together with the elasticity of the system, the jerk is decisive in introducing elastic deformation in the cam train. The results are vibration in the system with increasing noise emission, wear, and loss of adhesion. To minimize these negative consequences, valve lift curves are designed to be "low jerk."

Because the low-jerk requirement for cams is in conflict with maximization of the time-area integral of the valve, the design objective is to maximize the rigidity of the system and minimize the moving mass. This makes the valve train less sensitive to high-jerk values. The degree to which this objective can be achieved determines the limits for possible maximization of the time-area integrals.

~**Main cam.** The main cam determines the opening cross section for the charge cycle. It ends at a point corresponding to the cam nose.

~**Mass.** The mass of the camshaft should be kept as low as possible. For this purpose, camshafts are designed with a hollow shaft. As a matter of principle, the design of the entire valve train determines the load on the camshaft and, therefore, the required dimensions and mass.

~**Materials.** The dominating criteria for selection of the materials are the terminological characteristics and strength. For mass-produced camshafts, technical production aspects such as cost are of considerable significance. The terminological properties are decisive for the cam surfaces. The strength characteristics are significant particularly for shaft bodies subject to bending and torsion loads. A selection of the most important materials is shown in **Fig. C42**.

Material	Hardeness at shaft [Mpa]	Hardness of cam [HRC]	Suitability for sliding friction	Suitability for rolling friction
Case hardened steel	800...1,200	55...62	++	+++
Steel for induction hardening	800...2,100	50...62	+	+++
Gray cast iron with lamella graphite, induction hardened	100...300	40...55	+	++
Gray cast iron with lamella graphite, remelt hardened	100...300	40...55	+++	++
Clear shell casting with lamella graphite	100...300	40...55	+++	++
Clear shell casting with spheroidal graphite	500...700	40...55	+++	++
Gray cast iron with lamella graphite, induction hardened	500...700	40...55	+	++
Gray cast iron with spheroidal graphite, remelt hardened	500...700	40...55	+++	++
Malleable iron, inductive hardened	600...800	50...60	+	+++

Fig. C42 Summary of common camshaft materials

C

Fig. C43 Chill cast cam

Fig. C44 Inductive hardened cast iron cam

Figs. C43 and C44 show micrographs of two important materials.

The design of assembled camshafts allows these camshafts to be built using different materials. This allows individual elements of the camshaft to be produced using materials which are particularly suited for their specific loads.

~Number of camshafts. The number of camshafts required is determined primarily by the number of valves in addition to the design of the combustion chamber and ports as well as the required rigidity of the valve actuation. Ports designed for maximum flow and combustion chambers designed for optimum combustion require overhead valves arranged in a V shape: to maintain compact and rigid transfer elements with this valve arrangement, two overhead camshafts are required per cylinder bank. **Fig. C45** shows one version produced.

~Offset. With flat cylindrical tappets (e.g., bucket tappets), an offset is provided between the middle of the

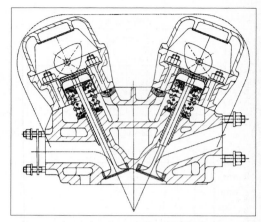

Fig. C45 Valve train with two camshafts and bucket tappets

cam and the middle of the tappet so that the tappet will turn. In combination with a slight conical curvature of the camshaft and a crowned tappet surface, the contact point between the cam and tappet is displaced from the middle of the tappet. The friction between the cam and tappet acts on an offset to the middle of the tappet, thereby rotating it. This basic arrangement is shown in **Fig. C46**.

~Production. The production method for the camshaft depends on the type of engine and the resulting loads.

Production of the cam profile requires maximum precision and is usually achieved by grinding. Demanding cam shapes require CNC (computer numerical control) machines with up to five axes.

~~**Assembled camshafts.** An assembled camshaft is a camshaft whose cams are produced separately from the body of the shaft and permanently connected with the shaft body at a later time (**Fig. C47**). Assembly can be accomplished by means of a friction fit or positive fit.

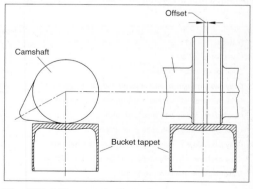

Fig. C46 Offset, cam/bucket tappet

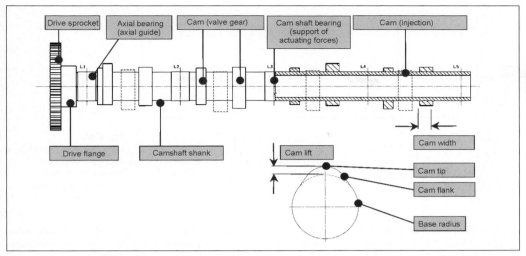

Fig. C47 Assembled crankshaft

The separation of the camshaft into cam and shaft allows selection of the material, production procedure, and heat treatment to be matched to the specific functions and loads. A large number of processes have been developed for producing assembled camshafts, which compete with the traditional process of casting and forging followed by machining.

Some assembled camshafts which have been introduced into mass-produced engines are described briefly below.

- *MWP-Süko*: Shaft body of tubing with the end ground off. Cams shrink-fitted in classic manner by heating. Cam material optionally hardened steel (e.g., 100 Cr 6) with subsequent final grinding or precision-fitted to final dimensions and heat-treated, without subsequent grinding.
- *Emishaft*: Cold-drawn precision tubing (e.g., St 52). Steel cams, forged and hardened. Cams are loosely slid over pipe, which is then widened to a defined dimension from the inside by applying high liquid pressure. Final processing by grinding bearings and cams.
- *Pressta*: Cold-rolled, ductile pipe (e.g., St 52). At mounting points for cams raised "bells" are formed by rolled circumferential grooves. Cams are hardened steel, provided with splined inner profile which provides a positive connection to the pipe when the cams are pressed on under controlled pressure.

~~Cast camshafts. Cast camshafts are characterized by their production process. Production of camshaft blanks by casting is widespread, and all common iron-based casting materials are used. Except for so-called clear chilled casting, most cast camshafts are given their required strength and terminological properties by a suitable heat treatment. The camshafts are cast hollow to reduce the weight.

~~Forged camshafts. The extremely high strengths of forged camshafts make them well-suited for uses requiring maximum bending and torsional loads. The wear resistance of the cams is increased by heat treatment.

~Radius of curvature. The radius of curvature of a cam running surface is an important factor for evaluation of the production feasibility and surface pressure. It plays an important role in the selection of the material and production process. Cam followers with small radii have negatively curved (concave) cam surfaces (**Fig. C48**) that require particularly small grinding wheels for production.

Fig. C36 shows the geometrical relationships on the cam.

~Scuffing. *See below*, ~Wear

~Surface pressure. *See above*, ~Cam width; ~Radius of curvature

~Variable camshaft adjustment. *See above*, ~Adjustment

~Wear. The wear between the cam and cam followers can be subdivided into primarily two types of wear:

- Scuffing
- Fatigue (pitting)

Scuffing occurs when the contact partners are not sufficiently separated by lubricant. The metal surfaces contact one another directly (mixed friction, dry friction), resulting in microscopic welding of the roughness peaks. This increases the roughness of the surface which further accelerates the process, leading to scuffing.

Because mixed friction cannot be completely eliminated during starting and at low engine speeds, it is important to select the materials for the friction pair so

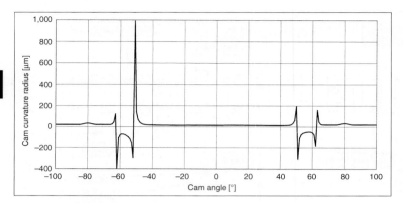

Fig. C48
Cam radius of curvature over cam angle for negatively curved acceleration sections

that they have a minimum tendency to bond to one another. To keep wear low, it is necessary to minimize the occurrence of mixed friction states in the design and operation of the engine. Measures leading to this goal include:

- Reduction of the contact forces
- Sufficient lubricant feed
- Favorable positioning and design of the lubrication point
- Avoiding states without lubrication

It is necessary to observe the following factors in terms of the properties of the sliding surfaces.

- Optimum selection and adaptation of the friction partners
- Favorable surface topography with high contact areas
- Suitable wear protection coating on friction surfaces for special applications

Fatigue wear results from local overloading of the material as a result of the periodic pressure stress. Fissures result in the area of the surface of the mating pair which leads to material erosion (so-called pitting).
Figs. C49–C52 show a number of types of wear.

Camshaft arrangement →Camshaft ~Arrangement

Camshaft base radius →Camshaft

Camshaft bearing →Bearings ~Bearing positions

Camshaft drive →Camshaft ~Drive; →Toothed belt drive; →Valve gear ~Gear components

Camshaft mass →Camshaft

Camshaft offset →Camshaft

Camshaft phasing →Camshaft; →Control/gas transfer; →Control/gas transfer ~Four-stroke engine ~~Timing; →Electronic open- and closed-loop control ~Electronic open- and closed-loop control, gasoline

engine ~~Functions; →Valve gear; →Variable valve control ~Operation principles, ~Systems with camshaft ~~Camshaft phasing systems

Camshaft phasing systems →Camshaft ~Adjustment; →Variable valve control ~Systems with camshaft

Fig. C49 Scuffing on hardened cast cam, precision cast valve lever

Fig. C51 Pitting after longer operating time, chill cast cams, steel tappets

Fig. C52 Flaking white layer, nitration-hardened steel cams, steel cam levers

Fig. C50 Scuffing due to insufficient oil, steel cams, chrome-plated steel rocker arms

CAN interface →Electronic/mechanical engine and transmission control ~Electronic components

Candle filter →Particle (Particulates) ~Particulate filter system ~~Particulate filter ~~~Filter media

Capability of achieving leaner mixtures →Air-fuel mixture; →Air-fuel ratio

Camshaft production →Camshaft

Capacitor discharge ignition (CDI) →Ignition system, gasoline engine ~High-voltage generation ~~High-voltage capacitor ignition

Camshaft radius of curvature →Camshaft

Camshaft vibration damper →Torsional vibrations ~Camshaft

C

Capsule →Engine acoustics ~Engine noise

Carbon dioxide →Combustion products

Carbon dioxide emissions →Fuel consumption

Carbon monoxide →Combustion products

Carbon residue (also, →Fuel, diesel engine ~Properties). Carbon residues are the accumulation of residues from partially combusted fuel and oil in the combustion chamber and on the valves and spark plugs. Deposits in the combustion chamber increase the tendency to knock—for example, by changing the compression ratio or increasing the thermal insulation. On the other hand, deposits of carbon residue can also lead to spontaneous ignition.

If carbon residues form on the valves (intake valves), these residues can act like a sponge, absorbing and releasing fuel in an uncontrolled manner. This can result in malfunctions. The fuels normally used today contain additives, namely detergents, that help prevent formation of such residues and deposits at critical points.

Carbureted engine →Carburetor

Carburetor. The task of the carburetor is to add fuel to the induced air of a gasoline engine in the desired appropriate mixing ratio depending on the operating point on the engine map, its desired change, and the operating condition. The carburetor also includes the throttle valve, which serves as an adjustment element for the flow of induced air.

Carburetors in the pure sense of the word existed during the early period of engine manufacturing. Ever since Wilhelm Maybach built the first spray carburetors in 1892/93 and the principle became accepted, they have been devices for metering fuel.

Carburetors work passively and without external drives. The energy needed for metering the fuel and feeding it inside the carburetor is taken from the airflow.

In terms of function, the carburetor of an engine, and particularly a branched intake manifold distributing the mixture produced by the carburetor to the cylinders, must be regarded as a unit. The engine operating behavior largely depends on the amount of care taken in developing the intake manifold together with the carburetor for uniform mixture distribution to all cylinders at all operating conditions.

Literature: A. Pierburg: Vergaser für Kraftfahrzeugmotoren, Vertrieb VDI Publishers, Düsseldorf, 1970. — K. Löhner, H. Müller: Gemischbildung und Verbrennung im Ottomotor, Vol. 6 in H. List: Die Verbrennungskraftmaschine, Springer Publishers, Vienna/New York, 1967. — H.P. Lenz: Gemischbildung bei Ottomotoren, Volume 6 in H. List, A. Pischinger: Die Verbrennungskraftmaschine, New Series, Springer Publishers, Vienna/New York, 1990.

~Acceleration enrichment. Mixture preparation starts in the carburetor. However, it is primarily prepared in the intake manifold for thermodynamic reasons and, after starting at low temperatures, also in the cylinder head or cylinder only. The fuel boiling curve and intake air heating plus intake manifold are both matched such that the ideal homogeneous mixture can be attained with the engine at its operating temperature. But even with the engine at operating temperature, there will be high-boiling fuel portions in the intake manifold still in liquid state in the form of film on the walls; this will be carried away by the air. The air reaches the cylinders faster than the fuel, and this results in temporary leaning of the mixture entering the cylinder; this is then compensated by injecting fuel.

~Actuation and design, second carburetor stage. With multistage carburetors, there always has to be fuel delivery corresponding to the airflow as the second stage is opened. Furthermore, mixture-flow redistribution has to be under control in the carburetor and intake manifold. There are several ways to achieve this actuation. It is necessary to distinguish between (a) mechanical and (b) pneumatic actuation.

a. Second stage mechanically actuated.
Second stage as fixed Venturi carburetor. Actuation is effected using a valve-lever linkage. The second stage cannot open until the first one is already half open. Both stages reach full opening at the same time. The design of the second stage corresponds widely with the first one. Transition systems and/or damping in the opening of the second stage facilitate the initiation of fuel delivery. The engine power peak is usually attained using enrichment.

Some carburetors contain an eccentrically pivoted additional flap valve downstream of the actual throttle valve, which is opened up by the intake manifold pressure and mixture flow for better transition into the second stage.

Second stage as constant depression stage. These two-stage carburetors are equipped with one-sided pivoted chokes in the inlet (**Fig. C53**). The throttle valves of the second stages are mounted on a common shaft and actuated through a valve-lever linkage. The chokes can only open then when the pressure inside the intake manifold has dropped such that a diaphragm cell releases a locking device holding them.

The second stages contain one constant pressure stage each and a transition system of a constant pressure stage.

b. Second stage, pneumatically actuated (**Fig. C54**). This type of carburetor has both its stages built like fixed Venturi carburetors. A diaphragm cell actuates the stage controlled by the throttle control lever of the second stage via connecting rod. The latter stage is, however, not independent but coupled with the first stage via a pivoted lever and cam disk. The roller of the pivoted lever abuts the cam disk with no play. Thus,

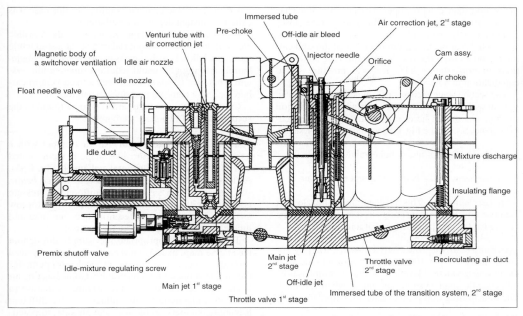

Fig. C53 Systems of a two-stage carburetor in downdraft arrangement, first stage as fixed Venturi carburetor with main system and idle system, second stage as constant pressure stage with transition system of a constant pressure stage; furthermore, premixture shutoff, recirculating air duct, and insulating flange

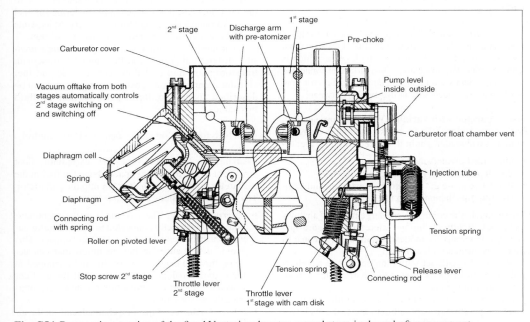

Fig. C54 Pneumatic actuation of the fixed Venturi carburetor second stage in downdraft arrangement

C

the second stage can only open when the first stage has opened nearly completely, and the second stage closes simultaneously with the first stage. The pressure against the rear of the diaphragm is taken through calibrated bores from the smallest Venturi tube cross sections of both stages. This leads to pressure signal equalization in both stages.

The second stage incorporates, in additon to the main system with normal structure, a second stage transition system, not drawn here, and immersed-tube enrichment, if necessary. See **Fig. C68** for the standard systems.

~Air choke. *See below,* ~Choke valve

~Air correction. *See below,* ~Compensating air

~Air correction jet. This is generally a nozzle that meters a flow of compensating air; it is specifically the nozzle for the compensating air of a main system.

~Altitude compensation. Mixture will enrich by approximately 5% per 1000 m increase of altitude with increasing geodetic elevation. Altitude compensators influence the main system fuel supply when executed. The transmitting unit is an evacuated barometric cell.

~Automatic choke. *See below,* ~Starting systems ~~Design types

~Bimetal spring. A bimetal spring consists of metal sheets with different thermal expansion coefficients bonded onto each other. In its flat form, the spring will deflect (cup) upon any change in temperature; in a spiral form, both ends twist against each other (*see* **Fig. C67**).

~Carburetor, mode of operation. A carburetor has a fuel accumulator—a carburetor float bowl in most cases—with a free fuel surface, the level of which is kept constant. The space above the surface is connected to the inlet duct via a breathing device (carburetor float bowl vent). It is necessary to distinguish between:

~~**Carburetor with variable Venturi cross section.** The constriction of the air intake channel is designed with a movable element in this case. Common are:

- One-sided pivoted choke
- Piston penetrating the channel orthogonally
- Rocker arm—e.g., pivoted downstream and constricting the channel from the side

These enable the control over a large range of airflows using but slightly changing differential pressure that causes comparatively modest carburetor resistance to fluid flow even at high airflows.

A conical needle jet with plunging pintle nozzle is linked for metering the fuel with the movable element for reasons of symmetry.

Running the engine under steady conditions at operating temperature, it is possible to meter the fuel over the whole range of airflows with the pintle nozzle, pro-

vided the movable element is idling. In this case it is called a constant depression carburetor.

A constant pressure stage occurs when the movable element does not move but rather rests at a stop while the engine idles. If such stage is incorporated in a single-barrel carburetor, it would require a separate idle system. This then is a hybrid between a fixed Venturi carburetor and a constant pressure carburetor.

Constant pressure stages are also to be found as the second stage of two-stage carburetors.

~~**Fixed Venturi carburetor.** An air trumpet with a fixed cross-section Venturi-like shape is incorporated as a constriction in the air intake channel (**Fig. C55**). At least one "main" jet is assigned to it (*see below,* ~Systems ~~Main system). The differential pressure generated with the Venturi tube remains small for low airflows. The pressure differential between inlet and intake manifold is used to meter the fuel.

To achieve the appropriate fuel supply throughout the engine map, fixed Venturi carburetors already need several additional systems and an accelerator pump for the engine at operating temperature. Compensating air is admixed to the fuel for premixture formation in order to compensate for the influence of the different Reynolds numbers on the fuel and air side, to practically neutralize the gravity effect of the fuel on its way to the carburetor channels, and to accelerate transportation in the channels. Carburetors are predominantly designed based on this principle.

~~**Single-barrel carburetor.** The aspirated airflow is conducted through a reduction in cross section (**Fig. C56**). A differential pressure is generated relative to the inlet duct and this is utilized for metering the fuel supply through a nozzle and for its delivery. At least one fuel jet is assigned to one reduction in cross section.

A carburetor is characterized by generating a pressure differential from an airflow and its immediate transformation into fuel flow. The air side and fuel side are identically designed in principle and can be described using the Bernoulli equation of fluid mechanics.

~Carburetor float bowl. *See below,* ~Level control

~Carburetor float bowl vent. The carburetor float bowl vent connects the gas space above the fuel in the carburetor float bowl with the inlet into the carburetor. The following versions are in use:

- *External venting.* The gas space in the carburetor float bowl is directly connected to the carburetor environment or sometimes to a carbon canister. This avoids disturbances at hot operation that can develop through evaporation of the fuel contained in the carburetor float bowl. Changes of the air-filter resistance due to contamination do, however, influence the mixture ratio.
- *Internal venting.* A pipe, or a channel, ends in the inlet or on the clean air side of the air filter. This type

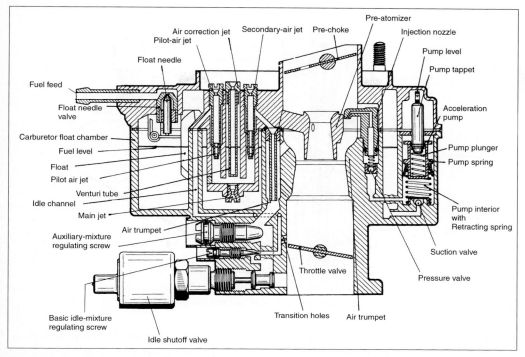

Fig. C55 Single-barrel carburetor as fixed Venturi carburetor in downdraft arrangement with mechanically actuated plunger-type dashpot pump, volume control system, and idle shutoff valve

of ventilation provides a system that is fluid-mechanically correct.

• *Switchover ventilation.* Switching over between internal and external venting is made using switchover ventilation. The switching is made either electrically with a solenoid valve or mechanically via an articulated cam follower on the throttle-valve actuation

~Carburetor float needle (float needle valve). The needle valve is the closing device of the float valve. It shuts off the flow of fuel into the carburetor and con-

trols the fuel level in the carburetor float bowl by interacting with the float.

~Carburetor stage. *See below,* ~Design types ~~Number of air intake channels ~~~Multistage carburetor

~Carburetor with variable Venturi cross section. *See above,* ~Carburetor, mode of operation

~Choke. *See below,* ~Starting systems ~~Functional systems

~Choke valve. *See below,* ~Starting systems ~~Functional systems

~Compensating air. Metered by separate jets, this air is added to partial fuel flows within a carburetor. In terms of their mass, compensating airflows are very small.

~Constant pressure carburetor. A Zenith-Stromberg CD carburetor is an example; SU carburetors are of similar design (**Fig. C57**).

~~Acceleration device. The hollow guide bore of the piston is filled with light oil into which a fixed damper piston plunges. This damper retards the opening of the piston with rapidly increasing airflow, causing temporary

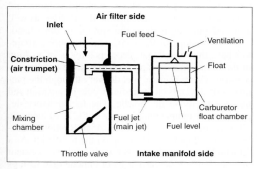

Fig. C56 Single-barrel carburetor

C

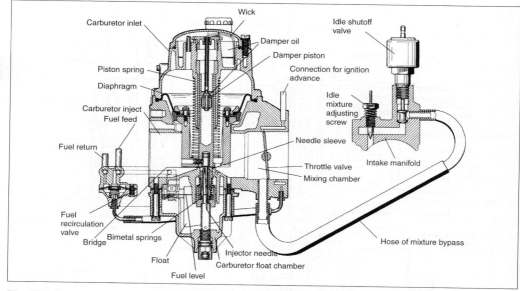

Fig. C57 Constant pressure carburetor with mixture bypass and fuel return

mixture enrichment. A relief valve in the damper allows the piston to drop immediately upon a decreasing air-flow. The carburetor cover is designed as an oil reservoir out of which oil is replenished into the hollow guide bushing through a wick.

With the engine at service temperature, it is possible to satisfy its fuel requirements with a single system owing to the existence of the damper and by using the intake system pulsations. Constant pressure carburetors are used in place of multistage carburetors. They avoid the weak points of the latter associated with the transitions between the systems. By contrast, the air-flow on constant pressure carburetors does not increase as quickly upon rapid opening of the throttle valve as is seen in fixed Venturi carburetors, which some people consider to be unresponsiveness.

~~Design *(also, see above, ~Carburetor, mode of operation).* In this case, a horizontal carburetor will be considered. The constricting element in the air intake channel is a circular cylindrical piston plunging from above and centrally guided. A spring along with the piston's own weight presses it against the bridge, filling the lower section of the air intake channel. A diaphragm closes the space above the piston, and the space is connected to the piston underside through the hollow piston. The underside of the diaphragm is ventilated toward the inlet. Because of the spring, a linearly increasing pressure differential (depression) is necessary for lifting the piston. The same pressure differential accelerates the induced air between inlet and the smallest point between piston and bridge so that the position of the piston indicates the value for the aspirated airflow. SU carburetors have no springs or diaphragm.

The upper part of the piston has a large diameter instead, sealing there against a piston in the carburetor cover with a slot. The pressure between the piston and the bridge actually remains constant.

The carburetor float bowl is located below the bridge. A needle jet is centrally fixed to the piston plunging into a pintle nozzle in the bridge, metering the fuel dependent on piston lift.

The piston hovers above the bridge during idle. At first, the fuel can be metered with the pintle nozzle following an airflow-dependent characteristic curve. The constant pressure carburetor is attached to the engine such that the piston will be fully open under full load and at about half maximum engine speed. In case of larger airflows at higher speed, the constant pressure carburetor works like a fixed Venturi carburetor. The mixing ratio of air and fuel can be convincingly designed across the entire engine map by utilizing the pressure pulsations in the intake system, which can be influenced via the intake manifold volume and cross section on the one hand and the air filter as well as air ducting before the carburetor on the other hand.

Some versions of these constant pressure carburetors have their needle jet placed positively in position in the pintle nozzle using a spring in the needle suspension system because fuel delivery through the pintle nozzle depends on the radial needle jet position and its eccentricity in the nozzle. The fuel delivery depends on the temperature in spite of the low kinematic viscosity of fuels because of the small Reynolds numbers of the flow in the ring gap on the needle jet. As a remedial measure, some versions of these constant pressure carburetors have a set of bimetal disk springs below the pintle nozzle.

~~**Starting systems** (*also, see below,* ~Starting systems). The starting systems for constant pressure carburetors correspond only in part to those for fixed Venturi carburetors. Normally, the throttle valve is opened to provide the increased flow rate of mixture for cold engine idle. Enrichment of the mixture is primarily geared to cold engine idle. The excess quantity of fuel or premixture is discharged from the outflow side of the bridge into the constant pressure carburetor using the prevailing differential pressure signal for metering, its signal being very similar to the one on the bridge. This is possible because, upon loading the engine when it is still cold, the higher viscosity of the cold oil in the damper leads to stronger acceleration enrichment than it would in an engine at operating temperature.

Given a manually actuated starting system, the throttle valve is set via a cam disk when the choke is pulled. At the same time, small holes in the jointly actuated slide valve open the path to the outflow side of the bridge for additional fuel.

The functional sequence of an automatic choke needs to be released before starting by single actuation of the accelerator pedal. The control unit for adapting to the engine temperature is a starter with a bimetal spring that can be heated electrically and that is surrounded by coolant flow.

~Constant pressure stage. *See above,* ~Carburetor, mode of operation; *see below,* ~Systems

~Dashpot pump. The dashpot pump (accelerator pump) serves for acceleration enrichment of the mixture supplied from the carburetor, primarily with the engine at its operating temperature.

In principle, dashpot pumps are volume-displacing pumps with a suction valve and a pressure valve. During the intake stroke, fuel flows from the carburetor float bowl into the pump interior through the suction valve. During the delivery stroke, fuel mostly flows through the pressure valve to an injection tube valve or calibrated injection nozzle leading to the air intake channel. The delivery stroke is initiated by a pump spring, which is tensioned in different ways. In most cases, a retracting spring is necessary, providing for the intake stroke.

There are various design types of dashpot pumps. The pump interior can be cylindrical with one piston, as shown in **Fig. C55**. Another common type is the flat pump interior with a flat or crimped diaphragm as the moving wall, as shown in **Fig. C58**. Another function of the dashpot pump can be as an opening damper for the throttle valve.

~Design types. Design types need to be distinguished according to the number of carburetor air intake channels and their spatial position.

~~**Number of air intake channels.** There may be up to four air intake channels in a carburetor.

~~~**Multistage carburetor.** A multistage carburetor contains two parallel air intake channels in a common carburetor housing designated as carburetor stages, where both work on an intake manifold or intake

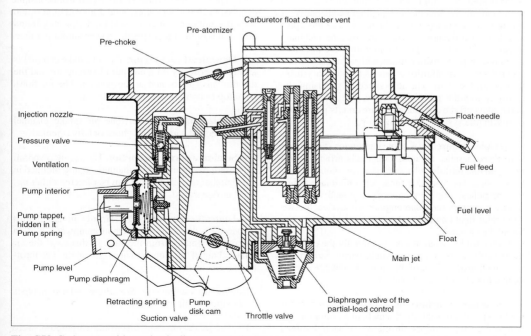

Fig. C58 Carburetor with mechanically actuated diaphragm dashpot pump and part-load control

C

manifold of one branch of the engine. Each throttle valve is opened in succession. Multistage carburetors have versions with one float bowl or with two carburetor float bowls.

The range of the differential pressure signal is very wide with the large span of the airflow conducted through a Venturi tube. Either the signal remains very small at small airflows in an operating range close to idling or maximum power output is limited because of the associated resistance of the carburetor to the fluid flow. As a way around this problem, the possible airflow through an engine is divided into two parts, with the first part using the first stage for smaller airflow rates, including idle and part load, and the second part—often larger in cross section—for reaching maximum power output only.

The first stage of a multistage carburetor is always designed like a single-barrel fixed Venturi carburetor with all the necessary systems. The dashpot pump and starting system are only required on the first stage, and this stage is directly actuated through the accelerator pedal. The multistage carburetor differs in actuation and design of the second carburetor stage (*see above*, ~Actuation and design, second carburetor stage).

~~~**Single-barrel carburetor.** A single-barrel carburetor has one air intake channel with one throttle valve. At the time it was first engineered, it had all the systems that were considered to be necessary for perfect engine operation. Single-barrel carburetors are the most common designs. **Fig. C55** shows a downdraft carburetor in a fixed Venturi carburetor as an example of the design. This single-barrel carburetor is equipped with a carburetor float bowl and a float needle valve. It incorporates a main system including Venturi tube, preatomizer, air correction jet with emulsion tube, and main jet, as well as a basic idle system with basic idle nozzle, basic air jet, transition holes, and basic idle mixture control screw. In addition, there is an auxiliary mixture control system. The idle speed is set using the auxiliary mixture regulating screw. The idle shut-off valve shuts off the auxiliary mixture and its parts from the basic idle system for engine stop.

~~~**Triple carburetor.** A triple carburetor is equipped with three parallel air intake channels arranged in a row with one throttle valve each in a common carburetor housing that works on three separate intake manifold branches. Triple carburetors designed as downdraft carburetors feed one engine cylinder each. Two carburetor float bowls are accommodated between the air intake channels.

Such a triple carburetor is functionally the embodiment of three fixed Venturi single-barrel carburetors into one housing. Each of the air intake channels is assigned to the same set of systems.

~~~**Two-barrel carburetor.** A two-barrel carburetor has two parallel air intake channels in a common carburetor housing with one throttle valve each that feed two separate intake manifolds or intake manifold branches. The throttle valves are activated synchronously. They may both be on one common shaft or on two shafts parallel to each other. The same applies to any choke valves.

A two-barrel carburetor represents the embodiment of two single-barrel carburetors in a single housing. Each of the air intake channels is assigned to the same set of systems. But normally there is only one carburetor float bowl and one dashpot pump. The necessary control units—for the starting system, for example—are provided in single form only.

~~~**Two-stage carburetor.** The two-stage carburetor is the structural embodiment of two multistage carburetors together with four air intake channels in a square arrangement with four throttle valves in a common carburetor housing; these supply two separate intake manifold branches. Downdraft carburetors designed as two-stage carburetors have two parallel throttle valve shafts, where one carries the throttle valves of the first stages and the second one those of the second stages. Both sides of these two-stage carburetors—that is, both multistage carburetors—are equally structured, aside from the fact that one carburetor float bowl will do.

Europe used such carburetors on large-displacement engines with six cylinders while their classic application in the United States was in mass-production V-8 engines.

~~**Position of the air intake channels.** The early spray carburetors were mostly updraft carburetors. Downdraft carburetors saw increasing use after the 1930s, only to prevail after World War II. Apart from constant depression carburetors, horizontal draft carburetors and semi-downdraft carburetors were predominantly found on engines with higher specific power output per liter.

~~~**Downdraft carburetor.** The air intake channel is arranged vertically on a downdraft carburetor and the throttle valve is located at the bottom. The air flows from top to bottom.

~~~**Horizontal carburetor.** The air intake channel on a horizontal carburetor is horizontally oriented.

~~~**Semi-downdraft carburetor.** The semi-downdraft carburetor has its air intake channel slightly inclined in the chamber and the airflow can be directed either upward or downward.

~~~**Updraft carburetor.** The airflow in an updraft carburetor is directed from the bottom upward. It often enters horizontally from the side. In this case, the air intake channel turns upward at a right angle. The throttle valve is located on top.

~Downdraft carburetor. *See above*, ~Design types ~~Position of air intake channels

~Electronic carburetor. Electronic carburetors were developed to save fuel in cases of uneconomical mechanical adaptation by refining the composition of the mixture to meet particular engine requirements, especially after cold start; lambda control was added later (**Fig. C59**). Electronic carburetors reduce back to the basic carburetor systems mechanically, but with additional features. Included in these is the throttle valve actuation that is effective in an operating range close to idling, mixture enriching intervention, sensors, and an electronic control unit.

The actual carburetor is a multistage, fixed Venturi carburetor with a pneumatically actuated second stage. The first stage comprises the main system and idle system with transition holes or transition slots, and the second stage consists of the main system, transition system, and immersed-tube enrichment. Throttle valve actuation of the first stage is made in an operating range close to idling, also using a continuously position-controlled throttle valve actuator.

A small electric motor applies a closing torque to the choke valve of the first stage for the mixture enrichment intervention. A tapping actuates a needle jet with the choke valve in the idle air jet designed as a pintle nozzle. The mixture delivered from the first stage can be continuously enriched over the entire range of airflows.

Engine stop and engine shutoff are indicated by closing the throttle valve or a valve jointly actuated by the choke valve servomotor that shuts off the fuel feed.

There is a throttle valve potentiometer on the throttle valve shaft of the first stage by which the throttle

valve position is determined and the change of position evaluated. Furthermore, a temperature pickup for the coolant, a temperature pickup on the intake manifold, an idle switch on the throttle valve actuator, and an engine-speed pickoff from the ignition system are added as sensors.

The electronic digital control unit processes the sensor signals and transforms them into first-stage throttle valve adjustments as well as choke valve adjustments. The following functions are available:

- Control of start and warm-up over all operating phases both on the air and fuel side
- Acceleration enrichment
- Lambda control
- Influence of the air-fuel ratio in the performance map
- Idle speed control
- Fuel shut-off while in overrun
- Engine stop
- Overspeed protection
- Function for preventing overheating of catalytic converter

Complementary functions are added, such as switching tasks for the PTC heating element in the idle system, early fuel evaporation system (EFE system), and exhaust gas recirculation and, furthermore, ignition timing diagnosis and control.

As in mechanical carburetors, starting with an electronic carburetor is done with the throttle valve in the

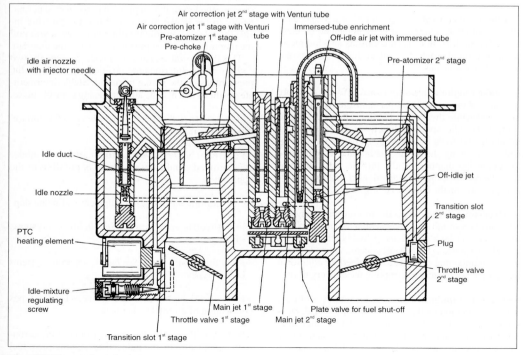

Fig. C59 Electronic carburetor.

C

first stage turned on and the choke valve completely shut. The idle speed control takes over throttle valve control using a set-point that depends on the engine coolant temperature. Mixture enrichment takes place with the aid of the choke valve, in which case basic steady state enrichment data dependent on the intake manifold temperature, already provided on a map, is used. This enrichment complements the acceleration enrichment as the engine is loaded, which depends on the intake manifold temperature, the starting point in the map, and the opening speed of the throttle valve. Lambda control, too, is achieved via the choke valve.

~Emulsion tube. *See below,* ~Systems ~~Main system

~Engine stop. *See below,* ~Equipment

~Enrichment. *See below,* ~Starting systems ~~Requirements; ~Systems

~Equipment. Carburetors have a number of devices that improve their function or make possible secondary functions that are desirable for other reasons.

~~**Engine stop.** There are several options in the case of fixed Venturi carburetors to avoid engine dieseling after cutoff—for example, by shutting off the idle fuel, idle premixture, or idle mixture. In addition, it is common to close the throttle valve so much that the orifices of the system delivering fuel during idle assume a position upstream of the throttle valve.

~~**Idle speed control.** Direct control of idle speed requires a suitable governor. It is implemented on electronic carburetors. Idle speed control is also indirectly possible through control of intake manifold pressure by mechanical means.

~~**Intake manifold pressure control.** When pressure is clearly dropping in the intake manifold with increasing speed while the engine is idling, it is possible to control the idle speed indirectly using mechanical means in the form of a throttle valve actuator (**Fig. C60**). Balancing the spring forces and compressive forces on the diaphragm, the throttle valve actuator will always open the throttle valve so that the adjustable pressure on the compression spring is maintained.

~~**Overrun.** Combustion will stop on many engines while in overrun and especially at high rotational speeds. This results in high emissions of hydrocarbons. There may also be noises in the exhaust system. A remedial measure is maintaining combustion by feeding additional mixture, or completely interrupting the delivery of fuel.

~~**Pressure tappings.** Pressures for purposes of control and actuation of the carburetor and other devices such as ignition distributor and exhaust gas recirculation valve are taken from the carburetor through bores. These holes can correspond to transition holes and can be covered by the throttle valve. A pressure signal dependent on airflow is obtained in the smallest cross section of a Venturi tube.

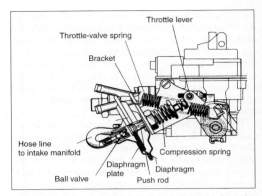

Fig. C60 Throttle valve actuator as intake manifold pressure control

~~**Throttle valve actuator.** A throttle valve actuator is a diaphragm cell, the compression spring of which is so strong that it can override the cutoff springs in the throttle control linkage and open the throttle valve it is acting upon (**Fig. C60**). A pushrod is fixed to the diaphragm mounted between two dishes pressing the compression spring against the throttle control lever until one dish rests in the housing. The space to the right of the diaphragm is connected to the ambient pressure and the space to the left of the diaphragm is connected permanently to the intake manifold. When the pressure inside the intake manifold is sufficiently low, it will pull the diaphragm into the position shown in the diagram.

This throttle valve actuator can only assume the two end positions. Intermediate positions result from the pressure inside the intake manifold. Stable intermediate positions can also be adjusted using control valves.

~Fixed Venturi carburetor. *See above,* ~Carburetor, mode of operation

~Flow bench. Device for testing carburetors under standardized conditions as close as possible to the application.

~Fuel shutoff. *See above,* ~Equipment ~~Engine stop

~Full-throttle enrichment. *See below,* ~Systems ~~Enrichment

~Fully automatic start. *See below,* ~Starting systems ~Design types

~Horizontal carburetor. *See above,* ~Design types ~~Position of air intake channels

~Hot operation. With the engine running, the carburetor is kept cool by the induced air and incipient fuel evaporation. However, when ambient temperatures are very high, there may be malfunctions due to postheating in the event that a hot-running engine continues to idle after the motor vehicle is parked, or when an engine stopped in a very hot condition is restarted after a short period or is driven away with high-power application after this hot start. Apart from the fact that the fuel pump will not deliver sufficient fuel, the causes also derive from excessively high temperatures in parts of the carburetor, which can lead to fuel evaporation. Particularly vulnerable are engines with the intake and exhaust manifolds arranged one on top of the other and screwed together for mixture heating.

Disturbances during hot operation strongly depend on the fuel boiling characteristics. Operation at high altitude enhances evaporation and susceptibility to these defects.

The susceptibility to such problems can be lessened by introducing supplementary air into the intake manifold, returning unused fuel to the tank, and retarding carburetor heating by shielding the carburetor or by using an insulating flange between the carburetor and the intake manifold.

~Icing. When parts of the fuel evaporate they draw energy from the aspirated air and the carburetor components; this can cause the moisture contained in the air to freeze, and the adherence of ice to parts of the carburetor may result in malfunctions. The maximum tendency to ice formation is at air temperatures around 5°C and in very high humidity atmospheres, particularly in fog.

~~Full-load icing. Full-load icing or Venturi tube icing takes place on older carburetors where the main system emulsion tube is centrally accommodated in the Venturi tube and the premixture passes into the air by a short route. The Venturi tube ices over, and this leads to a decrease in engine power or even engine stall as a result of reduced airflow and consequent over-enrichment of the mixture. Only a more suitable fuel can improve the situation if the intake air preheating cannot be enhanced.

~~Idle icing. Idle icing can occur when the main system is already delivering at low engine load and ice from the cooling air-fuel mixture deposits on the edge of the throttle valve. When the latter is closed, the engine may stall because of deficient air when idling.

Idle icing only occurs during a period of some minutes after starting when temperatures below freezing point occur because the air is not yet heated by the engine or not yet preheated. Idle icing will disappear by itself as the temperature of the system rises.

To avoid idle icing, preheating of the intake air, throttle valve, and endangered wall zones would be appropriate, as would changing the fuel to one with the right additives.

~Idle adjustment. *See below*, ~Systems ~~Idle systems with transition slot, ~~Volume control system

~Idle mixture control screw. Regulating screw for setting the fuel delivery via the idle system.

~Idle nozzle. The idle nozzle is the fuel jet in an idle system.

~Idle speed control. *See above*, ~Equipment

~Idle system. *See below*, ~Systems

~Intake manifold pressure control. *See above*, ~Equipment

~Jets. Jets are used to meter the flows of fuel, compensating air, and premixture and also the air in rare cases. It is possible to use holes for secondary purposes in place of jets. There is a variety of outer jet forms, with different shaping of the flow path. They have an inlet cone and often an outlet cone for shaping the flow and protecting the actual calibrating section. **Fig. C61** shows a typical main jet. It can be used for both flow directions.

Occasionally a pintle nozzle is used with a needle jet plunging into it that, in principle, is conically machined; for example, *see below*, ~Systems ~~Main system.

In rare cases there are orifices instead of jets.

~Lambda control. There are two solutions for implementing lambda control.

In fixed Venturi carburetors, an additional intervention has been integrated for lambda control. This solution uses valves similar to idle fuel shut-off valves incorporated into the main system and/or idle system, which influence the partial flows of fuel or compensating air and thus the mixing ratio supplied from the carburetor. These are switched in parallel or series depending on the carburetor.

Alternatively, another arrangement involves the part-load control, whereby a solenoid valve opens or shuts one channel to the intake manifold.

In electronic carburetors, the lambda control is part of the electronic control unit function volume. Variation of the mixing ratio is implemented via the choke valve servomotor and the choke valve.

~Level control. There are a variety of means to regulate the fuel level in a carburetor fuel accumulator.

~~Carburetor float bowl. The carburetor float bowl is a fuel accumulator for the carburetor. It usually contains a pivoted float immersed in the fuel contained in the carburetor float bowl. The float actuates a float

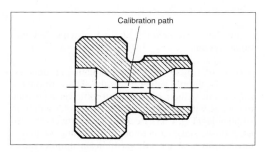

Fig. C61 Typical main jet

needle, which is the body that closes the float needle valve, and shuts off the fuel supply when the set level is reached. The float is either hollow or consists of foamed closed-cell plastic.

~~**Overflow carburetor.** An overflow carburetor contains a very small fuel accumulator into which fuel is constantly pumped from a carburetor float bowl that is separate from the carburetor and located below it. Unconsumed fuel runs through an overflow pipe into the carburetor float bowl which a further pump feeds from the tank.

~Main jet. The main jet is the fuel jet of a main system.

~Main system. *See below*, ~Systems

~Manual starting device. *See below*, ~Starting systems ~~Design types

~Mixing chamber. Space before the throttle valve where fuel and air mix.

~Mixture shut-off. *See above*, ~Equipment ~~Engine stop

~Multistage carburetor. *See above*, ~Design types ~~Number of air intake channels

~Overflow carburetor. *See above*, ~Level control

~Overrun. *See above*, ~Equipment

~Part-load control. *See below*, ~Systems

~Premixture. This is a mixture of a partial fuel-flow with a compensating airflow that is small in mass but volumetrically large. The premixture is so rich in fuel that it will not burn in the engine without strong air dilution.

~Pressure carburetor. A pressure carburetor is sealed pressure-proof to the outside for use on the supercharger pressure side. Also normal (vacuum) carburetors in pressure-proof housings have been used as an alternative.

~Pressure taps. *See above*, ~Equipment

~Pull-down. *See below*, ~Starting systems ~~Functional systems

~Recirculating air duct. A recirculating air duct bypasses the throttle valve. It sets the airflow through a stage at idle without causing the throttle valve position to change with respect to pressure tap holes and transition holes (**Fig. C53**). Recirculating air ducts are also used in two-stage carburetors for adjusting the unmetered airflows at the idle position on throttle valves mounted on a shaft.

~Semi-downdraft carburetor. *See above*, ~Design types ~~Position of air intake channels

~Single-barrel carburetor. *See above*, ~Carburetor, mode of operation

~Spray carburetor. Wilhelm Maybach built this carburetor as the first to work using the principle used in current carburetors. The fuel left the spray nozzle, which is the main jet in the modern device, that was mounted centrally in the Venturi tube and went directly into the flow of aspirated air. The spray carburetor did not have any compensating features at that time.

~Start carburetor. *See below*, ~Starting systems ~~Design types

~Starting systems. Engine control from the start to the end of the warm-up period is a complex task. Several phases of engine operation are passed through in succession in each case; these phases impose specific requirements on the carburetor. The important thing is to keep the composition of the mixture in the ignitable range and take it toward the lean operating limit without exceeding this limit.

There are several design solutions that cater to all phases of engine operation, which result in both simple and complicated starting systems when combined; fuel shutoff is combined with the starting system in some cases. The simpler the starting system, the more the requirements of the phases are fulfilled by more fuel addition than necessary, and this is uneconomical.

~~**Design types.** In this context, the designs of starting system used on fixed Venturi carburetors are discussed. The starting systems of constant pressure carburetors are discussed elsewhere (*see above*, ~Constant pressure carburetor).

Four basic designs are considered below: Automatic choke, fully automatic start, manual starting device, and start carburetor.

~~~**Automatic choke.** The mixture flow for starting and idle of a cold engine is ensured through opening the throttle valve (**Fig. C62**). Mixture enrichment is done by an eccentrically suspended choke valve, and a pull-down is provided. The temperature-dependent control unit is a bimetallic spring that is accommodated in the starter. The choke valve can be pulled open against the bimetallic spring.

Typical for such an automatic choke is that the connection between the choke valve and throttle valve positions is made using a stepped pulley. The functional sequence has to be initiated by depressing the accelerator pedal three times before starting. This causes the throttle control lever to lift from the stepped pulley. A cold bimetallic spring can shut the choke valve and the driving lever linked with the prechoke shaft rotates the stepped pulley into the start position. The throttle valve stop screw will rest on the top step of the stepped pulley—starting is now possible.

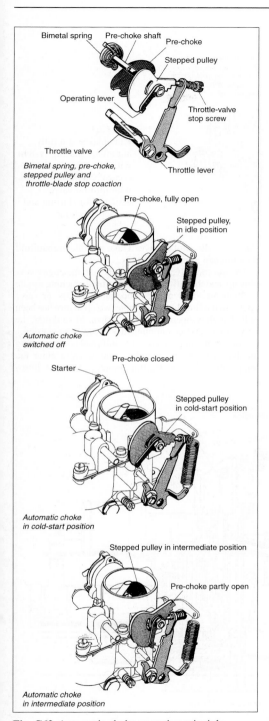

Bimetal spring, pre-choke,
stepped pulley and
throttle-blade stop coaction

Automatic choke
switched off

Automatic choke
in cold-start position

Automatic choke
in intermediate position

Fig. C62 Automatic choke operating principle

The pull-down opens the choke valve against the bi-metallic spring in the course of running-up and the very strong enrichment will decrease. Furthermore, the airflow can keep on pulling open the choke valve further.

The temperature of the bimetallic spring has to be adapted to the temperature condition of the engine. As the temperature increases, the bimetallic spring will open the choke valve and the mixture enrichment will be reduced. This is also possible when the engine is permanently coasted. The throttle control lever lifts from the stepped pulley on any action of acceleration. Its counterweighted shape causes the driving lever to drop and, while the bimetallic spring is heating, the throttle-valve stop screw will start resting on lower steps. The mixture flow for idle is reduced in a stepwise manner. An alternative to the counterweighted shape is a spring that presses a stepped pulley onto the driving lever.

With forced choke valve opening, the latter will be pulled open when the throttle valve is fully opened. The choke valve is open when the bimetallic spring is fully heated and the throttle control lever rests on its normal idle stop, which would be the lowest step of the stepped pulley in this case.

There is no separate mixture enrichment for idle in this example. There are carburetors on which the pull-down uses a piston releasing a channel through which compensating air is introduced into the idle system when the pull-down has opened. This way, the mixture provided from the idle system is enriched for starting.

~~~Fully automatic start. It is not necessary to initiate a fully automatic start before starting. The control of the flow rate of mixture for idling of the cold engine, and mixture enrichment on the systems that depend on the aspirated airflow are separate from each other.

A throttle valve actuator is employed for slightly opening the throttle valve for starting. The temperature-dependent control unit for mixture flow during idle is an expansion element that engine coolant flows around.

Mixture enrichment of the airflow-dependent system is done using a bimetallic spring that adjusts the choke valve—its operation is similar to that of an automatic choke. This also applies for the operating phases of running-up and continuous running. Together with complex throttle valve actuators, all desirable functions can be achieved to a great extent (e.g., engine stop and fuel shutoff while in overrun) except that acceleration enrichment needs to be supplied separately.

Idle mixture flow via the throttle valve. With this type of design, the entire flow rate of the mixture needed for idle is controlled by the throttle valve when the engine is cold (**Fig. C63**). The control unit is an expansion element. This type of design also includes a throttle valve actuator that adjusts the throttle valve for starting. After starting, the expansion element takes over throttle valve control for the duration of warm-up. After this the throttle valve can be completely shut with aid of the throttle valve actuator; this can also be applied for overrun fuel cutoff. Furthermore, this system can allow for the engagement of automatic transmission and/or air conditioning.

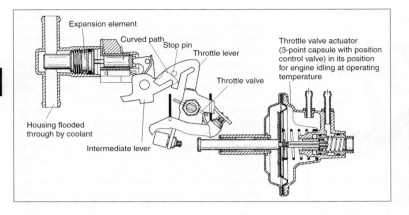

Fig. C63
Kinematics of a fully automatic starting system with idle mixture-flow control via throttle valve, with throttle valve actuator for starting and further functions

The throttle valve actuator is drawn in a position which it would take for the engine at operating temperature when idling, and the position control valve is operative. The diaphragm plate abuts the housing on the left side for starting and the throttle valve is adjusted.

Temperature bypass starting system. This system delivers additional mixture for idle and is always combined with a throttle valve actuator (**Fig. C64**).

The throttle valve is opened slightly by the throttle valve actuator, and the temperature-bypass air valve is shut, as is the choke valve. Thus, the intake manifold pressure can affect all systems.

The throttle valve actuator opens with the engine revving up, and the throttle valve shuts until reaching its idle position. In addition, the temperature-bypass air valve opens, bypassing the choke valve, which in turn has been pulled slightly open by the pull-down. In so doing, the very strong mixture enrichment is reduced for starting.

The thermostatic valve lets additional compensating air that is above its switching temperature enter the temperature bypass starting system. Thus, it intro-

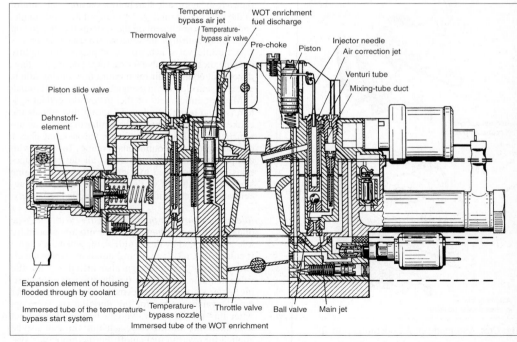

Fig. C64 Fully automatic start system with idle multimixture via the temperature bypass starting system, also compensating air part-load control in the main system

duces a temperature-independent stage into the mixture composition.

In the course of warm-up, the piston slide valve shuts off the temperature bypass starting system with growing coolant heating except for a small amount of leakage. The mixture composition for idling can be influenced through the temperature bypass starting system stage by increased addition of compensating air. In the example, the intake manifold pressure control partly compensates the change in load encountered by engaging automatic transmission and/or activating air conditioning.

~~~**Manual starting device.** Adjusting the throttle valve ensures the mixture flow for the starting and idle position of a cold engine. Enrichment is attained by shutting a choke valve (**Fig. C65**). The choke valve is pivoted one-sided in the inlet and can open under the influence of the aspirated airflow and independent from the choke lever position. In addition, a pull-down is usually provided as well. The throttle valve is often adjusted via a cam disk, thus creating latitude for engine adaptation.

A centrally pivoted choke valve contains a spring-loaded choke air valve that opens under the influence of the intake air.

Pulling the choke knob closes the choke valve for starting and at the same time the throttle valve is adjusted via the cam disk. The pressure from the intake manifold propagates across the throttle valve into the carburetor on starting and fuel can flow abundantly.

As soon as the pressure in the intake manifold drops while the engine is accelerating, a pull-down opens the choke valve and the very strong enrichment is reduced. Furthermore, a large airflow continues pulling open

the choke valve against a return spring. As the engine warms up and the idle speed seems to have risen sufficiently, the driver returns the choke knob back to its rest position. This also reduces the throttle valve adjustment, and the idle speed drops. The driver opens the choke valve a bit at the same time, thus decreasing mixture enrichment. The choke valve can open further under the influence of the airflow. There is no separate enrichment for the idle mixture; this also has to be done by the choke valve.

Certain pull-down versions allow separate consideration of running-up and continuous running. There are also manual starting devices with positive choke valve opening by which the choke valve is further pulled open against the return spring to avoid excessive mixture enrichment at wide-open choke when the throttle valve is fully opened.

~~~**Start carburetor.** Normally a start carburetor would not have its own carburetor float bowl and can be integrated into the carburetor or separately accommodated on the intake manifold. In the latter case it would usually get its fuel from the (main) carburetor. Start carburetors primarily control the additional mixture required for mixture enrichment at idle in a cold engine.

The differential pressure signal for fuel delivery is derived from the pressure differential between the inlet to the carburetor or environment and the intake manifold. Mixture enrichment on airflow-dependent systems cannot be addressed fully.

A start carburetor is normally actuated manually. More air and more fuel, or additional premixture, is usually controlled using a flat rotary valve, which releases the appropriate amount through holes in its sliding element depending on position. The choke knob is pulled into the position for starting and moved back during the course of warm-up. Often there is an air valve that is closed for the starting process; actuated by the intake manifold pressure, it will feed compensating air during running-up, thus reducing the very strong enrichment during startup.

~~**Functional systems.** The functional systems can be found on starting systems in different combinations. Several tasks are often fulfilled in parallel. These are:

~~~**Choke valve.** The choke valve (or air choke or choke) is a flap valve pivoted with a prechoke shaft in the inlet. When shut, the differential pressure signal of airflow-dependent systems will be increased, and the mixture is enriched.

Devices that influence flow rate:

- Through-flow valve. This contains an electrically heated expansion element that terminates the delivery of premixture from an auxiliary system on heating a needle jet. The initially very strong enrichment on idle is removed depending on the initial temperature and time after starting.
- Thermo-servomotor (**Fig. C66**). An electrically heated expansion element presses out the control pin depending on the initial temperature and time

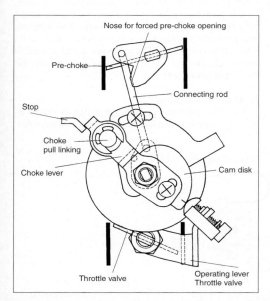

Nose for forced pre-choke opening

Pre-choke

Connecting rod

Stop

Choke pull linking

Choke lever

Cam disk

Throttle valve

Operating lever
Throttle valve

Fig. C65 Kinematics of a manual starting device in starting position

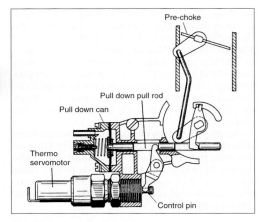

Fig. C66 Thermo servomotor

after starting. The pull-down can open the choke valve further and enrichment is decreased.

~~~**Pull-down.** Strong mixture enrichment generated by the choke-valve is reduced by slightly opening the choke using a pull-down as soon as the pressure drops in the intake manifold valve. There are several variants of this approach:

- *Simple pull-down.* This consists of a diaphragm cell with pull rod that opens the choke valve a bit through a lever on one end (**Fig. C66**). The strong enrichment of the mixture for starting is reduced.
- *Stepped pull-down.* The flow rate can be controlled in steps using a valve.
- *Dual pull-down.* There are carburetors with two pull-down cells with different lifts. One is active on running-up and the second one depends on a valve. Thus, the flow rate can be controlled in steps.

~~~**Starter.** In the narrower sense, the term "starter" is understood to mean the bimetallic spring carrier and components needed for it to function (**Fig. C67**).

A starter consists of the starter housing bolted to the carburetor and the starter cover closing the housing.

The housing bears the pull-down including the linkage, and, in the case of carburetors with automatic choke, it normally includes the stepped pulley as well as levers needed for its function. The starter cover accommodates the bimetallic spring, which transfers its torque with its outer end. If present, the resistance wire or PTC resistor (or two of them) for the electrical heating of the bimetallic spring is in the cover behind the bimetallic spring.

Provided the bimetallic spring is "heated" with coolant, the starter cover would have a coolant cover arranged on top which also bears the hose connections.

~~~**Thermal choke.** The running limit on the rich side can be exceeded if a manual starting device is not set back to its resting position in the course of engine warming—the engine can cut out in an extreme case.

This is avoided by a bimetallic spring arranged between the choke valve and its actuation device, such that the bimetallic spring opens the choke valve when it warms up.

~~**Requirements.** In principle, the requirements imposed on starting are the same at all temperatures. These requirements can be considered by looking at (a) the mixture flow for idle and (b) the mixture enrichment assigned to (c) the phase of engine operation, one by one.

a. *Mixture flow for idle.*

A cold engine requires more charge than at the operating temperature because of increased friction even at the same idle speed. Furthermore, the idle speed is often increased at lower temperatures to cover shortcomings in mixture distribution and mixture sequence.

It is possibly necessary to allow for the increased torques required when the air conditioning is switched on and particularly when automatic transmission is used under cold conditions. The reference variable used to assess whether the engine is warm is normally the coolant temperature.

b. *Mixture enrichment.*

The physical boundary conditions for mixture formation deteriorate considerably at lower temperatures.

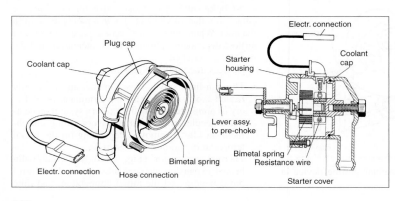

Fig. C67
Starter with bimetallic spring and heating element

The dew point for an air-fuel ratio of approximately $\lambda = 1$ is about 35°C at atmospheric pressure. Fuel cannot completely evaporate into the aspirated air if the suction region of the engine and the engine itself are colder.

Even when operating at steady state, the cold engine requires a richer mixture than at operating temperature to compensate for shortcomings in mixture sequence, mixture distribution, and homogenization. Furthermore, the acceleration enrichment has to be greater as the temperatures of the air and walls in the suction section are lower.

Using mechanical carburetors, the mixture, with few exceptions, is enriched so much that the lean operating limit is not reached, even under engine load.

Only electronic carburetors can allow mixture enrichment to be adjusted based on acceleration for the cold engine as well as for the warm engines. The main reference variable for enrichment is the temperature of the intake manifold.

c. *Phases of engine operation.*

- *Start-up and initial acceleration.* The starter cranks the engine during the actual start until its speed increases under the influence of the initially irregular and later regular combustion. Immediately after run-up, the engine assumes its idle speed corresponding to the given circumstances. The very strong enrichment has to be reduced to keep the mixture combustible.

  A strong air flow measured at the throttle valve under reference conditions with pressure ratio above the critical value effectively would support starting at all temperatures; however, the actual initial airflow is much less than required. An engine starts as soon as an ignitable mixture develops from the light fractions of the fuel under the conditions present, with excess fuel dispersed on the walls of the intake manifold. This process proceeds faster at high temperatures than at low ones. The reference airflow and enrichment do not need to depend on the initial temperature for starting.
- *Continuous running.* Continuous running is also called the post-starting phase, when the engine stabilizes after startup. Mostly the mixture flow for idle set for the end of running-up is maintained. However, additional steady-state mixture enrichment is needed depending on the initial temperature and time after starting. The acceleration enrichment must be higher in continuous running than during warm-up at the same intake manifold temperature.
- *Warm-up.* The running condition will continue to pass into the warm-up phase as the engine assumes its normal operating temperature. The mixture flow for idle can be controlled corresponding to the warming of the engine. The same applies for mixture enrichment dependent on the intake manifold temperature. Also decisive for acceleration enrichment are the engine load, which defines the starting point on the performance map, and the opening speed of the throttle valve.

~SU carburetor. *See above*, ~Constant pressure carburetor

~Synchronization (balancing). Synchronization (balancing) is understood to be the adjustment of carburetors on a system to give the same airflow and fuel delivery rate at idle.

~Systems (*also, see above*, ~Carburetor, mode of operation). A system is a fluid-mechanically coherent entity if it generates a differential pressure signal and transforms this into fuel flow. Carburetors are differentiated by the method of generation of the differential pressure signal as (a) systems depending on the pressure in the intake manifold and (b) systems dependent on the airflow.

a. *Systems that depend on the pressure in the intake manifold.* In fixed Venturi carburetors these supply the airflow zone of a stage in which the main system does not yet deliver fuel or not enough fuel.

~~**Idle system with transition slot.** The diagram shows the layout of the idle system of a modern carburetor (**Fig. C68**).

The idle nozzle meters the fuel taken downstream from the main jet. Compensating air is added through the idle air jet. The premixture flows through the idle duct and then it passes through the premixture shutoff valve. Further air is added at this stage, and part of the premixture passes the cone on the idle mixture control screw into the intake manifold. The rest enters the intake manifold via the transition slot section that is still below the control helix of the throttle valve. Meanwhile, air is also added through the section above. An alternative to transition slots are transition holes (**Fig. C55**).

At idle, there is a pressure ratio above the critical value at the ring gap between the edge of the throttle valve and the mixing chamber wall. As a result of this, the airflow through the stage depends on the position of the throttle valve to give the differential pressure signal for fuel metering. There will be proportionality between airflow and fuel flow in a first approximation. In the first instance, the idle speed is changed through the throttle valve position when setting idling and the mixture composition is controlled by the idle mixture control screw.

~~**Second transition system for fixed Venturi carburetors.** This system, shown on the right of the figure, improves the transition in a similar manner to the second stage of a fixed Venturi carburetor designed as a multistage carburetor (**Fig. C68**). It largely corresponds to an idle system. The example shows the differential pressure signal generated on the transition slot of the second-stage throttle valve.

~~**Volume control system.** This system delivers combustible mixture whose composition is independent of the quantity flow rate (**Fig. C55**). Adjustment of the auxiliary-mixture regulating screw suffices for changing the idle speed. This system supplements the classic

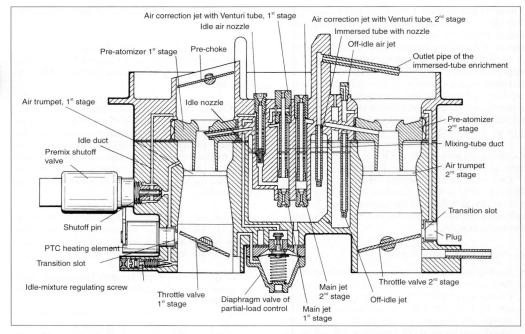

Fig. C68 Systems of a fixed Venturi-multistage carburetor in downdraft arrangement with main systems, idle system, part-load control, premix shutoff valve, PTC heating element, second-stage transition system, and immersed-tube enrichment

idle system, which is called the basic idle system in this connection.

Part of the airflow is branched off in the Venturi tube inlet and conducted to the intake manifold through a channel equipped with a Venturi tube that bypasses the throttle valve. A differential pressure develops in the gap between this Venturi tube and the pipe. This meters the partial flows of fuel and compensating air on the auxiliary nozzle and the secondary-air jet above, and the premixture thus generated at the small Venturi tube is delivered into the branched airflow.

b. *Systems that are dependent on the airflow.* There are two different systems—those with fixed Venturi carburetors and those with carburetors with variable Venturi cross sections:

Systems with fixed Venturi carburetors. It is necessary to consider:

~~Enrichment. The engine at operating temperature provides specific mixture enrichment, particularly for peak power. Enrichment works dependent of the airflow through the stage that the carburetor is currently in, yet at high airflows only, without interfering with the main system. Two enriching systems can be present on a single stage.

The dashpot pump can be employed for the second design function by opening the pressure valve and mechanically drawing fuel via the pump.

In case of immersed tube enrichment or full-throttle enrichment (**Fig. C68**), an immersed tube extends into the carburetor float bowl, where the lower end is formed into a nozzle.

Enrichment starts delivering fuel when the differential pressure signal developed by airflow contraction into the inlet lifts the fuel beyond the highest point of the outlet pipe. From this point, it delivers an increasing flow dependent on the airflow.

~~Main system. A main system works across the entire airflow range of a stage except for low airflows (**Fig. C68**). The differential pressure signal is generated from the airflow through a Venturi-type air trumpet inside the air intake channel. A pre-atomizer ending in the narrowest Venturi tube cross section amplifies the differential pressure signal, thus making the fuel distribution into the airflow and its atomization easier. Some carburetors have a further (smaller) pre-atomizer extending into the first one, which primarily affects fuel conditioning.

The pre-atomizer is channeled into an emulsion-tube channel that extends into the emulsion tube. The fuel-metering main jet is located in the inlet to the emulsion-tube channel, and the air correction jet is on top of the emulsion tube. Compensating air enters the emulsion tube through the air correction jet and moves onward to the emulsion-tube channel through the transverse holes, where it mixes with the ascending fuel.

Main systems without pre-atomizers are equipped with a discharge arm through which the premixture reaches the Venturi tube center from the emulsion-tube channel, which is laterally arranged in the wall. There are also many other ways of conducting the fuel and compensating air inside a main system and other ways of mixing them.

~~Part-load control. The mixture ratio supplied from a main system is normally adjusted for engine part load. This is enriched for full load and the region of the engine map close to full-load, among others, by intervention in the fuel delivery of main system and in case of multistage carburetors in that of the first stage. For the most part, the intervention depends on the pressure in the intake manifold.

Here, too, there are many versions. A second approach can be selected using a main jet bypass through which the fuel gets into the main system. **Fig. C68** shows a diaphragm valve, the back of which is open to the pressure in the intake manifold; this is closed at part load. It opens at high pressure inside the intake manifold so that additional fuel then flows into the emulsion tube. With a pintle-controlled main jet, a stepped needle jet whose position is dependent on the pressure in the intake manifold is immersed in the main jet. A needle jet immersing into the air correction jet has a reverse direction of action. Enrichment is an alternative or supplement to part-load control.

Systems of carburetors with variable Venturi cross section.

~~Constant pressure stage. This is the equivalent to the fixed Venturi carburetor main system. In the example depicted in **Fig. C53**, which shows a section through one side of a two-stage carburetor, there is an eccentrically pivoted air flap in the inlet of the second stage. The pressure in the intake manifold and flow of aspirated air can open the air flap when the throttle valve of the stage is open. There is a cutoff spring on one end of the air flap shaft, not shown here, countering the flap opening with torque and thus determining the differential pressure signal of this stage. The needle jet that plunges into the main jet from above through the air correction jet and the immersed tube is connected to the air flap via the cam assembly and the transfer lever. The needle jet contour controls the mixture ratio delivered from this stage depending on the position of the air flap and thus the air flow through this stage. Premixture from fuel and compensating air forms at the lower end of the immersed tube and flows toward the mixture discharge of this stage.

~~Transition system of a constant pressure stage. The transition system improves fuel delivery on second-stage start and is designed as a constant pressure stage **(Fig. C53)**. The transition system meets the closed air flap narrowly above its edge and starts when the edge of the air flap is passing over the outlet orifice. In principle, it uses the same differential pressure signal as

the constant pressure stage. A premixture is formed from the fuel and compensating air at the lower end of the immersed tube which flows through the immersed tube to the outlet.

Because the cross sections are constant on this transition system, its contribution to the mixing ratio delivered from the stage diminishes with increasing airflow through the second stage.

~Temperature bypass starting system. *See above,* ~Starting systems ~~Design types ~~~Fully-automatic start

~Thermal choke. *See above,* ~Starting systems ~~Functional systems

~Throttle valve. The throttle valve is used to set the flow quantity of the mixture.

~Throttle valve actuator. *See above,* ~Equipment

~Throttle valve potentiometer (*also, see above,* ~Electronic carburetor). This is a potentiometer fixed to the throttle valve shaft by which the position of the throttle valve is determined.

~Transition holes. *See above,* ~Systems ~~Idle system with transition slot

~Transition system of a constant pressure stage. *See above,* ~Systems

~Triple carburetor. *See above,* ~Design types ~~Number of air intake channels

~Two-barrel carburetor. *See above,* ~Design types ~~Number of air intake channels

~Two-stage carburetor. *See above,* ~Design types ~~Number of air intake channels ~~~Multistage carburetor

~Updraft carburetor. *See above,* ~Design types ~~Position of the air intake channels

~Venturi tube. The Venturi tube is a Venturi-type constriction mostly used in air intake channels that generates a differential pressure signal.

~Volume control screw. This is a screw for setting the flow rate of the auxiliary mixture.

~Volume control system. *See above,* ~Systems

~Warm-up. *See above,* ~Starting systems ~Requirements

~Zenith-Stromberg CD carburetor. *See above,* ~Constant pressure carburetor

Carburetor devices →Carburetor ~Equipment

Carburetor float bowl →Carburetor ~Level control

Carburetor float chamber vent →Carburetor

Carburetor float needle →Carburetor

Carburetor icing →Carburetor ~Icing; →Fuel, gasoline engine ~Additives; →Mixture formation ~Mixture formation, gasoline engine ~~Mixture cooling

Carburetor jets →Carburetor ~Jets

Carburetor stage →Carburetor ~Design types ~~Number of air intake channels ~~~Multistage carburetor

Carburetor starting devices →Carburetor ~Starting systems

Carburetor synchronization (balancing) →Carburetor ~Synchronization (balancing)

Carnot cycle →Cycle

Cast camshaft →Camshaft ~Production

Cast crankshaft →Crankshaft ~Blank ~~Cast

Cast iron with scaled graphite, alloyed, annealed →Piston ring ~Materials

Cast iron with scaled graphite, not annealed →Piston ring ~Materials

Cast iron with spheroidal graphite (spheroidal cast iron), alloyed, annealed →Piston ring ~Materials

Cast manifold →Exhaust system ~Exhaust gas manifold

Cast piston →Piston ~Blank production

Casting, connecting rod →Connecting rod ~Semifinished parts production ~~Casting

Casting process →Crankcase; →Cylinder head

Castor oil →Oil ~Body

Catalytic afterburning. During catalytic afterburning or the secondary reaction, the oxidizable constituents in the exhaust (primarily HCs, COs) are combusted or reoxidized to form H_2O and CO_2 by reducing the energy required for activation with a catalyst.

Catalytic coating →Particles (Particulates) ~Particulate filter system ~~Particulate filter ~~~Filter media ~~~~Regeneration of particulate filters

Catalytic converter. In motor vehicles, the exhaust gas converter is frequently called a catalytic converter. In the ready-to-install state, the catalytic converter consists of a housing containing the catalytic converter inside. The actual catalytic converter consists of the substrate (ceramic or metallic bed), the wash coat (coating), and a noble metal coating responsible for the catalytic effect. Catalytic converters not using noble metals are in development, but they are not currently capable of competition due to their lower activity and poor resistance to aging. At present, catalytic converters are the most effective system for treating the exhaust to reduce engine emissions. They reduce the emissions by up to 99%, thereby allowing even the strictest emission limits to be maintained.

~Activity. The magnitude of acceleration for chemical conversion with the aid of a catalyst is known as its activity. The activity can have different values for different chemical elements as well as different porosities (specific surface areas). Moreover, the activity does not remain constant, but decreases from aging and poisoning or contamination, thereby limiting the service life of the catalytic converter. While the decrease from aging continues more or less slowly over the service life of the catalytic converter, the activity can decrease very quickly to the point that the catalytic converter no longer functions at all as a result of poisoning or contamination. The activity of a catalytic converter is also very low at low temperatures, say below 200°C.

~Aging. Even a catalytic converter operating in a very pure gas flow (absolutely free of contaminants for catalytic converters) is subject to decreasing activity from aging. The cause for this aging is a sintering process of the noble metal crystals, decreasing the size of the active surface. This aging process depends on the temperature and proceeds more quickly at increasing temperature. The wash coat is also subject to the effect of aging. Here, the χ Al_2O_3 is converted to α Al_2O_3. This is accompanied by a change in the grain size leading to the pores in the wash coat closing and thereby reducing the inner surface area. Because noble metal (coating) is located in these pores as well, this process has a negative effect on the activity. **Fig. C69** shows typical changes in the specific surface area resulting from aging in relationship to the temperature. A sufficiently high conversion rate can be maintained in spite of increasing age by laying out the catalytic converter with an appropriately large active surface.

~Aluminum oxide. Aluminum oxide, Al_2O_3, plays a significant role as the wash coat or ceramic bed for heterogonous catalytic converters. Used as a substrate, the material can be present in the form of pellets for granulate-type catalytic converters or extruded elements.

Aluminum oxide is also used as a constituent of the ceramic catalytic converter substrate in the form of a compact extruded element interlaced with ducts (monolith). The material cordierite ($2\,MgO \cdot 2\,Al_2O_3 \cdot 5\,SiO_2$) is used preferably for the latter (ceramic substrate).

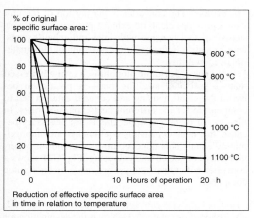

% of original
specific surface area:

Reduction of effective specific surface area
in time in relation to temperature

Fig. C69 Catalytic converter aging

~**Bypass catalytic converter.** Immediately after starting, the catalytic converter has no effect, because the temperature is still below the light-off temperature (*see below*, ~Light-off point/light-off curve). A small separate catalytic converter can be installed in parallel to the exhaust pipe close to the engine, through which the exhaust flows immediately after starting the engine in order to reach light-off temperature as quickly as possible. When the main stream catalytic converter has reached its light-off temperature, the catalytic converter located in the bypass is switched off and the exhaust is routed through the exhaust pipe to the main converter for normal operation. This is necessary to prevent operating the catalytic converter located close to the engine at impermissibly high temperatures. **Fig. C70** shows a typical layout for a bypass catalytic converter. Following a cold start, the shut-off valve is closed and the exhaust first flows through the start-up catalytic converter and then through the main catalytic converter. When the main catalytic converter has reached its operating temperature, the shut-off valve is opened and the exhaust flows directly through the main catalytic converter, except for a very small quantity which continues to flow to the start-up catalytic converter. The start-up catalytic converter can be shut off completely by installing a second shut-off valve.

~**Catalytic exhaust purification.** For catalytic exhaust purification, catalysts are used to convert toxic exhaust constituents into nontoxic compounds. Catalysts are substances which accelerate a chemical reaction without being used up themselves. Suitable substances which have a catalytic effect are noble metals. Already known for some time in the field of chemistry, catalysts were introduced for purification of exhaust gases from motor vehicles in the late 1960s. The challenge with the exhaust from motor vehicles was to convert different constituents in the exhaust (such as carbon monoxide, unburned hydrocarbons, oxides of nitrogen) simultaneously. Here, the following processes are accomplished:

1. Oxidation of pollutants and combustible constituents

    1.1 $CO + 1/2\ O_2 \Leftrightarrow CO_2$

    1.2 $C_mH_n + (m + n/4)\ O_2 \Leftrightarrow mCO_2 + n/2\ H_2O$

    1.3 $H_2 + 1/2\ O_2 \Leftrightarrow H_2O$

2. Reduction of nitric oxide

    2.1 $CO + NO \Leftrightarrow 1/2\ N_2 + CO_2$

    2.2 $C_mH_n + 2\ (m + n/4)\ NO \Leftrightarrow (m + n/4)\ N_2 + n/2\ H_2O + mCO_2$

    2.3 $H_2 + NO \Leftrightarrow 1/2\ N_2 + H_2O$

Processes 1 and 2 can be accomplished with good conversion rates in a single-bed catalytic converter (three-way catalytic converter) when a λ controller keeps the air-fuel mixture close to $\lambda = 1$ ("closed loop" three-way catalytic converter). With double-bed catalytic converters, two catalytic converter elements with various coatings are connected at the back of one another so that the oxidation and reduction processes are accomplished virtually separately from one another.

~**Cell density.** The cell density is usually specified in cells per square inch, or cpsi. On exhaust catalytic converters for motor vehicles, cell densities of 25 cpsi are achieved today for racing and up to 1200 cpsi for high-performance catalytic converters; 1200 cpsi corresponds to 186 cells/cm².

~**Ceramic bed.** Ceramic in the form of a cordierite monolith or as a packing—for example, in the form of small beads of aluminum oxide as granular-type catalytic converter—are used as the catalyst substrate. In addition to ceramic substrates, metal substrates are also used, and these are available only in the form of monoliths.

~**Cerium oxide.** A metal oxide with a capacity for adsorbing oxygen is added to the wash coat to compensate the control range of the lambda probe for HC surges. Cerium oxide has proven itself and is used primarily for this purpose. Alternating operation around $\lambda = 1$ results in a slight excess or deficiency of oxygen. This is compensated for by the oxygen storage capacity of the oxide film, which stores excess oxygen in the exhaust when available and releases it back into the exhaust during phases of oxygen deficiency in order to oxidize the HC and CO constituents. This results in the following process:

- *Oxidation:* Cerium (III) oxide absorbs oxygen from the exhaust. This is defined by the reaction equation:

    $2Ce_2O_3 + O_2 = 4CeO_2$.

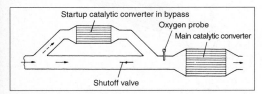

Fig. C70 Bypass catalytic converter

- *Reduction:* Here, a reducing agent is required for the cerium oxide to release the oxygen, and the CO in the exhaust acts as the reduction agent. For example, reduction with carbon monoxide gives

$$2CeO_2 + CO \rightarrow 2Ce_2O_3 + CO_2.$$

~Closed-loop. Normal operation of a three-way catalytic converter is designed as a closed loop when regulation of the air-fuel mixture has settled to a figure corresponding to $\lambda = 1$ and the temperature is above the light-off point.

~Closed-loop three-way catalytic converter. *See below,* ~Three-way catalytic converter

~Coating. The actual catalysts on motor vehicle exhaust purification systems are noble metals, such as platinum, palladium, and rhodium, which are applied in certain ratios to a substrate (ceramic bed or metal bed) in finely distributed form. For this purpose, the substrate is first coated with a wash coat consisting of Al_2O_3 to which the noble metals (catalysts) are then applied. A homogonous mixture of the wash coat and catalyst is also possible. Common ratios of noble metals as catalysts are 4:1–5:1 for platinum/rhodium or 1:14:1–1:28:1 for platinum/palladium/rhodium coatings (tri-metal coatings). Catalysts consisting of non-noble metals have not become generally accepted due to their low activity and low resistance to aging when used as catalytic converters for motor vehicles.

~Contamination. Contamination of the catalyst surface with extremely fine solid particulates can also be responsible for reduction of the activity. These clog the pores and thereby reduce the effective surface of the catalyst. In combustion engines, this can be caused by deposition of ash from the engine oil or metal abraded from the engine.

~Conversion. This designates chemical conversion of substances. In terms of the pollutants contained in the exhaust, we differentiate between oxidation and reduction processes.

~Conversion behavior. The conversion behavior of the individual constituent gases depends on a number of factors. The most important of these include:

- Efficiency of the catalyst for the specific process
- Composition of the gas—e.g., excess O_2 for converting CO to CO_2
- Temperature—most reactions are accomplished more rapidly at higher temperatures

~Conversion efficiency. This defines the magnitude of conversion of a substance present in the untreated gas. A conversion rate of 90% of the CO constituents, for example, means that 90% of the CO present in front of the catalytic converter is converted to CO_2 as it passes through the catalytic converter.

~Converter. The converter is the portion of the unit which provides catalytic exhaust purification. It consists of a steel housing with inlet and exhaust pipes, with a catalyst in the housing consisting of bulk material or monoliths.

In the case of ceramic monoliths, a resilient pressed mat—a so-called swelling mat—is positioned between the metal housing and ceramic catalyst bed to compensate for the different heat expansion rates. In the case of metal substrate, this is not necessary because the heat expansion coefficients are approximately equal. This honeycomb-type monolith is permanently connected to the housing. **Fig. C71** shows a converter with metal catalyst bed made by the Emitec company.

~DeNox catalytic converter (*also, see below,* ~NO$_x$ absorption-type catalyst). DeNox is the designation for conversion of oxides of nitrogen (present in combustion gases in the form of NO and NO_2) to nitrogen and CO_2 or H_2O. At $\lambda < 1$, this is accomplished by reaction with CO or C_mH_n (catalytic exhaust purification), while at $\lambda > 1$ with hydrocarbons (C_mH_n) or by reaction with NH_3 ($2NH_3 + 2NO + 1/2\ O_2 = 2N_2 + 3H_2O$) or urea. To date, addition of NH_3 or urea has been used only with stationary engines. Reactions using the hydrocarbons in the fuel increase the fuel consumption by 2–4%.

~Diesel engine catalytic converter. Although in diesel engines the percentage of CO, gaseous HC, and NO_x in the untreated emissions is lower than in gasoline engines, the problem of particulate emissions exists here in addition to the relatively high NO_x concentration still present. These particulates can be subdivided into solid (soot with deposited HCs and sulfates) and liquid particles (e.g., fuel constituents). Carbon monoxide and gaseous and liquid HCs can be converted to CO_2 and H_2O in oxidation-type catalytic converters.

The sulfur present in diesel fuel is combusted in the engine to form SO_2 and is converted with O_2 in the oxidation-type catalytic converter to form SO_3 which forms sulfuric acid in combination with the water present. This again leads to formation of sulfates. Therefore, the

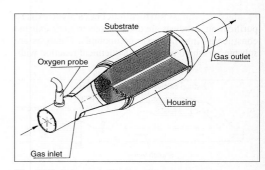

Fig. C71 Schematic layout of a converter

use of an oxidation-type catalytic converter increases the formation of sulfates compared to operation without a catalytic converter. This can be decreased by reducing the quantity of sulfur in the fuel, which is now being accomplished worldwide. A number of processes have been developed for separation of the solid particulates (soot):

- Electrostatic deposition of the soot particles and combustion in a separate combustion chamber using a burner with its own fuel supply. This process is technically complicated and is used only for large engines.
- Soot particles have been collected in a ceramic filter and combusted intermittently. Because the exhaust temperature is normally not high enough, it is necessary to fortify the fuel with additives (metals) to reduce the ignition temperature using a catalytic effect. The problem is that some of the metals may be entrained into the exhaust.
- When the diesel particulate filter is clogged with soot, the pressure loss and, therefore, the exhaust back pressure, increase, leading to poorer performance with higher fuel consumption and higher exhaust temperatures. This effect can be used to burn the soot out of the filter.
- Soot particles are collected in a ceramic or metal filter in the same manner as previously, however here they are burned out periodically by heating up with an electrical resistance-type heater.

Because diesel engines, unlike gasoline engines, are not operated at $\lambda = 1$, and because NO_x conversion cannot be accomplished in the lean range, it is necessary to reduce the NO_x in another manner. For conversion of the NO_x, it is necessary here to add reactive agents in the form of hydrocarbons, ammonia, or urea because the engine is operated with excess air, $\lambda > 1$ (*see above*, denox catalytic converter). Ammonia or urea is used only on larger engines due to the low quantity of NO_x emitted from diesel engines and the high technical and spatial requirements for the process.

~Dual-bed catalytic converter. Dual-bed catalytic converter is an old designation for a reduction catalytic converter and an oxidation-type catalytic converter connected parallel to one another using the so-called dual-bed system. Today, this type of catalytic converter has been superseded by three-way catalytic converters.

~Efficiency. The efficiency (conversion rate) indicates the percentage of the theoretical value achieved in a process based on per unit (1) or percentage (100%). The efficiency of a conversion process is the quantity of substance converted in relation to the total quantity of the substance in question—e.g., 90% CO conversion means that 90% of the CO present in the exhaust in front of the catalytic converter is converted to CO_2 and 10% remains in the exhaust in back of the catalytic converter. Efficiencies of more than 99% are achieved today for HC and CO oxidation.

~Exhaust back pressure increase. The exhaust back pressure has an influence on the volumetric efficiency of the engine. Particularly important here is the ratio p_l/p_g, where p_l represents the intake manifold pressure and p_g the exhaust back pressure. The higher the value of p_l/p_g, the higher the volumetric efficiency. The p_l value can be increased by supercharging (e.g., turbocharger) thereby increasing the volumetric efficiency. In the other direction, an increase in the exhaust back pressure reduces the p_l/p_g value and, therefore, the volumetric efficiency. The volumetric efficiency is defined as the ratio m_z/m_{th}. Here, m_z represents the mass of the intake charge without residual gas and m_{th} the theoretical charge quantity when the displacement chamber is filled completely with a fresh charge. The volumetric efficiency is proportional to the engine performance—that is, the engine output increases proportionally to the increase in the volumetric efficiency. Therefore, the objective is to counteract any increase in exhaust back pressure while reducing the pressure loss—that is, to the catalytic converter. Generally, a high volumetric efficiency also results in higher performance and, consequently, lower specific fuel consumption. **Fig. C72** shows the pressure loss resulting from catalytic converters from a cell density of 400 cpsi (cells per square inch) depending on the inflow rate w_A (i.e., gas flow rate in the empty space in front of the catalytic converter). For passenger car engines, the typical gas velocity at full load is around 30 m/s.

~Fuel mileage reduction. *See above*, ~DeNox catalytic converter; ~Exhaust back pressure increase

~Gasoline-engine catalytic converter. *See below*, ~Three-way catalytic converter

~Granular catalyst. *See above*, ~Ceramic bed

~Heater. The heater is used to heat the catalytic converter up to the light-off temperature following a cold start. This is accomplished by combustion of fuel in a

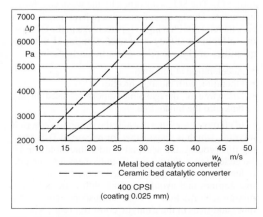

Fig. C72 Pressure loss through catalytic converter as function of inflow rate

burner system in the exhaust system or electrically by using the metal substrate in the catalytic converter as an electrical heating resistor or by installing an electric heater in front on ceramic monoliths.

When the catalytic converter is heated up by combustion of fuel, air and fuel are fed to a burner located separately in front of the catalytic converter. The increase in the temperature of the gas from this external combustion heats up the catalytic converter together with the engine exhaust.

On electrically heated catalytic converters, a portion of the catalytic converter at the inlet is designed as an electric resistance heater. The temperature of the heating resistor increases very quickly, so that the engine exhaust is heated up to the light-off temperature within approximately 10 s. Another possibility is to manipulate the engine parameters to heat up the catalytic converter (e.g., by retarding the ignition).

~Heat-up. Following a cold start, it is first necessary to heat up the catalytic converter to its light-off temperature before it starts operating effectively. This is accomplished, for example, in a passive catalytic converter system by using the engine exhaust. Active heating is possible to accelerate this heat-up process. This can be accomplished electrically or by additional combustion of hydrocarbons immediately in front of the catalytic converter. Consideration of technical feasibility and acceptable fuel consumption suggests a heat-up time of approximately 5–10 s is required to reach the future emission limits.

~Hydrogen sulfide (H_2S). Hydrogen sulfide (H_2S) can be formed in motor vehicles equipped with three-way catalytic converters. Even though the concentrations produced are far below the limits significant for health, hydrogen sulfide is annoying because it smells like rotten eggs.

H_2S is formed in the following process. The sulfur present in the fuel combusts in the engine to form SO_2. In the catalytic converter, the SO_2 combines with O_2 to form SO_3. The SO_3 is adsorbed by the wash coat forming sulfate. During operation with a rich mixture ($\lambda < 1$) and, therefore, with a high H_2 percentage in the exhaust, the sulfur in the sulfate is released, forming H_2S. Large quantities of H_2S are released particularly during:

- Acceleration at full load
- Operation with closing throttle
- Discharge of the gasoline vapors released in the tank into the intake line

In all three cases, the cause is the formation of a rich mixture, which is responsible for release of H_2S. During this process, the acceptable odor threshold can be exceeded for a short time. Countermeasures include reduction of the sulfur contained in the fuel for gasoline engines and avoiding peaks with extremely rich mixtures. Formation of H_2S can also be reduced by other additives in the catalytic converter (e.g., Ni).

With diesel engines, this effect does not occur, because they do not operate in the rich range.

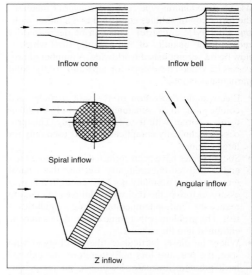

Fig. C73 Catalytic converter inflow sections

~Inflow. The cross section for holding the converter is significantly greater than that of the exhaust pipe, because this reduces the pressure loss and the limits of the length of the catalytic converter. It is therefore necessary to increase the cross section of the exhaust pipe to blend with that of the converter. The objective is to adapt the shape to keep the pressure loss as low as possible and also provide the most uniform possible speed over the cross section before the gas enters the body of the catalytic converter (monolith). **Fig. C73** shows a number of shapes used for the inflow sections.

~Intermediate layer. A wash coat is applied to the substrate (ceramic or metal bed) as an intermediate layer for application of the noble metal. The purpose of the intermediate layer is to increase the surface area for the layer of noble metal.

~Lambda probe (also, →Sensors). The lambda probe is a measuring device which determines the air-fuel ratio in the exhaust by measuring the partial pressure of oxygen and keeps it constant within a narrow range by means of a control system. A prerequisite for optimum operation of a three-way catalytic converter is that operation be maintained in the control range.

~Layout. A variety of catalytic converter models have been developed for different requirements in the motor vehicle sector. This variation is accomplished primarily by changing the four system components listed below.

- Substrate system (e.g., filler, ceramic, or metallic bed)
- Intermediate layer (wash coat)
- Catalytically active coating
- Location in vehicle

~Lean-burn engine catalytic converter. Lean-burn engine catalytic converters can be used in two areas: Either on diesel engines and on lean-burn gasoline engines or those with direct injection. Operation of the lean-burn engines currently in use is possible over wide ranges of air-fuel mixture (i.e., $\lambda > 1$ as well as $\lambda < 1$) even though the designation tends to indicate only lean operation with $\lambda > 1$. During operation of such engines, the HC contents in the exhaust decreases from $\lambda < 1$ to $\lambda = 1$ but increases again when λ increases. Operating with low HC concentrations in the exhaust can result in difficulties with the NO_x-reduction (catalytic exhaust purification, process 2). Because the catalyst has a certain capability of storing NO_x, this is stored in the minimum HC range. When operating with an HC increase (by changing λ), NO_x can then be released again. It reacts with HC, leading to a decrease in NO_x. If lean engines are operated primarily in the range $\lambda > 1$ (HC minimum), it is possible to add HC in front of the catalytic converter.

When used on diesel engines, the same processes are accomplished in principle; however, the entire temperature level of the exhaust is lower so that conversion of NO_x becomes increasingly difficult with increasing quantities of excess air. Moreover, the quantity of uncombusted hydrocarbons, which are used as a reduction agent, is too low to achieve reduction in the exhaust from diesel engines.

For this reason, it is necessary to add HC—for example, through subsequent injection, leading to an increase in the fuel consumption of 2–4%.

~Light-off point/light-off curve. The light-off point corresponds to the temperature at which the process begins. However, it has been determined that this temperature is slightly higher when heating up the gas flow than when cooling down. This hysteresis leads to the catalytic converter being effective slightly longer as the gas temperature decreases (**Fig. C74**).

~Light-off temperature (starting performance, light-off curve). Every catalytic converter has a certain temperature at which it becomes effective, the so-called light-off temperature. Below this temperature, the catalytic converter has little or no effect, and the reaction is accomplished only with very low conversion rates ($<50\%$). It is possible to allow hot gas to flow past the catalytic converter so that it reaches its light-off temperature more quickly or to equip the catalytic converter with an electric heater to heat up the catalytic converter itself and the inflowing gas to the light-off temperature. If the light-off temperature is defined as the temperature at which 50% of the pollutants are converted, the light-off temperature for operation with gasoline engines is approximately 250°C, depending on the type of pollutant.

For operation in diesel engines, the light-off temperature is below 200°C. However, the light-off temperature is not a constant value; it increases continuously with the age of the catalytic converter. Moreover, the light-off temperatures differ for the individual reactions and, therefore, for the different types of pollutants (**Fig. C86**).

~Location, catalytic converter. The catalytic converter can be located below the hood or further away from the engine below the body. Underfloor catalytic converters require a longer time to reach their light-off temperature, but operate at lower temperatures in continuous operation and, therefore, have a longer service life. Underhood catalytic converters are subject to higher thermal stress. The trend is more and more to use underhood catalytic converters that reach the light-off temperature more quickly after starting the cold engine and thereby to reduce the time before conversion of the emissions starts. This offers advantages particularly in reduction of the uncombusted hydrocarbons and carbon monoxide. A further reduction in the light-off time is possible by heating the catalytic converter. **Fig. C75** shows the times for achieving the light-off time for three typical catalytic converter locations. (*Also, see above,* ~Bypass catalytic converter; *see below,* ~Primary catalytic converter.)

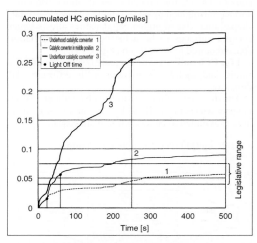

Fig. C75 Influence of location of catalytic converter on HC emission

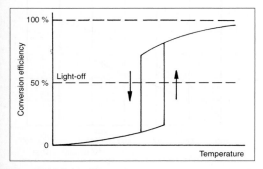

Fig. C74 Light-off curve

C

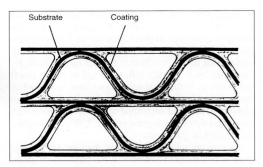

Fig. C76 Channels for metal substrate

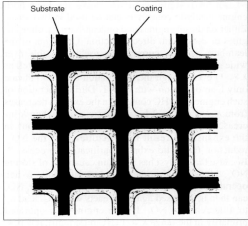

Fig. C78 Monolith channels for ceramic substrate

~Matrix. The matrix is the honeycomb element which serves as a bed for the catalyst in a monolith. The matrix can be constructed using a ceramic or metallic bed.

~Metal bed. The metal bed is a metal substrate in the form of a honeycomb element (converter). For automobile exhaust catalytic converters, iron CrAl steels are used in particular, in the form of foils as the initial product. These are shaped, wound, and brazed, depending on the shape of the catalytic converter. The advantage in comparison to ceramic monolith is particularly the lower pressure drop in the exhaust system due to the minimum wall thickness of the foil (**Fig. C76**).

~Monolith. The monolith is a honeycomb-shaped element of ceramic material (ceramic bed) coated with a wash coat (e.g., Al_2O_3) and then a layer of noble metal. The term monolith is frequently also used for the metal bed. In contrast to granular-type catalytic converters, monoliths have precisely defined passages in the form of channels (**Fig. C77**). These channels can have different shapes. **Fig. C78** shows the channels for a ceramic substrate in the form of squares, and **Fig. C76** shows channels for a metal substrate in the form of a surface surrounded by one straight and one sinusoidal-shaped line. Use of higher cell densities allows larger specific surfaces to be achieved or the installation space required to be reduced (**Fig. C79**).

Fig. C80 indicates the typical sizes for coated ceramic and metallic substrates with different cell densities.

The thickness of the substrate webs without coating is specified as the wall thickness. The coating thickness is assumed to be 0.025 mm in all cases.

~Noble metal layer. *See above*, ~Coating

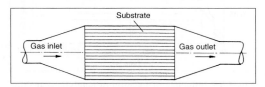

Fig. C77 Catalyst bed

~NOx absorption-type catalyst. The functional principle here is that during the "lean" phases, during which the engine operates with an equivalence ratio of $\lambda > 1$, the oxides of nitrogen produced are adsorbed, and during the so-called regeneration cycles, in which it is necessary to operate the engine rich ($\lambda < 0$), it can then be converted to N_2.

During the "lean" phase—i.e., $\lambda > 1$—a portion of the oxides of nitrogen are oxidized to form NO_2 on the noble metals (e.g., Pt) present in the catalytic layer and then adsorbed by the adsorption material (e.g., $BaCO_3$; **Fig. C81**). Here, the following reaction takes place: $2BaCO_3 + 4NO_2 + O_2 \rightarrow 2Ba(NO_3)_2 + 2CO_2$. In parallel with this, the carbon monoxide is oxidized and the uncombusted hydrocarbons are converted to CO_2 and H_2O. An NO_x sensor recognizes when the adsorption material is saturated and initiates a short (2- to 10-s) "rich" phase (regeneration phase). Then adsorbed NO_2 is released at λ values between 0.75 and 0.98 and converted to N_2

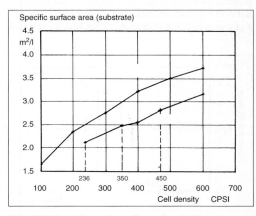

Fig. C79 Specific surfaces for various substrates

cpsi	Ceramic Bed Catalytic Converter				Metal Bed Catalytic Converter			
	Wall Thickness	Free Cross Section %	Geometric Surface	Heat Capacity J/K	Wall Thickness mm	Free Cross Section	Geometric Surface	Heat Capacity J/K
100					0.05	90.5	1.63	231
200					0.05	85.5	2.35	354
236								
300					0.05	83.1	2.74	415
350	0.14	73.84	2.48	526				
400	0.16	68.8	2.55	543	0.05	79.8	3.20	497
470	0.13	71.84	2.82	549				
500					0.04	80.2	3.50	462
600	0.11	71.24	3.17	529	0.04	78.8	3.72	499

C

Fig. C80
Typical values for ceramic and metallic catalyst substrates

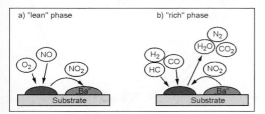

Fig. C81 NO$_x$ absorption-type catalyst

by the reduction agent present in the fuel (e.g., HC, CO). This is defined by the equation

$$2Ba(NO_3)_2 + 8CO \rightarrow 2BaCO_3 + O_2 + 6CO_2 + N_2.$$

The catalytic converter is regenerated and is then again ready to adsorb NO$_x$.

~Number of cells. The number of cells in a monolith can be calculated from the cell density in cpsi multiplied by the flow contact area in square inches (in^2) or with SI units as the cell density in cells per cm^2 multiplied by the flow contact areas in cm^2.

~Open-loop catalytic converter (also, see below, ~Three-way catalytic converter). Here, the λ values vary over a wide range above and below unity, due to the uncontrolled mixture formation (open loop). The conversion rates for CO and HC reactions decrease in the rich range and the rates for NO$_x$ decrease in the lean range. This is caused by the lack of oxygen for oxidation in the rich range and lack of reduction agents (hydrocarbons) in the lean range (**Fig. C88**). In all, this leads to significantly higher emissions in comparison to a "closed loop" three-way catalytic converter.

~Operation. See above, ~Catalytic exhaust purification

~Oxidation catalyst. See above, ~Dual-bed catalytic converter

~Oxygen storage capacity. It is necessary for the catalyst to be capable of adsorbing oxygen due to the con-

trol variations in closed-loop systems. All metal oxides, particularly cerium oxide, act as oxygen storage agents.

~Palladium (also, see above, ~Coating). Palladium is a noble metal used in combination with rhodium and also with platinum as a catalyst in three-way catalytic converters (coating).

~Platinum (also, see above, ~Coating). This is a noble metal used as a catalyst. In catalytic converters for motor vehicles, it is used together with rhodium, or with rhodium and palladium.

~Poisoning. The efficiency of a catalytic converter can decrease, or cease entirely, due to aging as well as poisoning. Typical catalyst poisons for platinum and palladium are the selenium, sulfur, and phosphorous frequently contained in the fuel or engine oil. These combine chemically with the catalyst making it ineffective. A strong poison for most motor vehicle exhaust catalysts is lead. For this reason, motor vehicles equipped with catalytic converters should never be operated with leaded gasoline.

~Pressure difference. This results from the flow resistance of the catalytic converter because of reduction of the cross section by the substrate walls as well as friction of the gas along the walls. The pressure difference should not be too great compared with the overall increase in the exhaust back pressure (**Fig. C72**). At the same mass flow, in relation to the cross section, monoliths and, here, metal substrates have a lower pressure difference than ceramic-type catalytic converters, due to the thinner cell walls.

~Pressure loss. See above, ~Exhaust back pressure increase; ~Pressure difference

~Primary catalytic converter (also, see below, ~Starter catalytic converter). The catalytic converter should be located close to the engine to achieve the light-off temperature quickly. If this is not possible for space reasons, a smaller primary catalytic converter is installed close to the engine to allow quick light-off whereby the main catalytic

111

converter is then located below the floor (**Fig. C70**). It is not necessary for a primary catalytic converter to be a by-pass catalytic converter. It can also be located in the main flow if the permissible temperatures are not exceeded.

~Recycling. Noble metals such as platinum, palla-dium, and rhodium (coatings) are used as the actual catalysts. For this reason it is not only ecological but also economically practical to reclaim these precious metals from used catalytic converters. Future require-ments are expected to give preference to products which can be recycled virtually completely—that is, in terms of reusing all materials.

When recycling ceramic bed catalytic converters, the monolith is removed from the housing and crushed. After crushing, it is necessary to separate the noble metal from the ceramic and wash coat. The reclaimed noble metal is then used to produce new catalytic con-verters. The steel housing can also be recycled.

With metal bed catalytic converters the entire con-verter—that is, the monolith and housing—are ground up together. The wash coat containing the noble metal is then separated from the steel. The wash coat, which contains a significantly higher concentration of noble metal as a result of the system used in this process, is then treated to reclaim the noble metal.

The steels in the ferritic steel of the metal substrate and the austenitic steel in the housing are separated with a magnetic cutter and fed to the steel smelting process separately. The degree of noble metal re-claimed is significantly higher with metal bed catalytic converters than with ceramic-type catalytic converters. The recycling flow chart is shown in **Fig. C82**.

~Reduction catalyst. *See above*, ~Dual-bed catalytic converter

~Regulations. Official regulations related to emission limits have been passed in the individual states and countries of the United States and Europe. **Fig. C83** shows a general summary of the emission limits for Europe and the United States. For individual applica-tions, the applicable laws are to be used as a basis in each case. In some cases, the quantity of pollutant is specified in g/km instead of g/test cycle. In this case,

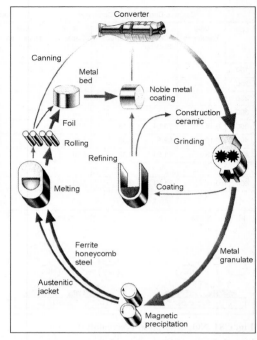

Fig. C82 Flow chart for catalytic converter recycling

test cycle 1 corresponds to a driven distance of 11.007 km in Europe or 17.9 km in the United States.

~Resistance. *See above*, ~Pressure difference

~Retention time. The specific retention time, t_s, in the empty space without catalyst is equal to the reciprocal of the space velocity, SV; that is, $t_s = 1/SV$. If V is the flow rate in m^3/s and V_0 the empty space in the catalytic converter, $t_s = V_0/V$. The effective retention time is then $t_w = V_k/V$, where V_k = the volume of all channels in the monolith.

~Rhodium (*also, see above*, ~Coating). Rhodium is a noble metal used as a catalyst (three-way catalytic

Regulation	Euro III		Euro IV		Regulation	USA LEV I		USA LEV II	
Effective date	1/2000–1/2001		1/2005–1/2006		Effective date	Up to 2004–2007 AS 2000		As of 2004	
Limits [g/km]	Gasoline, LGP, NG	Diesel	Gasoline, LGP, NG	Diesel	Limits [g/mi]	LEV	ULEV	LEV	ULEV
HC	0.2		0.1		HC	0.075	0.04	0.075	0.04
NOx	0.15	0.5	0.08	0.25	NOx	0.2	0.2	0.05	0.05
HC + NOx		0.56		0.30	HC + NOx				
CO	2.3	0.64	1.00	0.50	CO	3.4	1.7	3.4	1.7
PM		0.05		0.025	PM				
HCHO					HCHO	0.015	0.008	0.015	0.008
Note: LEV = Low Emission Vehicle, ULEV = Ultra Low Emission Vehicle									

Fig. C83 Emission limits for Europe and the United States

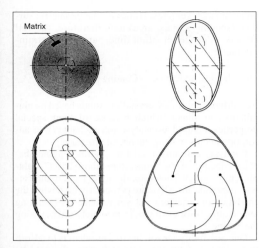

Fig. C84 Catalytic converter shapes

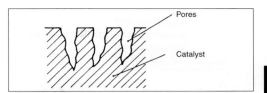

Fig. C85 Inner surface formed by pores

converters) in combination with platinum and palladium. Rhodium is particularly effective for reduction of NO_x.

~Selective catalyst. This is the designation for a catalyst which accelerates a desired process while simultaneously suppressing undesired secondary reactions.

~Shape. Here, reference is made to the shape of the cross section of the catalytic converter (monolith). The common shapes are round, oval (racetrack), elliptic, and irregular, whereby usually the installation space available in the vehicle dictates the cross section. **Fig. C84** shows different cross-section shapes, and the winding direction of the metal substrate in the matrix is illustrated in the figures for the oval and elliptic versions.

~Space velocity (also, see above, ~Retention time). The space velocity in the form of m³/h of gas per m³ of total catalytic converter volume is frequently used as a comparative value, and for transferring test values for laying out technical systems. This value expressed in units of m³/(m³h) or h⁻¹ is frequently called the SV (space velocity) value. Space velocities of up to 250,000 h⁻¹ are reached.

~Specific surface area. The specific surface area is the surface of the catalytic converter in contact with the gas in m² per liter of catalytic converter volume. Here, it is necessary to differentiate between the outer surface formed by the geometric surface (GO) and the inner surface of the wash coat, which includes the surface of all pores. This value is several times that of the geometric surface area. **Fig. C85** illustrates the inner surface formed by the pores.

~Starter catalytic converter. This is a small catalytic converter that serves as a primary or starter catalytic converter located close to the engine. The starter cata-

lytic converter can be located in the main flow, in front of the main catalytic converter, or in the bypass as a bypass catalytic converter (**Fig. C70**).

~Starting performance. The starting performance is closely associated with the light-off temperature. First it is necessary to reach the light-off temperature to start the reaction in the gas mixture. If the light-off temperature for a certain catalytic converter is relatively low, the chemical reaction begins immediately after starting at a low reaction rate. The reaction rate increases only when the temperature increases, thereby increasing the conversion rate. The typical starting performance for conversion of the most important constituents (CO, HC, NO_x) is shown in **Fig. C86**.

~Sulfate formation. See above, ~Diesel engine catalytic converter

~Swelling mat. When ceramic monoliths are installed in steel housings, a so-called swelling mat (e.g., mica compound) is used between the ceramic and the steel to compensate for their different heat expansion rates (resulting from the different heat expansion coefficients). **Fig. C87** shows such a design.

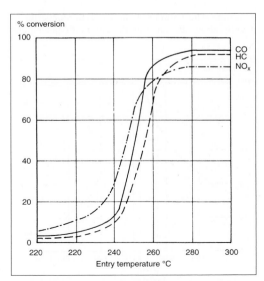

Fig. C86 Starting performance for catalytic converters used in gasoline engines

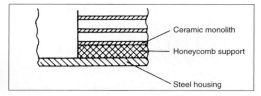

Fig. C87 Swelling mat in catalytic converter

~**Three-way catalytic converter.** A three-way catalytic converter must be effective for

• Conversion of carbon monoxide to carbon dioxide
• Conversion of hydrocarbons into carbon dioxide and water vapor
• Decomposition of oxides of nitrogen into nitrogen, water vapor, and carbon dioxide

With closed-loop three-way catalytic converters equipped with a lambda probe, the air-fuel ratio is kept around $\lambda = 1$ (λ window), and the conversion rates for all three of the processes specified above is greater than 90%.

Fig. C88 shows measured CO, HC, and NO_x conversion rates over a wide λ range. It can be seen that the greatest possible conversion of all three pollutants is achieved in the vicinity of $\lambda = 1$ (λ window).

With open-loop three-way catalytic converters without a lambda probe, the conversion rates are significantly lower due to the difference in the excess air figures, depending on how the engine is operated (resulting from the specific operating characteristics of the engine).

~**Two-bed catalytic converter.** *See above,* ~Dual-bed catalytic converter

~**Volume.** The volume of the catalytic converter is usually about the same as the engine displacement

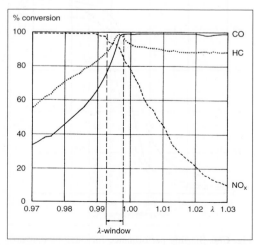

Fig. C88 Conversion in a three-way catalytic converter

on most stock vehicles. However, the geometric surface (specific surface area) and, therefore, the cell density—not the retention time—are the determining factors.

~**Wash coat.** *See above,* ~Coating

~**Zeolites.** Zeolites are crystalline solids based on aluminum and silicon which because of their special properties are used as catalyst, ion exchangers, or adsorption agents in many technical processes.

Catalysts currently used for gasoline engines become effective only at temperatures above 250°C (light-off temperature). Therefore, the hydrocarbon and carbon monoxide emissions are particularly high when the catalytic converter has not yet reached its operating temperature—for example, immediately after starting the engine.

One possibility for solving this problem is to adsorb the pollutants from the exhaust flowing through the system at relatively low temperature with the aid of a suitable adsorbent (e.g., zeolites) and then release these pollutants at higher temperatures (light-off temperatures)—when the catalytic converter is operating—and convert them in the catalytic converter. Recent publications deal with the development of adsorbing wash coats containing noble metals.

Catalytic converter aging →Catalytic converter ~Aging

Catalytic converter conversion ratio →Catalytic converter ~Conversion efficiency

Catalytic converter efficiency →Catalytic converter ~Catalytic exhaust purification, ~Efficiency

Catalytic converter heating →Catalytic converter ~Heater

Catalytic converter heat-up →Catalytic converter ~Heat-up

Catalytic converter inflow →Catalytic converter ~Inflow

Catalytic converter intermediate layer →Catalytic converter ~Intermediate layer

Catalytic converter layout →Catalytic converter ~Layout

Catalytic converter light-off temperature →Catalytic converter

Catalytic converter location →Catalytic converter ~Location, catalytic converter

Catalytic converter matrix →Catalytic converter ~Matrix

Catalytic converter poisoning →Catalytic converter ~Poisoning

Catalytic converter pressure difference →Catalytic converter ~Pressure difference

Catalytic converter resistance →Catalytic converter ~Pressure difference

Catalytic converter shape →Catalytic converter ~Shape

Catalytic converter volume →Catalytic converter ~Volume

Catalytic exhaust purification →Catalytic converter

Cavitation (*also,* →Bearings ~Operational damage; →Injection system (components) ~Diesel engine ~~Injection hydraulics). Cavitation can occur on any component in an internal combustion engine when a fluid flows through or around it. This phenomenon can be explained using the cylinder barrel in a diesel engine as an example. Damage from cavitation can occur on the water-cooled side as a result of high frequency vibration generated by the gas forces in the combustion chamber and the shock pulses from the piston as it moves from the thrust to antithrust surfaces. This vibration and these pulses result in high frequency vibration of the sleeve wall of the cylinder barrel which is transferred to the cooling water. Vapor bubbles, which implode when the pressure increases again, are created at points at which the pressure is lower than the vapor pressure of the liquid. This results in high forces at certain spots, which can lead to deterioration of the surface layer. Microscopic particles are detached, resulting in "pitting." The surface of the rear of the cylinder barrel may be damaged to such an extent that water leaks into the cylinder barrel or crankcase.

Factors influencing cavitation in this area are the engine speed, pressure variation in the combustion chamber, piston clearance and piston mass, piston axis layout, wall thickness of liner, material, dimensions of water gap, and coolant routing.

Cavitation can also occur in the fuel lines of diesel and gasoline engines. In fuel lines, the pressure can drop below the vapor pressure at spots, due to the high flow rates and to rapid rates of change of flow rate in the fuel. Vapor bubbles are formed, and the collapse of these bubbles results in cavitation damage in the lines. In such cases, increasing the system operating pressure helps.

Cavitation can also occur in the water pump impeller, oil pumps, injection pumps, and injection nozzles.

Cavitation/erosive corrosion →Coolant

CCMC →Oil ~Classification, specifications, and quality requirements ~~Classifications ~~~CCMC classifications

CEC →Oil ~Classification, specifications, and quality requirements ~~Institutions

Cell density →Catalytic converter

Cellulose paper →Filter ~Intake air filter ~~Filter media

Center electrode, spark plug →Spark plug ~Electrode gap

Center muffler →Exhaust system ~Muffler

Center offset →Valve guide ~Design

Central injection →Injection system, fuel ~Gasoline engine ~~Intake manifold injection systems

Central mixture formation →Mixture formation ~Mixture formation, gasoline engine

Centrifugal advance mechanism →Ignition system, gasoline engine ~Ignition ~~Spark control

Centrifugally controlled advance →Ignition system, gasoline engine ~Ignition ~~Spark control

Centrifuge, oil →Filter ~Lubricating oil filter

Ceramic bed →Catalytic converter

Ceramic monolithic cell filter →Particles (Particulates) ~Particulate filter system ~~Particulate filter ~~~Filter media

Ceroxide →Catalytic converter

Cetane index →Fuel, diesel engine ~Properties

Cetane number →Fuel, diesel engine ~Properties

Cetane number improver →Fuel, diesel engine

CFPP →Fuel, diesel engine ~Properties ~~Cold filter plugging point (CFPP)

CFV-CVS →Emission measurements ~Test type I ~~Calibration

Chain drive (*also,* →Cam-shaft; →Engine accessories). Timing chains on state-of-the-art engines drive

C

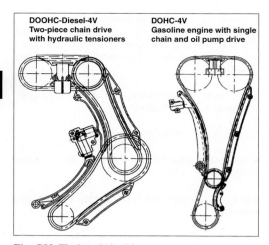

Fig. C89 Timing chain drives

the camshaft and frequently also drive other components such as the oil pump, water pump, and fuel injection pump (**Fig. C89**).

Because neither the camshaft nor the crankshaft rotates uniformly and because the fuel requirement for the injection pump is subject to extremely high periodic variations, the drive is subject to extremely complex dynamic loads. In the course of decades of experience, certain dimensions for roller chains and sleeve-type chains have proven to be particularly suitable for timing cases.

~Chain elongation. *See below*, ~Characteristic values

~Chain sprocket. The shape of the teeth on chain sprockets is standardized for roller chains, sleeve-type chains, and toothed-type chains (DIN 8196). Careful design of the tooth shape is just as significant for reliable operation of the timing chain drive as is the resistance of the chain to wear.

Usually, sprockets with maximum tooth space shape are used. This design allows the chain to run in and out without disturbance even at high chain speeds due to the low tooth crown heights and large tooth gap opening.

Disk-type sprockets or chain sprockets with a hub on one or both sides are used depending on the space available and type of application. The material used depends on the timing chain drive relationship, the operating conditions, and power transfer. For example, C10 material is used for super finish-punched sprockets, 16MnCr5 for machined sprockets, or D11 for sintered versions, with heat treatment appropriate to the material.

~Chain tensioner. The use of permanently acting tensioning and guide elements matched precisely to the specific engine allow the drive to be optimized so that its service life corresponds to that of the engine without special care beyond the specified service to the engine (**Fig. C90**).

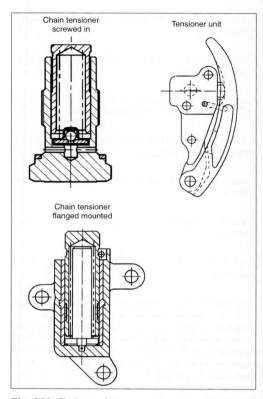

Fig. C90 Chain tensioner

The guide elements frequently consist of simple plastic rails or metal rails with plastic coating, which are straight or curved depending on the chain track.

The tensioner satisfies a number of purposes in the timing chain drive. On the one hand, the timing chain is tensioned with a defined force on the return side under all operating conditions even when wear elongation occurs during operation. A damping element utilizing either friction or viscous damping reduces vibrations to a permissible level.

~Chain wear. *See below*, ~Characteristic values

~Characteristic values. Three significant factors distinguish the application characteristics of control chains (**Fig. C91**):

• Resistance to breakage
• Dynamic strength
• Wear strength

Breakage can occur when the static or dynamic fracture load is exceeded.

The load on the chain is not usually uniform, and this is especially true for timing chain drives. The chain is subject to dynamic stresses due to the pulsating torque of the camshaft, the injection pump on diesel

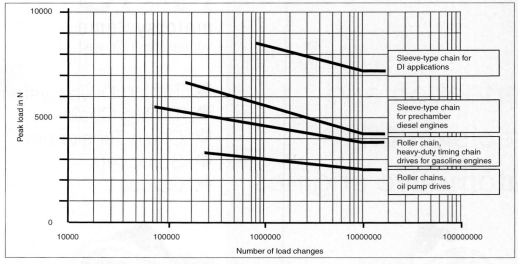

Fig. C91 Dynamic strength results for sleeve-type and roller chains

engines, the nonuniform rotation of the crankshaft, and the fluctuating chain elongation force caused by the polygon effect. These forces must not exceed the dynamic strength of the chain, because the number of such alternating loads during the service life of the engine is greater than 10^8 load cycles in any case.

The wear in the chain joints and the resulting elongation of the chain determine the permissible load for roller- and sleeve-type chains. On present-day engines with precision timing, low wear elongation figures of 0.2–0.5% of the chain length can be achieved for mileages up to 250,000 km.

~Damping. *See below*, ~Stiffness

~Designs. The rollers on a roller chain that rotate around the sleeves roll over the flanks of the teeth on the chain sprocket with minimum friction so that a different point on the circumference is always in contact (**Fig. C92**). The lubricant between the rollers and sleeves contributes to reduction of the noise and shock. With a sleeve-type chain, the teeth of the chain sprocket always make contact with the fixed sleeve at the same point. For this reason, proper lubrication of such drives is particularly important.

Sleeve-type chains have a larger joint surface than the corresponding roller chains with the same pitch and resistance to breakage. A larger joint surface allows lower joint surface pressure and, therefore, less wear to the joints. Sleeve-type chains have proven themselves especially for camshaft drives subject to high loads in high speed diesel engines. As soon as transfer of a given torque with a single chain would lead to less than 18 teeth at a certain maximum sprocket diameter, a multiple chain with smaller or the same pitch is preferable.

~Double roller chain. *See below*, ~Multiple chain

~Guide elements. *See above*, ~Chain tensioner

~Multiple chain. Multiple chains—for example, double roller chains—are used when transfer of the required torque is no longer possible with a sprocket with less than 18 teeth at the maximum possible sprocket diameter.

~Roller chain. *See above*, ~Designs

~Sleeve-type chain. *See above*, ~Designs

~Stiffness. A timing chain drive with mass, stiffness, and damping constitutes a vibrating system with a number of degrees of freedom (**Fig. C93**). This can cause resonance effects due to alternating effects in events corresponding to the excitation by the camshaft, crankshaft, injection pump, and so on, which lead to an extreme load for the timing chain drive.

The stiffness of the chain can be increased while maintaining the specific mass by particular design measures. This leads to a displacement of the resonant points to higher frequencies.

Chain drive computation →Calculation processes ~Application areas ~~Engine mechanics

Chain drive damping →Chain drive ~Stiffness

Chain drive designs →Chain drive ~Designs

Chain drive stiffness →Chain drive ~Stiffness

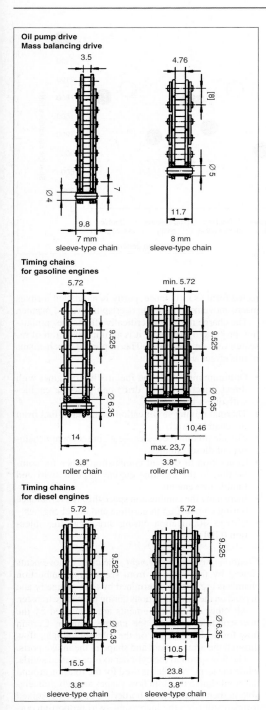

Oil pump drive
Mass balancing drive

7 mm
sleeve-type chain

8 mm
sleeve-type chain

Timing chains
for gasoline engines

3.8"
roller chain

3.8"
roller chain

Timing chains
for diesel engines

3.8"
sleeve-type chain

3.8"
sleeve-type chain

Fig. C92 Chain designs

Chain drive values →Chain drive ~Characteristic values

Chain elongation →Chain drive ~Characteristic values

Chain sprockets →Chain drive

Chain tension →Chain drive ~Chain tensioner

Chain tensioner →Chain drive

Chain values →Chain drive ~Characteristic values

Chain wear →Chain drive ~Characteristic values

Chamber volume →Engines ~Alternative engines

Championship Auto Racing Teams (CART) →Engine ~Racing engines

Change interval, oil →Oil ~Oil maintenance ~~Oil change interval

Characteristic curves. In contrast to maps, which illustrate a parameter—for example, specific fuel consumption as a function of two variables, engine speed and specific work—characteristic curves represent a parameter in relation to another parameter—for example, valve lift or crankshaft angle.

Characteristic values. With the aid of characteristic values or characteristics of an engine, it is possible to evaluate and compare engines. Characteristic values are values such as power, torque, mean effective pressure, efficiencies, fuel consumption, compression ratio, air-fuel ratio, pollution emissions, and so on.

Characteristic vector →Engine acoustics ~Resonant frequency

Characteristics →Characteristic values

Charge →Cylinder charge, ~Volumetric efficiency

Charge air →Supercharging

Charge air blow-off →Supercharging ~Blow-off valve

Charge air intercooler →Supercharging

Charge application. Charge application is the total fresh charge per combustion cycle.

Charge backpressure. Depending on the timing of the engine, a recharging effect occurs in the induction tract as a result of the kinetic energy in the medium, which leads to an improvement in the charge. At lower engine speeds, however, a very late intake closing leads to the opposite effect; that is, the charge is forced

119

C

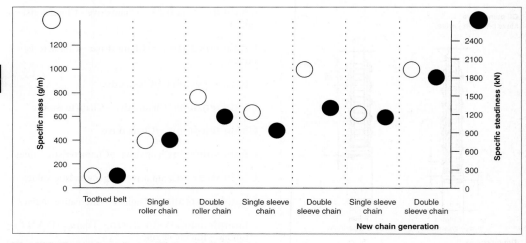

Fig. C93 Chain drive values, stiffness and dimensions

back into the inlet manifold. The resulting charge losses have a negative effect on performance and torque.

Charge control →Load ~Gasoline engine

Charge dilution. Pure fresh charge or fresh fuel mixture is characterized by a specific proportion of air and fuel. In a conventional gasoline engine these are typically in a ratio approaching the stoichiometric composition. For certain reasons (e.g., exhaust quality, load control) it is necessary to dilute the fuel mixture (the charge)—that is, to reduce the relative fuel proportion. This can be done in several ways. Either the air-fuel mixture is diluted with more air or it is mixed with exhaust gases so that primarily the nitrogen proportion is increased. The higher nitrogen proportion, which in the initial approach works as an "inert gas," and the resulting lower oxygen proportion affect the reaction process. The charge is diluted with exhaust gas, for example, by increasing the residual gas proportion in the cylinder by changing the timing or through exhaust gas recirculation.

Charge dilution by air is achieved in diesel engines through load control and in gasoline engines with direct injection. Exhaust gas charge dilution is used to reduce emissions of nitrous oxide in diesel and gasoline engines. The effect is based on the reduction of the combustion temperatures and can be increased further by cooling the recycled exhaust gas.

Charge movement. Powerful charge movements occur in the cylinders of combustion engines partly because of the inflow through comparably narrow valves—generating temporary and local high-speed gradients—and partly as a result of the number and geometry of the intake ports, as a result of the valves and through the formation of quench chambers. Additionally, folds are being placed increasingly in the intake tracts directly in front of the valves in order to achieve targeted charge movements. The air movements can exist in a nondi-

rected form as turbulence, partly in a directed form as macro-movements, and can overlap in a varied manner.

The charge movement critically affects the combustion process; overcoming it is an important aim of research and development. The important mechanisms of this effect are:

- Optimum mix of air and fuel in diesel engines with direct injection (spray drift) and gasoline engines with direct injection
- Thermal mixing stratification (removal of fuel from the chamber wall)
- Chaotic mixing processes—e.g., in secondary chambers of diesel engines
- Generation of defined combustion zones for combustion in the gasoline engine through swirl and tumble movements
- Increasing the combustion space through turbulence during combustion in gasoline and diesel engines
- Charge stratification during direct gasoline injection (GDI).

Creation of charge movement. The charge movements caused by the outflows from secondary combustion chambers are largely determined by their geometry and depend on the flow conditions during the charge transfer.

All other charge movements are generated by the flow conditions during the charge transfer. Certain flow forms are imposed on the internal cylinder flow through the arrangement and shape of the intake ports.

The flows around the cylinder axis are called swirls, while the term "tumble" is used for the rotation around the crankshaft axis. Both flow shapes are only conceivable in their pure form as borderline cases. The mixed forms result from the increasing swirl proportion corresponding to **Fig. C94** as follows: For a pure tumble without swirl components, there is a straight tumble axis (image A, **Fig. C94**). As the swirl components increase, the straight tumble axis is bent farther and farther by the

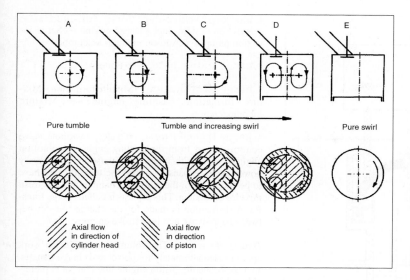

Pure tumble

Tumble and increasing swirl

Pure swirl

Axial flow in direction of cylinder head

Axial flow in direction of piston

Fig. C94
Swirl and tumble

C

rotation of the air that has already flowed into the cylinder (images B and C, **Fig. C94**) until the tumble vortex has turned fully about the Z axis to form a circle and becomes the toroid axis of a toroidal vortex (image D, **Fig. C94**). This flow type has been clearly confirmed in experiments in diesel engines with direct injection.

Measuring the charge movement. In research, it is possible and usual, but very expensive, to measure the charge movements using optical processes (LDA, light section method). For development and quality control measures, the stationary flow rate test is the standard procedure. Because of the great dependence of the emission levels on the charge transfer values, swirl measurements are increasingly taken in high-volume production.

Calculating the charge movement. Charge movements also can be calculated arithmetically using computational fluid dynamics (CFD). However, very powerful computers are required.

~High tumble. *See below,* ~ Swirl, ~ Tumble

~Laminar flow. A laminar flow occurs when the "flow particulates" arrange themselves such that they move in layers alongside each other. No mixing or saturation occurs. Laminar flows occur especially at low flow speeds. Laminar flows are rather rare in an engine. At best, they can be found at low airflow rates in the intake manifold; in the combustion chamber, by contrast, turbulent flow occurs almost exclusively.

~Quench flow. A quench flow is typically formed with help from quench areas, which are fitted in the cylinder head or on the piston.

In the simplest case, these are flat areas that force fuel mixture as the piston approaches top dead center. Quench areas are used in gasoline engines in order to accelerate the fuel mixture conversion and to reduce the risk of knocking.

~Swirl. Flows around the cylinder axis are called swirls. In order to generate the charge movement (swirl, tumble), spiral-shaped swirl channels are typical; tangential channels, in many forms as single and double channels, are typical for the valve axis. Distinguishing between the swirl channel and the charge port is not reasonable as these functions may in fact not be separated from each other.

~Swirl factor. With the stationary flow test, the angular momentum of the flow is measured as torque using a flow rectifier and is recorded as the swirl factor on the basis of the ram pressure. This is carried out using a spherical flow rectifier with a three-axis torque sensor for the angular momentum components of the three directions in space, whereby the simultaneous measurement of the swirl and tumble components is achieved (**Fig. C95**). The angular momentum figures calculated are the driving forces for the flow forms forming in the cylinder chamber. As the angular momentum factors are the only flow values able to contribute to the internal cylinder flow formed, their use in characterizing the charge movement is justified. This applies in particular to overall swirl factors. Because of the momentum qualities of the swirl factors measured for the various valve lifts, these can be determined through an expanded gas exchange calculation.

~Tumble. Flows whose rotational axes are in the same direction as the crankshaft axis are called tumble. Depending on the characteristics and intensity, they are described as "high tumble."

~Turbulent flow. If one increases the flow speed of a laminar flow field, the flow form changes above a critical

C

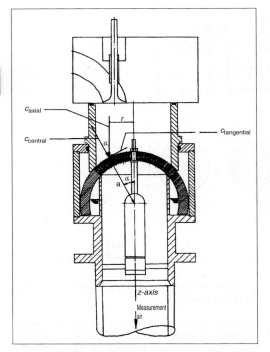

Fig. C95 Measurement set-up for determining swirl factors

value. This value, also known as the critical Reynolds number, depends on many parameters. Apart from the flow speed, factors such as density, dynamic viscosity and geometric dimensions are important. In a turbulent flow, the basic flow is influenced by disorganized, coincidental movement fluctuations in crosswise and lengthwise directions.

The flow in an engine's combustion chamber is generally highly turbulent.

~Variable swirl. Controls to adjust the charge movement in respect of load and speed can be achieved through changes in valve lift or timing and through the installation of variable elements that affect the flow — for example, folds in the swirl channels. This also applies to the tumble.

Charge port →Intake port, gasoline engine

Charge pressure →Supercharging

Charge pressure control →Electronic open- and closed-loop control ~Electronic open- and closed-loop control, gasoline engine ~~Functions; →Supercharging

Charge pressure control valve →Valve ~Exhaust gas control valves

Charge pressure ratio →Supercharging

Charge pressure-dependent full-load stop →Injection system, fuel ~Diesel engine ~~Serial injection pump ~~~Add-on modular assemblies

Charge status, battery →Battery

Charge stratification →Mixture formation ~Mixture formation, gasoline engine ~~ Mixture inhomogeneity

Charge transfer (also, →Cycle; →Variable valve control). Apart from the Stirling engine, combustion engines work with internal combustion. After every power cycle (expansion cycle) the burned gases have to be forced from the working chamber and the fresh mixture brought in. This process is called charge transfer. A distinction is made for the charge transfer between four-stroke and two-stroke engines.

Four-stroke engine. The volume change in the work space is used alternately for power and charge transfer. In a reciprocating piston engine, the control is preferably achieved through the intake and discharge valves. A combustion cycle comprises four cycles:

Induction – Compression – Expansion – Expulsion

Induction and expulsion are charge transfer cycles; compression and expansion are power cycles.

Fig. C96 shows the gas pressure development in the indicator diagram (p-V diagram) and in the p-a diagram for a four-stroke reciprocating piston engine.
Two-stroke engine. The charge transfer occurs between the compression and expansion power cycles by rinsing the exhaust gases with fresh gas or through a scavenge pump. In order to prevent fuel losses, in modern engines the exhaust gases are rinsed with fresh air, and then fuel is injected. Control is through intake and exhaust ports or valves. A combustion cycle comprises two cycles:

Compressing the charge – Expansion after ignition and discharge

Fig. C97 shows the gas pressure development in the indicator diagram (p-V diagram) and in the p-a diagram for a two-stroke reciprocating piston engine.

The quality of the charge transfer affects, for instance, the following parameters: power, mean pressure (torque), fuel consumption, exhaust, performance. In principle, the charge transfer must be designed such that as little power as possible is required for it so that fuel consumption can be reduced.

~Charge transfer computation (also, →Calculation processes ~Application areas ~~Thermody-namics). An experimental optimization of the charge transfer in combustion engines is time-consuming and cost-intensive. For this reason, design and optimization are achieved using computation models. Based on the engine geometry, flow mechanics, the condition of the medium in the inlet manifold, the engine's operating point (load and speed) and the timing, it is possible to

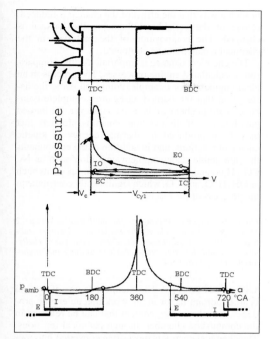

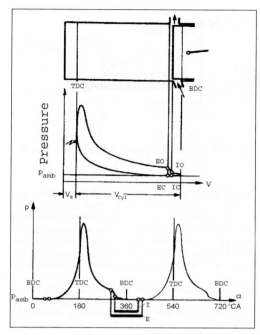

Fig. C96 Four-stroke process. (Source: F. Pischinger) E = Exhaust, I = Intake, α = Crankshaft angle, measured from gas transfer TDC up (°KW), V = Combustion chamber volume, V_h = Displacement, V_c = Compression volume, p_a = Ambient pressure, p_i = Ignition pressure, IO = Intake opens, EO = Exhaust opens, EC = Exhaust closes, TDC = Top

Fig. C97 Two-stroke engine. For descriptions and abbreviations, see Fig. C96. (Source: F. Pischinger)

calculate factors such as cylinder charge, pressure, and temperature of the charge as a function of the crankshaft angle. Further considerations include heat-exchange processes.

~Charge transfer devices (*also*, →Camshaft; →Control/gas transfer ~Four-stroke engine; →Valve). All elements are part of the charge transfer devices, which contribute to the periodic opening and closing of the intake and exhaust valves. In modern four-stroke engines, camshaft-controlled valves are used almost exclusively. The requirements for the charge transfer equipment are:

- Short times for opening and closing the valves
- Large opening time cross sections
- Impermeability under all conditions
- Variability in terms of time and duration.

~Charge transfer loss (*also*, →Variable valve control). Charge transfer losses (pressure losses) are charge losses of the combustion air or fuel mixture in the path to the cylinder through resistance in the intake system. Charge transfer losses occur as pressure losses—that

is, higher pressures—are required in order to transport the gas columns to and from the cylinder. This leads to increased induction and expulsion work, which has to be covered by all the energy available. Resistance includes, for example, throttle valve, air filter, air mass sensor, carburetor, inlet manifold, and cylinder head intake ports with inlet valves and their guides, insofar as they penetrate the intake port. In engines with charging and charge air intercoolers, other losses occur in the charge air intercooler and in the intake and discharge paths. In the gasoline engine, the throttle losses are primarily in the partial throttle as a result of load control resulting from throttling of the intake mixture. Those losses increase with increasing throttling (partial throttle) and result in a reduction of efficiency. Charge transfer losses can be overcompensated through charging. However, this is typically paid for with a higher exhaust counterpressure with the resulting negative concomitant phenomena. The greater the charge transfer losses, the worse the overall result of charging.

Charge transfer losses also occur on the exhaust side in the outlet port and in the exhaust system. Flow resistance also has to be minimized here.

~Charge transfer work. Charge transfer involves work in the form of the expansion and charging (induction or charging) and emptying (discharge) of the cylinder.

123

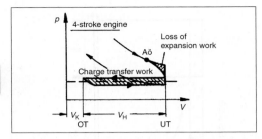

Fig. C98 Cylinder pressure during charge transfer (Source: A. Urlaub)

Therefore, it is necessary to develop the intake and exhaust more favorably for the flow. Cross sections must be optimized (maximized) and bends (changes in direction) in the induction and discharge tract have to be kept to a minimum. Intake and exhaust valves are often bottlenecks. For this reason, more and more engines are being fitted with several intake and exhaust valves.

Some manufacturers use three intake valves. This is especially sensible in long-stroke engines, as the relative cylinder diameter is smaller than in short-stroke engines.

Another important location is the throttle valve, which in conventionally operated gasoline engines is required for controlling the charge. The charge transfer power can be reduced here, for example, by using load-based valve timings. **Fig. C98** shows the cylinder pressure development and the charge transfer power required for the charge transfer in a four-stroke engine.

Charge transfer computation →Calculation processes ~Application areas ~~Thermodynamics; →Charge transfer

Charge transfer loss →Charge transfer

Charge transfer work →Charge transfer

Charging →Particles (Particulates) ~Particulate filter system ~~Particulate filter

Charging loss →Cylinder charge ~Volumetric efficiency

Charging process →Supercharging

Chemical balance. The initial materials do not vanish completely in chemical reactions involving a transition to a gaseous state; residuals remain, which means that a balance develops between all initial and end products. The balance moves from one side or other of the reaction equation depending on the boundary conditions. The general principle for the direction in which the balance of compounds moves when a variable is changed is as follows: The balance changes in such a way that the effect of the change is reduced, if one of the variables influencing the balance is changed. The composition of the materials in the chemical balance can be computed.

The chemical balance is important for the comparison of calculated engine processes. The process in an ideal engine is, for example, computed using the assumption that the burned gases after complete combustion are in chemical balance. However, the burned gas, because of dissociation, contains products that have been produced by thermal decays or kinetic chemical reactions, and in addition to the components that are stable under ambient condition such as N_2, CO_2, H_2O, and O_2, these can include components such as OH, H, O, and CO, which significantly contribute to the progress of the reaction.

Literature: Stephan, Mayinger: Thermodynamik, Volume 2 Mehrstoffsysteme und chemische Reaktionen, Springer Publishers, 1988. — W. Kleinschmidt: Untersuchung des Arbeitsprozesses und der NO-, NO₂-, und CO-Bildung in Ottomotoren, TH Aachen, 1974.

Chemical energy. The chemical energy of the fuel is an important factor in the work that an engine produces. It is the energy content that can be released in the combustion chamber through chemical reactions. The chemical energy is most often converted into mechanical or electrical energy.

The chemical energy of a fuel depends on the energies contained in the molecular bonds; these are dependent on its composition, which means on the ratio of the particular hydrocarbon components, because the commercially available gasoline and diesel fuels consist of hydrocarbon mixtures. The calorific value of the fuel can be determined by the energy released during the chemical reaction for certain processes.

Chemical reaction. Chemical reactions in the combustion chamber result in a conversion of energy—for instance, combustion processes convert the bond energies in the fuel into thermal energy, a part of which can be used as mechanically usable work.

The combustion process in the engine is based on oxidation of the fuel. For engines, this normally consists of hydrocarbon compounds. Gasoline and diesel fuels consist of a mixture of many different hydrocarbons, with oxygen attached to the hydrogen in some fuels—such as methanol, for example.

The change of state under the influence of chemical reactions can be related to time, for example, with the help of the rate kinetics of homogenous gas reactions. Rate kinetic analyses permit determination of the time dependency of reactions and of the associated components. The contribution of the individual elements is considered for the formation or dissociation of a component. The elemental reactions as a whole are characterized by the reaction scheme of the analyzed process. The conversion of compounds in the chemical processes is a result of an overlap of elementary reactions that run concurrently and consecutively. A certain level of activation energy is normally required to start the reactions, and this energy is supplied by the ignition spark in gasoline engines or by the energy of the compressed air in diesel engines.

In the combustion chamber of an engine, the fuel is converted through a chemical reaction into the combustion products, which in theory are only carbon dioxide and water. However, the burned gas contains a number of other products because of inhomogeneities in the process, the presence of nitrogen in the air, and temperature gradients.

These products include primarily unburned hydrocarbons, oxides of nitrogen (NO_x), carbon monoxide (CO), aldehydes, carbon (soot), hydrogen, oxygen, and nitrogen. The oxidation process uses a number of intermediate products or intermediate reactions, which can have short or long durability.

The composition of the gas and the concentration of its components change depending on the variation of pressure and temperature in the combustion gas.

Even today, it is still not possible to compute exactly the chemical process defining the combustion in the engine. The number of components in the fuel and the associated elementary reactions, which cannot be defined exactly, are only part of the problem. The kinetic reaction data for the decay and formation reactions are also not fully known.

Literature: H.D. Baehr: Thermodynamik, Springer Publishers, 1989. — F. Pischinger: Verbrennungsmotoren Lecture Reprint Volume 1, 1988. — Stephan, Mayinger: Thermodynamik, Volume 2 Mehrstoffsysteme und chemische Reaktionen, Springer Publishers, 1988.

Chemiluminescence detector →Exhaust gas analysis equipment

Chevy Indy V-8 →Engine ~Racing engines ~~IndyCar

Chilled casting →Crankcase ~Casting process, crankcase

Chip tuning. Chip tuning, through the exchange of electronic memory chips or through the change of circuits, can increase the performance of an engine when compared with values installed by the suppliers. These interventions in the engine control parameters for power increases can result in a number of negative consequences. They can especially affect the service life of the engine in addition to having an impact on fuel consumption and exhaust emissions.

Choke →Carburetor ~Starting systems, ~Systems

Choke valve →Carburetor ~Starting systems ~~Design types, ~~Functional systems

Choke valve servomotor →Carburetor ~Electronic carburetor

Chrome-plating →Piston ring ~Contact surface armor

Chromium-ceramic layer →Piston ring ~Contact surface armor

Circular plate cooler →Coolant ~Design

City driving cycle →Emission measurements ~Test type I

City test →Emission measurements ~Test type I

Classification, oil →Oil

CLD →Exhaust gas analysis equipment ~Chemiluminescence detector

Clean air line →Air intake system ~Thermodynamic air management system

Clean fuel vehicles →Emission limits

Clearance →Piston ring ~Gap

Clearance valve →Valve clearance

Close-coupled catalytic converter →Pollutant aftertreatment ~Aftertreatment concept with a three-way catalytic converter

Closed control loop →Catalytic converter ~Closed-loop

Closed deck →Crankcase ~Crankcase design ~~Cover plate

Closed loop →Catalytic converter

Closed-circuit cooling →Cooling circuit ~Force circulation cooling

Closed-loop control (*also*, →Electronic open- and closed-loop control ~Electronic open- and closed-loop control, gasoline engine ~~Functions). The three-way catalytic converter with closed-loop control has become established as the concept for the exhaust gas treatment of gasoline engines with external mixture formation. The control system ensures that the composition of the air-fuel mixture is maintained at the stoichiometric ratio ($\lambda = 1$) within a very narrow λ-range (λ-window) (**Fig. C99**). This is necessary to ensure efficient conversion of the pollutants CO, HC, and NO.

In closed-loop control, the air-fuel equivalence ratio is measured using the λ-probe mounted in the exhaust gas, this value is compared with the nominal value, and the fuel quantity is corrected as necessary. The air-fuel ratio must have a certain amount of fluctuation upstream of the catalytic converter in order to achieve the best possible oxidation of CO and HC

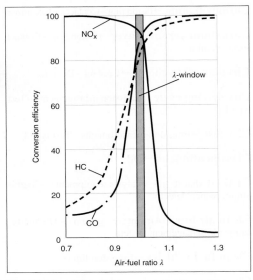

Fig. C99 Lambda window

and to reduce NO$_x$ as much as possible. This fluctuation means operation of the engine both in the excess air (weak) and in the deficient air (rich) ranges to prevent accumulation of oxygen molecules from deactivating the catalytic converter.

A distinction is made between binary and linear lambda probes.

The control algorithm (**Fig. C100**) for binary λ-control is based on a PI controller, where the P and I levels are stored in maps based on engine speed and load. With binary control, the stimulus for the catalytic converter (λ-fluctuation) occurs implicitly through the two-position control.

With linear closed-loop control (**Fig. C101**), a forced stimulus is required in order to suspend the λ-fluctuation. The diagram provides an overview of the structure of the linear λ-control, including forced stimulus and trim control.

Based on the difference between the actual λ and the nominal value (**Fig. C102**), the forced stimulus modulates a periodic deviation (λ-pulse) that optimizes the effectiveness of the catalytic converter.

The signal obtained feeds directly into the correction for the quantity of fuel as preregulation; the signal

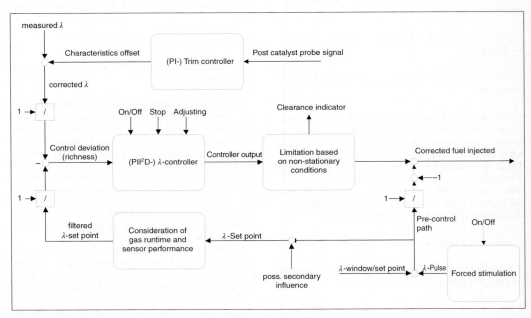

Fig. C100
Control algorithm for binary lambda probe (Source: Siemens)

Fig. C101 Closed-loop control for linear probe (Source: Siemens)

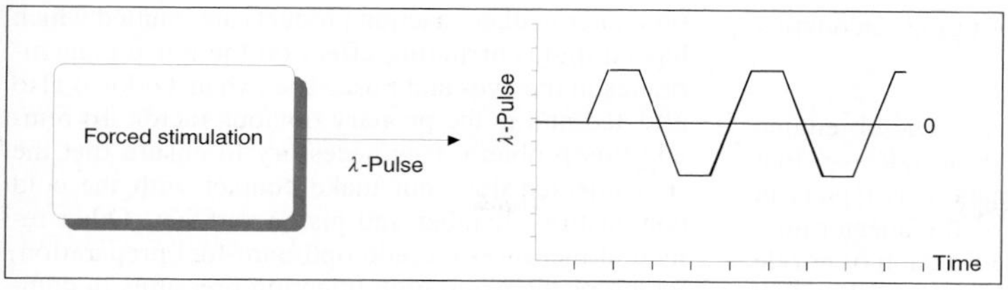

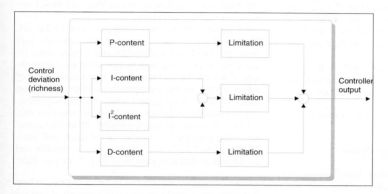

Fig. C102
Forced stimulus (Source: Siemens)

Fig. C103
Calculation of the control error (Source: Siemens)

also is loaded with a possible secondary factor and processed further as a difference between the filtered λ and nominal value that takes into account the travel time of the gas and the delay characteristic of the linear probe.

The signal from the linear λ-probe is converted into a λ-value using a stored curve. This curve can be corrected through the trim control. The trim control is a PI controller, which exploits the post-cat probe signal (ideally from a binary transition probe) that is subject to less interference. The control signal (error; **Fig. C103**; $= λ - 1$), is calculated from the corrected λ signal and the filtered nominal λ value, and then acts as the actual λ control value. This is designed as a PII^{2}D controller. The I^2 proportion serves to balance the oxygen load in the catalytic converter. The controller output also can be restricted under transient conditions. The correction of the injection quantity calculated in this way is included with the preregulation in the calculation of the quantity injected.

Linear closed-loop control offers the following advantages, compared with binary closed-loop control:

- Increase in the control dynamics and reduction in transient λ errors.
- Increased effectiveness of the catalytic converter through adjustable forced stimulus in the closed λ-control loop.
- Possibility to control at $λ ≠ 1$; as a result, controlled warm-up or controlled catalytic converter protection is possible.

Closing angle →Ignition system, gasoline engine ~Ignition

Closing pressure →Injection valves ~Diesel engine ~~Injection nozzle parameter

Closing time →Ignition system, gasoline engine ~Ignition

Closing time of valve →Variable valve control ~Load control

Cloud point →Fuel, diesel engine ~Properties

CNG →Fuel, diesel engine ~Natural gas

CO content →Emission limits; →Exhaust gas analysis

CO$_2$ content →Exhaust gas analysis; →Fuel consumption ~ECE/EUDC consumption

Coal dust engine fuel →Combustion process, diesel engine

Coating →Sealing systems ~Special seals

Coating catalytic converter →Catalytic converter ~Coating

Coaxial variable nozzle (CVN) →Injection valves ~Diesel engine ~~Vario nozzle

Coherence →Engine acoustics ~Noise excitation

Coil ignition →Ignition system, gasoline engine ~High-voltage generation

Coincidental frequency →Engine acoustics ~Threshold frequency

Coking (*also*, →Injection valves ~Diesel engine ~~Hole-type nozzle, ~~Pintle-type nozzle ~~~Flat pintle nozzle). Coking (coke formation) is typical in diesel engines. It usually appears on the injector nozzles in the injection valves and is a function of nozzle geometry, nozzle temperature, and, most notably, fuel grade. Coking is primarily promoted by the cracking of parts of the diesel fuel components at the end of the boiling range. Coking processes have not been subjected to significant study because they are very complex.

In gasoline engines, the main concern is coked valves and coked spark plugs. Additives in gasoline and lubricating oil can remedy these problems.

Cold filter plugging point →Fuel, diesel engine ~Properties

Cold operation →Cold start; →Warm-up

Cold sludge →Oil ~Classification, specifications, and quality requirements ~~Classifications ~~~CCMC classifications

Cold smoke →Cold start ~Diesel engine

Cold start (*also*, →Electronic open- and closed-loop control ~Electronic open- and closed-loop control, gasoline engine ~~Functions; →Fuel, diesel engine ~Additives, ~Properties; →Ignition system, diesel engine/preheat system). The term "cold start" means starting the engine when it is not at operating temperature. In this state, neither the engine nor fluids such as coolant and oil are at operating temperature. In order to achieve optimum engine operation—in terms of low-temperature operation, transient behavior, throttle response, starting characteristics, etc.—it is necessary to take measures to ensure good performance, particularly at temperatures far below the freezing point. These measures consist primarily, of adapting the injection timing, injection quantity, and ignition timing. Above all, however, it is necessary to ensure the necessary cranking speed for starting.

~Diesel engine. Diesel engines have fewer starting problems than gasoline engines due to the internal mixture formation in combination with the charge quality regulation and cold starting aids such as glow plugs. Regulation of idle speed is accomplished by changing the injection quantity. The injection quantity for starting at low temperatures is higher than that required when the engine is warm, because it is necessary to overcome additional friction losses. When starting at low temperatures, during the warm-up phase and in the low load range, generation of "cold smoke" can cause problems, leading to visible smoke in the exhaust. This results from incomplete chemical reaction, and uncombusted fuel is its cause. In addi-

tion, intermediate reaction products are emitted which have a highly irritating effect on the mucous membranes in the eyes and nose. The exhaust odor is also affected and is the primary obvious factor. To remedy this problem, it is necessary to ensure that the fuel injected does not make contact with the cold combustion chamber and piston surfaces. Other remedial measures include optimum fuel preparation, which results from high injection pressures in combination with small injection orifices on diesel engines with direct injection.

On diesel engines, white smoke can also be formed. This results from evaporation of the unburned fuel from the combustion chamber and piston surfaces; when this is expelled, high hydrocarbon emissions result which can be visible in the exhaust in the form of white smoke. As the temperatures of the combustion chamber walls increase, this white smoke disappears.

Cold starting problems can also occur when the diesel fuel contains too much paraffin. At low temperatures, this precipitates clogging of the filters and lines. Modern diesel fuels are, however, low in paraffin and are heated in critical engine areas such as the fuel filter.

~Gasoline engine. In contrast to diesel engines, where it is only necessary to increase the quantity of fuel for cold starts, with gasoline engines it is necessary to enrich the air-fuel mixture (cold start enrichment) and increase the mixture volume. The required enrichment is higher for carburetor engines and engines with central injection than with multipoint injection, because in the cold intake manifold, the fuel components with higher boiling point can condense out and then reach the combustion chamber in liquid form. This results in surges of hydrocarbons in the exhaust (hot spot heating). At low ambient temperatures, carburetor engines require an additional starting system, because at the low cranking speeds prevalent here, the vacuum in the intake manifold is too low for the fuel to evaporate.

With multipoint injection, an additional electromagnetically actuated injection valve for cold starting valve is opened; this supplies additional fuel to the intake manifold for a short time for starting at low temperatures. The systems commonly in use today solve the problem of increased fuel requirement by increasing the injection time.

Cold start parameters →Ignition system, diesel engine/preheat system ~Cold start ~~Influential parameters

Cold start test →Emission measurements ~Test type I

Cold starting aid →Air cycling valve ~Air cycling valve functions; →Ignition system, diesel engine/preheat system

Color, oil →Oil ~Properties and characteristic values

Combination governor →Injection system (components) ~Diesel engine ~~Mechanical control ~~~Control maps

Combination ignition systems →Ignition system, diesel engines/preheat system ~Cold starting aid

Combined regeneration process →Particles (Particulates) ~Particulate filter system ~~Particulate filter ~~~Regeneration of particulate filters

Combustion, diesel engine (also, →Combustion process; →Combustion process, diesel engine). Diesel engine combustion is characterized by the following processes: 1) compression of the air, 2) fuel injection, 3) mixture formation, 4) auto-ignition, and 5) combustion. These processes take place over a period of milliseconds in high-speed diesel engines. Fuel injection follows after the end of compression shortly before the top dead center. In indirect injection engines, the fuel is injected into the prechamber or swirl chamber; in engines with direct injection it goes directly into the main combustion chamber.

Distributor injection pumps and unit pumps as well as pump-nozzle injection systems and common-rail injection systems are used for the injection process. The injection pressures for direct injection diesel engines using distributor injection pumps are about 1400 bar, for pump-nozzle injection systems they reach 2000 bar, while 1800 bar is achieved for common-rail systems, although this is increasing all the time.

The injected fuel evaporates and mixes with the compressed hot air. The mixture that develops will auto-ignite when the conditions are right. Rapid injection and good atomization are required to achieve intensive mixture formation because there is only a short time available for mixture formation. **Fig. C104** shows qualitatively the processes into which diesel engine combustion can be subdivided.

The individual processes, particularly spray formation, droplet vaporization, and mixture formation are mutually connected with each other and, therefore, cannot be considered completely separate from one another. That is why the sequence of diesel engine mixture formation and combustion is extremely complex.

~Auto-ignition. The physical and chemical processes that occur during the ignition delay period are very complex. The most important physical processes are fuel atomization, evaporation, and mixing of fuel vapor with air until an ignitable mixture has developed. The chemical processes described in the following are the initial reactions in the mixture up to auto-ignition taking place at a local air-fuel ratio of $0.5 < 1 < 0.7$.

The start of hydrocarbon oxidation can be considered a branched chain-propagating process whose course is strongly dependent on temperature and which can be subdivided into three temperature ranges, according to [5] and as described in the following:

At high temperatures, $T > 1100$ K, chain branching

$$H^\bullet + O_2 \rightarrow O^\bullet H + O^\bullet$$

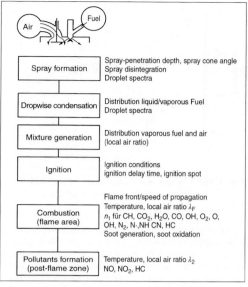

Fig. C104 Processes for mixture formation and combustion in a diesel engine

is dominant. This reaction is, however, strongly temperature-dependent and quickly loses significance at lower temperatures. In the middle of the temperature range, 900 K $< T < 1100$ K, the additional branches

$$HO_2 + RH \rightarrow H_2O_2 + R$$

$$H_2O_2 + M \rightarrow OH + OH + M$$

gain importance and the OH radicals will partly revert to the original HO_2 radical.

The decomposition of H_2O_2 is relatively slow in the low temperature range, $T < 900$ K, and degenerated branched-chain reactions gain significance characterized by the precursors of chain branching (e.g., RO_2) that would dissociate again at higher temperatures. This causes an inverse temperature dependence of the reaction rate, which can be described as a two-step reaction mechanism. This reaction mechanism, initially developed for the description of knocking combustion in the gasoline engine [5], leads to an extremely large reaction scheme because the developing remnant molecules can have many isomeric structures, giving some 6000 elementary reactions with about 2000 species for the auto-ignition of $n\text{-}C_{16}H_{34}$.

Several auto-ignition models have been developed for use in engine-related combustion simulations.

Today, a model with 30 elementary reactions and 21 species [6] is considered as the basic model in which the elementary character of the kinetics is maintained to a large extent.

Fig. C105 shows that results from this model agree very well with the measurements in stoichiometric mixtures of heptane and air at various pressures.

C

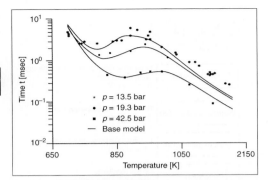

Fig. C105 Comparison of the basic model with auto-ignition measurements according to [6]

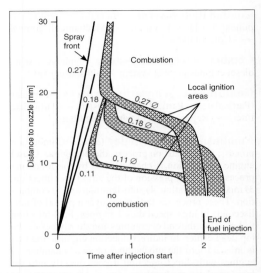

Fig. C106 Position of the local ignition areas against the time after start of injection for three different injector nozzles according to [7]

An eight-step reaction model containing a chain-propagating mechanism, known as the Shell Model, which is extended by a degenerated branched-chain process with two reaction paths and by two chain-terminating reactions, requires adaptation of 26 reaction parameters [12]. This model is used, for example, in the commercially available FIRE computer code.

A "relatively simple" five-step reaction mechanism describes the auto-ignition of mixtures made from n-heptane and isooctane very well in the mean temperature range [13]. This model is of great interest due to its simplicity for simulation calculations.

The above-mentioned complex models are often not absolutely necessary for diesel engine auto-ignition because this typically takes place at higher temperatures. In practice, a one-equation model based on only one Arrhenius equation can be used to describe the ignition delay satisfactorily:

$$\Delta t_{ZV} \; = \; A \cdot \frac{\lambda}{p^2} \exp\!\left(-\frac{E}{R \cdot T}\right).$$

It can be seen that the ignition delay, Δt_{ZV}, depends on pressure, temperature and air-fuel ratio [4]. In summary, **Fig. C106** shows the distance of the local ignition areas from the nozzle mouth, as a function of time after injection, for the various injector nozzles according to [7].

It can be seen that the mixture ignites after about one millisecond and that the smaller the nozzle hole diameter, the closer the ignition area is to the nozzle mouth.

~Combustion curve. *See below*, ~Combustion process

~Combustion process. The course of diesel engine combustion can be roughly subdivided into three phases. These are represented in **Fig. C107** and described in the following sections.

Fuel injected during ignition delay period mixes with the ambient air, forming a largely homogeneous and reactive mixture. This mixture will initially burn very quickly after the physical and chemical ignition delay. So this first phase, so-called premixed combus-

tion, is similar to gasoline-engine combustion. The combustion noise typical of diesel engines is caused by the high rate of increase in pressure during this premixed combustion. This noise can be influenced by changing the injection point. Advanced injection means "hard" (harsh) combustion and retarded injection "soft" combustion (**Fig. C108**). Combustion noise is significantly reduced by using a 5% pilot injection of the fuel quantity.

The mixture formation processes persist during main combustion period. The chemistry is fast in this second phase and the combustion sequence is diffusion-controlled; therefore, this phase is known as mixture-controlled diffusion combustion. The end of this main combustion phase is characterized by reaching the maximum temperature in the combustion chamber.

Pressure and temperature have dropped so much toward the end of combustion that the chemistry becomes slow compared to the simultaneously running mixing processes. Therefore, this third phase increasingly becomes diffusion combustion controlled by reaction-kinetics. A decisive factor in the thermodynamic quality of the overall combustion process is how the release of thermal energy progresses, given by:

$$\frac{dE_B}{d\varphi} \; = \; f(\varphi).$$

It results in the heating up of the air-fuel mixture in the cylinder, causing a substantial increase in pressure.

Fig. C109 shows the pressure and combustion curves (heat-release rate) for a high-speed, high-performance diesel engine with relatively retarded injection at full and part load.

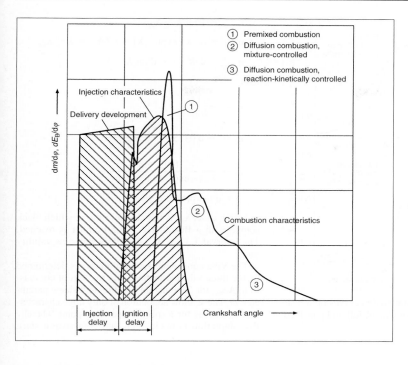

C

Fig. C107
Fuel delivery, injection,
and combustion curve for
the diesel engine [1], [4]

~**Equivalent rate of combustion.** A number of models have been developed for heat-release simulation; these are detailed in [1], [3], and [4]. Semi-empirical equivalent rates of combustion are quite often used in practice. However, these need to be adapted to the respective engine and, therefore, do not have extrapolation capabilities.

Starting from a triangular rate of combustion characteristic [8] and based on considerations of reaction kinetics, the relation

$$\frac{E_B}{E_{B,\,ges}} = 1 - \exp(-a \cdot y^{m+1})$$

can be obtained, where $E_{B,ges} = m_B H_u$, the maximum quantity of heat that can be released when burning all the fuel, and $y = (\varphi - \varphi_{BB})/\Delta\varphi_{BD}$ for the dimensionless crank angle, with $\Delta\varphi_{BD} = \varphi_{BE} - \varphi_{BB}$, defining the combustion duration. **Fig. C110** shows the rate of combustion (instantaneous heat release) curve

$$\frac{dE_B}{d\varphi} = f(\varphi, m)$$

and the cumulative combustion (heat release) curve

$$E_B = \int f(\varphi, m) \cdot d\varphi = F(\varphi, m),$$

showing the dependence on the dimensionless crank angle for various shape parameters, m.

With the combustion duration over—i.e., at $\varphi = \varphi_{BE}$ and, respectively, $y = 1$, $\eta_{U,ges}$—a certain percentage of the total energy injected with the fuel is supposed to be released. Hence follows the relation

$$\frac{E_B}{E_{B,\,ges}} = \eta_{U,\,ges} = 1 - \exp(-a)$$

and from there the numerical values in **Fig. C111**.

Betz [10] has stated the empirical relation for the conversion ratio

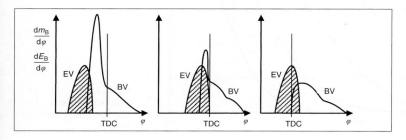

Fig. C108
Fuel injection and
combustion curves for
"hard" and "soft"
combustion [1], [4]

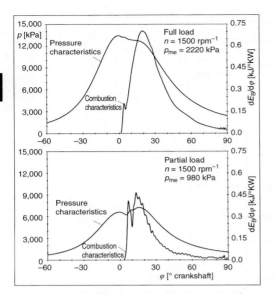

Pressure and combustion curves on a fast-running diesel engine at full and part load [1], [4]

$$\eta_{U, ges} = 1 \text{ for } \lambda > \lambda_{RB}$$

$$\eta_{U, ges} = a \cdot \lambda \cdot \exp(c \cdot \lambda) - b \text{ for } 1 \leq \lambda \leq \lambda_{RB}$$

$$\eta_{U, ges} = 0.95 \cdot \lambda + d \text{ for } \lambda \leq 1$$

with

$$c = -\frac{1}{\lambda_{RB}}$$

$$d = -0.0375 - \frac{\lambda_{RB} - 1.17}{15}$$

$$a = \frac{0.05 - d}{\lambda_{RB} \cdot \exp(-1) - \exp(c)}$$

$$b = a \cdot \exp(c) - 0.95 - d$$

in which case, λ_{RB} is the air-fuel ratio at which black smoke with a Bosch RB = 3.5 number is reached. The interval $1.17 < \lambda_{RB} < 2.05$ is stated as validity range.

The Vibe equivalent combustion rate is determined by the three Vibe parameters: φ_{BB} for start of combustion, $\Delta\varphi_{BD}$ for combustion duration and shape parameter, m. These allow adaptation of only three characteristic quantities for a specific operating point. Thereby the adaptation is made such that combustion start,

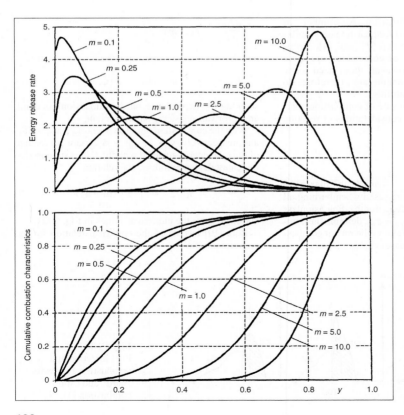

Fig. C110
Combustion (heat release) curves based on Vibe [8]

$\eta_{U,ges}$	0.999	0.990	0.980	0.950
a	6.908	4.605	3.912	2.995

Fig. C111 Factors for combustion characteristics

φ_{BB}, ignition pressure, p_Z, and mean effective pressure, $p_{m,i}$, agree with those measured in the real engine process. The adaptation is made through analysis of the pressure-volume diagram, which is described in detail in [4].

Adaptation of the Vibe parameters to any operating point is made using semi-empirical functions and depends primarily on the following variables: air-fuel ratio, λ; engine speed, n; power output; ignition delay, $\Delta\varphi_{ZV}$; and start of combustion, φ_{BB}, according to these:

$$\frac{\Delta\varphi_{BD}}{\Delta\varphi_{BD,0}} = \left(\frac{\lambda_0}{\lambda}\right)^{0.6} \cdot \left(\frac{n}{n_0}\right)^{0.5} \cdot \eta_{U,ges}^{0.6}$$

$$\frac{m}{m_0} = \left(\frac{\Delta\varphi_{ZV,0}}{\Delta\varphi_{ZV}}\right)^{0.5} \cdot \frac{p \cdot T_0}{p_0 \cdot T} \cdot \left(\frac{n_0}{n}\right)^{0.3}$$

$$\Delta\varphi_{ZV} = 6 \cdot n$$
$$\cdot 10^{-3}\left[0.5 + \exp\left(\frac{7800}{2 \cdot T}\right) \cdot \left(\frac{0.135}{p^{0.7}} + \frac{438}{p^{1.8}}\right)\right]$$

$$\varphi_{BB} = \varphi_{FB} + \Delta\varphi_{EV,0} \cdot \frac{n}{n_0} + \Delta\varphi_{ZV}$$

with the start of delivery φ_{FE} and injection delay $\Delta\varphi_{EV,0}$. Refer to [4] and [9] for further details.

~Flame front. *See above*, ~Combustion process

~Heat release. *See above*, ~Equivalent rate of combustion

~Spray propagation. The fuel sprays leave the injector nozzles at high velocity and break up into small droplets because of the high turbulence in the spray and the high relative velocity into the high pressure air, thereby becoming more atomized as the fuel penetrates into the combustion chamber. **Fig. C112** shows a qualitative sketch of the fuel spray leaving the injector nozzle.

The spray propagation in a diesel engine and thus the mixture formation are primarily determined by the injection system and the injection parameters, but they are also affected by the flow field (swirl and turbulence) of the compressed air in the combustion chamber. The turbulent kinetic energy introduced into the combustion chamber by the fuel spray is, however, at least one order of magnitude greater than the kinetic energy of the air, so that the flow field in the cylinder increases in importance only toward the end of injection when the spray has already strongly slowed down.

In case of very high injection pressures, breakup and evaporation of the liquid spray is already initiated by cavitation inside the nozzle bore. The vapor pres-

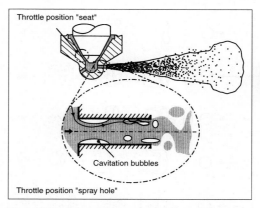

Throttle position "seat"

Cavitation bubbles

Throttle position "spray hole"

Fig. C112 Schematic representation of the spray propagation

sure is reached because of extremely high pressure drop along the injection jet, and the first vapor bubbles are formed. Implosion of these bubbles will then cause pressure pulsations, which accelerate both the breakup of the liquid spray core and primary droplet formation. Secondary droplet breakup will take place because of deformation of these primary droplets through their vibration characteristics and the surface forces caused by the high relative velocity between droplets and air in the combustion chamber. Droplets of varying size will form in the range of 10 to 100 μm in diameter.

This breakup of the secondary droplets is described by the Weber number

$$W_e = \frac{\rho_{Gas} \cdot d_T \cdot w_{rel}^2}{\sigma}$$

which represents the dimensionless ratio between the aerodynamic forces and surface tension force on the droplet. The dependence on the size of the Weber number of breakup mechanisms on the droplet is represented in **Fig. C113**:

- Vibration breakup (1): We < 12
- Bubble breakup (2) and bag-and-stamen breakup (3): We < 50
- Shear breakup (4): We < 100
- Catastrophic breakup (5): We > 100

In addition to single droplet breakup, there is also collision of several droplets in the fuel spray. The droplets rebound from each other depending on size, velocity, and impingement angle, breaking up into smaller ones or combining to form larger droplets (droplet coalescence).

At the edge of the spray, the fuel droplets mix with the hot air in the combustion chamber (air entrainment). This mixing heats up the droplets as a result of convective heat transfer and thermal radiation from the hot combustion chamber walls, and the droplets eventually evaporate. In addition to temperature, the

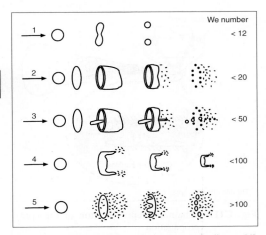

Fig. C113 Droplet breakup mechanism according to [4]

droplet evaporation rate is also influenced by fuel diffusion from the droplet surface into the droplet environment, so that heat transport and mass transport occur simultaneously.

Both spray propagation and mixture formation are qualitatively understood today and can be described relatively well using semi-empirical models for any of the partial processes.

Literature: [1] G.P. Merker, G Stiesch: Technische Verbrennung—Motorische Verbrennung, B.G. Teubner Publishers, Stuttgart/Leipzig, 1999. —[2] W.O.H. Mayer: Zur koaxialen Flüssigkeitszerstäubung im Hinblick auf die Treibstoffaufbereitung in Raketentriebwerken, Research Report DLRFB-93-09, 1993. —[3] J.R. Ramos: Internal Combustion Engine Modeling, Hemisphere Publ. Corp., New York, 1989. —[4] G.P. Merker, C. Schwarz, G. Stiesch, F. Otto: Verbrennungsmotoren - Simulation der Verbrennung und Schadstoffbildung. Teubner Publishers, Wiesbaden, 2004. —[5] J. Warnatz, U. Maas, R.W. Dibble: Verbrennung, 2nd edition, Springer Publishers, Berlin/Heidelberg, 1997. —[6] K. Fieweger, H. Ciezki: Untersuchung der Selbstzündungs- und Rußbildungsvorgänge von Kraftstoff/Luft-Gemischen im Hochdruckstoßwellenrohr, SFB 224 Research Report, 1991. —[7] E. Winkelhofer, B. Wiesler, G. Bachler, H. Fuchs: Detailanalyse der Gemischbildung und Verbrennung von Dieselstrahlen, Conference "Der Arbeitsprozess des Verbrennungsmotors," Graz, 1991. —[8] R.R. Vibe: Brennverlauf und Kreisprozess von Verbrennungsmotoren, VEB Publishers Engineering, Berlin, 1970.—[9] G. Woschni, F. Anisits: Eine Methode zur Vorausberechnung der Änderung des Brennverlaufs mittelschnelllaufender Dieselmotoren bei geänderten Betriebsbedingungen, MTZ 34, 4, 1973. —[10] A. Betz: Rechnerische Umsetzung des stationären und transienten Betriebsverhaltens ein- und zweistufig aufgeladener Viertakt-Dieselmotoren, Diss., TU Munich, 1985. —[11] C. Eigelmeier: Phänomenologische Modellbildung des gasseitigen Wandwärmeübergangs in Dieselmotoren, Diss., University of Hanover, 2000. —[12] M.P. Halstead, L.J. Kirsch, A. Prothero, C.P. Quinn: A mathematical model for hydrocarbon autoignition at high pressures, Proc. R. Soc. London A, 346, 1975, pp. 515–538. —[13] M. Schreiber, A. Sadat Sakak, A. Lingens, J.F. Griffith: A Reduced Thermokinetic Model for the Autoignition of Fuels with Variable Octane Rings; 25th Symposium (Int) on Combustion, 1994, pp. 933–940. —[14] G. Stiesch: Modelling Engine Spray and Combustion Processes, Springer Publishers, Berlin/Heidelberg, 2003.

Combustion, gasoline engine (*also,* →Combustion process). Gasoline engine combustion features these phases: 1, induction of air-fuel mixture or pure air; 2, compression (or fuel injection with mixture formation with direct injection); 3, ignition; 4, combustion. It is necessary to distinguish between two cases pertaining to fuel injection into the combustion chamber: homogeneous mixture formation with advanced injection to enable mixture formation and stratified charge. In the latter case, injection does not occur before compression, so that the stratification does not dissipate before combustion.

Both heat and exchange of material processes continue further combustion in the mixture after initiation by electric ignition. Two preconditions need to be met for normal flame expansion: First, the air-fuel ratio must be within the ignition limits, and second, no autoignition or glow ignition must occur.

~Combustion chamber shape. The combustion chamber design has an essential influence on knock characteristics, fuel consumption, and emission of pollutants. The following prerequisites are necessary to satisfy these variables:

1. Compact combustion chamber with central plug location for achieving short flame travel
2. Avoidance of hot spots shortly before end of combustion; therefore, plug(s) in the vicinity of the exhaust valves
3. High flow velocity in the combustion chamber, for example, induced through swirl or tumble for increasing flame speed
4. Least possible dead space (e.g., above top land).

Some combustion chamber designs are presented in the following:

- *Bathtub combustion chamber in the cylinder head*, **Fig. C114**. The valves along axis and parallel to each other; pronounced quench gap; spark plug lateral, inclined. Requirement 1 and 2 not met, requirement 3 met well.
- *Wedge-shaped combustion chamber*, **Fig. C115**. The valves parallel to each other and inclined to the piston axis; spark plugs lateral, inclined. Pronounced quench gap. Requirements 1, 2, and 3 complied with well.
- *Recessed combustion chamber in the piston*, **Fig. C116**. Valves parallel to each other in the cylinder head; pronounced quench gap; frequent air turbulence through the intake port that is designed as a swirl channel. Requirements 1, 2, and 3 complied with very well.
- *Hemispherical combustion chamber*, **Fig. C117**. Valves mounted V-shaped in the cylinder head; mostly without quench gap. Spark plug preferably

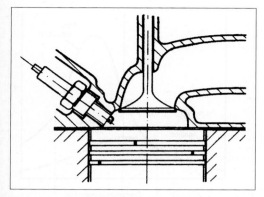

Fig. C114 Bathtub combustion chamber (Source: Pischinger)

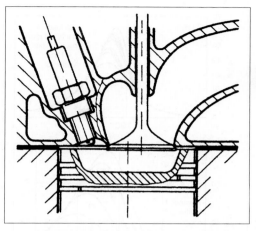

Fig. C116 Recessed combustion chamber in the piston (Source: Pischinger)

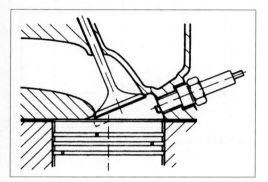

Fig. C115 Wedge-shaped combustion chamber (Source: Pischinger)

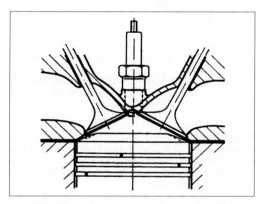

Fig. C117 Hemispherical combustion chamber (Source: Pischinger)

lateral, inclined. Requirement 1 met very well, requirements 2 and 3 complied with well.

Fig. C118 shows a special design of the hemispherical combustion chamber. This combustion chamber features a low tendency to knock because of the strong concentration of volume at the spark plug and a carefully machined quench gap. It also has a positive effect on the emission of unburned hydrocarbons.

~Cyclic variations. Relatively large fluctuations of the pressure-volume (p-V) diagram from cycle to cycle are typical in gasoline engine combustion. The causes of these cyclic variations are fluctuations of the turbulent velocity and mixture composition fields in the combustion chamber in terms of both time and location, but in particular in the area of the spark plug electrodes. The resultant cyclical variations in the ignition delay have a significant effect on the p-θ diagram in the combustion chamber and can lead to incomplete combustion.

Fig. C119 shows the effect of the cyclic variations (top) and the influence of the ignition advance on the p-θ diagram for methanol combustion. Comparison of the two diagrams shows that the cyclic variations have

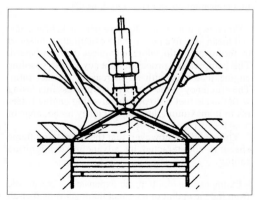

Fig. C118 Hemispherical combustion chamber with concentration of volume at the spark plug (Source: Pischinger)

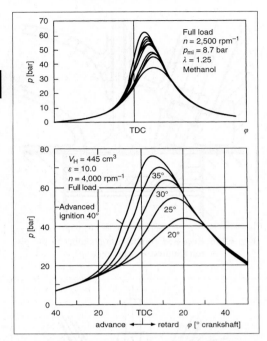

Fig. C119 Cyclic variations and influence ignition-timing on the p-θ diagram in the combustion chamber [7]

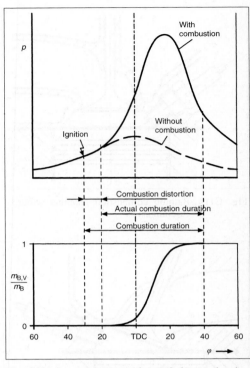

Fig. C120 Pressure curve and cumulative combustion curve against crank angle [7]

an impact similar to adjusting the ignition angle by approximately 15°CA. Therefore, the measured p-θ diagrams have to be averaged to obtain a thermodynamic evaluation of the tests, usually over 64 or 128 consecutive combustion cycles.

Reducing cyclic variation during gasoline engine combustion by optimizing mixture formation, ignition angle map, and flame expansion (dual ignition) is a promising aim for achieving lowering of the specific fuel consumption and HC emissions [2].

~Efficiency and mean effective pressure. Efficiency and mean effective pressure are essentially influenced by the air-fuel ratio, which influences the flame speed. The comparative process efficiency of the gasoline engine increases with an increasing air-fuel ratio. The efficiency of the comparative process increasing with the air-fuel ratio and the growing engine losses result in maximum internal efficiency in the range of $\lambda = 1.1$ to 1.3.

This requires, however, that the decreasing flame speed is counteracted by advancing the ignition timing.

~Flame expansion. If flame expansion is considered from the viewpoint of a perfect homogeneous fuel-air mixture, this would give the ideal case of a completely premixed flame. The chemical processes that take place in the flame front are slow compared to the ther-

mal and mass transport processes; so the gasoline engine combustion is chemically controlled.

Fig. C120 shows the pressure curve of the high-pressure cycle under both driven and firing engine operation. The terms "combustion duration," "ignition delay or combustion delay," and "effective combustion duration" are explained in **Fig. C120**. The end of combustion is defined as that time when the fuel is "completely" burned (99.9% as a rule). The illustration also represents the ratio of burned against available fuel, $m_{B,V}/m_{B}$ against the crank angle and this ratio is identical with the cumulative combustion curve, provided the fuel is completely burned.

In **Fig. C121**, the position of the flame front is represented at various crank angles for two combustion cycles with either intake manifold injection or with direct injection. From this it is evident that the differences are relatively minor and that they are within the range of the cyclical fluctuations described.

Instead of popular four-valve engines with centrally arranged spark plugs, three-valve engines with two spark plugs, arranged off-center (dual ignition), are used to obtain faster mixture burn-up [1, 2]. Looking at the flame expansion for the two engines represented in **Fig. C122**, it becomes apparent that in the three-valve engine, the mixture burns up faster because the combustion distances are shorter.

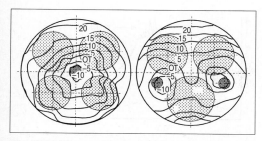

Fig. C121 Flame expansion with injection into the intake manifold in front of the intake valve and with direct injection during the intake cycle (operation at $\lambda = 1$) [7]

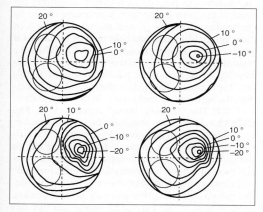

Fig. C122 Flame expansion in a four-valve engine with central ignition and in a three-valve engine with dual ignition [7]

~Ignition delay. The ignition delay has great significance for knocking. It depends on the fuel octane number, from the air-fuel ratio, and from pressure and temperature at the end of compression. The shorter the ignition delay of an air-fuel mixture, the greater the risk of auto-ignition and hence the risk of knock.

~Knocking. A relatively "soft pressure curve" can be observed with normal combustion, having maximum rates of increase in pressure of about 2 bar/°CA. By contrast, strong fluctuations of pressure occur in the air-fuel mixture with knocking combustion, which can be explained by the following process. With initiation of combustion by the ignition spark completed, the unburned residual mixture is further compressed by the expanding flame front and thereby additionally heated. Spontaneous auto-ignition will eventually start if the ignition limit of the mixture is exceeded during this process. This almost isochoric combustion of the residual gas leads to steep pressure gradients expanding in the form of shock waves in the combustion chamber, causing the notorious knocking or pinking noise.

Spontaneous auto-ignition is almost completely controlled by chemical kinetics. Tests with single-cylinder engines have demonstrated that knocking will occur at temperatures of approximately 1100 K. The elementary processes taking place thereby can be explained relatively well through chain branching of the HO_2 radical as in

$$HO_2 + RH \rightarrow H_2O_2 + R$$

$$H_2O_2 + M \rightarrow OH + OH + M.$$

However, the heat losses to the combustion chamber walls in multicylinder series engines are significantly higher so that knocking already occurs at considerably lower temperatures in the range between 800 and 900 K. Decomposition of H_2O_2 in this temperature range is relatively slow, and auto-ignition is described by the considerably more complex low-temperature oxidation.

Spontaneous auto-ignition essentially depends on the composition of the fuel. Reduced reaction mechanisms were developed in [3] and [4] to obtain qualitatively correct descriptions of the auto-ignition process in gasoline engines. While [3] describes a four-step mechanism for the auto-ignition of n-heptane, [4] developed a formally similar three-step mechanism for two-component gasoline fuels consisting of n-heptane and isooctane, corresponding to

$$F + \alpha O_2 \rightarrow P$$

$$F + 2O_2 \rightarrow I$$

$$I + (\alpha - 2)O_2 \rightarrow P$$

with the species involved

$$F = \frac{ON}{100}(iso - C_8H_{18}) + \left(1 - \frac{ON}{100}\right)(n - C_7H_{16})$$

$$I = \frac{ON}{100}I_8 + \left(1 - \frac{ON}{100}\right)I_7 + H_2O$$

$$P = \left[8\frac{ON}{100} + 7\left(1 - \frac{ON}{100}\right)\right]CO_2$$
$$+ \left[8\frac{ON}{100} + 8\left(1 - \frac{ON}{100}\right)\right]H_2O$$

the intermediate products

$$I_7 = OC_7H_{13}OOH$$

and

$$I_7 = OC_8H_5OOH$$

the stoichiometric coefficients of oxygen

$$\alpha = 12.5\frac{ON}{100} + 11\left(1 - \frac{ON}{100}\right)$$

and the fuel octane number (ON).

In addition to the composition of the fuel, the geometry of the combustion chamber has a decisive influence on the tendency to knock. Combustion chambers with low tendency to knock have:

C

- Short flame travel because of compact design and a centrally positioned spark plug
- No hot spots at the end of flame travel because the spark plug is positioned in the vicinity of the exhaust valve
- High flow velocities and thereby good mixture formation as a result of swirl, tumble and quench flows

The pent-roof combustion chamber represented in **Fig. C123** with its valves inclined by approximately 20–30° against the cylinder axis and the centrally arranged spark plug has turned out to be particularly advantageous for four-valve cylinder heads. Modeling of the detailed processes and 3-D simulation will contribute greatly to the optimization of combustion processes by taking combustion-chamber geometry into account.

A further kind of unwanted combustion sequence is glow ignition. It is caused by extremely hot zones, the so-called "hot spots" on the walls of the combustion chamber where temperatures of about 1200 K, significantly above the auto-ignition temperature, are reached. The most frequent "hot spots" are combustion residues that deposit, for example, as hot flakes on the walls, predominantly on the exhaust valve. **Fig. C124** sketches the qualitative pressure curve in knocking combustion with the start of both knocking combustion and glow ignition drawn in. Knocking combustion can only occur after the initiation of combustion by the spark plug, while glow ignition can take place earlier. The shock waves occurring in both processes can cause mechanical damage and, furthermore, partly melt zones on piston and cylinder head because of the increased thermal load.

See [5] for a detailed description of the reaction-kinetic processes on knocking. The many forms of glow ignition are elaborated on in [6].

~Mean effective pressure and fuel consumption.
The air-fuel ratio, λ, influences the flame speed quite significantly, and thereby the attainable mean effective

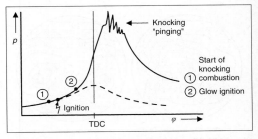

Fig. C124 Pressure curve at knocking combustion [7]

pressure and specific fuel consumption, over the course of combustion or pressure curve. With λ > 1.1, combustion will proceed increasingly sluggishly because of the lower combustion temperature caused by the heating of excess air and the resultant lowered flame speed. Minimum fuel consumption is reached at air-fuel ratios of approximately λ = 1.1. While the "calorific value" of the mixture increases with decreasing air-fuel ratio, the maximum mean effective pressure is already reached by about λ = 0.85. Therefore, the optimum air-fuel ratio is in the range 0.85 < λ < 1.1. The "fishhook curve" represented in **Fig. C125** shows the course of specific fuel consumption b_e against brake mean effective pressure, p_{me}, with various air-fuel ratios, λ, measured at constant engine speed and constant throttle-valve position. The employment of a three-way catalytic converter for emission clean-up on gasoline engines requires that the air-fuel ratio is controlled very precisely in the λ window 0.999 < λ < 1.002. **Fig. C125** emphasizes that this

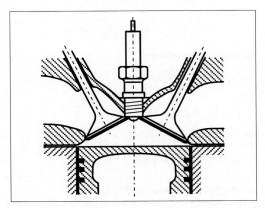

Fig. C123 Pent-roof combustion chamber of a four-valve cylinder head with centrally positioned spark plug [7]

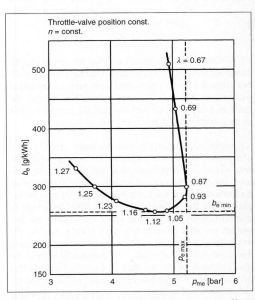

Fig. C125 Specific fuel consumption and mean effective pressure for various air-fuel ratios [7]

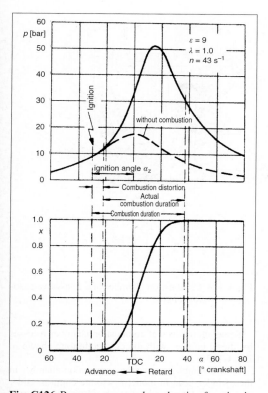

Fig. C126 Pressure curve and combustion function in a gasoline engine (Source: Pischinger)

value has to be rather precisely between $b_{e,\,min}$ and $p_{m,max}$.

~Pressure curve. The prerequisite for a normal pressure curve i.e., normal flame expansion—is no auto-ignition, no glow ignition, and the air-fuel ratio must be within the ignition limits.

Fig. C126 shows (upper diagram) a pressure curve against time that results from the variation of flame expansion and combustion-chamber volume as it changes over the piston travel. The bottom diagram shows the combustion function associated with the pressure-crank angle curve. As can be seen, combustion starts 8°CA after ignition and ends approximately 38°CA after TDC. Reactions for stable flame front formation between ignition and start of pressure rise take their course in the mixture volume between the spark plug electrodes. The heat released in the ignition zone and the generated reactive particles have to be sufficient to ignite the neighboring mixture particles.

~Rate of pressure rise. *See above*, ~Knocking

Literature: [1] H. Niefer, H.-K. Weining, M. Bargende, A. Walthner: Verbrennung, Ladungswechsel und Abgasreinigung der neuen Mercedes-Benz V-Motoren mit Dreiventiltechnik und Doppelzündung, MTZ 85 (1997), pp. 392–399.
— [2] M. Bargende, H.-K. Weining, P. Lautenschütz, F. Altenschmidt: Thermodynamik der neuen Mercedes-Benz 3-Ventil-Doppelzünder V-Motoren, in U. Essers (Ed.): Kraftfahrwesen und Verbrennungsmotoren, 2nd Stuttgart Symposium, Expert Publishers, Renningen-Malmsheim, 1997. — [3] U.C. Müller, N. Peters, A. Liñán: Global Kinetics for n-Heptane Ignition at High Pressures. Twenty-Fourth Symp. (Int.) on Combustion, The Combustion Institute, 1992, pp. 777–784. — [4] U.C. Müller: Reduzierte Reaktionsmechanismen für die Zündung von n-Heptan und iso-Octan unter motorrelevanten Bedingungen. Diss., RWTH Aachen University, 1993. — [5] J. Warnatz, U. Maas, R.W. Dibble: Verbrennung, 2nd edition, Springer Publishers, Berlin/Heidelberg, 1997. — [6] A. Urlaub: Verbrennungsmotoren, 2nd edition, Springer Publishers, Berlin/Heidelberg, 1994. — [7] G.P. Merker, G. Stiesch: Motorische Verbrennung, Teubner Publishers, Stuttgart/Leipzig, 1999.

Combustion air preheating →Intake air ~Air preheating

Combustion byproducts. A number of substances are produced by combustion in addition to the legally limited pollutant emissions such as unburned hydrocarbons, oxides of nitrogen, carbon monoxide, and particulates. Some of these are intermediate products, which decompose again quickly; however some persistent compounds remain in the exhaust gas. In spite of the generally low concentration of such byproducts, these can be a potential danger to health; the extent can only be determined with extremely complex studies. Moreover, some of these byproducts have an annoying odor. The byproducts include aldehydes, polycyclic aromatics, SO_2, olefins, paraffins, and hydrogen sulfide.

Combustion chamber. The conversion of the chemical energy in the fuel into mechanical energy is performed in the combustion chamber. The outputs of this process include combustion products and heat losses. The combustion chamber shape (geometry, position, valve arrangement, spark plug position, compression ratio, surface-volume ratio, position of the injection jet and of the glow plug, etc.) has a significant impact on the operating characteristics of the engine, for example, with respect to torque, power, acoustics, fuel consumption, and pollutant emissions. Generally, a compact combustion chamber provides the best efficiency for the gasoline engine.

The combustion chambers of engines have changed significantly over time. **Fig. C127** shows a flat combustion chamber with an extremely large surface-volume ratio and divided combustion chamber while **Fig. C128** shows compact multivalve combustion chambers as an example of the development of gasoline engines.

~Combustion chamber in cylinder head. The prechambers and swirl chambers are located in the cylinder head of a diesel engine that works as a chamber (indirect injection, IDI) engine (prechamber and swirl chamber), and they are connected to the main combustion chamber by narrow throats.

C

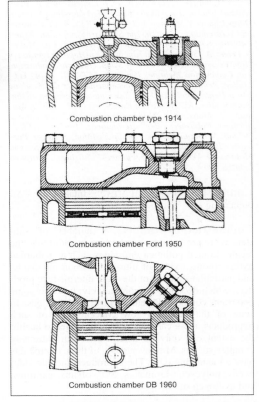

Combustion chamber type 1914

Combustion chamber Ford 1950

Combustion chamber DB 1960

Fig. C127 Combustion chamber evolution (Source Woschni)

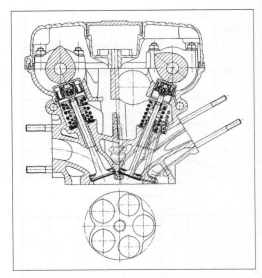

Fig. C128 Combustion chamber of a five-valve gasoline engine

Some alternatives of combustion chamber recesses for diesel engines with direct injection are shown in **Fig. C129**.

It is preferable to have the combustion chamber in the cylinder head in modern gasoline engines. Older designs also have combustion chambers which are, in large part or completely, inside the piston. A disadvantage is the large surface-volume ratio, which results in high HC emissions.

Literature: R. v. Basshuysen and others: Audi Turbodieselmotor mit Direkteinspritzung, Part 1 and Part 2. MTZ 50, 51 1989/1990. — W. Ebbinghaus and others: Der neue 1,9 Liter Dieselmotor von VW, MTZ 50 (1989) 12.

~**Main combustion chamber.** The partially combusted mixture flows from the prechamber or swirl chamber to the main combustion chamber during diesel engine combustion in an IDI diesel engine. A very rich mixture is present because the fuel is injected into the prechamber or swirl chamber. Air is present in the main combustion chamber and the rich air-fuel mixture flows into the chamber through the throat. This results

By contrast, the undivided combustion chamber in modern gasoline engines is located, completely or partly, in the cylinder head. For gasoline engines, the chamber can be designed as tub-shaped or wedge-shaped or as a hemispherical combustion chamber. Also common in gasoline engines are designs with part of the combustion chamber in the piston (e.g., 80% in the cylinder head, 20% in the piston as a shallow depression).

~**Combustion chamber in piston.** Diesel engines with direct injection into the cylinder have their entire combustion chamber in the piston bowl. The combustion chamber for four-valve engines is arranged symmetrically on the piston axis; however, in two-valve engines it is necessary to arrange it asymmetrically in relationship to the piston axis. A combustion chamber recess arranged with rotational symmetry in relationship to the piston axis has the advantage that no disturbance of the swirl occurs. It also permits central positioning of the injection jet.

The principle of a recess tucked in at the edges (reentrant bowl) results in lower NO$_x$ emissions.

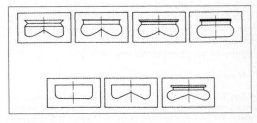

Fig. C129 Recessed shape in piston

in two-stage combustion with advantages in terms of the nitrogen oxide (NO_x) emissions and combustion noise, but with disadvantages for fuel consumption.

Combustion chambers in gasoline engines can also be divided into two parts. The spark plug is then positioned in a precombustion chamber containing a rich mixture. A typical example is the Honda CVCC process, which uses a stratification process. These processes have not become commercially important due to their flow and pressure losses.

~Multiple combustion chamber. *See below*, ~Subdivided combustion chamber

~Prechamber (*also*, →Combustion process). Diesel engines with a subdivided combustion chamber are generally called prechamber engines. The optimum prechamber volume in prechamber engines is approximately 35–40% of the main chamber volume. The oxygen share is not high enough for smokeless combustion if the prechamber is too small, and the HC, CO, and particle emissions will also increase. However, the NO_x emissions decline because the pressure increases and the resulting peak temperature also declines.

Of particular importance is the design of the throat passages. Oversized throat diameters result in low gas speeds, which means that there may possibly not be enough kinetic energy available in the gas flowing into the main chamber to ensure sufficient distribution and conditioning of the mixture. Undersized throat diameters can be constricted by contamination, with the effect that it may take too long before the content of the prechamber arrives in the main chamber, which then results in thermodynamic disadvantages.

Other important design features are the position of

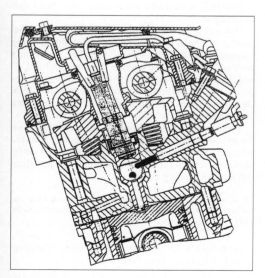

Fig. C130 Combustion chamber of a prechamber diesel engine

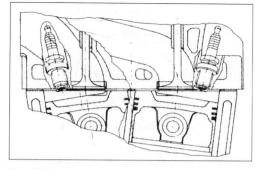

Fig. C131 Combustion chamber of a VR engine

the prechamber relative to the main chamber, the position and shape of the glow plug, and the angle of the injection jet. Four-valve technology supports a central position for the prechamber. **Fig. C130** shows a section through the cylinder head of a four-valve prechamber engine.

Prechamber diesel engines have now been largely replaced by diesel engines with direct injection.

Literature: M. Fortnagel, A. Peters, F. Thoma: Die neuen Vierventil-Dieselmotoren von Mercedes-Benz/Entwicklung von Verbrennung und Abgasreinigungssystem, MTZ 54 (1993) 9.

~Precombustion chamber. *See above*, ~Prechamber; *see below*, ~Swirl chamber

~Single combustion chamber. Single combustion chambers are used in conventional gasoline engines (**Fig. C128**), gasoline engines with direct injection, and diesel engines with direct injection (**Fig. C129**). They are generally compact and have advantageous surface-volume ratios, especially in gasoline engines and diesel engines with direct injection. **Fig. C128** shows an optimum combustion chamber for gasoline engines based on the example of a five-valve engine. **Fig. C131** shows a rather unfavorable combustion chamber using the example of a VR engine. The differences between the left and right combustion chamber shape, especially in terms of their geometry, result in significant design restrictions.

The combustion chamber of the Wankel engine is also single-stage. It has a large surface-volume ratio due to its particular geometry, which is unfavorable with respect to emission of unburned hydrocarbons. Additionally, the long flame travel, because of its narrow aspect ratio, results in unfavorable combustion conditions, which have a negative impact on fuel consumption. **Fig. C132** shows the diagram of such a combustion chamber.

~Subdivided combustion chamber (*also, see above*, ~Main combustion chamber). Subdivided combustion chambers are found in gasoline and diesel engines. Diesel engines primarily use prechamber and swirl chamber engines. Gasoline engines use primarily

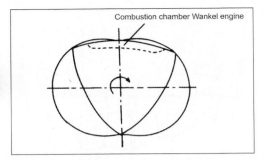

Fig. C132 Combustion chamber of a Wankel engine

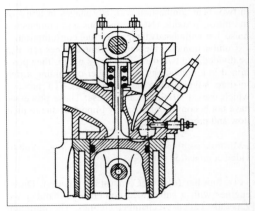

Fig. C133 Combustion chamber of a swirl chamber engine

stratified-charge arrangements such as the VW PCI, Porsche SKS, and Honda CVCC or the May-Fireball process. However, these are not used anymore because of their high fuel consumption.

Subdivided combustion chambers have advantages in terms of NO_x emissions and acoustics, but thermodynamic disadvantages due to flow losses from one chamber to the next and due to higher wall temperature losses.

~**Swirl chamber** (*also*, →Combustion process). The swirl chamber diesel engine has a divided combustion chamber similar to the prechamber engine. The volume ratio of the prechamber to the main chamber is approximately 1.0 in the ideal case. The ratio for real engines is usually between 1.07 and 1.15.

A significant reduction of the nitrogen oxide emissions with a small increase of the smoke emissions can be achieved if the volume of the swirl chamber is reduced. The swirl chamber geometry, in addition to the volume ratio, has a decisive impact on NO_x emissions.

Manipulation of the thermal characteristics is possible through varying the diameter, height, and throat cross-sectional areas. The thermal characteristics describe the released energy volume for each degree of crank angle. A reduction of NO_x generation is possible if energy conversion is delayed and reaches its maximum late in the process; this, however, is done at the expense of thermal efficiency. The shape and volume of the swirl chamber must be optimized to a swirl energy that is as high as possible to achieve favorable mixing action. Comparable to the prechamber engine, one main criterion is the position and geometry of the glow plug. A high swirl energy is achieved by using a thin glow plug positioned relatively far toward the back of the chamber to ensure that the flow processes are disturbed as little as possible. **Fig. C133** shows a schematic diagram of the swirl chamber combustion process. The optimum number of valves is not four, as for the prechamber engine, but three due to the eccentric position of the swirl chamber.

Combustion chamber charge →Cylinder charge

Combustion chamber charge stratification →Mixture formation ~Mixture formation, gasoline engine ~~Mixture inhomogeneity

Combustion chamber configuration →Combustion, gasoline engine ~Combustion chamber shape; →Combustion chamber

Combustion chamber deposits (*also*, →Fuel, diesel engine ~Additives; →Fuel, gasoline engine ~Additives). Combustion chamber deposits are primarily due to coking of oil and fuel. They can lead to increased wear. Additionally, the compression ratio is increased by the coking, which increases the knocking tendency in gasoline engines.

Deposits on the intake and exhaust valves can result in leakages.

Deposits at the intake valve can store fuel with characteristics similar to a sponge. This can result in leaner mixtures after a cold start and, therefore, to running problems with gasoline engines. Suitable fuel and oil additives can to a large extent prevent damaging deposits. **Fig. C134** shows, in addition to the combustion chamber deposits, additional problem areas regarding contamination in a gasoline engine.

Deposits in a diesel engine are primarily found on the injection jet, where they impact negatively on the mixture formation. Active materials, so called detergents, in fuel and oil can keep the injection jets clean and remove existing deposits.

Literature: K.G. Brand: Kraftstoffadditive–ein Beitrag zur sauberen Umwelt, AVL Conference Motor und Umwelt, Graz, 1991.

Combustion chamber documents →Sealing systems ~Cylinder head gaskets

Combustion chamber flow →Charge movement; →Variable valve control ~Charge movement

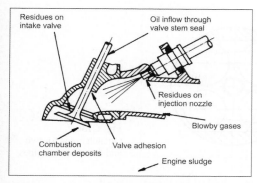

Fig. C134 Relevant problem areas for deposits in gasoline engines

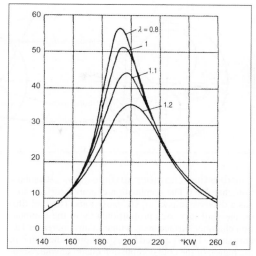

Fig. C135 Variation in pressure in the combustion chamber for different air-fuel ratios

Combustion chamber gasket →Sealing systems ~Cylinder head gaskets ~~Beadings

Combustion chamber in cylinder head →Combustion chamber ~Combustion chamber in cylinder head

Combustion chamber in piston →Combustion chamber ~Combustion chamber in piston

Combustion chamber internal wall temperature. The temperature of the internal wall of the combustion chamber is a quantity used, for instance, to calculate the heat transfer from working gas to the wall or through the wall to the cooling medium. Because the internal wall temperature is not locally constant, it must be calculated with an approximation to the mean internal wall temperature.

Cooling must be designed such that the salient properties—for example, strength factors—of the material are not exceeded for the particular wall materials and aluminum alloys used on modern engines.

Combustion chamber peak temperature →Combustion chamber temperature

Combustion chamber pressure. The combustion chamber pressure can reach 90 bar in supercharged gasoline engines. It is significantly higher for supercharged diesel engines. Supercharged diesel engines with direct injection achieve values of 160 bar with a trend toward increasing values in present car engines; large diesel engines achieve values of over 200 bar.

The pressure is limited primarily by the piston and crankcase strengths. Very high rates of pressure increase result in loud combustion noise. The pressure in the combustion chamber depends on the operating conditions in the engine. Peak values in gasoline engines are achieved for an air-fuel ratio of $\lambda \cong 0.8$; the peak pressures decline as lambda increases. The moment of ignition is another important factor, with early ignition tending to result in high combustion chamber

pressures. Volumetric efficiency, compression ratio, combustion chamber shape, the spark plug system, and supercharging rate also have a significant impact on the pressure in the combustion chamber. **Fig. C135** shows an example of the pressure variation with crank angle in the combustion chamber of a gasoline engine with external mixture formation for different air-fuel ratios.

Combustion chamber recess →Combustion, gasoline engine ~Combustion chamber shape; →Combustion chamber

Combustion chamber recirculation →Variable valve control ~Residual gas control

Combustion chamber sensor →Sensors ~Pressure sensors

Combustion chamber shape →Combustion, gasoline engine; →Combustion chamber

Combustion chamber squish areas. Squish areas in gasoline engines are designed to increase the charge movement in the combustion chamber. A displacement of the mixture occurs in these zones because when the piston approaches the cylinder head (**Fig. C136**), secondary flows (squish flows) are produced which support faster combustion of the mixture due to the more intense mixing. The size of the squish area can be approximately 10–15% of the piston surface area.

Combustion chamber squish flow →Combustion chamber squish areas; →Charge movement ~Quench flow

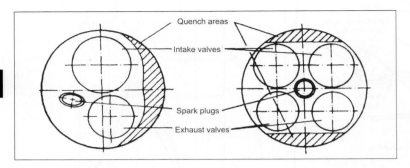

Fig. C136
Quench areas for two-
and four-valve engines

Combustion chamber surface (*also*, →Combustion chamber). The combustion chamber surface area has a decisive impact on HC emissions. The fuel film on the wall increases proportionately to the combustion chamber surface area, which in turn increases HC emissions. This is caused by extinction of the flame in the region of the cold combustion chamber wall. Combustion chambers with small surface-volume ratios should, therefore, be the target, which means compact combustion chambers. The heat losses also increase with an increasing surface, which in turn results in thermal efficiency reductions. However, NO_x generation is higher in compact combustion chambers due to the higher gas temperatures.

Combustion chamber surface volume ratio →Combustion chamber surface

Combustion chamber swirl →Charge movement ~Swirl

Combustion chamber temperature. The maximum gas temperature in the combustion chamber of gasoline engines is approximately 2600°C. The spatial mean value is approximately 2000°C for IDI diesel engines (supercharged swirl chamber engine), but higher peak temperatures may develop locally. Supercharged diesel engines with direct injection achieve maximum local gas temperatures of approximately 2600°C, which explains the higher NO_x generation as compared to that in chamber engines. A temperature reduction must be achieved by means of cooling to ensure that the maximum permissible component temperature, depending on the material used and the component involved (cylinder head, piston, gasket, etc.), will not be exceeded. The acceptable temperature for aluminum cylinder heads is approximately 240°C and for cast iron approximately 260–270°C. The surface temperatures of the piston in swirl chamber diesel engines reach approximately 300–350°C and valve seats (exhaust) in gasoline engines may reach 750°C.

Combustion chamber temperature reduction →Combustion chamber temperature

Combustion chamber turbulence →Charge movement ~Turbulent flow

Combustion chamber wall (*also*, →Flame quenching). The combustion chamber wall bounds the combustion chamber. The boundary on the top side is the cylinder head, and on the bottom side the piston crown, while the upper part of the cylinder liner completes the volume. The materials that line the combustion chamber must, among other things, transfer a proportion of the heat generated by combustion. This is why materials with high thermal conductivity are preferred; however, very high heat losses reduce efficiency.

Combustion curve (*also*, →Combustion, diesel engine ~Combustion process). The combustion curve describes the time-dependent conversion of the fuel or the air-fuel mixture. The air-standard cycles (ideal cycles) of diesel and gasoline engines assume time-dependent characteristics for heat release that are not feasible in practice. The reduction in efficiency caused by the real combustion process is described by the term "quality grade." Peak quality grade values are achieved when the centroid of the combustion curve is near top dead center (TDC); however, high peak pressures and peak temperatures are negative factors in this case because they increase the heat losses and, therefore, total efficiency. Nitrogen oxides as well as the stress in the components and noise increase as well. Measurements performed on actual engines show that the optimum position for the centroid of the combustion curve of diesel engines with direct injection is about 10–15° crank angle after TDC (ATDC), the optimum position of IDI engines is about 15–20°ATDC after upper dead point, and the optimum position of gasoline engines is approximately 10°(ATDC) after upper dead point.

Combustion cycle. The definition of the combustion cycle differentiates between two- and four-stroke engines. The combustion cycle for four-stroke engines requires two revolutions of the crankshaft. The individual cycles are intake, compression, expansion, and exhaust.

One combustion cycle per revolution occurs in the two-stroke engine.

Combustion delay →Combustion, gasoline engine ~Combustion chamber shape; →Combustion process, diesel engine; →Fuel, diesel engine ~Properties ~~Cetane number

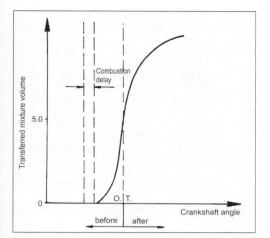

Fig. C137 Combustion curve and combustion duration for the conventional gasoline engine

Combustion duration. Combustion duration characterizes the time between the moment of ignition and the complete energy conversion of the mixture in the gasoline engine, as far as possible under conditions such as angle of ignition, valve timing, and air-fuel ratio (**Fig. C137**). However, it is primarily a function of the air-fuel ratio, which has a dominant influence on flame speed. Volumetric efficiency, residual gas content (exhaust gas recirculation), and compression ratio, among other things, also play a decisive role for a given fuel, but so do design parameters such as combustion chamber shape, spark plug position, and dual ignition.

The combustion rate of the fuel slows down as a result of an increase in the air-fuel ratio into the "rich" mixture regime. The combustion point then shifts farther into the expansion tract. The result is a lower flame speed due to temperature and pressure reductions, which in turn potentially result in a situation of incomplete combustion when the exhaust valve is opened. This results in increased fuel consumption and higher HC emissions.

The higher fuel consumption is also based on the fact that the best thermal efficiency is achieved during constant-volume combustion. Longer combustion durations move the process away from this optimum situation.

The combustion duration reaches values of approximately 60° crank angle at an equivalence ratio $\lambda = 1$ and average engine speed; with lean mixtures ($\lambda = 1.3$) durations of up to crank angles of 100° can be reached. Even leaner mixtures require special measures for stabilization of the combustion duration, such as charge layering, swirl, and tumble.

Fig. C137 shows the principal pressure characteristic, p, and the related converted mixture volume (i.e., the combustion duration) depending on the crank angle. Also shown are combustion duration and the combustion delay.

Combustion duration can be defined for the diesel engine as starting at ignition after injection time and lasting until the end of combustion. This is marked by phase 1 and phase 2 in **Fig. C138**. The injection characteristics m_{Fu} and the combustion curve $m_{FU\text{-}cons}$ are shown as time-related fuel conversion.

Combustion end pressure →Peak pressure

Combustion end temperature →Combustion temperature

Combustion engine. A combustion engine is characterized by the conversion of chemical energy of fuel into potential and/or kinetic energy via an oxidation process (combustion). The potential energy (pressure energy) of the medium can be further transformed—for example, into mechanical work (internal combustion engine). The kinetic energy of the medium can be translated into propulsion in jet engines or rockets.

Combustion (flame) speed →Combustion process, diesel engine, ~Flame propogation; →Combustion process, gasoline engine ~Flame propagation

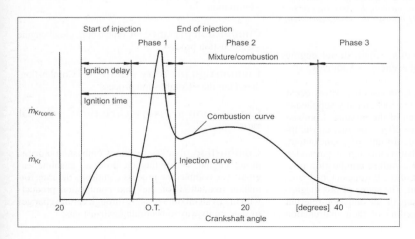

Fig. C138
Combustion curve for the diesel engine

Combustion function

Combustion function (heat release diagram).
The combustion function (heat release diagram) provides information about the burned fuel portion during the combustion process in the engine.

The combustion function $x(t)$ is principally identified by the following equation

$$x(t) = \frac{m(t)}{m_0}$$

where $m(t)$ is the already burned part and m_0 is the mass of the cylinder fresh charge. The combustion function has the value $x = 0$ at the moment of ignition and theoretically reaches the value $x = 1$ at the end of combustion. No theoretically justified approach exists for the combustion function. It is possible to calculate it from the measured time-dependent cylinder pressure characteristics (*p-V* diagram). Numerous models have been developed in the past, partly with the inclusion of chemical reactions which calculate the combustion function on the basis of the measured pressure characteristics in the cylinder.

Vibe attempted to develop a valid approach from combustion curve determined by tests. The resulting equation that bears his name is:

$$x = 1 - \exp\left(c \cdot \frac{t}{t_z}\right)^{m+1}$$

where

$x =$ the burned fuel part starting at the point in time t,
$t_z =$ the total mixture combustion duration,
$t =$ time,
$c =$ factor, calculated based on the burned fuel mass during the total combustion duration, and
$m =$ coefficient that defines the character of combustion.

The values depend on the engine operating parameters, especially on the air-fuel ratio and the engine speed, but the equation provides no means of evaluating these dependencies.

Literature: I.T. Vibe: Brennverlauf und Kreisprozess von Verbrennungsmotoren, Berlin, BEB Publishers Engineering, 1970. – W. Kleinschmidt: Untersuchung des Arbeitsprozesses und der NO-, NO₂- und CO-Bildung in Gasolinemotoren, Diss. TH Aachen, 1974.

Combustion improvers →Fuel, diesel engine ~Additives; →Fuel, gasoline engine ~Additives

Combustion misfire. Combustion misfires occur if neither ignition energy nor fuel quantity supplied is sufficient to ensure ignition of the mixture. Combustion misfires have negative effects on fuel consumption, torque, degree of nonuniformity of engine output, and the behavior of the exhaust emissions in particular. Therefore, detecting and avoiding combustion misfires is important and part of the European On-Board-Diagnosis (EOBD) system. The scope of this diagnosis requires the monitoring of all emission-relevant components and the recording of their malfunction

based on certain diagnostic thresholds. A central demand in connection with the EOBD is the detection of combustion misfires over the entire speed and load range; identification of the malfunctioning cylinder is necessary. Combustion misfires exert significant influence on the level of emissions of unburned hydrocarbons, and these would occur particularly at faults in the ignition characteristics. Increased HC and CO emissions, however, occur also at intermissions in injection processes (e.g., overrun) where there will be a temperature decrease of the combustion chamber walls and the exhaust gas. The oxides of nitrogen emissions can increase because of poor efficiency of the catalytic converter with excess air. In addition, the "wrong" lambda probe signal can affect mixture enrichment of the other cylinders.

There are several methods for detecting combustion misfire:

- Measurement of irregular running: The speed variations on the crankshaft or change of the angular acceleration are used as a measure of irregular running and hence as a measure of combustion misfire detection.
- Ionic current measurement: Ionization (chemically/thermally) of individual combustion components occurs during combustion. The electrons released by this generate a current, the so-called ionic current. The current flowing for only a few milliseconds can be taken as proof of combustion.
- Cylinder pressure indication: Using an appropriate sensor, the pressure, the pressure rise, or the pressure differential between the combustion p-θ diagram and compression p-θ diagram in the combustion chamber is used for misfire detection.
- Vibrations in the sealing gap between the cylinder head and cylinder block can be measured using sensors (e.g., piezo-electric) in the cylinder head gasket.

Combustion noise →Engine acoustics

Combustion particles →Particles (Particulates) ~Formation

Combustion pressure →Combustion, diesel engine ~Combustion process

Combustion pressure gradient →Combustion, diesel engine ~Combustion process

Combustion pressure variations →Cyclical variations

Combustion process (*also*, →Combustion process, diesel engine; →Combustion process, gasoline engine). The combustion process comprises the transformation (oxidation) of fuel into combustion products with the fresh mixture passing through certain changes in state on its way to becoming exhaust gas.

~Closed combustion process. The working medium has not changed, so it is running through a closed process. Combustion takes place outside the working chamber. Examples are the steam and Stirling engines.

~Constant pressure cycle →Cycle

~Constant volume cycle →Cycle

~Continuous combustion process. Combustion engines with external combustion normally feature a continuous combustion process. The working medium is continuously supplied with energy. Typical examples are Stirling and steam engines.

~Delayed combustion process. The real combustion process or combustion curve deviates from the theoretical isochoric-isobaric process. The isochoric combustion that would occur at a piston speed of zero cannot be attained in a combustion engine. Efforts are being made to keep the duration of combustion as short as possible for maximum thermal efficiency. It is, however, advisable for many other reasons, such as limiting the maximum rate of increase in pressure to allow a correspondingly long duration of combustion. Depending on the engine operating conditions, the combustion period can exceed the combustion time defined as optimal. This case is referred to as a delayed combustion process.

~Incomplete combustion. A combustion process, or combustion, is said to be incomplete when not all fuel components completely oxidize to CO_2, H_2O, SO_2, and so on. In case of incomplete combustion, the combustion products still contain combustible matters such as CO, which can yet be oxidized to CO_2.

~Intermittent combustion process. Intermittent combustion is typical for internal combustion piston engines in which high temperatures of the working medium occur for brief periods only. The combustion process takes place during a relatively short period of time toward the end of compression and is consistently interspersed with phases of compression, expansion, and charge transfer. This results in the positive consequence that the thermal load of the components is reduced.

~Internal combustion process. An internal combustion process or internal combustion is designated as a system in which the working medium is contained in a closed system—for instance, the combustion chamber, which is the working chamber.

~Knocking combustion process →Knocking

~Open combustion process. The working medium in the open combustion process is the air-fuel mixture or air and fuel. It is continuously renewed with the work done. An example of an open combustion process is the reciprocating piston engine.

~Process →Combustion, diesel engine ~Combustion process; →Combustion, gasoline engine ~Pressure curve

~Rate of increase in pressure →Combustion, diesel engine ~Combustion process

Combustion process, diesel engine. Diesel engine operation is based on burning the fuel injected into the combustion space by compression ignition in the aspirated air. Fuel is normally introduced into the combustion space using an injection system and multi-hole nozzle. There must be a sufficiently high air temperature for dependable compression ignition of the fuel. This is reached by using a correspondingly high compression ratio. The ensuing combustion necessitates an optimum mixing of the fuel and air in the combustion space.

The quantity of fuel injected, the charge movement in the combustion space, the combustion-chamber geometry, the temperatures of the cylinder charge, and the walls confining the combustion chamber determine the mixture formation. The mixture formation in the diesel engine is termed internal mixture formation, as opposed to that in the gasoline engine with intake manifold injection. The degree of homogeneity during fuel injection into the combustion space and the locally and time-based changes in the air-fuel concentration field (liquid and vapor) are a measure of the mixture formation attained. The local and temporal sequence of combustion, as well as the completeness and quality (pollutant formation), depend on the mixture formation. The pollutants emitted in the engine exhaust gas occur because of interplay between pollutant formation and reduction in the combustion space and exhaust system.

~Coal dust engine. Tests with coal dust as the fuel were carried out in the early days of diesel engine development. Even Rudolf Diesel looked into the subject with some random tests because it was not absolutely clear at that time what would be the most suitable engine fuel. Coal dust was an obvious choice only because coal was the main energy source at that time.

~DCCS (dilution controlled combustion systems). In the case of the HCCI (homogeneous charge compression ignition) technique (*also, see below,* ~Homogeneous combustion, diesel engine), it tried to effect combustion at low temperature and high air-fuel ratios. The consequence of this is that both the formation of NO_x and soot are largely suppressed. The DCCS system is to be seen as an alternative to that or as a variant of HCCI, whereby low combustion temperatures and low air-fuel ratios suppress soot and NO_x formation, too. Lowering of the combustion temperature to the necessary extent is achieved by charge dilution with extremely high EGR rates of up to 70%.

~Direct injection. A distinction is made between processes with direct and indirect fuel injection. Direct injection has an advantage in fuel consumption of 15–20% over indirect injection, and this is why it has gained worldwide acceptance in the field of automobile engines. The initial development problems with automobile engines such as untreated pollutant emissions and excessively loud combustion noise were mastered through decades of engineering work.

~~**MAN-FM technique** (*also*, →Combustion process, gasoline engine ~Stratified charge process) The MAN-FM technique is a further development of the MAN-M process. The letter F represents the German word for external ignition (*Fremdzündung*). The internal mixture formation, combustion chamber design, and load control were adopted from the M-method, but ignition is made using a spark plug as in a gasoline engine. The pressure development is nearly identical to that of the constant pressure cycle. The behavior of the emission of pollutants is a little bit more favorable that that in the M-process. The FM-process is attributed to the hybrid combustion process or stratified charge process because of the combination of the classic features of the diesel and gasoline processes.

~~**MAN-M process.** *See below*, ~~Single-spray process, wall-applied

~~**Multiple-spray method, air-distributing.** This method, as shown in **Fig. C139**, has prevailed worldwide in all diesel engines because it represents a better approach than the single-spray process. The recess arranged in the piston crown forms the actual combustion space. The piston recess comprises up to 80% of the clearance volume. Swirl is introduced in the induced air through appropriately designed swirl channels (swirl channel and/or tangential port) to produce

better mixture preparation in automobile engines. Modern four-valve engines work with controlled swirl; this allows reduction of the swirl at high speed and load for increasing the specific power, for instance. Multihole nozzles are used as injector nozzles—six to eight nozzle holes usually are used in passenger car engines. If the number of nozzle holes is reduced, the swirl requires enhancement, and vice versa. The injection pressure is between 1800 and 2300 bar depending on the particular injection system. Automobile engines require compression ratios between 16:1 and 19:1. Pencil-type glow plugs are required that need to be operated from 0°C and below to ensure a safe cold start. Today, these engines reach maximum speeds of 4500 rpm and have brake thermal efficiencies of approximately 44% at the best point on the characteristic map.

~~**Single-spray process, wall-applied.** This method (**Fig. C140**) did not gain acceptance, because the multiple-spray method has advantages over it: The specific power, maximum mean pressure, specific fuel consumption, and especially the hydrocarbons are significantly worse with the single-spray approach. It also must be remembered that it has less favorable starting response and an increase in offensive smell due to higher aldehyde concentrations in comparison with the multiple-spray method. Advantages of the wall-applied single-spray process lie particularly in smoother engine running. The fuel spray from the eccentrically positioned nozzle is directed against the wall of a spherical combustion-chamber recess in the piston. It is sprayed in the direction of the air turbulence to reduce atomization. A fuel film is formed on the combustion space recess using this approach, and only a small part

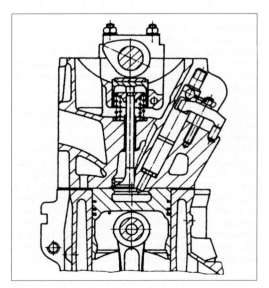

Fig. C139 Multiple-spray method

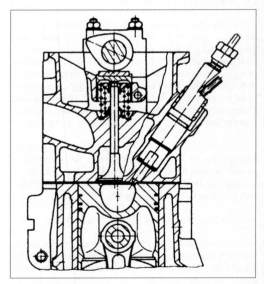

Fig. C140 Single spray process, wall-applied

of the fuel serves the purpose of ignition. This prevents a high ignition pulse from developing at the beginning of combustion. The air turbulence will then evaporate the fuel film from the combustion chamber wall after start of combustion. This technique is also called MAN-M process.

~~Wall-applied. *See above*, ~~Single-spray process, wall-applied

~Gas engine. In the gas engine, air is compressed as in a normal diesel engine. High-pressure gas fuel is blown into the compressed air. Local areas with random gas-air mixture are formed during the "injection" process. The auto-ignition temperature of the gaseous fuel used is generally higher than that of diesel fuel. A relatively large quantity of ignitable gas-air mixture can form during the extended ignition delay period caused by this, which then nearly deflagrates because of auto-ignition. This causes development of impermissibly high rates of pressure rise and maximum pressures in the combustion chamber. For this reason, a small quantity of diesel fuel (pilot fuel) is preinjected in these engines. Gas "injection" follows only after ignition of the pilot fuel quantity. Then the gas can burn at acceptable increases in cylinder pressure almost without any time delay.

~HCCI process (Homogeneous Charge Compression Ignition). *See below*, ~Homogeneous combustion, diesel engine

~Homogeneous combustion, diesel engine. Homogeneous diesel combustion offers the possibility of reducing NO_x and particulates at the same time, which is not possible using conventional combustion processes. The basis of this procedure is extensive homogenization of the mixture before initiation of combustion. The main effects are decreased peak combustion temperatures and avoidance of overenriched zones of mixture. The first leads to reduction of NO_x and the second to reduction of particulates. Combustion at temperatures of less than 2200 K and an air-fuel mixture ratio of $\lambda > 0.7$ would be optimal. **Fig. C141** shows the direction of development for combustion processes that make this possible.

Homogenization of the charge in the combustion chamber can be achieved in different ways; either advanced or retarded homogenization can be applied in the case of direct injection. Furthermore, fuel introduced via the intake manifold would also appear feasible.

The problems with homogeneous diesel combustion are:

- Controlled combustion at start of combustion and control of the combustion rate.
- Setting of the optimal λ-temperature window for combustion.
- Minimization of the HC and CO emissions. Normally, homogeneous diesel combustion leads to in-

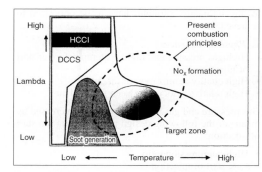

Fig. C141 Combustion process for low NO_x and particulate emissions (Source: Kamimoto)

creased HC and CO emissions (**Fig. C142**), which can be reduced with the help of an oxidation catalytic converter.
- Homogeneous mixture formation.
- Attaining sufficiently high mean effective pressure.

Literature: Brennverfahrensseitige Ansatzpunkte für Pkw-Dieselmotoren zur Erfüllung künftiger EU- und US-Abgasstandards, IAV GmbH, 5th International Stuttgart Symposium Feb. 2003.

~Hybrid combustion processes →Combustion process, hybrid engine

~Hybrid engine →Combustion process, hybrid engine

~Ignition delay. The physical and chemical processes that start at the beginning of fuel injection take some time to attain ignition conditions. This process, termed "ignition delay" is in the region of up to 2 ms depending on the conditions in the cylinder at the time of fuel injection. Engine development should be aimed at obtaining a short ignition delay because that keeps the combustion noise within reasonable limits (no abrupt initiation of combustion). The fol-

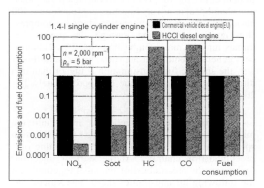

Fig. C142 State of development of homogeneous diesel combustion (Source: AVL)

C

lowing influential variables have to be optimized to achieve this:

1. *Physical influences*
- High gas pressure and high gas temperature at start of injection
- High quality atomization of the fuel
- High relative velocity between fuel and air

Gas pressure and gas temperature can be increased by means of the following design parameters at the start of injection:

- High compression ratio
- Late start of injection
- Supercharging
- High temperature of coolant
- Combustion chamber design (influence of the wall temperature)
- Employment of ignition aids (pencil-type glow plugs, intake air preheating)

2. *Chemical influences*
- High fuel ignition quality (high cetane number)
- High fuel temperature, high gas pressure, and high gas temperature at start of injection

~Mixture formation →Mixture formation ~Mixture formation, diesel engine ~~Mixture formation in single combustion chamber, ~~Mixture formation in divided combustion chamber

~Prechamber engine. As opposed to direct injection diesel engines, prechamber engines have a divided combustion space. One of the combustion chambers is the secondary combustion chamber and the other is the main combustion chamber. The secondary combustion chamber is designed as a prechamber and arranged in the cylinder head. The main combustion chamber is between piston and cylinder head, as in a conventional engine. Both combustion chambers are connected to each other via prechamber outlets, called "throats." The secondary combustion chamber is designed as a prechamber or swirl chamber (**Fig. C143**). In both cases, the fuel is injected into the precombustion chamber at relatively low pressure (maximum 400 bar). From this point, combustion starts by compression ignition continuing through the prechamber throats into the main combustion chamber. This process generates both heat and flow losses, and this is why the specific fuel consumption is higher than in a direct injection diesel engine by approximately 15–20%.

~~Air storage process. Injection is made into the main combustion chamber in this case. The injection spray is, however, directed against the opening of a precombustion chamber. At the start of injection, fuel enters the precombustion chamber together with the inflowing air and ignites there. The kinetic energy associated with the outflow process is utilized for mixture formation in the main combustion cham-

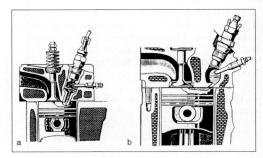

Fig. C143 Prechamber engines

ber. **Fig. C144** shows the principle of the air storage process.

~~**Ignition chamber.** The ignition chamber is understood as the prechamber and the swirl chamber.

~~**Lanova.** *See above*, ~Prechamber engine ~~Air storage process

~~**Prechamber.** *See above*, ~Prechamber engine

~~**Prechamber outlet.** *See above*, ~Prechamber engine

~~**Secondary chamber engine.** *See above*, ~Prechamber engine

~~**Swirl chamber.** *See above*, ~Prechamber engine

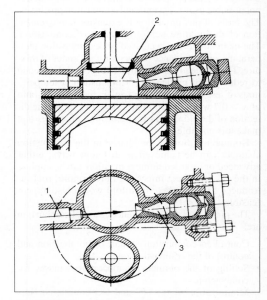

Fig. C144 Lanova air storage process

Combustion process, gasoline engine. Combustion in the gasoline engine is produced by external ignition using spark plugs. Conditioning of the air-fuel mixture can be achieved using different approaches. The power output is controlled by changing the amount of charge (quantitative control)—this is used when homogeneous mixture formation is employed. Control of the power output with stratified mixture formation is achieved by varying the air-fuel ratio (qualitative control), enabling throttle-free load control.

~Charge movement →Charge movement

~Combustion chamber design →Combustion chamber

~Cyclic variations. Fluctuations of the cylinder pressure characteristics from cycle to cycle are typical of combustion in gasoline engines. They are caused by fluctuations in the turbulent velocity field and the local charge composition—these influence the flame front propagation and hence the conversion of energy.

~External ignition →Ignition system, gasoline engine
~Ignition

~External mixture formation. The external mixture formation was produced in the past by means of a carburetor, and it is produced today using intake manifold fuel injection. The air-fuel ratio thus plays an important role in relation to fuel consumption, emission of pollutants, and combustion temperatures. The flame speed and hence the rate of combustion are the highest at an air-fuel ratio of $\lambda = 0.80$–0.85. However, an air-fuel ratio of $\lambda = 1$ is run for the largest part of the performance map because this allows more effective secondary exhaust treatment using a three-way catalytic converter.

~Flame propagation (also, →Flame expansion). Flame propagation should be fast for avoidance of knock and to improve the efficiency. This requires a central position for the spark plug, and four- and five-valve engines allow this most favorable central position. Two spark plugs should be employed in the case of three-valve engines. A further consequence of fast flame expansion is that there is a reduced tendency for flame quenching at the cylinder wall. This reduces the unburned hydrocarbons in the exhaust gas significantly. Furthermore, the fast conversion of energy reduces the cyclic fluctuations of gasoline-engine combustion.

~Flame speed →Flame expansion

~Homogeneous combustion, gasoline engine. In terms of fuel consumption, unthrottled lean-burn operation or stratified-charge operation is optimal for the gasoline engine. The charge dilution that can be reached is limited in conventional combustion processes in which ignition of the mixture is reached through an external ignition source (spark plug). The reasons for this are local ignition limits and incomplete flame expansion. Even

though further de-throttling reduces fuel consumption through charge stratification, the relatively narrow ignition limits in the central region of the combustion chamber cause higher NO_x emissions. These call for subsequent treatment of the NO_x in the lean mixture operating regions with the negative consequences of higher cost and use of low-sulfur or sulfur-free fuel.

Combustion producing much less NO_x emissions than can be achieved in the conventional spark ignition with its high combustion temperatures is required to ensure that these secondary exhaust treatment systems need not be employed. Certainly a further reduction of NO_x emissions is possible by applying further charge dilution, but the process is limited by mixture ignitability.

Homogeneous auto-ignition could remedy this situation by initiating combustion in the entire combustion space almost simultaneously. This does not happen as in the flame front common with spark ignition but takes place extensively over the whole volume. The extremely fast development of combustion with short combustion duration delivers peak temperatures, which remain far below those that produce high NO_x formation.

A safely functioning auto-ignition system under all operating conditions that can be influenced by, for example, storage temperature, residual gas content, cylinder pressure, degree of charge homogenization, and air-fuel ratio, is still a problematic proposition. Likewise, additional spark ignition—which, however, does not cause any flame front and, hence, no NO_x-producing temperatures—is possible because of the corresponding leaning and mixture homogenization processes.

Literature: A. Führhaper, W.F. Piock, E.M. Unger, G.K. Fraidl: CSI- Ein kostenoroentiertes Ottomotor-Gesamtsystem mit homogener Selbstzündung. — K.G. Fraidl, W.F. Piock, A. Führhapter, M. Unger, T. Kammerdiener: Homogene Selbstzündung—die Zukunft der Benzineinspritzung, MTZ 10/2002 63. — K.G. Fraidl, W.F. Piock: Otto Direkteinspritzung ohne DENOx-Kat, 23rd Intern, Vienna Engine Symposium, VDI Progress Reports Series 12, No. 490, April 2002.

~Hybrid engine →Combustion process, hybrid engine

~IFP Baudry process. This process with external mixture formation (**Fig. C145**) belongs to the group of stratified charge concepts with an undivided combustion chamber. The engine is supplied with lean mixture via an intake port. Rich mixture reaches the spark plug zone through a hollow stem valve via an auxiliary port. Precise charge stratification is, however, not possible with this method.

~Indirect injection. *See below*, ~Internal mixture formation

~Internal mixture formation. In comparison with the external mixture formation that occurs with intake manifold injection, the gasoline engine with internal mixture formation is believed to have a potential saving in fuel consumption of 15–20% when stratified combustion processes are applied. The causes are

C

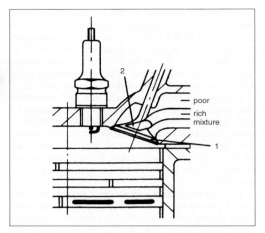

Fig. C145 IFP Baudry process

drastic decrease of the throttling losses, reduction of the wall-heat losses with the stratified combustion process, and higher effective and geometrical compression ratio enabled by internal mixture cooling and reduced tendency to knock.

It is possible to distinguish between three different combustion processes with charge stratification (**Fig. C146**).

- *Air-guided combustion process*. The fuel is transported from the injector nozzle to the spark plug by means of generated charge movement. Wetting of combustion chamber walls is excluded, provided this process is implemented. Exact timing of injection and stable charge movement are decisive for the quality of this process. The mixture formation supported by charge movement shows high mixture quality with dimensioning accordingly.
- *Wall-applied method*. The fuel is fed through a correspondingly contoured combustion chamber wall—the piston in this case—to the ignition spot. This technique brings about a high portion of fuel deposits on the combustion chamber walls because the evaporation effects until the moment of ignition usually cannot remove the whole fuel film. A stable process is achieved, however, because the procedure is based on uniform basic conditions.
- *Spray-guided method*. Introducing the fuel into the very vicinity of the ignition spot has the highest potential for fresh-charge stratification and thus the highest fuel consumption potential in comparison with all other techniques. The advantage in fuel consumption is negated in part by the unsatisfactory mixture quality on the spark plug at the moment of ignition. Ignitable mixture forms in part of the fuel spray only. Direct support of mixture preparation through charge movement cannot be applied in this case because there is the risk of the mixture spray drifting away from the ignition point. Permanent operation is not yet ensured, because the spark plug

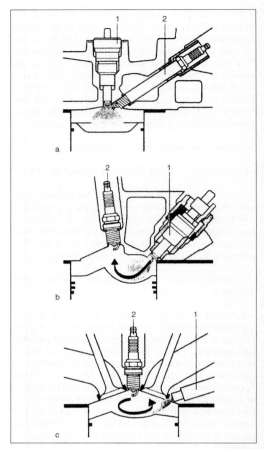

Fig. C146 Internal mixture formation on gasoline engines

is subjected to distinct alternating thermal load through occasional wetting by fuel. But in the future, this technology will be state-of-the-art due to the better fuel consumption obtainable.

Engines with internal mixture formation will become more prevalent in the future. The main engineering effort is going into secondary treatment of the pollutants. The engines can also be operated using a three-way catalytic converter if they are controlled at $\lambda = 1$. This, however, greatly reduces the advantage in fuel consumption.

~Knocking combustion →Knocking

~MAN FM process. Fuel from a single hole nozzle is sprayed against the wall of the combustion chamber in the piston, spreading there as a thin wall film. The spark plug is close to the combustion chamber wall, diametrically opposite the injection nozzle. Further mixture formation is obtained by evaporating the fuel

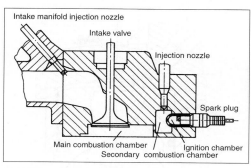

Fig. C148 Porsche SKS process

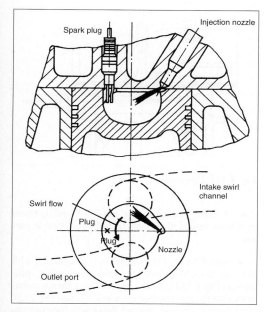

Fig. C147 MAN-FM process

from the wall. As opposed to the normal diesel engine, injection is made 10–15°CA earlier by means of the MAN-M process. The air density is even lower in this case so that increased air distribution of fuels or fuel components unwilling to ignite is prevented. Ignition is effected toward the end or even after termination of the injection period. The advantage of this extreme charge stratification is the ability to use high compression ratios (up to 16) together with pure qualitative control almost across the entire operating range of the engine. The fuel consumption figures achieved were close to those of a direct injection diesel engine. A specially designed spark plug with about 8 mm long parallel electrodes is needed to achieve this performance. **Fig. C147** shows the principle of the FM process.

~Mixture formation →Mixture formation ~Mixture formation, gasoline engine

~Porsche SKS. The Porsche SKS process (**Fig. C148**), is similar to the VW PCI process in its operating principle. Control requires additional air throttling, too. A high level of turbulence is generated in the charge through the very narrow connecting pipe to the combustion space for the purpose of rich mixture homogenization. The initiation of combustion takes place in an ignition chamber that is connected to a secondary chamber via throttling bores.

~Stratified charge process (*also, see above*, ~Internal mixture formation). In engines with stratified charging, the mixture in the vicinity of the spark plug is en-

riched to the extent that safe ignition is possible. However, combustion takes place, on average, with a strongly leaned mixture. The main objectives are better fuel consumption at part load and low NO_x and CO emissions, as in diesel engines.

~~**Deutz atmospheric pressure plasma process.** The Deutz atmospheric pressure process works using peripheral charge stratification and an undivided combustion chamber. The fuel is fed from two nozzle bores across the rotating airflow. The arrangement of the piston recess allows a larger free spray length so that the fuel will not come into contact with the bottom. This generates a mixture ring at the periphery of the combustion space. A part of the mixture flows into the area of the spark plug discharge gap and is ignited there. Ignition is initiated shortly after the start of injection. In addition, the course of combustion is determined by the mixing process and further progress of injection. This technique allows alcohol-operated commercial vehicle engines, in particular, to be operated nearly soot-free and with diesel-like efficiency. **Fig. C149** shows the principle of the procedure.

~~**Ford PROCO process.** →Combustion process, hybrid engine; and *see below*, ~~FPC-procedure

~~**FPC-procedure.** In this approach, fuel with a spray cone of approximately 100° is centrally sprayed into the air circulating in the piston combustion chamber, (**Fig. C150**). The fuel is transported from there into the outer zones of the combustion space by air rotation. A decreasing air-fuel ratio is established from the center of the combustion space toward the edges. The times of injection and ignition need to be coordinated very well because, on the one hand, a too-rich air-fuel mixture near the spark plug can cause problems with the ignition, whereas, on the other hand, too lean a mixture in the boundary zones can cause the flame to extinguish. To prevent the mixture from becoming too lean in the peripheral zone at part load, the intake air has to be throttled on falling below certain load conditions. This, however, leads to disadvantages in terms of the fuel consumption.

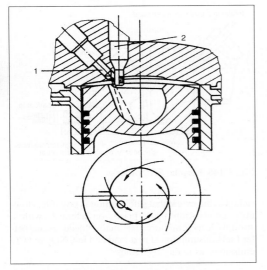

Fig. C149 Deutz atmospheric pressure process

~~**Hesselmann engine.** The Hesselmann engine, which dates back to 1934, belongs to the hybrid combustion processes as an air-compressing procedure. Its mode of operation is presented in **Fig. C151**. The air rotation around the axis of the combustion space is generated by an umbrella valve that makes sure that the injection spray directed into the piston recess will always deliver an ignitable mixture to the spark plug.

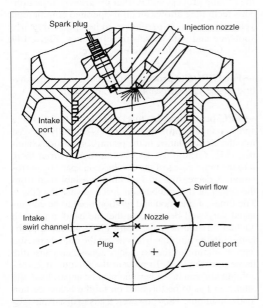

Fig. C150 Ford PROCO process

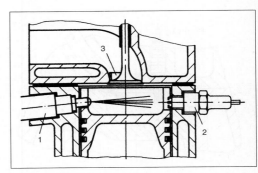

Fig. C151 Hesselmann engine

The compression ratio was only 7.5 and ignition problems forced the intake air to be throttled at part load.

~~**Honda CVCC.** This procedure is categorized with the mixture-compressing hybrid combustion processes (**Fig. C152**). A rich mixture flows from a carburetor into the prechamber via a hole (2) and valve (3). Simultaneously, lean mixture flows into the combustion space through intake valve (4). Spark plug (5) ignites the rich mixture in the prechamber toward the end of compression. The gases then blowing out of the chamber ignite the lean mixture in the main combustion chamber. This technique is characterized by low emissions of oxides of nitrogen.

~**Texaco CCS process.** In this process, a wide fuel jet is sprayed past the spark plug located on the periphery of the combustion space (**Fig. C153**). The ignition timing is set such that the first mixture cloud ignites. The rate of fuel injection determines the further course of combustion in the engine operated unthrottled in the whole of its part-load range. A zone develops near the spark plug, which is supplied with fresh fuel originating from the injection process and with fresh oxygen from the rotating air. The injection and ignition timing, the peripheral air speed, and the injection velocity have to be carefully coordinated to obtain a stationary flame front.

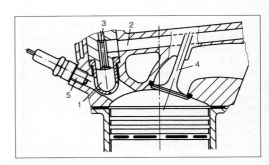

Fig. C152 Honda CVCC process

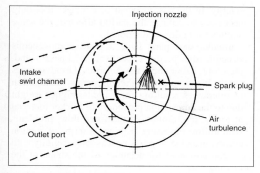

Fig. C153 Texaco TCCS process

~**Uncontrolled combustion process** →Glow ignition; →Knocking

~**VW PCI process.** This process is shown in **Fig. C154** and operates with both internal and external mixture formation. A very lean, homogeneous mixture in low part-load operation is introduced into the main combustion chamber (carburetor, injection system). Additional fuel is introduced into the swirl chamber for initiation of combustion by a spark. The initial phase of combustion is supported by the rotational motion caused by flow into the chamber during piston movement. The relatively lean mixture in the main combustion chamber is ignited by the hot high-speed gas jet leaving the chamber. Load control is achieved by throttling the induced air and changing the quantity of fuel fed to the main combustion chamber.

Combustion process, hybrid engine. Hybrid engines include engines or processes that combine characteristic features of both gasoline and diesel engines. They are proposed in an attempt to combine the advantages of the diesel engine—e.g., low specific fuel consumption—with the advantages of the gasoline engine—e.g., high specific power output and silent engine running. Essential characteristics of hybrid engines can be: mixture compression or air compression, homogeneous or nonhomogeneous mixture condition, external ignition or auto-ignition, quantitative or qualitative control, high or low compression ratio,

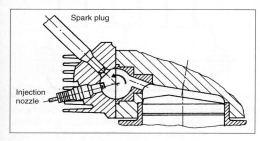

Fig. C154 VW PCI process

gasoline or diesel fuel whereby unusual combinations make up the hybrid engine. Furthermore, a large number of the processes are embraced by the term of stratified charge engine (e.g., Deutz-AD, Ford PROCO, FPC, MAN-FM, Hesselmann, PCI). Multifuel engines belong to the class of hybrid engines too. Strictly speaking, the gasoline engine with direct injection could also be counted among the hybrid combustion processes. This engine will be treated separately because of the importance of this technology.

~**Stratified charge process** →Combustion process, gasoline engine

Combustion process, multifuel engine (*also*, →Combustion process, hybrid engine). Multifuel engines are designed such that they are unexacting in terms of the ignition quality or antiknock properties of the fuel used. They work exclusively with internal mixture formation and retarded start of injection to exclude knocking. Multifuel engines have very high compression ratios because the ignition quality of fuel does not allow, or hardly allows, auto-ignition. Another option is ignition using a spark plug or glow plug. In this case, compression ratios, ε, of about 14:1 or 15:1 are used, so this is a value between gasoline and diesel engines. A further way of initiating ignition is the pilot injection process whereby ignition is ensured by a diesel-fuel spray of 5–10% of the full-load delivery of the diesel engine; this is injected directly into the combustion space. Multifuel engines have practically no importance at present.

Combustion process characteristics →Combustion process

Combustion process management →Combustion process; →Combustion process, diesel engine; →Combustion process, gasoline engine

Combustion products (*also*, →Emissions; →Exhaust gas analysis). The combustion products leaving the exhaust system of the engine propelled with conventional fuel are essentially unburned hydrocarbons, carbon monoxide, carbon dioxide, oxides of nitrogen, water, and particulates (solid/liquid). Furthermore, there are a multitude of low concentration components such as hydrogen, O-H-C compounds (preoxidized hydrocarbons), N-H compounds, and sulfur compounds.

Combustion residues. These develop during combustion of the fuel and lubricating oil in the combustion chamber and are deposited to some extent on the combustion chamber surfaces, which include the cylinder head, piston ring groove, spark plug, and valves. They consist of organic and inorganic compounds, since fuels and lubricating oils can form residues themselves, and also contain additives for a variety of purposes.

Combustion residues should be maintained as small

as possible because they cause a lot of negative effects. For instance, they can promote wear, obstruct the movement of the piston rings, and contaminate the spark plugs, and thus impede their function and affect gas transfer when on the valves. Also, in diesel engines the injector nozzles can coke, which would not occur with clean combustion. Above all, the compression ratio is increased (residue balance is usually reached after driving 5000–7000 km), which promotes knocking in gasoline engines.

Combustion temperature (*also*, →Combustion chamber temperature). The combustion temperature is a function of many parameters. Some of them are engine load, boost pressure, engine speed, the combustion process, and the fuel used. **Fig. C155** shows an example of the mass mean temperature curve for a gasoline engine and a diesel engine at full load. The diesel engine is supercharged in this case. Nevertheless, the diesel engine peak temperature is about 500°C lower than that of the gasoline engine.

Commercial vehicle crankcase (*also*, →Crankcase). The crankcase is the central, supporting section of the engine which contains and connects the functional groups. It represents the limit of the engine system because, together with the connected components, it forms the external boundary of the engine and the outlet for the working fluid—the coolant and lubricant—while preventing entry of dust, moisture, and contamination. In special cases, the crankcase also serves as a supporting structure for the machine to be driven.

~Crankcase construction. Large components of complicated structures such as the crankcase can be designed according to different principles:

- Differential design in which the individual part is divided up to the greatest possible extent into functional workpieces allowing simple production and assembly: A number of smaller parts are easier to process and handle during assembly, maintenance, and repair; however, they require more time for as-

sembly. In the case of wear or damage, it is only necessary to replace the part(s) affected; the same applies for faulty production.

Smaller cast parts have a more homogeneous structure due to the lower wall thicknesses and, therefore, better strength properties. Disadvantages are the higher machining requirements, the addition of tolerances resulting from joints, higher weight due to the connection procedure or thicker material at joints, interruption of the force flow, problems with uniform pressure at joints in terms of friction corrosion and leakage, as well as lower rigidity with its consequences for accuracy of the shape of the bearing bores.

- Integral design, in which a number of individual parts are consolidated into one component, results in a rigid construction with optimum force flow, particularly for castings with low overall weight. Casting difficulties increase with the size of the part. Greater wall thicknesses have a less homogeneous structure, and the values of strength are lower. Assembly and repair are more difficult due to the size of the component.

Because these are contradictory principles, the advantages for the one version are disadvantages for another. Depending on the type and size of the engine and the state of the art, crankcase designs have developed in the direction of integral design (monoblock housing), on the one hand, and, on the other, in the direction of differential design, with structures divided up to the greatest extent in individual sections (base plate, frame, intermediate frame, and individual cylinder). There are all types of intermediate stages between these two versions (**Fig. C156**).

~Crankcase design. In addition to function and concept, the effect of forces and moments are particularly decisive for the design of the crankcase (**Fig. C157**). The gas pressure in the cylinder acts on:

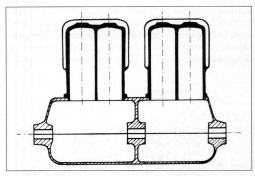

Fig. C156 Two-piece crankcase with crankshaft bearings on alternate throws and attached double cylinders, approximately 1910 (Source: Zima)

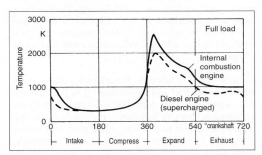

Fig. C155 Mass mean temperature curve for gasoline engines and diesel engines (Source: Pischinger)

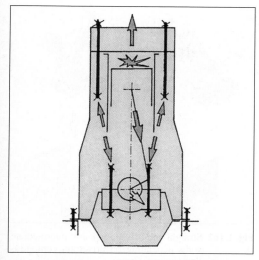

Fig. C157 Schematic: Force flow in crankcase intermediate wall (Source: Zima)

- The cylinder head, which for its part, acts on the (crankcase) intermediate wall through the cylinder head bolts
- The drivetrain on the (crankshaft) bearing caps, which also act on the intermediate wall through the bearing cap bolts

This completes the force flow, and the housing wall is subject to dynamic tension. This can be relieved by design measures to precompress the housing wall with tension anchor bolts so that it is first necessary to dissipate the operating loads before the tension becomes effective. These tension anchors can simultaneously be used for fastening the crankshaft bearing cap (**Fig. C158**).

The cylinder head bolts—four, six, or eight per head, depending on the engine size—are located around the cylinders, and the forces of the bolts are transferred directly to the area of the intermediate wall (**Fig. C159**).

The forces from the bolts in the area of the crank circle plane are transferred by means of special design features such as tie beams, ribs, and belts in the intermediate wall (**Fig. C160**).

The flow of force from the cylinder head to the crankcase can be influenced by the manner in which the cylinder head is mounted on the crankcase (**Fig. C161**).

The favorable "side-by-side rod" arrangement of the drivetrain in V-engines requires offset crankcase intermediate walls due to the offset of the connecting rod (**Fig. C162**).

The transfer of forces to the crankcase intermediate wall combined with the complicated cross-sectional forms result in additional stresses. Deformations resulting from assembly and operating forces in the area of the upper cylinder liner affect the piston bearing surface. Although the piston clearance is matched to the liner and its deformations, the rigidity of support for the inner and outer cylinders sometimes causes dif-

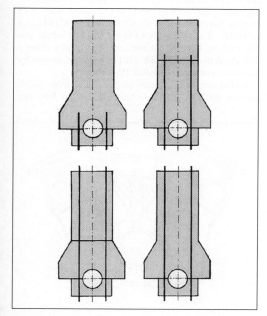

Fig. C158 Schematic: Development steps for tension rod tensioning of the crankcase (Source: Zima)

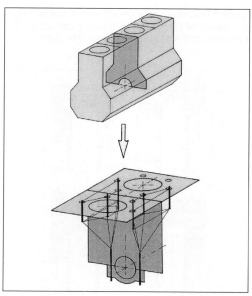

Fig. C159 Schematic: Transfer of forces from cylinder head bolts to crankcase intermediate walls

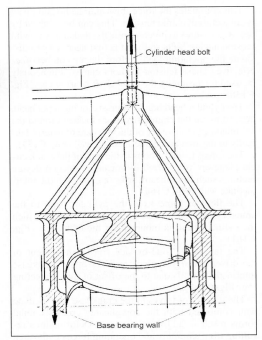

Fig. C160 Design of crankcase longitudinal wall for transfer of forces from cylinder head bolts to crankcase intermediate wall (Source: K. Groth; O. Syassen, *MTZ* 5/68)

ficulties. Deformations of the crankshaft bearing bores as a result of forces in the housing can affect the magnitude of the bearing clearances. The deformation

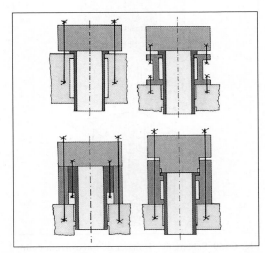

Fig. C161 Versions for fastening cylinder head (Source: Zima)

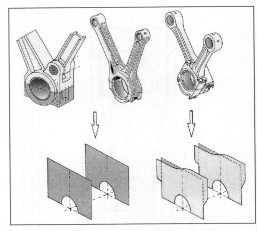

Fig. C162 Straight and offset crankcase intermediate walls depending on drivetrain configuration (Source: Zima)

characteristics of crankcases can be observed even during the development phase and, where applicable, influenced by designing with FEM calculations supported by measurements using elongation contour processes, optical strain measuring equipment, and strain gauges (**Fig. C163**).

In order to assemble the crankshaft, it is necessary to leave the crankcase open at the bottom end, which reduces the structural rigidity. A so-called tunnel housing can be obtained by using intermediate walls with bearing bores enclosed on all sides (**Fig. C164**). For assembly, it is necessary to insert the crankshaft axially into the bearing channel, which requires either a disk crankshaft (MTU BR 538) or a special design for the bearing inserts (Pielstick PA4-185).

Radial engines, aviation engines, and large diesel engines (historic engines) have drum-type housings

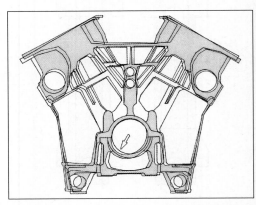

Fig. C163 Schematic representation of crankcase deformation under the effect of gas and inertia forces (Source: Zima)

C

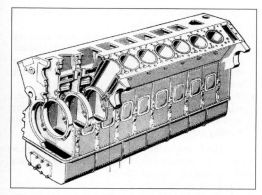

Fig. C164 Tunnel housing, version MTU BR 538 (Source: MTU publication)

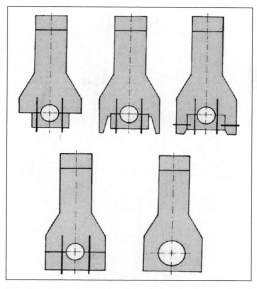

Fig. C166 Schematic: Development steps toward stiff crankcase (Source: Zima)

(**Fig. C165**) with two-piece design (allows single-piece main rods) or single-piece design (requires two-piece main rods).

The high lateral rigidity of tunnel housings can be achieved today with deep skirt-type crankcases in combination with lateral bracing of the crankshaft bearing caps and on small engines by integration of the bearing cap into a ladder frame construction (**Fig. C166**).

On marine engines, the base plate is designed to extend up beyond the center of the crankshaft. This had already proven to be necessary at an early stage for submarine engines in order to give their crankcase the necessary rigidity due to the "soft" hull of the ship (**Fig. C167**). Modern two-stroke, crosshead engines also have high base plates because when the bottom of the ship forces the engine to deform, the concave deflection of the crankshaft bearing exerts significantly more force on the crankshaft.

With V-engines, the main bearings each pick up the forces from two cylinders, and the lateral components are accordingly higher due to the inclination of the two cylinder banks. In order to securely take up

these lateral forces, crankcases were previously designed with a two-piece layout (**Fig. C168**) and the crankshaft was borne in the upper and lower sections of the crankcase.

However, difficulties occurred in bolting the crankcase halves to one another with sufficient rigidity to prevent deformation of the bearing bores on the one

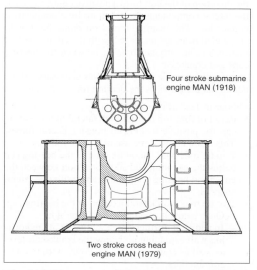

Four stroke submarine engine MAN (1918)

Two stroke cross head engine MAN (1979)

Fig. C167 Base plates with high sides: Submarine engine (1918) and two-stroke crosshead engine (1979) (Source: Körner, 2nd edition)

Fig. C165 Drum-type housing for radial engine, version BMW 801 (Source: BMW 801 Manual)

C

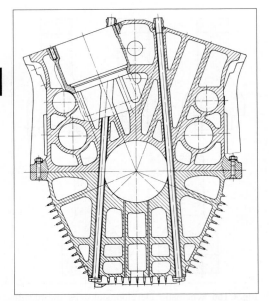

Fig. C168 Two-piece crankcase with crankshaft bearings between top and bottom section of crankcase. The two crankcase halves are braced by continuous tie rods (Source: Just/Zima: Aluminum 47 [1971] 8)

hand and to eliminate micromovements, which formed friction rust, on the other.

For this reason, crankshafts with suspended bearings were preferred, because the components acting in the direction of the bolt form a positive connection and the lateral components are taken up by a nonpositive connection. The bearing forces on crankshafts with suspended bearings increased, with the engine performance forcing stronger bracing between the bearing caps and crankcase.

Assuming two tie rod bolts, it was necessary to take additional measures as the bearing forces increased:

- Four instead of two tie rods.
- Relief of force on tie rods with lateral fit for bearing cap for positive locking to transfer lateral forces. This again allowed the use of two tie rods.
- Two tie rods with lateral fixation of the bearing cap below the cap joint and additional bolt between the cover and housing wall in this area with two lateral anchors.
- Two tie rods with lateral fixation of the bearing cap.
  —Once below the cap joint with lateral fit
  —In addition, further down, with a fit in the housing wall extending up below the crankshaft level as well as connection of the cover with two lateral anchors
- Four tie rods with lateral fixation of the bearing cap once below the cap joint and, in addition, further down with the housing wall extending down below

the crankshaft level as well as fastening the cover with two lateral anchors.

- Four tie rods with lateral fixation of the bearing cap once below the cap joint and, in addition, further down with the housing wall extending down below the crankshaft level as well as fastening the housing wall with four lateral anchors.
- Two tie rods and bolting of the cover to the housing wall, as well as an additional brace for the housing wall consisting of a long lateral anchor below the bearing cap.

The rigid connection achieved by lateral reinforcement of the bearing cap (**Fig. C169**) to the housing wall is called the "tunnel effect."

The bolts are hydraulically stretched with a clamping device so that the nuts can be loosened and tightened by hand to facilitate assembly and disassembly work on crankcases on large engines (**Fig. C170**).

It is necessary for the crankcase to hold functional parts and groups such as the gear drive, pumps, heat exchanger, filters, regulators, and so on. These are usually located in or on the engine in motor vehicles. In order to expand the range of application for large engines by means of design that is application-specific, the gear drive is installed in a removable gear box fastened to the final crankshaft bearing (where applicable with additional support) (**Fig. C171**).

In the case of high-speed, high-performance engines, entire function areas are consolidated into removable modules (**Fig. C172**). For example:

- In a turbocharging unit, exhaust turbocharger, charge air intercooler, induction air system, exhaust outlet air and coolant lines, and
- In a service block with auxiliary equipment, oil, coolant and fuel heat exchangers, oil and fuel filters, oil centrifuge and electronic boxes.

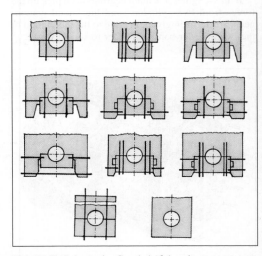

Fig. C169 Schematic: Crankshaft bearing cap mounts (Source: Zima)

C

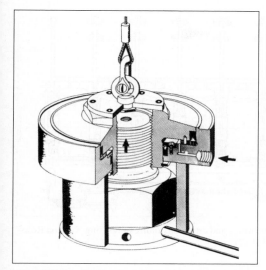

Fig. C170 Hydraulic clamping device, version MAN (Source: Scobel. Neuentwicklungen MAN-Großdieselmotoren für die Seeschifffahrt Hansa 102 [1965] 8)

On large marine and stationary engines, the accessories are preferably installed separately from the engine for reasons of logistics as well as better access for maintenance and repair.

~**Crankcase versions.** With the wide spectrum of engines—from small single-cylinder units up to large crosshead two-stroke engines—there are many different crankcase designs and versions. The design possibilities include:

- Design as divided crankcase: Single/multiple piece crankcase
- Layout of crankcase: Crankcase, engine block, base plate and frame or stands, and connection of these parts with bolts and/or tie rods.
- Location and number of separation planes
- Crankshaft bearing: Suspended, horizontal, in upper and bottom section of crankcase or tunnel crankcase
- Crankshaft bearing cap mounting: Cylinder studs, anchor bolts, lateral anchors
- Types of cylinder construction: Block cylinders, single cylinders, cylinder liners, cylinder liners cast into crankcase

These individual versions can be combined with one another—not in every manner, because some necessitate one another, others exclude one another. However, the variation in width is high. Examples include:

- Crankcase with block cylinders were the preferred design into the 1940s for inline and V aviation engines (Daimler-Benz, Junkers, Rolls-Royce, Allison, Hispano-Suiza); diesel engines (tank diesel

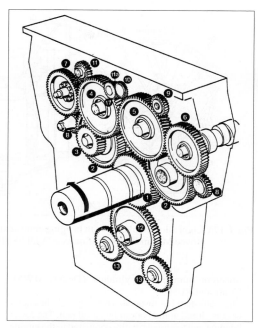

Fig. C171 Schematic: Removable gearboxes (Source: MTU publication)

W2) also have such crankcases with suspended crankshaft (**Fig. C173**), and flat crankcase connection with dry sump lubrication (*also*, →Engine ~Historic engines). The block cylinders—one on each side of the engine—contributed to reinforcing the engine structure.

- Air-cooled engines (**Fig. C174**) finned (individual) cylinders—attached to the crankcase—allowed a virtual module system for engines with

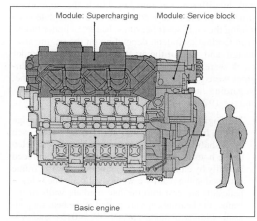

Module: Supercharging   Module: Service block

Basic engine

Fig. C172 Modular design of function areas, MTU BR 595 (Source: MTU publication)

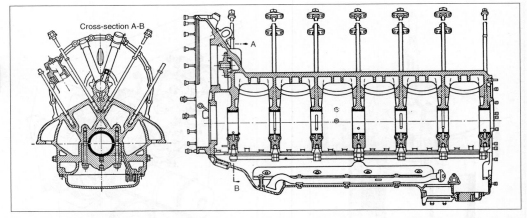

Fig. C173 Light alloy crankcase with hanging crankshaft bearings and cross tensioning of housing, version Rolls-Royce Merlin II (1940) (Source: ZVDI 88 [1944] 17)

different number of cylinders (Deutz, MWM, Continental).

- The crankshaft has suspended bearings and the crankcase is closed at the bottom by an oil pan. The lower bending and torsional rigidity compared to the engine blocks/crankcases of water-cooled engines was compensated for by reinforcing the affected cross sections.
- The most common design used today for water-cooled commercial vehicle engines is an engine block/crankcase with integrated cooling water, oil, and fuel lines; the coolant pump volutes from the oil cooler are also integrated into the housing. The crankshaft runs in suspended bearings, and the crankcase is covered at the bottom by an oil pan with deep oil sump to obtain reliable oil supply even when the vehicle is at an angle.

Large medium speed engines during the 1960s and 1970s had crankcases with horizontal crankshaft bearings in different versions, in which the supporting structure of the engine consisted of a base plate containing the crankshaft bearings, frame, cylinder block, or individual cylinders: depending on the engine size, design, and state of the art, the frame and/or cylinder block were again subdivided. Even at that time, engines were produced with a one-piece cylinder frame. Regarding large medium speed engines of the 1960s and the 1970s,

- The engine frame is reinforced at the top by bridge beams fitted laterally into the frame above each crankshaft bearing; the bridge beams are connected to the housing by long tie rods; the frame is braced laterally by elongating bolts on both sides. The frame is closed at the rear by a welded oil pan. The bearing pedestals are cast into the intermediate walls of the frame; the bearing caps are bolted to the bearing pedestal with two studs (M,4N VV 40154 and VV 52152) (**Fig. C175**). The box-shaped cylinder block is cast for three, four, or five cylinders, depending on

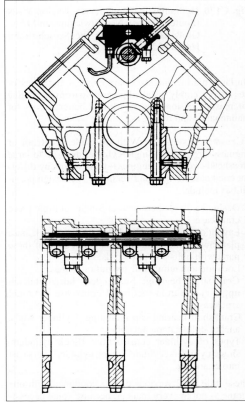

Fig. C174 Crankcase for air-cooled engine. The reinforced areas for compensating for the low bending and torsional rigidity of such crankcases are shaded (Source: H.-U. Howe: Der luftgekühlte DEUTZ-Dieselmotor FL 413 [SAE 700028])

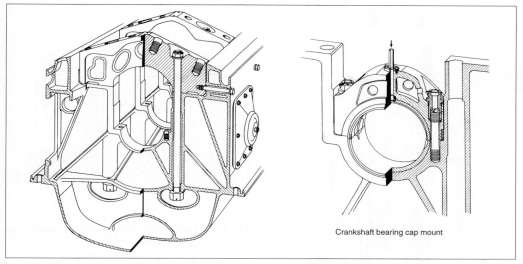

Crankshaft bearing cap mount

Fig. C175 Crankcase with horizontal crankshaft bearings, version MAN V 40/54 and V 52/52 (Source: MAN printed publication Mittelschnelllaufende Viertakt-Schwerölmotoren. Signet D 36 5128)

the number of cylinders in the engine. It holds the liners, the control shaft (camshaft), and the cylinder heads. The individual cylinder blocks are bolted to one another; they are connected to the bridge beams by long tie rods. The V-engines in these series have main connecting rods with attached auxiliary rod allowing straight intermediate frame walls.

- The V-engine frame extending up to the middle of the crankshaft is connected to the attached cylinder frame by long tie rods arranged at an angle. The

crankshaft, borne horizontally in the frame, is held from above by bearing caps (**Fig. C176**).

- The frame reaching far beyond the center of the crankshaft, with horizontal crankshaft bearings, is connected to the intermediate frame and cylinder block by long continuous tie rods (**Fig. C177**).

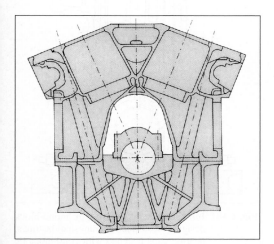

Fig. C176 Connection of base plate and cylinder frame with tie rods running at an angle: GMT 550 (Source: Pounder)

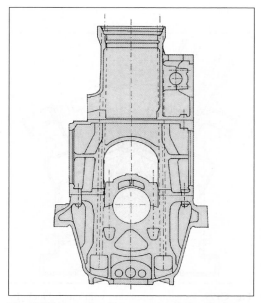

Fig. C177 Base plate, frame, and cylinder block connected by continuous tie rods: MaK (Source: MaK)

C

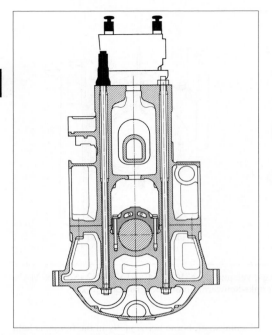

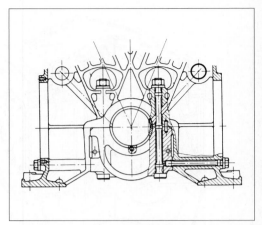

Fig. C180 Design of bearing cap bolts as continuous bolts (Source: MTZ 33 [1972] 3 p. 106)

Fig. C178 Engine housing with horizontal crankshaft bearings, base plate connected to engine frame by continuous tie rods, version MAN L 52/55B (1985) (Source: MAN printed publication D 2366135 st 7854)

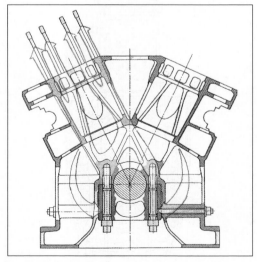

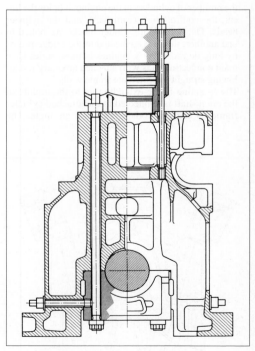

Fig. C179 V-engine crankcase with mono-block design with suspended crankshaft bearings, version MAN V 40/45 (1981) (Source: MAN printed publication D 2366107 st 10814)

Fig. C181 Inline engine crankcase with mono-block design with suspended crankshaft bearings, bearing caps fastened with continuous tie rods, version MAN L 40/54; L 48/60; L 58/64 (1994) (Source: MAN printed publication D 2366156)

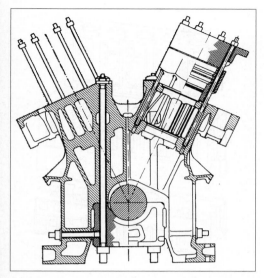

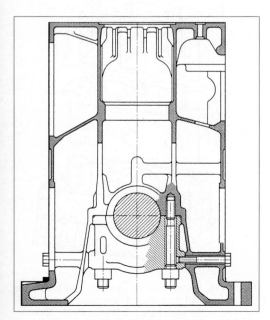

Fig. C182 Crankcase for V-engine with mono-block
design with suspended crankshaft bearings
and bearing caps fastened by continuous tire
rods, version MAN 48/60 (1994) (Source:
MAN printed publication D 2366178/1 st
6943 st 10814)

- Base plates with integrated crankshaft and frame are
tensioned with tie rods (**Fig. C178**).

The variety of types of large medium-speed diesel
engines has been reduced by progress in casting and
production technology and by operating experience to
the simple monoblock design. Since the 1980s, medium-speed engines, including large engines (up to
600 mm bore diameter and crankcase casting weights
of 100 t), have been designed with monoblock crankcases with attached individual cylinders and suspended crankshaft (**Fig. C179**) as well as an oil pan as
the bottom conclusion.

The crankshaft bearing caps are held by studs. In order to prevent notching tension by the thread, the bearing cap bolts were previously designed as continuous
bolts (**Fig. C180**).

Depending on the version and state of development,
bearing cap bolts were designed as tie rods extending
through the crankcase (**Figs. C181 and C182**). The
cases for smaller inline engines were provided with a
torsionally stiff rectangular cross section with integrated charge air guidance and camshaft bearings.

Single-piece cases (**Fig. C183**), have a favorable,
uninterrupted force flow from the cylinder head bolts

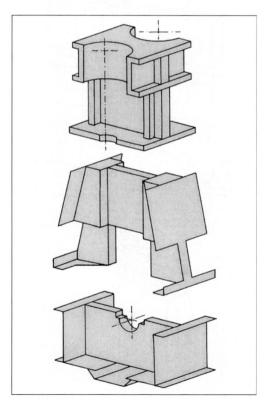

Fig. C184 Structural elements for engine housing for
two-stroke crosshead engine. Base plate,
stands, and cylinder block (Source: Zima)

Fig. C183 Mono-block crankcase, version MAN L
25/30 (Source: MAN printed publication)

to the crankshaft bearings, avoiding high peaks in stress. They are subject to less deformation, which is particularly good for the cylinders. The crankcases are no longer used for carrying the cooling water. This eliminates corrosion of the crankcase and the danger of water leaking into the lubricating oil. In total, the number of parts for the crankcase has been reduced.

The size of two-stroke crosshead engines requires a subdivision of the engine crankcase into individual parts. Due to its size and weight (up to 250 tons) alone, it is necessary for the crankshaft to run in horizontal bearings in a base plate. Stands are attached to the base plate—previously cast parts, today welded constructions (**Fig. C184**).

Individual stands with cylinder block and cylinder are subdivided depending on engine size and number of cylinders (**Fig. C185**). The base plate is designed as a welded composite construction consisting of two box-shaped longitudinal elements and cross members into which the cast steel bearing pedestals are cast. The thrust bearing housing is also integrated into the

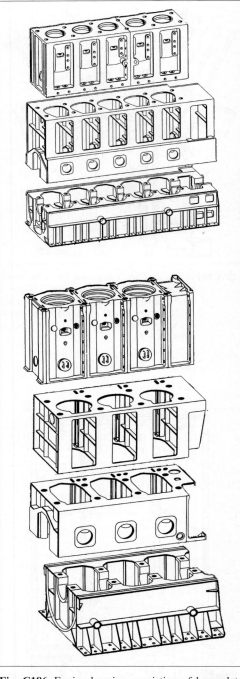

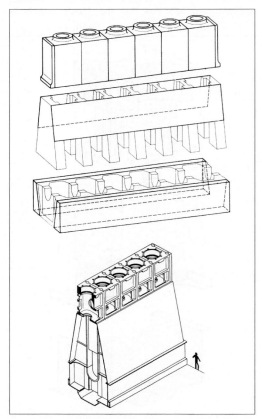

Fig. C185 Crankcase on crosshead two-stroke engine, consisting of base plate, stands, and cylinders (Source: MTU publication)

Fig. C186 Engine housing consisting of base plate and box-shaped longitudinal members, version MAN KSZ-C (1979) (Source: MAN printed publication D 2365226 st 10794)

base plate. The frame is covered at the bottom by a welded oil pan. The stands consist of box profiles with highly finned intermediate walls. The cylinders are cast individually and have cylinder liners inserted from the top. The cylinder jackets are bolted to a block with fitted bolts. Side openings with removable covers allow the pistons, piston rings, and ports to be checked.

In place of stands, box-shaped frames were also used, which were subdivided horizontally and vertically depending on the engine size and number of cylinders (**Fig. C186**). The engine housing consists of the box-shaped base plate with rack divided horizontally and vertically, depending on the engine size and number of cylinders, and high cylinder jackets.

The box design ensures high rigidity so that deformations transferred from the double floor of the ship to the base plate can be taken up slowly so that no abrupt curvature changes occur in the engine. This design ensures that the crankshaft bearing passage is deformed only slightly and continuously under internal as well as external forces. The frame with box design facilitates assembly and helps to prevent oil leakage. On "smaller" engines, the cylinders are cast into units of two or three.

~Crankcase weights and dimensions. As the heaviest part of the engine, the crankcase is particularly suited for saving weight by means of careful design and layout. This can be achieved primarily by reducing the crankcase

dimensions and less so by reducing the wall thicknesses or using specifically lighter materials (light alloys).

The dimensions, length, width, and height and, therefore, the end cross-sectional area, volume, and weight of the crankcase are determined by:

- The primary engine dimensions: Bore, stroke, and number of cylinders
- Engine concept: Arrangement of cylinder (inline, V, etc.), operating cycle, type of drivetrain (crosshead/ trunk pistons), type of cooling
- Crankcase design, which again depends on the engine concept, one or multiple piece, cast or welded construction, etc.
- Detailed design: Cylinder version, drivetrain configuration, location of control and accessories, etc.

~~**Height.** The height of the crankcase is determined primarily by the drivetrain and its arrangement. The height of crosshead engines ("cathedral engine") is nearly twice that of a comparable trunk piston-type engine, because the drivetrain length is increased by the length of the piston rod (**Fig. C187**).

On V-engines, large V-angles result in low overall height of the crankcase. Otherwise, the stroke/bore ratio and the connecting rod /crank ratio, are the primary factors influencing the height, depending on the cylinder bore. On high-speed engines, the overall height of the crankcase can also be reduced by shortening the compression height of the piston.

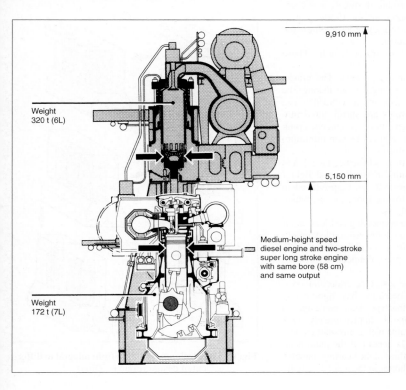

9,910 mm

Weight
320 t (6L)

5,150 mm

Medium-height speed
diesel engine and two-stroke
super long stroke engine
with same bore (58 cm)
and same output

Weight
172 t (7L)

Fig. C187
Size comparison,
two-stroke crosshead
engine and four-stroke
trunk piston engine
(Source: MAN printed
publication D 2366156)

~~Length. The length of the crankcase is determined primarily by the bore diameter, number of cylinders, and arrangement. The distance from the middle of the one cylinder to the next—the cylinder interval (cylinder spacing)—depends on the bore diameter, the cylinder design, and the drivetrain layout:

- On engine blocks/crankcases or engine blocks with suspended cylinder liners, and even more on engines with cast-in cylinder liners, the cylinders can be pushed closer together, particularly when the cylinders touch one another directly as is frequently the case for passenger car engines—i.e., are not separated by a water chamber.
- Individual cylinders cannot be located too close to one another because of their wall thicknesses:
  - it is necessary for them to transfer the forces from the cylinder wall to the crankcase so that the relative cylinder interval (cylinder interval/bore) on a series of medium-speed diesel engines (block cylinders with suspended cylinder liners) is 1.54, but for the subsequent series with individual cylinders it is between 1.7 and 1.75.
- The cylinder spacing on two-stroke engines is greater than that on comparable four-stroke engines due to the scavenging air force—on cross-scavenged and reverse-scavenged engines also due to the exhaust ports.

The cylinder spacing is also influenced by:

- The crankshaft bearings: When bearings are only located next to each alternate throw, as was common for passenger car engines in the 1950s—e.g., a three-bearing crank on a four-cylinder engine—the overall length can be reduced.
- Dimensions of crank throw (throw length). These are dependent on:
  - *The crankshaft production process.* The crankshafts for large two-stroke engines can no longer be produced in one piece; they are "assembled"—i.e., the main bearing journals are shrink-fitted into the webs of the separately produced crank throws. The shrink fit requires certain minimum dimensions for the webs.

    Recently, the crank throws have been welded to such shafts. This results in smaller and lighter shafts; a welded crankshaft can be produced in one piece instead of two or more pieces, which also shortens the crankcase.

    When the size of the connecting rod journal diameter is increased on large assembled crankshafts, the web thicknesses can be reduced without any effect to the press connection between the journal and web.
  - *Crankshaft design.* The disk-type crankshaft, particularly with roller bearings, has significantly shorter throw lengths than conventional designs; a disk crankshaft engine is approximately 18% shorter than a comparable web crankshaft engine.
  - *Crankshaft bearing.* The length of the plain bearings is one half to one third of the bearing diameter. Rolling bearings are shorter in the axial direction.

- *Drivetrain configuration.* On V-engines, the central connection between the rod journal and connecting rod of two cylinders opposing one another in a V-shape allows slightly shorter rod journal lengths.

~~Width. The arrangement of the cylinders (inline or V-engine) is primarily decisive for the weight; the diameter of the bore is secondary. The functional width of an inline engine results from the cylinder diameter and the so-called connecting rod motion diagram. On V-engines, the V-angle determines the width. The space on each side of the bank of cylinders or between the cylinder banks can be utilized optimally by clever arrangement of the control and engine accessories, especially for vehicle engines with accessories located on the engine. This applies even more for tank engines, in which extremely small external dimensions are required because the engine must be surrounded by armored plates. The double-acting MAN two-stroke engines for special marine applications and the Junkers two-shaft opposed piston engines (*also,* →Engine ~Historic engines), developed in the 1930s, were extremely narrow.

~Development of crankcase versions. It was quickly recognized—even with the first motor vehicle and aviation engines (→Engine ~Historic engines)—that it was necessary to protect the drivetrain from dust, water, and contamination (**Figs. C188 and C189**). These early engines were, therefore, provided from the

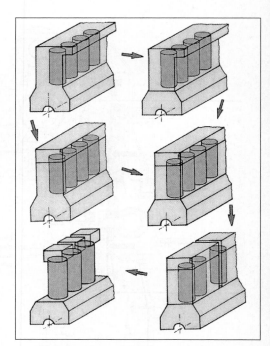

Fig. C188 Development stages from integral to differential design (Source: Zima)

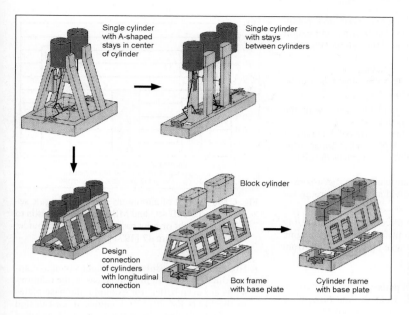

Single cylinder with A-shaped stays in center of cylinder

Single cylinder with stays between cylinders

Block cylinder

Design connection of cylinders with longitudinal connection

Box frame with base plate

Cylinder frame with base plate

Fig. C189
From single cylinder design to cylinder frame (Source: Zima)

very beginning with crankcases with individual or block cylinders. The crankcase was frequently laid out as a two-piece unit with the crankshaft located between the upper and lower sections.

The cylinders and cylinder cover on side-controlled engines were cast in one piece. This eliminated the cylinder head gasket, which posed problems at that time. The cylinders were then integrated into the crankcase, and the crankshaft was suspended to provide greater longitudinal and torsional rigidity. As production capabilities improved, operating experience increased, and engine size expanded, the engine block/crankcase was divided up into individual parts. Engines were developed in which the crankcase, cylinder block, and individual cylinder heads, and, finally, crankcases with individual cylinders and individual heads were all separate.

The first large engines did not have a crankcase. One or more individual A-shaped stands were located on the base frame in the plane of the crank circle, and these supported the cylinder(s). The disadvantage of the A stands was that they had to protrude extensively to provide sufficient clearance for the crank train because they were located in the crank circle plane and, therefore, required appropriately wide, heavy base plates. Moreover, the long lever alarms led to high bending moments with the corresponding bending load on the base plate. The main crankshaft bearings were easily accessible, but the drivetrain was not. In the crank circle plane, it was possible for the crosshead guide to be supported on one half of the stand, which allowed one-sided ("one rail") guidance of the crosshead, which was favorable for production and assembly. In multi-cylinder engines, the deformation of individual cylinders had no effect on the others. However, such engines only possess low longitudinal rigidity, which presented

problems on the soft foundations on-board ships. For this reason, the individual stands were connected with intermediate beams and were positioned between the cylinders instead of in the crank circle plane.

Because the stands were located between the cylinders, it was necessary to attach the slide rails for the crossheads between the stands, as a rule, in the form of a "four rail" guide. This also made the drivetrain easily accessible from both sides.

Upon introduction of pressurized circulating lubrication systems, it was necessary to cover the drivetrain to prevent lubrication and piston cooling oil from spraying into the surrounding area. This covering of sheet metal was the first step toward a box frame. The higher engine speeds (inertia effects) and outputs made it necessary to make the engines stiffer in the longitudinal and lateral directions. Box frames were constructed by screwing cast frame elements (stands) together to form a box.

These were connected with bolts to form a frame. In the beginning the cylinders were positioned singly on the frame, in order to prevent any interaction between them. The longitudinal connection of the stands to form box frames resulted in bolting the cylinders together to form blocks and provision of continuous supports to make the frame more rigid. A number of cylinders were also cast in a block—frequently in such a manner that the cylinders had a common water chamber. In engines with many cylinders, two or more blocks were connected with one another by flanges and bolts.

Another development was to have the cylinders and box frame cast as one part, which naturally was possible only up to certain engine sizes. The base plate—adapted to the specific frame construction—extended upward allowing it to better take up the lateral forces from the bearings. It also proved necessary to separate the cylinder lubrication from the bearing lubrication using a design which subdivided the cylinder chamber and crank chamber.

~Function. The functions described above had to be fulfilled under ancillary conditions:

- Good utilization of the available space with the smallest possible component masses
- Sufficient structural rigidity in terms of accuracy of the shape of the bearing bores and cylinder fit as well as leakage of attached parts, such as the oil pan and cylinder heads
- Greatest possible integration of engine accessories
- Good accessibility to individual function groups for maintenance and repair

The crankcase takes up the forces and moments acting in and on the engine and transfers them to the engine mounts; moreover, it must take up forces acting from the outside.

These include:

- Forces from the attached parts
- Radial and axial forces from the machine to be driven (supporting forces and axial thrust)
- Forces from the engine mount (e.g., deformation of the hull of the ship)
- Assembly forces
- Forces resulting from thermal expansion

~Materials and production. The following materials are used for large cases:

- Lamellar graphite cast iron (GGL)
- Vermicular graphite cast iron (GJV)
- Spheroidal graphite cast iron (GGG)

Today, casings subject to high loads are produced using spheroidal graphite cast iron (GGG), due to its higher strengths and expansion values.

The cases for aviation and tank engines were and are produced from aluminum/silicon alloys, which are capable of being hardened by heat treatment. The cases for engines for PT boats are also cast using aluminum alloys.

In the 1960s and 1970s, the crankcases for high-speed, high-performance diesel engines were produced using gray cast iron as well as antimagnetic light alloys (for mine sweepers). Due to the differences in the material properties, the wall thicknesses and resistance moments on aluminum crankcases are greater, the thread engagement depths are longer and the bearing overlap is higher. A comparison of the weights obtained when using different materials is shown in **Fig. C190**.

The crankcases for high-performance engines are designed as welded constructions of cast steel parts and segments. Welding offers significant advantages for large structures—assuming that the construction was designed for welding.

Initially, there were technical difficulties with welding, which have now long been mastered. The advantages of welded construction for large engines compared to casting are:

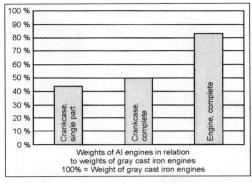

Weights of Al engines in relation
to weights of gray cast iron engines
100% = Weight of gray cast iron engines

Fig. C190 Weight of aluminum version of engine series MB 493 and MB 652 in comparison to weight of gray cast iron engines. (Source: Aluminum 47 [1971] 8)

- Higher strength, resulting in weight savings of approximately 15%. In individual cases, the relationships are even better: for example, the base plates of the MAN KZ 701120 A engine in cast design was 57 t, whereas the welded structure weighed only 31 t.
- Stiffer and more reliable construction, meaning less danger of breakage; moreover, fissures can be welded.
- No extremely expensive models—for large components—allowing the special requirements of shipbuilders to be satisfied with welded parts.
- Long-stroke engines have a tendency to allow lateral vibration; this has been compensated for by stiff welded constructions.
- Today, large engines are produced primarily on a license basis. Not all licensees are capable of casting large parts to the required dimensions; also shipbuilders prefer welding.

Disadvantages of welding are:

- The components cannot be designed with such an optimum force flow as cast parts.
- Welded parts are more susceptible to corrosion.

~Set-up. The crankcase design depends on:

- Size of engine
- Operating cycle (four/two-stroke)
- Cooling medium (water/air)
- Number of cylinders and configuration
- State of the art
- Technical "philosophy"

The design possibilities for crankcases are represented in the designations defined in DIN ISO 7967.

The crankcase encloses the crank chamber with crankshaft and holds the cylinder block (two or more cylinders cast or bolted together) or the single cylinder (assembly with or without cylinder sleeve). When the cylinders are integrated into the crankcase, this is called an engine block. At the bottom, the crankcase/engine block is terminated by the oil pan.

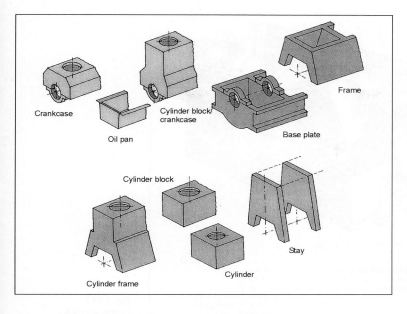

Fig. C191
Cylinder block/
crankcase for high-speed,
high-performance diesel
engine (Source: MTU)

The crankshaft is borne horizontally in a base plate—due to its mass—on larger engines. The base plate serves for supporting the engine in a ship, on a foundation, and so on: it can act as the oil pan. One of the following components is attached to the base plate (**Fig. C191**):

• The frame—a component surrounding the upper crank chamber without integrated cylinder
• The cylinder frame—a component surrounding the upper crank chamber with integrated cylinders
• The stands—components supporting the cylinder, cylinder jacket, or cylinder block

In principle, the crankcase or engine block consists of the intermediate walls, the upper cover plates, the longitudinal walls, and the end walls. The crankshaft and camshaft(s) are borne in the intermediate walls. Larger engines have installation openings with removable flaps, covers, or even doors in the longitudinal walls to allow access to the drivetrain for checking, for maintenance, and for repair work. Some of these covers are designed as explosion-protection flaps. On high-speed engines (**Fig. C192**), the water jacket is cast into the engine block, just as are the passages for the oil and coolant, and, where applicable, the charge air line.

At the bottom, the housing is closed by a base plate or oil pan. To increase the service life of cylinder blocks/crankcases, possibilities are provided for rebuilding—for example, at the cylinder liner collar contact point, at the locations for the cylinder liners, and, particularly, in the crankshaft bearing bores. The differentiating features of crankcases are:

• Number and location of housing separation planes
• Type of crankshaft bearing (suspended, horizontal)

• Interconnection of the individual housing parts
• Material and type of production (casting, welding)

Commercial vehicle cylinder designs →Commercial vehicle cylinders/liners

Commercial vehicle cylinder head. The cylinder head is the component which closes the combustion chamber with or without components for gas exchange (DIN ISO 7967). Colloquial language differentiates between cylinder heads (with elements for gas exchange) and cylinder covers (without elements for gas exchange).

The cylinder head terminates the working chamber; it holds the intake and exhaust valves, gas ports, injection nozzles, and, where applicable, the prechamber or auxiliary chamber, on large engines, it also contains the safety and decompression valve as well as the air starter valve. The valve actuation, including the valve turning device, is attached to the cylinder head.

~Design. Cylinder heads are laid out in a boxshape; they consist of the lower compression plate, the upper cover plate, and the side walls. The gas passages, the guides for the valves, and the injection nozzles, where applicable for the combustion chamber (pre- or swirl chamber), with intermediate walls and plate as well as fins are supported on the compression plate and reinforce the structure.

The coolant flow in the cylinder head is controlled by the intermediate plate, the connection of the pipes to the ducts (support ribs), and their passages (**Figs. C193–C195**).

The rigidity required for taking up the gas forces is achieved primarily by the height of the cylinder head.

171

C

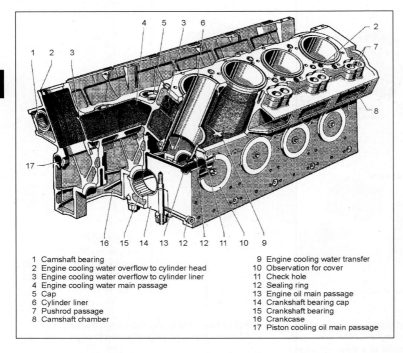

1 Camshaft bearing
2 Engine cooling water overflow to cylinder head
3 Engine cooling water overflow to cylinder liner
4 Engine cooling water main passage
5 Cap
6 Cylinder liner
7 Pushrod passage
8 Camshaft chamber

9 Engine cooling water transfer
10 Observation for cover
11 Check hole
12 Sealing ring
13 Engine oil main passage
14 Crankshaft bearing cap
15 Crankshaft bearing
16 Crankcase
17 Piston cooling oil main passage

Fig. C192
Designations for
crankcases (ISO 7967,
Part 1) (Source: Zima)

The contour of the individual cylinder heads is designed with a circular shape in the area of the cylinder head gasket for better transmission of force (**Fig. C196**).

On smaller engines, the cylinder heads for one engine bank are cast as a single block, allowing the camshaft (overhead camshaft) to be installed in the cylinder head, particularly on passenger car engines. Individual cylinder heads begin to be used at approximately 130 mm.

All told, single heads are more advantageous in terms of production and assembly; they are stiffer, which is advantageous for transfer of the force and sealing (**Fig. C197**). Design differences result from the number of valves and their arrangement. Small commercial vehicle engines have two or three valves, larger engines have four, and special versions even

have six valves. Each valve can open and close its own gas port, or two valves can be used for one port. The cross-flow arrangement is advantageous in terms of the flow; having the intake ports on one side and ex-

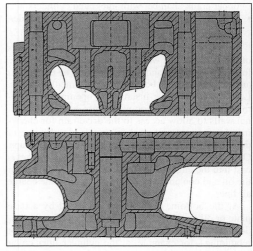

Fig. C194 Longitudinal cross section through the cylinder head (high-speed, high-performance diesel engine), approximately 230 mm, box-shaped layout with intermediate plate for coolant routing (Source: Zima)

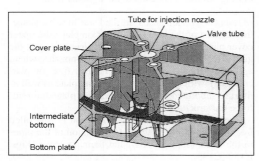

Tube for injection nozzle
Valve tube
Cover plate
Intermediate bottom
Bottom plate

Fig. C193 Schematic design of a cylinder head (Source: Zima)

C

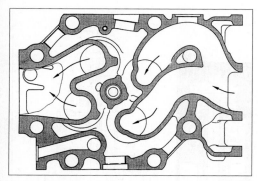

Fig. C195 Cross section of cylinder head for medium-speed diesel engine, version MaK M 332 C (Source: MaK printed publication)

haust ports on the other is more favorable in terms of supercharging, due to the shorter lines to the exhaust turbocharger. Depending on the design of the intake ports, either turbulence ducts are used, which swirl the air as it flows in, or turbulence-free charging ducts are used. Because the swirl required for certain combustion processes depends highly on the engine speed and also the load state of the engine, one of the two intake ports can be shut off at part load. The air then flows through one duct only—ensuring the required swirl rate.

Larger engines have overhead valves that are parallel to the cylinder axis. Angled overhead valves were used on passenger car engines and earlier primarily for aviation engines. Engines were also produced with in-block valves (side valves); this simple design with in-block valves was standard for passenger car engines into the 1950s. The valves are inserted through the valve guide into the valve port, so that they are centered on the seat. Simultaneously, the heat from the valves is transferred through the valve guide.

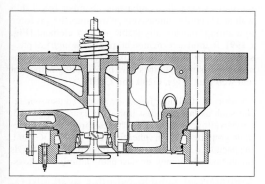

Fig. C196 Cylinder head with circular support on cylinder, version MAN V 32/36 (Source: MAN printed publication Signet D 2365211 st 4821)

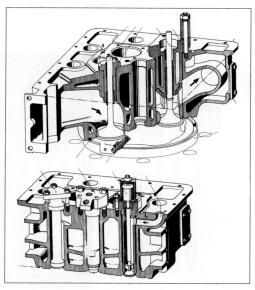

Fig. C197 Cylinder head for medium-speed, four-stroke diesel engine, version MAN L+V 40/54 (Source: MAN printed publication Mittelschnelllaufende Viertakt-Schweröl-motoren Signet D 36 5128)

The valves run directly in the cylinder head material or in special valve guide sleeves of gray cast iron or special bronze.

The injection nozzles are located either in a thin-walled sleeve or directly in a port in the head, so that the lubricating oil is not diluted even in the event of leakage. When the injection nozzles are positioned at an angle, it is necessary to locate them in a precise position to ensure a defined injection pattern. The coolant must be routed so that dead water (stagnant) areas or steam pockets cannot form: intermediate walls and support ribs are used for this purpose. The coolant is fed to and discharged from the cylinder head through corresponding passages in the crankcase, or, in larger engines, by separate cooling water lines. The intake valve seats are inductively hardened, and the exhaust valves are armored. On large engines, special valve seat rings are used for protection against oil carbon deposits, wear, and corrosion. The valve seats can be cooled, depending on the size, design, and version.

Exhaust valves, in particular those subject to high thermal loads and high-temperature corrosion, are located in valve cages, allowing them to be replaced quickly (**Fig. C198**).

The valves are provided with rotating devices (Rotocap) to prevent them from wearing into the seat. The exhaust valves on large engines are provided with a "propeller," which is driven by the exhaust gas and which rotates the valve.

C

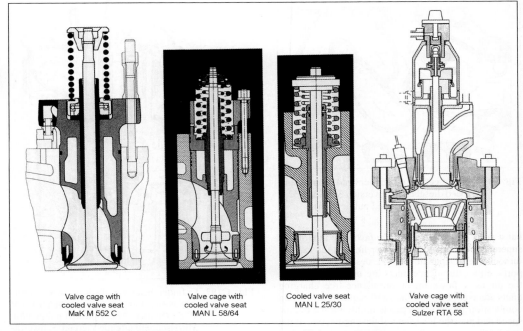

Valve cage with cooled valve seat MaK M 552 C

Valve cage with cooled valve seat MAN L 58/64

Cooled valve seat MAN L 25/30

Valve cage with cooled valve seat Sulzer RTA 58

Fig. C198 Valve seats and valve cages (Source: MaK, MAN, and Sulzer documents)

~Forces and load. The cylinder head is subject to mechanical loads resulting from the bolt forces and gas pressure. The high heat flows with nonuniform temperature distribution, in combination with a com-

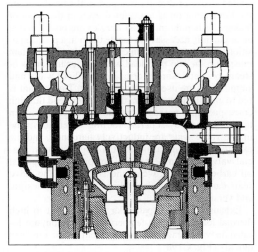

Fig. C199 Two-piece cylinder head on loop scavenged, two-stroke diesel engine, version MAN KSZ-CL (approximately 1979) (Source: MAN printed publication D 2365226 st 10794)

plicated cast structure and varying wall thicknesses resulting from the design, lead to high-temperature gradients, which are increased by the change in temperatures resulting from the alternating engine load. This combination of mechanical and thermal loads requires compensating measures to be taken in dimensioning: thin walls for high thermal loads and thick walls for mechanical loads. The heat conduction can be increased by using thin-walled flame plates on the combustion chamber side. The gas forces are taken up by a strong intermediate plate. Large, port-controlled, two-stroke engines have two-piece cylinder covers: the part closing the combustion chamber has a thin wall and is cooled intensively; it transfers the gas force to a strong, intrinsically stable support element (**Fig. C199**). To limit deformation in spite of the high gas pressures, slow running two-stroke engines and medium-speed four-stroke engines have bore cooling in the seat and web area (**Figs. C200 and C201**).

~Materials. Cylinder heads are produced using material such as lamellar cast iron—for example, GG 25, CroMo alloyed cast iron, ferritic cast iron with vermiculate graphite (GGV), spheroidal cast iron (GGG) and, in special cases, cast steel (GS). Finned cylinder heads on air-cooled engines are cast using aluminum alloys.

Commercial vehicle cylinder head forces/stresses →Commercial vehicle cylinder head

Commercial vehicle cylinder head layout →Commercial vehicle cylinder head

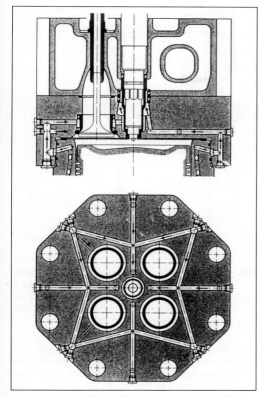

Fig. C200 Two-piece, bore-cooled cylinder head for medium-speed, four-stroke diesel engine, version Sulzer (Source: Sulzer Technical Review 1/1979)

Commercial vehicle cylinder head materials →Commercial vehicle cylinder head

Commercial vehicle cylinder layout →Commercial vehicle cylinders/liners ~Layout

Commercial vehicle cylinder liner versions →Commercial vehicle cylinders/liners ~Cylinder liner designs

Commercial vehicle cylinder materials/ machining →Commercial vehicle cylinders/liners

Commercial vehicle cylinders/liners. This engine component is called a cylinder because of its shape—a solid described by the edge of a rectangle rotated around the parallel edge as axis—and it forms a sliding pair together with the piston. According to DIN ISO 7967, it is necessary to differentiate between:

- Cylinder component in which a piston acts with or without a cylinder liner and with or without a cast on cylinder head

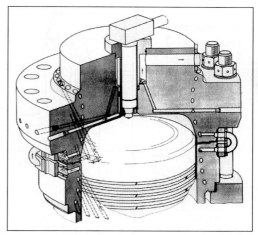

Fig. C201 Cylinder head (cylinder cover) on cross-flow scavenged, two-stroke diesel engine, version Sulzer RN 90 M (approximately 1980), with bore cooling (Source: Sulzer)

- Cylinder jacket: Component surrounding the cylinder containing the coolant and fastened to a frame or crankcase
- Cylinder liner: Liner inserted into the cylinder jacket with running surface for the piston

The cylinder liner fulfills a number of purposes: limitation of the working chamber, guiding the drivetrain, transferring the force to the crankcase, transferring heat to the coolant, and supporting the charge processes (on two-stroke engines). The cylinder jacket supports the cylinder liner and routes the coolant and lubricant, as well as holding engine accessories.

~Cylinder liner designs. Cylinder liners in medium-speed engines require a thick-wall design to provide the necessary inherent stability due to the high gas pressure, which has now reached up to 200 bar.

The cylinder liners are supported on the engine block by a support ring (water ring) (**Fig. C202**) so that it is no longer in contact with the coolant. This design positively eliminates housing corrosion and prevents water from leaking into the lubricant.

On high-speed diesel engines, this was achieved with a separate water jacket between the cylinder liner and crankcase (**Fig. C203**).

Larger engines are produced to an increasing extent using a single-cylinder design. The advantages are reliable piston operation, reduced wear, no mutual influence between cylinder deformations, and, all told, less deformation over the entire length of the engine. The cylinder liners are supported on the crankcase by a water jacket. Medium-speed engines using heavy oil as fuel are provided with cylinder lubrication (**Fig. C204**), with special lubrication oil fed through a system of individual bores in the cylinder liner.

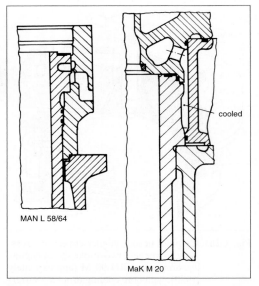

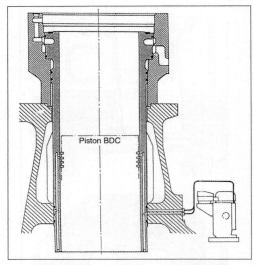

Fig. C202 Cylinder with water ring (Source: MAN-und MaK)

Fig. C204 Cylinder lubrication on medium-speed, four-stroke diesel engine, version MAN L 48/60 (Source: MAN B&W Three medium-speed engines—one design concept Signet D 2366156)

The cylinder liners are now cooled only in the upper area, whereby the cooling is intensified by ingenious routing of the coolant:

- By increasing the flow rate with a narrow gap, particularly on the exhaust valve side, to such an extent that equal temperatures are present over the entire circumference (**Fig. C205**), or
- With bore cooling (**Fig. C206**) using a system of angular bores in the upper, thick-wall area of the liners for specific dissipation of the heat

In two-stroke engines, ports in the cylinder liner are opened and closed by the piston to control the gas flows. While engines with uniflow scavenging have intake ports distributed uniformly around the circumference of the liner in the region of bottom dead center,

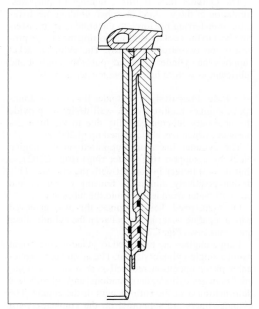

Fig. C203 Cylinder liner with water jacket, installation in engine block, version MMB MC 1060 (approximately 1968) (Source: MTZ 29 [1968] 5)

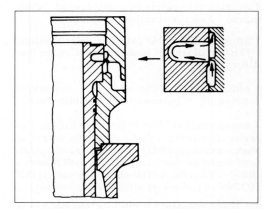

Fig. C205 Specific control of coolant flow MAN L 40/54 (Source: MAN)

C

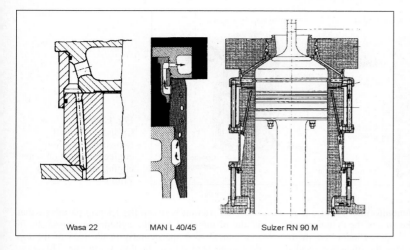

Wasa 22   MAN L 40/45   Sulzer RN 90 M

Fig. C206
Bore cooling on
medium-speed,
four-stroke and
slow-speed, two-stroke
engines (Source:
Company Publications,
Wasa, MAN, Sulzer)

the exhaust ports are located above the intake ports on engines with loop scavenging and across from the intake ports on engines with cross-flow scavenging.

An attempt was made to control deformation resulting from asymmetrical thermal loads by cooling the webs; the liner collar was also cooled (**Fig. C207**).

Small two-stroke gasoline engines have a cylinder with cast-on transfer ducts for scavenging (**Fig. C208**).

~Layout. The type and size of the engine, as well as the operating conditions, determine the design, materials, wall thickness, and layout of the collar section and cylinder cooling. On small engines (passenger car engines), the cylinders are cast into the cylinder housing; commercial vehicles and larger engines have cylinder liners, because these allow the use of materials best suited in each case for the individual functions/components; parts subject to wear and damage can also be replaced separately.

The cylindrical sleeves have a collar at the top with which they are supported axially directly on the top deck or in a machined fit in the engine block/crankcase (beam seat). The cylinder liner is free to expand downward (**Fig. C209**). The sleeves are held radially at the upper and lower sleeve collars by a sleeve fit in the engine block. On the outside, the cylinder sleeves are surrounded by coolant between the sleeve collars. They are provided with a corrosion-protection layer to prevent corrosion.

Blowby of combustion gases between the cylinder liner and cylinder head is prevented by the cylinder head gasket. The coolant water chamber is sealed at the top by the collar contact on the beam seat and at the bottom from the crank chamber by two (or three) sealing rings inserted into the cylinder liner. A relief groove is machined into the cylinder liner between the top and bottom sealing rings; a relief hole is located in the crankcase at the same height. If a sealing ring or relief groove is leaky, the leak water collects here and can flow out through the relief hole so that the leakage can be recognized.

A cast iron ring ("flame ring") (**Fig. C210**) inserted into a machined groove at the inside of the top of the cylinder liner and located slightly above the running surface reduces the piston clearance and thereby deposition of carbonized oil on the top land of the piston.

~Loads and deformations. Cylinder liners are subject to mechanical, thermal, terminological and chemical loads. High forces are applied to the liner collar when the cylinder head bolts are tightened. The cylin-

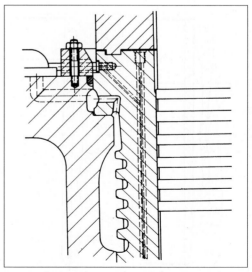

Fig. C207 Cylinder liner with cooled collar for loop scavenged two-stroke diesel engine, version MAN (approximately 1965) (Source: Scobel. Neuentwicklungen an MAN-Großdieselmotoren für die Seeschifffahrt. Hansa 102 [1965] 8)

177

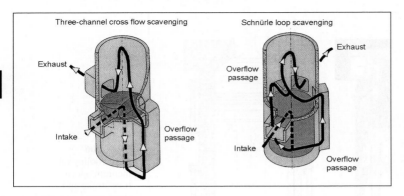

Fig. C208
Cylinder on small
two-stroke gasoline
engine with cast on scav-
enging ports (Source:
Zima)

der deforms radially and nonuniformly where it is located in the crankcase in the area of the cylinder head
bolts, leading to a reduction of the cylinder liner diameter; below the locating region, the liner is expanded
by the gas pressure. The change in the contact position
of the piston resulting from the movement between the
thrust and antithrust sides of the liner causes the cylinder liner to vibrate. The coolant cannot follow this vibration, which can result in the formation of bubbles in
the coolant (cavitation). High temperatures, large temperature gradients, and rapid temperature changes result in thermal stresses and deformation. Additional
stresses result as a consequence of unequal temperatures of the cylinders and cylinder head, which attempt
to compensate themselves through the gasket plane.

Mixed friction between the friction partners—that
is, the piston rings and the cylinder liner—results in
mechanical wear, particularly in the area of the top
dead center position of the first piston ring. This wear
is enhanced by corrosive components of combustion,
and sulfuric acid is formed in the combustion chamber
when the temperature drops below the condensation
point for SO_2. The bore wear at the TDC position of
the top ring is exacerbated by higher gas pressures and
poor fuels (heavy oil). Carbon particles from the oil in
the combustion chamber polish the cylinder liner
along the upper third of the piston stroke. Polished surfaces start to develop on the antithrust side and expand
over a wide area of the cylinder.

Bore polishing. The oil consumption and danger of
ring burning increases. Adhesive wear results when

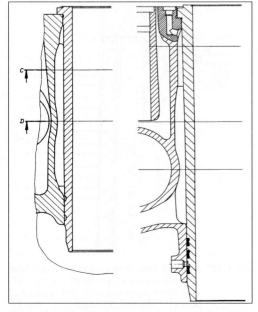

Fig. C209 Cylinder sleeve contact on high-speed diesel engines (Source: MTU/Zima)

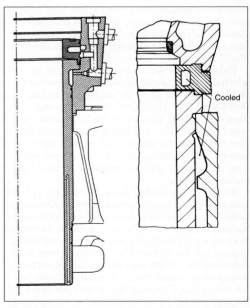

Fig. C210 Versions of flame rings (Source: MAN/MaK)

metal particles are torn out of the cylinder wall. Deformations and wear affect the operation of the piston and the functioning of the piston rings and finally limit the service life of the cylinder liners.

~Materials and processing. The terminological properties required for proper functioning of the cylinder liners are achieved by selection of the material and subsequent processing. Alloyed gray cast iron with pearlitic basic structure and a fine, hard phosphide network (support mesh effect) is used for the cylinder liners on diesel engines as follows:

- Cr alloy GGL (standard material)
- CrNi alloy GGL (higher resistance to wear and mechanical loads)
- High carbon CrMo alloy GGL (very good sliding properties)

The advantages of the various materials can be utilized by compound casting using lamellar cast iron (good running properties) and spheroidal cast iron (high strength). The cylinders for aviation engines were designed as thin-wall nitration-hardened steel liners.

The terminological characteristics for wear and running behavior were improved by:

- Hardening (induction hardening, nitriding, and laser hardening).
- Phosphatizing (coating): Phosphate crystals on the cylinder surface improve the adhesion of the oil and thereby the sealing effect which counteracts piston ring burning.
- Processing: The micro-geometry of the running surface has a decisive effect on the running characteristics of the piston rings and piston; for this reason, the cylinder running surfaces are honed.

Because engines react sensitively to so-called plate jacket formation with increasing output—during the honing operation, graphite lamella can be covered and scoring can occur due to the cutting pressure of the tool—a special honing process is used, known as a plateau honing. In this, the peaks are honed down, resulting in plateaus with reduced roughness, separated by deep scores (**Fig. C211**). The plateau structure is evaluated according to the Abbotts support curve (**Fig. C212**).

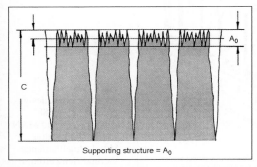

Fig. C211 Schematic: Plateau honing (Source: Gehring)

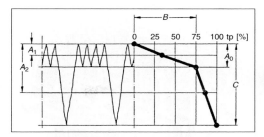

Fig. C212 Abbott support curve (Source: Gehring)

~Types of cylinder construction. Until the end of the 1920s, aviation engines were built using single-cylinder designs. Thin-walled steel liners with a water jacket and cylinder head were bolted to the crankcase as a single piece using a deep-seated collar on the cylinder liner. The cylinders for the PT boat diesel engine MTU 20 V 672 (formerly MB 518) were constructed in a similar manner: the light alloy cylinder head was bolted to the gray cast iron liner, an intermediate steel element, and the water jacket to form one unit (**Fig. C213**).

Diesel engines for rail propulsion, version Maybach GO6/G 6/GTO, have double cylinders consisting of two cylinders, cylinder heads, and water jacket cast as one piece (**Fig. C214**).

Four-stroke diesel engines for rail propulsion, version General Electric 7FDL, have closed cylinders with fitted steel cylinder heads (**Fig. C215**). The cylinder liner is inserted into the cylinder so that a chamber for the coolant is formed in between. The cylinder liner

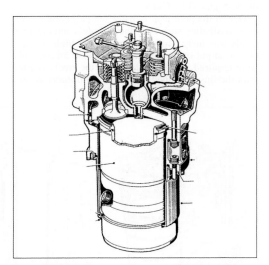

Fig. C213 Assembled cylinder: Cylinder head, intermediate element, cylinder liner, and water jacket of PT boat diesel engine, version MTU 20 V 672 (Source: MTU Operating Instructions)

C

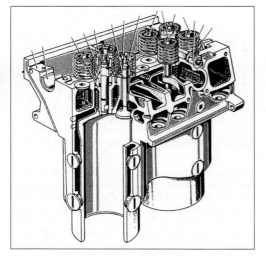

Fig. C214 Double block cylinder with cast-on cylinder heads for high-speed diesel engine, version Maybach GTO 6 (Source: Maybach company documents)

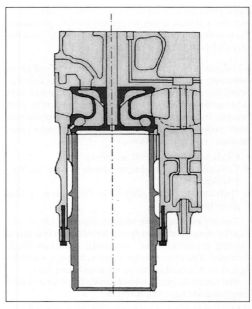

Fig. C215 Cylinder with fitted steel cylinder head and bolted cylinder liner, version General Electric 7FDL (Source: Pounder)

is bolted to the cylinder from the bottom, using a collar. Beginning in the 1930s, aviation engines used cylinder blocks with inserted cylinder liners. This design was also used for medium-speed diesel engines in the 1960s. Such cylinder blocks have one common water chamber for all cylinders. One or more blocks were bolted together to form one unit, depending on the size of the engine (**Fig. C216**).

Due to the poor heat transfer with air cooling, the heat-dissipating surface of the cylinders and cylinder heads was increased with fins. At the high specific power output of aviation engines, close fin intervals were required, and to achieve this, the steel cylinder with the fins machined out of the solid material was

connected to a cast light alloy cylinder head—in a separable or inseparable manner, depending on the version (**Fig. C217**).

A combination of aluminum and cast-iron materials provides optimum strength, running characteristics, and heat conduction:

- Light alloy finned element shrink-fitted onto gray cast iron liner.
- Compound casting: Aluminum finned cylinder with cast-in gray cast iron liner. Here, it is necessary to differentiate between:

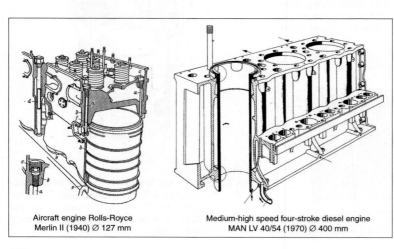

Aircraft engine Rolls-Royce
Merlin II (1940) ∅ 127 mm

Medium-high speed four-stroke diesel engine
MAN LV 40/54 (1970) ∅ 400 mm

Fig. C216
Cylinder block designs (Source: ZVDI 88 [1944] 17/18 and MAN printed publication Mittelschnelllaufende Viertakt-Schwerölmotoren Signet D 36 5128)

180

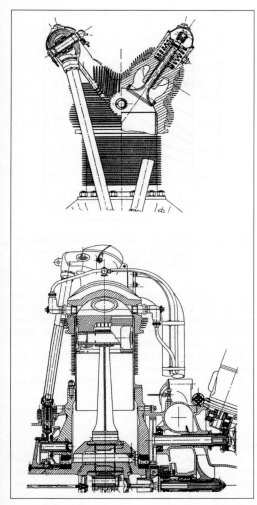

Fig. C217 Finned cylinder on air-cooled aircraft engine, version BMW 132 (Source: Katz: Der Flugmotor, p. 87, and BMW Operating Instructions)

— The AlFin process: Intermetallic connection layer between gray cast iron liner and aluminum cylinder
— Rough cast cylinder liner: The outer jacket surface of the thin-walled cylinder liner is roughed for a positive fit when clamped to various materials.

Commercial vehicle engines →Commercial vehicle crankcase; →Oil ~Classification, specifications, and quality requirements ~~Diesel engine oils

Commercial vehicle loads/deformations →Commercial vehicle cylinders/liners

Common rail →Injection system (components) ~Diesel engine ~~High-pressure rail; →Injection system, fuel ~Diesel engine, ~Gasoline engine ~~Direct injection systems

Common rail injector →Injection valves ~Diesel engine

Compact filter elements →Filter ~Intake air filter ~~Air filter elements

Comparative process →Cycle

Compatibility, bearing materials →Bearings ~Materials ~~Qualities

Compensating air →Carburetor

Complete measuring paths →Particles (Particulates) ~Particle measuring

Composite systems →Engine concepts

Composite technology →Crankcase ~Crankcase design ~~Cylinder

Composition, exhaust gas →Emissions; →Exhaust gas analysis

Composition, fuel →Fuel, diesel engine; →Fuel, gasoline engine

Compound →Supercharging ~Turbo-compound supercharging

Compound carburetor →Carburetor ~Design types

Compound charging →Supercharging ~Turbo-compound supercharging

Compressed natural gas →Fuel, gasoline engine ~Alternative fuels ~~Natural gas

Compressibility →Fuel, diesel engine ~Properties

Compression (*also*, →Compression ratio). The compression process is one of the four processes in the combustion engine cycle. It brings about an increase in the pressure and temperature of the working medium, thus enabling combustion with higher efficiency. The temperature increase of the compressed air to approximately 800 to 900 K creates the prerequisite in the diesel engine that the injected fuel will spontaneously ignite. **Fig. C218** shows the dependence of thermal efficiency on the compression ratio.

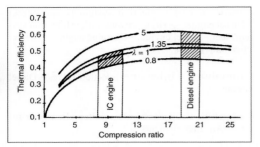

Fig. C218 Compression ratio and thermal efficiency

~Compression end pressure. The compression end pressure is a function of the intake air temperature of the given working medium, the initial pressure, and the compression ratio if heat transfer is not considered. The final compression pressure has been plotted against the charge pressure in **Fig. C219**.

~Compression end temperature. The final compression temperature is a function of the cylinder filling and the compression ratio of the given working medium and a given initial temperature of the medium if heat transfer is not considered. The compression temperature has been plotted against the intake air temperature in **Fig. C220**.

~Compression ratio. The compression ratio (geometrical) is defined as the quotient of the maximum and minimum cylinder volume. Maximum cylinder volume is given when the piston is at bottom dead center (BDC). When the piston position is at top dead center (TDC), the volume is its minimum value and is termed as clearance volume or dead volume.

The clearance volume is composed of the combus-

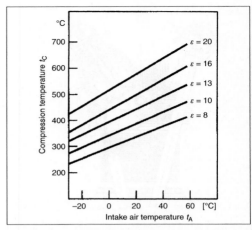

Fig. C220 Final compression temperature as function of intake air temperature and compression ratio (Source: Bosch)

tion-chamber volume of the cylinder head, the volume of the piston recesses, and piston relief, as well as the volume of the fire land up to the top compression ring. Clearance volume and displacement can be determined by volumetric measurement.

Fig. C221 shows a schematic representation of displacement and clearance volume.

Compression ratio, ε, for four-stroke engines is determined with

$$\varepsilon = \frac{V_h + V_c}{V_c},$$

where V_c = clearance volume.

The compression ratio of the gasoline engine is limited by knocking as well as glow ignitions. The compression ratio can be increased on direct injection gasoline engines because of improved internal cooling through internal mixture preparation. This results in an advantage in efficiency over the gasoline engine with intake manifold injection.

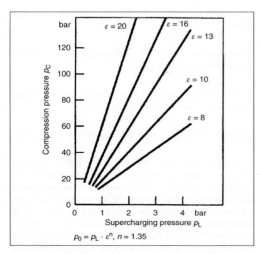

Fig. C219 Final compression pressure against charge pressure

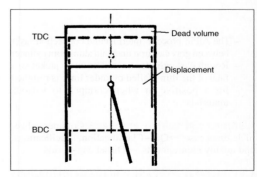

Fig. C221 Displacement and clearance volume

In the diesel engine, the compression ratio needs to be selected such that the engine will safely start at low temperatures. The required cold-start capability necessitates high compression ratios. Generally, the thermal efficiency rises with increasing compression ratio. An excessively high compression ratio will be at the cost of the effective efficiency caused by strongly increasing frictional forces. Independently of that, the peak pressure is limited by material strength and confines the practically realizable compression ratio.

NO_x and HC emissions increase with rising compression ratio. The oxides of nitrogen increase because of the increase of temperatures in the combustion space; the HC emissions increase because of the more cleaved combustion space (relatively higher portion of recesses), as well as increase of the combustion-chamber surface/combustion-chamber volume (surface-volume ratio). Hence, combustion chambers have to be designed to be as compact as possible. In addition, the exhaust gas temperature decreases with increasing compression ratio because of better efficiency, so that secondary reactions of unburned hydrocarbons and carbon monoxide in the exhaust are reduced. Increasing compression ratio normally would increase the capacity to achieve leaner mixtures, permitting more retarded ignition timing because of faster combustion. HC and NO_x emissions can be reduced through this. NO_x and particulates can be reduced by decreasing the compression ratio on the diesel engine.

In two-stroke engines with piston port control, it is necessary to distinguish between the geometrical compression ratio (ε) and the effective compression ratio (ε'). The effective compression will only start after the piston has shut the intake and exhaust ports. The effective compression ratio is calculated from

$$\varepsilon' = \frac{V'_h + V_c}{V_c},$$

where

$$V'_h = \frac{\pi d_K^2}{4} s'$$

V'_h = Residual volume above the ports

s' = Residual lift above the ports.

Fig. C222 represents the possible ranges of compression ratios for current engines.

Literature: [1] L. Bergsten: Saab Variable Compression SVC, MTZ 62 (2001) 6.

~**Effective compression ratio.** The effective compression ratio in gasoline engines decreases with advancing throttling at a constant geometrical compression ratio; this results in a decrease in efficiency. This decrease becomes even more drastic when a supercharged gasoline engine is considered. Considering higher knocking sensitivity near full load, the geometrical compression of a supercharged engine has to be reduced compared with that of a naturally aspirated engine. This causes a further decrease of efficiency during low part-load op-

Engine Type	ε	Limited by
IC engine (two-stroke)	7.5 to 10	Glow ignition
IC engine (two valves)	8 to 10	Knocking, glow ignition
IC engine (four valves)	9 to 11	Knocking, glow ignition
IC engine (direct injection)	11 to 14	Knocking, glow ignition
Diesel engine (chamber engine)	18 to 24	Efficiency loss at full load, component stress
Diesel engine (direct injection)	17 to 21	Efficiency loss at full load, component stress

Fig. C222 Compression ratios

eration. **Fig. C223** shows the effective compression ratio on a gasoline engine in the performance map.

~**Geometrical compression** →Compression ~Compression ratio

~**Isentropic compression** →Cycle

~**Isothermal compression** →Cycle

~**Variable compression.** Variable compression ratio is a way of increasing the efficiency of a gasoline engine, particularly since the engine's basic compression ratio is limited by the tendency of the gasoline fuel to knock at wide-open throttle. Increasing the compression ratio at part load will considerably improve the indicated efficiency. A 10% reduction in fuel consumption can be achieved in contrast to engines with fixed compression ratio in the range relevant to the CVS (constant volume sample) test. Even more important are improvements in the efficiency of supercharged engines with variable compression, since an additional gain through the shift in operating point would occur in this case. In a given case, the compression of a supercharged engine was raised to a compression ratio of

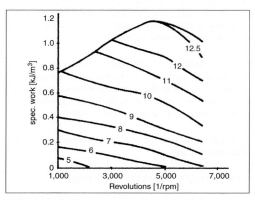

Fig. C223 Effective compression ratio on a gasoline engine at various operating points

$\varepsilon = 13.5$ at part load while the compression ratio was $\varepsilon = 8$ at full load. The gain in fuel consumption was more than 20% in the CVS test, with identical driving performance in this case.

Systems with variable compression ratios have not been accepted in series production because of their high expenditure. The following systems will be analyzed as examples:

- Pistons with variable compression height. The disadvantage is high piston mass.
- Combustion space enlargement or reduction by displacing—e.g., a cylinder in the cylinder head. The disadvantage is poorer combustion due to the cleft combustion chamber.
- Displacement of the crankshaft center line. This is very costly.
- Inclination of the cylinder head that is designed such that the joint between head and block is moved lower, that is, the block height is reduced in contrast to the conventional engine. This also is a very costly solution.

Compression chamber →Combustion chamber

Compression chamber/swept volume. Compression chamber or swept volume is the chamber above the piston and between the piston top land and the cylinder when it has reached upper dead center. It is thus characterized by the combustion chamber volume, which depends on the compression ratio. Compression chambers also can be found in engine components. A typical example is the compression chamber or the swept volume of an injection nozzle (blind-hole nozzle) for diesel engines, which affects the emissions behavior of the engine. The geometry of the compression chamber or swept volume is especially important in diesel engines because it negatively can influence the thermodynamic processes—it must, therefore, be minimized.

Compression (clearance) volume (*also*, →Compression pressure). The compression or clearance volume is the volume of the combustion chamber when the piston is at top dead center (TDC). It is also called the dead space or dead volume of the cylinder.

Compression end. The end of compression on a reciprocating piston engine is reached at the top dead center. Considering the sequence without ignition of the mixture, this would give the highest pressure and highest temperature at the end of compression.

Compression end pressure →Compression ~Variable compression

Compression height (CH) →Piston

Compression ignition →Combustion process, diesel engine

Compression pressure. The compression pressure is that in the combustion chamber when the piston is located at top dead center (TDC) of its compression stroke before ignition. The magnitude of the compression pressure is affected by the position of the throttle valve on gasoline engines and the geometric features of the combustion chamber, the compression ratio, and so forth, and depends on the type of combustion process. Its value is also dependent on the operating state of the engine—for example, component temperature and the status of the mixture or air in the combustion chamber. Generally, the curve of compression pressure against crank angle is approximately symmetrical about top dead center.

Compression ratio →Compression; →Fuel consumption ~Variables ~~Engine measures

Compression ring →Piston; →Piston ring ~Versions ~~Compression rings

Compression seal →Sealing systems

Compression temperature →Self-ignition engine

Compression volume →Injection valves ~Diesel engine ~~Hole-type nozzle

Compression wave →Engine acoustics ~Structure-borne sound

Compression work →Mean effective pressure; →Specific work; →Supercharging

Compressive creep strength →Sealing systems ~Special seals

Compressor air mass flow curve →Supercharging

Comprex →Supercharging ~Pressure wave supercharger

Comprex loader →Supercharging ~Pressure wave supercharger

Computation of evaporative emissions →Emission measurements ~Test type IV

Conditioning of the air-fuel mixture →Mixture formation ~Mixture formation, gasoline engine

Configuration of camshaft →Camshaft

Configuration of catalytic converter →Catalytic converter

Configuration of charge cycle valve →Valve arrangement

C

Configuration of cylinder →Engine concepts

Configuration of oil pump →Lubrication ~Oil pump

Configuration of spark plug →Spark plug ~Spark plug location

Conical springs →Valve spring ~Design types

Connecting rod (*also, see below,* ~Connecting rod ratio). The connecting rod (con-rod) is the connecting link between pistons and the crankshaft in a reciprocating piston engine. The connecting rod converts the oscillating motion of the piston into a rotating motion of the crankshaft. Weight and design of the connecting rod directly influence the performance and running smoothness of an engine. Therefore, a connecting rod with optimum weight is becoming increasingly important for engines with reduced emissions.

~Alternative materials. *See below,* ~Materials

~Bearing cap. The lower part of the connecting rod that is bolted with the shaft (connecting rod design).

~Bearing shell (*also,* →Bearings). The crankshaft performs a rotary motion inside the big end. To improve the sliding properties and to reduce wear, a partitioned bearing shell is inserted in the big end of the connecting rod prior to bolting it to the crankshaft. The bearing shell contains a guide element that is placed in a retaining groove to lock the bearing shell in position along the crankshaft axis. In automatic assembly, this guide element can be omitted.

~Bolt. (*also,* →Engine bolts). The shaft and cap are connected with connecting rod bolts. The inertial forces of the connecting rod shaft and piston as well as the transverse force, owing to the offset loading, act on the connecting rod bolt. The connecting rod bolt must prevent opening in the partitioned joint between the cap and shaft. During engine assembly, through tensile-strength-controlled tightening torque or by using the rotational angle method, a corresponding initial tension is applied in the bolt; its action opposes the effective inertial force.

Another function of the connecting rod bolt is centering the shaft and cap. Fitted bolts with collars in the partition plane are used for this purpose, and they prevent the shaft and cap from being displaced. When connecting rods are parted by means of cracking, the fitted bolts can be dispensed with because, in this case, the structured partition surface provides sufficient grip to counter relative motion between shaft and cap.

~Bolting (*also,* →Engine bolts). In order to fix it on the crankshaft, the connecting rod cap is bolted to the shaft by means of the connecting rod bolts. This bolted

connection must guarantee that the shaft and cap are exactly aligned with one another and secured against displacement. Several methods can achieve this:

- Guidance by means of fitted bolt or dowel
- Introduction of an indented joint in the partition plane
- Guidance through the fractured partition surface (cracking) (**Fig. C224**)

~Casting. *See below,* ~Semi-finished parts production

~Casting material. *See below,* ~Materials

~Connecting rod bearing pin. *See below,* ~Throw

~Connecting rod big end. *See below,* ~Design

~Connecting rod bolt →Engine bolts

~Connecting rod eye. *See below,* ~Design

~Connecting rod head. *See below,* ~Design

~Connecting rod motion diagram. (*also,* →Crankcase). The connecting rod motion diagram describes the space required for the connecting rod during one revolution of the crankshaft and is the locus of outer points on the connecting rod during motion (**Fig. C225**).

~Connecting rod ratio. The connecting rod ratio is a geometrical comparative size—formulated from the cranking radius, r, and the center distance, l (connecting rod length)—and is defined as

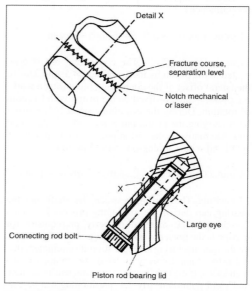

Fracture course, separation level

Notch mechanical or laser

Detail X

Large eye

Connecting rod bolt

Piston rod bearing lid

Fig. C224 Guidance in the fractured partition surface

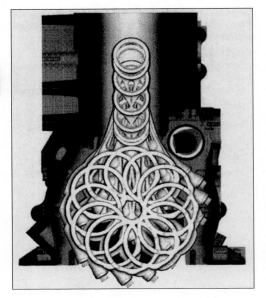

Fig. C225 Determining the connecting rod motion diagram

$$\lambda = \frac{r}{l}.$$

The connecting rod ratio normally lies between 0.23 and 0.33. With an increase in the connecting rod ratio, the lateral forces on the piston increase; with a decrease in the ratio, the design height of the engine increases. A compromise must be found between these two influences.

~Connecting rod shaft. *See below,* ~Design

~Connection to piston. The connecting rod is connected to the piston by means of the small end. Due to the lateral offset of the connecting rod during a working cycle, it must be fixed on the piston with freedom of rotation. During the machining process, a bearing bush carrying the piston pin is pressed into the small end. Alternatively, the piston pin is shrunk-fit into the small end with a shrunk-on connecting rod.

~Connection to the crankshaft. *See above,* ~Bolting

~Cracking. The parting between the shaft and the bearing cap during the machining process is termed "cracking." The material property prerequisite for parting is the presence of a coarse grain structure and, on the manufacturing side, partitioning equipment that exerts the required energy to "break" the component at a high speed. If the ratio between material tensile strength and yield point is nearly 2:1, partitioning can be carried out without large deformation. Raw parts in all

manufacturing processes (unfinished connecting rod production) can be partitioned by cracking.

Prior to the cracking process, notches or internal broaches are made with laser on the side surface of the big end to obtain a big notch effect at the desired partitioning plane (**Fig. C224**). The big end is placed on and clamped to a pair of the cracking punches. The cracking punch is spread at a high speed and the stress generated in the workpiece leads to a fracture that begins at the notches and then progresses radially outward. If the material and tools are well-matched, the roundness deviation after cracking is a maximum of 30 μm.

Above all, cracking is advantageous because it reduces the necessary machining steps. Conventional machining of the parted surface is dispensed with. The two halves can be accurately joined together after cracking and are protected against relative movement through their irregular fracture, so that no additional guiding elements are required. Another advantage is the application of a simplified connecting rod bolt, because the bolt must not fulfill the function of an assembly guide.

As such, fracture-partitioned connecting rods are cost-effective and technically a high-quality alternative to the conventionally partitioned connecting rods.

~Crankshaft hole. *See below,* ~Design

~Design. The connecting rod comprises a small end (or piston pin bore) for connecting the rod to the piston (link to the piston) and a big end (or crankshaft bore) for fastening the connecting rod to the crankshaft (link to the crankshaft) (**Figs. C226 and C227**). Because the connecting rod bores also assume the bearing function, they must have a stable design. To reduce the mass of the connecting rod, the small end can be made with a trapezoidal form upwards. Such a trapezoidal bore allows optimized design of the piston with regard to stress. Another variant is the stepped connecting rod (**Fig. C228**). Here, the load distribution between piston and connecting rod is further improved.

For the assembly of the connecting rod on the crankshaft, the big end is partitioned and held together with two bolts (connecting rod bolts); for commercial vehicle connecting rods, four screws are used in individual cases. The partition of the big end is normally perpendicular to the longitudinal axis of the connecting rod. Alternatively, to reduce the largest width of the connecting rod, the big end can also be partitioned in a skew plane.

This allows the diameter of the big end to be enlarged while the cylinder diameter remains unchanged. The disadvantage of a skew-partitioned connecting rod is that the blind hole for the bolt runs along the area with the greatest stress (connecting rod stress), and a high transverse force must be borne on the partitioned surface. Application areas for a skew-partitioned connecting rod are primarily in diesel engines that have a large connecting rod journal due to the load.

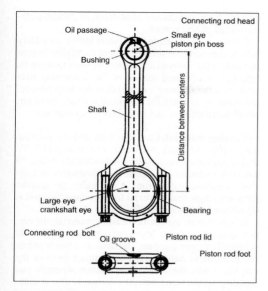

Fig. C226 Geometry of an evenly partitioned connecting rod

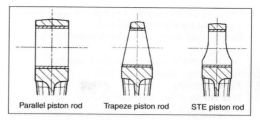

Fig. C228 Design of the connecting rod small end

The big and small ends are connected by the connecting rod shaft, which is designed as an "I" cross section. This lends itself to for the requirement of low weight while the section modulus (load) remains high.

Based on the position of the connecting rod inside the engine, the upper (piston side) part is termed "connecting rod head" and the lower (crankshaft side) part is termed "connecting rod foot." Elementary static equations are available for rough calculation of the re-

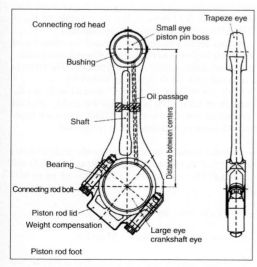

Fig. C227 Geometry of a skew-partitioned connecting rod

quired cross sections. Exact dimensional calculations of the connecting rod are today performed using FEM analysis. This allows design optimization and calculation of the deformation under dynamic loads; possible opening of the bolted connection is verified.

~Distance between centers. The distance between hole middle-points from the big and small connecting rod ends.

~Drop forging. *See below,* ~Semi-finished parts production

~Eye. *See above,* ~Design

~Force. *See below,* ~Stress

~Forged steel. *See below,* ~Materials

~Geometry. *See above,* ~Design

~Installation. To install the connecting rod in the engine, first the connecting rod head is fixed on the piston, and then this unit is inserted from the top through the cylinder onto the crankshaft. Then the two bearing half shells are inserted in the big end, and the connecting rod shaft and cap are bolted on the crankshaft.

~Lubrication. The connecting rod must guarantee sufficient sliding properties of the bearings in the small and big ends.

Grooves that facilitate oil supply can be provided in the face surface for lubricating the big end (**Fig. C226**). The piston pin bearing can be lubricated through a hole in the longitudinal axis of the shaft, through which the oil is fed from the big end (**Fig. C227**). As an alternative to the longitudinal hole in the shaft, one or several holes are placed on the piston side of the small end (**Fig. C226**).

~Machining. Raw parts (unfinished connecting rod during production) are machined to their finished size. In large series production, this is done on fully automatic machining lines. For smaller series production, machining centers with lower-level automation are used. The machining steps for partitioned connecting rods (cracking technique) are outlined in the following section:

- Grinding the face surfaces of big and small ends
- Preboring the big and small ends
- Drilling and screw-cutting of threaded holes
- Parting
- Screwing the bearing cap and rod together
- If necessary—inserting the bush
- Finish grinding
- Fine boring of big and small ends.

Subsequent to the machining process, the finished parts are weighed and assigned to classes. The connecting rods in a given weight class are then installed in an engine (selective assembly). If the raw part is already manufactured with a closer weight tolerance, classification into classes can be dispensed with. Alternatively, the nominal weight can be attained by milling off a portion to compensate for the weight difference.

~Manufacturing. See above, ~Machining; *see below,* ~Semi-finished parts production

~Mass. The inertial forces incurred during the acceleration-retardation motion within a working cycle of the reciprocating piston engine are influenced by the mass of the piston and the connecting rod. The mass of the connecting rod is divided into a rotational and an oscillatory mass component for simplified calculation of the forces while the overall mass and center of gravity of the rod are retained. The mass concentrated in the big end executes a rotary motion while the mass concentrated in the small end executes an oscillatory motion.

To calculate the mass components, first the center of gravity of the connecting rod is determined (s is the distance of the center of gravity from the big end). The mass component of the small end is derived from

$$m_{conrod,sm,end} = m_{conrod,total} \frac{s}{l},$$

with l as connecting rod length or middle distance between the connecting rod eyes. The difference from the total weight gives the mass of the big end.

The oscillating mass of the connecting rod (and piston with pin and rings) directly influences the running smoothness of the engine. Therefore, an effort is made to reduce the oscillating mass component. By making the small end a trapezoid or stepped connecting rod (connecting rod design), the oscillating masses of the small end, of the piston pin, and of the piston can be reduced.

~Mass distribution. See above, ~Mass

~Materials. Based on the application and occurring loads, different materials are applied.

~~Alternative materials. In addition to the following materials for connecting rods in batch production, alternative materials are being pursued to reduce the weight of the connecting rod while retaining the mechanical strength. To achieve this, carbon fiber-reinforced aluminum, spray-compacted aluminum, or carbon fiber-reinforced plastic is used.

Connecting rods made of titanium alloys are widely used in the racing sector; because of the higher fatigue strength of this alloy material, weight reduction can be achieved. A common feature of the connecting rods made of these alternative materials is that high production costs are justified for individual engines; however, broad distribution in serial engines is not common.

~~Casting material. Cast iron with nodular graphite (GGG-70) and black malleable cast iron (GTS-70) find application as connecting rod materials. GGG-70 has both technical and economic advantages in contrast to malleable cast iron: in particular, the specific fatigue strength that is crucial for the connecting rod is significantly higher in GGG-70.

GGG-70 is an iron-carbon casting material; the carbon content more or less exists in the structure as spherically shaped graphite. The basic structure is primarily pearlitic in nature, and the compact form of the graphite gives the material maximum strength and ductility. At the same time, the carbon is also responsible for the good casting properties. The required structural state is generated directly in the mold without additional thermal treatment.

Malleable cast iron is likewise an iron-carbon casting material. Its structure is established, however, only through thermal treatment downstream of the casting plant. In the process, the carbon is finally deposited in the form of flakes in the basic structure.

~~Forged steel. Most connecting rods are manufactured from steel using the drop-forging method. In this case, micro-alloy steels such as 27 MnVS6 BY or carbon-manganese steels such as C40 mod BY are used. For fracture-separated, drop-forged connecting rods (cracking technique), steel with high carbon content (C70 S6 BY) is used. Tensile strengths of $R_m = 1000$ MPa are attained with these materials.

Recent developments have pursued the objective of achieving higher material fatigue strength and guaranteeing suitability for the cracking process. The result is a material with the designation 36MnVS4.

With 34CrNiMo6 V, a steel material with tensile strength of 1200 MPa is available for highly stressed connecting rods. In this case, additional thermal treatment (hardening and tempering) is required.

~~Powder metal. For connecting rods made of powder metal, materials such as Sint F30 and Sint F31 are available. They attain tensile strengths of up to 900 MPa.

~Piston pin hole. See above, ~Design

~Powder metal. See above, ~Materials

~Semi-finished parts production. The semi-finished part can be produced in different ways, based on the particular application.

~~**Casting.** The starting point for production of semi-finished parts is a model made of plastic or metal, comprising two halves, which when combined depict a positive image of the connecting rod. Several identical model halves are combined on a model plate and connected with the model for the casting and gating system. In a multiple production process, the two model plates are molded by compacting the green sand. The resulting sand molds each depict a negative image of the corresponding model plate. When correctly placed over one another, they form a cavity in the shape of the connecting rod to be produced. This is then filled with liquid cast iron, which is smelted in a cupola or electric furnace with scrap steel as the charge. The metal solidifies slowly in the mold.

~~**Drop forging.** Source material for production of semi-finished parts is bar steel with a round or square cross section, which is heated to a temperature between 1250 and 1300°C. In a stretch-rolling process, the mass is first predistributed toward the big and small connecting rod eyes. As an alternative to stretch rolling, cotter rolling can be applied, whereby the accuracy of the preforming geometry can be improved.

The main forming occurs in a press or a hammer aggregate. The excess material flows into a fin that is removed in a subsequent operation. During the deburring process, the big end—and with commercial vehicle connecting rods also the small end—is concurrently perforated.

To attain the required structural properties based on the steel alloy, the connecting rod is cooled in a controlled airstream (BY = best yield cooling) or a separate heat treatment is carried out.

Finally, the scales on the raw part are removed by means of shot blasting, whereby residual compressive stresses of 300 MPa are generated in the near-surface area.

In most cases, the connecting rod and cap are forged together and separated during the machining process. Based on the connecting rod and plant size, double pieces—i.e., two connecting rods—are drop-forged concurrently, which leads to higher productivity from the plant.

~~**Sintering.** This production process begins with the servo-hydraulic pressing of alloyed powder to form a "green" pellet. Subsequent weighing ensures that the green pellet achieves a close weight tolerance of about 0.5%. The sintering process takes place in an electrically heated, continuous furnace in which the parts remain for about 15 minutes at about 1120°C. In the subsequent forging process, only the height reduction of the component is carried out up to the theoretical limit in order to increase the density of the component part. Finally, a state of residual compressive stress is established in the surface by means of shot blasting.

The forging process is costly with this production method. A new powder technology is in development, aimed at dispensing with this residual compressive stress state.

~Shaft. *See above*, ~Design

~Sintering. *See above*, ~Semi-finished parts production

~Skew-partitioned connecting rod. *See below*, ~Design

~Stress. The connecting rod is subjected to stress by gas forces inside the cylinder and the inertial forces of moving masses. The gas force acting on the connecting rod through the piston leads to compressive stress inside the connecting rod shaft. Bending stress occurring due to the lateral offset in the oscillation plane of the connecting rod can be neglected.

The acceleration-retardation cycle of the connecting rod and piston mass leads to tensile stress in the shaft; it is also transferred from the shaft to the big end. The connecting rod, as such, is subjected to alternating tensile and compressive stresses: the magnitude of the compressive force normally exceeds that of the tensile force. Due to this, protection against buckling must also be considered when dimensioning the connecting rod.

~Weight balance. To attain the nominal weight of a machined connecting rod, bosses can be provided on the raw small and/or big end of the component; these are milled during the mechanical machining process until the nominal weight is accurately attained. In modern manufacturing processes, the production parameters can be monitored accurately, so that raw parts can be fabricated with an adequate weight tolerance.

Connecting rod bearing pin →Crankshaft ~Throw

Connecting rod bearing shell →Bearings ~Bearing positions ~~Connecting rod bearings; →Connecting rod ~Bearing shell

Connecting rod bolt →Connecting rod ~Connection to piston; →Engine bolts ~Threaded connections

Connecting rod bushing →Bearings ~Bearing positions

Connecting rod casting →Connecting rod ~Semi-finished parts production

Connecting rod connection →Connecting rod ~Connection to piston; →Engine bolts ~Threaded connections

Connecting rod drop-forging →Connecting rod ~Semi-finished parts production

Connecting rod load →Connecting rod ~Stress

Connecting rod manufacture →Connecting rod ~Machining, ~Semi-finished parts production

C

Connecting rod motion diagram →Connecting rod; →Crankcase ~Connecting rod motion diagram

Connecting rod sintering →Connecting rod ~Semi-finished parts production

Connection to crankshaft →Engine bolts ~Threaded connections ~~ Connecting rod bolts

Connection to piston →Connecting rod

Constant bevel oil control ring →Piston ring ~Versions ~~Oil scraper rings ~~~Double beveled rings and constant beveled rings

Constant pressure carburetor →Carburetor

Constant pressure combustion (*also*, →Cycle). Constant pressure combustion is a theoretical ideal process that assumes that the pressure in the combustion chamber remains constant during the whole of the combustion process. This means the pressure reduction during expansion in the engine is compensated for by the converted chemical energy; this assumption is made in ideal cycles. This is called maximum pressure restriction.

Constant pressure cycle →Cycle

Constant volume combustion (*also*, →Cycle). The term constant volume combustion is used in connection with the ideal cycle for gasoline engines (the Otto cycle). With constant volume combustion the volume in the combustion chamber theoretically does not change during the combustion process. This would mean for the gasoline engine that the combustion would have to be completed in an infinitely short time, because the volume changes in the combustion chamber because of the piston movement. Constant volume combustion gives a higher theoretical efficiency than constant pressure combustion, which means that it is advantageous for reasons of thermal dynamics, to get as close as possible to constant volume combustion.

Constant volume cycle →Cycle

Consumption →Fuel consumption

Consumption calculation →Fuel consumption

Consumption concept →Supercharging

Contact breaker →Ignition system, gasoline engine ~Ignition timing sensor

Contact controlled coil-type ignition →Ignition system, gasoline engine

Contact pressure →Piston ring

Contact surface →Cylinder running surface

Contact surface hardening →Piston ring

Contact system →Electrical/mechanical engine and transmission control ~Requirements for mechanical systems and housing concepts ~~Plug connector

Continuous combustion process →Combustion process

Continuously variable length actuation →Intake system ~Intake manifold ~~Intake manifold design

Continuously variable stroke →Variable valve control ~ Variation parameters

Continuously variable valve actuation →Valve actuation

Contour, cylinder running surface →Cylinder running surface

Contour turning →Piston ring ~Shaping

Contra-rotating cylinder engines →Engine concepts ~Reciprocating piston engines ~~Single-shaft engines ~~~Radial engines

Control and safety components →Lubrication

Control maps →Injection system (components) ~Diesel engine ~~Mechanical/control

Control of throttle valve position →Actuators ~E-gas

Control piston →Piston ~Design

Control ports →Two-stroke engine ~Scavenging process

Control sleeve injection pump →Injection system, fuel ~Diesel engine ~~Serial injection pump

Control unit with inverter →Starter ~Starter-generators ~~Integrated starter-generator ISG (main drive)

Control units. In order to perform the charge exchange process, the inlet and exhaust openings of the engine and/or the working chambers inlet and exhaust openings must be periodically opened and/or closed by control units. The control units must fulfill certain tasks. The most important of these include:

- Availability of large opening cross-sectional areas
- Rapid opening and closing
- Low pressure loss during airflow
- Gas-tight closure

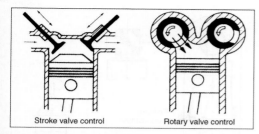

Fig. C229 Stroke (poppet) valve and rotary valve (Source: Pischinger)

In practice, three construction solutions have proven themselves for these tasks:

- Stroke (poppet) valves (**Fig. C229**)
- Rotary valves only in special designs (**Fig. C229**)
- Pistons as control unit for rotary pistons and/or two-stroke engines (**Fig. C230**)

Control/gas transfer (*also*, →Restrictor; →Variable valve control). The control enables the charge exchange of the engine—that is, filling of the working chamber with fresh gas and emptying of exhaust gas. The quality of the charge exchange has a major impact on the development of power by the engine. The intermittent operating mode of the piston engine demands intermittent control of fresh gas and exhaust gas flows. This can be achieved through:

- Oscillating poppet valves (movement of the shut-off unit in flow direction)
- Oscillating, vibrating, or rotating valves (shut-off unit will be moved in flow direction)

The gas transfer of two-stroke engines is partly or completely controlled by the working piston acting as a valve. Important for the charge exchange are the time-area integrals of the cross sections—this is the integral of the opened area and opening time.

~**Four-stroke engine.** The gas transfer in four-stroke engines is almost exclusively controlled with valves. Depending on the design, preferably two, three, four, and five valves per cylinder are used.

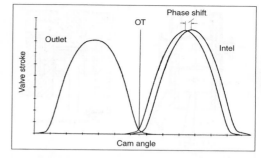

Fig. C231 Change of control time through phasing of the intake cams

~~**Open period.** The valve open period is the time in degrees crank angle or cam angle, during which the inflow and/or outflow of air/fresh mixture and/or exhaust gas from the cylinder takes place. The open period depends on the design—the open period is about 260°CA for inlet valves and about 220°CA for exhaust valves.

~~**Phasing.** Phasing is the offset and/or the valve spread angle of intake and exhaust cams. In case of the variable valve control, the phasing can be changed by rotating the intake cam relative to the crankshaft position depending on the engine speed (**Fig. C231**). This results in changed control times for the inlet valve and the valve overlap, which influence the torque curve, the fuel consumption, and the emission behavior.

~~**Slide control.** The first combustion engines were, based on steam engine technology, slide-controlled; then valves replaced slides (**Figs. C232–C234**). Because valve controls are noisy and the valves are sensitive to overheating and because hot exhaust valves resulted in spontaneous ignition and knocking in gasoline engines, different and better solutions were sought.

The double-slide control developed by Knight was used in Germany by Mercedes, in Belgium by Minerva, and in the United States by Willys before World War I; they ran very smoothly. Two concentric pipe or bushing-shaped sliders slid against each other in the

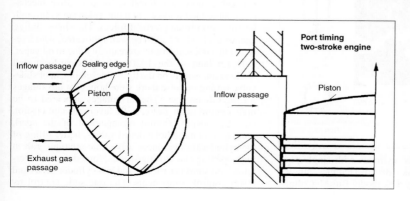

Fig. C230
Pistons as control unit
in rotary valves and
two-stroke engines

C

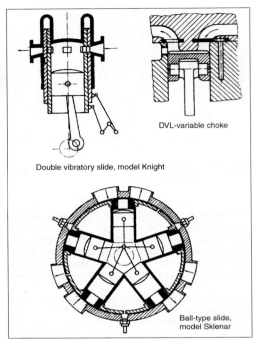

Double vibratory slide, model Knight

DVL-variable choke

Ball-type slide, model Sklenar

Fig. C232 Slide control double vibration slide, model Knight; DVL constant depression slide, ball-type slide, model Sklenar (Source: Dipl.-Arb. A. Kupfers)

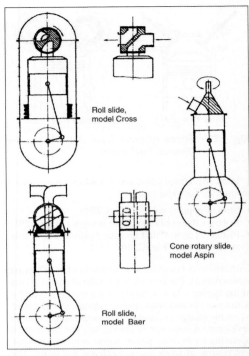

Roll slide, model Cross

Cone rotary slide, model Aspin

Roll slide, model Baer

Fig. C234 Slide control, centric slide, model Cross; cone-rotary valve, model Aspin; centric slide model Baer (Source: Dipl.-Arb. A. Kupfers)

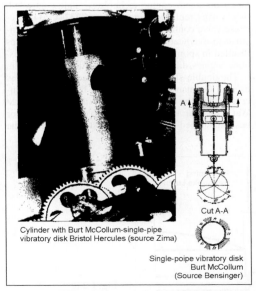

Cylinder with Burt McCollum-single-pipe vibratory disk Bristol Hercules (source Zima)

Cut A-A

Single-poipe vibratory disk Burt McCollum (Source Bensinger)

Fig. C233 Slide controls: cylinder with single-pipe vibratory disk, Bristol-Hercules single-pipe vibratory disk model BurtMcCollum

cylinder. Together with the piston in the cylinder head, the internal slide limits the combustion chamber. The slides are actuated via a crankshaft-shaped actuating shaft that rotates at half the speed of the crankshaft. Because this shaft is arranged parallel to the crankshaft axis, the sliders execute a pure stroke movement. If, at counter-rotating stroke movement, both slides cover each other, the gas routes for the inlet and/or the exhaust are opened. With further development of the engine, the disadvantages of the Knight slide control became more noticeable: poor heat transfer, complex drive mechanism, large inertia forces, and one-sided stress through lateral piston force (vertical force). The slides sliding against each other had to be well lubricated, which resulted in a reduction of mixture quality due to oil vapor.

In England in the 1920s, Harry Ricardo started working intensively on the further development of slide-valve timing because there were hopes for advantages for aircraft engines. The advantages appeared to be high compression ratios with favorable combustion chamber shapes, central arrangement of the spark plug, and no more hot exhaust valves; better filling of the cylinder through large valve time-area integrals at high opening and closing speeds, forced control operation, and short gas routes; and low cylinder heads and fewer control components. The one-slide control by

Timing in °KW	Ford 2.0 L	Porsche 911 Turbo	Audi V8, 4.2 L	VW V6
Inlet opens (IO)	4° b. OT	10° b. OT	25° a. OT	26° a. OT
Inlet closes (IC)	50° a. UT	20° a. UT	45° a. UT	1° b. UT–56° n. UT
Outlet opens (OO)	35° b. OT	41° b. UT	38° b. UT	29° b. UT–56° v. UT
Outlet closes (OC)	4° a. UT	9° b. OT	8° b. OT	4° b. OT–26° v. OT

Fig. C235 Gasoline engine timing

C

BurtMcCollum was preferred at that time. In order to get the control sequence necessary for the four-stroke procedure, the slide must make a sliding and rotating movement. The bushing-shaped slide slides between the cylinder and the piston. A crank vertical to the cylinder axis is engaged eccentrically on the slide circumference, so that the slide executes an elliptic curve. The vertical stroke determined the height, the rotation, the width, and/or the number of slots. When the slots in the slide cover the gas passages, the gas routes are cleared. The development of the BurtMcCollum control was costly but proved itself in English aircraft engines before and during World War II and in air-cooled radial engines and double radial engines of the Bristol Aeroplane Co. (Perseus, Aquila, Taurus, Centaurus, Hercules), as well as in water-cooled H engines by Napier (Sabre). Rolls-Royce developed slide-valve engines as well. In the Nordatlas transport planes, slide-controlled Bristol engines were used until the early 1970s.

In addition to the bushing-shaped vibrating slides, cylindrical (Cross, Baer), ring-shaped (Sklenar), and cone-shaped rotary valves (Aspin) were designed and built. The Bristol swan-plate engine had a ring-slide control. During World War II the Deutsche Versuchsanstalt für Luftfahrt (DVL) developed a constant-depression control for airplane engines ready for production. Irrespective of kinematic and aerodynamic advantages, the slide control could not replace the poppet valve. Poppet valve control is—despite the many single components—simpler and less problematic than slide control.

~~Spread angle. The spread angle is the distance in degrees CA of the maximum valve lift from top dead center for inlet and exhaust valve.

~~Timing. The valve timing of the engine defines the beginning and the end of the valve opening times, as well as the beginning and end of inlet and exhaust processes. Engine timing very much depends on the engine concept, especially the cylinder head design. The number of valves, the valve angle, the valve diameter, the valve lift, and the length of the intake manifold are only some of the parameters influencing the valve timing. In addition, operational parameters, such as ignition timing and engine speed, play an important role.

The determination of the timing is based on the following criteria:

- Torque
- Power/performance
- Fuel consumption

- Exhaust emissions
- Idle quality

Fig. C235 shows some gasoline engine timing examples.

~~~Exhaust closes. →Charge transfer

~~~Exhaust opens. →Charge transfer

~~~Inlet closes. (*also, see above,* ~Four-stroke engine ~~Open period). The inlet closing angle is more important than the outlet opening angle for the operation of the engine. It must be designed such that the charging with fresh mixture and/or air can reach a maximum—that is, the delivery ratio is as high as possible.

There are two settings:

~~~~Early inlet valve closure. Early inlet closing generally results in the volumetric efficiency increasing at low speeds and, thus, the engine torque decreasing at high speeds (**Fig. C236**).

~~~~Late inlet valve closure. Late inlet closing relates to the momentum of the inflowing air maintaining inflow to the cylinder even when the induction effect of the piston no longer exists (in the BDC range). If the inlet valve closure is too late, an undesired backflow of the fresh mixture and/or the air results, which leads to a lowering of the volumetric efficiency.

~~~Inlet opens. *See above,* ~Four-stroke engine ~~Open Period

~~Variable valve timing (*also,* →Variable valve control). Variable valve timing is the change of the start of

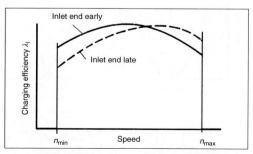

Fig. C236 Influence of the inlet closure on the volumetric efficiency

	Early Position	Late Position
Inlet opens	26° b. OT	26° a. OT
Inlet closes	179°–184° a. OT	231°–236° a. OT
Outlet open	231°–236° b. OT	209°–214° b. OT
Outlet closes	26° b. OT	4° b. OT

Fig. C237 Timing in °CA with variable adjustment

the opening or closing of the valves while the engine is operating. By expanding the degree of freedom of the valve control, clear improvements in the engine's thermodynamics are possible. These are:

- Increase of volumetric efficiency
- De-throttling
- Internal exhaust-gas recirculation
- Generation of charge movement

An example for the engine timing with variable adjustment can be seen in **Fig. C237**.

~~~**Valve clearance** → Valve; → Valve clearance

~~~**Valve lift** (*also*, → Valve lift curve). Valve lifts of current passenger car series production engines are at about 10 mm. With fully variable valve control, valve

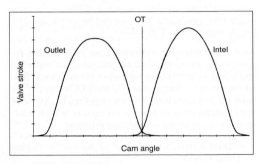

Fig. C238 Valve lift curve

lift can be varied between zero and the maximum valve lift. Exhaust valves have a smaller lift than inlet valves.

~~~**Valve lift curve.** The valve lift curve defines the valve lift as a function of crank and/or cam angle (**Fig. C238**). In general, the valve lift of the exhaust valve is smaller than that of the inlet valve.

~~~**Valve overlap.** Valve overlap is the time in °CA during which the inlet and the exhaust valves are open simultaneously (**Fig. C239**). In case of naturally aspirated engines, a valve overlap of 40–60°CA can be achieved, which guarantees the scavenging of the residual gas. In case of highly charged diesel engines, in which the intake manifold pressure is higher than the exhaust gas pressure, valve overlaps of up to 120°CA are possible. The size of the valve overlap depends on the torque curve, the fuel consumption, the exhaust pressure, and the idle quality. Often, the valve overlap helps to internally return the exhaust gas, which, in contrast to the external exhaust gas recirculation, cannot be controlled exactly. Moreover, the valve overlap influences the volumetric efficiency of an engine and defines the standard for the fresh charge remaining in the cylinder after the charge exchange is completed. In diesel engines, the valve overlap can also be used to cool thermally stressed engine components, especially in connection with supercharging.

~~~**Valve time-area integral.** Area A in **Fig. C240** is characteristic for the opening behavior of a valve: The larger that area A is, the greater will be the thermal stress on the valve; hence, slow valve openings should be avoided.

The areas B and D should be as large as possible— B can be smaller than D because the pressure differential between the combustion chamber pressure and exhaust is larger than that on the intake side.

Area C, the area of valve overlap, should not be chosen too large for the sake of smooth engine running at low speeds and/or at idle speed.

The torque parameters of an engine can be influenced by the size of area E. A large area E results in

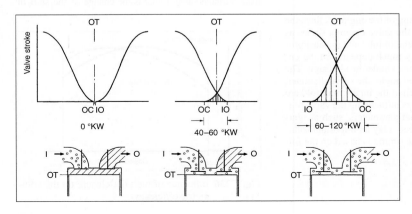

Fig. C239
Valve overlap (Source: Mollenhauer)

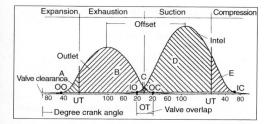

Fig. C240 Valve time-area diagrams, showing the time-area integrals

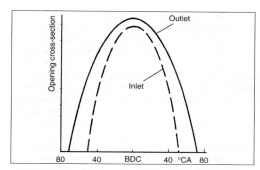

Fig. C242 Opening cross section for a two-stroke engine (Source: Pischinger)

good performance at high speeds but decreases the torque at low speeds, and vice versa.

~~~**Valve timing.** *See above*, ~Four-stroke engine ~~Charge transfer

~~**Valve cross-sectional area.** Not only is the open period decisive for the gas transfer, but also the valve cross-sectional area is. This is influenced by the cam contour, and the goal is to have steep valve lifting curves to improve filling, enabling fast opening to the maximum valve lift. This results in a fairly large time-area integral for the valve area. In contrast, the associated accelerations of the valve and the resulting stresses are high. Valve cross-sectional areas for different valve configurations are shown in **Fig. C241**.

~Two-stroke engine (*also*, →Two-stroke engine). The usual control of the gas transfer is done through slots in the cylinder of the two-stroke engine. This has the advantage that the complete drive to control the valves is eliminated. Lately, concepts have been created incorporating a valve control for two-stroke engines as well.

~~**Opening cross section.** The opening cross section for a two-stroke engine can be seen in **Fig. C242**.

~~**Piston port control** →Two-stroke engine

~~**Timing diagram.** The timing diagram depends on the scavenging process of the two-stroke engine. In **Fig. C243**, an asymmetrical timing diagram can be seen for a two-stroke engine.

~~**Valve control.** The advantages of control of the gas transfer process through valves—that is, with respect to the variability of the timing, exact separation of fresh mixture, and exhaust gas—have led to the use of gas transfer control for two-stroke engines. A combination of slot and valve control is possible—for example, slot control of the exhaust area and valve control for the inlet area.

~Valve control. Today, the charge exchange in four-stroke engines is exclusively controlled by poppet valves or mushroom valves; it is partly controlled by poppet valves or mushroom valves in two-stroke engines. Poppet valves—mechanically and thermally unfavorably stressed, with aerodynamic and kinematic disadvantages—have the advantage of being able to "help themselves." The arrangement of the valves in the cylinder has a mutually supportive effect and helps them to better fulfill their functions. The pressure against

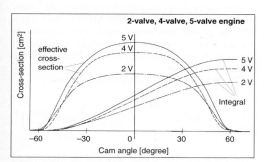

Fig. C241 Opening cross section for different valve configurations

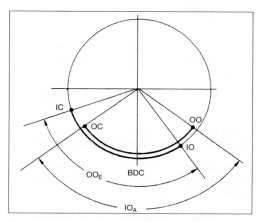

Fig. C243 Timing diagram for a two-stroke engine

which the valves must seal contributes to the sealing effect by pushing the valves deeper into their seat.

Poppet valves can be arranged horizontally or vertically (straight or inclined) in the combustion chamber, and they can influence the operating behavior of the engine. Vertical valves, directly driven by the camshaft(s), make control and construction easy but have the disadvantage—because they are located next to the cylinder—of requiring "sharp," flat combustion chambers. Gasoline engines with such valve arrangements have a tendency to knock. Valves arranged in a horizontal position lead to compact combustion chambers but result in a complicated construction with difficulties of control. Today, inclined or vertical valves are preferred. Large two-stroke engines with uniflow scavenging have a large central discharge valve; fast running two-stroke engines have up to four discharge valves. Four-stroke engines must control the inlet and the exhaust using valves, and, therefore, they must have at least one inlet and one exhaust valve. In the past, there were engines with a single joint inlet and exhaust valve. Larger engines, however, are built with two inlet and exhaust valves each (special engines have even three each). Passenger car engines have two inlet and one or two exhaust valves. The valves in passenger car engines are either directly actuated (bucket tappets) by the camshaft in the cylinder head or via rocker arms. For commercial vehicle engines, as is the case for all larger engines, the valves are actuated via pushrods and rocker arms. The speeds are lower, and in single-cylinder heads, the camshafts are mounted in a complex construction and are difficult operationally; the camshaft also must be disassembled for piston pulling. The valves are partially force-actuated; their dynamic equation is determined by the cam shape. Today, considering the mechanical stress of the control, smooth cams are preferred.

In the early days of engine building, automatic valves were used to control the inlet. These were opened and/or closed by the pressure differential between the gas in the cylinder and the atmospheric pressure. With increasing engine speed and—in the case of airplane engines with higher altitudes—this design proved to be less and less suitable, and the change was made to force-controlled valves.

Controlled glow system →Ignition system, diesel engine/preheat system ~Cold starting aid ~~Glow system

Controlled nonreturn valve →Air cycling valve ~Design

Controlled three-way catalytic converter →Catalytic converter ~Three-way catalytic converter

Controller Area Network →Electronic/mechanical engine and transmission control ~Electronic components ~~CAN interface

Conventional crankcase breather →Crankcase ~Crankcase ventilation

Conventional oils →Oil

Conversion →Catalytic converter

Conversion behavior →Catalytic converter

Conversion process →Catalytic converter

Conversion ratio →Catalytic converter ~Conversion efficiency

Converter →Catalytic converter

Convexity →Piston ring ~Running surface shapes

Coolant (*also*, →Radiator). The coolant is a liquid that circulates in the engine cooling system to absorb excess heat from the engine and keep the engine at the temperature level required for optimum efficiency. The coolant usually consists of 30–50% coolant concentrate and water softened in an ion exchanger or potable water.

~**Antifoam additive.** An antifoam additive is a component in the coolant concentrate to prevent foaming when the coolant is filled into and circulated in the engine cooling system. Fatty alcohols, fatty alcohol alkoxylates, polyglycols, and silicones are used individually or in mixtures in concentrations of 50–200 mg/kg in the coolant concentrate. The hydrophobic properties of the antifoam agent suppress formation of foam in the aqueous coolant. The foam formation can be tested according to ASTM Standard D 1881.

~**Antifreeze** (*also, see below,* ~Coolant concentrate). At temperatures below 0°C, it is necessary to protect the coolant against freezing. If the coolant freezes, it expands, resulting in an impermissibly large increase in the pressure. This can lead to formation of fissures (cracks) in the cylinder block, the cylinder head, the radiator, and other parts of the cooling system. The coolant concentrate contains primarily glycols, which provide antifreeze protection when present in sufficient quantities in the coolant.

~**Bitter compound.** The bitter compound is a constituent in the coolant concentrate that makes the otherwise pleasantly sweet taste of the MEG-based (monoethylene glycol) concentrate bitter. It is intended to prevent oral consumption by humans. Denatonium benzoate is used in the range of 25 to 70 mg/kg in the coolant concentrate.

Chemical structure of denatonium benzoate:

~**Boiling point.** The boiling point of an MEG-based coolant concentrate is approximately 170°C measured at normal pressure. The boiling point of the coolant containing 40% coolant concentrate in water is about 107°C (**Fig. C244**).

~Cavitation/erosive corrosion. Cavitation is a physical phenomenon that occurs in the cooling circuit at speeds that induce turbulent flow and high-pressure differences—these conditions lead to imploding vapor bubbles. When the implosion occurs on a metal surface, the pressure waves cause damage to the material. Conditions for cavitation can occur on the water pump impeller, in the coolant passages, in the cylinder block or head, and so on, or when the wet cylinder liners in diesel engines are subject to strong vibration.

~Coolant concentrate. The coolant concentrate consists of approximately 93% monoethylene glycol by weight and 5–7% corrosion inhibitors by weight. Small percentages of other additives such as silicate stabilizers, sequestering agents for taking calcium and magnesium ions out of hard water, antifoam additives, denaturants and dyes are also used. In some cases, monopropylene glycol is also used as one of the main constituents. Diethylene glycol (DEG) is also suitable for use as one of the primary components as a matter of principle; however, it is less effective in reducing the freezing point. Moreover, its availability and price limit its use to exceptional cases and to maximum concentrations of less than 20% in the coolant concentrate.

The coolant can be mixed ready for use with water— preferably softened water or potable water. The coolant concentrate protects the engine cooling system; this is achieved by corrosion and antifreeze protection. The coolant concentrate also increases the boiling point of the coolant and thereby increases the efficiency of the engine at the higher operating temperatures. A 40% solution of coolant concentrate in water should provide sufficient protection for the cooling system under most conditions.

~Coolant pump (*also*, →Coolant pump). The coolant flow through the engine cooling system is controlled primarily by the operation of the coolant pump. When the pump is driven by the engine, the flow rate is lowered at low engine speed and low heat generation and is higher at high engine speed when larger quantities of heat are developed. The flow rate can vary within this range from a few liters per minute up to more than 300 L/min. The coolant pump can also be driven according to map control and independent of the engine speed with the aid of an electric motor to reduce the effect its power requirement has on the fuel consumption.

The coolant pump usually consists of a cast aluminum housing and a pump impeller also made of an aluminum alloy, gray cast iron, steel, or plastic.

~Corrosion. Corrosion is damage to metal surfaces in the cooling system in contact with the coolant, usually caused by electromechanical reactions. Corrosion occurs in various forms, such as galvanic, hole, crevice, heat transfer, cavitation/erosion, stress, and cracking.

~Corrosion inhibitors. Corrosion inhibitors are substances that prevent the corrosion process. Corrosion inhibitors can protect the metal surfaces against corrosion by forming a protective film—consisting, for example, of a metal oxide or a metal phosphate. The active agents used are carefully matched to one another in the coolant concentrate, because different, partially mutually opposing effects or individual metals are present (corrosion inhibitor system). Examples of corrosive substances and corrosion inhibitors, including their mutually opposing effects, as well as a number of structural formulas are shown in **Figs. C245 and C246**.

~Corrosion inhibitor system. The corrosion inhibitor system is a combination of a number of corrosion inhibitors formulated to protect all metals present in the engine cooling system against corrosion. The individual inhibitors may protect one of the metals while having aggressive effects on other metals. Moreover, the concentration of the inhibitors is important.

Excessive concentration can be just as disadvantageous as when the concentration is too low. It is also

C

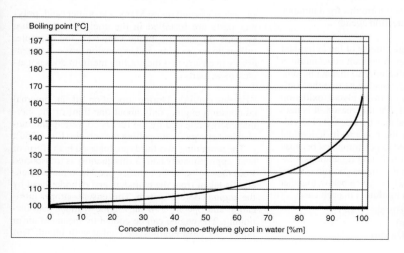

Fig. C244
Boiling points of monoethylene glycol/ water mixtures

Metal	Corrosion from	Corrosion protection by
Copper	Free amines, sodium nitrate	Benzotriazole, tolyltriazole, sodium niercaptobenzothiazole
Tin solder	Sodium nitrite, sodium nitrate, gluconic acid	Sodium benzoate, borax
Brass	Free amines, sodium nitrite	Benzotriazole, tolyltriazole, sodium mercaptobenzothiazole
Steel	—	Phosphate, sodium benzoate, sodium nitrite
Cast iron	Sodium benzoate, sodium nitrate, gluconic acid	Phosphate, sodium nitrite, salt of carboxyl acids
Aluminum	Sodium nitrite Borax	Sodium silicate, sodium nitrate, phosphates, sodium benzoate, benzotriazole, tolyltriazole

Fig. C245 Corrosive substances and corrosion protection agents

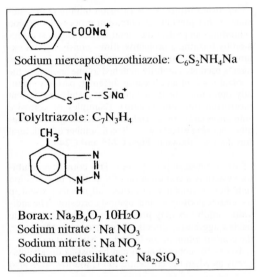

Sodium niercaptobenzothiazole: $C_6S_2NH_4Na$

Tolyltriazole: $C_7N_3H_4$

Borax: $Na_2B_4O_7\ 10H_2O$
Sodium nitrate : $Na\ NO_3$
Sodium nitrite : $Na\ NO_2$
Sodium metasilicate: Na_2SiO_3

Fig. C246 Structural and total formulas for a number of corrosion protection agents

necessary to take the synergic effects of the individual inhibitors into consideration. A coolant concentrate is formulated with a reserve alkalinity so that acidic substances either from exhaust, which gets into the cooling water unintentionally, or from glycol oxidation products can be neutralized. For this reason, it is necessary to ensure a proper balance between the individual corrosion inhibitors in the coolant. In individual cases, certain inhibitors need to be eliminated because of regional environmental protection regulations, toxicity, or to avoid deposits in the engine cooling system. Examples of corrosion causes are given below, together with the primary corrosion inhibitors in each case:

- Alkali metal phosphate/silicate (USA)
- Amine/phosphate (Japan)
- Silicate-free/phosphate (Japan)

- Benzoate/nitrite (Europe, USA)
- Nitrite/amine/phosphate-free, or NAP-free (Europe)
- Silicate-free (Europe, USA)

~Corrosion test. Corrosion protection places the highest requirements on a coolant concentrate. The ultimate test for a coolant is its performance characteristics in motor vehicles under operating conditions over extended periods of time. Such tests are, however, expensive and require a great deal of time. Therefore, they are unsuitable for evaluation of corrosion inhibitor systems during their development phase. For this reason, laboratory tests have been developed for corrosion inhibitor systems that allow a preliminary selection of candidate materials so that the products finally developed have good prospects of successful behavior in the subsequent motor vehicle tests. As described, the corrosion tests are performed in the following sequence:

- Laboratory tests
- "Circulation" tests simulating practical operating conditions
- Engine tests
- Vehicle tests

Fig. C247 shows the laboratory corrosion test apparatus and **Fig. C248** the corrosion "circulation" test.

~Crevice corrosion. Crevice corrosion is a galvanic process consisting of intensive local corrosion in material gaps and similar zones exposed to corrosive solutions. It appears frequently in combination with extremely slow-flowing solutions, typical for crevasses, holes, between touching metal surfaces, and below gaskets. Crevice corrosion in the radiator can have grave consequences when leaks occur.

~Density. Measurement of the coolant density allows a quick and easy check of the coolant concentrate when monoethylene glycol is used as antifreeze. **Fig. C249** shows the relationship between density and the strength of water/coolant concentrate mixtures.

~Elastomers/polymers. Many elastomer and polymer parts are installed in the cooling and heating systems on modern motor vehicles—for example, rubber hoses, gaskets, and expansion tanks. It is necessary to ensure that the cooling medium does not attack these parts.

~Engine cooling. During operation of an internal combustion engine, only about one-third of the thermal energy released by the fuel is converted into mechanical energy for propelling the vehicle (**Fig. C250**).

The remaining two-thirds are released in the form of heat; approximately half of this quantity is transported from the engine in the exhaust gas. The remaining energy is taken from the cylinders and cylinder head by the engine cooling system. This ensures proper lubrication and operation of the engine without malfunctioning. The engine can be cooled by means of air or liquid, but most engines are cooled with liquid. This

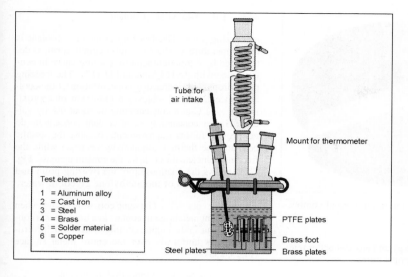

Test elements

1 = Aluminum alloy
2 = Cast iron
3 = Steel
4 = Brass
5 = Solder material
6 = Copper

Tube for air intake

Mount for thermometer

PTFE plates

Brass foot

Steel plates   Brass plates

Fig. C247
Apparatus for
laboratory corrosion test

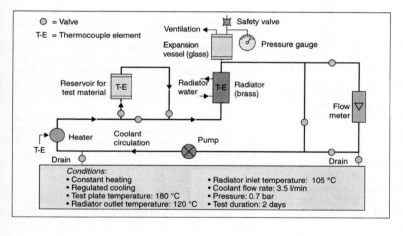

○ = Valve
T-E = Thermocouple element

Safety valve

Ventilation

Expansion vessel (glass)

Pressure gauge

Reservoir for test material   T-E   Radiator water   T-E   Radiator (brass)

Flow meter

T-E

Heater   Coolant circulation   Pump

Drain   Drain

Conditions:
• Constant heating
• Regulated cooling
• Test plate temperature: 180 °C
• Radiator outlet temperature: 120 °C
• Radiator inlet temperature: 105 °C
• Coolant flow rate: 3.5 l/min
• Pressure: 0.7 bar
• Test duration: 2 days

Fig. C248
Laboratory test rig for
corrosion analysis

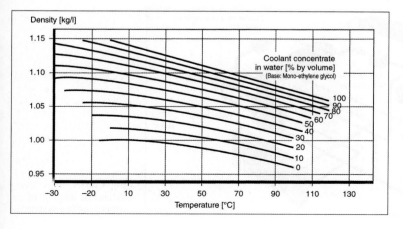

Density [kg/l]

Coolant concentrate
in water [% by volume]
(Base: Mono-ethylene glycol)

100
90
80
70
60
50
40
30
20
10
0

Temperature [°C]

Fig. C249
Density of coolant
concentrate/water
mixtures in relation to
temperature
(monoethylene
glycol base)

C

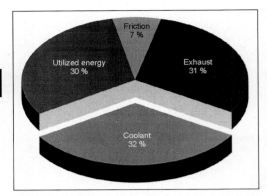

Fig. C250 Energy utilization in an internal combustion engine

generally requires less energy and ensures better regulation of the component temperatures.

~Engine cooling system. The engine cooling system, in which the coolant circulates, consists of the actual radiator with thermostat and fan, a cooling water pump, an expansion tank with cap, hoses, and a heater heat exchanger for heating the passenger compartment. The required passages for the cooling water are cast into the block and cylinder head. The metals present in the engine cooling system, which require corrosion protection, are aluminum alloys, gray cast iron, steel, brass, copper, and the soldering material. **Fig. C251** shows a schematic of the engine cooling system.

~Ethylene glycol →Radiator ~Coolant

~Expansion tank. The expansion tank is a part of the engine cooling system; it is usually produced of polypropylene and serves as a vessel for expansion of the coolant. It is provided with a filler fitting for the coolant, is usually translucent, and thus allows the coolant level to be checked easily.

~Flow rate →Radiator ~Coolant

~Freezing point. The freezing point of a coolant is the temperature at which crystals begin to form, as determined by a time/temperature cooling curve in conformance with ASTM Standard D 1177. The freezing point drops with increasing concentration of the coolant concentrate and achieves a minimum of approximately −53°C at a concentration of about 6% by volume. Higher concentrations of coolant concentrate do not provide further improvements, because the specific heat and the thermal conductivity decrease while the freezing point and the costs for the coolant increase. **Fig. C252** shows the relationship of the freezing point and the concentration of monoethylene glycol in water.

~Galvanic corrosion. Galvanic corrosion occurs when two different metals are in contact in a liquid. Such conditions occur in the engine cooling system. Such corrosion occurs uniformly over the entire metal surface (**Fig. C253**).

~Glycol. *See below*, ~Monoethylene glycol

~Heat transfer. Heat transfer occurs in the engine cooling system from the cylinder block and cylinder head to the coolant and from the coolant to the radiator. The heat transfer between the cooling system components and the coolant is a major factor in laying out the cooling system. Heat transfer corrosion can occur at the hottest points in the cooling system, usually in the vicinity of the exhaust ports in the cylinder head.

~Heat transfer corrosion. Heat transfer corrosion can occur on metal surfaces at high temperature, high heat conduction, and high heat transfer to the coolant. Heat transfer corrosion, which can penetrate deeply into the metal, occurs typically in aluminum alloy cylinder heads in the vicinity of the exhaust ports.

~Heater heat exchanger. The heater heat exchanger is a heat exchanger and the part of the engine cooling system that uses some of the heat in the coolant for

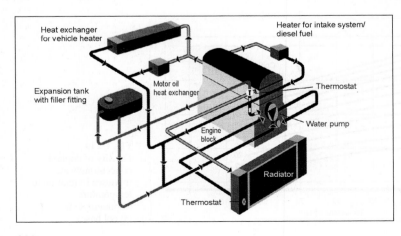

Fig. C251
Schematic diagram of engine cooling system

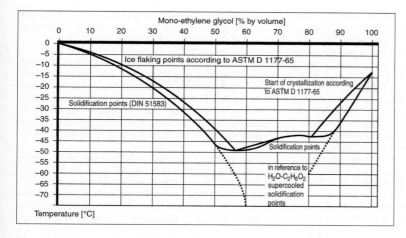

Fig. C252
Low-temperature
characteristics of
aqueous monoethylene
glycol solutions

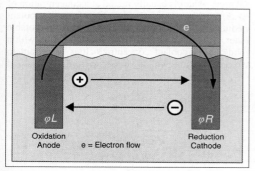

Fig. C253 Galvanic cell

heating the passenger compartment. It is usually soldered together using copper and/or brass parts or consists of aluminum alloys.

~**Monoethylene glycol.** Monoethylene glycol, also called ethandiol 1,2, is a polyhydric alcohol. It is also known under the designation MEG or simply glycol. Coolant concentrate consists of about 93% monoethylene glycol by volume.

~**Monopropylene glycol.** Monopropylene glycol, also called 1,2-propandiol, is also a polyhydric alcohol. It

is also known under the designation MPG and is less toxic than MEG. When used in coolant concentrate, it is also used at a concentration of approximately 93% by volume, the same as MEG. The primary data for coolant containing glycols are shown in **Fig. C254**.

~**pH level.** The pH level is an indication of the acidity or alkalinity of the coolant. A pH value of about 8 is usual (in Europe).

~**Pitting corrosion.** Pitting corrosion is a galvanic process consisting of intensive corrosion on a small area of the affected metal surface. This results in high corrosion with deep pitting and high penetration into the metal. Pitting corrosion can occur when the protective film on the metal surface is incomplete or has holes, and it can have grave consequences in an engine cooling system.

~**Pressure.** During engine operation, the pressure in the cooling system increases by 0.5–2 bar; this pressure is regulated by the radiator cap. The boiling point of the coolant is increased by the increase in pressure. This allows higher engine temperatures and more efficient engine operation.

~**Pump.** *See above*, ~Coolant pump

Total formula	$C_2H_6O_2$		$C_3H_8O_2$		$C_4H_{10}O_3$	
Structural formula	H_2C — CH_2 │ │ OH OH		H_3C — CH — CH_2 │ │ OH OH		H_2C — O — CH_2 │ H_2C │ OH	CH_2 │ OH
Density (20°C) [kg/m³]	1.113		1.036		1.118	
Melting point [°C]	−11.5		−60		−10.5	
Boiling point [°C]	198		189		245	
Specific heat (20°C) [kJ/(kg × K)]	2.407		2.460		2.307	

Fig. C254
Data on glycols suitable
for coolant

~Radiator cap. The entire cooling system is ventilated through the radiator cap. Two valves in the radiator cap provide for pressure compensation. The radiator cap ensures optimum operating pressure in the cooling system and prevents a vacuum from forming when the system cools down.

~Reserve alkalinity. The coolant concentrate is buffered so that a pH value in the range of 7.5 to 8.5 in Europe and up to 10 in the United States is achieved at a 33% concentration in demineralized water. The reserve alkalinity of the coolant allows neutralization of acidic constituents entering the coolant from the exhaust or as glycol oxidation products. The reserve alkalinity is measured according to ASTM Standard D 1121.

~Specification. The specification of a coolant concentrate establishes its quality and performance characteristics. It contains the requirements for protection against various forms of corrosion together with other physical and chemical properties such as freezing point, pH, reserve alkalinity, ash content, water content, and so on. The appropriate measured values are determined using standardized methods. A variety of specifications for coolant concentrates has been drawn up by

the producers of the concentrates and by vehicle and engine manufacturers as well as by standardization committees such as ASTM (**Fig. C255**).

~Specific heat (capacity). The specific heat of a coolant is its capacity to absorb heat and, therefore, its capability of removing heat from the engine. The higher the specific heat of the coolant is, the better the heat removal by the coolant. **Fig. C256** shows the change in the specific heat in relation to the temperature for various concentrations of monoethylene glycol in water.

~Thermostat. The thermostat is a part of the engine cooling system located at the coolant exit from the cylinder head. Its temperature-controlled function prevents circulation of the coolant through the radiator before the engine has reached its operating temperature of approximately 80–90°C. The thermostat is usually made of brass.

~Water. Water is usually mixed at a concentration of 30–50% with coolant concentrate to form the coolant medium. The best-suited quality of water is potable water, with moderate hardness in combination with a good-quality coolant concentrate. It is important to avoid deposition, which can occur from calcium and

Physical and chemical specifications		
Property	Specification values	ASTM test method
Density at 15.5°C in kg/dm³	1.110 to 1.145	D 1122
Freezing point in °C, 50% by vol. in distilled water	37 or less	D 1177
Boiling point in °C, undiluted	≥163	D 1120
• 50% by vol. in distilled water	≥107.8	D 1120
Attack on vehicle enamelling	no influence	D 1882
Ash Content in m%	≤5	D 1119
• pH level 50% by vol. in distilled water	7.5 to 11.0	D 1287
Chlorine content in mg/kg	≤25	D 3634
Water in m%	≤5	D 1123
Reserve alkalinity in ml	*	D 1121

* To be agreed between manufacturer and user

General specifications		
Property	Specification values	ASTM test method
Color	readily recognizable*	–
Influence on non-metallic materials	no adverse influences	–

* Preferred color: green or blue-green

Performance characteristics – specifications		
Property	Specification values	ASTM test method
Corrosion in laboratory test, weight loss in mg/test specimen		D 1384
• Copper	≤10	–
• Solder	≤30	–
• Brass	≤10	–
• Steel	≤10	–
• Cast iron	≤10	–
• Aluminum	≤30	–
"Rig" test with stimulated practice conditions – weight loss in mg/test specimen		D 2570
• Copper	≤20	–
• Solder	≤60	–
• Brass	≤20	–
• Steel	≤20	–
• Cast iron	≤20	–
• Aluminum	≤60	–
Corrosion of cast aluminum from surfaces with heat elimination	≤1.0	D 4340
Foaming		D 1881
• Volume in ml	≤150	–
• Time for foam collapse	≤5	–
Cavitation erosion, evaluating for pitting, cavitation or erosion of water pump	≥8	D 2809

Fig. C255
Specifications for coolant with monoethylene glycol base, ASTM Standard D 3306

C

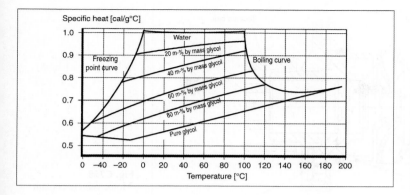

Fig. C256
Relationship of specific
heat to temperature and
various monoethylene
glycol/water mixtures

magnesium salts from hard potable water and certain
additives (e.g., phosphate) in the coolant concentrate.

The analysis values for the water should not exceed
the following limits:

- Water hardness: 0–20° GdH (0-3.6 mmpl/L)
- Chloride content: max. 100 ppm
- Sulfate content: max. 100 ppm

~**Water pump.** *See above*, ~Coolant pump

Coolant exchange heat accumulator →Heat
accumulator ~Design

Coolant flow rate →Coolant ~Coolant pump

Coolant preheating →Starting aid

Coolant pump (*also*, →Coolant). The coolant pump
circulates coolant in the cooling circuit as required

under all operating conditions and operating states of
the engine. A low failure rate, minimum drive power,
and cavitation-free operation must be ensured while at
the same time providing for minimum installation space
and low costs.

Presently, single-stage radial centrifugal pumps are
used in the vast majority of motor vehicle cooling circuits. The speed and, therefore, the pumping rate are
coupled to the engine speed by driving the pump directly from the engine crankshaft at a specific transmission ratio. The increasing requirement for coolant
pumping rates that are partly or completely independent
of engine speed can be satisfied with speed-regulated
coolant pumps or electric pumps. Because of the difference in the installation conditions dictated by this
requirement, as well as differences in the required
performance from the coolant pumps, different designs and production materials are used.

As shown in **Figs. C257 and C258**, the design of

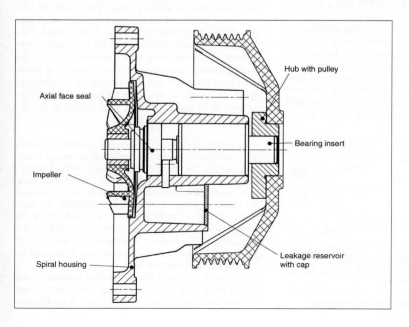

Fig. C257
Motor vehicle coolant
pump: attached pump
(Source: GPM GmbH)

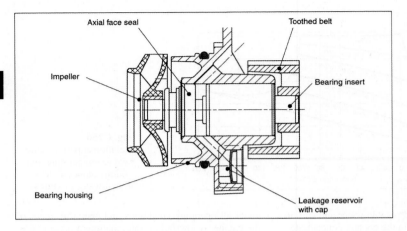

Fig. C258
Motor vehicle coolant
pump: insert pump
(Source: GPM GmbH)

a motor vehicle coolant pump consists primarily of a bearing housing, hub with pulley, impeller, axial base seal, bearing insert with integrated pump shaft, leak reservoir with cap, and, for attached pumps, housing with spiral channel.

Coolant pumps for engines can be differentiated into attached pumps (**Fig. C257**) and insert pumps (**Fig. C258**), depending on how they are mounted or attached.

With insert-type pumps, parts of the pump construction such as the spiral channel and inlet are located in the engine housing. Moreover, it is necessary to differentiate between various types of impellers and the location of the axial face seal on the intake or outlet side as well as the design of the drive. The relationship of the pump differential pressure, or pump head, and the required coolant flow rate and pump speed is illustrated in the coolant pump map. **Fig. C259** shows a typical map for a passenger car coolant pump.

The coolant pump map as shown in **Fig. C259** should be represented using the parameter pump head and should therefore be independent of the temperature and the mean flow composition—that is, the map applies to all coolant temperatures and percentages of antifreeze. By contrast, the curve using the pump parameter "differential pressure" would apply only for the mean flow states present during determination of the map.

In the case of the same impeller dimensions, the pump curve can be influenced by the number and shape of the impeller blades.

The intersection of the pump curves and the coolant circuit resistance indicate the operating points for the coolant pumps at the specific pump speeds. When the cooling circuit resistance changes by an action such as shutting off the heater or changing the thermostatic control, the operating points are changed accordingly. **Fig. C260** provides a schematic representation of the different cooling circuit resistances and operating points, AP, for idle and rated speed with the heater closed and open in the coolant pump map.

The coolant pump is usually laid out for a design point specified by the engine producer (pump differential pressure or pump head and flow rate) at the nominal pump speed or another pump speed between idle and cutoff speed and for a certain cooling circuit status—for example, thermostat open, heater closed. This design point should coincide with the operating points resulting from operation of the coolant pump in the cooling circuit, because at a different operating point, other cooling circuit flow rates would be present and, therefore, the partial flow rates through the circuit elements will be higher or lower than defined at the design point for the coolant pump.

~*Axial face seal.* An axial face seal prevents the coolant from leaking out to the outside of the pump. It can be located in the intake side or pressure side of the impeller. When axial face seals are located on the intake side, the frictional heat can be dissipated well, because the entire quantity of coolant flows around the seal. However, this can reduce the suction capability of the pump. When the axial face seals are located on the pressure side, the flushing flow rate across the impeller pressure side, axial face seal, and release holes must be sufficient to dissipate the heat.

The axial face seal friction pair is lubricated and cooled by the coolant. This ensures minimum leakage of vapor and liquid into the atmosphere. Such leakage

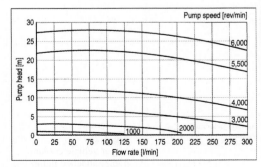

Fig. C259 Map for passenger car coolant pump

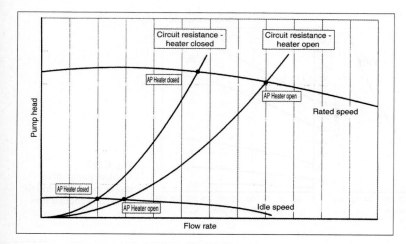

C

Fig. C260
Cooling circuit
resistances and
operating points in
coolant pump map

should be caught (**Fig. C261**) to eliminate the necessity of replacing a technically intact coolant pump.

~Cavitation. Cavitation results from the development and rupture of vapor bubbles in flowing liquids. Cavitation occurs when the pressure of the coolant is less than the vapor pressure at a point in the cooling circuit. Because of its low pressure level, the intake side of the coolant pump is particularly endangered by cavitation. Depending on the magnitude by which the coolant pressure is lower than the vapor pressure, the pump head of the coolant pump and, therefore, the cooling circuit flow rate are decreased. Moreover, material re-

moval can occur because of sudden bursting of the vapor bubbles in high-pressure areas—for example, in the spiral channel or engine block.

The pressure decreases occur locally at sharp edges and points where the direction changes as well as upon entry into the blade channels at high flow rates (i.e., high coolant flow rates at high pump speeds). Because of the difficulty in calculating and measuring these local pressure decreases in the rotating blade channel, cavitation tests are performed by the coolant pump manufacturers on a coolant pump test bench designed for this purpose, usually at the design point and at design speed. Here, the pressure on the intake side of the pump is decreased until a certain decrease in pump head is achieved—for example, 3%. The so-called NPSH (net positive suction head) value is then calculated from the suction pressure measured and the measuring conditions (mean flow temperature and composition) present during the bench tests. Because this NPSH value is independent of the mean flow composition and mean flow temperature—as is the pump head— it is possible to use it to calculate the pressure required on the intake side of the pump and, where applicable, the pressure required in the expansion tank or its connection point in the cooling circuit to avoid cavitation for the cooling circuit operating state to be considered (e.g., 40% antifreeze, coolant temperature at pump intake side = 108°C, heater closed). It is necessary to take into consideration that, when the coolant pump is operated in the coolant circuit, cavitation also can occur in the mid-temperature range in addition to very high coolant temperatures because of the usually high-pressure drop at the thermostat bypass plate and the low pressure in the expansion tank.

~Drive. The coolant pump may be driven by a toothed belt, a V-belt, or a poly V-belt. When the coolant pump is located outside of the belt plane(s), as a result of the engine design, it can also be driven by a stub shaft or gears.

~Drive power. **Fig. C262** shows the relationship among the pump drive power and the engine speed and

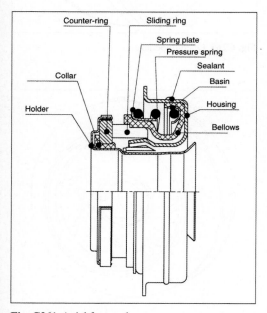

Fig. C261 Axial face seal

C

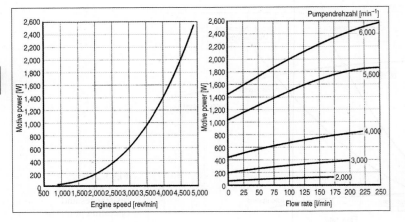

Fig. C262
Coolant pump drive
power in relationship to
speed and coolant flow
rate

pump flow rates (*also, see below,* ~Similarity relationships). The requirements of power to drive the pump at
the rated engine speed range lie between 500 W and
3.5 kW, depending on the required flow rate, pump
efficiency, and cooling circuit resistance. It should
be noted that the pump drive power can change at the
same engine speed, depending on the cooling circuit
operating state (e.g., heater open or closed; thermostat open, oscillating, or closed), because of the associated differences in the total resistance of the
cooling circuit.

~Impeller. Conversion of the mechanical energy fed
into the pump shaft through the drive pulley into pressure and speed energy is accomplished by the impeller.
With radial impellers, the coolant enters the impeller
axially and is forced radially to the circumference of
the impeller and into the impeller channel as a result
of centrifugal force. The flow is delayed after it exits
the impeller in the spiral channel.

Usually, radial impellers are used in motor vehicle
coolant pumps; however, axial impellers can also be
used for small impeller diameters and high pump
speeds. Radial impellers are produced using an open
design (**Fig. C263**) or closed design (**Fig. C264**).

A closed impeller is less sensitive to varying clearance gap sizes, resulting from the production processes,
than an open version. However, the production costs
are higher due to the additional cover plate.

The design layout of the impellers is accomplished
according to customer requirements after defining the
design point with a GPM specific calculation program.

~Production materials. Coolant pump housings are
produced using gray cast iron or aluminum; plastic
will be used increasingly in the future. Duroplasts or
thermoplasts, sheet metal, and aluminum are used primarily for impellers in the passenger car range and
gray cast iron in the commercial vehicle range.

~Similarity relationships. The pump curves for the
pump speed to be considered in each case can be cal-

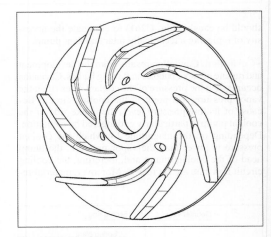

Fig. C263 Open impeller

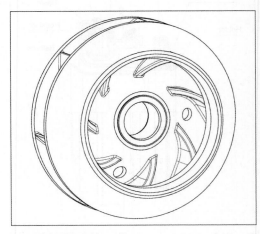

Fig. C264 Closed impeller

culated with the aid of similarity relationships when the pump speed deviates from the design speed. Here, the flow rate increases linearly; the pump differential pressure or pump head as well as the pump torque increase with the square of the pump speed; and the effective pump output increases with the cube of the pump speed. The power needed to drive the pump increases in a similar manner to the effective pump output (effective pump output is equal to the product of the pump differential pressure, according to Bernoulli, and pump flow rate), but it is also dependent on the efficiency curve. It is also important to consider that the similarity relationships apply only when the flow resistance of the cooling circuit remains constant. Since the cooling circuit resistance increases — at least in the engine low-speed range — because of incomplete hydraulic flow, the similarity relationships can be used only to a limited extent for conversion of the pump parameters for operation of the coolant pump in the cooling circuit — that is, only after reaching hydraulic rough flow.

~Speed control. The requirement for coolant flow rates independent of the engine speed can be met by using speed-regulated coolant pumps. In this manner, low flow rates can be achieved in the part-load range of engine operation to increase the coolant and oil temperatures and to reduce the engine frictional losses; in addition, high flow rates can be used at engine idle speed to support the requirements of the heater and for maximum heat transfer to the cooling system after operating at full load and at high engine torques in the engine low-speed range. Moreover, the coolant pump can be switched off during the engine warm-up phase to enable the engine to reach the operating temperature more quickly. Due to the saturation of the heater and radiator on the coolant side, the coolant flow rate is limited in the upper engine speed range, allowing a significant reduction in the power needed to drive the pump, as shown in **Fig. C262**.

These requirements are satisfied by an electric main water pump with coolant distribution function (**Fig. C265**).

The electric coolant pump shown in **Fig. C265** achieves high pump efficiency by using an axial impel-ler. Therefore, relatively low power is required to drive it. By integrating circuit control elements such as the thermostat, additional cooling circuit functions can be taken over economically by the electric coolant distribution pump and thus save space. The power to drive the pump and, therefore, the overall size of the electric motor as well as the generated load can be kept low by using flow optimized cooling circuits — for example, by reducing the main flow resistances and optimizing the cooling circuit network.

The use of electric coolant pumps frequently requires comprehensive modifications to the cooling circuit design. A switch-on and switch-off function for conventional coolant pumps is possible using on/off pumps, as shown in **Fig. C266**, without any design modifications to the engine or cooling circuit.

Complete integration of the clutch in the pump housing makes this on/off pump completely compatible and exchangeable with conventional, unregulated coolant pumps. The impeller can be switched off following a cold start and used to reduce the flow rate in the part-load range by switching it on and off dynamically.

~Spiral channel. The spiral channel of the pump (**Fig. C267**) or spiral housing can be considered to be a guide channel with a guide blade. The spiral corresponds to the angular cross section of the guide blade. Spiral housings are usually used for single-stage pumps. They can be located in the engine block (for insert-type pumps) or in the pump housing (for attached pumps). The spiral channel cross section increases in the direction of rotation of the impeller.

The same flow condition — that is, constant swirl — must be present in all parallel circuits in the spiral to ensure homogeneous flow symmetrical to the axis.

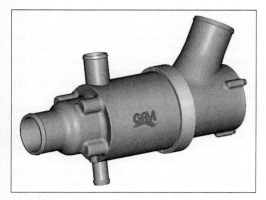

Fig. C265 Electric coolant distribution pump (Source: GPM GmbH)

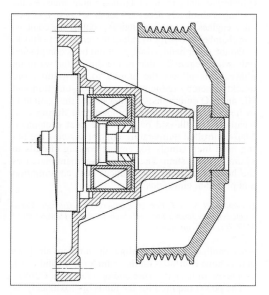

Fig. C266 On/off pumps (Source: GPM GmbH)

C

Fig. C267 Spiral channel

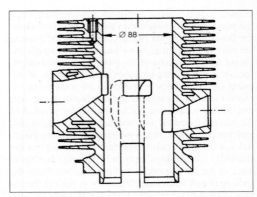

Fig. C268 Air cooling cylinder head on a two-stroke engine (Source: Pischinger)

Coolant specification →Coolant ~Specification

Coolant temperature →Radiator ~Coolant ~~Temperature

Coolant thermal capacity →Cooling, engine

Cooling, cylinder head →Cylinder head

Cooling, engine. To keep the thermal load to the engine and its components (e.g., the pistons with piston rings, cylinder heads, valves with guides, spark plugs, and cylinder) resulting from combustion within limits, engine cooling is absolutely necessary for diesel as well as gasoline engines. In gasoline engines, cooling the combustion chamber also helps to reduce the tendency to knock.

On the other hand, cooling should be accomplished only when required—that is, not during the warm-up phase. This allows the engine to warm up more quickly, thereby reducing the fuel consumption and HC and CO emissions. Another use of the coolant is to heat up the passenger compartment during cold weather with the aid of a suitable heat exchanger. Map-controlled temperature control systems are being used increasingly for engine cooling circuits to optimize the system. Thermal management increases the heating comfort and reduces fuel consumption and pollutant emissions.

Literature: R. Saur, P. Len, H. Lemberger, G. Huemer: Kennfeldgesteuertes Temperaturregelsystem für Motorkühlkreisläufe, MTZ 7/8, 1996, pp. 424 ff.

~**Air cooling.** With air cooling, the ambient air flows across finned surfaces on the cylinder head and block as a result of back pressure and/or a fan (**Fig. C268**).

The power required for the fan is approximately 3–4% of the engine output. The significance of air cool-

ing has decreased highly for diesel as well as gasoline engines because disadvantages are associated with this type of cooling. These are:

- A large fan is required with relatively high drive power and noise development.
- The absence of an acoustically insulating liquid jacket around the engine.
- Specific engine output is limited.
- A high delay occurs in vehicle heating, particularly during warm-up phase, because the cooling air heats slowly and the heat losses in the air lines are relatively high.
- There is low specific heat of air and low heat transfer.

Advantages of air cooling include:

- No liquid with antifreeze is required.
- No radiator is required.
- Simpler design and no leaky connections.
- No coolant pump is required.
- Greater heat transfer coefficient in comparison with air.

Literature: A. Urlaub: Verbrennungsmotoren, Springer Publishers, 1994.

~**Evaporative cooling.** Evaporative cooling (cooling by change of state) is based on the idea of providing a constantly high coolant temperature of 105°C in the cylinder head at ambient pressure. The basic design is shown in **Fig. C269**. This allows improvements in the specific fuel consumption of up to 5% in comparison to conventional cooling systems. **Figs. C270 and C271** show samples for a given case.

Evaporative cooling provides the following basic advantages:

- The engine can be operated at a boiling temperature of 105°C using a standard water/ethylene glycol mixture without an intrinsic system pressure.
- The temperature is independent of the density of heat flow and virtually the same everywhere in the engine.
- High local heat flow densities at thermally highly load points lead to only slight increases in local wall temperatures.

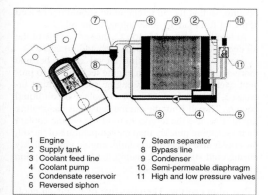

1 Engine
2 Supply tank
3 Coolant feed line
4 Coolant pump
5 Condensate reservoir
6 Reversed siphon
7 Steam separator
8 Bypass line
9 Condenser
10 Semi-permeable diaphragm
11 High and low pressure valves

Fig. C269 Basic layout of evaporation cooling system

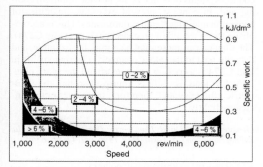

Fig. C270 Reduction of fuel consumption with evaporative cooling in steady state operation

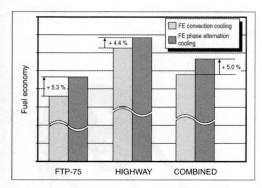

Fig. C271 Improvement of fuel economy in US test (simulated calculation)

- The coolant flow rate can be greatly reduced in comparison to conventional cooling. This allows use of a smaller coolant pump.
- The size of the radiator can be reduced.

Literature: P. Müller, E. Heck u. W. Sebbeße: Verdampfungskühlung—eine Alternative zur Konvektionskühlung MTZ (1995), 12.

~High-temperature cooling. High-temperature cooling is an excellent means of significantly reducing the specific fuel consumption and exhaust emissions, including NO_x. This requires use of a suitable high-temperature coolant with a high boiling temperature compared with water (e.g., ethylene glycol with boiling temperature above 170°C), whereby the system pressure can still be controlled at a temperature of 160°C in the cooling circuit.

The improvements in the fuel consumption result during the warm-up phase as well as in the part-load range with the engine running warmer and can be greater than 20% in diesel engines with direct injection. **Fig. C272** shows the results for a particular case. Here, WC means cooling with water as cooling medium and GC cooling with glycol as cooling medium. The indices $_{in}$ and $_{out}$ characterize the inlet and outlet temperature.

The following improvements were achieved for the exhaust emissions. The CO emissions were reduced by 35–45%, depending on the load and the HC emissions by 19–33%. By accepting a slight reduction in the fuel consumption improvement achieved of about 2%, compared with normal liquid cooling and by starting the fuel delivery later, it is possible to reduce the NO_x emissions by 54%.

The reason for improvement during the warm-up phase is the significantly faster warming of the engine (coolant and oil) due to the reduction in the heat transfer coefficient and lower heat capacity (for ethylene glycol, the heat transfer coefficient is approximately one-fifth, and the heat capacity about one-half that of water).

Countermeasures are necessary to prevent thermal overloading at full load. It is necessary to reduce the temperature by increasing the size of the radiator and/or increasing the coolant pump and radiator fan out-

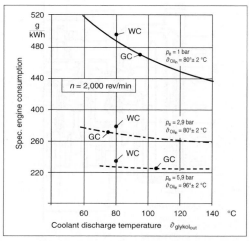

Fig. C272 Effect of coolant outlet temperature on effective specific fuel consumption at various mean effective pressures

C

puts. High-temperature cooling is a practical proposition, particularly for diesel engines. On gasoline engines, knocking problems occur relatively quickly when the coolant temperature is increased.

Literature: E. Mühlberger, W. Beßlein: Variable Heißkühlung beim Fahrzeug-Dieselmotor, parts 1 and 2. MTZ 1983, pp. 403ff and 505ff.

~Liquid →Coolant ~Coolant concentrate

~Liquid cooling. Liquid cooling uses 50–70% water and 50–30% antifreeze, usually ethylene glycol. Inhibitors also provide for corrosion protection. The percentage of antifreeze also increases the boiling temperature of the mixture, thereby allowing coolant temperatures up to 120°C at a system pressure of 1.4 bar, which also reduces the fuel consumption.

In the past, so-called thermo-siphon cooling was used, which was sufficient for the low engine loads at that time. Because it is less dense, the cooling liquid heated up in the engine rises from the upper part of the engine into the radiator where it is cooled, then floods downward under gravity and re-enters the engine at the bottom. Today's high-performance engines make coolant pumps in the cooling circuit necessary (**Fig. C273**).

~Oil cooling. When an oil cooler is not used, it is necessary for the heat in oil to be transferred to the surrounding air, particularly through the oil pan. For this purpose, sheet metal oil pans and aluminum oil pans sometimes have fins.

The increasing specific power and the associated high thermal load on the piston crown led to oil spray cooling, so that the oil temperature should not exceed 150°C at full load, to delay aging of the oil.

At temperatures above 150°C, the service life of the oil decreases by about half for each additional 10°C.

Specific oil cooling using an oil cooler became more and more frequent in engine technology to prevent overheating the oil as engine specific output increased. Oil coolers are oil/air coolers or oil/liquid coolers. The oil from the engine is used as the cooling liquid. Because the temperature of the engine coolant increases more quickly than the temperature of the oil, the cooler also heats up the entire engine more quickly during cold start, thus reducing the fuel consumption and HC and CO emissions.

~Precision cooling. The objective of precision cooling is to cool hot spots, particularly in the area of the cylinder head, more intensively than conventional cooling systems do—that is, to, provide optimum cooling and, in contrast, prevent excessive quantities of coolant in uncritical temperature zones. This is accomplished with the aid of precision passages (**Fig. C274**). The diameter of these may be 5 mm, for example, and they run next to the intake channels. The ring visible in **Fig. C274** is a cooling passage around the spark plug.

At full load, cooling is improved so that the volumetric efficiency and, therefore, the power output increase slightly.

Literature: T. Hütten, R. Duckworth: Der präzisionsgekühlte Zylinderkopf von Cosworth. MTZ 1996, vol. 11, pp. 636ff.

~Thermostat →Radiator ~Coolant

~Water →Coolant; →Radiator ~Coolant

~Water lubrication. Considerations are made continuously for lubricating the engine with water instead of mineral oil.

The advantages are:

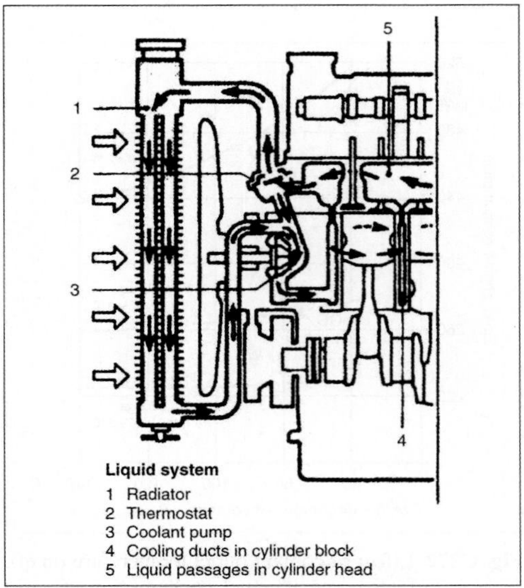

Liquid system
1 Radiator
2 Thermostat
3 Coolant pump
4 Cooling ducts in cylinder block
5 Liquid passages in cylinder head

Fig. C273 Liquid cooling system

Fig. C274 CAD model of cooling jacket in precision cooled cylinder head

- High heat capacity.
- Good heat conductivity.
- Environmentally friendly.
- Preserves natural resources in relation to use of mineral oil.
- Lubricating oil cannot reach the combustion chamber and pollute the exhaust gas and catalytic converter.
- Common cooling circuit with coolant imaginable. In this case, the oil cooler can be eliminated. Cooling can be accomplished with the main radiator.

The disadvantages are:

- Corrosion and antifreeze protection are required as with cooling water.
- Increased surface protection required for surfaces in contact with water.
- Permissible water temperature is less than 105°C (boiling temperature for water/ethylene glycol mixture).
- The possibility of cavitation cannot be excluded, for example, in the bearings or liquid pumps.

Cooling air fan balancing →Balancing of masses ~Balancing

Cooling air routing →Crankcase ~Crankcase construction ~~Air cooled

Cooling calculation →Calculation process ~Application areas ~~Fluid circulation

Cooling capacity →Radiator ~Liquid/air radiator

Cooling circuit (*also*, →Radiator). During combustion, peak temperatures of greater than 2000°C occur locally in the combustion chamber. Cooling is required to prevent thermal overload to the materials used for the cylinder head, valves, spark plugs, injector nozzles, prechambers, gaskets, pistons, piston rings, and cylinder liner.

In the EPA-city/highway driving cycle (**Fig. C275**) approximately 33% of the energy contained in the fuel is dissipated in the form of heat by the cooling system. This is accomplished by means of direct cooling (air) or use of a cooling medium (coolant), which can be de-

signed as evaporative cooling or preferably, today, forced circulation cooling.

As a rule, water with a coolant additive or in rare cases oil is used as the coolant. The volume of coolant, V_K, contained in the entire cooling circuit is about four to six times the engine displacement and is circulated by the coolant pump approximately 10 times per minute.

A desirable side effect of the water jacket is the noise attenuation. In addition to dissipation of heat, a primary objective is uniform distribution of heat to prevent thermal distortion from interfering with the engine operation. In addition, heat can be transported by a heat exchanger for heating the passenger compartment, and its rapid activation is important for comfort reasons.

A thermostat is responsible for the control of temperature (**Fig. C276**). The coolant path is opened and closed by a valve actuated by a wax actuator with high thermal expansion. In the cold state, the coolant circulation is initially limited to the circuit inside the engine to enable the engine to heat up quickly. The path to the radiator is opened only after the thermostat opening temperature is reached (*also, see below*, ~Map cooling).

Suitable additives to the coolant provide sufficient corrosion and cavitation protection of all surfaces as well as to reduce the freezing point (antifreeze) and increase the boiling point. This allows the cooling sys-

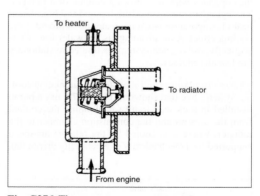

Fig. C276 Thermostat

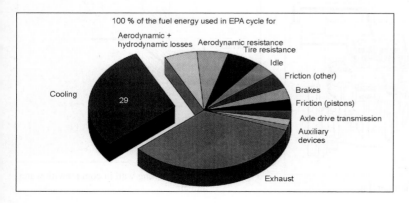

Fig. C275
Subdivision of fuel energy in EPA-city/highway driving cycle

C

Fig. C277 Air cooling

tem to be laid out as a closed continuous cooling system for operation throughout the year.

~Direct cooling system. Direct cooling with intensive airflow past the surfaces equipped with cooling fins is used only for small engines today (**Fig. C277**). Because of the low heat transfer rates with air, a high airflow rate is required (about 30–80 m³/kWh), and, as a rule, this can be achieved only with a single-stage axial fan.

~Evaporative cooling. Evaporative cooling (hot cooling) operates without a flowing coolant or a radiator. The cylinders are open at the top and the high latent heat of evaporation when the water boils provides the cooling effect. A disadvantage is the water loss. Consequently, use of such systems is limited to stationary (and small) engines.

~Force circulation cooling. A forced circulation cooling system, like that shown in **Fig. C278**, has a pump installed in a closed circuit for pumping the coolant from the area around the combustion chamber to the radiator, where it is cooled by the relative air-speed supported by a mechanically or electrically driven fan.

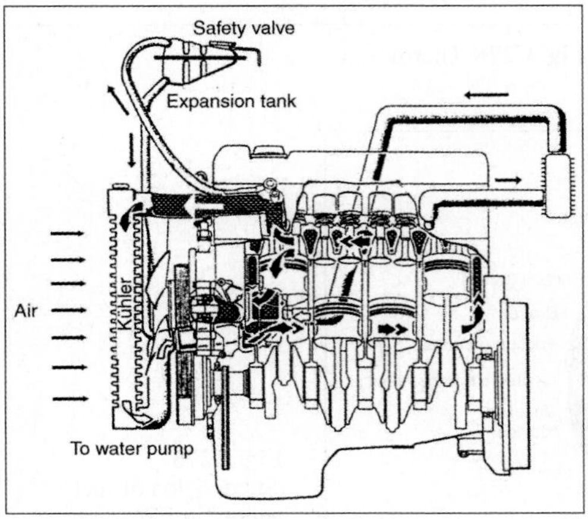

The forced circulation cooling is the main method of cooling currently used in vehicles.

~Heat flow. The dissipated heat flow can be calculated according to **Fig. C279** as follows:

$$Q = a_1 \cdot A \cdot (t_1 - t_1') = \frac{\lambda}{\delta} \cdot A \cdot (t_1' - t_2')$$

$$= a_1 \cdot A \cdot (t_2' - t_2),$$

from which the temperature differences

$$(t_1 - t_1') = \frac{\dot{Q}}{a_1 \cdot A} ; (t_1' - t_2') = \frac{\dot{Q} \cdot \delta}{\lambda \cdot A} ;$$

$$(t_2' - t_2) = \frac{\dot{Q}}{a_2 \cdot A} ;$$

and this results in

$$(t_1 - t_2) = \frac{\dot{Q}}{A} \cdot \left(\frac{1}{a_1} + \frac{\delta}{\lambda} + \frac{1}{a_2} \right)$$

or, finally,

$$\dot{Q} = k \cdot A \cdot (t_1 - t_2) = p \cdot \dot{V}_k \cdot c_p \cdot (t_1 - t_2) \approx P_e$$

where $\dot{Q}$ = heat flow, α = heat transfer coefficient, A = surface area, t = temperature, λ = thermal conduction coefficient, δ = wall thickness, k = heat transfer coefficient, ρ = density, c_p = specific heat capacity, $\dot{V}_k$ = coolant flow rate.

~Map cooling. An advance in thermostatically controlled forced circulation cooling is called map cooling. This allows the engine to receive variable cooling based on the cooling requirement. The thermostat with fixed opening temperature is replaced by a thermostat with electronic actuation. Depending on the cylinder charge, engine load, and vehicle speed, a higher temperature level can be programmed in the part-load range resulting in a reduction in the fuel consumption (about 1% per 10°C increase).

Fig. C278 Forced circulation cooling

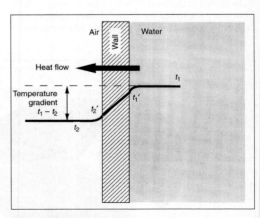

Fig. C279 Heat flow along wall in contact with water and air

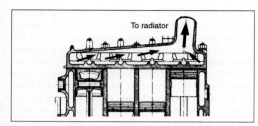

Fig. C280 Principle of thermo-siphon cooling

~Thermo-siphon cooling. Thermo-siphon cooling (also called self-circulating cooling) uses the differences between the densities of hot and cold water and does not require an additional pump. Because of its lower density, the water heated up around the combustion chamber rises to the radiator, which must be located at a higher level than the engine (**Fig. C280**). From there, the cooled water with lower density flows back because of a natural flow gradient. The disadvantage is the large radiator volume required and the overall height of the system. Moreover, the relatively low water flow rate results in a lower cooling effect than with forced circulation cooling with a water pump.

Cooling circuit cavitation →Cooling circuit

Cooling coil piston →Piston

Cooling duct piston →Piston ~Cooling

Cooling energy. The cooling energy of the engine that has to be dissipated is particularly dependent on the engine load in addition to the engine speed and combustion process. At full load, the energy to be dissipated by cooling is between 20 and 30% of the chemical energy contained in the fuel. Depending on the combustion process and whether the engine is supercharged, quotients between the quantity of the cooling energy dissipated, Q_K, and effective output of 0.45 result for supercharged diesel engines with direct injection to 1.5 for naturally aspirated gasoline engines.

Literature: A. Urlaub: Verbrennungsmotoren, Springer Publishers, Berlin Heidelberg New York, 1994, p. 344. — K. Mollenhauer: Handbuch Dieselmotoren, Springer Publishers, Berlin Heidelberg New York, 1997.

Cooling exhaust port →Exhaust port ~Exhaust port cooling

Cooling exhaust recirculation →Exhaust gas recirculation

Cooling fin design →Crankcase ~Crankcase construction ~~Air cooled

Cooling fins →Crankcase ~Crankcase construction ~~Air cooled, ~~Water cooled

Cooling intake port →Intake port, diesel engine ~Cooling; →Intake port, gasoline engine ~Cooling

Cooling losses →Heat flow

Cooling piston →Piston

Cooling system →Coolant ~Engine cooling system

Cooling thermostat →Radiator ~Coolant

Cooling water →Coolant ~ Water

Cooling water energy →Cooling energy

Cooling/heat dissipation →Oil ~Oil functions

Copper alloys →Bearings ~Materials

Copper plating →Piston ring ~Surface treatments

Corrosion →Bearings ~Operational damage; →Coolant

Corrosion effect on copper →Fuel, diesel engine ~Properties ~~Corrosive efect on metals

Corrosion inhibitor system →Coolant

Corrosion inhibitors →Coolant; →Oil ~Body ~~Additives

Corrosion protection →Oil ~Oil functions

Corrosion protection, coolant →Coolant

Corrosion protection, fuel, gasoline engines →Fuel, gasoline engine

Corrosion protection additives →Fuel, gasoline engine ~Additives

Corrosion test →Coolant

Corrosive wear →Piston ring ~Wear

Counterflow cylinder head (scavenging) →Cylinder head ~Cross flow

Counterpressure side →Piston ~Pressure side

Counterweight →Balancing of masses ~Counterweight; →Crankshaft ~Balance weights

Coupled 1-D and 3-D processes →Intake system ~Intake manifold ~~Computation process

Cover plate →Crankcase ~Crankcase design

Cover plate design →Crankcase ~Crankcase design ~~Cylinder ~~~Composite technology

C

Cracking →Connecting rod

Cracking carburetor. Cracking carburetors are systems to split fuel, mostly under endothermic conditions, to produce compounds that enable better mixture formation and, thus, engine combustion. Usually, liquid fuels, such as alcohol and benzene, were converted into gaseous fuels using a catalyst.

Currently, these processes are again connected to the fuel cell. The reformer, as a "cracking carburetor," uses the liquid fuels—for example, working as a benzene reformer—to produce the hydrogen necessary for fuel cell operation.

Crank angle →Crankshaft drive

Crank arrangement →Crank set

Crank sequence →Balancing of masses

Crank set (*also*, →Balancing of masses). The phasing (firing order) of a crankshaft—that is, the angle between the individual cranks—can be specified with the aid of a crank set. The crank set simultaneously indicates the direction of the effective lines of the rotating inertia forces. Crank sets for the first and second order reciprocating inertia forces can also be illustrated and used for graphical representation of the resulting inertia forces and their moments. The crank set of the second order forces results from doubling the crank angle of the first order crank set. **Fig. C281** shows this using four-cylinder inline engines with the same firing order as an example.

Crankcase (*also*, →Commercial vehicle crankcase). The crankcase is a part of the engine containing the cylinders, the cooling jacket, and the engine housing.

~**Acoustics** (*also*, →Engine acoustics). A focal point in acoustic development in motor vehicles and their drives is the achievement of legal noise regulations and fulfillment of product-specific noise requirements.

Smooth operation of an internal combustion engine is a function of many parameters and to a high degree is determined in advance by definition of the engine design and crankcase design.

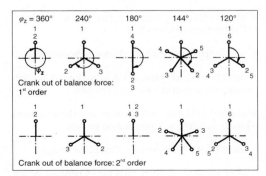

Crank out of balance force: 1st order

Crank out of balance force: 2nd order

Fig. C281 Crank sets of the first and second order

In addition to the crankcase functions, optimization of the acoustic characteristics of the crankcase structure is an important development objective, specifically the achievement of low noise radiation by avoiding resonant frequencies and the insulating of excitation vibrations.

The load on the crankcase (function) resulting from nonuniform torque curve at the crankshaft and from the free mass forces and mass moments leads to mechanical vibration. The excitation frequency of these loads has a certain relationship to the rotary frequency of the crankshaft corresponding to the excitation orders of the free gas and mass force effects. The mechanical vibrations caused by low excitation orders are low frequency and act primarily in the area of the main bearing block and crank chamber.

High-frequency vibrations in the crankcase walls are caused by the combustion process in part by pulse-shaped force transmission in the valve train and also by piston force excitations. The high frequencies are in the audible range and are called acoustic vibration.

Low- and high-frequency vibrations are transferred through the crankcase and engine bearings to the vehicle structure; these can be transferred to the vehicle regardless of the type of engine bearings. A portion of the high-frequency acoustic vibration is also radiated through the side walls of the crankcase.

The following factors must be taken into consideration for acoustic optimization of an engine:

- Causes of structure-borne sound excitation mentioned above
- Structure-borne sound paths in the cylinder head, cylinders, pistons, piston pins, connecting rods, and crankshaft
- Design of engine bearings and their connection to the crankcase or to other parts of the engine and drivetrain
- Structure of crankcase in connection with crankcase design

Modern crankcase development is accomplished using closed CAE (computed-aided engineering) process chains. Three-dimensional CAD (computer-aided design) illustrations and networking of the housing structure form the basis for FEM (finite element method) calculations for strength, rigidity, dynamics, and acoustics.

An experimental modal analysis on the finished crankcase provides additional information on its forms of intrinsic vibration.

Experience from more than 100 years of internal combustion engine construction and the design, calculation, and analysis capabilities available today provide basic information for crankcase design optimized from the point of view of noise: The objective is the stiffest possible crankcase and stiffest possible engine/transmission connection.

This can be achieved by measures independent of the crankcase design and by use of specific design advantages:

- A crankcase surface structure with bows and ribs reduces air-borne sound emission.

- Stiff cover plate and cylinder head bolts whose forces are applied at a point located deep below the cover plate lead to minimum distortion of the sealing surface and cylinder. The latter is a prerequisite for minimizing piston play and piston noise during operation.
- A stiff connection between the crankshaft and main bearing block ensures minimum bearing clearances.
- Stiff flanges to the oil pan and transmission are prerequisites for a stiff connection between the engine and transmission.

The various crankcase designs have different specific acoustic advantages:

- Closed deck design has a stiff deck surface with advantages for distortion of the sealing surface and cylinder in comparison to an open deck design. With the latter, cylinders cast together increase the rigidity of the cylinder barrel connection.
- A design consisting of an upper section and a lower section provides a stiff connection between engine and transmission, compared with a crankcase with side walls extending below the center of the crankshaft combined with separate main bearing caps. In the latter design, the rigidity can be increased by consolidating the individual main bearing caps into one ladder-type frame.
- On solid aluminum crankcases, consisting of an upper and a lower part, cast-in parts of gray cast iron at the main bearings reduce thermal expansion and, therefore, decrease the bearing play.
- The use of cast aluminum oil pans with a flange for the transmission, in combination with various crankcase designs, provides for a stiff connection between engine and transmission.

In addition to the functions and the acoustics, the design of the crankcase is also influenced by production quantity, costs, weight, company policies, and so on. Therefore, a variety of crankcase designs use the abovementioned design-independent rules for layout with regard to good accoustics.

~Air cooled. *See below*, ~Crankcase design

~Casting process, crankcase. Crankcases for motor vehicles are cast using cast iron or aluminum silicon alloys. Costs, quantities, and design are the primary criteria for selection of the casting process. The casting processes for crankcases include:

~~Chilled casting. Chilled casting can be subdivided into gravity flow and low-pressure casting.

~~Gravity feed chilled casting. A chilled mold is a metallic permanent mold of gray cast iron or hot-forming tool steel for production of castings using light metal alloys. As with sand casting, sand cores are placed in the mold. The mold is filled with the molten metal at atmospheric pressure by the force of gravity acting on the molten metal. The casting process is accomplished primarily on partly or fully automated casting machines.

The chilled casting process allows each mold to be used a number of times in contrast to sand casting. New sand cores are required for each molding operation (the so-called "broken cores"). The use of cores provides chilled casting with maximum freedom of design just as with sand casting; in contrast to die casting, undercuts in the casting are possible.

In contrast to sand casting, the molten metal is solidified quickly and specifically in the chilled mold. Specific cooling of the mold is possible and is used frequently. It is necessary to apply a parting compound (so-called release agent) to protect the mold from the molten light metal. In comparison to sand casting again, the chilled casting process provides castings with a finer structure, higher strength, higher dimensional accuracy, and better surface quality.

Double heat treatment is possible on chilled cast parts as with sand cast parts. In addition to specific control of the cooling of the casting in the mold as the first heat treatment, further heat treatment (artificial aging) is frequently performed.

~~~Low-pressure casting. The low-pressure casting process differs from the gravity-flow chilled casting process primarily by the method used for filling the metallic permanent mold with the molten light alloy metal: The molten metal is pressed into the mold from the bottom at a pressure of 0.2–0.5 bar and solidifies under this pressure. The nearly ideal solidification of the cast part achieved with this process is a primary reason for the high quality of low-pressure castings.

With the low-pressure casting process, double heat treatment is possible as with sand casting and chilled casting.

~~Die casting. Permanent molds of annealed hotforming tool steel are used for the die-casting process. Before each casting operation, the mold parts are treated with a parting compound. In contrast to sand casting and chilled casting, it is not possible to insert cores into the casting mold because the molten light alloy material is injected into the mold at high pressure and high speed.

The magnitude of the pressure depends on the size of the cast part and usually ranges from 400 to nearly 1000 bar. The pressure is maintained during solidification. For larger cast parts, the mold halves are cooled to provide specific solidification as well as quick cooling of the cast part.

After solidification of the cast part, the mold consisting of stationary and movable mold parts as well as movable mold slides is opened, and the cast part is ejected by means of ejector pins.

In comparison with sand casting and chilled casting, die casting allows the most precise reproduction of the hollow space in the mold and, therefore, of the cast part itself: thin-wall castings with close dimensional tolerances, high molding accuracy, and high surface quality as produced. Dimensionally precise casting of eyes, bores, and in many cases, fits and lettering without subsequent machining, as well as the casting in of

C

sleeves (including gray cast iron cylinder sleeves), pins, and other insert parts are possible.

The die-casting process has the highest productivity in comparison with sand, chilled, and low-pressure casting because all casting and mold motion sequences are automated to the greatest extent.

Disadvantages are the limited design freedom for the cast part, because undercuts are not possible. Possible air or gas pores in the casting prevent double heat treatment as with sand, chilled, and low-pressure castings.

Aluminum silicon alloy crankcases, particularly in combination with special cylinder sleeve technologies, are produced increasingly using the die-casting process.

~~**Lost-foam process.** This is a special form of sand casting. A foam model of the later casting is made by foaming EPS (expandable polystyrene) and, where applicable, gluing together individual segments. The foam model is coated with a water-based parting agent. After coating and drying, the model is molded into a casting container containing pure quartz sand without any binding agent by use of a vibration technique.

During the casting operation, accomplished within a very short time (15–20 seconds), the molten metal is allowed to flow against the foam model—so-called full-mold casting. The foam model is decomposed by the heat of the molten metal and its liquid and gaseous constituents are transferred to the mold sand. After cooling and removal from the mold, a burr-free casting is present.

The particular advantage of this method is that casting geometries not possible with conventional sand-casting processes can be produced as a result of the foam model production process described above, particularly the core gluing technique. The lost-foam process is suitable for production of gray cast iron as well as light alloy castings.

Presently, crankcase castings in mass production are not produced using the lost-foam process. Introduction of this process for aluminum silicon alloy crankcases is anticipated in the near future.

~~**Sand casting.** Models and core boxes of hard wood, metal, or plastic are used to form the shape of the later crankcase casting in the sand mold. As a rule, the molds are produced using quartz sand (natural sand, synthetic sand) and binding agents (synthetic resin, CO_2). Core shooters are used to shoot in sand to form the cores. The individual cores are assembled to form core packages, and the core packages and outer mold are assembled mechanically and fully automatically for production of medium quantities.

Complicated castings with undercuts can be produced by dividing the model, core, and mold up into various levels and inserting cores into the mold.

During the casting process, the hollow cavities between the outer mold and cores are filled with molten metal. After the casting process and solidification of the molten metal, the casting is removed from the sand

mold. The sand mold is destroyed during this process (the so-called "broken mold"). The casting is then cleaned: Dead head, gate marks, casting skin, and seams are removed. In mass production, the casting operation, removal of the casting from the sand mold, and cleaning are also automated. Double heat treatment is possible on Al-Si alloy sand cast parts—the first heat treatment consists of controlling the cooling process and time of the casting in the sand mold, and the second one consists of artificial aging by placing the casting in a furnace for a certain time at a certain temperature. Both heat treatments serve to increase the strength of the casting. The sand-casting process allows each mold to be used only once. Sand casting can have undercuts. By comparison, casting processes with permanent metal molds allow the molds to be used a number of times—as a rule, castings produced using these processes do not have undercuts.

Sand casting is the traditional casting method for mass production of gray cast iron crankcases. However, aluminum silicon alloy crankcases are now also being mass-produced using a precision sand-casting process.

A further application for the sand-casting process is production of prototypes and small lots of crankcases as well as other parts.

~~**Squeeze casting.** The squeeze-casting process is a combination of the chilled-casting process with especially the low-pressure casting and die-casting process. Permanent metal molds are filled from the bottom with the molten light alloy at a pressure of 0.2–0.5 bar. Then, they are allowed to solidify at a high pressure of approximately 1000 bar. The exceptionally good seal of the casting mold allows high-strength alloys that are hard to cast to be used also.

Solidification of the molten metal at high pressure results in a very fine structure in the casting. Slow mold filling and solidification of the molten material at high pressure result in a virtually core-free structure and, therefore, high permanent strength against cyclic loads and higher resistance to cyclic temperature changes compared to low-pressure casting and die casting.

As with die casting, it is not possible to use sand cores in the squeeze-casting process. This means that the same design restrictions exist for squeeze-casting parts in terms of producing undercuts as for die-cast parts.

In contrast to the die-casting process, double heat treatment is possible with squeeze cast parts due to the virtually pore-free structure. Therefore, squeeze casting combines the advantages of chilled casting/low-pressure casting and die casting.

~Chilled casting. *See above,* ~Casting process, crankcase

~Closed-deck. *See below,* ~Crankcase design ~~Cover plate

~Closed-deck design. *See below,* ~Crankcase design ~~Cover plate

~Composite technology crankcase. *See below,* ~Crankcase design ~~Cylinder

~Connecting rod motion diagram. The connecting rods complete an oscillatory motion each time the crankshaft rotates. The resulting envelope locus traced by the connecting rod has a characteristic shape similar to the outer contour of a violin (**Fig. C282**).

The design of the crankcase must ensure that sufficient clearance is present for the connecting rod motion.

The most important close points between the crankcase and the connecting rod are usually:

- The bottom edge of the cylinder on V, W, and horizontally opposed engines; also the opposing cylinder
- Crankcase sidewalls, particularly those with passages for oil return or crankcase ventilation located next to the connecting rod

As a rule, the clearance is 3.5~4.5 mm, which takes into consideration all the tolerances for the associated components, piston, connecting rod, big end bearing shell, crankshaft, crankshaft main bearing shells, and crankcase with main bearing caps, ladder frame construction or crankcase bottom section, depending on the crankcase design.

~Conventional crankcase ventilation. *See below,* ~Crankcase ventilation

~Cooling air routing. *See below,* ~Crankcase construction ~~Air cooled

~Cover plate. *See below,* ~Crankcase design

~~**Design of cover plate and cylinders.** *See below,* ~Crankcase design ~~Cylinder ~~~Composite technology

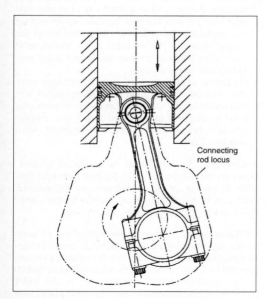

Connecting rod locus

Fig. C282 Connecting rod motion diagram

~Crankcase bottom section. *See below,* ~Crankcase design ~~Main bearing pedestal area

~Crankcase construction. Design of crankcases is characterized primarily by the type of crankcase as well as a number of other aspects.

~~**Air-cooled.** Air-cooled cylinders were used previously and are still common for motorcycle engines. Air-cooled cylinders are no longer used today on motor vehicle engines.

With air-cooled cylinders, the heat dissipation depends on the thermal conductivity of the cylinder fins and cylinder material, the design of the cooling fins, and the manner in which the cooling air is provided.

~~~**Cooling air routing.** Even today, air-cooled motorcycle engines are usually open and exposed to the relative wind velocity resulting from the speed of the motorcycle. However, there are also exceptions, such as motor scooters with engines cooled by a fan located under the body of the motor scooter.

In the case of air-cooled automobile engines, positive cooling with a fan was used exclusively. Here, the cooling air is blown from a fan housing over air guide plates (so-called baffles) surrounding the cylinders and cylinder heads so that it flows directly against and between the cooling fins. The better the flow conditions and heat transfer values, the lower the quantity of cooling air required and, therefore, the lower the power to drive the fan.

~~~**Design of cooling fins.** Cooling fins are located on the outer cylinder wall of air-cooled cylinders to increase the effective heat transfer area. Theoretically, fins with a triangular cross section are most effective. On cast cylinders, the casting process results in slightly trapezoidal shape fins with rounded edges which are nearly as good as fins with a triangular cross section.

A heat transfer at the cooling fins can be increased by:

- Increasing the fin surface area by increasing the number of fins by placing them more closely together or by greater fin height
- Increasing the flow rate of the cooling air
- Conversion from unguided to guided cooling airflow through the use of air baffles
- Use of a cylinder and cylinder fin material with the highest possible thermal conductivity—e.g., aluminum alloy instead of gray cast iron

~~**Primary dimensions.** Crankcases are characterized by the following primary dimensions, depending on the engine design—for example, inline, V, or horizontally opposed engine (**Fig. C283**).

- Length: Dimension from front edge of crankcase, also FFOB (front face of block), to engine/transmission flange, also RFOB (rear face of block)
- Width: Maximum overall width
- Height: Dimension from crankshaft center line to cover plate plane in cylinder axial direction

C

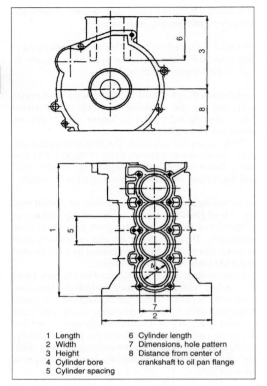

1 Length	6 Cylinder length
2 Width	7 Dimensions, hole pattern
3 Height	8 Distance from center of
4 Cylinder bore	crankshaft to oil pan flange
5 Cylinder spacing	

Fig. C283 Primary dimensions of crankcase

- Cylinder bore/cylinder diameter: Nominal inner diameter of cylinder
- Cylinder spacing: Dimension between two adjacent cylinders from center to center
- Cylinder offset: On V, W, and horizontally-opposed engines, interval between the centers of two adjacent cylinders in opposing cylinder banks
- Cylinder length: Dimension of cover plate to lower end of cylinder
- Dimensions of hole pattern for cylinder head bolts: depending on arrangement—e.g., four or six per cylinder
- Dimension from crankshaft center line to oil pan flange (*see also*, Crankcase design):
  —Equal to zero when oil pan separation plane is at crankshaft center line
  —Height of deep skirt on crankcase with long side walls
  —Height of crankcase bottom section

~~Ribbing. The ribbing on gray cast iron as well as aluminum crankcases fulfills a number of functions:

- Increased rigidity—e.g., of the crankcase side walls, thereby improving the noise emissions
- Optimization of the transmission of force from less rigid areas to supporting areas in the crankcase structure—e.g., by connecting casting eyes with one

another as well as with the supporting areas such as the cover plate, oil pan flange, engine/transmission flange
- Optimization of the casting process resulting from better flow of the molten material to points in the crankcase where unavoidable material accumulations may occur

In the past, the ribbing on a crankcase was designed empirically to a great extent. Today, additional state-of-the-art design methods such as FEM (finite element method) are available for calculation. These allow the ribbing to be optimized in terms of weight and function.

~~~Thermal conductivity of cylinders and fin material. The thermal conductivity of aluminum alloys is nearly three times as high as that of cast iron materials (*also, see above*, ~Casting process, crankcase; and *below*, ~Weight, crankcase). For this reason, cast iron cylinders have been replaced by cylinders using aluminum alloys now that suitable barrel technologies have been developed for aluminum cylinders.

The cylinder barrel technologies are basically the same as described → Cylinder running surface. Quasi-mono-metal designs using hypo-eutectic Al-Si alloys with nickel dispersion coating are used instead of a galvanic coating of chromium as was common previously, particularly for motorcycle cylinders.

For highly loaded air-cooled gasoline and diesel engines, the dimensional stability of the pure light alloy cylinder was occasionally insufficient. For this reason, cylinders were used in which a gray cast iron or steel liner was surrounded by a finned light alloy jacket. These so-called composite cast cylinders were produced using one of two processes:

- A finned jacket consisting of a light metal alloy was diecast around a gray cast iron cylinder liner with roughened outer wall or around a steel or gray cast iron liner on which a thin iron/aluminum coating was applied to the outer wall of the liner before the casting process. This resulted in an intermetallic connection between the liner and finned jacket providing uniform heat flow.
- On lightly loaded engines, cylinder versions were also used in which a light alloy finned jacket was cast around a gray cast iron liner with no particular bond to the light alloy finned jacket or a preprocessed light alloy finned jacket was shrunk-fit onto a cast iron liner.

~~Strength. The strength of the crankcase is determined by the material used, by the heat treatment possible depending on the casting process and material used, and by the design, characterized by the crankcase version, wall thicknesses, and ribbing.

~~Water-cooled. With very few exceptions, all automobile engines today are water-cooled. In contrast to air-cooled cylinders with cooling fins, the cylinders are surrounded by a water chamber, the so-called water jacket.

An important design characteristic is the water jacket depth, the dimension from the cover plate plane to the lowest point of the water jacket.

On earlier gray cast iron crankcases, this dimension amounted to up to 95% of the cylinder barrel length. On modern cast iron engine block designs, the water jacket ends in the area of the lower piston ring contact zone—that is, in the area between the first compression ring and the oil scraper ring when the piston is at bottom dead center.

On modern aluminum crankcases, the water jacket is even shorter. The water jacket depth is in the range of approximately one-third of the cylinder barrel length.

This is possible due to the higher thermal conductivity of the aluminum alloys compared with cast iron materials as well as to state-of-the-art piston design which uses smaller piston compression heights compared with earlier designs.

A shorter water jacket reduces the percentage of coolant in the engine and, therefore, the engine weight. The primary advantage of a lower quantity of coolant in the engine is that the engine warm-up phase is shortened; this results in a reduction of the pollutant emissions which are highest immediately after starting the cold engine.

~Crankcase design. The crankcase designs can be structured corresponding to the design in the area of the

- Cover plate
- Cylinder
- Main bearing pedestals

~~**Cover plate.** The crankcase cover plate is an important design feature. Its design influences the selection of the casting process. It is necessary to differentiate between closed-deck and open-deck designs for the cover plate.

~~~**Closed-deck.** Regardless of the design, the cover plate always has openings for the cylinders, the threaded holes for the cylinder head bolts, and, as a rule, bores

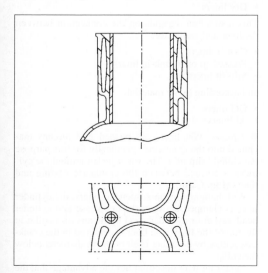

Fig. C284 Closed-deck design

and passages for pressurized oil, cooling water, oil return, and crankcase ventilation (**Fig. C284**).

With the closed-deck design, the crankcase cover plate is virtually closed in the remaining area around the cylinder. Here, the cover plate is interrupted only by small, defined openings for the coolant. These openings connect the water chamber (water jacket) surrounding the cylinders through matched passages in the cylinder head gasket and openings in the cylinder head combustion chamber plate with the water jacket in the cylinder head.

Production of the crankcase water jacket using the closed-deck design requires a sand core, because the water jacket is mostly closed at the top by the cover plate and, therefore, cannot be molded as a part of the outer mold for the upper section of the crankcase. It is necessary for the water jacket sand core to be held in the mold. These retaining points are generally present in the finished crankcase as casting eyes in the side walls of the crankcase. These core retaining eyes are plugged with sheet metal caps. On an assembled engine, where it not possible to see the cover plate, the presence of such core retaining eyes in the crankcase is an unmistakable indication of a closed-deck cover plate.

The advantage of the closed-deck design in comparison with the open-deck design is the greater rigidity of the cover plate. This has a positive effect on cover plate deformation, cylinder deformation, and noise. However, the crankcase design with a closed-deck cover plate limits the selection of casting processes. Because of the required water jacket sand core, the closed-deck design can be produced today using only sand casting, chilled casting or low-pressure casting processes.

Gray cast iron crankcases produced using the sand casting process virtually all have a closed-deck design.

Aluminum silicon alloy crankcases with closed-deck design have been mass-produced primarily using the chilled casting/low-pressure casting process, but have recently also been made using the sand casting process.

~~~**Open-deck.** With the open-deck design, the water jacket surrounding the cylinder is still open at the top (**Fig. C285**). In terms of casting, this means that a sand core and, therefore, core retainer eyes are not required for production of the water jacket core. The water jacket core can be molded without undercuts as a steel casting.

The water jacket is open at the top and allows better cooling of the hot upper area of the cylinder than is possible with the closed-deck design. The rigidity of the cover plate is lower for the open-deck design than with the closed-deck design. The resulting negative influence on deformation of cover plate and cylinder is compensated for today by using a metal cylinder head gasket. The lower creep behavior of this gasket, as compared to a conventional soft material cylinder head gasket, allows lower initial tension of the cylinder head bolts, thereby reducing deformation of the cover plate and cylinders.

Crankcases with open-deck design can be produced using all casting processes as a matter of prin-

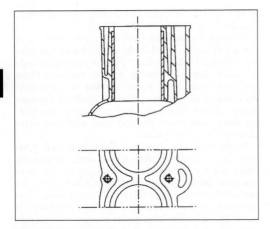

Fig. C285 Open-deck design

ciple. In individual cases, gray cast iron crankcases with open-deck design are produced using the sand casting process.

For crankcases of aluminum silicon alloys, the open-deck design opens up the possibility of production using the economical die-casting process; this process also allows realization of special cylinder/cylinder sleeve technologies.

~~Cylinder. The detailed parameters of the cylinders and cylinder liners are determined by various aspects of the design and materials. The two are related. The cylinder or crankcase design, in terms of material, is subdivided into:

- Composite technology
- Insert technology
- Mono-metal design

Details on the possible designs are given below for each case.

~~~Composite technology. Composite technology is a cylinder barrel technology that can be used only on aluminum alloy crankcases.

On composite technology aluminum crankcases, the cylinder barrel is produced using local material engineering in contrast to Al crankcases with mono-metal design. Cylindrical mold elements, so-called premolds, consisting of a composite of suitable materials, are placed in the mold and permeated by the molten aluminum alloy under high pressure during the casting process. This technology limits selection of the casting process to die casting or processes derived from die casting such as squeeze casting in the new die-casting process developed by Honda.

Design of the cover plate and cylinder with composite technology. The limitation to die-casting or casting processes related to die casting resulting from the technology allows for an open-deck design only.

Cylinders that are cast together as well as cylinders that are not cast together may be produced.

Honda MMC™ process. The Honda MMC-V process (metal matrix composite) already has been used in mass production for a number of years. This process is similar to the KS-Lokasil™ process in principle. Fiber premolds are inserted into the molding die. The premolds consist, among other things, of a composite of Al_2O_3 and carbon fiber and are permeated by the molten aluminum alloy during the new Honda die-casting process.

KS-Lokasil™ process. A highly porous cylindrical mold element consisting of silicon is permeated by the liquid aluminum alloy under high pressure using the squeeze casting process. The cylinder barrel is produced by three-stage honing. During the initial honing using diamond strips, many of the silicon crystals located at the surface are destroyed. This destroyed silicon crystal layer is removed by intermediate honing with silicon carbide. The third honing phase with an abrasive bonded resiliently in the strips exposes the silicon grain. These form a hard, wear-resistant cylinder barrel surface, similar to mono-metal design, in which the silicon crystals present in the hyper-eutectic alloy are exposed.

An iron-coated piston is required as the running partner. As a rule, the piston ring set usually used for gray cast iron cylinder barrels is satisfactory.

~~~Insert technology. Insert technology is used, as a rule, on cylinder liners in vehicle engines only in combination with aluminum crankcases. Liners produced of various materials are inserted into the crankcase in various ways.

A distinction is made according to function between:

- Wet liners
- Dry liners

This distinction depends on the connection between the liner and crankcase:

- Cast-in liners
- Pressed-in or shrunk-in liners
- Slip-fit liners

and according to the material

- GG liners
- Al liners.

Wet liners. Wet liners are pushed into mounts machined into the crankcase specifically for this purpose (so-called "slip fit"). The water jacket around the cylinder is formed between the crankcase casting and liners (**Fig. C286**).

With hanging or top-suspended liners, the cylinder liner is clamped between the crankcase and cylinder head gasket or cylinder head by means of a collar at the top of the liner. The liner is centered in the crankcase, either by the collar itself or by the diameter below the collar.

Centering with the collar has the advantage that the cylinder liner is cooled well in its uppermost section,

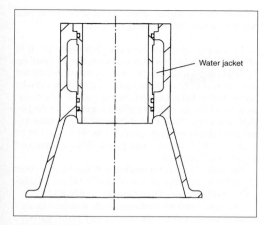

Fig. C286 Wet liners

where the thermal load is highest; a disadvantage is the high load on the hollow chamfer in the crankcase.

Centering the liner below the collar decreases the cooling effect at the upper end of the cylinder; however, it relieves the load on the hollow chamfer in the crankcase.

Wet, top-suspended liners are sealed at the top and bottoms by O-rings to prevent coolant and oil leakage (oil fog in the crank chamber).

With so-called upright wet liners, contact and centering is accomplished in the lower area of the liner. This liner design requires particularly careful layout to limit cylinder distortion.

Wet upright liners are sealed at the top by the cylinder head gasket and at the bottom with a flat gasket below the liner contact surface or with O-rings.

The alignment of wet cylinder liners in relation to the cover plate plane presents certain problems. Extensions or recesses have a negative effect on the surface pressure of the cylinder head gasket around the cylinder as well as on cylinder distortion. Therefore, it is necessary to limit liner extension or recess to the unavoidable minimum.

Extremely tight tolerances are required for the corresponding liner dimensions when the completely machined wet liners are inserted into the crankcase with a machined cover plate. Distance matching is also common for upright liners. Another possibility is final machining of the crankcase cover plate and inserted liners together. Wet liners are produced using gray cast iron (GG) as well as aluminum alloys. GG liners have the advantage of good running characteristics on the cylinder barrel; the disadvantages are higher weight and decreased heat transfer compared with aluminum liners.

Wet aluminum liners are produced using hyper-eutectic as well as hypo-eutectic Al-Si alloys. In the case of hyper-eutectic Al-Si alloys, the cylinder barrels are produced as described under the mono-metal design by means of chemical etching. Liners of hypo-eutectic Al-Si alloys are generally coated with a nickel disper-

sion coating, also as described under the mono-metal design.

Regardless of the design and material, wet sockets can in principle be combined with open-deck and closed-deck designs and with all casting processes commonly used for crankcases.

Wet liners are preferred in crankcases with open-deck designs produced using the die-casting process.

The advantages of using wet cylinder liners include freedom of selection of the liner material, production of various cylinder bores and, therefore, displacement by combining corresponding liners with the same crankcase, and simple interchangeability and repair. The disadvantages are the higher production costs in comparison to mono-metal design.

Wet GG sockets are used particularly in large diesel engines whose crankcases, however, are produced using cast iron materials because of ready interchangeability when repair is required.

Wet Al liners are used in light alloy engines, particularly in sports cars or racing cars.

Dry liners. Dry liners are pressed, shrink-fitted, or cast into the crankcase (**Fig. C287**). Pressing in or shrink-fitting is performed in the crankcase casting. For casting in, the liners are placed in the mold for the crankcase and the molten aluminum alloy is cast around them.

In contrast to wet liners, the water jacket is not located within the liner material and crankcase casting, but is rather a part of the crankcase casting as in a mono-metal design.

Therefore, no seal is required between the dry socket and the crankcase.

Protrusion of the dry, pressed-in, or cast-in liner in relation to the cover plate plane is eliminated by machining the installed liners and cover plate together. This presents problems in the case of different materials, such as GG liners and Al crankcases, in terms of the various cutting speeds required for the particular materials.

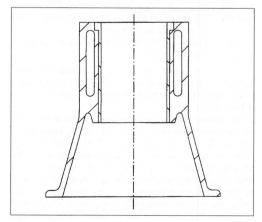

Fig. C287 Dry liners

C

Dry liners are produced as gray cast iron, Al alloys, or sintered liners using powder metallic materials. Dry GG and Al liners are designed with the same approach as wet liners in principle for the cylinder barrel and have the same properties as described there.

Regardless of the material, dry liners can be combined with open-deck and closed-deck designs and with all the casting processes commonly used for crankcases. Aluminum crankcases are mass-produced with the closed-deck design with permanent mold/low-pressure mold casting and pressed-in GG liners, as well as with an open-deck design, diecast and cast-in GG liners.

The advantages of dry-cylinder liners are freedom in the selection of the liner material; with GG liners, there is the simple possibility of repairing by honing out to oversize; with Al liners, there is the option of separate production of the cylinder barrels and combination with crankcases of a different Al alloy from the liner alloy.

Disadvantages are the higher production costs in comparison to mono-metal design as well as the poorer heat transfer between the cylinder barrel and water jacket.

~~~**Mono-metal design.** Typical representatives of mono-metal design are crankcases with cast-iron alloys. With such crankcases, the cylinder is a part of the crankcase casting. The surface quality required for cylinder function is created by processing in a number of steps (initial and fine machining, honing).

Mono-metal crankcases using aluminum silicon alloys (Al-Si alloys) are available in two versions:

1. *Production of crankcase casting from hypereutectic Al-Si alloy.* Al-Si alloys are called hyper-eutectic when the percentage of silicon is greater than 12%. The primary silicon precipitated in the casting is exposed following machining of the crankcase by means of a chemical etching process or by machining. This results in a hard, wear-resistant, so-called unreinforced cylinder barrel surface that requires an iron-coated piston as its running partner.

This mono-metal cylinder or crankcase version is mass-produced using the most economical casting process, depending on the type of cover plate:

- Hyper-eutectic Al-Si alloys and closed deck design in the low-pressure casting process
- Hyper-eutectic Al-Si alloy and open deck design in the die-casting process

Due to the higher percentage of silicon in hypereutectic Al-Si alloys, it is more difficult to machine parts consisting of these alloys than castings produced using hypo-eutectic alloys.

The primary silicon crystals precipitated in the casting are destroyed and splintered during machining. This results in undesirable short chip formation.

When hyper-eutectic Al-Si alloys are cast using the die-casting process, the primary silicon precipitates with significantly smaller grain sizes than with the low pressure casting process, and this significantly improves the machinability of the component. The smaller silicon crystals show less of a tendency to be destroyed or to splinter during machining, which can, therefore, be

accomplished at higher speeds while at the same time obtaining better cutting results.

2. *Production of crankcase using hypo-eutectic Al-Si alloy in combination with cylinder barrel coating.* With this quasi mono-metal crankcase design, a nickel dispersion coating is applied galvanically to the cylinder barrel. The nickel dispersion coating consists of a nickel layer in which the silicon carbide particles are distributed uniformly. Such coated cylinder barrels have extremely good running properties, low wear, and can be combined with pistons and piston rings of standard commercial materials.

The nickel dispersion coating is better known under registered trademarks such as Nikasil™ (Mahle), Galnical™ (Kolbenschmidt), or Gilnisil™ (BMW).

This cost-intensive coating process requires decalcification equipment for the preliminary baths, galvanic systems with vapor extraction, and washing equipment as well as disposal of the sludge.

The nickel dispersion coating can be combined with closed-deck as well as open-deck designs.

Because even microscopic porosities in the cylinder barrel result in coating problems in the form of separation of the coating, selection of the casting processes used for Al-Si alloy crankcases is limited. Conventional die casting—that is, without special measures such as vacuum support—is not possible.

Nickel-dispersion–coated cylinders were used frequently for single-cylinder engines on motorcycles and for the number of vehicle engines that are still produced with single cylinders. Multicylinder crankcases on motor vehicle engines mass-produced with nickel-dispersion–coated cylinders have been made only on a limited scale.

Cylinder layout with mono-metal design. Here it is necessary to differentiate between cylinders cast in a single block and those not cast together in the crankcase longitudinal axis.

Previously, crankcases with closed-deck as well as open-deck design and of gray cast iron as well as aluminum/silicon alloys were produced with the cylinders not cast together in the crankcase longitudinal axis. This was done in order to achieve the greatest possible uniformity of temperature distribution between the cylinders (resulting from the coolant being present between the cylinders) and minimized cylinder distortion (by preventing adjacent cylinders from influencing one another).

The overall engine length, determined primarily by the length of the crankcase, was not important to the extent it is today, due to the fact that the engines primarily were installed longitudinally in the vehicles, in the larger engine compartments used at that time.

Many years of experience and comprehensive studies have shown that even in cylinders cast together along the engine longitudinal axis, the temperature distribution in the cylinders is virtually uniform in spite of the absence of coolant between the cylinders. This means that no significant distortion problems occur, eliminating the resulting operating problems such as high oil consumption or blowby.

The advantages of cylinders cast together include high crankcase rigidity, shorter crankcase length, and resultant minimizing of the engine length and weight.

Today, the reduction of engine length is a dominant criterion for installing the engine transversely in the engine compartment, as well as continuous reduction of the installation space available for the drive assemblies.

Minimization of vehicle weight, in combination with optimization of the efficiency and weight of the drive assemblies, are prerequisites for reducing fuel consumption. A considerable reduction in overall length and weight is possible by designing the crankcase with the cylinders cast together, depending on the type of engine (inline engine, V-engine, horizontally opposed engine).

The limit for casting cylinders together is the thickness of the dividing wall between the cylinders (so-called cylinder web). Today, mass production engines are produced with cylinder dividing walls with thicknesses of less than 5.5 mm regardless of the crankcase material. This has been mastered functionally by the use of metallic cylinder head gaskets with minimum compression characteristics and, therefore, low initial tension requirements. In addition to proper sealing of the cylinder dividing wall, the cylinder deformation is reduced to a minimum as a result of the low initial tension for connecting the cylinder head and crankcase.

~~**Main bearing pedestal area.** The design of crankcases in the area of the crankshaft bearing, the so-called main bearing pedestal area, is a further characteristic of the crankcase design.

The crankcase design in the area of the main bearing pedestals can be structured according to the location of the separating plane between the crankcase and oil pan:

- Oil pan flange at crankshaft center
- Oil pan flange below crankshaft center

as well as corresponding to the design of the main bearing cap:

- Individual main bearing caps
- Ladder-type frame design
- Crankshaft bottom section

~~~**Crankshaft bottom section.** The crankcase bottom section (also called the girdle or bed plate) consolidates the individual main bearing caps into one part in the same manner as a ladder frame design. In contrast to a ladder frame, the crankcase bottom section is not inside the engine. The side walls of the bottom section form the outer edges of the crankcase. The flange for the oil pan forms the bottom plane of the crankcase bottom section.

In addition to the advantages and disadvantages of crankcase top section and crankcase bottom section described in ~Crankcase design, a crankcase bottom section offers the same design possibilities as described under ~~~Ladder-type frame design. Moreover, additional functions can be integrated into crankcase bottom sections, mass-produced almost exclusively using Al alloys with the die-casting process:

- Function of so-called oil baffle—i.e., radial stripping of the motor oil around the enveloping curves of the crankshaft counterweights and connecting rods (so-called connecting rod motion envelope)
- Parts of the oil circuit of the engine, such as oil intake channel between oil pump and oil sump, oil channel between oil filter flange and oil pump, oil filter flange itself, oil return passages, main oil channel and oil channels to the individual main bearings, partial integration of the oil pump housing
- Holding rear shaft seals and, where applicable, front shaft seals for sealing the crankshaft

As described above, crankcase bottom sections are used in mass-produced wholly aluminum engines and racing engines.

~~~**Ladder-type frame design.** With the so-called ladder-type frame design, the individual main bearing caps are consolidated into one part similar to the design with crankcase bottom section (**Fig. C288**). In contrast to the crankcase bottom section, the ladder frame design does not have a flange plane to the oil pan. The ladder frame is located inside the engine and is, therefore, covered by the oil pan on designs with the oil pan flange at the crankshaft center or by the long side walls in crankcases with a deep skirt design.

The ladder frame design has a number of advantages:

- Compared with individual main bearing caps, higher rigidity and, therefore, better acoustic properties, simpler and quicker to install than individual main bearing caps
- Virtually the same design freedom as the crankshaft bottom section in terms of integrating functions; however, more economical and lighter than a crankcase bottom section

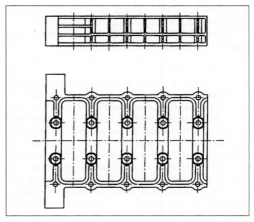

Fig. C288 Ladder frame design

Ladder frames can be die cast, particularly using Al alloys. This allows the finished part to be cast almost completely including integration of cast oil grooves for oil supply to the main bearings.

In the area of the individual bearing points, cast iron inserts with spheroidal graphite (e.g., GGG 60) can be cast into the individual bearing points. These have the same advantages as the combination of aluminum crankcases and gray cast iron main bearing caps: reduction of the operating bearing play of the crankshaft, increase in the rigidity of the ladder frame, and reduction of the noise radiated from the area of the main bearing pedestals.

Existing crankcase designs with individual gray cast iron main bearing caps can be replaced with a ladder frame design to increase the rigidity and improve the acoustic properties without having to completely redesign the crankcase. Compromise solutions between individual main bearing caps and a single integrated ladder frame are possible by connecting the individual bearing caps with one another by bolting them to a separate casting in the form of a ladder. The most widely used mass production design for crankcases with ladder frames is a combination of Al crankcase and Al ladder frame, where the crankcases are designed with the oil pan flange at the center of the crankshaft; some crankcases also have a deep skirt design.

~~~**Main bearing caps.** The main bearing caps are the bottom section of the main bearing pedestals to which they are bolted and have basically the same function: namely, taking up the forces and moments of the crankshaft bearing, holding the corresponding bearing (sliding bearing shells) on the fitted bearing, holding flanged bearings or, more rarely, bottom thrust plates on the last main bearing, and, where applicable, holding a radial shaft seal for sealing the rear end of the crankshaft.

The main bearing caps and pedestals in the crankcase are machined together and, therefore, are uniquely related to one another for all assembly processes following machining. The usual bearing cap anchorages consist of broached surfaces or holes for locating sleeves on the sides of the main bearing pedestals.

The main bearing caps are produced almost exclusively of cast iron and combined with crankcases of cast iron as well as Al alloys. The common machining of the Al main bearing pedestals and gray cast iron (GG) bearing caps does present problems in view of the different optimum cutting speeds for each material; however, this technique has been mastered and is the state of the art for mass production today. The combination of aluminum crankcases/aluminum main bearing pedestals and GG main bearing caps has advantages resulting from the GG material: The low thermal expansion coefficient of the GG main bearing cap limits the bearing play in the crankshaft bearing during operation. This reduces the oil flow rate through the crankshaft main bearings. Reduced main bearing play and higher material rigidity of the GG bearing cap (the modulus of elasticity of GG is higher than that of Al)

reduce noise development and its emission in the area of the main bearings.

The design used most frequently for mass production up to the middle of the 1990s used gray cast iron crankcases with separate gray cast iron main bearing caps. The crankcases were designed with the oil pan flange at the crankshaft center as well as crankcases with long side walls. The combination of Al crankcases and separate gray cast iron main bearing caps was used for V engines.

~~~**Main bearing pedestal.** The main bearing pedestal is the upper half of a crankshaft bearing point in the crankcase. Regardless of the design of the crankcase in the area of the crankshaft bearing, the main bearing pedestals are always integrated into the casting for the crankcase or crankcase upper section (**Fig. C289**).

The number of main bearing pedestals in a crankcase depends on the type of engine and particularly on the number of cylinders and the cylinder arrangement. Today, crankcases are produced almost exclusively with so-called full-bearing crankshafts. Full-bearing crankshafts have a main bearing journal between each crank throw. A full-bearing inline four-cylinder engine has five main bearing pedestals; full-bearing inline cylinder engines and opposing six-cylinder engines have seven main bearing pedestals; V-6 and V-8 engines have four and five main bearings, respectively, and so on.

The most important functions of the main bearing pedestals are to:

• Take up the forces and moments of the crankshaft bearings (radial and axial)
• Hold the upper bearing shells for the crankshaft radial bearings
• Hold the flanged bearing or through plate in a main bearing pedestal, the so-called thrust bearing for the crankshaft

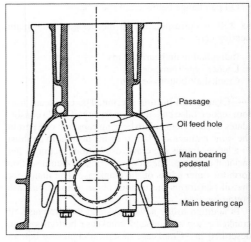

Fig. C289 Main bearing pedestal/main bearing cap

- Hold the thread for mounting the main bearing cap or ladder frame or crankcase bottom section
- Hold locating holes or aligner for locating main bearing cap or ladder frame or crankcase bottom section
- Almost always accept oil feed holes and oil grooves for supplying the crankshaft main bearing with oil
- Depending on the engine design, hold the radial shaft seal in the last main bearing pedestal to seal the rear end of the crankshaft

The main bearing pedestals frequently have passages for compensating the pressure in the individual chambers of the crankcase and thereby reduce losses resulting from internal engine friction.

Vertical holes or channels for oil return from the cylinder head or for crankcase ventilation are frequently present on the side of the main bearing pedestals.

This variety of functions necessitates great care in designing and laying out the main bearing pedestals and their combined parts, such as the main bearing caps or ladder frames or the crankcase bottom section. This process is facilitated by the design aids available today, such as FEM (finite element method) calculation.

~~~Oil pan flange at crankshaft center.
A design characteristic of crankcases still encountered today is the location of the separating plane between the crankcase and the oil pan at the middle of the crankshaft (so-called short skirt design) (**Fig. C290**). With this design, the upper halves of the crankshaft bearings are integrated into the crankcase casting as so-called main bearing pedestals. The lower halves of the crankshaft bearings are designed either as individual main bearing caps or as a ladder-frame construction.

The crankcase and oil pan are sealed between the flanges at the separating plane. The crankshaft is sealed at the front and rear ends in a specific manner corresponding to the engine design—for example, the front end of the crankshaft is sealed with radial shaft seal in the oil-pump housing or in the front cover; the rear end of the crankshaft is sealed with a radial shaft seal in the last main bearing pedestal or in a separate cover. Gray cast iron crankcases with the separating plane at the oil pan at the center of the crankshaft and with separate main bearing caps were used frequently up to the middle of the 1990s for small-displacement (up to about 1.8 L), inline, four-cylinder engines, as well as for some V-6 and V-8 engines. As a rule, such engines were initially designed 15–20 years previously.

The advantage of this design is the low production cost. The disadvantages of this design in comparison to crankcases with long side walls or with crankcase bottom sections are the lower rigidity and poorer acoustic properties.

~~~Oil pan flange below crankshaft center.
Two crankcase designs exist with the separating plane located between the crankcase and oil pan:

- Design with crankcase top section and bottom section (**Fig. C291**).
- Crankcase with long side walls (**Fig. C292**).

Design with crankcase top section and bottom section. With this design, the main bearing caps are consolidated into a single bearing housing, the so-called crankshaft bottom section. The separating plane between the crankcase top section and bottom section is at the center of the crankshaft. This means that the

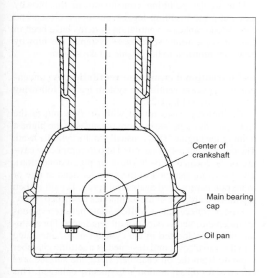

Fig. C290 Oil pan flange at center of crankshaft

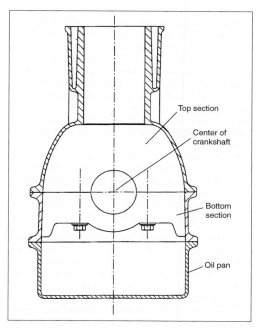

Fig. C291 Design with crankcase top section and bottom section

C

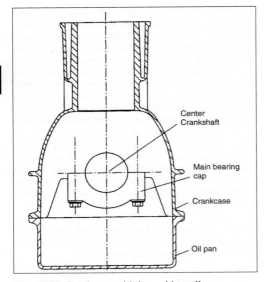

Center
Crankshaft

Main bearing
cap

Crankcase

Oil pan

Fig. C292 Crankcase with long side walls

component designated as the crankcase top section here corresponds to the crankcase design with oil pan flange at the crankshaft center.

The bottom plane of the crankshaft bottom section is the flange to the oil pan. The crankshaft is sealed depending on the engine design at the rear end by a radial shaft seal in the last main bearing pedestal and at the front by a radial shaft seal in the oil pump housing or end cover.

The advantages of this design are the high rigidity, good acoustic properties, and design possibilities — particularly for the body section of the crankcase as described above at ~~~Crankcase bottom section, and at ~~~Ladder-type frame design. An example is casting spherical graphite cast iron inserts into the area of the individual bearing points on aluminum alloy crankcase bottom sections produced by die casting. Disadvantages are higher production costs and, where applicable, slightly higher weight of individual main bearing caps.

This version is used in mass production with aluminum alloy crankcase upper and lower sections. Because racing engines frequently use the entire vehicle as a supporting part, racing engine crankcases are almost always produced using this design principle because of the higher rigidity required.

Crankcase with long side walls. With this design, the outer walls of the crankcase extend downward to below the middle of the crankshaft (so-called deep skirt design), where they terminate at the flange plane to the oil pan. The main bearing pedestals are still divided at the crankshaft center line for processing reasons.

Versions in production have individual main bearing caps as well as main bearing caps consolidated into a ladder frame construction. The advantages of construction with a ladder frame include similar high rigidity, similar good acoustic properties, and, also depending on the quantity, slightly lower production costs compared with designs with crankcase upper sections and bottom sections.

Gray cast iron (GG) crankcases on mass-produced engines are frequently designed with the deep skirt design and individual GG main bearing caps. Aluminum crankcases are also being mass produced with the deep skirt design and aluminum ladder frame construction.

~Crankcase ventilation (*also*, →Blowby). Crankcase ventilation returns blowby gases virtually free of lubricating oil to the engine intake system ensuring that virtually no pressure exists on the inside of the engine.

During operation of reciprocating piston engines, blowby gases escape from the combustion chamber into the area between the piston or piston rings (piston ring gap) and cylinder into the crankcase.

These blowby gases contain the same spectrum of emissions as the exhaust in addition to unburned fuel. The percentage of hydrocarbons (HC) in the blowby gas can be a multiple of the HC concentration contained in the exhaust, depending on the engine load level.

A speed-dependent pressure is developed in the chamber below the piston in the crankcase resulting from the quantity of blowby gases, depending on the engine speed and from the translateral piston motion. Because the crankcase is connected with the cylinder head through passages for the oil return, crankcase ventilation and (where applicable) timing chain staff, pressure is also present at these points on the inside of the engine.

The blowby gases, mixed with the lubricating oil, are present in the crankcase in the form of oil fog.

During the initial years of engine construction, engines were ventilated by releasing the blowby gas/lubricating oil mixture into the atmosphere without treatment.

Due to the polluting constituents in the blowby gases and the resulting legal regulations, closed and controlled crankcase ventilation systems have been in use for some time. With these systems, the blowby gases are returned to the engine intake system.

~~**Conventional crankcase ventilation.** A conventional crankcase ventilation system has the following basic function (**Fig. C293**).

The blowby gases mixed with oil flow out of the crankcase through one or more passages to the highest point of the engine. This is usually in the cylinder head. The passages may be integrated into the crankcase casting or be located outside (hose or pipe connection). The gases are returned to the cylinder head at one or more points protected against oil spray. The oil is then separated using one of the techniques described below.

First, it is necessary to differentiate between various arrangements and versions of oil separators. For example, some are located in the cylinder head, integrated into the valve cover and designed as a settling chamber separate from the oil chamber. Here, the surface area effective for oil separation is frequently enlarged. This was accomplished previously with steel mesh but it is

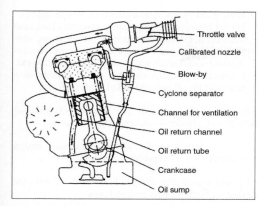

Fig. C293 Conventional crankcase ventilation

now done with expanded sheet metal. Another possibility is to locate the oil separator outside of the engine — for example, in the form of a cyclone oil separator.

The location and version depends on a variety of criteria, including the design of engine, the installation sites available (engine space package), and finally the design principles and the design philosophy of the engine or automobile manufacturer. For example, in V-engines, oil separators are located in the area between the cylinder banks, the "engine V," partially or completely integrated into the crankcase or located in this area as an external oil separator.

After separation of the oil, the blowby gases are returned to a suitable point at which, ideally, a vacuum is present in the engine intake system in nearly all engine operating states. As a matter of principle, this is the case in front of the throttle valve.

While the crankcase ventilation was usually discharged into the air filter housing earlier on carbureted engines, today the crankcase ventilation discharges into the engine intake system just in front of the throttle valve on engines with fuel injection to avoid contamination of the mass air sensor and idle charge control.

A connection between the intake system at back of the throttle valve and the crankcase produces a vacuum in the crankcase. This connection line contains a calibrated throttle to limit the effective vacuum.

An excessive vacuum inside of the engine in principle has the same effect on the engine sealing system as excessive pressure. It results in failure of the engine sealing system. If the vacuum is too high, unfiltered air is sucked into the engine. The result is accelerated aging of the oil from oxidation and formation of sludge. Engine oil leakage can occur if the pressure is too high.

This system satisfies the functions required for proper engine operation and maintenance of the legal regulations within acceptable component costs (oil separator, hoses, hose clamps, calibrated nozzle).

The functionality system is ensured by calibration of the nozzle for limiting of the vacuum in the engine, which is determined by optimization research and must be maintained during mass production. The blowby

gases contain fuel vapors and water vapor when the engine has not yet reached operating temperature. These vapors can freeze in the hoses, but this can be eliminated by suitable routing of the hose to eliminate siphoning and, where applicable, by other additional measures.

Even though the lubricating oil is separated from the blowby gases in front of the throttle valve before discharge, the constituents in the flow pose a hazard for contaminating the components present in modern engines with fuel injection, such as the air mass flow sensor and idle charge control.

~~Positive crankcase ventilation (PCV). With this system, fresh air cleaned in the intake air filter is fed into the engine in a controlled, continuous, or load-dependent manner. The mixture of blowby gas and lubricating oil is mixed with the fresh air. The system is controlled by tuned throttles and valves. The oil is separated in basically the same manner as in conventional or vacuum-regulated systems. The blowby gases are also fed to the engine intake system.

The water and fuel vapors contained in the blowby gases are picked up by the fresh air and continuously removed from the crankcase.

The disadvantage of PVC is that, in addition to the higher construction costs, accelerated engine oil aging from oxidation and black sludge formation can occur. Oxidation of the lubricating oil, which is always present, is reinforced by the oxygen contained in the circulating air. Although the fresh air is filtered to the greatest possible extent in the intake air filter, residual contamination can result in formation of oil sludge.

~~Vacuum-regulated crankcase ventilation. With this system, the blowby gas is discharged from the oil separator through a differential pressure valve into the intake system, downstream of the throttle valve — that is, into the intake manifold (**Fig. C294**). In comparison to the conventional system, this system eliminates the connection between the oil separator and intake system in front of the throttle valve as well as the vacuum line with an integrated throttle between the engine and intake system downstream of the throttle valve.

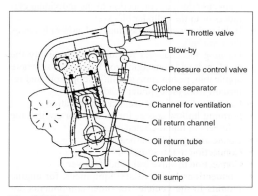

Fig. C294 Vacuum regulated crankcase ventilation

The differential pressure valve is a spring-loaded diaphragm-type valve with adjusted bypass. It regulates the vacuum inside the engine to a permissible maximum value under nearly all engine load states. Excessive vacuum in the engine could have negative effect as described for conventional crankcase ventilation systems.

With this system, a vacuum is possible in the crankcase over the entire engine map. In comparison to conventional crankcase ventilation systems, this system requires fewer components (hoses, hose clamps), and the danger of ice forming in the hoses is lower. Introduction of the blowby gases into the intake system downstream of the throttle valve virtually excludes contamination of the air mass flow sensor and idle charge control.

A more theoretical disadvantage is the possible failure of the diaphragm in the differential pressure valve. Such failure has not yet been encountered in practical operation.

~Crankcase with long side walls. *See above*, ~Crankcase design ~~Main bearing pedestal area ~~~Oil pan flange at crankshaft center

~Cylinder. *See above*, ~Crankcase design

~Design. *See above*, ~Crankcase design

~Design, cooling fins. *See above*, ~Crankcase construction ~~Air cooled

~Diecast. *See above*, ~Casting process, crankcase

~Dry liners. *See above*, ~Crankcase design ~~Cylinder ~~~Insert technology

~Function, crankcase. The primary functions of the crankcase are:

- Transferring the gas forces from the cylinder head to the crankshaft bearings
- Holding the drivetrain consisting of piston, connecting rod, crankshaft, and, as a rule, flywheel or torque converter drive plate
- Holding the cylinders, or with multiple section crankcase design, connection to the individual cylinders or to the cylinder block(s)
- Holding the crankshaft bearings and, now only rarely, the camshaft
- Forming the channels for lubricant to supply the crankshaft and connecting rod bearings and, where applicable, piston spray cooling, supply oil, and return oil from the cylinder head(s)
- Forming channels for crankcase ventilation
- Forming hollow spaces for coolant for cooling the cylinders in liquid-cooled engines
- Connection to cylinder head(s)
- Connection to transmission
- Connection to valve train and its cover
- Connections for auxiliary equipment, for engine mounts in the vehicle, for other components depending on engine concept such as coolant pre-

heating, oil/water heat exchanger, where applicable oil filter, where applicable external crankcase ventilation, external oil separator for crankcase ventilation, electric sensors such as oil pressure, oil temperature, crankshaft speed and knock sensors, etc.
- Connection of crank chamber to the outside, as a rule over the oil pan and radial shaft seals

~Honda MMC process. *See above*, ~Crankcase design ~~Cylinder ~~~Composite technology

~KS Lokasil process. *See above*, ~Crankcase design ~~Cylinder ~~~Composite technology

~Ladder-type frame design. *See above*, ~Crankcase design ~~Main bearing pedestal area

~Lost foam process. *See above*, ~Casting process, crankcase

~Low-pressure casting. *See above*, ~Casting process, crankcase ~~Chilled casting

~Magnesium. *See below*, ~Weight, crankcase ~~Reduction of specific weight of material

~Main bearing cap. *See above*, ~Crankcase design ~~Main bearing pedestal area

~Main bearing pedestal. *See above*, ~Crankcase design ~~Main bearing pedestal area

~Main bearing pedestal area. *See above*, ~Crankcase design

~Material specific weight reduction. *See below*, ~Weight, crankcase

~Materials. **Fig. C295** shows the most common crankcase materials—and also GGV (vernicular graphite cast iron) for comparison—as well as the most important material properties.

~Mono-metal design. *See above*, ~Crankcase design ~~Cylinder

~Oil pan. The oil pan is located at the bottom of most engines—that is, engines with wet-sump lubrication (**Fig. C296**). The most important functions of the oil pan include:

- To serve as a reservoir for holding lubricating oil
- To serve as a collection reservoir for lubricating oil flowing back from the engine
- To seal the crank chamber
- To hold a threaded element for the oil drain plug
- To hold the oil dipstick guide tube
- Where applicable, to hold an oil level sensor for indicating the oil level on the vehicle instrument panel
- To reinforce the engine/transmission assembly (so-called structural oil pan) by means of special oil pan design

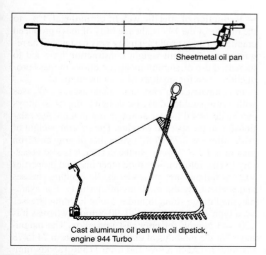

Material Group	Aluminum						Iron		
Material	AlSi6Cu4		AlSi17Cu4Mg		AlSi9Cu3		GG 25	GG 30	GGV
Remarks	Hypo-eutectic		Hyper-eutectic Heat Treated	Hyper-eutectic	Hypo-eutectic		Cast iron with Lamella Graphite	Cast iron with Lamella Graphite	Vermicular Graphite
Casting technique	Sand and chilled mold	Die cast	Sand and chilled mold	Die cast	Sand and chilled mold	Die cast			
Strength limit $R_{p0.2}$ (N/mm²)	100–180	150–220	190–320	150–210	100–180	140–240	165–228	195–260	240–300
Tensile strength R_m (N/mm²)	160–240	220–300	220–360	260–300	240–310	240–310	250	300	300–500
Strength at break A_g (%)	0.5–3.0	0.5–3.0	0.1–1.2	0.3	0.5–3.0	0.5–3.0	0.8–0.3	0.8–0.3	2–6
Brinel hardness HB	65–110	70–100	90–150	25	65–110	80–120	180–250	200–275	160–280
Fatigue strength under bending stress (N/mm²) NG = 25 – 108	60–80	70–90	90–125	70–95	60–95	70–90	87.5–125	105–150	160–210
Modulus of elasticity (kN/mm²)	73–76	75	83–87	83–87	74–78	75	103–118	108–137	130–160
Heat expansion coefficient. [20°–200°C] (10⁻⁴/K)	21–22.5	22.5	18–19.5	18–19.5	21–22.5	21	11.7	11.7	11–14
Thermal conductivity (W/mK)	105–130	110–130	117–150	117–150	105–130	110–130	48.5	47.5	42–44
Density (kg/dm³)	2.75	2.75	2.75	2.75	2.75	2.75	7.25	7.25	7.0–7.7

Source: Kolbenschmidt AG, Neckarsulm, Handbuch Alumnium - Gussteile, Heft 18, (Aluminum manual, cast parts, Book 18)
DIN 1691 Cast Iron With Lamella Graphite (Gray Cast Iron),
Porsche Technical Delivery Specifications 2002,
Vermicular graphite cast iron (GGV) – A new material for internal combustion engines, Aachener Kolloquium Fahrzeug- und
Motorentechnik 95, F. Indra, M. Tholl, Adam Opel AG. Rüsselsheim

Fig. C295 Crankcase materials

Sheetmetal oil pan

Cast aluminum oil pan with oil dipstick, engine 944 Turbo

Fig. C296 Oil pan

~~**Oil pan design.** On mass production engines, the oil pan is usually made of deep-drawn sheet metal. Usually only one layer of sheet metal is used. However, a new design with two layers of sheet metal with a plastic foil located in between is now being used to improve the engine noise levels.

On large volume production engines, Al-Si alloy oil pans are frequently used, which are manufactured using the chilled mold or die-casting process. Such oil pans can be and usually are designed as structural oil pans.

This design contributes significantly to stiffening the engine/transmission assembly and, therefore, to better acoustics as a result of the stiff design of the oil pan side walls and primarily by an integrated flange on the engine clutch end for connection of the transmission flange.

Al alloy oil pans are produced in single-piece as well as two-piece versions. Two-piece oil pans consist of a light alloy upper section and a sheet metal bottom section screwed to the former. This design has the advantage that the sheet metal oil pan bottom section deforms if the vehicle bottoms out in the area of the lowest point on the engine, while a single-piece oil pan would crack.

Today, this advantage is of only subordinate significance due to the increased use of underbody paneling on vehicles in this area of the engine.

~Oil pan flange at crankshaft center-line. *See above,* ~Crankcase design ~~Main bearing pedestal area

~Oil pan flange below crankshaft center-line. *See above,* ~Crankcase design ~~Main bearing pedestal area

~Open-deck. *See above,* ~Crankcase design ~~Cover plate

~Operating load. The crankcase is subjected to tension or compression, flexion, and torsion by:

- Gas forces transferred from the cylinder head to the crankshaft bearings
- Internal moment of inertia (bending moments) resulting from rotating and oscillating mass forces
- Internal torsional forces (tilting moments) between individual cylinders
- Crankshaft torque and the resulting reaction forces in the engine mounts
- Free inertia forces and inertia moments resulting from oscillating mass forces of the first and second order, which have to be taken up by the engine mounts.

The effect of the forces and the resulting moments on the inside of the crankcase, as well as on the outside (engine mounts, noise radiation produced by mechanical vibration; *see,* →Engine acoustics), depend on the design of the engine.

The primary parameters of the engine design that influence the crankcase load are:

- Number of cylinders
- Cylinder arrangement
- Crankshaft throw arrangement
- Firing order

The loads occurring in the crankcase have an effect on:

- The design of the crankcase and the crankcase layout in terms of sufficient strength, minimum deformation, low-cost production, recycling
- Noise radiation (acoustics)
- Crankcase weight and, therefore, the total engine weight

~Operating technology. *See above,* ~Crankcase design ~~Cylinder

~Permanent mold casting. *See above,* ~Casting process, crankcase ~~Chilled casting

~Positive crankcase ventilation. *See above,* ~Crankcase ventilation

~Primary dimensions, crankcase. *See above,* ~Crankcase design

~Ribbing. *See above,* ~Crankcase construction

~Sand casting. *See above,* ~Casting process, crankcase

~Squeeze casting. *See above,* ~Casting process, crankcase

~Strength, crankcase. *See above,* ~Crankcase design

~Thermal conductivity of cylinder and rib material. *See above,* ~Crankcase construction ~~Ribbing

~Thin-wall casting. *See below,* ~Weight, crankcase ~~Reduction of specific weight of material

~Vacuum-regulated crankcase ventilation. *See above,* ~Crankcase ventilation

~Water cooled. *See above,* ~Crankcase construction

~Weight, crankcase. Development targets for passenger cars include minimization of fuel consumption and emissions while providing attractive performance. Achieving these objectives requires implementation of consistent light-weight construction for all vehicle components in addition to other measures. One contribution to optimizing the weight of the entire drivetrain is reduction of the crankcase weight.

The crankcase weighs about 25–33% of the total engine weight (according to DIN 70020 A) depending on the size of the engine (displacement, number of cylinders), engine design (cylinder arrangement, gasoline or diesel engines) and crankcase design. Reduction of the crankcase weight significantly reduces the total engine weight.

The measures for reducing the crankcase weight can be subdivided into:

- Reduction of specific weight of material
- Weight reduction by optimization of the structure

~~**Reduction of specific weight of material.** Up to the middle of the 1990s, the majority of mass-produced crankcases were produced using gray cast iron. The requirements for light-weight construction have led to increased use of aluminum/silicon alloys in mass production, even for small displacement engines.

In comparison to cast iron crankcases, crankcases with a comparable crankcase design using Al-Si alloys can be designed lighter although not in a precise relationship to the specific weights. The specific weight of Al-Si alloys is 2.75 g/cm^3, while that of gray cast iron materials 7.2–7.7 g/cm^3. In the design, it is also necessary to take into consideration other material properties such as fatigue strength under cyclic bending stresses and particularly the modulus of elasticity. For example, the fatigue strength under cyclic bending stresses for a typical Al-Si alloy is 70–90 N/mm^2, whereas it is 120–145 N/mm^2 for gray cast iron alloy. The magnitudes for the modulus of elasticity range from 74 to 78 kN/mm^2 for Al-Si alloys and from 115 to 135 kN/mm^2 for gray cast iron alloys. The weight of aluminum crankcases can be reduced by 40–60% compared with gray cast iron crankcases, depending on the size of the engine.

On gray cast iron crankcases, reduction of the weight is possible by a combination of structural optimization and thin-wall casting. In the case of thin-wall casting, general wall thicknesses down to approximately 3 mm are possible, whereas the general wall thicknesses for cast iron crankcases are usually in the range of 4.0–5.5 mm.

The use of vermicular graphite cast iron (GGV) provides a cast material with high strength, and weight reductions up to about 30% can be achieved compared with crankcases of the same design using conventional cast iron material such as GG 25. Such weight savings require a crankcase design suitable for the GGV material.

The advantages of substituting GGV for GG 25 in crankcases are higher rigidity and better acoustic properties in addition to the possible weight savings. Disadvantages are the higher material costs, estimated to be 20–28% more per kg of cast material, which might be compensated for by the reduction in the weight as well as the longer service lives of the processing tools.

~~~**Magnesium crankcase.** Magnesium (Mg) alloy crankcases were used for a long time for air-cooled engines. Examples are the horizontally opposed four-cylinder engines for the VW Beetle and horizontally opposed six-cylinder engines for the Porsche 911, whose crankcases were produced using an Mg alloy from the end of the sixties to the beginning of the seventies. Today, Mg crankcases are produced only for racing engines. The primary advantage of using Mg alloys for the crankcase material is its low specific weight. Disadvantages in comparison to the Al-Si alloys used for mass production of light-weight crankcases today are the high material costs, lower material strength and lower resistance to corrosion.

The specific weight of Mg alloys is 1.8 g/cm^3; for Al-Si alloys the specific weight is 2.75 g/cm^3. Crankcases of comparable design using Mg alloys are not lighter than Al-Si alloy crankcases in proportion to the specific weights. It is necessary to consider the layout in terms of the stress and the differences in the material characteristics. For example, the 0.2% yield strength of a typical Al-Si die-cast alloy is 140–240 N/mm^2, while that of a typical Mg die-cast alloy is 140–160 N/mm^2. The values for the modulus of elasticity are 75 kN/mm^2 for Al-Si die-cast alloys and 45 kN/mm^2 for Mg die-cast alloys. In comparison to an Al-Si crankcase, weight savings of the order of 25% are possible using an Mg alloy with comparable crankcase design.

The lower strength and modulus of elasticity of Mg alloys can be compensated for in the design to the greatest possible extent by redesigning the structure to take this into account. For example, coolant ducts for cross-cooling integrated longitudinally into the Mg crankcase in racing engines increase the structural rigidity.

There are many reasons for not using Mg crankcases currently for mass production engines. Al-Si alloys have a significant cost advantage of about 3:1, compared with Mg alloys, due to the absence of an Mg recycling market. While Al-Si alloys are available economically in the form of secondary alloys from components that have been remolten, it is necessary to use expensive primary alloys for Mg alloys. However, the higher Mg alloy material costs must be compared from case to case with the lower costs resulting from weight reduction as well as shorter processing times and longer service lives of the die-casting molds and processing tools.

Without additional measures, the corrosion resistance of Mg alloy components is lower than that of components produced using Al-Si alloys whose natural cast surface/casting skin is already resistant to corrosion. It is necessary to differentiate between contact corrosion and surface corrosion.

Contact corrosion results when Mg alloy components are in contact with components consisting of other metals or metal alloys. This results from the different positions of the various metals in the electrochemical displacement series. Contact corrosion results, for example, at bolted connections and holes for fastening elements, such as alignment sleeves and alignment pins. The measures required for corrosion protection result in an increase in the costs, including use of washers of Al-Si alloy and special surface protection for bolts and alignment sleeves.

Surface treatment such as CrVI-free passivation before processing the part and application of wax or powder coating after processing are required to prevent surface corrosion on the outer surface of components of common Mg alloys. Components produced using some HP-Mg alloys (high purity) have sufficient protection against surface corrosion even without such surface treatments, but the sufficiency of protection must be considered within the context of the magnitude of the corrosion stress occurring.

For use of Mg in mass production crankcases, it is necessary to provide sufficient resistance to surface corrosion in the water jacket for the engine service life period required today, corresponding to a mileage of 160,000 km.

Due to the tribological properties of the Mg alloy, it is not possible to use Al-Si alloy pistons directly in Mg cylinders. Mg crankcases require development of a cylinder running surface technology that is compatible with the Mg alloy basic material, analogous to the cylinder running surface technologies for Al-Si alloy crankcases (using GG or Al sleeves, composite material engineering).

Mg alloy crankcases allow a further reduction in the weight of the drive equipment in comparison to Al-Si alloy crankcases as a prerequisite and contribution to further reduction in the fuel consumption of motor vehicles.

~~**Weight reduction by structural optimization.** The crankcase design has a considerable influence on the total crankcase weight. Design methods available today such as CAD (computer-aided design) and FEM (finite element method) calculations allow better specific optimization of all designs compared with previously.

With crankcases, these methods allow systematic optimization of the structure and ribbing to meet the load and function. This means that the wall cross sections required for the function and the exact position, number and geometry of fins, which serves to increase the rigidity and improve the acoustics, can be optimized in terms of weight today with the minimum use of material.

Cylinders cast together as well as integration of many functions in the crankcase also contribute to reduction of the total engine weight.

~Weight reduction by optimization of the structure. *See above,* ~Weight, crankcase

~Wet sockets. *See above,* ~Crankcase design ~~Cylinder

Crankcase bottom section →Crankcase ~Crankcase design ~~Main bearing pedestal area

Crankcase construction →Crankcase ~Crankcase construction

Crankcase design →Crankcase

Crankcase dimensions →Crankcase ~Crankcase construction ~~Primary dimensions

Crankcase materials →Crankcase

Crankcase stress →Crankcase

Crankcase ventilation, piston →Blowby; →Crankcase; →Crankcase ventilation systems

Crankcase ventilation systems (*also,* →Crankcase). During operation of an internal combustion engine, crankcase ventilation occurs for so-called blowby gas. The blowby gas escapes from the combustion chamber through the gaps between the cylinder wall and piston and between the piston and piston rings, through the ring gaps in the piston rings, and through the valve seals into the crankcase. In addition to products resulting from complete and incomplete combustion, the primary constituents in the blowby gas include water (vapor), soot, and fuel residues as well as engine oil in the form of microscopic droplets. This contaminated gas flow is forced out of the crankcase by a ventilation system with additional components such as pressure control devices, check valves, and so on and, in a closed ventilation system, returned to the combustion air sucked in by the engine (**Fig. C297**).

With open ventilation systems, the purified gas is discharged directly into the atmosphere. However, environmental legislation now permits use of open systems in exceptional cases only.

In turbocharged diesel engines and gasoline engines with direct fuel injection, the lubricating oil and soot

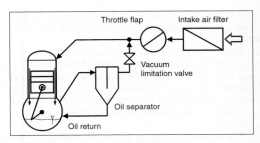

Fig. C297 Closed crankcase ventilation system on an internal combustion engine

contained in the blowby gas occasionally pose problems resulting from deposits on the turbocharger, in the charge air cooler, on the valves, on the injection nozzles, and, where applicable, in subsequent particulate filters (ash deposits from inorganic additives in the lubricating oil). These deposits can impair the function of the affected components. In addition to protecting various components, another important purpose is to reduce the oil consumption by limiting the quantity of engine oil taken out of the crankcase by the crankcase ventilation.

The lubricating oil contained in the blowby gas comes from the oil sprayed by moving engine parts, from the lubricating film on the cylinder wall, and from the condensing of evaporated lubricating oil for cooling the piston crowns. The mean droplet size and percentage of the entire spectrum differ depending on the source of the lubricating oil. The spray oil has a relatively coarse spectrum; the droplet size is partly in the millimeter range. This quantity can be influenced significantly by suitable selection of the removal point or simple measures inside the engine (shields, etc.). The droplets resulting from the lubricating films and particularly from the condensing of evaporated lubricating oil are primarily very fine. Here, the mean droplet size is between 0.5 and 1 μm. These extremely fine droplets are removed quantitatively from the engine chamber without suitable separation devices. Their percentage depends primarily on the operating conditions (engine load and speed).

The droplet size distribution shown in **Fig. C298** for the quantity of oil entrained in the gas shows that the mean droplet diameter is between 0.5 and 2 μm, depending on the type of engine. The droplet diameter decreases with increasing mean effective pressure in the combustion chamber and increasing oil temperature. The minimum droplet diameters are usually present in the load range close to the maximum engine torque.

~Oil separation systems. The crankcase ventilation system seldom consists of only one oil separator; it can be enhanced to achieve a more or less complex system, depending on the concept.

Oil return. This is achieved by a siphon or reservoir with valve located either below the oil level or without

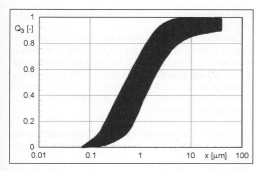

Fig. C298 Droplet spectrum of gas entrained oil in blowby gas. Aerodynamic diameter determined on various engines

connection to the oil sump. Direct return to the oil sump or through a siphon offers the advantage that even large quantities of oil can be returned without a problem. In the return flow to the sump below the oil level, it is absolutely necessary for reliable operation that the pressure difference when flowing through the oil separator is not greater than the static pressure of the oil column in the return line, resulting from the geodesic height even when the engine is at an angle. If this pressure difference is exceeded, the oil rises out of the oil sump through the return line and oil separator to the intake area. This can be prevented by a float valve which closes the return line when the oil level rises. If the separated oil is returned through a reservoir with a valve as shown in **Fig. C299**, this allows the separator to be installed in practically any position on the engine. Proper design of the reservoir and check valve allows oil to run back from the reservoir into the engine chamber to a limited extent, even when the engine is operating.

Preseparator. Because many oil separators cannot handle large quantities of oil, it may be necessary to inte-

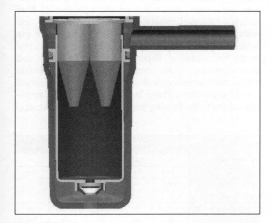

Fig. C299 Cyclone separator with reservoir and oil check valve

Fig. C300 Preseparator unit with louvers, oil return opposite inflow direction

grate an additional preseparator. The preseparator must be capable of separating and returning large quantities of oil with minimum pressure loss, so preseparators are usually volume-type separators with simple inserts (**Fig. C300**).

Vacuum limitation valve. A vacuum limitation valve or pressure regulation valve is used to prevent the vacuum in the crankcase from decreasing to that in the intake manifold (up to -80 kPa). Common versions for regulating the pressure use a spring-loaded diaphragm (**Fig. C301**). In addition, valves which regulate the flow rate are also in use. However, these require positive crankcase ventilation, because the pressure can be regulated only by the additional flow from this ventilation. Operating states at which the maximum flow rate which can flow through the flow control valve are exceeded, are critical.

Positive crankcase ventilation (PCV). PCV (**Fig. C302**) serves to reduce the concentration of condensable vapors (fuel, water) by using additional fresh air to such an extent that the vapors cannot condense out into the

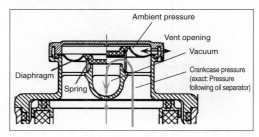

Fig. C301 Diaphragm-type vacuum limitation valve

C

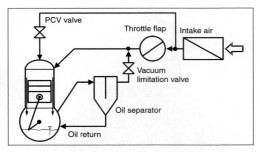

Fig. C302 PCV system with check valve (PCV valve)

crankcase even at low temperatures or at least only to a limited extent. At very low outdoor temperatures when the ambient air is accordingly dry, this is particularly successful. The quantity of water in the condensate can freeze at low temperatures and, in the worst case, the lubrication circuit can be interrupted, leading to complete destruction of the engine. Damage can also be caused by ice partially clogging the vent line, allowing the pressure in the crankcase to increase to a high level. At corresponding pressures, leakage can occur through the oil dipstick, radial shaft seals, or valve cover gaskets. It is necessary to integrate check valves into the system to prevent unpurified blowby gases from flowing back to the PCV line. These valves can act as one-way restrictors and also limit the flow rate and, in the event of a malfunction (e.g., vacuum limitation valve) also prevent uncontrolled flow through the crankcase.

Heating pipe. Formation of ice can be prevented at particularly exposed points in the ventilation system by using heated pipes and heating elements (**Fig. C303**). Heating is usually accomplished electrically with self-regulating positive temperature coefficient (PTC) elements that reduce the current upon reaching a steady-state temperature.

Check valves are required in positive crankcase ventilation systems (PCV) in the air feed line as well as in branched systems, in which the ventilation gases are fed to the intake system, depending on the pressure upstream of the exhaust turbocharger or downstream of the turbocharger and, where applicable, the throttle valve (gasoline engines).

~Oil separator. Various separating processes can be used for separating the oil contained in the blowby gas (**Fig. C304**).

Labyrinth-type separators are simple, relatively high-volume inertia-type separators with baffles in the direction of flow. The droplets precipitate in these baffles—that is, they deposit in the areas where the flow rate is decreased. Labyrinth-type separators are used primarily for separating large oil droplets and large quantities of oil. Separation of small droplets with sizes less than about 2 μm is practically impossible. Labyrinth-type separators are designed as permanent components for the service life of the engine.

Cyclone separators are also inertia-type separators in which the droplets are centrifuged out of the gas in a rotating flow field. An increase in the separation rate is possible when a number of smaller cyclones are connected in parallel to form multiple cyclone units (**Fig. C305**).

With the correct layout adapted to the engine conditions, droplets also can be separated around 1.5 μm to a significant extent. Cyclone separators require precise definition of the operating point, because the separation performance is linked closely to the pressure loss and, therefore, the flow rate. Cyclone-type separators are designed as permanent components for the service life of the engine, just as are labyrinth-type separators.

Fiber separators or coalescers are primarily diffusion-type separators, depending on the selection of fiber material and, particularly, the fiber diameter. Droplets with sizes less than 1 μm can still be separated out with this type of separator. The droplets are deposited in the fiber material, depending on the flow rate, and then flow out of the fiber material or agglomerate where they form larger drops which can be removed from the medium. These larger drops can then be removed from the flowing gas with simpler mechanical

Fig. C303 Electrically heated pipe (PTC heating elements)

	Labyrinth	Cyclone	Fiber separator	Centrifuges	Electrostatic separator
Separation rate	– to 0	0 to +	0 to ++	++	++
Pressure loss	+ to ++	0 to –	+ to –	++	++
Required space	0 to +	++	+	+	0
Sensitivity flow rate	0	– to 0	0	+	+
Service life of component	Yes	Yes	No	Yes	Yes
Additional auxiliary energy	No	No	No	Yes	Yes

Fig. C304 Evaluation of various oil separation systems

Fig. C305 Duo-cyclone with reservoir and check valve for five-cylinder CRD (common rail diesel) engine

separators. Because this process requires the use of extremely fine fiber mass (**Fig. C306**), which has a tendency toward clogged pores, particularly in the face of high quantities of soot, it is usually necessary to design fiber separators as a service part.

Centrifuges are assemblies that rotate at high speed, removing droplets from the gas by centrifugal force.

Due to the external drive, the degree of separation is independent from the pressure loss to a great extent. This allows the highest precipitation rates with moderate pressure losses. The centrifuges can be driven by an electric motor which is independent of the engine speed. Drive concepts which are dependent on the engine operating state are also possible using a mechanical coupling to the rotating engine parts, hydraulic oil repulsion drive, a Pelton turbine, or a pneumatic drive using compressed air from the compressor or turbocharger. Plate-type separators offer the maximum potential, because they achieve the highest separation rates with minimum installation space at acceptable speeds (**Fig. C307**). Centrifuge-type separators are designed as permanent components for the service life of the engine.

Electric separators utilize the forces acting on charged droplets or particles in a strong electric field for separation. As with centrifuges, this ensures maximum separation rates with minimum pressure losses. The high voltage required for satisfactory separation is about 5–15 kV. Problems frequently result from deposits on the electrodes from residues in the lubricating oil and blowby gas.

Crankcase weight →Crankcase ~Weight, crankcase

Crankcase with long side walls →Crankcase ~Crankcase design ~~Main bearing pedestal area ~~~Oil pan flange at crankshaft center

Cranking speed →Speed

Crankpin →Crankshaft ~Throw

Crankshaft. In internal combustion engines, the crankshaft converts the reciprocating motion of the pistons to rotary motion through the moments of the forces applied by the connecting rod. In combination with the periodic vibrations intrinsic to the system, this leads to significant stress on the crankshaft, for which it must be designed accordingly (*also, see below*, ~Stress resis-

Fig. C306 Coalescence or diffusion separator element

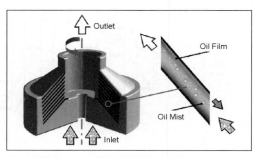

Fig. C307 Functional principle and design of plate-type separator for separation of oil vapor aerosols

C

tance). The main elements of a crankshaft are illustrated and identified in **Fig. C308**.

~Balance weights. Counterweights are usually attached to the crankshaft webs opposite the big end bearing journals to compensate for one-sided rotating masses such as the mass of the bearing journals plus the rotating percentage of the connecting rod mass (approximately two-thirds of connecting rod weight). This allows for compensation of external moments, minimizes internal moments, and reduces the vibration amplitudes and bearing loads.

The best compensation results when these are attached directly opposite each big end bearing journal and correspond to the rotating masses. The number of counterweights is frequently reduced to save costs and weight—currently, for example, only the outer and middle crank webs are offset at the circumference so that the external moments are compensated (Σ moments = 0). However, this increases the internal moments.

The counterweights can be cast on, forged, welded, bolted, or clamped on (e.g., with dovetail connections).

The maximum rotating radius is limited by the space available in the engine (passage/piston collar). Compensation can also be improved by reducing the material on the big end bearing journal side. This can be achieved with holes for weight reduction or by inserting heaving metal into the web on the bottom dead center side, which is relatively expensive. Another possibility is to design the crank webs with chamfers (**Fig. C308**).

A compensating moment can also be achieved by using counterweights at the ends of the crankshaft. Fine tuning of the counterweights in terms of the external moments is accomplished by means of dynamic balancing, usually by drilling holes in the counterweights to remove material. If the crankshaft is not compensated to the outside without rotating masses, appropriate substitute weights must be taken into consideration for balancing—for example, master weights on big end bearing journals.

~Big end bearing journal. *See below*, ~Throw

~Blank. There is a series of processes, described in brief below, for producing the blank.

~~**Cast.** The percentage is approximately 95% in the United States, 60% in Europe, and 25% in Japan for passenger car and commercial vehicle crankshafts not subjected to high loads. Low specific weight and differentiated shaping reduce the crankshaft weight, and "near finish" casting with minimum dimensions reduces the machining required. However, this does result in greater scatter of the stress resistance resulting from greater structural inhomogeneity.

~~**Die forged.** This process is used for crankshafts with unfinished weights of up to 2.5 tons and lengths of 4.2 m for small and medium batch sizes.

~~**Hammer forged.** This process is used for prototype and individual parts of all sizes.

~~**Press forged.** This process is used for passenger car and commercial vehicle crankshafts for large batches.

~~**Single stroke forging.** This process is used for production of large crankshafts with lengths of 3–12 m. The individual throws are forged one after another at the correct angular position. This allows many of the same tools to be used in sequence. Most versions are based on the TR process (Tadeusz Rut).

The production sequence, as shown in **Fig. C309**, is:

    I. Spindle inserted in the TR jig
   II. Distortion of the later crank webs
  III. Forging and shaping of the throw

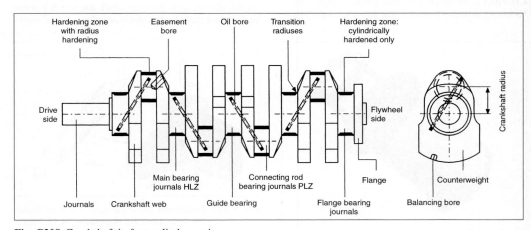

Fig. C308 Crankshaft in four-cylinder engine

C

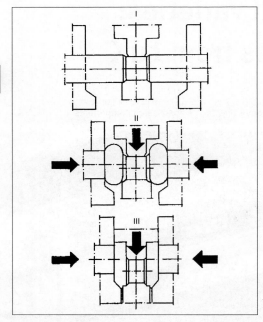

Fig. C309 TR forging process

~**Built-up crankshaft.** Larger crankshafts with a length greater than about 12 m are no longer produced in one piece. They are assembled from individual elements and by shrink fitting.

Even some very small crankshafts are built up, usually using serrations which are relatively expensive to produce. This allows the use of closed connecting rod eyes and self-aligned bearings on the big end bearing journals.

~**Crankshaft radius.** *See below*, ~Throw

~**Crankshaft web.** *See above*, ~Balance weights; *see below*, ~Stress resistance

~**Geometry.** The geometrical design of the crankshaft corresponds to the intended application and is determined primarily by the load. This applies particularly for the diameter and number of main and big end bearing journals, crank webs, crank radii, transition radii, and overall length.

The geometry of the bearing journals is of particular importance for the service life of the bearings. Here, the following elements are particularly important:

- Parallelism with one another
- Concentricity of main bearing journals with one another
- Surface line, frequently laid out with a crown shape to avoid edge supports
- Roundness/waviness (**Fig. C310**)
- Percentage of contact area/peak-to-valley height

The first two elements are influenced decisively by the curvature of the crankshaft (double curvature = "eccentric").

~**Guide bearing (thrust bearing).** The thrust bearing (**Fig. C308**) holds the crankshaft in the longitudinal direction and picks up axial forces. The thrust collars on the thrust bearing are larger in comparison to the thrust collars on the other main bearings and can be hardened (surface treatment) or burnished to reduce wear.

~**Machining.** This varies depending on layout, size, type of hardening, and batch size. An example of the machining for a four-cylinder passenger car crankshaft is given below. The crankshaft is inductively hardened, with a production rate of 500 parts/month.

- Centering
- Preliminary processing around center axis by turning, milling, and/or turn broaching
- Preliminary processing of the big end bearing journals by milling or turn broaching
- Boring of oil ducts
- Inductive hardening
- Flange and journal processing: boring, milling, turning
- Grinding of main bearing journals and end sections
- Grinding of big end journals

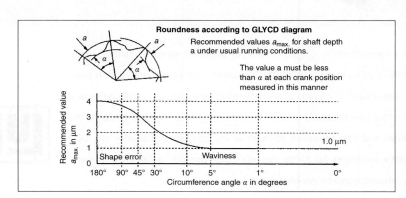

Fig. C310
Example for roundness or waviness requirements on passenger car crankshaft

C

- Magnet fluxing
- Manual work—e.g., rounding ends of oil holes
- Dynamic balancing
- Belt or stone finishing

Nitriding operations are accomplished after final grinding. Various types of machining can be accomplished in a fixture on rotary milling machines, a combination of lathes and machining center. This is interesting for prototypes and large crankshafts. Pendulum grinding has become well established for grinding the big end bearing journals. With this method, the crankshaft rotates around its center axis. This process requires the use of Cubic boron nitride (CBN) grinding wheels.

~Main bearing journal (*also, see below,* ~Throw). These are bearing points at the rotational center in the crankcase.

~Materials (*also, see above,* ~Blank). **Fig. C311** shows the materials used most frequently for crankshaft production and their mechanical properties (diameters range from 40 to 100 mm).

~Oil passage. The system of oil passages in the crankshaft is intended to supply all bearings with the required oil. The big end bearings are generally supplied with oil from the main bearings. However, separate supply of the connecting rod and main bearings is also possible, although more expensive.

Fig. C312 shows common oil passage systems: A) Most simple system with continuous passage from the surface of main bearing journal to surface of big end bearing journal. B) Additional cross-passages in main bearing journal. C) Cross-passages in main bearing and big end bearing journals connected with angular holes; moreover, these need to be closed. The many holes result in higher costs when it is necessary to deburr them on the inside.

A sufficient material to the transition radiuses at the critical cross section must be maintained for the corresponding connection passages. The angle of the hole in relation to the surface of the journal should be at least 40° to avoid fissures on hardened crankshafts. If the main bearing has only one oil groove in the bearing shell half at the supply point, it is necessary for the crankshaft to have a bearing cross passage (**Fig. C312B and C**).

~Production. *See above,* ~Blank; *see below,* ~Machining

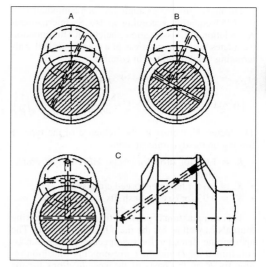

Fig. C312 System of oil passages

~Stress resistance. The crankshaft must be capable of enduring the permanent stress resulting during operation of the engine. This operating stress is composed of a bending load and torsional load in a different phase, resulting in a multi-axis stress condition. The torsion load results from the effective torque obtained by adding the rotary forces of the individual cylinders. The additional moments resulting from variations in the rotary force during an operating cycle as well as the stress from crankshaft deformation resulting from torsional vibration are superimposed on the effective torque.

The dynamic stress level in the crankshaft can be determined on a test bench (by recording component stress cycle diagrams for the bending and torsional stress) as well as calculated using empirically derived equations. These equations are based primarily on the results of vibration tests on crankshafts or material specimens. The equations attempt to take into consideration the influence of the crankshaft geometry (e.g., stress concentration factors based on journal diameters, transition radiuses, the crankshaft radius, and web design) and the effect of sizes and surfaces as well as the technological influence (resulting from the type of forging).

Material	Elongation Limit R_e [N/mm²]	Tensile Strength R_m [N/mm²]	Elongation at Break $A_{min.}$ [%]	Reduction in Area at Break $Z_{min.}$ [%]	Resistance to Impact $a_{k\,min.}$ (20°C)
Ck 45	370	630–780	17	45	25 DVM
38 MnS 6	550	850–1000	12	25	–
42 CrMo 4	650	900–1100	12	50	35 DVM
31 CrMoV9	>800	1000–1200	11	–	50 DVM
GGG70	>450	>700	>2	–	–

Fig. C311
Crankshaft materials

C

The most common example for such computation is the M53 equation according to IACS (International Association of Classification Societies). This equation calculates the dynamic stress of a crankshaft under alternating bending stress, in N/mm^2:

$$\sigma_{DW} = \pm K \cdot (0.42 R_m + 39.3) \cdot$$

$$\left(0.264 + 1.073 \cdot D^{-0.2} + \frac{785 - R_m}{4900} + \frac{196}{R_m} \cdot \sqrt{\frac{1}{R}} \right)$$

The factor, K, represents the influence of the type of forging on the dynamic strength.

- $K = 1.05$ for drop forge and HFH forged crankshafts
- $K = 1.0$ for hammer-forged crankshafts
- $K = 0.93$ for cast crankshafts

This computation is possible for the big end bearing journals as well as for the main bearing journals. The appropriate diameter for the specific journal can be substituted for D and the corresponding transition radius for R on the basis of the specifications for the strength certification for the big end bearing journal as well as main bearing journal. Moreover, it is clear that the dynamic strength of a crankshaft depends highly on the tensile strength, R_m, of the material, at least according to this equation.

In the other existing equation, this relationship to the tensile strength is also present; however, additional parameters are also included in the calculations. Other computation equations are:

- The "German Lloyd" equation takes into consideration the strength properties, 0.2 offset yield stress, and reduction in area on break.
- With the "FFV computation equation," the surface roughness at the transition radius is taken into consideration.
- In the "Crankshaft IV equation," the 0.2 offset yield stress has a relatively high influence.

In addition to the empirically determined equations, computer-aided strength calculations are being used more and more frequently today, which are frequently based on a 3-D FEM model.

There are basically two possibilities to increase the stress resistance of the crankshaft:

1. Design measures can reduce the stress concentration factors (e.g., by using larger transition radii) and the corresponding resistance moments can be increased (e.g., with greater web thickness, by increasing overlap between big end bearing journals and main bearing journals, by changing the shape of the webs, and by increasing the journal diameter). These design measures are intended primarily to reduce the magnitude of stresses present at points critical for fatigue stress.
2. The use of various surface treatment processes results in the formation of intrinsic compressive strains which counteract the tensile forces resulting from the load, particularly in the crankshaft areas with maximum stress concentration. Moreover, these processes increase the material strength of the crankshaft surface layer, which also has a positive effect on the stress resistance of the crankshaft. Therefore, the use of surface treatment processes is intended primarily to increase the stress resistance of the crankshaft without making it necessary to change the actual crankshaft dimensions. Because it is currently not possible to clearly calculate the increases in the stress resistance resulting from surface treatment preliminarily, dynamic component testing is necessary to determine the achievable improvement.

~**Surface treatment processes.** These are used to reduce wear as well as to increase the stress resistance of the crankshaft (**Fig. C313**). It is necessary to differentiate between mechanical and thermal or thermochemical processes.

Mechanical processes. These include burnishing (with press polishing rollers) for passenger car crankshafts as well as impact strengthening (with balls) on crankshafts for large diesel engines. Both processes result in semi-plastic cold deformation on the transition radii. This results in intrinsic compressive strain at

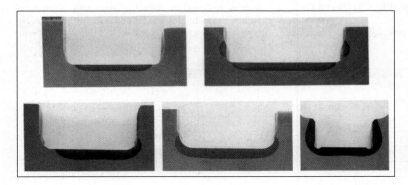

Fig. C313
Hardened zones produced by induction hardening

these points, partly increasing the strength. The dynamic stress resistance of the crankshaft is increased considerably by these two mechanisms.

Thermal processes. The primary thermal process is induction hardening, which is very significant for crankshaft production. The decisive advantage is that various hardening requirements can be achieved specifically in one operation with relatively simple means (e.g., resulting from the design of the inductors and selection of the process parameters). A selection of possible hardening zones is shown in **Fig. C313**. Here, the journal running surfaces, thrust bearing collars, sealing seat, and serrations are hardened with the intention of increasing the wear resistance. The stress resistance of the crankshaft is increased particularly by hardening at the transition radii. The increase in the stress resistance results from the intrinsic compressive strain produced and the partial increase in the strength (see *Mechanical processes*, above).

Induction hardening can be used on cast as well as forged crankshafts regardless of the size.

Thermo-chemical processes. With these processes, the chemical composition of the entire crankshaft surface is changed in the course of the heat treatment unless certain areas are specifically masked. In addition to nitrating procedures (e.g., gas nitrating, nitro carburization, plasma nitrating), case hardening (carbonization of the boundary layer) is still used in individual cases primarily for racing applications. The addition of nitrogen — with nitro carburization, carbon is added in addition — the strength of the crankshaft surface layer is increased partially and the material is subjected to compressive strain, increasing the stress resistance of the crankshaft. The improvement of the wear resistance is based on the formation of a hard compound layer ("white coating") on the surface.

The table in **Fig. C314** shows common examples of these processes.

The achievable surface hardness as well as the "layer depth" depend particularly on the material. The crankshaft geometry, dimensional relationships, and subsequent annealing processes also have an influence.

Case hardening (core hardness up to 1450 N/mm²) or plasma nitriding is also used for special purposes such as racing crankshafts.

Process	Material	Surface Hardness	"Treatment Depth"	Remarks
Induction hardening on running surfaces and transition radiuses	42 CrMo 4 CK 45 38 MnS 6 GGG 70	50–56 HCR	1.5–6 mm 1.5–4 mm	Depending on dimensions
Short-term gas nitriding (nitro carburetion) Time: 1.5 h–2 h	42 CrMo 4 CK 45	>450 HV 10 >350 HV 10	VS: 10–20μm DS: 0.2–03 mm	VS = Connecting layer DS = Diffusion layer
Long-term gas nitriding Time: 80 h–120 h	31 CrMoV 9	>700 HV 10	DS: 0.4–0.6mm	Depending on nitriding duration

Fig. C314 Surface treatment processes

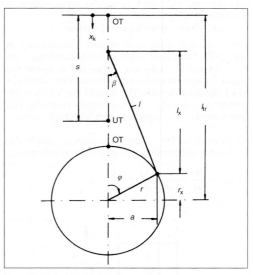

Fig. C315 Crankshaft drive

~Throw. The throw consists of the connecting bearing journals and crank web which are connected by the main bearing journals. The distance between the center of the crankshaft bearing journal and center of the crankshaft is known as the crankshaft radius (**Fig. C315**) and, together with the diameter and number of cylinders, determines the engine displacement. Crankshafts with one to ten throws and increments of 180°, 120°, 90°, 72°, 60°, 45°, 40°, and 36° are common for inline and V-engines. Generally, crankshafts have a bearing next to each throw. However, smaller crankshafts in gasoline engines, which are not subject to such high stresses, sometimes have bearings next to alternate throws. This results in a shorter engine and lower costs. On "split-pin" crankshafts (**Fig. C316**), the wider big end bearing journals (for V-engines) are offset in the circumferential direction, resulting in better firing intervals that thereby improve the operating smoothness of the engine.

The connecting rods can be guided laterally either by the piston or the crankshaft web. In the latter case, higher requirements are placed on the accuracy of the thrust collars and the width of the big end bearing journals.

Crankshaft bearing →Bearings ~Bearing positions

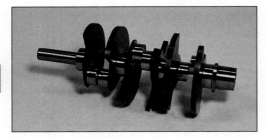

Fig. C316 Split-pin crankshaft

Crankshaft blank →Crankshaft ~Blank

Crankshaft drive. The crankshaft drive converts the reciprocating motion of the piston into a rotary motion. The primary elements of the crankshaft drive are:

- Piston (piston rings, piston pins), completing the reciprocating motion
- Connecting rod which completes reciprocating as well as rotating motion
- Crank, which completes rotary motion

A crankshaft drive is illustrated schematically in **Fig. C317**. The values required for the kinematics are the crank radius, r, the connecting rod or piston rod length, l, the crank angle, φ, and the angular velocity at which the crank rotates. The ratio of the length of the connecting rod to the crank radius is also relevant. The piston stroke, s, is equal to twice the crank radius.

~Acceleration. The piston acceleration, c_k, can be determined by differentiation from the piston velocity, a_k:

$$a_k = \frac{dc_k}{dt} = \frac{dc_k}{d\varphi}\frac{d\varphi}{dt} = \frac{dc_k}{d\varphi} \cdot \omega$$

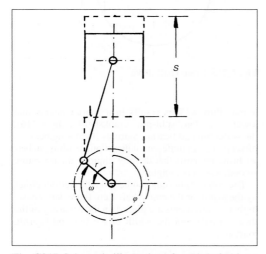

Fig. C317 Schematic illustration of crankshaft drive

But at constant angular velocity, ω, of the crankshaft, $d\varphi/dt = \omega = $ constant, therefore

$$a_k = r\omega^2\left[\cos\varphi + \left(\lambda + \frac{\lambda^3}{4} + \frac{15\lambda^5}{128}\right)\cos 2\varphi \right.$$
$$\left. - \left(\frac{\lambda^3}{4} + \frac{3\lambda^5}{16}\right)\cos 4\varphi + \frac{9\lambda^5}{128}\cos 6\varphi + ...\right]$$

where

dt = differential change in time
$d\varphi$ = differential change in crank angle
λ = connecting rod ratio r/l
r = crank radius
l = length of connecting rod
ω = angular velocity.

Because the connecting rod ratio, λ, is in the range of 0.17 to 0.3, it is a sufficient approximation to consider only the first two terms in the above equation. This results in:

$$a_k \approx r\omega^2(\cos\varphi + \cos 2\varphi)$$

Fig. C318 shows the basic curve for the piston acceleration in relation to the crank angle.

~Crank angle. The crank angle is the angle of the crank position measured from top dead center. The crank angle for one operating cycle on a four-stroke engine is 720° and on a two-stroke engine 360°.

~Crankshaft drive with articulated connecting rod. In contrast to engines with straight connecting rods, these engines have an additional connecting rod, the articulated connecting rod; **Fig. C319** shows a schematic representation. Articulated connecting rods are hardly ever used on engines today, because of the reduction in the overall space, and are significant only for radial engines. Similar compact designs can be realized with V-engines by using connecting rods located next to one another and offsetting the cylinder banks.

~Forces. The forces present in the crankshaft drive are caused by gas and inertia forces. The gas force acting on the piston is $F_G = p_G \cdot A_K$, with piston surface

$$A_K = D_K^2\frac{\pi}{4}\ (D_K = \text{piston diameter}),$$

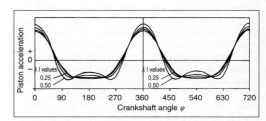

Fig. C318 Piston acceleration angles crank angle for various values of λ

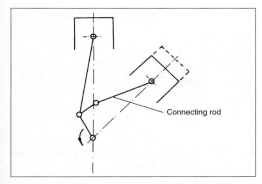

Fig. C319 Crankshaft drive with articulated connecting rod

and gas pressure $= p_G$. The curve of the gas forces is
proportional to that of the gas pressure acting on the
piston surface.

Inertia forces or mass forces act to move the propulsion parts in a reciprocating piston engine. They are
opposed to the direction of acceleration and correspond to the equation

$$F = -ma$$

where F = force, m = mass, and a = acceleration.

The masses of the piston with piston rings and piston pins and connecting rod and crankshaft are responsible for the mass forces. Just as with the gas forces,
the mass forces can be divided up into their components, where the mass force effective direction is defined by the crank angle.

For practical reasons, the engine parts are divided
up into reciprocating and rotating masses.

The force curve resulting from superimposing the
gas and mass force is responsible for the load on the
components and the torque available at the crankshaft.
To determine the load on the individual components,
the resulting force can be divided up into the following
components:

- Lateral force F_N
- Force along connecting rod F_P
- Tangential force F_T
- Radial force F_R

The force directions are shown schematically in **Fig.
C320**.

These forces and their effects are influenced primarily by:

- Number of cylinders (4, 5, 6, 8)
- Arrangement of cylinders (inline, V, horizontally
  opposed, radial)
- Firing order
- Throw sequence on crankshaft
- Design of components responsible for the mass forces
- Engine mounting

~Offset crank operation (*also*, →Piston ~Offset).

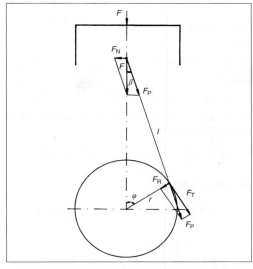

Fig. C320 Breakdown of forces resulting on crankshaft drive

With an offset crank, the center of the crankshaft is
offset in relation to the center-line of the cylinder (**Fig.
C321**). The offset dimension, σ, is defined as the offset, a, in relation to the connecting rod length, l,

$$\sigma = a/l$$

and is $\sigma < 0.05$ for common reciprocating piston engines. This means that the forces resulting from the
crankshaft drive are affected only to an insignificant
extent by the offset.

~Piston travel. The piston travel, x_k, results from the
geometries of the crankshaft drive, shown in **Fig. C315**.
The piston travel, x_k, results from

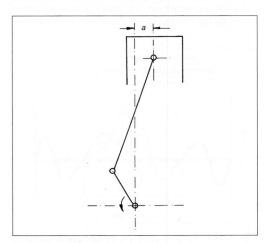

Fig. C321 Offset crank operation

$$x_K = \lambda + r - r_x - l_x.$$

If $r_x = r\cos\varphi$

and

$$l_x = \sqrt{l^2 - a^2},$$

then

$$x_K = \lambda + r - r\cdot\cos\varphi - \sqrt{l^2 - a^2}.$$

If substitution is made for $a = r\sin\varphi$ and $\lambda = r/\lambda$ in the equation, we obtain:

$$x_K = r\left(\frac{1}{\lambda} + 1 - \cos\varphi - \frac{1}{\lambda}\sqrt{1 - \lambda^2\sin^2\varphi}\right).$$

By series expansion:

$$x_K = r\left(1 - \cos\varphi + \frac{\lambda}{2}\sin^2\varphi + \frac{\lambda^3}{8}\sin^4\varphi\right.$$
$$\left. + \frac{\lambda^5}{16}\sin^6\varphi + \dots\right).$$

This defines the piston travel as a function of the crank angle, the crank radius, and the connecting rod ratio.

~Tangential force. The tangential force, F_T, changes periodically with the crank angle, φ. Therefore, the moment,

$$M = r\cdot F_T,$$

also changes periodically. **Fig. C322** shows the relationship for a single-cylinder engine. On multicylinder engines, the resulting tangential force can be obtained by adding together the tangential forces of the individual cylinders in the sequence of the crankshaft.

Because the torque (tangential force) changes with the crank angle, there is a certain degree of nonuniformity, which appears as variations in engine speed. These can be reduced by an appropriately dimensioned flywheel or by a large number of cylinders. The more cylinders an engine has, the lower the torque differences during one operating cycle and, therefore, the

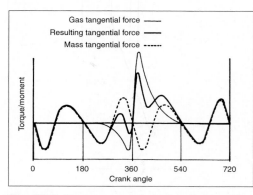

Fig. C322 Torque curve of single-cylinder engine
(Source: Küttner)

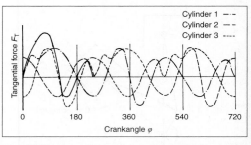

Fig. C323 Torque curve of a three-cylinder engine
(Source: Küttner)

lower the variation in the speed. **Fig. C323** shows this using a three-cylinder engine as an example.

~Velocity. The piston velocity, c_k, can be determined from the piston path, x_k, by differentiating with respect to time:

$$c_k = \frac{dx_k}{dt} = \frac{dx_k}{d\varphi}\cdot\frac{d\varphi}{dt}.$$

This results in:

$$c_k = r\cdot\omega\cdot\left[\sin\varphi + \left(\frac{\lambda}{2} + \frac{\lambda^3}{8} + \frac{15\lambda^5}{256}\right)\sin 2\varphi\right.$$
$$\left. - \left(\frac{\lambda^3}{16} + \frac{3\lambda^5}{64}\right)\sin 4\varphi + \frac{3\lambda^5}{256}\sin 6\varphi + \dots\right].$$

The first two terms in the series expansion are satisfactory for sufficiently precise description:

$$c_k \approx r\omega\left(\sin\varphi + \frac{\lambda}{2}\sin 2\varphi\right).$$

Crankshaft drive with articulated connecting rod →Crankshaft drive

Crankshaft drive with connecting rod articulation →Balancing of masses

Crankshaft journal →Crankshaft

Crankshaft main bearing →Bearings ~Bearing positions

Crankshaft production →Crankshaft ~Blank, ~Machining

Crankshaft radius →Crankshaft ~Throw

Crankshaft rigidity →Crankshaft ~Stress resistance

Crankshaft speed →Speed

Crankshaft star →Crank set

Crankshaft starter/generator →Fuel consumption ~Variables ~~Engine measures

Crankshaft throw →Balancing of masses ~Number of throws; →Crankshaft ~Throw

Crankshaft torsional vibrations →Torsional vibrations

Crankshaft vibration →Torsional vibrations

Crankshaft vibration damper. During operation of an internal combustion engine, the intrinsic resonance of the crankshaft is excited by the gas and inertia forces. This results in major torsion between one end of the crankshaft and the other, which can lead to noise and torsional fracture of the crankshaft. Today, torsion vibration dampers are usually used to reduce the load on the crankshaft. These are integrated into the pulley on the equipment at the free end of the crankshaft. Designers foresee eliminating the belt drive on some new generations of engine and driving the auxiliary equipment in another manner; however, the torsion vibration damper will still be required. The LuK Co. is planning to integrate the damper into the crank web without taking up additional space. The associated conditions of limited space allow only minimum moments of inertia of the vibrating masses. Sufficient reduction of the crankshaft torsional vibration is possible only by the use of steel pressure springs whose high energy absorbing capacity allows large pivot angles.

The advantages of an internal crankshaft damper are:

• Reduction of the crankshaft stress
• Lower noise development in the engine
• Less installation space
• Reduction of engine masses

Crankshaft web →Crankshaft ~Balance weights, ~Stress resistance

Crescent oil pump →Gear pump ~Internal gear pump

Crevice corrosion →Coolant

Cross-flow channel design →Cylinder head ~Cross-flow

Cross-flow cylinder head →Cylinder head ~Cross flow

Cross-flow head →Cylinder head ~Cross-flow

Crosshead →Crosshead engine

Crosshead bearings →Bearings ~Bearing positions

Crosshead engine. With the conventional design of high-speed passenger car engines, the pressure acting on the pistons is transferred directly to the crankshaft through the piston pin and connecting rod.

On large low-speed diesel engines and reciprocating compressors, an additional guide is frequently re-

quired for the piston. This is accomplished with the aid of a crosshead (**Fig. C324**).

Crosshead guides →Bearings ~Bearing positions

Cross scavenging →Two-stroke engine ~Scavenging process ~~Cross-flow scavenging

CRT system →Particles (Particulates) ~Particulate filter system ~~Particulate filter ~~~Regeneration of particulate filters

Crude oil →Oil

Cruise control →Actuators ~E-gas ~~Cruise control

Current flow →Electronic/mechanical engine and transmission control ~Electronic components ~~Block diagram

Cut-off intake port →Intake port, gasoline engine ~Cutoff

Cut-off speed →Speed ~Cut-off speed

CVCC →Combustion process, gasoline engine ~Stratified charge process ~~Honda CVCC

CVS (constant volume sampling) method →Emission limits; →Emission measurements ~Testtype I ~~Calibration

Cycle. The basic subject in thermodynamics is the description of the conversion of different forms of energy. This focuses particularly on the conversion of any form

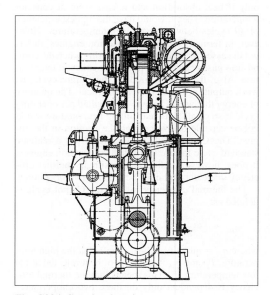

Fig. C324 Crosshead engine

C

of naturally occurring energy into mechanical energy. In the case of chemical energy (such as that in fuels), this process must occur continuously, not just once. Cycles can be used to describe such processes whose operating substances are subject to a series of changes of state.

When applied to internal combustion engines, the chemical energy in the fuel is converted to mechanical energy at the crankshaft during the engine operating process. Because of the complex interrelationships between the chemical reactions with superimposed flow processes and with substance and energy exchange operations, it is not yet possible to calculate the combustion process in an engine completely accurately.

For this reason, a series of ideal processes (standard cycles) is defined, which allows the real operations to be approached. These processes enable information on the operating characteristics to be obtained when parameters are varied.

The simplest models that reflect the theoretical engine process are closed, reversible cycles: The process of chemical conversion (combustion) is replaced by heat supply; the working fluid is returned to its initial point by removing the heat. This technique allows the thermal efficiency to be determined. Advanced processes are open standard cycles in which the chemical conversion is taken into consideration.

~Carnot process. The Carnot cycle is a reversible cycle consisting of two isentropic and two isothermal processes for the working fluid. Carnot recommended this cycle for a "heat engine." Heat (energy) is supplied and removed isothermally—the energy is absorbed at high temperature and released at low temperature. The efficiency of the Carnot cycle would be high only if heat absorption and release were at constant temperature. Then it would have the highest efficiency of all cycles between two given temperatures. However, in terms of internal combustion engines, this is not the case, because the heat occurring when the combustion gases cool is not present at constant temperature. Also, the area covered by the Carnot cycle, its work output, is too small to be practical. The quantity of energy that would have to be supplied per operating cycle is so small that the resulting lean mixture would not be capable of overcoming the friction in the engine. The Carnot cycle is, therefore, not a realistic standard cycle for use in internal combustion engines; it does, however, define the maximum thermal efficiency that can be attained between two temperatures.

The thermal efficiency, η_{th}, of the Carnot cycle is defined as:

$$\eta_{th} = \frac{q_{zu} - q_{ab}}{q_{zu}},$$

where q_{zu} = quantity of heat supplied at the high temperature, T_1, and q_{ab} = quantity of heat rejected at the low temperature, T_0. This means that the thermal efficiency is dependent only on these two temperatures and not on the operating substance,

$$\eta_{th} = 1 - \frac{T_0}{T_1}.$$

~Constant pressure cycle. The constant pressure cycle illustrated in **Fig. C325** represents one version of simple processes. Here, a maximum pressure is permissible, which can be achieved in a real engine—that is, within the strength of the components. The constant pressure cycle represents a first theoretical approach to the cycle in a diesel engine:

$$\eta_{th} = \frac{q_{zu} - q_{ab}}{q_{zu}},$$

where, q_{zu} = the quantity of energy supplied, and q_{ab} = the quantity of energy rejected.

If the quantity of energy supplied and rejected is defined using the constant specific heats and temperature differences, $T_3 - T_2$ or $T_4 - T_1$, and the isentropic exponent (ratio of specific heats) is κ, then

$$\eta_{th} = 1 - \frac{1}{\varepsilon^{\kappa-1}} \cdot \frac{1}{K}\left[\frac{(T_3 / T_2)^\kappa - 1}{T_3 / T_2 - 1}\right].$$

The thermal efficiency, η_{th}, is dependent only on the temperatures, T_3 and T_2—that is, on the quantity of heat supplied and the compression ratio, ε.

Fig. C326 shows the curve for the thermal efficiency as a function of the compression ratio for various quantities of energy supplied. At increasing values of ε, the efficiency increases; with increasing energy supply, the efficiency of the constant pressure cycle

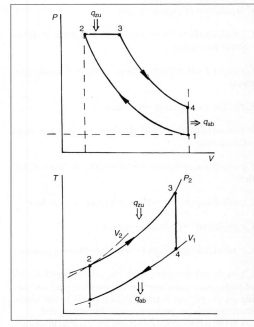

Fig. C325 Constant pressure cycle

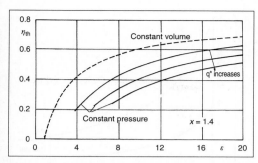

Fig. C326 Thermal efficiency (Source: Pischinger)

$$\eta_{th} = \frac{q_{zu} - q_{ab}}{q_{zu}}.$$

By substituting the compression ratio, ε, and taking the type of gas into consideration, defined here by its isentropic exponent (ratio of specific heats), κ, the following equation results after a number of operations:

$$\eta_{th} = 1 - \frac{1}{\varepsilon^{\kappa-1}}$$

The thermal efficiency, η_{th}, of the constant volume cycle increases with increasing ε (**Fig. C328**). The efficiency also depends on the type of gas, defined here by its isentropic exponent, κ, or the thermodynamic properties of the type of gas. Efficiency is independent of the energy supplied and rejected.

decreases; here q is a variable defining the energy supply.

The constant pressure cycle has a lower efficiency than the constant volume cycle, based on compression ratio and unlimited peak pressures.

~**Constant volume cycle.** The constant volume cycle is frequently considered as a cycle that reflects, in principle, the theoretical processes in a gasoline engine. This cycle is based on two isentropic and two isochoric changes of state (**Fig. C327**). The changes of state from 1 to 2 and from 3 to 4 are isentropic, those from 2 to 3 and 4 to 1 isochoric. The thermal efficiency can be defined by the ratio of the energy (heat) added to that rejected:

~**Dissociation.** Dissociation describes the decomposition of constituents in a substance such as a combustion gas. If a hydrocarbon reacts with air, according to the simplified reaction, only water and carbon dioxide are produced theoretically (nitrogen is considered to be an inert gas). Depending on the temperature and pressure of the mixture, the constituents will decompose into partly unstable intermediate products. For example, a mixture of CO_2, H_2O, O_2, and N_2 will break down into a mixture consisting of O, H, O_2, OH, H_2, N, N_2, CO, CO_2, H_2O, and so on. The percentage of the individual constituents depends primarily on the temperature.

Dissociation plays a role in calculating the engine cycle by means of open standard cycles. For example, it increases the overall thermal capacity of the gases that occur during combustion resulting in different final states, regardless of whether the energy involved in dissociation is taken into consideration.

~**Exergy loss.** Exergy defines all the available energies in a system that can be converted into power output by any process that is allowed by the Second Law of Thermodynamics.

In terms of internal combustion engines, this means that there are process losses which occur in the form of

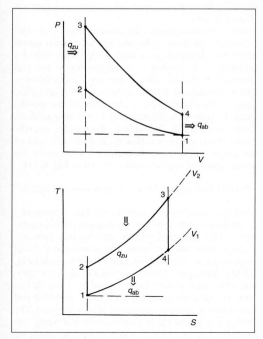

Fig. C327 Constant volume cycle

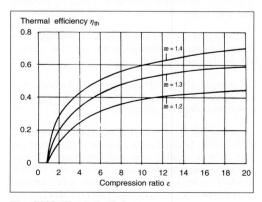

Fig. C328 Thermal efficiency

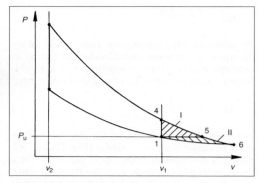

Fig. C329 Constant volume cycle

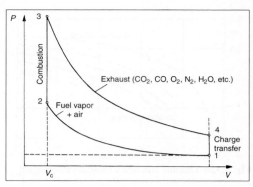

Fig. C330 Schematic representation of open standard cycle

exergy, which could be used if the process were controlled appropriately.

Fig. C329 shows the losses in the p-V diagram using the constant volume cycle as an example

A loss results because the process cannot be continued during the expansion phase down to ambient pressure (Point 5). The reason is that in a real engine, the exhaust valve opens before the combustion chamber has reached ambient pressure as a result of expansion (area 1-4-5-1). If the process were to continue all the way to the ambient temperature, T_1 (expanded), and the gas was then isothermally compressed back to the ambient pressure, the energy in area 1-5-6-1 could be utilized.

However, this type of process control is not possible because an extension of the expansion process would require a significantly larger combustion chamber. It would be necessary to extend the stroke.

~*Isentropic change of state (process).* An isentropic change of state is a theoretical process in which energy is neither supplied nor released, and it is also reversible—that is, the change in the entropy is zero.

~*Isobaric change of state (process).* An isobaric change of state is a theoretical process in which the pressure does not change.

~*Isochoric change of state (process).* An isochoric change of state is a theoretical process in which the volume does not change.

~*Isothermal change of state (process).* An isothermal change of state is a theoretical process in which the temperature does not change.

~*Open standard cycles.* With the cycles discussed up to this point, information analogous to the actual engine cycle is possible only with limitations. For this reason, a further approximation to the reality of open standard cycles is used. A frequently used standard cycle is that of a perfect engine. This is defined, for example, as a constant pressure cycle, illustrated sche-

matically in **Fig. C330**, with the following limiting conditions:

- Geometrically identical to the actual engine
- Pure charge, no residual gases in the cylinder
- Combustion sequence according to given definitions
- Same air-fuel ratio, λ, as in a real engine
- Loss-free gas exchange
- Adiabatic walls
- Combustion products in chemical balance.

Here, chemical balance means the state in which the number of reactions of formation is equal to the number of decomposition reactions—that is, viewed globally, the total composition of substances does not change in time.

This definition applies for constant volume as well as constant pressure cycles and also for a combination of the two. It represents a further approximation of a real engine. In particular, a relationship to the real combustion air-fuel ratio is present. Dissociation—that is, breakdown and recombination of combustion gases—also plays a role. Calculation of the perfect cycle is quite complicated and is not treated here. Under comparable limitations, it provides an efficiency that is always less than the thermal efficiency of the simple, ideal cycle. The efficiency curve for the perfect gasoline engine cycle is shown in **Fig. C331**.

~*Perfect engine. See above,* ~Open standard cycles

~*Seiliger (dual combustion) cycle.* This represents a combination of the constant volume and constant pressure cycles and theoretically corresponds approximately to the actual combustion process in a diesel engine just as does the constant pressure cycle (**Fig. C332**). Analogous to the previous cycles, the thermal efficiency results from the ratio of the difference between the quantity of energy supplied and that rejected, and the quantity of energy supplied. The energy is supplied partly isochorically and partly isobarically. For calculation, it is necessary to divide up the energy supplied:

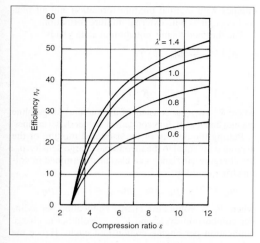

Fig. C331 Efficiency in relation to air-fuel ratio for perfect gasoline engine cycle

$$\eta_{th} = \frac{q_{zuv} + q_{zup} - q_{ab}}{q_{zuv} + q_{zup}}.$$

The following formula results from the relationships shown in **Fig. C332**:

$$\eta_{th} = 1 - \frac{1}{\varepsilon^{\kappa-1}} \frac{\left(\frac{T'_3}{T_3}\right)^{\kappa} \cdot \frac{T_3}{T_2} - 1}{\frac{T_3}{T_2} - 1 + \kappa\frac{T_3}{T_2} \cdot \left(\frac{T'_3}{T_3} - 1\right)}.$$

The efficiency of the Seiliger cycle (**Fig. C333**) is between that of the constant pressure and constant volume cycles and is dependent both on the way the energy supplied at constant volume and constant pressure is subdivided and on the compression ratio.

~Stirling cycle. The theoretical standard cycle for the Stirling engine is distinguished by isothermal compression, isochoric energy supply, isothermal expansion, and isochoric heat release.

In contrast to gasoline and diesel engines, the Stirling engine operates with external combustion.

Cyclical combustion variations →Cyclical variations

Cyclical variations (also, →Combustion, gasoline engine; →Combustion process, gasoline engine). Experience shows that different conditions in the combustion space at the beginning of ignition result in different maximum pressures in the cylinder from combustion cycle to combustion cycle. These fluctuations between successive combustion cycles can be statistically collected and are designated as cyclical variation. The causes are the time-based variation of different quantities such as turbulence, velocity pro-

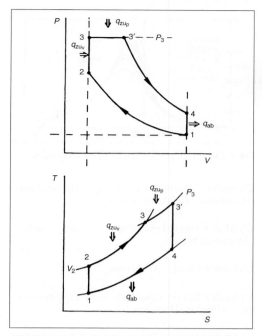

Fig. C332 Seiliger (dual combustion) cycle

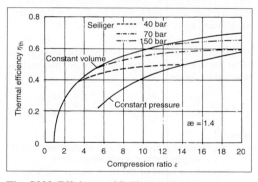

Fig. C333 Efficiency of Seiliger cycle

files, and mixture composition at the spark plug when ignition occurs.

These result, because of their different flame speeds, in different pressure curves in the cylinder, which in turn contributes to the degree of nonuniformity in the torque output. **Fig. C334** shows an example of the cyclical variation in a gasoline engine.

Cycloidal gearing →Gear pump ~Internal gear pump ~~Crescent-type oil pumps

Cyclone separator →Filter ~Intake air filter ~~Air filter elements ~~~Preliminary filter; →Lubrication ~Lubricating systems ~~Passenger car engines

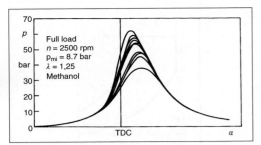

Fig. C334 Cyclical variation in a gasoline engine

Cylinder →Commercial vehicle cylinders/liners; →Crankcase ~Crankcase design

Cylinder barrel →Cylinder running surface ~Machining process

Cylinder block →Crankcase

Cylinder bore →Crankcase ~Crankcase construction ~~Primary dimensions

Cylinder charge. Engine power at equal air-fuel ratios is proportional to the air mass or mixture volume supplied to the engine, hence the charge trapped in the cylinder. Indexes λ_a and λ_l serve for defining the charging efficiency.

~Charging efficiency. The charging efficiency is the value for the fresh charge supplied to the engine. Hence the relation

$$\lambda_a = \frac{m_G}{m_{th}} = \frac{m_G}{V_h \cdot \rho_{th}} \quad \text{or} \quad \lambda_a = \frac{m_{Ggtot}}{V_H \cdot \rho_{th}},$$

where

$\quad m_G$ = the total mass of fresh charge supplied to a cylinder per combustion cycle

$\quad m_{Gtot}$ = total mass of the fresh charge supplied to the engine per combustion cycle

$\quad m_{th}$ = theoretical charge mass per combustion cycle (cylinder or whole engine)

$\quad \rho_{th}$ = theoretical charge density

The total of fresh charge supplied consists of:

$$m_G = m_K + m_L \quad \text{or} \quad m_{Gtot} = m_{Ktot} + m_{Ltot}$$

for the gasoline engine, and

$$m_G = m_L \quad \text{or} \quad m_{Gtot} = m_{Ltot}$$

for the diesel engine.

The theoretical fresh charge mass is determined from the geometric displacement and the ambient condition of the charge. In supercharged engines the thermodynamic condition before the intake valve is taken in place of the ambient condition. The induced charge is air in

engines with internal mixture formation and air plus fuel in engines with external mixture formation.

The thermodynamic equation of state yields:

$$p_u \cdot V_h = m_{th} \cdot R \cdot T_u$$

or

$$p_u \cdot V_H = m_{thtot} \cdot R \cdot T_u,$$

where $R = R_G$ (gas constant of the mixture) for gasoline engines and $R = R_L$ (gas constant of air) for diesel engines.

Putting the density of the aspirated mixture or the aspirated air equal to the theoretical charge density, ρ_{th}, the charging efficiency can also be determined in volumetric quantities:

$$m_G = V_G \cdot \rho_G \quad \text{or} \quad m_{Gtot} = V_{Gtot} \cdot \rho_G,$$

where V_G = volumetric charge application per cylinder combustion cycle and V_{Gtot} = volumetric charge application per engine combustion cycle. Hence the relation

$$\text{Gasoline engine: } \lambda_a = \frac{V_G}{V_h} \quad \text{or} \quad \lambda_a = \frac{V_{Gtot}}{V_H}$$

$$\text{Diesel engine: } \lambda_a = \frac{V_L}{V_h} \quad \text{or} \quad \lambda_a = \frac{V_{Ltot}}{V_H}.$$

To experimentally determine engine charging efficiency, the induced air volume on the air mass is measured. In addition, the pressure and air temperature as well as ambient conditions and fuel consumption in the gasoline engine need to be determined.

~Volumetric efficiency. The volumetric efficiency is defined as the amount of fresh charge remaining in the cylinder at the termination of charge transfer. Similar to the charging efficiency, the volumetric efficiency is based on the theoretical charge density.

$$\lambda_l = \frac{m_Z}{m_{th}} = \frac{m_Z}{V_h \cdot \rho_{th}} \quad \text{or} \quad \lambda_l = \frac{m_{Ztot}}{V_H \cdot \rho_{th}}$$

In the gasoline engine, the fresh cylinder charge, m_Z or m_{Ztot} is

$$m_Z = m_{ZL} + m_{ZK} \quad \text{or} \quad m_{Ztot} = m_{ZLtot} + m_{ZKtot}$$

and in the diesel engine,

$$m_Z = m_{ZL} \quad \text{or} \quad m_{Ztot} = m_{ZLtot},$$

where m_{ZL} = air mass in one cylinder, m_{Zltot} = total air mass in all engine cylinders, m_{ZK} = fuel mass in one cylinder, and m_{ZKtot} = total fuel mass in all cylinders.

The mass of charge remaining in the cylinder or in all engine cylinders cannot be directly determined or acquired by measurement. The following method is used to obtain an approximation:

A. Cylinder-pressure diagram in one or all engine cylinders.

B. Assumption that the temperature of the cylinder charge at "intake valve closes" is about equal to the temperature in the intake port just upstream of the intake valve (this temperature is measured by thermocouple).

C. Formulation of the gas equation at the time of "intake valve closes":

$$p_{ZEs} \cdot V_{Es} = m_z \cdot R \cdot T_{ZEs}.$$

R_G or R_L are again used for the gas constant, R.

The range of crank angle for valve overlap (time period over which both the intake valve and exhaust valve are simultaneously open) is relatively small for four-stroke gasoline engines. A good approximation for the small valve overlap is to set $\lambda_a \approx \lambda_l$.

For engines without supercharging, λ_a and λ_l are always less than one because the resistances to fluid flow prevent complete scavenging of the geometric cylinder displacement during induction and expulsion and, therefore, charging losses will occur. Engines with supercharging have operating conditions under which λ_a and λ_l can be greater than 1.

Diesel engines, and those with supercharging in particular, have large valve overlaps to give internal cooling and better expulsion of the residual gases from the combustion space. Here, λ_a can become « λ_l.

In piston-controlled engines, there is a significant difference between air efficiency and volumetric efficiency—this is because of the transfer losses. The quotient of volumetric efficiency and air efficiency indicates the capture rate, which is the value for the fresh charge that remained in the cylinder.

Cylinder charge losses →Cylinder charge

Cylinder crankcase →Crankcase

Cylinder design for mono-metal construction →Crankcase ~Crankcase design ~~Cylinder ~~~Mono-metal design

Cylinder head (*also*, →Commercial vehicle cylinder head).

The cylinder head is the main component for realizing the charge transfer of a reciprocating piston engine. It accommodates the charge transfer units and usually represents part of the combustion space. Moreover, the cylinder head has to fulfill the following general requirements:

- Sealing the combustion space
- Accommodating spark plug, pencil-type glow plug and injection nozzle, prechamber or swirl chamber
- Taking up the gas forces
- Lubricating the moving valve-gear components
- Dissipating the heat from combustion
- Mounting the camshafts.

There are generally two designs:

- One-piece connected with the cylinder **Fig. C335** shows a design with the cylinder head connected with two cylinders.

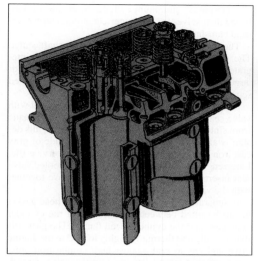

Fig. C335 Cylinder and cylinder head, one-piece, of a dual-block cylinder Maybach GTO engine

- Separate components for one cylinder or several cylinders (common today) (**Fig. C336**).

The combustion-chamber–oriented arrangement of intake and exhaust valves, precombustion chambers, spark plugs, and so on, and the accommodation of usually four cylinder-head fastening screws make high demands on the design of the cylinder head. The layout arrangement must additionally pay attention to material castability, machining, assembly, thermal loading, cooling, mechanical loading, weight, and cost.

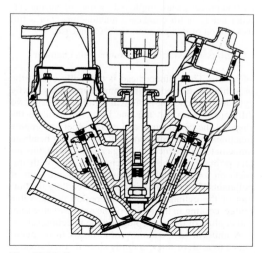

Fig. C336 Cross section through conventional cylinder head

C

"One-piece" aluminum cylinder heads have established themselves for passenger-car engines with overhead camshafts.

The upper flange surface is then on camshaft center. The mounts for valve actuation are preferably integrated. Cast-iron cylinder heads for truck engines or large engines have their control components mounted separately.

Aluminum cylinder heads need to be equipped with valve seat inserts and guides that are primarily made from sintered materials and pressed in. Today, this design is also selected for cylinder heads made of gray cast iron because of the demands for high service-life. Large engines are partly fitted with liquid-cooled valve seat inserts that can be removed from above together with the valve and guide (caged valves).

In separate cylinder heads, the cylinder head gasket has to be fitted between the surfaces of the cylinder crankcase and the cylinder head flange. The gasket is dynamically and thermally highly loaded in the engine drive and has to seal against high combustion pressures, pressurized oil flowing into the valve control, low pressure oil return and coolant flows. Disturbance of these functions can quickly result in engine failure.

Separate sealing elements are employed for the various sealing functions on large diesel engines.

The intake manifold or charge air tube and the exhaust manifold are flange-mounted on the long sides of the cylinder head.

The coolant outlet neck is located at the highest point of the cylinder head.

Vehicle engines have aggregates mounted at the front, at the back, and at the side that are driven from the camshaft.

~Bolt →Engine bolts

~Casting process. As a matter of principle, the same casting processes are employed for manufacturing cylinder heads as for crankcase manufacture. For a full description of this process, *see* →Crankcase.

~Combustion chamber →Combustion chamber

~Cooling. The cylinder head is thermally one of the most highly stressed engine components. That is why cylinder head cooling is of great significance from two different aspects: on the one hand, to stabilize the shape of the component, and on the other, the engine functions that the temperature of the cylinder head can immediately influence. In the gasoline engine, this refers primarily to the knock characteristics with impacts on torque, fuel consumption, and emission of pollutants. Essential in this context is cooling the bottom of the cylinder head in the region of the valve bridge, cooling in the region of the exhaust valve seat, and cooling of the spark plug and the injection valves.

A uniform temperature profile needs to be maintained to achieve shape stability since thermal stresses result from locally different temperatures. The maximum limit of temperature should be around 250°C

since the modulus of elasticity of aluminum-alloy drops sharply above this temperature. Likewise, the thermal conductivity of the aluminum alloy is of importance, which is in the region of 150 W/mK.

This renders optimized coolant flow in the cylinder head mandatory. It is possible to differentiate between air cooling and liquid cooling in principle, and liquid cooling can be made using "cooling water" and oil. Today, liquid cooling with cooling water on high-speed engines is common. The cooling water consists of a mixture of water with antifreeze and corrosion inhibitors.

~~**Air cooling. Fig. C337** shows cross-sectional and longitudinal section views of an air-cooled industrial engine—the cylinder head is die-cast aluminum and connected with the housing by four cylinder studs. Typical of this design is the air-ducting housing and the cooling fins in the upper region of the cylinder. The fin spacing for the cooling fins is generally around 5–8 mm and the fin height is about 70 mm.

~~**Water cooling.** The water flow in cylinder heads can be either crosswise or lateral in direction. **Fig. C338** shows the cylinder head cooling system of a state-of-the-art four-cylinder, four-valve engine. The

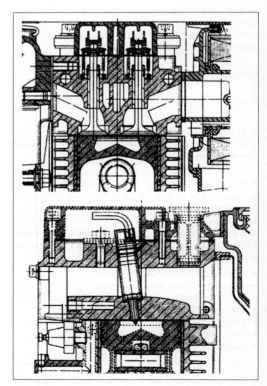

Fig. C337 Air-cooled cylinder head of a Hatz diesel engine (Source: *MTZ*)

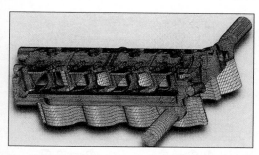

Fig. C338 Internal engine cooling system of a BMW four-cylinder engine with Valvetronic system

principle is a quench flow in the cylinder head with a nonforced flow through the crankcase. The coolant is directly supplied to a distributing duct in the cylinder head via an inflow channel. Starting from this channel, the coolant flow is adjusted through individual ribbing in the webs between the exhaust valves of the individual cylinders. The critical web temperatures between the cylinders can thus be decreased to 200°C.

The objective is to reduce the power requirement of the coolant pump in addition to decreasing the peak temperatures and achieving a uniform temperature distribution—for example, by minimizing the flow resistances.

Oil cooling of the cylinder head can be employed for special applications (**Fig. C339**).

This allows cooling of particularly critical zones through specific bore systems and the supply of oil just as it is performed using general bedplate cooling. **Fig. C339** shows an oil-cooled cylinder head. In this configuration the bedplates for each cylinder are surrounded by ring grooves that overlap each other. This renders longitudinal flow possible in the cylinder head. The oil flow can be adjusted for each cylinder individually by design of the corresponding cross sections.

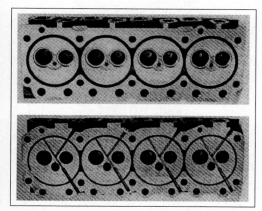

Fig. C339 Oil-cooled cylinder head of a Deutz diesel engine (Source: *MTZ*)

~Counterflow. *See above*, ~Cross flow

~Cross flow. One can classify cylinder heads into cross-flow (parallel flow) and counterflow systems pertaining to flow of gases through the engine. When the intake and exhaust valves are on the same side of the cylinder head, the system is counterflow. With cross- or parallel flow, the fresh gas flows in on one side of the cylinder head and exhaust gas escapes from the other side of the cylinder head. State-of-the art passenger-car engines nowadays exclusively use the cross-flow design. This helps in subjecting the flowing-in air or mixture to less thermal load.

~Cylinder head cover. The cylinder head cover closes the top side of the cylinder head. It is made of sheet metal, light-alloy casting, or plastic material. Vehicle engines frequently have the oil filler cap and the engine ventilation line connected there. Where feasible, the cover contains an oil separator space for engine ventilation. Engine design and noise emission additionally influence the design of this component. State-of-the art gasoline engines also have their ignition coils arranged in the cylinder head cover.

Because of the fuel lines, electric cables, noise emission, and so on, an extra plastic cover is frequently used.

~Design type. There are many types of design for cylinder heads depending on the engine application. Depending on engine design and application it is possible to distinguish between cylinder heads for:

- Two-stroke engines, with and without valves
- Four-stroke engines
- Gasoline engines and diesel engines
- Small, passenger car, truck, and stationary engines
- Air or liquid cooling
- Valve gear with in-block camshaft or overhead camshaft
- Individual heads or multiple heads

Today, light and cheap air-cooled small engines with loop scavenging mostly run using the two-stroke principle and large "low-speed" and durable liquid-cooled marine engines with uniflow scavenging. Moreover, there are two-stroke engines with valves. **Fig. C340** shows the cylinder head longitudinal and cross-sectional view of a two-stroke diesel engine with uniflow scavenging.

Currently passenger car engines and truck engines are exclusively liquid-cooled four-stroke engines. Intake and exhaust ports, valves, and valve control for gas transfer are integrated into the cylinder heads of these engines forming the upper wall of the combustion chamber. In gasoline engines the spark plugs and the injection valve—if direct injection is fitted—are additionally built in. Depending on the combustion system, the precombustion chamber or swirl chamber—or in direct injection, the injection valve and glow plug—are arranged in the diesel engine cylinder head. **Fig. C341** shows an example of a modern aluminum cylinder head for a direct injection diesel engine.

C

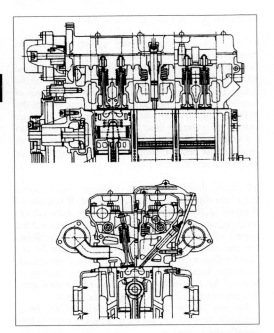

Fig. C340 Passenger car two-stroke diesel engine, AVL

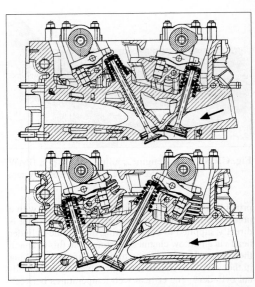

Fig. C342 Sectional views of a VR cylinder head with four valves

![photo]

Fig. C341 Two-part aluminum cylinder head of a direct injection engine (Ford Duratorq) with fur valves and central injection nozzle

Fig. C336 shows an example of a modern four-valve cylinder head with central spark plug position. The cylinder head for a V-engine with small V-angle (RV-engine) represents a special constructional variant. When two camshafts are used, the required asymmetry results in different length intake or exhaust ports and different length valves for various inclinations to the cylinder axis. Cylinder heads of this design type are very broad. **Fig. C342** shows an example of such a design.

~Dimensions. The key dimensions of a cylinder head (**Fig. C343**) include:

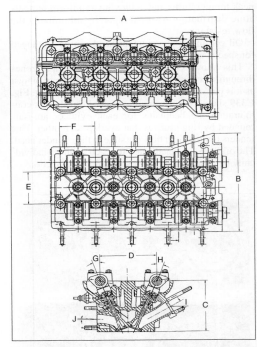

Fig. C343 Key dimensions on a cylinder head

- Number of cylinders: layouts of up to six cylinders (V-12 engine, W18 engine) are used for passenger-car engines.

- Cylinder spacing: The cylinder spacing is primarily determined by the minimum permissible web width.
- Bore diameter.
- Valve angle: The larger the valve angle on four-valve cylinder heads the greater the camshaft spacing and the wider the cylinder head.
- Number of valves.
- Position and number of camshafts: Cylinder heads with central overhead camshaft require relatively little mounting space. Twin overhead camshaft cylinder heads require a rather bulky cylinder head.
- Camshaft drive: The channel for the camshaft drive chain in integrated into the cylinder head in some designs.
- Position and design of the gas transfer ports.
- Valve actuation: Engines with bucket tappets for valve actuation normally require less cylinder-head clearance height than cam followers, for instance.

~Gas transfer ports →Exhaust port; →Intake port

~Injection-nozzle arrangement. The arrangement of the injection nozzle on the conventional gasoline engine is made in the intake manifold. The injection nozzle is mostly arranged laterally in the cylinder head of a direct-injection gasoline engine, as shown in **Fig. C344**. Depending on the combustion system, the diesel engine is differentiated between direct injection and indirect injection, in which the fuel is injected into a precombustion chamber. The precombustion chamber can be a swirl chamber or prechamber. Today, engines with direct injection are used almost exclusively on grounds of better fuel consumption. Dependent on the injection system selected, the following options are used:

- fuel-injector mount for a block pump
- pump jet
- injection valve for common rail

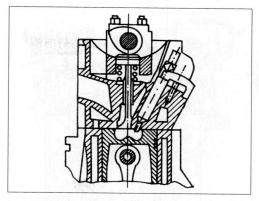

Fig. C345 Direct injection diesel engine and fuel-injector mount

Fig. C346 Direct injection diesel engine and pump-nozzle as well as built-in glow plug in the cylinder head

Figs. C345–C347 each show a cross section of the cylinder head of a diesel engine with direct injection, with **Fig. C346** showing a system with pump-nozzle and **Fig. C347** representing a common rail system.

Large engines are operated with inline fuel pumps, pump jet, individual pumps, and common rail.

~Inlet duct →Intake port

~Materials. Two groups of materials for cylinder heads have gained acceptance in principle: gray cast iron and light-metal alloys. Al-Si alloys predominate.

- Gasoline engine: Aluminum silicon alloys with 5–10% silicon and 2–4% copper or, for example, AL-SI10 Mg(Cu) wa (artificially warm-aged for strength enhancement)
- Diesel engine: Aluminum silicon alloys with 6–9% silicon, copper, and magnesium constituents, capable of being artificially aged; GG25; GG26 with alloy constituents Mn, Cu, Ni, Cr, Mo; compacted cast iron GGV
- Large engines: GGG40

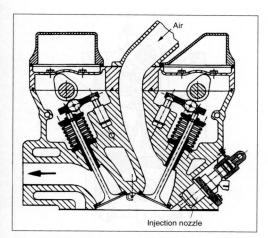

Fig. C344 Cross section through a cylinder head for a gasoline direct injection engine

C

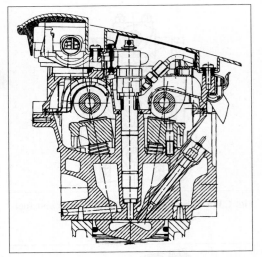

Fig. C347 Direct-injection diesel engine with common rail system

Mechanically low-stressed cylinder heads for gasoline engines are primarily manufactured from fusible aluminum alloys. Cylinder heads of precombustion chamber diesel engines are thermally very highly loaded in the region of the combustion space, hence requiring casting from primary alloys with a particularly high degree of purity and additional heat treatment to cope with the demanded durability. These materials are also advantageous for direct-injection supercharged engines with their additional high mechanical stresses.

Aluminum cylinder heads for series production engines are manufactured by low-pressure/gravity permanent mold casting processes. The density and microstructure in the region of the combustion space and the associated strength values need to be ensured using careful process control, particularly for highly stressed diesel engine cylinder heads.

Cylinder heads for truck engines and large engines are manufactured by sand casting from ferrous alloys with good overall heat-transfer coefficients and material strength values. Wall thicknesses and wall shapes need to be selected such that the assembly and operating stresses do not exceed the permissible values of material endurance limit. Because the strength characteristics of material decrease sharply with increasing temperature, a crown thickness that is as small as possible would be to the best advantage; mechanical durability has to be ensured by the overlying cool parts.

~Outlet port →Exhaust port

~Parallel flow →Cylinder head ~Cross flow

~Sealing →Sealing systems ~Cylinder head gaskets

~Spark plug location. Depending on the combustion system, one or two spark plugs are screwed into the cylinder heads of gasoline engines to initiate combustion.

The spark plug is located in the center of the combustion space in cylinder heads with four valves. Two spark plugs are useful on cylinder heads with three valves for reducing the combustion distances; the spark plugs, however, are eccentrically arranged.

~Swirl chamber and prechamber. The cylinder head concept for passenger car diesel engines with indirect injection is essentially determined from the arrangement of the precombustion chamber with its injection valve and glow plug. It is necessary to differentiate between swirl chambers and prechambers in cylinder heads with a precombustion chamber. **Figs. C348 and C349** show cross sections of such designs.

~Two-stroke engine. Charge transfer on conventional two-stroke engines is controlled via ports that are located in the cylinder liner. The cylinder head is, therefore, a rather simple construction. It commonly contains the injection nozzle and its associated drive on diesel engines (**Fig. C350**). The cylinder head can be air-cooled or liquid-cooled.

Fig. C351 shows the cylinder head of a piston-controlled, reverse-scavenged two-stroke diesel engine with a two-piece design, made by MAN. On these large engines, the cylinder head is a two-part design with a relatively thin-walled section that is intensively cooled confining the combustion space. The applied gas force is transferred to a dimensionally stable supporting body.

The variant of a bore-cooled cylinder head for controlled cooling in the seat and web region is shown in **Fig. C352**.

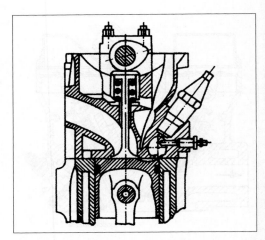

Fig. C348 Swirl chamber cylinder head of a two-valve diesel engine

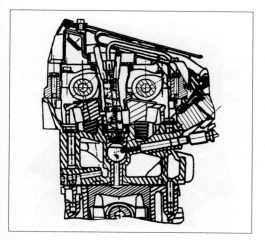

Fig. C349 Two-valve diesel engine prechamber cylinder head

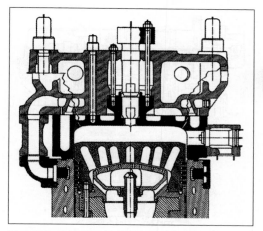

Fig. C351 Two-piece cylinder head of a two-stroke diesel engine with reverse scavenging (MAN)

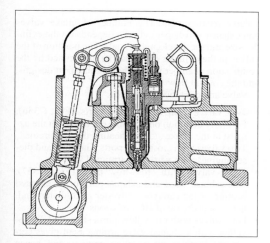

Fig. C350 Cylinder head of a GM two-stroke diesel engine with Roots supercharger and direct injection through pump-nozzle with in-block camshaft for pump drive (1936)

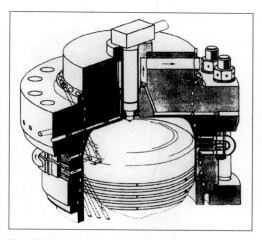

Fig. C352 Cylinder head of a Sulzer two-stroke diesel engine with cross-flow scavenging and bore-cooling

Later developments have the gas transfer carried out either with valves or a mixture of valves and ports. The expenditure on the head design is similar to that for a four-stroke engine if valves are employed. A camshaft drive has to be provided including camshaft(s), valve actuation, and valves. The cylinder head has an increased height requirement.

Depending on the approach and field of application there are advantages for controlling the exhaust with ports and the intake with valves, or vice versa. If the application should favor the diesel engine, advantages could accrue from achieving the intake swirl through intake valves, for instance.

If valves are employed for intake and exhaust control, this would require head loop scavenging in two-stroke engines, which would normally lead to low scavenging efficiency and an increase in the charge transfer work. **Fig. C353** shows the longitudinal and cross-sectional views of a passenger car engine with uniflow scavenging as an example of a valve-controlled two-stroke engine.

~**Valve actuation.** Today, gasoline engines and diesel engines for passenger cars are built exclusively with overhead camshafts. Camshaft actuation via chain, drive belt, or gears influences the cylinder head design. Gasoline engines have one or two camshafts. Valve ac-

C

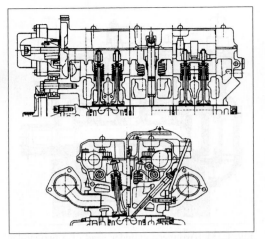

Fig. C353 Longitudinal and sectional view through the cylinder head with uniflow scavenging of a passenger car two-stroke diesel engine

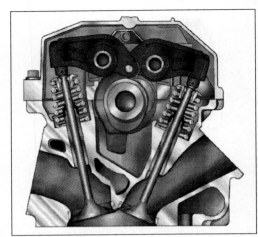

Fig. C355 Four-valve cylinder head gasoline engine with rocker arm

Fig. C354 Roller-type finger with hydraulic supporting element (Source: INA)

tuation is effected using bucket tappets (**Fig. C336**), roller finger followers (**Fig. C354**), or rocker arms (**Fig. C355**). The levers increasingly are fitted with rollers for minimizing frictional losses. Four-valve diesel engines frequently have two camshafts because of the space required for the injection valve.

Valve clearance is mechanically adjusted or automatically hydraulically for the sake of easier assembly and maintenance.

The camshaft of truck engines and large engines with low speed and a demand for a high service-life is arranged en bloc in the cylinder housing and driven by gears. The valves are actuated via tappets, pushrods, and rocker arms.

~**Valve arrangement.** The number of intake valves and exhaust valves per cylinder depends on the cylinder size and the development goals. The position of the valves in the cylinder head may be determined by the combustion chamber shape, the possible port designs, and the coolant flow.

Valve and duct layouts include:

- Two valves with counterflow ports (**Fig. C356**): Formerly common on gasoline engines with the exhaust at the intake side for heating the carburetor.
- Two valves with cross-flow ports: Gasoline and diesel engines with vertically arranged valves
- Three valves with cross-flow ports (**Fig. C357**). One exhaust valve and exhaust port for accelerated heating of the catalytic converter at cold start and, hence, a decrease of HC emissions
- Four valves with cross-flow ports (**Fig. C358**): Together with V-shaped valve arrangement gives better gas transfer on gasoline engines and hence higher power output and lower fuel consumption. Preferably vertical arrangement on a diesel engine because of combustion space for better gas transfer, variation of air swirl in the cylinder and optimum central position of the injection valve
- Five valves with cross-flow ports for gas transfer improvement

See **Fig. C359** for an illustration of an engine with three intake valves and two exhaust valves. The valve diameter, the valve shape, and the associated ports are optimized by means of gas-transfer calculations and laboratory measurements on models. The aim of this design and development process is to produce as little resistance to fluid flow in the ports as possible, which minimizes the charge-exchange losses and has a positive impact on fuel consumption. A characterizing quantity of resistance to fluid flow is the loss coefficient, represented in **Fig. C360** as function of valve lift.

C

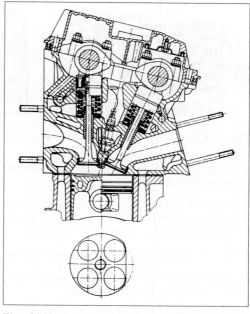

Fig. **C358** Four valves with cross-flow ports and asymmetric valve angles

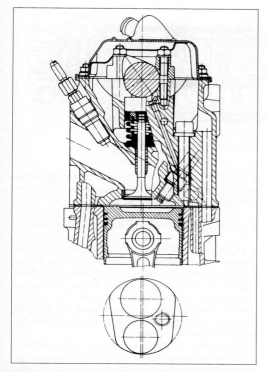

Fig. **C356** Two-valve gasoline engine with cross-flow ports (exhaust port and intake port) on the same side as in the cylinder head

One port on direct-injection diesel engines is generally designed as the swirl port. **Fig. C361** illustrates potential port shapes and port layouts.

Cylinder head bolt →Engine bolts

Cylinder head cooling duct →Cylinder head ~Cooling

Cylinder head cover →Valve cover

Cylinder head cover modules →Sealing systems ~Modules ~~Valve umbrella module

Cylinder head dimensions →Cylinder head ~Dimensions

Cylinder head gaskets →Sealing systems

Cylinder head gaskets with metal layers →Sealing systems ~Cylinder head gaskets

Cylinder head lubrication →Lubrication ~Lubricating systems ~~Passenger car engines

Cylinder jacket →Crankcase ~Crankcase design ~~Cylinder ~~~Insert technology

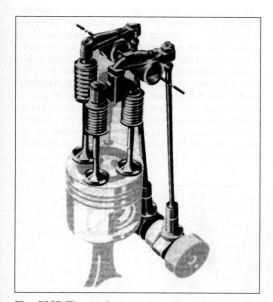

Fig. **C357** Three-valve arrangement

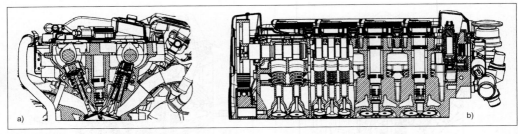

Fig. C359 Engine with five valves (a) cross-sectional view, (b) longitudinal section view

Cylinder layout →Engine concepts ~Reciprocating piston engines

Cylinder liner →Commercial vehicle cylinders/liners; →Crankcase ~Crankcase design ~~Cylinder ~~~Insert technology; →Cylinder running surface

Cylinder liner surface →Cylinder running surface

Cylinder material →Crankcase ~Crankcase design ~~Cylinder ~~~Insert technology

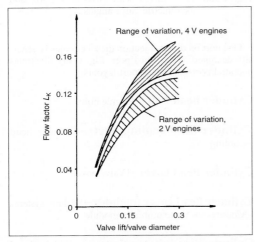

Fig. C360 Flow coefficients for the intake port of gasoline engines

Cylinder mixture movement →Mixture formation ~Mixture formation, gasoline engine ~~Mixture movement ~~~In the intake manifold

Cylinder pressure →Cylinder pressure curve

Cylinder pressure curve (*also*, →Combustion, gasoline engine; →Peak pressure). A time-based pressure curve arises depending on expansion of the flame and the variation of the volume of the combustion chamber that occurs with piston travel. The cylinder pressure curve is strongly dependent on the filling of the combustion space. Supercharged engines have the highest cylinder pressure.

Because the cylinder pressure curve reflects the effect of many variables on the operating behavior of the engine, it is perfectly suitable as a reference variable for engine control. However, this presupposes dependable and cost-effective acquisition of the cylinder pressure as well as real-time processing of the volume of data acquired.

~Diesel engine. The pressure curve for the diesel engine depends strongly on the combustion principle (**Fig. C362**). In comparison with direct injection diesel engines, the prechamber system shows a more moderate pressure rise that causes less combustion noise, and the maximum pressure gradient is also low. The precombustion chamber system does a particularly good job in this respect, but diesel engines with prechambers or swirl chambers are, however, no longer in series production for reasons of fuel consumption. One or several pilot injections are provided for reducing the rate of pressure increase in direct injection engines.

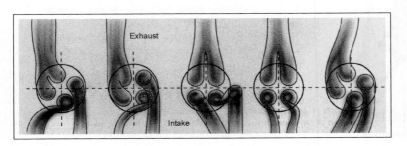

Fig. C361
Port shapes for a four-valve engine

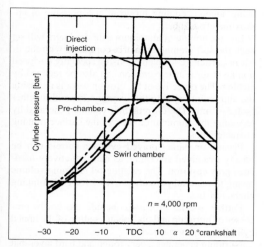

Fig. C362 Cylinder pressure curves for different combustion principles (diesel engine)

~Gasoline engine. Combustion in gasoline engines is initiated using spark plugs. The energy released during combustion causes the temperature and pressure of the cylinder charge in the combustion space to rise.

However, analysis of the cylinder-pressure curve indicates a time delay after ignition initiation (**Fig. C363**), which is due to local heating of the mixture to ignition temperature in the immediate vicinity of the spark plugs and is approximately 1 ms, independent of the engine speed.

The burn rate can be determined from the combustion curve (**Fig. C363**): the figure shows the ratio of the mass of fuel burned to the total mass of fuel as a function of the crankshaft angle. Through this, the start of combustion and its duration as well as its thermodynamic effect can be evaluated. The optimum location of the "center" of combustion for a homogeneous mixture is at approximately 8°CA after TDC and the effective combustion duration 30°–50°CA depending on the operating point and the combustion process.

The flame front starts from the spark plug and spreads across the cylinder in an engine with intake manifold injection and normal combustion at about 20–25 m/s. Reduction of the combustion duration gives efficiency advantages because the process comes closer to isochoric (constant volume) conversion of energy, and this can be reached by the following measures:

- Fast flame front speed achieved by higher charge movement (swirl, tumble, or quench flow)
- Shorter flame travel through compact combustion chamber design with centrally arranged spark plug or several spark plugs

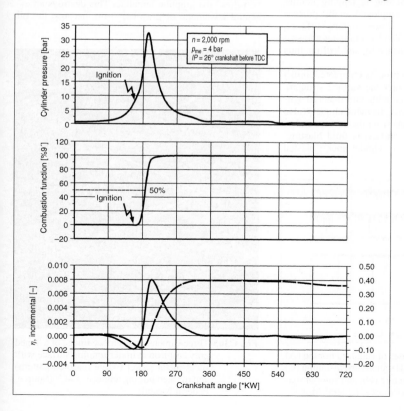

Fig. C363
Pressure curve and analysis of the cylinder pressure diagram

Cylinder running surface (*also*, →Crankcase ~Crankcase design ~~Cylinder). The cylinder running surface of combustion engines is tribological running mate and sealing face for pistons and piston rings.

~Machining process. Standard honing generates a normally distributed surface structure by single-stage or multistage machining—that is, there will be as many depressions as peaks in the roughness profile.

Plateau honing uses an extra processing step, however, to cut off the rough peaks, thus creating a plateau-like sliding surface with deep oil-retaining scores.

Short-stroke honing is a further development of plateau honing. This method primarily differs from plateau honing by having less roughness (peak roughness particularly) and a very large honing angle of 120°–150° for the deep scores. A very even surface roughness is reached by special honing strips following the bore shape.

Laser structuring enables a nearly free surface design by controlled erosion of material by laser. The cylinder running surface is structured in the TDC region and the rest is designed smoothly.

Structures such as spiral-shaped slots and pockets as well as pot-shaped forms are possible in addition to uniform conventional crisscross scored structures. **Fig. C364** shows the roughness profiles of various honing techniques.

A complex variant of honing is lapping honing wherein free abrasive grit is employed. In this process loose grit is used to give the cylinder running surface a chaotic high-low structure. The hard lapping abrasive is partly pressed into the surface by solid strips, thus creating a plateau surface.

With the standard honing process completed, brush honing rounds off and deburrs the surface structure with the aid of a brush coated with hard material. Another technique used to remove the metal chiplets (also called the sheet-metal jacket) from the surface and to flush existing pores in the surface is fluid blasting. With this technique, the whole cylinder liner is blasted

using aqueous cooling lubricant at a pressure of approximately 120 bar.

Erosive honing of aluminum cylinder liners will set back the soft aluminum matrix compared with the fiber or particle reinforcement using specially designed honing strips. Particle erosion can also be realized by etching. The objective of the erosion work is recessing the aluminum bonding that is prone to weld together by 0.5–1 mm. The oil-retaining volume generated by recessing the aluminum improves the surface running characteristics.

Plasma or flame-sprayed cylinder barrels can be ideally machined smoothly, as can inductively hardened gray cast iron. The existing oil retaining volume owing to material porosity provides for good running characteristics.

Further special processes include the costly processes of nitriding and phosphate coating of the honed cylinder running surfaces.

Nitriding generates a very rough and hard layer that cannot be used as cylinder running surface without additional treatment, such as phosphate coating, for instance. Phosphate coating is also employed without nitriding, and both has a smoothing effect and acts as a solid lubricant.

Fig. C365 shows the surface with a honed gray cast iron cylinder running surface.

The honing process on gray cast iron cylinder liners "crushes" the graphite lamellae. This destroys an essential advantage of gray cast iron, because the oil

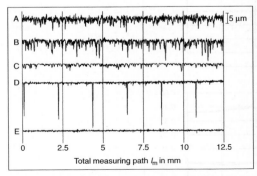

Fig. C364 Roughness profiles of standard honing (A), plateau honing (B), short-stroke honing (C), laser pocket structure (D), and smooth standard honing (E) (Source: Federal-Mogul).

Total measuring path l_m in mm

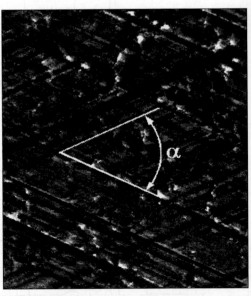

Fig. C365 Three-dimensional surface image of a honed gray cast iron cylinder running surface with sheet-metal jacket (white marbled pattern) and marked honing angle, $\alpha = 47°$ (Source: Federal-Mogul)

storage capacity of the "crushed" graphite lamellae is limited. A new method, which re-exposes the graphite lamellae, evaporates the top layer using a laser. In additon to exposing the graphite lamellae, the method has the advantage that a nano-crystalline layer with a high proportion of nitrogen is formed that provides ceramic properties at the contact surface. The method, soon to be series launched by one engine manufacturer, shows wear advantages of up to 90% over conventional honing and reduced oil consumption by up to 75%.

~Surface. The surface finish of the cylinder running surface contributes significantly to the buildup and distribution of an oil film between the running mates. There is a strong connection between cylinder roughness and engine oil consumption and wear. Cylinder roughness with Ra < 0.3 μm is state of the art.

Finish-machining of cylinder running surfaces is realized by means of fine milling or turning and subsequent honing treatment. A rotary and an alternating translatory motion overlap to combine with the cutting motion. This yields a cylindrical shape of below 10 μm and uniform surface roughness. The honing scores created by the cutting motion include the so-called hone angle.

Machining (**Fig. C366**) should preferably be handled gently on the material to prevent break-outs, crushing of the boundary zone and burr formation. Material cutting is carried out using honing stone retainers under aqueous cooling lubricant or special honing oil. Material erosion of 100 μm diameter at specified contact pressure or forward feed is achieved in less than one minute.

~Wear. The cylinder running surface is subjected to abrasive wear even under favorable tribological conditions; this is caused by contact with the sliding components such as the piston/piston rings. Moreover, secondary corrosive wear also occurs. Wear caused by the piston primarily derives from support at inclined connecting rod positions or piston secondary motion. On the other hand, both of these phenomena cause comparatively little contact surface wear.

The piston rings exert a much more significant influence, particularly at the dead center positions. The loss of speed at these positions results in the loss of the hydrodynamically built-up lubricant film, as a result of which mixed lubrication and high levels of local wear occur. The pressure in the combustion chamber causes the tangentially prestressed piston ring to press additionally against the cylinder wall. This generates the so-called "bore wear at top ring TDC" and oil consumption increases. The permissible value for bore wear at top ring TDC strongly depends on the pairing of materials. Gray cast iron contact surfaces are normally more critical than aluminum contact surfaces. **Fig. C367** shows formation of bore wear at top ring TDC at the upper ring reversal point as an example.

Literature: G. Flores: Grundlagen und Anwendungen des Honens, Vulkan Publishers, Essen, 1992. — U. Klink: Laser-Honing für Zylinderlaufbahne, MTZ Motortechnische Zeitschrift 58 (1997), Vieweg Publishers/GWV Speciality Publishers, Wiesbaden. — A. Robota, F. Zwein: Einfluss der Zylinderlaufflächentopografie auf den Ölverbrauch und die Partikelemissionen eines DI-Dieselmotors, MTZ Motortechnische Zeitschrift 60 (1999), Vieweg Publishers/GWV Speciality Publishers, Wiesbaden. — U.-P. Weigmann: Neues Honverfahren für umweltfreundliche Verbrennungsmotoren, in Werkstatt und Betrieb (Workshop Business), Year 132 (1999), Carl Hanser Publishers, Munich. — P. Trechow: UV-Laser lässt Motoren aus Grauguss lange leben, in VDI Newsletter 7 March 2003.

Cylinder running surface topography →Piston ring ~Lubrication

Cylinder shut-off →Displacement; →Electronic open- and closed-loop control ~Electronic open- and closed-loop control, gasoline engine ~~Functions; →Fuel consumption ~Variables ~~Engine measures, Air cycling valve ~Functions, air cycling valve; →Variable valve control

Cylinder sleeve →Crankcase ~Crankcase design ~~Cylinder ~~~Insert technology

Cylinder spacing →Crankcase ~Crankcase construction ~~Primary dimensions

Cylinder stud (*also*, →Cylinder head ~Cooling ~~Air cooling). Cylinder (head) studs are fasteners

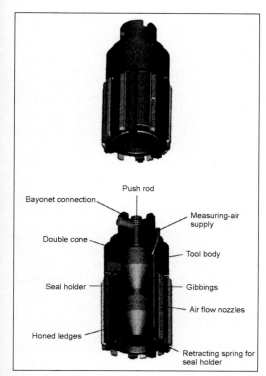

Bayonet connection
Push rod
Measuring-air supply
Double cone
Tool body
Seal holder
Gibbings
Air flow nozzles
Honed ledges
Retracting spring for seal holder

Fig. C366 Multiple-strip honing tool with air measurement system (Source: Nagel)

C

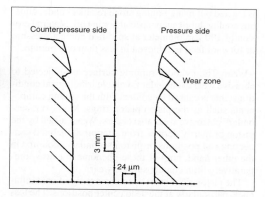

Fig. C367 Example of bore wear formation at top ring TDC (Source: Köhler)

for connecting the cylinder head, cylinder block, cylinder head gasket and, if necessary, preferably cylinder liners that are inserted into the crankcase and tensioned by cylinder studs. Cylinder studs would mostly have their point of load application in the lower crankcase area. If only the cylinder deformation is considered, a threaded connection below the water jacket is good. There are also designs in which the cylinder head and main bearing housings are screwed against the cylinder crankcase using straight-through cylinder studs.

Cylinder-selective knock control →Knocking

Cylindrical spring →Valve spring ~Design types

Cylinder volume →Displacement

Cylindrically asymmetric spring →Valve spring ~Design types

D

Damping →Engine acoustics ~Intake noise ~~Intake silencing

Dashpot →Actuators ~E-gas ~~Load-reversal damping; →Throttle dashpot

DC motor →Actuators ~General Purpose Actuator (GPA) ~~Electric motors

DCCS (Dilution Controlled Combustion Systems) →Combustion process, diesel engine

DC/DC converter ~Starter ~Starter-generators

Deactivation. The engine is deactivated if the permissible maximum engine speed is exceeded in a gasoline or diesel engine. This is normally done by interrupting the fuel feed in the diesel engine and not by interrupting the ignition as normally done in the gasoline engine. This prevents unburned fuel from getting into the catalytic converter, where it would react and damage the catalytic converter due to the high temperatures.

Dead center, or dead point (if not on a reciprocating engine). The dead center characterizes the extreme positions of the pistons, for which the piston speed equals zero.

~Bottom dead center. The bottom dead center (BDC) defines the point at which the direction of piston travel reverses after expansion and/or induction. Just as with top dead center, this can serve as a marker for engine control.

~Intersection dead point. The intersection dead point defines the smallest chamber volume between the casing and the rotor flank in a rotary piston engine.

~Top dead center. The top dead center (TDC) defines the point at which the direction of piston travel reverses after compression and/or expulsion. It serves as a marker for engine control—that is, as a relative point for the calculation of ignition timing and injection parameters.

Decentralized fuel mixture formation →Mixture formation ~Mixture formation, gasoline engine

Decibel (dB) →Engine acoustics

Decoupled analysis →Engine acoustics ~Acoustic transfer function

Decoupled exhaust system →Exhaust system

Decoupling systems →Sealing systems ~Elastomer sealing systems ~~Structure-borne noise decoupling

Deep-groove ball bearings →Bearings ~Materials

DEF →Oil ~Classification, specifications, and quality requirements ~~Institutions

Deflection jockey pulley →Toothed belt drive ~Camshaft drive ~~Design criteria

Degree of absorption →Engine acoustics

Degree of balance →Balance shaft; →Balancing of masses ~Residual imbalance

Degree of noise reduction →Engine acoustics

Degree of nonuniformity (cyclic variation). The degree of nonuniformity, δ, can be derived from the nonuniformity. It is defined as:

$$\delta = \frac{\omega_{max} - \omega_{min}}{\omega_{min}},$$

where ω_{min} = minimum angular velocity and ω_{max} = maximum angular velocity. The degree of nonuniformity can be minimized by increasing the size of the flywheel or the number of cylinders. When dimensioning the flywheel, the transient engine behavior must be considered.

Degree of protection →Electronic/mechanical engine and transmission control ~Requirements for mechanical and housing concepts

Degree of variation →Engine acoustics ~Psychoacoustic parameters

Delay angle. A delay angle on gasoline engines is defined as the angle—for example, in crankshaft degrees—from start of fuel injection to intake valve opening. Mixture preparation can be influenced by the delay angle.

Delayed combustion process →Combustion process

Delayed inlet closing →Control/gas transfer ~Four-stroke engine ~~Timing ~~~Inlet closes

Delivery characteristics, injection pump →Injection system (components) ~Diesel engine ~~Injection hydraulics, diesel engine

Delivery degree (*also*, →Volumetric efficiency). Delivery degree or volumetric efficiency is a measure of the quantity of charge in the cylinder. The delivery degree, λ_1, is a measure of the fresh charge in the cylinder after completion of the charge exchange process. It is defined by

$$\lambda_1 = \frac{m_z}{m_{th}} = \frac{m_z}{V_H \cdot \rho_{th}},$$

where m_z = mass of the fresh cylinder charge, m_{th} = theoretical mass of charge at certain specified conditions, V_H = cylinder displacement, and ρ_{th} = theoretical density, often based on inlet manifold or ambient conditions. The fresh cylinder charge, m_z, for gasoline and diesel engines, respectively, is

$$m_z = m_{z,L}, \quad \text{and} \quad m_z = m_{z,L} + m_{z,Kr},$$

where $m_{z,L}$ = mass of air in the cylinder and $m_{z,Kr}$ = mass of fuel in the cylinder.

For four-stroke engines with a small valve overlap, the delivery degree increases and approaches the same values as the volumetric efficiency, a. The aim of engine development is to maximize the delivery degree because the power output of the engine can be influenced directly through the relationship to the mass in the cylinder.

Using the wave action in the intake and exhaust tracts, values greater than unity can be achieved if corresponding control times are selected for the delivery degree. Values greater than unity also can be achieved for supercharged engines, depending on the definition used.

Demand development fuel →Fuel, diesel engine

Demulsifying capability →Fuel, gasoline engine

DeNox →Catalytic converter ~DeNox catalytic converter

Density, air →Intake air

Density, fuel →Fuel, diesel engine ~Properties ~~Density; →Fuel, gasoline engine ~Density

Density recovery →Electronic/mechanical engine and transmission control ~Requirements for mechanical and housing concepts

Deposits →Combustion chamber deposits; →Fuel, gasoline engine ~Additives

Desmodromic valve operation →Valve gear

Detergents →Fuel, diesel engine ~Additives; →Fuel, gasoline engine ~Additives; →Oil ~Body ~~Additives

Deterioration factor →Emission measurements ~Test type I ~~Mass determination, emissions

De-throttling. The operating characteristics of the engine can be improved by de-throttling of the intake and exhaust paths. De-throttling reduces the work that the engine has to perform to induce or to exhaust the air or the mixture. Higher performance and lower fuel consumption are the result. Air filters, the intake manifolds, the throttle valve and the intake ports are the main flow resistors on the intake side. The resistances from these can be minimized by optimizing the design, the cross-sectional areas, the surface roughness, the deflections, and so on.

The flow resistance of the manifold (junctions), the catalytic converter, and the muffler are of importance on the exhaust gas side. However, the exhaust gas routing also has an impact on the throttling rate. A junction of the exhaust gases from the individual cylinders (firing order, crankshaft throw) with good flow characteristics can also result in de-throttling, as, for example, the design of the intake and exhaust of the catalytic converter with good flow characteristics.

Deutz atmospheric pressure process →Combustion process, gasoline engine ~Stratified charge process

Diagnosis (*also*, →Electronic/mechanical engine and transmission control ~Electronic components). The engine control components are subject to failure monitoring during operation. The reasons for monitoring are to increase the reliability by switching to limp-home features, to monitor relevant exhaust gas components (onboard diagnostics, OBD), and to identify and localize defects during service work.

Diagnosis generally includes a check of sensors, processes, and actuators. Tests of the software function are performed in addition to functional component tests.

Identified defects are stored in a permanent memory in the form of code for readout when required.

Literature: C. Rätz, B. Adam: Innovative Lösungen für die Kfz-Diagnose, in Automotive Electronics 11/2002, Special Edition.

Diagonally split connecting rod →Connecting rod

Diameter port →Exhaust port ~Exhaust port design; →Intake port, gasoline engine

Diametrical force →Piston ring

Diaphragm pump →Diaphragm valve; →Reed valve

Die casting →Crankcase ~Casting process, crankcase

Diesel comparison process →Cycle ~Seiliger (dual combustion) cycle

Diesel control →Electronic open- and closed-loop control ~Electronic open- and closed-loop control, diesel engine; →Injection system, fuel ~Diesel engine ~~Mechanical control

Diesel engine (*also*, →Cold start; →Combustion process, diesel engine; →Electronic open- and closed-loop control ~Electronic open- and closed-loop transmis-

sion control, diesel engine; →Performance characteristic maps; →Exhaust gas analysis; →Injection system
~Diesel engine; →Injection system, fuel ~Diesel engine; →Load). The diesel engine is a machine with
auto-ignition of the injected fuel in the air in the combustion chamber; the air is heated by the high compression ratio ($\varepsilon = 17$–19 for car diesel engines with direct
injection). The air in the combustion chamber is
sucked in naturally or is supplied under pressure by
turbocharging. Normally diesel engines are reciprocating piston engines; other designs, such as the
Wankel engine, have not met with acceptance.

The method of formation of the internal mixture
used in the diesel engine produces a nonhomogeneous
mixture. The range of air-fuel ratio is between $\lambda = 1.05$ and 8. The "rich" side is limited by the smoke
number, and the "lean" side by the mechanical losses.
The load is controlled by changing the volume of fuel
injected, while the volume of air is kept more or less
constant; this is called qualitative control. The air-fuel
ratio is approximately $\lambda = 1.05$–1.2 at full load and increases accordingly with reduced load.

Different types of engines include direct injection
and an injection into a swirl or prechamber, also called
a precombustion chamber (**Fig. D1**). The diesel engine
has disadvantages in efficiency when designed as a
prechamber engine, compared with one with direct injection. These disadvantages are the result of higher
heat losses through the wall due to a larger, more irregular combustion chamber surface and flow losses as
the mixture transfers from the prechamber into the
main chamber. This is why the diesel engine with direct injection has an efficiency that is about 15% better
than the efficiency of prechamber engines and why it is
not only preferred in large diesel engines but also for
turbocharged engines in car diesel engines. A charge
air intercooler can be used in addition to increase efficiency and performance on turbocharged engines.

Diesel engines are offered in the power range of
about 2–40,000 kW and the turbocharged diesel engine is the standard version.

Four-cycle and two-cycle diesel engines are available, and single-cylinder, inline, and V-engines are offered. Diesel engines are normally water-cooled, but
air-cooled engines are also in production.

Diesel engines have the highest efficiency of all reciprocating piston engines—it is more than 50% for
large engines, and car engines already achieve efficiencies of 45%. The best fuel consumption values are
for large diesel engines about 145 g/kWh and for car
diesel engines about 200 g/kWh.

The untreated exhaust gas of the diesel engine contains less carbon monoxide, unburned hydrocarbons,
and NO_x than a comparable gasoline engine. However,
it has a higher particulate content, although this has
been reduced by the use of the low sulfur fuels available today. The lean total fuel-air mixture of the diesel
engine prevents the effective use of a controlled three-
way catalytic converter.

This is the reason why oxidation catalysts, lean catalytic converters (DeNox), and, increasingly, particu-

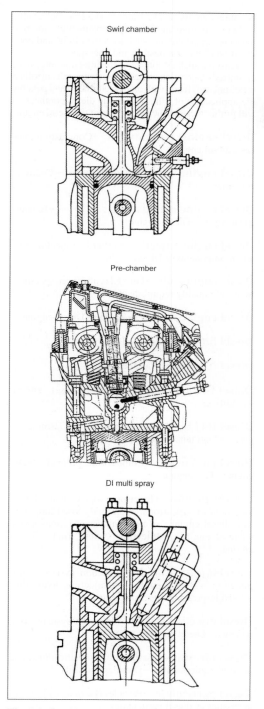

Fig. D1 Combustion process for the diesel engine

late filters are used. The Euro-4 values are achievable without a particulate filter by implementing internal measures in the engine, such as cooled EGR and optimization of the combustion and injection.

Measures such as control of the injection rate, swirl optimized combustion, variable injector nozzle, modified (reformulated) fuels (with oxygen attached), and possibly the application of a homogeneous diesel combustion will further increase the advantages of the diesel engine.

Diesel engine, prechamber →Combustion process, diesel engine

Diesel engine catalytic converter →Catalytic converter

Diesel engine exhaust gas analysis →Exhaust gas analysis ~Diesel engine

Diesel engine injection system →Injection system (components) ~Diesel engine

Diesel engine oils →Oil ~Classification, specifications, and quality requirements

Diesel engine torque →Torque ~Diesel engine

Diesel filter →Filter ~Fuel filters

Diesel fuel →Fuel, diesel engine

Diesel fuel detergents →Fuel, diesel engine ~Additives

Diesel fuel tank →Injection system ~Gasoline engine ~~Fuel tank systems

Diesel gas engine →Combustion process, diesel engine ~Gas engine

Diesel injection →Electronic open- and closed-loop control ~Electronic open- and closed-loop control, diesel engine; →Injection system (components) ~Diesel engine; →Injection system, fuel ~Diesel engine

Diesel injection pump →Injection system (components) ~Diesel engine ~~Injection hydraulics ~~~High-pressure injection pump

Diesel injection system →Injection system (components) ~Diesel engine

Diesel oil →Oil ~Classification, specifications, and quality requirements ~~Diesel engine oils

Diesel particulate →Particles (Particulates) ~Characteristics of diesel particulates

Diesel prechamber engine →Combustion process, diesel engine ~Prechamber engine

Diesel process →Combustion process, diesel engine; →Diesel engine

Dieseling. Dieseling is a process which is important for engines with carburetors. It occurs when fuel is sucked into the combustion chamber after the engine is switched off, where it ignites on hot components (spontaneous ignition). This causes the engine to continue to run on for a few seconds after the ignition is switched off. One possible remedy is to switch off the fuel feed electrically.

Dieseling is practically impossible on modern engines with electronically controlled injection systems that switch off the fuel feed.

Differential gear →Balancing of masses; →Mass balancing mechanism

Differential hall sensors →Sensors ~Speed sensors ~~Active speed sensor

Differential pressure sensors →Sensors ~Pressure sensors

Differential pressure valve →Lubrication ~Control and safety components

Diffusion. If a closed container contains two or more gases that are initially separated from each other by a wall, when the wall is removed complete mixing of the gases will take place after awhile purely by diffusion.

This process is important, for example, for combustion in an engine. Highly reactive components are carried into the air-fuel mixture by the diffusion of components and intermediate products of the burned gases, such as atomic or molecular hydrogen, atomic hydroxyl, and other radicals. These components support the start and acceleration of the reactions.

Diffusion field →Engine acoustics

Digital engine development →Virtual engine development

Digital interface →Electronic/mechanical engine and transmission control ~Electronic components ~~Serial interfaces

Dilution tunnel. The dilution tunnel is used to measure the particulate emissions on engine test benches for vehicles (**Fig. D2**). It is possible to distinguish between partial-flow and full-flow systems. When using the full-flow system, the entire mass of exhaust gas flows into the dilution tunnel and is then diluted with air. In the partial-flow system, part of the exhaust gas mass flows through the dilution system. The EPA has issued regulations relating to the design of and sampling from tunnels with full-flow systems for measuring particulate emissions from diesel engines.

The exhaust gas (partial flow or full flow) is generally diluted with filtered air. The dilution factor is determined by measuring the CO_2 concentration in the diluted exhaust gas. The diluted exhaust gas is passed over a filter. Apart from the weighting factors, the deposited particle mass, sample gas mass, dilution ratio, the entire exhaust gas mass flow, and the engine performance are elements from which the specific particle emission can be determined, in g/kWh for instance.

Fig. D2 shows the schematic representation of a full-flow dilution tunnel according to EPA regulation.

Dilution-controlled combustion processes →Combustion process, diesel engine ~DCCS

Dimensioning oil pump →Lubrication ~Oil pump

Diolefin content →Fuel, gasoline engine

Diolefins →Fuel, gasoline engine ~Diolefin content

Direct cooling system →Cooling circuit

Direct injection →Combustion process, diesel engine; →Combustion process, gasoline engine; →Fuel consumption ~Variables ~~Engine measures; →Injection functions ~Gasoline engine; →Injection system (components) ~Diesel engine; →Injection valves ~Gasoline engine

Direct injection process →Combustion process, diesel engine; →Combustion process, gasoline engine; →Fuel consumption ~Variables ~~Engine measures; →Injection functions ~Gasoline engine; →Injection system (components) ~Diesel engine; →Injection valves ~Gasoline engine

Direct injection systems →Injection functions ~Gasoline engine; →Injection system, fuel ~Diesel engine, ~Gasoline engine

Direct sound →Engine acoustics

Direct starting. The so-called direct starting of gasoline engines with direct injection has been under analysis for several years now. By means of targeted injections and ignitions, it is possible to accelerate the engine from standstill to its idle speed in a few revolutions without the help of an electric starter. This system is, however, not yet ready for production.

Direct valve actuation →Valve gear ~Actuation of valves

Dirt collection capacity →Filter ~Filter characteristics

Dirt particles →Filter ~Filter characteristics

Discharge voltage, battery →Battery

Disk cooler connecting rod →Radiator ~Design ~~Fin

Dispersing agents →Oil ~Body ~~Additives

Displacement. The displacement is the product of the stroke and the cross-sectional area of the bore. It is the most important variable next to speed in determining the performance of an engine, and it must be selected in such a way that the required maximum performance can be achieved. The displacement in gasoline and diesel automotive engines is primarily in the range of 1 to 3 liters, preferably distributed over

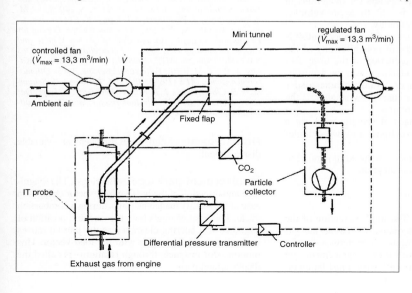

Fig. D2
Full-flow dilution system

three to six cylinders. The quantity of air induced in gasoline engines must be reduced by throttling for part-load operation (quantitative control), and it must be reduced by the mass of fuel in diesel engines (qualitative control). Throttling generates losses in gasoline engines, which means that the filling is not equivalent to that possible based on the displacement. There are several alternatives to reduce the throttling losses, such as variable displacement, cylinder shutoff, and downsizing. The charge transfer losses can be reduced with these alternatives because the amount of throttling is reduced due to reduced displacement for a given load point. Downsizing enables the required performance to be achieved through a smaller displacement and turbocharging.

The displacement is an important variable for assessing the torque of an engine especially at low speeds. A larger displacement delivers a higher torque, but it has the disadvantage that it increases the fuel consumption.

Small capacity diesel engines (approximately 1 liter displacement) with direct injection are currently under development as part of the effort to reduce fuel consumption. These extremely small combustion chambers represent problems for the optimal design of the combustion chamber and the injection parameters. To counter these problems, the displacement in these engines is distributed over a smaller number of cylinders—for example, three.

~Cylinder shutoff. Cylinder shutoff is a good method for adjusting the apparent displacement of the engine to meet its current power requirements and to reduce the charge transfer losses significantly at part loads. This can result in fuel consumption savings of 16% to more than 30%, depending on the number of cylinders and the load point. The reduction in fuel consumption occurs because engines with smaller displacements must be operated in higher load ranges to achieve the same performance. Cylinder shutoff is a solution that offers one of the largest potentials for reductions in fuel consumption, which is why there are vehicles in production with this technology.

One of the major problems of the cylinder shutoff is the worsening of the degree of nonuniformity of the crankshaft speed and the reduction of the firing frequency. This results in disadvantages for the acoustics as well as for the passenger comfort characteristics (vibration). These disadvantages grow smaller as the number of cylinders is increased.

Fig. D3 shows the ranges of specific work for a production engine in four-cylinder and two-cylinder operation.

The emission of pollutants is reduced in parallel to the reduction of the fuel consumption.

~Downsizing →Downsizing

~Variable displacement. The implementation of the variable stroke is one alternative to reduce the gas transfer work in gasoline engines. The engines must be operated at higher load levels if the displacements get reduced while the same specific work must be maintained. This also reduces the intake manifold vacuum and the throttling losses. The effect on the reduction of the gas transfer work results in the desired improvement of the fuel consumption.

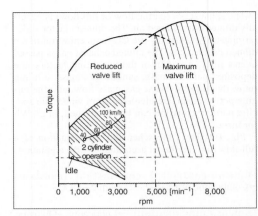

Fig. D3 Two- and four-cylinder operation of a gasoline engine (Source: K. Hatano and others)

Design solutions for a continuous variation of the piston stroke have been known for a long time and are an optimum solution with respect to a reduction of the gas transfer work, because in extreme cases they can eliminate the throttle valve completely.

Tests with variable stroke volumes have been conducted; however, the technical solutions were too expansive. One solution was, for example, shifting the crankshaft laterally, which reduces the stroke and, therefore, the displacement. A disadvantage was the reduced bore-stroke ratio due to the shift.

Literature: R. van Basshuysen: Zylinderabschaltung und Ausblenden einzelner Arbeitszyklen zur Kraftstoffersparnis und Schadstoffminderung, MTZ 54 (1993) 5, S 240 ff. — M. Schwaderlapp, S. Pischinger, K.I. Yapici, K. Habermann, C. Bolling: Variable Verdichtung—eine konstruktive Lösung für Downsizing-Konzepte, Aachen Colloquium on Vehicle and Engine Technology, Oct. 2001. — L. Guzella, R. Martin: Das SAVE-Motorkonzept, MTZ (1998) 10. — K.G. Fraidl, P. Kapus, W. Piock, M. Wirth: Fahrzeugklassenspezifische Ottomotorenkonzepte, MTZ (1999) 10. — M. Fortnagel, J. Schommers, R. Clauß, R. Glück, R. Nöll, C. Reckzügrl, W. Treyz: Der neue Mercedes-Benz-Zwölfzylindermotor mit Zylinderabschaltung, MTZ (2000) 05.

Displacement, variable →Displacement ~Variable displacement

Displacement path (*also*, →Bearings). The magnitude and direction of the bearing load vector changes in case of transiently loaded bearings during a combustion cycle. This also changes both the size and position of the narrowest bearing clearance as the journal moves in the bearing bore in line with the load vector. This motion—for instance, through rotation—is called the displacement path.

Displacement scavenging →Two-stroke engine ~Fresh mixture losses

Dissipation. Dissipation describes the irreversible processes in a system. A major process is the conversion of work into heat through friction.

Dissociation (also, →Chemical reaction; → Cycle). Dissociation describes the decomposition of a compound into other components; for example, CO_2 can disassociate into O_2, CO into O. This decomposition occurs at combustion temperatures of about 1500°C and presents intermediate products in high temperature combustion. The extent of dissociation, which means the displacement of the reaction to the one side (O_2, CO, O) or the other side (CO_2) of the reaction depends primarily on the pressure and temperature of the gas during the process. A chemical balance is achieved, for example, when the reaction described above causes as many CO_2 molecules to be produced as to decompose.

Distance between centers, connecting rod →Connecting rod

Distance sensor →Sensors

Distillation residue →Fuel, gasoline engine

Distortions →Calculation processes ~Application areas ~~Stability

Distributor contact breaker points →Ignition system, gasoline engine ~Ignition timing sensor

Distributor injection pump, edge-controlled →Injection system, fuel ~Diesel engine ~~Distributor pumps

Distributor pumps →Injection system, fuel ~Diesel engine

Divided combustion chamber →Combustion chamber ~Subdivided combustion chamber

DMF (dual mass flywheel) →Engine acoustics ~Transmission rattle; →Flywheel; →Torsional vibrations ~Flywheel

Double stopper →Sealing systems ~Cylinder head gaskets

Double-pulse holography →Engine acoustics ~Laser holography

Downdraft carburetor →Carburetor ~Design types

Downsizing (also, →Displacement; →Fuel consumption ~ Variables ~~Engine measures). Downsizing describes a process that allows engines with small displacements to achieve the same performance values on the road as engines with larger displacements. The rea-

son for adopting this approach is to reduce the fuel consumption and as a result the CO_2 emissions. A particular propulsive power can be delivered in a higher speed and load range by reducing the displacement size in gasoline engines and, therefore, operating the engine in a "de-throttled" mode; it means a shift from the operating points that were often used with a large engine to an operating point with a lower specific fuel consumption. Additionally, engines with smaller displacements have a smaller friction loss.

However, engines with small displacements have the disadvantage of a lower torque, especially at lower speeds, and, therefore, have reduced dynamic characteristics, such as flexibility. This must be compensated for to a large extent by a combination of different technologies. A downsizing concept must orient itself toward the following objectives:

- Reduction of fuel consumption and CO_2
- Reduction in weight, compared with engines with the same performance but larger displacements
- Smaller displacement and possibly lower number of cylinders if compared with conventional comparable engines
- Comparable performance by increasing the mean effective pressure
- Compensation of the torque weakness, compared with a conventional engine with larger displacement, e.g., through supercharging

These objectives can be achieved in gasoline engines by the following measures:

- Supercharging, especially the dynamic characteristics in the low-speed range must be improved. The approaches for solving include the electrically supported turbocharger and the e-booster; mechanical supercharging is also helpful.
- Fully variable valve drive.
- Variable compression ratio for the gasoline engine.
- Pulse supercharging through an air cycling valve.

Literature: C. Balis, P. Barthelet, C. Morreale: Elektronisch unterstützte Turboaufladung, Einfluss auf Downsizing und Übergangsmoment, MTZ (2002) 9. — M. Schwaderlapp, S. Pischinger, K.I. Yapici, K. Habermann, C. Bolling: Variable Verdichtung—eine konstruktive Lösung für Downsizing-Konzepte. Aachen Colloquium on Vehicle and Engine Technology, Oct. 2001. — H. Drangel, L. Bergsten: Der neue Saab SVC-Motor—Ein Zusammenspiel zur Verbrauchsreduzierung von variabler Verdichtung, Hochaufladung und Downsizing. Aachen Colloquium on Vehicle and Engine Technology, Oct. 2000. — P. Wolters, B. Biermann, K. Habermann: Downsizing—Möglichkeiten und Grenzen, Engineering Academy of Esslingen, 1998. — S. Pischinger, K. Habermann, K.I. Yapiei, H. Baumgarten, H. Kemper: Der Weg zum konsequenten Downsizing, MTZ (2003) 5, p. 398.

DP/DPC/DPCN/DPS distributor-type fuel injection pump (Delphi/Lucas) →Injection system, fuel ~Diesel engine ~~Distributor pumps ~~~Radial piston distributor injection pump

DPF (diesel particulate filter) →Particles (Particulates) ~Particulate filter system ~~Particulate filter

Draining test →Friction ~Friction measuring techniques

Drive belt design →Toothed belt

Drive-by noise →Engine acoustics ~Exterior noise

Drive-by-wire →Actuators ~E-gas

Driving cycle →Emission measurements ~FTP-75, ~Test type I

Drop forging →Connecting rod ~Semi-finished parts production

Droplet breakup. In order to improve the ignitability of the air-fuel mixture, the fuel droplets must break up quickly. Factors that affect this process are heat as well as flow and the inertia force acting on the droplet. Droplet breakup occurs when the inertia forces get bigger than the surface forces on the droplet due to internal flow turbulences.

Droplet radius. The droplet radius depends on the relative speed of the intake air. For a gasoline engine with carburetor, the maximum droplet radius is dependent on the relative speed, as shown in **Fig. D4**. (*See also*, droplet breakup.)

Droplet size →Injection system, fuel ~Gasoline engine ~~Direct injection systems ~~~Air-"blast" direct injection

Dry cylinder liner →Crankcase ~Crankcase design ~~Cylinder ~~~Insert technology

Dry sleeves →Crankcase ~Crankcase design ~~Cylinder ~~~Insert technology

Dry sump lubrication →Lubrication

DTC memory →Electronic/mechanical engine and transmission control ~Electronic components ~~Diagnostics

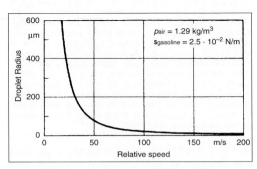

Fig. D4 Maximum droplet radius depending on the relative speed in the carburetor (Source: F. Pischinger)

DTC register →Electronic/mechanical engine and transmission control ~Electronic components ~~Diagnostics

Dual contour turning →Piston ring ~Shaping

Dual ignition →Valve arrangement ~Number of valves ~~Three valves ~~~Three valves and two spark plugs

Dual inline engines →Engine concepts ~Reciprocating piston engines ~~Multiple-shaft engines ~~~Twin-shaft engines

Dual-bed catalytic converter →Catalytic converter

Dual-branch exhaust system →Exhaust system

Dual-mass flywheel (DMF) →Engine acoustics ~Transmission rattle; →Flywheel; →Torsional vibrations ~Flywheel

Dual-mass torsional vibration damper →Torsional vibrations ~Dual-mass flywheel

Dual-roller chain →Chain drive ~Multiple chain

Dual-spark ignition coil →Ignition system, gasoline engine ~High-voltage generation ~~Ignition coil

Dual-spring nozzle holder →Injection valves ~Diesel engine ~~Fuel-injector body

Dummy head test →Engine acoustics

Duplex chain →Chain drive ~Multiple chain

Dust capacity →Filter ~Intake air filter ~~Filter characteristics

Dust passage →Filter ~Intake air filter ~~Filter characteristics

Dust-proof sealing →Electronic/mechanical engine and transmission control ~Requirements for mechanical and housing concepts

Dynamic charging →Air cycling valve ~Air cycling valve functions ~~Torque increase; →Supercharging

Dynamic sealing gap pulsations →Sealing systems ~Cylinder head gaskets

Dynamic seals →Sealing systems ~Flat seal designs

Dynamic stiffness →Engine acoustics

E

Early intake closing →Control/gas transfer ~Four-stroke engine ~~Timing ~~~Inlet closes

Easy-change filter →Filter ~Lubricating oil filter ~~Oil filter change; →Lubrication ~Filter ~~Filter designs

eBooster →Supercharging ~Electrically supported turbocharging

Eccentricity →Bearings ~Computation

ECE/EUDC consumption →Fuel consumption

Effective compression →Compression

Effective compression ratio →Compression ~Compression ratio

Effective fuel consumption →Fuel consumption

Effective power output (*also*, →Power output). Power is divided into indicated and brake power outputs. The following equation describes the indicated (or internal) power P_i of an engine:

$$P_i = p_{mi} \cdot V_h \cdot n \cdot i \cdot z,$$

where

> n = engine speed (rpm),
> i = combustion cycles per revolution,
> p_{mi} = indicated mean effective pressure,
> V_h = displacement of a cylinder, and
> z = number of cylinders.

The brake power output is calculated as:

$$P_e = p_{me} \cdot V_h \cdot n \cdot i \cdot z.$$

The difference between the brake power and the indicated power is the friction loss, P_r.

$$P_e = P_i - P_r$$

The power of an engine also depends on the ambient conditions. The physical variables of the ambient conditions—that is, pressure and temperature—have the highest impact. The filling of the cylinders and, therefore, the mixture or air mass in the engine are lower; therefore, the power is lower if the air temperature, for example, is very high. There is a similar effect for the air pressure.

Certain reference and standard conditions have been defined to achieve better comparability between test results. The reference conditions in accordance with DIN 70020 are pressure, $p_0 = 1013$ mbar, and the tem-perature, $T_0 = 20°C$. This results in the following correction equation for the power P_0 at standard state:

$$P_0 = P \cdot \frac{1013}{p} \cdot \left(\frac{273 + T}{273 + T_0} \right)^{0.5}$$

where p is the air pressure (mbar) and T is the air temperature (°C). Humidity is not included, even though it has an impact on the power characteristics (high water content in the air means a power loss of 1–2%).

Effective pressure →Mean effective pressure

Effective specific work →Mean effective pressure

Effective work. Effective work is the actual work that the engine performs. Effective work is measured at the engine output shaft to the transmission, in Nm or Joule.

E-gas →Actuators

EGR →Actuators ~Actuators for exhaust gas recirculation; →Exhaust gas recirculation

EGR bypass valve →Actuators ~Actuators for exhaust gas recirculation ~~Exhaust gas recirculation valve

EGR cooling →Exhaust gas recirculation

EGR radiator →Exhaust gas recirculation; →Radiator

EGR rate →Exhaust gas recirculation

EGR valve →Valve ~Exhaust gas control valves

Elastomer compatibility →Fuel, diesel engine ~Properties; →Fuel, gasoline engine; →Oil ~Properties and characteristic values

Elastomer lip seals →Sealing systems ~Cylinder head gaskets

Elastomer materials →Sealing systems ~Elastomer sealing systems

Elastomer sealing systems →Sealing systems

Elastomer seals →Sealing systems ~Elastomer sealing systems

Electric EGR valve →Actuators ~Actuators for exhaust gas recirculation ~~Exhaust gas recirculation valve

Electric fuel pump →Injection system (components) ~Gasoline engine ~~Fuel-feed unit ~~Fuel feed pump

Electric machine →Starter ~Starter-generators ~~Integrated starter-generator ISG (main drive)

Electric motor →Actuators ~General Purpose Actuator (GPA); →Engine accessories

Electric regeneration while stationary →Particles (Particulates); →Particulate filter system ~~Particulate filter ~~~Regeneration of particulate filters

Electric starter →Starter

Electric variable valve control →Variable valve control ~Operation principles, ~Systems without camshaft

Electrical conductivity →Fuel, diesel engine ~Properties

Electrically heated catalytic converter (EHC) →Pollutant aftertreatment ~Aftertreatment concept with a three-way catalytic converter

Electrically heated particulate filter →Particles (Particulates) ~Particulate filter system ~~Particulate filter ~~~Regeneration of particulate filters

Electrically supported turbocharger →Supercharging

Electrically triggered injection valve →Injection valves ~Gasoline engine ~~Intake manifold injection

Electrode design →Spark plug

Electrode gap →Spark plug

Electrode wear →Spark plug

Electrodes →Spark plug

Electrohydraulic transmission control →Electronic/mechanical engine and transmission control ~Requirements for mechanical and housing concepts

Electrohydraulic valve actuation →Variable valve control ~Operation principles

Electrohydraulic valve control →Variable valve control ~Operation principles

Electrohydraulic variable valve control →Variable valve control ~Operation principles, ~Systems without camshaft

Electromagnetic variable valve control →Variable valve control ~Operation principles, ~Systems without camshaft

Electromechanical valve actuation →Variable valve control ~Operation principles, ~Systems without camshaft

Electromechanical valve control →Variable valve control ~Operation principles, ~Systems without camshaft

Electromechanical variable valve control →Variable valve control ~Operation principles, ~Systems without camshaft

Electronic accelerator pedal →Actuators ~E-gas ~~Drive-by-wire; →Electronic open- and closed-loop control ~Electronic open- and closed-loop control, diesel engine ~~Functions, ~Electronic open- and closed-loop control, gasoline engine

Electronic carburetor →Carburetor

Electronic components →Electronic/mechanical engine and transmission control

Electronic control/open-loop control gasoline engine →Electronic open- and closed-loop control

Electronic engine and transmission control →Electronic open- and closed-loop control

Electronic ignition system →Ignition system, gasoline engine

Electronic open- and closed-loop control

~Electronic open- and closed-loop control, diesel engine. The control functions for the diesel engine are provided by an electronic engine control system.

The sensors are only slightly different from those in the gasoline engine; however, the actuators are adapted to the particular injection systems (distributor pump, serial pump, pump-line nozzle, pump nozzle, common rail) and to the particular engine design (exhaust turbocharging with electronic charge pressure control, exhaust gas recirculation, controlled coolant circulation).

~~Diagnostics. Diagnostics are classified as follows:

- End-of-line diagnostics are performed at the startup of the engine. This is done by implementing special modes of operation in the control instrument.
- Offboard diagnostics are primarily performed on an external computer—in most cases a workshop tester. These include special test procedures for the analysis of the engine, its sensors, and actuators, based on existing information in the error memory

or through interaction with the electronic engine system.

- Onboard diagnostics are integrated in the engine electronics and are active during engine operation. These are also called self-diagnostics. The exhaust gas-related diagnostics, which are required by law, are also called onboard diagnostics (OBD).

~~~**Electronic accelerator pedal.** Electronic engine control systems in vehicles acquire the demand from the driver by an electronic accelerator pedal, called E-gas. For safety reasons, the acquisition of the accelerated pedal position is design with multiple channels. The signals of several sensors are transmitted through separate wires to the respective control instrument.

~~~**End-of-line programming.** Engines and vehicles are delivered by the vehicle manufacturers in many different options. An effort is being made to build the injection system with uniform components to reduce the number of options.

It is possible to implement the adaptation with electronic engine control systems by programs and data that are stored in the control instrument. The control instrument is programmed with the data at the end of the vehicle production process. The data reflect the actual features and the legal regulation for the respective country of use. Changes by workshops are also possible. For example, fine adjustment of the idle speed can be performed with the help of a workshop tester. The programming is done through self-diagnostic interfaces and is protected against unauthorized tampering.

~~~**Engine brake.** An engine brake is used in commercial vehicles in addition to the mechanical brakes. This brake can be enabled by the driver, and the actual activation is performed by a control instrument to protect the engine brake from overload.

The engine brake is integrated in the total concept in vehicles with ABS and ASR systems (antilock braking system and antislip control). The engine control instrument is again connected through a data connection to the control instruments of the brake system and a possibly available vehicle control instrument.

~~~**Exhaust gas recirculation (EGR).** Exhaust gas recirculation is used in diesel engines to reduce the emissions of oxides of nitrogen. However, the increase of the quantity of inert gas in the cylinder has its limits — for example, the emissions of soot are increased. The first exhaust gas recirculation systems consisted of valves between exhaust pipe and intake port with relatively simply control systems, but the requirements for the EGR valves increased with the tightening of the exhaust gas emissions legislation, and this resulted in increasing the functions for the control of these systems. In most cases, current exhaust gas recirculation systems have a position sensor to adjust the position of the EGR valve with great accuracy. The physical interaction between reciprocating engines and the turbo-

charged engines (turbocharger) and the resulting variation in the pressure drop between the exhaust pipe and intake port together with all relevant boundary conditions must be taken into consideration to achieve the increased accuracy required from the calculation of the exhaust gas recirculation rates.

~~~**Fuel consumption gauge.** The instantaneous fuel consumption can easily be calculated by using the information available about quantity of fuel injected, engine speed, and driving speed. This calculation can be performed by the engine control instrument and by a board computer. The fuel consumption is normally displayed by the board computer on a display instrument. The information exchange is performed via a data connection.

~~~**Fuel temperature correction.** All known injection systems and metered fuel injected processes control the opening times. However, the injected fuel quantity is the important variable for the combustion process. The control time is calculated by using injection specific performance maps with the conventional input values of engine speed, quantity, start of injection (pump-nozzle), or pressure (common rail). The fuel temperature, measured by a sensor, is used to include changes of density and viscosity of the fuel that are dependent on the temperature. Additional system specific effects, such as compression losses and amounts of fuel leakage, can be compensated for by a correction of performance map established in a test.

~~~**Idle speed control.** As for the mechanically controlled car diesel engines, the electronic engine controller in a car implements control of the idle speed. Similar to the mechanical controller, the idle speed of an electronic controller is controlled to a constant value. However, the idle speed in an electronically controlled engine can be changed very easily by changing the set point. The idle speed set point is increased to improve the engine smoothness depending on temperature and load — for example, after a cold start or when the air conditioner is switched on.

~~~**Immobilizer.** The vehicle insurance companies require that these are electronically coded and self-activating and that they cannot be disabled by switching or short-circuiting two wires.

Immobilizers consist of three parts: the operating unit, the immobilizer control instrument, and the engine control instrument. A key code stored and generated in the operating unit is transmitted in remote controls with the help of infrared or radio or, if transponders are used, by electromagnetic near-field connections to the immobilizer control instrument. This is where the key code is compared with a stored code and transmits a release code on a bi-directional data line to the engine control unit if there is concurrence. The engine control unit performs a target-actual comparison and releases the engine start only when there is concurrence.

E

~~~**Injector control.** The control of the injection system (pump nozzle, common rail, etc.) in a modern electronic injection system is centralized in a separate module. This module computes the specified quantity of fuel injected and actuates the respective driver stages under consideration of the particular characteristics of the injection system and the ambient conditions (e.g., fuel temperature, fuel pressure). The driver stages in connection with the actuators may feature underlying control loops to increase accuracy and robustness.

~~~**Intelligent actuator.** An actuator with integrated electronics is called an intelligent actuator. This type of actuator is primarily used in components with an underlying position control loop. This means that the component comprises the actuator, a sensor, and analysis electronics including functions. They are also in parts equipped with components that permit communication with the higher ranking control instruments through a bus system (e.g., CAN).

~~~**Intermediate speed control.** The engine in commercial vehicles is often used as a drive for power take-offs such as cable winches or ladders. The engine speed in this application must be adjustable and maintained constant independent of the load. This is done with an additional control function, which is switched on through an operation unit.

~~~**Limitations.** The engine operation is limited by its boundary conditions. These include the mechanical stress limits, based on

- Acceleration forces — primarily due to the speed of the engine and its auxiliaries and on the exhaust turbocharger;
- Pressure forces, defined by the cylinder peak pressure;
- Load transfers in the drivetrain and also in the mechanically driven supercharger.

The thermal stress limits of the individual components depend directly or indirectly on

- the exhaust gas temperature,
- the charge air temperature,
- the coolant temperature,
- the lubricating oil temperature, and
- the emission limits for pollution and noise.

The adherence to these limits is guaranteed by the determination of the steady state engine operating range (**Fig. E1**). Exhaust turbocharging expanded this concept by introducing a fill limitation dependent on the charge pressure during transient operation of the engine.

Due to the use of an electronic engine control system, the operating limits of the engine can now be used more accurately than with the mechanical control instrument available in the past. The robust and accurate acquisition of the engine operating parameters and of the ambient conditions and their inclusion in suitable functions is the basis for this application.

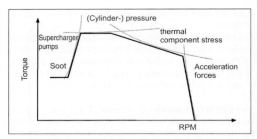

Fig. E1 Steady-state engine map of a supercharged vehicle engine with the activity areas of individual operating limits

The engine operation may also be limited if the diagnostics identify an error and the control instrument is switched to limp-home operation to enable the journey to continue safely.

The limits imposed by other control instruments installed in the vehicle, which have an impact on the engine operation, ideally intervene via the torque structure interfaces.

~~~**Load reversal damping.** Electronically controlled injection systems can change the injected fuel quantity extremely quickly over a very large range of quantity. This can lead to severe, abrupt load changes if this potential is realized by abrupt changes of the set point — for example, by fast movements of the accelerator pedal. The resulting intense and uncomfortable movement of the engine and its mountings and the severe load changes in the drivetrain must be avoided. This is why the load reversal damping system filters out high load gradients in the set point by using suitable algorithms and without limiting the dynamics of the engine.

~~~**Metered fuel control.** The performance request by the driver through the accelerator pedal was converted in older electronic diesel control instruments, through injection system specific performance maps, into a set point for the quantity of fuel injected. This set point was adjusted to comply with the available air mass, the engine and fuel temperature, and the engine speed. The thermal and mechanical limits in this system also had a direct impact on metered fuel computation. The quantity injected for startup was increased depending on temperature and time.

The turbocharging characteristics, especially in transient operations, are also included directly through the performance maps for the calculation of the maximum quantity of fuel.

The engine requirements for engine controls with integrated torque structure are coordinated primarily within this torque structure. The quantity of fuel injected is computed from the torque for one or more injections and for the associated injection times. The quantity of fuel injected will subsequently be converted into corresponding control signals by using control algorithms for the particular injection system (pump-line nozzle, pump nozzle, common rail, etc.)

~~~Offboard diagnostics *See above*, ~~Diagnostics

~~~Onboard diagnostics (OBD). The complexity of fault-finding in case of a dysfunction increases significantly as the high number of components and connection wires in the electronic engine control systems increase. This is why monitoring and replacement functions are implemented within the electronic engine and transmission system. These monitor the components and functions during operation to identify errors and store them in an error memory. The stored information can be read through a diagnostics output with a workshop tester.

Each signal has a permissible operating range. The signal range monitor is activated if the measured signal is outside this range. The signal can be tested for plausibility if several signals deliver the same or similar information.

Short circuits and breaks in contact in connecting wires to actuators are detected by intelligent semiconductor output stages. Control loops are monitored for permanent control deviations.

Errors within a control instrument—for example, in the computer part—are detected by the watchdog timer or by time monitoring. Especially critical points are designed with redundant structures.

Onboard diagnostics used as a monitoring system for the injection system detects changes in the exhaust gas-related parameters in the injection system based on malfunctions due to aging. The driver will get the information displayed if these changes exceed specified limits. The monitoring is performed in the control instrument.

Legal regulations determine the limits, the type of driver information, and the report obligations of the vehicle manufacturer to the authorities if the number of malfunctions exceeds certain limits for a vehicle type.

~~~Speed control. Diesel engines need a controller to stabilize the speed of the engine because of their inherent instability. This used to be implemented by using the centrifugal force in mechanical or hydromechanical governors. The speed for commercial vehicles is controlled over the total engine operating range (*see also*, →Injection system ~ Diesel engine ~~ Mechanical control ~~~Control maps). However, it is common for car engines to stabilize only the idle speed with an idle speed controller and control the maximum speed by a speed limiter or maximum speed controller. Other than this, the engine is operated without control or it is controlled by the driver.

~~~Speed limit. In some countries, the maximum speed of commercial vehicles must be legally limited to a fixed speed. The vehicle speed for powerful cars is also limited to a responsible speed.

This function can be easily and accurately implemented with electronic engine control systems through a limitation on the engine speed. The driving speed can be measured by an additional sensor and acquired by the engine control instrument or by the communication interface—for example, antilock brake system (ABS).

~~~Start of injection control. The set point for the start of injection is evaluated from the speed, the quantity of fuel injected, and the engine temperature. The control signals stored in characteristic curves/performance maps determine the start of injection. Closed-loop control is possible if the actual start of injection in the injection nozzle is also detected by a sensor. The control of start of injection can be performed with an accuracy of $\pm 0.5°$ cam angle. Open-loop control of the start of injection is significantly poorer than closed-loop control and depends on the particular system.

~~~Torque structure. The engine control system is networked with other control systems. The control systems for automated transmissions or automatic transmissions, the brake system, vehicle stability control system, and the air conditioning are important for the power required by the vehicle. The interface for the drivetrain related communication in diesel engine control instruments is integrated in the torque structure. The required engine torque is computed within the torque structure by using all torque requirements and the activated limits. Included are also load reversal damping, active surge control, and, depending on the system, the constant velocity control of the cylinder (engine smoothness control). The engine torque of the diesel engine that needs to be adjusted is converted at the output of the torque structure into one or more quantities of fuel injected and the required start of injection.

~~~Transmission control. The control of an automatic transmission or of an automated manual transmission is achieved by means of an independent transmission control instrument. For optimization of the control strategies, the transmission control instrument exchanges speed, torque requirements, limits, and error status (diagnostic results) information with the engine control instrument through a data line (e.g., CAN), normally directly, but also through a vehicle control instrument in some cases.

~~Functions. Functions (**Fig. E2**) are the algorithms implemented by software for the control and monitoring of the combustion engine.

The core functions of a diesel engine control system include the functions for calculating the quantity of fuel injected, control of the start of injection, the idle speed control, and the maximum speed limitation or deactivation. The other basic functions required for the operation of an electrically controlled injection system and of a fuel supply system are added.

The electronic engine control instrument offers the opportunity to add other functions for improving the

E

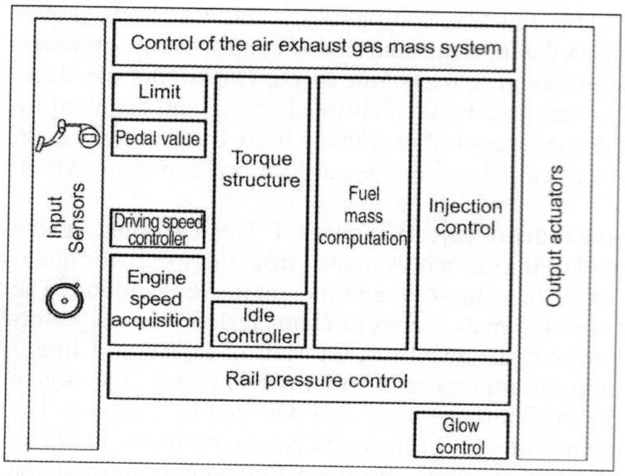

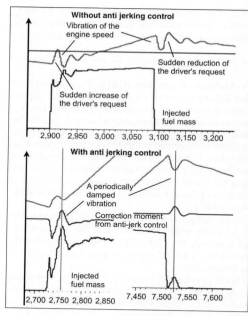

Fig. E2 Modular design of the functions for a turbocharged car diesel engine with common rail

Fig. E3 Effect of the active surge control

emissions, engine performance, road performance, and driving comfort as well as safety, availability, and maintainability of the engines.

Compliance with the ever more stringent exhaust emission regulations is only possible with the use of an electronic engine control system. On the other hand, the increasing complexity of the engine control functionality requires new processes for operation and maintenance of these systems.

~~~**Active surge control.** The drivetrain of a vehicle is a vibratory mechanical system. If this system gets stimulated by the inertia changes that develop during start or rapid acceleration, an unpleasant vehicle surge is experienced (**Fig. E3**). It also can develop at low speed if too high a gear is chosen. The active surge control is designed to detect these vibrations or, even better, predict them and reduce them to an acceptable or defined level by suitable opposite phase changes to the quantity of fuel injected.

~~~**Charge pressure control.** The turbochargers used in vehicles are designed to give good vehicle response, which means the effect of the so-called turbo lag is minimized. However, this requires that the achievable charge pressure gets limited for the higher engine speeds and performance range. In the past, this was implemented with the help of purely pneumatically controlled bypass valves or waste gates. The electronic engine control system does not only offer the opportunity to limit the charge pressure, but it can also react flexibly to existing boundary conditions, especially during acceleration. This is achieved by using appropriate actuators such as an electropneumatic controlled waste gate or a variable area turbine. The electronic engine and transmission control system also effectively permit the interaction of the charge pressure controller with the exhaust gas recirculation system and possibly existing secondary exhaust treatments (e.g., a diesel particulate filter).

~~~**Cruise control.** This controls the driving speed to a freely selected set point. The determination of the target speed is performed through it own operator unit. The driver can increase the driving speed at any time by accelerating—for example, for passing maneuvers—and can then continue with the originally selected speed without changing the cruise control. The cruise control is switched off by using the operating unit or by using the brake or clutch pedal.

~~~**Cylinder balancing.** Tolerance and wear specific differences between individual engine cylinders and individual cylinder components of the injection system can result in different torques from each cylinder. This results in speed fluctuations, which are seen in an irregular engine operation especially in the idle speed and low speed ranges. This roughness of engine operation is detected by the appropriate algorithms, and it is compensated for by an adjustment of the quantity of fuel injected for each individual cylinder in low performance ranges (**Fig. E4**).

The nonuniformities are normally not compensated if they exceed a defined range, but they are recognized by the self-diagnostics as a malfunction in the system.

~~~**Cylinder shutoff.** The fuel supply to the engine cylinders in diesel engines gets shut off during overrun. Individual cylinders also can be switched off selectively for solenoid valve controlled injection systems in lower load ranges to reduce the fuel consumption and the exhaust emissions in this range.

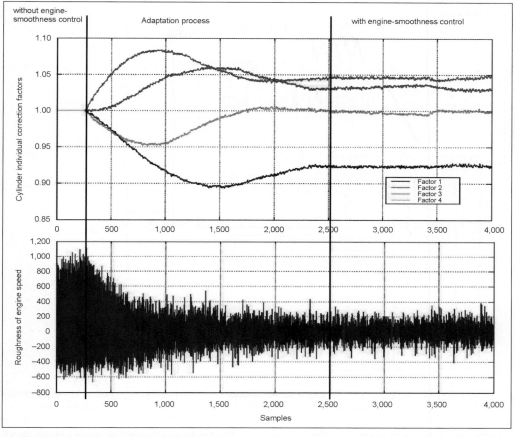

Fig. E4 Effect of the engine smoothness control

E

~Electronic open- and closed-loop control, gasoline engine

~~**Actuators.** A large number of actuators are required for the operation of a modern engine management system. The following lists some examples:

• Electronic throttle control (ETC)
• Injection valve
• Injection system (ignition coil and spark plug)
• Camshaft phasing valve
• Exhaust gas regeneration valve
• Relay for fuel pump control
• Relay for main relay control
• Intake manifold switch valve

~~**Electronic accelerator pedal.** The engine performance demanded by the engine management system is primarily controlled by the electronic accelerator pedal (**Fig. E5**). The request from the driver is determined by the sensing pedal value at the foot levers (accelerator pedal), and this is transmitted with a redundant analog control signal to the engine control unit. This drive-by-

wire technology permits a flexible design of the engine performance characteristics or the torque delivery of the engine through a performance map. This technology makes it possible for typical characteristics for the vehicle performance to be generated, such as good drivability and a high level of driving comfort—for example, free of drivetrain influences during switching processes.

~~**Functions**

~~~**Automatic stop-start.** The engine is shut off during long periods at idle speed, taking into consideration the actual engine and catalytic converter temperatures, to reduce fuel consumption.

Restarting is performed by the driver actuating the accelerator pedal. Several release conditions must exist to successfully perform a restart—for example, the clutch pedal for manual transmissions or the brake pedal for automatic transmission must be actuated.

The use of this technology represents an additional load on the starter battery because of repeated restarts

E

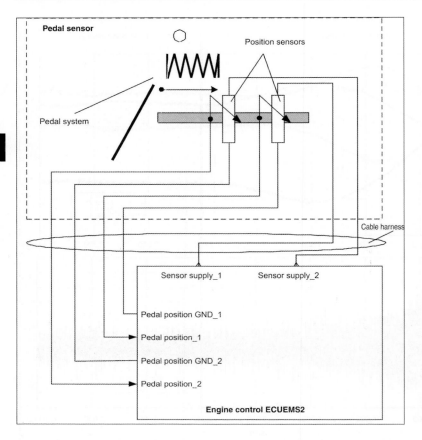

Fig. E5
Overview of electronic
accelerator pedal

in normal operation. The customer acceptance of this feature is significantly limited due to the limited number of permissible charge and discharge cycles for current starter batteries and the normally long start times for restarts combined with a high level of noise. This has resulted in very limited use of this technology for everyday operation, despite the potential fuel savings existing.

~~~**Camshaft phasing** (*also*, →Valve gear). The change in angle of the camshaft relative to the crankshaft (camshaft phasing) in a car gasoline engine can have a positive influence on the maximum performance, the torque characteristic, the exhaust gas characteristics, and the idle speed stability.

Camshaft phasers are produced today for two discrete angle positions and also for variable change of the angle position. **Fig. E6** shows the adjustment options of two continuously variable camshaft phasers.

The currently used designs of camshaft phasers are divided as follows:

- Two-position phasers (with helical teeth or with chain run shift)
- Continuous phasers (with helical teeth or based on the slewing motor principle).

Fig. E6 shows a camshaft phaser as a binary phaser with shift of the chain run and continuous phase in accordance with the slewing motor principle.

The adjustment of the camshaft position relative to the crankshaft is performed by electrohydraulic proportional valves driven by the engine oil pressure. The end position of the camshaft is the "early" position for the exhaust camshaft and the "late" position for the

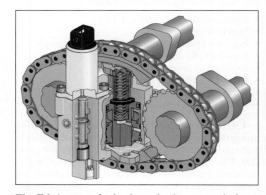

Fig. E6 Actuator for intake and exhaust camshaft

intake camshaft. The valve overlap is a minimum in these positions, which eventually results in smaller residual gas content in the cylinder. The phasing is moved away from the end position by actuating the proportional valves through the engine control system.

The position is controlled by the engine control system in continuous camshaft phasers by measuring the camshaft position relative to the crankshaft and analyzing the corresponding camshaft or crankshaft sensors.

~~~**Charge (boost) pressure control** (*also,* →Supercharging ~Supercharging pressure control turbocharging). Exhaust turbochargers are used to improve the power output and torque characteristics of gasoline engines. Relatively small turbochargers are used to ensure a positive impact on the torque characteristics even at low engine speeds. The turbochargers need a relief valve (waste-gate) to avoid high charge pressures at high loads and speeds, and this is achieved by routing part of the exhaust gas flow around the turbine (**Fig. E7**).

The position of the waste gate valve is controlled by the engine control system through an electropneumatic pulse valve and a pneumatic actuator. The charge pressure, p_L, is in most applications the control parameter. The relief valve will also be opened in certain part-load ranges because this reduces the pumping work required from the engine against the turbine back pressure. This also has the effect that the pressure gradient at the throttle valve and the pressure and temperature at the turbocharger exit are reduced, which results in a reduction of fuel consumption at part load.

~~~**Cold start.** Cold start occurs when the engine is started after standing for a long time and cooling down to the ambient temperature (this is in contrast to hot starting or repeated starts). The condensation of fuel on the cold components (the intake manifold and combustion chamber walls) must be taken into consideration for the control of injection and ignition during a cold start. The parts of the condensed fuel do not take part in the combustion and are released as unburned hydrocarbons (HC) to the exhaust port. The quantity of fuel delivered to the engine must be increased and the ignition timing adjusted to compensate for the above effects during the start. The measured air mass flow signal cannot be used, because the intake manifold must first be emptied. The control instruments provide fixed injection times based on the cooling water temperature. The injection duration can be evaluated only based on the load acquired after the start has been completed, which means after the speed threshold, which is dependent on the temperature of the cooling water being exceeded.

A short start time is initiated with a preinjection after a speed signal is detected. The sequential injection to individual cylinders, which is based on selective injection in accordance with a certain camshaft position, can only be performed after synchronization of the engine and transmission control system has been completed. The control of the injection duration depends on the number of rotations completed since the start (cycle counter), the engine speed, and the cooling water temperature. This means that the injection valves do not inject the quantity of fuel required for the wall moistening (to ensure an acceptable mixture in the cylinder) longer than necessary. The further time-dependent progress of the quantity injected, depending on the number of revolutions since the start, compensates for the decrease in the additional quantity required by the engine and depends on the energy converted since the start.

The first ignition pulse is issued after synchronization has been completed. The ignition timing and the injection durations are adjusted to the cold engine using its speed and the cycle counter. A band of spark ignition is used to improve the ignition at low starting temperatures—several ignition pulses are produced per combustion cycle during this phase.

~~~**Cruise control.** Cruise control is a function of the engine control system. A target speed specified by

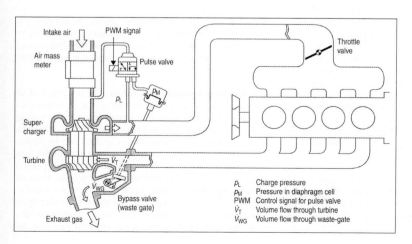

Fig. E7
Charge pressure control
(system arrangement)

Symbol	Description
p_L	Charge pressure
p_M	Pressure in diaphragm cell
PWM	Control signal for pulse valve
$\dot{V}_T$	Volume flow through turbine
$\dot{V}_{WG}$	Volume flow through waste-gate

the driver can be controlled with the cruise control. The target speed is adjusted by the driver through a switch on the steering wheel. Typical switch operation functions are listed in the following table:

Switch 1:	SET+	Set/accelerate
Switch 2:	SET−	Delay
Switch 3:	OFF	Control switch off
Switch 4:	Resume	Resume
System main:	Mainswitch	System switch on
Switch:		Switch off

A target torque requirement for the engine is calculated based on the specified target speed. This target torque is adjusted by the torque control system and the electronic throttle valve control system.

The driver can always increase the driving speed by accelerating.

This may, for example, be required during overtaking maneuvers. The previously adjusted target speed is returned to after the additional acceleration has been completed.

Just as the driver can increase the driving speed at any time, the driver can also cut off the cruise control at any time. This can be accomplished by actuating the brake or the switch operation unit.

~~~**Cylinder shutoff.** Engines at part load are operated in a range with unfavorable specific fuel consumption. If cylinder shutoff is used, the remaining cylinders must work at a higher load because of the switching off of individual cylinders; this moves them to a range of lower throttle losses and, therefore, lower fuel consumption. The remaining cylinders are switched on again only when the performance demand cannot be met anymore by the reduced number of cylinders.

The intake and exhaust valves must be switched off as long as the cylinder switch-off is active to reduce the charge transfer losses of the switched-off cylinders. This can be done by using switching elements in the valve gear. Known switching elements are, for example, the switching of the bucket tappet or the rocker arm with a switch-off function (**Fig. E8**). The switching function is normally actuated hydraulically and is coordinated by the engine and transmission control. This requires that the injection switch-off is triggered at exactly the same time as the intake and exhaust valve switch-offs.

The engine and transmission control actuates the throttle valve position (E-gas) via the torque model and, for a short time, the ignition timing controller, to make the switching process torque-neutral and, therefore, comfortable for the driver. The engine control system makes the decision when certain cylinders or cylinder groups should be switched off depending on many operating parameters, such as torque demand, engine temperature, and speed.

~~~**End-of-line programming.** Engines and the associated electronic engine control systems are used in different vehicle platforms. Variations between the requirements in different countries and, therefore, variations to adapt the engine system configuration to the

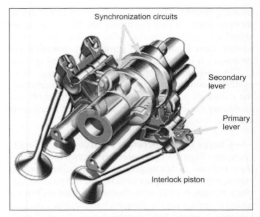

Fig. E8 Example of a DC 12-cylinder engine with hydraulic valve shutoff

respective market requirements—e.g., fuel qualities—also increase the number of options.

The data and program segments that are required for the operation of the control system are often programmed at the end of the line to reduce the number of options of the engine control systems delivered by the subsupplier. Another option to reduce the options is the coding of the control instrument at the end of the line. To achieve this, all data and program segments are normally provided in an engine option. The end-of-line programming selects the required dataset or even the program segment.

This process is also used for the activation of the speed control function integrated in the engine control system. This eliminates the hardware option of the control instrument for this function.

End-of-line programming also permits the adjustment of the idle speed in accordance with the transmission or drivetrain option—workshop testers also can activate it. This enables the service workshop to implement adaptation to the idle speed or to the adjusted standard fuel.

The end-of-line program has special access and protection mechanisms to avoid unauthorized access to the programming interface through the diagnostics interface, which is normally used for end-of-line programming.

~~~**Engine smoothness control.** The operation of an engine with a lean mixture can result in better efficiency when the engine is designed accordingly, as long as a certain limit of engine smoothness is not exceeded. This is why engine smoothness control can be implemented as an alternative to the $\lambda = 1$ control for lean operation at part load. The cylinder pressure diagram is used to assess the engine smoothness. Analysis of the signal from a combustion chamber sensor delivers a value for the engine smoothness; the value from one cylinder is adequate. The engine speed can also be analyzed, as an alternative to cylinder pressure, for

detecting engine smoothness, and it can now be used for limiting the air-fuel ratio or the cylinder uniformity.

~~~**Exhaust gas recirculation control** (*also*, →Exhaust gas recirculation). Exhaust gas from the engine exhaust system is routed through an exhaust gas recirculation valve (**Fig. E9**) to the intake tract of the engine. This approach increases the amount of inert gas in the fresh mixture (inert gas = gas that will not take an active part in the combustion) and, therefore, reduces the flame speed and the peak temperatures in the combustion chamber.

The process reduces the generation of thermal NO, and the de-throttling of the engine that results leads to increases in efficiency up to a certain limit. However, too much exhaust gas results in increases in emissions of HC and fuel consumption as well as a reduction of the engine smoothness. Exhaust gas recirculation is normally controlled by an electropneumatic converter and a pneumatic exhaust gas recirculation valve, although electrical exhaust gas recirculation valves are also used. The control function is stored as a target performance map in the engine and transmission control system—e.g., as a function of the required torque and the engine speed. Because the exhaust gas recirculation is an important factor in terms of the pollutants in the exhaust gas, appropriate diagnostics of the functional chain by the OBD are required.

~~~**Fuel consumption gauge.** The actual fuel consumption can be evaluated from the vehicle speed and integration of the injection times. The computed actual fuel consumption is transmitted by a bus system—for example, by CAN (Controller Area Network) bus, or by an output pin to the combination display.

~~~**Generator load control.** The increasing number of electrical loads in the vehicle results in an increased demand for electrical energy; this increases the demand for engine energy for operating of the engine accessories. The generation of 1 kWh of electrical energy typically requires more than 2.5 kWh of energy by the engine because of losses in the generators and the belt drives. Current generators can make 2–3.5 kW of electrical energy available. This means an additional load on the engine in the idle speed range. A reduction of the motive power load is important for small engines for drivability and to achieve good acceleration values. This is why the generator commutation may be limited for certain times depending on the energy available from the engine. It also includes the battery charge status. The generator load is increased and decreased with torque control.

~~~**Idle speed control.** The internal torque of the engine must overcome the "friction" torque at idle speed. The friction torque of the engine is determined by many factors, such as changing oil viscosity and energy consumers, which change their load (electrical generator, automatic transmission, air-conditioning, etc.). This is why an adequate control of the idle speed is required.

The control is performed by means of the torque, and it is divided into a fast and a slow path. The slow path influences the airflow to the engine at idle speed with the help of a throttle valve (**Fig. E10**). The fast path controls the efficiency of the combustion process through the position of the ignition timing relative to the top dead center. The ignition timing in this application is adjusted later if overspeed is detected and earlier to increase the internal torque if the speed is under the set value.

The current position of the throttle valve is adapted through an algorithm and stored in a nonvolatile memory in the engine control system, to include the long-term effects on the idle speed control, such as the contamination of the throttle valve.

~~~**Ignition timing/determination.** The instantaneous ignition timing is evaluated from a basic igni-

Fig. E9 Exhaust gas recirculation valve

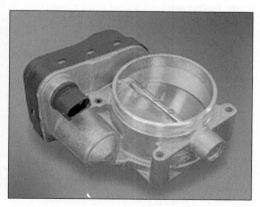

Fig. E10 Throttle valve actuator, E-gas

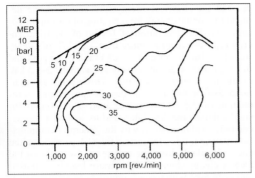

Fig. E11 Basic ignition angle performance map

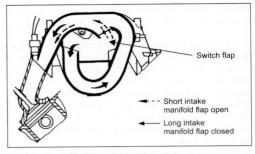

Fig. E12 Intake manifold switch over

tion angle stored in a performance map (**Fig. E11**) by including the current engine operating parameters. The actual ignition timing can be significantly different from the basic ignition timing depending on the ambient conditions. This requires special start ignition timings or late ignition timing for rapid heating of the catalytic converter.

~~~Immobilizer. Immobilizers are used in electronic engine and transmission control systems to meet the requirements of vehicle insurance companies. The requirement for an electronically coded and self-activating system must be met, which must not be disturbed from the outside by short-circuiting or rewiring.

Systems in current use consist of at least three components:

- Immobilized control instrument
- Engine control instrument
- Operating unit

A key code, which is stored in the operating unit, is normally transmitted via the electromagnetic near-field coupling of transponders. The immobilizer control instrument compares the key code received with the stored value. A bidirectional data protocol with the engine control system is started and a release code is transmitted to the engine control system if the key code is identical to the stored value.

The engine control system performs a check on the release code. Ignition and injection are released and a normal engine operation is started if the release codes are identical.

Increasing requirements for future immobilizer systems will be met by improved safety mechanisms and the expansion of the release code to other electronic control instruments in the vehicle — e.g., dynamic driving control instruments.

~~~Intake manifold switchover. The resonance conditions in the intake manifold are designed to use the charging effects optimally to improve the cylinder contents. This requires a compromise with respect to the requirement for short intake manifolds with large di-

ameters for the high speed range and long intake manifolds with relatively small diameters for low speed. This compromise stems from the requirements for the highest possible torques at low speeds and maximum filling at high speeds.

The resonance effects that exist in the manifold are used by the intake manifold geometry for these requirements to increase the cylinder fillings for low speeds. The intake manifold switchover offers the option to switch between two or more intake manifold lengths or intake manifold geometries, thereby reducing the conflict between high- and low-speed needs. Plastic valve systems are primarily used in the intake manifold for switchover of the length, and this is actuated by a vacuum motor (**Fig. E12**) that operates the valve systems.

Another option to increase the filling of the cylinders is the resonance charging in six-cylinder engines. In most cases, it is combined with the intake manifold switchover and represents a subgroup of the intake manifold switchover.

~~~Knock control. Knocking is uncontrolled self-ignition in the normally stable cylinder contents. It is accompanied by high flame speeds in the range of the speed of sound and high peak pressures. Permanent knocking combustion leads to engine damage, especially on the piston, cylinder head gaskets, and cylinder head.

The knock control segment of the engine control system keeps the engine at its knock limit by using closed-control loop when the engine is in the critical knock operating ranges. This is performed for engines without a turbocharger by altering the ignition timing and for turbocharged engines by altering both the boost pressure and the ignition timing.

The knock control uses the vibration generated by the pressure pulsations in the combustion chamber by measuring the structure-borne signals using a knock sensor. A seismic mass in the knock sensor acts on a piezo-ceramic device and induces a charge proportional to the structure-borne vibration. The vibration is normally found in a frequency range of 5 to 15 kHz, and it develops as a resonance frequency of the engine structure to the high frequency parts of the cylinder pressure, which are generated during knocking by the turbulent flame speeds in the combustion chamber.

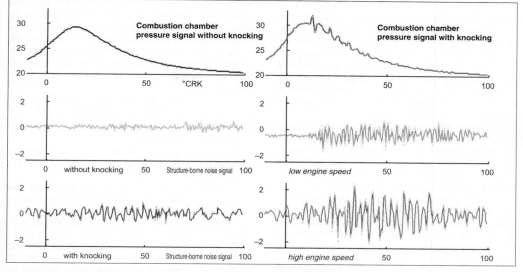

Fig. E13 Cylinder pressure diagram and structure-borne sound detection

Fig. E13 shows a typical cylinder pressure diagram and the structure-borne signal for normal and for knocking combustion.

The engine controller evaluates the knocking from the electrical knock signal by first formatting the raw signal in an integrated circuit (IC; **Fig. E14**). The formatted raw signal is now processed by the microprocessor. A knock event is present in a cylinder if the formatted raw signal exceeds the previously administered knock limit during engine operation. This analysis is performed in a knock window, which is specified for each cylinder by the crank angle position of the engine. The energy contained in the knock signal determines in a further step the size of the correction to the ignition timing.

In four-cylinder engines, one knock sensor is normally positioned on the crankcase between cylinders 2 and 3 on the intake side. This makes the knock vibration from all four cylinders detectable. Two knock sensors are used for six-cylinder inline engines, and the same number is also used for V-6– and V-8–cylinder engines (one knock sensor per cylinder block).

~~~**Lambda control.** The λ controller ensures that pollution components such as CO, HC, and NO are optimally converted in the catalytic converter. This re-

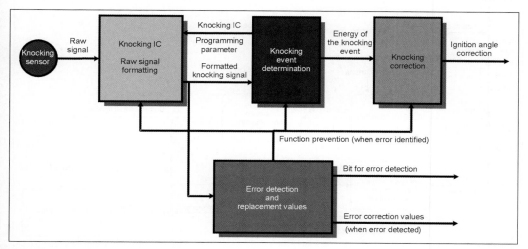

Fig. E14 Knock signal processing

quires that the air-fuel equivalence ratio ($\lambda = 1$) is maintained at the stoichiometric composition so that the three-way catalytic converter operates in a very narrow λ range of 0.99 to 1.00 (λ window). The air-fuel equivalence ratio, λ, is measured in a closed loop by the λ sensor in the exhaust pipe, the actual air-fuel equivalence ratio is compared with the set point, and the quantity of fuel is adjusted as required. Linear λ sensors (broadband sensors) are also used in addition to binary λ sensors (transition sensors).

The control algorithm for binary λ control is based on a PI controller, and the P and I parts are stored in performance maps based on engine speed and torque. The stimulation of the catalytic converter (λ fluctuation) is an implicit result of the binary control—the amplitude of the λ fluctuation is adjusted to 3%.

A forced stimulation is required to adjust the λ fluctuation for the linear lambda control.

Fig. E15 gives an overview over the structure of the linear λ control system including forced stimulation and trim control. The forced stimulation modulates a periodic deviation (λ pulse) onto the actual λ set point to optimize the efficiency of the catalytic converter. The resulting signal is integrated directly into correction of the quantity of fuel as a precontrol signal; a secondary influence is potentially added to the signal and it is processed by including the travel time and the delay characteristics of the linear sensor as a filtered λ set point.

The signal of the λ sensor is converted into a λ value in accordance with a stored characteristic curve. This characteristic curve can be corrected by the trim controller. The trim controller is a PI controller and uses the signal after the catalytic converter (preferably binary transition sensor), which has less cross-sensitivity.

The control deviation is computed as richness ($= \lambda -1$) by using the λ signal and the filtered λ set point,

and it is used as input for the actual I-controller. This controller is designed as a PII^{2}D controller and is shown in **Fig. E15**. The I^2 part is used for balancing the oxygen supply to the catalytic converter. The control output can be limited when transient conditions are present. Together with the precontrol, the computed injection quantity correction is used for the fuel injection calculation.

Linear lambda control has the following advantages, compared with the binary lambda control:

- Increased speed of the control dynamics and reduction of the transient λ error
- Increased efficiency of the catalytic converter through adjustable forced stimulation in a closed λ control loop
- Option to control $\lambda = 1$; this permits controlled warmup or controlled catalytic converter protection

~~~**Limp-home control.** When a vehicle engine malfunction occurs, the injection and the ignition availability are maintained through replacement values and emergency functions from the time of occurrence to the time when a repair workshop can be reached, so that the car can be operated with limited comfort. The controller replaces the missing information or provides a replacement value when a malfunction has been identified. Independent of the malfunction type, individual limp-home measures are taken in the case in which an actuator malfunctions. For example, injection to the affected cylinder is switched off to avoid damage to the catalytic converter if a malfunction is detected in the ignition loop.

~~~**Maximum speed governor.** This protects the engine from overspeed damage, which is why the engine speed is limited to a maximum value. The same inter-

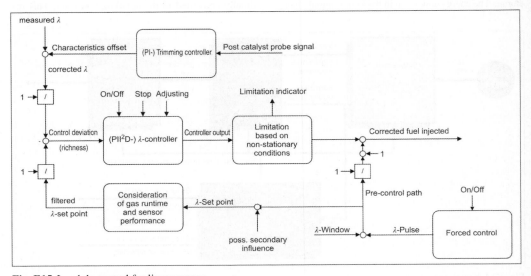

Fig. E15 Lambda control for linear sensor

vention, when used in conjunction with a vehicle speed signal, can also be used to control vehicle speed limits.

The engine torque is initially reduced with the help of the electronic throttle valve when the engine speed reaches the maximum value specified in the engine control system. The fuel injection is interrupted long enough for the speed to fall below the threshold in case the throttle action does not reduce the speed sufficiently.

~~~**Onboard diagnostics.** The onboard diagnostics include the monitoring of all exhaust gas-related vehicle components and the systems for their functioning related to the normal driving operation. The defective component should be identified as accurately as possible and defect type and location should be stored in memory if a defect is identified. The driver must be informed by a signal lamp on the vehicle instrument panel (check engine light) and must be asked to visit a repair shop if the defect exceeds specified exhaust gas values. Additional measures must be taken to maintain driving safety, to permit continued driving, and to avoid consequential damage. The repair shop must have the facilities to read the defect memory to be able to quickly identify the defect and facilitate repairs.

This is why California as the first state developed a law requiring the onboard diagnostics of engine control systems starting with the model year 1988. This law initially only included monitoring all components that were connected to the electronic control instrument in the engine control system. Expanded onboard diagnostics, called OBD II, were legally required starting with model year 1994. This law required, for the first time, the monitoring of all exhaust gas-related vehicle components and systems.

The following components or systems must be monitored:

- Catalytic converter system
- Lambda sensors
- The complete fuel system, which comprises the injection valves, the fuel pressure controller, the fuel pump, and the fuel filter
- Secondary air system
- Exhaust gas recirculation system
- Tank venting system consisting of activated charcoal filter and tank venting valve
- Other exhaust gas-related systems not directly controlled by the engine control system, such as transmission control for automatic transmissions that is equipped with elevation of the gear shifting point to improve the heating of the catalytic converter

Engine misfires must also be identified. A standardized malfunction lamp control and a standardized tester interface, which permits the reading of the defect memory in the repair workshop, are required in addition to system monitoring.

A second conventional lambda sensor (monitoring sensor) must be installed downstream of the catalytic converter, in addition to the lambda sensor upstream of the converter (control sensor), to monitor the functioning of the three-way catalytic converter. The lambda

control oscillation is increasingly damped over the length of the catalytic converter, because a catalytic converter that is functioning stores oxygen. This damping declines with increasing age and the monitoring sensor identifies lambda oscillations downstream of the catalytic converter. The direct comparison of the lambda signal upstream and downstream of the catalytic converter can, therefore, indicate the aging status of the catalytic converter and can switch on a diagnostic lamp in the case of a malfunction.

~~~**Overrun fuel cutoff.** The fuel injection is switched off while the ignition continues to operate until the upper speed for the idle speed control has been achieved, if the engine is operated in overrun, which means the engine actually delivers an engine torque of zero (noncombusted operation) or even a negative braking torque due to pumping.

The complete switch-off of the injection is interrupted to avoid cooling of the catalytic converter. The transition to overrun and the return are operated to guarantee a transition without jolt by targeted torque increases or decreases.

~~~**Speed restriction.** A further increase of the speed is prevented in the engine management system after a defined speed limit has been achieved by interference in the injection and ignition systems and the air path. This avoids a further increase in torque and, therefore, an increase in speed. The interference is done in several stages. The intensity of the interferences depends primarily on the actual engine operating status, the vehicle status, and the gradient of the speed increase.

The torque is reduced in the first stage by interference with the ignition. This results in an increase of the exhaust gas temperature and an increase of the component temperatures on the exhaust side. This is why the interference is performed primarily by disabling the injection in individual cylinders in the firing order for a fast reduction of the torque. Comfort reduction is avoided by a slow adaptation of the air path. The throttle valve angle is reduced to reduce the torque and to prevent exceeding the maximum permissible speed limit by throttling. Limiting the speed by interferences in the filling of the cylinders is performed adaptively, corresponding to the existing ambient conditions.

~~~**Warm-up control (catalytic converter heat-up).** Direct control of warm-up is unnecessary due to the use of modern multipoint injection systems. The buildup of wall films to ensure that an ignitable mixture exists is no longer required because the injection is close to the intake valve; the wall film is rather undesirable due to the additional HC emissions.

Special warm-up control through adapted ignition, injection, and cylinder charging is only used to achieve a defined threshold of engine temperature to compensate for the increased torque demand after the engine cold start. However, special catalytic converter heating is required to warm up the catalytic converters after a cold start. The rapid heating of the catalytic converter with late ignition timing and the resulting high exhaust gas temperatures are different from the very rich mixtures and the injection of secondary air into the exhaust pipe upstream of the catalytic converter.

~~**Sensors** (*also*, →Sensors). A modern engine management system typically consists of the following sensors:

- Crankshaft position sensor
- Camshaft position sensor
- Air mass flow sensor
- Intake manifold pressure sensor
- Accelerator pedal sensor for sensing the driver demands
- Engine-cooling water sensor
- Intake air temperature sensor
- Knock sensor
- Lambda sensor in front of and behind the catalytic converter
- Throttle valve position feedback (two redundant sensors)

Literature: R. van Basshuysen, F. Schäfer (Ed.): Handbuch Verbrennungsmotoren, Wiesbaden, Vieweg Publishers, 2002.

~Electronic open- and closed-loop control/transmission control

~~**Diesel engine.** The control instrument for the automatic transmission or the automated transmission can compute control signals for managing the switching transition or the change of gear ratio by using the operating data stored in the electronic control system of the diesel engine. The two control instruments exchange the engine speed and torque information. An engine control intervention is performed during the switching transition and the engine torque is specified.

During the overlap phase, the transferable torque is transferred from a clutch that switches off to a clutch that switches on in transmission systems without traction interruption. The synchronization of the rotating engine masses is supported by a reduction of the engine torque during the subsequent peak phase. The reduction of the engine torque by changing the quantity of fuel injected and the start of injection is performed in coordination with the clutch control. The engine intervention enables a smooth switching transition in conjunction with high-level dynamics.

In transmission systems with traction interruption, the drivetrain is "interrupted" during the switching transition by means of a separation clutch. The engine is controlled by the torque or speed set point during the switching transition. It is required to reduce the engine torque in a controlled manner to "zero" to execute the switching transition. The demanded gear is synchronized and engaged after the separation clutch has been opened. The drivetrain is closed because of the controlled closing of the clutch, and the engine is guided to the calculated demand speed. The specified engine torques are adjusted by the electronic diesel control sys-

tem through the quantity of fuel injected and start of injection.

The switching strategy in the control instrument determines the demanded gear ratio and the gear which should be engaged. Different switching programs (performance- or consumption-oriented) are available through programmable switching maps. The switching programs are selected by "intelligent" driving or switching strategies based on a classified driving situation. The driving style selected by the driver (performance- or consumption-oriented) and the driving situation of the vehicle by recognition of the environment (road inclines, curves, or low road friction values) can also be classified. The adaptation to driving situations ensures that the engine works as closely as possible to its optimum fuel consumption range, and drivability, driving comfort, and driving safety are also taken into consideration.

~~**Gasoline engine.** The control instrument for the automatic transmission or the automated transmission can compute the control signals for the management of the switching transition or the change of gear ratio by using the operating data of the electronic engine and transmission control system for this application. The two control instruments exchange the engine speed and torque information. An engine control intervention is performed by the transmission control system during the switching transition and the engine torque is specified.

During the overlap phase, the transferable torque is transferred from a clutch that switches off to a clutch that switches on in transmission systems without traction interruption. The synchronization of the rotating engine masses is supported by a reduction of the engine torque during the subsequent peak phase. Because of the dynamic requirements, the engine torque is reduced by adjusting the ignition angle in conjunction with the clutch control system. The engine intervention enables a smooth switching transition in conjunction with high-level dynamics.

In transmission systems with traction interruption, the drivetrain is "broken" during the switching transition through a separation clutch. The engine is controlled by the torque or speed set point during the switching transition. It is required to reduce the engine torque in a controlled manner to "zero" to execute the switching transition. The demanded gear is synchronized and engaged after the separation clutch has been opened. The drivetrain is closed because of the controlled closing of the clutch, and the engine is guided to the calculated demand speed. The adjustment of the specified engine torques is performed in the gasoline engine by the E-gas system through the electronic engine and transmission control system.

The switching strategy in the control instrument determines the demanded gear ratio and the gear which should be engaged. Different switching programs (performance- or consumption-oriented) are available through programmable switching maps. The switching programs are selected by "intelligent" driving or switching strategies based on a classified driving situation. The driving style selected by the driver (performance- or consumption-oriented) and the driving situation of the vehicle by recognition of the environment (road inclines, curves, or low road friction values) can be classified. The adaptation to driving situations ensures that the engine works as closely as possible to its optimum consumption range, while the drivability, the driving comfort and the driving safety are also taken into consideration.

Literature: Braess/Seiffert (Ed.): Vieweg Handbuch Kraftfahrzeugtechnik, Vieweg, Braunschweig/Wiesbaden, 2000.

Electronic open- and closed-loop transmission control →Electronic open- and closed-loop control

Electronic slip →Flywheel

Electronic spark control →Ignition system, gasoline engine ~Ignition ~~Spark control

Electronic stability control →Actuators ~E-gas

Electronic/mechanical engine and transmission control

~Electronic components

~~Angle sensor →Sensors ~Position sensor

~~Arbitration. *See below*, ~~CAN interface

~~Bare die. Electrical components, which in contrast to conventional components are not delivered in a plastic housing but which are installed without housing, are called bare die or bare chip (**Fig. E16**). They are fixed on the substrate being used with an appropriate jointing method (e.g., gluing of the chip with its back to the substrate). The electrical connections are established with the help of so-called bonding. Bonding means the connection of the connection pads on the upper side of the die with the appropriate pads onto

Fig. E16 Two bare dies with additional components on a switching carrier

the substrate with the help of very thin gold or aluminum wires with diameters of 25–300 μm.

~~Binary O_2 sensor interface. It is used for the acquisition of the O_2 value. The output delivered by the O_2 sensor assumes a voltage of approximately 200 mV ($\lambda > 1$) or approximately 800 mV ($\lambda < 1$) depending on the oxygen concentration in the exhaust gas. This voltage is read directly by the microcontroller of the engine and transmission control system. The gasoline/air mixture for the engine must be either enriched or weakened depending on the measured value, and this results in permanent control of the exhaust gas composition appropriate $\lambda = 1$. A short duration current is injected into the sensor to control the sensor temperature. This results in a temperature-dependent voltage increase.

Binary O_2 sensors are currently used in almost all gasoline engines.

~~Block diagram. The block diagram (**Fig. E17**) is an abstract way of showing the primary circuit elements of an electronic control device for gasoline and diesel engine control systems and for transmission control systems. The signal flow from the sensors through the input filter structures to the microcontroller is shown.

The microcontroller analyzes these and computes the important output variables in the form of magnitude and time information. The microcontroller then transfers this information through output stages to the actuators. Additionally, a power supply is required to provide power to the circuit components and the extensive reset logic to guarantee orderly functioning of the device. **Figs. E18–E20** show examples of gasoline, diesel, and transmission control systems.

~~Bonding. *See above*, ~~Bare die

~~Bridge output stage. *See below*, ~~Power transistor

~~CAN interface (controller area network, CAN). This interface is primarily used for rapid data exchange between control units (e.g., engine and transmission control systems) and intelligent sensors or actuators. The maximum data transfer rate is 1 MBit/s. This international interface is described in ISO 11898. The bus protocol was specially designed for relevant safety applications in vehicles. Transfer defects are reliably identified by the sender. All CAN participants can transfer information or can sample the bus. The object-oriented signals or messages contain information such as engine speed, vehicle speed, and temperatures and are available to all components at the same time. Based on the transmitter identifier, each receiver decides whether this message is meant for it. The CAN bus has multimaster characteristics, which means each component can start sending with equal rights; this requires an arbitration process. The arbitration of the bus devices is priority-controlled by its identifier. The arbitration and the intrinsic safety require error frames, and that is why the bus should only be utilized up to 15% with useful messages.

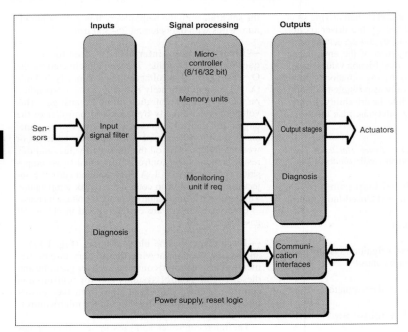

Fig. E17
Signal flow in electronic
control instruments

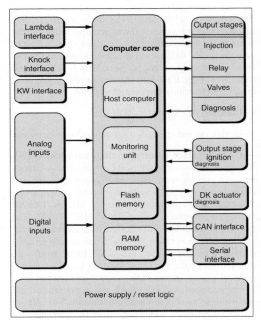

Fig. E18 Microcontroller for gasoline engine control
systems

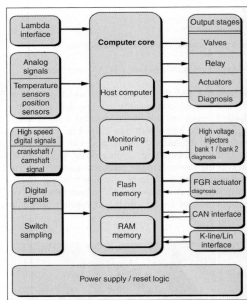

Fig. E19 Microcontroller for diesel engine control
systems

~~**Computer unit.** The computer unit consists of
the host computer, the read-only memory (ROM) for
the program code and characteristics, the variable data
storage, and possibly a monitoring unit.

~~**Controller Area Network.** *See above*, ~~CAN
interface
~~**Diagnostics.** Extensive diagnostics are included in
the control instruments to monitor the orderly functioning

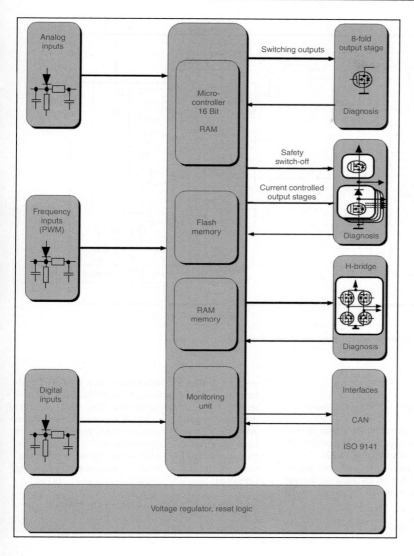

Fig. E20
Microcontroller for
transmission control
systems

of the device. Logical and functional correlations (software) decide whether an orderly function is present or not. These diagnostics are applied to the input as well as to the output signals. The input signals are tested with the help of plausibility analysis and combinatorial analysis of the signals with each other to find out whether the new status or the acquired change of the signal is possible or not. The output signals are monitored in a similar manner. The basis for the functional decisions is the logic integrated in the respective components. This logic monitors whether overload (high current, short circuit) or no function (low current or idle) is present in the corresponding output. In many cases a thermal overload of the component can also be monitored (temperature monitoring). This information is registered in an error register

or error memory and can be called up by the microcontroller. It can be analyzed and can be stored for later workshop diagnosis. (Note: An additional logical interconnection with information and analyses from other functions may be required in many cases.) **Fig. E21** shows the example of a fourfold output stage. It shows that analysis logic is connected to the actual output stage transistor, which transfers signals to an error register. The information is transferred serially to the microcontroller through a connected sliding register.

~~**Digital interface.** *See below*, ~~Serial interfaces

~~**Distance sensor** →Sensors ~Position sensor

~~**DTC register.** *See above*, ~~Diagnostics

291

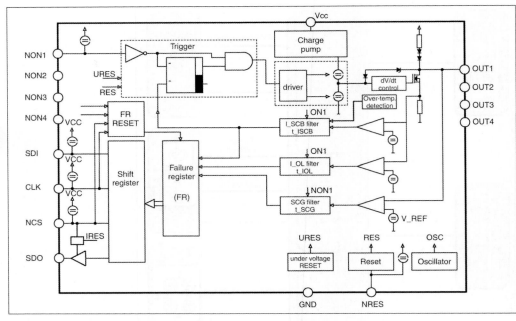

Fig. E21 Quadruple output stage with integrated diagnostics logic

~~**DTC storer.** *See above*, ~~Diagnostics

~~**Enable logic.** *See below*, ~~Voltage regulator

~~**Flip chip.** These are components in a bare die format. The component is installed face down on the substrate and is fastened by soldering (**Fig. E22**).

Special connection geometry—for example, soldering bumps (**Fig. E23**)—must be installed on the surface of the component to enable an electrical connection to be established.

~~**Function blocks.** *See above*, ~~Block diagram

~~**Heat conductive adhesive.** A heat conductive adhesive is used to glue the ceramic substrates to the corresponding control instruments. This establishes the mechanical connection in the equipment, and it also guarantees a good discharge of the power dissipated from the electronic components based on the special characteristics of the adhesive on the substrate. It prevents overheating of the electronic components during operation.

~~**High currents.** *See above*, ~~Diagnostics

~~**High-side driver.** *See below*, ~~Power transistor

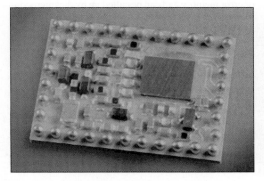

Fig. E22 Flip chip on a transmission frame

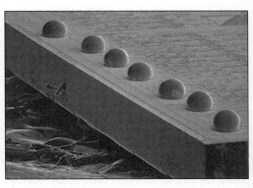

Fig. E23 Soldering bumps on an IC

~~**Host computer.** *See below,* ~~Microcontroller

~~**Identifier.** *See above,* ~~CAN interface

~~**Inductive load.** *See below,* ~~Power transistor

~~**Input signals.** Input signals are conditioned by suitable input filter structures for the microcontroller in the control instrument. Input signals can be analog (e.g., from sensors) or digital (e.g., from switches or other control instruments). The stochastic signal generated by the knocking sensor is conditioned by the knocking IC input filter component (**Fig. E24**), and it is transmitted in a digital form through a serial interface to the microcontroller for analysis and further calculation.

~~**Interference signals.** These can develop in vehicle networks because of a variety of influence factors. The variety of electrical consumers, which induce high positive or negative voltage peaks during the on and off switching processes, can force malfunctions in connected electronic instruments. A low charge status in the vehicle battery, aged generators, or corroded contacts are all reasons why the vehicle electrical system may drop to a voltage below 5 V when a powerful electrical consumer (e.g., starter, auxiliary heating, or horn) is switched on, and this can result in malfunctions of the control instruments. Different types of test pulses are defined in ISO/DIS 7637-2.2 for testing the operating safety of electronic equipment, and test samples are subjected to them during a test run to test their operability. A classification of the requirements for the electronic control instruments is established in the description of the desired functional states.

~~**Interference voltages.** *See above,* ~~Interference signals

~~**K-line interface.** This interface is defined in ISO 9141. It is a diagnostics interface for the communication between electronic control instruments in the vehicle or with workshop testers. The data are transferred through single wires with a transmission speed of about 12 kbit/s.

~~**Knock IC.** *See above,* ~~Input signals

~~**Knock sensor.** *See above,* ~~Input signals; *also,* →Sensors

~~**LIN interface.** LIN (Local Interconnect Network) describes a cost-effective communication standard in the vehicle for controlling the comfort electronics, intelligent sensors and actuators, and nonsafety-critical engine and transmission control components. The communication is based on a bit-serial single wire communication in accordance with the SCI (UART) data format. The maximum transmission speed is 20 kbit/s. The individual nodes are synchronized without stabilized time basis. The specifications are similar to ISO 9141. Collision of the messages on the data bus is avoided by organizing into a master and one or more slaves (only the master can initiate a communication).

~~**Linear O$_2$ sensor interface.** This is used for the continuous acquisition and analysis of the O$_2$ value for the whole measuring range. An integrated component controls the voltage to 450 mV through a Nernst cell in the sensor by controlling the pump current for the O$_2$ sensor in accordance with the oxygen concentration in the exhaust gas. The current is, therefore, equivalent to the measured O$_2$ value. Linear O$_2$ sensors have a reference cell (oxygen concentration approaches infinity), which is filled with oxygen by a small constant current. The measurement of the internal resistance of the

E

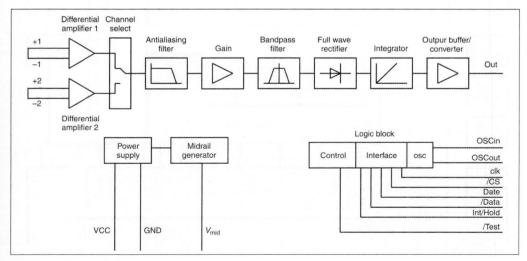

Fig. E24 Signal conditioning for a knocking IC

sensor permits the sensor temperature to be determined. The optimum operating point of the sensor is approximately 750°C. A temperature controller is, therefore, required in the engine and transmission control system to achieve high accuracy for the O_2 measurement. The sensor is electrically heated depending on the exhaust gas temperature. The production tolerances of the sensor must be compensated to meet the high accuracy specifications for the input signal. Each individual sensor is trimmed by the sensor manufacturer to the behavior of the standard sensor by adjusting a resistor integrated into the sensor. The interface permits measurement of the O_2 values between about $\lambda = 0.65$ and 4.9. Linear O_2 sensors are used primarily for especially severe specifications of exhaust gas composition (e.g., Euro 5) or for gasoline engines with direct injection.

~~Low-side driver. *See below,* ~~Power transistor

~~LTCC. *See below,* ~~Substrate, ~~Transmission frame

~~Main controller. *See below,* ~~Voltage regulator

~~Microcontroller. Specially designed microcontrollers are used for many applications in the automotive industry (**Figs. E25 and E26**). Components with a bus bandwidth of 8, 16, or, in new applications, 32 bits are used.

These components combine high computing power in the computer cpu with high integration of the peripheral components that are required for analysis of the input signal and for control of the output stages. One of the microcontrollers must be defined as master in the case where a control instrument uses several microcontrollers.

~~O$_2$ sensor →Sensors

~~Output stages. *See above,* ~~Block diagram; *see below,* ~~Power transistor

~~Over-temperature. *See above,* ~~Diagnostics

~~Peripheral components. *See above,* ~~Microcontroller

~~Power semiconductor. *See below,* ~~Power transistor

~~Power transistor. The power transistors are electronic switches that can switch a current range from a few milliamps to several thousand amps. They are divided into bipolar technology with NPN and PNP types and the power MOSFETs (N-channel and P-channel versions). They are used in electronic control instruments for the control of electrical consumers. In vehicles, these are normally electrical valves, relays, and heating elements.

Two switch methods exist—the low-side driver and the high-side driver (**Fig. E27**)—to operate the inductive and resistive loads of the consumers.

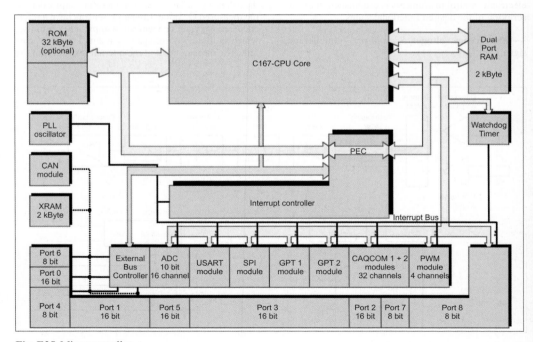

Fig. E25 Microcontroller

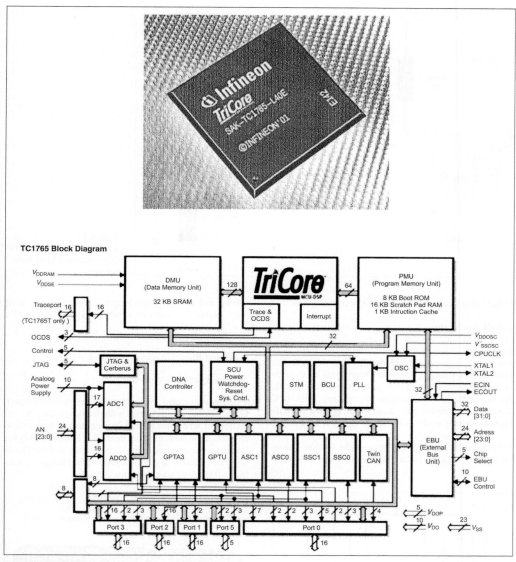

TC1765 Block Diagram

Fig. E26 Microcontroller

Bridge circuits or bridge output stages (H-bridges) are used for the control of electrical DC motors, which must execute forward and reverse rotations (**Fig. E28**). The direction of rotation is determined by appropriate control of the power transistors.

~~**Pressure sensor** →Sensors

~~**Printed circuit board.** *See below,* ~~Transmission frame

~~**Quadruple output stage.** This is the integration of four low-side or high-side drivers in one inte-grated component. Channel currents of up to 3 A can flow permanently when a housing with suitable performance is used. The diagnostic characteristics (diagnostics) of the component permit the detection of a wire break as well as an external short circuit to the battery voltage or to ground. These components also have an over-temperature fuse, which interrupts the current flow at a silicon temperature of about 150°C.

~~**Reset logic.** *See above,* ~~Block diagram; *see below,* ~~Voltage regulator

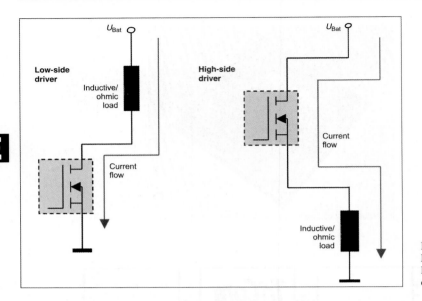

E

Fig. E27
Principle description for low-side and high-side drivers

~~**Resistive load.** *See above,* ~~Power transistor

~~**Self-protection.** *See above,* ~~Diagnostics

~~**Sensors.** These constitute the interface to the outside world (vehicle environment). They deliver the input signals that are required for analysis by the microcontroller. Sensors are used to measure temperatures, travels, rotational speeds, angles, or pressures. The knock sensor used in engine and transmission control systems requires special analysis by the so-called knock IC.

~~**Serial interfaces.** These are single or dual wires for the transmission of digital information between a sender and a receiver in electronic systems. The most used interface in vehicles is currently the CAN bus.

~~**Short circuit.** *See above,* ~~Diagnostics

~~**Signal flow.** *See above,* ~~Block diagram

~~**Speed sensor** →Sensors

~~**Substrate.** (*also, see below,* ~~Transmission frame). Conventional components with housings can be installed on substrates (also called ceramic transmission frames). However, in many cases it is advantageous for space-saving reasons to use components without housings, so-called bare dies. Another option for space saving is the use of so-called flip chips.

~~**Temperature sensors** →Sensors

~~**Test pulses.** *See above,* ~~Interference signals

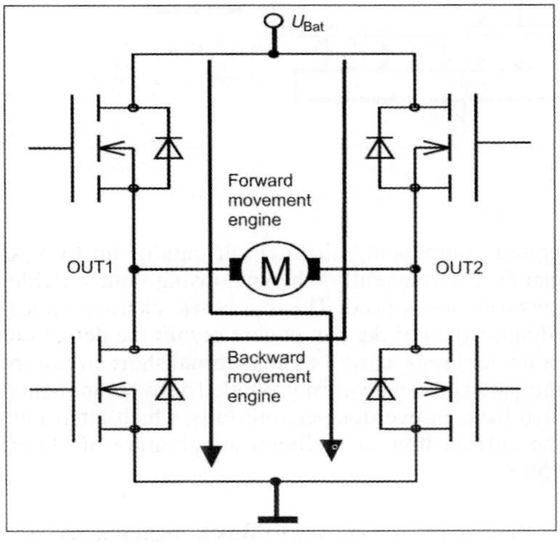

Fig. E28 Example of an H-bridge circuit

Fig. E29 Transmission frame

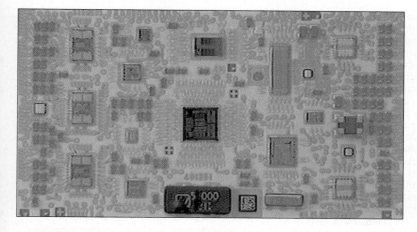

Fig. E30
Transmission frame

~~Transmission frame. Transmission frames are printed circuit boards which can be used for the installation of electronic components. The transmission frame establishes electrical connections between components. The materials used for their manufacture are FR4 (**Fig. E29**) or ceramics. Thick-film technology or LTCC (low temperature co-fired ceramics) are used for ceramic transmission frames (also called substrates) (**Fig. E30**).

~~Voltage regulator. The power for electronic control instruments is supplied by the 14 or 28 V vehicle electrical systems. The voltage can vary between 6 and 26 V for cars and between 5 and 48 V for trucks, depending on the load of the electrical system and the charge status of the battery. The voltage regulator (**Fig. E31**) converts this variable primary voltage into the uniform fixed value (e.g., 5 V, 3.3 V, or even lower) that is required for the operation of the electronic control instruments. The switching voltage controller and the longitudinal controller are the switching methods used most frequently. The secondary voltage for the switching controller is generated from the higher input

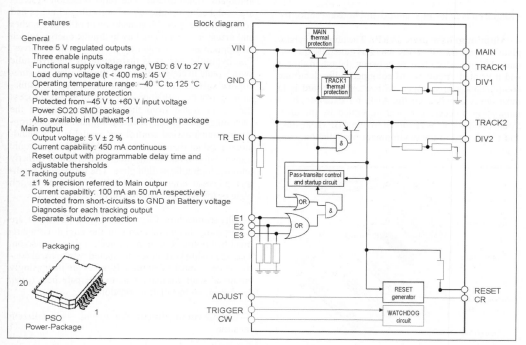

Fig. E31 Example of a modern voltage regulator (longitudinal controller) with integrated reset logic, watchdog, and tracking outputs

voltage by a switched transistor together with an inductance and a freewheeling diode. A capacitor provides for smoothing and buffering of the voltage. The control resistor in the longitudinal controller acts as an adjustable resistor that adjusts the secondary voltage by its voltage drop. A capacitor is used to buffer the voltage. Modern integrated voltage regulators have many auxiliary functions. There is a so-called power-on reset, which uses its internal rest logic (release logic) to release the computer system only after a defined minimum voltage has been reached. It can detect whether the secondary voltage inadvertently drops below a safe operating level, and this results in a low-voltage reset, which resets all integrated circuit components of the control instrument to a defined status. A watchdog function also can be integrated, which expects a switch signal from the host computer at frequent time intervals. The host computer is reset to a defined starting condition if this signal is late or missing. Low-power voltage regulators (so-called tracking outputs) with their typical 5-volt level, derived from the main regulator and, therefore, are ratiometric and short-circuit–resistant and integrated in the same component—and can deliver a current of approximately 100 mA. It is used for the supply of external electronic components such as potentiometers or active sensors.

~~**Voltage supply.** *See above,* ~~Voltage regulator

~~**Watchdog function.** *See above,* ~~Voltage regulator

~Requirements for mechanical and housing concepts

~~**Aluminum base plate (ABP).** The aluminum base plate is a heat transfer body which is generally the primary heat transfer element between the environment and a printed circuit board equipped with electronic components. An equipped printed circuit board is in most cases installed on the ABP, for example, by a lamination process. The heat generated by the electronic components is absorbed by the ABP (heat sink) and transferred to the environment by radiation, con-

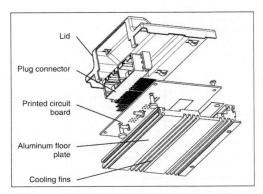

Fig. E32 Control housing concept for stand-alone products

vective processes (cooling fins), or heat conduction (heat bridge). The ABPs are also installed on large engine and transmission housings, which in turn act as a heat sink and dissipate and transfer the absorbed heat.

Fig. E32 shows the general design of a transmission control (here a stand-alone product) with its main components. The ABP is equipped with cooling fins in this example and this increases the heat transfer performance significantly.

~~**Assembly extension.** *See below,* ~~Installation location

~~**Auxiliary fan.** *See below,* ~~Electronics box

~~**Connection sealing.** *See below,* ~~Plug connector

~~**Contact system.** *See below,* ~~Plug connector

~~**Dustproof.** Depending on the installation space in a vehicle, dust-proofing is a requirement for housings of electronic control equipment. A housing is dustproof if only a small amount of dust, which does not impact the function and the safety of a control instrument, penetrates.

~~**Electrohydraulic transmission control.** This is the total control mechanism, including control and switching strategy, for the activation of ratio selection in an automatic transmission. The ratio selection systems are switched hydraulically by activating clutches and actuators—these switch gearsets on or off. The clutches and actuators are actuated by hydraulic energy, which is supplied or discharged by pressure-control valves. The hydraulic energy is distributed by the hydraulic switch plate in conjunction with the installed pressure-control valves. The electronics performing transmission control drive the pressure-control valves, process the actual signal from the sensor, and convert the switching strategy into set point signals. The electronics for transmission control are available as a stand-alone product or mounted as an MTM (mechatronic transmission module, an integrated product) directly onto the hydraulic switch plate. The former has no direct physical connection with the hydraulic switch plate or the pressure-control valves.

~~**Electronics box.** Electronics boxes (E-boxes) are waterproof, dustproof boxes in the engine compartment that house control instruments, relays, and so on. The increasing heat losses dissipated by electrical control systems means that these E-boxes are increasingly equipped with auxiliary fans to dissipate heat (*also, see below,* ~~Installation location).

~~**Engine compartment.** *See below,* ~~Installation location

~~**Environmental conditions.** Environmental conditions are all of those physical, chemical, and mechanical factors that influence a product from the outside. Physical factors are (permanent) temperature, pressure, and moisture. Chemical influence factors include the components of the local medium such as fuel, oil, and detergent additives, and also salt sprays (medium compatibility). Mechanical factors that might influence a component are (permanent) the impact of vibration stresses and other forces, and also (protection type, impermeability) the impact vapor and water sprays and dust in the atmosphere. The environmental conditions are a major impact factor for defining the housing concept for stand-alone and integrated products and their design and fastening, for selection of the material, and for the protection type classification. Integrated products are normally subjected to significantly harder environmental conditions than stand-alone products.

~~Heat sink. A heat sink is a thermodynamic environment or, as a design term, a heat transfer body which has a high heat capacity and, therefore, can maintain a quasi-stationary temperature condition because of its high capacity to absorb heat. A heat sink is the opposite of a heat source, which can continuously supply heat to achieve a quasi-stationary temperature state. The efficiency of a heat sink as a component is characterized by high heat capacity and high conductivity of the material. An aluminum base plate is an example, which absorbs and transfers heat from the connected printed circuit board or electronic components to maintain a quasi-stationary temperature.

~~Heat transfer bodies. These are components which are generally integrated in a heat transfer system. They are characterized by their basic effect—for example, heat conductivity (heat bridges), heat radiation, heat absorption capability (heat sink), and convective heat exchange capability (cooling fins). Heat transfer bodies can be applied in a targeted manner to produce defined thermal conditions in a system as part of thermal management.

~~Housing concept. This defines the designed philosophy of the housing for an electronic control system in terms of installation location, installation class, thermal class, protection class, and ambient conditions.

~~Hydraulic switch plate. The hydraulic switch plate is the central distribution component for hydraulic energy in an electronically controlled transmission. The hydraulic switch plate, in most cases manufactured from two aluminum castings, is positioned in the oil pan at the lowest place in the transmission housing to guarantee a reliable oil supply for the hydraulic functions under all operating conditions. The hydraulic switch plate transmits the hydraulic energy in the form of oil pressure and volume through a network of channels, lines, and openings to switch cylinders, hydraulic intensifiers, and stores, from where the hydraulic effects can be applied to clutches, actuators, and switches. The flows of hydraulic energy are normally controlled by electromagnetic pressure-control valves, which in turn

Fig. E33 Integrated electrohydraulic control instrument for a continuously variable transmission (CVT)

are controlled by integrated or stand-alone transmission electronics (stand-alone products, integrated products). MTMs are normally installed directly on the hydraulic switch plate. This establishes a direct electrical contact to the pressure-control valves already installed on the hydraulic switch plate and to the pressure measurement orifice; it also establishes a heat sink for the control electronics mounted on the aluminum base plate.

Fig. E33 shows the assembly of a complete electrohydraulic transmission control system for a CVT (continuously variable transmission) with hydraulic switch plate (HSP), MTM, and pressure-control valves.

~~Impermeability. Impermeability defines the protection against penetrating contaminants and materials, and it is classified into protection types. The impermeability of a housing for the existing ambient conditions is achieved by specific design activities such as the use of dispersing seals or insertion gaskets or welding or gluing of housing elements. The materials used for the sealing must reliably withstand all temperature, vibration, and pressure loads and must be compatible with the media.

~~Installation classes. *See below, ~~* Installation location

~~Installation location. The installation location (**Fig. E34**) for electronic controls in a vehicle is divided into the following four installation classes:

- Passenger compartment or electronics box
- Engine compartment (chassis installation)
- Added accessories
- Integration in engine

All installation spaces can have an associated value for the local environmental conditions (**Fig. E35**)—for example, thermal class, vibration, and protection against certain materials (unpressurized and pressurized fluids, solid materials, etc.).

~~Integrated products. In contrast to stand-alone products, integrated products are system modules (e.g., control instruments) that are physically integrated in the control system or engine, as a result of which these

E

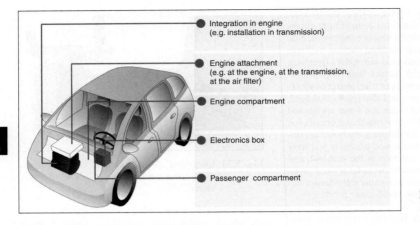

Fig. E34
Installation spaces

	Passenger compartment/ electronics box	Engine compartment	Engine attachment (e.g. at the engine, at the transmission, at the air filter)	Integration in engine (e.g. installation in transmission)
Vibration value (depending on frequency)	... 5 g	... 16 g	... 28 g	... 40 g
Thermal class (ambient temperature)	... 80 °C	... 105 °C	... 125 °C	... 140 °C
Impermeability	Dustproof	Dustproof, spray waterproof	Dustproof, steam jet impermeability	Transmission oil leak-tightness

Fig. E35
Environmental requirements

integrated products are subjected to the same local conditions, such as permanent vibration and permanent temperatures, as the engine itself. The local conditions and specific engine characteristics, such as installation location, available installation space, and orientation, are the conditions for the design of the integrated products and in particular their thermal management.

Integrated products—such as, for example, the MTM shown in **Fig. E33** (transmission controls installed in the transmission oil pan)—must, in addition to the existing specific local conditions, meet design requirements under all local conditions, such as impermeability of the electronics box, the sensors housed at the periphery, and the total compatibility of materials—for example, against hot transmission oil.

~~**Lamination process.** Printed circuit boards are glued on the carrier with the help of the lamination process. The carrier is normally a die-casting or bent sheet metal part. An adhesive film in the shape of the printed circuit board is attached to the carrier. The printed circuit board is put onto the film in a second step of the process; the glue is cured under pressure and heat. High vibration resistance and good heat dissipation are advantages of the laminated printed circuit board.

~~**Leak tightness to sprayed water.** Depending on the installation space in a vehicle, leak tightness to spraying water is a requirement for housings of electronic control equipment. A housing is spray watertight if it shows no damage after a high-pressure water jet is directed at the housing from all directions.

~~**Long-term/extended vibration loads.** *See below,* ~~Vibration stresses

~~**Long-term temperature loads.** *See below,* ~~Thermal stresses

~~**Mechatronic transmission module (MTM).** In addition to the electronic control of the switch components, a transmission control system includes the acquisition of peripheral values through travel, angle, rotational speed, temperature, and pressure sensors. In addition to the electronic control and peripheral sensor functions, a mechatronic control module (MTM) integrated in the transmission system combines the following subfunctions, ideally in a separate complete module: potential and signal distribution, contact to the switch components, and other hydraulic and electronic interface devices. In contrast to the stand-alone transmission control units (stand-alone products), the control instruments integrated into the transmission (integrated products) deliver the highest potential of mechatronic functionality, because all important input and output components are directly installed on the transmission. The primary advantages of an MTM are a significant reduction in the number of contact points, the precalibration and tests of the sensor and control functions, and the simplified installation and logistics, which are important for the customer. A maximum level of integration can be achieved if the transmission and the integrated control system are designed together during the concept phase, because the arrangement of all required components with respect to location, orientation and technology can be optimized. **Fig. E36** shows an example of an overview for the arrangement of MTM components.

~~**Medium compatibility.** Housing concepts including the peripheral sensors and the control components, which are subjected to aggressive local conditions, must be compatible with the medium. Compatibility with the medium is achieved by using the appropriate and resistant materials, alloys, and surface coating processes, which meet the lifespan requirements of the module with respect to integrity, function, impermeability, and optics. Fuels, engine and transmission oils, detergents, hot steam, and salt sprays are materials to which the stand-alone products are subjected and for which a medium compatibility must be achieved under all conditions. Integrated products, such as MTMs, are always surrounded by aggressive transmission oils and require an especially careful analysis of medium compatibility for all the components in contact with the oil.

~~**MTM.** *See above*, ~~Mechatronic transmission module

~~**Passenger compartment.** *See above*, ~~Installation location

~~**Plug connector.** Electronic controls are connected to the vehicle with plug systems. A plug system consists of the plug connector and the plug socket. The contact system is the core of all plug systems. It defines the connection of each individual plug pin with the receiving socket. Different contact systems are available worldwide; however, the vehicle producers normally use only one contact system. The routing of the cable harness into the socket housing is sealed with sealing elements. The most commonly used elements are single-loader seals and collection seals. In single-loader sealing applications, all wires are individually sealed with a sealing element against the socket housing. In collection sealing applications, the individual wires are routed through holes in a common rubber seal. The advantage of single loader against collection seals is the higher impermeability class; however, the costs are higher.

The plug is defined by its

- Impermeability,
- Number of pins,
- Number of modules (chambers), and
- Direction of the plug connector (vertical or parallel to the printed circuit board).

Other criteria include the electrical performance, plug forces, and vibration resistance.

Requirements for different numbers of pins can be implemented with the same modular connection systems (**Fig. E37**).

~~**Pressure loads.** Pressures in hermetically sealed housings increase or decrease with a change in the operating temperature (ambient conditions), and these can in some cases exert significant forces on the housing walls, seals, and fastening elements. These pressure loads must be contained with highly stable materials, so that a reliable and safe operation is guaranteed for the total life of the equipment (housing).

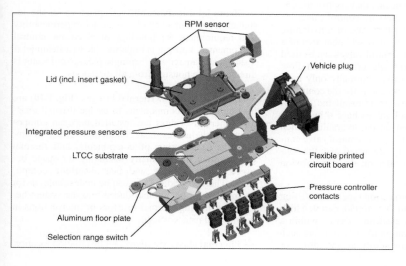

Fig. E36
Main components of a modern MTM

E

water, water sprays, water splashes, high pressure steam sprays, and also permanent immersion) and against contact by these media. The housing concept of a stand-alone product is significantly influenced by the protection type.

~~**Selection range switch/sensor.** This is a passive position acquisition component which senses the drive range or driving position of a transmission.

The axial position of the main control pistons, actuated by hydraulics or electric motors, which perform the distribution of the hydraulic energy to the respective clutches and actuators, is measured by a travel sensor, and it is then analyzed for the following control steps of the transmission control electronics. The selection range switch is an integral component in integrated products such as MTMs, and it is correctly positioned and calibrated during installation on the hydraulic switch plate. **Fig. E33** shows an integrated selection range switch and all its design relationships mounted on an aluminum base plate.

~~**Silicon gel.** Silicon gel is a two-component silicon product, which is filled into an electronics box in the low viscosity primary curing phase to protect electronic components and contact points from humidity and corrosion. The sensitive bond wires and connections are protected and stabilized against impact and vibration loads with the good damping characteristics of the cured gel.

~~**Single supercharger.** *See above*, ~~Plug connector

~~**Stand-alone products.** Electronic control systems are divided into stand-alone products and integrated products (**Fig. E39**). Stand-alone products are engine and transmission controls that are installed in vehicles as separate units, in contrast to the integrated products, which are combined with another functional unit (e.g., transmission).

~~**Steam jet impermeability.** Depending on the installation space in a vehicle, steam jet impermeability is a requirement for housings of electronic control equipment. A housing is impermeable to a steam jet if it shows no damage after strongly pressurized water is directed at the housing from all directions.

~~**Thermal bridges.** Thermal bridges (**Fig. E40**) are heat-conducting components to bridge thermal resistors. Thermal bridges are installed between components and the nearest heat conduction surface to prevent the overheating of components, and thermal bridges span the thermally insulating air space between heat-producing and heat-absorbing components. Thermal bridges also can be undesirable, if, for example, local heat is introduced into the system because of missing thermal resistors or interruptions in the thermal resistor chain.

~~**Thermal class.** *See above*, ~~Installation location

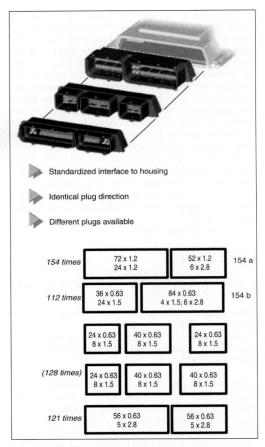

Fig. E37 Modular connection system

~~**Pressure-control valves.** These are electromagnetic switch and control elements that control the energy in a hydraulic system. Pressure-control valves influence the pressure and the flow rate of a hydraulic system by changing the flow cross-sectional area of a valve. Proportional pressure-control valves can be used to adjust and maintain variable cross-section flow areas, whereas switching valves normally only have "fully open"/"fully closed" positions for the control of the hydraulic energy status. To transmit the control commands, pressure-control valves have a hydraulic interface which is mounted on the hydraulic switch plate, and an electrical interface, the control valve connection, which is electrically connected with the contacts of the pressure controller on the transmission control side (**Fig. E38**).

~~**Protection type.** The protection type is an existing standard in accordance with DIN 40050, part 9, which classifies the safety of an electronic device within a housing against the penetration of contaminants such as dust and water (impermeability against drops of

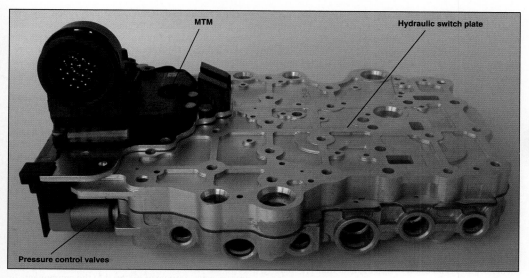

MTM

Hydraulic switch plate

E

Pressure control valves

Fig. E38 Electrohydraulic control instrument for CFT 23 transmission

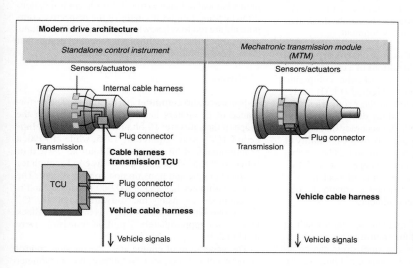

Modern drive architecture

| Standalone control instrument | Mechatronic transmission module (MTM) |

Sensors/actuators

Internal cable harness

Plug connector

Transmission

Cable harness transmission TCU

TCU

Plug connector
Plug connector

Vehicle cable harness

↓ Vehicle signals

Sensors/actuators

Plug connector

Transmission

Vehicle cable harness

↓ Vehicle signals

Fig. E39
Description of
stand-alone and
integrated equipment

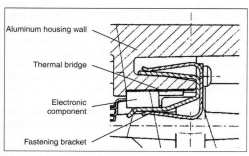

Aluminum housing wall

Thermal bridge

Electronic
component

Fastening bracket

Fig. E40 Thermal bridge

~~**Thermal management.** Thermal management is the use of technical and physical options for targeted heat dissipation/heat supply to generate defined temperature conditions at selected locations. The heat generated by an electronic control system must be dissipated from the electronics even in a warm environment to guarantee the integrity of the component, connections, and contacts and, therefore, the lifespan of the equipment and its control reliability.

The technical and physical possibilities of heat transfer mechanisms consist of conduction, radiation, and convection phenomena for quasi-stationary conditions. Also used for the thermal management of transient conditions are the capacitive characteristics of the material. Thermal Cu or Al bridges are designed with cooling fins for their task as heat conducting components for housing areas, to increase the convective heat transfer to the environment. Aluminum or steel base plates are used as heat sinks to smooth out large temperature swings in the attached electronics. The efficiency of the thermal management in a system of heat transfer bodies is computed during the design phase by simulating the thermal behavior of the component.

~~~**Cooling fins.** Cooling fins are design elements made from heat conducting components, which provide increased convective and radiative heat transfer through their large surface. **Fig. E32** shows a configuration of a cooling fin on the underside of an aluminum base plate for a stand-alone transmission control system. Cooling fins significantly increase the heat transfer due to their numbers and their design features (height, length, and thickness), and they also influence the thermodynamic impact factors of the surrounding (heat absorbing) medium so that the heat transfer characteristics at the interface cooling/environment are increased. The design of the cooling fins is preceded by a thermal simulation, which determines the dimensions depending on the amount of heat to be dissipated from the thermal and design environment.

~~**Thermal simulation.** This is an analytical process, in which the temperature conditions of a component or structure are computed using particular computing methods (e.g., Finite Element Method [FEM] analyses) that take into consideration the geometrical and dimensional specifications and the defined thermodynamic boundary conditions. The accuracy of a thermal simulation is determined primarily by the reliability of the thermodynamic boundary conditions. The thermal simulation is used in thermal management, which is based on the thermal dimensioning of heat transfer bodies such as cooling fins, heat sinks, radiation bodies, and the thermodynamic environment.

~~**Thermal stresses.** Thermal stresses are a result of the local conditions and of the control instrument itself, because heat and, therefore, thermal stress can be induced by either. Thermal stresses have a significant impact on the housing concept, because thermal management requires very specific physical and technical solutions. Peak temperatures that are only present for a short time can normally be reduced by heat sinks and the inherent heat capacities of the system. On the other hand, high thermal stresses have a significant impact on the lifespan of the electronic components in an electronic control instrument.

~~**Transmission oil leak-tightness.** Impermeability against transmission oil is a requirement for housings for electronic controls when they are mounted at corresponding installation locations. The housing is leak-tight with respect to transmission oil if the control system is not damaged when the particular oil remains at its location.

~~**Underhood installation.** The underhood installation of an electronic control instrument is done on or at an assembly, which in turn is connected to the engine (e.g., air filter). This results normally in lower local requirements (temperature) than direct installation on the engine (installation location).

~~**Vibration stresses.** The stimulation spectra from components in the vehicle—for example, the engine or the suspension—are transmitted to the fastening interfaces of electronic control systems; the shape of the received vibration spectrum depends on the location of the control system on the vehicle. The transferred stimulation spectrum results in a characteristic vibrational stress of the housing and all of the assemblies physically connected to it. Vibration stresses result in oscillating forces with particular frequencies, which occur as peak resonant values or as permanent values (permanent vibration stress). If the peak resonant stresses are higher than the fracture strength of the stimulated component or the mechanical connection, the component will break. The permanent vibration stresses cause fatigue of material of the components or of their connections. Permanent vibrations also result in mechanical damage to the electronic control system if the bending/fatigue strengths of the respective components are too low. Knowledge of the local conditions makes it possible to establish a mechanical design that takes these vibration stresses into account.

~**Software**

~~**Software and computer performance.** The performance of computers has increased in line with the scope of the software: 16-bit processors—such as Infineon C167—were required by 1995, while in 1990, a programmable 8-bit processor—Motorola 68HCll or Infineon 8OC517, for example—was good enough for a typical engine and transmission control system. The former computer has about 500,000 transistors and is programmed by using the software language "C." Thirty-two-bit processors—such as Motorola Black Oak—with approximately 7 million transistors were used in 2000.

The generation of engine and transmission control systems currently under development has a computer with about 40 million transistor functions.

A significant part of the software (currently more than 50%) in a modern engine and transmission control system is not used for the "actual" function (control of the engine) but provides tasks such as diagnostics (engine and transmission control system, sensors, actuators, OBD II), communication (gateway function in vehicle, communication to the outside or workshop diagnostics) and self-monitoring (drive-by-wire, E-gas).

~~**Software architecture.** The software architecture can be implemented by a distinctive layer model (**Fig. E41**).

The relationship between the layers is generally determined as follows:

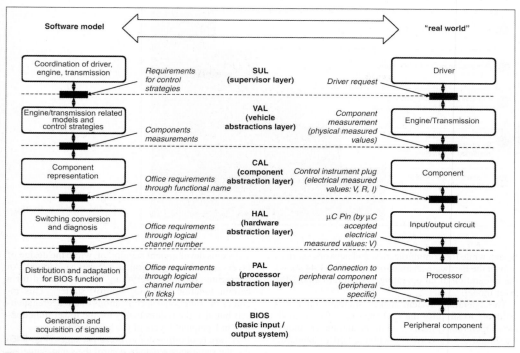

Fig. E41 The six layers of drivetrain software

- A layer can use all services of the layer below, but not the services of the layers above.
- A layer can exchange data with the layer below, but cannot jump layers.
- A layer can have control flows (without data transmission) to other layers; the layer may not know the receiver of control flows for control flows to higher layers.
- Only certain data types are permitted for each layer.
- Each layer must be designed independently from hardware, processor, and compiler. An exception is the two lowest layers, but these should also follow the rules as far as possible.
- The six software layers correspond to the respective abstraction levels of the "real world."

The following targeted advantages are combined with this software layer model:

- A new microcontroller or a new input/output component impacts only the BIOS (basis input/output system) and PAL (Processor Abstraction Layer) and possibly also the HAL (Hardware Abstraction Layer).
- A function can assume the required data as given — the input and output of data can be implemented independently or they can be re-used.
- Externally developed functions can be integrated with minimum effort, if they accord with the interface conventions.
- The hardware independence and the encapsulation represent a major step toward "software as a product," which means a step toward the sales and purchase of individual software functions.
- The model also permits an increase of the re-use rate, which reduces the exponential increase of the software scopes and the development effort required by new software.

~~**Software development process.** Efficient and quality-oriented software development in larger teams requires the use of a suitable well documented development process, which may, for example, be in accordance with the Capability Maturity Model (CMM) (**Fig. E42**).

The SPICE model (ISO/IEC TR 15504), which is currently being discussed in detail in the automotive industry, should also be mentioned in this context. The requirements for the CMM and the SPICE model can be compared, so that organization of the development process can be achieved that satisfies both models. The extensively used V-model is a possible option for the software development process (**Fig. E43**) because it describes the interaction of the requirements analysis, the abstraction levels, and the corresponding tests especially well. The V-cycle is completed for each software release.

~~**Software for operating systems.** The operating system is a standard software module that generates functions from the different layers available. The services of the operating system are also available for the upper

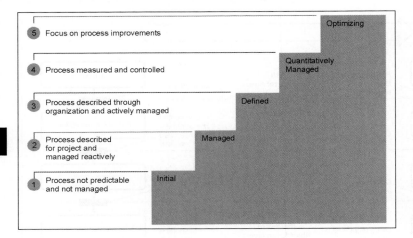

Fig. E42
CMM model with five
maturity levels

layers, while the communication and network services impact the hardware abstraction layer. **Fig. E44** shows the integration of an operating system into a possible layer model.

The operating system makes standard interfaces available to simplify the exchange of software modules. Operating systems used in the automotive industry should be in accordance with the OSEK standard.

~~Software function packages. The establishment of functional packages (aggregates) is important from a systems development standpoint, while the design of layers has advantages from a software standpoint. An aggregate combines everything that is required for a certain function (e.g., O_2 control, starter). This includes the functional specifications of the aggregate itself but also the requirements for the lower, functionally independent layers of the layer model.

A variety of different functional packages (**Fig. E45**) use the same sensor data or control the same actuators, and this is made possible by the layer model which makes standardized interfaces for this application available.

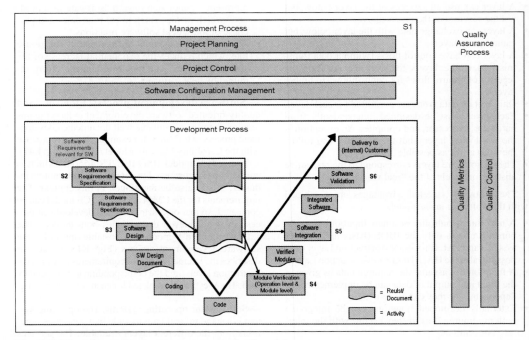

Fig. E43 The software development process in the V-model

E

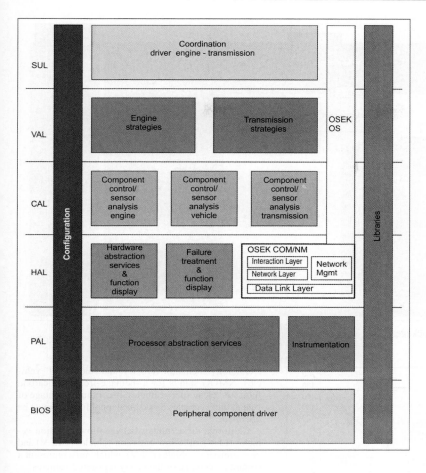

Fig. E44
Integration of an operating system into the six software layers

Modules, which have certain aggregates or sensors/actuators allocated to them on a software layer, now represent the granular resolution that is preferred for re-use.

~~~Software scopes. The scope (**Fig. E46**) of the software for a typical engine and transmission control instrument has doubled approximately every three years in the past. It is assumed that this tendency will continue when planning current projects.

~~~Software team sizes. The increasing scopes of software and reduction of development times result in a large increase in the sizes of the software team. In 1990 it was possible to find two developers on a project. In 2003, twelve developers were required for a new project. This size of team requires strict adherence to a defined development process with extensive quality controls.

~~Software requirements. The importance of the software and also the requirements for the software for vehicle electronics have increased dramatically during the last 20 years, especially for control of the engine and transmission. Functions that previously were implemented by mechanical or electronic solutions now are better realized and more cost-effective. Also, the option of a programmable computer opens the way to new and previously unrealized functions.

Included in these are the extensive self-diagnostic capabilities of modern control instruments and the fine-tuning of the combustion process enabled by software, which minimizes emissions and fuel consumption.

Electronics box →Electronic/mechanical engine and transmission control ~Requirements for mechanical and housing concepts

Electropneumatic EGR valve →Actuators ~Actuators for exhaust gas recirculation ~~Exhaust gas recirculation valve

Emission limits (*also*, →Engine acoustics ~Exterior noise). Air pollution by exhaust gases has increased during the past few decades because of the ever-increasing number of vehicles.

The exhaust gas components must be measured to determine and control the emission limits for vehicles.

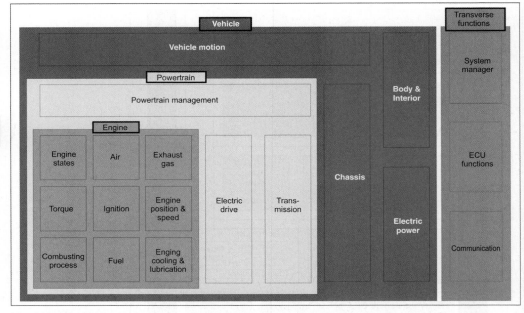

Fig. E45 Functional packages

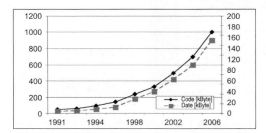

Fig. E46 Software scopes of a six-cylinder engine and transmission control system in Kbytes

This must include the fact that the emission or air polluting materials are dependent on the driving conditions. Several countries have, therefore, defined driving cycles that must be executed on a driving performance test stand. The emitted exhaust gases are collected in bags, and then specific instruments are legally required for the analysis of the concentrations of the particular components.

The first emission limits were defined in California in 1960, and have since been gradually tightened. The particulate emissions of automotive diesel engines were limited in 1982. The fleet average fuel consumption was limited to 27.5 miles per gallon (US) in 1985.

Fig. E47 shows the emission standards for the United States and for California. Additional information can be found in the current laws. The limits shown in the table are applicable to several vehicle types. These include vehicles with improved exhaust

technology, methanol vehicles, and "clean fuel" vehicles. NMOG stands for a methane-free hydrocarbon limit for "clean fuel" vehicles. A certain percentage of "clean fuel" vehicles based on alcohol fuel use must be operated.

California differentiates between low emission vehicles (LEV), ultra-low-emission vehicles (ULEV) and zero-emission vehicles (ZEV), which must represent a certain percentage of the newly registered vehicles.

Some Japanese emission standards are shown in **Fig. E48**.

The exhaust gas legislation in the United States created an increasing demand for limitations of the exhaust gas emissions of cars in Germany and other European countries. The emissions of carbon monoxide and hydrocarbons were first limited by a guideline from the Economic Commission for Europe (ECE)—R 15/00 for the detoxification of car exhaust gases in Germany—on October 1, 1971.

In the early 1980s, guideline 15/04 further reduced the limits for carbon monoxide and combined the limits for oxides of nitrogen and hydrocarbons. At the same time, the internationally accepted FID process was made mandatory for the measurement of hydrocarbons.

Beginning in 1985, a mandatory annual exhaust gas test was legally required in Germany for cars with gasoline engines.

The European Union limits were tightened in several steps, which meant that, almost exclusively, only gasoline engine cars with a controlled three-way catalytic converter were newly registered in Germany be-

Operating time	Emission category	THC	NMHC	NMOG	CO	NO	PM	HCHO
					g/mile			
5 years	Tier 0	–	0.39	–	7.0	0.4	0.08	0.015
50,000	Tier 1	–	0.25	–	3.4	0.4	0.08	0.015
Miles	TLEV	–	–	0.125	3.4	0.4	–	0.015
	LEV	–	–	0.075	3.4	0.2	–	0.015
	ULEV	–	–	0.040	1.7	0.2	–	0.008
	ZEV	0.00	0.00	0.000	0.0	0.0	0.00	0.000
10 years	Tier 0	–						
100,000	Tier 1	–	0.31	–	4.2	0.6	–	–
Miles	TLEV	–	–	0.156	4.2	0.6	0.08	0.018
	LEV	–	–	0.090	4.2	0.3	0.08	0.018
	ULEV	–	–	0.055	2.1	0.3	0.04	0.011
	ZEV	0.00	0.000	0.000	0.0	0.0	0.00	0.000

Fig. E47
Emission standards for the United States (Source: Basshuysen and Schäfer)

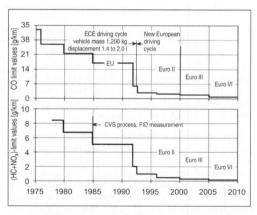

E

Vehicle weight	Emission category	CO max	CO mean	HC max	HC mean	NOₓ max	NOₓ mean	PM max	PM mean
						[g/km]			
Diesel	1997	2.7	2.1	0.62	0.40	0.55	0.40	0.14	0.080
<1265 kg	2002a	–	0.63	–	0.12	–	0.28	–	0.052
Diesel	1997	2.7	2.1	0.62	0.40	0.55	0.40	0.14	0.080
>1265 kg	2002a	–	0.63	–	0.12	–	0.30	–	0.056
Gasoline	1997	2.7	2.1	0.62	0.25	0.48	0.25	–	–
	2002a	–	0.670	–	0.08	–	0.08	–	–

a) Proposal for 2002

Fig. E48
Japanese emissions limits for cars. The maximum limits are valid for production volumes of fewer than 2000 vehicles per year, the mean limits for larger production volumes (Source: Basshuysen and Schäfer)

ginning in 1990. Guideline 91/441/EEC represents the current status. **Fig. E49** shows the development of exhaust gas standards over time. **Fig. E50** shows current exhaust gas values and the emission categories of the European Union.

The exhaust gas test for vehicles with gasoline engines and catalytic converters but without lambda-controlled mixture preparation has been mandatory since December 1, 1992. Exhaust gas tests for vehicles with gasoline engines and catalytic converters and with lambda-controlled mixture preparation and for vehicles with diesel engines have been mandatory since December 1, 1993.

Literature: R. van Basshuysen, F. Schäfer (Eds.): Handbuch Verbrennungsmotor, 2nd edition, Wiesbaden, Vieweg Publishers, 2002.

Emission maps →Performance characteristic maps

Emission measurements (*also*, →Engine acoustics). Emission measurements can be performed under steady-state or transient conditions on the engine test stands and directly on the vehicle. A certain legally mandated process has to be followed in measuring emissions to guarantee a uniform basis of valuating the emissions measurements; the main part comprises the regulations for determining the vehicle emissions. This is demonstrated for the most important structures in the following sections—more detailed analyses and execution requirements can be found in the legal regulations.

Fig. E49 Development of the exhaust gas standards over time (Source: Basshuysen and Schäfer)

Drive	Emission category	CO	HC	HC + NO	NO	PM
				[g/km]		
Diesel	Euro 1	0.64	–	0.56	0.50	0.050
	Euro 2	0.50	–	0.30	0.25	0.025
Gasoline	Euro 3	2.30	0.20	–	0.15	–
	Euro 4	1.00	0.10	–	0.08	–

Fig E50 Current limits in the European Union (Source: Basshuysen and Schäfer)

309

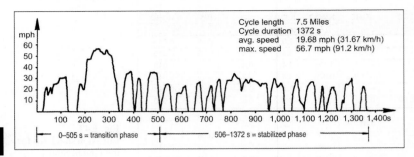

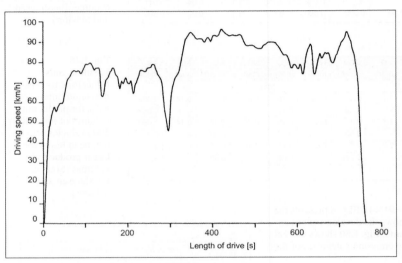

Length of drive [s]

Fig. E52
Driving cycle II

~FTP-75. The American driving cycle consists of two driving tests, driving cycle I (**Fig. E51**), and driving cycle II (**Fig. E52**), on a driving performance test stand, during which time the quantities of pollutant gases and particulates are measured.

~~**Dynamometer adjustment.** The dynamometer adjustment is determined by the driving resistance of the vehicle. It can either be determined by the energy conversion during deceleration or by the torque at constant speed. The road used must be horizontal and dry. The vehicle must be in a normal driving and adjustment state, must be loaded up to a reference mass, and must be at the operating temperature.

The measurement of deceleration is performed in such a way that the vehicle is accelerated to a speed of more than 10 km/h above the selected test speed, V. The transmission is shifted to neutral. The delay time, t_1, of the vehicle from the speed $v_2 = (v + \Delta v)$ km/h to $v_1 = (v - \Delta v)$ km/h is measured, where $\Delta v = 5$ km/h. Correspondingly, t_2 is determined in the other direction. The mean value, T_1, is computed by using t_1 and t_2. The test is repeated until the statistical accuracy, p, for the mean value, T, is the same or smaller than 2% ($p \leq 2\%$). The mean value is computed as follows:

$$T = \frac{1}{n}\sum_{i=1}^{n} T_i .$$

The performance can be calculated with the following equation:

$$P = \frac{M \cdot V \cdot \Delta v}{500 \cdot T}.$$

where

P = performance in kW,
M = reference mass in kg
V = test speed in m/s (selected 80 km/h = 22.222 m/s),
Δv = deviation from speed V in m/s (5 km/h = 1.389 m/s), and
T = time in seconds.

The same procedure is used on the test stand. The equivalent inertial mass, I, is adjusted and the vehicle and test stand are brought to the operating temperature. M is replaced by the equivalent inertial mass, I, in the performance equation.

~~**Equivalent inertia masses.** A flywheel mass is specified from **Fig. E53** in accordance with the refer-

Pr (kg)	I (kg)	Pr (kg)	I (kg)
Pr ≤ 480	450	1390 < Pr ≤ 1450	1420
480 < Pr ≤ 540	510	1450 < Pr ≤ 1500	1470
540 < Pr ≤ 600	570	1500 < Pr ≤ 1560	1530
600 < Pr ≤ 650	625	1560 < Pr ≤ 1620	1590
650 < Pr ≤ 700	680	1620 < Pr ≤ 1670	1640
700 < Pr ≤ 780	740	1670 < Pr ≤ 1730	1700
780 < Pr ≤ 820	800	1730 < Pr ≤ 1790	1760
820 < Pr ≤ 880	850	1790 < Pr ≤ 1870	1810
880 < Pr ≤ 940	910	1870 < Pr ≤ 1980	1930
940 < Pr ≤ 990	960	1980 < Pr ≤ 2100	2040
990 < Pr ≤ 1050	1020	2100 < Pr ≤ 2210	2150
1050 < Pr ≤ 1110	1080	2210 < Pr ≤ 2320	2270
1110 < Pr ≤ 1160	1130	2320 < Pr ≤ 2440	2380
1160 < Pr ≤ 1220	1190	2440 < Pr ≤ 2610	2490
1220 < Pr ≤ 1280	1250	2610 < Pr ≤ 2830	2720
1280 < Pr ≤1330	1300	2830 < Pr	2940
1330 < Pr ≤ 1390	1360		

Pr = Reference mass of vehicle
I = Equivalent inertial mass

Fig. E53 Inertial mass in relationship to the reference mass of the vehicle.

ence mass of the vehicle (mass empty plus 136 kg), and this determines the total inertia of the rotating masses. The next highest mass is used if the associated mass is not available at the test stand. It cannot have a difference of more than 120 kg from the reference mass.

~~**Mass determination, emissions.** The masses of the emissions are determined as follows: To replace the listed equations for the calculation of emitted quantities of gaseous pollutants, the following equations for the calculation of the emitted masses of gaseous and solid air polluting materials in drive cycle I (city part) and drive cycle II (highway part) are used:

Drive cycle I:

$$M_i = 0.43 \frac{m_{iCT} + m_{iS}}{S_{CT} + S_S} + 0.57 \frac{m_{iHT} + m_{iS}}{S_{HT} + S_S}$$

where

M_i = emitted quantity of component i in g/km,
m_{iCT} = the quantity of component i (m_i) in g emitted in phase 1,
m_{iS} = the quantity of component i (m_i) in g emitted in phase 2,
m_{iHT} = the quantity of component i (m_i) in g emitted in phase 3,
S_{CT} = measured driving distance of phase 1 in km,
S_{HT} = measured driving distance of phase 3 in km, and
S_S = measured driving distance of phase 2 in km.

Drive cycle II:

$$M_i = \frac{m_{iHW}}{S_{HW}}$$

where

M_i = emitted quantity of component i in g/km,
m_{iHW} = emitted quantity of component i (m_i) in g/km, and
S_{HW} = measured driving distance in km.

The masses emitted in the test phases are calculated as

$$m_i = 10^{-6} \cdot V_{dil} \cdot \zeta_i \cdot C_i \cdot k_H,$$

where

m_i = emitted quantity of the gaseous air pollutant in g/test phase,
V_{dil} = volume of the diluted exhaust gases corrected for nominal conditions (273.2 K, 101.33 kPa) in l/test phase,
ζ_i = relative density of the gaseous air pollutions under nominal conditions (273.2 K, 101.33 kPa) (*see* Quantification emissions),
C_i = concentration of gaseous air pollution in the diluted exhaust gases, shown in ppm and corrected with its concentration in the dilution air (*see* Quantification emissions), and
k_H = humidity correction for the calculation of the emitted quantity of oxides of nitrogen (for HC and CO no humidity correction permissible) (*see* Quantification emissions).

The other calculation equations can be found in quantification of the emissions. The calculated emission quantities are multiplied by legally regulated deterioration factors. The emission quantities obtained this way need to be below the limiting valves as amended from time to time if type approval is expected to be granted.

~~**Performed driving cycles.** This test is performed on a rolling road dynamometer (engine test bench). The driving speed is recorded as a function of time to be able to judge the validity of the test. The driven distance is determined separately for each quantity of test gas collected.

The absolute humidity, H, in the test room or the engine intake air must be in accordance with the following conditions: 5.5 g ≤ H ≤ 12.2 g H_2O/kg of dry air. A fan is used to keep the engine and vehicle temperatures at the normal values for street-driving operations—the vehicle hood is open during this procedure.

A proportional part of the flow is taken from the exhaust gas; it is diluted by the ambient air, and then routed into a collection bag. The gas flow of the sample for drive cycle I is routed into a separate collection bag for each of the three phases. The gas flow of the sample for drive cycle II is routed into a collection bag. Ambient air samples are collected in parallel to determine the impact of the ambient air concentration. The measurements of the gas flow sample are adjusted to the desired values, but must not be below 5 L/min.

Drive cycle II is executed twice. The first run is used for conditioning, and the emission quantities are measured only during the second run.

E

311

The speed deviation can only be about 3.2 km/h from the highest or the lowest speed at the time of about 1 second. Higher deviations must not last longer than 2 seconds.

Special equipment is used for measurements on vehicles with self-ignition engines. The particulates are extracted on separate filter pairs during the test with driving cycle I. The flow through the particulate filters is adjusted in such a way that it is almost constant, with deviations of approximately 5%. The extracted particulate quantity should be between 2 and 5 mg per filter pair. The particulate emissions are not measured during drive cycle II. The HC concentrations are measured and integrated with a heated FID (HFID) for each phase or driving cycle. The sample gas flow for the HFID must be at least 2 L/min. The analysis is performed as for test type I.

~~**Vehicle preparation.** The test vehicle must be in good order. It must be driven long enough to stabilize the exhaust emissions. However, it should not have been driven for more than 6,400 km before the test. The ambient temperature during the test should be between 20°C and 30°C. The vehicle is filled with test fuel.

The temperature of the tank content is measured with a temperature sensor in the middle of the fuel for vehicles with external ignition, once the tank is filled up to 40% of the nominal tank volume.

For preconditioning, drive cycle I is executed without the parking phase and the third driving phase.

~Low-temperature test. *See below,* ~Test type VI

~Test type I. This is a test of the exhaust emissions after a cold start. For this test, the vehicle is driven on the European driving cycle on an engine test bench—this simulates vehicle resistance and inertia mass. It consists of four basic city driving cycles and one highway driving cycle (**Fig. E54**). The results of each test are multiplied by the appropriate deterioration factors. The calculated emission quantities must be below the corresponding emission limits.

~~**Analysis.** The analysis of the gases in the bag is completed within 20 minutes of the conclusion of the driving cycle. The loaded particulate filters are brought into the chamber within one hour of the emission test. They are conditioned for 2–36 hours, and they will subsequently be weighted. The analyzer will now be adjusted in accordance with the calibration curves by using calibration gases; nominal concentrations should be between 70% and 100% of the upper range value of the corresponding scale. The zero adjustment is tested afterward. The process must be repeated if the measured value deviates by more than 2% of the upper range value from the previously adjusted zero adjustment. The samples are analyzed subsequently.

~~**Calibration**

a. *Engine test bench.* The calibration of the engine test bench determines how much power is absorbed by friction and how much by the dynamometer.

The vacuum method for dynamometer adjustment can be used, for example, for vehicles with external ignition and test stands with a fixed load curve. The performance dynamometer can also be adjusted in a simpler manner so that the power at a speed of 80 km/h is equivalent to the power on the powered wheels shown in **Fig. E55**.

The performance values shown in **Fig. E55** must be multiplied by a factor of 1.3 for cars with a reference

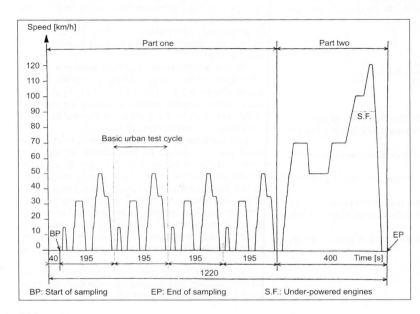

Fig. E54
ECE driving cycle
(city and highway)

BP: Start of sampling   EP: End of sampling   S.F.: Under-powered engines

Reference Mass of Vehicle, Pr (kg)	Equivalent Inertial Mass, Pa (kW)
Pr ≤ 750	4.7
750 < Pr ≤ 850	5.1
850 < Pr ≤ 1020	5.6
1020 < Pr ≤ 1250	6.3
1250 < Pr ≤ 1470	7.0
1470 < Pr ≤ 1700	7.5
1700 < Pr ≤ 1930	8.1
1930 < Pr ≤ 2150	8.6
2150 < Pr ≤ 2380	9.0
2380 < Pr ≤ 2610	9.4
2610 < Pr	9.8

Fig. E55 Absorbed power in relationship to the reference mass

mass of more than 1700 kg or for vehicles with a permanent all-wheel drive. This means that the characteristic value of the engine test bench can be calculated as

$$P_a = K \cdot v^3,$$

where

K = characteristic value of the engine test bench,
P_a = received power, and
v = speed.

The inertia masses are adjusted accordingly; the test stand is accelerated to a speed of 80 km/h. Subsequently, the speed is increased to 90 km/h. The time to decelerate from a speed of 85 km/h to 75 km/h is recorded. This process is repeated with other dynamometer loads until the performance range of the street is covered.

The deceleration work is divided by the time in the following equation to calculate absorbed power:

$$P_a = \frac{M_i \cdot (v_1^2 - v_2^2)[kg \cdot m^2]}{2 \cdot t \ [s^2 \cdot s]} \cdot \frac{1 \ [kW]}{1000 \ [W]}$$

where

P_a = absorbed power in kW,
M_i = equivalent inertial masses in kg (the inertia mass of the idle running rear roll is not included),
v_1 = starting speed in m/s (85 km/h = 23.61 m/s),
v_2 = final speed in m/s (75 km/h = 20.83 m/s), and
t = time (s) for the deceleration of the rolls from 85 km/h to 75 km/h.

b. Venturi pipe (CFV). Calibration of the CFV is based on the flow equation for constant volume sampling (CVS) with critical flow:

$$Q_s = \frac{K_v \cdot p}{\sqrt{T}},$$

where

Q_s = flow,
K_v = calibration coefficient,
p = absolute pressure (kPa), and
T = absolute temperature (K).

The following describes calibration process used to determine the value of the calibration coefficient for measured values of pressure, temperature, and airflow. The adjustment of the flow control valve is changed in steps, in such a way that at least eight measurements are performed in the critical flow range of the Venturi tube.

The values for the calibration coefficient, K_v, are calculated for each measuring point with the equation

$$K_v = \frac{Q_s \cdot \sqrt{T_v}}{P_v},$$

where

Q_s = flow quantity (m³/min at 273.2 K and 101.33 kPa),
T_v = temperature at the inlet of the Venturi pipe (K), and
p_v = absolute pressure at the inlet of the Venturi tube (kPa).

This makes it possible to record the curve of K_v related to the pressure at the inlet of the Venturi tube.

c. Positive displacement pump (PDP). Calibration is performed for each drive speed of the pump. The flow can be calculated from the calibration equation by measuring the pump characteristic values (PPI, PPO, n) in a normal test for the determination of the exhaust emissions.

The following measurement accuracies must be maintained for this test.

• Air pressure (corrected) (P_B): ±0.03 kPa
• Ambient temperature (T): ±0.2 K
• Air temperature at the LFE (ETI): ±0.15 K
• Vacuum in front of LFE (EPI): ±0.01 kPa
• Pressure drop due to LFE nozzle (EDP): ±0.0015 kPa
• Air temperature at the intake of the CVS pump (PTI): ±0.2 kPa
• Air temperature at the outlet of the CVS pump (PTO): ±0.2 kPa
• Vacuum at the intake of the CVS pump (PPI): ±0.22 kPa
• Pressure at the outlet of the CVS pump (PPO): ±0.22 kPa
• Pump speed during the test (n): ±1 rpm
• Duration of the test (t, min. 250 s): ±0.1 s

A volume flow rate of air, Q_s, in m³/min is determined for each test point from the measured values of the flowmeter. The volume flow rate of air will now be converted to the pump flow, V_0, in m³ per revolution at absolute temperature and absolute pressure at the pump intake:

$$V_0 = \frac{Q_s}{n} \cdot \frac{T_p}{273.2} \cdot \frac{101.33}{P_p},$$

where

V_0 = pump flow volume at T_p and P_p in m³/revolution,
Q_s = airflow volume at 101.33 kPa and 273.2 K in m³/min,
n = pump speed in rpm,
T_p = temperature at the pump intake in K, and
P_p = absolute pressure at the pump intake in kPa.

313

Reference Mass of Vehicle, Pr (kg)	Equivalent Inertial Masses, I (kg)
Pr ≤ 750	680
750 < Pr ≤ 850	800
850 < Pr ≤ 1020	910
1020 < Pr ≤ 1250	1130
1250 < Pr ≤ 1470	1360
1470 < Pr ≤ 1700	1590
1700 < Pr ≤ 1930	1810
1930 < Pr ≤ 2150	2040
2150 < Pr ≤ 2380	2270
2380 < Pr ≤ 2610	2270
2610 < Pr	2270

Fig. E56 Inertia mass related to the reference mass of the vehicle.

~~**Equivalent inertia masses.** The inertia mass is supposed to simulate the total inertia of the rotating masses of the vehicle. It should be equivalent to the reference mass (the mass of the roadworthy vehicle minus the reference mass of 75 kg for the driver and plus a reference mass of 100 kg) of the vehicle. **Fig. E56** shows the corresponding values.

The total inertia mass of the rotating parts (including the simulated inertia mass if required) must be within ±20 kg of the equivalent inertia mass class.

~~**Mass determination, emissions.** The emitted quantities of gaseous pollutants are calculated from the equation

$$M_i = \frac{V_{mix} \cdot Q_i \cdot k_H \cdot C_i \cdot 10^{-6}}{d},$$

where

M_i = emitted pollutant quantity, i, in g/km;
V_{mix} = volume of the diluted exhaust gases, in L/test (total driving cycle), and corrected to nominal conditions (273.2 K; 101.33 kPa);
Q_i = density of the pollutant, i, in g/L, at nominal conditions:
 Carbon monoxide (CO) = 1.25 g/L
 Hydrocarbons ($CH_{1.85}$) = 0.619 g/L
 Nitrogen oxides (NO_2) = 2.05 g/l;
k_H = humidity correction factor for the computation of the emitted pollutant quantities (no humidity factors are available for HC and CO);
C_i = concentration of the pollutant, i, in the diluted exhaust gases, in ppm, and corrected by the pollutant concentration, i, in the diluted air; and
d = the distance driven in accordance with the driving cycle.

The volume of the diluted exhaust gases at the withdrawal system with positive displacement pump is calculated with the equation

$$V = V_0 \cdot N,$$

where

V = volume of the diluted exhaust gases (before correction), in L/test;
V_0 = gas volume pumped by the positive displacement pump under test conditions, in l/revolution; and
N = revolutions of the pump during the test.

The volume of the diluted exhaust gases is corrected by the following equation to arrive at nominal conditions:

$$V_{mix} = V \cdot K_i \cdot \frac{P_B - P_l}{T_p}$$

where

K_i = 273.2 K/101.33 kPa = 2.6961 K/kPa;
P_B = air pressure in the test room, in kPa;
P_l = pressure difference between the vacuum at the intake of the positive displacement pump and the ambient pressure, in kPa; and
T_p = mean temperature, in K, of the diluted exhaust gases at the intake into the positive displacement pump during the test.

The corrected concentration of the pollutants in the collection bag is computed

$$C_i = C_e - C_d\left(1 - \frac{1}{DF}\right),$$

where

C_i = concentration of the pollutant, i, in the diluted exhaust gases, in ppm, and corrected by the pollutant concentration i in the diluted air;
C_e = measured concentration of the pollutant, i, in the diluted exhaust gases in ppm;
C_d = measured concentration of the pollutant, i, in the air used for dilution, in ppm; and
DF = dilution factor.

The dilution factor is computed with the equation

$$DF = \frac{13.4}{C_{CO_2} + (C_{HC} + C_{CO})10^{-4}},$$

where

C_{CO_2} = CO_2 concentration in the diluted exhaust gases in the collection bag, in percent by volume;
C_{HC} = HC concentration in the diluted exhaust gases in the collection bag, in ppm of hydrocarbon equivalent; and
C_{CO} = HC concentration in the diluted exhaust gases in the collection bag, in ppm.

To correct the impact of the humidity on the results achieved for the nitrogen oxides,

$$k_H = \frac{1}{1 - 0.0329 \cdot (H - 10.71)},$$

where

$$H = \frac{6.211 \cdot R_a \cdot P_d}{P_B - P_d \cdot R_a \cdot 10^{-2}}$$

and, in both equations,

H = absolute humidity, in grams of water per kilogram of dry air;

R_a = relative humidity of the ambient air, in percent;

P_d = saturation vapor pressure at ambient temperature, in kPa; and

P_B = air pressure in the test room, in kPa.

The mean HC concentration is calculated to determine the mass of the HC emissions for compression ignition engines:

$$C_e = \frac{\int_{t_1}^{t_2} C_{HC} \cdot dt}{t_2 - t_1},$$

where

$\int_{t_1}^{t_2} C_{HC} dt$ = integral of the values measured by HFID during the test time $(t_2 - t_1)$,

C_e = HC concentration, measured in diluted exhaust gas in ppm for C_i, and

C_i replaces C_{HC} directly in all appropriate equations.

Based on the different weights of the particulate masses deposited on both filters, the particulate mass, m, is determined (m_1 = mass in first filter, m_2 = mass in second filter).

- If $0.95\,(m_1 + m_2) \leq m_1$, then $m = m_1$.
- If $0.95\,(m_1 + m_2) > m_1$, then $m = m_1 + m_2$.
- If $m_2 > m_1$, then test is invalid.

The particulate emission, M_p, in g/km, is computed with the following equation if the gas samples are routed out of the tunnel,

$$M_p = \frac{(V_{mix} + V_{ep}) \cdot m_e}{V_e \cdot d},$$

and if the gas samples are routed back into the dilution tunnel,

$$M_p = \frac{V_{mix} \cdot m_e}{V_e \cdot d}.$$

In both equations,

V_{mix} = standard volume for the diluted exhaust gas,

V_{ep} = standard volume for the exhaust gases that flowed through the particulate filters,

m_e = mass of the particulates deposited on the filters,

d = distance of the driving cycle in km, and

M_p = particulate emission in g/km.

All measured emission quantities are multiplied by deterioration factors and can, therefore, be compared with emission limits. The deterioration factors in **Fig. E57** can be used if the emissions quantities must be determined before the conclusion of the type V test.

~~**Particulate filter preparation.** The particulate filters are conditioned for at least 8 hours but not more than 56 hours prior to the test in an open bowl, protected against dust deposition, in an air-conditioned chamber for measurement with vehicles with self-

Engine Type	Deterioration Factor (D.E.F.)		
	Co	HC + Nox	Particulates*
External ignition engine	1.2	1.2	–
Compression ignition engine	1.1	1.0	1.2
*For vehicles with compression ignition engine			

Fig. E57 Deterioration factors

ignition engines. The unused filters are now weighed. The filter should be removed from the chamber not more than one hour before the test.

~~**Perform driving cycle.** The vehicle is on an engine test bench for the test, and the vehicle resistance and inertia mass are simulated. The temperature of the test room must be between 293 and 303 K (20 and 30°C) during the test.

The vehicle hood is open during the test. A fan is used to maintain a normal engine temperature.

For measurements of vehicles with self-ignition, the flow through the particulate filters must be constant, with maximum deviations of ±5%, and must be adjusted in such a way that the filtered particulate mass is between 1 and 5 mg if 47-mm (diameter) filters are used.

A proportional partial flow is taken from the exhaust gas during the test, and it is diluted by the ambient air, then routed into a collection bag. Ambient air samples are collected in parallel to determine the effect of the ambient air concentration.

The engine is operated at idle speed for 40 seconds before the first cycle is started. The driving cycle consists of part 1—the city driving cycle, which is executed four times—and part 2, the highway cycle.

General data of the city cycle
- Average speed: 19 km/h
- Actual operating time: 195 s
- Equivalent driving distance for 4 cycles: 4.052 km

General data of the highway cycle
- Average speed: 62.6 km/h
- Actual operating time: 400 s
- Theoretically driven distance per cycle: 6.955 km

Maximum speed: 120.0 km/h

~~**Total system check.** A known quantity of polluted gas is supplied to the system during normal operation to determine the total accuracy of the CVS withdrawal system and the analyzers. This is followed by an analysis.

~~**Vehicle preparation.** The vehicle must be in proper mechanical condition, run-in, and must have been driven for at least 3000 km. The adjustment of the systems must be in accordance with the information provided by the manufacturer. The vehicle is filled with a reference fuel—for example, indolene.

For measurement of the particulates, the part 2 of the driving cycle is executed for a maximum of 36 hours or a minimum of 6 hours for vehicles with a compression-ignition engine. Three consecutive cycles are performed.

All vehicles, independent of the ignition process, are conditioned before the test in a room with the temperature between 293 and 303 K (20 and 30°C). The conditioning must be performed for at least 6 hours and the temperature of the engine oil and the coolant must be within ±2 K of the room temperature.

~Test type II. The test type II includes the emission test for carbon monoxide at idle speed in vehicles with an external ignition engine.

Test type II must be executed immediately after the last driving cycle of test type I at engine idle speeds without the use of cold start equipment. A basic driving cycle (part 1) of test type I is performed immediately before further measurements of the carbon monoxide content. Measurement with the adjustments used for the test type I is performed at the start of the test. The carbon monoxide content in the exhaust gases is subsequently measured in all possible positions or at the determined positions of the adjustment equipment.

The possible positions of the adjustment equipment are limited by the engine speed.

On one hand, by the higher of the two values—the lowest engine speed at idle speed or at the idle speed recommended by the manufacturer minus 100 rpm; on the other, by the lowest of the three following values—the highest engine speed, which can be achieved by interfering with idle speed equipment, the recommended idle speed by the manufacturer plus 250 rpm, or the cut-in speed for automatic transmissions.

The CO (CCO) and CO_2 (CCO_2) concentrations are determined by using the respective calibration curves from the display values or from records of the measuring instruments.

The equation for the corrected CO concentration for four cycle engines is

$$C_{Co,\,corr} = C_{CO}\frac{15}{C_{CO} + C_{CO_2}} \ (\text{Vol.\%}) .$$

The carbon monoxide content at idle speed for the adjustment used for test type I cannot exceed 3.5% by volume and it cannot exceed 4.5% by volume for the adjusted range specified above.

~Test type III. This test includes a test of the crankcase emissions.

The crankcase gases consist primarily of unburned and cracked hydrocarbons from the fuel and can be a significant part of the total HC emissions in vehicles with external ignition engines. The crankcase emissions in vehicles with diesel engines are negligible, since only pure air is compressed during the compression stroke.

Test type III is executed after the vehicle has completed test types I and II.

The crankcase ventilation is checked for faultless functionality for the operating conditions shown in **Fig. E58**. The aeration and ventilation openings are left unchanged for this check. The pressure in the

Operating Conditions No.	Vehicle Speed km/h	Power Absorbed by the Brakes
1	Idle	None
2	50 ± 2 (in 3rd gear or "drive")	In accordance with adjustments test type 1
3	50 ± 2 (in 3rd gear or "drive")	In accordance with operating conditions No. 2, multiplied by factor 1.7

Fig. E58 Operating conditions for test type III

crankcase is measured in the opening for the oil dipstick and cannot exceed the atmospheric pressure for any measuring condition during the measurement. An additional process can be executed if the crankcase pressure exceeds the atmospheric pressure under any operating condition. An impermeable bag for the crankcase gases, with a capacity of about 5 liters, is attached to the opening for the oil dipstick for 5 minutes per operating condition.

The vehicle is in accordance with the regulations when no visible filling of the bag can be detected at any operating condition.

~Test type IV. Test type IV includes the determination of the evaporative emission for vehicles with external ignition engines, known as the sealed housing for evaporative determination test (SHED). The evaporative emissions, which consist primarily of hydrocarbons, are generated by the background emissions—for example, by plastic parts of the interior furnishings and the filler neck of the fuel tank.

The evaporative emissions of a vehicle with self-ignition are negligible because the fuel system is completely closed.

~~Background emission. This test is designed to detect whether the chamber materials emit hydrocarbons.

~~Calculation of evaporative emissions. The evaporative losses from venting of the tank and the hot switch-off phase are computed with the equation

$$M_{HC} = k \cdot V \cdot 10^{-4} \cdot \left(\frac{C_{HC,f} \cdot P_f}{T_f} - \frac{C_{HC,i} \cdot P_i}{T_i}\right),$$

where
M_{HC} = the emitted hydrocarbon quantity (grams) during the test phase,
H/C = ratio hydrogen/carbon,
$k = 1.2 \ (12 + H/C)$,
V = net volume of the cabin, corrected by the vehicle volume with open windows and open luggage compartment; a volume of 1.42 m^3 is deducted if the volume of the vehicle has not been determined,
C_{HC} = the hydrocarbon concentration measured in the cabin (ppm [volume] C_1 equivalent),

f = the final value,
P = air pressure in kPa,
i = the starting value, and
T = temperature of the ambient air in the cabin, K.

A value of 2.33 is assumed for H/C for tank ventilation losses; a value of 2.20 is assumed for H/C for hot switch-off losses.

The total quantity of the emitted hydrocarbons is computed as

$$M_{total} = M_{THE} + M_{HS},$$

where

M_{total} = the total quantity of the vehicle emissions (grams),
M_{THE} = the quantity of the hydrocarbon emissions during the tank heating (grams), and
M_{HS} = the quantity of the hydrocarbon emissions during hot switch-off (grams).

The evaporative emission must be less than 2 g/test.

~~**Calibration.** The following gases are used for operation and calibration: cleaned synthetic air with an oxygen content between 18 and 21% by volume, FID combustion gas (40 ± 2% hydrogen), and propane.

~~**Evaporative emissions—hot switch-off.** The hood is closed completely at the end of the driving cycle. The vehicle will subsequently be driven into the measuring chamber. The engine is switched off before any part of the vehicle enters the measuring chamber.

The doors of the chamber are closed gas tight within two minutes after the engine has been switched off and within seven minutes after the driving cycle is completed.

The hot switch-off phase with a duration of 60 ± 0.5 minutes begins as soon as the chamber is closed. The hydrocarbon concentration, the temperature, and the air pressure are measured, and these are used as initial values $C_{HC,i}$, P_i, and T_i for the hot switch-off test. These values are used for the computation of evaporative emissions. The ambient air temperature, T, in the chamber must not be below 296 K and not above 304 K during hot switch-off, which lasts 60 minutes.

The hydrocarbon concentration, the temperature, and the air pressure in the chamber are measured at the end of the test. These are the final values $C_{HC,f}$, P_f, and T_f for the hot switch-off test. This concludes the test process for the determination of the evaporative emissions.

Fig. E59 shows the evaporative emissions due to hot switch-off of low-emission vehicles with activated carbon traps. They represent the average emissions of a sampling of the most common vehicle types. The diagram includes the equation for the approximation curve determined by nonlinear regression.

~~**Hydrocarbon retrieval.** The net internal volume can be controlled and leakages can be measured with the test of the chamber for hydrocarbon retrieval.

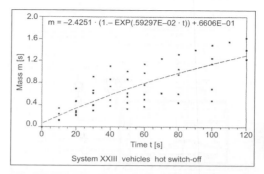

System XXIII vehicles hot switch-off

Fig. E59 Evaporative emission due to hot switch-off, low-emission vehicles (Source: UBA, RWTÜV)

~~**Tank venting evaporation emissions.** A tank emptying and refill process starts between 9 and 35 hours after the completion of the preconditioning driving cycle.

The fuel can be heated to the starting temperature of 289 ± 1 K. The fuel tank cap is put into place and the chamber is closed gas tight as soon as the temperature reaches 287 K. The hydrocarbon concentration, the air pressure, and the temperature are measured as soon as the fuel temperature reaches 289 ± 1 K, and these measurements are used as $C_{HC,i}$, P_i, and T_i for the tank heating test. The linear heating starts now.

The final hydrocarbon concentration, the air pressure, and the temperature in the cabin are measured, once the fuel heat-up is completed. These are the final values $C_{HC,f}$, P_f, and T_f for the tank heating test.

A city driving cycle with cold start and a highway driving cycle are performed on an engine test bench.

Fig. E60 shows the evaporative emissions due to the tank venting of low-emission vehicles with activated carbon trap. They represent the average emissions of a sampling of the most common vehicle types. The diagram includes the equation for the approximate curve determined by nonlinear regression.

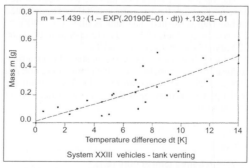

System XXIII vehicles - tank venting

Fig. E60 Evaporative emission due to tank venting, low-emission vehicles (Source: UBA, RWTÜV)

317

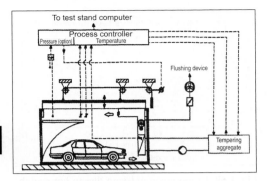

Fig. E61 SHED chamber

~~**Test equipment.** The vehicle is in a gas-tight SHED chamber in accordance with **Fig. E61** for the measurement of the evaporative emissions. The volume of the SHED chamber is about 70 m³. At least one of the surfaces is made from an elastic, impermeable material to compensate for pressure changes due to small temperature changes.

The hydrocarbon concentration is measured by a flame ionization detector.

The fuel in the tank must be heated by 14 K from 289 K within 60 minutes, and it must be within approximately 1.5 K of the required temperature.

The temperature in the chamber is recorded at two points by temperature sensors. The hydrocarbon concentration in the chamber is changed to the concentration of the ambient air. The air in the chamber is mixed by ventilators or fans to achieve a uniform temperature and hydrocarbon concentration in the chamber. The vehicle must not be affected by a direct airflow from the ventilators or the fan.

~~**Test preparation.** The vehicle must be in proper mechanical condition and must have been broken in for at least 3000 km before the test. The fuel tank is equipped with a temperature sensor, which is positioned in such a way that the temperature is measured in the center of the fuel at a tank filling of 40%.

The vehicle is transported into the test room, which has a temperature between 293 and 303 K. The activated carbon trap is flushed. The tank is filled up to 40% ± 2% with a test fuel, which has a temperature between 283 and 287 K (between 10 and 14°C).

A linear heat-up by 14 ± 0.5 is started over a time frame of 60 ± 2 minutes when the temperature reaches 289 ± 1 K. The temperature of the fuel during the heat-up process must be within ±1.5 K of the following equation:

$$T_r = T_0 + 0.2333t,$$

where

T_r = required temperature (K),
T_0 = starting temperature of the container (K), and
t = time from the start of the heat-up in minutes.

The fuel emptying and filling process must be started within one hour as it is described above.

A second tank heating-up process, as described above, is started within two hours after the conclusion of the first tank heating-up period.

One driving cycle (part 1) and two driving cycles (part 2) are performed with the vehicle on an engine test bench within one hour after the conclusion of the second tank heating up. No exhaust emissions are sampled during the tests.

The vehicle will now be switched off for 10–26 hours. The temperature of the engine oil and the coolant must now be within ±2 K of the temperature of the switch-off range.

~**Test type V.** The type V test includes an aging test to check the durability of the emission-reducing equipment of vehicles with external ignition engines or self-ignition engines for a distance of 80,000 km. A deterioration factor is determined by this test.

The test is performed on a test track or on a roller dynamometer. The driving cycles, which must be considered in the drive, consist of 11 cycles of 6 km each.

The deterioration factor, DF, used for the exhaust emission correction is computed for each pollutant as follows:

$$DF = \frac{Mi_2}{Mi_1},$$

where

Mi_1 = mass emission of the pollutant in g/km, interpolated at 6400 km, and
Mi_2 = mass emission of the pollutant in g/km, interpolated at 80,000 km.

~**Test type VI.** This test shows the average exhaust emissions of carbon monoxide/hydrocarbons at low ambient temperatures after a cold start. The test consists of four basic city drive cycles of the type 1 test. The low temperature, which is applied for a total of 780 seconds, is performed without interruption and begins with the starting of the engines at an ambient temperature of −7°C. The vehicles must be uniformly coordinated before the start of the test. The exhaust gases must be diluted during the test and proportional samples must be collected. The diluted exhaust gases are analyzed for carbon monoxide and hydrocarbons. The measured quantities must be below the limits shown in **Fig. E62**.

Test temperature 7 °C			
Vehicle class	Type–	CO[g/km]	HC [g/km]
M₁[1]	I	15	1,8
N₁	II	15	1,8
N₁[2]	II	24	2,7
		30	3,2

1) Except vehicles with more than six seats and vehicles with a maximum mass of over 2.500 kg
2) And the vehicles of class M1 listed in [1]

Fig. E62 Limits of the low-temperature test

Emission value →Engine acoustics ~Exterior noise

Emissions (*also*, →Emission limits; →Emission measurements; →Engine acoustics; →Variable valve control ~Effect of fully variable valve control).

Emissions are the pollutants released by engines from the combustion process, the hydrocarbons released from the fuel tank or during refilling, and also the noise radiation.

The pollutant components of main importance are carbon monoxide, CO, unburned hydrocarbons, C_mH_n, and oxides of nitrogen, NO_x. Particulates are also present, especially in diesel engines, and hydrocarbons can attach to them. Other pollutant components that are emitted are, for example, sulfur, sulfur compounds, and aldehydes. These exhaust gas components all have an impact on the environment, which is why the legislators in many countries have regulated emission values.

Emissions control →Pollutant aftertreatment

Emissions during regeneration →Particles (Particulates) ~Particulate filter system ~~Particulate filter ~~~Regeneration of particulate filters

Emissions legislation →Injection system (components) ~Gasoline engine ~~Fuel tank, ~~Fuel tank systems

Emissions test →Exhaust gas analysis

Emulsions →Fuel, diesel engine

Enable logic →Electronic/mechanical engine and transmission control ~Electronic components ~~Voltage regulator

Encapsulation →Engine acoustics ~Engine noise

End of boiling →Fuel, gasoline engine ~Boiling characteristics

End of intake →Control/gas transfer ~Four-stroke engine ~~Timing

End-of-line programming →Electronic open- and closed-loop control ~Electronic open- and closed-loop control, gasoline engine ~~Functions

Energy (*also*, →Fuel, diesel engine ~Properties; →Fuel, gasoline engine ~Properties). Internal combustion engines are heat engines which convert energy internally to produce external work. A number of energy forms are generated during the process.

~Balance. In an engine, only a part of the energy, supplied in the form of chemical energy contained in the fuel, is converted to mechanical energy at the crankshaft. The conversion is defined by the brake thermal efficiency or by the effective efficiency,

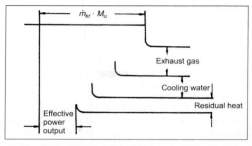

Fig. E63 Energy flows in an engine

$$\eta = \frac{P_e}{\dot{m}_{Kr} \cdot H_u},$$

where

P_e = effective, or brake, power,
$\dot{m}_{Kr}$ = the flow rate of the fuel, and
H_u = the lower calorific value of the fuel.

By far the largest part of the chemical energy of the fuel is transmitted unused into the environment. Gasoline engines have maximum brake thermal efficiencies of approximately 30% and diesel engines with direct injection have brake thermal efficiencies up to 45%.

A schematic description of the energy flows is shown in **Fig. E63**.

It is necessary to analyze a steady-state engine operating point to evaluate the energy balance. The total system can be seen as a steady-state thermodynamic flow process. A system boundary is defined and the inflowing and out-flowing energy and material flows are balanced; this is shown schematically in **Fig. E64**.

The following flows cross the system boundaries: P_e = effective, brake, power output; $\dot{Q}_{Rest}$ = heat transfer to the environment based on the radiation, conduction, and convection; $\dot{H}_{air}$ = enthalpy flow rate of the air; $\dot{H}_{Kr}$ = enthalpy flow rate of the fuel; $\dot{H}_{cwi}$ = enthalpy flow rate of the cooling water at inlet; $\dot{H}_{cwo}$ = enthalpy flow of the cooling water at outlet; and $\dot{H}_{exh}$ = enthalpy flow rate of the exhaust gas.

A balance of the material and the energy flows that enter the control volume or leave it results in

$$\dot{H}_{Kr} + \dot{H}_{air} + \dot{H}_{cwi} = \dot{H}_{cwo} + P_e + \dot{Q}_{Rest} + \dot{H}_{exh, T2}.$$

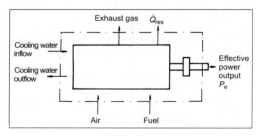

Fig. E64 Material and energy flows in a combustion engine

The introduction of the calorific value of the fuel results in

$$\dot{m}_{Kr} \cdot H_u = \Delta \dot{H}_{CW} + P_e + \dot{Q}_{Rest} + \Delta \dot{H}_{exh}$$

where $\Delta \dot{H}_{exh}$ is the enthalpy difference between exhaust gas at the exhaust gas temperature and the temperature at which the fresh mixture was supplied.

The above description shows that the energy supplied by the fuel is divided into the effective, brake, power output, the heat transfer, and the enthalpy difference of both the cooling water and exhaust gas.

The enthalpy difference of the cooling water can be computed by using the flow rate of the cooling water and the temperature difference. The effective, brake, power output is determined by measuring the torque and the speed at the power dynamometer, the enthalpy difference of the exhaust gas in the mass flow (air + fuel), and the temperatures and the specific heat of the exhaust gas. The heat transfer can be determined because the calorific value of the fuel is known.

~Internal energy. The internal energy of a body is the sum of all the quantities of energy supplied to the body, starting with an initial state until a final state is reached, and which are, therefore, contained in it. The internal energy is related to the matter of the body—it is a property of the system. It is higher if the quantity of matter analyzed is greater.

The internal energy is a state variable that is important for the computation of comparative processes (diesel and gasoline engine). It is independent of the path (process) that was used to get the body to the state to be analyzed. It has a certain value for each state of the matter.

~Kinetic energy. Kinetic energy is the form of energy that is present as energy of movement (speed energy). The kinetic energy of the exhaust gas and of the intake air are important for the engine—for example, during the pulse turbocharging.

High kinetic energy at the inflow into the cylinder supports the mixing of air and fuel.

~Potential energy. Potential energy is the form of energy that normally exists depending on a position or pressure. The position energy is not of great importance under engine boundary conditions; however, the pressure energy is a major factor. The energy conversion in an engine starts from the chemical energy of the fuel, which gets converted into pressure energy (in the combustion chamber), and this energy in turn gets converted into mechanical energy at the flywheel.

Energy balance →Energy ~Balance

Energy density →Fuel, diesel engine ~Properties; → Fuel, gasoline engine ~Properties

Energy of turbulence →Combustion chamber ~Swirl chamber

Energy storage →Battery; →Flywheel; →Heat accumulator

Engine (also, →Engines). The engines treated in this volume are internal combustion engines, which differ in terms of design, control, type, area of use, and so on. The common feature is that they convert the chemical energy in the fuel into mechanical work. For this purpose, liquid fuels are usually used. The two most frequent types of engine are called gasoline engines and diesel engines; their specific differentiating characteristics are described in detail within this reference work.

Here, a distinction is made according to:

- Alternative engines
- Current engines
- Historic engines
- Aviation engines
- Stationary engines
- Racing/sports engines

~Alternative engines. Alternative engines are propulsion systems which have significant differences from the internal combustion engines in common use today, both in terms of design and function. Moreover, they can also be different in terms of the operating fluid and the fuel conversion. However, the only systems considered here are still designated as combustion engines.

~~Hybrid powertrain. Hybrid propulsion systems are generally distinguished by at least two different energy "sources" and energy converters. The energy converters are preferably electric motors and combustion engines; the energy "sources" are preferably conventional fuels and batteries. The use of one or the other system or parallel operation allows the advantages of the particular system to be used, depending on the requirements placed by the vehicle operating characteristics.

This leads to advantages in terms of:

- Fuel consumption
- Emissions
- Noise

These are offset by disadvantages in terms of costs and space required for installing the second propulsion system.

It is necessary to differentiate between

- Parallel hybrids. Here, electrical only, combustion engine only, or combined propulsion are the possibilities. The advantages include operation in areas where emissions and noise are critical factors. Moreover, electric motors can assist the combustion engine during acceleration and the combustion engine can be shut off in operating ranges with particularly low efficiency.
- Serial hybrids. With this system, an electric motor is always used for propulsion, and the combustion engine generates the electrical power with the aid of a generator; this is stored in a battery. The advantages are particularly in the area of pollutant emissions.

IRON MEN

ROBUST. RELIABLE. PROFITABLE.

MTU SERIES 4000 WORKBOAT EDITION
www.mtu-online.com

The Toyota Prius, which can be called a mixed hybrid, is presently in mass production. With a series hybrid, the combustion engine can be connected through to the wheels, for example.

Literature: R. van Basshuysen, F. Schäfer: Handbuch Verbrennungsmotor, Wiesbaden. Vieweg Publishers, 2002. — H.H. Braess, U. Seiffert (Eds.): Handbuch Kraftfahrzeugtechnik, Wiesbaden, Vieweg Publishers, 2001. — I. Harada: Entwicklung eines neuen Toyota Hybrid-Fahrzeuges, Proceedings Motor und Umwelt, AVL Graz 7/8 Sep. 2000. — A. Friedrich: Gegenüberstellung von Pkw mit Verbrennungskraftmaschinen, Hybridantrieben und Brennstoffzellen aus Umweltsicht, VDI Report 1418, 1998.

~~Steam engine (*also*, →Rankine cycle). Newer developments for a modern steam engine as a propulsion engine for passenger cars go back to the 1970s. The starting point was the discussion regarding emissions legislation. Primary advantages were considered to be:

- Low exhaust emissions without secondary exhaust treatment, particularly low NO_x emissions
- Lower requirements in terms of type and quality of fuel
- Favorable torque curve (**Fig. E65**) with the possibility of eliminating the transmission system

These are offset by the following disadvantages:

- Poor thermal efficiency and, therefore, higher fuel consumption than diesel engines with direct fuel injection (**Fig. E66**)
- Height/weight
- Deficits in terms of spontaneous operation
- Poor control

The advancement in materials and use of electronic circuitry for control and regulation purposes resulted in a completely new situation in considering the steam engine. Any alternative propulsion method to the reciprocating piston engine must fit into current consumer demand, as well as meeting the environmental constraints as on efficiency and, therefore, fuel consumption. This simultaneously defines the limits on the CO_2

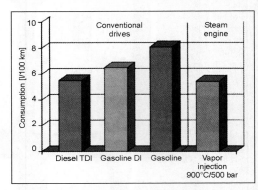

Fig. E66 Fuel consumption in US FTP 75 cycle, steam engine

emission. Conventionally designed steam engines are based on the Rankine cycle.

Alternatives to these are steam engines with isothermal process control or steam injection, because here the operating capacity is higher at the same maximum temperature. Basic research has shown that even at high process temperatures (900°C) and pressures (500 bar), the fuel consumption with steam injection in the US FTP 75 cycle is higher than that of a diesel engine with direct fuel injection.

Characteristic data for a steam engine are shown in **Fig. E67**.

Literature: G. Buschmann, H. Clemens, M. Heotger, B. Mayr: Der Dampfmotor—Entwicklungsstand und Marktchancen, MTZ 62 (2001) 5. — G. Buschmann, H. Clemens, M. Heotger, B. Mayr: Zero Emission Engine—Der Dampfmotor mit isothermer Expansion, MTZ 61 (2000) 5.

~~Stirling engine (hot gas engine). In 1818, the principle of the Stirling engine was described by the Scottish inventor R. Stirling. The engine operates with external, continuous combustion in a closed regenerative thermal power cycle. The thermal energy is transferred to the operating medium by a heat exchanger. The operating medium is moved back and forth between chambers with constant high or low temperatures. This results in periodic pressure varia-

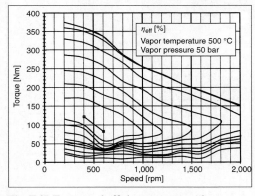

Fig. E65 Torque and efficiency, steam engine

Number of cylinders	3
Engine capacity	992 cm
Bore/stroke	90/52 mm
Maximum burner output	108 kW
Rated power	50 kW
Rated speed	2000 rpm
Maximum speed	2500 rpm
Rated torque between 200 and 1500 rpm	300 Nm
Maximum torque	500 Nm

Fig. E67 Characteristic data for a steam engine

tions which are converted to mechanical energy by an operating piston and crank drive.

This cycle is distinguished by the following changes of states:

- Isothermal compression with heat dissipation
- Isochoric (constant volume) heat supply
- Isothermal expansion with heat supply
- Isochoric heat dissipation

The thermal efficiency of the Stirling cycle is the same as that of the Carnot cycle—that is, dependent only upon the maximum and minimum cycle temperatures achieved. The work output is, however, at technically realizable pressures and compression ratios, comparable to the work used as the basis for the idealized cycles for diesel or gasoline engines.

The principal processes in the cycle are shown in **Fig. E68** in the form of a *p-v* diagram for full and part load.

The advantages of the Stirling engine include:

- Use of any desired heat source
- Low pollutant emissions
- Very good maximum achievable efficiencies
- Good acoustic properties

These must be contrasted to the disadvantages in terms of

- Vehicle response
- Load control
- Overall size
- High production costs

These disadvantages make the Stirling engine unsuitable for mobile applications such as motor vehicles. **Fig. E69** shows characteristic values for a Stirling engine.

~~Wankel engine. The Wankel or rotary piston engine is a special version of the reciprocating piston engine and is an intermediate step between the classic crankshaft engine and other types of drives. It operates exclusively on the four-stroke principle. The trochoidal piston contour controls the intake and exhaust ports, which can take two different forms, requiring

Parameter	Value
Specific value	100–500 W/kg
Maximum efficiency	approx. 40%
Part-load efficiency	approx. 30%
Specific power output	50–500 W/L
Costs	50–1500 D/kW
Service life	>11,000 hours of operation

Fig. E69 Characteristic values for a Stirling engine

different control mechanisms. If the ports are designed in the housing jacket, this is called circumferential control; if the ports are located in the side plates, it is called side control.

Fig. E70 shows the layout and functioning of a Wankel engine, which has the advantages of ideal balancing, compact design, elimination of a valve train, and good torque up to high engine speeds of approximately 8000 rpm. The disadvantages are the unfavorable combustion chamber shape (based on the geometric of the mating components), the negative quench effect, relatively high HC emissions, high fuel and oil consumption values and limited possibility of production in diesel engine form. Nevertheless, after its first use in the NSU Spider and then in the RO-80, Mazda has continued use of this design in mass production up to today. Tests have shown that the long, thin combustion chamber is very good for operation with hydrogen, because the Wankel engine combustion process is intrinsically longer than in a reciprocating piston engine.

~Current engines. A complete description of all contemporary engines would go beyond the scope of this work. For this reason, the description here is limited to a few significant, characteristic types of engines, that reflect the present state of the art in engine technology. The

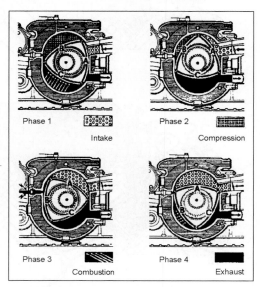

Fig. E70 Principle of Wankel engine

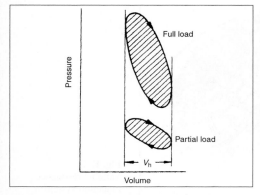

Fig. E68 Stirling engine process (Source: Pischinger)

development and optimization goals in engine development are characterized primarily by the requirements for:

- Improvement of driving performance
- Minimization of fuel consumption and CO_2 emissions
- Fulfilling standards for pollutant emissions such as EU4, EU5, ULEV
- Improvement of passenger comfort
- Minimization of weight
- Minimization of costs

This requires development of modules, systems, and assemblies capable of satisfying the partly contradictory requirements mentioned above. In most cases, the

development steps derived from the definition of the objective usually represent a compromise.

Modern engines for personal transport are distinguished by the following primary features:

Diesel engines.

- Engines with direct fuel injection.
- Multihole injection nozzles and air distributing combustion systems.
- Highpressure fuel injection systems with common-rail fuel injection at injection pressures in the range of 200 bar and pump nozzle systems with over 2000

Design	[–]	V-8 with 90°-V angle	Engine weight DIN	[kg]	189
Displacement	[dm³]	4.172	Intake valves	[–]	3
Stroke/bore	[mm]	82.4/84.5	Exhaust valves	[–]	2
Compression ratio	[–]	11.0	Camshaft adjustment range	[°CA]	22 in advanced direction
Cylinder spacing	[mm]	90	Specific power output	[kW/l]	59.0
Cylinder offset	[mm]	18.5	Max. mean effective pressure	[bar]	12.9
Crankshafts	[–]	90° throw	Firing order	[–]	1–5–4–8–6–3–7–2
Connecting rod length	[mm]	154	Max. torque	[Nm]	430 at 3500 rpm

Fig. E71
Technical data on Audi V-8 engine

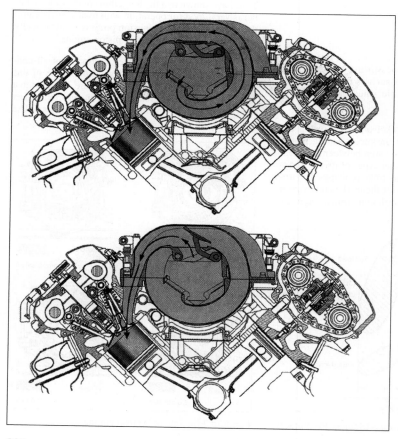

Fig. E72
Variable resonance induction system (Source: *MTZ*)

bar: these two fuel injection systems dominate. The percentage of distributor injection pumps is decreasing.
- Primarily four valves per cylinder with centrally located injection nozzle.
- Swirl intake systems, with a move in the direction of variable swirl and port switch-off.
- Electronic diesel control.
- Aluminum as material for cylinder head and, to an increasing extent, for cylinder block as well.
- Secondary exhaust treatment system using catalytic converter and NO_x accumulator-type catalyst to an increasing degree with particulate filter.
- Exhaust turbocharging with variable turbine geometry and charge air cooler.
- Cooled exhaust gas recirculation.
- Predominantly three-, four-, six-, and eight-cylinder engines, where six-cylinder engines are laid out as both V- and inline engines, while eight-cylinder engines are V-engines.

Gasoline engines.

- Engines with intake manifold and direct injection. Direct injection is increasing for reasons of fuel economy.
- Primarily naturally aspirated engines. Use of turbocharged engines will increase.
- Aluminum as material for cylinder head and cylinder block.
- Variable camshaft timing systems right up to completely variable timing, mechanical, electromechanical, or electrohydraulic.
- Regulated three-way catalytic converter; not on lean engines.
- Predominantly four valves per cylinder.
- Single ignition coils and cylinder selective spark control.
- Cylinder selective injection, cylinder shutoff on large volume and multiple cylinder engines.

- Heat management for optimization of cooling and warm-up.
- Variable resonance induction system.
- Preferred four, six, or eight cylinders, whereby six-cylinder engines are produced as V-engines and inline engines, and eight-cylinder engines as V-engines.

A number of examples of current diesel and gasoline engines are listed below without any pretense of completeness.

Audi V-8, 4.2. Some of the technical data on the 4.2-liter, V-8 engine is shown in **Fig. E71**. The engine is a five-valve engine with identical cylinder heads, installed in mirror-image formation.

The crankcase consists of a hyper-eutectic aluminum alloy ($AlSi_{17}Cu_4Mg$) produced using the low-pressure chilled casting process. The silicon crystals exposed by etching form the bearing partner for the iron-coated pistons. The five-bearing crankshaft is forged using $42CrMoS_4$ material. Cranked steel connecting rods allow positive fixation and, therefore, centering of the connecting rod bearing cap. The engine has a die-cast magnesium variable resonance induction system with intake manifold "tuned" lengths between 285 and 705 mm (**Fig. E72**). The exhaust manifold is a high heat-resistant air gap insulated sheet metal header-type manifold with pulse optimized re-unification and optimized inflow into the underhood, multiple-stage catalytic converter which achieves its conversion temperature quickly after starting the engine due to catalytic converter heat-up measures. A longitudinal section and a cross section of the engine are shown in **Fig. E73**.

VW W12–6.0 l. This W-type engine is a combination of two V-6 engines to form a V-engine with a cylinder bank angle of 72° and one common crankshaft. The ar-

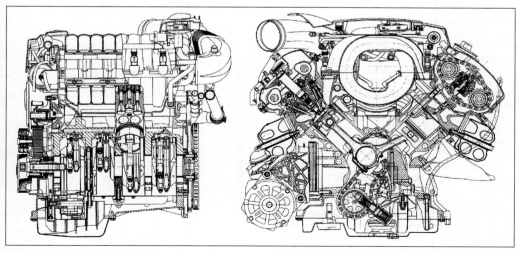

Fig. E73 Longitudinal section and cross section

Design	[–]	W	Rated power	[kW]	309
Displacement	[dm³]	5.998	Max. torque	[Nm]	550/3000 rpm
Stroke/bore	[mm]	84.0/90.168	Length/width/height	[mm]	513/710/715
Compression ratio	[–]	10.75	Bank angle	[°]	72°
Connecting rod length	[mm]	168.5	V angle	[°]	15°
Crankshaft bearing	[–]	7	Cylinder spacing	[mm]	65
Adjustment range camshaft	[°CA]	Intake 52/exhaust 22	Firing order	[–]	1-12-5-8-3-10-6-7-2-11-4-9-

Fig. E74
Technical data for
W12–6.0 l

E

rangement in combination with a V-angle of 15° results in a very compact, torsionally rigid design. The primary technical data are given in **Fig. E74**.

The crankcase with closed deck design consists of an Al-Si alloy with wear resistant cylinder liners.

The crankshaft bearing cross member, with integrated oil slot, consists of AlSI Cu₃ with main bearing caps of GGG50. The primary geometrical relationships for the crank drive are shown in **Fig. E75**.

The four-valve cylinder heads for the V-6 engine serve as the basis for the W-engine. Modifications to the cylinder heads attached to the specific cylinder banks with one intake and one exhaust camshaft each were related to the oil return and water cooling. Each cylinder bank has one intake and exhaust camshaft adjuster.

The ferrostan-coated pistons, which are identical pistons for both cylinder banks, have an inclined quench area due to the V-angle and are selectively installed to achieve a piston clearance between 20 and 40 μm. The 675 g connecting rods are forged from 42CroMo₄ material.

An underhood catalytic converter concept ensures fulfillment of the Euro 4 standard. This concept includes four underhood catalytic converters for startup into which the exhaust flows from insulated head pipes and two underbody catalytic converters. An exhaust manifold with insulating air gap is also used. **Fig. E76** shows the map of the specific fuel consumption for the W-engine.

BMW-R6 diesel engine. This supercharged three-liter, inline, six-cylinder engine is a four-valve diesel engine with direct injection of the fuel. The injection nozzles are positioned upright on the axis of the cylinder, and it has two intake ports per cylinder: one is designed as a swirl port and the other as a charging port. The ports are distinguished by a high swirl level at low flow rates for the swirl port and high charging at high flow rates for the charging port. The pistons have cooling ducts with an axially symmetrical recess. The two exhaust ports are connected together in the cylinder head.

Injection is accomplished using a common rail injection system with maximum injection pressures of 1600 bar.

A newly developed microblind hole nozzle with six holes provides for optimum spray preparation even at low quantities. Multispray injection in certain operating ranges ensures appropriate noise improvements. The maximum permissible pressure in the combustion chamber is 180 bar.

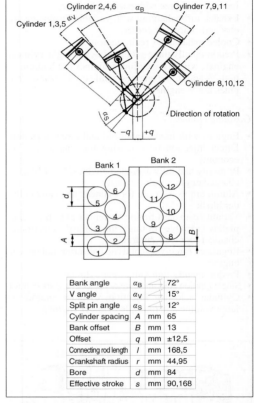

Bank angle	α_B		72°
V angle	α_V		15°
Split pin angle	α_S		12°
Cylinder spacing	A	mm	65
Bank offset	B	mm	13
Offset	q	mm	±12,5
Connecting rod length	l	mm	168,5
Crankshaft radius	r	mm	44,95
Bore	d	mm	84
Effective stroke	s	mm	90,168

Fig. E75 Relationships on crankshaft drive

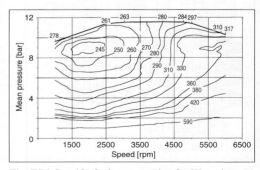

Fig. E76 Specific fuel consumption for W-engine

Design	[–]	Inline six cylinder	Power-weight ratio	[kg/kW]	1.3
Engine mass	[dm³]	2.993	Engine mass DIN	[kg]	208
Stroke/bore	[mm]	90/84	Mean effective pressure	[bar]	20.1
Connecting rod length	[mm]	136	Specific power	[kW/liter]	53.5
Cylinder spacing	[mm]	91	Cutoff speed	[rpm]	4900
Compression ratio	[–]	17	Max. power	[kW]	160 at 4000 rpm
Number of valves/cyl	[–]	4	Max. torque	[Nm]	500 at 2000 rpm

Fig. E77
Engine data on a
BMW six-cylinder,
inline engine

Design	[–]	Four-cylinder, inline engine	Four-cylinder, inline engine
Combustion process	[–]	Port injection	Direct injection
Firing order	[–]	1-3-4-2	1-3-4-2
Engine mass	[dm³]	1.796	1.796
Stroke/bore	[mm]	85/82	85/82
Cylinder spacing	[mm]	90	90
Compression ratio	[–]	8.7 to 10.2 depending on performance version	10.5
Rated power	[kW]	105 to 141	125
Max. torque	[Nm]	220 at 2500 rpm to 260 at 3500 rpm	250 at 3000 rpm
Engine weight	[kg]	158	163
Charging process	[–]	Mechanical supercharging	Mechanical supercharging
Camshaft adjustment range	[°CA]	Exhaust 40°CA (retarded) Intake 30°CA (advanced)	Exhaust 40°CA (retarded) Intake 30°CA (advanced)

Fig. E78
Technical data on a
four-cylinder,
supercharged engine

The exhaust turbocharger has adjustable turbine geometry in the form of adjustable turbine vane unit, and it also has charge air cooling. The exhaust gas flows to the turbocharger through an exhaust manifold with air gap insulation. **Fig. E77** shows the most important engine data.

Literature: F. Steinparzer, W. Mattes, W. Hall, C. Bock: Die Dieselantriebe der neuen BMW 7er-Reihe. MTZ 63 (2002) 10, p. 790. – F. Anisits, K. Borgmann, H. Kratochwill, F. Steinparzer: Der neue BMW Sechszylinder Dieselmotor, MTZ 59 (1998) 11.

Four-cylinder supercharged engine from Daimler-Chrysler. These four-cylinder engines with mechanical supercharger are available in three versions based on the combustion process: Intake manifold or port injection and a version with direct gasoline injection. The basic engine has a displacement of 1.8 liters and four valves per cylinder. Further technical data on the individual versions are given in **Fig. E78**.

The second order vibration excitations resulting from the unbalanced inertia forces are eliminated by a Lancaster harmonic balancer, which runs at twice the engine speed.

The intake valves are driven by cam followers and double camshafts.

A camshaft varying system allows the phases of the intake and exhaust camshafts to be shifted.

The crankcase is a die-cast version ($AlSi_9Cu_3$) with deep skirt design and cast-in gray cast iron cylinder liners. The two-piece, four-valve cylinder head is produced using $AlSi_{10}Mg$ with the chilled molding process. It can be used for the version with port injection as well as that with direct injection.

The exhaust system consists of a dual exhaust manifold with insulating air gap and integrated underhood catalytic converter as well as an underbody catalytic converter. The underbody catalytic converter with a total volume of 1.6 liters and 400 cells/inch² is coated with platinum/rhodium. The exhaust system and engine layout ensure fulfillment of the Euro4 values and the ULEV limits for the US market. **Fig. E79** shows a partial cross section of the four-cylinder engine with mechanical supercharging and port injection.

Fig. E79 Partial section of four-cylinder engine with mechanical supercharging (Source: *MTZ*)

Literature: L. Miculic, B. Heil, M. Mürwald, K. Bruchner, A. Pietsch, R. Klein: Neue Vierzylinder-Ottomotoren mit Kompressoraufladung, MTZ 63 (2002) 6.

VW 1.9-liter TDI. The 1.9-liter TDI with an output of 110 kW is a two-valve engine with pump/nozzle high-pressure injection system with high performance and torque potential as well as low fuel consumption and emission values.

The cylinder head is the same as on the 85 kW engine, but the pump/nozzle element has been adapted to the increased requirements. The primary characteristics of the pump/nozzle system are improvement of the geometry of the nozzle holes, which are rounded by hydroerosion and optimized for maximum flow; a pre-injection quantity over the entire map, and injection pressures of up to 2050 bar. The pistons, connecting rods, crankshafts, and crankcase are adapted to the increased performance level to withstand the mean effective pressures of 22 bar and combustion pressures with values greater than 170 bar. Optimum piston cooling, reinforced piston pins, and top land height of 12 mm as well as geometrically and material-optimized connecting rods (42 $CrMo_4$) and crankshaft (42 $CrMoS_4$) were necessary to increase the output.

EGR cooling as well as an oxidation-type catalytic converter are used to reduce the exhaust emissions and satisfy the Euro 3 standard.

A signal from the crankshaft is used in the calculation of the start of injection and injection period. A second sensor on the camshaft senses the phase position (i.e., which stroke) to allow the engine to be started even during the first rotation. Other sensors are used to sense the charge pressure, start of injection, EGR, and so on. **Fig. E80** shows a summary of the inter-relationships for engine control.

Further development of this version with four valves will result in a further increase in the performance of the engine.

Fig. E81 shows a cross section of the engine, while **Fig. E82** shows the primary engine data.

Literature: 25 Jahre Dieselmotoren von VW, MTZ extra, 2001.

DaimlerChrysler CDI diesel engine. The technical data for the four- and six-cylinder engines in the CDI series are shown in **Fig. E83**. A cross section through the engine cylinder head is shown in **Fig. E84**.

The engines are equipped with an exhaust turbocharger with electrically actuated guide vane adjustment as well as a charge air intercooler. The common rail injection system with central injection nozzle operates with injection pressures of 1600 bar, and a regulated high-pressure pump is used for charging this as required. The seven-hole nozzle used has conical, flow-optimized nozzle holes for highly efficient mixture preparation.

The four-cylinder engine is equipped with a balancing system that compensates out annoying second order out-of-balance forces and adapts the smoothness of the engine to the comfort required. The system is equipped with two masses rotating in opposite directions, driven by gears with a ratio of 2:1 and attached to the crankcase from below.

The recycled exhaust is cooled with the aid of the cooling water to reduce the NO_x emissions and improve the NO_x/particulate tradeoff (**Fig. E85**). A catalytic converter system consisting of an underhood catalytic converter and an additional underbody catalytic converter provides for secondary exhaust treatment.

Literature: R. Klingmann, W. Fick, H. Brüggemann: Die neuen Common-Rail-Direkteinspritz-Dieselmotoren in der modellgepflegten E-Klasse—Motorkonstruktion und mechanischer Aufbau, Part 1, MTZ 60 (1999) 7/8. — D. Naber, K.H. Hoffmann, A. Peters, H. Brüggemann: Die neuen Common-Rail- Direkteinspritz-Dieselmotoren in der modellgepflegten E-Klasse-Verbrennung und Motormanagement. MTZ 60 (1999) 9. — H. Brüggemann, R. Klingmann, W. Fick, D. Naber, K.H. Hoffmann, R. Binz: Dieselmotoren für die neue E-Klasse, MTZ 63 (2002) 4.

Ford Duratec HE. These engines are distinguished by aluminum cylinder heads with four valves and integrated

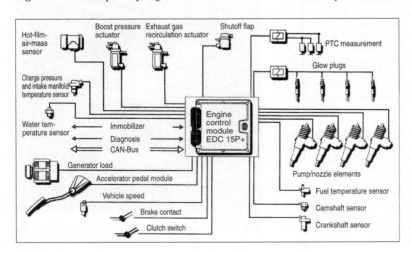

Fig. E80
Engine control
(Source: *MTZ*)

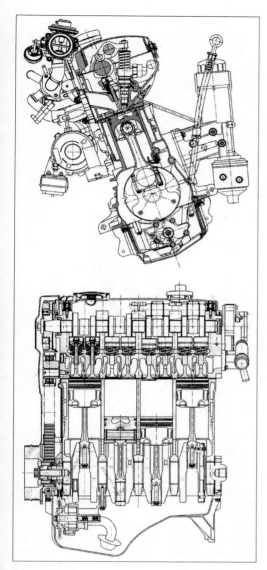

Fig. E81 Engine longitudinal section and cross section

Design	[–]	Water-cooled, four-cylinder, inline engine
Stroke/bore	[mm]	95.5/79.5
Cylinder spacing	[mm]	88
Displacement	[dm³]	1896
Valve train	[–]	Overhead cam with toothed belt drive
Compression ratio	[–]	18.5
Capacity	[kW]	110 kW at 4000 rpm
Spec. power output	[kW/liter]	58
Max. torque	[Nm]	320 at 1900 rpm
Exhaust standard	[–]	EU 3
Brake mean effective pressure	[bar]	22

Fig. E82 Technical data for 1.9-liter TDI

The crankshaft, produced from spheroidal graphite cast iron, has four counterweights; the connecting rods are produced using a sintering casting process and are broken for economic fixation of the bearing cap and connecting rod.

The exhaust manifold consists of welded steel pipes with a four-into-one junction. Classification in EU 4 was possible without the use of an underhood catalytic converter.

The engine is controlled by a Visteon-Levanta engine management system.

Fig. E86 shows a longitudinal section and cross section of the engine; **Fig. E87** shows the primary technical data for the engines.

Porsche 911 Turbo. The Porsche 911 Turbo is a 3.6-liter engine with horizontally opposed cylinders with water cooling and four-valve head design. The primary features include a vertically divided crankcase with wet cylinder liners; cylinder head and camshaft housing as separate assembly as well as dry sump lubrication. The $AlSi_{10}Mg$ crankcase is artificially aged and produced using the chilled casting process with inserted sand cores. The cylinder liners are pressed in, coated with Nikasil, and clamped into the crankcase at the top as well as centered at the bottom. The water jacket is sealed at the cylinder head and side of the crankcase with O-rings.

The cylinder heads are also produced using the chilled casting process with the high-temperature alloy Rolls Royce 350. The valve timing is controlled by the VarioCamPlus which operates using a combination of valve stroke switching and camshaft phasing. The adjustment range of the phase shift for the intake side is 30°CA; the valve stroke varies from 3 to 10 mm. The hollow exhaust valves are sodium-cooled.

The engine has two turbochargers connected in parallel, which provide for high cylinder charging and short acceleration lag times when combined with a low intake manifold volume and short exhaust manifold, as well as charge air intercooling. **Fig. E88** shows some of the technical data for the engine.

EGR duct. The AlSi alloy cylinder head is produced using the low-pressure casting process and subsequently heat-treated for structural homogenization and increased strength.

The valves are actuated by mechanical bucket tappets and the camshaft is driven by a chain with hydraulic tensioners on the loose side. The timing case with oil pump drive is closed at the front by an engine cover. The aluminum crankcase is cast as a precision sand casting with cast-in gray cast iron cylinder sleeves and a closed-deck construction with a deep skirt design.

Design	[–]	Inline 4	Inline 6
Valve train	[–]	4-valve DOHC	4-valve DOHC
Engine mass	[dm³]	2.148	3.222
Stroke/bore	[mm]	88.34/88	88.34/88
Cylinder spacing	[mm]	97	97
Connecting rod length	[mm]	149	149
Compression ratio	[–]	18	18
Max. Brake mean effective pressure	[bar]	15.8	19.5
Max. ignition pressure	[bar]	155	155
Rated power	[kW]	90 at 4200 rpm	150 at 4200 rpm
Max. torque	[Nm]	270 at 1400 rpm	500 at 1800 rpm
Injection system	[–]	Common-rail direct injection	Common-rail direct injection

Fig. E83
Technical data for
four- and six-cylinder
engines

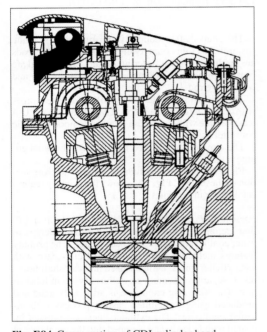

Fig. E84 Cross section of CDI cylinder head

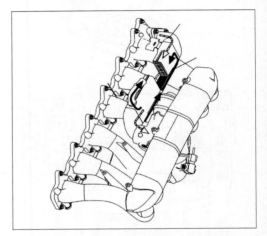

Fig. E85 Recirculated exhaust gas cooled by cooling
water for a six-cylinder, inline CDI engine

Lupo-FSI (fuel stratified injection) engine. The combustion process used by the FSI engine is direct injection with stratified charging. While a mixture capable of ignition is present in the area of the spark plug, pure air is present almost exclusively in the remainder of the combustion chamber. This process provides better fuel economy than intake manifold injection, particularly in the low load range.

The basic engine is a 1.4-liter engine with four-valve head, aluminum crankcase with plasma-coated steel and molybdenum cylinder liners honed using a special process. The engine has map-controlled cooling allowing higher coolant temperatures in the part-load range.

The EGR discharges into the intake system with its

integrated tumble system allowing 35% of the exhaust gas to be recycled. The camshaft phasing allows an adjustment range of 20°CA.

The high-pressure injection system consists of a high pressure pump, fuel distribution rail, pressure sensor and high-pressure injection valve. A primer pump supplies fuel to the high pressure pump at a pressure of 3 bar. The fuel pressure can be adjusted in the range of 20 to 120 bar. The high-pressure injection valve has a spray angle of 70°, and the spark plug is located in the center of the head. The intake port is divided into an upper and lower half by a tumble plate and designed as a charge port with moderate tumble characteristics.

The exhaust system is distinguished by an under-hood three-way preliminary catalytic converter as metal bed catalytic converter integrated into the exhaust manifold. The NO_x accumulator-type catalytic converter used is an underbody catalytic converter with NO_x sensor. The system is cooled for better operation of the NO_x accumulator and minimization of the thermal load. **Fig. E89** shows the technical data for the engine.

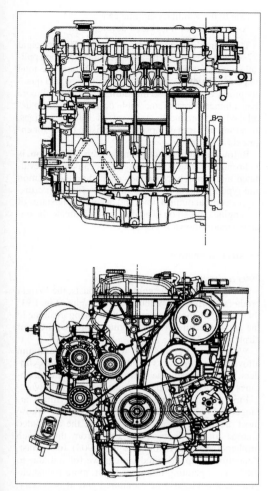

Fig. E86 Longitudinal section and cross section of Duratec engine

Design	[–]	Horizontally opposed engine
Number of cylinders	[–]	6 in two banks
Engine mass	[dm³]	3.6
Stroke/bore	[mm]	76.4/100
Cylinder spacing	[mm]	118
Connecting rod length	[mm]	127
Compression ratio	[–]	9.4
Camshaft adjustment	[–]	30°CA
Rated power	[kW]	309 at 6000 rpm
Max. torque	[Nm]	560 at 2700 rpm
Max. charge pressure	[bar]	1.80
Max. brake mean effective pressure	[bar]	19.2
Cooling	[–]	Water-cooled

Fig. E88 Technical data for Porsche 911 Turbo

2.2-liter ECOTEC Opel. The upper part of the two-piece cylinder block for the Ecotec engine is produced using 319 T5 aluminum in the lost foam process with a number of styrofoam sections glued together. The bottom section, which almost completely surrounds the crankshaft, is produced using 319 F aluminum in the low-pressure casting process. The open-deck design with cooling ducts between the cylinders is equipped with pressed-in gray cast iron cylinder liners with wall thickness of 1.5 mm, which are produced using the rotation molding process.

The cylinder head consists of the same material as the crankcase and is also produced by the lost foam process. The cylinder head contains two inductively hardened camshafts of cast iron with spheroidal graphite driven by a simplex roller chain.

The connecting rod is sinter-forged from C70 S6 material. The spheroidal cast iron crankshaft has eight counterweights. Two cast iron balance shafts running in opposite directions rotate at twice the speed of the crankshaft to compensate for the second-order inertia forces.

The intake manifold is produced using nylon shells with 30% fiberglass by means of friction welding.

Design	[–]	Four-cylinder, inline engine
Engine capacity	[dm³]	1.999
Stroke/bore	[mm]	83.1/87.5
Cylinder spacing	[mm]	96
Connecting rod length	[mm]	146.25
Compression ratio	[–]	10.8
Valve train	[–]	DOHC
Rated power	[kW]	107 at 6000 rpm
Max. torque	[Nm]	190 at 4500 rpm
Secondary exhaust treatment	[–]	Underbody three-way catalytic converter
Max. valve lift intake/exhaust	[mm]	8.8/7.7
Max. brake mean effective pressure	[bar]	12.0

Fig. E87 Technical data for two-liter Duratec engine

Design	[~]	Four-cylinder, inline engine
Engine capacity	[dm³]	1.39
Combustion process	[~]	Direct injection
Stroke/bore	[mm]	75.6/76.5
Valves per cylinder	[~]	4
Compression	[~]	11.5
Capacity	[kW]	77 at 6200 rpm
Max. torque	[Nm]	130 at 4500 rpm
Max. EGR rate	[%]	35
Max. injection pressure	Bars	120

Fig. E89 Technical data for Lupo-FSI engine

Low resistance air guidance, cylinder selective feed, and high uniformity of charge distribution to all cylinders, particularly the crankcase ventilation and uniform routing of the intake cross sections, distinguish the induction module.

The exhaust manifold is manufactured from cast iron with a high percentage of silicon/molybdenum. The exhaust system includes an underhood metal bed start-up catalytic converter of the oxidation-type catalytic converter and an underbody catalytic converter with two lambda probes. **Fig. E90** shows some of the technical data for this engine.

Literature: Motortechnische Zeitschrift, Vieweg Publishers, Wiesbaden.

~Historic engines. Historic engines are engines that are no longer built today but which did play an exceptional role in the history of engine development or engineering, or were used for military and commercial purposes. The special features of historic engines can be explained on the basis of their physical/technical features, the state of the art at the particular time, and the particular conditions under which, and on the basis of which, they were developed. During a period of more than a hundred years, the many engines produced differ in terms of their basic concept (engine concepts). Engines are significant in terms of both engineering history and general history:

- As individual models
- As the final point of a development point
- As the peak of a one-sided development aimed at special characteristics
- As the exponent that led to the development of a standard
- As a typical representative of certain development trends
- As examples for a certain development status
- As examples for the requirements and conditions of a certain time
- Due to their economic success — e.g., due to large quantities produced

- Due to their military significance
- As examples of misguided development

The individual criteria may overlap. The principal layout and design details are determined by the application of the engines and the state of the art. The total range of the engines, in terms of size, design, and quantity produced, is large, so that during the course of more than a hundred years remarkable designs have been developed and produced. A survey of historic engines, therefore, provides glimpses into decisive trends for engine development.

Below, subdivided according to their application, is a chronology. The performance data can only be evaluated in general, because they depend on the duration and type of output (e.g., for aircraft: climbing, cruising, or combat output), load and development status of the engine, and so on. The data given indicate the engine version shown in each case.

~~Aircraft engines

1903 to 1918
Using motor vehicle engines as a basis, the Wright brothers, Orville and Wilbur, built an engine (**Fig. E91**) that was light enough to propel their flying machine in the air. Powerful but light engines were required for airplanes.

Weight could be reduced with the new aluminum alloy, higher output was achieved by using more cylinders, and higher engine speeds resulted. Aviation engines began to develop into a category of their own.

First fan-shaped and then star-shaped cylinder arrangements resulted in short, light engines, whose most distinguished representatives are the air-cooled Gnome rotary engines (**Fig. E92**) from the French brothers Louis and Laurent Seguin (1908). In contrast, Zeppelin airships required more powerful and, due to the longer operating times per flight, more robust engines that could also be monitored and, if necessary, repaired with the Zeppelin in the air. One example of this design is shown in **Fig. E93**.

National version	[–]	United States	Europe
Design	[–]	Four-cylinder, inline engine	Four-cylinder, inline engine
Engine capacity	[dm³]	2.198	2.198
Stroke/bore	[mm]	94.6/86	94.6/86
Max. power	[kW]	103 at 5600 rpm	108 at 5800 rpm
Max. torque	[Nm]	195 at 4200 rpm	203 at 4000 rpm
Weight (DIN)	[kg]	138	138
Compression ratio	[–]	9.5	10.0
Valve train	[–]	DOHC roller cam levers	DOHC roller cam levers
Valve angle	[–]	Intake 18/exhaust 16	
Specific power output	[kW]	47	49
Antiknock control	[–]	Global	Cylinder selective
Secondary exhaust treatment	[–]	3-way catalytic converter	3-way catalytic converter/EGR
Emissions legislation	[–]	LEV	EU4

Fig. E90
Technical data for Ecotec engine

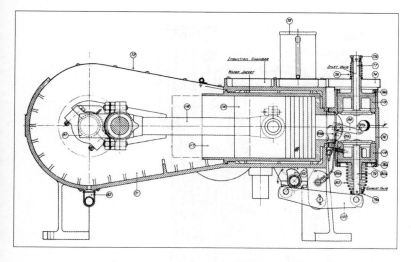

Fig. E91
Wright aircraft engine
(USA, 1903)

World War I resulted in refinements in the types of aviation engines, which eliminated rotary cylinder engines and eggbeater-type engines. The field was then dominated by water-cooled single- and multicylinder, inline engines.

The Hispano-Suiza V-8 engine (**Fig. E94**), developed by Swiss engineer Marc Birkigt and produced in large quantities in France, England, and the United States, was considered to be one of the best aviation engines during World War I.

The German aviation engines were heavier than those used by the Allies, but more robust and reliable.

The Mercedes D IIIa engine (**Fig. E95**) had welded steel cylinders, a design that proved itself and was adopted by many aircraft engine manufacturers.

A highlight of aviation engine development during World War I was the American Liberty engine (**Fig. E96**). It was developed, designed, and built as a test engine and tested on a test bench within a few weeks in a combined effort by designers from various companies. Three months after the design work was begun, the 12-cylinder version completed a 50-hour endurance test.

1918 to 1955

Following World War I, water-cooled single- and multicylinder inline and air-cooled single- and multicylinder radial engines competed with one another. High-performance water-cooled engines were built as 12-cylinder versions, as three-row engines (W-engines) (**Fig. E97**) and as two-row engines (V-engines) (**Fig. E98**).

The W-engine was complicated in terms of design, but rigid, an advantage for the torsional vibration characteristics of the engine. Torsional vibration in six-throw crankshafts on (single) inline and V-engines

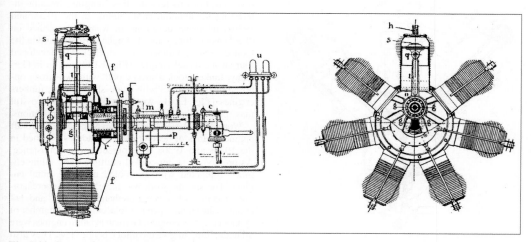

Fig. E92 Gnome Omega rotary engine (France, 1908)

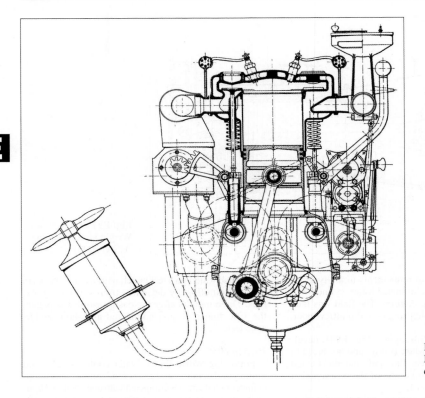

Fig. E93
Maybach CX Zeppelin
engine (Germany, 1913)

became a grave problem in the 1920s, a problem which was later solved by using torsional vibration damper.

As the result of the Treaty of Versailles, Germany did not participate in development of large air-cooled radial

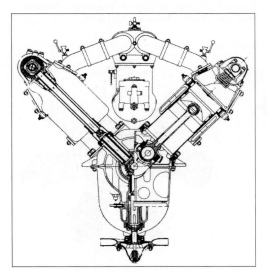

Fig. E94 Hispano-Suiza engine (France, 1915)

engines such as were produced by Bristol in England and Pratt & Whitney and Wright in the United States. This deficiency was made up by obtaining licenses for production of foreign engines. BMW built the Pratt & Whitney Wasp, Siemens the Bristol-Jupiter (**Fig. E99**).

Although diesel engines are heavier than corresponding gasoline engines, if the fuel required for long-distance travel is also considered, this relationship is reversed. For this reason, diesel engines became interesting for aviation applications. In the 1920s and 1930s, intensive development was accomplished on diesel engines for aircraft. In England, Bristol, and in the United States, Guiberson and Packard, developed four-stroke diesel engines (**Fig. E100**), while in Germany, Junkers took a special path with two-stroke opposed piston engines with a two-shaft design. Junkers engines (**Fig. E101**) were the only truly successful mass-produced diesel engines for aviation applications.

World War II also hastened development of aircraft engines. Lines of development already established in the 1920s and 1930s were pursued with persistence. As previously, water-cooled multiple bank inline engines competed with air-cooled multiple radial engines. The specific work was increased by supercharging, injection of a water-methanol mixture, and hot cooling; the engine speed, displacement, and number of cylinders were increased. German aviation engines were all underslung to give the pilot a better view. They also operated with fuel injection, which offered advantages

E

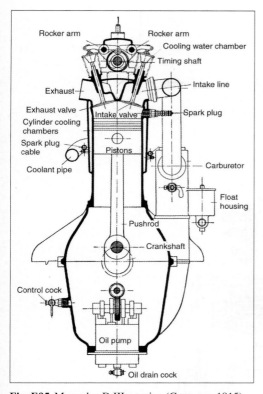

Fig. E95 Mercedes-D IIIa engine (Germany, 1915)

Labels in Fig. E95:
Rocker arm — Rocker arm — Cooling water chamber — Timing shaft — Intake line — Exhaust — Exhaust valve — Intake valve — Spark plug — Cylinder cooling chambers — Spark plug cable — Pistons — Coolant pipe — Carburetor — Float housing — Pushrod — Crankshaft — Control cock — Oil pump — Oil drain cock

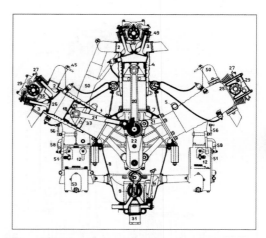

Fig. E97 12-cylinder W-engine, Lorraine design (France, 1926)

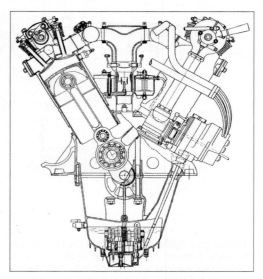

Fig. E98 12-cylinder V-engine, BMW design (Germany, 1929)

in extreme flying positions. A typical representative of this design is the Daimler-Benz DB 601 — a standard engine used by the German air force (**Fig. E102**).

The English Rolls-Royce Merlin — a product of long, complicated, evolutionary development — was considered to be the best aviation engine during World War II (**Fig. E103**). It was built in many versions into the 1950s with a total production of over 168,000. Installed in aircraft such as the Vickers-Supermarine Spitfire fighter and Lancaster bomber, it contributed significantly to English air supremacy.

A masterpiece of British design was the 24-cylinder H-engine Napier-Sabre I with slide valve timing (**Fig.**

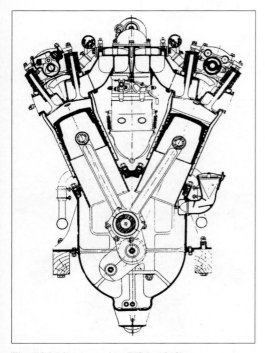

Fig. E96 Liberty engine (USA, 1918)

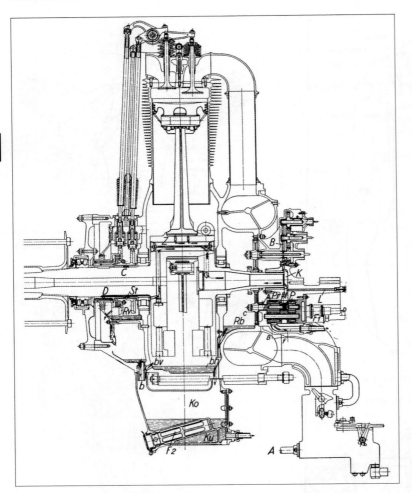

Fig. E99
Bristol-Jupiter
nine-cylinder radial
engine (England/
France, 1929)

E104); it gave fighter planes such as the Hawker Typhoon and Tempest a speed capable of allowing their pilot to keep up with the Vl flying bombs. The American air-cooled Pratt & Whitney R 2800 radial engine (**Fig. E105**) was used for military and civilian applications up to the end of the piston engine era. In this engine, the mixture was formed in the supercharger; moreover, the engine also operated with water-methanol injection in civilian applications.

~~Diesel engines, large. The diesel engine was a result of the attempt of Rudolf Diesel to create a rational combustion engine with significantly better efficiency than the steam engine. The breakthrough was achieved in 1897 with the third test engine built and improved by MAN: it had an output of 13.1 kW (17.8 PS), and the brake thermal efficiency was 26.2%. Due to lack of any other possibilities, the fuel was "blown" into the cylinder with highly compressed air, which required a multistage compressor with intermediate cooling.

Fig. E100 Packard aviation diesel engine (USA, 1928)

The compressor was heavy, expensive, and prone to malfunctioning; moreover, it required 10–15% of the engine power. Nevertheless, diesel engines were accepted astonishingly quickly for marine purposes: first for river and inland navigation, then for maritime ships, and not only due to their good efficiency. The performance and speed range corresponded to the requirements of nautical operation; it was possible to drive the ship propeller directly, which, however, required reversibility. The output could be regulated to meet the instantaneous requirements. The entire machine required less space than an equivalent steam engine because it had no boiler or condensers—thus fewer personnel—no heaters and trimmers. Operation with liquid fuel facilitated storing and trimming; the fuel supply, stored in double-wall tanks, did not take up room required for cargo.

In contrast to steam engines, internal combustion engines were ready to start immediately at any time and could also be shut off even for short stops. All this made diesel engines ideal for marine applications. In 1911/12, the Danish shipyard and machine factory Burmeister & Wain built the first ocean-going motor ship, the MS *Selandia*, with two 919 kW (1250 hp) four-stroke crosshead diesel engines (**Fig. E106**) developed by the company itself.

Its use in ships had consequences for the design of the diesel engine. Due to the possible deformation of the hull of the ship, it was necessary to replace the single-stand design taken over from the steam engine with a continuous, stiff crankcase. At that time, it was not yet clear whether the four- or two-stroke cycle would be better suited for marine engines. Two-stroke

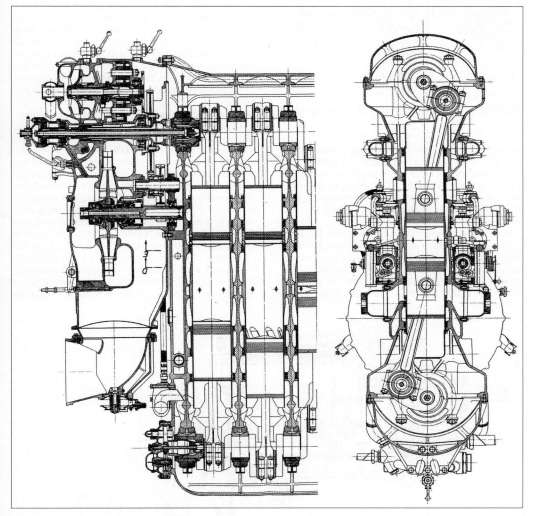

Fig. E101 Junkers aviation diesel engine Jumo 205 (Germany, 1941)

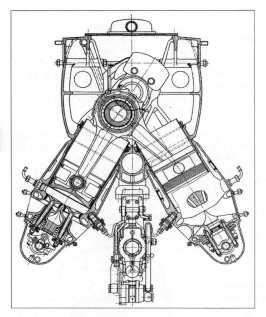

Fig. E102 Daimler-Benz DB 601 A/B (Germany, 1939)

engines offered higher output from a unit—about 75–90% more than comparable four-stroke engines—and control of the exhaust process by the piston, which eliminated the difficulties of thermally overloaded exhaust valves. For this reason, the two-stroke cycle quickly became favored in the higher output range.

Navies also recognized the advantages of the diesel engine. In 1909, the German Naval Office commissioned the MAN Nuremberg plant, which was experienced in the construction of large gas engines, to develop a six-cylinder diesel engine with an output of 8823 kW (12,000 PS) (**Fig. E107**) at a time when it

was just possible to obtain 735 kW (1000 hp) from a single engine.

The diesel engine obtained its real significance in application in submarines. On the one hand, high power output, and on the other, a space-saving, rigid design was required. For this reason, the engines were conceived with a plunger piston design with a strong continuous crankcase. At Vickers Ltd. in England, technical director James McKenchie was the first to achieve direct injection of liquid fuel into the combustion chamber ("compressorless injection"), and he introduced this application in English submarine engines (**Fig. E108**). In Germany, the fuel continued to be blown in by air. With respect to submarine engines, the Germania shipyards (Krupp) concentrated on two-stroke engines, while MAN used the four-stroke process (**Fig. E109**).

Engine development was pushed on by World War I. Power outputs increased considerably, and, although this was obtained at a high price, valuable operating experience was collected under extreme conditions. Basic knowledge was gained in all areas of engine technology; one example is knowledge about the torsional vibration characteristics of combustion engines.

It is no wonder that the diesel engine continued to gain ground in marine applications during the 1920s, while four-cycle and two-cycle engines continued to compete.

To achieve the specific power output of two-cycle engines, four-cycle engines were built so that they were double acting (**Fig. E110**). Because it was also possible to design two-cycle engines as double-acting engines, double-acting four-stroke engines did not survive in the long run. The transition to compressor-less injection proceeded for large diesel engines during the first half of the 1920s. The primary differentiating characteristic of the individual two-stroke engine designs is the scavenging, called "uniflow," loop, and cross-flow scavenging. It was necessary to "pay for" the technical advantages of uniflow scavenging in terms of the design: either with single-shaft opposed piston engines (Doxford), control of the exhaust with a control piston (Burmeister & Wain; Cooper-Bessemer, Fairbanks-

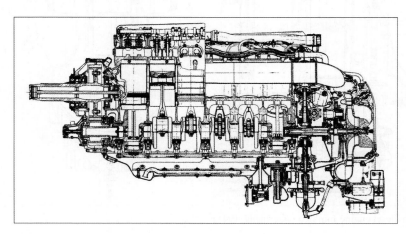

Fig. E103
Rolls-Royce Merlin
(England, 1939)

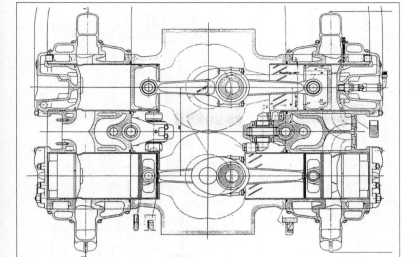

Fig. E104
Napier-Sabre II
(England, 1943)

Morse), or with exhaust valves (Burmeister & Wain). For this reason, the simpler cross-flow scavenging mechanism (Sulzer, Krupp, Fiat) and loop scavenging (MAN) were preferred.

Due to Allied regulations, it was only possible for the German empire to build warships of very limited size ("pocket battleships"), which forced designers to achieve high engine outputs with low engine mass and

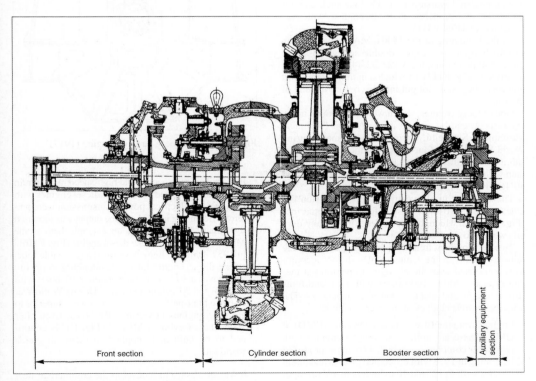

Front section Cylinder section Booster section Auxiliary equipment section

Fig. E105 Pratt & Whitney R 2800 (USA, 1943–1955)

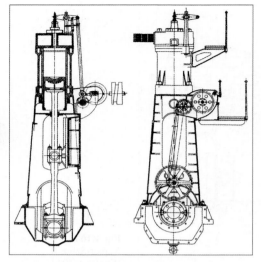

Fig. E106 B&W engine from MS *Selandia* (Denmark, 1912)

minimum installation space. For this purpose, MAN developed special double-acting two-stroke engines. The armored ship *Deutschland* was the first purely engine-powered warship in 1933; four such engines drove the two propeller shafts through a "volcano" transmission (**Fig. E111**).

At the beginning of the 1940s, MAN developed an extremely light, 24-cylinder, double-acting two-stroke diesel V-type engine, the V12Z 32/44, for propelling destroyers (**Fig. E112**), of which a number were built; however, they were not put into use before the end of the war.

One of these engines, which were remarkable in every respect, can be viewed at the Auto + Technical Museum in Sinsheim, Germany. In Germany during World War II, only four-stroke engines were used as submarine engines, first as naturally aspirated engines, then with exhaust turbocharging (Krupp F 46; MAN V 40/46) (**Fig. E113**).

The US navy used two-stroke opposed piston engines with a two-shaft design for their submarines (**Fig. E114**), in which the newly developed heavy-duty oils were also developed.

Two development trends can be distinguished in the railroad sector. In the United States, heavy locomotives were used with diesel engines operating at medium speeds (700 to 1000 rpm), with two- and four-stroke design and electric power transmission. The principal representatives of this trend include:

- Engines from the Electromotive Division (EMD) of GMC. These are uniflow scavenged engines with two Roots compressors, which were mass-produced beginning in 1938 with oil-cooled pistons and pump nozzles with 6, 12, and 16 cylinders in a V arrangement (**Fig. E115**).

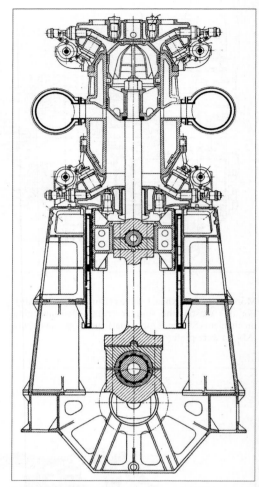

Fig. E107 Nuremberg large oil engine (1917)

- In Germany, rail coaches and high-speed rail car train sets with high-speed four-stroke diesel engines with hydraulic or electrical power transmission were preferred. Beginning in 1934, engine output was increased by exhaust turbocharging; an example here is the Maybach GO6 four-stroke diesel engine (**Fig. E116**).
- In 1937, Daimler-Benz developed a reversible four-stroke diesel engine for the Hindenburg Zeppelin, which was used in a marine version in PT boats with 12-, 16-, and 20-cylinder engines. During World War II, these high-performance engines were superior to gasoline engines in German PT boats. Even after 1945, the 20-cylinder MB518 (**Fig. E117**) continued to be built and supplied to many shipyards worldwide.

It was necessary to redevelop the commercial fleet destroyed during the war, so there was a large demand

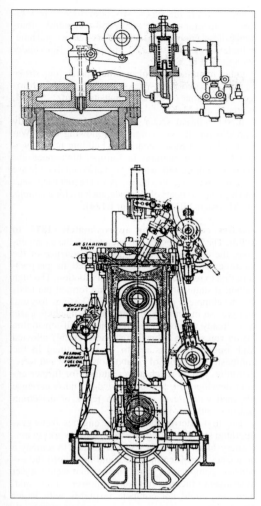

Fig. E108 Vickers compressorless submarine diesel engine (England, 1914)

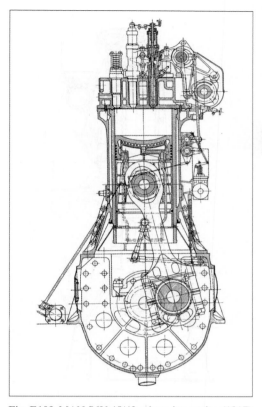

Fig. E109 MAN S6V 45/42 submarine engine (1917)

for marine engines, which was addressed to an increasing extent by slow-running, single-acting, two-stroke diesel engines. The diesel engine continued to displace piston-type steam engines as well as steam turbines in the medium output range, particularly because it was possible to convert the two-stroke engines for use with heavy oil.

Double-acting two-stroke engines offer high outputs per unit (75–80% more than conventional engines), without the engine mass and dimensions increasing to the same extent. However, their weak points include the piston rod packing gland, whose sensitivity to malfunction and damage drive up operating costs. Exhaust turbocharging provided a means of obtaining high output with fewer problems, so that construction of double-acting engines was discontinued by the mid-

1950s. However, it was not simple to supercharge two-stroke engines, because of the higher specific airflow rates compared with four-stroke engines and because of the lower exhaust temperatures, so that the bottoms of the pistons also were adapted to provide the charge.

To investigate the possibilities of exhaust turbo-charging, MAN constructed a test engine as a four-stroke crosshead engine at the end of the 1940s; this was designed especially for the stresses encountered with high pressure supercharging (**Fig. E118**). The supercharging was 170%, and a minimum fuel consumption of 190 g/kWh was measured.

The increasing award of licenses forced manufacturers of large engines to take the production possibilities suggested by their licensees into consideration in the design. Because the latter were capable of casting high quality large castings only to a limited extent, the crankcase was designed as a welded construction; this also reduces the mass. An early large engine with welded crankcase was the Sulzer RDS (**Fig. E119**).

Doxford, an English company, started building opposed-piston engines (**Fig. E120**) with a Junkers license in 1913 and continued to produce large two-stroke marine diesel engines with single-shaft opposed-piston design up until the 1970s.

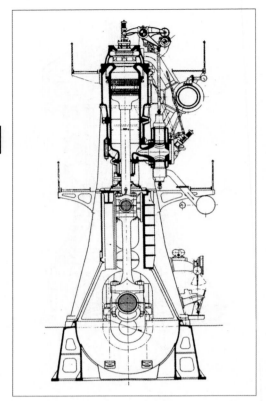

Fig. E110 Double-acting four-stroke engine, design Werkspoor (Holland, 1925)

The size (water displacement) of the ships increased continuously throughout this period, and tankers, in particular, were constructed as continually larger units making it necessary to increase the engine outputs because single machine systems—that is, single-engine systems with direct propeller drive—were preferred. In 1967, MAN produced a large engine, the KZ 105/180, with an output per cylinder of 2941 kW (4000 hp) (**Fig. E121**).

With the PC 2 (**Fig. E122**), the designer Gustav Pielstick created a medium-speed, four-stroke diesel engine in France during the 1950s based on the V40/46 submarine engine that he had developed while at MAN. This engine set standards for this class of engines.

PC engines—further developed to PC3 and PC4—covered a large portion of the medium-speed engine market in the 1960s and 1970s.

Four-stroke engines gained a great deal of ground in marine construction; medium-speed, turbocharged, and—heavy oil-compatible—four-stroke engines penetrated into performance ranges previously reserved for large slow-running engines. The low overall height and the capability of application as multiple engine systems were primary advantages for ferries and pas-

senger ships. In the 1960s, MAN developed a new series of medium-speed, heavy oil-compatible four-stroke diesel engines consisting of six-, eight-, and nine-cylinder inline and 10-, 12-, 14-, 16-, and 18-cylinder V-engines (**Fig. E123**).

Other manufacturers of medium-speed, four-stroke engines included Stork-Werkspoor, Sulzer, GMT, Mirrlees Blackstone, Allen, Daihatsu, Deutz, MWM, Wärtsilä, and MaK.

In the railroad sector, the trend that had already started before the war continued following World War II. The United States continued building heavy two- and four-stroke engines; in Europe, high-speed engines were preferred, which were also suitable for use in ships, working machines, power generating sets, and high-speed boats of all types, such as, for example, the Maybach-MD engine (**Fig. E124**).

~~**First engines (from approximately 1875 to 1900).** The basic design was dictated by the steam engine; the crank drive controlled the sequence of the thermodynamic processes and converted the gas pressure first to reciprocating, then rotary motion. The high technical status of the steam engine formed the basis for development of other engines. Casting, forging, and precise machining of complicated machine parts was already mastered. The one-piece, self-tensioning piston rings from John Ramsbottom (1854) allowed high operating pressures to be maintained in the combustion chamber of combustion motors and were, therefore, just as much a prerequisite for further engine development as the knowledge and experience obtained with steam engines in terms of drivetrain bearings and their lubrication.

The first internal combustion engines were gas-operated stationary engines for driving working machinery of all types; they were intended to remedy a deficiency. Small companies could not afford the expensive, complicated steam engines that were subject to stringent official regulations as power sources and, therefore, could not compete favorably with larger companies.

This resulted in development of such drives at various points. Nikolaus August Otto successfully introduced his engine based on the four-process cycle already described by the Frenchman Beau de Rochas (**Fig. E125**). The decisive advantage of this compared to the gas engines designed by the Frenchman Jean Joseph Etienne Lenoir was the preliminary compression of the mixture. The British engineer Dougald Clerk "shortened" the four-stroke cycle to obtain the two-stroke cycle by eliminating the strokes for exchanging the charge (**Fig. E126**).

Gottlieb Daimler, Wilhelm Maybach and Karl Benz produced lightweight, high-speed engines which provided the basis for motor vehicles operating without rails (**Figs. E127 and E128**).

Diesel's "rational thermal engine," (**Fig. E129**) was also used initially for stationary applications; naturally, this also applied for its predecessors, the Brayton (**Fig. E130**) and Akroyd engines (**Fig. E131**).

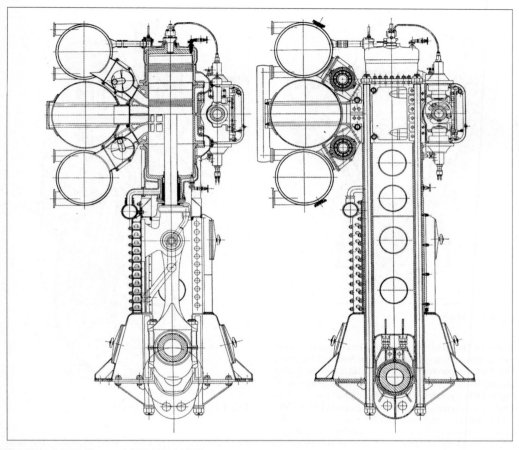

Fig. E111 MAN (Source: MTZ 42/46 1933)

~~Stationary engines. The stationary gas engine was the "engine for small companies." City or coal gas served as fuel; where this was not available, so-called Dowson gas, produced by passing water and air over a glowing bed of coal, could also be used. The resulting mixture of carbon monoxide, hydrogen, and nitrogen was captured and fed to the engine after purification in a scrubber. There was considerable demand for such a small stationary engine, so that by 1900 the Deutz Co. offered an entire series with versions from 4.4 kW to 10.3 kW (6–14 hp) (**Fig. E132**). These were all horizontal single-cylinder engines with a large flywheel to provide the required uniformity of speed.

The energy present in the large quantities of blast furnace gas available from the steel industry led to the development of large horizontal engines produced by Oechelshäuser (single-shaft opposed piston engines), by Ehrhard & Sehmer, the Elsass machine construction company, John Cockerill (Belgium), and the MAN plant in Nuremberg (**Fig. E133**).

The experience gained with such large machines was later used for diesel engines. While initially stationary gasoline as well as diesel engines provided the basis for development of motor vehicle engines, the trend now reversed: "light-weight engines" derived from motor vehicle engines and with reduced speed and output saw increasing use as stationary engines. Large engines were used as marine as well as stationary engines. In 1926, Blohm & Voß built the largest diesel engine (MAN version) at that time to cover the peak load at the Neuhof power plant owned by the Hamburg electric works. This engine consisted of a double-acting two-stroke diesel engine with an output of 11,029 kW (15,000 hp) (**Fig. E134**).

Small stationary engines were all gasoline engines, because diesel engines could not be constructed for small outputs at that time due to the compressor required for air injection. A solution was offered by the glow-head engines whose particular advantage was their simplicity: any blacksmith could repair them. Production of small diesel engines in the range of 5.9 to 2.9 kW (8 to 4 hp) began approximately in the middle

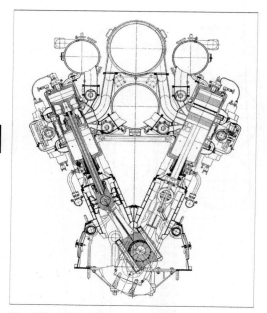

Fig. E112 MAN V12Z 32/44

of the 1930s. Stationary "small diesel engines" were
built as two- or four-stroke upright or horizontal engines.
The engines operated with indirect injection. At the beginning of the 1940s, Hatz introduced a cross-flow scavenged twin-cylinder, two-stroke diesel engine with
crankcase scavenging and direct injection (**Fig. E135**).

Gas engines—stationary engines, per se—were designed as spark ignition or self-ignition-type gasoline
engines in many versions beginning in the 1930s: in
Europe this was for reasons of independence and in the
United States to utilize natural gas as a cheap energy
source. The American Nordberg Co. produced large
11-cylinder radial engines with vertical crankshafts as
dual engines for (natural) gas or diesel operation during the 1940s and 1950s (**Fig. E136**). These engines
were used for driving generators.

~~Vehicle engines

1900 to 1918

The most difficult problem on the early engines was
ignition of the charge.

The nonelectrical ignition processes—flame ignition
(Otto) and uncontrolled hot tube ignition (Maybach/
Daimler)—represented a hurdle for engine development which was first overcome with the electrical ignition processes: low voltage magneto ignition (Otto),
trembler ignition (Benz), and finally, high voltage
magneto ignition (Bosch).

The next problem was to improve the mixture formation in terms of quality and quantity. Only highly
volatile constituents in the gasoline (final boiling point
approximately 100°C) could be used with wick, sur-

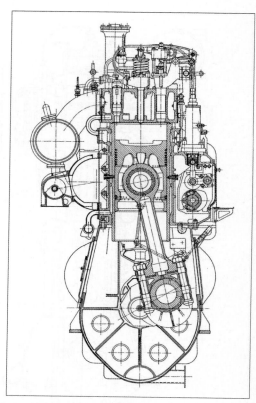

Fig. E113 Krupp F 46 (1942)

face, and brush-type carburetors; moreover, the individual fuel constituents did not evaporate simultaneously. With the injection nozzle carburetor from
Wilhelm Maybach, the fuel was no longer "vaporized,"
but rather atomized. This made it possible to use heavy
gasoline (final boiling point around 200°C) which
could be provided in nearly any desired quantity—a
primary prerequisite for further increases in performance. Carburetors with automatic auxiliary air regulation from Krebs, Claudel (Zenith), and Menesson
and Goudard (Solex) improved the operating characteristics of the engines and reduced the fuel consumption.
With increasing output, it was also necessary to dissipate
more heat in the cooling water—made possible by a
honeycomb-type radiator from Wilhelm Maybach.

Once these engine prerequisites were established,
motor vehicles began to develop rapidly. More and
more companies started producing motor vehicles and
engines. In order to increase the performance and improve the smoothness of running, the number of cylinders was increased—from one to two and then to four
as with the Mercedes-Simplex engine (**Fig. E137**).

In other countries—France, Italy, England, and the
United States—production of motor vehicles and engines was also under way, initially still on the basis of

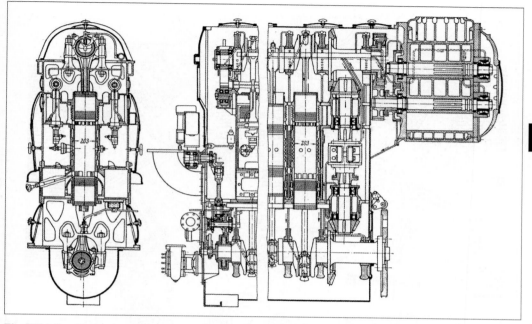

Fig. E114 Two-stroke opposed-piston engine, Fairbanks-Morse 38 D (USA, 1941)

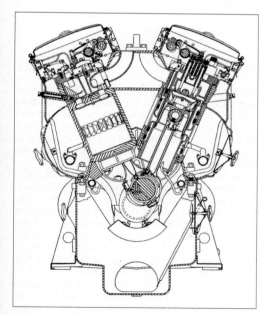

Fig. E115 Locomotive engine GMC 567 (USA, 1938)

the German pioneers; however, new designs quickly appeared, such as the Panhard & Levassor engine (**Fig. E138**) and the Napier 15hp engine (**Fig. E139**).

Of particular note are the twin-cylinder Gobron-Brilliee engines (**Fig. E140**) with single-shaft opposed piston design (four pistons). A Gobron-Brilliee racing car was successful in the Gordon-Bennett races.

In addition to passenger cars, commercial vehicles were also in production (**Fig. E141**).

Six-cylinder engines were developed from four-cylinder engines. Self-actuated (lifted by the cylinder vacuum) inlet valves were abandoned in favor of mechanically actuated valves. Nevertheless, the design, layout, and production of the valve control mechanisms were far from perfect, so that the Knight valve control was superior in terms of smoothness.

Knight valve engines (**Fig. E142**) were built in England by Daimler Co., in Belgium by Minerva, and in Germany by the Daimler engine corporation.

The shift in passenger cars from a recreational pleasure for the rich to a basic commodity was occurring in the United States even before World War I. In 1909, Henry Ford started production of the Model T ("Tin Lizzie"); 15 million of these vehicles were produced by 1927. **Fig. E143** shows the engine of the Ford Model T.

1918 to 1945

In the 1920s, combustion knocking became the criterion which limited the performance of gasoline engines. In 1921, Midgley and Boyd in the United States discovered the efficacy of tetra ethyl lead (TEL) as an antiknock agent.

Moreover, it was known that certain fuels such as benzene were less sensitive to knocking than standard commercial gasoline. In England, Harry Ricardo rec-

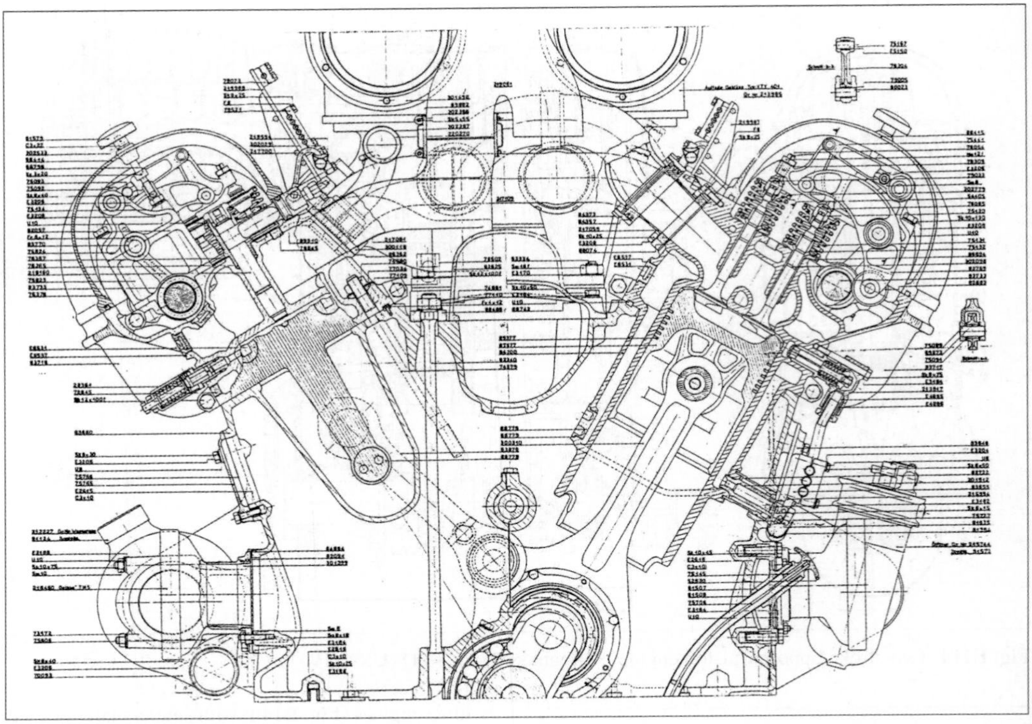

E

Fig. E116 Maybach GO 6 four-stroke diesel engine with exhaust turbocharger (1935)

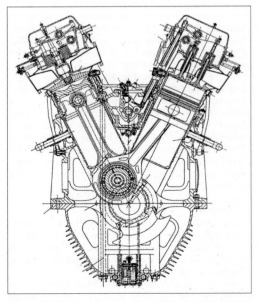

Fig. E117 PT boat diesel engine MB 518 (1938–1975)

ognized that engines with compact combustion chambers, such as those with overhead valves, had less tendency to knock (aviation engines commonly had overhead valves); nevertheless, flat-head engines with valves in the block (side-valve engines) continued to be produced for passenger cars up to the middle of the 1950s for reasons of cost.

The transition from gray cast iron to light alloy pistons allowed higher compression ratios and, therefore, higher performance and lower fuel consumption. The steel/lead/bronze bearings developed by the American aviation engine manufacturer Allison provided engine bearings capable of withstanding high loads. At the Heinrich Lanz AG Agricultural Machine Factory, A. Diefenthaler and K. Sipp discovered that pearlite iron was an excellent material for the cylinders.

In the relatively affluent United States, people could afford vehicles with powerful engines and fuel costs played no particular role. For this reason, large-displacement six- and eight-cylinder inline engines dominated (**Fig. E144**) and 12- and even 16-cylinder V-engines were also built, all as water-cooled four-stroke gasoline engines with comparatively high compression ratios (due to the TEL additive in the fuel) and low operating speeds. Six- and eight-cylinder engines were equipped with vibration dampers.

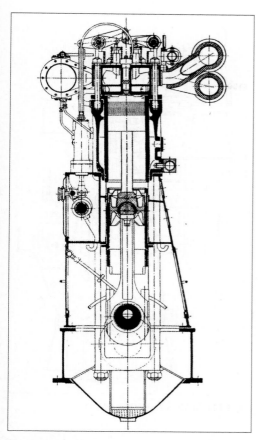

Fig. E118 MAN test engine K6V 30/45 (1951)

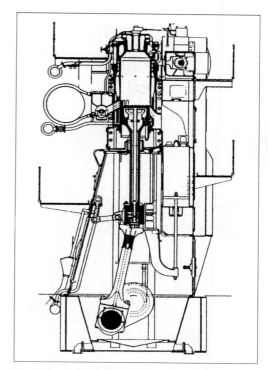

Fig. E119 Sulzer RSD 75 engine (Switzerland, 1953)

In Europe, the development was not uniform. Large passenger car engines (**Fig. E145**) were built for large, luxurious vehicles, such as Mercedes, Maybach, Horch, Stöhr, and so on; however, compact and miniature cars such as the Opel (Laubfrosch) and the compact car from Hanomag known as the "bread loaf" (**Fig. E146**)—due to its unusual appearance—predominated for sale to the middle class.

Modern engine concepts began to be adopted in automotive construction in the 1930s. The V-8 90° engines (**Fig. E147**) increasingly replaced inline eight-cylinder engines, although the latter continued to be produced in the United States up to the end of the 1940s. Engines with a short stroke allowed higher engine speeds; compact combustion chambers and anti-knock control led to improvements in mixture formation, combustion, and the charge process (**Fig. E148**).

The crankshaft had bearings next to each crank throw. Even in Germany, engines were now equipped with air filters. Higher speeds, better roads, and in Germany particularly the new autobahns (freeways) required larger radiators and, in many cases, oil coolers as well.

Two-stroke engines.

The possibilities for this operating cycle were studied. Two arguments militated for the two-stroke principle, which were mutually exclusive in the end: high specific power output on the one hand, and simple design on the other hand. Two-stroke engines with crankcase scavenging—that is, without valves—offered advantages for motorcycles and small passenger cars. The Schnürle reverse scavenging process from DKW (**Fig. E149**) was a decisive advance compared to cross-flow scavenging, because it allowed better scavenging of the cylinder and also made it possible to replace the thermally highly stressed wet pistons with flat pistons.

The high power density made the two-stroke cycle attractive for automotive diesel engines. However, at their (relatively) high speeds, only the complex design using uniflow scavenging provided prospects of success. In Germany, Junkers had developed single-shaft two-stroke opposed piston engines (**Fig. E150**), in which the bottom piston controlled the exhaust port and the upper piston the intake port. The intake and scavenging pistons were consolidated in terms of design.

These engines were produced under license by Krupp in two-, three-, and four-cylinder versions.

In the United States, General Motors Corporation (GMC) developed a series of uniflow scavenged two-stroke engines with intake ports and exhaust valves, Roots supercharger, and inline pump nozzles consisting

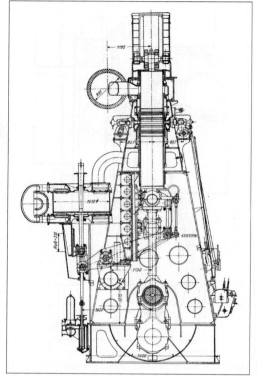

Fig. E120 Two-stroke opposed piston engine Doxford 76 J (1964)

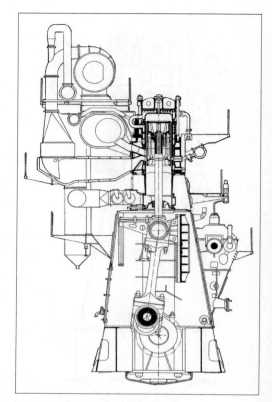

Fig. E121 MAN KZ 105/180 (1965)

of three-, four-, and six-cylinder engines (**Fig. E151**). Later, another small and another large series were added. The GMC two-stroke engines were developed with a purpose and at great expense to become one of the most successful engine series ever; they were produced by Detroit Diesel into the 1990s.

A different type of two-stroke diesel engine was the glow-head engine, such as that produced by Heinrich Lanz AG (**Fig. E152**) in Mannheim for propulsion of agricultural tractors (Lanz-Bulldog) and road tractors. These were low-compression ratio engines ($\varepsilon = 6{:}1$), into which the fuel was injected at relatively low pressure (30–40 bar) and processed in the red glowing auxiliary chamber ("glow-head," semi-diesel engines). Glow-head engines are simple and insensitive to fuel. Used particularly in agricultural applications as tractor engines, these engines were the backbone of German agriculture during the war and the deficit period thereafter.

Fuel injection using compressed air ("air injection") was a hindrance for use of diesel engines in motor vehicles. At the beginning of the 1920s, intensive work was accomplished by various parties on "compressorless" injection. Based on the preliminary work before and during World War I (L'Orange, Leissner), motor

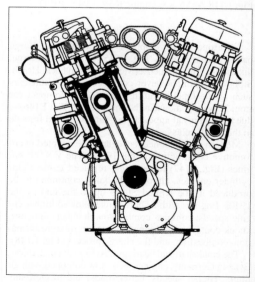

Fig. E122 Four-stroke diesel engine Pielstick PC 2 (France, 1961)

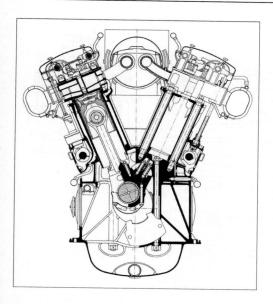

◀ **Fig. E123** Four-stroke diesel engine MAN V 40/54 (1968)

vehicle diesel engines without compressors were developed in Germany by MAN (**Fig. E153**), Benz (**Fig. E154**) (later Mercedes-Benz), and Junkers.

Based on the Acro patents, the Robert Bosch Co. developed complete injection systems for automotive diesel engines. The injection pumps had port-and-helix control and spill regulation. Indirect injection (prechamber and swirl chamber, air buffer) was preferred because of the difficulties involved with direct injection on motor vehicle engines with their large speed range. The principles for modern direct injection were established by the Swiss company, Adolph Saurer, at the beginning of the 1930s with the double swirl combustion engine (**Fig. E155**).

The first commercial vehicle diesel engines failed after a short service life as a result of wear to the ring

▼ **Fig. E124** Maybach MD engine (1950 to present)

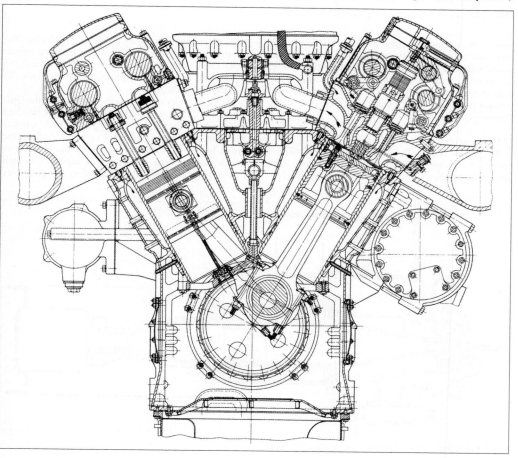

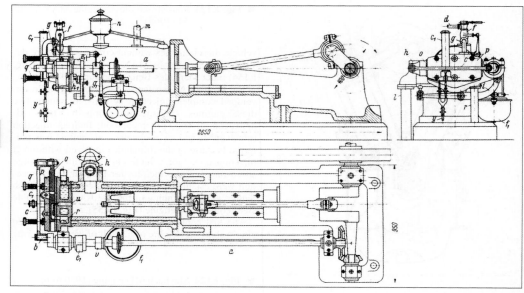

Fig. E125 Otto's four-stroke engine (Germany, 1876)

grooves in the aluminum pistons. Ernst Mahle was the first to provide a remedy with ring carrier pistons. The diesel engine (**Fig. E156**) proved itself well in heavy commercial vehicles and was used increasingly in light commercial vehicles and finally even in passenger cars (Mercedes-Benz, Hanomag, Oberhänsli, Colt, Cummins, etc.).

Military engines

World War II interrupted the development of engines for civilian applications. Engines which had proved themselves in civilian vehicles were used in large quantities for military commercial vehicles. The simply designed six-cylinder, inline gasoline engine from GMC was installed in over one half million 2.5-t trucks and helped to provide supplies to the Allied troops on all fronts.

On the German side, it was the engine from the Opel Admiral which powered the Opel "Blitz," the standard truck used by the German military (**Fig. E157**). It was necessary for tank engines to be compact, because the armor has a high influence on the vehicle mass. With the German tank engines—gasoline engines with high power densities—this was achieved by using a roller

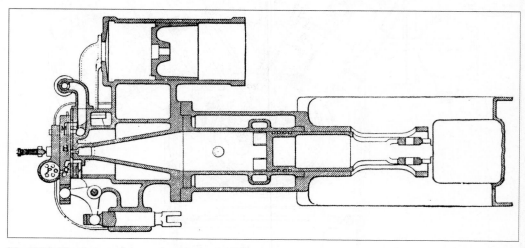

Fig. E126 Clerk's two-stroke engine (England, 1885)

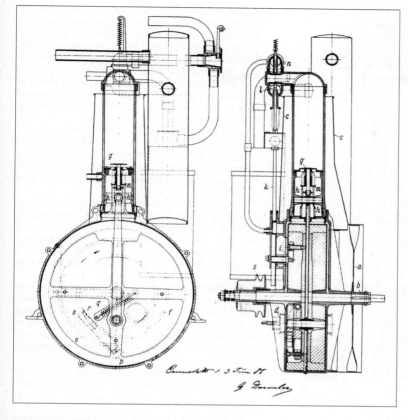

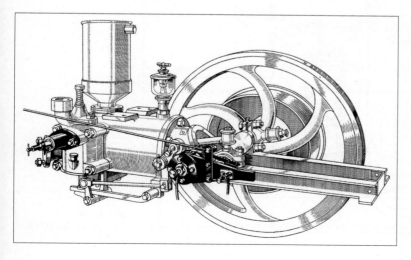

Fig. E127
Daimler's 1/2 hp motor
cycle engine (Germany,
1885)

bearing sectional crankshaft with short cylinder offset
(**Fig. E158**).

The Soviet Union was the only nation involved in the
war to use a high-performance diesel engine—type W2—
in its T14 and heavy KW tanks (**Fig. E159**). This engine
continued to be advanced and produced into the 1980s.

Air-cooled engines

Air cooling was significantly more difficult for
motor vehicle engines than for aviation engines due to
the low vehicle speed and unfavorable operating conditions for cooling. Nevertheless, attempts were
made to cool motor vehicle engines directly with air. A

Fig. E128
Benz's first automobile
engine (Germany, 1885)

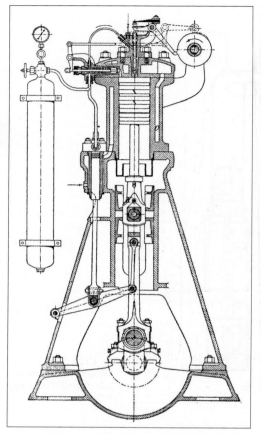

Fig. E129 Diesel's third engine (Germany, 1897)

pioneer in this area was the American company Franklin Mfg. Co., which produced an air-cooled, six-cylinder inline engine even before World War I (**Fig. E160**). The flywheel was designed as a fan with drum impeller, and it sucked the cooling air through the engine. The gray cast iron cylinder had 56 steel fins arranged parallel to the cylinder axis. Later, the fan was placed in front of the engine and the cooling fins were arranged in a radial pattern; the cooling air was pushed through the engine.

Air-cooled motor vehicle engines were developed and produced during the 1920s and 1930s in Europe as well, by Krupp and Phaenomen (**Fig. E161**), for commercial vehicles, for passenger vehicles from Tatra, and by Ferdinand Porsche for the new Volkswagen.

The air-cooled horizontally opposed "boxer" engine (**Fig. E162**) from Volkswagen—first used in off-road and amphibian vehicles, and later in the "Beetle"—became a synonym for reliability and sturdy design.

Up to today, the VW Beetle retains the world production record for passenger cars.

The advantages of air cooling induced the military weapons office to commission a development association under the auspices of the Deutz Co. to develop an air-cooled commercial vehicle diesel engine in 1942. After the war, this engine (**Fig. E163**), was used in commercial vehicles, tractors, and construction machinery.

Similar considerations were the reason the American army used air-cooled engines for their medium-heavy tanks, first as gasoline and later as diesel engines.

The Continental Co. built V-12 engines (**Fig. E164**) intended as large air-cooled engines for agricultural vehicles.

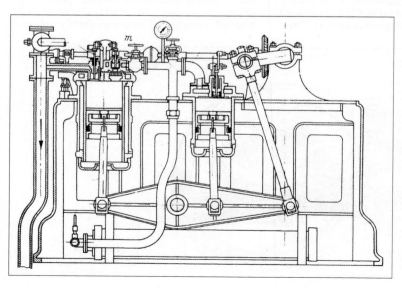

Fig. E130
Brayton engine (USA, 1872)

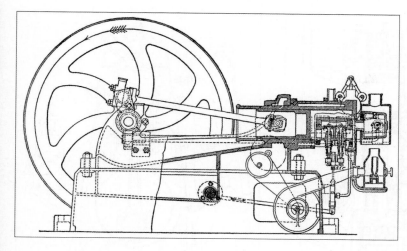

Fig. E131
Hornsby-Akroyd engine
(England, 1892)

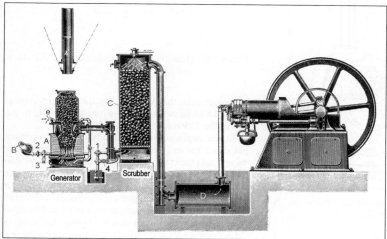

Fig. E132
Stationary small
gas engine, Deutz
version (Germany,
circa 1900)

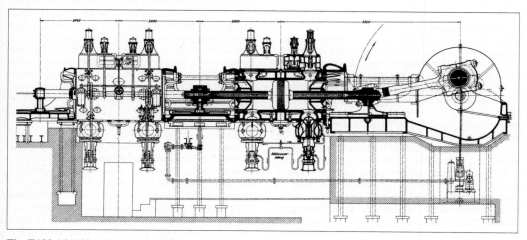

Fig. E133 MAN large gas engine (Germany, 1910)

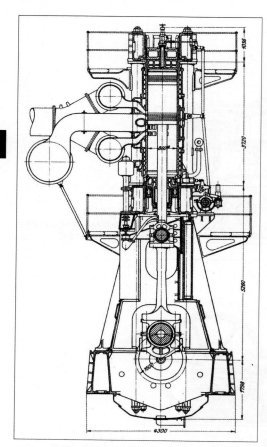

Fig. E134 Double-acting two-stroke engine MAN D9Z 86/150 (Germany, 1926)

After 1945

Passenger car engines

Initially, production of the prewar engines was resumed. New developments began production at the beginning of the 1950s. In the United States, large volumes of engines were built, most as V-8 engines (**Fig. E165**).

In Europe, a large number of compact and miniature cars were driven with air- and water-cooled, two- and four-stroke engines: to mention a few from Germany: Gutbrod, Lloyd, Goliath, and DKW. In order to avoid the disadvantageous scavenging loss of fuel with two-stroke gasoline engines, the Gutbrod and Goliath engines (**Fig. E166**) were equipped with a gasoline injection system.

Some four-stroke engines were equipped with a gasoline injection system, particularly to achieve increased performance; one example is the 3-liter engine from Mercedes-Benz. In the course of the "economic miracle," the demand for compact cars decreased and with it, the use of two-stroke engines in passenger cars decreased; in East Germany, however, two types of

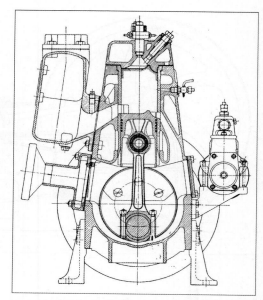

Fig. E135 Hatz two-stroke diesel engine A2 (Germany, 1941)

passenger cars, the Wartburg and the Trabant, continued to be equipped with two-stroke engines up to the end of the 1980s (**Fig. E167**).

At the beginning of the 1950s, many four-stroke passenger car engines still had in-block valves (side valves), and the crankshafts had bearings after every two throws. However, this began to change, and new engines were designed according to the state of the art with higher displacements, higher engine speeds, and higher performance. Mercedes-Benz again participated successfully in racing; the Silver Arrow engines had positive valve control (desmodromic control) and a gasoline injection system derived from aviation engines (**Fig. E168**).

In view of the obvious disadvantages of the reciprocating piston engine, attempts were made to design other engine configurations. Of the many such attempts, such as swash plate engines, cam plate engines, and angular plate engines, only the rotary piston engine (**Fig. E169**), Wankel version, was successful enough to reach mass production. However, it was not successful commercially due to its unfavorable combustion chamber shape.

Commercial vehicle engines.

Influenced by the success of the GMC two-stroke engine, many companies in Europe designed two-stroke diesel engines, some with uniflow scavenging (Commer, Krupp) (**Fig. E170**), others with loop scavenging (Krauss-Maffei, Ford/Köln, Hanomag, Alfa-Romeo, Foden, and Gräf & Stift); however, none of these engines was successful. Production of two-stroke commercial vehicle diesel engines was abandoned in Europe by the end of the 1950s and the beginning of the

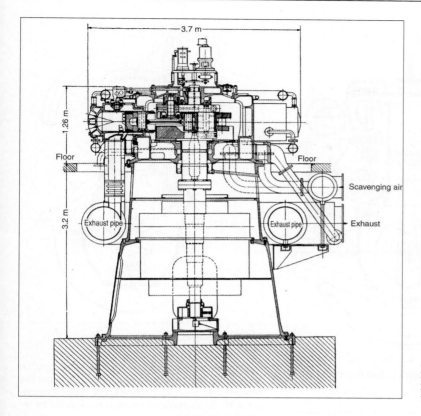

Fig. E136
Nordberg Radial
Engine (USA, 1950)

1960s. From the current point of view, MAN was on the "right path" in 1952 when it introduced a turbo-charged, four-stroke diesel engine (D 1246 GT).

Exhaust turbocharging became popular only later in Germany because in Europe traffic conditions require frequent load—that is, gear—changes, an operation that still presented problems for exhaust gas turbo-charging at that time. In contrast, in the United States, with its great distances and other conditions favorable for smooth layout of engines and transmissions, exhaust gas turbocharging increased rapidly.

At the beginning of the 1960s, commercial vehicle diesel engines were changed over to direct injection for fuel economy reasons.

On the one hand, the difficulty of mixture formation over the wide speed range for the commercial vehicle engines had been mastered and, on the other hand, the piston problems, such as thermal overload resulting from the compact combustion jet from the precham-ber, became more and more pressing.

~**Racing engines.** Racing engines are internal com-bustion engines for a special application with spe-cific outputs of up to 500 kW/L. In principle, they do not differ from mass-produced motor vehicle en-gines. In detail, the following global differences can be listed:

- Higher engine speeds and greater speed range, with continuous operation at speeds of over 19,000 rpm—for example, Formula 1—are made possible prima-rily by the use of pneumatic valve actuators (air springs) (**Fig. E171**).
- Least possible throttling in the intake path.
- Power optimized components such as intake mani-fold, exhaust manifold, and exhaust system.
- Large valve lift and four valves per cylinder—the valve material is usually titanium.
- Improved cooling of cylinder head.
- Dry sump lubrication to cope with extreme accelerations.
- On gasoline engines, highly knock-resistant com-bustion chambers requiring relatively small valve angles and moderate compression ratios and pistons with crown as even as possible.
- Due to higher thermal and mechanical load, adapta-tion of structure, materials, and connection ele-ments (bolts) to meet the increased requirements.
- Lowest possible mass for components (e.g., tita-nium, carbon fiber-reinforced plastics).
- Service life adapted to racing operation, special measures for quality assurance of parts installed (individual testing).

~~**American Le Mans series/Le Mans.** In the Amer-ican Le Mans series (ALMS), the same technical

E

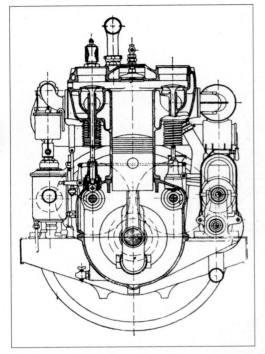

Fig. E137 Mercedes-Simplex engine (Germany, 1902)

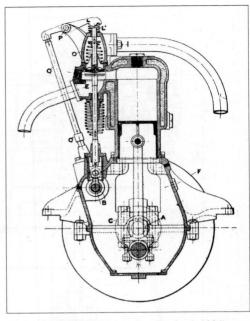

Fig. E139 Napier 15-hp engine (England, 1904)

regulations of the Automobile Club de l'Ouest (ACO) apply as for the 24-hour Le Mans race. The races last between 2:45 hours and 12 hours. The IMSA (International Motor Sports Association) is responsible for organizing the racing.

~~~**Engine.** Naturally aspirated engines with a maximum displacement of six liters and turbocharged engines up to a maximum displacement of four liters are allowed in the LMP 900 class (**Figs. E172–E174**). The number of cylinders is not regulated. The maximum width is 2 meters, the maximum length 4.65 meters, and the tank capacity is limited to 90 liters. Only one type of fuel is permitted (Euro Premium 98).

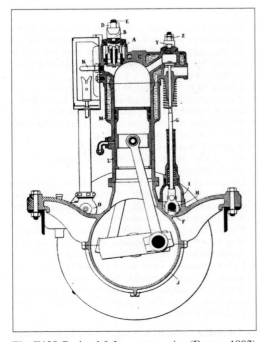

Fig. E138 Panhard & Levassor engine (France, 1902)

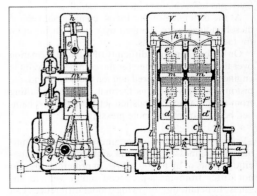

Fig. E140 Gobron-Brilliee engine (France, 1902)

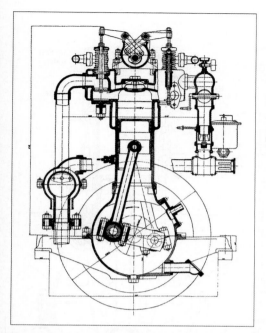

Fig. E141 Büssing commercial vehicle engine (Germany, 1903)

To reduce the speeds at Le Mans and the ALMS, smaller air quantity restrictors were generally specified for all vehicles for the 2003 season. On the V-8, twin-turbo Audi engine, the air intake in front of the two turbochargers was reduced from the previous 32.4 mm to a diameter of 30.7 mm. This reduces the engine output by nearly 10%. Instead of the 610 PS produced previously, the 3.6-liter, V-8, twin-turbo engine now has an output of approximately 550 hp. The advantages of the R8 engine are the low fuel consumption and good "drivability." Both result from the direct injection system used by Audi since the middle of the 2001 season.

~~~**Regulations.** A Le Mans prototype (LMP/LM GTP) (**Figs. E175–E177**) is a racing car for which a minimum production volume is not required, provided that it

- Satisfies the FIA safety regulations and
- Fulfills the technical regulations published by the Automobile Club de L'Quest (ACO)

A Le Mans prototype (LMP) is an open vehicle with a minimum weight of 900 kg. A Le Mans GT prototype (LM GTP) is a closed vehicle with a minimum weight of 900 kg.

The Audi R8 is an example of the type of car belonging in the category of large Le Mans prototypes (LMP 900), the "top class" for sports cars. This category allows engineers a great deal of design freedom. A variety of options are available, especially for the engine.

These range from small turbocharged engines to large displacement naturally aspirated engines. Air quantity restrictors with different diameters and, for turbocharged engines, different maximum surcharging pressures ensure an "equal" opportunity for all competitors.

The Le Mans prototypes differentiate between closed coupes (LM GTP), such as the Bentley EXP Speed 8, and open roadsters, such as the Audi R8 (LMP 900). As a matter of principle, the same regulations apply for both—with minor differences. For example, the open Le Mans prototypes are allowed to start with wider, larger tires; however, the engine is subject to more restrictions and, therefore, has slightly less output than the coupes.

In addition to the "top class," the LMP 900, there is also an LMP 675 version. Technically, the two classes are nearly identical. The numbers in the designation refer to the different minimum weights: 900 kg for the LMP 900 class, 675 kg for the smaller LMP 675 version, which also has smaller engines.

~~**Championship Auto Racing Teams (CART).** This is a formula racing series in the United States. The vehicles are single-seat—so-called monopostos—free wheels, rear engine with displacement of 2.6 liters and an open cockpit. The V-8 engine has an output of more than 800 hp, is equipped with a turbocharger, and operates with methanol. The vehicles achieve top speeds of 400 km/h. The races are run on ovals as well as on street courses.

The races have been held by the American Automobile Association (AAA) since the beginning of the twentieth century. As the accident figures increased dramatically in 1955, AAA disassociated itself from the series for image reasons and in 1978 transferred all rights to the USAC (United States Auto Club) as the umbrella organization.

The similarities between the champ car racing cars and current Formula 1 cars is of a purely optical nature. The American series differs significantly in terms of its culture as well as its financial and technical structure.

~~~**Regulations.** The standard drive consists of a 90-degree, V-8 Cosworth XFE (**Figs. E178–E180**), which has four camshafts and a displacement of 2650 cm^3. None of the systems or components on the Cosworth engines can be modified or replaced without written approval from Cosworth. The engine speed was limited to 12,000 rpm for the 2003 season. CART reserves the right to continue to adapt this limit for safety reasons. All engine adjustments must be made by the driver onboard or during a box stop, and telemetry between the car and pits is prohibited.

Exhaust system. The exhaust manifolds must have an outer diameter of 5.08 cm (2 inches) and a length of 50.8 cm (20 inches). Variable geometry exhaust systems are prohibited. The only material allowed is steel with a minimum wall thickness of 0.12 cm (0.049 inches). The exhaust systems must be designed so that accidents involving fire are excluded to

E

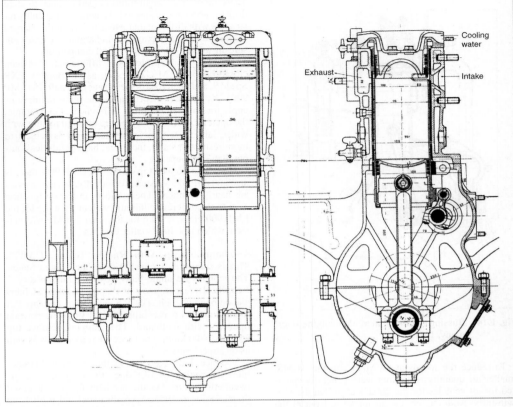

Cooling
water

Exhaust

Intake

Fig. **E142** Daimler engine with Knight valve control (England, 1910)

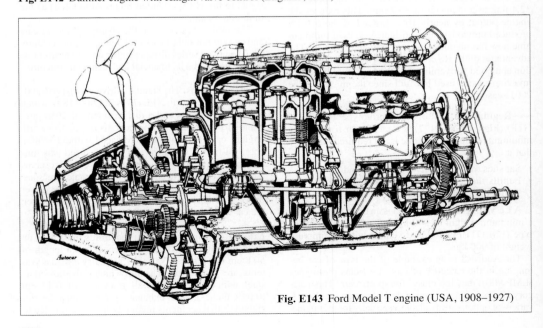

Autocar

Fig. **E143** Ford Model T engine (USA, 1908–1927)

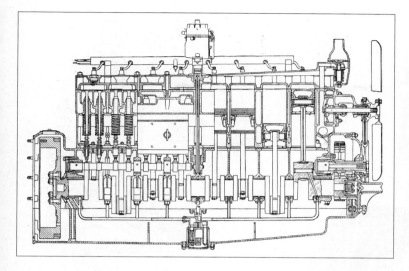

Fig. E144
Eight-cylinder inline
Packard engine
(USA, 1923)

the greatest extent possible and hazards are avoided for other competitors. If the exhaust system and diffuser are consolidated, the end of the exhaust may extend 1.27 cm (0.5 inches) beyond the end of the diffuser. It is only permissible to use the air filter supplied by the manufacturer. Modifications are not permitted.

Material. Moving masses may not consist of carbon, ceramic, or metal matrix material. The use of materials such as ceramic, metal matrix, or composite materials or of engine parts other than those mentioned above must be approved by the technical director of the series.

Titanium aluminide, lithium-aluminum, beryllium, and aluminum-beryllium are not permissible for engine components. Beryllium-copper is allowed. The

use of ceramic, metal matrix, and composite materials is prohibited unless otherwise decided by CART.

Oil and cooling system. A dry sump system is prescribed for oil supply. The oil and cooling system must have a catch tank with a minimum capacity of 3.785 liters (4 quarts). Only the engine oil approved by Cosworth may be used.

Accessories. The turbocharger supplied by Cosworth may not be modified or replaced. Installation of a waste gate valve controlled pneumatically by the intake pressure is prescribed. It is not permissible to actuate this valve hydraulically or electrically. Charge air intercoolers or similar devices are prohibited.

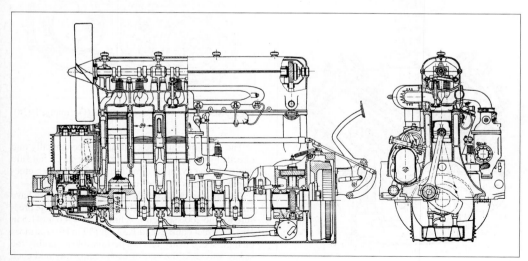

Fig. E145 Mercedes compressor engine (Germany, 1928)

E

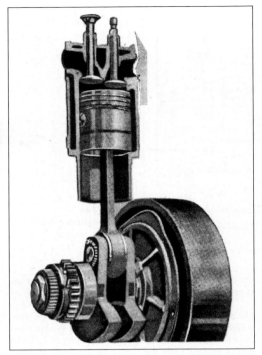

Fig. E146 Engine for the Hanomag car, called *Komissbrot* or (pan loaf) (Germany, 1926)

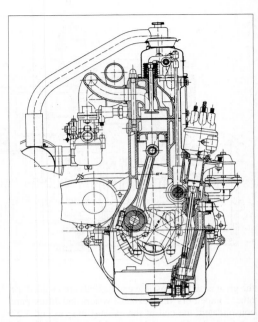

Fig. E148 Head controlled BMW engine (Germany, 1938)

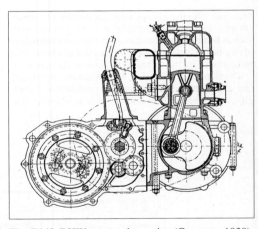

Fig. E149 DKW two-stroke engine (Germany, 1938)

The gas flow must be controlled mechanically and directly by the driver's foot. The gas must have a fail-safe device in the event of a malfunction at "open"; electronic control is not permissible.

Racing cars with turbocharged engines must have an approved safety valve. The permissible charge pressure for the 2003 season was limited to 1.4 bar (41.5 in Hg) for street courses and 1.32 bar (39 in Hg) for races on high-speed ovals. It is not permissible to modify the safety valve. At the end of each racing day and each racing event, the valves are collected by the event

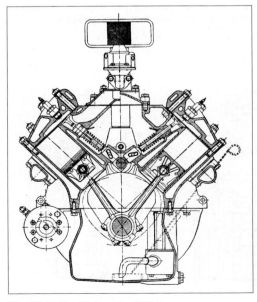

Fig. E147 Ford V-8 engine (USA)

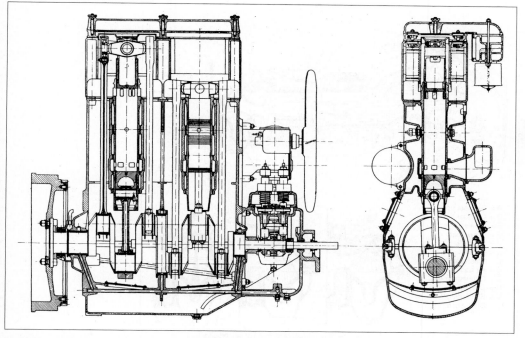

Fig. E150 Junkers opposed piston engine HK 80 (Germany, 1928)

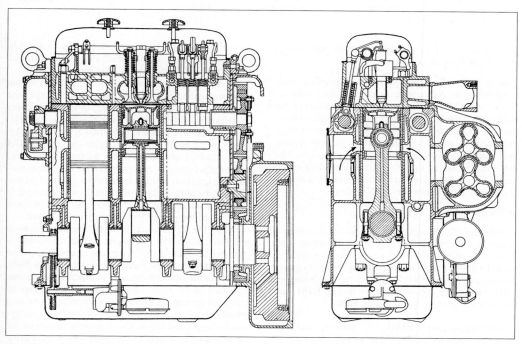

Fig. E151 GMC 71 (USA, 1938)

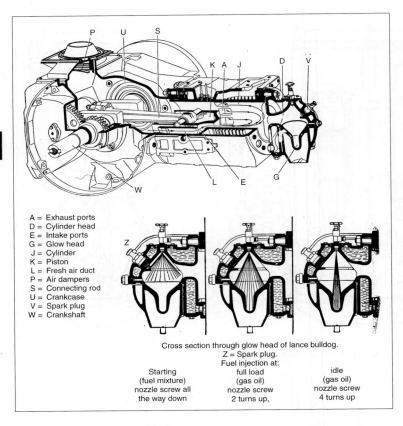

A = Exhaust ports
D = Cylinder head
E = Intake ports
G = Glow head
J = Cylinder
K = Piston
L = Fresh air duct
P = Air dampers
S = Connecting rod
U = Crankcase
V = Spark plug
W = Crankshaft

Cross section through glow head of lance bulldog.
Z = Spark plug.
Fuel injection at:

Starting (fuel mixture) nozzle screw all the way down	full load (gas oil) nozzle screw 2 turns up,	idle (gas oil) nozzle screw 4 turns up

Fig. E152
Tractor "Lanz-Bulldog" engine HR 5 (Germany, 1928–1935)

steward, checked for conformity with the regulations, and kept until the next event.

State of the art. The 2003 season also specified Ford-Cosworth as the only engine supplier for the CART series with its XFE V-8 engine. The new version of the successful XF engine now carries the official designation Ford-Cosworth XFE V-8. A number of design modifications were required to satisfy the new regulation changes in terms of output and service life. The 2003 XFE is laid out for 700–750 hp, depending on the route. This performance must be guaranteed over a distance of 1931 kilometers (1200 miles).

~~Formula 1

~~~Design.
The design of the engine for next season begins in January and February of the preceding year. This means that 20 of the 220 total employees working on an F1 project in a Formula 1 team begin designing the engine for the 2003 season at the beginning of 2002. The engine control is shown in **Fig. E181**, and the technical data are given in **Fig. E182**.

~~~Engine, example BMW.
The primary development objectives for an F1 engine are maximum power output, stability, low specific fuel consumption, low overall weight, good drivability, and quick throttle response with a service life which still could be limited to the distance of one race in 2003 (now changed).

In terms of internal combustion engines, this means maximizing the work through maximum charging of the cylinders in the charging process, minimizing losses from friction, and optimizing thermal efficiency in combustion. The time factor is taken into consideration by achieving the highest possible engine speed. **Fig. E183** shows the F1 engine from BMW for the 2002 season.

Formula 1 engines are optimized for maximum output. The question of acceptable torque is based primarily on drivability. When sufficient torque is present to accelerate out of slow curves, this facilitates optimum metering of the power development for the driver.

During the 2002 season, the BMW engine, designated internally as the P82, set impressive standards with a peak engine speed of 19,050 rpm and nearly 900 hp. The speed and output were increased even further in 2003. In addition to further increases in the performance, one objective of the team was to optimize the center of gravity of the engine. Reduction of the weight was not a primary development target; the distribution of weight was the important factor. Every gram is important for the reduction of weight in the upper area of the engine block, while further reduction of the weight at the bottom of the engine block does not provide any advantages. A portion of the vehicle ballast is already located there.

~~~**Engines from previous year.** A further advantage of the planned changes in the regulations is that independent teams will no longer have to use engines from the previous year. In 2004, it will not be possible to use an engine from the previous year, even though it only had to withstand one race in 2003. The same applies for 2005 and 2006. As of 2006, the number of engines required for racing and tests will already be so small that no significant efforts will be required for a large team or engine manufacturer to supply current engines to an independent team. The same applies to chassis and drivetrain parts.

~~~**Quality control.** A racing engine has nearly 5000 individual parts, and of those, 1000 different parts must be checked before assembling a racing engine in an average of 80 working hours. Although stability is decisive, preservation is not important for Formula 1. In this high-tech sport, the goal is to achieve the maximum that is physically achievable.

The limit can only be established when one is prepared to also exceed the limit.

~~~**Regulations.** An excerpt from the FIA regulations for Formula 1 engines specifies that only four-stroke piston engines are permitted. The displacement may not exceed 3000 cubic centimeters. Supercharging is prohibited. The engines must have 10 cylinders, each cylinder must be round. The engines may not have more than five valves per cylinder.

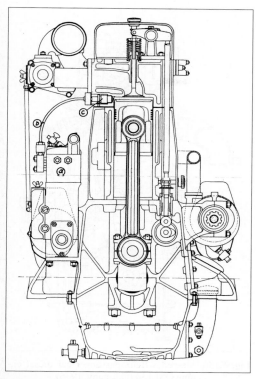

Fig. E153 MAN W4V 11/18 (Germany, 1925)

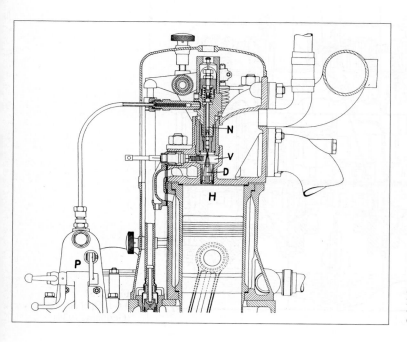

Fig. E154
Mercedes-Benz OM 5
(Germany, 1926)

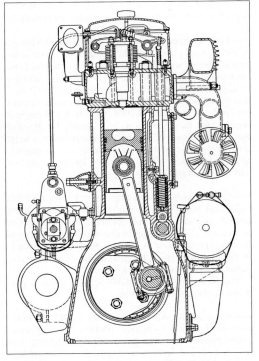

Fig. E155 Saurer twin swirl engine (Switzerland, 1934)

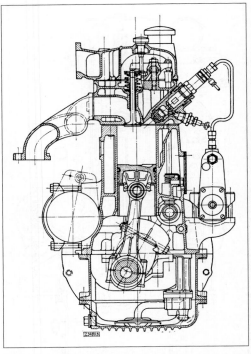

Fig. E156 Mercedes-Benz passenger car diesel engine OM 138 (Germany, 1936)

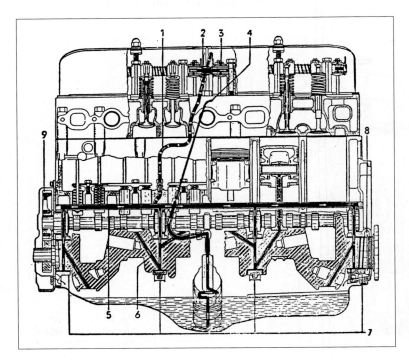

Fig. E157
3.6-L Opel engine
(Germany, 1939)

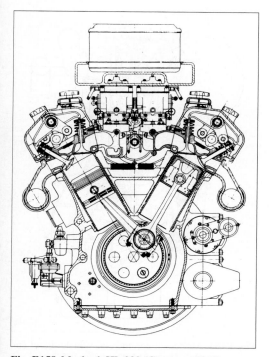

Fig. E158 Maybach HL 230 (Germany, 1943)

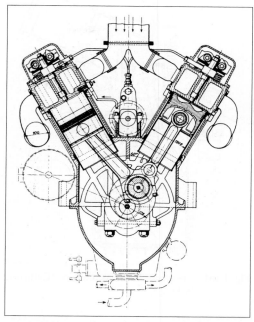

Fig. E159 Soviet tank diesel engine W2 (USSR, 1941)

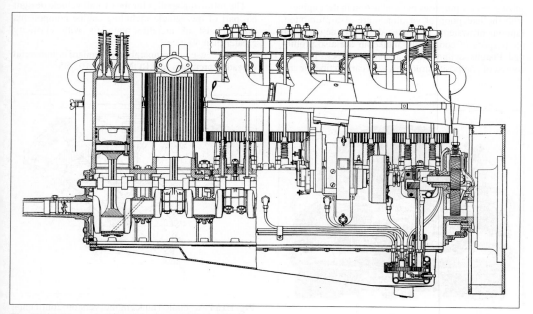

Fig. E160 Air-cooled Franklin engine (USA, 1914)

E

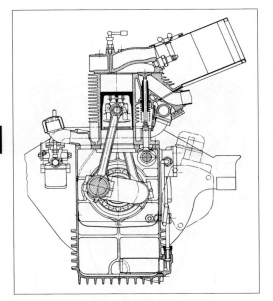

Fig. E161 Air-cooled Phaenomen engine (Germany, 1931)

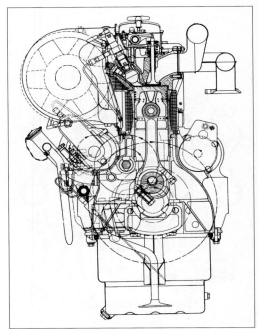

Fig. E163 Air-cooled commercial vehicle diesel engine, Deutz 514 (Germany, 1952)

Drives other than the 3-liter, four-stroke and 10-cylinder engine described above are not permitted. The grand total of renewable energy and energy stored in the vehicle may not exceed 300 kJ.

Variable exhaust systems, injection of water or other substances for purposes of combustion in the engine—with the exception of gasoline—are also prohibited. Equipment, systems, processes, designs or devices with the purpose of cooling the intake air are also prohibited. Pistons, cylinder head and engine blocks may not consist of composite structures with carbon fiber or ara-

mide reinforcements. The basic structure of the crankshaft and camshaft must be steel or cast iron.

The following applies for the overall vehicle design: No part of the vehicle including engine components may consist of metallic materials with elasticity greater than 40 g/cm^3.

The fuel used is comparable to standard commercial unleaded premium fuel RON/MON: 102/85

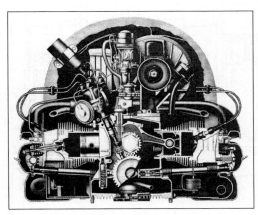

Fig. E162 VW "Beetle" engine (Germany, 1961)

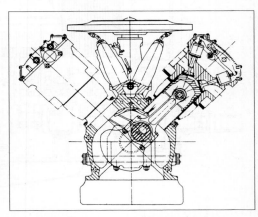

Fig. E164 Air cooled tank engine, version Continental (USA, 1957)

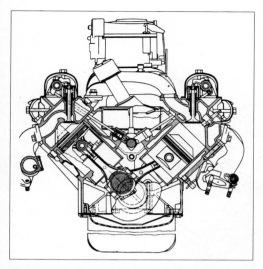

Fig. E165 Buick V-8 engine (USA, 1955)

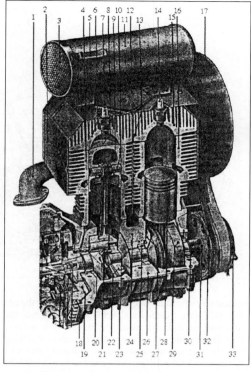

Fig. E167 Two-stroke engine P 60 (Germany, to the end of 1980s)

~~~~**Planned changes to regulations as of 2004.**
From the viewpoint of the year 2003, the engine is the greatest possible accessibility for independent teams. Assuming a realistic price of approximately 20 M (US) for the engine equipment for an independent Formula 1 team in 2003, this figure should be less than $10 million (US) in 2004, because then a team will only require one engine per GP weekend and vehicle, instead of two or three as in 2003. For the 2004 season, it will only be permissible to use one engine for the entire Grand-Prix weekend according to the applicable regulations. This increases the required service life from previously 400 to 800 km.

If the recommendations made to the teams by the FIA are accepted, the costs should be closer to $5 million (US) from 2004 on, because the engines will then

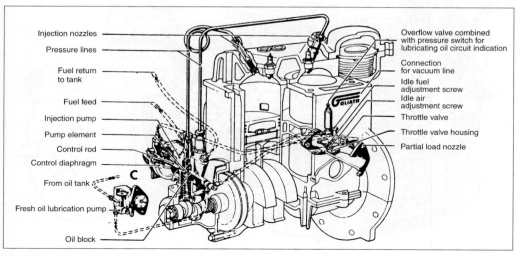

Injection nozzles
Pressure lines
Fuel return to tank
Fuel feed
Injection pump
Pump element
Control rod
Control diaphragm
From oil tank
Fresh oil lubrication pump
Oil block

Overflow valve combined with pressure switch for lubricating oil circuit indication
Connection for vacuum line
Idle fuel adjustment screw
Idle air adjustment screw
Throttle valve
Throttle valve housing
Partial load nozzle

Fig. E166 Goliath two-stroke gasoline engine with fuel injection (1960s)

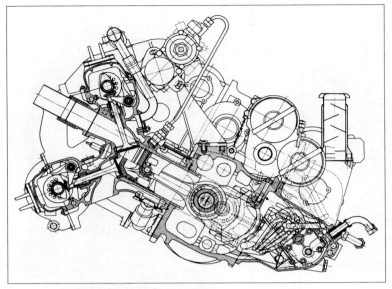

Fig. E168
Mercedes-Benz racing
engine (Germany, 1954)

be used for two racing weekends. From 2006 on, the engines will have to stand six racing weekends, and the costs will be reduced to an estimated $1.6 million (US). During the next three years, the costs resulting from the engines alone would be reduced from $20 per team to $1.6 million (US) per team or $100,000

(US) per race. This recommendation should provide new impulses for Formula 1 racing and provide a realistic cost basis for private teams.

Comparatively speaking, if the costs for the design and development of an F1 engine are not considered, an engine supplier could supply engines for the entire

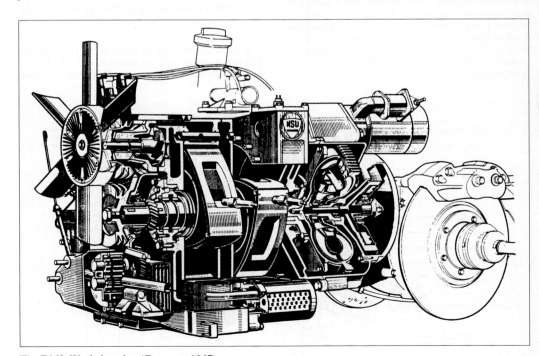

Fig. E169 Wankel engine (Germany, 1967)

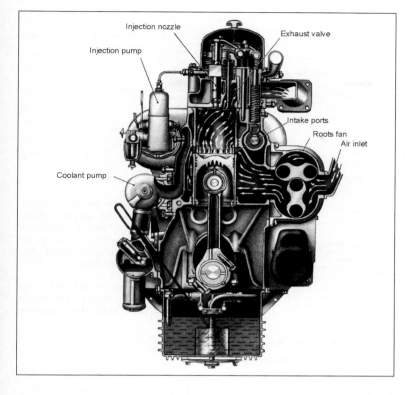

Fig. E170
Krupp two-stroke diesel
engine D 688 (SW 6)
(Germany, 1958)

Formula 1 field in 2006 at the same cost as required to equip a single F1 team with high performance engines in 2003.

In the opinion of the experts, engines designed for six racing weekends will still be capable of providing over 700 hp at more than 16,000 rpm. This performance corresponds to that of the best engines during the 1996 season. It is expected that the development costs for the longer-life engines will only be marginally higher than for the current "instant engines." A further positive point would be that the new technical challenges would be significantly closer to the basic business of engine manufacturers than designing an engine capable of surviving only 400 km or, in extreme cases, a qualification engine good only for 50 km.

Materials. The costs for F1 engines can be reduced even further by prohibiting additional exotic and expensive materials. This also changes the technical

Fig. E171 Pneumatic valve actuation (Source: Indra)

Fig. E172 Le Mans engine 2002, the successful Audi R8 with direct gasoline injection

Category: LMP 900
Engine: V-8 engine with turbocharging
90-degree cylinder angle
4 valves per cylinder
2 Garrett turbochargers
Air quantity limitation to 2 ~ 30.7 mm in conformance with regulations
Charge pressure limitation to 1.67 bar absolute
Direct gasoline injection FSI
Engine management: Bosch MS 2.9
Engine lubrication: Dry sump, Shell oil
Displacement: 3600 cm^3
Output 550 hp
Torque: more than 700 Nm
Drivetrain: Rear-wheel drive
Clutch: CFK clutch
Transmission: Sequential 6-speed racing transmission, Partner Ricardo
Drive shafts: Constant velocity plunging tripod joint shafts
Differential: Multiple disk limited slip differential
Tank capacity: 90 liters

Fig. E173 Technical data on Audi R8 (2003)

Category: LMP-01
Engine: V-8, Elan Power Products V-8 aluminum engine block and cylinder heads
Valve train: 2 overhead valves per cylinder
Displacement: 6 liters
Output over 600 bhp at 7000 rpm
Torque: more than 678 Nm (500 lb-ft)
Maximum speed: 7250 rpm
Injection system: Elan Cross-Ram gasoline injection
Fuel: ELF Sportsman 100 unleaded

Fig. E174 Panoz LMP 01

challenge for the engine designers, because currently, teams with good financial backing can purchase expensive, exotic materials from aeronautic suppliers or special laboratories, thereby obtaining an edge on the competition. Moreover, this form of technical challenge is much closer to mass production.

~~~**Stability.** While dimensions, weight, and center of gravity are defined in the concept and design phases, the primary focal point during further develop-

Minimum weight : 900 kg							
Normally aspirated engines			Supercharges engines				
Displacem. (cm^3)	Restrictors diameter (mm)		Displacem. (cm^3)	Restrictors diameter (mm)		Boost pressure (mmb)	
	1	2		1	2	2 valve	3+ valve
			3 valve engines and more				
6000	42.5	30.3	4000	43.0	30.7	1700	1500
5500	43.0	30.7	3800	43.0	30.7	1790	1580
5000	43.4	31.0	3600	43.0	30.7	1900	1670
4500	43.9	31.4	3400	43.0	30.7	2010	1770
4000	44.4	31.7	3200	43.0	30.7	2130	1880
3500	44.9	32.0	3000	43.0	30.7	2270	2000
3000	45.3	32.4	2800	43.0	30.7	2440	2150
			2600	43.0	30.7	2630	2310
			2400	43.0	30.7	2840	2500
			2200	43.0	30.7	3100	2730
			2000	43.0	30.7	3410	3000

Fig. E175
Le Mans prototype 900

	Normally aspirated engines	Turbocharged engines
Weight minimum	875 kg	
Displacement max.	3,400 cm^3	2,000 cm^3
1 restrictor	41.7 mm	40.8 mm
Boost maximum	–	2,500 mbar

Fig. E176
Le Mans prototype 675

Minimum weight : 900 kg							
Normally aspirated engines			Supercharges engines				
Displacem. (cm^3)	Restrictors diameter (mm)		Displacem. (cm^3)	Restrictors diameter (mm)		Boost pressure (mmb)	
	1	2		1	2	2 valve	3+ valve
			3 valve engines and more				
6000	44.4	31.7	4000	43.9	31.4	1910	1680
5000	45.1	32.2	3600	43.9	31.4	2130	1870
4000	45.8	32.7	3200	43.9	31.4	2390	2100
3500	46.5	33.2	2800	43.9	31.4	2730	2400
			2400	43.9	31.4	3180	2800
			2000	43.9	31.4	3820	3360

Fig. E177
Le Mans prototype GTP

Fig. E178 90-degree Ford-Cosworth XFE V-8—
standard drive for CART series

Official designation	Ford-Cosworth XFE
Design	V-engine
Displacement	2.65 L (161.7 in³)
Number of cylinders	8
Valves	32
Output	750 bhp at 1.4 bar charge pressure on road courses
	700 bhp at 1.32 bar on ovals
Torque	475 Nm (350 lb.-ft.) at 11,000 rpm
Maximum speed	12,000 rpm
Service life	1931 km (1200 miles) maximum
Engine block	Aluminum
Crankshaft	Steel
Piston	Aluminum
Engine management	Cosworth Electronics
Lubrication system	Dry sump
Fuel	Methanol

Fig. E180 Technical data on Ford-Cosworth XFE

Fig. E179 A CART engine must have an "engineered" lifetime of just 2000 kilometers.

Fig. E181 Engine control on BMW Formula 1 engine

ment and test work is initially the stability. Here, maximum attention is required particularly for quality control. **Fig. E184** shows a cylinder head mold.

~~~**Telemetry.** This means the wireless transfer of all relevant readings from the racing car to the service park, where they are recorded and analyzed by engineers (**Fig. E185**). It includes the following parameters: engine speed; water temperatures; oil temperatures in the engine, transmission, and differentials; oil pressures; accelerator pedal position and output; fuel consumption; transmission gear and vehicle speed; and all dynamic driving values such as lateral acceleration, acceleration, deceleration, and many more.

~~~**Weight.** An F1 engine has a weight of less than 100 kg.

Design	10-cylinder V naturally aspirated engine
Bank angle	90 degrees
Displacement	2998 cm³
Valves	4 per cylinder
Valve train	Pneumatic
Engine block	Aluminum
Cylinder head	Aluminum
Crankshafts	Steel
Oil system	Dry sump lubrication
Engine control	BMW

Fig. E182 Technical data on BMW P83

Fig. E183 BMW P 82, Formula 1 engine for 2002 season

Fig. E185 Telemetry: engineers monitor the technical values of Formula 1 vehicles with the aid of over 200 sensors.

~~German touring car masters (DTM)

~~~**Engine.** There are no restrictions regarding the engine as long as it meets the following conditions: Only four-stroke gasoline engines are permitted. Hybrid engine systems are prohibited (**Figs. E186 and E187**).

Eight cylinders are prescribed. The maximum displacement is 4000 cm³.

The axis of rotation of the crankshaft must be parallel to the longitudinal axis of the vehicle when viewed from above. Moreover, the engine must be installed so that the middle of the crankshaft is 1075 mm in front of the center of the wheelbase in relation to the longitudinal axis of the vehicle.

The minimum weight of the engine without engine liquids and without exhaust manifold must be 165 kg. The following are also taken into consideration in determining the engine weight: Auxiliary equipment, the complete induction system including air restrictors, the starter, the engine wiring harness, and all parts of the mixture preparation system fastened to the engine.

Only V-8 engines are permitted. Only aluminum alloys may be used as the material for the engine block. The cylinder bank angle must be 90°. The cylinder interval must be at least 102 mm. The bore must be cylindrical and have a diameter of at least 93 mm.

The crankshaft must consist of steel and have certain minimum dimensions. A central power takeoff in the drivetrain is not permissible. One connecting rod, including all fastening elements, must have a minimum weight of 450 grams. One piston, including the piston pin, piston rings, and all mounting parts, must have a minimum weight of 350 grams. The piston pin must consist of steel and have a minimum diameter of 19 mm. The minimum height of the piston rings used must be 1.1 mm.

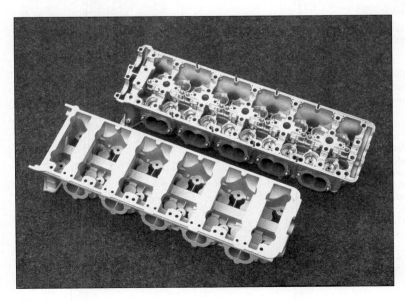

Fig. E184
Mold for F1 engine cylinder head.

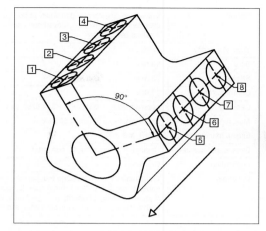

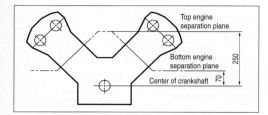

Fig. E186 and E187 Only V-8 engines with 90° bank angle are permitted in the German touring car masters

Cylinder head. Only aluminum alloys may be used as the material for the cylinder head. A maximum of four valves per cylinder is permissible. The valves must be actuated exclusively by means of bucket tappets and coiled springs. All devices and measures which allow variable engine timing and/or variable valve stroke are prohibited. The camshaft may be driven only by chains or toothed belts.

Valves. Valves must be made of steel or titanium. The valve stem diameter must be at least 6 mm over 50% of the entire length of the valve. The valves must not be hollow.

Intake system. With the exception of the throttle valve(s) or throttle slide, every device or measure allowing a variable cross section in the intake tract is prohibited. Only a direct mechanical connection is permitted between the accelerator pedal and engine. Electrical, hydraulic, and pneumatic systems which have an effect on the mixture quantity fed to the engine are prohibited under all circumstances. Any device or measure which allows a variable length in the intake tract is prohibited.

The intake system must be equipped with two air restrictors, each of which may have a maximum diameter of 28 mm over a minimum length of 3 mm. The entire quantity of air sucked in by the engine must flow through these air restrictors. The air restrictor must be produced of metal or a metal alloy.

Supercharging. Supercharging is prohibited. A pressure reservoir in the intake system is prohibited.

Exhaust system. The exhaust system must consist completely of steel. The minimum wall thickness for the pipe material used is 1 mm. Any device or measure allowing a variable length or variable cross section in the exhaust system is prohibited.

The vehicle must be equipped with an emission con-

trol system capable of proper functioning at all times and containing one or more catalysts approved by the German Motor Sport Bund (DMSB) or the Federation Internationale de l'Automobile (FIA). All exhaust gases must flow through the emission control system.

The monolith in the catalytic converter must have the following minimum dimensions:

- Diameter: at least 88.0 mm
- Length: at least 74.5 mm
- Cell density: at least 50 cpsi (cells per square inch)

Noise generation. 120 dB(A), measured at 3800 rpm using the near-field measurement method. It is not permissible to exceed this value. There are restrictions regarding the mixture preparation. Only one injection nozzle per cylinder is permissible. Injection of the fuel into the combustion chamber is prohibited (direct injection).

Ignition system. Only one (injection nozzle) spark plug per cylinder is permissible.

Engine cooling. Coolant pumps must be mounted directly on the engine and driven in a purely mechanical manner by the engine. Radiators must be installed in front of the central axis of the front wheels in the longitudinal vehicle direction. All heat-carrying substances are prohibited, with the exception of ambient air, water, oil, corrosion protection agents, and antifreeze. Radiators and reservoirs for coolants are independent components and must not be a part of the engine.

Lubrication system. Engine oil pumps must be mounted directly on the engine and be driven in a purely mechanical manner by the engine. Coolers for engine oil must be installed in front of the central axis of the front wheels in the longitudinal vehicle direction. It is not permissible for an oil tank to be located in the cockpit. No part of the vehicle containing lubricating oil may be located more than 600 mm from the vertical plane through the longitudinal axis of the vehicle.

Design	V-8 engine, installed longitudinally at front, cylinder bank angle 90°, cylinder interval 102 mm, 4 overhead camshafts, driven by timing chain, 4 valves per cylinder
Engine capacity	3998 cm³
Bore	93 mm
Stroke	73 mm
Compression	13.0:1
Capacity	approximately 340 kW/462 PS at 6750 rpm
Max. torque	approximately 510 Nm at 5250 rpm
Engine management	Bosch MS 2.9
Exhaust purification	Regulated 3-way catalytic converter
Lubrication	Dry sump
Drivetrain	Longitudinal transaxle racing transmissions (DTM unit part; manufacturer: either X-trac or Hewland) with 6 speeds, straight gears, unsynchronized. Sequential shifting pattern, fixed gear ratio, variable overall transmission ratio, carbon fiber clutch, mechanical limited slip differential, rear-wheel drive

Fig. E188
Technical data on Opel
Astra V-8 coupe DTM

Engine. Example: Opel (**Fig. E188**). Under the conditions of the DTM regulations, engineers concentrated particularly on measures to reduce friction when optimizing the DTM engines. Optimization potential is also present in the area of the charging processes as well as in the continuous improvement of the combustion process. Measures for reducing the internal friction are made more difficult by the regulations due to precise specification of the size and weight. Developers have, therefore, concentrated on reducing the friction resulting from the pistons and auxiliary systems.

Generally, the durability of the engines plays a decisive role. The engine is sealed for the entire season and may be opened only in exceptional cases to adjust the valve clearance or to eliminate damage resulting from accidents. The service life for the DTM V-8 engines is greater than 5000 km.

The lubrication system for the Opel V-8 engine consists of a dry sump system with four pumps driven by a common shaft. Each pump supplies one crankshaft chamber.

~~~Regulations. Every competition vehicle must have a stock vehicle as its basis. The stock vehicles must have been produced in a quantity of at least 10,000 identical units for 12 months in succession. In terms of the number of seats, the stock vehicle must be approved by EEC or KBA as a four-passenger vehicle.

The stock vehicle must have a fixed sheet metal roof which cannot be removed, with or without sliding roof.

~~IndyCar. The Indy Racing League (IRL), promoter of the familiar American racing series, also specifies the design appearance of the engine and vehicle in detailed regulations. Considered superficially, the monopostos—that is, single-seat racing cars with uncovered wheel suspension and driver seat on the vehicle longitudinal axis—resemble the Formula 1 cars. However, the technical differences are considerable.

~~~Design features. A current Indy engine as designed for the new Chevy-Indy V-8 for 2003, has a displacement of 3.5 liters, four-valve aluminum cylinder heads, two overhead camshafts per bank, and sequential electronic fuel injection. According to GM specifications, it has an output of more than 675 (SAE) hp and weighs just 127 kg. The costs remain within limits: the Chevy Indy V-8 can be purchased for approximately $120,000 (US).

From the concept phase up to the test bench, the engineers required approximately nine months of development time.

The camshafts on the Chevrolet Indy V-8 are driven by gears; the engine has two injection nozzles per cylinder and exchangeable, wet cylinder liners of light alloy material.

The prescribed cylinder diameter of 93 mm requires a crank throw of 64.4 mm to provide a displacement of 3.5 liters. This stroke leads to a piston speed of 133.65 meters (4385 feet) per minute and a maximum acceleration corresponding to 7000 times the force of gravity at an engine speed of 10,300 rpm. The engines must be laid out to withstand this stress over a 500-mile race, which may require full load over the greatest portion.

Five main bearings are supplied with oil from the oil pan. The engine is a dry deck design using piston rings between cylinder and engine block and O-rings at the top of the cylinder instead of conventional cylinder head gaskets.

Peak output is particularly important for races on high speed ovals. The requirement of maximizing the output of the V-8 engine, assuming a top engine speed of 10,300 rpm determines the timing and valve sizes as well as the intake and exhaust volumes. The cylinder heads in the case of the Chevy Indy V-8 engine consist of an aluminum alloy machined on CNC machines. Four camshafts actuate the 32 titanium valves directly over bucket tappets. The valves are closed by conventional steel springs: pneumatic valve drive is not allowed.

The Indy V-8 engine (**Fig. E189**) has an electronic ignition system. The engine speed is limited to the prescribed 10,300 rpm by a temperature-resistant governor, which is programmed and checked by the IRL inspectors. The governor was introduced to keep the

Displacement	3.5 liters (214 in³)
Output (SAE-PS)	675 hp at 10,300 rpm
Fuel	Methanol
Top engine speed	10,300 rpm (regulation)
Compression ratio	15:1
Piston diameter	93.0 mm (3.66 in)
Crankshaft throw	64.4 mm (2.53 in)
Cylinder angle	90 degrees
Valve train	Dual overhead camshafts
Valves per cylinder	4
Camshaft drive	Gear
Crankcase material	Aluminum
Cylinder liner material	Wet, light alloy
Cylinder head material	Aluminum
Crankshafts	180 degrees
Fuel system	Sequential gasoline injection
Lubrication system	Dry sump

Fig. E189 Technical data on Chevy Indy V-8 engine

Fig. E191 Indy V-8 engine

costs low and the safety at the races at a high level. The majority of the IRL V-8 engines (**Fig. E190**) have four valve heads with titanium-coated valves. In contrast to this, the Chevy SB2 "Small Block," which powers the Monte Carlo team at the NASCAR Winston Cup, as well as the Chevy LS 1 "Small Block" for the C5-Rs Le Mans Corvette have only two valves per cylinder. Generally, a four-valve head provides better charging of the cylinder. Generally, four small valves per cylinder are also lighter than the two large valves in a two-valve engine, making the entire valve train more resistant to high speed. A four-valve head has a considerably higher airflow rate than a two-valve layout. The airflow rate in a Chevy Indy V-8 engine amounts to approximately 28 m³.

~~~**Indy V-8 engine.** State of the art and affordability are the main points in the specifications for the Indy V-8 engine (**Fig. E191**).

~~~**Methanol as fuel.** The IRL engines run on methanol. This fuel was introduced during the 1960s for safety reasons, because a methanol fire can be extinguished with water, while a fire fed by the petroleum base fuels common today requires chemical extinguishing agents. Moreover, methanol allows higher

compression ratios. In order to achieve maximum output figures comparable to those of a gasoline-driven internal combustion engine, methanol, however, requires a fuel/air mixture that is twice as rich.

The octane rating for methanol is normally approximately 10 points higher than gasoline so that this fuel is less susceptible to auto-ignition. Methanol allows higher compression and thereby improves the engine efficiency. Methanol-operated engines run relatively cool, a further decisive advantage for a 500-mile race. This cooling effect increases the air density and thereby the engine output.

The disadvantages of methanol as a fuel are its corrosive characteristics on aluminum, magnesium, and the other materials commonly used in racing engines. Moreover, the alcohol can attack the rubber and other synthetic materials in the fuel system. In order to avoid these negative effects, conservative IRL teams allow the engines to run at the end of the day with regular gasoline until the methanol has been completely displaced in the fuel system. Moreover, methanol has highly toxic properties — as do many other fuels. These risks require extremely careful handling of the fuel.

~~~**Power limitation.** To ensure a sufficient safety level and keep the costs within affordable limits, the maximum output is limited for many race series. This can be accomplished by different means, including displacement limits, limited gasoline feed, and pressure valves on turbochargers. The regulations of the IRL

	2002 Chevy Indy V-8	2003 Chevy Indy V-8	Difference
Engine mass	3.5 liters	3.5 liters	nil
Output (SAE-PS)	675	675	No information
Weight	143 kg	127 kg	−16 kg
Speed (regulation)	10,700 rpm	10,300 rpm	−400 rpm
Injection nozzles	8	16	+8
Camshaft drive	Chain	Gear	
Cylinder liners	Gray cast iron	Wet, light alloy	

Fig. E190
Comparison of Indy V-8 engines, 2002 and 2003

375

specify an electronically limited top engine speed of 10,300 rpm. In fact, the output of the Indy V-8 engine is actually restricted by three governors. The two "hard" governors shut off the ignition as does an on/off switch upon reaching the top permissible speed. The so-called soft governor progressively interrupts the ignition current as the engine speed approaches the specified limit.

The IRL governor is supplied and serviced by the sponsor. Moreover, it measures the speed in real time; the sponsors, therefore, have the possibility of checking the engine speed at any time during the race.

The actual IRL governor has a very "hard" effect. For this reason, the IRL teams install an electronic governor in the engine control software, which also determines the engine speed accurately and is set to engage "softly" just below the official top speed of 10,300 rpm. This eliminates abrupt complete interruption of the ignition for the drivers. The so-called soft governor switches off individual cylinders instead of the ignition for the entire engine. The engine continues to provide power and allows the driver to operate closer to the maximum performance limits.

For IRL teams, it is necessary to carefully balance the transmission ratio, aerodynamics, and track conditions to maximize the vehicle speed without coming too close to the top engine speed. The fourth, fifth, and sixth gears are so close to one another that the driver can keep the Indy engine, which begins to reach its maximum output only in the higher speed range, in this optimum range at all times without exceeding the prescribed maximum speed. In spite of all attempts by the teams to maintain the specified maximum speed, it may still be exceeded for short periods due to external circumstances. Examples of these are spinning rear wheels, slipstreaming behind a car ahead, or tailwind. When racing at night, the cooler ambient air can have an effect on the vehicle speed and, therefore, the engine speed.

~~~**Regulations.** The maximum displacement is 3.5 liters (**Fig. E192**). The regulations permit only 90° V-engines with eight cylinders, maximum bore of 93 mm, a maximum of four camshafts driven by chain or gears, and a maximum of four valves per cylinder. The engines must not be supercharged, have variable timing, or have variable length or volume features in the intake or exhaust system, and only steel valve springs are permitted. Hydraulic or pneumatic valve spring systems are prohibited. The dry weight of the engine without radiator, clutch, engine control module or ignition box and filter must be at least 128 kg. A minimum size is also prescribed.

A maximum of two injectors are permitted per cylinder and the maximum permissible injection pressure is 150 lb_f-in^2. Only one spark plug per cylinder is permitted. The engine speed is limited by a governor to 10,300 rpm (400 revolutions per minute less than up to 2002 and limiting the output to approximately 650 (SAE) hp.

The primary differences in relation to European race series are in the area of the fuel. The Indy engines use methanol, which, in comparison to gasoline, has less than half the specific heat and burns stochiometrically at approximately 6:1 air-fuel ratio.

The IRL does not require engines to be based on any stock engines. This regulation, therefore, allows manufacturers without their own V-8 with dual overhead cams in the model line to participate in the series—the engines are then designed specifically for this purpose.

Since 2000, the IRL specifies 3.5-liter engines instead of the 4-liter engines used up until that time. One of the reasons for this change was to increase the reliability. In 2003 and 2004, three manufacturers were approved by IRL as engine suppliers: Chevrolet, Honda, and Toyota.

~~~**Software.** The functional principle of engine management system for racing cars corresponds in principle to that of normal highway vehicles. The fuel supply is regulated by the injection nozzles, which themselves are controlled by the engine management system.

At the maximum engine speed of 10,300 rpm, the injectors on the Indy V-8 engine open and close more than 82,000 times per minute. The pause width—that is, the time interval during which the injector allows fuel to flow—determines the quantity of fuel injected. The engine management synchronizes the quantity injected with the firing order.

In contrast to other racing series, the use of electronics is subject to stringent regulations in IRL. Modifications to the engine calibration must be accomplished manually by the driver. For this purpose, IRL drivers have a control knob with eight different settings. Usually, tailor-made programs are programmed here for maximum output, fuel saving mode, or yellow flag phases. Telemetry is prohibited.

One of the few automatic settings permitted by IRL is the "pit lane speed limiter," a device which automatically limits the vehicle speed in the pit lane to the permissible limit in each case. This system can also be programmed so that it switches off again a certain distance after leaving the pit without the driver doing anything.

One of the main components in the Indy V-8 engine is the 180-degree crankshaft. While the four connecting rods are installed with a 90° offset on conventional V-8 crankshafts, they are mounted with an offset of 180° in this engine. A 180° crankshaft may be lighter and also more rigid than its 90° counterpart. The primary dis-

Type	3.5-liter, V-8, 32-valve, dual-overhead-cam (DOHC), naturally aspirated engine (turbo not allowed)
Size	3.5-liter displacement, Max. bore 93 mm four camshafts, 4 valves per cylinder
Weight	Dry weight at least 128 kg (without radiator, clutch; engine control module: ignition box, filter)
Engine speed	Maximum 10,300 (governor) rpm
Output	Approximately 650 (SAE) hp
Fuel	Methanol
Injection	Electronic
Manufacturer	General Motors, Honda, Toyota

Fig. E192 IndyCar engine regulations

Fig. E193 NASCAR Winston Cup vehicles

Official designation	e.g., GM NASCAR Winston Cup Engine
Fuel supply	Carburetor
Configuration	V-8
Displacement	5.8 liters
Number of cylinders	8
Valves	16
Output	850 (SAE) hp
Torque	550 lb-ft at 7200 rpm
Maximum engine speed	9500 rpm
Service life	600 miles (engines are overhauled after each race)
Engine block	Aluminum
Crankshafts	Steel
Piston	Aluminum
Engine management	By team
Oil supply	Dry sump
Fuel	110 octane (leaded)

Fig. E194 NASCAR engine figures

advantage is the imbalance, because in reality a V-8 with 180° crankshaft is nothing more than two four-cylinder engines connected by a common crankshaft.

Indy V-8 engines are also subjected to hard bench testing: for example, the engineers at GM racing simulate a complete race.

The data for the racing simulators are taken directly from the racing rounds. The engine is accelerated and decelerated in 55 exactly defined stages. These stages are intended to simulate the aerodynamic and mechanical forces acting on the vehicle. The cycle corresponds exactly to a 41-second round at an average speed of approximately 355 km/h.

~~NASCAR. NASCAR (National Association for Stock Car Auto Racing) has various classes which even include a class for commercial vehicles (**Fig. E193**). Only the top class, the so-called NASCAR Winston Cup, is considered here.

~~~Engine. The engine was presented to the NASCAR officials, for the first time, in October 1995. The V-8 engine was approved for the NASCAR Winston Cup (**Fig. E194**), for the 1998 season.

The SB2 is an engine package combining optimized reliability and simple maintenance with reduced production and maintenance costs.

In spite of all advances, the engine is to continue to be operated with carburetors and its valves driven by the crankshaft over push rods. The carburetors and combustion chambers of the SB2 are positioned so that they form a straight line.

The NASCAR regulations prescribe a maximum compression ratio of 12:1. It was necessary to reinforce the engine block to reach this limit. This design simplifies service at the race track because the intake can be removed without draining the cooling water or removing the carburetors.

~~~Regulations. All engines must be based on production engines whose primary parts such as the engine block, cylinder head, or crankshaft can also be purchased by retail customers. All significant parts must be approved by NASCAR officials before participating in competition. Only one engine may be used per race, including training, qualifying, and running. It is not permissible to remove the engine from the vehicle without the permission of NASCAR officials.

All engine parts must be steel or aluminum, but the crankcase itself must not be aluminum. Only so-called "Small Blocks" are permitted—that is, V-8 engines with displacements of 5.8 liters. The cylinder diameter must not exceed 10.6299 cm. The displacement is calculated according to the following equation: $V_{cyl} = 0.7854 \times Bore^2 \times Stroke$. This figure multiplied by the number of cylinders is the displacement of the engine.

The maximum compression ratio may not exceed 12:1.

Only round aluminum pistons with a maximum weight of 400 g are permitted. The piston pin must be steel or titanium and must weigh at least 70 g. Only solid steel piston rods with a minimum weight of 525 g are permitted.

The cylinder heads must be approved by NASCAR officials. Only steel or titanium is permitted for the valve materials. The valve springs must be steel. Only two valves per cylinder are permissible.

~~World Rally Car (WRC). This is run under generous technical regulations that no longer require production of 2500 highway vehicles of the same basic design (**Figs. E195 and E196**). In other words, a WRC does not require a street version with turbocharged engine and all-wheel drive, as described for Groups A and N. The minimum weight is 1230 kg.

~~~Homologation. This is the official registration process of a Rally car with the FIO racing authority. All decisive technical peculiarities are defined and recorded meticulously in the homologation, because they serve as the basis for all technical inspections before and during the rally by the technical inspectors in the future.

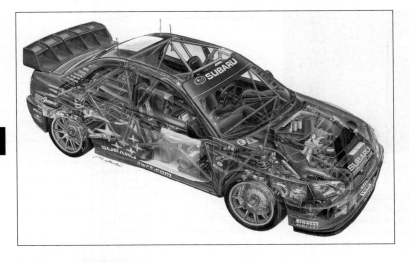

Fig. E195
Subaru WRC
four-cylinder horizontally
opposed engines for low
center of gravity

Engine	Four-cylinder, 16-valve horizontally opposed engine with IHI turbocharger (34 mm restrictor) 1994 cm³. Programmable electronic Subaru engine management system. Approximately 220 kW (300 hp) at 5500 rpm. Max torque at 480 Nm and 4000 rpm. Acceleration (0–100 km/h) in approximately 4.5 s
Transmission	Electrohydraulically controlled six-speed transmissions with joystick. Permanent all-wheel drive 50/50. Three electrohydraulically controlled differentials
Suspension	MacPherson struts. Bilstein shock absorbers All-round internally ventilated Alcon/Prodrive disk brakes (305 mm standard/366 mm for asphalt rallies)
Dimensions	Length 4405 mm, width 1770 mm, height 1390 mm, wheelbase 2535 mm. WRC minimum weight 1230 kg
Accessories	Three microprocessor controllers for engine. Differentials and data logging with CAN bus link. LCD color monitor with eight selectable data displays for passenger. Kenwood communication systems. Pirelli tires with OZ magnesium rims. NGK spark plugs. 80-liter safety fuel tank

Fig. E196 Technical data on Subaru WRC 2003

~~~**Regulations.** The number of cylinders in the engines is limited to eight. The displacement is permanently defined for:

- Naturally aspirated engines
- With a maximum of two valves per cylinder to a maximum of 3 liters,
- With more than two valves per cylinder to a maximum of 2.5 liters
- Turbocharged engines

The displacement is limited to a maximum of 2.5 liters. The turbocharging system must be that specified in the vehicle homologation, and its output is limited by a restrictor.

The total airflow used by the engine for combustion must flow through the restrictor. The maximum inner diameter of the restrictor is 34 mm. The restrictor must be installed on the pivot charger housing. The outer di-

ameter of the restrictor must be at least 38 mm at the smallest point and over a length of at least 5 mm to each side. On diesel engines, the inner diameter must be at least 35 mm and the outer diameter at least 41 mm.

The designers are free to select the material and number of camshafts as desired. The material and shape of the valves can also be selected freely.

The specific power output of the WRC engines is very close due to the airflow restrictor specified in the regulations. Increases in performance can, therefore, be achieved only by continuously increasing the efficiency of the engine and turbocharger. Generally, four-cylinder inline engines are used; an exception here is the Subaru with a horizontally opposed engine. The boost pressure is approximately 3.5 bar.

Bang-bang system. The bang-bang system is an automatic circulating air system with which an electrically operated compressor maintains the boost pressure of the rally car even when the accelerator is released, thereby eliminating power losses resulting from so-called turbocharger lag.

Other regulations pertaining to the World Rally Car.

- The tips of the turbine blades reach speeds of approximately 2000 km/h (0.0345 mm radius at 150,000 rpm)
- The engine sucks in 0.21 kg/s of air, equal to the flow rate of air required for 1500 humans (air consumption per human approximately 10,000 liters per day)
- In the maximum torque range, the force acting on the piston is equivalent to the weight of 10 compact cars.

~~~~**Group A.** These already quite generous technical regulations require that 2500 basic models of a potential rally car be built within 12 months in sequence. Group A superseded the so-called Group B Rally monster in 1987 and led to interesting highway cars such as the Lancia Delta Integrale, the Mitsubishi Lancer Evo

I through VII (current status), as well as the Ford Sierra and Escort RS-Cosworth. In the middle of 2001, the last factory team, Mitsubishi, changed from a group A car to a modern World Rally Car.

~~~~**Group N.** Technical regulations for near-stock rally cars. As for Group A, construction of 2500 identical homologation vehicles is required within 12 months.

~Naturally aspirated engine. A naturally aspirated engine is one which, in contrast to a supercharged engine, sucks its air or air-fuel mixture from the atmosphere primarily on the basis of the piston motion as it moves to bottom dead center. The density of the charge is approximately equal to the density of the ambient air if the supercharging effect resulting from the wave action and resonance in the induction system, etc., is disregarded.

A contribution to achieving high specific output is minimization of the throttle losses during the charge cycle (induction and exhaust) by reducing the resistances in the intake and exhaust tracts. A supercharged engine should be compared with a naturally aspirated engine. In this, an operating device provides precompressed air to the cylinder.

The specific output of naturally aspirated gasoline engines is presently about 30 kW/L for small, economical engines, more than 50 kW/L for good midsize engines and up to 80 kW/L for applications in sports vehicles. For naturally aspirated diesel engines, the current values are approximately 25–27 kW/L. Diesel engines with turbochargers and charge air intercooling provide figures around 70 kW/L of displacement.

~Supercharged engine →Supercharging

Engine accessories. Auxiliary assemblies are devices that are installed as attachments to the engine such as the alternator, power steering pump, vacuum pump for the brakes, air-conditioning compressor, secondary air for exhaust gas treatment, fan, and possibly also the oil pump and the coolant pump. The coolant pump and the oil pump are usually integrated into the engine and are, therefore, normally not included among the auxiliary assemblies. Different assembly drive concepts for these devices are required to address very different tasks. The oil pump in production engines is normally directly powered by the crankshaft through gears or chains. The coolant pump is normally integrated in the coolant circuit of the engine block and it is, therefore, often powered by the camshaft drive. The first electrically powered coolant pumps are now in production. The other auxiliary assemblies are normally powered by a belt.

Gear drives have been used in special applications (races)—for example, for the alternator drive and for drives by the cardan shaft at the rear axle. The engine accessories are often powered on two levels, which means two V-belts or multirib belts.

The main requirement for the drives is slip-free operation under all load conditions.

~Chain drive. Chains are normally used to power the oil pump. Single row chains are usually acceptable. Duplex chains are often used for camshaft drives.

~Drive belt. Drive belts are rarely used to power engine accessories. The only exception is the water pump, which is powered by drive belts in connection with the camshaft drive. In special cases, the oil pump is also powered with a drive belt directly from the crankshaft. Drive belts are primarily used to power the camshaft.

~Electric motor. All auxiliary assemblies could in principle be driven by electric motors. This is, however, only being done in exceptional cases because of the comparatively low total efficiency factor, cost, and weight. The secondary air pump is normally powered electrically because of the accuracy required in the on-off switching functions. Shortage of space and the desire or requirement for engine speed-independent control is another reason to opt for an electrical drive for engine accessories.

~Fan belt. The fan belt is the classic drive element used for engine accessories. This is an endless belt made of fabric and rubber with a cross-section in the shape of a triangle or trapezium, which runs over pulleys with a corresponding shape. The fan belt transfers the forces through the belt flanks. The belt tip must not sit on the groove base of the pulley, because this relieves the flanks and the force-transferring surfaces become too small. The result is belt slippage and squeaking due to reduced force transfer.

The contact angle influences the maximum possible force transfer for the design of the V-belt drive. Driving several devices with one fan belt is only a limited option because the maximum tension load is soon reached and because the contact angles become too small.

The fan belts are often cross-slit, which gives them a toothed belt profile to permit a better adaptation of the fan belt to pulleys of different sizes. A deflection of the back of the fan belt is not possible with fan belts as it is done for multirib belts or drive belts due to the negative influence on its service life.

Transmission ratios can be adjusted through the different diameters of powering and powered pulleys. A deviation of auxiliary assembly speeds from the engine speed is not possible without significant additional expenditure.

~Fan drive. Fans can be powered electrically or directly by the engine through the auxiliary assembly drive. A viscous clutch is installed for direct drives between drive wheel and fan wheel, because the fan power is not required for all engine operating conditions. The fan wheel is not operated when the viscous oil is cold and the speed of the auxiliary assembly drive is low. The temperature of the oil increases with increasing speed and when the fan wheel is switched on. The temperature dependence of the oil represents a disadvantage because it changes the cut-off speed depending on the temperature. Electromagnetic clutches, with temperature dependent control, are used in addition to the viscous clutches.

E

~Gears. In special cases, the oil pump is powered by the crankshaft directly with the help of gears. Gears are also used to power camshafts.

Literature: J. Hadler, K. Blumensaat, W. Nederkorn, A. Kracke, P. Urban: Der neue Fünfzylinder-Dieselmotor von VW, MTZ (2004) 1, p. 8.

~Multirib belt. The powering of engine accessories in modern machines is currently achieved with multi-rib belts, which have four to eight ribs and can have ribs on both sides.

The multirib belt permits the integration of several auxiliary assemblies; it can transfer higher power levels and has the advantage of requiring less space for complex engines. The state of the art today is the integration of alternator, water pump, steering booster pump, and air-conditioning compressor. These drives are designed as serpentine drives with multirib belts.

The multirib belt consists of a fiber-reinforced rubber mixture, the cord, and a back fabric and a rubber coating. The contact angles can be increased by deflection pulleys on the back of the belt, for example. Belt tension (spring loaded, hydraulically damped) is often used in the return strand for damping and for constant belt pressure. The design must be done carefully, because this approach increases friction loss. In contrast to the fan belt, a correctly designed multirib belt rarely requires replacement during the life cycle of the engine. A large dust load, high temperatures, and frequent and strong deflections over the back of the belt have a life-cycle–reducing effect. In contrast to the V-belt drive, the powering of engine accessories with a multirib belt permits smaller transmission ratios.

The maximum transferable torque is approximately 30 Nm, the maximal transferable power is approximately 20 kW, and the temperature load is a maximum 100°C. **Fig. E197** shows an example of an auxiliary assembly drive.

Literature: H. Niggemeyer and others: Die neuen BMW Sechzylinder-Vierventilmotoren, MTZ 51 (1990) 3. — F. Nau-

mann, D. Voigt, H. Deutsch: Der neue VR6-Motor von VW. MTZ 52 (1991) 3. — M. Arnold, M. Farrenkopf, S. McNamara: Zahnriementriebe mit Lebensdauer für künftige Motoren, MTZ 62 (2001) 2. — R. van Basshuysen, F. Schäfer: Handbuch Verbrennungsmotor, 2nd Edition, Wiesbaden, Vieweg Publishers, 2002.

~Variable engine. The required power of the auxiliary assembly is not always proportional to the engine speed. The most important example is the alternator, which is heavily used while idling, especially in winter, by power consuming devices such as the lights, the rear window defroster, the fan, etc.; however, it generates only a small amount of power due to the low idle speed. The situation changes at high engine speeds, when it can deliver more current than necessary. A variable engine speed ratio would be desirable, because it could increase the alternator speed significantly during engine idle, and it would not overload the mechanical parts at high speeds. This, however, is an expensive solution, which also requires more space and is heavier. Pulleys with an adjustable distance between the side walls, which accept the fan belt, are feasible. The belt runs on a larger radius if the side walls are closer together and on a smaller radius if they are farther apart.

The avoidance of large slip and maintaining optimum control are difficult while maintaining a constant belt pressure. Another disadvantage is that powering several devices at the same time is not possible, because this requires extra space, which is often not available in modern engine compartments.

The disadvantages of variable auxiliary assembly drives—price, weight, space—are currently greater than the advantages, which means that variable auxiliary assembly drives are not yet used in high-volume automobile production.

Engine acoustics. Engine acoustics have become an important factor in the development of engines and motor vehicles due to increasing demands for passenger comfort and legal regulations for noise emissions.

~Absolute silencer. This is a large-volume auxiliary silencer which fully eliminates the induction noise or the exhaust pipe tip noise. Absolute silencers are used for measurement purposes in vehicle tests to determine whether the induction noise or exhaust pipe tip noise still contributes to the total vehicle noise. The difference between the noise measured with and without the absolute silencer indicates the improvement potential for optimization of the intake silencer or exhaust muffler system. The development target is reached when this difference is zero.

~Absorption-type muffler. This is a volume present in pipe systems or pieces of pipe that is provided with an acoustically effective lining in which the sound waves are attenuated by friction between the oscillating gas molecules and the absorbing material such as mineral wool or foam (*also, see below*, ~Noise absorption). Absorption-type mufflers are used in exhaust systems as the primary or middle muffler; the air filter

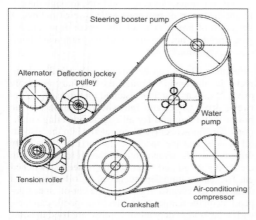

Fig. E197 Example of an auxiliary assembly drive

in the intake system also acts as an absorption-type silencer (*also, see below,* ~Intake noise).

~Acceleration passing. Acceleration passing is the driving state of a motor vehicle during the legal exterior noise test (*also, see below,* ~Exterior noise ~~Legal measurement).

~Accessory noise. The noises from the accessories and their drives are associated with the engine just as is the noise from the valve train and its drive; these usually have a negative influence on the acoustic pattern of the engine—that is, the noise quality. The most important contributions to the noise include: generator noise, power steering pump noise, and air-conditioning compressor noise.

~Acoustic holography. *See below,* ~Sound field analysis

~Acoustic impedance. Acoustic impedance, analogous to mechanical impedance, is the ratio between noise velocity and acoustic pressure:

$$\tilde{z} = \frac{\bar{v}}{\bar{p}}.$$

~Acoustic power. The power of sound waves passing through an imaginary surface—that is, the integral of the noise intensity—is given by

$$P = \int_s I ds.$$

The entire acoustic power radiated by an acoustic radiator is an important value for evaluating its noise emission.

~Acoustic power level. *See below,* ~Decibel

~Acoustic pressure. Acoustic pressure is the scaled acoustic field value for airborne and liquid-borne sound, which does not describe the absolute pressure, but rather the pressure variations in relation to the steady ambient pressure. The acoustic pressure level, where applicable, divided up into its various frequencies and evaluated serves as an important basis for evaluating the noise perceived by humans.

~Acoustic pressure level. *See below,* ~Decibel

~Acoustic short circuit. *See below,* ~Hydrodynamic short circuit

~Acoustic transfer function. Analogous to the mechanical transfer function, the acoustic transfer function is the frequency-dependent ratio of noise absorbed at an observation point, j (e.g., driver's ear position) and excitation magnitude at introduction point, i (e.g., engine mount). If the transfer path includes a solid body at the input point, a force, F, is usually used as the excitation magnitude. In the case of pure airborne sound transmission paths, however, volumetric sources, q (velocity times surface area) are usually used:

$$\tilde{H}_{ij} = \frac{\tilde{P}_j}{\tilde{F}_i} \quad \text{or} \quad \tilde{H}_{ij} = \frac{\tilde{P}_j}{\tilde{q}_i}.$$

In the case of more than one input and/or observation point, these are compiled to form a transfer matrix. The concept of the transfer function assumes linear system properties.

~Acoustically dead room. An Anechoic chamber. *See below,* ~Free field.

~Acoustics test bench. Acoustics test benches must satisfy two primary requirements in comparison to "normal" engine and vehicle test benches (**Fig. E198**).

It is necessary to prevent reflections from the walls to the greatest extent possible with the aid of absorbent coverings in order to provide free field conditions (an exception is the reverberation chamber).

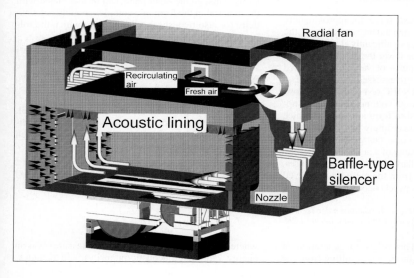

Fig. E198
Acoustic vehicle dynamometer (Source: Akustikzentrum GmbH)

Extraneous noises and vibrations (from cooling fan, chain drive, brakes, adjacent machine shops, etc.) must be reduced to a sufficient degree by use of appropriate measures (e.g., acoustic insulation and damping, vibration insulation).

~Active engine mount. *See below*, ~Engine/accessory mount ~~Active engine mount

~Air-borne sound. *See below*, ~Structure-borne sound

~Air-borne sound radiation. *See below*, ~Sound radiation

~Air-conditioning compressor noises. Vibratory excitation and transmission are basically similar to noise from the power steering pump, but the level of pressure pulsation generated is considerably lower. Nevertheless, air-conditioning compressor hoses under pressure represent a problem for engine acoustics because, with their high rigidity, they form a structure-borne sound bridge between the engine and vehicle body— that is, transfer engine noise. The air-conditioning compressor can increase the engine hum indirectly because of its inertia when it resonates in the frequency range of an unbalanced engine order due to an insufficiently rigid mount.

~Alternator noise. *See below*, ~Generator noises

~Annoyance. Subjective annoyance is, similar to noise quality, not only a question of the acoustic pressure level. Pulse-type noises such as diesel "ping" are perceived to be particularly annoying, whereas a continuous uniform noise is perceived to be less annoying. In terms of measuring technology, these differences in the character of the noise can be recognized more clearly by considering the time curve from the microphone signal than from the spectrum itself (*also, see below*, ~Psychoacoustic parameters).

~Assessment curves. These serve to approximately correct for the frequency-dependent sensitivity of the human ear, which decreases significantly at very low and very high frequencies—the lower the absolute acoustic pressure, the greater the range of variation. Measured acoustic pressure spectra are assessed with the assessment curves defined in DIN IEC 651 (Curve A for low, B for medium, and C for high volumes) and, where applicable, then consolidated to form a total level which can then be represented accordingly (e.g., Index A). It must be noted that assessment curves do not provide any information on the quality of a noise and, therefore, the assessed level, particularly for narrow band noises, can deviate significantly from the actually perceived loudness (*also, see below*, psychoacoustic parameters).

Literature: M. Heckl, H.A. Müller: Taschenbuch der Technischen Akustik, 2nd Edition, Springer Publishers, Berlin, 1994.

~Balancing of masses. This refers to generation of counterforces to oppose the forces resulting from the reciprocating motion of the pistons and connecting rods and thereby reducing or completely obliterating engine block vibration. The piston motion can be broken down into frequency components at the crankshaft rotation frequency and multiples (or orders) of it. It is possible to balance the first, second (twice crankshaft frequency), or higher order of the components—called first- or second-order balancing.

~Capsule, encapsulation. *See below*, ~Engine noise

~Characteristic vector. *See below*, ~Resonant frequency

~Combustion noise. Combustion noise originates from the pressure increases in the combustion chamber and unsteady flow processes. The resulting dynamic forces excite the engine structure, and its surface deformations radiate the combustion noise as airborne sound or corresponding vibrations of the engine and transmission block that are transferred to the body through the engine mounts as structure-borne sound. A second, primary component of combustion noise results from superimposition of the pressure pulses in the induction system and the exhaust manifold, whose characteristics are determined to a significant extent by the noise characteristics of the engine (noise quality). These noises reach the human ear by secondary radiation from the piping system and exhaust noise. The most familiar examples are diesel knock resulting from rapid rates of pressure rise in the combustion chamber or the typical combustion noise resulting from the firing sequence of the two cylinder banks on V-8 engines.

~Complex modulus of elasticity. This is a mathematical aid for treating attenuation in vibration problems. In the time response of a system, attenuation manifests itself as a lag in the motion behind the exciting force (hysteresis). In the frequency range such a phase shift can be produced very easily by complex rigidity. This, on the other hand, can be associated with the material in the form of a complex modulus of elasticity (or also by a shear modulus), for which the following notations are commonly used:

$$\tilde{E} \ = \ E' + iE'' \ = \ E(1 + i\eta) \ = \ E\sqrt{1 + \eta^2}\,e^{i\vartheta} \ ,$$

where E' = storage modulus, E'' = loss modulus, η = loss factor, and ϑ = phase angle.

~Decibel (dB). This is a dimensionless unit (1/10 bel, named after Alexander Graham Bell), and is principally a random value for logarithmic scaling, which can be converted to the level L:

$$L_x \ = \ 20\log_{10}\!\left(\frac{x}{x_0}\right)\!dB$$

or

$$L_x \ = \ 10\log_{10}\!\left(\frac{x}{x_0}\right)\!dB,$$

where x is a field value (e.g., acoustic pressure), X is an energy value (e.g., acoustic intensity), and x_0 and X_0

are their reference values. This value is frequently used in acoustics because it is often necessary to represent, compare, or evaluate values over several orders of magnitude and because the human ear hears approximately "logarithmically." For the acoustic pressure, $p_0 = 2 \times 10^{-5}$Pa (effective value); for the acoustic power, $P_0 = 10^{-12}$W; and for the acoustic intensity, $I_0 = 10^{-12}$Wm^{-2}. (*Also, see above, ~*Assessment curves.)

~Degree of absorption. This refers to the ratio of noise intensity absorbed to noise intensity present:

$$\alpha = \frac{Iabsorb}{Ipresent} \cdot a$$

This is used, among other things, as a characteristic value for absorbing materials, linings, and so on.

~Degree of noise reduction. The difference in acoustic pressure level in dB between the front and back of the insulating component is called the degree of noise reduction. It is the dimension for the noise reduction of the component or material specimen.

~Degree of variation. *See below, ~*Psycho-acoustic parameters

~Diffusion field. This is an acoustic field which has the same amplitudes everywhere in the chronological middle and no energy flow. This can be approximated in reverberation chambers by a very large number of reflections.

~Directional characteristics. *See below, ~*Directional effect

~Directional effect. Direction effect is a property of acoustic transducers (acoustic radiators, microphones) which do not have the same effect or same sensitivity in all directions. This is particularly noticeable when the dimensions of the transducer are greater than the airborne wave lengths (e.g., high-frequency engine block vibration). A polar diagram of the directional dependency is called the directional characteristics.

~Direct sound. This is the percentage of an acoustic field which reaches the observer along a direct path—for example, without reflection.

~Dissipation. Dissipation is the conversion of mechanical or electrical energy (e.g., acoustic energy) to heat.

~Double-pulse holography. *See below, ~*Laser holography

~Dual-mass flywheel (DMF). *See below, ~*Transmission rattle

Literature: W. Geib (Ed.): Geräuschminderung bei Kraftfahrzeugen, Braunschweig, Vieweg Publishers, 1988. — H. Klingenberg: Automobil-Messtechnik, 2nd Edition, Volume A, Acoustics, Springer Publishers, Berlin, 1991.

Fig. E199 Dummy head acoustic test for recording and evaluating noise quality (Source: Head Acoustics)

~Dummy head test. This is a stereophonic noise recording based on human hearing in which the two recording microphones are located in the ears of a dummy head (**Fig. E199**).

~Dynamic rigidity. Dynamic rigidity is the quotient of resulting vibratory force (spring force plus damping force) and vibration path, analog to static rigidity consisting of force per unit extension of a spring.

~Engine harmonics. *See below, ~*Harmonic.

~Engine noise (*also, see below, ~*Exterior noise, ~Interior noise, ~Noise generation in engine, ~Noise source, engine).

~~Encapsulation, acoustic effect. Reduction of the exterior noise, which is the objective of engine encapsulation, is limited by the size and number of openings required for cooling the vehicle. Development of complete engine encapsulation is, therefore, more a problem of cooling than acoustics. Development vehicles with completely encapsulated engine compartments achieve spectacular public effects; however, they are usually far from being capable of withstanding a trip over a mountain path with a trailer or a test in a hot country. In the case of the figures measured in dB(A), which quantify the noise reduction measured by encapsulation, it is necessary to differentiate:

- dB(A) value for reduction of radiated engine noise
- Reduction of external noise emission values—i.e., the overall vehicle noise measured in the legal pass-

by test. Example: During the passing test, a value of 74.8 dB(A) is measured for a passenger car, where the total energy content consists of 70 dB(A) tire noise and 73 dB(A) engine noise. Through extensive engine encapsulation, the acoustic energy radiated from the engine can be reduced to one half — i.e., to an acoustic pressure level of 70 dB(A). Tire noise and engine noise of 70 dB(A) each when added together still result in a total level of 73 dB(A). The reduction in the exterior noise emission value through encapsulation is, therefore, 1.8 dB(A).

~~**Engine noise calculation.** *See below,* ~~Simulation

~~**Measurement, procedures.** The legal exterior noise measurement and interior noise measurement while driving on the road are the most important of the standardized measuring procedures for considering various types of operation or aspects of noise emission, such as standing noise measurement or exhaust tip noise measurement. The latter is performed preferably in second or third gear or in drive stage 2 and represented as the level in relation to the engine speed. Measurements in second or third gear usually differ only slightly due to the higher contribution of wind and road noises in third gear. However, the most important parameter is the engine load. For this reason, three tests are usually performed at part load—full load and deceleration and represented separately. Here, the part-load acceleration is usually defined at a vehicle acceleration rate that is kept constant during the test — e.g., 1 m/s² — which is indicated by a simple pendulum-type instrument as the speed increases and is maintained by the driver.

~~**Measurement, types of representation.** While the dB(A) level or psychoacoustic parameter serves as the dimension for the subjectively perceived volume, only frequency analysis of the measured microphone signal provides information on the type of noise and its cause. The transfer mechanism with which the engine noise reaches the passenger compartment is frequency-dependent: structure-borne sound below approximately 500 Hz, airborne sound above approximately 1000 Hz, and both structure-borne and airborn sound between 500 and 1000 Hz may be involved (**Fig. E200**). In the structure-borne sound range, the low-frequency engine noise consists of the sum of sinusoidal vibrations, whose frequency is based on the speed of the crankshaft, the so-called engine orders. All reciprocating piston engines generate a noise spectrum with their combustion cycles consisting of half orders of the crankshaft speed, corresponding to the firing frequency of each individual cylinder once every two revolutions of the engine. The best overview of an order analysis results from a 3-D representation of the noise level in relation to frequency and speed, either in the form of a "waterfall" diagram or of a so-called Campbell diagram, in which the color or brightness indicates the magnitude of the level (**Fig. E201**). In these diagrams, the engine orders are recognizable as "tracks" increasing at an angle with constant speed/frequency ratio. The magnitude and distribution of the orders distinguish the source of noise excitation such as the engine; in contrast, increases at constant frequency indi-

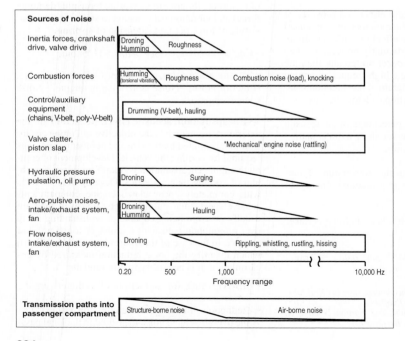

Fig. E200
Noise sources, engine
(Source: AFT Atlas
Fahrzeugtechnik)

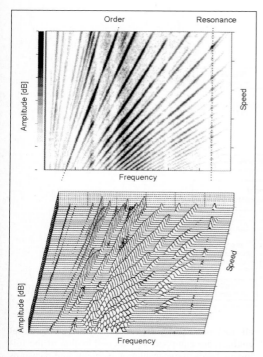

Fig. E201 Campbell diagram (top) and waterfall representation (bottom) (Source: AFT Atlas Fahrzeugtechnik)

cate resonances in the transfer path. These summary representations are less suitable for quantitative representation of the level or comparisons of the level. For this purpose, diagrams of the most important order over the engine speed, so-called "order curves," are commonly used, as shown in **Fig. E202**. The highest orders in the noise spectrum are always the greatest uncompensated mass forces and moments and the firing frequency—on

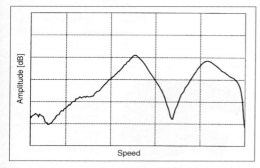

Fig. E202 Order curve in relation to speed, cross section of **Fig. E203** (Source: AFT Atlas Fahrzeugtechnik)

a four-cylinder inline engine, the second engine order, on five-cylinder engines, the second and 25th engine order, etc. In the higher frequency airborne sound range ("mechanical engine noise"), it is usual to subdivide the noise into frequency bands and record the octave level or major third level as a curve in relation to the engine speed. An increase in this level can, for example, indicate deficiencies in the acoustic insulation between the engine compartment and the passenger compartment.

~~**Simulation.** Preliminary calculation of vibration and radiated airborne sound level of an engine is still in the development stage and continues to be a demanding assignment today (**Fig. E203**). The starting point is a dynamic FE model of the engine/transmission block, which is fine enough to consider the complex forms of vibration even in the high-frequency range. However, the most complicated and difficult part is to calculate the forces which act on the block forces during operation and to determine the magnitude of the vibration amplitudes of the engine surface. For this purpose, it is necessary to couple a model of the crankshaft drive with the model of the block structure ("bottom-end code")—the coupling conditions are highly nonlinear because of the hydrodynamic oil film in the crankshaft bearings and on the cylinder running surfaces. It is necessary either to take the gas forces on the crank drive from a measured indicator diagram or to calculate them with a complicated gas dynamic and thermodynamic simulation program.

The initial results obtained are the vibration amplitudes of the surface of the entire unit. This data can be read into a program that can calculate the airborne sound radiation—BE (boundary element) models instead of FE (finite element) modules are usually used today because they also can calculate effectively the propagation of the sound waves in free space as required for predicting the exterior noise. When calculating the vehicle interior noise, it is necessary to differentiate between the low-frequency noise <1000 Hz—that is, the "humming" transferred in the form of structure-borne sound—and the high frequencies >1000 Hz, which are transferred as airborne noise into the interior and perceived subjectively as "mechanical engine noise." For the structure-borne sound, the vibration amplitudes of the engine supports can be considered to be so-called path excitation of the rubber mounts for mounting the engine. The contribution of structure-borne sound to the total noise level can be calculated for each mount using the dynamic stiffness of the engine mounts and the acoustic transmission factor of the body at the mounting points. These contributions can be added together for all mounts and all directions of vibrations. The acoustic transfer functions of the body may be present in the form of measured data or as the result of dynamic FE calculations of the body structure and hollow cavities FSI (fluid structure interaction). The airborne sound path over which the high frequencies are transferred from the engine compartment to the interior through the bulkhead, in spite of its acoustic

E

E

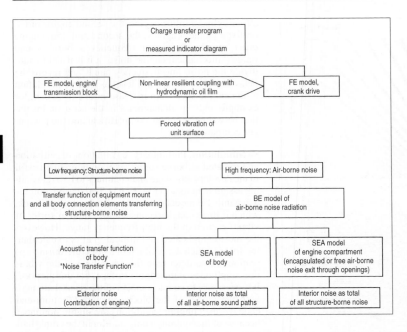

Fig. E203
Engine noise simulation process

lining, can also be calculated today. Because the forms of body vibration at these high frequencies are too complex and numerous to determine—calculate with an FE model, for example—it is more practical to select the static mean value for the vibration rate over a subsurface as the calculation variable and to consider the flow of vibration energy along the path from the engine compartment to the interior; the SEA (statistical energy analysis) method is used for this purpose. In practical applications, it is normal to depart from the idea of an acoustically "complete vehicle model" which can be used to calculate the entire interior noise of the vehicle, because of the unrealistically large amount of time required to create the model. The trend is frequently toward so-called hybrid models—that is, a combination of assemblies measured technically or input variables with computation models for new components. This solution is particularly good for new developments based on existing vehicle floor assemblies or engine families already present.

Literature: O. v. Estorff, G. Brügmann, A. Irrgang, L. Belke: Berechnung der Schallabstrahlung von Fahrzeugkomponenten bei BMW, ATZ 96 (1994) 5, pp. 316–320.

~~Underfloor engine noise encapsulation. This term refers to auxiliary parts mounted on the body that create a virtually closed chamber around the engine compartment and from which only minimum engine noise escapes to the outside. In addition, the peripheral surfaces are lined on the inside with acoustic absorbing materials such as foam rubber or cotton matting, which is protected in the lower areas against oil and moisture and is noncombustible, to reduce the increased noise level in the engine compartment. Nearly

all stock passenger cars, particularly those with diesel engines, are equipped with a combination of the following types of encapsulation elements and measures (**Fig. E204**):

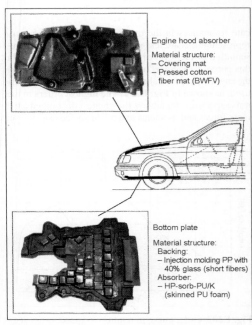

Engine hood absorber

Material structure:
— Covering mat
— Pressed cotton fiber mat (BWFV)

Bottom plate

Material structure:
Backing:
— Injection molding PP with 40% glass (short fibers)
Absorber:
— HP-sorb-PU/K (skinned PU foam)

Fig. E204 Sound absorber in engine compartment
(Source: HP-Chemie Pelzer)

- Absorbing engine hood lining consisting of foam material, fiber matting, frequently quilted at intervals to the sheet metal to achieve increased absorption at low frequencies as the result of the plate resonator effect, or created with a honeycomb pattern which acts as Helmholtz resonators.
- Underbody: Plastic or metal shells which close the engine compartment at the bottom. These usually have a cutout for the oil pan to allow greater ground clearance and often stop in front of the bulkhead. In special cases, on acoustically critical vehicles, the front section of the transmission tunnel is also closed at the bottom as long as the resulting cooling problems can be mastered. The underfloor covers are also usually jacketed with an acoustic absorbing material on the inside, thereby protecting against saturation with oil, fuel, etc., which has priority over optimum acoustic absorption, hence, material such as hard foam is used.
- Closing the bridges at the side to the wheel arch for the tie rods and, where applicable, driveshafts for the front wheels with resilient covers such as rubber bellows, etc.
- Closing the cooling air inlet at the front of the vehicle—e.g., with thermally controlled louvers which are closed when the engine is started cold and thereby reduce the annoyance for the neighbors resulting from the cold knocking noise of the diesel engine.

~~Underhood engine noise encapsulation. This term refers to components fastened resiliently to the engine block which have the acoustic function of an attached shell and reduce the sound radiated from the surface of the engine. However, complete underhood encapsulation is almost impossible to achieve on mass-produced passenger cars because:

- The surface of the engine is virtually covered with mounts for auxiliary equipment, hoses, and electrical lines, which all have to be closed.
- Cooling problems occur.
- Space requirements in engine compartment limit underhood encapsulation.
- Vibrational problems resulting from resilient mounting of the encapsulation. Practical applications provide for resilient suspension or vapor coating of the attached parts present, whose noise radiation is thereby reduced—e.g., valve covers, oil pans, intake manifold, timing case cover, or covers on fuel injection lines and wiring harnesses in the area of the head, that are provided for visual reasons.

~Engine order. *See below,* ~Order

~Engine/accessory mount. Engine and transmission mounts consist of rubber mounts that are intended to prevent the transmission of vibration and structure-borne sound to the body. The physical function is the same as for any vibration insulating foundation, whereby the engine/transmission block represents the foundation mass and the body console is the floor of the building. Acceptable vibration insulation starts at just above 1.4 times the resonant frequency of the spring/mass system and increases approximately with the square of the frequency. For this reason, the objective from an acoustic point of view is a mount with the lowest possible resonant frequency—that is, a rubber mount with low spring rigidity. On passenger car engines these values are between 6 and 15 Hz, which is equivalent to 100–300 N/mm stiffness measured statically. However, the main criterion for a rubber bearing is the dynamic stiffness—that is, the stiffness measured under vibration excitation—and this is a function of the frequency and can be calculated from the total of the spring force and damping force. Because the damping force increases approximately linearly with the frequency, rubber mixtures with high material attenuation are unsuitable for engine mounts because of the associated "dynamic hardening" of the mount at high frequencies. Damped mounts are, therefore, designed as hydraulically damped equipment mounts, short hydro-bearings, in which the attenuation is effective only at low frequencies below the audible range.

~~Active engine mount. The term "active engine mount" describes a mount in which a dynamic counterforce is produced within a rubber mount—for example, as a result of built-in electrodynamic excitation and a computer-controlled circuit—that extinguishes the structure-borne sound otherwise acting on the body. Active engine mounts differ from active noise control systems only in that structure-borne sound, not airborne noise, is generated as antinoise. The advantage is that the engine noise as well as vibrations transferred through the engine mounts are eliminated to the greatest extent. The high costs for active noise control systems are an impediment to their use in mass-produced passenger cars and commercial vehicles.

Literature: R. Freymann: Passive und aktive Systeme zur Dämpfung von Hohlraumeigenschwingungen, ATZ 91 (1989) 5, pp. 285–292. — A. Felske, H.J. Gawron, K. Schaaf: Aktive Innengeräuschreduzierung bei Kraftfahrzeugen, ATZ 92 (1990) 1, pp. 6–12. — D. Föller: Antischall—Chancen und Grenzen, ATZ 94 (1992) 2, pp. 88–93.

~~Hydraulically damped engine mounts. Contrary to widespread opinion, hydraulic damping in engine or transmission mounts does not serve only to improve the acoustic properties, because from the point of view of pure structure-borne sound insulation, an undamped mount would be optimum. The requirement for hydraulic damping results exclusively from the requirements for suspension comfort on uneven roads where the resonance of the engine mass in soft mounts would lead to unpleasant reverberation. The front seat passengers in a car, especially, perceive this so-called shaking as poor suspension comfort and instinctively associate it with the front axle, which has frequently led to the criticism "front axle shakes" in test reports.

The advantages of using hydraulically damped mounts compared to damping using conventional vibration dampers such as those used for the suspension, in addition to the more favorable costs, are particularly the fact that they eliminate the problem of friction between the pistons and cylinders of the damper, and also the hydraulic damping is increased precisely at the resonant frequencies between 6 and 15 Hz, where it is required by means of special tuning, and is decoupled above this frequency, where it would be disadvantageous acoustically.

This principle is shown clearly in **Figs. E205 and E206**. **Fig. E205** shows the spring/mass vibration model; **Fig. E206** schematically illustrates the technical execution and a cross section of the device. The greater vibration amplitude of the engine in the range of 1 to 2 mm on uneven roads is absorbed by displacing liquid from one chamber to another. During this process, it is necessary for the hydraulic fluid to pass through a ring-shaped overflow channel in which it is accelerated to a high flow rate.

At a certain frequency, to which the hydro-bearing can be tuned in a manner similar to the tuning of an intake manifold, this flow rate and, therefore, the damping effect, reach a maximum. As the fluid flows into the second chamber, the flow velocity is decreased by turbulence—that is, the damping effect is not dependent on the viscosity of the fluid. For this reason, water with antifreeze is used as the hydraulic fluid.

The vibration model separates the integrated functions of the rubber element into two springs: the left spring, also called the support spring, supports the static weight of the engine and represents the behavior of the bearing without hydraulic fluid, where only shear deformation of the rubber element occurs. As soon as it is necessary to displace fluid from the first chamber, the increase in the pressure leads to distention of the chamber. In terms of forces, this acts like a second spring connected in parallel, the so-called distention spring in the vibration model.

The corresponding force or pressure increase in the chamber accelerates the column of water in the

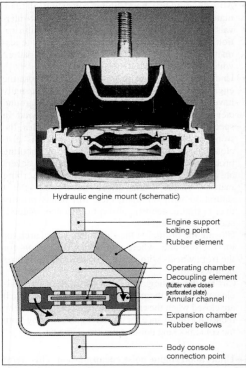

Hydraulic engine mount (schematic)

Engine support bolting point
Rubber element

Operating chamber
Decoupling element (flutter valve closes perforated plate)
Annular channel

Expansion chamber
Rubber bellows

Body console connection point

Fig. E206 Hydraulic engine mount (Source: AFT Atlas Fahrzeugtechnik)

annular channel, which acts as an inertia mass. The distortion spring with stiffness, c_B, and water mass, m_W, form a spring/mass system with a resonant frequency of

$$\frac{1}{2\pi}\sqrt{\frac{c_B}{m_W}}.$$

At this frequency, the damping effect is the greatest, due to the subsequent kinetic energy of the turbulence in the fluid. The second characteristic component in a hydro-bearing is the so-called uncoupling diaphragm, which acts as a flutter valve to close the chamber. This has a clearance range of approximately 0.1 mm and decouples the hydraulic effect at low structure-borne sound amplitudes.

Literature: D. Bösenberg, J. Van den Boom: *Motorlagerung im Fahrzeug mit integrierter hydraulischer Dämpfung—ein Weg zur Verbesserung des Fahrkomforts,* ATZ 81 (1979) 10, pp. 533–539. — K. Holzemer: *Theorie der Gummilager mit hydraulischer Dämpfung,* ATZ 87 (1985) 10, pp. 545–551. — H. Bathelt, J. Bukovics, D.J. Young: *Die Entwicklung der Aggregatelagerung des Audi V8,* ATZ 91 (1989) 2, pp. 93–102. — G. Kern, T. Großmann, D. Göhlich: *Computergestützte Auslegung von hydraulisch gedämpften Gummilagern,* ATZ 94 (1992) 9, pp. 462–472. — M. Hermanski: *Prüfstand zur Bestimmung der Übertragungssteifigkeit,* ATZ 97 (1995) 1, pp. 38–44.*

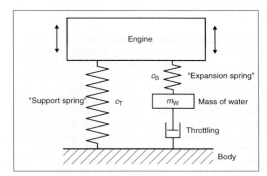

Fig. E205 Spring/vibration model for hydro-bearing (Source: AFT Atlas Fahrzeugtechnik)

E

~~**Semi-active engine mounts.** Semi-active engine mounts allow the stiffness or damping to be controlled to match certain factors such as the engine speed. The first major application consisted of the engine mounts for the Audi TDI, on which the distinctive spring stiffness (*see above*, ~~Hydraulically damped engine mounts) and hydraulic damping of the hydro-bearing are decoupled at idle and low-engine speeds by turning a taper shaft with an actuator. This prevents the high vibrations and rough operation of the DI engine in the lower speed range from being transferred to the body.

Literature: R. van Basshuysen, G. Kuipers, H. Hollerweger: Akustik des Audi 100 mit direkteinspritzendem Turbo-Dieselmotor [Acoustics of Audi 100 with Direct Injection Turbo Diesel Engine], ATZ 92 (1990) 1, pp. 14–21.

~Enveloping surface method. *See below*, ~ Sound measurement

~Exhaust noise. *See below*, ~Exhaust system noise

~Exhaust pipe tip noise. This is the airborne noise coming from the open end of the exhaust pipe.

~~**Measurement.** The exhaust pipe tip noise is measured on highway vehicles with a microphone fastened to the rear of the vehicle. The position of the microphone is usually the same as that used for measuring the stationary noise: at a lateral distance of 0.5 m from the end pipe and at an angle of 45° to the rear (**Fig. E207**). The dB(A) level curve measured with this microphone at full-load acceleration indicates the acoustic efficiency of the muffler system. There are no legally defined or generally applicable limits for the magnitude of this level; the limit curve results from the requirement that the exhaust pipe tip noise must not provide any measurable contribution to the total vehicle noise, either to the level measured during the legal exterior noise test or to the interior noise level. This can be checked by comparing the measurement with an absolute silencer.

~Exhaust system noise. This is a combination of the noise present at the tip of the exhaust pipe and secondary exhaust system noises.

~~**Secondary exhaust system noise.** The airborne sound radiated from the surfaces of the exhaust system is called secondary radiation, and it is generated when the sheet metal surfaces on the mufflers are excited by structure-borne vibration; this is then radiated as airborne noise (it is similar to the intake noise caused when airborne vibration is transferred to the interior of the vehicle).

Literature: F. Lehringer, D. Kattge: Schallabstrahlung von Abgasanlagen, ATZ 87 (1985), No. 10, pp. 559–563.

~Exterior noise. The operating noise of a motor vehicle radiated to the outside is perceived as traffic noise in the environment. In contrast to the interior noise, whose level is determined by the low-frequency noise, the exterior noise is dominated by the middle and high frequencies between 1 and 10 kHz. The primary factors that contribute to exterior noise when driving are the tire noise and the engine noise. The exterior noise is measured quantitatively with the legally prescribed method for exterior noise measurement. The general operating permit and, therefore, the approval for a motor vehicle to be driven on public roads requires that the exterior noise generated by a motor vehicle not exceed a maximum dB(A) level, determined according to the vehicle category and known as the exterior noise emission limit.

~~**Emission limits.** Emission limits are the maximum dB(A) levels permissible during type approval in order to achieve a general operating permit for the vehicle model. In the European Union, a value of 74 dB(A) applies for passenger cars as of 2003. In the future, further reduction to 71 dB(A) is in discussion in the European Union, which would correspond to reducing the acoustic power emitted by one half. However, this is also to be combined with a modification of the test conditions, whereby the tire noise is to be given a higher priority in comparison to the engine noise at

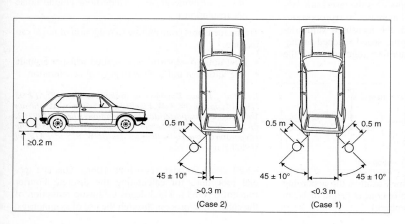

Fig. E207
Measuring setup for stationary noise measurement

full load, which better corresponds to reality, particularly at high speeds.

~~**Emission value.** This term refers to the dB(A) level determined in the legal exterior noise test.

~~~**Interpretation of emission value.** The obvious conclusion "low-emission value = quiet vehicle" cannot necessarily be concluded by testing one single vehicle in a driving state lasting less than 2 s. The following must be taken into consideration for differentiated examination:

- A low exterior noise emission value can result from an engine that is actually quiet, as well as poor "throttle response" of the vehicle during the first second—i.e., the vehicle starts to get louder only after it has exited the measuring path).
- The measured A-classified noise level (*see above*, ~Assessment curves) is not a value of the perceived loudness of subjective annoyance from the noise for many types of noise. Particularly noises containing pulses are perceived to be more annoying and louder by up to 20 dB(A) than indicated by the dB(A) measurement. A familiar example is a clattering motor scooter with defective or manipulated muffler that produces a noise which is perceived to be extremely loud even though the dB(A) level is comparatively low.

~~**Legal measurement.** Measurement of the exterior noise is standardized in ISO R 362. This internationally recognized measuring procedure is simple to perform; however, its informational value is limited. The principal procedure is illustrated in **Fig. E208**: The vehicle must approach a 20-meter–long measuring path at a constant speed of 50 km/h, and the throttle valve must be suddenly opened completely at the beginning of the measuring path. The noise level is measured during the subsequent full-load acceleration lasting for approximately 1.3 s; therefore, this is also called "accelerated passing," and this measurement is accomplished with two microphones—one each at a lateral distance of 7.5 m on either side of the vehicle path.

The maximum dB(A) value that is registered by either the left or right microphone is the test result. Because the engine noise level increases overproportionally with the engine speed, the measured value depends primarily on the engine speed reached during the test. An abundance of auxiliary regulations for the

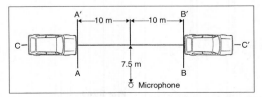

Fig. E208 Exterior noise measuring setup according to ISO 362, microphone at height of 1.2 m (Source: Klingenberg)

individual vehicle categories, therefore, specifies the particular gear for vehicles with manual transmission or the drive stage on vehicles with automatic transmission in which the vehicle is to accelerate as it drives past (for example, *see below*, ~Lex Ferrari). In a passenger car with five-speed manual transmission, this test is accomplished in second and third gears and the average value of the two measurements is the test result. On cars with automatic transmissions, the measurement is accomplished in drive stage "D" with the "kick-down" function switched off.

Literature: P. Ehinger, H. Großmann, R. Pilgrim: Fahrzeug-Verkehrsgeräusche. Messanalyse- und Prognose-Verfahren bei Porsche, ATZ 92 (1990) 7/8, pp. 398–409. — W. Betzel: Einfluss der Fahrbahnoberfläche von Geräuschmessstrecken auf das Fahr- und Reifen-Fahrbahn-Geräusch, ATZ 92 (1990) 7/8, pp. 411–416. — H. Klingenberg: Automobil-Messtechnik, Volume A, Berlin, Springer Publishers, 1991, p. 225ff.

~~**Reduction.** Certain development goals for the reduction of exterior noise affect the engine. These are:

- Reduction of engine speed. However, this leads to lower noise emission in practical traffic applications only when the car can actually be operated at the higher gear ratio at low engine speeds by increasing the torque available from the engine.
- Optimized combustion. A "softer" combustion process provides a more pleasant combustion noise, particularly with diesel engines.
- Vibration-optimized housing structure with high dynamic rigidity—e.g., with ribs.
- Vibration decoupling and coating of parts on the engine surface and attached parts such as resiliently mounted cylinder head cover, coated oil pan.
- Noise optimization of accessories such as the radiation fan and generator.

Modifications to vehicles for reduction of noise can include:

- Noise optimized tires. The contribution of the tires to the total noise is approximately 50% even at the low speeds in the legal exterior noise measurement.
- Encapsulation of the engine compartment (*also, see above*, ~Engine noise ~~Underfloor engine noise encapsulation.)

Traffic measures contributing to reduction of noise can include:

- Low-noise road surfaces, so-called whisper asphalt
- Regulation of traffic flow to prevent acceleration

Literature: R. van Basshuysen: Motor und Umwelt, ATZ 93 (1991) 1, pp. 36–39. — W. Eikelberg, G. Schlienz: Akustik am Volkswagen Transporter der 4. Generation, ATZ 93 (1991) 2, pp. 56–66. — J. Albenberger, T. Steinmayer, R. Wichtl: Die temperaturgesteuerte Vollkapsel des BMW 525 tds, ATZ 94 (1992) 5, pp. 244–247.

~**FFT (Fast Fourier Transformation).** This is a special procedure for calculating the discrete Fourier transform (DFT), which leads to drastic reduction of the calculation process through the use of symmetries.

The FFT is restricted to certain values for the number of support points, usually powers of two.

~Filter. *See below*, ~Frequency filters

~Fluid-borne sound. The primary difference between airborne sound and fluid-borne sound is the significantly higher characteristic acoustic impedance of liquids resulting in better transformation to structure-borne sound. As a rule, this means that due to the high degree of interaction, the frequencies of the solid bodies and their radiation can no longer be treated separately. Moreover, acoustic insulation is made more difficult.

~Fourier analysis (frequency analysis). Decomposition of a time signal into its frequency components— that is, determination of a spectrum—can be achieved with the aid of Fourier transforms. Fourier analysis is the most important tool for diagnosis of vibration and acoustic problems and serves as the basis for many other analysis procedures (*also, see below*, ~Signature analysis).

~Fourier transformation. This is integral transformation on the basis of trigonometric functions. The Fourier transform converts a "time function," $f(t)$, into a spectrum, $F(\omega)$ (frequency range), and vice versa:

$$F(\omega) = \int_{-\infty}^{\infty} F(t) \cdot e^{-i\omega t} \cdot dt \leftrightarrow f(t)$$

$$= \frac{1}{2\pi} \int_{-\infty}^{\infty} F(\omega) \cdot e^{i\omega t} \cdot d\omega$$

In addition to valuable information on the frequency distribution of a signal, Fourier transformation is particularly useful because many operations (integration, convolution, etc.) can be performed considerably more simply in the frequency range than in the time range. In practice, Fourier transformation is performed numerically on the basis of discrete support points (discrete Fourier transformation, or DFT). This is used very widely, particularly because the computation required was reduced drastically by development of the fast Fourier transformation (FFT). In terms of pure measurement technology, a spectrum can also be determined with the aid of frequency filters.

~Free field. The term refers to sound field without reflections. Free field conditions are frequently the prerequisite for precise and reproducible acoustic measurements and are, therefore, generated approximately in so-called low reflection chambers ("anechoic" chambers).

~Frequency analysis. *See above*, ~Fourier analysis

~Frequency band. *See below*, ~Octave analysis, ~Spectrum

~Frequency evaluation. *See above*, ~Assessment curves

~Frequency filters. These are systems whose characteristic property is that they change the spectrum of an input signal in a certain manner. The primary applica-tion is signal analysis, where frequency filters are introduced in the form of analog electronic circuits or digital filters. Frequent applications include:

- Low-pass: Reduces frequencies above a specified limit (e.g., anti-aliasing filters)
- High-pass: Reduces frequencies below a certain limit (e.g., for suppression of low-frequency interference)
- Band pass: Combination of low- and high-pass filters with specified pass range (e.g., octave or major third filter). Mechanical/acoustic systems are also used for frequency filters with certain characteristics (e.g., exhaust system).

~Frequency range. Frequency range is (a) a spread of frequencies and (b) the designation for a method of observing primarily stationary (steady state) signals based on their frequency contents and not on their time characteristics (time range), expressed as frequency-dependent amplitude and phase (spectrum). The basis for this is the Fourier transformation.

~Frequency spectrum. *See below*, ~Spectrum

~Fundamental frequency. Frequency generated by periodic operation (e.g., crankshaft speed).

~Gear meshing noises. *See below*, ~Transmission noise

~Generator noises. The generator/alternator can contribute to the overall engine noise directly in three ways:

- The flow noise from the connected fan impeller, which consists of a mixture of background noise and a periodic noise at the basic frequency, equals rotational speed times number of blades.
- So-called magnetic howling or whistling is produced by the magnetic forces, as a result of electric load, which excite the generator structure into mechanical vibrations that, for their part, are the cause of the noise radiation. The basic frequency of this purely periodic sound results from the rotational speed and the number of poles. This noise occurs to an increased degree as the electric load is increased and can, therefore, be detected more easily in the vehicle passenger compartment by switching on high electric loads.
- So-called electric howling or whistling results from the frequency of the electric current produced, which then causes other electronic components in the vehicle passenger compartment to vibrate at high mechanical frequencies.

The generator can also be associated, indirectly, with two other types of noise:

- The inertia mass of the generator together with the stiffness of the mounting represent a spring/mass system fastened to the engine block which can increase the engine vibration at its own resonant frequency.
- Due to its high speed, the generator is the most important rotating mass in the V-belt drive for the accessories. The resonance of this rotary vibration system consisting of the vibrating mass of the gen-

erator and the longitudinal rigidity of the V-belt can lead to V-belt flutter and corresponding "drumming noises" in the speed range close to idle.

~**Hammer excitation.** *See below,* ~Pulse hammer excitation

~**Harmonic.** This term refers to that portion of vibration with an integral multiple of the fundamental frequency (*also, see below,* ~Order).

~**Harmonic third analysis.** *See below,* ~Octave analysis

~**Helmholtz resonator.** This is a system capable of oscillation, consisting of a closed chamber with an opening, similar to a bottle, in which the air mass at the opening oscillates and the volume of compressible air in the container acts as a spring.

~**Heterodyne (beat).** Heterodyne is the periodic increase and decrease in the volume of a harmonic noise—for example, engine hum. This results from the superposition of two sinusoidal vibrations with slightly different frequencies. In the case of engine noise, this occurs particularly when the excitation resulting from a harmonic of the crankshaft drive—that is, an engine order—and the unbalanced excitation from an accessory have nearly the same amplitude and same frequency. As a result of belt slip or slight differences in speed, the phase position of the two excitations changes continuously so that they alternately reinforce or cancel one another. As a remedy, ratios close to 1:1 or integral ratios are avoided in laying out an accessory drive.

~**Hydraulic mount.** *See above,* ~Engine/accessory mount ~~Hydraulically damped engine mounts

~**Hydrodynamic near field.** *See below,* ~Hydrodynamic short circuit

~**Hydrodynamic short circuit.** A hydrodynamic short circuit is the local obliteration of sources when the distance from areas vibrating in the opposite phase is less than the length of one airborne sound wave. The air molecules are moved back and forth in a hydrodynamic near field without compression and dilation, whereby the radiation is very low.

~**Impedance.** *See above,* ~Acoustic impedance; *see below,* ~Input impedance

~**Input impedance.** This is defined as the ratio of vibration rate at the effective force point in direction of force and active force

$$\tilde{z} = \frac{\tilde{v}}{\tilde{F}}$$

(generally complex and dependent on frequency). It is the measure of the local resistance of a structure to excitation by an external vibration. The input frequency is inversely proportional to the actual portion of the input impedance.

~**Insertion attenuation.** This refers to differences in level before and after inserting acoustic insulation:

$$D_e = L_{o.D.} - L_{m.D.}$$

~**Intake noise.** The intake noise is caused by the pressure pulsations in the induction system. Its basic frequency, f, is determined by the number of opening and closing operations of the intake valves in a four-stroke engine; therefore, one-half the number of cylinders, z, times the crankshaft frequency:

$$f = \frac{zn}{2 \times 60} \text{ Hz,}$$

when the engine speed, n, is in rpm—that is, in a four-stroke gasoline engine equal to the ignition frequency. Subjectively, the intake noise is perceived as low-frequency humming or droning. In the part- and full-load range, the intake noise is difficult to separate from the engine combustion noise; it can be heard clearly in diesel engines when operating under overrun conditions, because they do not have a throttle valve. For test purposes, it is also possible to make the intake noise more clearly perceptible even in gasoline engines by opening the throttle valve on overrun with the ignition switched on.

~~**Intake silencer.** The components in the induction system in front of the intake manifold are designed as an intake silencer. Here, the air filter casing acts as a reflection-type silencer and the paper filter provides a slight absorption effect. Considerably high absorption of the intake noise is achieved by lining a longer intake snorkel with open-pore foam material or fiber matting. A certain acoustic effect can also be achieved by placing holes in the wall of the intake snorkel.

~~**Intake silencing.** It is possible to reduce intake noise by using: air filter casing with large volume, thick walls, and specific ribbing; intake snorkel with large length, thick walls, and absorbing inner jacket, and, where applicable, rubber supports or bellows with thick walls.

~~**Secondary intake noise.** A proportion of the intake noise is radiated by the surface of the intake system, which causes the pulsating airflow in the interior to vibrate. This radiated noise depends on the mass and wall stiffness of the intake system, particularly the air filter casing.

~**Intake opening noise.** A portion of the intake noise consists of pressure pulsations which reach the atmosphere through the open end of the intake snorkel (primary intake noise).

~**Interior noise.** The interior noise in a vehicle is determined primarily by three main components: engine noise, road noise, and wind noise. For these three, a certain level is accepted as unavoidable regardless of state or class of the vehicle. Noises from other components occurring beyond this, such as transmission noise, fan noise, noises from auxiliary equipment, and noises from interior equipment such as crackling, rattling, etc., are considered to be malfunctions and are not accepted by demanding customers. For this reason, the term "interior noise" always pertains only to the three main components.

Literature: W. Geib: Schallfluss im Automobil, ATZ 93 (1991) 9, pp. 562–576. — Q.H. Vo, W. Sebbeße: Entwicklung eines subjektiv angenehmen Innengeräusches, ATZ 95 (1993) 10, pp. 508–519.

~~**Measurement.** Measurement of interior noise is accomplished by recording the noise with two to four microphones when driving on a road or, when applicable, on an acoustic dynamometer. The four microphone positions are defined in ISO 5128 and are located in the middle of the seat and at the height of the head of a passenger of average height. However, because practical experience shows that the subjective impression of noise is determined more at the higher level at the outer ear, located closest to the side window from which the noise comes, many automotive developers use microphone positions, "standardized" within the company, and which deviate from the ISO standard. Here, the microphones are usually located at the position of the outer ear. When the subjective noise perception is stored in addition to exact measurement of the acoustic pressure level and the noise quality analyzed, a dummy head measuring system with stereophonic recording of signals at both ears is used so that the question of the ear position is eliminated. Because it is not possible to recognize from the A-rated interior noise level measured in the direct gear in relation to the vehicle speed whether the noise comes from the engine, the road, or the wind, such measured results are usually only suitable for use in a press release. Acoustic experts always attempt to analyze only one type of noise depending on the noise source and select the driving cycle so that the other two noise sources are eliminated to the greatest possible extent. This means that wind noises are measured at high vehicle speeds on a road with quiet pavement surface and with the engine shut off or idling; road noises, in contrast, are recorded at medium speeds on critical, defined surfaces such as rough asphalt, cobblestones, or individual obstacles such as cross joints in the highway, manhole covers, etc. The engine noise is measured at low vehicle speeds, usually in second gear or in drive stage 2 with automatic transmissions, which makes the engine speed the most important parameter, and in the operating states of full load, part load, and deceleration. Moreover, the dB(A) level is usually not taken into consideration, but rather the frequency ranges typical for the specific types of noise are filtered out, the level of which then represents a quantitative criterion for comparison and evaluation of vehicles of the same class.

~**Intrinsic shape.** *See below*, ~Resonant frequency

~**Laser Doppler vibrometry.** This is an optical process for measuring surface velocities. It is based on the frequency shift of laser light reflected from a moving surface (Doppler effect). In contrast to laser holography, the speed is measured at a point; that is, it is necessary to "stand" the part to determine the vibration forms (scanning vibrometer); this requires steady-state operation. Because the speed can be measured continuously and in phase, it is possible to determine the spectrum and transfer functions. The advantage of the optical process is, among other things, its suitability for measuring hot or rotating surfaces (exhaust manifolds, flywheel).

~**Laser holography (holographic interferometry).** This is an optical process which can be used to make very slight deformations in components such as the vibration of sound-radiating housing surfaces visible, as shown in **Fig. E209**.

For this purpose, the geometry of the surface is illuminated with laser and the image is stored photographically in a hologram (**Fig. E209a**) by superimposition of two holograms recorded at an interval of approximately 1 ms apart (double pulse holography) results in interference lines, similar to contour lines, which represent the deformation at this moment (**Fig. E209b**).

In contrast to laser Doppler vibrometry, transient vibration images can also be recorded. The photographically stored interference images are then digitized and converted to deformation images in a computer by precise evaluation of the graduation of the gray value between the interference lines (**E209c**). Computer processing of the holograms to obtain descriptive 3-D deformation images (**E209d**), usually also in color and animated, allows this procedure to be used for industrial development work. This process is being replaced to an increasing degree by laser speckle interferometry (*also, see below*, ~Laser speckle interferometry).

~**Laser speckle interferometry.** Like laser holography, laster speckle interferometry is also based on the interference of laser light; however, the photographic storage of a hologram is eliminated. With this procedure, **Fig. E209c** can be generated directly in a digital camera and further processed in the same manner. This makes the process much easier to use in practice.

~**Legal regulations for noise emission.** *See above*, ~Exterior noise ~~Emission limits, ~~Legal measurement

~**Level.** *See above*, ~Decibel

~**Lex Ferrari.** Lex Ferrari is a legal regulation for measurement of exterior noise according to ISO R 362, according to which sports cars or other passenger cars with particularly high acceleration capability must complete the acceleration passing test in third gear only, in contrast to normal passenger cars with five-speed transmissions that are measured in second and third gears. This takes into consideration the fact that the prescribed measuring process would be disad-

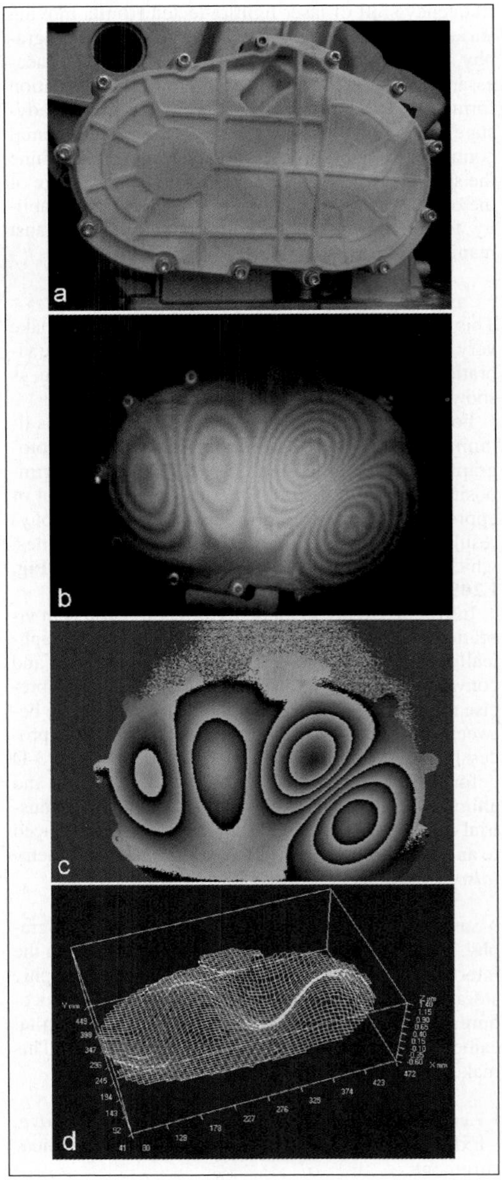

Fig. E209 Vibration distribution on surface of transmission case (Source: Laser Laboratory, FHT Esslingen)

vantageous for such vehicles in an unrealistic manner (*also, see above*, ~Exterior noise, ~~Interpretation of emission value).

~Loss factor. The loss factor is the value for quantifying damping, generally dependent on frequency. It is defined as the quotient of energy dissipated per vibration cycle and the reversibly stored energy:

$$\eta = \frac{W_d}{2 \pi W_r}.$$

The loss factor is used as a value for a specific material as well as for characterization of the damping of compound systems. It is closely related to the concept of the complex modulus of elasticity and can, therefore, only be used in the frequency range under restricted terms.

~Loudness. *See below*, ~Psychoacoustic parameters

~Low reflection chamber. Acoustic measuring chamber for obtaining approximately free field conditions. This is achieved by use of absorbing coverings (porous materials, membrane absorbers). The characteristic value for a low-reflection chamber is its fundamental frequency below which "complete" absorption is no longer achieved. With porous coverings (e.g., foam rubber wedges), this depends primarily on their depth (*also, see above*, ~Acoustics test bench).

~MDOF. *See below*, ~Multiple degree of freedom system (MDOF)

~Modal analyses. Modal analysis is the determination of the resonant frequencies and intrinsic forms (modes) of a system. In terms of calculation, this is accomplished by solving an intrinsic value problem, usually on the basis of a finite element model. Experimental modal analysis is based on the fact that (under simplified assumptions for the damping distribution) any vibratory response can be represented as the weighted sum of all its modes. Here, the weighting factors are frequency-dependent functions with poles located at the corresponding resonant frequencies. The information required to determine the poles and the modes can be obtained by measuring a sufficient number of transfer functions (vibration replies) at various locations in the system. The resonant frequencies and intrinsic forms can be determined from this by comparison with the modal sum representation with the aid of so-called "curve-fitting processes" (approximations). Pulse hammer excitation is the preferred method of measuring the transfer functions.

Literature: D.J. Ewins: Modal Testing, Theory and Practice. Research Studies Press Ltd., Letchworth, 1984.

~Modes. *See above*, ~Modal analyses

~Modulation. Modulation refers to constant, frequently periodic variation of amplitude (amplitude modulation) and/or frequency (frequency modulation) of a harmonic "carrier wave." The frequently observed periodic modulation, which can also be understood as an additive superimposition of vibrations, manifests itself in the spectrum with so-called side bands—that is, peaks to the left and right of the carrier frequency (and, where applicable, its harmonics) at the interval of the modulation frequency. This can be used to identify mechanical damage (bearings, gears) or production inaccuracies (eccentricity). Moreover, modulations have an important influence on the quality of a noise (e.g., exhaust tip noise, psychoacoustic parameters).

~Muffler rattle. This is the high-frequency noise resulting from turbulent flow through the restrictions or bends in the exhaust pipe.

~Muffler system. This term was previously the common designation for the exhaust system before emission clean-up functions were included.

~Muffling. Muffling is the dissipation of acoustic energy. An attenuation effect going beyond the inner damping always present is achieved for airborne noise particularly by using porous materials and for structure-borne sound by the use of highly insulating materials, insulating coverings, or discrete dampers (also resonance absorbers) as well as the use of joints.

~Multiple degree of freedom system (MDOF). This is a mathematical/mechanical model used to describe the vibratory characteristics of systems which can be represented (at low frequencies) in the form of discrete elements, such as masses, moments of inertia, springs, and dampers (e.g., piston/connecting rod/crankshaft, valve train). The model can also be used, among other things, to calculate resonant frequencies and intrinsic shapes.

~Narrow-band analysis. Frequency analysis with higher resolution of the frequencies is typically provided by FFT analyzers. In contrast to octave or major third analysis, narrow-band analysis enables the individual excitation frequencies, harmonics, and side bands to be identified, and it is the prerequisite for applications in vibration diagnosis.

~Near field. A near field is the zone around an acoustic radiator in which the acoustic field does not yet have the character of a smooth wave—that is, the acoustic pressure and acoustic velocity are not in phase. The acoustic field here contains reactive energy which cannot be radiated. The near field is important for measurements of acoustic power which do not generally have to be taken (or should not be taken) outside of the near field.

~Noise absorption. Noise absorption is the reduction of the noise level by converting the vibration energy into heat that results from friction. The absorption materials used in practice consist of textile materials, glass matting, or open pore foam in which the oscillating air molecules are slowed down by friction with the fibers and throttling losses in the pores. A further possibility for converting the energy consists of transferring the vibratory energy in the air to another vibratory system in which it can then be decreased by material damping or contact friction. For this purpose, the vibratory system must be capable of being excited easily

by the airborne sound wave—that is, its own resonant frequency must be tuned to the frequency of the incident noise. It is common to speak of resonance absorbers or, in reference to the design shape, of plate absorbers, when the connected vibrational system is a smooth wall or plate whose flexural stiffness is tuned to the frequency range to be absorbed. Another version of resonance absorption is a reflecting wall with a large number of small Helmholtz resonators. For example, the intake zone of many jet engines is designed with a perforated plate which provides a field of hollow chambers with a "bottleneck" in combination with the honeycomb sandwich construction located behind.

~Noise emission. *See below*, ~Sound emission

~Noise excitation. Noise excitation is the idealized excitation for measurement of mechanical and acoustic transmission functions in which a frequency generator is fed with a noise signal. On the one hand, this has the advantage that all frequencies are excited simultaneously (reduction of testing time) and, on the other, that the noise provides information on nonlinearities in the system with the aid of coherency. In practice, band-limited noise is used in order to concentrate the excitation power in the frequency interval of interest. Depending on the application, white noise (constant spectral power output) or pink noise (spectral power output proportional to $1/f$) is used, where the latter emphasizes the low frequencies to a greater extent.

~Noise generation in engine. Various physical phenomena and various engine components are responsible for engine noise, depending on the frequency range (**Fig. E200**):

- *Inertia forces on crankshaft drive*: With the exception of the rotary acceleration of the connecting rod and crankshaft, all these forces act along the cylinder axis and cause the portion of the engine block vibration that is independent of load (*see below*, ~Noise source, engine).
- *Combustion noises*: These act at low frequency and are responsible for the nonuniform rotation of the crankshaft drive and the reaction of a corresponding torsional vibration of the crankcase. Both are highly *dependent* on the load, and their effects on the acoustics and vibrations in the vehicle are, therefore, subjective and most perceptible at the point of changeover from unloaded to full-load operation. In particular, the rate of pressure rise in the cylinder is responsible for the high-frequency combustion noise—in diesel engines this is the familiar diesel knock.
- *Inertia forces of valve gear*: These result as a reaction to the acceleration of the valve masses and principally have the same effect as the inertia forces from the piston motion. Even though lower in magnitude, they can contribute to a humming noise in engines with good balance—that is, six- and eight-cylinder engines.
- *Valve gear drive*: The noise of the timing gear—timing chain howling, toothed belt howling—is in

the medium frequency range corresponding to the engagement frequency of the teeth on the chain or toothed gear sprockets. This noise is generated in a manner similar to that of gears (*see below*, ~Transmission noise) and results from the periodic loading and unloading of the teeth or chain links at the beginning of engagement. These howling noises are particularly audible at idle and in the lower speed range where they are not yet covered by the increasing mechanical and combustion noises of the engine. Moreover, *low-frequency noises* can result from vibration of the timing chain or toothed timing belt.

- *Valve striking noise*: This determines the high-frequency "mechanical" engine noise to a high degree and is frequently associated with the hard sound from multiple valve engines.
- *Piston slap*: Piston slap comes from the piston "slapping" against the cylinder wall as it moves from the thrust to the antithrust side.
- *Hydraulic pressure pulsations of engine oil pump*: These appear acoustically, particularly when external oil coolers are used which are caused to vibrate by these pulsations and then introduce structure-borne sound into the mounting elements.
- *Pressure pulsations in the intake air*: See above, ~Intake noise.
- *Pressure pulsations of exhaust*: See above, ~Exhaust pipe tip noise, ~Exhaust system noise.
- *Flow noises in manifold and exhaust system* (e.g., exhaust rattle, head pipe whistle).

~Noise immission. *See below*, ~Sound immission

~Noise intensity. This is a vector value indicating the mean power transported per surface unit. The intensity vector, I, is parallel to the velocity vector, v. It can be calculated from $I = \overline{p(t)v(t)}$ (time range averaging) or in the frequency range from

$$I = \frac{1}{2}Re\{\tilde{p}\tilde{v}^*\}$$

(p = acoustic pressure). The vector character of the acoustic intensity can be used with measurements for locating sources in complicated acoustic fields as well as for "robust" sound measurement.

~Noise level. *See above*, ~Decibel

~Noise quality. This is a concept for describing the subjective perception of noise and is determined by significantly more parameters than the measured acoustic pressure level or noise intensity (*also, see below*, ~Psychoacoustic parameters).

~Noise reduction. This refers to keeping acoustic energy away from a part of the overall system by reflection of the sound waves on impedance offsets. For reduction of airborne sound, this is accomplished with "walls" (bulkheads, doors, encapsulation, attached shells, etc.) Reduction of s tructure-borne noise is ac-

complished with resilient intermediate elements and/or blocking masses (vibration insulation). Acoustic energy is not "destroyed" by insulation; for this reason, acoustic attenuation is always required.

~~**Measurement of noise reduction.** The noise reduction of materials and components in automotive construction is measured on a so-called window test bench or sound intrusion test bench (**Fig. E210**). It consists of an excitation chamber, in which a high airborne sound level is generated artificially, and an opposing reception chamber which is insulated as perfectly as possible from the excitation chamber: the two are connected by a window. The specimen to be measured—that is, "through which the sound is to be transferred"—is installed in this window. The reduction in the level measured, compared to that through the open window, is the degree of noise reduction in decibels.

~Noise source, engine. The engine acts as a noise source in two physically different manners:

- By vibration of the engine/transmission block which transfers structure-borne sound to the body through all mechanical connections such as the accessory mounts, driveshafts, hydraulic hoses, accelerator control cable, shift linkage, etc. This excitation is responsible for the low-frequency interior noise, the "engine hum" radiated from the stiff metal body surfaces to the vehicle passenger compartment (**Fig. E211**).The vibratory acceleration values are measured at the engine supports to determine the value of this excitation and as a quantitative acoustic quality criterion for the specific engine.
- From the radiated airborne sound of the vibrating surfaces of the engine/transmission block. This airborne sound determines to a primary degree the exterior vehicle noise as well as the high-frequency share of the interior noise, the "mechanical engine noise." The technical measuring criterion for the engine as a source of airborne noise is the radiated

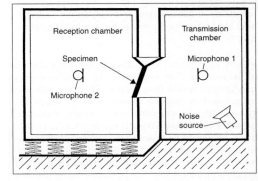

Fig. E210 Sound intrusion test bench for measuring noise reduction (Source: Klingenberg)

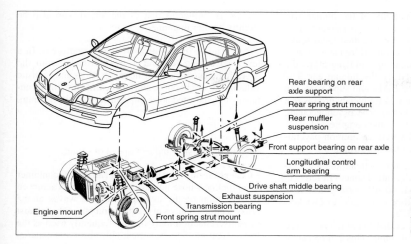

Rear bearing on rear axle support

Rear spring strut mount

Rear muffler suspension

Front support bearing on rear axle

Longitudinal control arm bearing

Drive shaft middle bearing

Exhaust suspension

Transmission bearing

Engine mount

Front spring strut mount

Fig. E211
Structure-borne sound transfer points on body (Source: *ATZ*)

acoustic power as a function of load and engine speed. The noises from the auxiliary equipment are frequently associated with the engine.

~Octave analysis. Octave analysis refers to coarse frequency categorization when measuring acoustic pressure and acoustic power (one octave corresponds to doubling of the frequency), achieved classically by using analog octave filters. The interval limits for the octaves are standardized in DIN 45401. The determination of the octave spectra is included in many measuring regulations; however, for diagnosis of acoustic problems, they are hardly suitable because of their coarse frequency resolution. The same also applies for major third analysis, which subdivides each octave again into three intervals.

~Operating mode vibration analysis. This is the visualization of the forms of vibration of a system (e.g., engine/transmission unit) that occur during operation—for example, acceleration measurements animated with a wire lattice model or laser holography. In contrast with experimental modal analysis, the intrinsic form can be determined only approximately, if at all; however, the expenditures are frequently considerably lower and the excitation corresponds to "reality."

~Order. Spectra of nonlinear, oscillating systems (e.g., crankshaft drive) contain harmonics in addition to the fundamental frequency. If the fundamental frequency is variable (e.g., engine speed), the frequencies of the harmonics vary at a constant ratio to the fundamental frequency as well as to one another. They are then designated as orders, whereby the number of the order indicates the factor in relation to the fundamental frequency. For example, the second engine order is the frequency curve corresponding to twice the engine speed. In contrast to harmonics, multiples and fractions in the system can be represented by nonintegral orders (e.g., half intervals on four-cycle engines).

~Order analysis. *See below,* ~Signature analysis

~Phase spectrum. *See below,* ~Spectrum

~Power steering pump noise. Vane pumps or radial piston pumps are used as power-steering pumps in passenger cars. Due to their design, they generate pressure pulsations at the basic frequency (engine speed times number of vanes, or number of pistons) and, therefore, the noise spectrum output consists of this fundamental frequency and integral multiples of this frequency. These harmonics determine the "sawing" character of the noise which is perceived as unpleasant. The average customer associates this noise with the engine, because it changes with speed in the same manner as the engine noise, and evaluates its sound quality accordingly. The practiced ear of a specialist distinguishes this noise from the engine noise because it becomes stronger when the load on the power steering is increased by turning the wheels. The power steering pump noise reaches the vehicle passenger compartment in three possible ways:

- Hydraulic transfer through the high-pressure hose. The pressure pulsations cause the hose as well as the load—i.e., the steering—to vibrate so that all mounting points for hoses and steering on the body represent potential structure-borne sound transfer points.
- Structure-borne sound transfer through engine and engine mounts. The vibration from the power steering pump is transferred by its mounts to the entire engine block and from there through the engine mounts to the body.
- Airborne sound radiation in the engine compartment. The surface of the pump and the adjacent engine structure are caused to vibrate by the inner forces and radiate airborne noise into the engine compartment.

~Psychoacoustic parameters. These serve for theoretical evaluation of acoustic signals in terms of improved approximations to the sensitivity of the human ear. The basis for many psychoacoustic parameters is a modulated frequency scale, the so-called tonal scale, which is based on the nonlinear local frequency transformation of the Basilar diaphragm and, therefore, the natural frequency perceived by the human ear. The tonal scale uses units called Barks and subdivides the audible frequency range into 24 frequency groups in a nonlinear manner (0–24 Barks). The psychoacoustic parameters are calculated by some complicated algorithms from the measured spectra. Common psychoacoustic parameters include the following:

- Loudness: Linear value for evaluation of a perceived volume using units called sones (Reference: 1 kHz sinusoidal sound, 40 dB corresponds to 1 sone). A calculation method (according to Zwicker) is standardized in ISO 532.
- Volume: Level value for evaluation of the perceived volume using "phones" as units. This can be calculated approximately from the loudness.
- Sharpness: Evaluation that accentuates high frequencies that constitute the sharpness of a noise (unit: 1 acum).
- Variation intensity: Evaluation of extremely low frequency (<20 Hz) modulations of the signal level which are usually perceived as annoying.
- Roughness: Evaluates modulations in the frequency range of 20 to 300 Hz, which makes a sound appear "rough," not necessarily a negative property.
- Tonality: Serves for classification of noises in terms of their percentage of pure tones in relation to the noise.

~**Pulse hammer excitation.** This is used particularly for experimental modal analysis. A primary advantage is the extremely wide and uniform spectrum of a pulse, which is wider the shorter the duration of the pulse (pulse duration "adjustable" with hardness of hammer tip). In this manner, all relevant frequencies can be excited with literally one impact which means a great time savings in comparison to sinusoidal excitation at various frequencies. A further advantage is that no special devices (e.g., mount for vibration generator) are required for excitation. A disadvantage is that the excitation energy cannot be bundled to a single-frequency range and, therefore, very high forces may be required, which can lead to local damage and nonlinearity.

~**Radiation loss factor.** Ratio of sound energy, P, actually radiated from a surface, S, to that of a large (significantly larger than the sound wavelength) contraphase vibration plate with the same mean square velocity, $S\bar{v}^2$, density, ρ, and sound velocity, c:

$$\sigma = \frac{P}{\rho \cdot c \cdot S \cdot \bar{v}^2}$$

where σ is the value for the efficiency of a radiating surface. If the contraphase vibrating areas in the surface are significantly larger than the wavelength of the sound, $\sigma \approx 1$. In the opposite case, $\sigma < 1$. In the transition area where the "wavelength" of the radiating

surface is similar to that of the air, σ can also be slightly greater than one.

~**Rated noise impedance (surge impedance).** In a level wave field (e.g., at a great distance from the acoustic radiator) the acoustic pressure, $\tilde{p}$, and acoustic velocity, $\tilde{v}$, are in phase and can be linked with one another using the rated noise impedance, Z, of the medium:

$$Z = \frac{\tilde{p}}{\tilde{v}} = \rho c ,$$

where ρ = density, and c = speed of sound.

~**Reflection muffler.** In principle, an empty container in a pipeline whose noise deadening effect is achieved by reflection of the acoustic pressure waves at the cross section changes at the inlet and outlet. In exhaust systems, reflection mufflers are frequently used as the final or rear mufflers because highly absorbent material which could become saturated with liquid is not desired here due to the possible formation of condensate.

~**Resonance.** This is present when the frequency of an external excitation coincides with the resonant frequency of the excited system. In this case, the amplitudes of response of the system are limited only by internal and external damping as well as by nonlinearity and are, therefore, frequently very high. The associated intrinsic form of vibration then dominates when the system is excited.

~**Resonant frequency.** This is the characteristic frequency at which a system continues to oscillate without external excitation following an initial excursion. Systems with a number of degrees of freedom generally have just as many resonant frequencies as they have degrees of freedom; continuous systems have an infinite number of resonant frequencies whose density generally increases with increasing frequency (e.g., oil pan). A characteristic intrinsic shape is associated with each resonant frequency: this can be represented as a characteristic vector for discrete systems, which must be present as an initial excursion to excite the associated resonant frequency. Any form of vibration can be built up using the intrinsic shapes (modal analysis). In acoustics—knowledge of resonant frequencies and intrinsic shape—is very interesting because systems react particularly strongly when an outside excitation is present whose frequency is close to one of the resonant frequencies (resonance). In this case, the associated intrinsic shape then dominates.

~**Reverberation chamber.** This is an acoustic measuring chamber with smooth, acoustically "hard" walls articularly designed to give the least possible absorption. The objective is to produce a quasi-diffused acoustic field by means of multiple reflections, for which simple relationships apply between input and dissipated acoustic power (absorption) as well as the measured mean acoustic pressure (formula from Sabina). Characteristic values of a reverberation chamber are its volume and (largest possible) reverberation time (time for 60 dB decrease in acoustic pressure level). Angular walls as well as fixed and moving diffusers are used to reduce the influence of chamber resonances (standing waves). Reverberation chambers are used primarily for measuring acoustic power and determining degrees of absorption for lining materials (so-called α chambers).

~Reverberation chamber process. *See below*, ~Sound measurement

~Reverberation field. A reverberation field is part of an acoustic field dominated by reflected sound waves.

~Reverberation radius. This is the distance from the radiator at which the energy density of the direct sound and the reverberation field are equal.

~Rigid body modes. These are the lowest intrinsic forms of solid bodies which are not restrained or are only weakly restrained in certain directions of motion. Solid bodies perform lateral or rotational vibrations without distortion (e.g., cadence vibrations of engine/transmission unit).

~Road noise. When discussing the road noise of a vehicle, it is necessary to differentiate between the exterior noise, which is usually perceived by the environment as an annoyance and the interior noise, which influences the comfort of the passengers.

~Roughness. *See above*, ~Psychoacoustic parameters

~SDOF (single degree of freedom system). SDOF is a single-mass oscillator.

~Secondary radiation. In contrast to direct radiation, secondary radiation occurs when airborne sound results after conversion through this sequence: airborne sound, structure-borne sound, airborne sound (*also, see above*, ~Exhaust system noise ~~Secondary exhaust system noise).

~Sensitivity. This is the frequency-dependent ratio between the observed value and excitation value: it is frequently used for the acoustic transfer function.

~Sharpness. *See above*, ~Psychoacoustic parameters

~Side bands. Side bands are peaks in the spectra that occur at a slight interval next to the primary frequency. They result from superposition of vibrations or periodic operations with different frequencies and can be an indication of imbalance or bearing damage (*also, see above*, ~Modulation).

~Signature analysis. The spectra of an acoustic source with variable frequency (e.g., engine) change continuously with the frequency generator (e.g., engine speed, vehicle speed) as a general rule. If the spectra

are plotted in relation to the frequency parameter, the "signature" of the acoustic source is obtained. This is accomplished in 3-D form as a so-called waterfall diagram, or in the form of a so-called Campbell diagram with different colors coded to the ranges of levels (**Fig. E201**). Signatures are very useful because they allow recognition of whether an increase in the spectrum is the result of increased excitation (variable frequency peaks, order lines) or resonance in the transmission path (constant frequency). Moreover, it is possible to recognize how many orders were generated and which of these are significant. If only a few orders are relevant, it is possible to limit the study to analysis of a single order for data reduction. Here, only the level of the order lines is charted in relation to the engine speed (**Fig. E202**) and data reduction can be achieved even during measurement by using inline (variable frequency) filters.

~Sound emission. Sound emission refers to radiation of acoustic waves. Generally, the acoustic power level, L_W, serves as the value for noise emission.

~Sound field analysis. This is a measuring procedure which allows for two-dimensional or three-dimensional analysis of an acoustic field and which in practice is used frequently for localization of highly radiating areas such as a side of the engine. The technique uses one of the following methods:

- Noise intensity scanning, in which a noise intensity probe guided by a positioning robot over a matrix of "dots" provides a two-dimensional "intensity map" of the radiating surface in combination with the corresponding evaluation software, with which radiation hot spots can be recognized. This process is simple and robust and provides information on the absolute source strengths; however, it requires steady conditions over a long measuring period.
- Array/beam forming process, in which the acoustic pressure is recorded in parallel using a series of specially located microphones (array) at a certain distance from the source of the sound (short measuring time, transient processes). The software allows evaluation of the directional dependency of the source of the acoustic pressure, allowing a "source distribution" to be drawn up for the plane of the acoustic radiator surface.
- Spatial transformation sound field (STSF)/acoustic holography, with which the complete acoustic field can be determined from measured acoustic pressure signals with the aid of computation models; this also allows measurement directly at the surface of the radiator. The acoustic pressure signals are recorded by an array of microphones located at a distance from the acoustic radiator; additional reference sensors are used, if required, for separation of independent sources. High performance systems allow short measuring times. Newer versions of this process allow transient acoustic events to be analyzed.

Literature: M. Quickert, O. Andres: Moderne Verfahren zur Ortung von Schallquellen am Beispiel schwerer Nutzfahr-

zeugdieselmotoren, In H. Tschöke, W. Henze (Ed.): Motor- und Aggregate-Akustik, House of Engineering Technical Manual, Volume 25, Expert Publishers, Renningen, 2003. — Acoustic Holography, Noise Source Identification and Quantification, LMS Application Note, Publication #4.0/2052/A20/ 07.96, Leuven (BE). — J. Hald: Non-Stationary STSF, Brüel & Kjær Technical Review No. 1-2000, Nærum (DK) 2000, ISSN-0007–2621.

~Sound immission. Incidents of sound waves at a point of observation. Generally, the acoustic power level, L_W, serves as the value for acoustic pressure level, Lp.

~Sound intrusion test bench. *See above*, ~Noise reduction ~~Measurement of noise reduction

~Sound measurement. The acoustic power radiated serves as the basis for evaluating the noise emission of an acoustic radiator; the following processes are common for measurement:

- Enveloping surface acoustic pressure measurement. The acoustic pressure is measured at discrete points on an imaginary surface that envelops the radiator, and the acoustic power is calculated from these measurements. This is based on the assumption that the acoustic pressure and acoustic velocity are coupled uniquely with one another by means of the rated noise impedance (level wave propagation) and, therefore, the acoustic power requires sufficient distance from the radiator as well as free field conditions (e.g., low-reflection measuring chambers). If these prerequisites are violated, it may be possible to work with corrective terms.
- Reverberation chamber. The acoustic power is determined from the averaged acoustic pressures measured at various points in the diffuse field of a reverberation chamber. The basis of this method is that in a steady state the acoustic power input and the acoustic power that is dissipated must be at equilibrium, hence the dissipated acoustic power can easily be determined for a reverberation chamber from its known absorption (or reverberation time).
- Measurement with comparative noise source. A measurement according to procedure/method 1 is accomplished for the acoustic radiator to be studied as well as for a comparative noise source with known acoustic power, positioned at the same location. The actual acoustic power can then be concluded in consideration of ambient reflections by comparing the results.
- Enveloping surface intensity measurement: Procedure is the same as in method 1; however, instead of the acoustic pressure, the acoustic intensity is measured (generally with a double microphone probe). As a result of the vector character of the noise intensity, the acoustic flow is taken into consideration with the correct signs, whereby superimposed extraneous (incoherent) noise sources as well as reflections stand out. This procedure can also be used under "normal" ambient conditions, which do not satisfy the requirements of a measuring chamber.

Literature: H. Henn, G.R. Sinambari, M. Fallen: Ingenieura-kustik, 3rd Edition, Braunschweig/Wiesbaden, Vieweg Publishers, 2001.

~Sound particle velocity. This is the velocity of the local vibratory motion of particles in a wave. Because the vibratory motions have a direction, depending on the shape of the wave, the particle velocity is a vector value. If this is not specified particularly, usually its quantity or the main direction is meant.

~Sound radiation. In the closer sense, this is generation of sound waves by a vibrating solid structure (e.g., engine block), in contrast to aeropropulsive or aerodynamic sound sources (e.g., exhaust pipe tip noise, flow noises). In practice, the radiated acoustic power is usually important. If the areas of the acoustic radiator oscillating in opposite phase are significantly larger than the airborne sound wavelengths, it is proportional to the mean square of the velocity of the radiator surface. The other case is considerably more complicated to consider, however, because here, the radiated acoustic power generally decreases with the number of opposite phase areas present and the closer they are together.

~Spectral power output. *See below*, ~Spectrum

~Spectrum. Subdivision of a signal into its various frequencies (frequency analysis) provides its spectrum.

Because the individual frequencies may have different phase positions in relation to one another, a spectrum is usually a complex variable—that is, it can be split into a sum spectrum and a phase spectrum. Energy and power spectra, in contrast, are purely real components and represent the frequency distribution of the energy or power. In practice, spectra are formed either by filter banks (series of band pass filters, usually octave or harmonic third filters) or with the aid of discrete Fourier transformation (DFT). The resolution of spectra is therefore limited, so that the discrete values of a spectrum represent the contents of a finite frequency band including overlaps in each case (important for continuous spectra—e.g., noise). The spectral power density (PSD), in contrast, indicates the power per unit of frequency band, which is independent of the resolution.

Literature: R.B. Randell: Frequency Analysis, Rev. Sept. 1987, Brüel & Kjaer DK BT 007-11, ISBN 87 87355078.

~Speed of sound. This is the rate at which sound waves propagate. The speed of sound is a characteristic value for a transmission medium and, in the case of structure-borne sound, for a type of wave as well. For the most important types of waves, the speed of sound can be calculated as follows:

- for gases and liquids: $c = \sqrt{K / \rho}$
  K = bulk modulus, ρ = density
- for longitudinal waves: $c_L = \sqrt{E / \rho}$
  E = modulus of elasticity, ρ = density
- for torsion waves: $c_T = \sqrt{T / \theta'}$
  T = torsional rigidity, θ' = polar moment of inertia
- for flexural waves: $c_B = \sqrt[4]{\omega^2 B / m'}$
  B = flexion or plate stiffness, m' = mass (per unit

length or surface area), ω = angular frequency. The flexural wave speed is, therefore, independent of the frequency.

~Standing wave. In a spatially limited acoustic field, so-called standing waves are formed by harmonic excitation because the sound waves are reflected in a regular manner and are then superimposed to obtain a stationary "wave pattern." This becomes particularly obvious in the case of excitation at a resonant frequency so that the waves that are superimposed on themselves are in phase everywhere and develop a dominant intrinsic form (e.g., interior resonance in a passenger compartment).

~Stationary noise measurement. This is the procedure described in DIN (ISO) 5130, in which the engine is allowed to idle and then revved up to the rated speed and the dB(A) level measured at the end of the exhaust pipe (**Fig. E207**). The maximum value is entered in the vehicle papers; however, it has no significance for registration of the vehicle.

~Statistical energy analysis (SEA). This is the method for estimating the interaction of coupled vibratory systems (e.g., body/passenger compartment) at high frequencies. Under the "statistical prerequisite" of high modal density (many resonant frequencies located very close to one another), it is possible to assume simplified relationships for the power transmission in which the masses, modal densities, and loss factors of the subsystems and transfer points are used as the determining values.

~Structure-borne sound. This refers to noise in solid bodies. In contrast to acoustic waves in airborne and liquid-borne sound, which consist of a pure compression wave, acoustic waves in solid bodies are propagated in many different forms (even simultaneously) due to their capability of absorbing shear stress. A few examples include: extending waves (valve shaft), flexion waves (body sheet metal), torsion waves (camshaft).

Literature: L. Cremer, M. Heckl: Körperschall, Springer Publishers, Berlin, 1995.

~Structure-borne sound transfer function. (*also, see above*, ~Acoustic transfer function). This is the frequency-dependent ratio between vibratory response, x (movement, velocity, or acceleration), at one point, j, in relation to excitation value (usually one force) at input point, i:

$$\tilde{H}_{ij} = \frac{\tilde{x}_j}{\tilde{F}_i}.$$

~Sum spectrum. *See above*, ~Spectrum

~Surge impedance. *See above*, ~Rated noise impedance

~Test conditions, noise emission. *See above*, ~Exterior noise

~Threshold frequency. Threshold frequency is the frequency used to designate frequencies at which the characteristic behavior of a system changes when the frequency exceeds or drops below the threshold designated. In acoustics, it is used, among others, for: coincidental frequency, frequency above which certain wedge shapes are no longer capable of propagation in pipelines (e.g., exhaust system); cutoff frequencies of frequency filters; and range limits of instruments.

~Timing chain howl. The periodic load on the chain links when they engage with the teeth on the sprocket results in a periodic noise with a fundamental frequency equal to the number of teeth times the speed — that is, frequency of tooth engagement. This excitation in the circumferential direction is in principle the same process that occurs when two gears in a transmission engage (*see below*, ~Transmission noise). In addition, the so-called polygonal effect on the chain links results in momentarily radial acceleration when running onto the chain sprocket which is known as the "contact impact." As a countermeasure, the chain links in the valve train of high-quality passenger car engines are rubber-coated.

~Toothed belt whine. In addition to the processes described above (~Timing chain howl), this term also includes the local elongation and bending of the teeth when they engage with the sprocket teeth.

~Transfer function. *See above*, ~Acoustic transfer function, ~Structure-borne sound transfer function

~Transmission degree. This is the ratio between transmitted noise intensity (allowed to pass through) and incident noise intensity:

$$\tau = \frac{I_{trans}}{I_{incident}}.$$

It is used as a characteristic variable for the specific effect of insulating measures.

~Transmission grinding. Grinding transmission noises are usually a sign of mechanical defects such as pitting on the teeth flanks or on the surfaces of the bearings.

~Transmission noise. Every transmission with gears generates noise at the tooth engagement frequency equal to the number of teeth times shaft speed. This is generated by the periodic load on each individual tooth which has to take up a circumferential force proportional to the torque transferred at the beginning of the tooth engagement; this load is relieved when the next tooth engages. The resulting sinusoidally periodic noise and its harmonics are known as "transmission whine" at low speeds in the vehicle — for example, "differential whine" — and as high-frequency "transmission singing" or "transmission whistle" at high speeds.

The most important variables in terms of design are the contact ratio of the gear (straight or helical gears), the resistance of the shaft to bending, and the stiffness of the housing structure. Also, in terms of manufacture, the pitch and other geometrical factors and the surface quality of the teeth flanks have an effect, as do the operating parameters—for example, the load and oil viscosity (i.e., oil temperature). A further noise, which also occurs even on mechanically intact transmissions, is transmission rattle.

~Transmission rattle. Transmission rattle occurs in manual transmissions as a result of the faces of the teeth on idle gears that are rotating at the same speed and hitting against one another. This is caused by the nonuniform rotation of the engine, which produces the corresponding alternating accelerations on the transmission shaft so that the gears not under load at that time "rattle" within their backlash tolerances: the excitation from the engine is greater at high load and low engine speed. It is necessary to differentiate between two frequencies for the resulting, nonharmonic noise. The teeth faces hit against one another at a frequency of twice the nonuniformity of the engine speed, each time the circumferential acceleration changes—that is, the pulse sequence frequency is twice the firing frequency. The noise pulses themselves are, however, high frequency and wideband in the range of 1 to 4 kHz: the noise is said to be "modulated" at twice the firing frequency (*also, see above,* ~Modulation). The primary variables of influence on the design side are all measures which determine the nonuniform rotation of the transmission input shaft: the torque diagram of the engine, the size of the flywheel, and torsion spring rigidity and friction of the torsion damper. The last two of the most important tuning parameters for the drivetrain—a soft torsion spring that decouples the transmission from the nonuniform rotation of the engine—however, can, in interreaction with other soft rotational elements in the drivetrain, result in bucking or buildup of low-frequency vehicle vibration, the so-called bonanza effect, as a result of the undesired windup effect during load changes. The torsion spring, together with the rotating masses of the transmission, forms a resonant system which is responsible for transmission rattle at a certain speed which differs slightly depending on the gear engaged—and is usually particularly distinguishable between 1000 and 2000 rpm. It is necessary to suppress this resonance with friction in the torsion damper. The best solution for vibration insulation of the nonuniform rotation is the so-called dual mass flywheel (**Fig. E212**), which has particularly low torsion-spring rigidity, and shifts a proportion of the flywheel mass to the transmission side.

This approach shifts the resonant frequency of the system to far below the idle speed, and it is used particularly on high-quality and critical engines such as diesel engines with direct injection in passenger cars. Countermeasures on the transmission side are naturally to keep the gear backlash as low as possible and to optimize the transmission casing for minimum noise radiation. On the vehicle side, measures include acoustic insulation, particularly in the area of the transmission tunnel, because the noise is trans-

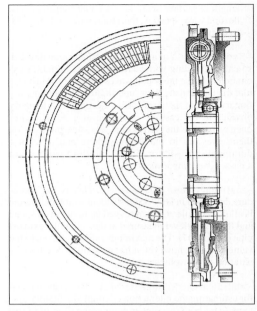

Fig. E212 Dual mass flywheel (Source: LuK)

ferred into the vehicle interior in the form of airborne sound.

Literature: H. Finsterhölzl, W. Keller: Leerlaufqualität in der Fahrzeugbeurteilung, ATZ 92 (1990), 5, S 268–282. — G. Weidner, G. Lechner: Klapper- und Rasselgeräusche in Fahrzeuggetrieben, ATZ 92 (1990), 6, pp. 320–326.

~Transmission whistle. *See above,* ~Transmission noise

~Underwater sound. *See above,* ~Fluid-borne sound

~Velocity. *See above,* ~Sound particle velocity

~Vibration insulation. Vibration insulation is used when there is a difference of vibration level at the front and the back of a resiliently mounted component.

~Volume. *See above,* ~Psychoacoustic parameters

~Window test bench. *See above,* ~Noise reduction ~~Measurement of noise reduction

~Windowing technique. This is an experimental procedure for identification and evaluation of partial surfaces of an acoustic radiator. The acoustic radiator (engine, transmission) is first surrounded completely by a highly insulating, close-fitting jacket (e.g., mineral wool and lead plate). Then, a small "window" is removed from the jacket and the influence of its removal on the radiation measured. This suppresses the interaction with other partial surfaces: the procedure is valid only for uncoupled partial radiators, strictly speaking (no airborne wavelengths relative to radiator, incoherent partial sources).

Engine air mass flow curve →Supercharging

Engine and transmission control →Electronic/mechanical engine and transmission control ~Electronic components, ~Requirements for mechanical systems and housing concepts

Engine arrangement →Engine acoustics ~Order

Engine assemblies and components →Engine damage

Engine block →Crankcase

Engine bolts. A basic state-of-the-art engine has between 250 and 320 bolted connections, held by 80–160 different types of bolts. The number of bolts is primarily dependent on the design—for example, whether the engine is an inline four-cylinder or V-6 engine—and is less dependent on the combustion process (diesel or gasoline). In comparison to European engines, engines developed in Japan have approximately 15% more bolted connections with simultaneously less variety of types. The size of the bolt increases in proportion to the displacement.

Automation in final assembly has been introduced to a very high degree by all European automobile manufacturers in mass production since the middle of the 1980s. For this purpose, it was necessary to design bolts to facilitate automatic feed and assembly.

Engine construction requires high-precision production of components such that the production tolerances for the basic parts (e.g., cylinder block and cylinder head) are extremely low and the positioning accuracy of the operating equipment and robots is less than 0.5 mm. For this reason, engine production has always been highly automated in comparison to other production units. If only a small number of engine versions are produced on a production line, full automation is practical; partial automation is more practical for a larger variety of types. In the latter case, the connecting elements are fed manually and the bolts are only tightened by a single or by multiple screwdrivers in an automatic tightening station—to facilitate compensation of the reaction torque.

With the advancement of control systems and ergonomic design, manual screwdrivers with integrated electronic circuitry (torque and angular rotation sensors) are being used increasingly for monitoring or controlling the tightening operation. This reduces the investment and maintenance costs for the assembly line while increasing the flexibility in the direction of "joint production systems."

For automation, the use of high-quality bolts is important. If bolts from different manufacturers are installed on a mixed basis, experience shows that this results in difficulties if precise specifications in terms of material, yield strength ratio, and friction values have

not been made from the very beginning. Manufacturers find that after changing suppliers, it is frequently necessary to readjust the system.

When manufacturers select the tightening and assembly procedures, they need to remember that passenger car engines are usually produced in larger quantities while commercial vehicle engines are produced frequently in small quantities or as individual products.

~Threaded connections. Generally, five critical bolted points are present on the engine; they are described below. The bolted engine connections and flanges are not considered here, with the exception of the threaded connections for magnesium components. In these components high-strength bolts in the diameter range beginning with M6 are usually used; thus, these are primarily standardized or similar versions.

~~**Connecting rod bolts.** The connecting rod connection is a typical example of a bolted connection subject to high dynamic load. The dimensioning range for passenger car engines is from M 7 to M 9, for commercial vehicle engines from M 12×1.25 to over M 14×1.5 up to M 16×1.5, for the extremely large diesel engines used in construction vehicles or for stationary application, up to UNJF 5/8^{00}-1 8 Gg and even UNJ 1 1/1 6^{00}-1 6 Gg or the corresponding metric sizes (UNJF and UNJ are designations from the "Unified Screw Threads" standard used in countries using Imperial units).

Predecessor engines or engines of a similar type and size are used for correct dimensioning of the connecting rod bolt. In the case of connecting rod housing bolts, the operating load for the bearing housing is known from the simulation of the crank drive (mass and gas forces).

Initially unknown are the operating loads resulting from the size, direction, and position in relation to the bolt axis at the separating joint plane, which are introduced into the individual bolt to determine the deformation and stresses for the bolt. The specialized literature gives various analytical procedures by which the axial force FA, the lateral force FQ (calculated value from the friction at the separating joint), and the eccentric distance, a, of the axial force from the bolt axis can be evaluated, depending on the design parameters of the connecting rod housing. If these values are available, it is possible to use a PC program such as "KABOLT" (bolt calculation according to VDI 2230) to calculate the required initial tension to prevent partial lifting and lateral displacement of the bolted joint and, therefore, the corresponding thread dimensions and strength class for the bolt. The bolt tightening specifications can then be determined on the basis of the values measured. After analytical layout of the connecting rod connection is completed, the durability can be checked on the entire connection rod with pulsing tests. The calculation and laboratory results are then confirmed in field tests. The calculation parameters for connecting rod bolts are given in **Fig. E213**, where the example is based on a four-cylinder gasoline engine with a displacement of 1996 cm^3.

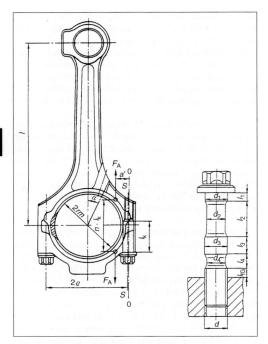

Fig. E213 Relationships for connecting rod connection

Subject characteristic:

- Linear computation formula (can also be calculated nonlinearly: Rectangular separating joint area, eccentric force application)
- Bolt with reduced shank
- Thread start rolled
- Washer not present
- Surface phosphated (→Friction value torque tightening; →Tightening factor)

KABOLT definition:

Other bolts with cylindrical shank

M $8\times1\times43$–11.9 ($R\rho_{0.2} = 1035$N/mm^2)

Bolt section

No. 1 Diameter	$d_1 = 7.75$ mm
Length	$l_1 = 3.0$ mm
No. 2 Diameter	$d_2 = 6.5$ mm
Length	$l_2 = 14.0$ mm
No. 3 Diameter	$d_3 = 8.3$ mm
Length	$l_3 = 5$ mm
No. 4 Diameter	$d_4 = 7.35$ mm
Length	$l_4 = 3.55$ mm
Free thread length	$l_f = 3$ mm

Loads and characteristic data

Operating force	$F_{Ao} = 10600$ N
Operating force	$F_{Au} = 0$ N

Initial residual clamping force: $F_{Kerf} = -$
Yield point $\quad\quad\quad\quad\quad\quad R\rho_{0.2} = 1035$ N/mm^2
Tightening factor $\quad\quad\quad\quad \alpha_A = 1.181$
Friction, thread $\quad\quad\quad\quad\quad \mu_G = 0.101$
Friction under head $\quad\quad\quad\quad \mu_K = 0.14\,1$
Flange material $\quad\quad\quad\quad\quad\quad = C\,45$

Geometry

Nominal bolt diameter	$d = 8.00$ mm
Thread pitch	$P = 1.00$ mm
Head contact diameter	$d_w = 12.70$ mm
Friction diameter	$d_{km} = 10.60$ mm
Clamped sleeve	$D_A = 13.18$ mm
Clamping length	$l_K = 28.55$ mm
Force introduction factor	$n = 0.501$
Force introduction factor	$n_{red} = 0.451$
Bore diameter	$d_h = 8.50$ mm
Bore bezel diameter.	$d_F = 9.20$ mm
Eccentricity, bolt	$s = 0.48$ mm
Eccentricity, bolt	$a = 2.70$ mm
Length	$c_T = 13.20$ mm
Length	$c_B = 12.70$ mm
Width	$b = 33.10$ mm
Gauge for connecting rod hole	$l = 132.00$ mm
	$l_1 = 31.50$ mm
	$l_2 = 100.50$ mm
Connecting rod bearing eye radius	$r_m = 27.00$ mm
Distance between bolt centers	$2\rho = 62.00$ mm
Clamping angle	$\alpha = 23°$

The connecting rod bolt design depends primarily on the loads and the manner in which the connecting rod is assembled. The bolts are provided either with a head shape with force transfer elements or rotation prevention elements, depending on whether a nut is used. The two halves of the connecting rod are centered either by means of a knurled section, fitted collar, or separate pin; in the case of large connecting rods on commercial vehicles, splines are frequently used according to the tongue-and-groove principle. In the case of cracked connecting rods, centering is accomplished by means of the resulting fracture surfaces.

When making new investments, procurement of a complete production line is frequently not economical for connecting rods produced only in medium quantities. Here, the use of sintered connecting rods is more economical. While in the case of conventional production it was necessary to cut the large connecting rod eye with a machine in order to insert the connecting rod bearing shell after machining, "cracking" has become widespread during the past three years on sintered, cast, and forged connecting rods. Here, what later becomes the bearing cap is cracked off in a device by a defined force applied externally to an intended breaking point on the connecting rod. In addition to eliminating the cutting operation, the advantage is that both halves of the connecting rod center themselves by means of the fracture surfaces when assembled later (for machining the bearing shell seats). For this reason, a precision fitted shank on the connecting rod bolt

With more than six decades of solving complex fastening

applications, KAMAX has gained superior product knowledge and

KAMAX fastener solutions worldwide

unparelleled experience in the fastener industry. Our extensive

knowledge includes not only

high-strength fasteners but a wide range of other unique

fastening solutions to support the worldwide automotive industry.

KAMAX provides value by offering total technical support to

ensure an optimal product design for your specific application

needs while also meeting budgetary requirements. The global

reach and overwhelming success of our quality products has

allowed KAMAX to become the leader in fastening solutions.

KAMAX

info@kamax.de · www.kamax.com

is not required for cracked connecting rods. The bolt diameter can have a tolerance of up to 0.1 mm.

Each connecting rod is installed twice after cutting. The first time it is installed for machining the bearing shell seat: here, the initial assembly tension must be similar to that later in use so that similar deformation results for the connecting rod bearing housing. For this reason, the bolts are tightened either to a specified torque or rotation angle up to just below the yield point or directly by means of yield point control. After this machining, the connecting rods are disassembled (for inserting the bearing shells) and then assembled on the crankshaft. Here, either an angle-controlled tightening procedure, which tightens the bolts beyond the elastic limit, or the yield point-controlled tightening process is used. If angle-controlled tightening is used, complex laboratory tests are required to have established previously the tightening recommendations. For yield point-controlled tightening, it is sufficient to define the so-called "green window" with a few tightening tests.

The question arises as to which bolts are best suited for yield point-controlled tightening, particularly because the connecting rod bolts have to be tightened twice in the yield point range as a result of the connecting rod production process.

When dimensioning bolted connections, consideration must be given that the bolt will break at its weakest cross section if subjected to excessive stress from static tension forces. This is generally the case in the free loaded section of the thread or in the area of the reduced shank. With the recently developed multiple waist bolts, the fracture is also located in the area of the waist. The connecting rod bolts shown in **Fig. E214** are particularly capable of being tightened beyond their elastic limit.

In the case of bolts with shanks (similar to DIN EN 4014), at least six free-bearing thread turns should be present in order to distribute the plastic elongation over a larger area and, therefore, avoid the danger of premature constriction.

Bolts with reduced shanks have the best tightening characteristics in the range beyond their elastic limit. This is also the case for bolts with threads down to just below the head (similar to DIN EN 4017). The measured flexibilities place multiple waist bolts between bolts with reduced shank diameter and bolts with threads down to just below the head.

The durability of bolted connections is determined exclusively by the value of the local stress concentrations. As a rule, the breaking strength of a notched rod in comparison to a smooth rod should be greater than 1 for the bolt material—that is, the material must have sufficient ductility. The additional dynamic forces, which have to be taken up by the bolt when dynamic operating loads are applied, decrease in inverse proportion to the initial tension level. This also speaks in favor of a tightening procedure beyond the elastic limit.

~~**Cylinder head bolts.** The function of cylinder head bolts is to provide a secure connection between the entire system consisting of the cylinder head, cylinder head gasket, and crankcase to give long-term operation when the maximum possible ignition forces are considered. The objective is particularly to achieve low, uniform component loads and prevent leakage of the combustion gases, lubricant, and coolant.

Although it was previously necessary to retighten cylinder head bolts up to twice to compensate for settling, cylinder head connections which do not require retightening are now state of the art.

This was made possible by the use of reduced shaft diameter or threaded elongating bolts with high elasticity, limited tolerances of tensile strength, and friction characteristics with cylinder head gaskets with minimum settling (e.g., full metal gaskets) and a tightening procedure with minimum dispersion of the initial tension. Angle-controlled tightening in the range beyond the elastic limit has become the most widely accepted tightening procedure. The increase in requirements for lightweight construction and the resulting lower component rigidity of the block and cylinder head is usually compensated for by reducing the maximum bolt strength. Maintenance of the minimum required bolt tension can be achieved only by drastic limitation of the tolerances for the tensile strength and friction values. It is frequently necessary to take thermal influences into consideration in laying out the cylinder head bolts. It is possible for the cylinder head bolts to heat up less during the warm-up process of the engine than the parts which they clamp together: the cylinder head and crankcase. This can lead to a considerable increase in the initial tension when these components consist of materials with high coefficients of thermal expansion such as aluminum. The use of bolts with reduced shank diameter or threaded "stretch" bolts, as shown in **Fig. E215**, can be advantageous in this aspect, because the increase in the load on the bolts is significantly lower due to the lower increase in the spring curve.

The thermal expansion characteristics of steel differ considerably from those of aluminum or magnesium. For this reason, the latest developments provide for cylinder head bolts of austenitic materials whose coefficient of thermal expansion is similar to that of aluminum or magnesium.

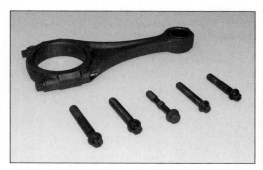

Fig. E214 Connecting rod bolts for tightening beyond the elastic limits

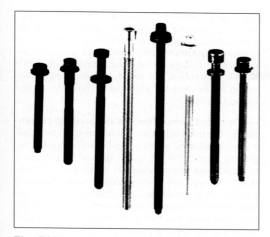

Fig. E215 Bolts with reduced shank diameter or "stretch" bolts for fastening cylinder heads

The increasing necessity of cost reduction has been implemented by optimizing cylinder head bolts in two areas:

- Use of threaded "stretch" bolts as a viable compromise between sufficient elasticity and reduced production costs in comparison to bolts with reduced shank diameter with a considerably more expensive production process.
- Replacement of the washer segment, even on aluminum cylinder heads, by integrating it into the bolt head in the form of a bolt with flanged head. However, for this purpose, a great deal of experience is required for defining, producing, and assuring the quality of the geometry of the contact area below the head to prevent galling, which can occur frequently, when installing the bolts. This includes surface treatment with extremely low variance of the friction values and excellent adhesion to the base material such as the process of quasi-amorphous thin layer phosphatization.

~~Flywheel bolt. A relatively small reference circle is provided by the design on the crankshaft. During assembly, it is necessary to ensure that sufficient clearance is present between the bolts for the tightening tools. The bolts are tightened under yield point control simultaneously by a multiple spindle unit. This is also necessary because short clamping lengths are present here (e.g., 7 mm). The head heights are less than the standard due to the extremely small interval between the crank pins and flywheel. To ensure that the required torque is applied, a twelve-point or six-point head or a head with internal serrations for a spline wrench is used. When lubricating oil is supplied through passages in the crankshaft, the bolts are provided with micro-encapsulated sealing glue or a circumferential nylon coating to seal against oil leakage.

Some engine manufacturers still tighten flywheel bolts with torque control, which are then manually "retightened."

With the dual mass flywheels, which are being used more and more recently, the entire module is delivered to the vehicle manufacturer complete with flywheel bolts, where it is installed as a unit. Here the bolts are tightened using a multiple spindle screwdriver through holes in the clutch plate spring and clutch plate.

~~Main bearing cap bolts. These are used to connect the main bearing cap on the crankshaft bearings to the crankcase.

Generally, two bolts are used for each main bearing cap, and these are designed as completely threaded collar bolts and, where applicable, used with washers. **Figs. E216 and E217** show the measuring setup for determining the initial tension as well as the force flow in the area of the main bearing cap bolts.

The primary problem in laying out these bolts is the limited installation space available, in most cases, for the bolt head. This means that it is necessary to ensure very precise maintenance of the permitted surface pressure for the contact surface at the bottom of the bolt head and the opposing surface. Each main bearing cap bolt is installed twice: first, when the bearing shell seats are machined to the fitted dimension and then, after installation of the crankshaft, when the bearing cap is installed. During the second installation procedure, the threads can "gall" (e.g., have impact points)

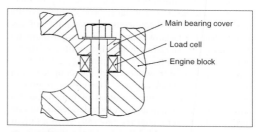

Fig. E216 Measurement of force on main bearing cap bolts.

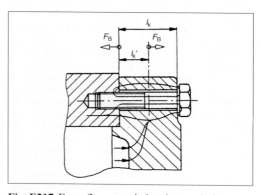

Fig. E217 Force flow at main bearing cap bolt

if the bolt is damaged in the area of the end/thread start. This is particularly the case for female aluminum threads.

As a preventive measure, the end is designed optimally and the drop heights during production are kept as low as possible (maximum 300 mm).

The drop height is the height that the still-untempered bolt drops into the catch container when discharged from the machine. The end design includes beveling the end of the bolt shank before cutting the thread to ensure that the thread does not break out when cut. In order to increase the rigidity of the crankcase, so-called ladder frames are being installed more and more frequently in engines. These provide a connection between the individual main bearing caps. This allows the lower engine area to be designed so that it is more resistant to torsion. Usually, the bearing caps in the "ladder frame" ("bed plate") are cast using light alloy material. In this case, the entire unit is bolted on with the main bearing cap bolts.

Yield point-controlled or angle-controlled tightening procedures are used for assembly.

~~**Pulley bolts.** Pulleys are fastened with a bolt in the center.

While only the generator or alternator was previously driven by a pulley, today such drives are used for auxiliary equipment such as the air conditioner. In addition to the V-belt pulley, a gear for driving the oil pump and possibly a vibration damper are bolted to the crankshaft. The inner bore of the pulley is positioned on the crank journal. The large bore diameter of the pulley allows a good connection between the bolt and pulley, using a large washer or larger collar diameter. Frequently, an M 12 bolt is provided with a washer or collar with a diameter up to 38 mm (e.g., gasoline engine) or an M 18 bolt with collar diameter up to 65 mm (e.g., 2.5-liter diesel engine). The pulley is pressed separately onto the crank journal or pulled into the crankshaft by a bolt at a previously defined tightening torque. Commercial engines generally have a dimension range of up to M 24×1.5 and the washer is installed just before assembly. On large commercial vehicle engines, the pulley is attached with a vibration damper and bolted directly to the crankshaft with six or eight bolts or studs (e.g., M 10).

Earlier, pulley bolts were tightened to a specified torque only. Today, the angle-controlled tightening process is used. The torque is applied through a joining moment until all joining surfaces are pressed snugly against one another. The bolt is then turned further, controlled by measuring the angle of rotation. Extremely high final tightening torques can be achieved in this manner. Torques of up to 260 Nm can be achieved for an M 12×1.5–10.9 bolt, whereby the theoretically calculated final torque is between 120 and 150 Nm. The large dispersion in the final torque results from the large head contact area, which begins to gall when canted even slightly. A yield point-controlled tightening process is not possible for pulley bolts with an extremely large collar diameter or when

a number of parts are clamped together, which results in a large number of separation joints between the parts to be clamped. Due to inaccuracies in production resulting from unavoidable contaminations, this situation results in high settling characteristics or such a flexible connection that the yield point is not determined by the bolt but rather, impermissibly, by the connection itself.

~~**Threaded connections in magnesium components.** The trend to lightweight construction in the automotive industry, which has continued since the beginning of the 1990s, requires the use of alternative materials such as magnesium in addition to optimization of components of proven construction materials such as steel, aluminum, and synthetics. The advantage of magnesium is its relative stiffness, even on designs with extremely thin walls, while in terms of density it is comparable to plastics.

In the area of the engine, this material is used only for auxiliary components such as the cylinder head covers for encapsulated engines or air filter intake fittings, where magnesium replaces plastic. In the engine block itself, the thermal loads are too high for it to be used in the overall connection: steel bolts can be used in magnesium connections only at room temperature due to the creep and relaxation characteristics. For this reason, engine assemblies have heat-treated aluminum bolts using: AL 6056 in combination with the die-cast magnesium alloy AZ 91, AS 21 used up to temperatures of 120°C (peak temperature: 150°C). When magnesium components are used in combination with steel or aluminum bolts it is necessary to take into consideration the contract corrosion characteristics.

Literature: S. Jende: Robotergerechte Schrauben–Hochfeste Verbindungselemente für flexible Automaten, Techno TIP 12/ 84. Würzburg, Vogel Book Publishers. — N.N.: Industrie-werkzeuge–Montagewerkzeuge, Company Catalogue 2000– 2001 of Atlas Copco Tools GmbH, Essen. — Technical Specification ISO/TS 16949: QM–Systeme: Besondere Anforderungen bei Anwendung von ISO 9001: 2000 für die Serien- und Ersatzteilproduktion in der Automobilindustrie, VDA-QMC, Frankfurt, 2002. — K.H. Illgner, D. Blume: Schraubenvademecum Herausgeber: Textron Verbindungstechnik, 6th Edition, 2001. — O.R. Lang: Triebwerke schnelllau-fender Verbrennungsmotoren, Design Books No. 22, BerlinHeidelberg, Springer Publishers, 1966. — H. Grohe: Otto- und Dieselmotoren: Arbeitsweise, Aufbau u. Berechnung von Zweitakt- u. Viertakt-Verbrennungsmotoren. Kamprath–Series kurz und bündig, Technik (Concise Engineering), 6th Edition, Würzburg, Vogel Book Publishers, 1982. — VDI: Systematische Berechnung hochbeanspruchter Schraubenverbindungen, VDI Guideline 2230 (1986) sowie (2002). — S. Jende: KABOLT–ein Berechnungsprogramm für hochfeste Schraubenverbindungen, Beispiel: Die Pleuelschraube, VDI-Z 132 (1990), No. 7, July, pp. 66–78, Düsseldorf, VDI Publishers GmbH. — PC-Bolt '98 (Schraubenberechnungsprogramm), Institute for Mechanical Engineering/Design Engineering, Technical University of Berlin, 1998. — K.H. Kübler, G. Turlach, S. Jende: Schraubenbrevier, 3rd Edition, 1990 (KAMAX-Werke Rudolf Kellermann GmbH & Co. KG, Osterode am Harz). — W. Scheiding: Verschrauben von Magnesium braucht mehr als Alltagswissen, in Konstruktion und Engineering, 04/01, Landsberg/Lech: Publishers Moderne Industrie. — K.H. Kübler, W. Mages: Handbuch der hochfesten

Schrauben. Herausgeber: KAMAX-Werke, 1. Auf l. Essen, Publishers W. Girardet, 1986. – S.J. Engineer, H. Finkler, H. Gulden, H. Köhler, K. Sartorius: Stähle mit 1% Chrom und Bor für hochfeste Schrauben, Stahl und Eisen 112 (1992) vol. 2 (Publishers Stahleisen mbH, Düsseldorf). – H. Köhler, G. Roth, E. Walper: Hochfeste Verbindungselemente aus alternativen Werkstoffen, Ingenieurwerkstoffe 4 (1992) No. 5, pp. 36–39 (VDI Publishers, Düsseldorf). – H. Köhler: Hochfeste Verbindungselemente aus Dualphasen–Stahl ohne vergütende Wärmebehandlung, ATZ 100 (1998), vol. 10. – K. Westphal: Keine Patentlösung in Sicht. Neue Cr6-freie Zinklamellensysteme für Verbindungselemente der Automobilindustrie. In: Metalloberfläche, 56th year, Munich, Hanser Publishers, May 2002. – K. Westphal: Verschraubung von Magnesiumkomponenten, in, Metall, 56th year, vol. 1–2 Isernhagen, Giesel Publishers, 2002. – H. Köhler, K. Westphal: Chromatfreie Korrosionsschutzsysteme für Verbindungselemente, Conference House of Engineering "Korrosionsschutz am Automobil" (Essen, Sept. 2003).

~Tightening procedure

~~Angle-controlled tightening.

In the case of angle-controlled tightening beyond the yield strength, the average initial tensioning torque is 25–30% higher than for torque-controlled tightening. While for torque-controlled tightening, the initial tension varies by ±25% (practically to the same degree as the friction), the variation in the initial tension for angle-controlled and yield strength-controlled tightening is only ±10%. For angle-controlled tightening, the initial tension control depends primarily on the friction only within the range up to the joining torque. The joining torque is the force which must be applied until the surfaces of all separating joints are "snugly" pressed against one another by elastic and plastic deformation as a result of tightening the connection. The dispersion results primarily from the different actual elongation strengths of the bolts, assuming that the required repetition accuracy is achieved in moving to the set angle. This is guaranteed with the pulse sensors used today. Moreover, the curves above the yield point indicate that angle control has only a subordinate effect on the preliminary assembly tension (**Fig. E218**): a torque monitoring feature is used to ensure the quality of the connection.

A further advantage of angle-controlled tightening beyond the elastic limit is that determination of the tightening angle for bolts with a length of greater than $2 \times d$ is virtually uncritical and tolerances of even 20 degrees can be accepted. For example, with a bolt with a pitch of $P = 1.5$ mm, for example, turning the bolt beyond the yield point by 30 degrees results in plastic elongation of approximately 0.125 mm. In relation to a bolting length of 60 mm, this results in permanent deformation of 0.21%. This value is generally recognized as safe. Today, bolts are tightened a number of times beyond the elastic limits. The rule of thumb is that, in such cases, a maximum permanent deformation of 1% in relation to the clamping length is permissible. However, it is necessary to consider that multiple tightening operations can damage the head contact surface as well as the section of the thread bearing the load, and therefore such bolts have a tendency to "gall." In these cases, the required initial tension is no longer reached.

Also advantageous is the reproducibility of angle-controlled tightening with simple tightening tools, so that it is preferred for initial tightening on the production line as well as for service.

~~Torque-controlled tightening.

Torque-controlled bolt tightening is usually used for bolts of secondary importance and only in special cases for high-quality applications (e.g., installing pulleys) on automated assembly lines (bolts of secondary importance do not require a precisely defined minimum initial tension). This will also continue to be used in the service sector. The problem is that the specified torque must be selected so that the initial tension—that is, the force present on the connection after completion of assembly—does not exceed the yield strength of the bolt even in unfavorable cases (low friction coefficient) and that the value for this is far below the minimum initial tensions required (high friction coefficient).

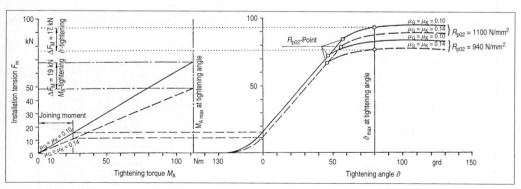

Fig. E218 Tightening curves for a bolt DIN EN ISO 4014–M12×1.5×70–10.9 for torque-controlled (*left*) and angle-controlled and yield-point–controlled tightening (*right*), illustrating influences of thread and head friction as well as bolt strength

The expected friction coefficients agreed upon between the bolt manufacturers and automotive industry are between $\mu_{tot} = 0.08$ and $\mu_{tot} = 0.14$. They are part of the quality agreement and can be checked on a random sample basis on a friction value test bench for each batch of bolts.

A special form of torque tightening is the combination with so-called retightening—that is, after conclusion of the bolting operation, the connection is tightened again with a measuring pipe torque wrench. This process is used in mass production at the remaining manual assembly points and in small batch production for all critical connection elements. Manual tightening is accomplished with a torque-controlled pneumatic screwdriver up to the set torque, after which final tightening is accomplished with a measuring-type torque wrench; such points are usually then color-coded. The torque required until the bolt starts to turn again is the retightening torque (final tightening torque).

"Retightening" is accomplished according to experience to a value slightly above the value set on the torque wrench, so that this frequently results in an indirect angle-controlled tightening operation.

~~Yield point-controlled tightening process. In

comparison to angle-controlled tightening, this process has the advantage that the actual yield point of the bolt installed is reached in each case.

The permanent elongation of the bolt in the tightening operation is between 0.1 and 0.2%, depending on the sensitivity of the screwdriver systems, still below the 0.2% permanent elongation limit: unacceptable permanent elongation of the bolt beyond the 0.2% limit is virtually impossible. In comparison to the angle-controlled tightening process, the average initial tension level is 4–7% lower. Quality control of the connection is accomplished by monitoring the "green window," which defines the switch-off point for the screwdriver in the yield point range of the bolt defined by the minimum and maximum angle and torque specifications.

Engine brake (*also*, →Electronic open- and closed-loop control ~Electronic open- and closed-loop control, diesel engine ~~Functions; →Variable valve control). When a vehicle with reciprocating engine is decelerated by "releasing the throttle" with the engine connected to the drive wheels, the friction and work required for the gas exchange process decelerates the vehicle. This provides a portion of the braking energy required. This is advantageous because the vehicle brakes are not designed for continuous operation and can fade in the face of thermal overload (e.g., on long downhill grades). In extreme cases, the braking system may fail completely. This applies particularly for commercial vehicles with a heavy load. In order to increase the effect of the engine brake, two types of so-called retarder systems are used in addition to the conventional braking system:

- Engine brake with exhaust valve. Up to the present, this version has been used most frequently. Increasing the exhaust back pressure results in a braking force of 14–20 kW/L (**Fig. E219**).
- Engine brake with constant throttle in addition to exhaust valve (**Fig. E220**). Use of a small throttle valve in the bypass to the exhaust valve allows an additional increase in the engine braking performance. Actuation of this valve is accomplished with compressed air in the same manner as the actuator for the engine braking valve. During operation of the engine brake, this valve is open continuously, thereby providing a constant cross section.

Engine camshaft →Camshaft ~Drive

Engine capacity characteristics (*also*, →Supercharging). The capacity curve of an engine can be represented as a function of the pressure ratio and airflow rate. At constant engine speed, the airflow rate increases with increasing pressure ratio. However, an internal combustion engine must be considered as a displacement machine whose throughput increases to a

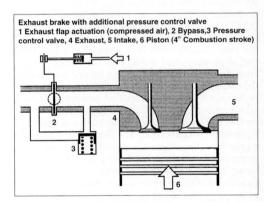

Exhaust brake with additional pressure control valve
1 Exhaust flap actuation (compressed air), 2 Bypass, 3 Pressure control valve, 4 Exhaust, 5 Intake, 6 Piston (4ᵗʰ Combustion stroke)

Fig. E219 Engine brake with constant throttle (Source: Bosch)

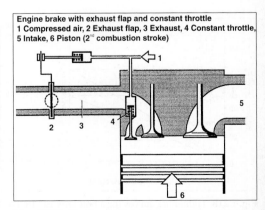

Engine brake with exhaust flap and constant throttle
1 Compressed air, 2 Exhaust flap, 3 Exhaust, 4 Constant throttle, 5 Intake, 6 Piston (2ⁿᵈ combustion stroke)

Fig. E220 Engine brake with third valve (Source: Bosch)

far greater extent with increasing speed than with pressure ratio.

Engine combustion process →Combustion process

Engine compartment →Engine concepts

Engine compartment covers →Sealing systems
~Modules ~~Beauty cover

Engine compartment temperature. In spite of cooling by the relative air stream and cooling fan, temperatures considerably higher than the ambient level are present in the engine compartment due to the heat transfer from the internal combustion engine. It is necessary to ensure that these temperatures are below the temperature required for proper functioning of the components installed in the engine compartment. Here, the specifications for the permissible thermal load on the corresponding components are important. These include particularly synthetic and electrical and electronic components. Endangered components exposed to a direct heat source, such as the exhaust system, are frequently protected by insulating materials which also reflect radiation (e.g., aluminum coated insulating mats). When considering the relative air stream flowing through the engine compartment, it is necessary to balance the size of the openings at the front of the vehicle which increase the aerodynamic drag (c_w value) with the cooling effect achieved. The critical vehicle states in terms of the engine compartment temperature are the maximum speed on level road, driving up mountainous roads at high speeds, and driving up mountainous roads at low speed with a trailer.

Engine concepts. The engine concept—i.e., the basic design of an engine—is influenced by many factors, some of which are absolutely necessary, some of which have limitations, and some of which are freely selectable. Some important factors are the operating process (two-stroke or four-stroke), the operating principle (diesel or gasoline), function (simple or double-acting), type of cooling (water or air), performance rating, number and arrangement of cylinders, engine configuration, bearings (slide/friction bearings), crankcase design (differential or integral design with all intermediate stages), type of control (valve control, turbocharging), location of balance shafts, etc.

Engine concepts are determined particularly by the application of engine, its output as well as auxiliary conditions; they are also related to the state of the art, so that over the course of time, very many and highly differing engine concepts have been developed. The most important criterion for an engine is its application (**Fig. E221**); this determines the output and the conditions under which the power must be provided (e.g., duration of power output). In the performance equation for power, P,

$$P = z \cdot \frac{\pi}{4} \cdot d^2 s \cdot w_e \cdot \frac{n}{i}$$

due to higher order regularities of similarity mechanics, the individual factors are not independent of one another, so that the operating process and procedure and type of cooling play a role in addition to output, displacement, and engine speed (**Fig. E222**). In the previous equation, z = number of cylinders, d = bore diameter, s = stroke, w_e = specific work, n = engine

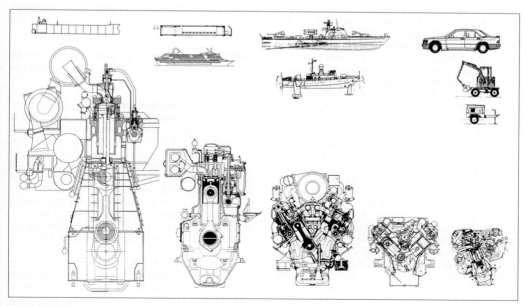

Fig. E221 Applications of engines and engine size (Source: Zima)

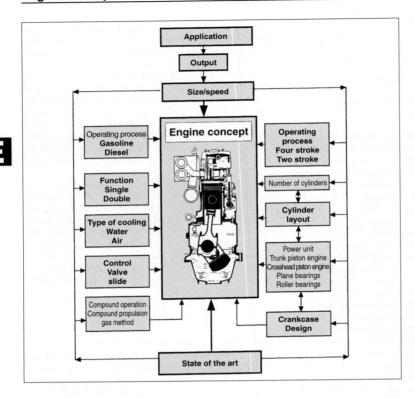

Fig. E222
Engine concept:
influences,
interrelationships,
and interactions
(Source: Zima)

speed, and i = factor for two-stroke or four-stroke process. High absolute power outputs can only be achieved with large cylinder dimensions (bore, stroke); in contrast, high specific outputs (power/displacement; power/engine mass) can be represented with the frequency of the combustion cycle and specific work. The two-stroke process for increasing the working frequency is only used for extremely large and extremely small engines today—with a few exceptions—and double acting (alternate loading of the bottom and top of the piston with the pressure of the working gas) is not used at all today. The problems encountered in engines increase with the engine speed. Inertia effects are difficult to control, as are thermal loads, and the charge exchange cycle is more difficult so that limits are set for using an increase in the speed as a means of increasing the power output.

Because the output and size of engines are determined by the speed range, engines are divided up into low-speed engines (60–200 rpm), medium-speed engines (>200–1000 rpm) and high-speed engines (>1000 rpm). Each of these groups of engines has its own characteristic features. With the exception of passenger cars, motorcycles, and small working machines, all engines operate according to the diesel principle.

The basic design of engines is dictated in principle by the type of operation. To date, reciprocating piston engines have proven to be superior to all types of engines without crankshafts. This applies to control of the ther-

modynamic process as well as to the efficiency of the conversion of the reciprocating motion to rotary motion.

~Cam engines. *See below*, ~Engines without crankshaft

~Composite systems

~~**Free piston engines.** Free piston engines can be subdivided into free piston compressors and fuel gas processes.

~~~**Free piston compressors.** A free piston compressor is a combination of two-stroke diesel engine and piston compressor whereby the crankshaft drive is eliminated (**Fig. E223**). In each cylinder, two groups of pistons operate in opposite directions, consisting of an engine piston and compressor piston in the first and second stages. The engine stage operates as a uniflow-scavenged two-stroke diesel engine. The group of opposing pistons is synchronized by a pair of racks, which simultaneously drive the fuel and lubricating pumps—connected by a gear. Free piston compressors run extremely smoothly, due to their total internal balance.

~~~**Fuel gas process.** The fuel gas process divides up the output even further by taking the entire effective output from the axial turbine system (conversion of energy in the working substance into mechanical work); the piston engine, a two-stroke diesel engine

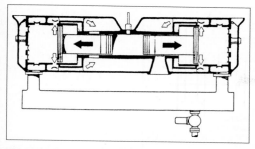

Fig. E223 Free piston compressor (schematic) (Source: Junkers Operating Instructions)

without crankshaft, serves only to generate the fuel gas (working substance preparation and combustion; **Fig. E224**). The diesel stage consists of a two-stroke cylinder operating with uniflow scavenging and two adjacent compression chambers on each side. Two operating pistons run in opposite directions in the cylinder.

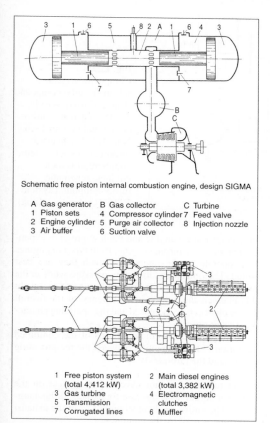

Schematic free piston internal combustion engine, design SIGMA

A Gas generator	B Gas collector	C Turbine
1 Piston sets	4 Compressor cylinder	7 Feed valve
2 Engine cylinder	5 Purge air collector	8 Injection nozzle
3 Air buffer	6 Suction valve	

1 Free piston system (total 4,412 kW)	2 Main diesel engines (total 3,382 kW)
3 Gas turbine	4 Electromagnetic
5 Transmission	clutches
7 Corrugated lines	6 Muffler

Fig. E224 Fuel gas process schematic free piston combustion engine, version SIGMA (Source: Cernea). Power plant for tourist ship "Fritz Heckert" (Source: Cernea)

The pistons are synchronized by a mechanism. One compressor piston is rigidly connected to each working piston. During the piston power stroke, air is sucked into the compressor chambers on the same side as the compressor pistons; on the other side, air is compressed, which provides a buffer (air spring) for the piston motion and then—as it expands—it again forces the pistons against one another (rebound stage). At the end of the power stroke, the one piston opens the exhaust ports in the working cylinder, the other opens the intake ports—slightly later—for the usual two-stroke process. The compressed air is routed into a gas turbine together with the exhaust gas. The gas turbine is designed as a pure power turbine whose entire output is then available as effective power.

The high temperatures required for good efficiency are obtained in the diesel stage—where they can be controlled, that is—and then the temperature from the mixing of exhaust and compressed air is low enough to be withstood by the steels in the gas turbine under continuous operation. The working gas transfers the compression work directly to the air through pistons. A crankshaft is not required either on the engine or the compressor side, and it is not necessary for the gas turbine to output any mechanical work to the axial compressor; finally, it is possible for a number of diesel units to provide the working gas for one turbine. Such a concept certainly promises advantages. However, the diesel stage, gas turbine, and transmission required for reduction of the high turbine speed make the system very complicated compared to a reversible two-stroke diesel engine acting directly on the propeller shaft.

During the 1940s and 1950s, numerous free-piston internal combustion engines were put into operation for driving locomotives—e.g., Renault-SIGMA with up to 1765 kW; in the United States by Baldwin-Lima-Hamilton with outputs up to 2350 kW; as well as in the Union of Soviet Socialist Republics. Although these free-piston locomotives operated well—even in scheduled service—they were not able to displace the "classic" diesel engine. Here, marine applications appeared to offer better possibilities. Different types of ships were equipped with free piston drives. High-speed marine vehicles, fishing boats, freight ships—for example, the Liberty ship *William-Patterson* and passenger ships such as the FDGB tourist ship *Fritz Heckert*. Free-piston engines were also built as generator systems and tested as power plants for motor vehicles.

In the 1950s, General Electric developed a composite system (Orion) consisting of a gas turbine and air-cooled two-stroke diesel engine in which the diesel stage served only as a gas supplier for the gas turbine and for driving the scavenging and cooling fan. The exhaust from the engine, as well as the cooling air, were banked at high pressure—in back of each cylinder—and expanded together in the gas turbine. The advantage is that the heat transfer from the cylinder wall, which is picked up by the cooling air, was also used.

~~Turbo-compound systems (composite systems)
(*also,* →Supercharging ~Turbo-compound supercharging;

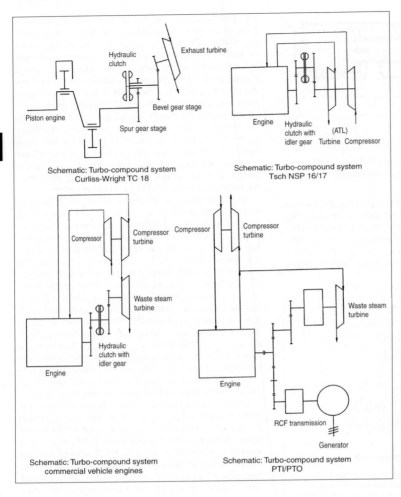

Schematic: Turbo-compound system
Curliss-Wright TC 18

Schematic: Turbo-compound system
Tsch NSP 16/17

Schematic: Turbo-compound system
commercial vehicle engines

Schematic: Turbo-compound system
PTI/PTO

Fig. E225
Composite systems:
schematic turbo-compound
system Curtiss-Wright
TC 18; schematic turbo-
compound system Tsch
NSP 16/17; schematic
turbo-compound system,
commercial vehicle
engines; schematic
turbo-compound system
PTI/PTO (Source: Zima)

Figs. E225 and E226). Turbo-compound systems or composite systems are a combination of reciprocating engines and turbines in which both types of engines participate directly in developing the power—that is, the overall output. Approximately a third of the chemical energy present in the fuel is not utilized in the reciprocating engine due to incomplete expansion of the working fluid. The cylinder volume that would be required for complete expansion of the working fluid cannot be achieved in a practical technical manner on reciprocating engines. It would either be necessary to use a significantly longer stroke or to expand the fluid in two stages—similar to the proven principle used in steam engines. In either case, the gain in effective work would be absorbed by friction and heat losses, and this is aside from the complicated design required for such solutions. For this reason, the individual changes of state in the thermodynamic process are divided up so that the high-pressure processes—for example, combustion and initial expansion—take place

in a reciprocating engine and the low-pressure expansion process in a turbine engine. Both types of engines then operate in the range for which they are best suited. The otherwise "lost" low pressure work in the reciprocating engine can be used to a great extent. In this manner, the output and efficiency of the overall power plant can be increased and the operating characteristics improved.

There are two extremes in terms of the combined operation of the interaction between the reciprocating engine and the turbine(s):

• The turbine does not have a direct effect on the power output, as is the case for exhaust turbochargers. The entire mechanical output from the exhaust turbine is used by the compressor.
• In the combustion process, the entire effective output of the system is produced by the turbine. The subdivision of the specific pressure increase or pressure decrease between the reciprocating engine and

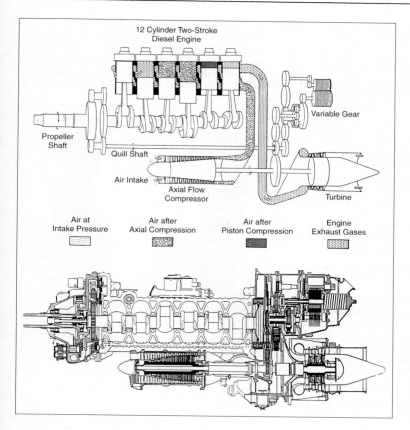

Fig. E226
Composite systems:
schematic turbo-
compound system
Napier-Nomad
(Source: Chatterton)

turbine depends on the engine concept and the state of the art.

An imbalance in the performance between the turbine and compressor on turbocharged engines can be compensated in certain operating ranges by mechanically coupling the turbocharger to the engine. When the output from the turbine is insufficient to drive the compressor, the turbocharger is driven by the engine when excess power is available from the turbine; mechanical power is transferred to the engine. This solution was used for medium-speed two-stroke diesel engines for locomotives and four-stroke diesel engines for PT boats. Hydraulic couplings will damp the torsional vibrations in the gear train between the engine and turbocharger; torsion rods in the powertrain act as predetermined breaking points for protection of the gear drive in the event of moments of extreme deceleration such as those that can occur when a piston seizes up.

In the 1940s and 1950s, aircraft which operated at very high altitude with low atmospheric pressures made use of the high expansion ratio available by using turbo-compound engines such as the double-radial Curtiss-Wright TC 18 engine. Its 18 cylinders were divided up into three groups of six each, each of which fed an exhaust turbine. The three exhaust turbines transferred their power to the crankshaft through two-stage reduction gears and hydraulic clutches. In contrast, the two-stage turbochargers were driven mechanically.

A 12-cylinder V-180 design of two-stroke diesel aircraft engine called the Nomad, produced by Napier, was supercharged by a twelve-stage axial compressor which was driven through direct gearing by a three-stage axial turbine, so that the compressor operated in the range of good efficiency and constant charging pressure, regardless of the load on the engine. The turbine had a maximum output of 1654 kW, of which 1352 kW was required to drive the compressor and 302 kW was output to the crankshaft. The output was increased even further by injecting fuel into the exhaust line—possible due to the excess air resulting from the high scavenging flow. Developed at the end of the 1940s for long-range reconnaissance planes and airfreight traffic, this initially promising development was made obsolete by the appearance of jet engines.

In the meantime, as a result of the high exhaust turbocharger efficiencies, the entire exhaust gas energy is not required for compressing the charge; a proportion of this energy was fed to an operating turbine and the turbine output connected to the engine or used to drive generators. This has been used for slow-running two-stroke, medium-speed four-stroke, and commercial vehicle diesel engines; there are various possibilities for guiding the gas flow and utilizing the effective output:

- The turbine driving the compressor (VT) and working turbine (NT) are connected in series on the gas side; usually the working turbine is located downstream of the compressor turbine or both turbines operate in parallel (also used on large two-stroke diesel).
- The working turbine acts through various coupling and gear configurations on the crankshaft of the main or an auxiliary engine (power-take-in, PTI).
- The working turbine operates in parallel to an RCF transmission with a generator via a gear on the crankshaft (power-take-off/power-take-in [PTO/PTI]).
- The working turbine drives a generator via a clutch and transmission.
- The working turbine drives a generator together with a steam turbine fed from an exhaust boiler, thereby increasing the generator output.

The fuel consumption savings resulting from turbo-compound operation amount to 4–7 g/kWh.

~~~**Propelling nozzle.** The exhaust energy can also be utilized in another way on aviation engines: the exhaust system can be shaped in the form of a jet nozzle (nozzle cross-section area, approximately 60% of exhaust pipe area) to obtain additional propulsion from the exhaust gases. At high-flying speeds, the output can be increased by 10–15% with propulsion amplifiers (propelling nozzles).

~Cylinder layout. *See below*, ~Reciprocating piston engines

~Dual inline engines. *See below*, ~Reciprocating piston engines ~~ Multiple-shaft engines ~~~ Twin-shaft engines

~Engines without crankshaft. Attempts have continually been made to replace the crankshaft with other, supposedly more advantageous, mechanisms. The motives for this included:

- The desire for higher output and power-to-weight ratio, making it necessary to increase the number of cylinders. The limits were soon reached in terms of the powertrain mechanics, so that only cylinder configurations other than the conventional drivetrains were required—also in consideration of a small front surface area for aviation engines.
- Improvement of the mechanical efficiency when converting reciprocating motion to rotary motion.
- Internal speed reduction in the engine for driving aircraft and marine propellers.
- Changing the motion of the pistons to improve the thermo-dynamic process.

In the case of Fairchild-Caminez engines, the rotary axis of the cam is positioned perpendicular to the plane of the cylinder axes; on curved path engines, the cylinder axes and direction of rotation of the driveshaft are parallel (**Fig. E227**).

~~**Curved path engines.** Two cylinder blocks each with four, six, or eight cylinders located coaxially in a circle are bolted together to form one unit; each pair of opposing pistons is bolted together and acts on the sinusoidal shape curved path with rollers (**Fig. E228**). This turns under the force of the pistons

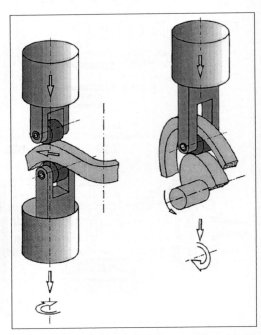

Fig. E227 Fairchild-Caminez engines and curved track engines: drive torque direction of effect (Source: Zima)

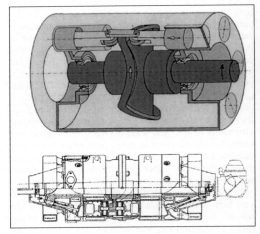

Fig. E228 Curved path engines, schematic (Source: Zima). Curved path engine, version Herrman (Source: Judge, p. 295)

driving the output shaft. Coaxial concentrically arranged cylinders provide compact high-output drives with low front surface area. In the 1920s and 1930s, such drivetrain configurations were considered to be a promising alternative to conventional aviation engines.

~~Fairchild-Caminez engines. The piston rods act directly on a cam instead of on a crankshaft, causing the cam to turn **(Fig. E229)**. The number of protrusions on the cam allows the output speed to be reduced relative to the frequency of the combustion cycles; the motion of the pistons can be varied—within limits—by the shape of the protrusions.

The pistons in the Fairchild-Caminez engines (1926) acted on a figure-eight–shaped cam path; two combustion cycles resulted in one rotation of the driveshaft. Because the two opposing systems run in the opposite

direction, the masses were balanced internally. Built and tested in the United States in the middle of the 1920s, this aviation engine was never put into mass production. On the German Michel engine, a two-stroke diesel engine, two pistons are arranged in a radial shape with one common combustion chamber; they act on the cam disk through connecting rods and rollers. The Michel engine was built and tested in the beginning of the 1920s.

~~Inclined plate or angular plate engines. A plate is fastened to the driveshaft at a certain "skew" or "inclined" angle **(Fig. E230)**. The pistons are arranged circumferentially around this shaft and act on the plate through piston rods and articulated or socket joints; this causes the plate to turn under the effective axial load. Swash plate engines are limited in terms of

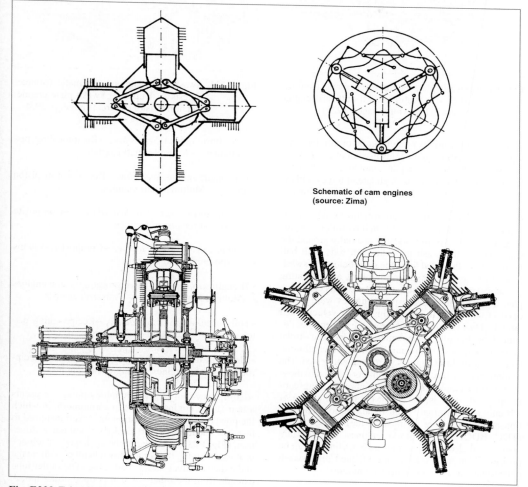

Schematic of cam engines
(source: Zima)

Fig. E229 Fairchild-Caminez engines, schematics (Source: Zima). Fairchild-Caminez engines, version Fairchild-Caminez (Source: *Der Motorwagen* 1927, p. 692)

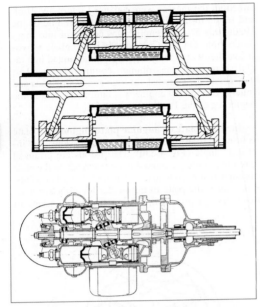

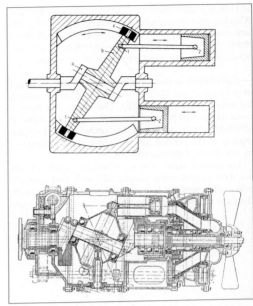

Fig. E230 Swash plate engines schematic (Source: Thesis A. Kupfers). Ali-outboard engine (Source: Der Motorwagen 1927)

Fig. E231 Swash plate engines schematic (Source: Thesis A. Kupfers). Swash plate engine, version Bristol (Source: Judge; p. 293)

speed (sliding velocity) due to unfavorable terminological relationships in the transmission of power from the piston rod to the swash plate; moreover, high piston friction is present due to the high lateral forces. Otherwise, the same applies as for swash plate engines.

~~Swash plate engines. A nonrotating swash plate is supported on a V-shaped offset shaft by roller bearings (**Fig. E231**). The pistons located coaxially around the shaft act on the swash plate through rods with ball heads. Under this axial load, the shaft below the swash plate turns so that the latter completes an oscillating (gyratory) motion; a torque support prevents the swash plate from turning. Swash plate engines were developed in the 1930s for propelling motor vehicles and aircraft. They became known from the bus engines designed by Bristol Motors, nine-cylinder four-stroke gasoline engines, with annular slide valve timing. Only a few of these engines were built and tested. The motion relationships of the pistons correspond to those of a normal crank drive with an infinitely long connecting rod—that is, the pistons complete a pure or virtually pure harmonic (sinusoidal) motion. The motion relationships are identical at the reversal points (TDC, BDC). On two-stroke engines, the ports are opened and closed in a shorter time. Cylinders can be positioned on both sides of the inclined plate/swash plate to increase the power-to-weight ratio.

~Fairchild-Caminez engines. *See above,* ~Engines without crankshaft

~Five-shaft engines. *See below,* ~Reciprocating piston engines ~~Multiple-shaft engines

~Four-shaft engine. *See below,* ~Reciprocating piston engines ~~Multiple-shaft engines

~Free piston compressor. *See above,* ~Composite systems ~~Free piston engines

~Fuel gas process. *See above,* ~Composite systems ~~Turbo-compound systems

~H engines. *See below,* ~Reciprocating piston engines ~~Multiple-shaft engines ~~~Twin shaft engines

~Horizontal engines. *See below,* ~Reciprocating piston engines ~~Location of cylinders

~Horizontally opposed engines. *See below,* ~Reciprocating piston engines ~~Single-shaft engines

~Humphrey pump. The Humphrey pump represents a four-stroke engine reduced to a minimum, in which the power from the combustion gases is transferred to the flow medium (water) directly—without a drivetrain (**Fig. E232**). One leg of the U-shaped "machine" is closed forming the combustion chamber with intake and exhaust valves and a spark plug. The outlet tube extends from the other—which is open at the top. A gas/air mixture is forced into the combustion chamber by a compressor to start the engine. This mixture is

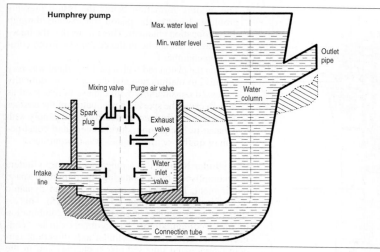

Fig. E232
Humphrey pump
(Source: Zima)

ignited, and the pressure generated by combustion presses a water column into the other leg of the U tube. A proportion of the water exits through the outlet tube; the other proportion is lifted and then oscillates back, forcing the exhaust gases out of the combustion chamber. When the system oscillates back again, a vacuum is created in the combustion chamber, which sucks in a fresh charge and water. The next time the column of water swings back, the mixture is compressed, and so on. Two of these pumps used in the Cobdogla irrigation system on the Murray river in southern Australia still function as technical monuments.

~Inclined plate or angular plate engines. *See above*, ~Engines without crankshaft

~Inline engines. *See below*, ~Reciprocating piston engines ~~Single-shaft engines ~~~Single-bank engines

~Inverted engines. *See below*, ~Reciprocating piston engines ~~Location of cylinders

~Multiple shaft engines. *See below*, ~Reciprocating piston engines

~Number of cylinders. *See below*, ~Reciprocating piston engines

~Opposed-piston engines. *See below*, ~Reciprocating piston engines ~~Single-shaft engines

~Opposed-piston engines, two-shaft design. *See below*, ~Reciprocating piston engines ~~Multiple shaft engines ~~~Twin-shaft engines

~Position, crankshaft. *See below*, ~Reciprocating piston engines

~Position, cylinders. *See below*, ~Reciprocating piston engines

~Propelling nozzle. *See above*, ~Composite systems ~~Turbo-compound systems

~Radial engines. *See below*, ~Reciprocating piston engines ~~Single-shaft engines

~Radial inline engines. *See below*, ~Reciprocating piston engines ~~Single-shaft engines ~~~Radial engines

~Radial rotating piston engines. *See below*, ~Reciprocating piston engines ~~Single-shaft engines ~~~Radial engines

~Reciprocating piston engines. (*also*, →Engine). The frequency of the combustion cycles determines the specific power output of the engine. Depending on the number of strokes, a two-stroke engine has an output—at least theoretically—which is twice that of a four-stroke engine, because it has a power stroke during each crankshaft revolution. The output could be doubled (theoretically) again by using the bottom of the piston for an additional power stroke—using a double-acting piston—similar to a steam engine. Four-stroke engines could also be double-acting, and they would then have the output of a two-stroke engine (theoretically). However, the force delivered by double-acting four-stroke engines is not as uniform as with two-stroke engines, because it is always necessary for two power strokes to occur in sequence. Double-acting engines are built only as diesel engines; tests to design gasoline engines as double-acting engines have all failed.

While construction of double-acting engines using the four-stroke process was given up in the 1930s, large double-acting two-stroke engines continued to be built for marine and stationary operation up to the middle of the 1950s. The peak of double-acting two-stroke engines (**Fig. E233**) was the 24-V-engine MZ 32/44 developed by MAN for the German Navy (historic engine).

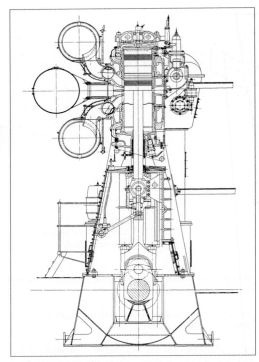

Fig. E233 Double-acting two-stroke crosshead diesel engine, MAN DZ 53/800. $d = 530$mm, $s = 800$ mm, cylinder output 596 kW, $n = 214$ rpm (Source: Mayr, 2nd Edition, p. 422)

There are many reasons why double-acting engines have not been successful:

- The engine is subject to extremely high mechanical and thermal loads. The heat dissipation from the pistons is difficult, making the design of the piston very difficult—and, therefore, sensitive.

- The mixture formation below the piston presents problems because the piston rod extends through the combustion chamber. This means that the injection nozzle cannot be in the centered (cloven combustion chamber.)

- Sealing the piston rod with a packing gland was very sensitive to malfunction and damage. This difficulty increased considerably with the use of heavy oil as fuel. Because the specific power output of the engines could be increased more effectively and simply with exhaust turbocharging, development and construction of double-acting engines was discontinued.

~~**Cylinder layout.** The arrangement of the cylinders is accomplished by considering the minimum space requirement, low specific mass (m/P), and dynamic mechanical behavior (**Figs. E234 and E235**). This results in a large number of possibilities of combination (cylinder configuration): The cylinders can be "connected" in series—in the cylinder axis direction—tandem engines and opposed-piston engines and in the crankshaft direction; and inline engines.

When the cylinders are arranged concentrically on one crank throw, the result is a radial engine. In the case of swash plate engines, the cylinders are located in a circle with their axes in parallel. Rows of cylinders in a polygon with a number of crankshafts result in three-, four-, and five-shaft engines.

Combining one or more inline engines results in a V-, W-, or X-engine with one crankshaft as well as double or H-engines with two crankshafts; by combining two or more "radials," double or multiple radial engines are obtained as well as radial inline engines. The configuration of the cylinder and drivetrain are closely related. Aspects for the arrangement of the cylinders include:

- The installation conditions limit the engine dimensions: in the case of locomotive engines, both the height and width may be limited due to the interior space profile that has to be maintained; when cylinders are installed transversely in a passenger car, the

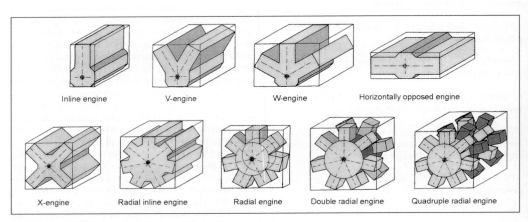

Fig. E234 Cylinder layout 1 (Source: Zima)

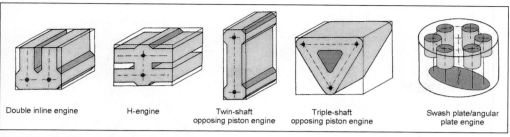

Fig. E235 Cylinder layout 2 (Source: Zima)

engine length is limited; in the case of underfloor engines, there are limitations in height; and for aviation engines, the front surface area is limited due to the aerodynamic resistance; finally, for tank engines it is necessary to consider the weight of the armor for the overall volume (height × width × length).

- Engine dynamics: The sensitivity of crankshafts to torsional vibration increases with the number of throws.
- State of the art: The output can be increased most easily technically by adding engine units, which has led to double engines.

~~**Location of crankshaft.** With few exceptions, all engines have horizontal crankshafts. In special cases— for example, driving generators, as well as special military applications—engines with vertical crankshaft have been built (**Fig. E236**) particularly in the United States by General Motors and Nordberg.

~~**Location of cylinders.** In most engines, the cylinders are positioned upright (**Fig. E237**). The pistons act "from top to bottom" on the horizontal crankshaft.

If the crankcase were rotated around the crankshaft, the cylinders would first be inclined at a certain angle, then horizontal, and finally inverted.

~~~**Horizontal engines (underfloor engines).** Even lower overall heights can be obtained by turning the engine through 90°—that is, installing it lying on its side—or by using V-engines with a V-angle of 180° (**Fig. E238**). Applications for such flat engines include small vans, trucks, buses, and rail motor coaches.

In the 1950s and 1960s, the Büssing company made a name for itself with such underfloor engines.

Horizontal engines were also constructed with vertical crankshafts, as were used in the United States for engines driving generators (Nordberg, Cleveland Diesel Engine Division) (*see also* location of crankshaft).

~~~**Inverted engines.** In aircraft, engines have also been installed upside down (**Fig. E239**). This design allows installation of larger—that is, slower rotating— propellers with better efficiency without increasing the length of the fuselage. Moreover, this provides the pilot with a better view, and service work from the ground can be facilitated.

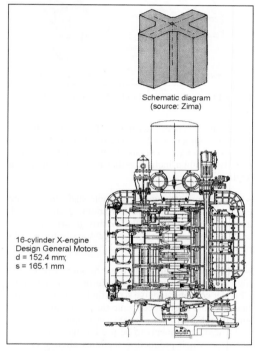

Schematic diagram
(source: Zima)

16-cylinder X-engine
Design General Motors
d = 152.4 mm;
s = 165.1 mm

Fig. E236 Engines with crankshaft (Source: Zima), 16-cylinder X-engine, General Motors version (Source: *MTZ* 13 [1952] 9, p. 228)

The inverted design was preferred in German aviation engine construction by companies such as Junkers, Daimler-Benz, Hirth, and Argus. Inverted engines have dry sump lubrication with one or two pumps which suck oil out of the crankcase and into a collection tank, from where it is pumped to the lubricating circuits.

~~~**Vertical engines.** This design with upright cylinders is the "natural" layout (**Figs. E240 and E241**). The control devices and drivetrain are easily accessible, and the engine oil collects at the lowest region of the oil pan, from where it is sucked up and returned to the lubrication circuit.

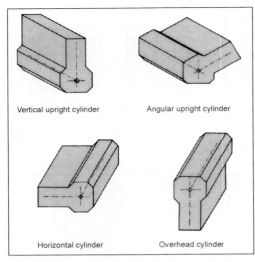

Fig. E237 Orientation of cylinders (Source: Zima)

Because it is necessary to keep the height of the engine compartment low to reduce the drag coefficient on modern passenger cars, passenger car engines are frequently installed inclined—that is, tilted to a certain angle.

~~Multiple-shaft engines. Multiple-shaft engines are engines with two or more drivetrains, each of which acts on its own crankshaft. The drivetrains can be located parallel, inclined, or concentrically to one another or in a polygonal shape.

~~~Five-shaft engines. In order to quickly provide a high-performance tank engine during World War II, Chrysler consolidated five six-cylinder commercial inline gasoline engines concentrically to obtain one engine (**Fig. E242**). This allowed a large number of tried-and-tested components to be used as interchangeable parts for which production facilities already existed (Chrysler Multibank).

~~~Four-shaft engines. A design highlight in opposed piston engines was the Jumo 223 developed by

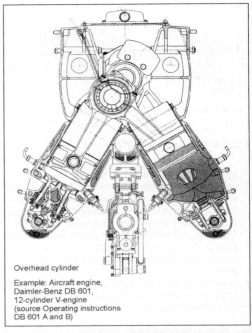

Overhead cylinder

Example: Aircraft engine,
Daimler-Benz DB 601,
12-cylinder V-engine
(source Operating instructions
DB 601 A and B)

Fig. E239 Aviation engine, Daimler-Benz DB 601 12-cylinder V-gasoline engine (Source: Operating instructions DB 601 Au. B)

Junkers during World War II, with four crankshafts and 24 cylinders ("square engine") (**Fig. E243**). The development of this 2200 kW two-stroke diesel engine was abandoned in 1942 due to other priorities. Another development project which never advanced beyond the test phase was the four-shaft group engine developed by FKFS consisting of four 12-cylinder V-engines in a concentric arrangement using the Hirth HM 512 aviation engine as a basis.

~~~Three-shaft engines. At the beginning of the 1920s, Michel-Motorengesellschaft developed a two-stroke diesel engine with three shafts as an "inverted" three-radial four-cylinder inline engine, in which each

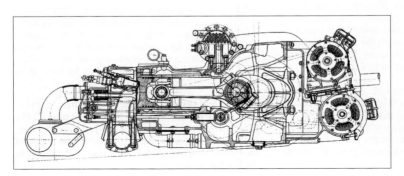

Fig. E238
Horizontal cylinders,
Büssing underfloor
engine U 15 (Source:
MTZ 23, 1962)

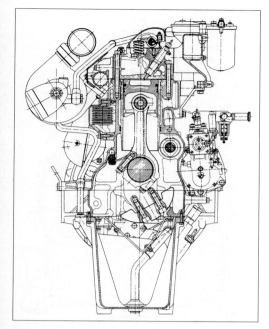

Fig. E240 Vertical upright cylinders, Commercial vehicle diesel engine MAN D 2855 (Source: MAN)

cylinder acted on its own crankshaft based on a common combustion chamber for a total of three crankshafts (**Fig. E244**). This engine was never put into mass production. Following World War II, Napier in England developed the Deltic engine in a triangular format, which was put into mass production: in this, three two-stroke opposed-piston diesel engines were

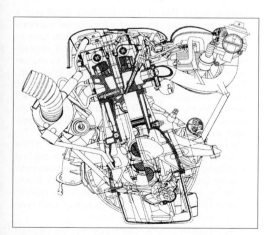

Fig. E241 Inclined cylinders, Mercedes-Benz OM 611 passenger car diesel engine (Source: *MTZ* 58 [1997] 11)

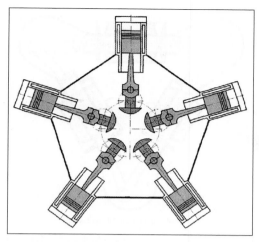

Fig. E242 Five-shafts engine, Chrysler Multibank engine (Source: Zima)

consolidated to form a single unit in the form of an equilateral triangle. Built with 9 and 18 cylinders, the Deltic was used in PT boats as well as for driving locomotives.

~~~**Twin-shaft engines.** With twin-shaft engines, it is necessary to differentiate between dual inline engines and H-engines.

~~~~**Dual inline engines (or parallel and twin engines).** Two engines with complete powertrains are consolidated to form one unit (double engines). In spite of an unchanged basic design, this allows the output to be doubled, the development time shortened,

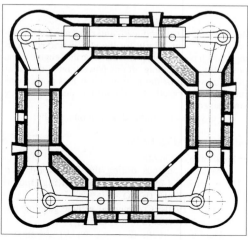

Fig. E243 Four-shaft engine, Junkers Jumo 223 (Source: Thesis A. Kupfers)

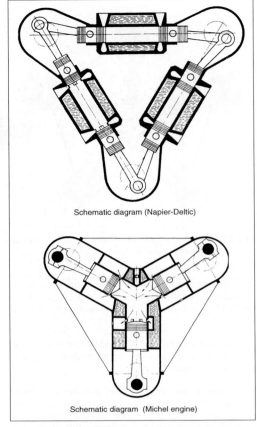

Schematic diagram (Napier-Deltic)

Schematic diagram (Michel engine)

Fig. E244 Three-shaft engine schematic, Napier-Deltic and Michel engine (Source: Thesis A. Kupfers)

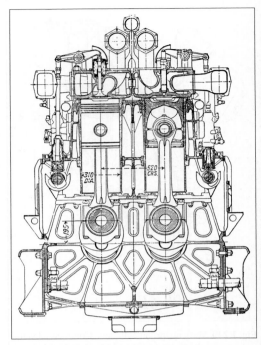

Fig. E245 Dual inline engine, 2×6-cylinder, Sulzer. $d = 310$ mm, $s = 390$ mm, $P = 1618$ kW, $n = 700$ rpm

development costs reduced, and risks avoided in development—particularly for the connecting rod bearings, which posed problems at that time in V-engines. The possibility of switching off one powertrain in the event of malfunction or damage or when less power was required increased the reliability and efficiency simultaneously.

The engines are arranged next to one another.

Inline engines (**Fig. E245**)

• Parallel to one another
 —Aviation engines during the first world war: Adler 2×4-cylinder (Germany); King-Bugatti 2×8-cylinder (USA)
 —End of 1920s: Commercial vehicle gasoline engines Henschel (2×6-cylinder)
 —During the 1930s (2×6-cylinder), locomotive diesel engines: MAN, Sulzer; Aviation gasoline engines: Menasco (USA)

• Inclined at a certain angle to one another
 —Aviation engines: Maybach (1918)
 —Commercial vehicle engine: Büssing (during 1930s)

Double V-engines (**Fig. E246**)

• During World War II, Daimler-Benz developed a double engine, the DB 606, for long-range bombers, consisting of two DB 601 engines placed next to one another. These two twelve-cylinder V-engines were arranged inverted at an angle of 44° to one another and drove the propeller through an idler shaft. Each subengine could be operated separately. Double engines were also derived from the DB 605 model (double engine DB 610) and DB 603 (double engine DB 613). In the United States, Allison developed such a double engine (V3420A).

• The engines were located next to one another. At the beginning of the thirties, Fiat selected this method to achieve the power required to meet the competition for the Schneider cup, a coveted trophy in aviation, by combining two V-12 engines as a tandem system (Fiat AS 12). The two subengines were connected to one another by a gearbox.

• They drove two contrarotating propellers through a transmission on two driveshafts extending through the engine saddle and running inside one another.

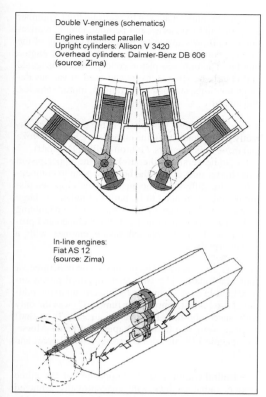

Double V-engines (schematics)

Engines installed parallel
Upright cylinders: Allison V 3420
Overhead cylinders: Daimler-Benz DB 606
(source: Zima)

In-line engines:
Fiat AS 12
(source: Zima)

Fig. E246 Double V-engines: Allison and Daimler-Benz DB 606 (Source: Gerdorf/Grassmann, 3rd Edition, p. 111). Inline version, Fiat AS, $P = 2059$ kW (Source: Handbuch Motor- und Segelflug)

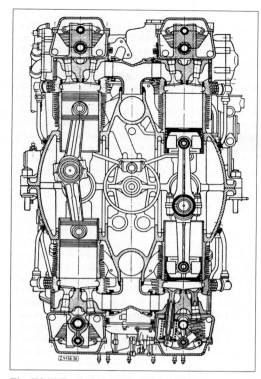

Fig. E247 Dual-shaft H-engine, Napier Dagger. $P = 588$ kW, $n = 4000$ rpm (Source: *ZVDI* 82 [1938] 12)

- At the beginning of the 1950s, Krupp Südwerke designed a series of uniflow-scavenged two-stroke diesel engines for commercial vehicles, which were combined as two three-cylinder engines to obtain the D 688 (SW 6) six-cylinder engine. It was the most powerful commercial vehicle engine in Germany at that time, with an output of 154 kW.

~~~~H-engines. H-engines are based on the principle of two identical V-1800 drivetrains in a common crankcase, acting on one output shaft through an idler shaft (**Fig. E247**). This design was preferred, particularly in England, for aviation engines; specifically by the Napier Company, which brought out a 16-cylinder dual-shaft engine with vertical cylinders (Rapier) in 1929.

The Dagger was then followed with a 16-cylinder engine as well and finally—as a highlight of aviation engine development—the 24-cylinder Sabre with slide-valve timing. Rolls Royce built the Eagle, and Pratt & Whitney built the H-3730, both with 24 cylinders. In the 1970s, MTU developed, in cooperation with Amiot (France), a 40-cylinder H PT diesel engine (40 H 672),

in which half the engine could be switched off, depending on the power requirement.

~~Number of cylinders. The basic form of an engine is the single-cylinder engine. In terms of output (absolute and specific), uniformity of torque and speed, as well as starting characteristics, the majority of engines have a number of cylinders.

The maximum number of cylinders is limited by production costs, maintenance expense, and the torsional vibration characteristics of the drivetrain. Today, vehicle engines have from 1 to 12 cylinders (motorcycles, 1 to 4; passenger cars, 3 to 16; commercial vehicles, 4 to 12), aviation engines in mass production have up to 28 cylinders, and special and test versions have as many as 48 cylinders. High-speed, high-performance diesel engines have 6 to 20 cylinders (special versions have had up to 56); medium-speed, four-stroke engines between 6 and 18 cylinders, and two-stroke crosshead engines 4 to 12.

~~Single-shaft engines. Single-shaft engines are reciprocating piston engines with one crankshaft.

~~~Horizontally opposed engines. A horizontally opposed engine is an engine with the cylinders arranged

E

in one plane and with two rows of cylinders located opposite one another (DIN 1940) (**Fig. E248**). These are 180° V-engines in which the opposing pistons and connecting rods act on their own throw. Because the motion of the pistons is reminiscent of boxers exchanging punches, these are also called boxer engines. The cylinders of adjacent crank throws are located opposite one another; they can, therefore, be located closer to one another in the crankshaft direction. This is the reason that such horizontally opposed engines, in which each cylinder acts on its own throw, are shorter than inline engines with the same number of cylinders.

Horizontally opposed engines are extremely flat. In the 1930s, Krupp developed air-cooled four-cylinder horizontally opposed engines for commercial vehicles as four-stroke gasoline and diesel engines. In Czechoslovakia, Tatra developed a four-cylinder air-cooled gasoline engine for the passenger car called the Tatraplan. During the second half of the 1930s, horizontally opposed engines were also built by Daimler-Benz (12-cylinder: OM 807), Deutz (12-cylinder: A 12 M 319), DWK (8-cylinder: 8 V 19; 12-cylinder: 12 V 19), and MAN (12-cylinder: W 12 V 13/19). Following World War II, horizontally opposed engines were also developed for motor vehicles by companies such as MAN. In the United States, Lycoming developed a 12-cylinder air-cooled four-stroke gasoline horizontally opposed engine for aircraft during the 1940s and 1950s and a series with 4, 6, 8, and 12 cylinders. Familiar horizontally opposed engines are the air-cooled twin-cylinder motorcycle engine from BMW and particularly the engine in the VW Beetle—as well as the Porsche.

~~~**Opposed piston engines.** Two pistons operate in the opposite direction in a single cylinder, and the pistons drive one or two crankshafts (**Fig. E249**). The advantages of opposed piston engines are the high spe-

cific power output achieved by consolidating a number of engine units in series and parallel, simple realization of uniflow scavenging with initial opening of the exhaust port and mutual compensation of the engine inertias. The opposed piston concept was propagated particularly by Hugo Junkers and used in various designs for ships, stationary operations, motor vehicles, and aircraft.

Single-shaft design. One piston in the cylinder acts directly on the crankshaft, the other through a cross yoke and drive rods (Junkers, Gobron-Brillie, Doxford).

On some versions, the stroke of the upper piston was shorter than that of the bottom piston to compensate for the different component masses. Opposed piston engines continued to be built in England as large marine diesel engines by Doxford up to the beginning of the 1970s. Commer and Sulzer developed and produced single-shaft opposed piston engines with a rocker arm drive.

Multiple-shaft design. The pistons act on two, three, or four crankshafts. The twin-shaft opposed-piston engine built in large quantities by Junkers up to the middle of the 1940s (Jumo 205 or Jumo 207) was the only truly successful diesel engine for aircraft. Twin-shaft engines were also built by Napier, Fairbanks-Morse, Compagnie Lilleoise des Moteurs, Rolls-Royce, and Leyland.

~~~**Radial engines.** As the "opposite" of the inline engine, radial engines with concentrically arranged cylinders result in a very short and light engine with, however, a large frontal area. This property—considered by itself—is disadvantageous for aviation engines due to the high aerodynamic resistance. However, the cylinders located in the shape of a star can be cooled

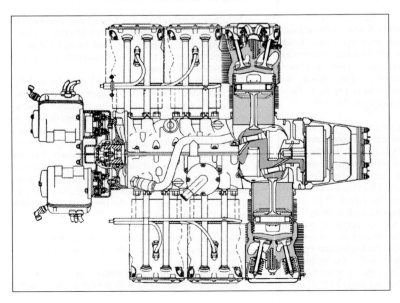

Fig. E248
Six cylinder horizontally opposed engine, Lycoming GO-435. $d = 123.8$ mm, $s = 98.4$ mm, $P = 191$ kW, $n = 3400$ rpm (Source: Mackerle)

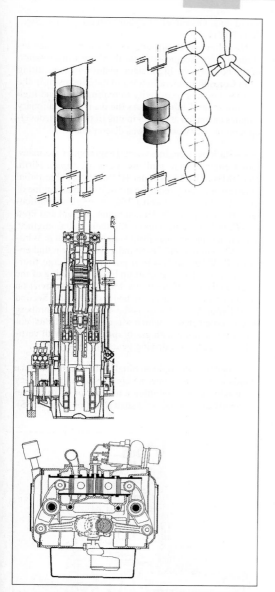

Fig. E249 Opposed piston engines, schematic (Source: Zima); rocker arm engine, Commer (Source: Newton/Steed: *The Motor Vehicle*, p. 223); two-stroke diesel engines, Doxford (Source: Pounder B 92)

well with air due to the large free-flow cross sections. In order to ensure proper cooling of the rear cylinder banks on two- and four-bank radial engines, the cylinders were offset in relation to one another. Air-cooled engines do not require a water jacket, a water pump, coolant or, above all, a radiator, which reduces the en-

gine mass and susceptibility to malfunction. Air-cooled radial engines were, therefore, a preferred design for aircraft for some time. In order to achieve a uniform firing order, the number of cylinders on four-stroke engines was odd. It is necessary to differentiate between radial so-called rotary engines and basic radial engines.

~~~~**Counter-rotating cylinder engines.** An attempt was made to compensate for the disadvantages of rotary cylinder engines with counter-rotating cylinder engines, in which the crankcase and crankshaft rotated in opposite directions (**Fig. E250**). In this manner, the housing rotated more slowly—while maintaining the high average piston velocity required for the output—thereby reducing the ventilation losses. Moreover, the gyrostatic moments cancelled one another. Counter-rotating cylinder engines were built before and during World War I by Siemens (Type Sh 1 to Sh 3).

At the beginning of the 1920s, even a motorcycle was built with a counter-rotating cylinder engine in which the crankshaft rotated at five times the engine speed in the opposite direction to the crankshaft connected to the front wheel (Megola).

~~~~**Radial inline engines.** The maximum concentration of power was offered by radial inline engines, which were all water-cooled because it was no longer possible to sufficiently cool the rear cylinders with air (**Fig. E251**). Junkers developed a 24-cylinder engine as a six-radial four-cylinder inline engine, the Jumo 222, at the beginning of World War II; in the United States, Lycoming developed a nine-radial four-cylinder inline engine, the XR-7755.

While these aviation engines were produced only in small quantities or as test engines, a series of liquid-cooled radial inline engines was built in large quantities in the Union of Soviet Socialist Republics during the 1960s and 1970s as seven-radial six-cylinder and eight-cylinder radial inline engines (with a total of 42 or 56 cylinders) for high-speed boats (Tsch SNP 1611). These were four-stroke compound-type diesel engines.

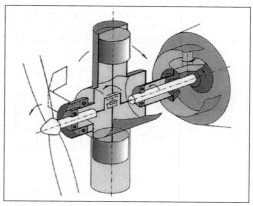

Fig. E250 Schematic of counter-rotating cylinder engine, Siemens Sh 1 (Source: Zima)

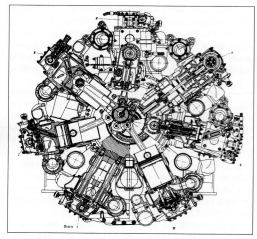

Fig. E251 Radial inline engine, 42- or 56-cylinder four-stroke diesel engine, Tsch NSP 16/17 (Source: Technical description)

~~~~**Rotary engines (radial rotating cylinder engines).** Rotary engines reverse the usual kinetic principle of engines; the crankcase rotates around the crankshaft axis while the pistons and piston rods rotate in a circle around the crank pins (**Fig. E252**). The crankshaft remains stationary and takes up the reaction forces. By rotating the housing, the engine itself provides the cooling air required, which was a decisive advantage in early engine engineering, because at that time, problems were encountered with the cooling air for stationary engines. Advantages of rotary engines were: their power-to-weight ratio was only two-thirds of that of conventional engines (stationary engines); and they have practically no free mass effects and, therefore, ran exceptionally smoothly, which was important for the lightweight aircraft before World War I. The rotating crankcase also provided a "flywheel effect," which gave an extremely uniform torque curve.

Familiar producers of such engines were the French companies Gnome, Rhone, and Clerget. Rotary engines were produced under license in Germany (Mo-

torenfabrik Oberursel), England (Bentley), and Sweden (Thulin).

The principal disadvantages of rotary engines include the gyroscopic moment resulting from rotation of the housing, the high lateral piston forces resulting from Coriolis acceleration, ventilation losses from the rotating housing, sensitivity to malfunction, and high consumption of lubricant. As the design of stationary engines advanced at the beginning of the 1920s, development of rotary engines was discontinued.

~~~~**Stationary engines (radial engines).** In Germany, radial engines were built by Siemens, Argus, Hirth, and BMW. The 14-cylinder BMW 801 double-radial engine was considered to be the best German engine of this type (**Figs. E253 and E254**). In England, particularly Bristol and in the United States, Wright and Pratt & Whitney became known for their radial engines. The compound engines from Wright and Pratt & Whitney (18-cylinder and 28-cylinder four-bank radial engines R-3350 and R-4360) powered the large four-engine transatlantic aircraft up to the beginning of the jet era at the end of the 1950s. Modified radial aviation engines were also used in tanks during the Second World War. A peculiarity is the 11-cylinder Nordberg water-cooled engine, which was designed as a dual engine for operation with gas or diesel fuel—with vertical crankshaft—for driving generator sets.

~~~**Single-bank engines.** Single-bank engines are the standard version of internal combustion engines; they are produced by combining a number of cylinders sequentially in the axial crankshaft direction (**Fig. E255**).

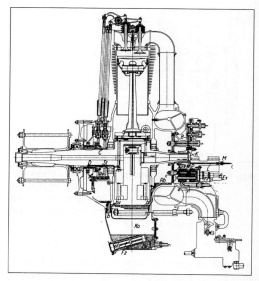

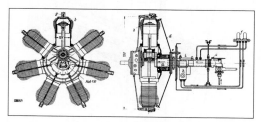

Fig. E252 Seven-cylinder rotary engine, Gnome. $d = 105$ mm, $s = 110$ mm, $P = 31$ kW, $n = 1200$ rpm (Source: *ZVDI* 84 [1940] 7, p. 117)

Fig. E253 Nine-cylinder radial engine, Siemens Jupiter (License Gnome & Rhone). $d = 146$ mm, $s = 190$ mm, $P = 375$ kW, $n = 2100$ rpm (Source: Siemens Jupiter Motor, Handbuch)

Fig. E254
14-cylinder BMW 801
double radial engine
(Source: BMW)

The crankcases on inline engines have a very simple design; inline engines are also very easy to service and repair.

Today, motor vehicle engines are usually designed as inline engines with up to six cylinders: larger four-stroke engines with up to nine cylinders are also built. Two-stroke crosshead engines are constructed only as inline engines today—with up to 12 cylinders—because of their size.

~~~**Single-cylinder engines.** Single-cylinder engines are built for low output with small cylinder dimensions and are usually air-cooled (**Fig. E256**). As gasoline and diesel engines, they are used to drive small working machines and generator sets, as gasoline engines for driving light motor bikes, motor scooters, and manually operated working machines. Special measures are required to compensate for the inertia effects and to smooth the torque (flywheel, balancing gears).

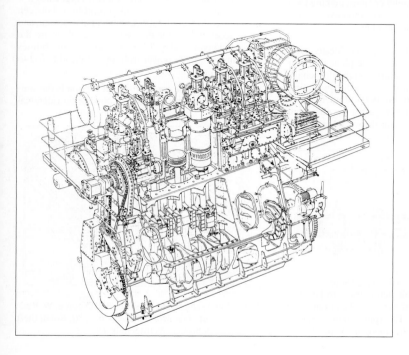

Fig. E255
Single-bank MAN B&W
two-stroke crosshead
diesel engine (Source:
Two-stroke drive systems
for ships MAN B&W,
1, 1991)

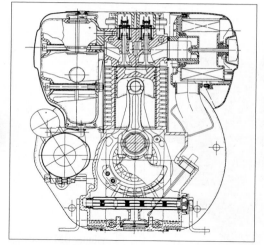

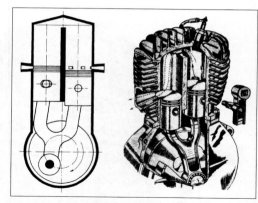

Fig. E258 Twin-piston engines: schematic (Source: Thesis Kupfers) and Puch version (Source: *ATZ*)

Fig. E256 Air-cooled single-cylinder diesel engine, Hatz 1B20. $d = 69$ mm, $s = 62$ mm, $P = 3.4$ kW, $n = 3600$ rpm (Source: Hatz Company Publication)

~~~**Tandem engines.** The pistons in two cylinders located back-to-back in the cylinder axial direction are connected to one another by a piston rod and act together on one crank throw (**Fig. E257**). This design, which was taken over from steam engines, was used for large stationary gas engines and diesel engines into the 1930s. Due to their intrinsic length, tandem engines are designed as horizontal engines, making them easily accessible for service work; moreover, it was possible to keep the machine hall low.

~~~**Twin-piston engines.** Two cylinders located close to one another in the crankshaft circle plane of a two-stroke engine have one common combustion chamber; they simultaneously form a "U" (U-type engine) (**Fig. E258**). The pistons in the two cylinders (twin-piston engine) operate either together on one crank throw or each on its own crank throw, whereby one

piston slightly leads the other. The leading piston controls the exhaust, the trailing piston the intake port. This allows advantageous uniform scavenging to be achieved with minimum design expenditure. Twin-piston engines are built particularly for motorcycles and were particularly successful in motor vehicle racing.

~~~**V-engines.** V-engines (older designation: fork engines) are a structural consolidation of two banks of cylinders inclined to one another at a certain angle, the V-angle, whose pistons act on a common crankshaft—in each case, two pistons moving in cylinders in the opposite banks act on one crank throw (**Fig. E259**).

V-engines represent a good compromise between high power-to-weight ratio, compact basic design, and good accessibility; for this reason, V-engines are the preferred layout today. There are various possibilities for transferring the power from the connecting rod to the crank pin:

- The side-by-side connecting-rod version is the simplest solution in terms of the powertrain (identical

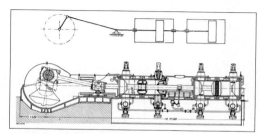

Fig. E257 Tandem engines: schematic (Source: Zima); and Skoda double-acting four-stroke diesel engine. $d = 850$ mm, $s = 1250$ mm, $P = 1100$ kW, $n = 125$ rpm (Source: Special VDI booklet, *Diesel engines* V, p. 41)

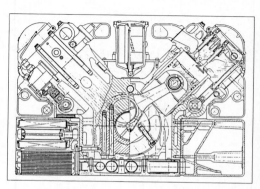

Fig. E259 V-engine, MTU series 880 (Source: W. Rudert: *Die Motorenbaureihe 880*, Sonderheft Jahrbuch Wehrtechnik 14)

connecting rods and identical bearings). However, due to the axial offset of the drive point on the crank pin (connecting rod offset), offset intermediate walls are required.

- Straight intermediate walls with different connecting rods and bearings and with a concentric connecting rod action can be achieved with:
  —Fork and blade connecting rods
  —Main and pivoting (articulated) connecting rods

Due to the low supporting force of the earlier bearing materials, the *side-by-side rod* layout was preferred because of its greater bearing width—which also resulted in longer crank pins and, therefore, longer engines; however, the power flow in the offset intermediate walls also presented difficulties. For this reason, the concentric arrangement was preferred for high-performance engines (gasoline as well as diesel engines). Today, only the side-by-side rod layout is used. Due to the higher lateral components in the bearing forces in V-engines, fastening the crankshaft bearing cap to the crankcase is more complicated than in inline engines.

On small engines, V-engines require two cylinder heads (one cylinder head for each engine bank), as well as two sets of camshaft bearings. The design of the intake manifolds is more complicated and there are two "hot" sides of the engine. Turbocharging is also more complicated, and V-engines are less favorable than inline engines in terms of their inertia effects. Even six-cylinder engines can be installed transversely in passenger vehicles when designed as V-engines; in commercial vehicles, the narrow crankcase waist of the V-engine fits well into the limited space in the front region of the vehicle. The space between and below the engine banks can be used to install the engine accessories (injection pump, turbocharger, filter, etc.) to obtain a compact drive unit. High-speed, high-performance engines are produced with six or more cylinders, and medium-speed engines are manufactured with 12 or more cylinders in a V shape.

~~~~**V-angle.** The criteria for selection of the V-angle (fork angle) are: firing intervals, inertia effects, torsional vibration characteristics, supercharging, and, above all, limitation of the engine dimensions in terms of height and width (**Fig. E260**). Other aspects include whether an engine version is to be constructed as a gasoline and/or diesel engine, the number of cylinders in an engine series, as well as the production facilities available.

Practically all V-angles are used between the extreme values of 0° (inline engine) and 180° (boxer engine).

Small V-angles require longer connecting rods (smaller connecting rod ratio $l = r = l$), to ensure sufficient clearance of the connecting rods in the cylinders. This results in longer crankcases but lower lateral

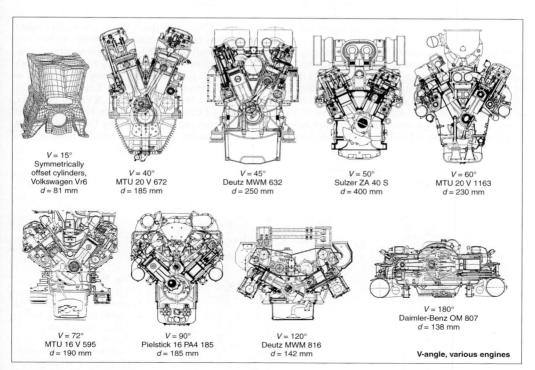

| $V = 15°$ Symmetrically offset cylinders, Volkswagen Vr6 $d = 81$ mm | $V = 40°$ MTU 20 V 672 $d = 185$ mm | $V = 45°$ Deutz MWM 632 $d = 250$ mm | $V = 50°$ Sulzer ZA 40 S $d = 400$ mm | $V = 60°$ MTU 20 V 1163 $d = 230$ mm |

| $V = 72°$ MTU 16 V 595 $d = 190$ mm | $V = 90°$ Pielstick 16 PA4 185 $d = 185$ mm | $V = 120°$ Deutz MWM 816 $d = 142$ mm | $V = 180°$ Daimler-Benz OM 807 $d = 138$ mm |

V-angle, various engines

Fig. E260 V-angle of various engines (Source: Zima)

piston forces as a result of the smaller pivot angle executed by the connecting rod. Equal firing intervals can be obtained by selecting the V-angle with $\delta = 720°$/number of cylinders (four-stroke engine) or $\delta = 360°$/number of cylinders (two-stroke engine).

A V-angle of 90° is preferred for motor vehicle engines and high-speed diesel engines, because this allows complete compensation of the first order reciprocating inertia forces by means of rotating counterweights. On eight-cylinder 90° four-stroke V-engines, the V-angle corresponds to the (uniform) firing interval. When the number of cylinders and V-angle do not correspond, equal firing intervals can still be obtained by "spreading" the crank pins by the difference between the V-angle and firing interval (split-pin crankshaft, offset crank pins, offset stroke). Today, six-cylinder passenger car and commercial vehicle engines are built with V-angles of 90° (e.g., Audi, Deutz, Mercedes-Benz), 60° (Ford) and even 54° (Opel), requiring a total throw offset of 30°, 60°, and 66°, respectively. Larger engines (for locomotives and ships) require narrower V-angles of 60°, 50°, or 45° because of the installation conditions. Because such engines are usually built as a series with between 6 and 20 cylinders, it is only possible for the V-angle to be "correct" for one specific number of cylinders. The crank pins are not spread for reasons of stability; unequal firing intervals ("long" [$\delta + 360°$] or "short" firing sequence [δ]) are tolerated, particularly because such engines are all equipped with torsional vibration dampers. A special case here is an eight-cylinder V-45° diesel engine with 90° throw offset whose offset crank pins are connected by an intermediate web (SKL).

~~~**VR-engines.** The overall length of inline engines can be shortened by "moving the cylinders apart" in the crankshaft rotation plane and then "pushing them together" in the crankshaft longitudinal direction; in this manner, it is possible to obtain a V-engine with a very small V-angle (**Fig. E261**). During the 1920s, Lancia used this design, which became known as the Lancia principle (Lancia Lambda: 15° V-angle; Lancia

Dilambda 24°). At the beginning of the 1990s, Volkswagen developed a VR6 engine for transverse installation in a passenger car based on this principle with a V-angle of 15°. The clearance for the drivetrain—with a medium-length connecting rod—was achieved by moving the cylinder banks apart (while maintaining the same V-angle), so that the cylinders actually intersected below the center of the crankshaft (offset crank drive). The advantages of this design include a short engine length achieved by "packing" the cylinders closer together, which results in a lower engine mass, only one cylinder head, and low inertia effects. However, the different lengths for the intake and exhaust paths, unequal spark plug lengths in the two cylinder banks, and different fire land height resulting from the oblique angle of the pistons result in less favorable relationships for the gas exchange cycle, combustion, and emissions.

~~~**W-engines.** W-engines—that is, engines with three common banks of cylinders acting on one crankshaft—were constructed as aviation engines before and during World War I and in the 1920s (**Fig. E262**). Tests were also performed with W-engines in motor vehicle construction; the Rumpler teardrop car had a W-6 engine at the beginning of the twenties.

During the 1950s Mitsubishi developed a 24-cylinder-W two-stroke diesel engine for PT boats (**Fig. E262**). At the beginning of the 1990s, Audi developed a twelve-cylinder W-engine, which, however, was never put into production.

~~~**X-engines.** With the X arrangement, four banks of cylinders each at a certain angle to the others act on one crankshaft (**Fig. E263**). The angle between the engine banks can be equal or it may be different. Such engines were built before and during World War II as gasoline engines for aircraft: in England by Rolls-Royce, Exe (24-cylinder); Eagle XVI (16-cylinder); Vulture and Crecy (24-cylinder); in France by Clerget (16-cylinder) and in Germany by Daimler-Benz DB 604 (24-cylinder). They were also used as diesel engines

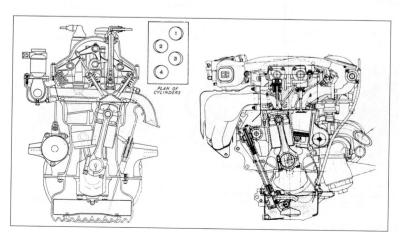

Fig. E261
VR-engines: Lancia principle: Lancia Lambda engine (Source: A. W. Judge: *Automobile Engines*, p. 234) Volkswagen VR5 engine (Source: 59 *MTZ* [1998] 1)

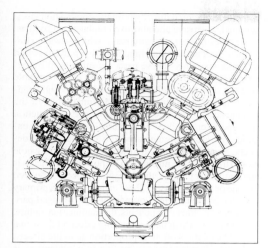

Fig. E262 Mitsubishi 24 WZ 24-cylinder W-engine, two-stroke diesel. $d = 150$ mm, $s = 200$ mm, $P = 2647$ kW, $n = 1800$ rpm (Source: *MTZ* 34 [1973] 11, p. 371)

for high-speed PT boats (for finding and destroying submarines) in the United States by GMC (16-cylinder: model 16–184 A). At the end of the 1950s, Fiat developed a 32-cylinder X-diesel engine for PT boats whose cylinder banks were arranged in the vertical engine direction at an angle of 130° each and in the horizontal plane at an angle of 50° (type 560). In the 1950s, the Cleveland Diesel Engine Division of GMC built a 16-cylinder X-engine for driving generators with a vertical crankshaft for operation with either diesel fuel or gas (model 16–358), as well as a small diesel version for special marine applications.

~Rotary piston engine →Engine ~Alternative engines ~~Wankel engine

~Single-bank engines. *See above*, ~Reciprocating piston engines ~~Single-shaft engines

~Single-cylinder engines. *See above*, ~Reciprocating piston engines ~~Single-shaft engines

~Single-shaft engines. *See above*, ~Reciprocating piston engines

~Stationary engines (radial engines). *See above*, ~Reciprocating piston engines ~~Single-shaft engines ~~~ X engines

~Swash plate engines. *See above*, ~Engines without crankshaft

~Tandem engines. *See above*, ~Reciprocating piston engines ~~Single-shaft engines

~Three-shaft engines. *See above*, ~Reciprocating piston engines ~~Multiple-shaft engines

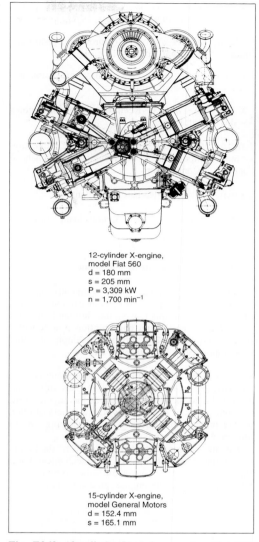

12-cylinder X-engine, model Fiat 560
d = 180 mm
s = 205 mm
P = 3,309 kW
n = 1,700 min⁻¹

15-cylinder X-engine, model General Motors
d = 152.4 mm
s = 165.1 mm

Fig. E263 12-cylinder X-engine, version Fiat 560 (Source: *MTZ* 22 [1961], 6, p. 210); 16-cylinder X-engine, version General Motors (Source: *MTZ* 13 [1952], 9, p. 228)

~Twin-piston engines. *See above*, ~Reciprocating piston engines ~~Single-shaft engines

~Twin-shaft engine *See above*, ~Reciprocating piston engines ~~Multiple-shaft engines

~U-type engine. *See above*, ~Reciprocating piston engines ~~Single-shaft engines ~~~Twin-piston engines

~V-engines. *See above*, ~Reciprocating piston engines ~~Single-shaft engines

~Vertical engines. *See above*, ~Reciprocating piston engines ~~Location of cylinders

~VR engines. *See above*, ~Reciprocating piston engines ~~Single-shaft engines

~W engines. *See above*, ~Reciprocating piston engines ~~Single-shaft engines

~X engine. *See above*, ~Reciprocating piston engines ~~Single-shaft engines

Engine controller →Engine acoustics; →Injection system (components) ~Diesel engine ~~Mechanical control ~~~Control maps

Engine cooling →Radiator ~Coolant

Engine cooling circuit →Cooling circuit; →Radiator

Engine cooling fan →Fan

Engine cooling system →Coolant

Engine curve. Engine maps frequently have a characteristic oval appearance, similar to the contours of a sea shell. For this reason, the German term for a map of the specific fuel consumption is a "conchi-form" curve. **Fig. E264** shows an example of a fuel consumption map for a gasoline engine.

Engine damage. Internal combustion engines are sophisticated machines, but damage does still occur. This is due to the characteristics of the engine itself, the complexity of its functions with a variety of often hardly manageable influences, and varying—often unfavorable—operating conditions.

The internal combustion engine converts the fuel in energy into "heat" and, partly, into mechanical work. Because this energy conversion—the combustion—takes place inside the engine, precision parts of the

system are charged directly with the combustion gases and their corrosive and abrasive combustion products.

Engines consist of many parts, which—when connected structurally and functionally—must function together smoothly. These parts are influenced by numerous effects that, in many cases, are not foreseeable and can only be marginally influenced by design engineers in the production plants. Malfunctioning of the engine accessories, the transmission, and construction and operational failure of the vehicle all affect the engine. Therefore, engine damage is often a sign of defects, weaknesses, and malfunctioning of the whole powertrain and the machine to be powered. Damage is the result of a process, the damaging process, triggered by one or several causes. These causes can be a primary damage in the affected component, the damaged part, or damage in other components, the malfunction, failure, or damage of which has a damaging effect on the actual part. The cause is often disproportionate to the effects.

Engine damage, just as any other form of damage, is undesirable, but it is part of the learning experience and development. Therefore, damage is induced on purpose in engine development to find the processes, causes, and influential variables of characteristic damage and how to avoid them.

Independent of the cause, the event, their symptoms, and the type of affected component, functional group, or machine, damages are the result of numerous physical, chemical, and electrochemical processes, which are collected in VDI regulation 3822. All damage to machine parts can be traced back to the following:

- Damage due to mechanical use
- Overload breakage
- Vibration breakage
- Damage due to corrosion in aqueous mediums
- Corrosion without mechanical stress
- Corrosion types with additional mechanical stress
- Hydrogen-induced corrosion
- Corrosion due to microbiological processes

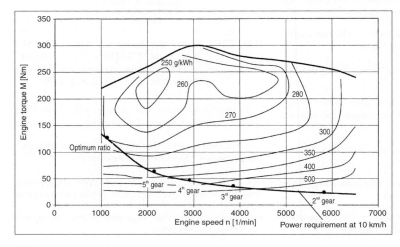

Fig. E264
Engine map

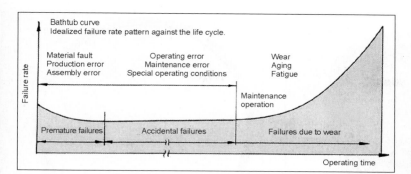

Fig. E265
Bath tub curve

E

Damage due to thermal stress
- Thermal breakage
- Thermal rupture
- Thermal surface damage
- Functional damage due to deposits
- Damage due to diffusion processes
- Damage due to tribological stress
- Sliding abrasion
- Rolling abrasion
- Vibration abrasion
- Abrasive wear
- Flow abrasion

Causes of damage include:

- Wear and tear
- Technical defects (product fault)
- Design defects
- Material defects
- Production defects
- Operational defects
- Overload
- Changes in operational conditions
- Operating defects

The "normal" damage behavior during the life of the engine is described by a bathtub curve (**Fig. E265**). In this diagram, the curve with a reducing slope describes the break-in phase with accidental failures—that is, due to faults in material and in production or assembly processes. The second part, at a constant level, describes the area of use in which the failure rate is constant; these failures are due to operating mistakes, dirt, vibrations, loosening, water impact, or maintenance faults. The third part—progressively rising—describes the increase in damage frequency due to wear and tear, aging, or vibration breakage (fatigue). Using the correct maintenance measures, this final area can be shifted to the right to delay the onset of damage and limit its exacerbation.

Engines, just as any other technical entity, are subject to wear and tear. This can be simple wear, aging, corrosion, biological material damage, or breakage. Wear and tear cannot be avoided during operation of the machine, and the rate of wear—that is , the progression of wear per time unit—is the decisive factor.

~Engine assemblies and components

Damage to the engine transmission unit

a. *Pistons*

The combined dynamic mechanical, quasi-static thermal, and tribological stresses exert a lot of load on the pistons so that even small operational irregularities can impair the function of the pistons and cause damage (**Figs. E266–E268**).

Such impairment of function and damage can appear as:

- Hard surface appearance, wear, friction points, seizing
- Initial cracks, cracks, breakages, initial fused areas
- Holes
- Erosion, corrosion, and oil-derived deposits

All areas of the piston are involved:

- Stem: Seizing due to overloads, oil shortage, insufficient clearance, faulty geometry, dirt in the oil, fuel flooding
- Piston crown and piston-pin boss: External forces, such as foreign bodies in the combustion chamber due to breakage or induction; overheating through faults in the mixture formation; hub and gap breakage in the area of the piston-pin boss; piston-land breakage; ring-groove wear
- Piston crown area: Holes due to faults in the mixture formation and the ignition timing, recess wall cracks,

Fig. E266 Piston seizing through overheating after coolant shortage

E

Fig. E267 Combustion on a diesel engine piston due to poorly ejecting injection valve

Fig. E268 Burning and hole in piston crown

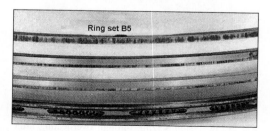

Ring set B5

Fig. E269 Burnt piston rings

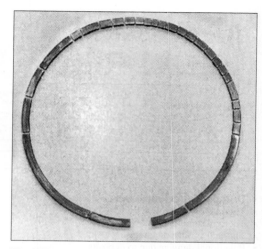

Fig. E270 Piston ring broken into pieces

mechanical damage through valve wear, or foreign bodies in the combustion chamber, erosion

b. *Piston rings*

Assembly faults (too much enlargement on fitting, damage to the cylinder liner), loss of bias voltage, wear, scorch marks (partial seizing; **Fig. E269**)

Ring flutter, ring breakage (**Fig. E270**)

Ring clogging and high oil consumption

c. *Piston pin*

Piston pins are dynamically stressed and, therefore, sensitive to production and assembly faults—insufficient bias voltage of storage trays and bushings leads to frictional corrosion. Fretting can trigger vibration breakage (fatigue breakage) in the bearings and the piston pin. If screws are not tightened according to specifications, the large connecting-rod eye deforms irregularly, which can lead to bearing seizure. Liquid lock and connecting-rod bolt breakage can cause heavy damage to the piston pin, which often leads to overall engine damage (**Fig. E271**).

d. *Crankshafts*

The primary damage to crankshaft is one or more of the following:

- Bending fatigue breakage, mostly starting at highly stressed fillets (**Fig. E272**)
- Torsional fatigue breakages (**Fig. E273**)

Less common, secondary crankshaft damage is often due to bearing damage as well as the failure or the malfunctioning of vibration dampers.

e. *Bearing damage*

In most cases, bearing damage is caused by the breakdown of the lubricating film. When the bearing runs in the mixed and boundary friction range—locally

Fig. E271 Heavily deformed connecting rod due to hydraulic lock

Fig. E273 Torsional fatigue breakage of a crankshaft

Fig. E274 Crankshaft bearing of a diesel engine with pronounced contact pattern (polishing)

and in large areas—first a polished surface appears (**Fig. E274**), then rubbing (**Fig. E275**), and, finally, seizure (**Figs. E276 and E277**). Seizure often develops into complete destruction of the bearing, which makes determination of the cause difficult, if not impossible. Often, bearing damage is caused by events outside the bearing itself. Those causes are: particles in the oil—dirt, abrasion, oil-derived deposit, dust, fibers, etc.—installation errors, deformation of the bearing bore, bearing track, crankshaft, and shape and bearing deviations of the engine transmission unit parts, as well as lubrication deficiencies.

DIN 31 661 summarizes possible changes and damages to plain bearings, giving an overview of the multitude of potential damage to plain bearings and their causes:

Changes and damage to the bearing material

- Dirt, abrasion through ingress of particles
- Wear through mixed friction
- Overheating
- Fatigue, surface disruption
- Washout, cavitation erosion
- Tribochemical reaction, corrosion
- Coating formation

Fig. E272 Flexural fatigue breakage of a crankshaft

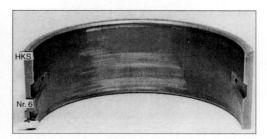

Fig. E275 Crankshaft bearing of a diesel engine with friction points (rubbing)

Changes and damage to the bearing seat

- Microgliding of the bearing bushing
- Surface condition
- Housing bore location
- Bearing support breakage

Fig. E276 Heavy seizing of a diesel engine crankshaft bearing. The bearing was completely flattened in the crankshaft fillet.

Fig. E277 Signs of heavy seizure of a liner with aluminum adhesion from the piston due to piston seizure caused by coolant shortage

Damage to the cylinder head, cylinder, and crankcase

a. *Cylinder head*

Cylinder heads are thermally and mechanically subject to high stress, so that cracks and breakages can occur at critical points, especially when deposits have formed at these points. Valve bridges are especially susceptible. High flow speeds can cause erosion. Valve seats and valve guides are subject to wear. Overheating, due to shortage of coolant and improper tightening of the cylinder head bolts, can cause an offset of the cylinder head face so that the cylinder head gasket can no longer seal properly. The result is blowby of the combustion gases.

b. *Cylinder*

As the gliding partner of piston rings and pistons, the cylinder often incurs wear. This primarily affects the reversal point of the top piston ring: cylinder-bore wear. The mechanical wear through adhesion and abrasion is worsened by corrosion if sulfurous acid is created when the temperature of the sulfur dioxide falls below its dew point. When piston action is affected, cylinders are damaged by furrows and seizing. Hard alternations of the piston contact between the thrust and antithrust sides may cause cavitation on the coolant side, and this is intensified by corrosion (**Fig. E278**).

c. *Crankcase*

In the event of unfavorable stresses, cracks and breakage can occur in the crankcase. These are caused not so much during production or due to assembly errors, but mostly by the effects of forces resulting from piston rod damage. Cavitation and corrosion at the surfaces in contact with the coolant can be counteracted through piston offset (in smaller engines), suitable coolant additives, and construction measures (in larger engines).

Fig. E278 Cylinder liner of a diesel engine with signs of corrosion on the coolant side

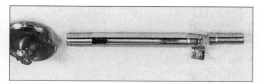

Fig. E279 Double valve breakage: Fatigue breakage due to fretting through valve collets, then breakage of the valve head resulting from touching of the valve on the piston.

Fig. E281 Camshaft of a diesel engine after long running period, with noticeable signs of wear

Damage to control elements

a. *Valves*

In the event of overload due to deficiencies in engine operating parameters (mixture formation, ignition timing), valves—especially the exhaust valve—can be thermally overloaded. Elastic deformation caused by high gas pressure leads to micromotion of the valve and cracking, which is intensified by hard combustion residues.

If carbonized oil accumulates in the valve guide, the stem clearance can decrease to the point that the valves jam. If the kinetics of the valve are affected, breakage results when the valve bottoms on the piston (**Fig. E279**).

Fretting at the contact surfaces of the valve seats causes cyclical bending fatigue of the head (**Fig. E280**). Valves of engines in heavy-oil operation are endangered by heat corrosion.

b. *Intake valve spring*

Intake valve springs are under a lot of dynamic stress. Therefore, even insignificant material and surface faults can trigger vibration breakages.

c. *Camshaft and cam follower*

Typical forms of damage to these components are wear, fatigue, micro-pitting, fretting rust and bearing seizure (**Fig. E281**).

Damage to the injection system (diesel engines)

Typical damage to the injection pump and injector nozzle includes cavitation, corrosion, wear, and seizure (**Fig. E282**). Due to the extremely limited clearances, the pump and the nozzle are at risk from contamination in the fuel. Cavitation can occur in injection lines (**Fig. E283**).

Damage to the spark plugs

The functioning of the spark plug is affected by the various deposits that can be found on it: soot, oil, glaze, and so on. Thermal overload—that is, resulting from wrong ignition point—leads to initial fusing of the electrodes.

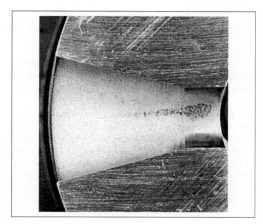

Fig. E282 Cavitation at the intake bore of an injection pump

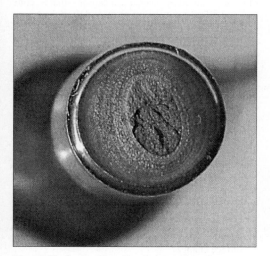

Fig. E280 Cyclical bending fatigue of a valve

Fig. E283 Cavitation groove in an injection line

Damage to the heat exchanger

The heat exchanger must be constructed for use in the normal service conditions of the engine. Especially the case of medium- and low-speed engines with different installation conditions, damage may result from improper design and installation. These errors include corrosion due to potential differences between different materials in the cooling circuit and vibration excitation by the engine. In case of unfavorable operating conditions and maintenance faults, heat exchangers may be contaminated; this causes a narrowing of cross sections and a decrease in the cooling power as well as corrosion of the ventilation elements.

Damage to the exhaust turbocharger

Exhaust turbochargers are at risk from foreign bodies if the air is not perfectly filtered. Crankcase ventilation into the intake line leads to carbon deposits from the oil onto the blades of the impeller in the event of high oil consumption. On the turbine side, the rotor is at risk from foreign bodies in the combustion chamber, produced from piston, piston ring, or valve damage. As is the case for all components under high stress, fatigue failures can develop on the blades of impellers and turbines. Damage may also occur in event of insufficient cooling—that is, after the engine is shut off. Temporary oil cooling after engine shut off is recommended.

Literature: Greuter/Zima: Motorschäden. Würzburg, Vogel Publishers, 2000.

~Stress. Differentiation is made between mechanical, thermal, and tribological stresses—as well as stress through corrosion.

~~**Corrosion in watery media.** Corrosion is an electrochemical reaction of the material and its environment during which an anodic metal–metal ion reaction follows electrolytic metal abrasion (DIN 50900, part 1). Corrosion causes metal abrasion, which impairs the functioning of the components and shortens the life of the material.
Corrosion is determined by:

- The material
- The attacking media
- The electrolytes

Corrosion is encouraged by:

- Oxygen in the air
- Amount of dirt, soot, and salt in the air

~~**Mechanical stress.** This is stress imposed from outside the component or/and internal tension. This can cause cracks, which—more or less quickly—cause the breakage of the component.
Decisive for the degree of breakage are the following stress types:

- Force/pressure
- Thrust/shear
- Bending
- Torsion
- Surface pressure
- Surface of a hole
- Stress condition: single axis/multiple axes
- Stress-time curve: Static/dynamic (resting/swelling, changing, shock-like)
- Speed of application of stress
- Temperature-strength behavior = f (temperature)

A distinction is drawn between:

- Forced breakages through thrust or shock movements and/or overload
- Vibration breakages (fatigue breakages), which are the characteristic failure type for dynamically stressed engine components

The development of vibration failures depends on:

- Previous damage (initial crack)
- Stress amount
- Duration of load exposure

Vibration failures on engine components are primarily caused by forces, pressures, cyclical bending, and torsional load.

~~**Thermal stress.** Thermal stress leads to:

- Thermal failure
- Thermal cracking
- Thermal deformation
- Thermal surface damage
- Malfunctioning due to deposits

~~**Tribological stress.** A distinction is drawn between solids wear and wear from mixed friction and boundary friction.
Wear is the progressive material loss at the surface of a solid, caused by mechanical causes—that is, contact and relative movement of a solid, liquid, or gaseous body close to it [DIN 50 320].
Crack mechanisms include:

- Surface disruption
- Abrasion
- Adhesion
- "Tribochemical" reactions

In engines, cracks are primarily caused by:

- Boundary and mixed friction (incomplete separation of basic and proximate bodies)
- Liquid friction (incomplete separation of basic and proximate bodies)
- Cavitation
- Erosion
- Ablation

The effects of wear are:

- Sectional weakening
- Surface change
- Impairment of function through cyclic expansion

- Reduction of contact ratio
- Impairment of geometry and kinetics

Possible effects of wear:

- Increased friction
- Seizing
- Force and fatigue breakages

In the engine, wear has an effect on:

- Transmission of movement (engine transmission unit, actuation)
- Movement limitation (vibration damper)
- Force transmission (engine transmission unit, actuation, gear drive, press-fit connections, and gear, chain, and belt drives)
- Material transportation (injection lines, coolant lines)
- Information transfer (electric contacts)

Engine design →Engine concepts

Engine dimensions, primary (*also*, →Engine concepts). The operating characteristics of an engine are determined significantly by the primary dimensions in addition to the combustion process. Moreover, these also are the first approximation for the mass of the engine and the installation space required in the vehicle. The following can be defined as primary dimensions:

Cylinder bore

- Stroke: The stroke multiplied by the cylinder cross-sectional area is the displacement and is a primary factor determining the overall height of the engine.
- Engine length: This is primarily a function of the cylinder bore, number of cylinders, distance between the cylinders, the cylinder arrangement, and location of accessories.
- Engine width: The engine width is determined on nearly all engines by the arrangement of the cylinders—for example, V, RV, inline or by the number of camshafts. On inline engines with a number of camshafts, the distance between the camshafts is primarily responsible for the engine width. The type and location of the intake and exhaust ducts as well as the location of accessories also affect the width.
- The target of engine development is to keep the overall size and, therefore, mass as low as possible. This has positive effects on the overall installation in the vehicle (e.g., cooling) and, in terms of the engine mass, advantages with regard to fuel consumption.

Engine drag torque. For special engine tests (e.g., friction loss), the engine is run without firing (i.e., motored) using an electric motor or similar. The size of the torque required depends on the engine configuration and speed. This torque is called the engine drag torque, and it is used for overcoming the engine friction.

Engine efficiency. The efficiency of a combustion engine defines utilization fuel energy. In this case, generally speaking, the "collectable" energy is engine performance, and energy input to the engine is through the fuel (chemical energy). It is also necessary to consider how completely the chemical fuel energy, for example, is transformed into collectible energy, and to distinguish between various efficiencies depending on approach.

~**Effective (brake thermal) efficiency.** Effective efficiency is defined as

$$\eta_e = \frac{P_e}{\dot{m}_{Kr} H_u},$$

where P_e = effective (brake) power output measured on a dynamometer, H_u = lower calorific value of fuel, and $\dot{m}_{Kr}$ = fuel mass flow.

~**Ideal engine.** The efficiency of the ideal engine is represented as the quotient of the difference of the internal energy at the initial and final conditions of the process and the chemical fuel energy.

~**Indicated (thermal) efficiency** →Engine efficiency
~**Internal efficiency**

~**Internal efficiency.** Internal efficiency is defined as

$$\eta_i = \frac{P_i}{\dot{m}_{Kr} H_u},$$

where P_i = indicated power output evaluated from an indicator diagram, H_u = lower calorific value of fuel, and $\dot{m}_{Kr}$ = fuel mass flow.

~**Mechanical efficiency.** The mechanical efficiency evaluates the effective (brake) power output as a proportion of the internal (indicated) power and primarily allows for the frictional losses. Mechanical efficiency is defined as

$$\eta_e = \frac{P_e}{P_i}.$$

~**Thermal efficiency** (*also*, →Cycle). The thermal efficiency defines the ratio between collectible (brake) energy and energy involved in an ideal closed heat engine cycle. As an example, the efficiency of the Carnot cycle is

$$\eta_{th, C} = 1 - \frac{T_1}{T_3},$$

where T_1 is the temperature of energy rejection and T_3 is the temperature of energy addition.

Engine elasticity. The term "engine elasticity" is important in the context of the vehicle. The engine elasticity refers to the acceleration characteristics of a vehicle, particularly in higher gears. Here, a measurement is taken of the time required to accelerate a vehicle from 40 to 120 km/h in fourth or fifth gear on a flat road. This time is influenced by a number of factors. In addition to the engine, the vehicle weight and transmission

E

layout are significant. The prerequisite on the engine side for high elasticity is high torque and high mean effective pressure even at low engine speeds. Engines with turbochargers can have particularly high elasticity, depending on the layout. Engines with good elasticity are also economical in terms of fuel consumption because they can be operated at lower engine speeds.

Engine exhaust →Combustion cycle; →Exhaust system

Engine harmonics →Engine acoustics ~Harmonic

Engine installation. Various installation positions are possible and necessary for the engine, depending on the design of the vehicle.

~Longitudinal (axial) installation. Longitudinal installation was used frequently in the past. However, inline engines with a number of cylinders have the disadvantage that their overall length requires a vehicle with a large wheel base, or a large "overhang" results. This has disadvantages for the overall length and the handling of the vehicle. Moreover, the front end of the engine prevents "streamlining" by lowering the hood, which has a negative effect on the aerodynamic resistance. Longitudinal installation presents problems for front wheel drive, particularly on inline engines with more than four cylinders.

~Mid-engine. In mid-engined vehicles, the drivetrain is located in the middle of the vehicle. Today, mid-engines are hardly ever used in passenger cars because too much of the space required for the passengers and luggage is lost. For this reason, this layout is only used for sports cars. However, it does provide advantages in terms of the weight distribution.

~Rear engine. In a vehicle with a rear engine, the engine is located at the rear of the vehicle. Depending on the type of drive (front-wheel or rear-wheel drive), this design can offer advantages: however, there are disadvantages in terms of engine cooling. Moreover, little space is available for the exhaust system. The weight distribution in the vehicle is unfavorable, because it is concentrated at the rear.

~Transverse installation. In order to avoid the problems of longitudinal installation, engines are frequently installed in the transverse direction today. This makes it possible to install engines in a given engine compartment without changing the dimensions of the vehicle while maintaining the hood contour, front end, and so on, and while retaining a low drag coefficient. Moreover, advantages can result in terms of the location of the accessory equipment, routing of the exhaust, and so forth. This type of installation usually requires a special layout for the transmission and the exhaust system due to the motion of the engine.

Engine intake →Intake

Engine layout →Engine concepts

Engine map →Engine curve; →Performance characteristic maps

Engine measures →Fuel consumption ~Variables

Engine mechanics →Calculation processes ~Application areas

Engine noise →Engine acoustics

Engine octane rating →Fuel, gasoline engine ~Octane number

Engine oil →Oil

Engine oil cooler →Radiator ~Oil cooler

Engine operating point. The operating point of an engine is defined by the engine speed and engine load and is represented by the torque or by the mean effective pressure present. The engine, transmission, and vehicle must be matched to each other so that in addition to the driving performance, an optimum engine operating point is present over wide ranges — for example, an engine operating point providing high efficiency or minimum fuel consumption.

Engine performance →Power output

Engine power loss. Real engine processes are subject to losses compared to the ideal theoretical cycles. The basic reference value is the chemical energy present in the fuel, and if it were possible to convert all of this into mechanical work, the efficiency would be equal to 100%. However, because the engine only has an efficiency of approximately 25–43%, depending on the combustion process, this energy difference is dissipated by "losses." These are:

- Losses resulting from deviations in the cylinder charge (e.g., filling)
- Losses resulting from deviation from ideal combustion
- Losses resulting from leakage
- Heat losses
- Charge transfer losses
- Frictional losses

The effective work and the losses for a passenger car prechamber diesel engine are shown in **Fig. E284**. This is also illustrated for a passenger car gasoline engine in **Fig. E285**. **Fig. E286** categorizes the individual types of losses and the remaining effective work for a passenger car gasoline engine.

Engine service life. The engine service life is the period of time during which an engine can be used for its intended application. The primary causes that limit the engine service life are wear and aging of the materials

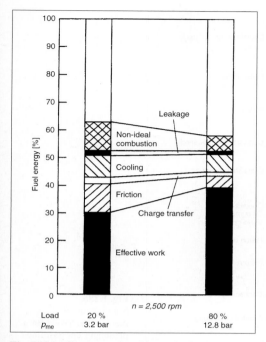

Fig. E284 Effective work and losses in a passenger car diesel engine (Source: Pischinger)

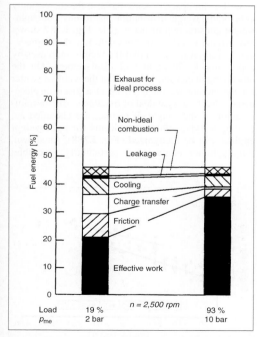

Fig. E285 Effective work and losses for a passenger car gasoline engine (Source: Pischinger)

used. The desired engine service life depends on the application and is developed depending on the requirements placed on the product. Unnecessary reserves should be avoided, because these increase weight and cost.

The engine service life ranges from a few hours (Formula 1 engines) to decades (large stationary engines).

It is necessary to lay out the same basic engine differently depending on the specific application. A diesel engine for use in a passenger vehicle requires a service life of approximately 2000 to 3000 hours; if it is necessary to lay out the same engine for an application such as a heat pump, the service life is in the range of 20,000 hours—that is, 10–15 times longer, which makes the engine considerably more expensive.

The requirement for passenger car engines is, as mentioned, 2000 to 3000 hours or 150,000 to 300,000 km and differs depending on the individual manufacturer. Two-stroke motorcycle engines are designed for approximately 50,000 km, four-stroke motorcycle engines for approximately 100,000 kilometers. The service life of two-stroke engines is limited because it is not permissible for the piston rings to rotate due to the gas exchange ports, and the gas exchange ports interrupt the lubricating film.

Commercial vehicle engines are developed for mileages of 1.5 million kilometers and, in the meantime, even 1.0 million miles.

Engine smoothness →Operating limit; →Variable valve control

Engine smoothness control →Electronic open- and closed-loop control ~Electronic open- and closed-loop control, gasoline engine ~~Functions

Engine speed →Speed

Engine stop →Carburetor ~Equipment

Engine temperature (*also,* →Piston; →Valve). The temperatures in and on the engine have a significant influence on the operating characteristics and service life. Important temperatures include: water temperature (coolant), oil temperature, temperature of intake air, combustion temperature, exhaust temperature, and various component temperatures.

All parts in contact with the combustion gas from the engine are subject to extremely high thermal loads; these include the electrode area of the spark plug, the piston, combustion chamber (cylinder head), exhaust valves, exhaust system, valve guides, and cylinder head gasket.

Moreover, the temperatures required on a number of components differ between gasoline engines and diesel engines. For example, to prevent knocking, the temperature of the cooling water and the cylinder head of a gasoline engine at full load must be slightly lower (80° to 95°) than on a diesel engine, where higher temperatures are desirable because this can reduce the fuel consumption and emissions.

E

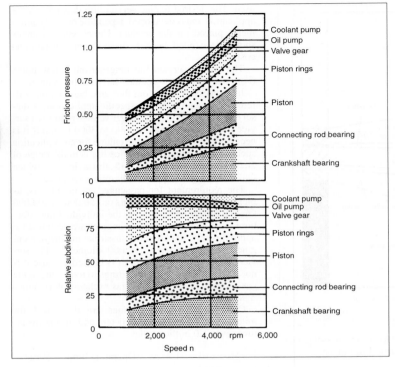

Fig. E286
Subdivision of
mechanical losses for a
passenger car gasoline
engine (Source:
Pischinger)

Rapid warm-up of the catalytic converter is of fundamental importance in terms of minimizing pollutant emissions, and the legal requirements in this area can only be fulfilled by this means. Rapid warm-up of the engine to its operating temperature following a cold start is also important for minimization of fuel consumption.

Engine test bench →Emission measurements ~FTP-75 ~~Performed driving cycles

Engine weight (*also*, →Weight-to-power ratio). Low engine weights and low requirements for space are becoming more important for reasons of fuel economy,

as well as for space reasons and also the increasingly unfavorable weight distribution that results from light-weight construction of the vehicle. **Fig. E287** shows how engine weight varies in relation to the displacement. Here, it is shown that the weight increases by approximately 50 kg/liter of displacement. On the other hand, it can also be seen that the width of scatter for the engine weight is 50 kg at a given displacement—that is, a great deal of freedom exists for minimization of the weight. If engines are classified according to types, the first concepts for light-weight engines can be recognized (**Fig. E288**). As shown, three-cylinder engines are particularly favorable,

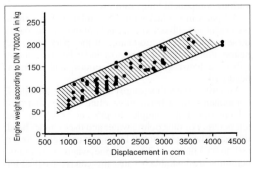

Fig. E287 Engine weight and displacement

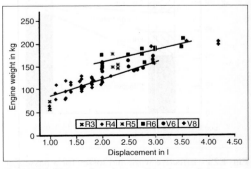

Fig. E288 Types of engines and engine weight

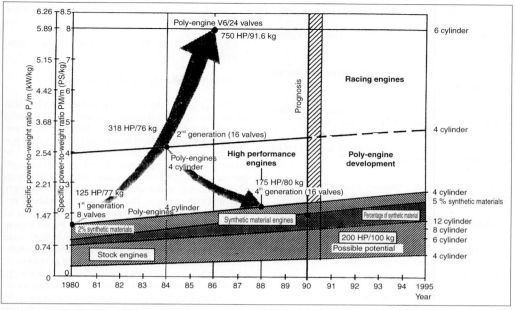

Fig. E289 Specific power/weight ratio

whereas inline six-cylinder engines are particularly unfavorable in this area. However, the production costs of inline six-cylinder engines are 15–20% lower than comparable V-6 engines. **Fig. E289** shows the development of the weight-to-power ratio from 1980 to 1995, and this development has continued.

Engines. The combustion engine is, next to the electric motor, the most frequently used device to power machines, vehicles, and so on. Two basic applications can be described:

- Engines for mobile units such as cars, commercial vehicles, marine engines, and aircraft engines
- Engines for stationary applications such as emergency generators, block heating, and power plants

The use of gasoline, diesel, or special engines is possible depending on the application. Cost effectiveness, availability, energy consumption, cost and power requirements are crucial for the selection.

Engines without crankshaft →Engine concepts

Enrichment →Air-fuel mixture; →Carburetor ~Starting systems ~~ Requirements, ~ Systems; →Mixture formation ~Mixture formation, gasoline engine ~~Mixture metering

Entry of foreign matter →Oil ~Power during operation

Enveloping surface method →Engine acoustics ~Sound measurement

Environmental conditions →Electronic/mechanical engine and transmission control ~Requirements for mechanical and housing concepts

Environmental effects →Particles (particulates) ~Effects on the human organism

EP additives →Oil ~Body ~~Additives

EPIC distributor injection system →Injection system, fuel ~Diesel engine ~~Distributor pumps ~~~Radial piston distributor injection pump

Equivalent combustion rate →Combustion, diesel engine; →Combusion process, diesel engine

Equivalent diameter →Filter ~Intake air filter ~~Filter characteristics

Equivalent inertial masses →Emission measurements

ESP →Actuators ~E-gas ~~Electronic stability control

Ethanol →Fuel, gasoline engine ~Fuels containing ethanol

Ether →Fuel, gasoline engine ~Alcohol (and ether) content, ~Alternative fuels

Ethylene glycol →Coolant ~Monoethylene glycol

445

European driving cycle →Emission measurements

Evaporative cooling →Cooling, engine; →Cooling circuit

Evaporative emissions →Emission measurements ~Test type IV

Evaporative heat →Fuel, diesel engine ~Properties, →Fuel, gasoline engine; →Vaporization, fuel

Evaporative loss →Fuel, gasoline engine; →Oil ~Properties and characteristic values

Evaporative losses →Emission measurements ~Test type IV ~~Calculation of evaporative emissions

Evaporative residue →Fuel, gasoline engine

Even mixture distribution →Mixture formation ~Mixture formation, gasoline engine ~~Mixture preparation

Excess air. With an air-fuel ratio of $\lambda = 1$, the air and fuel masses are stoichiometrically balanced—that is, each carbon and hydrogen atom in the fuel molecule theoretically has an atom of oxygen available for combustion. With excess air there is more oxygen than is theoretically necessary to burn all the fuel.

If gasoline engines are operated with excess air, this leads to low fuel consumption but also to lower power output at full load. With too much excess air, the ignition threshold of the mixture can be exceeded and misfires will occur, which, among other things, lead to an increase in HC emissions.

In diesel engines, HC emissions are low even at high air-fuel ratios because there are almost always zones of stoichiometric mixtures of air and fuel in the combustion chamber.

Combustion in (pre-) chamber diesel engines starts with extreme air deficiency in the chamber and with high excess air in the main combustion chamber. Considerably lower NO_x emissions are generated as a result, compared to engines with direct injection.

Excess pressure valve →Lubrication ~Control and safety components

Exergy loss →Cycle

Exhaust →Combustion cycle; →Exhaust system

Exhaust backpressure →Exhaust gas back pressure

Exhaust cam →Camshaft

Exhaust closes →Control/gas transfer ~Four-stroke engine ~~Timing

Exhaust control →Control/gas transfer ~Four-stroke engine ~~Timing

Exhaust emissions →Emission measurements; →Emissions

Exhaust flow rate →Exhaust system

Exhaust gas. Exhaust gas includes materials that are present after combustion, and were either present in the fuel or have been generated during combustion. Conventional fuel, which is normally a mixture of hydrocarbons, reacts with the constituents of the air. The major resulting components are carbon dioxide, carbon monoxide, nitric oxide, water, unburned hydrocarbons, oxidized hydrocarbons, and particles (primary only in diesel engines). The unburned and oxidized hydrocarbons and the particulates are a conglomerate of different compounds with different toxicity for human beings and the environment. Legal regulations have been established to minimize the effect of these materials, and there are legal limits to the emission of pollutants.

The exhaust gas contains a certain amount of energy (thermal, chemical, potential, kinetic), depending on load and engine speed; this is often unused and is released into the atmosphere. However, the energy can also be used for power increases in exhaust gas turbocharging.

Exhaust gases are often added to the fresh mixture or to the intake air (exhaust gas recirculation) to reduce the oxides of nitrogen.

Exhaust gas afterburning →Pollutant aftertreatment

Exhaust gas analysis. Exhaust gas analysis shows the composition of the exhaust gas of combustion engines. The most important exhaust gas components are:

- Carbon monoxide (CO)
- Carbon dioxide (CO_2)
- Unburned hydrocarbons ($C_m H_n$)
- Nitrogen oxides (NOx)
- Aldehydes (HCO compounds)
- Oxygen (O_2)
- Nitrogen (N_2)
- Water (H_2O)
- Hydrogen (H_2)
- Sulfur dioxide (SO_2)
- Particles (sulfates, soot, combined water, metals, hydrocarbons with a high boiling point, etc.)

The concentration of components in the exhaust gas depends among other things on the fuel used, on the engine oils combusted in the combustion chamber, on the combustion process, and on the engine operating conditions. CO, CO_2, N_2, H_2O are generally present at levels in percent by volume, while the $C_m H_n$, NOx, and H_2 components are found in the ppm (parts per million) range. Particles are normally defined in mg/m^3. The quantity of oxygen in the exhaust gas depends on the air-fuel ratio. Also, additional components in the fuel—for example, sulfur in diesel fuel—generate pollutants.

~Diesel engine. As in the gasoline engine, the air-fuel ratio for the diesel engine represents important parameters, which has a significant effect on the oxygen concentrations.

For each power cycle, the average air-fuel ratio in the combustion chamber is significantly higher for the diesel engine than for the gasoline engine. The diesel engine, therefore, delivers significantly lower CO concentrations than the gasoline engine.

The total concentration of unburned hydrocarbons is lower in the untreated emissions in the diesel engine than in the gasoline engine. The NOx concentration is lower than in the gasoline engine, while the NO_2 content in NOx is slightly higher, at between 5 and 15%. However, comparison of a gasoline engine with a controlled three-way catalytic converter and a diesel engine with an oxidizing catalytic converter shows lower NOx concentrations for the gasoline engine. Sulfur compounds depend only on the sulfur content in the fuel. A mass emission of 2.6 g gaseous SO_2 would be obtained for the combustion of one liter of diesel fuel with 0.15% sulfur and a 100% conversion of the sulfur into sulfur dioxide (SO_2). However, the result would be 10 g (with approx. 1.3–1.5 g water per g of sulfate) including the water bound to the sulfates, if all the sulfur was converted into sulfate (SO_4).

These sulfates are part of the particulate emissions, which are approximately 16–20% of the particulate emissions in FTP-75. The reduction of the sulfur content or the existence of sulfur-free diesel fuel in some countries may mitigate these issues.

The particulates consist of solid-organic-insoluble and fluid-organic-insoluble phases. **Fig. E290** shows an example of the particulate composition.

The solid phase consists of:

- Soot as amorphous carbon, ash, oil additives, and corrosion and abrasion products
- Sulfates and their molecularly bound water

The fluid phase consists of:

- Portions of fuel and lubrication fluids primarily combined with soot. The hydrocarbons are still primarily gaseous in the hot exhaust gas and transfer into a fluid, organically soluble phase (particles) after cooling and turbulent mixing with air.

Soot or particulates are only generated in significant amounts in diesel engines, and under certain circumstances in gasoline engines with direct injection based on locally restricted extremely oxygen-deficient air. They consist primarily of carbon. However, certain substances (hydrocarbons) adhere to the soot.

Some of these accumulated substances are present in mg/km orders of magnitude. They are collectively called polycyclic aromatic hydrocarbons (PAH), and will be removed in part by catalytic converters.

~Gasoline engine. A typical composition of the exhaust gas is shown in **Fig. E291**. The exhaust gas was analyzed for a European test cycle run with the reference fuel specified for this test.

Approximately 1% of the total emitted exhaust gas mass can be classified as pollutants in the European test cycle, if CO_2 is excluded (it is one order of magnitude lower with a catalytic converter). The CO content in exhaust gas depends primarily on the air-fuel ratio, and high CO concentrations occur in rich mixture regions which are deficient in air ($\lambda < 1$). The CO concentration can be reduced by increasing excess air.

The composition of the unburned hydrocarbons is variable. The main components are usually aromatics (e.g., benzene, toluene, ethyl benzene), but also polynuclear aromatics, which are said to be carcinogenic, and olefins (e.g., propane, ethylene) and paraffins (e.g., methane, pentane).

NO contributes approximately 90–98% of the NOx emissions during engine operations. Aldehydes are hydrocarbons with additional oxygen atoms. These so-called O-H-C compounds are especially created during the combustion of fuels that have a high oxygen content (e.g., alcohols). The odor of some aldehydes is very strong. An important representative of this pollutants group is formaldehyde (HCHO), which is already legally restricted in California.

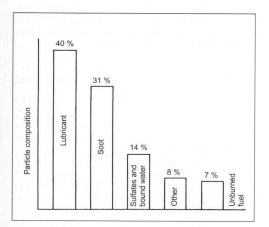

Fig. E290 Particle composition in diesel engine exhaust gases

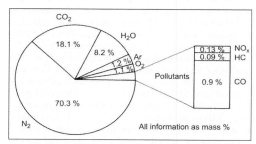

Fig. E291 Average exhaust gas composition in one catalytic converter in ECE test

The sole cause of lead emissions from gasoline engines was the use of lead fuel additives. These were normally antiknock additives based on chlorine or bromine compounds and they are now prohibited.

Literature: H. Friedl: Katalytische Nachbehandlung von Dieselmotorenabgas unter besonderer Betrachtung der Partikel und Schwefelemission, Progress Reports VDI, Series 12, 162, 1991. — M. Houben, G. Lepperhoff: Der Kraftstoffeinfluss auf die Partikelemission von Dieselmotoren, Technical Working Conference Hohenheim, 1990. — Presentation. — F. Pischinger: Lecture manuscript TH Aachen, 1994. — K. Engeljeringer: Probenahme und Messung von Komponenten im Abgas, MTZ 64 (2003) 6, p. 493.

Exhaust gas analysis equipment. The exhaust gases collected in accordance with legally regulated processes must be analyzed. This process also often collects unregulated exhaust gas components such as CO_2 and O_2 in addition to the restricted components. The concentration of the gaseous components is determined with the help of gas analyzers. The individual measuring methods for the gases are tuned selectively and specifically to the particular gas. Particles that are generated during engine combustion will be separated with the help of filters and their total mass determined gravimetrically by analyzing the loading of the filter (blackening).

The following analyzers are normally used for the measurement of the legally limited exhaust gas components:

- Chemiluminescence detector (CLD) for oxides of nitrogen (NOx)
- Flame ionization detector (FID) for hydrocarbons (THC)
- Nondispersive infra-red spectroscopy (NDIR) for carbon monoxide (CO) and carbon dioxide (CO_2)
- Dilution tunnel and filter for particulate measurement. Other processes for the determination of particulate emissions are currently being tested.

The following are the most important measuring methods and instruments, which are currently used to analyze gasoline and diesel engines:

~Calibration. Each measuring range must be calibrated before each measurement. The calibration curve of the measurement instrument is determined by at least five calibration points. The nominal concentration of the calibration gas with the highest concentration must at least be 80% of the upper scale value. The calibration curve is calculated in accordance with the least squares method. It cannot deviate more than 2% from the nominal value of each calibration gas. The calibration test is performed with a zero gas, normally N_2, and a calibration gas, whose nominal value is between 90 and 95% of the value to be analyzed. Legislation requires the calibration of a flame ionization detector (FID) with propane (C_3H_8). The analysis is valid if the difference between the two measurements is <2%.

The pure gases used for the calibration and the operation of the instruments must meet the following conditions:

- Purified nitrogen (purity ~1 ppm C, ~1 ppm CO; ~400 ppm CO_2, ~0.1 ppm NO)
- Purified synthetic air (purity ~1 ppm C, ~1 ppm CO; ~400 ppm CO_2, ~0.1 ppm NO), oxygen content between 19 and 21% by volume.
- Purified oxygen (purity ~99.5% by volume O_2)
- Purified hydrogen (and mixtures with hydrogen content) (purity ~1 ppm C, ~400 ppm CO_2).

The actual concentration of a calibration gas must be within ±2% of the nominal value.

~Chemiluminescence detector (CLD). The CLD is used for the determination of the concentrations of the exhaust gas components nitric oxide (NO) and nitrogen dioxide (NO_2), which are normally combined and called oxides of nitrogen, NOx. Such an instrument is shown schematically in **Fig. E292**.

The CLD mode of operation is based on a chemical reaction, which causes a light emission that is measured by a photocell. The following processes are performed:

- The nitric oxide is oxidized to nitrogen dioxide by means of ozone (O_3). This results in molecules with a stimulated electron condition. The molecules, which are at a higher energy conditions, return to their initial electrochemical state by the emission of photons (light quanta, hν).
- This process is called chemiluminescence. The emitted photon current is proportional to the nitric oxide concentration. It is amplified and displayed as parallel flow. The reaction equation for this process is:

$$\text{Efficiency (\%)} = \left(1 + \frac{a-b}{c-d}\right) \cdot 100.$$

In this equation, h is Plank's constant and ν is the frequency of the emitted light. The exhaust gas flow is guided directly into the reaction chamber to measure the NO content. The concentration of NOx is measured by guiding the exhaust gas first into a converter, which converts the NO_2 into NO at about 650°C. The exhaust gas, which now contains only NO, subsequently flows to the reaction chamber.

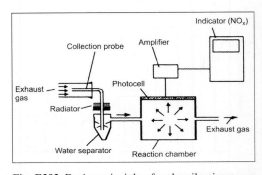

Fig. E292 Design principle of a chemiluminescence detector

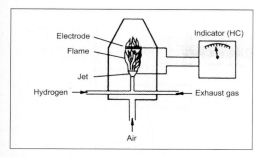

Fig. E293 Flame ionization detector

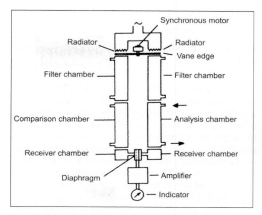

Fig. E294 Infrared absorption analyzer

~**Flame ionization detector (FID).** The FID measures the concentration of the total hydrocarbons (THC) in the diluted exhaust gas. **Fig. E293** shows the schematic design of an FID.

A hydrogen flame burns in pure air in a combustion gas oven, which is free of hydrocarbons. Voltage is applied between the burner nozzle and an electrode within the flame. Practically no ions are generated as long as only pure hydrogen burns. An ionic current is generated in the exhaust gas as a result of the hydrocarbons in the exhaust gas introduced into the flame. It is proportional to the volumetric concentration of hydrocarbons in the exhaust gas. The ionic current does not only depend on the concentration of the hydrocarbons, but also on the number of carbon atoms in the molecule.

A feed line heated up to 190°C must be used for this measurement to prevent hydrocarbons with a higher boiling point from accumulating on the walls of the line, which could falsify the results. Due to the wide spectrum of composition of the hydrocarbon compounds, which are composed of nonoxidized or partly oxidized fuel components, it is necessary to calibrate the analyzer with a typical component present in the exhaust gas.

~**Nondispersive infra-red absorption analyzer (NDIR).** This instrument measures the CO and CO_2 components. The measuring principle of the NDIR is based on the absorption of electromagnetic radiation. Each gas has a radiation absorption for each wave length range, which is used to detect the existence of the corresponding gas. **Fig. E294** illustrates the measuring principle.

A wire spiral is used as the radiation source. Both paths of rays are interrupted periodically by an aperture wheel. One of the two paths of rays contains the analysis chamber with the diluted exhaust gas probe, the other contains a comparison chamber filled with an IR inactive gas. A filter chamber filled with interference gas is placed in front of each chamber. This prevents cross-sensitivity, because the share of the radiation that would be absorbed by the interference gas is filtered out. The receiver is filled with a mixture of argon (Ar) and the pure gas to be analyzed (e.g., CO), and, therefore, can only be heated by the heat radiation of the wavelengths absorbed by CO.

The exhaust gas probe extracts a part of the CO specific radiation from its bundle of rays if the probe contains CO. Only a residue remains for the corresponding receiver part. It will be heated less, and the gas volume inside this receiver part will expand less than in the other receiver part. A metal membrane used as a separation wall then bends accordingly. The degree of bending is converted into a corresponding signal and serves as a measure of the concentration of the gas under investigation.

~**NOx converter.** This converter reduces NO_2 to NO. The effectiveness of the converter must be checked frequently to exclude measuring errors. An ozone generator is used for this application and the following process is executed.

The analyzer is calibrated to the measuring range used most. The calibration gas must have a NO content of about 80% of the upper scale value. The NOx analyzer is adjusted to a NO operation, which means that the calibration gas does not get into the converter. The concentration displayed will be recorded.

The gas flow is then continuously fed with oxygen or synthetic air until the displayed calibration concentration is reduced by 10%. The concentration, c, displayed is recorded. The ozone generator is shut off during this process.

The ozone generator is then switched on to produce enough ozone to ensure that the NO concentration is reduced to 20% (minimum 10%) of the calibration concentration. The concentration, d, displayed is recorded.

The analyzer is then switched to operation status and the gas mixture, consisting of the components NO, NO_2, O_2, and N_2, now flows through the converter. The concentration, a, displayed is recorded.

The ozone generator is then switched off. The gas mixture flows through the converter into the measuring device. The concentration, b, displayed is recorded.

The efficiency of the NOx converter can now be calculated using the following formula:

$$\eta = 1 - \frac{a - b}{c - d}$$

The efficiency of the converter cannot be lower than 0.95.

~Oxygen analyzer. This measuring principle is based on the paramagnetic characteristic of oxygen. Paramagnets generate a strong inhomogeneous magnetic field inside the measuring chamber while the measured gas flows through it. The oxygen molecules are pulled into the inhomogeneous magnetic field due to their paramagnetic behavior if oxygen is present in the measured gas. Different partial pressures of oxygen are generated due to the strong inhomogeneity of the magnetic field. The partial pressures are higher at locations with high field strength than at locations with low field strength. The differential pressure between the partial pressure of the measured gas and the partial pressure of a comparison gas is a measure of the oxygen concentration in the exhaust gas.

~Particle measurement. **Fig. E295** shows the particulate measuring technology normally used in the car industry for the analysis of engine exhaust gas. The dilution tunnel is one of the main elements of the measuring technology for this measuring principle. The particulate emissions are calculated based on the exhaust gas mass, the particle mass collected on the particulate filter, the gas volume fed over the particulate filter, and the NOx dilution ratio.

~Smoke measuring instrument. The soot or smoke emission is measured by a diesel smoke meter. The soot content in the exhaust gas of diesel engines can be measured photo-electrically with this analyzer with the help of the filter paper method. For this measurement, a specified exhaust gas volume is extracted from the engine exhaust gas with a collection probe and sucked through filter paper with a defined surface area. The exhaust gas soot particles blacken the filter paper. The level of blackening is measured with a reflection

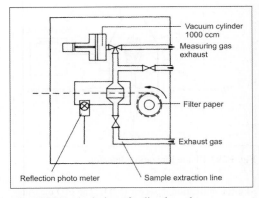

Fig. E296 Basic design of a diesel smoke meter

photometer and assessed with reference to a blackening chart. Complete reflection is equivalent to the blackening number zero; no reflection is equivalent to the blackening number ten. **Fig. E296** shows the principle design of such a system.

Exhaust gas back pressure The exhaust gas back pressure is the pressure in the exhaust gas pipe, normally measured after the joining of the exhaust gas branches of the cylinders.

The exhaust gas back pressure should be as small as possible to allow for an optimal gas exchange in the engine. Back flows from the exhaust gas to the cylinder can develop in the area where the valves overlap, which reduces the filling of the cylinder, leading to torque and power losses. An internal exhaust gas recirculation may be intended in some cases to reduce oxides of nitrogen. The additional work of expulsion against the higher exhaust gas back pressure leads to efficiency reductions and, therefore, increases specific fuel consumption.

A negative scavenging loop exists if the exhaust gas back pressure is higher than the pressure on the intake

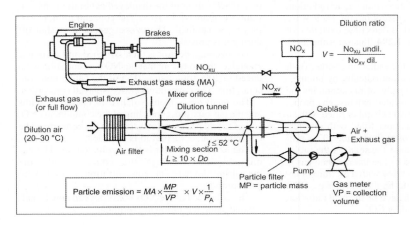

Fig. E295
Diagram of particulate measuring technology

side—particularly in supercharged engines. This results in filling losses due to a high proportion of residual gas and also a higher fuel consumption.

The pressure in the exhaust gas pipe is increased by installations such as catalytic converter, exhaust gas turbocharger, particulate filter, and so on. For supercharged engines and depending on the charge level, exhaust gas back pressures of up to several bars are possible upstream of the turbine.

The exhaust gas back pressures at full load and high engine speeds are approximately 300 mbar for naturally aspirated engines.

Exhaust gas back pressure increase →Catalytic converter; →Filter; →Particles (Particulates) ~Particulate filter system ~~Particulate filter

Exhaust gas cleaning →Pollutant aftertreatment

Exhaust gas components →Exhaust gas analysis

Exhaust gas composition →Emissions; →Exhaust gas analysis

Exhaust gas concentration →Exhaust gas analysis

Exhaust gas control valve →Valves

Exhaust gas cooling →Exhaust gas recirculation ~EGR cooling, ~EGR radiator

Exhaust gas energy utilization (*also*, →Exhaust gas heat). In addition to turbocharging, the thermal portion of the exhaust gas energy can, for example, be used for the charging of a heat accumulator system, air or mixture preheating, or heating of the passenger area. The kinetic portion in the form of pulsation can be used by the air suction system for afterburning.

Exhaust gas exergy. The exhaust gas exergy is that portion of the exhaust gas energy that can be converted into other forms of energy by means of a heat engine. The exergy of the exhaust gas can, for example, be used partly for supercharging. Exergy losses in the exhaust gas are, for example, caused by heat transfers. The anergy is the opposite of the exhaust gas exergy. This is the portion of the energy that cannot be converted to exergy.

Exhaust gas filter →Particles (Particulates) ~Particulate filter system ~~Particulate filter

Exhaust gas "heat" (*also*, →Exhaust gas energy utilization). The exhaust gas "heat" represents the part of the energy in the fuel that is released through the exhaust gas. This is approximately one third of the energy in the fuel. A direct use of the exhaust gas energy to improve the engine performance is not possible in practical applications. There are, however, some opportunities for using the exhaust gas heat. These are, for example:

- Exhaust gas turbocharging. This only uses a relatively small part of the thermal energy. The differential pressure of the turbine contributes the largest part to the power of the exhaust gas turbine.
- Compound supercharging. A part of the exhaust gas turbine power is directed at the engine shaft.
- Heating. A part of the exhaust gas heat is used to heat the inside of the vehicle.
- Catalytic converter. The thermal energy of the exhaust gas is used for the catalytically supported reactions for exhaust gas cleaning.
- Particle filter. The exhaust gas heat is used for the regeneration of the filter just as for the catalytic converter.
- Heat accumulator to improve the cold start characteristics.
- Preheating of air and mixture.

Exhaust gas loss. The chemical energy of the fuel, the product of fuel flow rate and thermal value, is converted in the engine into the following: effective power output, heat transfer into the coolant, heat flow into the environment, and enthalpic exhaust gas flow. The latter represents the exhaust gas loss when the exergetic portion of the exhaust gas energy is not used, for example, by turbocharging. This loss can be up to 35% of the energy in the fuel depending on the combustion process and the engine operating point.

Exhaust gas maps →Performance characteristic maps ~Emission maps, ~Exhaust temperature maps

Exhaust gas measurement →Emission measurement

Exhaust gas measuring equipment →Exhaust gas analysis; →Exhaust gas analysis equipment

Exhaust gas monitoring check. Exhaust gas monitoring check is used to determine the exhaust gas characteristics of vehicles in traffic with spark-ignition or compression-ignition engines in accordance with §47 a StVZO (German Road Traffic Regulations). The checks must be performed at 12-month intervals for cars with spark-ignition engines with or without catalytic converter but also without lambda controlled mixture conditioning. The time intervals for new cars with a catalytic converter and lambda controlled mixture conditioning and for cars with a compression-ignition engine are 36 months for the first check and 24 months for further checks.

~**Exhaust gas measuring equipment** (*also*, →Exhaust gas analysis equipment). The exhaust gas measuring equipment used must meet the legal calibration regulations. Exhaust gas measuring equipment with operator control and systems for storage and data output in the form of a test certification are preferable.

The systems described under exhaust gas analysis equipment including a lambda meter are used for exhaust gas measurement of the relevant exhaust gas components in external ignition engines.

~~Opacimeter. The opacimeter is used for exhaust gas measurements in compression-ignition engines.

The measuring principle is based on the attenuation (increasing opacity) of a light source by soot particles in the exhaust gas. The attenuation of the light can be displayed as opacity coefficient (m^{-1}) or opacity level (%). A value of 0 means no light absorption and a value of 100% means total absorption. The measurement pipe is heated to a constant temperature to avoid the impact of condensate drops due to the influence of temperature.

~Monitoring methods. Monitoring methods are shown in **Fig. E297**.

~~Compression-ignition engine. A smoke gaseous opacity test using the partial flow method under free acceleration of the engine is used for vehicles with compression-ignition engines. The test uses commercially available diesel fuel.

A visual inspection is performed first. It includes the components that are subject to pollution, including the exhaust system, for presence, completeness, tightness, and damage. It also includes a full-load phase with the accelerator pressed to the floor.

A check of the idle speed, the cutoff speed and the exhaust gas behavior follows after conditioning in accordance with the manufacturer's information. This is done by acquiring the peak value of the smoke gas opacity at free acceleration.

The check is performed for an engine at operating temperature. The engine temperature is recorded during the test (min. 80°C oil temperature).

While idling the accelerator pedal is depressed to the floor in a continuous motion. This position is then maintained long enough after achieving the engine cut-off speed.

The accelerator pedal is subsequently released until the idle speed is achieved again. This procedure is carried out at least four times. The opacity peak value is recorded after the second pass. The peak values of the smoke gaseous opacity, which are the basis for the check, must be within a bandwidth of $0.5\ m^{-1}$. The peak value determined by this procedure must not exceed the maximum opacity level specified by the vehicle manufacturer for this vehicle. An opacity coefficient of $2.5\ m^{-1}$ is used as the upper opacity limit if the vehicle manufacturer did not specify an opacity value. The opacity value can only be exceeded in special cases. The peak values to be analyzed must be within a bandwidth of $0.7\ m^{-1}$ for opacity coefficients greater than $2.5\ m^{-1}$.

~~Spark ignition engine. Spark ignition engines are available both with and without lambda control.

a. *Vehicles with or without catalytic converter, without lambda control*

A visual inspection is performed first. It includes the components that are subject to pollution, including the exhaust system: for presence, completeness, tightness, and damage. If required, it also includes the constricted tank filler neck or the tank fill label.

This is followed by a check of the pollution relevant service data to ensure conformity with the values specified by the vehicle manufacturer for the vehicle. This check is carried out by following the guidelines of the vehicle manufacturer. The analysis is performed for an engine at operating temperature. The engine oil temperature must be at least 80°C. The closing angle for contact controlled ignition systems, the moment of ignition (spark timing), the idle speed, and the CO content in the exhaust gas during idling are tested. The operability of the exhaust gas recirculation system, the secondary air system, and the effectiveness of the catalytic converter, if present, are also tested. The test of the catalytic converter is performed by measuring the CO content at an increased idle speed, which ensures operation at a lean mixture ($\lambda > 1$), if this is possible without adjusting the mixture conditioning. The lambda value is calculated by the measuring instrument.

The adjustment is considered to be in accordance with the state of the art if the pollution emissions are minimized with reliable engine function if the vehicle manufacturer did not provide specifications.

The CO content while idling cannot exceed 3.5% by volume, unless it also cannot be met when the engine and the pollution-relevant components are in proper condition.

Test Process	Vehicle Group		
	Vehicles with External Ignition Engine without Cat and with U-cat	Vehicles with External Ignition Engine with G-cat	Vehicles with Compression-Ignition Engine
Preparation	• Data vehicle certification • Target data • Conditioning (engine warm-up)	• Data vehicle certification • Target data • Conditioning (engine warm-up)	• Data vehicle certification • Target data • Conditioning (engine warm-up)
Visual inspection	• Exhaust system • Tank filler neck if necessary • Catalytic converter if necessary • Pollutant-relevant components, etc.	• Exhaust system • Catalytic converter • Lambda sensor • Tank filler neck • Pollutant-relevant components, etc.	• Exhaust system • Full-load stop • Pollutant-relevant components, etc.
Check and functional inspection	• Moment of ignition • Closing angle • Idle speed • CO value (idle speed) • CO value (increase idle speed) if necessary • Lambda value if necessary	• Moment of ignition • Idle speed • CO value (idle speed) • Lambda value • CO value (increased idle speed) • Lambda control loop (through feed-forward control acc. to veh. supplier)	• Idle speed • Cut-off speed • Peak value of smoke gas opacity during free acceleration (no load)
Documentation	• Complete/print test certificate	• Print test certificate	• Print test certificate

Fig. E297 Test runs, exhaust gas monitoring check

b. *Vehicles with catalytic converter and lambda control*
This test also starts with a visual inspection. The tank filler neck and the service data are checked. The check is performed for an engine at operating temperature. The engine oil temperature must be at least 80°C. The moment of ignition (spark timing) and the idle speed are tested. The control loop is tested for operability at the test speed, based on simple up and down variation of an interference parameter provided by the vehicle manufacturer, by determining the lambda characteristics. In addition, the value for lambda is tested with an allowable deviation of ±2% at increased idling speed. The speed at increased idling must be at least 2500 rpm and cannot exceed 2800 rpm and must be maintained for at least 30 seconds before measurement.

The permissible λ value is 1 ± 3%, unless the manufacturer specifies a different value. Otherwise, the specified minimum value of −0.02 and the specified maximum value of +0.02 are the limits.

The upper limit valid for the CO content at idling is 0.5% by volume and at increased idling 0.3% by volume. Different limits can be used in special cases.

Exhaust gas opacity meter →Exhaust gas monitoring check ~Exhaust gas measuring equipment

Exhaust gas oxidation →Pollutant aftertreatment

Exhaust gas pulsation →Exhaust system ~Muffler

Exhaust gas recirculation (EGR) (*also*, →Air cycling valve ~Air cycling valve functions; →Electronic open- and closed-loop control ~Electronic open- and closed-loop control, diesel engine ~~Functions). Exhaust gas recirculation is the supply of exhaust gas to the fresh mixture or to the air sucked into the cylinder. The use of exhaust gas recirculation is primarily aimed at the reduction of NOx emissions for gasoline and diesel engines. The NOx reduction is primarily caused by the following factors:

• The heat capacity (c_p) of the recirculated exhaust gas is higher than the heat capacity (c_p) of the air. This leads to lower temperature increases for the same amount of energy released by combustion
• Reduction of the O_2 partial pressure and, therefore, lower oxygen mass inside the cylinder, because a portion of the combustion air is replaced by exhaust gas with a lower oxygen content
• Reduction of the combustion speed and, therefore, lower temperature increase

Literature: F. Schäfer, R. van Basshuysen: Schadstoffreduzierung und Kraftstoffverbrauch von Pkw-Verbrennungsmotoren, Vienna/New York, Springer, 1993.

~EGR cooling (*also, see below,* ~EGR radiator). The recirculated exhaust gas can be cooled to further reduce the NOx emissions. A reduction of up to 50% is possible for radiators that are designed accordingly.

~EGR radiator (*also*, →Radiator). Cooling of the recirculated exhaust gas is one way to reduce the pollution emissions without significant additional fuel consumption. NOx generation is reduced by lowering the gas temperature in the combustion chamber. To achieve this, a specially developed EGR radiator is installed, for example, between the EGR valve and the intake manifold entry point for the exhaust gas (**Fig. E298**). The EGR radiator can be used for cooling the exhaust by using a pipe bundle design. Swirl generating elements are implemented in the radiator in the form of bosses, for example, to intensify the heat transfer. The arrangement of these swirl generating elements depends on the requirements for heat transfer, pressure loss, and contamination. In addition, the resistance to corrosion is an important factor, which is the reason why austenitic stainless steels are used. The cooling requirement is in the range of 2 kW for cars and up to 40 kW for trucks.

~EGR rate. The exhaust gas recirculation rate represents the volumetric percentage of the exhaust gas in the fresh mixture. An EGR rate of 5–10% is normal for gasoline engines, although engines with good mixture generation (rich mixture in the region of the spark plug) permit significantly higher exhaust gas recirculation rates, sometimes exceeding 20%: these make a potential consumption reduction of up to 7% and reduction of the HC and NOx pollution emissions by approximately 35%.

Recirculation rates of more than 60% are possible for diesel engines with direct injection (DI) and up to 40% for indirect injection (IDI) engines.

Newer combustion processes, currently in the development stage, such as homogeneous diesel combustion, enable even higher EGR rates.

~EGR valve →Valve ~Exhaust gas control valves

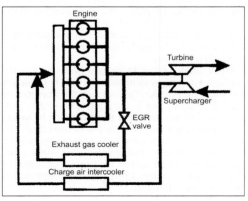

Fig. E298 Design of an EGR radiator (Source: *MTZ* 1999 7/8)

E

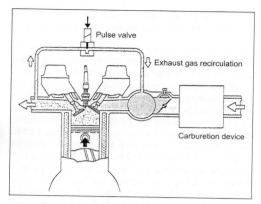

Fig. E299 Diagram of external exhaust gas recirculation

~**External exhaust gas recirculation.** The external exhaust gas recirculation (**Fig. E299**) works with the supply of exhaust gas in the intake cycle by using, for example, an EGR cycle valve.

The EGR is only useful in the part-load range for gasoline engines because at full load the engine is operated at $\lambda < 1$ for performance reasons; therefore, NOx is only generated in small concentrations.

The EGR cycle valve is normally a diaphragm valve that is electrically cycled. It releases or closes during the desired time frame, supplying the desired exhaust gas amount to the intake system. The valve can be regulated by injection maps. Gasoline engines normally use an exhaust gas recirculation rate of 5–10% (5–10% of the volume of the total intake mixture is exhaust gas), which makes possible a significant reduction in NOx emissions.

Diesel engines can use a higher exhaust gas recirculation rate than gasoline engines due to the larger fuel ignition limits. Diesel engines with direct injection allow up to twice the amount possible with indirect injection engines ($>60\%$ of the volume).

Excessively high residual exhaust proportions, especially around the idling range, result in misfiring and irregular engine function with corresponding emissions of unburned hydrocarbons.

Today, the exhaust gas recirculation is implemented in many engines to enable them to meet stringent exhaust gas regulations.

When implemented in engines with large valve overlaps, exhaust gas recirculation results in inefficient flushing of the cylinder. This increases the proportion of residual exhaust—that is, unburned gas—in the cylinder. A quantitative control of this proportion by means of valve overlap is only possible to a limited extent. The use of valve timing phase transformers is one way to regulate the process. For example, the intake camshaft can be rotated against the exhaust camshaft to change valve overlap. Fully variable valve drives are even more effective.

Exhaust gas recirculation control →Electronic open- and closed- loop control ~Electronic open- and closed-loop control, diesel engine ~~Diagnostics ~~~Exhaust gas recirculation, ~~~Intelligent actuator

Exhaust gas recirculation cooler →Exhaust gas recirculation ~EGR radiator; →Radiator

Exhaust gas recirculation initiation →Intake system ~Intake manifold ~~Intake manifold design

Exhaust gas recirculation rate →Exhaust gas recirculation ~EGR rate

Exhaust gas recirculation rate map →Performance characteristic maps

Exhaust gas recirculation valve →Actuators ~Actuators for exhaust gas recirculation (EGR); →Valve ~Exhaust gas control valves

Exhaust gas routing →Exhaust system

Exhaust gas secondary reaction →Pollutant aftertreatment

Exhaust gas sensor →Sensors ~Lambda sensors

Exhaust gas temperature. The exhaust gas temperature depends among other things on the combustion process, the load, and the engine speed. Gasoline engines achieve exhaust gas temperatures of more than 1050°C, while supercharged diesel engines with direct injection achieve maximum exhaust gas temperatures of about 800°C. The temperature of the exhaust gas is important for emissions control. The light-off performance of the catalytic converter and the reaction of the lambda probe also depend significantly on it. The conversion efficiency of the catalytic converter is reduced if the exhaust gas temperature goes below 250–300°C.

The exhaust gas temperature is also an indirect measure of the combustion quality. A high compression ratio and hence expansion ratio in the gasoline engine results in a low exhaust gas temperature (high thermal efficiency); late ignition angles move the combustion partly to the exhaust tract and increase the exhaust gas temperature; this results in a faster light-off of the catalytic converter during the warm-up phase. A disadvantage is the resulting increase in fuel consumption.

The exhaust gas temperature in diesel engines can be increased from time to time by intake air throttling and intake air preheating, to support, for example, the regeneration of soot filters. Later start of injection and postinjection also help increase the exhaust gas temperature.

Exhaust gas temperature maps →Performance characteristic maps

Exhaust gas test →Emission measurements; →Exhaust gas analysis

Exhaust gas thermal reactor. Postcombustion reactions, especially involving the CO and HC components in the exhaust gas, take place in the engine exhaust manifolding. The thermal reactor creates better conditions to support this reaction. High temperatures and sufficient oxygen are required for the reaction; this oxygen can be added in the exhaust port or in the reactor by adding air.

Consideration has to be given to the retention time required to reduce the CO and HC contents significantly. This process takes place in the reaction chamber, which is thermally insulated to achieve high temperatures. The reaction chamber is mounted immediately downstream of the engine exhaust manifold if possible.

A secondary air supply is required if the engine is operated with rich mixtures. A high conversion ratio is achieved with temperatures above 700°C in the thermal reactor. Significant HC and CO reductions can be achieved, especially with rich mixtures. However, the thermal reactor has not been accepted in the market. The reasons are:

- Additional cost of high temperature reactor materials and insulation, large additional mass and secondary air pump required.
- Larger installation space required and high weight.
- The effect is large with rich mixtures, but these cause high fuel consumption, at least under partial loads.
- Danger of burnout during ignition failure (ignitable air-fuel mixtures get directly into the reactor).
- Conversion efficiency at low temperatures (part load) is not adequate.
- NOx cannot be influenced.

Exhaust gas treatment →Pollutant aftertreatment

Exhaust gas turbine →Supercharging ~Exhaust gas turbocharging

Exhaust gas turbocharging →Supercharging

Exhaust manifold →Exhaust system ~Exhaust gas manifold

Exhaust muffler →Exhaust system ~Muffler

Exhaust noise →Engine acoustics ~Exhaust system noise

Exhaust odor. The exhaust gas has a varying quantity of constituents causing odor depending on the different components in the fuel used, the combustion process, and the engine conditions. The significant odor content materials in the exhaust gas are: Hydrocarbons (olefins, acetylenes, aromatics, etc.), aldehydes (usually fuels with oxygen content), sulfur dioxide (SO_2), soot with hydrocarbons attached. The unpleasant odor of these strong-smelling materials can be significantly reduced by treatment of the exhaust gas (catalytic converter, soot filter). This is especially true for hydrocarbon compounds and for aldehydes.

Exhaust opens →Control/gas transfer ~Four-stroke engine ~~Timing

Exhaust pipe →Exhaust system ~Exhaust pipe

Exhaust pipe insulation →Exhaust system ~Insulation

Exhaust pipe outlet noise →Engine acoustics

Exhaust plenum chamber →Supercharging ~Ram induction

E

Exhaust port. The burned gases flow from the combustion chamber of the engine through the exhaust port into the exhaust gas manifold. The exhaust port represents resistance to the flow, which results in losses (work of expulsion). The resistance comes primarily from redirection of flow, surface impacts, separation and turbulence of the flow as well as the partial blocking of the cross-sectional area by the valve and the valve guide. The latter are inherent, but the other resistance can be minimized.

~**Exhaust port cooling.** The hot exhaust gases from the combustion chamber (higher than 1000°C in gasoline engines) would melt the materials of the exhaust port. This is why the exhaust port must be cooled and it should be as short as possible. The cooling is performed generally with coolant, a mixture of alcohols for freezing point reduction (glycol), water, and corrosion protection. The composition depends on the desired antifreeze properties. The intensity of the cooling depends primarily on the design of the liquid chamber, the liquid volume, and, therefore, on the flow speed and the local wall thicknesses.

The design calculations of the exhaust port cooling are difficult due to the complicated design of the liquid chamber and the transient heat flow, presenting uncertainties.

~**Exhaust port design.** Diameter and length and basic design of the exhaust port are significant for the exhaust gas flow resistance. A lower flow resistance is required to further empty the combustion chamber of exhaust gas. This reduces the work of expulsion, and the engine can, therefore, achieve its best power at low fuel consumption.

The following criteria for the design of an exhaust port must be considered:

- No sudden cross-section changes but steady cross-section increases around the valve guide and the valve stem
- Large radius of curvature
- No port offset against seat ring or manifold
- Ports as hot as possible and, for multiple valve engines, a short junction region to minimize the heat transfer at the cylinder head

The mechanical flow optimization of the exhaust port for a maximized flow rate or minimized resistance

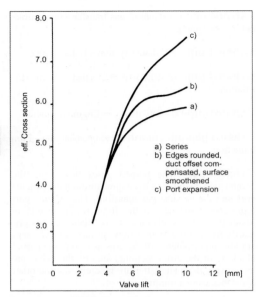

Fig. E300 Exhaust port optimization

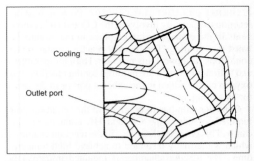

Fig. E301 Exhaust port with cooling

can, for example, be assessed with steady flow measurements. An effective flow cross section will be determined for this application.

Fig. E300 shows the example of the results of a port optimization. It shows the effective flow cross-sectional area (not the geometrical) as a function of the valve lift for two optimization steps. Based on the cast mass production port (a), an optimization with respect to interfering edges — for example, due to port offsets during casting — rough surfaces, and casting unevenness (b) up to the port expansion (c) has been implemented. A clear improvement can be seen from (a) to (c).

The design-dependent curvature of the port and, therefore, the redirection of the flow also have a significant influence on the effective cross-sectional area. The port curvature is normally larger on the exhaust side than on the intake side. The reason is the required minimization of the heat removing surface of the exhaust port and cooling by appropriate water chambers (**Fig. E301**).

The space conditions in a cylinder head are normally very tight, which means that the exhaust port cannot be designed to achieve its ideal shape and diameter. Additionally, the valve stem also disturbs the flow in the front part of the exhaust port. Therefore, a port as short as possible is therefore desired. The flow resistance is small in this case and the heat transfer from gas to the cylinder head is minimized (exhaust port cooling).

Air (secondary air) can be forced into the exhaust port after the cold start to reduce pollutants. This can be introduced through ports in the cylinder head or through pipes from the manifold into the intake port or through port liners especially designed for this pur-

pose. The closer the secondary air is brought to the exhaust valve, the stronger is the after-reaction.

~Exhaust port diameter. *See above*, ~Exhaust port design

~Exhaust port length. *See above*, ~Exhaust port design

~Exhaust port shape. *See above*, ~Exhaust port design

~Port liner. These are normally ceramic inserts, which are cast or inlaid into the exhaust port to prevent a direct contact of the hot gases with the cylinder head. The reduced heat removal from the exhaust gas supports the oxidation of the HC emissions by after-reactions, especially if secondary air is supplied, and additionally it unburdens the cooling circuit. The good insulation of the ceramics prevents, in addition, the excessive heating of the cylinder head. This process is expensive and, therefore, rarely used, but it is used in some high-output engines for mass production.

Exhaust port feedback →Variable valve control ~Residual gas control

Exhaust port manifold →Exhaust system ~Exhaust gas manifold

Exhaust pressure →Exhaust gas back pressure

Exhaust side. The exhaust side is the engine side which supplies the exhaust gases to the exhaust port manifold. Modern engines separate the exhaust and intake side and, therefore, have cross-flow cylinder heads. This has the advantage that the fresh mixture supplied is not heated by the exhaust gas heat, which would result in fill losses.

Exhaust system. The exhaust system, or exhaust, consists generally of the exhaust gas manifold, head pipe (bifurcated front pipe), catalytic converter or catalytic converters (startup catalytic converter, underfloor catalytic converter), particulate filter, exhaust gas pipe, muffler (absorption and/or reflection muffler) and rear exhaust pipe. Its purpose is to release the exhaust gas from the engine with minimal losses and noise, whereby exhaust gases must also be aftertreated or cleaned. A low level of heat radiation is another objective. The exhaust gas manifold must receive the exhaust gas of the cylinders and must combine the exhaust gas optimally depending on the number of cylinders and the firing order. It is recommended to combine the individual cylinders as late as possible depending on the requirements— for example, power, torque, and space. The length and the cross-sectional area of the pipes also have a significant influence on engine behavior. The exhaust system has a significant influence on power, acoustics, and fuel consumption and, therefore, must be tailored to the engine design, for instance in terms of exhaust flow rate, exhaust gas back pressure, and firing order.

The exhaust flow rate, $\dot{m}_A$, is calculated with $\dot{m}_A = \dot{m}_L + \dot{m}_{fuel}$, whereby the air mass $\dot{m}_L$ is determined by

$$\dot{m}_L = \rho V_H n i \lambda_a ,$$

where $\dot{m}_{fuel}$ = the fuel mass flow, ρ = the air density, V_H = the displacement, n = the engine speed, i = the number of combustion cycles per revolution, and λ_a = the volumetric efficiency.

The dimensions (cross-sectional areas) of the exhaust gas carrying parts, and as a result the pressure loss of the system, are primarily determined by the exhaust flow rate.

The exhaust system is often insulated in the exhaust gas manifold and head pipe areas to ensure a faster warm-up of the catalytic converter.

There is a high thermal demand on the exhaust system. Exhaust gas temperatures of up to 1050°C are achieved in gasoline engines and up to 850°C in diesel engines upstream of the turbocharger turbine. Inner parts of the resonator reach temperatures of about 700°C and surface temperatures of about 550°C. Maximum temperatures between 250 and 500°C are achieved at the end of the exhaust system depending on the muffler type: absorption or reflection.

Surface treated (aluminum coated) metal sheets are used for cost reasons, in addition to high-alloyed steel sheets to achieve corrosion resistance (combustion water with acid content) and to withstand the high temperature requirements.

Both single and double flow systems are used, depending on the engine in the vehicle. Double flow systems are preferred for V-engines.

The exhaust gas branches can be combined in the head pipe (one lambda probe) or can be individually directed to the parallel mounted catalytic converters (two lambda probes). The double flow can be continued to the end of the exhaust system to increase the power.

~Absorption-type muffler. *See below*, ~Muffler

~Cast manifold. *See below*, ~ Exhaust gas manifold

~Catalytic converter (*also*, →Catalytic converter). A startup catalytic converter plus an underfloor catalytic converter or only an underfloor catalytic converter is generally used for gasoline engines. Oxidizing catalyst and NOx accumulator-type catalytic converters can be used for the diesel engines in addition to diesel particulate filters.

~Central muffler. *See below*, ~Muffler

~Decoupled exhaust system. The exhaust system is often decoupled from the exhaust gas manifold to improve the vibration behavior (fatigue fracture). This is especially important for transversally installed engines to enhance buffering of the tilting moments across the exhaust system.

Flex couplings and flexible flange couplings can be used as decoupling systems (**Fig. E302**).

~Dual branch exhaust system. Two exhaust gas pipes will be installed at the rear of the vehicle for two or multipass exhaust systems. These can have corresponding mufflers or catalytic converters.

~Exhaust flow rate. The exhaust flow rate depends primarily on the displacement, the engine speed, the intake manifold pressure, and the fill of the engine. It is identical to the mass of the intake air and the fuel used, if leakage losses are disregarded.

~Exhaust gas manifold. The exhaust gas manifold is responsible for the aerodynamic discharge of the exhaust gas in the cylinders from the cylinder head. Depending on the number of cylinders and the firing order, combining of the individual exhaust gas branches is required to achieve optimal efficiency. This ensures that the pulsation of one cylinder does not affect the outflow of the other cylinder, but rather supports it with low pressure waves. The combustion gases of the individual cylinders reach the exhaust gas manifold at different times depending on the firing order. This means that significantly large pressure and vibration differences are present, which influence engine behavior (e.g., mean pressure, power, exhaust gas). Joining of the individual exhaust gas branches as far as possible downstream is, therefore, an important element in performance efficiency.

The manifold can be cast, welded, or pressed using sheet metal (high performance manifold). Tubular heaters are often jacket manifolds with two jackets. The gas carrying parts are pipes and the carrying parts are

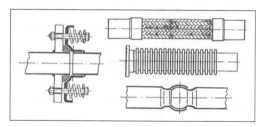

Fig. E302 Decoupled exhaust system

sheet metal jackets. Due to the high load, the internal pipes are made from high temperature steel with a wall thickness of approximately 1 mm. They are connected to each other by close sliding fits to compensate for the thermal expansion. Tubular heaters have a smaller mass and consequently a smaller heat capacity than die manifolds. They permit faster warm-up and, therefore, an early light-off of the catalytic converter. The outside temperature of the manifold is approximately 500°C for gasoline engines depending on the insulation material and exhaust gas temperature compared to a wall temperature of approximately 800°C for the cast manifold. This results in a reduction of the engine compartment temperature. Insulated jacket manifolds are up to 40% lighter than cast manifolds.

Die manifolds made from materials with nickel content are preferably used for high temperature applications (e.g., for gasoline turbo engines). They normally have better acoustic characteristics than tubular heaters.

~Exhaust gas pipe. *See below*, ~Exhaust pipe

~Exhaust manifold. *See above*, ~Exhaust gas manifold

~Exhaust noise. *See below*, ~Muffler

~Exhaust pipe. The connection elements between manifold, catalytic converter, and mufflers are pipes, the so-called exhaust pipes. The length of the pipes is determined by the position of the individual elements. The pipe between engine and muffler can be seen as a pipe closed on one side with respect to vibration behavior. The other pipes behave like pipes that are open on both sides.

For acoustic reasons, systems with only one muffler should be designed in such a way that the pipe length behind the muffler is longer than the pipe length in front of the muffler. The first muffler should be placed as close as possible to the engine if two mufflers are used to achieve an early damping of the exhaust gas pulsations. The selection of the pipe lengths depends on the frequency of the engine (number of cylinders and engine speed), to achieve optimal silencing. Insulated head pipes reduce the start-up time of the catalytic converter.

Developments for future systems are aimed at the elimination of the exhaust noise. Sound waves will be produced in such a way that they mirror and overlay the wave of the unwanted noise (counterwave) to eliminate the unwanted noise (active sound compensation). This is implemented with loudspeakers and microphones. Smaller exhaust gas back pressure, partial elimination of the large mufflers, and the resulting weight reduction are advantages of this technology.

~Exhaust tail pipe. The last part of the exhaust system is called the exhaust tail pipe.

~Exhaust tract. This is the name of the engine part through which the exhaust gases flow.

~Head pipe. *See above*, ~Exhaust pipe

~High performance manifold. *See above*, ~Exhaust gas manifold

~Insulation. Insulation of parts of the exhaust system is recommended so the catalytic converter will kick in earlier. This applies in particular to the exhaust gas manifold and the head pipe. The insulation can, for example, be implemented via air gaps for multilayer designs, or insulation material can be attached to the outside of the exhaust gas pipes.

~Jacket manifold. *See above*, ~Exhaust gas manifold

~Manifold. *See above*, ~Exhaust gas manifold

~Muffler (*also*, →Engine acoustics). Mufflers are used to reduce the exhaust noise. The acoustic pressure level at the exhaust valve is between 150 and 60 dB(A). This value must be reduced to the legally regulated value. The operation of the muffler (silencer) is based on the principle of reflection and/or absorption sound insulation. By combining the two operations, insulation can be achieved in the relevant range of 50 to 8000 Hz.

Exhaust noise is part of the vehicle noise, which is legally limited to 74 dB(A). A further reduction to 71 dB(A) is under consideration for all cars within the EU, whereby, however, the measuring method may also be changed. A reduction of 3 dB(A) reduces the acoustic power by half.

A certain minimum volume for the reflection or absorption area or several mufflers (front, central, and rear muffler) is required depending on the engine concept (displacement, power, supercharging, multiple valve engine, number of cylinders, etc.).

The gas flow is guided through the muffler if an absorption-type muffler is used. The gas carrying pipe is perforated in the muffler area and the chamber next to it is filled with absorption material (steel wool, basalt wool, or fiberglass). The pulsating gas flow can expand through the perforation into the chamber filled with absorption material. Friction reduces much of the vibration energy and converts it into heat. The gas flow that leaves the muffler is, therefore, free of pulsations, for the most part. The absorption-type muffler is especially effective for the frequency range above 500 Hz.

The attenuation in the reflection-type muffler (interference muffler) is implemented by diversions, cross-sectional area changes and partitions inside the muffler. The corresponding chamber and cross-sectional area changes, must be exactly matched. Interference occurs if the sound waves eliminate each other (e.g., 180° phase shift) after traveling two paths of different lengths. This principle is especially effective below 500 Hz.

Literature: N.N.: Walker Deutschland GmbH, technical information. — N.N.: Abgasanlagen für Kraftfahrzeuge, Schalldämpfer für Kraftfahrzeuge, Die Bibliothek der Technik (Library of Engineering), Volumes 47 and 83, 1990.

~Multipass exhaust system. *See above*, ~Dual branch exhaust system

~Pipe. *See above*, ~Exhaust pipe

~Power loss. The resistance in the exhaust system normally results in power losses. Length, cross-sectional area, and integration of the individual exhaust gas branches are among the factors with a significant impact on the power or torque characteristics of the engine, and suitable designs or implementations can result in a small supercharging effect (resonance induction) within a narrow engine speed range.

~Rear muffler. The last muffler is called the rear muffler (see above, ~Muffler).

~Reflection muffler (silencer) →Engine acoustics

~Rigid exhaust system. See above, ~Decoupled exhaust system

~Secondary air injection →Pollutant aftertreatment ~Aftertreatment concept with a three-way catalytic converter ~~Secondary air system

~Soot filter →Particles (Particulates) ~Particulate filter system ~~Particulate filter

~Thermal reactor →Exhaust gas thermal reactor

~Trap (HC) →Pollutant aftertreatment ~Aftertreatment concept with a three-way catalytic converter ~~HC absorber; →Pollutant retention system

~Tubular heater. See above, ~Exhaust gas manifold

~Variable exhaust gas pipe length. Similar to the intake systems, the lengths of the exhaust systems can be changed, for example, by flaps (valves) in the area of the head pipes. This may improve the torque characteristics by influencing the exhaust gas vibrations.

Exhaust system noise →Engine acoustics

Exhaust tract. This is the name of the engine part through which the exhaust gases flow.

Exhaust valve (Port in two-stroke engines) (also, →Exhaust port; →Gas transfer; →Valve ~Gas transfer valves; →Variable valve control ~Variation parameters). Exhausts are openings which are used to transport the exhaust gases into the exhaust port and from there to the exhaust gas manifold. The residual gas volume in the combustion chamber is controlled by, among other things, the opening times of the exhaust valve. The exhaust must be open long enough to ensure that the burned mixture is removed, preferably completely, from the combustion chamber, so that it can receive a large volume of fresh charge. The exhaust opening must be closed before the start of the intake period to prevent repeated intake of the burned gases. A defined exhaust residual proportion can, however, be intended in modern engines to influence the pollutant formation (internal exhaust gas recirculation rate). A rapid reduction pressure develops in the cylinder and the remaining potential energy will not be transferred to the piston if the exhaust opens too early.

Most engines today have poppet valves for control functions, despite the fact that these, from a flow-mechanical standpoint, cannot be ideally designed and are very complex with respect to control. The special advantage is the simple sealing of the valve head on the valve seat. Exhaust valves represent an especially high resistance for the flow, because the exhaust gases must flow first around the valve head.

A number of slide valve gears are available in addition to the poppet valves. These are systems which are used to let the air-fuel mixture, the air, or the exhaust gas flow into or out of the combustion chamber under defined conditions. Slide valve gear can perform a steady or a nonsteady movement. Combinations of valve and slide valve gears are also possible. The slide valve gears can be divided into pipe valves, rotary valves, and constant depression valves. The external slide valve gears have not prevailed for production engines, especially for four-stroke engines, because a lasting sealing of the combustion chamber to the outside has not been achieved. Valves, in contrast, are pressed against the seat by the combustion pressure, which means that the sealing is easily achieved.

The two-stroke engine is often controlled by a slide valve gear, and the piston is used as the control slide. The sealing is achieved by the piston rings, which are required anyhow. In this case, the combustion gases are released directly from the cylinder into the exhaust manifold without additional moving parts. However, the almost ideal simplicity of this design does not permit the clean separation and control of fresh mixture or exhaust gas and this results in disadvantages regarding engine efficiency and emissions.

The simple exhaust port systems permit only small variations of the opening times during the engine operation. Therefore, there is currently little chance for further development.

Exhaust valve control time →Control/gas transfer ~Four-stroke engine ~~Timing

Exhaust valve spring →Valve spring

Expansion. The expansion is a change of state (a process), which must be performed in many machines to achieve an energy transformation. For example, expansion of the working medium always takes place in a turbine. The expansion in engines is, for example, part of the four-cycle or two-cycle process. It can run independently from the analyzed machine under different thermodynamic aspects, such as isothermal, isentropic, adiabatic, and polytropic.

The expansion is considered to be an isentropic expansion in theoretical comparisons of diesel and gasoline engines.

~Adiabatic. Adiabatic expansion is in simple terms a change of state (process) requiring no heat exchange. Adiabatic systems can in principle execute only processes, from a given state, that maintain the entropy constant. It is, therefore, not possible to reach every possible end state.

~Isentropic. Isentropic expansion is a change of state (process) for which heat is neither supplied nor discharged. However, pressure, specific volume, and temperature can change. It is, in contrast to the adiabatic change of state, a reversible change of state, therefore taking place at constant entropy.

~Isothermal. Isothermal expansion is a change of state (process) for which the temperature remains constant. Pressure and specific volume change in a certain context. The pressure is inversely proportional to the specific volume. Heat must be supplied for an isothermal expansion.

~Polytropic. Polytropic is a general term for a change of state (process)—for example, an expansion or compression. Isentropic and isothermal expansions are special cases of a polytropic expansion. The general equation of such a change of state is

$$pv^n = \text{constant}.$$

Depending on the selection of the exponent, the above equation will deliver the relationship for isentropic ($n = \kappa$) or isothermal ($n = 1$) processes, with κ being the isentropic exponent.

The polytropic change of state (process) between these limits requires that heat must be supplied or discharged, depending on the direction, while the temperature at the same time increases or decreases.

By approximation, such processes are present in the engine. The conditions of isothermal or isentropic changes of state cannot be achieved in an engine.

Expansion chamber (also, →Engine ~ Alternative engines ~~ Stirling engine). The work chamber (hot reservoir) of a Stirling engine is called the expansion chamber. The medium is pressed from a compression chamber into the expansion chamber and it transfers work to a piston. Subsequently, the medium flows through the cooler, the regenerator, and the heater.

Expansion loss. The expansion loss in a real engine stems from the fact that the exhaust valve opens before bottom dead center is reached. From the start of exhaust-opening until bottom dead center, the expansion does not follow the progression of change of state that would exist for a closed exhaust valve. This results in a larger reduction in pressure in the combustion chamber and could theoretically be avoided if the exhaust valve opened infinitely fast at bottom dead center.

A form of expansion loss that is different in principle stems from the fact that the expansion in an engine is not performed all the way down to the ambient pressure. A further loss due to incomplete expansion could be avoided if expansion were continued to ambient pressure and afterward also to the ambient temperature, with a subsequent isothermal compression to the ambient pressure.

However, such an extension of expansion is not feasible from a technical standpoint because of the considerable structural expenditure required.

Expansion ratio. The expansion ratio represents the pressure ratio, from the maximum cylinder pressure to the pressure in the cylinder, which is present when the exhaust valve is open. The pressure in the cylinder is higher than the ambient pressure at the time of exhaust. Thermodynamic advantages would be experienced if the expansion were continued until the ambient pressure is reached (efficiency). However, this requires significant additional construction efforts.

Explosion limits →Fuel, gasoline engine

External combustion. Engines are divided into those with internal and external combustion. Engines with internal combustion are, for example, gasoline, diesel, and Wankel engines. External combustion is employed in the Stirling engine. It is based on a closed-loop regenerative thermal power process with continuous combustion. This combustion type has advantages with respect to noise and pollution emissions. A disadvantage is the inferior controllability. This is one of the reasons why the Stirling engine is primarily applicable for hybrid powertrains or for stationary usage only (see also, →Engine ~Alternative engines ~~Stirling engine).

External exhaust gas recirculation →Actuators ~Actuators for exhaust gas recirculation; →Exhaust gas recirculation ~ External exhaust gas recirculation

External fuel mixture generation →Combustion process, gasoline engine; →Mixture formation ~ Mixture formation, gasoline engine

External gear pump →Gear pump

External ignition →Ignition system, gasoline engine ~ Ignition

External ignition engine →Combustion process, gasoline engine; →Exhaust gas monitoring check ~ Monitoring methods

External noise →Engine acoustics

External regeneration →Particles (Particulates); →Particulate filter system ~~ Particulate filter ~~~ Regeneration of particulate filters

External tank venting systems →Injection system (components) ~ Gasoline engine ~~ Fuel tank systems

External valve spring →Valve spring

External ventilation →Carburetor ~ Carburetor float bowl vent

Externally opening →Injection valves ~ Diesel engine

Externally opening injection valves →Injection valves ~ Diesel engine, ~ Gasoline engine ~~ Direct injection

F

F1 engine →Engine ~Racing engines ~~Formula 1

Facility change →Piston ~Pressure side

Fan (*also*, →Cooling, engine→Supercharging). A high cooling capacity has to be provided in order to ensure sufficient engine cooling, even at low vehicle speed and, for example, when towing a trailer uphill. To achieve this, the cooler (radiator) must be power-ventilated, and the drive power can be up to 15 kW. The fans used can be single-piece injected plastic fans with riveted metal blades or fully plastic fans. The drive can be through a belt from the engine, an electric motor, or the fan can be fitted directly on the crankshaft. When designing the fan, its acoustic behavior must be considered because at high speeds it can cause considerable noise emissions.

The control of the fan is important because the operating modes, for which ram pressure is mostly sufficient for ventilation, comprise up to 95% of the operating time depending on the vehicle type and type of use; hence, fuel energy does not have to be consumed by the fan in these regimes. The control of the region of application of the fan takes place through a two-position controller—the fan is actuated through an electric temperature switch only above a certain coolant temperature.

~Visco clutch. The fluid-friction clutch (visco clutch) has become established as a mechanical drive for the fan in commercial vehicles and in automobiles (**Fig. F1**).

It primarily comprises the primary driven disk, the secondary driven part, and the controller. The torque is transferred to the walls through the internal friction of the high-viscosity fluid—there is always a slip between the drive and the output. A scraper, which revolves with the secondary part, constantly moves working fluid into the storage chamber, to which it flows as a result of the centrifugal force through a

valve that opens into the work chamber. As the ambient temperature drops, the control unit uses a bimetallic strip to close the valve so that the fluid collects in the storage chamber and the working chamber is emptied. The clutch is deactivated except for a negligible residual torque. Depending on the temperature of the bimetallic strip in the cooling airflow, the desired speed is adjusted smoothly over the entire range of control.

Fan characteristics →Supercharging ~Turbo supercharger

Fan drive →Engine accessories

Fan type →Supercharging ~Fan

Fatigue strength →Bearings ~Materials ~~qualities

Fatigue strength (pitting) →Camshaft ~Wear

FEM →Calculation processes ~Numerical processes ~~Finite element method; →Finite element method

Ferrocene. Soot is collected and burned in pollutant retention systems (particulate filters) in diesel engines. However, soot only oxidizes at temperatures above 550°C and these temperatures are only rarely present in diesel engine operation. This is where oxidation materials come in, which reduce the ignition temperature of soot down to 250°C. Ferrocene is an oxidation material that is based on iron and is used for the regeneration of particulate filters. It is normally added to the fuel in a separate container. Other oxidation materials are produced based on Cu, Ca, Mn, and Cr. A medium-size car needs approximately 1 L per 10,000–15,000 km driving distance; the size of the dosage is 10–20 ppm, but the trend is for the amount to reduce.

Ferrotherm piston →Piston ~Design

FFT (Fast Fourier Transformation) →Engine acoustics ~FFT, ~Fourier transformation

Ficht injection systems →Injection system, fuel ~Gasoline engines ~~Direct injection systems

FID →Exhaust gas analysis equipment ~Flame ionization detector (FID)

Film evaporation in combustion chamber →Wall film ~Diesel engine

Filter. Filters are used to remove solid contaminants from the fluid flows and circuits in engines.

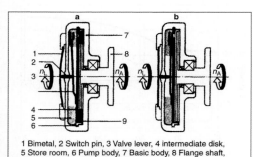

1 Bimetal, 2 Switch pin, 3 Valve lever, 4 intermediate disk, 5 Store room, 6 Pump body, 7 Basic body, 8 Flange shaft, 9 Work room, n_A Drive speed, n_L Fan speed

Fig. F1 Visco clutch

The separation of water from the fuel (coalescence) is an additional task of the filter in diesel engines.

~Air filter. *See below*, ~Intake air filter

~Filter characteristics. Other special characteristics will be covered at the applicable text passages.

~~**Dirt collection capacity.** The dirt collection capacity is the mass of solid particles that are absorbed by the filter up to a defined level of flow resistance.

~~**Dirt particles.** These are solid contaminants that need to be removed by the filter. Dirt particles come from the environment, are small wear parts from the engine, or they are combustion products such as soot.

~~**Durability.** The durability is the time of use of a filter until it needs changing after the completion of a service interval.

~~**Dust retention capacity** →Filter ~Intake air filter ~~~Filter characteristics

~~**Filter mesh.** The filter mesh value is defined as the capability of the filter to separate particles of a certain size. The filter mesh value can be determined by gravimetric processes or processes with particle size determination. Test standards in accordance with ISO, SAE, JIS, or DIN are used to determine the filter mesh value depending on the filter type (air, oil, fuel).

~~**Flow resistance.** The flow resistance is the static pressure difference between the intake and outflow sides of the filter (also defined as raw and clean side).

~Fuel filters. An acceptable fuel quality is needed to protect modern injection systems for gasoline and diesel engines from wear. The primary task of the fuel filter is to separate and remove safely the contaminants in the fuel, such as solid inorganic and organic components.

Additionally, the separation of free or emulsified water from diesel fuels is required for modern diesel injection systems to prevent corrosion damage in the injection system (e.g., high-pressure pump). The maximum dirt load and the maximum water content are regulated in applicable standards (diesel, DIN EN 590; biodiesel, DIN V 51606; gasoline fuel, DIN EN 228 a.o.). The maximum water value in diesel fuels is normally 200 mg/kg, but it can also reach values of up to 500 mg/kg [1]. Diesel fuels in accordance with DIN EN 590 must have particulate contents of under 24 mg/kg. The World-Wide Fuel Charter [1] calls for 2002 values of <10 mg/L. Diesel fuels offered in central Europe normally have particulate contents of under 10 mg/L. Significantly higher particulate concentrations can often be found outside this region. The size distribution is important in addition to the total content of dirt particles. One liter of diesel fuel can contain more than 5×10^4 particles in the >15-μm size class (coarse fraction) and more than 5×10^5 par-

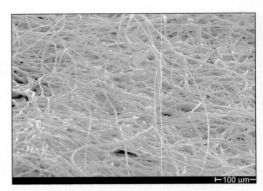

Fig. F2 REM image of a fully synthetic layer of fuel filter medium

ticles in the >5-μm size class (fine fraction) [2]. Particle sizes in the range of 3–10 μm are responsible for the wear of modern injection systems. The results of insufficient fuel filtration can lead to the loss of an injection system. Injectors of the injection system can, for example, be worn by the dirt particles so that the specified injection quantities change. This results in significant loss of performance from the vehicle because of injection and combustion that are not optimal anymore. Not only injectors must be protected against wear, but also all other components of the injection system, such as high-pressure pumps in common rail systems.

Completely new filter media have been developed in the last five years to meet the increased requirements [4, 5]. Filter media with integrated gradient structures and multilayer filter media with fully synthetic layers (**Fig. F2**), are currently available [6].

These innovative filter media are high-pass filters, which means that the dirt particles are separated inside the filter medium. Approximately 10-fold durability is achieved in comparison to simple screen filters (surface filtration), so it is possible to achieve higher separation performance and at the same time extend the exchange intervals (**Fig. F3**).

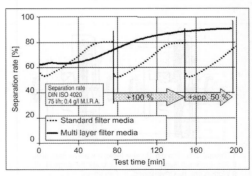

Fig. F3 Separation rate and durability of multilayer media in comparison to standard media

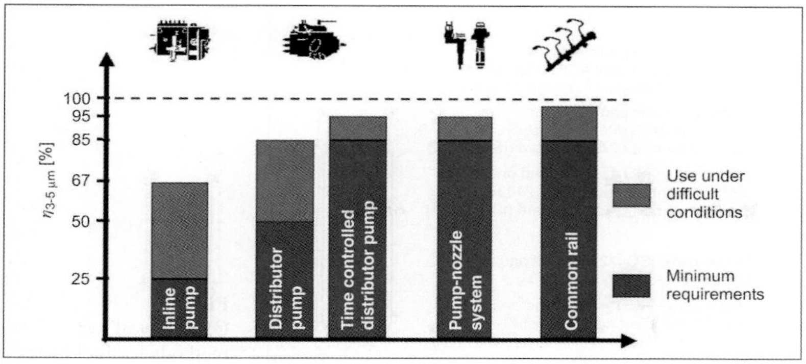

Fig. F4
MANN+HUMMEL recommendation of initial separation rates for diesel fuel filter under normal and severe conditions

	Initial separation rate in accordance with ISO/TR 13353:1994 $\eta_{3-5\mu m}$ [%] 1996	Initial separation rate in accordance with ISO/TR 13353:1994 $\eta_{3-5\mu m}$ [%] 2003	Useful life [tkm] 1996	Useful life [tkm] 2003
Diesel fuel filter car	25–45	85–95	30–90	60–90
Diesel fuel filter truck	25–45	85–99	40–100	60–150
Gasoline filter car	20–45	25–67	40–120	90–Lifetime (>240)

Fig. F5
Change in the performance profile of fuel filters from 1996 to 2003

The required filter mesh is generally specified by the injection system manufacturer or by the respective requirements of the injection system. The operating conditions of the vehicle are also significant for the selection of the required filter mesh. The filter mesh must be adjusted for severe conditions or for difficult territories (e.g., dirt roads) or for use outside Europe, Japan, or the North American Free Trade Agreement (NAFTA) zone. The fuel filter must ensure that an increased number of particles get removed because of the increased dirt presence under these "severe conditions." The filter mesh increases, therefore, in contrast to "normal conditions" (**Fig. F4**).

The significantly improved filter performance of the above-mentioned filter media makes it possible to meet the requirements for higher filter meshes and also for increased durability. **Fig. F5** shows the development of filter performance in the last few years.

The radial V-folding technique of the filter medium has been accepted as the standard (**Fig F6**) to achieve a high packing density and a homogenous inflow.

The folded bellows is supported by a pressure-resistant center tube. The through-flow is radial from the outside to the inside. Fuel filters are also offered as spiral V-shaped filters, which consist of concentrically positioned paper filter pockets made from conventional filter media. Spiral V-shaped filters are losing their market share. The fuel filters are normally

Fig. F6 Diesel exchange filters with separate water collection space; element with radial V-folding

F

Separation rate:		
■ ISO/TR 13353:1994 / Part 1	Single-pass and particle count, initial separation rate; ACFTD (air cleaner fine test dust), old sensor calibration standard (ISO 4402)	
■ ISO/TR 13353	Single-pass and particle count, initial separation rate; test dust ISO 12103-A3; new sensor calibration standard (ISO 11171)	
■ ISO 19438	Multi-pass (≈ ISO 4548-12) and online particle count; test dust ISO 12103-A3; new sensor calibration standard (ISO 11171)	
Useful life:		
■ ISO 4020	Single-pass; ISO M2 test dust and organic test dirt	
■ ISO 19438	Multi-pass (≈ ISO 4548-12) and online particle count; test dust ISO 12103-A3; new sensor calibration standard (ISO 11171)	

Fig. F7
Overview of test methods for fuel filters

positioned between the fuel feed unit and the high-pressure pump.

Different standardized test processes are used for testing filter performance and they are continuously upgraded to represent the state of the art. Modern test processes measure the particle numbers online before and after the fuel filter and, therefore, the resulting separation rate. **Fig. F7** shows an overview of the test processes.

~~Diesel filter. Current time-controlled diesel injection systems generate injection pressures of up to 2000 bar. The technical design is, therefore, appropriately precise and can only be realized by very accurate and extremely small gap dimensions (less than 100 μm). The fuel filtration of diesel engines has to meet significantly higher requirements than that of modern gasoline injection systems (DI; **Fig. F4**). The injection system must also be protected for a longer vehicle life span (up to more than 1 million km for heavy trucks). The diesel fuel filter has also a second important function in addition to the high fine particulate separation. The diesel filter must guarantee the separation of emulsified and free water to prevent corrosion damage in flow zones with little turbulence and to prevent cavitation damage at fuel pumps and injection nozzles [3]. ISO 4020 requires water separation of more than 93% at nominal flows for time-controlled injection systems. The water content for diesel fuels is generally below the regulated limit, so that the water separation in automobile applications can be eliminated in part. The water separation is performed by coalescence at the filter medium. The separated water is accumulated in

Fig. F8 Preliminary filter for water separation with transparent water collection space

a water collection space integrated in the filter housing. The water level is monitored either by optical or transparent components (bowl; **Fig. F8**) or by conductivity sensors because water has a conductivity different from that of diesel fuel. The water will now be discharged either manually by a discharge screw or through suction devices. Fully automatic systems for water discharge are currently under development, but they are not yet ready for volume production. Addi-

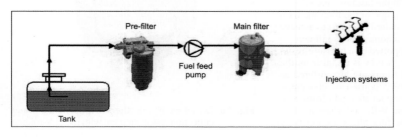

Fig. F9
Two-stage concept for diesel filtration

Fig. F10 Diesel filter with preheat valve (recirculation controlled) and water aspiration

Fig. F11 Diesel filter (inline)

tional heating systems (supplied by the 24-V board power supply) are integrated, especially for certain water separators in commercial vehicles, to ensure that the separated water does not freeze at low ambient temperatures. Filter systems with a preliminary filter (**Fig. F9**) adapted to the fine filter (main filter) are available for especially severe wear protection and water separation requirements.

Paraffins are also separated, and this results in clogging of diesel filters if the fuel temperature falls below a certain temperature limit. However, the temperature limit (cold filter plugging point, or CFFP) of -20–$-25°C$ provided by the manufacturer does not provide information about the filtration capabilities of the diesel fuel by using a modern diesel filter (**Figs. F10 and F11**).

Depending on the additives in the diesel fuels, paraffins are separated significantly before the CFFP, which means that fine filters for winter diesel fuel in Central Europe can already clog at -8–$-12°C$ if no additional measures are taken. Electrical heating systems or intake or recirculation controlled fuel preheaters are used to prevent this clogging. **Fig. F12** shows the principle of recirculation controlled fuel heating.

This already recirculates a very warm or hot recirculation volume from the injection system directly back to the fuel filter (small circulation) on cold start of the vehicle. Mixing of the warm fuel with the large, cold residual volume in the tank does not take place (large circulation). An appropriate valve will open once a certain temperature is reached in the small circulation system and the excess thermal energy will be supplied to the large circulation system. Other functions, such as pressure sensors for monitoring the status of the load, hand pumps for the filling of the system, or fuel cooling at high temperatures, can be integrated into diesel fuel filters (**Fig. F13**). Aluminum or plastic inline filters are increasingly used in addition to conventional diesel fuel filters as exchange filters. Housing filters, which require the exchange of the filter element, are used if different additional functions must be implemented. Housing and other components remain in the vehicle during the exchange.

~~**Gasoline filter.** The requirements for gasoline filtration were increased in parallel with the transformation of the gasoline injection systems into direct injection systems. Multilayer filter media is used currently, while the carburetor engines in the past used primarily inline filters with standard filter media. Additionally,

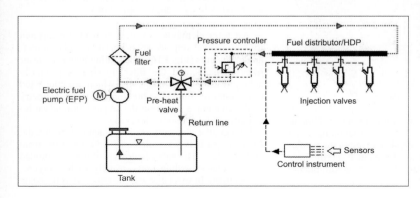

Fig. F12 Principle drawing preheating

Fig. F13 Housing fuel filter for motor cars with heating, cooling

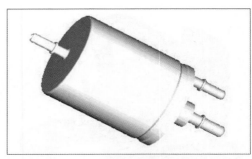

Fig. F15 Gasoline filter (inline) with pressure control valve

the exchange intervals were drastically increased, so that lifetime filters (for more than 240,000 km) are increasingly used, which means that the filters cannot be exchanged anymore, or can only be exchanged with great effort. The chemical stability of the filter media and filter materials had to be adjusted to the lifetime requirements and to the gasolines used. Gasolines in the NAFTA area may contain, for example, ethanol and methanol (E22, M15). Lifetime filters are generally positioned in the fuel tank and are installed together with other in-tank modules (**Fig. F14**). The integration of the gasoline filter in the tank is also of great importance for the reduction [7] of the permissible hydrocarbon emissions. Current legislation permits a value of 2 g (test three-day diurnal). More severe limits in the range of 0.5 g and lower (0 g) are planned for the future. The previous connections between the fuel line and gasoline filter are eliminated or are now inside the tank and no longer contribute to the development of emissions. Gasoline filters are always installed on the pressure side. The above

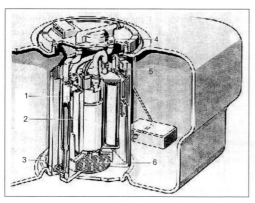

Fig. F14 In-tank module with lifetime filter element: 1, lifetime filter; 2, electric fuel pump; 3, jet pump (controlled); 4, fuel pressure regulator; 5, tank fill level sensor; 6, preliminary filter. (Source: Bosch) Picture credit: picture UMK 1439-1Y

mentioned inline filters and lifetime filters are available on the market as design types. The pressure sensors for the control of the system pressure to a constant differential pressure in relationship to the ambient pressure are in part also integrated into the gasoline filters (**Fig. 15**).

Literature: [1] World-Wide Fuel Charter, Broschüre der Int. Verbände der Automobilindustrie: ACEA, Alliance, EMA, JAMA, December 2002. — [2] M. Durst, G.-M. Klein, N. Moser: Filtration in Fahrzeugen. Grundlagen und Beispiele zur Luft-, Öl- und Kraftstofffiltration, Die Bibliothek der Technik Volume 228, Publishers Moderne Industrie, 2002. — [3] U. Projahn, K. Krieger: Diesel-Kraftstoffqualität—Erkenntnisse aus Sicht des Einspritzsystemlieferanten, Proceedings 9th Aachen Colloquium on Vehicle and Engine Technology, S. Pischinger (Ed.), Aachen 2000, pp. 929–944. — [4] J. Reyinger, U. Weipprecht, G.-M. Klein: Common Rail Diesel Filters, New Filtration Concepts, Proceedings 3rd Int. Conf. Filtration in Transportation, Stuttgart, L. Bergmann (Ed.), 2001, pp. 28–32. — [5] G.-M. Klein: Changes in Diesel Fuel Filtration Concepts, Proceedings 2nd International Conference "Filtration in Transportation," LaGrange, USA, 1999, pp. 45–49. — [6] G.-M. Klein: Verschleißschutz moderner Diesel-und Ottokraftstoff-Einspritzsysteme: Anforderungen und filtrationstechnische Lösungen, House of Engineering Technical Manual Volume 17, Expert Publishers, 2002. — [7] The California Low-Emission Vehicle Regulations (as of May 30, 2001).

~Intake air filter. Intake air filters are used to protect the engine from the solid contaminants that are part of the air induced from the environment. This avoids wear and failures in engine operations.

~~**Air filter elements.** Air filter elements for cars and for commercial vehicles are primarily circular or rectangle elements (**Fig. F16**). The sealing between the air filter elements and the filter housing is primarily achieved using elastomer seals, which are connected to the filter medium. The preference for the geometry of the element is based on the construction space requirements for the individual case.

Rectangular elements are manufactured from a flat folded paper bellows, which is glued tight at the front cutting edges. The paper bellows will now be molded into an elastic sealing mass, usually PUR (polyurethane) foam. In special cases, elastomer seals based on

Fig. F17 Compact air filter element

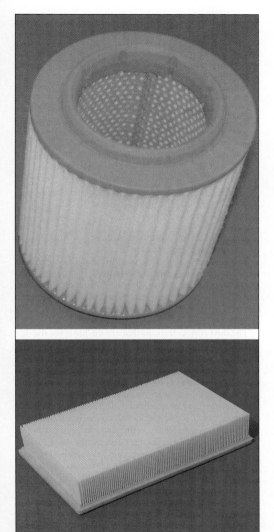

Fig. F16 Standard circular or rectangular air filter element

silicone or PVC (polyvinyl chloride) are used. Elements with staged bellows heights or trapezoid shapes are used for special construction spaces.

The filter medium is assembled to a closed bellows for circular filter elements. This bellows will now be sealed on the front side. Elements with axially acting seals will be fitted with perforated sheet metal or plastic elements for axial stabilization. Circular filter elements are the preferred solution for large air volume flows in commercial vehicles.

~~~**Compact filter elements.** Compact filter elements are inline filters with axial flow (**Fig. F17**). The special

design makes it possible to install a maximum paper surface into a given volume. The compact and mechanically stable design permits the use of special media.

~~~**Cyclone separator.** *See below,* ~~~Preliminary filter

~~~**HC absorber.** After the engine is shut off, gases with hydrocarbon content flow from the crankcase ventilator or from the intake manifold against the normal flow direction through the air filter element into the environment. These HC emissions can be reduced significantly by integrating additional absorbing layers. The loaded absorbent is regenerated by using fresh intake air during driving operation.

~~~**Intake air temperature control.** The intake air temperature control is used to preheat the combustion air. The required warm air is normally drawn from near the exhaust pipe and added to the fresh air by an electrically, mechanically, or pneumatically operated flap mechanism. Icing problems can be reduced especially for engines with exhaust gas recirculation, and the warm-up phase will be shorter due to the induction of warm air.

~~~**Maintenance indicator.** Mechanical or electrical pressure indicators signal that a certain differential pressure across the air filter element has been reached. Maintenance indicators permit requirement-related maintenance of the air filter elements for vehicles that are highly stressed.

~~~**Oil bath air filter.** The particles are separated by a pool of oil and a knitted steal mesh. Oil bath air filters can be cleaned and do not require changing during maintenance. These air filters are only used in special applications because of their significantly lower separation rate.

F

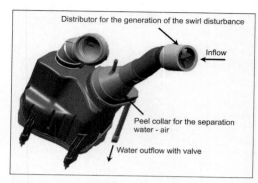

Fig. F18 Air filter system Picoflex with Multizyklon preliminary filters for separation of coarse dust, compact filter element, and safety element

Fig. F19 Automobile air filter housing with upstream centrifugal water separator

~~~**Preliminary filter.** Preliminary filters are used to increase durability of the main filter, especially in commercial vehicle applications with high dust loads in the ambient air. These centrifugal or cyclone separators (**Fig. F18**) separate coarse dust from the intake air and, therefore, reduce the dust load on the downstream filter element. In addition to dust separation, preliminary filters are also used as water separators. Designs with a vane ring around the circular filter element or with a preliminary multicyclone unit are currently in use.

The installation of a preliminary filter increases the flow resistance of the total system, which means that it is only used if absolutely necessary.

~~~**Safety elements.** Safety elements are installed behind the main filter element in the clean air region. They consist of (oil soaked) fleeces or papers. They are used when it has to be guaranteed that no dust can get into the clean air region during filter maintenance or when a minimum driving capability has to be provided, even though the main element is defective.

~~~**Water separation.** When driving in rain, significant quantities of water can be induced into the intake system, depending on the location of the air intake. The induced water can also get to the clean air side of the filter because the air filter element cannot protect against water intake. This can lead to malfunctioning or damage to the hot-film air mass flowmeter (HFM). The time for water induction can be significantly extended by taking certain measures upstream.

Preliminary filters with suitable water separation capabilities are used in this case, or a combination of measures including the design of the housing with additional water separation fleeces mounted on the filter element might be used. **Fig. F19** shows a centrifugal water separator with water separation capabilities, which is mounted upstream of the air filter housing.

~~**Design of air filter elements.** The intake volume flow, the engine performance, the engine use, and the required durability are the main factors for the design of air filter elements.

The design is based on two criteria: the durability and the separation rate. The durability is determined by the weighted average volume flow and the dust content of the intake air. It must then be checked whether the maximum volume flow still provides the required separation rates.

The dust concentrations that are the basis for the design vary over a wide range. The average dust concentration on paved roads is approximately 0.3 mg/m³, and the values on construction sites or in agricultural environments can get to more than 6 g/m³. The particle size distribution of the environmental dust also covers a wide range: from approximately 0.1 to up to more than 100 μm. High particle concentrations normally have larger particle diameters in the particle distribution.

The position of the intake point on the vehicle also influences the dust load on the filter elements. The particle concentration declines with increasing height above the road; however, the size distribution shifts to smaller particle diameters.

The correlation between the dust capacities measured with standardized test dusts in laboratories and the durability can only be determined in extensive practical tests.

~~**Filter characteristics.** The most important filter characteristics are defined below.

~~~**Dirt capacity.** *See above,* ~Filter characteristics

~~~**Durability.** The durability is the time of use of the filter until maintenance or exchange of the filter element. The information is typically given as driving distance (km) or as time of use (h) for stationary engines.

~~~**Dust passage.** Quantity of particles that are not separated by the filter medium.

~~~**Dust retention capacity.** The dust retention capacity or the dirt capacity is the mass of solid contaminants that can be absorbed by the filter element or the

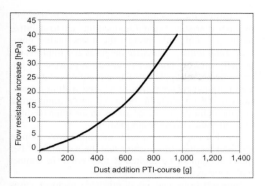

Fig. F20 Dust retention capacity of an air filter element as a function of the flow resistance increase

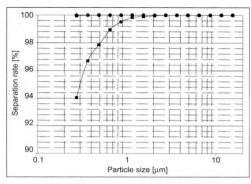

Fig. F21 Fraction separation rate of an air filter medium after a flow resistance increase of 1 (■) or 5h Pa (♦)

filter system up to a defined flow resistance or up to a defined time. **Fig. F20** shows the dust retention capacity of air filter elements as a function of the flow resistance.

Standard test dusts are used in laboratory tests for comparative analyses to determine the dust retention capacity. The measured dust retention capacity depends on the test dust used; fine dusts result in low dust retention capacities, coarse dusts result in higher dust retention capacities.

~~~**Equivalent diameter.** The equivalent diameter is the diameter of a sphere that has the same physical characteristics (projected area, volume, settling rate, light diffusing characteristics) as the analyzed particles with irregular shapes.

~~~**Filter mesh.** *See above*, ~ Filter characteristics

~~~**Flow resistance.** *See above*, ~ Filter characteristics

~~~**Fraction separation rate.** The fraction separation rate $\eta(x_i)$ indicates the separation rate for the specified particle size or particle size ranges. It is defined for each particle size, or each particle size range, x_i, as the ratio of the separated particles (number n or mass m or concentration c) in relationship to the supplied particles:

$$\eta(x_i) = \frac{n_{G(x_i)}}{n_{A(x_i)}} = \frac{m_{G(x_i)}}{m_{A(x_i)}} = \frac{c_{1(x_i)} - c_{2(x_i)}}{c_{1(x_i)}}.$$

As with the separation rate, the fraction separation rate is also dependent on the loading of the filter elements, which means the associated loading status must also be defined here. **Fig. F21** shows the fraction separation rate of an air filter medium for two loading states.

The fraction separation rate is also called the filter mesh. The average filter mesh describes the particle size at which 50% of the particles will be separated; the absolute filter mesh describes the particle size at which 100% will be separated.

The measuring process used and the particle characteristics have an impact on the result of the particle size measurements (equivalent diameter), and this is why the measuring process used and the test conditions must be stated.

~~~**Initial separation rate.** The initial separation rate describes the separation performance of the filter element or filter system at the start of the loading—that is, when it is new. The initial separation rate is defined as the ratio of the retained contamination mass, $m_{G,0}$, to the supplied contamination mass, $m_{A,0}$:

$$\eta_0 = \frac{m_{G,0}}{m_{A,0}}.$$

The initial separation rate for air filter elements is defined in ISO 5011.

~~~**Separation rate.** The separation rate, η, describes the current separation performance of a filter or a filter system. It is defined as the ratio of the separated particle mass, m_G, or the particle mass concentration ($c_1 - c_2$) in relationship to the supplied particle mass, m_A, or the particle mass concentration, c_1:

$$\eta = \frac{m_G}{m_A} = \frac{c_1 - c_1}{c_1}.$$

The total separation rate is often used for the integral description of the filtering performance, because the separation rate in storage filters changes in most cases with the dust load of the filter element.

~~~**Total separation rate.** The total separation rate η_G is an indication for the separation performance of the filter or the filter element. The total separation rate is defined as the ratio of the total retained contamination mass, m_G, in relationship to the supplied contamination mass, m_A,

$$\eta_G = \frac{m_G}{m_A}.$$

469

The total separation rate is normally related to a certain admissible increase in pressure loss during the flow through the filter element—for example, after 2000 Pa or 4000 Pa.

~~Filter media.
Fiber materials based on cellulose, plastics, or a combination of these materials are used worldwide for air filtration. The papers (cellulose fibers) or fleeces (nonwoven synthetic fibers) have a defined composition and fiber structure corresponding to the required filter performance (filter characteristics).

Depending on the application, different filter medium requirements exist with respect to separation rates.

- Commercial vehicles, >99.9%
- Motor cars with diesel engines, >99.8%
- Motor cars with gasoline engines, >99.5%

The different requirements are based on the different mileages required from the engines and the different clearances in the area of the cylinder liners.

The duration requirements have continuously increased with the development of new generations of vehicle. Current values are between 40,000 and 120,000 km for motor car applications. The duration is specified as a combination of kilometer performance and time because the materials that are used to produce air filter elements degenerate over time.

~~~Cellulose paper.
Cellulose-based papers are currently still the most commonly used media for intake air filters. These filter media are produced with a conventional paper manufacturing process. Cellulose fibers with a diameter of approximately 20–50 μm are used as raw material (**Fig. F22**). The selection of the fibers and the process control can result in different characteristics.

The papers must receive impregnations to stabilize the fibers in the compound and the folds after the imprinting process. The stability of the media can be optimized by adding synthetic fibers and glass fibers to the cellulose fibers, and the stability of the folded medium in the filter element especially under wet environmental conditions (driving in the rain) can be improved significantly.

~~~Impregnations.
Impregnation is used to improve the characteristics of the filter medium and to stabilize the fiber compounds, protect the fibers against environmental impacts, stabilize the folds and imprints especially in wet conditions, introduce processing improvements, and apply new flame-proofing compounds. Cellulose papers are primarily treated with impregnations based on phenolic resins or acryl compounds, and the cellulose paper becomes an air filter medium only after the correct impregnation.

Another important aspect is the flame protection of the filter media—this ensures that the flammable fiber materials cannot be ignited if they come in contact with a cigarette stub, for example.

Impregnations can be eliminated, in most cases completely, for applications with synthetic fiber nonwovens. Water repellant materials in the impregnations improve the wet stiffness of the media and at the same time permit to some extent the separation of water from the filter element.

~~~Synthetic fiber, nonwoven.
High-performance synthetic nonwoven fibers are used for the most severe requirements because of their higher filter performance. They exist in most cases as synthetic meltblown fibers with diameters from <1 to 20 μm (**Fig. F23**). As with the cellulose paper, the fiber compound must also be stabilized. The possibility for a thermal "welding" of thermoplastic fibers exists because of edge melting of the fibers at the contact point, and the use of chemical binders (impregnations) is, therefore, not required.

~Lubricating oil filter.
Lubricating oil filters remove and reduce particles from the engine oil, which otherwise would lead to damage and wear in the lubricant circulation. The particles can become enriched when the filtration is insufficient and the wear will be accelerated, because the engine oil circulates in the oil circuit continuously. Lubricating oil filters remove particulate contamination from the lubricating oil. Liquid or soluble components, such as water, additives, or age-dependent catabolites of the oil will not be separated.

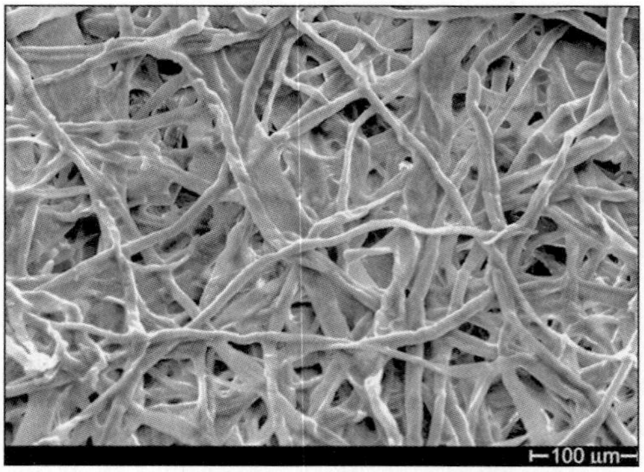

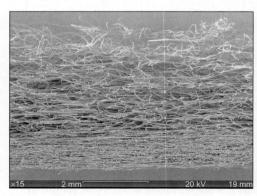

Fig. F22 REM image of cellulose fiber filter media

Fig. F23 REM image of synthetic fiber, nonwoven

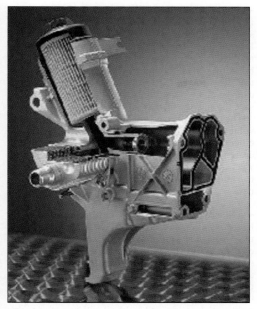

Fig. F25 Oil filter change with antidrain membrane and bypass valve

Fig. F24 Oil filter housing with metal-free filter element and accessories

The number and size of the particles determine the wearing effect of the particles in the oil circulation system. A typical spectrum of particle sizes in the engine oil is in the range of 0.5 to 500 μm. The filter mesh of the lubricating oil filter is, therefore, specifically adapted to the requirements of the particular engine.

Lubricating oil filters are in principle available in two important design types: as easy-change filters and as housing filters. The filter element in the exchange filter is positioned in a housing that cannot be opened and it is connected through a threaded socket to the engine block. The complete exchange filter is exchanged during maintenance.

The housing filter consists of a housing that is permanently connected to the engine block and that can be opened, and an exchangeable filter element (**Fig. F24**). Only the filter element will be exchanged during maintenance because the housing is a lifetime component. The filter elements in modern engines are produced without metal parts.

In most cases, both filter designs have, in addition to the filter element, a filter bypass valve that opens at high differential pressures and guarantees the lubrication of the required points in the engine. Typical opening pressures are in the range of 0.8 to 2.5 bar. High differential pressures are generated by high oil viscosities, for example, during cold starts in winter or when the filter element is very exhausted. Both filter designs can also have a clean oil side and a raw oil side antidrain valve depending on the requirements of the engine (**Fig. F25**). These valves prevent the emptying of the oil filter housing after the engine is switched off. This

ensures immediate lubrication of the engine after starting, because otherwise the volume of the filter housing would have to be filled first before the oil pressure would get to the lubrication points, and this would result in noise and wear.

~~Bypass oil filter. A bypass oil filter (**Fig. F26**) is used for fine filtration of engine oils. This filter removes significantly finer particles from the oil than does a full-flow oil filter. The smallest abrasive particles are removed to improve wear protection, and soot particles are separated to decrease the increase in viscosity. The maximum permissible soot concentrations are 3–5%. Higher concentrations result in severe increases in oil viscosity, and this reduces the functionality

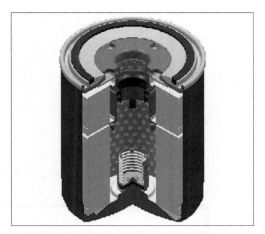

Fig. F26 Combination filter with full-flow and bypass unit

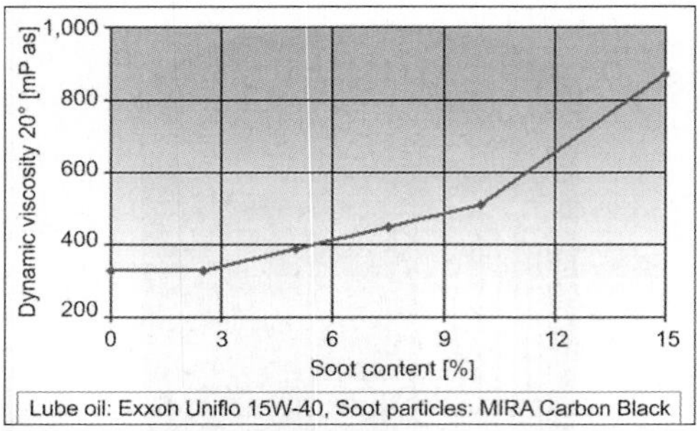

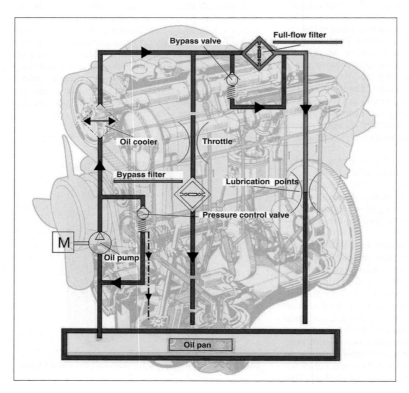

Fig. F27 Lubrication oil viscosity increase in relation to the soot concentration

Lube oil: Exxon Uniflo 15W-40, Soot particles: MIRA Carbon Black

F

of the oil (**Fig. F27**). Bypass filters are, therefore, almost exclusively used in diesel engines. Higher pressure losses are required to achieve the high filter units. Only a part of the volume flow of oil, which is delivered by the oil pump, is, therefore, routed through the bypass oil filter. Typically, 8–10% of the volume flow is recirculated to the oil pan after it passes through the filter. The bypass oil filter removes particles mechanically from the circulating oil, but it does not remove dissolved oil catabolites and, therefore, it cannot extend the oil change intervals caused by the chemical decay of the oil during driving operations.

~~**Full-flow oil filter.** All modern vehicles have a full-flow oil filter. **Fig. F28** shows the schematic design of a lubrication oil circulation system with full-flow oil filter. This method of oil filtration routes the total volume of oil flow, which is transported to the lubrication points, through the filter, which guarantees that all particles that could contribute to significant damage and wear are separated during their first passage through the filter.

The filter surface is primarily determined by the oil volume flow and the dirt absorption capacity. Specific surface loads between 0.8 and 1.5 L/cmh are design guidelines. For these surface loads and an oil viscosity of 20 mm/s, the differential pressure across the filter element is approximately 0.1–0.3 bar when new.

~~**Oil extractor.** Centrifuges are used in addition to high-pass filters for the separation of fine particles in the bypass flow (**Fig. F29**). The centrifuges are powered by the oil pressure without an external energy supply, and the oil exits through open jet nozzles that are tangentially positioned at the rotor. This means that the rotor achieves speeds of up to 10,000 rpm. The exiting oil is collected and returned to the oil pan. The solid particles contained in the oil will be moved to the rotor wall by the centrifugal forces and are separated there as a solid layer. The separation performance of the centrifuge depends on the difference between the

Fig. F28
Oil circulation with full-flow and bypass oil filter

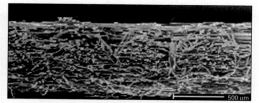

Fig. F30 Fully synthetic filter medium with gradient structure

Fig. F29 Oil bypass centrifuge with fully incinerable plastic rotor

density of particles and the oil as well as on the rotational speed and the design of the rotor. The advantage of the centrifuge is its compact design in relation to the storage volume for the dirt.

~~Oil filter change. The oil filter change is always performed at the same time as the oil change. The filter absorbs dirt from the oil during operation and stores this in the three-dimensional high-pass filter matrix. The differential pressure of the filter increases quickly if the dirt absorption capability is exhausted. The filter must, therefore, be exchanged before it is completely blocked. If the oil filter is not changed together with the oil, the separated dirt can be stripped from the filter by the new oil and its unused additives (dispersants und dispergents). The second parameter requiring the exchange is the decay and the aging of the filter medium and the sealing materials during operation. The temperature, time, oil used, fuel used, operating conditions, driving pattern, and other factors all play a role in determining the deteriorating conditions. The materials are designed especially for the service intervals and depend on the engine and the application. Skipping the oil filter change or exceeding the time interval can result in permanently open bypass valves and damage to the oil filter component due to decay. The intervals for automobiles are currently between 15,000 and 50,000 km, and they range from 60,000 to 120,000 km for commercial vehicles. Developments are aimed at even longer service intervals. This will require improved filter materials such as fully synthetic oil filter media.

~~Oil filter media. Different high-pass filter media are used for oil filtration. These are fiber-based filters, which are available in different combinations. Flat media are used in most cases. They are mostly pleated, and for bypass oil filters they are sometimes also wound or used in the form of fiber packings. Cellulose is the most commonly used material for this application. Additional plastic or glass fibers can be added to the cellulose in almost unlimited proportions. These filter media are subjected to a synthetic resin impregnation to guarantee their resistance against oil. Synthetic fiber filter medium is increasingly used because it has a significantly better chemical resistance, which results in longer service intervals. Additionally, it offers better alternatives for the development of a three-dimensional fiber matrix to optimize the separation and dirt absorption performance. **Fig. F30** shows a synthetic fleece with a gradient structure for oil filtration.

The filter unit ranges of the full-flow and bypass oil filter media are shown in **Fig. F31**.

Literature: M. Durst, G.M. Klein, N. Moser: Filtration in Fahrzeugen, Publishers Moderne Industrie, 2002. — M. Kolczyk: Economy and Ecology—Innovative Filter Solutions for Customers Needs, Proceedings 4th International Filtration Conference, 2001. — K. Knickmann, M. Kolczyk: Fluid Management with Oil and Diesel Filter Systems, Proceedings 5th International Filtration Conference, 2002. — A. Samways: The Development of an Improved Self-Powered Centrifugal Oil Cleaner for Engine Applications, Proceedings 5th International Filtration Conference, 2002. — K. Knickmann: Flüssigkeitsmanagement an modernen Verbrennungsmotoren, Filtersysteme im Automobil, Expert Publishers, 2002. — ISO 4548, Schmierölfilterprüfverfahren. — J. Spanke, P. Müller: Neue Ölwechselkriterien durch Weiterentwicklung von Motoren und Motorölen, MTZ Motortechnische Zeitschrift (International Trade Journal on Engine Technology) 58 (1997) 10. — W. Dahm, K. Daniel: Entwicklung der Ölwechselintervalle und deren Beeinflussbarkeitdurch Nebenstromfeinstölfilterung. MTZ Motortechnische Zeitschrift 57 (1996) 6.

~Maintenance *See above*, ~Lubricating oil filter ~~Oil filter change

~Particulate filter →Particles (Particulates) ~Particulate filter system; →Pollutant retention systems

Filter, service →Filter ~Intake air filter ~~Air filter elements ~~~Maintenance indicator

Filter change →Filter ~Filter characteristics ~~Durability

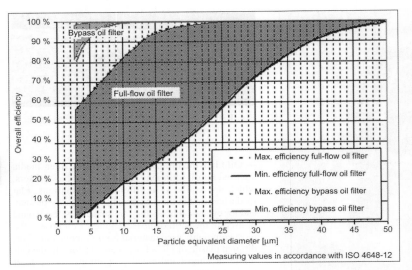

Fig. F31
Filter mesh of full-flow and bypass filters (according to ISO 4548-12)

Filter characteristics →Filter; →Filter ~Intake air filter

Filter contamination →Filter ~Intake air filter ~~Filter characteristics ~~~Dust retention capacity

Filter designs →Lubrication ~Filter

Filter felts →Particles (Particulates) ~Particulate filter system ~~Particulate filter ~~~Filter media

Filter impact →Filter ~Intake air filter ~~Filter characteristics ~~~Fraction separation rate

Filter materials →Filter ~Intake air filter ~~Filter media

Filter media →Filter ~Intake air filter; →Particles (Particulates) ~Particulate filter system ~~Particulate filter

Filter mesh →Filter ~Filter characteristics

Filter papers/filter felts →Particles (Particulates) ~Particulate filter system ~~Particulate filter ~~~Filter media

Filter resistance →Filter ~Filter characteristics ~~Flow resistance

Filter service life →Filter ~Filter characteristics, ~Intake air filter ~~Filter characteristics

Filter systems →Oil ~Oil maintenance

Filter/catalytic converter combinations →Particle (Particulates) ~Particulate filter system

Filtering →Filter

Final boiling point →Fuel, diesel engine ~Properties; →Fuel, gasoline engine

Finite element analysis →Sealing systems ~Development methods ~~Calculation of the connection between components

Finite Element Method (FEM) (*also*, →Calculation processes ~Numerical processes). FEM is currently an indispensable development tool for the engine technology. Complex loading patterns and geometries with free-form surface areas are divided into fields, which can be used in the FEM to simulate stresses due to static and dynamic forces and thermal stresses. The geometry of the structure, which is in most cases available as a CAD data file, can be reproduced by space or shell elements. Deformations and stresses can be determined at each point of the analyzed component by using the defined forces and force application points at a particular load. Application areas of FEM in engine development are, for example:

- Component dimensioning (piston rod, crankshaft, crankcase, etc.)
- Vibration problems (acoustic behavior due to gas forces and oscillating masses)

Literature: R. van Basshuysen, F. Schäfer (Eds.): Handbuch Verbrennungsmotor, 2nd Edition, Wiesbaden, Vieweg Publishers, 2002. — E. Köhler: Verbrennungsmotoren. Motormechanik, Berechnung und Auslegung des Hubkolbenmotors 3rd Edition, Wiesbaden, Vieweg Publishers, 2002.

Finite elements →Calculation processes ~Numerical processes

Finite volumes →Calculation processes ~Numerical processes

Finned valve →Reed valve; →Two-stroke engine ~Charge transfer, ~Scavenging fan

Fire land →Piston

Fire land height →Piston ~Fire land

Firing order (*also*, →Ignition system, gasoline engine ~Ignition). The sequence in which the cylinders of an engine fire is termed the firing order. For example, the usual firing orders for inline engines are:

4 cylinders: 1–3–4–2 or 1–2–4–3
5 cylinders: 1–2–4–5–3
6 cylinders: 1–5–3–6–2–4 or 1–2–4–6–5–3, etc.
8 cylinders: 1–6–2–5–8–3–7–4 or 1–3–6–8–4–2–7–5, etc.

For V-engines:

4 cylinders: 1–3–4–2
6 cylinders: 1–2–5–6–4–3 or 1–4–5–6–2–3
8 cylinders: 1–6–3–5–4–7–2–8 or 1–8–3–6–4–5–2–7, etc.

The firing order can have considerable impact on mixture distribution if central carburetion devices are used. Distributor injection pumps with only one delivery element for all cylinders have to supply the injection nozzles of the individual cylinders according to the firing order. In the case of multicylinder engines, the effects of forces inside and outside are influenced by, among other things, the firing order. Moreover, design, demand for equal ignition intervals, and crankshaft design and its operational stress all have a significant influence on the firing order.

First engines →Engine ~Historic engines

Fitted bolt →Piston pin

Five valves →Valve arrangement ~Number of valves

Five-shaft engines →Engine concepts ~Reciprocating piston engines ~~Multiple-shaft engines

Five-valve technology →Valve arrangement ~Number of valves

Fixed venturi carburetor →Carburetor ~Carburetor, mode of operation

Fixed-tension pulleys →Toothed belt drive ~Belt tensioning systems

Flame expansion (*also*, →Combustion process, gasoline engine). Diesel and gasoline engine combustion are different in their basic process. This has an impact on the conversion of the fuel and the expansion of the flame. The distribution of air and fuel is extremely inhomogeneous in a diesel engine, while as a first approximation a homogeneous mixture is present in a conventional gasoline engine. This is also true for a gasoline engine with direct injection. The air-fuel ratio varies from $\lambda = 0$ to $\lambda = \infty$ in the area of the injected fuel spray in a diesel engine and for a gasoline engine with direct injection. A mixture that delivers optimum output and minimum exhaust gas emissions is targeted for diesel engine combustion. The flame expansion process, or combustion, can be divided into three phases for the diesel engine. Phase one includes injection and ignition. By the time of the ignition a large part of the fuel will have been converted into a gaseous phase and has been mixed with the air, and ignition takes place at an air-fuel ratio that is significantly below $\lambda = 1$. The actual condition of the mixture is important for the combustion immediately after ignition. The injected spray "burns" similarly to a layered mixture cloud.

The injected fuel is further mixed with air and combustion gas during the second phase. A third combustion phase follows after injection has ended. It is characterized by a further leaning of the mixture due to the mixing.

The expansion of the flame in a diesel engine is primarily influenced by the temperature at the end of compression, the injection pressure, the spray pattern, the number of injection sprays, the distribution of the fuel quantity during the injection, the combustion chamber geometry, and the swirl of the air.

The flame front in gasoline engines with conventional combustion processes and direct injection is initiated by the electrical ignition at the spark plug and the subsequent heat and mass transfer processes. The mixture volume between the electrodes of the spark plug, which must be within the ignition limits, results in the development of a stable flame. Enough energy must be released to generate reactive that can ignite the surrounding zones of mixture.

The ignitable mixture must be transported in a controlled manner and it must be repeatable for processes with charge stratification. Zones with incombustible lean mixtures must be avoided, as must zones with mixtures that are too rich.

The flame in gasoline engines should run through the air-fuel mixture as quickly as possible, so that knocking regions cannot develop.

Additionally, low speeds of flame propagation result in a delayed combustion process with potentially high levels of unburned hydrocarbons in the exhaust gas and a large deviation from the ideal, efficient constant volume combustion. The flame speed depends strongly on the air-fuel ratio and the movement of the mixture. The maximum value of 20–25 m/s is achieved in a homogeneous mixture with an air-fuel ratio of $\lambda = 0.85$.

The best conditions for fast combustion are offered by combustion chambers with a small surface:volume ratio. Additionally, the spark plug must be positioned centrally to minimize the flame travel distances. This can be implemented in multivalve engines (e.g., four or five valves per cylinder) when the spark plug is positioned in the middle of the combustion chamber. Two spark plugs per cylinder are recommended for two- and three-valve engines to minimize the flame travel distances.

F

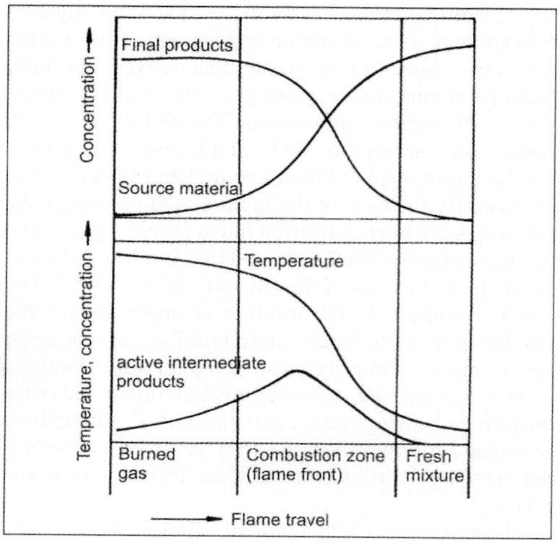

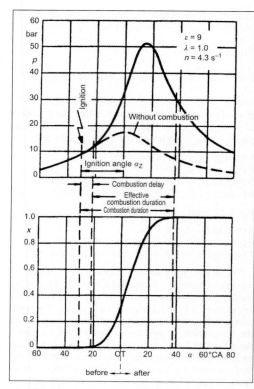

Fig. F32 Processes in the combustion zone (Source: F. Pischinger)

Fig F33 Pressure curve p and combustion function x in gasoline engines (Source: Pischinger)

The flame path to the exhaust valves must be short; otherwise self-ignition of the mixture may occur in the area of the hot exhaust valves, which in turn will result in knocking.

Fig. F32 shows the principal processes in the combustion zone. Preconditions for a normal expansion of the flame are an air-fuel ratio that is within the ignition limits of the mixture and the absence of self- or glow ignition. A combination of the above features and the variation of the combustion chamber volume by the motion of the piston result in a cylinder pressure curve as shown in **Fig. F33**.

The terms in **Fig. F33** are:

- Ignition angle a_z: Time of the electrical spark in degrees crank angle (°CA) before or after top dead center
- Combustion duration: Time between ignition and end of combustion (°CA)
- Effective combustion duration: Time between the first, just measurable pressure increase relative to the compression curve and the end of the combustion (°CA)
- Combustion delay: Time between the ignition and the just measurable pressure increase (°CA)

The lower half of the diagram shows the derived combustion function. The first fuel gets visibly converted 5–10°CA after ignition, and the combustion is in this case completed approximately 38°CA after top dead center (TDC).

The expansion of the flame front is subject to cyclical variations on consecutive combustion cycles. These are based on time-dependent differences in local turbulence and mixture composition—for example, at the spark plug and in the combustion chamber. The cyclical variations are smallest for small air-fuel ratios of approximately $\lambda = 0.85$ because of the high flame

speeds in this strength mixture. The mixture burns more slowly and the cyclical variations increase the closer the process gets to the ignition limits. **Fig. F34** shows the impact of the cyclical variations on the pressure curve.

The flame speed is the sum of the combustion speed and the gas velocity, which in turn are the result of the turbulence or other mixture movements (the mixture movement must be included vectorially in relationship to a location). The transport speed is influenced by the piston movement, the resulting quench flow, and the

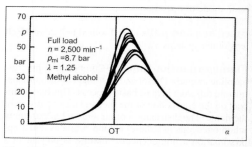

Fig. F34 Cyclical variations in a gasoline engine (Source: Pischinger)

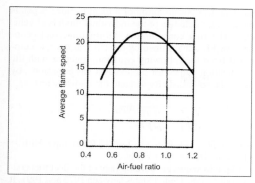

Fig. F35 Average flame speed

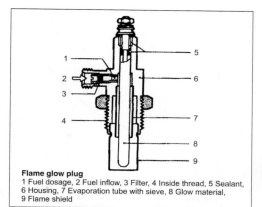

Flame glow plug
1 Fuel dosage, 2 Fuel inflow, 3 Filter, 4 Inside thread, 5 Sealant, 6 Housing, 7 Evaporation tube with sieve, 8 Glow material, 9 Flame shield

Fig. F36 Flame glow plug (Source: Bosch)

intake process of the mixture. Swirl and turbulence have a significant impact on this process. The average flame speed w_F is used because swirl and turbulence have strong local variations and are time-dependent. It is calculated based on the distance and the travel time between two points in the combustion chamber. The average flame speed depends on:

- The air-fuel ratio
- The combustion chamber shape
- The spark plug location
- The mean piston speed
- The piston position or the crank angle

Fig. F35 shows the impact of the air-fuel ratio on the flame speed. It can be seen that the combustion speed and, therefore, also the flame speed are highest for an air-fuel ratio of approximately $\lambda = 0.85$.

The combustion chamber shape and the plug location determine the distance that the flame has to travel to cross the chamber. Compact combustion chambers with centrally positioned spark plugs are desirable to achieve short combustion duration. The combustion duration decreases as the flame speed increases, which means that the combustion is also completed faster. This is the way to get close to the ideal constant volume combustion, which is desirable for its low fuel consumption.

The impact of the mean piston speed is also of importance. The mean piston speed increases in proportion to the engine speed. However, the flame speed does not increase proportional to the piston speed, and this prolongs the combustion period in terms of crank angle as the engine speed increases, which depends on the combustion duration. The moment of ignition must be advanced with increasing engine speeds to achieve optimum efficiencies, because the peak pressure should be achieved shortly after top dead center for thermodynamic reasons.

Literature: R. van Basshuysen, F. Schäfer (Eds.): Handbuch Verbrennungsmotor, 2nd Edition, Wiesbaden, Vieweg Publishers, 2002.

Flame front →Combustion, diesel engine ~Combustion process; →Combustion process, diesel engine; →Flame expansion

Flame front speed →Flame expansion

Flame glow plug (*also*, →Starting aid). The flame glow plug (**Fig. F36**), heats the intake air by the combustion of fuel. The fuel feed unit for the injection system normally delivers the fuel via a solenoid valve to the flame glow plug. The contact pin of the flame glow plug includes a filter and a dosage device. It releases a fuel quantity tuned to the particular engine. The fuel evaporates in an evaporation tube positioned around the pencil-type glow plug and mixes with the intake air. The mixture ignites in the front part of the pencil-type glow plug, which has a temperature of more than 1000°C.

Literature: Bosch (ed.): Kraftfahrtechnisches Taschenbuch, 22nd Edition.

Flame ionization detector (FID) →Exhaust gas analysis equipment

Flame path →Flame expansion

Flame propagation →Combustion process, gasoline engine

Flame propagation speed →Flame expansion

Flame quenching. Flame quenching describes the extinguishing of the flame at the relatively cold combustion chamber walls of engines. Flame quenching is, therefore, responsible for a significant part of the emissions of unburned hydrocarbon. This is why compact combustion chambers are desirable, because they have a small surface:volume ratio. The cold combustion chamber surface reduces in line with the reduction of the specific surface (**Fig. F37**).

It is also important for gasoline engines with intake manifold injection that optimum quench surfaces are available in the combustion chamber. These result in the generation of the desired mixture movement by displacement of the mixture when the surfaces are

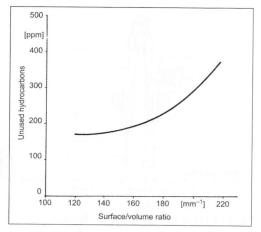

Fig. F37 HC concentration in relationship to the surface:volume ratio

moved toward each other during the upward movement of the piston. This produces a turbulent flow with intense mixing, and even part of the hydrocarbon particulates that are detached from the wall surface participate in the oxidation processes.

Flame radius. The flame expands after ignition in the combustion chamber of the gasoline engine in a spherical manner. Flame radii of different lengths and, therefore, different durations of combustion, result from different combustion chamber shapes and spark plug positions (i.e., centrally or at the edge of the combustion chamber). The flame travel distances should be minimized to achieve optimum efficiencies (e.g., constant volume combustion). This leads to combustion chambers with central spark plug positions. Multivalve engines are best for this application. A shortening of the flame travel distances reduces knocking in gasoline engines; a higher compression ratio and, therefore, better thermal efficiencies become possible.

Flame speed →Flame expansion

Flashpoint →Fuel, diesel engine ~Properties; →Oil ~Properties and characteristic values

Flat pintle nozzle →Injection valves ~Diesel engine ~~Pintle-type nozzle

Flat seals →Sealing systems ~Flat seals

Flat tappet →Valve gear ~Gear components ~~Pushrod

Fleet consumption. The US government has established consumption limits for vehicles. However, individual vehicles do not have to meet the consumption limits, but the total fleet of a manufacturer must do

so. The fleet consumption of a manufacturer depends on the sales volumes of the individual models of vehicles. The fuel consumption information (fuel economy, or FE), determined by the US regulator, is computed from the following equation, together with the certification values for the exhaust gas emissions, by using the so-called city test and the highway test:

$$FE = \frac{1}{\dfrac{0.55}{CFE} + \dfrac{0.45}{HFE}} \text{ mpg,}$$

where CFE = FE from city test (FTP 75), mpg, HFE = FE from highway test, mpg.

Fig. F38 shows the development of the fleet fuel consumption in the United States between 1975 and 1991. It can be seen that the reduction of the fuel consumption has slowed significantly in recent years. It is approximately 28 mpg (US) for a fleet of newly licensed vehicles.

The large potential for reductions in fuel consumption can only be utilized if measures are taken on the vehicle, the transmission, and the engine. In addition to engine-related measures, reductions in vehicle weight, new transmission concepts, and improvements in the aerodynamics are significant.

Additional consumption reductions are possible with measures such as multiple valve technology, turbocharging, improvements in fuel injection, optimization of direct injection for gasoline and diesel engines, cylinder shut-offs, reductions in friction, and an increase in the compression ratio for gasoline engines. All measures for the reduction of fuel consumption also result in a reduction of the greenhouse gas CO_2.

The focus for diesel engines is further development of the engine with direct injection, additional optimizations of the combustion processes, increases in injection pressure, variable injection nozzles, and turbochargers with variable turbine geometry—prechamber engines are a thing of the past.

Forecasts for the expected fuel consumption vary widely. For example, a fleet consumption of 45–75 mpg (US) is possible in the United States for the model year (MY) 2010. **Fig. F39** shows that this could be realistic because it highlights a few vehicle prototypes with especially low levels of fuel consumption.

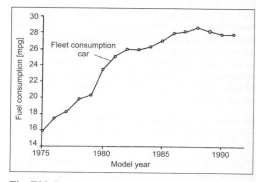

Fig. F38 Development of fuel consumption

	Power [KW]	Fuel consumption [mpg]	Engine
General Motors	28	61 city 74 hwy	Gasoline
Volkswagen	38	74 city 99 hwy	Diesel
Renault	37	63 city 81 hwy	Diesel
Renault	20	78 city 107 hwy	Gasoline
Peugeot	37	55 city 87 hwy	Diesel
Peugeot	21	70 city 77 hwy	Gasoline
Ford	29	57 city 92 hwy	Diesel
Toyota	41	89 city 110 hwy	Diesel
Daihats THS	–	142 (Japanese urban fuel economy)	GDI Hybrid
Honda IMAS	–	96	Hybrid

Fig. F39 Fuel consumption of vehicle prototypes

Vehicles with a consumption of three liters per 100 km already exist, while vehicles with a fuel consumption of less than three liters per 100 km are available as prototypes.

Literature: H. Barske: Entwicklungspotential des Kraftstoffverbrauchs von Antriebskonzepten mit Verbrennungsmotor, 3rd Aachen Colloquium on Vehicle and Engine Technology, 1991. – F. Piech: 3 l/100 km im Jahr 2000, Symposium Energieverbrauch im Straßenverkehr, Vienna, 1991, VDI Publishers No. 158. – N.N.: Automotive Fuel Economy. How Far Should We Go? National Research Council, National Academy Press, Washington, DC, 1992. – L.D. Bleviss: The New Oil Crisis and Fuel Economy Technologies, Quorum Books, Greenwood Publishing, New York, 1988.

Flexible fuel vehicles. These vehicles have engines that can use alternative fuels, such as ethanol, methanol, biofuels, etc. The vehicles are used to protect the environment from additional emissions of CO_2 and sulfur compounds.

Flexible valve actuation →Valve gear

Flip chip →Electronic/mechanical engine and transmission control ~Electronic components

Float →Carburetor ~Level control ~~Carburetor float bowl

Float needle valve →Carburetor

Floating piston pin →Piston pin

Flow →Charge movement; →Variable valve control ~Charge movement

Flow behavior →Oil ~Oil functions, ~Properties and characteristic values ~~Viscosity

Flow bench →Carburetor

Flow cross section. When the inlet valve opens, a ring-shaped geometric cross section results. Because of separation and constriction of the air or mixture plume, the flow is constricted. This results in the flow cross section being smaller than the geometric cross section.

~**Isentropic flow cross section.** If an isentropic flow is assumed through the slot cross section created by the inlet valve, a theoretical speed results in the flow cross section, c_{is}. The real speed is less than the theoretical speed due to friction influences. Thus, the isentropic flow cross section can be defined as

$$A_{is} = \varphi \cdot \psi \cdot \delta : \delta_{is} \cdot A,$$

where A = geometric cross section, φ = friction coefficient, ψ = coefficient for flow constriction, δ = actual density, and δ_{is} = density with isentropic flow. With this defined isentropic flow cross section, it is possible to assess the quality of flow passages through valves.

Flow improver →Fuel, diesel engine ~Additives

Flow inlet port →Intake port, gasoline engine

Flow losses →Intake system

Flow rate control →Lubrication ~Oil pump ~~Pump loss of performance

Flow resistance →Filter ~Filter characteristics

Flowmeter intake air (*also*, →Sensors). The air volume or the air mass must be measured in gasoline engines with fuel injection, or a proportional signal of these values must be available. The required fuel amount is calculated based on this information in stoichiometric mixture ($\lambda = 1$) engines for example, and it is delivered through electrically or mechanically actuated injection valves. The maximum air mass to be measured for a medium-size car is approximately 600 kg/h. The ratio of maximum airflow for idle speed or full load is about 1:100, which means that only systems with a high resolution can be used.

The instruments are divided into air volume and air mass meters. Air volume meters need additional variables, such as air pressure and temperature, to determine the mass, because they cannot measure density changes in relation to the geodetic altitude. This can be corrected with the help of an absolute pressure meter (barometric cell) and the intake air temperature.

~**Hot-film air mass flow sensor.** Hot-film or hot-wire anemometry offers the possibility of fluid and gas mass flow measurements. It is based on the effect of the heat transfer between a heated body and the surrounding material. The heat release depends primarily on the flow velocity, the thermal conductivity, the specific thermal capacity of the material, and the temperature difference between the fluid and the basic sensor.

The resistor for the hot-film air mass sensor is placed as a thin resistive film on the heating resistor. The sensor will not be heated directly and is, therefore,

F

F

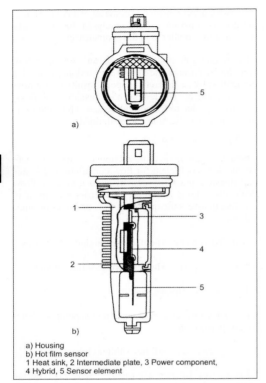

a) Housing
b) Hot film sensor
1 Heat sink, 2 Intermediate plate, 3 Power component,
4 Hybrid, 5 Sensor element

Fig. F40 Hot film air mass sensor (Source: Bosch)

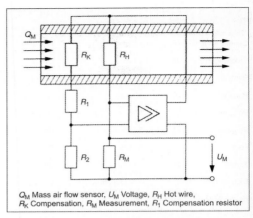

Q_M Mass air flow sensor, U_M Voltage, R_H Hot wire,
R_K Compensation, R_M Measurement, R_1 Compensation resistor

Fig. F41 Hot-wire air mass sensor (Source: Bosch)

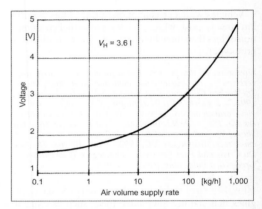

Fig. F42 Characteristic curves of a hot-wire air mass
flow sensor

less susceptible to contamination—as opposed to hot-wire sensors, which must be cleaned by burning from time to time. The film is protected against external impacts by a crystal coating. The measuring principle is based on the sensor resistor being held at a constant temperature that is significantly higher than the material flowing past it. A current control function keeps the sensor at a constant temperature. The current is proportional for the energy transferred into the fluid that flows by, and it is, therefore, proportional to the air mass that passes it. **Fig. F40** shows the design of such an air mass sensor.

~Hot-wire air mass flow sensor. The hot-wire air mass flow sensor uses the same principle as the hot film air mass flow sensor described above. A constant overtemperature (compared with the air temperature) is generated in a thin platinum wire. The required heating current is proportional to the airflow. Air backflows and pulsations—possible for engines with a small number of cylinders—must be avoided, because this would otherwise result in measurement errors. This is the reason a sensor has been integrated into the intake or castor offset. **Fig. F41** shows the design of such a hot-wire air mass flow sensor. **Fig. F42** shows the characteristic curves of an air mass flow sensor for a 3.6-L naturally aspirated engine.

~Karman Vortex flowmeter. This type of flowmeter measures the volume flow rate. Air vortices are generated in the intake airflow with the help of a flow obstruction and their frequency is proportional to the volume flow rate. The volume flow is measured by sending ultrasound waves across the intake airflow. The propagation speed of these waves is impacted by the vortices, and this variation is detected by an ultrasound receiver and analyzed in a control instrument (**Fig. F43**).

~Load-speed flow measurement. Using this flow measurement, sensors detect the engine speed and the intake manifold pressure. The two values are proportional to the intake air mass. The control instrument calculates the fuel injection duration based on the two variables. This is proportional to the fuel volume to be injected.

~Ram pressure flow measurement. This is an air volume sensor. A movable sensor flap opens a variable

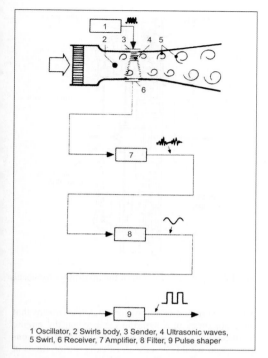

1 Oscillator, 2 Swirls body, 3 Sender, 4 Ultrasonic waves,
5 Swirl, 6 Receiver, 7 Amplifier, 8 Filter, 9 Pulse shaper

Fig. F43 Karman Vortex flowmeter (Source: Bosch)

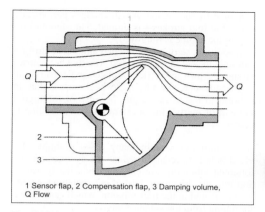

1 Sensor flap, 2 Compensation flap, 3 Damping volume,
Q Flow

Fig. F44 Sensor flap air volume meter (Source: Bosch)

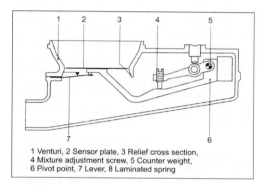

1 Venturi, 2 Sensor plate, 3 Relief cross section,
4 Mixture adjustment screw, 5 Counter weight,
6 Pivot point, 7 Lever, 8 Laminated spring

Fig. F45 Rotameter principle air volume meter (Source: Bosch)

cross section depending on the ram pressure. A potentiometer measures the flap angle, which is proportional to the volume flow. Errors develop during rapidly changing loads when the sensor flap cannot follow fast enough because of its inertia. A compensation flap is connected to the sensor flap to eliminate the impact of vibrations caused by pulsations in the air—the impact of these is balanced because the pressure vibrations act on the sensor flap and on the compensation flap with the same effect.

A temperature sensor detects the change in air density because of temperature changes and corrects this error.

A force created by the flow of the intake air acts on the sensor flap, and this force is opposite to the reset force of a spring. The contour of the measurement channel and the deflection of the flap are harmonized. The open cross section increases with increasing air volumes. A logarithmic relationship has been found to be advantageous, because the resolution of the air volume sensor is high for small airflows, which require a high degree of accuracy. This system is used in the L-Jetronic. **Fig. F44** shows the principle of a ram pressure flow measurement.

~Rotameter principle. This is based on the principle that a floating device—for example, a sensor plate—is moved out of its resting position when air flows around it. The sensor plate is positioned in a Venturi nozzle, and its contour opens a certain cross section (**Fig. F45**).

The movement of the sensor plate is transferred through a lever system to a control piston, which allocates the required fuel mass. This type of air measurement is used for mechanical injection—that is, K-Jetronic.

Flow-through components →Calculation processes ~Application areas ~~Thermodynamics

Fluid circulation →Calculation processes ~Application areas

Fluid friction →Bearings ~Friction loss conditions

Fluid sound →Engine acoustics

Flywheel (also, →Balancing of masses; →Torsional vibrations). The flywheel is a mechanical energy storage system with which kinetic energy can be stored. The stored energy can be calculated from the moment of inertia and the angular velocity. Using a flywheel, the total moment of inertia of the crank drive and flywheel is increased. Therefore, the irregularity of the crankshaft rotational speed—produced by the torque

F

481

generated by the engine, which changes periodically across the four strokes—is limited. The smooth running of the engine is improved.

In addition, startup is facilitated and the control response of the engine is improved. The flywheel carries the starter crown and carries the clutch and/or parts of it.

A disadvantage of a high flywheel mass is that, in addition to the extra weight, because of the higher moment of inertia, the acceleration of the engine is slower.

~Dual-mass flywheel. In many cases a simple flywheel is not sufficient to reduce the cyclic variation. Torsional vibrations are continued via the transmission into the drivetrain. Disturbing snatching sounds can be created in the transmission, which can cause a drumming from the vehicle body.

To dampen these effects, a dual- or triple-mass flywheel is used. With the dual-mass flywheel, the total flywheel mass is divided into two partial masses. With this, the natural vibration of the drivetrain and the transmission is subject to the natural vibration from the engine. As a result, the engine and the transmission cannot excite the drivetrain to develop noise any longer, or at least it is excited at a reduced rate. Design and force distribution of the dual-mass flywheel can be seen in **Figs. F46 and F47**. In very critical cases, a triple-mass flywheel can help. Here, the total flywheel mass is divided into three partial masses.

~Electronic slip. Specific slip of the clutch (about 2%) can result in an improvement of the degree of irregularity of engine speed. The slip is created by an electronic control motor and is regulated using speed sensing on the inlet and outlet side of the flywheel. A disadvantage is a small increase in the fuel consumption. Therefore, electronic slip should only be used in critical map areas.

~Energy storage system. The flywheel acts as an energy storage system to dampen out the effects of the

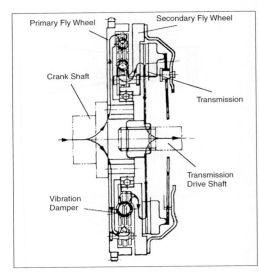

Fig. F47 Force distribution of the dual-mass flywheel

periodic variation of engine torque, which means that there are times in a working cycle when the engine produces work and others when it absorbs it. The engine speed would decrease drastically during times in which no work is performed. The flywheel as energy storage system results in a more apparent uniformity of torque output.

~Flywheel mass. The flywheel mass essentially determines the degree of cyclic variation of engine speed through its effect on the total moment of inertia. The moment of inertia, Φ, is proportional to mass, m, and the square of the distance, r, from the axis of rotation. For a cylindrical flywheel:

$$\Phi = \frac{1}{2}m \cdot r^2.$$

~Flywheel wobble. Both gas and inertia forces in the engine lead to bending stresses in the crankshaft. Additional bending stresses in the crankshaft are caused by flywheel wobble (**Fig. F48**).

~Triple-mass flywheel. *See above*, ~Dual-mass flywheel

Flywheel balancing →Balancing of masses ~Balancing

Flywheel bolt →Engine bolts ~Threaded connections

Flywheel class. The emission of pollutants and the fuel consumption of a vehicle are partly dependent on its acceleration and, thus, the vehicle mass. Because this must be considered when performing mandatory exhaust gas tests and driving programs, a flywheel

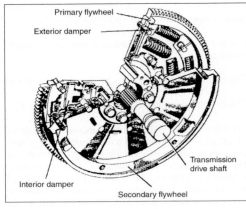

Fig. F46 Dual-mass flywheel

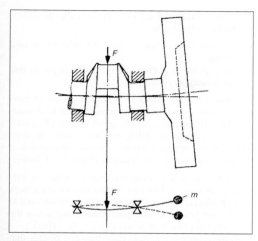

Fig. F48 Flywheel wobble (Source: F. Pischinger)

class is defined proportional to the vehicle mass. This flywheel class is set during operation on a roller dynamometer. Based on this, it is possible to adapt the translationally moved vehicle mass. Therefore, the flywheel class is chosen at which a total inertia of the equivalent "rotating mass" of the vehicle is reached, and this equals the reference mass of the vehicle according to the values indicated in **Fig. F49**. The reference mass of the vehicle consists of the empty weight and the defined payload.

Flywheel mass →Flywheel

Vehicle Reference Mass, Pr (kg)	Equivalent Flywheel Mass, I (kg)
Pr ≤ 480	455
480 < Pr ≤ 540	510
540 < Pr ≤ 595	570
595 < Pr ≤ 650	625
650 < Pr ≤ 710	680
710 < Pr ≤ 765	740
765 < Pr ≤ 850	800
850 < Pr ≤ 965	910
965 < Pr ≤ 1080	1020
1080 < Pr ≤ 1190	1130
1190 < Pr ≤ 1305	1250
1305 < Pr ≤ 1420	1360
1420 < Pr ≤ 1530	1470
1530 < Pr ≤ 1640	1590
1640 < Pr ≤ 1760	1700
1760 < Pr ≤ 1870	1810
1870 < Pr ≤ 1980	1930
1980 < Pr ≤ 2100	2040
2100 < Pr ≤ 2210	2150
2210 < Pr ≤ 2380	2270
2380 < Pr ≤ 2610	2270
2610 < Pr	2270

Fig. F49 Vehicle reference mass and equivalent flywheel mass

Flywheel wobble →Flywheel

FM process →Combustion process, gasoline engine ~MAN FM process

FMVSS 124 →Actuators ~E-gas

Foaming →Oil ~Properties and characteristic values ~~Foaming characteristics

Foaming characteristics →Oil ~ Properties and characteristic values

Follower →Friction ~Friction measuring techniques

FON (Front Octane Number) →Fuel, gasoline engine ~Front octane number

Forced circulation cooling →Cooling circuit

Forced piston oil cooling →Piston ~Cooling ~~Forced oil cooling

Ford PROCO process →Combustion process, gasoline engine ~ Stratified charge process ~~FPC-procedure

Forged camshaft →Camshaft ~Production

Forged crankshaft →Crankshaft ~Blank

Forged piston →Piston

Forged steel →Connecting rod ~Materials

Form matching capability →Piston ring

Formaldehyde. Formaldehydes—for example, HCHO—are part of the strong-smelling, toxic compounds in the exhaust gas of gasoline engines. This is why the US regulator, as an extension of the Clean Air Act, requests the introduction of reformatted gasoline to reduce the release of formaldehyde, benzene, and polycyclic organic compounds.

A limit of 0.015 g/mile of formaldehyde for TLEV (transitional low-emission vehicles) and LEV over an operating period of five years or 50,000 miles is set for clean-fuel cars. The same limits also are set for NLEV (US national low-emission vehicle) standards. Formaldehyde is not regulated for the US Federal Certification Exhaust Emission Standards.

For TLEV and LEV, 0.018 g/mile are valid for an operating period of 10 years or 100,000 miles. ULEV (ultralow-emission vehicles) values are between 0.008 and 0.011 g/mile, depending on the operating period.

Formula 1 (F1) →Engine ~Racing engines

Four valves →Valve arrangement ~Number of valves

Four-stroke engine. The change in volume of the combustion space in a four-stroke engine is alternately used for work and charge transfer. The processes are controlled by the intake and exhaust valves. One combustion cycle includes two crankshaft rotations consisting of four work or charge transfer processes: (1) induction, (2) compression, (3) expansion, (4) exhaust.

Friction. Frictional effects can basically be distinguished into two types:

- Components in contact moving relative to one another.
- Components in contact not moving relative to one another; this is called static friction.

Only systems in which the components are in contact and moving relative to one another will be considered here. It is possible to distinguish between frictional bodies in contact according to their relative motion. There are two relative motions of particular importance to engine operation, which can be defined as follows:

- Sliding, as a tangential traversing motion of surfaces in contact. This can involve frictional partners with relative motion from different directions and at different velocities. A typical case in engines is the motion of pistons or piston rings in the cylinder.
- Rolling, as a theoretical case between bodies that have point or linear contact with one another. In the process, at least one of the bodies executes a rotary motion about the axis of rotation lying in the contact area. Typical examples in engines are roller-type cam followers.

The components in contact are more or less separated by means of a lubricant. Several different frictional states are defined in this way:

- Solid friction, whereby the boundary layers of the materials in contact determine the friction
- Viscous friction, whereby a liquid substance between the frictional partners determines the friction
- Static friction, whereby the adhesive layers on the frictional partners determine the friction
- Mixed friction, in which all the above types occur side-by-side (solid, viscous, and static friction)

In addition to heat and charge-transfer losses, as well as leakages and process-related losses, friction is a very important component in the overall losses of an internal combustion engine. The increased requirements of maximizing overall engine efficiency make it increasingly necessary to reduce friction in the engine. All mechanical losses are usually combined under the general term "friction losses." A suitable variable for evaluating the so-defined frictional losses is the friction mean effective pressure. It is defined as the specific work that must be done in order to overcome mechanical losses or friction in an engine.

The variation of the friction mean effective pressure of gasoline and diesel engines is shown in **Fig. F50**.

The distribution of the components that are primarily responsible for mechanical losses in a gasoline engine are shown in **Fig. F51**. In addition to clear dependence of frictional losses on engine speed, it is recognizable that, based on maximum mean effective pressure at full throttle of about 10–12 bar, the friction mean effective pressure is about 10% of the brake mean effective pressure. In the part-load range, at high speeds, the relative share is significantly higher. Moreover, the piston group as well as engine bearings (connecting

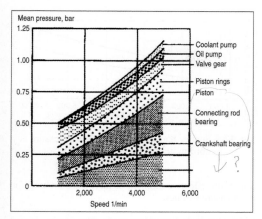

Fig. F50 Friction mean effective pressure of gasoline and diesel engine (Source: Pischinger)

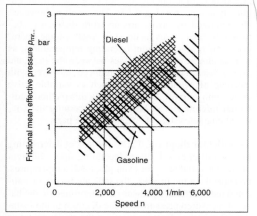

Fig. F51 Proportion of friction contributed by individual components (Source: Pischinger)

rod and crankshaft) contribute the greatest shares of losses.

As an example, **Fig. F52** shows the distribution of mechanical losses (friction) in a two-liter automotive direct injection diesel engine at full "throttle."

It is apparent that the piston assembly, consisting of compression rings, oil control ring, and piston, contribute the greatest share of mechanical losses.

If the losses caused by pistons and piston rings (**Fig. F53**), are further broken down, it can be seen that, particularly at low speeds, the oil control ring makes the greatest contribution to friction.

This low-speed effect occurs because hydrodynamic lubrication prevails more at high than at low speeds; conversely, the share of piston friction increases because of the higher inertial forces at high speed.

The following factors basically affect piston friction: the temperature of the cylinder liner and the oil; the piston clearance and piston skirt; the piston pin offset; the piston mass; and deformation of the cylinder liner.

~Friction loss →Fuel consumption ~Variables ~~Engine measures; →Power output

~Friction mean effective pressure (*also*, →Mean effective pressure). The friction mean effective pressure p_{mr} is defined as

$$p_{mr} = p_{mi} - p_{me},$$

where p_{mi} = indicated mean effective pressure, and p_{me} = brake mean effective pressure.

~Friction measuring techniques. The following processes are used to measure overall friction in the engine:

• Indexing
• Dragging, whereby it is possible to switch off individual cylinders or assemblies that contribute to friction loss

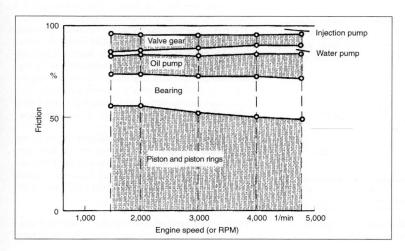

Fig. F52
Distribution of mechanical losses of a direct injection diesel engine (Source: Affenzeller/Glaeser)

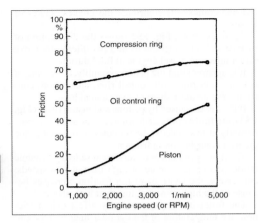

Fig. F53 Distribution of friction within the piston assembly (Source: Affenzeller/Glaeser)

- Coast-down tests of engines or assemblies
- Shut-down technique
- Willians line
- Stripping method

An overview of the results of the friction mean effective pressure, as determined with the individual methods, is shown in **Fig. F54**. It must be recognized that, based on the method applied, different elements are incorporated in the value defined as the friction mean effective pressure—for example, charge exchange losses are in some but not others.

~~**Coast-down tests.** In this case, the rate of deceleration is determined as a function of time after turning off the engine at a stable operating point. The frictional torque results from the Newton's equation of rotary motion:

$$M_r = J\frac{d\omega}{dt}$$

where J = total moment of inertia of engine rotating parts, ω = angular velocity, and t = time. Disadvantages of this method are that load dependencies cannot be determined; charge exchange losses and the varia-

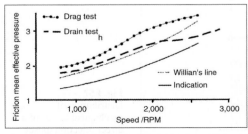

Fig. F54 Method of determining the friction mean effective pressure (Source: FVV)

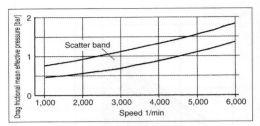

Fig. F55 Scatter band of mean effective pressure according to the drag method

tion of the effective moment of inertia are included in the measured values.

~~**Driving technique.** Determining the friction losses by driving the whole engine is a very common method. The motive power required to drive the engine is determined in unfired operation—for example, on the engine dynamometer (**Fig. F55**). The result is, however, characterized by errors arising from differences between the heat losses incurred in "real" engine operation, which give different operating temperatures, as well as charge exchange losses between the fired and driven cases. The latter can be reduced, however, with the help of suitable systems, such as variable valve drives with zero lift. Likewise, using this method, individual inferences can be drawn concerning the frictional characteristics of specific engine parts.

~~**Indicator diagram.** This is the task of determining the overall engine friction depending upon different operating variables—for example, load, temperatures, and speed. In the process, the indicated work or indicated mean effective pressure is determined on the basis of measured pressure-volume diagram in the cylinder. The brake effective work or the brake effective mean effective pressure is determined from data measured on the engine test rig. The friction mean effective pressure is defined as the difference between indicated and brake mean effective pressures.

The process is very cumbersome, because a pressure transducer must generally be installed inside the combustion chamber and many variables have an effect on the calculated result (e.g., exact determination of the top dead center, temperature dependence of the sensing element, and calibration of quartz pressure sensor). Indicator diagrams can be measured in both driven (cylinder switch-off) as well as in fired (cylinder firing) operation.

~~**Shut-down technique.** This method is used for multicylinder engines. By switching off the fuel supply to one cylinder, the latter is driven by the fired cylinders. By measuring the difference between torques, prior to and after fuel cut-off, the frictional loss can be determined.

~~**Strip-down method.** The engine is dismantled step-by-step on a drag dynamometer. From the difference

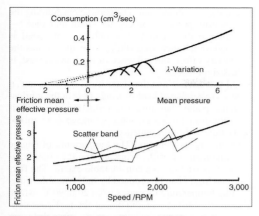

Fig. F56 Willian's line (Source: FVV)

between the measured values—that is, with and without the individual engine components—an inference is drawn concerning the friction of such components. The overall friction loss is the result of the adding together of all the losses incurred by individual components.

~~Willians line. The Willians line method is used for determining the friction mean effective pressure (**Fig. F56**). The fuel consumption is plotted at constant speed over a range of mean effective pressures (loads). If this curve is extended back to the intersection point with the negative mean effective pressure axis, the magnitude of the ordinate represents the friction mean effective pressure. This method produces only an indicative value for mean effective pressure and cannot be considered as an absolute one. The dependence on engine load cannot be depicted; likewise, the charge exchange losses are included in the measured value. The method is inaccurate—that is, affected by broadly scattered measurement results because the dependence of specific fuel consumption on the mean effective pressure is not linear and, as such, it is difficult to determine the exact tangent for projecting the curve.

~Friction reduction. The reduction of friction is of particular importance for modern engines, for which development is focused on minimizing the fuel consumption. Often, all possible mechanical losses are comprised by the term "friction," which includes plunging losses, oil shearing, and internal friction in operating fluids. As shown in **Fig. F51**, the piston assembly, bearings, and valve drives are the main sources of friction losses. Therefore, measures are taken to improve these components by focusing on their frictional characteristics. However, the application of, for example, roller bearings for reducing friction in the crankshaft bearings has not been successful with the four-stroke engines. The primary focuses for reducing friction loss are the optimization of material pairs and coatings of sliding partners, effective lubrication systems, optimized engine oils, quick attainment of the

operating temperature, and the design of components—for example, roller-type cam followers instead of bucket tappets.

Also, generally, lowering the engine speed also leads to a reduction of friction losses.

~Frictional energy. The frictional energy is the difference between the indicated and brake powers. Depending on the definition used, it also includes the power required by the ancillary equipment, in additon to the friction losses. With the help of the definition for friction mean effective pressure, p_{mr}, the friction loss, P_r, can be calculated from

$$P_r = i \cdot n \cdot p_{mr} \cdot V_H,$$

where i = number of work cycles per revolution, n = engine speed, p_{mr} = friction mean effective pressure, and V_H = displacement volume.

~Valve gear friction. There is a series of contradictory requirements with respect to the efficiency and optimum design of valve gears. For instance, two such contradictory requirements are the possibility of attaining a high volumetric efficiency—for example, by extreme valve timing and the resulting high acceleration, high force levels and surface pressures in the cam-follower pair—and the pursuit of minimum friction. It is necessary to minimize valve gear losses because their share in the overall engine friction increases, in particular at low speeds. The starting point for the analysis of the loss characteristics of valve gear is determining the torque curve on the drive wheel of the camshaft(s). An example of such a torque curve is shown in **Fig. F57**.

Fig. F58 shows, for example, the shares of valve gear components in the friction mean effective pressure. **Figs. F59 and F60** give an overview of the frictional characteristics of camshaft drives with two and four valves or the difference when different valve-operating elements are used.

~Variables (*also*, →Oil). Essential variables for friction characteristic are determined by:

* Break-in status of the engine. Break-in describes the process of matching the friction partners, thus eliminating surface unevenness. Wear takes place and is accompanied by an increase in friction loss.

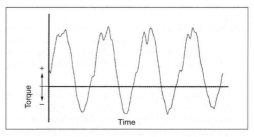

Fig. F57 The torque band of valve gear (unfired)

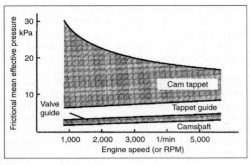

Fig. F58 Variation of friction mean effective pressure of valve gear with bucket tappets against speed, for a gasoline engine, at an oil temperature of 100°C (Source: Affenzeller, Glaeser)

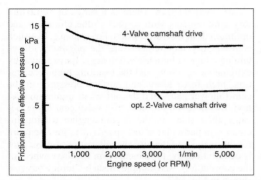

Fig. F59 Friction mean effective pressure of optimized four-valve and two-valve versions with roller tappets for a gasoline engine. (Source: Affenzeller, Glaeser)

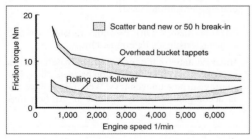

Fig. F60 Friction torque curve of the valve gear with roller-type cam followers and bucket tappets, for a gasoline engine, at oil temperature of 120°C (Source: Affenzeller, Glaeser)

Depending on the friction partners, an engine is broken in after 20 to 30 hours. After break-in, a more or less constant level of friction sets in until further wear causes the durability limit of the components to be reached.

- Engine operating point. The parameters defining the operating point of the engine—load and speed—influence the friction. The influence of speed is based on the increased sliding velocity and has the primary effect; the influence of load, in contrast, is less significant.
- Oil viscosity. When the viscosity of the oil is low, it leads to a reduced load-bearing effect at the lubrication gap and, therefore, to a reduction of the lubrication-film thickness, with an increase in mixed friction as the consequence.
- Operating temperature. The temperature of engine components or of the lubricant and the coolant influences friction to a considerable extent. The frictional losses can be reduced by half when the operating temperature increases from 20 to 90°C, which leads to substantial reduction of the fuel consumption between the operating points after the start/warm-up phase and when the engine has attained its operating temperature.

Literature: J. Affenzeller, H. Glaeser: Lagerung und Schmierung von Verbrennungsmotoren, Die Verbrennungskraftmaschine; New Series Volume 8, Springer Publishers, 1996. — R. Pischinger, G. Kraßnig, G. Taucar, T. Sams: Thermodynamik der Verbrennungskraftmaschine, Die Verbrennungskraftmaschine, New Series Volume 5, Springer Publishers, 1989. — F. Koch, F. Haubner, M. Schwaderlapp: Thermomanagement beim DI Ottomotor—Wege zur Verkürzung des Warmlaufs, 22nd International Vienna Engine Symposium, Vienna, 26.04.–27.04.2000. — F. Koch, F.-G. Hermsen, H. Marckwardt, F.-G. Haubner: Friction Losses of Combustion Engines—Measurements, Analysis and Optimization. Internal Combustion Engines Experiments and Modeling, Capri, Italy 15–18.09.1999 — A. Haas: Aufteilung der Triebwerksverluste am schnelllaufenden Verbrennungsmotor mittels eines neuen Messverfahrens, Dissertation RWTH Aachen University, 1987. — R. van Basshuysen, F. Schäfer: Handbuch Verbrennungsmotor, Wiesbaden, Vieweg Publishers, 2002. — S. Pischinger: Lecture Reprint Verbrennungsmotoren, 21st Edition, 2000.

Friction damper →Torsional vibrations

Friction loss →Friction ~Friction reduction; →Fuel consumption ~ Variables ~~Engine measures; →Power output

Friction loss conditions →Bearings

Friction mean effective-pressure →Friction; →Mean effective pressure

Friction reduction →Friction

Front octane number →Fuel, gasoline engine

Front pipe →Exhaust system

FTP-75 →Emission measurements

Fuel →Fuel, diesel engine; →Fuel, gasoline engine

Fuel, diesel engine. Fuels for diesel engines are mixtures of hydrocarbons. Different percentages of various hydrocarbons in the fuel result in different effects on the operating characteristics (power, mileage, acoustics, emissions).

~Additives. Additives should not be used by consumers because they could interfere with the effect of the additives used by the manufacturer, which are matched precisely to the specific fuel. Moreover, incompatibilities between the various substances are possible, resulting in filter problems or counteracting the positive characteristics of individual constituents.

In the face of unsatisfactory low-temperature behavior, it was necessary in the past to add gasoline (preferably regular gasoline) to the diesel fuel or, even better, kerosene, to dilute the paraffins. However, the addition of regular gasoline entails some disadvantages. If mixing is done outside the vehicle tank, it is necessary to ensure that storage and transport are accomplished according to the hazard class A1 specified for gasoline. The cetane number of the mixture decreases. This makes cold start more difficult and increases the pollutant emissions. The viscosity drops and the wear protection decreases. Bubbles forming in the fuel system may lead to rough engine operation.

For these reasons, gasoline should not be added, particularly because the low-temperature behavior of the diesel fuel is adapted throughout Europe to the anticipated temperatures depending on the season and region.

Possible additives that could be used by customers after conferring with the fuel supplier include biocides — use of which can be advantageous following infection of storage tanks by microbes, to prevent re-infection after cleaning and drying the tank.

Additives for diesel fuels are usually in the ppm concentration range and significantly improve the fuel properties and behavior of the fuels. **Fig. F61** shows the most important commercial additives and their effects.

Ignition accelerators allow a cost-effective increase in the cetane number with a corresponding positive effect on the combustion and exhaust emissions (**Fig. F62**). The effective ingredients are organic nitrates; in particular, ethylhexyl nitrate is used as an ignition accelerator in commercial applications. Lubricity additives (to improve the lubricating capability) provide the required wear protection — for example, in distributor injection pumps and pump/nozzle systems when desulphurized fuels are used. Experience with additives for aviation turbine fuels and extremely desulphurized diesel fuels in Sweden has ensured that suitable additives are available in Europe for introduction of diesel fuels containing low quantities of sulfur (<500 ppm of sulfur; **Fig. F63**).

Residues in the injection nozzles (particularly on throttling pintle nozzles) resulting from carbonizing of the fuels can lead to a delay in the start of injection in the pilot injection phase, reducing the quantity of fuel injected at the beginning of combustion. This leads to a steeper increase in the pressure with higher noise and exhaust emissions. Diesel fuel detergents can significantly reduce such residues so that the carbon deposits on the nozzles and, therefore, the fuel consumption and emissions, remain within acceptable limits (**Figs. F64 and F65**). A series of substances is suitable for use as detergents in diesel fuels such as amines, imidazolines, amides, succinimides, polyalkyl succinimides, and amines as well as polyether amines.

Additives also improve the efficiency and compensate for the various, frequently mutually contradictory, requirements. For example, flow improvers and wax antisettling additives in winter allow the use of paraf-

Diesel fuel additive constituents	Active ingredient		Improvement of
Ignition accelerators	Organic nitrates, e.g. ethyl-hexyl nitrate		Fuel cetane number
			Engine cold start, exhaust white smoke, engine noise, exhaust emissions, fuel consumption
Detergents	Protection of fuel system (supply network and vehicle) e.g. amines, amides, succinimides, polyalkyl-succinimides, polyetheramines		Nozzle cleanliness
Flow improvers	Ethyl-vinyl-acetates		Vehicle operating reliability at low temperatures, use of paraffin constituents with high cetane number
Wax anti-settling additive	Alkyl-acryl-amides		Fuel storage at low temperatures
Lubricity improver	Fatty acid derivates		Injection pump wear, particularly for hydrogen-treated, low-sulfur fuels
Antifoam additive	Silicone oils		Convenient fueling (less overflow)
Corrosion protection additives	Allyl succinic acid esters or amine salts of alkenic succinimide acids		Protection of fuel system (Supply network and vehicles)

Fig. F61 Overview of diesel fuel additives, effective agents, and their influence on engine behavior

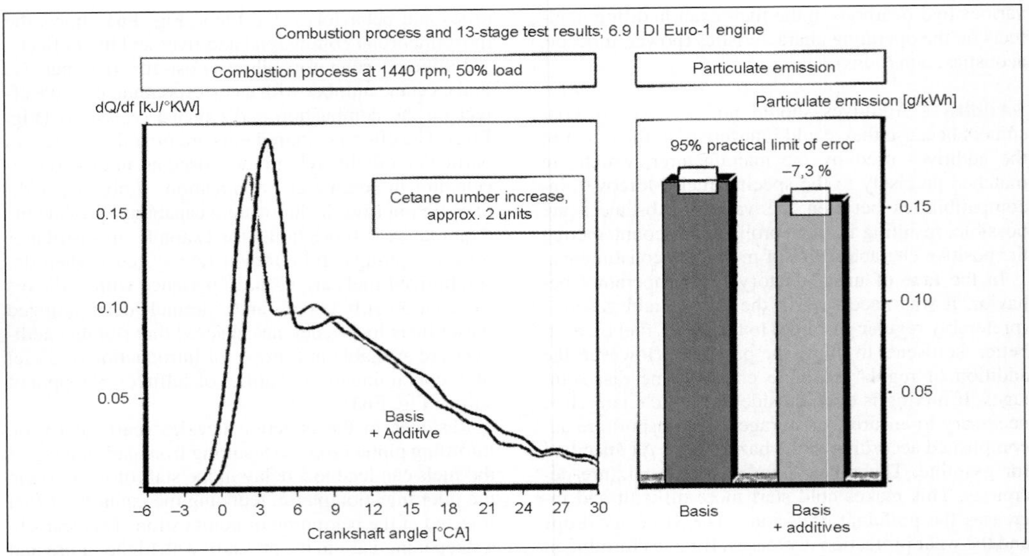

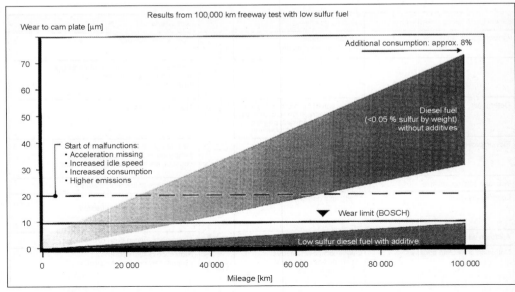

Fig. F62 Effect on combustion and particulate emissions of using ignition accelerators to increase cetane number

finic constituents with high cetane number but limited low-temperature behavior. Flow improvers cannot completely prevent formation of paraffin crystals; however, they can reduce their size and prevent agglomeration.

Typical of the compounds used as improvers are ethylvinyl acetate (EVA). Reduction of the CFPP (cold filter plugging point) by more than 10°C below the cloud point is generally only possible with flow improvers. A further reduction of the CFPP is possible by additional use of wax antisettling additives (WASA). This simultaneously prevents sedimentation of paraffin crystals during long-term storage below the fuel cloud point.

Corrosion protection additives provide sufficient protection for the fuel system even under critical conditions such as long vehicle standstill periods with formation of condensed water. Polar molecular groups of

Fig. F63 Lubricity additives reduce the wear to the injection pump with low-sulfur fuels

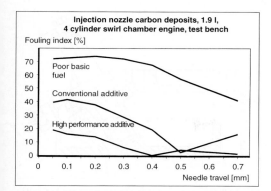

Fig. F64 Detergents reduce the residues on the injection nozzles

esters or alkenic succinimide acids form a monomolecular protective coating on the metal surfaces and prevent contact with water and acids.

Antifoaming agents reduce the foam when fueling and thereby shorten the time required to fill the tank and reduce overflowing and, therefore, contamination of the ground with diesel fuel.

Other additives, used occasionally and as required, include anti-oxidants for stabilization of the cracked constituents, biocides for destruction of micro-organisms in storage tanks, antistatic additives for improving electrical conductivity, "pipeline drag reducers" for reducing the pumping power required for transporting the fuel to pipelines, and deodorants for covering up the unpleasant odor of diesel fuel. The latter have become less important because reduction of the sulfur content in diesel fuels has resulted in reduction of the odor.

All additives currently in commercial use are based on organic compounds—that is, they burn to form water vapor, CO_2, and nitrogen. The only exceptions are antifoaming agents, whose effective substance, silicone oil, gives traces of silicon compounds in the exhaust gas. Organo-metallic additives such as those based on barium, calcium, or iron, use of which would

be possible for reduction of the soot and particulate emissions, are not used commercially because of undesirable side effects on the environment, incompatibility with exhaust treatment systems, and possible exhaust toxicity.

~Alcohols. *See below*, ~Alternative fuels

~Aldehydes. *See below*, ~Alternative fuels ~~Vegetable oil

~Alternative fuels. Alternative fuels are fuels not produced from a petroleum base under usual refinery conditions. The most important types are alcohols, vegetable oils, and ethers.

~~Alcohols. Alcohols (methanol and ethanol) can also be used as alternative fuels in diesel engines. However, compared to conventional diesel fuel, they have basic disadvantages (e.g., poor ignition performance, high latent heat of evaporation, low lubricating effect, high volatility, and a high tendency to promote corrosion) which require significant modifications to the engine and the fuel system concepts.

In all cases, adaptation of the injection system to the low volumetric energy (heat) of reaction of alcohol—that is, an increase in the volume injected—is required to obtain a maximum performance comparable to that achieved using diesel fuel. Moreover, adaptation of the fuel or engine is required because of the poor tendency to self-ignite. A comparison of energies of reaction and cetane numbers is shown in **Fig. F66** for alcohols.

On the otherwise unmodified engine, an increase in the ignition performance is achieved by adding organic nitrates. While concentrations of far below 1% are common for diesel fuels, concentrations of commercially available ethylhexyl nitrate of greater than 5% are required for ethanol and concentrations greater than 10% for methanol. More effective nitrates such as diethylene glycol dinitrate cannot be used for safety reasons because they are explosive in the unmixed state or after evaporation of the alcohol.

Another possibility for using alcohols in diesel engines is to use a mixture of alcohol and diesel fuel. Because methanol and ethanol are practically not miscible with diesel fuel at ambient temperatures, this concept requires the simultaneous use of large quantities of solubilizers—for example, ethyl acetate (**Fig. F67**). The use of ignition accelerators, as well as solubilizers, reduces efficiency.

Adaptation of diesel engines for operation with pure alcohol requires use of a partial operating second

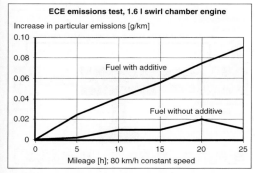

Fig. F65 Detergents prevent an increase in the particulate emissions at increasing mileage

		Diesel fuel	Methanol	Ethanol
Calorific value H_u	[MJ/kg]	approx. 42.5	19.66	26.77
Calorific value H_u	[MJ/l]	approx. 35.7	15.63	21.25
Ignition quality (cetane number)		approx. 50	3	8

Fig. F66 Energies of reaction and cetane numbers of alcohols in comparison to diesel fuel

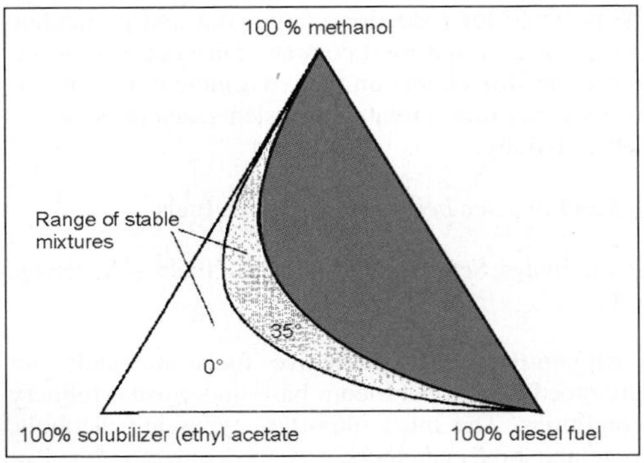

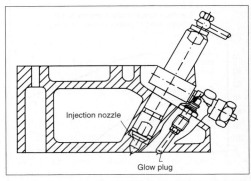

Fig. F67 Three-phase solubility diagram for alcohol/diesel fuel with ethyl acetate as the solubilizer

Fig. F69 Cross section through cylinder head of a diesel engine with injection nozzle and glow plug for operation with methanol

injection system for dual fuel operation—that is, diesel fuel for cold starting, warm-up and idle operation, as well as alcohol to an increasing extent at increasing load and speed (**Fig. F68**) or use of glow plugs (**Fig. F69**) or spark plugs.

The advantage of using alcohols in diesel engines is particularly a reduction of the soot (particulates) and NOx emissions.

The concepts mentioned, as well as others, have been tested technically. They are not, however, economical with the present cost structure and taxation.

~~**Biological fuels (Biofuels).** These are fuels from renewable raw material resources. Ideally, biofuels emit only as much carbon dioxide (CO_2) when burned as they took from the atmosphere while growing—for example, this would allow a closed CO_2 "cycle" which did not increase the CO_2 concentration in the atmosphere (**Fig. F70**). In practice, since it is necessary to

use other types of energy for the production of the plants and their harvesting, transport, and preparation, only a gradual CO_2 reduction is possible (**Fig. F71**). However, the CO_2 balance can also be negative (e.g., ethanol produced from potatoes and grain). Of the practical alternatives in Germany—ethanol, methanol, and vegetable (rapeseed) oil ester—the last represents the best alternative. In terms of cost, other alternatives for reducing the CO_2 emissions (greenhouse effect) (e.g., thermal insulation, production of energy using

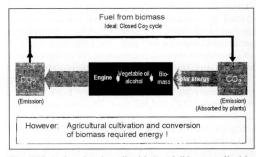

Fig. F70 A closed carbon dioxide "cycle" is not realizable

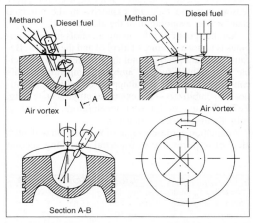

Fig. F68 Combustion chambers for "dual fuel" concepts with injection nozzles arranged for variable mixed operation with methanol diesel fuel

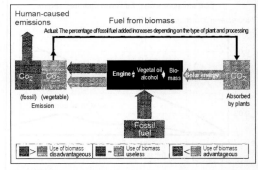

Fig. F71 Agriculture, harvesting, and processing require additional energy

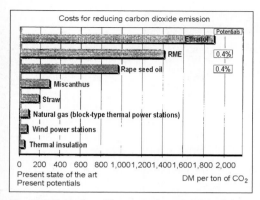

Costs for reducing carbon dioxide emission

0   200 400 600 800 1,000 1,200 1,400 1,600 1,800 2,000
Present state of the art
Present potentials   DM per ton of CO_2

Fig. F72 Fuels from biomasses for reduction of CO_2 are very cost intensive

wind turbines) are considerably more effective at present (**Fig. F72**).

The suitability of biofuels for engine operation has been proven in comprehensive research programs. However, adaptation of the fuel systems and the engines as well as the engine oils used is required in most cases. The expenditure required for design modifications on engines to accommodate biofuels and their derivates increases with increasing demand for low exhaust emissions.

The EU directive "for promotion of the use of biofuels and other replenishable fuels in the motor vehicle sector" requires attempting to define a minimum percentage of biofuels in the total fuel consumption.

Indicative reference values are 2% for 2005 and 5.75% for 2010. The actions taken for promotion, resources used, and percentages of the market reached are to be reported annually to the commission. The European Commission reports to the European Parliament and the Council, every two years enclosing an evaluation of the cost effectiveness, the life cycle prospects including the influence on reducing CO_2 emissions, as well as the economical aspects and environmental effects of further cultivation. In Germany, plans currently provide for reduction of mineral oil taxes on fuels proven to contain biofuels. Vegetable oil methylester is particularly interesting as a mixing constituent for diesel fuels. Pure vegetable oil methylester has already established itself on the market.

~~**Ethyl alcohol (ethanol).** *See above,* ~~Alcohols

~~**Fatty acid methylester.** FAME is produced by transesterification of vegetable oils with methanol (CH_3OH; **Fig. F73**). Animal fats can also be used as the initial product. The intrinsic low-temperature characteristics and viscosity of the vegetable oils and their thermal stability are improved by transesterification. In addition, undesirable byproducts are removed. For this reason, fatty acid methylester is significantly better as a fuel for diesel engines than pure vegetable oils. The required properties of methylester are listed in the standard draft prEn 14214. It is particularly important to maintain the limits for glycerin and glycerides to ensure acceptable elastomer compatibility (**Fig. F74**).

The most familiar product in this group is rapeseed oil methylester (RME; **Fig. F75**). Additional energy is required for the transesterification process, and, in comparison to rapeseed oil, for which approximately 65% of the energy contained in the product is required for its preparation, this means that RME is even poorer with approximately 77% being needed. This value could be improved to approximately 30% by use of the theoretically imaginable energy in all byproducts (meal, glycerin, straw; **Fig. F76**).

RME requires significant subsidies to enable it to be offered at filling stations at the same price as diesel fuel. Naturally, this difference in the cost decreases as

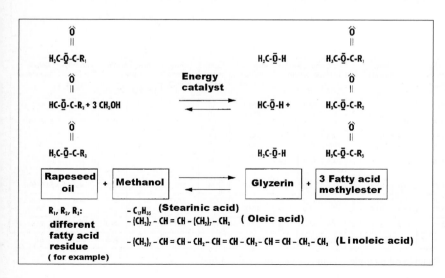

Fig. F73 Transesterification of a vegetable oil to obtain the corresponding fatty acid methylester

493

Properties	Units	Limits min.	Limits max.	Test procedure
Ester content	% m/m	96.5		EN 14103
Density at 15 °C	kg/m^3	0.860	0.900	EN ISO 3675 EN ISO 12185
Viscosity at 40 °C	mm^2/s	3.5	5.0	EN ISO 3104
Flash point	°C	120	–	pr EN ISO 3679
Limit for filterability1) 15. 04 to 30. 09, class B 01. 10 to 15. 11, class D 16. 11 to 28. 02, class F (in leap years, until 29. 02) 01. 03 to 14. 04, class D	°C °C °C °C		0 –10 –20$^{2)}$ –10	EN 116
Sulfur content	mg/kg		10	pr EN ISO 20846 pr EN ISO 20884
Carbon residue (percentage by weight) of 10% distillation residue	%		0.30	EN ISO 10370
Ignition quality (cetane number)	–	51		EN ISO 5165
Ash content (sulfate ash)	% m/m		0.02	ISO 3987
Water content	mg/kg		500	EN ISO 12937
Total contamination	mg/kg		24	EN 12662
Corrosion effect on copper (3 h at 50 °C)	Degree of corrosion		1	EN ISO 2160
Oxidation stability 110 °C	Hours	6.0	–	EN 14112
Acid number	mg KOH/g		0.5	EN 14104
Iodine index	g iodine/100g		120	EN 14111
Percentage of linolenic acid methyl ester	% m/m		12	EN 14103
Percentage of fatty acid methyl ester with more than 4 double bunds	% m/m		1	
Methanol content	% m/m		0.2	EN 14110
Monoglyceride content Diglyceride content Triglyceride content Percentage of free glycerin	% m/m % m/m % m/m % m/m		0.80 0.20 0.20 0.020	EN 14105 EN 14105 EN 14105 EN 14105 EN 14106
Percentage of total glycerin	% m/m		0.25	EN 14105
Percentage of alkali metals (Na + K)	mg/kg		5.0	EN 14108 EN 14109
Percentage of alkaline earth metals (Ca + Mg)	mg/kg		5.0	pr EN 14538
Phosphorous content	mg/kg		10	pr EN 14107

$^{1)}$ Valid values for Germany from table for temperate climate

Fig. F74
Minimum requirements placed on methylesters from vegetable and animal oils (FAME) for use as diesel fuel in conformance with EN 14214

Methyl Ester from Fatty Acids in	Number of Double Bonds	Total Formula			Composition (% by weight)			Molecular Weight (g/Mlol)	Iodine Index (g/l)	Density at 15°C (kg/m^3)	Lower Calorific Value in		Viscosity at 40°C (mm^2/s)	Cetane Number
		C	H	O	C	H	O				(MJ/kg)	(MJ/l)		
Coconut oil	0.08	13.4	26.7	2	73.3	12.2	14.5	220.4	10	872	35.3	30.8	2.7	62.7
Palm nut oil	0.13	13.9	27.7	2	73.7	12.3	14.1	227.6	15	–	35.5	–	–	–
Palm oil	0.59	18.0	34.9	2	76.3	12.4	11.3	283.7	52	872–877	37.0	32.4	4.3-4.5	64.3–70.0
Peanut oil	1.10	19.0	35.0	2	77.0	12.2	10.8	296.3	94	–	37.2	–	–	–
Cotton seed oil	1.20	18.5	34.6	2	76.9	12.1	11.1	289.2	105	–	37.0	–	–	–
Rape seed oil (rich in erucic acid)	1.25	21.0	39.5	2	77.8	12.3	9.9	324.2	98	–	37.7	–	–	–
Rape seed oil (low quantity of erucic acid)	1.33	19.0	35.4	2	77.2	12.0	10.8	296.0	114	882	37.2	32.8	4.2	51.0–59.7
Sun flower seed oil	1.49	18.9	34.8	2	77.2	11.9	10.9	294.1	129	885	37.1	32.8	4.0	61.2
Soja beam oil	1.51	18.8	34.6	2	77.2	11.9	10.9	293.0	131	–	37.1	–	–	56
Linseed oil	2.12	18.9	33.6	2	77.5	11.6	10.9	292.1	183	891	37.0	33.0	3.7	52.5
Diesel fuel, typical	–	1	1.85	0	86.6	13.4	0	120–320	–	830–840	42.7	35.5	2-3.5	51

Fig. F75 Data on vegetable oil methylesters in comparison to diesel fuel

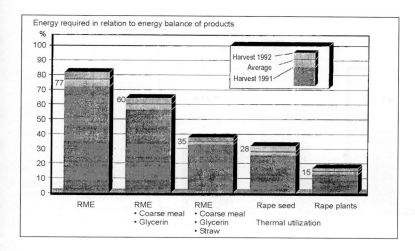

Fig. F76
Energy required to
produce RME and
byproducts for use as
energy

the price for crude oil increases. The EU commission has specified the additional costs for "biodiesel" based on rapeseed oil at 300/1000 liters at a crude oil price of $25/barrel, while at a crude oil price of $35/barrel, the additional costs would be 200/1000 liters. The specified values assume optimum production of RME in large plants, include the additional consumption of RME in the engine, and assume a typical price level for the byproducts such as glycerin.

Used in vehicles adapted for it (elastomer compatibility), RME has lower particulate, PAH, HC, and CO emissions; however, it has a tendency to higher levels of NO_x and aldehyde emissions, together with higher volumetric fuel consumption and lower engine performance (**Fig. F77**).

~~Hydrogen →Fuel, gasoline engine ~Alternative fuels

~~Liquefied petroleum gas (LPG). *See below,* ~Liquefied petroleum gas (LPG). *Also,* →Combustion process, diesel engine

~~Methyl alcohol (methanol). *See above,* ~~Alcohols

~~Natural gas. *See below,* ~Natural gas. *Also,* →Combustion process, diesel engine

~~Rapeseed oil. *See below,* ~~Vegetable oil

~~RME. *See below,* ~~Vegetable oil

~~Vegetable oil. On the basis of its physical data, vegetable oil is suitable for use as a fuel in diesel engines as a matter of principle. However, the availability is limited by the areas available for agriculture. The high fuel costs in comparison to diesel fuel and the competition between fuel and foodstuffs allow substitution of diesel fuel with vegetable oils to a limited extent only.

Technical problems are posed particularly by the high viscosity of vegetable oils, which, in addition to making operation at low temperatures more difficult, also produces problems for atomization of the injected fuel. This leads to poorer combustion and higher exhaust emissions; this is seen particularly on high-speed (passenger car) diesel engines. Other problems result from lack of stability, natural impurities, contamination by fungus and bacteria and, outside of tropical and subtropical regions, poor resistance to low temperatures. These limitations also apply when adding small concentrations of vegetable oils to diesel fuel. Some useful experiential data have been gathered with commercial utilization of coconut oil in Asia.

Studies in Germany, at the request of the Federal Ministry for Research and Technology, have shown that the diesel engines used widely and available on the market are not suitable for use with the rapeseed oil available in Germany. In addition, measurements of the exhaust emissions from a four-cylinder passenger car with swirl chamber engine showed that with the exception of NOx emissions, all legally limited pollutants as well as the emission of aldehydes and polycyclic aromatic hydrocarbons increase with rapeseed oil (**Figs. F78 and F79**).

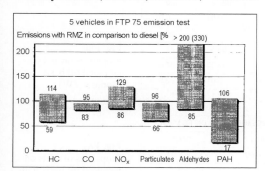

Fig. F77 Comparison of pollutant emissions resulting from engine operation with rapeseed oil methylester or diesel fuel

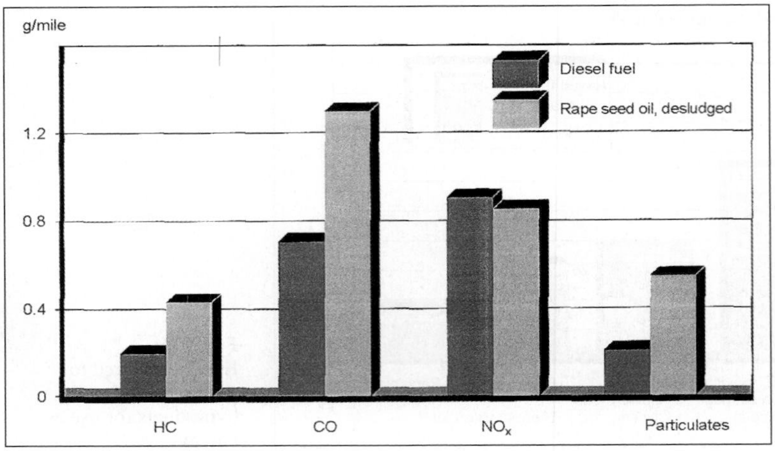

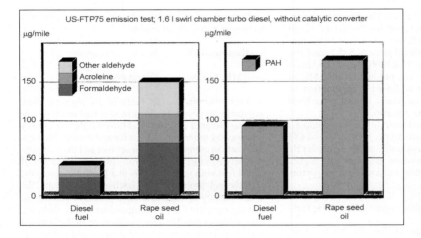

Fig. F78
Rapeseed oil reduces the NOx emission slightly but significantly increases the other pollutants

Fig. F79
Aldehyde emissions and emissions of polycyclic aromatic hydrocarbons (PAHCs) with diesel fuel and pure rapeseed oil

Parameter	Unit	Pure Diesel Fuel	DK-R10	DK-R20	DK-R30	Pure Rape Seed Oil
Density at 15°C	kg/m³	841.5	835.7	830.5	824.9	920
Sulfur content	g/100g	0.19	0.13	0.09	0.04	0.01
CEPP	°C	−9	−7	−5	−2	16
Cetane number	—	54.5	59	63	66.5	41
Calorific value H_u	MJ/kg	42.82	42.98	42.84	43.23	37.4
Viscosity 20°C	mm²/s	4.90	4.99	5.01	5.01	73.5

Fig. F80
Properties of fuel on rapeseed oil basis (Source: VW)

Parameter	Unit	DK-R10/MDE	DK-R20/MDE	DK-R30/MDE
Density at 15°C	kg/m³	836.7	832.1	827.5
Sulfur content	g/100g	0.13	0.09	0.04
CEPP	°C	−5	−4	−2
Cetane number	—	58	63	69
Calorific value H_u	MJ/kg	42.92	43.06	43.11

Fig. F81
Properties of fuels on rapeseed oil basis with transformation through refinery process (Source: VW)

In addition, there is also the familiar unpleasant odor of the exhaust gases, which, however, can be reduced with the use of catalytic converters. For the reasons mentioned, conversion of vegetable oils is generally required before they can be used as fuel. This can be accomplished either by transesterification (vegetable oil methylesters) or by hydrocracking at a refinery after mixing with hydrocarbon refinery products.

Fig. F80 shows the properties of fuel rapeseed oil and diesel fuel as well as three diesel fuels produced by adding rapeseed oil to the distillate gas oil followed by hydration in the hydrocracker. The figures for mixed fuels refer to the percentage originating from rapeseed oil in the final product.

Rapeseed oil can also be added in the middle distillate desulfurization plant in an analogous manner. **Fig. F81** shows the properties of such fuels, where the figures in the fuel designation represent the quantity originating from rapeseed oil in the final product.

~**Aromatic content.** See below, ~Composition

~**Aromatics.** See below, ~Composition

~**Ash content.** See below, ~Properties

~**Biogas.** This is obtained from the digestion tanks in sewage treatment plants and is suitable, with reservations, for driving gas engines. The primary constituent is methane CH_4 (approximately 60%), the remainder is CO_2. The byproducts are oxygen, O_2, as well as hydrogen sulfide H_2S. To avoid engine corrosion damage, the percentage of H_2S should not be greater than 0.15% and the total quantity of sulfur should not be greater than approximately 2.000 mg/m^3. The usual treatment of sewage gas is limited to drying and precipitation of the condensate.

In all cases, careful adaptation is required to match the lubricating oil (gas engine oils) to the quality of the biogas used.

~**Biological fuel.** See above, ~Alternative fuels

~**Boiling characteristics.** See below, ~Requirements

~**Boiling curve.** See below, ~Properties

~**Carbon residue.** See below, ~Properties

~**Cetane index.** See below, ~Properties

~**Cetane number.** See below, ~Properties

~**Cetane number improver.** Organic nitrates are particularly suitable for increasing the ignition performance of diesel fuels. However, relatively good results have been achieved in tests with peroxides. In addition to good efficiency and low production costs, the safe storage and handling of ignition accelerators in concentrated form (i.e., before mixing with the fuel) is an

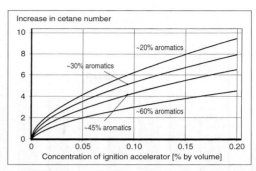

Fig. F82 Ignition accelerators have a greater effect in fuels containing low quantities of aromatics

important criterion for commercial use. For this practical reason, peroxides and a number of highly effective organic nitrates cannot be used. The use of EHN (ethylhexyl nitrate) has proved suitable in common commercial fuels. Typical concentrations are less than 0.1% and usually between 0.01 and 0.1%.

Paraffinic fuels react better than aromatic fuels to the addition of ignition accelerators. The cetane number enhancing effect decreases when the added quantity of improver decreases (**Fig. F82**). At a concentration of 0.05%, the ignition performance of common commercial fuels is improved by approximately two to four cetane numbers (**Fig. F83**).

Fuels in which the cetane numbers have been improved by an ignition accelerator show approximately the same positive effect on the engine behavior and particularly the HC and CO emissions as fuels in which an increase in the cetane number has been achieved with refinery constituents (**Figs. F84 and F85**).

~**CFPP.** See below, ~Properties

~**Cloud point.** See below, ~Properties

~**Cold filter plugging point.** See below, ~Properties

~**Cold start.** See below, ~Properties

~**Composition.** The composition of a fuel determines its primary characteristics. Commercially available diesel fuels consist of a mixture of various hydrocarbons.

~~**Aromatic content.** See below, ~~Aromatics

~~**Aromatics.** Aromatics are ring-shaped hydrocarbon compounds in which the carbon atoms are alternately connected by double bonds, such as in benzene (a compound not present in diesel fuel because of its low boiling point of 80°C).

It is common to differentiate between mono-aromatics, di-aromatics, tri-aromatics, and tri-plus–aromatics according to the number of aromatic ring systems. **Fig. F86** shows a number of examples relevant to diesel fuel.

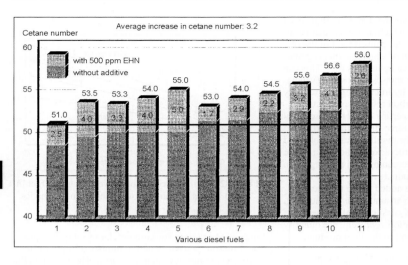

Fig. F83
Increase in cetane
number of standard
commercial diesel fuels
with ignition accelerator
(ethylhexyl nitrate)

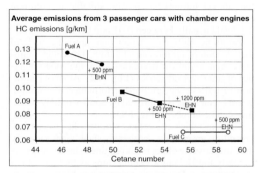

Fig. F84 Increase in the cetane number with EHN ignition accelerator leads to lower hydrocarbon emissions

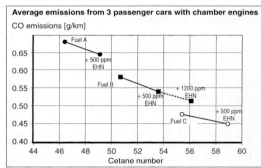

Fig. F85 Increase in the cetane number with ignition accelerator leads to lower carbon monoxide emissions

Because of the boiling range of approximately 180–370°C, diesel fuel contains a variety of highly differing aromatic compounds. The aromatic content is measured with the aid of high-pressure liquid chromatography (HPLC). This analysis technique takes into consideration the entire molecule containing the single or multiple aromatic ring compounds and includes the side chains. The method developed especially for diesel fuels is described by the British Institute of Petroleum under the designation IP 391 ("backflushing"). This method categorizes the concentrations of mono-, di-, tri-, tri-plus–, and higher polyaromatics into groups. Backflushing—that is, reversal of the flow direction at the end of analysis—is used particularly to determine the percentage of higher polyaromatics (tri+) more precisely. Exact analysis of the individual aromatics in diesel fuel, in contrast to gasoline, is possible only with extremely complicated procedures because of the many compounds that may be present. Conventional diesel fuels contain primarily mono-aromatics with variously located side chains (approximately 15–25%

by weight), the di-aromatics are less than approximately 5% by weight, and a total of primarily tri-aromatics and higher polyaromatics with percentages of usually less than 1% by weight.

The characteristics of mono-aromatics with paraffin side chains vary. While aromatic characteristics (e.g., high solubility of other products and low cetane number) dominate for molecules with relatively short side chains, molecules with long side chains behave more as paraffins. In order to study the effects of aromatics on factors such as exhaust emissions, the requirement was adapted: rather than studying the molecules with aromatic constituents, only the percentage of cyclic hydrocarbons contained in diesel fuel should be analytically determined. (In contrast to the values listed above, the cyclic compounds including the side chains are categorized as aromatics.) For research purposes, other methods are available, including nuclear magnetic resonance (NMR) spectroscopy, which is best suited for determining the cyclic hydrocarbons actually contained in a fuel.

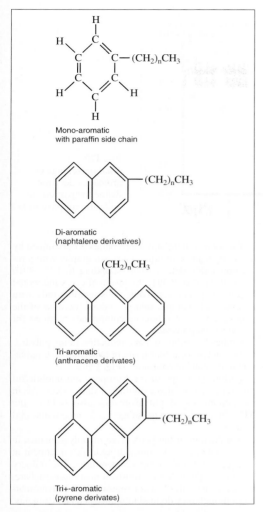

Mono-aromatic
with paraffin side chain

Di-aromatic
(naphtalene derivatives)

Tri-aromatic
(anthracene derivates)

Tri+-aromatic
(pyrene derivates)

Fig. F86 Examples of aromatics in diesel fuel

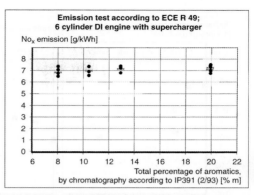

Emission test according to ECE R 49;
6 cylinder DI engine with supercharger

NO_x emission [g/kWh]

Total percentage of aromatics,
by chromatography according to IP391 (2/93) [% m]

Fig. F87 Total aromatic content has no effect on exhaust emissions (example NO_x)

As expected, NMR spectroscopy indicates significantly lower aromatic contents (aromaticity) than the liquid chromatography method mentioned above. However, because of the relatively complicated apparatus required, this method is used only when such accurate values are required for specific research purposes.

In addition to the methods mentioned, there are others that either provide insufficient precision or are too complicated (Concawe Report 94/58). The influence of the aromatic content in diesel fuel on the combustion and exhaust emissions was controversial for some time. Frequently, intercorrelation of the fuel constituents made analysis of test results difficult. Generally, increasing aromatic content was associated with increasing density and decreasing cetane number.

Therefore, it was practically impossible to analyze the effect of the aromatic content separately from other product characteristics. It is only recently that test fuels have been successfully produced which allow variation of the individual parameters independently of one another. Evaluation of the results for these fuels indicates that mono-aromatics have no recognizable influence on particulate and NO_x emissions (**Figs. F87 and F88**). In contrast, an increase in the percentage of polyaromatics is accompanied by an increase in particulate emission (**Fig. F89**). In the European standard, EN 590, the percentage of polycyclic aromatic hydrocarbons is limited to 11% (m/m). The definition stipulates that this means the total percentage of aromatics according to EN 12916 minus the mono-aromatics— that is, it includes only the di-, tri-, and tri-plus aromatics. The percentages of polycyclic aromatics typical in Germany is considerably below this limit with an average value of barely 3% (m/m).

~~**Naphthenes.** Naphthenes are cyclic saturated hydrocarbons (single bond between the carbon atoms in the molecule). Naphthenes are present in various quantities in crude oil and are also produced by hydration (treatment with hydrogen) of aromatic middle distillates. Naphthenes in diesel fuel result in good low-temperature behavior, but only medium cetane numbers (however, they are higher than in aromatic hydrocarbons).

~~**Olefins** (*also*, →Fuel, gasoline engine). Olefins are mono-unsaturated or poly-unsaturated chain or branch chain hydrocarbons. In comparison to n-paraffins they have lower, although still relatively high, cetane numbers; for example, n-paraffin cetane, $C_{16}H_{34}$, has a cetane number of 100; the single unsaturated olefin with the same number of carbon atoms, 1 cetane, $C_{16}H_{32}$, has a cetane number of 84.2. In comparison to the short chain olefins used in gasoline, the single double bond in the long chain olefins found in diesel fuel has only a slight influence on physical characteristics and combustion properties.

F

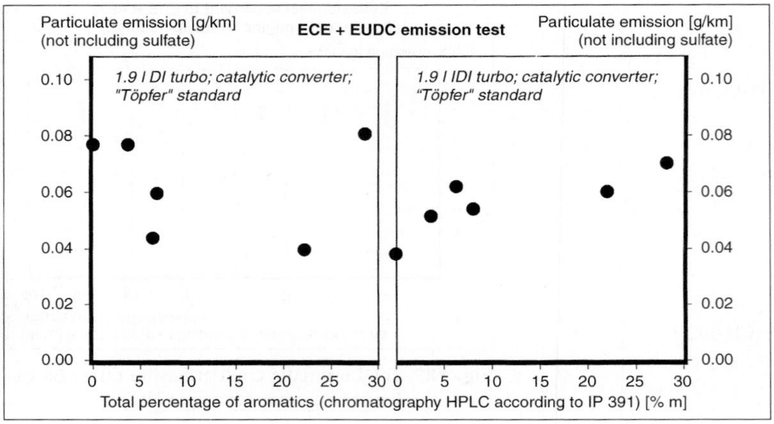

F

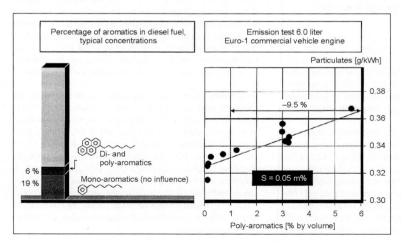

Fig. F88
Total percentage of aromatics does not define the particulate emission for passenger cars

~~Paraffins. Paraffins are saturated hydrocarbons. Normal chain paraffins are particularly suitable for diesel fuel. They are distinguished by good ignition performance (high cetane number), but limited low temperature properties. Paraffins represent the greatest percentage of hydrocarbons in diesel fuel. Molecule combinations of mono-aromatics with long paraffin side chains are also present and show behavior similar to that of paraffins.

~~Polyaromatic content (*also, see above,* ~~Aromatics) This is the percentage of inorganic foreign matter in diesel fuel. In European specification EN 590, the percentage of polyaromatics is limited to a maximum of 11% (m/m). Tri-aromatics (three rings) begin to boil at >300°C, four-ring aromatics at 385°C. The concentration of dual nucleus aromatics (di-aromatics) in diesel fuels is less than 5% by weight. The total content of polyaromatics with three and more rings in diesel fuel is generally less than 3% by weight. Here the concentration decreases as the number of rings increases.

The total content of polyaromatics is determined by using high-pressure liquid chromatography with a refraction index detector corresponding to EN 12916 (previously also IP 391). Inclusion of every individual polyaromatic is not possible or is possible only with exceptionally complicated procedures because of the nonavailability of reference compounds required for the detection process.

Studies have shown that a reduction in the polyaromatic content of a fuel leads in particular to a reduction in particulate emissions (**Fig. F90**).

Within the scope of the European automobile/oil program, it was also established that a decrease in the percentage of polyaromatics also reduces the NO_x emissions from passenger cars and commercial vehicles.

A reduction in the percentage of polyaromatics in diesel fuel is possible through hydrogen treatment at high pressure and higher temperatures. The refinery facilities required for desulfurization of diesel fuel are, however, not sufficient for removing all polyaromatics, although they do tend to reduce such constituents.

Fig. F89
Polyaromatics in fuel increase the particulate emissions

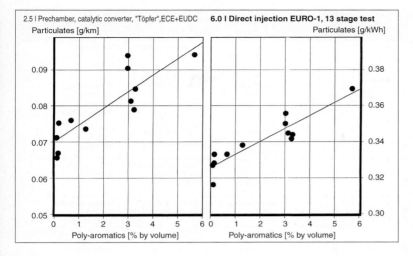

Fig. F90
Reduction of percentage
of polyaromatics in fuel
reduces the particulate
emissions

~~**Reformulated.** Reformulated fuel is fuel with a modified composition and/or modified physical parameters with the purpose of reducing emissions. In complex studies performed by individual companies, research associations, and more recently throughout Europe in the inter-industry European automobile oil program, all significant diesel fuel parameters were studied to determine their effect on exhaust emissions. An increase in the cetane number is, for the most part, advantageous.

With the exception of economic disadvantages such as cost and availability, some of the possible measures have contradictory effects on the individual emission levels or do not provide uniform results in passenger cars and commercial vehicles (**Fig. F91**).

Emission improvements can, therefore, only be achieved by integrating consideration of the effects on air quality models with corresponding priorities for air pollutants. It is also necessary to consider the effects

on other characteristics when modifying fuels to improve emissions (**Fig. F92**).

For example, reduction of the final boiling point also reduces the cetane number because paraffins with long chains and a higher boiling point have good ignition performance. By comparison, the response characteristics of low-temperature additives are reduced so that adaptations are required to ensure sufficient resistance to low temperature. When the sulfur content is reduced to values <10 ppm, it is usually necessary to use a lubricity additive to ensure that components— the distribution injection pump in particular—are lubricated properly.

~~**Sulfur content** (*also, see below,* ~Production). The natural sulfur content in diesel fuel depends on the sulfur content of the crude oil used and would be in the range of approximately 0.2 to >1% without desulfurization. The sulfur content can be determined using

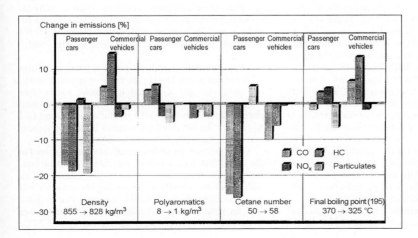

Fig. F91
Various effects of fuel
modifications on the
emission behavior of
passenger cars and
commercial vehicles

Fuel parameters	Effect on exhaust emissions					Possible refinery process	Efficiency	Technical effects
	CO	HC	NOx	Part.	SO₂			
Sulfur reduction to 0.05 % by volume	–	–	–	⇓	⇓	Hydro-desulfurization	Additional costs	Tendency to reduced lubricity
Reduction in density	⇕	⇓	⇓	⇓	–	Unlimited constituent selection	Additional costs, low availability	Higher consumption, less power, reduced cetane number
Reduction of mono-aromatics	–	–	–	–	–	Stringent hydro-treater	Considerable additional costs	Reduced lubricity, low density
Reduction of poly-aromatics	⇕	⇕	⇓	⇓	–	Milder hydro-treater	Additional costs	
Increase in cetane number	⇓	⇓	⇓	⇓	–	Additives, stringent hydro-treater	Low additional costs, considerable additional cost	See aromatic decrease
Reduction in final boiling point	⇕	⇑	⇕	⇓	–	Unlimited boiling section	Additional costs, low availability	Reduced cetane number, lower density
Diesel fuel/water emulsion	⇑	⇑	⇓	⇓	–	Mixing system/emulsifier availability	Additional costs from emulsifier and deionized water	Limited storage stability, low-temperature resistance
Diesel fuel/alcohol mixture	⇑	⇑	⇓	⇓	⇓	Constituent tanks	Considerable additional costs for alcohol and solubilizer	Tendency to corrosion, cetane number decrease, limited storage stability

Fig. F92 Reformulated diesel fuel: possible measures and consequences

various methods such as combustion of the fuel or x-ray fluorescence analysis. Methods specified for diesel fuels are described in EN 24260, EN ISO 8745, and EN ISO 14596. The limits for the maximum permissible sulfur content have been reduced continuously in the past.

The designations "low sulfur" and "sulfur-free" are now used frequently instead of the actual limits and refer to a maximum sulfur content of 50 and 10 mg/kg respectively.

Since 2000, the maximum sulfur content in diesel fuels has been limited to 350 mg/kg by the European standard. Since January 1, 2003, the quantity of sulfur in German fuels is less than 10 mg/kg (sulfur-free, according to definition) with practically no exceptions because of the tax advantage. A Europe-wide reduction of the sulfur content to a maximum of 50 mg/kg was approved for the year 2005. The update of the EU fuel guidelines provides for full availability of sulfur-free fuels in all 15 European Union countries by 2005 as well as overall changes in the requirement to specify sulfur-free fuels as of 2009. The sulfur is present in chemically bound form and is burned to primarily gaseous SO₂ (>95%) during combustion of the fuel; the remainder is contained for the most part in the exhaust particulates (**Fig. F93**). The increasing limitations on particulate emissions and the use of exhaust treatment systems has made a corresponding reduction of the sulfur content in fuels necessary (**Fig. F94**).

The particulate mass formed from sulfur in the fuel cannot be influenced by design changes (with the exception of fuel consumption) and depends solely on the quantity of sulfur contained in the fuel used. The relative percentage of the particulate mass formed by the sulfur in the fuel, therefore, increases as the overall levels of engine particulate emissions decrease (**Fig. F95**).

In the presence of an oxidation-type catalytic converter, the percentage of sulfur in the fuel converted to SO₃ increases depending on the type of catalytic converter and catalytic converter temperatures (engine load and speed). In this manner, the particulate mass can increase to impermissible levels, rendering the catalytic converter less than useless. Therefore, the advantages for particulate emissions resulting from a catalytic converter concept for use in commercial vehicles require especially adapted catalytic converters and/or the extremely low sulfur fuels now available (**Fig. F96**).

~~**Water content.** Crude oil contains small quantities of dissolved water. During the refining process, washing processes are also used to remove undesirable water-soluble constituents. Water vapor is also used during desulfurization in the production of diesel fuel to remove the intermediate product hydrogen sulfide, and this is naturally followed by drying.

The maximum possible quantity of water dissolved in the diesel fuel decreases with decreasing aromatic content and fuel temperature: A diesel fuel saturated with water becomes cloudy at decreasing temperature and water is ultimately precipitated. In comparison to gasoline, the water absorption capacity is significantly lower in the range of approximately 50–100 mg/kg at +20°C.

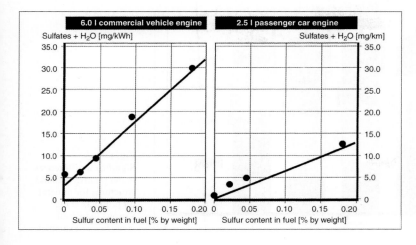

Fig. F93
Sulfur reduction in fuel decreases the particulate sulfates

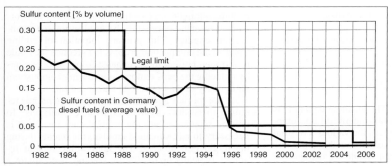

Fig. F94
Development of sulfur content in diesel fuels in Germany

A maximum limit of 200 mg/kg is permitted in the quality standard for diesel fuels EN 590. Commercially available fuels contain less than 100 mg/kg of water. The water content can be determined analytically by titration using the Karl Fischer method in conformance with EN ISO 12937. However, any infiltration of water into the diesel fuel, particularly in winter, should be avoided because the ice crystals forming could accelerate filter clogging in combination with the paraffin crystals that could precipitate.

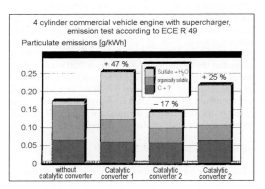

Fig. F95 Sulfur in fuel leads to the same absolute particulate emissions in various types of engine designs; however, in engines with low emission levels, the relative percentage of particulate mass increases

Fig. F96 Conversion of sulfur dioxide in exhaust catalytic converters to sulfates can increase the particulate emissions; therefore, modification of the catalytic converter is required

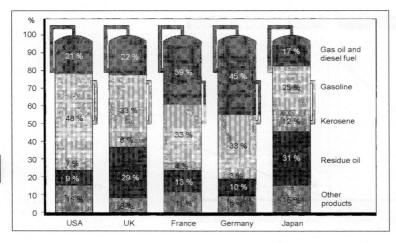

Fig. F97
Significant differences in the relative demand structures for mineral oil products using various industrial countries as examples

~**Compressibility.** *See below,* ~Properties

~**Corrosion effect on copper.** *See below,* ~Properties

~**Density.** *See below,* ~Properties

~**Detergents.** *See above,* ~Additives

~**Development of demand.** The absolute demand, as well as the percentage of total mineral oil consumption, differs significantly between the various industrial countries depending on the vehicle numbers and distances driven/loads transported. While the demand for diesel fuel in the United States amounts to only 8% of the total mineral oil consumption, it is significantly higher in France (22%) and Germany (16%) (**Fig. F97**).

The predicted increase in cargo traffic on the road will result in an absolute as well as a relative increase in demand for diesel fuels in Germany with significantly decreasing demand for gasoline (**Fig. F98**).

The increase in demand results from transport of larger quantities of goods, and a larger percentage of diesel passenger cars is expected to exceed the possibilities of savings resulting from more efficient engines and vehicles as well as from optimum logistic planning. The demand for diesel fuel in Germany, which exceeded the demand for gasoline in 2003 with a figure of 28.5 Mtonne, is expected to continue to increase to approximately 31 Mtonne by 2007. The additional requirement can be covered without quality losses because of the decrease in demand for light fuel oil and further improved flexibility of the refineries.

~**Diesel fuel detergents.** *See above,* ~Additives

~**Elastomer compatibility.** *See below,* ~Properties

~**Electrical conductivity.** *See below,* ~Properties

~**Emulsions.** Emulsion fuels are a mixture of hydrocarbons and water. Mixtures of diesel fuel, gasoline,

naphtha (straight-run gasoline), and other constituents with water can conceivably be used in internal combustion engines. The primary intention of adding water to the combustion process is to reduce the oxide of nitrogen present in the exhaust from diesel engines by reducing the peak temperatures during combustion, particularly in the upper load and speed range.

An appropriate agitator or ultrasonic generator is required to produce hydrocarbon-water emulsions. To prevent separation into the original constituents, it is necessary to add so-called emulsifying agents during the production process. The high water content required, typically up to 30% by volume, necessitates a series of additional measures. These include addition of corrosion protection and wear protection additives to protect the fuel system and engine as well as use of low-temperature protective additives to prevent the water from freezing at low temperatures and clogging the fuel system (e.g., fuel filter).

Studies indicate that the NO_x can be reduced by approximately 13–20% in state-of-the-art engines by using emulsion-type fuels. The particulate emissions can also be reduced by approximately 10–15%. However, this is only possible at the cost of increasing the

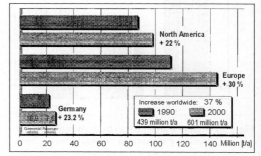

Fig. F98 The increase in the worldwide demand for diesel fuels

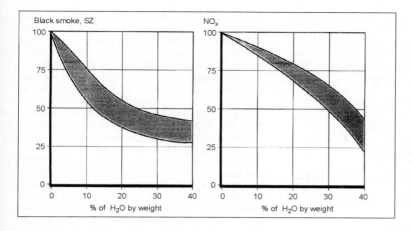

Fig. F99
Reduction of black smoke and nitrogen oxide emissions as a function of the percentage of water in diesel/water emulsions.

emissions of unburned hydrocarbons (HC), carbon monoxide (CO), and carbon dioxide (CO_2). In previous engines not optimized in terms of emissions, the advantages as well as disadvantages of diesel fuel-water emulsions were greater (**Figs. F99 and F100**).

The significantly lower heating (calorific) value results in additional volumetric consumption of up to 60%, which leads to a corresponding reduction in the range if the volume of the vehicle fuel tank remains the same.

In contrast to commercial vehicles, passenger cars are operated primarily in low load ranges, so that for this vehicle segment, the many disadvantages mentioned would not be offset by the minor advantages in terms of NO_x reduction that would result because the NO_x emissions are already low in this range. In the meantime, other highly promising solutions for reduction of NO_x have also been developed for commercial vehicles. For example, the NOx emissions can be reduced to a minimum by the use of $DeNO_x$ catalytic converters with carbamide.

When emulsion fuels are used in vehicle engines, a uniform percentage of water is injected into the combustion chamber over the entire load and speed range. A variable composition of the fuel-water emulsion depending on the operating point of the engine or a separate and appropriately controlled water injection system would be required for optimum reduction of the oxides of nitrogen while simultaneously reducing the disadvantages of increasing the other pollutant emissions as described above. Better conditions exist for use of emulsion-type fuels in stationary engines, preferably operated at defined operating points. Aquazole, an emulsion fuel for operation of stationary diesel engines and inter-urban public transportation and supply services, has been available since 1998.

In addition to engine problems, a series of other technical application problems are posed by diesel/water emulsions because of the higher viscosity, the low long-term stability of the emulsion (particularly at low-temperatures), and potential infection by microorganisms. Alternate use of conventional diesel fuels

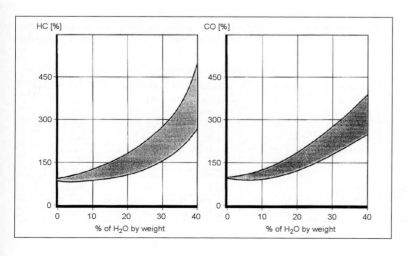

Fig. F100
Increase in HC and CO emissions as a function of the water content in diesel/water emulsions

and emulsion-type fuels is not possible or is possible only after completely cleaning the fuel system.

The problems described and the higher fuel price resulting from the required emulsifying agents have prevented extensive use of diesel/water emulsion fuels to date.

~Energy. See below, ~Properties

~Energy density. See below, ~Properties

~Ether. See above, ~Alternative fuels

~Ethyl alcohol (ethanol). See above, ~Alternative fuels

~Final boiling point. See below, ~Properties

~Flashpoint. See below, ~Properties

~Flow improvers. See above, ~Additives

~Fuel vapor lock. See below, ~Vapor lock

~Gas oil. Gas oil is the designation for hydrocarbon middle distillates (refinery products) in the boiling range of diesel fuel/light fuel oil (HEL). It is sold in lots (tankers, fuel depots) either as mixed constituents or as final products. The analytical data for the gas oil provide information on the suitability for use as diesel fuel, fuel oil, or mixed constituents. In customs and tax law, categorization of a mineral oil product in the class of gas oils requires a volatile percentage of at least 85% by volume at 350°C.

~Gross heating (calorific) value. See below, ~Properties

~Heat of evaporation (latent heat of vaporization). See below, ~Properties

~Heating (calorific) value. See below, ~Properties

~Heavy oils. Heavy oils are used as fuels for most slow-running marine engines. They are generally residues from the refinery process, and their composi-

tion is influenced to the greatest extent by the crude oil. In addition to the hydrocarbons typical for diesel fuels, they usually also contain asphaltenes, which are kept in a preflowing state by aromatic solvents. They are classified under the designation "marine fuel oil" according to viscosity and density. The viscosities can vary from 30 to 700 mm²/s at 50°C with maximum densities of 970–1010 kg/m³. In addition to the specifications of worldwide fuel producers—for example, Shell (**Fig. F101**) specifications from neutral organizations have also been in use for approximately 20 years (e.g., ISO 8217, CIMAC, BS-MA 100).

The remaining specification values are primarily the flashpoint (safety aspect for storage) and limitations for impurities (water, acids/metals, sulfur). Oils with high viscosity can contain up to 5% sulfur, 1% water, and 600 mg/kg vanadium. The cetane number of these fuels is not specified. Another primary quality characteristic is storage stability, which is tested at increased temperatures with a special filtration test. Before heavy oils are used, it is necessary to heat them up and separate mechanical impurities. The high percentage of sulfur requires highly alkaline oils for lubricating the cylinder running surfaces.

In contrast, fuels for high speed marine engines (marine gas oils and marine diesel fuel) have similar specifications to diesel fuel for passenger cars and commercial vehicles, albeit usually with lower cetane numbers, higher densities and less efficient low-temperature behavior.

~Hydrogen →Fuel, gasoline engine ~Alternative fuels

~Ignition accelerators. See above, ~Additives

~Ignition delay. See below, ~Properties ~~Cetane number

~Ignition performance. See below, ~Properties ~~Cetane number

~Landfill gas. Landfill gas is obtained from garbage (refuse) dumps and is suitable within certain limits for operation of gas engines. However, such engines must

			Designation			
Properties	Unit	Test method	MFO 30	MFO 100	MFO 380	MFO 700/1010
Viscosity at 50°C	mm²/s	ASTM D445	Max. 30	Max. 100	Max. 380	Max. 700
Density at 15°C	kg/m³	ASTM D1298	Max. 970	Max. 991	Max. 991	Max. 1010
Flash point	°C	ASTM D93	Min. 60	Min. 60	Min. 60	Min. 60
Pour point	°C	ASTM D97	Max. 0/6/24	Max. 30	Max. 30	Max. 30
Carbon residue	%m	ASTM D524	Max. 10/14	Max. 15/20	Max. 18/22	Max. 22
Sulfur content	%m	ASTM D4294	Max 3.5	Max. 5	Max. 5	Max. 5
Water content	%vol.	ASTM D95	Max 0.5	Max. 1.0	Max. 1.0	Max. 1.0
Total Acid No (TAN)*	mg KOH/g	ASTM D974	Max. 3	Max. 3	Max. 3	Max. 3
Ash	%m	ASTM D482	Max. 0.10	Max. 0.10/0.15	Max. 0.15/0.20	Max. 0.20
Vanadium	mg/kg		Max. 150/300	Max. 200/500	Max. 300/600	Max. 600

Fig. F101 Specification values of a number of marine fuel oils (MFO)

operate on the spark-ignition principles used in gasoline engines with external ignition.

The primary constituent in landfill gas is methane CH_4 (approximately 40–50%), but in addition, the gas contains large quantities of nitrogen and carbon dioxide. Undesired byproducts include organic sulfur and chlorine and fluorine compounds, as well as ammonia, dust (silicon), condensate from higher hydrocarbons, and water. The composition of landfill gases usually differs depending on the type of material in the dump and the season of the year.

In addition to environmental pollution, undesirable byproducts can also significantly reduce the service life of the engine—resulting, for example, from corrosion damage, formation of residues on the intake valves and in the combustion chamber, as well as increased wear. To add to these difficulties, it is preferable to operate gas engines with engine oils containing a low quantity of ash and, therefore, limited alkaline reserves, in order to reduce combustion chamber residues.

For this reason, engine manufacturers have established limits for the above-mentioned byproducts to reduce engine damage; typical values are listed in **Fig. F102**. In all cases, careful adaptation of the engine is required to match the oil for the gas engine to the quality of landfill gas used.

~**Light oil.** Light oil is an oil the concept of which is derived from customs and tax law, and it contains mineral oil products, which are more volatile than diesel fuels. Light oils have volatile percentages of greater than 90% at a temperature of 210°C (e.g., gasoline fuels)

~**Liquefied petroleum gas (LPG)** (*also*, →Fuel, gasoline engine ~Alternative fuels). Liquefied petroleum gas (LPG) is a mixture consisting primarily of propane and butane (properties are standardized in EN 589).

In gas engines with external ignition and exhaust catalytic converter (gasoline engine principle), liquefied petroleum gas provides a potential for significant reduction of particulate and nitrogen oxide emissions, particularly in interurban traffic. However, higher costs and a limitation of the vehicle range limit its possible use to vehicle fleets such as urban public transportation.

Liquefied petroleum gas can only be used in permanent dual fuel operation in diesel gas engines because

of its poor ignition performance. To overcome this, a small quantity of diesel fuel is injected (pilot injection) to initiate combustion.

~**Lower heating (calorific) value.** *See below*, ~Properties

~**Low-temperature behavior.** *See below*, ~Properties

~**Lubrication additive.** *See below*, ~Properties ~~Lubricity

~**Lubricity.** *See below*, ~ Properties ~~Lubricity

~**Lubricity additives.** *See above*, ~Additives

~**Methyl alcohol (methanol).** *See above*, ~Alternative fuels

~**Natural gas** (*also*, →Fuel, gasoline engine). Natural gas (primarily methane, CH_4) is used in gas engines that operate almost exclusively according to the spark-ignition engine principle with external ignition. Because of the low ignitability of natural gas, its use in diesel engines requires injection of a small quantity of diesel fuel to initiate combustion (permanent dual fuel operation, also called pilot injection). Although technically possible and tested, this combustion concept is not commonly used commercially. Natural gas for driving gas engines is used in facilities for gas supply (increasing pressure for intermediate storage and transport) as well as block-type thermal power stations.

~**Other fuels.** *See below*, ~Synthetic diesel fuels

~**Oxidation stability.** *See below*, ~Properties

~**Polyaromatic content.** *See above*, ~Composition

~**Polycyclic aromatic hydrocarbons.** *See above*, ~Composition

~**Production.** Conventional diesel fuel is produced primarily by distillation of crude oil followed by desulfurization (**Fig. F103**). The quantity of middle distillates (diesel fuel and light fuel oil) achievable through distillation depends on the source of the crude oil. Typical yields for several types of crude oil are shown in **Fig. F104**.

Relative humidity (water)	Max. 60–80%
Relative humidity (C_mH_n vapors)	Max. 60–80%
Dust	Max. 25–50 mg/m³
Sulfur (in form of H_2S)	Max. 2000–2200 mg/m³
Hydrogen sulfide	Max. 0.15%
Chlorine	Max. 100 mg/m³
Fluorine	Max. 50 mg/m³
Ammonia	Max. 55 mg/m³

Fig. F102 Typical limits for byproducts in landfill gases (additional limitations apply to engines with catalytic converters)

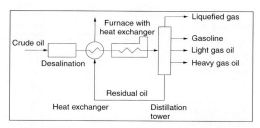

Fig. F103 Simplified schematic of crude oil distillation

Type of crude oil	Middle East Arabian Light	Africa Nigeria	North Sea Brent	South America Maya
Liquefied gas	<1	<1	2	1
Naphtha (gasoline)	18	13	18	12
Middle distillate	33	47	35	23
Residual oils	48	39	45	64

Fig. F104 Percentage yield from distillation of various crude oils

Product designation	Density kg/m^3	Boiling range °C	Cetane number
Hydrocrack constituents	860.0	170–400	52
Thermally cracked constituents (from Vis breaker)	857.0	180–400	40
Light cycle oil from catalytic cracker	953	195–410	40

Fig. F107 Properties of refinery constituents from cracking processes for production of diesel fuel

Intensive distillation to separate the middle distillate into light and heavy constituents is required for production of diesel fuel according to specification. Typical refinery constituents for the production of diesel fuel are designated according to their boiling range and the production facilities; their approximate characteristics are shown in **Fig. F105**.

In addition to these components, usually produced by distillation, other products can be obtained from cracking processes. **Fig. F106** shows a survey of the cracking processes.

The constituents obtained from these processes have approximately the characteristics shown in **Fig. F107**.

The constituents for production of the diesel fuel are selected according to the refinery equipment, the physical data of the constituents, their quantity balance, and the quality requirements to be obtained. The primary criteria for mixing include the density, boiling characteristics, flashpoint, low-temperature behavior, and cetane number required for the fuel. It is necessary to increase the percentage of kerosene and light gas oil and reduce the percentages of heavy gas oil and vacuum gas oil to produce winter fuel. As shown by the

constituent data, this is accompanied by a tendency to reduce ignition performance.

More or less intensive desulfurization of the diesel fuel constituents is required depending on the availability and use of the various types of crude oil. Sulfur is present in crude oil in a chemically bound form, and the quantity of sulfur is highly dependent on the source of the crude oil (**Fig. F108**).

Reduction of the sulfur content is accomplished by treatment with hydrogen at high pressure and high temperature in the presence of a catalyst. The refinery facilities required depend on the crude oil and the desired final sulfur content. **Fig. F109** shows a simplified schematic of the process for a refinery desulfurization plant.

The hydrogen required for desulfurization is obtained from the catalytic reformer used to refine gaso-

Product designation	Density kg/m^3	Boiling range °C	Cetane number
Kerosene	805,0	150–260	45
Light gas oil	840,0	210–320	55
Heavy gas oil	860,0	200–400	55
Vacuum gas oil	870,0	250–400	56

Fig. F105 Constituents from distillation of diesel fuel indicating typical properties

Source	Designation	Sulfur content
North Sea	Brent	0.4 % by weight
Middle East	Iranian heavy	1.7 % by weight
Arabian light		1.9 % by weight
Arabian heavy		2.9 % by weight
Africa	Libyan light	0.4 % by weight
Nigeria		0.10.3 % by weight
South America	Venezuelan	2.9 % by weight
Russia		1.5 % by weight
North Germany		0.622 % by weight

Fig. F108 Typical sulfur contents in several types of crude oil

	Vis breaker	Coker	Catalytic cracking	Hydro-cracker
Processing characteristics	Mild	Severe	Severe	Severe
	Thermal cracking	Thermal cracking	Catalytic cracker	Catalytic cracking in hydrogen atmosphere
Initial product	Atmospheric residue	Vacuum residue	Vacuum distillate	Vacuum distillate
Yield* Gases Gasoline Middle distillate HS constituents Coke	2 % 5 % 13 % 80 % –	7 % 20 % 27 % 17 % 29 %	21 % 47 % 20 % 7 % 5 %	18 % oder 7 % 55 % oder 28 % 15 % oder 54 % 12 % oder 11 % – –
Subsequent treatment of conversion products	yes	yes	yes	no
Flexibility of plant	low	low	medium	high
* In relation to initial material, source: BP				

Fig. F106
Overview of conversion plant

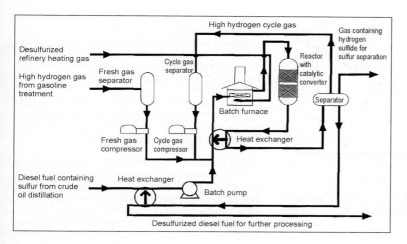

Fig. F109
Simplified schematic for
the process of diesel fuel
desulfurization

line. The separated sulfur in elemental form is used for further processing in the chemical industry.

If there is a general requirement for diesel fuels with a reduced percentage of aromatics (tri-aromatics, di-aromatics, and possibly mono-aromatics), further treatment with hydrogen together with additional energy is required in complicated refinery facilities.

If the aromatic and sulfur content of the fuel is decreased, this is accompanied by a decrease in the density (**Fig. F110**). This causes an increase in the volumetric consumption and a decrease of the fuel lubricity (lubricating capability).

The use of low-temperature additives in winter is practical for the production of diesel fuels to meet the standards. This also allows the use of heavy components with high cetane number, although only to a limited extent. The use of ignition accelerators may also be required depending on the crude oil.

The remaining additives are specific to the market and are added automatically when the tank car is filled in order to increase the quality of the diesel fuel to beyond the standard quality.

~**Properties.** These are the combined physical and chemical parameters for diesel fuels. The properties relevant for transport, storage, suitability for engine operation, and the effect on the environment are defined in European specification EN 590. They can be influenced by selection of the crude oils to be processed and even more by the refinery process used, as well as by additives.

Engine design and calibration and the fuel properties must be matched to one another. Standard adjustment fuels (CEC reference fuels), each of which represents the average fuel quality, are available to engine manufacturers to achieve optimum operating and emission characteristics.

~~**Ash content.** This is the quantity of inorganic foreign matter in diesel fuel. It must not be greater than 0.01% (m/m). The ash content is measured by burning a fuel sample (reducing it to ash) in conformance with EN ISO 6245. In typical fuels on the market, the ash content is generally below the detection limits. Higher quantities are found only in mechanically contaminated fuel (dust, etc.).

~~**Boiling curve.** Diesel fuels consist of mixtures of hydrocarbons that boil in the range between 170°C and about 380°C. Boiling apparatus and distillation conditions (among others, variable energy supply for constant distillation rate of 4–5 ml/min) are defined in EN ISO 3405. This method does not specify any precise physical boiling point but rather approximate boiling characteristics under the practical conditions of rapid evaporation. During this process, some volatile constituents are held back at increasing product mixture temperatures and, on the other hand, constituents with higher boiling points are already entrained. The exact physical initial boiling point or final boiling point is, therefore, lower or higher than specified. However, the standard method is well suited for evaluation of diesel fuels.

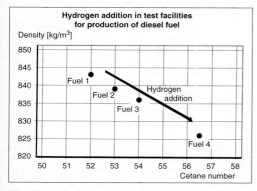

Fig. F110 Hydrogen treatment of diesel fuel for reducing content of sulfur and aromatics increases the cetane number and reduces the density

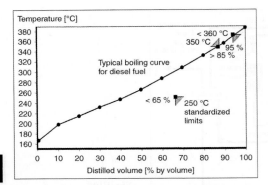

Fig. F111 Boiling curve for typical diesel fuel with basic values from requirement standard EN 590

In principle, hydrocarbons outside the usual boiling range also are suited for combustion in diesel engines—that is, machine engines operate with fuels at a significantly higher boiling point. A series of marginal conditions (e.g., viscosity, flowability at low temperatures, density, ignition performance, flash point) limit the permissible boiling range of fuels for motor vehicle operation including the layout of fuel systems.

For this reason only three points defining the fuel in the range of the middle to final boiling point are specified in the requirement standard EN 590 (**Fig. F111**).

Because diesel fuel is partly obtained directly from distillation of crude oil, a limitation of the boiling range and, in particular, reducing of the final boiling point means a reduction in the quantity of diesel fuel available (**Fig. F112**).

~~Carbon residue. This is determined by carbonizing the last 10% of the distillation quantity of diesel fuels at low temperature (method described in EN ISO 10 370). The carbon residue contains primarily organic as well as inorganic constituents. It provides

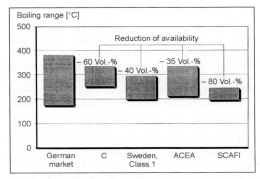

Fig. F112 Limiting the boiling range for diesel fuel reduces the availability

information on the tendency of fuels to deposit carbon residues on the injection nozzles.

Because some diesel fuel additives (e.g., ignition accelerators) increase the carbon residue and would, therefore, lead to erroneous interpretations, determination of this factor is only practical in diesel fuels without additives.

The carbon residue should not exceed 0.3% according to requirement standard EN 590. The carbon residue in commercially available fuels is significantly lower and averages approximately 0.03%.

~~Cetane index. The cetane index corresponding to EN ISO 4264 can be calculated from the fuel density and boiling characteristics as a substitute for evaluation of the ignition performance of conventional diesel fuels without measuring the cetane number. The empirical equation developed for typical fuels on the market is based on evaluation of approximately 1200 diesel fuels. The primary assumption, with a series of corrections, is that the cetane number decreases with increasing density (increasing percentage of cracked products with double bonds) and that the cetane number increases as the high boiling constituents increase (larger molecule chain length).

The formula has already been changed a number of times corresponding to the long-term changes in refinery structures and the constituents in diesel fuel. It is only suitable with reservations for some fuel constituents. In the same manner, it cannot be used to represent the cetane number for fuels with ignition accelerators. The latest calculation formula for the cetane index, CI, is:

$$CI = 45.2 + 0.0892 T_{10N} 3 (0.131 + 0.901B) \cdot T_{50N}$$
$$+ (0.0523 - 0.420B) \cdot T_{90N}$$
$$+ 0.00049 (T_{10N}^2 - T_{90N}^2)$$
$$+ 107B + 60B$$

where

T_{10} = temperature (°C) for 10% evaporation by volume,

T_{10N} = $T_{10} - 215$,

T_{50} = temperature (°C) for 50% evaporation by volume,

T_{50N} = $T_{50} - 260$,

T_{90} = temperature (°C) for 90% evaporation by volume, and

T_{90N} = $T_{90} - 310$.

B = $[e^{(-0,0035 D_N)}]$

D = density at 15°C in kg/m³ and

DN = $D - 850$,

A minimum value of CI = 46.0 is specified in the requirement standard EN 590. Correlation between the cetane index and measured cetane number for German diesel fuels from the market analysis is shown in **Fig. F113**.

The deviations between the measured cetane number and calculated cetane index result from the inaccuracy of the empirical equation (particularly its failure

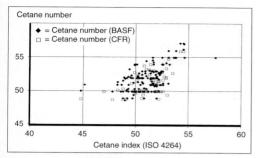

Fig. F113 Correlation between measured cetane number of German diesel fuels from market analysis in 1995

to consider fuels containing ignition accelerators) as well as the relatively wide range of deviation for cetane number measurements.

~~Cetane number. The cetane number is an indicator for the ignition performance of diesel fuels. A minimum cetane number of 51 is specified in the European requirement standard for diesel fuel EN 590. German diesel fuels have a cetane number of 52 with a tendency to higher values in summer fuels and lower values for winter fuels because of the partial reduction of high boiling paraffins to ensure sufficient resistance to low temperatures. The cetane number is determined in a standardized single cylinder test engine operating with constant ignition lag between fuel injection and start of combustion. In the CFR engine, specified in EN ISO 5 165, this is achieved independently of the fuel by changing the compression ratio (high fuel ignition performance requires reduction of the compression ratio). In the BASF test engine used as an alternative (DIN 51 773), a constant ignition delay is achieved by variable throttling of the intake air. Cetane and α-methylnaphthalene with a defined cetane number of 100 or 0 are used as reference fuels (**Fig. F114**). A fuel that requires the same engine setting in the test engine as a mixture of 52% cetane and 48% α-methylnaphtha-

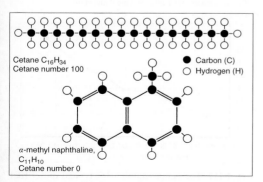

Fig. F114 Reference fuels for cetane number measurement

lene, for example, is defined as having a cetane number of 52.

The comparability, R, and repeatability, r, obtained from these tests require improvement. The variation depends on the value of the cetane number. In the range of 52–56 — above the defined minimum value of 51 — the following apply:

CFR engine: $R = 4.3$–4.8

$$r = 0.9\text{-}1.5$$

BASF engine: $R = 2.5$

$$r = 1.0$$

International research is underway to improve this situation. The BASF engine evaluates higher values of the ignition performance for the specimens than does the CFR engine (on the average, approximately 1.5 CN units higher), and the cetane numbers measured in the BASF engines are corrected accordingly.

Fuels with poor ignition performance lead to higher ignition lag, resulting in higher gas and noise emissions (**Fig. F115**). Paraffins have a high ignition performance (cetane number), while hydrocarbons with double bonds, particularly aromatics, have low ignition performance. The cetane number increases with increasing chain lengths of the paraffin — for example, high molecular weight and boiling point (**Figs. F116 and F117**).

The ignition performance of conventional diesel fuels is characterized satisfactorily by the cetane number. This also applies for fuels whose ignition performance has been increased with ignition improvers. By comparison, alcohol fuels containing high quantities of ignition accelerators behave differently than expected based on the cetane number in full engines.

The starting and noise characteristics improve with increasing cetane number (**Fig. F118**), and the gaseous exhaust emissions of HC and CO, in particular, decrease (see cetane number improvers), as well as NO_x (**Fig. F119**). Studies with standard engines have shown that the operating and emissions characteristics deteriorate significantly when using fuels with cetane numbers in the range of 40 and below but that only gradual improvements occur at cetane numbers above 60 (i.e., the behavior is nonlinear).

An increase in the cetane number of refinery constituents is possible through hydrogen treatment at high pressure and high temperatures. The appropriate equipment, which also leads to high energy consumption at the refinery, is not generally available today. Synthetic diesel fuels, which are available only in low quantities, generally are distinguished by extremely high cetane numbers.

~~Cloud point. This is the temperature at which clouds of paraffin crystals first appear when diesel fuel is cooled. The cloud point provides information on the fuel pump and filtration capability only when flow improvers (additives) have not been added. It cannot be influenced to any great extent by additives. However,

F

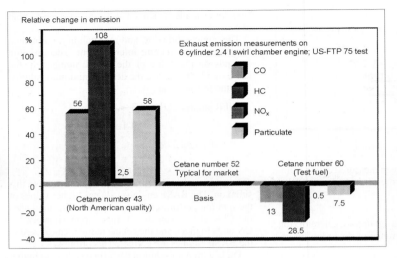

Fig. F115
Low cetane numbers
lead to high exhaust
emissions in passenger
car swirl chamber
engines (relative change
based on typical fuels on
the market)

the cloud point is an aid for fuel producers in determining the quantity and type of flow improvers to be used. Because of the limited information regarding the low-temperature behavior of diesel fuels in vehicles, it is not included in the requirement standard. Typical values for winter fuels are less than –7°C, and for summer fuels, about 0°C. Higher cloud points allow the use of greater quantities of paraffinic constituents with higher cetane numbers. The cloud point is determined in conformance with EN 23015.

~~Cold filter plugging point (CFPP). This is the laboratory method for predicting the lowest temperature at which a diesel fuel continues to flow without interruption and remains capable of filtering. This is measured with a screen with a mesh size of 45 μm and wire thickness of 32 μm and a cooling rate of approximately 1°C/minute in conformance with EN 116.

With fuels without low-temperature additives, the CFPP is only slightly below the cloud point—that is, the point at which paraffins begin to precipitate. Low-temperature additives reduce the size of the paraffin

crystals to such an extent that CFPP values can be lowered to more than 20°C below the cloud point depending on the type and quantity of additive. CFPP values down to approximately 15°C below the cloud point can be achieved by flow improvement additives alone. Further reductions are possible by combining flow improvers and wax antisettling additives.

The requirement standard for diesel fuel (EN 590) requires various resistance to low temperatures corresponding to the ambient temperatures expressed with the CFPP. The following values were selected for Germany:

Winter	(Nov. 16 to Feb. 28)	max. –20°C
Spring	(Mar. 1 to Apr. 14)	max. –10°C
Summer	(Apr. 15 to Sep. 30)	max. ±0°C
Fall	(Oct. 1 to Nov. 15)	max. –10°C

The operating reliability of the vehicles is described approximately by the CFPP. Because the fuel below the cloud point and in the range of the CFPP contains small paraffin crystals, filters should be installed in the

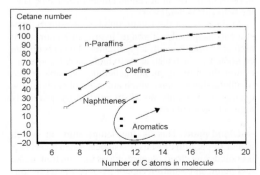

Fig. F116 Increase in cetane numbers with increasing molecular size

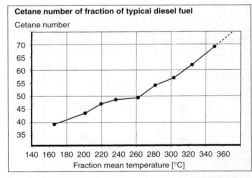

Fig. F117 The cetane number for typical diesel fuel fractionation products increases with the boiling temperature

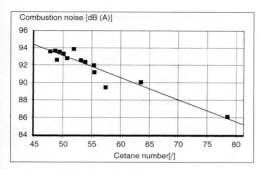

Fig. F118 Increasing cetane number reduces the combustion noise

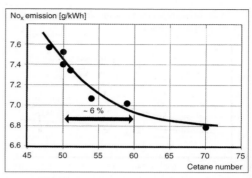

Fig. F119 Increasing cetane number decreases NOx emissions from commercial vehicles

fuel system only, where they can be heated by the heat from the engine or other measures after starting the engine. Fuels with actively heated filters allow reliable operation even at temperatures lower than indicated by the CFPP. On vehicles with filters heated by the engine heat, the operating reliability corresponds approximately to the CFPP of the fuel. At very low CFPP values, the operating reliability, however, can actually be at slightly higher temperatures—that is, the CFPP gives the impression of better low-temperature characteristics than actually occur. For this reason, various alternative processes for CFPP are in development.

~~**Compressibility.** Compressibility is a fuel parameter, significant for layout and operation of the injection system. With increasing molecule size and decreasing temperature, the paraffins primarily present in diesel fuel have decreasing compressibility.

~~**Corrosive effect on metals.** Diesel fuel encounters moisture in the air and oxygen as a matter of course during transport, storage, and use in motor vehicles. For this reason, corrosion can result on lines and in containers—for example, under exposure to alternating temperatures and water condensation. The corrosion products formed can damage the distribution chain and the vehicle fuel system.

~~~**Corrosion effect on copper.** Corrosion of materials containing copper in contact with the fuel (e.g., pump components at filling stations) presents problems for two reasons: On one hand, the components themselves are damaged, and on the other, the dissolved copper is catalytically active, leading to the formation of contaminants with high molecular weight in the fuel. The corrosiveness of a fuel depends primarily on the water content, compounds containing oxygen, the type and quantity of sulfur compounds, and, naturally, the corrosion protection additive used. A polished copper strip is brought into contact with diesel fuel at 50°C for three hours to test the corrosion value defined according to the European requirement standard (EN ISO 2160). Additives provide for far-reaching protection of all metals being exposed to the fuel, even under severe conditions.

~~~**Corrosive effect on steel.** The corrosion characteristics are tested in accordance with DIN 51 585. Here, a round steel rod is immersed for 24 hours at 60°C in a 10:1 mixture of fuel and distilled water (version A) or synthetic sea water (version B). After conclusion of the test, the formation of rust is evaluated visually. This method is used, for example, for testing the effectiveness of corrosion protective additives.

~~**Density.** This is a value for determination of the fuel mass/volume and is specified in kg/m³ according to the standards (EN ISO 3675, EN ISO 12185) at 15°C.

The density of fuels is traditionally measured with hydrometers (densimeters) or, today, with vibration-measuring instruments. For this purpose, a small quantity of the fuel to be tested is filled into a vibrating tube. The change in the vibration frequency of the tube enables the density of the fuel to be determined by means of calibration data.

With increasing carbon content in the diesel fuel—that is, increasing chain length of the paraffin molecules and increasing percentage of double bonds (aromatics, olefins), the density increases; whereas an increasing percentage of hydrogen in the diesel fuel reduces its density. The permissible density range for diesel fuel is specified in the requirement standard EN 590 to be 820–845 kg/m³ (800–840 kg/m³ for arctic fuels produced without long-chain paraffins). Hydrogen treatment of the diesel fuel during its production (to reduce the sulfur content and/or percentage of aromatics) has a tendency to reduce the fuel density.

The density of German diesel fuels is usually between about 827 and 841 kg/m³ in summer and tends to be between 821 and 840 kg/m³ in winter because it is necessary to use fewer high-density components because of their poor low-temperature characteristics.

At the given maximum diesel engine injection volumes, set by the vehicle manufacturers, increasing fuel density reduces the volumetric fuel consumption and simultaneously increases soot and particulate emissions (**Fig. F120**). Contrary to this, decreasing density decreases the emission, albeit at the cost of reducing

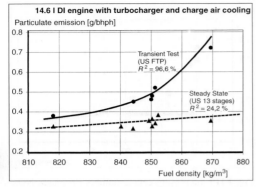

Fig. F120 The effect of fuel density on particulate emissions

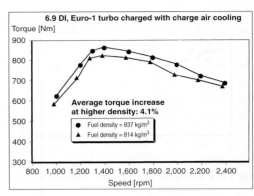

Fig. F121 The effect of fuel density on engine torque

maximum possible output (**Fig. F121**), and causes an increase in volumetric fuel consumption.

~~**Elastomer compatibility.** Diesel fuels with the compositions common on the market to date—that is, with an aromatic content of approximately 20–30% without addition of alternative constituents (fatty acid methylester)—are virtually neutral in relation to the elastomers used in the fuel system.

In individual cases, elastomer shrinkage, which can lead to corresponding fuel leakage, have been noted with fuels containing only small quantities of aromatics—for example, Swedish diesel fuel (Class 1).

By comparison, rapeseed oil methylester can lead to impermissible swelling and damage to elastomers based on acryl-nitrile in mixtures with diesel fuel. The use of such fuels can require the use of special elastomers—for example, silicone rubber.

~~**Electrical conductivity.** It is necessary to ensure that the fuel has a sufficiently high electrical conductivity to prevent electrostatic charging at high fuel pumping rates. The "prevention of ignition hazards resulting from electrostatic charges" is regulated in BGR 132 dated March 21, 2002.

Measured conductivities for individual paraffins are in the range of 0.01 to 0.02 pS/m (pico-Siemens/meter). The natural conductivity of diesel fuels has, in the past, proved sufficient for the common pumping rates for handling diesel fuels. Reduction of the sulfur content has a tendency to reduce the electrical conductivity of the basic fuel, because polar fuel constituents are lost during the desulfurization process. This effect can be compensated by adding so-called conductivity improvement agents.

~~**Energy.** The energy contained in diesel fuel by volume or weight that can be released during combustion is measured as the heating (calorific) value (energy of reaction). For practical operation, the lower heating (calorific) value—that is, before condensation of the water vapor—is relevant. For diesel fuels, lower ener-

gies of reaction of approximately 43 MJ/kg or approximately 36 MJ/L are common.

Conversion of the chemical energy present in the fuel to mechanical energy for propulsion of the vehicle is accomplished with diesel engines considerably better than with gasoline engines because of combustion with excess air and lack of a throttle in the air intake. Nevertheless, commercial vehicle engines with direct injection are only capable of converting a maximum of 43% and passenger car prechamber engines approximately 36% of the chemical energy contained in the fuel into mechanical energy at the crankshaft at the best point in the engine map. The remaining portion is lost in the form of heat transferred to the cooling water, energy in the exhaust, and that radiated directly by the engine into the ambient air. Various proportions of the energy potentially available are also lost by driving auxiliary equipment; the quantity depends on the operating point of the vehicle. By definition, 0% of the energy contained in the fuel is used for propulsion of the vehicle when the engine is idling.

In the so-called energy chain for overall vehicle operation, it is also necessary to consider the effective load ratio—that is, the percentage of payload in relation to the total vehicle, as well as the efficiency of the engine and transmission. In addition, the energy required for production and distribution of the fuels is taken into consideration. The energy chains shown for passenger cars refer to vehicles operated in the ECE city driving cycle (**Figs. F122 and F123**).

Although diesel engines, particularly engines with direct injection, have significantly better performance values than gasoline engines, the low effective energy conversion in relation to the payload moved is initially surprising. However, the balance is significantly better for commercial vehicles fully loaded in long-haul operation.

Optimum utilization of the energy in the fuel, with values of 84–86%, can be achieved when diesel fuels are used in stationary engines with combined cycles (e.g., mechanical energy used for operation of a heat pump) and use of the energy in the cooling water and exhaust gas for heating purposes.

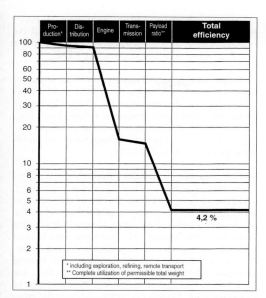

Fig. F122 Energy chain using regular gasoline from petroleum as an example

~~Energy density. The energy density is the energy contained in a fuel in relation to its weight or volume.

While the energy density in terms of weight is usually used for scientific purposes, in practical applications the energy density per unit of volume is decisive, because fuels are metered and sold on a per-unit-of-volume basis (**Fig. F124**).

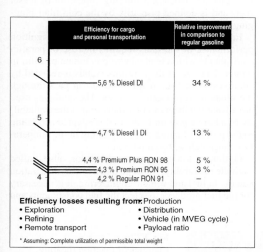

Efficiency losses resulting from: Production
• Exploration
• Refining
• Remote transport
• Distribution
• Vehicle (in MVEG cycle)
• Payload ratio

* Assuming: Complete utilization of permissible total weight

Fig. F123 Propulsion with diesel engines has a higher overall efficiency than propulsion with gasoline engines (basis: fuels from petroleum in passenger cars)

	Calorific value H_u	
	MJ/kg	MJ/l
Diesel fuel	approx. 43	approx. 35.7
Gasoline (Premium)	approx. 41	approx. 30.8
Methanol	19.66	15.63
Rape seed oil	37.2	32.7

Fig. F124 Energy density (heating [calorific] value H_u) of diesel fuels in comparison to other liquid fuels

Diesel fuels have an approximately 15% higher energy density, related to the volume, compared with gasoline—that is, even at the same thermal efficiency for the engine, this would already result in a correspondingly lower volumetric consumption for the diesel engine. In comparison to RME, an advantage of approximately 9% is present on a volume basis.

~~Final boiling point (*also, see above*, ~~Boiling curve). Maximum temperature indicated at the time of evaporation of the last liquid from the bottom of the flask in the boiling apparatus.

The final boiling point for diesel fuels frequently cannot be determined precisely because cracking processes may already start in the temperature range above 350°C when the last residues of diesel fuel evaporate. As with gasoline, the final boiling point determined according to the standard method is not the exact physical final boiling point of the diesel fuel, but rather an approximate value under practical conditions. Traces of fuel constituents with higher boiling points are already entrained prematurely at the specified distillation rate. The exact physical boiling is, therefore, higher than the boiling point measured using the boiling analysis method in conformance with EN ISO 3405. The final boiling point of diesel fuels is not specified in the requirement standard EN 590, because of the inaccuracy mentioned above.

~~Flashpoint. The flashpoint of a liquid is the lowest temperature at standard (atmospheric) pressure at which vapors develop in a closed vessel in quantities such that a vapor/air mixture results which could be ignited by an external source (determination according to EN 22719, ISO 2719). The laws on storage and transport of combustible liquids (VbF) classify fuels in various hazard classes depending on their flashpoint. The safety equipment required for the tank, etc., is laid out for the various fuel hazard classes. Diesel fuels have a flashpoint of >55°C according to this categorization and, therefore, are classified in hazard class AIII (gasoline fuels with flashpoints <21°C are classified in hazard class A1). Even small quantities of gasoline, which may be present in diesel fuel when transported alternately in the same tank chambers, can decrease the flashpoint of "diesel" fuel to below the limit of 55°C, resulting in a safety risk. Studies from previous years have shown that poor supervision during transport has frequently led to flashpoints below 55°C for diesel fuel (**Fig. F125**). For this reason, a transport

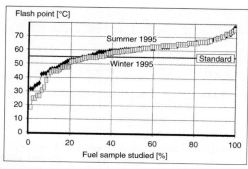

Fig. F125 Flashpoints of diesel fuels in Germany

system has been developed for brand-name fuels that excludes even slight intermixing. In the production of diesel fuel, concern about the flashpoint limits the use of highly volatile components.

~~Gross heating (calorific) value (*also, see below,* ~~Heating (calorific) value). Quantity of energy released per kilogram of fuel during complete combustion including the latent heat of condensation of the water vapor occurring (DIN 51900, Part 2; the water formed during combustion of the compounds in the fuel is condensed from the products).

~~Heating (calorific) value (*also, see above,* ~~Energy density). A differentiation is made between the upper heating (calorific) value (H_o) and lower heating (calorific) value (H_u).

The upper heating (calorific) value H_o, or gross heating (calorific) value, is determined in a calorimeter (combustion bomb) by complete combustion in an oxygen atmosphere at a pressure of 30 bars. Carbon dioxide and any sulfur dioxide present in the products remains in gaseous form following combustion but the resulting water vapor condenses. Because the water vapor does not condense when combustion is accomplished in an engine, the gross heating (calorific) value is too high for practical evaluation of fuels. For this reason, the hydrogen content is determined by analyzing the elements; this is used to calculate the latent heat of condensation of the water vapor, which is subtracted from upper heating (calorific) value to finally obtain the lower heating (calorific) value (method described in DIN 51900).

During fuel production, the heating (calorific) value is a parameter that results from the density, boiling characteristics, and fuel composition, which is not measured for the purpose of quality control. Measurement is required only for fuels used for certain research and development purposes. Values for a number of typical diesel fuels are given in **Fig. F126**.

~~Latent heat of vaporization. This value is important for production of fuel at the refinery and for mixture preparation after injection. It is affected by the hydrocarbon constituents contained in the diesel fuel. The latent heat of vaporization decreases with increasing molecule size. The typical range for diesel fuel is approximately 240 to 260 kJ/kg.

~~Lower heating (calorific) value (*also, see above,* ~~Energy density, ~~Heating (calorific) value). Quantity of energy released per unit of mass (or volume) of fuel during complete combustion, when the water resulting from combustion is present in gaseous form. The lower heating (calorific) value is the heating (calorific) value relevant for evaluating engine combustion. It can be calculated from the gross heating (calorific) value (upper heating [calorific] value; see DIN 51900, Part 2), but in this case the latent heat of evaporation for the water is subtracted from the gross value.

General equation:

$$H_u = H_o - L\, m_F/m_P$$

where H_u = lower heating (calorific) value, H_o = gross heating (calorific) value (upper heating [calorific] value), L = latent heat of vaporization for water, m_F = mass of water resulting from analysis of the combustion process, and m_P = mass of fuel specimen

~~Low-temperature behavior. This describes the flow characteristics and filtering capability of diesel fuels and is defined primarily by the cold filter plugging point.

Economical fuel production and sufficient ignition performance of the fuels requires the use of paraffin constituents with a tendency to precipitate paraffin in winter. The size of the paraffin crystals is reduced by additives so that the major portion can pass through the filters used in motor vehicles. The additives used are adapted to the particular fuels so that the availability of additive present in the fuel is adapted according to the profile of paraffins precipitating in the temperature range below the cloud point.

Optimum low-temperature behavior, also called "vehicle operability," must be taken into consideration equally in fuel and vehicle engineering. It is necessary for the fuel producer to select the fuel constituents and

| | Density 15°C (kg/m³) | Organic analysis | | H_o MJ/kg | H_u MJ/kg | H_u MJ/l |
		C (m%)	H (m%)			
Diesel fuel	829.8	86.32	13.18	45.74	42.87	35.57
A	837.1	85.59	12.7	45.64	42.9	35.91
B	828.3	86.05	13.7	46.11	43.12	35.72
C						
Average value	831.7	85.99	13.19	45.84	42.96	35.73

Fig. F126
Energies of reaction (heating values) and elementary analysis of commercially available diesel fuels (Source: DGMK, Hamburg)

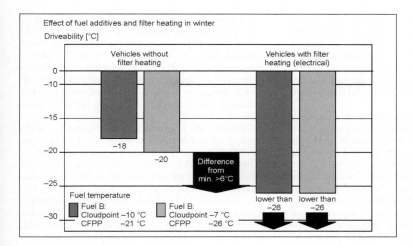

Effect of fuel additives and filter heating in winter

Driveability [°C]

Fig. F127
Improved operating reliability in winter with fuel additives and filter heating

additives to ensure sufficient filtering capability at low temperatures. In addition, it is also necessary for vehicle manufacturers to provide for the possibility of heating the fuel filters (by convection in the engine compartment or, even better, with active heaters) so that the paraffin crystals that are deposited to a limited extent after a cold start melt during operation (**Fig. F127**).

Various degrees of operating reliability at low temperatures are obtainable for given low-temperature fuel behavior depending on the design of the fuel system. Vehicles with actively heated filters allow operation at lower ambient air temperatures than indicated by the CFPP of the fuel. In contrast, vehicles without filter heaters may have limited operability in winter in spite of fuel additives.

~~Lubricity (lubricating capability). The lubricity of diesel fuel is reduced by desulfurization of the diesel fuel with the aid of hydrogen at high pressure and temperature. With the refinery procedures required for desulfurization, the polar molecules, which are natural constituents of fuel, are simultaneously reduced. This can reduce the lubricity of the desulfurized diesel fuels to such an extent that significant wear problems can result on distributor injection pumps and pump/nozzle systems. With increasing desulfurization of the fuels, their lubricity generally also decreases. Fuels which have been desulfurized to an extremely large extent — such that they contain only a few ppm of sulfur (according to Swedish Class 1 specification) — already result in high levels of pump wear even after a short engine operating time. Fuels in which the sulfur has been reduced to a less severe limit of 0.05% do not generally lead to high short-term wear, although their long-term lubricity is limited.

In addition to the relatively expensive endurance test on pump testing benches, other mechanical laboratory tests have been developed for testing the wear characteristics of diesel fuel. It was possible to use experience gained in wear tests on fuels for aircraft en-

gines. The HFRR (high frequency reciprocating wear rig) was developed and selected uniformly by vehicle manufacturers, pump manufacturers, and the mineral oil industry for evaluation of fuels corresponding to ISO 12156-1.

The HFRR test simulates friction wear in the injection pump by rubbing a ball (diameter 6 mm) at constant contact pressure against a polished steel plate under liquid at a test temperature of 60°C (**Fig. F128**). The flattening of the ball resulting after 75 min is measured as the test result (average wear diameter in μm). The requirement standard for diesel fuel allows a maximum wear diameter of 460 μm.

The lubricity of desulfurized diesel fuels can be improved with additives. With sufficient additives, the same lubricity can be achieved in sulfur-free fuels (<10 ppm) as in conventional fuels containing sulfur.

~~Oxidation stability. During long periods of storage (storage times >1 year — e.g., for strategic stocks), fuels can partially oxidize and polymerize, leading to the formation of insoluble constituents and, therefore,

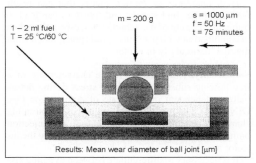

Fig. F128 Schematic of laboratory device for determination of fuel lubricity (HFRR lubricity test)

to filter clogging. The chemical mechanism causing this is the splitting of hydrogen and attachment of oxygen, particularly to unsaturated olefin fuel molecules. The continuous process of oxidation and polymerization, as intermediate products resulting from the formation of so-called "free radicals," can be prevented and effectively interrupted by the use of antioxidants (additives).

To measure the oxidation stability in the laboratory, the fuel is aged at an accelerated rate according to EN ISO 12205 for 16 hours at a temperature of 95°C in an open vessel aerated with pure oxygen (3 L/h). During this process, it is not permissible for more than 25 g/m of soluble and insoluble resinous substances to be produced. The quantity of resinous substances measured according to this method in standard commercial diesel fuels is significantly lower, generally less than 1 g/m. This method can be used only with restrictions for safety reasons because of the addition of oxygen. Alternative laboratory tests using air as the oxidation medium and temperatures close to those encountered in practice require several weeks and are, therefore, not suitable for checking the quality of fuels.

~~**Solidification point.** This is similar to the pour point. The solidification point is the temperature limit below which a diesel fuel no longer flows. The solidification point is below the cold filter plugging point (CFPP) relevant for "flowability" and filterability. Therefore, it provides little information on the operating characteristics of diesel fuels and is not included in the requirement criteria.

~~**Total contamination.** In accordance with requirement standard EN 590, the total of all undissolved foreign matter in fuel (rust, sand, undissolved organic constituents) must not exceed 24 mg/kg of fuel. Values obtained with standard commercial fuels are generally less than 10 mg/kg of fuel.

The quantity is determined according to EN 12662. The fuel is filtered with a 0.8-μm diaphragm-type filter. After the filter is washed with n-heptane and then dried, the total of the remaining undissolved foreign matter is determined by weighing.

High quantities of foreign matter that can result from improper fuel storage and transport can lead to frequent clogging of the vehicle filter, especially when in combination with paraffin crystals precipitated from the fuel, particularly in winter.

~~**Viscosity.** The viscosity is the characteristic of a free-flowing substance to absorb stress during deformation, dependent only on rate of deformation (see DIN 1342). The fuel viscosity influences the pumping characteristics of the fuel in the fuel and injection pumps as well as the atomization of the fuel by the injector nozzle.

It is necessary to differentiate between dynamic viscosity, μ (Pas = Ns/m^2), and kinematic viscosity, $v = \mu/\rho$ (mm^2/s). The kinematic viscosity is the quotient of the dynamic viscosity, μ, and density, ρ.

For diesel fuels, the kinematic viscosity is measured with an Ubbelohde capillary viscometer (EN ISO 3104). During this procedure, the time that a 15-ml specimen takes to flow through a defined capillary tube at a temperature of 40°C is measured. In the requirement standard EN 590, a viscosity range of 2.0 to 4.5 mm^2/s is required for diesel fuels. Lower minimum values down to 1.2 mm^2/s apply to arctic fuels.

Commercially available fuels have viscosities in the range of 2.0 to 3.6 mm^2/s at the specified temperature of 40°C. Viscosity is normally not a primary criterion in fuel production but rather results from the other fuel parameters. The viscosity increases at decreasing temperature and increasing pressure. For example, the viscosity of diesel fuels approximately doubles when the temperature decreases from 40 to 20°C or when the pressure increases to approximately 600 bars (**Fig. F129**). The viscosity influences the flow and pumping characteristics of the fuel in the fuel system as well as deformation of the injection jet in the combustion chamber by the injector nozzle. Exceptionally high viscosity will limit the pumpability of the fuel at low temperatures, and this leads to cold starting problems, while exceptionally low viscosity makes it difficult to start the engine when hot and leads to losses in performance at high temperatures in addition to pump wear.

~Rapeseed oil. *See above*, ~Alternative fuels

~Rapeseed oil methylester (RME). *See above*, ~Alternative fuels ~~Vegetable oils

~Reformulated. *See above*, ~Composition

~Requirements. The minimum requirements for diesel fuels are specified for all of Europe in the DIN EN 590 standard. Significant parameters are the density, ignition performance, boiling characteristics, flashpoint, viscosity, low-temperature behavior, and sulfur content, as well as a series of parameters describing the stability, corrosion effects, and concentration of extraneous substances.

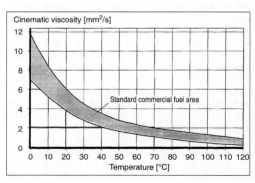

Fig. F129 Viscosity/temperature diagram for diesel fuel

	Units	Minimum requirements according to EN 590	Values obtained/ average) (2003)	Test procedure
Density at 15 °C	kg/m^3	820–845	830	EN ISO 3675 EN ISO 12185
Cetane number, cetane index		min. 51 min. 46	54 53	EN ISO 5165 EN ISO 4264
Distillation: Total evaporated quantity up to 250 °C up to 350 °C 95% point	% by volume % by volume °C	max. 65 min. 85 max. 360	42 96 352	EN ISO 3405
Flash point	°C	min. 55	60	EN 22719
Viscosity at 40 °C	mm^2/s	2.00–4.50	2.50	EN ISO 3104
Filtrability/CFPP 15. 04 to 30. 09 01. 10 to 15. 11 16. 11 to 28. 02 01. 03 to 14. 04	°C °C °C °C	max. 0 max. –10 max. –20 max. –10	–3 [2] –28 [2]	EN 116
Sulfur content	mg/kg	max. 350 as of 2005 max. 50	7	EN ISO 8754 EN 24260 EN ISO 14596
Carbon residue	m/m %	max. 0.30	0.03 [3]	EN ISO 10370
Ash content	m/m %	max. 0.01	[1] [3]	EN ISO 6245
Cu corrosion	Degree of corrosion	1	1 [3]	EN ISO 2160
Oxidation stability	g/m^3	max. 25	0.4 [3]	EN ISO 12205
Total contamination	mg/kg	max. 24	6.5 [3]	EN 12662
Water content	mg/kg	max. 200	< 100 [3]	EN ISO 12937
Lubricity HFRR [4]	μm	max. 460	407	EN ISO 12156-1
Polyaromatics	m/m %	max. 11	0.3	EN 12916

[1] Below detection limit [2] Not determined [3] Values from 1996, newer data not available
[4] Corrected "wear scar diameter" at 60 °C

Fig. F130
Quality of diesel fuels: minimum requirements according to EN 590 and average values obtained

Fig. F130 compares the minimum requirements with the average values obtained in Germany.

~RME. *See above*, ~Alternative fuels

~Sewage gas. *See above*, ~Biogas

~Solidification point. *See above*, ~Properties

~Sulfur content. *See above*, ~Composition

~Summer fuel. *See above*, ~Properties ~~Cetane number

~Synthetic diesel fuels. In addition to conventional diesel fuel, other synthetically produced substances are suitable for combustion in diesel engines. To date, significantly higher costs and limited availability have been limiting factors for all synthetic fuels.

A very old process is the Fischer-Tropsch synthesis technique (developed in 1925), in which synthetic gas is first obtained from coal or natural gas. With appropriate catalysts, this synthetic gas can be converted to hydrocarbons, which can then be refined to obtain diesel fuel or gasoline. The very low efficiency of this process means that it is hardly ever used today, with the exception of the SASOL process in South Africa.

Two synthetically produced hydrocarbon constituents have obtained limited practical significance, although their present availability is very low. These are "SMDS" diesel fuel (Shell middle distillate synthesis)

obtained from natural gas and "XHVI" diesel fuel (extra high viscosity index) which occurs as a byproduct in small quantities during production of synthetic lubricating oils. Both substances are characterized by very high cetane numbers but have lower densities than conventional diesel fuels and, therefore, tend to give poorer engine performance and increased volumetric fuel consumption. In the same manner as conventionally desulfurized fuels, their unmixed use requires the addition of lubricity additives to ensure lubrication of the injection system components; specially adapted flow improvers are also required. An advantage of their exclusive use would be the significantly lower emissions of pollutant. Because of the higher costs and low availability, these two products are currently used exclusively as mixing components for conventional diesel fuels with the exception of test operation (**Fig. F131**).

In addition to these synthetic liquid hydrocarbons, other substances are suitable as diesel fuels for use in

	XHVI	SMDS
Cetane number [/]	Approx. 70	70–80
Boiling range [°C] (T10–T90)	240–350	260–330
Sulfur content [% by weight]	0.004	0.000
Paraffin content [% by weight]	66	95

Fig. F131 Physical data on synthetic diesel fuels

519

		DME	DMM
Cetane number	[/]	55	50
Density (at 20 °C as liquid)	[kg/m³]	660	860
Boiling point	[°C]	−25	43
Calorific value H_u	[MJ/l]	18,7	−

Fig. F132 Data for dimethylether and dimethoxymethane

compression ignition engines. Possible compounds are, for example, dimethylether, DME (CH_3-O-CH_3) and dimeth-oxymethane, DMM (CH_3O-CH_2-OCH_3) (**Fig. F132**).

These two products have the required cetane number level, and limited tests with these fuels have shown significantly lower particulate and NO_x emissions. A disadvantage of both fuels is the comparatively low volumetric energy density requiring approximately 1.8 times the injection quantity compared to diesel fuel. Therefore, with such fuels it is necessary to accept either lower vehicle ranges or use larger fuel tanks and to design an adapted injection system. DME has the further disadvantage that it is present in gaseous form at normal ambient temperatures and, therefore, has to be carried in pressurized gas containers (similar to liquefied petroleum gas).

In contrast, liquid synthetic diesel fuels have the advantage of being able to use the existing infrastructure and require only minor modification of the injection system.

~Total contamination. *See above*, ~Properties

~Vapor lock. Usually, this is not a problem when standard diesel fuels are used; however, it may occur when gasoline is added to the diesel fuel to improve its low-temperature behavior. Formation of vapor in the injection system can lead to rough engine operation, loss of power, and hot starting problems.

~Vegetable oil. *See above*, ~Alternative fuels

~Vegetable oil methylesters. *See above*, ~Alternative fuels

~Viscosity. *See above*, ~Properties

~Water content. *See above*, ~Composition

~Winter fuel. *See above*, ~ Properties ~~Low-temperature behavior

Fuel, gasoline engine. As in fuels for diesel engines, fuels for gasoline engines consist of mixtures of hydrocarbons obtained primarily from crude petroleum.

~Acetone. *See below*, ~Alternative fuels

~Additives. Additives are substances added to a fuel in concentrations normally considerably less than 1%

that bring about significant improvement of the characteristics and behavior of the fuel in the engine.

Generally, the use of additives by consumers is not recommended, because the fuel system is matched precisely by the producer and could be impaired by such additives. Incompatibilities between the various substitutes are possible, which could result in filter problems, or the positive properties of individual constituents could be lost.

The only exception was an additive for protecting the exhaust valve seats on some old vehicles when unleaded fuels were used. Vehicles with "soft" valve seats—for example, gray cast iron—require particular wear protection in the fuel when the engine is operated at high load and speed (primarily freeway driving). In the past, this protection was provided by the lead compounds added to the fuel. The effective mechanism was the formation of lead oxides that were embedded into the lattice in the upper layer of the gray cast iron. Organic potassium or sodium compounds soluble in fuel were developed to ensure sufficient protection of the soft exhaust valve seats. These provided reliable protection, even at significantly lower middle concentrations, compared to lead.

The best-known additives include the antiknock products containing lead, which are no longer used because of their incompatibility with the exhaust treatment systems and for toxicological reasons. Meanwhile, improved refinery technology is capable of achieving the required octane values for fuels in the production process without lead additives. In contrast, detergents that prevent most deposits in the injection system, which can lead to poorer operating characteristics and exhaust emissions, are absolutely necessary for state-of-the-art engines.

Detergents have completed an approximately 40-year development phase. Initially, additives (e.g., alkyl amines, alkyl phosphates, oleyl amides, imidazolines) prevented residues resulting, in part, from gases from crankcase ventilation only in the carburetor area. Modern detergents act over the entire intake system, particularly on the intake valves and injection nozzles.

Active ingredients include poly-isobutene amines, poly-isobutene polyamides and polyether amines. A particularly good effect with these detergents is usually obtained in combination with temperature-stable synthetic carrier oils such as polypropylene oxides or polyalphaolefins. In addition to fuel additives, high quality engine oils also have a positive effect on keeping the valves clean (**Fig. F133**).

In addition to preventing deposits in the intake system, state-of-the-art high performance detergents are also distinguished by their capability of reducing residues that have already formed—resulting, for example, from the use of fuels without additives.

Due to the various design features of engines and the wide range of operating characteristics, it is necessary for detergents to function efficiently over a wide temperature range, even in contact with burned gas flowing back into the intake system. Moreover, it is necessary to ensure that they do not lead to an increase

Residues on intake valves; results from 5000 km road tests; 4 cylinder engine, 1.3l, 55 kW		
Fuels	**Motor oils**	**Typical residue weights per valve (mg)**
Standardized fuels without additives	Multi-grade oil simple quality API SE	400
Standardized fuels without additives	Synthetic heavy duty oil API SH	150
Standardized fuel with effective detergents	Synthetic heavy duty oil API SH	10

Fig. F133
Effect of fuel additives and motor oils on valve residues

in combustion chamber residues (octane value requirement) or sticking of the intake valve stems. The use of detergents with performance characteristics proven in standardized engine tests (ACEA fuel charter) is promoted by the European engine industry.

Moreover, a variety of other engine and driving tests is required for development of such additives — for testing the effectiveness of additives in all important engines in vehicle production, their long-term characteristics, and their compatibility with engine oils, for example.

In addition to detergents, antioxidants and corrosion protection additives are also commonly used. Antioxidants are added during production of unstable crack constituents to prevent formation of polymer residues (gum) that can result with di-olefins and dienes during fuel storage.

Corrosion protection additives have a particular significance for new vehicles that are transported over great distances before reaching the customer; these sometimes have long standstill times with the tank nearly empty. In cases in which fuels are produced with corrosive alcohol constituents, corrosion protection additives are used to coat the metal surface to be protected with their polar molecule groups (carboxylate, ester, or amino groups), thus forming a protective coating and thereby keeping corrosive constituents away from the metal (**Figs. F134 and F135**).

Anti-icing additives for protection against carburetor icing or throttle valve icing in central injection systems have become less important because of the thermostatically controlled intake air temperatures used to achieve low exhaust emissions. Agents are in use for reducing the freezing point (alcohols, glycols) or surface-active substances, which prevent ice crystals from forming on metal surfaces. Additives for reducing the freezing point offer the best protection under extreme conditions. While particular agents for reducing the freezing point, such as isopropyl alcohol or dipropylene glycol, were used in earlier automotive engineering to effectively prevent throttle valve icing, surface effective detergents are sufficient with modern engine technology, and special anti-icing additives are no longer required.

Additives were required to reduce combustion chamber residues when antiknock agents containing lead were used. The best-known lead compounds were tetraethyl lead and methyl containing i-dichloroethane and dibromoethane (so-called scavengers). Undesirable lead oxide was converted into highly volatile lead chloride or lead bromide, most of which left the combustion chambers in gaseous form. Studies have shown that these scavengers are no longer required with the low lead contents common today. For this reason, scavengers were not used in low lead fuels in Germany. Other additives, including phosphorus (e.g., tricresyl phosphate) and boron compounds, were used for converting combustion chamber and particularly spark plug deposits containing lead. Use of such additives prevented the misfiring that resulted after a certain period of time using fuels with a high lead content. The elimination of antiknock additives containing lead has made agents for converting the combustion chamber residue of the type described superfluous. Due to their unacceptable side effects (toxicity of combustion products and incompatibility with catalytic converters), they are undesirable in modern fuels anyway.

Although organic additives for the prevention of combustion chamber residues and hence an increase in the required octane rating have proven to have a certain effect in specific engine tests, comprehensive fleet tests under European driving conditions with the present vehicle population have not shown any generally convincing reduction in the octane rating requirement.

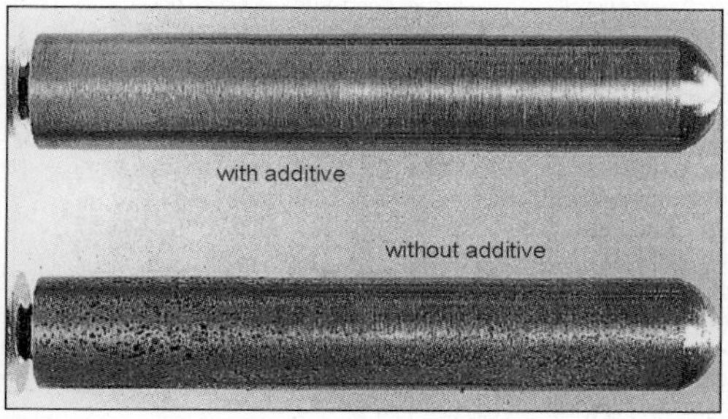

Fig. F134 Effect of corrosion protection additives in fuel/laboratory test

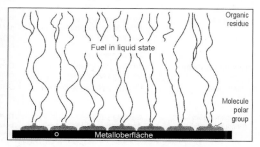

Fig. F135 Effect of corrosion protection additives

Organic potassium and sodium compounds have shown extremely good protective characteristics for "soft" exhaust valve seats as a substitute for lead. The concentrations required are significantly lower than the previously required minimum limit for lead (0.07 g/liter of fuel), established for protection of the exhaust valve seats. The potassium and sodium-based additives developed were also generally recognized as being safe in terms of toxicity in the form of the additive itself as well as its combustion products. Having been introduced on the market as an additive for fuel in older vehicles affected by legislation, they also opened up the way to complete elimination of leaded fuels. Because old vehicles with soft exhaust valves are practically no longer driven, such additives have lost their significance.

In addition, potassium acts as a stabilizer and accelerator for the start of combustion initiated by the spark plug (it is a "spark aider"). This reduces the cyclic variations in the combustion cycle and accelerates the overall combustion process slightly. However, because of general reservations regarding all additives containing metals, products containing potassium and sodium are not used widely. In Germany, they are prohibited by the leaded gasoline law.

Friction reducers (friction modifiers) are known from lubricating oil engineering; they can also contribute to a reduction of the friction as a fuel additive and thereby reduce the fuel consumption. When used in fuel, friction reducers decrease the friction between the piston rings and cylinder liners (particularly in the area of the upper piston ring, which is primarily in contact with the fuel). **Fig. F136** gives a summary of gasoline additives. Metering of the additives in commercially available fuels to increase the fuel quality to beyond the requirements of the normal minimum quality is accomplished for specific brands by computer-controlled metering systems when the tank trucks are filled.

~Alcohol (and ether) content. The type and maximum possible concentration of permissible alcohols (and ethers) and their mixtures are defined for standardized commercial fuels in EU Directive 85/

Additive constituents			Active ingredient	Improvement of	Remarks
Antiknock agent					
	Lead		Tetraethyl lead Tetramethyl lead	Octane rating; exhaust valve seat wear	required for medium + high concentrations additional chlorine and bromine compounds as scavengers, highly toxic
			Iron pentacarbonyl	Octane rating	Problems with combustion residues, engine wear
	Other metals		Ferrocene		No undesired side effects, however, only small increase in octane rating
			MMT (containing manganese)		Compatibility with catalytic converters questionable
		Organic	Aniline + derivates		Low effectiveness and therefore not economical
Antioxidants			Paraphenylene diamines, sterically impeded aklyphenoles	Storage stability, prevention of polymerization (formation of residues)	Used for stabilization of crack constituents
Metal deactivators			N,N-disalicylide diamine derivates	Storage stability, stops catalytic effect of metal ions	Used for stabilization of crack components
Corrosion protection additives			Carboxyl, ester, amine compounds	Corrosion protection	Also frequently effective for non-ferrous metals, usually used together with detergents in specific brands
Anti-freeze protection			Freezing point reducer such as alcohols and glycols; surface active substances such as amines, diamines, amides	Throttle valve icing in moist, cold weather; operating characteristics, idle, mileage	Highly effective, however no longer required with modern vehicle engineering; surface active detergents also provide sufficient protection against throttle valve icing
Detergents			Polyisobutenamine, Polyisobutenpolyamide, Langkettige Carboxyl-säureamide, Polyetheramine	Poly-isobutene amines, poly-isobutene polyamides, long chain carboxyl acid amides, polyether amines	Engine intake and fuel system cleanliness, operating characteristics; exhaust emissions
Residue converter			Tricrysyl phosphate boric acid metal ester	Conversion of deposits containing lead	No longer required
"Spark aider"			Organic calcium compounds	Stabilization of firing phase	Only limited use due to general reservations regarding additives containing metal
Friction modifier (Friction Modifier)			Carboxyl-, Ester-, Aminverbindungen	Reduction of friction between piston rings/cylinder running surface	Supports effect of fuel consumption reducing motor oils

Fig. F136 Summary of gasoline additives

Constituents containing oxygen	Maximum concentration corresponding to European Standard for gasoline fuels EN 228 in % by volume
Methanol (stabilizers required)	3
Ethanol (stabilizers may be required)	5
Iso-propyl alcohol (IPA)	10
Iso-butyl alcohol (IBA)	10
Tert.-butyl alcohol (TBA)	7
Ether, C5 or more C-atoms	15
Other organic compounds containing oxygen with final boiling points not higher than 210 °C	10
When a number of the constituents listed are used, it is necessary to ensure that the oxygen content in the fuel does not exceed 2.7% m/m.	

Fig. F137
Maximum permissible use of components containing oxygen in gasoline

F

Average alcohol and ether content in German fuels in % by volume (market analysis 1986, 1995, 1998)						
	Regular gasoline		Premium fuel		Premium Plus	
	Alcohol	Ether	Alcohol	Ether	Alcohol	Ether
1986	2.4	0.4	2.7	1.0	Type not available	Type not available
1995	0.1	0.5	0.3	1.2	0.4	6.3
1998	0.2	0.3	0.3	1.0	0.2	6.6
2003	0.0	0.8	0.1	2.3	0.0	9.7

Fig. F138
Average quantity of alcohol and ether in German fuels

536EEC. This directive is a part of the European standard for gasoline (EN 228; **Fig. F137**).

The individual alcohols/ethers and the maximum oxygen content were limited to exclude undesirable side effects—for instance, changes in the materials in the fuel system (elastomers) and exceptionally lean mixtures with poor operating characteristics, particularly in vehicles without lambda control or in unregulated operation (e.g., cold-start characteristics).

The individual alcohols/ethers are determined by gas chromatography according to EN 1601 and prEN 13132. The total oxygen content is determined as described in the same standard, with an oxygen-specific flame ionization detector (O-FID) and the associated equipment (crack reactor, hydration reactor).

While alcohols are hardly ever used today for economic reasons (earlier methanol/TBA mixtures were common), MTBE (methyl tert-butyl ether) is frequently used today up to the maximum possible concentration of 15% by volume, particularly in premium plus fuel (**Fig. F138**). Ether, as well as MTBE, allows a high octane rating for the fuel while simultaneously limiting the aromatic and benzene content.

Blended fuels containing gasoline and alcohol, with higher quantities of alcohol, require special modification of the engines. In Germany, for example, a blended fuel containing 15% methanol (designation M 15), has been researched intensively.

~Alcohol content →Fuel, diesel engine ~Requirements

~Alcohols. Alcohols are hydrocarbon compounds which have an OH group (hydroxyl group) in the molecule instead of a hydrogen atom (**Fig. F139**).

~~Production.

Methyl alcohol (Methanol). This is obtained almost exclusively from synthesized gas using one of the following gross equations:

$$CO + 2H_2 \Leftrightarrow CH_3OH - 91 \text{ kJ/mol}$$

$$CO_2 + 3H_2 \Leftrightarrow CH_3OH + H_2O - 50 \text{ kJ/mol}$$

Practically all substances containing hydrocarbons, such as natural gas, refined petroleum products, and coal as well as wood and agricultural products, are suitable for production of the synthesized gas. The synthesized gas is produced according to various processes depending on the initial substance. Common to all processes is partial oxidation of the carbon and generation of hydrogen from the water vapor added previously.

Technically, the gas is synthesized primarily from natural gas, refined petroleum products, and coal. Schematic illustrations of methanol production are shown in **Figs. F140 and F141**.

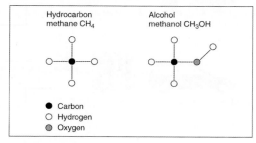

Fig. F139 Example for alcohols

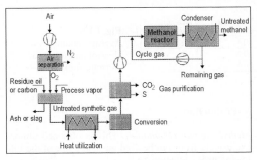

Fig. F140 Schematic representation of methanol production from natural gas

Containing sugar	Containing starch	Containing cellulose
Sugar cane	Grain	Wood
Sugar beets	Corn	Wood scraps from forest
Sugar millet	Manioc	Quick growing trees or shrubs
	Potatoes	Annual plants
		• Hemp
		• Gamboge fiber
		Agricultural residues
		• Cane trash
		• Straw
		• Stems, pods, shells
		Municipal waste
		• Old paper
		• Wood residues

Fig. F142 Initial vegetable products for production of ethanol (Source: Menrad)

F

[Fig. F141 image]

Fig. F141 Schematic representation of methanol production from coal or oil residues

Ethyl alcohol (ethanol). In addition to industrial production—for example, from ethylene—ethanol can be produced on a large scale by fermentation of agricultural products. In principle, all products containing sugar, starch, and cellulose can serve as initial products (**Fig. F142**).

Glucose is converted to alcohol by yeast according to the following equation:

$$C_6H_{12}O_6 \Leftrightarrow 2[C_2H_5OH] + 2[CO_2] - 234 \text{ kJ/mol}$$

Today, the production of ethyl alcohol from sugar cane for the propulsion of motor vehicles is most notable as an economic factor in Brazil.

To increase the yield and reduce competition between production of foodstuffs and fuels, the use of plants containing primarily cellulose would be more advantageous. However, fermentation would have to be preceded by a process of converting the various percentages and types of cellulose into glucose, depending on the type of plant. Cellulose hydrolysis with acids has been tested technically (**Fig. F143**). However, in the long term, hydrolysis using enzymes (bacteria, molds) also appears possible (**Fig. F144**).

In principle, alcohols are well suited as fuels for propulsion of motor vehicles with spark-ignition (gasoline) engines. The technologies for production are known and mature. Transport, storage, and distribution could be accomplished in practical terms using the present systems. In the vehicle, a nonpressurized tank system such those in use today is possible and the energy density, although lower, is still within the same order of magnitude as that of hydrocarbons. For combustion in gasoline engines in particular, the high octane rating of alcohols is an advantage as well as the greater increase in the air-fuel mixture following evaporation of

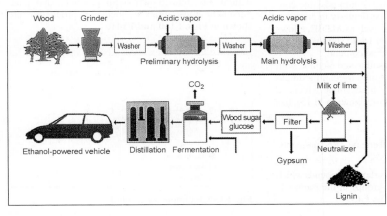

Fig. F143 Schematic illustration of ethanol production from wood using acid hydrolysis

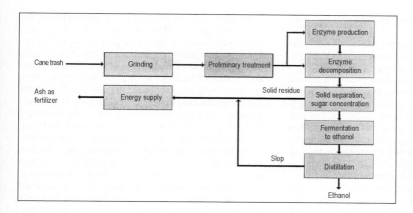

Fig. F144
Schematic representation of ethanol production by means of enzyme hydrolysis

the fuel, which allows higher mean effective pressures compared to operation with gasoline. However, other properties of alcohols, particularly methanol and ethanol, do require adaptations. These include:

- Low calorific value and low air requirement (stoichiometric air-fuel ratio) necessitate a higher fuel delivery rate. A larger tank capacity is also required to obtain the same vehicle range.
- The higher rate of combustion (flame speed) requires adaptation of the ignition map.
- A higher tendency to preignition, particularly with methanol when used in engines, requires engine modifications (heat range of spark plugs, etc.).
- A higher latent heat of vaporization requires adaptation of the preheating system for the mixture, as well as the measures for cold start and cold operation. However, the latent heat also increases the volumetric efficiency and internal cooling of the engine.
- The attack on metal and elastomers (resulting partially from traces of byproducts) requires the use of particular materials and corrosion protection additives.

For blends with hydrocarbons, it is also necessary to observe the following:

- Azeotropic mixing characteristics (constant boiling point of mixture) with hydrocarbons requires special adaptation of the fuel formulation.
- Hygroscopic (water absorbing) properties require the use of stabilizing higher alcohols such as isopropyl alcohol.

The most important physical characteristics for alcohol compared to gasolines are listed in **Fig. F145**.

Alcohols can be used in pure form as well as in blends with other hydrocarbons, whereby low concentrations are possible without any modifications to the vehicle (*also, see below*, ~Requirements).

However, higher concentrations — for example, 15% methanol in gasoline — do require adaptations. While methanol and ethanol can be used in pure form as well as in blends, higher alcohols can only be used as blending components in gasoline fuels — for instance, as stabilizers for methanol and ethanol;

Designation	Total formula	Melting point °C	Boiling point °C	Density at 20 °C kg/m³	Calorific value H_u MJ/l	Reid vapor pressure kPa	Octane rating RON	Octane rating MON	Oxygen content in % by weight	Latent heat of evaporation KJ/kg
Methanol (methyl alcohol)	CH₃OH	−93.9	64.7	791.2	15.7	32	114.4	94.6	49.93	1100
Ethanol (ethyl alcohol)	C₂H₅OH	−117.3	78.5	789.4	21.2	18/16	111.4	94	34.73	910
Isopropyl alcohol (2-propanol) (IPA)	C₃H₇OH	−89.5	82.4	785.5	23.6	14	118	101.9	26.63	700
sec butyl alcohol (2-butanol) (IBA)	C₂H₅CH-(OH)CH₃	−114.7	100	808.0	27.4		110.4	90.1	21.59	
Tert. butyl alcohol (tert. butanol) (TBA)	(CH₃)₃COH	+25.6	82.80	788.7	26.8	7			21.59	544
Premium gasoline (for comparison)	Multi-fuel mixture	−	20–215 °C	725–780 °C	ca. 32	Summer: 35–70 Winter: 55–90	approx. 96	85	0–2.8	380–500

* Density at 15 °C

Fig. F145 Important physical characteristics for alcohols compared to gasoline

for cost reasons, this is done primarily in low concentrations.

Methanol and ethanol have been used at various times and at various locations during motor vehicle history—for example, ethanol from cane sugar in Brazil.

Recently, methanol has also been used as a fuel for fuel cells because of the simple storage compared to hydrogen. Hydrogen is first generated in an intermediate step in the vehicle, and it is then fed to the fuel cell.

Methanol and fuels containing methanol have been studied in a cooperative project by the German motor vehicle and mineral oil industry sponsored by the federal government to determine all aspects of its use in motor vehicles with internal combustion engines. However, the price and cost structure for alcohols compared to petroleum products currently prevent widespread commercial use in industrialized countries. But, methanol is used as a product in production of methyl tertiary butyl ether (MTBE), which has optimum properties as a blending component for gasoline fuels.

Gasolines containing methanol have been tested in the research programs mentioned above, primarily in two versions:

- Methanol fuels containing >90% methanol
- Methanol/gasoline-blended fuel containing 15% methanol.

Other blending ratios are also possible, but then it is necessary to consider the danger of separation at decreasing methanol content in the presence of water (**Fig. F146**).

~Alternative fuels

~~Acetone. Acetone, CH_3COCH_3, is usually used as a solvent. Due to its high octane rating, RON 114*, MON 104* (*blended octane ratings), it is also suitable as a component in gasoline fuels at least in principle. However, its use is counterbalanced by to aggressiveness with elastomers and paints. Common fuels on the market do not contain acetone.

~~Biofuels. The EU directive "for promotion of the use of biological fuels and other renewable fuels in the motor vehicle sector" provides for promoting and defining a minimum percentage of biological fuels within total fuel consumption. The reference values are 2% for 2005 and 5.7% for 2010. The actions taken for promotion, resources used, and percentages of the market reached are to be reported annually to the commission. The EU commission reports to the European Parliament and the Council every two years, and it includes an evaluation of the cost effectiveness, the life-cycle prospects (including the influence on reduction of CO_2 emissions), and the economical aspects and environmental effects of further culture. In Germany, plans currently provide for reduction of mineral oil taxes on fuels proven to contain biological fuels. Of primary interest for gasoline fuels is the use of bio-ethanol and bio-ETBE (ethyl tert-butyl ether) as blending constituents.

~~Butane. Butane is a natural constituent of petroleum and occurs as a component in the processing of petroleum during distillation and cracking. It is a saturated hydrocarbon (paraffin), and assumes a gaseous state at normal pressure and temperature.

- Empirical formula: C_4H_{10}
- Boiling point: ~0.5°C
- Octane rating MON: 89.6
- Calorific value: 45.8 MJ/kg

Butane is marketed primarily as a liquefied petroleum gas (LPG), either alone or blended with propane. Moreover, butane is used in low concentration as a constituent in gasoline-based fuels to adjust the vapor pressure.

~~Ether. Ethers are hydrocarbon compounds containing oxygen in which the CH_2 group has been replaced by an oxygen atom. The ethers suitable for fuels in motor vehicle engines have at least five carbon atoms and are distinguished by their high octane rating (RON > 100, MON nearly 100), low vapor pressure, and good miscibility with hydrocarbons without azeotropic increase of the volatility with low sensitivity to water. Due to the low oxygen content compared to methanol (approximately 15%), reduction of the calorific value is within acceptable limits compared to conventional hydrocarbon fuel constituents. Large-scale technical production is accomplished by reaction of alcohols with branched olefins (**Fig. F147**).

In addition to MTBE, which is produced on a large scale technically and used in a series of countries as a fuel constituent, other ethers are suitable as constituents in fuel as a matter of principle. Because of the comparatively high production costs, they are, however, hardly used (exceptions are ETBE = ethyl tert-butyl ether and TAME = tertiary amyl methyl ether, which still have achieved a certain practical significance). The designation, composition, and most important properties of the ether compounds possible for gasolines are shown in **Fig. F148**.

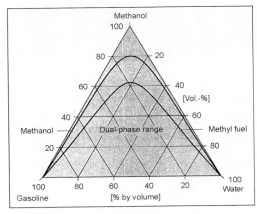

Fig. F146 Stability of methanol/gasoline/water mixtures

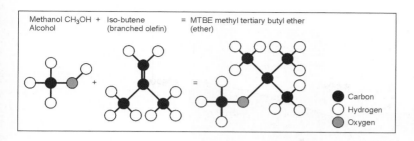

Methanol CH₃OH + Iso-butene = MTBE methyl tertiary butyl ether
Alcohol (branched olefin) (ether)

● Carbon
○ Hydrogen
◐ Oxygen

Fig. F147
MTBE from methanol
and isobutylene

F

The EC directive 85/536/EEC, which regulates use of constituents containing oxygen in gasoline fuels, allows ethers with five or more hydrocarbon atoms in the molecule with boiling points below 210°C at a maximum concentration of 15% by volume.

MTBE has proven to be particularly suitable in premium plus quality gasoline as an octane rating-increasing constituent, because its use achieves a high octane rating without significantly increasing the percentage of aromatics. The slightly higher fuel consumption (approximately 1–2%) and slight increase in the concentration of aldehyde in the exhaust are minor disadvantages compared to fuels produced using hydrocarbons exclusively. However, aldehydes can be broken down by catalytic converters in the same manner as hydrocarbons. The advantage by comparison is a reduction of the CO as well as HC emissions. MTBE, therefore, represents an attractive constituent for gasoline. Its partial water solubility, relatively poor biological degradability, and the resulting disagreeable taste of polluted water (even in very low concentrations) led to prohibition of the use of MTBE in 2003 in California because of complaints about drinking water. In Germany, the contamination risk is minimized by the high safety standards (double wall storage tanks, leak indicators) and MTBE continues to be used because of its other positive properties. At present, in German it is added to premium plus gasoline in the amount of approximately 6–7% by volume.

~~Hydrogen. Hydrogen is the most promising alternative fuel for the long term because the basic material, water, is available without limitation, and water is produced again during combustion (closed cycle). Breakdown of the water would be accomplished preferably using renewable energy such as solar power or water power because of the high energy requirement of the processes. Utilization of bacteria is being studied as a further possibility for production of hydrogen. In hydrogen production, the quantity is not limited, as with the direct utilization of solar energy using biomass. Theoretically, there are three possibilities for transport and storage in the distribution system, as well as in the vehicle itself. These are:

- High-pressure tank
- Metal hydride accumulator
- Liquid storage

For safety reasons, high-pressure tanks are expensive, heavy, and unsuitable for the transport of large quantities of hydrogen or for installation in passenger cars; however, their use in commercial vehicles cannot be fully excluded. Metal hydride accumulators, in which the hydrogen is attached to metal alloys, offer a high safety standard; however, the storage quantity is limited in spite of the status of development already reached. A further method of storage is low-pressure liquid storage at −235°C, in combination with high efficiency insulation. However, the high energy requirement for this extremely low temperature cooling of hydrogen decreases the energy balance significantly, because it cannot be reclaimed. Also, when the vehicle stands still for longer periods of time, small quantities of hydrogen are lost due to blow-off. Outdoors, this

Designation	Total formula	Boiling point °C	Density kg/m³ at 20°C	Vapor pressure hPa	Octane rating RON	Octane rating MON	Calorific value H_u MJ/kg	Oxygen content % by weight
MTBE/Methyl tertiary butyl ether	CH₃OC(CH₃)₃	55	740	480	114	98	26.04	18.15
ETBE/Ethyl tertiary butyl ether	C₂H₅OC(CH₃)₃	72	742	280	118	102	26.75	15.66
DIPE/Di-Isopropyl ether	(CH₃)₂CH-OCH(CH₃)₂	68	725	240	110	100	26.445	15.66
TAME/Tertiary amyl methyl ether	H₃C-O-CH₂-C(CH₃)₃	85	770	160	111	98	27.905	15.66
PTBE/Isopropyl tertiary butyl ether	(CH₃)₂CH-OC(CH₃)₃	88,5	740	200			27.461	13.77
Premium gasoline for comparison		25–215	approx. 750	500–900	96	85	approx. 41	0–2

Fig. F148 Ethers are suitable as constituents in gasoline fuels as a matter of principle

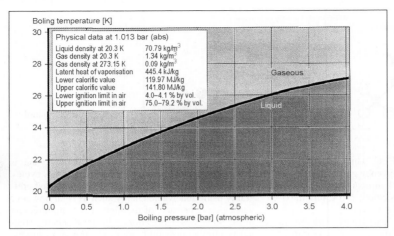

Fig. F149
LH2 boiling curve
and physical data of
oxygen (Source:
Solar-Wasserstoff-
Bayern GmbH)

does not pose a safety problem, because hydrogen (15 times lighter than air) rises immediately in contrast to hydrocarbon gases and distributes in the atmosphere resulting in low, noncritical concentrations.

The boiling curve and physical characteristics of hydrogen are shown in **Fig. F149**. In relation to its mass, liquid hydrogen contains approximately three times as much energy (lower calorific value) as hydrocarbon fuels and significantly wider ignition limits in the atmosphere. The caloric value of a stochiometric hydrogen/air mixture is, in contrast, slightly lower than that of hydrocarbon fuels (**Fig. F149**).

It is necessary to place high requirements on the fueling operation for the vehicle, so that practically only automatic fueling without human manipulation is feasible. The low temperature and complete evacuation of moisture and air require new technologies, and these must be sufficiently robust for everyday operation. The first hydrogen fueling facilities have already proven

their capability for proper operation. **Fig. F150** shows the basic schematics of one of the first such systems.

Fuel cells with an electric motor as well as gasoline engines could conceivably convert the energy of the hydrogen into power in the vehicle. The fuel cell is considered to have the greatest potential for the future, because it is capable of converting the energy contained in hydrogen at high efficiency and with minimum emissions in a "cold combustion process." The first stock vehicles with fuel cells are already in use.

In gasoline engines, mixture preparation and control of combustion requires adaptation to the properties of hydrogen. Combustion in the very lean range is optimum in terms of control of the combustion process as well as the exhaust emission (e.g., NO_x; **Fig. F151**), although it does involve power losses. Air/hydrogen ratios with a higher energy content—that is, a minimum of excess air—can be controlled only by injecting additional water into the intake manifold at the

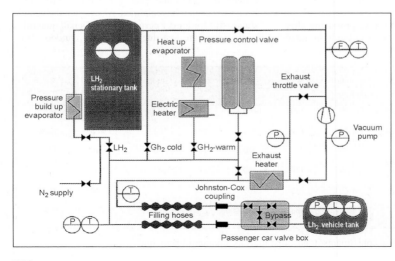

Fig. F150
Basic schematic of an
LH2 fueling facility

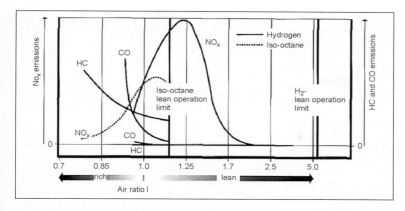

Fig. F151
Exhaust emissions without exhaust treatment, and lean limits for use of hydrogen in gasoline engines compared to iso-octane

present status of development, because without water injection, backfiring occurs in the intake system. Work is currently being done on injecting the liquid water directly into the combustion chamber to eliminate this problem. The tendency of hydrogen to ignite easily makes it impossible to use high compression engines—that is, the thermodynamic efficiency is lower.

Hydrogen is the optimum fuel for vehicles with fuel cells and electric motors. Because of the technical problems involved in carrying hydrogen in motor vehicles, methanol is now being used. The methanol is converted to hydrogen before it is fed to the fuel cell. It is preferable to use saturated hydrocarbons instead of methanol, and this could also be satisfied. Both versions produce natural carbon dioxide, so that completely CO_2-free combustion is not achievable by these methods.

~~**Liquefied petroleum gas (LPG).** Liquefied petroleum gas is an alternative gaseous fuel at normal pressure and temperature consisting primarily of the refinery gases propane and butane. LPG provides advantages compared to gasoline and particularly to diesel fuel in terms of the exhaust emissions (CO, HC, particulate, ozone formation potential). Because of the increasing use of natural gas as an alternative to gasoline, the significance of liquefied petroleum gas has decreased considerably.

The quality requirements for liquefied petroleum gas are standardized in the Europe-wide EN 589. The primary requirements are shown in **Fig. F152**.

In addition, it is necessary to ensure that a characteristic unpleasant odor is perceptible by 20% of the lower explosion limit. To ensure cold starting ability, it is necessary to maintain a vapor pressure of at least 250 kPa in the various classes, A–D, at the defined

Properties	Units	Min. limits	Max. limits	Test procedure
Resistanced to knocking MON		89		1)
Percentage of dienes	Quantity of substances in %		0.5	EN 27941
Hydrogen sulfide		Negative, passed		EN ISO 8819
Percentage of total sulfur (after odorizing)	mg/kg		100	EN 24260 ASTM D 3246
Corrosive effect on copper (1h at 40°C)	Degree of corrosion			EN ISO 6251
Evaporation residue	mg/kg		100	EN ISO 13757
Vapor pressure (pressure gauge vapor pressure) at 40 °C	kPa		1550	EN ISO 4256 EN ISO 8973
Vapor pressure (pressure gauge vapor pressure) min. 150 kPa at temperature of for class A for class B for class C for class D	°C		−10 −5 0 +10	EN ISO 8973
Water content		Free of undissolved water at 0°C		
Methanol content	mg/kg	--	2000	ISO 8174
Odor		Unpleasant and specific at 20% of lower explosion limits		

Fig. F152 Quality requirements for liquid petroleum gas

temperature, between -10 and $+10°C$. The corresponding class must be defined nationally, depending on the ambient temperature. On the fuel side, the vapor pressure can be manipulated by changing the propane/butane ratio in the mixture. The amount of dienes (diolefins: i.e., double unsaturated hydrocarbons—e.g., 1,3 butadiene) is limited to ensure the storage stability and to prevent formation of residues in the engine intake system.

Taxation in various countries has highly influenced the use of liquefied petroleum gas in vehicles. In Germany, in contrast to The Netherlands and Italy, liquefied petroleum gas was subject to taxation at a level similar to that of gasoline. The costs for installation of an additional tank and an evaporator in the vehicle amounting to several thousand Euros could not be recovered during the course of vehicle operation. Additional limitations resulting from less space in the vehicle trunk and prohibitions regarding the use of underground parking facilities and parking buildings, plus the lack of a complete network of filling stations around the country, have also impeded the use of liquefied petroleum gas.

"Dual fuel" operation with either gasoline or liquefied petroleum gas also prevented optimization of the engines—for example, the compression ratio could not be raised for operation with LPG. This meant that advantages of liquefied petroleum gas that were technically feasible could not be fully utilized.

~~**Natural gas.** Natural gas is a gas consisting primarily of methane (CH_4) used worldwide for heating and power production in industry and private households. It has various compositions and properties depending on the location of the source. In addition to the primary constituent, methane, heavier hydrocarbons (ethane, propane, butane) as well as carbon dioxide and particularly nitrogen can be present. The calorific value of natural gas decreases depending on the percentages of CO_2 and N_2, and it is differentiated into H gas (high calorific value) and L gas (low calorific value; **Fig. F153**). LL gas with even lower calorific value, is hardly

	Lower calorific value H_u		
	MJ/m³	MJ/l	MJ/kg
Natural gas, H-gas, North Sea	39.1	–	46.8
Natural gas, H-gas GUS	35.9	–	49.1
Natural gas, L-gas, Netherlands	33.4	–	40.3
Gasoline, Premium	–	30.3–31.5	40.7–41.6

Fig. F153 Typical calorific values of various types of natural gas compared to premium gasoline

suited for use in passenger car engines. After production, natural gas is desulfurized and cleaned; otherwise, its composition is not changed. However, local gas suppliers frequently have the possibility of compensating peak requirements by adding liquid petroleum gas and air. Although this has no significant effect on the calorific value, it is disadvantageous for use in motor vehicle engines, because the octane rating decreases.

The long-term availability, existing infrastructures, and good combustion properties for gasoline engines have made natural gas one of the most promising alternative fuels; it has replaced the other alternative fuel, liquid petroleum gas, in many areas. More than one million natural gas-propelled vehicles are already in operation in Argentina, the United States, and Italy. The percentage of such vehicles is also increasing in Germany, primarily in commercial fleets.

Natural gas can be stored either in gaseous form at a pressure of 200 bar (CNG = compressed natural gas) or in liquid form at $-160°C$ and 2 bar (LNG = liquefied natural gas). The latter storage is technically very complicated (*also, see above*, ~~Hydrogen) and the energy required for liquefaction cannot be reclaimed. For this reason, only the gaseous form of storage is used for commercially used vehicles. At special CNG gas filling stations, after the gas is dried, a compressor and high pressure tank are used to store the natural gas from the municipal gas line for use in motor vehicles (**Fig. F154**). Generally, the advantages of operation with natural gas are lower untreated emissions, particularly carbon monoxide, air toxics, and

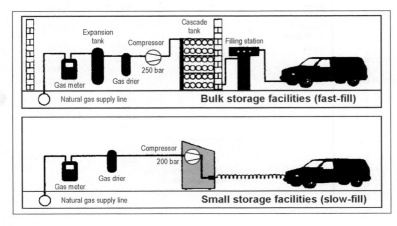

Fig. F154
Layout of natural gas tank facilities

the ozone-formation potential of the exhaust gases. Use in city buses with gasoline engines also provides the advantage of virtually no particulate emissions compared to operation with diesel engines. In gasoline engines converted to solely natural gas operation, the significantly higher octane rating of 115–130 allows the compression ratio to be increased; therefore, the thermal efficiency is improved. Because methane does not react as quickly as gasoline, it is necessary to adapt the combustion process by advancing the ignition timing and modifying the catalytic converters.

A further disadvantage compared with gasoline and diesel fuels is the lower storage density, resulting in lower range with the same fuel tank volume, as well as reductions in performance of passenger car engines because of the lower filling of the combustion chambers and significantly higher consumption (in city buses approximately 25% greater compared to diesel fuel). In passenger cars with dual natural gas/gasoline operation, the additional costs and space requirement for the second fuel system must also be considered.

~~Propane. Propane is a natural constituent of crude oil just as butane is, and it is produced as a product during distillation and cracking of petroleum. Propane is the main constituent in liquefied petroleum gas.

Propane is a saturated hydrocarbon (paraffin) that is in a gaseous state at normal pressure and temperature.

- Chemical formula: C_3H_8
- Boiling point: $-42°C$
- Octane rating: MON 96.0
- Calorific value: 46.3 MJ/kg

Because of its high volatility, propane is not preferred for use in gasoline and, at most, only traces are present.

~Antiknock additives. Antiknock additives improve the knock behavior (increase the octane rating) of gasolines. Antiknock additives were particularly significant before it was possible, or only possible to a limited extent, for refinery processes to increase the octane rating of fuels. The most familiar product here was tetraethyl lead. The use of antiknock additives permitted a significant increase in the octane rating and, hence, the compression ratio (**Fig. F155**) as well as significant improvement of the specific engine performance and a corresponding reduction of specific fuel consumption.

The incompatibility of lead with catalytic converters and increasing concern regarding the health and environmental influences of lead emissions, in combination with the "scavengers" based on chlorine and bromine that are usually used to prevent combustion residues, have led to legal measures to prohibit lead and other organo-metallic antiknock additives in most countries. In Germany, only unleaded fuel has been sold since 1996.

Other organo-metallic antiknock additives include primarily:

- Iron pentacarbonyl
- Manganese pentacarbonyl

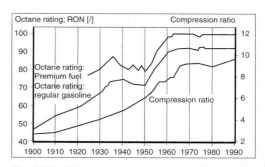

Fig. F155 High octane ratings of fuels allow higher compression ratios in engines

F

- MMT (methylcyclopentadienyl manganese tricarbonyl)
- Ferrocenes (dicyclopentadienyl iron).

However, because of the side effects and reservations regarding their toxicity, these products are not used worldwide or their use is limited (e.g., MMT in Canada in low concentrations of 0.018 g Mn/L of fuel, giving an octane rating increase of approximately 1–2 units). **Fig. F156** shows the effectiveness of some such compounds.

The use of iron pentacarbonyl results in deposits of iron oxide in the combustion chamber; these led to spark plug problems and high engine wear when used during the 1930s. In contrast to iron pentacarbonyl, ferrocene is intended for use in low concentrations (10–30 ppm). At such concentrations, the RON can be increased by more than one unit and the MON by just one unit.

In Germany, organo-metallic additives are prohibited by the leaded gasoline law unless they have been proven to be safe in terms of their toxicological effects after many years of testing; such tests have been successfully completed for ferrocene.

Unfortunately, the comprehensive search for effective organic antiknock agents that can be produced economically, that are soluble in gasoline, and that are free of undesirable side effects has been unsuccessful to date.

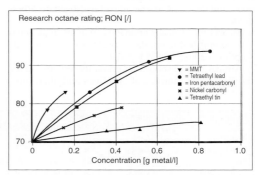

Fig. F156 Increase in octane rating with organo-metallic antiknock agents

Constituents	RON	MON
3.4-dimethyl aniline	370	320
3.5/-dimethyl aniline	340	210
p-toluidine	340	305
p-ethyl aniline	320	300
Diphenyl amine	310	300
Aniline	310	290
p-tert.-butyl aniline	300	260
N-methyl aniline	280	250
Idoline	300	150
N,N-dimethyl aniline	95	84

Fig. F157 Blending octane numbers for organic compounds containing nitrogen (values determined by adding 2% by weight to premium fuel) (Source: Ullmanns Encyclopedia of Industrial Chemistry 1991)

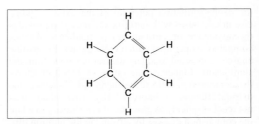

Fig. F158 Chemical structure of benzene

The reason for this could be their rapid decomposition at the beginning of combustion. The best known products are organic nitrogen compounds (aniline and aniline derivatives). However, the concentrations to be used were in the percentage range for octane rating increases of 1–2 units. In addition to the costs, other disadvantages included possible formation of residues and an increase in oxides of nitrogen (NO_x) in the exhaust. **Fig. F157** shows blending octane numbers for compounds containing nitrogen measured at a concentration of 2%.

~Antiknock properties. *See below*, ~Octane number

~Antioxidants *See above*, ~Additives; *also*, →Fuel, diesel engine ~Additive

~Aromatics →Fuel, diesel engine ~Aromatic contents

~Aromatics contents. Aromatics are organic compounds containing one or multiple benzene aromatic bond systems. The typical "aromatic" odor is characteristic. Aromatics are already present in crude oil, although they are produced primarily by catalytic reformers (*see below*, ~Production), releasing hydrogen (required for desulfurization). They are the hydrocarbons containing the lowest ratio of hydrogen to carbon atoms.

Measurement of the aromatics contents in gasoline is accomplished using the fluorescence indicator adsorption method ASTM D 1319.

Aromatics in gasolines are primarily responsible for the high octane ratings of gasolines and, therefore,

allow high compression ratios in the engines with high specific power and low specific fuel consumption. Because of the higher octane rating of premium fuel, it also contains more aromatics than regular gasoline.

The average aromatics contents in German fuels (from market analysis in 2003) are:

- Regular gasoline   31% by volume %
- Super   35% by volume %
- Premium plus   37% by volume %

In contrast to diesel fuel, only single-ring aromatics are present in gasoline, due to the lower final boiling point of maximum 210°C (only traces of multiple ring aromatics can be measured in the ppm range). The simplest aromatic compound is benzene C_6H_6 (**Fig. F158**).

Because of its known toxicity, benzene lost its significance as a gasoline constituent for increasing the octane rating some time ago and is minimized or removed in the refinery constituents for production of gasoline (*see below*, ~Benzene content). The aromatics remaining in gasoline are shown in **Fig. F159**.

With these aromatics, one or more of the hydrogen atoms in the benzene molecule is replaced by a methyl group (CH_3) or a longer hydrocarbon chain (e.g., toluene; **Fig. F160**).

Only slight concentrations of aromatics with more than 10 carbon atoms are present due to the specified fuel boiling curve and the final boiling point. **Fig. F161** shows typical percentages of aromatics in German gasolines.

During combustion in the engine, the aromatics behave similarly to other hydrocarbons—that is, greater than 99% is burned, depending on the operating condition. The catalytic converter then converts the remaining operating fuel, amounting to <1%, with an efficiency of greater than 90% at constant operating conditions and nearly 100% under the conditions of the exhaust emissions test. Aromatics (e.g., toluene) and olefins (e.g., ethene) contribute significantly to

Product	Formula	Boiling point/range	Mixed octane rating RON	Mixed octane rating MON
Toluene	C_7H_8	110 °C	124	112
Ethyl benzene	C_8H_{10}	136 °C	124	107
Xylenes	C_8H_{10}	138–144 °C	120–146	103–127
C_9 aromatics	C_9H_{11}	152–176 °C	118–171	105–138
Small quantities of C_{9+} aromatics	$C_{10}H_{12}$ $C_{11}H_{13}$	169–210 °C	114–155	117–144

Fig. F159
Aromatics in gasoline

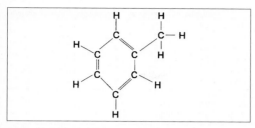

Fig. F160 Chemical structure of toluene

Concentration of specific aromatics in German fuels				
		Regular	Premium	Premium plus
Toluene	% by volume	9.3	13.0	9.3
Xylene	% by volume	9.5	9.8	9.5
C_{8+} aromatics	% by volume	11.3	13.4	11.1

Fig. F161 Percentage of aromatics in German fuels
(average, winter 2003)

reduction of NO_x in the catalytic converter when the air-fuel ratio is adjusted slightly into the rich range. Alkanes, such as methane, are only capable of converting the NO_x in the catalytic converter under richer

conditions (**Fig. F162**). Emission tests with various fuels accordingly showed that fuels with higher percentages of aromatics led to lower NOx emissions. However, the carbon monoxide and hydrocarbon emissions were, at the same time, higher. Studies performed by the automotive and mineral oil industry on the influence of the fuel on exhaust emissions (EPEFE program) have confirmed these results (**Fig. F163**). Further studies have shown that the remaining unburned aromatics in the exhaust have a tendency toward dealkylation. In an attempt to further reduce the unburned aromatics remaining in the exhaust downstream of the catalytic converter (including benzene), the percentage of aromatics in gasolines has been limited to a maximum of 42% by volume since the year 2000. The next step is to limit this value to a maximum of 35% by volume for 2005. For premium and premium plus fuels, this level requires construction of new refinery facilities for the production of isoparaffins.

~Autogas. *See above,* ~ Alternative fuels ~~ Liquefied petroleum gas

~Basic gasoline. Gasoline conforming to the standard before the addition of quality-improving additives.

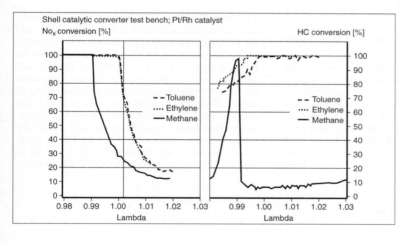

Fig. F162
Aromatic support
conversion of NOx

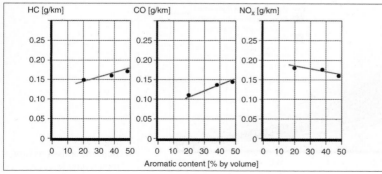

Fig. F163 Reduction in NO_x emissions and increase in HC and CO emissions in the exhaust brought about by aromatics in fuel (ECE + EUDC cycle, average: 16 vehicles with catalytic converter)

~Benzene. *See below,* ~Benzene content

~Benzene content. Benzene is a natural constituent in crude oil and is the basis for all aromatic compounds. It has a boiling temperature of 80°C and is, therefore, within the boiling range of gasoline. Benzene is also formed in small quantities from other fuel constituents during the combustion process in the engine. Because of its high octane rating (RON and MON each greater than 100) and its availability from hydration of coal, benzene was previously used as an active blending constituent in gasoline. After introduction of the catalytic reformer in the 1950s, the blending in of benzene lost significance in Germany, and, following recognition of the health risks involved with handling benzene, it was then eliminated as an additive. Since the year 2000, the maximum permissible concentration of benzene has been 1% by volume. However, even before this limit became effective, benzene was minimized in fuel and its constituents. Various procedures were used for this purpose:

- Maximization of constituents containing little benzene but with acceptable octane ratings, such as butane, isomerisate, light "cat crack" gasoline, polymer gasoline, alkylate, and MTBE
- Reduction of the benzene content in catalytic reformat—by increasing, for example, the initial boiling point of the product added
- Minimization of benzene "producers" in the reformer product
- Extraction of benzene and processing in the chemical industry

The percentage of benzene in German gasolines has been reduced by the measures described to an amount below the legally specified limit of 1% by volume.

For high octane super plus fuel, this limit has been maintained on a voluntary basis since 1995 because this fuel was initially used frequently in vehicles without catalytic converters and with higher exhaust emissions. This has resulted in a significant reduction in the relatively high benzene emissions in the

	Regular	Premium	Premium plus	Premium leaded
1986	2.4	2.8	–	2.8
1988	2.2	2.6	–	2.8
1990	2.2	2.8	2.6	2.7
1992	1.8	2.5	2.4	2.5
1994	1.6	2.1	2.0	2.3
1996	1.5	2.1	0.8	–
1998	1.6	2.0	0.8	–
2000	0.8	0.7	0.6	–
2003	0.7	0.7	0.6	–

Fig. F164 Development of benzene content in German gasolines (the limit, up to 1999, was a maximum of 5% per volume; after 2000, it was a maximum of 1% by volume)

exhaust from such vehicles. Technically, this minimization in premium plus was possible because this fuel is produced in relatively low quantities. **Fig. F164** shows the development of the benzene content in German gasolines.

Together with the rapid penetration of vehicles with catalytic converters and further improvements in motor vehicle engineering, the decreasing benzene content in the fuel has led to a continuous and significant reduction of benzene emissions from motor vehicles (**Fig. F165**).

~Benzine. Benzines are fuels with low density and a low final boiling point. They were the only usable fuel for the first engines because initially only so-called surface carburetors were available, and these were capable of forming mixtures capable of ignition only with extremely volatile benzine. In such carburetors, the fuel/air mixture was produced by allowing the air sucked in to pass over the fuel present in the flow chamber with a certain turbulence. In this type of mixture preparation, it was necessary for the fuel to be significantly more volatile than it is now. Typical densities of benzine were significantly below 700 kg/m³ with final boiling points less than 100°C. For example,

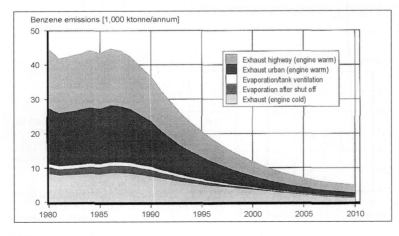

Fig. F165
Benzene emissions from motor vehicles

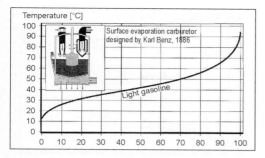

Fig. F166 Surface carburetor and boiling curve for benzine

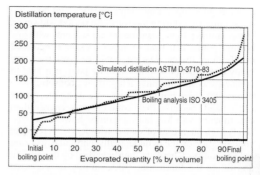

Fig. F168 Boiling curves for gasolines evaluated by different methods

in 1907 the US Navy required a maximum fuel density of approximately 670 kg/m³ (**Fig. F166**).

Although as early as 1983, Maybach invented a spray nozzle carburetor which was capable of processing fuels with today's boiling characteristics, fuel densities and final boiling points developed slowly (up to approximately 1920) to the level now common: this occurred in unison with the improved carburetor engineering.

~Biogas →Fuel, diesel engine ~Biogas

~Biomass. *See below*, ~Fuels containing ethanol

~Blended fuels. *See below*, ~Methanol fuel

~Boiling characteristics. The boiling characteristics are an important evaluation criterion for the quality of gasoline. Unsuitable boiling characteristics inhibit starting and operation of gasoline engines (e.g., diesel fuel in a gasoline engine). The variety of hydrocarbons contained in gasoline (up to 400 different substances) does not allow specification of a specific boiling point determined with the standardized boiling apparatus (EN ISO 3405), but rather a boiling curve with initial boiling point at approximately 30°C and final boiling point between 195 and 210°C (**Fig. F167**).

According to the boiling analysis method specified in EN ISO 3405, the fuel specimen used is evaporated at variable heat supply with a defined temperature increase of 1°C/minute and then condensed (constant distillation rate of 4 to 5 ml/min). The resulting boiling curve provides a great deal of information for application-related criteria. Because of the high distillation rate in this standardized analysis method, however, the boiling analysis is not exact physically because more highly volatile substances in the fuel are held back while less volatile substances are entrained prematurely. A significantly more precise description of the

boiling characteristics, not required for practical evaluation of fuels, is the so-called true boiling point distillation method according to ASTM Standard D 2892 and "simulated distillation," according to ASTM Standard D 3710, which is simulated from gas chromatographic analysis of the fuel. Because of their complexity, however, the last two methods mentioned are not suitable for standard quality control of fuels. A comparison between the boiling curves using the standard apparatus according to EN ISO 3405 and simulated distillation indicates that the exact initial boiling point is below -20°C and the final boiling point over 250°C and that the boiling curve has the shape of a finely incremented stepped curve depending on the individual hydrocarbons in the mixture (**Fig. F168**).

The boiling characteristics influence primarily the starting and operating behavior of the engine as well as the emissions. The significance of the curve and the individual standard boiling curve ranges for engine operation are shown in **Fig. F169**.

Well-balanced boiling characteristics are a primary prerequisite for operation of automotive gasoline engines under all the operating conditions encountered. Light components—that is, with low boiling point—ensure that the cold engine can be started quickly and

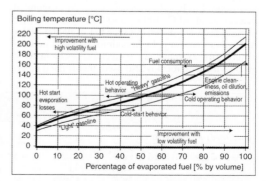

Fig. F169 Gasoline boiling characteristic—significance for engine operation

Final boiling point	Regular	Premium	Premium plus
Range (°C)	180–206	181–203	185–201
Average (°C)	192	189	192

Fig. F167 Final boiling point for German gasolines, standard value: max 210°C (winter 2003)

provide good operating characteristics and low exhaust emissions during the warm-up period. However, in the summer, high quantities of constituents with low boiling points can lead to vapor lock and increased evaporation losses. During cold, wet weather, too many constituents with low boiling point can lead to carburetor or throttle valve icing. Heavy constituents — that is, with high boiling points — are desirable, on the other hand, because they contain more energy than constituents with low boiling points; however, excessively high quantities of such constituents can result in condensation on the cylinder walls, particularly during cold operation. These are then absorbed by the oil film and dilute the lubricating oil, resulting in increased wear and higher exhaust emissions.

Fig. F170 shows different boiling curves for fuels; a balanced boiling curve for a standard commercial fuel is illustrated in the middle. The fuel with boiling gaps, with too few constituents in the medium boiling range, leads to poor operating characteristics such as "jerking during acceleration." A so-called plateau fuel (aviation fuel Avgas 100 LL) consists primarily of constituents that boil at about 100°C (iso-octane in alkylate used).

In addition, a methanol-blended fuel, "M 15," with 15% methanol, is shown. The boiling curve indicates the azeotropic boiling characteristics for this fuel in the lower boiling range. The higher latent heat of vaporization of the methanol, which leads to poorer cold starting characteristics, is not indicated in the illustrated dependency of the temperature and the evaporated percentage of the fuel, nor can it be recognized from the boiling curve, because of the automatic adaptation of the energy supplied to maintain the specified temperature increase during distillation.

In the European requirement standard for gasoline, six volatility classes are defined, of which the four highly volatile classes are subdivided again in order to take into consideration geographic and seasonal changes in the ambient temperatures. According to this method, the permissible volatility ranges for the gasolines are defined taking into account the local vapor pressure, which depends on the country and season.

Typical production values for the primary distillation values E_{70}, E_{100}, E_{180} (in each case, the evaporated percentage at 70°C, 100°C, 180°C, respectively) for German fuels are shown in **Fig. F171** using the winter fuel type as an example. In addition to the E nomenclature — for example, E_{70} (percentage evaporated at 70°C) — the reciprocal is also used; that is the temper-

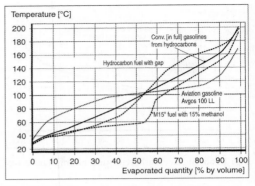

Fig. F170 Boiling curves of various gasolines

ature at which 10%, 30%, and 50% of the fuel is evaporated, or T_{10}, T_{30}, T_{50}.

~**Boiling curve.** *See above*, ~Boiling characteristics

~**Butane.** *See above*, ~Alternative fuels; *also*, →Fuel, diesel engine ~Liquefied petroleum gas (LPG)

~**Calorific value.** It is necessary to differentiate between the gross calorific value, H_O, and the lower calorific value, H_U. The gross calorific value is determined in a combustion bomb (calorimeter) by complete combustion in conformance with DIN 51900. Following combustion, carbon dioxide is present in gaseous form while the resulting water vapor has condensed. However, the water vapor does not condense during combustion in the engine, and the gross calorific (heating) value is unrealistically high for practical evaluation of fuels and engine performance. For this reason, the hydrogen content is determined by means of gravimetric analysis, the energy of condensation of the water vapor is calculated using this figure, and this is then subtracted from the gross calorific value. The result is referred to as the lower calorific value. The calorific value of fuel is one of the values resulting from fuel composition during its production. It is not measured as a quality criterion; however, it is necessary to determine this value for certain research and development work. **Fig. F172** shows calorific values for regular, premium and premium plus fuels. The composition of the fuels is shown in **Fig. F173**. It is interesting to note that, on the basis of mass, regular gasoline has a higher

		Regular	Premium	Premium plus	Standardized range (EN 228) Class D Germany Winter
E70 (evaporated percentage at 70 °C)	Average	37	35	34	
	range	26–46	28–47	24–48	22–50
E100 (evaporated percentage at 100 °C)	Average	58	55	54	
	range	50–68	48–63	45–61	46–71
E180 (evaporated percentage at 180 °C)	Average	93	97	96	–
	range	95–99	93–99	95–99	
E150 (evaporated percentage at 150 °C)	Average	87	87	87	min. 75
	range	81–96	84–95	84–91	

Fig. F171
Distillation values for German gasolines in winter 2003 (percentage by volume)

No.	Type	Density 15 °C (kg/m³)	Chemical analysis			Calorific value, H$_o$ (MJ/kg)	Calorific value, H$_u$	
			C (m%)	H (m%)	O (m%)		(MJ/kg)	(MJ/l)
1	Regular unleaded	730.6	86.85	12.51	0.23	44.62	41.89	30.61
2	Regular unleaded	739.1	87.81	11.98	0.05	43.79	41.18	30.43
3	Regular unleaded	735.6	87.75	11.94	0.12	44.09	41.49	30.52
	Average value	735.1	87.47	12.14	0.13	44.17	41.52	30.52
4	Premium unleaded	745.2	88.22	11.34	0.11	43.14	40.67	30.30
5	Premium unleaded	755.3	88.65	11.04	0.36	43.10	40.69	30.73
6	Premium unleaded	746.7	88.65	11.40	0.04	43.42	40.93	30.56
7	Premium unleaded	756.9	88.06	11.32	0.44	44.11	41.64	31.52
	Average value	751.0	88.40	11.28	0.24	43.44	40.98	30.78
8	Premium plus	753.1	87.16	11.40	1.47	42.80	40.31	30.36
9	Premium plus	772.6	88.40	10.15	1.29	42.30	40.09	30.97
10	Premium plus	748.3	86.87	11.71	1.17	42.74	10.19	30.07
11	Premium plus	770.1	88.09	10.50	1.39	42.93	40.64	31.30
	Average value	761.0	87.63	10.94	1.33	42.69	40.31	30.67

Fig. F172
Calorific values and basic analysis data of common commercial fuels

F

gravimetric calorific value due to its higher paraffin content and, therefore, its higher percentage of hydrogen. However, when the volumetric calorific values relevant for practical operation are compared, the premium fuels have an advantage due to their higher density. Although premium plus fuel has a slight advantage over regular gasoline in terms of volumetric calorific value, it does not quite reach the level of premium fuel because of the oxygen content resulting from the use of MTBE (methyl tertiary butyl ether). The calorific values alone do not provide any information on fuel consumption. The design of the engine (e.g., compression ratio) and its adaptation to the fuel quality selected by the manufacturer are also significant.

~Carburetor icing. *See above*, ~Additives

~CNG (compressed natural gas) →Fuel, gasoline engine

~Composition (*also, see below*, ~Production). Gasoline consists of a variety of individual hydrocarbons with different structures and molecular sizes belonging to three major groups.

Alkanes. Collective designation for saturated hydrocarbons with maximum possible quantity of hydrogen. It is possible to differentiate among normal paraffins

(carbon atoms without branches in a chain), isoparaffins (branched chains), and naphthene hydrocarbons in which the carbon atoms are arranged in the shape of a ring (cyclic).

All n-paraffins and isoparaffins have the chemical formula C_nH_{2n+2}, while naphthenes have the chemical formula C_nH_{2n} (**Fig. F174**).

Alkenes (olefins). Collective designation for unsaturated hydrocarbons. Characteristics include one or more double carbon bonds in a carbon chain or carbon ring and olefins (**Fig. F175**).

Aromatics. Compounds containing one or multiple double aromatic benzene bonds and with side chains (**Fig. F176**).

In a study recently performed at the request of DGMK (German Coal Science and Technology Society for Petroleum, *See above*, ~Alternative fuels ~~Natural gas), the individual hydrocarbons in a number of standard commercial fuels were determined by means of complex gas chromatographic analysis. About 200 hydrocarbons and compounds containing oxygen with mass percentages of at least 0.1% were identified and analyzed.

~Corrosion protective additives. *See above*, ~Additives

~Corrosive effect on metals

No.	Type	FIA analysis (% by volume)			MTBE (% by volume)
		Aromatics	Olefins	Paraffins	
1	Regular unleaded	24.5	20.5	55.0	–
2	Regular unleaded	32.0	19.5	48.5	–
3	Regular unleaded	29.5	23.0	47.8	–
4	Premium unleaded	34.0	14.6	48.5	–
5	Premium unleaded	41.0	8.3	48.4	–
6	Premium unleaded	36.0	12.0	52.0	–
7	Premium unleaded	–	–	–	–
8	Premium plus	39.9	5.9	44.9	9.3
9	Premium plus	51.6	5.1	35.5	7.7
10	Premium plus	38.2	7.4	47.5	6.8
11	Premium plus	–	–	–	9.8

Fig. F173
Composition of fuels listed in **Fig. F172**

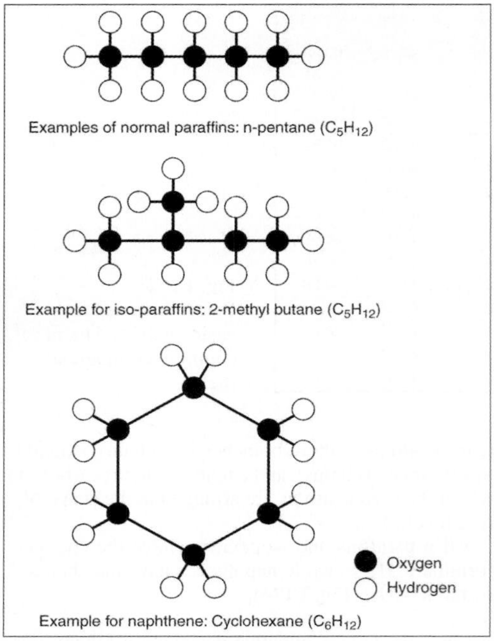

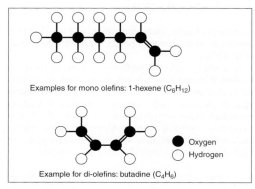

Examples of normal paraffins: n-pentane (C_5H_{12})

Example for iso-paraffins: 2-methyl butane (C_5H_{12})

● Oxygen
○ Hydrogen

Example for naphthene: Cyclohexane (C_6H_{12})

Fig. F174 Example for alkanes (saturated hydrocarbons)

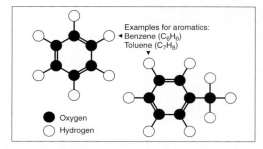

Examples for mono olefins: 1-hexene (C_6H_{12})

● Oxygen
○ Hydrogen

Example for di-olefins: butadine (C_4H_6)

Fig. F175 Examples of alkenes (unsaturated hydro-carbons)

Examples for aromatics:
◄ Benzene (C_6H_6)
Toluene (C_7H_8)
▼

● Oxygen
○ Hydrogen

Fig. F176 Examples of aromatic hydrocarbons

Corrosive effect on steel. Gasoline comes into contact with moisture in the air and oxygen as a matter of course during transport, storage, and use in motor vehicles. For this reason, corrosion can occur on lines and tanks—for example, under the influence of changing temperature and the formation of condensed water. The corrosion products formed can cause damage in the distribution chain and in the fuel system of the vehicle.

The corrosive characteristics are tested according to DIN 51 585. For this purpose, a round steel rod is immersed for 24 hours at 60°C in a 10:1 mixture of fuel and distilled water (version A) or synthetic seawater (version B). After conclusion of the test, the formation of rust is evaluated visually. The method is used, for example, for testing the effectiveness of corrosion protective additives.

Corrosive effect on copper. Corrosion of materials containing copper that come into contact with fuel (e.g., components in gasoline pumps) poses problems for two reasons. On the one hand, the components are damaged, and on the other, the dissolved copper is catalytically active, leading to formation of high molecular weight impurities in the fuel. The corrosiveness of a fuel depends primarily on the water content, the compounds containing oxygen, the type and quantity of sulfur compounds, and, naturally, the corrosion-protection additive used. To test the corrosion limit established in the European specification, a polished copper strip is brought into contact with gasoline at a temperature of 50°C for three hours (EN ISO 2160). Additives provide protection for virtually all metals coming into contact with the fuel, even under severe conditions.

~Demulsifying capability. The demulsifying capability is the ability of a fuel to precipitate water absorbed during improper transport and storage. The demulsifying capability decreases with increasing amounts of aromatics and/or MTBE, particularly in the presence of surface active substances. "Demulsifiers" or "dehazers" can be added as a remedy. The primary effect of these additives is to reduce the surface tension between the water droplets in suspension, thus allowing them to agglomerate to form larger drops, which precipitate more readily.

~Density. The density is a value for determining the mass of a particular volume of fuel and is specified in kg/m^3 corresponding to the analysis standards (EN ISO 3675, EN ISO 12185) at 15°C.

The density of fuels traditionally has been measured with hydrometers (aerometers); however, today vibration-type measuring instruments are predominant. For this purpose, a small quantity of the fuel to be tested is filled into a vibrating tube. The change in the vibration frequency of the tube enables the density of the fuel to be determined using calibration data.

The density increases with increasing carbon proportion in the gasoline—that is, increasing percentage of double bonds (aromatics, olefins), etc. Premium fuels,

Density of common commercial German gasolines (kg/m³), (Winter '03)			
	Regular	Premium	Premium Plus
Range	725–760	726–762	732–759
Average	735	741	749

Fig. F177 Density of standard commercial German gasoline fuels (winter 2003)

therefore, have higher densities than regular gasoline. Accordingly, increasing hydrogen content (isoparaffins, normal paraffin) reduce the density (reformulated gasoline).

The range of densities is specified uniformly for all three unleaded gasoline qualities—regular, premium, and premium plus—in DIN EN 228 as 720 to 775 kg/m³. **Fig. F177** shows average values and density ranges for commercially available German fuels.

Increasing the density of the fuel leads, as a matter of principle, to higher volumetric energy content, decreases volumetric fuel consumption. According to experience, an increase in the density of 1% results in a volumetric reduction of consumption of approximately 0.6%.

~Deposits. *See above*, ~Additives

~Detergents. *See above*, ~Additives

~Diesel fuel →Fuel, diesel engine

~Diolefin content. Diolefins are double unsaturated olefinic hydrocarbon compounds—such as, for example, 1,3-butadiene $H_2C = CH-CH = CH_2$.

They are particularly unstable and, therefore, undesirable in gasoline. Fuels containing diolefins show a greater tendency to form residues in the intake system. Diolefins are usually present in thermally cracked products. They can be removed by partial hydration (mild hydrogen treatment); the slight residues remaining (<1% by volume in final fuel blend) are stabilized sufficiently with antioxidants.

The high reaction tendency of diolefins does, however, result in high rates of combustion; therefore, they are of certain interest as constituents in racing fuels in spite of the described disadvantages and relatively low octane rating.

~Diolefins. *See above*, ~Diolefin content

~Distillation residue. Distillation residue is the quantity of liquid remaining in a cooled distillation flask following distillation. According to the boiling analysis conforming with EN ISO 3405, this should not be more than 2% by volume. Higher distillation residues

can indicate mixing—with, for example, lubricating oils or diesel fuel.

~Elastomer compatibility. Gasolines without alcohol additives lead to only slight swelling of all elastomers and synthetic materials used in fuel systems. If, however, alcohols (even in the permissible low concentrations) are used as blending constituents, the swelling tendency for elastomers increases considerably, particularly in the case of acrylonitrile-based elastomers. On the fuel side, the combination of methanol containing high quantities of aromatics has proven to be particularly critical. For this reason, the use of suitable elastomers—for example, fluorocaoutchouc (caoutchouc: unvulcanized rubber)—was required on the vehicle side before EU approval of alcohol additives for gasolines. Test liquids with an increased quantity of alcohol and aromatics are available for testing elastomers for their resistance to possible commercially available fuels (DIN 51604, Parts 1–3).

The swelling characteristics of various elastomers in relation to the aromatic and methanol content are shown in **Fig. F178**.

~Energy. The quantity of energy present in gasoline per unit of volume or weight (energy density) which can be released during combustion is measured as the calorific value. The lower calorific value (before condensation of the water vapor) is relevant for practical operation.

Typical values for conventional fuels are slightly above 40 MJ/kg, which is equivalent to slightly above 30 MJ/L. Due to the low density, the volumetric calorific value of gasoline is significantly lower than the value for diesel fuel.

The overall conversion into power of the energy contained in the fuel is less efficient in gasoline engines than in diesel engines because it is necessary to throttle the intake air for the former in the part-load range. Moreover, gasoline engines in comparable vehicles are operated more frequently in the unfavorable part-load range than are diesel engines because they frequently have a higher maximum output. For this reason, only approximately 4.2% of the energy contained in the fuel is used over the operating cycle for regular gasoline and 4.4% for premium plus for moving the payload in the ECE driving cycle (energy chains, fuel, diesel engine).

~Ethanol. *See above*, ~Alcohols

~Ethanol content. Ethanol, C_2H_5OH, may be contained in gasoline up to a maximum concentration of 5% by volume in conformance with EC directive 85/536/EEC. Because of the high production costs of

F

Fuel	Percentage of aromatics	30 % by volume		40 % by volume	
	Percentage of methanol	0 % by vol.	3 % by vol.	0 % by vol.	3 % by vol.
Type of elastomer	Acryl nitrile butadiene caoutchouc	20 %	28 %	27 %	34 %
	Flouro caoutchouc	1 %	2 %	2 %	3 %

Fig. F178
Elastomer compatibility depending on fuel in percent of swelling (increase in volume)

ethanol compared with gasoline, high subsidies were required for production from biomass in pilot facilities in Germany. For this reason, ethanol is currently no longer blended with gasoline in Germany. In the future, ethanol could again become significant as a constituent in gasoline fuels because the EU Commission has plans to promote biological fuels (including blending constituents). Ethanol as a blending constituent has achieved greater significance, but also has high subsidies in Austria, France, and areas of the United States and as pure alcohol fuel in Brazil. Problems with contamination (water and acids) resulted, particularly in Brazil, from locally decentralized production from sugar cane with insufficient quality control.

~Ethanol fuels. *See below*, ~Fuels containing ethanol

~Ether. *See above*, ~Alternative fuels

~Evaporation residue. It is necessary to differentiate between "unwashed evaporation residue" and "washed (existing) evaporation residue" washed with n-heptane. The unwashed evaporation residue primarily contains pure additives and their carrier oils as well as undesirable resinous fuel residues. In contrast to this, in washed (existent) evaporation residue, the additives are specifically washed out with n-heptane, leaving only the insoluble constituents with resinous properties (gum).

Both types of evaporation residues are measured in conformance with EN ISO 6246 by blowing air at a temperature of 160–165°C over a 50-ml fuel specimen for 30 minutes.

A maximum value of 5 mg/100 ml of fuel is specified in the quality standard for gasoline fuels EN 228. Typical fuels on the market have washed evaporation residues in the range of 0 to 2 mg/100 ml. Higher amounts of washed evaporation residue levels can be an indication of undesirable sticking in the intake systems on gasoline engines (carburetor, injection nozzles, intake valves). Such problems can occur particularly when fuels without additives are used.

~Evaporative heat (latent heat of vaporization). This is the amount of energy that must be added to the fuel to change it from the liquid to the gaseous phase without changing the temperature. It is significant for mixture preparation and internal cooling and is approximately 350 kJ/kg. The evaporative heat for methyl tertiary butyl ether (MTBE) as a blending constituent is of the same magnitude (322 kJ/kg). In contrast, alcohols have significantly higher values of latent heat. However, at the maximum permissible concentration for alcohols in gasoline in the lower percentage range, the latent heat of the fuel is not influenced significantly.

~Evaporative loss. Evaporative losses, particularly of light fuel constituents, can occur over the entire chain of fuel production, storage, and distribution up to the vehicle tank during vehicle operation and standstill. A series of legal regulations has been established to limit

hydrocarbon emissions resulting from evaporation. The implementation of these regulations has significantly reduced the evaporative losses (primarily through complete gas reclamation procedures). In refineries and tank storage facilities, constituents and mixing and storage tanks for gasolines are equipped with solid roofs and gas reclamation lines. Connections for gas reclamation are also present for filling the oil tankers as well as for emptying them at the service stations. When the fuel tanks at the service station are filled, the displaced fuel vapors are caught in the oil tanker and fed into a vapor reclamation system at the oil tanker filling facility. Such systems have been installed at all refineries and tank storage facilities. This has made it possible to reduce the evaporative losses in the vicinity of the refineries from about 0.05% of the initial product to be processed (e.g., crude oil) to about 0.01%. In a second stage, evaporative losses during vehicle fueling have been reduced by approximately 70–80% with so-called vapor recovery booths on the filling station pumps. This has reduced the displacement losses while fueling from approximately 1.5 g/liter of fuel to approximately 0.3 g/liter of fuel (about 0.04%; **Fig. F179**). This entire package of measures in the distribution chain for gasoline has reduced the evaporative emissions by nearly 90%. About 90% of the fuel vapors released during fueling are primarily low boiling fractions, such as butanes and pentanes.

Evaporative losses from vehicles are limited by law in the same manner as exhaust emissions. The limit for passenger cars, after shutting off the hot engine, is determined by using the SHED test (sealed housing for evaporative determination) and is currently 2 g of hydrocarbons per test. The tests are performed using reference fuels with characteristics based on those of typical fuels on the market. Among other measures, the introduction of "charcoal canisters" in motor vehicles has significantly reduced hydrocarbon emissions from evaporation, even though fuels in the vehicle tank continue to lose light constituents, albeit only through the charcoal canisters and engine. Although the evaporative losses are slight during normal operation, relatively high evaporative losses can occur when the tank is nearly empty, particularly under extreme operating

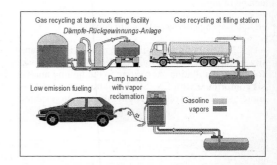

Fig. F179 Fuel vapor recycling and reclamation during fuel handling

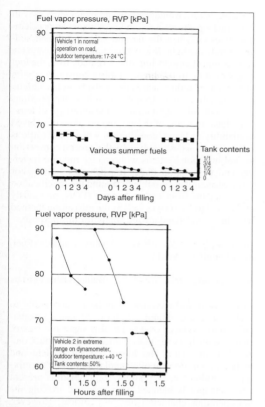

Fig. F180 Evaporative loss in a vehicle illustrated by a change in fuel vapor pressure in the vehicle tank

conditions (e.g., freeway operation with high system temperatures followed by stop-start motoring; **Fig. F180**). However, these emissions do not reach the atmosphere, but rather are fed to the engine for combustion either directly or through the charcoal canister. Factors that accelerate evaporation of light constituents on the vehicle side, in addition to high temperatures in the fuel system, also include high fuel pumping rates, which result in a rapid increase in the tank temperatures. Improvements are possible by means of requirement-controlled pumping capacity of the fuel pump and good thermal insulation of the fuel system to prevent heat transfer from the engine and exhaust system.

Naturally, on the fuel side, evaporation losses can be prevented by reducing the acceptable vapor pressure and producing fuel that is not particularly volatile. The maximum permissible vapor pressure has, therefore, been reduced for summer fuels from 70 to 60 kPa. However, further reduction of the fuel volatility would be disadvantageous for cold starting, cold operating characteristics, the HC and CO exhaust emissions when the engine is operating cold, and the CO_2 emissions (the light hydrocarbons have a lower percentage of

carbon), as well as the ozone formation potential of the unburned hydrocarbons remaining in the exhaust.

Analysis of fuel vapors from vehicles shows that, in addition to butane, the vapors contain primarily pentanes and, in lower concentrations, heavier hydrocarbons; however, the latter are in decreasing concentration with increasing boiling point (**Fig. F181**).

~Explosion limits. The so-called explosion limits mark the concentration range (percent of fuel by volume in the gas phase) within which an air-fuel vapor mixture can be caused to explode by an external ignition source. It is necessary to differentiate between a lower explosion limit (low fuel vapor level) and upper explosion limit (high fuel vapor level) in air. At fuel vapor concentrations below the lower explosion limit and above the upper explosion limit, the mixture does not explode when ignited.

Gasoline/air mixtures have explosion limits of approximately 0.6–1% by volume of fuel in air (lower explosion limits) and approximately 6–8% by volume (upper explosion limits).

When gasoline is stored, an air-fuel mixture usually forms above the fuel with a high percentage of fuel vapor far above the upper explosion limit—that is, in the safe rich range. Studies in a vehicle fuel tank under simulated operating conditions showed, however, that it was possible to drop below the upper explosion limit with fuels that had the previously acceptable minimum vapor pressure according to the requirement standard EN 228 and low volatility in combination with low ambient temperatures; then it would have been possible to ignite the fuel vapor/air mixture in the vehicle tank (**Fig. F182**). For this reason, the minimum possible vapor pressure values, which practically never occur anyway, were increased.

~Ferrocene. *See above*, ~Antiknock additives

~Final boiling point (*also, see above*, ~Boiling characteristics). The final boiling point is the temperature at which all the fuel, with the exception of the distillation residue, has evaporated in the boiling analysis performed according to EN ISO 3405. In the quality

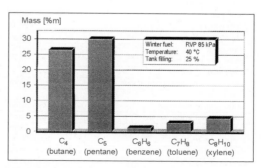

Fig. F181 Composition of fuel vapors in the vehicle tank, primarily light fuel constituents

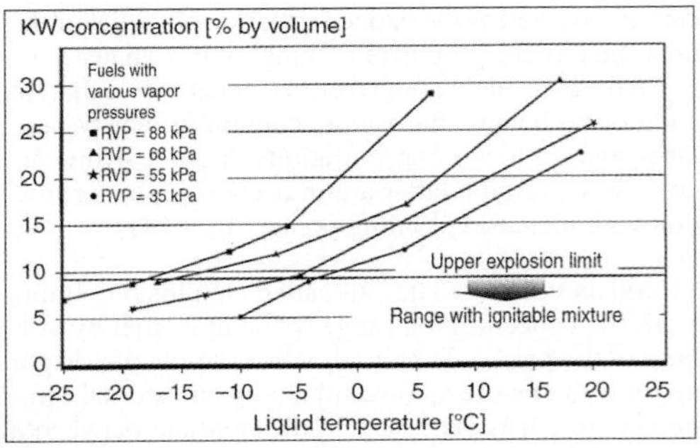

KW concentration [% by volume]

Fuels with various vapor pressures
- RVP = 88 kPa
- RVP = 68 kPa
- RVP = 55 kPa
- RVP = 35 kPa

Upper explosion limit

Range with ignitable mixture

Liquid temperature [°C]

Fig. F182 Hydrocarbon concentration in motor vehicle tank as a function of liquid phase

standard EN 228, the maximum final boiling point is specified to be 210°C. Utilization of the final boiling point allows the use of constituents with higher density and usually also higher octane rating (e.g., C_9 aromatics), which provide for low volumetric fuel consumption. However, dilution of the engine oil by condensing fuel can increase as the final boiling point increases, particularly during city driving in the winter. An increased final boiling point indicates mixtures with diesel fuel or lubricating oil.

~FON. *See below,* ~Front octane number (FON)

~Front octane number (FON). The front octane number is the octane number (RON) of the percentage of the fuel which distills out up to 100°C—that is, the highly volatile constituents. In vehicles with carburetors, this number provides information on the tendency of the fuel to knock during acceleration. The light constituents of the fuel, with the exception of butane, generally have a limited octane rating level (distillate, light reformat). This results in a lower octane rating for the light half of the fuel compared to the total fuel. This is particularly disadvantageous in carburetor engines with long intake paths resulting in knocking under acceleration, because when the throttle valve is opened suddenly, only the light, low-octane constituents reach the combustion chambers initially. A remedy from the fuel side would be to use high volatility constituents with higher octane rating such as isomerisate, light catalytic crack gasoline, and alcohols as well as the previous use of high volatility tetramethyl lead instead of tetraethyl lead. In terms of design, the problem of delayed availability of heavy constituents when the throttle is suddenly opened was solved by the general conversion to fuel injection systems with precise mixture metering and preparation under transient conditions. For this reason, it was possible to rescind the specification values for the front octane number (FON, previously also called RON 100) as well as the measuring method introduced and standardized solely for this purpose.

~Fuel additives. *See above,* ~Additives; *also,* →Fuel, diesel engine ~ Additives

~Fuel boiling curve. *See above,* ~Boiling characteristics

~Fuels containing ethanol. Similar criteria apply to fuels containing ethanol as to those containing methanol, but to a reduced extent. The low vapor pressure of ethanol does, however, make cold starting difficult. Compared to methanol, the lower oxygen content and higher air requirement of ethanol require a richer mixture. Simultaneously, a slight azeotropism is present when ethanol is blended with gasoline, making the hot-operating characteristics less critical. However, with ethanol from biomass it is particularly necessary to monitor the water content, the acids, and other extraneous constituents; **Fig. F183** shows specifications for such ethanol fuels. The ethanol fuels used in Brazil did not always fulfill these specifications and, therefore, led to malfunctions during operation.

In addition to ethanol, other alcohol fuels are pos-

Ethanol	[% by volume]	Min. 94
Water	[% by volume]	Max. 6
Denaturant		e.g., 0.76% methyl ethyl ketone
Molecular weight		46.07
Boiling point	[°C]	78.1
Melting point	[°C]	−117.3
Density (94 ~% by weight) at 20°C	[°kg/l]	0.806
Flash point	[°C]	12
Ignition temperature	[°C]	425
Evaporation index		8.3
Maximum workplace concentration	[ppm]	1000
Hazard class		B
Lower ignition limit in air	[% by volume]	3.28
Upper ignition limit in air	[% by volume]	18.95
Odor threshold	[ppm]	350
Evaporation residue	[mg/100 ml]	Max. 10.0
Residue on ignition	[mg/100 ml]	Max. 1.0
Acid as acetic acid	(mg/100 ml)	Max. 1.0
Aldehyde as ethanol	[mg/100 ml]	Max. 4.0
Ester as Ethyl acetate		Max. 100

Fig. F183 Examples of specifications for ethanol containing water (94%)

sible from special production processes—for example, blends consisting (primarily) of methanol and higher alcohols (methyl fuel).

~Gasoline. *See below,* ~Premium gasoline, ~Premium plus, ~Regular gasoline, ~Synthetic gasoline

~Gasoline requirements. *See below,* ~Requirements

~Gross calorific value →Fuel, diesel engine ~Properties

~Heat capacity (specific heat capacity). The specific heat capacity of fuels is usually not measured during fuel analysis because it is not a primary criterion for the behavior of fuel during transport, storage, and combustion. It is approximately half that of water with values just slightly over 2 J/kg K (J/kg° C). Increasing temperature and decreasing fuel density lead to an increase of the specific heat capacity. The specific heat capacity of paraffin is slightly higher than the values for naphthenes and aromatics. For this reason, the specific heat capacity of gasoline and diesel fuel is approximately the same because the influences of density and types of hydrocarbons approximately counterbalance each other. Diesel fuel has a higher density (lower heat capacity); however, it has a higher paraffin content (= higher heat capacity) than gasoline.

~Hot start. *See below,* ~Vapor lock

~Ignition accelerators. Ignition accelerators as used in diesel fuels are undesirable in gasolines, because neither combustion before ignition by the spark plug nor multiple ignition points following ignition are desirable in the gas not yet ignited in front of the flame front (antiknock properties, octane rating). However, additives with a potassium base are known that can accelerate the flame front. This can lead to shorter and more compact combustion periods with a positive effect on the engine thermal efficiency (**Fig. F184**). The law in Germany prohibiting lead in gasoline excludes the use of such additives containing metals.

~Initial boiling point. *See above,* ~Boiling characteristics

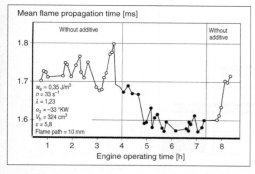

Fig. F184 Reduction of flame propagation time with additives

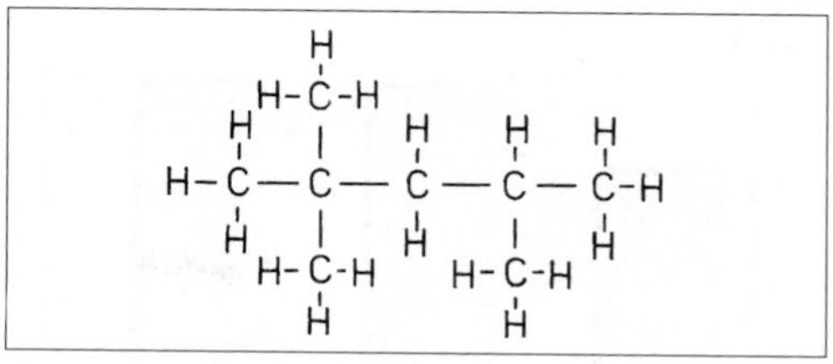

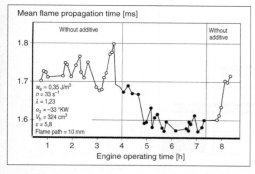

Fig. F185 Chemical structure of iso-octane (2,2,4-trimethyl pentane)

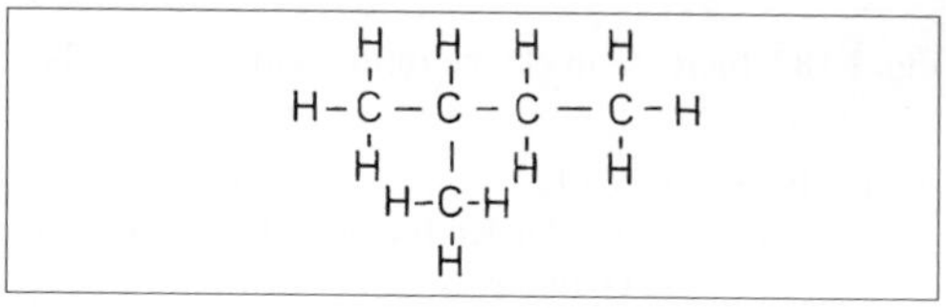

Fig. F186 Structural formula for iso-pentane

~Iso-octane. (2,2,4-trimethyl pentane) is a saturated, chain-shaped hydrocarbon with branches C_8H_{18} (**Fig. F185**). It is used as a high octane reference fuel (octane rating 100) for determination of the octane number. Iso-octane is a constituent of alkylate, a refinery product consisting primarily of isoparaffins.

~Isoparaffins. Isoparaffins are saturated chain-shaped hydrocarbons with branches such as iso-octane. Products containing isoparaffin are valuable constituents in fuel because they ensure a high octane rating (RON and MON) in the lower boiling range. Isomerisate is a widely used blending product containing large quantities of isopentane and isohexane—for instance, iso-pentane C_5H_{12} (constituent in isomerisate; **Fig. F186**).

~Knocking. *See below,* ~Octane number

~Knocking under acceleration. *See below,* ~Octane number

~Lead additives. *See above,* ~Additives

~Lead content. At the beginning of commercial utilization of tetraethyl lead (TEL: $[C_2H_5]_4$ Pb) in 1929, concentrations in the magnitude of 1 g of lead/L were usual in the United States. This allowed the octane rating of gasoline, consisting primarily of distillate and cracked products, to be improved by more than 15 units. The concentration figures were first specified on a volumetric basis—that is, mL of tetraethyl lead/liter of fuel. After introduction of tetramethyl lead (TML: $[CH_3]_4$ Pb) in 1960 with TML's higher lead content per unit volume, it proved more practical to specify the lead concentration in grams of lead per liter of fuel. The maximum limits for German gasolines were:

up to 1971 0.63 g/L (corresponding to 1 mL of TEL/l of fuel)

as of 1972 0.40 g/L

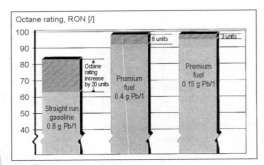

Fig. F187 Increase in octane rating with lead alkyls

as of 1976 0.15 g/L
1996 Sale of leaded fuel was discontinued in
Germany
2000 Leaded fuel prohibited in Europe

The average octane rating increase for these limits is shown in **Fig. F187**.

Because of increasing consumption of gasoline, the lead consumption in Germany initially increased continuously following the Second World War. In 1976, with the limitation to 0.15 g/L, the lead concentration in gasoline was reduced by more than half. However, the overall consumption of lead decreased markedly only after introduction of unleaded fuels began in 1985 and then the discontinuation of sales of leaded regular gasoline in 1988.

The introduction of high octane premium plus fuel in 1988 accelerated the reduction of lead consumption. In 1996, sale of leaded premium fuel was also discontinued in Germany. This ended the era of leaded gasoline fuels in Germany (**Fig. F188**).

~Lead emissions. *See above*, ~Antiknock additives

~Lead-free additives (*also, see above*, ~Antiknock additives). Tetraethyl lead and tetramethyl lead, in ad-

dition to increasing the octane rating, also have the property of protecting the valve seats against wear, particularly the relatively soft gray cast iron frequently used in the past. Following the introduction of unleaded fuels, wear problems occurred on exhaust valve seats, frequently leading to burning of the valves on vehicles with "soft" exhaust valve seats when operated at high load and speed.

Substitute substances based on organic potassium and sodium compounds were found for protecting the exhaust valve seats from wear; these provided sufficient protection for the valve seats at concentrations of only a few ppm (**Fig. F189**). Compounds with these materials were offered to owners of older vehicles as gasoline additives; they did not form any toxic constituents in the exhaust gas. Because the vintage vehicles still in use are rarely operated over long distances at full load and speed, such additives are no longer required.

~Liquefied gas. *See above*, ~Alternative fuels

~Lower calorific value. *See above*, ~Calorific value

~Methanol content. The methanol content in gasoline may not exceed a maximum of 3% by volume without special marking according to EEC directive 85/836 when suitable solubilizers are used in addition (higher alcohol, usually tertiary butyl alcohol or TBA). The methanol content in fuels is measured by means of gas chromatography (alcohol content).

Before widespread usage the middle of the 1980s, it was necessary to ensure that components such as the fuel hoses and carburetor floats were not sensitive to attack by using appropriate elastomers. The combination of aromatics and methanol in particular can lead to considerable swelling of nitrile caoutchouc. During production, transport, and storage of fuels containing methanol, it is necessary to take great care to prevent water from getting into the fuel; otherwise phase separation can occur (water content). Currently, the use of methanol in fuels has become uneconomical for most

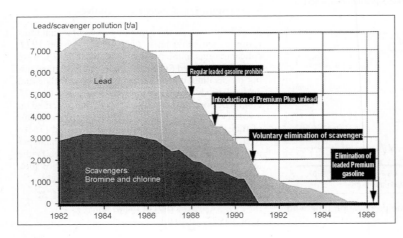

Fig. F188
Development of lead and scavenger pollution from 1982 to 1996 in Germany

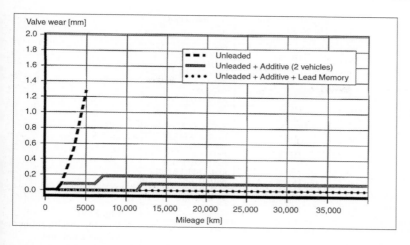

Fig. F189
Fleet test with "lead substitute" additive

	Regular gasoline	Premium gasoline	Premium plus
1988 (% by volume)	1.2	1.5	–
1998 (% by volume)	0.0	0.1	0.0
2003 (% by volume)	0.0	0.0	0.0

Fig. F190 Average methanol contents of German fuels (market analysis data)

refineries and for this reason is hardly ever used as a blending constituent (**Fig. F190**).

~Methanol fuel. This is the designation for fuel containing primarily methanol, which is, however, not offered commercially at present. Methanol fuels are being technically tested, but they are not currently economical. They can be used in gasoline as well as in diesel engines. The addition of small quantities of light straight-run gasoline and butane as blending constituents has proven itself well for use in gasoline engines. They improve the limited cold-start and cold-operating properties of pure methanol resulting from its low vapor pressure and high heat of evaporation. As with gasoline, the vapor pressure can be adjusted to match the season with different combinations of the two gasoline constituents. Simultaneously, the addition of hydrocarbons adds sufficient color to the virtually invisible flame of pure methanol in the event of unintentional ignition. The comparatively low octane rating of straight-run gasoline (approximately 60–70) reduces the octane rating considerably when it is blended with methanol.

In addition to the limited cold-start and cold-operating properties of methanol, it contributes to corrosion resulting from low acid and water contents. In practical tests, unprotected zinc and aluminum surfaces in tanks, fuel tankers, gasoline pumps, and vehicle components formed gelatinous corrosion products (metal hydroxides) that frequently led to filter clogging in the vehicles. This corrosion problem was solved by using a corrosion protection additive in the fuel; eliminating zinc, tin, and other nonferrous metals in the fuel system; and protecting the surfaces of components.

Moreover, when methanol is used as fuel, special lubricating oils are required because the ash-free dispersing agents usually used in these oils lead to sticky residues when they come into contact with methanol. After development of special lubricating oils also capable of controlling cylinder wear resulting from corrosion, it was possible to increase the service life of methanol engines to the usual level. The advantages of methanol fuel—high octane rating, high rate of combustion and large volume expansion of the air-fuel mixture—have significantly reduced the increase in the volumetric fuel consumption expected on the basis of volumetric calorific value. It was possible to reduce the theoretical additional consumption based on the difference in the calorific value in relation to gasoline from a factor of two to approximately 1.7 by optimization of the engines for methanol. **Fig. F191** shows the possible specifications for a methanol fuel.

In contrast to methanol fuels with high quantities of methanol, engine technology designed for gasoline can be used for blended methanol fuels with low percentages of methanol.

The addition of methanol is currently limited to 3% by volume in commercially available gasoline. This low level was specified in consideration of older vehicles, because higher methanol contents lead to problems with some materials in the fuel system (plastics and elastomers) and to impermissibly lean mixtures with operating problems, particularly during cold operation. Vehicles with catalytic converters, oxygen probe, and adaptive injection quantity which adapts to the fuel—even for cold operation—could be operated without problems even with fuels containing higher quantities of methanol (alcohol), whereby the upper limit for the possible alcohol content in the fuel is limited by the achievable, maximum injection quantity of the nozzles, among other things. It is necessary to consider, in addition to the required stoichiometric mixture preparation, the higher latent heat of vaporization for methanol and the azeotropic boiling characteristics (**Fig. F192**). The latter together with an increase in the

		Summer	Winter	Test method
Methanol	[%m]	Min. 82	Min. 82	GC
Hydrocarbons HC total*	[%m]	Min. 10–Max. 13		GC
Butane C4	[%m]	Max.1.5	Max. 2.5	GC
Density d15	[kg/m³]	770–790		DIN 51757, ASTM D 941
Vapor pressure RVP (dry)	[mbar]	550–700**	750–900**	DIN 51754, PREN 12, ASTM D 323
Water content	[ppm]	Min. 2000–Max. 5000***		DIN 51777, ASTM D 1744
Higher alcohols	[%m]	Max. 5		GC
Formic acid	[ppm]	Max. 5		
Total acid (calculated as acetic acid)	[ppm]	Max. 20		ASTM D 1613
Evaporation residue	(mg/kg)	Max. 5		DIN EN5, washed with MeOH
Chlorine	[ppm]	Max. 2		DIN 51408, Part I ASTM D3120, mod. & ASTM D 2988
Lead	[ppm]	Max. 30		ASTM D C2S7
Phosphorus	[ppm]	Max, 10		ASTM D 3231
Sulfur	[ppm]	Max. 100		ASTMD 3120
Additive	[%]	Max. 1		

* Type of hydrocarbon, boiling characteristics and quantity, depending on cold start and reliability requirements
** Example for Central Europe, other values possible according to local gasoline specifications
*** With corrosion inhibitor

Fig. F191
Possible specifications
for methanol fuel

gasoline vapor pressure by approximately 15–20 kPA depending on the measuring process (**Fig. F193**) requires virtually complete elimination of butane in the fuel. Since introduction of the dry measuring method (DVPE) for determining the vapor pressure, the de-scribed azeotropic vapor pressure increase is taken into consideration fully in the measured value, and the admixture of small quantities of methanol to the fuel has, therefore, become uneconomical. Moreover, because of the low carbon content and high octane rating, butane has become a preferred constituent that should be used in fuel production if possible.

The octane-increasing effect of methanol in blended fuels is hardly noticeable at low concentrations (maximum 3%), but it becomes quite noticeable at concentrations of 15% and higher (**Fig. F194**).

Other properties of fuels containing alcohol, such as elastomer compatibility and corrosion characteristics, have already been taken into consideration in the materials used in modern vehicles. The advantage of methanol in terms of energy conversion is maintained even as a blending constituent. In fleet tests with methanol blended fuel (15% methanol) approximately 1000 participating vehicles showed only an approximately 5.5% increase in the volumetric fuel consumption. This means that the energy in the fuel is utilized approximately 2.6–3% better.

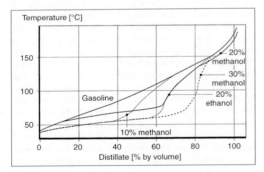

Fig. F192 Azeotropic boiling characteristics of gasoline/methanol blends (Source: Menrad)

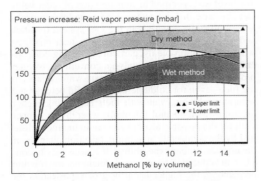

Fig. F193 Increase of vapor pressure of gasoline upon addition of methanol

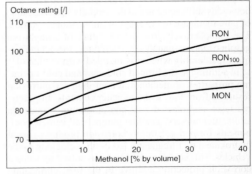

Fig. F194 Effect of methanol on the increase in octane number of gasoline (Source: Menrad)

~Methanol/benzine blended fuel. *See above*, ~Alcohols

~Methyl tertiarybutyl ether (MTBE). *See above*, ~Alternative fuels

~Motor octane number (MON). *See below*, ~Octane number

~Octane number (RON, MON). The octane number or rating is a value that indicates the resistance of gasoline to detonation (knock)—that is, its capability of preventing uncontrolled combustion in the remaining gas not yet combusted before arrival of the flame front. Knocking operation with a rate of combustion approximately 10 times higher than normal is only permissible for a very short time. If knocking continues for too long, the results are steep pressure peaks, high rates of pressure variations, and high temperatures in the combustion chamber that can damage the spark plug, pistons, cylinder head gaskets, and valves, particularly when knocking causes preignition.

The tendency of a fuel to undesired preignition can differ greatly from its antiknock properties. For example, benzene and methanol have very high octane ratings, but they also show a tendency to preignition.

It is necessary to differentiate between research octane number (RON) and motor octane number (MON). Both are measured in so-called CFR single cylinder engines, which have been standardized worldwide according to EN 25164 and EN 25163 (previously so-called BASF test engines were also used as an alternative primarily in Germany). In the test engines, the compression ratio can be varied smoothly through the measurement process so that it can be adjusted to the same knocking intensity for any fuel. The knocking intensity is measured by a pressure sensor located in the combustion chamber, and it is indicated as an average value over a number of cycles by means of an electronic time-lag circuit.

The operating conditions for the engines differ when measuring each octane number, and the motor octane number (MON) has the more severe conditions (**Fig. F195**). This is why the RON is generally higher than the MON. The difference between the two values is known as the sensitivity.

Reference fuels are mixtures of iso-octane C_8H_{18} (2,2,4-trimethyl pentane; **Fig. F196**) with the assigned octane number of 100 and normal heptane C_7H_{16} with the assigned octane number of 0 (**Fig. F197**). The octane number of a reference fuel mixture by definition corresponds to the volumetric contents of iso-octane, according to the two methods RON and MON. The octane number of a fuel to be measured is determined by "bracketing" with two different reference fuel mixtures with lower and higher octane numbers (not more

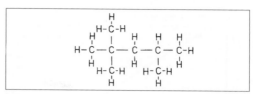

Fig. F196 Chemical structure of iso-octane

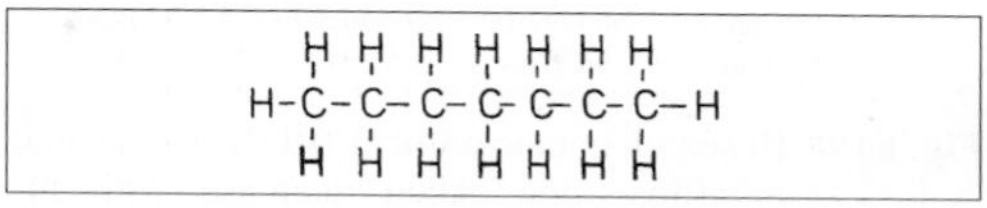

Fig. F197 Structural formula for normal heptane

than two octane numbers apart) and by linear interpolation of the compression ratio measured in each case for the same knocking intensity.

The two different test methods, RON and MON, are justified by the difference in the behavior of stock engines under different operating conditions. For acceleration at full throttle and low speed, most stock engines evaluate the fuel according to its RON (acceleration knocking). At higher speed and full load, the MON becomes more significant (high speed knocking, which is usually less easily perceived acoustically and is generally more dangerous for the engine).

The increasing significance of the MON at higher load and speed applies particularly to four-stroke and two-stroke engines with throttle intake systems. The various hydrocarbon compounds are evaluated differently by RON and MON; particularly aromatics and olefins are evaluated more stringently and devalued by the MON:

- Normal paraffins have the same octane rating in RON and MON. At increasing molecule sizes, the octane rating tends to decrease greatly so that, in terms of the antiknock properties, only butane C_4H_6 represents a usable constituent (**Fig. F198**).
- Isoparaffins have approximately the same RON and MON value. With increased branching, the octane rating increases while the number of carbon atoms remains constant (**Fig. F199**).
- Aromatics have very high RON and high MON values; with the xylenes, the different arrangement of the methyl groups in the benzene ring also has an effect on the octane rating (**Fig. F200**).
- Light olefins have a medium RON and a low MON value. With increasing molecule size (chain length), the octane ratings decrease and the reduction of the MON decreases (**Fig. F201**).

	Engine speed	Intake air temperature	Mixture preheating	Ignition timing	Compression ratio
RON	600	51.7 °C	–	13 ° BTDC	4 : 1 to 16 : 1 variable
MON	900	38 °C	variable 285 °F to 315 °F	variable 14 ° to 26 ° BTDC	4 : 1 to 16 : 1 variable

Fig. F195
Operating conditions for determining RON and MON

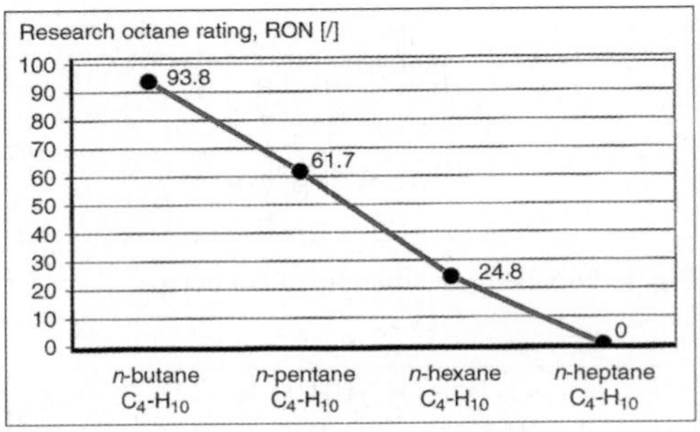

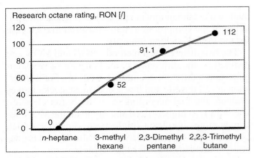

Fig. F198 Research octane number (RON) for normal paraffins — note steep decrease with increasing molecule size

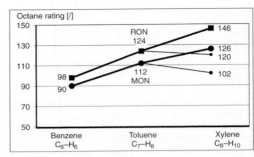

Fig. F200 Octane ratings of a number of aromatics: RON and MON mixed octane numbers in 60-40 mixture of iso-octane and n-heptane

Research octane rating, RON [/]

Fig. F199 Research octane number (RON) for isoparaffins

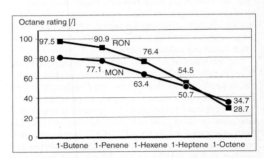

Fig. F201 Octane ratings of a number of olefins; note that as chain lengths increase, the reduction resulting from the MON decreases (although this is only of theoretical interest because it is at a very low level of octane rating)

Refinery constituents have different octane rating profiles corresponding to their different compositions, which result from the production processes. In combination with boiling characteristics, the octane rating of the fuels is the dominant criterion for their usability. Because of the production process, standard commercial fuels usually have only very slight reserves in terms of their MON. This applies particularly to premium and premium plus fuels. For this reason, the MON for nearly all fuels is just slightly above the limit of 85 or 88, while the RON has a larger range (**Fig. F202**).

~Octane rating requirement. This is the minimum octane rating required by an engine for operation without knocking. The octane rating requirement of a vehicle depends on design parameters, operating conditions, and atmospheric conditions as well as the fuel:

- Design parameters that result in a high octane rating requirement include a high compression ratio, a high volumetric efficiency, an advanced firing point, jagged combustion chambers, low charge movement, long combustion paths (e.g., spark plug not located in center), so-called hot spots (e.g., hot exhaust valves), and high engine chamber temperatures (**Fig. F203**).

- Operating conditions can affect the octane number requirement; in particular, slow operation over longer periods of time increases the octane rating requirement because combustion chamber residues increase under such conditions. This results, first, in an increase in the effective compression ratio and, second, in poor conduction from the combustion chamber, resulting in higher combustion temperatures (**Fig. F204**).

- Atmospheric conditions that require higher octane ratings include high atmospheric pressure, high atmospheric temperatures, and low humidity
  In countries with extensive high-altitude plateaus, the octane rating requirement is several units lower at the low ambient pressures, so that fuels with lower octane rating can also be used (e.g., in East Africa: octane rating on the coast is 95, octane rating on high altitude plateau is 93).

- Fuel influences the octane rating requirement. Generally, the specified reference fuels for measuring the octane rating — i.e., blends of the paraffin constituents iso-octane and n-heptane — provide the lowest octane rating requirement. Standard com-

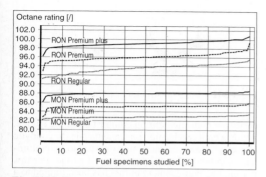

Fig. F202 Octane rating, market analysis of gasolines, summer 1995

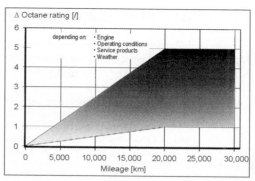

Fig. F204 Increase in octane rating requirement with increasing operating time

mercial fuels, which generally contain aromatics and olefins, result in a higher octane rating requirement when the measured octane rating requirement is based on the research octane number according to definition (**Fig. F203**).

The octane rating requirement with standard commercial fuels increases with engine speed as well as temperature and percentage of residual gas in the combustion chamber. In other words, these are the conditions under which the MON (motor octane number) of a fuel gains significance. Extreme examples of this are two-stroke engines with mixed scavenging, which load fuels even more stringently than the MON evaluation. This can result in such engines not being capable of operation without knocking, even with premium fuel, even though the octane rating requirement measured with iso-octane/n-heptane is at the level of regular gasoline.

Generally, the octane rating of the fuel used should be higher than the octane rating requirement of the engine to exclude knocking and possible engine damage. Occasional operation in the knocking region—for example, during full-load acceleration at low speeds or also part-load acceleration—does not result in engine damage, but knocking at high

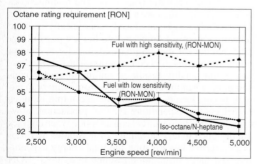

Fig. F203 Octane rating requirements depending on fuel and engine speed at full load; example: V-6 engine, displacement 2.5 L, compression ratio 9.7:1

speeds over long periods of time is critical. Here, high temperature increases may occur on the spark plugs and other components, resulting finally in pre-ignition on these hot surfaces. This generally leads to destruction of the cylinder head gasket, the valves, or the piston.

Previously, the octane rating requirement was measured primarily with the aid of a stethoscope attached to the engine; today, this is accomplished with knock sensors attached to the engine or with special spark plugs equipped with a pressure sensor for sensing vibration in the combustion chamber. However, it is necessary to ensure that the measured values are not affected by gas vibrations in the spark plug channel. Recognizable pressure vibrations in the oscillogram generally do not cause engine damage because the component stress also depends on the intensity of the knocking.

On vehicles with knock control, the ignition map adapts automatically to the fuel in the tank. At low octane rating—for instance, using regular instead of premium gasoline—retardation of the ignition prevents knocking but also results in a loss in power, high fuel consumption, and higher exhaust temperatures (with concomitant strain on the catalytic converter). If an appropriate control circuit is present for higher octane ratings—for example, premium plus instead of premium—an advanced ignition setting can increase the output and reduce fuel consumption.

Optimum utilization of the energy contained in the fuel requires careful tuning between the engine and fuel. Better engine efficiency can be achieved theoretically at the highest possible compression ratio with very high octane rating requirement. However, production of such fuels requires higher energy expenditure at the refinery. In contrast, the lowest energy expenditure for fuel production is for straight run gasoline with a modest octane rating in the range of 60 to 70. But, from a practical viewpoint, the requirements for engine and fuel production have found an optimum compromise. Experience has shown that, for a number of reasons, compression ratios greater than 12:1 are no longer

549

Olefin content	Regular	Premium	Premium plus
Group	1–18	1–14	1–10
Average value	7	5	3

Fig. F205 Ranges and average values of olefin content in German gasolines (from market analysis)

F

optimal for standard gasoline engines (high surface/volume ratio in combustion chamber, increased wall heat losses, increased HC and NO_x emissions). On the other hand, the requirements of the fuel market favor the use of catalytic crackers as well as the hydrogen requirement for desulfurization, which requires the use of catalytic reformers. Both of these refinery facilities provide high octane constituents, so that the octane ratings of the fuels normally used today, together with the engines adapted for this purpose, result in roughly optimum energy utilization from the crude oil in premium gasoline.

~Olefin content. Olefins are found particularly in catalytically cracked gasoline. They have a medium to high research octane number (RON) but only a limited motor octane number (MON). They are, therefore, particularly suitable for regular and premium gasoline but are less suitable for premium plus fuel. Accordingly, the last fuel usually has a low percentage of olefins. Olefin contents of between 0% and just below the maximum limit of 18% by volume are present, depending on the refinery configuration and type of fuel. The olefin content in gasoline is determined by means of fluorescence indicator analysis (FIA), ASTM D1319 (**Fig. F205**).

~Olefins. Olefins are simple or multiple unsaturated (double carbon bonds) straight-chain or branched-chain hydrocarbons. A differentiation is made into mono-, di-, and tri-olefins, depending on the number of double carbon bonds. Of these, the mono-olefins with only one double carbon bond are particularly suitable for use in gasoline. The low boiling species—for example, 1-hexene C_6H_{12} (**Fig. F206**)—provide good research octane numbers (RON).

Olefins with multiple double bonds, particularly conjugated diolefins with alternating double/single bonds such as butadiene (**Fig. F207**), are less suitable for gasoline because of their low stability and tendency to form residues and are, therefore, converted in the refinery process through partial hydration. The remaining mono-olefins are usually stabilized with oxidation inhibitors (additives) to improve the storage stability of the fuel. Olefins are reactive constituents that

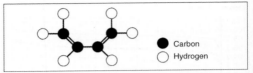

Fig. F207 Chemical structure of butadiene

easily are converted during combustion in the engine, giving a high combustion rate, and then in the catalytic converter. However, they are suspected to be preliminary substances for the formation of ozone. For this reason, the olefin content in fuel has been limited to a maximum of 18% by volume since the year 2000.

~Oxidation stability. The oxidation stability (resistance to oxidation) is an indicator of the storage stability of a fuel. This can be tested in a rig: The fuel is oxidized in a pressure vessel (according to EN ISO 7536), which is filled at the beginning with pure oxygen at a pressure of 7 bar, then maintained at a temperature of 100°C. Possible reactions between the oxygen and unstable fuel constituents lead to a pressure loss. The test is terminated when the pressure loss is more than 0.14 bar within 15 minutes. The time until the pressure decrease of 0.14 bar is reached is the value for the oxidation resistance. Fuels with untreated crack products require antioxidants (additives) to achieve the oxidation stability required in DIN EN 228 of >360 minutes.

~Premium fuel. *See below*, ~Premium gasoline, ~Premium plus

~Premium gasoline. This gasoline is standardized throughtout Europe in EN 228. Requirements with octane numbers:

- Min. 95.0 RON
- Min. 85.0 MON

Introduction was accomplished all over Europe in 1985 and was the prerequisite for the introduction of vehicles with catalytic converters. The minimum octane ratings specified are a compromise between types of fuels that can be produced economically in large quantities all over Europe and the highest possible octane number for the design of engines for maximum efficiency and low fuel consumption. The specified octane rating quality results in minimum energy consumption when the entire process from refinery to engine is considered. Increasing octane rating requires higher energy expenditure for fuel production; however, it results in lower fuel consumption when engines are designed to take advantage of it. Studies by European industry investigating the total energy expenditure and fuel consumption resulted in a "Pool" RON (average octane rating of various types of gasoline in consideration of the quantity relationships) of 95 as the optimum for the engine and refinery technology at that time (**Fig. F208**).

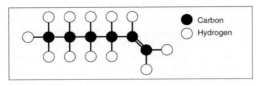

Fig. F206 Chemical structure of 1-hexene

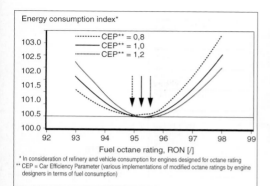

Energy consumption index*

* In consideration of refinery and vehicle consumption for engines designed for octane rating
** CEP = Car Efficiency Parameter (various implementations of modified octane ratings by engine designers in terms of fuel consumption)

Fig. F208 Optimum fuel pool octane rating for low fuel consumption

~**Premium plus.** This is an unleaded gasoline with the same octane rating (RON 98/MON 88) as previously leaded premium, and it was introduced in 1988, eight years before leaded premium fuel was taken off of the market in 1996. Premium plus is available everywhere in Europe (with, however, octane levels for the specific country, levels which are occasionally less than the values common in Germany—98 RON/88 MON) and is produced to the greatest extent with high octane ether (methyl tertiary butyl ether—MTBE) as an additional blending constituent. At the initiative of the German mineral oil companies, the percentage of benzene in this fuel was already limited to 1% by volume in 1995. The low benzene content particularly reduces the benzene emissions from old vehicles without catalytic converters, which were originally designed for operation with leaded premium. Premium plus ensures maximum engine efficiency and power/torque in engines designed accordingly. It also ensures proper operation of older engines developed for use with leaded premium fuel without changing the engine

tuning. Potassium-based additives are available for drivers to add to the fuel themselves when additional wear protection is required for the exhaust valves.

~**Production.** Gasoline is produced from crude oil in refineries. Although production from coal and natural gas as well as other raw materials containing carbon and hydrogen is technically feasible (synthetic gasoline), such production methods are not economical due to the present low price for crude oil.

Fig. F209 shows a simplified flow chart of a mineral oil refinery.

In an oil refinery, the crude oil consisting of hundreds of different hydrocarbons is initially cleaned coarsely (desalinated) and then separated physically into light and heavy hydrocarbons in one or more distillation systems (**Fig. F210**). Because the requirement for hydrocarbons with low boiling points is greater than the quantity of such hydrocarbons present in crude oil, the heavy hydrocarbons are "cracked" to produce hydrocarbons with higher volatility.

After removal of the undesirable sulfur compounds, intermediate products suitable for production of gasoline are treated in so-called reformers. In these facilities, the molecular structure of the hydrocarbons is "reformed" using various procedures to obtain high-quality, high-octane constituents for production of the gasoline regardless of the particular crude oil used.

In contrast to diesel fuel, which is produced primarily from distillates, gasoline is composed almost exclusively of refinery products which have been subjected to further treatment. In addition, small concentrations of high octane alcohols and ethers are permissible (alcohols).

The products used depend on the conversion and treatment facilities available in the refinery in question. The products have various compositions which supplement each other mutually, allowing fuels to be blended to meet the specifications.

F

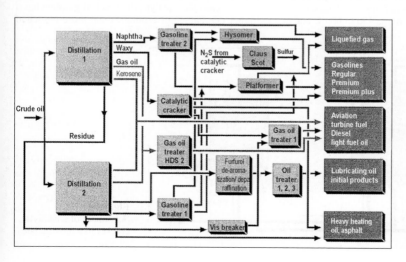

Fig. F209
Simplified schematic of an oil refinery

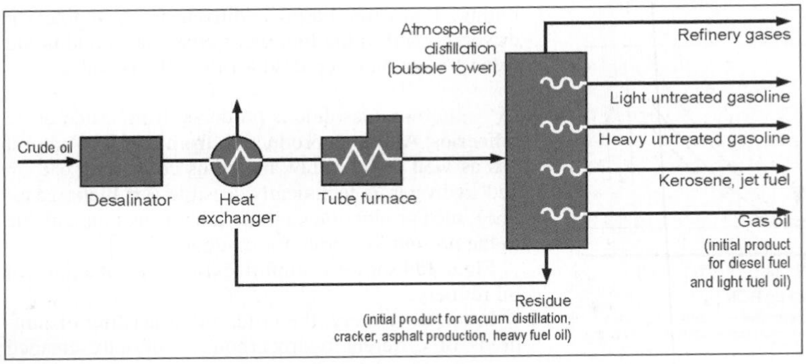

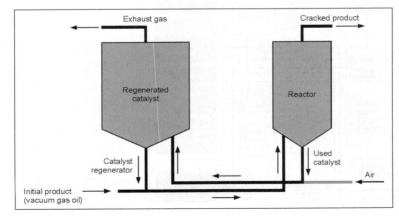

Atmospheric distillation (bubble tower)

Refinery gases

Light untreated gasoline

Heavy untreated gasoline

Kerosene, jet fuel

Gas oil
(initial product
for diesel fuel
and light fuel oil)

Crude oil

Desalinator

Heat exchanger

Tube furnace

Residue
(initial product for vacuum distillation,
cracker, asphalt production, heavy fuel oil)

Fig. F210
Simplified schematic of
a distillation plant

The majority of constituents in gasoline are produced using the following methods:

- Cracking method: thermal cracking, catalytic cracking, and catalytic cracking with hydrogen, or hydrocracking
- Reforming: Platforming with platinum catalyst
- Alkylating
- Isomerization
- Polymerization

Because of limitations on the content of aromatics and olefins in the fuel, the processes for production of high octane isoparaffins, alkylation, and isomerization have gained significance.

The cracking processes at high pressure and high temperature and, in part, in the presence of suitable catalysts and by the addition of hydrogen allow the heavy (long chain, high carbon) crude oil constituents to be used to produce light products with high octane rating suitable for gasoline. The cracking processes provide the prerequisites for producing products in the qualities and quantities required by the demand structure from various crude oils:

Thermal cracking. Conversion of heavy hydrocarbons to obtain lighter products is accomplished exclusively at high temperatures in the range 450 to 600°C and at pressures of 20–40 bar. In the meantime, thermal crackers have been replaced, to a great extent, by other cracking processes. This also applies for modified processes such as "visbreaking," with extremely short reaction times and the increased version, "coking," with extended reaction times of up to 24 hours. The thermal cracking processes initially lead to breaking down of the hydrocarbon molecules between the carbon bonds. The resulting free radical—that is, incomplete—hydrocarbon compounds form light olefins and, in part, also diolefins and paraffins. Parallel to this, naphthalenes in the initial product are converted to aromatics.

Catalytic cracking. The catalytic cracking process is more common than thermal cracking (**Fig. F211**).

In the refining plants, heavy gas oil from atmospheric or vacuum distillation is usually brought into contact with an aluminum silicate (zeolite) catalyst in a moving bed reactor at 450–500°C.

The catalyst is continuously regenerated by burning out the carbon deposits.

In contrast to thermal cracking, the molecule is broken down by an ionic mechanism. This then forms olefins and also branched hydrocarbons (through isomerization) as well as aromatics. Compared to thermally cracked gasoline, catalytically cracked gasoline has a higher percentage of aromatics and more branched

Exhaust gas

Cracked product

Regenerated catalyst

Reactor

Catalyst regenerator

Used catalyst

Air

Initial product
(vacuum gas oil)

Fig. F211
Schematic of a catalytic
cracker in a moving bed
system

hydrocarbon bonds and fewer dienes, and it is, therefore, more suitable for combustion in engines.

Hydrocracking. In hydrocracking, an advancement in catalytic cracking hydrogen is introduced in addition to the initial product (vacuum gas oil or also other middle distillates) at a temperature of 320–420°C and a pressure of 100–200 bar. The hydrogen is available from the catalytic reformer. The desired hydrogen addition to the cracked molecules requires a special catalyst. As with catalytic cracking, this consists of an aluminum silicate with finely distributed molybdenum, tungsten, cobalt, palladium, or nickel.

Products from the hydrocracker are practically 100% saturated hydrocarbons. For use in gasoline, it is, therefore, usually necessary to treat them in subsequent facilities to increase the octane rating.

Reformers. With reformers, hydrogen is separated from the molecules at high temperatures and high pressures (20 to 50 bar) in the presence of catalysts to form, primarily, aromatic compounds with high octane rating. Heavy benzene from distillation or from the hydrocracker is used as the initial product. The production flow runs over a line of platinum catalytic converters at temperatures around 500°C. This results in the formation of aromatics from saturated ring-shaped hydrocarbons (naphthalenes) by means of cyclization followed by dehydration. During this process, the formation of benzene can best be prevented by preliminary distillation treatment of the initial product.

The released hydrogen can be used for the desulfurization facilities as well as for the hydrocrackers, if present. The catalyst in the reformer is sensitive to metallic impurities, sulfur, and nitrogen compounds, and so the desulfurization system, which is required anyway to maintain low sulfur concentrations, precedes the reformer (**Fig. F212**).

Desulfurization. Sulfur is a natural constituent of crude oil, but its proportion varies greatly depending on the source of the crude. The initial product is treated with hydrogen in the presence of a catalyst at a pressure of approximately 80 bar and at approximately 350°C to separate the sulfur, which is primarily bonded chemically. The sulfur is detached from the hydrocarbon molecules under the processing conditions. The resulting hydrogen sulfide is separated from the product in the next system and converted to pure elemental sulfur in the so-called Claus process. Constituents containing nitrogen and oxygen are also removed in parallel in this type of system.

Alkylation. Alkylation is the production of isoparaffins by reaction of isobutene with C_3 to C_5 olefins under the influence of sulfuric acid or hydrofluoric acid as catalyst. Due to its high resistance to knocking, high calorific value, low vapor pressure, and favorable boiling range, alkylate gasoline is excellent as a constituent in fuel and is also the primary constituent of aviation gasoline.

Isomerization. Isomerization is the production of light isoparaffins from normal paraffins. In this process, high volatility straight-chain paraffins with low octane

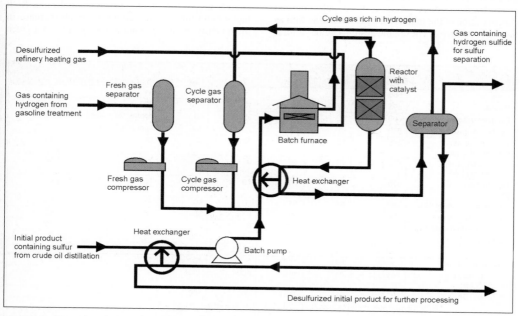

Fig. F212 Simplified process schematic of a desulfurization plant

rating are converted to branch chain isoparaffins with significantly higher octane rating using a platinum catalytic converter on zeolite basis. The initial product can be light distilled gasoline, which can be used as automotive fuel to a limited extent only because of its low octane rating.

Polymerization. Polymerization is a process with minimum practical significance in which primarily C_6 to C_{12} olefins are produced from propane or butane by means of oligomerization.

The products for the production of the basic fuel are selected according to availability (configuration of refinery facilities), the physical data of the constituents, and the quality requirements to be fulfilled. **Fig. F213** shows the primary basic data for main components in gasoline and their typical composition.

The primary criteria for blending formulas are the octane rating, the boiling characteristics, and the vapor pressure of the constituents. Brand-specific additives to improve the properties are added automatically to the completed and already standardized gasoline when the fuel tanker is filled. The quality of the fuel depends decisively on the type and quantity of additives used.

~*Propane. See above,* ~Alternative fuels

~*Properties.* These are described in terms of the physical and chemical values for gasoline.

The characteristics and minimum requirements relevant for transport, storage, suitability for engine operation, and influence on the environment are defined in the European requirements EN 228. They can be influenced by the refinery processes used, selection and concentration of the individual constituents, as well as the use of additives. The engine design, engine calibra-

tion, and fuel properties must be matched. For this reason, standardized reference fuels (CEC reference fuel with designation CEC-RF) representing the average fuel quality on the market in each case are available to engine manufacturers, the aim being to achieve optimum operating and emissions behavior.

~*Racing fuels.* There are a number of ways to increase the performance of engines by means of modifying the fuels used or by optimization of the fuel, depending on the regulations for motor vehicle racing competitions.

Basic prerequisites for the suitability of constituents of fuel for racing are sufficient safety for storage, transport, and handling (explosion protection) as well as acceptable toxicity. Moreover, the fuel volatility should allow virtually complete evaporation in the intake system. The classic indirect method of increasing engine performance through the fuel is to increase the octane rating either by selection of the constituents or by addition of antiknock agents. (In the past, high concentrations of tetraethyl or tetramethyl lead were common). High octane ratings allow high compression ratios and, therefore, high specific engine output. Hydrocarbons with high octane ratings suitable for racing fuels are shown in **Fig. F214**. Refinery constituents with appropriate composition include alkylate (isooctane and other isoparaffins) and catalytic reformulated fuel (toluene, xylene, and other ring-shaped hydrocarbons). Other possibilities for increasing the octane rating, when permissible according to regulations, consist of using high octane alcohols, such as methanol and ethers—for example, MTBE or TAME.

A direct increase in the output by means of the fuel is possible by selecting constituents with high energy content and/or low stochiometric air-fuel ratio. Both factors and their combinations allow an increase in the quantity of energy input to the cylinder at a given

Constituents	Density 15 °C	Octane rating		E 70*	E 100**	Paraffins	Olefins	Aromatics
	(kg/m^3)	RON	MON	(% by volume)	(% by volume)	(% by volume)	(% by volume)	(% by volume)
Distillate	680	62	64	70	100	94	1	5
Butane	595	87–94	92–99	100	100	100	–	–
Pyrolysis gasoline	800	82	97	35	40	approx. 20	approx. 10	ca. 70
Light coker gasoline	670	69	81	70	100	approx. 57	approx. 40	ca. 30
Light catalytically cracked gasoline	685	80	92	60	90	61	26	13
Heavy catalytically cracked gasoline	800	77	86	0	5	29	19	52
Light hydrocracked gasoline	670	84	90	70	100	100	0	0
Full range reformulate 94	780	84	94	10	40	45	–	55
Full range reformulate 99	800	88	99	8	35	38	–	62
Full range reformulate 101	820	89	101	6	20	29	1	70
Isomerisate	625	87	92	100	100	98	–	2
Alkylate	700	90	92	15	45	100	–	–
Polymer gasoline	740	80	100	5	10	5	90	5
Methyl tertiary butyl ether (MTBE)	745	98	114	100	100	–	–	–
Methanol/TBA (1:1)	790	95	115	50	100	–	–	–
* E 70 = Evaporated percentage of 70 °C, ** E 100 = Evaporated percentage of 100 °C								

Fig. F213 Density, octane rating, boiling characteristics, and typical composition of important gasoline constituents

Hydro-carbon	Summer formulas	Octane rating	
		RON	MON
Iso-octane	C_8H_{18}	100	100
Triptane (2.2.3-trimethyl butane)	C_7H_{16}	112	101
Iso-decane	$C_{10}H_{22}$	113	92
Cyclo-pentane	C_5H_{10}	101	85
Toluene	C_7H_8	120	109

Fig. F214 Hydrocarbons with high octane rating for racing fuels

engine air supply volume. A typical example of such a compound is nitromethane, which has a significantly lower calorific value than gasoline but allows more than twice the amount of energy to be supplied because of its much lower stoichiometric air-fuel ratio (**Fig. F215**). However, aside from the regulations, the use of nitromethane is limited by the thermal and mechanical load on the engines.

Hydrocarbon constituents, which allow a significantly more modest but still measurable direct increase in performance, include stretched hydrocarbons such as quadri-cyclane (C_7H_8) or diolefines such as di-iso-butylene (C_8H_{16}). Such constituents were considered unsuitable according to conventional evaluation methods because of their usually lower octane ratings; however, due to their significantly higher combustion rates when adapted appropriately to the ignition map in racing engines, they are significantly less sensitive to knocking than would be expected on the basis of their octane rating (**Fig. F216**). A further advantage of high combustion rates is the tendency to displace the energy conversion to the region of top

dead center, thereby improving the thermal efficiency at the usually high operating speeds for racing engines. This advantage also applies to conventional gasoline cracked constituents (containing olefins).

The partly contradictory characteristics of the fuel constituents make it necessary to carefully match the fuel and engine in all cases and particularly to consider the current regulations. Moreover, it is also necessary to consider the availability and, frequently, the significantly higher costs for special fuels in addition to the cost of engine adaptation.

~Reformulated fuel. The term "reformulated fuels" originated in the United States. There, the Clean Air Act legally prescribed the use of especially formulated fuels for areas with extremely high summer smog (ozone) and/or CO pollution. The term has since been extended, and now reformulated fuel generally refers to a change in the composition and/or physical characteristics with the objective of reducing the exhaust and fuel evaporation emissions. Although mechanical modifications on the vehicle such as regulated three-way catalytic converters and charcoal canisters have a considerably higher effect than a change in the fuel composition, the latter can still contribute to reduction of emissions in the existing vehicle population.

In complex studies performed by individual companies, research associations, and the inter-industry, European automobile oil program (EPEFE), all significant gasoline parameters were studied to determine their effect on exhaust emissions.

The possibilities and consequences are shown in qualitative form in the table below. With the exception of the economic disadvantages, several of the possible

F

Properties	Nitromethane	Methanol	Iso-octane
Total formula	CH_3NO_2	CH_3OH	C_8H_{18}
Mol weight	61	32	114
Oxygen content (%m)	52.5	49.4	0
Latent heat of evaporation (MJ/kg)	0.56	1.17	0.27
Stochiometric air/fuel ratio	1.7 : 1	6.45 : 1	15.1 : 1
Lower calorific value (MJ/kg)	11.3	19.9	44.3
Specific energy at stochiometric air/fuel ratio*	6.7	3.1	2.9
* Quotient from lower calorific value and stochiometric air/fuel ratio			

Fig. F215
Nitromethane and methanol lead to higher energy input at stoichiometric air-fuel ratio than iso-octane because less oxygen is required for complete combustion

	Di-iso butylene	Quad recyclane	Iso-octane	Toluene
Quad recycling	C_8H_{16}	C_7H_8	C_8H_{18}	C_7H_8
Density (kg/m³)	719	919	699	874
Boiling temperature (°C)	101	109	99	111
Octane ratings				
RON	98*	54*	100	124*
MON	78*	19*	100	112*
Lower calorific value (MJ/kg)	44.6	44.1	44.3	41.0
Stochiometric air/fuel ratio	14.7 : 1	13.4 : 1	15.1 : 1	13.4 : 1
Specific energy at stochiometric air/fuel ratio*	3.0	3.3	2.9	3.1
* Mixed octane ratings				
** Quotient from lower calorific value and stochiometric air/fuel ratio				

Fig. F216
Quadri-cyclane has a higher specific energy factor than toluene and iso-octane; di-iso-butylene improves the engine efficiency, particularly because of the high combustion rate

Fuel parameters	Limited emissions			Unlimited emissions			Possible refinery process	Economic disadvantages	Technical disadvantages
	HC	CO	NOx	Benzene	Aldehydes	Butadiene			
Sulfur reduction	⇓	⇓	⇓	–	–	–	Plant expansion Hydro desulphurization	Additional costs	Higher CO_2 emissions at refinery
Reduction of aromatics	⇓	⇓	⇑	⇓	⇑	⇑	Considerable investments (isomerization alkalization units)	Considerable additional costs	Octane rating level endangered, higher CO_2 emissions at refinery
Reduction in benzene	–	–	–	⇓	–	–	Modification distillation process	Moderate additional costs	–
Higher fuel volatility in medium boiling range	⇓	⇓⇑	⇑	⇓–	–	⇑	Investments required only in part	Low additional costs	MON level endangered with present equipment
Use of ethers as fuel constituents	–	⇓	⇓⇑	⇓	⇑	–	(Additional) constituent tank required	High product costs	Slightly higher volumetric consumption

Fig. F217 Reformulated gasoline fuel: possible measures and consequences

measures have contradictory effects on the individual emissions values.

The only measure that reduces all pollutants in the exhaust is reduction of the sulfur content in the gasoline. A reduction of the aromatic content reduces the concentration of unburned hydrocarbons (HCs) and carbon monoxide (CO); however, it increases the emissions values for oxide of nitrogen (NO_x). The achievable reduction in the emissions for a vehicle is, however, counterbalanced by an increase in the CO_2 emissions during production of the fuel. Highly volatile fuels in the medium range of the boiling curve have a tendency to reduce the HC emissions and to increase the NO_x emissions. In terms of the CO emissions, a minimum was achieved at an evaporated percentage of 50% at 100°C, which is a typical value for standard commercial fuels. **Fig. F217** shows the data for and consequences of possible reformulated gasolines. Common evaluation of the results by the automotive and mineral oil industry, the European Commission and the EU Parliament led to changes in the gasoline specifications which will be implemented in two stages (**Fig. F218**).

Reduction of the maximum vapor pressure in the summer in Germany from 70 to 60 kPa is intended to reduce evaporation emissions.

The newly defined limitation of the aromatic content to initially 42% by volume and later 35% by volume and the reduction of the maximum permissible benzene content to 1% by volume were both intended to reduce the benzene emissions. However, the effects are less than would be suspected from the change in the specifications, particularly because the values generally significantly exceed the previously permissible maximum benzene content of 5% by volume, and a maximum of 1% benzene by volume was already observed on a voluntary basis for premium plus. The limitation of the aromatic content to 35% by volume planned for the year 2005 will increase the requirement for ethers (MTBE) as well as isoparaffins and require expansion of the corresponding production facilities.

The newly introduced limitation of the olefin content to a maximum of 18% by volume is intended to reduce the ozone formation potential from unburned hydrocarbons as well as those not oxidized by the catalytic converter and those from evaporative emissions. This measure has made reformulation of a portion of the regular gasoline produced necessary. Constituents containing olefins such as light and heavy cracked gasoline will have to be replaced by constituents with pure olefins, such as reformulated fuels.

	EN 228 Status 1999	Specification changes to minimize emissions	
		2000	2005
Sulfur [ppm]	500	150	50
Benzene [% by volume]	5.0	1,0	
Aromatics [% by volume]	not specified	42	35
Olefins [% by volume]	not specified	18	
Vapor pressure (summer) [kPa]	max. 70	max. 60	

Fig. F218
Modification of specifications for gasolines to reduce emissions

Reduction of the sulfur content primarily increases the efficiency of exhaust catalytic converters. Moreover, it also reduces the problems of formation of hydrogen sulfide in the catalytic converters. The low limits require, in particular, the desulfurization of light and heavy cracked gasoline.

The minor reduction in the maximum oxygen content from 2.8 to 2.7% m/m is of only subordinate significance. The maximum value for MTBEs of 15% by volume is still possible approximately even at a maximum oxygen content of 2.7%.

~Regular gasoline. Regular gasoline is an unleaded gasoline specified in European requirement standard EN 228 whose octane rating is, however, established locally (specifically for each country). In Germany, this value is at least 91.0 for RON and at least 82.5 for MON.

Leaded regular gasoline was prohibited in Germany in 1988. Regular gasoline is the least expensive type of gasoline with a market percentage of nearly 40% in Germany (1994).

Because of its low octane rating, in comparison to premium gasoline, regular gasoline can only be used in engines with low compression ratios or engines with knock sensors and retarded ignition. In combination with its lower density and, therefore, lower volumetric calorific value, it has a lower thermodynamic efficiency during combustion, resulting in higher volumetric fuel consumption of approximately 5–8% in comparison with premium gasoline.

~Requirements. Unleaded premium fuel is standardized throughout Europe in EN 228, whereby various specification classes have been created which are used, depending on the location and season, for climate-dependent criteria (vapor pressure and basic values for description of boiling characteristics, evaporation quantity at 70°C, as well as a characteristic value from the two parameters). Even though premium plus and, particularly, regular gasoline are not available in all European countries, they are also standardized in the national appendix—for example, DIN- EN 228. A European standard for leaded fuels, which are no longer on the market, has not been created.

Generally, fuels are standardized primarily in terms of their characteristics, not their composition. Excep-

	Test procedure		Unleaded gasoline according to EN 228		
			Premium Plus	Premium	Regular
Density at 15 °C	EN ISO 3675 EN ISO 12185	kg/m^3	720–775		
Resistance to knocking (octane rating) RON MON	EN ISO 25164 EN ISO 25163		min. 98.0 min. 88.0	min. 95.0 min. 85.0	min. 91.0 min. 82.5
Lead content	EN 237	mg Pb/l	max. 5		
Boiling curve: Total evaporated percent by volume up to 70 °C (E 70)	EN ISO 3405				
Summer		% by volume	20–48		
Winter		% by volume	22–50		
up to 100 °C (E 100)					
Summer		% by volume	46–71		
Winter		% by volume	46–71		
up to 150 °C (E 150)		% by volume	min. 75		
Boiling point (FBP)		°C	max. 210		
Distillation residue in percent by volume		% by volume	max. 2		
Flüchtigkeitskennziffer (Vapor lock index) VLI = 10 · VP + 7 · E 70 Transition period summer/winter winter/summer			max 1150		
Vapor pressure (DVPE)	EN 13016-1				
Summer		kPa	45–60		
Winter		kPa	60–90		
Evaporation residue, washed	EN ISO 6246	mg/100 ml	max. 5		
Benzene content	EN 238 EN 12177	% by volume	max. 1		
Olefins	ASTM D 1319		max. 18		
Aromatics	ASTM D 1319		max. 42		
Sulfur content	EN 24260 EN ISO 14596 EN ISO 8754	mg/kg	max. 150		
Oxidation stability	EN ISO 7536	Minutes	min. 360		
Corrosive effect on copper	EN ISO 2160	Degree of corrosion	max. 1		
Oxygen content	EN 1601 EN 13132	m %	max. 2.7		

Fig. F219 Gasolines: requirements and properties corresponding to EN 228 for Germany

tions are the upper limits of constituents to ensure environmental compatibility for benzene, sulfur, aromatics, and olefins. In addition, there are upper limits to ensure proper functioning with alcohols and ethers, which are permitted in low concentrations (alcohol content). All three unleaded types of gasoline are listed in the German version of the Quality Standard DIN-EN 228 (**Fig. F219**).

Details on the individual criteria are given under the keywords in each case.

~Research octane number (RON). See above, ~Octane number

~Road octane number. This describes the antiknock properties of a gasoline—not, however, in the standardized single cylinder test engine, but rather in a stock vehicle. The road octane number evaluates the antiknock characteristics of fuels in typical vehicles taken from the vehicle population. This was required before defining obligatory quality values for gasolines with simultaneous introduction of new refinery procedures, which generated fuel blends not yet tested. In contrast to the standardized single-cylinder CFR test engine (octane rating), the ignition timing—but not the compression ratio—is changed in the stock engine to evaluate the fuel (because a change in the compression ratio is not practical). The tests are performed either under full-load acceleration or at constant load and speed (on a rolling-road dynamometer).

In comparison to the CFR test engine, the measuring range is highly limited because informative values can only be measured within a crank angle range of approximately 10–15° around the ignition timing defined by the vehicle manufacturer. This coincides approximately to a "bandwidth" of five to six octane numbers. For the test, a comparative curve of the timing point at which knocking starts (trace knock) is completed with a number of iso-octane and n-heptane blends. Then the firing point for trace knocking is measured with the fuels to be evaluated. **Fig. F220** shows

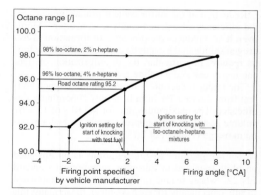

Fig. F220 Determination of road octane number

how the road octane numbers can be read off with the aid of the iso-octane/n-heptane comparison curve.

The road octane number for a fuel is usually between the RON and MON. At low-engine speeds, it tends to be closer to the RON and, at higher speeds and higher residual gas percentages, closer to the MON. With carbureted engines and road octane number measurement under acceleration conditions, it was also possible to reduce the road octane number by means of nonuniform octane number distribution over the fuel boiling range. (Light fuel constituents that reach the combustion chambers without delay during acceleration have a lower octane rating; *see above,* ~Front octane number.)

The use of excessive quantities of blending constituents with a low MON is excluded by defining the motor octane number MON in the requirement standard DIN EN 228 at a high level (considered relatively high for all types of fuels). Moreover, the sensitivity of the engine to nonuniform octane number distribution over the fuel boiling range has been eliminated by the introduction of fuel injection. This generally removes the

Constituent Example	Constituent Properties		Effect on Road Octane Index	
	Octane Index	Boiling Characteristics	Acceleration (Carburetor Vehicles)	Constant High Load and Speed
	RON	MON			
Light distillate	Low	Low	Volatility	Negative	Negative
Butane i-pentane i-hexane	High	High	Volatility	Positive	Positive
Light cracked gasoline	High	Low	Volatility	Positive	Negative
Heavy reformulated	High	Medium/High	Volatility	Negative	Positive
Heavy cracked gasoline	Medium	Low	Volatility	Negative	Negative
Tetraethyl lead			Volatility	Negative	Positive
Tetramethyl lead			Volatility	Positive	Positive

Fig. F221
Influence of a number of fuel and lead constituents on the road octane number

necessity for measuring the road octane number, because it is hardly ever used today for the evaluation of fuels. **Fig. F221** shows the influence of a number of fuel and lead constituents on the road octane number.

~Sewage gas →Fuel, diesel engine ~Biogas

~Sulfur. *See below*, ~Sulfur content; *also*, →Fuel, diesel engine ~Composition

~Sulfur content. Sulfur is a natural constituent in crude oil, where it is present in a chemically bonded form (**Fig. F222**). The sulfur content is highly dependent on the source of the crude oil.

The sulfur is removed from the fuel to a great extent by special refinery processes. While the sulfur content of gasoline in the past depended highly on the crude oil used and the production process as well as selection of the components relative to production, the differences between the fuels have decreased because of the low sulfur limits required since the year 2000.

The sulfur content is measured using the energy dispersive X-ray fluorescent method corresponding to EN ISO 8754, the wavelength dispersive X-ray fluorescent method in EN ISO 14596, or the Wickbold combustion method as outlined in EN 24260.

Sulfur compounds are undesirable in gasoline. Comprehensive studies in the American as well as European auto oil research program have shown that the efficiency of catalytic converters decreases with increasing sulfur content in the fuel and that the HC, CO, and NO_x emissions increase accordingly (**Fig. F223**). The limits for the maximum permissible sulfur content have been reduced continuously in the past. Fuels that do not contain any catalytically cracked gasoline, such as most premium plus fuels, contain very low quantities of sulfur. The designations "low sulfur" and "sulfur-free," used frequently in lieu of the actual limits, refer to a maximum sulfur content of 50 or 10 mg/kg respectively.

Since 2000, the maximum sulfur content has been limited to 150 mg/kg by the European requirement standard. Since January 1, 2003, the percentage of sulfur in German fuels has been less than 10 mg/kg ("sulfur-free," according to definition) with practically no exceptions because of the tax advantages available. A Europe-wide reduction of the sulfur content to a max-

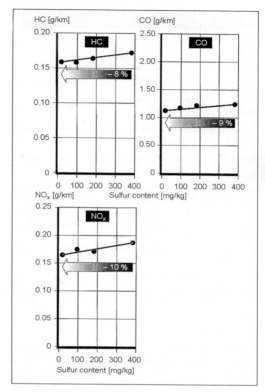

Fig. F223 Decrease in sulfur content in fuel reduces exhaust emissions

imum of 50 mg/kg was approved for 2005. The update to the EU fuel guidelines requires availability of sulfur-free fuels over the full area of the 15 EU countries by 2005 as well as changeover to entirely sulfur-free fuels by 2009 (**Fig. F224**).

~Synthetic gasoline (*also*, →Fuel, diesel engine ~Synthetic diesel fuels). In principle, gasolines (as well as other hydrocarbons) can be produced from other raw materials containing hydrocarbons, in addition to crude oil, by using various processes. However, the conversion processes require a great deal of energy so that the overall efficiency for the production of synthetic fuels is significantly lower than that for processing crude oil (**Fig. F225**).

Production of synthetic gasoline from coal has certain significance at times when the availability of crude oil is limited. However, because of the high carbon and low hydrogen content of coal (H/C atom ratio

Sulfur content	Regular	Premium	Premium plus
Range (mg/kg)	2–18	2–37	2–12

Fig. F224 Typical sulfur content in German gasoline fuels (winter 2003)

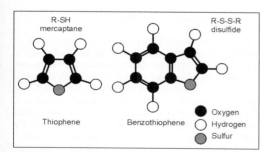

Fig. F222 Typical sulfur compounds in crude oil

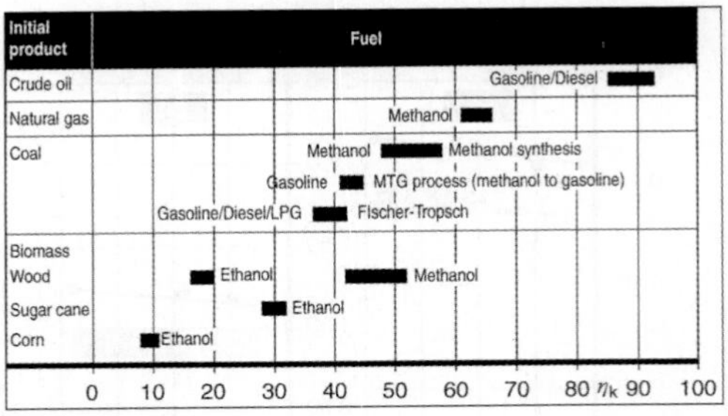

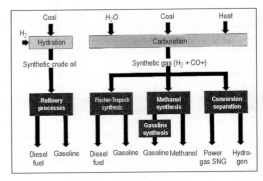

Fig. F225 Thermal efficiencies of fuel production

Fig. F226 Synthesis of fuels from coal

approximately 0.8 in comparison to crude oil with approximately 2.0), the addition of hydrogen is necessary. This can be accomplished directly by means of coal hydrogenation or by common gasification of coal with water vapor with the intermediate step of synthetic gas production (**Fig. F226**). These processes are described below:

1. *Hydrogenation.* During hydrogenation, the coal is decomposed and cracked in a number of stages and then enriched with hydrogen. This process is exothermic, corresponding to the following basic reaction:

$$n(C + H_2) = C_nH_{2n} - n1.04 \ kj/mol.$$

High pressure hydrogenation at a pressure of approximately 200 bar and temperatures of approximately 450–520°C is preferred. Gasoline and diesel fuels can be obtained in a final conventional refinery process.

2. *Gasification.* During gasification, coal is initially processed with water vapor to obtain synthetic gas. The primary reaction is the "heterogeneous water gas reaction,"

$$C + H_2O = CO + H_2 .$$

This reaction is highly endothermic. The energy requirement is provided in the present state of the art by burning approximately a third of the coal used. Gasification is accomplished at temperatures above 800°C usually without pressure or at relatively low pressures. The following reactions occur:

$$C + H_2O = CO + H_2 + 118 \text{ kJ/mol}$$
$$C + O_2 = CO_2 - 394 \text{ kJ/mol}$$
$$CO + H_2O = CO_2 + H_2 - 42.3 \text{ kJ/mol}$$
$$CO + 3H_2 = CH_4 + H_2O - 206 \text{ kJ/mol}$$

The synthetic gas resulting during gasification is cleaned and then used in one of the following synthesis processes:

- Fischer-Tropsch synthesis
- Methanol synthesis
- Gasoline synthesis

a. *Fischer-Tropsch synthesis*
Conversion of the conditioned synthetic gas (CO and H_2) on an iron catalyst to obtain liquid hydrocarbons. The basic reaction is:

$$n(CO + 2H_2) = (-CH_2-)_n + nH_2O - n159 \text{ kJ/mol}$$

b. *Methanol synthesis*
Conversion to methanol using a catalyst (e.g., Cu, Zn, Cr). The resulting methanol has a high degree of purity and can either be used directly in internal combustion engines or further processed using a gasoline synthesis process. The basic reaction is

$$CO + 2H_2 = CH_3OH - 90.0 \text{ kJ/mol.}$$

c. *Gasoline synthesis (MTG process)*
Reversible dehydrogenation of the methanol to obtain dimethyl ether is accomplished using catalysts. Methanol and dimethyl ether are initially converted to light olefins and then to paraffins and aromatics by condensation, transposition, and polymerization. The basic reaction is

$$n(CH_3OH) = (-CH_2-)_n + nH_2O - n50.9 \text{ kJ/mol}$$

Because of the low efficiency and higher costs compared to refining crude oil, production of synthetic fuel has been used only when the availability of crude oil is limited (Germany, World War II; South Africa, during commercial embargo) (Source: Pischinger RWTH Aachen University).

~Tetraethyl lead. *See above*, ~Antiknock additives

~Tetramethyl lead. *See above*, ~Antiknock additives

~Throttle valve icing. *See above*, ~Additives

~Toluene. *See above*, ~Aromatics contents

~Vapor lock. Vapor lock is premature evaporation of highly volatile fuel constituents in the fuel system. This can result in delays during hot starts or the engine missing following a hot start due to insufficient fuel. Under particularly critical conditions, vapor lock can also lead to malfunctions during normal operation — for example, stop-start operation in the summer or in mountainous terrain. In the same manner, excessively rich air-fuel mixtures were made possible in carbureted engines when vapor lock occurred in the float chamber, allowing fuel to be pressed into the intake system in an uncontrolled manner. Vapor lock can also occur when fuel is being pumped.

On the fuel side, vapor lock can be prevented by reducing the highly volatile constituents (e.g., butane, isopentane). This can be achieved by low limits for vapor pressure and the evaporated quantity at 70 and 100°C in summer. In terms of design, the tendency to vapor lock can be reduced by limiting heat-up of the fuel and by higher pressure in the fuel system. This is assisted by favorable location of the fuel pumps and mixture generation components.

Vapor lock can be described by various fuel parameters in sensitive vehicles. On vehicles with a carburetor, a dimensionless parameter has proven useful for this purpose based on E_{70} (evaporated quantity up to 70°C). The value determined from this fuel data is known as the vapor lock index (VLI). The maximum possible limits for the vapor pressure and E_{70} corresponding to the requirement standard for gasolines cannot be utilized fully for maintaining the maximum VLI; that is, at maximum utilization of the vapor pressure, the quantity evaporated at 70°C is reduced and vice versa.

Other fuel parameters—for instance, the vapor/liquid ratio at a given temperature (or the reverse: the temperature for a given vapor/liquid ratio)—correlate in similar form to the vapor lock in the fuel system on vehicles with gasoline engines (i.e., also with injection systems).

The vapor pressure measurement at high temperatures (Grabner method) has been developed for evaluation of fuels in terms of hot-start and hot-operation problems. This method shows particularly the azeotropic increase in the vapor pressure of gasolines containing methanol in the temperature range above 38°C (the previously defined temperature for measuring vapor pressure).

An exact correlation between the specified fuel parameters and the hot-start and hot-operation characteristics of the entire vehicle population is not feasible due to design differences and the wide variety of operating conditions; therefore, the response characteristics of the vehicles differ to a considerable extent. However, for sensitive vehicles, it can generally be said that a decrease in vapor pressure and lower evaporated fuel

quantity at 70°C (E_{70}) reduce problems resulting from vapor lock (**Fig. F227**). Fuel mixtures for this purpose have, however, led to poorer cold starting characteristics, cold operating characteristics, and an increase in the exhaust emissions, particularly at low temperatures (**Fig. F228**). Moreover, the carbon content in the fuel increases slightly, also increasing the CO_2 emission.

~**Vapor pressure.** The vapor pressure is the pressure caused by evaporation of a liquid in a closed container in relation to its temperature. The fuel vapor pressure, in part in combination with other fuel volatile characteristics, influences the hot- and cold-starting characteristics of the engine, cold operating characteristics at low temperatures, evaporation losses, and formation of the vapor phase above the liquid level in storage tanks. It is influenced primarily by highly volatile fuel constituents (e.g., butane, isopentane).

Various methods are used to characterize gasolines. Previously, Reid vapor pressure, RVP, was used. The test temperature was 37.8°C (100°F) at a vapor/liquid ratio of 4:1.

Measurement inaccuracies resulting from various quantities of moisture in the air in the vapor chamber of the measuring instrument were prevented by flushing the vapor chamber with water. However, when alcohol, particularly methanol, was present in the fuel, a portion of the alcohol went into solution in the water phase, so that a slightly higher vapor pressure was indicated than in a completely dry system. Nevertheless, because all fuels contain traces of water, the wet Reid method is sufficiently accurate for practical applications.

In modern engines with fuel injection, the fuel is subjected to higher temperatures, particularly upstream of the injection nozzles. For this reason, a laboratory method was developed for measuring the fuel vapor pressure in another temperature range (the standard specifies the range of 40 to 100°C) (EN 13016-2).

The measurement here is accomplished in relation to a vacuum; the fuel is air-saturated (ASVP: air-saturated vapor pressure). Exact determination of the partial pressure of the gas dissolved in the fuel also allows

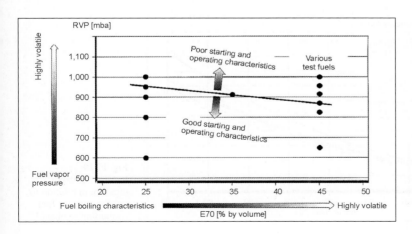

Fig. F227
Hot-start and operating characteristics in relation to fuel vapor pressure and boiling characteristics

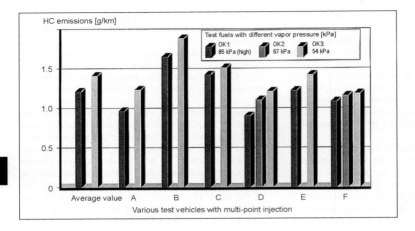

Fig. F228
Higher HC emissions at
low fuel vapor pressure

correction of the total vapor pressure. The vapor/liquid ratio for this method is 3:2.

With this method, fuels containing methanol show a considerably higher increase in the vapor pressure, particularly at high temperatures, while fuels containing alcohols with higher boiling points—TBA or ether, for instance—have the usual increase in the vapor pressure for pure hydrocarbon fuels (**Fig. F229**). The significant increase in the vapor pressure of fuels containing methanol is the result of formation of an azeotrope. Pure alcohols themselves have only a relatively low vapor pressure.

In the requirement standard for gasoline, the Reid vapor pressure was replaced, beginning February 1, 2000, by a computed DVPE (dry vapor pressure equivalent) corresponding to EN 13016-1. The DVPE can, for example, be calculated from the vapor pressure measured according to Grabner with the dry ASVP method described above at 37.8°C. For this purpose, the following formula was determined by multiple ring tests: DVPE = (0.965 ASVP) − 3.78.

With this method, the formation of the methanol-hydrocarbon azeotrope is taken into consideration and,

at the same time, the unsatisfactory imprecision of the Reid method used previously is eliminated with the modified dry measurement.

In addition to the measuring method for vapor pressure relevant to practical application, the "real" vapor pressure at 50°C is known. It is listed in the transport regulations and is defined by the pressure at which a vapor:liquid ratio of about 0:1 develops. The real vapor pressure is intended to eliminate variations in the reading, resulting from changes in the liquid phase when a portion of the fuel has evaporated as present with the other methods. A method generally used in the laboratory for direct measurement of the real vapor pressure is not available. The values are determined by computation of the vapor pressure according to Reid (RVP) and distillation data.

The specification standard EN 228 for European gasoline fuels specifies various volatility classes for fuels depending on the ambient air temperature and also contains various limits for the evaporated quantity of fuel at 70°C. This is in addition to the vapor pressure with and without the VLI (vapor lock index) calculated from the two values. For Germany, the following data were selected:

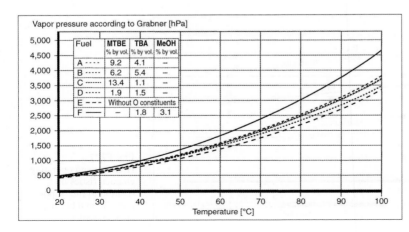

Fig. F229
Increase in vapor
pressure in relation to
temperature for fuel
containing methanol

May 1 to September 30: Class A (without VLI), summer products

November 16 to March 15: Class D (without VLI), winter products

October 1 to November 15 and March 16 to April 30: Class D1 (with VLI) (**Fig. F230**)

Correlation of vehicle and engine behavior as well as evaporation losses with the fuel vapor pressure even in combination with other fuel parameters is sufficiently good to be used for individual vehicles under defined test conditions. Because of the large variety of designs and the various conditions under which the vehicles are used and operated, however, only general trends can be established for the entire vehicle population. A sufficiently high fuel vapor pressure is required for good cold-starting at low temperatures as well as for proper acceleration when the engine is cold. The formation of an air/vapor mixture above the upper explosion point when the fuel is stored (i.e., in the "safe" rich range) also requires a sufficiently high vapor pressure. In contrast, limitation of the evaporation losses and elimination of problems when the engine is started or operated when already hot require limitation of the vapor pressure. The standardized vapor pressure ranges, therefore, represent a compromise between the various requirements (i.e., cold and hot operation). The maximum permissible vapor pressure for summer fuels was reduced from 70 kPa previously to 60 kPa beginning in the year 2000 to reduce the evaporation losses.

~Volatility. The volatility of a fuel characterizes its tendency to evaporate. It is described by the vapor pressure and boiling curve. The requirement standard EN 228 specifies six different fuel volatility classes, A–F, for summer and winter and different temperature conditions that take geographical differences into consideration. The highly volatile classes C–F are each subdivided again; classes C1–F1 have an additional limitation of the maximum potential volatility consisting of a characteristic value derived from the vapor

pressure and evaporated quantity at 70°C, the VLI (vapor lock index; **Fig. F230**). Adaptation of the gasoline volatility to the current outdoor temperatures is required to ensure optimum cold- and hot-operating characteristics as well as minimum exhaust and evaporation emissions. The fuel volatility is generally calculated using a dimensionless parameter from the vapor pressure (DVPE) and evaporated percentage of the fuel at 70°C (E_{70}), which is designated as the vapor lock index (VLI = 10 DVPE + 7E70).

The limits for volatility were changed in the year 2000. The minimum values for the vapor pressure were increased to prevent explosive mixtures in storage and vehicle fuel tanks at low temperatures (*also, see above,* ~Explosion limits). The upper values for summer fuels were decreased to prevent emission of the evaporated fuel.

~Water content. To prevent water from collecting in the vehicle fuel system, gasoline should not be cloudy or contain any undissolved water. The water absorption capacity of gasoline and, therefore, the potential water content are notably higher than for diesel fuels because of the higher polarity of the molecules (aromatics, alcohols, ethers). In addition to the chemical composition, temperature plays a decisive role for the capability of the fuel to physically dissolve water—a decrease in the temperature (e.g., sudden onset of winter) can lead to cloudiness. Normally, the fuel clears up relatively quickly as the droplets drop to the bottom of the tank. So-called dehazers can be used to support this process (*see above,* ~Demulsifying capability). The water collecting at the bottom of storage tanks as a result of this process is pumped out at regular intervals.

The water content is measured by titration using the Karl-Fischer method (DIN 51 777). A coulometric process also exists as an alternative (EN ISO 12937). Typical values for the water content in standard commercial gasolines (without alcohols) are approximately 80 mg/kg for regular gasoline and 120 mg/kg for premium fuel.

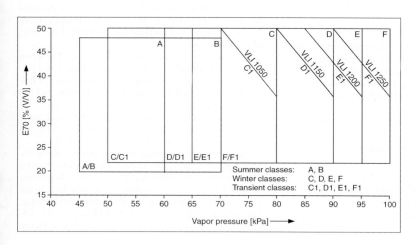

Summer classes: A, B
Winter classes: C, D, E, F
Transient classes: C1, D1, E1, F1

Fig. F230
Volatility classes for European gasoline fuels; vapor pressure and evaporated quantity at 70°C in conformance with EN 228

The water content in fuels containing alcohol requires special attention. Dissolved water can lead to precipitation of water-alcohol mixtures, with phase separation at decreasing temperatures. Such water-alcohol phases can also result at the bottom of a tank when a distribution system is changed over from alcohol-free fuels to fuels containing alcohol and water is present at the bottom of the storage tank.

Higher alcohols such as tertiary-butanol (TBA) can be used as solubilizers to significantly increase the water absorption capacity and provide for homogeneity in fuels containing alcohol. After the tank is cleaned, any water residues still present even in the vehicle tanks can then be absorbed reliably by the fuel without cloudiness or phase separation.

Fuel additive →Fuel, diesel engine ~Additives; →Fuel, gasoline engine ~Additives

Fuel atomization →Injection valves ~Diesel engine ~~Injection nozzle parameter ~~~Injection spray, ~Gasoline engine ~~Intake manifold injection ~~~Spray preparation

Fuel boiling curve →Fuel, gasoline engine ~Boiling characteristics

Fuel bubble formation →Fuel, diesel engine ~Vapor lock; →Fuel, gasoline engine ~Vapor lock

Fuel cell. The fuel cell is a possible future alternative to the internal combustion engine for use as a vehicle engine. The reasons for development of the fuel cell as a powerplant include:

- Low total pollution emissions, especially CO_2 emissions, by using hydrogen that is produced by a regenerative process. The most promising technology today for the use in mobile systems is the PEM (polymer electrolyte membrane). The polymer membrane is a proton conductive polymer film with a high power density and a working temperature below 100°C.

The disadvantages of fuel cells are:

- Expensive stainless steel catalytic converters, containing platinum, result in high purchase cost.
- Specific power density of cell is relatively small.
- Fuel storage and refueling is very cost-intensive.
- Warm-up times are currently still very high.
- Cold-start consumption is high.
- Cost and space requirements are currently not controllable.

The following materials are possible energy sources:

- Hydrogen (H_2)
- Methyl alcohol (MeOH) (CH_3OH)
- Ethyl alcohol (EtOH) (C_2H_5OH)
- Dimethyl ether (DME)
- Diesel (reference mixture $C_{12.95}H_{24.38}$)
- Modified gasoline

The utilization of fuel cells depends primarily on whether they can exceed the efficiency of conventional combustion engines and especially that of the future diesel engines with direct injection. The best values for fuel cells operated with water are currently at a maximum of about 40% and are, therefore, no better than the values for diesel engines.

Fuel composition →Fuel, diesel engine ~Composition; →Fuel, gasoline engine ~Composition

Fuel compressibility →Injection system (components) ~Diesel engine ~~Injection hydraulics

Fuel consumption (*also*, →Combustion, gasoline engine ~Mean effective pressure and fuel consumption; →Variable valve control ~Effect of fully variable valve control ~~Fuel consumption). The fuel consumption of motor vehicles has been one of the central focal points of vehicle development during the last 20 years. This is primarily the result of public awareness of the greenhouse effect (because the formation of CO_2 increases proportionally to the fuel consumption), limitation of the natural crude oil reserves, and an increase in the cost of fuels.

~**Absolute fuel consumption.** This is the fuel consumption of a vehicle in liters/100 km or in miles/gallon (mpg). In this case, it is necessary to note the difference between the calorific values of diesel fuel and gasoline. Different variables have an influence here: the vehicle/transmission/engine combination; the vehicle weight; vehicle equipment; aerodynamic drag coefficient; rolling resistance; engine performance and engine friction as well as the individual driving characteristics, including the operating profile, ambient conditions, and traffic conditions. An objective comparison between vehicles in actual traffic is, therefore, only feasible in a limited sense.

Better possibilities for comparison result from driving cycles on defined dynamometers. Using the so-called one-third mix, the fuel mileage is determined for constant operation at 90 km/h and 120 km/h and in the ECE test cycle, which are then combined at a ratio of one-third each.

The new measuring methods in the European Union, the United States, and Japan are based on a carbon balance from the emission data. The basis for this is the ECE and EUDC test cycle in Europe and the FTP test cycle in the United States.

The methods just mentioned allow different vehicles to be compared, but the results do differ from the fuel mileage of an "actual" vehicle on the road.

Currently, the best real road mileage figures are achieved by vehicles equipped with a turbocharged diesel engine with direct injection (magnitude approximately 6 L/100 km for medium-sized vehicles). In the future, the fuel consumption will continue to decrease. For example, fuel mileage figures of less than 3 L/100 km can be achieved with extremely light, small vehicles,

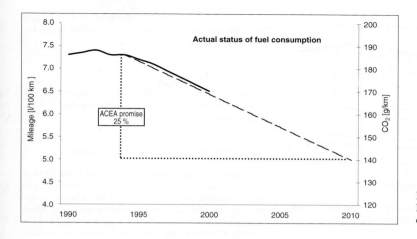

Fig. F231
Development of fleet
consumption

equipped with a three-cylinder engine (direct injection diesel or gasoline engine).

In spite of the continuous increase in vehicle weights, it has been possible to significantly reduce the fuel consumption of motor vehicles and, therefore, the CO_2 emissions in the past few years.

The association of European automobile manufacturers ACEA has agreed with the European Union to reduce the CO_2 emissions of their vehicle fleets from the present average of 180 g/km to 140 g/km by 2008 (**Fig. F231**).

~Carbon dioxide emissions. The CO_2 mass emissions from a vehicle are directly related to its fuel consumption and can be calculated using the following formula according to [19]:

$$m_{CO_2} = \frac{0.85 m_{Kr} \cdot 0.429 CO - 0.866 HC}{0.273}$$

where

m_{Kr} = fuel mass,
CO = carbon monoxide emission factor, and
HC = emission factor for unburned hydrocarbons.

Therefore, every measure to reduce fuel consumption contributes directly to reduction in the CO_2 emissions. For example, a reduction in the weight of the vehicle from 1500 to 1300 kg reduces the CO_2 emissions by approximately 20 g/km.

~ECE/EUDC consumption. The CO_2 and fuel-related emissions are measured in a test cycle. The fuel consumption is calculated from the carbon (C), carbon monoxide (CO), and carbon dioxide (CO_2) emissions. Limits are not specified.

~Effective fuel consumption. The effective fuel consumption, b_e (brake specific fuel consumption BSFC), is an engine-related specific value determined on engine test benches and expressed in maps in g/kWh in relation to the load and speed. It is defined as:

$$b_e = \frac{H_u \dot{m}_{Kr}}{P_e},$$

where H_u = 32,000 kJ/dm^3 for gasoline and 36,000 kJ/dm^3 for diesel, and where H_u = lower calorific value of fuel, $\dot{m}_{Kr}$ = fuel mass per time, and P_e = effective engine output.

Typical fuel consumption maps for an MPI gasoline engine and DI diesel engine are shown in **Figs. F232 and F233**. The lines of constant specific fuel consumption are also called engine graphs. The minimum value for the specific fuel consumption (so-called best value) is located in the lower speed range at high load. The best values for current passenger car engines are approximately 200 g/kWh, for diesel engines with direct injection; and approximately 230 g/kWh for gasoline engines.

The specific fuel consumptions for various engine concepts are generally compared with one another at a brake mean effective pressure of 2 bar and engine speed of 2000 rpm. The current status of various concepts is shown in **Fig. F234**.

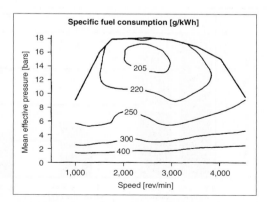

Fig. F232 Typical fuel consumption map for an MPI gasoline engine [2]

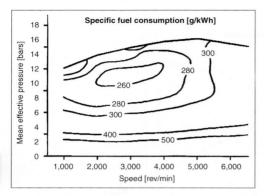

Fig. F233 Typical fuel consumption map for a direct injection diesel engine [2]

~**Fuel consumption calculation.** In the United States, manufacturers are required by law to maintain specified fleet fuel consumption limits. The individual vehicles sales are calculated into the balance weighted according to the numbers sold. The fuel consumption in countries using the US emission technique according to the carbon balance method is calculated from the results of the emissions test. This assumes that all exhaust constituents from a hydrocarbon fuel (HC, CO, CO$_2$) exhausted from the engine must have been input into the engine in the form of fuel, whereby HC, CO, and CO$_2$ are specified in g/mile and ρ_{Krst} in kg/L.

The method for calculating the fuel consumption is also used in the ECE/EUDC test cycles.

~**Indicated fuel consumption.** *See below,* ~Internal fuel consumption

~**Internal fuel consumption.** The internal or indicated fuel consumption, b_i (indicated specific fuel consump-

tion, or ISFC), is a specific engine-related value that does not take into consideration the friction in the engine. It is defined as

$$b_i = \frac{H_u \dot{m}_{Kr}}{P_i},$$

where m_{Kr} = fuel mass per unit of time, P is determined from the indicator (p-V) diagram, and P_i = indicated engine power output.

~**Specific fuel consumption.** *See above,* ~Effective fuel consumption, ~Internal fuel consumption

~**Variables.** A certain amount of energy in the form of fuel is required to overcome the road resistance. The fuel consumption of a vehicle can be reduced by improving the efficiency of the drive source and drivetrain, as well as by reducing the road resistances of the vehicle. Generally, the fuel consumption of a vehicle can be calculated as follows:

$$B_e = \frac{\int b_e \cdot \frac{1}{\eta_{\ddot{u}}} \left[\left(m \cdot f \cdot g \cdot \cos\alpha + \frac{\rho}{2} c_w \cdot A \cdot v^2 \right) + m \cdot (a + g \cdot \sin\alpha) \right] \cdot v \cdot dt}{\int v \cdot dt}$$

where

$m \cdot f \cdot g \cdot \cos\alpha$ = rolling resistance (**Fig. F235**),

$\frac{\rho}{2} c_w \cdot A \cdot v^2$ = aerodynamic drag,

$m \cdot a$ = acceleration resistance, and

$m \cdot g \cdot \sin\alpha$ = climbing resistance.

~~**Driving behavior.** The driving behavior of a driver has a significant influence on fuel consumption, particularly on vehicles with manual transmission. The fuel consumption map (**Fig. F233**), for an internal combustion engine shows that the range with the best specific

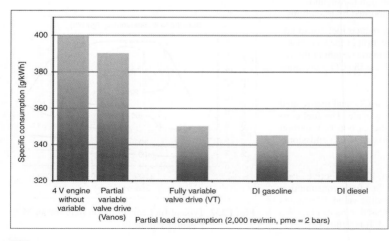

Fig. F234 Comparison of specific fuel consumption for various engine concepts

Size		Unit
B_e	Mileage	g/km
b_e	Specific consumption	g/kWh
$\eta_\ddot{u}$	Drive train efficiency	–
α	Road inclination angle	°
p	Air density	kg/m³
c_w	Drag coefficient	–
A	Bulkhead area	m²
m	Vehicle mass	kg
f_R	Rolling resistance coefficient	–
g	Gravitational acceleration	m/s²
v	Vehicle speed	m/s
e_1	Additional factor for rotating masses in gear i	–
a	Longitudinal acceleration	m/s²
t	Time	s

Fig. F235 Values and units [2]

fuel consumption is located in a narrow speed and load range: at high load and medium-to-low engine speed; the driver should try to operate the vehicle in this range. When the vehicle is driven in the part-load range, the propulsion force required to maintain a constant speed is available in a number of gears. It is possible for the driver to select the gear for best fuel consumption without having to accept disadvantages when driving at a constant speed. In detail, this means:

- Shifting up at the lowest possible engine speed
- Accelerating with high load at low engine speed
- Driving with foresight uniformly in the highest possible gear, using "Engine braking"
- Using only 70–80% of the maximum vehicle speed
- Shutting off the engine during longer idle phases

The savings potentials in daily rush-hour traffic by utilizing low shifting speeds are shown in **Fig. F236**.

Shutting off the engine during longer idling phases using an automatic start/stop feature when the engine is at operating temperature reduces the fuel consump-

tion. However, it is necessary to ensure the vacuum supply to the power brake is maintained as well as take into consideration the power used by vehicle air conditioning and the charging of the battery. OEM systems that do this have been offered in the compact car sector for a number of years. The fuel consumption savings achieved amount to 0.1–2 L/100 km. The discussed introduction of integrated crankshaft mounted starter-generators could allow wider application of automatic start/stop devices. However, this is currently still prevented by the considerable additional costs and significant additional weight of these.

~~**Engine measures**

~~~**Compression ratio** (*also*, →Compression). In internal combustion engines, the thermal efficiency can be improved by increasing the compression ratio. In gasoline engines with quantitative control of the combustible mixture, the effective compression ratio is very low in the part-load range (e.g., at 2000 rpm, BMEP [brake mean effective pressure] = 2 bar, 2.0 liters displacement, ε_{eff} = 3). When the geometric compression ratio is increased to about 15 (**Fig. F237**), there is an increase in the effective compression ratio at part load and, therefore, to a significant improvement in fuel consumption. The compression ratio at full load is limited to ε ~ 12 to avoid knocking combustion. If the fuel consumption is to be optimized, a variable compression ratio dependent on speed and load is desirable. On turbocharged gasoline engines, fuel enrichment at full load is reduced or higher supercharging is possible, thereby improving the thermal efficiency.

~~~**Crankshaft starter-generator.** A synchronous or asynchronous generator driven directly by the crankshaft can be used as a crankshaft starter/generator. This design reduces the noise when starting, because there

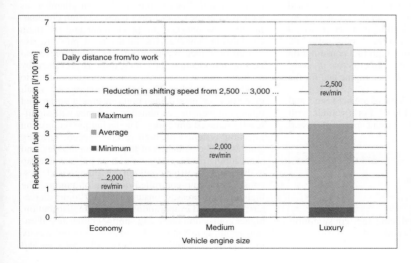

Fig. F236
Reduction in fuel consumption by reducing shifting feeds according to [18]

567

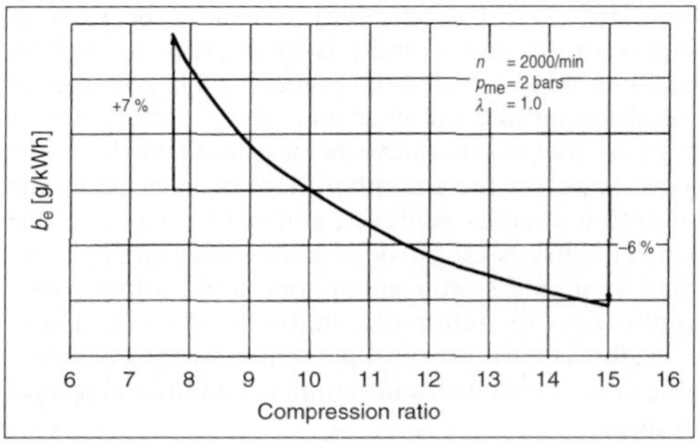

F

Fig. F237 Effect of compression ratio on specific fuel consumption [6]

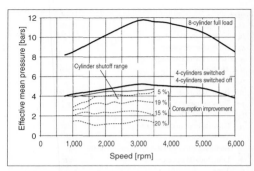

Fig. F239 Fuel consumption advantage resulting from cylinder shutoff on an eight-cylinder engine according to [7]

is no gear noise. Such machines can also be used to reduce torsional vibrations; however, they still require too much power today.

During the last ten years, the electrical power requirement for electric motors, auxiliary heaters, control units, appliances (radio, CD player, GPS system, etc.), and electric actuators has decreased dramatically. Crankshaft-type generators can easily provide such outputs with significantly higher efficiency than can conventional generators. This results in a major enhancement of the fuel consumption advantages when the system is also utilized as a start/stop feature or for reclamation of the energy (regeneration) used for braking (**Fig. F238**).

~~~**Cylinder shutoff.** The possible advantage in terms of fuel consumption by cylinder shutoff is based on shifting the load onto the cylinders still operating — that is, in the part-load range. The torque of the individual cylinders still firing is increased to achieve a higher load level, with better fuel consumption, for these cylinders. On passenger cars with large displacement engines, high power output, and torque, only a fraction of the available power from the engine is used in city traffic. The higher the power output and torque of an engine, the lower its load level in the part-load range, leading to high losses at the throttle valve and low efficiencies and, therefore, high fuel consumption figures. **Fig. F239** shows the mean effective pressure curve for a complete eight-cylinder engine and for the cylinder shutoff range.

The fuel consumption advantages are between 5 and 20% depending on the load level. For a vehicle with an eight-cylinder engine, at constant speeds of 90 and

120 km/h, the improvement in the fuel consumption was determined to be 15 and 13% respectively; in the NETZ, an improvement of 6.5% was achieved.

With cylinder shutoff, it is necessary to ensure that the temperature of catalytic converter does not drop to below its operating point when cold air is pumped through the engine and that the shutoff procedure does not result in unburned fuel reaching the catalytic converter.

~~~**Direct injection** (*also*, →Injection systems, fuel ~Diesel engine ~~Common rail, ~Gasoline engine ~~Direct injection systems ~~~Common rail). Since the introduction of direct injection in passenger car diesel engines at the end of the 1980s, prechamber and swirl chamber engines have been eliminated almost completely; new passenger car diesel engines all have direct injection.

The 15 to 20% lower specific fuel consumption of direct injection engines compared to the chamber engines results among other factors from the lower heat losses by not dividing the combustion chamber, as well as from the lower losses obtained from eliminating the flow between the prechamber and the main combustion chamber. The high rate of pressure rise caused by direct injection (noises), which prevented general introduction for a long time, is prevented by multiple injections with injection pressures of up to 2000 bar and higher. This preinjection of very small quantities, or pulsed injection, ensures a softer and more homogeneous combustion sequence with significantly reduced emissions. The technique requires an injection system with high reaction speed, virtually constraint-free actuation of the injection nozzles, and exact tuning of the provision of high pressure fuel up to the nozzle by using common rail, pump/nozzle systems, or solenoid-controlled distribution pumps. The high injection pressure ensures smaller droplet diameters, more rapid evaporation and mixture distribution, higher conversion rates, and shorter combustion duration, resulting in better soot oxidation and significantly lower particulate emissions with smaller particles and

Function/Property	Total savings potential
Start-Stop (ECE cycle)	
Increase in efficiency/42 V electrical system	15 ... 25 %
Brake energy reclamation	
Booster operation	

Fig. F238 Potential for reducing fuel consumption through use of crankshaft starter-generator [14]

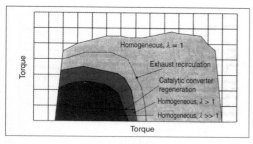

Fig. F240 Strategies for direct injection on gasoline engines [2]

lower fuel consumption. However, it is necessary to find the best compromise between the fuel consumption advantages resulting from the high injection pressures and the additional fuel consumption required for driving the high pressure pump. The common rail system has the lowest possible maximum drive power.

Direct injection has been used in gasoline engines for mixture preparation since around 1995. The gasoline engine is operated in stratified mode with $\lambda \gg 1$ or homogeneously with $\lambda = 1$, depending on the engine load and speed (**Fig. F240**). In stratified operation, the engine load is controlled by means of the injection quantity (as with diesel engines) with the throttle valve open.

This throttle-free load control and the extremely lean mixture results in reduced fuel consumption. Further fuel advantages result from internal mixture cooling and fuel consumption optimized firing point as well as increasing the isentropic exponents of the working fluid. In the meantime, engine development departments are working on the third generation of injection systems — the so-called jet-controlled method — after previously using the wall and air-controlled processes (**Fig. F241**).

With the wall-controlled combustion process, mixture formation of the injected fuel is initially brought about

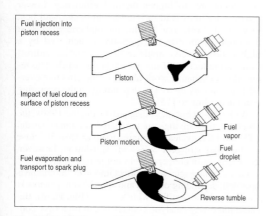

Fig. F241 Wall-controlled reverse tumble-combustion process [16]

because the majority of the fuel is deposited on the combustion chamber wall, from which it then evaporates.

In contrast, the mixture formation and formation of the charge stratification with jet-controlled combustion is based primarily on the characteristics of the fuel jet. With this process, the motion of the charge is not used for specific support of the mixture formation process. Positioning the injection nozzle in the center of the cylinder head and sharply focusing the fuel jet in the vicinity of the cylinder axis results in a favorable concentration of the fuel/air mixture for part-load operation with a thermally insulating sleeve of air or air/residual gas mixture.

Fuel consumption advantages of about 8% are currently achieved using the wall- and air-controlled processes in the European, American, and Japanese test cycles commonly used today. In contrast, the jet-controlled process provides a potential of 15% and higher.

~~~Downsizing. An increase in the mean effective pressure of the engine — that is, by supercharging — leads to an increase in the effective engine output from the same displacement. This has made it possible for engines with small displacement to achieve the same output data as engines with larger displacements. For example, a turbocharged four-cylinder engine with a displacement of 1.8 liters is capable of delivering the same power as a six-cylinder naturally aspirated engine with a displacement of 2.6 liters. Smaller engines have lower absolute friction values. A higher overall efficiency is possible at high mean effective pressure with simultaneous shift of the operating point to lower speeds. Fuel consumption savings between 10 and 20% [13] are possible compared to the savings from comparable naturally aspirated engines, depending on the driving cycle selected.

With diesel engines, the increase in the injection pressure from 600 to 1000 bar leads to an increase in the mean effective pressure of 17% with the same specific fuel consumption according to [12]. Today, injection pressures of 2000 bar are used, whereby the increase in the mean effective pressure is utilized to increase the output.

A downsized engine has lower vehicle package requirements reducing the vehicle weight. This leads to additional reduction of the vehicle fuel consumption.

~~~Friction loss (*also*, →Friction). The mechanical efficiency, η_m, of an internal combustion engine is defined as the ratio of the brake, p_{me}, and the indicated, p_{mi}, mean effective pressure,

$$\eta_m = \left(\frac{p_{me}}{p_{mi}}\right) = \left(\frac{p_{mi} - p_{mr}}{p_{mi}}\right),$$

where p_{mr} = mean friction pressure.

The way in which the engine friction has changed over time is shown for four-cylinder gasoline engines in **Fig. F242**.

During the last ten years, the friction characteristic particularly has been improved considerably. Statisti-

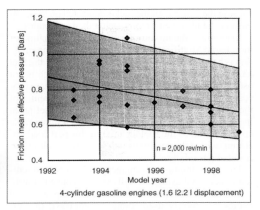

Fig. F242 Development of friction in four-cylinder gasoline engines according to [4]

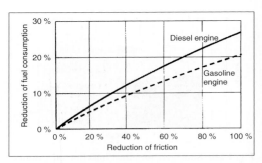

Fig. F243 Effect of reducing friction on fuel consumption at 2000 rpm [5]

cally speaking, the friction in a 2.0-liter engine has been reduced by approximately 20%. The reduction in fuel consumption by reducing the friction in the engine at operating temperature at a speed of 2000 rpm is shown in **Fig. F243**. At a speed of 2000 rpm, the friction losses amount to as much as 10% of the indicated output at full load. In the part-load range, the mechanical efficiency drops so that the effect of the friction in the fuel consumption increases even more. Therefore, a reduction in friction continues to be a primary development target.

~~~Fuel optimization. Fuels for gasoline engines should have the lowest possible inclination to self-ignition to prevent knocking. The octane number (ON) is the value for this tendency to knock.

When fuels with a high octane rating are used, it is possible to set the ignition timing to optimum values for reducing the fuel consumption. Modern engines are equipped with knock control, which changes the firing point precisely, depending on the engine speed, load, and temperature. On engines with knock control, it is possible to increase the compression and operate at more favorable firing angles.

This provides a potential for reducing the fuel consumption by 2–4%. Gasoline and diesel fuels can be improved considerably by specific development. For example, sulfur content has a direct effect on the fuel consumption in gasoline engines with direct injection because it is necessary to operate the engine with a rich mixture to remove the sulfur from the NO_x catalyst.

~~~Ignition. At high mean effective pressure and high thermal efficiency, rapid, uniform ignition and combustion rates are required for burning the air-fuel mixture. The combustion process and location of the centroid of the combustion depend on the angle of ignition and air-fuel ratio, λ (**Fig. F244**). If the ignition timing is advanced, this leads to lower specific fuel consumption and higher power output. However, knocking combus-

tion, excessive exhaust temperatures, and excessively high piston temperatures, for example, in the first ring groove, make it necessary to retard the ignition angle.

The thermal efficiency can be increased at critical operating points, depending on the combustion chamber geometry, by using two spark plugs when, for example, more uniform combustion and more rapid burning are achieved at high exhaust gas recirculation rates. In standard operation, the fuel consumption was reduced by 2% using this approach, compared to that with a single spark plug arrangement according to [9].

~~~Lean concepts (*also*, →Lean-burn engine).When gasoline engines are operated in the part-load range, the specific fuel consumption is reduced by operation with excess air ($\lambda > 1$, homogeneous)—that is, in the lean range (**Fig. F244**). During lean operation, the total charge mass increases because of the excess air, thus also increasing the effective compression ratio resulting in increased temperatures and pressures following compression. The quantity of heat released during combustion is absorbed by a greater charge mass, which reduces the mean process temperature. Both effects lead to an increase in the isentropic exponents of the charge and, therefore, to higher thermal efficiency. Lower mixture densities lead to a reduction of the wall heat losses in the wall area. The higher total charge mass at the same mean effective pressure is achieved by a higher throttle valve angle leading to partial dethrottling and reduction of the gas exchange work. The reduced gas exchange work increases the effective overall efficiency by up to 4%, depending on the operating point according to [15].

With the test engines designed during the 1980s, the potential fuel savings in the part-load range (usual FTP and ECE cycles) amounted to up to 15%. Because of the most stringent emissions legislation, however, these lean-burn concepts were not pursued further.

On gasoline engines with direct injection, the engine is operated, depending on load and speed, homogeneously lean ($\lambda > 1$) or stratified with $\lambda >> 1$ (**Fig. F240**). Because three-way catalytic converters can no longer convert the NO_x emissions in these ranges, it is necessary to buffer the NO_x emissions in a storage catalytic converter.

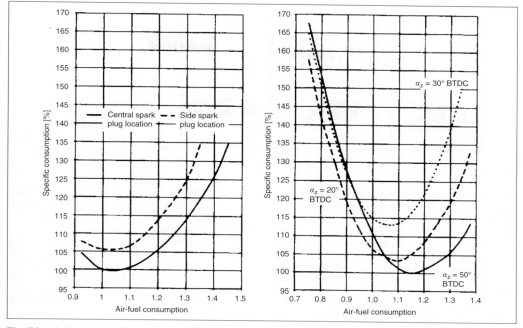

Fig. F244 Influence of spark plug position (left), preignition, and air-fuel ratio on fuel consumption according to [3]

F

~~~**Lubricant optimization.** Multiple grade oils with SAE ratings of 0W or 5W at their low temperature viscosity are LL oils (light viscosity lubricating oils) or FE oils (fuel economy oils). Two measures result in a reduction in the fuel consumption:

- Reducing the viscosity in the full lubrication range (hydrodynamic lubrication)
- Additives for reducing friction in the boundary lubrication range (mixed friction range)

The reduction of the viscosity has the greatest influence. Reduction of the friction caused by high viscosity oils was significant in engines not optimized in terms of friction. With modern engines with rolling friction in the valve drive and temperature management with which the viscosity is already controlled by the lubricating oil temperature, the effect has been reduced. On such engines, the fuel consumption advantage is in the range of 1 to 1.5%.

~~~**Supercharging (turbocharging).** Supercharging is already a standard feature in modern passenger car diesel engines. In contrast, this feature is limited to niche products for gasoline engines. Achieving the maximum possible output data is the primary reason for using turbochargers on most gasoline engines. On turbocharged gasoline engines, the compression ratio has to be reduced to 8.5 from 9.5 and the mixture enriched in the fuel load range to prevent knocking—both measures lead to higher fuel consumption. A comparison between the fuel consumption maps for a naturally aspirated gasoline engine and a turbocharged version shows significantly higher specific fuel consumption values in the lower part-load range [1].

~~~**Thermal management.** Thermal management ensures that the lubricating oil is heated as quickly as possible to about 90°C. This reduces the engine friction, particularly in the piston assembly and valve drive, resulting from the associated reduction of the oil viscosity, and the catalytic converter is heated to its light-off temperature as quickly as possible. This is achieved by using regulated heat exchangers or water pumps with regulated flow rate. Electrically driven water pumps can be switched off during the warm-up phase and the flow rate can be controlled according to the load. This allows the fuel consumption to be reduced by between 1 and 3%.

~~~**Thermodynamic optimization.** The heat conducted through the walls of the combustion chamber results in a significant energy loss. The combustion chamber surface area is minimized by making the combustion chamber shape as close as possible to spherical. With a basically long stroke engine layout, the ideal combustion chamber shape is approached. However, this also results in an increase in the piston velocity and the friction.

Moreover, smaller piston diameters lead to smaller valve cross sections resulting in higher pressure losses and, therefore, more gas exchange work.

~~~**Variable valve timing gear.** Variable valve timing gear—that is, with variable opening time, variable

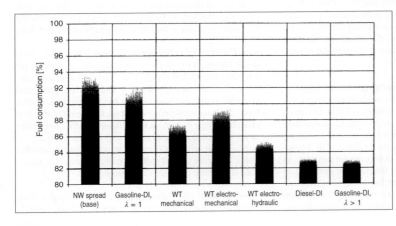

Fig. F245
Gas exchange:
conventional timing,
early intake valve closing,
and late intake valve
closing.

F

valve lift, and/or variable open period—reduce the fuel consumption in gasoline engines through their influence on the gas exchange work, the mixture preparation, and the combustion.

By varying the period on the intake and/or exhaust side, the residual gas can be controlled in the part-load range using internal exhaust gas recirculation, and the engine efficiency can be increased by dethrottling and improving the mixture preparation.

If the engine load is controlled by variable timing gear with variable valve lift and open period, the intake valve is closed at precisely the moment the required mixture quantity is in the cylinder. Because the throttle valve can then be eliminated or the throttle fully opened, the region around the intake valve is at virtually ambient pressure and the temperature of the mixture is reduced as a result of expansion. The potential fuel savings by reducing the gas exchange work is in the range of 6 to 8% for a 2.0-liter engine. In the part-load range and at full load at low engine speed, the engine is operated with minimum valve lift values. At these valve lifts, the mixture flows through the valve gap at sonic speed. This ensures extremely fine distribution of the fuel droplets and significantly improves mixture preparation. According to [10], this effect alone can result in fuel consumption savings of up to 4%.

Throttle-free load control is usually accomplished by closing the intake valves early (**Fig. F245**). Cylinders can also be disabled by completely closing the intake valves for a number of cycles.

Newer variable valve arrangements also allow one or two intake valves to be shut off. In this manner, the effects of variable valve control and direct injection can be combined. The potential improvement in fuel consumption from various concepts is compared to the present standard state with variable camshaft spread in **Fig. F246**.

~~**Road resistances.** A certain amount of energy (i.e., fuel) is required to overcome the road resistances.

~~~**Aerodynamic drag.** The aerodynamic drag of a vehicle increases with the square of the vehicle speed (with the air flowing in the longitudinal direction).

Aerodynamic drag

$$F_L = c_W \cdot A \cdot \frac{\rho_L \cdot v^2}{2},$$

where c_w = drag coefficient, ρ_L = air density, v = vehicle speed, and A = lateral transverse cross section.

The c_w value is a function of shape, and the transverse cross section A is a measure of size, which can be

Fig. F246
Comparison of potential
savings in fuel
consumption for various
engine concepts, data
according to *MTZ* 3/2000

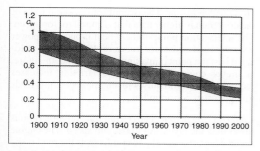

Fig. F247 Development of drag coefficient, c_w, since 1900 [2]

influenced to a great extent by the design. The development of the c_w value since 1900 is shown in **Fig. F247**, and the effect on the maximum speed and fuel consumption is shown in **Fig. F248**.

A reduction in the c_w value is limited by design trends, the flow of air through the engine compartment and passenger compartment, wheel clearances, measures to prevent lift at the two axles, flow of air through the wheel well to cool the brakes, flow of air through the radiator to cool the engine and for vehicle air conditioning, and flow of air around the exhaust system and by attached parts such as mirrors, windshield wipers, antennas, and door handles.

~~~**Rolling resistance.** The rolling resistance results from the work required to change the shape of the tires and the roadway.

Rolling resistance

$$F_R = f \cdot F_G,$$

where f = rolling resistance coefficient and F_G = vehicle weight.

It is necessary to take into account the resistance caused by the subsurface changing shape only for off-

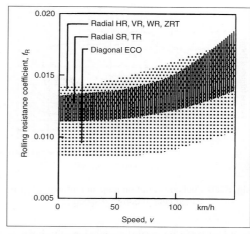

Fig. F249 Rolling resistance as a function of vehicle speed [11]

road applications. On paved roads, the rolling resistance is determined primarily by the losses resulting from tire flexing, which, for their part, are influenced by the compression of the suspension, wheel load, tire inflation pressure, and vehicle speed. New tires with low rolling resistance achieve coefficients of 0.008 in the low-speed range (**Fig. F249**).

Because F_R is defined along the longitudinal axis, it is necessary to take into consideration the additional load resistance resulting from the lateral force (toe-in resistance) in curves.

~~**Transmission adaptation.** The torque from the engine is converted to the propulsion force required for the vehicle drive wheels by a transmission system with individual gear ratios and the ratio of the subsequent differential (**Fig. F250**). In manual transmissions, the individual gears can be shifted manually by

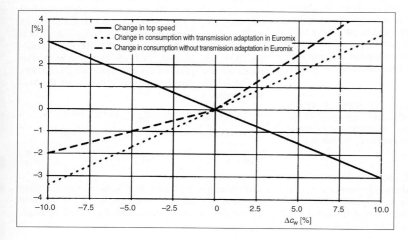

Fig. F248
Effect of c_w value on top speed and fuel consumption [2]

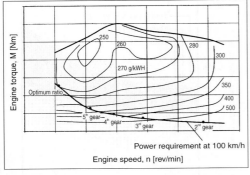

Power requirement at 100 km/h

Engine speed, n [rev/min]

Fig. F250 Fuel consumption map and influence of selected gear at constant speed [2]

the driver, but in automatic transmissions, this is determined by a hydraulically actuated shifting program. The following factors are, therefore, effective for the total gear ratio between the engine and wheels:

- Gear ratio of gear engaged i_{Gi}
- Differential ratio i_D

Total gear ratio of drivetrain in gear engaged, i_A:

$$i_A = i_{Gi} \cdot i_D$$

The magnitude of the differential gear ratio as well as the individual gear ratios and number of gears influence the fuel consumption (**Fig. F250**).

Introduction of a six-speed transmission made it possible for one vehicle manufacturer to reduce the fuel consumption in the ECE cycle by 4%. In laying out the gear ratios, selection of the high gear and overall ratio are of particular significance.

When engines with spur gear transmissions are installed in the longitudinal direction, it is possible to lay out one of the gears as a direct gear (ratio = 1:1). With this direct drive, none of the gear pairs is engaged under load, resulting in higher transmission efficiency. In order to fully use this advantage to achieve fuel savings, the gear used most frequently in operation is selected for the direct ratio. In passenger cars, this is normally the highest gear that can have a service life of above 80%. A transmission efficiency of 98% is achieved in the direct gear in comparison with 95–96% in the lower gears.

Total gear ratio in highest gear. The lowest transmission ratio in the drivetrain affects the top speed of the vehicle, the excess force, and, therefore, the dynamic agility of the vehicle, the noise emitted, engine gear, as well as fuel consumption.

Layout for maximum speed. The overall transmission ratio is selected so that the driving resistance curve on level ground (wheel resistance + air dynamic) intercepts with the maximum wheel drive power at this speed. The engine runs at the rated speed at the rated power point.

Overrevving layout. The overall transmission ratio of the drivetrain is higher than that required for top speed. The intersection point of the wheel propulsion power with the road resistance curve on level ground is located past the point of maximum output and, therefore, at higher engine speed. Because friction increases more than proportionally with engine speed, this leads to higher fuel consumption.

Underrevving layout. The overall transmission ratio is lower than the layout for maximum speed. The engine speed where the wheel power intersects the road resistance curve is below the rated output speed and, therefore, in the range of lower fuel consumption. However, the maximum possible speed is not achievable, and the excess power at lower speeds is lower.

With a highly underrevving layout, the transmission is shifted down at increasing road resistance because the excess power available is not efficient. This cancels the fuel consumption advantages of such a low-revving design. An improvement in specific fuel consumption of 16% is shown in **Fig. F251** for a highly underrevving layout at the achievable top speed in each case when the most extreme high and low ratios in the highest gear are compared.

~~**Vehicle equipment (energy loads).** The increase in the demand for convenience during the past few years has resulted in additional vehicle equipment, such as cruise control systems, electrohydraulic steering, power windows, electrically operating sliding roofs, automatic distance warning devices, brake assistance systems, active suspension, and parking assistant systems. These have led to a high increase in electrical consumption. **Fig. F252** shows that the electrical energy required in vehicles results in an average fuel consumption increase of more than 1.0 L/100 km. An increase in the electric power by 100 W leads to additional fuel consumption of about 1% for the customer. **Fig. F253** shows a few examples of electric loads.

~~**Vehicle mass.** The mass of the vehicle has a large influence on the vehicle dynamics and fuel consumption. Approximately 20 years ago, an additional fuel consumption of 1.0 L/100 km was calculated for an additional vehicle weight of 100 kg. Today, the effect is calculated using a figure of 0.4–0.6 L/100 km, depending on the vehicle model and type of engine. In the Euromix test, this influence amounts to about 0.2 L/100 km.

The vehicle mass has a linear influence on the road resistances.

Climbing resistance: $F_a = m \cdot g \cdot \sin\alpha$

Acceleration resistance: $F_a = e_i \cdot m \cdot a$

where $e_i = \dfrac{\Theta_{\text{red}}}{m \cdot R_{\text{dyn}}^2} + 1$,

and where $g = 9.81$ m/s^2, $m =$ vehicle mass including cargo, $a =$ road inclination angle, $\Theta_{\text{red}} =$ reduced rotating moment of inertia, and $R_{\text{dyn}} =$ tire dynamic radius.

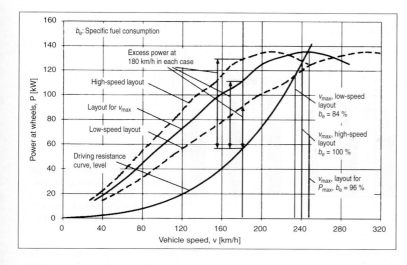

Fig. F251
Different layouts of
overall transmission ratio
in highest gear [2]

Today, the weight of the average vehicle is increasing as a result of the increased demands for convenience features such as electrical actuators for windows; sliding roof, mirrors, and seats; air conditioning; seat heaters; power steering; automatic transmissions, and advanced safety equipment such as traction and brake control systems, dynamic driving control, active bumpers and lateral stabilizers, air bags, and seat belt tensioners. The trend to high performance engines and the associated heavier components in the drivetrain as well as the larger brakes also tends to increase the weight, as does the increasing use of diesel engines. Also the motor consumption measures, the introduction of more and more catalytic converter devices (precatalyst and underbody catalyst), as well as new body structure developments to improve crash safety, all increase the vehicle mass. The development of the weight in the various vehicle classes during the past few years is shown in **Fig. F254**.

The increase in weight requires higher performance and, therefore, as a rule, heavier engines to provide the same or improved driving performance. A number of ideas and measures exist to reverse this spiral of increasing weight. Here, a large percentage of the total weight of the body and wheel suspension systems can be reduced by using intelligent light-weight structures such as aluminum rear axles or front-end structures of aluminum alloy. Replacement of steel by light alloys, plastics, or magnesium in the body area and in the interior also contributes to this weight reduction. In the area of the engine, the use of aluminum instead of gray cast iron for the crankcase could reduce the weight considerably. The introduction of magnesium/aluminum composite structures in this area opens up additional possibilities for reducing the vehicle weight. The replacement of large swept volume internal combustion engines with small capacity turbocharged engines (downsizing) could also contribute to a reduction in the weight.

Literature: [1] F. Pischinger, Verbrennungsmotoren, Lecture Reprint — [2] R. van Basshuysen, F. Schäfer (Eds.): Handbuch Verbrennungsmotoren Grundlagen, Komponenten, Systeme, Perspektiven, ATZ/MTZ Technical Manual, Braunschweig/Wiesbaden, Vieweg, 2002. — [3] Robert-Bosch-GmbH (Eds.): Ottomotor-Management, Braunschweig/Wiesbaden, Vieweg, 1998. —[4] M. Schwaderlapp, F. Koch, C. Bollig, F.G. Hermsen, M. Arndt: Leichtbau und Reibungsreduzierung Konstruktive Potenziale zur Erfüllung von Verbrauchszielen, 21st International Vienna Engine Symposium, Vienna 04–05.05.2000.

Fig. F252 Percentage of fuel used by electrical system [17]

Loads	Power consumption
Rear window heater	0.1 kW
Windshield heater	0.7 kW
Wiper motor	0.1 kW
Exterior lighting	0.16 kW
Power supply to control modules	0.2 kW
Fuel pump	0.15 kW
Dashboard display	0.15 kW
Sound system	0.2 kW
Car computer	0.15 kW
Ventilation fan	0.1 kW
ABS/FDR pumps	0.6 kW
Total	2.91 kW

Fig. F253 Power consumption of electrical loads in a passenger car [2]

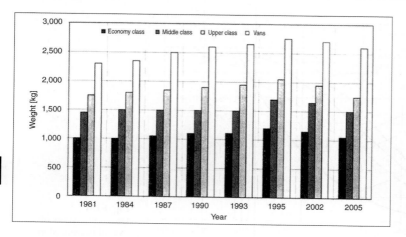

Fig. F254
Development in vehicle weight since 1981 and weight prognosis for various vehicle classes. Data according to [8]

— [5] F. Koch, U. Geiger: Reibungsanalyse der Kolbengruppe im gefeuerten Motorbetrieb, GfT Tribologie-Fachtagung, Göttingen 05–06.11.1996. — [6] C. Bollig, K. Habermann, H. Marckwardt, K.I. Yapici: Kurbeltrieb für variable Verdichtung, MTZ 11/97. — [7] M. Fortnagel, G. Doll, K. Kollmann, H.K. Weining: Aus Acht mach Vier, Die neuen V8-Motoren mit 4,3 und 5l Hubraum, in: ATZ/MTZ Yearbook 1998, Braunschweig/Wiesbaden, Vieweg, 1998. — [8] 20th International Vienna Engine Symposium, VDI 1999, Düsseldorf. — [9] 18th International Vienna Engine Symposium, VDI 1997, Düsseldorf. — [10] W. Wolfram, D. Hieber, J. Pietsch, H. Tschöke: Einfluss des Einlassventilhubes auf Gemischbildung, Kraftstoffverbrauch und Abgasemissionen von Ottomotoren. MTZ Motortechnische Zeitschrift 61 (2000) 7/8. — [11] Robert-Bosch-GmbH (Eds.): Kraftfahrzeugtechnisches Taschenbuch, 22nd edition, VDI 1995, Düsseldorf. — [12] U. Essers (Ed.): Dieselmotorentechnik 98, Expert, 1998, Renningen. — [13] H.D. Erdmann: Mehrventiltechnik und Aufladung als Verbrauchskonzept. Symposium Entwicklungstendenzen auf dem Gebiet der Ottomotoren, Esslingen, 1994. — [14] Kraftfahrwesen und Verbrennungsmotoren, 4th International Stuttgart Symposium, Expert, Renningen. — [15] H. Carstensen: Systematische Untersuchung der Konstruktions- und Betriebsparameter eines Zweiventilmagermotors auf Kraftstoffverbrauch Schadstoffemissionen und Maximalleistung, MTZ Motortechnische Zeitschrift 3/2000. — [16] Kume, Iwamoto, Jida, Murakami, Akishino: Combustion Control Technologies for Direct Injection SI Engines, SAE-Paper 960600. — [17] H.H. Braess, U. Seiffert (Eds.): Handbuch Kraftfahrzeugtechnik, 1st Edition, Braunschweig/Wiesbaden, Vieweg, 2000. — [18] Aral AG (Ed.): Kraftstoffe für Straßenfahrzeuge, Fachreihe Forschung und Technik. — [19] J. Abthoff, C. Noller, H. Schuster: Möglichkeiten zur Reduzierung der Schadstoffe von Ottomotoren, Technical Library, Daimler-Benz, 1983.

Fuel consumption gauge →Electronic open- and closed- loop control ~Electronic open- and closed-loop control, gasoline engine ~~Functions

Fuel conversion →Combustion function

Fuel cooler →Radiator ~Liquid/liquid radiator

Fuel dilution →Oil ~Power during operation ~~Contamination by foreign matter

Fuel economy (*also*, →Fleet consumption). Fuel economy is a fuel consumption value, which is calculated based on the certification method for the exhaust emissions and a carbon balance. Fuel economy is described in mpg (miles per gallon).

Fuel effect on particles →Particles (Particulates) ~Formation ~~Variables for particulate formation

Fuel efficiency maps →Performance characteristic maps

Fuel energy →Fuel, diesel engine ~Properties; →Fuel; gasoline engine

Fuel enrichment →Mixture formation ~Mixture formation, diesel engine, ~Mixture formation, gasoline engine

Fuel evaporation →Fuel, diesel engine ~Properties ~~Latent heat of vaporization, →Fuel, gasoline engine ~Volatility

Fuel feed →Injection system (components) ~Diesel engine, ~Gasoline engine ~~Fuel-feed unit

Fuel feed unit →Injection system (components) ~Gasoline engine

Fuel film →Mixture formation ~Mixture formation, gasoline engine ~~Wall film

Fuel film evaporation. A portion of the fuel deposits on the wall during mixture formation. This deposition can occur inside the combustion chamber (wall-applied direct injection) or outside the combustion chamber in the intake manifold (carburetor or central injector). This leads to a thin film of fuel that evaporates more or less quickly, depending on the physical conditions. The evaporation rate depends primarily on the evaporation heat (latent heat of vaporization) of the fuel and the rate at which energy is supplied. This process leads to uncontrolled variations in the air-fuel ratio, particularly with carburetors or central injection systems.

Fuel filter →Injection system (components) ~Diesel engine, ~Gaso-line engine

Fuel gas process →Engine concepts ~Composite systems ~~Free piston engines

Fuel heat capacity →Fuel, gasoline engine ~Heat capacity (specific heat capacity)

Fuel injection →Injection system, fuel

Fuel jet →Carburetor, ~Idle nozzle, ~Main jet; →Injection system (components) ~Diesel engine ~~Injection hydraulics

Fuel latent heat of vaporization →Fuel, gasoline engine ~ Evaporative heat; →Mixture formation ~Mixture formation, gasoline engine ~~Mixture cooling

Fuel line. The fuel line is used to transport the fuel to the injection valves, fuel distribution rail, or carburetor. To keep the fuel from vaporizing, the fuel line should be located in the engine compartment so that it is subjected to minimum heating effects. Moreover, basic safety requirements must be taken into consideration in the event of a collision. With high-pressure systems, such as those used with diesel engines, the elasticity and length of the fuel line affect the injection parameters and, therefore, the operating characteristics of the engine. For those reasons, it is practical to keep the length of the lines as short as possible and equal for all cylinders.

Fuel line pressure →Lubrication ~Oil line, ~Oil pump

Fuel lubricity →Fuel, diesel engine ~Properties ~~Lubricity

Fuel mass →Cylinder charge ~Volumetric efficiency

Fuel optimization →Fuel consumption ~Variables ~~Engine measures

Fuel preheating →Starting aid

Fuel preparation →Mixture formation ~Mixture formation, diesel engine, ~Mixture formation, gasoline engine ~~Mixture preparation

Fuel pressure damper →Injection system (components) ~Gasoline engine

Fuel pressure regulator →Injection system (components) ~Gasoline engine

Fuel production →Fuel, diesel engine ~Production; →Fuel, gasoline engine ~Production

Fuel pump →Injection system (components) ~Diesel engine, ~Gasoline engine

Fuel rail →Injection system (components) ~Gasoline engine

Fuel rail direct injection →Injection system (components) ~Gasoline engine

Fuel requirement development →Fuel, diesel engine

Fuel requirements →Fuel, diesel engine; →Fuel, gasoline engine

Fuel return →Carburetor ~Hot operation; →Injection system (components) ~Diesel engine

Fuel savings →Oil

Fuel shut-off →Carburetor ~Equipment; →Electronic open- and closed- loop control ~Electronic open- and closed- loop control, gasoline engine ~~Functions ~~~Overrun fuel cutoff, ~~~Cylinder shut-off; →Fuel consumption ~Variables ~~Engine measures ~~~Cylinder shut-off

Fuel supply pump →Injection system ~Diesel engine, ~Gasoline engine ~~Fuel feed unit

Fuel systems →Calculation processes ~Application areas ~~Fluid circulation

Fuel tank →Injection system (components) ~Gasoline engine

Fuel temperature. The fuel temperature affects the fuel density and thereby the mass of fuel supplied as well as the performance characteristics of the engine. Moreover, a high fuel temperature for gasolines leads to increased fuel evaporation in the atmosphere, which can be minimized with the aid of a fuel evaporation retention system. Further, vapor can form in the fuel system, leading to misfiring and hot-starting difficulties. With diesel engines, a change in the fuel temperature also changes the quantity in terms of the composition. This results in a noticeable effect on the engine parameters such as output, torque, and fuel consumption.

Effective measures to prevent excessive fuel heat-up include:

- Adaptation of the fuel requirements to the consumption (low return quantities)
- Insulation of the tank against heat transfer—e.g., from the exhaust system
- Shielding the fuel lines against heat transfer
- High fuel pump efficiency

Some vehicles are equipped with a fuel cooler.

Literature: D.B. Forster, W. Jung: Einfluss der Kraftstoff-Eingangstemperatur auf die Abgasemissionen von Dieselmotoren, MTZ (2003) 4, p. 274.

Fuel vapor pressure →Fuel, gasoline engine ~Vapor pressure

Fuel viscosity →Fuel, diesel engine ~Properties ~~Viscosity

Fuel wall deposition →Fuel film evaporation

Fuel-air mixture →Air-fuel mixture

Fuel-injector mount →Injection valves ~Diesel engine

Fuels containing ethyl alcohol →Fuel, gasoline engine

Fuel-water emulsion →Fuel, diesel engine ~Emulsions; →Water injection

Full slipper skirt piston →Piston

Full-flow burner →Particles (Particulates) ~Particulate filter system ~~Particulate filter ~~~Regeneration of particulate filters

Full-flow centrifuge →Filter ~Lubricating oil filter ~~Oil Extractor; →Lubrication ~Filter ~~Filter designs

Full-flow filter →Filter ~Lubricating oil filter ~~Full-flow oil filter

Full-flow oil filter →Filter ~Lubricating oil filter; →Lubrication ~Filter

Full-load/wide-open throttle →Load

Full-skirt piston →Piston

Full-skirt strut piston →Piston ~Design

Full-throttle enrichment (*also*, →Carburetor ~Starting systems ~~Requirements, ~Systems). Full-throttle enrichment is applied on gasoline engines to enable an increase in power output. It is used for "internal cooling" by decreasing combustion temperature and exhaust gas temperature. Disadvantages are the higher emissions of pollutants, because the three-way catalytic converter cannot work at its optimum operating point with a rich mixture. Fuel economy disadvantages are also observed.

Full-throttle fuel delivery (*also*, →Injection system, fuel ~Diesel engine ~~Serial injection pump ~~~Boost pressure-dependent full-load stop). Full-throttle fuel delivery is the fuel quantity that must be injected to obtain full-load power.

Full-throttle injection quantity →Injection functions ~Gasoline engine ~~Intake manifold injection

Fully automatic start →Carburetor ~Starting systems ~~Design types

Fully synthetic oils →Oil ~Body ~~Basic oils

Fully variable valve gear →Variable valve control

Fully variable valve timing →Variable valve control

Function blocks →Electronic/mechanical engine and transmission control ~Electronic components ~~Block diagram

Fundamental frequency →Engine acoustics

Fuses → Piston ring ~Gap

G

Galvanic corrosion →Coolant

Gap dimension →Charge movement ~Quench flow; →Combustion chamber squish areas

Gap width →Piston ring

Gas engine →Combustion process, diesel engine

Gas engine oils →Oil ~Classification, specifications, and quality requirements ~~Special engine oils

Gas force (*also*, →Balancing of masses) The combustion of the mixture of fuel and air generates a pressure inside the combustion chamber and, therefore, a gas force on the piston surface area. This is transferred to the crankshaft through the piston, the piston pin, and the piston rod. The gas force will then appear as torque outward on the crankshaft, and from there, to outside the engine. In addition, inertia forces also act on the crankshaft drive. The gas and inertia forces generate gas torques and couples (due to the inertias) through a lever (the throw of the crankshaft).

Fig. G1 shows the forces, which act on the crankshaft drive. F is the net force on the piston generated by adding the gas force, F_{Gas}, and the inertia force, F_M, from the reciprocating masses.

The most important components are:

Net piston force: $F = F_{Gas} - F_M$

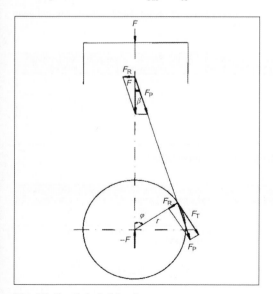

Fig. G1 Crankshaft drive forces

Net vertical force: $F_N = F \cdot \tan\beta$

Force along connecting rod: $F_P = \dfrac{F}{\cos\beta}$

Tangential force: $F_T = F \cdot \dfrac{\sin(\varphi + \beta)}{\cos\beta}$

Gas oil (light oil) →Fuel, diesel engine

Gas pressure diagram →Balancing of masses; →Cylinder pressure curve

Gas temperature. The temperature of the gas in the combustion chamber can locally reach 3000 K. The highest value of the mass averaged temperature is achieved shortly after the top dead center. It is about 2500 K in gasoline engines and 2000 K in diesel engines (**Fig. G2**). The thermal conductivity of the gas and the heat transfer at the combustion chamber wall result in steep temperature gradients in the gas in the combustion chamber.

Gas transfer →Charge transfer

Gas transfer control →Control/gas transfer; →Variable valve control ~Load control

Gas transfer loss →Charge transfer ~Charge transfer loss

Gas transfer valves →Valves

Gas transfer work →Charge transfer

Gas work. A gas enclosed in a cylinder with a moving piston is analyzed as it undergoes work caused by the volume change. The volume changes by the increment dV if the piston is moved by the distance dx. The

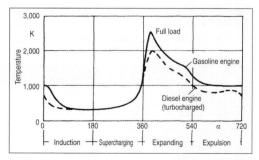

Fig. G2 Mass average temperature for gasoline and diesel engines (Source: Pischinger)

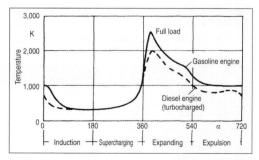

579

gas pressure on the piston must be compensated by a pressure of the same value from the piston on the gas if the system is in stable equilibrium. Incremental work dW is performed when the force acts on the piston, F, which is the product of the pressure p and the piston surface area, by the distance dx. The work done by the gas is, therefore, computed as:

$$W = \int p \cdot dV$$

Gaseous fuels →Fuel, diesel engine ~Biogas, ~Natural gas; →Fuel, gasoline engine ~Alternative fuels ~~Butane

Gaseous hydrocarbons →Fuel, diesel engine ~Biogas, ~Natural gas; →Fuel, gasoline engine ~Alternative fuels ~~Butane

Gasoline →Fuel, gasoline engine

Gasoline burner →Pollutant aftertreatment ~Aftertreatment concept with a three-way catalytic converter

Gasoline engine →Cold start; →Combustion process, gasoline engine; →Electronic open- and closed-loop control ~Electronic open- and closed-loop control/transmission control; →Emission maps; →Injection system, fuel ~Gasoline engine; →Load; →Piston ring; →Wall film

Gasoline engine exhaust gas analysis →Exhaust gas analysis

Gasoline engine injection functions →Injection functions ~Gasoline engine

Gasoline engine injection system →Injection system (components)

Gasoline engine torque →Torque

Gasoline filter →Filter ~Fuel filters

Gasoline fuel pump →Injection system (components) ~Diesel engine ~~Fuel-feed pump, ~Gasoline engine ~~Fuel-feed unit ~~~Fuel-feed pump

Gasoline injection →Injection system, gasoline engine

Gasoline requirements criteria →Fuel, gasoline engine ~Production

Gasoline tank →Injection system (components) ~Gasoline engine ~~Fuel tank systems

Gasoline-alcohol mixture →Fuel, gasoline engine ~Methanol fuel, ~Requirements

Gear meshing noise →Engine acoustics ~Transmission noise

Gear pump (also, →Lubrication ~Oil pump). Gear pumps on automobile engines are primarily employed as oil pumps. Selection criteria include the space required, the cost, and efficiency in the particular application. The only designs in use are so-called external or internal gear pumps.

~**Characteristics.** Important characteristics can be gathered from **Figs. G3 and G4**.

~**Cycloidal gear.** *See below*, ~Internal gear pump ~~Oil pumps without crescent or annular gear pumps

~**Delivery rate, oil pump.** *See below*, ~External gear pump

System	Max. drive speed [rpm]	Actual displacement at 1500 [rpm]	Permissible working pressure [bar]	Permissible operating temperature [°C]	Kin. viscosity [mm²/s]
Internal gear pump with filler piece	1,200 ... 5,000	5,6 ... 576	63 ... 250	−20 ... +80	20 ... 100
Internal gear pump (annular-gear pump) w/out filler piece	1,500 ... 1,800	4 ... 50	120	−10 ... +80	16 ... 150
External gear pump	800 ... 3,000	6,5 ... 280	120	−15 ... +80	22 ... 90

Fig. G3 Typical characteristics of pump systems from the literature (Source: SHW)

System	Max. engine speed [rpm]	Actual displacement at 1,500 [rpm]	Typical working pressure [bar]	Typical side-to-side width [mm]	Permissible operating temperature [°C]	Kin. viscosity [mm²/s]
Internal-gear pump with filler piece	650 ... 6,500	5,6 ... 576	1 ... 13	8 ... 14	−35 ... +160	5 ...
Internal-gear pump (annular-gear pump) w/out filler piece	600 ... 7,500 (Crankshaft) 350 ... 5,000 (Power take off)	4 ... 50	1 ... 13	8 ... 14 (Crankshaft) 20 ... 32	−35 ... +160	5 ...
External gear pump	800 ... 3,000	6,5 ... 280	1 ... 13	25 ... 60	−35 ... +160	5 ...

Fig. G4 Typical values of oil pumps for combustion engines (Source: SHW)

Fig. G5 External gear pump (Source: SHW Automotive GmbH)

Fig. G6 Involute teeth (Source: SHW)

G

~External gear pump. The external gear pump consists of two or more externally toothed gears. The mode of operation can be described as follows: The nonengaged teeth of the driven and the driving external gear act as positive-displacement devices, delivering from the suction space to the pressure space (**Fig. G5**). These pumps have the problem of having high pressure peaks applied to the base of their teeth that can be reduced by employing suitable relieving grooves. This design offers advantages for larger displacement volumes.

~Gerotor pump. *See below*, ~Internal gear pump ~~Oil pumps without crescent or annular gear pumps

~Injection pump cavity. *See above*, ~External gear pump

~Internal gear pump. The family of double-rotor systems includes the internal-gear pump because two meshing elements—internal and external gear—do the rotary motion. In this case, the internal rotor rotates inside the outer rotor, whereby the outer rotor is positioned eccentrically to the internal rotor by half the tooth height. The system is mostly driven by the internal rotor in such a way that the outer rotor is put into rotation by gear teeth. There are design types both with and without crescents.

~~**Crescent-type oil pumps.** A crescent generating a sealing area over several teeth serves as the sealing element on this kind of gearings. Its advantage is higher pump pressures but its disadvantage is the higher space requirement. Two gearing systems are implemented: one with involute teeth (**Fig. G6**), and a second

using a profile designed with trochoidal gear (**Fig. G7**). These pumps are exclusively employed in case of a directly driven crankshaft. They are mostly accommodated in the engine front control cover. Space requirements and producibility of the crescent do not allow its employment in the power takeoff (PTO)—for example, the oil sump.

~~**Oil pumps without crescent or annular gear pumps.** These pumps normally have gear ratios $z_I = z_O - 1$: typical tooth counts are between 4/5 and 13/14 teeth. Three gear-tooth geometries are commonly used:

- Gear teeth, gerotor, designed in the form of linked up circular arcs (**Fig. G8**).
- Gear teeth, duocentric, designed in the form of non-linked up circular arcs (**Fig. G9**).

Fig. G7 Trochoidal gear (Source: SHW)

Fig. G8 Gerotor pump (Source: SHW)

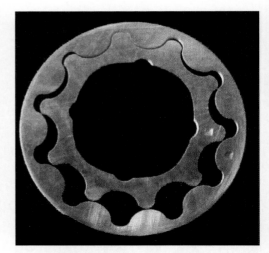

Fig. G10 Duocentric IC pump (Source: SHW)

• Cycloidal gearing. This is a gearing designed from hypocycloids and epicycloids, duocentric IC (internal combustion; **Fig. G10**). The advantage of this gearing is lower noise emissions.

These pump systems can be found both on the crankshaft and on the power takeoff—for example, in the oil sump.

~Internal rotor. *See above*, ~Internal gear pump ~~Oil pumps without crescent or annular gear pumps

~Involute teeth. *See above*, ~Internal gear pump ~~Crescent-type oil pumps

~Oil sump. *See above*, ~Internal gear pump ~~Crescent-type oil pumps

~Pump pressure chamber. *See above*, ~External gear pump

~Trochoidal gear. *See above*, ~Internal gear pump ~~Crescent-type oil pumps

Literature: R. van Basshuysen, F. Schäfer (Eds.): Handbuch Verbrennungsmotor, 2nd edition, Vieweg Publishers, 2002.

Gear pump characteristics →Gear pump

Gear scuffing →Camshaft ~Wear

Gears →Engine accessories

General Purpose Actuator (GPA) →Actuators

Generator →Alternator

Generator load control →Electronic open- and closed-loop control ~Electronic open- and closed-loop control, gasoline engine ~~Functions

Generator noises →Engine acoustics

Generator operation →Supercharging

Geometrical compression →Compression ~Compression ratio

Geometry, crankshaft →Crankshaft

Geometry, piston rod →Connecting rod ~Design

German Touring Car Masters (DTM) →Engine ~Racing engines

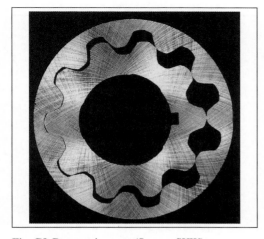

Fig. G9 Duocentric pump (Source: SHW)

Glow ignition (*also*, →Premature ignition; →Thermal ignition). Glow ignition is an effect that is found only in gasoline engines. It is caused by reactions that initiate combustion that has not been started by a spark provided by the spark plug. Such ignition is triggered by overheated components, such as the electrodes of spark plugs; the exhaust valves; sharp, overheated edges in the combustion chamber; or overheated combustion chamber walls. Deposits in the combustion chamber, which are generated by combustion residues from the fuel and oil, can also be the reason for glow ignitions. Temperatures of at least 1100–1200°C are required to initiate glow ignitions. These temperatures are significantly higher than the auto-ignition temperatures of the mixture. Glow ignitions that are initiated at lower temperatures are extinguished by the cooling effect of the neighboring mixture and the relatively cold combustion chamber walls.

As stated previously, glow ignitions can be initiated by the spark plug and its electrodes without a spark. This means that the thermal value of the spark plug is too low. The thermal value can be adjusted to the thermal conditions of the combustion chamber by designing a suitable geometry (heat absorptivity). However, the thermal value of the spark plug must not be too high, because this would make the temperature of the spark plug too low for part-load operation, and this would no longer permit burn-off of the residues on the spark plug electrodes, resulting in misfiring.

The heat transfer from the exhaust valves must be improved if the exhaust valves are too hot. This requires efficient heat transfer at the valve seats. Other supporting actions are rotation devices for the exhaust valve. Hollow stem valves are used for highly stressed valve applications. The hollow space in the stem gets close to the valve head, and it is filled to 50% by sodium. This becomes liquid at 97°C. The shaker effect of the liquid sodium removes the heat absorbed by the valves through the stem and the valve guide.

Overheating of the combustion chamber walls can occur through malfunctions of the cooling system or through lack of coolant, and this can also result in glow ignitions.

The most frequent reasons for glow ignitions are incineration ash from the oil and fuel residues. The combustion residues attach in a shingle fashion to the combustion chamber walls and the piston crown. They can flake and move as hotspots within the combustion chamber.

The following describes the manifestations that can cause glow ignitions and the effect of glow ignitions:

- Early glow ignition happens before the ignition spark. This results in higher gas temperatures and pressures than in a normal combustion, which can cause knocking glow ignition. Spontaneous combustion results, and the knocking process cannot be influenced, changing the ignition timing.
- Late glow ignitions occur simultaneously with the initiation of the normal combustion by the spark plug or even later.

- Accelerated glow ignitions can develop from late glow ignitions. These continue to advance to earlier ignition times in subsequent glow ignitions and can develop into early glow ignitions.
- Infrequent glow ignitions develop due to infrequent occurrences, such as the moving of hot deposited particles in the combustion chamber.
- Rumbling glow ignition happens when the ignitions occur in each power cycle at several locations in the combustion chamber. The resulting rapid energy conversions generate high rates of pressure rise and subsequently high gas pressures. They result in large loads on the powertrain and this in turn results in rumbling noises with low frequencies in the range of 80 to 100 Hz. Knocking noises, in contrast, are at approximately 7000 Hz.
- Dieseling glow ignitions can develop after switching off the ignition of a carburetor engine that was previously highly stressed. Mixing is still induced in carburetor engines even though the ignition is switched off. The declining speed increases the exposure time of the mixture to hot combustion chamber locations, resulting in the dieseling of the engine due to glow ignitions.

Fuels with higher octane numbers lower the susceptibility to glow ignition or even prevent it for all glow ignitions. Glow ignitions can result in engine damage such as burn-out of a piston or exhaust valve. **Fig. G11** shows the effect of glow ignitions and knocking on the pressure curve.

Glow plug →Ignition system, diesel engine/preheat system ~Cold starting aid ~~Glow system

Glow system →Ignition system, diesel engine/preheat system ~Cold starting aid

Glycol →Coolant ~Monoethylene glycol, ~Monopropylene glycol

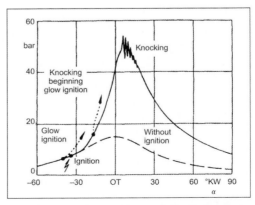

Fig. G11 Impact of knocking engine operation on the pressure curve (Source: Pischinger)

Glycol entry →Oil ~Power during operation ~~Contamination by foreign matter

GM SB2 engine →Engine ~Racing engines ~~NASCAR

GPA subgroups →Actuators ~General Purpose Actuator (GPA)

Granulate catalyst →Catalytic converter

Gravity die casting →Crankcase ~Casting process, crankcase ~~Chilled casting

Gross calorific value →Fuel, gasoline engine ~Calorific value

Ground electrode →Spark plug ~Electrode gap

Group A →Engine ~Racing engines ~~World rally car, or WRC ~~~Regulations

Group efficiency factor →Supercharging

Group N →Engine ~Racing engines ~~World rally car, or WRC ~~~Regulations

G-supercharger →Supercharging ~Rotary-piston supercharger

Guide bearing →Crankshaft

Guide elements →Chain drive ~Chain tensioner

H

Hall field →Engine acoustics

Hall sensor →Sensors ~Speed sensors ~~Active speed sensor

Hammer drop-forged crankshaft →Crankshaft ~Blank ~~Die forged

Hammer excitation →Engine acoustics ~Pulse hammer excitation

Hammer forged crankshaft →Crankshaft ~Blank ~~Hammer forged

Harmonic cam →Camshaft

Harmonics →Engine acoustics

HC absorber →Filter ~Intake air filter ~~Air filter elements; →Pollutant aftertreatment ~Aftertreatment concept with a three-way catalytic converter

HC concentration →Exhaust gas analysis; →Exhaust gas analysis equipment ~Flame ionization detector

HC emissions →Emission measurement ~FTP-75 ~~Mass determination, emissions

H/C ratio, fuel →Fuel, diesel engine ~Composition; →Fuel, gasoline engine ~Composition

HC trap →Pollutant aftertreatment ~Aftertreatment concept with a three-way catalytic converter ~~HC absorber

Heat accumulator. Heat accumulators collect the "waste heat" from the engine that is dissipated to the coolant and store it by efficient thermal insulation. The stored energy, in the form of heat, is delivered at the next cold start with a power of up to 80 kW within two to three minutes. The heat accumulator directs hot air to the windshield defroster nozzle, while simultaneously heating the engine. Comfort in the passenger-compartment is achieved at the same time as environmental benefits in the form of lower cold-start emissions and reduced fuel consumption. The benefits are further enhanced when a door switch-actuated electric pump initiates coolant circulation in the cooling circuit when the motor vehicle is opened. Hot air from the defroster nozzle instantly flows when the ignition, in this case, is switched on. Starting the engine is also improved and cold-start emissions of HC and CO are drastically reduced, particularly at low temperatures.

~Applications. The applications are for cab heating and engine heating at cold start, when there are short breaks in operation, or in stop-and-go traffic, when the emission of pollutants is also reduced.

~~Cab heater. In the search for fuel economy in modern engines, engines with direct injection in particular, many motor vehicles suffer from heating deficits which affect both the comfort of passengers and the defrosting of the windshield. Heat accumulators are well suited for defrosting windshields quickly and make the defrosting easier. Passenger comfort is also enormously enhanced by fast availability of hot air. **Fig. H1** shows the temperature profile at the defroster nozzle with a rapid rise in temperature a few seconds after cold startup to approximately 55°C when the usual coolant (flow rate 600 L/h) flows through the accumulator.

Because the thermal energy stored in the accumulator (contents: 9 L coolant) is quickly "emptied" in the process, the defrost temperature will drop to about 42°C for a short time only, then steadily rise in parallel with the basic heating. A depleted level that lasts approximately 4 min followed by a steady rise in temperature generated by the engine that has heated up in the meantime will develop, and this results in temporary deactivation of the coolant flow from 600 to 150 L/h.

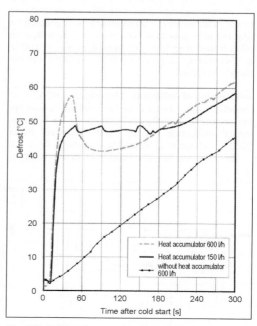

Fig. H1 Cab heating at cold start, defrost temperatures

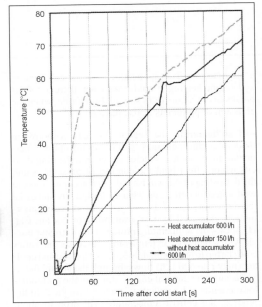

Fig. H2 Engine heating at cold start, coolant temperature at engine exit

Fig. H2 shows the effect of a coolant exchange-type heat accumulator used for engine heating at idle. In this case the coolant temperature at the engine exit is considered to be the representative variable. The temperature level of 55°C, corresponding to the point when the accumulator is discharged, is reached after less than 60 s with a coolant flow of 600 L/h and after 180 s with a coolant flow of 180 L/h. This heating-up process would take 240 s without the accumulator.

~~**Emission of pollutants.** The influence of heat accumulators on exhaust emissions and fuel consumption was very carefully analyzed by the US Environmental Protection Agency (SAE Technical Paper 922244 — Hellmann, Piotrowski, Schaefer).

The following improvements were measured in the first exhaust-gas bag of the FTP test on a gasoline-powered vehicle with a latent heat accumulator:

> Latent heat accumulator without preheating:
> HC 35%, CO 54%, MPG 9%

> Latent heat accumulator with preheating:
> HC 69%, CO 76%, MPG 14%

The FTP composite emission reductions were almost at the same level while the fuel economy (MPG) was merely down to 1%.

The results of the +24°C test were:

> Latent heat accumulator without preheating:
> HC 9%, CO 2%, MPG 2%

> Latent heat accumulator with preheating:
> HC 12%, CO 42%, MPG 4%

Here also, the percentage reduction of the HC and CO emissions in the FTP composite test was almost identical with bag 1, and the fuel economy advantage was nearly zero.

~Design. Three types of design of heat accumulators for vehicle installation have become generally used. The most popular are the latent heat accumulators offered by VW and BMW as optional equipment as well as a coolant exchange heat accumulator, offered by DaimlerChrysler and VW as dealer installations. Heat accumulators for coolant transfer have only been used experimentally.

~~**Coolant exchange heat accumulator.** The coolant is the storage medium, and hot coolant is pumped into the heat accumulator at engine cutoff and kept warm—the cold contents of the accumulator flow into the engine (coolant exchange). The hot coolant in the accumulator is exchanged with cold coolant at cold start (**Fig. H3**). The accumulator is separated from the cooling circuit during the heating-up phase, and an integrated control unit switches depending on temperature. The internal and external containers and their system of connections can be adopted unchanged from the latent heat accumulator (**Fig. H4**). The internal structure of the coolant exchange heat accumulator is very simple. All that is important is that intermixing of the cold and warm coolant is prevented, at least when the accumulator is discharged. For this purpose, the internal container is subdivided by horizontal deflectors

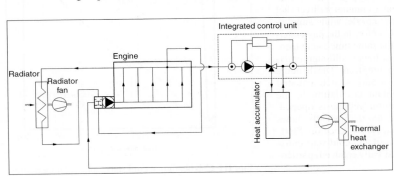

Fig. H3
Cooling circuit,
coolant exchange heat
accumulator

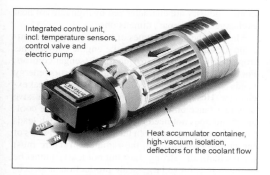

Integrated control unit, incl. temperature sensors, control valve and electric pump

Heat accumulator container, high-vacuum isolation, deflectors for the coolant flow

Fig. H4 Coolant exchange-type heat accumulator

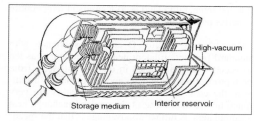

High-vacuum

Storage medium Interior reservoir

Fig. H6 Latent heat accumulator (Fritz Werner Präzisionsmaschinenbau GmbH)

at 20-mm intervals. The coolant flows by a meandering path from the bottom upward. Thus, the intermixing effect is low when the accumulator is discharged because the cold coolant flows in at the bottom and pushes the warm coolant upward.

~~**Heat accumulator for coolant transfer.** The coolant is the storage medium as well. Warm coolant transfer into the empty accumulator will follow when the engine is stopped and the coolant volume in the engine decreases accordingly.

Transfer of the coolant that has been kept warm from the heat accumulator to the engine is affected when the engine is started. This procedure is very advantageous thermodynamically and also from the point of view of weight but practically of no relevance in spite of good experimental results. Drainage and filling would be too difficult with many engines.

~~**Latent heat accumulator. Fig. H5** shows the integration of a latent heat accumulator into a vehicle cooling circuit in its plainest form. This presupposes a latent heat accumulator with low internal thermal resistance. The coolant will, therefore, constantly flow through the latent heat accumulator, and the charging and discharging of the accumulator are self-regulating. The location of the latent heat accumulator in front of the thermal heat exchanger helps cab heating without substantially reducing the heating of the engine.

Fig. H6 shows the layout of a latent heat accumulator. The storage medium, barium hydroxide octahy-

drate, is stacked in thin-walled copper elements, and spacers designed as turbolators form the flow path for the circulating coolant. The solid storage medium melts when the coolant temperature exceeds the transition temperature of 78°C. The "latent heat" absorbed remains stored during engine standstill—overnight, for example. The cold coolant will flow from the engine through the accumulator on engine cold start—for example, next morning—and the storage medium heats up the coolant. With the latent heat used up, the storage medium solidifies and its temperature drops below 78°C.

The salt bags are stored in an internal container together with the spacers, and this vessel is then placed in an external container. There is a high vacuum in the empty space between both containers that would suppress thermal conduction between both containers to a great extent. The walls of the vacuum space are silver-plated to minimize heat radiation. The mounting forces between both containers are generated by thin, high-strength stainless steel tensioning bolts to minimize the thermal conduction through the mounts.

The third fundamental losses between the internal and external container are caused by the tubes that deliver the coolant into and out of the internal container and from the coolant contained in these tubes. The tubes are run vertically in the vacuum space and are connected to the bottom of the cold external container and on the top of the warm internal container. Thus, the coolant forms stratified layers because of gravity, and convection is prevented and the heat losses are limited to thermal conduction in the coolant and the tubes. With heat accumulators designed in this way, thermal losses per day can be limited to less than 2 W at 80°C

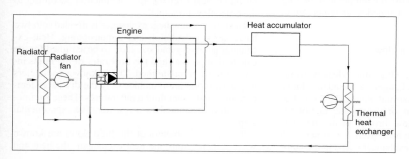

Radiator Radiator fan

Engine

Heat accumulator

Thermal heat exchanger

Fig. H5 Coolant circuit with latent heat accumulator

internal temperature and $-20°C$ ambient temperature in series production vehicles.

~Operating experience. Operating experience with standard production latent heat accumulators has led to acceptable deficiencies in quality that stem from both production deficiencies and design flaws. In addition, the costs of latent heat accumulators are very much higher than those of the coolant exchange heat accumulators. A further disadvantage is that the speeds for charging and discharging are lower and depend on the thermal conductivity and transition temperature of the latent storage medium. It can, therefore, be assumed that latent heat accumulators will not be used in vehicular applications in future.

It is to be expected that the coolant exchange heat accumulator will have an interesting future. Simple design and safe operation in all situations combined with its low cost make this accumulator an ideal solution for future cold-start problems with engine and cab heating. These problems are the consequences of increasing pressure to reduce fuel consumption and emissions.

The heat accumulators for coolant transfer will not be considered, because engine filling and draining is considered to be too difficult and unsafe.

~Requirements. The greatest requirements are imposed on volume, weight, charging/discharging efficiency, and thermal insulation. Thereby, the most important variables are the properties of the heat accumulator medium (latent heat of salts or tangible coolant heat), the concept of the heat exchanger, and the type of thermal insulation.

Heat accumulator, requirements →Heat accumulator ~Requirements

Heat accumulator, type →Heat accumulator ~Design

Heat accumulator operating experiences →Heat accumulator

Heat accumulators for coolant transfer →Heat accumulator ~Design

Heat capacity (*also*, →Fuel, gasoline engine). Heat capacity or specific heat (capacity) is that quantity of energy (heat) required to raise the temperature of 1 kg of substance by 1 K. It is a function of temperature, and the SI unit is J/kgK.

Heat conduction →Heat flow

Heat exchanger (*also*, →Radiator). Heat exchangers, often called heat interchangers, are used to achieve specific changes in the thermal state of a fluid. The heat exchangers cool, heat, or change the state of aggregation of fluids that flow through them—heat exchangers always transfer heat from a warmer to a colder medium. The heat exchangers used in the auto-

motive industry are characterized by high rates of heat exchange and compact dimensions.

Many of the various types of heat exchangers in automobiles are used to protect the engine components from overheating or to provide for passenger comfort. For this purpose, a great deal of the combustion energy needs to be released to the ambient air via the heat exchanger. Heat exchangers should not exert excessive flow resistance on the coolant to ensure that the coolant pump does not need too much power, which leads to pressure change between the outlet and inlet sides of the pump. Furthermore, the radiator structure must be mechanically strong. Last but not least, it must be possible to achieve large-scale production at an economic price.

Heat gets from the hotter fluid to the cooler heat exchanger wall through convection. It is then transported by thermal conduction through the material of the heat exchanger walls to the opposite surface from where it is also released into the cooler fluid through convection. So the material of the radiator wall must have good heat-conducting properties or have thin walls so as to cause as little resistance to heat flow as possible. The quantity of heat transferred is determined through the heat-transfer coefficients on the surfaces involved, the "wetted" area, and temperature difference between wall surfaces and the fluids. The fluid flow rates and thermo-physical properties are additional basic variables. Because gases generally have low heat-transfer coefficients, the surface which they contact must be inevitably very large. Cooling fins and guide fins are used to enlarge these surfaces, thus enhancing the heat exchanger performance.

The heat exchangers can be categorized into roughly three classes based on the flow of the fluids:

- Parallel flow
- Counterflow
- Cross-flow

In this context, the temperature profiles representative of counterflow heat and parallel flow heat exchangers with two fluids (indexed without dash [fluid 1], with dash [fluid 2]; inlet with 1, outlet with 2) are shown in **Fig. H7**. The cross-flow heat exchanger represents a mixture between the parallel and counterflow principles. The counterflow heat exchanger has the highest efficiency of all three types of design. In most applications, heat exchangers in vehicles (**Fig. H8**) are air/liquid heat exchangers that function according to the cross-flow principle.

Design of a new heat exchanger is divided into two basic tasks: calculation and dimensioning. Heat exchanger geometry and size have to be assumed for the calculation process and the flow rates of the fluids involved must be predefined. The unknown parameters in practical operation include only the discharge temperatures and mean fluid temperatures. These parameters can be evaluated by a few iterative experimental steps.

The inlet parameters of the fluid flows are known when dimensioning the heat exchanger; the heat and

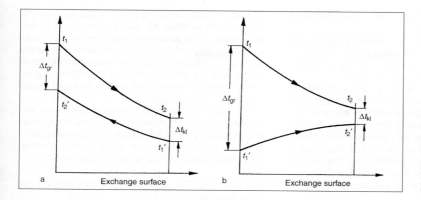

Fig. H7
Schematic representation of the fluid temperatures for (a) counterflow principle and (b) parallel flow principle (Source: Dubbel)

pressure losses are also specified so that the design engineer can evaluate the size of the heat exchanger. The basic design, however, is significantly more complex than the calculation as such because a large number of decisions have to be made in the preliminary stages. The distribution of flow, the material to be employed, and the geometry need to be specified. The specifications and their parameters are very complex and do not follow conventional decision-making.

The main difficulty is selecting the surface geometry of the heat exchanger. It is very expensive to develop an optimal heat exchanger for a new vehicle.

Vehicle heat exchangers can additionally be subdivided into five groups by the type of fluids used in the heat exchange processes:

1. Gas/gas heat exchangers
2. Gas/two-phase heat exchangers
3. Gas/liquid heat exchangers
4. Liquid/liquid heat exchangers
5. Liquid/two-phase heat exchangers

1. Gas/gas heat exchangers are primarily used in charge-air cooling in turbocharged engines. In a charge air cooler, the gas compressed and heated in the compressor of the turbocharger is cooled down by the ambient air that acts as the cooling medium. This cool-

ing increases the density of the charge air, and a larger mass of air is available for the combustion process. The high air requirement and good efficiency from a powerful engine need low pressure loss in the charge air cooler. The available space, however, usually sets narrow confines to dimensioning the component. Air/air charge air coolers are mostly designed as flat-tube heat exchangers (*see below*, ~Design types).

2. Gas/two-phase heat exchangers are used in the air-conditioning system, for instance. The air-conditioning coolant changes its state from liquid to gas in the evaporator. In doing so, it removes heat from the air flowing through the heat exchanger into the passenger compartment. Because the evaporator must be integrated into the air ducting of the passenger-compartment heating, there is very little space available—in most cases, the evaporator is a round pipe or plate-type heat exchanger (*see below*, ~Design types). The coolant is compressed in the compressor of the air-conditioning system and, from here, it transports the absorbed energy to the condenser. Here, the coolant transfers its heat to the air flowing through, and the coolant then condenses. Today, condensers mostly are designed as flat-tube heat exchangers.

3. Gas/liquid heat exchangers. Air/liquid heat exchangers are the most common heat exchangers in automobiles because there is no other coolant readily available to the overall motor vehicle system but ambient air. The commonest heat exchanger is the radiator. It dissipates the waste heat of the combustion engine to the ambient air. An example will illustrate how large its cooling capacity is: At top speed, a coolant volume of about 11,000 L/h flows through the radiator of a motor vehicle with a 2.0-L engine and a power output of 130 kW. The radiator has to dissipate 70–100 kW in this case. The most common types of design are round-tube/flat-tube heat exchangers (*see below*, ~Design types). The thermal heat exchanger, too, belongs to the group of air/liquid heat exchangers. It utilizes part of the engine waste heat in the cooling water to heat the passenger compartment. Like the radiator, it is designed as a round-tube/flat-tube heat exchanger. Further uses in vehicles for air/liquid heat exchangers are:

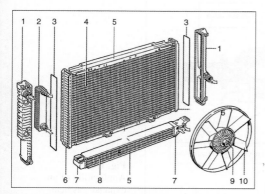

Fig. H8 Passenger car cooling (Source: Bosch)

- Radiators that prevent overheating of the engine oil to avoid engine damage.
- Fluid coolers for automatic transmission systems to decrease the temperature of the transmission fluid.
- Steering servo coolers to cool the hydraulic steering fluid, ensuring a reliable steering boost.
- Fuel coolers to reduce the fuel temperature and prevent vapor lock. Modern injection systems work using fuel circulation, and a considerable proportion of the delivered fuel quantity flows back into the tank. This fuel is used to cool the injection system components and dissipate the heat into the tank. Fuel temperatures rise significantly, especially when the quantity of fuel remaining in the tank is low. The absence of heat exchangers may damage system components.

Today, all qualified gas/liquid heat exchangers are designed as round-tube/flat-tube heat exchangers (*see below*, ~Design types). Other types of design can be employed for engine exhaust gas, for instance. In this case allowance must be made for high temperatures, a tendency for contamination by exhaust gas constituents, and increased anticorrosion requirements.

The types of heat exchanger in use also include shell-and-tube heat exchangers *see below*, ~Design types).

- EGR radiators are heat exchangers that cool part of the flow of the hot exhaust gas from the exhaust system before it is added to the fresh air in the engine intake. This enables improvements in exhaust emissions.
- Full-flow exhaust-gas heat exchangers (**Fig. H9**), take off heat from the entire exhaust gas which, for instance, is fed to the engine coolant. This energy can be used to warm up the air inside the passenger compartment faster and get the engine to operating temperature more quickly. Shorter cold phases reduce fuel consumption, improve exhaust emission figures, and reduce wear.

4. Liquid/liquid heat exchangers are primarily used to cool the lubricating oil and the transmission oil. Owing to their compact design, they are installed in space directly on the engine block, on the gearbox casing, or on the radiator collector. The engine oil cooler, for example, is located in the main flow of the oil circulation between oil filter and engine block. The heat is passed on to the engine cooling circuit and dissipated via the radiator. Both engine oil and transmission oil coolers can be manufactured as plate heat exchangers (*see below*, ~Design types).

5. Liquid/two-phase heat exchangers. This type of heat exchanger has not yet been introduced for series application. However, deployment for operation as a heat pump in the air-conditioning system using the engine waste heat in the coolant as heat source is being analyzed.

~Design types. The main types of heat exchanger designs found at present are:

- Round-tube heat exchanger (mechanically constructed)
- Flat-tube heat exchanger (brazed)
- Plate heat exchanger (brazed)
- Shell-and-tube heat exchanger (brazed).

The automotive industry is widely using round-tube heat exchangers for stationary cooling systems. In essence, these consist of a pack of fins and circular tubes with 180-degree deflections soldered onto the tube ends. The pack of fins is composed of individual thin plates, and air flows along the plates, thereby taking up the heat. In the manufacturing process, the tubes are put through circular cutouts in the pack of fins. A heat-conducting connection is only made through expansion of the mechanical tube from inside using a mandrel that is drawn through the tube. The distribution of flow through the heat exchanger can be determined by any interconnection of individual tubes or tube assemblies. This generates high flexibility which is important for the possible cooling capacity of the aggregate. Most round-tube heat exchangers (**Fig. H10**) consist of light alloy to reduce the weight and make production easier.

Flat-tube heat exchangers (**Fig. H11**) also frequently are used in the automotive industry. In es-

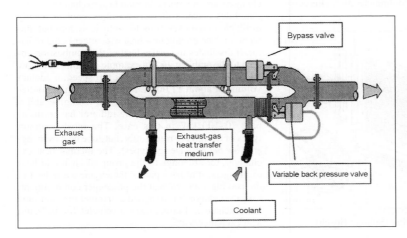

Fig. H9
Application of an exhaust-gas heat exchanger in full flow

Fig. H10 Example of a round-tube heat exchanger (Source: Visteon)

Fig. H11 Example of a flat-tube heat exchanger (Source: Visteon)

sence, these consist of fins, flat tubes, and the collecting mains. The fins are wave-shaped, folded ribs positioned at right angles to the direction of airflow. The individual fin bars can additionally be slotted, producing the so-called gills that improve the heat transfer efficiency.

The flat tubes either consist of extruded aluminum sections or folded sheet aluminum that is bent into flat tubes by folding and welding together on the seam. The collecting mains have one or several rows of stamped duct holes for the flat tubes. The flat tubes, packs of fins, and collecting mains are put together as an assembly during production. Subsequently, the heat exchanger is soldered in a brazing oven—this thermal joining ensures good heat conduction.

Plate heat exchangers consist of a pack of mostly rectangular plates. Embossing the individual plates forms gaps in which the heat-exchanging fluids flow. The individual plates are interconnected via ducts such that one heat-dissipating fluid at a time flows between two hollow spaces with the heat-absorbing fluid and vice versa. The plates seal against each other. Plate heat exchangers are often built into deep-drawn aluminum housings or sheet steel housings with integrated interconnections.

Shell-and-tube heat exchangers are also widely deployed in systems technology. Stainless steel tubes are mostly used for reasons related to the possibility of corrosion and high temperatures in exhaust-gas heat exchangers. Exhaust gas constituents are less prone to build up heat-transfer–obstructing layers in the circular tubes, provided that sufficiently high flow velocities are maintained.

Heat flow. The heat that flows in the engine is developed primarily from the energy released by burning the fuel. The temperature rise of the working fluid that occurs at this time has the disadvantageous effect that energy is drawn from the working gas through the adjacent components into the cooling water and from these components and the cooling water into the atmosphere (**Fig. H12**). A further proportion of heat losses is taken to the environment by the exhaust gas because its temperature is higher than the ambient temperature. In principle, engines with internal combustion do not require the flow of heat outside the engine boundary, and it may even be harmful. An adiabatic process is, however, not feasible.

Heat flow resulting in energy transfer can happen in three ways:

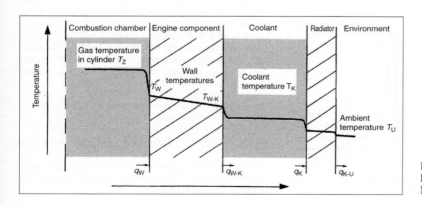

Fig. H12
Heat flow (Source: Mollenhauer)

- Convection: heat transfer in a flowing medium
- Heat conduction: e.g., through solid matter due to a temperature difference
- Radiation: by electromagnetic waves

~Heat conduction. The gas inside the working chamber transfers heat to its environment, to cylinder walls, or to cylinder head. There are temperature fluctuations in the wall because the heat transfer process is a transient one. The magnitude of these temperature fluctuations decreases from the inside of the combustion space toward the outside of the containing wall and depend on wall thickness. This process, called heat conduction, can be calculated using the Fourier equation for thermal conduction. The calculation result is, for example, the temperature variation in the cylinder wall.

~Heat flow density. The flow of energy from the gas to the combustion chamber walls is primarily caused through convection. Thus, the heat transfer is influenced by operating conditions (e.g., load and speed) and the particular engine design (e.g., combustion chamber design, intake manifold configuration). The density of heat-flow (W/m²) can be defined as heat flow per unit of area of combustion chamber surface. It is constant neither with time nor position so that calculations are normally performed using a mean heat-flow density for the whole combustion chamber surface. **Fig. H13** shows the heat-flow density as a function of crank angle or time.

~Heat losses through the wall →Adiabatic combustion chamber; →Heating law

~Heat release →Combustion curve; →Heat flow

~Heat removal →Combustion curve; →Heat flow

~Heat transfer. The heat flow density depends on the temperature difference—for example, between mean gas temperature and combustion chamber wall—on the contact side at a certain location in the combustion space, as well as the local heat-transfer coefficient at this point. The local heat-transfer coefficient cannot be determined exactly, because the flow conditions at this

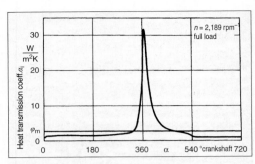

Fig. H14 Variation of the heat-transfer coefficient with crank angle (Source: Pischinger)

point, in particular, exercise major effects on the value. Therefore, some approximations to the heat-transfer coefficient based on the laws of similarity mechanics have been made. This results in relationships of varying complexity into which parameters are entered, including: characteristic lengths, such as cylinder diameter; characteristic speed, such as mean piston speed; and the gas temperature and pressure in the combustion space. The corresponding coefficients for the equations can be determined experimentally. Other approaches allow for the temporal or positional variations of pressure and temperature in the combustion space, the transient thermal conduction, as well as chemical reactions.

Fig. H14 shows an example of the variation of heat-transfer coefficient as a function of crank angle.

Heat interchanger →Heat exchanger

Heat losses through the wall →Adiabatic combustion chamber; →Heating law

Heat release →Combustion, diesel engine ~Equivalent rate of combustion; →Combustion curve; →Heat flow

Heat release characteristics. Heat release characteristics describes the net release of heat (thermal energy) used to heat the working medium. The net heat release becomes negative if the gas temperature exceeds the wall temperature during the compression process and if the initially released fuel energy does not yet compensate the heat transfer. Based on the measured pressure characteristics, the combustion characteristics can be calculated through the heat release characteristics.

Literature: A. Urlaub: Verbrennungsmotoren, Springer Publishers, Berlin/Heidelberg, 1994.

Heat removal →Cooling circuit; →Cycle; →Heat flow

Heat sink →Electronic/mechanical engine and transmission control ~Requirements for mechanical and housing concepts

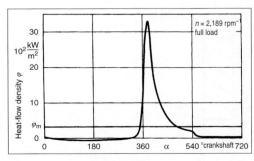

Fig. H13 Variation of heat-flow density with crank angle (Source: Pischinger)

Heat supply →Cycle

Heat transfer →Coolant; →Heat flow

Heat transfer bodies →Electronic/mechanical engine and transmission control ~Requirements for mechanical and housing concepts

Heat transfer corrosion →Coolant

Heat-conducting adhesive →Electronic/mechanical engine and transmission control ~Electronic components

Heat-flow density →Heat flow

Heating flange (*also*, →Diesel engine; →Ignition system, diesel engine/preheat system ~Cold starting aid). Heating flanges sometimes are used to heat the intake air of combustion engines during their warm-up phase—they consist of heated glow wires or glow coils. They are preferred in combustion engines and are installed between air filter and cylinder head. Another alternative for heating the intake air is the use of flame starting systems.

Heating law. The time-dependent conversion of the fuel in the combustion chamber is defined by the heating law. The heating law covers only that part of the fuel quantity that can be identified in the measurable pressure increase, while the combustion law includes all parts. The difference between combustion law and heating law is primarily the heat losses through the wall. The heating law, $x(\varphi)$, is described as

$$x(\varphi) = \frac{m_{B,\text{converted}}(\text{increase} - \text{pressure})}{m_{B,\text{mass-per-cycle}}}$$

where $m_{B,\text{converted}}$ = the converted fuel mass at crank angle, φ, and $m_{B,\text{mass-per-cycle}}$ = the fuel mass converted for each combustion cycle.

A slow increase of the fuel conversion up to the middle of the combustion process is desirable, followed by a continued decline to zero. This reduces the mechanical load of the powertrain, and it reduces combustion noise. However, the best efficiency is achieved with a fast fuel conversion (approximation to constant volume combustion).

The heating law in diesel engines is strongly influenced by:

- Type and time dependent progression of the fuel injection
- Type of mixture formation—e.g., wall or air distributed
- Pressure and temperature at the end of compression
- Combustion chamber geometry (air movement)
- Start of injection
- Air-fuel ratio

The heating law in gasoline engines is influenced by:

- Air-fuel ratio
- Pressure and temperature after combustion
- Residual gas content
- Combustion chamber geometry (mixture movement)
- Ignition timing

Heating the catalytic converter →Catalytic converter ~Heat-up

Heavy gasoline →Fuel, gasoline engine ~Production

Heavy oil →Fuel, diesel engine

Height-offset balance shaft →Balancing of masses ~Balance shaft

Helical gear pumps →Oil pump

Helmholtz resonator →Air intake system ~Acoustics ~~Acoustic elements of tubing systems; →Engine acoustics

Hemispherical combustion chamber →Combustion chamber ~Combustion chamber in cylinder head

H-engine →Engine concepts ~Reciprocating piston engines ~~Multiple-shaft engines ~~~Twin-shaft engines

Hertzian stress →Camshaft

Hesselmann bowl (*also*, →Combustion chamber ~Combustion chamber in piston). The complete combustion chamber in diesel engines with direct injection is located in the piston. Some alternative combustion chamber geometries are shown **Fig. H15**. The Hesselmann bowl is the recess in the middle. It also is called a "Mexican hat" because of its characteristic shape.

Hesselmann engine (*also*, →Combustion process, gasoline engine ~Stratified charge process). The Hesselmann engine, which was first designed in 1934, is a hybrid engine. The objective of the design was to combine the advantages of the diesel engine with those of the gasoline engine. This included an air compression process with heterogeneous mixture formation, external ignition, quality control at full load and quantity control at partial load and during idle speed. Quantity control is required because ignition problems would otherwise be generated. **Fig. H16** shows the mode of operation.

An injection nozzle (1) and a spark plug (2) on opposite sides of the cylinder protrude into the combustion chamber through windows in the piston—they reach the circumference of the piston recess. Air swirl

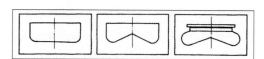

Fig. H15 Hesselmann bowl (middle)

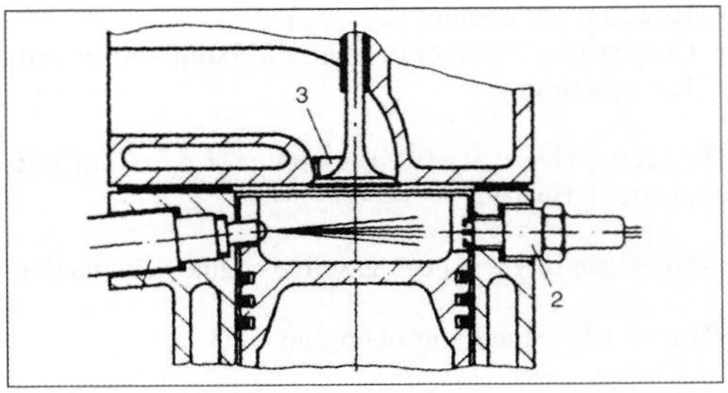

Fig. H16 Hesselmann engine (Source: F. Pischinger)

around the combustion chamber axis is generated by an umbrella valve (3). This is why the injected spray, in combination with the swirling air, always results in an ignitable mixture at the spark plug—the spray is aimed at the piston recess. The compression ratio of the engine was 7.5.

Heterogeneous air-fuel mixture →Air-fuel mixture ~Homogeneous mixture

Heterogeneous mixture →Air-fuel mixture ~Homogeneous mixture

HFID (hydrocarbon flame ionization detector) →Emission measurements ~FTP-75 ~~Performed driving cycles

High tumble →Charge movement ~Tumble

High-compression engine →Compression

High-pass filter →Particles (Particulates) ~Particulate filter system ~~Particulate filter ~~~Operating mode of particulate filters

High-performance manifold →Exhaust system ~Exhaust gas manifold

High-pressure generation →Injection system (components) ~Diesel engine ~~Injection hydraulics

High-pressure injection →Injection system (components) ~Diesel engine

High-pressure pump →Injection system (components) ~Diesel engine ~~Injection hydraulics ~~~High-pressure injection pump

High-pressure rail →Injection system (components) ~Diesel engine

High-pressure sensors →Sensors ~Pressure sensors

High-pressure supercharging →Supercharging

High-pressure system →Injection system (components) ~Diesel engine ~~Injection hydraulics

High-side driver →Electronic/mechanical engine and transmission control ~Electronic components ~~Power transistor

High-speed engines. The term "high-speed engine" is used mostly for diesel engines. These are usually passenger car engines with a maximum engine speed currently of about 4200–4600 rpm. In contrast, there are low-speed engines (large ship engines) with speeds around 80–100 rpm and medium-speed engines with speeds of about 500 rpm.

High-speed knocking →Knocking ~High-speed knocking

High-temperature cooling →Cooling circuit ~Evaporative cooling

High-volatility fuels →Fuel, gasoline engine ~Vapor lock

High-voltage capacitor discharge ignition (CDI) →Ignition system, gasoline engine ~High-voltage generation

High-voltage distribution →Ignition system, gasoline engine

High-voltage generation →Ignition system, gasoline engine

Highway driving cycle →Emission measurements ~Test type I

Highway test →Emission measurements ~FTP-75

Historic engines →Engine

Hole corrosion →Coolant

Hole diameter →Injection valves ~Diesel engine ~~Hole-type nozzle

Hole length →Injection valves ~Diesel engine ~~Hole-type nozzle

Hollow chamfer →Valve ~Gas transfer valves

Hollow stem valve →Valve ~Gas transfer valves ~~Valve designs ~~~Hollow stem valves

Hollow valves →Valve ~Gas transfer valves ~~Valve designs

Homogeneous air-fuel mixture →Air-fuel mixture ~Homogeneous mixture

Homogeneous charge compression ignition (HCCI) process →Combustion process, diesel engine ~Homogeneous combustion, diesel engine

Homogeneous combustion →Combustion process, diesel engine; →Combustion process, gasoline engine

Homogeneous mixture →Air-fuel mixture

Homogenization →Air-fuel mixture ~Homogeneous mixture; →Combustion process, gasoline engine ~Homogeneous combustion, gasoline engine

Homologation →Engine ~Racing engines ~~World rally car, or WRC

Honda CVCC →Combustion process, gasoline engine ~Stratified charge process

Honda MMC process →Crankcase ~Crankcase design ~~Cylinder ~~~Composite technology

Honeycomb support →Catalytic converter

Honing, cylinder contact surface →Cylinder running surface ~Machining process

Horizontal carburetor →Carburetor ~Design types ~~Position of the air intake channels

Horizontal engines →Engine concepts ~Reciprocating piston engines ~~Location of cylinders

Host computer →Electronic/mechanical engine and transmission control ~Electronic components ~~Microcontroller

Hot air engine →Engine ~Alternative engines ~~Stirling engine (hot gas engine)

Hot bulb ignition →Thermal ignition

Hot film air mass sensor →Flowmeter intake air; →Sensors ~Air mass flow sensor

Hot gas engine →Cycle; →Engine ~Alternative engines ~~Stirling engine (hot gas engine)

Hot operation →Carburetor; →Oil ~Power during operation ~~Effect of different operating conditions

Hot spot heating. The fuel in mixture formation systems, such as carburetors, single-point injection, and central injection, which are rarely used today, was only supplied to the intake air at one point. This results in problems with multicylinder engines because it is difficult to supply the same mixture composition to all cylinders.

Many fuel droplets of different sizes are transported with the air because the fuel only evaporates gradually, and part of the fuel does not evaporate until it reaches the cylinder. Recondensation can result in deposition of the fuel at bends in the intake manifold, and the now

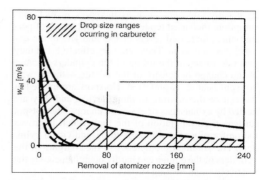

Fig. H17 Relative speed between air and fuel in relationship to intake manifold length (Source: A. Urlaub)

liquid fuel can enter the cylinder by flowing along the wall of the intake manifold. In extreme cases up to 50% of the fuel supplied by the mixture-forming device can be in a liquid state. The cylinder then receives different quantities of fuel depending on the detailed design of the intake manifold (intake manifold length, diameter, bends, changes of cross-sectional area, etc.) and depending on the firing order and, therefore, on the intake sequence.

Fig. H17 shows the relative speed, w_{rel}, between the air and the fuel for a given case depending on the length of the intake manifold. The curves show the relationship for droplet sizes, d_T, between 10 and 100 μm— the hatched area shows the region of operation of a carburetor. **Fig. H18** shows possible distributions of the mixture in a four-cylinder engine, depending on the firing order.

The different air-fuel ratios are the result of possible asymmetries in the intake manifold. Intake manifold preheaters are used to improve the nonuniform distribution to the individual cylinders and thereby reduce

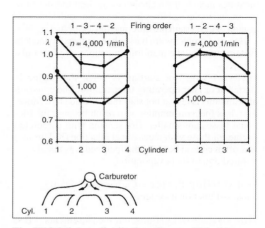

Fig. H18 Mixture distribution (Source: Urlaub)

its negative impacts, such as increases in fuel consumption, increased tendency to knock, losses of performance, increased levels of CO and HC in the exhaust gas, and so on. These are very effective, but they generate losses in the filling of the cylinders because of the change in density of the mixture—this is why hot-spot heating is preferred. Hotspots are located at the exit of the mixture-forming device, and they are heated by exhaust gas or electrical energy. The hotspot furthers the evaporation of the fuel, which improves the uniformity of distribution to the individual cylinders. The volumetric efficiency or the charging of the cylinders of the engine can hardly be influenced by the relatively small heating surfaces.

Hot start (*also*, →Fuel, gasoline engine ~Vapor lock). The hot start of an engine is defined as the restarting of a warm engine. Hot-start problems exist in the gasoline engine only because of vapor locks in the fuel, which are related to the boiling characteristics and, therefore, the composition of the gasoline fuel. Hot-start problems occur when the initial boiling point is similar to the temperature of the intake air.

The vapor lock problem in carburetor engines can occur in the carburetor float chamber. The lightly volatile components of the gasoline engine enter the intake manifold and enrich the mixture because of the evaporation even when the engine is not running. This can result in starting problems and hydrocarbon peaks in the exhaust gas and out-of-center idle speed.

Vapor bubbles can occur at any point in the fuel feed unit, and especially in the fuel feed pump. A fuel retention valve on the carburetor is the solution for these cases. It lets a certain fuel quantity recirculate back to the fuel tank during idling and part load. This prevents overheating of the fuel by cooling it with the relatively cold fuel from the fuel tank, and this suppresses vapor locking. Vapor lock in engines with injection systems occurs mainly in the fuel lines and in the injection valves. The solution here is also cooling of the fuel lines with additional fuel, which is recirculated to the tank. Other measures include the reduction of the heat transfer from the engine to the fuel lines by using nonmetallic lines, elevation of the system pressure, and the use of fuel coolers.

Hot-start problems are basically eliminated in modern systems with intake manifold injections and fuel rails.

Hot starting. Hot starting is when the engine is started at or near operating temperature. The emission of pollutants is less at hot start than at cold start. Moreover, less fuel consumption will result because the frictional losses are smaller. Hot starting with carbureted engines or those with mechanical injection (K-Jetronic) was often critical because the fuel feed was not always ensured due to fuel evaporation.

Hot starting device →Carburetor ~Starting systems ~~Functional systems

Hot switch-off →Emission measurements ~Test type IV ~~Calculation of evaporative emissions

Hot wire anemometer →Flowmeter intake air; →Sensors ~Air mass flow sensor

Housing concept →Electronic/mechanical engine and transmission control ~Requirements for mechanical and housing concepts

Housing filter →Filter ~Lubricating oil filter

Housing materials →Sealing systems ~Modules

Humidity →Carburetor ~Icing; →Intake air

Humphrey pump →Engine concepts

Hybrid combustion processes →Combustion process, hybrid engine

Hybrid engine →Combustion process, hybrid engine

Hybrid powertrain (*also*, →Engine ~Alternative engines). Hybrid powertrains are a combination of at least two different drive systems. These can include, for example, a combination of combustion engine and electric motor. An objective of hybrid powertrains includes the use of the "pollutant free" electrical drive in city traffic and the use of the combustion engine for intercity and highway traffic.

The main features of a hybrid powertrain are:

- Use of different sources of energy (gasoline fuel, diesel fuel, electrical energy, solar energy, etc.)
- Interim energy storage or energy recuperation (braking energy, downhill driving, etc.)
- Combination of engines in a system depending on the requirements

Vehicles with a hybrid powertrain generally have a higher vehicle weight and, therefore, a lower performance and a smaller cruising range for the same engine power. Hybrid powertrains often consist of a combination of electrical motor and gasoline engine. Current vehicles have installed electrical powers of around 50 kW and diesel engine powers of 60 kW, depending on the application—the trend is toward increased power. Europe is currently focusing on the development of diesel engines, while Japan and the United States are concentrating on the hybrid powertrains with gasoline engines as the combustion engine.

The biggest disadvantage of hybrid powertrains is the cost.

Literature: U. Seiffert: Perspektiven der zukünftigen Automobilentwicklung, 3rd Aachen Colloquium on Vehicle and Engine Technology, 1991. — B. Harbolla, W. Buschhaus: Konzeptauswahl und Entwicklung eines Pkw Hybridabtriebes, 3rd Aachen Colloquium on Vehicle and Engine Technology, 1991. — T. Takaoka, Y. Kobayashi, T. Nishigaki: Ein hocheffizientes emissionsarmes Hybrid-Fahrzeug, 9th Aachen Colloquium on Vehicle and Engine Technology, 2000. — L.R. Gordon: I-MoGen Ein Mild-Hybrid-Antriebskonzept für künftige Dieselfahrzeuge, ATZ (2002) 11. — R. van Basshuysen, F. Schäfer (Eds.): Handbuch Verbrennungsmotor, 2nd Edition, Wiesbaden, Vieweg Publishers, 2002.

Hydraulic flow →Injection valves ~Diesel engine ~~Hole-type nozzle

Hydraulic medium →Oil ~Oil functions

Hydraulic mount →Engine acoustics ~Engine/accessory mount ~~Hydraulically damped engine mounts

Hydraulic oil cooler →Radiator ~Oil cooler

Hydraulic switch plate →Electronic/mechanical engine and transmission control ~Requirements for mechanical and housing concepts

Hydraulic valve actuation system →Variable valve control ~Operation principles

Hydraulic valve clearance compensation →Valve gear

Hydraulic valve shut-off →Valve shutoff

Hydraulically damped engine/accessory mount →Engine acoustics ~Engine/accessory mount

Hydrocarbon analyzer →Exhaust gas analysis equipment

Hydrocarbon concentration →Emission measurements; →Raw emission

Hydrocarbon emissions →Emission measurements; →Injection system (components) ~Gasoline engine ~~Fuel tank, ~~Fuel tank systems; →Raw emission

Hydrocarbon recovery →Emission measurements ~Test type IV

Hydrocarbons →Exhaust gas; →Fuel, diesel engine; →Fuel, gasoline engine

Hydrodynamic near field →Engine acoustics ~Hydrodynamic short circuit

Hydrodynamic short circuit →Engine acoustics

Hydrodynamics →Bearings ~Function

Hydrogen →Fuel, diesel engine ~Alternative fuels; →Fuel, gasoline engine ~Alternative fuels

Hydrogen engine oils →Oil ~Classification, specifications, and quality requirements ~~Special engine oils

Hydrogen sulfide →Catalytic converter

Hyperbaric supercharging →Supercharging

I

Icing →Carburetor; →Mixture formation ~Mixture formation, gasoline engine ~~Mixture cooling

Ideal balancing of masses →Balancing of masses

Ideal engine →Engine efficiency

Ideal process →Cycle

Identifier →Electronic/mechanical engine and transmission control ~Electronic components ~~CAN interface

Idle →Load

Idle actuator (gasoline engine) →Actuators ~E-gas ~~Idle control (gasoline engine)

Idle adjustment →Carburetor ~Systems

Idle air jet →Carburetor ~Electronic carburetor

Idle control (gasoline engine) →Actuators ~E-gas

Idle injection volume →Injection functions ~Gasoline engine ~~Intake manifold injection

Idle jet →Carburetor

Idle mixture control screw →Carburetor

Idle mixture shutoff valve →Carburetor

Idle shutoff speed. Idle shutoff speed is the speed at which the engine speed controller is activated or deactivated—it is typically around 1200 rpm. If the engine speed falls below this value, the controller is activated so that the engine does not stall. If this value of speed is then exceeded again, the controller is deactivated.

Idle speed →Speed

Idle speed control →Actuators ~E-gas; →Carburetor ~Equipment; →Electronic open- and closed-loop control ~Elec-tronic open- and closed-loop control, gasoline engine ~~Functions

Idle speed controller →Injection system (components) ~Diesel engine ~~Mechanical control ~~~Control maps

Idle speed reduction. Fuel consumption can be decreased by reducing the idle speed. In a warmed-up engine, stable idling is possible at lower speeds than during the warming phase due to good mixture prepa-

ration. Limits on the reduction in speed are created by the accessories being driven by the engine, such as the alternator, the power steering pump, and the air-conditioning, all of which require a certain minimum speed, and also by the behavior of the engine in terms of comfort at extremely low speeds. The lowest idle speeds today are approximately 500 rpm.

Idle switch →Carburetor ~Electronic carburetor

Idle system →Carburetor ~Systems

Idle valve →Actuators ~E-gas ~~Idle control (gasoline engine) ~~~Idle actuator (gasoline engine)

Idle volume (*also*, →Injection functions ~Gasoline engine ~~Direct injection). Idle volume refers to fuel consumption at idle; it is a function of the engine speed, friction loss, and the combustion efficiency. The fuel consumption at idle is around 0.5–2 L/h and depends, among other things, on the combustion process (gasoline/diesel), engine displacement, engine temperature, compression ratio and the power consumption of the accessories.

To assess the injection system, it is important to know the idle volume per combustion cycle or power stroke. For example, for diesel engines it is 5–7 mm^3 of fuel/stroke.

IFP Baudry process →Combustion process, gasoline engine

Ignitability →Ignition conditions

Ignition →Fuel consumption ~Variables ~~Engine measures; →Ignition system, gasoline engine

Ignition accelerators →Fuel, diesel engine ~Additives, ~Cetane number improver; →Fuel, gasoline engine

Ignition chamber →Combustion process, diesel engine ~Prechamber engine

Ignition characteristics. The ignition characteristics of a fuel are marked by its ease of ignition. Gasoline fuels need to have low ignition quality so that the fuel is not prone to knocking. The octane number is the measure for ignition quality (antiknock properties).

Diesel fuel requires ready ignition, and low ignition quality would lead to the accumulation of large quantities of fuel vapor-air mixture before auto-ignition takes place. This would result in a rapid pressure rise, resulting in harsh combustion noise. The measure of the ignition quality in a diesel engine is the cetane number.

Ignition coil →Ignition system, gasoline engine ~High-voltage generation

Ignition conditions (also, →Ignition system, gasoline engine ~Ignition). Ignition of the air-fuel mixture requires provision of satisfactory ignition conditions. This is attained on the direct-injection diesel engine through the high compression ratio that generates temperatures in the combustion space that ignites the fuel. Ignition aids (e.g., pencil-type glow plugs) are required for special operating conditions—cold start, for instance. Ignition of the homogeneous mixture in the case of the conventional gasoline engine is initiated by a spark within relatively narrow air-fuel ratio limits. Ignition conditions in the case of the direct-injection gasoline engine are determined both by the local air-fuel ratio and by the energy of the spark.

Ignition delay →Combustion, gasoline engine ~Combustion chamber shape; →Combustion process, diesel engine; →Fuel, diesel engine ~Properties ~~Cetane number

Ignition distributor (conventional) →Ignition system, gasoline engine ~High-voltage distribution

Ignition energy →Ignition system, gasoline engine ~Ignition

Ignition failure detection →Ignition system, gasoline engine ~Ignition

Ignition improver →Fuel, diesel engine ~Properties ~~Cetane number

Ignition interval. The ignition (firing) interval, as an example, is the value in degrees crank angle at which the cylinders fire one after the other according to the firing order. It is beneficial to achieve equal ignition intervals so as to obtain an even torque diagram and low cyclic speed variation. For instance, the ignition intervals of inline engines are dependent on the number of cylinders:

• Three-cylinder engines: 240° crankshaft angle
• Four-cylinder engines: 180° crankshaft angle
• Five-cylinder engines: 144° crankshaft angle
• Six-cylinder engines: 120° crankshaft angle

Ignition lead →Ignition system, gasoline engine

Ignition limits (also, →Inflammation). The ignition limits represent the limits of mixture ignitability for air and fuel on both the lean and rich side of stoichiometric. For gasoline fuels, it is 1–8% by volume in air, and in the engine, this results in lambda values between 0.4 and 1.4. The flame speed is much slower at lean or rich mixtures reaching zero at the ignition (flammability) limits. Furthermore, considerable emissions of unburned hydrocarbons occur when the ignition limits are approached.

Ignition limits between lambda 1.05 and 8 are utilized in the engine power cycle when diesel fuels are used.

Ignition map →Performance characteristic maps ~Ignition and injection maps

Ignition pressure →Peak pressure

Ignition quality →Fuel, diesel engine ~Properties ~~Cetane number

Ignition spark →Ignition system, gasoline engine ~Ignition

Ignition speed →Combustion, gasoline engine

Ignition system →Ignition system, diesel engine/preheat system; →Ignition system, gasoline engine

Ignition system, diesel engine/preheat system

~Auto-ignition. A number of interconnected physical and chemical processes (such as spray formation, evaporation, air-fuel intermixing) and initial chemical reactions (such as chain branching of the injected hydrocarbons) start when diesel fuel is injected into the hot, compressed air of the combustion space. With the ignition delay time over, these processes lead to local auto-ignition.

The combustion process progresses from the early noise, typical of diesel knock (rattling), which is caused by the spontaneous combustion of already premixed areas, to the predominantly diffusion-controlled combustion of the injected fuel. The courses of ignition and combustion are affected by largely fixed parameters, such as the properties of the injection system, the geometry of the combustion-chamber, the charge movement, and the fuel composition; however, each cycle is particularly dependent on the temperature and pressure of the fresh gas in the combustion space. The quantity of fuel involved in premixed combustion can be reduced by using pilot injection. This makes the combustion process less spontaneous and, therefore, less noisy.

The smoothness of combustion and the fuel consumption of diesel engines were very significantly improved by using flexible high-pressure injection systems (common rail or pump-nozzle injection systems) and improved combustion processes using direct-injection diesel engines.

Literature: H.-G. Schmitz: Zündung Dieselmotor, Handbuch Verbrennungsmotor (Basshuysen/Schäfer), pp. 471–479, Vieweg Publishers, 2002. — G. Henneberger: Elektrische Motorausrüstung, Braunschweig/Wiesbaden, Vieweg Publishers, 1990. — Bernd Rau: Versuche zur Thermodynamik und Gemischbildung beim Kaltstart eines direkteinspritzenden Viertakt-Dieselmotors, Diss., Technical University of Hanover, 1975. — Th. Heinze, Th. Schmidt: Fuel-Air Ratios in an Injection Jet, Determined between Injection and Autoignition by Pulsed Spontaneous Raman Spectroscopy, SAE Int. Fuels & Lubricants Meeting, SAE 892102, Baltimore, 25/28 September

1989. — T.A. Baritaud, Th. Heinze, J.-F. Le-Coz: Spray and Self-Ignition Visualization in a DI Diesel Engine, SAE Int. Congr. and Exposition, SAE 940681, Detroit, Feb. 28–March 3, 1994. — Rudolf Petersen: Kaltstart- und Warmlaufverhalten von Dieselmotoren unter besonderer Berücksichtigung der Kraftstoffrauchemission, VDI Progressbericht, Series 6, No. 77 (1980). — Uwe Reuter: Kammerversuche zur Strahlausbreitung und Zündung bei dieselmotorischer Einspritzung. Diss. RWTH Aachen University, 1990. — Alles über Glühkerzen: Technische Information No. 04, Beru AG. György Sitkei, Kraftstoffaufbereitung und Verbrennung bei Dieselmotoren. Warnatz, Technische Verbrennung, Springer Publishers, 1996. — Max Endler: Schlanke Glühkerzen für Dieselmotoren mit Direkteinspritzung, MTZ 59 (1998) 2. — Hans Houben, Günther Uhl, Heinz-Georg Schmitz, Max Endler: Das elektronisch gesteuerte Glühsystem ISS für Dieselmotoren, MTZ 61 (2000) 10. — Rolf Merz: Elektrische Ansaugluft-Vorwärmung bei kleineren und mittleren Dieselmotoren, MTZ 58 (1997) 4. — Magnus Glavmo, Peter Spadafora, Russell Bosch: Closed Loop Start of Combustion Control Utilizing Ionization Sensing in a Diesel Engine, SAE paper 1999-01-0549.

~Cold start. The ignition quality of the mixture of diesel fuel and air depends decisively upon the temperature and pressure in the combustion space. The conditions for auto-ignition fall off particularly when a cold engine is cranked—so much so that satisfying the criteria for ignition down to low temperatures is not achieved without additional measures. The quality of starting would deteriorate above average at falling temperatures to a point at which the engine cannot be started at all.

~~Criteria for evaluation of starting. The passenger car diesel engine is expected to feature dependable starting followed by stable and smooth engine operation. Evaluation of the quality of starting is based primarily on the potential perception by the driver of noises or emissions smells, visible exhaust fumes (soot, blue smoke, or white smoke), vibrations, waiting time till start, the starting time itself, and engine response. Cold-start quality can be assessed by measuring the noise level, the smoke density, and further exhaust emissions—HC in particular—as well as by evaluating idle-speed variations and speed increase as response to a jump in quantity of fuel injected.

~~Influential parameters. Fig. I1 represents the major parameters influencing starting performance and the parameters linking them. The representation considers further development of cold-start components and injection systems with several degrees of freedom. Representation of further links such as direct temperature influence on charge losses (gap dimensions/oil film) or final compression temperature was set aside for clarity reasons.

Low temperatures also reduce battery performance and increase engine friction, so these would result in lower attainable starter speeds due to the higher torque requirement. The engine rotational speed decreases in the range of the compression point or ignition top-dead center at very low ambient temperatures such that the long dwell time of the hot compressed charge in the combustion space causes a significant decrease of temperature and charge pressure due to heat transfer. This results in a dramatic deterioration in the conditions for mixture formation and auto-ignition, in which case the temperature has a considerably greater influence than does pressure on starting quality.

Higher starter speeds are needed with decreasing temperature to ensure a safe cold start. The required minimum cranking speed and, therefore, the limiting cold-starting temperature can be lowered enormously enabling starts at temperatures of around −20°C and below (Fig. I2).

The combination of physical and chemical ignition delays—that is, the time from injection initiation to ignition—is decisive for any good engine start. The

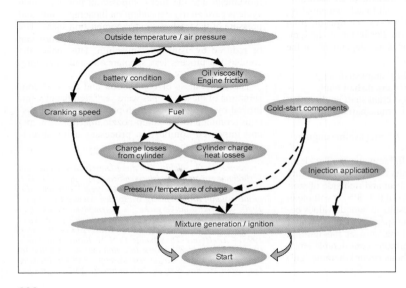

Fig. I1
Important cold start parameters

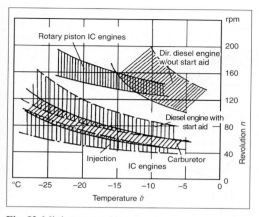

Fig. I2 Minimum cranking speed

ignition delay increases exponentially with decreasing charge temperature. As far as cranking speed is concerned, the curves of physical and chemical ignition delay are opposed (**Fig. I3**). The physical ignition delay decreases with the cranking speed because of better mixture preparation, and the chemical ignition delay in degrees of crank angle rises in proportion to speed. This results in a minimum at an average cranking speed of about 200 rpm.

The extended ignition delay under cold-start conditions cannot be compensated for by arbitrarily advancing the fuel injection. Fuel injected prematurely deposits on the combustion chamber walls and is, therefore, no longer available for auto-ignition.

~Cold starting aid. Conditions for fast auto-ignition and complete combustion of the injected diesel fuel deteriorate with decreasing temperature. Without a cold-starting aid, the starting quality would decrease so much that starting will last an extremely long while or will not be possible at all with temperatures below $-10°C$. Procedures using cold-starting aids improve the conditions for auto-ignition and complete fuel mixing to the extent of allowing dependable cold engine starts down to $-30°C$.

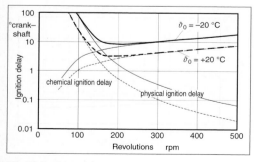

Fig. I3 Ignition delay

~~**Combined systems.** The combination of glow plugs as tried and tested cold-starting aids and the use of the heating flange for suppressing emissions of smoke during cold-start or improving the smooth running of the engine is suitable for fast starting with minimum start emissions and optimized smooth running. This has particular significance for complying with future tightening of emissions legislation but also for an increasing number of cylinders or increasing displacement.

~~**Glow system.** The most important cold-starting aid on diesel engines is the glow system, which uses glow plugs as active heating elements in the combustion space. Modern passenger car diesel engines are equipped with glow plugs as standard. The glow plug is also indispensable for engines with divided combustion chamber (prechamber or swirl chamber) for ensuring starting within the frequently occurring temperature range of 10 to 30°C.

Because of the drastic deterioration of starting quality below freezing point, the glow plug also is used as a cold-starting aid for direct-injection diesel engines.

The glow plug is typically located close to the injection nozzle but is not directly positioned in the injected spray and extends about 3–8 mm into the combustion space for locally transferring and applying its relatively small heat output away from the hot surface as effectively as possible. Depending on design and size, the power input is 30–150 W when in a state of equilibrium; at that the glow plug reaches temperatures around 800°C up to 1100°C on its surface. The physical and chemical ignition delay is reduced in the vicinity of the hot tip of the pencil-type glow plug through accelerated evaporation of the fuel droplets and the faster initial reactions at higher temperatures, thereby resulting in ignition of the mixture. The glow plug acts as an indirect local ignition aid. Energy for igniting the remaining main part of the fuel then has to come from the independently running combustion.

The glow plug needs to be further energized (postheated) for up to three minutes after starting, depending on engine temperature, to ensure favorable ignition conditions also during the engine warm-up phase. With quickly heating glow plugs, good starting qualities can now be achieved without additional preheating, in contrast to older concepts.

The glow-plug filament and regulating filament are series-connected in a self-regulating sheathed-element glow plug to form a common resistance element (**Fig. I4**).

The glow-plug filament is made of high-temperature resistant material, the electrical resistance of which is largely temperature-independent. It forms the heating zone together with the front part of the pencil-type glow plug. The regulating filament is attached to the current-carrying terminal stud. Its resistance has a large positive temperature coefficient. This renders a quickly heating, self-regulating pencil-type glow plug, and at the same time overheating of the glow plug is prevented (**Fig. I5**). Both filaments are coated with compacted, electrically insulating ceramic powder with very good thermal conductivity throughout the

601

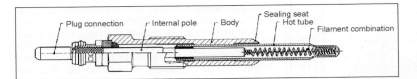

Fig. I4
Glow plug layout

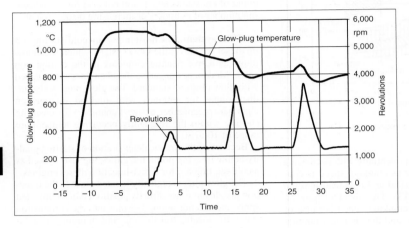

Fig. I5
Start with
self-regulating glow
plug

entire hot tube. The powder is compacted by mechanical compression such that the filaments are seated as they were cast to permanently withstand vibrational stresses from the engine.

The space available for the glow plug, especially in the case of modern engines in four-valve technology with pump-nozzle elements or injectors, is limited so that the glow plugs need to be designed as slim as possible. On the other hand, glow plugs also need a certain robustness because they often are meant to last the life of the engine due to a very poorly accessible installation position.

~~~**Controlled glow systems.** Electronically controlled systems will eventually replace the self-regulating glow plugs. Regulated systems that do not need complex calculation of control power against engine parameters are currently under intensive development. In this case, the glow plug requirements have to be transmitted from the overriding engine control unit to the preheating controller as target values and the controller in turn controls the necessary voltage on the glow plug accordingly.

Special glow plugs were developed for this purpose, plugs which also feature fast response at low voltages and high heating reserves and which are able to return an easily analyzed and stable temperature signal to the preheating controller.

~~**Heating flange.** Heating flanges are used primarily on diesel engines in commercial vehicles for intake air preheating. They ensure good engine starting down to extremely low temperatures as the result of improved mixture preparation and reduced ignition delay, and at the same time they very effectively reduce HC cold-

start emissions such as visible white smoke (**Fig. I6**). They are increasingly gaining importance, particularly with regard to compliance with new emission limits. With growing requirements for the reduction of emissions during cold start and the improvement of ride comfort, electrical heating flanges are also gaining interest for passenger car applications.

Heating flanges are employed with thermal capacity starting from 0.5 kW up to 2.5 kW. Two variants are offered: An unregulated simple variant with standard heating materials and a self-regulating variant in the form of a PTC (positive temperature coefficient) resistor heating flange.

The heating flange heats the flowing air in the intake tract by approximately 50 K. In this phase, the heating element has a temperature of around 600°C (**Fig. I7**).

Using engine compression, the temperature rise of the supplied air is multiplied up to the end of compres-

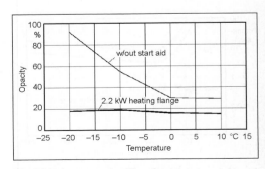

Fig. I6 Exhaust opacity 30 s after start

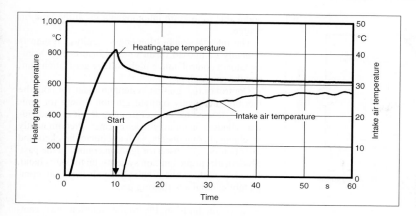

Fig. I7
Heating flange
heating-up behavior

sion (**Fig. I8**). For instance, the increase of the intake air temperature through a heating flange of $D_{T1} = 50$ K will result in a rise $D_{T2} = 147$ K at a compression ratio $\varepsilon = 18.5$.

The electric heating flange is installed in the intake-air duct as an intermediate piece about 20 mm long and has a power connection and current feed for the internally positioned heating element. It accommodates the power electronics and the heating element together with its insulation. The heating element consists of one or more metallic strips that are typically run in ceramic insulation meandering with some five loops, which are connected on one side to the frame for the grounding connection (**Fig. I9**).

~~Ionic-current measurement. A cost-effective technique of ionic-current measurement in the combustion space of a diesel engine uses special glow plugs that are electrically insulated from their threaded body. Ions generated in the combustion process can be col-

lected at a favorable location inside the combustion space using the pencil-type glow plug as electrode so insulated (**Fig. I10**).

Potential application areas of ionic-current measurement inside the diesel engine combustion space are:

- Compensation for different quality fuels.
- Cylinder equalization relating to start of combustion, compensation of tolerances in the injection and intake system, etc.
- Detection of combustion misfires
- Direct feedback from the combustion space for OBD applications

Separate grounding of the heating pin to the cylinder head is necessary due to the insulation of the heating pin against the cylinder head, which can be interrupted for ionic-current measurement. A corresponding switching circuit is integrated into the glow plug so that there will be no change in the external design of the glow plug.

Ignition system, gasoline engine. The internal combustion engine requires external ignition of the air-fuel mixture. The ignition energy is locally supplied (usually from a spark plug) with the aid of the ignition system to a suitable central location inside the combustion

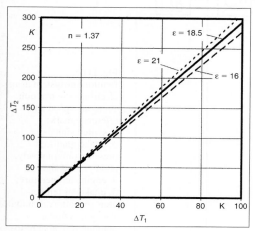

Fig. I8 Increase in final compression temperature with intake air preheating

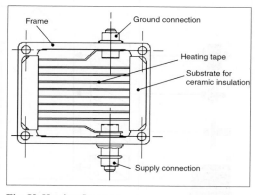

Fig. I9 Heating flange

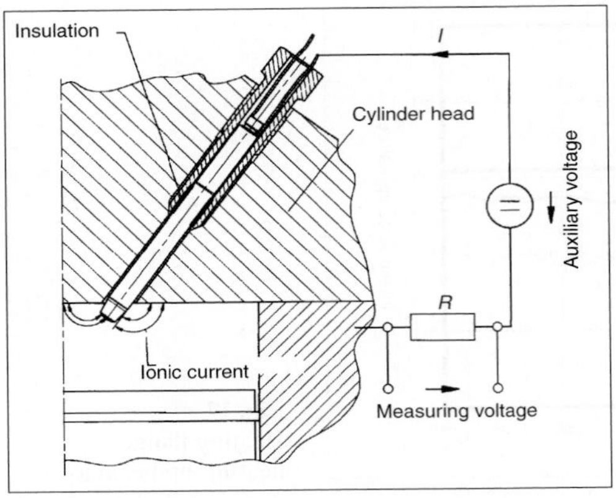

Insulation

Cylinder head

Auxiliary voltage

R

Ionic current

Measuring voltage

Fig. I10 Principle of ionic-current measurement

space in the final phase of the compression cycle. The ignition energy produces a strong, local temperature rise of several thousand degrees Kelvin in the gas mixture. A flame kernel develops from which the mixture is ignited in radial direction by a self-preserving flame front. Good engine-related operating behavior requires dependable ignition, preferably of the complete mixture in the combustion space. The ignition system always will have to provide sufficient ignition energy to achieve this, even in the case of lean-mixture fluctu-

ations due to mixture inhomogeneities at the ignition source.

The internal combustion engine ignition system is typically composed of an intermediate energy accumulator, a control unit, a high-voltage generation unit, and the spark plugs.

Control of the ignition system provides for both the thermodynamically optimal ignition timing (or ignition angle) and the correct firing order of the individual cylinders on multicylinder engines. Predominantly mechanical components, such as contact breakers and distributors, were used for ignition control in the past. Modern electronic ignition systems, on the other hand, function with static high-voltage distribution completely without any moving parts.

~Electronic ignition. Electronic ignition has completely taken over the control functions and high-voltage generation of the ignition system over the course of time; it has replaced mechanical ignition components. Electronic ignition has meanwhile been integrated into the electronic engine control unit (**Fig. I11**). Together with the ignition output stage, it takes over the switching functions in the primary circuit, such as charging the ignition coil and breaking the coil current, so as to initiate external ignition of the mixture in a certain cylinder at the desired moment.

Above all, the optimal ignition timing depends on engine speed and engine load. Other variables that possibly influence the ignition process are engine management

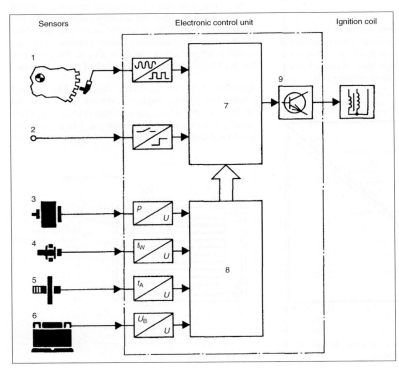

Sensors   Electronic control unit   Ignition coil

Fig. I11
Ignition subsystem in an electronic control unit:
1, engine speed and reference mark;
2, switch signals;
3, intake manifold pressure (alternatively, another load signal);
4, engine temperature;
5, intake air temperature; 6, battery terminal voltage; 7, microcomputer; 8, analog-to-digital converter;
9, ignition output stage

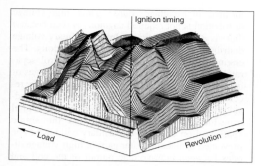

Fig. I12 Optimized ignition angle map for electronic ignition timing

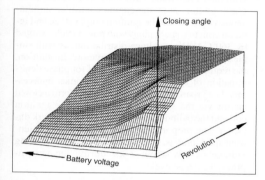

Fig. I13 Dwell-angle map

Fig. I14 Chopper disk as pulse generator wheel for the crankshaft speed signal with inductive pickup

functions such as knock controls and emission controls. Information required for control of ignition timing are stored in multidimensional performance maps (**Fig. I12**) or are calculated via numeric models.

The integrated control of the closing angle provides for sufficient energy (or current) in the ignition coil at the desired moment. The primary current of the ignition coil is switched on by a dwell-angle map (**Fig. I13**) that allows for the relevant physical dependencies such as vehicle system voltage, speed, and the electrical data of the ignition components.

Time-based synchronization of the control of ignition with the engine crankshaft position is made through speed sensors that sense pulse generators (**Fig. I14**) on the crankshaft and camshaft.

~**High-voltage distribution.** Electronic static high-voltage distribution or predominantly mechanically rotating high-voltage distribution was employed in the past for high-voltage distribution to the spark plugs on the different cylinders of multicylinder gasoline engines.

~~**Ignition distributor (conventional).** The (ignition) distributor (**Fig. I15**) is the main component of a mechanical ignition system for multicylinder engines.

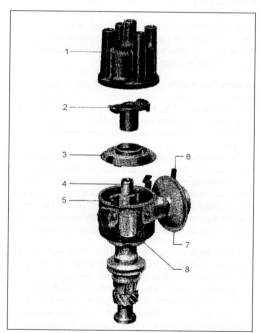

Fig. I15 Ignition distributor with ignition-advance adjusting device: 1, distributor cap; 2, distributor rotor with electrode E; 3, dust cover; 4, distributor shaft; 5, contact-breaker cams; 6, port for vacuum hose; 7, vacuum unit; 8, ignition capacitor (centrifugal advance device and the vacuum advance device are inside the cup-shaped distributor housing)

In fully mechanical ignition systems, this component contains, in addition to purely high-voltage distribution to the spark plugs according to firing order, the complete function of ignition breaking and advance of spark timing.

The distributor (rotor) arm rotates in the distributor housing for ignition voltage distribution. The rotor diode is laterally attached to the distributor arm. The rotor arm is supplied with high voltage from the ignition coil via a spring-loaded carbon element through the central tower of the distributor cap. At the time of ignition, the rotor arm faces a certain fixed electrode in the distributor cap according to the firing order, thus passing high voltage to the spark plug in a particular cylinder via the ignition lead located in the appropriate outer tower in the distributor cap. This requires a small air gap between rotor electrode and fixed electrode, which is bridged by the distributor spark. At the same time as the onset of ignition coil charging, the distributor spark gap represents a booster gap that suppresses undesired switch-on sparks at the spark plug. On four-stroke engines, the distributor arm is either rotated by the crankshaft by way of a 1:2 reduction gear or is rotated directly from the camshaft and thus is placed into the correct cylinder position relative to the internal fixed electrodes of the distributor cap.

The distributor is not required anymore because of fully electronic ignition systems with static high-voltage distribution of the ignition. In a transitional phase, it was used on partly electronic ignition systems with rotating high-voltage distribution.

~~Rotating high-voltage distribution.
In the case of rotating high-voltage distribution, a mechanical (ignition) distributor (**Fig. I15**) was employed to assign the high voltage generated in the ignition coil to only one exclusively designated spark plug according to the firing order.

The distributor with its internal, rotating distributor arm is usually locked to the camshaft, thus rotating in phase with the corresponding gasoline engine work cycles. The rotating high-voltage distributor is employed for mechanical ignition systems and partly electronic transistorized ignition systems. Fully electronic ignition systems employ static high-voltage distribution.

~~Static high-voltage distribution.
High-voltage delivery on static high-voltage distribution to the spark plugs is made without mechanical moving parts. The designation "static" is used because the mechanical ignition distributor with its rotating distributor arm is not required anymore. Elimination of mechanical moving parts results in simplifications of the engine design, less wear, and reduction of noise. In addition, avoidance of open spark gaps on contact members significantly reduces the emission of electromagnetic interference. Static high-voltage distribution can be achieved both with direct-fire ignition coils and with dual-spark ignition coils:

- Static high-voltage distribution with single spark coils uses one ignition coil for each spark plug as an intrinsic high-voltage source. Thus, high-voltage distribution is not required anymore as such because there is a shift to static electronic "primary voltage distribution" to the respective ignition coils. The engine control unit takes over this control separately for each direct-fire coil with output stages of its own (power transistors). The engine controller receives the correct phasing relationship for the individual power cycles from the camshaft via a signal transducer. Switch-on sparks from the single spark coil on charging the primary coil have to be prevented by using booster gaps or high-voltage diodes between high-voltage output of the coil and the spark plug, because there may be a low pressure ignitable mixture in the cylinder that only requires a low ignition voltage.

- With static high-voltage distribution employing twin-spark ignition coils, the engine controller simultaneously ignites, via the ignition output stage and ignition coil, two spark plugs with power cycles shifted by 360°CA. High-voltage distribution, as with single spark coils, is not required anymore. In addition, the requirement of camshaft signals for phase determination of control of ignition becomes unnecessary. Both spark gaps of the spark plugs form a closed circuit via the secondary winding of the ignition coil and the cylinder head for the period of spark duration. The proportional ignition energy of the unused spark plug is, however, low because the arcing voltage is only about 2 kV during the uncompressed exhaust cycle.

~High-voltage generation.
Ignition of the compressed air-fuel mixture by spark plugs requires a high voltage between 10 and 30 kV. For this purpose, the available vehicle voltage is transformed upward using the ignition coil or a fast low-loss ignition transformer. At the moment of ignition, the energy stored in the magnetic field of the ignition coil or in a capacitor is transferred as secondary high-voltage ignition pulse to the spark plug by means of the high inductance (number of turns) of the secondary coil of the ignition coil or the ignition transformer.

~~Coil ignition.
Coil ignition systems use an auto-transformer-switched ignition coil as the energy accumulator and for generating a high voltage. In the case of the conventional coil ignition systems, the primary circuit of the ignition coil is switched on through the ignition switch from one side (terminal 15) and through a series resistor to the positive terminal of the motor vehicle battery. The second terminal of the ignition coil primary circuit (terminal 1) is grounded through the distributor contact breaker points and thus connected to the negative terminal of the motor vehicle battery when the distributor contact breaker points are closed. Thus, DC charging current flows through the primary winding (the ignition output stage takes over the function of the distributor contact breaker points in case of electronic ignition). The ohmic resistance of the primary winding limits the maximally attainable

charging current. The magnetic field connected with the flowing current contains stored energy W_{mag}, which is

$$W_{mag} = \frac{1}{2}L_1 i_1^2 \ ,$$

where L_1 is the inductance of the primary winding and i_1 is the current in the primary coil.

The charging current is interrupted by the ignition breaker, or ignition output stage, at the moment of ignition. This causes an abrupt breakdown of the magnetic field. This change of the magnetic field induces a voltage into the secondary winding, the slewing rate of which is determined by the inductance of the secondary winding and the capacity of following line elements. In electronic ignition systems, the output stage of the ignition system has to be protected from voltage induced into the primary winding at the same time by appropriate steps.

The ignition spark is generated as soon as the induced (secondary) voltage reaches its flashover voltage at the spark plug gap. Subsequently the secondary voltage drops to the spark voltage of the spark plug. The ignition coil discharges its stored energy in the form of an arc discharge followed by a glow discharge on the spark plug. As soon as the voltage is no longer sufficient to maintain the glow discharge, the spark will break down and the residual energy will decay into the secondary circuit of the ignition coil. The induced (secondary) voltage in the ignition coil is negatively polarized against vehicle mass at the time of spark breakdown to the center electrode. Vice versa, the ignition voltage requirement would be somewhat higher. The duration of the spark from coil ignition systems is relatively long, and hence coil ignition systems ensure high reliability for ignition of turbulent and/or nonhomogeneous lean mixtures.

~~**High-voltage capacitor ignition (capacitor discharge ignition, CDI).** High-voltage capacitor ignition systems are used only in special cases for high-performance and motor sport engines because they offer only small advantages over modern and almost equally fast coil ignition systems.

High-voltage capacitor ignition uses a capacitive energy accumulator. The ignition energy is supplied from the electric field of a capacitor with capacitance, C, and charging voltage, U, and lead-to-energy content, W_{el}, according to the formula

$$W_{el} = \frac{1}{2}C U^2 \ .$$

Fig. I16 illustrates the schematic circuit diagram of a high-voltage capacitor ignition system. The ignition energy storage capacitor is connected via a charging and control circuit to an ignition transformer that generates a high voltage. The charging circuit charges the capacitor to approximately 400 V. At the moment of ignition, a thyristor switches the capacitor to ground on the charging side. For this reason, the high-voltage capacitor ignition is also termed thyristor ignition. On the primary side, the discharge current abruptly flows through the ignition transformer. The rapid change of the magnetic field in the secondary winding of the ignition transformer induces the required ignition voltage.

The very low internal resistance of the CDI produces a very steep rise in the secondary voltage (some kV/ms), which, in contrast to conventional coil ignition systems, generates the ignition spark more quickly.

Therefore, high-voltage capacitor ignition systems are highly resistant to parallel connections caused by spark plug coatings. A disadvantage with nonhomogeneous mixtures is the very short ignition duration of only 100 ms, which may cause misfiring, and the high spark current leading to rapid erosion of the spark

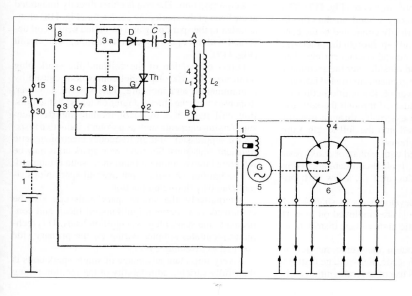

Fig. I16 Circuit diagram of the breakerless CDI with inductive pickup system in the ignition distributor: 1, battery; 2, ignition-and-starter switch; 3, switching device; 3a, charging device; 3b, control stage; 3c, pulse shaper; D = diode; C = storage capacitor; Th = thyristor with gate connection; 4, ignition transformer; L1 = primary winding; L2 = secondary winding; 5, inductive pickup; 6, ignition distributor

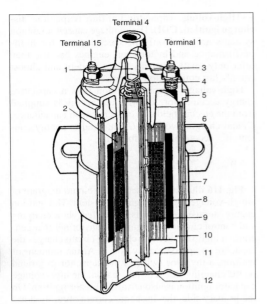

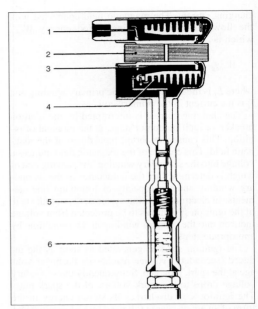

Fig. I17 Conventional ignition coil as sectional view: 1, outer high-voltage terminal; 2, winding layers with insulating paper; 3, insulating cap; 4, internal high-voltage terminal with spring contact; 5, housing; 6, mounting clamp; 7, magnetic metallic shield; 8, primary winding; 9, secondary winding; 10, sealing compound; 11, insulating body; 12, iron core

Fig. I18 Single-spark coil: 1, outer low-voltage terminal; 2, laminated iron core; 3, secondary winding; 4, primary winding; 5, internal high-voltage terminal via spring contact; 6, spark plug

plug, thus resulting in relatively high levels of spark plug wear.

~~Ignition coil. A standard cylindrical ignition coil is built around a laminated soft-iron core (**Fig. I17**). The secondary winding is directly wound on the core in central position and electrically connected to the center tower in the ignition-coil cap through the otherwise insulated installed iron core and contact springs. The primary winding is wound outside over the secondary winding. The low-voltage connections are led laterally through to the center tower on the ignition-coil cap. The magnetic flux is conducted through the soft iron core inside the coils and externally through metallic shields, which are inserted into the full-length, cup-shaped, outer steel-metal housing.

The inner body is sealed with asphalt or epoxy resin for insulation and fixing. Typical coil data are: Turns ratio, 1:100; primary inductance L_1, a few mH; primary resistance, 0.2–3 Ω.

Further ignition coil developments as single-spark or twin-spark ignition coils are employed on modern ignition systems with static high-voltage distribution.

~~Single-spark coil. Modern electronic multicylinder ignition systems with static high-voltage distribution require simultaneous use of several ignition coils.

In the case of ignition systems with single spark coils, each cylinder is ignited with an ignition coil of its own. Such ignition systems are suitable for engines with any number of cylinders, particularly odd cylinder numbers. When using single spark coils without an ignition lead, the ignition coil and the spark plug connector are fitted as a common unit. The coil is either directly integrated at the head of the spark plug connector (**Fig. I18**) or integrated in the spark plug as a co-called pencil coil designed with an open, wide-stretched magnetic circuit (**Fig. I19**).

Direct connection of the coil and the spark plug minimizes the risk of parallel connections caused by contamination and moisture. This system, however, dispenses with the shielded ignition leads, so the ignition coil itself has to accommodate interference-suppression elements such as a wound, inductive resistor for suppression of high-frequency electromagnetic transient emissions. Several single spark coils may be integrated into a common ignition cassette if required. The ignition cassette is put over all spark plugs simultaneously during production.

Alternatively, the single spark coils can also be combined as a common ignition-coil block and connected to the spark plugs via ignition leads. This technique facilitates simpler wiring on the primary side and makes assembly easier.

A very important advantage of single spark coils is their fully variable adjustability of the ignition timing

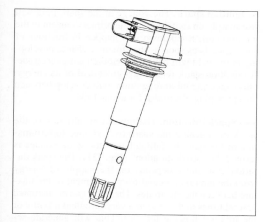

Fig. I19 Pencil coil

up to highest engine speed, because there is always enough time available for charging the coil. In particular, overlapping the closing times of several coils is practicable.

~~Twin-spark ignition coil. Electronic ignition systems with static high-voltage distribution for gasoline engines with an even number of cylinders frequently use twin-spark ignition coils (or dual-firing coils). In contrast to direct-fire coils, twin-spark ignition coils represent a space-saving alternative: Half the power output stages compared with the direct-fire coil solution are sufficient for ignition control.

In the case of the twin-spark ignition coil (**Fig. I20**) the two secondary winding connections with two spark plugs each are series-connected for cylinders whose firing orders are offset to each other by 360°CA. Thereby the twin-spark ignition coil generates an igni-

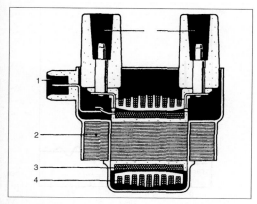

Fig. I20 Twin-spark ignition coil for static high-voltage distribution: 1, low-voltage terminal; 2, iron core; 3, primary winding; 4, secondary winding; 5, high-voltage terminals

tion spark on the two spark plugs simultaneously. The first spark is in a cylinder filled with a compressed air-fuel mixture, and the second spark is in a cylinder just bending its exhaust process. A strong main spark develops in the highly compressed medium, while a weak supporting spark with low additional voltage requirement occurs in the uncompressed one.

One of the two spark plugs ignites with positive high voltage because of the series connection and the other spark plug ignites with negative high voltage. Electrode burning on the spark plugs is highly asymmetric because of the different polarities of the ignition voltage. Various designs for ignition with dual-firing coils can be realized. The combination of the dual-firing coils into one ignition module and connection to the spark plugs via ignition leads is possible. Electric contact can be applied on the ignition coil directly as an alternative and connected to the spark plug of the corresponding cylinder via an ignition lead.

~Ignition. Spark plugs and other features that are of importance in the context of the byword "ignition" in gasoline-engine combustion are discussed under this generic term.

~~Closing angle. The closing angle results from engine speed multiplied by energy accumulator charging time (the ignition coil primarily). Charging time is converted into crank angle rotation during this period. "Closing angle" is originally derived from closing the mechanically operated distributor contact breaker points of contact-breaker controlled coil ignition.

~~Closing (dwell) angle control. The closing (dwell) angle (or closing time) has to be sufficiently long to charge the ignition coil with the necessary ignition energy but must not become too long at low speed and increased vehicle voltages so that the ignition coil will not be overheated. Control of the closing angle with speed and vehicle voltage is termed "closing angle control"—electronic ignition systems driven by engine controllers use dwell-angle maps for this purpose.

~~Closing (dwell) time. Closing (dwell) time is the time needed to charge the energy accumulator (primary coil of the ignition coil). It corresponds to division of the closing angle by the speed.

~~External ignition. The gasoline engine requires external ignition for ignition of the compressed air-fuel mixture. External energy in the form of an electrical spark is introduced into the mixture, thus heating it up locally by several thousand Kelvin. There has to be ignitable mixture in the region of the spark, and required minimum ignition energy must be coupled in so that the mixture can complete self-sustained combustion.

~~Firing order. Firing order in multicylinder engines is the sequence in which the individual cylinders are fired. With mechanical ignition distribution, the mechanical coupling of the distributor automatically established assignment to the "correct" cylinder. With static high-voltage distribution, the engine control unit provides the correct order at the ignition-coil controller by using chopper disks and engine speed sensors.

~~**Ignition advance range.** The ignition angle can be selected in degrees crank angle within the ignition advance range according to the respective engine operating state before or after TDC as needed.

~~**Ignition conditions.** In gasoline engines, all parameters that influence ignition sparks and combustion of the mixture are counted among the ignition conditions. The status of the mixture in the region of the spark plug electrodes, in addition to ignition system properties (e.g., ignition energy and condition of the spark plugs), is decisive for successful ignition that is followed by complete combustion of the cylinder charge. There has to be a combustible mixture at the spark gap that can be ignited by the ignition energy. An independent combustion process with complete combustion of the existing mixture will only develop from the ignition source if conditions suitable for ignition and flame propagation are available.

~~**Ignition energy.** Under optimal conditions, a stoichiometrically homogeneous air-fuel mixture requires ignition energy of approximately 0.2 mJ for ignition. The energy demand will rise to more than 3 mJ in richer or leaner mixtures. The ignition energy required to ensure reliable inflammation of the mixture in a real engine is considerably more. The turbulent, nonhomogeneous mixture in the region of the spark plug as well as transfer losses and heat losses at feed lines and electrodes significantly increase the ignition energy demand. A long spark duration and a large spark gap improve the possibility of ignition and decrease the ignition energy demand.

Conventional ignition systems would provide the spark plug with about 40 mJ during approximately 1 ms of spark duration for ensuring ignition. With electronically controlled ignition systems, the secondary available energy is about 60–70 mJ. Human contact with parts that carry high voltage involves a life-threatening risk.

~~**Ignition failure detection.** Ignition failure detection is a partial function of vehicular self-diagnosis (onboard diagnostics [OBD]) for complying with ruling emission limits. It monitors the ignition system for misfiring. For instance, crankshaft acceleration is measured through crankshaft torque contributions via the so-called engine smoothness monitoring system. Nonfired cylinders can be detected with the aid of appropriate computing algorithms and shutoff from fuel.

~~**Ignition pressure.** Ignition pressure is the maximum pressure developed during combustion of the air-fuel mixture.

~~**Ignition spark.** The voltage at the spark plug will rise sharply on switching off the primary current in the ignition coil. Breakdown will happen in fractions of nanoseconds as soon as the streamer charge developing in the field will have reached the facing electrode. An ignition spark releasing a great deal of its energy, thus triggering mixture ignition, will develop between the spark plug electrodes for about 1 ms.

~~~**Spark duration.** The time from initiation of the spark on reaching the spark-over voltage to termination of the spark by falling below the spark voltage is termed the "spark duration" (**Fig. I21**). The spark duration essentially depends on the supply of energy from the ignition coil and from movement of the mixture between the electrodes. The spark can be disrupted several times in the case of a very turbulent mixture or the mixture flowing too fast. This will generate sequential sparks the total duration of which would be sufficient for safe inflammation of the ignited mixture provided there is still enough residual energy in the coil. Depending on the ignition system and the available closing time, the spark duration can drop to 0.6 ms at high speeds.

~~~**Spark energy.** Secondary energy available at the ignition coil during the ignition process is only partially fed into the plasma between the spark plug electrodes. About half the energy stored in the coil is lost at the electrodes due to thermal dissipation, shunts, and capacitive loads. The residual energy actually getting into the mixture via the spark is termed "spark energy." Safe inflammation of the mixture requires the spark energy to be significantly above the minimum ignition energy under all engine operating conditions.

~~~**Spark gap.** The spark gap defines the path the spark takes between the electrodes. This is termed

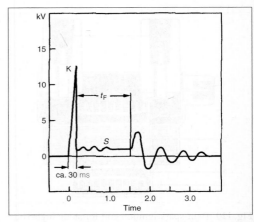

Fig. I21 Voltage characteristic between the spark plug electrodes during ignition process: K = spark head; S = spark tail; t_F = spark duration

the spark air gap or spark surface gap, depending on the electrode arrangement. It is called a spark air gap when the shortest path between the electrodes is open and the ignition spark takes the almost direct path through the air-fuel mixture. The second variant is the spark surface gap. Drawing back the ground electrodes can result in the top of the insulator nose being located at or in the direct path between the electrodes. The spark will then flash to the insulator nose, passing it and thereby forming the so-called spark surface gap. This has the advantage that soot deposits are prevented on the insulator nose through the surface-gap ignition spark, which results in less parallel connections on the spark plug. The disadvantage is, however, energy loss through heat release to the insulator and possibly reduced mixture accessibility.

~~~Spark head.
Spark head is understood to be the first very short phases of the spark—that is, the breakdown—and arc phases, in which a highly excessive spark current flows into the plasma and the electrode potential breaks down to the spark voltage at the spark plug.

~~~Spark position.
The spark position describes the position of the spark gap relative to the combustion space at the front of the spark plug screw-in base (**Fig. I22**). The spark should be delivered as deep as possible into the combustion space for safe ignition of the mixture. In this context it is necessary to differentiate between spark positions extended and recessed.

Modern engines would normally have an advanced spark position of approximately 3 mm, and the spark plugs would need to be designed for a higher thermal load.

In the case of engines with very high specific power (racing cars and special engines), spark plugs with a recessed spark position are used. The spark gap will be within the spark-plug shell in this case. Extended low-load operation could cause soot deposits on recessed spark plugs because the self-cleaning temperature is not reached.

~~~Spark tail.
The spark tail (**Fig. I21**) describes the third and last phase of the spark. During this phase, the spark burns as a glow discharge under almost constant spark voltage and at relatively low current of less than about 100 mA. The spark tail describes the ignition spark phase that is the longest by orders of magnitude, thus determining the ignition spark duration in essence and the main part of energy transfer at the same time. It immediately follows the spark head. The spark tail ends when the voltage falls below the spark voltage, the spark breaks, and the spark gap again ceases to be conducting.

~~~Switch-on spark.
Switching on the charging current also causes a strong change in the magnetic field to develop inside the secondary coil of the ignition coil. Voltage is thereby induced at the firing end (spark plug) that can lead to switch-on sparks at low pressures. The switch-on spark can produce unwanted premature ignition of the mixture and is, therefore, suppressed by means of a booster gap or a high-voltage diode.

~~Ignition timing.
The ignition timing corresponds to the end of the ignition coil closing time—that is, the time of primary ignition-coil current switch-off, which would immediately initiate the spark and, thus, ignition of the air-fuel mixture. The ignition timing is stated in degrees crank angle before or after TDC.

~~Ignition voltage.
The ignition voltage is applied to the electrodes of the spark plug to ignite the spark between the electrodes through gas ionization and to subsequently maintain the spark over the necessary combustion duration. The ignition voltage will drop to the considerably lower spark voltage of the glow discharge after spark-over.

The ignition voltage varies depending on the operating point of the engine and according to spark phase can be between a few, 100 V, and 20 kV; it can even reach 30 kV at cold starting.

~~Ignition voltage requirement.
Fig. I23 shows the minimum and maximum ignition voltage requirements (spark-over voltages), depending on engine speed. These were determined from a mix of overland driving and circular track test driving with a high acceleration component. The comparison of new against old shows a significant rise in the voltage requirement over the running period. Burning encountered by the electrodes widens the electrode gap and thereby the ignition voltage required.

The ignition voltage requirement needs to be kept below a certain limit because further increase of the voltage supply would cause problems with the high-voltage handling capability of the feed lines and would result in increased electrode burning due to the higher ignition energy. The limitation of the ignition voltage requirement can be achieved through suitable spark plugs (with smaller electrode cross sections, for instance) and assurance of an ignitable air-fuel mixture in the electrode region.

The difference between the high-voltage supply provided by the ignition coil and the necessary ignition

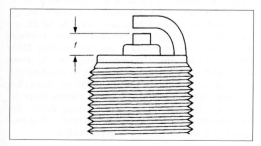

Fig. I22 The spark position reference point for measure f is the front side of the screw-in base

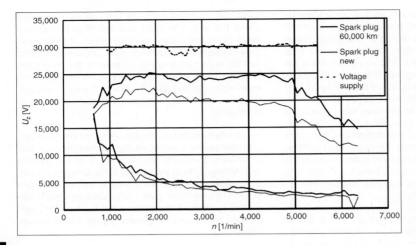

Fig. I23
Ignition voltage
requirement (min/max)
and voltage supply

voltage required for safe inflammation mixture defines the voltage reserve (**Fig. I23**). Therefore, the voltage reserve has a large effect in determining the maximum possible lifespan (field time) of the spark plug.

~~Misfiring. Misfiring occurs when the ignition spark fails between electrodes and spark plug. The causes for this may also be attributed to spark plug parallel connections or an ignition system defect.

Combustion missing, in which case the mixture will not ignite in spite of successful ignition, must be differentiated from misfiring,

In both cases, significantly higher HC emissions will result at a relatively low percentage of total power cycles. Extended operation with misfiring or missing combustion can damage or destroy the catalytic converter because of overheating.

~~Spark control. The ignition timing has a decisive influence on the fuel consumption, performance, and exhaust emissions of gasoline engines (**Fig. I24**) because the spark initiates the combustion process. The ignition timing has to be set for the maximum generation of mechanical work using automatic spark control system such that the combustion pressure will reach its peak value preferably in the crank angle range between 15 and 20°CA after TDC.

The control of the spark has to be dependent on speed and load for the following reasons:

- The course of combustion varies at varying engine load and hence at variable cylinder filling because the flame propagation speed of the compressed mixture depends on its properties.
- Less time is available for combustion per degree crank angle as the engine speed is increased. Combustion has to be initiated ever earlier relative to TDC with increasing engine speed so that the peak combustion pressure will be in the optimum crank angle range at all engine operating points.

There are mechanical and electronic methods for controlling the spark. Using mechanical spark control with a centrifugal ignition advance device and vacuum advance device allows only limited timing characteristics (spark advance curve) to be achieved.

Electronic ignition systems that are controlled from a microcomputer-based engine controller operate with numeric maps of ignition angle, thereby replacing the mechanical advance characteristics; they also offer better adaptation possibilities. In addition, further influencing variables such as engine or intake-air temperature can be entered into the computation. The current ignition angle, in combination with knock control, transmission control, catalyst heating function, traction control system, or electronic engine-power control is additionally influenced by the requirements of these subsystems that are dependent on operating point.

The engine controller, furthermore, computes the optimal charging time (closing "dwell" time) of the ignition coil, making allowance for engine speed and battery terminal voltage, and automatically initiates the charging in time. With the engine stopped, the current flow through the ignition coil is automatically interrupted so as to prevent coil overheating.

~~Spark current. Spark current is understood to be the current that flows in the secondary circuit of the ignition coil via the spark plug electrodes during the ignition sparks. It strongly depends on the respective phase that the ignition spark is in at this moment (**Fig. I25**). For a very short time, the spark current rises to more than 100 A in the nanosecond range after reaching the flashover voltage. Subsequently, in the arc phase, for less than one millisecond the spark current drops to some few amperes. In the final sparking phase, the glow discharge, the spark current will further drop to a current below 100 mA for about one millisecond.

Literature: M. Adolf: Zündkerzen, in Handbuch Verbrennungsmotor, R. van Basshuysenm, F. Schäfer (Eds.), pp. 465–470,

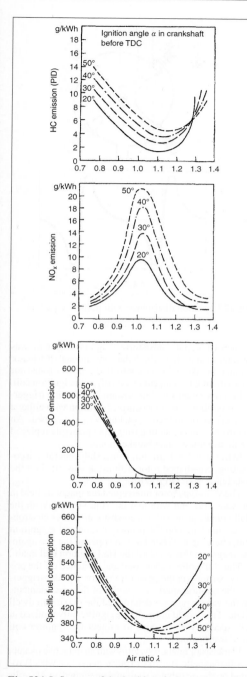

Fig. I24 Influence of the ignition timing on emissions and fuel consumption

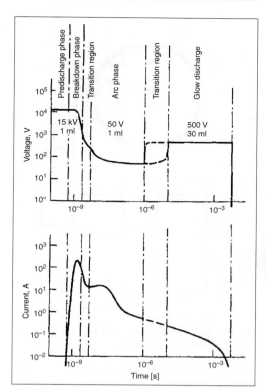

Fig. I25 Variation of current and voltage of transistorized coil ignition over time. Typical values for occurring voltages and energy transfer during the individual spark phases

ternal Combustion Engine Fundamentals, McGraw Hill, New York, 1989. — Autoelektrik, Autoelektronik am Ottomotor, Bosch, VDI Publishers, 1987. — H. Albrecht, R. Maly, B. Saggau, E. Wagner: Neue Erkenntnisse über elektrische Zündfunken und ihre Eignung zur Entflammung brennbarer Gemische, in Automobil-Industrie (4): 45–50, 1977. — R. Maly, M. Vogel: Ignition and Propagation of Flame Fronts in Lean CH4-Air Mixtures by the Three Modes of the Ignition Spark, Proc. of 17 Int. Symp. on Combustion pp. 821–831, The Combustion Institute, 1976. — M. Schäfer: Der Zündfunke, Diss., Univ. Stuttgart, 1997. — P. Hohner: Adaptives Zündsystem mit integrierter Motorsensorik, Diss., Univ. Stuttgart, 1999. — R. Herweg: Die Entflammung brennbarer turbulenter Gemische, Diss., Univ. Stuttgart, 1992.

~**Ignition leads.** Variants of ignition leads are employed in gasoline engines: either resistance ignition leads with a wire resistor or carbon resistor in the lead or copper-cored ignition leads with interference-suppression resistors incorporated in the plug connectors.

In wire resistor leads, stainless steel wire is wound around a ferromagnetic silicone carrier, subsequently fitted with a first silicone insulation, reinforced by fabric, and again fitted with high-temperature–resistant silicone sheathing.

Vieweg Publishers, 2002. — H. Klein: Zündkerze, Shell Lexikon Verbrennungsmotor, R. van Basshuysen, F. Schäfer (Eds.), Series 86/87, Vieweg Publishers, 2003. — J.B. Heywood: In-

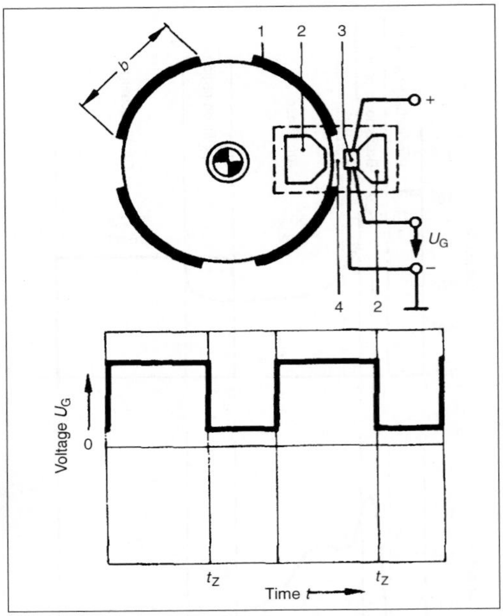

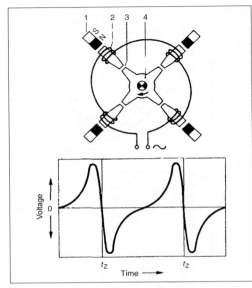

Fig. I26 Ignition triggering by Hall generator in the ignition distributor

Copper ignition cables have high-voltage–resistant copper cores and are provided with glass-fiber–reinforced, soap-stoned silicone sheathing. Both variants are highly flexible and can carry high voltages up to 40 kV.

~Ignition timing sensor. Ignition timing sensors are used for synchronizing the ignition system with the engine operating process. In ignition coil engine-type ignition systems, older systems use cam-controlled mechanical techniques as the ignition timing sensor. Ignition systems using noncontact methods with Hall generators or inductive pickups are used in electronic ignition systems (**Figs. I26 and I27**).

~~**Distributor contact breaker points.** The distributor contact breaker points switch the ignition coil primary circuit on and off in the ignition distributor. The ignition distributor is actuated by the cam on the rotating distributor shaft. This direct-switching principle is characterized by low performance and a limited life span (through mechanical and electrical load) and results in expenditures by the owner on maintenance and repairs.

~~**Transistor ignition.** The transistor ignition was developed to improve the life of distributor contact breaker points before sensor-based electronic engine controls were available on motor vehicles. It features a higher ignition voltage than coil ignition with distributor contact breaker points, thus leading to better cold- and hot-start performance. In addition, it is more robust to spark-plug wear and spark-plug fouling. The

Fig. I27 Ignition distributor with inductive pickup

breaker-triggered transistorized ignition system was introduced at first, in which the conventional distributor contact–breaker contact controlled a power transistor for switching the current on and off in the ignition coil.

In further decisive steps toward nonwearing triggering of ignition, the transistorized ignition with either a Hall effect or inductive pickup was developed. An electronic, noncontacting triggering principle replaced the distributor contact breaker points.

In transistorized ignition with a Hall generator, a co-rotating, pot-shaped rotor with mask was fitted on the distributor shaft (**Fig. I26**).

The vane-type rotor interrupted the magnetic field in the vane area and with that the Hall voltage. With the Hall voltage switched off, downstream power electronics could switch on the primary circuit of the ignition coil, charging it. When the vane opened the gap again, the magnetic flux reached the Hall ignition coil again.

The Hall voltage increased and switched off the primary current via the power transistor, thereby triggering the ignition spark. The width of the vane elements on the vane-type rotor determined the maximum dwell angle. This would normally be significantly reduced at low speed by electronics to protect the ignition coil from overheating.

In transistor ignition systems with an inductive pickup, a soft-magnetic star rotor (pulse generator wheel) rotated on the distributor shaft and generated an AC voltage in externally fixed coils (**Fig. I27**). This voltage was additionally amplified by means of downstream permanent magnets. The electronic ignition system generated from that a square-wave signal for triggering the spark. The dwell angle is controlled through voltage evaluation at a low-value resistor in the ignition transistor emitter line and intervention at the trigger level of the inductive pickup signal.

Ignition temperature (*also*, →Ferrocene; →Inflammation). The ignition temperature is a physical property of the various fuels. It is the lowest temperature at which auto-ignition of fuel takes place in an open vessel. This value is not the immediate determining factor for engine combustion. Its importance is in the field of safety-related aspects. The ignition temperature of gasoline fuels is at 220°C . . . 300°C . . . 450°C and that of diesel fuels is around 230°C.

Ignition timing →Ignition system, gasoline engine ~Ignition

Ignition timing map →Performance characteristic maps ~Ignition and injection maps

Ignition timing sensor →Ignition system, gasoline engine

Ignition timing/determination →Electronic open- and closed-loop control ~Electronic open- and closed-loop control, gasoline engine ~~Functions

Ignition voltage →Ignition system, gasoline engine ~Ignition

ILSAC →Oil ~Classification, specifications, and quality requirements ~~Institutions

Immissions. Immissions are negative environmental effects, primarily from anthropogenic sources. They can include air pollution, noise, radiation, heat energy, and so on. Immissions are the factors that have an effect on the environment (humans, animals, plants, materials, etc.). Emissions are the cause of immissions— for example, air pollutants that are emitted by a facility and that are transmitted by transport mechanisms (e.g., the atmosphere) to the acceptors (humans, animals, plants, etc.).

Immobilizer →Electronic open- and closed-loop control ~Electronic open- and closed-loop control, gasoline engine ~~Functions

Impedance →Engine acoustics ~Acoustic impedance, ~Input impedance

Impermeability →Electronic/mechanical engine and transmission control

Impregnations →Filter ~Intake air filter ~~Filter media

In situ process for aerosol characterization →Particles (Particulates) ~Particle measuring

Included angle between cylinder banks →Engine concepts ~Reciprocating piston engines ~~Single-shaft engines ~~~V-engines ~~~~V-angle

Incomplete combustion →Combustion process ~Incomplete combustion

Inconvenience →Engine acoustics

Indicated efficiency →Engine efficiency ~Internal efficiency

Indicated fuel consumption →Fuel consumption ~Internal fuel consumption

Indicated power →Power output

Indicated specific work →Mean effective pressure

Indicating →Friction ~Friction measuring techniques

Indicator diagram. The indicator diagram is the pressure curve in the working chamber of an engine in relation to the crank angle or the displacement. It is possible to determine the indicated engine efficiency with the help of this diagram. The indicator diagram is normally determined by measurements with a pressure sensor, which measures the pressure in the combustion chamber. A relationship between the crank angle and the corresponding pressure is established at the same time. The actual volume of the cylinder is related to the crank angle.

Indirect drive →Valve gear ~Actuation of valves

Indirect injection →Injection system, fuel ~Gasoline engine ~~Intake manifold injection systems

Indolene →Emission measurements ~Test type I ~~Vehicle preparation

Induction tract →Intake system ~Intake manifold ~~Intake ports

Inductive load →Electronic/mechanical engine and transmission control ~Electronic components ~~Power transistor

Inductive sensors →Electronic/mechanical engine and transmission control ~Electronic components ~~Sensors

Industrial engines →Lubrication ~Lubricating systems ~~Truck and smaller industrial engines

Indy Car →Engine ~Racing engines

Indy engine software →Engine ~Racing engines ~~Indy Car

Inertia →Balancing of masses

Inertia effect →Balancing of masses

Inertia force →Balancing of masses; →Crankshaft drive; →Piston rings

Inertial force balance →Balancing of masses; →Crankshaft ~Balance weights

Inflammation (*also*, →Air-fuel mixture). The inflammation (combustion) of the air-fuel mixture in gasoline engines is started by the ignition. The spark-over caused by an electric spark in the mixture concerned leads to first thermal reactions between the electrodes of the spark plug at temperatures between 3000 and 6000°C. Only a small part of the reaction is generated by the impact of heat; the bulk of the impact comes from molecular stimulation and ionization. In addition, the air-fuel ratio must be within the ignition limits. Current gasoline engine fuel mixtures are ignitable with an air-fuel equivalence ratio of 0.6:1.6 (lower and upper ignition limits). The ignition limits for hydrogen, however, are between 0.2 and 9. A short duration spark is unfavorable for the inflammation of lean mixtures because of the low probability that a combustible mixture will be present between the electrodes of the spark plug.

Diesel engines usually work with locally excess air (in the proximity of the fuel droplets and, therefore, the ignition source), but with a similar air-fuel ratio and ignition limits as gasoline engines. The inflammation after ignition requires a finite time, called the ignition delay. The inflammation is defined as the first appearance of a visible flame or the start of the pressure increase above the polytropic compression progression in the combustion chamber. The ignition delay is defined as the time, or the crank angle, from the time of ignition to the controlled inflammation of the mixture.

The reasons for the ignition delay in a spark-ignition engine are as follows: The spark generates reactions. These accelerate after a certain time in such a pronounced manner that a significant mixture conversion occurs. **Fig. I28** shows the impact of different spark durations on the ignition delay for a given air-fuel ratio. Chemical energy is released during the ignition delay period by the conversion of the mixture within and on the surface of the first flame. Heat conduction, radiation, and convection withdraw energy from the flame core at the same time. The excess energy guarantees the development and propagation of the flame. The flame speed increases proportionally to the energy released. The initial flame gets extinguished if the energy losses are too high.

Literature: A. Urlaub: Verbrennungsmotoren, 2nd Edition, Springer Publishers, 1994.

Infrared analyzer →Exhaust gas analysis equipment ~Nondispersive infra-red absorption analyzer (NDIR)

Initial boiling point →Fuel, gasoline engine ~Boiling characteristics

Initial reaction →Combustion, diesel engine ~Auto-ignition

Initial separation rate →Filter ~Intake air filter ~~Filter characteristics

Injection (air) →Mixture formation ~Mixture formation, gasoline engine ~~Mix-ture injection

Injection angle →Injection valves ~Diesel engine ~~Injection nozzle parameter ~~~Spray angle

Injection curve →Combustion, diesel engine ~Combustion process

Injection functions. Specific parameters must be tuned to operate the engine and to satisfy the specification in terms of performance, torque, noise, fuel consumption, emissions, and so on.

~Diesel engine →Electronic open- and closed-loop control ~Electronic open- and closed-loop control, diesel engine

~Gasoline engine. The category of gasoline engines can be divided into engines with direct injection and with intake manifold injection.

~~Direct injection

~~~Idle volume. The idle volume is the injection quantity required by the unloaded engine at idle speed. The flow through the injection valve must be designed in such a way that the idle volume can be injected in the linear range of operation of the valve, under consideration of the tank venting volume.

~~~Injection duration. The injection duration is the period in degrees of crank angle or in ms during which the injection valve is open and the fuel gets injected into the combustion chamber. The injection duration is calculated by the engine control system based on the required volume to be injected in relationship to the flow of the injection valve [mg/ms] and the injection pressure. The high-pressure injection valve must be designed for static flow in such a way that the required fuel volume for the peak engine performance can be injected during the maximum injection duration (start minus end of injection). The injection duration must be

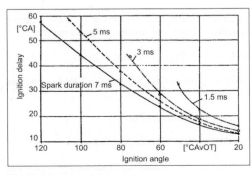

Fig. I28 Ignition delay

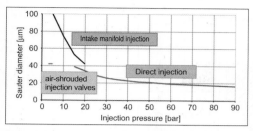

Fig. I29 Injection pressure and atomization quality

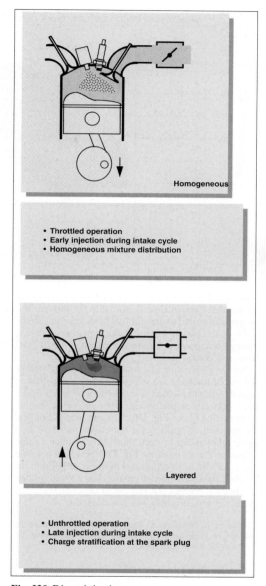

Homogeneous

• **Throttled operation**
• **Early injection during intake cycle**
• **Homogeneous mixture distribution**

Layered

• **Unthrottled operation**
• **Late injection during intake cycle**
• **Charge stratification at the spark plug**

Fig. I30 Direct injection process

adjusted according to a quadratic function (flow equation, Bernoulli), if the injection pressure is changed.

~~~**Injection pressure.** The fuel atomization process improves with increasing injection pressure. **Fig. I29** shows the atomization quality of a high-pressure injection valve with swirl atomizer in comparison to a low-pressure injection valve and an air-encompassed injection valve.

Variable injection pressures in the range of 5 to 12 MPa are used for wall and air guided combustion processes. Injection pressures of up to 20 MPa are used for spray-guided combustion processes. The injection pressure must be optimized in the performance map of the engine to achieve stable engine operation without misfires and low exhaust emissions. The motive power of the high-pressure pump also increases with increasing injection pressure, which increases the "friction" losses of the engine and in turn the fuel consumption.

~~~**Metered fuel injection.** Metered fuel injection is differentiated as either homogenous or stratified engine operation (**Fig. I30**). Similar to intake manifold injection, the volume of the cylinder at the time of injection for homogenous engine operation is filled with fresh charge in such a way that a homogenous exhaust gas is present in the exhaust system for an optimum conversion of the exhaust gases of the three way catalytic converter. The fresh cylinder charge is calculated in such a way that the required torque can be generated at the crankshaft. This load control process is called quantitative control and is also used for engines with intake manifold injections. The injection volume for the stratified engine operation is calculated from the required torque at the crankshaft. This load control process is called qualitative control and is also used in diesel engines. Lean catalytic converters are required to reduce the oxides of nitrogen from gasoline engines to the legally required emission limits. Engine and transmission control systems with torque-based functional structures are used to guarantee the transition between homogenous and stratified engine operation without a change of torque at the crankshaft.

~~~**Start and end of injection.** For direct injection, both values depend on the engine operating type.

The start of injection for preparation of a homogenous mixture is selected in such a way that enough time is available for the mixing of fuel and air. The injection is performed during the intake cycle of the engine (**Fig. I30**).

The homogenous distribution of fuel and air without attachment of fuel to a combustion wall is the target for good emission values and low soot emissions. The start of injection for engines with direct injection must be performed after closing the exhaust valve to avoid scavenging losses of unburned fuel into the exhaust

hegma 025.299 PP466-150V 3,000 1/min 0.4 kJ/dm³ single inj. GFK-2mm

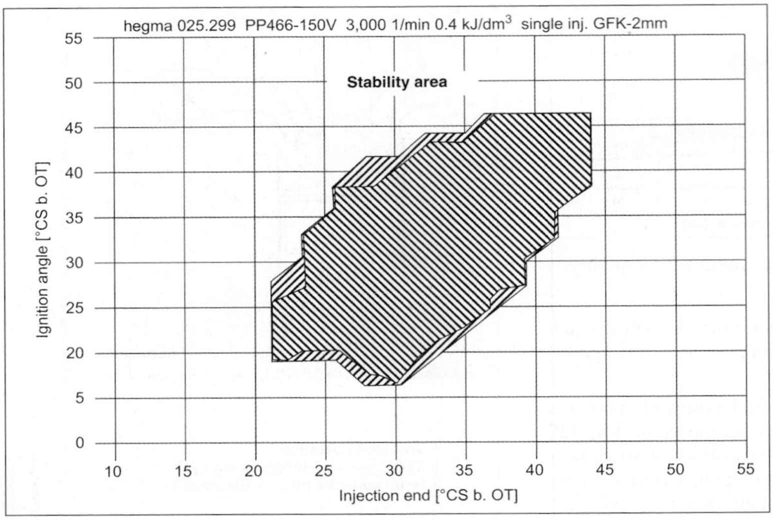

Stability area

Ignition angle [°CS b. OT]

Injection end [°CS b. OT]

Fig. I31
Relationship between
end of injection and
ignition angle

manifold. The end of injection must be early enough to ensure that enough time is available for mixture homogenization and to prevent fuel from attaching to the walls of the piston surface—soot emissions increase during homogeneous engine operations if the fuel adheres to a combustion wall.

For the preparation of a stratified mixture the start or the end of injection are selected in such a way that an ignitable mixture cloud is available around the spark plug. Injection is performed at the end of the engine compression cycle (**Fig. I30**). The end of injection and the moment of ignition for the engine must be in a close relationship to each other to guarantee engine operation free of misfires. **Fig. I31** shows the possible range for end of injection and moment of ignition for engine operations free of misfires for a spray-guided combustion process (1). The end of injection is a result of start of injection and injection duration.

Literature: K. Fröhlich, K. Borgmann, J. Liebl: Potenziale zukünftiger Verbrauchstechnologien, 24th International Vienna Engine Symposium, 2003.

~~**Intake manifold injection**

~~~**Full-throttle injection quantity.** The required full-throttle injection quantity, Q_{max}, depends primarily on the following parameters:

n_{max} = Maximum engine speed
ρ = Fuel density [g/cm²]
b_e = Specific engine fuel consumption [g/kWh]
P = Power [kW]
z = Number of cylinders
n = Engine speed [min⁻¹]
Q_{max} = Full-throttle injection quantity

$$Q_{max} = \frac{P \cdot b_e}{60 \cdot \rho \cdot n \cdot z}$$

~~~**Idle injection volume.** The idle injection volume is designed for the unloaded engine at target idle speed. Optimum raw exhaust emissions and a constant engine speed due to high accuracy and repeatability of the metered injection quantity are most important for this design step. Good mixing of the intake air with the fuel is necessary for the achievement of optimum values.

Stringent requirements for minimum metered quantity and its controllability are the result of specifications requiring the lowest possible fuel consumption level and concomitantly low injection quantities due to reduction of the target idle speed.

~~~**Injection duration.** The injection duration is determined by the metered injection volume. The required (to be adjusted) injection duration for the most common solenoid valve controlled injection systems is determined by the design of the injection cross section and the defined or adjusted system injection pressure between rail and intake manifold.

~~~**Injection pressure.** A minimum fuel pressure compared with that in the intake manifold is used for the design of the mostly fixed injection pressure to avoid hot-start problems due to degassing of fuel. The mixture may become leaner because of the degassing of the fuel. This can in turn result in hot-start problems. Injection pressures of more than 3 bar differential pressure compared to that in the intake manifold are adequate in most cases for currently available fuel qualities.

~~~**Injection quantity.** The required injection quantity depends on the engine operating point, and the adjustment of the desired air-fuel mixture is determined by the specifications for the control time for the solenoid valve required for the injection quantity.

618

The calculation is based on the injection valve performance map.

The required injection quantity depends primarily on the engine operating point (e.g., engine speed and load) and the engine operating status (e.g., engine warm-up), as well as the adjusted injection pressure.

In addition, the control time for the solenoid valve required for the adjustment of the desired injection quantity is defined by the system characteristics of the solenoid coil of the injection valve depending on the control frequency.

All these influences are calculated in the engine management system and the desired fuel quantity is metered in accordance with the operating point.

The absolute minimum required fuel quantity is taken into consideration for the selection and the design of the required injection quantities.

~~~**Start of injection.** The start of injection is primarily determined by the intake valve timing. Injections into the open period of the intake valve should be avoided to achieve low HC emissions. This is especially important during cold starting of the engine. The position of start of injection is otherwise not critical for the intake manifold injector.

Injection hydraulics, diesel engine →Injection system (components) ~Diesel engine

Injection law, diesel engine →Combustion, diesel engine ~Combustion process

Injection line →Injection system (components) ~Diesel engine

Injection maps →Performance characteristic maps ~Ignition and injection maps

Injection period →Injection functions ~Gasoline engine ~~Direct injection, ~~Intake manifold injection; →Injection valves ~Diesel engine ~~Fuel-injector body, ~~Injector nozzle parameter ~~~Opening pressure

Injection point →Injection functions ~Gasoline engine ~~Direct injection ~~~Start and end of injection; →Injection system (components) ~Diesel engine ~~Injection timing device

Injection pressure →Injection functions ~Gasoline engine ~~Direct injection, ~~Intake manifold injection

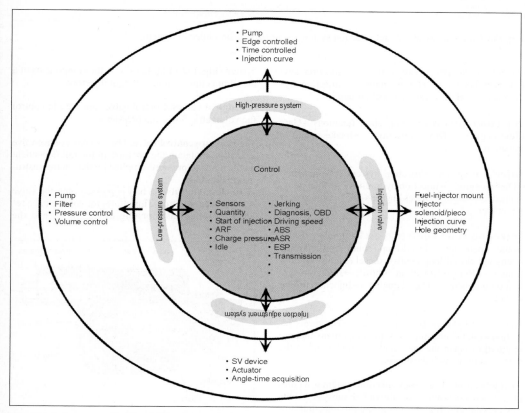

Fig. I32 Subsystems of a diesel injection system

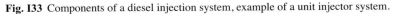

A Fuel supply (low pressure part)
1 Fuel container
2 Fuel filter
3 Fuel pump with return valve
4 Pressure limiting valve
5 Fuel cooler

B High-pressure part
6 Unit Injector

C Electronic diesel control (EDC)
7 Fuel temperature sensor
8 Control instrument
9 Accelerator pedal sensor
10 Driving speed (inductive)
11 Brake contacts
12 Air temperature sensor
13 Camshaft speed sensor (Hall sensor)
14 Intake air temperature sensor
15 Charge pressure sensor
16 Intake manifold flap
17 Hot film air mass sensor
18 Engine temperature sensor (coolant)
19 Crankshaft speed sensor (inductive)

D Peripherals
20 Instrument cluster with signal output
   for fuel consumption, speed, etc.
21 Glow time control instrument
22 Pencil-type glow plug
23 Clutch switch
24 Controls for cruise control (FGR)
25 Air-conditioning compressor
26 Controls for air-conditioning compressor
27 Driving switch (glow start switch)
28 Diagnosis interface
29 Battery
30 Turbocharger

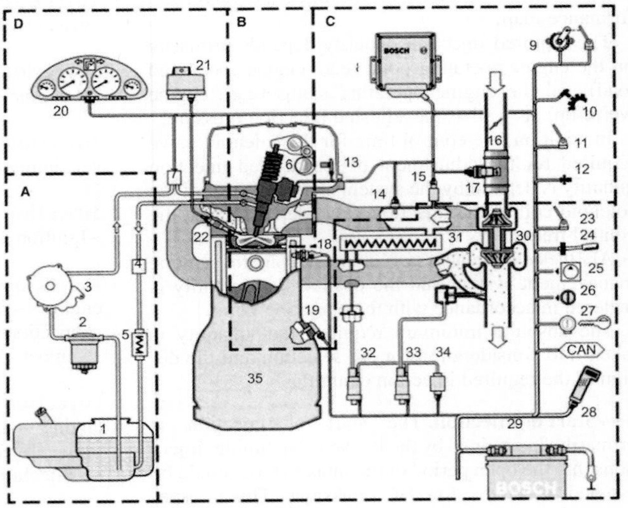

31 Exhaust gas recirculation cooler
32 Exhaust gas recirculation actuator
33 Charge pressure actuator
34 Vacuum pump
35 Engine
CAN CAN Controller Area Network
   (serial data bus in vehicle)

Fig. I33 Components of a diesel injection system, example of a unit injector system.

Injection pump, diesel engine →Injection system (components) ~Diesel engine ~~Injection hydraulics ~~~High-pressure injection pump

Injection quantities →Injection functions ~Gasoline engine ~~Direct injection, ~~Intake manifold injection

Injection spray →Injection system (components) ~Diesel engine ~~Injection hydraulics ~~~Fuel spray; →Injection valves ~Diesel engine ~~Injection nozzle parameter

Injection start control diesel engine →Electronic open- and closed-loop control ~Electronic open- and closed-loop control, diesel engine ~~Functions ~~~Start of injection control; →Injection system (components) ~Diesel engine ~~Injection timing device

Injection system (components). The term "injection system" encompasses all components that are required to inject the fuel, correctly timed and in the correct volume, into the engine.

~Diesel engine. The diesel injection system is a complex system consisting basically of five subsystems: the low pressure, the high pressure, and the injection timing systems, as well as the injection valve and a control

system (**Fig. I32**). **Fig. I33** shows the components of a diesel injection system with unit injectors.

~~Control valves. Control valves are used to control the volumetric flows and pressures.

~~~Pressure control valve. The pressure control valve is used to maintain the pressure in the rail dependent on the demand pressure supplied by the control instrument (**Fig. I34**).

To close the valve, a ball is pressed onto the valve seat by solenoid force. This allows for volume to be released through fast opening of the valve, and the

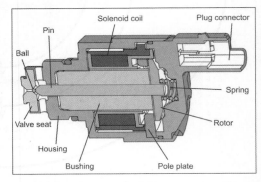

Fig. I34 Pressure control valve

pressure can be lowered in this way. This provides a safety function to guard against overpressure in the rail, and also the pressure in the pump or the rail can be quickly reduced. This function is used, for example, during the transition from load operation to overrun of the engine.

Both valves are also used to switch the engine off, which is achieved by closing the volumetric flow control valve and opening the pressure control valve. This results in a situation in which the pump cannot build up high pressure anymore and the existing pressure in the rail is subsequently reduced.

~~~Volumetric flow control valve. Throttle slides are actuated by a solenoid. The changed cross section creates a volume flow changeable in a range of 20 to 100%, which is supplied to the high-pressure part of the CR pump. The design goal is to achieve a defined characteristic curve with low hysteresis (**Fig. I35**).

~~Diesel control →Electronic open- and closed-loop control ~Electronic open- and closed-loop control, diesel engine

~~Fuel filter (*also*, →Filter). The diesel fuel filter is used to remove harmful impurities from the fuel. Modern diesel injection systems need very clean fuel to satisfy the challenging requirements with respect to pressure generation and injection accuracy for the high-pressure pump and the injection valves. The filter removes particle contamination and free water content from the fuel. This prevents wear and corrosion in the injection system and ensures a long product life. Particles with a size of only a few micrometers can generate significant wear. This is why the use of filter paper with initial separation rates of almost 95% for particle sizes of 3–5 μm in accordance with ISO TR 13353 (1994) is mandatory for diesel common rail injection systems.

~~Fuel injection line. The fuel injection line is used to transport the injection volume from the injection pump to the injector nozzle. The length of the injection line causes delays in the start of injection relative to the start of delivery depending on the speed (injection timing device).

The injection lines for multicylinder engines must have the same lengths to achieve the same injection

conditions for all cylinders. Length and internal diameter of the injection line are important design parameters. Pressure losses due to restrictions result from internal diameters that are too small, and losses due to compression effect come from diameters that are too large.

Careful design of the line length can generate an injection pressure at the injector nozzle that is up to 300 bar higher than the pressure at the injection pump. This effect is based on the wave effects in the pressure of the injection volume and its reflection at the injector nozzle; this pressure difference increases with increasing speed. It is, therefore, especially pronounced in systems for high-speed diesel engines. The effect increases the effective fuel preparation differential pressure at the nozzle without an additional load on the injection pump. Injection lines can safely be designed for 2000 bar peak pressure and the important parameters for the design are the quality of the inside surface (SN < 20 μm) and the quality of the material (St52 quality P DIN 2391).

It becomes increasingly difficult to adjust the injection hydraulics to avoid dribbling of the fuel, cavitation, and hydraulic instabilities due to temporary cavities as the length and volume of the line increase [3].

~~Fuel return. Depending on the injection system, the low-pressure primer pumps and the high-pressure pumps at times deliver significantly more fuel than required for injection. For example, for vane-type fuel pumps in distributor injection pump systems, a part of the excess supply (a multiple of the injection quantity) is used for cooling and lubrication of the high-pressure pump and for the adjustment of the injection timing device. This fuel is normally returned to the tank, but it can also be supplied to the hydraulic system in areas that are close to the pump. For very high pressures in the unit injector and in some common rail applications in cars, the fuel can be heated to such an extent by the relief process that it must be cooled with a special fuel cooling circuit before it is returned to the tank. Other return quantities (comparably small) are generated in sealing gaps in high-pressure pumps, nozzles, and injectors. This quantity will also be returned to the low-pressure circuit or into the fuel tank.

~~Fuel-feed pump. The fuel-feed pump is used to supply the high-pressure pump with the required fuel quantity and to prepressurize it. It uses a vane-type pump and is integrated into the high-pressure pump. This is why it is also called an "integrated primer pump." It sucks the fuel from the fuel tank through the connection lines of the low-pressure system and the fuel filter. The excess fuel quantity that is not accepted by the high-pressure pump is returned through a pressure-limiting valve to the suction side of the fuel-feed pump. The pump limiting valve produces the required prepressure for the high-pressure pump. The installation of an electrical primer pump is required for high delivery rates to supply the integrated primer pump with a defined overpressure.

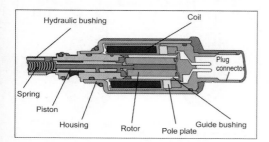

Fig. I35 Volumetric flow control valve

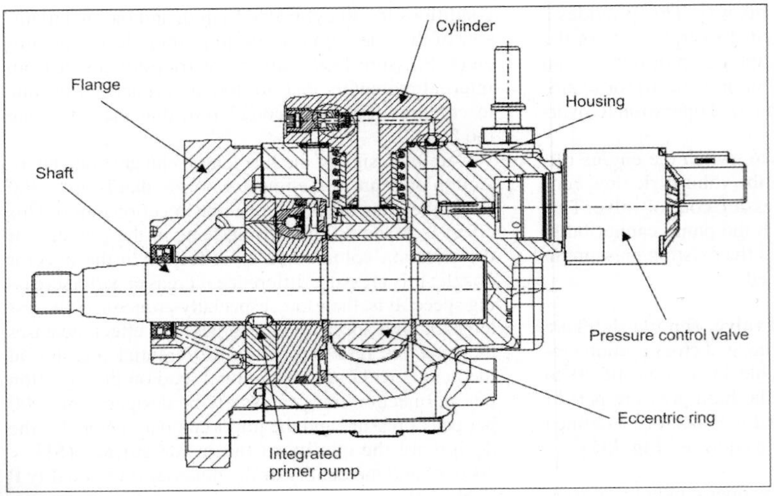

Cylinder

Flange

Shaft

Housing

Pressure control valve

Eccentric ring

Integrated
primer pump

Fig. I36
Design of a
high-pressure pump

~~High-pressure pump. The high-pressure pump for common rail systems must provide the high-pressure rail with diesel fuel in adequate volumes and at an adequate pressure, currently up to 1500 bar (**Fig. I36**).

It is powered by the engine through belts, chains, gears, or transmission. The normal transmission ratios n_{pump}/n_{engine} are 1/2 or 2/3. The pump normally gets lubricated by the fuel.

The pump consists of three pumping cylinders that are positioned on a radius, one eccentric ring, one integrated primer pump, two control valves (volume flow and pressure-control valves), and a housing.

The integrated primer pump sucks the fuel from the tank and is supplied by an electrical in-tank pump. This produces pressures of up to 7 bar.

A pressure-limiting valve is installed to prevent the pressures in the primer pump from becoming too high. The pre-pressurized fuel volume is controlled in the downstream volumetric flow control valve from 20 to 100% of the maximum volume. The advantages of this concept are found in the partial delivery range, because this process supplies only the required volume to the rail of the high-pressure cylinders, which results in a high total efficiency of the pump and in lower motive powers or in lower driving torques in this important operating range.

The suction valve sucks the fuel into the cylinder during the downward movement of the piston (**Fig. I37**).

The pressure valve, with rail pressure on its secondary side, is closed for this part of the process. The fuel is pressurized during the upward movement and is delivered through the pressure valve into the collection bore of the pump housing. The suction valve is closed for this part of the process. The pressure valve is also positioned in the collection bore of the housing.

The three-cylinder concept of the high-pressure pump delivers a relatively constant volumetric flow as the result of the overlapping suction and pressure

phases for the individual high-pressure cylinders. This in turn results in little fluctuation of the drive torque for the high-pressure pump. Although the torque fluctuations increase in the partial delivery range, they are significantly less critical than in cam-controlled pumps.

The goal for the pump design is high efficiency, which consists of the mechanical and the volumetric efficiency.

The mechanical efficiency is determined by the friction losses in the bearings and bearing surfaces of the eccentric ring (typical value is greater than 80%). The volumetric efficiency (typical value is greater than 85%) is a result of the closing times from the suction and pressure valves of the high-pressure cylinder and the leakage between piston and cylinder bore.

This is the reason why the pistons, which have no piston rings for sealing, are paired in the cylinder with a clearance between 2 and 4 μm.

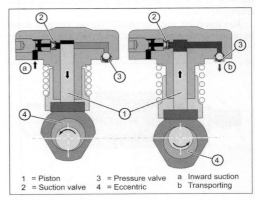

| 1 = Piston | 3 = Pressure valve | a Inward suction |
| 2 = Suction valve | 4 = Eccentric | b Transporting |

Fig. I37 Cylinder function

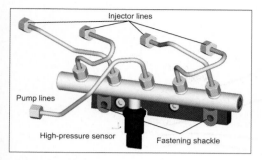

Fig. I38 High-pressure rail

~~**High-pressure rail.** This rail has the task of storing the fuel under high pressure and damping the pressure fluctuations as effectively as possible (**Fig. I38**). The pressure fluctuations are introduced into the rail by the pumps and the injectors. The pressure should be maintained as constantly as possible and it should follow the pressure demand from the control system as closely as possible. The rail has an inflow connection to the high-pressure pump, connections for the injector lines, and a rail pressure sensor. The rail volume is an application variable. An approximate reference value would be 20 cm^3 for a conventional 2-L engine. V-engines normally use one rail for each cylinder bank to achieve lines from the rail to the injectors that are as short as possible.

~~**Injection hydraulics.** The term "injection hydraulics" describes the processes on the high-pressure side of a diesel injection system.

~~~**Cavitation.** Cavitation is the development of vapor bubbles in hydraulic systems by local pressure changes and their subsequent implosion; these are connected to erosion effects in the system (cavitation wear).

Many opportunities exist in high-pressure injection systems for the development of time dependent and "local" vacuums (control process, shutting of valves, pump processes between moving gaps, low-pressure waves in bores and lines). These gas bubbles implode in the subsequent high-pressure phases. The high-energy density as the bubbles collapse can result, after a time, in surface erosion (cavitation wear) if the bubbles are close to a wall. This process is shown schematically in **Fig. I39**. The cavitational impact is not always found at the point where the bubble was formed, because the gas bubbles are transported by the flow; cavitation can often be found in "dead water zones." Cavitations can be prevented to a limited extent by improving the material quality or hardening and the surface. The goal must be to avoid the generation of gas bubbles and to avoid their negative impact through optimization of the flow conditions.

Gas and air bubbles can result in dynamic irregularities during the volume metering. These so-called hydraulic instabilities are normally generated by over-relief of the line system and the pressure valve [3].

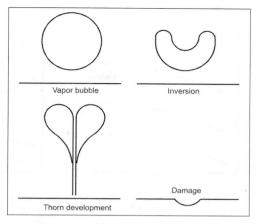

Fig. I39 Imploding of a cavitation bubble [3]

~~~**Compression volume.** Compression volume is the sum of the volume on the high-pressure side, which must be compressed during each injection, and the volume when it is relieved again at the end of injection.

The compact dead volume has an impact on the pump or nozzle side, but it is different from the filiform volume in the injection line. This will only be compressed locally because of the dynamic processes due the pressure wave. The hydraulic efficiency of the injection volume decreases if the compression volume increases.

The progress of injection in a system with large compression volume gets delayed by the compression losses. Similarly, the progress of injection follows the speed changes of the camshaft better in a system with small compression volume (e.g., unit injector).

The compression volume in a common rail system has no significant impact on the injection process as described above.

~~~**Delivery characteristics.** The delivery characteristic of an injection pump is the variation of the injected fuel volume in relation to the speed at constant target volumes (e.g., for constant control rod travel, constant control sleeve position, or constant timing of a solenoid valve). This results in a map of volume against speed with constant target volume setting being the other parameter.

A pump with volume measurement (edge controlled) through a control edge and constant volume relief valve generally has a high-pressure pump with constant pressure release with increasing speed, but declining volume characteristics (**Fig. I40** [1], [2]).

The most important parameters having an impact on the delivery characteristics are:

- Leakage losses (decline strongly with speed)
- Compression losses (increase strongly with speed [pressure])
- Drive elasticity (increases strongly with speed [pressure])

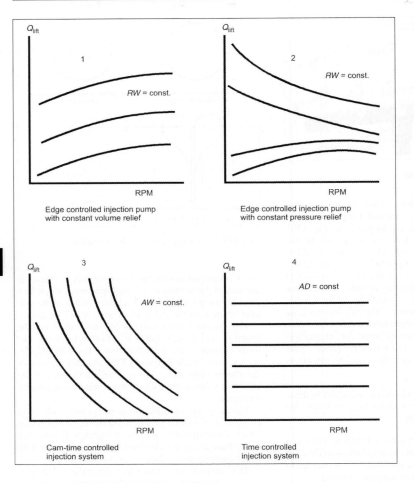

Fig. I40
Delivery
characteristics [3]

- Pre/post delivery (increases with speed)
- Constant volume relief (declines with speed)
- Constant pressure relief (increases strongly with speed in full-load range).

For constant control angles (control time declines with increasing speed), a cam-time controlled injection system provides delivery characteristics with hyperbolically reducing characteristic curves (**Fig. I40** [3]).

Characteristics that are independent of speed are the result of solely time-controlled injection systems (pressure constant, common rail) (**Fig. I40** [4]).

~~~**Fuel compressibility.** The compressibility of diesel fuel must be taken into consideration for the pressure ranges of present-day high-pressure injection systems. The following applies:

$$\Delta V = V_0 \cdot \Delta p \cdot 1 / E_k$$

and

$$c_s = \sqrt{\frac{E_k}{\rho_k}},$$

where ΔV = volume change through compression of the initial volume, V_0, Δp = pressure differential, E_k = compressibility of fuel $\approx$ 20,000 bar (pressure and temperature dependent), c_s = speed of sound in diesel fuel $\approx$ 1500 m/s (pressure and temperature dependent), and ρ_K = fuel density = 0.88kg/dm³.

For example, a difference volume of $\Delta V = 5$ mm³ is "stored" at a differential pressure of $\Delta p = 100$ bar in a volume of $V_0 = 1$ cm³.

The pump must be designed larger than its delivery rate to compensate for the compression losses. The compression losses are converted to heat during the relief, and they have a significant impact on the hydraulic efficiency.

~~~**Fuel spray.** The fuel spray develops after the fuel exits the nozzle. A single conical, fan-shaped spray develops in a pintle-type nozzle. The number of cone-shaped sprays in orifice nozzles for engines with direct injection is equivalent to the number of bores (**Fig. I41**). The fuel spray disintegrates into small droplets when it enters the combustion chamber. The spray dis-

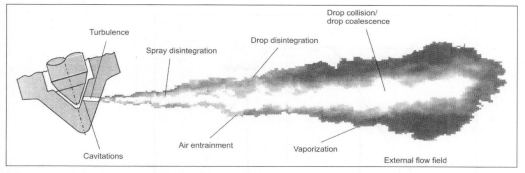

Fig. I41 Spray development in an orifice nozzle [4]

integration that begins at exit from the nozzle bore and immediately thereafter is called primary disintegration.

Even finer droplets develop under the influence of the surrounding fluid (air entrainment), and they evaporate under pressure and temperature; in addition, collisions and coagulations occur based on the high droplet density. The interaction between the droplets and the air results in a pulse exchange, which reduces the spray momentum and increases the gas movement. **Fig. I42** shows the penetration depth of the fluid and vaporized spray tip in relation to the injection pressure, measured in a high-pressure chamber. The spray propagation during the injection process is shown in **Fig. I43**. The development and propagation of the spray can be computed with the help of CFD (computational fluid dynamics) simulations; they can be analyzed in injection/combustion chambers or measured with laser optics in so-called transparent engines. The spray momentum can be measured with a force sensor, which is moved, together with a deflector plate, directly into the fuel spray. The spray momentum is a result of the integral of the force with time, as

$$I = \int F\,dt.$$

The spray momentum is also defined as the product of the mass flow and its velocity, $I = mv$, which means the exit speed of the spray can be determined from the momentum if the mass is known [5].

~~~**High-pressure injection pump.** Piston pumps are used in general for the generation of high injection pressures, because only they have a sufficiently tight piston fit (2–5 μm) and the required tolerances to achieve the required hydraulic efficiency connected with high pressures. This is the case irrespective of the different injection systems.

The piston pump must deliver a flow level adequate to generate the high pressure necessary to achieve the desired injection pressure independent of the required injection quantity and injection duration by subsequently throttling the injector nozzle cross section.

The delivery flow is calculated from

$$\dot{Q} = A_K \cdot v_K,$$

where A_K = piston surface area and v_K = piston speed.

The camshaft of the high-pressure injection pump is optimized to the highest possible piston speed during the delivery phase to achieve the highest possible delivery rate in cam-driven piston pumps. High-pressure pumps are generally designed with long strokes and relatively small piston diameters to enable control of the engine forces despite high peak pressures.

A good hydraulic efficiency is of great importance. It requires:

- Good tightness, which means small clearances, small deformations, long and fast overlapping of the control edges
- Minimum compression volume
- Low throttling inside the flow route

Relatively high peak driving torques are generated (maximum 100–600 N·m for direct injection) by high-pressure pumps independent of the cam design (pressure) and the cam stroke.

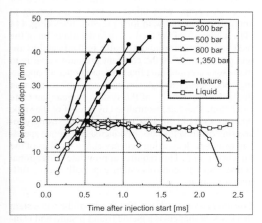

Fig. I42 Penetration depth of the fluid and vaporized spray tip in relation to the injection pressure, chamber temperature 900 K, chamber pressure 50 bar [4]

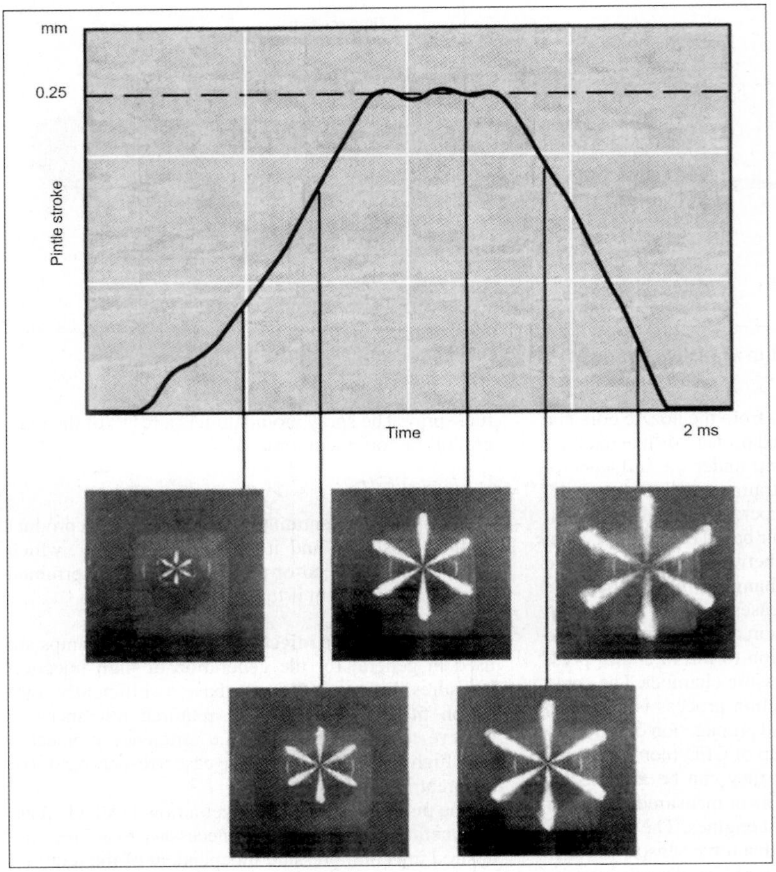

Fig. I43
Fuel spray generation correlated to pintle stroke [2]

High-pressure pumps for injection systems that do not need the injection pressures to be synchronous with the crankshaft position and the injection process, as in common rail systems, are generally designed with eccentrics instead of cams. A high piston speed is of little importance in this case (note: leakage), and the maximum driving torques are reduced to only 20–40 N·m.

~~~**Pressure valves.** A pressure valve is normally added between the injector pump and the jets in diesel injection systems with longer pressure lines. It is positioned directly at the pump outlet. The pressure valve is used to bring the pressure in the line and the nozzle chamber down to a certain pressure level, the so-called static pressure, and to separate the fuel high-pressure circuit between the pressure line and the pump pistons during the pause between deliveries. The pressure relief results in a fast and precise closing of the nozzle and prevents an undesired after-dribble of the fuel.

However, this particular function requires different solutions depending on the application (injection pressure, engine speed range, length of fuel line, and dead volume on the high-pressure side).

Constant pressure valve (CPV). The CPV is used for injection pumps with injection pressures starting at about 1000 bar and with long pressure lines (**Fig. I44** [1]). It consists of the feed valve in the direction of the feed and a pressure-holding valve in the backflow direction. The task of the constant pressure valve is to maintain a constant fuel line pressure under all operating conditions. Cavities can have a negative effect on the hydraulic stability of the system and can generate cavitation. These are neutralized by a relief of the static pressure to about 30–120 bar in the line. A strongly increasing overrelief occurs with increasing engine speeds and results in decreasing delivery characteristics of the injection pump. This requires appropriate correctional actions from the control system, especially during full-load operation.

Constant volume valve (CVV). A part of the valve stem in a CVV is shaped as a piston (relief piston) and is inserted with a small clearance into the valve carrier (**Fig. I44** [2]). This relief piston enters the valve carrier after the end of delivery and takes a certain volume from the pressure line. As a result, the pressure level in

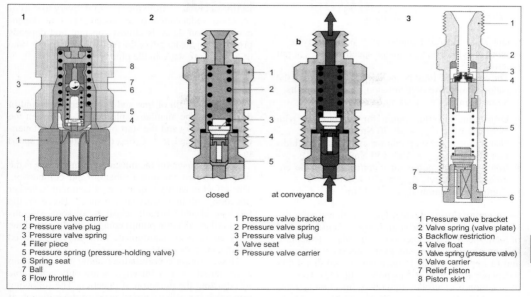

1 Pressure valve carrier
2 Pressure valve plug
3 Pressure valve spring
4 Filler piece
5 Pressure spring (pressure-holding valve)
6 Spring seat
7 Ball
8 Flow throttle

1 Pressure valve bracket
2 Pressure valve spring
3 Pressure valve plug
4 Valve seat
5 Pressure valve carrier

1 Pressure valve bracket
2 Valve spring (valve plate)
3 Backflow restriction
4 Valve float
5 Valve spring (pressure valve)
6 Valve carrier
7 Relief piston
8 Piston skirt

Fig. I44 Pressure valve designs [2]: 1, constant pressure valve; 2, constant volume valve; 3, constant volume valve with backflow restriction

the line is reduced, as required for a rapid end to the injection process and to avoid dribbling of the fuel. The valve seat separates the pressure line from the high-pressure chamber of the pump. The relief volume must be designed individually for each application. The application limit of a sole CVV is at injection pressures of about 700 bar; a design without overrelief is difficult to implement for higher pressures — that is, for lower engine speeds. Torque controlled valves are used in special cases to enable the desired delivery rate characteristics. They have an additional ground section at the relief piston. Constant volume valves are often combined with backflow restrictors (**Fig. I44** [3]).

Backflow restrictor (BR). The BR is often used in addition to the constant volume valve and is used to damp and neutralize the returning pressure waves that are generated during the closing of the nozzle (**Fig. I44** [3]). This reduces or prevents wear and tear and cavitation (cavity generation) in the high-pressure chamber. The BR is positioned in the upper part of the pressure valve holder — that is, between the constant volume valve and the line. The BR valve body has a small hole adapted to the particular application. It produces the desired damping of the returning pressure wave and, therefore, avoids further wave reflection to a large extent. The valve opens with its full stroke in the direction of delivery. It provides no damping nor does it influence the delivery characteristics in the forward flow direction. The BR increases the application limit of the constant volume relief to about 800–1000 bar.

Diesel engines with direct injection and injection pumps with high delivery rates (e.g., distributor injec-

tion pump, VP 44) have a backflow restrictor without an additional pressure valve. They are called "valveless pumps." The relief effect is comparable to a "leaking" pressure valve in the return direction. The lack of a return valve in the delivery direction can be a disadvantage with respect to the charging of the high-pressure valve. The pressure line must, therefore, be filled during the first startup, and special provisions must be implemented to avoid an emptying of the tank. The advantages of this solution are the low installation effort, the low pressure losses, and the low compression volume.

~~~**Pump drive power.** The following simplified derivation is valid for the pump drive power of cam-powered piston pumps [3]:

$$P = \left(\frac{V}{E_k} \cdot \frac{p2_{Pmax}}{2} + V_E * \frac{p_{Pmax}}{2} \right) \cdot np * z$$

where V = total compression volume on the high-pressure side of the injection system — i.e., volume content of the chamber filled with fuel, V_E = injection quantity, p_{Pmax} = max peak pressure on the pump side, E_K = compressibility of the fuel, n_P = pump speed, and z = number of cylinders.

The leakage was not included in the above approximate formula. Edge-controlled, high-pressure injection pumps have a drive efficiency of only about 30–40% due to the high leakage and deactivation losses, and the power needed to drive them can be several kW.

The following pump power is required for individual fuel displacement (incompressible) into the storage of a common rail injection system:

$$P = \frac{\pi}{4} \cdot D_P \cdot e \cdot z \cdot n_p \cdot p_{Rail}$$

where D_P = piston diameter, e = eccentricity, z = number of cylinders, n_p = piston speed, and p_{Rail} = rail pressure.

The peak torque must be included in the design of the injection pump drive (transmission, gears, drive belts). Some examples for peak torques:

- Distributor injection pump for car engines with swirl chamber: d_{max} = 30–70 Nm
- Distributor injection pump for car engines with direct injection: d_{max} = 100–150 Nm
- Inline injection pump for commercial vehicle engines with 2 L/cyl.: d_{max} = 350–600 Nm
- Common rail pump: d_{max} = 150 Nm.

~~Injection timing device. The injection timing device is used to adjust the start of injection independent of load, speed, and temperature. The goal is to achieve the optimum emission values for exhaust gas and noise and to improve the fuel consumption and drivability independent of the operating point (**Fig. I45**): The start of injection is controlled in modern injection systems (*also*, →Electronic open- and closed-loop control). The start of delivery (SD) is often used in inline systems (start of the fuel volume delivery in the pump) instead of the start of injection because it is actually easier to determine SD. This means that the SD actuator must also be able to compensate for the so-called in-

jection delay (time from the start of delivery to start of injection) independent of the speed. The time-dependent injection delay is almost constant and depends primarily on the length of the fuel line. The transit time of a pres- sure wave in the pipe is

$$T_W = \frac{L}{v},$$

where L = length of line and v_s = speed of sound in diesel fuel. The angular displacement φ between the start of delivery and the start of injection is, independent of the speed, $\varphi = T_W 6 n_p$, if n_p, is the pump speed in rpm.

Another reason for the injection timing device is the time-controlled mixture formation and combustion. This effect normally requires an adjustment to earlier injection with increasing engine speed. However, the mechanical and hydraulic elasticity of some systems (e.g., inline injection pump) compensates a part of the early adjustment requirement with increasing speed (increasing pressure).

These complex correlations show that optimum characteristic values for engines are only achievable by adjusting the initiation of injection. The following principles can achieve adjustments of the initiation of delivery and injection:

The angle between the engine output shaft and injection pump camshaft is adjusted by a flyweight-controlled adjuster of initiation of delivery. However, this robust and solely mechanical solution permits only a fixed, speed-dependent adaptation, with the exception of expensive mechanical-hydraulic solutions. The injection adjuster as a phase adjuster is positioned between the drive of the injection pump on the engine side and the pump itself. It is used in conventional inline injection pumps.

The initiation of delivery (changeable prestroke) in a control sleeve injection pump can be adjusted with low forces and independent of the volume adjustment by a control sleeve containing the control bore. This permits electronic control of start of injection. This principle permits the adjustment of start of injection even when the pump is at a standstill, and this is important for a start free from white smoke. Because of the changed pressure level, the volume adjustments must be corrected depending on the cam design.

The start of delivery or injection for cam-time controlled systems (solenoid valve-controlled systems) can also be adjusted by changing the initiation of control of the solenoid valve. This basic concept is similar to the mechanical control sleeve solution. It has advantages if only one actuator (solenoid valve) is required for the fuel volume and the injection adjustments—for example, unit injector and unit pump (*also*, →Injection functions). The prestroke adjustment at the cam is often not adequate for high-speed engines (e.g., in passenger cars) with line systems, so that an additional injection timing device is required (*also*, →Distributor injection pump).

The injection adjustment in solely time-controlled systems (common rail) is exclusively realized by controlling the actuator (solenoid valve or piezo).

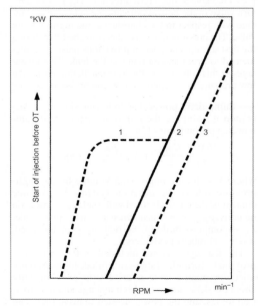

Fig. I45 Start of injection in relationp to speed and load in car engines at cold start and operating temperature (example) [2]: 1, cold start (<0°C); 2, full load; 3, part load

~~Injection valve (*also*, →Injection valves ~Diesel engines). The injection valve or the injector nozzle is the "interface" of the injection system with the combustion chamber of the diesel engine. It ensures that the fuel is injected into the combustion chamber in the form required; this depends on the combustion process, the combustion chamber geometry, the air movement, and the operating point (*also*, →Fuel spray; →Injection valves ~Diesel engine). The injection valve is normally installed with a fuel-injector mounting or directly into the injector (unit injector or common rail injector), which in turn is installed in the cylinder head of the diesel engine. Another task of the injection valve is the sealing of the fuel system against the combustion chamber during pauses in the injection process.

~~Mechanical control. Each diesel engine needs a speed controller (governor), because the performance map of the edge-controlled injection pumps (fuel quantity as a function of the speed for a specified load adjustment) does not correspond to the target performance map of the engine. This is especially true for idle speed and full load. A stable idle speed requires an increase in fuel quantity with declining speed or increasing load. However, the basic characteristics of most performance maps of mechanically controllable injection systems are not totally acceptable, and the speed controller compensates for this (delivery characteristics). The smoke limitation is, in addition to other influences, especially important during full load, which means that combustion must not generate visible smoke when mixture enrichment increases, and it is necessary that it must comply with the maximum allowable blackening number.

The diesel engine also requires reliable control of maximum engine speed; otherwise, due to its unthrottled operation and the delivery quantity characteristics of the edge-controlled pump, it could accelerate to a level of mechanical destruction. The processing of correction variables and application-specific functional values must also be performed by the diesel controller.

~~~Control maps. A number of different target performance maps are the result of different diesel engine applications independent of the specific injection systems.

~~~~Combination governor. The combination governor has a mixed performance map consisting of a variable speed governor and an uncontrolled range comparable to the final-idle speed controller. End speed control and idle speed control are of course available. The intermediate speed control can be in the upper or the lower speed range.

~~~~Engine controller. The engine controller (governor) is used for power generation systems. A simple final speed controller or an idle speed controller with a design for a defined speed—for example, the generator speed—is often adequate. The target speed for power generation systems must be maintained within narrow limits and can change only in accordance with the tolerance in DIN 6280, even under load changes.

~~~~Final-idle speed controller. The final-idle speed controller (also called the vehicle controller) controls the final speed in addition to the idle speed (**Fig. I46**). Its use is preferred in road vehicles, and it is also qualified for automatic transmissions due to its direct pick-off of a load signal at the control lever (e.g., by way of a potentiometer).

~~~~Variable speed governor. The variable speed governor (also called full-load controller) controls the engine speed over the whole speed range in addition to the maximum and the idle speed (**Fig. I47**). In addition to being used in road vehicles, this controller is also used in vehicles that need to maintain a certain engine speed or driving speed independent of the load—for

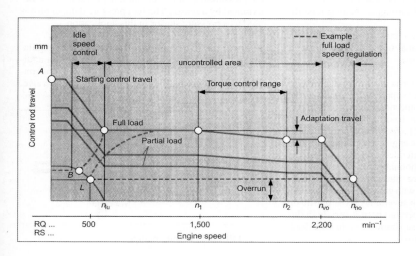

Fig. I46
Control map of a final-idle speed controller with adaptation of the full-load characteristic curve for commercial vehicles [2]

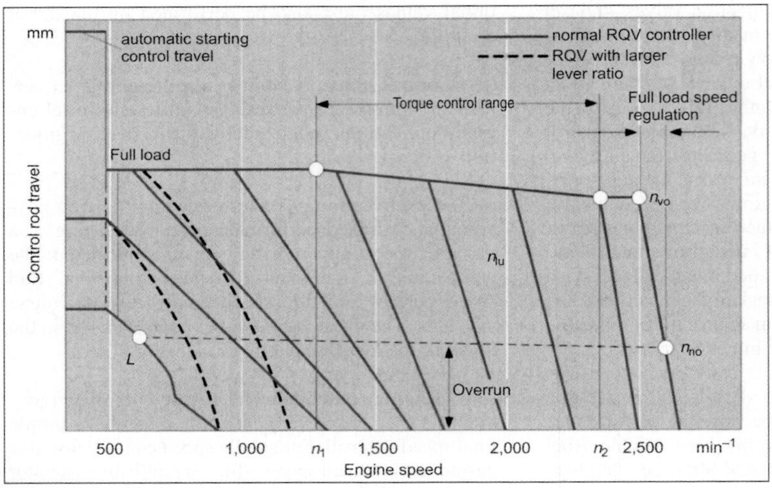

mm | automatic starting control travel — normal RQV controller / RQV with larger lever ratio

Torque control range — Full load speed regulation

Full load

Control rod travel

n_{vo}

n_u

L

n_{no}

Overrun

500   1,000 n_1 1,500   2,000 n_2 2,500   min⁻¹

Engine speed

Fig. I47
Control map of a
variable speed governor
RQV for commercial
vehicles [2]

example, in tractors, sweepers, boats, or vehicles with power takeoffs.

~~Primer pump. The electrical fuel primer pump is used to supply the fuel delivery pump in the high-pressure pump with a defined overpressure. It must increase the pressure level in the fuel volume required to one that prevents cavitation in the high-pressure pump. The electrical primer pump is positioned inside the swirl pot in the fuel tank and is a part of the fuel delivery unit. The pressure increase of the electrical primer pump is in a range of 0.5 to 1 bar.

~Gasoline engine

~~Diesel fuel tank. Diesel fuel shows almost no tendency to fuel fumigation. Emissions that develop during the refueling of a diesel fuel tank comprise the gas content, which is replaced in the tank by the fuel.

These gases are routed through a venting tube directly to the environment.

Ambient air flows into the tank during operation of the vehicle. It is proportional to the volume of diesel fuel consumed by the engine.

~~Fuel filter. The fuel filter is used to protect important components of the fuel supply system, such as the fuel pressure regulator and injection valves, from dirt particles and resulting wear; in diesel applications it is also used for water separation. The fuel filter is positioned between the fuel pump and the engine and justifies its presence by contributing to extended vehicle durability.

In addition to the conventional paper filter, newly developed high performance plastics are increasingly used, especially in engines with gasoline direct injection.

They are laminated with several layers of plastic on the conventional filter paper and permit very high dirt absorption and an initial separation rate of over 90% for particle sizes of 3–5 μm.

~~Fuel pressure damper. A constant, possibly intake-manifold–related flow pressure in the fuel line is required in injection systems for gasoline engines. Rapid pressure changes, by load changes or pulsations generated by the injection valves, may not be controlled by the fuel pressure controller. Pulsations also can result in acoustic disturbances because of the connection of the system to the car body. The fuel pressure damper reduces the above-mentioned effects (**Fig. I48**).

Fluids are basically incompressible, which is why they cannot contribute to the storage of pressure energy and to damping, and this is why systems with pressure-dependent storage volumes, normally fitted with a diaphragm and spring, are used in vehicles. At the target pressure, the diaphragm-spring system is positioned in the middle of the working range to be able to absorb pressure loads as well as pressure relief.

For electronically controlled fuel-supply systems, the pressure-dependent diaphragm position can be determined by a sensor. The fuel pressure damper also is used as a pressure sensor in this application.

~~Fuel pressure regulator. The fuel pressure regulator (**Fig. I49**) is added to the fuel circuit as an additional bypass valve, and it is used to maintain a constant fuel differential pressure at the injection valve.

Fig. I48 Fuel pressure damper

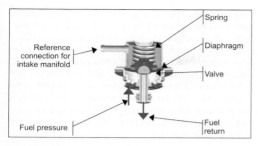

Fig. I49 Fuel pressure regulator

The technical principle is equivalent to that of a proportional controller. The absolute fuel pressure is changed by reference to the absolute intake manifold pressure, measured by sampling line. The design of the fuel controller is based on a valve loaded with a spring and diaphragm. The design shown in **Fig. I49** is typically used for working ranges of up to 5 bar.

~~Fuel rail. The fuel is supplied to the individual injectors with the help of a fuel rail (**Fig. I50**). This prin-

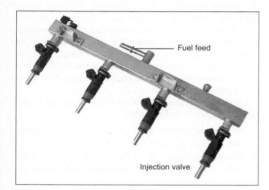

Fig. I50 Fuel rail

ciple has also been called "common rail" in recent publications. The fuel rail is also used in gasoline engines with intake manifold injection to locate the injection valves and possibly other components, such as fuel pressure regulator, fuel pressure damper, and cables. The components are normally made of coated steel or stainless steel to achieve high robustness. For appropriate engine designs and markets with controlled fuel quality, plastic designs are feasible if the special material characteristics are taken into consideration. High specifications of cleanliness are required for the fuel rails and for all other fuel-carrying components to guarantee faultless functioning of the injection valves. Entrained particles (e.g., metal shavings, casting burrs, plastics, glass fibers) and uncontrolled production materials should be avoided.

~~Fuel rail direct injection. The fuel rail is used to provide a uniform supply of fuel for injection. It is also used to house the components of the fuel system. A major requirement is how leak-proof the fuel system is under all engine operating conditions. The volume of the fuel rail is determined during the design of the fuel system and is a compromise between the pressure drop during injection and the required time for the pressure change during unsteady state operating conditions of the engine (i.e., changing speed or load) and during a high-pressure start.

The pressure drop during injection is calculated as

$$\Delta p = \frac{\beta \cdot MFF}{V_{rail} \cdot \rho},$$

where Δp = pressure drop, MPa, β = fuel compressibility, MPa (800 MPa for cold fuel and 600 MPa for hot fuel); MFF = injection quantity, kg, V_{rail} = volume fuel rail, m³, and ρ = fuel density, kg/m³.

Fig. I51 shows the pre-assembled fuel module with all components and electrical connections. It is fully tested before it is delivered to the production line of the engine manufacturer. This permits very short engine assembly times. O-rings are used to seal the connection

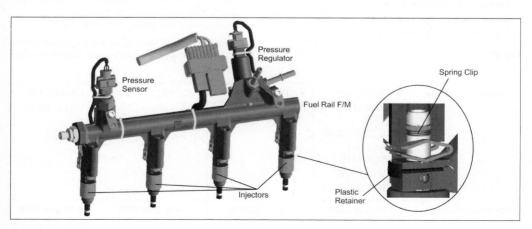

Fig. I51 Injection module

Fig. I52 Stainless steel rail

between injection valve and fuel rail as well as to compensate for tolerances in the longitudinal direction between fuel rail and cylinder head. The injection valves are pressed against the cylinder head by a spring. The fuel rail is made of forged aluminum, and these fuel rails can be used with a maximum concentration of alcohols and ethers and their mixtures in accordance with the European standard for gasoline fuels EN228.

Stainless steel fuel rails with metallic seals are used for worldwide fuel qualities with higher concentrations of alcohols and to make the fuel system leak-proof down to −40°C (**Fig. I52**). Similar to diesel common rail systems, the assembly of the individual components of the fuel system is performed at the end of the production line.

Europe
European Union
ECE R34
70/220/EEC
Official Journals
United States
California Air Resources Board (CARB)
California Code of Regulation (CCR), title 13
Environmental Protection Agency (EPA)
Code of Federal Regulation (CFR), title 40, part 86

Fig. I53 Emissions legislation

~~**Fuel tank.** The specification for a large volume tank and optimal use of the available vehicle construction space results in complex fuel tank shapes, such as single-chamber and multichamber tanks. The geometries require demanding production processes and also tank functions such as fuel-feed and tank venting systems. The selection of the materials for the tank is the responsibility of the vehicle manufacturer, and it is made taking into consideration the emissions legislation in the respective markets. The tanks are made of either plastic or metal.

~~~**Emissions legislation.** The emissions legislation defines the limits for the emission of hydrocarbons by motor vehicles into the environment. The most important regulations are shown in **Fig. I53**. Starting in 1980, product design has been significantly influenced by the need to reduce hydrocarbon emissions (**Fig. I54**).

~~~**Hydrocarbon emissions.** These emissions can be the solid, liquid, or gaseous materials or compounds that are emitted into the environment by a fixed or locally changing source or product. One goal for the design of new fuel tank systems is the reduction of hydrocarbon emissions in accordance with the emissions legislation. Emissions are divided into diurnal losses, running losses, and hot soak losses. The emissions are created by permeation and/or leakage. The connections of lines (flow, external bleeder line) to the tank and service openings are sources of hydrocarbon emissions in plastic fuel containers as well as metal ones.

Permeation is the penetration of one material by another. It happens in four steps:

1. Adsorption: Attachment of the permeation molecule on the internal wall
2. Absorption: Solution of the absorbed gas or fluid molecules into the plastic wall; penetration into the molecule structure
3. Diffusion: Movement of the molecules in the plastic wall in the direction of a negative concentration gradient (in the direction of the outside)
4. Desorption: Evaporation at the exit surface

The measurement of the hydrocarbon emissions from the total vehicle is performed using a flame ionization

Emission legislation	• CARB (1980) • CARB LEV I (1995) • EU 2000 (1996) • EPA LEV I (2001) • EU III (2003) • EU IV (2005)	• EPA Tier II (2004)	• CARB LEV II (2004)	• CARBZero Evap (2005)
Threshold emission	2.00 g HC/24h	0.95 g HC/24h	0.50 g HC/24h	0.35 g HC/24h

Fig. I54 Development of emission limits for motor cars

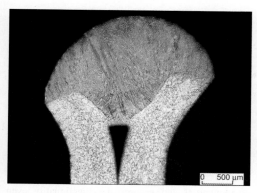

Fig. I55 Metallographic cross section through a MAG weld seam of a MFT

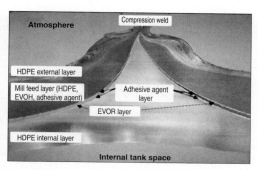

Fig. I56 Description of a compression weld area of a COEX tank with six layers (Source: Kautex)

detector (FID) in a SHED (sealed housing for evaporative determination) chamber.

~~~**Metal fuel tank (MFT).** The basic materials of metal fuel containers are fire aluminized sheets (FAL) and stainless steel. In contrast to working with plastics, completely different processing or shaping characteristics (deep drawing) as well as factors such as corrosion and material stability must be taken into consideration. The half-shells are normally connected using the seam welding process through MAG or plasma welding (**Fig. I55**). The selection of the stainless steel material (normally 1.403) is based on the emission requirements for the US market.

~~~**Plastic fuel tank (PFT).** The majority of tanks are made from the high density polythene plastic, PE-HD. This material can be used to produce complex tank geometries. Mono-layer tanks (i.e., tanks made from a single layer) are made exclusively from the material PE-HD. Several processes are available to reduce permeation—the most important ones are fluorination and co-extrusion. The internal container space is treated with an F_2/N_2 mixture for fluorination, which reduces the permeation of hydrocarbons (hydrocarbon emissions) in comparison to the untreated mono-layer tank. The barrier characteristics of PFT are improved by an additional barrier layer such as B. EVOH (ethylene vinyl alcohol; **Fig. I56**). The COEX tanks are known as multilayer tanks.

The extrusion blow-molding production process is used for mass production. In this process, the tank raw material inside the mold is blown into the desired shape. The tank can be removed as a formed component after it is cooled down. In a few cases, tanks are also produced using the thermo-forming process.

In this process, plastic plates are heated, deep drawn, and welded. The process is still new in tank production and is now in serious production.

~~**Fuel tank systems.** The supply of the engine with fuel requires a fuel tank, which normally is located in the region of the rear axle of a vehicle. It is the part of the fuel tank system that provides for a number of tank functions, including fuel storage, fuel supply (fuel-feed pump, fuel filter, pressure controller), operational and fill venting (tank venting system), fuel filling (filler tube), fill-level limitation (dip tube or float valve), and fill-level measurement (fill level sensor). The tank system provides the engine with fuel through the flow line within a defined start time and at a defined pressure and volume flow independent of the driving conditions and ambient temperature.

Tank systems are designed for gasoline fuels or diesel fuels. Depending on the market and the emissions legislation, the tank systems are divided depending on the flow routing of the gases (fuel vapor air mixture, hydrocarbon emissions) that develop during fuel filling (refueling) and during operation. Emissions legislation has made the most progress in Europe and the United States, and this development has had a significant influence on the development of tank systems (**Fig. I57**).

~~**Fuel-feed unit.** Fuel-feed units (**Fig. I58**) are divided into conventional systems and systems with a higher degree of integration. Conventional systems are systems with a return. This means that the fuel-feed unit delivers a constant fuel volume to the engine through a flow line and an external filter. The excess fuel is routed through the return line into the tank and, therefore, back to the fuel circuit, because the engine does not always use the total fuel volume supplied.

New systems have a higher degree of integration. This means that the fuel filter and pressure regulator are integrated into the feed unit and, therefore, in the fuel tank. This eliminates the fuel return line and the hot return from the engine to the tank. As a result, requirements for the tightness of fuel supplies from the viewpoint of emissions are better satisfied. The fuel temperature must be kept low, especially for the plastic tanks currently in use.

The following is a functional description of an integrated system: Gasoline flows through a valve to the swirl pot for a first filling. A fuel-feed pump delivers

633

Fuel	Gasoline fuel tanks			Diesel tanks
Market	Europe	USA		Europe/USA
Tank identification	ECE tank	ORVR tank		Diesel tank
Tank option	—	Mechanical seal	Liquid seal	—
Flow routing of emissions during filling	Suction at fuel nozzle	Activated charcoal filter	Activated charcoal filter	Environment
Flow routing of emissions during driving operation	Activated charcoal filter	Activated charcoal filter	Activated charcoal filter	

Fig. I57 Classification of tank variables according to emission routing

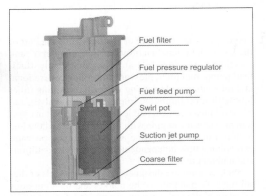

Fuel filter

Fuel pressure regulator

Fuel feed pump

Swirl pot

Suction jet pump

Coarse filter

Fig. I58 Fuel-feed unit without return but with integrated pressure controller

fuel from the swirl pot to the fuel filter. The filter has a sufficiently large surface to separate from the fuel the dirt that is generated in the fuel system during the life of the vehicle—therefore, it does not have to be replaced. The filtered fuel flows through the fuel pressure regulator, which keeps the fuel pressure in the flow line constant. The excess fuel is routed into a suction jet pump, which ensures that the swirl pot is always filled.

This has the advantage that the supply is always guaranteed, even if the tank has only a small fuel volume. A coarse filter protects the feed unit from coarse

contamination and ensures the functionality even under the most difficult circumstances.

~~~Fuel-feed pump. This in-tank pump delivers the fuel continuously from the fuel container into the injection system—it is installed directly into the tank. The in-tank pumps used today are generally integrated into the feed units. The feed volume is larger than the maximum fuel demand of the engine to guarantee the required fuel pressure under all operating conditions. The engine and transmission control unit switches the fuel-feed pump.

The electric motor and the pump stage are located in a combined housing, and fuel always flows through them. **Fig. I59** shows the example of such a pump in which fuel flows through on the inside. The flow on the inside delivers good cooling of the electric motor. An ignitable mixture cannot be generated in the pump because of the lack of oxygen; hence, there is no danger of explosion. A nonreturn valve prevents the return from the fuel line.

Fig. I60 shows several functional principles that are used in fuel pumps, either individually or in combination. The positive displacement pumps (gerotor principle) are better qualified for higher pressures, while the pumps using the flow principle (side-channel design) are less critical with respect to noise generation in the vehicle.

~~~Suction jet pump. Suction jet pumps are used during operation for the filling of the swirl pot and in multichamber tanks for the extraction of the fuel from

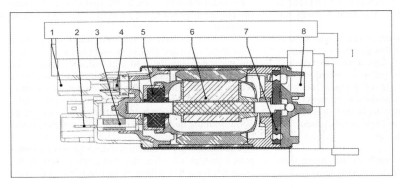

Fig. I59
Section through the main pump, side-channel principle:
1, pressure-pipe tube;
2, electrical connection;
3, shield unit; 4, return valve; 5, commutator;
6, rotor; 7, pump wheel (side-channel pump);
8, induction pipe

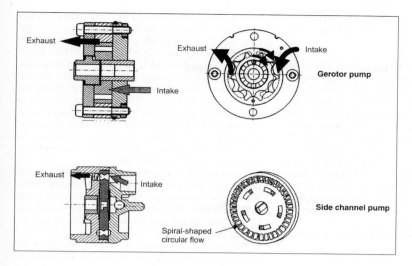

Fig. I60
Fuel-feed pumps/
functional principles

the tank side without feed unit. The function is based on a venturi nozzle (named for the Italian scientist G.B. Venturi), which operates on a suction process without moving parts. The fuel spray is used to "drive" the device, and it must exit the nozzle with adequate speed to carry an additional volume flow generated by internal friction and turbulent mixing. One of the possible tasks of a suction jet pump is that it must deliver the fuel through the tank saddle to the fuel-feed pump into the fuel-feed unit. The fuel-feed pump must also deliver the fuel spray of the suction jet pump in addition to the engine consumption. Another option is the use of the return volume from the pressure controller as the fuel spray.

~~~**Swirl pot.** The swirl pot is part of the fuel-feed unit, and it houses the fuel-feed pump. It is used as a fuel reservoir th7at guarantees delivery of the fuel even if only a small fuel volume is in the tank, particularly when the vehicle tilts and during cornering.

~~**Gasoline tank.** Gasoline tanks are called ECE (Economic Commission for Europe) tanks in Europe

and ORVR (onboard refueling vapor recovery) tanks in the United States.

The gases that develop during refueling in an ECE tank are removed by the refueling system. Only the fuel vapors that develop during driving operations are cleaned and removed by an activated charcoal filter.

The cleaning of the hydrocarbon vapors in an ORVR tank during refueling and driving operations is performed by the activated charcoal filter, which is two to four times larger than the one in the ECE tank. The gap between the fuel nozzle and the fuel pipe is sealed to prevent the exit of hydrocarbons. Two processes are used here: a liquid seal and a mechanical seal. The liquid seal system generates a vacuum caused by the fuel flow in the fuel pipe during refueling (fluidic sealing; **Fig. I61a**). Air from the atmosphere is sucked into the tank system in this process. The principle of recirculation is often used to minimize the volume of gas flow through the activated charcoal filter. It sucks a part of the vapors through an additional line from the fuel tank to the fuel pipe and returns it to the fuel tank with the air taken from the atmosphere. This circuit reduces the volume of fuel vapor during refuel-

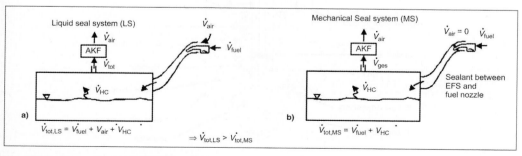

Fig. I61 Sealing processes for ORVR tanks

ing. The mechanical sealing system (**Fig. I61b**) seals the gap between the fuel nozzle and the plug neck with a gasket.

~~**High-pressure pump (direct injection).** The high-pressure pump is used to increase the fuel pressure of the electrical fuel delivery pump to 0.38–0.5 MPa, the required pressure for the gasoline direct injection. Injection pressures of 5–12 MPa are required for wall and air-guided combustion processes. Spray-guided combustion processes require a fuel pressure of 20 MPa. Additionally, pressure reductions in the fuel rail during injection should be as low as possible and a constant drive torque from the high-pressure pump should be available. The power consumption of the high-pressure pump (e.g., a radial-piston pump; **Fig. I62**) should be as small as possible to avoid an increase in friction losses in the engine.

The power consumption of the high-pressure pump is calculated as

$$P = \frac{\Delta p \cdot Q}{\eta},$$

where Δp = pressure differential between inflow and output of pump [Pa], Q = delivery flow of the high-pressure pump [m³/s], and η = overall efficiency.

The compressing elements are powered by an ec-centric pump shaft. Three pistons and three cylinders are positioned around the eccentric—the pistons are fixed and the corresponding cylinders move. A shoe transmits the eccentric movement to the cylinder, which is pressed onto the shoe by a spring. Fuel enters the space between piston and cylinder through a suction valve. The cylinder is moved against the piston by the eccentric and pressurizes the fuel. The suction valve is closed and the outlet valve is opened. The outlet valve is closed and the suction valve opens when the eccentric is at the upper dead center point during the downward cylinder movement. The fuel leaves the compression unit through the outlet valve.

Three compressor units with a phase angle of 120 degrees are used to develop an almost constant delivery of flow because the compression maxima are at different times. The radial piston pump is connected by a flange to the end of the camshaft to receive actuation.

The radial piston pump is used for pressure-controlled fuel systems. In this application, the pump always delivers a fuel volume that is proportional to its speed. The target fuel pressure is controlled by an electrically controlled pressure controller. The power requirements for such a fuel system are relatively high due to the return line from the fuel pressure controller to the intake of the high-pressure pump. The friction

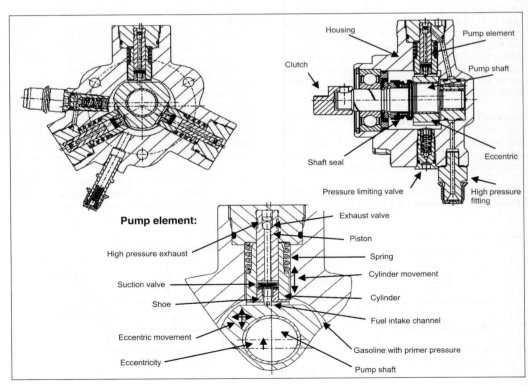

Fig. I62 Cross section of a three-cylinder radial piston pump

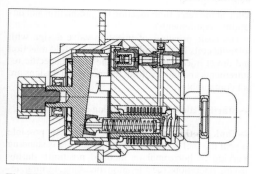

Fig. I63 SAH three-cylinder axial piston pump

loss of the engine is increased at higher fuel system pressures during part load. The control concept results in nearly constant propulsion torque requirements for the high-pressure pump and in low levels of variation of the fuel pressure.

The motive power for the pump is optimized for the SAH three-cylinder axial piston pump shown in **Fig. I63**. The fuel circuit is separated from the oil circuit of the pump by a steel diaphragm. All moving pump parts are lubricated by oil, which minimizes wear and friction losses. The swash-plate disk is rotated by a mechanically driven shaft and generates the alternating movements by the three pistons. The oil moves at the lower dead center via a groove in the swash plate from the oil reservoir through the piston into the pump chamber. The intake valve on the fuel side is closed through the movement of the pump piston from the

lower to the upper dead center and the outlet valve opens subsequently. The fuel is compressed to the pressure level in the fuel rail. The pressure level in the oil chamber and the fuel chamber are almost equal. The fuel pump is capable of running dry, which permits volume control through a three-way valve that is integrated on the intake side of the pump. This control system permits a fast pressure reduction in the fuel rail.

The housing, which is in contact with the fuel, is made from stainless steel to accommodate worldwide fuel qualities.

~~~**Control valves.** A three-way valve is used for flow control of the axial piston pump (**Fig. I64**). The flow cross section opened by the throttle slide increases in proportion to the current through the solenoid coil. An additional pressure control valve connected to the high-pressure circuit of the pump is integrated because the throttle slide valve does not permit zero delivery. Only the leakage amount will flow through the throttle slide when small currents flow through the solenoid coil, and this permits pressure control. It also permits a fuel flow higher than the leakage amount from the high-pressure part of the pump to the low-pressure side, and this enables a fast pressure reduction in the fuel rail.

A fuel controller (**Fig. I65**) is used for pressure-controlled fuel systems.

The electromagnetic pressure control valve is a two-way bypass valve. The valve keeps the pressure in the rail constant, independent of the pump delivery and the volume of fuel injected. The bypass is open or closed depending on the required injection volume.

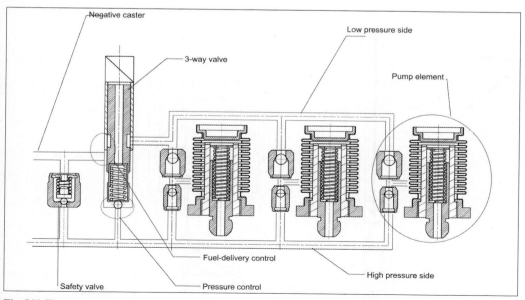

Fig. I64 Three-way valve

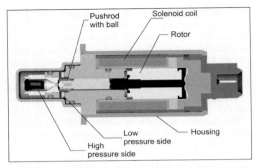

Fig. I65 Pressure controller for direct injection

The electromagnet must be strong enough to exceed the maximum forces generated by the fuel pressure on the ball seat of the needle.

~~Injection valves. Injection valves (**Fig. I66**) are used for fuel metering with the help of electrically controlled solenoid coils. The opening coil in a linear design has been accepted as the standard version. Other possible designs (e.g., opening and closing coils) cannot achieve the same price-performance ratio with the same high robustness. The solenoid circuit switches quickly and has fast dynamic characteristics. On-off low side drivers have been accepted as the standard control output stages for intake manifold injection. Current controlled output stages are used in racing car engines or they are used for direct injections in combination with high side drivers.

The mass of the injection valve should be kept as small as possible to achieve high overall dynamics, while at the same time meeting the production and durability requirements.

The basic principle of injection valve design with solenoid coils is identical for intake manifold injection and direct injection. However, the engine specific requirements and the resulting technological designs are not comparable.

~~Starting aid systems →Starting aid

~~Tank venting system. It must always be possible to vent the fuel container when the vehicle is standing (parking in horizontal or inclined position), during driving operations (acceleration, breaking, turning, incline or decline roads), and during refueling. Tank systems are divided into external and internal tank venting systems.

~~~External tank venting systems. Conventional tanks normally have two expansion tanks (operations and refueling venting) at the fuel pipe (**Fig. I67**).

One of them is used as a reserve volume for the thermal expansion of the fuel and as a separator for fuel drops, which are carried by the gas volume flow during venting. The fuel flows from the container back to the tank because of the difference in height between the container and the tank. The removal of the gases during refueling normally starts through a line, which has an open end (dip tube) or which is connected to the float valve. The position in the fuel tank guarantees a flooding of the line when the specified fill volume has been reached, and venting is now interrupted. This is valid for horizontal positions and inclines of up to 5°; at the gas station. It results in an increase of the internal pressure in the tank. An increase of the fuel column is the result,

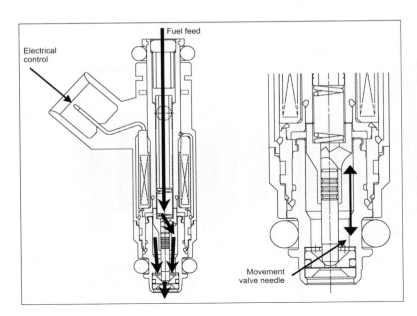

Fig. I66
Injection valve

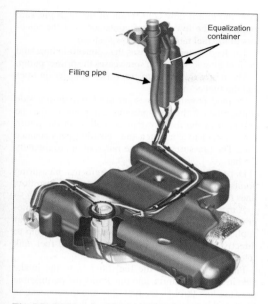

Fig. I67 ECE tank with external venting system, PFT

and the fuel nozzle interrupts the fueling process (pressure-induced fuel level limitations).

~~~Internal tank venting systems. More stringent emissions legislation requirements have led to the external venting systems being integrated into the tank (**Fig. I68**). The number of outside pipe interfaces, which are a source of hydrocarbon emissions, are reduced by this design. The high height difference required for the independent emptying is missing now because of the relocation of the expansion tank. Fuel that gets into the expansion tank during driving operations because of swashing in the fuel tank is transported through a suction jet pump back to the tank. Fuel level limits are controlled either by pressure-induced systems (external tank venting systems), by a multifunction

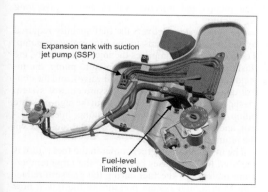

Fig. I68 ORVR tank with internal venting system, MFT

valve in the venting line, or with a float-controlled fuel level limit valve at the end of the fuel pipe.

Injection system, air (*also,* →Mixture formation ~Mixture formation, gasoline engine ~~Mixture injection). Systems for air injection are used to oxidize components, such as hydrocarbons and carbon monoxides, by supplying oxygen into the exhaust system and performing an after-reaction.

Systems such as fans, turbochargers, and so on that are used for turbocharging and, therefore, for increases in performance can also be called injection systems.

Another possibility is mixture injection, which is generated outside the combustion chamber. The principle of the mixture injection was tested with two-stroke engines. In this test, fuel was injected into an initially closed mixing chamber and mixed with hot air. This provided more time for fuel preparation, which then can be used for direct injection because the fuel can be injected in an evaporated—and, therefore, partly gaseous—state, with injection pressures of up to 20 bar. This results in advantages regarding fuel consumption and pollutant emissions. **Fig. I69** shows the diagram of a mixture injection.

Literature: G. K. Fraidl, R. Knoll, H. P. Hazeu: Direkte Gemischeinblasung am Zweitakt-Ottomotor, VDI Conference on Vehicle Engines, Dresden, 1993.

Injection system, fuel. Of the many options possible for injection systems, the main ones used in gasoline engines are direct injection and multipoint injection into the intake manifold. Direct injection systems are used for car diesel engines.

~Diesel engine. The fuel for diesel engines is supplied by an injection pump or by a pressurized fuel reservoir.

The qualification of the different injection systems for both basic combustion processes—direct injection into the main combustion chamber or indirect injection into the prechamber or swirl chamber—depends primarily on the magnitude of the injection pressure achievable for a given injection quantity.

Indirect injection requires a maximum of 450 bar at

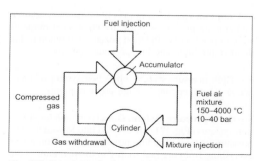

Fig. I69 Principle of mixture injections (Source: AVL)

the nozzle, while the currently dominant direct injection process requires, depending on the application, a maximum of 1400 bar to 2000 bar (at the nozzle). The available pressure range is listed for the injection systems described below (*also*, →Mixture formation ~Mixture formation, diesel engine).

In addition to the injection pressure that can be attained at high and low speeds, the following parameters are also important: the number of cylinders, the capability of making multiple injections and the injection curve, and, finally, the packaging and cost.

The following system rankings are used currently:

- Systems with injection synchronizer, which means pressure generation in relationship to the engine stroke. These include:
  - Inline injection pumps, which include a number of pump-cylinder units equivalent to the number of engine cylinders. They are positioned serially in a housing and are actuated by their own pump camshaft. The pump-cylinder units are connected to the fuel-injector body in the cylinder head through pressure lines.
  - Distributor pumps. The generation of the high pressure and the metering of the fuel injected are performed in only one central location. The individual pump outlet or fuel-injector bodies are controlled by a rotating distributor.
  - Single-injection pumps (pump-nozzle units with separate controller and unit pumps without separate camshaft).
  - Unit injector. The high-pressure pump and the injector nozzle are installed as a unit into the cylinder head for this application. The pump pistons are actuated directly by the camshaft in the cylinder head or by pushrods, and a separate unit is needed for each cylinder.
- Systems with a central pressure storage (common rail). These systems provide the required injection quantity through injection valves (injectors), which are assigned to the cylinders, under the control of a management system.

The latter systems are called modular injection modules (unit injector, unit pump, common rail) in contrast to the serial or distributor pump designs. **Fig. I70** shows the currently used injection systems.

~~Common rail. In common rail injection systems, a fuel storage plenum is normally positioned immediately in front of the injection nozzles. Common rail is divided into high- and low-pressure systems.

~~~High-pressure system →Injection system (components) ~Diesel engine ~~High-pressure pump

~~~Low-pressure system. Diesel common rail injection systems (**Fig. I71**) are divided into low- and high-pressure systems. The low-pressure system comprises everything from the fuel supply in the fuel tank to the inlet of the low-pressure system (feed system) and

from the return flow connection of the high-pressure pump and the low-pressure connections at the injectors to the fuel tank (return flow system).

The feed system comprises the connection lines and the fuel filter as well as in some cases the primer pump.

The return flow system comprises connection lines and the fuel cooler.

Defined pressure ranges on the low-pressure side connections of the high-pressure components must be observed in the feed system as well as in the return flow system to guarantee the specified performance data. The pressures are in the range of approximately 2.5 bar_{abs}.

Depending on the system requirements, measuring points can be integrated into the feed and return flow lines (e.g., fuel return temperature sensor).

A mechanical fuel feed pump has already been integrated in the high-pressure pump for the majority of applications. This enables the high-pressure pump to directly take fuel from the swirl pot in the fuel tank (feed unit).

Fuel flows through the fuel filter from the intake side in this application, and this limits the permissible pressure loss through the fuel filter and the connection lines. In this case, the input pressure of the high-pressure pump is lower than the ambient pressure. Cavitation danger exists for the fuel feed pump if the intake pressure is too low. This is why positioning on the intake side is limited to the engines with small displacements (low maximum feed rates). Engines with large volumes are normally supported by an electrical primer pump in the fuel tank, which can be used to establish a defined prepressure for the fuel delivery pump in the high-pressure pump. This prevents cavitations and it can at the same time increase the pressure loss permissible through the fuel filter. The fuel filter will be operated in a position on the pressure side in this application. The pressure increase of the electrical primer pump is in a range of about 0.5–1 bar. The fuel filter removes particle contamination and any contained free water from the fuel to avoid wear and corrosion in the intake system. It is required to remove these contaminants from the fuel to meet the demanding requirements of the high-precision components in high-pressure pumps and injection valves and to give long service life.

The feed system delivers the fuel quantity required by the injection systems to the high-pressure pump. The fuel quantity comprises the injection quantity, the flushing quantity for the high-pressure pump, the injector control quantities (injection valve), and the control quantity required by the pressure control system via the pressure control valve (PCV) in the high-pressure pump. The fuel used for pump flushing, injector control, and PCV control are returned to the fuel tank via the return flow system.

The installation of a fuel cooler in the fuel return is required in many cases because of the large temperature increase of the fuel during the pressure reduction from a high system pressure to the return flow pressure.

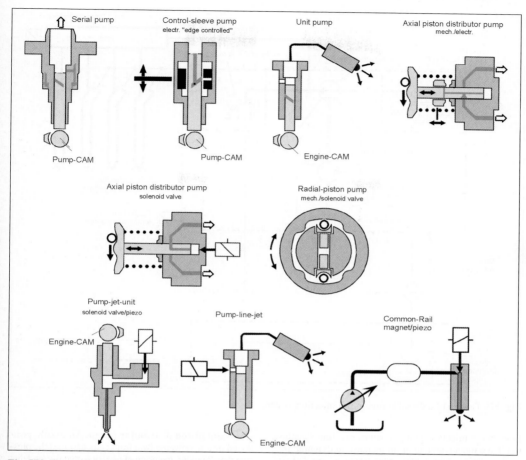

Fig. I70 Currently used injection systems: *top*, edge-controlled systems; *center*, solenoid valve–controlled compact systems, six cylinder; *bottom*, single-cylinder systems, solenoid valve– or piezo-controlled

~~Control sleeve injection pump. The control sleeve injection pump is a serial injection pump for commercial vehicles (**Fig. I72**). The quantity of fuel delivered and the start of delivery can be adjusted independently from each other in this design. The control sleeve is a new part, as compared with a conventional design. The prestroke of the pump and, therefore, the start of injection, change if the control sleeve is moved in the longitudinal direction of the piston.

A later start of feed results in a larger prestroke and vice versa. The fuel feed starts when the cross bore in the piston is completely covered by the control sleeve during the upward movement. The end of feed occurs when the control edge of the pump piston dips into the suction bore of the control sleeve. **Fig. I73** shows the phase characteristics of the piston movement for a control sleeve serial injection pump.

Adjustment of the fuel quantity is performed in a similar manner to adjustment in conventional serial pumps—that is, by rotation of the pump piston. The

control sleeve injection pump is only supplied as a system with electronic diesel control. The rotation of the pump piston (quantity) and the height adjustment of the control sleeve (start of delivery) are each performed by a linear magnet, which in turn is controlled by the control device (electronic open- and closed-loop control, diesel engine). The control sleeve injection pump is designed for engines with an output up to 70 kW/cyl and maximum injection pressure on the nozzle side of approximately 1350 bar.

~~Distributor pumps. Distributor pumps of different designs are used for high-speed diesel engines with three, four, five, and six cylinders and for light trucks and tractors with cylinder performances of up to 45 kW. All known distributor pumps work with fuel lubrication, in contrast to serial injection pumps.

Distributor pumps are available with mechanical or electronic control. Two design types are available: the axial and the radial piston distributor injection pump.

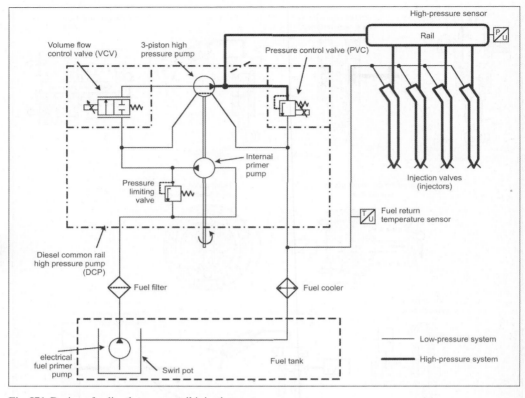

Fig. I71 Design of a diesel common rail injection system

Distributor injection pump systems are line systems that have the advantage of an increase in dynamic injection pressure at higher speeds (*also, see below*, ~~Serial injection pump).

~~~**Axial piston distributor pump.** An axially positioned piston, powered by a front cam, assumes the pressure development, metering of injection quantity, and distribution to the outlets, which means the engine cylinders.

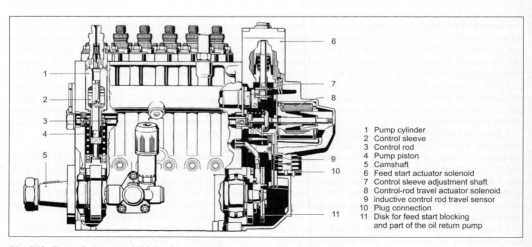

1 Pump cylinder
2 Control sleeve
3 Control rod
4 Pump piston
5 Camshaft
6 Feed start actuator solenoid
7 Control sleeve adjustment shaft
8 Control-rod travel actuator solenoid
9 inductive control rod travel sensor
10 Plug connection
11 Disk for feed start blocking
    and part of the oil return pump

Fig. I72 Control sleeve serial injection pump for commercial vehicles [2]

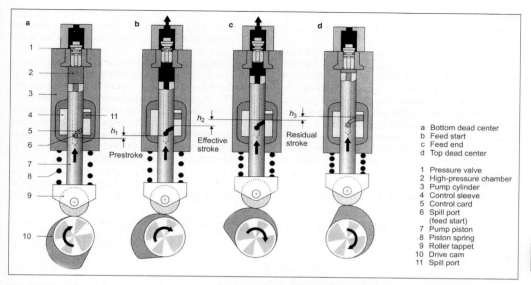

Fig. 173 Functional description of the control sleeve serial injection pump [2]

a Bottom dead center
b Feed start
c Feed end
d Top dead center

1 Pressure valve
2 High-pressure chamber
3 Pump cylinder
4 Control sleeve
5 Control card
6 Spill port (feed start)
7 Pump piston
8 Piston spring
9 Roller tappet
10 Drive cam
11 Spill port

~~~~**Add-on modular assembly for VE distributor fuel injection pump, mechanically controlled.** Depending on the engine requirements, the edge- and mechanically-controlled VP can be completed in a modular manner with several adaptation devices. The most important are:

- supercharging pressure–dependent full-load stop
- hydraulically actuated adaptation for naturally aspirated engines
- load-dependent start of feed
- atmospheric pressure–dependent full-load stop
- cold start accelerated
- multistage or adjustable start quantity
- temperature-dependent idle speed increase

- electrical shut-off valve
- pressure sensor in the central screw plug as feed signal sensor

~~~~**VE distributor pump (Bosch), edge-controlled.** **Fig. 174** shows the mechanically controlled basic version.

The VE pump is primarily used in cars for engines with indirect injection and engines with direct injection up to approximately 30 kW/cyl. The range of injection quantity achieves a maximum of about 125 mm³/injection and the injection pressures on the nozzle side are between 350 and 1250 bar.

The distributor injection pump is powered in such a way that the drive shaft runs parallel to the crankshaft

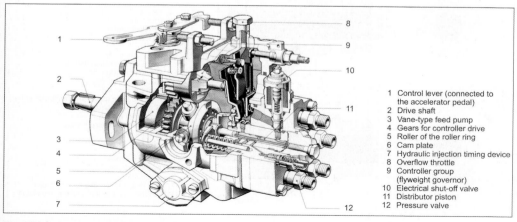

1 Control lever (connected to the accelerator pedal)
2 Drive shaft
3 Vane-type feed pump
4 Gears for controller drive
5 Roller of the roller ring
6 Cam plate
7 Hydraulic injection timing device
8 Overflow throttle
9 Controller group (flyweight governor)
10 Electrical shut-off valve
11 Distributor piston
12 Pressure valve

Fig. 174 Axial piston distributor injection pump VE with mechanical flyweight governor [7]

of the engine. This imposed drive arrangement is implemented using drive belts, plug-in pinions, gears, or chains.

The vane-type feed pump is positioned on the drive shaft of the distributor injection pump. This supports the roller ring, which is not connected to the drive shaft and which is also anchored in the pump housing. The cam plate, which is supported by the rolls of the roller ring and which is powered by a cross coupling (for longitudinal and angle compensation), creates a rotary-stroke movement, which is transmitted to the distributor piston. **Fig. I75** shows the stroke and rotating phases of the axial piston distributor injection pump based on the Bosch design.

The pump piston has a clearance of 3–5 mm when positioned in the distributor head. The electrical shut-off device for the interruption of fuel supply, the central screw plug with bleeder screw, and the pressure valves with pressure valve brackets are installed in the distributor head.

Similar to the mechanical controllers of the serial injection pumps, different controller options are available for VE pumps (e.g., variable speed governors; **Fig. I76**) and the associated add-on modular assemblies.

The driveshaft powers the controller group through a pair of gears. The control groups have flyweights and governor sleeves. The control lever assembly, consisting of adjustment lever, start level, and tension lever, is supported in the housing and can rotate. The position of the slide control on the pump piston is influenced by it. The control spring, which is connected through the adjustment lever shaft with the externally positioned

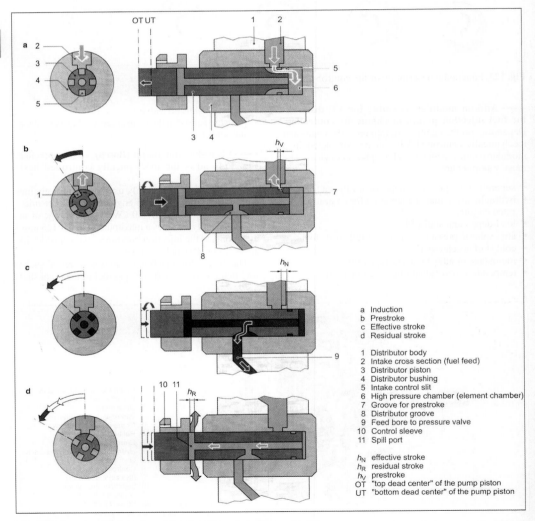

a Induction
b Prestroke
c Effective stroke
d Residual stroke

1 Distributor body
2 Intake cross section (fuel feed)
3 Distributor piston
4 Distributor bushing
5 Intake control slit
6 High pressure chamber (element chamber)
7 Groove for prestroke
8 Distributor groove
9 Feed bore to pressure valve
10 Control sleeve
11 Spill port

h_N effective stroke
h_R residual stroke
h_V prestroke
OT "top dead center" of the pump piston
UT "bottom dead center" of the pump piston

Fig. I75 Functional principles of the distributor injection pump [7]

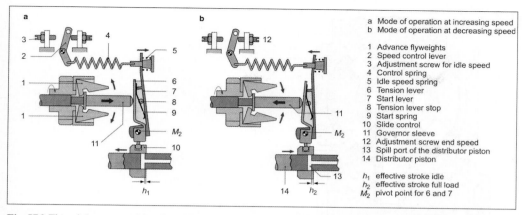

a	Mode of operation at increasing speed
b	Mode of operation at decreasing speed

1 Advance flyweights
2 Speed control lever
3 Adjustment screw for idle speed
4 Control spring
5 Idle speed spring
6 Tension lever
7 Start lever
8 Tension lever stop
9 Start spring
10 Slide control
11 Governor sleeve
12 Adjustment screw end speed
13 Spill port of the distributor piston
14 Distributor piston

h_1 effective stroke idle
h_2 effective stroke full load
M_2 pivot point for 6 and 7

Fig. I76 Flyweight governor for distributor injection pumps, variable speed governor in load operation [7]

adjustment lever, engages on the upper side of the controller components. The adjustment lever shaft is supported in the controller cover, and the pump function is influenced through the adjustment lever.

The hydraulic start of feed or adjustment of start of injection is integrated on the bottom of the VE pump at right angles to the longitudinal axis of the pump. Its function is influenced by the internal chamber pressure, which is determined by the feed pump and the pressure control valve. The adjustment range for the start of feed is a maximum of 24°CA. **Fig. I77** shows how the hydraulically actuated injection timing device functions.

~~~Radial piston distributor injection pump. In contrast to the axial piston distributor pump, the pressure is generated in this application by two, three, or four small pump pistons, which are positioned radially and fit into a rotary distributor. The pistons are supported by a fixed cam ring and are actuated by the internal cam elevation at the cam ring. The rotary distributor and the pump piston are powered at half the engine speed by a driver.

The cam ring is adjustable within a limited angle sector for the definition of the injection adjustment effect. The individually known designs of radial piston distributor injection pumps are based on either the metering or the control principle.

~~~~DP/DPC/DPCN distributor-type fuel injection pump (Delphi/Lucas) [8]. In contrast to the DPS model, a direct volume control of the start and full-load volume is available in the DP/DPC distributor pump in addition to the control of the injection quantity of the basic map through an intake throttle. This limits the free stroke of the pump piston through an adjustable mechanical stop. For full load, this mechanical stop can be adjusted by a diaphragm actuator as a function of the boost pressure or in naturally aspirated engines as a function of the atmospheric pressure. The DP200 distributor injection pump is, for example, also offered

with an electronic controller. It is applicable for cylinder displacements of 1.3 L. The DPC is used for cars and light trucks up to about 2.5 L total displacement and for indirect injection engines.

The DPCN version is based on the DPC distributor injection pump and works with partial electronics and control instruments. The hydraulic pressure of the injection actuator control piston is modulated by an electronically controlled actuator. The feedback of start of injection is given by a pintle motion sensor.

~~~~DPS distributor-type fuel injection pump (Delphi/Lucas) [8]. The DPS distributor-type fuel injection pump is the basic model of the Lucas radial piston distributor pumps with mechanical flyweight controller and hydraulically controlled injection adjuster. The hydraulic injection adjuster is based on the rotation of the cam ring relative to the drive shaft. The metered fuel injected is based on the intake throttle principle.

~~~~EPIC distributor injection system (Delphi/Lucas). The EPIC system consists of an electronically controlled radial piston pump for engines with indirect and direct injection (**Fig. I78**).

The metering of the fuel injected (volume control) is performed by direct control of the pump piston power stroke. The distributor rotor with the radial pump piston is moved axially in this application by the controlled hydraulic pressure on the front side.

The pump piston stroke is limited externally by a stop on the roller shoes. The stroke travel is terminated when the tapered ramps at the shoes touch the corresponding ramps of the drive shaft. The axial shift of the rotor, therefore, creates a change in volume through changing the stroke limit of the pistons. The control pressure on the front side of the rotor is varied through the use of two electrohydraulic actuators. The control pressure acts against a retractor spring to change the axial position of the rotor.

Closed-loop control is achieved with a sensor on the rotor. The sensor measures the axial rotor posi-

645

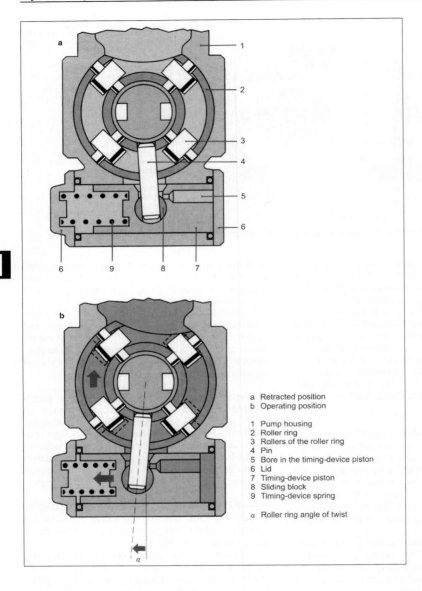

a Retracted position
b Operating position

1 Pump housing
2 Roller ring
3 Rollers of the roller ring
4 Pin
5 Bore in the timing-device piston
6 Lid
7 Timing-device piston
8 Sliding block
9 Timing-device spring

α Roller ring angle of twist

Fig. 177
Injection adjuster for
axial piston distributor
injection pump [7]

tion and creates the control loop through the control instrument.

The start of injection of the EPIC injection pump is also controlled. The method is similar to the common DPS and DPC technology, but it was modified for electronic control. The cam ring is rotated with the help of an electrohydraulic actuator. It acts with variable pressure on a piston, which is connected to the cam ring, and the piston acts against a prestressed spring. The cam position is determined by an inductive sensor and is compared with the set point in the control instrument [9, 10]

A fuel-injector body with pintle motion sensor can be used when more accurate control of the dynamic start of injection is required.

~~~~**VP 44 radial piston distributor pump (Bosch).**
A maximum injection pressure of more than 1300 bar is required on the nozzle side for performance enhanced and emission optimized automotive diesel engines with direct injection. This is where the radial piston design has advantages (balanced radial load of the cam ring, no piston retraction spring, improved tightness due to larger useful strokes).

The VP 44 achieves injection pressures of 2000 bar maximum in automotive applications and up to

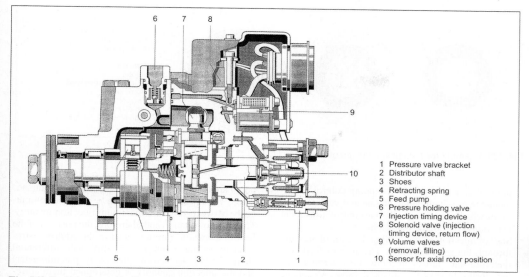

1	Pressure valve bracket
2	Distributor shaft
3	Shoes
4	Retracting spring
5	Feed pump
6	Pressure holding valve
7	Injection timing device
8	Solenoid valve (injection timing device, return flow)
9	Volume valves (removal, filling)
10	Sensor for axial rotor position

Fig. I78 Radial piston distributor injection pump EPIC design Delphi/Lucas [9]

approximately 1500 bar in commercial vehicle applications. Cylinder outputs of up to 45 kW can be achieved.

Fig. I79 shows the design principle with the following special features:

- Highly dynamic metering of fuel quantity through a central high-pressure solenoid valve (metered fuel injected), which includes a pre-injection option. To determine the exact quantity, the actual closing time of the high-pressure solenoid valve is determined from the electrical current progression during the control.

- Angle/time control of the solenoid valve by an increment system installed in the pump, also called angle of rotation sensor (DWS system), that also can detect local and time dependent speed variations during metering.

- The integration of the pump control instrument into the pump housing is required to permit model specific performance map balance to compensate for system tolerances (tolerances in the solenoid valve,

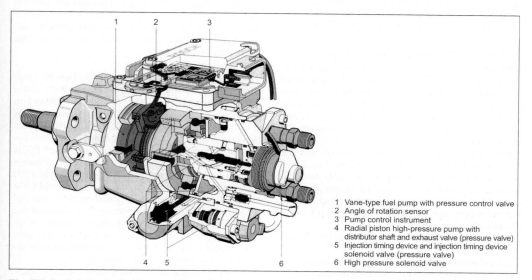

1	Vane-type fuel pump with pressure control valve
2	Angle of rotation sensor
3	Pump control instrument
4	Radial piston high-pressure pump with distributor shaft and exhaust valve (pressure valve)
5	Injection timing device and injection timing device solenoid valve (pressure valve)
6	High pressure solenoid valve

Fig. I79 Radial piston distributor injection pump VP44 [11]

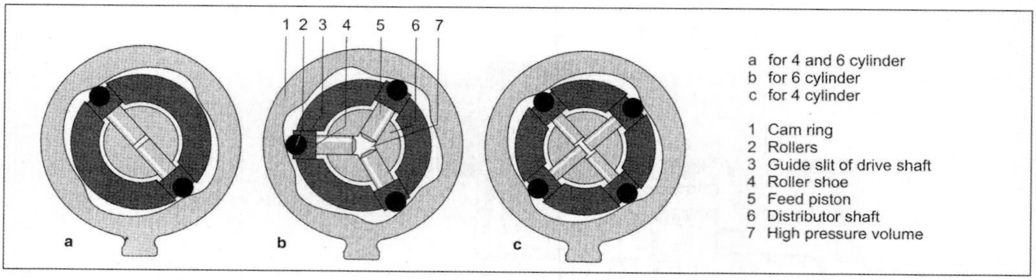

	a	for 4 and 6 cylinder
	b	for 6 cylinder
	c	for 4 cylinder
	1	Cam ring
	2	Rollers
	3	Guide slit of drive shaft
	4	Roller shoe
	5	Feed piston
	6	Distributor shaft
	7	High pressure volume

Fig. I80 Example for feed piston design of the VP44 [11]

pump part, and control). The pump control instrument can also include the engine and vehicle functions, which means that only one control instrument is required for engine management.

- Start of injection by controlling the control pressure through the solenoid valve. The start of feed and, therefore, the start of injection are determined by the closure time of the solenoid valve. A separate pintle motion sensor is not necessarily required. The injection adjuster piston will be controlled by a control sleeve positioned in the middle of the former. A control piston, which acts on the integrated control sleeve, is adjusted to the set point by the pressure, which is modulated by the solenoid valve. This permits a fast injection adjustment with low hysteresis, because the pulsed solenoid valve must only make small changes on the control piston. The range of adjustment for start of feed can be up to 40°CA.

The feed-rate of the VP 44 is significantly higher than that of the axial piston distributor pumps due to the larger number of pump pistons and due to the demand depending diameters. **Fig. I80** shows three examples for the feed piston arrangement in the VP 44.

~~~VP15, VP37 distributor pumps (Bosch), edge-controlled. When electronic diesel control (EDC) was introduced, the VE (Bosch)-designed edge-controlled distributor pump received electromagnetic actuation in place of the mechanical controller with its many add-on modular assemblies. An enhanced processing of variables is possible with the control instrument, the sensors, and the actuators.

The VP15 is used for indirect injection and the VP37 is used for direct injection engines. The differences between the designs are the materials used, the coatings, the dimensioning, and the machining tolerances. The basic function of metering the fuel injected is equivalent to the one in the VE distributor pump (Bosch), edge-controlled.

Fig. I81 shows the design of the electronically and edge-controlled VP.

The solenoid actuator (rotation actuator) engages by way of a shaft at the control sleeve. The deactivation sections of the mechanically controlled injection

pump are opened earlier or later depending on the position. The injection quantity can be adjusted continuously between zero and maximum injection quantity. The angle of rotation and, therefore, the position of the control sleeve is communicated to the control instrument by an angle sensor (currently: half differential short circuit ring sensor, previously potentiometer) and the injection quantity is determined there corresponding to the speed and other sensor signals related to fuel quantity through performance maps and characteristic curves. The retractor springs adjust a "zero" fuel feed quantity if no current is flowing.

Similar to the mechanical injection timing device, the speed related internal pump pressure has an effect on the timing-device piston. For the control of start of injection, an electromagnetic pulse valve with variable pulse ratio (ratio of time open to the total duration of a combustion cycle of the solenoid valve needle), which can be changed continuously by the electronic control instrument, is integrated into the injection timing circuit. This valve behaves as a variable nozzle and can, therefore, change the control pressure on the adjustment piston, so that the roller ring can get into any position between "early" and "late" feeding. A permanently open solenoid valve (pressure reduction) gives late delivery; a fully closed valve (pressure increase) delivers early start of injection.

~~~VP29, VP30 distribution pump (Bosch), cam-timing controlled. Further development of the cam-/edge-controlled VE distributor pumps resulted in solenoid-valve–controlled distributor injection pumps (VE-MV). The cylinder performances are comparable to those with edge-controlled pumps. The VP29 is designed for indirect injection engines, while the VP30 is designed for direct injection engines and achieves maximum injection pressures of approximately 1500 bar on the nozzle side. A high-pressure solenoid valve is used for the metered fuel injected replacing the mechanically actuated control sleeve. This increases the flexibility of the fuel metering and adjustment of the start of injection. The delivery of the fuel starts after the closing of the solenoid valve; the shut-off time determines the quantity of fuel injected.

Short switching times, which require small masses, of the high-pressure solenoid valve are necessary for

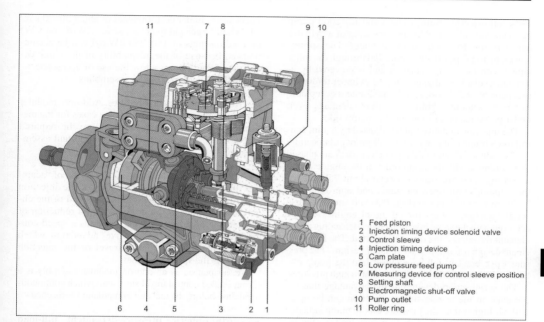

1 Feed piston
2 Injection timing device solenoid valve
3 Control sleeve
4 Injection timing device
5 Cam plate
6 Low pressure feed pump
7 Measuring device for control sleeve position
8 Setting shaft
9 Electromagnetic shut-off valve
10 Pump outlet
11 Roller ring

Fig. I81 Axial piston distributor injection pump with electromagnetic actuator [7]

accurate metering of the fuel injected. Large valve cross sections are designed for a fast and complete filling of the high-pressure chamber of the pump. **Fig. I82** shows the VP29/30. The pressure is generated, similar to the one in the VE distributor injection pump, edge controlled by an axial moving piston.

An angle-of-rotation sensor (incremental angle-time sensor, or IWS) integrated into the pump assumes the delivery of the angle of rotation and the cylinder synchronization. The connection of the sensor to the roller ring enables the correct correlation of the signals

to the cam stroke position, even when the cam ring is rotated because of an injection adjustment. The signals of this sensor are transmitted to and analyzed in the control instrument on the top side of the pump. The high-pressure solenoid valve and the injection adjustment pulse valve are controlled by this control instrument. It only controls the pump, which requires an additional control instrument for the engine and vehicle functions (*see above*, ~~~Radial position distributor injection pump ~~~~VP 44 radial position distributor pump [Bosch]).

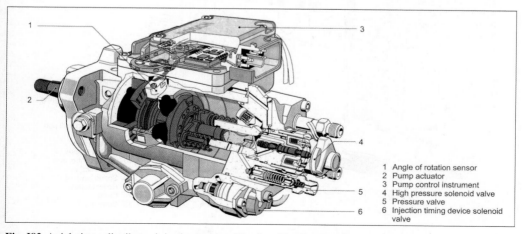

1 Angle of rotation sensor
2 Pump actuator
3 Pump control instrument
4 High pressure solenoid valve
5 Pressure valve
6 Injection timing device solenoid valve

Fig. I82 Axial piston distributor injection pump with solenoid valve quantity metering [2]

~~**Serial injection pump.** The serial injection pump has a so-called pump element consisting of a cylinder and a piston for each engine cylinder. The spring-loaded piston is positively connected through the roller tappet with the pump camshaft and is powered by it. The pump piston is aligned with the cylinder through a precision fit of 2–5 μm, and this guarantees acceptable tightness without additional sealing elements even under pressures of 1150 bar on the pump side.

The injection quantity can be adjusted by a joint control rod through a pinion gear or ball/groove connection.

The individual cams of the pump camshaft are phased in accordance with the firing order of the engine. Maximum injection pressures between about 750 bar for the A-pump with open elements and approximately 1350 bar for the P-pump (**Fig. I83**) and control sleeve serial injection pumps for DI-diesel engines can be achieved with the serial injection pump, depending on the element design and the quantity injected. The flange element design is used for the two latter designs. The pressure valve bracket will be screwed into the pump element, which makes the system resistant to high pressures.

The serial pump design has the advantage that the pressure on the nozzle side at injection can be up to 250 bar higher than the pressure on the pump side due to the additive nature of the pressure waves—the increase in pressure depends on the length of the injection line and the engine speed.

The serial pump is normally designed as a single unit together with a mechanical controller or an electro-mechanical actuator. Also included in the pump controller unit is a fuel feed pump and sometimes an injection timing device. The serial pump is a flexible solution for different engine classes from 2 to 12 cyl-inders and cylinder performances of about 30–70 kW/cyl. When larger pumps (e.g., series ZWM and CW) are used, outputs of 150–250 kW/cyl can be accommodated because of the adaptability of the pump design, and larger pumps allow the use of accessories—the so-called add-on modular assemblies.

~~~**Add-on modular assemblies.** Add-on modular assemblies may be required in some cases for the mechanical flyweight governor to achieve the required adaptation for an engine. The most important assemblies are:

~~~~**Adaptation.** Adaptation is the process of "adapting" the natural feed characteristics of the injection pump in the full-load range to the demand of the engine with respect to torque, power, and reduction of smoke. The adaptation device works as a speed controller in the mostly used "positive" adaptation, which means it reduces the control travel or the injection quantity as the speed increases.

The adaptation as an add-on modular assembly is a spring-loaded control travel stop, with which continuous regulation before the real end of regulation is reached.

~~~~**Atmospheric pressure-dependent full-load stop.** In countries with large variations in altitude, the quantity of fuel injected must be adapted—which means reduced—at certain higher altitudes to take into account the deterioration in the air quantity in the engine cylinders. The air mass in the engine cylinder is reduced at high altitudes because of the reducing density. The black smoke limit would be exceeded if this quantity of fuel was not reduced.

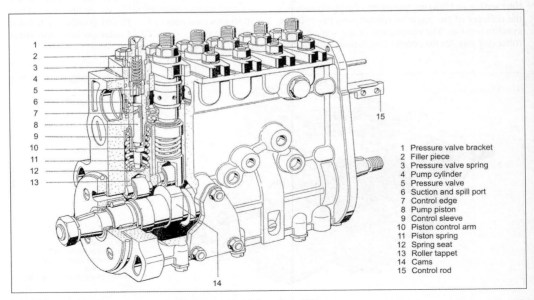

1 Pressure valve bracket
2 Filler piece
3 Pressure valve spring
4 Pump cylinder
5 Pressure valve
6 Suction and spill port
7 Control edge
8 Pump piston
9 Control sleeve
10 Piston control arm
11 Piston spring
12 Spring seat
13 Roller tappet
14 Cams
15 Control rod

Fig. I83 Serial injection pump series P for commercial engines [6]

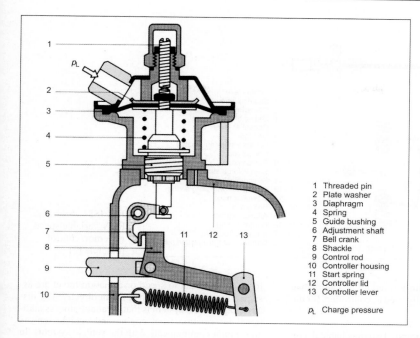

1 Threaded pin
2 Plate washer
3 Diaphragm
4 Spring
5 Guide bushing
6 Adjustment shaft
7 Bell crank
8 Shackle
9 Control rod
10 Controller housing
11 Start spring
12 Controller lid
13 Controller lever

p_L Charge pressure

Fig. I84
Boost-pressure–
dependent full-load
stop [6]

The atmospheric pressure-dependent full-load stop is a control rod stop which takes the full-load stop of the control rod back in accordance with the increasing altitude; its position is related to the decreasing air pressure.

~~~~**Boost-pressure–dependent full-load stop.** The full-throttle fuel delivery for a turbocharged engine is adapted to the increased air mass as a result of the boost pressure (**Fig. I84**). However, the boost pressure is less at low speeds and this reduces the air mass drawn into the engine cylinders. The full-throttle fuel delivery must, therefore, also be adapted to the reduced air mass in a corresponding ratio. The boost-pressure–dependent full-load stop is used for this purpose. It is normally a spring-loaded diaphragm cell, which will be charged with the boost pressure from the exhaust turbocharger, and it is used as a control rod stop. The control rod will be moved in the direction of the stop at declining boost pressures. The boost-pressure–dependent full-load stop can be unlocked once during the start process through a manual or electromagnetic intervention. This start release must be relocked safely during the first ramp-up of the engine to avoid excessive smoke development and a thermal overload of the engine. Careful adjustment of the boost-pressure–dependent full-load stop is required especially for dynamic load changes (e.g., acceleration). On the one hand, it is desirable to have an increase in fuel delivery that is as fast as possible to support the acceleration, and on the other hand, the required quantity cannot rise "faster" than the increase in boost pressure from the turbocharger, because this would result in unacceptable smoke development.

~~~~**Shut-off device.** The diesel engine can only be shut off if either the airflow or the fuel injection gets interrupted. The latter method is normally used. Mechanical intervention into the control linkage, pneumatic capsules, or solenoids may be used for this purpose. They all are used to bring the control rod into the stop position.

An electrohydraulic cutoff valve for the fuel feed in electronically controlled serial pumps is an alternative solution. This cutoff method is used as a redundant safety cutoff for electromagnetic actuators. The normal cutoff is performed in this case by magnet actuation through the control rod. Switch-off through an intake airflow cutoff is often used for large diesel engines as a safety cutoff.

~~~~**Temperature-dependent starting stop.** In modern diesel engines with direct injection, an increased fuel quantity at start is only required at low ambient temperatures and for a cold engine.

The injection of an increased quantity at start must be kept at the absolute minimum to fulfill the environmental protection requirements. This is why the controller has a device that only releases the required control travel at the start.

The demanded quantity during warm start is limited through the injection pump control rod travel with the help of an expansion element or a temperature-controlled solenoid; these are controlled by the ambient temperature and the warm start quantity demand, and the device is adjusted to the ambient engine temperature.

~~~**Controller for diesel injection pumps.** Mechanical flyweight governors (measuring instrument) or

651

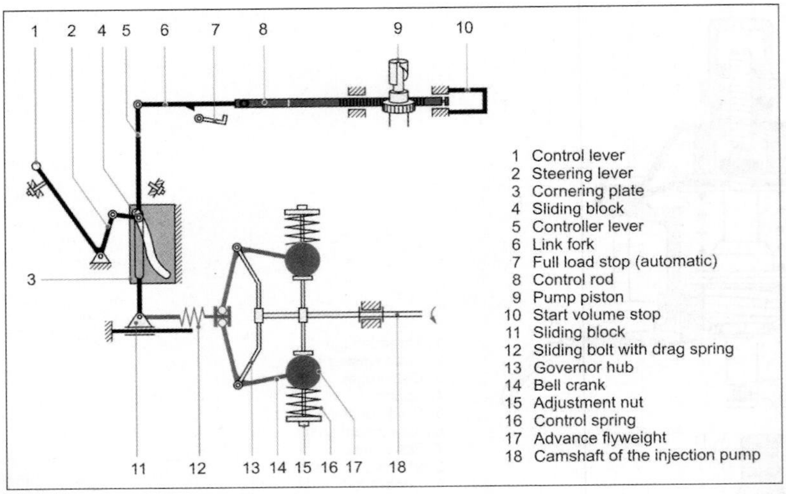

1 Control lever
2 Steering lever
3 Cornering plate
4 Sliding block
5 Controller lever
6 Link fork
7 Full load stop (automatic)
8 Control rod
9 Pump piston
10 Start volume stop
11 Sliding block
12 Sliding bolt with drag spring
13 Governor hub
14 Bell crank
15 Adjustment nut
16 Control spring
17 Advance flyweight
18 Camshaft of the injection pump

Fig. I85
RQV variable speed
governor, control
principle [2]

electronic controllers with a solenoid actuator (magnetic actuator) are currently used as controllers for diesel injection pumps.

~~~~**Measuring instrument.** Two mechanical controller designs have been proven [3].

R-RQV/RQVK controller family. This works with flyweights and an integrated revolving control spring (**Fig. I85**). It has the advantage that the controller linkage is kept free of high frictional forces and results in a high dynamic response and low hysteresis.

The individual abbreviations mean the following:

RQ = Idle end speed controller or aggregate controller

RQV = Variable speed governor, step controller

RQVK = Variable speed governor with cam-controlled, full-load characteristics (**Fig. I86**).

RSF/RSV controller family. This works with folded weights and external, fixed control springs (**Fig. I87**). This results in compact dimensions and advantages for adjustment through simpler interaction with the control spring. These controllers can be used only for smaller serial injection pumps because of their higher controller friction and the small controller work potential.

RSF = Idle end speed controller for automobile serial injection pumps

RSV = Variable speed governor especially for tractors, machines, and power generators. The RSV controller also is called a spring-loaded controller because of the control spring, which is adjusted through the adjustment lever.

~~~~**Solenoid actuator for diesel injection pump.** In mechanical applications for controlling speed, the functions of speed measurement and conversion into a

force proportional to speed or actuation travel are assumed by the measuring instrument directly. The disadvantage of this system is that it cannot process additional information from engine and vehicle sensors and cannot communicate with the vehicle systems. In a solenoid actuator, which is installed in a serial injection pump instead of the mechanical controller, a linear magnet moves the control rod. The control sensor, which is also integrated into the actuator, communicates the control rod position to the control instrument, and a demand/actual comparison is performed here. The speed sensor can be integrated in the solenoid actuator, and it measures the actual pump speed with the help of a pulse wheel. In other cases, the speed sensor is attached to the engine, and it samples the rotational markings on the starter ring gear of the engine (**Fig. I88**).

~~**Single-injection pump.** Single-injection pumps do not have their own camshaft. They are used for single- and multicylinder engines. They are mechanically or electromechanically controlled by a controller that is integrated into the engine. The injection quantity is transferred through short injection lines to the fuel-injector bodies (**Fig. I89**).

Applications for the single-injection pumps can be found in small single- and multicylinder engines for construction machines, aggregates, and tractors (4–76 kW/cylinder), as well as in large engines for ships, locomotives, and stationary engines (75–1000 kW/cylinder). Powerful options also have been developed for these single-cylinder engines; the upper limits of the injection peak pressure are between 800 and 2000 bar, depending on the size of pump and fuel used.

The highest-pressure pumps are designed for the use of heavy oil and for a long service life. These pumps use so-called blind-hole elements to minimize deformation. The cylinder body of this design is closed by a base, and the high-pressure chamber is connected

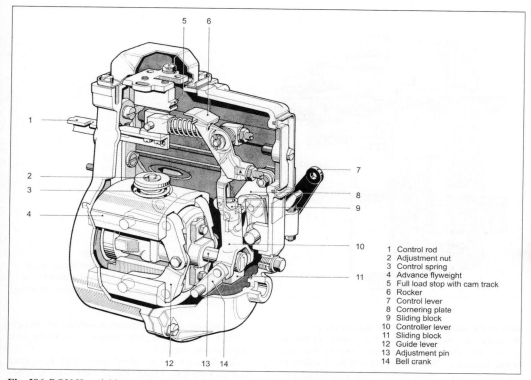

Fig. I86 RQV-K variable speed governor with cam-controlled, full-load adaptation [2]

1 Control rod
2 Adjustment nut
3 Control spring
4 Advance flyweight
5 Full load stop with cam track
6 Rocker
7 Control lever
8 Cornering plate
9 Sliding block
10 Controller lever
11 Sliding block
12 Guide lever
13 Adjustment pin
14 Bell crank

to the pressure chamber valve through an exhaust bore only.

~~**Unit injector (UI).** The electronically controlled injection system with UI has a modular design. The UI is a single-cylinder injection pump with integrated so-lenoid valve or piezo actuator (control valve) and integrated injector nozzle. It is directly installed in the cylinder head of the diesel engine. The UI is powered by rocker arms or in some cases by pushrods through the upper engine camshaft. The injection is performed by switching of the control valve (**Fig. I90**). The closing

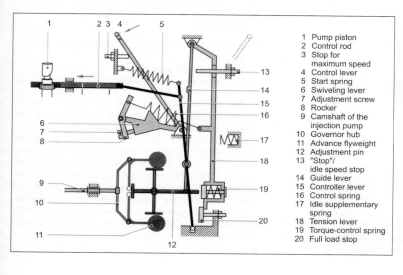

1 Pump piston
2 Control rod
3 Stop for maximum speed
4 Control lever
5 Start spring
6 Swiveling lever
7 Adjustment screw
8 Rocker
9 Camshaft of the injection pump
10 Governor hub
11 Advance flyweight
12 Adjustment pin
13 "Stop"/ idle speed stop
14 Guide lever
15 Controller lever
16 Control spring
17 Idle supplementary spring
18 Tension lever
19 Torque-control spring
20 Full load stop

Fig. I87
RSV variable speed governor, control principle [2]

653

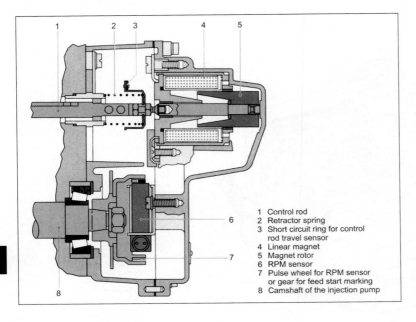

1 Control rod
2 Retractor spring
3 Short circuit ring for control
rod travel sensor
4 Linear magnet
5 Magnet rotor
6 RPM sensor
7 Pulse wheel for RPM sensor
or gear for feed start marking
8 Camshaft of the injection pump

Fig. I88
Solenoid actuator for
electrical control of a
serial injection pump [2]

time of the valve determines the start of injection, and the closed duration determines the quantity of fuel injected. The control valve is activated by the electronic control instrument, which can be installed on the engine (*also*, →Electronic open- and closed-loop control ~Electronic open- and closed-loop control, diesel engine).

A part of the pressure losses caused by compression are eliminated because the UI does not need a pressure line. Also eliminated are the shifts in start of injection which are dependent on speed because of the injection line (injection delay). In contrast to systems with pres-

sure lines, the small compression volume has the effect that the injection characteristics follow the cam shape almost exactly.

The electronic valve control provides the UI system with the option to realize multiple injections or injection curve shaping. UI systems are available up to 2000 bar for cars and up to 1800 bar for commercial vehicles. **Fig. I91** shows the design of a unit injector for cars with integrated hydraulic/mechanical pre-injection through a storage or bypass piston.

~~Unit pump (UP). The unit pump system has a modular design and is only used for car engines (**Fig. I92**). Each engine cylinder is supplied by its own injection module, which consists of the following components:

- High-pressure unit pump with fast-switching sole-noid valve
- Very short high pressure line
- Nozzle bracket combination

The high-pressure unit pump is powered by the associated injection cams through a roller tappet. The engine camshaft with the injection cam is normally located at the side or below. The injection is performed by the switching of the solenoid valve. The closing time of the valve determines the start of injection and the closed duration determines the quantity of fuel injected. **Fig. I93** shows the design of a unit pump with a short injection line and nozzle bracket combination.

Unit pump systems have a lot of similarities with the unit injector system and have the same control options (electronic open- and closed-loop control diesel engine).

1 Pressure valve
2 Bleeder screw
3 Pump cylinder
4 Pump piston
6 Control sleeve
7 Guide bushing

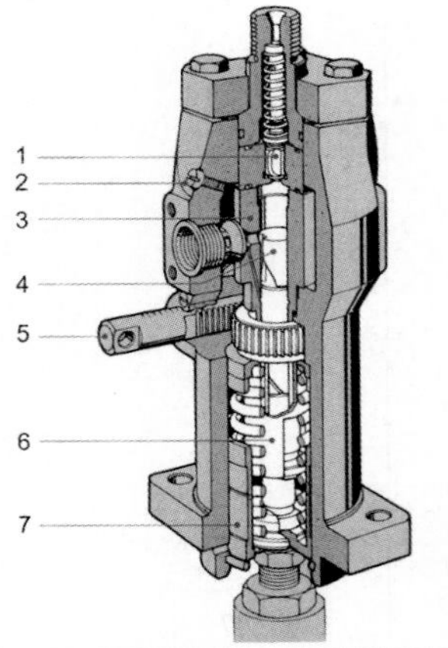

Fig. I89 Single-injection pump with mechanical controller for large engines, design PF1D [2]

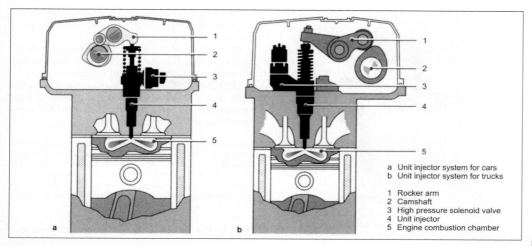

Fig. I90 Design of a unit injector in the cylinder head [2]

a Unit injector system for cars
b Unit injector system for trucks

1 Rocker arm
2 Camshaft
3 High pressure solenoid valve
4 Unit injector
5 Engine combustion chamber

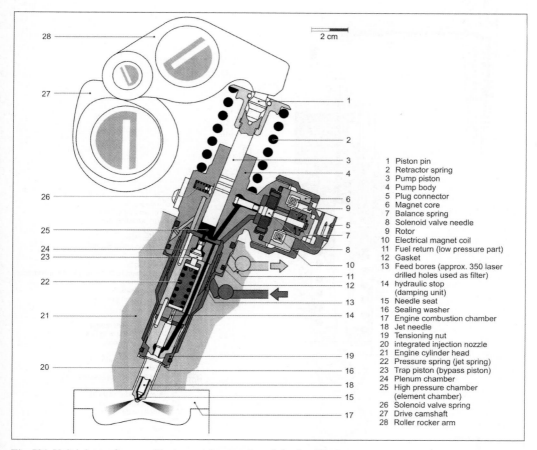

2 cm

1 Piston pin
2 Retractor spring
3 Pump piston
4 Pump body
5 Plug connector
6 Magnet core
7 Balance spring
8 Solenoid valve needle
9 Rotor
10 Electrical magnet coil
11 Fuel return (low pressure part)
12 Gasket
13 Feed bores (approx. 350 laser
    drilled holes used as filter)
14 hydraulic stop
    (damping unit)
15 Needle seat
16 Sealing washer
17 Engine combustion chamber
18 Jet needle
19 Tensioning nut
20 integrated injection nozzle
21 Engine cylinder head
22 Pressure spring (jet spring)
23 Trap piston (bypass piston)
24 Plenum chamber
25 High pressure chamber
    (element chamber)
26 Solenoid valve spring
27 Drive camshaft
28 Roller rocker arm

Fig. I91 Unit injector for car with storage piston and pre-injection [2]

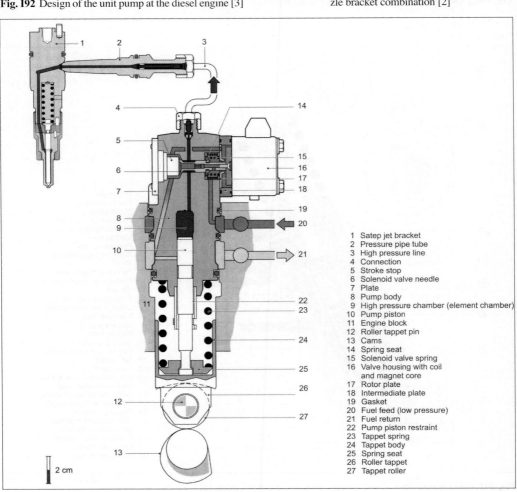

The UP system has the advantage of a simple installation concept and it can be handled in a simpler manner during service.

The same good results are also achieved with respect to the injection characteristics and controllability of injection quantity and start of injection.

The UP system is used for commercial vehicle engines with 1–2 L/cylinder, and the achievable peak pressures on the nozzle side are 1600–1800 bar, similar to those of the unit injector system.

Literature: [1] Robert Bosch GmbH (Ed.): Diesel-Einspritzsysteme, Unit Injection System/Unit Pump System, Technical Instruction, Order No. 1 987 722 056, Stuttgart, 1999. – [2] Robert Bosch GmbH (Ed.): Dieselmotor-Management, 3rd Edition, Wiesbaden, Vieweg, 2002. – [3] R. Schwartz: Einspritztechnik, in Shell-Lexikon Verbrennungsmotor, Supple-

1 Camshaft	4 Engine combustion chamber
2 Unit pump	5 Jet bracket combination
3 High pressure solenoid valve	6 Short high pressure line

Fig. I92 Design of the unit pump at the diesel engine [3]

▼ **Fig. I93** Design of the unit pump for commercial vehicles with short injection line and nozzle bracket combination [2]

1	Satep jet bracket
2	Pressure pipe tube
3	High pressure line
4	Connection
5	Stroke stop
6	Solenoid valve needle
7	Plate
8	Pump body
9	High pressure chamber (element chamber)
10	Pump piston
11	Engine block
12	Roller tappet pin
13	Cams
14	Spring seat
15	Solenoid valve spring
16	Valve housing with coil and magnet core
17	Rotor plate
18	Intermediate plate
19	Gasket
20	Fuel feed (low pressure)
21	Fuel return
22	Pump piston restraint
23	Tappet spring
24	Tappet body
25	Spring seat
26	Roller tappet
27	Tappet roller

2 cm

ment to ATZ und MTZ in 90 issues (04/1995 to 05/2003), Is-
sues 7 to 12, Wiesbaden, Vieweg, 1995. − [4] T. Pauer:
Laseroptische Kammeruntersuchungen zur dieselmotorischen
Hochdruckeinspritzung−Wirkkettenanalyse der Gemischbil-
dung und Entflammung, University of Stuttgart: Dissertation,
2001. − [5] L. Komaroff, K. Melcher: Messung der Strahl-
kraft und Strahlbewegungsgröße zur Beurteilung der Zer-
stäubungsgüte von Einspritzstrahlen, Bosch Techn. Reports,
vol. 6 (1971) No. 6. − [6] Robert Bosch GmbH (Ed.): Diesel-
Reiheneinspritzpumpen, Technical Instruction, Order No. 1
987 722 012, Stuttgart, 1998. − [7] Robert Bosch GmbH
(Ed.): Kantengesteuerte Diesel-Verteilereinspritzpumpen,
Technical Instruction, Order No. 1 987 722 014, Stuttgart,
2002. − [8] Delphi (Troy, Michigan, USA): Company Papers.
− [9] G.R. Lewis: Das EPIC-System von Lucas, MTZ 53
(1992) 5, pp. 224–229. − [10] M. Norman: Aktuelle Trends
und zukünftige Entwicklungen der Dieselkraftstoff-Einspritz-
technologie, MTZ 54 (1993) 2, pp. 80–85. − [11] Robert
Bosch GmbH (Ed.): Diesel-Radialkolben-Verteilereinspritz-
pumpen, Technical Instruction, Order No. 1 987 722 053,
Stuttgart, 1998.

~Gasoline engine. Gasoline engines are divided into
direct and intake manifold injection systems.

~~**Direct injection systems.** Direct injection systems
for gasoline engines have only been established during the last few years and are in addition to the intake
manifold injection systems. Mechanical gasoline direct injection systems were first used in aircraft engines in 1938 to avoid carburetor icing and to increase performance.

Current direct injection systems are used to guarantee
economical, environmentally friendly, and powerful engine operation and to improve the engine performance.

The reduction in fuel consumption compared with
the intake manifold injection system is achieved by

- Reduction of the throttling losses at the throttle
  valve during stratified charge operation,
- Improved thermal efficiency through a higher compression ratio, and
- Reduced wall heat losses

~~~**Air-shrouded direct injection.** Air-supported
processes with direct injection, such as the OCP™
(Orbital Combustion System, a registered trademark
of Orbital; **Fig. I94**), exist in addition to the liquid
high-pressure processes with direct injection in accordance with the common rail method. Stable combustion processes with good stratification performance and
compatibility with high exhaust gas recirculation rates
are possible with the air-supported direct injection process based on the special mixture preparation quality.
The main component of air-supported direct injection
is a design of an electromagnetic fuel injection valve
and an electromagnetically actuated air injection valve,
which is used to inject atomized fuel into the combus-

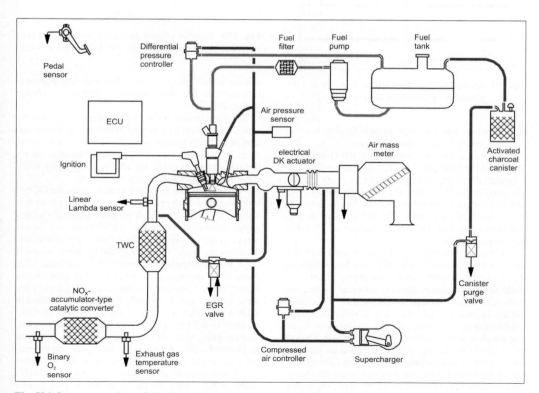

Fig. I94 System overview of the air-supported Orbital direct injection system

Fuel

MPI fuel injector

Air

Air injector

Injected mixture cloud

Control pulse sequence

Fuel injection pulse

Direct injection pulse

Ignition

| 720° TDC | 360° TDC | 180° | 0° TDC |

Fig. I95 Arrangement of fuel and air injector and the phased progression of the control pulses

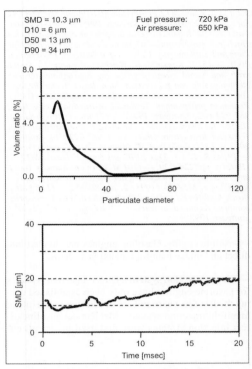

SMD = 10.3 μm D10 = 6 μm D50 = 13 μm D90 = 34 μm

Fuel pressure: 720 kPa Air pressure: 650 kPa

Volume ratio [%]

Particulate diameter

SMD [μm]

Time [msec]

Fig. I96 Volume related and drop size distribution with respect to time for the Orbital process

tion chamber. The injection system is also feasible for spray-guided combustion processes.

It is divided into two subsystems, the compressed air path and the fuel path. The required compressed air is produced by a supercharger driven by gears or belts. The pressure level is adjusted to the set point by a mechanical pressure controller, and the fuel is delivered by an electrical fuel pump. The fuel pressure is controlled to a constant differential pressure of 1.5 bar to the compressed air.

The fuel is metered using a conventional injection valve for intake manifold injection. The fuel is injected into a mixing chamber in the air injection valve. A finely atomized mixture cloud is delivered to the combustion chamber through an air injector. This cloud can be ignited immediately, and a spray-dependent combustion process with low raw emissions is therefore possible with stratified charge operation. Correct coordination of phasing (i.e., start of the air injector) and ignition guarantees that optimum air ratio of mixture in the area of the spark plug can be maintained to achieve stable combustion under all operating conditions. The spatial arrangement of fuel and air injector

and the phased progression of fuel metering and mixture injection into the combustion chamber are shown in **Fig. I95**.

The compressor in the system offers a new and interesting solution for flushing of the activated charcoal canister. The air volume from the compressor is routed over the activated charcoal canister. This also allows the application of the stratified charge operation without disadvantages for the raw emissions.

The difference in the fuel preparation for the OCP™ injector to the high-pressure injection valve is as follows. The spray in a high-pressure injector disintegrates primarily because of turbulences and inertia forces in the fuel spray itself; a distance of about 10–50 times the outlet diameter of the nozzle is required until the disintegration process is completed. The spray in an air-supported injector disintegrates when aerodynamic forces exceed the surface tension of the fluid. The pressure level in the air injector is selected in such a way that the critical pressure ratio at valve opening is not exceeded during the injection process. The resulting speed of sound of the airflow results in large aerodynamic forces at the fuel spray and the major part of the atomization process is already completed immediately at the valve exit. Other thermodynamic effects, such as the evaporation, play a role especially during the fuel injection into a medium with

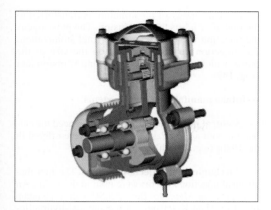

Fig. I97 Cross section of the air compressor

increased temperature. This interaction of the fuel spray with the cylinder contents is the interface between fuel system and combustion system. **Fig. I96** shows the atomization quality of an OCP™ injector. The Sauter mean diameter is 10.3 μm, and only a very small part of the volume has a diameter larger than 40 μm.

The required air mass flow for the OCP™ air injector relative to the total air quantity acquired by the engine varies from 15% for homogenous idle operation to 1.5% at full load. This results in about 5–9 mg of air per injection pulse for a 1.5-L four-cylinder engine. The pressure at the air rail is preferably adjusted to 6.4 bar. The compressed air is produced by a water-cooled piston compressor, which is driven by the engine

through gears or belts. **Fig. I97** shows the cross section of a compressor with a displacement of 30 cm³, which is designed for a 1.5-L four-cylinder engine. A fuel pressure range of 7.2 to 10 bar is achieved with an air pressure of 6.5 bar and a differential pressure in the range of 0.7 to 3.5 bar.

The air-supported direct injection "Orbital" is produced for two-cycle boat engines and scooters.

Literature: Motorrad, vol. 23/94, pp. 28–30.

~~~Common rail. In contrast to the direct injection systems of the past, which used mainly mechanical fuel injection pumps, the fuel in this application is injected directly into the combustion chamber under high pressure through an electronically controlled injection valve from a fuel distributor rail ("common rail" method).

Fig. I98 shows a system overview of a pressure-controlled common rail injection system. A speed-related feed volume is delivered from the high-pressure pump to the fuel distributor rail. A pressure regulator controlled by the engine control system controls the pressure to the pressure set point independent of the fuel quantity injected into the engine. The return from the pressure controller goes directly to the intake of the high-pressure pump. A pressure sensor is used for pressure measurements. A pressure relief valve is integrated into a high-pressure circuit for safety reasons; it limits the maximum fuel pressure. The electrical pressure controller can basically be replaced by a mechanical pressure controller if the fuel pressure in the distributor rail is kept constant over the total engine speed and load range.

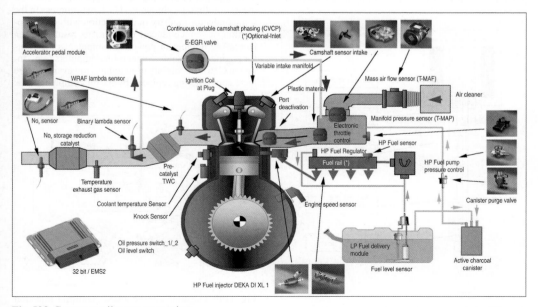

Fig. I98 Common rail system overview

An engine with direct injection requires the use of an electronically controlled throttle valve for the different operating conditions under homogeneous loading and stratified charge. The mixture control for lean mixtures with stratified charge requires a linear λ sensor, which can also guarantee this parameter for the homogeneous operation with λ = 1. The lean mixture for combustion during stratified operation causes a problem with the secondary exhaust treatment because conventional three-way catalytic converters cannot reduce the NO$_x$ emissions. A special secondary treatment of the NO$_x$ emissions is required to fulfill the emission limits despite the reduced raw emissions NO$_x$ because of exhaust gas recirculation rates of up to 30%. So-called NO$_x$ accumulator-type catalytic converters are used currently in addition to selective NO$_x$ reduction catalysts, whose thermal stability is still low.

The accumulator-type catalytic converters absorb/adsorb the NO$_x$ emissions during lean operation and convert them into N$_2$ and CO$_2$ under stoichiometric operation. An extensive engine management function controls this process. Accumulator-type catalytic converters have a tendency to be poisoned by sulfur and, therefore, need fuel with low sulfur content. An NO$_x$ sensor is used for monitoring and control of the lean catalytic converter.

The effective fuel consumption savings decline because of the requirement for secondary exhaust treatment.

~~~**Ficht injection system.** This system is an electromagnetic pump-nozzle unit, which is used preferably in two-cycle engines. The injection point and the injection quantity are determined by a actuation so-lenoid. The energy of the armature is transferred to the fuel after it hits the feed piston. The pulse opens the injection valve and is used for fuel atomization. The injection quantity depends on the size of the pulse, which can be varied by a control instrument (**Fig. I99**).

~~**Intake manifold injection systems.**

~~~**Central injection.** Systems that replaced the carburetor with a central injection valve were offered in the 1980s (e.g., Bosch Mono-Motronic).

~~~**Mechanical fuel injection systems.** The first mechanical injection systems were used in the 1950s. An example is the Mercedes Benz 220SE with serial pump injection. The K-Jetronic, a mechanical-hydraulic-controlled system, was developed in the 1960s.

~~~**Multipoint injection.** An electrical multipoint injection system was first described in a 1956 patent by Bendix (**Fig. I100**).

First volume production of electronic systems developed in the 1960s (Bosch D-Jetronic). Systems with direct air volume metering and lambda control followed. The analog circuits were gradually replaced by digital-controlled circuits using microcomputers. In the following years, the mechanical and electrical parts of the engine and transmission control were increasingly combined to an electronic unit, which comprised all operating parameters and the required control interactions with the help of electronically controlled actuators.

In the following years, additional control functions were introduced into engine development because of the use of digital technology. The large progress in the semiconductor industry in conjunction with mass production of semiconductor components resulted in an increasing number of powerful computer generations for engine electronics. Sixteen-bit computers (e.g., Infineon C167) are mainly used today. However, they are increasingly being replaced by 32-bit computers. The

| Drive magnet |
| Feed piston |
| Feed valve |
| Pressure line |
| Injection nozzle |

Fig. I99 Ficht direct injection two-cycle system

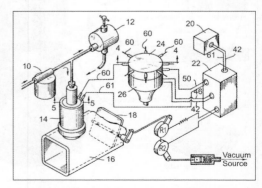

Fig. I100 Gasoline engine intake manifold injection system, Bendix

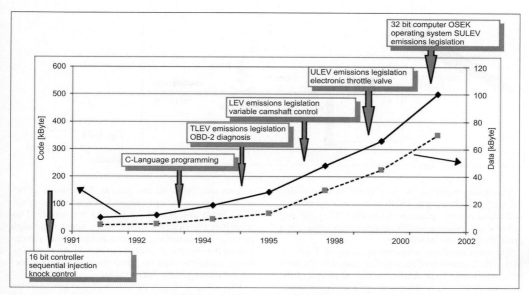

Fig. I101 Development of functional scope of engine and transmission control

increase in computer performance resulted in an increase of the functional scope.

Multipoint systems are currently equipped with a multitude of electronically controlled functions (**Fig. I101**; *also*, →Electronic engine and transmission control).

Injection valve →Injection system (components) ~Diesel engine, ~Gasoline engine; →Injection valves

Injection valve construction →Injection valves ~Gasoline engine ~~Intake manifold injection

Injection valve curve →Injection valves ~Gasoline engine ~~Intake manifold injection

Injection valve design →Cylinder head ~Injection-nozzle arrangement

Injection valves. Injection valves are used to control the fuel metering dependent on volume and time.

~Diesel engine. The injection valves, installed on an associated fuel-injector body or injector, are used for the following tasks:

- Fuel metering into the combustion chamber with a defined timing
- Fuel atomization
- Forming of the injection curve
- Sealing of the combustion chamber

The combustion processes and the geometry of the combustion chamber must be taken into consideration

for the geometrical design. The shape and the direction of the spray, the fuel spray pulse, the atomization of the fuel spray, the injection time, and the injection quantity must be optimized.

The dimensions and the principal design features of the nozzles and the fuel-injector bodies are standardized in accordance with the following standards: ISO 2697, ISO 2699, ISO 2974, ISO 3539, ISO 7026, ISO 7030.

The injection nozzles are divided into two groups depending on the combustion process. Pintle-type nozzles are used for processes with divided combustion chambers (preswirl and swirl chamber engines — i.e., indirect injection engines). Orifice nozzles with a different number of holes are used in engines with direct injection and undivided combustion chambers.

~~Common rail injector. The core of the injector is a piezo actuator, which permits the use of relatively low electrical voltages and which reliably fulfills vehicle-related requirements relative to temperature and vibration. **Fig. I102** shows the design principle of the common rail piezo injector.

The actuator can open or close the servo valve in less than 100 μs. The opening speed of the nozzle and, therefore the injection curve and the minimum injection quantity, which is determined by the shortest possible control time, can be influenced together with the coordinated intake and outflow throttle to the control volume above the needle jet. These processes are basically performed without dead time. This shows that the piezo technology permits high repeatability of injection.

Fig. I103 shows an enlarged example of such an injector.

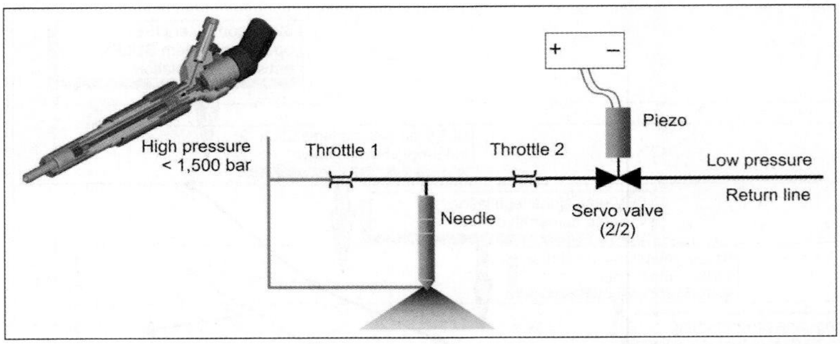

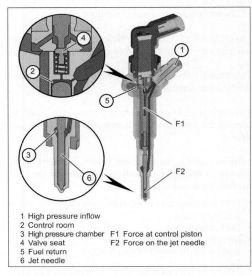

Fig. I102
Design of a CR-piezo
actuator

The piezo actuator (4) is a stack in a so-called multilayer technology, which connects a number of individual ceramic plates to each other. They are prestressed in a housing—temperature compensation is a problem that must be solved.

The required large temperature range of a vehicle makes the expansion of the housing caused by the temperature change large relative to the expansion of the ceramic plates when voltage is applied. Temperature compensation is implemented by selecting the appropriate materials for the housing around the piezo stack together with a prestressed spring and with clever determination of the idle stroke of the actuator. On the one hand, the injector should not stay open (idle stroke too small), which could result in material damage; but on the other hand, very small control times should not lead to nonopening (idle stroke too large), which increases the combustion noise significantly because of the missed pilot injection. The servo valve is another specialty, which opens to the inside of the high-pressure chamber, in contrast to the valve actuated by a

magnet that opens to the outside. The reason is because the piezo not only expands when voltage is applied, but it also provides a force to the outside. This simplifies the opening of the valve against the high pressure in contrast to the piezo and simplifies the design of the injector in contrast to movement in the opposite direction when the piezo receives a current during the closing of the servo valve.

This actuator design permits a 40-μm stroke of the control valve for the total temperature range of −30 to +140°C in a vehicle engine. **Fig. I104** shows the functionality of this design.

Fuel at a high rail pressure is present in the injector control chamber (2) and in the high-pressure chamber (3) of the nozzle if the injector is not controlled. The bore to the fuel return (5) is closed by a valve cluster (4) with the help of a spring. The hydraulic force (F1) applied by the fuel high pressure to the needle jet (6) in the control chamber (2) is larger than the hydraulic force (F2), which is applied to the nozzle tip, because the area of the control piston in the control chamber is

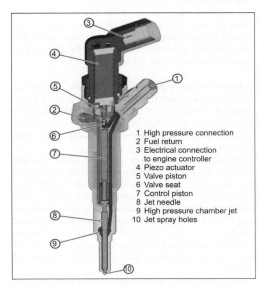

1 High pressure connection
2 Fuel return
3 Electrical connection
   to engine controller
4 Piezo actuator
5 Valve piston
6 Valve seat
7 Control piston
8 Jet needle
9 High pressure chamber jet
10 Jet spray holes

Fig. I103 Cross section of a piezo injector

1 High pressure inflow
2 Control room
3 High pressure chamber   F1 Force at control piston
4 Valve seat   F2 Force on the jet needle
5 Fuel return
6 Jet needle

Fig. I104 Piezo actuator, not controlled

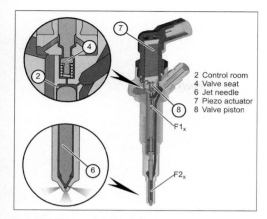

2 Control room
4 Valve seat
6 Jet needle
7 Piezo actuator
8 Valve piston

Fig. I105 Piezo actuator actuated

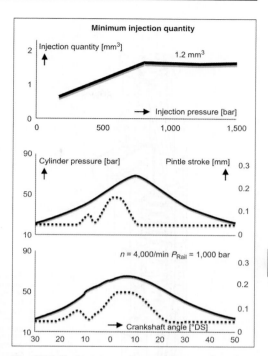

Fig. I106 Performance of a second-generation piezo injector

larger than the free area underneath the needle jet. The nozzle of the injector is closed.

The piezo actuator (7) presses the valve piston (8) and the valve cluster (4) and opens the bore, which connects the control chamber (2) with the fuel return, if the injector is actuated (**Fig. I105**).

This results in a pressure drop in the control chamber, and the hydraulic force, which is applied to the tip of the needle jet, is larger than the force on the control piston (F1) in the control chamber. The needle jet (6) moves upward, and the fuel is transferred through the spray holes into the combustion chamber of the engine.

The valve, which connects the control chamber to the fuel return, and the injector nozzle are closed by spring force when the engine stops. **Fig. I106** shows the performance of a second generation piezo injector.

Minimum injection quantities of below 1.5 mm at up to 1600 bar can be achieved for engine sizes with cylinder volumes of 0.5 liters based on a total adjustment of the injector, which means with sufficiently fast opening and closing edges. At the same time, the start of injection distance for pre- and main injection can be selected to be very small. The smallest distance for a speed of 2000 rpm is below 6°CA, and a distance of approximately 12°CA is still achievable at a speed of 4000 rpm. Larger starts of injection distances have no limits.

Both diagrams show that preinjection is possible over the whole area of the performance map, which means over the total pressure range as well as the total speed range.

~~**Externally opening.** For the externally opening diesel engine injection nozzles, the needle jet leaves the jet body partly in the direction of the combustion chamber during the opening of the injection openings. The needle sealing seat is positioned downstream of the main throttle point relative to the flow direction (e.g., to the spray holes). The low pollutant volume is

of advantage. The disadvantage is the thermal stress of the nozzle in the area of the seat and the resulting increased danger of coking.

~~**Fuel-injector body.** The fuel-injector body is used for the pressure-tight installation of the nozzle in the cylinder head. It contains the pressure spring and other components for stable long-term adjustment of the opening pressure. The assembly of nozzle and fuel-injector body is called fuel-injector body combination. The internal pressure resilience depends on the highest injection pressure. This pressure is currently 350–500 bar for prechamber engines with pintle-type nozzles or 900–2000 bar for engines with orifice nozzle direct injection (**Fig. I107**).

The installation space in the cylinder head is normally very limited, which means that a stable design for the high pressure is challenging. The fuel-injector body for the orifice nozzle is, therefore, designed for different peak pressure ranges.

A sealing washer (e.g., made of copper) is placed between the fuel-injector body and cylinder head to seal the cylinder chamber to the outside. The fuel-injector body is preferably guided at the shaft diameter, in special cases also at the nozzle nut. The installation dimensions must be maintained precisely to guarantee flawless functioning of the injection nozzle.

Fixing of the fuel-injector body in the cylinder head is required in addition to the fixing of the rotary

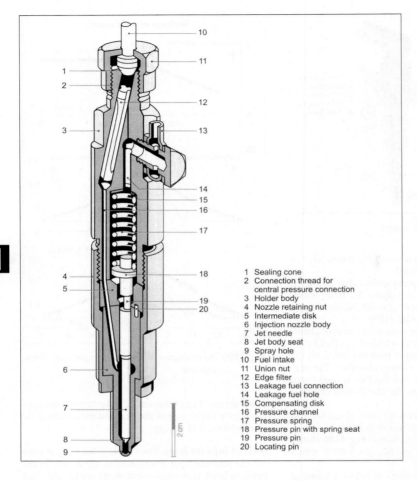

1 Sealing cone
2 Connection thread for
  central pressure connection
3 Holder body
4 Nozzle retaining nut
5 Intermediate disk
6 Injection nozzle body
7 Jet needle
8 Jet body seat
9 Spray hole
10 Fuel intake
11 Union nut
12 Edge filter
13 Leakage fuel connection
14 Leakage fuel hole
15 Compensating disk
16 Pressure channel
17 Pressure spring
18 Pressure pin with spring seat
19 Pressure pin
20 Locating pin

Fig. I107
Standard fuel-injector
body combination for
direct injection engines

position of the fuel-injector body to guarantee correlation of the fuel sprays to the combustion chamber. Damping of the needle jet to prevent overshooting of the needle and to improve the metering quality is used for the smallest injection quantities—for example, preinjection. **Fig. I108** shows a hydraulic damping unit in the fuel-injector body.

Literature: Robert Bosch GmbH (Ed.): Dieselmotor-Management, 3rd Edition, Wiesbaden, Vieweg, 2002.

~~~Dual-spring fuel-injector body. Dual-spring fuel-injector bodies are derived from the standard body for throttling pintle nozzles and orifice nozzles (**Fig. I109**). The needle jet opens initially only the first stroke stage (prestroke h1) during the injection process. This transfers only a small fuel quantity into the combustion chamber (**Fig. I110**).

The needle jet is now opened to the full stroke and the main quantity will be injected if the pressure in the fuel-injector body increases further. This two-stage injection curve results in "smoother" combustion with a significant reduction in noise and often even NO_x, because the peak combustion temperature declines. At high speed when the pressure is developed very quickly, the level of hydraulic force is high enough to overcome the higher opening pressure of the nozzle immediately.

The injection is now comparable to a normal main injection without a two-stage needle stroke curve.

~~~Pintle motion sensor, fuel-injector body with pintle motion sensor. A start-of-injection sensor is required for a closed start of injection control loop for electronically controlled line systems such as serial injection pump and distributor injection pump. **Fig. I109** shows a dual-spring fuel-injector body with pintle motion sensor. Motion of the needle jet induces a speed- (not stroke-) related signal voltage because of the change of the magnetic flux in the coil, and the signal will subsequently be processed by an analysis circuit. **Fig. I111** compares the characteris-

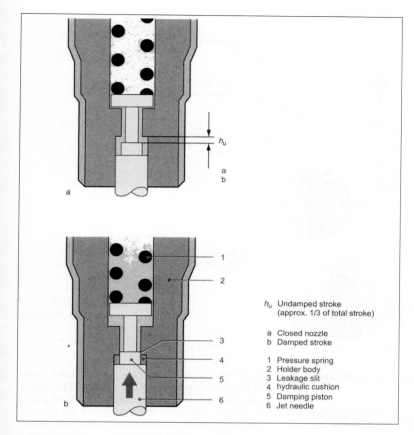

h_u Undamped stroke
(approx. 1/3 of total stroke)

a   Closed nozzle
b   Damped stroke

1   Pressure spring
2   Holder body
3   Leakage slit
4   hydraulic cushion
5   Damping piston
6   Jet needle

Fig. I108
Needle jet damping

tics of the pintle stroke with the signal of the pintle motion sensor.

When the threshold voltage is exceeded, this signal is used by the analysis circuit to determine the start of injection.

~~Hole-type nozzle. Hole-type nozzles are used in diesel engines with direct injection (**Fig. I112**). They are divided into the following types dependent on the internal design of the nozzle tip:

* Blind-hole nozzles
* Seat-hole nozzles

Different design sizes can be used depending on the application, and they differ from each other by their outside dimensions and/or their needle diameter:

* P-type with 4-mm needle guidance diameter
* S-type with 5- or 6-mm needle guidance diameter

Hole-type nozzles normally have five and eight spray holes, which form a uniform circle of holes on the surface of a cone inside the nozzle. A symmetrical distribution of spray relative to the nozzle axis is the result for four-valve engine applications. The spray distribution for two-valve designs is asymmetrical,

which means that the height angles of the individual spray holes are different. The number of spray holes and their shape, diameter, and position (spray distribution) result from:

* The installation position
* The required injection quantity
* The combustion chamber shape (especially the depression geometry)
* The swirl in the combustion chamber

~~~Blind-hole nozzle. The spray hole for this design starts in a blind-hole below the seat plug, which can be cylindrical as well as conical (**Fig. I112**). Conical blind-holes produce a lower compression volume, compared with the cylindrical blind-holes, and are often called mini-blind-hole nozzles. The nozzle wall thickness in the area of the spray holes must be at least about 0.7 mm to meet stability requirements, which means that the external tip shape of the nozzle is shaped in accordance with the blind-hole shape and is also conical.

Blind-hole nozzles produce very symmetrical spray distributions in the combustion chamber because of the favorable flow conditions at inlet to the spray holes. Radial motions of the needle jet during the opening

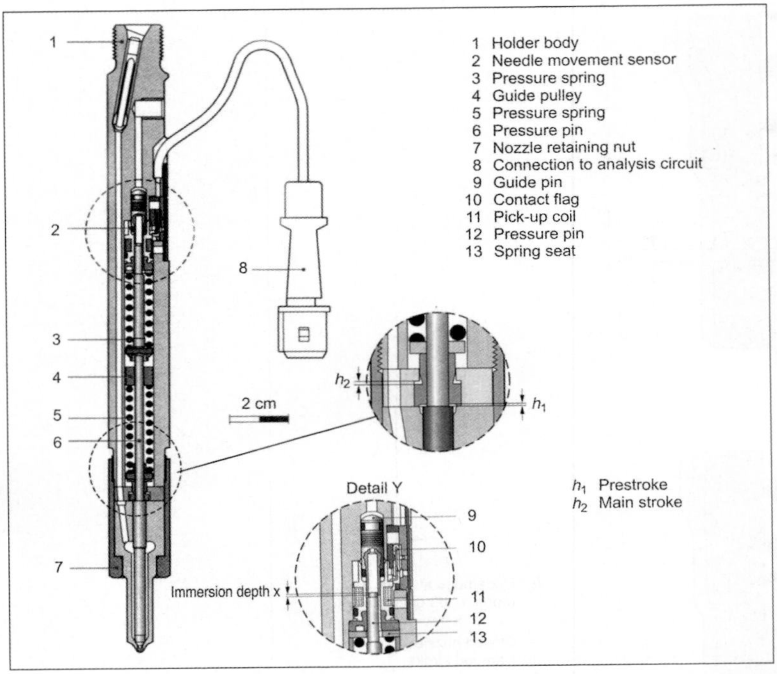

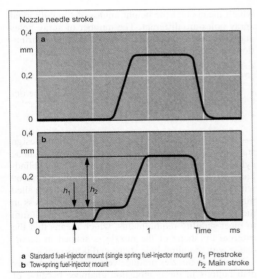

1 Holder body
2 Needle movement sensor
3 Pressure spring
4 Guide pulley
5 Pressure spring
6 Pressure pin
7 Nozzle retaining nut
8 Connection to analysis circuit
9 Guide pin
10 Contact flag
11 Pick-up coil
12 Pressure pin
13 Spring seat

2 cm

Detail Y

Immersion depth x

h_1 Prestroke
h_2 Main stroke

Fig. I109
Dual-spring
fuel-injector body
combination with needle
motion sensor for direct
injection engines

phase do not result in a significant disturbance of the intake flow.

~~~Coking. The hydraulic flow (HD) in orifice- and pintle-type nozzles changes when the spray holes be-

come coked. This leads to a late injection balance point in pressure-controlled injection systems with slightly increased pressure and, therefore, changed engine conditions. The result is a reduction of the injection quantity for needle stroke-controlled injection systems (e.g., common rail systems) and, as a consequence a loss of engine performance, these effects are proportional to the reduction in HD. High temperatures support coking (injection nozzle diesel engine, externally opening). Poor quality or contaminated fuels also result in a coking of the spray holes.

Nozzle needle stroke

a Standard fuel-injector mount (single spring fuel-injector mount) h_1 Prestroke
b Tow-spring fuel-injector mount h_2 Main stroke

Fig. I110 Comparison of the pintle stroke curves for standard and dual spring fuel-injector bodies

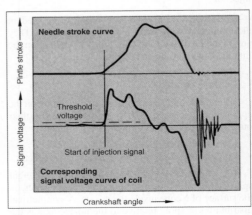

Needle stroke curve

Threshold voltage

Start of injection signal

Corresponding signal voltage curve of coil

Crankshaft angle

Fig. I111 Pintle stroke and signal from the needle travel sensor

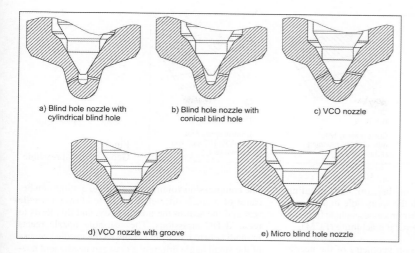

Fig. I112
Nozzle types

a) Blind hole nozzle with cylindrical blind hole

b) Blind hole nozzle with conical blind hole

c) VCO nozzle

d) VCO nozzle with groove

e) Micro blind hole nozzle

~~~**Compression volume.** The volume below the pintle seat is filled with fuel after injection has been completed. This fuel can get into the combustion chamber, where it can increase the HC emissions (unburned hydrocarbons) because it is poorly conditioned. The harmful volume must be reduced to minimize the HC emissions. Minimizing the compression volume can be achieved in a seat-hole nozzle by a VCO nozzle (valve-covered orifice; **Fig. I112**). Design options are available for blind-hole nozzles to reduce the compression volume (e.g., mini blind-hole nozzle, micro blind-hole nozzle).

~~~**Hole diameter.** The diameter of the spray hole together with the number of holes determines the hydraulic flow and the atomization characteristics of injection nozzles. The design of the holes is a compromise: The diameter must be large enough to achieve adequately short injection times at maximum injection pressures for maximum engine performance but as small as possible to achieve good atomization at engine performance curve points that have an impact on exhaust emissions. These opposing requirements result in an optimum diameter of the holes.

The diameters are between 0.11 and 0.23 mm for present day hole-type nozzles. The number of holes and the size tapering and intake edge rounding of the spray holes also have an impact on the hydraulic flow.

~~~**Hole length.** The length of the hole is directly equivalent to the tip wall thickness of the injection nozzle. A minimum wall thickness is required to achieve an acceptable stress on the tip. It is greater than about 1.0 mm for seat-hole nozzles and greater than approximately 0.7 mm for blind-hole nozzles. The length of the hole has an impact on the atomization of the injection spray: A short hole results in an injection spray with a larger spray angle than obtained

with a longer one. The volume of the spray hole is directly proportional to the length of the hole and acts as compression volume.

~~~**Hydraulic flow (HD).** This is defined as volume flow in cm³ over 30 seconds at an oil temperature of 40°C when using test oil below 100 bar in accordance with ISO 4113. Tolerances of ±1–2% for hole-type nozzles with radiusing of the entries of spray hole are typical.

~~~**Micro blind-hole nozzle.** This nozzle is a compromise between compression volume and spray pattern symmetry (**Fig. I112**). Minimizing of the volume below the seat reduces the hydrocarbon emissions compared with the blind-hole nozzle. The relatively undisturbed fuel flow to the spray holes results in a significantly more symmetrical spray pattern in comparison to the seat-hole nozzle.

~~~**Number of holes.** Single-hole and multihole nozzles are different. Single-hole nozzles are used, for example, for the obsolete MAN-M process. Between five and eight holes are currently used for the combustion processes depending on the swirl in the combustion chamber. In general, the higher the swirl level is, the lower the number of holes.

~~~**Rounding (radiusing) of the inlet edge of the spray hole.** This increases the hydraulic flow (**Fig. I113**).

The inlet edge of the spray hole is radiused in production to avoid rounding during operation that is caused by abrasive particles in the fuel. The efficiency of the injection nozzle increases as a result. Additionally, the tolerances for the hydraulic flow can be reduced as a result of the rounding process and the use of volume flow measurements. The degree of rounding is described as a percentage of the flow increase during the rounding process. So-called hydro-erosive radiusing

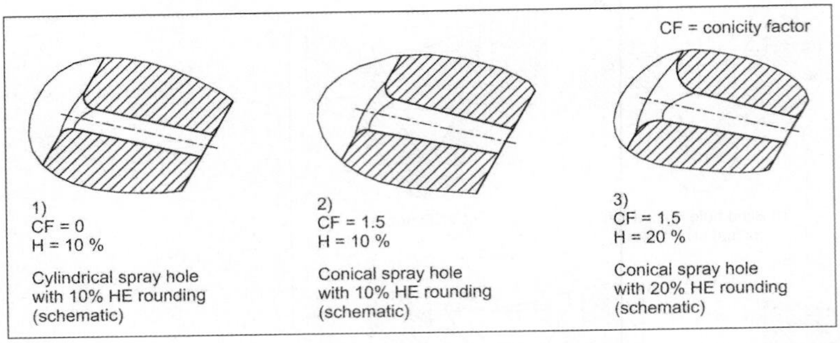

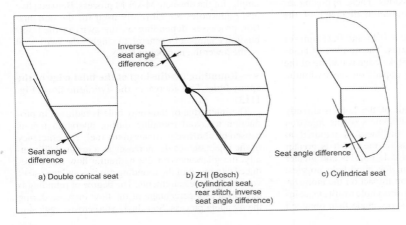

1)
CF = 0
H = 10 %

Cylindrical spray hole
with 10% HE rounding
(schematic)

2)
CF = 1.5
H = 10 %

Conical spray hole
with 10% HE rounding
(schematic)

3)
CF = 1.5
H = 20 %

Conical spray hole
with 20% HE rounding
(schematic)

CF = conicity factor

Fig. I113
Geometry of spray hole

(HE) is primarily used; in this, oil with grinding "paste" is pressed through the spray hole in a process similar to fluid lapping. Other processes include Extrude-Hone (EH) and electrochemical machining (ECM).

~~~**Seat geometry.** The seat geometry of the needle can be designed in different ways (**Fig. I114**). The main geometries are:

- Duplex cone
- Cylindrical seat with inverse seat angle and undercut (ZHI)
- Cylindrical seat with/without groove

Twin needle nozzles are mainly used in conventional injection systems, while the ZHI geometry and the cylindrical seat are used in common rail systems. High-precision production tolerances on the seat diameter, with a significant impact on the opening characteristics of the injectors, are common for the geometries of the latter designs.

Literature: D. Potz, W. Christ, B. Dittus: Diesel Nozzle—The Determining Interface between Injection System and Combustion Chamber, THIESEL 2000 Thermofluidynamic Processes in Diesel Engines.

~~~**Seat-hole nozzle.** The spray holes have their inlet directly in the seat cone of the nozzle body and are covered by the needle when it is closed (**Fig. I114**).

The compression volume is reduced to a minimum because of a small difference in angle between needle seat and the seat in the nozzle body, and this leads to reduced HC emissions. The seat-hole nozzle reacts immediately to radial deflections of the needle because of the small angle difference between needle and nozzle body, which means the spray distribution in the combustion chamber shows significant imbalance in flow, especially for small needle strokes. One solution is a groove in the needle seat positioned close to the spray holes. It guarantees a uniform fuel supply to the spray holes even when the needle is deflected.

The stress at the spray holes in seat-hole nozzles is significantly greater than in blind-hole nozzles and requires minimum wall thicknesses of approximately 1 mm for conical external tip shapes.

~~~**Spray hole.** This is the injection opening for orifice nozzles. Spray holes are either drilled with helical drills or eroded. The former production technology can only be used for spray holes larger than 0.15 mm.

~~~**Taper, injection hole.** A positive taper means that the diameter of the inside of the hole in the nozzle is larger than its diameter hole on the outside (**Fig. I113**). A positive taper of the spray hole results in an increase in the efficiency of the flow (flow coefficient)

Inverse
seat angle
difference

Seat angle
difference

a) Double conical seat

b) ZHI (Bosch)
(cylindrical seat,
rear stitch, inverse
seat angle difference)

Seat angle difference

c) Cylindrical seat

Fig. I114
Seat geometry

and a reduction of the danger of cavitation in the flow through the spray hole. Most modern automobile diesel engines have nozzles with tapered spray holes.

~~**Injection nozzle parameter.** The design can have a significant impact on the performance and emission parameters of the engine. The nozzle design must take the combustion process and the potential of the injection system into consideration and includes the following significant parameters:

~~~**Closing pressure.** Closing and opening pressures interact directly. The same spring force is applied to the needle for opening and closing. The closing pressure is always below the opening pressure because the fuel pressure is applied to the total needle diameter during closing, while the fuel pressure is only applied to the ring surface between the needle diameter and needle seat at the start of the needle opening process. It must definitely be above the maximum combustion chamber pressure so that the combustion gases do not flow back into the nozzle. A high closing pressure results in a high acceleration of the needle during closing, and this causes a large impact on closure. The maximum permissible force must not be exceeded.

The closing pressures in common rail systems are almost equal to the rail pressures and, therefore, to the injection pressures. The full rail pressure is applied to the nozzle from the end of injection. Depending on the pressure, the closing of the needle jet is started by an electrical control signal.

~~~**Injection spray.** The fuel that exits the injection opening of an injection nozzle into the gaseous environment in a diesel engine is called the injection spray. The spray from orifice nozzles consists of a dense liquid core, which is surrounded by fluid ligaments and drops. One spray exits each spray hole. Pintle-type nozzles produce a hollow tapered spray. The pressure energy in the fluid is converted to flow energy while flowing through injection opening. The turbulence of the flow and the forces created by the implosion of cavitation bubbles will result in a primary disintegration of the spray. Further atomization is created by momentum interchange with the surrounding gas (secondary disintegration). An increase in the injection pressure increases the degree of atomization. The quality of the atomization also depends on the temperature and density of the gas in the combustion chamber and also on the physical fuel characteristics. Average drop diameters of 3–30 μm are achieved under combustion chamber conditions, depending on the injection pressure.

~~~**Opening pressure.** The opening pressure of the injection nozzle at the start of injection influences the injection pressure and therefore the atomization of the injected spray for pressure-controlled injection systems. The injection nozzle remains closed when the fuel pressure is lower than the opening pressure due to the force of a spring that acts on the needle jet.

The needle jet is lifted once the fuel pressure increases and the opening pressure is exceeded. The surface area subjected to the pressure increases once the pressure underneath the needle seat increases and an additional force acts on the needle jet against the spring force.

The opening pressure for pintle-type nozzles is 120–170 bar. It is 200–400 bar for orifice nozzles. It must definitely be above the maximum combustion chamber pressure so that the combustion gases do not flow back into the nozzle. It is connected directly to the closing pressure via the spring.

The opening pressure in the common rail system is equal to the rail pressure and, therefore, to the injection pressure. The full rail pressure is applied to the nozzle from the start of injection. Depending on the pressure, the opening of the needle jet is started by an electrical control signal.

~~~**Pintle seat (pintle sealing seat).** The cone of the needle jet is smaller than the cone in the nozzle body by one seat angle difference. This ensures that needle and nozzle body have only a line contact when the needle is closed without any force applied. This line is the needle seat or the sealing edge. A closing force applied to the needle changes the line contact to a surface area contact because of the elastic deformation. However, the closing force should not exceed the range of elastic deformation at the contact surface areas. Bad design of the needle closing force results in increased wear or in rupture of the nozzle tip.

~~~**Pintle stroke.** The maximum lift of a needle jet is called the threshold stroke. The threshold stroke is preferably adjusted to a level that ensures that throttling of the flow at the needle seal seat is much lower than throttling at the injection openings. This guarantees that the total injection pressure is used for good atomization. The hydraulic flow as a function of the needle lift represents the performance map of an injection nozzle. The function of an injection system can be judged based on the variation of lift with time. For this application, test injectors are equipped with a sensor that produces a signal for carrier frequency amplifiers; this delivers a signal proportional to the needle lift. The signal delivers information regarding start of injection, needle jet opening speed, needle jet closing speed, and end of injection. The injection characteristics can be approximated from the performance map of the nozzle and the needle lift characteristics.

~~~**Spray angle.** The cone angle of an individual injection spray is called spray angle. The spray angle in pintle-type nozzles can be adjusted by changing the cone of the pintle. The spray angle in orifice nozzles is influenced by the geometry of the spray hole and the injection pressure.

~~~**Spray cone.** The spray cone must be designed in such a way that each individual spray hits the piston

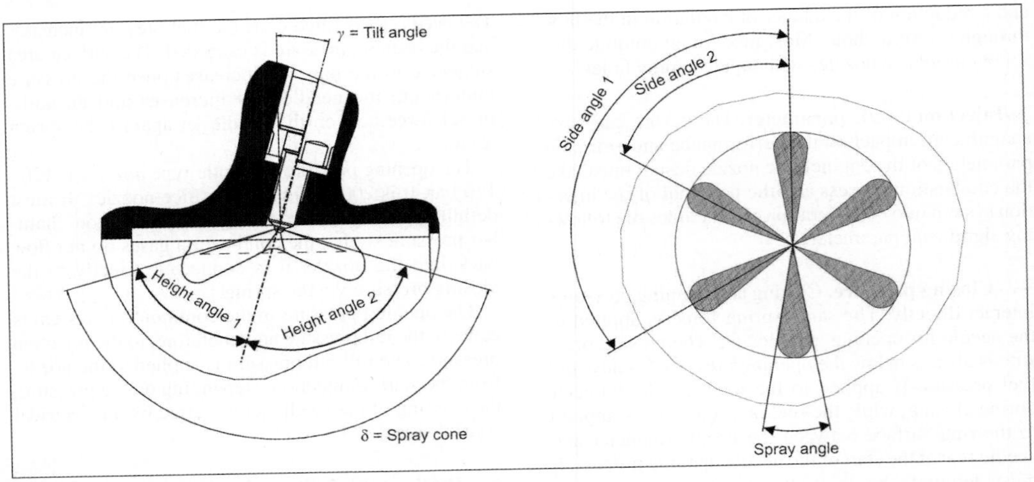

Fig. I115 Spray distribution in the combustion chamber

bowl at the same height (**Fig. I115**). The actual spray direction differs more or less from the direction of the axis of the spray hole depending on the type of nozzle — e.g., hole-type perforated seat nozzle or blind-hole nozzle — therefore, appropriate corrections must be taken into consideration during design.

~~~**Spray distribution.** Optimized spray distributions are becoming much more important because of the increasingly severe emission limits. Small deviations during the distribution can have a negative impact on black smoke emission, performance and fuel consumption. Uniform distribution of the fuel in the combustion chamber is the goal. This distribution is achieved by having the same arc length between the individual sprays relative to the impact point on the surface of the piston bowl. The angle of each individual spray on the impact surface of the piston bowl is called the side angle (**Fig. I115**). It is measured relative to a reference angle.

The term height angle describes the angle between nozzle axis and spray axis of an individual spray (**Fig. I115**). All sprays have the same height angle in centrally installed nozzles (four-valve technology), as long as the nozzle is positioned centrically to the piston bowl. The height angles of sprays from tilted nozzles (two-valve technology) are all different, but they form a symmetrical spray cone in the piston bowl.

The ignition spray, which is injected in direct proximity to the glow plug, is of special importance for the spray distribution. The cold start characteristics and the idle noise of the engine when cold are directly influenced by the ignition spray. The ignition spray can be injected tangentially on the side or tangentially below the glow plug into the combustion chamber. The flow direction of the incoming air has a special impact on the position of the ignition spray, which is related to the shape of the piston bowl and the air swirl.

~~**Installation position.** The fuel spray must be tuned to the piston bowl (combustion chamber) and the air movement by selecting the correct installation position. The design of the installation position and, therefore, the spray distribution are two of the most important elements that contribute to optimization of the combustion process.

The following combustion processes exist:

• The installation position for prechamber engines with pintle-type nozzles must be selected in such a way that the pencil-type glow plug is touched by the injection spray. The spark positions, which can be across, horizontal, vertical, or downstream relative to the air motion and the surface of the cylinder head gasket, are the options.
• The nozzle should ideally be positioned centrally and vertically relative to the combustion chamber for engines with orifice direct injection. This can be achieved easily for four-valve assemblies. For functional reasons, the offset of the nozzle position should not be more than 25% of the combustion chamber radius, and the tilted position should not be more than 15° from vertical for cylinder heads with two valves (**Fig. I116**). The nozzle injection time or the nozzle tip should penetrate the combustion chamber only a little to minimize the thermal stress.

Literature: Robert Bosch GmbH (Ed.): Dieselmotor-Management, 3rd Edition, Wiesbaden, Vieweg, 2002.

~~**Internally opening.** The needle jet of the internally opening injection nozzle inside the nozzle body in a diesel engine moves away from the combustion chamber during the opening of the injection orifices.

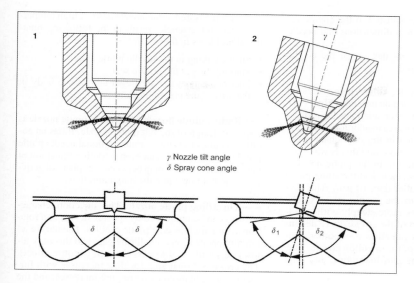

γ Nozzle tilt angle
δ Spray cone angle

Fig. I116
Installation positions of
the fuel-injector body
assembly: 1, central
position (four-valve
direct injection engine);
2, tilted position
(two-valve direct
injection engine)

Relative to the flow direction, the needle sealing seat is positioned upstream of the injection orifice. In contrast to the outside nozzle, the lower thermal stress of the needle sealing seat due to combustion in the engine is an advantage. The higher pollutant volume compared with the externally opening injection nozzles is a disadvantage. The hole-type nozzles and pintle-type nozzles currently used are internally opening nozzles.

~~**Pintle-type nozzle.** Pintle-type nozzles are used for prechamber and swirl chamber engines. The fuel in these engines is primarily conditioned by the turbulence of the air movement. The cone-shaped injection spray supports the process of fuel conditioning. **Fig. I117** shows the standard design of a pintle-type nozzle. The needle jet of the pintle-type nozzle has a so-called needle pintle at its end, which extends into the spray hole of the nozzle body with a small clearance.

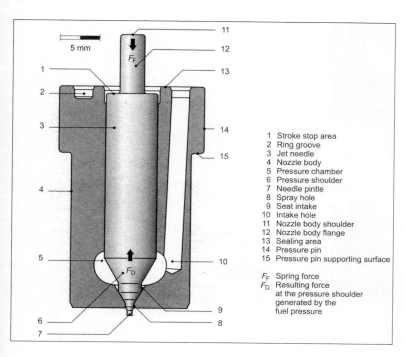

5 mm

1 Stroke stop area
2 Ring groove
3 Jet needle
4 Nozzle body
5 Pressure chamber
6 Pressure shoulder
7 Needle pintle
8 Spray hole
9 Seat intake
10 Intake hole
11 Nozzle body shoulder
12 Nozzle body flange
13 Sealing area
14 Pressure pin
15 Pressure pin supporting surface

F_F Spring force
F_D Resulting force
      at the pressure shoulder
      generated by the
      fuel pressure

Fig. I117
Standard pintle-type
nozzle

The injected spray can be adjusted to the engine requirements by using different dimensions and needle pintle designs.

Two options with different designs of the needle pintle area are normally used.

~~~**Flat pintle nozzle.** The throttling pintle nozzle for the flat pintle nozzle is beveled on one side (**Fig. I118**). The resulting surface opens (for small needle strokes) another flow cross-sectional area in addition to the ring gap. Deposits are prevented in the flow channel because of the higher volume flow, and this is why coking is lower and more uniform in flat pintle nozzles. The ring gap between spray hole and throttling pintle nozzle is very small (smaller than 10 μm). The beveled surface is parallel to the needle jet axis. The flow gradient can increase more steeply with an additional tilt in the flat part of the stroke curve dependent on the flow, and this can allow a smoother transition to the full opening of the injection nozzle. This additionally improves the part-load noise and road performance.

Literature: Robert Bosch GmbH (Ed.): Dieselmotor-Management, 3rd Edition, Wiesbaden, Vieweg, 2002.

~~~**Heat insulation for pintle-type nozzles.** Several measures are used to provide heat insulation (intermediate disks, protective sleeves between nozzle nut and cylinder head) to reduce the direct heat transfer from the combustion chamber to the nozzle. These measures are required when a high temperature (above 220°C) is measured at the nozzle base. A high temperature can, for example, result in the following functional problems in the nozzle:

- Strong coking in the needle pintle area
- Stability loss at the nozzle seat
- Fuel deposits at the needle jet and nozzle body in the area of the pressure chamber

~~~**Radius pintle nozzle.** The radius pintle nozzle is a special design of the flat pintle nozzle. It has an additional radius transition or an additional needle pintle profile, which provides an even smoother transition of the flow performance map between the sphere of activity of the flat pintle and the maximum flow.

~~~**Throttling pintle nozzle.** The injection characteristics can be influenced by specific design and rotation symmetrical needle pintle profiles. The needle jet initially opens a very small ring gap, and this only allows very little fuel to pass through (throttle effect). The cross-sectional area for flow increases with further opening (caused by a pressure increase), and the main quantity will be injected at the end of the needle stroke.

The combustion and, therefore, the engine operation become smoother because of these injection characteristics because the pressure in the combustion chamber increases slowly. The NO_x emission is also reduced, because the peak temperature of combustion decreases or because there is not enough oxygen in the chamber to develop oxides of nitrogen.

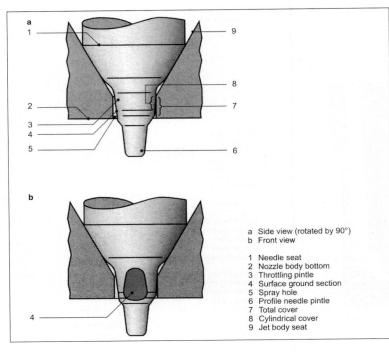

a Side view (rotated by 90°)
b Front view

1 Needle seat
2 Nozzle body bottom
3 Throttling pintle
4 Surface ground section
5 Spray hole
6 Profile needle pintle
7 Total cover
8 Cylindrical cover
9 Jet body seat

Fig. I118
Flat pintle nozzle

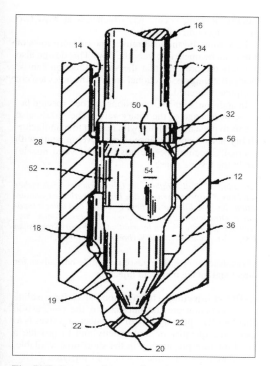

Fig. I119 Rate shaping nozzle

~~**Rate shaping nozzle (RSN).** The rate shaping nozzle (**Fig. I119**) is an orifice nozzle which is used for shaping the injection curve of pressure-controlled injection systems. It has a nozzle gap between needle jet and nozzle body, and the width of the gap depends on the needle lift. This reduces the nozzle flow at small needle lifts in contrast to the conventional orifice nozzles. The nozzle gap is not effective at full needle lift. The injection rate is similar to the one for the dual-spring nozzle holder.

~~**Vario nozzle.** The dimensioning of the spray holes presents a conflict for the injection nozzles of current diesel engines. Relatively small injection quantities are required for the part-load range preinjection or postinjection. These should ideally be atomized through small spray holes to achieve a reduction of the exhaust pollutants.

The best fuel atomization can be achieved if the total pressure reduction and, therefore, the conversion into kinetic energy takes place completely in the spray hole. However, small spray holes present the problem that the required fuel quantity for full load cannot be injected during the specified time available. This means that a compromise must be found for the design of the diameter of the spray holes, which fulfills on the one hand the requirements for the exhaust emissions and on the other hand also enables maximum engine performance. The required minimum injection quanti-

ties are achieved in certain ranges of the performance maps by not opening the needle jet completely, limiting the flow in the needle seat range. However, this results in a throttling of the pressure losses in front of the spray holes, so that the full pressure energy is no longer available for conversion into kinetic energy. The term vario nozzle describes a nozzle design in which this conflict has been addressed with variable or staged injection cross-section areas. Several design solutions are available.

~~~**Coaxial vario nozzle (KVD).** The most promising solution for the vario nozzle is the KVD (**Fig. I120**). The nozzle tip in this design has two rows of spray holes that are controlled separately by an internal and an external needle.

The minimum quantities required for pre- and postinjection and for idle speed and part load can be implemented by a few small spray holes on the upper spray hole level. A second row of spray holes with more and larger spray holes is opened for the full-load range.

~~~**Injection nozzle efficiency.** The efficiency of the injection nozzle is equivalent to the flow coefficient in the Bernoulli equation based on volume flow. The injection nozzle efficiency describes the ratio of the actual volume flow relative to the theoretically possible volume flow, which is based on differential pressure and fluid density. The injection nozzle efficiency is proportional to the calculated exit speed of the injection spray. The hydraulic flow measured with the needle jet is normally used to determine injection nozzle efficiency.

~~~**Twin needle nozzle.** The twin needle nozzle is a special form of the vario nozzle. The needles and the rows of spray hole in this design are implemented next to each other and not inside each other. However, this option cannot be used in automobiles because of the large installation requirement.

~~~**Variable orifice nozzle (VON).** The VON (**Fig. I121**) is a possible design for the vario nozzle. It has an internal rotary valve in front of the spray holes. The rotary valve is powered by a stepper motor.

Different throttle levels can be achieved depending on the angle of the rotary valve. However, the inflow of the spray holes is very asymmetrical and the spray

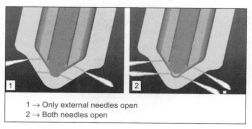

1 → Only external needles open
2 → Both needles open

Fig. I120 Coaxial vario nozzle

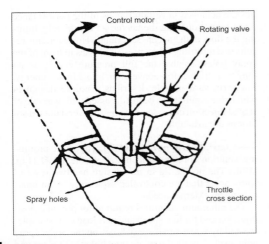

Fig. I121 Variable orifice nozzle

development is, therefore, critical. Manufacturing for volume production and wear characteristics also seem to be problematic. However, the detected turbulence in inlet to the spray hole could result in improved atomization.

~~~Variable two-stage nozzle. The variable two-stage nozzle (**Fig. I122**) can be seen as another implementation of the vario principle. The main feature is the ex-

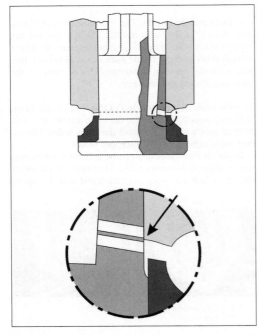

Fig. I122 Variable two-stage nozzle

ternally opening needle jet, which has at least two rows of injection holes.

Depending on the load range, one or more rows of holes can be opened to vary the flow. This design also presents the problem that the spray direction can be changed if the cross-sectional areas of the holes are not completely opened.

In addition, high HC emissions can be expected because of leakage of the rotary valve. The nozzle also seems to present problems with respect to manufacturability and wear resistance. New developments with internally opening needle jets (coaxial vario nozzle) are in the testing phase.

Literature: E. Kull: Einfluss der Geometrie des Spritzloches von Dieseleinspritzdüsen auf das Einspritzverhalten, Dissertation, 2003. — Further reading: B. Bonse, and others: Innovative Dieseleinspritzdüse—Chancen für Emissionen, Verbrauch und Geräusch, 5th International Stuttgart Symposium 2003, "Kraftfahrwesen und Verbrennungsmotoren."

~Gasoline engine. The gasoline engine has valves for direct and intake manifold injection.

~~Direct injection. The injection valves for gasoline direct injection are the interface to the combustion chamber. The primary task of the injection valves is to meter the quantity of fuel exactly and to provide a good mixture preparation in the short time available. Compared with the intake manifold injection valves, the time for the fuel injection is shorter by a factor of four, and the fuel pressure is higher by a factor of 25.

Internally and externally opening injection valves are used for gasoline injection.

~~~Air shrouding. →Injection system, fuel ~Gasoline engine ~~Direct injection systems ~~~Airshrouded direct injection

~~~Externally opening. This type includes a ring gap for mixture preparation and metering of the fuel injected. This concept for mixture preparation produces a rotationally symmetrical hollow cone. The angle of the spray cone depends on the back pressure. It qualifies these injection valves well for spray-guided combustion processes.

The mixture injection injector for the "Orbital" air-supported direct injection is also an externally opening injector. This injector has a high-resistance solenoid coil and is of similar design to an intake manifold injection valve. It is controlled by a single output stage similar to the one in intake manifold injection valves, and because the combustion chamber pressure and the injection pressure are very similar, the actuation forces are small. The injection quantity for this system is not metered by a mixture injection valve but by its own injection valve. An intake manifold injection valve can be used for fuel injection, because the same time is available for the fuel injection as in intake manifold injection applications.

Delphi has developed an externally opening dual-coil valve (2; **Fig. I123**). This design requires that the

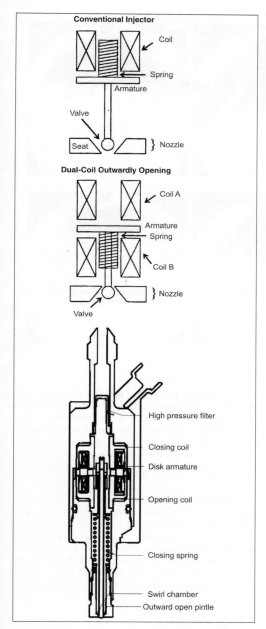

Conventional Injector

Coil

Spring

Armature

Valve

Seat

} Nozzle

Dual-Coil Outwardly Opening

Coil A

Armature

Spring

Coil B

} Nozzle

Valve

High pressure filter

Closing coil

Disk armature

Opening coil

Closing spring

Swirl chamber

Outward open pintle

Fig. I123 Externally opening dual coil injector compared with the operation of a single coil system

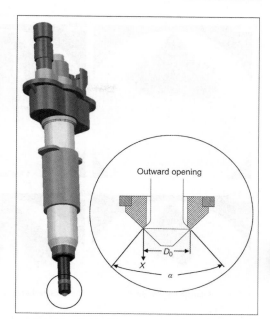

Outward opening

D_0

X

α

Fig. I124 Externally opening piezo injector

spring force is dimensioned in such a way that the injector is safely sealed for the fuel system design pressure. The opening coil (A) must produce the differential force between spring force and the force from the fuel pressure. The closing of the injection valve is sup-

ported by a closing coil (B). Both coils are actuated by a control voltage of 12 V. The low cost for the electrical control is offset by the additional cost for the second solenoid coil.

Externally opening injection valves have a very high static flow rate. A piezo-electric control system is especially qualified to fulfill the requirements for a minimum injection quantity inside the injector without using a throttle point. An externally opening injector with ring gap metering and piezo-actuation is shown in **Fig. I124**.

The valve needle for this injector is powered directly in contrast to the currently used diesel common rail injectors. This also permits changes of the needle stroke during operation by changing the control voltage of 160 V and by multiple injections in very high frequencies. The 20 MPa pressure in the total fuel system is available for atomization. Minimum injection times of 150 μs are required to fulfill the requirements for the minimum injection quantity. By avoiding a pocket volume behind the sealing seat and by providing a high fuel system pressure, the mixture preparation concept is also very resistant to deposits and, therefore, fulfills the requirements for long-term stability of the spray patterns for spray-guided combustion processes (**Fig. I125**).

The penetration depth is small compared with injection valves with swirl atomizer or multihole valves, which reduces the danger of wetting the piston wall with liquid fuel. The quantity of fuel injected was kept constant for this comparison. The penetration depth of the piezo injector is smallest at 0.6 MPa back pressure (**Fig. I126**).

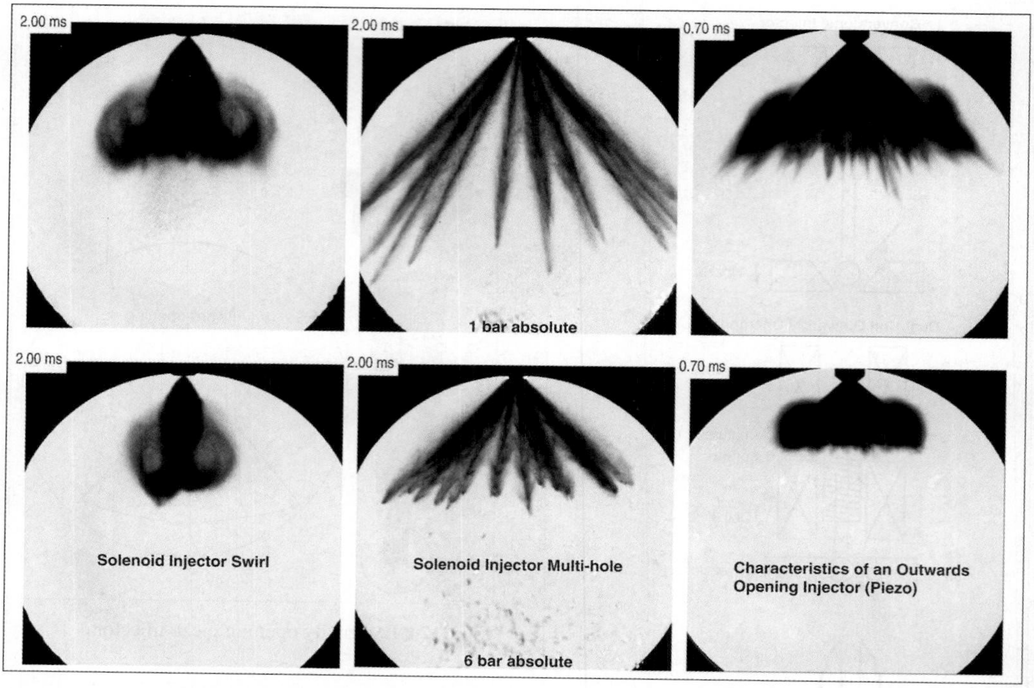

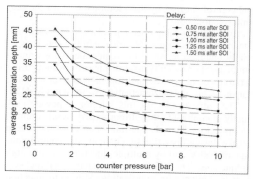

Fig. I125 Spray patterns of different direct injectors

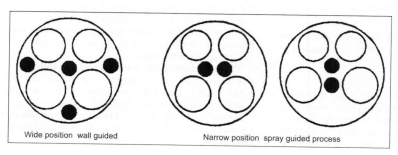

Fig. I126 Penetration depth of the injection spray relative to the back pressure

~~~**Installation positions. Fig. I127** shows the possible arrangements of injection valve and spark plug in the combustion chamber. Different installation positions are required for particular combustion processes. It must be taken into consideration that a good and uniform cooling in the cylinder head is guaranteed because of the installation position of spark plug and injection valve.

~~~~**Air-guided combustion processes.** The injection valve is also positioned on the side for this application, but, in contrast to the wall-guided combustion process, the fuel is injected into the air in the center of the combustion chamber. The fuel will be transported to the spark plug by the internal cylinder flow. This requires air movement that is generated by a variable tumble motion. It is more difficult to achieve a stable

Fig. I127
Possible arrangement of injection valve and spark plug

combustion process than with wall-guided combustion processes because of the strong cyclical variations of the cylinder flow.

~~~~Homogeneous combustion processes.
Homogeneous combustion processes provide an ignitable mixture in the whole combustion chamber. As for intake manifold injection, load control is performed by a throttle valve that changes the density of the trapped mixture. The injection valves can either be arranged centrally in the combustion chamber or at the side in the combustion chamber between both injection valves. The piston shape is designed for an optimal combustion and for maximum performance.

~~~~Spray-guided combustion processes.
These processes have the highest potential for lean engine operation and, therefore, the highest potential for fuel savings. The injection valve is positioned centrally in the combustion chamber and the spark plug is positioned in close proximity tilted to the side. This prevents the fuel from contacting the piston or the combustion chamber walls. This combustion process requires the highest performance for spray preparation in the injection valve. Finely atomized fuel must be present in the area of the spark plug to ensure safe ignition and to avoid coking of the spark plug, and the spray cone angle cannot change significantly as a result of the changing pressure inside the combustion chamber. The spark plug may not be moistened with liquid fuel to reduce the thermal stress of the spark plug.

Fig. I128 shows the possible operating range of spray-guided combustion processes compared with wall and air-guided combustion processes [2]. The significantly larger range with stratified charge and the performance and torque potential with stratified charge are shown. This process also has the highest consumption potential.

~~~~Stratified charge combustion process.
In gasoline engines, as in diesel engines, the load control in the lower range is performed by changing the "quality" of the mixture. This enables the engine to be operated unthrottled. For the gasoline engine, an ignitable mixture must be transported at the right time to a fixed ignition location, namely the spark plug. A cloud of stratified charge with an ignitable mixture must be arranged around the spark plug at the time of ignition. Ideally, no fuel is present outside of this stratified charge cloud.

This can be accomplished by the following:

- Wall-guided combustion processes
- Air-guided combustion processes
- Spray-guided combustion processes

Fig. I129 provides an overview of the combustion processes.

~~~~Wall-controlled combustion processes.
The injection valve is positioned on the edge of the chamber. Injection valve and spark plug are widely separated in relation to each other. A large part of the fuel is sprayed onto the piston base in stratified operation. The injected fuel is transported to the spark plug by the shape of the piston bowl and the type of airflow. Depending on the geometry of the intake port, the airflows are divided into so-called reverse tumble or tumble processes and swirl processes with channel switch-off (1). A very stable combustion process is possible due to the routing of the fuel along a combustion wall. However, the heavy piston and the unfavorable combustion chamber shape require compromises for full load. This is why the fuel consumption advantage is limited due to the wall losses, the high exhaust emissions, and the strongly reduced operating range with stratified charge.

Literature: G.K. Fraidl, M. Wirth, W. Piock: Direkteinspritzung im Ottomotor—Brennverfahren und Entwicklungsrichtungen, Conference on Direct Injection in the Otto Engine in the House of Engineering Essen, 1997. — K. Fröhlich, K. Borgmann, J. Liebl: Potenziale zukünftiger Verbrauchstechnologien, 24th International Vienna Engine Symposium, 2003.

~~~Internally opening.
These can be designed as swirl valves or as multihole valves for the mixture

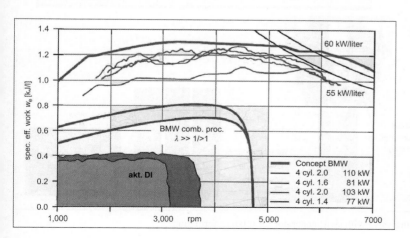

Fig. I128
Spray-guided combustion process (BMW concept) compared with wall and air-guided processes

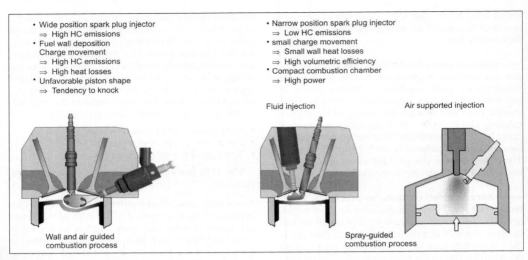

Fig. I129 Overview of combustion processes

preparation. **Fig. I130** shows the cross section of an internally opening direct injection valve. The injection valve consists mainly of housing, valve seat, valve needle, cutoff spring, and its solenoid coil for the generation of a magnetic field. The magnetic field lifts the valve needle from the valve seat against the spring force and the fuel pressure once the solenoid coil receives electric energy. The spring force must be dimensioned in such a way that the injection valve cannot be opened during the combustion and that the

requirements for the injection quantity are met. The solenoid coil of the injection valve is designed with low electrical resistance to guarantee a fast rise in the magnetic field.

A current-controlled output stage with a voltage rise to 77 V during the valve opening phase is integrated into the control instrument. A DC/DC converter generates the 77 V in the control instrument. A peak current of 11.5 A will be switched to a holding current of 3 A, which will be drawn from the battery. This makes a

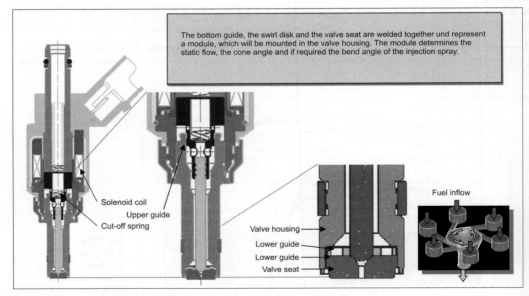

Fig. I130 Internally opening valve with swirl generator for mixture formation

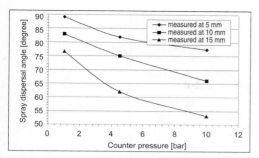

Fig. 1131 Reduction of the spray angle with increasing back pressure

minimum injection time of shorter than 0.5 ms during the linear range of the injection valve possible. The maximum injection time is limited due the time available from the time the exhaust valve closes to the compression stroke.

The spray is generated by a swirl insert. The fuel is rotated by the use of tangential channels, and a fin is generated when the fuel exits the valve. This fin disintegrates because of the aerodynamic forces and forms small fuel droplets. The pressure distribution inside the fuel spray has a declining gradient in the direction of the spray axis, which means that a secondary flow with vortices is generated. This process transports air and small fuel drops in the direction of the spray axis. An increase of the back pressure, which is present during injections in the compression stroke, results in a reduction of the spray cone angle (**Fig. 1131**).

Injection valves with swirl generators are not suitable for spray-guided combustion processes, because these combustion processes require spray cone angles independent of back pressure. A prespray with a large penetration depth is generated in swirl injection valves by the atomization concept. The volume that is not put into rotation, and which therefore produces the pre-

spray, can be adapted to the requirements of the combustion process if the volume between the swirl disk and sealing seat is varied.

Tilted spray patterns can be generated if the exit hole for the fuel has a bend angle relative to the injection axis. This makes spray cone angles of 35–70° and bend angles of up to 25° possible.

The coil injectors can be equipped with a multi-hole disk or a slotted disk instead of a swirl generator (**Fig. I132**). The mixture is prepared immediately after fuel exit based on the aerodynamic viscous forces. As in diesel injectors, the atomization quality is a function of the exit speed of the spray or the pressure of the fuel. The spray cone angle depends on the back pressure. However, the penetration depth is higher than it is with swirl valves, which increases the risk of combustion chamber wetting with liquid fuel. The advantage of multihole valves is the simple adaptation of the spray shape to the requirements of the combustion process; the disadvantage is the risk of valve tip coking.

Literature: Y. Sonoda, J. Harada, K. Norota, H. Nihei, T. Kawai: Entwicklung eines neuen DI-Otto-Motors für den europäischen Markt, 9th Aachen Colloquium on Vehicle and Engine Technology, 2000. – D.L. Varble, M. Xu, P. VanBrocklin: Entwicklung eines außen öffnenden Benzin-Direkteinspritzventils für Schichtladebrennverfahren, 7th Aachen Colloquium on Vehicle and Engine Technology, 1998.

~~~Spray dispersal angle. *See above,* ~Diesel engine ~~Injection nozzle parameter

~~Intake manifold injection. The injection valve for the intake manifold injection is positioned either in the intake manifold or in the cylinder head. The injected spray is directed at the intake valve for both options.

~~~Characteristic curve. The characteristic curve (**Fig. I133**) of the injection valve is distinct in defined limits. With a constant feed voltage, the metered fuel mass is directly proportional to the duration of the voltage pulse. The slope of the characteristic curve is equivalent to the flow rate through a permanently open injection valve, and the flow is determined by the effective cross-section area for a given differential pressure. The mass of fuel injected during the electrical pulse is called the dynamic flow.

The characteristic curve of the injection valve is undefined at the lower and upper ends of the directly proportional work range. The reason for this is the inertia of the spring mass system as well as the time-dependent characteristics for establishing or discharging a sufficiently strong magnetic field. Modern engine and transmission controls have nonlinear characteristics of the injection valve at the end of its range stored in the characteristic curve in the engine control system, so that parts of these ranges can still be used.

~~~Design. A design with one internally opening valve and two defined valve positions has been established as the industry standard. The drive is a linear drive, and it is characterized by one opening coil, one rotor in pot

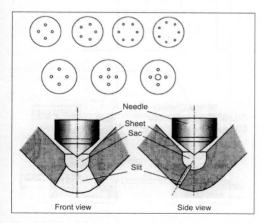

Fig. 1132 Multihole injector with different multihole disks

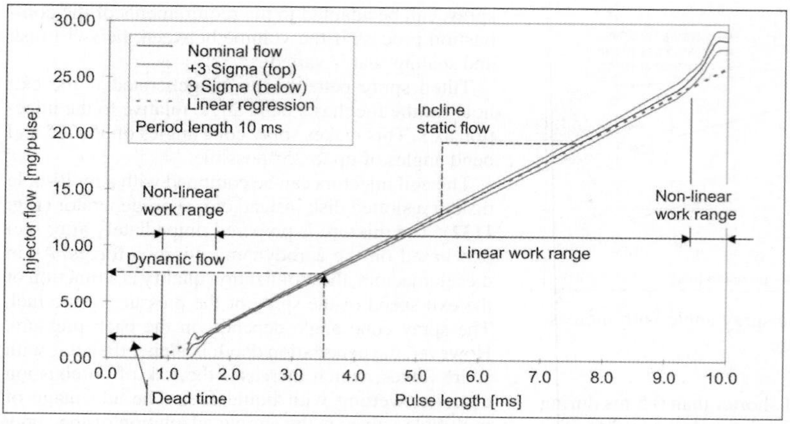

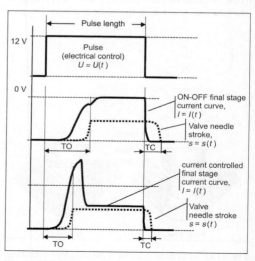

Fig. I133
Injection valve
characteristic curve

design, and a spring to close the valve. The injection valve is one of the fastest-closing solenoid valves available. The fuel is measured and the spray is shaped by a dosage plate which is located downstream of the valve seat. **Fig. I134** shows a cross section of a valve.

~~~**Electrical control.** The electrical control (**Fig. I135**) of the injection valves is performed by low-side drivers (on–off switching of the ground wire) with the help of the vehicle voltage of 12 V. This type of control has developed into an international industry standard. The coil is typically produced with resistance values between 10 and 14.5 ohm. The switching characteristics of the electrical output stage have a significant impact on the dynamic characteristics of the injection valve. The selection of the cutoff voltage in the output stage has a significant impact on the elimination of the stored magnetic field energy during closing of the valve.

The cutoff voltage is typically 75 V. However, this value can be reduced to 55 V under consideration of the engine requirements, which makes it possible to select a small silicon surface for the output stage chip. Also, all electrical resistors in the electric circuit (e.g., internal resistance of the output stage, cable harness resistor, connector transition resistors) must be considered as well.

The use of injection valves with low electrical resistance (typically 2.4 ohm) and switched or current-controlled output stages should also be mentioned. They are currently exclusively used for motor sports applications and are not used for the general market in the automotive industry because their costs are significantly higher.

~~~**Functionality.** The functionality of an injection valve can be summarized with the following three terms:

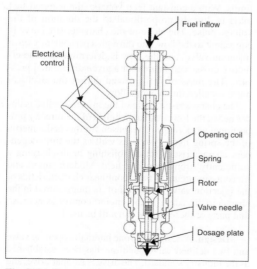

Fig. I134 Cross section of injection valve

Fig. I135 Injection valve control

- Accurately timed metering of the required fuel mass
- Tightness against uncontrolled flow of fuel (e.g., valve seat sealing function)
- Spray preparation to support the mixture formation

The product must, in addition, also fulfill a number of other requirements, such as guarantee of the functionality during the specified product life and environmental and quality requirements.

~~~**Set-up.** The injection valve consists of two main components:

- The linear drive with opening coil
- The valve group

The function of the linear drive is based on the conversion of the electrical current into a magnetic field, which provides the force for the "rotor" on the valve needle and which lifts it against the fuel pressure and the spring (**Fig. I136**).

The axial magnetic force applied to the rotor must be higher than the sum of the forces acting on the valve needle itself—such as the spring force, the friction force, and the hydraulic closing force—to ensure that the valve needle connected to the rotor is lifted during static calculations. The hydraulic closing force is the result of the fuel pressure and the resulting effective cross-sectional area in the valve seat. The dynamic calculation includes the inertia forces of the moving valve mass. The switching characteristics of an injection valve are shown in **Fig. I137**).

The spring force must be larger than the remaining magnetic force and the frictional force to close the valve when the dynamic influences from the fuel flow are disregarded.

The valve group consists primarily of the valve needle and the connected rotor, the valve seat body, and the dosage plate. Lifting the valve needle (typically 80 μm) opens an adequate cross section in the seat area so that the fuel can flow to the dosage plate with the lowest

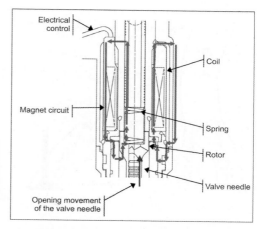

Fig. I136 Injection valve solenoid circuit

possible pressure losses. The differential pressure is completely converted to kinetic energy at the dosage plate. The dosage plate, in interaction with the upstream shapes for flow formation, has the task of optimizing the spray formation and turbulence generation to improve the spray breakup after the fuel exits from the dosage plate.

~~~**Spray formation.** The spray formation or the spray preparation is performed by the dosage plate. Two- or four-hole options are normally used. The dosage plate normally has thin walls. The coking characteristics increase with increasing wall thickness of the dosage plate, which results in strong reductions in flow. The type of spray shape is selected depending on the installation and on the cylinder head geometry of the engine. The focus of development is on avoiding depositing of a wall film in cold-areas of the intake

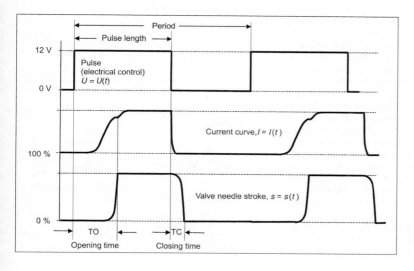

Fig. I137
Switching
characteristics

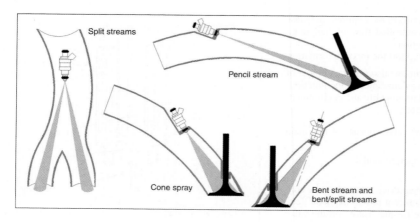

Split streams

Pencil stream

Cone spray

Bent stream and
bent/split streams

Fig. I138
Injection valve spray
shapes

tract, and in optimizing the spray design (shaping and preparation) with respect to the cold-start characteristics of an engine. Typical spray geometries are shown in **Fig. I38**. New developments in production technology and further developments for the simulation of flow regimes and spray disintegration mechanisms will result in future designs that are better optimized for the particular characteristics of the engine.

~~~**Spray preparation.** The spray preparation for intake manifold injection is almost exclusively defined by the disintegration mechanism of the primary spray; the disintegration of the secondary spray (interaction fuel-air, depending on the relative speed between the air and fuel) has only little impact. However, the disintegration of the secondary spray plays a major role for direct injection systems as well as for diesel engines. The effect of the disintegration of the secondary spray to reduce the droplet size was utilized for the intake manifold injection with air-shrouded injection valves. The high relative speed between air and fuel at the valve tip resulted in a significant reduction in the droplet size. The air-shrouded injection valves are not the preferred solution from an engine standpoint, because good spray preparation can only be achieved with a sufficiently high differential pressure at the air shrouding, and this is only achieved in the lower part-load range of the engine.

Different measuring methods with different objectives are used to determine the quality of the spray prepared by an injection valve. The laser diffraction technology (frequently used measuring equipment: Malvern) delivers a fast and readily repeatable measurement of drop size distributions within the defined volume of a spray. Phase Doppler anemometry (PDA) is a powerful research tool which is used to locally measure individual drops in small spray volumes and to determine their size, speed, and direction. However, measurements with a PDA are time-consuming. The difference among the measuring methods is in the way in which the data are measured.

Fig. I139 shows the spatial arrangement and the sizes of the measuring volumes. The Malvern analyzer computes a histogram of the drop sizes, which would

produce a theoretical pattern such as the detected diffraction pattern. However, the PDA permits measurement of the actual individual drops in the measurement volume. Both measuring processes can only include spherical drops in the measurements. Any drops that are not spherical due to the spray pattern or which were deformed by aerodynamic effects within the measurement volume will not be included.

The values measured by both measurement methods are strongly dependent on the measurement parameters selected and on the calibration of the measuring equipment. Comparisons of different laboratories show significant deviations between the measured values at different laboratories. Therefore, a comparison of the quality of spray preparation between different injection valves is only possible under the condition that all the measurements are performed on the same measuring equipment in the same laboratory. The Sauter mean diameter (SMD), named after the German scientist Sauter, is normally used to describe the quality of spray preparation (**Fig. I140**). The SMD (or the D32) characterizes the evaporative characteristics of a spray adequately.

$$\text{SMD} = D_{32} = \frac{\sum\limits_{i=1}^{n} D_i \cdot \pi \cdot D_i^2}{\sum\limits_{i=1}^{n} \pi \cdot D_i^2} = \frac{\sum\limits_{i=1}^{n} D_i^3}{\sum\limits_{i=1}^{n} D_i^2}$$

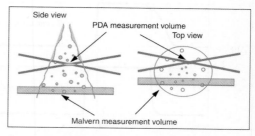

Side view

PDA measurement volume

Top view

Malvern measurement volume

Fig. I139 Spray preparation quality measurement

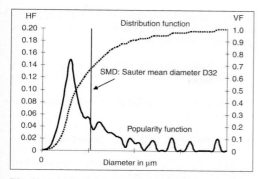

Fig. I140 SMD droplet size distribution

Injector →Injection valves ~Diesel engine

Injector nozzle →Injection valves

Injector nozzle parameter →Injection valves ~Diesel engine

Inlet opens late (SE) →Variable valve control ~Load control

Inline engine →Balancing of masses ~Inertia forces ~~Inline engine; →Engine concepts ~Reciprocating piston engines ~~Single-shaft engines ~~~Single-bank engines

Inner valve spring →Valve spring ~External valve spring

Input impedance →Engine acoustics

Input signals →Electronic/mechanical engine and transmission control ~Electronic components

Installation bending stress →Piston ring

Installation classes →Electronic/mechanical engine and transmission control ~Requirements for mechanical and housing concepts

Installation conditions and ambient conditions →Electronic open- and closed-loop control ~Electronic open- and closed-loop control, diesel engine

Installation length, valve guide →Valve guide ~Design

Installation position, injection valve →Injection valves ~Diesel engine ~~Installation position, ~Gasoline engine ~~Direct injection ~~~Installation positions

Installed insulation factor →Engine acoustics

Institutions →Oil ~Classification, specifications, and quality requirements

Insulation exhaust system →Exhaust system ~Insulation

Insulator →Spark plug

Intake (*also*, →Combustion cycle; →Exhaust system; →Variable valve control ~Variation parameters). "Intakes" are the valves or openings for admitting the fresh mixture into the gasoline engine or the air into the gasoline engine with direct injection and into the diesel engine. The filling of the cylinder and, therefore, the power and the torque are mainly determined by the opening time or by the cross-sectional area-time diagram. The time of intake closing exerts a major influence on this process.

The intake and the exhaust of the working fluid into and from the combustion chamber can be implemented with poppet valves or throttle slides.

Intake air. The thermodynamic condition of the intake air has a significant influence on the operating characaterstics of the engine. The density, ρ, is defined by the temperature, T, the intake pressure, p, and the gas constant, R. The state equation gives the relationship

$$\rho = \frac{p}{R \cdot T}.$$

~Air density. The fill of the engine increases with the density of the air. The density declines with increasing geodetic altitude, and this requires an adjustment of the fuel mass. The density can be increased significantly by supercharging.

~Air filtering →Filter ~Intake air filter

~Air pollution →Filter ~Intake air filter

~Air preheating. The mixture conditioning, which means nebulization and vaporization of the fuel, is not optimal during the warm-up of gasoline engines, especially for engines with central, single-point injection, due to the cold intake pipes and the low flow speed. A large part of the fuel attaches to the intake pipe walls as a film or condensates in the combustion chamber. Intake air preheating resolves this problem. The exhaust gas energy or electrical energy is used to preheat the intake air for this application.

Intake air preheating results in fill losses in the cylinder and must therefore be switched off after the intake pipe has warmed up, for example, by control systems with thermostats. The intake air preheating is primarily used in engines with carburetors or central injection, which also use intake air preheating to avoid throttle icing due to condensation of the humidity in the air.

~Air pressure →Trapped pressure

~Air temperature. The air density, and as a result the fill of the engine and its power, are reduced with increasing air temperatures. Therefore, strong heating of the air on the way to the combustion chamber of the engines should be avoided. The air is significantly heated (up to 120°C) due to the compression in supercharged engines, so it is recommended that the air be subsequently cooled with a charge air cooler.

~Humidity. Increased water vapor content displaces a significant part of the dry air, about 1–2% if the humidity is high (close to 100%). The oxygen content supplied to the engine is reduced by this amount and the power is, therefore, also reduced.

Intake air filter →Filter

Intake air filtering →Filter ~Intake air filter

Intake air pollution →Air intake system ~Thermodynamic air management system ~~Clean air line; →Filter ~Intake air filter ~~Filter characteristics ~~~Dust retention capacity

Intake air pressure →Trapped pressure

Intake air swirl →Charge movement; →Intake port, diesel engine; →Intake port, gasoline engine ~Swirl

Intake air temperature control →Filter ~Intake air filter ~~Air filter elements

Intake air throttling →Load ~Load control

Intake cams →Camshaft ~Cam shape, ~Intake camshaft

Intake closing →Control/gas transfer ~Four-stroke engine ~~Timing

Intake control →Control/gas transfer ~Four-stroke engine ~~Timing

Intake cross section →Intake port; →Intake port, diesel engine; →Intake port, gasoline engine

Intake cycle →Charge transfer

Intake manifold →Intake system

Intake manifold gasket →Sealing systems ~Modules ~~Intake manifold module

Intake manifold injection →Injection functions ~Gasoline engine; →Injection valves ~Gasoline engine

Intake manifold mixture flow →Mixture formation ~Mixture formation, gasoline engine ~~Mixture movement ~~~In the intake manifold

Intake manifold module →Sealing systems ~Modules

Intake manifold preheating →Hot spot heating; →Intake system

Intake manifold pressure →Intake resistance

Intake manifold pressure regulation →Carburetor ~Design types

Intake manifold resonance system →Intake system ~Resonance system

Intake manifold switch-over →Electronic open- and closed-loop control ~Electronic open- and closed-loop control, gasoline engine ~~Functions

Intake module →Modules

Intake noise →Air intake system ~Acoustics ~~Orifice noise; →Engine acoustics

Intake noise emission →Air intake system ~Acoustics ~~Noise generation; →Engine acoustics

Intake opens →Control/gas transfer ~Four-stroke engine ~~Timing

Intake orifice →Air intake system

Intake port (*also*, →Cylinder head; →Intake port, diesel engine; →Intake port, gasoline engine. The intake port or the intake ports are used to supply the mixture or the air to the engine. The port must be designed particularly carefully with respect to flow to achieve higher volumetric efficiencies and lower flow losses, which are required for optimum engine operation. The total intake cross section can be increased by 40% with a change from one intake port per cylinder to two intake ports (two intake valves). A total cross-section area that is 60% larger is achieved if three intake ports per cylinder (three intake valves) are used. A 10–15% steady jet-shaped narrowing of the intake port in the direction of the intake valve has shown advantages. The fine tuning and fluid-mechanical optimization of the intake ports are normally performed on so-called flow test stands.

Intake port, diesel engine (*also*, →Intake port). The intake port design is especially important for diesel engines with direct injection. This is why the following refers primarily to this engine.
The intake port in a diesel engine is an important variable for conditioning the fuel, because it can influence the swirl of the incoming air. Sustaining the air swirl

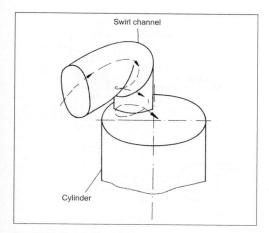

Fig. I141 Swirl channel

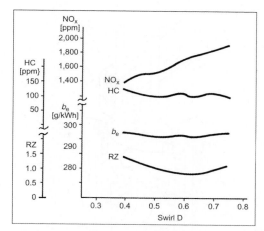

Fig. I143 Impact of swirl on exhaust emissions and fuel consumption

determines the flow conditions in the combustion chamber. Two design options are available for the intake ports: the swirl channel (**Fig. I141**) and the tangential port. In the swirl channel, a swirl is generated by a spirally shaped channel around the intake valve just before inlet to the cylinder, while air is supplied tangentially in the tangential port design. The filling of the cylinder is slightly higher if a tangential port (**Fig. I142**) is used.

For diesel engines with direct injection and four valves (two intake and two exhaust valves), it is recommended that the intake port be designed as a swirl channel for low filling of the cylinder at load and speed and the second channel as a tangential port for better filling at high engine speed.

New developments prefer processes with low swirl, because this reduces the NO$_x$ emissions (**Fig. I143**) and maintains high torque and high power. However, the smoke number increases if the swirl gets too low. High injection pressures, which permit a larger number

of injection sprays per injector nozzle, improve the conditions at low swirls with respect to NO$_x$ emissions.

Literature: R. van Basshuysen, D. Stock, R. Bauder: Part 1, Grundsatzentwicklung der dieselmotorischen Brennverfahren mit direkter Einspritzung zur Konzeptauswahl, MTZ 10/89.

~**Cooling.** The heating of the intake port in the cylinder head must be avoided if possible, so that the combustion air is not significantly heated on its way into the combustion chamber, because this would result in charging losses. A short channel is, therefore, advantageous.

~**Design.** *See below*, ~Tangential port; *also*, →Intake system ~Intake manifold ~~Intake manifold design

~**Diameter** →Intake port, gasoline engine

~**Flow** →Intake port, gasoline engine

~**Length** →Intake port, gasoline engine ~Design

~**Spiral port.** *See below*, ~Swirl channel

~**Swirl.** The flows in the combustion chamber can be divided into axial and radial components. The radial flow has an impact on the mixture formation only in a very limited crank angle range. The axial speeds, which are important for an optimum mixture formation, are relatively low if no special measures are taken. This cylindrical rotation (swirl flow) can be intensified in undivided combustion chambers by appropriately designed intake ports. The diesel engine with direct injection requires special attention in respect to the swirl. The optimum swirl depends, among other things, on the combustion processes, the combustion chamber shape, the piston design, and the injection conditions. It must, therefore, be adapted to individual cases.

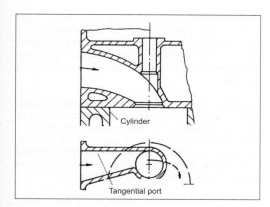

Fig. I142 Tangential port (Source: Urlaub)

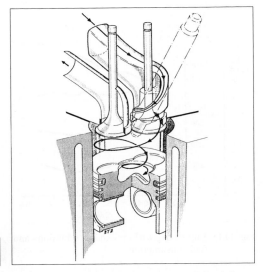

Fig. I144 Swirl channel and guidance of the airflow

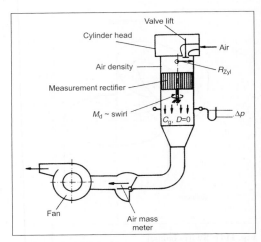

Fig. I145 Swirl measurement test stand.

~**Swirl, variable.** It is particularly advantageous to use two intake ports in diesel engines with direct injection. The swirl in the engine can be increased with the help of a swirl channel, if the second intake port is shut off at low airflows, and this has advantages for the mixture preparation. The switching on and off of channel is called swirl control. This achieves optimum conditions for low and high airflows. The second channel is designed as a charge port, and in most cases it is designed as a tangential port.

~**Swirl channel.** The swirl or spiral port is designed to produce a desired swirl to optimize the mixture preparation. A possible design is shown in **Fig. I144**. The swirl channel requires more installation space than the tangential port; however, it is less sensitive to production tolerances.

~**Swirl factor.** This is defined as the ratio of tangential speed to the axial speed. Several methods are available to determine the swirl factor. One method is the rectifier swirl measuring method according to Tippelmann (**Fig. I145**).

This measurement method requires the air to be sucked in through the swirl channel and a cylinder of corresponding diameter to the engine bore. The air rotation is rectified by a grid, and the resulting torque reaction is measured and is converted in accordance with the following equation to the swirl, D:

$$D = \frac{M_d \cdot R_{Cyl}}{\dot{V}^2 \cdot \rho_L},$$

where M_d = torque reaction, R_{Cyl} = cylinder radius, $\dot{V}$ = volumetric airflow, and ρ_L = air density in the cylinder.

Fig. I143 shows the impact of swirl on emissions and specific fuel consumption at a representative oper-

ating point (n = 3000 rpm, w_e = 0.4 bar). Low charge rotational speeds are desired to reduce NO_x and increase the filling of the cylinder. An air supply free of swirl or with low swirl is, therefore, the target for the future. However, this can only be realized if it is possible to improve the mixture preparation, for example, by even higher injection pressures, without increasing the smoke number (SN) to an inadmissible value.

~**Tangential port.** The tangential port is an alternative to the swirl channel (**Fig. I142**). It requires less combustion space, which is important if several intake ports per cylinder are used, and it permits slightly higher filling of the cylinder (1–2%).

Intake port, gasoline engine (*also*, →Intake port). The intake port for gasoline engines is the channel that is cast into the cylinder head between the intake manifold and the intake valve(s). The intake port in a gasoline engine is partially divided only if two or three intake valves are present.

~**Charge port.** In engines with two intake ports (four-valve technology), one port can be used for the supply of rotational energy, and one port can be used for the optimum filling of the cylinder. The charge port must be designed with as little flow resistance as possible.

~**Cooling.** The air or the mixture is heated up during the flow through the intake port, and this reduces its density, which means the power output is also reduced. Heating of the intake air can be reduced by appropriate routing of the cooling water flow or the cooling airflow in the engine compartment and by avoiding heating in the intake port by external heat sources.

~**Cutoff.** If a cylinder has several intake ports (multivalve engines), it is possible to cutoff one channel at

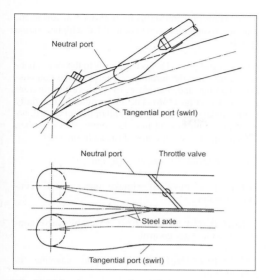

Fig. I146 Intake port configurations

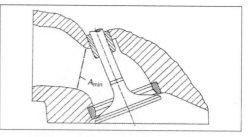

Fig. I148 Section through the intake port of a two-valve engine

and tests. **Figs. I147 and I148** show sections through intake ports of four- and two-valve engines that are currently in production. Modern four-valve engines have a port with small cross-sectional changes, and only slight redirection is performed during the flow of the mixture into the combustion chamber (trap channel). This results in small flow or charge transfer losses.

~Diameter. The relationships between the port diameter, the diameter of the cylinder bore, and the intake valve are major factors affecting the air or mixture masses flowing into the cylinder. However, the intake port normally has an ideal cross section at very few places due to the narrow space available for routing it through the cylinder head. An equivalent "diameter" at all cross sections down the port can be defined by using the hydraulic diameter, which is the diameter that is calculated if the real cross-sectional area is recalculated in a circular form.

Fig. I149 shows sections cut through different positions of the intake port.

~Flow. The air or mixture flow and, therefore, the air or mixture mass supplied to the cylinder determine the engine power. The flow is determined by the cross section of the port, the port geometry, the port surface, valve timing, pulsations in the gas flow, and so on.

The result of good port design is a large flow rate for a given inlet port. The flow number is a relationship

low loads and engine speeds. This improves the mixing and flow conditions and can, for example, produce a swirl or tumble. This can be implemented, for example, through a throttle valve in the channel (**Fig. I146**) or by closing one intake valve in multivalve engines.

~Design. The design, including cross section, shape, length, and position of the cylinder axis, is the primary determinant of the flow resistance of the incoming fluid and also influences the pulsation characteristics in the intake branch. The redirection for the flow in the combustion chamber is minimized if design produces a steep port, which means that the angle between the port center line and the cylinder axis is minimum.

Port design has a major impact on the engine power, but it depends strongly on the space available in the cylinder head and must be optimized by calculation

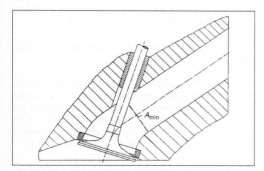

Fig. I147 Section through the intake port of a four-valve engine

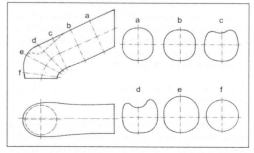

Fig. I149 Sectional plane and cross-section shapes at an intake port

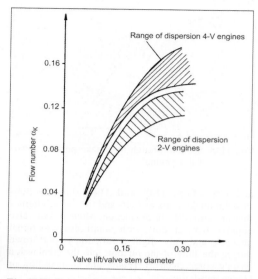

Fig. I150 Range of variation of the flow numbers in an intake port

between the actual and ideal flows that might be achieved and gives a specific value that is a measure of the quality of the airflow. **Fig. I150** shows the range of variation for the flow number for two- and four-valve engines.

The flow number, α_K, is defined as:

$$\alpha_K = \frac{\dot{m}_{real}}{\dot{m}_{theor}}$$

where $\dot{m}_{real}$ = the measured airflow and $\dot{m}_{theor}$ = the theoretical airflow (i.e., evaluated as isentropic flow through the port).

~Length. The length of the intake port should be as short as possible for flow reasons and to avoid heating up the air on its passage to the combustion chamber.

~Swirl (*also*, →Intake port). A good mixing of fuel and air is required for optimum combustion, which produces minimum emission of pollutants.

The mixing is optimized by the generation of rotation of the air in the intake port with the help of the intake port guidance. One channel can be designed as a swirl or tangential port and the second as a charge port in engines with two intake valves.

~Swirl channel →Intake port, diesel engine; →Intake port, gasoline engine ~Swirl

~Swirl variable. The generation of distinctive swirl at high flow rates results in poorer filling of the cylinder. The best design solution is to incorporate variable swirl, because intake swirl has advantages at certain operating ranges of the engine. This approach enables

the mixture preparation to be improved at lower engine speeds and fill losses to be avoided at higher engine speeds.

~Tangential port. This is used, as is to the swirl channel, to increase the turbulence energy in the combustion chamber and, therefore, to improve the mixture preparation (**Fig. I146**). The air is routed tangentially to the intake valve so that the fluid or the mixture flows with a rotating motion into the combustion chamber. This design is a competitor of the swirl channel.

~Trap channel. *See above*, ~Design

Intake port cooling →Intake port, gasoline engine ~Cooling

Intake port recirculation →Variable valve control ~Residual gas control

Intake port shut-off →Intake port, gasoline engine ~Cutoff

Intake ports in cylinder →Intake port, diesel engine; →Intake port, gasoline engine

Intake pressure loss →Trapped pressure

Intake resistance →Intake system ~Intake manifold ~~Intake manifold design, ~Intake resistance; →Trapped pressure

Intake sequence →Firing order

Intake side →Exhaust side

Intake silencer →Engine acoustics ~Intake noise

Intake silencing →Engine acoustics ~Intake noise

Intake stroke. The stroke of the piston from top dead center to bottom dead center for the intake of air and/or the air-fuel mixture.

Intake swirl →Charge movement ~Swirl factor; →Intake port, diesel engine ~Swirl channel

Intake swirl generation →Intake port, diesel engine ~Swirl channel

Intake system (*also*, →Air intake system). The intake system primarily supplies the engine with air. Intake systems contain elements for intake silencing and for measuring and regulating the air mass delivered to the engine, a device for controlling the supply of recirculated gas and blowby gases, and activated charcoal filters.

~Air distributor (plenum volume). The air distributor is a primary tuned intake runner to the plenum volume, whose task it is to distribute the induced air

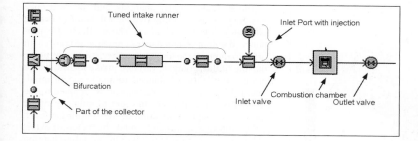

Fig. I151
1-D calculation model
of an intake module

volume as evenly as possible to all cylinders. The quality of distribution achieved by the air distributor depends on its position in the manifold and the component geometry as well as the design of the transitional area between the manifold and the tuned intake runner.

~Intake manifold

~~Computation process. Computation processes aim to evaluate the flow behavior in the intake manifold. This allows for a derivation of design criteria for the design of the intake manifold.

~~~Coupled 1-D and 3-D procedures. The dynamic gas behavior and the resistance to flow in very complex components can only be evaluated with a degree of uncertainty using a 1-D calculation process. In contrast, dynamic 3-D flow calculations (three-dimensional flow simulation, unsteady state) deliver more exact results, although the interaction with the total powertrain is missing. The direct coupling of both calculation processes uses their respective advantages to obtain a very realistic simulation of the intake system.

In such a simulation, components with complex geometry can be inserted three-dimensionally into the 1-D calculation model. At the model interfaces, the boundary values will be passed on to the other parts of the program.

~~~One-dimensional charge transfer calculation (1-D). The 1-D charge transfer calculation constitutes the time-dependent simulation of the engine process with integrated gas dynamics in the intake and exhaust-gas flow system defined as one-dimensional components—that is, the geometry is described using a 1-D model consisting of elements such as pipes, volumes, and orifices (**Fig. I151**). In order to create a suitable model, these elements are connected and calculated by considering their resistance coefficients. The necessary data are attained through tests or 3-D calculations. Because the calculation is based on the assumption that the state variables pressure, density, and gas velocity are constant across the cross section of the manifold element, these 1-D processes can only depict the flow behavior in complex three-dimensional geometries to a limited degree.

In order to evaluate intake manifold concepts, the effect of dynamic charging (tuned intake runner sys-

tem, resonance system) can be simulated; this depends on the nature of the pressure pulsations in the intake and exhaust system. Pulsation amplitudes and phase relation determine the charging effect, which is manifested by the output quantities of volumetric efficiencies, torque, and power.

~~~Three-dimensional flow simulation (3-D), dynamic. Steady-state 3-D calculations deliver insufficient information for complex procedures, such as the addition of recirculated exhaust gas (EGR supply) to the fresh air entering the engine. These secondary processes are dominated by the transient gas transfer processes occurring in the intake tract. In order to perform a dynamic computation in contrast to a steady-state one, the boundary conditions have to be defined in a time-dependent manner. These boundary conditions can be determined using a 1-D calculation of the total engine gas exchange system (one-dimensional charge transfer calculation) or measurements.

Integral results allow for an evaluation of the equal distribution of the recirculated exhaust gas, which delivers time-dependent output quantities such as gas flow speeds and direction, and also enable additional statements regarding the optimization potential of the flow guidance in the intake system, as shown in **Fig. I152**.

~~~Three-dimensional flow simulation (3-D), steady-state. In most cases a steady-state calculation of

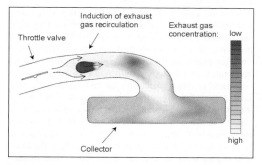

Fig. I152 Dynamic flow calculation: distribution of the recirculated exhaust gas in an intake system at a certain point in time in the engine cycle

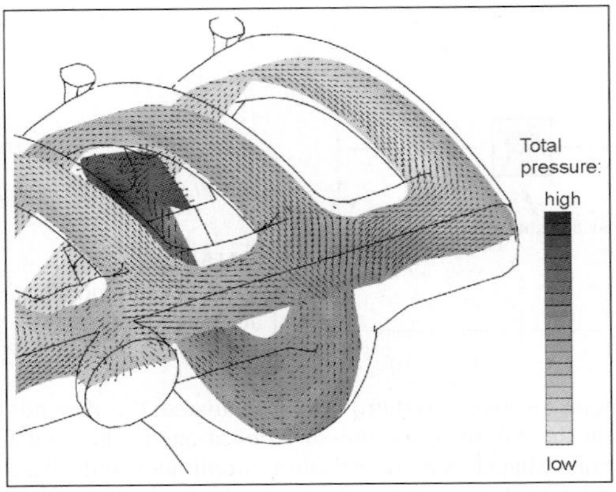

Fig. I153 Velocity field and pressure distribution in the intake manifold

the flow rate in the component with a three-dimensional flow model is sufficient. This type of model is defined in an analogous way to a finite element calculation method using a grid. It solves the differential equations of fluid mechanics that apply in each cell of the domain (momentum, energy, and mass conservation, as well as the gas equation and turbulence models). The output quantities are spatial pressure and velocity fields in the intake system. Important quantities that have an effect on the dimensioning of the manifold are minimum pressure loss as well as an optimization of the flow distribution in the intake to the individual engine cylinders (**Fig. I153**).

~~Intake manifold design. The design of the intake manifold depends on the technical specifications, the design space, the acoustic requirements, and the construction material used. The performance and torque requirements for the engine determine the geometry of the intake manifold—in particular the diameter, the length, and the plenum volume of the air distributor; from the plenum volume the individual intake manifolds branch off to the cylinders.

In intake manifolds for gasoline engines with a central mixture generating device, symmetrical manifold configurations that could heat up quickly have been preferred, and this favors thin wall cast designs to reduce the fuel wall film. This requirement has become less important for modern multipoint injection systems (manifold or direct injection systems), because the surface wetted by fuel is almost independent of the intake manifold design. This allows the choice of a design favorable to the flow of the whole intake system and the introduction of optimized gas dynamic charge effects. In order to realize a design that is favorable to the flow, a manifold with large radii of curvature, few deflections, and clean intake sections and transitions as well as a smooth interior surface without cross-sectional steps or disturbances is needed to make good use of the dynamic charging effects.

Tuned intake runner charging and the resonance charging are used to improve the fill factor (charging efficiency) of the cylinders. These effects are pronounced within certain narrow engine speed ranges. In order to improve the torque characteristic of the engine outside of these ranges, variable intake manifolds are often used. With a balanced torque curve in mind, manifold lengths and volumes can be adapted to the respective circumstances in order to obtain a torque increase through application of the charging effects. Other improvements can be made by means of external measures, such as supercharging or pulse-charging systems.

~~~Blowby supply (*also*, →Crankcase ventilation systems). The introduction of gases from crankcase ventilation usually takes place into the clean-air line, the collector, or the individual tuned intake runners for selected cylinders. This leads to the requirement for the material to be resistant to the aggressive components of the blowby gases and the blowby condensate. The introduction site is heated electrically in arrangements that allow this in order to prevent freezing of the ventilation line because of the water in the blowby.

~~~Continuously variable length actuation. A simple, switchable two-stage intake manifold can only be a compromise, since there is an optimum vibration system for each speed. In case of a fixed manifold diameter, the manifold can be adapted to this by variation of the manifold length. The technical conversion can be accomplished through telescopic manifold segments or flexible helical tubes or by adding lengths (**Fig. I154**). The general difficulty with all of these concepts is the time-consuming and complicated sealing of the individual manifold parts as well as the adherence to required shifting times during dynamic driving. In general, optimum gas vibration for charge exchange can be attained using suitable combinations of manifold diameter and manifold length.

~~~Design. The design of the intake system is determined by the engine installation and the spatial arrangement of the engine accessories (e.g., routing of the exhaust system, positioning of accessories). The manifold cross sections themselves are round, portal-shaped, or square.

In tuned intake runners with straight manifold runs or large-curve radii, the pulsation effects of the gas column are very pronounced.

Curved manifolds with narrow radii lead to a damping of the pressure pulsations through reflections off the walls of the bends and the more pronounced influence of wall friction, thus lowering the usable charging benefits. In the case of very tight-bend radii further losses occur through flow separation.

The inflow to the intake manifolds at the plenum volume benefits from the inlet horn. The result is small flow contraction and, thus, a large effective manifold cross section.

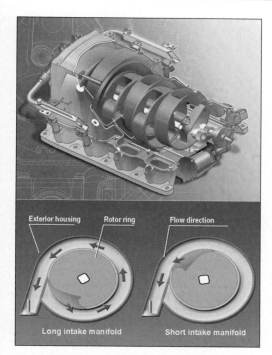

Fig. I154 Continuously variable intake system with variable lengths (Source: BMW)

~~~**Diameter.** The design of intake manifolds is done according to flow-dynamic aspects, usually using one-dimensional calculation of gas dynamic characteristics of the air column in the intake manifold. The length and the diameter of the manifolds are adapted to reach the maximum charge. Large diameters create a high torque at high speed through good dynamic charging, whereas small manifold cross-sectional areas enable good charging at low speeds. However, small diameter manifolds also increase flow pressure losses through the increased influence of wall friction; the effects of the manifold diameter on the torque can be seen in **Fig. I155**.

~~~**Exhaust gas recirculation initiation.** Exhaust gas recirculation reduces the emissions of oxides of nitrogen (NO_x) from gasoline and diesel engines by reducing their formation in the combustion chamber by lowering the temperature of the combustion gases. In gasoline engines, exhaust gas recirculation is also used for de-throttling the engine in the part-load range; recirculation rates can amount to about 30%.

The initiation of exhaust gas recirculating in the intake tract must be accomplished so that homogenous mixing with the induced fresh air is obtained and also so that it is distributed equally to the individual cylinders. Especially in modern diesel engines with exhaust gas recirculation rates of up to 60%, this requirement must be adhered to in the part-load range. In many cases, intake air throttles are used to reach the required mass flow of exhaust gas in order to realize the necessary pressure gradient.

The blending of the flow streams is accomplished by electrically or pneumatically driven EGR valves discharging the exhaust gases into the collector or the clean air line. These are often integrated with exhaust gas coolers in the intake manifold in order to further increase the efficiency of reduction of NO_x by the EGR by lowering the exhaust gas temperature. The discharging of exhaust gases into the intake system places higher demands on the materials of the parts conducting the exhaust gas in the intake system, because this is where corrosive condensates containing sulfuric acid are generated. In engines without exhaust gas cooling featuring plastic intake manifolds, effective thermal decoupling of exhaust gas-leading parts from the intake components must be realized. Recent research has examined the possibility of layered EGR supply directly at the inlet valve in the cylinder head, which can lead to a further increase of the achievable EGR rate.

~~~**Geometry.** →Intake system ~Intake manifold ~~Intake manifold design

~~~**Length.** In general, good filling of the cylinders is accomplished at low engine speeds with long tuned intake runners. The intake system generates a high maximum torque within a low speed range. Shorter manifold lengths lead to improved torque at higher speeds and, thus, to higher maximum performance of the engine. The torque maximums lie over a broader range, but are on the whole at a lower level than in long tuned intake runners. The effects of different lengths are shown in **Fig. I156**.

Thus, for each engine speed, the intake manifold length can be determined at which, through better timing of the gas exchange, optimal filling of the combustion chamber can be achieved.

Adaptations for larger speed ranges are realized through variable intake manifolds.

~~~**Materials.** The selection of the material for the manifold is becoming more important because of the necessity to reduce weight as a way to obtain savings in fuel consumption and cost. Generally speaking, plastics, aluminum, and magnesium are being used. The mate-

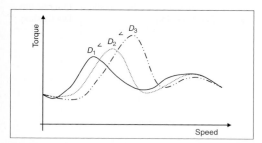

Fig. I155 Torque as a function of the manifold diameter

691

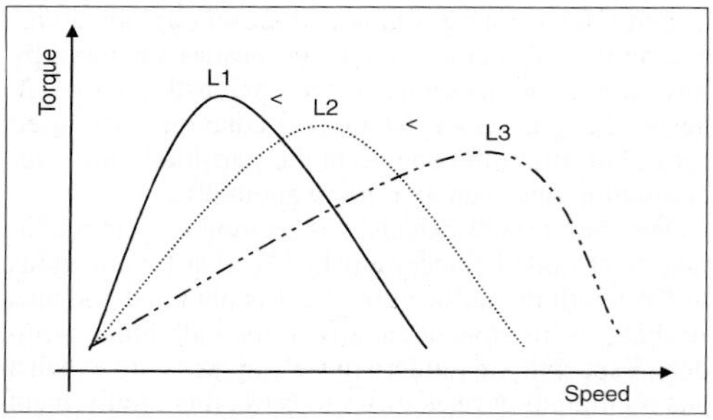

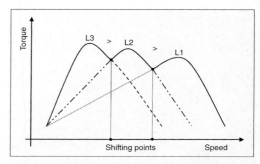

Fig. I156 Torque as a function of the intake manifold length

Fig. I158 Characteristic torque curve for three-stage intake manifold

rial selection is determined by the manifold design and the characteristics of the material; the evaluation of production processes and materials can be seen in **Fig. I157**.

~~~**Three-stage intake manifold.** The three-stage intake manifold (**Figs. I158 and I159**) allows for a "generous" torque curve for the engine through the overlaying of three torque curves. This represents the transition to continuously variable length actuation.

~~~**Torque boost.** A "generous" torque curve enables a "longer" transmission ratio with a higher gear. This allows the use of lower engine speed, which leads to fuel consumption advantages while driving. A balanced torque curve requires maximum filling of the cylinder chamber with air over the entire speed range of operation, which is significantly influenced by intake manifold parameters such as length, diameter, shape, and the respective volumes. In order to balance specific disadvantages of a fixed geometry configuration, often more complex intake manifolds are used that vary the intake manifold geometry to the current operating condition, thus achieving the highest possible torque.

~~~**Two-stage intake manifold.** In the case of two-stage intake manifolds, a differentiation is made between a short intake length for a good filling and a long intake length for good torque behavior at medium speeds. The switching is done by an electric or pneumatic actuator.

The length variation can be accomplished by the opening of bypass flaps or through the connection of manifold segments. Two-stage intake manifolds are often used for switching from resonance charging to tuned intake runner switching (*see* →Intake system ~Resonance system). **Fig. I160** shows the effects on the torque. In **Fig. I161**, some principles for switchable intake manifolds are shown.

~~~**Variable intake manifold.** Torque, fuel consumption, and exhaust gas emissions of engines can be positively influenced through the use of variable intake manifolds. The primary aim is to achieve a high volumetric efficiency and good mixture preparation over the whole engine map. A short intake manifold length is favorable for good performance at the higher speeds; long intake manifold lengths optimize the torque behavior in the low and medium speed range (resonance tube/tuned intake runner system). In order to improve the charge transfer over the whole engine map, switchable elements in the intake manifold contribute to a targeted change of the geometry of the intake system and, thus, a switch between two or more different gas dynamic tuned systems.

In order to generate charge movement on demand in the combustion chamber with low air mass flow rates, intake ports are throttled and switched off. The additional swirl or tumble results in greater running stability, a capacity for leaner operation, and a greater toler-

Material Manufacturing Process	Plastic Injection Molding	Meltable-Core Technique	Aluminum Sand Casting	Die Casting	Magnesium Die Casting
Surface Quality	+ +	+	−	+ +	+ +
Weight	+ +	+ +	−	+	+ +
Residues	+ +	0	0	+ +	+ +
Molding life	+ +	+ +	−	+	+ +
Price per unit	+ +	0	−	+ +	−
Design freedom	+	+ +	+ +	0	0
Manufacturing automation	+ +	+	-	+ +	+ +
Material evaluation: + + advantageous . . . − disadvantageous					

Fig. I157 Materials and production processes

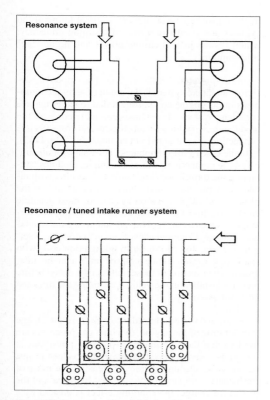

Resonance system

Resonance / tuned intake runner system

Fig. I159 Example of variable intake systems

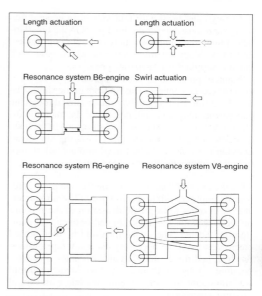

Fig. I161 Principles for switchable intake systems

ance of the engine for recirculated exhaust gas due to the improved mixture.

~~Intake manifold mass. The mass of the intake manifold is primarily dependent on the material used. If magnesium is used, weight reductions of up to 40% can be obtained compared to cast intake manifolds made of aluminum. Since the beginning of the 1990s, plastic intake manifolds are often used; these are man-

ufactured by fusible core or injection-molding sandwich element technology and offer additional potential for light-metal engine design. Plastic intake manifolds allow for weight reductions of up to 50% compared to conventional aluminum intake systems (**Fig. I157**). Additional weight advantages can be realized through additional integration of functional modules into the intake system (modular design).

~~Intake ports. In order to influence the charge motion in the cylinder from outside, the individual intake manifold may be split up into two or three ports. By throttling these ports, the charge motion in the combustion chamber (swirl and/or tumble) can be controlled. The control elements are powered by electric or pneumatic actuators.

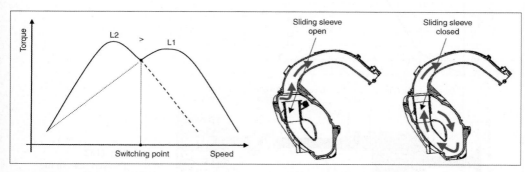

Fig. I160 Characteristic torque curve for two-stage intake manifolds and example of a gearshift sleeve concept (Source: Mahle)

In a cylinder head with four valves, in general two designs are possible for the inlet ports in the tuned intake runner: the single-tube port shape for both valves, and twin (Siamese) inlet ports, in which case each valve is supplied via a separate port.

Siamese ports can be guided separately to the inlet valve or can be connected just in front of the valve. The increased wall friction can be compensated through the corresponding design and increase of the manifold diameter. When one port is shut off, the speed in the open manifold branch increases; this allows the air stream entering the cylinder to be forced into a swirling motion, thus improving the mixture formation.

~Intake manifold preheating. Intake manifold preheating improves mixture formation at low air temperatures. In order to avoid condensation of the mixture on the cold walls, the intake air and/or the intake manifold are heated. The intake manifold is often heated electrically until the necessary coolant temperature is reached, and the cooling circuit then takes over temperature regulation. Preheating of the intake manifold was important for earlier engine models with central mixture generation.

~Intake resistance. The intake resistance is the amount of resistance exerted against the intake process brought about by the downward movement of the piston. It influences the actual airflow. All the components in the intake tract, which includes the intake pipes and volumes, filters, air-mass flow probes, and throttle valve, affect the intake resistance through their flow geometry as well as their surface structure (**Fig. I162**).

The theoretical maximum induced volume of air entering the engine results from the product of swept volume and the engine speed. The throttling of the airflow

by the intake resistance prevents optimal filling of the cylinder and results in a loss of charging efficiency, which results in less engine power and higher fuel consumption.

~Modular design, intake system. The modular design of an intake system is the union of several components in the engine compartment that together make a complete system. The consistent adherence to the modular design allows for an efficient and cost-effective connection of several individual functions (**Fig. I163**).

In the case of the components necessary for air supply, these components are, for example, modules containing the throttle valve actuator, the air filter with intake silencing, the air-mass flow sensor, as well as the blowby and EGR initiation system. In most gasoline engines the injection valves necessary for mixture formation as well as the fuel rail are integrated into the intake module. Components such as actuators (e.g., setting elements for throttle shifting), ignition coils with spark plugs, and engine control units can be integrated. The function of these units can be tested before they are installed in the vehicle, reducing the time and costs for assembly.

~Resonance system. In order to minimize the charge transfer time of the engine, dynamic gas processes are used so that the amplitudes of pressure pulsations at the intake valve attain their maximum values at just before valve closure, thereby increasing the mass of fresh charge supplied to the cylinder. In the case of the resonance charge, the charging effect is created by a "vibrating" container-manifold system. The periodic intake cycles of the individual cylinders create pressure pulsations in the container via short intake manifolds, pulsations which increase the pressure differential

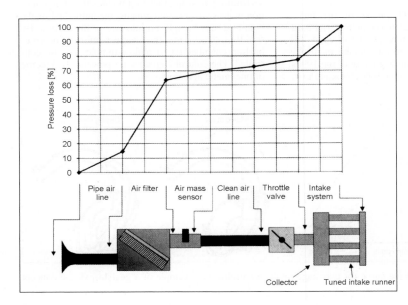

Fig. I162
Pressure loss in the intake system against ambient pressure

Fig. I163
Intake module for
BMW I-6 engine
(Source: Mahle)

between the inlet port and the combustion chamber at the beginning and the end of the induction phase.

These pressure pulsations, which cause a significant increase in the charging efficiency, have a distinct maximum when the excitation through the cylinders matches the intrinsic angular frequency of the container-manifold system.

An optimal prerequisite for vibration excitation is the offset of the individual intake phases by about 240°CA—that is, three cylinders each per resonance container.

The resonance intake system has a special meaning in combination with turbocharging to balance and improve engine torques in the lower speed range. Furthermore, a combination of tuned intake runner charging and resonance charging lends itself to 6- and 12-cylinder engines. In this case, the resonance vibration in the plenum chamber at low speed is used, because short intake manifolds contribute as tuned intake runner systems to increase the volumetric efficiency at the high-speed range. **Fig. I164** shows a schematic diagram of a resonance system.

The adaptation between operating modes is realized through the opening or closing of the resonance valve. In the torque position the resonance control valve is closed so that two three-cylinder intake systems with coupling are effective via the resonance tube and the resonance collector. The resonance control valve is open in the power position, and the intake module acts as tuned intake runner system for all six cylinders, which are fed from the main collector with short tuned intake runners. The cross sections and length were calculated and optimized to obtain these effects by using one-dimensional computational processes. The control of air-mass flow is achieved with a central throttle valve. The torque gain in such a system is shown in **Fig. I165**.

~Tuned intake runner system. In order to minimize energy expenditure for charge exchange, gas-dynamic processes are applied—air pulsation occurs in the connecting pipe between the intake manifold collector and the cylinder. The piston moving downward initiates a

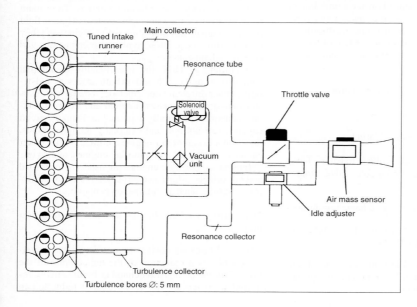

Fig. I164
Diagram of an intake
system for an I-6 engine
(Source: Mahle)

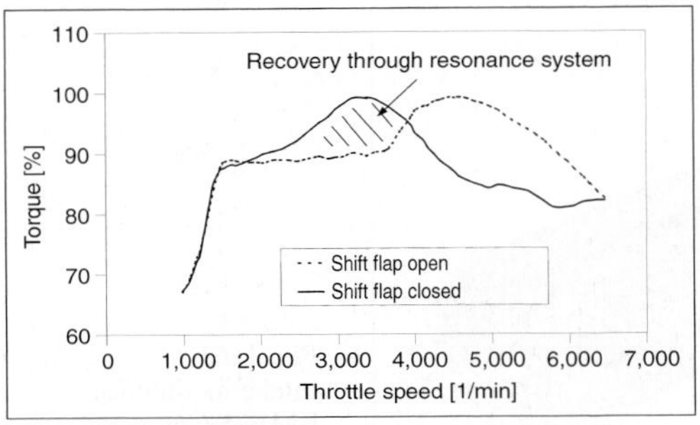

Fig. 1165 Torque curve for a six-cylinder gasoline engine with resonance system (Source: Mahle)

vacuum pulse, which runs against the flow direction through the tuned intake runner to the intake manifold collector. There, the vacuum initiates the beginning of the inflow process of the air into the tuned intake runner and the cylinder. Later, despite the upward movement of the piston, the momentum of the air-mass flowing at high speed initiates another inflow into the cylinder, which increases the filling process. Synchronization of the processes leads to the pressure impulse arriving at the cylinder shortly before the inlet closes. This synchronization is primarily accomplished through the transit time of the pressure waves in the intake manifold. Thus, the length of the tuned intake manifolds has a dominant effect.

~Turbulence system. A turbulence pipe system integrated into the intake manifold results in low emissions and low fuel consumption in idle state and in the lower part-load range.

The turbulence system has extremely small cross-sectional areas and is introduced off the center line of the intake port to produce a specific charge movement. The air supply is regulated using a separate throttle element, which adds air to the throttle valve in the bypass after the air mass sensor (→Intake system ~Resonance system).

Intake throttling →Load ~Load control

Intake valve →Valve ~Gas transfer

Intake valve spring →Valve spring

Intake valve timing →Control/gas transfer ~Four-stroke engine ~~Timing; →Variable valve control ~Variation parameters

Intake work →Charge transfer ~Charge transfer loss

Integrated electronic control unit →Actuators ~General Purpose Actuator (GPA)

Integrated products →Electronic/mechanical engine and transmission control ~Requirements for mechanical and housing concepts

Integrated starter generator (ISG) →Starter ~Starter-generator, ~Integrated starter-generator

Integrated starter-generator, or ISG (main drive) →Starter ~Starter-generators

Intelligent sensor →Sensors

Interference signals →Electronic/mechanical engine and transmission control ~Electronic components

Interference voltages →Electronic/mechanical engine and transmission control ~Electronic components ~~Interference signals

Interior noise →Engine acoustics

Intermediate cooling →Supercharging ~Multistage supercharging

Intermediate wheel bearing →Bearings ~Bearing positions

Internal bending moment →Balancing of masses

Internal bevel ring →Piston ring ~Versions ~~Compression rings ~~~Ring with inside bevel or inside angle on bottom flank

Internal combustion engine (also, →Combustion process, diesel engine; →Combustion process, gasoline engine; →Combustion process, hybrid engine). Combustion engines (also called internal combustion, or IC, engines) are heat engines that convert the chemical energy of a fuel into mechanical energy by means of a combustion process.

Diesel and gasoline engines are among the most widely used internal combustion engines. Their main features are that they are open systems (i.e., a fluid flows through an internal combustion engine) and they have mixture generation, ignition, combustion, and charge exchange processes.

According to DIN, the classification of internal combustion engines is based on both processes and designs. Other differentiations include, for example, two- and four-stroke processes; internal and external mixture generation; naturally aspirated and supercharged engines; self-ignition and externally ignited engines; injection and carburetor engines; inline, V-, and boxer engines; single-fuel and multifuel engines, and so on.

Internal combustion engines are used to power road and rail vehicles, ships, and airplanes as well as steady-state facilities such as emergency generators.

Internal combustion process →Combustion process

Internal cooling (mixture cooling). The internal cooling is a reduction of the temperature of the mixture in the combustion chamber by the energy required to evaporate the fuel. A large reduction in temperature can be experienced if the latent heat of evaporation of the fuel is high (e.g., for alcohols). A reduction in temperature increases the density of the mixture and, therefore, the mass trapped in the cylinder. The increase in performance is proportional to the increase in the mass of mixture of the same energy content in the cylinder. The internal cooling can also be used for reducing NO_x.

Internal cylinder flow →Calculation processes ~Application areas ~~Thermodynamics

Internal energy →Energy

Internal exhaust gas recirculation →Variable valve control ~Exhaust gas recirculation, internal

Internal gear pump →Gear pump

Internal mass balancing (test oriented) →Balancing of masses

Internal mixture formation →Combustion process, gasoline engine; →Mixture formation

Internal opening →Injection valves ~Diesel engine

Internal opening injection valves →Injection valves ~Diesel engine, ~Gasoline engine

Internal power →Power output

Internal tank venting systems →Injection system (components) ~Gasoline engine ~~Fuel tank systems, ~~Tank venting system

Interpretation emission value →Engine acoustics ~Exterior noise ~~Emission value

Intersection dead point →Dead center, or dead point (if not on a reciprocating engine)

Intrinsic shape →Engine acoustics ~Resonant frequency; →Torsional vibrations ~Vibration damper

Inverted engines →Engine concepts ~Reciprocating piston engines ~~Location of cylinders

Inverter →Starter ~Starter-generators ~~Integrated starter-generator ~~~Control device with inverter

Ionic current measurement →Ignition system, diesel engine/preheat system ~Cold starting aid

Irregular cam drive →Variable valve control ~Systems with camshaft

Isentropic change of state →Cycle

Isentropic compression →Cycle ~Isentropic change of state (process)

Isentropic enthalpy gradient →Supercharging

Isentropic expansion →Expansion ~Isentropic

Isentropic flow cross section →Flow cross section

Isobaric combustion →Cycle ~Constant pressure cycle, ~Seiliger (dual combustion) cycle

Isochoric combustion →Cycle ~Constant volume cycle

Iso-octane →Fuel, gasoline engine

Iso-paraffin →Fuel, gasoline engine

Isothermal change of state →Cycle ~Carnot process; →Expansion ~Isothermal

Isothermal compression →Cycle ~Carnot process

Isothermal expansion →Cycle ~Stirling cycle; →Expansion ~Isothermal

J

Jacket manifold →Exhaust system ~Exhaust gas manifold

Jamming →Bearings ~Operational damage

Jet carburetor →Carburetor ~Jets

K

k factor →Piston ring ~Parameters

K line interface →Electronic/mechanical engine and transmission control ~Electronic components

Karman Vortex flowmeter →Flowmeter intake air

Kerosene →Fuel, diesel engine ~Production

Keystone ring →Piston ring ~Versions ~~Compression rings

Kinematics, crankshaft drive →Crankshaft drive

Kinetic energy →Energy

Knit fiber filter →Particles (Particulates) ~Particulate filter system ~~Particulate filter ~~~Filter media

Knock control (*also*, →Electronic open- and closed-loop control ~Electronic open- and closed-loop control, gasoline engine ~~Functions). The purpose of a knock controller is to protect the engine from the deleterious effects of knocking combustion caused by fuel with a low octane rating or engine influences. For many engine operating points, an ignition point can be used at which the knock limit is beyond the optimum ignition point in terms of efficiency. At other engine operating points, the optimum firing point cannot be set because it is prevented by the knock limit—that is, knocking combustion occurs before the engine reaches the optimum ignition setting. In such cases, it is most practical for minimization of the fuel consumption and maximization of the output to operate the engine as close as possible to the knock limit without being damaged. This can be accomplished by changing the ignition timing.

Because the knock limit does not represent a sharp separation line but rather changes under various influences such as temperature, engine age, and fuel quality, adaptation of the timing is required. This is accomplished by knock control. Knocking combustion generates characteristic noises in the frequency range between 4 and 20 kHz, which can be detected with the aid of a suitable sensor, the knock sensor, and fed to an electronic control module. Comprehensive measurements during the application phase are required to find the suitable point for attaching the sensor on the engine in order to detect knocking in all cylinders.

Measurement of the knocking noise is performed within a certain time-window during the operating cycle of each individual cylinder to separate the knocking noise from other mechanical engine noises which might interfere with it.

When knocking is recognized, the knock controller readjusts the ignition timing to quickly terminate knocking operation in the corresponding cylinder. When knocking is recognized, the ignition point for the affected cylinder is retarded. After expiration of a certain number of ignition cycles, the timing is advanced again in a number of increments.

If knocking occurs repeatedly at certain engine operating points, the basic data for the ignition map can be changed with an adaptive controller (knock controller).

On turbocharged engines, the knock controller also decreases the charging pressure in addition to retarding the ignition.

~Adaptive knock controller →Knock control; →Knocking

~Cylinder selective knock control →Knock control; →Knocking

Literature: R. Sloboda, W. Häming, W. Fischer: Tool Chain for Effective Application of Knock Control Systems, MTZ 65

(2004) 1, p. 26. — M. Fischer, M. Günther, K. Röpke, M. Lindemann, R. Placzek: Knock recognition in gasoline engines, MTZ 64 (2003) 3, p. 186.

Knock limit →Knocking

Knock sensor →Electronic/mechanical engine and transmission control ~Electronic components ~~Input signals; →Sensors

Knock-IC →Electronic/mechanical engine and transmission control ~Electronic components ~~Input signals

Knocking (*also*, →Combustion, gasoline engine; →Fuel, gasoline engine ~Octane number). Knocking or pinging is a significant phenomenon of combustion in gasoline engines. Under normal conditions, the combustion is started by an ignition using a spark plug. The flame front moves three-dimensionally from the spark plug, or spark plugs, through the combustion chamber until the entire air-fuel mixture has been oxidized. In the case of knocking, a portion of the mixture combusts prematurely at the hot point of the combustion chamber (e.g., in the vicinity of the exhaust valve, due to the increased combustion pressure and the resulting higher mixture temperature) before the flame front has reached these areas (knocking point). This results in sudden combustion with a steep pressure increase up to 8 bars/ °CA. This corresponds to about 50,000 bars/s.

The combustion speed increases from approximately 20 to 250–300 m/s. This results in high local pressures, shock waves, pressure variations, and, above all, increased heat transfer, which can cause engine damage. For example, the pistons and valves can be burned and the crankshaft damaged. **Fig. K1** shows the flame propagation for knocking combustion. **Fig. K2** shows schematically the effect of knocking in the pressure diagram in relation to the crankshaft angle α. The variations in the pressure result from gas vibration due to local pressure differences and are perceived as a knocking or pinging noise.

Knocking must be prevented because it can lead to

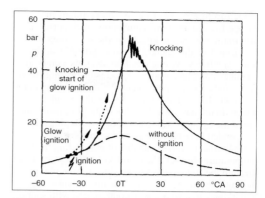

Fig. K2 Pressure curve for knocking combustion (Source: F. Pischinger)

destruction of the engine. Knocking limits the compression ratio as well as the ignition timing capability, thereby influencing the efficiency of gasoline engines. The relationship between thermal efficiency, η_{th}, and compression ratio, ε, for the idealized constant volume cycle is expressed in the following equation (where κ is the isentropic exponent):

$$\eta_{th} = 1 - \frac{1}{\varepsilon^{\kappa-1}}.$$

The more the compression ratio is limited by knocking, the lower will be the thermal efficiency. Knocking can be influenced positively by modifying the engine combustion chamber as well as by using knock-resistant fuels. In terms of the damage caused, it is necessary to differentiate between knocking duration and knocking intensity. Knocking duration is distinguished by the number of operating cycles during which knocking occurs. Knocking intensity is distinguished, for example, by the pressure gradients.

In general, low-intensity knocking does not necessarily lead to damage, even with a longer knocking duration.

Low speed negatively influences the tendency to k nocking because the mixture can be heated up for a longer period of time. However, high speed knocking can also occur during prolonged operation in the maximum performance range; the high combustion chamber temperatures are responsible for this. High-speed knocking is particularly dangerous because it is not audible over the driving noise level and has a high intensity.

The following factors also promote knocking:

- High compression ratio
- High thermal load in the combustion chamber
- High ignition advance
- Deposits in the combustion chamber
- High air density
- Insufficient engine cooling or excessive coolant temperatures
- Poor location of spark plug resulting in long flame paths

It is necessary to design the combustion chamber according to certain criteria to realize high compression

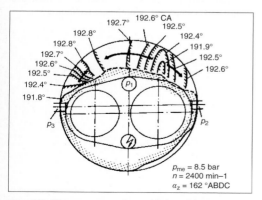

Fig. K1 Flame propagation with knocking combustion (Source: F. Pischinger)

ratios (up to $\varepsilon = 12$) to achieve maximum fuel efficiency. In a combustion chamber optimized in terms of minimum tendency to knocking, the spark plug should be located in the center, which is possible in three-, four-, and five-valve engines. Moreover, the combustion chamber must be compact. This results in short flame paths, and knocking points can form only at higher compression. In engines with two or three valves and which do not have space for the spark plug in the middle of the cylinder head, use of two spark plugs is recommended to reduce the knocking tendency.

Sodium core valves capable of conducting the high temperatures by their shaker effect from the valve head through the stem to the valve guide are suitable for reducing the temperature of the hot exhaust valves. Knocking can also occur in combination with spontaneous ignition.

Knock sensors are used to enable the engine to be operated as close as possible to the knocking limit. The sensors retard ignition until the knocking ceases. With the aid of knock control, it is also possible to use fuels with different octane ratings (RON and MON) because the electronic engine controller automatically adapts the ignition timing to the octane rating of the available fuel. These are called adaptive knock control systems.

~Acceleration knocking. Acceleration knocking occurs during full-load acceleration from low engine speeds (e.g., driving uphill in high gear). This is easily perceptible acoustically and seldom leads to engine damage due to the usually short acceleration phases.

~High engine speed knocking. *See below*, ~High-speed knocking

~High-speed knocking. High-speed knocking or high engine speed knocking occurs at high engine speeds and high loads (e.g., freeway driving) This is seldom perceptible due to the driving noises and almost always leads to severe engine damage.

~Knock limit. The knock limit usually characterizes the firing point at which knocking occurs. It is a relative value and is dependent on a large number of pa-

rameters (fuel quality, combustion chamber shape, location of spark plug, engine load, engine speed, compression ratio, cooling, etc.). A factor initiating knocking combustion and therefore affecting the knock limit is the intensity of the knocking—that is, the magnitude of the pressure vibrations.

If it is possible to set the ignition timing at the particular operating point to the maximum achievable torque without knocking occurring, then the knock limit does not have a significant for engine tuning. However, this is not possible if the ignition point for maximum efficiency is "behind" the knock limit—that is, in the knocking range; then knock control usually is activated to keep the point of ignition as close as possible to the knock limit.

~Knocking point. Knocking points are regions in the combustion chamber which offer favorable prerequisites for knocking operation. These can be deposits of fuel or oil residues, ash, insufficiently cooled points, hot residual gases, and so on. When developing an engine, it is necessary to ensure that such knocking points are not present.

~Knocking tendency. The knocking tendency describes the sensitivity of the engine to knocking. This can have various causes such as use of a fuel with a tendency to knock (low octane number), early ignition, high compression, and so forth.

Knocking combustion →Knocking

Knocking during acceleration →Fuel, gasoline engine ~Octane rating requirement; →Knocking

Knocking intensity →Fuel, gasoline engine ~Octane number

Knocking point →Knocking

Knocking tendency →Knocking

KS Lokasil process →Crankcase ~Crankcase design ~~Cylinder ~~~Composite technology

L

Laboratory tests →Sealing systems ~Development methods

Ladder frame design →Crankcase ~Crankcase design ~~Main bearing pedestal area

Lambda (λ) →Air-fuel ratio

Lambda map (also, →Performance characteristic maps). The lambda map is a representation of the specific work output over the engine speed range with air-fuel equivalence ratio as an additional parameter.

Lambda meter. The air-fuel equivalence ratio, λ, can be calculated either using, for example, C and O figures from the exhaust components (with one calculated figure coming from the air and the other from the fuel) or using a probe (e.g., lean lambda sensor).

The air-fuel equivalence ratio is calculated from the voltage of the sensor when the sensor ceramic temperature is taken into account. The sensor ceramic temperature can be measured indirectly from the resistance of the sensor to an internal alternating current.

Lambda probe (also, →Sensors ~Lambda sensors). The lambda probe is a measuring device which determines the air-fuel ratio in the exhaust by measuring the partial pressure of oxygen, and keeps it constant within a narrow range by means of a control system. A prerequisite for optimum operation of a three-way catalytic converter is that operation be maintained in the control range.

Lambda probe aging. The lambda probe is designed for very high long-term stability. However, incorrect operation can lead to premature aging of the probe and its failure.

Lambda range →Air-fuel ratio

Lambda sensors →Sensors

Lambda window (also, →Catalytic converter). Three-way catalytic converters for vehicles are used to convert all three pollutants (CO, HC, and NO_x) simultaneously to an extent sufficient to meet legislation in a very narrow range of air-fuel equivalence ratio (**Fig. L1**). This range of approximately λ = 0.99 to –1.00 is called the lambda window. **Fig. L1** shows the lambda window, with the dashed curves showing the emissions without a catalytic converter. The solid lines indicate the emissions with catalytic aftertreatment.

Laminar charge movement →Charge movement

Laminar flow →Charge movement

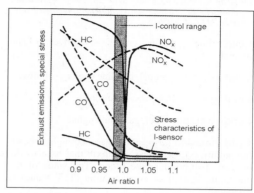

Fig. L1 Control range of the lambda probe (lambda window) and exhaust emissions in front of and behind the catalytic converter

Lamination process →Electronic/mechanical engine and transmission control ~Requirements for mechanical and housing concepts

Lancaster harmonic balancer →Balance shaft; →Mass balancing mechanism

Landfill gas →Fuel, diesel engine

Lanova process →Combustion process, diesel engine ~Prechamber engine ~~Air storage process

Large diesel engines →Engine ~Historical engines ~~Diesel engines, large

Large engines →Lubrication ~Lubricating systems

Laser holography →Engine acoustics

Laser-Doppler-Vibrometry →Engine acoustics

Latent heat accumulator →Heat accumulator

Law of combustion. The actual time-dependent process of fuel energy conversion is covered by equations known as the law of combustion, the combustion curve, and the heat release law. The law of combustion defines the conversion of the fuel mass used for each combustion cycle. The heat release law covers only the part that is equivalent to the pressure increase in the combustion chamber. The differences comprise mainly the heating losses, which are approximately 10%. The heat release law, $x(\varphi)$, is described as

$$x(\varphi) = \frac{m_{B,conv.}}{m_{B,comb.cycle}},$$

where $m_{B,conv.}$ is the combustion mass converted at the crank angle, φ, and $m_{B,comb.cycle}$ is the converted fuel mass for each combustion cycle.

Literature: Woschni: Beitrag zum Problem des Wärmeübergangs im Verbrennungsmotor, MTZ 26, 1965. — Woschni: Elektronische Berechnung von Verbrennungsmotoren-Kreis–prozesse, MTZ 26, 1956.

Lead →Fuel, gasoline engine ~Additives, ~Lead content

Lead additives →Fuel, gasoline engine ~Additives

Lead alloys →Bearings ~Materials

Lead content →Fuel, gasoline engine

Lead emission (*also*, →Fuel, gasoline engine ~Lead-free additives). Lead is a cytotoxin that reduces the oxygen absorption in the blood. The value of the maximum workplace concentration (TLV value) is 0.1 mg/m^3.

The lead emissions from combustion engines have been significantly reduced since the introduction of lead-free gasoline. The trend for lead addition is now moving toward zero. **Fig. L2** shows the stepwise reduction of the lead content in fuels as determined by the regulator in Germany. No fuel with lead content has been marketed since 1984.

The components with a lead content that improve the antiknock properties of gasoline are mainly lead alkyls—for example, tetraethyl lead (TEL) and tetramethyl lead (TML). These have been replaced by antiknock fuel components (e.g., hydrocarbons with circular or short chains).

Leaded fuel→Fuel, gasoline engine ~Lead content

Leaded gasoline→Fuel, gasoline engine ~Lead content

Leaded normal gasoline→Fuel, gasoline engine ~Regular gasoline

Lead-free additives→Fuel, gasoline engine

Lead-free gasoline→Fuel, gasoline engine ~Lead content

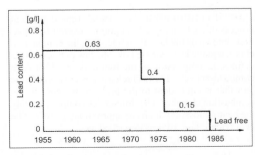

Fig. L2 Lead content reduction

Leakage losses (*also*, →Blowby). Leakage losses occur at various positions in the engine. For example, in turbochargers they occur in the compressor and turbine. They also occur in mechanical superchargers—for instance, in Roots blowers.

Leakage losses from the combustion chamber are particularly important; these are called blowby losses. Leakages from the exhaust system are also important, and leaking exhaust lines in front of the lambda probe can conceal the danger that air is entering the system as a result of pressure pulsations. The result of this is that mixture formation is affected, because the signal obtained from the lambda probe is false.

Lean air-fuel mixture →Air-fuel mixture ~Capability of achieving leaner mixtures

Lean catalytic converter →Catalytic converter

Lean concepts →Fuel consumption ~Variables ~~Engine measures; →Lean-mixture engine

Lean lambda sensor (*also*, →Sensors ~Lambda sensors). Lean lambda sensors measure values of λ very much less than 1.00. If necessary, the output can be used for controlling the λ value. Lean lambda sensors can be used, for example, for lean-burn and lean-mixture engines or for gasoline engines with direct injection.

Lean mixture →Air-fuel mixture

Lean operating limit →Operating limit

Lean operation →Lean-mixture engine

Lean run capability →Operating limit

Lean run O$_2$ sensor →Piston ~Materials; →Pollutant aftertreatment ~Pollutant aftertreatment lean concepts; →Sensors

Lean (weak) mixture →Air-fuel mixture ~Capability of achieving leaner mixtures

Lean-burn engine (*also*, →Catalytic converter). In contrast to lean-mixture engines, lean-burn gasoline engines are tuned preferably so that $\lambda \gg 1$, except when they are at full load. The objective is to burn a lean mixture through mixture formation measures. Lower untreated pollutant emissions result if it is possible to maintain the combustion even at mixtures of λ to 2. Further weakening of the mixture to make the use of a catalytic converter superfluous would be desirable. The problem here is the flammability of extremely lean mixtures.

A typical example of a lean-burn engine is the direct injection gasoline engine. A large proportion of the part-load range is tuned to a lean mixture (up to $\lambda = 4$) to minimize the fuel consumption; however, operation is around $\lambda = 1$ at full load.

Lean-burn engine catalytic converter →Catalytic converter

Lean-burn engine management →Electronic open- and closed-loop control ~Electronic open- and closed-loop control, gasoline engine ~~Functions ~~~Lambda control

Lean-mixture engine (*also*, →Catalytic converter). A lean-mixture engine is a gasoline engine in which the mixture formation is tuned so that it is rich in the full-load range, λ = 1 in certain part-load ranges, and in other part-load ranges has λ values much greater than one. This results in high torque at full load and minimal pollutant emissions (HC, CO, and NO$_x$) in the part-load range depending on the operating points and minimum fuel consumption. Minimization of the emissions can only be accomplished using catalytic converters. The extremely lean range that is desirable in terms of fuel consumption is, however, associated with higher emissions because the three-way catalytic converter does not have good conversion rates in this range. The problem with tuning is to keep the emissions below the values specified by law in spite of the extensive region in which λ ≠ 1.

Legal limits →Emission limits; →Engine acoustics; →Particles (Particulates)

Legal regulations →Air intake system ~Acoustics; →Emission limits; →Engine acoustics ~Exterior noise ~~Legal measurement; →Particles (Particulates)

Legislation, vehicle noise →Air intake system ~Acoustics; →Engine acoustics ~Exterior noise ~~Legal measurement

Length, connecting rod →Connecting rod ~Design

Length, exhaust port →Exhaust port ~Exhaust port design

Length, inlet manifold design →Intake system ~Intake manifold ~~Intake manifold design

Length, inlet port →Intake port, diesel engine; →Intake port, gasoline engine ~Design

Length compensation, valve →Valve ~Gas transfer valves ~~Valve length compensation

LEV →Emission limits

Level control →Carburetor

Lex Ferrari →Engine acoustics

Light oil →Fuel, diesel engine ~Gas oil

Light-metal crankcase →Crankcase ~Weight, crankcase ~~Reduction of specific weight of material

Light-off curve →Catalytic converter ~Light-off temperature

Light-off-point →Catalytic converter

Limp-home air mass flow →Actuators ~E-gas ~~Limp-home feature

Limp-home control →Electronic open- and closed-loop control ~Electronic open- and closed-loop control, gasoline engine ~~Electronic accelerator pedal

Limp-home feature →Actuators ~E-gas

Limp-home position →Actuators ~E-gas ~~Limp-home feature

Linear O$_2$ sensor →Sensors ~Lambda sensors

Linear O$_2$ sensor interface →Electronic/mechanical engine and transmission control ~Electronic components

LIN interface →Electronic/mechanical engine and transmission control ~Electronic components

Liquefied gas →Fuel, gasoline engine ~Alternative fuels

Liquefied petroleum gas →Fuel, gasoline engine ~Alternative fuels ~~Liquefied petroleum gas

Liquid cooler →Radiator ~Liquid/air radiator

Liquid cooling →Cooling, engine

Liquid sealing →Sealing systems

Liquid/air intercooler →Radiator

Liquid/liquid intercooler →Radiator

Load. Load is a measure of the torque of an engine between no load (e.g., idle), part load, and full load and also trailing throttle.

The desired load (load control, load regulation) on a conventional gasoline engine is set through the mass of mixture in the cylinder (quantitative control). The mass of mixture is in turn prescribed by a control unit, which is typically a throttle valve. However, it can also be achieved by piston port control or a rotary valve. The load on a diesel engine is determined exclusively through the mass of fuel injected. No throttle valve is required to control the load—that is, the mass of air induced is always the same on a naturally aspirated engine (this is called qualitative control). In contrast to gasoline engines, there is no or only limited throttle loss in diesel engines.

In a concept engine with hybrid combustion processes, the load is frequently set through a combination of qualitative and quantitative control. The quantitative control in this case occurs at idle and in the lower part-load range because with qualitative control the mixture preparation and ignition and combustion would be incomplete. Part and full load are achieved using qualitative control.

A measure for the load is the specific work (mean effective pressure) that the engine produces.

If additional loads occur at idle in a gasoline engine—for example, loading from an alternator, the power steering, or an activated air-conditioning system—this tends to lead to a reduction in the engine idle speed. This reduction in speed can be compensated through, for example, an idle control device (adjuster), which adds a greater mass of air or mixture according to the engine speed—and, sometimes, also temperature—thus increasing the "no load" indicated power. The increased friction loss after a cold start can also be compensated for by means of this technique.

A similar approach is used in a diesel engine, in which the air is enriched with fuel instead of the air or mixture mass being increased.

~Diesel engine. With a diesel engine, the device for measuring the fuel (e.g., control rod of the distributor pump, injection nozzles) is influenced by a signal from the accelerator pedal. The mass of air induced is almost unchanged for a constant engine speed. Load control is realized by varying the amount of fuel injected. This is called qualitative control, because the quality of the mixture in the combustion chamber is influenced (from approximately $\lambda = 1.05$ to 8).

~Full load. Full load determines the maximum torque and, at the corresponding engine speed, the maximum engine performance. When designing the full-load point of an engine, a number of conditions have to be considered, such as sufficient cooling, knocking in gasoline engines, fuel consumption, load on components. In gasoline engines, full load is characterized by a rich air-fuel equivalence ratio ($\lambda < 1$) and in diesel engines by a fuel-air ratio of approximately $\lambda = 1.05$.

~Gasoline engine. In a conventional gasoline engine, the accelerator pedal controls the throttle valve to generate different effective diameters for the throttle valve to control the flow of the air-fuel mixture. As the engine load depends on the induced mass, load control is possible by throttling the mixture induced (charge control). The composition of the mixture is almost constant to a first approximation. This is called quantitative control. Engines with stratified charge or lean-mix engines are operated with a combination of qualitative and quantitative control.

~Idle. Idle is part of the no-load operating state, when the engine speed is as low as possible. This results in low fuel consumption and low noise emissions. However, a stable idle is important; that is, engine speed fluctuations at idle should be as small as possible. Idle engine speeds of about 500–800 rpm are typical, but higher values occur in diesel engines: the reason for this is the larger irregularity of the engine operation as a result of the higher compression ratio.

~Load control. Load control in conventional engines is performed through the accelerator pedal.

In a conventional gasoline engine, the quantity of mixture is varied by the throttle valve; the air-fuel equivalence ratio remains almost constant to first approximation at λ equals 1; in diesel engines and gasoline engines with direct injection, the quantity of fuel is altered by controlling the injection pump or with a common-rail through the valve opening time.

~Load regulation. *See above*, ~Load control

~Load transfer. Load transfer takes place when the position of the accelerator pedal is changed rapidly. As this is accompanied by a sudden change in torque, the torque reactions in the vehicle appear as a shuddering deceleration or acceleration, which the driver experiences as "juddering." Principally in the idle range, this juddering can build up into what is known as the "Bonanza effect," a self-increasing jerking that takes over the entire vehicle. Modern vehicles minimize these effects to a large extent—for example, through design of the engine mounting and intervention in the engine or gears steering.

~No load. No load is the operating state of an engine which, starting from idle, covers the entire engine speed range without load.

~Part load. Part load is an engine operating state between no load and full load. Close to full load is called upper part-load operation and close to idle is called lower part-load operation. The behavior of an engine at part load is important because the cumulative load during the life of an engine occurs overwhelmingly in this range.

~Trailing throttle. Trailing throttle is a load state in which negative work is performed—that is, the engine does not produce work but consumes it. Trailing throttle occurs in a gasoline engine when the throttle valve is suddenly closed; the engine speed is greater than the idle speed so that the vehicle mass is decelerated through the consumption of work by the engine—also called engine braking. In diesel engines and frequently also in gasoline engines, trailing throttle is achieved by switching off the fuel supply.

~Transient engine operation. Transient engine operation is characterized by the rapid change of speed and/or load.

Load control →Load

Load regulation →Variable valve control

Load transfer →Load

Load-speed flow measurement →Flowmeter intake air

Local curves →Balancing of masses

Locomotive oils →Oil ~Classification, specifications, and quality requirements ~~Special engine oils

Longitudinal engine installation →Engine concepts; →Engine installation

Long-life spark plug →Spark plug

Long-stroke →Bore-stroke ratio

Long-stroke engine →Bore-stroke ratio

Long-stroke motor →Bore-stroke ratio

Long-term/extended temperature loads →Electronic/mechanical engine and transmission control ~Requirements for mechanical and housing concepts ~~Thermal stresses

Long-term/extended vibration loads →Electronic/mechanical engine and transmission control ~Requirements for mechanical and housing concepts ~~Vibration stresses

Loop scavenging →Two-stroke engine ~Scavenging process

Loss factor →Engine acoustics

Lost foam process →Crankcase ~Casting process, crankcase

Loudness →Engine acoustics ~Psychoacoustic parameters

Loudness perception →Engine acoustics ~Psychoacoustic parameters

Louver →Radiator

Louver JASO →Oil ~Classification, specifications, and quality requirements ~~Institutions

Low-emission vehicles (also, →Emission limits). Low-emission vehicles are vehicles complying with certain legal emission standards. They have tax advantages.

Lower calorific value →Fuel, diesel engine ~Properties; →Fuel, gasoline engine ~Calorific value

Low-jerk cams →Camshaft

Low-pressure casting →Crankcase ~Casting process, crankcase ~~Chilled casting

Low-pressure injection →Injection functions ~Gasoline engine ~~Direct injection ~~~Injection pressure

Low-pressure system →Injection system, fuel ~Diesel engine ~~Common rail

Low-reflection room/anechoic chamber →Engine acoustics ~Free field

Low-side driver →Electronic/mechanical engine and transmission control ~Electronic components ~~Power transistor

Low-speed motor. Low-speed motors are engines with low nominal speeds of about 100 rpm and compression ratios of approximately 12:1, and they are mostly used as diesel engines for ships. Medium-speed engines have nominal speeds of about 500 rpm with compression ratios from 12:1 to 14:1. High-speed engines are automobile diesel engines with prechambers or engines with direct injection. They have compression ratios of 17:1 to 23:1 and nominal speeds of 4000–5000 rpm—the higher values of the compression ratio and nominal speeds apply to prechamber engines. Compression ratios that are too high result in increased peak pressures, which can exceed the permissible component load. In addition, excessively high compression ratios lead to increased wall heat transfer and friction losses. The increased combustion temperatures with excessive compression ratios also lead to the formation of NO_x and more combustion noise.

The lower compression ratios found in low- and medium-speed engines are brought about by the high levels of charging efficiency typical today.

L

Low-swirl combustion →Combustion process, diesel engine ~Direct injection ~~Multiple spray method, air distributing; →Intake port, diesel engine

Low-temperature behavior →Fuel, diesel engine ~Properties

Low-temperature behavior, oil →Oil ~Power during operation ~~Effect of different operating conditions ~~Cold operation

Low-viscosity oil →Oil ~Properties and characteristic values ~~Viscosity

LPG →Fuel, gasoline engine ~Alternative fuels ~~Liquefied petroleum gas

L-shaped compression ring →Piston ring ~Versions ~~Compression rings

LTCC →Electronic/mechanical engine and transmission control ~Electronic components ~~Substrate ~~Transmission frame

LTV →Air cycling valve

Lubricant →Oil

Lubricant additive →Fuel, diesel engine ~Properties

Lubricant optimization →Fuel consumption ~Variables ~~Engine measures

Lubrication (*also*, →Oil). The lubricating system must essentially fulfill the following functions:

- Supply the main, connecting rod, and piston pin bearings with engine oil
- Cool the piston and lubricate the piston, piston rings, and cylinder liners with engine oil
- Supply engine oil to the camshaft bearings, the valve clearance compensation elements, and other bearing surfaces
- Supply bearing points in the area of the valve gear and gear drives
- Supply the bearing points of the support accessories (injection pump, cooling fan, viscous fan, tension roller, chain tensioner, etc.)

Fig. L3 shows the lubricating system of the four-stroke passenger car diesel engine.

It consists mostly of the oil pan, the intake line with strainer, oil pump, control valves for the oil pressure, oil filter, oil cooler, bypass valves, safety devices (pressure-differential switch, oil pressure-holding valves), injection devices, covers, and nozzles.

The oil pump sucks the oil from the oil pan and delivers it to the bearings on the engine through an oil filter. The filter has the task of filtering foreign particles out of the engine oil to keep wear and tear at the bearings as low as possible. The oil filter housing usually contains the pressure regulator valves and/or the overflow valves and the bypass valves. Pressure regulator valves are necessary to keep the supply pressure to the engine bearings as near to constant as possible. The excess oil fed by the oil pump runs from the pressure relief valve back to the oil pan. The remaining oil volume flows into the main oil passage of the engine.

The bypass valves reliably supply the engine with oil, even when the filters are heavily soiled. The bypass valve makes bypassing of the filter possible,

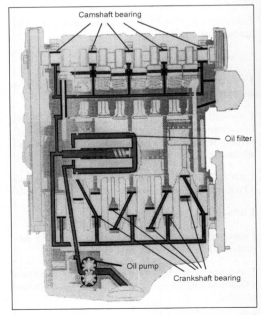

Fig. L3 Lubricating system of a four-stroke passenger car engine

which, however, leads to a high-risk operating state with high wear and tear on the bearing points.

The main oil passage supplies the base bearings, and the connecting rod bearings supply the crankshaft and, if necessary, the small end bearing as well as the piston with oil. Oil is fed from the main oil passage to the cylinder head to lubricate the camshaft and the valve actuating elements. In case of a supercharged engine, part of the oil is also used to lubricate the turbocharger. The oil used for lubricating the turbocharger may not be fed through a bearing surface, in order to avoid carrying abraded parts. Highly stressed engines require an oil cooler, because the oil not only needs to lubricate, but must also transport heat from the bearings and/or from the piston and the cylinder liners. The temperature in the lubrication gap as well as on the piston should not exceed certain limits, depending on the engine type. Furthermore, the aging of the oil depends on the oil temperature level. An increase of the oil temperature in the oil pan—that is, from 120 to 130°C—results in a halving of the oil change interval.

~**Bypass valves**. *See below*, ~Control and safety components

~**Control and safety components**. The prerequisites for the efficiency of the pressure control valve are throughput and discharge cross sections large enough to avoid exceeding the desired pressure.

Behind the pump, a safety valve is arranged to protect from the overload that can occur when a cold engine is started. In passenger cars, this safety valve is the pressure-regulating valve. In truck engines and/or medium-speed engines, separate pressure-regulating valves are usually arranged at the end of the main oil passage and in the cylinder head. The safety valve is in most cases adjusted to between 800 and 1200 kPa.

Oil-regulating valves, which also can be used as safety valves and pressure differential valves, are shown in **Fig. L4**. The valve seat is either conical or level and,

based on experience, the opening cross-sectional area of the valves should be at least 1.4–1.5 times the cross-sectional area of the pressure line.

Piston valves must be sufficiently long, and the sliding surfaces and the valve seat must be hardened and ground. At the lower end of the piston shaft, several small holes must be arranged so that the oil enclosed behind the piston valve can be reached during the vertical motion.

SBy arranging the discharge opening from the valve slightly higher than the surface of the valve seat, a quick opening of the discharge cross section and a soft bottoming during closing of the valve is achieved under the triggering pressure. In order to avoid seizure of the valve due to contamination, the diameter of the piston shaft above the seat is about 0.4–0.8 mm smaller than the guiding diameter.

With piston valves, soft valve springs are recommended to attain large cross-section openings to control the oil pressure uniformly. Piston valves set at either side of the oil pressure are used as differential pressure valves.

Usually, flat- or conical-seat valves are used as pressure-relief valves (**Fig. L4**). Flat-seat valves, each of which has one pressure spring, are used as differential pressure valves in filters. **Fig. L4** shows a conical valve for pressure regulation of the spray nozzles to cool the pistons. These valves open at excess pressures between 100 and 150 kPa and ensure that the main bearings are supplied with sufficient oil, even at low engine speeds.

For small oil capacities, balls are used as valves to regulate pressure. Ball valves tend to seize the springs and balls, and—due to the usually short, hard springs—tend to "chatter"; this often causes the seat to leak. In the case of pump or filter housings made of light metal or gray cast iron, valve seats made of steel must be pressed in for ball valves (**Fig. L5**). Because of their vulnerability, ball valves are now being used less frequently in lubricating systems.

As additional control valves, differential pressure valves and thermostatic valves (oil cooler) are also

L

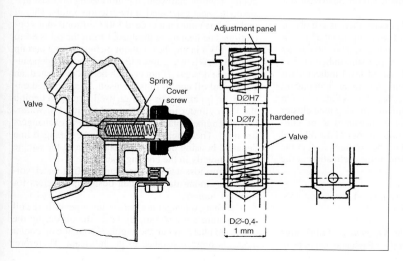

Fig. L4
Lubricating oil
pressure-regulating valve

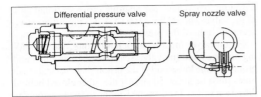

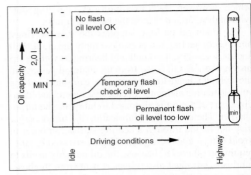

Fig. L5 Conical valve

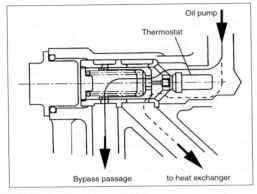

Fig. L6 Thermostatic valve

Fig. L7 Responsiveness of the oil level indicator

~Differential pressure valve. *See above*, ~Control and safety components

~Dry-sump lubrication. Dry-sump lubrication is used in engines that are subject to extreme use, such as acceleration during long, fast cornering. This prevents the lubricating oil from being moved away from the entrance of the intake pipe to the oil pump because of high centrifugal force. This would be possible in the case of wet-sump lubrication with an oil pan and oil container unit—the oil supply could not be guaranteed. In the case of dry-sump lubrication, the oil supply is not saved in the oil pan but in a separate oil tank. The oil returning from the engine is sucked from the very flat oil pan into this oil tank. Another pump or another pump level or both suck the oil from there and transport it to the lubricating points on the engine.

Due to the small depth of oil pan, there is also more ground clearance with dry-sump lubrication, which is important for off-road vehicles and/or sports cars and racing cars.

~Engine cooling (*also*, →Radiator). In addition to reducing friction and wear, the lubricating oil must dissipate the heat generated at the friction points. The oil in the oil pan must not exceed 130°C in continuous operation. The heat to be dissipated from the oil is 84 to 125 kJ/kWh in engines without piston cooling and up to 250 kJ/kWh in engines with piston cooling (depending on the design). If not enough heat is dissipated, an additional oil cooler must be used. In air-cooled engines, an oil air cooler is often used, arranged behind the last cylinder in the cooling-air route. For water-cooled engines, plate or pipe-bundle heat exchangers are used. These heat exchangers can be arranged in a separate housing or in the cylinder area in the engine block, directly in the cooling water.

Pressure loss and the necessary cooling water volume are among the important design parameters for such oil coolers.

The heat exchanger not only reduces peaks in the oil temperature but also quickly heats the oil during the warm-up phase. When the engine is cold, the oil cooler presents more resistance to the lubricant. Therefore,

present in engines. **Fig. L6** shows a thermostatic valve for a four-cylinder, 3.3-liter, direct-injection diesel engine. At low temperatures, the oil flows through the bypass opening to the engine. At higher temperatures, the oil is transported to the engine through the oil cooler. In order to prevent draining of the oil from the lines (e.g., cylinder head with hydro push rods), oil nonreturn valves are used. These are ball valves without springs or suspension valves. Differential pressure valves protect the filter and the cooler from pressure levels that are too high, which would otherwise destroy the components.

At the end of the main oil passage there is usually an oil-pressure indicator sensor. The arrangement of this oil pressure sensor must be chosen to avoid possible leaks into the oil stream from the oil pressure indicator.

It is necessary to provide an oil level indicator for monitoring the oil cycle. Examples are the dynamic oil flow sensors used by Mercedes-Benz (**Fig. L7**). On these, a warning lamp illuminates in the gauge cluster when the dynamic oil level is measured to be constantly below 60 s. Another way to control the oil level is the familiar dipstick. It must be arranged so that the oil level shown is independent of the inclination of the vehicle.

~Cyclone separator. *See below*, ~Lubricating systems ~~Passenger car engines

~Cylinder head lubrication. *See below*, ~Lubricating systems, ~Lubricating systems ~~Passenger car engines

the oil flow bypasses the cooler via a flow valve and flows directly into the engine. The circulating oil reaches its operating temperature faster that way. Only during slow warm-up does a part of the oil flow into the cooler. The valve must be adjusted so that all of the lubricant flows through the cooler when the engine has reached operating temperature.

For the design of air-cooled oil coolers, a differential between the oil and the air temperature of 70–80°C is assumed, and for water-cooled air coolers, it is 30–40°C.

~Filter (also, →Filter ~Lubricating oil filter). Filter systems for the engine oil are important components of modern engines. Along with the current development status of modern engine lubricating oils, they add to nominal engine life. Oil filters must also fulfill this requirement with even further reduction in their installation space.

There are two requirements for the oil filter:

• Reduction of wear through removal of abrasive particles in the size range dangerous for the lubricating gap
• Limitation of the concentration of fine solid contamination in order to prevent failures in the oil cycle

The lubricating oil in the engine is contaminated particularly by metallic abrasion, dust from the induced air, formation of water through condensation, and dilution with fuel, soot, and corrosion products. This readily results in damage to the bearing surfaces and, thus, to a progressive increase in wear. Furthermore, wear particles influence the ease of pumping the viscosity and/or sludging of the engine oil—the last results in increased wear at high concentrations and/or in a deterioration of cold-start behavior. Filtering the engine oil removes these particles, reducing engine wear.

Filters are classified as either full-flow filters or partial-flow filters. **Fig. L8** shows filter switching, full-flow filters, and partial-flow filters.

The total volume of engine oil flows through the full-flow filter. Up to 30% of the total oil quantity flows through the partial-flow filter; this is not returned via the bearings but directly into the oil pan. Partial-flow filters have finer filter materials and reduce the fine contamination in the oil.

Almost all full-flow filters in engines for road vehicles have inserts made of phenolic resin-impregnated cellulose paper. The filter inserts in partial-flow filters, which must reduce fine contamination in the oil, consist of either microporous paper, cotton, or fiberglass.

The filtration performance of the oil filter depends on the fineness of the filter element. When evaluating the filter fineness, it must be observed that in cyclic operation, as in the engine lubricating system, the degree of filtration is shifted to much smaller values after several cycles.

Therefore, in full-flow filters, compromises must be made when choosing the fineness. The primary selection criterion for filter fineness is the required degree of wear protection from the reliable filtering out of wear particles and large and hard particles and, thus, favorable downtime behavior under varying and extreme operating conditions. A contrary requirement is to guarantee sufficient circulation volume of oil, especially during a cold start.

The partial-flow filter elements consist primarily of cotton fibers. These filters have proven themselves in practice, especially with regard to specific values of soiling capacity. With slow flow speeds (penetration), a higher percentage of the fine contamination can be filtered out in partial-flow filters, which is not filtered by the full-flow filter.

~~**Filter designs.** In smaller engines, such as passenger car engines, replacement filters are used as full-flow filters (**Fig. L9**). The housing and cartridge are firmly connected and are replaced as a unit during maintenance. The filter housing also contains the

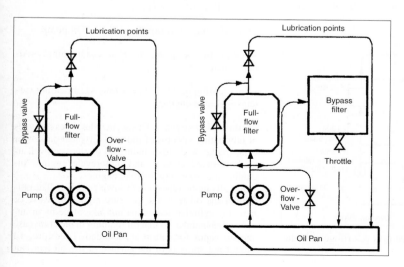

Fig. L8
Connection diagram for full-flow and partial-flow filters

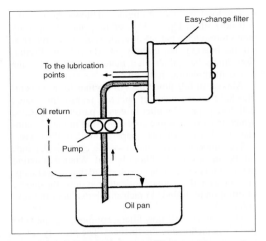

Fig. L9 Full-flow replacement filter

bypass valve and, depending on the design, the return lock valve, on either the inflow or the outflow side or on both sides. The lock valves prevent the filter from emptying when the engine is shut off.

In trucks and larger engines, housing filters with exchangeable inserts are used as full-flow filters. The filter cartridges are replaced during maintenance. Full-flow and partial-flow filters are used to achieve long oil change intervals. These can be arranged separately or in combination in a single housing.

Instead of partial-flow filters, full-flow centrifuges can also be used (**Fig. L10**). Essentially, they consist of a housing and a rotor. The oil, branched off into a partial-flow line downstream from the full-flow filter, flows into the centrifuge and through the central hollow shaft into the rotor.

After flowing through the rotor, the oil under pressure is directed to the drive nozzle mounted under the rotor. The reaction force created by the oil jets gives

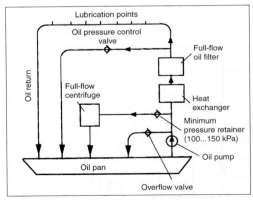

Fig. L10 Position of the free-flow centrifuge in the lubricating system

the rotor a high torque dependent on the oil pressure. The centrifugal force created by the rotation throws the contamination from the oil onto the interior wall of the stator, where it adheres as a homogeneous layer. The growing layer easily can be removed. The clean oil flowing from the drive nozzles into the centrifugal housing flows without pressure from the housing into the engine oil pan or into a separate oil container.

Filters have a life of about 30,000 km in the gasoline engines fitted in passenger cars and 60,000 km for diesel engines in trucks. Depending on the lubricating oil used, its quality, and the operating mode of the engine, the paper loses mechanical strength after longer periods of use. The limits for this bursting strength are around 80 kPa. The flow-through resistance with the normal flow-through volume without contamination is 20–40 kPa at about 95°C.

~Flow rate control. See below, ~Oil pump ~~Pump loss of performance loss

~Free-flow centrifuges. See above, ~Filter ~~Filter designs

~Fuel gallery. See below, ~Oil pump ~~Spiral gear pumps

~Fuel line pressure. See below, ~Oil line

~Full-flow filter. See above, ~Filter

~Gear pump. See below, ~Oil pump

~Intake manifold. See below, ~Oil pump ~~Arrangement

~Lubricating oil gallery. See below, ~Lubricating systems ~~Large engines

~Lubricating oil hole. See below, ~Lubricating systems ~~Passenger car engines

~Lubricating oil pump. See below, ~Oil pump

~Lubricating oil sump. See below, ~Oil pump ~~Arrangement

~Lubricating oil tank. See below, ~Lubricating systems ~~Large engines

~Lubricating systems. This section describes the types of lubricating systems of individual typical passenger cars, trucks, and small industrial engines as well as medium-speed engines and motorcycle engines including high-speed two-stroke engines.

For all engine types, the following applies: The outflow of the oil from the pressure-regulating valve or from the cylinder head must not take place in an uncontrolled manner, so that oil foaming in the crankcase and oil vapor formation are avoided. Therefore, oil should not spray onto the surface of the oil pan, but should drain off along the walls or be fed in under the oil level.

Depending on the engine type, the oil pumped from the oil pan, as mentioned before, is circulated via the oil pump at rated engine speed four to six times per minute in passenger car and truck engines and twice to four times in medium-speed engines. The degree of oil foaming determines whether the higher or lower value applies. The air content of the lubricant should be under 7%, especially if hydraulic valve clearance compensation elements are used.

In order to fulfill this requirement, oil foaming must be kept at a minimum when designing the oil cycle. During the design of the crankcase, individual functions such as oil separation, crankcase ventilation, oil return, and the possibility of oil movement in the oil pan must be considered as a complete system. The engine oil flowing out of the cylinder heads should be returned via large cross sections in the engine block and, from there, through deflectors into the oil pan. At the same time, it must be recognized that the blowby gases from the crankcase can rise through holes or shafts into the highest points of the cylinder head. Return openings for the oil and the rising possibilities for blowby gases should, therefore, be arranged separately.

~~**Large engines. Fig. L11** shows the schematic of the engine oil cycle for an MTU engine. This engine is used for heavy trucks, locomotives, and power generators as well as for marine and speedboat operations. At the front of the engine, a gear pump for circulating the oil is driven by the crankshaft. From there, an oil line in the region of the oil pan is taken to the two oil coolers and oil filters on the side, on the right-hand side of the cross section.

The two oil filters and oil cooler units are fed separately from the oil duct in the oil pan, but there is a joint return. The oil first enters the oil cooler and is advanced to the full-flow filters and from there to the transversally positioned disc filter and to the engine block.

From the second unit near the flywheel, oil is fed via the front unit to the return-flow bore. The oil filter on each unit can be locked separately. The lubricating oil flows from these oil cooler filter units into the main oil

gallery. From there, oil is branched off to cool the piston. This oil is sprayed toward the piston through nozzles which have a spring-loaded valve. The piston itself has collector nozzles to receive the oil necessary to cool the piston. After cooling the piston, the oil is used to lubricate the small end bearing. From the main oil gallery, a vertical bore branches off to the main bearings, which lubricates the main bearings and the big end bearings via the crankshaft. To supply oil to the camshaft and the wear parts, such as rocker shafts and contact points between the pushrod and the rocker arm on one side and the valve and rocker arm on the other side, oil is fed per cylinder via a bore hole from the main oil gallery through the camshaft bearing to the cylinder head, where the blind hole of the camshaft bearings has a groove. The pushrod itself is a roller tappet. From the front, last main bearing, the hydraulic vibration damper is supplied with engine oil. This last bearing is not an engine bearing with bearing contact, but rather it limits radial movements to the vibration dampers. The locating hole for the installation of the gearwheel operation also has supply lines and discharge lines for vibration damper lubricant. The main bearings of such engines have a groove in the upper shell but none in the lower shell. Part of the oil flowing into the main oil passage is branched off for lubricating the turbocharger, the injection pump drive, and the cylinder shut-off. Furthermore, oil is mixed with fresh air after the intercooler to protect the valve seats.

Fig. L12 shows the schematic of the lubricating oil cycle of a six-cylinder Mitsubishi engine. Depending on the temperature level, which is controlled by a thermostatic valve, the lubricating oil is pumped from the oil pan via a gear pump directly or via an oil cooler to the oil filters. There is a pressure regulating valve in front of the oil filters, which limits the pressure to ~50 kPa. The lubricating oil reaches the main oil passage from the four oil filters arranged in a row. From there, the main bearings and big end bearings of the crankshaft are supplied with lubricating oil. The big end bearing has a lubricating groove in the upper shell. This is supplied with lubricant by the crankpin, and two cross-holes in the upper shell lead to a hole in the short connecting rod blade, whose axis is slightly bent toward the screw axis. This hole extends into the longitudinal hole in the connecting-rod shank to supply the small end bearing and via the piston pin to cool the built-up piston.

From the main oil passage, lubricant is branched off to lubricate the camshaft, on which the valve cams and the injection cams are arranged. The camshaft itself is hollow and receives lubricant from the seventh camshaft bearing (in a six-cylinder engine) on the flywheel-end of the engine to lubricate the other camshaft bearings.

The lubricant hole to the camshaft extends into the cylinder head in order to guarantee the supply of oil to the rocker-arm shaft. The hollow rocker-arm shafts on each cylinder are connected with lines. From the rocker-arm shaft the valve bridges and/or pushrod contact points are lubricated via the rocker-arm.

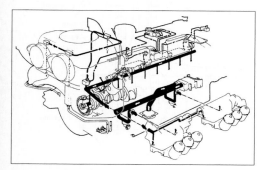

Fig. L11 Lubricating oil cycle of the 16V-396T-4-MTU engine

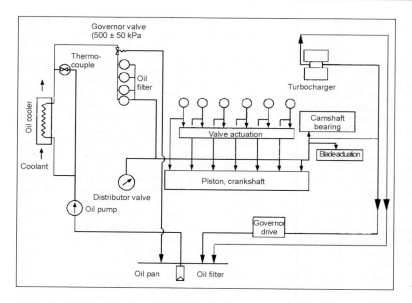

Fig. L12
Lubricating oil
cycle schematic of a
six-cylinder Mitsubishi
engine

The oil capacity in the oil pan of this engine is 340 liters. The oil pump is a gear pump with a speed ratio to the crankshaft of 1:1.32. It pumps 550 L of lubricating oil through the engine at a speed of 1200 rpm— that is, the lubricating oil is circulated 1.6 times/min at an oil pressure between 500 and 700 kPa.

The larger the engine, the more sophisticated is the lubricating system. An example of a lubricating system for a medium-speed engine operated on heavy fuel is shown in **Fig. L13**.

Large engines, especially marine engines, are often two-stroke engines. **Fig. L14** shows a cross section of a two-stroke Mitsubishi marine engine. This engine has two separate oil cycles. The cleaning system for the lubricant via a secondary cycle is shown in **Fig. L12**. The primary oil cycle of the engine lubricates the crank components and the crosshead and cools the piston. Furthermore, the injection pump and the exhaust valve control as well as the turbocharger are supplied

with lubricating oil from this oil cycle (**Fig. L15**). A second oil cycle lubricates the cylinders. It also has an oil pan to collect the lubricant flowing from the engine, which flows off into the lubricating oil tank.

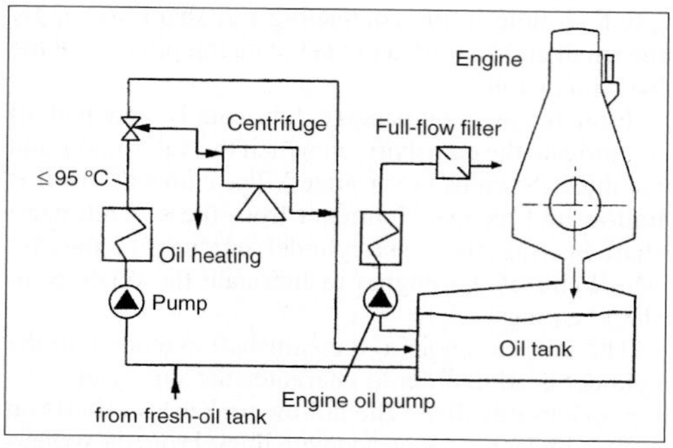

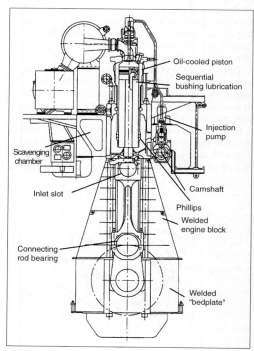

Fig. L14 Cross section of a two-stroke Mitsubishi marine engine

Fig. L13 Principle of a lubricating oil cycle of a medium-speed engine

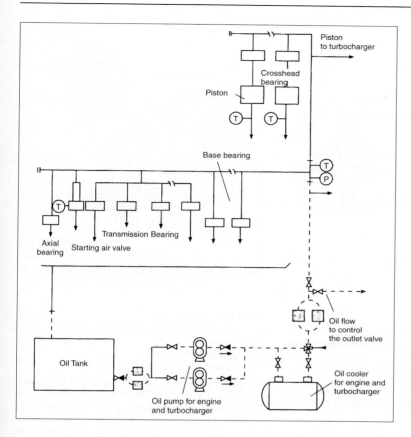

Fig. L15
Schematic of the
lubricating oil supply of a
two-stroke Mitsubishi
marine engine

The lubricant cleaning system via a secondary cycle is shown in **Fig. L13**.

Engine lubrication is realized by means of two oil pumps, and lubricating oil is taken from the lubricating oil tank via a strainer. Pressure regulator valves are arranged in the lines after the pumps. The lubricant flows through a thermostatic valve and, depending on the level of the temperature, through an oil cooler or directly to the engine. From the lubricating oil line to the engine, lubricating oil is branched off for the actuation system of the exhaust valves to lubricate the turbocharger as well as the injection pump elements. The main oil passage supplies the main bearing and the large end bearing as well as the camshaft with lubricant. The thrust bearing in this engine is lubricated separately. Lubricant flows from the supply line to the camshaft to the pneumatic starting valves. The pressure in the lubricating oil line for the camshafts is 350–450 kPa.

The crosshead area is supplied with lubricating oil by a separate line. From the crosshead, lubricating oil is guided to the piston via a pushrod, and this cools the piston through the shaker effect.

SAE 30 oil is used for engine lubrication; the pressure at the bearings and the crosshead is 150–250 kPa.

The second oil cycle lubricates the cylinders, as shown in **Fig. L16**.

The cylinder lubrication takes place in the upper third of the stroke; lubricating oil is guided to the cylinder via the bay. The lubricant is fed via a separate, electrically driven lubricating oil pump.

From a high-level tank, oil flows to the pumps via an intermediate tank.

Each cylinder of this two-stroke engine has eight oil lubrication points arranged on one level so that the lubrication of the piston and the piston rings is guaranteed. The oil supply for lubrication takes place when the piston, with the piston ring pack, strikes the oil supply holes on the cylinder.

~~**Motorcycle engines.** As an example of a lubricating system for a motorcycle engine, a Yamaha engine is described below. High demands are placed on the lubricating system of this engine because of the high speed of 8000 rpm attained by it and its high specific performance of 150 kW/L. **Fig. L17** shows a schematic of the lubricating system. **Figs. L18 and L19** show individual steps for better assignment and design of lubricating holes, filter arrangements, and the oil pump position.

1. Rocker arm
2. Pushrod

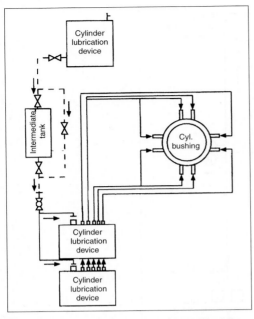

Fig. L16 Cylinder lubrication of a two-stroke Mitsubishi marine engine

3. Protective tube
4. Hydraulic start filling and/or engine protection automatic
5. Crankshaft
6. Crankshaft bearing
7. Hydraulic belt tensioning device
8. Lubricating oil pump
9. Oil pan
10. Plug
11. Threaded pin
12. Oil suction strainer

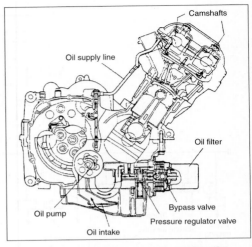

Fig. L18 Assignment of the lubricant holes, filter arrangement, and oil pump position

13. Mass balancer shaft
14. Lubricating oil filter with bypass valve
15. Oil cooler
16. Oil pressure switch
17. Oil pressure relief valve

The oil pump sucks oil from the oil pan and pushes it via the oil cooler and the oil filter into the main oil distribution line in the engine. Right in front of the oil cooler is a branching point to the transmission and/or the drive axle. Depending on the oil temperature, the oil cooler may be bypassed through a bypass valve. The oil filter also has a bypass valve, so that sufficient oil can get to the engine even if there is too much filter resistance. The arrangement of the oil pump, the oil cooler, and the filter can be seen in **Fig. L18**. There is a pressure valve in the line in front of the oil cooler. The oil pump is a trochoidal pump, arranged between

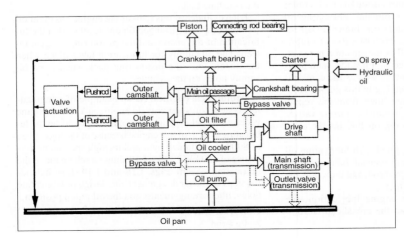

Fig. L17
Schematic of the lubricating system for a Yamaha motorcycle engine

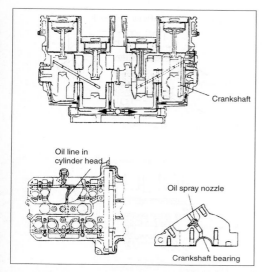

Fig. L19 Design of the lubricant holes, filter arrangement, and oil pump position

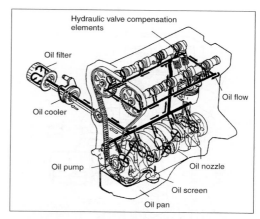

Fig. L20 Oil cycle of the BP DOHC, four-cylinder, 1.9-L Mazda engine

the transmission and the crankshaft axle. The five bearings are lubricated from the main oil passage. The four connecting-rod bearings are supplied with lubricant through holes from the second and fourth bearings. The output takes place in the center of the engine block. As can be seen in **Fig. L19**, the pistons are sprayed and cooled via a spray nozzle on the main bearings, which have a groove. In addition to lubricating the bearings of the crankshaft, the oil axially locates the crankshaft between its two ends and the closing covers—the oil cushions there represent the thrust bearings. Moreover, oil is guided toward the starter and the clutch from the crankshaft bearings. From the main oil passage, an oil hole is branched off between the crankcase and the clutch bell in the transmission in an upward direction and transports oil via an oil line to the cylinder head on the intake cam side. On both the right and left sides in the cylinder head, there is a longitudinal hole to supply the camshaft bearing and the hydraulic valve-clearance balancing elements with oil. The two longitudinal holes are connected to an oil line between the cylinders (**Figs. L18 and L19**).

~~Passenger car engines. Fig. L3 shows the lubricating system of a four-cylinder, 1.3-L VW diesel engine. The gear pump for the oil, which is driven by a chain, is a unit underneath the bracket bow in the oil pan. The pressure relief valve is arranged to limit the oil pressure in the pump housing.

The lubricating oil is passed via the pressurized oil passage to the main oil filter, which is an easy-change filter with a differential-pressure valve incorporated. From the filter, the oil reaches the main oil passage, from which the main bearing and, via the crankshaft, the big end bearings are lubricated. Oil is directed

from the main oil passage through a vertical hole in the block to the cylinder head gasket and from there via the hollow space between the cylinder head screw and the cylinder head screw whistle into the oil passage in the cylinder head. From this passage both the camshaft and the bucket tappet are supplied with pressurized oil. The oil pressure switch is in the oil distribution hole in the cylinder head. Oil dosage to the cylinder head is performed by a throttle in the cylinder head gasket. From this lubricating oil hole, the vacuum pump is supplied with oil, and the oil return is on the flywheel side.

Fig. L20 shows the oil cycle of the BP DOHC, four-cylinder, 1.9-L Mazda engine. On this engine, the oil pump is a trochoidal pump at the front on the crankshaft. This pump absorbs the oil from the oil pan and pumps it first through a hole in the side to the oil filter and the oil cooler. The oil filter is a replaceable unit and consists of a main filter and/or a full-flow and partial-flow filter. The pressure-differential valve opens when the pressure decrease via the cooler and filter is 80–120 kPa. The oil cooler is cooled with cooling water.

From the filter the oil is pumped into the main oil passage, then to the main bearings and via the crankshaft to the connecting-rod bearings, and/or from the partial-flow filter into the oil pan. The piston of this engine is spray-cooled with oil. There is an oil hole between the third and fourth cylinders to the oil distribution line in the cylinder head to enable the supply of oil to the two camshafts. In the cylinder head, the lubricant is fed to a longitudinal hole, and the front and rear of the second longitudinal hole are connected to the first one. From these, the camshaft bearings and the bucket tappets are lubricated. The oil return takes place via four return passages in the cylinder head and/or in the crankcase. Moreover, from the main oil passage, lubricant is branched off to the vacuum pump.

Figs. L21–L23 show a cross section of the DOHC, VTEC, four-cylinder, 1.6-L Honda engine as a schematic and as a cross section. This engine with four

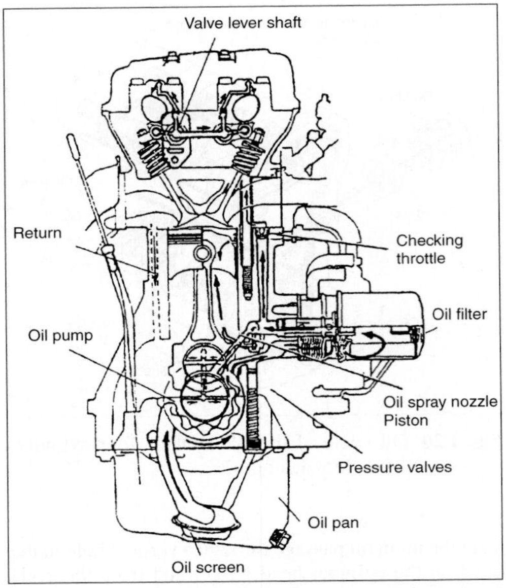

L

![Fig. L21]

Valve lever shaft

Return

Oil pump

Checking throttle

Oil filter

Oil spray nozzle
Piston

Pressure valves

Oil pan

Oil screen

Fig. L21 Oil supply of a DOHC, VTEC, four-cylinder, 1.6-L Honda engine

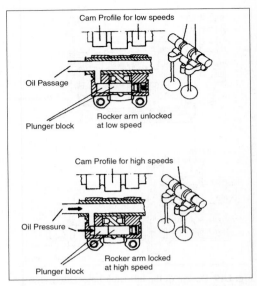

Cam Profile for low speeds

Oil Passage

Plunger block

Rocker arm unlocked at low speed

Cam Profile for high speeds

Oil Pressure

Plunger block

Rocker arm locked at high speed

Fig. L22 Lubricating system of the variable valve control of the DOHC, VTEC, four-cylinder, 1.6-L Honda engine

valves per cylinder is interesting because it has a variable valve control which is controlled by the oil pressure. The oil pump is located at the front end of the crankshaft as a trochoidal pump that pumps oil from the oil pan (**Fig. L21**). The oil is transported via a main filter (replaceable unit) and flows from there via an oil cooler to the main oil passage. From the main oil passage, the main bearings and/or, via holes in the crankshaft, the connecting-rod bearings are supplied with oil. From the main oil passage, oil is branched off to spray nozzles for piston cooling. A passage leads from the main oil passage to the cylinder heads.

The oil passage to the cylinder head in the center of the block has a throttle in the area of the upper deck. In the cylinder head, the oil is transported through the hollow space between the cylinder screw whistle and the cylinder head screw. From there, oil flows in the longitudinal direction along the engine to two oil passages located near each camshaft. From there, camshaft bearings and cams as well as cam followers are primarily lubricated. As can be seen in **Fig. L23**, an oil hole branches off the main oil passage and leads via a control valve that regulates the pressure level into a second longitudinal hole in each camshaft to control the valve actuation device.

Fig. L24 shows a schematic of an oil cycle for a V-6 Audi engine. The oil pump, which is a trochoidal pump, is at the front end of the crankshaft. The pressure regulating valve is arranged on the left side next to the oil pump. Oil is transported to the oil filter via a short line arranged on the left side of the oil pan. The passage for distributing pressurized oil and the flange for the oil filter and the cooler are integrated into the oil pan casting.

From there, oil is transported via a hole in the crankcase to a V-shaped oil passage in the center. From there, the crankshaft is supplied with oil, and from the main oil passage a hole between cylinders 1 and 2 leads to the cylinder head shown on the right side of the figure and between cylinders 3 and 4 to the cylinder head shown on the left side of the figure.

There are nonreturn valves in the rising line to prevent idling of the oil passages and/or the bucket tappets. There is also a throttle in the cylinder head gasket for the calibration of the oil flow so that the cylinder head lubrication is hydraulically decoupled from the crankcase lubrication cycle.

In the cylinder head, the bucket tappets and/or the camshaft bearings are supplied with oil. Moreover, the oil pressure relief valves are located in the cylinder heads, limiting the oil pressure to a maximum of 270 kPa. These are arranged at the end of the distribution lines. Through the calibration device in the cylinder head gasket, the full pressure in the oil system of the crankcase is preserved at full oil pressure in the cylinder head.

Fig. L25 shows the oil cycle of the VR6 2.8-L VW engine. The V angle of the engine is 15°. The engine has two valves per cylinder. In this engine, a gear pump is driven by a vertical shaft on the side, which is located under the crankcase of the oil pan. This component pumps oil to a longitudinal hole in the crankcase via a connecting line and from there to the replaceable filter. From there, oil flows via an oil cooler into the actual main oil passage, shown on the right of the cross section in the figure. From this main oil passage the main bearings and, via the crankshaft, the connecting-rod bearings are supplied. Moreover, spray

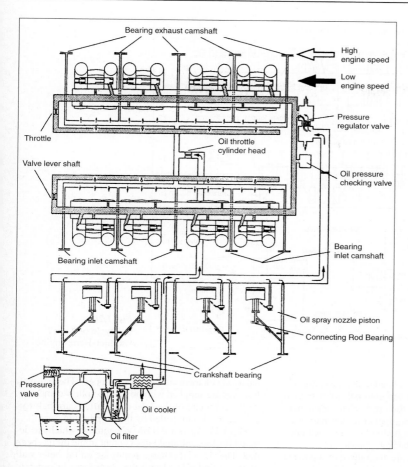

Fig. L23
Lubricating system of
the DOHC, VTEC,
four-cylinder, 1.6-L
Honda engine

nozzles are supplied with oil through a groove in the
area of the bearing pedestal to cool the pistons.

These spray nozzles have a spring-loaded ball valve,
which opens at a set pressure level (100–150 kPa) so
that the oil pressure in the oil system can build up quickly

when the engine is started. From the main oil passage
on the left of the figure, oil flows to a secondary main oil
passage underneath the oil pump drive. From this longitudinal hole, oil is supplied from the vertical shaft
and the lower chain tensioner. From the connecting
passage between the two longitudinal holes for oil distribution, a rising hole leads via a nonreturn valve to

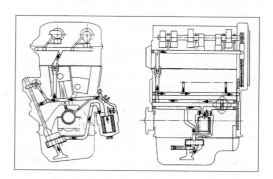

Fig. L24 Schematic of oil cycle of the V-6 Audi engine

Fig. L25 Oil cycle of the VR6 2.8-L VW engine

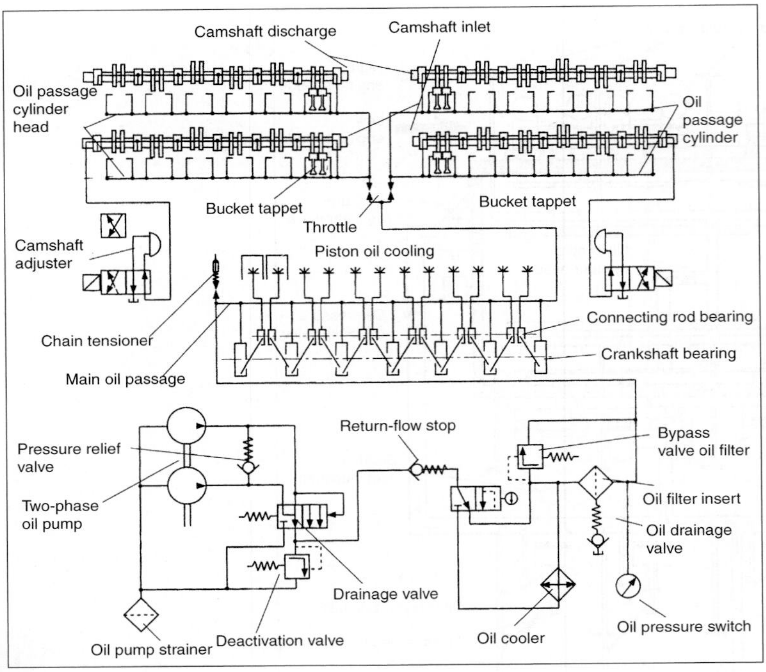

Fig. L26
Oil cycle of a V-12
Daimler-Benz engine

the cross-hole in the cylinder head under the camshaft. From this cross-hole, four longitudinal holes branch off; the outer ones supply the upper chain tensioners with oil, and the center ones supply the camshaft bearings with oil as well as the hydraulic valve clearance balancing elements in the bucket tappets.

Fig. L26 shows a schematic diagram of the oil cycle of a V-12 Daimler-Benz engine. A chain drives the register oil pump with a primary and a secondary pump. Both pumps deliver oil over the whole speed range. The secondary pump delivers oil via a pressure relief valve to the discharge bore of the primary pump when the pressure is low. If the oil pressure is sufficient, the oil discharge takes place directly via the deactivation valve from the secondary pump. The primary pump then transports the oil through the engine. This guarantees sufficient oil supply at low speeds, and at higher speeds the power loss from the oil pump is minimized.

The oil is transported directly via the deactivation valve, the nonreturn valve, and a temperature-regulating valve to the oil cooler or directly to the oil filter. From there oil is transported to the main oil passage, from where it is supplied to the bearing points of the crankshaft as well as the big end bearings with the spray nozzles to cool the pistons. On the flywheel side, there is a rising passage to supply the four camshafts in the cylinder heads. There are throttles in these rising passages to distribute the lubricant evenly and under constant pressure to the two cylinder heads. These oil passages also supply the hydraulic valve-clearance-

balancing elements in the bucket tappets. The camshaft bearings are supplied with oil via the hollow camshafts. At the front end of the inlet camshaft, a camshaft torsion device is controlled by the oil pressure.

Fig. L27 shows a V-8 BMW engine in which the demand for less entrained gas content in the oil has been met. The six oil discharge points are on the outer wall of the cylinder head and the crankcase. In order to avoid foaming of the oil, the oil discharge passages lead to a separation level between the crankcase and the oil pan. Then the returning oil flows via guide plates in the oil pan to the engine sump. The V space contains an arrangement of eight ventilation holes through which the blowby gases reach the highest points of the cylinder heads.

This reduces the pumping loss in the crankshaft drive. For the separation of the blowby gases and the engine oil, there is a cyclone separator in the chain space which can be seen in the right-hand cylinder row of the figure.

Condensate and fuel particles exist in the cyclone separator along with the blowby gases. They are directed almost oil-free to the pressure-regulating valve and from there into the intake manifold for combustion in the engine. The pressure regulating valve in the collection lid regulates the pressure in the crankcase throughout the whole operating range of the engine to values under 2 kPa vacuum.

This arrangement ensures that the ventilation process is not prone to ice formation, because the temperature in the cyclone is the same as the engine temperature.

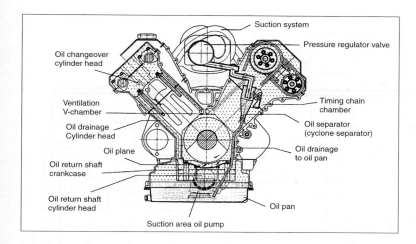

Suction system

Pressure regulator valve

Oil changeover
cylinder head

Ventilation
V-chamber

Timing chain
chamber

Oil drainage
Cylinder head

Oil separator
(cyclone separator)

Oil plane

Oil drainage
to oil pan

Oil return shaft
crankcase

Oil return shaft
cylinder head

Oil pan

Suction area oil pump

Fig. L27
Crankcase ventilation on
a V-8 BMW engine

In all oil cycle designs only full-flow filters with or without integral oil cooler are used. The pump types used most frequently are the latest versions of trochoidal pumps and/or gear pumps with a gear train to the crankshaft. This minimizes power losses because of the oil pumps.

~~Truck and smaller industrial engines. Fig. L28 shows a cross section of a six-cylinder, 12-L Mitsubishi truck engine, and **Fig. L29** shows a schematic of the oil supply. The gear pump attached underneath the crankshaft on the flywheel side pumps the oil into the crankcase. First, the oil flows through the full-flow filter, and then some of the oil flows through the partial-flow filter back into the oil pan. The oil through the full-flow filter is sent either through the oil cooler via a bypass valve or directly into the main oil passage.

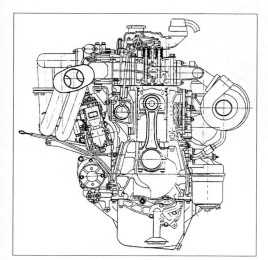

Fig. L28 Cross section of a six-cylinder, 12-L Mitsubishi truck engine.

From the oil cooler the oil for lubrication is branched off into the compressor. The oil cooler is a disc oil cooler in the region of the water jacket. From the main oil passage, oil is branched off via the pressure valves for oil-spray cooling of the pistons. Some holes guide oil to each main bearing. The small end bearing is supplied with oil through a hole in the connecting rod that comes from the large end bearing.

From the main bearings on the flywheel side, oil is branched off to lubricate the wheel drive on the flywheel side. From the main bearings, a cross-hole extends to a longitudinal hole in order to lubricate the camshaft, and from here oil is taken to lubricate the injection pump. Oil is transported through holes from the camshaft bearing to the cylinder heads to lubricate the rocker-arms and their wear points. This engine has rocker-arm shafts and their associated bearings for each cylinder; the hollow rocker-arm shafts are connected by lines. The pressure is kept constant at the end of the main oil passage by means of a pressure regulator valve. From this main oil line, an oil line is taken to the compressor.

Fig. L30 shows the oil supply schematic of a four-cylinder, air-cooled industrial engine. This engine has an air oil-cooler, which is supplied with the air from the A/C blower. First, oil is guided from the oil pump to the alternating-current filter and then into the longitudinally attached oil cooler. From there, oil flows into the main oil passage, which supplies the crankshaft and the spray nozzle bearing for the piston cooling. From the central main bearing, an oil hole extends to the camshaft. From this supply line, a longitudinal hole branches off to supply the camshaft bearings with oil. The rocker arms and the valve guides are lubricated via the pushrods. At the front end of the engine, oil flows from the first camshaft bearing to the camshaft gear drive shafts to lubricate the gearwheel drive.

~~Two-stroke engines. Two-stroke engines for motorcycles, passenger cars, and industry often have crankcase scavenging. In these, ball bearings and/or

Lubrication

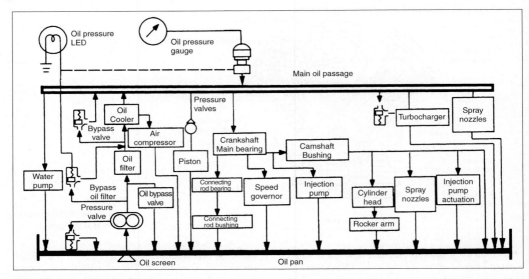

Fig. L29 Oil supply schema

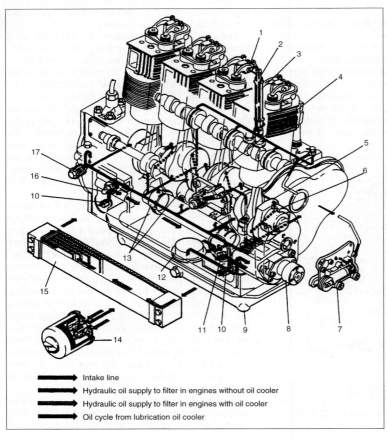

→ Intake line

→ Hydraulic oil supply to filter in engines without oil cooler

→ Hydraulic oil supply to filter in engines with oil cooler

→ Oil cycle from lubrication oil cooler

Fig. L30
Lubricating oil cycle of
the 4L40C-Hatz diesel
engine

needle roller bearings are used; and in order to lubricate these bearings, oil is either mixed with the fuel or, downstream of the carburetor, a fuel-air mixture is added. If oil is added through a controlled pump to the fuel or to the fuel-air mixture, it is possible to vary the fuel:oil ratio if necessary. The variables that are used to control the fuel-oil ratio are the throttle position and the engine speed. The fuel:oil ratio usually varies between 20:1 and 50:1, but in modern engines, it can be as high as 100:1.

The intake air is not the best medium to transport the lubricating oil because there is not much air where the lubricant is needed most. Based on experience, the lubrication of a crankcase-scavenged two-stroke engine with a fuel/oil mixture or a separate oil supply downstream of the carburetor is sufficient. Looking at emissions regulations shows that minimizing the oil consumption still further seems to be necessary to fulfill future requirements. In order to minimize the oil consumption, it is necessary to supply the bearing points with oil directly—that is, to supply the oil above the slits in the area between the piston rings in the upper dead-center position of the piston as well as to lubricate the bearing points directly. In order to guarantee lubrication throughout the stroke, an oil pump can be used. This metering pump delivers a certain amount of oil during a revolution of the crankshaft to lubricate the piston in the cylinders and/or to lubricate the bearings. In order to deliver enough oil to the connecting-rod bearings in addition to the main bearings and to drain off the oil being discharged from the bearings, certain structural measures are necessary.

~Lubrication points. Lubrication points are all points of the engine which must be reached by engine oil so that the components there can fulfill their function. Important lubricating points are the crankshaft bearings and the camshaft bearings, the big and small end bearings, guideways, pistons/cylinders, bucket tappets, valve drives, turbochargers, accessories, and so on.

~Main oil passage. See above, ~Lubricating systems

~Oil capacity. See below, ~Oil demand

~Oil circulation. See above, ~Lubricating systems

~Oil cooler See above, ~Lubricating systems ~~Large engines; also, →Coolant

~Oil demand. Using calculation models, answers to the following questions regarding lubricating systems can be derived:

- Does the lubricating system have too much hydraulic resistance at certain points?
- How sensitive is the lubricating system to manufacturing tolerances at the oil exit points and/or bearings?
- Is the dimensioning of the oil pump sufficient for low engine speeds?

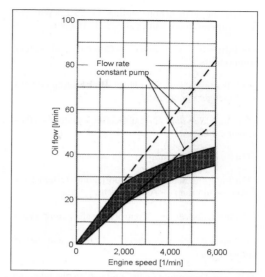

Fig. L31 Oil demand as a function of the engine speed for two 2-L engines

Fig. L31 shows the oil demand of two typical two-liter passenger car engines at an oil temperature of 95°C.

First, the oil flow increases at low speeds up to a speed of about 1500–2500 rpm because the engine needs a certain amount of oil to avoid wear and tear at low speed. Then, both the lubricant pressure and the oil capacity are limited by the pressure regulator valve, or the oil discharge flow is limited through self-regulation by the lubricant pump. If the oil flow is divided into the individual areas that consume oil, the following capacities result according to AVL: main bearing, 15%; big end bearing, 20–25%; cylinder head (total valve drive and hydraulic valve clearance compensation), 35–50%; others, 6–10%. These values are derived from typical passenger car engines without piston cooling but with hydraulic valve clearance compensation. For the calculation of the oil discharge flow, different programs are used that consider the individual lubricating points, bearings, valve drives, and so on.

The calculation of the oil capacity when designing the feed pump is based on an engine speed of 1200–1500 rpm for passenger car engines and between 1000 and 1200 rpm for commercial vehicles; the pressure base is 3 bar (43.5 lb$_f$/in^2). It is necessary that the oil capacity for piston cooling, the gearwheels, and valve lubrication are included in the calculation. In addition to shaft bearings, piston cooling is important for the oil pump design. If the oil is used to cool the piston, the oil capacity for sufficient oil cooling is 4–6 L/kWh. For the design of a piston with cooling ducts in which the oil flows through the duct to cool the piston, in case of built-up pistons or articulated-skirt pistons, an oil capacity of 4–5 L/kWh must be planned.

Pragmatic values for the oil pump design for truck or stationary engines without piston cooling are 40–48

L

L/kWh and with piston cooling 44–55 L/kWh. For truck engines, these values are 28–35 L/kWh.

~Oil dipstick. *See above*, ~Control and safety components

~Oil drain passage. *See above*, ~Lubricating systems ~~Passenger car engines

~Oil duct. *See above*, ~Lubricating systems ~~Passenger car engines

~Oil flow rate. *See above*, ~Oil demand

~Oil foaming. *See above*, ~Lubricating systems; *see below*, ~Oil pump ~~Arrangement;

~Oil level. *See above*, ~Control and safety components

~Oil level gauge. *See above*, ~Control and safety components

~Oil line. Depending on the engine type, the lubricant present in the oil pan is circulated by the oil pump four to six times per minute in small engines and twice to four times in larger engines at rated speed. The pressure lines in the engine should be as small as possible. Larger hollow spaces, from which oil can flow during standstill, should be avoided. Cast lines should only be used up to the filter, to avoid sand residues in the bearings that do not have filtration. If, however, lost-foam technology allows a 100% clean lubricant line, oil lines in the engine block or in the cylinder head may be cast after the filter. The diameter of the holes in the main oil gallery to the bearings should be between 6 and 8 mm, which leads to oil speeds of 2.5 m/s at rated engine speed. In general, the oil cycle must be cleaned of any dirt, such as processing chips or sand residue.

In addition to the individual bearing points, as mentioned above, oil is also used for piston cooling. To make sure that enough pressure can build up for bearing lubrication, spring-loaded valves are used, opening at a line pressure of 100–150 kPa.

~Oil pan →Crankcase

~Oil pressure gauge. *See above*, ~Control and safety components

~Oil pressure switch *See above*, ~Control and safety components

~Oil pump. For the supply to the lubricating points, gear-wheel pumps or trochoidal pumps are used to pump the oil; these work according to the displacement principle.

~~Arrangement. Because the best position in the oil sump for the gear pump often cannot be realized for construction reasons, the following requirements must be considered for the location of the lubricating oil pump:

The oil sump gear pump should be as deep and as close as possible to the oil level in the oil pan. The marginal requirement is that the system is designed in such a way that the pump cannot run empty and long intake lines are avoided. If this is not possible, a valve should be installed at the end of the suction manifold.

It should be noted that the oil flowing out of the pressure deactivation valve must not spray uncontrollably, in order to avoid oil foaming and oil vapor formation in the crankcase.

The oil temperature in the oil pan must not exceed 150°C.

~~Dimensioning. For the determination of the oil pump capacity, the oil flow in the engine must be evaluated at a certain supply pressure as well as a specified engine speed. In order to determine the oil flow, it is necessary to analyze the whole lubricating system, taking into account the holes, clearances, flow resistance, and oil requirement in the bearings of the engine transmission unit and the cylinder head. For that purpose, not only the cold clearance but also the operational clearance must be determined, because bearing clearances influence the necessary oil capacity. It is possible that if the movement of balance elements to hydraulically compensate for valve clearance is too large, the oil requirement in the engine may be 20% higher than for normal clearances.

~~Gear pump. A usual gear pump consists of two outer toothed wheels (**Fig. L32**, cross section A-A), of

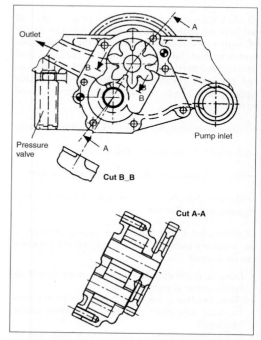

Fig. L32 Gear pump

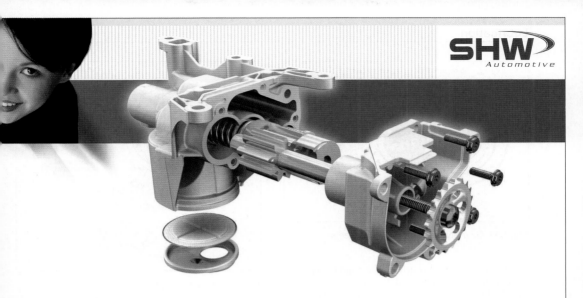

We've got the
Future of the Oil Pump Under Control.

SHW is synonymous with oil-pump innovation. As the developer and producer of the world's first map-controlled oil pump for combustion engines, we are the partner for the automotive and utility vehicle industry wherever a power-saving and reliable oil supply is required — because our production experience is based on the manufacture of more than 6 million pumps each year.

With know-how based on 25 years of pump development, plus control concepts for internal gear pumps, external gear pumps and vane pumps, we'll have your application under control, too.

SHW Automotive GmbH. Wilhelmshütte Plant · Enzisholzweg 11 · D–88427 Bad Schussenried
Phone: +49 7583 946-0 · Fax: +49 7583 946-211 · wi@shw.de · www.shw.de

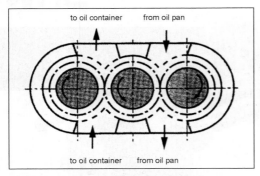

Fig. L33 Gear pump with three gearwheels

which one gear wheel is driven. The liquid volume in each tooth gap is advanced by the turning of the wheels along the housing wall to the pressure side. The gear engagement creates the sealing between the pressure and the suction side. The peripheral speed of the gear teeth is usually under 10 m/s. The ratio of the teeth widths and the reference diameter is smaller than 2 and the module mostly between 2 and 7.

The gear wheels are shrunk onto friction-bearing shafts in the housing. The clearance of these bearings is about 1 in relation to the shaft diameter.

The inflow speed of the oil to the pump should be between 1.5 and 2.5 m/s. The speed in the pressure line should not exceed the maximum speed of 4.5 m/s.

To relieve the pump bearings in passenger cars, trucks, and large engines, an alternative must be provided for the oil pushed out in the gear engagement between the teeth. This so-called relief is carried out by grooves or cast recesses in the housing.

In addition to the usual gear pump design with two gear wheels, there are models with three gear wheels as can be seen in **Fig. L33**. Such pumps are used for dry sump engines, such as underfloor engines. On one side, oil is advanced from the pan to the oil container, and on the other side, from the oil container via the lubricating points to the oil pan.

~~Pump loss of performance. At the rated speed of an engine, oil pumps contribute to up to 10% of the mechanical "losses" in the engine and, thus, increase fuel consumption. Therefore, it is necessary to keep the feed performance of oil pumps as low as possible. This can be achieved through a reduction of the oil pump speed as well as the reduction of the outer diameter. The use of pumps with regulation of the feed flow are currently being developed.

~~Spiral gear pumps. Another example for the design of an oil pump is shown in **Fig. L34**: a pump for a medium-speed V-12 engine with a bore diameter of 190 mm. Such pumps usually have spiral-toothed wheels. Because there is high axial pressure with spiral teeth, the teeth are flattened, as can be seen in cross section C-C and view X. This flattening absorbs the

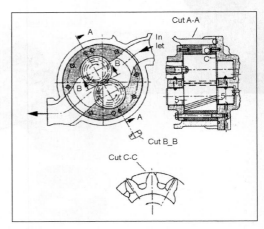

Fig. L34 Spiral gear pump

axial pressure between the gearwheels and the pump housing via a lubricating film and avoids possible seizure or too much wear on the housing. Here, discharge grooves are present as well. The engine oil is branched off to the pressure gallery for lubricating the bearings and is then transported centrally to the bearings. The oil discharging from the bearings toward the pump gear wheels is collected over a large radius and returned to the fuel gallery. On the side of the pressure-regulating valve, the oil discharged from the bearings is advanced via these discharge openings.

~~Trochoidal pumps. In passenger car engines, trochoidal pumps are often used as oil pumps, and they are usually attached to the crankshaft. The pump consists of an inner and an outer toothed wheel and can have a sickle-shaped part, in which case the pump is called a trochocentric pump. This sickle-shaped part assures very small feed and pressure fluctuations.

Fig. L35 shows a trochoidal pump.

~Oil pump drive. *See above*, ~Oil pump

~Oil return flow. *See above*, ~Lubricating systems
~~Passenger car engines

~Oil separation. *See above*, ~Lubricating systems

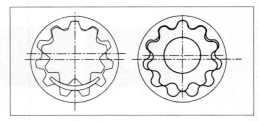

Fig. L35 Trochoidal pump

~Oil speed. *See above*, ~Oil line

~Oil sump. *See above*, ~Oil pump ~~Arrangement

~Oil temperature stress. The oil temperature must not exceed 150°C. Oil life is reduced by 50% per 1°C increase in temperature.

~Oil vapor. *See above*, ~Lubricating systems

~Oil vapor formation. *See above*, ~Lubricating systems

~Oil volume sensor. *See above*, ~Control and safety components

~Overflow valves. *See above*, ~Control and safety components

~Partial-flow filter. *See above*, ~Filter

~Pressure control valve. *See above*, ~Control and safety components

~Pressure line. *See above*, ~Lubricating systems

~Pressure relief valve. *See above*, ~Control and safety components

~Replacement filter. *See above*, ~Filter

~Strainer →Lubrication

~Thermostatic valve. *See above*, ~Control and safety components

~Valve lubrication. *See above*, ~Oil demand

Lubrication, connecting rod →Connecting rod; →Lubrication

Lubrication, crankshaft →Crankshaft; →Lubrication

Lubrication pocket →Bearings ~Design solutions ~~Lubricating elements

Lubricity →Fuel, diesel engine ~Properties

Lubricity additive →Fuel, diesel engine ~Additives

Lysholm compressor →Supercharging ~Miller method

L

M

Magnesium →Crankcase ~Weight, crankcase ~~Reduction of specific weight of material

Magneto ignition →Ignition system, gasoline engine ~High-voltage generation ~~Coil ignition

Main bearing →Bearings ~Bearing positions ~~Crankshaft main bearings

Main bearing block →Crankcase ~Crankcase design ~~Main bearing pedestal area

Main bearing block area →Crankcase ~Crankcase design

Main bearing cover →Crankcase ~Crankcase design ~~Main bearing pedestal area

Main bearing cover bolt →Engine bolts ~Threaded connections

Main cams →Camshaft

Main chamber →Combustion chamber ~Main combustion chamber

Main combustion chamber →Combustion chamber

Main controller →Electronic/mechanical engine and transmission control ~Electronic components ~~Voltage regulator

Main dimensions crankcase →Crankcase ~Crankcase design

Main fresh air supply →Two-stroke engine

Main jet →Carburetor

Main oil passage →Lubrication ~Lubricating systems

Main system →Carburetor ~Systems

Maintenance indicator →Filter ~Intake air filter ~~Air filter elements

Malleable iron connecting rod →Connecting rod ~Materials ~~Casting material

MAN FM process →Combustion process, diesel engine ~Direct injection; →Combustion process, gasoline engine ~Stratified charge process

MAN M process →Combustion process, diesel engine ~Direct injection ~~Single spray process, wall-applied

Manifold →Exhaust system ~Exhaust gas manifold

Manual starting device →Carburetor ~Starting systems ~~Design types

Manufacture, camshaft →Camshaft

Manufacturing process piston ring →Piston ring

MAP (manifold absolute pressure sensor) →Sensors ~Pressure sensors

Map cooling →Performance characteristic maps

Marine engines →Oil ~Classification, specifications, and quality requirements ~~Diesel engine oils; →Piston ring ~Ring set

Mass airflow sensor →Flowmeter intake air; →Sensors

Mass balancing mechanism (*also*, →Balance shaft; →Balancing of masses). A basic disadvantage of reciprocating piston engines is the generation of mechanical vibrations due to the mechanical/kinematic aspects of the crank drive. Depending on the number of cylinders and layout, these vibrations can lead to forces and couples which result in noise and vibration in the vehicle. While the forces resulting from rotating masses can be compensated for by counterweights, it is necessary to integrate more or less extensive compensation elements or balance gears into the engine to compensate for the reciprocating forces. These usually consist of balancer shafts that are designed in the appropriate arrangement and rotate at the appropriate speed. The shaft(s) can be located in the oil pan or in the crankcase. Toothed belts, chains, and gears are used as the drive elements. The advantages of improved acoustics, reduction of assembly vibrations, and, therefore, reduced vehicle vibration with balancing mechanisms, are counteracted by the disadvantages of higher power losses, greater engine masses, larger numbers of parts, and, hence, higher assembly costs.

Literature: H. Neukirchner, O. Arnold, A. Dittmar, A. Kiesel: Die Entwicklung von Massenausgleichseinrichtungen für Pkw-Motoren, MTZ 5/2003.

Mass determination, emissions →Emission measurements ~FTP-75

Mass distribution, connecting rod →Connecting rod ~Mass

Mass flow rate. The mass flow rate is the quantity of air and fuel flowing through the engine. The magnitude of the mass flow rate depends, to a first approximation, on the cylinder displacement, engine speed, type of engine (two- or four-stroke), charging mechanism, intake air temperature, air:fuel ratio, and in gasoline engines, the position of the engine throttle valve. Moreover, atmospheric conditions such as ambient pressure, temperature, and humidity in the air have an effect on the mass flow rate.

Mass reduction →Balancing of masses

Mass substitution system →Balancing of masses

Material behavior →Calculation processes ~Application areas ~~Stability

Material compatibility →Electronic/mechanical engine and transmission control ~Requirements for mechanical and housing concepts

Material models. The calculation of the operational stability by using service life software requires a detailed description of the material characteristics.

Material-specific weight reduction →Crankcase ~Weight, crankcase

Maximum full-throttle speed. Maximum full-throttle speed is the speed at which the engine is operated with the throttle valve fully open (quantitative control) or with maximum available enrichment (qualitative control)—that is, at maximum specific work of the speed present in each case.

Maximum pressure restriction →Cycle ~Constant pressure cycle, ~Seiliger (dual combustion) cycle

Maximum speed →Speed

Maximum speed governor →Electronic open- and closed-loop control ~Electronic open- and closed-loop control, gasoline engine ~~Functions

MDOF (multiple degree of freedom) →Engine acoustics ~Multiple degree of freedom system (MDOF)

Mean effective pressure (*also,* →Combustion process, gasoline engine; →Peak pressure). Mean effective pressure (or, effective specific work) is a dimensional value allowing the "load" on engines with different displacements to be compared. It represents the work per operating cycle in relation to the displacement and, seen physically, has the units of pressure.

The higher the mean effective pressure of comparable engines, the better the efficiency and, therefore, the lower the fuel consumption. The objective of engine

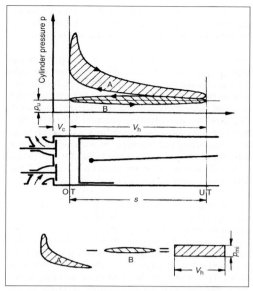

Fig. M1 Determination of mean effective pressure

development is, therefore, to achieve the highest possible mean effective pressure. A high mean effective pressure, even at low engine speeds, results in a "flexible" engine, which also allows adaptation of the transmission ratio to further improve the fuel efficiency.

Fig. M1 shows how mean effective pressure is measured. The work represented by the *p*-*v* diagram is determined by integration. The area "B" (work required for the charge cycle) is subtracted from area "A" (the work output). The difference is converted into a rectangle based on the same base length as the displacement and covering the same area as the original diagram. The height of this rectangle then corresponds to the mean effective pressure.

Three definitions of mean effective pressure are common in engine engineering: brake, friction, and indicated mean effective pressure.

~Brake mean effective pressure. The brake mean effective pressure is the difference between the indicated mean effective pressure and frictional mean effective pressure. It is an important value for evaluating an engine, because it can be determined simply from the displacement, V_H, the measured brake power, P_e, and the engine speed, n. The factor i takes the type of engine into consideration ($i = 1$ for two-stroke, $i = 2$ for four-stroke).

$$p_{me} = \frac{P_e}{i \cdot n \cdot V_H}$$

Maximum brake mean effective pressure valves for naturally aspirated gasoline engines are 12–13 bar, which means 1.2–1/3 kJ/dm³ for the brake specific work. On supercharged engines, the achievable mean effective pressure depends on the degree of super-charging. Mean effective pressures of greater than 20 bar are achieved with multiple stage supercharging.

~Friction mean effective pressure. The friction mean effective pressure is the difference between the indicated and brake mean effective pressures. It is dependent on the speed of the engine and reflects the losses resulting from friction of the individual components. The friction mean effective pressure is highly influenced by the piston and piston rings, crankshaft bearings, connecting rod bearings, oil and coolant pumps, valve train, and so on. Common values for the friction mean effective pressure are between 0.5 and 1.2 bar, depending on engine speed and design.

~Indicated mean effective pressure. The indicated mean effective pressure can be determined by measurements on the engine. For this purpose, it is necessary to measure the pressure in the combustion chamber and the associated specific crank angle. The work produced in the cycle can be determined using the geometrical parameters of the crank drive, and integration gives the indicated mean effective pressure as described above.

~Specific work. Mean effective pressure, or, effective specific work, is a result of the gas work. The incremental change of the gas work on the piston is calculated based on the cylinder pressure, p, the bore diameter, D, and the incremental change of the stroke, ds:

$$dw_K = p \cdot D^2 \cdot \frac{\pi}{4} \cdot ds.$$

The gas work, w_K, is calculated as follows:

$$w_K = \oint p \cdot dV_h,$$

where p = cylinder pressure and V_h = swept volume of a cylinder.

This describes the cylinder work delivered for each combustion cycle. The following equation describes the delivered work for an engine with z cylinders.

$$W_K = w_K \cdot z$$

This work is also called indicated (or internal) engine work.

The thermodynamic processes, which go clockwise on the p-V diagram, are rated as positive and produce positive delivered work, whereas counterclockwise processes are rated negative and produce negative work. This definition (an opposite definition is also possible) means that the high-pressure loop for the four-cycle engine encloses a positive area, while the low-pressure loop encloses a negative area (**Fig. M2**). Areas in the pressure-volume diagram represent the delivered work. The sum of both areas is a measure of the indicated work delivered by the engine, assuming the positive and negative convention is followed.

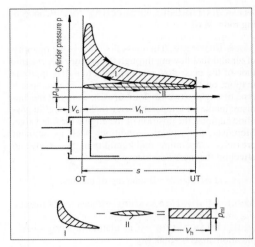

Fig. M2 Determination of the indicated specific work (four-stroke process)

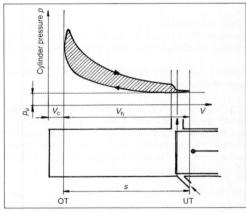

Fig. M3 Determination of the indicated specific work (two-stroke process)

The p-V diagram for the two-stroke engine is shown in **Fig. M3**. If the resulting area is converted to a rectangle, with the swept volume, V_h, as the baseline, the "height" represents the mean effective pressure, p_{mi}, or the specific work, w_i.

This variable is of prime importance in engine technology, because it enables the comparison of engines with different displacements. The engine with the higher specific work is the better engine for the analyzed operating point. This work is called indicated specific work, w_i, or indicated mean effective pressure, p_{mi}, because this work represents the energy exchanged between the working fluid and the piston.

The brake mean effective pressure or the brake specific work is among the most important variables for engines. Comparisons of engines with different displacements are possible similar to the comparisons that can be made using the indicated mean effective pressure. The engine with the higher brake specific work or the higher brake mean effective pressure has a higher torque for the same displacement.

Mean piston speed →Piston

Measuring estimate →Injection system, fuel ~Diesel engine ~~Serial injection pump ~~~Controller for diesel injection pumps

Mechanical efficiency →Engine efficiency

Mechanical fuel injection systems →Injection system (components) ~Gasoline engine ~~Direct injection systems

Mechanical losses →Engine power loss

Mechanical regulation/control, diesel engine →Injection system (components) ~Diesel engine

Mechanical supercharging →Supercharging

Mechanical valve clearance adjustment →Valve clearance

Mechanical valve control →Valve gear

Mechanical valve gear →Valve gear

Mechanical valve shutoff →Valve gear

Mechanical variable valve timing →Variable valve control ~Operation principles, ~Systems with camshaft, ~Systems without camshaft

Mechatronic transmission module →Electronic/mechanical engine and transmission control ~Requirements for mechanical and housing concepts

Medium speed. Medium-speed engines are generally diesel engines, which almost exclusively today operate using the four-stroke principle. The speed range is between 300 and 1200 rpm. The cylinder diameters range from 200 to 600 mm. The mean piston speed is around 10 m/s, and the brake mean effective pressure achieves values around 25 bar corresponding to a specific work of 2.5 kJ/dm³. This represents the upper limit for single-stage supercharging. Based on these values, performances in the range of 100 kW/cylinder to 1500 kW/cylinder are obtained. Medium-speed engines are used particularly as marine and stationary engines for driving generators.

Metal beaded gaskets →Sealing systems ~Special seals

Metal bed catalytic converter →Catalytic converter ~Converter

Metal content in oil →Oil ~Power during operation ~~Contamination by foreign matter

Metal deactivators →Oil ~Body ~~Additives

Metal fuel tank →Injection system (components) ~Gasoline engine ~~Fuel tank

Metal hydride accumulator (*also,* →Fuel, gasoline engine ~Alternative fuels ~~Hydrogen). Among all the alternative fuels, hydrogen assumes a special position. However, in addition to adaptation of the engine, the problem of storage of hydrogen in the vehicle poses particular problems. One possibility is to store it using metal hydrides. Some metals or metal compounds such as iron—titanium or magnesium—and nickel are capable of binding nitrogen or releasing it under certain conditions. This is accomplished by energy obtained from the engine exhaust.

Metal-elastomer cylinder head gaskets →Sealing systems ~Cylinder head gaskets

Metal-elastomer gaskets →Sealing systems ~Elastomer sealing systems

Metal/soft material cylinder head gaskets →Sealing systems ~Cylinder head gaskets

Metal/soft material gaskets →Sealing systems ~Special seals

Metered fuel injected →Electronic open- and closed-loop control ~Electronic open- and closed-loop control, diesel engine ~~Functions ~~~Metered fuel control; →Injection functions ~Gasoline engine ~~Direct injection

Methane →Fuel, gasoline engine

Methanol →Engines ~Racing engines ~~IndyCar; →Fuel, diesel engine ~Alternative fuels, →Fuel, gasoline engine ~Alcohols

Methanol engine oils →Oil ~Classification, specifications, and quality requirements ~~Special engine oils

Methanol fuel →Fuel, gasoline engine

Methyl alcohol/gasoline fuel blend →Fuel, gasoline engine ~Alcohols

Methyl tertiary butyl ether (MTBE) →Fuel, gasoline engine ~Alcohols, ~Alternative fuels

M

Micro-blind hole nozzle →Injection valves ~Diesel engine ~~Hole-type nozzle

Micro-controller →Electronic/mechanical engine and transmission control ~Electronic components

Micro-welding →Piston ring ~Wear

Mid-engine →Engine installation

MIL specification →Oil ~Classification, specifications, and quality requirements ~~Classifications ~~~MIL classification

Miller process →Supercharging; →Variable valve control

Mineral oil recycling →Oil ~Safety and environmental aspects ~~Used oil ~~~Recycling of mineral oil

Mineral oils →Oil ~Body ~~Basic oils

Misfire. Misfires are the stochastic variation of the air-fuel mixture ignition. They are usually caused for two reasons:

- Defects in the ignition system
- No ignitable mixture is present at the spark plug at the time of ignition

Misfires result in a sudden drop of power, higher fuel consumption, and higher pollution emissions, especially HC emissions, if they occur in one or more cylinders.

Damage to the exhaust gas treatment system can occur in case of an ignition failure, because a very high exothermic reaction can occur in the catalytic converter, which can result in the destruction of the system. This results in a dramatic increase in pollution emissions.

Misfire limit →Air-fuel mixture

Misfiring →Ignition system, gasoline engine ~Ignition

Mixed friction →Bearings ~Friction loss conditions

Mixing chamber →Carburetor

Mixing ratio →Air-fuel ratio; →Radiator ~Coolant

Mixture →Mixture formation ~Mixture formation, diesel engine; →Mixture formation ~Mixture formation, gasoline engine

Mixture calorific value. The calorific value of the mixture is actually the calorific value of the air-fuel mixture and depends on the calorific value (energy content) of the fuel and on the air-fuel ratio. The calo-

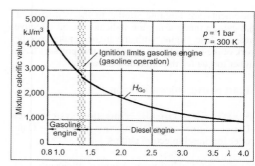

Fig. M4 Mixture calorific value as a function of the air-fuel ratio

rific value of the ignitable air-fuel mixture is the determining factor for engine performance and not the calorific value of the fuel. A fuel with a low calorific value requires large fuel quantities to achieve an appropriate-mixture calorific value. **Fig. M4** shows the change of the mixture calorific value in relationship to the air-fuel ratio of the mixture.

Mixture composition →Electronic open- and closed-loop control ~Electronic open- and closed-loop control, gasoline engine ~~Functions ~~~Lambda control

Mixture control →Electronic open- and closed-loop control ~Electronic open- and closed-loop control, gasoline engine ~~Functions ~~~Lambda control

Mixture cooling →Mixture formation ~Mixture formation, gasoline engine

Mixture density →Mixture formation ~Mixture formation, gasoline engine

Mixture dilution (*also*, →Exhaust gas recirculation). The dilution of the mixture is achieved by a reduction of the proportion of fuel in the mixture. It can be achieved in two ways:

- By reducing the fuel content while keeping the air mass constant
- By adding exhaust gas to the fresh mixture

The latter is also called exhaust gas recirculation (EGR). EGR rates of significantly more than 50% are used, for example, to reduce the NO_x emissions from diesel engines with direct injection.

Mixture dilution is used for load control in diesel engines but also for a large region of the performance curve in gasoline engines with direct injection.

Mixture distribution →Mixture formation ~Mixture formation, gasoline engine ~~Mixture preparation

Mixture enrichment →Carburetor ~Starting systems ~~Requirements, ~Systems; →Mixture formation ~Mixture formation, gasoline engine ~~Mixture metering

Mixture formation. The objective of the mixture formation and preparation is the generation of an ignitable mixture based on air and fuel available for combustion in an engine. The quantity of oxygen required for the combustion of the fuel in a combustion engine can be supplied to the combustion chamber in two ways:

- In a bonded form as a component of the fuel or an additional oxygen-releasing compound
- In an unbonded molecular form (O_2) as a component of the intake air

The advantage of the second alternative is that oxygen does not have to be kept in the vehicle, and this reduces significantly the weight of the fuel. As a result of this, combustion engines usually use the oxygen in the air to satisfy their oxygen demands.

Additives to the fuel of oxygen-carrying components such as methanol are used not necessarily to increase the oxygen content of the combustible mixture but because of their positive impact on the antiknock properties of the gasoline fuel and the reduced emissions of pollutants that they bring about. They also help to reduce the amount of soot emitted by diesel engines.

The ideal fuel should have a mass that is as small as possible, a small volume, and a high calorific value, and it should have the capability to be easily and safely stored. It must also be reactive with the free molecular oxygen in the ambient air, so that the transport of an additional oxidation material is not necessary.

~Mixture formation, diesel engine. The mixture formation in a diesel engine results in the formation of an internal heterogeneous mixture with auto-ignition properties. The trapped air in the cylinder is compressed to approximately 50 bar and heated to approximately 800°C during the compression cycle. This results in auto-ignition of the fuel, which is injected shortly before the end of compression near the top dead center. The subsequent combustion and the utilization of the induced combustion air depend, in heterogeneous processes, primarily on the processes forming the mixture. The injection of fuel is performed by following an appropriate injection curve (volume of fuel injected per time unit or per degree crank angle) that depends on the combustion process. This results in the formation of an ignitable mixture, consisting of the induced air and the injected fuel, in the cylinder. At the time and point of injection, the fuel is injected into the highly compressed hot air.

The highest injection pressures are achieved with direct injection systems; these can be up to more than 2000 bar depending on the injection system, and, therefore, good atomization of the fuel occurs (*also*, →Combustion process, diesel engine).

Fig. M5 shows ideal injection curves, which are different depending on whether the engine is opti-

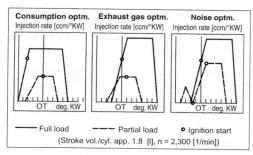

Consumption optm.	Exhaust gas optm.	Noise optm.
Injection rate [ccm/°KW]	Injection rate [ccm/°KW]	Injection rate [ccm/°KW]

——— Full load   - - - - Partial load   ○ Ignition start
(Stroke vol./cyl. app. 1.8 [l], n = 2,300 [1/min])

Fig. M5 Ideal injection curves (Source: W. Haas)

mized for fuel consumption, exhaust gas emissions, or noise.

It can be an advantage to design the system to give one or more small preinjections before the main injection process—these reduce the combustion noise and the levels of NO_x. The reason for this is that it reduces the ignition delay for the main injection quantity and, therefore, the rate of pressure rise of the combustion (reduced combustion noise).

The initially compact injection spray broadens, and fuel droplets are stripped off its edges. The spray disintegrates into individual droplets during the continued injection curve, and these evaporate because of the high temperature.

Fig. M6 shows how ignition zones and local air conditions are formed around the injection spray. It can be seen that the ignition in a diesel engine also occurs at a local range of air-fuel ratio around 1, which means with a stoichiometric mixture. However, the diesel engine, in contrast to the gasoline engine, has many ignition points where the ignition can start.

~~**Air movement** →Intake port, diesel engine; →Intake port, gasoline engine

M

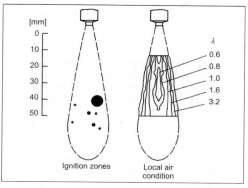

Fig. M6 Ignition zones in the injection spray (Source: F. Pischinger)

~~**Fuel atomization** →Injection valves ~Diesel engine ~~Injection nozzle parameter ~~~Injection spray, ~Gasoline engine ~~Intake manifold injection ~~~Spray preparation

~~**Fuel evaporation** →Fuel, diesel engine ~Properties; →Fuel, gasoline engine ~ Volatility

~~**Internal mixture formation.** →Injection system, fuel ~Diesel engine

~~**Mixture composition** →Air-fuel mixture

~~**Mixture control diesel engine** →Electronic open- and closed-loop control ~Electronic open- and closed-loop control, diesel engine

~~**Mixture formation in divided combustion chamber** (not used very much at present). Prechamber and swirl chamber engines are possible; in these, the injection is performed into a secondary combustion chamber, which is separated from the main combustion chamber by a passage. The ignition in the precombustion chamber is performed with deficient air (i.e., a rich mixture), because the total quantity of fuel is injected into only a part of the total quantity of trapped air. This is why only part of the combustion of fuel can take place here. The burning mixture then flows under high pressure through appropriate openings or passages from the precombustion chamber to the main combustion chamber where it burns completely. Based on the large surface area in engines with divided combustion chambers, the temperatures achieved at the end of compression are often so low when the engine is started that the injected diesel fuel cannot be ignited. It is, therefore, generally necessary to arrange glow plugs in prechamber so that the injected fuel is ignited under cold-start conditions, and this initiates the combustion of the remaining fuel. Glow plugs are only required in engines with direct injection for temperatures below 0°C or for pollutant reduction.

The glow plugs are switched off after the engine has started and combustion has become stable. However, the glow plugs are often not switched off immediately, which has a positive impact on the acoustics and the pollutant reduction.

~~**Mixture formation in single combustion chamber.** Direct injection is characterized when the formation of the mixture and the subsequent combustion in a diesel engine are performed in a single chamber, as is customary today. Engines with direct injection have significantly lower fuel consumption than do engines with divided combustion chambers, because the surface area of the combustion chamber and, therefore, the heat losses, are smaller. Additionally the transfer losses between the chambers are eliminated.

~~**Mixture formation systems, diesel engine** →Injection system, fuel ~Diesel engine

~~**Mixture ignition limits** →Ignition limits

~~**Mixture movement.** The air supplied to the cylinder is forced into a swirl or tumble motion by appropriately formed intake ports. The air movement continues in a reduced form in the combustion chamber even after the closing of the intake valves. This is required to achieve good mixing of the injected fuel with the air, with the result that the oxygen existing in the air can be used optimally; the highest possible performance is achieved with this approach. In addition, the pollutant components can also be minimized.

~~**Mixture preparation.** *See below*, ~Mixture formation, gasoline engine

~Mixture formation, gasoline engine. The formation of the mixture is used to generate an ignitable mixture consisting of air and fuel; the latter can be supplied in a liquid or gaseous state. The ignition limits depend on the fuel and the temperature of the mixture as well as on its composition. The upper ignition limit is $\lambda > 1$ (lean mixture) and the lower ignition limit is $\lambda < 1$ (rich mixture).

An ignitable, homogenous mixture must be present in the cylinder of the gasoline engine at the start of external ignition at the spark plug. To meet the above requirements, high-quality atomization and vaporization characteristics must be available if the fuel is supplied in a liquid and not a gaseous state. Fuels with high volatility characteristics are of significant advantage, but they also come with the disadvantage of generating vapor locks in the fuel lines and in the mixture-forming devices.

The compression of the air-fuel mixture in the cylinder, which takes place after the induction stroke, results in heating up of the fuel vapors. In gasoline engines, the self ignition temperature of the mixture must be higher than the temperature at the end of compression, because the ignition should only be started by the spark from the spark plug. It also must be higher than the temperature achieved by the last quantity of mixture burned due to compression by the expanding flame front, because this would result in knocking.

Some countries have established standards for minimum requirements for liquid gasoline fuels to achieve a unified design and a worldwide operation of gasoline engines. These include specification of the density, antiknock properties, vapor pressure, boiling characteristics, and distillation and vaporization residue as well as the contents of sulfur, benzene, oxygen-carrying components, and so on.

The formation of the mixture in gasoline engines is divided into metering, preparation, transport, or distribution of the fuel to individual cylinders. Extensive evaporation of the liquid fuel and its uniform mixing with the air are the objectives of mixture formation in gasoline engines. Uniform distributions of air and fuel to the individual cylinders in multicylinder engines is also required (*also*, →Combustion process, gasoline engine).

~~**Central mixture formation.** This is formation of the mixture using carburetors or central injections. The required air-fuel mixture is generated centrally for all cylinders of the engine and then supplied to the individual cylinders. Generally one of these mixture-forming devices is fitted on each engine. However, if carburetors are used (they are not used anymore in automobile engines), the following is the situation: Two carburetors are required for engines with more than four cylinders because one carburetor cannot supply a uniform mixture for combustion with more than four cylinders. These carburetors are typically designed as two-barrel carburetors in a single housing and are positioned centrally because the synchronization (balancing) of several carburetors for uniform air and fuel flow rates creates problems. A large installation space would be required in addition to complicated actuation mechanisms.

Central mixture formation has the disadvantage that long distances are required from the location where the mixture is formed until it enters the cylinder. Auxiliary equipment is used to prevent problems in transient operation and during the warm-up phase. The problems are due to the condensation of the fuel on the cold intake manifold walls.

Central injection is used as an alternative design to the carburetor for central formation of the mixture (Monojetronic, Single-Point-Injection).

~~**Charge stratification.** Charge stratification occurs when the combustion chamber is not filled with a homogeneous mixture. In this case, some zones develop that have a richer or a leaner combustion mixture than others. An objective of charge stratification is to place a richer mixture near the spark plug for better ignition and a leaner mixture farther from the spark plug to achieve better fuel economy. This measure should achieve lower part-load fuel consumption and reduced NO_x and CO emissions.

~~**Condensation.** Condensation of evaporated fuels from the mixture can develop at low temperatures with central mixture-forming devices. This means fuel that condenses is transported by the airflow in the form of droplets or it becomes a film on the manifold wall. **Fig. M7** shows the influence of the air-fuel ratio and the intake manifold pressure on the so-called mixture saturation temperature. Condensation of the evaporated fuel must be expected below this temperature (*also, see below, ~~Wall film*).

~~**Decentralized fuel mixture generation.** The decentralized generation of the fuel-air mixture has one mixture formation system for each cylinder. The injection into the intake port before the intake valve is typical, and this has one injection nozzle for each cylinder. The advantages of this arrangement include a more uniform distribution of fuel to the individual cylinders and short distances for the supply of the fuel to the cylinder. Other advantages include design flexibility for the intake pipe and the utilization of the motion of intake valve to improve the supply rate.

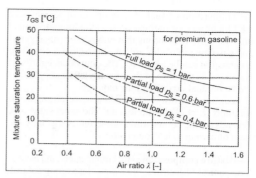

Fig. M7 Effect of air-fuel ratio and intake manifold pressure on mixture saturation temperature

~~**External mixture formation.** External mixture formation in a gasoline engine describes a process by which the fuel is delivered into the intake pipe, for example, in the direction of the intake valve. External generation of the mixture is currently the process most commonly used in gasoline engines. It is characterized by a ready supply of fuel, in contrast to the timed supplies in processes with direct injection. Therefore, the supply need not be synchronous with the crank angle. The external formation of the mixture can be implemented in two ways:

- By carburetors (not used anymore in automobile engines).
- By injection, which can be performed intermittently (currently prevailing) or continuously. This type of external formation of mixture can also be divided into central (no longer used) and decentralized injection.

~~**Fuel atomization** →Injection valves ~Diesel engine ~~Injection nozzle parameter ~~~Injection spray, ~Gasoline engine ~~Direct injection, ~Gasoline engine ~~Intake manifold injection ~~~Spray preparation

~~**Fuel evaporation** →Fuel, diesel engine ~Properties; →Fuel, gasoline engine ~Volatility

~~**Internal mixture formation.** Internal formation of the mixture occurs when the fuel is injected directly into the combustion chamber. A stratified charge is required for extremely lean operations because of the narrow ignition limits of the gasoline fuel, and this cannot be achieved using intake manifold injection.

The advantages of operating with internal mixture formation are better charge-air–cooling, the opportunity to achieve lean burn concepts, the reduction of the throttle losses, and improved thermal efficiency. These are balanced by the disadvantage of a short preparation time for the mixture (Source: D. Scherenberg).

~~**Mixture.** A certain limited mass ratio in the intake manifold or the combustion chamber is required for

M

the combustion of fuel. A mixture of approximately 14.7 kg air per kg fuel is called a stoichiometric mixture ($\lambda = 1$). This means that enough air (or oxygen) and gasoline molecules are present so that a complete combustion can be achieved. The equivalence ratio, λ, is defined as the ratio of the quantity of air supplied to the theoretical quantity of air required.

Lean mixtures ($\lambda > 1$) contain more air, and rich mixtures ($\lambda < 1$) contain less air. The exact value of λ depends on the composition of the fuel, especially the ratio of carbon to hydrogen atoms.

~~**Mixture composition** →Electronic open- and closed-loop control ~Electronic open- and closed-loop control, gasoline engine ~~Functions ~~~Lambda control

~~**Mixture compression.** In contrast to diesel engines, which compress air, gasoline engines work with compression of the air-fuel mixture, which means the mixture, usually generated outside the cylinder, is compressed.

~~**Mixture control** (*also*, →Closed-loop control). This control loop is used to generate a predetermined fuel-air ratio in the mixture supplied to the combustion chamber. In contrast to open-loop mixture control, closed-loop control of the mixture guarantees the desired air-fuel ratio under all operating conditions. An appropriate sensor is required for closed-loop control of the mixture (lambda sensor or air ratio sensor). The sensor must be capable of determining the corresponding air ratio by measuring the residual oxygen in the exhaust gas and then controlling the strength of the mixture through a change of the supplied fuel quantity through the injection valves or the electronically controlled carburetor.

~~**Mixture cooling.** The evaporation of the liquid gasoline fuel on the way to the combustion chamber results in cooling of the mixture (evaporative heat). The cooling can be 8–15°C depending on the volatility of the fuel. The water in the intake air, which is present in the form of vapor, condenses at intake air temperatures of 0–10°C and can, therefore, result in icing of the main jet in engines with carburetors—for example, on the throttle valve. Countermeasures include additives to the fuel to lower the freezing point, such as a 2–3% addition of alcohol or additives that have a surface impact on the throttle valve so that no ice can stick to it. Another alternative is the heating of the throttle valve with the help of electrical energy (heating coil around the throttle valve edge) or with coolant from the cooling circuit, which is heated fast enough for this purpose; air preheating also has been used.

The cooling effect of the mixture also has a positive impact because the filling of the cylinder is increased by the internal cooling. This means that the volumetric efficiency and, therefore, the performance of the engine are increased.

The total reduction of temperature in engines with direct injection is used for internal cooling. Racing en-gines often use high-volatility fuels to further increase the internal cooling.

Gasoline fuels have a latent heat of evaporation of less than 250 kJ/kg; however, methanol has a value of 1190 kJ/kg. This results in a theoretical reduction in temperature of approximately 14°C for an equivalence ratio of 1 when using methanol, and this can increase the performance significantly. A special problem exists with the formation of the mixture when alcohols are used. The physical limits of mixture formation make a stoichiometric air-fuel ratio possible down to temperatures that are significantly below –20°C with gasoline, but the methanol limit is 17°C, and it is 29°C for ethanol. This means that the vast majority of the fuel quantity must be supplied in liquid form, resulting in additional requirements for heating for the cold start and the warm-up phases of the engine when using these fuels.

~~**Mixture density.** The mixture density defines the density of the air-fuel mixture. It changes with the air-fuel ratio and, therefore, with the quantity of fuel in the air. The mixture density is also affected by the air pressure and the temperature.

~~**Mixture distribution.** *See below*, ~~Mixture preparation

~~**Mixture enrichment.** *See below*, ~~Mixture metering

~~**Mixture formation processes** →Combustion process, gasoline engine; →Injection system, fuel

~~**Mixture formation systems** (*also*, →Carburetor; →Injection system, fuel ~Gasoline engine). Mixture formation systems are used to supply the engine with the required air-fuel mixture for each operating point. Different mixture formation systems are used depending on the requirements for the operating characteristics of the engine. They are divided into central and decentralized mixture forming devices.

~~**Mixture forming device** →Carburetor; →Injection system, fuel ~Diesel engine, ~Gasoline engine

~~**Mixture inhomogeneity.** Mixture inhomogeneity is generated by a nonuniform distribution of the fuel molecules with the air molecules in the air-fuel mixture supplied for combustion. A targeted inhomogeneity of the mixture is planned with so-called charge stratification. This process provides a rich mixture near the spark plug to better support the ignition, and a lean mixture at a distance from the spark plug in order to reduce fuel consumption.

~~**Mixture injection.** In this application, air-fuel mixtures are injected into the cylinder under pressure (special format of the mixture formation—e.g., for hydrogen engines, two-cycle engines).

~~**Mixture leaning.** *See below*, ~~Mixture metering,

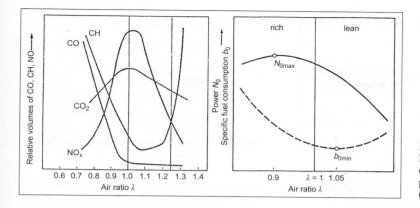

Fig. M8
Effect of air-fuel ratio on the most important pollutant components (Source: Bosch)

~~**Mixture metering.** Mixture metering is the metering of the fuel into the air used for combustion. A stoichiometric mixture or an air-fuel ratio (air number) of $\lambda = 1$ exists, if the combustion air is supplied with sufficient fuel, so that all fuel molecules can react chemically with the available oxygen. The air-fuel ratio of $\lambda = 1$ is equivalent to a mass ratio of approximately 14.7 kg air/1 kg fuel. The actual value of this ratio depends on the composition of the fuel, especially the H/C ratio and whether it is oxygenated (e.g., ethanol).

Fig. M8 shows the effect of the air ratio on the generation of individual components of pollutants such as oxides of nitrogen (NO_x), carbon monoxide (CO), unburned hydrocarbons (HC), and the combustion product carbon dioxide (CO_2). An air ratio smaller than 1 means a deficiency of air or an excess of fuel; this is called a rich mixture or mixture enrichment. An air ratio larger than 1 means that, based on the chemical reaction equation, less fuel is available than required for the complete combustion of the available air—this is called a lean mixture. **Fig. M8** shows also the effect of the air ratio on the fuel consumption and the maximum performance achievable. Lean mixtures result in a better fuel consumption than stoichiometric mixtures, but they have a tendency to misfire when the mixture becomes too lean, and this leads to high HC emissions. Rich mixtures, to a certain degree, generate the highest performance.

~~**Mixture movement.** Mixture movement is the transport of the air required for the combustion and the fuel in the intake manifold or in the cylinder.

~~~**In the cylinder.** The mixture enters the cylinder after the transport of the mixture in the intake manifold, and it is distributed either uniformly or nonuniformly (charge stratification) in the cylinder, depending on the geometry of the intake port(s) (two- or multivalve engines). The movement of the charge in the cylinder is primarily determined by the intake flow characteristics during the intake phase.

The movement of the charge in the cylinder can be divided into a swirl movement and a tumble movement. The swirl is a rotational movement around the cylinder axis, and tumble is a "vortex" motion along the cylinder axis. Appropriate measuring procedures deliver characteristics for the swirl and tumble behavior of the charge movement that is generated by the port geometry of the cylinder head. The so-called swirl and tumble numbers are defined by the ratio of the circumferential to the axial speed in the cylinder in relationship to the corresponding axes—both are measured with appropriate impellers.

Tumble flows are primarily generated in multivalve engines with single-branch intake systems (**Fig. M9**). In contrast, two-branch intake systems, with a port that can be switched off, generate swirl flows at low speeds. The different forms of charge movement have a strong impact on the flame speed. Tumble flows generate an increase of the turbulent flame speed, but the effect is less significant with swirl flows.

The advantage of the increased turbulent flame speed is that a leaner mixture can be used, and this reduces the specific fuel consumption (**Fig M10**).

It is important for the formation of mixture in the cylinder that an ignitable mixture is present at the spark plug at the time of ignition. This is especially neces-

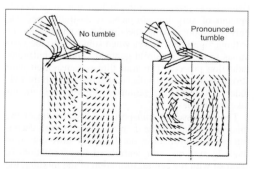

Fig. M9 Flow structure for tumble flow (Source: FEV)

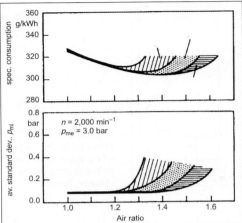

Fig. M10 Impact of the tumble flow on specific fuel consumption (Source: FEV)

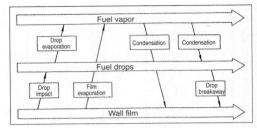

Fig. M11 Different fuel transport alternatives (Source: H.P. Lenz)

sary with stratified charge engines, which normally only burn a lean mixture (charge stratification). The movement of the mixture should not be too strong at the spark plug, because this can extinguish the developing flame.

~~~**In the intake manifold.** It is important to differentiate between centralized and decentralized formation of the mixture. Fuel and air are transported through the intake manifold if a centralized mixture-forming device is used. The intake manifold basically transports only air, and it is in only the last part of the intake duct that air and fuel are carried if a decentralized mixture-forming device is used—that is, the injection nozzles are positioned close to the intake valve.

The fuel can be transported in three different conditions:

• As fuel vapor
• As fuel droplets
• As wall film (**Fig. M11**)

Extensive evaporation of the fuel, or at least the development of very fine droplets, is desirable, because only fuel vapor and fine fuel droplets can follow the air without significant delays when subjected to the variations in intake manifold geometry and the pulsating induction process. The wall film is transported with strong delays, and it is distributed unevenly to the cylinders because of the geometric differences in the intake manifolds.

The processes of transport and distribution of the mixture that take place in intake systems with decentralized fuel formation are basically different from those in systems with centralized fuel formation. In contrast with the central formation of the mixture, the major parts of the intake port are in contact with only

air due to the late supply of the fuel into the intake air because, in most cases, the fuel will be injected only just before or directly at the intake valve. This means that many of the problems of multiphase flows are eliminated, such as phase segregation, nonuniform phase distribution, wall film storage, and so on, and this results in significantly improved design flexibility for the intake system with respect to the utilization of dynamic gas effects.

~~**Mixture preheating.** It may be an advantage with the central mixture formation systems to preheat the mixture of fuel and air during cold start and warm-up to prevent fuel condensation and generation of a wall film. This is required because a strong wall film ratio results in a slow and nonuniformly distributed transport of the fuel to the individual cylinders, a situation that necessitates extra enrichment of the mixture so that every cylinder gets an ignitable mixture. Preheating the mixture enables starting and warm-up with lower enrichment, and this generates less pollutants (HC and CO).

Mixture preheating can be achieved by preheating the induced air, which, for example, can be routed over the exhaust manifold, and also by using hotspots in the intake manifold, which evaporate a part of the wall film. Electrically heated honeycombs in the intake tract for mixture heating are also possible.

A further benefit of preheating the mixture is that icing of the throttle valve is avoided.

~~**Mixture preparation.** Mixture preparation is the atomization, the evaporation, and the mixing of the fuel with air. The objectives include fine atomization and extensive evaporation of the fuel, because ignition is only possible if the fuel is evaporated and mixed with air. The fineness of atomization of the fuel is important for the evaporation of the fuel, because a finely atomized fuel evaporates faster than a less finely atomized one. The fineness of atomization of the fuel is also important for uniform and fast transport from the mixture-forming device (injection valve or carburetor) through the intake manifold to the cylinder.

Fuel drops that are smaller than 10 μm are basically transported with the airflow without any delay, through all the curves along the port and into the cylinder. Larger droplet diameters create a centrifugal effect in the curves of the intake manifold, which results in a

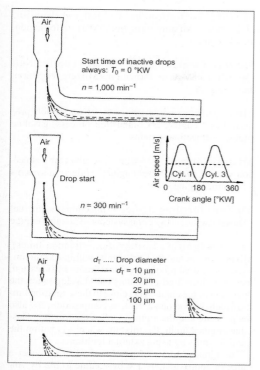

Fig. M12 Impact of droplet diameter and speed on droplet trajectory (Source: H.P. Lenz)

wall film. This in turn causes a delay in the transport of fuel to the cylinders and a nonuniform distribution to the individual cylinders.

Fig. M12 shows the effect of the droplet diameter and engine speed (in relation to the air speed in the intake manifold) on the trajectory of the droplets for a

downdraft carburetor with 90° bend in the intake manifold. It can be seen that only droplets that are smaller than 10 μm follow the bend in the manifold. All the other droplets are separated out from the airstream by centrifugal forces depending on the speed in the bend. **Fig. M13** shows the effect of the droplet diameter on the droplet speed. It can be seen that the droplets with diameters of 10 μm basically follow the airflow without delay, while droplets with larger diameters show significant delays in the airflow, and that they get to the respective cylinder only after the intake valve has been closed. They are, therefore, only available for the next power cycle.

~~Mixture saturation temperature. *See above,* ~~Condensation

~~Mixture stratification. *See above,* ~~Mixture inhomogeneity

~~Mixture temperature. The temperature of the mixture has a significant impact on the performance of the engine. As a first approximation, the mass of mixture trapped in the cylinder of a gasoline engine is proportional to it. Its impact on the performance of turbocharged engines is also significant, which is why most turbocharged engines have charge air cooling: It permits performance increases of up to 15%.

~~Mixture transport. *See above,* ~~Mixture movement

~~Mixture volume →Electronic open- and closed-loop control ~Electronic open- and closed-loop control, gasoline engine ~~Functions ~~~Lambda control

~~Wall film. The wall film is that part of the fuel which is not transported to the cylinder in vaporized form or as droplets and which occurs especially when central mixture-forming devices are being used. The fuel film is deposited on the intake manifold wall by centrifugal force and flows slowly and uncontrollably to the cylinders.

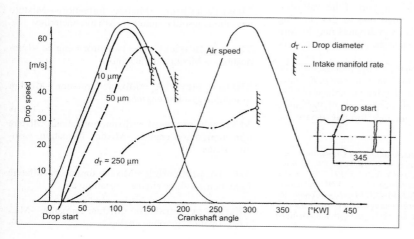

Fig. M13
Impact of droplet diameter on droplet speed (Source: H.P. Lenz)

Inadequate fuel preparation results in the development of droplets and, combined with abrupt changes in the direction of the flow of mixture, these attach to the intake manifold wall and generate a film. The formation of a fuel film can be up to 50% of the total quantity of fuel supplied during full-load operation, and the proportion can be even higher during a cold start.

Generation of a wall film is even possible during cold starts and the warm-up phases of an engine—that is, when the mixture has not been heated adequately—even when very good fuel preparation has been performed in the mixture forming device—for example, complete evaporation. In this case, part of the fuel condenses on the cold intake manifold walls if the mixture saturation temperature of the fresh charge in the intake manifold is too low.

The wall film attached to the wall of the intake manifold is moved in the direction of the cylinder by the shearing strain on the film surface caused by the air flowing over its surface. The shearing strain on the surface of the film is transferred to the other levels of the fluid in the wall film by internal friction. The average speed of the film is significantly lower than the speed of the intake air (**Fig. M11**; *also*, →Mixture formation ~Mixture formation, gasoline engine ~~Mixture movement ~~~In the intake manifold).

In addition to condensation of the wall film and its transport, more or less effective evaporation of the wall film takes place in the intake manifold of engines with central mixture formation, depending on the load point and the engine operating status. The turbulent flow of mixture with low partial pressures over the film surface allows part of the film to evaporate again. The film in the intake manifold is fractionated and more parts with higher boiling temperatures accumulate in the film, because the more volatile parts of the fuel in the film evaporate first. A quantitative and uniform distribution of the fuel must be achieved in addition to the uniform distribution of the fuel components because of the different antiknock properties of these components compared to the initial fuel. In addition to basic transport of the wall film, surface waves are generated due to the shearing strain on the film surface, and droplets may be torn out of the film surface when the gas speeds are very high.

The effect of gravity on the transport of the wall film is significantly higher than on the transport of droplets. Significant changes in the distribution of the wall films can be achieved by small inclinations of the intake manifold when central mixture forming device are used.

Literature: H.P. Lenz: Gemischbildung bei Ottomotoren, Die Verbrennungskraftmaschine, New Series, Volume 6, Springer Publishers, Vienna, 1990, ISBN 3-211-82193-7. — Bosch (ed.), Kraftfahrtechnisches Taschenbuch, VDI Publishers GmbH, Düsseldorf 1, ISBN 3-18-419114-1 1. — Conference "Der Arbeitsprozess des Verbrennungsmotors" 18/19.9.1995, TU Graz, vol. 67, Institut für Verbrennungskraftmaschinen und Thermodynamik.

Mixture formation in divided combustion chamber →Mixture formation ~Mixture formation, diesel engine

Mixture formation in one-piece combustion chamber →Mixture formation ~Mixture formation, diesel engine

Mixture formation process →Combustion process, diesel engine; →Combustion process, gasoline engine; →Injection systems

Mixture formation systems →Carburetor; →Injection system, fuel ~Diesel engine; →Injection systems ~Gasoline engine

Mixture forming device →Injection system, fuel ~Diesel engine; →Injection systems ~Gasoline engine

Mixture ignition limits (*also*, →Ignition limits). Mixtures of air and fuel only ignite within certain limits, the so-called mixture ignition limits. Misfires and, therefore, increased pollutants can develop when the operation of the engine gets close to these ignition limits, which depend on the fuel, its composition, and also the pressure and the temperature of the mixture. Gasoline engines ignite the mixture when it is within the ignition limits by using external ignition.

The ignition limits of gasoline fuel in air cover a range of approximately 1% by volume to 8% by volume, and the ignition limits of the engine in relationship to lambda are between $\lambda = 0.4$ and $\lambda = 1.4$. Similar values are found for diesel engine combustion. Gaseous fuels—for example, hydrogen—have ignition limits in a range of $\lambda = 0.5$ to 10.5.

Mixture inhomogeneity →Mixture formation ~Mixture formation, gasoline engine

Mixture injection →Mixture formation ~Mixture formation, gasoline engine

Mixture leaning →Mixture formation ~Mixture formation, gasoline engine ~~Mixture metering

Mixture lubrication →Two-stroke engine ~Lubrication ~~Mixture lubrication

Mixture metering →Mixture formation ~Mixture formation, gasoline engine

Mixture movement →Mixture formation ~Mixture formation, diesel engine, ~Mixture formation, gasoline engine

Mixture preheating →Mixture formation ~Mixture formation, gasoline engine; →Starting aid

Mixture preparation →Mixture formation ~Mixture formation, gasoline engine

Mixture saturation temperature →Mixture formation ~Mixture formation, gasoline engine ~~Condensation

Mixture shut-off →Carburetor ~Equipment

Mixture stratification →Mixture formation ~Mixture formation, gasoline engine ~~Mixture inhomogeneity

Mixture temperature →Mixture formation ~Mixture formation, gasoline engine

Mixture transport →Mixture formation ~Mixture formation, gasoline engine ~~Mixture movement

Mixture volume →Electronic open- and closed-loop control ~Electronic open- and closed-loop control, gasoline engine ~~Functions ~~~Lambda control

Modal analysis →Engine acoustics

Model construction →Calculation processes ~Numerical processes

Modular intake system design →Intake system ~Modular design, intake system

Modularization, intake system →Air intake system

Modulation →Engine acoustics

Modules (*also*, →Sealing systems). Modules are functional assemblies built up from subsystems and components which perform certain functions. Adjacent components are integrated into one common pre-assembled component. This approach also provides for mutual interaction—for example, in terms of thermal, flow, and structural mechanical behavior. The advantages of modular design are:

- Single parts are matched to one another in a module.
- Integration of a number of functions reduces variety of parts.
- Mutual interactions can be taken into consideration.
- Better understanding of overall system if possible.
- Simulation of the overall system is easier.
- The system is the responsibility of one company.
- Weight can be reduced.

Examples of modules are:

- Valve cover module—e.g., including functional features such as oil separation, blowby guidance, acoustic insulation, mount for ignition module, cable guides, hose holder, sealing, and acoustic coupling functions.
- Cooling modules—e.g., with functional features of heat transfer, fan operation, fan drive, air guidance, space utilization, cooling capacity.
- Oil modules—e.g., with integrated oil filter, oil pump, oil/water heat exchanger.

Fig. M14
Multifunction module
(Source: ATZ)

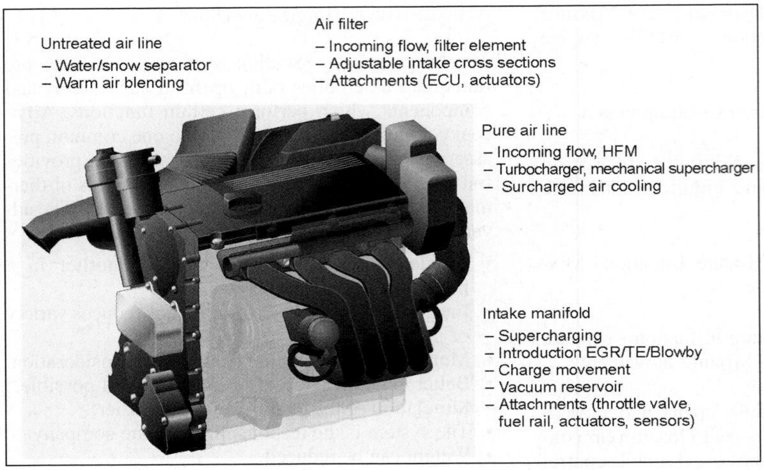

Air filter
– Incoming flow, filter element
– Adjustable intake cross sections
– Attachments (ECU, actuators)

Untreated air line
– Water/snow separator
– Warm air blending

Pure air line
– Incoming flow, HFM
– Turbocharger, mechanical supercharger
– Surcharged air cooling

Intake manifold
– Supercharging
– Introduction EGR/TE/Blowby
– Charge movement
– Vacuum reservoir
– Attachments (throttle valve,
  fuel rail, actuators, sensors)

Fig. M15
Air guidance modules
for combustion engine
(schematic)

- Induction module—e.g., with integrated variable resonance induction system with valves and actuating elements.
- Intake module: This module is widely used because it provides for the air guidance for the majority of the engine and is suitable due to its size for attachment to highly varied components such as air baffles, filter, intake scoops, air mass meter, sensors, etc. It is not necessary for these components to be a part of the air guidance system. One example is installation of the engine control in the air filter, where the air flowing past is used to cool the electronic circuitry.
- Multifunction module for fuel injection and tumble valves (Lupo FSI engine; **Fig. M14**) consisting of: tumble valve arrangement, vacuum actuation for tumble valve, potentiometers for indication of tumble valve position, high-pressure fuel distribution rail with connection to high-pressure pump, pressure sensor and pressure control valve, mount and fuel supply for high-pressure injection valves.

Fig. M15 shows a schematic representation of the air guidance system for a four-cylinder engine with the most important functions and a number of attachment parts.

Moisture →Intake air ~Humidity

Molybdenum coating →Piston ring ~Running surface shapes

Moment of ignition →Ignition system, gasoline engine ~Ignition

MON →Fuel, gasoline engine ~Octane number

Mono-ethylene glycol →Coolant

Monolith →Catalytic converter

Mono-metal design →Crankcase ~Crankcase design ~~Cylinder

Mono-metal valves →Valve ~Gas transfer valves ~~Valve designs

Mono-propylene glycol →Coolant

Motorcycle →Oil ~Classification, specifications, and quality requirements ~~Gasoline engine oils, four-stroke, ~~Gasoline engine oils, two-stroke

Motorcycle engines →Lubrication ~Lubricating systems

Mounting location electronics/mechanics →Electronic/mechanical engine and transmission control ~Requirements for mechanical and housing concepts

Movement of combustion chamber charge →Charge movement

MTM →Electronic/mechanical engine and transmission control ~Requirements for mechanical and housing concepts ~~Mechatronic Transmission Module

Muffler →Engine acoustics ~Intake noise ~~Intake silencer; →Exhaust system

Muffler rattle →Engine acoustics

Muffler system →Engine acoustics

Multibody simulation (MBS) →Calculation processes ~Numerical processes

Multicylinder engine →Engine concepts ~Reciprocating piston engines ~~Number of cylinders

Multifuel engine →Combustion process, multifuel engine

Multigrade oils →Oil ~Properties and characteristic values ~~Viscosity classes for vehicle transmission oils

Multimass torsional vibration damper →Torsional vibrations ~Dual-mass flywheel

Multiple combustion chamber →Combustion chamber ~Subdivided combustion chamber

Multiple ignition coil →Ignition system, gasoline engine ~High-voltage generation ~~Ignition coil

Multiple shaft engine →Engine concepts ~Reciprocating piston engines

Multiple valve arrangement →Valve arrangement ~Number of valves

Multiple weight modulator →Engine acoustics

Multiple-spray injection →Combustion process, diesel engine ~Direct injection

Multiple-spray method →Combustion process, diesel engine ~Direct injection

Multipoint injection →Injection system, fuel ~Gasoline engine ~~Direct injection systems ~~~Intake manifold injection systems

Multistage supercharging →Supercharging

Multistage turbocharging →Supercharging

Multivalve engine →Variable valve control ~Valve stoppage

Mushroom valve lifter →Valve gear ~Actuation of valves ~~Direct valve actuation

N

Naphthalene →Fuel, diesel engine ~Composition

Napier ring → Piston ring ~Versions ~~Oil scraper rings

Narrow band analysis →Engine acoustics

NASCAR →Engine ~Racing engines

Natural gas CNG (Compressed Natural Gas) →Fuel, diesel engine ~Alternative fuels

Naturally aspirated engine →Engine ~Current engines

NDIR analyzer →Exhaust gas analysis equipment

Near field →Engine acoustics

Needle jet →Carburetor ~Jets

Needle lift sensor (*also,* →Injection valves ~Diesel engine ~~Fuel-injector body ~~~Pintle motion sensor, fuel-injector body with pintle motion sensor). Hole-type nozzles are used on diesel engines with direct injection (injection systems). The motion of the needle in the injection nozzle is used to obtain information on the start and end of injection, and so on. These values are measured by a needle lift sensor (**Fig. N1**). Fuel in-

jector mounts with an integrated electrical coil, which plunges into an ion core, are used for this purpose. This induces an electrical voltage which is proportional to the needle lift. This motion provides information on the pintle opening and closing processes, which are used for controlling the beginning of injection, for diagnosis, and for determination of the fuel consumption and engine speed.

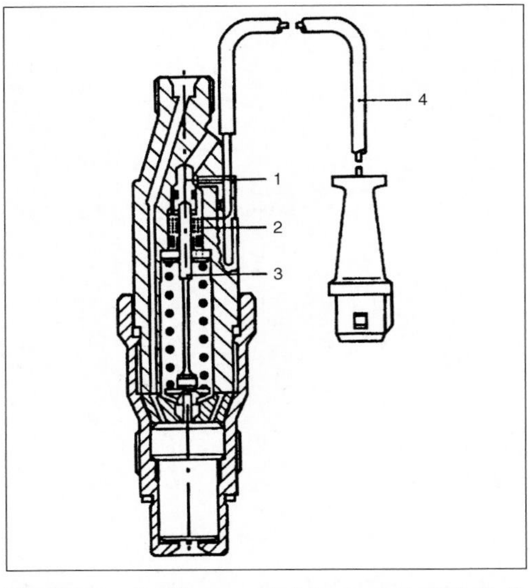

Needle lift sensor, nozzle holder with needle lift sensor →Injection valves ~Diesel engine ~~Fuel injector body

Net power →Power output ~Effective (brake) power output

Neutralization capacity →Oil ~Oil functions

Neutralization of combustion products →Oil ~Oil functions, ~Properties and characteristic values

Nitration →Oil ~Power during operation ~~Aging of oil

Nitration hardening and nitro-carburizing →Piston ring ~Surface treatments

Nitric oxide →Exhaust gas analysis

Nitrogen dioxide →Exhaust gas analysis

Nitrous oxide emission →Emission measurements ~Test type I ~~Mass determination, emissions

Nitrous oxide sensor →Sensors

Nitrous oxides →Emission measurements ~Test type I; →Exhaust gas analysis

NMHC (nonmethane hydrocarbons) →Emission limits

NMOG (nonmethane organic gases) →Emission limits

Noise →Engine acoustics ~Engine noise ~Exterior noise, ~Noise generation in engine, ~Noise source, engine; →Piston

Noise absorption →Engine acoustics

Noise emission →Engine acoustics ~Sound emission

Noise emission limit →Emission limits; →Engine acoustics ~Exterior noise ~~Emission limits

Fig. N1 Needle lift sensor: 1. Adjustment bolt; 2. Coil; 3. Pressure pin; 4. Cable with connector (Source: Mollenhauer)

Noise excitation →Engine acoustics

Noise field analysis →Engine acoustics

Noise generation →Air intake system ~Acoustics; →Engine acoustics

Noise immission →Engine acoustics ~Sound emission

Noise intensity →Engine acoustics

Noise level →Engine acoustics ~Decibel (dB)

Noise quality →Engine acoustics

Noise reduction →Engine acoustics

Noise reduction measures →Engine acoustics ~Exterior noise ~~Reduction

Noise source, engine →Engine acoustics

Nondispersive infrared analyzer →Exhaust gas analysis equipment

Nonengine particulates →Particles (Particulates) ~Particulate emissions

Nonhomogeneous air-fuel mixture →Air-fuel mixture ~Homogeneous mixture

Nonstationary operation →Load ~Transient engine operation

Nonuniformity (cyclic variation) (*also*, →Torsional vibrations). The gas forces from combustion acting on the piston vary with crank angle. If these are overlapped with the periodically varying inertia forces of the engine parts, the resulting force, which produces the torque, results. The variation in torque over a working cycle leads to a cyclically varying rotational speed on the crankshaft. The resulting speed variation can be defined as nonuniformity.

NOₓ accumulator-type catalytic converters →Pollutant aftertreatment ~Pollutant aftertreatment lean concepts

NOₓ concentration →Emissions

NOₓ converter →Emission limits; →Exhaust gas analysis equipment ~Chemiluminescence detector (CLD)

NOₓ emission →Injection system, fuel ~Gasoline engine ~~Direct injection systems ~~~Common rail

NOₓ particulate trade-off. This describes the inter-relationship between particulate emissions and NO_x emissions on diesel engines. Engine measures to reduce NO_x emissions generally result in an increase in the particulate emissions and vice versa.

The objective of engine development is to create degrees of freedom whereby one of the two constituents—for instance, the particulate emissions—is to be reduced by measures not associated with the engine, so that the other constituent can be reduced by modifying the engine operation without the particulate emissions increasing. An example of this is the use of a particulate filter which compensates for increased particulate emissions resulting from measures in the engine which reduce the NO_x.

Fig. N2 shows the inter-relationships of the individual emission stages and possible potentials of the techniques for reduction of emissions.

Greater homogenization of the mixture by means of appropriate combustion processes shifts the trade-off in the direction of lower NO_x and particulate emissions, whereby the untreated emissions are reduced.

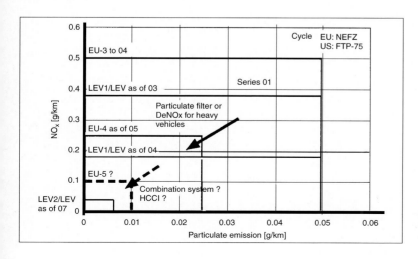

Fig. N2
Emission stages and techniques for reduction of emissions

Number of camshafts →Camshaft ~Number of camshafts

Number of cells →Catalytic converter

Number of cylinders →Balancing of masses; →Engine concepts ~Reciprocating-piston engines

Number of holes →Injection valves ~Diesel engine ~~Hole-type nozzle

Number of throws →Balancing of masses

Number of valves →Valve arrangement

Numerical optimization →Calculation processes ~Numerical processes

Numerical process →Calculation processes

N

O

OBD (On-board diagnosis) →Electronic open- and closed-loop control ~Electronic open- and closed-loop control, gasoline engine ~~Functions

Octane number (RON, MON) →Fuel, gasoline engine ~Octane number (RON, MON)

Octane rating requirement →Fuel, gasoline engine ~Octane number (RON, MON)

Octave analysis →Engine acoustics

Offset →Piston

Offset crank operation →Crankshaft drive; →Piston ~Offset

Oil. The development of the combustion engine is closely associated with the development of materials required for its operation. Therefore, both the fuel and the engine oil are decisive "design elements" for reliable operation of the engine.

The first engines were equipped with a total-loss lubrication system. In these engines, the engine oil had only to cater once for the lubrication point, then it left the engine and was exhausted into the environment. Very soon demands were made for better oils that would be capable of more than just lubricating: they should be used for cooling, sealing, and collecting "dirt" from the combustion residues. Primarily due to cost, it became necessary to collect the oil after its application at the lubrication point and make it available to the engine again: this required an oil circulation system. Consequently, the question arose: How long could oil be used before a change of engine oil became necessary?

In 1949, engine oil would deteriorate to such an extent that it had to be changed after a distance of only 1500 km; by 1972, the possible distance prior to oil change was already 5000 km. Widespread introduction of additives in engine oils and the production of multigrade oils contributed to the longer distance covered prior to oil change. Only 20 years later in 1992, the oil change interval had been increased threefold again (15,000 km). Today, lengthening oil change interval is still a central theme for joint-venture work between engine manufacturers and the mineral oil industry.

Nevertheless, mere lengthening of the oil change interval no longer stands alone in the forefront of development today; rather attempts are being made to accurately determine specific loads on engine oil at the factory of manufacturers and on this basis to define individual oil lifespans. Gathering and interpreting vehicle-specific operating parameters plays a central role, and in this process, driving conditions are very significant. The joint-venture work between engine manufacturers and the mineral oil industry becomes particularly clear at the interface between industrial data collection and interpretation of data. Thus, operating conditions generated by particular customers—for example, frequent short-distance operation or extreme motorway operation—put the engine oil under significantly more stress than operation under moderate conditions. The vehicle manufacturer undertakes data-gathering with sensors (engine speed, load, oil temperature, water temperature, etc.) that are installed in the vehicle; the mineral oil industry then helps with the evaluation of the oil stress arising from these operating conditions on the basis of extensive test-stand and field results. The information jointly gathered here is used in further processes for defining the respective oil-change interval.

Determining the oil-change interval "based on requirement"—that is, to change the oil when it can no longer fulfill its function—will be a determinant theme of development work even in the future. Here, common efforts are necessary with a greater scope than before in order to provide modern vehicle engineering with the correct oil specification. Engine and oil developers must work together to define the possibilities and limits of further expansion of oil-change intervals and assess the partly contradicting requirements. At this point the main questions pertain to fuel consumption, exhaust gas emissions, and about the health risks when people come in contact with severely contaminated waste oils.

In the process, essential requirements will be imposed on the quality of engine oil; specifically, synthetic high-performance oils provided with very good additives will permit a longer oil-change interval under comparable driving conditions. At the same time it must be ensured that fuels (both diesel and gasoline) are compatible with the engine oil formulations and that no harmful interactions occur between the basic fuel components and the fuel additives; ideally they should also complement one another in the case of a four-stroke engine. This requires competent development engineers who consider lubricant, fuel, and engine as a single entity.

Based on these assessments, in the future there will be possibilities of formulating lubricants that can fulfill these requirements and have the greatest possible benefit to drivers.

~Body. Nearly all lubricants are formulated on the basis of two "pillars": basic oils—in the actual sense the liquid "oily" components—and the additives which give the basic lubricant certain desired properties. In comparison with all other lubricants, engine oils exhibit the highest proportion of these additives. The quality of any lubricant and thus, also, of engine oils is,

Basic oils	1 Mineral oil	2 Semi-synthetic	3 Fully synthetic
Additives (concentration without solvent carrier)	A Normal alloy ca. 4–7%	B High alloy ca. 6–9%	C Highest alloy ca. 8–12%
Viscosity class (typical)	X Single range oil 15W-40 15W-50	Y 10W-30 10W-40 5W-40	Z 5W-40 5W-30 5W-20 0W-40 0W-30
Demand (e.g., for passenger cars)	Light	Medium	High

Fig. O1
Overview of quality classes of engine oils

therefore, always determined by the quality of these two components together and their reciprocal matching (**Fig. O1**).

For example, combination of Basic Oil 2 with Additive C set at Viscosity Y is technically suitable for the highest demands; that is, the additive package is more decisive than basic oil as far as the quality of engine oil is concerned. From this consideration two important rules are obtained for assessing the quality of engine oils:

- The viscosity class is not a quality criterion; it says little about the suitability of an oil type for a certain application purpose.
- Fully synthetic engine oils do not automatically exhibit good quality, because the popular designation — synthetic oil — relates only to basic oil.

~~Additives. Modern combustion engines require high performance from their lubricants, which goes beyond mere reduction of friction and prevention of wear. To satisfy these requirements, properties are demanded which nonpremium mineral oils or synthetic liquids do not exhibit at all or only exhibit with lower performance. This is the reason why engine oils are blended with chemically and/or physically acting additives that today reach (including their solvent carriers) metering rates of 20% and more in fully formulated engine lubricants. Additives can improve existing properties (e.g., the flow behavior) or give oils properties that they do not have yet (e.g., detergent effect). The primary areas of application for additives are protection of engine components against wear, dirt, and corrosion and improvement of the viscosity-temperature behavior of the oil. Nonetheless, not all properties of engine oil can be influenced by additives. For instance, thermal conductivity, viscosity-pressure dependency, solubility of gas in oil, and air removal capacity either cannot be or can only be influenced to a small extent.

The common types of engine oil additives which find application in modern high-performance engine oils are compiled in **Fig. O2**. In principle, the types of basic-additive blending of synthetic oils or low-viscosity oils do not differ from the type of basic-additive blending of conventional — that is, mineral oil–based — high-performance oils. However, with the selection of corresponding additives, different response of the additives ("additive response") in different basic oil types has to be considered. Engine oil additives were developed for optimum effectiveness in conventional solvent-neutral oils (mineral oils) — that is, matched to their solvent properties.

The solution properties of many synthetic basic oils differ from those of solvent-neutral oils, so that these covalent basic oils (to which the so-called poly-alpha-olefines [PAO] belong) with stronger polar components (e.g., mineral oils or esters) become necessary or the application of modified additive becomes necessary. Conventional engine oils are, in most cases, naturally mixable with synthetic oils; however, this mixture of oils of different quality classes is not advantageous because the properties of high-performance oils become "diluted" and can therefore not be fully effective.

In addition to the manufacture and selection of possibly environmentally neutral basic oils, the required additives must naturally also be assessed, optimized, and applied with regard to their environmental acceptance. For example, chlorine-containing additives are no longer used. Chlorine contents in some engine oils that are still in use today originate exclusively from undesired production-based impurities with chlorine compounds.

Additives and basic oils must be carefully matched during the formulation of an engine oil with regard to the application purpose and the desired behavior of the lubricant. The desired properties can seldom be achieved with only one additive; in general, mixtures of different additives are used.

Additives act on their own or through their decomposition and conversion products or through mutual interaction among themselves; these can occur in a synergetic or antagonistic manner. When these interactions are not carefully matched, negative effects will occur on the metal surfaces of the engine to be protected. For instance, if the proportion of detergent is too high, the high alkalinity would neutralize the acidic components formed from zinc dithiophosphates for protection against wear and would result in mechanical wear. Conversely, if the proportion of wear-protection additives is too high and the basic components in the oil are too low, this would lead to excessive acidic components with chemical wear as a consequence. It is necessary to choose additives that are suitable for several requirements. For example, zinc dithiophosphates act both as wear protection and as oxidation inhibitors.

Type	Example	Function
Basic detergents	Calcium or magnesium sulfonate, phenolate or -salicylate	Neutralization of acids; prevention of lacquer formation; Reinigung des Motors
Ash-free dispersing agents	Polyisobutene–succinimide, PIB-Mala-Penta	Dispersing agents (to keep floating) for soot, aging products, and other foreign substances; prevention of soot deposits, foreign substances; prevention of acquer formation
Corrosion inhibitors	Calcium or sodium sulfonate, organic amine	Prevention of corrosion
Metallic deactivators	Complex organic sulfur and nitrogen compounds	Prevention of oil oxidation (aging) and oil thickening
Oxidation inhibitors (substances that inhibit aging)	Zinc dithiophosphates, phenoles, amines, metal salicylate	Prevention of oil oxidation and oil thickening
Pour point improver (stock point improver)	Polymere methacrylates	Improvement of flow properties at low temperatures
Friction reducer (friction modifier)	More mild EP-additive, fatty acids and fatty acid derivatives, organic amines	Improvement of friction; minimization of frictional force losses
Antifoaming agent	Silicone compounds	Prevention of foam formation
Wear reducer/ high-pressure additives EP (extreme pressure) additives	Zinc dithiophosphates, organic phosphates, organic sulfur compounds	Prevention of wear
Viscosity index improver	Polymeric compounds of esther type (polymethacrylates) or carbohydrate type (PIB, copolymere)	Improvement of relative viscosity/temperature dependence

Fig. O2
Classes of engine oil additives

Most additives are consumed when the oil is used; the "additive level" drops and their effect is reduced. When the additive content falls short of a minimum level, an oil change is due even if the demand on the oil is low due to dirt and foreign substances.

The essential additive types used in engine oils are discussed in detail below.

~~~**Antifoaming agents.** Air or other gases can exist in lubricating oil in a dissolved state as finely distributed bubbles or in the form of surface foam. The amount of actually dissolved or dispersed air in the oil depends upon the entry of air through the swirling inside the crankcase and upon the pressure and temperature; this can hardly be influenced positively through additives. Also the dwell time (measured as "air removal capacity") cannot be influenced positively through additives, but it can be prolonged by high viscosities and impurities in oil.

The surface foam can be brought to a state of collapse by reducing the surface tension between oil and air. To act as antifoaming agents, these additives must be generally insoluble in the mineral oil and possess a lower surface tension than the mineral oil. Today this happens mostly with the help of silicone oils (polydimethylsiloxanes), which exhibit good effects even in very low concentrations (e.g., 0.01 g/kg oil). Their structure can be seen in **Fig. O3**.

When silicone oils are used, however, it must be observed that air bubbles dispersed in oil are frequently stabilized by the silicone film floating on the oil. The air removal capacity can also deteriorate.

~~~**Antifriction agents.** Reduction of friction increases the overall efficiency of the engine and as such serves to save fuel or to increase power. There are two possibilities of reducing mechanical friction losses with the help of lubricating oil: in the area of liquid friction this happens through lowering the viscosity; in the area of mixed friction it happens by adding friction reducers (reducer of the coefficient of friction, friction modifier). It should be noted that by lowering the viscosity, the range of mixed friction is increased from a certain minimum viscosity. This disadvantageous effect must again be compensated for by the friction-reducing additives.

Friction modifiers are mostly polar molecules which are adsorbed on metal due to adsorption forces. In the process the polar end of the molecule is aligned toward the metal while the long hydrocarbon chains are aligned toward the oil. Based on the strength of the polarity, several layers can result whereby the long chains are aligned perpendicular to the metal surface. The films that occur reduce the friction between the surfaces that move relative to one another (**Fig. O4**).

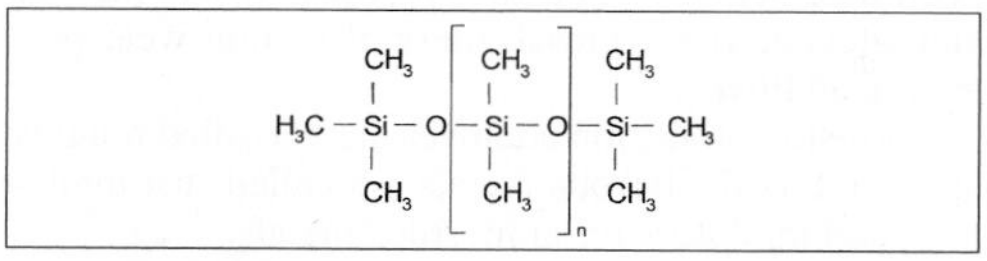

$$H_3C - \underset{\underset{CH_3}{|}}{\overset{\overset{CH_3}{|}}{Si}} - O - \underset{\underset{CH_3}{|}}{\overset{\overset{CH_3}{|}}{Si}} - O - \underset{\underset{CH_3}{|}}{\overset{\overset{CH_3}{|}}{Si}} - CH_3$$

Fig. O3 Silicone oil

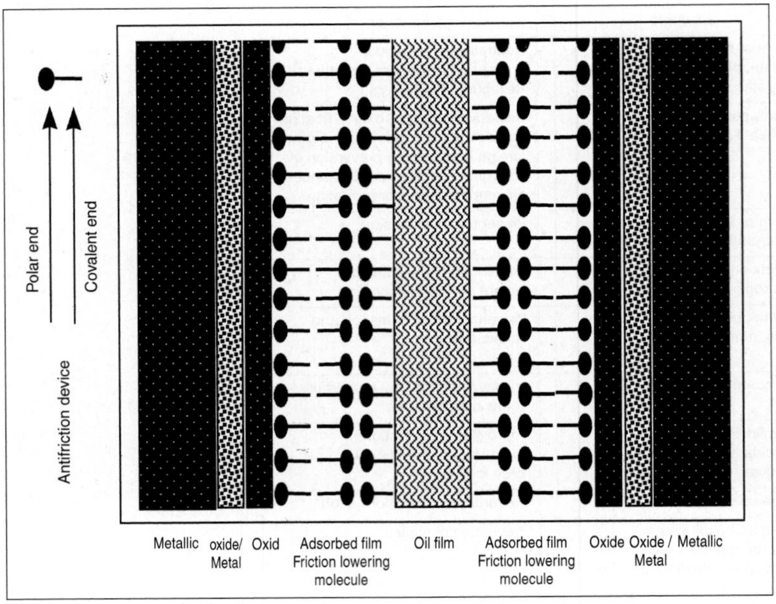

Fig. O4
Alignment of the
antifriction additive
between metal surface
and oil film, idealized

In addition to the solid substances and metallo-organic compounds mentioned under antiwear additives, organic additives can also be used for building up friction-reducing absorption and reaction layers often in the form of friction polymers. Among others, ester, saturated fatty acids, and dicarboxylic acids are suitable for this purpose.

~~~**Corrosion inhibitors.** Corrosion is an electrochemical process by which the attacked metal gets oxidized and the attacking medium is reduced. The iron and nonferrous metal surfaces in the engine must be protected against corrosion both in operation at high temperatures as well as during standstill periods. Special demands on corrosion protection are made, on the one hand, by marine diesel engines that are operated with heavy, sulfur-rich residual oils and must, as such, be protected against sulfuric acids and their derivatives; and, on the other hand, gasoline two-stroke engines at standstill, because the metal parts in the engine are protected only by a very thin oil film against the corrosive effect of moisture in combination with combustion residues.

Because nonpremium synthetic basic liquids and mineral oils only have a weak corrosion protection behavior, the oils are mixed with additives. They protect the surfaces against the corrosive attack of air and oxygen, neutralize acidic combustion products of engine oils and alleviate the aggressiveness of certain wear protection additives.

Corrosion protection additives are classified roughly as inhibitors for ferrous metals (so-called rust inhibitors) and inhibitors for nonferrous metals:

• *Rust inhibition additives*: Corrosive substances are water, metal chloride, and metal bromide as well as sulfur, nitrogen oxides, organic and inorganic acids, and oxygen. Compounds that carry a strong polar group on a long chain alkyl molecule are suitable as rust protection additives. Compounds of this type can accumulate on the metal surface and form dense hydrophobic (water repellent) protection films.

Whereas for applications in engine oils the inhibitor must also exhibit neutralization capacity in addition to effective surface protection (the entry of acidic combustion products is possible), surface protection is of decisive importance for rust inhibitors in transmission oils. The neutralization effect only plays a secondary role here.

Among the most important representatives of the rust inhibitors are 2-(1-alkenyl)diethyl succinate, the alkali and alkaline earth salts of high molecular sulfonic acid, and petrol-sulfonate (**Fig. O5**).

• *Corrosion protection additives for nonferrous metals*: The corrosion of nonferrous metals primarily concerns the effect of acids on bronze, nonferrous metals, and white metals. Such corrosion affects plain bearings (e.g., in engines and transmissions) as well as clutches and synchronization components. Corrosion protection additives form a passivation film on the metal surface. Important among the additives are zinc dithiophosphates, which act as antioxidant agents and wear protection additives.

When selecting a corrosion inhibitor, one must choose an additive that is optimally effective but that by itself does not lead to corrosion due to excessive reaction capacity. The best-known example is the corrosive effect of sulfur compounds on nonferrous metals, such as copper.

2-(1-alkenyl) diethyl succinate

$$C_{10}H_{21} - CH = CH - CH - COOH$$
$$| $$
$$CH_2 - COO - CH - CH_2 - OH$$
$$| $$
$$CH_3$$

Alkaline and alkaline earth salts of high molecular sulfonic acid

$$\left[R - (CH2)_n COO \right]_{1-2} Me$$

Petrol-sulfonates

$$R - SO_3 Na \quad (R - SO_3)_2 Me \quad R - SO_3 - Me - O - COOH$$

Fig. O5 Structure of the most important rust protection additives (Me = metal [primarily calcium or magnesium], R = alkyl aryl-remainder with molecular weights of approximately 300–400)

~~~**Detergents.** As can be seen from the definitions in **Fig. O2**, detergents and dispersing agents have different duties. The differentiation between the additive types to be classified under the terms detergents and dispersion agents is certainly often blurred or even opposite; often these terms are also used synonymously and referred to as "DD-additives" in summary. For the following discussion:

- Detergents are basic, metalliferous additives.
- Dispersion agents are defined as ash-free, thus non-metalliferous additives.

The term detergent (Latin *detergare:* cleaning, washing off) already defines the cleaning task of this class of substances, also known as "soap." Detergents have two primary functions: They should prevent or remove carbon and lacquer deposits on hot parts of the engine, and provide sufficient alkalinity to neutralize the acidic combustion products. Additionally, detergents can counteract oil aging, and, depending upon type, they contribute to friction reduction in varying degrees.

The detergents normally contain oil-soluble organo-metallic, ash-producing compounds, the so-called metallic salts. The most important representatives are metallic sulfonates, metallic alkyl phenolates, and metallic salicylate (**Fig. O6**). The metallic component is mostly calcium or magnesium; in the past barium was used, but it is rarely in use today for toxicological reasons.

The above-described neutral detergents—that is, salts of weak organic acids with calcium or magnesium—are not sufficient for the neutralization of stronger acids which are produced during the combustion of fuels with high sulfur content, such as residual oil for ship engines.

The so-called overbasic oil-soluble detergents, which contribute to more alkalinity in the additive in the form of carbonates or hydroxides, were developed for the above-described purpose and consequently used in nearly all other engine applications.

The strongly dispersing effect of neutral compounds (**Fig. O6**) enables this additional basic calcium or

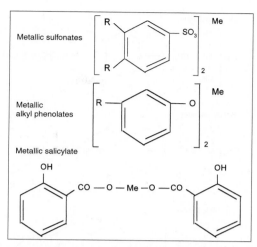

Metallic sulfonates

Metallic alkyl phenolates

Metallic salicylate

Fig. O6 Structure of typical detergents for engine oils (Me = metal [mostly calcium or magnesium], R = hydrocarbon rest)

magnesium carbonate to be provided in the finest distribution surrounded with the above-mentioned detergents as so-called micelles in oil-soluble form (**Fig. O7**). Such micelles are extremely alkaline compared to the surrounding neutral detergent. Overbasic detergents can even neutralize the acids produced in very thin oil films from the combustion process before the acids can reach the metal surface through the oil film.

Different quantities and types of detergents are utilized in oils based on the application of the oil. Because they do not burn free of ash, due to the required metal content, residue formation in the combustion chamber must be taken into consideration when they are used in gasoline engine oils. This especially applies to the frequently cold-running two-stroke boat engines,

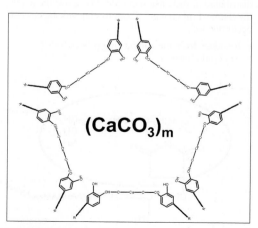

$$(CaCO_3)_m$$

Fig. O7 Calcium salicylate overalkalinized with calcium carbonate

Mono-succinimides

Bi-succinimides

PIP Mala Penta

Structures of important
dispersing agents

in which case an excessive ash content of the oil will be formed and either no or only little detergent of the above type can be used due to the danger of residue formation on the spark plugs. During application in the engine, the detergents are consumed as determined. Replenishment is possible during topping up, depending on the amount of engine oil consumed. That is why the engines that do not consume any oil or that only consume a small volume of oil contaminate the oil more severely, and this makes an earlier oil change desirable.

~~~**Dispersing agents.** Dispersing agents should primarily keep oil-insoluble foreign substances finely distributed in suspension in order to remove them during the next oil change. These foreign substances can occur due to:

- Residues from the combustion process (soot)
- Material abrasion

- Road dust
- Aging of the lubricant

The dispersion agents are often ash-free additives such as mono- and bi-succinimides or polyisobutenes — maleic acid anhydrides and combinations of the same (**Fig. O8**).

Because the combustion soot occurs primarily in very small particles (<0.5 mm), the dispersing agents can keep these particles in suspension and bond them "firmly" in the oil such that they cannot be filtered out. Even extremely fine filtering does not lead to success. These foreign substances can only be removed from the engine through an oil change. A better illustration of the structure of a typical dispersing molecule is given in **Fig. O9**.

Without dispersing agents, oil-insoluble particles would collect on the floor of the oil sump and form the so-called sludge. This sludge blocks passages in the engine and deteriorates the heat dissipation of the oil.

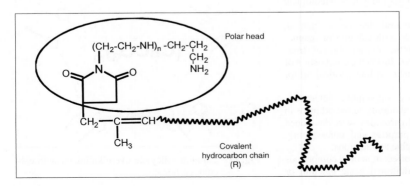

Fig. O9
Polybutene
succinimide (molecular
weight 1000–3000)

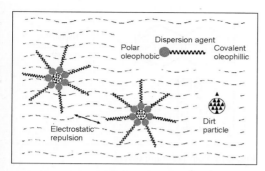

Fig. O10 Action of dispersing agents, schematically

In the presence of dispersing agents, the dirt particles are encapsulated by additive molecules so that further growth is prevented through steric inhibition. At the same time, the covalent oleophilic remainder of the dispersing agent molecule, through electrostatic repulsion, prevents the formed dispersion colloids from joining together (**Fig. O10**).

~~~**Metal/elastomer gaskets.** Metal surfaces or metal salts dissolved in oil encourage oil aging via catalytic means. Metal deactivators encapsulate metals or metal ions, which stops their undesired catalytic effect during the oil aging process. Particularly complex organic sulfur and nitrogen compounds are well suited for this. These products form a passivation film on the metal surface similar to corrosion protection additives and, thus, serve as a chemical separation from oil. Typical products are illustrated in **Fig. O11**. A further type of metal deactivators compensates for the metals already dissolved in oil.

~~~**Oxidation inhibitors.** Oxidation of lubricating oils is also known as aging. It is triggered, for example, by increased temperatures and is a chemical reaction of hydrocarbons with the oxygen in the surrounding air. This reaction is influenced by the presence of metals and certain chemical compounds which can act as catalysts.

In the oxidation process, acids finally occur after intermediate stages under the formation of organic per-

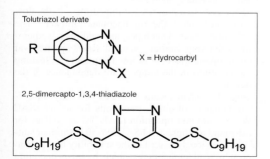

Fig. O11 Example of film-forming metal deactivators

R• + A-OH	⟶	RH + A-O•
R-OO• + A-OH	⟶	R-OOH + A-O•
R• + A-O•	⟶	A-O-R
2 A-O•	⟶	(AO)$_2$

Fig. O12 Effect of primary antioxidants (free-radical scavengers), in the example (R = rest of oil molecule, A = rest of antioxidants, • = radical)

oxides, and in advanced stages, oil-insoluble polymerisates occur which lead to lacquer and sludge-like deposits. Oxidation of oils can be avoided or reduced in a different manner. The so-called primary antioxidants interrupt the chain reaction that leads to oil aging products by intercepting the radicals (**Fig. O12**).

Structures of important primary antioxidants follow:

Phenols

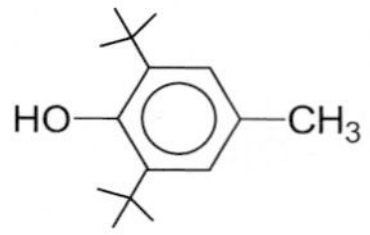

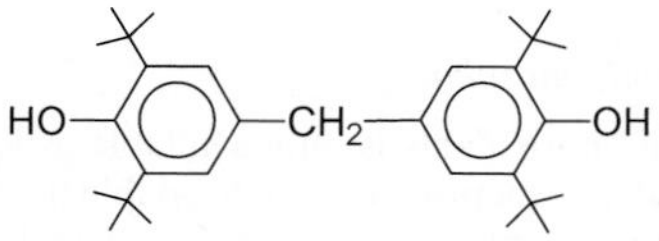

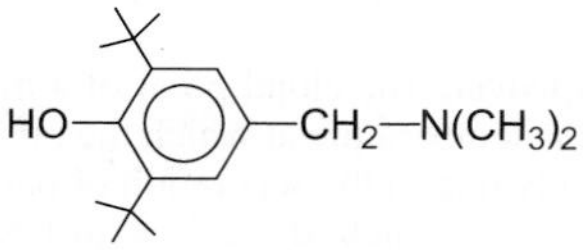

Phenols are used primarily in hydraulics, turbines, and insulating oils, but are also used in engine oils in combination with other classes of oxidation inhibitors. They are most effective at low temperatures (temperature ranges below 150°C).

Alkylated diphenylamines

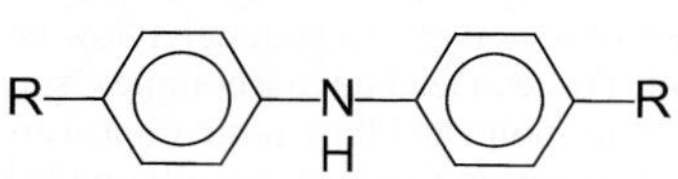

Several amines are still effective at temperatures above 150°C. They are used also in lubricating grease and synthetic diester oils for aviation.

The so-called secondary antioxidants, on the other hand, do not react with the radicals directly, but rather decompose hydroperoxides (**Fig. O13**, line 2) that are formed in the course of oil oxidation, which

(1) R-OOH $\longrightarrow$ HO• + RO•

(2) R-OOH + A-S-A' $\longrightarrow$ ROH + A-SO-A'

Fig. O13 Effect of secondary antioxidants (peroxide decomposers), in the example (R = rest of oil molecule, A = rest of antioxidants, • = radical)

would otherwise lead to further chain reactions (line 1). They are also still effective at temperatures above 150°C.

Structures of important secondary antioxidants are:

Sulfur–phosphor compounds

(R) P (S) . . . S . . . R

They find application in engines, transmissions, and hydraulic oils. The most famous representative is zinc dithiophosphate. In addition to antioxidant effects, this group has wear protection properties.

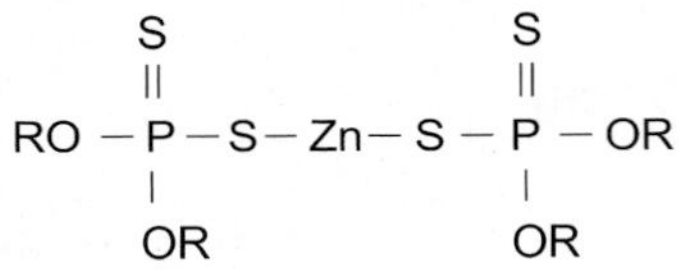

$$RO-\underset{\underset{OR}{|}}{\overset{\overset{S}{\parallel}}{P}}-S-Zn-S-\underset{\underset{OR}{|}}{\overset{\overset{S}{\parallel}}{P}}-OR$$

Dialkyl zinc dithiophosphate

The effect of antioxidants is only available as long as the antioxidants themselves are available in oil, even if in minimum quantities. After they have been consumed, the oil begins to age very quickly.

~~~**Pour point improver.** The cloud point of a mineral oil designates the temperature at which the oil begins to become cloudy due to the separation of paraffin. The temperature at which the oil is just still capable of flowing is termed the pour point. The solidification point is the temperature at which the oil is no longer capable of flowing.

During the cool-down process of lubricating oils in certain temperature ranges, crystallization and separation of paraffin occur. Individual crystals can join together to form a paraffinic structure through mutual hook-up (cross-linking). This "sponge" is capable of binding the rest of the oil and impeding its flow below the cloud point. This can lead up to a complete stagnation of the oil. Pour-point and flow-point improvers act against this stagnation in that they adsorb on the paraffin crystals at the beginning of separation and impede their cross-linkage to form dense and stable structures. More compact crystals are formed, crystals that no longer can cross-link but rather can move freely in oil.

Pour-point improvers influence neither the quantity of the precipitated crystals nor the cloud point of oil. Thus, they do not stop crystallization.

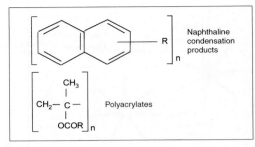

Fig. O14 Pour point improvers

The most important group of pour-point improvers is naphthalene condensation products and polymethyl acrylates (which at the same time can act as viscosity index improvers; **Fig. O14**).

~~~**Viscosity index improver ("VI improver").** Modern and high-performance engine oils for cars and trucks require full functional capability over a wide temperature range due to prolonged oil change intervals and high mobility: from winter to summer, from north to south. This means

- Sufficiently low viscosity for a safe start at low temperatures without underlay and fuel-saving lubrication in cold-winter operation in the north, and,
- At the same time, sufficiently high viscosity for wear-protecting lubrication during typical summer motorway driving in the south.

Natural viscosity–temperature behavior of oils, even that of synthetic basic oils with a higher viscosity index, is not adequate for these requirements; that is why the application of viscosity index improvers is required.

A series of long-chain polymers, in which the solubility in the oil at high temperatures is better than at low temperatures, is suitable as VI-improvers (**Fig. O15**). At low temperatures the applied long-chain polymers shrink together and increase the viscosity of basic oil to a relatively lesser extent. With increasing temperature they gradually swell out. This leads to a state in which their spatial expansion relative to that of basic oil gradually increases.

This leads to a progressive impediment to the flow of the molecules; the oil becomes "thicker." The thickening effect of such polymer additives is, therefore, relatively higher—several times greater—at high temperatures than at low temperatures, leading to a reduction in the temperature dependence of the viscosity.

Fig. O16 schematically shows the viscosity–temperature behavior of single range oils for winter (SAE 5W) and summer operation (SAE 30) as well as for year-round multigrade oil SAE 5W-30. "Summer suitability" obviously describes the suitability for application at high oil temperatures which, based on operating conditions, can obviously also occur in the winter months.

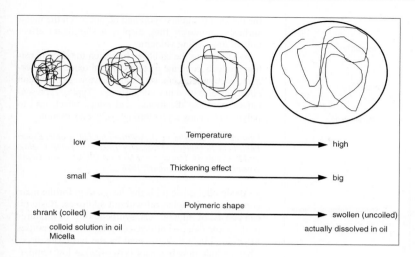

Fig. O15
Schematic diagram of
the effect of viscosity
index improvers

When selecting and metering the VI-improvers, one must pay attention to the prevention of undesired side effects. Very high dosing can lead to residue formation in the engine and in the intercooler. With the application of extremely long-chain polymers (with molecular weights >200,000), there is danger at high temperatures that the swollen and, thus, very long polymer chains may be cut at narrow gaps and edges and thereby lose their thickening effect. Through this permanent destruction of viscosity index improvers, the viscosity properties of the oil are persistently impaired, making problematic effects on engine operation unavoidable. Therefore, shear stability is a criterion in engine oil specifications. At narrow lubrication gaps, on the other hand, orderly alignment of long polymer chains can occur, whereby they lose their viscosity-improving effect temporarily. This temporary viscosity loss, however, must not be disadvantageous, because it generally does not lead to impairment of the lubrication safety but, rather, has a positive effect on fuel savings.

The most important representatives of VI-improvers are polyolefins, polyisobutene, styrene/olefins copolymers, and polyacrylates (**Fig. O17**).

~~~**Wear reducer, EP additives.** These additives find application where complex states of friction lead to mixed friction between sliding engine parts.

Here the task of additives is to form both partitioning as well as slippery layers. This prevents fusion between highly stressed lubrication points, and the wear is also reduced. This can be brought about physically (absorption), chemically (reaction), or through the additive of solid lubricants (graphite, molybdenum disulfite).

EP-additives are chemically reactive high-pressure additives—for example, organic phosphor, sulfur, and chlorine compounds—that react with the metal surface. Locally increased frictional heat and plastic deformations can lead to activation of the chemical reaction at the extremely stressed points. Mostly solid layers occur on the metal, layers which stand in chemical equilibrium through abrasion and new formation.

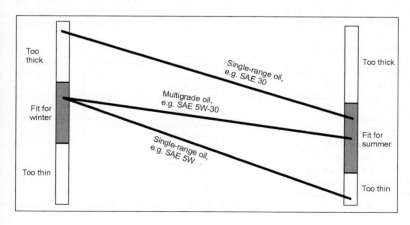

Fig. O16
Comparison of single
and multigrade oils

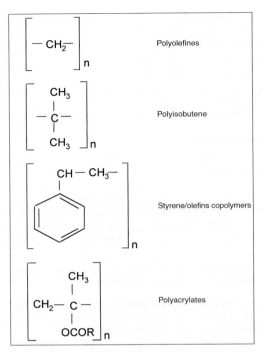

- $-CH_2-$ $\Big]_n$ Polyolefines

- CH3 | -C- | CH3 $\Big]_n$ Polyisobutene

- CH-CH3- (benzene ring) $\Big]_n$ Styrene/olefins copolymers

- CH3 | CH2-C- | OCOR $\Big]_n$ Polyacrylates

Fig. O17 Structure of the most important VI-improvers

The most important representative is zinc dialkyl dithiophosphate. Decomposition products originating from zinc dialkyl dithiophosphates are the actual components that become active as wear-protection additives, among others, and contain the following:

Disulfide, dithiophosphates $(RO2)P(S)SR$,
Thiophosphates $(RO)2P(S)OR$,
Thiophosphates $(RO)2P(S)OR$,
Phosphates $(RO)3P(O)$, $(RO)2P(O)O$—,

For an engine oil to have wear protection over the entire temperature range, a combination of different zinc dithiophosphates with different decomposition temperatures is usually required. Zinc dialkyl dithiophosphates are applied both as primary alkyls as well as in the form of secondary alkyls. Zinc diaryl dithiophosphates are also common. The application of specific components depends strongly on the purpose of the oil—that is, for gasoline or diesel engines.

Wear reducers are physically adsorptive additives. They cling to metal surfaces due to their polarity and, thus, form protective and reactive layers.

Graphite and molybdenum disulfite (MoS_2) are those primarily used as solid lubricants. They are particularly characterized by their emergency running properties in the event of failure of the oil supply system and also work at high temperatures. Solid lubricant additives in engine oils are, nonetheless, mostly problematic, because they are kept in suspension by

the dispersing agents in oil and do not reach the metal surfaces; moreover they display a significant effect only at low sliding velocities.

In addition to the additives by which the decomposition products build the protective metal oxide layers, organic friction reducers, which also have wear protection effect, are also used in some instances. Typical representatives are some dicarbonic acid esters, which lead to polymer surfacing layers through polycondensation.

Literature: W.J. Bartz, et al: Additives for Lubricants, Expert Publishers, Renningen-Malmsheim, 1995. — J. Falbe, U. Hasserodt: Catalysts, Tensides and Mineral Oil Additives, Georg Thieme Publishers, Stuttgart, 1978.

~~**Basic oils.** Basic oil is the designation for the main component of engine oil without additives. Basic oil can consist of mineral oils, synthetic oils, or mixtures of the same (semi-synthetic oils). Essential properties of engine oils are determined by the basic oils used. They include flow behavior (viscosity) at low temperature ranges and evaporation loss at high temperatures. Different basic oil types respond differently to additives. Therefore, changing basic oils or their composition, or production from crude oils of different origin and other types, would require extensive engine tests for proof of engine compatibility, even though the oils retain the same physical data.

ATIEL (Association Technique de l'Industrie Européenne des Lubrifiants) has introduced a rough classification of basic oils in five groups (**Fig. O18**).

In addition to these quoted data, an entire series of other criteria determines the suitability of a basic oil component for use as engine basic oil. These properties are discussed in the following sections.

~~~**Castor oil.** Castor oil is a vegetable oil that is obtained by pressing oil from the seeds of the herbaceous perennial castor plant from the plantation regions in Brazil and India. It is composed of 80–85% glyceride of ricinoleic acid and the rest from glycerides of other organic acids. In the past, castor oil was introduced without mixing or as a component with good lubrication properties (adhesive properties) in racing engines. With mechanically problematic engines, castor oil provided immediate lubrication certainty without piston seizure when this was no longer possible using

Group	Composition	Sulfur Content	Viscosity Index
I	<90% saturated hydrocarbons	>0.03%m	80–120
II	>90% saturated hydrocarbons	≤0.03%m	80–120
III	>90% saturated hydrocarbons	≤0.03%m	>120
IV	Polyalphaolefines		
V	all other basic oil components (e.g., ester)		

Fig. O18 ATIEL classification of basic oil components

engine oils based on mineral oil. However, disadvantages include a lack of oxidation stability and hence inadequate aging behavior.

~~~**Fully synthetic oils.** Synthetic basic oils are used in engine oils if high-performance multigrade oils with low oil consumption, little residue formation, and long oil-change intervals are required and costs do not have top priority. The synthetic basic oil components especially applicable to engine oils have, in particular, lower temperature-dependent viscosity changes and lower evaporation loss at high temperatures than comparable mineral basic oils. This is based on their chemically uniform structure. While mineral oils consist of a variety of different molecules according to their separation by distillation and subsequent refinement, synthetic components are generally composed of uniformly structured molecules depending upon the manufacturing process.

Among the variety of available synthetic oils, synthetic hydrocarbons—namely isoparaffin and poly-alpha-olefins as well as organic ester and dicarbonic acid esters—in particular are suitable as basic oil components for engine oils.

The manufacture of isoparaffin and poly-alpha-olefins requires different processes, but in either case the process starts from mineral oil. The manufacture of synthetic XHVI™ basic oil using the Shell process and the manufacture of poly-alpha-olefins is illustrated schematically in **Fig. O19**.

The feedstock for XHVI™ basic oils is paraffin: slack wax, which is obtained during deparaffinization of lubrication oils. Isoparaffin is formed through catalytic hydro-isomerization. After subsequent vacuum distillation to separate light components and MEK (methyl ethyl ketone) solvent extraction to separate unconverted long-chain normal paraffin, the basic oil component retains a very high content of isoparaffin

with a viscosity index of about 150. The viscosities of this XHVI™ oil can be set by selecting the feedstock and the process conditions.

Poly-alpha-olefins are manufactured in several process steps from ethylene, a cracking product of raw gasoline. As an intermediate product, alpha-olefins with greater chain length, preferentially 1-decene (chain-shaped hydrocarbon with 10 hydrogen atoms and double bonds between the first two hydrogen atoms) are formed. Poly-alpha-olefins are synthesized from here primarily through oligomerization (linking of several 1-decene molecules around double bonds so that the end product comprises two up to four) of these initiator (molecules). Due to linking, double bonds disappear so that isoparaffin with a different number of equally long side chains occurs. The designation "poly-alpha-olefin" strictly refers to the initial product. Subsequent aftertreatment brings about the separation and conversion of undesired components. From among the many different esters, dicarbonic acid esters have proven themselves especially useful in engine oils. They are synthesized from an organic acid, a C_{12}-dicarbonic acid, for example, and a branched propylene or butylene alcohol:

$$2\ R{-}OH + HOOC{-}(H_2C)_x{-}COOH$$

$$\rightarrow ROOC{-}(H_2C)_x{-}COOR + 2\ H_2O$$

The basis for the manufacture of organic acids is ethylene from the cracking of raw gasoline, just as it is the basis for the manufacture of poly-alpha-olefin.

In comparison with mineral oil components, dicarbonic acid esters have a high-viscosity index and low evaporation loss. They are especially suitable as substitutes for low-viscosity mineral basic oils if high demands are put on these properties, such as extreme temperature stress. The application of esters, however, results in their more pronounced inclination to corrode

O

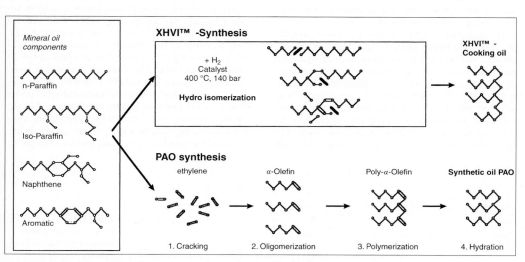

Fig. O19 Manufacture of XHVI™ basic oils and of poly-alpha-olefins by the Shell process

nonferrous metals, frequently problematic compatibility with elastomers, and possibly negative influence on piston cleanliness at high concentrations (which nonetheless are hardly considered for cost reasons). Dicarbonic acid esters have proven to be excellent basic oil components when used, for example, to improve the solubility of synthetic oils for additives.

~~~**Mineral oils.** Basic oils based on crude oil contribute the overwhelming share of all lubricants used as engine oils as well as in most areas of lubrication. This is on the one hand due to their comparatively low cost and high availability in comparison with synthetic lubricants, but also due to the excellent properties which can be attained with the modern manufacturing processes. There is also the fact that the given properties of natural components, such as low sulfur content or aromatic compounds, will already to some extent have beneficial properties such as pressure resistance or aging stability, so that the effect of the additives is assisted.

Mineral basic oils are manufactured in plants which are normally affiliated with refineries for the manufacture of fuels, heating oil, and bituminous materials. Through the selection of special crude oils, it is guaranteed that, after separation of volatile components in the atmospheric distillation plant (for fuel and heating oil production), a suitable "residue"—the so-called "long residue"—will become available for lubrication oil production. In contrast to fuels that can be manufactured in most of the modern refineries from a variety of different crude oils, special crude oils are selected for manufacturing basic oils.

Fig. O20 shows the composition of several crude oils suitable for basic oil production. Mineral basic oils are manufactured from the above-mentioned "long residue" as follows: At first oils of different boiling ranges (and, thus, different viscosities) are separated from heavy residue, mostly bituminous, in the vacuum distillation plant. The distillate goes through several stages of aftertreatment.

The manufacturing process of separation by distillation described below and subsequent refinement (extraction and/or conversion through mild hydrogen treatment) of undesired components leave the molecular structure of oil more or less unchanged. Although, in the past, both paraffinic as well as naphthenic basic oils were used for engine oils, paraffin-based oils are used today, due in particular to the low temperature dependency of viscosity and low residue formation. For this reason, only selected paraffin-based crude oils (with

Middle East	Sulfur content %	Paraffinic hydrocarbons %
Arabian light	1.7	43.5
Iranian light	1.4	43.5
Oman	0.9	45.0
Kuwait	2	47.4
Qatar Marine	1.5	37.8
North Sea Brent Blend	0.3	39.2
South America Lagomar	1.3	49.3

Fig. O20 Crude oils suitable for the manufacture of lubrication basic oils (selection)

high proportions of paraffinic hydrocarbons) are suitable for manufacturing basic oils for engine oils.

For the manufacture of engine oils, based on the desired viscosity (for multigrade oils the viscosity in low temperature range), a series of mineral basic oils are available corresponding to the type of separation by distillation, thus with different viscosities—from the thin spindle oil up to the viscous bright stock (**Fig. O21**).

A mixture of at least two basic oil components is required for adjusting the desired viscosity and other properties for the manufacture of an engine oil. The low-viscosity spindle oil is now hardly ever used, even in low-viscosity engine oils, due to the high evaporation loss. In contrast to this, the high-viscosity bright stock (viscosity above SAE 50) in low concentration makes for a stable lubrication film, even in low-viscosity oils.

Manufacture of mineral oil–based basic oils

Vacuum distillation

For the manufacture of basic oils, the residue (long residue) from atmospheric distillation is used in so-called vacuum distillation (**Fig. O22**). With vacuum distillation, the feedstock is exposed to a reduced pressure. Relative to the atmospheric pressure of 1013 mbar, the boiling temperature falls at a pressure of 10 mbar by about 150°C. The feedstock is heated in a tube furnace to between 360 and 400°C and fed into a vacuum column. The vaporous hydrocarbons under these pressure and temperature conditions rise in the vacuum column, cool down in the process, and condense according to their boiling ranges on, for example, a bubble-cup tray; the liquid portion remains on

	A	B	C	D	E	F
Viscosity at 100°C mm²/s mm²/s	3.7	5.1	7.6	12.2	22.5	32
VI	97	98	96	96	96	96
Noack evaporation loss (mm²/s mm²/s)	30	15	7	3	1	0
Flashpoint COC °C	200	215	240	265	310	300
Pour point °C	−18	−15	−9	−9	−6	−9

Fig. O21
Properties of typical mineral oil–based basic oils for engine oil manufacture

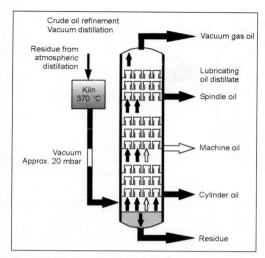

Crude oil refinement
Vacuum distillation

Residue from atmospheric distillation

Kiln 370 °C

Vacuum Approx. 20 mbar

Vacuum gas oil

Lubricating oil distillate

Spindle oil

Machine oil

Cylinder oil

Residue

Fig. O22 Vacuum distillation plant

the tray of the column. Therefore, vacuum distillation is a pure, physical separation; the structure of hydrocarbons is not changed during the distillation process.

In vacuum distillation the following products are extracted:

- Vacuum gas oil
- Spindle oil distillate
- Machine oil distillate
- Cylinder oil distillate
- Short residue (further processing, e.g., to make bituminous compounds)

Solvent refinement

With regard to solvent refinement, undesired impurities still present in the lubricating oil distillate, especially the aromatic compounds such as compounds with a sulfur content and olefins are extracted with a solvent, generally furfural. The lubrication basic oils thus manufactured are designated as solvent raffinates or solvates.

Furfural (Latin *furfur*, bran) is technically manufactured from degrained maize cobs, rice, and cereal husks. It is liquid, toxic, flammable, and has a penetrating smell. Furfural possesses a greater dissolving power relative to ring-shaped hydrocarbons than to paraffin and isoparaffin and has a higher density than mineral oil. If it is mixed with lubricating oil distillates (containing aromatic and paraffinic compounds), a heavy aromatic-rich furfural extract phase forms in the lower part of the mixing tower and in the top part of the mixing tower a lighter paraffinic oil phase with furfural content forms. Both layers are drawn off separately; furfural will be won back through evaporation and fed back into circulation. After distillation of the furfural, an aromatic extract, and the solvate, a nonaromatic raffinate with mainly paraffinic compounds

that exhibits a good viscosity–temperature characteristic is obtained. The selectivity of the process can be controlled through dwell time, temperatures, and change of the furfural amount. As such, an increase of the viscosity index (VI) can also be achieved by increasing the selectivity, obviously at the expense of the yield of solvate.

Hydro-refining

Hydro-refining has the purpose of reducing the aromatic content through hydrogenation of the aromatic compounds, sulfur reduction, and saturation of further unsaturated compounds, such as, olefins. Furthermore, the nitrogen and oxygen content is reduced (for neutralizing acidic distillates among other reasons), and an increase of the viscosity index is achieved through the increased yield of paraffin. Hydro-refining is a modern refinement method with the advantages that no exploitable residues and byproducts occur.

Through hydrogenation (hydrogen adsorption), unsaturated compounds can be converted into saturated ones. Hydrogenation involves a chemical conversion process by which the molecules change in their structure. Inside a Luboil–Hydrotreater (LHT), a high-pressure reactor, the product to be improved is mixed with hydrogen in the presence of a catalyst at a high temperature (350–370°C) and a high pressure (140–170 bar). The hydrogen saturates the aromatic compounds and other unsaturated compounds, whereby more stable hydrocarbon compounds occur. Moreover, elements — such as sulfur, nitrogen, and oxygen — that are harmful to the quality of basic oil are removed. In case of excessive hydrogen, the sulfur from sulfur compounds is converted to H_2S (hydrogen sulfide) and, thus, removed from the oil.

A clear characteristic of hydrogenated oils is that they are lighter in color than the solvates treated with solvent. This effect is not important for most lubricating oils, whereby more emphasis is placed on light color in the case of a number of the other oil types produced.

Deparaffinization

Paraffin-based raffinate and solvates contain a high proportion straight-chained paraffins of high molecular weight, which are dissolved in oil at higher temperatures, but readily crystallize out at ambient temperatures. They agglomerate in the process and bring the oil to stagnation. At low temperatures these precipitated paraffins lead to clogging of filters in lubrication systems. Removal of paraffin from the oil leads to better flow behavior and a lower solidification point or pour point.

Precipitation of undesired paraffin components from oil can be attained through simple cool-down; separation of precipitated solid paraffin through filtration, however, causes difficulties due to the high viscosity of cooled oil. Therefore, a solvent mixture — e.g., methyl ethyl ketone and toluene — is used in deparaffinization plants (**Fig. O23**) for lowering the viscosity. Methyl ethyl ketone favors the precipitation of paraffin,

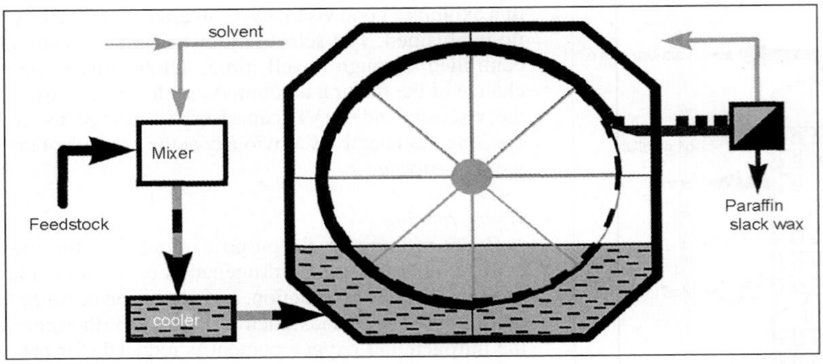

Fig. O23
Operation of a
deparaffinization plant

toluene retains the oil in the solution at the deparaffinization temperature. The oil is added to a mixture of both solvents and then cooled depending upon the type of oil. The mixture is cooled down in the so-called scraped wall chillers, consisting of double-walled tubes, in which the coolant flows in the jacket side and in the walls of the inner tubes; then scrapers driven from outside scrape the mixture, enriched with paraffin, from the wall.

Paraffin from the oil-solvent mixture is removed from the oil through filtration with a rotary drum filter. The oil-paraffin mixture is sucked through the filter fabric by means of vacuum into a chamber in the rotary drum, whereby the paraffin extract is precipitated on the fabric. This extract is then released into one of the chambers through excess pressure. The paraffin content separated from the oil-solvent mixture through filtration is called slack wax.

It can serve as feedstock for the manufacture of synthetic basic oils or also be processed further to make paraffin waxes. By washing the filtered-out paraffin with solvents to remove the oil still contained in it and by renewed crystallization, oil-free paraffin can be produced. This is used as a primary source product in the petrochemical industry and is used there for the manufacture of candles, floor polish, and so on. Paraffin wax is also used as an insulation material and in the manufacture of paraffin-treated paper.

The solvent is won back from the filtrate (deparaffinized oil) and from slack wax by distillation.

The target product is deparaffinized, solvent-free lubrication oil fraction. These deparaffinized oils exhibit a substantially lower pour point and a somewhat lower viscosity index than the feedstock.

Fig. O24 shows an example of the change of the product properties through solvent refinement and deparaffinization.

~~~**Semi-synthetic oils.** Engine oils in which the basic oil component consists of a mixture of mineral oils and synthetic oils are often designated and marketed as semi-synthetic. This approach is reasonable both technically and economically for oils developed for medium loads as far as the manufacturer is successful in achieving a good compromise between the advantageous technical properties of fully synthetic basic oils and the cost-effective mineral oils. For instance, through partial application of synthetic components the evaporation behavior (e.g., evaporation loss; oil consumption) of engine oils can be improved.

~~**Changing from mineral to synthetic oils.** Mineral oils and synthetic oils are basically mixable (semi-synthetic oils). For this reason, especially during oil change, in which only small amounts of residues from

	Distillate	Solvate	Deparaffinized Oil Solvate	
Viscosity at 50°C	120	49	68	mm²/s
VI	Approx. 60	Approx. 100	95	
Density at 15°C	0.94	0.88	0.88	g/ml
Flashpoint COC	270	270	270	°C
Pour point	+40	+47	−12	°C
Overall aromatic content	Approx. 50–60	Approx. 20–22	24	%

Fig. O24
Effect of solvent
refinement and
deparaffinization

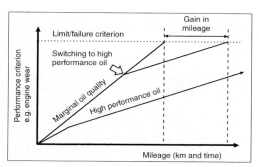

Fig. O25 Improving the engine's service life, schematically

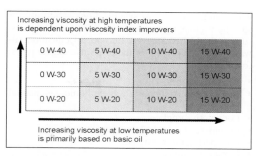

Increasing viscosity at high temperatures is dependent upon viscosity index improvers			
0 W-40	5 W-40	10 W-40	15 W-40
0 W-30	5 W-30	10 W-30	15 W-30
0 W-20	5 W-20	10 W-20	15 W-20

Increasing viscosity at low temperatures is primarily based on basic oil

Fig. O26 Setting the viscosity class

the previous oil remain inside the engine, changing from mineral-based to synthetic oil presents no problems. Generally speaking, it is advisable to change to higher-quality engine oils, irrespective of the vehicle's age and the state of engine wear, because from the time of the change onward the wear rate and residue deposit levels will be reduced due to the improved oil quality. **Fig. O25** shows the influence of two engine oils of very different quality classes on, for example, the engine wear in a schematic and simplified manner. With a new engine the wear is obviously zero, and with increased mileage it rises, depending on the type of oil quality used, at different rates. Influenced by many parameters (e.g., engine design, driving conditions, fuel quality, etc.), the wear can become so great that engine failure may occur. If all other conditions are assumed to be the same, this limit will obviously be reached much later with a high-performance oil providing very good wear protection and dispersion agent/detergent performance.

Therefore, consistent application of very good oils is the ideal way to retain engine value, but also changing over from an engine oil with only marginal wear protection will lead to a significant extension of operating life. Naturally, wear that has already occurred cannot be reversed, but the significantly reduced wear increase after the change to a high-performance oil can lead to a significant gain in mileage.

Even when refilling engine oil between the oil changes, there are no incompatibilities between mineral oils and synthetic oils. Here the general rule applies that the combination of different additive systems does not lead to the same good result that an optimized additive system alone achieves. In any case, it is better, if the oil in the engine is not available to refill, to use another oil of comparable quality standard than to dispense with refilling entirely.

~~**Production/mixing oil.** The decisive factors influencing the quality of engine oil include both the properties of the basic oils and those of the additives. For example, **Fig. O26** shows how the viscosity level of a fully formulated engine oil is influenced both by the

viscosity of basic oil components as well as by the properties of the additives—here, the viscosity index improvers.

Engine oils are manufactured and filled according to their specified formulation in special mixing facilities (**Fig. O27**). A description of the manufacturing process and discussion of its many and diverse details would exceed the scope of this lexicon. Reference will be made here only to the importance of production and final quality control prior to release of the products, generally based on chemical analysis. For instance, because of the "sensitivity" of the oils to contamination (e.g., with the liquids for automatic transmission), the final release can only be done after analysis of a sample taken directly from the delivering tanker.

~**Classification, specifications, and quality requirements.** There are a great many different specifications for the quality and performance characteristics of engine oils, reflecting the wide variety of application conditions. These specifications define minimum requirements for the quality of engine oils. They are in some cases formulated in cooperation with engine manufacturers, oil manufacturers, and relevant consumer organizations and continuously adjusted to new developments. There are specifications with world-

Fig. O27 Mixing boiler with feed lines for basic oils and additives

wide validity as well as specifications valid only for specific regions with specific application conditions, such as the United States or Europe. Specifications always refer to oil types for the respective engine types, for example,

- Aircraft engines
- Gas engines (stationary, vehicle liquid gas, natural gas)
- Locomotive engines
- Commercial vehicle diesel engines
- Car spark ignition engines
- Car diesel engines
- Marine engines
- Two-stroke SI engines (motorcycles, chain saws, outboards)

In addition there are also special oil specifications for special application areas—for instance, for methanol engines or hydrogen engines. These are, however, only of limited general interest because there is still little or no commercial application of engines using these types of fuels.

The specifications describe physical properties of engine oils, such as viscosity, evaporative loss, and shear stability; performance characteristics in engine tests, such as wear protection and cleanliness; and in some cases influence on fuel consumption as well as the changes observed in the engine oil during operation, such as viscosity changes (thickening).

The most famous specification is the SAE viscosity specification for engine oils (*see below*, ~Properties and characteristic values ~~ Viscosity classes for engine oils). The performance characteristic of engine oils is determined in the classifications of API (American Petroleum Institute) and ACEA (Association des Constructeurs Européens d'Automobiles) and in specifications of military consumers as well as, especially, by individual renowned engine manufacturers like BMW, DaimlerChrysler, VW, and Volvo. The variety of tests adapted to different engine designs, materials, and operating conditions makes it impossible to formulate oils that satisfy all specifications, because requirements for additive effectiveness differ in some cases. In addition to meeting the classifications, many manufacturers require additional proofs for the performance capability of an oil type—for example, successful testing of newly developed engine oils in practical operation. This reflects the fact that the development of new engine oils takes several years. In the following passage, an overview is provided of the most important requirements and terms. There are also special oils for particular application areas such as motor sports, the requirements for which are also discussed in this section.

~~Classifications. A number of classification systems exist; some of these are listed in the following sections.

~~~ACEA classifications. The ACEA (Association des Constructeurs Européens d'Automobiles) was founded in March 1991 as the successor organization for CCMC that initially took over the engine oil specifications from CCMC in unrevised versions under the old designations.

On January 1, 1996, the CCMC classifications were replaced by revised ACEA classifications that at present represent the most current standards for European engines by a European association of automotive manufacturers. The new ACEA classifications define minimum requirements for vehicle engine oils, which must be verified in both chemical and physical laboratory tests and in test-stand tests in modern engines. A link exists between American engine oil classifications and ACEA classifications in that American engine tests are prescribed in the ACEA test sequences. The designation structure for new specifications is similar to the CCMC classifications. It comprises a character and a digit as well as the year of publication (e.g., ACEA B2-96).

ACEA classifications are listed with the respective engine oil types.

~~~API classification. The first classification was introduced in 1947 by the American Petroleum Institute (API) for the then large passenger car market in the United States. This classification for car engine oils for decades was "the measure of all things" worldwide in terms of the quality of engine oil. Since that time, a number of revisions have been introduced.

A distinction is made between two classes of oil: the "S" class oils, where S stands for service, used in gasoline engine oils, and the "C" class oils, where C stands for commercial, is the identification character for diesel engine oils (for commercial vehicle operation).

The API classification system is open-ended, that is, new classes with more stringent requirements can be added if necessary. This is then done using the next letter in the alphabet.

API classes were the dominant quality scale for engine oils worldwide up to the beginning of the 1980s, and they are still important today in Europe and particularly in Germany, although they no longer set the trends. When an oil type satisfies "only" the API requirements, it will no longer meet the requirements of modern European engines, because API classes are adapted to the requirements for American engines and traffic circumstances. For instance, the speed limit in the United States (88 km/h) results in thermal stress on American car engines that is lower than in their European counterparts. This has led to significant design differences. The higher demands made on oils by European engines (with smaller swept volume, higher compression, higher rotational speeds, higher output, higher average speed, and, above all, top speed) require an oil type with properties different from those prescribed in the API classes.

Engine oils are classified according to the requirements they have to meet under different operating conditions, and according to their design features. An oil type is qualified for a certain API class by fulfilling standardized laboratory and engine tests in which its different properties are determined.

CCMC G1	This defines minimum requirements for engine oils in cars with gasoline engines. The quality level corresponds to API-SE, complemented by engine tests in three European engines. *Invalid since April 1989.*
CCMC G2	In contrast to G1, it has been raised to the API-SF quality level; it also includes engine tests in three European engines. Applies to conventional engine oils, not low-viscosity ones. *Was replaced in 1989 by CCMC G4.*
CCMC G3	Applies to low-viscosity oils; the quality level is above that of API-SF. Includes engine tests in three European engines. *Was replaced in 1989 by CCMC G5.*
CCMC G4	Comparison with API classification feasible in part only (G4 exceeds SF and fulfills SG to a large extent); G4 represents conventional multigrade oils. It has higher requirements than previous ones with respect to evaporation losses, elastomer compatibility, antifoaming, resistance to aging, residue formation at high temperatures including "piston ring seizure," cold sludge formation, and wear in the valve drive area.
CCMC G5	G5 is the CCMC class for low-viscosity oils. All the higher requirements mentioned under G4 obviously also apply to G5. Furthermore, G5 oils must pass a test comprising 30 cycles in a Bosch diesel injection pump at 100°C without significant viscosity loss. The requirements regarding evaporation losses are stricter than in G4. In G5 it is required that a low-viscosity oil lose a maximum of 13% of its mass due to the evaporation of lighter components after one hour at 250°C.

Fig. O28
CCMC classifications
for gasoline engine oils

API classifications are listed with respective engine oil types.

~~~CCMC classifications. As already described for API classes, the requirements for the American and European engines have been continuously drifting apart since the end of the 1970s. Oils in compliance with API and MIL have become less and less qualified for use in European engines.

The European automobile manufacturers, therefore, developed a series of tests for testing oils for engines of European origin; this was established by the Comité des Constructeurs d'Automobiles du Marché Commun (CCMC) in cooperation with the Coordinating European Council (CEC), to which both the automobile and the mineral oil industries belong. CCMC classifications have since been replaced by ACEA classifications.

With the CCMC classifications, the character G stands for gasoline, D for diesel, and PD for passenger car diesel. **Figs. O28 and O29** show CCMC classifications for gasoline and diesel engine oils, which were valid from January 1, 1990, to January 1, 1996.

~~~DHD-1 emission standards. The worldwide diesel engine oil specification Global DHD-1 has been valid since 2001. It was formulated jointly by ACEA, EMA (Engines Manufacturers Association, USA) and JAMA (association of Japanese Automobile Manufacturers). DHD-1 was developed to keep up with the trends in worldwide distribution of engine technologies.

~~~MIL classification. Since 1941, the US Army has published specifications for engine oils that are continuously adapted to new developments. The term "HD" (for heavy duty) appeared in the first engine oil speci-

CCMC D1	This defines the minimum requirements for diesel engine oils for use with light commercial vehicles using naturally aspirated diesel engines. The quality level is the same as API-CC, supplemented by two German diesel engine tests. *CCMC D1 has been invalid since 1989.*
CCMC D2	Defines engine oils for commercial vehicles with naturally aspirated and turbocharged diesel engines; quality level corresponds to API-CD, likewise supplemented by two German diesel engine tests. *Became invalid in 1989 and replaced by CCMC D4.*
CCMC D3	Describes high-performance oils for commercial vehicles with supercharged diesel engines and prolonged oil change intervals. Quality level clearly higher than API-CD owing to two additional tests in German diesel engines. *Was withdrawn in 1989 and replaced by CCMC D5.*
CCMC D4	Class D4 goes beyond the requirements for CD and CE. Fulfills the engine requirements of Mercedes Benz operating materials regulation, Sheets 227.0 and 227.1.
CCMC D5	Class D5 lists only requirements for oils for commercial diesel engines. No API class with similarly high requirements exists. A D5 oil type also fulfills the engine requirements of Mercedes Benz operating materials regulation, Sheet 228.3.
CCMC PD1	For passenger car diesel engines including turbocharged diesels. Quality level corresponds to API-CD/SE. Contains the same requirements as those demanded by Mercedes Benz Sheet 227.0/1.
CCMC PD2	PD2 is the class for passenger car diesel engine oils including turbocharged diesels. PD2 also comprises engine tests for naturally aspirated passenger cars and turbocharged diesel engines and it can be fulfilled only by high-performance multigrade oils.

Fig. O29
CCMC classifications
for diesel engine oils

MIL-L-46152 A up to MIL-L-46152 E	These military specifications have now been withdrawn without replacement. Engine oils that qualified according to this specification were suitable for use in American gasoline and diesel engines. The last valid specification MIL-L-46152 E, which was replaced in 1991, corresponded in terms of performance characteristics to an oil of type API SG/CC.
MIL-L-2104 C	This specification classifies high-premium engine oils for gasoline engines as well as naturally aspirated and turbocharged diesel engines from US manufacturers.
MIL-L-2104 D	This specification covers the requirements of MIL-L-2104 C and contains an additional engine test in a highly charged Detroit two-stroke diesel engine; moreover, the requirements of Allison C-3 and Caterpillar TO-2 are also covered.
MIL-L-2104 E	Contains the requirements according to MIL-L-2104 D; the gasoline engine tests are updated and include more stringent testing procedures (Seq. III E/Seq. VE).

Fig. O30
MIL specifications

fication issued by the US Army in 1941 and designates the transition from the pure mineral oils used until that time — without additives (nonpremium oils) — to the premium engine oils which could only meet the requirements of the US Army with the help of additives. Although directed only to the military sector, MIL specifications (American military specifications) have found worldwide application alongside the API classification in the civilian sector.

Military authorities demand that engine oils fulfill certain specifications which contain, in addition to physical and chemical data, standardized engine tests. **Fig. O30** shows the MIL specifications valid today as well as several that have since been replaced and a few that are still applicable. Whereas the first ones were limited to tests in diesel engines, the specifications that followed included tests in gasoline engines. MIL specifications quickly acquired international importance beyond the borders of the United States — their continuously updated editions accounted for the continually rising quality requirements for engine oils due to the development of more stringent standards and inclusion of newer engine tests.

They are still used today by engine manufacturers and consumers, in their most recent valid version, to describe the required or desired engine oil quality. Nevertheless, their main importance today, as in the beginning, is for commercial vehicle diesel engine oils of American manufacture. The importance of MIL specifications to the German market has dropped off sharply.

~~Diesel engine oils

~~~**Car diesel engines.** Whereas in Europe passenger cars powered by diesel engines have become important, the proportion of passenger cars with diesel engines is still very low in the United States. This is why there is no classification for oils for passenger car diesel engines in the United States. In Europe, the development of engines pushed ahead in parallel with the development of special passenger car diesel engine oils.

B1-02	Fuel-economy, low-viscosity engine oils Changed requirements compared with B2-98 for HTHS — viscosity (>2.9 mPas) Proof of a defined fuel-consumption advantage in M111FE. Oils of this type are normally unsuitable for engines not expressly designed and approved for them.
B2-98 Issue 2	STANDARD engine oils Requirements in: • Evaporation loss • HSHT viscosity (~3.5 mPas) • Cleanliness • Wear • Oil thickening • Oil consumption
B3-98 Issue 2	PREMIUM engine oil Enhanced requirements in the items: • Evaporation loss • HSHT viscosity (~3.5 mPas) • Cleanliness • Wear • Oil thickening • Oil consumption
B4-02	STANDARD engine oils Additional requirements compared with B2 in a direct-injection diesel engine car, in piston cleanliness.
B5-02	Low-viscosity engine oils with long-drain properties to improve fuel-economy; compared with B1-98 Changed requirements for HTHS viscosity (>2.9 and <3.5 mPas) Proof of a defined fuel-consumption advantage in M111 FE. Enhanced requirements regarding soot carrying-capacity and piston cleanliness. Due to their viscometric properties, oils of this type are normally unsuitable for engines not expressly designed and approved for them.

Fig. O31
ACEA quality classes for passenger car, diesel engine oils

Test Method (engine)	Test Conditions	Test for
VW 1431 TC/IC	Hot	Piston cleanliness Piston ring seizure
PSA XUD11 BTE	Hot	Piston cleanliness, oil thickening Dispersing capacity
OM602A	Cold/hot	Wear Cleanliness Oil oxidation Oil consumption
VW DI	Hot	Piston cleanliness Ring seizure
M111	Cold/warm	Fuel consumption

Fig. O32 ACEA engine tests for passenger car diesel engines

In contrast to commercial diesel engines, the high engine speed under heavier load in the valve train and low temperature operation under city traffic conditions in particular demand an additive formulation that is different from that applied to diesel engine oils for commercial vehicles.

In oils for passenger cars, a high level of dispersion capacity for soot particles is required without the oil thickening too much. The aim is to achieve oil change intervals equal to those of gasoline engines, even though the oil volume based on engine power is smaller than in commercial vehicles. **Fig. O31** gives an overview of the ACEA quality classes for car diesel engine oils. The physical characteristic data of fresh oils are specified by the ACEA according to the same criteria as for gasoline engine oils (**Fig. O32**). Four special, standardized car diesel engines have been designated for the tests in the engine. Only the influence on fuel consumption is investigated in the test engine designated for gasoline engine oils.

~~~**Commercial vehicles.** Diesel engines with direct injection, exhaust turbocharger, and intercooler for greater efficiency have established themselves worldwide in commercial vehicles. The operating conditions cover a wide range, including mainly low-speed engines that operate under variable load for vehicles used by the municipal waste disposal companies, fixed route buses in city traffic, and continuous operation at high

API CA	Oils for gasoline and self-priming diesel engines per MIL-L-2104 A. Suitable for engines manufactured up into the 1950s.
API CB	Oils for gasoline and self-priming diesel engines since 1949. Offer protection against high-temperature deposits and bearing corrosion.
API CC	Oils per MIL-L 2104 B for self-priming diesel engines. Protect against low-temperature sludge, rust, corrosion, and high-temperature deposits. Beginning in 1961.
API CD	Oils for turbocharged diesel engines with improved protection against wear and deposits if different diesel fuels are used. Cover the requirements of Caterpillar Series 3. For engines beginning in 1955.
API CD II	In addition to CD, the requirements for two-stroke diesel engines of American design are covered; they provide enhanced protection against wear and deposits.
API CE	Engine oils for highly turbocharged, commercial vehicle engines of American design manufactured since 1983 and operated under changing conditions. Improved protection against oil thickening and wear, control of oil consumption, and improved piston cleanliness.
API CF 4	In contrast to CE oils, these exhibit improved performance characteristics in American high-performance diesel engines based on a new engine test (CAT IK) and the higher requirements in the Cummins NTC 400 test. Beginning in 1990.
API CF	Valid since 1994. Meant for indirect-injection diesel engines and others that can use a wider range of fuels (also fuels with sulfur content >0.5%). Covers the requirements of API CD.
API CF-2	Valid since 1994. For two-stroke diesel engines. Cover the requirements of API CD-II; do not cover the requirements of API CF or API CF-4.
API CG-4	Valid since 1994. For highly loaded diesel engines. For quick-running, four-stroke diesel engines under heavy operation on the road (sulfur content in the fuel <0.5%) and off-road vehicles (sulfur content in the fuel >0.5%). These oils are especially suitable for engines that are designed to comply with the US exhaust emission standards of 1994. Cover the requirements of API CD, API CE, and API CF-4.
API CH-4	Valid since 1998. For highly loaded diesel engines. More stringent requirements than API CG-4; tailored to current American commercial vehicle engines.
API CI-4	Valid beginning in 2002. For heavily loaded diesel engines that fulfill rigorous emission limit values, e.g., by means of exhaust gas recirculation.

Fig. O33
API classification for commercial vehicle engine oils

763

E1-96	STANDARD OILS for naturally aspirated engines with normal requirements (Invalid with the introduction of ACEA E5-99) (Invalid with the introduction of ACEA E5-99)
E2-96 Issue 4	STANDARD—European commercial vehicle oils for nonturbocharged and turbocharged engines at normal oil-change intervals
E3-96 Issue 4	PREMIUM—European commercial vehicle engine oils, with enhanced demands as to: • Bore polishing • Piston cleanliness • Cylinder barrel wear • Sludge-carrying capacity • Oil consumption • Viscosity increase in case of high soot content in oil, comparable with Mercedes Benz Sheet 228.3
E4-99 Issue 2	PREMIUM—European commercial vehicle engine oils cancelled on the basis of E3-96 test OM 364 A and replaced with OM 441 LA, so that higher requirements are met: • Wear • Piston cleanliness • Turbocharger residues All oils approved according to MB Sheet 228.5 and MAN standard 3277 fulfill ACEA E4-98.
E5-02	STANDARD/PREMIUM—Engine oils for combined requirements of American and European commercial vehicle diesel engines. Reduced requirements relative to ACEA E4-98. Engine requirements for Mack T8, Mack T9, and Cummins M11 tests were taken over from API-CH-4.

Fig. O34 ACEA classifications for commercial vehicle diesel engines

speeds and heavy loads in long-haul traffic. Due to the technical framework conditions derived from these conditions, from the economic viability of operation, and, obviously from environmental compatibility, special requirements for commercial vehicle engine oils come into play which distinguish them from passenger car oils. The focus of the quality requirements is as follows:

- High level of wear protection for cylinder liners, at the same time avoiding bore polishing
- High capacity of detergent activity for long-term piston cleanliness

- High capacity for dispersing soot to produce less oil thickening during operation, thus guaranteeing low fuel consumption over the entire period between oil changes. This particularly applies to low-emission engines with delayed injection
- High additive content to achieve long oil-change intervals
- Low levels of residue formation in turbochargers and intercoolers

These quality features are especially fulfilled by oils that conform to the highest API and ACEA quality classes (**Figs. O33–O36**).

DHD-1 emission standards

The worldwide diesel engine oil classification Global DHD-1 has been in effect since 2001. It was formulated jointly by ACEA, EMA (Engine Manufacturers Association, USA) and JAMA (Association of Japanese Automobile Manufacturers). DHD-1 was developed to keep pace with the trends in the worldwide distribution of engine technologies. Exhaust requirements are likewise being standardized globally, and these provide a further reason for defining a high standard engine oil type with worldwide availability. **Fig. O37** illustrates the constantly growing requirements as defined by emissions legislation in the regions of North America, Japan, and Europe.

~~~**Marine (ship) engines.** The requirements that marine engine oils must meet depend on engine type and the fuel used. Slow-running, crosshead engines are basically lubricated with two different types of oil. High-alkaline oils are used for cylinder lubrication and low-alkaline oils for the drive gear. Also, in medium-speed and high-speed plunger engines, the quality of oil is based on the fuel used. In a number of engine types, just as in the crosshead engines, additional cylinder lubrication is applied. The drive gear oils for crosshead engines and engine oils for high-speed engines are also used for lubricating the transmission and propeller shaft, if necessary.

Cylinder lubricating oils in the first place must ensure that sulfuric acid created during the combustion

API	ACEA	Test Purpose	Testing Method
X	X	Optimum viscosity	SAE viscosity classification
X	X	Prevention of oil thickening due to loss of lighter oil components	Noack evaporation loss at 250°C
X	X	Reliable oil supply by limitation of foaming	Foaming/anti-foaming at 24, 94, and 150°C
X	X	Viscosity loss through shearing	Shear stability: Viscosity loss at 100°C after shearing
	X	Viscosity loss through shearing high-temperature characteristic	Viscosity at high-shear gradient and at high temperature
X		Corrosion protection	Corrosion protection for nonferrous metals
	X	Compatibility with sealing materials	Swelling tests with standardized sealing materials
	X	Prevention of combustion chamber deposits	Sulfate ash

Fig. O35
Overview of physical characteristic data of engine oils.

API	ACEA	Test Method (engine)	Test Conditions	Test for
X		L38	Hot	Bearing corrosion; oil oxidation
X		Sequence IIIE	Hot	Oil oxidation
X		Caterpillar 1 K/1 N/1 R	Hot	Piston cleanliness; wear pattern; oil consumption
X	X	Cummins M11		Valve drive wear; sludge
X		Cummins M11 EGR		Soot/EGR indicated valve drive wear
X	X	Mack T8	Hot	Dispersing capacity
X	X	Mack T9		Cylinder wear
X		Mack T10	With EGR	Cylinder wear; bearing corrosion
X		GM 6.2 Liter	–	Valve drive wear
	X	OM 364LA	Hot	Piston cleanliness Cylinder wear Sludge Oil consumption
	X	OM 602A	Cold/Hot	Wear Cleanliness Oil oxidation Oil consumption
	X	OM 441LA	Hot	Piston cleanliness Cylinder wear Turbocharger cleanliness

Fig. O36
Engine tests for oil specification for commercial vehicle diesel engines (selection)

of residue oils is neutralized. The oils thus protect the cylinders and pistons from corrosion. With declining fuel quality—that is, rising sulfur content—higher alkalinity is required in the engine oil. In addition, a high level of detergency is required to prevent the buildup of residues, provide aging stability, and give a low coefficient of friction. In the past, naphthalene-based raffinates were generally used as basic oils because they form readily soluble, soft residues. Such nonpremium cylinder oils were no longer sufficient for lubrication when residual fuel oils with high sulfur content came into use. Engine oils with correspondingly high-alkaline detergents were not available. The required high level of alkalinity was achieved with so-called two-phase oils—an emulsion of oil and water with alkaline substances. Well-known stability problems with such emulsions led to the development of highly alkaline oil-soluble detergents, so that, beginning in 1959, single-phase oils with higher reserves of alkalinity were available. The viscosities of cylinder oils correspond approximately to the viscosity class

Region	Year	Tested Accord	NO$_x$ Limit Value	PM Limit Value
United States	1990	None	6.0 g/bhph	0.6 g/bhph
	1998	None	4.0 g/bhph	0.1 g/bhph
Europe	1992	ESE R-49	8.0 g/kWh	0.612 g/kWh
	2000	ESC ELR	5.0 g/kWh	0.1 g/kWh
Japan	1994	13-step test	6 g/kWh (DI)	0.96 g/kWh
			5 g/kWh (IDI)	
	1997–1999	13-step test	4.5 g/kWh	0.25 g/kWh

Fig. O37 Requirements of emission legislation (International)

SAE 50. Over-basic calcium, barium, or magnesium salicylate, sulfonate, or phenates are used as additives.

With this additive technology, a very high oil-soluble alkalinity is attained with total base number (TBN) values of 70 up to 100 mg KOH/g. The storage stability and homogeneity of basic oil-additive mixtures could clearly be improved through the development of the above-mentioned additives.

In the past, viscosity class SAE 30 mineral oils, without additive content, were used initially as the drive gear oils in crosshead engines. In the meantime, however, they have also been improved with additives, especially in the form of oxidation and corrosion inhibitors. Special demands on drive gear oils come from oil-servicing systems, centrifuges, and separators. Because the addition of water in oil has proven itself for centrifuging foreign and aging substances, a good demulsification capacity is demanded of the engine oils. At the same time it must be ensured that the additives remain in the oil phase and are not washed away.

The engine oils for medium-speed plunger engines must have high alkalinity with TBN values in the range of 30 to 40 mg KOH/g if they are operated with heavy residual fuel oil. For additional cylinder lubrication, the same quality of oil is frequently used for both cylinder lubrication and drive gear lubrication. Technically speaking, however, nonpremium oils are also used for drive gear lubrication if possible. The oils must exhibit a good demulsification capacity, just as the drive gear oils for crosshead engines do. Mostly, paraffin-based solvates in the SAE viscosity classes 30 or 40 range find application as basic oils.

When higher-value distillates (gas oils) are used as fuel, engine oils with a lower alkalinity reserve and with TBN base numbers ranging from about 10 to 20 are adequate. Such oils are also designed for lubricating high-speed auxiliary engines, transmissions, the propeller shaft, and other auxiliaries. It thus turns out that, in addition to the demands made on engine oils, a high load-bearing capacity is also a quality criterion.

Typical quality data of oils for marine engines are shown in **Fig. O38**. The development of new marine diesel engine oils is expensive and protracted due to the size of experimental engines and the long-term testing required. Prior to engine testing on the dynamometer, a series of special laboratory tests must be passed. These are, among others, tests of the corrosion and oxidation properties, water separation during centrifuging, load-bearing capacity under drivetrain strain, the "caterpillar" coefficient of friction, and the swelling of sealing materials. After successful testing in the laboratory, expensive testing follows, partly in large engines on the dynamometer; finally, long-term testing that can take several years begins in practical operation, involving both engine manufacturers and ship owners.

~~Gasoline engine oils, four-stroke. Oils for vehicles operated with standard gasoline fuel—that is, commercially available gasoline—are considered in this section. Oils for four-stroke gasoline engines operating with other fuels—for example, gas engines, aviation engines, and so on—is described under →Oil ~Classification, specifications and quality requirements ~~Special engine oils.

Although motorcycle and car engines present many similar requirements for the engine oil, there are also many clear differences which require specific formulation.

~~~Car. The majority of cars are powered by gasoline engines. The specifications for gasoline engines have, therefore, attained a special importance. While primary attention was paid to resistance to high temperatures over a long period, in spite of the cold start capacity of the oils, quality aspects at a lower oil temperature have become more important following the increase in traffic density and the growing proportion of lower-speed driving. Parallel to this, environmental protection and cost reduction for vehicle operators have had clear effects on the development of quality requirements. This means that there is a special class

of "fuel-economy" oils and all gasoline engine oils have had to be adjusted for catalytic converter operation. During the stoichiometric combustion process during catalytic converter operation, a high content of nitrogen oxides is produced in the combustion chamber. The engine oils must be especially protected against chemical attack by oxides of nitrogen (NO_x).

The car engine oil classifications of API and ACEA are shown in **Figs. O39 and O40**.

The properties specified in **Fig. O41** are mainly for fresh oils. Substantial differences exist between current European requirements according to ACEA and the former requirements according to CCMC. **Fig. O42 and O43** once again show a comparison of different engine tests for gasoline engine oils in cars.

~~~Motorcycle. Motorcycle engines present special requirements for engine oils in comparison with car engines due to their high specific power, higher piston speeds, and integrated transmissions and, above all, due to the engine-oil-moistened "wet" clutches. The high specific engine power requires high-quality oils. Because of the integrated transmission, the oil must exhibit particularly high shearing stability. This is achieved through the application of synthetic basic oils with a high viscosity index, restriction of the multigrade characteristic to the required range (e.g., 10W-40 or 20W-40), and hence limitation of the required quantity of viscosity index improvers. A requirement of wet clutches is that they not slip or stick; therefore, special care is needed when selecting the additives in order to insure smooth operation—it is necessary to more or less dispense with friction reducers.

~~Gasoline engine oils, two-stroke. The design principle of a conventional two-stroke gasoline engine with precompression of the air-fuel mixture in the crankcase and charge exchange in the cylinder through the transfer port and exhaust port requires a different lubrication system compared to the four-stroke engine. Because forced-feed circulation lubrication, with spraying of the cylinder liners, is not possible from the crankcase due to the gas movement there, engine oil in low concentration is mixed in advance with the fuel or added to the air-fuel mixture based on the load and engine speed (loss lubrication).

Oil type	Application	Viscosity (SAE class)	Alkalinity Reserve (TBN, mg KOH/g)
Cylinder oil	Cylinder lubrication of crosshead engines in heavy oil operation	50	70–100
Drive gear oil	Circulation oil low alkalinity for crosshead engines, transmission, and auxiliary aggregates except engines	30	5
Engine oils, high alkalinity	Plunger engines in heavy oil operation	30 or 40	30 to 40
Engine oils, medium alkalinity	Plunger engines in gas oil operation, transmission, auxiliary aggregates	30 or 40	9 to 18

Fig. O38
Typical data of marine engine oils

API Classifications of Car Gasoline Engine Oils	
API-SA	Nonpremium mineral oils; antifoaming agents and improvers of solidification point may be part of the content.
API-SB	Oils with low wear, aging, and corrosion protection additives. Since 1930.
API-SC	Oils with enhanced protection against gear scuffing, oxidation, and bearing corrosion. Additional additives against cold sludge and rust. Covers the requirements of US automotive manufacturers from 1964–1967.
API-SD	Improved oil quality compared to SC. The requirements of US automotive manufacturers were covered until 1971.
API-SE	Improved oil quality in comparison with SD for higher requirements of US automotive manufacturers were covered from 1972 to 1979.
API-SF	Compared to SE enhanced oxidation stability, improved wear protection, better engine cleanliness, and reduction of cold sludge formation. Fulfills the requirements of US automotive manufacturers of the 1980s.
API-SG	Includes SF. Also protects against black sludge and oxidation as well as improvement of the wear characteristic. For US automobiles from 1988.
API-SH	Valid since 1992. Includes SG; additional requirements for evaporation loss, foaming behavior.
API-SJ	Enhanced requirements for evaporation loss, foaming, and sludge formation.
API SL	Introduction of new exhaust-optimized engine as of 2001.

Fig. O39
API classifications

Despite clear reductions in oil volume as related to the fuel flow rate (petroil ratio) in the course of engineering development from 1:10 to 1:25 to 1:100 to 1:150, oil consumption by two-stroke engines is still many times higher than that of a four-stroke engine of comparable power. Part of the engine oil is still finely distributed in the combustion gas and in the combustion chamber due to its transport with the fuel or via the same path as the air-fuel mixture. Therefore, combustion residues from the oil can easily be deposited on the spark plug of a two-stroke engine.

Different two-stroke engine designs, special methods of lubricant supply, and the particular environment involved (e.g., outboard engines) clearly require a lubrication technology that differs from that of the four-stroke engines.

The essential criteria are:

- Enhanced corrosion protection, particularly for longer standstill periods, because crank drive and bearings are in contact with the ambient air. This applies in particular to outboard engines in a seawater environment.
- Less residue formation of engine oils during combustion to prevent deposits on the spark plugs and exhaust ports.
- Reliable lubrication (fretting corrosion protection) for piston rings and piston skirt.
- Low-smoke combustion as well as low levels of environmental pollution from emitted oil or its combustion products.
- Ready biodegradability, especially in the case of application in sensitive environments (water, forests).
- Good solubility in fuel.

In contrast, some of the requirements for four-stroke engine oils, such as dispersing capacity, wear protection, and viscosity-temperature characteristic (with multigrade oils) fade into the background.

Physical characteristic data of outboard engine oils is shown in **Fig. O44**, and these are reflected both in the specification of characteristic values and in engine tests.

~~~**Boat drives. Figs. O44 and O45** show the specifications of engine oils for two-stroke outboard en-

Service-Fill Oils for Gasoline Engines	
A1-02	Fuel economy low-viscosity engine oils compared with A2-96, changed requirements for HTHS viscosity (>2.9 mPas) Proof of defined fuel consumption advantage in M111FE
A2-96 Issue 3	STANDARD engine oils Requirements: Evaporation stability HSHT viscosity (~3.5 mPas) Wear Cleanliness Black sludge Oil oxidation
A3-02	PREMIUM engine oils Enhanced requirements: Shear stability HSHT viscosity (~3.5 mPas) Wear Cleanliness Black sludge Oil oxidation
A4-nn	Classification of oils for direct injection gasoline engines (in preparation)
A5-02	Fuel economy low-viscosity engine oils with long-drain properties. HTHS viscosity >2.9 and <3.5 mPas. Proof of a defined fuel consumption advantage in M111 FE. Same engine requirements as ACEA A3-02. Oils of this type are generally unsuitable for engines not expressly designed or approved for them due to their viscometric properties.

Fig. O40 ACEA classifications

O

API	ACEA	Test Purpose	Test Method
X	X	Optimum flowing ability	SAE viscosity classification
X		Flow behavior at low temperatures	Gelling index
X		Filter permeability	GM filter test
X	X	Prevention of the oil thickening effect due to the loss of lighter oil components	Noack evaporation loss at 250°C
X		Prevention of the oil thickening effect due to the loss of lighter oil components	Simulated distillation at 371°C
X	X	Reliable oil supply through antifoaming	Foaming/antifoaming
X	X	Viscosity loss through shearing	Shear stability: viscosity loss at 100°C after shearing
	X	Viscosity loss through shearing	Viscosity at high shear-gradient and high temperature
X		Corrosion protection	Nonferrous metal corrosion protection
X		Safety during storage	Flash point
X		Mixability with other oils	Mixing with reference oil
X		Oxidation stability	Chrysler high temperature test
	X	Compatibility with sealing materials	Swelling tests with (5) standardized sealing materials
	X	Combustion chamber deposits	Sulfate ash

Fig. O41
Specification of physical characteristic data of engine oils for gasoline engines (overview)

gines according to the NMMA (National Marine Manufacturing Association).

Two-stroke engine oils for application in boat drives with exposure of oily exhaust to water provided the initial incentive for the introduction of biodegradable lubricants. Biodegradable two-stroke engine oils were intended to stop the accumulation of hydrocarbons in lakes that in some cases also serve as drinking water res-

API, ILSAC	ACEA	Test Method (engine)	Conditions	Testing for
X		L38	Hot	Bearing corrosion Oil oxidation
X		Sequence IID	Cold	Corrosion
X	X	Sequence IIIE	Hot	Wear Cleanliness Oil oxidation
X		Sequence IIIF	Hot	Piston cleanliness Viscosity increase, Wear
X		Sequence IVA (Nissan KA24)	Cold/Hot	Valve drive wear Wear
X	X	Sequence VE	Cold/Warm/Hot	Sludge Piston cleanliness Wear
X	X	Sequence VG	Cold/Warm	Sludge Engine lacquer formation
X		Sequence VIA	Cold/Warm	Fuel consumption
X		Sequence VIB	Cold/Warm	Fuel consumption
X		Sequence VIII		Bearing wear
	X	PSATU 3MS	Cold/Hot	Valve drive wear
	X	PSATU 3MH	Hot	Piston cleanliness Oil oxidation
	X	PSATU5 JP	Hot	Piston cleanliness Oil oxidation
	X	M111 Sludge	Cold/Hot	Sludge (Wear)*
	X	M111(FE)	Cold/Warm	Fuel consumption

Fig. O42
Engine tests for specifying engine oils in car gasoline engines

Engine Test	ACEA A1, 2, 3, (4), 5	CCMC G–4, 5
L-38		yes (or Petter W1)
Seq. IID		yes
Seq. IIIE	yes	yes
Seq. VE	yes	yes
Seq. VG	yes	
Petter W1		yes (or L-38)
TU3M VTW	yes	yes
TU3M HT	yes	
TU5JP	yes	
Fiat 132		yes
M102 E		yes
M111 FE	yes	
M111 Sludge	yes	

Fig. O43 Comparative requirements according to ACEA and CCMC

Test Purpose	Testing for
Rust protection	Corrosion in salt water solution
Miscibility in the fuel	Mixing at $-25°C$
Adequate flow behavior at low temperature	Dynamic viscosity according to Brookfield method: $= <7500$ cP at $~25°C$
Filter pass (gelling) for mixtures with water and potassium (contamination with other engine oils)	Filter test, reduction of flow rate ≤20% compared with unmixed fresh oil
Compatibility with other engine oils	Homogeneity test with two defined reference oils after mixing and 48 hours of storage

Fig. O44 Physical characteristic data of outboard engine oils

Test Purpose	Test for
Adequate viscosity at operating temperature	Min. viscosity at $100°C ~6.5$ mm^2/s
Safety during storage and transport	Flash point according to national legislation
Operation safety of spark plugs (prevention of fouling causing bridging or whiskering)	Limiting of sulfate ash content: JASO max. 0.25% ISO max. 0.18%
Exclusion of four-cycle engine oils, lifespan of oxidation catalytic converters	No phosphorus (only JASO)

Fig. O46 Physical characteristic data

ervoirs. Synthetic esters in combination with ash-free additives result in high-performance outboard engine oils. The biological degradability is 80% after 21 days.

~~~**Motorcycle. Figs. O46 and O47** show the specification of engine oils for two-stroke motorcycle engines according to the Japanese (JASO) and international (ISO) quality requirements.

~~**Institutions.** The most important international standardization institutions are:

~~~**ACEA.** Association des Constructeurs Européens d'Automobiles (Association of European Automobile Manufacturers) which, among other things, determines the quality requirements for engine oils reflecting European engine technology and operating conditions in particular.

~~~**API.** American Petroleum Institute, which together with SAE determines internationally applicable quality requirements for automobiles and commercial engine oils.

Test Purpose	Test Method, Engine, Test Conditions	Test Criterion
Reliable lubrication, (prevention of piston seizure)	Yamaha CE 50 S (50 cm^3), 12 hours of testing, fuel/oil mixing ratio 150:1	Torque decrease compared to reference oil
Prevention of glow ignition (preignition) when leaded fuel is used	Yamaha CE 50 S (50 cm^3), 100 hours testing period, mixing ratio of leaded fuel/oil 20:1	Rare occurrence of preignition compared to reference oil
Engine cleanliness	Mercury 2-cyl. 262 cm^3 100 hours of testing, mixing ratio fuel/oil 100:1	Piston-ring sticking, piston cleanliness, piston seizure, freedom of motion of needle bearing
Engine cleanliness	OMC 2 cyl.737 cm^3 100 hours of testing, mixing ratio fuel/oil 100:1	Piston-ring sticking, piston cleanliness, ring zone, skirt, underbody
Engine cleanliness	OMC 3 cyl. 913 cm^3 100 hours of testing, mixing ratio fuel/oil 50:1	Piston-ring sticking, piston cleanliness compared to reference oil

Fig. O45 Engine tests

Test Purpose	Test Method, Engine, Test Conditions	Test Criterion
Reliable lubrication, (prevention of piston seizure)	Honda DIO AF 27 full-throttle at 4000 rpm mixing ratio of fuel/oil 50:1	Torque directly after cold start and at operating temperature
Engine cleanliness (piston rings, piston skirt, combustion chamber residues)	Honda DIO AF 27 full throttle at 6000 rpm testing duration: JASO 1 h, ISO 3 h	Evaluation of engine parts at the end of testing
Limitation of smoke in exhaust	Suzuki SX 800 R partial load and 0-load at 3000 rpm mixing ratio fuel/oil 10:1	Evaluation of visible smoke
Cleanliness of exhaust ports	Suzuki SX 800 R load alterations for exhaust temperature between 330 and 370°C at 3600 rpm mixing ratio fuel/oil 10:1	Limit value of negative pressure in the intake area

Fig. O47
Engine tests

~~~**CCMC.** Comité des Constructeurs d'Automobiles du Marché Commun (a cooperation of European automobile manufacturers) which in the past defined quality requirements for engine oils. Its tasks are now carried out by the ACEA.

~~~**CEC.** Coordinating European Council for the Development of Performance Tests for Lubricants and Engine Fuels, a professional committee of European automobile and mineral oil manufacturers that defines test methods (e.g., engine tests) for engine lubricants and fuels.

~~~**DEF.** Abbreviation for military specifications (Ministry of Defense) for the United Kingdom.

~~~**ILSAC.** The International Lubricant Standardization and Approval Committee has generally adopted American (US) quality standards for engine oils.

~~~**JASO.** Japan Automobile Standards Organization, whose quality requirements for two-stroke gasoline engine oils are used worldwide.

~~~**MIL.** The abbreviation for American (US) military specifications for engine oils, among other things.

~~~**SAE.** Society of Automotive Engineers, association of American (US) automotive engineers that introduced the worldwide viscosity classification for engine and transmission oils (SAE classes).

~~**Special engine oils**

~~~**Aircraft engine oils.** The proportion of aircrafts with reciprocating engines has constantly decreased since the introduction of gas turbine engines. They are only used in historic aircraft, sports aircraft, and in aircraft for special purposes — such as pest control. In spite of the relatively low demand for special aviation engine oils, this kind of oil has been further developed nonetheless. Two classes of engines are used in aircraft with reciprocating engines: engines developed specifically for air traffic, including radial engines or engines derived from passenger cars. Whereas the latter are partly lubricated with passenger-car engine oils, special air traffic engines impose special demands on oil. This is reflected in the specifications, the composition, and the properties of aviation oils.

Since the Second World War, US military specifications have been used worldwide for aviation engine oils without additives and later with additives. They were later replaced by the SAE specifications — SAE J-1966 — for oils generally without additives, and SAE J-1899 for oils with additives. The aviation engine oils must also meet the requirements of the aviation engine manufacturers.

Whereas multigrade oils have been in use in passenger car engines for a long time, the application of viscous, single-range oils of SAE classes 30–60 predominates in aircraft engines. The reason for this is partly the relatively large bearing clearance, particularly for radial engines. Multigrade oils — in the viscosity range of SAE 15W-50, for example — with high content of synthetic basic oils have come to prevail only recently.

Single-range oils without additives are always still recommended for new and overhauled engine break-in; in contrast, high-quality oils with additives according to SAE J-1899 show improved

- Oxidation stability,
- Dispersion of abrasion and combustion products,
- Anti-foaming,
- Corrosion protection, and
- Wear protection.

The formulation of aircraft engine oils is, however, quite different from that of oils for four-stroke gasoline engines in cars. Due to the special materials used in some engine types, it is necessary to dispense with zinc dithiophosphates that are otherwise proven as wear-protection additives for bearing shells. Furthermore, it has become clear that the interactions between the highly leaded fuels, common in flight operation, with the organo-metallic detergents in engine oils can lead to undesirable combustion chamber residues and hence to misfiring. For this reason, organo-metallic detergents are dispensed with, so that aircraft engine oils are practically ash-free. Car engine oils with ash content are, therefore, not suitable for special aircraft engines; conversely, aircraft engine oils do not fulfill the requirements for four-stroke gasoline engines in passenger cars.

~~~**Break-in engine oils.** In the past, it was common to use special break-in engine oils for new engines and to change the oil after a short duration in order to remove the extra content of metallic abrasion products. To accelerate the break-in effect, oils with low viscosity were frequently applied with a higher content of corrosion protection additives (for internal conservation of the engines during export) and otherwise with low levels of additives. With such oil types, however, it was possible to expose the engines only to limited stress. However, the improved surface quality of new engines led to reduced metal abrasion during the break-in phase. At the same time, due to improved oil technology, a slight increase of foreign substances can easily be kept suspended in the oil, thus ensuring reliable corrosion protection. It was, therefore, possible to dispense with both the special break-in oils and shorter oil-change intervals for new car and commercial vehicle engines after the break-in period.

Increased levels of dirt, such as sand, are occasionally found in the first oil filling, even today. Therefore, it is recommended that the oil be changed after approximately 500 to 1000 km, especially to protect highly loaded engines such as those in high-performance sports cars.

~~~**Gas engine oils.** There is a variety of gas engine applications: engines derived from gasoline or diesel engines, stationary or mobile applications, liquid or compressed gas applications, natural gas, landfill gas or petroleum gas application, and so on. Special requirements for different engines in different applications lead to a situation in which a mineral oil manufacturer must provide about a dozen different oil types in order to cover all gas-engine applications. The following discussion can, therefore, cover only a few examples.

Mobile CNG applications. The essential differences between operation with compressed natural gas (CNG) and operation with gasoline are:

- Natural gas burns at a higher temperature. Because natural gas exhibits higher resistance to engine knock than gasoline fuel, it is possible to obtain higher engine output, especially with modern injection and ignition systems. This generally causes a higher temperature level in the engine.
- Natural gas is drier and does not smear. Even the liquid gases contained in varying proportions in natural gas boil at very low temperatures, so they cannot form a liquid film.
- Natural gas produces more water during combustion. The higher proportion of hydrogen in natural gas (some 25%) in contrast to gasoline fuel (about 13%) automatically leads to the formation of twice the amount of water during combustion. Increased water content in the engine oil can lead to more corrosion and intensified wear; this must be avoided by using an engine oil type formulated accordingly. The lifespan of an engine can be impaired in particular if the vehicle is operated almost exclusively in short-haul traffic and in the process reaches only low oil temperatures, so that the water cannot "boil out." Therefore, the engine oil should be changed once a year, if possible in spring, in vehicles frequently used for short-haul operation.

These specific natural gas application problems obviously occur only in single-fuel operation, which is with exclusive use of natural gas. When natural gas is used in dual-fuel operated vehicles—for instance, in combination with gasoline fuel—the intake valves are reliably cleaned by standard brand gasoline fuels.

From these considerations, it is clear that the requirements imposed on natural gas operation can be fulfilled much better with special engine oils. When natural gas is used in gasoline engines, for instance, the following additional requirements arise:

- Increased combustion temperatures intensify the tendency to form deposits in the combustion chamber and on the pistons. Also, these deposits may be particularly hard. What is required, therefore, is a low-ash engine oil formulation that shows very little tendency to form deposits.
- The dryness of natural gas, or rather the absence of solvent properties in natural gas, has the consequence that natural gas cannot wash out the deposits that form in the intake area in every engine, and in particular on the intake valves. The composition of engine oil can accommodate this problem to a certain degree, similar to the manner in which combustion chamber deposits are prevented.
- Because dry natural gas (and, in comparison to gasoline, every quality of natural gas is dry) does not have any lubricating capacity, engine oil alone must be used to lubricate the valves. For this purpose, the oil formulation requires adjustment.

Specific problems of gas-engine oils include contradictory requirements for low-ash content (thus low level of additive mixture) to prevent combustion chamber residues and the longer durability that makes the oils economical.

Stationary applications. Stationary gas engines must "digest" a wide variety of gas qualities depending on the particular application. Stationary gas engines work mainly with external (spark) ignition based on the gasoline combustion process. The engine oils used for this purpose are approved by the manufacturer according to the analytical data and on the basis of successful long-term engine tests. Oil quality and oil change intervals are matched to the reactivity of the fuels used, (e.g., mainly neutral natural gas compared to aggressive landfill gas). The oil change intervals are frequently determined by a combination of oil quality monitoring during the operation period by the operator, the oil manufacturer, and the engine manufacturer. To this end, relevant analysis data of oil, such as viscosity increase, water content, flash point, alkalinity and acid numbers, oxidation products, organic nitrates, metal abrasion, and silicon are determined. The required oil change interval is determined as needed according to changes in these parameters and compliance with limit values based on empirical data. Special oils with the following criteria have proven qualities for the lubrication of stationary gas engines:

- Single-range oils based on mineral oil of SAE viscosity classes 30 or 40. Earlier, naphthalene-based basic oils were used. In gas operation, these oils prevent the formation of undesirable hard residues. Through further development in the additives technology, it is today also possible to use "common" paraffin-based basic oils.
- Synthetic oils without viscosity index improvers, but with higher oxidation stability for longer oil change intervals.
- Lower content of sulfate ash for the reduction of combustion chamber residues.
- Admixed additives for corrosion and wear protection, for oxidation stability, and for neutralizing acidic combustion products.

Gas engines for commercial vehicles developed based on diesel engines also impose special requirements on engine oils. Additional oil specifications for these applications are also written by the respective engine manufactures. In principle, their requirements are similar to those of the above-mentioned specifications for stationary gas engines; on the other hand, single-range oils cannot be used.

~~~**Hydrogen engine oils.** Parallel to the development of hydrogen-fueled combustion engines, suitable engine oils have been developed in a cooperative effort involving engine manufacturers and several mineral oil firms. Operation with hydrogen, compared to gasoline or diesel fuel, involves differences in the operating conditions of the engine and in the exhaust composition. Both may influence the requirements for the engine oil being used. Due to a higher water vapor content in the exhaust and possible water spray, more

water-in-oil emulsion can be expected in the oil. This raises the requirements for corrosion protection and dispersing capacity. The water admixed with the oil must be kept suspended before it evaporates once an adequate oil temperature is attained to prevent deficient lubrication. On the other hand, combustion residues and deposits are very low with hydrogen and hence lower oil contamination can be expected, especially in terms of cleaning behavior (detergency).

The starting point when developing specific oils is first to test high-quality commercial engine oils, then to perform continuous analytical monitoring at shorter intervals during operation with the oil. Based on possible peculiarities of used-oil data, adjustments can be made in a timely manner by changing the formulation of the oil before engine damage occurs. The development of suitable oils for hydrogen engines and their broad commercial application is still in the initial phase, so at present no conclusive evaluation is possible.

~~~**Locomotive oils.** Locomotive diesel engines are commonly lubricated with high-performance engine oils that correspond to the upper quality classes for commercial vehicles—that is, the US-API and the European ACEA classifications. A classification system issued by the American "Locomotive Maintenance Officers Association" (LMOA) is still on record. In addition to suitability (proved by means of analytical investigation) and engine tests (according to the above classifications), a positive long-term test in a limited number of locomotive engines is normally required prior to oil type approval by the engine manufacturer. In this process, special emphasis is put on long-term behavior to facilitate prolonged oil change intervals. Additional quality requirements are issued by the American manufacturers General Electric (GE) and General Motors Electromotive Division (EMD). GE requires low friction coefficients for combinations of steel shafts and bronze bearings. Oils for EMD locomotive engines must show neutral behavior toward bearing shells. The demand for oils without zinc dithiophosphate as a wear-protection additive is a logical consequence.

~~~**Methanol engine oils.** Within the scope of a research project to study the suitability of methanol as a fuel for road transport sponsored by the Federal Ministry for Research and Technology (Germany), engine oils for use in methanol-operated engines were also investigated by the participating firms from the automotive and mineral oil industries. Special requirements for engine oils used in methanol engines were defined based on long-term road tests and additional investigations using dynamometers:

- Low residue formation on pistons reduces requirements for detergents and dispersing agents.
- Incompatibility between methanol and long-chained organic viscosity index improvers and dispersing agents leads to substantial residue formation in the intake system due to the contact between fuel and

engine oil vapor from crankcase ventilation. Maximum exclusion of VI improvers and reduction of detergent content (possible due to reduced piston soiling) results in intake system cleanliness.

- High inclination to corrosion in cold operation requires a higher level of corrosion protection from engine oils.
- Higher levels of cylinder wear in cold operation require special additives. The cause is presumably the reduced wetting ability of the engine oil on metal surfaces in the presence of methanol.

~~~Racing engine oils. There is no "universal oil" for all engines; likewise, no "racing oil" for every racing engine and competition exists. Engine oil must be matched to the engine, the applicable regulations, and the type of competition if optimum performance is to be obtained. Some of the criteria include engine type (e.g., a Formula-1 engine or a commercial vehicle diesel engine for "truck racing"), regulations about refilling oil during the competition, and the type of competition (e.g., short distance in summer or long distance—rally—in winter).

Changes have come about due to the development of engines and new technologies for basic oils and additives in the historic development of racing oils. In the past, racing engines were frequently supercharged series engines; and naturally aspirated engines in non-rated series with insufficient mechanical strength and rigidity for increased performance. For such sensitive engines, castor oil partly mixed with synthetic esters was used because it allowed for an increase of lubricating reliability plus prevention of piston seizure, albeit at the expense of very restricted oxidation stability, which often required immediate oil drain and engine cleaning after the race. With the design of special racing engines with performance-adapted mechanical elements, engine oils no longer needed to be conceived for the emergency operation, but rather for an optimum overall result. The essential criteria and oil properties are listed here:

- Generally: Consistent oil properties over the entire competition distance, such as shear stability and good oxidation stability. "Chemical" wear-protection through additives for high stress and boundary-film lubrication. Due to high engine speed and oil circulation, a good antifoaming effect is required.
- Short distance with maximum performance: viscosity as low as possible while insuring hydrodynamic lubrication. The application of oils with the lowest possible viscosity in contrast to oils with higher viscosity leads to a power increase.
- Long distance with limited oil refill: omission of low-viscosity basic oil components to reduce evaporation losses and oil consumption. The viscosity is set somewhat higher than the minimum required, insuring additional wear safety.
- Engines that suffer from oil-enriched operation under fuel dilution: high dispersing capacity of the

oil is required to prevent centrifugal removal of foreign substances and additives from oil diluted with fuel.

In addition to these special requirements, "normal" criteria such as elastomer compatibility and low toxicity must also be fulfilled. However, the requirements are reduced when it comes to "year-round suitability"—that is, long-term operation and cold-start suitability.

~~~Tractor oils. Diesel engines for tractors can be lubricated with engine oils for commercial vehicle diesel engines, with the quality level determined by the manufacturer (specifications for commercial vehicle diesel engine oils). To simplify warehousing in agricultural operations, universal oils for agriculture have been developed which can be used concurrently in the engine, in the transmission, in the hydraulic system, and in the oil-cooled, so-called wet brakes. These so-called "Super Tractor Oils Universal" (STOU) oils must concurrently fulfill relevant specifications for all oil-lubricated tractor components. A special challenge during the development of these oils was the combination of the diverse properties required:

- Engine performance characteristics
- Wear protection for high-pressure stress and shear stability in the transmission and differential gearbox
- Adaptation of the coefficient of friction properties for smooth and effective braking
- Requirements for hydraulic oils (viscosity-temperature characteristic, filtering ability, stability in contact with moisture, air removal capacity)
- Compatibility with all materials used

Although nearly all of the currently required specifications can be fulfilled by the STOU oils, compromises are obviously required; and optimized performance similar to that of special oil—that is, optimized for individual tractor components—is unattainable. In particular, the requirements for high-performance engine oil are difficult to reconcile with some other demands. For the majority of tractors currently in operation, this is meaningless because good "Super Tractor Oils Universal" oils clearly surpass their quality requirements. The performance level of STOU oils will not remain adequate in future for tractors with more powerful, low-emission engines that will also be equipped with more demanding components. More powerful tractors with low exhaust emissions are increasingly using top-quality engine oil for commercial vehicles in combination with universal oil in the transmission, axles, hydraulics, and brakes—the so-called "Universal Tractor Transmission Oil," or UTTO.

~Fuel economy. Engine oils that cause less friction and lead to reduction of fuel consumption or increases in power are frequently termed low-viscosity oils or fuel-economy oils. Compared to conventional multigrade oils, this characteristic is achieved through low viscosity and special additives, the so-called friction reducers. In addition, the requirements for lubricating

O

reliability in the upper temperature and load range must also be fulfilled. Engine wear, cleanliness, and oil consumption may not deteriorate and a significant part of the fuel-saving properties of the oil must be retained over its entire useful life (i.e., low viscosity increase). Particularly with diesel engine oils, which are usually susceptible to thickening—due to soot particles, for instance—this requires a special additive that reduces the viscosity even in the presence of higher contents of solid foreign substances. High quality, fuel-saving oils can only be manufactured from basic oils with a high viscosity index, low evaporation loss, and high aging stability.

Obviously, no "miracle" should be expected to result from application of fuel-saving oils. Even if the energy losses through friction, based on engine designs, may be slightly different (e.g., the energy loss in the valve gear is apparently dependent upon the number of cylinders and valves), the maximum fuel-saving potential that can be attained by means of oils in the drivetrain is approximately 4% (compared to conventional oils).

Although the viscosity level alone is not decisive for the fuel-economy potential of a given oil type, the choice of the right viscosity class is an essential aspect. Depending on the operating conditions to be expected, the focus when lowering the viscosity must be on different temperature ranges. Gasoline engines that are frequently run under cold conditions require the viscosity to be lowered over the lower or the entire temperature range (e.g., 5W-30 instead of 10W-40); in contrast, commercial diesel engines that are run primarily under hot conditions require the viscosity to be lowered in the upper temperature range (e.g., 10W-30 instead of 10W-40). Measurements on a turbocharged, commercial vehicle diesel engine have resulted in the reduction of drag torque by 17% with low-viscosity oils (**Fig. O48**).

Significantly faster oil supply after cold start was measured using low-viscosity oils (both fresh and used) in the same engine type (**Fig. O49**). The slight thickening effect of oil during use is a direct consequence of the oil quality used. Particularly impressive

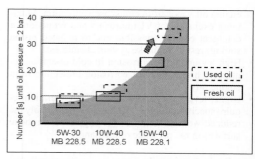

Fig. O49 Pumping suitability in an OM 441 LA engine at 0°C

savings in fuel consumption are achieved if the low-viscosity concept is also applied to lubricants for the transmission, rear axle, and wheel bearings. Experiments with fully loaded commercial vehicles have resulted in consumption reductions above 5%, depending on speed and operating conditions (**Fig. O50**). In general, the magnitude of fuel savings always depends on the operating conditions. In transit bus operation, for example, savings of some 3% were achieved by changing from a 15W-40 standard engine oil type (MB 228.1) to a 5W-30 high-performance oil type (MB 228.5); the same products are also illustrated in **Fig. O49**.

Road tests are a highly suitable but extremely expensive way of demonstrating the fuel-savings aspect. This is due to the large number of vehicles required to obtain reliable data that will factor the influence of different operating conditions (in particular different drivers). Dynamometer tests in standard engines or standard driving cycles on the roller dynamometer exhibit relatively good levels of reproducibility of results; nonetheless, they only rarely provide information on the fuel-saving aspects under varying operating conditions because of their predetermined cycles. For instance, in the tests on roller dynamometers, reproduction of the starting behavior of heavy commercial vehicles or buses is bound to be imperfect. Therefore, driveline dynamome-

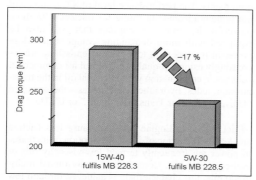

Fig. O48 Reduction of friction in OM 441 LA (Euro II engine)

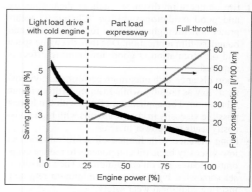

Fig. O50 Consumption reductions with a commercial vehicle (compiled from field test results)

ter specialized for measurement of fuel economy was developed by Shell, in cooperation with some commercial vehicle manufacturers. This dynamometer combines far-reaching flexibility of operating conditions with extraordinary accuracy.

~Oil functions. In addition to the lubrication, which is discussed in detail below, the engine oil has further important functions to fulfill, without which a combustion engine would not be functional:

- Sealing the combustion chamber against the crankcase between the piston and cylinder—that is, supportive fine-sealing between the piston rings and the cylinder liner. Sealing the intake and exhaust ducts via the valve guides against the valve drive.
- Hydraulic power transmission, for example, in hydraulic tappets and chain tensioners.
- Corrosion protection of the engine parts threatened by aggressive combustion products through formation of protective coats on the metal surface.
- Neutralization of aggressive combustion products through chemical conversion.
- Keeping the engine clean by removing combustion residues (and aging products of engine oil) with oil soluble soaps.
- Lubrication—that is, separation of metal surfaces sliding relative to one another—either by forming a load-bearing hydrodynamic lubrication film or by changing the metal surface for operating states and components that do not allow hydrodynamic lubrication (wear protection).
- Dispersing solid foreign substances, dust, wear, combustion products such as soot or ash through dispersion agents, so that these cannot be deposited in the oil circulation system or in the crankcase sump, but are rather filtered out, centrifuged, or removed during oil changes.
- Protection against wear of engine parts that move relative to one another such as bearing journals/bearing bushes, piston/cylinder, and valve drive.

To summarize, it is the task of the engine oil to sustain the operating safety of the engine, whereby wear protection and cleanness have top priority.

Some of the tasks of engine oil are described in detail below.

~~**Cooling/heat dissipation.** Cooling of engine parts that are not directly accessible to the coolant, mainly piston and crankshaft, is another task of engine oil. Due to the high circulation rate of the oil and the intensive movement, a sufficient cooling effect can be achieved despite the significantly lower cooling effect in comparison to a coolant based on a water/glycol mixture. A comparison of the specific heat and the heat conductivity with water and water/monoethylene glycol mixture is shown in **Figs. O51 and O52**.

~~**Corrosion protection.** Corrosion is an electrochemical process by which the attacked metal is oxidized

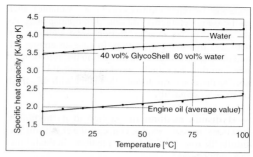

Fig. O51 Specific heat capacity of engine oil in comparison to water and a water/monoethylene glycol (MEG) mixture of 60/40

and the attacking medium is reduced. The iron and nonferrous metal surfaces in the engine must be protected against corrosion both in operation at high temperatures as well as during standstill periods.

Special demands on corrosion protection are made, on the one hand, by (ship) diesel engines that are operated with heavy, sulfur-rich residual oils and must, therefore, be protected against sulfuric acids and their derivatives, as well as by gasoline two-stroke engines during downtime, because the metal parts in the engine are protected only by a very thin oil film against the corrosive effect of moisture in combination with combustion residues.

Furthermore, increased corrosion protection is important for longer transport journeys (export) and longer downtime periods, for example, putting the vehicle out of operation during the winter period, because the crankshaft drive and bearings remain in contact with the ambient air. This applies especially to outboard engines in a marine environment. Prior to a longer standstill period, an oil change is especially advisable. On the one hand, the corrosive combustion products that had accumulated in the oil will be removed; on the other hand, the concentration of the corrosion protection additives will be refreshed. The effect of the corrosion protection additives is further

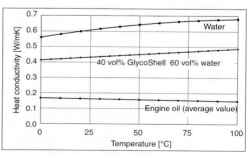

Fig. O52 Heat conductivity of engine oil in comparison to water and a water/monoethylene glycol (MEG) mixture of 60/40

enhanced by other additives (so far as available), such as the basic detergents and zinc dithiophosphates. As mentioned under wear protection, such additives are activated by engine operation. In case an engine oil, and not a special corrosion protection oil, is to be used when putting the vehicle out of operation, it is reasonable to drive several dozen kilometers with this fresh oil prior to standstill.

~~**Hydraulic medium.** In addition to its traditional functions, in newer engine concepts the engine oil also fulfills the role of a "hydraulic medium" in the following components:

- Hydraulic valve clearance compensation
- Hydraulic timing chain tensioner
- Hydraulic camshaft adjusting unit

In these components as well, lubricant viscosity plays an important role. If the viscosity is too low, the oil will flow out of the pressure chambers of the mentioned components and the components can no longer fulfill their functions. The driver will notice this functional failure either as a loud noise component caused by rattling hydro tappets or strong timing chain vibration or as a power drop if the camshaft adjustment does not work.

A very high gas content in the engine oil has a similar effect on the hydraulic components mentioned. If the gas content is too high, the engine oil becomes compressible and, thus, cannot fulfill the power transmission function assigned to it. This may also result in increased noise development, power loss, or even engine damage. To lower the foaming tendency of engine oils, so-called antifoaming additives (mostly based on silicone oil) are added to the oil.

~~**Keeping clean.** Keeping the engine clean from deposits of any kind is achieved primarily by uptake of combustion residues and other solid and/or liquid foreign substances in finely distributed form by oil and is also due to the fact that deposits already formed are dissolved by the oil for the most part. This is, for instance, of special importance for the cleanliness of the piston ring grooves; the deposit of "carbon" or combustion residues would lead to ring riding or ring seizure with the consequence of increased oil consumption and performance loss or even to severe engine damage.

The dirt and foreign substances dissolved in the oil are then eliminated by changing the oil.

~~**Lubrication.** Lubrication has the function of reducing friction between two surfaces. Friction is the resistance of a surface to move relative to another surface; the direction of the force that must be applied to move the body is thereby oriented opposite to the friction. Friction, therefore, always leads to energy loss.

Even smooth surfaces exhibit roughness; whenever two surfaces are in contact, the contact takes place in reality only at the peaks of this roughness (**Fig. O53**). Because the actual contact surface is very small, very

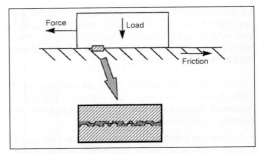

Fig. O53 Basics about friction

high pressures occur, potentially even resulting in welding of the surfaces.

The friction between two bodies is independent of the contact surface and dependent on the load one body exerts on the other (**Fig. O54**).

In first approximation, the friction μ is described by

$$\mu = \frac{Force}{Load}.$$

However, the coefficient of friction is not constant, but rather depends on the loads acting on the system: The coefficient of friction is dependent on the surface texture, load duration, temperature, sliding velocity, and the surface.

Separation of the metal surfaces sliding over one another is one of the main functions of engine oil. It is facilitated by the formation of a load-bearing hydrodynamic lubricant film as well as by changes of the metal surfaces in operating states and components that are inimical to hydrodynamic lubrication (wear protection).

Variations in formation of the lubricant film and, thus, in the kind of lubrication and wear protection, depend on design and operating parameters as well as on the viscosity of the lubricant. The three typical areas— boundary-film lubrication, mixed-film lubrication and hydrodynamic lubrication—are characterized by the so-called Stribeck curve (**Fig. O55**). The formation of an oil film that is load-bearing and fully separates the metal parts requires a sufficiently high relative velocity of the sliding parts, sufficiently high oil viscosity, and adequately low specific surface load.

Boundary lubrication

Zone A of the curve in **Fig. O55** corresponds to the conditions at low velocities (e.g., when starting the

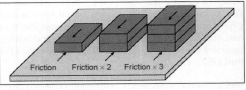

Fig. O54 (Sliding) friction

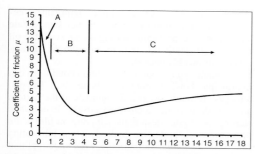

Fig. O55 Different types of lubrication (Stribeck curve)

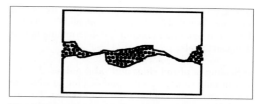

Fig. O57 Thin-film lubrication or mixed friction

engine) and (independent of the engine speed) the conditions at the top and bottom dead center of the piston (i.e., at reversal of the direction of motion).

Under these circumstances, lubricant film thickness is practically zero (i.e., the surfaces of the frictional partners are only covered with absorbed lubricant molecules or the load is born by only a very thin lubricant film; **Fig. O56**), and surface contact occurs elsewhere. The effectiveness of lubrication here depends primarily on the chemical interactions between the lubricant and the surface and, thus, on the chemical composition of the lubricant. The factors viscosity and load have practically no influence on the coefficient of friction μ under these conditions.

Thin-film lubrication or mixed friction

In Zone B in **Fig. O55**, the thickness of the lubrication film corresponds to the surface roughness; the load is borne by both the lubricant film as well as the surface contact (**Fig. O57**).

A steep curve shape is characteristic of the range of mixed-film lubrication: a relatively small velocity drop or increase in load effectuates a strong increase of the friction coefficient μ and the transition in the range of boundary friction.

Hydrodynamic lubrication

Zone C in the curve **Fig. O55** corresponds to the conditions of higher sliding velocities and sufficiently high viscosity (**Fig. O58**).

The surfaces are fully separated by the lubricant film in this case. To achieve full separation of the surfaces, an adequately high pressure buildup in the lubrication gap is required as a precondition (e.g., a wedge-shaped constriction of the lubrication gap).

The thickness of the lubrication film increases with an increase in viscosity and velocity or with a decrease in load. This is the ideal state to be targeted for reliable lubrication; design constraints do not allow this everywhere in engines.

The lubrication film can either be maintained hydrodynamically or hydrostatically. Hydrodynamic means that through the motion of bodies, a natural pressure buildup occurs between the surfaces so that they do not touch. If the pressure between the body surfaces is maintained with the help of a pump, this effect is termed hydrostatic.

Elastohydrodynamic lubrication

A special type of hydrodynamic lubrication is elastohydrodynamic lubrication at the borderline to thin-film lubrication. It occurs when highly loaded surfaces move relative to one another in a rolling contact (e.g., in transmissions).

If the surface shape and the type of motion are suitable, lubricant can be briefly retained between the surfaces, where it can be exposed to very high pressures (up to 30,000 bar).

This high pressure has two important effects:

- The viscosity of the lubricant rises substantially, and with it the loadbearing capacity.
- The corresponding surfaces become slightly deformed, so that the load is spread over a larger area.

~~Neutralization of combustion products. The neutralization capacity of an engine oil is its capability to neutralize acidic combustion products that penetrate into the engine oil and the acidic aging products of the engine oil, thus preventing wear and residue formation by virtue of its alkaline additive. Also to be neutralized are the reaction products produced by entry of nitrogen oxides (blowby gases). The gradual

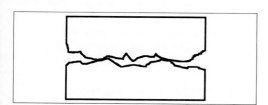

Fig. O56 Boundary lubrication.

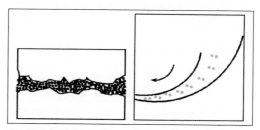

Fig. O58 Hydrodynamic lubrication

exhaustion of the neutralization capacity of the engine oil during operation due to the "consumption" of the additives is a significant criterion for determining the oil change intervals.

~~**Sealing of piston rings and valve guides.** An important duty of engine oil is fine-sealing of different spaces inside the engine from one another. For instance, the combustion chamber has to be sealed under all operating conditions against the crankcase and against the valve compartment in the cylinder head area. Rough sealing of these spaces is achieved on the one hand by the piston ring assembly and on the other hand by the valve stem seals, which are fixed on the valve guides inside the valve compartment. Fine-sealing in these areas is realized by the engine oil. Here the oil must fulfill its sealing function under very high temperatures (up to 300°C in the first piston ring groove) and under very high pressures in the finest sealing gaps, without leaving the sealing space, which in the most inconvenient case would lead to increased oil consumption and, thus, to increased oil-based exhaust emissions.

To guarantee reliable sealing , the choice of a suitable viscosity is of decisive importance. If the viscosity is too low, this may under certain circumstances lead to rupturing of the sealing oil film, so that in the end the combustion gases enter the crankcase and lead to an increased blowby effect. If the viscosity is too high, it can lead to internal friction in the lubricant and, thus, to a drop in performance or an increase in fuel consumption.

The volatility—that is, the evaporation tendency of the engine oil—is of great importance in the piston ring zone in particular. A very high evaporation tendency of the engine oil leads, especially in this area in which the oil fulfills a sealing function, to an increase in oil consumption as well.

~~**Wear protection.** One of the essential functions of engine oil is wear protection of engine parts that move relative to one another, such as bearing journals/ bearing bushings, cylinder liners, piston/piston rings, and valve drive (**Fig. O59**). In the process, the friction partners nonuniformly shaped in the micro range are separated from one another by the intermediate medium—engine oil. The engine oil must guarantee this under all possible loads. If the size of the frictive surfaces formed by the impinging forces is great enough—the width of the bearings, for instance—the wear protection is ensured mainly under the conditions of hydrodynamic lubrication through the choice of suitable engine oil viscosity. A temporary pressure increase in the lubrication gap makes for a clear increase of the loadbearing capacity of the lubricant. A suffi-

cient minimum viscosity at a higher temperature ensures wear protection even under higher steady-state loads. For wear protection during a cold start, the oil must be sufficiently thin in order to reach the lubrication points quickly enough.

Optimum hydrodynamic lubrication conditions can in principle, however, not always be reached. The pistons pass through low-speed segments at the bottom and top dead center areas respectively. Even when starting the engine, despite the adhering oil film at the beginning, the protective hydrodynamic lubricating film is not available. Furthermore, even under optimal lubrication conditions, contact between the rough salients of the friction partners cannot be ruled out. A clear improvement in the wear protection of engine oils is achieved through additives, which guarantee a separation of the friction partners under critical conditions through the formation of protective layers. Most common are organic zinc phosphorous compounds of different compositions, mainly zinc dialkyl dithiophosphates and zinc diaryl dithiophosphates (wear protection additives).

These so-called EP (extreme pressure) additives react with the metal surfaces upon attaining certain reaction temperatures and, thus, prevent contact between the metal surfaces that would otherwise lead to wear or welding. A balance must be attained between adequate EP-effectiveness and other undesired chemical reactions (e.g., corrosion).

The EP demand depends on

• Driving style,
• Engine design, and
• Oil consumption.

These EP additives are activated only after they reach certain reaction temperatures (wear protection additives); thus, generally after initial engine warmup. In practice, it has, however, become clear that the effect of these activated additives is a function of time. If the activation (in the case of short-distance operation, possibly insufficient) of these agents has taken place some time before, flawless functioning of engine components moving in the mixed friction range cannot always be ensured.

~Oil maintenance

~~**Filter systems.** A significant extension of the oil change intervals by means of an oil filter that is additionally fitted in the vehicle is only feasible to a limited extent. Experiments with commercially available bypass filters showed that the greatest proportion of impurities remained in the oil:

• Solid foreign matter, approximately 90%
• Liquid foreign matter, approximately 95%

Concentration too low = Mechanical wear	Optimal EP–add. concentration = Safe for moderate driving style	Concentration too high = Chemical wear

Fig. O59
Principle of EP additives' activity

- Aging products, approximately 90%
- Chemical reaction products, approximately 80%

The bulk of solid foreign matter is present in the oil in particulate sizes of 0.1–0.5 μm. The size of filter pores is much larger than this. Even deep-acting filters are hardly capable of removing all the solid foreign matter from the oil, because the dispersant effect of oil additives is greater than the adsorbing force of the filter medium. The fuel quantities dissolved in oil are fully soluble in the oil and hence cannot be filtered; the same applies to low-molecular-weight aging products. The increase in the volume of the oil in the engine due only to the bypass filter leads to a lower contamination level (kW/L) of the oil, and as such it becomes possible to extend the oil-change interval proportionally.

According to the investigations sponsored by the Federal Environment Agency, Germany, it was found that the use of retrofitted bypass filters and dispensing with oil changes led to unacceptable deterioration of the engine oil; engine wear increased, fuel consumption rose, and operating costs increased (W. Dahm and K. Daniel, Einfluss Nebenstromölfilter auf Ölwechselintervalle, *MTZ* 57 (1996) 6).

The Federal Environment Agency summarized in its press release no. 46/95: "No extension of engine oil durability, no advantages for the environment. Fitting these filters in vehicles has no influence on the suitability of engine oils for use and would not lead to the reduction of waste oil quantity." This assessment is fully justified in view of the results of this study:

- Clear increase in fuel consumption (0.4 liter per 100 km in cars)
- Engine damage (preliminary stage for piston seizure)
- Increased wear
- Overall increase in operating costs

To sum up:

- Bypass filters cannot slow the exhausting of the engine oil; that is, they cannot slow down the inevitable consumption of additives.
- Extended oil-change intervals are paid for with increased fuel consumption.
- Bypass filters increase the proportion of hazardous waste.
- Extension of the oil-change interval is possible, proportional to the increase in oil volume, by means of installation of a bypass filter. A commercial vehicle manufacturer allows an equivalent extension of the oil change intervals either through a bypass filter or an additional 10-liter oil tank.

~~Oil change interval. Oil changes are required because engine oils suffer from irreversible physical and chemical changes during operation; essentially, this also includes solid foreign matter from both unburned fuel components and condensate water. The formation of aging and nitration products as well as the creation of additives—that is, the rest activity of oil—also influences the distance that can be travelled before chang-

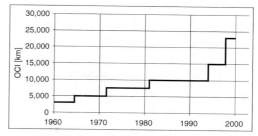

Fig. O60 Development of oil change intervals of gasoline engines for cars over 40 years (from German manufacturer)

ing the oil. These changes reduce the performance of oil; in turn, the operating characteristic of the engine will be poorer (difficult cold-starting), the engine lifespan is shortened, and environmental pollution increases through higher fuel consumption. The continuous quality improvement of engine oils through additives, but also through further development of basic oils, has led to a significant lengthening of the oil change interval despite increasing oil contamination (**Fig. O60**).

Oil change intervals are usually determined by engine manufacturers and are based on the period of operation or mileage by considering the prescribed oil quality, the specific oil contamination, the oil consumption and additionally with the help of engine and vehicle tests; even under inconvenient operating conditions, the quality of oil reserves is retained. Alternatively, the required oil change can be indicated indirectly by monitoring the quality of the oil in the engine. In this case, engine speed, number of cold starts, distance covered, and further relevant operating parameters are considered and processed in the vehicle and output as an "oil change indicator."

Recent joint-venture developments of the vehicle and mineral oil industries also allow direct quality monitoring of engine oil with the help of oil sensors in the vehicle.

Additional monitoring of oil quality in operation together with flexible determination of oil change intervals offers a series of advantages. The acquisition of operating data defines the maximum possible oil mileage dependent upon the operating conditions, as described in **Fig. O61**, by means of the continuous line. Measuring the oil quality then modifies the result derived from the operating data, as depicted in the gray range: Lowering the oil quality overproportionally leads to reduction in the mileage. This guarantees operating safety. A persistently high oil quality (via refilled oil) extends the mileage. This gives the owner of the vehicle a corresponding cost advantage.

Nevertheless, even with improved oil monitoring today, it is not possible to acquire all the relevant data on used oil that would be required for an "absolutely" safe definition of the oil change deadline. The continuous, analytical oil monitoring system, with an oil change according to current analysis data (used oil anal-

O

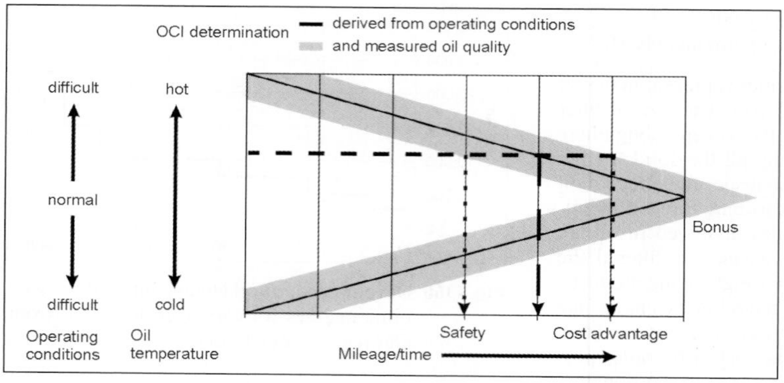

Fig. O61
Requirement-based oil
change intervals

ysis) as practiced for large engines, is a technical objective: such a system would allow the oil quality to be
utilized in an optimum manner.

~~**Used oil analysis.** Analysis of the used oil is an indispensable aid for field tests and engine development.
It helps the engine manufacturer determine the oil-
change intervals dependent on engine condition and
oil quality. In the event of damage, it helps to clarify
the causes. With large engines (e.g., ship diesel engines), used-oil analysis is applied in routine oil and
engine monitoring. In large fleets with specific operating conditions, the used-oil analysis can help test potential longer oil-change intervals following switch-
overs to better oil qualities. This solution is, however,
not practicable for passenger cars and commercial vehicles because the cost of oil analysis is not proportional to the potential cost savings. A good alternative
for vehicle fleets, in cooperation with engine manufacturers and oil suppliers, with fixed operating conditions (e.g., public transport companies) is analytical
oil monitoring on representative vehicles and engine
oils and determination of the appropriate oil-change
intervals based on the results obtained. In contrast, individual used-oil analysis for passenger cars and commercial vehicle engines is in most cases less meaningful because it only shows the actual state and not the
trend. In such cases, "equipment condition monitoring" is preferred because it provides indications of
possible engine problems (e.g., coolant ingress into
the oil). Exceptions are devices for limited "onboard"
oil analysis (dielectric constants) installed in several
vehicle types as an aid for determining the requirement-based oil-change interval.

The condition of engine oil and its change of properties compared with fresh oil allow conclusions to be
drawn about the state of the engine and its operation.
The most common tests are:

- Infrared analysis
- Viscosity measurement
- Base number
- Flashpoint
- Solid foreign matter

- Metal abrasion
- Water and glycol content
- Dispersing power

The assessment of oil change interval requires experience in typical cases based on the type of operation
and engine. For some parameters (viscosity, base
number, dispersing power), it is necessary to also
make sure that no mixing with other oils takes place.
Under certain circumstances, this can be determined
by means of infrared analysis. Spectroscopic infrared
analysis according to DIN 51 451 mainly, qualitatively, detects a variety of compounds contained in engine oils. These are basic oil components and additives
as well as aging products and foreign substances such
as glycol, fuel, and water. The infrared spectrum of
used oil, for instance, in **Fig. O62** facilitates identification of the engine oil being used and provides clear evidence of its changes due to operation in the engine by
comparison with known spectra of fresh oils in many
cases. Specific investigations regarding aging and foreign matter are frequently carried out downstream of
the instrument based on the qualitative results of the
infrared analysis.

The quantitative determination of abrasive elements
provides hints on the state of wear. In general, spectro-
metric methods are used to determine them depending
on the method used, covering abrasive particles up to a
size of about 15 μm. The following measuring methods are used to prove the presence of metals:

- Atomic absorption spectrometry (AAS) according
  to DIN 51 396-3, DIN 51 397-1, and DIN 51 397-2:
  this method is seldom used today because it requires a high scope of sample preparation.
- X-ray fluorescence spectroscopy according to DIN
  51 396-2: for elements in a concentration of about
  1 mg metal per kg oil, the presence of metals can be
  determined with this method quantitatively up to a
  percentage range. The advantages of this method
  are concurrent proof from a simple sample of several elements side by side and the omission of working steps when preparing a sample.
- Optical emission spectrometry with inductively coupled plasma (ICP OES) according to DIN 51 396-1,

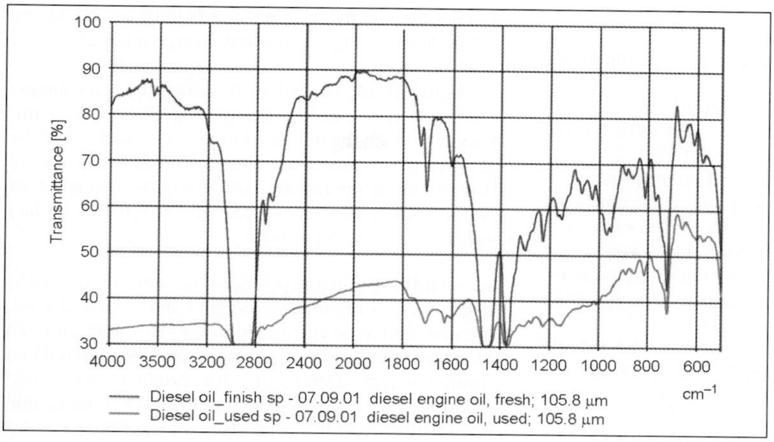

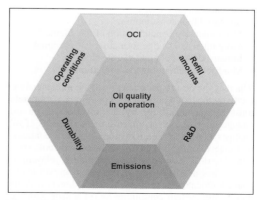

Fig. O62
Infrared spectrum of used oil (below) in comparison with fresh oil (above)

all elements to be determined in the content range of approximately 0.01 mg/kg up to about 10,000 mg/kg can be determined concurrently from one sample.

Several essential used-oil data can also be determined "on site" without expensive analysis and allow a limited evaluation of oil quality. This includes viscosity change (thickening or dilution), alkalinity, foreign matter content, dispersing power, as well as water and fuel contents. The manual methods described cannot replace the precision analysis methods required for oil and engine developments; however, they do enable a rough evaluation of engine oil in the field when a decision about an oil change is needed without expensive analysis. However, the evaluation does require some experience.

Changes of kinematic viscosity can be evaluated on an inclined plate by comparing the droplet flow velocity of used oil with fresh oil and with reference oils with a known viscosity value.

The alkalinity content in used oil can be evaluated by matching an existing additive in special engine oil against an indicator liquid; the color change of the oil sample on the filter paper is decisive.

Foreign matter content, dispersing power, and water and fuel content can be evaluated by means of the so-called oil drop test on special filter paper. A uniformly stretched oil drop without a sharp edge, with uniform color, indicates whether sufficient dispersing power is present. In the case of insufficient dispersing power and high content of foreign matter, a concentration of foreign matter (black coloration) is evident in the middle of the droplet. Water in oil is revealed by an irregularly jagged droplet edge; fuel is revealed by a light yellow edge outside the oil droplet.

~Power during operation. The oil quality in operation, or engine oil load and the resulting aging process, will obviously depend on the design parameters of the engine and vehicle (**Fig. O63**). At the same time, oil quality is also influenced by other factors, or it exerts effects on other factors: length of oil-change interval, other maintenance measures, oil refill quantities, demand for or guarantee of fuel economy, and minimization of emissions and engine protection to maximize the lifespan of the vehicle. The operating conditions are obviously decisive for oil quality in operation.

Over the past forty years, great progress has been made in the development of engines and lubricants. Changes in engine design have brought about large changes in specific outputs, engine speeds, and oil temperatures — and hence contamination. At the same time, oil consumption and oil sump contents have been reduced and the oil change intervals prolonged. In sum, these factors have led to much higher contamination levels in engine oils.

The features that increase contamination of the oil and that, therefore, necessitate higher oil qualities (or shorter oil-change intervals) — subdivided into design, service-relevant, and operational influences — are:

Fig. O63 Influence factors and effects of the oil quality in operation

Design parameters

- High specific output of the engine (output/liter swept volume)
- Lower-volume oil content of engine
- Lower oil consumption levels (i.e., no refreshing of oil through refill quantities)
- High blowby volumes at high cylinder pressures
- High oil-foaming due to missing seals in the crankcase
- Omission of oil cooling in highly loaded engines
- Inadequate ventilation of interior engine space
- High oil contamination on individual engine components (noncooled pistons, narrow bearings, high surface pressures in the valve gear)
- Engine size not optimally adapted to the vehicle (engine too large = too much cold operation; engine too small = too much hot operation)

Service-relevant parameters

- Prolonged oil change interval
- Dirty or defective air filter
- Oil level below minimum, without refill
- Application of low-quality oils
- Wrong operating temperature (e.g., because of defective cooling water thermostats)

Operational parameters

- Frequent operation at high loads and speeds
- Frequent cold starts and cold start operation, mainly short-distance operation
- "Police operation" (i.e., frequent city operation with short load peaks)

"Extreme" operating conditions—that is, inconvenient combinations of the above parameters—occur regularly in a significant proportion of all vehicles and lead to the following definitive changes in engine oils:

- Viscosity increase or viscosity decrease
- Disturbance in additive equilibrium
- Minimization of dispersing capacity

Potential consequences include:

- Increased engine wear
- Increased fuel consumption
- More intensive contamination of the engine
- More intensive sludge formation in the engine
- Impairment of cold and warm start
- More severe environmental pollution

Whereas the design parameters may be influenced directly by the manufacturer, the operational parameters must be accounted for by the instructions on oil change intervals. Clear improvements in terms of oil changes as needed (i.e., reflecting the actual contamination level of the oil) are made possible by collective consideration of the operating conditions relevant to oil changes and processing of this data to create an oil-change indicator. Better yet would be a direct oil status monitoring system combining an oil sensor with the engine operating data. Processing of the operating data may run into certain difficulties (e.g., high and low

temperatures), and rotating speeds must be considered when determining service and change intervals.

~~**Aging of oil.** The oil in its function as an engine "waste dump" is exposed to various kinds of contamination through the input of foreign substances. Not the least of these effects are chemical changes of the oil during engine operation, particularly the formation of liquid (low molecular) and solid (high molecular) aging and reaction products.

~~~**Additive consumption.** Some engine oil additives, such as alkaline detergents and wear reducers, are intended to be consumed in use. However, in fresh oils, these additives can generally only be metered to a limited extent. Detergents, for instance, form ash; overdosing an oil with them would lead to undesirable combustion chamber deposits. From a different point of view, the additives compete for the metal surface areas in the engine; an excess of wear protectors, for instance, would practically "block" the corrosion inhibitors from accessing the metal surface.

The decline of additive concentrations over the application period of the oil is, therefore, an important criterion for assessing the remaining capacity and the possible necessity of changing the oil.

Two examples are given for this case:

- The gradual exhausting of the neutralizing potential in an engine can be followed, for example, by means of the TBN. A decline of the "alkalinity reserve" to about 50% of the original value in fresh oil is generally harmless. Only below this value does the oil lose the capacity to keep engine parts cleared of combustion residues and oxidation products; it also loses the capacity to neutralize the acidic combustion products, so that corrosion can occur, dirt can accumulate in the engine, and the oil supply can "collapse."
- The wear reducers are chemically reactive, high-pressure additives that react with the metal surface. Locally increased heating due to friction and plastic deformation may activate chemical reactions at highly stressed points. The decomposition products resulting from the process are the actual components that act as wear-protecting additives. Under extreme stress or in case of impermissibly prolonged oil change intervals, these additives may be completely consumed, and this leads to severe engine damage.

Modern engines often consume less than 100 ml of oil per 1000 km, so that oil is generally not refilled within the normal oil change interval of 15,000 km. The recommendation, especially for highly stressed engines, is to refill the missing amount of oil after half the oil change interval, because this not only reduces the oil temperature due to the larger amount of oil, but also counteracts the effect of consumed additives.

On the other hand, concentrated additives sometimes offered for refreshing engine oil, instead of an oil change, cannot be recommended. Of course the decline in additives can be stopped to a certain extent, but the dirt in oil, oil thickening, and oil dilution cannot be compensated in this way. Such products make little technical sense and put the engine at risk.

~~~**Nitration, oil sludge.** The formation of oil sludge in gasoline engines is becoming a new problem even with oils hitherto adequately mixed with additives; this is one result of the increasing use of engines with stoichiometric or lean combustion. The cause is the higher content of oxides of nitrogen (NO_x), and these enter the crankcase past the piston rings. They are converted into nitrogen dioxide (NO_2) during the dwell time in the crankcase, either in the gas phase or through reactions with components in the engine oil:

$$NO + RO_2^{\cdot} \Leftrightarrow NO_2 + RO^{\cdot}$$

($^{\cdot}$ designates a radical: a chemically highly reactive state).

The reactive nitrogen dioxide, NO_2, finally reacts with polar oil components to form organic nitrates, which in turn lead to undesirable deposits, similar to oil sludge, in both the crankcase and valve gear and can lead to premature failure of the engine oil. The content of organic nitrates in used oils is, therefore, a good indicator for its state and suitability for further application. It can be determined with the help of infrared spectroscopy according to DIN 51454.

Antisludge oils can be obtained with nonpolar basic oils such as highly refined mineral oils or synthetic basic oils, but also with special, admixed additives or by combining both measures. Phenolic antioxidants and zinc dithiophosphates act, for example, as NO_2 scavengers providing adequate protection against the formation of organic nitrates and oil sludge formation as long as these additives still remain in sufficient concentration in the oil.

Testing of oils for anti-oil sludge characteristics can be done in the laboratory by simulating a nitration process under concurrent addition of oxygen in the presence of reactive fuel components. Ultimately, testing in the ACEA test engine M111 and fleet tests under critical conditions and with prolonged oil change intervals provide the required information.

~~~**Oxidation.** Oxidation of lubricating oil occurs in the engine through the reaction of oil components with oxygen in the air; high temperatures encourage this process. High-quality basic oils exhibit high resistance to oxidizing attacks. However, at temperatures $>150°C$, oxidation of engine oil takes place particularly in the presence of incompletely burned fuel components. Acids and insoluble products occur in the process.

In the oxidation process, acids finally occur after intermediate stages, accompanied by formation of organic peroxides and, in the advanced stages, oil-insoluble polymerizates, leading to an increase in viscosity and finally to deposits resembling lacquers and sludges.

$$RH \rightarrow R^{\cdot} + H^{\cdot},$$

$$R^{\cdot} + O_2 \rightarrow ROO^{\cdot},$$

$$ROO^{\cdot} + RH \rightarrow ROOH + R^{\cdot},$$

where R = rest of the oil molecule and $^{\cdot}$ = radical.

The presence of some metals substantially increases the oxidation rate. Some of the oxidation products that are formed attack certain metals, the dissolved compounds of which then further accelerate the aging process. Oxidation inhibitors interrupt this chain reaction by scavenging the reactive intermediate products, the so-called radicals.

~~~**Viscosity drop.** The dilution of engine oil accompanied by a decrease in viscosity during engine operation can have several causes. The most well-known process is oil dilution with fuel or water, which can admix with the engine oil via the cylinder liners especially under persistent cold-engine operation (i.e., due to condensed fuel components and water vapor from the combustion process). Low-emission engines with little mixture enrichment even in cold operation are less inclined to oil dilution. Oil dilution with fuel can also occur due to engine damage in multiple-cylinder engines, for instance, if incomplete combustion occurs in one of the cylinders or if fuel that is leaner than intended for stoichiometric operation is used in a racing engine for maximum output and interior cooling. Furthermore, oil dilution can also be caused by the formation of low-molecular-weight aging products (oxidation). A third possible indirect cause of oil dilution is reduction of viscosity through shearing of the viscosity index improvers in multigrade oils or temporary viscosity loss at a high shear gradient.

In the case of severe oil dilution, the lubrication of highly stressed engine components, such as bearings and camshaft, is compromised because the lubricant film collapses due to low viscosity under heavy load and metal surfaces are then exposed to severe wear. A high content of fuel in the oil also spoils the effectiveness of the additives (e.g., their dispersing capacity) so that, for instance, residues are deposited inside crankshaft oil drillings and lead to bearing damage. Oil dilution is, therefore, one of the main criteria for determining oil change intervals, especially for vehicles that operate primarily over short distances and achieve a low accumulated mileage, and, therefore, for which a timed oil-change period applies. Oil dilution may also be concealed by opposing effects (oil thickening). The dilution of the oil caused by the fuel can be measured according to DIN 51 565 by means of a distillation process. Due to the lower cost and high accuracy, however, a gas chromatographic analysis technique is now used most frequently. The viscosity drop through shearing of viscosity index improvers is determined by different methods of viscosity measurement. Some examples of the potential extent of oil dilution in vehicles with gasoline engines in short distance operation are provided in **Fig. O64**.

~~~**Viscosity increase.** Oil thickening—that is, viscosity increase of engine oils during operation—can have different causes: evaporation of low-boiling oil

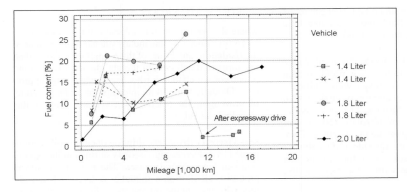

Fig. O64
Fuel dilution under
extreme short-distance
operation

components, increase of content of solid foreign substances, oxidation, and polymerization of oil components through oil aging.

An increase in viscosity can occur particularly in long-term operation, under heavy load, and at high speed. Oil thickening jeopardizes the cold starting of the engine due to increased drag; in addition, the oil supply to sensitive engine components such as the rocker arm is delayed after a cold start and fuel consumption rises. Oil thickening is, therefore, a criterion for determining the oil change interval. When engine oils are formulated, the inclination to thickening can be reduced by both the choice of basic oils and the additives. Basic oils with a narrow boiling point range give lower evaporation losses and hence less thickening than those with a wide boiling point range or formulations composed of a mixture of volatile low-viscosity and high-viscosity components. Accordingly, synthetic basic oils, due to their uniform structure, have low evaporation losses and, thus, have advantages with regard to thickening. Among the additives contained in engine oils, antioxidants and dispersion agents contribute essentially to the reduction of oil thickening. In used engine oils, thickening and diluting phenomena may overlap. To insure lower levels of oil thickening, the maximum permissible thickening for approval of engine oils is currently being determined in a series of international engine tests and also special engine tests conducted by some vehicle manufacturers. Oil thickening in practical driving operation, under heavy load and with longer oil change intervals, is shown in **Fig. O65**. Top quality oils thicken by approximately one SAE viscosity grade, whereas low quality products thicken more.

As explained in the discussion of fuel economy, the choice of a suitable viscosity grade is a necessary, albeit not the only, condition for selection of a fuel-economy oil. Therefore, only products that still ensure a significant fuel consumption advantage at the end of the oil change interval can be considered as "real" fuel-economy oils—for example, oils that do not show a pronounced viscosity increase during the oil change interval.

The factors that can lead to viscosity changes in engine operation are listed in **Fig. O66**. It is important to note that the thickening and diluting phenomena can overlap; the correct viscosity of used oil alone is, therefore, not an indication of good oil quality.

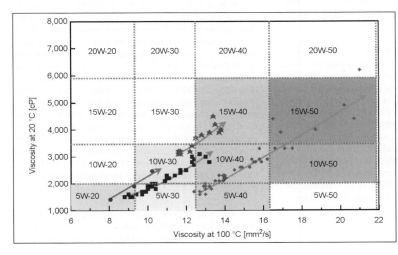

Fig. O65
Oil thickening under
heavy load, plotted over
30,000 km of oil
mileage; results for
several vehicles

Viscosity Change in Used Oil Through		
Decrease		Increase
←		→
←	Shear stress	→
←	Structure viscosity	
	Soot, foreign matter	→
	Evaporation loss	→
	Basic oil oxidation	→
←	Polymeric oxidation	
←	Fuel	
	Pressure	→

Fig. O66 Summary of viscosity changes in operation

~~**Contamination by foreign matter.** The engine oil is contaminated with different types of foreign matter:

- Solids: soot that causes abrasion
- Liquids: fuel, water, glycol (antifreeze)
- Gases: oxides of sulfur, oxides of nitrogen, oxygen

The gases speed up the aging process of the oil, while the solid and liquid foreign matter must be kept suspended by the oil: filter systems are of little value in this process.

~~~**Dilution.** The dilution of oil with fuel has already been discussed as the main reason for a drop in the viscosity. *See above,* ~~Aging of Oil ~~~Viscosity drop.

~~~**Glycol contamination.** Glycol from engine coolant is only present in the oil if a sealing system is defective or corrosion occurs on wet cylinder liners in water-cooled engines. The performance characteristic of engine oil is significantly lowered by its glycol content, in particular by residue formation increases in the engine. The glycol content is determined according to DIN 51 375 by pouring it out with water and subsequently performing a gas-chromatographic analysis.

~~~**Metal content.** Mechanical and corrosive wear of the engine leads to a gradual increase of metal content in engine oil. During the break-in phase of a new or an overhauled engine, the rate of increase in metal content is relatively high; however, it reduces by the end of the break-in phase and then increases gradually again with increasing engine age. However, due to improved production methods, the previously very high rates of metal abrasion during the engine break-in phase have been significantly reduced. High-quality oil providing high levels of wear and corrosion protection lowers the quantity of metal content in oil because of the low rate of wear.

In addition to oil quality, the engine itself and the operating conditions play a decisive role in terms of abrasion and metal content. Short-distance driving with many cold-starts generally leads to a high rate of fuel admixture with the engine oil, and the resulting oil dilution increases the wear and hence the metal content. During engine and oil development, continuous monitoring of the metal content in oil and of its composition can give important indications about the development of wear on individual components (bearing, cylinder liners) and about the reduction of wear and corrosion protection additives in the oil. Analysis of used oils generally reveals the following contaminants: silicon and sodium from the environment via the air filter; aluminum, iron, and chromium from the combustion chamber; copper, lead, and tin as bearing material; and zinc, phosphor, potassium, and magnesium as additive components in the oil being used. Based on the content of these individual elements and their increase with time, conclusions can be drawn about wear on individual engine components.

An increase of the contents of foreign metals accelerates oil aging due to the catalytic effect of the metals. As such, they are passivated by means of special additives—metal deactivators—and kept suspended by dispersion agents in the oil. Their removal from the engine oil through the main or bypass filter is hardly possible because their particulate size is generally smaller than the size of filter pores. In addition, the ratio of oil-fill amount to engine output is also relevant, so that cross-comparisons between different engine types are only possible within limits.

~~~**Methods of measuring foreign matter**

DIN 51365 (centrifuge method). A sample is dissolved in a solution and then centrifuged. In further steps, the oil in the centrifuged residue is removed ("washed") with a solvent. In the standard procedure, this is done with n-heptane as solvent. The solvent method can detect soot, abrasion, foreign solid particles such as dust and sand, and aging products already present in the oil as solid particles. In addition, the aging fall-out products are detected by means of the n-heptane solvent.

The method can be applied to diesel and gasoline engine oils (and industrial oils); nevertheless, with oil samples from modern diesel engines (Euro III technology), the difficulty gradually arises that the soot particulates are becoming too small to be adequately centrifuged from the oil. Therefore, the meaningfulness of this method has recently come to appear debatable.

If the method is carried out with n-heptane and then with toluene as solvent, the difference between the results of the two processes can be used as a criterion for evaluating the oil-aging effect; this is because toluene exhibits a good solvent characteristic with regard to the products of aging. A serious disadvantage of this method is the relatively high time required for the test. Manual handling is intensive in this process, and its precision is often poorer in reality than the standards specify. Discussions about removal of this method from the DIN standards is ongoing.

Filtration through a membrane filter. This process is also conducted by dissolving a sample in a solvent and then filtering it through a membrane filter. The filter cake and membrane filter are then washed free of oil using a solvent.

The content is determined according to the withdrawn DIN 51592 regulation. As standard procedure, a 0.45-μm cellulose acetate membrane filter is used, with toluene as the standard solvent.

The method is well suited to all industrial oils. It is not applied as a standard procedure for engine oils because the filter is often fully clogged when dirty samples are used. As for samples with soot content, part of the soot is also not held back by the filter.

The lack of precision of the method was one of the reasons why it was withdrawn as the DIN standard. In addition to soot, all kinds of filterable solid dirt are measured by this method, just as with the centrifuge method. Analogous to the method described above under the certrifuge method 1, the solvent influences the magnitude of the result.

Determining the soot content by means of IR (DIN 51452). This method is very popular and widely accepted in the mineral oil industry. When considered on the basis of its principle, it is highly suitable for determining the soot content used in diesel engine oils up to approximately 4%. With content levels >4%, it is also possible to evaluate higher soot contents by diluting the sample. However, in contrast to determining the soot content by means of the centrifuge or filtration method, not all the dirt is evaluated with this method, but only the soot content. Therefore, it is suitable neither for gasoline engine oils nor for industrial lubricants.

With this method, the soot content is determined on the basis of correlation factors reflecting the reduction in the transparency caused by the soot content. The weakness of this method is that these correlation factors were actually defined for other methods of determining foreign matter content. For instance, the correlation factor used in the DIN method is too low, because according to data from Shell it is based on the filtration method. Therefore, the correlation factor mentioned above finds application in the fundamental studies conducted on this method by Severin.

It would be more correct, although not very user-friendly, if only the absorption of light per unit thickness of layer (e.g., A/cm) were used instead of soot content in mass percent.

The method is easy to apply and can be readily automated with modern IR devices coupled with an exchanger for the samples and appropriate evaluation programs. A disadvantage is that the spectrum is required for the particular fresh oil to compensate for the extinction caused by the oil itself.

Determining soot content by means of UV. This method (IFP 303/CEC-L-82-A-97) works similarly to the IR method: that is, it is based on the absorption of light by the soot. In contrast to the IR method, the oil is dissolved in toluene so that the detection limit can be compensated from the beginning due to a very high concentration of soot and the associated total absorption of light. Experience shows that the correlation coefficients used in this method to calculate the soot content are realistic, and the results correspond to the results obtained with the IR method when the factor formulated by Severin is applied.

This method is not as widespread in the mineral oil industry as the IR method.

Heated blotter spot test (Shell method). This photometric method is also a process that is based on light reduction by means of the soot content. A defined drop of the sample is applied to a likewise defined filter paper. After storage of the filter paper for over one hour at 80°C, the reduction of light transmitted through the "oil spot" on the filter paper is measured, whereupon the factored content of foreign matter on it can be calculated.

This method is quite suitable for diesel engine oils with foreign matter contents of approximately 3–4%, but not for gasoline engine oils and industrial oils. The process delivers results comparable to the IR method, and it is comparable with the UV method. The advantage of this method is that it is less time-consuming and that a large number of samples can be tested in a short time.

~~~Soot, solid foreign matter. The sum of all insoluble components contained in used engine oils is designated as solid foreign matter. This primarily consists of soot particulates from the combustion process; solid products from the engine oil aging process; metallic abrasion particles; and dust that enters into the oil with the engine intake air (measured as silicon). Furthermore, solid lubricants such as graphite and molybdenum sulfide, if admixed with the engine oil, are also included in this amount of foreign matter.

An increased foreign-matter content of the engine oil leads to thickening of the oil and speeds up the oil aging process. The content of solid foreign matter is, therefore, one of the main criteria for an oil change or for determining the oil change interval. The increase of foreign matter content in the engine oil depends on both the type of engine and the operating conditions. For instance, the foreign matter content in prechamber diesel engines is higher than in direct injection diesel engines and lowest in gasoline engines. Operation in the upper load range increases the foreign matter content through an increase in soot formation during combustion (diesel engines) and through an intensified oil aging process. Dispersion additives in the engine oil keep the foreign matter in suspension and prevent it from joining to form larger particulates and then falling out in the oil circulation system and leading to clogging and deposits. Efforts are made through the choice and concentration of dispersion agents to keep the size of foreign matter as small as possible to prevent engine damage through foreign matter.

If necessary, a far-reaching analysis of the foreign matter can provide information on its composition. This may, for instance, be necessary in the development process for both engines and engine oils. Effective reduction of the content of foreign matter during engine operation, by means of oil filtration, is possible neither through the main nor through the bypass filter,

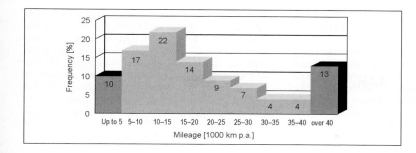

Fig. O67
Annual mileage for the cars of a German manufacturer (Source: SAE paper 951035)

because the particulate size is generally less than the size of the filter pores.

~~~**Water.** Used engine oils contain small amounts of water from condensed exhaust that admixes with the oil, particularly during frequent short-distance operation. Due to damage to sealing elements or corrosion of wet cylinder liners, large amounts of water can mix with oil, and it is common for the water content to be between 0.1 and 1% of the oil. Water is an undesirable component in engine oil, because it promotes the formation of acids, corrosion, and oil aging; at higher concentrations, the lubricating properties are obviously also impaired. The disadvantageous effects of low water concentrations are neutralized through the additives used in oil.

The water in the oil can be measured using different methods. Simple qualitative methods, in the case of high water content, are a visual evaluation test for emulsion formation. At low concentrations, the so-called crackle sample is used, whereby the oil sample is heated in a test tube and the presence of water is detected through the crackling noise. Quantitative methods include the distillation method with addition of xylene according to DIN ISO 3733, the titration method with methanol according to Karl Fischer based on DIN 51 777, and the so-called FINA test that uses calcium hydride as a reaction agent and is commonly used due to its comparatively low cost and accuracy.

~~**Effect of different operating conditions.** More than 20% of the car output of a particular German

manufacturer operates under unusual conditions (**Fig. O67**). Ten percent of all vehicles have a mileage under 5000 km/year. These cars are mostly used as second cars, thus mostly in urban short-distance operation, with frequent cold starts and often under "stop-and-start" conditions. On the other hand, 13% of all vehicles exhibit annual mileages of over 40,000 km. It is assumed that these vehicles are frequently used at high speed on the expressways.

~~~**Cold operation.** Short distances of travel with frequent cold starts lead to very low oil temperatures, so that large amounts of fuel may accumulate in the engine oil. This dilution of the oil minimizes its lubricating capacity and causes it to become more reactive due to the increase in aromatic components, so that an increased rate of engine wear can occur.

Some examples of the spectrum that fuel entrainment in oil can take are shown in **Fig. O64**; this applies to vehicles with a gasoline engine in short-distance operation. Fuel content levels far above 10% have been observed in oil after only a few thousand km, in a fleet-test framework under typical second-car conditions. For one of the vehicles, the values have been plotted to show how the fuel content returns to normal values after an expressway stage.

The effect of strongly diluted oils on engine wear is illustrated in **Fig. O68**. A rapid rise in the iron content in the engine oil is a clear indication of the engine wear processes. This abrupt rise was observed in all vehicles, although at different mileage levels due to the different models. Engine evaluation after 10,000 km, in short-distance operation, indicated that, under these

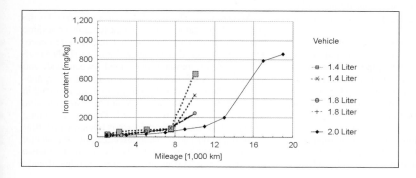

Fig. O68
Increase of metal content levels under extreme short-distance operation

operating conditions, significant wear had occurred on the cylinder liners, piston rings, valve gear, and bearings.

~~~**Hot operation.** High speeds, especially under heavy load leads (typical for cars with mileages above 40,000 km/year) result in high oil temperatures with an increase in nitration, oxidation, and viscosity of the oil as the consequence. This increases the risk that the wear protection additives in the engine oil will be prematurely exhausted in the case of long oil-change intervals. Excessive increases in the level of the viscosity of the oil, which frequently occur under these operating conditions, jeopardize the functioning of the engine and lead to increased fuel consumption.

~~**Oil consumption.** The oil consumption by combustion engines is influenced by design parameters, state of wear, operating conditions, and engine oil. In four-stroke engines, oil is consumed by a number of different mechanisms: piston rings in the combustion chambers, valve guides in the intake and exhaust ducts, and evaporation or atomization in the crankcase with voiding through the crankcase ventilation. Oil consumed in this manner is generally burned or oxidized in the catalytic converter. In addition, oil loss may occur through damaged sealing elements. Oil consumption has been continuously reduced through improved production methods. For passenger cars, after the engine break-in phase, consumption levels below 150 g/1000 km are common. In the break-in phase of a new engine, however, substantially greater oil consumption, up to about 350 g/1000 km, can occur. In commercial vehicle engines, after the break-in phase, oil consumption levels of up to 350 g/1000 km are common.

Oil consumption can be reduced by means of the particular oil, including formulations with lower evaporation loss, achieved by omitting low-viscosity mineral basic oil components. It is still possible to produce low-viscosity oils by using alternative synthetic basic oils.

If little or no oil consumption is noticed during the operation of the vehicle, this may be an indication of detrimental oil dilution—for example, caused by unburned fuel in short-distance travel with a cold engine. With constant oil consumption necessitating topping-up between oil changes, the oil is refreshed by the re-filled amount; that is, the consumed additives are partly replaced and the contaminants diluted. With regular refills of large amounts, the oil change interval can be extended.

In two-stroke gasoline engines, with total-loss lubrication, the oil consumption is proportional to the amount of fuel consumed: for example, with a mixing ration of 1:100 and a fuel consumption of 5 L/100 km, the result is oil consumption of 0.5 L/1000 km. This oil consumption, in spite of the low dosing rate quoted here, is many times greater than is the case in four-stroke engines of the same output power.

The cylinder liners in slow-running two-stroke diesel engines for ships are likewise lubricated with total-loss lubrication. The required amount of oil can be set based on the fuel quality. When residual oils with high sulfur content are used as the fuel, a dosing rate in the range of 0.8 to 1.4 g/kWh is required in order to protect the cylinder liners against corrosion and wear. With low lubricant dosing, even with highly alkaline oils, no uniform distribution of sufficient alkalinity is possible along the cylinder circumference, so that wear strips can form in the areas with a thin oil film. Medium-speed ship diesel engines with separate cylinder lubrication systems require dosing rates roughly similar to the cylinder oil, about 1.0 g/kWh.

~**Properties and characteristic values.** The methods of determining physical and chemical properties of engine oils have been continuously developed and rendered more precise in the course of more than 100 years of experience with the requirements and behavior of engine oils. Viscosity metering is a good example. At the beginning oil was tested only to determine whether oil flowed from the oil can in a cold environment; now oil viscosity is determined in a complex and globally standardized measuring process, reflecting as far as possible the operating conditions in the engine both during a cold start and at full throttle steady-state operation. The objective of all applied laboratory methods is the determination of characteristic data of an oil and their evaluation for the suitability of application in engines. Because engines and fuels are continuously changing, the requirements for engine oils and the analytical methods used to test them are changing as well. In addition to upholding standardized testing methods and their improvement through newly available measuring techniques, a running verification of whether individual characteristic data will still result in relevant deductions about the behavior of the oil in the engine, or whether new methods are required, is a constant concern.

The following passages explain the most important methods for investigating fresh engine oils. Despite modern and extensive analysis technique, one cannot dispense with engine operation test in the development of new engine oils, in test-stand experiments, and in road tests. Such tests, therefore, comprise an important element in the specification of engine oils (*see above*, →Oil ~Classification, specifications, and quality requirements). Laboratory investigations are applied in evaluation of waste oils.

~~**Ash content.** The ash content of an engine oil is the portion of inorganic materials that remains as residue after reduction to ash (burning out and incandescence; **Fig. O69**). This is determined according to DIN 51575 at 775 C. A differentiation is made between oxide and sulfate ash. Noninflammable components of organometallic additives as well as impurities are included in ash content; in waste oils, silicon (dust) and metallic abrasion are considered as well. When determining sulfate ash content, the oxide ash is treated with sulfuric acid and burned to incandescence until the weight remains constant. In engine oils, the common practice is to determine sulfate ash because sulfates already

Application	Fuel	Type of engine	Sulfate ash
Outboard engines	SI	Two-stroke	0–0.05
Motorcycle	SI	Two-stroke	0.1–0.2
Car	SI	Four-stroke	1.0–1.3
Gas engines (stationary)	SI	Four-stroke	0–2.0 (dependent upon engine type and gas composition)
Motorcycle	SI	Four-stroke	0.9–1.2
Car	Diesel	Four-stroke	1.1–1.5
Truck	Diesel	Four-stroke	1.5–2.0
Ship (piston lubricant)	Diesel	Four-stroke	Up to >10

Fig. O69 Typical contents of sulfate ash in qualitative, high-quality, European engine oils

occur during engine operation through the contact with fuel with sulfur content. If sulfate ash is determined both from fresh oil and waste oil, it is possible to make deductions about dust and abrasion.

Basic oils without additives do not contain any ash. Engine oils mixed with additives have different ash contents depending on the type and concentration of the organometallic additives used. These reflect specific application requirements. For instance, two-stroke boat engine oils may not have ash (formation of bridges to the spark plugs) and cylinder oils for ship engines must have a high ash content for neutralization of the sulfurous acid resulting from combustion due to calcium and magnesium carbonate. Within individual groups of engine oils and with a knowledge of the organometallic additives, the ash content can provide limited information about the oil quality. Due to the reduction of sulfur content in fuel and development of highly effective ash-free organic additives, a tendency to low ash content has been evident in recent years.

~~**Color.** The color of an engine oil is not a quality feature. Only in the case of fresh oil can it be used for quality control purposes, together with other criteria. Fresh oils are already dark due to the additives. In general, oil becomes darker the more agents it contains. Used oils, especially from diesel engines, are already black after a short period of running, without any impairment of performance.

~~**Corrosion protection.** Corrosion protection testing of engine oils involved various different test pro-

cesses. In principle the test methods are very similar. Generally speaking, standard test pins or plates are moistened with the oil to be tested, then exposed at somewhat increased temperatures to a variety of corrosive mediums, such as seawater.

~~**Elastomer compatibility.** Engine oils of different composition have different effects on the sealing materials used in the engine as fresh and waste oil. An imperative condition for safe engine sealing over its entire life cycle and prevention of environmental pollution through leakage oil is the compatibility between elastomers and engine oils. Higher oil temperatures owing to powerful engines have in the course of time led to nearly exclusive application of fluoro-rubber as a sealing material in high-load sealing elements. Parallel to this, the compositions of engine oils have also changed. Aromatic basic oils which can lead to swelling with subsequent brittleness, especially of nitrile rubber, have in many cases been replaced by strongly hydrogenated paraffinic or synthetic oils. These oils are less inclined to cause swelling of sealing materials. **Fig. O70** provides an overview of materials for radial shaft seal rings and their thermal resistance and qualitative compatibility with basic oils.

There are a variety of test methods, different in detail, which are continuously undergoing further development by engine manufacturers in particular. In principle, the test methods are very similar. In general, standardized qualities of different elastomer types are stored in the oil to be tested at raised temperatures and afterwards checked for changes, for instance volume change, for crack formation, and for changes in breaking strength.

In addition to basic oils, the additives used in engine oils also influence the sealing materials; the indispensable dispersion agents in particular may lead to brittleness in fluoro-rubber. This presumably involves an amine-base catalyzed after-reaction of the fluoro-rubber. Oils aged inside the engine affect elastomers much less critically even after a short running period, so that the impairment—for example, the breaking strength of the fluoro-rubber—remains within tolerable limits. Therefore, testing of elastomer compatibility should always take place in a full engine under realistic operating conditions. When this occurs, instead of the sealing elements, special test bodies which are fitted inside the oil sump of the engine can also be used. Long-term investigations in engines have confirmed that the combination of fluoro-rubber with synthetic engine oils guarantees flawless functioning of the sealing elements over the lifespan of the engines.

Requirements	NBR (nitrile rubber)	ACM (acrylate rubber)	VMQ (silicon rubber)	FPM (fluoro-rubber)
Thermal resistance	Limited up to approx. 100°C	Good, up to approx. 130°C	Very good, up to 150°C	Very good, up to approx. 150°C
Compatibility with basic oils	Limited strongly swelling in aromatically rich mineral oils and esters	Limited strongly swelling in aromatically rich mineral oils and esters	Good but strongly dependent upon the viscosity	Very good Only very few restrictions (swelling in some ester types)

Fig. O70 Suitability of different elastomers as radial shaft seal ring materials

~~Evaporative loss. Evaporative losses of low-boiling engine oil components can occur due to high temperatures, for instance at the piston rings and on the piston underbody. They lead to undesired oil thickening and to increased oil consumption.

Selection of suitable basic oils can reduce evaporative losses. Oil formulas manufactured without low-viscosity minera-oil–based basic oils (and/or basic oils with a wide boiling range) show little evaporative losses, as do synthetic basic oils. The previous solutions of additives — that is, their oil components (structure, manufacture) — must also be included in the evaporative losses.

The evaluation of an engine oil with regard to the evaporative loss can be done by means of the so-called Noack laboratory test, DIN 51 581 — a component part of many specifications. In the Noack test, a sample of engine oil is maintained at a temperature of 250°C over a period of one hour. The oil vapors that form are continuously extracted with a negative pressure of 20 mm water head. Maximum permissible weight loss, depending on the engine oil specification, is 13–15%. **Fig. O71** shows the Noack evaporative losses of commercial type engine oils according to their quality grade.

In vehicle applications, high evaporative losses of engine oils are observed with engines frequently operated in the upper load range and at high speed.

~~Flashpoint. The flashpoint is the lowest temperature at which vapor develops from an oil sample in amounts such that if the oil vapor/air mixture is exposed to a flame it ignites and explodes weakly. With fresh oils the flashpoint is influenced by low-boiling oil components. It is covered in a series of specifications on storage safety. The flashpoint of fresh oils is normally measured according to DIN ISO 2592 in an open crucible.

In used oils, the flashpoint sinks due to fuel components that may penetrate into the oil especially due to frequent cold operation of the engine, due to overenrichment of the air-fuel mixture, or due to interrupted combustion owing to engine damage. The flashpoint of used oils is normally determined according to DIN

EN 22719 in a closed crucible. With spark ignition engine oils, this parameter is less meaningful, because even small amounts of fuel result in a pronounced lowering of the flashpoint. For diesel engine oils there are derivative tests that can be applied quickly, for instance at fixed temperatures of 160 or 190°C.

~~Foaming characteristics. Due to turbulence inside the crankcase, engine oil absorbs gas that is partly dissolved but partly coexists in the form of bubbles. Gas bubbles in oil and on its surface should actually be considered separately, but are considered together here under the term "foam." With declining viscosity of oil, a reduction of the surface tension, and an increase of the turbulence, gas absorption increases: that is, thinner oils, increasing oil temperature, the application of dispersing additives, and increasing engine speeds encourage the foaming effect. In addition, water and glycol (in case of penetration into the oil) encourage foaming: that is, used oils mostly have a poorer foaming characteristic than do fresh oils. Foaming can be additionally encouraged if vibrations occur inside the crankcase at critical speeds and in case of sudden pressure relief.

Foam in oil can, in extreme cases, lead to reduced oil flow due to the pressure relief associated with it and, thus, to engine damage. Due to a high gas content, the engine oil will also be compressible and can then no longer safely fulfill its hydraulic tasks.

With special additives, the antifoaming agents, the surface tension of oil is increased and the surface foam significantly reduced. On the other hand, high concentrations of antifoaming agents reduce the so-called air removal capacity of oil in an unfavorable manner. Because the mutual influence of basic oils and additives with regard to foaming characteristic is not yet completely clarified, foaming characteristic must be optimized on the basis of test data gathered during the development of an oil type.

The foaming characteristic of oils can be assessed by means of laboratory tests. The method described in ASTM D 892 is applied mostly by introducing air in the oil through a sinter brick. Both the magnitude of the forming foam and its disintegration time can be determined with this method. With ASTM D 6082 a method for determining foaming characteristic is now available that should correspond more closely to the behavior of the oil under engine conditions measured at high temperatures. The air removal capacity is measured according to DIN 51 381. Transferability of these static methods to the behavior of oils under dynamic conditions in the engine is, however, limited. A significant improvement was achieved through the dynamic RUB (Ruhr University Bochum) foam test.

~~Neutralization capacity. The neutralization capacity of an engine oil is its capacity to neutralize acidic combustion products that penetrate into the engine oil and the acidic aging products of the engine oil, thus preventing wear and residue formation by virtue of alkaline additives. The gradual exhaustion of the neutral-

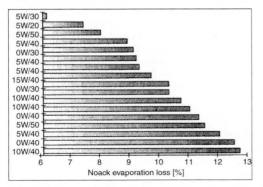

Fig. O71 Noack evaporative losses of commercially available engine oils

ization capacity of the engine oil during the operation through the consumption of the additives is a significant criterion for determining oil change intervals.

The neutralization capacity of engine oils is determined by different methods. Characteristic data of these methods include the total base number (TBN) and the strong base number (SBN), which specify the concentration of the entire and the stronger bases, and the total acid number (TAN) as well as the strong acid number (SAN) for quantitative specification of all acids and stronger acids. Because in used engine oils acidic combustion products and alkaline additives are concurrently present, in the range of values 4 to 9 pH both TBN and TAN are measurable. This also applies to a lesser extent to nonacidic fresh oils, because amphoteric additives (e.g., zinc dithiophosphates) react both with the acidic and also with the basic reagent for determining the TBN and TAN. A further characteristic variable for specifying the neutralization capacity of oils according to DIN 51558 is the neutralization number (NN). It is generally not specified for engine oils because a color change occurs during titration that is not recognizable in dark used oils.

~~~**TAN.** The total acid number describes the content of weak and strong acids in engine oil. It is determined with the help of electrometric titration with tetramethyl ammonium hydroxide according to ASTM D 664 and DIN EN 12634. It is specified in mg KOH/g—that is, the equivalent amount of potassium hydroxide that would be consumed to neutralize the acids. In principle, the increase of acid content in engine oil during the operation period is a measure of oil aging. Because the TAN, however, covers both the weak organic—and, thus, for the engine uncritical acids—as well as the strong acids (sulfuric acid that has formed from the sulfur content in the fuel), its meaning is limited. In addition, even in the case of nonacidic fresh oils with additives, TAN is measured because, for example, zinc dithiophosphates (amphoteric substances) react with both acids and basic reagents. A typical TAN curve with increased mileage of oil due to the conversion of zinc dithiophosphates is characterized first by a drop and subsequently by a gradual increase. The method for determining TAN is normally applied to used oil analysis of gas engine oils with low ash content. The stronger acids that are dangerous to the engine are covered by SAN (strong acid number).

~~~**TBN.** The total base number describes the alkaline reserve of engine oil and, thus, its content of effective alkaline additives (detergents, rust protection). The detergents determining alkalinity (soaps) are required to guarantee engine cleanliness. Furthermore, the alkaline additives neutralize the acids occurring during the combustion process. TBN is specified in mg potassium hydroxide/g engine oil (mg KOH/g) —that is, the equivalent amount of KOH, which corresponds to the alkalinity of different alkaline additives contained in oil.

TBN is determined by means of titration according to a variety of methods, such as with perchloric acid according to ISO 3771 or with hydrochloric acid according to ASTM D 664 or IP 177. The TBN is determined from the admixed acid quantity at the transition point between basic and acidic behavior. Different methods result in different values for TBN; for instance, the method according to ISO 3771 generally results in higher values. Based on the application of engine oil and sulfur content of the fuel, widely differing concentrations of the alkaline detergents and rust protection additives are required and, therefore, also different TBN levels. While fresh spark ignition engine oils, depending upon quality level, exhibit a TBN of 5–12 mg KOH/g, the TBN of cylinder lubricating oils for large diesel engines which are operated with residue oil as fuel lie in the range of 70 to 100 mg KOH/g.

The decline of TBN during the usage period of the oil is an important criterion for assessing the remaining capability and a required oil change if necessary; in this case, due to the "alkalinity" reserve contained in fresh oil, a decline to about 50% of the fresh oil value is generally unobjectionable. Only below that value does the oil lose its capacity to keep the engine clean within the lubrication range from combustion residues and oxidation products and to neutralize the acidic combustion products.

~~**Residue formation.** There is no generally recognized and standardized test method for determining the tendency of engine oils to form residues, especially in turbochargers or in the intercooler system, in the combustion chamber, and on intake valves. The formation of coking residues at high temperatures can be simulated in the laboratory according to Conradson, DIN 51 551, or Ramsbottom, ASTM D 524. For today's engine oils with high additive content, both methods are nonetheless not very useful. They are hardly used at all for engine oils.

Only road tests with long mileage provide a degree of certainty. Only a few vehicle manufacturers prescribe such fleet tests as a basis for approval of oils in the commercial vehicle sector. Responsible engine manufacturers conduct such tests "voluntarily."

~~**Viscosity.** The viscosity of an engine oil describes its thickness or thinness, at defined temperatures. High viscosity means thickness with high load-bearing capacity of a lubrication film but also a large force requirement for moving the metal surfaces separated with an oil.

~~~**Basic terms, units.** To define viscosity, it is appropriate to illustrate the conditions in a lubrication gap in a simplified and magnified manner (**Fig. O72**).

A surface, A, moved by a permanent action of force is held at a gap by an oil film on a fixed surface, B. The force to be applied based on the size of the surface is defined as shearing τ,

$$\frac{\text{Force}}{\text{Area}} = \text{Shearing stress } \tau .$$

The unit for the shearing stress is pascal (Pa).

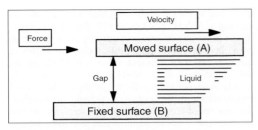

Fig. O72 Conditions in the lubrication gap

The behavior of the oil in a lubrication gap can be depicted as movement of a multiple number of thin liquid layers which are shifted relative to one another between a moved and a fixed plate. In this case, the liquid layer on the moved plate also exhibits its velocity. The liquid layer on the fixed plate does not move. Within the lubrication gap there is, therefore, a velocity gradient in the oil film, which increases with the increase in velocity of the moved plate and reducing gap (thin lubrication film).

$$\frac{\text{Velocity}}{\text{Gap}} = \text{Velocity gradient } D$$

Because in this equation of velocity and lubrication film thickness the gap cancels out, the remaining unit for the velocity gradient is s^{-1}.

For ideal liquids, whose properties will be presented below, the required shearing stress increases directly proportional to the velocity gradient, whereby the dynamic viscosity of the liquid between the moved surfaces is the proportionality factor.

$$\text{Shearing stress } \tau = \text{dyn.viscosity } \eta \cdot D$$

This relationship was discovered by Isaac Newton (Newtonian liquids). If this equation is resolved for the dynamic viscosity, η,

$$\eta = \frac{\tau}{D}.$$

The dimension for the dynamic viscosity is then

$$\eta = \frac{N \cdot m \cdot s}{m^2 \cdot m} = \frac{N \cdot s}{m^2} = 1 Pa \cdot s.$$

If an oil flows under the influence of the force of gravity—for example, from an oil can or in a capillary tube—the mass that acts on the flow process must also be considered. The mass is linked with the density via the relationship mass = density × volume. Because the volume under normal conditions during the flow process does not change, the dynamic viscosity can be correlated with the density. From the relationship between dynamic viscosity and density one can derive kinematic viscosity:

The dimension obtained for kinematic viscosity is m^2/s; it is, however, normally specified in terms of the unit mm^2/s.

In an engine, kinematic viscosity is only of subordinate importance; it is, however, a variable that can

be determined in a relatively simple viscosity measuring process and is, therefore, very good for the classification of oils, for their quality control, and for used-oil investigations in engines, and it is applicable in oil development.

In contrast to the SI units, outdated viscosity values were (and still are) common in viscosimetry (Poise for dynamic viscosity and Stokes for kinematic viscosity):

Conversion of viscosity specifications

1 Pa	=	10 dyn cm^{-2} ($= 10^{-5}$ bar)
1 Pa s	=	10 Poise (P)
1 mPa s	=	1 Centipoise (cP)
1 m2 s^{-1}	=	10^4 Stokes (St)
1 mm2 s^{-1}	=	1 Centistoke (cSt)

~~~**Flow behavior.** The flow behavior of engine oils requires special consideration. In an (ideal) Newtonian liquid, the shearing stress and velocity gradient are directly proportional to one another; that is, their viscosity is constant and independent of the velocity gradient. The majority of liquids exhibit a flow behavior that is, however, divergent:

- Liquids that begin to flow from minimum shearing stress (Bingham bodies)—e.g., fats of toothpaste. At very low temperatures, engine oils may also exhibit this behavior.
- Liquids whose viscosity increases with increasing velocity gradient (dilatant), among them high-viscosity suspensions.
- Liquids whose viscosity decreases with increasing velocity gradient (non-Newtonian): multigrade oils for instance and often paints as well. Measured viscosities of non-Newtonian: oils are termed pseudo-viscosity.

Viscosity curves of these different liquid types are illustrated in **Fig. O73**.

~~~**Measuring method.** Viscosities of engine oils are measured in viscosimeters. Because engine oils are

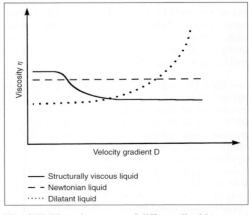

Fig. O73 Viscosity curves of different liquid types

not Newtonian (ideal) liquids and change their viscosities based on the operating conditions, several measuring methods are required to determine viscosity behavior under conditions similar to those in a cold or hot engine at low and high speed. The development of these measuring methods stretches through nearly the entire past century.

Orifice method for measuring kinematic viscosity. At first, the flow behavior of engine oils was defined under the influence of the force of gravity. The outflow time for a quantity of oil from a funnel or through a tube (capillary tube) was measured. The relatively simple outflow beaker (**Fig. O74**), does, however, not provide exact kinematic viscosities as measured values because a high proportion of the energy of the liquid to be tested is consumed through turbulences around the constrictions of cross-sectional areas. "Viscosities" thus measured allow only comparisons between different oils under the specification of the measuring device, but are not defined using the units Newton, meter, and second on which viscosity is based according to the viscosity equations. This also applies to the earlier used short capillary viscosimeter, because also in these devices the viscosity values are falsified by the inlet turbulence.

Typical examples of such viscosimeters are Engler, Saybold and Redwood beakers. Significant improvements came in the form of capillary viscosimeters with long, thin capillary tubes and length/diameter ratios >50:1, in which the inlet losses are sufficiently small and can be negligible. These devices include viscosimeters according to Ubbelohde and Cannon Fenske, which are still used today (**Fig. O75**). With these the measuring method according to Canon Fenske with a rising liquid column in the range of the measuring marks has advantages for dark (used) engine oils, because it is more easily visible when the measuring mark is reached by the liquid. In both devices, however, the flow velocity and with it the velocity gradient changes during the measuring process, because the height of the liquid column decreases. For this reason, measured results for structurally viscous liquids, whose viscosity decreases with the increase of the velocity gradient (e.g., engine oils) are only interpretable

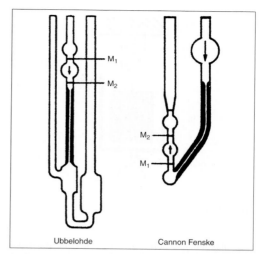

Fig. O75 Different versions of capillary tube viscosimeters

to a limited extent. Common temperatures for computing the viscosity index, for comparative evaluation of engine oils as well as for determining the viscosity classes, are the measurements of kinematic viscosity at 40 and 100°C.

Kinematic viscosity at 40°C does not by itself constitute a quality characteristic for fresh (i.e., unused) engine oil. It is, however, frequently used as a quick measuring method for comparing a used oil value with fresh oil value. This applies to the evaluation of used oils in engine tests such as the sequence III E test (API, ACEA) or the VW T4 test (PV 1449-VW 502.00), in which the viscosity increase at 40°C is used as a measure of oil oxidation as well as for quick estimation of the state of the used oil in the laboratory, showing, for instance, the first hints of dilution with fuel and water or of thickening.

The main application of kinematic viscosity at 100°C derives from the definition of the viscosity classes for engine oils. It also serves as a starting value for determining shearing stability based on shear rate.

Rotating viscometer for measuring dynamic viscosity. The limited significance of kinematic viscosity, measured in the traditional way, in relation to the dynamic stress to which oil is exposed in the engine led to the development of viscometers (sometimes called viscosimeters, particularly in Germany), which work under more practical conditions—that is, at higher flow velocities and, thus, higher shearing rates—and over a more extended temperature range than the traditional apparatus does. Such rotating viscometers (**Fig. O76**) can measure dynamic viscosity but clearly cost more than capillary viscometers. It must be ensured in particular that the temperature of the oil in the measuring gap remains constant during the measuring process. Two measuring

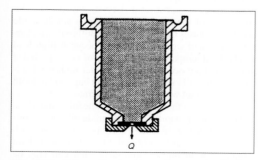

Fig. O74 Schematic drawing of an outlet beaker (Source: Haake)

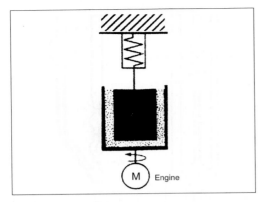

Fig. O76 Rotating viscometers

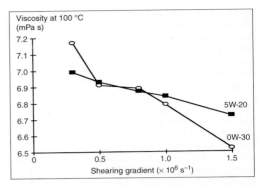

Fig. O77 Temporary viscosity loss of multigrade oils at high-shear gradients

principles are generally possible: either the rotational speed of the rotor is specified and the torque is measured as a variable for viscosity or vice versa. Both principles are used in viscosity measurement of engine oils.

Three different approaches are adopted to cover the whole operating range:

- The cold-cranking simulator for measuring "cold-start viscosity" (for adequate starter speed)
- The mini-rotary viscometer, also used to measure cold start viscosity (for adequate flow behavior in the suction section of the oil pump)
- Different high temperature high-shear viscometers for measuring viscosity in full-throttle continuous operation

In the cold-cranking simulator (CCS), the apparent viscosity of engine oil is determined according to ASTM D 5293 at temperatures in the range -30 to $-5°C$ by measuring the rotor speed.

The speed gradient in CCS is relatively high (approximately 10^5–10^6 s^{-1}); it is not checked but rather adapts itself. The CCS measurement is applied in a simulation of the cranking process during a cold start and correlates with the starter speed at low temperatures.

With the mini-rotary viscometer (MRV), the apparent viscosity of the engine oil is determined according to ASTM D 4684 at temperatures between -40 and $-10°C$. Measurement takes place at very low speed gradients ($<10^2$ s^{-1}), which correspond to the flow of the engine oil to the oil pump. To satisfy the viscosity class specification, the pumping limit of an engine oil must lie at least $10°C$ below the starting limit determined in CCS to guarantee that the oil can be pumped during a cold start. Determining the intrinsic viscosity with the MRV is a protracted process, because a 48-hour cool-down time is required for the oil.

In the devices for high temperature high shear (HTHS) viscosity, the dynamic viscosity is measured at a temperature of $150°C$. Measurement takes place at a fixed shear gradient of 10^6 s^{-1}. Both the shear gradient and the temperature correspond approximately to the conditions under full-throttle continuous operation

in crankshafts, in connecting-rod bearings, and between piston rings and the cylinder wall. Under these conditions, the so-called temporary viscosity loss of multigrade oils in particular becomes apparent due to the high shear gradient (narrower lubrication gap, high speed), as does the viscosity decrease due to the rise in temperature (its dependence on temperature). **Fig. O77** shows that at extremely high-shear gradients a higher-viscosity oil (0W-30) according to the SAE classification at $100°C$ exhibits a lower viscosity than the low-viscosity oil (5W-20). The designation "temporary" is used because this viscosity loss is reversible—that is, the original viscosity is measured again after a relatively brief exposure to shear stress at low velocity gradients. Nonetheless, the viscosity increase in the engine due to the high oil pressure is not taken into consideration by this measuring method. This applies analogously to the other measuring methods.

~~~**Shear stability.** In contrast to temporary viscosity loss under stress of brief duration, a strong shear stress over a long period can lead to permanent viscosity loss. The shear stability of an engine oil describes its resistance to permanent viscosity loss caused by "shearing" of the long-chain polymer additives, the viscosity index improvers (VI improvers) used in multigrade oils that cover several viscosity classes. The viscosity loss of the oil due to shearing is not desirable because it causes the loss of some multigrade oil characteristics, and wear damage may occur at high load and high oil temperatures. Shearing of viscosity index improvers may occur especially under high mechanical stress in oil in narrow lubrication gaps and at tooth flanks in contact with high oil temperatures, under exposure to which the VI improver molecules uncoil.

The shearing stability of multigrade oils can be improved by using special VI improvers, those with limited molecular weight and, thus, shortened chain lengths, for instance. Obviously, this can also be achieved by using basic oils with a very high viscosity index, such as synthetic oils, because the required

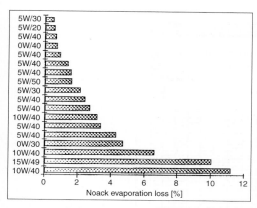

Fig. O78 Shear stability of commercially available engine oils after 30 cycles in the Bosch Injector Test (CEC-L-14-A-88)

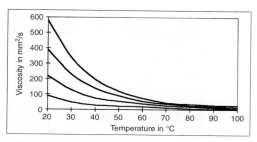

Fig. O79 Viscosity-temperature representation in a linear coordinate system

concentration of the polymer VI improvers can be lowered with their help. The notion that spectacular multigrade characteristics of engine oils over several viscosity classes are not technically necessary has also come to prevail. This means the required amount of VI improvers can be limited as early as the oil formulation stage.

Shear stability of multigrade oils is a component in the specifications that determines the characteristic performance of engine oils. It is measured in the laboratory with the help of a test apparatus comprising a diesel fuel injection pump and injection nozzle, in which engine oil is pumped repeatedly through circulation cycles at 100°C according to DIN 51 381 or CEC-L-14. In extensive experiments with car fleets it has been proved that this method correlates very well with the shear stability of the oil in the vehicle.

As can be seen from **Fig. O78**, substantial quality differences exist between different oils on the market.

A higher level of shear stress prevails, however, in the transmission and axles than in the engine. To cover this higher shearing load, a more exacting tapered roller bearing shearing stability test is used, based on DIN 51 350.

~~~**Temperature dependency.** In addition to the influence of shear gradient on viscosity, pressure and temperature also have significant effects. The viscosity rises with increases in pressure and falls significantly with increasing temperature. The temperature-dependent change in viscosity (VT behavior) exhibits a logarithmic characteristic; with decreasing temperature the viscosity is disproportionately higher (**Fig. O79**).

Using adapted scales in conventional viscosity-temperature diagrams according to Ubbelohde-Walther for illustrating kinematic viscosity, the temperature-dependent viscosity characteristic of basic oils can be depicted as a straight line (**Fig. O80**).

Fig. O80 shows the viscosity-temperature characteristic of different basic oil components (M = mineral

oil base, S = synthetic) as well as the approximate representation of 5W-30 (W) engine oil (shown by a dashed line). Engine oil characteristics cannot be represented correctly even in an Ubbelohde-Walther diagram, which is why there is the curved characteristic between the specified borderline points for a typical 5W-30 oil (viscosity at 100°C about 11.5 mm²/s, at −30°C about 7200 mm²/s) for the viscosity classes of engine oil.

It can be seen that the drop in viscosity with increasing temperature approximately corresponds to that of a synthetic basic oil. For engine operation, however, it would be desirable if the viscosity was constant under hydrodynamic conditions because that guarantees good lubrication over the entire temperature range. Success in this area has been achieved with optimized basic oils with a high viscosity index and additives (VI improvers) by formulating engine oils with small viscosity margins that provide for perfect engine operation throughout the year and under all operating conditions—from cold start at −30°C up to highway operation in summer. Nonetheless, even with the best multigrade oils, engine power loss in cold operation is comparatively high due to the viscosity increase at low temperatures.

~~~**Viscosity index.** The viscosity index (VI) describes the temperature-related kinematic viscosity change of a basic oil or of a ready-for-use formulated oil, but not its actual viscosity. Oils with a low VI show more pronounced temperature-dependent viscosity changes than those with a high VI. The numerical value of VI is based on a conventional scale in which two different American (US) oil types with clearly divergent viscosity–temperature behavior were classified as having VI = 0 and designated LVI (Low Viscosity Index; large change of viscosity with temperature) and a VI = 100, designated HVI (High Viscosity Index; little change of viscosity with temperature). From comparison of particular viscosities with these two reference qualities, the VI of an oil can be specified. In accordance with the common measuring method used for kinematic viscosity at the time this definition was formulated in the United States, viscosity was determined in terms of the "Saybolt Universal Second" unit at 100 and 210°F. In DIN 51563 the

O

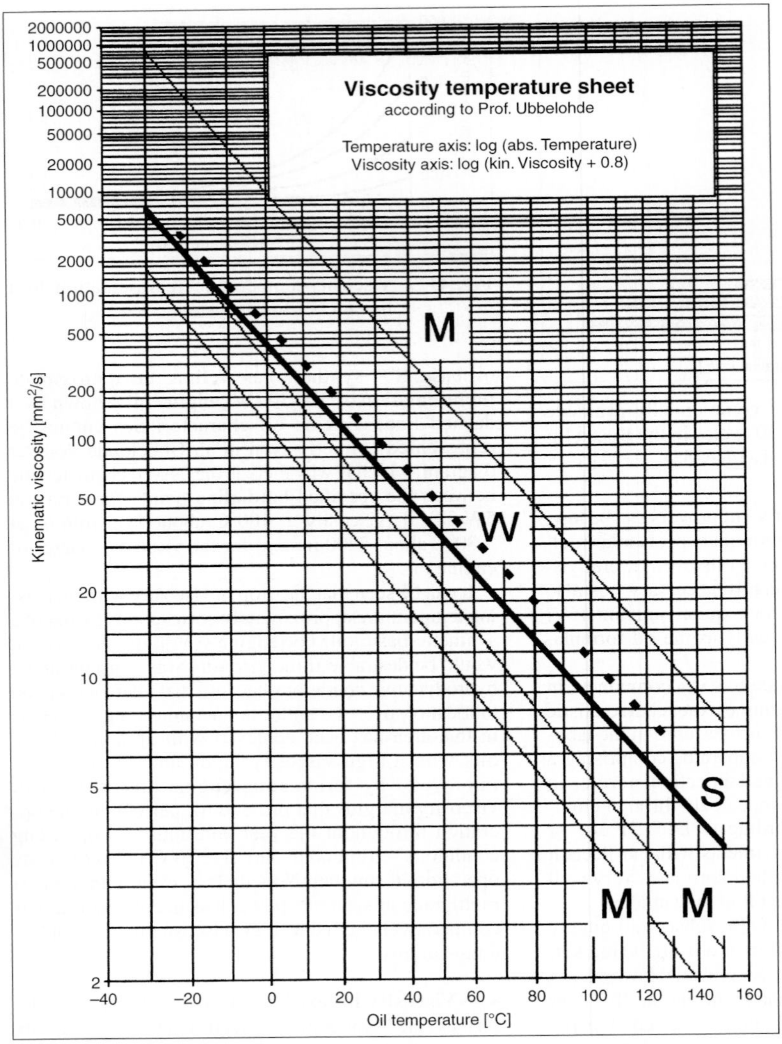

Fig. O80
Viscosity temperature
representation in an
Ubbelohde-Walther
diagram

values of both reference oil series are specified in the standard unit of kinematic viscosity.

Because the temperature-related viscosity change should be as low as possible for efficient engine operation, basic oils have subsequently been developed with a significantly higher VI than the previous maximum of 100 as a result of new manufacturing methods. In **Fig. O79**, the viscosity indexes of typical basic oil components are compared, and in **Fig. O80**, a selection of these oils is depicted in a V-T diagram (viscosity plotted against the temperature). Furthermore, the VI of engine oils can be increased with additives, so-called viscosity index improvers.

~~**Viscosity classes for engine oils.** Viscosity classifications for engine oils have developed parallel to the

expansion of knowledge about rheological relationships, the development of measuring methods, and increasing driver expectations concerning vehicle operating safety.

In the early development of engines, a visual test of flow behavior from an oil can apparently sufficed (**Fig. O81**).

As early as 1911, however, the SAE classification that is still binding today after a number of revisions

Engine oil recommendation
about 1910
"It must still comfortably flow from a can in the cold winter climate."
"It must have a high degree of lubricity."

Fig. O81 Engine oil recommendation in approximately 1910

was introduced in the United States by the Society of Automotive Engineers (**Fig. O82**). In the current specification for fresh oils, a total of twelve classes are listed, six each for "winter" (0W to 25W) and "summer" operation (20 to 60). The low temperature classes are specified according to two methods (viscosity measuring method) in relation to a given starter speed in the cold cranking simulator and flow behavior in the intake section of the oil pump with the minirotary viscometer. In this case it must be ascertained that the oil flows without apparent limit from a minimum 10°C below the temperature at which starting engine rotation is certain.

For 10W oil, the maximum viscosity in the cold cranking simulator for the starting characteristic is 7000 mPa·s at −25°C and maximum viscosity in the minirotary viscometer for oil pump supply is 60,000 mPa at −30°C. The values of the respective viscosity limits are based on practical experience. Just as in the low-temperature classes, the viscosities of high-temperature classes are specified with two different methods, the kinematic viscosity at 100°C and the high temperature high shear (HTHS) viscosity at 150°C to ensure lubrication characteristics under heavy load and at high speed in connection with temperatures clearly above 100°C, as these frequently occur in modern engines. The kinematic viscosity at 100°C used for high-temperature classes with the specified maximum values for individual viscosity classes has nonetheless only secondary importance in practical operation. It expresses little about the operating characteristic of the oil in the engine and serves effectively only for classifying the viscosity classes and providing adequately thin liquid in the case of single range oils. For commonly used multigrade oils, the specifications of low-temperature classes provide for this.

The ways in which the characteristics of fresh oil

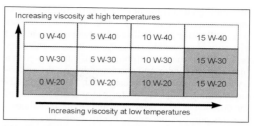

Fig. O83 Viscosity classes of multigrade oils (uncommon or nonsaleable settings are shaded in gray)

change during operation (thickening, thinning), in addition to the viscosity data, are obviously decisive factors in the engine operation. The maximum permissible viscosity change of used oil is specified in a series of engine tests (ACEA specification).

Low-viscosity oils must be especially manufactured for lowering fuel consumption such that their fuel-saving properties are retained over the time they are used in the engine—that is, only little thickening may occur due to aging and foreign substances. The viscosity of gasoline (SI) engine oils is uniformly classified worldwide by the SAE grades. The engine or vehicle manufacturer can select suitable viscosity grades according to climatic conditions (**Fig. O83**).

~~~**Multigrade oils.** Multigrade oils have a lower temperature-related viscosity change than single-range oils, thus a higher viscosity index. In contrast to single-range oils, they are suitable for year-round use in vehicle engines operated at varying load levels and over a wide temperature range. The designation "multigrade oil" refers to the viscosity classification SAE J300, with the low-temperature classes 0W to 25 and the high-temperature classes 20 to 60. A partic-

SAE Viscosity Class	CCS Viscosity [mPa·s] at Temperature °C (maximum low temperature viscosity)	Borderline Pumping Viscosity [mPa·s] at Temperature °C	Kinematic Viscosity [mm²/s] at 100°C		HTHS Viscosity [mPa·s] at 150°C and 10⁶ s⁻¹
	max.	max.	min.	max.	min.
0W	6200 at −35	60,000 at −40	3.8	−	
5W	6600 at −30	60,000 at −35	3.8	−	
10W	7000 at −25	60,000 at −30	4.1	−	
15W	7000 at −20	60,000 at −25	5.6	−	
20W	9500 at −15	60,000 at −20	5.6	−	
25W	13,000 at −10	60,000 at −15	9.3	−	
20	−	−	5.6	<9.3	2.6
30	−	−	9.3	<12.5	2.9
40	−	−	12.5	<16.3	2.9*
40	−	−	−12.5	<16.3	3.7**
50	−	−	16.3	<21.9	3.7
60	−	−	21.9	<26.1	3.7

* For 0W, 5W, 10 W
** For 15W, 20W, 25W, and single-range oils

Fig. O82
SAE viscosity classes for engine oils (SAE J300, binding from July 2001)

ular multigrade oil might concurrently fulfill the viscosity requirements of several viscosity classes (e.g., 10W-40 or 5W-30) due to low temperature-related viscosity change. Generally, the multigrade character is achieved through a combination of basic oils with a high viscosity index and special additives, the viscosity index improvers. The designation "multigrade oil" says nothing about the performance characteristic of the oil, such as wear protection and engine cleanliness. For this purpose, the other specification data are required (classification). The low-temperature viscosity of a multigrade oil is mainly determined by the basic oil. High-temperature viscosity results from the viscosity index of basic oil and the thickening effect of the VI improver.

Application of polymers as viscosity index improvers changes the rheological properties of an oil type. For instance, multigrade oils are structurally viscous — that is, their viscosity reduces with rising shear gradient (shear stability). This effect can lead to both advantages and disadvantages in the operating behavior of multigrade oils. The formulation of multigrade oil, therefore, requires considerable experience. By means of suitable selection and coordination of the individual additives, the frictional resistance, as one example, can be lowered and fuel consumption reduced without disadvantages in terms of lubrication reliability.

In addition to temporary viscosity loss, a permanent viscosity loss may become apparent through "shearing" of the polymers. By introducing suitable test methods and boundary values and applying shear-stable viscosity index improvers, these factors can be influenced in a controlled and focused manner.

Application of thinner (low-viscosity) mineral basic oils with high concentrations of VI improvers can also lead to increased evaporative loss and, thus, to associated oil thickening. For instance, based on inappropriate formulation, a 20W-40 can develop from a 10W-50 oil in engine operation. This is because the shearing of VI improvers reduces the high-temperature viscosity (e.g., from SAE 50 to SAE 40), while thickening due to evaporation increases the low temperature viscosity (from 10W to 20W). Application of synthetic basic oils with better evaporative characteristics may, on the other hand, significantly reduce the evaporative losses. Owing to the higher viscosity index of these basic oils and omission of a bandwidth in the overlap of viscosity classes (e.g., 5W-30 instead of 10W-50) that is in most cases technically unnecessary, the concentration of VI improvers can at the same time be lowered to a significant extent.

Viscosity index improvers may also be sensitive to oxidation, whereby deposits can form in the engine due to decomposition products of unstable VI improvers. Modern multigrade oils have achieved a very high quality level with long-term consistency and have become indispensable for modern cars and commercial vehicle engines, not least due to the development of temperature and shear-stable viscosity index improvers.

~~~**Single-range oils.** Single-range oils show a highly temperature dependent viscosity change and, thus, a low viscosity index. They are, therefore, suitable only for engines operated under generally constant conditions and at nearly constant temperatures — that is, engines for power generation. The designation "single-range oil" refers to the viscosity classification SAE J300, with the classes 0W to 60. Single-range oils satisfy viscosity requirements for only one class, such as SAE 30 with a kinematic viscosity between 9.3 and 12.5 mm^2/s at 100°C and a high-temperature highest viscosity of at least 2.6 mPa·s at 150°C. Single-range oils in vehicle engines must be changed according to the season and ambient temperatures according to manufacturers' instructions. For the manufacture of single-range oils, basic oils with a low viscosity index but without viscosity index improvers are used. The designation "single-range oil" says nothing about the performance characteristic and quality level of an oil type. Application restrictions due to the narrow range of permissible operating temperatures have greatly reduced the importance of single-range oils in the vehicular sector.

~~**Viscosity classes for vehicle transmission oils.** An overview of viscosities of vehicle transmission oils is provided in **Fig. O84** for the purpose of completeness and better clarity at this point. Whereas the SAE classes of engine oils bear the designation SAE 0W to SAE 60, the designations SAE 70W up to SAE 250 apply to transmission oils. The higher values for the transmission oils do not, however, indicate that they have higher viscosity levels; these specifications apply only within the respective engine oil and transmission oil specifications. Generally, used transmission oils of classes 70W to 90 have viscosities similar to engine oils of classes 10W to 50, which is apparent from a comparison of the viscosities at 100°C (**Figs. O84 and O81**).

~**Safety and environmental aspects.** Mineral oil, as a mixture of different hydrocarbon compounds, is a natural product of the earth. Once it is brought from great depths to the surface, it leads to familiar environmental problems due to its incompatibility with the

SAE Viscosity Class	Temperature (°C) for a Viscosity of 150 Pa·s	Kinematic Viscosity at 100°C (mm²/s)	
		Min	Max
70W	−55	4.1	
75W	−40	4.1	
80W	−26	7.0	
85W	−2	11.0	
90		13.5	24.0
140		24.0	41
250		41.0	

Fig. O84 Viscosity classes for vehicle transmission oils

waters of the biosphere. Although the market for lubricants only accounts for about 1% of the entire mineral oil consumption in Germany, there is a strong potential of environmental damage from this sector. If 1 liter of lubricant mixes with water, it can render several thousand cubic meters of water unpotable. Legal regulations for preventing environmental damage through lubricants are, therefore, based primarily on the control of water pollution and the effect on living organisms.

Recently, the carcinogenicity and disposal of used oil have also been covered by legislation. The most important legal basics in Germany are set out in the water ecology act, in waste recycling and disposal laws, in legislation regarding chemical pollution, and in the hazardous substance ordinance.

The effects of lubricants on inappropriate surroundings (environment) include a variety of factors. A distinction is drawn between primary and secondary effects. Primary effects are the direct influences that a lubricant exercises when it enters an inappropriate environment (e.g., due to its water-threatening effect). Secondary effects are those the lubricant has on its environment by virtue of its technical application (energy economy by lowering friction values).

Total-loss lubricants that affect the environment continuously, albeit in small amounts and fine distribution (two-stroke engine oils), must be evaluated according to their primary effects such as biodegradability and toxicity. Lubricants used in closed systems can make a more effective contribution toward environmental protection by extending the oil-change interval or saving energy than through improved biodegradability. The most effective environmental protection is and will remain the avoidance of leakages and total-loss lubrication.

In addition to the manufacture and selection of possible environmentally neutral basic oils, the required additives also have to be assessed and optimized with regard to their environmental acceptance. Frequently, undesirable production-based impurities (e.g., chlorine compounds) are involved, and these could be reduced significantly with new manufacturing processes.

~~Secondary environmental compatibility.
The environment-friendly characteristic of a lubricant is not exclusively expressed in typical toxicological or biological properties and effects. Because the lubricants exist, environmental protection is secondarily also determined by criteria based on the application, for example:

- Safe transport and handling of lubricants
- Appropriate use or application
- Optimization of container size
- Design features of lubricated points
  —Total-loss lubrication or lubrication at the depot
  —Quantity minimization in the case of total-loss lubrication
  —Leakage prevention through better sealing materials
  —Lifetime lubrication

—Reduction of power loss by decreasing the friction
—Reduction of splashing work
- The maintenance
  —Extension of oil-change intervals
  —Consumption control
- Recycling
- Waste disposal

~~Toxicology and effect on biological organisms.
Testing the lubricant for toxicity in the sense of hazardous material regulations, especially their carcinogenic nature, is an important aspect. Unused engine oils are not classified as hazardous matter in the sense of hazardous substance ordinances. During operation, various contaminants accumulate in used engine oils: for instance, the carcinogenic polycyclic aromatic hydrocarbons (PAH) change the toxicological properties of engine oils disadvantageously. Used engine oils must as such be classified as carcinogenic under the German *Gefahrstoffverordnung* (hazardous substance law). Skin contact with used engine oil should, therefore, be absolutely avoided.

For biodegradable oils—for example, total-loss lubricants such as two-stroke engine oils for use in boat drives—their effect on microorganisms, animals, and plants is tested: in the first approach, the tests involve acute toxicity—that is, the short-term effect on algae (producers of organic substance), small crabs (primary consumers), fish (secondary consumers and final member of the aquatic nutritional chain) as well as bacteria (decomposers that bring about the degradation of organic substances). In addition, the effect on organisms living in the soil must also be considered. These are the mammalian toxicity (rats), earthworm toxicity, and growth inhibition of higher plants.

The evaluation is based on objective test criteria, according to which the substances must be classified and designated based on their acute or long-term effect in aquatic (waters) or nonaquatic systems (soil, air). Acute toxicity is defined quantitatively for fish, daphnia (water flea), and algae. The biodegradability of the substance is included in the evaluation with different methods. The evaluation criteria for danger to the environment in nonaquatic systems have not been further defined because of a lack of suitable measuring methods. The following R designations have already been established:

R 54, toxic for plants
R 55, toxic for animals
R 56, toxic for soil organisms
R 57, toxic for bees
R 58, can have long-term harmful effect on the environment
R 59, dangerous for ozone layer

~~~Water ecology act [Germany].
The water ecology act has practical effects above all on the storage, production, and marketing suitability of potentially dangerous substances and is of great importance to the mineral oil industry even in the case of accidents with

Water Hazard Classification (WHC)	Classification	Examples
WHC 1	Weak—water threatening substances	Lubricants and basic oils without additives
WHC 2	Water threatening substances	Lubricants with additives
WHC 3	Strong—water threatening substances	Used oils

Fig. O85
Water hazard
classification

"oil contamination." The Advisory Board of the Federal Ministry for Environment compiled a catalogue of water-threatening substances in order to register the potential danger of a substance in water. This catalogue classifies the substances in three classes: from "strongly hazardous to water" up to "weakly hazardous to water." The classification is based on the values determined through tests for the biodegradability and toxicity (poisonous effect). The biodegradability and the possibility of accumulation and persistence of the lubricant to be evaluated are considered in a bonus-malus process during the classification process (**Fig. O85**).

~~**Used oil**

~~~**Basic liquids.** Biodegradability and the toxicological behavior of a liquid depend on the molecular structure of the basic liquids, the degree of refinement, and the additive content. Three different groups of suitable basic liquids have established themselves in practice. These groups differ substantially among themselves, and are also different from well-known mineral oil–based basic oils for lubricants: polyethylene glycols, vegetable oils, and synthetic esters. The development objective is to improve environmentally friendly liquids so that their reaction with oxygen and water becomes harmless. Therefore, it cannot be ruled out that low-viscosity poly-alpha olefins and mineral oils from XHVI synthesis technology will be used in certain future applications.

Polyethylene glycols are water-soluble but are insoluble in mineral oil. The latter is a substantial disadvantage of polyglycols, becuase the conversion of technical plants in operation is made difficult and the risk of making a mistake is increased. Solubility in water has to be considered from two sides: on the one hand, it

presumably encourages quick biodegradation in surface water; whereas, on the other, with the help of water the product quickly gets into deep soil layers that are poor in bacteria and oxygen inhibit biodegradation. Market experience shows that machine manufacturers and users consider the disadvantages mentioned above very carefully and hence prefer vegetable oils and synthetic esters as basic liquids (**Fig. O86**).

Rapeseed oil is preferred among the group of *vegetable oils* (**Fig. O87**). Rapeseed oil is derived by pressing or extraction from the seeds of rape; the proportion of oil is 35–45%. Just as with other vegetable oils, rapeseed oil essentially comprises glycerin esters of higher fatty acids: that is the three OH groups of glycerin are esterified with different fatty acids: they are also called *triglycerides*.

Synthetic esters (**Fig. O88**) are reaction products from different alcohols and carbonic acids (fatty acids). The results are *carbonic acid esters* that form by splitting of the water. The most important representatives for the production of environmentally friendly lubricants are *dicarbonic acid esters (diesters)* and *polyol esters*.

~~~**Biodegradability.** If an ecological system is not destroyed by lubricant leakage—for instance, if the quantities and/or durations of action are short and the lubricant is finely distributed—the environment is capable of withstanding this attack. The most important criterion for this is biodegradability. It determines the lifetime of the foreign substance in the environment. The faster the biodegradation, the lower the probability of accumulation and possible toxic effects.

Biodegradation is enzymatic oxidation of hydrocarbons caused by microorganisms (e.g., bacteria, algae,

Advantages	Disadvantages
• Very good lubricating properties (except Al, Al alloys/steel) • Good corrosion protection • Good aging stability • Shearing stability • Wide range of application temperature • Extremely good VT behavior • Practical experience over long-term	• More expensive than mineral oil • Cavitation and suction problems due to high density • Attacks lacquer and plastics • Elastomeric incompatibility • Not mixable with mineral oil • Paper filter not applicable • Waste disposal as Cat. II or III • Water soluble

Fig. O86 Advantages and disadvantages of polyglycol-based basic liquids

Advantages	Disadvantages
• Very good lubricating properties • Not water soluble • Eco-toxicologically harmless • Good elastomeric compatibility • Good VT characteristic • Good load-bearing capacity • Shearing stability • All filter types are suitable • Mixable with mineral oil • Thermally very stable	• More expensive than mineral oil • Poor aging stability (oxidative and hydrolytic) • Waste disposal as Cat. II or III

Fig. O87 Advantages and disadvantages of basic liquids based on natural ester oils

Advantages	Disadvantages
• Very good lubricating properties • Not soluble in water • Good corrosion protection • Very good aging stability (thermal and oxidative) • Good VT characteristic • Wide temperature app. range • Shearing stability • All filter types are suitable • Mixable with mineral oil • Lower evaporation loss • Good low temperature characteristic	• Significantly more expensive than mineral oil • Elastomeric compatibility ester-dependent (to be tested case by case) • Hydrolytically more unstable than mineral oil

Fig. O88 Advantages and disadvantages of basic liquids based on synthetic ester oils

lower fungi), which convert the hydrocarbon compounds during intake of food into energy, water, carbon dioxide, and biomass (propagation). The degradation can be distinguished into aerobic splitting (i.e., splitting with oxygen) and anaerobic splitting that occurs under exclusion of oxygen. For hydrocarbons, the latter is of less importance.

Ease of biodegradation alone does not make a lubricant environmentally friendly, although it is an essential consideration. Ecological toxicity has to be considered in addition to biodegradation. The primary criteria of environmental compatibility—such as degradability, water hazard classification, and toxicity—are especially important for total-loss lubricants that are continuously, albeit in small amounts and fine distribution, released into the environment:

• Chain saw oil
• Railway switch lubricant and wheel flange lubricant
• Commercial vehicle lubricants
• Formwork oil/forming oil
• Two-stroke engine oils

An ecologically high potential danger is posed by some 7000 tons of chainsaw oil that annually contaminate forest soil; also lubricant for railway switches and wheel flanges or the central lubricant supply for commercial vehicles used on construction sites and on roads must be included. In all cases, small amounts of lubricants are distributed over wide areas and released into the environment. All told, these quantities of lubricants add up to many thousand tons per annum. These applications are, therefore, particularly predestined for readily biodegradable products. Suitable lubricants are on the market.

Two-stroke engine oils for application in boat drives, due to the proximity of oily exhaust gas to water, were the initial incentive for introduction of biodegradable lubricants. Biodegradable two-stroke engine oils should help prevent an increase of hydrocarbons in drinking water. Synthetic esters in combination with ash-free additives result in high-performance outboard engine oils. The biodegradability is 80% after 21 days.

For lubricants finding application in closed systems (recirculating lubrication), questions about biodegradability are of less importance in principle. In the event

of leakage, though, they do not pose a less serious potential danger (e.g., hydraulic oils). Hydraulic systems on mobile working equipment in the building industry are typical of such large-volume, closed oil-systems that work under high pressure in the open. The agility of these pieces of equipment requires flexible hydraulic lines that pose real risks in rugged construction site operation.

From the comparison with hydraulic oils, it is clear that lubricants for four-stroke engines pose less of a problem due to the generally closed oil systems. The environmentally friendly features of four-stroke engine oils are found in the intended performance characteristics rather than in biodegradability:

• Reduced fuel consumption
• Reduced oil consumption
• Improved elastomeric compatibility for preventing leakages
• Extension of oil change intervals
• Improvement of exhaust quality

The difference between four-stroke engine oils and all previously mentioned lubricants is that the latter are released into the environment practically unchanged in all relevant aspects of biodegradability—that is, practically in the original fresh oil state. Engine oils, in contrast, are subjected to significant changes during use, for example, through the absorption of combustion residues. These substances that get stored in the oil and the aging products from oil oxidation are generally not biodegradable and may even inhibit the biodegradation of basic liquids. Original, biodegradable, fresh, four-stroke engine oil has the same effects in terms of environmental relevance in the event of leakage as conventional engine oil. Consequently, there is no test method for determining biodegradability of four-stroke engine oils for application in land-based vehicles. Occasionally there are four-stroke engine oils on the market that are designated as biodegradable; they are normally tested according to the test methods for boat engines. It remains to be seen whether an end consumer ignorant of the facts will be misled to carry out an oil change "in the forest"—which would at the very least be a violation of legal regulations in Germany.

~~~**Recycling of mineral oil.** Used lubricants must be collected and disposed of in a professional manner, so that after their proper use they do not contaminate the environment. In contrast to simple thermal—for example, as fuel in the cement industry—material exploitation, recycling back to basic oils for lubricants, is gradually gaining attention as a result of environmental policies. About half the amount of waste oil was reprocessed into secondary raffinates in 1997 (**Fig. O89**).

A number of contaminants, especially the residues from fuel combustion, accumulate in engine oils during operation. The enrichment of the carcinogens of polycyclic aromatic hydrocarbons (PAH) in particular changes the toxicological properties of engine oils disadvantageously. Skin contact with used engine oil

Type of Waste Disposal	Amount of Used Oil (in 1000 tons)
Second refinement	213
Black pump (gasification)	13
Steel industry (reduction means)	19.5
Cement industry	170
Special waste incineration	2.3
	435.8

Fig. O89 Used oil disposal as waste in 1997 (Source: Federal German government response to an inquiry from the SPD (Social Democratic Party) in the Bundestag dated 20 May, 1998, BT-Drs. 13/10773)

should, therefore, be avoided. Direct application of used oils as refinery raw material is, therefore, practically impossible. To eliminate undesirable components in addition to the PAH from used oils, such as other oxidation products, low-boiling fuel residues, esters, metals and organometallic compounds, salts, halogen compounds, and so on, a series of special processing steps is required. These are partly different from those required for the production of basic oils from crude oil.

The sulfuric acid/clay treatment of used oils was one of the first methods applied in used-oil treatment. This process no longer represents the state of the art due to its high level of environmental contamination from process-based byproducts (e.g., acidic resin) and due to the inadequate quality of basic oils. A modern alternative comprises the removal of water and lighter components, followed by vacuum distillation or thin-film evaporation that separates the lubricating oil fractions from the high-boiling-point components and other byproducts. The distillates thus obtained can be processed to high-quality basic oils for lubricants by means of hydrogenation or through a combination of solvent extraction and hydrogenation.

Whereas earlier basic oils of this type exhibited high values for PAH and halogens, hydrogenated recycling oils can, in particular, exhibit an extremely low content of such undesirable impurities and they hardly differ, in general properties, from mineral oils obtained by conventional production.

Nevertheless, recycled oils cannot attain the quality level of lubricants based on synthetic basic oils: The quality of such top products is in fact determined to a greater extent by means of special basic oils defined with precise tolerances—basic oils for making recy-

cled oils, in contrast, at their best, automatically correspond to the mineral oils.

However, there are still fully formulated engine lubricants on the market that do not conform to these modern requirements, but rather have significantly higher PAH content levels, sometimes even reaching or exceeding the PAH extract content levels (**Fig. O90**, samples A and B). In Germany, these extracts must be labeled with a skull and crossbones and are not freely marketable.

~~~Test methods

Testing for ready biodegradability. Testing for ready biodegradability is carried out under inconvenient degradation conditions. Microorganisms are only given a short period for adjustment. The only nutrient source is the substance being tested, the concentration of which is low in order to rule out toxic effects, but also to keep the formation of biomass down.

If a test substance is degraded in one of the test methods, then quick and complete degradation is expected in the environment as well. The test substance is then considered to be readily biodegradable.

Water is poured into a container, and then nutrient salt and a microorganism mixed culture (inoculum)—frequently from a communal sewage treatment facility—are added. Adequate oxygen concentration is insured and the container is stored at a predetermined temperature in a dark place for a certain period.

The following analysis methods for observing biodegradation are expensive: Oxygen consumption, carbon dioxide formation. Dissolved organic carbon (DOC), total organic carbon (TOC), loss of substance.

Test methods for determining potential degradability. Even if a substance does not fulfill ready biodegradability criteria, biodegradability is still not ruled out. It will possibly take longer to occur. More conducive testing conditions are, therefore, chosen to assess the degradability characteristic. The testing principle, however, is comparable with those for determining easy biodegradability. The microorganisms are certainly given more time to adapt to the provided nutrients (adaptation), and higher concentrations of microorganisms are used.

Test methods for determining whether the oil is degradable in principle (simulation test). If a substance shows no degradation or a very low rate of degradation in the above tests, a simulation test can be carried out

		A	B	C	Extract	Primary Raffinate
Viscosity Grade		15W-40	10W-40	15W-40	1997	15W-40
Sample Procured on		April 97	April 97	April 97		1997
PAH						
Total PAH (GC-MS)	mg/kg	400	880	17	240	nn
Sum of Grimmer PAH	mg/kg	270	570	12	160	nn
Benzopyrene	mg/kg	11	7	nn	12	nn
nn = not detectable, i.e. below the level that can be detected by the applied method.						

Fig. O90 Content of PAH in engine oils (different proving methods; samples A–C are marked on the drum as secondary raffinate)

to estimate the potential danger to the environment. When such a test is conducted, the most probable entry path into the environment is considered, and environmentally relevant testing conditions are chosen. These tests are extraordinarily expensive. Furthermore, proof must be provided that the simulation is in accordance with the considered environmental compartment.

Meaningfulness and transferability to practical conditions. The testing methods for determining the biodegradability are based on experiential data gathered with substances under real environmental conditions. The laboratory tests serve to distinguish different substances under reproducible conditions. Test modifications without appropriate security can lead to a wrong assessment.

The results of laboratory tests allow only assessments concerning the real degradation characteristic. A concrete conclusion about the duration of degradation under real conditions cannot be drawn. The processes in nature depend on numerous different parameters such as the composition of the microorganism population, the temperature, available nutrients and salts, the moisture, and the concentration of the contaminant. Therefore, the statements occasionally made stating that after a leakage of a readily biodegradable lubricant, it will be fully eliminated within 21 or 28 days are not correct. Rather, the biodegradation process in the soil is drastically slowed due to the magnitude of the concentration after an oil accident; the lack of oxygen in deep lying soil layers and lack of moisture are the reasons for this.

~~~**Used oil collection.** Engine oils with domestic sales of 420,900 tons per annum (figures for 1993) had a share of about 37% of the overall sales of lubricants in Germany. Of these, some 380,000 tons were used in vehicles, the rest in other areas. About 40% of the fresh oil amount is burned (the oil consumption of vehicle engines was calculated for 1993 with some 166,000 tons) and is, therefore, not collectable as waste oil. In 1993, the amount of waste oil from engines was 217,000 tons, reaching a proportion of 51.6% of fresh-oil sales and allowing for a difference of approximately 6,900 tons between initial filling and drained amounts from immobilized vehicles (partly due to vehicle exports) and (calculative) losses of approximately 9,300 tons. Due to extended oil change intervals and the lowering of the rate of oil consumption by engines, there is a decline in fresh oil sales despite the increase in the number of vehicles in operation (**Fig. O91**). A decline is therefore also expected in the volume of waste oil.

The reduction of waste oil volume is also possible through simple pragmatic measures without the necessity of making high-tech efforts, such as reducing containerized waste in oil sales.

~~~**Waste disposal.** Waste disposal of used oil is regulated in the Waste Recycling and Disposal Act dated August 27, 1986, and in the Recycling of Used Oil dated October 27, 1987. According to these laws, used

	Engine Oils (in 1000 tons)	Vehicle Transmission Oils (in 1000 tons)
1991	444.3	63.9
1992	414.2	61.4
1993	420.9	61.8
1994	419.1	61.7
1995	418.3	63.9
1996	405.5	63.4
1997	411.5	66.9
1998	381.6	64.0

Fig. O91 Domestic sales of engines and vehicle transmission oils (Source: Federal Agency for Economics)

oil is first classified as waste, and the owner is then responsible for its appropriate disposal. An assessment determines whether the quantity of used oil should be recycled or must be disposed of as costly waste. This automatically leads to the classification of the entire waste in different categories. "Burned" engine oils correspond to Category I—that is, the used oils that can be recycled. In order not to disturb the recycling process of engine oils, it is necessary to collect vegetable oils separately. These same vegetable oils, well known for their quick biodegradability, are the materials which, in the most inconvenient case, do not fulfill the standards for used recyclable oils as prescribed by law and which must then be treated as special waste.

In 1997, a total of 435,800 tons of used oil (including engine oils) were subjected to waste oil disposal (Source: BT-Drs. 13/10773). **Fig. O89** shows the waste disposal routes.

The law allocates to the owner the responsibility for appropriate waste disposal of used oil. The legislature has enacted special regulations to help used-oil owners with proper waste disposal. The latter are mostly private end-consumers, handling small amounts of engine and transmission oils.

In Germany, whoever supplies engine or transmission oil to the end consumer has the following obligations to the buyer:

• Provision of information about proper waste disposal
• Take-back obligation (free of charge)
• Proof of professional oil change and take-back of waste oil

Oil additives →Oil

Oil aging →Oil ~ Power during operation

Oil bath air filter →Filter ~ Intake air filter ~~Air filter elements

Oil change interval →Oil ~Oil maintenance

Oil characteristics →Oil

Oil circuit →Oil circulation

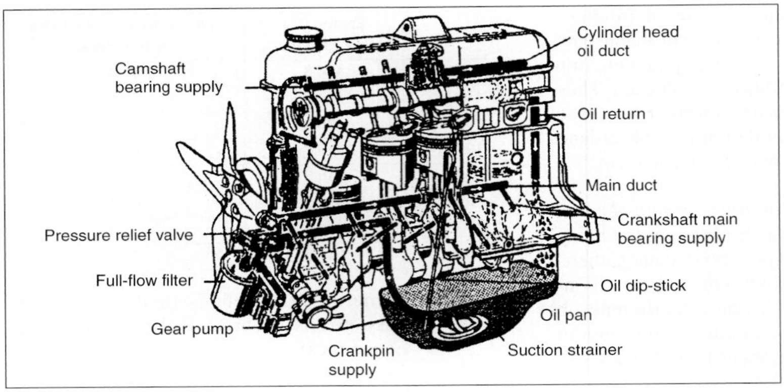

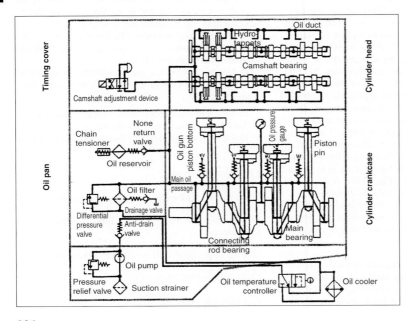

Camshaft
bearing supply

Cylinder head
oil duct

Oil return

Main duct

Crankshaft main
bearing supply

Pressure relief valve

Full-flow filter

Oil dip-stick

Oil pan

Gear pump

Crankpin
supply

Suction strainer

Fig. O92
Oil flow in the engine

Oil dip-stick

Oil circulation (*also*, →Lubrication ~Lubrication systems; →Oil). The oil circulation systems in modern engines are designed for forced-feed lubrication in a closed system. Due to increased needs/requirements, in contrast to older designs, many other hydraulic requirements must be fulfilled (**Fig. O92**).

First, all engine components that move relative to one another are lubricated, then the friction is reduced, and then the hydraulic drives are filled. What is decisive for wear prevention and operation after a long standstill period is the time taken by the lubrication system before the last consumer point is supplied with oil.

In addition to lubrication and friction reduction, further essential duties for the oil circulation system are heat removal and oil distribution. Following these are the fine sealing of piston rings, removal of combustion residues, and transportation of wear particles.

A schematic illustration of a possible oil circulation system is shown in **Fig. O93**.

The oil is sucked from the oil sump by means of the oil pump. This is generally designed as an internal (**Fig. O95**), or external axle gear-wheel pump and is driven by means of gears, chain, or shaft from the crankshaft, or it is mounted directly on the crankshaft (**Fig. O94**).

Directly driven oil pumps have advantages due to low manufacturing costs; indirectly driven pumps, because of the possibility of lowering the speed, have the advantage of better efficiency.

Original dirt and larger chips from the sump are separated in a sieve on the suction strainer at the beginning of the suction tube. The maximum oil pressure is limited approximately 3–5 bar by means of a pressure relief valve; this is mounted in parallel to the oil pump,

Fig. O93
Schematic illustration
of the oil circulation
system

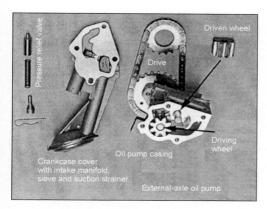

Fig. O94 Elements of an oil pump with external axle

and the bypassed oil is returned to the suction side of the pump. The valve and the gradual shut-off cross sections are dimensioned such that, even under inconvenient circumstances, the oil filter will neither burst nor be lifted off.

The oil pressure indicator on the dashboard derives its signal from a sensor in the main duct. If the system pressure falls short of the minimum oil pressure, an oil pressure control lamp lights, triggered by the oil-pressure switch also located in the main duct.

The oil pressure switch is triggered generally at about 0.8 bar. This oil pressure is absolutely inadequate at high speeds because the pressure at the last outlet point in the cylinder head is even lower.

Antidrain valves prevent the engine from idling after switch-off, and also after introduction of air into the system after restart.

If extensive cooling functions are required from the oil circulation system, an oil cooler (controlled) is installed in the circuit. The latter is either configured under the oil filter as a water/oil device or via pipes in the engine compartment as air/oil heat exchanger.

Engines without oil coolers reach oil temperatures >150°C under certain driving and ambient conditions. When coolers are used, the temperatures can be limited to 130°C maximum.

In case of emergency and to ensure that the engine is

not endangered in the case of heavy dirt buildup and filter blockage, the full-flow oil filter can be bypassed via a differential pressure valve.

If smaller particles and dust have to be filtered off from the circulation system of special-purpose diesel engines—for instance, in construction site applications—then a partial amount of approximately 5–10% of the volumetric flow can be diverted from the full-flow filter to a bypass filter and fed back to the oil sump without pressure.

After the filter, the full-flow oil duct in the engine block is supplied. From this duct, oil passages lead to individual main bearings of the crankshaft. The connecting rod bearings are supplied with oil via inclined passages in the crankshaft, from the crankshaft journals through the webs into the crankpin. From the latter, oil for lubricating the piston pins is tapped through passages within the connecting rods.

An intermittent oil gun for noise suppression can be provided via an additional lateral passage in the big connecting rod end. To lower the piston temperatures, oil gun nozzles can be supplied from the full-flow duct (permanent oil gun) that opens at a predetermined minimum oil pressure.

From the full-flow duct, another duct leads to the cylinder head; also the oil for the chain tensioner is derived from this duct. In the cylinder head itself the oil is fed to the camshaft bearings, either through individual holes or from one bearing through the camshaft. If the engine is provided with hydraulic valve clearance compensation (**Fig. O96**), the corresponding elements must also be supplied with oil from this cylinder head duct. Other consumers of oil are hydraulic cam adjusters as well as valve switch-off elements and elements for variable valve lift.

After all consuming elements have been supplied with oil, the oil must flow back to the sump under all circumstances; to ensure this, the oil must also be separated from crankcase blowby. Furthermore, the gas content in oil may not exceed the critical value for the particular engine.

If automatic oil return into the oil sump is not possible from all parts of the engine, the oil must be

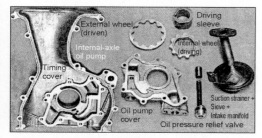

Fig. O95 Elements of an oil pump with internal axle

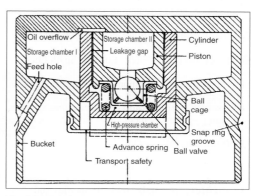

Fig. O96 Cup for hydraulic valve clearance compensation

sucked from different points (suction oil pumps) and pumped into a reservoir, from where the pressure pump draws the oil. In this case, the sum of suction volume flows must always be greater than the pressure volume flow. In a case in which no oil volume is left in the oil sump and the latter is replaced with a cover, the term used is "dry-sump lubrication."

During the development of the engine, the oil-sump content and oil-fill amount were steadily reduced, but the rate of circulation was increased. This leads to a higher rate of oil contamination, but on the other hand also to greater problems with the degassing of oil. Nevertheless, it was also possible to substantially prolong oil-change intervals through focused development. Covering a 30,000-km distance between oil changes is no longer a problem.

Oil circulation systems were earlier optimized in extensive driving tests; today, dynamic swivel arm test stands are used, which allow precise and quicker optimization.

Oil composition →Oil ~Body

Oil consumption →Lubrication ~Lubricating systems; →Oil ~Power during operation; →Piston

Oil coolant →Coolant

Oil cooler →Lubrication ~Lubricating systems; →Radiator ~Liquid/air radiator

Oil cooling →Cooling, engine

Oil dipstick →Lubrication ~Control and safety components

Oil drain passage →Lubrication ~Lubricating systems ~~Passenger car engines

Oil extraction →Lubrication ~Lubricating systems

Oil extractor →Filter ~Lubricating oil filter

Oil filter →Filter ~Lubricating oil filter; →Oil ~Oil maintenance; →Oil circulation

Oil filter change →Filter ~Lubricating oil filter

Oil filter media →Filter ~Lubricating oil filter

Oil flow rate →Lubrication ~Oil demand, ~Oil line

Oil foaming →Lubrication ~Lubricatiing systems, ~Oil pump ~~Arrangement

Oil heat capacity →Oil

Oil heater →Lubrication

Oil incineration ash →Oil ~Properties and characteristic values ~~Ash content

Oil level →Lubrication ~Control and safety components

Oil level gauge →Lubrication ~Control and safety components

Oil level sensor →Lubrication ~Control and safety components; →Sensors

Oil level switch →Sensors

Oil maintenance →Oil ~Oil maintenance

Oil measuring method →Oil ~Properties and characteristic values ~~Viscosity

Oil oxidation →Oil ~Power during operation

Oil pan →Crankcase; →Lubrication

Oil pan design →Crankcase ~Oil pan

Oil pan flange below crankshaft center →Crankcase ~Crankcase design ~~Main bearing pedestal area

Oil pan flange on crankshaft center →Crankcase ~Crankcase design ~~Main bearing pedestal area

Oil pan module →Sealing systems ~Modules

Oil passage →Lubrication ~Lubricating systems ~~Passenger car engines

Oil passage crankshaft →Crankshaft

Oil pipe →Lubrication

Oil pressure →Lubrication ~Lubricating systems

Oil pressure gauge →Lubrication ~Control and safety components

Oil pressure switch →Lubrication ~Control and safety components

Oil properties →Oil

Oil pump →Lubrication ~Lubricating systems

Oil pump drive →Lubrication ~Oil pump

Oil quantity →Lubrication ~Oil demand, ~Lubricating systems

Oil requirement →Lubrication

Oil return →Lubrication ~Lubricating systems

Oil return flow →Lubrication ~Lubricating systems ~~Passenger car engines

Oil ring groove →Piston ~Ring ~~Ring groove

Oil scraper ring →Piston ring ~Versions

Oil scraper ring with expander spring →Piston ring ~Versions ~~Oil scraper ring

Oil separation →Lubrication ~Lubricating systems

Oil separation modules →Sealing systems ~Modules

Oil separator →Lubrication ~Lubricating systems

Oil service →Oil ~Oil maintenance

Oil specifications →Oil ~Classification, specifications, and quality requirements

Oil spray →Lubrication ~Lubricating systems ~~Motorcycle engines; →Piston ~Cooling

Oil spray cooling →Piston ~Cooling ~~Cooling duct piston, ~~Spray oil cooling

Oil spray nozzle →Piston ~Cooling ~~Cooling duct pistons, ~~Spray oil cooling

Oil temperature →Lubrication ~Oil pump ~~Arrangement

Oil temperature stress → Lubrication ~Lubricating systems

Oil vapor →Lubrication ~Lubricating systems

Oil vapor formation →Lubrication ~Lubricating systems

Oil volume →Lubrication ~Lubricating systems

Olefin content →Fuel, gasoline engine

Olefins →Fuel, gasoline engine ~Olefin content

On-board diagnosis →Electronic open- and closed-loop control ~Electronic open- and closed-loop control, gasoline engine ~~Functions

On-board diagnostics →Electronic open- and closed-loop control ~Electronic open- and closed-loop control, diesel engine

One-dimensional charge transfer calculation/computation (1D) →Intake system ~Intake manifold ~~Computation process

Opacimeter →Exhaust gas monitoring check ~Exhaust gas measuring equipment

Open air test side →Engine acoustics

Open combustion process →Combustion process

Open standard cycle →Cycle

Open-deck design →Crankcase ~Crankcase design ~~Cover plate

Opening duration →Control/gas transfer ~Four-stroke engine; →Variable valve control ~Variation parameters

Opening time cross-section →Control/gas transfer ~Two-stroke engine; →Variable valve control ~Variation parameters

Operating characteristics. The operating characteristics of an engine can be divided into a steady-state and nonsteady or transient behavior.

The steady-state behavior defines, for example, the power, torque, supercharging (boost) pressure, emissions, and so on. for a given operating point. The point is generally identified by an engine speed and a load which are kept constant for the analyzed point in time.

The nonsteady behavior provides information about dynamic effects, such as during the acceleration phase. The differences between operating points are significant, for example, for an engine with exhaust gas turbocharging. An engine that is accelerated from low engine speed achieves lower loads for certain engine speed ranges than an engine that is operated under steady conditions at one given engine speed.

The operating characteristics of the engine should be tuned in such a way that the engine behaves optimally under all conditions (e.g., altitude, ambient temperature, engine temperature, acceleration, delay, idle) with regard to power, torque, acoustics, consumption, emissions, and so on.

Operating conditions, seals →Sealing systems

Operating limit. In a gasoline engine, successful ignition of the mixture is determined by the flammability limits. There is a flammability limit on both the rich and lean side of a stoichiometric mixture. The flammability limit is particularly important for the lean side. Deficiencies in the combustion process and misfires occur to an increasing extent before this is reached. They lead to incomplete combustion resulting in increasing HC emissions and uneven, irregular engine operation.

The operating limit is defined as the permissible amount of irregular operation. A measure for irregular operation is the additional fluctuation of the angular velocity of the flywheel. Irregular operation can be assessed by measuring the change in the angular velocity of the flywheel. If it exceeds the permitted extent, the air-fuel mixture can be enriched by operating a limit controller. The operating limit can be made leaner through charge stratification.

Lean operation is necessary in gasoline engines with direct injection. Air-fuel equivalence ratios up to

807

about $\lambda = 4$ are achieved here in part load. Lean operation requires specific measures to stabilize combustion but offers the benefit of lower fuel consumption (to 15%, compared with conventional gasoline engines).

With controlled engines cooperating at stoichiometric mixture of $\lambda = 1$, there is also a risk of misfiring when exhaust gas recirculation is used.

Another aspect of the operating limit in addition to ignition is the friction properties of the engine. If the mixture is too lean, the energy content is no longer large enough to overcome the friction work of the engine, and the operating limit of engine is reduced. This represents the limit of the air-fuel ratio of diesel engines at idle.

Operating materials (*also*, →Coolant; →Fuel, diesel engine; →Fuel, gasoline engine; →Oil). Operating materials are the materials required to operate an engine. These include fuel, lubrication fluid (oil), and coolant. A detailed description of the operating materials can be found under →Fuel, →Oil, and →Coolant

Operating modes of the starter alternator →Starter ~Starter-generators

Operating point →Engine operating point

Operating technology →Crankcase ~Crankcase design ~~Cylinder

Operation with methyl alcohol/methanol →Fuel, gasoline engine

Operational stress →Engine damage

Opposed-piston engine →Engine concepts ~Reciprocating piston engines ~~Single-shaft engines

Opposed-piston engine, two-cell design →Engine concepts ~Reciprocating piston engines ~~Multiple shaft engines ~~~Twin-shaft engine

"Orbital" air-based direct injection →Injection system, fuel ~Gasoline engine ~~Direct injection systems

Orbital combustion system (OCP) →Injection system, fuel ~Gasoline engine ~~Direct injection systems

Order →Engine acoustics ~Order

Order analysis →Engine acoustics ~Signature analysis

Order/order number →Balancing of masses

Orifice noise →Air intake system ~Acoustics

Orifice viscosity measurement →Oil ~Properties and characteristic values ~~Viscosity ~~~Measuring method

Other fuels →Fuel, diesel engine ~Alternative fuels, ~Synthetic diesel fuels; →Fuel, gasoline engine ~Alternative fuels

Out-of-balance couple →Balancing of masses

Output per piston surface area →Piston ~Specific output per (piston) surface area

Output stages →Electronic/mechanical engine and transmission control ~Electronic components ~~Block diagram, ~~Power transistor

Oval piston →Piston

Oval piston rings →Piston ring

Over square stroke-bore ratio →Bore-stroke ratio ~Oversquare

Overall efficiency (*also*, →Engine efficiency). The overall efficiency is an efficiency that combines at least two other efficiencies—it is the product of the subefficiencies. The overall efficiency is usually the product of the indicated efficiency of the cycle and the mechanical efficiency of the engine. The overall efficiency is always smaller than the smallest individual efficiency.

Over-enrichment. The air-fuel ratio of the gasoline engine with external mixture formation is in a limited range around $\lambda = 1.0$. In many regions of the performance map, the engine is operated with a mixture close to the stoichiometric composition. In order to reach full power and high torque at full load, the mixture is enriched—that is, $\lambda > 1.0$. That applies to gasoline engines with intake manifold injection as well as gasoline engines with direct injection. The term over-enrichment is used as well. Also, over-enrichment can be necessary for internal cooling or to avoid knocking in the region of full load.

Overflow →Electronic/mechanical engine and transmission control ~Electronic components ~~Diagnostics

Overflow carburetor →Carburetor ~Level control

Overflow valves →Lubrication ~Control and safety components

Overhead bucket tappet →Valve gear ~Actuation of valves ~~Direct valve actuation

Overhead valves →Valve arrangement

Overheating →Bearings ~Operational damage

Overlap →Control/gas transfer ~Four-stroke engine ~~Variable valve timing; →Variable valve control ~Variation parameters

Overrun fuel cut-off →Electronic open- and closed-loop control ~Electronic open- and closed-loop control, gasoline engine ~~Functions; →Load

Overtemperature →Electronic/mechanical engine and transmission control ~Electronic components ~~Diagnostics

Oxidation (*also*, →Chemical reaction; →Oil ~Power during operation ~~Aging of oil). Oxidation is a chemical reaction with oxygen. The fuel, essentially a mixture of different hydrocarbons, is oxidized with the oxygen in the air. This converts its chemical energy into heat energy, which is partly converted into mechanical work in the engine. Through oxidation, oxidation products such as oxides of nitrogen, carbon monoxide, and reduced hydrocarbons occur based on the reaction conditions (e.g., temperature, pressure, air-fuel ratio).

Because the exhaust of the engine still contains components that are not fully oxidized, postoxidation is possible to minimize emissions. This can happen through the supply of additional oxygen via a secondary air valve or with the help of a catalyst.

Oxidation catalyst →Catalytic converter ~Dualbed catalytic converter; →Particles (Particulates) ~Formation ~~Variables for particulate formation

Oxidation converter →Pollutant aftertreatment ~Pollutant aftertreatment lean concepts

Oxidation inhibitors, oil →Oil ~Body ~~Additives

Oxidation stability →Fuel, diesel engine ~Properties; →Fuel, gasoline engine ~Properties

Oxidized hydrocarbon →Combustion products

Oxidizing agent →Ferrocene

Oxygen analyzer →Exhaust gas analysis equipment

Oxygen concentration. The oxygen concentration in the air is about 21% by volume and 23% by mass. Because of a high residual gas content and/or exhaust gas recirculation, the concentration of oxygen in the air available for combustion in the combustion chamber of an engine is reduced. This leads to a lower nitrogen concentration and, depending on the residual oxygen concentration, a power loss.

On the exhaust side, the oxygen concentration is used to measure the air-fuel ratio. Using a suitable probe, the λ-probe, the oxygen level in the exhaust is measured and used to regulate the composition and/or the air-fuel ratio of the mixture.

Oxygen concentration probe →Lambda (λ) probe

Oxygen sensor →Lambda (λ) probe

Oxygen storage capacity →Catalytic converter

Ozone. Ozone (O_3) is the three-atom form of oxygen and exists in the stratosphere up to a content of approximately 100 ppm. Here, it is in equilibrium with the two-atom oxygen (O_2) that is also contained in the air we breathe. The short wave UV radiation of the sun constantly breaks down ozone atoms to form the dual-atom oxygen. This process absorbs "hard" UV radiation in the stratosphere that is harmful to organic life, and hence, this radiation does not reach the area of life in the biosphere.

In terms of effects of emissions in the biosphere, there are two phenomena to be distinguished regarding ozone:

- Ozone rarefaction ("ozone hole")
- Ozone stress

Whereas ozone rarefaction has a global character, ozone stress occurs near the ground and is preferentially also widespread near emitters or in metropolitan areas.

A feature common to these effects is that they are primarily caused by anthropogenic emissions (industrial facilities, house firing, and traffic). These anthropogenic emissions of trace gases and aerosols have increased along with industrialization since the beginning of this century. The emissions reach the atmosphere, where they are changed through physical and chemical conversion processes. The conversion processes with worldwide importance are those based on photochemical reactions, resulting in the so-called photochemical smog, and the processes are subsumed under the term "greenhouse effect." In additon to automobile exhaust, a number of other emitters are involved, such as industry, house firing and energy generation, and others. These emissions include halogenated hydrocarbons (CFCs), which are responsible for stratospheric ozone damage, as well as the SO_2 acidic rain and the greenhouse effect (CO_2). Whereas the processes that cause acidic rain are mainly based on reactions of sulfur dioxide and dioxides of nitrogen, in photochemical smog the NO_x, CO, and HC components, among others, are involved. Through solar radiation, substances known as photochemical oxidants are formed. Ozone (O_3) deserves special attention in this regard for two reasons:

- Ozone is the most important component based on its percentage.
- Ozone is extraordinarily climatically effective and also contributes substantially to the greenhouse effect in the atmosphere.

The production processes of ozone in the troposphere are different from those in the stratosphere. Atmospheric ozone has a very short lifetime in the atmosphere in contrast to the other climatically relevant trace gases. Furthermore, it is important because its concentration in the troposphere, especially in the northern hemisphere, is increasing; in the stratosphere, however, it is being reduced worldwide (ozone depletion or "hole").

At high concentrations, ozone causes irritation of the mucosa and the eyes as well as respiratory distress in humans, especially when the body is under stress. According to information issued by the Federal Ministry of Health (Germany), the maximum acceptable ozone concentration is 120 mg/m³, and, up to this, disadvantageous health effects can be ruled out with certainty. Peak values of 130–150 mg/m³ and beyond occur relatively frequently in northern latitudes in the summer months.

In other zones, particularly in Los Angeles, significantly higher values occur (hourly mean values between 600 and 700 mg/m³).

The portion of ozone in the troposphere depends on transport from the stratosphere and formation in the troposphere. During this formation, the oxides of nitrogen NO and NO_2 equally play a role just as the compounds CO, CH_4, and higher hydrocarbons do. The ratio NO_x:O_3 is of special importance in the process.

Ozone occurs during the reaction of molecular oxygen with atomic oxygen. The atomic oxygen, for instance, is generated by means of NO_2 photolysis:

$$NO_2 + hv \Leftrightarrow NO + O$$

The most efficient reduction occurs through the reaction:

$$NO + O_3 \Leftrightarrow NO_2 + O_2$$

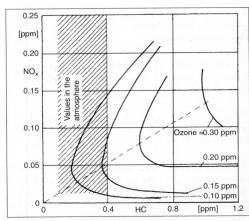

Fig. O97 Influence of HC and NO_x on ozone formation

These reactions alone do not exclusively influence the formation of ozone.

Investigations show that the presence of both oxygen components, hydrocarbons without methane content, and NO_x is important for the formation of ozone in the near-ground layers. Under certain circumstances, one-sided reduction of NO_x emissions would lead to an increase of ozone formation (**Fig. O97**). Therefore, in metropolitan areas it is necessary to decrease the amounts of both HC and NO_x components.

O

P

PAH (polycyclic aromatic hydrocarbons) →Fuel, diesel engine ~Composition

Palladium →Catalytic converter

Paraffins →Fuel, diesel engine ~Composition; →Fuel, gasoline engine ~Isoparaffins

Parallel engines →Engine concepts ~Reciprocating piston engines ~~Multiple-shaft engines ~~~Twin-shaft engines

Partial coating →Sealing systems ~Cylinder head gaskets

Partial flow burner →Particles (Particulates) ~Particulate filter system ~~Particulate filter ~~~Regeneration of particulate filters

Particle aging →Particles (Particulates) ~Characteristics of diesel particulates

Particle measurement instrument →Exhaust gas analysis equipment ~Particle measurement

Particle measuring →Exhaust gas analysis equipment ~Particle measurement; →Particles (Particulates) ~Particle measuring

Particle number →Particles (Particulates)

Particle size distribution →Particles (Particulates) ~Characteristics of diesel particulates

Particle surface →Particles (Particulates) ~Characteristics of diesel particulates

Particles (Particulates) (Latin *particula*). These are small particles that appear in a dispersed manner. Particles acquire the character of pollutants if their sizes are so small that they tend to remain suspended in gases and hence can access the respiratory system. The upper limit for total suspended particulates (TSP) is 57 μm, with a settling velocity of 10 cm/s; the theoretical settling velocity of a spherical particle of 100 nm mobility diameter and a density of 1 g/cm^3 (corresponds to a diesel particle) is 8.6×10^{-3} cm/s.

Particle-charged gases are multiphase mixtures or aerosols; the gas is the continuous phase in which the particles are dispersed. Based on the type of particles, they are referred to as dust (solid particles, fiber dust), smoke (solid matter from a thermal process), or fog (liquid substances, condensate).

Particles occur through abrasion, size-reduction, erosion, and corrosion as well as through condensation and nucleation from the gas phase, especially with incomplete combustion. These processes create typical size distributions and particle structures. Particles are viewed as a particularly critical health risk if they penetrate into the lungs, if their half-life in the lungs is long, or if they contain toxic substances and/or adsorb toxic substances to their surface. In the case of vehicles, abrasion particles from tires, brakes and clutches, corrosion particles, and fiber particles from exhaust pipes all play a role in addition to the combustion particles. Particles can change through agglomeration, aggregation, adsorption of gases and liquids, and chemical change—for example, under the influence of sunlight, moisture, and further pollutants in the atmosphere (sulfates, nitrates). Based on their origin, a distinction is made between primary particles and secondary particles.

The characterization of particles by means of particle measuring techniques provides information about the surface, number, size distribution, and particle ingredients. The following information is of concern above all for particles from engine combustion processes. Also, the particles from typical vehicular abrasion processes involving engine, tires, clutch, and brakes are gradually becoming important.

Pollution from automotive traffic contributes only about 0.5% of the global anthropogenic dust emissions. Automotive traffic's share of the fine particulate emission (PM2.5) in the inner cities can, however, attain values greater than 25%.

Literature: F. Loeffler: Staubabscheiden, Georg-Thieme Publishers, 1988. — W.C. Hinds: Aerosol Technology, John Wiley, 1989. — Aerosole - Stäube, Rauche und Nebel, MAK, 24th installment 1997. — Ultrafeine (Aerosol) Teilchen und deren Agglomerate und Aggregate, BIA Working Porfolio, 21st installment X/1998. — Ch. Cozzarini: Quellen der Partikel-Emissionen, 20th Vienna Engine Symposium, May 1999. — TLV und BAT-Werte-Liste der DFG, Report 32/1996 "Pruefung Gesundheitsschaedigende Arbeitsstoffe," VCH Publishers Weinheim, 1994.

~**Braided fiber filter.** *See below*, ~Particulate filter system ~~Particulate filter ~~~Filter media

~**Candle filter.** *See below*, ~Particulate filter system ~~Particulate filter ~~~Filter media

~**Catalytic coating.** *See below*, ~Particulate filter system ~~Particulate filter ~~~Regeneration of particulate filters

~**Ceramic monolithic cell filter.** *See below*, ~Particulate filter system ~~Particulate filter ~~~Filter media

~**Characteristics of diesel particulates.** Diesel particulates occur in general as agglomerates (**Fig. P1**) of the primary particles with a mobility diameter of

811

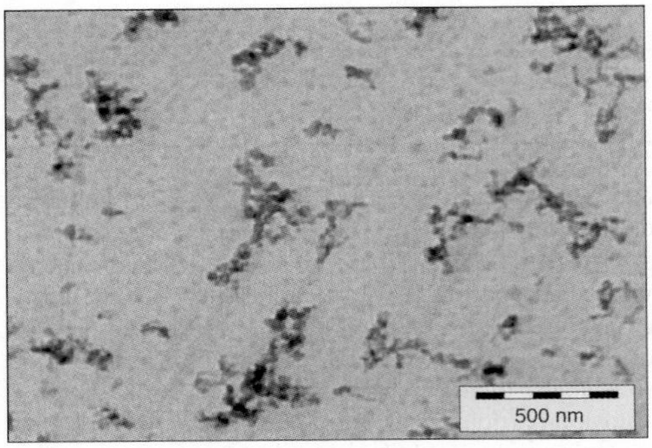

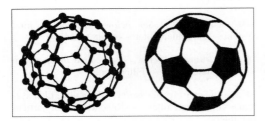

Fig. P1 Diesel particle agglomerate (Source: H.U. Franke)

10–30 nm (1 nm = 10^{-9} m) formed in the combustion chamber. The agglomerates typically have a particle size of about 20–300 nm; in extreme cases, they can grow up to 10 µm, usually in the case of adsorption to surfaces (exhaust pipes, mufflers, particulate filters).

Diesel engine primary particles consist of radial, symmetrically layered, curved layered carbon (bulb-like structure), with a grid layer gap of about 0.35 nm. Individual crystallite packets have thicknesses of up to 2 nm. The interplanar crystal spacing curvature is created through alternating formation of hexagonal and pentagonal structures; in an ideal case, they are formed by so-called fullerenes (**Fig. P2**). This is the third form, after diamond and graphite for the occurrence of carbon.

The specific surface area of soot particles determined according to the BET method ("BET" for Brunauer, Emmet, and Teller, the developers of the calculation) is about 50–150 m^2/g and can be derived from primary particles of about 10–30 nm. The density of primary solid matter particles is about 1.8 g/cm^3; the density of the agglomerate is 0.02–0.06 g/cm^3; in the deposited soot cake, it is up to 0.4 g/cm^3. Soot is practically inert, odorless, and insoluble in water and organic solvents. It is highly adsorbent for hydrocarbons, aldehydes, and adoriphores with oxygen content. The composition of the soot strongly depends on the generation process as well as on the sampling conditions.

Literature: H.U. Franke et al.: Otto von Guericke University of Magdeburg, 3D-Morphology of diesel soot particles and RME soot particles, published in the 2nd ETH Symposium on Nanoparticle Measurement, 7 August 1998. – E. Pauli: Regenerationsverhalten monolithischer Partikelfilter, Diss., RWTH Aachen University, 1986. – K.H. Zierock: Charakterisierung der partikelförmigen Emissionen von Dieselmotoren, Staub - Reinhalt, Luft 43 (1983). – N. Pelz: Rußpartikel, FVV Project No. 261/1985. – N. Metz et al.: Characterization of Particulate Matter Emissions from Modern Diesel Passenger Cars, 8th International Symposium "Transport and Air Pollution," Graz, 1999. – A. Zeilinger: Einsteins Schleier, die neue Welt der Quantenphysik, C. & H. Beck Publishers, Muenchen, 2003, BET: Oberflaechendefinition nach Brunauer, Emmet und Teller, see VDI-66131.

~~**Particle aging.** Particles can change after leaving the engine combustion process—through agglomeration with other particles, adsorption of gases and liquids and chemical changes, and especially through oxidation because of solar radiation. They can grow or shrink because of such effects and even harden in the process. The question of time and distance of the measuring location from the source always plays a major role when particles are considered. Boundary cases consider the practically unaged particles in the canyons between high buildings a few seconds after their occurrence, and, on the other hand, the particles in the free atmosphere, for example, at high altitudes months after their occurrence.

• *Agglomeration (coagulation)* (**Fig. P3**), simplified according to Hinds:

$$dN/dt = \alpha \, N^2,$$

where N = particle number concentration, α = 10^{-9} $cm^{3/s}$, a constant, and α is size-dependent; for instance, small particles are adsorbed to big particles

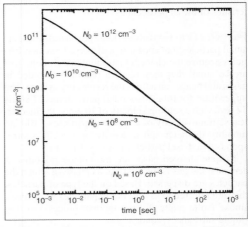

Fig. P3 Time sequence of the coagulation of different initial concentrations (Source: Hinds)

Fig. P2 Fullerenes have the same structure as a soccer ball on which a carbon atom is attached at every corner of a pentagonal or hexagonal patch. The illustrated fullerene molecule consists of 60 carbon atoms. Fullerenes were first discovered in 1985 by Kroto, Smalley, and Curl (Source: A. Zeilinger)

Eberspächer Exhaust Technology

rth America: The Corvette Z06
ieves a rated 505 bhp. A chal-
ging "Exhaust Manifold" under
mal timing constraints, this
duct was designed, developed,
idated and delivered into mass
duction in less than 12 months
n OEM sourcing.

ime mover for the automotive industry

erspächer is numbered among the leading manufactur-
world-wide of exhaust technology and is the largest
plier acting completely independent of the stock
change and holding companies. As development
tner of the passenger car and commercial vehicle
ustry, Eberspächer designs and produces mufflers,
alytic converters, SCR systems, particulate filters,
nifolds, pipes as well as complete exhaust systems.
ound 6,000 employees on 4 continents – more than
00 of these in the exhaust technology division
ontribute to mastering the technological challenges of
ay and tomorrow with innovative ideas.

ith innovative methods to new concepts

ovation is a corporate maxim at Eberspächer: More
n 300 highly qualified engineers and technicians are
rking within the company, or frequently also on-site
he customer's location, on new ideas shaping the
ure of the automobile. Around seven percent of the
es flow directly into research and development. When
eloping new technologies, Eberspächer uses the
ne modern CAD systems as the customers – supple-
nted by simulation and calculation programs. And the
est methods are also standard in production – whether
n technique, adaptive canning, laser welding or
ernal high-pressure forming.

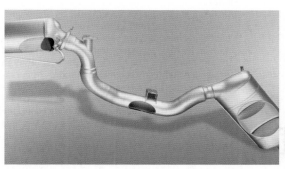

Future-oriented nitrogen oxide reduction for passenger cars:
Between the particulate filter (left) and the special SCR converter,
a nozzle injects urea, where a specifically designed pipe allows
optimal mixing with the exhaust gas. Particulate filters and nitrogen
oxide reduction strengthen the position of the diesel engine.

An active exhaust after treatment without engine modifications
is immediately possible with the fuel processor (in red). Converters and particulate filters can then become more efficient, while
the engine controls can be focused entirely on fuel economy and
performance. The predevelopment project is being tested now on
test vehicles.

more efficiently than small ones coagulate with one another.

Upon leaving the engine, the particle density is usually about 10^8 $1/cm^3$ or higher. At this level, rapid agglomeration takes place so that the concentration sinks within a few seconds by one to two orders of magnitude. Concentrations of about 106 $1/cm^3$ are, therefore, mostly measured in the exhaust. High-proportion rapid dilution takes place and stops further agglomeration so that after the exit of the gas through the exhaust pipe into the atmosphere — during which dilution factors of >1:1000 occur after a few meters — hardly any more agglomeration effects are expected. A very low level of dilution, as common in the measuring process applied in the CVS-dilution (constant volume sampler) tunnel with 1:5 to 1:10, does not rule out further agglomeration.

- *Adsorption of gases*: Primarily limited by the BET surface. For soot, this surface is in the range of 50 to 150 m^2/g and is thus sufficiently large enough to bind significant volumes of gases in a mono-molecular arrangement. In the case of combustion aerosols, additional highly toxic substances such as polycyclic aromatic hydrocarbons (PAH) may be bonded to solid particles and brought deep inside the lungs.
- *Adsorption of liquids*: Some particles are hygroscopic and can, therefore, rapidly adsorb water or diluted acids in humid atmospheres (exhaust temperature under the dew point), thus growing larger and greatly increasing their weight. Because of high surface tension (caused by a small radius of curvature of the particles), some degree of super-saturation is required. If the vapor pressure of the first adsorption layer is reduced by dissolving salts, the growth takes place much faster. This process also occurs in the respiratory system and leads to the situation whereby hygroscopic particles rapidly become deposited in the external respiratory passages. Soot particles are not primarily hygroscopic (Weingartner); they retain their size and penetrate up to the alveoli. They can develop a certain hygroscopic nature only after a longer period of solar radiation and then serve as a condensation core for droplet formation in the sky or influence the formation of secondary particles (sulfates, nitrates).

Literature: M. Kasper: Aerosole–Definitionen und physikalische Gesetze, House of Engineering, Munich, 12 October 1999; Diss. ETH 12F25/1998. — Weingartner/PSI, Modification of Combustion Aerosols in the Atmosphere, PSI Report No. 96/July 96; Envir. Sci Technol. 29/2982, Klingenberg, Wandlung von Dieselpartikeln laengs des Abgasstranges, MTZ 58 (1996).

~~**Particle composition.** **Fig. P4** shows the mean values of measurements on passenger cars as an example of the composition of combustion particles. According to this, the particles mainly consist of elementary carbon (soot), hydrocarbons, condensed sulfuric acid, and metal oxides. The form of the values shown in **Fig. P4**

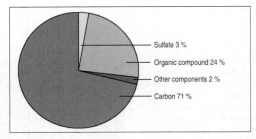

Fig. P4 Material composition of diesel particles (Source: Klingenberg)

can strongly vary depending on the engine, operating point, fuel, lubricating oil consumption, and exhaust aftertreatment system.

With a commercial vehicle engine that is operated at higher loads than a private car, elementary carbon is predominant; the share of hydrocarbons can be much higher from light-load operation of a car. The proportion of condensed sulfuric acid in the cooled exhaust mostly depends on the conversion of SO_2-SO_3, which can be strongly accelerated by oxidation catalytic converters. Metal oxides increasingly occur because of the application of corresponding additives in lubricating oil or also in fuel (for the purpose of filter regeneration). In addition, polycyclic aromatic hydrocarbons (PAH) and highly toxic trace substances such as dioxins, furans, and nitro-PAH can be adsorbed to particles. In the pure gas downstream from an efficient particulate filter, practically no solid matter (soot, metal oxide, ash) can be found in the exhaust. On the other hand, if the corresponding dew points are exceeded during sampling, sulfuric acid aerosols with bound water, hydrocarbon droplets and, at very low temperatures and low dilution, even condensed combustion water are found.

Common methods of determining particle composition are:

- Separation into soluble and insoluble components. In a simple analytical process, the particle mass found on the measuring filter is frequently subdivided into the following groups through a chemical extraction process and differential weighing:
  —SOF = soluble organic fraction: Hydrocarbons removed through solution in dichloromethane.
  —Water soluble particles: sulfate particles removed through the solution in diluted isopropanol.
  —INSOF = insoluble fraction: this residue is interpreted as soot and ash components.

According to the industrial medicine guidelines (TLV), either only the elementary carbon (EC) or the so-called total carbon (TC) is valid as the indicator substance for diesel engine particle emission (DME). TC comprises the elementary carbon and organic carbon: TC = EC + OC; it is determined either by a two-step coulometric method or by the thermographic method based on VDI 2465.

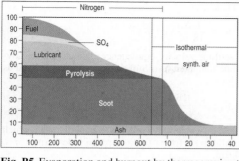

Fig. P5 Evaporation and burnout by thermogravimetric analysis (TGA; Source: ACEA)

- Thermogravimetry: By heating up a sample under nitrogen in a thermobalance, the volatile components can be determined by means of gravimetric analysis depending on their evaporation temperature; subsequent addition of oxygen provides information about the proportion of elementary soot; the residue is ash. **Fig. P5** shows a typical thermogram for the soot emitted by modern car diesel engines.
- More accurate determination of particle composition based on filter residues is possible with GC–MS (gas chromatography–mass spectrometry), a method for generally categorizing the chemical substances with very high accuracy but involving significant expenditure.
- When applying the ICP–MS method (inductively coupled plasma–mass spectroscopy), the detection limit is only a few nanograms per filter sample, depending on the chemical categorization—i.e., consideration of the entire material content of the sample for many elements (Ulrich).
- The PIXE (proton-induced X-ray emission) method is also applied for atmospheric dusts. It also can attain a similarly high resolution to the ICP–MS method, although it can only provide information about the composition of near-surface layers.

- Detailed information about the composition of particles directly after their formation is provided by the aerosol time-of-flight mass spectrometer, or ATOFMS.
- Categorization of surface composition according to elements using the REM (reflection electron microscopy) technique, under the microscope, is possible with the EDX (energy dispersion X-ray) analysis; with this, it is also possible to identify certain substances.
- The chemical character of the soot surface is of greater importance for the reactivity of the soot particles in the organism. A chemical code (fingerprint) can be obtained from the photoelectric yield (Source: Burtscher, Siegmann). This can be measured on-site by the PAS method directly on the particles in the aerosol. The emission of electrons under ultraviolet radiation depends largely on the substance on their surface. The metal oxide particles and the condensate emit more weakly than the combustion soot particles. The cause is seen in typical PAH coating of combustion aerosols. The photoelectric PAS method is also suitable as a dynamometer measuring technique, even under transient driving cycles.

For other particles, such as wear particles, fiber particles, and so on; *see below*, ~Particulate emissions ~~Nonengine particles.

Literature: H. Klingenberg et al.: Rußpartikel im Dieselmotorabgas—Entstehung und Messung, aekologische Forschung GSF/BMFT 1992. — ACEA Programme on the Emissions of Fine Particles from Passenger Cars, ACEA, December 1999. — VDI 2465: Chemisch-analytische Bestimmung des elementaren Kohlenstoffs ZH1/120.44, anerkannte Analyseverfahren für krebserzeugende Arbeitsstoffe, Karl Heimanns Publishers, Koeln, 1990. — D. Dahmann: Ringversuch zur Bestimmung von elementarem Kohlenstoff, Gefahrstoffe Reinhaltung der Luft, Maerz, 1999. — M. Kasper et al.: NanoMet, a New Instrument for On-line Size- and Substance-Specific Particle Emission Analysis, SAE 2001-01-0216. — A. Ulrich, A. Mayer et al.: Retention of Fuel Borne Catalyst Particles by Diesel Particle Filter Systems, SAE 2003-01-0287.

~~Particle size. Fig. P6 compares the size of soot particles with numerous other airborne particles.

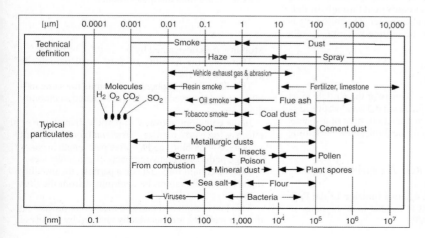

Fig. P6
Magnitude ranges of different particles
(Source: Cozzarini)

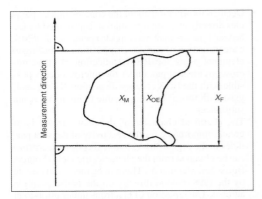

Fig. P7 Statistical particle dimensions (Source: Loeffler)

Particles generally have no regular geometrical shapes (such as sphere, cube, etc.) through which they can be described. "Dispersion sizes" that are related to particle size are termed according to DIN 66 141 "fineness feature." The following definitions are common:

a. *Statistical particle dimensions* (**Fig. P7**): If the particles can be "seen" through a microscope (light microscopy, scanning electron microscopy [SEM] , transmission electron microscopy [TEM]), they are described by:

X_F = Feret diameter: Distance of the tangent perpendicular to the measuring direction

X_M = Martin diameter, surface area dissecting chord, parallel to the measuring direction

X_{OE} = Long chord, parallel to the measuring direction

b. *Geometrically equivalent diameter*: Dimension of particles of regular shape with the same geometrical properties—for instance, diameter of a circle with the same projection of surface areas.

c. *Aerodynamic diameter*: Particles with the same aerodynamic characteristic such as the same settling velocity are considered as equivalent through the definition of an aerodynamic diameter. The following is the general equation for the settling velocity, w_s, of a sphere:

$$w_s^2 = \frac{4}{3}\frac{1}{c_w(Re)} \cdot \frac{\Delta\rho}{\rho_f}gd,$$

where $\Delta\rho = \rho_p - \rho_r$, ρ_p = density of the particle, ρ_f = density of the fluid, g = gravitational acceleration, d = particle diameter, c_w = coefficient of drag of a sphere, μ = kinematic viscosity of the fluid, and w_s = settling velocity.

$$Re = \frac{w_s \cdot d \cdot \rho_f}{\mu} = \text{Reynolds number}$$

The coefficient of drag, c_W, is a function of the Reynolds number. For Re $\leq$ 0.25 (that is in the range of Stokes' laws) is $c_w = 24/Re$.

Hence,

$$w_s = \frac{1}{18}\frac{\Delta\rho}{\mu} \cdot g \cdot d^2.$$

A settling velocity equivalence diameter or aerodynamic diameter, also Stokes' diameter, is obtained as follows:

$$d_{Aero} = \sqrt{\frac{18\mu \cdot \varphi_\sigma}{g \cdot \Delta\rho}}$$

where d_{Aero} = aerodynamic equivalent diameter, which is valid in air for particles of magnitude 1 μm $< d <$ 50 μm.

The aerodynamic equivalence diameter, therefore, depends upon the density. Generally (in contrast to DIN 66 112) the density difference $\Delta\rho = 1$ g/cm³ is used as a basis.

Conversion for other densities:

$$d_{eff} = d_{St} \sqrt{\Delta\rho_{St}}$$

Transitional range: With particle diameters below 1 μm, the settling velocity law must be corrected as a result of the effect of Brownian motion:

$$w_s = \frac{1}{18}\frac{\Delta\rho}{\mu} \cdot g \cdot d^2 \cdot \text{Cu}$$

Cu, the Cunningham correction, changes strongly with the particle size:

d (μm)	0.01	0.1	0.5	1.0	5.0	10.0
Cu	22.3	2.9	1.33	1.16	1.03	1.01

d. *Mobility diameter*: With very small particles— small in comparison with the free path length λ in carrier gases—the impact forces transmitted through the Brownian motion of molecules on the particles outweigh the shearing forces or viscosity forces and the inertial forces.

In this so-called "molecular field," the force, F, transferable to the particle is proportional to the cross-sectional area of the particle ($\sim d^2$), and force and drift velocity, v, are mutually proportional:

$$F \sim \frac{d^2 \cdot v}{\lambda}.$$

Mobility, b, is designated by $b = v / F$.

Therefore, the mobility of very small particles is proportional to $1/d^2$.

The diameter of a sphere that exhibits the same mobility as the observed particle is designated as mobility diameter.

In the equations above, d = particle diameter, F = effective force on a particle, F = frictional force, v = drift velocity, b = mobility, and λ = free path length in gas = average path between two impacts of gas molecules.

When a certain force acts on a particle, the mobility diameter can, therefore, be determined from the drift velocity.

This law is exploited during a measuring process, in that the particles are electrically charged by means of diffusion charging—for instance, in a passage through a coronal stretch, then exposed to an electric field. The force, F, is then known; the mobility is observed, and the mobility diameter can be determined from it.

The mobility diameter determined in this way is independent of the mass and the density of the particle.

All particles with the same mobility (independent of their density) are assigned the same mobility diameter.

The mobility, d, is directly linked with the diffusion constant, D, via the Stokes-Einstein relationship:

$$b = \frac{D}{k \cdot T} \sim \frac{1}{d^2},$$

where b = mobility, d = particle diameter (mobility diameter), D = diffusion constant, k = Boltzmann constant, and T = absolute temperature.

The behavior of very small particles (e.g., precipitation in the respiratory organs) is substantially determined by diffusion. The mobility diameter is, therefore, a reasonable characteristic variable for small particles. From the quadratic connection, it is also plausible that small particles, the mobility of which rises in proportion to the square of the decrease in diameter, very rapidly diffuse on the surfaces of the inner respiratory passages and can as such become biologically effective.

e. *Optical fineness features*: The intensity of scattered light emitted from the particles can likewise be used as a measure of particle dimensions—the scattered-light–equivalent diameter. In particular, ultrafine and nanoparticles are invisible to a great extent because they exhibit diameters below the wavelength of visible light.

When calculating the scattering characteristic, the Mie theory is applied because of the nonnegligible expansion of the particles. The scattering characteristic of agglomerates conforms in good approximation to the sum of the scattered signals of the primary particle from which they originate. If the optical equivalence diameter is to be compared with other diameters, its dependence on the wavelength of the light being used and the material of the particles, particularly its refractive index, are important.

Different particle diameters are, therefore, derived from different measuring methods. For instance, from the impact method, the aerodynamic diameter is derived; from the electrical mobility method, the mobility diameter; from the scattered-light method, the scattered-light–equivalent diameter.

For the evaluation of the small particles that are deposited preferentially in the alveolar area of the lungs, the mobility diameter is decisive because the deposition takes place primarily through diffusion, not through sedimentation and inertial separation.

Literature: Ch. Cozzarini: Quellen der Partikel-Emissionen, 20th Vienna Engine Symposium, May 1999. — F. Loeffler: Staubabscheiden, Georg Thieme Publishers Stuttgart, 1988. — W.C. Hinds: Aerosol Technology, BIA Working Portfolio, October 1998, Erich-Schmidt Publishers; DIN 66141, DIN 66 111, DIN EN 481, ISO 7708.

~~**Particle size distribution.** Particles occur preferentially in three areas according to the formation processes, as assigned in **Fig. P8**. A three-modal distribution is frequently referred to: nucleus mode, accumulation mode, large particles:

- Directly after their formation from the gas phase, particles are referred to as nuclei of the respective primary particles. Their mobility diameter mostly lies in the range of 2 to 3 nm, with a maximum of 10 nm.
- Agglomerate: Loose connection (van der Waal forces) in the exhaust, typically around 100 nm, after adsorption to arbitrarily large surfaces, frequently up to 10 nm (these are called aggregates if the connections are essentially more solid or sintered).
- Large particles form in a very different formation process—that is, mechanically, through granulation or wear.

The following classes are defined, based on this appearance:

- TSP (total suspended particulate matter): Airborne particles with settling velocity <10 cm/s, aerodynamic diameter <57 μm.
- Large particles: >2.5 μm.
- Fine particles: <2.5 μm.
- Ultrafine particles: <0.1 μm—i.e., <100 nm.
- For the ultrafine particles, the term "nanoparticle" (1–999 nm) is also becoming established.
- PM10: Particles precipitated through a filter, whose fractional collection efficiency has a 50% passage (50% of the mass) at 10 nm. In this fraction, even larger particles—i.e., about 1% of the particle mass with diameters of 30 nm—are included.
- PM2.5, PM1, PM0.1, analogously.
- Respirable particles: Subdivided according to respirable fraction, thoracic fraction and alveole intrusive fraction (DIN EN 481 9/93; **Fig. P9**).

To determine particle size distribution (**Fig. P10**), the particles must be classified according to measuring techniques and assigned to the various size classes. This is mostly done with impactors, electric mobility analyzers, diffusion batteries, cyclones, or centrifuges. Different distributions are obtained, depending on whether the particle number, particle surface area, or particle mass is illustrated as a function of the particle size.

This typical particle size distribution from a direct injection diesel engine shows a large number concentration for particles of small mass in the nm range and a negligibly small number for large mass in the μm range. Correlation of number and mass is, therefore, not always possible. The few large but heavy particles are agglomerates, as shown, for instance, in **Fig. P11** from a fiber filter.

Very different distributions occur even when the reference diameter is chosen differently—for instance, according to aerodynamic diameter instead of the mobility diameter. The size distribution of soot particles at the engine exit is generally very uniform and generally gives log-normal distributions around 80–100 nm (**Fig. P12**).

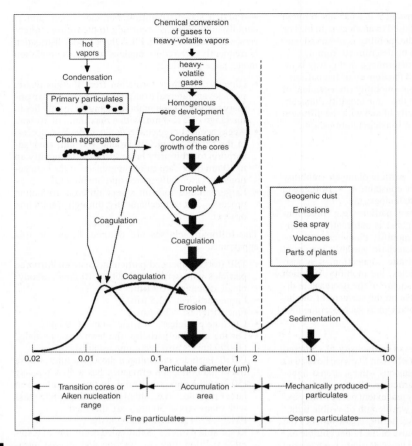

Fig. P8
Formation processes and size distribution of particles (Source: Peters)

P

Other particles, such as condensate, which also occur in the exhaust in addition to the soot particles, can distort the distribution and generate bimodal forms.

Two examples are shown in **Figs. P13 and P14**. In **Fig. P13**, a case that frequently occurs is shown; this case occurs when the dew point is not reached during the sampling process. The forming condensate (mostly sulfuric acid condensate, possibly also hydrocarbon droplets) is generally not differentiated from carbon particles through the common particle-counting pro-

cess. If the gas sample is heated, however, these droplets evaporate again and the typical diesel-soot distribution profile becomes apparent. The distribution pattern for the dew point lowered step-by-step (thermogram) provides information on the material composition of the aerosols.

Fig. P13 shows heavy-metal oxide particles in the range of 20 to 30 nm due to the addition of a regeneration additive with iron content (satacen from OCTEL) for fuel. In this case, nothing changes—however, the

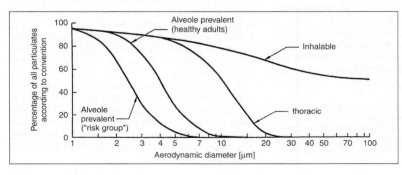

Fig. P9
Size distribution conventions (DIN EN 481)

Fig. P10 Size distribution of the same soot aerosol, depicted according to the number, surface area, and mass (Source: Matter). Key: —— number, –·– surface area, ···· mass/volume

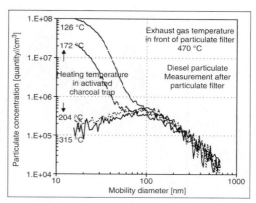

Fig. P13 Formation of condensate by cooling the exhaust during the sampling process. This type of illustration is termed thermogram (Source: Matter)

exhaust sample is heated—because the particles have a solid matter character.

Literature: Peters et al.: Staub und Staubinhaltsstoffe/Feine und ultrafeine Partikel, Handbuch Umweltmedizin, 14th Suppl. Installment 10/98. — ACEA Programme on the Emissions of Fine Particles from Passenger Cars, ACEA, December 1999. — Standard DIN EN 481, Festlegung der Teilchengrössenverteilung zur Messung luftgetragener Partikel, September 1993. — Matter et al.: Volatile and Nonvolatile Particles in Exhaust of Diesel Engines with Particulate Traps, 3rd ETH Workshop on Nanoparticle Measurement, 1999. — Mayer et al.: Particulate Traps for Retro-Fitting Construction Site Engines VERT: Final Measurements and Implementation, SAE 1999-01-0116.

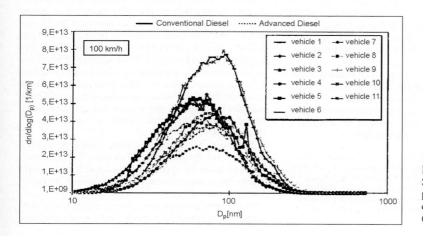

Fig. P11 Soot particles precipitated on a filter fiber (fiber diameter, 10 μm) with precipitated fine particles around 100 nm and very large dendritic-structured agglomerates (Source: J. Mayer, ETHZ)

~~Particle surface. Because the particles mostly do not exhibit geometrically clean—such as spherical—shapes but can assume any arbitrary shape (chain form or aciniform), the surface area cannot be directly

Fig. P12
Size distribution of solid particles for 11 modern car diesel engines (Source: ACEA)

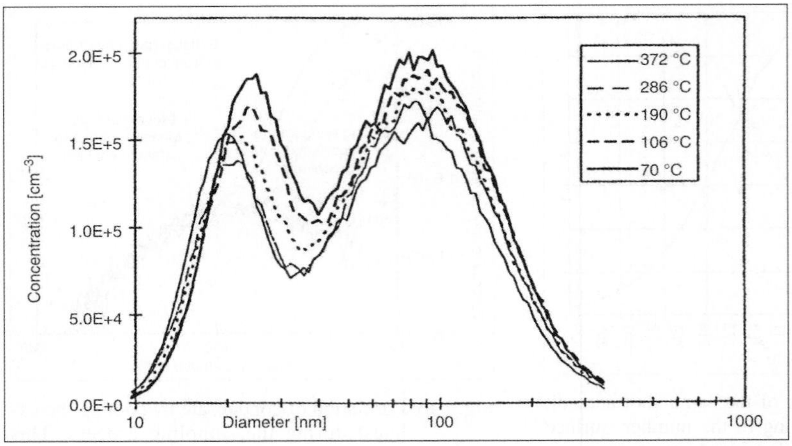

Fig. P14
Bimodal distribution
during the application
of a fuel additive with
heavy metal content
(Source: Matter)

derived from the diameter. Measurements on combustion particles frequently show that the particles become quasi-flat or chain-shaped during the agglomeration process—that is, the effective surface area quickly increases with the mass in the measuring process, corresponding to fractal dimension 2.

Because the specific surface area is of great importance for the adsorption of other substances and all biological processes, definitions and measuring processes are required:

- BET surface area: Using the gas-adsorption method according to Brunauer, Emmet, and Teller (DIN 66 131), the surface area is determined via the volume of a monomolecular moistening layer through a gas, mostly nitrogen. With soot surfaces, the range is from 50 to 150 m²/g.
- Active surface: The active surface is determined based on the adsorption of radioactive atoms or from ions of plasma. The active surface area is inversely proportional to the mobility (mobility diameter). For particles <100 nm—the typical diesel particle range—it is proportional to d².
- Mobility diameter and active surface areas are equivalent and characteristic for the behavior of the particle in the respirable range, because the diffusion constant D is also proportional to d. This active surface is also termed, by N. A. Fuchs, the Fuchs surface area.
- It is obvious that the variable, which is easily determined in aerosols and physically exactly defined (particle measuring technique), is also decisive for the chemical and biological processes occurring on the surface, and as such is generally recommended for characterizing the respirable particles.

Literature: Keller et al./ETH: Evaluation of the surface properties of nanoparticles, Third International ETH Workshop on Nanoparticle Measurement, 9 August 1998, Zurich. — Brunauer, Emmet, Teller: Adsorption of Gases in Multimolecular Layers, J. Am. Chem. Soc., 60th year (1938), p. 309. — Fuchs: The Mechanics of Aerosols, Pergamon Press, Oxford, 1964.

~Charging. See below, ~Particulate filter system ~~Particulate filter

~Collection method for aerosol characterization. See below, ~Particle measuring

~Combined regeneration process. See below, ~Particulate filter system ~~Particulate filter ~~~Regeneration of particulate filters

~CRT system. See below, ~Particulate filter system ~~Particulate filter ~~~Regeneration of particulate filters

~Deep-bed filter. See below, ~Particulate filter system ~~Particulate filter ~~~Regeneration of particulate filters

~Diesel particles. See above, ~Characteristics of diesel particulates

~Effects on climate. The effects of fine particles on climate are distinguished physically according to their direct effects—that is, the solar radiation is either absorbed from the aerosols or scattered back into space—and their indirect effects—the influence of the aerosols on cloud formation. According to IPCC (Intergovernmental Panel on Climate Change/1995), both of the above effects can show significant regional differences in particular but are much more difficult to estimate than the effects of trace gases because very different particles are involved and aerosol concentrations may be subject to substantial spatial and chronological fluctuations.

Natural airborne dust emissions exceed the anthropogenic ones by about sixfold in mass, but anthropogenic particles are smaller and therefore have a significantly larger surface area and, for this reason, probably have a greater effect.

Soot particles, moreover, absorb sunlight very efficiently, in contrast to mineral dust (weathering and erosion) or sea salt—the two major sources of natural aerosols.

Recent research by Jacobson concluded that the greenhouse relevance of diesel soot, based on mass, exceeds that of CO_2 by a factor of more than 500,000. Moreover, the advantage of the diesel engine, based on its CO_2 emissions, due to better thermodynamic efficiency, only plays a role if its particulate emission levels are reduced by more than 99% according to current standards. A reduction of particulate emissions would be effective within a few years, whereas the reduction of CO_2 will be not be felt for many decades.

Literature: Ch. Cozzarini: Quellen der Partikel-Emissionen, 20th Vienna Engine Symposium, May 1999. — Climate Change 1995, The Science of Climate Change, IPCC, Cambridge University Press, 1996. — Mark Z. Jacobson: Control of Fossil-Fuel Particulate Black Carbon and Organic Matter, Possibly the Most Effective Method of Slowing Global Warming, in Journal of Geophysical Research, accepted April 12, 2002.

~Effects on the human organism. The place of entry for air pollutants is the lung. The respiratory tract branches out in some 20 stages up to the bronchioles and ends in the 500 million alveoli (pulmonary alveolus) through the walls (one cell thickness, about 1 μm) where gas exchange occurs (**Fig. P15**). The air velocity is thereby constantly reduced; the dwell time is correspondingly long—that is, the deeper a particle penetrates into the respiratory tract, the longer it remains on the walls. The probability of pulmonary intrusion and disposition are, therefore, of great importance for evaluating pollutants.

In the course of evolution, efficient defense mechanisms against natural dusts have formed in the respiratory tract system: dust is separated on moistened surfaces; the mucus layer is constantly moved toward the throat by means of fine cilia; and a warning system with sensitive chemical sensors and diverse measures such as coughing and sneezing help keep the lungs clear. In addition, movable macrophages are located in the alveoles and the respiratory tracts that collect the particles, digest (phagocytize) them, and carry them away.

Technical dusts, especially particles from engine combustion, however, exhibit grain sizes that are up to 100 times smaller than natural dusts. When exposed to such ultrafine particles, the defense mechanisms of the lungs are no longer effective: these tiny particles advance through the protective zone into the bronchioles, in which the cilia are missing, then finally reach the alveoli and remain there for months or even years (**Fig. P16**). Therefore, there is a risk that they may get inside the system of blood vessels or lymph and into the nerve cells, spreading throughout the entire organism— or they may be permanently deposited in the lung tissue, where they can cause continuous irritation.

Every single particle that contacts the cell surface of the lung epithelium triggers a reaction process—either allergic or inflammatory in nature. Understandably, this process is more pronounced the more the cells react—that is, the more the particles are in contact with the cell surfaces in the alveolar section. Short-term effects show up in inflammatory states, light coughing, bronchitis, asthma, and severe allergic reactions.

After these particles penetrate into the vascular system, acutely damaging effects may be triggered, for instance, because of increased blood viscosity or an influence on the cardiac rhythm. Epidemiological studies indicate a potential for cardiopulmonary and cardiovascular damage in addition to carcinogenic damage, because of particles from combustion engines.

According to the "Monetization study" (1996), each year air pollution in Switzerland leads to 3800 premature deaths (seven times more than accidental deaths), 53,000 cases of bronchitis in children, and 791,000 days of work disability.

Investigations by Heinrich show the carcinogenic potential of inert particles; Denissenco verified the causality of tumor formation through benzopyrene, which

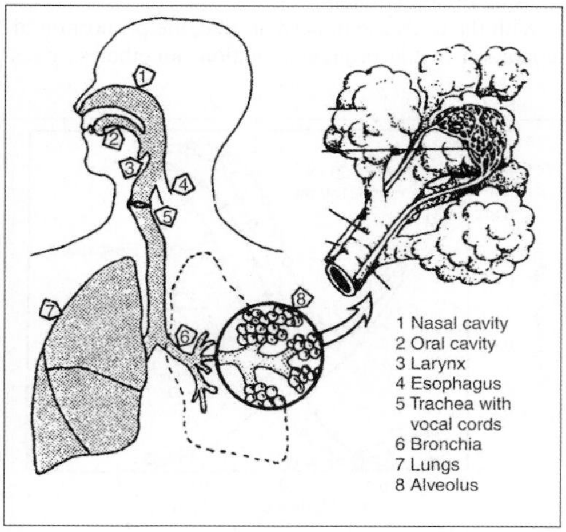

P

Fig. P15 The lung, gateway for air pollutants (Source: Birgersson)

1 Nasal cavity
2 Oral cavity
3 Larynx
4 Esophagus
5 Trachea with vocal cords
6 Bronchia
7 Lungs
8 Alveolus

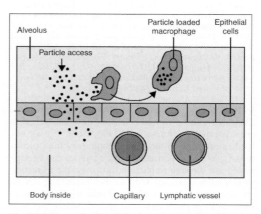

Fig. P16 The path of particles into the organism (Source: Donaldson)

is known to be adsorbed by combustion particulates. According to Wichmann, the health risk increases with decreasing particle size.

Diesel soot particulates were classified as "probably carcinogenic" on the basis of animal experiments with cell cultures and epidemiological findings as early as 1987 by the International Agency for Research on Cancer (IARC). This evaluation was confirmed after testing of further findings in 1999 by the IARC. In Germany, diesel soot particulates are classified in *MAK* (threshold limit value, TLV) class III A 2 and are as such subject to the minimization proposal. In 1998, the California Air Resources Board, ARB, classified diesel exhaust as a toxic air contaminant (TAC) based on its particulate content. The German Federal Agency for the Environment (*Umweltbundesamt*) evaluated diesel exhaust in terms of carcinogenic potential, according to the 1999 standard, as being 18 times higher than that of gasoline engine exhaust based on conclusions arrived at by a research project conducted by the Fraunhofer Institute. According to the 2005 standard, it is still evaluated as being 15 times greater. With the introduction of particulate filters, however, it is practically equivalent. According to industrial medical criteria, particles are considered particularly health-threatening if they are pulmonary-intrusive and hardly soluble in the organism; both apply to soot particulates from engine combustion.

Literature: Chemie und Gesundheit Birgersson, VCH Publishers, 1988. — Monetarisierung der verkehrsbedingten Gesundheitskosten der Schweiz GVF Report 272/5, 96. — N.N.: Quantification of Effects of Air Pollution on Health in the UK, ISBN 0 11 322102 9/1998. — N.N.: Preferential Formation of Benzo(a)Pyrene Adducts at Lung Cancer Mutational Hotspots in P53, Denissenco et al., in Science, 274 (1996) pp. 430–432. — Heinrich et al.: Influence of Number and Size of Particles on the Health Risk from Diesel and Otto Engines, Engineering Academy of Esslingen, Fuels 99. — H.E. Wichmann, Respiratory Effects Are Associated with Ultrafine Particles, in American Journal of Respiratory and Critical Care Medicine, Nov. 1996. — D.W. Dockery et al.: Association between Air Pollution and Mortality in 6 U.S. Cities, in New England Journal of Medicine, Dec. 1993. — Heinrich et al.: HEI Diesel Workshop, 7–9 March 1999. — UBA-Forschungsvorhaben 216 04 001/1, Risikovergleich Dieselmotoremissionen/Ottomotoremissionen, Fraunhofer Institut, December 1998. — Donaldson et al.: Ultrafine (Nanometre) Particle Mediated Lung Injury, J. Aerosol Sc. 5.6.1998. — William S. Beckett: Occupational Respiratory Diseases, In: New England J. of Medicine 342 (2000) pp. 406–413. — N. Metz: Bewertungskriterien fuer die Wirkung von Dieselruss, 4th International Vienna Engine Symposium, 2003.

~~Pulmonary intrusion, pulmonary deposition. The lungs can be viewed as a filter in which particles are separated in diverse ways (**Fig. P17**). The air velocities are high in the upper respiratory tract (nose, throat), so that in particular large particles can be separated out through impaction; however, the influence of diffusion increases with further branching out of passages and reduction of the air velocity. In the innermost areas of the lungs—the alveoli, where the flow stagnates—the diffusion and, thus, the mobility diameter of the particle is decisive.

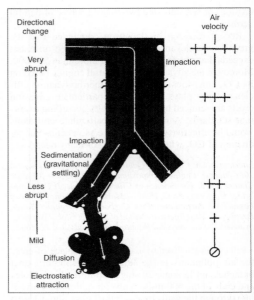

Fig. P17 Schematic illustration of the respiratory tract and the mechanism of particle deposition (Source: BUWAL)

Fig. P18 shows the deposition characteristics of fine particles in different areas of the lungs, depending on volume breathed (750 and 2150 ml/min):

The lung has an area, as does every filter, in which impaction no longer works and diffusion still does not work, so that the minimum size of separation is about 300 nm. About half of the particles in this area are breathed out again. Larger particles are efficiently separated in the nose and larynx area, and finer particles are separated in the alveoli—more strongly with increasing breath volume.

With the decrease in particle size, the proportion of separation in the bronchial section nonetheless rises

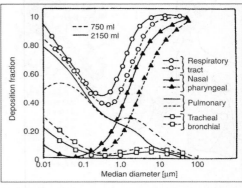

Fig. P18 Separation of fine particles in the nose, bronchi, and alveoli (Source: Hinds)

due to the increase in mobility of the particle. Significantly, smaller particles would not reach the alveoli at all, as the reversing curve already indicates 20 nm for 750 ml/min.

Of greater importance is also the question of whether or not the inhaled particles are hygroscopic. Hygroscopic particles rapidly increase in size in moisture-saturated segments of the respiratory tract and are separated out at an early stage, whereas non-hygroscopic particles can advance into the alveoli; diesel soot is hydrophobic if it has not remained for very long in sunlight in the atmospheric.

Compounds that are not soluble in water and, thus, not absorbed by the mucus layer in the area of the upper respiratory tracts, must obviously be classified as being more critical. This applies to gases and fine dusts: they have a low diffusion pressure on the mucus surface, and can as such advance to the alveoli.

Literature: BUWAL, PM10: Fragen und Antworten zu Eigenschaften, Immissionen und Auswirkungen, January 1998. — W.C. Hinds: Aerosol Technology, John Wiley, 1989. — Aerosole—Staeube, Rauche und Nebel, MAK, 24th installment, 1997. — Chemie und Gesundheit Birgersson, VCH Publishers, 1988.

~Electrical heating during operation. *See below,* ~Particulate filter system ~~Particulate filter ~~~Regeneration of particulate filters

~Electrical standstill regeneration. *See below,* ~Particulate filter system ~~Particulate filter ~~~Regeneration of particulate filters

~Emissions during regeneration. *See below,* ~Particulate filter system ~~Particulate filter ~~~Regeneration of particulate filters

~External regeneration (interchangeable filter). *See below,* ~Particulate filter system ~~Particulate filter ~~~Regeneration of particulate filters

~Filter papers/filter felt. *See below,* ~Particulate filter system ~~Particulate filter ~~~Filter media

~Formation. In the range of magnitude that allows pulmonary intrusion, particles are created by abrasion (e.g., of tire, clutch, brake, engine), by nucleation of solids from the gas phase, through condensation (acid droplets, hydrocarbon condensate), through incomplete combustion (soot particles, diesel particulates), and by chemical compounds such as $CaSO_4$ (gypsum) from calcium in lubricating oil and sulfur in both lubricating oil and fuel. Furthermore, particles can be brought into the engine process (suspended road dust) and then they reappear in the exhaust aerosol in a changed form.

Particle formation and decomposition can be accelerated through catalytic effects. Catalytic converters—mostly noble metals or oxides of transition metals coated on substrates or introduced as additives in fuel—can form particles by themselves, additive particles, and oxide particles.

~~Combustion particles in a diesel engine. In the case of older engines with low injection pressures, larger droplets, and a larger proportion of wall adsorption, particulates can occur because of carbonization processes; in the range of lower excess-air amounts, more diesel smut is produced because of these processes. These effects characterize the diesel soot produced by older engines as dense smoke clouds that have an offensive smell and that reduce visibility.

Modern diesel engines avoid wall adsorption, generate extremely fine, well-distributed droplets, offer abundant excess air, and guarantee nearly complete burnout of the particles.

Nevertheless, soot occurs mostly in the form of very small particles, mainly in the invisible range around 100 nm.

These soot particles are not carbonization residues from the injection droplets as with older engines, but they occur on a molecular basis—the reason for their small size. The cause is certainly always inhomogeneous combustion—that is, even under optimum mixture formation, microscopic and temporary oxygen deficits prevail.

With all the assumptions for modeling soot formation, pyrolysis of fuel occurs at the beginning, whereby the radicals and smaller molecules and, above all, the acetylene are formed. Also, aromatic rings can occur back in the gas phase because of their extraordinary stability.

Different opinions are held with respect to nucleation of the particles:

- *Polycyclic hypothesis*: According to this theory, large polycyclic hydrocarbons form in the gas phase (PAH). As a result of the van-der-Waal forces, the PAHs condense to clusters that also accumulate to form primary soot particles. What is characteristic of this model is that the PAHs are considered to be precursors of the particles and that the typical carbon-rich core of the soot particle occurs only in the second step through dehydration.

- *Radical hypothesis* (Siegmann). According to this model, the radicals condense to form the first cores, whereby chemical bonds are made; this, therefore, involves chemical condensation (**Fig. P19**). The first cores initially consist of chain-shaped aggregates with high mobility. Benzene rings also exist already, but they are still connected by means of flexible hydrogen chains. During the aging process of these cores in the heat of the combustion zone, further aromatic rings form, which finally lead to graphitization of the particles. The PAHs are synthesized on the surface of the particles, which is more efficient than the synthesis in the gas phase. First—that is, in the heat of the combustion zone—the PAHs evaporate in the gas phase; later, while the exhaust is cooling, the larger PAHs condense again on the surface of the particles and are carried out with the latter. This radical model can be verified in detail for the simplest fuel: methane.

~~Combustion particles in a gasoline engine. Conventional gasoline engines show very low particle

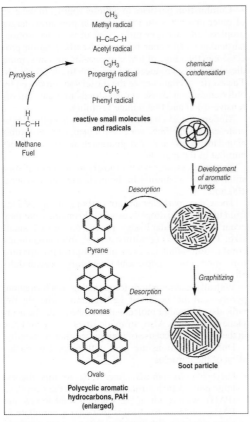

Fig. P19 Occurrence of soot particles (Source: Siegmann)

emission in the part-load range (typical driving cycles). However, as the load increases and particularly upon enrichment of the mixture, particle concentration is measured that generally corresponds to that of the diesel engine pertaining both to particle concentration and particle characteristic (**Fig. P20**).

With the direct-injection gasoline engine, the situation is similar to that of the diesel engine. Above all, high concentrations of fine particles are detected in a stratified operation. Efforts are being made to lower particulate emissions based on the combustion method and its further development.

Literature: S. Aufdenblatten et al.: Partikelemission von modernen Verbrennungsmotoren, MTZ (2002) 11, p. 962. – K. Siegmann, H.C. Siegmann: Die Entstehung von Kohlenstoffpartikeln bei der Verbrennung organischer Treibstoffe, Siegmann, Seminar Feinstpartikelemissionen von Verbrennungsmotoren, House of Engineering, Muenchen, 12 October 1999. – M. Woelfle: Schadstoffbildung im Zylinder eines DI-Dieselmotors, Diss., RWTH Aachen University, 1994. – FVV Report 594/996. – L. Jing: Charakterisierung der dieselmotorischen Partikelemission, Diss., University of Bern, 1997. – Gaskow, Kittelson, et al.: Exhaust Particulate Emissions from a Direct Injection Spark Ignition Engine, SAE 1999-01-1145. – ACEA Programme on the Emissions of Fine Particles from Passenger Cars, December 1999.

~~**Variables for particulate formation.** Many aspects of engine operation, including the materials used, have an influence on particulate formation.

~~~**Fuel.** Whereas the soot particles in the submicron range can hardly be influenced by reformulation of the diesel fuel (Sieverding), other important side effects such as sulfate formation, ash particle formation, and secondary emission formation are substantially influenced by fuel composition. Significantly, a stronger

P

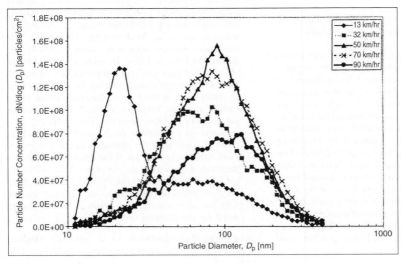

Fig. P20
Particle size distribution for a gasoline engine with direct injection (Source: Gaskow)

influence is exercised when the diesel fuel is mixed with oxygen-rich substances (Lutz) and when the so-called organic (bio) fuels such as rapeseed methylester (RME) are used.

- Sulfur: Sulfur in fuel is oxidized to SO_2 during engine combustion; from exhaust temperatures above $450°C$, a shift toward SO_3 becomes apparent because of the increase in atomic oxygen that is available; thus the formation of sulfate ions as well as sulfur condensate H_2SO_4 occurs in the cooled exhaust upon exceeding the dew point. This effect is reinforced when catalytic measures, through fuel additives or exhaust post-treatment, accelerate the conversion from SO_2 to SO_3, which can be the case especially with oxidation catalytic converters. This undesirable effect is most easily avoided through elimination of the sulfur content in fuel.
- Sulfates in the fuel can form calcium sulfate (gypsum) with calcium in the lubricating oil and with this solid particles; these can coat a downstream particulate filter to the extent that it cannot be regenerated.
- Water emulsions: Addition of water to fuel, simply in the form of emulsions with 13% water, as offered on the market, reduces the particulate mass by approximately 30% (and for older engines with poor injection characteristics even more); emissions of oxides of nitrogen are reduced by about 15%.
- Oxidants: Addition of methanol or other oxygen-rich components to the diesel fuel reduces the particle mass and also influences the particle number concentration in a favorable manner (Stommel, Fuels 99; Lutz).
- Fuel additive: Organometallic compounds can be used as fuel additives to reduce the particulate mass by 20–30%; however, they form their own fraction of fine ash particles at around 20 nm so long as they are not adsorbed to soot particles, which necessitates use of particulate filters suitable for this purpose (Mayer).
- Aromatics: The reduction of aromatics in fuel contributes to the reduction of aromatics in the exhaust, whereby the PAH coating on the particle can also be favorably influenced.
- Organic fuels: e.g., rapeseed oil methylester (RME). Organic fuels have a convenient effect on combustion, thanks to their increased oxygen content. Frequently, a reduction of the entire particulate mass is noticed; however, the composition changes generally in that the soluble portion increases. The boiling points of these soluble substances are mostly higher than those of diesel fuel (Walter). This different composition has to be considered during exhaust post-treatment; a combination of catalytic converter and filter is required in most cases.

Literature: Engineering Academy of Esslingen, Fuels 99, Contributions from Sieverding, Stommel, Mayer. — M. Gairing et al.: Einfluss von Kraftstoffeigenschaften auf die Abgasemissionen moderner Dieselmotoren, MTZ 55 (1994) 10. —

Bostel: Einfluss der Kraftstoffzusammensetzung auf die PAH-Emissionen eines modernen Dieselmotors, Gefahrstoffe Reinhaltung der Luft, 1999. — Th. Lutz et al.: Charakterisierung der Partikelemission von modernen Verbrennungsmotoren, MTZ (2002) 11. — F. Tort: Influence of water emulsions on nanoparticle emission characteristic, 4th Intern, ETH Conference on Nanoparticle Measurement, 7 August 2000. — Th. Walter: Untersuchung des Emissionsverhaltens eines Nutzfahrzeug-Motors bei Betrieb mit Rapsoel-Methylester, EMPA Duebendorf, Research Report 133439, 2001.

~~~Lubricating oil. The proportion of soot particles formed from the ingredients of lubricating oil can range from 20 to 30%. The lubricating oil particularly influences the formation of particles because of its high sulfur content (mostly >5000 ppm), calcium, and other metallic additives (such as zinc). Even the chlorine content (mostly >100 ppm) can change the chemical character of the particle, for example, through the formation of dioxins and furans. The ash content in the lubricating oil, which can be much higher than the ash content in the fuel, forms inert particles that can be deposited on the particulate filter and as such block it after a long period; at high temperatures, it can possibly even attack the filter materials with a disadvantageous effect or even become sintered with materials. Because the composition of lubricating oil is not subject to general limit values to date, valid quantitative inferences cannot be generally drawn from it.

Modern developments have had an effect on the influence of lubricating oil: its consumption has been reduced significantly (<0.1% of the fuel consumption), and the use of additives is increasingly aimed at promoting good health.

~~~Oxidation catalytic converter. The oxidation catalytic converters used in vehicles are normally axial, open-cell structures with a wall coating (**Fig. P21**). All reactions proceed in contact with the catalytic converter on the walls. Clear distance of the channels is around 1 mm; the dwell time of the gaseous medium traveling through is several hundredths of a second. For gas molecules, this time is sufficient to reach the surface and to react because of the strong diffusion pressure. Particles, however, have a very low diffusion

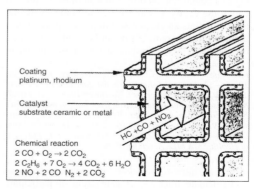

Coating
platinum, rhodium

Catalyst
substrate ceramic or metal

Chemical reaction
$2\ CO + O_2 \rightarrow 2\ CO_2$
$2\ C_2H_6 + 7\ O_2 \rightarrow 4\ CO_2 + 6\ H_2O$
$2\ NO + 2\ CO \rightarrow N_2 + 2\ CO_2$

Fig. P21 Oxidation catalytic converter

velocity at room temperature, according to Hinds, or only about 30 mm/s—that is, 100 to 1000 times less than the molecules. During a dwell time of 0.01 s in a catalytic converter, a particle can approach the wall only by a maximum of 0.3 mm. Precipitation, therefore, does not occur to a large extent. Case-adsorption effects occur to only a very small extent—solid particles are, as a result, practically not reduced through oxidation catalytic converters; thus, their distribution is not changed.

The soot deposited inside the oxidation catalytic converter is generally only converted slowly through catalytic means (NO_2 effect). In inappropriate cases, however, there is the danger of removal of the catalytic coating that can thereby become ineffective for the conversion of pollutants from the gas phase.

The high activity of the oxidation catalytic converter in converting gaseous substances, however, leads to the formation of SO_3 from SO_2 and, thus, to a strong increase of the entire particulate mass, depending on fuel composition.

NO is further converted in greater quantities through the oxidation catalytic converter to NO_2, which, according to TLV (threshold limit value) classification, is a significantly more toxic component of the oxides of nitrogen.

The conversion of hydrocarbons and carbon monoxide, while it is a positive aspect, is generally not required for a modern diesel engine.

Due to the high priority for minimizing particulate emissions owing to their carcinogenicity and the problem caused by NO_2, the oxidation catalytic converter is, therefore, not acceptable as the sole emission reduction measure for the diesel engine, and this method is strictly rejected for certain applications (mines, closed rooms).

Oxidation catalytic converters are seeing increased use in particulate filter systems due only to the increased formation of NO_2, with the help of which the soot precipitated in a filter can be converted at temperatures starting at 230°C—significantly lower than is the case with oxidation using oxygen. This process—known as CRT (continuous regeneration trap), according to a patent obtained by Johnson Matthey Inc., in which the catalytic converter is combined with the filter—depicts an elegant solution for a passive particulate filter system, but it can be used only where the increased NO_2 emission does not pose a problem.

Literature: W.C. Hinds: Aerosol Technology, John Wiley, 1989. — Treatment of diesel exhaust gas, European Patent 0341832/Johnson Matthey Inc., November 1989. — H. Friedl: Katalytische Nachbehandlung von Dieselmotorenabgas unter besonderer Betrachtung der Partikel- und Schwefelemissionen, VDI Progress Reports 162, 1992. — Diesel Emission Control—Sulfur Effects (DECSE) Program, US Department of Health/EMA MECA, August 1999.

~Full-flow burner. *See below,* ~Particulate filter system ~~Particulate filter ~~~Regeneration of particulate filters

~Knit fiber filter. *See below,* ~Particulate filter system ~~Particulate filter ~~~Filter media

~Legal limit values. In all industrial countries there are still no values for limits of particle abrasion from tires, brakes, and clutches, as well as from internal engines, mufflers, and catalytic converters; but the limit values for particle emission by combustion engines, as well as for exposure to emissions in the workplace and in the air that is breathed, do exist and are being developed in similar forms worldwide. For example, historically, one of the first limits of diesel smoke emission was VDI guideline 2281 (1961) with a full-load blackening number (BN) = 6.5 at 50 kW, reducing to 5 at 250 kW.

The limit for particulate emissions was introduced in the United States in 1982, with a gravimetric method for cars; in 1988, it was extended for trucks and buses.

The following limited emission values are gravimetrically defined, but apply to different test cycles and sampling methods. They are as such comparable to restrictions.

USA	Truck (g/kWh)
1990	0.8
1991/93	0.34
1994/97	0.13
2007	0.013
Europe	Cars (g/kWh)
1992	0.36
1995	0.15
2000	0.10
2007	0.02

- EU guideline for construction machines and agricultural tractors (class 75-130 kW)
  Stage 1 (1998) 0.7 g/kWh
  Stage 2 (2002) 0.3 g/kWh
  Stage 3b (proposal valid only for construction machines) 0.04 g/kWh
- Stationary diesel engine LRV 98 (CH): 5 mg/m³
- Immission limit values:
  Exposure at workplace:
  Switzerland, EC 100 µg/m³
  TRK Germany 100 µg/m³
  USA: Proposal 95 160 µg/m³
  Proposal 98 50 µg/m³
- Guide value for inner city in Germany 98 (soot) 8 µg/m³
- PM10 Switzerland LRV 98 20 µg/m³
- US EPA, reference concentration 5 µg/m³

Literature: M. Walsh: Global Trends in Diesel Emissions Control, 1999 update, SAE 1999-01-0107, TRGS 55 4. — EU Construction Machine Directive 97/68 EC 2/98, Switzerland. — Clean Air Ordinance LRV 98, Identification of Diesel Exhaust as Toxic Air Contaminant TAC, CARB, Feb. 98.

~~**Test cycles for particulate filter.** In order to check whether a particulate filter is functioning, it is necessary to measure not only its separation rate (fractional collection efficiency) dependent on the particle size, but also its characteristics under different charges and during regeneration. Whereas, in general, a steady-state test under maximum throughput and maximum temperature reached in application (worst case) is sufficient for checking the separation characteristic, charge and regeneration lead to transient states that have to be checked additionally. The influence of charging can be determined by recording the maximum permissible charge; the regeneration cycles with stepwise increase of load (temperature) at constant throughput, as introduced by DEGUSSA, is suitable for evaluating regeneration.

Suitability tests for evaluating particulate filter systems were first formulated with the so-called UBA measuring and charging cycle in the scope of the large-scale UBA soot filter experiment in 1989 and adopted during the introduction of TRGS 554 for the workplace in 1994. With the VERT suitability test, a detailed test method was introduced in Switzerland in 1998, which also comprised evaluating the separation characteristic based on particle size as well as verifying secondary emissions.

The Swiss Federal Agency for Environment, Forest and Landscape (BUWAL) publishes a list of all particulate filter systems approved in accordance with this testing method in the VERT Filter List.

Literature: B. Engler et al.: Catalytically Activated Diesel Particulate Traps—New Developments and Applications, DE-GUSSA AG, SAE 860007. — H. Bluemel: Russfilter-Grossversuch, UBA Berlin, Traffic Engineering, 31 (1990) 7/8. — Information zur Einfuehrung des Partikelfilter-Obligatoriums, Swiss Accident Insurance Institute, 24 Jan. 2000. — Mayer et al.: VERT-Particulate Trap Verification, SAE 2002-01-0435. — VERT Filter List: www.umwelt-schweiz.ch/Buwal.

~**Lubricating oil influence on particulate formation.** *See above*, ~Formation ~~Variables for particulate formation ~~~Lubricating oil

~**Partial-flow burner.** *See below*, ~Particulate filter system ~~Particulate filter ~~~Regeneration of particulate filters

~**Particle measuring.** It is necessary to characterize the particles physically and chemically, based on the particular problem. Among the physical characterizations required are variables such as mass, size, number, density, and possibly also shape, structure, and surface finish. Chemical characterization comprises element analysis as well as defining the chemical compounds in the entire particle sample, and possibly certain sizes as well.

Particle measuring poses considerable problems for the sampling process, to avoid the loss of particles (precipitation, electrostatic forces), occurrence of particles (condensation), or changes in particle composition (agglomeration, chemical reactions). Therefore, sampling must be done, if possible, using short lines while avoiding electrostatic and thermophoresis effects, under rapid dilution, and falling below the dew point of volatile ingredients in the measured gas. Isokinetic sampling is required for large (greater than 1 μm) particles; with sampling of smaller particles, this can be dispensed with because of the low inertia.

The measuring processes can be subdivided into two basic classes: the collective method and the in situ method.

With the collective method, the particles are first deposited on a filter, weighed, and finally analyzed with all the conventional methods (microscopy, electron microscopy, spectroscopy, wet chemistry). However, real-time measurements are mostly not possible, and the information about particle size and particle number is lost, posing a risk of artifacts caused by condensation, evaporation, or chemical change.

With in situ methods, the particles are characterized in their environment, thus in the aerosol. The artifacts mentioned above can be avoided and real-time measurements are possible with time resolutions less than 1 second; however, conventional analysis of a sample can no longer be applied.

In various extensive international measuring programs—for instance, the particulate measurement program (PMP), the UNECE (United Nations Economic Commission for Europe)/GRPE (Working Party on Pollution and Energy)—in situ methods have been investigated with respect to their applicability for the certification and field monitoring of diesel vehicles. For the medium term, the legislatures have planned an amendment or even a replacement of the limiting value for particle emissions, which is today determined purely gravimetrically by means of in situ measuring methods.

~~**Collective process for aerosol characterization.** The following methods are common:

- *Determining the entire particle mass* (PM; particle emission). According to the legally prescribed, so-called gravimetric methods (particle emission), a sample is taken from a dilution tunnel, cooled down to less than 325 K and separated on a glass fiber filter of a defined separation grade. The sample is conditioned with respect to temperature and humidity and then weighed. Measurement provides information about overall weight but no information about chemical composition and particle-size distribution; real-time information is not possible. Long collection periods, based on the particle concentration, are necessary in order to attain sufficient accuracy. Depending on exhaust gas composition, the dew points of volatile ingredients can already be exceeded at 52°C, so that the result is falsified by condensation. In the course of a worldwide effort to increase accuracy, the US Environmental Protection Agency (EPA 2007) is proposing important parameters such as that the filter temperature be strictly limited during sampling, a system of preseparation of coarse particles, and precise weighing techniques.

P

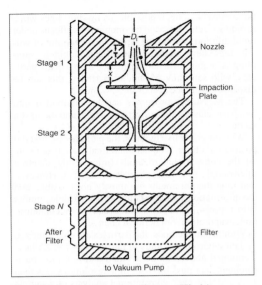

Fig. P22 Cascade impactor (Source: Hinds)

- *Impactors* (**Fig. P22**). By exploiting inertial effects, particles of a certain size fraction are aerodynamically separated. The fractions are collected on plates connected on a cascade; finally, they can be weighed and analyzed (Andersson impactor, Berner impactor). At low pressures, the size classification range can be extended downward (Berner low-pressure impactor). If the particles are electrically charged, the separated electric charge delivers a directly evaluated real-time signal (ELPI = electric low pressure impactor).
- *TEOM* (tapered element oscillating microbalance). Particles are collected on a vibrating filter. The change in resonance frequency is measured. This is a very sensitive process, frequently applied in environmental aerosols with the advantage of real-time information.
- *Beta absorption.* The method consists of measuring the absorption of beta radiation by a collective filter, from which information about the mass can be derived. In addition, this method is frequently applied to aerosol samples. Moreover, it also facilitates instantaneous information concerning the change of mass present on the filter.
- *Light absorption* (blackening). These processes are based on the absorption of visible light through a filter sample and are a measure for the soot that is separated. Thanks to the light-absorption properties of soot, the signal correlates well with the elementary carbon content of the filter sample (MIRA correlation). With the Bosch process (blackening number), the reflection of light is evaluated; with the aethalometer, the attenuation of a penetrating light ray is measured. Also, *see above*, ~Characteristics of diesel particulates ~~Particle

composition, for subsequent analysis of the filtered samples.

~~Complete measuring paths. The complete measuring path of an aerosol measurement comprises sampling, dilution, volatile/solid separation, size classification, and sensors, where the sensors should give information about the number, mass, surface finish, and chemical indicators as well.

~~In situ method of aerosol characterization.

- *Extinction measurement* (scattering plus absorption; **Fig. P23**).The opacity of a light ray penetrating the sample gas—that is, the sum of scattering and absorption—is the measured signal. In the Raleigh range, the absorption is about proportional to the mass. Because absorption dominates in the case of soot, this process is suitable for application. Corresponding devices are known as opacity measuring devices or opacimeters and are used in the so-called smoke test according to 72/306 EEC for field inspection of diesel vehicles. This method allows real-time measurement—in the free exhaust pipe stream or by means of sampling, for instance—and is a widespread method of measuring emission and immission. Because the light absorption reduces with the third power of the particle size and the scattering even more so, this method is insensitive for very small particles
- *Scattered-light measurement.* Methods that evaluate only the scattered-light signal, such as nephelometers and tyndallometers, are above all suitable for larger particles. Such instruments are also available in a large variety of hand-operated devices.
- *Laser-induced incandescence.* The particles are heated by a laser beam of a high intensity to about 4000 K. The mass concentration in the sample volume can be derived from the maximum intensity of the light radiated by the particles, and the rate at which this subsides gives information about the average primary particle diameter. This method is primarily used for measuring carbon particles.
- *Photo acoustics.* By means of a laser, the particles are strongly heated in a short time, such that the air volume surrounding the particle expands explosively. This pressure wave can be recorded as an acoustic signal, which allows the carbon mass of the sample volume to be determined.

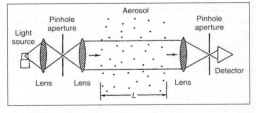

Fig. P23 Principle of the extinction measurement (opacimeter, turbidimetry; Source: Hinds)

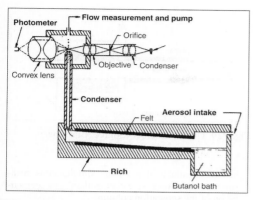

Fig. P24 Optical condensation nucleus counter (Source: TSI)

- *Particle counting* (**Fig. P24**). Particle size (several μm) can be optically counted. If the particles are too small, they are guided in a condensation nucleus counter through saturated butanol vapor, where they serve as condensation nuclei that rapidly expand (to 10 μm), rendering them available for optical counting. The measuring principle is the intensity of the light scattered from individual particles. For large particles, it is even possible to determine an "optical diameter" from this signal (optical fineness features).
- *Diffusion charge* (**Fig. P25**). The particles acquire an electric charge through adsorption of ions (which are generated by a corona discharge) that they release on the measuring filter after precipitation. The number of charges corresponds to the active surface

of the particle and, thus, to its mobility. The electrometer signal provides real-time information on the overall surface of collective particle resolution over periods of seconds. As for the preceding classification, the signal can also be evaluated for number, and with knowledge of the density and the fractal dimension of the particle (for soot, good assumptions) an estimate of mass is possible. Because ion adsorption is independent of the substance of the particles, all particles are evaluated equally (soot particles plus droplets and ash particles).

- *Photoelectric charging* (**Fig. P26**). When the electric charging of the particles is undertaken via photo emission through UV radiation, the entire charging of a particle depends on its chemical composition in addition to its active surface: soot particles show a high photoelectric yield due to their PAH setting, ash particles only show a small yield, liquid droplets do not react at all. In combination with the diffusion charging method, an analysis is possible concerning the nature of the aerosol and the changes in composition under real-time conditions.
- *Time of flight—mass spectrometry.* Particles are brought from the gas stream into a vacuum, then (partially) evaporated under laser bombardment. The ions and ionized compounds that evaporate from the particle surface are accelerated in an electric field. From the time of flight, the mass can be derived very accurately and the composition thus determined. The method is expensive and is suitable only for the laboratory or research department operation. In addition to determining the properties of the particle,

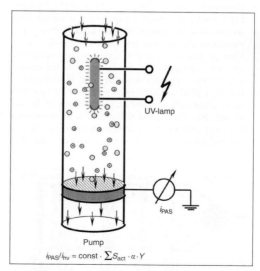

Fig. P26 Photoelectric charging. i_{PAS}, signal current; i_{hv}, photon stream ($hv = 6.0$ eV, $1 = 207$ nm); S_{act}, medium active surface per particle; a, light absorptions coefficient; Y, photoelectric yield (Source: Burtscher)

Fig. P25 Diffusion charge. i_{DC}, signal current; i_0, discharge current; S_{act}, medium active surface area per particle (Source: Kasper)

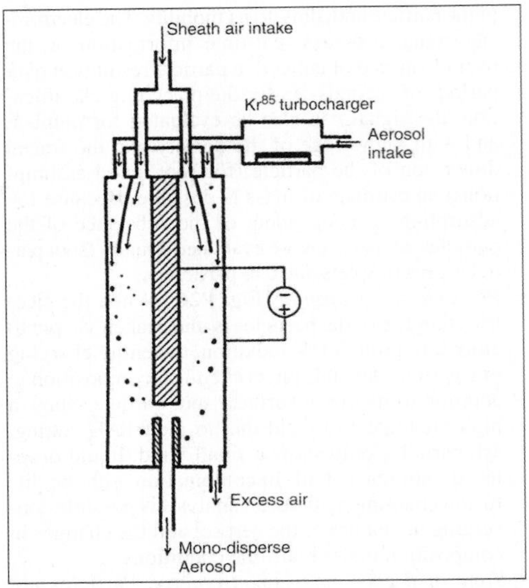

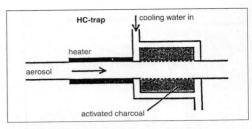

Fig. P27 Differential mobility analyzer (Source: TSI)

Fig. P29 Thermodesorber (Source: Burtscher)

become available that enable simultaneous measurement of several size classes thanks to segmentation of the separation column.

- *Diffusion battery* (**Fig. P28**). During the passage through structures with large surface areas, such as sieve packets, and based on the dwell time and pore structure, particles of a certain size are separated and the larger particles pass through. Through the series connection of several of such packets of different structures, classification according to mobility diameter in the range of fine particles <200 nm is feasible by simple means.

~~**Separation of solid and volatile particles.** By using a so-called thermodesorber (**Fig. P29**), volatile substances can be removed from the aerosol, so that only solid substances are fed into the actual measurement process (e.g., counting).

The thermodesorber comprises the connection in series of a heating section and a cooled activated carbon adsorption section. By varying the heating temperature, the content of volatile substances can be scanned according to the measure of its evaporation temperature by means of a thermogram.

Instead of the active carbon case, a heated dilution system can be used, which, given a sufficient dilution factor, can prevent recondensation of volatile exhaust components (thermodiluter).

in situ size classification is an important element in the particle measuring process. For classification according to the aerodynamic diameter, flow-dynamic systems such as impactors, cyclones, and centrifuges are among the possibilities. For classification according to mobility diameter, which has greater importance in the range of the combustion aerosol, the following two methods are used:

- *Differential mobility analyzer* (**Fig. P27**). Electrically charged particles experience a radial drift motion in a cylindrical electric field. Through the choice of the voltage and flow rate, only a very specific narrow-band size class—a mono disperse aerosol that can be fed, for instance, to the counting process—reaches the outlet slot. This device—usually called an SMPS (scanning mobility particle size) determiner—enables rapid scanning of, for instance, 100 size classes within two to three minutes for large spectra and is suitable for particles in the submicrometer range. Recently, devices have also

Literature: H. Burtscher: Tailpipe Particulate Emission Measurement for Diesel Engines, GRPE-PMP, Study CH1, Swiss Agency for the Environment, Forests and Landscape (BUWAL), 2001. — W.C. Hinds: Aerosol Technology, New York, 1998. — K. Willeke et al.: Aerosol Measurement–Principles, Techniques and Applications, Van Nostrand Reinhold, New York, 1993. — A.L. Nichols (Ed.): Aerosol Sampling Guidelines, The Royal Society of Chemistry, Cambridge, 1998. — D. Kittelson: Review of Diesel Particulate Matter Sampling Methods, University of Minnesota, January 1999. — M. Kasper et al.: NanoMet, a New Instrument for On-line Size- and Substance-Specific Particle Emission Analysis, SAE 2001-01-0216. — D.P. Moon, J.R. Donald: UK Research Programme on the Characterisation of Vehicle Particulate Emissions, ETSU-R98; Partikelmessverfahren Nfz-Motoren, FVV Report 466/1990. — A. Schuetz et al.: Feinstaub, Definition Messverfahren, Staub, in Reinhaltung der Luft 34 (1974) 9. — W. Schindler: Eine neue Messmethodik der Bosch-Zahl mit erhöhter Empfindlichkeit, MTZ 54 (1993) 1. — H. Hardenberg: Grenzen der Russmassenbestimmung aus optischen Transmissionsmessungen, MTZ 48 (1987) 2. — N.N.: EU Directive 20/220/EEC (Engineering Specifications for Gravi-

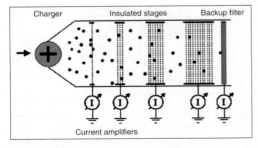

Fig. P28 Diffusion battery, principle design (Source: Burtscher)

metric Measurement). — M. Mohr, U. Lehmann: Comparison Study of Particle Measurement Systems for Fututre Type Approval Application, Swiss Material Testing Institute (EMPA) and Swiss Agency for the Environment, Forests and Landscape (BUWAL), 2003.

~Particle number. The number of particles per unit of volume of the aerosol carrier gas is a common measure for characterizing aerosols. The number can be measured with high accuracy in the condensation core counters. Normally, concentrations up to 1 million/cm³ can be registered by optical counting methods. For a higher particle density, it is necessary to resort to the scattered-light method (particle measuring technique).

The measurement of ultrafine particles in the invisible range is possible, in that the particles are magnified by condensing butanol on their surface until they can be counted visually. With a higher particle density, the particle number is a valid piece of information only momentarily because it can change by magnitudes within seconds due to agglomeration processes.

~Particulate emissions. The term "particulate emissions" of an engine, according to the legal definition on which the prototype test is based, is understood to mean the total particulate mass emitted in the exhaust. This particulate mass is defined using the measuring method. According to EU Guideline 88/77/ECC, the total solid and liquid exhaust components are combined under the term "particulate mass"; these are extracted from the exhaust, then diluted with ambient air (dilution 1:5–1:10), cooled down to <325 K, and then precipitated on a (respectively, two serially connected) defined sample filter. The mass is determined by weighing after conditioning as prescribed (temperature, humidity). The particulate total mass, PM, is determined according to this method, which is valid worldwide and is not specified further with regard to chemical composition and size distribution. It can be significantly influenced further through condensation during the sampling process if the sample falls below the acidic dew point, which is above 100°C, in the presence of sulfur in fuel. It is, therefore, not possible to classify the health-relevant parameters such as particle size, particle surface, and particle substance.

Recent efforts (GRPE-PMP project) are aimed at improving this method in order to obtain information about particle number in the range of magnitude that can be inhaled and for chemical composition in the real-world aerosol state.

The scale could still be a substance-specific mass similar to the EC industrial medicine definition, a particle number concentration (BIA proposal), or a statement about the total surface area of the aerosol, always linked with information about the ingredients. Another objective of these efforts is to harmonize the definitions and measuring methods in the emission range and the definitions and measuring methods in the inhalation range with one another.

Literature: Hauschulz et al.: Emissions- und Immissionsmesstechnik im Verkehrswesen, Publishers TÜV Rheinland, 1995.

— EU Directive 88/77/EWG. — Suva: Grenzwerte am Arbeitsplatz 2001 in der Schweiz. — Ultrafeine (Aerosol) Teilchen und deren Agglomerate und Aggregate, BIA Working Portfolio, 21st installment X/1998. — M. Mohr: Swiss Contribution to GRPE Particle Measurement Programme, EMPA/Duebendorf, Report 2002.779, May 2003. — GRPE-PMP, www.unece.org/trans/main/wp29/wp29wgs/wp29grpe/grpeinf46.htlm.

~~Nonengine particles. In addition to the particles from the combustion process, vehicles also emit particles from other sources, especially metallic particles from engine internal abrasion, wear processes in the clutch and brakes, as well as tire particles of different composition, fiber particles from muffler stuffing, corrosion particles, and particles released from the catalytic converters. Furthermore, vehicles are significantly involved in resuspension of road dust.

All these particles are accounted for in the inhalation measurement of airborne particles in the respiratory system and can be assigned to their sources using modern analytical methods. Based on their respiration nature and the toxicity of their ingredients, these particles are gaining importance alongside combustion particulates.

Tire abrasion: With levels of some 70 mg/km for cars and 1 g/km for trucks, tires are major emitters of particles. Tire particles have a mean diameter of 15 μm and contain a certain proportion of fine particles. Tires contain up to 30% soot, the composition of which is similar to diesel soot; it also exhibits the same PAH proportions, and contains sulfur and heavy metals (Source: Israel, Mayer).

Clutch abrasion: about 3 mg/km for cars, median grain size about 5 μm, also a certain portion of fine particles, particularly under high temperatures. The abrasion particles contain heavy metals, ceramic fiber fragments, and a variety of organic substances such as bonding agents.

Brake dust: About 20 mg/km for cars, median particle size about 10–15 μm, very different chemical composition, above all metals as well.

Catalytic converter: With modern monolithic carriers (in contrast to the pellet catalytic converters used earlier), emissions are still only about 100 ng/m³ (Inacker), with an extremely small particle size of about 5 nm. As far as coating material — platinum, palladium, and rhodium — crystallites are concerned, several μm are emitted in the case of fragments from the wash coat.

Engine abrasion: For cars, 1–10 mg/km, for a median particle size 1 μm, a fine portion is also observed after passing through the combustion process (Israel). These particles, heavy metals, or their oxides are frequently embedded in combustion particulates.

Fiber particles from mufflers: Fibers are subject to special requirements, according to TRGS 905, with regard to material solubility, diameter, length, and length/

P

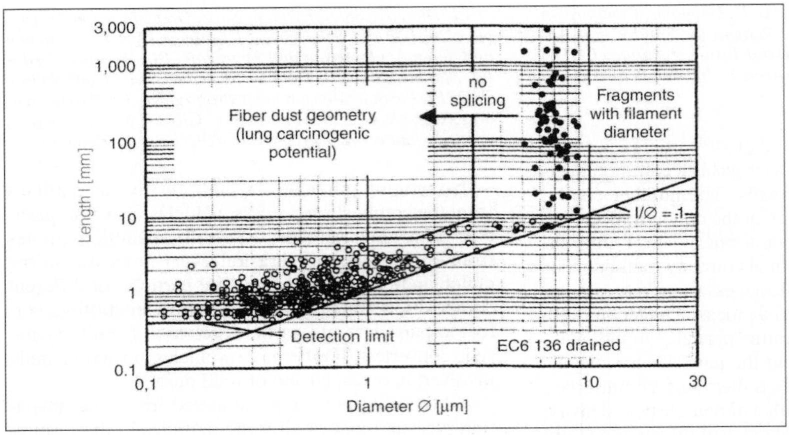

Fig. P30
Fragment geometry
distribution for a
glass-fiber yarn
(Source: Oser)

diameter ratio—derived from the asbestos problem. Muffler fibers that generally used to exhibit diameters of around 3 μm are now thicker in order not to fall into this range (Source: Oser).

It is apparent, however, that during the fracture of fibers (a normal process in mufflers to dampen vibrations), the glass fibers investigated here give off very fine fragments in addition to the fragments with filament diameters, which can fall in the critical range (**Fig. P30**). Therefore, the muffler stuffing must also be included among critical particle sources in vehicles.

Literature: A. Mayer: Emission Factors for Fine Dust from Road Traffic, 2nd ETH Workshop on Nanoparticle Measurement, August 1998. — M.P. Oser: Fasergestricke aus Glasfilamentgarnen für die Abgasreinigung bei Kleinemittenten, Eigenschaften und Biokompatibilitaet, Diss. ETH Zurich No. 12654, 1998. — G. Israel et al.: Rußemission in Berlin, VDI Progress Reports No. 152. — R. Inacker: Malessa, Experimentalstudie zum Austrag von Platin aus Automobilabgas-Katalysatoren (VPO 03). — BMBF/GSF: "Bericht Edelmetallemissionen," Hannover, October 1996.

~~~Secondary emissions through particulate filter technique. Particulate filters are high-surface-area structures that are frequently enriched with catalytically effective substances and can, therefore, be considered to be chemical reactors. After being embedded with soot, the particulate filter of a truck engine can attain an active surface area of 50,000 m^2 or more, so that exhaust substances can be embedded there, which provide long dwell times for chemical reactions. In the presence of copper, the formation of dioxins and furans was observed in extraordinarily high concentrations (an increase of three to four orders of magnitude) (Heeb), and in the presence of noble metals, increased generation of NO$_2$ and SO$_3$ was observed (Mayer). In addition, other secondary emission formations such as nitro-PAH were verified (Heeb); regulations demand that, going beyond the general requirements (USA, Federal Register, O.S. Clean Air Act Section 2002 [a]), increased verification of secondary-emission for-

mation should take place for particulate filter systems (BUWAL, EJPD).

Literature: N.V. Heeb: Influence of Particulate Trap Systems on the Composition of Diesel Engine Exhaust Gas Emissions, Swiss Material Testing Institute, Report 172847, 1998. — A. Mayer, M. Wyser-Heusi: Particulate Traps Used in City Buses in Switzerland, SAE 2000-01-1927, June 2000. — N.N.: Geprüfte Partikelfilter-Systeme für Dieselmotoren, Swiss Agency for the Environment, Forests and Landscape BUWAL, Switzerland, 1.3.2003 and EJPD Directive 8/1990.

~~Primary particles. Primary particles are generally understood to be the particles in the source state—that is, at short chronological and spatial distance from the source. In a strict definition, this means immediately after nucleation and prior to the beginning of agglomeration processes. The term is often more broadly interpreted, for instance, for particles in the emitted state at the end of the exhaust pipe.

~~Secondary particles. Particle-like substances formed though gaseous emission into the atmosphere are designated as secondary and include, above all, sulfate and nitrate particles (**Fig. P31**).

Literature: Ch. Cozzarini: Quellen der Partikel-Emissionen, 20th Vienna Engine Symposium, May 1999.

~Particulate filter system. A particulate filter system suitable for vehicular engines comprises the actual particulate filter (filter element, filter medium), the regeneration device and a control device, for monitoring the function.

~~Filter/catalytic converter combinations. In addition tothe series connection of filters and catalytic converters for the purpose of gas-phase reaction and regeneration support, the particulate filter can also act as a carrier for the catalytic converter. Because of their function, particulate filters have very high surface areas, whether configured as porous ceramic or fibrous structures, and are, therefore, suitable for the combination

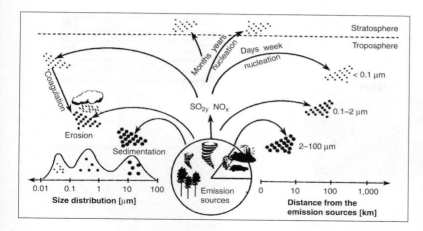

Fig. P31
Emission–
transmission–
inhalation and
formation of
secondary particles
through nucleation
and coagulation
(Source: Cozzarini)

of filtration and catalysis processes. The catalysis can be oxidation for the conversion of gaseous components or solid components such as soot, or reduction for converting NO$_x$. The medium may be coated on the inlet side, within the porous filter structure, and on the outlet side, so that the different effects can be achieved.

~~Location of particulate filter on engine.
To minimize the supply of external energy for regeneration, it is recommended that the particulate filter be located as near as possible to the engine; however, this is frequently not done because of construction limitations or high mechanical stress caused by vibrations. Modern filter systems with a small design volume and high mechanical strength for the filter medium are attractive for this purpose. With a turbocharged engine it is possible to locate the filter in the high-pressure section— upstream of the turbocharger turbine. This position not only offers the advantage of the highest possible exhaust temperature, but also simultaneously reduces the negative effects of filter counter-pressure on the latter. The principal disadvantage is that the filter medium acts as a heat sink, which delays the response of the turbocharger, but this can be minimized with modern filter media—the efficiency of the turbocharger is raised by the effect it has in compensating the exhaust pulsations (Mayer).

Literature: K. Oblaender: Entwicklung der Abgastechnologien, Rueckblick–Stand–Ausblick, VDI Reports 559. — A. Mayer: Pre-Turbo Application of the Knitted Fiber Diesel Particulate Trap, SAE 940459.

~~Particulate filter.
The technical requirements for a diesel particulate filter (DPF) are high (**Fig. P32**). The filter must withstand high temperatures and rapid temperature changes and exhibit a high separation rate, even for the finest particles in the 10–500 nm range. In addition, the lowest possible pressure drop, long lifespan, and a low cost level are demanded. From an overview of the technically available filter mediums, it is apparent that only structures with very large surface

area: volume ratios can be used: these include ceramic or metallic sintered or fiber structures.

Therefore, the separation rates for particles required in the typical diesel particulate range of magnitude cannot be achieved with cyclones, gas washers, and electrofilters.

~~~Charging.
The particulate filter can be loaded with precipitated soot, and the pressure loss will increase due to the charging effect. The separation rate increases because of the formation of soot cake in some filter types (surface-type filters); in other filter types, it has a tendency to reduce (deep-bed filter).

When a limit charging level is reached, usually defined by back pressure, the filter must be regenerated or cleaned.

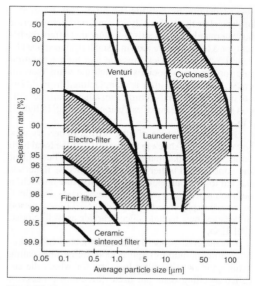

Fig. P32 Separation rate of technical filter systems as a function of particle size (Source: Zievers)

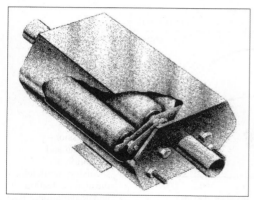

Fig. P33 Particulate filter with ceramic yarn—wrap-
around filter elements (Source: Engelhard/
3M)

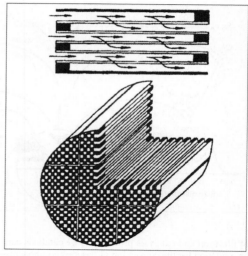

Fig. P34 Ceramic monolithic cell filters, so-called
"wall flow" filter

~~~**Filter media.** The following filter media are
common:

~~~~**Braided fiber filter.** High-temperature fibers are
also offered as braided fabrics and can be fixed over
metallic carrier structures for filtration. Such systems
were developed by Hug and 3M.

~~~~**Candle filter.** Yarns made of high-temperature
fibers (from mullite, product name Nextel from 3M)
are wound by a special winding technique that creates
a rhombic duct structure on a perforated carrier tube.
Filter cartridges of this type have been developed by
3M and Mann & Hummel (**Fig. P33**).

~~~~**Ceramic monolithic cell filter.** This is similar
in design to the cell catalytic converters; however, it
has alternately closed cells, a feature that offers the filter type a large surface area in a small design space (1–
3 m^2/L), thus low back pressure and high separation
rate for low gas velocities through the walls (several
cm/s). This type of filter is manufactured from different ceramic materials (cordierite, silicon carbide, silicon nitride) by means of extrusion (Corning, IBIDEN;
Fig. P34).

Originally, this type of filter proved to be very sensitive to thermal shock in the presence of the local high
temperature peaks that resulted from fast burnout of
the soot cake. Further development of the materials
has led to thermal-shock resistant structures, which in
addition to the improved cordierite variants, for which
extensive experience has been gathered over a period
of decades, now include silicon carbide as well (Notox,
IBIDEN).

The material properties and brittleness of the ceramic filter monoliths require good storage and sealing
against the metallic filter housing, for which the so-
called swelling mats are used; the entire process of
packing the ceramic components in the metallic structure is called "canning."

~~~~**Filter papers/filter felts, filter sponges.** Paper
filters, which are made similar to the vacuum cleaner
filters, are only used if the exhaust temperatures can be
reliably held low: papers are available that withstand
application temperatures of about 300°C (Donaldson)
(**Fig. P35**).

In principle, these papers are fiber filters with short
fibers; the fibers are arranged irregularly and fixed
in their structure by means of bonding agents. For
higher temperatures, felts made of ceramic fibers can
also be used, just as they have already been used for
a long time in application in industrial hot-gas filtration (BWF).

Numerous other filter systems are in development/
testing, such as fleeces made of ceramic or metallic fibers and ceramic and metallic sponge structures.

Electrostatic effects can improve the filter performance, but flow dynamic effects do not give a benefit
due to the small mass of the diesel particles. Electric
filtration has also been applied occasionally in the

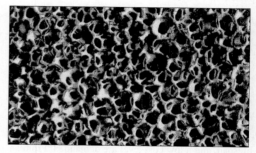

Fig. P35 Ceramic sponges as filter medium (Source:
Alusuisse)

P

course of these developments, but it has not led to solutions suitable for series production.

~~~~Knit fiber filter. Circular knit elements made of ceramic yarns are processed to make deep-bed structures by pleating. The fiber surface area:volume ratio typically reaches 200 m²/L. Buck developed this filter type, and it is offered with catalytic coating and an electric internal heating element. The flow through it is preferably from the inside outward; the flow rate is delayed and the deep-bed effect is, therefore, improved.

~~~~Sintered metal filter. A cell filter of similar design was developed by SHW using metallic materials by depositing a sinter structure in a steel mesh carrier structure. These filters are relatively heavy, but very robust compared with ceramic; they also exhibit naturally good thermal conductivity. HJS and Purem have further developed these metal-sintered filters as accordion-type bellow structures.

~~~Operating mode of particulate filters. The separation of particles, solids, or droplets with such small particle diameters occurs not as a result of blocking effects (sieve effects) but either through inertial separation (impaction) or through trapping from boundary particle trajectories or diffusion (**Figs. P36 and P37**).

The filtering effect requires, in addition to the capacity to separate, also the capacity to retain the separated particles—reduction of so-called blow-off effects.

The probability that a particle will be trapped or adsorbed to the surface depends on the filter structure, the pore size, the available internal surface area, the flow rate, and the dwell time. The bonding of the particle to the surface generally occurs because of van der Waals forces, which are high for dry substances in the presence of liquid films but can be reduced rapidly.

Particles that are deposited on the surfaces agglomerate and can grow to form large dendritic structures (**Fig. P38**).

Such loosely deposited soot increases the inner surface area of the filter and improves the separation rate but also increases the pressure loss at the same time.

Filters in which the deposited soot forms the actual soot cake are called surface filters (**Fig. P39**). With the growth of the soot cake, the separation rate is strongly improved ("soot filters soot"), but the back pressure

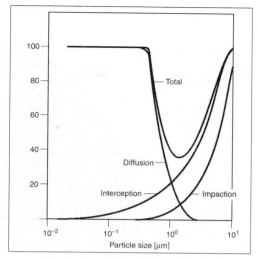

Fig. P37 Separation effects in a filter medium as a function of particulate size (Source: Hinds)

rises progressively. The combustion of concentrated soot deposits leads to heat generation ("exothermic") and can, if the heat dissipation is insufficient, result in a hot spot and the subsequent failure of the filter structure due to thermal shock or even melting. Deep-bed filters (**Fig. P40**) do not form soot cakes. The soot is finely distributed and deposited in the structure, whereby every filter layer has the same separation rate and the deposited quantity reduces from the inflow side toward the outflow side in a logarithmic manner. The salient variables in the design of these filters are filter depth, fiber diameter, pore size, and velocity distribution (delay in the flow direction is advantageous). From a certain charge level onward (saturation), large dendrites can break loose and exit the filter through the relatively large pores (blow-off). Fine particles are then still always separated out reliably (filter = agglomerator). The back pressure then reaches a limiting level.

Whereas surface filters need a certain charge in order to attain their top efficiency, deep-bed filters have their best values at low back pressures in the low charge range.

Most technical filters are mixed types that function as deep-bed filters with low charges, forming filter cakes with higher charges.

Literature: Emissionsminderung Automobilabgase—Diesel-motoren, VDI Report 559, 1999. — H. Hardenberg: Russeigen-schaften und Filtertechnologie, in Der Nahverkehr 5/86. — B. Wiedemann et al.: Diesel-Partikelfiltersystem mit additiv-gestuetzter Regeneration, ATZ 91 (1989) 12. — H. Houben: Optimierte Dieselmotoren für Gabelstapler, VDI Reports 33. —P.N. Hawker: Diesel Emission Control Technology Platinum Metals Review, January 1995, Vol. 39, No. 1. — G. Hueth-wohl, C. Kohberg: Ein neues Partikelfiltersystem fuer Diesel-motoren, MTZ System Partners 98. — A. Buck et al.: Passive

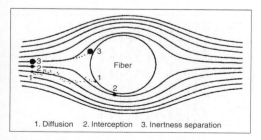

1. Diffusion 2. Interception 3. Inertness separation

Fig. P36 Filter mechanism on an individual fiber

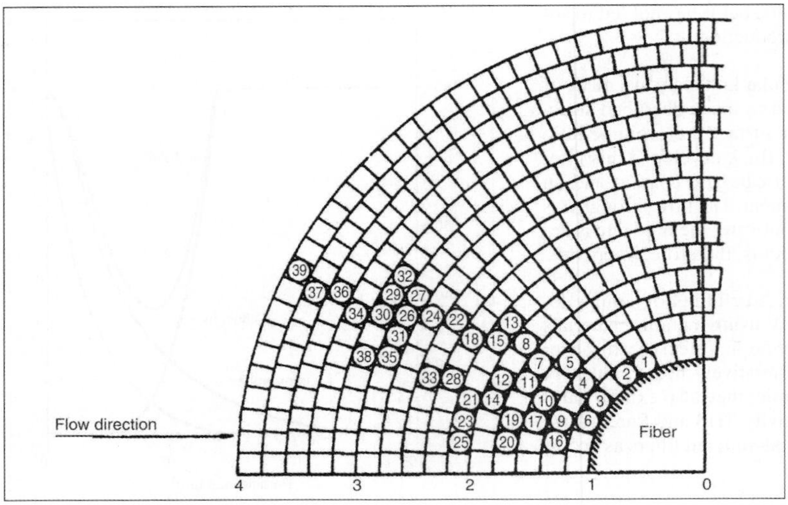

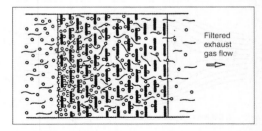

Fig. P38
Dendrite growth,
model calculation
according to Kanaoka
(Source: Jodeit)

Regeneration of Catalyst Coated Knitted Fiber Partic-
ulate Traps, SAE 960138. — B. Stanmore: The Ignition and
Combustion of Cerium Doped Diesel Soot, SAE 1999-01-
0115. — Eastwood: Critical Topics in Exhaust Gas, Wiley and
Sons, England. — E. Thomas et al.: Non-Thermal Plasma Af-
tertreatment of Particulates, SAE Paris, June 2000. — H.
Schulte et al.: Particulate Filter Technology for Diesel En-
gines, VDA Congress 1999. — C.M. Fleck: An Electrical Soot
Trap for Solid and Condensed Nanoparticles with a Continu-
ous Electrochemical Conversion of Soot and Hydrocarbons,
ETH Zurich, Nanoparticle Measurement Workshop, 1998.
— A. Mayer: Engine Intake Throttling for Active Regenera-
tion of Diesel Particle Filters, SAE 2003-01-0381.

~~~**Pore size.** The pore size of these technical filter
mediums lies in the range of 10 to 30 μm, but for deep-
bed filters it is significantly larger. However, the pores
are at least 100 times larger than typical diesel particulates. Obviously, blocking or sieving effects are not
applicable to this filter so long as the soot layers
formed are not compact. The total pore volume or the
empty space of a filter can reach magnitudes greater
than 90%.

~~~**Pressure loss.** The flow through the fine porous
structure of the diesel particulate filter is generally
laminar, because the Reynolds numbers are below one,
and the pressure loss is that appropriate to the flow type.

The pressure loss is related to the flow by the equation for flow around a cylindrical fiber (according to
Jodeit):

$$\Delta p = K'z\left(\frac{1-\varepsilon}{\varepsilon}\right)\cdot\left(\frac{v_0\ \mu}{d^2}\right) + K''\left(\frac{1-\varepsilon}{\varepsilon^2}\right)\cdot\left(\frac{\rho v_0^2}{d}\right),$$

where z = filter depth, v_0 = inflow rate, ε = porosity,
μ = dynamic viscosity of the exhaust, ρ = density of
the exhaust, d = characteristic flow variable (fiber diameter), and K', K'' = constants.

In the flow range of the microscopic structures of the
filter media, the second — namely, the turbulent element —
becomes negligibly small, leading to what is known as
Darcy's law:

$$\Delta p \sim \frac{\mu \cdot v_0 \cdot z}{d^2}.$$

The most important feature with laminar flow is that,
in contrast to the turbulent case, the pressure loss does
not depend on the density of the gas and increases only

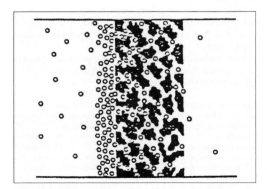

Fig. P39 Surface filter

Fig. P40 Deep-bed filter

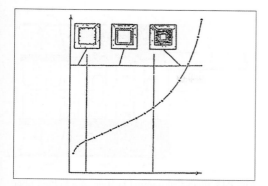

Fig. P41 Pressure increase due to charging for a ceramic cell filter (Source: NGK)

in a linear manner with the flow rate. On the other hand, it is strongly influenced by the fine structure of the cell and the dynamic viscosity of the gas, which is independent of pressure but strongly dependent on temperature.

The way in which the pressure-loss characteristic increases with age depends on the filter type as well as on the rate of charging (**Fig. P41**). After an initial phase, when only particles are embedded in the structure, a proportional increase of pressure loss with charging is evident. As soon as the filter cake is formed, a progressive increase follows. For deep-bed filters, which deposit all the material in the filter structure, a regressive increase characteristic is typical, explained through the so-called fiber-growth model (**Fig. P42**)—that is, during charging of a filter, the fiber dimensions grow, moderating the pressure-loss increase according to Darcy's law.

The total pressure-loss characteristic of a real filter generally exhibits a disproportionate flow-rate increase. This is because, in addition to pressure loss in the fine porous filter media, which obeys the laminar characteristic throughout, there are portions with turbulent flow characteristics: these are the pressure-loss components of ducts and pressure loss terms for inflow and outflow in pipe sections.

Literature: H. Jodeit: Untersuchungen zur Partikelabscheidung in technischen Tiefenfiltern, VDI Progress Reports

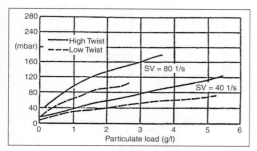

Fig. P42 Pressure increase for knit fiber filters (Source: Buck)

No. 108. — M. Baraket: Das dynamische Verhalten von Faserfiltern fuer feste und fluessige Aerosole, Diss. ETH Zurich No. 9738/1992. — H. Hardenberg: Rußeigenschaften und Filtertechnologie, in Der Nahverkehr 5/86. — A. Mayer et al.: Particulate Traps for Retro-Fitting Construction Site Engines, VERT: Final Measurements and Implementation, SAE 1999-01-0116. — Diesel Exhaust Aftertreatment 1999, SP-1414, March 1999. — H. Schulte et al.: Particulate Filter Technology for Diesel Engines, VDA Congress, Frankfurt, 1999.

~~~**Regeneration of particulate filters.** When connected downstream of a standard diesel engine, the filter is charged with soot up to a maximum permissible back pressure within hours and must be "regenerated" frequently. Regeneration occurs more or less without residue by burning the stored soot; dry soot burns sufficiently quickly at temperatures above 550°C. Because the temperatures in diesel exhaust mostly do not reach this level, an auxiliary system must be made available. Such systems can be differentiated between:

- Active and passive regeneration. Passive exists if the regeneration automatically sets in once certain operating states are reached. Active occurs if the process is triggered through an automatic control system that monitors the back pressure. This trigger signal either releases energy to raise the exhaust temperature (burner, electric heater, delayed injection, catalytic combustion) or reaction products are added, which lower the reaction temperature (oxidants, catalytic converters).
- Vehicle-dependent systems and interchangeable filter. Vehicle-dependent (onboard) systems are those in which all functions occur without the removal of the filter from the vehicle. Interchangeable filters must be removed for the purpose of regeneration. For this purpose, quick-release systems are provided in most cases.
- Permanent filters and push-on filters. Push-on filters are used for only a short time, for particular operating periods—for example, loading operations of trucks in indoor areas and/or in garage operation.

The following regeneration systems are used in most cases:

~~~~**Catalytic coating.** With transition metals, it becomes possible to reduce the ignition temperature of soot to below 400°C. Further reduction of the ignition temperature is possible through the additional application of noble metals, whereby the reaction of the soot with NO_2 as described in the CRT process is used; by this means, light-off temperatures below 300°C are attained.

When these temperature ranges are attained frequently and long enough during operation, no further measures are necessary—this is called a purely passive regeneration method. It has been observed that regeneration on a catalytically coated wall proceeds slowly, with an advantage as regards the completeness of the conversion, but requiring a longer period of availability of higher exhaust temperatures (Engelhard, DCL, ECS).

~~~~**Combined regeneration processes.** Combinations of different regeneration methods are also possible. The combination of fuel additives with heating systems or catalytic coating with heating appears especially interesting, because in this case the energy for heating the exhaust is reduced to a fraction of the basic requirement.

As soon as the possibility of a modern engine management in the filter/regeneration system can be integrated, new possibilities arise: the missing energy can be provided by post-injection of fuel in the expansion phase. This post-injection can be implemented in an elegant manner if a common rail direct injection system is used. Such systems achieve both injection process formation and injection process distribution. Other possibilities are to convert the injected fuel on an oxidation catalytic converter located upstream of the filter in order to raise the temperature or to throttle the intake airflow, which is in fact permissible under part load during the short regeneration period (Source: Mayer).

~~~~**CRT (continuous regeneration trap) system.** This system applies the property of a noble-metal–coated oxidation catalytic converter. When fitted in an engine exhaust, it generates more NO_2 from NO. In a soot filter connected downstream, simply expressed, the reverse process occurs, and the oxygen atom that has become free burns the carbon at very low exhaust temperatures (from about 230°C).

This is a fully passive system. It requires, nonetheless, use of generally sulfur-free fuel (less than 10 ppm), because otherwise unacceptably high sulfate emissions occur and simultaneously reduce the conversion NO → NO_2 and the catalytic converter sustains permanent damage. The increased NO_2 content in the exhaust is not problematic in many applications where high levels of dilution can be assumed. However, it can lead to high NO_2 concentrations in closed rooms, depending on the number of air exchanges. What is essential for the function of the CRT method is that the ratio of soot/NO_2 should at least be double that of the stoichiometric ratio of approximately 15.

~~~~**Electric heating.** The electrical heating of the entire exhaust stream and the filter system as well as the regeneration temperature require more energy than can be provided by the systems on current vehicles (**Fig. P43**). This method is only applied if the regeneration temperature is substantially lowered by means of catalytic measures (additives, coating). Systems with sequential regeneration—that is, heating of individual filter cartridges or filter sections at intervals—are under development. Even here, the performance requirements are high if the regeneration temperature is to be guaranteed for a sufficiently long regeneration period; in most cases, this is prohibitive in vehicle applications unless the flow rate is otherwise throttled during regeneration or the regeneration temperature is lowered by catalytic effects. Such methods are currently available with electrically conductive filter media such as metal fiber felts from Bekaert or conductive silicon carbide from Heimbach.

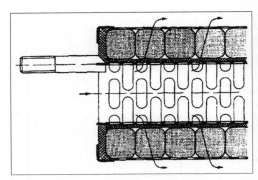

Fig. P43 Electrical internal heating of a knitted fiber filter cartridge with special heat transfer elements and catalytic coating (Source: Buck)

Another approach comprises igniting the soot cake electrically under conditions in which it can then continue to burn without further energy supply. Such a system was developed by Gillet with FEV, but it was not marketed.

~~~~**Electric standstill regeneration (onboard).** If the manner of operating the vehicle or machine allows, the regeneration can be carried out at standstill under well-controlled conditions (**Fig. P44**). Either the engine is then operated at idling speed and the low quantity of the oxygen-rich but relatively cold exhaust is heated to the required temperature by means of an installed electric heater that is now connected to the electrical mains or the airflow rate is provided by means of an additional air pump. Because the gas volumes are low, a longer period is generally required for heating up the system—that is, the process normally takes several hours, which has the advantage that the regeneration occurs in a very controlled manner and that the filter is not exposed to high stress.

~~~~**Emissions during regeneration.** When burning the deposited soot, substantial emission peaks can occur. Particles should not be released in the process.

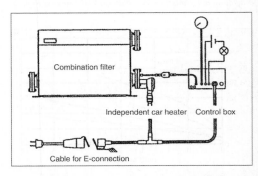

Fig. P44 Electrical regeneration at standstill (Source: UNIKAT)

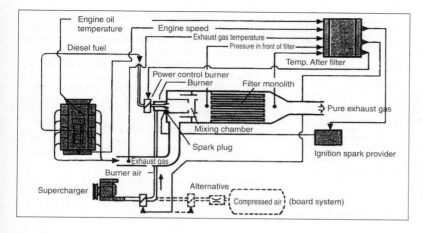

Fig. P45
Particulate filter system with full-flow burner (Deutz)

With rapid regeneration cycles, however, it is unavoidable that a part of the carbon will be oxidized not to CO_2, but to CO and a part of the adsorbed hydrocarbons will be transformed into the vapor phase and exit the filter without complete combustion to harmless substances. Moreover, the release of large amounts of water vapor is often observed (white plume), especially during a cold start after a long stationary period or longer idling periods.

~~~~**External regeneration (replacement filter).** If it is possible to remove the filter from the engine at regular intervals—for instance once daily—to perform the regeneration procedure, the possibility of external regeneration is suitable; this is then carried out by electrical means or with burners in a well-controlled process. Such an interchangeable filter with external regeneration is frequently used on forklift trucks, where the filter is then mounted on the exhaust system by quick-coupling devices.

Systems in which the charged filter cartridge is disposed of in a manner similar to an air filter (one-way filter) are also used. Such systems require either a high storage capacity or they can be applied only where small amounts of soot are produced. An especially interesting application of this type is found in coal mines, where the whole exhaust gas system has to be very effectively cooled to prevent explosions.

In addition, other ideas for regeneration are under development, such as the application of cold plasmas, microwave heating, heating of electrically conductive filter materials through passage of parallel flow, and the application of aerodynamically reverse-pulsed systems, as is common in plant engineering.

~~~~**Full-flow burners.** In the inflow section of the particulate filter is a burner that is dimensioned such that it can heat the exhaust at arbitrarily different operating states of the engine up to the regeneration temperature (greater than 700°C) when fueled with diesel fuel (**Fig. P45**). The burner is switched on when the back pressure has reached a limiting value and switched off when the

filter has been self-cleaned and is free of soot. The process takes approximately 10 minutes (Deutz, ArvinMeritor).

The burner is equipped with the necessary air and fuel supply as well as with an ignition device. The burner and burning capacity are controlled electronically based on the pressures and temperatures measured upstream and downstream of the filter. The control of burner output is necessary in order to avoid destruction of the filter due to excessively high temperatures.

~~~~**Partial-flow burner.** This involves twin filter systems, whereby the filter to be regenerated with low flow rates can be operated under controlled conditions while the second filter assumes the exhaust cleaning functions (**Fig. P46**).

These systems are less demanding in terms of burner control than full-flow burners; but in terms of design size and expense they are hardly competitive and have not been used much.

~~~~**Regeneration additives.** Numerous substances have the property of lowering the soot ignition temper-

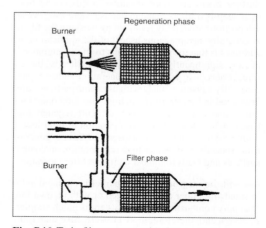

Fig. P46 Twin filter system with flap control

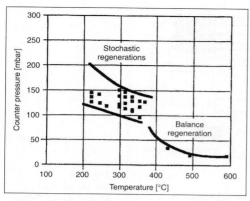

Fig. P47 Regeneration through fuel additives (Source: Rhodia)

ature to around and below 400°C (**Fig. P47**). Examples are cerium, iron, copper, and strontium as well as platinum. These catalytically active substances are generally added to the fuel in the form of organometallic compounds by way of dosing systems. The concentration is usually kept below 25 ppm, frequently around 10–15 ppm. With this method, which is also termed "passive regeneration," positive experiential data have been collected for many applications (Octel, Rhodia, CDT, Lubrizol).

It is apparent that regeneration at acceptable back pressures only occurs at temperatures above about 350°C.

Surprisingly, regeneration is also observed at low temperatures when using additives; they are probably triggered by the stochastic ignition of the SOF (soluble organic fraction), are hardly controllable, and can endanger the strength of the filter in case of high filter charging due to excessive development of heat.

A basic disadvantage of additives is that the additive substances are deposited in the oxidized form and are thus inert and gradually raise the filter back pressure. Surface filters are more sensitive in this connection than deep-bed filters that can deposit more ash. These ash portions are readily removed by washing the filter. A generally advantageous aspect of the application of additives is that the raw emissions from the engine are already significantly lowered and the filter is, therefore, relieved. In addition, regeneration additives act especially advantageously with ultra-fine particles; are also suitable for retrofitting in high-emission older engines; and cope reliably with difficult regeneration cases such as long periods of operation at light loads with massive soot cake formation. Additive regeneration proceeds much faster than regeneration on coated surfaces and leads to almost complete filter cleaning.

~~~~**Stationary burner.** The burner is operated only at standstill, either at idling speed or with its own fan to supply combustion air. The system is much simpler in design and, therefore, more cost-effective than full-flow burners.

~~~**Separation rate and penetration.** The effectiveness of a filter is described in terms of the separation rate or penetration. The separation rate is the ratio of the quantity of the material removed to the total material that was fed to the filter (**Figs. P48 and P49**).

The separation rate can be defined in terms of the mass or the particle number or the elementary carbon content (EC) of the total mass. The following definitions are commonly used:

Particle mass—separation rate PMAG

$$PMAG = \frac{PM_{beforePF} - PM_{afterPF}}{PM_{beforePF}} \ (\times\,100\%)$$

Particle number—separation rate PZAG

$$PZAG = \frac{PZ_{beforePF} - PZ_{afterPF}}{PZ_{beforePF}} \ (\times\,100\%)$$

The separation rate for elementary carbon, ECAG, is defined analogously

Penetration = 1—separation rate

$$= \frac{PM_{afterPF}}{PM_{beforePF}} \quad \text{or} \quad \frac{PZ_{afterPF}}{PZ_{beforePF}}$$

Good particulate filters attain separation rates of greater than 99% of solid particles—and for only these can the filter efficiency be defined; with modern developments, rates greater than 99.9% are often attained. The separation rates can be similar for transient states, even for the extreme case of free acceleration.

If, however, the separation rate is—physically incorrect—defined via the total particle mass (PM), according to the legally prescribed method, an apparently very small separation rate can be measured due to the effect of dew point. With high conversion rates of SO_2 to SO_3, which are the result of catalytic effects within the filter system, negative separation rates can occasionally be found—that is, a significantly higher filterable mass is found with a filter than without one. The mass is then largely composed of condensed and water-diluted sulfuric acid.

~~~**Storage capacity.** The storage capacity defines how much material a filter can deposit before a given pressure loss is registered. This limiting value of the storage capacity specifies when the filter must be self-cleaned through regeneration or when the inert components must be cleaned out.

The storage capacity depends on the substances to be filtered, and it is also dependent on their densities as well as whether uniform or concentrated deposition occurs on certain filter parts. The storage capacity for soot is different from the storage capacity for ash components. For pure soot, the storage capacity is in the range of 10 to 20 g per liter of filter volume. This corresponds to 20–50% loosely deposited clogging of the filter, depending on the density and the space present.

~Pulmonary intrusion. *See above*, ~Effects on the human organism

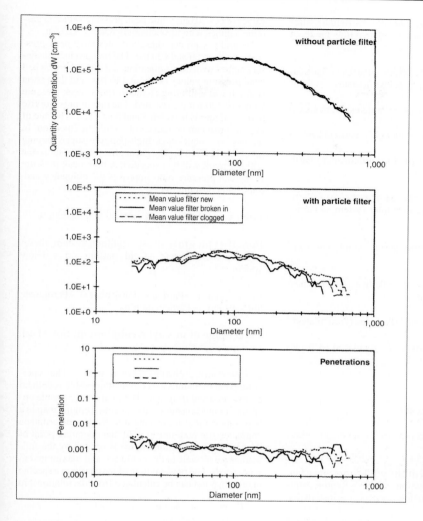

Fig. P48
Separation rate of a ceramic cell filter on a commercial vehicle DI diesel engine, defined according to the particle number (Source: Mayer)

P

~Regeneration additive. *See above*, ~Particulate filter system ~~Particulate filter ~~~Regeneration of particulate filters

	PMAG	PZAG	ECAG
Filter 1	77.5	97.8	97.2
Filter 2	76.5	95.4	94.0
Filter 3	12.4	99.9	99.3
Filter 4	54.7	99.0	98.1
Filter 5	13.0	99.9	98.5
Mean values	46.8	98.4	97.4

Fig. P49 Separation rate of five filters after a filter test of approximately 2000 h, determined for the mass (PMAG), the number of solid particles (PZAG), and the elementary carbon content of the sample (ECAG) (Source: SAE 2002-01-0435).

~Secondary emissions of the filter. *See above*, ~Particulate emissions ~~Nonengine particles

~Secondary particulates. *See above*, ~Particulate emissions

~Sintered metal filter. *See above*, ~Particulate filter system ~~Particulate filter ~~~Filter media

~Soot. *See above*, ~Characteristics of diesel particulates

~Soot formation. *See above*, ~Characteristics of diesel particulates

~Soot particulate. *See above*, ~Characteristics of diesel particulates

~Storage capacity. *See above*, ~Particulate filter system ~~Particulate filter

~Surface-type filter. *See above*, ~Particulate filter system ~~Particulate filter ~~~Operating mode of particulate filters

Particulate characteristics →Particles (Particulates) ~Characteristics of diesel particulates

Particulate composition →Particles (Particulates)

Particulate emission →Particles (Particulates)

Particulate filter →Particles (Particulates) ~Particulate filter system

Particulate filter mode of operation →Particles (Particulates) ~Particulate filter system ~~Particulate filter

Particulate filter preparation →Emission measurements ~Test type I

Particulate filter system →Particles (Particulates) ~Particulate filter system

Particulate formation →Particles (Particulates)

Particulate size →Particles (Particulates) ~Characteristics of diesel particulates

Part load →Load

Part-load range →Load

Passage →Combustion process, diesel engine ~Prechamber engine

Passenger compartment →Electronic/mechanical engine and transmission control ~Requirements for mechanical and housing concepts ~~Installation location

Passenger-car oil →Oil ~Classification, specifications, and quality requirements ~~Diesel engine oils ~~~Car diesel engines

Passive speed/engine speed sensor →Sensors ~Speed sensors

Peak combustion pressure →Peak pressure

Peak combustion temperature →Combustion temperature

Peak pressure. The peak pressure, combustion pressure, or maximum gas pressure is essential for the evaluating the mechanical load on the crank mechanism, the bearings, the crankcase, and the crankshaft. High peak pressures result especially in charged engines. Naturally aspirated gasoline engines have peak pressure values of 60–70 bar; while peak pressure values in gasoline engines with turbochargers reaches 120 bar.

Naturally aspirated diesel engines have peak pressures between 80 and 100 bar. The first production diesel engines with direct injection and turbochargers had peak pressure values of about 140 bar, and the piston strength was a limiting element. Today, peak pressure values of 170 bar are reached. The rate of pressure rise in the cylinder is between 4 and 8 bar/°CA. The rate of pressure rise can be reduced by multiple injection. In the case of single-stage high-pressure supercharging (about 24 bar and higher mean effective pressure), the 200-bar mark will be exceeded. A high peak pressure to median pressure ratio improves the efficiency and, thus, fuel consumption.

Peak temperature →Combustion temperature

Pencil-type glow plug →Ignition system, diesel engine/preheat system ~Cold starting aid ~~Glow system

Penetration →Particles (Particulates) ~Particulate filter system ~~Particulate filter

Percentage of methyl alcohol/methanol →Fuel, gasoline engine

Performance characteristic maps. The operating point of an internal combustion engine is defined by its speed and its torque. The total of all possible operating points constitutes the so-called engine map in a two-dimensional representation. In the performance characteristic map in **Fig. P50**, the operating point of the internal combustion engine is limited by the full-load curve as well as the minimum and maximum engine speeds. The power output of the engine at a specific operating point can be calculated from the relationship

$$P_e = 2 \cdot \pi \cdot M \cdot n .$$

Lines representing the same power are drawn as power hyperboles in the engine map. The specific values for load in relation to displacement, mean effective pressure, and specific work are frequently used instead of the torque to allow engines with different displacements to be compared.

The engine map is used to display certain engine characteristics in relation to the operating point. This representation consists of specifying discrete values with individual points. When a large number of individual values are available for the engine operating range, the specific engine characteristics, the ISO curve, can be calculated from these individual values by interpolation of the curves. The most common type of map representation is specific fuel consumption, ISO curves of which are visible in the map as so-called engine graphs.

In addition to the engine characteristics, the characteristics of the vehicle and its drivetrain can also be

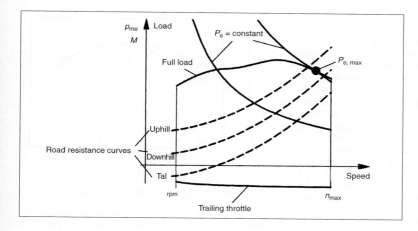

Fig. P50
Engine map

shown in the characteristic map. This is usually accomplished with driving performance curves. These curves illustrate the interrelationship between the engine speed and the torque transferred to the drivetrain for each gear when driving at constant speed on a level surface. Driving resistance curves displaced parallel to, one another result when accounting for driving uphill or downhill.

If the operating point of the engine is above the driving performance curve, the vehicle is accelerated; if it is lower, it is decelerated. The excess power available for acceleration results from the current engine speed and excess torque, corresponding to the interval between the driving performance curve and full-load curve. Shifting gears changes the torque at the same vehicle speed by changing the engine speed while maintaining approximately the same power requirement—that is, the operating point is moved along the power hyperbola to the point of intersection with the driving resistance curve corresponding to the gear change. In this manner, the changes in the operating or emission characteristics can be evaluated from the engine map in relation to particular vehicle and operating conditions.

For vehicle states with operating points at low vehicle power requirements, as occur within wide ranges of the emission cycles for vehicle type approval or in urban traffic, operating points with low to medium speed/load combinations are more relevant. In contrast, load loci typical of driving on a freeway are located in the upper right segments of the engine map.

Characteristic maps are used for documenting the operating parameters such as the ignition timing, injection timing, or fuel-air ratio for clarification of the operating strategy as well as for evaluation of the resulting measure and calculated values such as emissions, fuel mileage, or temperatures. In **Fig. P51**, a gasoline engine with direct injection is used as an example of how the engine map can be used to represent the operating strategy for the engine.

Characteristic areas in the map are marked differently to indicate the operating strategy. In the present

example, the engine is operated below a load of $p_{me} = 4$ bar and up to a speed of 3500 rpm by injection of a stratified fuel/air mixture during the compression stroke with high excess air. In the remaining map, injection of the fuel is accomplished during the intake phase, with the result that the longer time for mixture preparation results in a more homogeneous mixture. Even in homogeneous operation, a range with excess air is present in the lower load range for speeds between 3500 and 4500 rpm. Over the remaining load and speed range, with stratified and homogeneous lean operation, this engine operates in a similar manner to a conventional gasoline engine with a stoichiometric mixture. As full load is approached, particularly at higher engine speeds, the mixture is enriched.

This general map indicates that external exhaust gas recirculation has been accomplished over the entire stratified range and also over a portion of the stoichiometric range. Other characteristics of the operating strategy of the engine can be represented in the operating map in a similar manner. This includes, for example, coordination of camshaft timing or map cooling. The

P

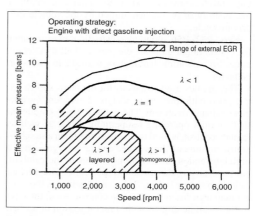

Fig. P51 General map

engine map represents a highly compact source of information enabling experienced engineers to evaluate the performance of the engine in question. In comparative evaluation of characteristic maps, it must be remembered that, based on experience, design criteria such as the displacement, stroke/bore ratio, compression ratio, or layout and arrangement of the injection valves only make minor differences to the parameters. In contrast, operating conditions such as the specific tuning of variable systems (variable resonance induction systems, camshaft adjusters) and the engine control system, as well as measures for exhaust treatment (catalytic converter systems, thermal insulation of exhaust system up to catalytic converter) cause significant differences in the operating characteristics for similar engines. Examples of this are gasoline engines with direct injection on the Japanese market, which have characteristics similar to the engine described previously, while the tuning of the same basic engine for the European market has no stratified or homogeneous lean range in the entire map. This clearly shows that measures for exhaust treatment for different markets or even for more stringent emission certification levels show more significant differences in the engine map than would normally be expected on the basis of the specific manufacturer or design differences.

*~EGR map. See below, ~*Emission maps

~Emission maps. As a rule, emission maps are concerned with the untreated emissions of the legally limited polluting substances—that is, hydrocarbons, oxides of nitrogen, and carbon monoxide. These are usually illustrated in the form of specific values in relation to the output (g/kWh), in the form of concentration values (% or ppm), or as mass flows (g/h). For diesel engines as well as gasoline engines with direct injection, the particulate emission maps are also significant. In addition to the maps for untreated emission, the emission values downstream of the catalytic converters are also frequently specified. These allow, on the one hand, evaluation of the conversion ratio in the catalytic converter and, on the other, estimation of the quantity of pollutants emitted by the vehicle in a driving cycle.

Figs. P52–P56 show characteristic emission maps for conventional gasoline engines as well as selected maps for operation parameters relevant to emission behavior.

The engines used as a basis for the maps shown are equipped with three-way catalytic converters for efficient exhaust treatment. The maps apply to engines at their normal operating temperature. This, as well as lambda control, is used to precisely maintain a stoichiometric mixture on the selected engines and guarantee a high conversion rate for all pollutants in the catalytic converter, using the three-way principle. The air-fuel ratio map shown in **Fig. P52** clearly indicates the largest map area for active lambda control. As in the case of the gasoline engine with direct injection described previously, the mixture is enriched in the full-load

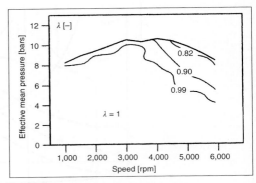

Fig. P52 Air-fuel ratio (MPI gasoline engine)

range as well as at high engine speed here also. The minimum air-fuel ratios, with values around $\lambda = 0.80$, are calibrated in the range of the rated output.

The reduction in the concentration of CO emissions is primarily a function of the air-fuel ratio, as shown clearly in the corresponding map in **Fig. P53**. In the map area with active λ control, the concentrations are usually in a noncritical magnitude between 0.5 and 0.8% by volume. At full load, combustion is accomplished with an oxygen deficiency due to enrichment of the mixture. At the maximum enrichment rates in the rated output range, the maximum CO concentration is 7.5% by volume. This high dependency on the air-fuel ratio expressed on the CO map can be considered critical for state-of-the-art gasoline engines with high specific output.

For untreated NO_x emissions, possibilities exist for influencing the level even in stoichiometric operation by tuning the operating parameters. When tuning the engine in the map, retardation of the ignition timing is selected partially for this purpose. However, this measure also causes decreases in the efficiency, which must also be taken into consideration in evaluating the fuel economy map. In contrast, exhaust gas recirculation (EGR) in the part-load range offers a significant

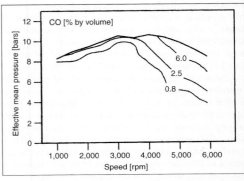

Fig. P53 CO emissions upstream of catalytic converter (MPI-gasoline engine)

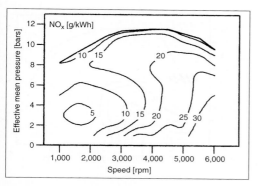

Fig. P54 Specific NO$_x$ emissions (MPI gasoline engine)

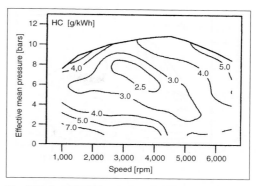

Fig. P56 Specific HC emissions (MPI gasoline engine)

potential for reducing the untreated NO$_x$ emissions while simultaneously providing efficiency advantages resulting from de-throttling the engine.

Recirculation of the exhaust gas can be accomplished either externally with a valve or in the form of internal exhaust gas recirculation using camshaft (valve timing) adjusters. The map of the specific NOx emissions from a gasoline engine with external exhaust gas recirculation is shown in **Fig. P54**, and the associated map of the EGR rates calibrated using the EGR valve shown in **Fig. P55** provides an example of a practical application of exhaust gas recirculation. The minimum NO$_x$ emission level of 2.3 g/kWh is achieved at the operating point with the maximum EGR rate of 18%. The typical behavior for NOx emissions is present outside the map area for external EGR. The large decrease in NO$_x$ emissions seen at full load as well as at high engine speeds results from enrichment of the mixture.

With internal EGR now also used for mass production engines, frequently with continuously acting camshaft adjusters, additional advantages are present in full-load operation, in addition to the described effects for the part-load emission and fuel economy characteristics. The volumetric efficiency advan-

tages can be maximized, resulting in an improved torque curve by optimizing the timing as a function of the engine speed.

In contrast to the NO$_x$ and CO emissions, the level of the untreated HC emissions is influenced to a considerably higher degree by design parameters. The first factor here is the shape of the combustion chamber, whereby the ratio of the surface area/volume represents an important parameter. Although the HC emissions are sensitive to operating conditions with the engine at operating temperature, this sensitivity is of subordinate significance in the usual variation ranges. Internal EGR with variable timing can have a positive effect on the HC emissions, because the HC peak typically present at the end of the exhaust stroke is fed back for recombustion. A typical HC map for a gasoline engine with single-stage intake camshaft adjustment is shown in **Fig. P56**.

The emission maps not shown here are those downstream of the catalytic converter which are distinguished by virtually complete conversion of the pollutants in state-of-the-art gasoline engines with three-way catalytic converters. Deviations from the extremely low emission level result in substoichiometric operation, when the catalytic oxidation of the HC and CO constituents is limited due to the lack of oxygen.

The carbon monoxide and hydrocarbon concentrations typical of diesel engines are significantly lower than with gasoline engines because of combustion with excess air (**Figs. P57 and P58**). Further reduction of these emissions is possible using oxidation-type catalytic converters due to the residual oxygen always present in the exhaust from diesel engines.

However, the untreated NO$_x$ emissions are more critical for diesel engines, as shown in **Fig. P59**. Because subsequent catalytic treatment poses a problem in the presence of excess air, the primary focus is on limiting formation of NO$_x$ by influencing the combustion process. As with gasoline engines, the measures used for this purpose include exhaust gas recirculation as well as retardation of the injection process, which is virtually the equivalent of retarding the ignition on gasoline engines.

In diesel engines, to increase the efficiency of EGR

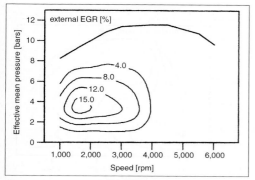

Fig. P55 Exhaust recirculation rate (MPI gasoline engine)

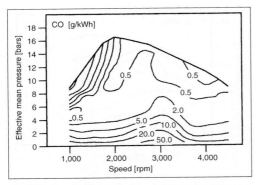

Fig. P57 Specific CO emissions (DI-TCI diesel engine)

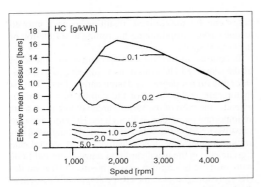

Fig. P58 Specific HC emissions (DI-TCI diesel engine)

in terms of reducing the emission of nitrous oxides, the recirculated exhaust gas is cooled. The map of the EGR rates (**Fig. P60**) indicates that, in the present example, the exhaust gas recirculation is calculated primarily for the map region relevant to the emissions. The exhaust recirculation rate reaches values of up to 50%, which is significantly higher than in gasoline engines. In contrast to gasoline engines, the possibility of exhaust gas recirculation is not limited by the problem

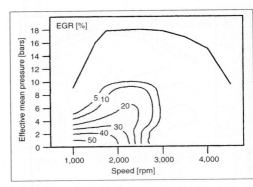

Fig. P60 Exhaust recirculation rate (TI-TCI diesel engine)

of misfire. Here, the combustion is accomplished with a high quantity of excess air and the oxygen concentration in the exhaust still as high as 15% by volume.

For diesel vehicles, the particulate emissions are also limited by law in Europe. A common parameter for evaluation of the particulate emissions from diesel engines is the Bosch smoke number. The increased smoke values in the emission-relevant map area, **Fig. P61**, indicate a relationship between particle formation and EGR, and this highlights the familiar conflict of objectives between NO_x and particulate emissions.

Outside of the map area tuned with EGR, the smoke numbers are relatively low. However, they increase significantly in the range close to full load, particularly at low engine speeds, due to the predominantly low quantities of air in the air-fuel mixture.

Particle formation can be counteracted by good preparation of the injected diesel fuel. For this reason, high-pressure injection with high atomization quality is a primary development trend for state-of-the-art diesel engines. As the particulate emission limits become more stringent, the use of particulate filter systems will allow significant progress, in addition to measures that can be performed within the engine itself. In spite of the relatively high costs, this technology is already

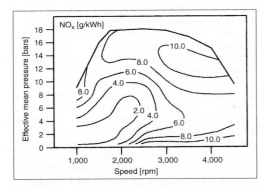

Fig. P59 Specific NO_x emissions (DI-TCI diesel engine)

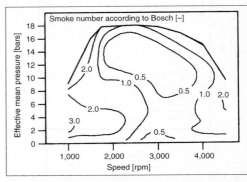

Fig. P61 Particulate emissions (DI-TCI diesel engine)

being used in mass production vehicles today. In a deviation from the steady-state calibration documented in the engine maps, intermittent regeneration of the particulate filter requires manipulations of engine calibration, which serve for temporary increase of the exhaust temperatures in certain map areas in order to promote combustion of the particles that have collected on the surface of the filter.

~Exhaust temperature maps. The variation of exhaust temperatures for a gasoline engine at different operating conditions is shown in **Fig. P62**. The rate of increase in the exhaust temperature at high loads requires specific measures to protect the exhaust catalytic converter against thermal aging or even destruction. This can be accomplished using design measures as well as calibrating the engine operating parameters. For engines with exhaust turbocharging, the gas temperature at the turbine inlet is also critical for protection of the component. In gasoline engines, enrichment of the fuel/air mixture is accomplished as described previously in the critical exhaust temperature map range as an effective measure for component protection.

In contrast, it is necessary to avoid excessively low exhaust temperatures when operating at low load points to prevent the catalytic converter from cooling down. For this reason, relatively high retardation of the ignition timing may be required.

In addition to these measures recognizable in the steady-state maps, the ignition timing and EGR rates are usually modified after a cold start so that the light-off temperature required for converting the untreated emissions into harmless constituents is achieved quickly.

The exhaust temperature map of a direct-injection turbocharged (DI-TCI) diesel engine is shown in **Fig. P63**. The measuring point for the exhaust temperature is located here at the inlet into the exhaust turbine. Analogous to gasoline engines, where it is necessary to protect the exhaust catalytic converter against ex-

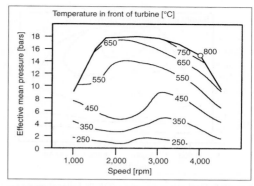

Fig. P63 Exhaust temperature map at turbine inlet (DI-TCI diesel engine)

cessive temperatures, here it is necessary to avoid excessive thermal load for the exhaust turbine. This applies particularly to the design with variable turbine geometry used widely today, in which it is necessary to protect the adjustable guide vanes against overheating. Measures for influencing the exhaust temperatures during operation are, on the one hand, appropriate tuning of the excess air by suitable layout of the turbocharger and, on the other hand, limitation of the injection quantity and, therefore, the full-load torque.

~Fuel efficiency maps. A typical fuel efficiency map for conventional gasoline engines with intake manifold injection is shown in **Fig. P64**.

As was previously mentioned, the curves of the constant specific fuel consumption are also called engine graphs. The minimum specific fuel consumption is achieved in the lower engine speed range at high load. The variation of specific fuel consumption is relatively flat over a large range around the minimum consumption point. At low loads, the gradient shows more rapid increases. The primary factors for this are increasing throttle losses at low load, as well as the friction becoming an increasing percentage of the effective

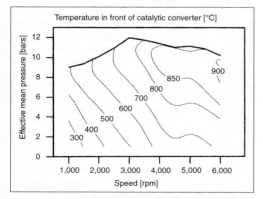

Fig. P62 Exhaust temperature map at inlet to catalytic converter (MPI gasoline engine)

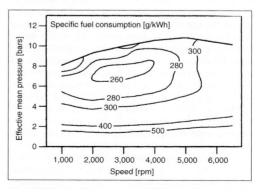

Fig. P64 Fuel efficiency map (MPI gasoline engine)

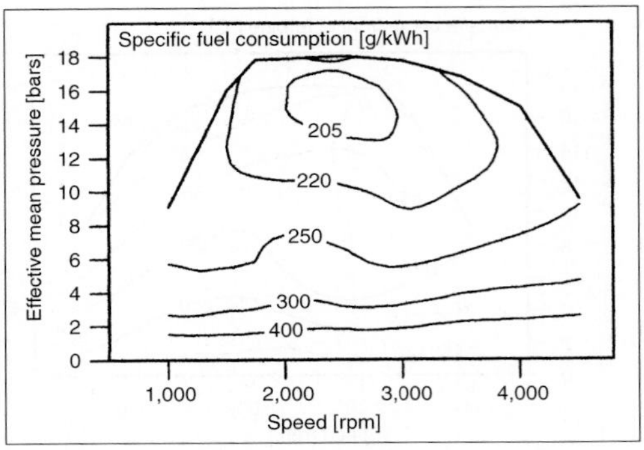

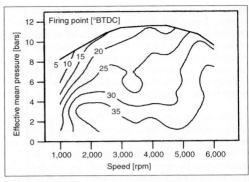

Fig. P65 Fuel efficiency map (DI-TCI diesel engine)

Fig. P66 Firing point (MPI gasoline engine)

torque output. These two factors lead to a visible increase in the consumption at constant load and increasing engine speed. In the full-load range, it is necessary to compensate for the tendency of the engine to knock and to keep the exhaust temperatures below the limit critical for catalytic converter aging by enriching the mixture. This also results in an increasing gradient for the consumption increase.

Fig. P65 shows a typical fuel efficiency map for a diesel engine with direct injection and turbocharging. The smaller increase in consumption at decreasing load is conspicuous, because the mixture quality control for diesel engines is not associated with throttle losses. In spite of the significantly more favorable part-load consumption values in comparison to gasoline engines, the consumption values obtained by vehicle calibration are higher than those for consumption by optimized tuning, particularly in the region of the map relevant to the European driving cycle. The reason for this is that the injection timing has to be retarded to achieve the permitted NO_x and particulate emissions.

The fuel efficiency map for an engine can also be used for calculating the fuel mileage for the vehicle when primary specific vehicle data are known, such as the driving resistances and transmission ratios.

For computation of the fuel consumption (mpg) in the dynamic test cycle, the driving curve is divided up into a series of steady-state operating points as a function of the specific vehicle parameters. Each of these steady-state points is characterized by engine speed and torque. The load points are then weighted by their time duration in accordance with their relevance for the driving cycle for computation of the average cycle fuel consumption (mpg). The models required for exact computation of the fuel consumption take into consideration additional consumption-relevant processes such as engine warm-up as well as gear shifting and other unsteady effects in addition to the specific vehicle data. With the aid of such models, the influence of the vehicle on the consumption and emission characteristics of the engine can be estimated. Examples of applications for this procedure are transmission tuning or the strategy for control of a continuous transmission (CVT).

~Ignition and injection maps. Calibration of the ignition timing on conventional gasoline engines with lambda control is highly dependent on the operating points. As a rule, the ignition timing is calibrated in the range of optimum performance at the mean part load. **Fig. P66** shows that the basic tendency is that increasing ignition advance is required at increasing engine speed as well as decreasing load. Other effects are superimposed on this behavior. In the lower load range, a significant ignition advance is already clearly recognizable at low engine speeds. For the engine shown, the external exhaust gas recirculation is calibrated in this range. The exhaust gas recirculated acts as an inert gas and retards the combustion curve; therefore, the point of ignition must occur earlier. Moreover, a retardation of the ignition can also be seen close to full load at engine speeds around 4500 rpm. This behavior results from the knock limitation frequently noted in the range of maximum volumetric efficiency. According to the present state of the art, the disadvantages resulting from such measures can be minimized by use of dynamic knock control systems.

These allow optimized ignition advance in terms of torque without the danger of damage to the engine from knocking combustion.

In diesel engines, the combustion is controlled primarily by the injection process. The start of delivery, therefore, has significance comparable to the ignition timing on gasoline engines. After the introduction of direct injection predominant today, rapid combustion with steep cylinder pressure curve gradients has led to engine acoustics problems. An effective measure for reduction of the cylinder pressure gradients with modern electronic diesel control (EDC) of the engine is pilot injection. With pilot injection, combustion is first initiated with injection of a smaller quantity of fuel. The remainder of the diesel fuel is then added to the process during the main injection period. **Figs. P67 and P68** show the maps for the start of delivery for the pilot and main injection on a state-of-the-art diesel engine for passenger cars. As shown, pilot injection is limited by the EDC to a certain speed range.

P

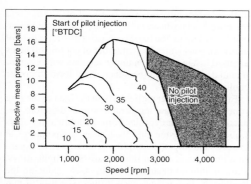

Fig. P67 Pilot injection start of delivery (DI-TCI diesel engine)

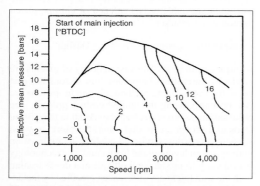

Fig. P68 Main injection start of delivery (DI-TCI diesel engine)

The relatively late start of delivery for main injection also results from the use of the previously described measures for reduction of the NO_x emissions. In contrast, the main injection operation is advanced in the map area without pilot injection.

~Injection maps. *See above*, ~Ignition and injection maps

Performance (during operation) →Oil

Phase adjuster →Variable valve control ~Operation principles, ~Systems with camshaft ~~Camshaft phasing systems

Phase shifter →Control/gas transfer ~Four-stroke engine ~~Timing

Phase spectrum →Engine acoustics ~Spectrum

Phasing →Camshaft; →Control/gas transfer ~ Four-stroke engine ~~Phasing; →Variable valve control ~Variation parameters

Phosphate coating →Piston ring ~Materials; ~Surface treatments

Photoelectric charging, particle →Particles (Particulates) ~Particle measuring ~~In situ method of aerosol characterization

Piezoelectric sensor →Sensors ~Knock sensors

Pilot injection process. The pilot injection process is characterized by a two-stage mixture or fuel feed into the combustion space, and this is classed among the hybrid combustion processes. In the first phase, a lean homogeneous mixture is generated and aspirated outside the combustion space. The second phase is characterized by direct fuel injection. Ignition is effected by autoignition in the heterogeneous mixture zone.

Pinging →Knocking

Pintle displacement path →Bearings ~Computation ~~Displacement path

Pintle nozzle →Carburetor ~Constant pressure carburetor ~~Design

Pintle seat →Injection valves ~Diesel engine ~~Injection nozzle parameter

Pintle stroke →Injection valves ~Diesel engine ~~Injection nozzle parameter

Pintle-type nozzle →Injection valves ~Diesel engine

Piston. The piston is the first element in the chain of power-transferring parts in a piston engine. Its purpose is to convert thermal energy into mechanical work. However, its purpose is not limited to simply transferring the effective forces. With its sealing elements, the piston rings, it seals the combustion chamber from the crankcase and passes the heat transferred to it to the coolant and cylinder wall. The lubricating oil present on the contact surfaces supports this sealing effect. These various types of functions result in some contradictory requirements for the piston, making it more difficult to design the individual cross sections of the piston geometry.

As part of the crankshaft drive, the piston is subjected to the gas force, the inertia force, the lateral guidance force, as well as the torque in relation to the piston center of gravity. The gas force results from the periodic sequence of the pressure change in the combustion chamber, the maximum pressure being related primarily to the combustion process, the fuel, and the compression ratio.

In addition to the piston, the piston ring, piston (gudgeon) pin, and a portion of the connecting rod contribute to the inertia forces of the oscillating engine parts. Because the force of inertia changes with the square of the engine speed, it is necessary to keep the weight

P

of the piston including piston pin and rings as low as possible, particularly in high-speed engines.

The deflection of the connecting rod resulting from the crankshaft drive results in a lateral force on the piston perpendicular to the cylinder axis. In many cases, the piston pin is not located on the central axis of the piston to prevent piston noise and reduce the thermal load on the piston ring groove. This causes a force that attempts to rotate the piston around its center of gravity, which is supplemented by the friction in the pin bearing and by the friction force on the cylinder wall.

Literature: M.D. Röhrle: Kolben für Verbrennungsmotoren, Publishers Moderne Industrie AG, Landsberg/ Lech, 1994.

~Articulated skirt piston. *See below,* ~Design

~Assembled pistons. Assembled pistons are not used today for passenger car engines because they are too heavy. They consist of a crown of forged steel (piston top section) and a piston bottom section of pressed piston alloy, consisting of spherolithic cast iron or forged steel (**Figs. P102 and P69**). Both parts are bolted together with high-quality connection elements. Hollow spaces through which oil flows during operation are located between the piston crown and piston bottom section. The temperatures in the area of the ring zone and on the piston crown can be influenced, depending on the cooling oil feed over a cooling chamber from the outside to the inside or from the inside to the outside. Common ignition pressures on standard commercial vehicle diesel engines today are in the range of 180 bar. The parts are tested at an overload of 200 bar. On Monotherm pistons for Euro IV classification, the

tests are accomplished at an overload of 240 bar, and for Euro V classification, the pistons are tested at an overload of up to 260 bar. For reasons of strength, steel upper sections with a large number of cooling oil holes are used instead of piston upper sections with circumferential outer cooling chamber. They form the outer cooling area.

Effective cooling, particularly in the area of the ring zone, results in low temperatures and, therefore, reduction of carbon deposits from the lubricating oil as well as lower piston ring wear. Low temperatures at the fire land, together with the minimum heat expansion of the steel crown in comparison to light alloys, allow minimum clearances in this area, reducing the load on the piston rings even further. The user of materials with high specific weight increases the overall weight in comparison to an aluminum alloy piston by approximately 20–50%. The piston ring grooves in the steel area are inductively hardened or chrome-plated to reduce the wear.

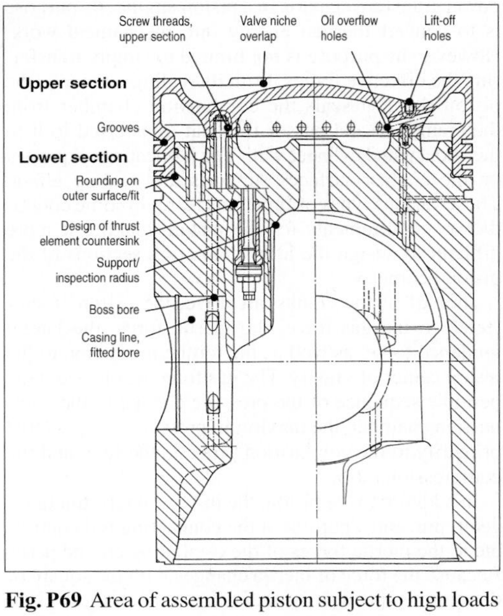

~Autothermic piston. *See below,* ~Design

~Blank production. The vast majority of pistons are produced by casting. Gray cast iron pistons are cast using the sand casting method, aluminum pistons are cast primarily using the chilled casting process. The rapid solidification of the liquid material in the casting leads to a fine grain structure and, therefore, to good strength properties. Optimum cooling in combination with optimum layout of the feeder and gate technology is necessary to produce a fault-free and dense casting with the wall thickness differences specified by the design from thin skirt to thick piston crown to achieve the required solidification. Multiple piece casting molds and casting cores allow a wide range of freedom for laying out the piston geometry so that undercuts—for example, in the inner shape of the piston—can be realized. Expansion regulating strips, ring carriers, cooling hollow chambers, and other design features can be cast in. The cooling hollow chambers for positive oil cooling are formed by casting in pressed salt cores which are then dissolved with water. Small lots are cast in manual molds. In contrast, when machine casting processes are used, all steps including pouring in the metal are accomplished automatically. Here, multiple molds and casting robots are used for mass production to meet the high requirements for quality and economy. The squeeze casting process (Liquostatik) is used for pistons with ceramic fiber inserts. This squeeze casting process differs from gravity mold casting in terms of the pressure applied to the molten material (100 Mpa and higher), which is maintained until the casting has completely solidified. The extraordinarily good contact between the solidifying molten material and walls of the mold results in very quick solidification. A fine structure advantageous for the material strength is achieved in this manner. Pistons reinforced locally on the piston crown in the groove or boss area with ceramic fibers or porous metallic materials can be produced using the squeeze

Fig. P69 Area of assembled piston subject to high loads

casting method. These cast-in parts are penetrated completely by the piston alloy as the result of the pressure applied to the molten material.

~Blowby (*also*, →Blowby). One of the primary purposes of the piston is to seal the combustion chamber under pressure against the crankcase in interaction with the piston rings. Due to the operating clearance between the piston and the cylinder, combustion gases can pass into the crankcase (blowby) during the kinematics motion sequence. However, gas leakage in the form of blowby not only means a power loss, it also poses a hazard for the piston and ring lubrication by displacing and contaminating the lubricating film and finally all of the engine lubricating oil. The compression rings provide the primary seal to prevent blowby. In comparison to the engine exhaust, crankcase gases can contain many times the concentration of hydrocarbons. While in the past these gases were allowed to escape into the atmosphere without treatment, today they are returned to the engine intake system. On naturally aspirated engines, the blowby quantity does not exceed a maximum of 1% of the theoretical intake volume; on turbocharged engines, this maximum figure is 1.5%.

~Cast pistons. *See above*, ~Blank production

~Coating. *See below*, ~Surface coatings

~Compression height (CH). The compression height is defined as the distance from the center of the piston pin to the top of the fire land. Protuberances and recesses are not specified separately. The compression height affects the overall height of the engine, in addition to the piston weight itself. The number and height of the piston rings, the required ring lands, the piston pin diameter, and the fire land height result in a minimum value for the compression height. This is approximately 33% of the piston diameter for three-ring pistons and, due to elimination of one piston ring and one ring land, is approximately 28% for two-ring pistons.

Reduction of the compression height is also associated with disadvantages. Although changeover from three-ring pistons to two-ring pistons by eliminating one piston ring offers a possibility for reducing the compression height, possible disadvantages can be expected in terms of the blowby quantity, oil consumption, blue smoke, or difficulties starting the engine after longer periods of operation. The short distance between the inner piston crown and boss bore and associated higher incidences of heat can lead to cracking occurring in the boss bore at high outputs and ignition pressures, which can be prevented only by piston cooling or other expensive modifications to the pistons (shape bores in the pin bore).

~Contact change. *See below*, ~Pressure side

~Contact pattern. A so-called contact pattern becomes visible on the skirt of pistons which have been in operation, due to abrasion on the surface peaks on

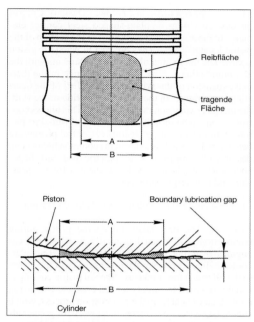

Fig. P70 Boundary lubrication gap between piston and cylinder and piston contact pattern

the machined skirt surface. This influences the regulating effect of the piston skirt under alternating operating conditions, the piston or skirt design, and the width of the contact surface available on the basis of the so-called free circumferential length, as well as its support resulting from the construction, due to the operating clearance and the resulting concave oval skirt contour (**Fig. P70**). Generally, the contact surfaces formed by the pressure/counterpressure side should not be too narrow to ensure that the mean surface pressure between the piston and the cylinder is kept within acceptable limits. The transition to the nonsupporting skirt areas should be laid out with a flowing design to prevent hard contact points.

~Contact pressure →Piston ring

~Control elements (*also, see below*, ~Control piston). Control elements are ring- or plate-shaped steel inserts cast into the area of the upper piston skirt on control pistons for reducing the thermal expansion in the pressure/counterpressure direction of the piston. Control elements cause elastic deformations as a result of the different thermal expansion of steel and Al-Si piston alloys during casting, which reduces the expansion of the piston by this amount when heated. This allows the undesired expansion of the piston in the running direction during operation to be diverted to the direction of the piston pin axis. The skirt running clearance can be kept virtually equal in the cold and warm state in this manner.

P

~Control piston (*also*, *see below*, ~Design). Control pistons have ring- or plate-shaped steel control elements in the area of the piston skirt or the section of the piston located above. They are cast in when the piston is cast. The interaction of these steel elements with the aluminum silicon alloy leads to a change in the thermal expansion of the skirt under the effect of the operating temperature. In the pressure direction of the skirt, the thermal expansion rate is reduced. This allows control pistons to be installed in the form of full-skirt pistons with low operating clearance. Control pistons are being used less and less today in the common lightweight design with symmetric or asymmetric oval skirt shapes and, where applicable, wall thicknesses for the pressure and counterpressure sides.

~Cooled ring carrier piston. *See below*, ~Design

~Cooling. The heat taken up by the piston during combustion is dissipated primarily in passenger car engines over the piston rings to the air- or water-cooled cylinder and over the inner surface of the piston to the oil fog in the crankcase. Special piston cooling is usually required for highly loaded engines. The piston cooling takes heat from the combustion process, which should be converted to mechanical work for greater efficiency. For this reason, piston cooling should be used sparingly—that is, only at high loads.

The influence of the piston diameter results from the circumstance that the surface area increases with the square of the geometric enlargement; however, the volume increases with the cube of the diameter. It is necessary to take up and dissipate a larger quantity of heat per unit of combustion chamber surface per time unit.

With the simplest type of heat dissipation—spray oil cooling—oil is taken from the lubricating oil circuit and sprayed against the inner surface in the area of the piston crown through a nozzle located at the bottom of the cylinder liner in the crankcase.

~~**Cooling duct pistons.** Pistons with cooling duct allow greater reduction of the temperature level in comparison to pistons which are sprayed with oil only locally on the bottom of the piston. Pistons with positive oil cooling with cooling duct are produced with a hollow cavity in the piston filled with a salt core—these are also on electron beam welded pistons. The cooling duct is filled only partially with oil to achieve the desired shaker effect—that is, intensive mixing of the cooling oil and thereby good heat dissipation. Cooling duct pistons are supplied with oil continuously from a stationary nozzle in the housing through a vertical feed duct in the piston. One or more bores on the inside of the piston usually opposite to the feed allow the oil to drain freely into the crankcase. The dimensioning of the feed, drain, and quantity of cooling oil fed in is balanced to achieve optimum fill levels with values of 30–50%. The oil oscillate in the cooling chamber (shaker effect) and its turbulent flow leads to high heat transfer rates. In this manner, more heat is dissipated than would be possible if the duct was completely filled with oil. The quantity of oil reaching the cooling duct is called the capture rate or feed efficiency of the piston oil to the piston.

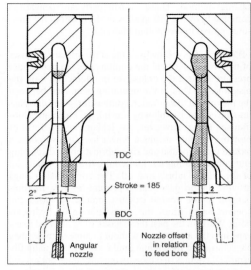

Fig. P71 Effect of various nozzle positions on cooling oil feed

~~**Forced oil cooling.** For higher cooling requirements, the cooling ducts are located in the area of the ring zone in the piston (positive oil cooling). As with spray oil cooling, the cooling oil is sprayed from a nozzle. This nozzle sprays the cooling oil into the funnel-shaped feed bore in the piston (**Fig. P71**).

On larger pistons, the cooling oil is frequently taken from the lubricating oil cushion in the crankshaft journal bearing and fed through the connecting rod and piston pin to the cooling duct in the piston.

~~**Spray oil cooling.** The simplest method of removing additional heat from the piston when required is spray cooling on the inside of the piston with an oil jet. In high-speed motor vehicle engines, the oil feed is accomplished either through a bore from the small connecting rod eye or, as is preferable with motor vehicle gasoline engines, from the large connecting rod eye (**Fig. P72**). The effect is limited to the area in contact with the oil jet. Due to the short retention time of the oil, heat transfer is limited at this point. In order to achieve positive effects, larger quantities of oil (up to 8 L/kWh) are required. A more effective method is to spray the piston with a stationary nozzle located at the bottom of the cylinder. Instead of an oscillating, intermittent oil jet inside the connecting rod pivot angle, the cooling oil spray is continuous but is dependent on the specific position of the piston in terms of quantity. The retention time for the oil can be increased by using oil catch grooves cast or screwed to the inside of the piston, resulting in more intensive heat dissipation (shaker pistons).

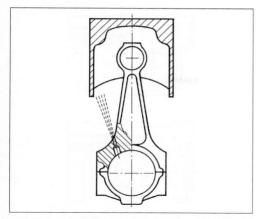

Fig. P72 Spray cooling from large connecting rod eye on a gasoline engine for motor vehicle

~**Counterpressure side.** *See below,* ~Pressure side

~**Crown.** The piston crown is a part of the combustion chamber. On pistons for gasoline engines, it can be flat, raised, or recessed. On pistons for diesel engines, the combustion chamber recess is usually located in the piston crown. Its shape is affected by the specific combustion process (prechamber, turbulence chamber, or direct injection method). The geometry of the piston crown is frequently also affected by the location and number of valves (**Fig. P73**). The thickness of the piston crown (crown thickness) is determined by the combustion pressure and the quantity of heat to be dis-

sipated. The piston crown is the section of the piston subject to the highest thermal load.

~**Deformation.** The gas, mass, and guidance forces acting on the piston cause deformations and, therefore, tension. The piston crown is bent inward by the effect of the gas pressure in the combustion chamber. Because the piston pin box is supported on the piston pin in the area of the skirt, the skirt is bent downward in the pressure/counterpressure direction and deformed ovally at its open end. Its diameter becomes greater in the piston pin direction and smaller in the pressure/counterpressure direction.

The deformation pattern becomes even more complicated due to deformation of the pin (flattening and sagging) and by the effect of the contact forces in the cylinder, which also tend to deform the piston. Deformations resulting from the temperature field present on the piston are superimposed on these piston deformations resulting from mechanical forces. This leads to bulging of the crown and an increase in the diameter from the bottom of the skirt to the fire land in comparison to the cold state when the pistons are heated up to operating temperature. The deformations are in the magnitude of the installation clearances of the pistons. At high continuous loads in positions having thin wall skirts, permanent deformation can occur which can have a negative influence on the operating characteristics of the piston. Therefore, it is necessary to ensure that the wall thicknesses of the piston are thick enough, on the one hand, to prevent fracture and permanent deformation from lateral forces, but on the other hand, resilient enough in a certain range to allow the piston to yield to deformation resulting from the outside—for example, from the cylinder. Deformations and tensions

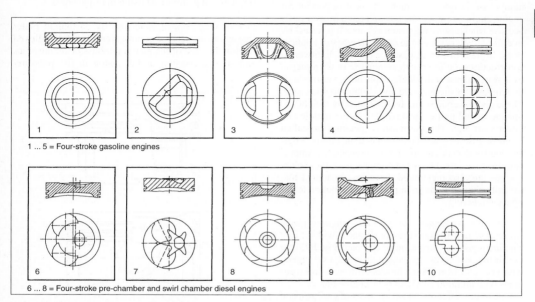

1 ... 5 = Four-stroke gasoline engines

6 ... 8 = Four-stroke pre-chamber and swirl chamber diesel engines

Fig. P73 Examples of piston crowns for various gasoline and diesel engines

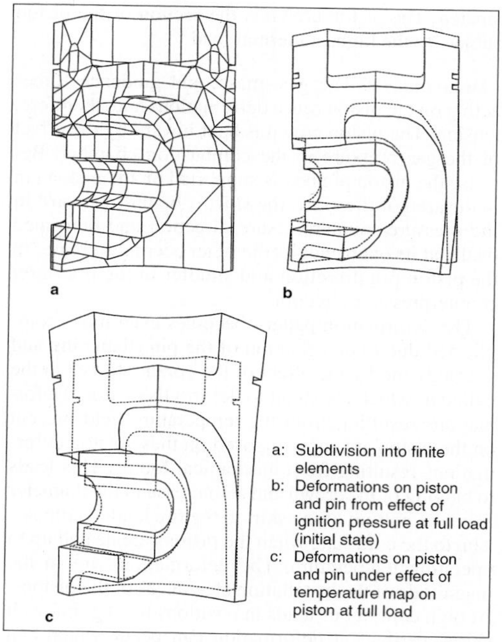

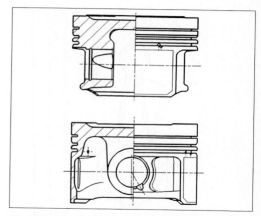

Fig. P75 Cast single metal piston (full-skirt piston)

a: Subdivision into finite elements
b: Deformations on piston and pin from effect of ignition pressure at full load (initial state)
c: Deformations on piston and pin under effect of temperature map on piston at full load

Fig. P74 Computations of deformations on a full-skirt piston (qualitative illustration)

resulting from the effect of impressed forces and temperatures can be calculated using the finite element method (**Fig. P74**).

~Design. Piston development has produced a large number of designs, of which the most important for engine engineering will be introduced as proven types because when discussing piston versions it is necessary to differentiate between the different types of applications. In state-of-the-art gasoline engines, lightweight constructions are used with symmetric or asymmetric oval skirt shapes and, where applicable, various wall thicknesses for the pressure and counterpressure side. These piston versions are distinguished by weight optimization and particular flexibility in the middle and lower skirt area. For the reasons specified, so-called "control pistons" which dominated in engine engineering for decades are becoming less and less important. However, for the sake of completeness, these older designs are also mentioned, because considered historically, they provide an important contribution to engine engineering.

Pistons for passenger car gasoline engines (four-stroke and two-stroke). Previously used widely, cast aluminum pistons without single cast part (**Fig. P75**) are used in the classical design today only in low-load engines with gray cast iron engine blocks or in two-stroke engines with aluminum cylinders or engine blocks.

Due to the different expansion rates of cast iron and aluminum, greater piston installation clearances are required in cast iron blocks, which lead to higher noise development, particularly at idle and in the partial-load range. With aluminum cylinders or aluminum engine blocks, however, the heat expansion does not present problems. In two-stroke engines, single metal pistons such as window pistons or full-skirt pistons are standard. However, in the meantime—in a deviation from the classical piston shape—pistons that are highly reduced overall, have reduced skirts, and have shorter piston pin bearings are used in four-stroke gasoline engines which do not offer sufficient space for single cast parts. Due to the production process, pressed pistons are always single metal pistons (**Fig. P76**). Due to the single piece extrusion die, the design possibilities for the inner shape are highly limited in comparison to single metal pistons cast using multiple cores.

Pistons with control elements. By casting in plate-shaped or ring-shaped control elements of nonalloyed steel in the area of the piston pin boss or on the skirt, undesired expansion of the piston in the pressure/

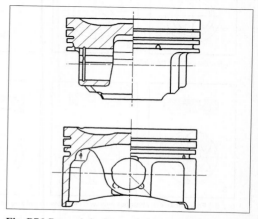

Fig. P76 Pressed single metal piston (full-skirt piston)

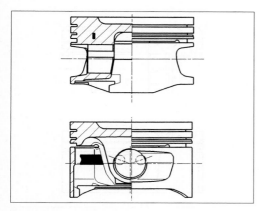

Fig. P77 Autothermic piston

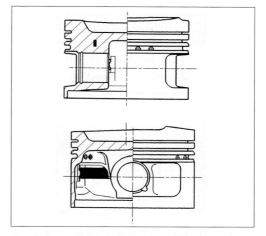

Fig. P79 Autothermic piston

counterpressure direction is prevented during operation, and the pin direction is deviated. This, in combination with a skirt profile designed according to hydrodynamic aspects, allows minimum installation clearance for the piston with all its advantages in all operating states. Autothermic pistons or hydrothermic pistons (**Fig. P77**) and ring belt pistons (**Fig. P78**) are used worldwide in passenger car gasoline engines. The cast-in control element and the slot between the piston crown and piston skirt in the area of the oil ring groove allow particularly smooth operation and long service life with these pistons.

A similar design was used for this piston—without a slot in the oil ring groove—operating according to the same control principle (autothermic piston or hydrothermic piston [**Fig. P79**] segment belt pistons, etc.). With these pistons, the heat transfer from the piston crown to the skirt is not disturbed. However, the transition cross sections are dimensioned to prevent any significant interference to the effect of the steel belt. Such pistons are preferable for use in high performance gasoline engines.

A modern piston version, the *Asymdukt* piston, without control element (**Fig. P80**), is distinguished by extremely low weight, optimized support, and box-type oval skirt layout. It is excellent for use in state-of-the-art passenger car gasoline engines and is equally well suited for aluminum engine blocks as well as gray cast iron engine blocks. The flexible skirt design allows compensation of various heat expansion rates between the gray cast iron block and aluminum piston extremely well in the elastic range. The systems can be cast as well as forged. The forged version is used particularly in highly loaded sport engines or highly loaded turbocharged gasoline engines.

The Ecoform design shown in **Fig. P81** presently offers the greatest potential for reducing the rate of four-stroke gasoline engines for stock application.

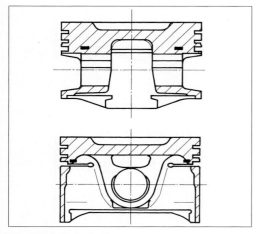

Fig. P78 Ring belt pistons

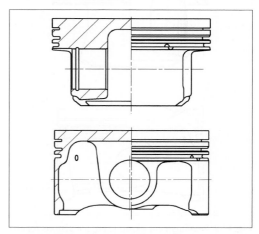

Fig. P80 *Asymdukt* piston for passenger cars

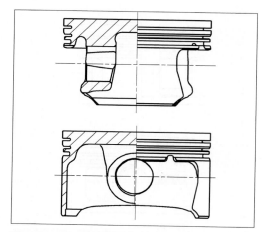

Fig. P81 Passenger car Ecoform piston

These pistons are characterized by a particularly distinguished high casting in combination with angular skirt walls and are used particularly in combination with trapezoidal support. Particularly designed crown recesses make this piston excellent for gasoline engines with direct injection (**Fig. P82**). This also gives them a tendency to have a slightly higher overall weight than conventional pistons for gasoline engines. The combustion recess in the piston crown is usually machined on cast or pressed pistons. Pistons for racing applications are all special designs. The compression height is very low and the pistons are extremely optimized, all totaled in terms of weight. Only forged pistons are used. **Fig. P83** shows an example of a piston for a Formula 1 racing engine. Here, the weight optimization and piston cooling are decisive criteria for the layout of the piston. In Formula 1 applications, engine speeds of greater than 18,000 rpm are common. The service life is matched to the extreme conditions. Window or full slipper skirt pistons are preferred for this design. However, electron-beam–welded cooling duct pistons are also used for racing engines (**Fig. P84**).

Pistons for passenger car diesel engines. Due to the higher combustion pressures and temperatures, primarily pressed hypereutectic single-metal pistons were used widely in passenger car diesel engines due to the higher combustion pressures and temperatures. As demands on passenger car diesel engines increased in terms of ring groove wear, operating smoothness, and service life, full-skirt strut pistons of eutectic aluminum/silicon alloy with Niresist ring carriers in the first ring groove were used more and more (**Fig. P85**).

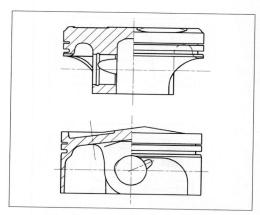

Fig. P83 Formula 1 piston

P

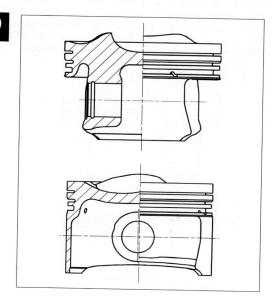

Fig. P82 Passenger car gasoline engine pistons for engines with direct injection

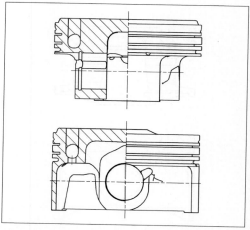

Fig. P84 Electron beam welded piston for racing

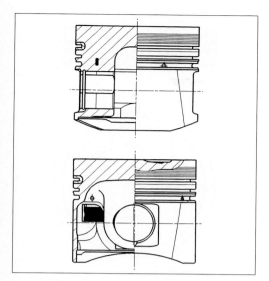

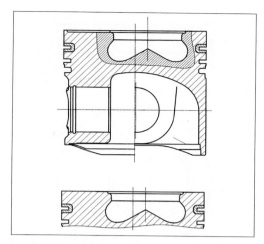

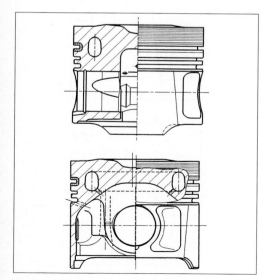

Fig. P85 Passenger car diesel autothermic piston with ring carrier

Fig. P87 Pistons for passenger-car engines with direct injection

Today, this is the standard design for engines of this type and it is used in vans as well as passenger cars. Salt core cooling duct pistons are used as a standard feature in diesel engines for passenger cars (**Fig. P86**).

Fig. P87 shows the typical shape of a piston for engines with direct injection in the passenger-car sector. A primary characteristic is the shape of the piston recess, which is arranged asymmetrically on engines with two valves. Various recess shapes for pistons for

passenger-car engines with direct injection are shown in **Figs. P88–P91**.

Pistons for commercial vehicle diesel engines. For some time, the standard version of pistons for commercial vehicle diesel engines were ring carrier pistons produced using the chilled casting method on which the first and in some cases also the second ring groove was located in the metallically bound ring carrier (**Fig. P92**). Cast full-skirt pistons without ring carrier were used only in engines with short life expectancy.

On supercharged commercial vehicle diesel engines, salt core cooling duct pistons with ring carrier are used (**Fig. P93**). When salt core cooling ducts cannot be positioned between the ring carrier and combustion recess for reasons of space, pistons with cooled ring carrier represent a new version with positive oil cooling. For this purpose, a hollow element, shaped from sheet metal, is welded to the ring carrier (**Fig. P94**).

P

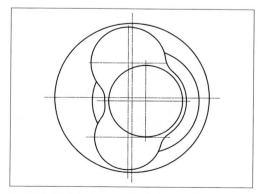

Fig. P86 Salt core cooling duct pistons with ring carrier for passenger-car diesel engines

Fig. P88 Piston recess for passenger-car engines with direct injection

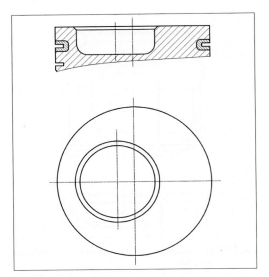

Fig. P89 Piston recess for passenger-car engines with direct injection

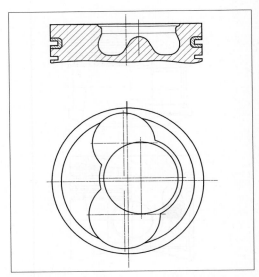

Fig. P91 Piston recess for passenger-car engines with direct injection

Cooled ring carrier pistons are used in the commercial vehicle and diesel passenger car sector. They allow significantly improved cooling of the first ring groove and the thermally highly stressed recess edge on the combustion recess. Intensive cooling of the first ring groove makes it possible to replace the double keystone ring usually used with a rectangular ring. Frequently shaping measures such as shape boring, relief pockets, or oval layout of the pin bore are no longer sufficient to improve the strength of the center bore, which is one of the piston areas subject to the

highest loads on highly loaded diesel pistons. For this reason, the piston pin bore is armored with a shrink-fit bushing of higher strength material (e.g., CuZn31Si1)

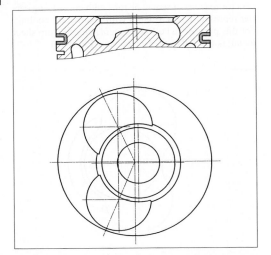

Fig. P90 Piston recess for passenger-car engines with direct injection

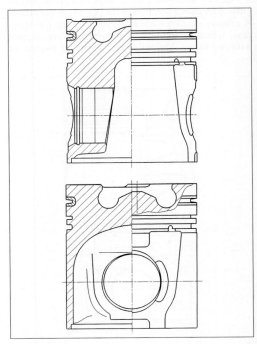

Fig. P92 Commercial vehicle full-skirt piston with ring carrier

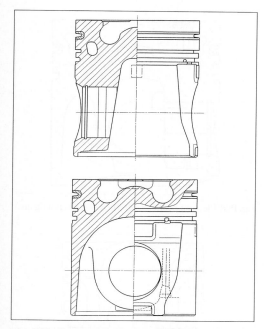

Fig. P93 Commercial vehicle salt core cooling duct piston with ring carrier

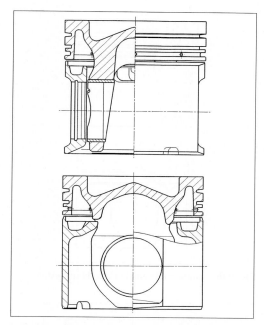

Fig. P95 Ferrotherm piston (articulated skirt piston)

particularly for high piston pin box temperatures. Two-piece systems (articulated skirt pistons, Ferrotherm pistons) are used for particularly high mechanical and thermal loads (**Fig. P95**). On such pistons, the guide and sealing functions of the piston are separated. The

Fig. P94 Passenger car diesel piston with cooled ring carrier

piston crown, primarily of forged steel, assumes the sealing function together with the piston rings. The gas forces are transferred to the cranktrain over it. The piston skirt is connected to the piston crown by the piston pin, allowing it to move. It provides for guidance and transfers the lateral forces to the cylinder.

Because iron materials conduct heat less well than does aluminum, a greater quantity of cooling oil is required due to the higher temperatures resulting on the surfaces in order to prevent the oil from aging. Oil catch grooves in the lower section of the piston or spring plates on the bottom of the cooling chamber in the piston crown (**Fig. P96**) also allow closed cooling chambers to be created in the piston crown, providing a shaker effect and longer retention time for the oil fed in there with spray nozzles. The first ring groove, machined in the iron piston head, guarantees low wear rates. The reduced fire land clearance ensures constant low oil consumption over long operating periods and has a positive effect on the exhaust emissions. An advancement of the Ferrotherm piston led to single-piece pistons of forged steel, called Monotherm pistons (**Fig. P97**). These pistons are optimized in terms of their weight. The compression height can be kept low and it is possible to approach the weight of a comparable aluminum piston by machining the inside of the piston boss. The outer cooling chamber is closed by two spring plate halves to improve piston cooling. Monotherm pistons are used in highly loaded commercial vehicle engines.

On locomotives, stationary, and marine diesel engines,

P

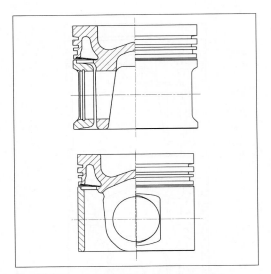

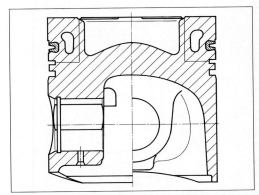

Fig. P98 Electron-beam–welded piston with cooling duct

Fig. P96 Commercial vehicle Ferrotherm piston

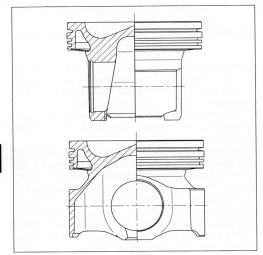

Fig. P97 Commercial vehicle Monotherm piston

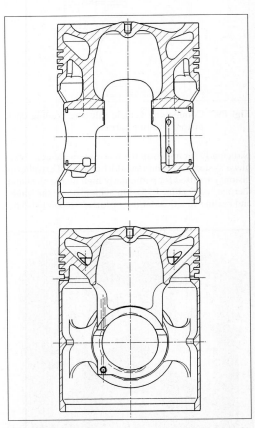

Fig. P99 Single-piece nodular cast iron (monoblock) piston

cast, pressed, electron-beam welded pistons (**Fig. P98**), single-piece nodular cast iron pistons (monoblock pistons; **Fig. P99**), and assembled pistons (**Figs. P100–P103**) are used. While assembled pistons with aluminum skirt limit the range of application for reasons of strength and assembled pistons with cast nodular iron pistons present problems for casting and machining and require a high degree of testing, assembled pistons with forged steel skirts (micro-alloy 38 Mn Vs 6) allow higher loads (**Fig. P103**). The expenses for processing and quality control are lower. The skirts can be produced using preform forging or on mechanical or hydraulic presses.

~Dimensions. The primary dimensions of the piston are dictated to a great extent by the primary dimensions of the engine and its components, such as the engine block, crankshaft, and connecting rod. The primary

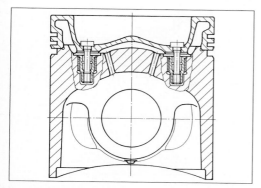

Fig. P100 Assembled piston with steel crown and aluminum bottom section

parts of the piston are the piston crown, the ring area with fire land, the piston pin boss, and the skirt (**Fig. P104**). The piston design is distinguished by additional functional elements such as control elements and ring carriers. The piston rings, the piston pin, and the piston pin retainers are also considered part of the piston assembly.

In addition to the piston diameter (D = cylinder diameter), after the installation clearance, the most important design dimension for the piston is the compression height (CH), which is the distance between the center of the piston pin and the top of the fire land (not taking into consideration protuberances and recesses). The bottom length (BL) is the dimension from the middle of the piston pin bore to the end of the skirt, so that the total length (TL) = CH + BL. Another important dimension is the distance between the eyes (AA) and the pin diameter (BO).

~Ferrotherm piston. *See above*, ~Design

~Fire land. The distance between the edge of the piston crown and the upper flank of the first piston ring groove in the piston ring zone is called the fire land. Subsequent lands between the piston ring grooves are called "ring lands." The dimensioning of the fire land results from the requirement that the uppermost piston ring must be located in a temperature range suitable for its function. This, for its part, is highly dependent on the overall design of the engine, the combustion process, the cylinder layout, and other factors.

~~Fire land height. On gasoline engines, the fire land height is 6.5–8% of the piston diameter with decreasing tendency for reducing the hydrocarbon emissions with gaps. On passenger car diesel engines with indirect injection, these values are between 10 and 15% of the piston diameter. On commercial vehicle diesel engines with indirect injection, we differentiate between a classic fire land height of 15–20% of the piston diameter and a tendency toward "short" fire

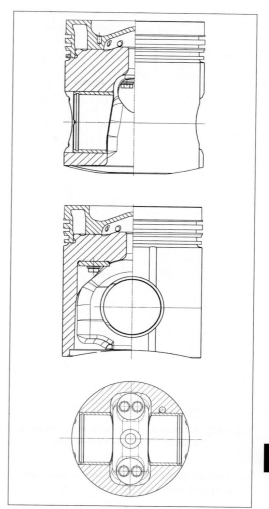

Fig. P101 Assembled piston with steel upper section and aluminum bottom section, bolted together as a twin.

lands of 10–15% of the piston diameter for optimization of the fuel consumption and exhaust quality by minimizing the dead volume in the gap above the first piston ring. An extreme form of this design is pistons with so-called head land ring, on which the ring carrier is located extremely high.

~Friction. The significant values for the friction loss in the piston assemblies in wide ranges of the applicable Newton law of friction losses are the supporting piston skirt surface area, the friction loss coefficient, and the thickness of the oil film between the piston and the cylinder. Measures to influence the piston friction can frequently result in contradictory results. For example, maximum piston clearance is advantageous for

P

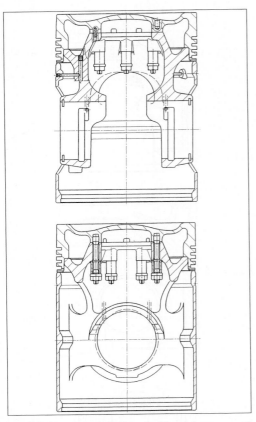

Fig. **P102** Assembled piston with steel crown and nodular cast iron bottom section

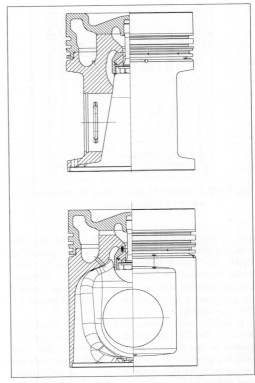

Fig. **P103** Ferrocomp piston, assembled piston with steel upper section and bottom section, bolted at center.

low mechanical friction; however, in terms of piston flap and piston noise, it has a negative effect.

~Full-skirt pistons. On cast and pressed full-skirt pistons, the piston crown, ring zone, and skirt form one sturdy, uniform element. The outer design of the piston skirt can be laid out as a full-skirt version (round skirt without spring backs), as window pistons for versions with short piston pin, or as pure slipper skirt pistons. Full-skirt pistons are used in engines of all types.

~Full-slipper skirt piston. In contrast to pistons with round skirt, full-slipper skirt pistons or window pistons are recessed in the skirt area in the pin direction. The piston is stiffer in the skirt area. A shorter piston pin can be used.

~Installation clearance (*also, see below,* ~Skirt). In order to ensure malfunction-free engine operation, the piston requires a slight clearance to its friction partner, the cylinder, in the warm and cold states. The installation clearance is the difference between the cylinder diameter and the maximum piston diameter in the area

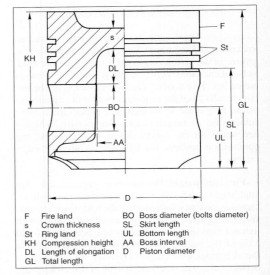

F	Fire land	BO	Boss diameter (bolts diameter)
s	Crown thickness	SL	Skirt length
St	Ring land	UL	Bottom length
KH	Compression height	AA	Boss interval
DL	Length of elongation	D	Piston diameter
GL	Total length		

Fig. **P104** Important piston dimensions

of the skirt. It includes dimensional and shape tolerances resulting from production. Common installation clearances for light alloy pistons for vehicle engines for passenger car control pistons in cast iron blocks amount to 0.03–0.05% of the nominal diameter and 0.07–0.13% for pistons for commercial vehicle diesel engines. In aluminum engine blocks, the installation clearances are lower.

~Lubrication. For the piston and cylinder materials in contact with one another with their specific structure, hardness, strength, elasticity, and heat conductivity, the type and condition of lubricant and the lubricating oil film thickness play a significant role to prevent "seizing" or "galling" of the sliding surfaces. Other influential factors are the geometrical shape, the accuracy and peak-to-valley height of the sliding surfaces, as well as external conditions of surface load, sliding velocity and temperature, and heat dissipation.

~Mass (also, see below, ~Weight). The pistons together with the piston rings, piston pin, and retainers, and a percentage of the connecting rod are the main components of the reciprocating masses. These masses have a notable effect on the bearing loads, the required counterweight, and, therefore, the overall motor weight, particularly at high speeds. Lightweight pistons, therefore, offer a technical advantage that, however, can be obtained only by reduction of the heat flow cross section, rigidity, and strength. For this reason, materials with high heat conductivity and strength combined with low specific weight are particularly suitable for pistons.

~Materials. The materials used for pistons must satisfy many different types of requirements. In addition to the requirements for high strength under alternating temperature loads, other factors which play a significant role are low density, good heat conductivity, favorable wear characteristics, and the thermal expansion properties. The inertia forces of the reciprocating engine parts can be kept low with materials which satisfy these requirements, the temperature can be influenced positively by the heat conductivity, and high resistance to deformation and high fatigue strength can be obtained by the thermal strength properties of the materials.

Aluminum/silicon alloys are primarily used for pistons in internal combustion engines. Introduced in 1926 as standard alloys with 12% Si, the eutectic aluminum silicon alloy has not been replaced by any other alloys which better satisfy all requirements to date. In addition to aluminum, this multiple material alloy contains 11–13% silicon and approximately 1% each of copper, nickel, and magnesium. The material Mahle 124, which is also used for aluminum cylinders, belongs to this alloy group used most frequently for engine construction. It offers an ideal combination of mechanical, physical, and technological properties for most applications. The alloy Mahle 142 with larger quantities of Cu and Ni was developed especially for

use at high temperatures. It is distinguished by improved and significantly higher thermal stability. A further advance is the virtually eutectic Mahle alloy 174.

By increasing the percentage of silicon, hypereutectic alloys with approximately 18% Si and approximately 24% Si were produced for reduction of the thermal expansion and wear, but at the cost of the strength. Aluminum copper alloys, once used for pistons due to their good thermal stability, are now used only rarely.

Cast iron pistons were used in the first internal combustion engines. When the inertia is not of primary significance—for instance, in large, slow-running, two-stroke engines—iron pistons are still used today. Spherolithic types of cast iron (cast iron with spherical graphite) with pearlitic base material are used primarily today. Due to the poorer heat conductivity, higher temperatures occur at the piston crown and in the ring zone on cast iron pistons. In order to assure proper function, they nearly all require specific cooling. Single-piece cooled pistons of this type are used for diesel engines operating at medium speeds with cylinder diameters between 150 and 250 mm.

The requirements for production of blanks for such complicated cast parts are high. On assembled pistons, the crowns usually consist of pressed, thermally stable types of steel—for example, standard annealed carbon steel 40 Mn 4, quenched and drawn steel 42 CrMo 4V, or valve steel X 45 CrSi9. The alloy constituents chrome and molybdenum promote formation of carbide and increase the capability for hardening all the way through. Moreover, the thermal stability is increased. Previously, the bottom sections were usually pressed using eutectic piston alloys; however, today they are being cast increasingly using nodular cast iron. On Ferrotherm pistons,[1] an advancement to the articulated skirt system, the piston head consists of high strength materials such as nodular cast iron, cast steel, or forged steel.

When the requirement profile of a material no longer satisfies the requirements, composite materials can provide a remedy. Short ceramic fibers consisting of aluminum oxide (Al_2O_3) or whiskers cast into the part locally can significantly increase the strength of the aluminum silicon materials. Whiskers consist of extremely thin single-crystal treads with a high apparent yield point and elasticity, which form materials with extremely high tensile strength when entwined with one another. The same properties can be obtained by casting in porous metal elements. As with fiber materials, squeeze molding is also required here.

In addition to the classic piston materials, attempts have continuously been made to produce pistons using other materials. However, such experiments usually remained in the test phase or were used only in specialty areas, such as racing. Worthy of mention in this context are magnesium alloys, powder metallurgical aluminum alloys, and, recently, unreinforced fine grain carbon (Mesocarbon). Frequently the price is still too high today for rapid introduction, in spite of some excellent properties.

P

863

~Mean piston speed. The mean piston velocity v_m is affected by the engine speed in rpm and the mean stroke/bore ratio s/D. In gasoline engines for motor vehicles, the mean piston speed is as high as 17 m/s, on rating engines up to 22 m/s.

~Monothermic piston. *See above*, ~Design

~Noise. The parameter is dictated by the piston design. Geometry, inertia, operating clearance, and skirt elasticity influence the course of the piston secondary motion and, therefore, the piston noise. At the end of the piston secondary motion, the piston slaps against the cylinder wall on the pressure side or the counterpressure side of the cylinder, transferring a pulse to the cylinder. However, noise can also be caused by the motion of the piston in the longitudinal engine direction — resulting, for instance, from component asymmetries on the piston and connecting rod from production. For development of noise it is assumed that the pulse initiated by the piston leads to vibration excitation of the cylinder and creation of structure-borne noise. The structure-borne noise is transferred through the structure of the cylinders to the cylinder surface and radiated as air-borne noise, depending on the transmission function and degree of radiation.

Even with quiet pistons, piston noises can occur sporadically at individual characteristic points which frequently occur only during instationary vehicle operations. These noises are annoying subjectively, even when they are not dominant in terms of the overall noise.

~Offset. Offset is a displacement of the pin axis from the center axis of the piston to the pressure or counterpressure side to provide a favorable effect on the secondary motion of the piston occurring as a result of the clearance between the piston and cylinder in terms of the noise characteristic or thermal load. Noise offset — that is, displacement of the piston pin for the pressure side — is frequently used in gasoline engines for passenger cars. In contrast, with diesel engines, displacement of the pin is accomplished more frequently for the counterpressure side — that is, thermal offset. The magnitude of the offset depends largely on the location of the combustion recess in the piston crown.

~Oil consumption. For operational reliability, it is necessary to ensure that the pistons, piston rings, and cylinder are lubricated sufficiently. However, it is also necessary to ensure that the oil does not overflow, because otherwise some of the oil increases with HC emissions in the form of uncombusted hydrocarbons.

Oil consumption in an internal combustion engine cannot be avoided completely. The term "oil consumption" refers to the quantity of lubricating oil passing into the combustion chamber, including the exhaust manifold, per unit of time, where it is combusted partially or completely. The losses resulting from leakage and evaporation are not taken into consideration.

Maintenance of oil consumption limits ($\leq$0.5 g/kWh) depends on the design of the piston, the piston rings

and the cylinder surface. Oil consumption limits must also be maintained because they are related to HC emissions. On supercharged commercial vehicle diesel engines, cylinder polishing represents a particular problem because this results in an increase in the oil consumption over the operating time and can lead to long-term piston seizing. Cylinder polishing can be avoided and low oil consumption achieved with a larger fire land clearance ($\approx$2% of piston diameter). On pistons with low fire land clearance, which allows low fuel consumption and low emissions, it is necessary to keep the formation of oil carbon on the fire land within closed limits to prevent cylinder polishing. Stepped fire lands represent a compromise solution between high and low fire land clearance.

Numerous methods have been developed to determine the oil consumption; these can be subdivided into conventional weighing methods (operating discontinuously as well as continuously) and marking procedures with indicating substances (radioactive and nonradioactive indicating substances). In contrast to consumption measurement, in which the oil quantities before and after a certain period of engine operation are compared (basic principle for all weighing procedures), the marking procedures are based on the principal of measuring the quantity of lubricating oil contained in the exhaust.

~Oil ring groove. *See below*, ~Ring ~~Ring groove

~Output per piston surface area. *See below*, ~Specific output per (piston) surface area

~Oval pistons. Usually, pistons are round, disregarding the deviation (ovality) from the circular shape in the microrange intended in the design. With oval pistons, the smaller diameter is the same as the diameter of comparable circular-shaped pistons; however, the large diameter is approximately 1.3 times this value. This allows the displacement to be increased by approximately 40% while maintaining the same overall engine block length and the same cylinder interval. Production of the pistons does not present problems; however, machining of the cylinder barrel poses greater difficulties. As with two-stroke pistons, the piston rings are fixed and cannot rotate.

~Ovality (out-of-round). Generally, pistons — particularly control pistons — have a slightly greater diameter in the pressure/counterpressure direction than in the piston pin direction (ovality).

The oval shape of the head and skirt provides a variety of layout possibilities. Due to the oval shape of the skirt (**Fig. P105**), the thermal expansion is diverted in the direction of the piston pin axis. The oval shape can be varied to provide a uniform, sufficiently wide contact pattern. In addition to so-called simple ovality, double ovality is also possible. This is designated as positive or negative depending on whether the local piston diameter is larger or smaller than with simple ovality. The contact pattern is widened by double

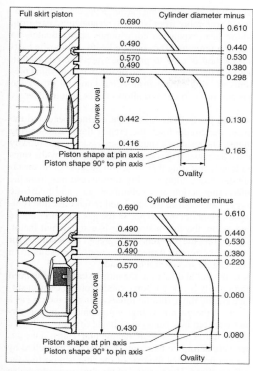

Fig. P105 Piston shape and mirror image of a full-skirt and autothermic piston

positive ovality and reduced by double negative ovality. Generally, the ovality is 0.3–0.8% of the piston diameter.

~Piston acceleration →Balancing of masses

~Piston clearance. *See above,* ~Installation clearance; *see below,* ~Thermal expansion

~Piston cooling. *See above,* ~Cooling

~Piston crown. *See above,* ~Crown

~Piston diameter. *See above,* ~Dimensions

~Piston head. The piston head is the upper section of the piston with piston crown and ring zone. It is the part of the piston subject to the highest thermal and mechanical loads. On assembled pistons, it is called the piston top section.

~Piston installation clearance. *See below,* ~Thermal expansion

~Piston materials. *See above,* ~Materials

~Piston pin →Piston pin

~Piston pin retainer →Piston pin ~ Piston pin retainer

~Piston recess. *See above,* ~Crown

~Piston ring →Piston ring

~Piston ring flutter. *See below,* ~Ring ~~Ring flutter

~Piston ring groove. *See below,* ~Ring ~~Ring groove

~Piston ring zone. *See below,* ~Ring

~Piston secondary motion. *See below,* ~Secondary motion

~Piston shape. *See below,* ~Shape

~Piston skirt. *See below,* ~Skirt

~Piston slap. *See below,* ~Secondary motion

~Piston speed. *See above,* ~Mean piston speed

~Pressed pistons (*also, see above,* ~Design). Pistons subject to high load are produced as blanks by forging or using the hot-flow pressing method. The reshaping process results in dense, high-tensile–strength piston material. The structure is more homogeneous and has a finer grain than chilled cast material. Sections of continuous cast bars are used as the initial material. Shaping is accomplished in a number of stages in hydraulic and mechanical presses. Pressed pistons are generally completely annealed. The advantages are the long-term strength, resistance to temperature change, and absence of porosity. The disadvantage is that it is not possible to produce pistons with control inserts and ring carriers. A further possibility is the use of semifinished products consisting of materials produced by means of spray compacting or powder metallurgy. Extremely heat-resistant materials which cannot be produced using melting metallurgical techniques can be produced for pistons subject to extremely high loads (racing) using this production technology.

~Pressure side. The kinematics of the cranktrain in a reciprocating engine lead to multiple changes in the contact between the piston and the cylinder wall. The skirt surface of the piston coming into contact with the cylinder wall after top dead center under the combustion pressure is called "offset." The side of the piston skirt opposite to this is called the "counterpressure side."

~Production. *See above,* ~Blank production, ~Pressed pistons

~Ring (*also,* →Piston ring). In combustion engines, the purpose of piston rings is to seal the moving part of the combustion chamber—the piston—off from the crankcase. They also support transfer of the heat from the piston to the cylinder wall and regulate the oil by distributing it on and wiping it off of the cylinder wall.

The various piston rings are subdivided into compression rings and oil control or oil scraper rings, depending on their function.

The rings have the shape of an open, annular spring to produce the required contact pressure against the cylinder wall. The ring is given the desired radial pressure distribution for its function in the cylinder after removing the section corresponding to the free gap by means of dual contour turning, simultaneous machining of the flanks of the ground black on the inside and outside on a contour lathe. This process ensures that the ring presses against the cylinder at the given radial pressure and seals so tightly against the cylinder wall that not even a light slit is present. The effective radial spring force present in the installed state is increased significantly during engine operation by the gas pressure also present behind the ring. The axial contact at the piston groove flank is produced primarily by the pressure of the gas against the ring flank.

~~Compression ring. *See below,* ~~Ring groove

~~Compression rings. These assume the sealing function and dissipate heat to the cylinder wall. They also contribute to control of the lubricating oil consumption. Rings with a rectangular or trapezoidal cross section and convex, asymmetric convex and conical running surface have proven themselves well as compression rings.

~~Oil scraper ring or oil control ring. This ring regulates and limits the oil on the cylinder wall. It scrapes excess lubricating oil off of the cylinder wall and returns it to the cylinder chamber. Oil scrape rings usually have two separate contact surfaces. Their contact pressure can be varied with spiral-type expanders. Stepped rings or stepped taper faced rings are hermaphrodites. They act as compression as well as oil scraper rings.

Piston rings are usually produced of high-quality cast iron with lamellar graphite interstratifications and nodular graphite. In addition, special materials laid out for high strength and low wear and various types of steel are used for compression and oil scraper rings as well as the spiral expanders for increasing the tangential tension. To increase the service life, the running surfaces of piston rings are galvanically chrome-plated, filled with molybdenum using the flame-spraying or plasma-spraying process, or coated over the whole surface or sprayed with metallic, metal ceramic, and mixed ceramic coatings. Service treatment is accomplished on the entire ring to improve break-in as well as to reduce wear on the flanks and contact surface. Processes such as phosphate coating, nitrogen case hardening, Ferro oxidation, copper-plating, and tin-plating as well as other processes are used.

~~Ring carrier. On diesel pistons, which are subject to significantly higher combustion pressures than are pistons in gasoline engines, the requirement for a first ring groove resistant to friction and impact wear is sat-isfied by casting in a ring carrier. Ring carriers are produced preferably from Niresist, an austenitic cast iron, whose heat expansion corresponds approximately to that of aluminum. The so-called Alfin composite casting method is used to produce a metallic connection between the ring carrier and piston material which prevents the ring carrier from being knocked off by the gas and mass forces and allows better heat transfer.

~~Ring flutter. The higher the engine speed and the lower the engine load, the greater the hazard of ring flutter. The result is a sudden increase in the gas blowby and a decrease in the output. This is caused by the rings no longer fulfilling their function because of an unstable state; they "flutter." This motion resulting from the alternating action of acceleration, gas, and hydraulic forces can occur in the axial and radial direction (collapse). Here, the difference in the gas pressures above and below the first piston ring is a significant factor. The smaller the difference, the greater the tendency for the ring system to flutter. This condition results from high engine speeds and low loads.

In the radial direction, the motion can be influenced particularly by the gas forces acting on the contact surface and by the hydraulic forces acting on the ring as a result of the oil film. Rings with conical running surface (tapered face rings) used in the first groove have a lower flutter speed than rectangular rings. The gas pressure acting on the contact surface is cancelled by the gas pressure at the rear of the ring, supporting the ceiling effect. The position of the flutter speeds is also dependent on the magnitude of the taper angle.

~~Ring groove. The piston ring zone consists generally of three ring grooves which hold the piston rings and of the two ring lands located in between. They are a part of the piston sealing system (gas sealing and control of lubricating oil consumption). It is, therefore, necessary for the rings and ring lands to have an extremely high-quality surface. Poor sealing leads to blowby of the combustion gases into the crankcase, to heating up of the surfaces exposed to the hot gas flow, and to destruction of the vital oil film on the running surfaces of the sliding and sealing partners. When pressed into the groove, down to the outer diameter of the piston, the piston ring should not hit against the bottom of the groove in the piston. It requires adequate radial clearance.

Today's lubricating oils allow groove temperatures of 200°C and higher in pistons in gasoline engines and from 200 to 280°C in pistons in motor vehicle diesel engines without residues forming in the groove or the rings sticking. In large high-speed diesel engines running at uniformly high load over longer periods, the ring groove temperature should not exceed 180°C. In order to increase the wear resistance of the second piston ring groove on diesel pistons subject to high thermal and mechanical loads, or the first piston ring groove on high-performance gasoline engines without increasing the piston weight, selective surface treatment—without also coating the adjacent area such as the

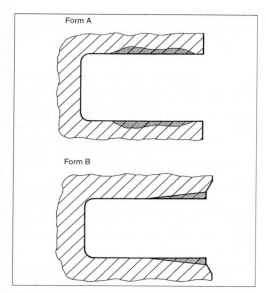

Fig. P106 Hardened groove flanks on assembled pistons

piston crown and fire land—was introduced. While coating is generally accomplished over the full surface in open baths, the selective technique applies the chemical only where it is required on the component.

On assembled pistons for operation with heavy oil, the grooves are hardened or chrome-plated on the surface to increase the resistance to wear (**Fig. P106**). Usually the inductive hardening process is used for hardening.

~~Ring land. The ring land is the section of the ring zone of a piston located between the two grooves. Above all, the first ring land—which is subject to high pressure from the combustion chamber—must be dimensioned sufficiently to prevent ring land fracture. On pistons for gasoline engines, the ring land height is 4.5–5% of the piston diameter; on passenger car diesel engines, 6%; on turbocharged passenger car diesel engines, 7–8%; and on commercial vehicle diesel engines with exhaust turbocharging and charge air cooling, 8–9% of the piston diameter. The second or remaining ring lands can have smaller dimensions due to the lower pressure to which they are subjected.

~~Ring land height. *See above*, ~~Ring land

~~Ring zone. The ring zone, usually three ring grooves with two ring lands located in between, holds the two compression rings and theoil scraper ring. The land located above the first ring groove is also called the fire land. The ring zone is subject to high thermal and mechanical loads resulting from the combustion gases and combustion pressure.

~Ring belt piston. *See above*, ~Design

~Ring groove wear. *See above*, ~Ring ~~Ring groove

~Running characteristics. (*also, see below*, ~Surface coatings). This collective term includes the significant factors decisive for the operating characteristics of the pistons: wear resistance, friction resistance, tendency to seize in the face of overload or poor lubrication, and the break-in characteristics.

~Salt core cooling duct pistons. *See above*, ~Design

~Secondary motion. This designates the motion of a piston in the cylinder of an internal combustion engine which is superimposed on the actual, reciprocating motion in the direction of the cylinder axis (primary motion) within the scope of its clearance and is called the secondary piston motion. It is composed of a translation in the connecting rod oscillation plane perpendicular to the cylinder axis and a rotation around the piston pin axis. The secondary motion can result in the piston slapping against the running surface of the cylinder, particularly in the top dead center area, and can result in the transfer of a pulse to the cylinder (piston click, piston slap).

~Shaker piston. *See above*, ~Cooling ~~Spray oil cooling

~Shape. The running surfaces of a piston in the pressure and counterpressure direction should not be too narrow, so that the specific pressure between the piston and the cylinder remains low. The skirt shape between the running surfaces and skirt without contact requires soft transitions to avoid hard contact points. The piston is reduced slightly at the top and frequently at the bottom end of the skirt as well in order to aid in forming the supporting wedge of lubricating oil (**Fig. P107**). The deviations from the nominal shape are known as shape tolerances. On pistons for passenger car and commercial vehicle engines, the tolerances are approximately ± 7 mm in the area of the skirt and 10–15 μm in the ring area. The dimensional tolerances resulting from processing are also added to these shape tolerances. These are between 16 and 22 μm, depending on the piston diameter. The reference level on the piston is designated as D_N and is located preferably in an area with stable shape or at the point of closest clearance tolerance between the piston and the cylinder.

~Single metal piston. *See above*, ~Design

~Skirt. The piston skirt, more or less the zone surrounding the lower section of the piston, ensures that the piston is guided straight in the cylinder. It can fulfill this function only when it has sufficient clearance in the cylinder. So-called piston slap—which occurs when the piston contact changes from one side of the cylinder wall to the opposite side (secondary piston motion)—is kept low by providing sufficient skirt length and closed guidance.

For diesel pistons, the so-called sliding skirt piston with its closed skirt interrupted only in the area of the

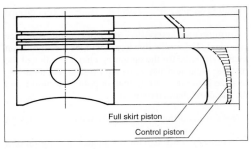

Fig. P107 Piston shape

pin bore dominates. The types of piston skirts for pistons in gasoline engines are more versatile. For reasons of weight due to the higher speeds, their skirt shape is limited only to relatively narrow skirt areas, which has led to full slipper skirt pistons, window pistons, and asymmetric pistons with different contact surface widths.

In terms of the strength and, therefore, the design, it is necessary for the skirt to fulfill a number of requirements. On the one hand, it must pick up the lateral forces without deformation or cracking, and on the other, it must adapt resiliently to deformation of the cylinder. Because the piston skirt is not cylindrical but rather a section of the piston reinforced locally by the two pin bosses, the transitions must be particularly soft. This is especially important because the piston crown deforms downwards under the load of temperature and ignition pressure, and the skirt deforms to an oval shape in the pressure and counterpressure direction, leading to an increase in the diameter in the direction of the piston pin and to a decrease in diameter in the pressure/counterpressure direction. However, permanent skirt collapse resulting from plastic deformation should be avoided. Remedial measures for endangered pistons include greater wall thicknesses, oval piston interior shapes, and small circumferential skirt lengths.

~Skirt shape. *See above,* ~Skirt

~Specific output per (piston) surface area. An illustrative factor for the total load to be withstood by a piston is the specific output per piston surface area. This consists of the engine output in kW, produced per cm^2 of projected piston crown area. It is a dimension for the quantity of energy passing through the engine and, therefore, allows a good approximation for comparing the piston load and the requirement for specific heat dissipation in various types of engines. Because this factor treats the mean pressure and speed equally — that is, it makes no difference whether a certain cylinder output is generated by high mean pressure or high speed — the comparison can be used for a wide range of engine types.

~Spray cooling. *See above,* ~Cooling

~Spray oil cooling. *See above,* ~Cooling

~Stroke. The stroke is the distance which the piston moves between bottom dead center and top dead center.

~Stroke/bore ratio. →Bore-stroke ratio

~Surface coatings. The running characteristics of pistons are affected highly by the surface treatment on the friction partners: the piston skirt and the cylinder running surface. The peak-to-valley height of the friction partners is decisive for breaking in the pistons and preventing seizing in the face of deficient lubrication. This can be accomplished by machining the surfaces of the pistons with diamond tools and honeying the cylinders.

The surface coatings used on pistons can be classified into four categories: coatings for improving the sliding characteristics, coatings for increasing the wear resistance, coatings for improving the thermal characteristics, and coatings for increasing the resistance to knocking.

Coatings for improving the sliding characteristics. Formation of a hydrodynamic lubricating film between the piston and cylinder is impaired by the reciprocating motion of the piston stopping at top and bottom dead center. This results in mixed friction states, particularly during the break-in phase. Even when the aluminum silicon alloys used for production of pistons have good emergency running characteristics, an additional running layer applied before the break-in process can be advantageous. Thin metallic as well as graphite coatings are suitable for this purpose. Lead and tin have good running properties and these can be applied using solutions of lead or tin salts without electric current by means of ionic exchange with the aluminum surface. Because both metals are nobler than aluminum in the electrochemical displacement series, they are deposited on the aluminum surface. The aluminum is dissolved until a closed lead or tin coating has formed. The resulting 1–2-μm thick metal coatings are used to a great extent for pistons in gasoline engines due to their good emergency running characteristics.

The Grafal synthetic resin graphite coating consists of fine graphite, bonded with phenol resol resin. It is approximately 10–20-μm thick and is baked at high temperature after application by spraying or screen printing. Improved adhesion characteristics can be achieved by previously applying a thin coat of metallic phosphate (bonder coat) prior to the Grafal. Grafal coatings are being used increasingly for pistons for gasoline and diesel engines in the passenger-car sector, particularly for large pistons, due to their advantageous surface characteristics for oil and good emergency running properties as well as for increasing wear resistance.

Coatings for increasing wear resistance. Pistons running in nonarmored, hypereutectic aluminum cylinders require wear-resistant surfaces because aluminum does not have good running properties when running against aluminum. For this reason, the running surface

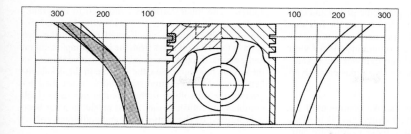

Fig. P108
Operating temperatures on pistons in passenger car engines at full load (schematic)

is coated with a wear-resistant layer of iron or chrome, which is then covered again with a thin layer of tin for break-in. This is called inversion. The thickness of the iron coating is 10–20 μm. These pistons are called Ferrostan pistons.

The piston ring grooves in the steel crowns of assembled pistons can be coated with an approximately 0.2-mm-thick layer of hard chrome to increase the wear resistance. This reduces the wear to the flanks of the ring grooves drastically, particularly when fuels containing large quantities of sulfur are used.

Coating for improvement of thermal characteristics. Hard anodized coatings are used to prevent thermal and mechanical overloads, particularly on the piston crown, as well as fissures at the edge of the recess and crown. The surface is converted to an approximately 40–80-mm-thick ceramic coating of aluminum oxide in special electrolytes. Because the coating is not applied but rather the base material on the surface is converted, no problem exists with adhesion. The favorable effect of this coating on the resistance of the piston material to fissures and cracking in the area of the piston crown results from the tension forces in the coating which are superimposed on the pressure tension forces from the thermal and mechanical load on the piston crown.

Coatings for increasing the resistance to knocking. With increasing utilization of the fuel up to the knock limit in combustion engines, it is necessary to provide antiknock coatings on the piston crowns in individual cases in order to avoid material erosion on the piston crown, on the fire land, and in the first ring groove resulting from knocking combustion. For such applications, chrome and a chemical nickel coating—a nickel coating applied without electric current with a thickness of 10–15 mm—are particularly suitable. These can be applied to conform precisely to the contour and dimensions to form a smooth surface even in the frequently bearing narrow first ring groove.

~Temperatures. The extremely rapid conversion of a portion of the energy contained in the fuel to heat leads to a considerable increase in the temperature and pressure during combustion. The heat temperatures of the gases present in the combustion chamber can result in temperatures between 1800°C and 2600°C. This temperature increase is influenced by the fuel, by the gasoline or diesel process, by the combustion process—

for example, prechamber or direct injection on diesel engines—and by the type of gas change—two-stroke or four-stroke process. The various operating procedures result in different compression and air-fuel ratios. Although heat is transferred to the surfaces forming the combustion chamber during the combustion cycle, the exhaust gas still has temperatures between 500 and 900°C, depending on the combustion process. A three-dimensional temperature field (**Fig. P108**) results on the piston, which can also be calculated using finite element programs, with the aid of marginal values.

~Terms (designations on piston). Function areas (**Fig. P109**) on pistons are the piston crown, the ring portion with fire land, the pin boss, and the skirt. Additional function elements such as control elements and ring carriers distinguish the type of piston. The piston assembly also includes the piston rings, the piston pin, and, depending on the layout, the piston pin retainer.

~Thermal expansion. Close installation clearances between pistons and cast iron cylinders require that the thermal expansion rates of the two materials be as close as possible to ensure proper clearance over the entire engine operating temperature range. The thermal expansion coefficient for cast iron is approximately 9×10^{-6} 1/K and that of pure aluminum is 24×10^{-6} 1/K. The thermal expansion coefficient of aluminum piston materials can be reduced to 17×10^{-6} 1/K with suitable alloying elements, particularly silicon. The remaining difference in relation to cast iron can frequently be compensated for on pistons for gasoline engines by casting in steel control strips (control

P

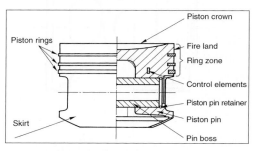

Fig. P109 Important terms on pistons

pistons). The thermal expansion of the piston and cylinders still has a slight influence, even on the full aluminum engine blocks used more and more today for reasons of weight, as well as on engine blocks with cast in, hypereutectic aluminum cylinder liners (alusil) and aluminum engine blocks with plasma-sprayed metallic surface coatings.

~Two-stroke piston. On two-stroke pistons used for speeds up to 16,000 rpm, the thermal load is particularly high due to the fact that these pistons are subjected to twice the amount of heat that pistons for four-stroke engines are subjected to. Moreover, the cooling face present in a four-stroke engine for intake of the fresh mixture is not present. Furthermore, it is responsible for covering and exposing the intake, exhaust, and transfer ducts during the reciprocating motion of the cylinder—that is, it controls the gas exchange. This results in a high thermal and mechanical load. Two-stroke pistons for small engines with diameters in the range of 30 to 60 mm are produced as cast and pressed single metal pistons. They can be equipped with one or two piston rings, and, in terms of their design, they vary from open-window-type pistons to full-skirt pistons. This depends on the design of the transfer ducts (long ducts or short handle-type ducts). The pistons are usually produced using hypereutectic aluminum silicon alloys. They are produced from eutectic and hypereutectic piston alloys. The typical shape of a two-stroke window piston is shown in **Fig. P110**.

~Variable compression ratio. Pistons with variable compression ratio allow the compression ratio to be varied for partial-load and full-load operation, which is practical for improving the efficiency of gasoline engines.

Such pistons consist of two main parts, a pot-shaped body (outer jacket and crown) and a piston-type inner carrier, which are capable of moving in relation to one another. The inner carrier is connected to the crankshaft in the usual manner by a connecting rod and piston pin. Oil chambers are located between the two parts, into which oil flows from the lubricating oil circuit in the engine through bores and spring-loaded valves. As the engine load increases, the volume in the oil chamber decreases (low compression ratio); as the load decreases, the oil volume in the chamber increases (high compression ratio). To date, mass production has been contradicted by the high weight and additional costs.

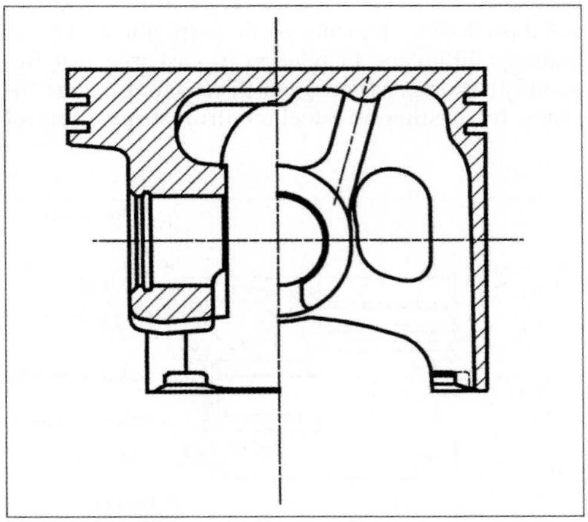

~Weight. The piston weight depends on many factors but particularly on the geometry of the piston and the primary dimensions. In addition to the piston diameter, the primary dimensions include particularly the compression height and piston boss spacing as well as the piston pin dimensions. For the geometrical layout, the type of piston and the shape of the piston crown—with recess or protrusion, with or without valve recesses—are influential. Categorization of the engine oscillating masses (100%) for a passenger car gasoline engine provides approximately the following pattern: Naked piston (50–60%), piston pin (15–20%), connecting rod (20–25%), piston rings (4–6%). Modifications of the shape in the skirt area frequently appear spectacular but contribute little to reduction of the weight. **Fig. P111** shows piston weights, G_N (without piston rings and piston bolts), for passenger car engines in relation to the piston diameter. The weight factor, X, has proven to be a good value for comparing different types of pistons. The specific weight of the piston material for common Al-Si piston alloys is approximately 2.7 g/cm^3; for cast iron, 7.2–7.3 g/cm^3; and for steel, 7.6–7.8 g/cm^3.

~Window piston. *See above*, ~Design

Piston acceleration →Balancing of masses; →Crankshaft drive ~Acceleration

Piston acceleration curve →Balancing of masses ~Piston acceleration

Piston balancing →Balancing of masses ~Balancing

Piston bolt →Piston pin

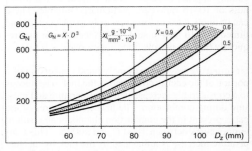

Fig. P110 Two-stroke window piston

Fig. P111 Weights of pistons for passenger cars

Piston boss →Piston pin

Piston installation clearance →Piston ~Installation clearance, ~Thermal expansion

Piston coating →Piston ~Surface coatings

Piston compressor. Piston compressors are used to compress gaseous fluids. However, they are not currently used for charging engines because of their complicated design, the increased irregular operation, their size, and, last but not least, because their cost is higher than that of a radial compressor. In addition, the optimum use of the piston compressor is in applications where very high pressures are required, and this is not the case for the charging of engines.

Piston construction →Piston

Piston contact change →Piston ~Pressure side

Piston contact pattern →Piston ~Contact pattern

Piston contact pressure →Piston ring ~Contact pressure

Piston control elements →Piston ~Control elements

Piston cooling →Piston ~Cooling

Piston crown →Piston ~Crown

Piston deformation →Piston ~Deformation

Piston design →Piston ~Design

Piston designations (designations on piston) →Piston ~Terms

Piston diameter →Piston ~Dimensions

Piston friction →Piston ~Friction

Piston head →Piston

Piston installation clearance →Piston ~Installation clearance, ~Thermal expansion

Piston knock →Piston ~Noise

Piston lubrication →Piston ~Lubrication

Piston mass →Piston ~Mass

Piston materials →Piston ~Materials

Piston noise →Piston ~Noise

Piston ovality →Piston ~Ovality

Piston pin. The piston pin represents the connection in the power flow between the piston and the connecting rod. It is subject to high loads in opposing directions due to the reciprocating motion of the piston and superimposed gas and inertia forces. Unfavorable lubrication conditions are present because of the slight rotary motion at the bearing point between the piston, the piston pin, and the connecting rod. Piston pins must have low weight, high rigidity, sufficient strength and endurance, high surface quality, geometrical accuracy, and high surface hardness.

The standard design is the tubular piston pin with "floating" bearing, which is prevented from lateral migration by the piston pin retainer. On pistons for gasoline engines, the piston pins can also be arrested in the connecting rod by means of a shrink-fit.

On large pistons, piston pins with cooling oil bores are used because the cooling oil is transferred from the connecting rod to the piston through the piston pin. When the piston pin extends all the way through the piston, solid piston pins are bolted to the connecting rod.

~Fitted pins. Fitted pins are piston pins on which the outer diameter is reduced slightly by plunge-cut grinding in the area of the inner edge of the piston center bore (**Fig. P112**). This allows softer adaptation of the piston pin and center bore under load and significantly reduces the tension forces in the center bore.

~Piston pin retainer. If the pin is not held in the connecting rod by a shrink-fit, it must be secured to prevent lateral motion out of the center bore and hitting against the cylinder wall. For this purpose, spring steel retaining rings which clamp from the inside are used almost exclusively in the grooves on the outer edge of the center bore. For small piston pin diameters, coiled rings of round wire are used primarily. The ends may be bent to the inside, forming a hook to facilitate installation in somewhat slow-running engines. Retaining rings for racing engines frequently have a hook bent toward the outside to prevent rotation. If the pins are subject to high axial thrust, retaining rings which clamp from the outside are used in grooves on the ends of the pins in individual cases.

Piston pin boss →Connecting rod ~Connection to piston

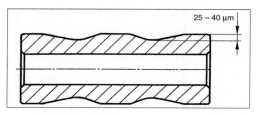

Fig. P112 Fitted pins

Piston pin bushing →Bearings ~Bearing positions

Piston pin retainer →Piston pin

Piston port control →Combustion process; →Two-stroke engine

Piston production →Piston ~Blank production

Piston pump →Carburetor ~Dashpot pump

Piston recess →Piston ~Crown

Piston ring (*also*, →Piston). Piston rings are metallic seals which seal the combustion chamber off from the crankcase, support heat transfer from the pistons to the cylinder wall, and regulate the oil supply for the engine, particularly with the oil scraper rings. It is necessary to ensure that the pistons and piston rings are lubricated without interruption at minimum oil consumption.

For this purpose, it is necessary for the outer circumference of the piston rings to make tight contact with the cylinder wall and the flank of the ring to make tight contact against the flank of the piston groove (**Fig. P113**). Radial contact is ensured by the intrinsic spring force, although additional springs may be used with oil scraper rings. The gas pressure in the combustion chamber reinforces the radial as well as axial contact, whereby the axial contact can alternate between the upper and lower piston groove flank due to the effect of inertia and friction forces.

Literature: Kolbenringhandbuch (Piston Ring Manual), Federal Mogul GmbH, New edition, 2003.

~**Abrasive wear.** *See below*, ~Wear

~**Adhesive wear.** *See below*, ~Wear

~**Axial play.** *See below*, ~Versions ~~Compression rings ~~~Keystone rings

~**Breaking.** Broken piston rings can no longer fulfill their sealing function; the malfunctions that results are

significant. Blowby combustion gases heat up the piston, leading to seizing.

Ring breakage is caused by:

* Improperly dimensioned gap clearance (too low or too high)
* Axial piston ring groove clearance too high
* Piston ring groove flanks not flat
* Worn grooves
* Axial ring wear
* Extreme ring flutter
* Extremely high increase in combustion pressure dp/da
* Local ring sticking
* On two-stroke engines, compression of the piston rings in the scavenging ports of the cylinder, resulting from improper radial pressure characteristics of the piston rings
* Incorrect ring installation (pull-over flexural stress too high)

Broken oil scraper rings lead to extremely high oil consumption. Oil carbonization leads to the rings sticking, resulting in ring breakage.

~**Bushing casting material.** *See below*, ~Production process

~**Cast iron with scaled graphite, alloyed, annealed.** *See below*, ~Materials

~**Cast iron with scaled graphite, not annealed.** *See below*, ~Materials

~**Cast iron with spheroidal graphite (spheroidal cast iron), alloyed, annealed.** *See below*, ~Materials

~**Characteristics.** *See below*, ~Designations

~**Chrome-ceramic coating.** *See below*, ~Contact surface armor

~**Compression ring.** *See below*, ~Versions ~~Compression rings

~**Contact pressure.** The contact pressure is the decisive piston ring variable for its sealing function. It is the pressure with which the piston ring presses against the cylinder wall. The pressure can be constant around the entire circumference or it can have intentionally variable, radial pressure characteristics (**Fig. P121**).

The contact pressure value is determined by the dimensions of the ring and the modulus of elasticity of the ring material. At constant pressure distribution, the tangential force F_t can be calculated according to the equation:

$$p = \frac{2 \cdot F_t}{d \cdot h} \ [\text{N/mm}^2] ,$$

where p = contact pressure, d = nominal diameter, and h = ring height

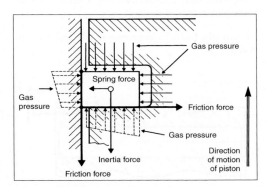

Fig. P113 Forces on piston ring

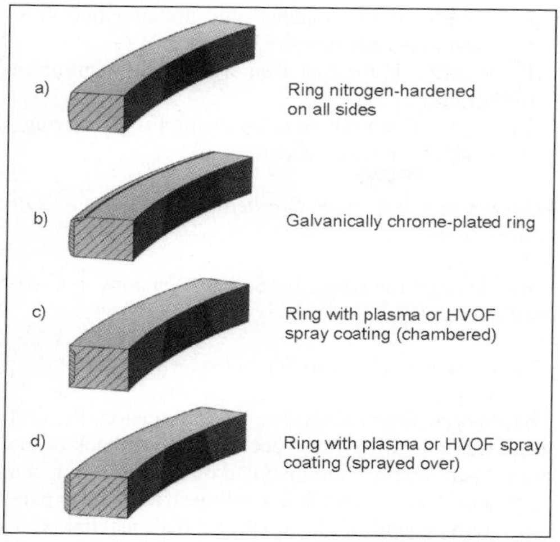

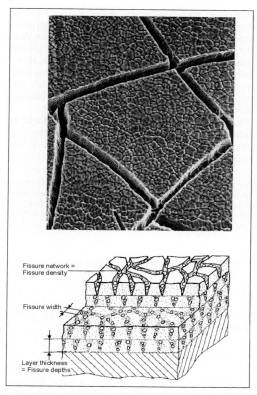

a) Ring nitrogen-hardened on all sides

b) Galvanically chrome-plated ring

c) Ring with plasma or HVOF spray coating (chambered)

d) Ring with plasma or HVOF spray coating (sprayed over)

Fig. P114 Contact surface armor

~Contact surface armor. The ring running surfaces are armored with wear-reducing protective coatings to increase the service life of the piston rings and cylinders (**Fig. P114**).

~~**Chrome plating.** Hard chrome coatings applied to the running surfaces galvanically are distinguished by their extremely high resistance to abrasive as well as corrosive wear. Moreover, they are less sensitive to scorch marks than are untreated running surfaces and running surfaces with treated surface. According to experience, a chrome-plated ring in the first piston ring groove reduces the wear to the ring assembly in question by approximately 30% in comparison to rings without chrome plating. The wear to the cylinder running surface is reduced by 50%.

For this reason, engines with high service life expectancies—commercial vehicles and marine and stationary diesel engines—are equipped with more than one chrome-plated piston ring.

Literature: W.H. Charlsworth, W.L. Brown: Wear of Chromium Piston Rings in Modern Automotive Engines, SAE Paper 670 042 (1967).

~~**Chromium ceramic coating (CKS®).** The good wear characteristics of hard-chrome finishes have been improved with chromium ceramic coatings.

By inclusion of ceramic particles (Al_2O_3) in the galvanically deposited chrome layer, the wear resistance is improved over the entire service life of the finish as is its resistance to thermal stress—that is, resistance to forming scorch marks. The galvanically deposited chrome (GDC) coating was developed for extremely high engine loads.

Fig. P115 shows the schematic structure of the chromium ceramic layers. It shows, among other items, that the finish is produced in a multiply repeated process to

Fissure network = Fissure density

Fissure width

Layer thickness = Fissure depths

Fig. P115 Schematic illustration of CKS chrome plating

ensure distribution of the ceramic particles over the entire thickness of the layer.

Chromium ceramic finishes are used with great success in diesel automotive engines and stationary engines, including heavy oil engines. Piston rings with chrome, CKS, and GDC contact surface armor finishes can be produced by chrome plating as well as chambered on one side.

Literature: U. Buran: Chrom-Keramik-Kombinationsschichten für Kolbenringe, Special Edition, MTZ/ATZ "Werkstoffe im Automobilbau 1996."

~~**Molybdenum coating.** In spite of its high resistance to scorch marks, thermally sprayed molybdenum coatings are only used in individual cases due to their comparatively low resistance to wear. Coatings with a molybdenum-based and self-flowing alloys such as MP 43 are used for practical purposes only in gasoline engines.

~~**Plasma spray coating.** The technologies for plasma spray coating and high-velocity oxy-fuel (HVOF) spraying make it possible to produce metallic and particularly metallic-mixed coatings whose basic materials have a particularly high melting point. The wear

P

protection coatings achieved in this manner have an even higher resistance to wear than molybdenum coatings and a greater resistance to scorch marks than chrome finishes. Due to its high melting capacity, plasma spraying is a very economical universal process. However, it is less suitable for production of advantageous hard metallic structures. The HVOF process was developed for this purpose. In the Goetze ring coating line, such coatings have the designation MKJet.

Plasma or HVOF spray coatings can be produced in chambered form (**Fig. P114c**), which allows sharp ring running edges, and can also be sprayed over the entire surface (**Fig. P114d**).

Literature: U. Buran, Chr. Mader, M. Morsbach: Plasmaspritzschichten für Kolbenringe, Stand und Einsatzmöglichkeiten. Specialist Publication K 35, Goetze AG, 1983. — [9] C. Herbst-Derichs, F. Münchow: Modern Piston Ring Coatings and Liner Technology for EGR Applications, SAE Paper 2002-01-0489.

~Convexity. *See below,* ~Running surface shapes

~Corrosive wear. *See below,* ~Wear

~Designations. The designations relevant for piston rings are indicated in **Fig. P116.**

~Designs. *See below,* ~Gap

~Diametrical force. The diametrical force F_d is a measure of the spring force of a piston ring when the ring is compressed diametrically to the nominal diameter in a plane located 90° to the ring gap. This method is preferred for mechanical ring testing. The following empirically determined relationships exist for various ring materials in terms of the diametrical force and tangential force, F_t:

- $F_d = 2.05 \cdot F_t$ for annealed and nonannealed gray cast iron materials (empirical)
- $F_d = 2.15 \cdot F_t$ for cast iron with nodular graphite (empirical)
- $F_d = 2.21 \cdot F_t$ simplified relationship for piston rings in usual layout range (theoretical)

~Double beveled rings. *See below,* ~Versions ~~Oil scraper rings

~Double keystone rings. *See below,* ~Versions ~~Compression rings

~Dual contour turning. *See below,* ~Shaping

~Expansion flexural strain. The expansion flexural strain is the bending stress occurring at the back of the rings when the rings are installed on the pistons; it can be greater than the installed bending stress. The expansion strain depends highly on the type of installation—tangential expansion or expansion sleeve.

On piston rings produced according to DIN/ISO standards, this strain is taken into consideration in designing the rings by using mathematical equations developed especially for this purpose; careful design of this parameter is critical for the installation.

~Flank wear. *See below,* ~Wear

~Flutter. Flutter is an unstable state of the piston ring which can occur through interaction of inertia, gas, and hydraulic forces at high engine speeds. Low engine loads increase the tendency of the rings to flutter because the difference in the gas pressures above and below the first compression ring is relatively low. The lower this difference, the more sensitive the ring system is to flutter. Piston ring flutter leads to sudden high gas blowby and power loss. The rings flutter in the axial direction between the flanks of the piston ring grooves and can even collapse in the radial direction.

This interference is supported by the gas forces acting on the running surface, reducing the forces pressing behind the ring. Compression rings with a conical running surface, therefore, have lower flutter limit speeds than rectangular rings.

Piston ring flutter can be influenced positively by reducing the inertia forces (axially low rings) and suitable selection of the radial pressure distribution.

Literature: P. Dykes et al.: Piston Ring Movement during Blow-by in High Speed Petrol Engines Transact, Institute of Mechanical Engineers Vol. 2, 1997. — H. Steinbrenner: Messung zur Erfassung des Ringflatterns in schnelllaufenden Kolbenmaschinen, MTZ 7/22 (1961). — G. Wachtmeister, K. Zeilinger: Einfluss der Druckanstiegsgeschwindigkeit auf die Bauteilebelastung, FVV Research Report 413, 1988.

~Form-matching capability. This is the property of a piston ring to adapt to even out-of-round cylinders.

A high form-matching capability—that is, great flexibility—ensures proper sealing against gas and oil

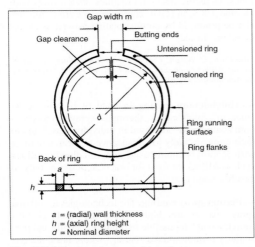

Fig. P116 Designations on piston rings

and, therefore, high performance and economical oil consumption for engines.

The form-filling capacity at which contact between the ring and cylinder wall is just achieved with a radial pressure of p = 0 can be calculated with U_i as the radial deformation of the cylinder to the i order:

$$U_i = \frac{k \cdot r}{(i^2 - 1)^2} = \frac{1}{8} \frac{(d - a)^3}{(i^2 - 1)^2} \cdot \frac{F_t}{E \cdot I},$$

where a = ring wall thickness, k = piston ring parameter, d = nominal diameter, E = modulus of elasticity of system ring material, I = planner moment of inertia of ring cross section $h \cdot a^3 / 12$, F_t = tangential force, and h = ring height.

Because the form-filling capacity increases to a power of four at increasing ordinal number, i, the cylinder distortion factors of a higher order are particularly critical for the function of the piston rings.

The gas pressure and spring elements behind the ring improve the form-filling capacity to obtain

$$U_{iz} = U_i \left(1 + \frac{p_z}{p}\right)$$

and

$$U_{ij} = U_i \left(1 + \frac{p_f}{p}\right),$$

where U_{iz} = radial deformation of cylinder to the i order in consideration of the gas pressure, p_z; U_{ij} = radial deformation of cylinder to the i order in consideration of the contact pressure of the spring-tensioned piston ring; p_z = gas pressure behind ring; p = contact pressure of piston ring, and p_f = contact pressure of spring-tensioned piston ring.

Literature: B.I. Gintsburg: Splittles-Type Piston Rings Russian Engineering Journal, Volume XLVI, No. 7. — A. Mierbach: Berechnung der Radialdruckverteilung in verzogenen Zylinderbuchsen, Specialist Publication K 15 der Goetze AG, 1973.

~Friction. Piston ring friction results in a loss in power similar to that caused by the valves and bearings and increases the fuel consumption.

The piston assembly contributes up to 40% of the total friction on the engine, of which the piston rings are responsible for approximately half. Worldwide studies have not clearly shown consistent results, probably because of variations in the test conditions, the definitions of objectives, as well as varying emphasis on piston ring friction.

Piston ring friction is determined by

- The surface pressure (spring force and gas pressure at the bottom of the groove),
- Ring height (width of running surface),
- Running surface shape (convexity), and
- Friction coefficient of ring running surface (contact surface coating).

The proportion of power loss from the ring assembly is approximately 20% of the total friction loss in the engine.

Surface pressure is the primary cause for ring friction; the oil scraper rings alone produce nearly 60% of the total, while 40% is attributable to the number of compression rings. The shape of the running surface influences the lubrication of the rings in the new state because of formation of an oil film. It is, therefore, influential only during the break-in phase. After breaking in, rings have an intrinsic convexity for the specific engine, resulting in a friction state which no longer can be influenced.

The friction coefficient of the ring running surface has a significant effect in the range of mixed friction (that is, starting), when the emergency running characteristics of the friction partners (rings/cylinders) are required. Because of the low piston velocities in this range, this is of only minor significance in terms of friction loss.

Measures taken to reduce the piston ring friction should not interfere with the functional characteristics of the piston rings. It is necessary to maintain the sealing effect of the ring assembly both for combustion gases and lubricating oils.

Literature: R. Jakobs: Zur Reibleistung der Kolbenringe bei Personenwagen-Ottomotoren, MTZ 49 (1988) 7/8 and Specialist Publications K 34, K 39, K 40 der Goetze AG, 1988. — S. Pischinger: Verbrennungsmotoren, Volumes I and II, Lecture at RWTH Aachen University, 2001. — R. Lechtape, G. Knoll: Kolbenringreibung I, FVV Research Report vols. 570-1 and 570-2, 1994.

~~Friction force. The piston ring friction force in the axial direction consists of the spring force of the ring, the gas force, and the force produced by the effective friction coefficients acting against the motion of the piston. It reduces the tendency of the ring to flutter. The piston ring friction force in the radial direction is determined by the gas pressure and the friction coefficient effective on the ring flank. It opposes the radial piston motion, therefore promoting possible seizing of the ring in the piston ring groove.

~Friction force. See above, ~Friction

~Gap. The ring gap is the gap formed between the ends of the ring, which is required for compensation of the thermal expansion of the piston ring. Wide gaps result in increased gas losses (blowby); narrow gaps can lead to the ends of the ring pressing together as the result of ring expansion, leading to ring breakage.

~~Clearance. The gap clearance is the size of the remaining gap between the ends of the rings in the installed state. It allows for thermal expansion of the ring and is dimensioned according to standards for a temperature difference between ring and cylinder of at least 100°C for compression rings and 80°C for oil scraper rings.

~~Designs. Generally, ring gaps are designed to be straight (**Fig. P117a**). Angular gaps (**Fig. P117b**) and overlapping gaps (**Fig. P117c**) are used only in hydraulic

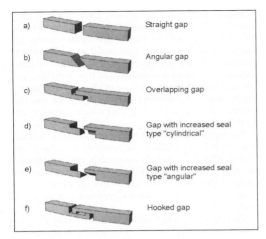

Fig. P117 Ring gap designs

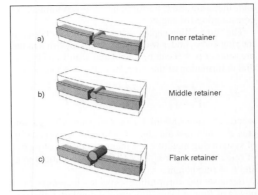

Fig. P118 Ring gap retainers

systems and in compressors. They do not provide any advantages in terms of sealing.

Single gaps, as shown in **Fig. P117d and e**, improve sealing of the ring in comparison to a straight gap. These gap designs are recommended for two-ring piston designs and have been tested with varying success. Because of the high production costs, no applications are known in mass production.

Single gaps are still used to varying extents in medium and low-speed engines. Hooked gaps (**Fig. P117f**) serve only to facilitate installation of rectangular rings in automatic vehicle drivetrains to seal hydraulic pressures.

~~**Retainers.** Piston ring retainers (**Fig. P118**) prevent the rings from turning in two-stroke engines. This prevents the ends of the ring from springing into the scavenging ports in the cylinder and breaking. The retaining effect is provided by a pin located in a permanent position in the piston ring groove and extending into the additional recesses at the ends of the ring. On inner and middle retainers (**Fig. P118a and b**), the pin is located at the bottom of the piston ring groove; on flank retainers (**Fig. P118c**), they are located in the bottom and top groove flank.

~**Gap width.** The gap width is the opening in the ring in the untensioned state, measured in the neutral chamfer. It is occasionally used as the value for the spring force, therefore serving as the contact force for piston rings in slow-running diesel engines.

~**Inertia force.** This is the force resulting from the mass of the piston ring and the piston acceleration. In interaction with the gas and friction forces, it affects the position of the piston ring in the ring groove (**Fig. P113**). High inertia forces resulting from high engine speeds and/or large piston ring masses (ring dimensions) can lead to ring flutter and promote axial wear.

~**Inside bevel rings.** *See below,* ~Versions ~~Compression rings

~**Installation.** In mass production applications, piston rings are automatically installed by machines. Here, the rings are widened to the dimensions at which they will slip over the piston by means of so-called expander sleeves or spreading claws. Here, the resulting expansion flexural strain (and stress) is taken into consideration for rings dimensioned according to standard.

~**Installation bending stress.** The installation bending stress is the flexural strain on the piston ring in the installed state in the cylinder. The maximum tension is present at the rear of the ring and can be calculated as follows:

Rectangular rings: $\sigma_b = \dfrac{a \cdot E}{d - a} \cdot 2k$ [N/mm^2],

Slotted oil control rings: $\sigma_b = \dfrac{x_1 \cdot E}{d - a} \cdot 2k \cdot \dfrac{I_m}{I_s}$ [N/mm^2],

where a = ring wall thickness, d = nominal diameter, E = ring material modulus of elasticity, k = piston ring parameter, $I_m = I_u + I_s / 2$, I_u = planner moment of inertia of unslotted cross section, I_s = planner moment of inertia of slotted oil scraper ring, and x_1 = double distance from center of gravity to outer diameter.

~**k factor.** *See below,* ~Parameters

~**Keystone ring.** *See below,* ~Versions ~~Compression rings

~**L-shaped compression rings.** *See below,* ~Versions ~~Compression rings

~**Lubrication.** Proper lubrication of the ring and the cylinder running surfaces is essential to reduce operational malfunctioning of the piston rings.

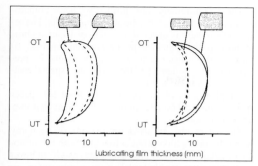

Fig. P119 Effect of convex running surface and ring height on lubricating film thickness

The following factors determine the quality of lubrication:

- Quality of oil used
- Aging of oil used
- Running surface shape of ring (convexity)
- Topography of ring running surface
- Topography of cylinder running surface
- Cylinder out-of-roundness

~~Contact surface design of piston ring. The shape of the running surface of the piston ring influences the thickness and distribution of the oil film over the length of the piston stroke. Convex rings provide a thicker oil film than do rectangular rings (**Fig. P119**).

The ring height also has a significant influence, as shown in **Fig. P119**. Rings with greater axial height create a thicker lubricating film than rings with low axial height. However, in practical applications, it is also necessary to consider the increased friction loss, the greater tendency to flutter, and, under certain circumstances, higher oil consumption.

~~Topography of cylinder running surface. As with "special lapping" of the piston ring running surfaces, the cylinder running surfaces can be designed to prevent galling by special honing, referred to as plateau honing.

Literature: G. Dück: Zur Laufflächengestaltung von Zylindern und Zylinderlaufbuchens. Specialist Publication K 9, Goetze AG, 1973. — Honatlas Goetze AG, Beurteilungskriterien für die Honung von Zylinderlaufflächen (unpublished).

~~Topography of ring running surface. Attempts to minimize oil consumption in engines lead to the tribological limits of the lubricating conditions. These limits can be extended by special design of the piston ring surface to prevent the friction partners from galling.

"Special lapping" of chrome-plated ring running surfaces (**Fig. P120**) prevents exceptional roughness resulting from plateaus and valleys and forms sealing surfaces with good bearing properties containing

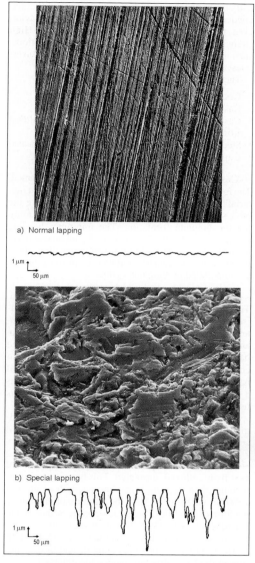

a) Normal lapping

b) Special lapping

Fig. P120 Topography of chrome-plated piston ring running surfaces

micro-sized oil reservoirs, which allows the tribological limit conditions to be extended.

Literature: H.W. Plankert, F. Stecher: Oberflächengestaltung von Kolbenringlaufflächen—ein Ergebnis tribologischer Untersuchungen, Specialist Publication K 24, Goetze AG, 1979.

~Marine and stationary diesel engines →Piston ring
~Ring set

~Materials. The piston ring materials used today are characterized by the requirements for good standard and emergency running characteristics, good thermal conduction properties, and good resilient behavior, even under the effect of high temperatures. High strength is required when extremely severe conditions, such as high speeds or high combustion pressure rise rates, $dp/d\alpha$, are present on the engine side.

The following materials are used:

- Cast iron with scaled graphite, not annealed
- Cast iron with scaled graphite, annealed
- Cast iron with spheroidal graphite (spheroidal cast iron), annealed
- Steel containing high qualities of chromium, annealed
- Carbon steel (spring steel)

~~Cast iron with scaled graphite, alloyed, annealed. The low structural strength values of the "standard material" can be increased markedly by annealing the basic structure (martensitic) and, simultaneously, refining the graphite formation. The minimum bending strength is 650 MPa. Because of the increased hardness of 35–49 HRC, resulting from annealing, these materials are particularly suitable for piston rings for chrome-plated or hardened cylinders.

~~Cast iron with scaled graphite, not annealed. This material, designated as a "standard material," satisfies all normal conditions placed on piston ring materials. The graphite formation ensures good break-in and emergency running properties; the unannealed basic structure (pearlite) provides satisfactory wear characteristics when it is free of ferrite. The material hardness is 200–290 HB, and the bending strength has a relatively low minimum value of 350 MPa, as is usual for gray cast iron.

The "standard material" is only used in modern engines for rings in the second ring groove and for oil scraper rings.

~~Cast iron with spheroidal graphite (spheroidal cast iron), alloyed, annealed. This type of cast iron is distinguished by a high bending strength of at least 1300 N/mm, resulting from the presence of graphite in the form of spherules. The basic structure is annealed to a hardness of 310–430 HB (martensitic). For some applications, it is annealed to a hardness of 390–470 HB.

The spherical shape of the graphite and its distribution in the basic structure reduce the emergency running properties of spheroidal cast iron in comparison to those of cast iron with scaled graphite. For this reason, it is necessary to provide piston rings produced of spheroidal cast iron with a wear-protection coating on the running surface. Spheroidal cast iron is preferred for rings in the first piston ring groove because of its high flexural strength.

~~Steel. Steel is used for piston rings wherever high resistance to breakage is required—for instance, rings with extremely low axial dimensions ($h < 1.2$ mm) for high speeds in gasoline engines and diesel engines with high pressure increase rates $dp/d\alpha$; it is also used for steel lamellas and spacer springs for oil scraper rings.

These requirements are fulfilled by steels containing high quantities of chrome and spring steels. The resistance to wear is ensured by finely distributed chromium carbide contained in the annealed matrix. To improve the resistance to wear, these steels are preferred in nitrogen-hardened forms or with a special coating on the running surface.

Simpler types of carbon steel are used for spacer springs and steel lamella in oil scraper rings, but they are also annealed.

Steel may also be used for reasons of economy when the production process together with the armor coating required in any case (nitriding) results in lower costs than rings of spheroidal cast iron.

~~Synthetic materials. The synthetic materials tested to date, such as Teflon and PBI plastic, have not proven suitable for use in piston rings in internal combustion engines. The reasons for this are, in particular, the high temperatures in the piston ring zone, as well as combustion residues which cause premature wear and burning of the synthetic material.

In compressors, piston rings of synthetic material have been used successfully in special cases.

The general development in the area of synthetic materials does not exclude the possibility that plastics could be developed for piston rings in internal combustion engines that would satisfy the material requirements of the ring, but currently such materials are not available.

~Molybdenum coating. *See above,* ~Contact surface armor

~Nitriding and nitrocarburizing. *See below,* ~Surface treatments

~Number of rings. The number of rings per piston has an effect on the friction losses in the engine. Their weight (mass) is a portion of the reciprocating inertia forces, particularly at high speeds, and is decisive for the bearing load and the size of the required counterweights—that is, the piston rings are an important decisive factor in the overall weight of the engine.

These facts explain the trend to use as few rings as possible per piston. Generally, three rings are used per piston—two compression rings and one oil scraper ring. Two rings per piston are also used in mass production to reduce the friction losses. In such cases, it is necessary to consider the risk that the sealing effect of the rings would be lost completely if one ring were to break. Moreover, a portion of the reduction in the friction losses is counteracted by the fact that with only two rings, it is necessary to increase the spring force.

Medium- and slow-running marine and stationary engines were equipped with four and more rings in the past. In the future, there is likely to be a high tendency to use three rings.

Here, weight and construction space as well as piston ring friction do not play such a significant role.

~Oil scraper ring. *See below*, ~Versions

~Oil scraper ring with expander. *See below*, ~Versions ~~Oil scraper rings

~Oil scraper steel strap ring. *See below*, ~Versions ~~Oil scraper rings

~Operating characteristics. The operating characteristics of the piston rings—that is, proper sealing against the combustion gas and lubricating oil—depend on the design of the engine, the thermal and combustion dynamic load, and the cylinder design and machining, which affect ring and cylinder wear as well as the formation of scorch marks, just as does the lubricating oil used. The quality of the rings themselves is significantly related to their operating characteristics.

~Out-of-roundness (*also*, →DIN/ISO 6621 Testing quality characteristics). Out-of-roundness, or ovality, is the difference in the diameters (measured across ring gap minus diameter measured at 90° to ring gap plane) of a ring pressed together to eliminate the gap clearance in a flexible tensioning strap. It is used as the value for the radial pressure characteristics and can be positive (four-stroke out-of-round), negative (two-stroke out-of-round), or 0.

~Parameters. The parameter, k, is a value that characterizes the resilient properties of the piston ring. For rectangular rings, it is defined as follows:

$$k = 3 \cdot \frac{(d-a)^3}{h \cdot a^3} \cdot \frac{F_t}{E}$$

using the tangential force, F_t, or,

$$k = \frac{2}{3 \cdot \pi} \cdot \frac{m}{d-a}$$

using the gap width.

~Passenger car diesel engines. *See below*, ~Ring set

~Phosphate coating. *See below*, ~Surface treatments

~Piston ring groove. →Piston

~Plasma spray coatings. *See above*, ~Contact surface armor

~Pliers. Special pliers are required for manual installation of the rings to prevent the ring from being subjected to excessive stress when pulled over the piston. Expansion of the ring is limited to a minimum dimension, and the force required for this purpose is supported by additional flexural torque introduced at the ends of the ring.

~Production process. The performance and service life of modern internal combustion engines can only be ensured with components which satisfy the highest quality requirements.

For piston rings, the variables decisive for satisfying these requirements are the material and shaping as well as the coatings of the ring running surfaces and flanks.

~~**Bushing casting material.** This procedure for producing cast bushings using gravity casting or centrifugal casting is seldom used for piston rings with nominal diameters up to 200 mm. It is preferred for larger rings—for instance, for marine diesel and stationary engines.

The piston rings are sliced off individually from the bushings and further processed as round flanks using the "thermal process" or double contour turning as with individual castings.

~~**Single casting material.** Piston rings of cast iron are generally produced using the single casting process—that is, each ring is shaped and cast individually on mold plates according to a certain mathematical model.

Generally, this process is accomplished according to the required material specifications so that the basic structure is already achieved during the casting process. Subsequent thermal treatment (annealing) is seldom used, due to the changes possible in the shape of the blanks.

~~**Steel material.** Cold drawn profile steel with various specifications is preferred for piston rings. Here, in addition to simple profiles for compression rings, special profiles with holes for oil scraper rings are also used.

~Protective coatings. *See above*, ~Contact surface armor; *see below*, ~Surface treatments

~PVD. *See below*, ~Surface treatments

~Quality. The quality characteristics are defined in ISO 6621 "Testing quality characteristics."

~Radial pressure characteristics. The radial pressure characteristics of a piston ring are defined by the shape of the distribution of the contact pressure of the ring around its circumference. **Fig. P121** shows three typical shapes as they are generally selected, depending on the application.

The pressure characteristics in piston rings for four-stroke engines (**Fig. P121a**) show that the radial pressure is increased at the ends of the ring, acting primarily to "dampen" piston ring flutter, which generally starts in this location. However, there has only been evidence of this effect at higher engine speeds (above 6500 rpm) to date. Rings with these characteristics have a tendency to wear more at the gap and have a lower form-matching capability than do pistons with constant pressure distribution.

For this reason, rings with constant radial pressure characteristics are preferred for high-speed diesel engines that do not reach extremely high speeds, as shown in **Fig. P121b**.

On the piston rings for two-stroke engines (**Fig. P121c**) the radial pressure is reduced greatly at the

P

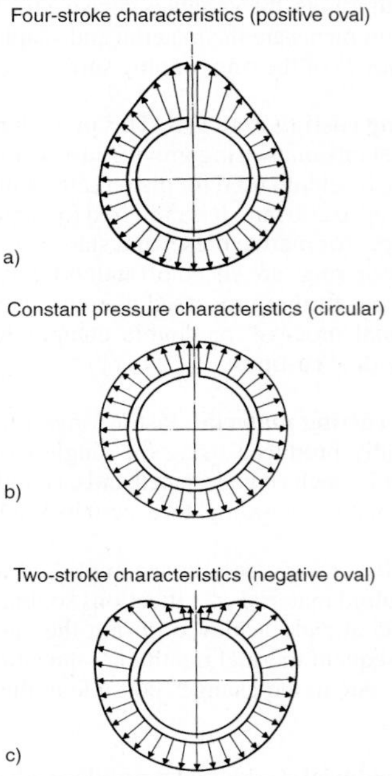

a)

Four-stroke characteristics (positive oval)

b)

Constant pressure characteristics (circular)

c)

Two-stroke characteristics (negative oval)

Fig. P121 Radial pressure characteristics of piston rings

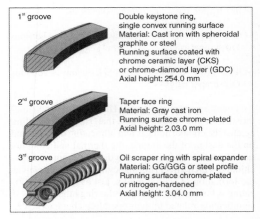

1ˢᵗ groove	Double keystone ring, single convex running surface Material: Cast iron with spheroidal graphite or steel Running surface coated with chrome ceramic layer (CKS) or chrome-diamond layer (GDC) Axial height: 254.0 mm
2ⁿᵈ groove	Taper face ring Material: Gray cast iron Running surface chrome-plated Axial height: 2.03.0 mm
3ʳᵈ groove	Oil scraper ring with spiral expander Material: GG/GGG or steel profile Running surface chrome-plated or nitrogen-hardened Axial height: 3.04.0 mm

Fig. P122 Piston ring set for commercial vehicle diesel engines

ends of the ring. This pressure reduction can decrease to zero or even assume negative values to prevent the ends from extending into the scavenging ports on these engines; it also can reduce the wear near the gap on four-stroke engines.

The characteristics illustrated in **Fig. P121** are achieved by contour turning (copy turning) and represent ideal shapes that can be affected by the surface treatments and variations in the ring material.

Literature: H. Arnold, F. Florin: Zur Berechnung selbstspannender Kolbenringe von konstanter Stärke, in Konstruktion 1 (1949) vol. 9.

~Rectangular ring. *See below*, ~Versions ~~Compression rings

~Retainers. *See above*, ~Gap

~Ring breakage. *See above*, ~Breaking

~Ring flutter. *See above*, ~Flutter

~Ring repair kits. Ring repair kits are specially developed ring systems for overhauling used engines. These rings reduce the increased oil consumption that results from ring and cylinder wear by using an increased oil scraping effect.

The various ring systems available on the market are all equipped with highly flexible oil scraper rings, slotted oil control rings with spiral-type expanders (**Fig. P131d**), or steel strap rings (**Fig. P131f**). Slotted oil

control rings with spiral-type expanders are preferred in repair kits for diesel engines, while steel strap rings are used in gasoline engines.

~Ring set. The ring set is the entire ring assembly installed on the piston. The ring set is determined by the design of the engine (lightweight, stiffness), by the operating economy (fuel and oil consumption), by the regulations on pollutant emissions, and by the production costs for the rings themselves. Generally, it represents a compromise between these conditions.

~~**Commercial vehicle diesel engines.** The high service life expectancy of these engines of more than one million kilometers can only be fulfilled by ring sets with high wear resistant properties. **Fig. P122** shows a typical ring set for long service life.

~~**Gasoline engines.** The piston ring set show in **Fig. P123** provides an example of the trend for gasoline engines.

~~**Marine and stationary diesel engines.** Although the pistons in these engines were equipped with four and more piston rings in the past, the trend today is to use three rings. To ensure extremely long service lives for these medium and slow-running engines, piston rings are used which retain their full function for up to 20,000 hours of operation in spite of a lower number per piston.

Today, chrome ceramic coated rings with an additional chrome finish on the flanks for use with heavy oil are standard for use in the first groove.

Fig. P124 shows an example of a ring set for a marine diesel engine with a 430-mm cylinder bore: (1) Groove 10 mm high with chrome-ceramic-coated rectangular ring; running surface, asymmetric convex; flanks chrome plated. Material: lamellated gray cast iron. (2) Groove 10 mm high with chrome-ceramic-coated rectangular ring; running surface asymmetric

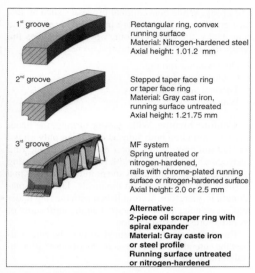

1ˢᵗ groove	Rectangular ring, convex running surface Material: Nitrogen-hardened steel Axial height: 1.01.2 mm
2ⁿᵈ groove	Stepped taper face ring or taper face ring Material: Gray cast iron, running surface untreated Axial height: 1.21.75 mm
3ʳᵈ groove	MF system Spring untreated or nitrogen-hardened, rails with chrome-plated running surface or nitrogen-hardened surface Axial height: 2.0 or 2.5 mm
	Alternative: **2-piece oil scraper ring with spiral expander** **Material: Gray caste iron or steel profile** **Running surface untreated or nitrogen-hardened**

Fig. P123 Piston ring set for gasoline engines

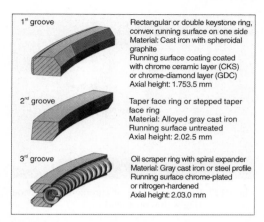

1ˢᵗ groove	Rectangular or double keystone ring, convex running surface on one side Material: Cast iron with spheroidal graphite Running surface coating coated with chrome ceramic layer (CKS) or chrome-diamond layer (GDC) Axial height: 1.753.5 mm
2ⁿᵈ groove	Taper face ring or stepped taper face ring Material: Alloyed gray cast iron Running surface untreated Axial height: 2.02.5 mm
3ʳᵈ groove	Oil scraper ring with spiral expander Material: Gray cast iron or steel profile Running surface chrome-plated or nitrogen-hardened Axial height: 2.03.0 mm

Fig. P125 Piston ring set for passenger car diesel engines

convex. Material: standard cast iron. (3) Groove 10 mm high with chrome-plated spiral expander ring with profile ground running lands. Material: standard gray cast iron.

~~Passenger car diesel engines. Fig. P125 shows a typical ring set for long service life.

~Ring vibration. See above, ~Flutter

~Ring with inside bevel. See below, ~Versions ~~Compression rings

~Ring with inside bevel or internal angle on bottom flank. See below, ~Versions ~~Compression rings

~Rotation. In addition to the axial motion of the piston ring resulting from the interaction of inertia, gas, and friction forces, the piston ring is also subjected to a rotary motion on the circumferential direction. This is also caused by the above forces and is very impor-

tant for the function of the ring. It consists of keeping the ring "free"—that is, reducing the tendency of the ring to stick.

On high-speed two-stroke engines with large scavenging ports in the cylinders, however, it is necessary to prevent rotations, because otherwise the ends of the rings would extent into the scavenging ports and break. For such applications, the piston rings are prevented from turning in the piston ring groove by the use of gap retainers. In low-speed two-stroke engines, due to the scavenging ports, the rings are provided with radial pressure characteristics decreasing to zero at the gap ends, allowing them to turn.

~Running surface shapes (also, see above, ~Lubrication). The shapes of the running surface have a great influence on the running characteristics of the piston rings. Rings with a rectangular running surface are seldom used today in the first ring groove, but they are sometimes used in the second groove.

Symmetrically convex, asymmetrically convex, and optimized asymmetrically convex shapes are currently the state of the art (**Fig. P126**). They have proven

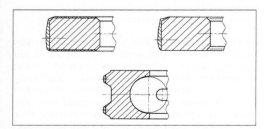

Fig. P124 Piston ring set for marine diesel and stationary engines

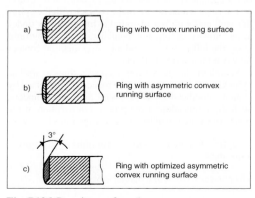

a)	Ring with convex running surface
b)	Ring with asymmetric convex running surface
c)	Ring with optimized asymmetric convex running surface

Fig. P126 Running surface shapes

Fig. P127 Running surface of chrome-plated piston ring with scorch marks

themselves in practice and in many tribological tests. With optimized asymmetric geometry, the upper third of the running surface is conical at an angle of 3°, so that the gas pressure in front of the ring acts against the pressure behind the ring, preventing an exceptionally high surface pressure on the running surface (**Fig. P126c**); moreover, this minimizes the amount of oil film pushed into the combustion chamber.

Optimized convexity is used only for piston rings in medium- and low-speed engines.

~Scorch marks. The term "scorch marks" was originally defined by the appearance of the running surface of piston rings (**Fig. P127**). The discolorations and so-called "paving stone appearance" on the chrome surface are indications of thermal overload. If the concept of scorch marks is expanded to include other materials, all types of appearances indicating thermal overload on the running surface of the piston rings are included. There is no definite point for distinguishing scorch marks from scouring. In the event of heavy scoring, heavy scratches and/or material transfer is recognizable on the surface of the piston rings. Scorch marks and scouring are caused by metallic contact with the friction partner, occurring due to lack of lubricating oil or exceptionally high piston ring edge pressure. Loose piston rings which do not make contact with the cylinder wall and with out-of-round cylinders allow undesired blowby of the hot combustion gases burning away the lubricating oil and allowing metallic contact between the friction partners.

Scorch marks and scoring on piston rings as well as cylinder running surfaces can be reduced or prevented by suitable modifications to the piston rings and cylinders, such as optimization of the pressure distribution of the piston rings and use of tribologically optimized surfaces.

~Sealing rings. *See below*, ~Versions ~~Compression rings

~Secondary wear. *See below*, ~Wear

~Shape winding. *See below*, ~Shaping ~~Winding

~Shaping. While conventional processes such as surface grinding and lapping are used for processing the ring flanks, the outer contour, which determines the characteristics of the piston rings, is created by

- Dual contour turning,
- Contour turning,
- Thermal stressing, and
- Winding.

~~**Contour turning.** With contour turning, the blank already ground on both sides is machined only on the outside (on copy lathe). Then the section of the ring corresponding to the gap width is cut out, and the ring is pressed together to nearly the nominal diameter in a bushing and rimmed out on the inside. The inner machining is, therefore, accomplished with the ring under tension, which can have an effect on the uniformity of its radial pressure distribution.

The copying cam is designed so that the ring realizes the intended radial pressure distribution after installation in the cylinder.

~~**Double contour turning.** This production procedure is usually used for correct shaping of the piston ring.

In this process, the ground blank is machined simultaneously inside and outside on a copying lathe. The cutting forces on the inside and outside cancel each other mutually, and a more uniform wall thickness is achieved over the circumference of the ring.

After removing the section of the ring corresponding to the gap width, the ring has the untensioned form; the desired radial pressure distribution is realized after installation in the cylinder. The shape of the copying cam is based on a mathematical principle laid out especially for the specific radial pressure characteristic of the ring.

The cam-controlled contour lathes are being replaced increasingly by CNC-controlled processes— that is, "electronic" cams.

~~~**Shape winding.** Shape winding is another production method for steel rings in which each ring is wound separately and is thereby immediately given the shape of an open ring instead of a round ring. The ring is then cut off in consideration of the gap width so that a free, open ring is produced which already has its functional shape. The tensions resulting from winding are annealed out in a thermal treatment, because otherwise, they would be relieved during operation of the engine, which could lead to undesired distortion of the ring.

~~**Thermal stressing.** This production process was originally used for producing piston rings. It is still used in a few cases for rings with diameters greater than 450 mm. Such rings are cast round, ground axially on both sides, turned round on the outside and rimmed out on the inside. They are cut and spread to the gap width. The spreading tension is then provided by a heat treatment so that the rings retain their open shape.

Installed in the cylinder, thermally stressed rings can achieve practically only a constant radial pressure distribution when they are in contact with the cylinder wall, leaving only a light gap.

~~**Winding.** Winding of piston rings is used for steel rings. The profile-drawn steel wire is wound to a round shape; the resulting spiral package is cut longitudinally, separating the rings, which are then drawn and shape-annealed on a shaping pin. The shape of the pin corresponds to the outer contour of an open, tension-free ring, laid out for certain radial pressure characteristics.

The open rings are pressed into a bushing when required, rimmed out on the inside, and then machined round on the outside.

On steel oil scraper rings, spreading is not required because these do not have a gap width and, therefore, have no intrinsic tension; it is only necessary for them to be round.

~Single casting material. See above, ~Production process

~Single side keystone rings. See below, ~Versions ~~Compression rings ~~~Keystone rings

~Slotted oil control ring. See below, ~Versions ~~Oil scraper rings

~Special lapping. See above, ~Lubrication ~~Topography of ring running surface

~Spiral-type expander. Spiral-type expanders for piston rings are cylindrical helical compression springs of round wire used to increase the contact pressure of oil scraper rings. They are dimensioned so that their spring travel corresponds to approximately the width of the gap in the oil scraper ring.

The spring force is determined by the wire diameter and winding pitch. Installed in a round or V-shaped groove in the oil scraper ring, the spiral-type expander exerts a uniform pressure over the entire circumference of the ring.

If required, spiral-type expanders are ground centerless or coiled using flat wire to reduce the secondary wear between the ring and spring. The contact surface in the area of the ring gap can be increased by winding the coils more tightly in this area, thereby reducing the specific pressure and hence the wear even further.

~Spring force. As an open ring spring, the piston ring exerts the force necessary for its sealing function on the cylinder running surface in the installed, tensioned state.

The spring force is measured by determining the tangential force. As a measuring alternative, the diametrical force covers only one-half of the piston ring. The spring force can be reinforced by additional spring elements, such as spiral expanders and expander springs.

~Spring-tensioned, spring-supported oil scraper rings. See below, ~Versions ~~Oil scraper rings

~Standard. Piston rings are standardized in DIN/ISO 6621-6627.

~Steel material. See above, ~Production process

~Steel ring. See above, ~Materials ~~Steel

~Stepped ring. See below, ~Versions ~~Oil scraper rings

~Stepped taper face ring. See below, ~Versions ~~Oil scraper rings

~Sticking. Ring sticking is caused by exceptionally high piston temperatures in the ring zone and by combustion residues.

The lubricating oil carbonizes at high temperatures, preventing free motion of the ring in the piston ring groove; the ring then sticks and loses its sealing function. The combustion gases blow past the piston, heating up the ring zone to an even greater extent. This can result in ring and piston seizing as well as ring breakage.

The same type of malfunctions also occur when there are extremely high deposits of combustion residues in the piston ring groove.

~Stress. In the installed state, piston rings are subject to continuous stress from the installation flexural strain caused by pinching the ring down to the diameter of the cylinder.

In addition, dynamic loads are present, which can be caused by the alternating effect of the gas, inertia, and friction forces (**Fig. P113**). In critical cases, this can cause ring flutter and ring vibration, which can lead to ring breakage.

The rings are subjected to extremely high loads as a result of oil carbon deposits in the piston ring groove, which can lead to sticking and ring breakage.

During assembly, the ring is subjected to flexural strain as it is pulled over the piston, which can be greater than the installation flexural strain when not standardized.

~Surface treatments. Surface treatments can improve the break-in properties of piston rings as well as reduce the wear of the running surface and flanks.

~~**Nitriding and nitrocarburization.** Nitriding/nitrocarburization is used primarily to improve the operating properties of piston rings in gasoline engines. The formation of a nitrate layer and the associated significant increase in the hardness (900 HVO, 1), and particularly improve the resistance to abrasion and adhesion as well as scorch marks. This is a thermochemical treatment with addition of nitrogen/carbon in the temperature range of 450 to 585°C to form a multiple-phased nitrated steel layer.

P

Selection of the piston ring material to be subjected to this nitration treatment has a decisive influence on the characteristics of the layer formed. Steels containing high quantities of chromium are primarily used in nitrate-hardened compression rings as well as in oil scraper rings (steel strap rings, two-piece oil rings).

The thickness of the nitrate-hardened layer is defined by means of the hardening curve in the diffusion layer as the so-called nitrating depth (see DIN 50190, Part 3).

Literature: H.J. Neuhäuser et al.: Steel Piston Rings: State of Development and Application Potential, T& N Symposium 1995, Paper 16.

~~**Phosphate coating.** The surface of the piston ring is changed to phosphate crystals by chemical treatment. This layer of phosphate is softer than the base material of the ring and, therefore, wears down more readily, accelerating the break-in process for the rings; it also prevents or reduces the formation of scorch marks. Zinc phosphate as well as manganese phosphate coatings can be used.

~~**PVD coatings (physical vapor deposition).** With the PVD method, hard material coatings can be deposited reactively from the vapor phase on the surface of piston rings. During this process, thin and thick coatings with a high resistance to wear are formed on the piston rings, conforming precisely to the contour. The high resistance to abrasion and adhesion results from the high hardness of the layer of 2000–3000 Vickers as well as the coating structure with an extremely smooth surface resulting from the low flaw density.

Because of the characteristics of the ceramics, these coatings also provide a very high resistance to scorch marks.

PVD coatings preferred for used on piston rings are based on the CrN system.

A disadvantage of the layers is the relatively low coating thickness, with a maximum of 50 μm, which results from increasing intrinsic tension. Currently, this limits applications in engines with extremely high service life requirements. However, improved coating systems promise more extensive application in the future.

~~**Tin-plating and copper-plating.** Both coatings are applied using galvanic processes. They act more or less as lubricants because of their low hardness, but are less susceptible to scorch marks. Frequently, they are also used as corrosion protection for storage, which is particularly important for ring repair kits.

~Synthetic ring. *See above,* ~Materials

~Tangential force. The tangential force, F_t, is the force required at the ends of the ring at the outer diameter to compress the piston ring to the gap clearance (**Fig. P128**). As a test parameter, it is the value for the spring force of the piston ring.

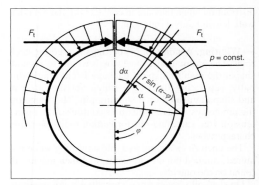

Fig. P128 Force relationships at constant contact pressure

It is the decisive value for determination of the contact pressure of the piston ring on the cylinder wall, which can be calculated as

$$p = \frac{2F_t}{d \cdot h} \ [\text{N/mm}^2],$$

where d = ring diameter and h = ring height.

For technical and mechanical reasons, the radial force rather than the tangential force is measured for mechanical testing of the spring force.

The contact pressure of the compression rings resulting from their intrinsic tension is generally in the range of 0.12 to 0.25 N/mm² for rings conforming with DIN/ISO. In addition to the intrinsic pressure defined above, the gas pressure behind the ring increases the contact pressure many times over.

~Taper face ring. *See below,* ~ Versions ~~Compression rings

~Thermal stressing. *See above,* ~Shaping

~Tin-plating and copper-plating. *See above,* ~Surface treatments

~Top rings. Certain types of piston rings require installation with a certain orientation on the piston for technical reasons; these include, for example, taper face rings, stepped rings, inside bevel rings, and inside rings as well as constant beveled oil control rings.

The upper flank, which must point toward the piston crown in the installed state, is marked with the word "top."

~Topography of cylinder running surface. *See above,* ~Lubrication

~Topography of ring running surface. *See above,* ~Lubrication

~Truck diesel engines. *See above,* ~Ring set

~Types. *See below,* ~Versions

~Versions. The versions differ according to their purpose:

- Oil scraper rings remove excess lubricating oil
- Compression rings seal in the combustion gases

~~Compression rings (Fig. P129)

~~~Keystone rings (DIN ISO 6624)

~~~~Double keystone rings. These are compression rings with trapezoidal cross section (Fig. P129c). The conical ring flanks automatically increase the axial clearance of the piston ring during radial motion of the piston, preventing the ring from sticking as a result of combustion residues to a high degree. The ring continuously frees itself from carbonization and combustion residues.

~~~~Single side keystone ring (DIN V 76624). These are compression rings with angled flank at top only (Fig. P129d). These are preferred for high-speed engines. The standard allows these rings in the form of cast iron rings up to a cylinder diameter of 80 mm and as steel rings up to a cylinder diameter of 100 mm.

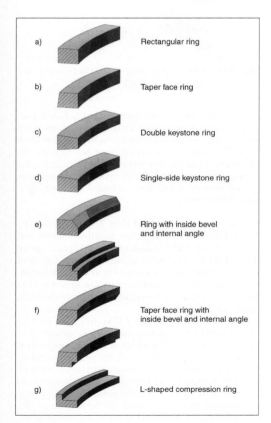

Fig. P129 Compression rings

As with double keystone rings, the angled flank increases the axial clearance for radial piston motion.

As rings with reduced height, they are frequently installed particularly close to the piston crown to minimize the damaging fire land space, which has a positive effect on the hydrocarbon emission values.

Single side keystone rings are preferred for use in the first groove on passenger car and commercial vehicle diesel engines, when a rectangular ring is no longer sufficient due to ring sticking and a double keystone ring is not yet required. A further application is two-stroke gasoline engines such as are used in snowmobiles and ultralight aircrafts.

~~~L-shaped compression rings. These are used in piston port controlled two-stroke engines for improving the control helix (Fig. P129g). The vertical L leg ends at the top edge of the piston crown, providing more exact timing.

In individual cases, the L-shaped compression ring is used in diesel engines for motor vehicles to minimize dead spaces in the combustion chamber.

~~~Rectangular rings (DIN-ISO 6622). These are piston rings with a rectangular cross section, used for sealing under normal operating conditions (Fig. P129a).

~~~Ring with inside bevel or inside angle on bottom flank. These rings are also called negative torsion rings (Fig. P129f). The interruption in the cross section at the lower ring flank results in negative twisting of the ring in the installed state.

~~~Ring with inside vessel or inside angle. The interruption in the cross section of the inside vessel or inside angle is achieved by the fact that the ring distorts to form a plate when installed, resulting in a conical running surface on the cylinder wall (Fig. P129e). As with tapered face rings, this results in an oil scraping effect. The ring is pressed flat by the gas pressure. This results in an additional dynamic load.

~~~Taper-faced rings. Taper-faced rings have a conical running surface which is shortened by the break-in time for the rings (Fig. P129b).

At the beginning, only linear contact is present with high pressure between the ring and cylinder running surface. The gas forces acting on the running surface at the beginning result in a certain pressure relief. Tapered-face rings are also used to support control of the engine oil consumption due to their oil scraping effect.

~~Oil scraper rings. These rings are of decisive importance for economic engine oil supply. They can be categorized into:

- Self-tensioning cast iron rings (DIN ISO 6623, DIN ISO 6625). These are used for support of the oil scraper function of the ring system on high-speed engines and, in some cases, operate alone as oil scraper rings.

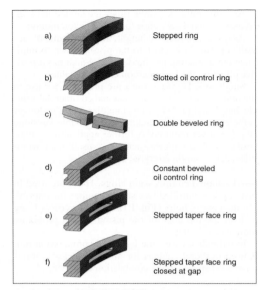

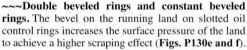

Fig. P130 Oil scraper rings (self-tensioning cast iron rings)

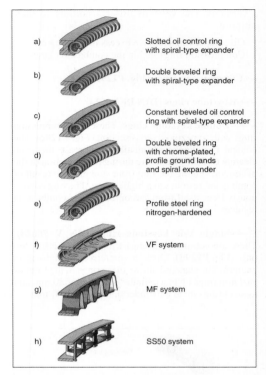

Fig. P131 Oil scraper ring (spring-tensioned and spring-supported)

- Spring-tensioned or spring-supported oil scraper rings (DIN ISO 6626), which are used in passenger-car gasoline and diesel engines as well as in medium-high speed engines.
- Spring-tensioned oil scraper rings of profile steel are being used increasingly in gasoline and diesel engines for motor vehicles.

~~~**Double beveled rings and constant beveled rings.** The bevel on the running land on slotted oil control rings increases the surface pressure of the land to achieve a higher scraping effect (**Figs. P130e and f**).

On constant beveled oil control rings, the bevel is machined in parallel on the upper edge of the running land. This results in an increased directional oil wiping effect in the direction of the crankcase.

~~~**Oil scraper rings with spiral expander.** Oil scraper rings are rings whose radial force and form-matching capacity are reinforced by cylindrical coiled pressure springs, so-called spiral expanders (**Fig. P131a–e**). The springs located in a round or V-shaped groove in the cast iron or steel ring act uniformly around the entire circumference increasing the high flexibility of this ring design.

To reduce ring land wear at high surface pressures, the running land on these rings is chrome plated or, in certain types of applications, provided with a sprayed metal coating (plasma or flame spray coating; **Fig. P131d**).

Profile steel oil control rings (**Fig. P131e**) are generally nitrogen-hardened on all sides in order to minimize secondary wear between the ring and spring. The ring is provided with punched holes for oil drainage.

Profile grinding of the running land makes it possible to maintain extremely close tolerances, allowing the specific contact pressure to be kept within close limits. To avoid high secondary wear between the spiral expander and the cast iron ring, the spiral expander is frequently coiled by use of rectangular wire or is ground centerless and also protected by covering it with Teflon tubing.

~~~**Oil scraper steel strap ring.** These ring designs are preferred in passenger car gasoline engines and consist of two steel lamellas and a steel spacer spring (**Fig. P131f–h**). The spacer spring is used in a variety of shapes whereby the secondary wear as well as economical production play a decisive role. The spacer springs and steel lamellas are frequently nitrogen-hardened to minimize secondary wear.

Designs as shown in **Fig. P131f and g** are particularly suitable for rings with low axial heights of up to 2 mm. The spring curve for these rings is steeper than that for spiral-type expander oil scraper rings, leading to greater reduction of the surface pressure of the land over the service life of the engine and, therefore, to reduced oil scraper effect as the running surface wears. On multiple-piece steel strap oil scraper rings, the scraping effect is supported by the radial motion of the steel lamellas acting independently of one another.

~~~**Slotted oil control rings.** The high surface pressure of the two running lands provides for its oil scraping effect (**Fig. P130d–f**).

The slots around the circumference facilitate flow of the excess lubricating oil scraped off toward the rear where it can flow out through the oil bores in the piston ring groove to the crankcase.

In modern mass-production passenger-car engines, single-piece oil scraper rings are generally no longer used because these types of rings no longer satisfy today's functional requirements.

~~~**Spring-tensioned and spring-supported oil scraper rings.** These are highly flexible sealing rings, which can adapt to major cylinder distortion and ensure particularly low oil consumption values for engines. Their high form-matching capability is achieved by small cross sections (resistance moments) and the contact force of the support spring (**Fig. P131**).

~~~**Stepped rings and stepped taper-faced rings.** Practically speaking these are compression rings with an oil-scraping effect (**Fig. P130a–c**). In the same manner as taper-faced rings and inside bevel rings, they deform to the shape of a plate in the installed state, and the bottom running edge of the lug makes contact with the running surface of the cylinder. The lug back-off ensures that the oil scrapped off is routed on by the running edge, preventing the oil from backing up at this point, which would reduce the scraper effect. Stepped taper-faced rings have a closed gap so that they prevent blowby better than the two other versions.

~**Wear.** The service life of the piston ring seal is determined by its wear. High radial wear results in rapid decrease of the contact pressure, whereas axial wear, usually to the ring and piston ring groove, causes an increase in the axial clearance. Wear decreases the sealing effect right up to the state at which it becomes zero.

The overall tribological system is extremely complex because nearly all usual types of wear—such as abrasive, adhesive, and corrosive wear—occur to various extents. The materials as well as the surface treatment and design (topography) of the friction partners, the lubricating oil used, and the fuel play an important role in the wear of the ring.

Fig. P132 shows a systematic illustration of the wear at top dead center in the cylinder, the ring running surface and ring flank wear, as well as the groove wear in the piston.

~~**Abrasive wear.** Abrasive wear is caused by foreign particles (dust, dirt, and combustion residues) between the friction partners. It can best be minimized with effective air and oil filters. Also, combustion residues can be influenced by good combustion, high quality fuels, and the lubricating oil. Naturally, cleanliness during assembly is a primary prerequisite to minimize abrasive wear.

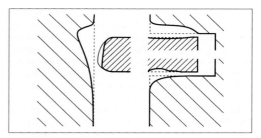

Fig. P132 Systematic illustration

~~**Adhesive wear.** Adhesive wear results when the lubricating film between the ring and cylinder running surface is interrupted, resulting in metallic contact. In extreme cases, this leads to scorching and galling of the piston rings. During the break-in phase, the adhesive wear can be influenced by special break-in coatings (e.g., phosphating) or surface treatment (special ring lapping, plateau honing in the cylinders) to improve later wear characteristics.

Wear protection layers on the rings such as chrome, plasma, or HVOF coatings have a general tendency to reduce adhesive wear when compared to untreated surfaces. Chrome ceramic coatings containing embedded ceramic particles have proven particularly satisfactory for increasing the resistance to wear and scorching. A further increase in the resistance to scorching can be achieved with HVOF and PVD coatings.

Literature: L. Wiemann: Die Bildung von Brandspuren auf den Laufflächen der Paarung Kolbenring-Zylinder in Verbrennungsmotoren, Special Printing der Goetze AG, MTZ 32 (1971).

~~**Corrosive wear.** Corrosive wear occurs frequently in diesel engines because the sulfur contained in the fuel forms acids (H_2O_3 and H_2SO_4) during combustion, which cause corrosion. Moreover, the lubricating oil oxidizes in engines subject to high thermal loads, which also leads to the formation of sulfur-based acids.

Of the various wear-resistant coatings for piston rings, galvanically deposited chrome is the most resistant to corrosion. Therefore, it is frequently applied to the ring flanks.

The percentage of corrosive wear is particularly high in heavy (residual) fuel engines because the sulfur contained in these oils is considerably higher (greater than 3.5%) than in lighter diesel fuels. The corrosion is frequently recognizable on the edges of the rings.

Literature: J.J. Broeze, A. Wilson: Sulphur in Diesel Fuels, Automobile Engineer, March 1949.

~~**Micro-welding.** "Micro-welding" designates a special form of damage to the piston ring flanks and piston groove flanks. This occurs on the top piston ring consisting of ferrous materials or on the associated

P

groove in aluminum pistons, even after relatively short engine operating times. Local material bonding, including material breakout and material deposition from the piston to the edges of the ring, is characteristic of this type of damage. This destroys the seal between the ring and piston so that the operating characteristics of the engine are impaired, leading in extreme cases leading to engine failure.

In addition to specific modifications to the piston, the piston rings can be protected by reducing the edge roughness and/or applying suitable armor coatings to the edges (such as phosphating, tinning, PVD coating, and chemical treatment of the edges [CPS]). Another method used is to round off the ring impact edge on the flank to prevent "micro-welding."

~~Secondary wear. Secondary wear is wear resulting from the relative motion between the ring and spring (spiral expander) or between the spacer spring and steel lamellas (steel strap rings). The spring digs into the ring, eliminating its freedom of motion, which has a negative effect on the function of the oil scraper ring. Because the relative motion is greatest near the gap in the piston ring, the greatest wear also occurs here.

A lower specific surface pressure and, therefore, lower wear can be achieved by increasing the contact surface of the spring. This can be accomplished by using springs wound with flat wire, centerless ground springs, and springs wound more closely in the area of the ring gap. The greatest protection against secondary wear is offered by a Teflon tube installed over the spiral expander. This solution is a standard one in high-speed diesel engines for passenger cars.

The secondary wear on multipiece steel oil-scraper rings occurs on the lamella as well as on the spacer springs. Here, the supports for the spacer springs dig into the lamella on certain systems, preventing their motion, which is important for the oil scraping effect. This secondary wear can be counteracted by nitrating the lamina and spacer springs.

~~Side wear. Side wear is observed most on the compression rings in the first piston ring groove. In gasoline engines, it has a greater tendency to occur at high engine speeds because of ring flutter. Uneven groove flanks and rings that twist in the installed state (stepped rings with internal bevel/internal angle) are reasons for additional ring motion, which increases the frictional stress to the sides of the ring. In the same manner, excessive axial clearance in the grooves allows the rings to "hammer" between the edges of the groove in the piston.

The side wear in diesel engines is caused by high ring groove temperatures, high fire land clearances, combustion residues, poor supply of lubricating oil, and high rates of pressure rise, $dp/d\alpha$. Rings as well as groove flanks are armored with wear protection coatings to prevent extreme side wear. The sides of the rings are chrome-plated or nitrogen hardened; the piston groove flanks are also chrome-plated or protected with Niresist ring carriers or, in steel pistons, by hardening.

Literature: F. Grunow: Schutz der Kolbennut durch Oberflächenbehandlung des Kolbenringes, Special Printing by ATZ and MTZ, Werkstoffe im Automobilbau, 1998/99.

Piston ring breakage →Piston ring ~Breaking

Piston ring characteristic parameters →Piston ring ~Designations

Piston ring clearance →Piston ring ~Gap

Piston ring contact surface design →Piston ring ~Lubrication

Piston ring designations →Piston ring ~Designations

Piston ring flutter →Piston ring ~Flutter

Piston ring gap →Piston ring ~Gap

Piston ring groove →Piston ~Ring

Piston ring lubrication →Piston ring ~Lubrication

Piston ring material →Piston ring ~Materials

Piston ring motion →Piston ring ~Rotation, ~Flutter

Piston ring ovality →Piston ring ~Out-of-roundness

Piston ring parameters →Piston ring

Piston ring retainers →Piston ring ~Gap

Piston ring rotation →Piston ring ~Rotation

Piston ring set →Piston ring ~Ring set

Piston ring shaping →Piston ring ~Shaping

Piston ring spring force →Piston ring ~Spring force

Piston ring standard →Piston ring

Piston ring sticking →Piston ring ~Sticking

Piston ring stress →Piston ring ~Stress

Piston ring support →Piston ~Ring ~~Ring carrier

Piston ring types →Piston ring ~Versions

Piston ring wear →Piston ring ~Wear

Piston ring zone →Piston ~Ring

Piston rings, number of →Piston ring ~Number of rings

Piston running characteristics →Piston ~Running characteristics

Piston secondary motion →Piston ~Secondary motion

Piston shape →Piston ~Shape

Piston skirt →Piston ~Design, ~Skirt

Piston skirt shape →Piston ~Skirt

Piston slap →Piston ~Secondary motion

Piston speed →Crankshaft drive ~Velocity; →Piston ~Mean piston speed

Piston spray cooling →Piston ~Cooling ~~Spray oil cooling

Piston spray oil cooling →Piston ~Cooling

Piston stroke →Piston ~Stroke

Piston surface coating →Piston ~Surface coatings

Piston temperature →Piston ~Temperatures

Piston travel →Bore-stroke ratio; →Crankshaft drive ~Piston travel; →Piston ~Stroke

Piston weight →Piston ~Weight

Pitting →Camshaft ~Wear

Plain bearing →Bearings

Planned regulation changes starting in 2004 →Engine ~Racing engines ~~Formula 1 ~~~Regulations

Plasma jet ignition system →Ignition system, gasoline engine

Plasma spray coatings →Piston ring ~Contact surface armor, ~Surface treatments

Plastic fuel tank →Injection system (components) ~Gasoline engine ~~Fuel tank, ~~Fuel tank systems

Plastic modules →Sealing systems ~Modules ~~Housing materials

Plastic ring →Piston ring ~Materials

Plate cooler →Coolant ~Design ~~Plate

Platinum coating →Catalytic converter ~Coating

Plenum chamber →Air distributor

Plenum volume →Intake system

Pliers →Piston ring

Plug →Spark plug

Plug connector →Actuators ~General Purpose Actuator (GPA); →Electronic/mechanical engine and transmission control ~Requirements for mechanical and housing concepts

Plug face →Spark plug

Plug location →Spark plug ~Spark plug location

Pneumatic (air) valve control systems →Variable valve control ~Operation principles, ~Systems without camshaft

Pneumatic (air) valve spring →Valve spring

Pollutant aftertreatment. Due to constant tightening of emission limits (**Fig. P133**), pollutant emissions (HC, CO, NO_x, and particulates) from vehicles have been reduced by about 95% since laws limiting exhaust gas emissions were introduced. This is due both to a continuous improvement of the combustion process and to the introduction of electronic engine control; these have considerably improved the untreated emissions from the engine. **Fig. P134** uses HC as an example to show that untreated emissions have

P

All values in g/km	Gasoline		Diesel		
	HC + NO_x	CO	HC + NO_x	CO	PM
EU I (1993)	0.97	2.72	0.97 (DI)	2.72	0.14
			1.36 (DI)		0.2 (DI)
EU II (1997)	0.50	2.20	0.70 (DI)	1.00	0.08
			0.90		0.1 (DI)
EU III * (2001)	0.2 + 0.15	2.30	0.56	0.64	0.05
			$NO_x = 0.50$		
EU IV * (2006)	0.1 + 0.08	1.00	0.30	0.50	0.025
			$NO_x = 0.25$		
* with new MVEG-B testing cycle					

Fig. P133
European emission limits

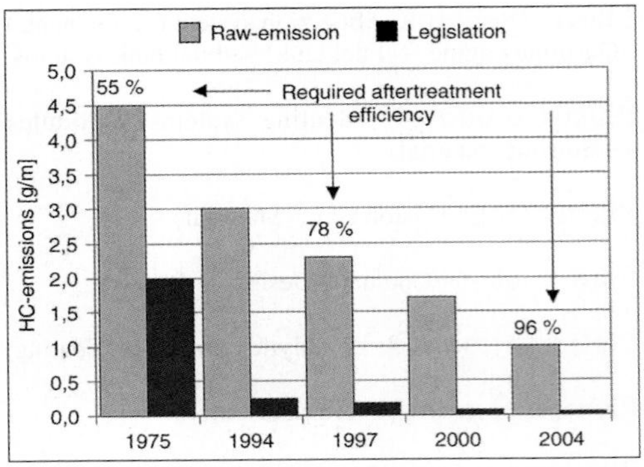

Fig. P134 Development of limits for untreated HC emissions and exhaust gases, with California as an example

been reduced by 80% since 1975. However, the gap between untreated emissions and the legislated limits is growing, so that more and more effective secondary systems for treating exhaust gas are necessary. Without the introduction of the catalytic converter technology, the current low emissions levels would not have been feasible. In gasoline engines with stoichiometric combustion, conversion rates of over 99% can be reached with modern catalytic converters. However, this only applies to catalytic converters at operating temperatures, starting at about 300°C until they reach the light-off time. Depending on the exhaust gas system, up to 90% of all emissions are emitted during this process. Therefore, in stoichiometrically fueled gasoline engines, all emission-reducing measures aim at a reduction of the converter light-off time.

In the case of lean-burn gasoline engines and diesel engines, the main problem is the catalytically supported reduction of oxides of nitrogen (NO_x) in an oxidizing environment. The potential reducing agents, such as HC and CO, preferably oxidize with the existing oxygen instead of the oxides of nitrogen. Here, special secondary treatment systems must be developed; these are described below. For diesel engines, particulate emissions are reduced to adhere to the emission limits. In smaller engines, a simple oxidation catalyst usually suffices. In larger engines and, especially, commercial vehicles, a much more complex particulate filter system is needed.

~Aftertreatment concept with a three-way catalytic converter. Aftertreatment concepts for the reduction of the light-off time in stoichiometrically operated catalytic converters include:

- Close-coupled (catalytic) converter
- Electrically heated catalytic converters
- Gasoline burners
- HC absorbers
- Reduced thermal capacity
- Secondary air systems

~~Close-coupled (catalytic) converter, or CCC. Prevention of heat loss just in front of the converter is achieved by direct connection of the converter to a compact manifold. Often, the converter is welded directly to the manifold without the use of an intermediate flange. The light-off time of the converter is reduced to a few seconds, which leads to potential emission advantages in HC and CO of up to 70%.

In this close-coupled position, exhaust gas temperatures of nearly 1050°C and accelerations of up to 75 G are reached, presenting extreme requirements for catalyst design. Because of consequent development, the catalytic coating now can stand up to the thermal stress. Another development focus of the CCC is optimization of the inflow to the converter using measurements or CFD (computational fluid dynamics) calculations. This has improved the efficiency and aging behavior of catalytic converters. Ever-increasing restrictions on emission limits tend to lead to larger-volume catalytic converters, which hardly can be accommodated in current engine compartments. Furthermore, such converters emit a lot of heat, which then must be vented from the engine compartment. As a compromise, divided systems have been developed, whereby part of the converter volume is located close to the engine and part of the volume is located in the underbody area of the vehicle. The volume of the close-coupled catalytic converter should not be less than 40% of the engine capacity.

In close-coupled catalytic converters, the negative influence on engine performance and, in particular, on the engine torque curve must be considered. This requires careful tuning of the total manifold/CCC system, which must compensate for the effect of the converter as far as possible through optimization of the intake system.

Despite the increased development costs and the problems described, the close-coupled catalytic converter is the most effective device today for complying with the EU4 and CARB standard.

~~Electrically heated catalytic converters (EHC). The goal of this concept is to reduce the light-off time of the catalyst by supplying electrical power (**Fig. P135**). The system consists of a heating element (using the principle of the electric resistance heater), the secondary primary catalyst, and the main catalyst.

During a cold start, the engine operates rich in order to provide sufficient fuel in the form of HC and CO to light the converter. Furthermore, the supply of sufficiently large amounts of necessary secondary air is by means of an electrically driven pump. The heating element is powered by the 12V electric system and is limited to about 1–1.5 kW. The operating principle of the heating catalyst is similar to that of a cigarette lighter. The heater represents the ignition spark, the rich mixture the combustion gas, and the flame is created from the catalytically supported exothermal reaction of the combustion gas with the secondary air in the primary catalyst. This "flame" reduces the heating time of the catalytic converter to less than 12 seconds, reducing cold start emissions by up to 70%. It must be

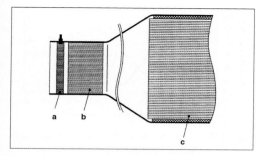

Fig. P135 Schematic diagram of electrically heated catalytic converter (EHC): (a) heating element, (b) primary catalyst, (c) main catalyst

considered, however, that the largest share of emission reduction results from the above-mentioned secondary effect; the additional advantage from heating is relatively small. In view of the complexity of the heating catalyst itself—a control unit, a power switch, the electric lines and connections, and possibly an additional battery—the effectiveness of the heated catalyst system is questionable. The use of heated catalysts is, therefore, limited to very few niche applications.

~~**Gasoline Burners.** Parallel to the electrically heated catalytic converter, concepts were developed in which the additional energy necessary to heat the catalytic converter more quickly is provided by a gasoline-fueled burner. Such burners have a much higher heat output than electric heaters. Disadvantages are structural expense, cost, and possible secondary emissions from improperly tuned burners. Due to a lack of acceptance of such burner systems from safety concerns, further development of such systems was discontinued.

~~**HC absorber.** The idea of simply storing HC emissions during the cold-start period led to the development of a variety of HC absorber systems. Originally, so-called activated charcoal filters were checked for suitability for vehicle applications. In these filters, the HCs are stored in the micropores of the carbon grid, and the efficiency was very good within narrow, well-tuned temperature limits. When the temperature increases, the filter desorbs the stored HCs, which are then reconverted in the secondary catalyst. At 250°C, about 90% of the HCs are desorbed. However, the activated charcoal starts oxidizing at temperatures above 270°C, and the efficiency quickly decreases. Starting at 500°C, the active carbon ignites and burns. Tests to bypass the thermal stability problem through special positioning of the active carbon filter (i.e., in the area of the rear muffler) resulted in the filter not being completely desorbed in the usual emissions test. Another problem was the sensitivity of the filter to the exhaust gas concentration always present during a cold start. The filter also stores the water, which uses up storage

space and significantly decreases the efficiency of the HC filter.

Some of the problems described above were resolved by the subsequent development of ZSM5 zeolite. Thermal stability was increased to 600–700°C. Desorption was quicker, but also at lower temperatures (90% at up to 180°C). Unfortunately, the zeolites proved to be hydrothermally unstable. The efficiency quickly decreased at higher temperatures when exhaust gas condensates were present. Furthermore, zeolites show poor storage efficiency for short-chain HCs with fewer than four C atoms (e.g., methane and ethane).

All HC adsorbers have the general problem of a temperature gap between the desorption (180–250°C) and reaching the operating temperature of the secondary catalyst (~300°C), which leads to clear HC peaks and a worsening of overall system efficiency. Tests to solve this problem using complicated designs (e.g., Low Hydrocarbons Emissions System [LHES] by Engelhard or PUMA industrial emission control system by Corning) failed because of their complexity and costs, so that HC absorbers were no longer considered for vehicle applications.

~~**Reduced thermal capacity.** A simple measure to speed up the heating of the catalytic converter is to reduce the heat loss from the exhaust gases on their way to the catalytic converter. This is achieved by reducing the wall thicknesses and, by this means, the heat capacity of the gas-carrying components (manifold and head pipe). For reasons of strength, in view of the considerable thermal and mechanical stress, the thin-wall gas-carrying pipe must be supported by a thicker outer pipe. Ideally, an insulating air gap is created between the two pipes, which reduces heat loss to the outside and, therefore, increases the temperature upstream of the catalytic converter. Another advantage of the air-gap method is the significantly decreasing insulating effect (heat radiation) at high temperatures, which prevents thermal overload of the catalytic converter. Depending on the particular application, HC and CO emission advantages of up to 30% are possible. With extremely thin internal pipes, the construction problems are much greater than the advantage gained, so that wall thicknesses of less than 0.6 mm no longer make any sense.

~~**Secondary air system.** A very effective way to speed up the heating of the catalytic converter is to supply energy through highly exothermic free gas reactions (**Fig. P136**). In this case, the engine is operated at a very rich mixture with an air ratio of $\lambda = 0.6$–0.7. In a gas phase reaction, large amounts of the emitted HC and CO are burned with additional air, which is blown directly into the manifold. Because free gas reactions take place at about 600°C, the air should be supplied as close to the exhaust valve as possible.

Usually the air is delivered via an electrically driven pump, either single-phase or multiple phase. The optimum effect is obtained at an exhaust gas lambda, λ, of between 1.1 and 1.2, necessitating control of the

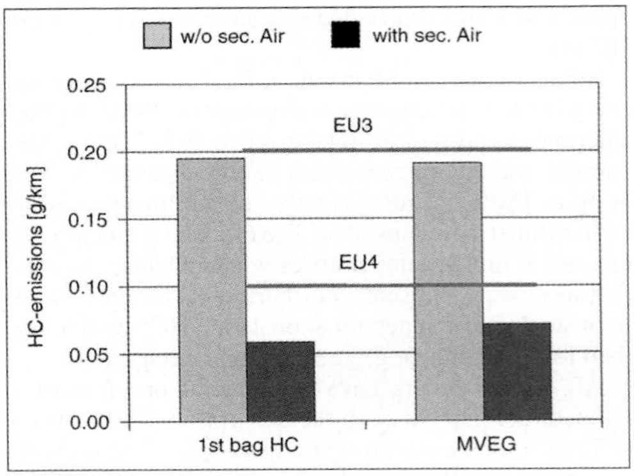

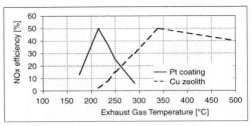

Fig. P136 HC-emissions advantage with secondary air system (2.0-L automobile with underbody catalytic converter).

volume of air in the exhaust gas process. If too much air is supplied, the exhaust gas cools down, eliminating part of the emission advantage.

A secondary air system makes improvements in HC and CO emissions of up to 70% possible. This technology enables compliance with the EU IV standard. Because of the complexity of the system (pump, injection device, etc.) and the related costs, cheaper solutions with similar effectiveness are still being sought.

~Particulate filter →Particles (Particulates) ~Particulate filter system

~Pollutant aftertreatment lean concepts. Exhaust gas aftertreatment systems for diesel engines and lean burn gasoline engines.

~~HC-SCR (selective catalytic reaction) concepts for NO$_x$. Operating a diesel engine with excess air reduction of NO$_x$ by the classical three-way principle is not possible (Fig. P137). Conventional oxidation catalytic converters directly can reduce NO$_x$ with hydrocarbons in the range of 180 to 250°C.

The efficiency of this reaction can be improved through modification of the wash coats. The maximum possible efficiency depends on the hydrocarbon-NO$_x$ ratio as well as the type of hydrocarbon available from the combustion. The steady-state efficiency is up to 50% and it is about 5–15% in dynamic driving operation. For current and future limits, this concept is usable for light- and medium-weight passenger cars. The layout of the exhaust system is determined by the raw NO$_x$ emissions in city and highway traffic. If a significant NOx reduction in both driving ranges is necessary to reach the emissions limit, two separate catalytic converters are used which, depending on the temperature profile in the exhaust system and the operating range of the coating, are positioned near the engine and in the underbody area. For lower efficiency levels, a catalytic converter can be installed at a temperature-optimized position in the exhaust system.

Fig. P137 NO$_x$ conversion efficiency at steady-state (N)SCR activity

Increased HC emissions for increased efficiencies can be achieved through engine modification—for example, with postinjection. It is important here that the possible separation of the oil film from the cylinder liner is avoided. A problem is the provision of the reducing agent HC for the second converter. There are several possibilities:

- Overdosage of fuel so that enough reducing agent gets through the first converter for the second converter
- Areas near the boundary of the first converter are not coated in order to enable a hydrocarbon bypass
- Using a valve, additional fuel is injected into the exhaust system as a reducing agent after the first converter

This preserves the oxidizing function of the catalytic converter over the whole operating range.

~~NO$_x$ catalytic converters with hydrocarbon trap. In gasoline engines with direct injection (GDI), the exhaust system for current and future limits consists of an underhood primary catalytic converter and a downstream catalytic converter with hydrocarbon trap (Fig. P138).

A wide-range lambda probe, positioned upstream of the primary catalytic converter, regulates and controls lambda = 1.0 as well as the lean operation of the engine. The primary catalytic converter after this, with its three-way coating, is necessary to adhere to the emissions limits during a cold start. Its function as well as the thermal management of the downstream exhaust system is controlled by a thermocouple.

The catalytic converter with a hydrocarbon trap deposits oxides of nitrogen in the storage component during lean-burn driving operation. Once the catalytic converter with hydrocarbon trap has reached the allowed filling capacity, the stored oxides of nitrogen are desorbed again during short-term lean burn driving operation. These oxides of nitrogen are reduced again after the coating process according to the well-known three-way principle.

The operating range for this storage function is at exhaust gas temperatures between 250 and 450°C.

This determines maximum lean-burn driving range of the vehicle. Above this exhaust gas temperature, the catalytic converter with hydrocarbon trap works as a three-way catalytic converter. Its maximum operating

P

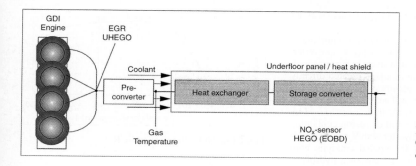

Fig. P138
Gasoline engine with
direct injection system

temperature is about 800°C. Together with the engine control, the exhaust gas system must regulate the temperature balance for the chemical function between the primary catalytic converter and the catalytic converter with hydrocarbon trap. On one hand, the necessary temperature protection can be accomplished in the engine alone with increased fuel consumption. On the other, this can also be accomplished through expansion of the lean-burn driving range via support of so-called exhaust gas heat exchangers. If the negative effect on the warm-up behavior of the system is too great, the exhaust gas system is equipped with a switching element to switch between a short pipe in the warming-up phase and the heat exchanger during driving operation. The current weakness of the system is the sulfur content of the fuel, which poisons the storage components even at low concentrations. This is partially possible under conditions of rich engine exhaust gas at high temperatures of 700–750°C. The downstream NO_x probe, or the lambda probe, controls the functioning of the catalytic converter as well as the onboard diagnosis.

This system also can be used for diesel vehicles. In most cases, the catalytic converter with hydrocarbon trap is located close to the engine. More important here is the possibility to regenerate sulfur as well as the achievement of the required rich exhaust gas composition.

The storage capacity, with its dependence on different parameters, is shown in **Fig. P139**.

~~**Oxidation converter.** Oxidation catalytic converters convert the hydrocarbons and carbon monoxide

contained in the exhaust gas into carbon dioxide and water. They also oxidize part of the volatile hydrocarbons condensed on the nucleus of the particle.

Oxidation catalytic converters are used as retrofit components for older diesel engines to obtain better taxation classification because of improved pollutant emissions. They also are used optionally in commercial vehicles and buses. It must be considered in such cases that more particulate emissions can occur because of sulfate production resulting from use above 350°C of a highly sulfurous diesel fuel. In view of current and future limits in passenger cars, its use is limited to small vehicles. The volume, as well as the Pt level, are adapted to suit the application. In order to optimize cold starts, additional zeolites can be integrated as temporary storage for hydrocarbons. The efficiencies for steady-state conversion are shown in **Fig. P140**.

~~**SCR catalytic converters.** This exhaust gas aftertreatment system is derived from the reduction of oxides of nitrogen with ammonia as used in power stations. In contrast to power station construction, the ammonia used as the reducing agent is created directly in the vehicle. Available reducing agents are solid carbamide, aqueous 32% carbamide solution, and ammonia carbamate.

The solid carbamide comes as a reducing agent in the form of pellets or sticks, the aqueous carbamide solution is in a tank, and ammonia carbamate is an oily emulsion. These are transported on the vehicle as secondary operating agents in a reactor. The metered addition of the carbamide solution is done by com-

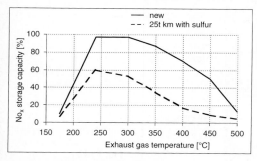

Fig. P139 NO_x storage capability

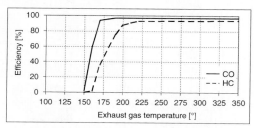

Fig. P140 Efficiency during steady state HC/CO
conversion

pressed air, or it can take place through a valve without compressed air.

The conversion of the carbamide into ammonia is done through hydrolysis in the catalytic converter, downstream from the exhaust system. In systems with ammonia carbamate, the ammonia is created directly in the reactor and added gradually into the exhaust gas system via temporary storage. The reduction of NO_x is then carried out at the SCR catalytic converter (**Fig. P141**). This can consist of a full extrusion containing V_2O_5, a coated catalytic converter with V_2O_5, or metal-exchanged zeolites. With the coated catalytic converter, the converter system consists of a hydrolysis catalytic converter for generation of NH_3, an SCR catalytic converter for reduction of NO_x, and a lock catalytic converter to avoid ammonia slip. With full catalytic converters, all three functions are integrated onto one substrate. The efficiency is directly dependent on the ammonia quantity used in relation to the amount of oxides of nitrogen, and can be as much as 90–98% in steady-state operation. Dynamically, they can reach 60–80% in the testing cycle. The operating range is, depending on the SCR catalytic converter type, 180 to 530°C for V_2O_5, or 300 to 550°C for zeolites. These ranges are determined by the ammonia creation, the SCR reaction, and the direct ammonia oxidation. To improve the low-temperature activity of the SCR catalytic converter, a primary catalytic converter can be added to the system to generate NO_2. The use of primary catalytic converters and/or coated catalytic converters requires the use of low-sulfur diesel fuel. SCR systems (**Fig. P142**). These devices are suitable for commercial vehicles, buses, transporters, and heavy passenger vehicles to guarantee adherence to future passenger car and commercial vehicle emissions limits.

Pollutant aftertreatment concepts →Pollutant aftertreatment ~Aftertreatment concept with a three-way-catalytic converter

Pollutant aftertreatment lean concepts →Pollutant aftertreatment

Pollutant emissions →Emissions; →Heat accumulator ~Applications ~~Emission of pollutants

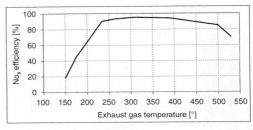

Fig. P141 NO_x efficiency in SCR

Pollutant formation →Emissions; →Particles (Particulates)

Pollutants (exhaust gas emissions). Pollutants are generally substances emitted into the environment that change the existing natural composition of the living space. In the case of the combustion engine, these are gaseous substances such as exhaust emissions and emissions from the fuel tank as well as liquid components and solids, subsumed under the term "particulates."

Gaseous exhaust emissions contain, among other things, the compounds HC (hydrocarbon), CO, and NO_x, which are limited by law. The emissions from the fuel tank and the crankcase ventilation system consist almost exclusively of hydrocarbons, and they are usually generated during the combustion process.

Carbon dioxide, which also causes the greenhouse effect, is not considered a pollutant with respect to vehicle emissions.

Pollution retention system →Particles (Particulates) ~Particulate filter system; →Pollutant aftertreatment

Poly V-belt →Engine accessories

Poly V-belt drive →Engine accessories

Poly V-belts →Engine accessories

Polyaromatic content →Fuel, diesel engine ~Composition

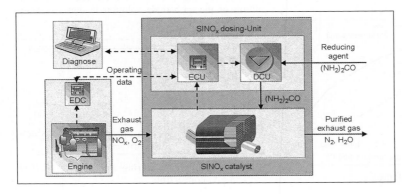

Fig. P142
SCR system

Polycyclic aromatic hydrocarbons (PAH) →Fuel, diesel engine ~Composition

Polytropic expansion →Expansion ~Polytropic

Poppet valve →Valve

Pore size →Particles (Particulates) ~Particulate filter system ~~Particulate filter

Porsche SCS. The Porsche SCS (stratified charge chamber system) method is a stratified charging method for gasoline engines: the SCS method applies a combustion chamber that is divided into two. Furthermore, direct injection in the precombustion chamber is combined with intake manifold injection (**Fig. P143**). Testing was carried out to implement charge stratification based on this design and operating conditions. The ignition occurs in the precombustion chamber, which receives a rich mixture; the combustion in the main chamber is initiated through a shot of flame from the precombustion chamber. Due to the stratification operation with a higher level of stratification and a higher overall air-fuel ratio is possible, with the advantages of low emissions of NO_x and HC.

Port line →Exhaust port

Port switch-off →Intake port, gasoline engine

Position, camshaft →Camshaft

Position, crankshaft →Engine concepts ~Reciprocating piston engines

Position, cylinder →Engine concepts ~Reciprocating piston engines

Position, glow plug →Ignition system, diesel engine/preheat system ~Cold starting aid ~~Glow system; →Starting aid ~Glow plug

Position, spark plug →Spark plug

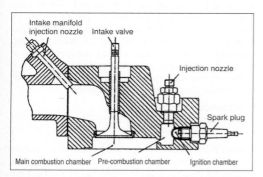

Fig. P143 Porsche stratified charge chamber system

Position sensors →Actuators ~General Purpose Actuator (GPA); →Sensors

Positive crankcase ventilation →Crankcase ~Crankcase ventilation

Postignition (*also*, →Spark plug). The terms and definitions for uncontrolled ignition of air-fuel mixtures have been established according to an international agreement (ISO 2542-1972) for adaptation of the heat values for spark plugs. Auto-ignition means ignition independent of a spark. If ignition occurs before the electric moment of ignition, this is called preignition. If ignition occurs after the moment of ignition, it is called postignition. Postignition is not critical for engine operation. However, preignition can result in severe damage. In the case of preignition, it is necessary to differentiate between incipient preignition and runaway preignition—actual spontaneous ignition.

It is, therefore, necessary to adapt the spark plugs so that preignition does not occur even at full load. Greater value is placed on the heat range reserve value, which describes the interval up to the beginning of preignition with further increasing of the thermal load on the spark plug. This heat range reserve is expressed in degrees of crank angle, the amount by which the standard set firing point can be advanced without preignition occurring. Retardation of the firing angle in the postignition range results in a linear increase in the mean combustion chamber temperature in the cylinder as well as the equilibrium temperature at the tip of the insulator base of the spark plug. The value of this temperature increase is three to ten degrees per degree of crank angle, depending on the engine. Ion flow measurements offer the possibility of adapting the heat range of spark plugs to any engine as well as measuring the heat range in a test engine. Moreover, the ion flow measuring process (**Fig. P144**), allows postignition and its percentage of the blanking rate to be followed by increasing the combustion chamber temperature (advancing the firing angle) by blanking out the ignition spark at certain intervals.

A change in the ion flow diagram on the oscilloscope (**Fig. P145**) allows the transition from postignition to incipient preignition to be determined exactly even without blanking out the electric ignition spark. This makes this measuring method an additional aid in evaluating individual design parameters in terms of their tendency to produce spontaneous ignition at high loads.

Poststarting phase →Carburetor ~Starting systems ~~Requirements

Potential energy →Energy

Pour point improver →Oil ~Body ~~Additives

Power control. In combustion engines, there are two ways to control power:

- by changing the engine speed
- by changing the load

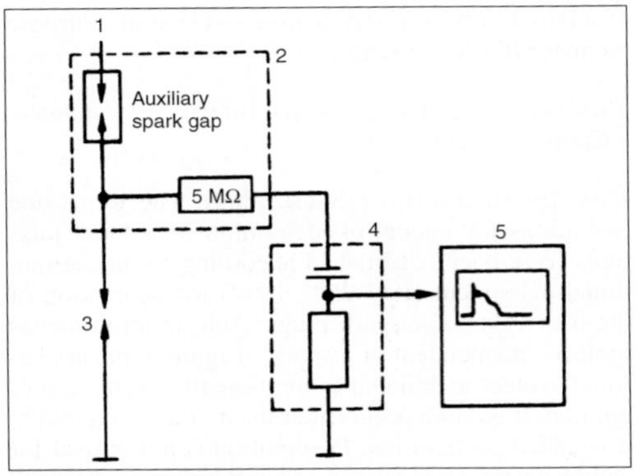

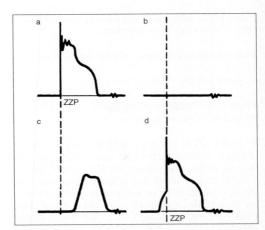

Fig. P144 Schematic diagram of ion flow measurement: (1) from ignition distributor, (2) ion flow adapter, (3) spark plug, (4) ion flow instrument, (5) oscilloscope (Source: Bosch)

Fig. P145 Characteristics oscillograms: (a) normal operating state, (b) ignition blanked out without postignition, (c) ignition blanked out with postignition, (d) preignition (Source: Bosch)

When the load is changed, a distinction is made between quantitative control (also charging, quantity, or throttle control) and qualitative control (mixture control).

Quantitative control is applied in conventional gasoline engines as the ignition limits of the mixture require it. The mass of air and fuel trapped in the cylinder per combustion cycle changes according to the requirement of the load.

Qualitative control means that the air-fuel ratio of fuel and air changes according to the requirement of the load. The mass of air trapped in the cylinder per combustion cycle remains almost constant. Qualitative control is applied in diesel engines and gasoline engines with direct injection. The advantage, compared with quantitative control, is less gas transfer work, which results in a considerable reduction in fuel consumption, especially at part load.

Power cycle →Combustion cycle

Power increase (*also*, →Supercharging; →Variable valve control). The effective power output, P_e, of an engine is calculated from the equation

$$P_e = V_H \cdot w_e \cdot n \cdot i,$$

where V_H = displacement of the engine (m³), w_e = specific work (Nm/m³), n = engine speed (rps), and i = combustion cycles per revolution. This shows that an increase in power can be achieved through an increase in the displacement, the specific work, and/or the engine speed. A two-stroke engine would have twice the power output of a four-stroke engine if all other parameters remained the same, which is not currently the case with the specific work.

The specific work, and, thus, the power output, can be increased substantially through supercharging.

Power limitation →Engines ~Racing engines ~~IndyCar ~~~Design features; →Power output

Power loss (*also*, →Friction). Power loss can be caused by engine components and processes and by external components and processes. Together with, for example, the combustion process, power control, cooling system, viscosity of oil, friction, mixture formation, and the intake and exhaust system, accessories such as power steering pump and air-conditioning system can cause power losses.

Friction causes a significant proportion of any power loss, with the friction mean effective pressure being up to 2.5 bar—this represents up to 20% of the brake mean effective pressure.

Power loss exhaust system →Exhaust system

Power mass →Weight-to-power-ratio

Power output (*also*, →Variable valve control ~Effect of fully variable valve control). The energy contained in the cylinders is the main parameter that determines the torque from a combustion engine and, with the corresponding engine speed, also its power output. A measure for the energy content of the cylinders is the mass of air (mass of oxygen) with the accompanying mass of fuel. The air and, as a result, the fuel mass, can be raised considerably by supercharging.

Factors that have an impact on the trapped mass in the cylinder for a given displacement are

- Air temperature,
- Air pressure, and
- Humidity

Moreover, all significant operating parameters such as the air-fuel ratio, the volumetric efficiency, the combustion efficiency, engine cooling, and so on affect the power output.

The engine power output is calculated as follows:

$$P = M \cdot \omega = M \cdot 2 \cdot \pi \cdot n,$$

where M = torque (Nm), ω = angular velocity (rad/s), and n = engine speed (rps). The power output is typically given in kW (formerly hp).

~**Effective (brake) power output.** The effective or brake power output is the useful or net power output at the driveshaft or flywheel, which is available at the corresponding engine speed. The gross power output of the engine is reduced by the loss of power used for driving ancillary drives, such as the alternator, the injection pump, scavenge and cooling fans, the cooling water pump, and superchargers. The result is the net power output, given by P_e, which is

$$P_e = V_H \cdot p_{me} \cdot n \cdot i,$$

where V_H = displacement (m³), p_{me} = effective mean pressure (N/m²), n = engine speed (rps), and i = number of combustion cycles per revolution.

~**Friction loss** (*also*, →Friction). The friction loss of an engine is defined as the friction loss in the engine itself—for example, from the pistons, piston rings, camshaft bearings, connecting rod, and crankshaft bearings, and also the power required to drive the ancillary devices and accessories, such as water and oil pumps.

~**Indicated power.** The indicated power or internal power output is calculated from the indicated mean effective pressure measured from the p-V diagram in the combustion chamber.

$$P_e = V_H \cdot p_{mi} \cdot n \cdot i,$$

where V_H = displacement (m³), p_{mi} = effective mean pressure (N/m²), n = engine speed (rps), and i = number of combustion cycles per revolution.

~**Internal power.** *See above*, ~Indicated power

~**Net power.** *See above*, ~Effective (brake) power output

~**Power limitation.** Power limitation can be achieved by restricting the engine speed or charge. Both measures serve, for instance, to protect the engine from becoming mechanically overloaded. Air restrictors are fitted to sports engines in the form of baffle plates (air restrictors).

~**Power loss.** The power loss, P_r, is that lost through friction, flow losses and accessory drives. It is calculated by subtracting the effective power, P_e, from the indicated power, P_i:

$$P_r = P_i - P_e.$$

~**Power output per liter/cubic inch.** *See below*, ~Specific power output per liter

~**Rated power.** The rated power is the maximum net power output of the engine. It is linked to the accompanying engine speed.

~**Specific power** (*also*, →Weight-to-power ratio). The specific power, P_s, of a combustion engine is defined as the power, P_e, per unit engine mass, m_M. This produces

$$P_s = \frac{P_e}{m_M} \text{ [kW/kg].}$$

In series-production gasoline engines that are made from the same materials, engines with the smallest number of cylinders have the highest specific power output—that is, the specific power output falls as the number of cylinders increases. V-engines with six cylinders have a higher specific power output than inline, six-cylinder engines as they are more compact, and turbocharged engines have the highest specific power output of all. Modern naturally aspirated engines with four- or five-valve technology constructed of aluminum with variable valve timing and variable geometry induction systems have specific power outputs of 0.8–0.9 kW/kg; the potential value in the near future is 1.5 kW/kg. Formula 1 engines have specific power outputs of up to 4.6 kW/kg, primarily because of the high engine speeds of up to around 19,000 rpm.

~**Specific power output per liter.** The specific power output per liter or power output per unit volume is the power output per liter (or cubic inch) of displacement of the engine. It is also called the specific power output:

$$P_H = \frac{P_e}{V_H}$$

where V_H = displacement (l), P_H = specific power output per liter (kW/L), and P_e = brake power output (kW).

The specific power output per liter of naturally aspirated diesel engines is currently about 25–30 kW/L at engine speeds of 4000–4500 rpm for chamber diesel engines and engines with direct injection. An exception is four-valve prechamber engines, which have very high speeds (about 5000 rpm) with a specific power output per liter of approximately 33 kW/L. The high engine speeds have a negative effect on fuel consumption and noise, however.

The specific power output per liter for direct injection diesel engines with turbocharging is more than 80 kW/L at engine speeds of 4000–4500 rpm and the trend is rising.

Modern naturally aspirated gasoline engines with four- or five-valve technology are at around 50–60 kW/L at an engine speed of 6000 rpm. Turbocharged gasoline engines achieve 80 kW/L at 6000 rpm and more.

Naturally aspirated Formula 1 racing engines, with a displacement of 3 liters, achieve specific power outputs of more than 300 kW/L at about 19,000 rpm.

Power output per liter. →Power output ~Specific power output per liter

Power semiconductor →Electronic/mechanical engine and transmission control ~Electronic components ~~Power transistor

Power steering pump. The power steering pump is an accessory or other device on the engine that is used for power-assisted steering. It is driven by the engine, typically using the belt drive for other accessories, and provides pressurized oil for the hydraulic power steering system through a high-pressure, vane-type gear or radial piston oil pump. The delivery rate is up to 10 L/min at pressures of approximately 60–100 bar. The average power required depends on the movement of the steering. In the future, power steering pumps will all be electrically driven because that is more efficient.

Power steering pump noise →Engine acoustics

Power stroke →Bore-stroke ratio; →Charge transfer

Power take-off (PTO) →Engine accessories

Power transistor →Electronic/mechanical engine and transmission control ~Electronic components ~~Power transistor

Power-driven machine. Power-driven machines accept work. Powertrains, on the other hand, deliver work (combustion engine, turbine). Power-driven machines are used in connection with combustion engines for supercharging. These are, for example, superchargers, fans, etc.

Powertrain →Engine

Powertrain engineering →Engines

Prechamber →Combustion chamber; →Combustion process, diesel engine ~Prechamber engine

Prechamber diesel engine →Combustion process, diesel engine

Prechamber engine →Combustion process, diesel engine

Precision cooling →Cooling, engine

Precombustion chamber →Combustion chamber

Precombustion process →Combustion process, diesel engine

Precompression →Two-stroke engine ~Crankcase precompression

Precontrol map. Data are stored in the precontrol map, such as values for the injection fuel quantity required at this point on the performance map. The data—for example, obtained using sensors and possibly modified depending on the correction of the fuel delivery rate—are then used to compute the injection quantity at the particular point on the performance map together with the precontrol values. The precon-

trol map can be generated for many parameters, such as ignition angle, idle characteristics, and exhaust gas recirculation quantity.

Preheating →Intake air ~Air preheating; →Starting aid

Preheating, coolant →Starting aid

Preignition (*also*, →Postignition). Preignition starts in gasoline engines when ignition occurs before the spark.

Preliminary catalytic converter →Catalytic converter ~Starter catalytic converter

Preliminary filter →Filter ~Intake air filter ~~Air filter elements

Premature ignition. The term premature ignition defines a moment of ignition relative to the top dead center and has the following characteristics:

- Premature ignition in gasoline engines is ignition by the spark plug significantly before top dead center, which can result in, among other things, knocking combustion. Early moments of ignition in general result in a better efficiency and reduce the exhaust gas temperature, but they also result in higher levels of oxides of nitrogen. Early moments of ignition are not without justification, but must be oriented toward the operating conditions of the engine—e.g., efficiency.
- Premature ignition, called preignition in this case, also results if the thermal range of the spark plug is too small. This results in a situation in which the mixture ignites at the hot spark plug before the ignition spark starts the proper ignition process. The thermal range of the spark plug describes the maximum operating temperature that is achieved by the spark plug in equilibrium between heat loss and heat gain.
- Preignition can also be generated at hot points in the combustion chamber or at the hot exhaust valve. The results may include piston burn-out, resulting in engine destruction. This ignition type is also called glow ignition.

Premium fuel →Fuel, gasoline engine ~Premium plus, ~Road octane number

Premium gasoline →Fuel, gasoline engine

Premium plus →Fuel, gasoline engine

Premixture →Carburetor

Pressure carburetor →Carburetor

Pressure control valves →Electronic/mechanical engine and transmission control ~Requirements for mechanical and housing concepts; →Injection system (components) ~Diesel engine ~~Control valves, ~~Injection hydraulics; →Lubrication ~Control and safety components

Pressure coolant →Coolant ~Pressure

Pressure diagram (*also*, →Combustion, gasoline engine ~Knocking; →Cylinder pressure curve). The cylinder pressure diagram shows the pressure variation in the cylinder in relationship to the crank angle. The maximum pressure is reached, in case of maximum efficiency, at 8–10° crank angle after top dead center in gasoline engines. The pressure diagram provides, for example, information about the combustion process, the mixture conversion, and efficiencies. A typical variation of the cylinder pressure in relation to crank angle is shown in **Fig. P146** for one combustion cycle. The abbreviations are: P, cylinder pressure; P_U, ambient pressure; OT, top; UT, bottom dead centre; A, exhaust, and E, intake.

Pressure differential catalytic converter →Catalytic converter

Pressure fluctuations →Cyclical variations

Pressure gauge →Pressure measuring instrument; →Sensors

Pressure gauge oil →Lubrication ~Control and safety components

Pressure gradient →Combustion, diesel engine ~Combustion process; →Combustion, gasoline engine ~Knocking; →Cylinder pressure curve

Pressure increase combustion →Cylinder pressure curve

Pressure intake air →Trapped pressure

Pressure line →Injection system ~Diesel engine ~~Injection hydraulics, diesel engine; →Lubrication ~Lubricating systems

Pressure loads →Electronic/mechanical engine and transmission control ~Requirements for mechanical and housing concepts

Pressure loss →Catalytic converter ~Pressure difference; →Exhaust gas back pressure; →Particles (Par-ticulates) ~Particulate filter system ~~Particulate filter; →Trapped pressure

Pressure oil →Gear pump ~Characteristics

Pressure radiator →Radiator ~Coolant ~~Pressure

Pressure sensor →Pressure measuring instrument; →Sensors

Pressure side →Piston

Pressure switch oil →Lubrication ~Control and safety components

Pressure valves →Injection system (components) ~Diesel engine ~~Injection hydraulics

Pressure waves (Comprex) →Supercharging ~Pressure wave supercharger (Comprex)

Pressure-measuring instrument (*also*, →Sensors). A large number of pressure-measuring instruments are used in engines during testing and in production. They are divided between piezo-electric and piezo-resistive pressure sensors. The surface of piezo-electric sensors is charged electrically when the crystal gets stressed mechanically. It is used for rapidly changing pressures such as the cylinder pressure in the combustion chamber of engines. The piezo-resistive effect is based on the elastic bending of a single crystal silicon membrane when pressure is applied to one side. A Wheatstone bridge diffused into the sensor membrane is brought into a state of imbalance because of the application of pressure.

Piezo-electric vibration sensors are primarily used in engines to determine knocking in gasoline engines. This is achieved by structure-borne noise measurements on the external skin of the crankcase.

Pressure-measuring instruments are also found in the intake system if, for example, the pressure is required to be a value proportional to the engine load to control the engine.

Pressure sensors (piezo-resistive) are used in diesel engines for the measurement of the high-pressure side pressure during electronic injection. Pressure measurement in the cylinder of production engines with state-of-the-art crystals (higher temperature resistance and higher measurement accuracy) may in the future become a new control variable for engine management systems. This is because the cylinder pressure variation contains a large amount of information about the combustion process. The measurement of cylinder pressure variation provides, in connection with the respective calculations, information about internal efficiencies, mixture conversion, friction loss, rate of combustion and combustion duration. These variables allow, among other things, judgments concerning the quality of combustion.

Literature: 1st Darmstadt Indexing Symposium, 17/18 May, 1994, Joint Presentation by AVL Germany with Prof. Hohenberg, TH Darmstadt.

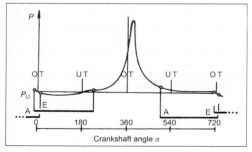

Fig. P146 Pressure diagram of a four cycle engine

Pressure-wave supercharger →Supercharging

Primer pump →Injection system (components) ~Diesel engine

Printed circuit board →Electronic/mechanical engine and transmission control ~Electronic components ~~Transmission frame

Process →Combustion process

Process control →Combustion process

Process peak temperature →Combustion temperature

Process progress →Combustion process

Production, fuel →Fuel, diesel engine; →Fuel, gasoline engine

Production/mixing, oil →Oil ~Body

Progression (throttle valve) →Actuators ~E-gas

Propane →Fuel, gasoline engine ~Alternative fuels

Propeller curve (*also*, →Supercharging ~Propeller operation). For the application of an engine for driving a fluid, adjustment of the torque via the rotational speed is important in order to match the characteristic of the torque curve of the fluid to be driven. A propeller curve is depicted in **Fig. P147**. Along this curve, the torque is proportional to the square of the speed. A typical application is the drive for a ship propeller.

Propeller operation →Propeller curve; →Supercharging ~Propeller operation

Propelling nozzle →Engine concepts ~Composite systems ~~Turbo-compound systems

Properties, fuel →Fuel, diesel engine ~Properties; →Fuel, gasoline engine ~Properties

Protective coating →Piston ring ~Contact surface armor, ~Surface treatments

Protective side layers →Piston ring ~Wear

Psycho-acoustic parameters →Engine acoustics

Pull down →Carburetor ~Starting systems

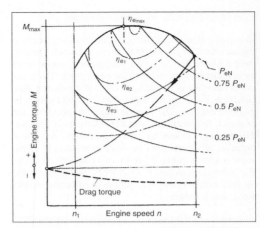

Fig. P147 Illustration of engine torque with propeller curve (Source: Mollenhauer)

Pulley screw →Engine bolts ~Threaded connections ~~Pulley bolts

Pulse →Supercharging ~Pulse turbocharging

Pulse converter →Supercharging ~Pulse supercharging

Pulse hammer excitation →Engine acoustics

Pulse supercharging →Actuators; →Air cycling valve ~Air cycling valve functions ~~Torque increase; →Supercharging ~Pulse turbocharging

Pulse turbocharging →Supercharging

Pulse versions →Piston ring ~Gap

Pump limit →Supercharging ~Pulse converter

Pushrod →Connecting rod; →Valve gear ~Indirect valve gear

Pushrod control →Valve gear

p-V diagram (*also*, →Cycle). The progress of the processes in an internal combustion engine can be reasonably depicted in a *p-V* diagram. In this diagram, the pressure in the combustion chamber is shown as a function of the cylinder volume for the phases of the cycle—for example, with a four-stroke engine, induction–compression–expansion–exhaust.

Q

Quadruple output stage →Electronic/mechanical engine and transmission control ~Electronic components

Qualification, emissions →Emission measurements ~Test type I

Quality control →Load

Quality specifications for fuels →Fuel, diesel engine; →Fuel, gasoline engine

Quality specifications for oil →Oil ~Classification, specifications, and quality requirements

Quench area →Combustion chamber squish areas

Quench edge →Combustion chamber squish areas

Quench effect. During the combustion of hydrocarbons in the engine, unburned hydrocarbons occur in the exhaust in relatively high concentrations. These hydrocarbons primarily originate from zones that are not fully covered by combustion or flame. These are mostly areas near the chamber walls, in the case of gasoline engines, in which, owing to the high dissipation of heat to the wall, the combustion rate is greatly reduced or the flame extinguishes. This process is known as the "quench effect." In particular, the flame extinguishes in gaps that exist in the combustion chamber of gasoline engines—for example, in the section around the cylinder head gasket and toward the ring section. Therefore, to minimize the hydrocarbon emissions, compact combustion chambers must be developed.

Quench flow →Charge movement; →Combustion chamber squish areas

Quench gap →Combustion chamber squish areas

Quench zone →Quench effect

Quicker passing →Engine acoustics

Q

R

Racing engine oils →Oil ~Classification, specifications, and quality requirements ~~Special engine oils

Racing engines →Engine

Racing fuels →Fuel, gasoline engine

Radial bearing →Bearings ~Function

Radial compressor →Supercharging ~Exhaust gas turbocharging

Radial engine →Balancing of masses ~Inertia forces; →Engine concepts ~Reciprocating piston engines ~~Single-shaft engines

Radial inline engines →Engine concepts ~Reciprocating piston engines ~~Single-shaft engines ~~~Radial engines

Radial piston distributor injection pump →Injection system, fuel ~Diesel engine ~~Distributor pumps

Radial rotary piston engines →Engine concepts ~Reciprocating piston engines ~~Single-shaft engines ~~~Radial engines

Radiation loss factor →Engine acoustics

Radiator (also, →Heat exchanger). Radiators are a part of the engine cooling circuit. These consist of heat exchangers in which the energy is transferred, because of the temperature difference, from a substance flowing through at high temperature (e.g., coolant, oil, charge air, etc.) to one with a lower temperature (e.g., ambient air, coolant) as each flows through the device.

In addition, a radiator in a motor vehicle must also fulfill technical requirements in terms of stability and resistance to corrosion. Liquid/air, air/air, and liquid/liquid radiators are used, depending on the mediums to be cooled. Exhaust/coolant radiators (EGR radiators) also exist for cooling the recirculated exhaust gases.

A portion of the energy from the combustion process is transferred specifically to the atmosphere to cool the engine components (e.g., piston, cylinder head, cylinder liner). The widely used liquid cooling is accomplished using a closed coolant circuit. Here, the excess heat is initially transferred from the engine to the coolant and then transferred to the ambient air in a coolant/air radiator. Evaporation of phase change cooling is a special type of liquid cooling, still in the development phase, in which flow, whose circulation is reduced considerably, evaporates on the hot engine components and is then returned to the liquid phase in an air-cooled condenser.

Fig. R1 Charge air intercooler for passenger car

With air cooling, the ambient air is blown directly against the engine components to be cooled.

~Air/air cooler (also, →Supercharging ~Charge air cooler). The meshes in these coolers are generally made of aluminum; they are very similar to those in coolant radiators (**Fig. R1**). The cooler tubes are provided with fins on the inside and outside to increase the surface for heat transfer (**Fig. R2**). When used in heavy commercial vehicles, the air housings are usually made of chilled cast aluminum and are welded to the bottom of the block. Because of the lower-charge air temperatures and pressures, plastic cases made of polyamide can be used in passenger-car applications. On lighter commercial vehicles, air housings of temperature-resistant plastic materials are also used today to reduce costs and weight.

~Charge air intercooling. Charge air intercooling is a suitable means of increasing the performance of

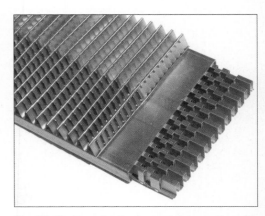

Fig. R2 Cooling tubes in charge air intercooling with fins on inside and outside

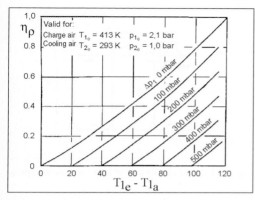

Fig. R3 Density recovery η_p in relation to charge air intercooling $\Delta_t = T_{1e} \sim T_{1a}$ and charge air pressure loss Δp_1

supercharged engines and simultaneously reducing pollutant emissions.

The density "recovery" is an indication of the quality of charge air intercooling (**Fig. R3**). The more the charge air is cooled, the lower the pressure loss on the charge air side and the higher the density recovery.

The temperatures of the charge air at the inlet to the cooler range up to 150°C on passenger car engines and up to 220°C on commercial vehicle engines.

The absolute pressure of the charge air is between approximately 2 bar (passenger car) to 3 bar (commercial vehicles).

The density recovery, η_p, can be considered to be an indicator for the performance increase of an engine resulting from charge air intercooling. It is defined as follows:

$$\eta_p = \frac{\Delta\rho_1}{\Delta\rho_{1max}} = \frac{\dfrac{T_{1e}}{T_{1a}}\left(1 - \dfrac{\Delta p_1}{p_{1e}}\right) - 1}{\dfrac{T_{1e}}{T_{2e}} - 1}, p$$

where $\Delta\rho_1$ = actual change in density of charge air, $\Delta\rho_{1max}$ = maximum possible change in density at given operating state, Δp_1 = pressure decrease in charge air over entire intercooler, T_{1e} = temperature of charge air at cooler inlet, T_{1a} = temperature of charge air at cooler outlet, T_{2e} = temperature of cooling air at cooler inlet, and p_{1e} = absolute pressure of charge air at cooler inlet.

The density recovery is high at high charge air intercooling and low pressure loss on the charge air side.

~Coolant (*also*, →Coolant). Coolants consist of mixtures of water (potable water quality) and 30–50% antifreeze by volume. The basic substances used as antifreeze agents are glycols, primarily ethylene glycol, with corrosion protection additives. In addition to dissipating the heat, the coolant also performs other functions:

• It prevents parts of the entire cooling system from freezing.

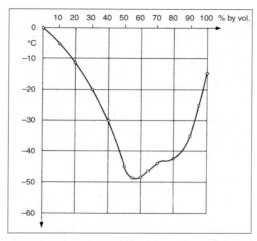

Fig. R4 Freezing point of coolant in relation to quantity of glycol added

• It protects the inside of radiators, circuit components, and the engine from corrosion.
• It increases the boiling temperature in the cooling system.

Fig. R4 shows that up to about 55% of antifreeze by volume, the freezing point of the mixture decreases; at higher percentages it begins to increase again.

The boiling temperature of cooling mixtures is higher than that of pure water. **Fig. R5** shows the vapor pressure in relation to the temperature and concentration.

~~**Flow rate.** Depending on the performance class of the engine, between 4000 and 18,000 L/h of coolant are pumped through passenger car engines and about 8000 to 32,000 L/h in commercial vehicle engines.

~~**Mixing ratio.** The mixing ratios common in Europe are in the range of 35 to 50% glycol by volume.

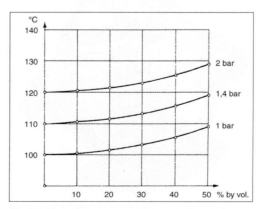

Fig. R5 Vapor pressure curves for water/glycol mixtures

~~**Pressure.** The maximum permissible pressures in the cooling system for passenger cars are currently in the range of 1.3 to 2.0 bar, and for commercial vehicles, 0.5 to 1.1 bar. Increasing the pressure in the cooling circuit allows the coolant temperature to be increased to achieve a higher effective temperature gradient without the coolant boiling. In gasoline engines, this increase is limited—otherwise the tendency to knock increases.

~~**Temperature.** The maximum permissible coolant temperatures, depending on the vehicle manufacturer and operating state, are 100–120°C for passenger car engines, and 98–105°C for commercial vehicle engines.

~~**Thermostat.** Coolant thermostats are used in all vehicle cooling surfaces (**Fig. R6**). They control the flow of coolant so that it begins to flow through the radiator only after it has reached a certain temperature at which the thermostat opens a valve. With so-called map thermostats, it is possible to adapt this opening temperature to engine operation. This allows the coolant temperature to be increased in the part-load range, which is less critical thermally, in order to reduce friction losses in the engine.

~Coolant circuit (*also*, →Cooling circuit; →Cooling, engine). Due to the continuous increase in the specific engine output and the associated higher thermal load on the cylinder walls and cylinder heads in internal combustion engines, the layout of the cooling circuit is becoming more and more important (**Fig. R7**).

The key points are:

- Coolant must not be lost during warm-up and after shutting off the engine (hot shutdown).
- It is necessary for gas and vapor to separate out of the coolant.
- Sufficient reserve quantities of coolant must be available for operation and for minor losses from leakage.

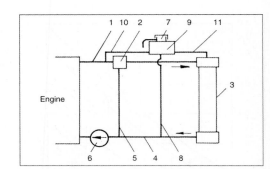

Fig. R7 Coolant circuit with expansion tank in bypass

- A vacuum should not be created on the intake side of the coolant pump (danger of cavitation in the pump).

Today, primarily coolant circuits with an expansion tank in the bypass are used to satisfy these requirements. The design of such a coolant circuit is shown in **Fig. R7**. The most important components are the feed line (1) from the engine to the radiator, the coolant thermostat (2) installed in the front return line, the radiator (3), the return line (4) from the cooler to the coolant pump (6), as well as the bypass line (5) leading from the thermostat to the intake side of the pump. The coolant reservoir (9) with pressure and vacuum valve (7) provides for a ventilation function as well as storage. In order to ensure continuous ventilation even when the thermostat is closed, a first ventilation line (10) branches off in the flow direction in front of the thermostat to the expansion reservoir. It is also practical to have a second vent line (11) branching off to the expansion tank from the coolant reservoir in the radiator. It is necessary to keep the flow rate in the coolant tank very low to allow separation of gases escaping into the coolant through cylinder head gaskets, hose connections, and coolant pump gaskets. This is achieved by selecting a smaller cross section for the connection line (8) than for vent lines (10) and (11).

~Coolant temperature →Radiator ~Coolant

~Cooling capacity →Radiator ~Liquid/air radiator

~Design. Plate, fin, and disk-type radiators are used in motor vehicles.

~~**Disks.** Disks are a special version of a liquid/liquid radiator (usually oil coolers). The disks consist of two semi-shells inserted into the turbulence plate. The cooling capacity and pressure drop can be adapted to the requirements by stacking the disks on top of one another (**Fig. R8**). The coolant flows through the smooth surface without fins or through the finned outer side of these disks. For this reason, disk-type radiators require a separate housing or a suitable installation space in the engine block or filter housing.

Fig. R6 Coolant thermostat

R

Fig. R8 Commercial vehicle oil/coolant radiator with disk design

~~Fin. Fins are used particularly on the air or exhaust gas side of radiators. As a rule, these consist of thin strips of material that are connected to the radiator tubes in a manner that promotes heat conduction. In addition to the associated increase in the surface available for transferring the heat (depending on the fin thickness), such fins allow the airflow to be given a specific turbulence (gills on soldered radiators, turbolators, or gills on mechanically-connected radiators), thereby increasing the heat transfer to the air. Two production processes are possible:

- Rolling of corrugated fins
- Stamping of fins

Usually, the rolling processes allow higher production rates than stamping the fins.

~~Plate. *See above,* ~~Disks

~~~Stacked disks. Stacked disks are a special form of disk radiators with disks shaped like a pan (**Fig. R9**). The design of the disks allows a circumferential seal from one disk to the next on the oil side as well as the coolant side. Stacked disk radiators, therefore, do not require an additional housing. The oil and coolant sides are equipped with either separate turbulence inserts or surface structure on the disk for creating turbulence.

~Disk cooler. *See above,* ~Design

Fig. R9 Different versions of stacked disk oil coolers

~EGR radiator →Exhaust gas recirculation

~Engine cooling circuit →Cooling circuit

~Exhaust/coolant radiator →Exhaust gas recirculation ~EGR radiator

~Fin cooler. *See above,* ~Design

~Fuel cooler. *See below,* ~Liquid/liquid radiator

~Liquid cooling. *See below,* ~Liquid/air radiator, ~Liquid/liquid radiator

~Liquid/air radiator. The individual components of a coolant/air radiator are shown in **Fig. R10**. These consist primarily of the radiator network consisting of tubes and fins, the coolant box, the side sections, the radiator bottom, and the rubber seal between the coolant box and the bottom. On mechanically joined coolant/air radiators, a rubber seal is also present between the connection between the tube and the bottom. The illustration also shows a cooler for transmission oil, for installation in the coolant box.

The cooling capacity and the radiator weight are influenced significantly by the materials used and the design of the radiator network. For this reason, aluminum alloys are used to a great extent today to replace the previously common copper and brass radiator materials, particularly in Europe. In comparison to a copper/brass radiator with the same cooling capacity, an aluminum radiator for passenger cars has a weight advantage of 20–30%. Moreover, the aluminum radiator is very stable under pressure. Replacement of brass coolant reservoirs with fiberglass reservoirs has also contributed to reducing the total weight of aluminum radiators. The fiberglass coolant reservoirs produced by injection molding are connected mechanically to the bottom together with the rubber gasket inserted into the bottom.

Aluminum radiators are separated, depending on the production process, into mechanically joined networks (**Fig. R11**) and brazed networks (**Fig. R12**).

R

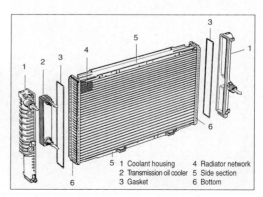

1 Coolant housing	4 Radiator network	
2 Transmission oil cooler	5 Side section	
3 Gasket	6 Bottom	

Fig. R10 Structure of passenger car radiator

Fig. R11 Mechanically joined radiator network with flat tubes

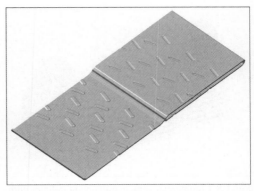

Fig. R13 Radiator tube with winglet structure

The mechanically joined radiator network consists of seamless drawn round or oval tubes and attached, stamped fins. The fins are slotted crosswise to the direction of the airflow in the form of gills to improve the heat transfer. In some cases, simple corrugated fins are used. The tubes are connected to the fins (press connection) by expanding the tubes by 0.3–0.5 mm. Improvement in the expansion process has led to the use of continuously thinner oval tubes, thereby increasing the performance of these heat exchangers.

On brazed flat-tube/corrugated-fin systems, the network is formed of welded flat tubes solder-plated on the outside and rolled corrugated fins with gills crossways to the direction of airflow. Networks with a row

of tubes at the bottom offer advantages to networks consisting of a number of rows of tubes, particularly in terms of cost.

The performance of radiators can be increased by using turbulence inserts. In the case of mechanically joined radiators, the turbulence inserts consist of coils, strips bent to a wave shape, or other specially developed aluminum or plastic structures into which the tubes are inserted. On brazed radiators, stamped aluminum strips are used.

Another more economical method of increasing the performance is the use of structured tube surfaces. Such a structure is shown in **Fig. R13**. The half-opened drain tube has so-called winglets that can be embossed into the surface of the tube from the outside; this winglet structure also increases the turbulence in the flow.

Fig. R14 provides a summary of the capacity of various radiator systems. The specific heat capacity is based on the inlet temperature gradient between the two substances flowing through the radiator.

On brazed aluminum oil/air radiators, flat tubes with turbulence inserts and extruded profiled tubes are used. This is required on the one hand to increase the stability under pressure; on the other hand, it also significantly increases heat transfer on the oil side.

~Liquid/liquid radiator. This type of radiator is very versatile in terms of design and allows high power densities to be achieved. Nearly all designs use brazed turbulence inserts that increase the internal pressure loss and improve the heat transfer in the flow through the tubes by increasing the turbulence. The oil flow or fuel flow is separated from the coolant by tightly brazed plates, disks, or flat tubes.

When high requirements are placed on the stability of the oil cooler (e.g., use in commercial vehicles), steel is used as the material for the turbulence inserts and stainless steel for the plates. The components are held in place by copper brazing. Aluminum is also being used increasingly for such applications.

The use of oil/coolant coolers offers the advantage that the lubricating oil is warmed more quickly during

R

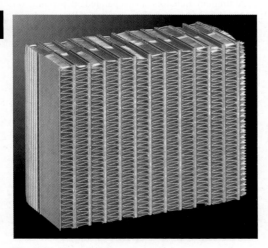

Fig. R12 Brazed radiator network

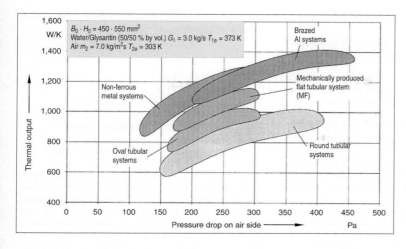

Fig. R14
Performance of various
coolant radiator systems

the warm-up phase by the coolant, the temperature of which rises more quickly than that of the engine, thereby warming the engine more quickly to operating temperature.

Transmission oil coolers are frequently constructed in the form of aluminum flat tubular coolers in a receptacle for the coolant radiator. However, because of the increasing power requirements, stacked plate-type oil coolers are being used more and more often.

~Louvers. Regulated cooling air louvers (**Fig. R15**) take into consideration that the cooling air flowing into the engine compartment is not required and not necessarily even desirable under all operating conditions.

They fulfill three different functions:

- When completely closed, they reduce the noise emission from the engine compartment to the front during the warm-up phase. This contributes to reducing the exterior noise, particularly on diesel engines.
- In the closed position, the louvers also prevent heat loss from the engine to the cooling air, thereby accelerating the warm-up phase, which contributes to a reduction in the fuel consumption. At low ambient temperatures, they also increase the thermal comfort in the vehicle by reducing the heat losses.
- Closed or partially opened louvers reduce the aerodynamic drag of the vehicle, thereby reducing fuel consumption, particularly at high speeds.

Radiator louvers with a rectangular shape that are installed between the coolant radiator and fan are the best-known form of louver. However, intake louvers located directly behind the radiator grille or circular fan louvers behind the fan impeller are also used.

Actuation is possible with devices consisting of expanding material, vacuum boxes, stepping motors, or the relative wind (for reducing noise emissions while standing still).

~Materials. The properties of aluminum such as low-specific weight (density), high thermal conductivity, strength and corrosion resistance, as well as excellent formability and easy handling make it an ideal material for modern radiators. In combination with plastic collection reservoirs, typically PA66 with fiberglass, weight savings of up to 50% are possible for commercial vehicles and up to 30% for passenger cars when compared with copper and brass radiators of the same size.

While aluminum alloys have replaced conventional copper and brass radiator materials in Europe, soldered radiators with brass tubes and copper finned

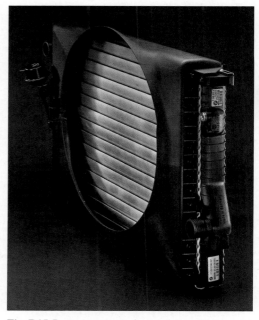

R

Fig. R15 Passenger car coolant radiator with louvers

tubes are still used to a greater extent in other automobile-producing countries.

Pb-Zn solder is used on soldered copper/brass radiators, while aluminum radiators are brazed.

~Oil cooler (*also, see above,* ~Liquid/air radiator). Oil coolers are used in vehicles particularly for the engine, transmission, and hydraulic equipment.

~~Engine. The compact design and high engine outputs of current gasoline and diesel engines cause higher component temperatures. This results in increased oil temperatures, which must be reduced by specific measures to cool the engine oil. This is usually essential in supercharged engines and engines with additional piston cooling.

Moreover, oil cooling is considered necessary to increase the service life of the engine oil and extend oil-change intervals.

The thermal capacity of engine oil coolers is in the range of 6 to 50 kW, whereby the maximum oil temperatures occurring at the inlet to the cooler can be as high as 155°C. Oil flow rates reach 30–60 L/min (60–200 L/min in commercial vehicles), and the operating pressures are in the range of 6 to 12 bar.

~~Hydraulic system (power steering). Hydraulic oil coolers are manufactured in the form of simple oil/air coolers by placing a finned or unfinned bank of tubes in the cooling airflow, particularly in front of other heat exchangers. For higher cooling requirements, brazed oil/air coolers with extruded, profiled tubes are also used.

~~Transmission. Transmission coolers in the form of double copper tube coolers or aluminum flat tubular coolers are frequently installed in a receptacle in the coolant radiator. Another possibility for transmission cooling is attachment of an oil/coolant cooler directly on the transmission, where a stacked plate-type oil cooler is usually used. In addition to limiting the maximum temperatures, it is also desirable to warm up the transmission oil quickly to its operating temperature to minimize friction losses. For this reason, modules consisting of an oil/coolant cooler and a mixed thermostat are used to equalize the temperature of the transmission oil. It is possible to increase the transmission oil temperature during the warm-up phase by allowing warm coolant to flow through the oil/coolant cooler. After reaching the operating temperature, the system is switched over to cooling. If a higher cooling capacity is required, oil/air coolers can also be used; the operating pressure range is up to 6 bar. The quantity of oil circulating is about 5–15 L/min, whereby cooling capacities of 3–15 kW are achieved.

~Plate cooler. *See above,* ~Design

~Pressure system. *See above,* ~Coolant ~~Pressure

~Temperature gradient →Radiator

~Transmission oil cooler. *See above,* ~Liquid/air radiator, ~Liquid/liquid radiator

Radiator design →Radiator ~Design

Radiator fan →Fan

Radiator fan drive →Cooling circuit

Radiator material. →Radiator ~Materials

Radiator tank. →Radiator ~Liquid/air radiator

Radicals. The reaction of the fuel-air mixture in the engine involves physico-chemical processes. In addition to the reactive atoms such as H, O, and N, radicals such as OH, CH, CH_3, and C_2H_5 are formed during combustion, and these contribute decisively to the reaction process. The concentration of radicals is a function of the temperature and pressure of the mixture and whether it is in a state of chemical equilibrium or nonequilibrium.

Radius pintle nozzle →Injection valves ~Diesel engine ~~Pintle-type nozzle

Ram →Ram effect; →Supercharging

Ram effect. By tuning the intake and exhaust systems, a ram effect can be achieved with the aid of the pulsating gas columns that increases the volumetric efficiency and, therefore, the performance and torque of the engine. Here, it is important that a pressure wave be present at the intake valve toward the end of the intake cycle just before the valve closes, providing a ram effect as a result of the higher mixture density.

The ram effect is dependent on many parameters, including the length and diameter of the intake manifold. Moreover, the engine speed and intake flow rate are important. For this reason, tuning is required in the specific speed range in which the desired ram effect is to occur. In order to include wide areas of the speed range, variable resonance induction systems are used, which change the length of the manifold depending on the speed. Variable resonance induction systems have two or three different lengths. In the meantime, infinitely adjustable variable resonance induction systems are also in mass production. Short manifold lengths have a positive effect for ram at high engine speeds, while long intake manifolds are more favorable at low engine speeds. **Fig. R16** shows a variable resonance induction system on a V-6 engine that is located in the engine V, and the effect on the torque is shown in **Fig. R17**.

On engines of two or more intake valves per cylinder, the total cross-section area and length of the intake manifold can be varied with load and speed by switching off a port (**Fig. R18**). The effect of this procedure on the cylinder charging is shown in **Fig. R19**. Such measures influence the mixture preparation and the lower speed range in addition to the cylinder charge. The efficiency in the part-load range is improved par-

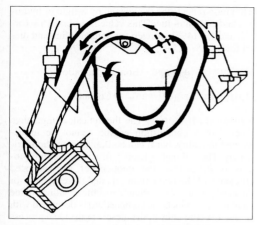

Fig. R16 Variable resonance induction system

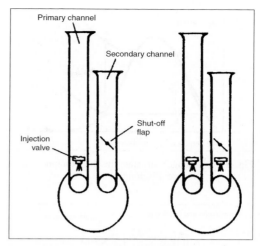

Fig. R18 Intake ports on multiple valve engines with port shutoff

ticularly in lean operation by specific charge movement—for instance, a swirl. Ram effects can be supported by appropriate layout of the exhaust system. A vacuum wave in the exhaust system just before the exhaust valve closes is of practical assistance in obtaining the maximum fresh charge in the cylinder.

Ram induction →Supercharging

Ram on supercharged engines →Air cycling valve ~Air cycling valve functions ~~Torque increase

Ram pressure flowmetering →Flowmeter intake air

Ram process →Ram effect; →Supercharging

Rankine cycle (*also*, →Engines ~Alternative engines ~~Steam engine). The Rankine cycle was origi-

nally applied to steam engines. In recent times, the Rankine cycle has been reconsidered in the search for alternative power cycles for vehicles. The changes of state in the Rankine cycle are illustrated in **Fig. R20**. **Fig. R21** shows a schematic illustration of a steam engine that operates on the Rankine cycle.

The system essentially consists of:

- A boiler for evaporating the working fluid
- A reciprocating machine for expanding the steam
- A regenerator for transferring heat
- A condenser
- A pump

The real cycle, in contrast to the ideal one, is affected by a number of losses—for example, in the evaporator and during condensation (flow losses).

Currently, steam engines running on the Rankine cycle do not present an alternative to gasoline or diesel engines in vehicle applications.

Rapeseed oil →Fuel, diesel engine ~Alternative fuels

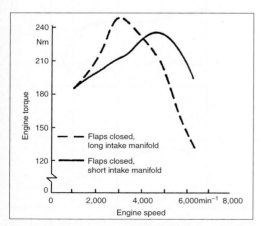

Fig. R17 Effect on the torque of a variable resonance induction system

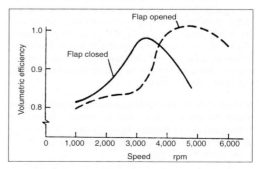

Fig. R19 Effect on the volumetric efficiency of port switch-off

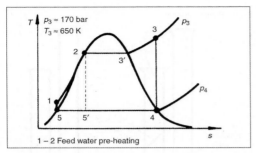

Fig. R20 Changes of state in the Rankine cycle (Source: Pischinger)

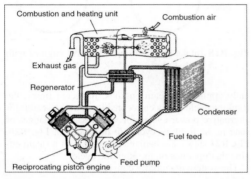

Fig. R21 Schematic illustration of a steam engine according to the Rankine cycle (Source: Pischinger)

Rate of pressure rise →Combustion, diesel engine ~Combustion process; →Combustion, gasoline engine ~Knocking

Rate shaping nozzle (RSN) →Injection valves ~Diesel engine

Rated noise impedance →Engine acoustics

Rated power (*also*, →Power output). The rated power for a given displacement, fuel calorific value, and selected combustion process (two-stroke, four-stroke) depends primarily on the engine speed, the volumetric efficiency, and the overall cycle efficiency. The latter values are a function of the charging processes, whose maximization during engine development leads to the maximum rated output.

Rated speed. The rated speed is the engine speed at which the engine achieves its maximum output. For high-speed diesel engines, this is usually in the range of 4000 to 4500 rpm and for gasoline engines in the range around 6000 rpm.

Ratio, hydrogen/carbon (HC) (*also*, →Fuel, gasoline engine ~Synthetic gasoline). The hydrogen/carbon (HC) ratio reflects the percentage by weight of hydrogen and carbon in fuel. This value is around 86% carbon and 13–14% hydrogen for both gasoline and diesel fuels.

Ratio, stroke/bore →Bore-stroke ratio

Ratio, water/antifreeze →Coolant

Rational efficiency. The theoretical, comparative process of a combustion engine cannot be implemented in reality, but it can be implemented approximately. The rational efficiency of a real working process is defined as the quotient of the indicated efficiency of the real engine, η_i, and the efficiency of the ideal process. The efficiency of the gasoline engine ideal process (isochoric combustion) is described with η_v, and the efficiency of the diesel engine ideal process (Seiliger combustion) is described with η_{vp}.

The resulting rational efficiency, η_g, is

$$\eta_g = \frac{\eta_i}{\eta_v} \text{ for the gasoline engine}$$

(isochoric [constant volume] combustion) and

$$\eta_g = \frac{\eta_i}{\eta_{vp}} \text{ for the diesel engine (Seiliger process)}.$$

The achievable rational efficiency depends on the actual working process, such as the gasoline or the diesel process. The operating parameters are the variables. The air-fuel ratio is of significant importance. **Fig. R22** shows this correlation for a gasoline and a diesel engine.

Gasoline engines have a maximum equivalence ratio of 0.8. The air-fuel ratio maximum for diesel

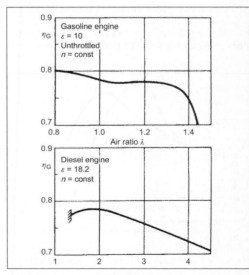

Fig. R22 Variation of rational efficiency for diesel and gasoline engines as a function of the air-fuel ratios (Source: Pischinger)

engines is at much higher values. The rational efficiency declines continuously with increasing air-fuel ratios or declining loads, but it is still relatively high due to the qualitative regulation in the diesel engine. The diesel engine has a limit on the rich side due to soot generation. The following rational efficiencies are available:

- Combustion rational efficiency
- Combustion curve rational efficiency
- Calorific value rational efficiency
- Charge transfer rational efficiency

Literature: A. Urlaub: Verbrennungsmotoren, Grundlagen-Verfahrenstheorie-Konstruktion, 2nd Edition, Springer Publishers, 1994.

Raw air pipe →Air intake system ~Thermodynamic air management system

Raw emission (*also*, →Emissions). Raw emissions are the emissions that come directly from the engine. Their levels are higher by a factor of approximately ten than the concentration prescribed for treated exhaust gas within the limitations framework for exhaust gas components in Europe. The design and tuning of the engine must be such that raw emissions are reduced as far as possible, this being a vital prerequisite for optimum exhaust gas treatment. The potential for further lowering of the raw emissions from gasoline and diesel engines is considerable, which is the reason such a great deal of effort is being made in this direction.

Reaction speed (*also*, →Chemical reaction). This term designates the rate at which a substance is formed or consumed during a chemical reaction. The higher the rate of reaction during combustion in the engine, the quicker the combustion process is completed. In theory, this means an increase in thermal efficiency, because the real cycle becomes increasingly close to an ideal constant volume one. At the same time, pressure and temperature in the combustion chamber generally increase, with increased formation of oxides of nitrogen and higher wall heat losses as consequences.

Reactor →Exhaust gas thermal reactor

Rear engine →Engine installation

Rear exhaust pipe →Exhaust system

Rear muffler →Exhaust system ~Muffler

Recess shape →Piston ~Design

Reciprocating engine (*also*, →Crankshaft drive). The reciprocating engine transforms the gas force created during combustion through a piston, a connecting rod, and a crankshaft into an effective torque. Normally, several structural versions and models are used. They differ with regard to position and arrangement in the cylinder. The basic principle of the slider-crank

mechanism, in which a connecting rod attached eccentrically to the crankshaft transforms the back and forth movement of the piston into a rotary movement, remains unchanged.

Reciprocating inertial force →Balancing of masses ~Inertia forces

Reciprocating masses (*also*, →Balancing of masses ~Inertia forces). The reciprocating masses in a combustion engine are all the masses of engine components that execute back-and-forth motion. These are, for example, valves, pistons, piston pins, piston rings, and parts of the connecting rod. These masses lead to inertial forces during their acceleration, based on number of cylinders or the cylinder arrangement, and to out-of-balance forces and couples, and should therefore be as light as possible due to reasons of material stress and comfort.

Reciprocating piston engine →Engine concepts ~Reciprocating piston engines

Recirculating air duct →Carburetor

Recirculating lubrication →Lubrication

Rectangular ring →Piston ring ~Versions ~~Compression rings

Recycling →Catalytic converter

Recycling of mineral oil →Oil ~Safety and environmental aspects ~~Used oil

Reduction catalyst →Catalytic converter ~Dual-bed catalytic converter

Reduction speed →Speed ~Speed reduction

Reed valve. Reed valves, also known as diaphragm valves, are primarily used on two-stroke engines for controlling gas transfer in the intake port (**Fig. R23**). These valves have elastic reeds that rest on a basic body. The reeds are self-acting and open when vacuum builds up in the engine crankcase so that air or fresh mixture can enter freely. As the pressure increases in the crankcase up to the pressure in the intake manifold, the reeds automatically close, thus preventing backflow of the induced charge. This enables long intake opening times and optimum utilization of gas-dynamic effects at high engine speeds without showing the typical shortcomings in the lower speed range.

Reed valves are primarily employed on two-stroke motorcycle engines. Tests have also been carried out on four-stroke engines, and these have resulted in more than 30% torque improvement. However, the large space requirement remains problematical.

Reed material today is fiber-reinforced plastic. Reeds made from spring steel previously were used, but they

R

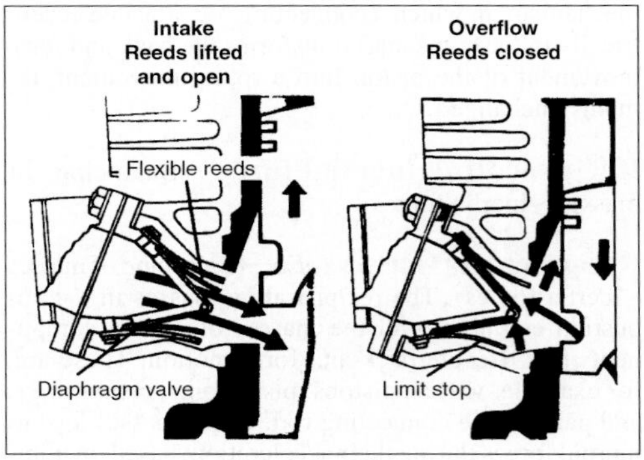

Intake
Reeds lifted
and open

Overflow
Reeds closed

Flexible reeds

Diaphragm valve

Limit stop

Fig. R23 Diaphragm valve for two-stroke engines (Source: Stoffregen)

were susceptible to breakage, leading to expensive engine damage.

Reference fuel →Fuel, diesel engine ~Properties

Reference sizes →Emission measurements ~FTP-75 ~~Equivalent inertia masses

Reflection muffler →Exhaust system ~Muffler

Reflections (*also*, →Intake system). When a gaseous medium is subjected to pressure disturbances, it spreads out like waves. Gas flow in a pipeline, for instance as an intake or exhaust manifold, can be depicted as a superposition of waves moving back and forth. In such a process, different geometries, diameters, and lengths of the pipe have considerable influence on the behavior of the waves. It is possible to distinguish the following cases:

- Manifold with a throttling point at the end
- Manifold with closed pipe end
- Manifold with open pipe end
- Manifold with cross-sectional steps

Different reflection conditions arise depending on the type of design. These reflection conditions can be used to influence engine filling and, thus, to influence the torque and power, primarily through intake tuning, or to optimize the exhaust scavenging process through tuning on the exhaust side.

With a four-stroke engine, the filling process can be improved if high pressure waves occur toward the end of the intake cycle. This results in a supercharging effect. What is vital is that the length and diameter of the intake manifold and any resonance bodies in the intake duct are correctly tuned. The length of the manifold that is correct for a high volumetric efficiency depends on the engine speed, so that this positive effect is achieved only within a limited engine-speed range. Basically, short-intake manifolds are advantageous at high speeds and long-intake manifolds at low speeds as far as the filling effect is concerned. This is the rea-

son multipath-intake manifolds are often used, because these simulate short-or long-intake manifold lengths, depending on the engine speed.

It is important to observe the reflection behavior in a conventional two-stroke engine with port-controlled scavenging because the gas change and, thus, the filling process can influence the exhaust behavior, fuel consumption, and so on. During the scavenging process, a negative pressure wave on the exhaust port improves the scavenging effect. Fresh gas is prevented from leaving the cylinder shortly before the "intake" port is closed by means of an excess-pressure wave.

Reformulated fuels →Fuel, diesel engine ~Composition ~~Reformulated; →Fuel, gasoline engine; →Vaporization, fuel

Regeneration of particle filters →Particles (Particulates) ~Particle filter system ~~Particulate filter

Regular fuel →Fuel, gasoline engine

Regular gasoline →Fuel, gasoline engine

Replacement system →Valve gear ~Elastic valve gear

Research octane number →Fuel, gasoline engine ~Octane number (RON, MON)

Reserve alkalinity →Coolant

Residual gas. During the gas transfer in engines, the burned air-fuel mixture is exhausted from the combustion chamber through the exhaust system to the atmosphere. In this process, it is theoretically possible to fully scavenge the residual gas to facilitate more complete subsequent fresh-gas charging of the cylinder. In practice, however, a certain amount of residual gas is desired in the cylinder. The quantity of pollutant substances in the exhaust—NO_x and HC—can be reduced by the presence of some residual gas in the cylinder. In addition, the combustion temperature is lowered, thus facilitating less formation of NO_x. The higher proportion of HC causes the HC emissions to be reduced through the post-reactions taking place in the subsequent work cycle compared to the mean concentration of the scavenged exhaust. In four-stroke engines, the volume of residual gas (internal exhaust gas recirculation) can be controlled by means of the charge transfer devices, and internal exhaust gas recirculation is especially interesting for part-load operation. At full throttle, proper scavenging of the residual gas is desirable, because this increases the fresh mixture charged into the cylinder and leads to increased power. With high proportions of residual gas, combustion misfire may occur and the engine will not run smoothly.

In port-controlled two-stroke engines, the residual gas scavenging process is undefined and, therefore, more difficult to implement in a controlled manner. Long cylinders help to improve the situation.

R

Residual gas content →Exhaust gas recirculation ~External exhaust gas recirculation; →Variable valve control ~Residual gas control

Residual gas control →Variable valve control

Residual unbalance →Balancing of masses

Residue formation →Oil ~Properties and characteristic values

Resistance, catalytic converter →Catalytic converter ~Pressure difference

Resistance to corrosion →Bearings ~Materials ~~Qualities

Resonance chamber →Supercharging ~Resonance induction

Resonance charging →Supercharging

Resonance flap (also, Intake system; →Supercharging ~Resonance induction, ~Resonance tube). Resonance flaps regulate the routes through multipath-intake manifolds for increasing the maximum achievable torque as a function of speed. This makes it possible to adjust the length (mostly in discrete steps) of the intake manifold as a function of the speed.

Resonance speed →Speed

Resonance supercharging →Supercharging

Resonance tube →Supercharging ~Resonance induction

Resonant frequency →Engine acoustics

Resonant frequency computation →Calculation processes ~Application areas ~~Structure dynamics

Resonator →Supercharging ~Resonance induction

Resonator chamber →Intake system ~Resonance system; →Supercharging ~Resonance induction

Respirability →Particles (Particulates) ~Effects on the human organism

Restrictor. Restrictors are used to control the engine load by air or mixture changes in conventional gasoline engines. "Restrictor" also is called "quantitative control," as opposed to "qualitative control," which is used in diesel engines.

One throttle valve for each cylinder or a double-flow throttle valve component is used, depending on displacement and degree of supercharging. One throttle with a small cross section opens up to a certain load or speed. A second throttle valve with a larger cross section is opened if larger flow rates are required.

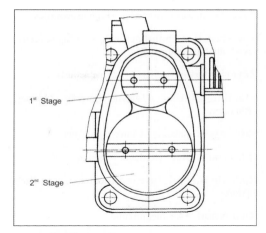

1ˢᵗ Stage

2ⁿᵈ Stage

Fig. R24 Dual-branch throttle valve

The throttle valve can have a progression (spherical zone) to control the air volume better at lower flow rates. A dual-branch throttle valve component is shown schematically in **Fig. R24**.

Rotary and throttle valves also are used in rare cases for engine control in addition to the throttle valve. The gas transfer valves (i.e., intake and exhaust) also can be used as restrictors—for example, **Fig. R24** shows dual-branch throttle valve parts with free variation of the valve lift or the valve opening durations. The conventional throttle valve can be eliminated if the intake valve is used for throttle valve control.

The qualitative control in diesel engines permits the engine to suck the total air volume into the combustion chamber without restriction by the throttle valve.

Throttling losses are generally caused by restrictors, and these reduce the engine efficiency.

Resultant inertial force →Balancing of masses ~Inertia forces

Resultant moments of inertia →Balancing of masses ~Out-of-balance couples

Retarded ignition. Retarded ignition is a term used in connection with gasoline engines. In order to assure effective and optimum combustion, the trapped mixture must be ignited at a certain time, depending on its composition and the operating state of the engine. For certain reasons, such as emissions behavior and warm-up, the most effective moment of ignition can be shifted toward TDC. Then the term retarded ignition or delayed moment of ignition is used.

Retarder →Starter ~Starter-generators ~~Operating modes of the starter-generator ~~~Alternator and retarder

Retention time, catalytic converter →Catalytic converter

R

Reverberation chamber →Engine acoustics

Reverberation chamber process →Engine acoustics ~Sound measurement

Reverberation radius →Engine acoustics

Rhodium coating →Catalytic converter ~Coating, ~Rhodium

Ribbing →Crankcase ~Crankcase design

Rib-type cooler →Radiator ~Design

Rich air-fuel mixture →Air-fuel mixture ~Rich mixture

Rich cloud →Air-fuel mixture

Rigid body dynamics →Calculation process ~Application areas ~~Structure dynamics

Rigid body shapes →Engine acoustics

Ring →Piston ring

Ring belt piston →Piston ~Design

Ring breakage →Piston ring ~Breaking

Ring carrier →Piston ~Ring

Ring flutter →Piston ~Ring ~~Ring flutter; →Piston ring ~Flutter

Ring groove →Piston ~Ring

Ring groove wear →Piston ~Ring ~~Ring groove

Ring land →Piston ~Ring

Ring land height →Piston ~Ring ~~Ring land

Ring portion →Piston ~Ring

Ring running surface topography →Piston ring ~Lubrication

Ring seal →Piston ~Ring; →Piston ring

Ring vibration →Piston ring ~Flutter

Ring with inside bevel (IB) →Piston ring ~Versions ~~Compression rings ~~~Ring with inside vessel or inside angle

Ring with internal angle (IA) →Piston ring ~Versions ~~Compression rings ~~~Ring with inside vessel or inside angle

RME (rape seed methylester) →Fuel, diesel engine ~Alternative fuels ~~Vegetable oil

Road noise →Engine acoustics

Road octane number →Fuel, gasoline engine

Road performance →Fuel consumption ~Variables

Rocker arm →Valve gear ~Indirect valve gear ~~Rocker arm valve train

Rocker arm actuation →Valve gear ~Activation of valves ~~Indirect valve activation

Rocker arm bearing →Bearings ~Bearing positions

Rocker arm control →Valve gear ~Actuation of valves ~~Indirect valve actuation

Rocker arm valve drive →Valve gear ~Indirect valve gear

Roller (rolling road) dynamometer (*also*, →Emission measurements). Roller dynamometers are used for measuring components and systems in vehicles. The drive wheels of a vehicle are put on a roller, which is then driven by the vehicle drive wheels. The advantage of measuring torque on a dynamometer is that the values are reproducible and the desired parameters can be set without being exposed to the variable conditions encountered in a road test. Frequent applications of these dynamometers include measuring the power, the braking, and the emission characteristics of vehicles.

The emission measurements, for instance, are based on legally prescribed general conditions that ensure that comparable results can be obtained under the same general conditions. The results determined are compared with the corresponding legally prescribed values.

For this purpose, driving programs more or less derived from typical driving characteristics in road traffic have been developed. These driving programs, in the form of speed–time characteristic curves, are simulated with a stationary vehicle on the roller dynamometer. The roller dynamometer comprises one or a number of rollers that have contact with the drive wheels of the vehicle and a flywheel that is used to simulate the vehicle inertia; in addition, there are devices for setting the vehicular resistance. The resistances, essentially the rolling and air resistances, are supposed to replicate the running characteristics of the vehicle on a plane surface.

The principle design of a roller dynamometer for the purpose of exhaust analysis is shown in **Fig. R25**.

Roller chain →Chain drive ~Designs: →Valve gear

Roller tappets →Valve gear ~Gear components ~~Pushrod

Rolling resistance →Fuel consumption ~Variables ~~Road resistances

RON →Fuel, gasoline engine ~Octane number

R

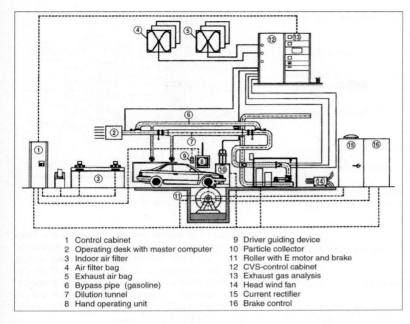

1 Control cabinet
2 Operating desk with master computer
3 Indoor air filter
4 Air filter bag
5 Exhaust air bag
6 Bypass pipe (gasoline)
7 Dilution tunnel
8 Hand operating unit
9 Driver guiding device
10 Particle collector
11 Roller with E motor and brake
12 CVS-control cabinet
13 Exhaust gas analysis
14 Head wind fan
15 Current rectifier
16 Brake control

Fig. R25
Roller dynamometer for
exhaust analysis

Roots compressor →Supercharging ~Rotary-piston supercharger

Roots supercharger →Supercharging ~Rotary-piston supercharger

Rotameter principle →Flowmeter intake air

Rotary-piston blower →Supercharging ~Rotary-piston supercharger

Rotary-piston engine →Engine ~Alternative engines ~~Wankel engine

Rotary-piston supercharger →Supercharging ~Rotary-piston supercharger

Rotary sleeve valve. The charge transfer periods for gas exchange are induction and exhaust. These processes are carried out by the effect of piston movement. The cylinder intake and exhaust have to be periodically opened and closed by means of control units. The control units should be characterized by:

• Maximum cross-sectional areas for gas flow
• Rapid opening/closing processes
• Aerodynamic design
• Dependable sealing effect
• A lifespan corresponding to the life of the engine

Fig. R26 shows two types of control unit designs for four-stroke engines. Poppet valves (left) allow simple and dependable sealing. The pressure in the cylinder enhances the sealing effect. The high valve accelera-

tions lead to considerable operating stresses on the valve gear due to inertia forces. Furthermore, it has to be ensured that contact between the components does not get lost at high speeds (valve bounce).

Rotary valves (right) have shorter opening/closing times. The inertia forces are also eliminated. Sealing and operational durability at high temperatures are problematical, however. The components could seize up and jam, which is why current engines are designed with poppet valves.

Rotary valve →Control/gas transfer ~Four-stroke engine ~~Slide control

Rotary vane →Supercharging ~Pressure wave supercharger (Comprex)

Rotary voltage distribution →Ignition system, gasoline engine ~High-voltage distribution ~~Rotating high-voltage distribution

R

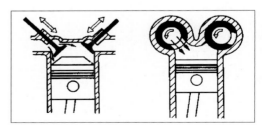

Fig. R26 Design types of control units (Source: Pischinger)

915

Rotary-piston compressor →Supercharging
~Rotary piston supercharger

Rotating device, valve →Valve rotating devices

Rotating high-voltage distribution →Ignition
system, gasoline engine ~High-voltage distribution

Rotating inertial force →Balancing of masses
~Inertia forces

Rotating masses →Balancing of masses

Rotating valve →Valve rotating devices

Rotating viscometer →Oil ~Properties and characteristic values ~~Viscosity ~~~Measuring method

Rotation →Speed

Rotational uniformity →Torsional vibrations
~Nonuniformity

Roto cap →Valve rotating devices

Rounding of the spray hole inlet edge
→Injection valves ~Diesel engine ~~Hole-type nozzle
~~~Blind-hole nozzle

Running gear (*also*, →Supercharging ~Exhaust
gas turbocharging). The term running gear is used for
the rotating parts of a turbocharger. It comprises the

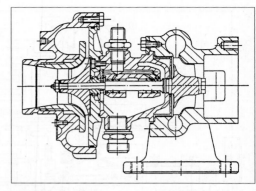

Fig. R27 Cross section of an exhaust turbocharger

compressor and turbine rotor and the shaft that links
these components together. **Fig. R27** shows the design
of an exhaust turbocharger. The turbine rotor should
have a low moment of inertia (GD^2), and this is
achieved by compact construction and light materials—for instance, aluminum for compressors and ceramic for turbines. The transient response of a vehicle
is improved with "lighter" rotors.

RV engine →Engine concepts ~Reciprocating piston engines ~~Single-shaft engines ~~~VR engines

RV-6 engine →Balancing of masses ~Inertia forces

R

S

SAE. This is the abbreviation for the Society of Automotive Engineers (USA), which defines itself as the Engineering Society for Advancing Mobility Land Sea Air and Space.

SAE viscosity classes →Oil ~Classification, specifications, and quality requirements

Safety wheel (two-stroke engine). The safety wheel, α, is the volume of fresh charge present in the cylinder, L_z, divided by the sum of the volumes of fresh charge and short-circuited charge. It is a measure for charging losses caused by short-circuiting of flow through the cylinder. The short-circuited quantity is the volume that escapes through the exhaust ports or the exhaust valves during the charging process before the intake is closed; it therefore does not participate in the combustion. The safety wheel, therefore, is

$$\alpha = \frac{L_z}{L_z + L_A}.$$

The values L_z and L_A are typically expressed in cubic meters.

Literature: K. Groth: Grundzüge des Kolbenmaschinenbaus 1, Verbrennungskraftmaschinen, Vieweg Publishers, 1994.

Sampling →Exhaust gas analysis equipment

Sand casting →Crankcase ~Casting process, crankcase

Scavenging →Two-stroke engine ~Scavenging process

Scavenging air (*also*, →Two-stroke engine). Scavenging air is especially needed for two-stroke engines to flush the residual gases from the cylinder and to fill it with fresh mixture. To accomplish this, a scavenging pump is necessary. Types that are possible for this purpose include piston pumps on the underside of the engine piston (crankcase scavenging), rotary piston blowers (Roots blowers), and centrifugal blowers.

Scavenging efficiency. Scavenging efficiency is especially important in two-stroke engines. It indicates the amount of fresh charge in the cylinder. It cannot be selected in crankcase scavenged engines, but depends on the compression ratio in the crankcase. In blown engines, the charging efficiency can be chosen, but an increase above a certain level does not result in an increase of the scavenging efficiency and requires unnecessarily high power input (gas transfer work).

Scavenging fan →Two-stroke engine

Scavenging losses. Scavenging losses can occur if there is a large valve overlap. In this case, part of the charge flows through the cylinder without contributing to the combustion process. This worsens the efficiency in mixture-inducing engines. However, the advantages are good scavenging of residual gases, better filling of the cylinder, and higher engine performance.

Scavenging pressure →Two-stroke engine ~Scavenging fan

Scavenging pump →Two-stroke engine ~Scavenging fan

Scavenging slope →Supercharging ~Exhaust gas turbocharging, ~Charging process

Schnuerle loop scavenging →Two-stroke engine ~Scavenging process

Scorch marks →Piston ring

Scotch yoke engine →Balancing of masses

SCR catalysts →Pollutant aftertreatment ~Pollutant aftertreatment lean concepts

Screw, connecting rod →Engine bolts

Screw, cylinder head →Engine bolts ~Threaded connections

Screw connections →Engine bolts

Screw-type connector →Supercharging

Scuff resistance, bearing →Bearings ~Materials ~~Qualities

SDOF →Engine acoustics

Seal designs →Sealing systems

Seal rings →Piston ring ~Versions ~~Compression rings

Sealing systems. A variety of elements are used as seals in engines to guarantee safe sealing of the different materials (oil, coolants, fuels, combustion gases) from the outside and also from each other. The sealing materials can act as force transfer elements between engine components and also, depending on the application, can take on additional functions. With the exception of dynamic seals, the seals used in engines are mainly flat, and they can be divided into cylinder head gaskets and special seals. Different seal designs are used depending on the area of application, such as cylinder head/crankcase, intake, liquid, or exhaust gas areas.

~Compression seal. This is so-called because of the surface compression at the seal area (here: interface sealing—component), which seals a pressurized material against the surroundings. A compression seal is often applied in the form of line compression (defined in N/mm), for example through seal beadings (*see below*, ~Cylinder head gaskets ~~Beadings) or elastomer lip seals.

~Cylinder head gaskets. The cylinder head gasket provides the sealing between crankcase and combustion chamber for the combustion gases, the coolant, and the pressurized and return oil. It also acts as a force transmission element between the two components and therefore has a significant influence on the force distribution within the total tensioning system and the resulting elastic deformations of the components. Cylinder head gaskets with metal layers are preferred in car engines today—metal/soft material cylinder head gaskets also may be used for spare-part applications. Metal/elastomer cylinder head gaskets are mainly used in commercial vehicles.

~~**Beadings.** The sealing function of cylinder head gaskets with metal layers is mainly determined by the beadings in the spring steel layers. A beading works like a classical spring, which builds the required sealing force depending on its deformation.

The deformation characteristics allow plastic adaptation to the rigidity of the components and also a high resilient capability to compensate for dynamic vibrations in the sealing gap and thermal deformation of the components. The line pressures required for the sealing are achieved by the use of half beadings ("step-shaped" geometry) in the fluid areas and in general full beadings (semi-circle geometry) for the combustion chamber (**Fig. S1**).

For metal/elastomer cylinder head gaskets, the beading also attempts to compensate for the different rigidities of components (**Fig. S2**).

~~**Combustion chamber gasket.** *See above*, ~~Beadings

~~**Combustion chamber pad rings.** Welded to plastically moldable beadings or supporting positions, combustion chamber pad rings are used to increase the sealing force or concentrate it for metal/elastomer cylinder head gaskets (**Fig. S3**).

~~**Cylinder head gaskets with metal layers.** This type of sealing system can be adapted to the specific requirements of the engine because of its modular design (**Figs. S1 and S4**). It consists of beaded, elastomer-coated spring steel layers and can be designed as single or multiple layer depending on the application. The sealing is done with full beadings (combustion gas) or half beadings (coolant, pressurized, and return oil) in the functional layer (*also, see above*, ~~Beadings).

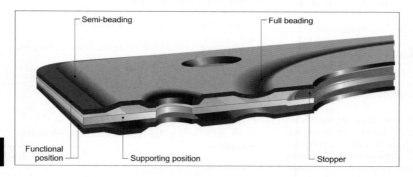

S

Fig. S1
Design of a cylinder head gasket with metal layers

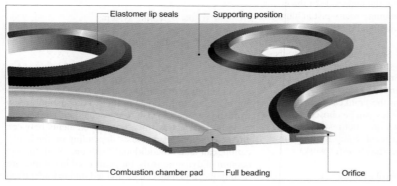

Fig. S2
Design of a metal /elastomer cylinder head gasket

**Experience Mobility.
Into the Future with the Power of Innovation**

Millions of vehicles all over the world – on the road with ElringKlinger technology. With cylinder head gaskets and specialty gaskets, cam covers and thermal shielding we are the leading development partner and system supplier for almost every automobile manufacturer worldwide when it comes to achieving ambitious goals. With our engineering competence and drive for innovation, we speed things up from the original idea to the start of production and set the standards for the engine development of the future.

Dedicated to economical and environmentally friendly mobility.

ElringKlinger AG | D-72581 Dettingen/Erms | www.elringklinger.de

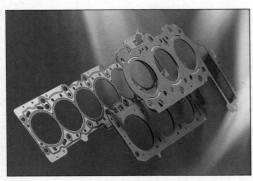

Fig. S3
Metal/elastomer
cylinder head gaskets
with combustion
chamber pad

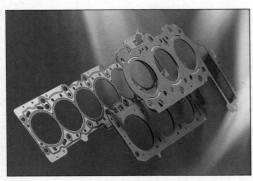

Fig. S4 Cylinder head gaskets with metal layers for car engines

The elastomer coating of the metal layers is used for microsealing. The technical strength of the cylinder head gaskets with metal layers is especially obvious in high-performance gasoline and diesel engines with direct injection, which is currently the leading application internationally.

~~Double stopper. This design consists of one stopper in front of and one behind the combustion chamber beading (**Fig. S5**). By using a double stopper, especially for the bushing design, the admission of force into the bushing is defined in such a way that plastic deformation and a lowering of the bushings is avoided.

~~Elastomer lip seals. These act as the coolant and oil seals in metal/elastomer cylinder head gasket applications. Material and geometry are adapted to the respective engine.

~~Firing pressure. The increase in pressure that develops in the combustion chamber during the combustion is called firing pressure.

~~Metal/elastomer cylinder head gaskets. The robust and highly durable metal bed sealing system with the elastomer profiles attached by vulcanization is currently the leading technology for commercial vehicle engines (**Figs. S2 and S6**). Different functions, such as ignition pressures (especially in engines without bushings and those with slip-fit bushings), low screw forces, and low number of screws can be handled dependably. The separation of duties between the combustion chamber and fluid seals is typical. Metal beadings (*also, see above,* ~~Beadings) are used to harmonize the distribution of the high sealing pressures in the combustion chamber area, while the sealing of the cooling water and oil passages is done with elastomer lip seals, which require a seal with only a low compression pressure.

~~Metal/soft material cylinder head gaskets. This carrier panel sealing design with soft material attached to both sides was primarily used in the early 1990s and is currently still important for spare parts. The limits of the system were shown at that time especially by engines with high thermal stress, small web widths, and large sealing gap vibrations. These limitations subsequently led to the development of higher-performance systems.

~~Minimum seal pressure. This is the surface pressure required to achieve reliable sealing.

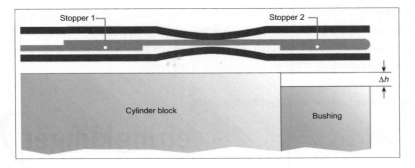

Fig. S5
Double stopper design
for aluminum crankcase
with cylinder liner

Freudenberg Dichtungs-
und Schwingungstechnik

Your Technology Specialist

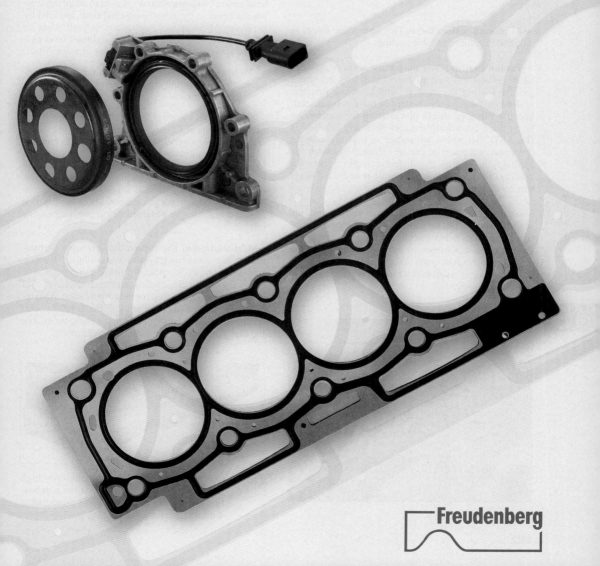

Freudenberg

Fig. S6 Metal/elastomer cylinder head gaskets with directly scorched elastomer profiles for commercial vehicle engines

~~**Minimum sealing force.** This force (line force) is required to achieve reliable sealing.

~~**Partial coating.** Only the surface areas of cylinder head gaskets with metal layers that are important for the sealing are coated, which means that the sealing surfaces in the coolant or oil are not coated, because this guarantees that coating separation can be avoided even under critical boundary conditions (**Fig. S7**). An added advantage is that the coating thickness and material can be selected based on the application, so that the different requirements for coatings in the combustion chamber and fluid areas can be accommodated.

~~**Sealing gap vibrations.** Dynamic vibrations of the sealing gap include, for example, high frequency relative movements between cylinder head and crankcase in the engine area caused by the firing pressure. In the case of vibrations in quasi-static sealing gaps, this relative movement is caused by thermal expansion and is of low frequency.

Fig. S7 Partial coating

~~**Stopper.** Cylinder head gaskets with metal layers normally have one stopper which tensions the engine components at the circumference of the combustion chamber. This stopper results in a reduction of the vibrations of the sealing gap caused by the gas force and, at the same time, it prevents unacceptable deformation of the full beadings (*see above*, ~~Beadings). The stopper is built by folding, impressing, or welding-on of a supporting ring, and the normal stopper thicknesses are between 100 and 150 µm.

The stopper also can be built with a height profile to adapt to the specific stiffness conditions in the engine if required (**Fig. S8**).

Under certain conditions, a stopper may not be necessary in gasoline engines, especially when light-metal crankcases are used.

~~**Supporting layer.** The supporting layer in cylinder head gaskets with metal layers is used to adapt the sealing thickness to the installation conditions required by the design. Micro-alloyed steels, stainless steels, or spring steels are used for metal/elastomer cylinder head gaskets to achieve an appropriate beading characteristic.

~~**Web width.** Web width describes the distance between the combustion chamber bores of neighboring cylinders.

~Development methods. Engine test runs are still a major component of leakage tests. Calculation of the connection between components using finite element analysis and laboratory tests under engine conditions have become more important because tests of a running engine on a test stand are expensive and time-consuming, and because the trend is toward shorter development times. Significant information about the function of seal design can be obtained before the real engine test.

~~**Calculation of the connection between components.** The tensioning system consisting of flanges, seals, and screws is calculated using a computer. The geometry of the components and the characteristics of the materials are required for the computation. Also required for the computation of the cylinder head gasket are the temperature distribution in the components and the firing pressure in the combustion chamber.

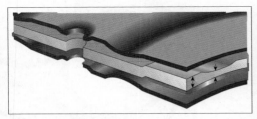

Fig. S8 Compensation of stiffness variations by high profiling of the carrier panel at the combustion chamber by using a cylinder head gasket with metal layers

VICTOR REINZ® – you need a strong drive to get ahead.

Motor components by VICTOR REINZ® — choose top quality and development and service competence on the highest level.

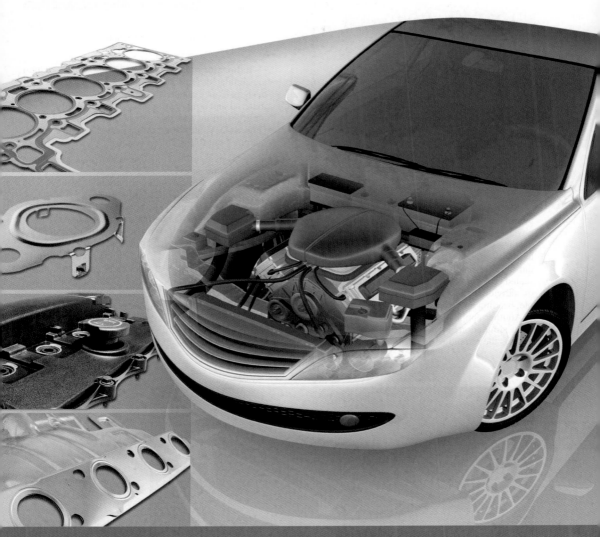

Sealing, shielding and plastic components

- Mulitlayer metall cylinder head gaskets with partial coating or full coating
- Singlelayer/multilayer flat gaskets as well as soft material gaskets
- Valve cover modules with integrated oil separation
- Shielding systems for thermal and acoustic insulation
- Shielding systems with integrated exhaust manifold gasket

VICTOR REINZ®

REINZ-Dichtungs-GmbH
Reinzstr. 3-7 89233 Neu-Ulm
Tel. +49 731 7046-0
Fax +49 731 71 90 89
www.reinz.com

Original Equipment
Original Service Parts
Industrial Applications
Fuel Cell Components

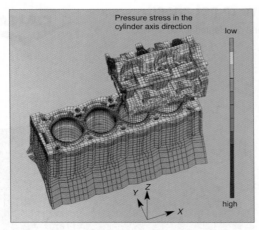

Fig. S9 Nonlinear structural analysis of the sealed connection: determination of the pressure distribution at the combustion chamber under ignition pressure and temperature

Fig. S11 Simulation of dynamic internal pressure with original engine components

~~**Finite element analysis.** The finite element computational method (FEM) is used for the calculation of physical problems such as statics, dynamics, fluid mechanics, chemistry and electromechanics. FEM is used to analyze the structure of the total system (**Fig. S9**) during the development of gaskets, including the sealed connection. It is also used for detailed analysis of individual functional parts (**Fig. S10**)—for example, beading and stopper.

Another application area for the FEM computation is the simulation of complex manufacturing processes, such as the remolding of elastic seal beadings.

~~**Laboratory tests.** The actual stresses in operation are simulated in the laboratory and depend on the seal designs—for example, tests for the determination of material and temperature resistance, durability, adaptability, settling characteristics, and sealing efficiency (**Fig. S11**). Established test processes for the determination of durability are based on

- Servo-hydraulic testing machines for the simulation of dynamic stress;
- The hydraulic simulation of combustion pressures for testing cylinder head gaskets;
- Shaker and temperature chamber for the evaluation of valve umbrella modules, for example; and
- Hot gas generator for the simulation of thermal stresses in the exhaust branches.

~Dynamic seals. Radial shaft seals are used for sealing shafts guided through bores in housings to protect against lubrication leakage and dirt. They are composed primarily of elastomers or PTFE (polyetrafluorethylen/Teflon). They are mainly applied to the crankshaft and the camshaft, which have to be sealed on one or two ends. Radial shaft seals are also used in the axis

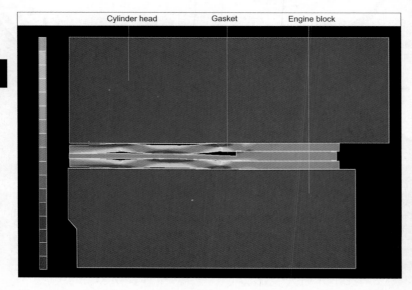

Fig. S10
Detailed evaluation: cross section of a sealed connection with local pressure and tensile stresses in the individual layers of a cylinder head gasket with a metal layer

areas. Dynamic seals with an axial sealing function are also used, for example, for valve stem gaskets or transmission piston gaskets in automatic transmission; these are also normally made of elastomers or PTFE.

~Elastomer sealing systems. Engine components are increasingly being made of plastics or magnesium to reduce weight and improve functionality. The result is reduced component rigidity and subsequent high deformation during tensioning, which must be compensated for by the sealing system. Elastomer sealing systems—with or without a metal bed—are extremely well-qualified for this task because they have the following advantages: safe sealing with low compression seals, compensation for large component tolerances, and acoustic component decoupling.

Elastomer sealing systems include, on the one hand, metal/elastomer cylinder head gaskets and, on the other hand, elastomer seals and metal/elastomer gaskets in the special sealings field (*see below*, ~Special seals).

~~**Decoupling systems.** *See below*, ~~Structure-borne noise decoupling

~~**Elastomer materials.** These rubber-elastic materials, which vulcanize under the impact of temperature, consist of polymers, fill materials, vulcanization agents, antioxidant agents, pigments, and so on. The temperature range is between -80 and $+250°C$ depending on the polymer used. Examples are: FPM (fluorine rubber), MVQ (silicone rubber), ACM (poly-acrylate rubber), AEM (ethylene acrylate rubber), FMVQ (fluorine silicone rubber), EPDM (ethylene propylene diene rubber) and HNBR (hydrated nitrile rubber).

~~**Elastomer seals.** These special seals made from elastomer materials can be used for all seals in oil and coolant applications of the engine (**Fig. S12**). Suitable seals are selected depending on the materials to be sealed, their temperatures, and the requirement profiles. A groove or a step must be machined into the components to be sealed to get the elastomer seal into the force bypass. The component-specific profile geometries ensure maximum functionality for seals with minimum compression.

Examples of applications are: intake manifold seals (for plastic intake systems), cylinder head gaskets (for metal or plastic covers—optionally including a decoupling system), oil pan seals, timing-case seals (seals between engine housing and the timing-case cover, which covers the timing-chain drive), and water pump seals.

~~**Metal/elastomer gaskets.** The sealing material of this design, which consists of elastomer profiles scorched onto a metal bed, is located in the force bypass. The tensioning force of the components is mainly transferred by the metal components. Metal/elastomer gaskets, shown in **Figs. S13 and S14** (*also, see below*, ~Special seals), are especially qualified for the use in force-carrying connections such as intake manifold seals, cylinder head gaskets (for aluminum

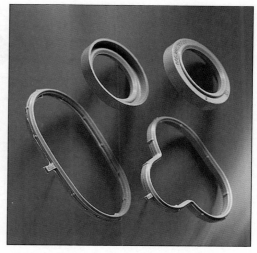

Fig. S12 Elastomer seals for intake manifold and spark plug gasket

covers—optionally also with decoupling systems), oil pan seals, timing-case seals (sealing between engine housings and the time-case covers that cover the chain drives), and water pump seals. Additional functions are often integrated, such as the calibration of fluid flows, exhaust gas recirculation, and cable passage.

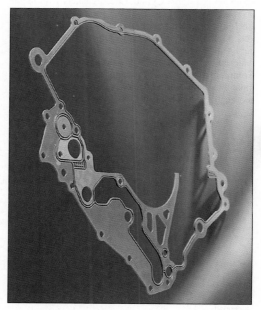

Fig. S13 Metal/elastomer gaskets for crankcase, consisting of different metal segments that are connected through the elastomer sealing lip

S

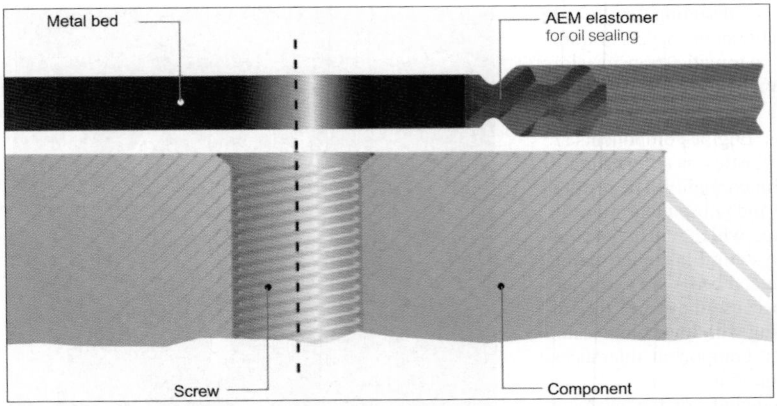

Metal bed

AEM elastomer
for oil sealing

Fig. S14
Design of a metal/
elastomer gasket

Screw

Component

Metal/elastomer cylinder head gaskets are primarily used in commercial vehicles.

~~Structure-borne noise decoupling. The radiation of engine noise can be reduced effectively by elastically decoupling resonant bodies such as valve umbrellas, intake manifolds, and oil pans. The component that needs to be decoupled is mounted elastically between elastomer seals, which must provide sealing forces and a decoupling element to ensure efficient damping. The decoupling system for valve cover modules (**Fig. S15**) might consist of seals, decoupling elements, screws, and spacer sleeves.

~Flat seal designs. Cylinder head gaskets with metal layers made from beaded, elastomer-coated spring steel layers currently dominate the cylinder head gaskets applied to car engines; metal/soft material cylinder head gaskets made from a metallic carrier with soft material on both sides may also be used for spare parts. Commercial vehicles use mainly metal/elastomer cylinder head gaskets made from metal carriers with a scorched elastomer profile.

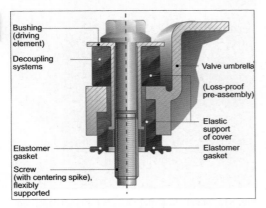

Bushing
(driving
element)

Decoupling
systems

Valve umbrella

(Loss-proof
pre-assembly)

Elastic
support
of cover

Elastomer
gasket

Elastomer
gasket

Screw
(with centering spike),
flexibly
supported

Fig. S15 Example of valve umbrella: elastomer seals and decoupling systems

Special seals can be divided into metal beaded gaskets based on elastomer-coated and uncoated metal carrier materials, soft material seals, metal/soft material gaskets made from a metal carrier with soft material attached to both sides, elastomer seals, and metal/elastomer gaskets made from metal carriers with elastomer profiles attached to them by vulcanization.

~Flat seals. Flat seals are used to seal statically fixed components (*also, see above*, ~Cylinder head gaskets, and *see below*, ~Special seals) as opposed to dynamic seals for moving components such as rotating or axial moving shafts.

~Modules. It is important that the sealing system for an optimally functioning sealed connection not be analyzed in isolation, but that the complex interaction of all individual parts be considered. Assembly-ready, multifunctional modules—for example, valve umbrella and oil pan modules and "beauty covers" (engine compartment covers)—with complex sealing and connection technology and accessories are replacing individual components in increasing numbers. Depending on the requirements, different combinations of housing materials and sealing systems are being implemented. Many additional functions can be integrated, depending on engine design and customer requirements.

~~Beauty cover. A beauty cover is used to make the engine compartment optically pleasing. It can also have additional functions such as noise reduction and targeted heat transfer to avoid heat buildup (**Fig. S16**). Plastic parts are generally used for this application, which may have design elements such as emblems or logos.

~~Cylinder head cover modules. *See below*, ~~Valve umbrella module

~~Engine compartment covers. *See above*, ~~Beauty cover

~~Housing materials. Housing modules can be made from plastics, aluminum, magnesium die casting, or

S

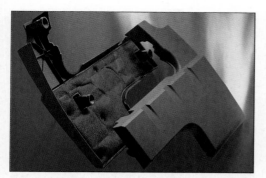

Fig. S16 Beauty cover with decoupled bearing and noise-reducing special foam

Fig. S17 Oil sump module from hot-galvanized, cathodic dip-painted sheet steel, with clipped-on, removable metal/elastomer gasket as well as oil drain plug

steel sheet metal. Plastics are divided into thermoplastics (polyamide PA 6/PA 6.6 with glass and mineral fill materials) and duroplastics (polyesters [BMC, SMC], and phenolic resins [PF]). The particular application (material, temperature, forces, stiffness, etc.) determines which material is used.

~~~**Plastic modules.** Numerous additional functions can be integrated economically due to the special processing characteristics of plastics. These include, for example, in the case of the valve umbrella modules: integration of the oil separation (blowby gas); integration of the pressure-control valve of the crankcase ventilation; preassembly of seals; decoupling elements and oil cap; decorative surface design—for example, by graining or printing. Another major advantage of the use of plastics is weight reduction.

~~**Intake manifold module.** This module consists of an intake manifold with a number of other add-on parts such as seals, fastening system, thread inserts, throttle valves, exhaust gas recirculation, switching systems for air volume control, sensors, and so on.

~~**Oil separation modules.** The oil separator is currently placed as a separate component in the engine compartment. It is used to remove the oil in the blowby gas, which comes from the crankcase, and return it to the crankcase; the gas is then introduced into the intake air. New developments integrate the oil separator into the valve umbrella. The housing material is primarily plastic (*also, see above*, ~~Intake manifold module ~~~Plastic modules).

~~**Oil sump module.** These consist of oil pans, fastening systems, and integrated sealing systems; all established housing materials can be used (**Fig. S17**). The stress in the component must be taken into consideration (e.g., stone impact) when plastics are used, because of the poor mechanical characteristics of this material. Several additional functions can also be integrated, such as those involving filters, oil return lines, oil drain plug, swirl bodies to steady the oil, and oil intake manifold.

~~**Valve umbrella module.** All established housing materials can be used for the housing (**Fig. S18**). The following additional functions can be achieved (*also, see above*, ~~Intake manifold module ~~~Plastic modules): integration and preassembly of the sealing system; preassembly of the decoupling system consisting of decoupling elements, sleeves, and secured preassembled screws (*also, see above*, ~Elastomer sealing systems ~~Structure-borne noise decoupling); integration of the oil separator; integration of the pressure control valve of the crankcase ventilation; and preassembly of the oil cap.

~**Operating conditions.** The functional impact parameters in the sealing systems, such as temperature, system pressure, compression seal, dynamics, surface topography, materials, and others, are encompassed by this term.

~**Sealed connection.** A sealed connection is the interaction of the sealing, the surrounding components, and the threaded connections. For example: cylinder head— cylinder head gasket—crankcase—cylinder head bolts.

S

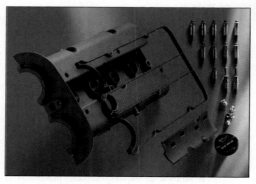

Fig. S18 Individual components and complete plastic valve umbrella module I

~Special seals. These seals are used for a number of sealing tasks in the engine, transmission, exhaust system, and accessories — for example, supercharger and pumps. Both the material and design are optimized for the specific requirements. Special seals can be divided into metal beaded gaskets, soft material seals, metal/soft material gaskets, elastomer seals, and metal/elastomer gaskets.

~~Coating. The coating on metal beaded gaskets, primarily elastomer, with coating thicknesses of 5–100 μm, assumes "microsealing," which means the sealing of the flange roughness. The coatings on soft material seals are generally used to improve the surface seals. Additional antistick coating improves the handling during assembly and simplifies the disassembly of sealing connections.

~~Compressive creep strength. A standardized test determines the thermal and mechanical sealing settings. Special specifications for metal/soft material gaskets and soft material seals are often agreed upon between suppliers and customers.

~~Elastomer seals. *See above*, ~Elastomer sealing systems

~~Metal beaded gaskets. This very reliable and cost-effective sealing system consists of a metal bed (cold-rolled strip, spring steel, aluminum) and is, in the majority of cases, elastomer coated (**Fig. S19**). The functional principle is based on line sealing over a beading. Almost all applications in an engine can be sealed because of a large number of combinations that can be achieved from base materials, the different elastomer materials, and the variable beading geometries. Examples of applications include intake system seals, cylinder head gaskets, oil pan seals, crankcase seals, water pump seals (**Fig. S20**), and exhaust seals. Metal beaded gaskets can also assume additional functions, such as the integration of oil deflectors or sensors for even more efficient engine management. It is possible to achieve preassembly solutions (**Fig. S21**) such as fastening clips and centering elements.

~~Metal/elastomer gaskets. *See above*, ~Elastomer sealing systems

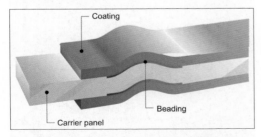

Fig. S19 The sealing system for metal beaded gaskets is aligned to the requirement profile by a variable combination of the parameters of carrier panel, coating, and beading

Fig. S20 Water pump sealing for gasoline engine

Fig. S21 Sealing with preassembly elements

~~Metal/soft material gaskets. These differ from soft material seals by metal inserts in the material center, which increase the tensile strength and the dimensional stability. The performance can be further increased by additional part-coating with elastomer. Metal/soft material gaskets are primarily used in coolant, oil, and fuel applications, and the materials used are adapted exactly to the particular requirements.

~~Sectional tightness. Penetration of the medium into the sealing material is prevented by selecting a suitable composite material (*also, see below*, ~~Soft material seals) that meets the requirement of the medium to be sealed, the temperature of the application, and the tensioning conditions. This achieves sectional tightness (*also, see above*, ~Development methods ~~Laboratory tests).

~~Setting characteristics. Seals that are tensioned in force transmission can settle. This process has considerable impact on the sealing function for composite material seals, especially under heat exposure.

S

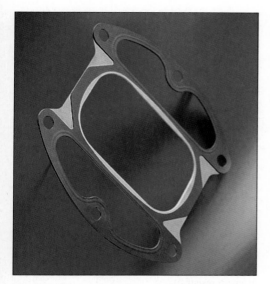

Fig. S22 Exhaust pipe with coolant channels

~~Soft material seals. The basic components of the composite sealing material are normally fibers, binding agents, and fill materials. The performance of the seal can be increased by additional applications of partial elastomer layers. The range of applications of these seals is broad due to the large number of material qualities available for their manufacture. The applications are limited by the high thermal stress to which the seals are exposed.

~Temperature resistance. The following temperature resistance is normally required for the seals used in automotive applications: in the coolant area, from -40 to $+120°C$; for the oil sealing, from -40 to $+150°C$; and in the exhaust gas area, up to $900°C$ (**Fig. S22**).

~Tensioning system. This is a system with several components that are pretightened with screw connections. An example is an exhaust manifold, which is connected to the cylinder head with gasket and screws.

Literature: R. van Basshuysen, F. Schäfer (Eds.): Dichtsysteme, Handbuch Verbrennungsmotor, 2nd Edition, Vieweg, 2002, pp. 272–290. — Elring Klinger AG, 72581 Dettingen, Germany, Fachdokumentation: Zylinderkopfdichtungen, Spezialdichtungen, Module- und Elastomer-Dichtsysteme. — A. Diez, Dr. U. Maier, G. Eifler, M. Schnepf: Integrierte Drucksensorik in der Zylinderkopfdichtung, MTZ (2004) 1, pp. 22–25. — A. Diez, T. Gruhler: Dichtung mit Profil, Automobil Industrie Special Edition—Mercedes-Benz E-Klasse, May 2002, p. 60. — E. Griesinger: Kompaktes Design, vielfältige Funktionen—Ventilhaubenmodule von ElringKlinger, MTZ No. (2003) 6, pp. 504–507. — G. Walter, E. Griesinger, Kunststoffmodule—Funktion und Ästhetik, ATZ /MTZ System Partners, 2002, pp. 32–37.

Seat geometry →Injection valves ~Diesel engine ~~Hole-type nozzle

Seat hole nozzle →Injection valves ~Diesel engine ~~Hole-type nozzle

Seat reinforcement →Valve ~Gas transfer valves, ~Valve seat ~~Seat hardening

Seat ring →Valve guide

Second carburetor stage actuation →Carburetor ~Actuation and design, second carburetor stage

Secondary air (*also*, →Pollutant aftertreatment). Secondary air is usually added to the exhaust gas system to improve reactions, especially the oxidation of CO and HC, which results in lower HC and CO emissions. It is helpful to position the supply point close to the engine outlet, so that the high temperature can improve oxidation. The supply can either be controlled or can take place via a self-actuated valve based on the pressure changes in the exhaust gas system.

Secondary air system →Pollutant aftertreatment ~Aftertreatment concept with a three-way catalytic converter

Secondary chamber engine →Combustion process, diesel engine ~Prechamber engine

Secondary circuit →Ignition system, gasoline engine ~High-voltage generation ~~Coil ignition

Secondary combustion chamber →Combustion chamber ~Prechamber, ~Swirl chamber

Secondary emission particulate filter →Particles (Particulates) ~Particulate emissions ~~Non-engine particles

Secondary exhaust gas system noise →Engine acoustics ~Exhaust system noise

Secondary intake noise →Engine acoustics ~Intake noise; →Oil ~Safety and environmental aspects

Secondary particles →Particles (Particulates) ~Particulate emissions

Secondary radiation →Engine acoustics

Secondary wear →Piston ring ~Wear

Seiliger process →Cycle

Selection range switch/sensor →Electronic/mechanical engine and transmission control ~Requirements for mechanical and housing concepts

Self-ignition →Combustion, diesel engine; →Combustion process, diesel engine; →Ignition system, diesel engine/preheat system; →Thermal ignition

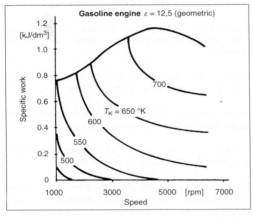

Fig. S23 Compression temperatures for swirl-chamber diesel engine

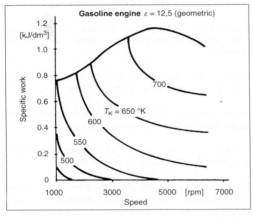

Fig. S24 Compression temperatures for gasoline engine

Self-ignition engine (*also*, →Exhaust gas analysis). Self-ignition (or compression-ignition) engines are the designation for engines in which ignition is accomplished by injecting fuel into the precompressed and, therefore, hot air. The compression leads to a rise in the temperature of the air in the combustion chamber, depending on the compression ratio, which is sufficient for ignition of the fuel injected. Possible compression temperatures for a swirl-chamber diesel engine and a gasoline engine are show in **Figs. S23 and S24.** The compression temperature for direct-injection diesel engines is slightly lower than that for chamber-type engines because of the lower compression ratio.

Self-protection →Electronic/mechanical engine and transmission control ~Electronic components ~~Diagnostics

Semi-active engine mount →Engine acoustics ~Engine/accessory mount

Semi-downdraft carburetor →Carburetor ~Design types

Semi-synthetic oils →Oil ~Body ~~Basic oils

Semi-variable valve control →Variable valve control

Sensors (*also*, →Electronic open- and closed-loop control ~Electronic open- and closed-loop control, diesel engine) Sensors, which also are called probes or sensing elements, measure physical or chemical values— it is preferred that the value is provided as an electrical signal (current or voltage). In most cases, the signals are transferred to electronic controls (control devices) and processed by those devices. Sensors influence technical systems; they are used for monitoring, diagnosis, and documentation of processes or to inform the user. The value to be measured (such as temperature, pressure, travel, angle, humidity) is converted by the sensor into an electrical signal using a physical effect (such as inductive, piezo-resistive, piezo-electric, or magneto-resistive).

The sensor itself can include the electronics needed to process the signal (amplifying, filtering, digitization).

The signals from analog sensors usually reflect the measured value through electrical voltage, current, or frequency (**Fig. S25**).

~Active sensor. An integrated electronic system for signal processing transfers standard signal levels; these are used in the directly connected electronic control unit.

~Air mass flow sensor. The hot film anemometer, or HFA (**Fig. S26**), uses a heated housing that loses energy to the surrounding air. The heat transferred depends on the air mass flow, and this can be used as the measured parameter.

Two temperature-dependent metal film resistors (R_S and R_T) are arranged in the intake airflow. These two resistors are connected in a bridge connection in combination with R_1 and R_2.

R_S is more or less cooled by the aspirated airflow. The electronics regulates the necessary heating current

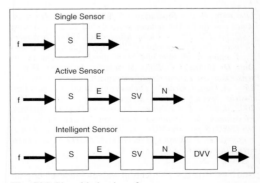

Fig. S25 Signal behavior of sensors

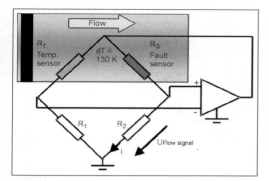

Fig. S26 Principle of a hot film air mass sensor

Fig. S27 Knock sensor with integrated plug

so that there is always a constant temperature difference (e.g., 130 K) for the air temperature measured at R_T. The heating current is converted into a voltage signal on resistor R_2.

The resistors R_S and R_T are tuned so that the characteristic output curve is independent of the air temperature. The HFA signal is almost independent of temperature, pressure, and contamination.

Because diesel engines usually do not have a throttle and, therefore, the pressure in the intake manifold cannot be used to measure the induced mass flow of fresh air, an HFA must be used.

~Exhaust gas sensors. These are mounted directly after the manifold and regulate the fuel injection system and, thus, the volume of fuel injected in order to reach an optimal conversion rate of the catalyst. If they are attached after the catalyst, they monitor its functionality and enable the fulfillment of the OBD (onboard diagnosis) requirements.

All currently used probes consist of ZrO_2, which is oxygen conducting above 350°C; they are built in several layers and use the so-called Nernst equation—the voltage, which is obtained via a ZrO_2 layer, depends only on the difference of the partial pressure of oxygen on both sides of the layer.

~Intelligent sensor. If more complex signal processing steps are performed by the electronics in the sensor (such as self-calibration, disturbance variable compensation), then it is also called a smart sensor.

In bus-compatible smart sensors, the measured signals are digitized in the sensor and (pre)processed by microprocessors. The information is then transferred via serial bus systems, such as CAN (controller area network), Profibus, InterBus.

~Knock sensors. Broadband knock sensors (**Fig. S27**) measure frequency between 3 and more than 20 kHz. They are mounted at a suitable position on the engine block to measure the vibrations created by the combustion process. In order to recognize the knocking of each individual cylinder, several knock sensors are used in

multicylinder engines, for instance two sensors for six cylinders or four sensors for eight cylinders.

The principle of operation of knock sensors is based on a piezo-ceramic ring that converts engine vibrations into electrically usable signals using a seismic mass.

The sensor sensitivity is expressed in mV/g or pC/g and is almost constant across a wide frequency range. The behavior of the knock sensor can be adapted to the requirements of the engine by the selection of the seismic mass.

If the seismic mass is reduced the resonant frequency can be increased. The sensitivity tolerance band is about ±30%. When tuning the engine control device, limit sample sensors are used.

Knock sensors are sometimes used in diesel engines today to control the start of injection from the injection nozzles (also, → Electronic open- and closed-loop control ~Electronic open- and closed-loop control, diesel engine ~~Functions ~~~Onboard diagnostics, ~Electronic open- and closed-loop control, gasoline engine ~~Functions ~~~Onboard diagnostics).

~Lambda sensors

~~Binary lambda probe. In the case of the binary lambda probe (**Fig. S28**), the Nernst voltage is measured between a catalytically active, exhaust-gas side electrode and a reference electrode in the air; the voltage changes suddenly at $\lambda = 1$.

Binary probes allow control of the air-fuel ratio around the stoichiometric point $\lambda = 1$ and, thus, set the supply of fuel for optimal conversion in the three-way catalyst.

~~Linear lambda probe. With the linear lambda probe, the air-fuel ratio in a sensor-internal chamber applies a regulated current—also called a pumping current—at the $\lambda = 1$ Nernst voltage. The reference air is created either via a channel in the ceramics or by a constant oxygen supply in a cavity. The pumping current is the measuring signal and depends on the value of lambda in the exhaust gas.

S

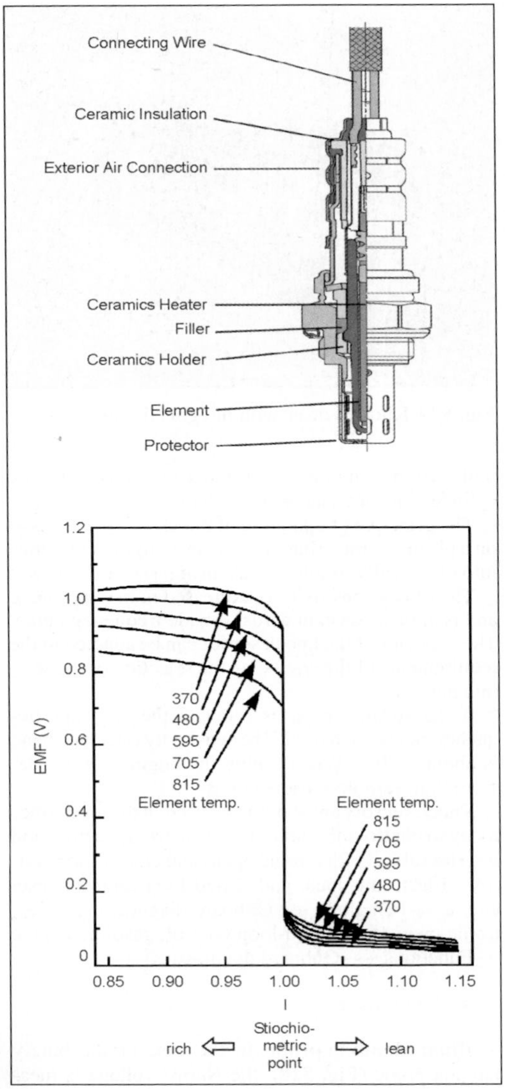

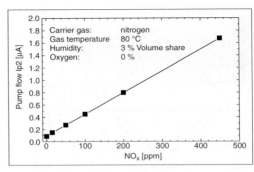

Fig. S29 Typical characteristic curve of NO$_x$ sensor

requirements for checking the three-way catalyst for low-emission concepts (SULEV, LEV 2).

The work principle of NO$_x$ sensors is based on the dissociation of oxides of nitrogen by a catalytically active electrode. The amount of oxygen produced is measured by the linear lambda probe.

The design of the multilevel ZrO$_2$ ceramics sensor has two chambers. In the first chamber, the oxygen in the exhaust gas is reduced (lean exhaust gas) or increased (rich exhaust gas) to a constant partial pressure of tens of parts per million by applying a pumping current. The necessary current is proportional to the reciprocal of the air-fuel ratio. The reduction of the NO$_x$ on the measuring electrode takes place in the second chamber. The current necessary to keep the surroundings of the electrode free of oxygen is proportional to the concentration of oxides of nitrogen, and this is the measuring signal.

~~**Smart NO$_x$ sensor.** The smart NO$_x$ sensor is equipped for complete control of the heating and regulation of the pumping current as well as digital communication and engine control (**Fig. S30**).

~**Oil level sensor (absolute measures).** The oil level is measured using a temperature-dependent wire resistor. The wire resistor is heated, and the existing oil ca-

Fig. S28 Structure and characteristic curve of a binary lambda probe

Linear probes measure the air-fuel ratio between the rich mixture and the air constantly and are especially suited for control of lean-burn engines, such as gasoline engines with direct injection.

~**Nitrogen oxides sensor.** The NO$_x$ sensor allows direct measurement of the concentration of oxides of nitrogen in the exhaust gas of gasoline and diesel vehicles. A typical characteristic curve is shown in **Fig. S29**. It makes possible the optimal control and diagnosis of NO$_x$ catalysts through engine control (i.e., NO$_x$ storage, SCR catalysts) and the fulfillment of the OBD

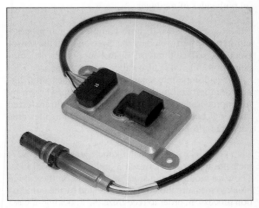

Fig. S30 Smart NO$_x$ sensor with control electronics

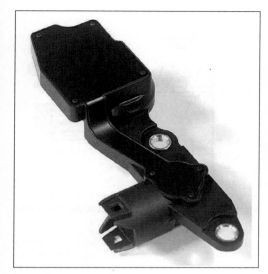

Fig. S31 Angle sensor for variable valve stroke control

Fig. S32 View of a typical high pressure sensor

Applications and measuring range:

- Fuel tank leakage detection: -30 to $50+$ mbar
- Variable valve lift (VVL) systems, inlet under pressure regulation: -150 to 50 mbar
- Diesel particulate filter, filter check: 0 to 1000 mbar
- Exhaust gas recirculation (EGR) measurement: 0 to 250 mbar

pacity determines the degree of cooling. The resulting resistance is a measure of the quantity of oil. It is used in engine control and transmission control.

~Oil level switch. These have reed contacts actuated by a float with a magnet. They allow limit value measurements.

~Position sensor. A position sensor (**Fig. S31**), using potentiometers, measures the position of the throttle, the gas pedal, the exhaust gas recirculation valve, the helical flap, the clutch position, and the transmission position. Contactless sensors enable goniometry (measurement of angles) with Hall sensors and magnetoresistive sensors. Travel measurements can be made with analog Hall sensors, IMS sensors, or PLCD sensors.

~Pressure sensors

~~**BAP (barometric absolute pressure) sensor.** This is used to determine the ambient pressure. The resulting information is used to compensate the air pressure in different levels. The measuring range is between 0.5 and 1.1 bar.

~~**Combustion chamber sensor.** The combustion chamber sensor measures the pressure curve in the cylinder, enabling the combustion curve to be evaluated. Direct measuring combustion chamber pressure sensors have contact with the combustion chamber and are exposed to temperatures of up to 600°C. Indirect measuring combustion chamber sensors can be integrated into a spark plug or a fuel injector.

~~**Differential pressure sensors.** These measure the pressure difference between the measuring chamber and a reference chamber or the ambient air.

~~**High-pressure sensors.** High-pressure sensors operate above 100 bar (**Fig. S32**). The normal design is for a hexagonal head section with an M12 bolt-thread connection.

The primary application areas are:

100–200 bar, HPDI (high-pressure direct injection) system
200–280 bar, brake pressure sensors
1300 bar–2000 bar, common rail diesel injection systems

There are a number of concepts for media separation with hermetic separation of the measuring element through a steel housing and a steel membrane or a thin shaft membrane. The design space for the measuring element (silicone pressure sensor) is filled in a vacuum with the silicone oil pressure transfer medium.

The mentioned concepts for media separation use thick-layer and thin-layer strain gauges on the membrane.

~~**MAP (manifold absolute pressure) sensor.** This is used as an intake air pressure sensor to determine the pressure in the intake manifold after the throttle. The typical measuring range is from 0.2 to 1.1 bar (**Fig. S33**).

When the MAP sensor is used together with the intake air temperature, the air mass can be calculated. Therefore, an integrated temperature sensor is used often to reduce installation effort.

~~**Turbo-MAP.** The sensor for measuring manifold absolute pressure of boosted engines determines the charge pressure of engines with turbochargers or superchargers. The typical measuring range is from 0.5 to 2.5 bar.

The engine controller optimizes the combustion parameters using the charge pressure information. Charge pressure information is also used to regulate the turbocharger.

Fig. S33 MAP without and with integrated temperature sensor

~Single sensor. The electric signal is transferred directly to the primary controller without further electronic processing.

~Speed sensors

~~**Active speed sensor.** Active speed sensors have integrated electronics for signal processing. Therefore, active sensors transfer standard signal levels that are used without additional signal processing in the electronic control device.

~~~**Differential Hall sensors.** Differential Hall sensors are the most commonly used devices for measuring speed. Ramp change of a ferromagnetic trigger wheel results in a difference of the magnetic field on the differential Hall sensors. The sensors, which work on the differential principle, are largely insensitive to disturbances such as temperature changes and external magnetic fields, and are, therefore, very accurate. With the differential principle, sensors capable of sensing speed as low as 0.1 rpm (zero speed) can be realized. Because of the differential principle (**Fig. S34**), these sensors can be used in only one installation position.

~~~**Hall sensor.** Active sensors based on the Hall effect are the most common sensors.

~~~**Single element Hall sensor.** These are used for a static function. These sensors allow the recognition of teeth or gaps without the trigger wheel (true power on). Through the arrangement of single elements, any orientation between the Hall sensor and the trigger wheel is possible.

~~**Passive speed sensor.** Inductive sensors are also called, and used as, passive speed sensors and variable reluctance (VR) sensors.

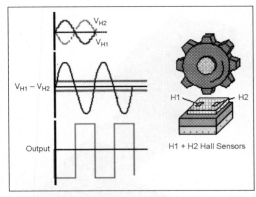

Fig. S34 Measuring principle of a differential Hall sensor

Inductive sensors essentially consist of a coil wound around a magnetically preloaded core. If the inductive sensor is close to a moving ferromagnetic trigger wheel, voltage is induced. This voltage is evaluated in an electronic control device. Each flank of the trigger wheel induces an electrical voltage. In inductive sensors, the level of the induced voltage depends on the speed. Therefore, there is a limit to the low speed/frequency at which the inductive sensor will function.

~Temperature sensors. Most temperature measurements in vehicles use the temperature-dependence of electrical resistance material with a negative temperature coefficient (NTC). Due to the drastic nonlinearity of these devices, a large temperature range can be covered (**Fig. S35**).

For applications at very high temperatures (exhaust gas temperatures of up to 1000°C) platinum sensors are used.

The change in resistance is converted into an analog voltage through a voltage divider connection with optional parallel resistance for linearization.

Separate starter-generator (SSG) power takeoff →Starter ~Starter-generators

Separation of solid and volatile particles →Particles (Particulates) ~Particle measuring

Separation rate →Filter ~Filter characteristics, ~Fuel filters; →Particles (Particulates) ~Particulate filter system ~~Particulate filter

Serial interfaces →Electronic/mechanical engine and transmission control ~Electronic components

Service life F1 reliability →Engine ~Racing engines ~~Formula 1 ~~~Engine, example BMW

Service strength →Calculation processes ~Application areas ~~Stability

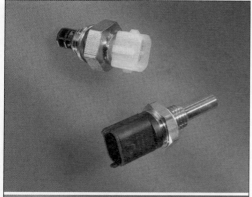

Fig. S35 Typical design of different temperature sensors

Setting characteristics →Sealing systems ~Special seals

Sewage gas →Fuel, diesel engine ~Biogas

Shaft seal →Throttle valve location

Shaker piston →Piston ~Cooling ~~Spray oil cooling

Shape winding →Piston ring ~Shaping ~~Winding

Sharpness (sound) →Engine acoustics ~Psychoacoustic parameters

Shear stability →Oil ~Properties and characteristic values ~~Viscosity

SHED (Sealed Housing for Evaporative Determinations) test →Emission measurements ~Test Type II

Shell-and-tube cooler →Radiator ~Design ~~Fin

Shift of operating point →Downsizing

Short circuit →Electronic/mechanical engine and transmission control ~Electronic components ~~Diagnostics

Short stroke →Bore-stroke ratio

Short stroke engine →Bore-stroke ratio; →Engine

Shut-off device radiation loss factor →Injection system, fuel ~Diesel engine

Side bands →Engine acoustics

Side electrode →Spark plug ~Electrode design

Side layer wear →Piston ring ~Wear

Side valves →Valve arrangement

Side-draft carburetor →Carburetor ~Design types ~~Horizontal carburetor

Signature analysis →Engine acoustics

Silencer →Engine acoustics ~Engine noise ~~Encapsulation, acoustic effect

Silicon gel →Electronic/mechanical engine and transmission control ~Requirements for mechanical and housing concepts

Simulation process (*also*, →Calculation processes). Simulation processes in the development of an engine are used to predict the behavior of process flows, functions, components, engine modules, and, in the future, the behavior of the whole engine. Simulation processes are, thus, an important element of virtual product development. Their use improves the whole creation process for the product, such that early in the development phase, predictions can be made regarding the behavior of the construction and design base of the system. Thus, in addition to minimizing development risks early on, simulation processes can optimize product functions. This results in a reduction in the number of prototypes as well as cost and time savings. In the last few years, several simulation processes in engine development have been established. Important processes include, for example:

- Disturbance simulation to optimize the intake tract, the cooling cycle, the cylinder head through-flow, the oil cycle, etc.
- Finite element methods for structural, deformation, and strength analysis, such as crankcase strength, acoustic optimization, and crankshaft calculation.
- Simulation of dynamic processes, such as valve drive, dynamic model of cylinder head, crank drive.
- Combustion simulation considering superimposed flow requirements.

Simultaneous engineering. This is the simultaneous processing of tasks within the framework of development, for example, of a new engine. The goal of simultaneous engineering is the reduction of development time and, thus, a shorter time to market. This results in potential savings of up to half of the usual development time. It also means high requirements for cooperation between all participating functions in the company, and can include external developers in simultaneous engineering. In addition to shorter development times, the quality of the development process can be improved.

Single combustion chamber →Combustion chamber

Single element Hall sensor →Sensors ~Speed sensors ~~Active speed sensor

Single injection, gasoline engine →Injection system, fuel ~Gasoline engine ~~Intake manifold injection systems ~~~Central injection

Single injection pump →Injection system, fuel ~Diesel engine

Single range oils →Oil ~Oil functions, ~Properties and characteristic values ~~Viscosity classes for engine oils

Single sensor →Sensors

Single supercharger →Electronic/mechanical engine and transmission control ~Requirements for mechanical and housing concepts ~~Plug connector

Single-bank engines →Engine concepts ~Reciprocating piston engines ~Single-shaft engines

Single-barrel carburetor →Carburetor ~Carburetor, mode of operation, ~Design types

Single-bed capability →Bearings ~Materials ~~Qualities

Single-cylinder engines →Engine concepts ~Reciprocating piston engines ~Single-shaft engines

Single-hole nozzle →Injection valves ~Diesel engine ~~Hole-type nozzle ~~~Number of holes

Single-layer materials →Bearings ~Materials

Single-metal piston →Piston ~Design

Single-shaft engines →Engine concepts ~Reciprocating piston engines

Single-sided keystone ring →Piston ring ~Versions ~~Compression rings ~~~Keystone rings

Single-spark coil →Ignition system, gasoline engine ~High-voltage generation

Single-spray nozzle →Injection valves ~Diesel engine ~~Hole-type nozzle ~~~Number of holes

Single-spray process, wall-guided →Combustion process, diesel engine ~Direct injection

Single-stage combustion chamber →Combustion chamber ~Single combustion chamber

Single-stage supercharging →Supercharging

Single-stroke forged crankshaft →Crankshaft ~Blank ~~Single stroke forging

Sintered metal filter →Particles (Particulates) ~Particulate filter system ~~Particulate filter ~~~Filter media

Sintering →Connecting rod ~Semi-finished parts production

Sinusoidal disk engines →Engine concepts ~Engines without crankshaft

Six valves →Valve arrangement ~Number of valves

Skirt piston →Piston ~Skirt

Sleeve type chain →Chain drive ~Designs

Slide valve →Control/gas transfer ~Four-stroke engine ~~Slide control

Slide valve gear →Control/gas transfer ~Four-stroke engine; →Valve gear

Slotted oil control ring →Piston ring ~Versions ~~Oil scraper rings

Smart NO$_x$ sensor →Sensors ~Nitrogen oxides sensor

Smog (photochemical). Photochemical smog is created through the emission of contaminants into the atmosphere. Automobiles are a major contributor to this. The main influencing factors are traffic density, traffic type, and road routing. Additional factors are climatic effects, solar radiation, and wind speed and direction.

In addition to the emission of contaminants, there are chemical reactions from solar radiation resulting in so-called smog. For example, there is the Los Angeles smog, which is primarily a result of sulfur compounds from domestic fuel and industry. Vehicle exhaust gases are only a small contributor to these sulfur compounds: there are not many diesel vehicles on the road in that location.

Substance	t_m	ψ_m
CH_4	Approx. 7 years	1.6 ppm
Other HC	Few hours to few days	
CO	Approx. 60 days	0.05 to 0.2 ppm
CO_2	2 to 4 years	0.033%
NO	3 to 30 hours	<0.2 ppb (pure air)
NO_2	1 to 2 days	<0.2 ppb (pure air)
SO_2	Approx. 5 days	<0.2 ppb (pure air)
O_3	35 to 40 days (pure air). few hours (impure air)	20 to 40 ppb

Fig. S36 Mean dwell time, tm, and mean volume share in the troposphere

Smog reduction can be accomplished either through expulsion by precipitation or through chemical reactions. **Fig. S36** shows the median dwell time of contaminants and the particles in the troposphere. Carbon dioxide (CO_2) and methane (CH_4) are reduced slowly. It is assumed that CO_2 causes global warming because of increased absorption of the infrared radiation reflected by the earth. Since 1900, the concentration of CO_2 in the atmosphere has increased by about 15%. The contaminants CO, NO, and SO_2 reduce quickly because of chemical reactions.

The formation of photochemical smog can be seen in **Fig. S37**. It shows that HC and NO_x, emitted by vehicles, are major contributors to smog.

Smoke density →Black smoke; →Particles (Particulates)

Smoke limit. The smoke limit for a diesel engine defines the opacity level of the smoke at which impermissibly high emission of soot occurs. The emission of soot from a diesel engine increases approximately with the (mean) air-fuel ratio because a homogeneous mixture does not prevail. This is also the reason why a limited amount of fuel can be applied at full-throttle in a diesel engine and why the torque cannot be raised without limit by enriching fuel-air ratio in supercharged diesel engines (**Fig. S38**).

In Europe, visible smoke is limited by the ECE-R24 regulation, which is based on the engine air throughput (**Fig. S39**).

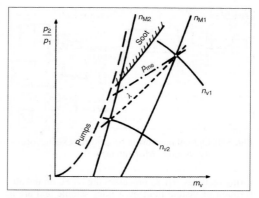

Fig. S38 Limits of torque increase on supercharged diesel engines (Source: Pischinger)

Smoke measurement →Smoke measuring instrument

Smoke measuring instrument (*also*, →Exhaust gas analysis equipment). Basically, two possibilities exist for analyzing gas loading with foreign bodies:

- Collection process—e.g., through separation on a filter
- In situ process—for determination of gas loading with, for example, soot

Smoke measuring devices usually function on the basis of the in situ process and optical analyses—for instance, by extinction measurements or light-scattering methods. They apply the property that the intensity of radiation is weakened by scattering or absorption of the light upon striking the solid contents of the gas—such as particles. This characteristic is used to determine exhaust blackening, which is primarily caused by soot particles or exhaust opacification.

Commonly used measuring devices either optically determine the opacity of gas or draw a defined quantity of exhaust gas through a filter and determine the blackening of the filter by photocells to measure the opacity of the gas (Bosch filter process). **Figs. S40 and S41** are schematics that show the principles of these mea-

S

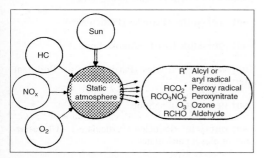

Fig. S37 Formation of photochemical smog (Source: F. Pischinger)

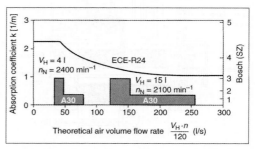

Fig. S39 Smoke limit curves according to ECE-R24 and A30 Sweden (Source: Mollenhauer)

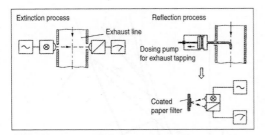

Fig. S40 Optical measuring process for exhaust opacity (Source: Mollenhauer)

suring methods. These processes are based on the absorption of visible light by a filter sample and are a measure of the amount of deposited soot.

The light absorption properties of soot ensure that the opacity measurement correlates well with the elemental carbon content on the filter sample (MIRA correlation). The reflection of light from a sooty filter is evaluated by the Bosch process (blackening number), but an aethalometer measures the attenuating effect on a penetrating light ray. (For subsequent analysis of the filtered samples →Particles (Particulates) ~Particulate filter system.)

The opacity of a light ray penetrating through the sample gas—that is, the sum of scattering and absorption—is equivalent to the measured signal. In the Raleigh range wavelength, the amount of light absorbed approximately corresponds to the mass of soot. Because absorption dominates in the case of soot, the process is suitable for application. The corresponding devices are known as opacity-measuring devices or opacimeters, and they are used in the so-called smoke test according to 72/306 EEC for inspection of diesel vehicles in the field.

Smoke number →Black smoke; →Particles (Particulates)

Smoke opacity →Black smoke; →Particles (Particulates)

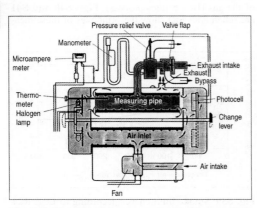

Fig. S41 Hartridge exhaust opacimeter (Source: DC)

Sodium-filled valves →Valve ~Gas transfer valves

Soft material seals →Sealing systems ~Special seals

Software →Electronic/mechanical engine and transmission control

Solenoid actuator →Injection system, fuel ~Diesel engines ~~Serial injection pump ~~~Controller for diesel injection pumps

Solid exhaust system →Exhaust system ~Decoupled exhaust system

Solid valve train →Valve gear

Solidification point →Fuel, diesel engine ~Properties; →Oil ~Body ~~Additives ~~~Pour point improver

Solidification point depressant →Oil ~Body ~~Additives ~~~Pour point improver

Sommerfeld number →Bearings ~Computation ~~Lubricating film pressure

Soot (*also*, →Particles [Particulates]). Soot, which is caused by combustion of the highly inhomogeneous structure of the mixture, is a typical exhaust component of diesel engines. It occurs during combustion under extreme lack of air. The formation of soot takes place in several stages, beginning with thermal cracking of the fuel molecules in a mixture with deficient oxygen. This leads to splitting of hydrogen into subhydrous structures. Macromolecules containing more and more carbon are produced via acetylene and polymerization, which then agglomerate to form soot particles similar to graphite. These seed particles measure on the order of 1 to 10 nanometers in magnitude. In later phases, primary soot particles join together to form larger units and agglomerate up to the order of magnitude of 100 nanometers.

An exact clarification of the kinetic reaction processes during the formation of soot has not been provided to date for the full magnitude range of soot particles. The range of soot formation and soot oxidation is shown for a diesel engine in **Fig. S42**.

Soot emission →Particles (Particulates)

Soot emission limit →Smoke limit

Soot filter →Particles (Particulates); →Pollutant aftertreatment

Soot generation →Soot

Soot particle →Particles (Particulates) ~Characteristics of diesel particulates

Soot separator →Particles (Particulates); →Pollutant aftertreatment

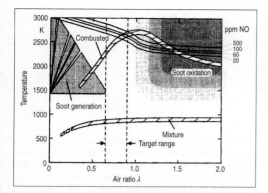

Fig. S42 Soot formation (Source: Pischinger)

Sound →Engine acoustics

Sound capsule →Engine acoustics

Sound intrusion test stand →Engine acoustics
~Noise reduction ~~Measurement of noise reduction

Sound particle velocity →Engine acoustics

Sound radiation →Engine acoustics

Sound spectrum →Engine acoustics

Spark →Ignition system, gasoline engine ~Ignition

Spark plug. The spark plug carries the electrodes that generate the ignition spark in the combustion space of the gasoline engine (**Fig. S43**). It delivers the electrical energy supplied from the ignition system locally to the air-fuel mixture at the moment of ignition in order to get externally controlled ignition. The mechanical and electric properties of the plug need to be designed accordingly. The spark plug has to ignite the mixture under all operating conditions and be stable against influences of temperature and aging.

The area of the spark plug facing the combustion chamber, therefore, has to quickly reach the self-cleaning temperature of about 400°C and ought not to exceed a peak temperature of approximately 850°C at wide-open throttle.

~Electrode design. Mixture accessibility, required ignition voltage, electrode wear, and heat dissipation all depend on the design of the electrode. It is possible to make a basic differentiation among the electrode designs used (**Fig. S44**) of the top electrode (or face electrode; **Fig. S44a**) and the side electrodes (**Fig. S44b**). The design determines the kind of spark gap produced.

Top electrodes are used for a pure spark air gap. Improved mixture accessibility can be achieved in this layout by advancing the position of the gap or reducing the electrode cross sections. The latter additionally brings about enhancement of the electric field strength and a reduced ignition voltage requirement through electrode surface area reduction. Thinner electrodes, however, require suitable materials, such as silver for better thermal conduction, because of the higher field-strength design measures that include platinum anchoring positions on the electrodes to give less spark erosion.

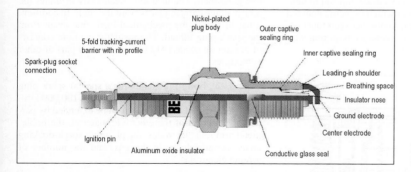

Fig. S43
Spark plug design

Fig. S44
Electrode designs (a, top electrode; b, side electrodes; c, surface-ignition-type electrodes)

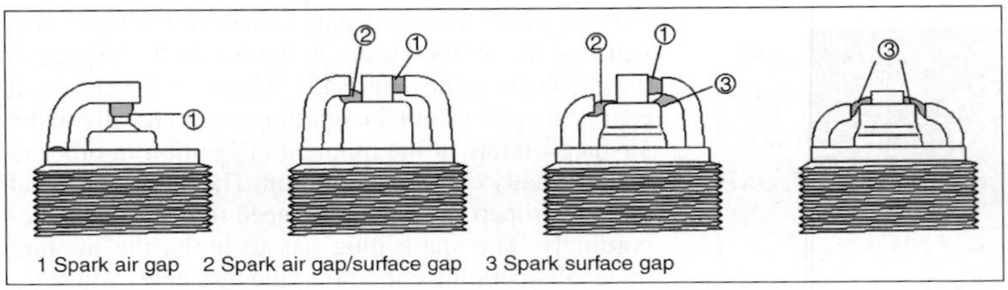

Fig. S45

1 Spark air gap   2 Spark air gap/surface gap   3 Spark surface gap

Fig. S45
Various paths of spark

Side electrodes are employed for spark surface gap concepts or mixed concepts comprising spark air gap and spark surface gap. Depending on the number and position of the side electrode(s), the ignition spark selects the ideal spark run for the operating state of the engine (**Fig. S45**). This can be a pure spark surface gap, a pure spark air gap, or a combination of both. Good side electrodes excel with low wear and eliminate the need to readjust the electrode gaps.

For all electrode types, the following applies: The electrodes are subjected to high thermal loading. Therefore, nickel-based special alloys are normally used as electrode material. Enhancement of the thermal conductivity of the electrode can be achieved by composite design with an inner copper core (**Fig. S46**).

~Electrode gap. The electrode gap is the shortest distance between the central electrode and the ground electrode(s). The voltage requirement for ignition increases linearly with electrode gap and gas pressure. For safe ignition of a flowing and/or turbulent air-fuel mixture, it is very important to activate the greatest possible volume with the spark by using the largest possible electrode gap. The gap is, however, limited by the ignition voltage supplied by the ignition system. That is why definition of the electrode gap and dimensioning of the ignition system with its specific demand for voltage supply are a compromise between function, ignition system costs, and maintenance requirements.

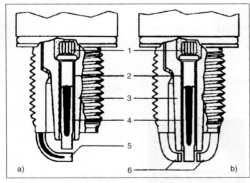

Fig. S46 Spark plugs with compound electrodes: 1, conductive glass; 2, air gap; 3, insulator nose; 4, composite center electrode; 5, composite ground electrode; 6, ground electrodes

~Electrode wear. The spark plug electrodes wear because of thermal loads, corrosion, and spark erosion. The thermal loads due to engine combustion give rise to hot-gas corrosion and electrode scaling. Aggressive gases and deposits, resulting from the high temperatures and continuous changes in temperature, develop on the electrodes under the influence of the fuel, additives, and lubricating oil, producing oxidation of the electrode surface area. When exposed to spark erosion, the electrode material is partially melted and eroded by evaporation due to the high temperatures in the plasma channel.

All wear mechanisms lead to rounding of the electrode edge and enlargement of the gap because of metal erosion. This leads to an increase in the required ignition voltage by several kV, and this must be supplied as a minimum ignition reserve in the ignition system voltage supply. The ignition system voltage supply thus determines the maximum possible lifespan (time of use) of the spark plug.

~Insulator. The insulator contains the connector pin carrying the high voltage along with the exterior high-voltage terminal and the central electrode. Both are connected in the insulator via an electrically conductive glass melt. The insulator electrically insulates the high-voltage carrying parts from the shell. In addition, it has to absorb the mechanical loads that occur from the spark plug thread. The insulator ceramic consists of aluminum oxide (Al_2O_3) with a small part of other admixtures.

~Long life. Long life stands for a special spark plug designed to achieve a long lifespan (e.g., 100,000 km). The useful life of a spark plug can be extended by particular selection of the electrode material, the noble-metal armored electrodes, the platinum spark-docking areas, certain electrode designs, and the number of ground electrodes.

~Plug face. Insulator nose, electrodes, and the front part of the spark plug hole are continuously exposed to the engine combustion process. This causes deposits that are characteristic of the particular engine operation, engine condition, and engine tune-up. The so-called plug face can be considered in this context (**Fig. S47**).

The following classes of plug faces primarily appear:

1. Normal spark plug appearance: Low electrode burning and a gray-white/gray-yellow to fawn discolored insulator nose.

Fig. S47
Plug faces

2. Leaded: The insulator nose shows brown-yellow glaze here and there that also can turn greenish.
Cause: Additives in fuel and lubricating oil form ash-type deposits.
Effect: The additives would liquefy and become electrically conductive under too sudden application of full engine load.

3. Sooted: Insulator nose, electrodes, and spark plugs are covered with velvety, black soot.
Cause: Faulty mixture setting—mixture too rich, air filter strongly contaminated, defective cold-start device. Predominantly used in short-distance traffic. Thermal range of the spark plug too high.
Effect: Leakage currents cause poor cold-start performance and misfiring. Unburned fuel can enter into the catalytic converter and damage it.

4. Oil-wetted: Insulator nose, electrodes, and spark plugs are covered with blackish oil film.
Cause: Too much oil in the combustion chamber, oil level too high, badly worn piston rings and cylinders and/or valve guides.
Effect: Misfiring or even shorted spark plug, total failure.

5. Glaze formation: The insulator nose shows brown-yellow glaze here and there that also can turn greenish.
Cause: Additives in fuel and engine oil form ash-type deposits.
Effect: The additives would liquefy and become electrically conductive under too sudden application of full engine load.

6. Deposits: Strong deposits of oil and fuel additives on insulator nose and ground electrode. Slag-type deposits (oil-derived deposit).
Cause: Alloy constituents, particularly from oil, can form residues that deposit in the combustion space and on the spark plug.
Effect: Can result in glow ignitions with power loss and engine damage.

~Spark plug location. The location of the spark plug in the combustion space has a decisive influence on the combustion diagram. The distance of flame travel to the combustion chamber wall should be equally long, as far as feasible, to obtain a combustion time as short and stable as achievable and, thus, bring operation close to the thermodynamically ideal constant volume combustion process as far as possible.

In the case of four-valve engines, only a central plug location between the valves would be suitable. In the case of two-valve engines, various spark plug locations, depending on the shape of the combustion space, are conceivable. As a precaution against knocking (autoignition of the mixture), when determining the plug location it should be ensured that the flame front ignites the knock-endangered hot zone of the exhaust valve as quickly as possible. This results in an optimal spark plug position close to the exhaust valve. In addition, the selected spark plug location also has to enable sufficient thermal dissipation from the plug.

~Thermal range. The thermal range of a spark plug is mainly determined by the length of the insulator nose and electrodes (**Fig. S48**). The more these extend into the combustion space, the more heat they absorb. At the same time, and depending on design, the thermal connection to thermal dissipation is less favorable, resulting in a further rise in temperature.

When operating, a spark plug should reach its self-cleaning temperature of more than 400°C as quickly as possible. At the same time, a peak temperature of 850°C should not be exceeded, to avoid glow ignition and increased thermal loads on both engine and spark plug. The thermal range of a spark plug describes how much energy the spark plug absorbs from the combustion space and to what extent the absorbed quantity of energy will result in temperature rise (heating up) in the insulator nose and electrodes (**Fig. S49**).

Engines with different maximum temperatures in their combustion chambers need spark plugs with a specifically adapted thermal range.

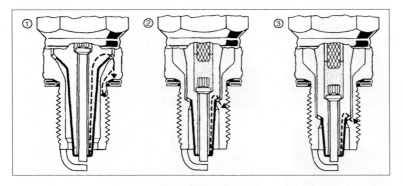

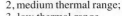

Fig. S48
Influence of the
spark-plug geometry on
the thermal range:
1, high thermal range;
2, medium thermal range;
3, low thermal range

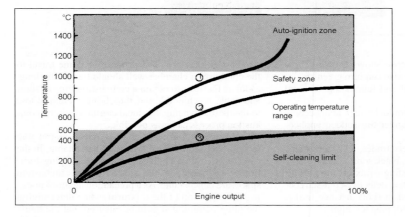

Fig. S49
Spark plug response to
temperature changes with
different heat range code
number

~Type formula. The type formula characterizes the spark plug type by a combination of letters and numbers. These describe the thermal range and the electrode material as well as mechanical features such as internal thread diameter and type of sealing seat.

Literature: M. Adolf: Zündkerzen: In R. van Basshuysen, F. Schäfer (Eds.), Handbuch Verbrennungsmotor, Vieweg Publishers, 2002, pp. 465–470. — H. Klein: Zündkerze: In R. van Basshuysen, F. Schäfer (Eds.), Shell Lexikon Verbrennungsmotor, Series 86/87, Vieweg Publishers, 2003. — S. Pischinger, J.B. Heywood: Einfluss der Zündkerze auf zyklische Verbrennungsschwankungen im Ottomotor, MTZ 52 (1991) 2. — Y.G. Lee, D.A. Grimes, J.T. Boehler, J. Sparrow, C. Flavin: A Study of the Effects of Spark Plug Electrode Design on 4-Cycle Spark-Ignition, Engine Performance, SAE, 2000-01-1210. — J. Geiger, S. Pischinger, R. Böwing, H.-J. Koß, J. Thiemann: Ignition Systems for Highly Diluted Mixtures in SI-Engines, SAE, 1999-01-0799. — Th. Kaiser, A. Hoffmann: Einfluss der Zündkerzen auf das Entflammungsverhalten in modernen Motoren, MTZ 61 (2000) 10. — Alles über Zündkerzen, Technische Information No. 02, Beru AG, 2000.

Spark plug electrode →Spark plug ~Electrode design

Spark plug face →Spark plug ~Plug face

Spark plug location →Spark plug ~Spark plug location

Spark plug position →Ignition system, gasoline engine ~Ignition ~~Ignition spark

Spark plug temperature →Spark plug ~Thermal range

Spark tail →Ignition system, gasoline engine ~Ignition ~~Ignition spark

Spark voltage →Ignition system, gasoline engine ~Ignition ~~Ignition spark, ~~Ignition voltage

Special engine oils →Oil ~Classification, specifications, and quality requirements

Special lapping →Piston ring ~Lubrication ~~Topography of ring running surface

Special sealing →Sealing systems

Special valves →Valve ~Gas transfer valves

Specific bearing load →Bearings ~Computation ~~Specific load

Specific catalytic converter surface →Catalytic converter ~Specific surface area

S

Specific fuel consumption →Fuel consumption ~Effective fuel consumption

Specific heat →Coolant

Specific piston area output →Piston ~Specific output per (piston) surface area

Specific power →Power output; →Weight-to-power ratio

Specific power output →Weight-to-power ratio

Specific power output per liter →Power output

Specific work (*also*, →Mean effective pressure). Specific work is the work performed by an engine related to the swept volume—that is, it has the dimension kJ/m³ or bar. It represents a reference parameter of different engines that is independent of the stroke volume of the engine.

A differentiation is made between indicated and effective (brake) specific work. While the indicated specific work is calculated from the p-V diagram in the combustion chamber, the effective (brake) specific work relates to the work output at the crankshaft.

Spectral power output →Engine acoustics ~Spectrum

Speed. In a four-cycle engine, the maximum engine speed, n, is determined by the displacement, V_H, the mean effective pressure, p_{me}, and the power output, P_e, by the following equation:

$$P_e = p_{me} \cdot V_H \cdot n \cdot i,$$

where i is the number of combustion cycles per rotation. In a four-cycle engine, $i = 0.5$.

The higher the speed, the higher is the power if everything else remains the same. The power increases in proportion to the speed if the mean effective pressure and displacement are kept constant. The speed for production engines is limited to control fuel consumption, noise, and cost. The maximum speed for gasoline engines, apart from exceptions, is approximately 6000 rpm.

A reduction in speed results in a reduction of the gas-exchange losses; it also reduces the friction losses and, therefore, increases the mechanical efficiency. The highest engine speeds are found in Formula 1 engines, and their purpose is to achieve increased power; speeds of more than 19,000 rpm are reached.

Diesel engines tend to have lower speeds. Prechamber diesel engines have a maximum speed of about 5000 rpm, while those with direct injection reach about 4000–4500 rpm. The limitation of the maximum speed of engines with direct injection is based on the available injection systems and on the duration of fuel conversion during the mixing and combustion processes.

~Cranking speed. Engines must be started externally because they cannot start on their own. This is achieved with external energy. The crankshaft is rotated with the help of a starter until a minimum speed is achieved.

The minimum speed must ensure that the signals for the start of the ignition and injection are present during electronic control of the gasoline engine.

The cranking speed in a diesel engine must ensure, in addition to the ignition assistance at low temperatures (pencil-type glow plug), that the temperature of the charged air is high enough to initiate combustion. Cranking speeds in diesel engines are higher than in gasoline engines. The minimum speed required to start an engine is different for each engine type, depending on ambient temperatures and mixture formation systems. It is about 60–100 rpm for gasoline engines and 80–200 rpm for diesel engines.

~Cutoff speed. The engine normally gets switched off at its highest permissible speed to prevent mechanical damage. The engine is switched on again once the speed falls below the highest speed. The switch-on and switch-off function in gasoline engines was usually handled in the past by interrupting the ignition. This allowed unburned fuels to enter the catalytic converter, which could destroy it, and the fuel supply was then interrupted.

~Idle speed. The idle speed should be as low as possible to minimize fuel consumption at idle speeds. However, auxiliary drives such as alternators, steering booster pumps, and air-conditioning compressors must deliver the required power at idle speeds as well. The minimum idle speeds are currently 550–750 rpm for gasoline engines and 800–900 rpm for diesel engines. The valve overlap must be kept as small as possible to achieve a stable idle speed. The starting performance of the catalytic converter is reduced by a low idle speed, because the exhaust gas temperature is also lower. This can be compensated for, if the idle speed is elevated during the warm-up phase.

~Maximum speed. The maximum speed is the highest allowable constant engine speed.

~Resonance speed. Resonance vibration develops when the frequency of the stimulation is equal to the resonance speed. This means that the stimulation forces act to accelerate the system in phase with the instantaneous direction of movement and as a result increasingly amplify the vibration. The amplitudes are now only limited by the damping, which transforms the mechanical energy into heat energy. A subcritical characteristic is present if the stimulation frequency is lower than the resonant frequency, and it is called supercritical if the stimulation frequency is higher than the resonant frequency.

The relative maximum deformation and stress values in the engine are achieved at the resonant frequency. This is the reason why high resonant frequencies

S

should be specified, so that the operating speeds are below the resonant frequency.

Small inertias and high crankshaft rigidity should be defined to achieve the desired torsional resonance frequency characteristics. Other important parameters are, for example, the type of design, the firing order, and the crankshaft arrangement. Balance weights or balance shafts increase the inertia, reducing the resonant frequency.

~Speed measurement. Speed measurement is required for engine control; this is normally performed by a pulse generator sensing the passing of the flywheel teeth.

~Speed range. The speed range indicates the limits of the usable speed of an engine. It is defined as the difference between the allowable highest speed and the idle speed. The "elasticity" of an engine increases with the increase of the usable speed range, and this means that a vehicle can be operated at a lower noise level.

~Speed reduction. The reduction of the maximum speed has advantages especially for noise and fuel consumption: It reduces the friction and gas transfer losses. In addition, it provides cost advantages and improves durability. A reduction in idle speed also increases the comfort and reduces fuel consumption while keeping the idle speed quality. However, there are limits for the speed range in the ability of the accessories (alternator, steering booster pump, etc.) to function.

~Threshold speed. The threshold speed is the maximum allowable engine speed. It depends on many factors. The inertia forces, the injection period in engines with direct injection, the speed limit or the speed range of the accessories, and the thermal and mechanical stability of components all play a major role. A power increase can be achieved with an increase in the threshold speed, because power increases in proportion to speed.

Literature: Kraftfahrtechnisches Taschenbuch, VDI Publishers, 19th Edition, 1984. — Autoelektrik/Autoelektronic am Ottomotor, VDI Publishers, 1987.

S

Speed, piston →Piston ~Mean piston speed

Speed limit →Electronic open- and closed-loop control ~Electronic open- and closed-loop control, diesel engines ~~Functions

Speed measurement →Speed

Speed of sound →Engine acoustics

Speed range →Speed

Speed reduction →Speed

Speed regulation →Electronic open- and closed-loop control ~Electronic open- and closed-loop control, gasoline engine ~~Functions

Speed restriction →Electronic open- and closed-loop control ~Electronic open- and closed-loop control, gasoline engine ~~Functions; →Speed ~Cutoff speed

Speed sensor →Sensors ~Speed sensors

Speed shut-off →Speed ~Cutoff speed

Spherical cap (throttle) →Actuators ~E-gas ~~Progression (throttle)

Spiral expander →Piston ring

Spiral port →Intake port, diesel engine ~Swirl channel

Spiral-type supercharger →Supercharging ~Rotary-piston supercharger

Splash lubrication →Lubrication

Sports engines →Engine ~Racing engines

Spray angle →Injection valves

Spray carburetor →Carburetor

Spray cone →Injection valves ~Diesel engine ~~Injection nozzle parameter

Spray cooling →Piston ~Cooling ~~Spray oil cooling

Spray dispersal angle →Injection valves ~Diesel engine ~~Injection nozzle parameter

Spray distribution →Injection valves ~Diesel engine ~~Injection nozzle parameter

Spray formation →Injection valves ~Gasoline engine ~~Intake manifold injection

Spray hole →Injection valves ~Diesel engine ~~Hole-type nozzle

Spray preparation →Injection valves ~Gasoline engine ~~Intake manifold injection

Spray propagation →Combustion, diesel engine

Spray shape →Injection valves ~Gasoline engine ~~Intake manifold injection ~~~Spray formation

Spray-guided combustion process →Injection valves ~Gasoline engine ~~Direct injection ~~~Installation positions

Spread angle →Control/gas transfer ~Four-stroke engine; →Variable valve control ~Variation parameters

Spring →Valve spring

Spring bending tension →Piston ring

Spring designs →Valve spring ~Design types

Spring loading →Valve spring ~Load on the valve spring

Spring map →Valve spring

Spring production →Manufacture, valve spring; →Valve spring

Spring resonance frequency →Valve spring ~Load on valve spring

Spring seat. The spring seat is used to fix the valve spring and the drivetrain between cylinder head and valve end (**Fig. S50**). It ensures that the intake and exhaust valves sit securely and gas-tight on the valve seat at the end of the cam lift. The spring seat also transmits the rotational motion of valve rotating devices (rotocap, rotovalve, rotocoil) that are positioned between valve spring and cylinder head. Rotating valves are required to compensate for local temperature differences at the valve seat. This avoids wear at the valve seat.

Spur gears →Valve gear

Sputter bearing →Bearings ~Materials

Squeeze casting →Crankcase ~Casting process, crankcase

SSG →Starter ~Starter-generator

Stacked plates →Radiator ~Design ~~Plate

Stage →Carburetor ~Design types

Stainless steel catalytic converter →Catalytic converter

Stainless steel coating →Catalytic converter ~Coating

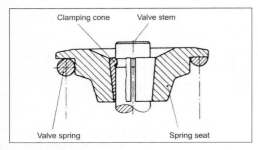

Fig. S50 Valve head

Stand-alone products →Electronic/mechanical engine and transmission control ~Requirements for mechanical and housing concepts

Start →Carburetor ~Starting systems ~~Design types

Start and end of injection →Injection functions ~Gasoline engine ~~Direct injection

Start carburetor →Carburetor ~Starting systems ~~Design types; →Starting aid ~Starting fuel injection

Start of injection →Injection functions ~Gasoline engine ~~Direct injection ~~~Start and end of injection; →Injection system, fuel ~Gasoline engine ~~Intake manifold injection systems

Start of injection timing →Electronic open- and closed-loop control ~Electronic open- and closed-loop control, diesel engine ~~Functions ~~~Start of injection control

Start of intake →Control/gas transfer ~Four-stroke engine ~~Timing

Start pilot →Starting aid

Starter. The task of the starter is to ensure that the necessary starting speed (minimum speed) for the combustion engine is achieved even in unfavorable conditions. From the moment of the first successful work stroke onward, the starter must support the running-up of the engine to the minimum self-sustaining speed. The minimum speed necessary for the starting process is 60–100 rpm for gasoline engines and about 80–200 rpm for diesel engines, depending on the ambient temperature.

The starter pinions and the wheel sprocket on the engine flywheel make up a transmission system, the ratio of which (about 1:10 to 1:17) converts the high start-up speed with low torque of the starter motor into a low engine speed with high torque.

To start combustion engines, electric motors (direct, alternating, and three-phase current engines; **Fig. S51**) and hydraulic and pneumatic motors are used. The electrical parallel flow series motor, in which field winding and rotor resistance are connected in series, is especially suitable as a starter motor because it develops the necessary high initial torque to overcome the start-up torque and to accelerate the inertia of the powertrain.

The excitation field necessary for the electric motor can be produced by the field winding and also by permanent magnets (**Fig. S52**). The connected low rotor torque must be balanced through a larger total ratio between rotor and crankshaft at a much higher rotor torque. The increase of the ratio can be reached using a countershaft gearbox integrated into the starter.

S

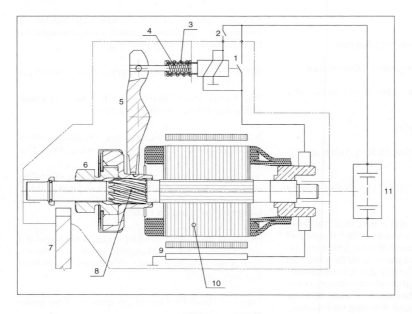

Fig. S51
Starter with permanent
magnet excitation

A starter consists of the following main components:

- Electric motor
- Engaging relay
- Piston-engaging drive

At startup, the pinion must engage the wheel sprocket. After the startup, the combustion engine can accelerate quickly to high speeds. A roller-type freewheel deactivates the transmission of force between the pinion and the rotor to protect the rotor from overspeeding with its associated centrifugal forces.

Depending on the system of engagement, a differentiation must be made among the following types:

- Bendix-type starter

- Sliding-gear starter
- Pre-engaged drive starter

The pre-engaged drive starter (**Fig. S53**) is the most common starter type for automobiles nowadays.

~**Starter battery** →Battery

~**Starter-generators.** Starter-generators offer a high potential for hybrid vehicles with regard to the constantly increasing demand for energy and the need to save fuel. A differentiation is made between the following levels of hybridization:

- *Minimum hybrid* (μHybrid), mostly as a separate starter-generator and in connection with the usual

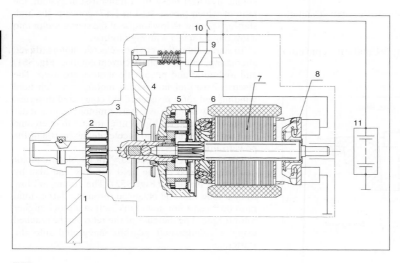

Fig. S52
Starter as parallel flow
series motor

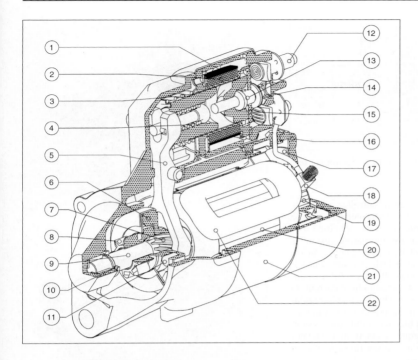

Fig. S53
Pre-engaged drive starter

14-V electric system, offers the possibility for start-stop operation.

- *Mild hybrid*, as a separate and integrated starter-generator together with an efficient 42-V electric system (or dual-voltage electric system (14 V/42 V) through bidirectional DC/DC converter), allows other functions such as boost and brake energy recovery in addition to start-stop operation.

- *Full hybrid*, as a separate or integrated starter-generator on the drivetrain side, in connection with a 42-V or a high-voltage electrical system, enables additional motor functions including pure electrical drive.

Another differentiation as far as starter-generators are concerned is made depending on the system design type and/or their arrangement in the drivetrain. **Fig. S54** shows an overview of the different systems.

~~Integrated starter-generator, or ISG (main drive)

~~~Control device with inverter. To regulate and/or control the starter-generator, a suitable electronic control device is needed. A typical block diagram of this component can be seen in **Fig. S55**. The often integrated CAN interface is the connection with the pri-

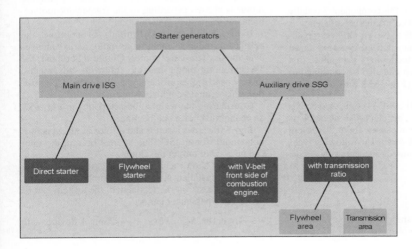

Fig. S54
Classification of the starter-generators

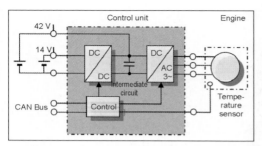

Fig. S55 Block diagram of a control device for starter-generators

mary system, such as the engine controller or battery management. This allows the starter-generator to be integrated in the torque structure of the drivetrain. For that, the necessary three-phase alternating current is produced from the parallel flow by means of an inverter, which then provides the desired torque. The so-called field-based regulation is often used for the controller that sets the train voltage and phase current on the stator of the electric motor. This method of regulation requires a high computing power from a microcontroller with at least 16 bits and a dynamic system behavior (time constant torque: about 10 ms) and very good system efficiency, which in the ISG can exceed 80% in wide operating ranges.

The phase current of the electric motor and the drive cycle defined by the automobile manufacturer, which reflects the duration of the current load in the inverter at defined environmental conditions, are central design criteria for the control device.

An example of a control device for a starter-generator can be seen in **Fig. S56**.

~~~DC/DC converter. The proposed architectures in the dual 14-V/42-V electrical systems mostly have a DC/DC converter, which is almost always executed bi-directionally, meaning that energy is transported in both directions.

The desired performance of this component is usually between 0.5 and 2.5 kW. The maximum efficiency is usually expressed as 90 to 95%. New multilevel topologies for the converter are characterized by a very small current ripple. The design technology is equal to that of the inverter, with which it can also be combined to make an economical unit (**Figs. S55 and S56**).

~~~Electrical machine. The ISG is connected coaxially with the crankshaft. The electrical machine is characterized by a large diameter of about 24–32 cm and a short length of about 4–9 cm. The following types of electric motors are usually used for energy conversion:

- Asynchronous motor
- Synchronous motor
- Reluctance motor
- Axial flow motor

The starting torque and the generator performance determine the performance of the motor. If the design

Fig. S56 Control device for a starter-generator

space is limited, necessary flows increase unfavorably. Typical performance requirements are:

- Starting torque: >200 Nm
- Battery power load: <500 A

Fig. S57 shows a rough overview of the characteristics of different types of motors offering advantages as well as disadvantages in practical use. A suitable motor is chosen according to the framework requirements of the system specification.

If the design space is limited, for example, the permanent-magnet synchronous motor has advantages over the asynchronous motor. On the other hand, the asynchronous motor has the following advantages: lower costs, higher efficiency at high torque, no position sensor, and recognized sturdiness.

Considering the weight characteristics in **Fig. S57**, an economic choice can be made.

Fig. S58 shows a stator and a rotor in an asynchronous motor for an ISG. The internal bore of the rotor can be used to mount the clutch.

Fig. S59 shows an example of a starter-generator system in installed state on the combustion engine. Here, after tests on the test bench, the system can be proved in a passenger car.

~~~Inverter. *See above*, ~~~Control device with inverter

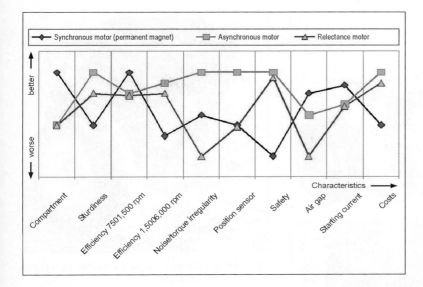

Fig. S57
Comparison between synchronous, an asynchronous, and a reluctance motor

~~Operating modes of the starter-generator

~~~**Alternator and retarder.** When the combustion engine is at idle or just faster, the starter-generator can produce energy for the electric system. Through implementation of the brake energy recovery function (in connection with a suitable storage media), additional fuel consumption advantages are possible, because the kinetic energies of the engine and vehicle, which are lost when the vehicle brakes are used, is converted into collectible energy.

The required braking torque is converted into energy, which can be fed into the electrical system on the vehicle. Energy recovery in the braking phases can only occur when the battery and/or the storage media are not at their maximum allowed charge status.

When using all functions, a reduction of the fuel consumption of up to about 20% is possible in standard driving cycles.

~~~**Boost function.** In this operating mode, the combustion engine is supported by the electric motor, especially in the lower speed range. The operating range is between idle speed and medium speeds. This allows the selection of an engine with smaller displacement for the same driving performance in a given vehicle application during the concept stage (engine downsizing). The maximum possible torque support also depends on the type and the charge status of the electrical energy supply (i.e., battery).

~~~**Retarder.** *See above*, ~~~Alternator and retarder

~~~**Starter.** This replaces the function of the usual shift starter, but it can tow-start the combustion engine, depending on the system design, up to the idle speed. This enables a low-emission, quiet, and quick start in about 300–500 ms.

~~~**Start-stop function.** The main interest when using a starter-generator is use of the system in the so-called start-stop operating mode. The electronic system of the starter-generator can, under certain circumstances, stop the combustion engine (i.e., driving speed ≈0 km/h, gear lever in neutral position) and restart the engine again via access to the engine control (i.e., CAN interface, clutch is engaged, shifting). By stopping the engine in periods when the vehicle is not in motion (in which the engine would usually run at idle), noticeable fuel and contaminant reduction of the combustion engine (about 5–10% in standardized drive cycles) can be reached.

~~**Separate starter-generator, or SSG (auxiliary drive).** The separate starter-generator is an electric motor based on the classic generator. Complex inverter electronics (**Fig. S60**) allow the motor to be used in generator mode as well as that of a motor (starter) and, thus, to realize both functions.

The separate starter-generator does not require large construction changes in the engine compartment. The motor is installed at the same location as the usual generator.

The electronics can be placed in a separate housing at a suitable location—that is, in the free space behind the radiator grill—where the environmental conditions provide adequate cooling of the components in the case of air-cooled electronics. There are separate starter generators for the usual 14-V electrical system architecture, as well as for 42-V and/or twin-voltage electric systems (with DC/DC converter).

In generator mode, a magnetic field is created in the rotor of the starter-generator, which is controlled by the electronics depending on the necessary electrical power output. The alternating current produced through electromagnetic induction is converted into parallel flow by the electronics, whereby the

S

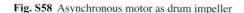

Fig. S58 Asynchronous motor as drum impeller

transistors of the inverter can also transport the current against the conventional direction and, thus, replace the usual diode rectifier. The significantly lower power loss of the switched transistors in the conducting state, compared with diodes, has a positive effect on efficiency. For this operation, which also is called synchronous rectification, it suffices to continuously determine the sign of the voltage on the preceding individual switches and to switch the transistors accordingly.

In starter mode, a magnetic field is produced in the rotor. In the stator, the electronics develop a rotating magnetic field depending on the position of the rotor relative to the stator (determined by position sensors), which, in turn, puts the rotor into rotary motion based on the induction principle. Because the transistors work in so-called continuous operation, costly switching for pulse generation is not necessary, because the transistors are each 180° electrically conductive.

Fig. S59 Installation example for an ISG on a combustion engine

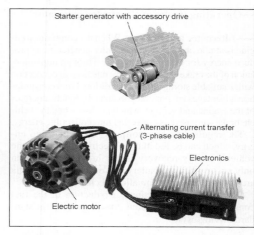

Fig. S60 Separate starter-generator with electronics

This leads to significantly reduced switching losses in the inverter, because the switching frequencies that occur are relative low.

By using the starter-generator as motor (starter), the combustion engine can be started via the belt of the accessory drive. The high torque produced, in conjunction with the high speeds the starter-generator can reach in starting mode, mean that relatively short starting times can be obtained. In comparison to starting with a classical starter motor, the start through the belt using a separate starter-generator is almost silent. The high torques to be transferred by accessory drive must be considered by adapting its design. Alternatively, the separate starter-generator can be used directly in the powertrain, such as in the flywheel area or in the transmission area.

Literature: Alfred Krappel, et al.: Kurbelwellenstartgenerator (KSG), Basis für zukünftige Fahrzeugkonzepte, 2nd edition, Expert Publishers, 2000, ISBN 3-8169-186-9. — Cédric Plasse, et al.: L'alterno-démarreur, du Stop&Go au groupe motopropulseur hybride, Société dés Ingénieurs Automobile, Palais de Congrès de Versailles, 13–14 November 2001. — Heinz Schäfer, et al: Integrierter Starter-Generator (ISG)–Das multifunktionale Bindeglied zwischen Bordnetz und Antriebsstrang im Kraftfahrzeug, Expert Publishers, 2001, ISBN 3-8169-1964-4. — H. Schäfer: Startergenerator mit Asynchronmaschine und feldorientierter Regelung, Special Edition on Automotive Electronics, ATZ/MTZ, January 2000. — F. Renken, V. Karrer: Leistungselektronik für den Integrierten Starter-generator, HdT Conference, 22, Conference Elektronik im Kraftfahrzeug, Dokumentation: Kfz-Elektronik, Conferences. No. E-H030-06-056-2, House of Engineering, Stuttgart, 2002. — E.C. Lovelace, et al.: Interior PM Starter-Alternator for Automotive Applications, Proceedings ICEN '98, 2–4 September 1998, Istanbul, Turkey. — J.M. Miller, et al.: Starter-Alternator for Hybrid Electric Vehicle: Comparison of Induction and Variable Reluctance Machines and Drives, Proceedings IEEE '98. — G. Altenbernd, et al.: Present Stage of Development of the Vector Controlled Crankshaft Startergenerator for Motor Vehicle, Proceedings speedam 2000, 13–16 June 2000, Ischia, Italy. — K.S. Rasmussen, P. Thogersen: Model Based Energy Optimiser for Vector Controlled Induction Motor Drives, Trondheim, EPE, 1997. — H. Späth: Steuerverfahren für Drehstrommaschinen, Springer Publishers, Berlin/Heidelberg/New York, 1983. — Peter Skotzek, et al.: High Performance Power Electronics for Integrated Starter/Generator Systems, Electronic Systems For Vehicles, 9th VDI Congress, Baden-Baden, Germany, 5–6 October 2000. — F. Renken, V. Karrer: Leistungselektronik für den Integrierten Starter-generator, HdT Conference, 22, Conference Elektronik im Kraftfahrzeug, Dokumentation: Kfz-Elektronik, Conferences. No. E-H030, 2-056-06, House of Engineering, Stuttgart, 2002.

Starter balancing →Balancing of masses ~Balancing ~~Accessory equipment ~~~Starter

Starter catalytic converter →Catalytic converter

Starter glow plug →Ignition system, diesel engine/preheat system ~Cold starting aid ~~Glow system; →Starting aid ~Glow plug

Starter valve. When using the L-Jetronic system as a carburation device in gasoline engines, additional fuel was injected into the intake manifold, depending on the temperature, via a separate injection valve called the starter valve. This was necessary to guarantee a safe cold start. The deactivation of the amount of fuel up to a predefined operating temperature took place by changing the cycle time of the starter valve, controlled via a thermo-time switch.

Starter-generator →Starter

Starting aid. Starting aids are necessary for diesel engines particularly when the temperature of the intake air and the motor speed are low. The starting speed can be increased through increasing the power of the starter motor and the battery power. New starter-generator systems increase the starting speed to idle speed, which significantly improves the starting of the engine. Low-viscosity engine oils are also helpful and, thus, necessary.

~Air preheating. *See below*, ~Intake air preheating

~Coolant preheating. In order to quickly heat up an engine to its operating temperature, it is recommended that the coolant be preheated. This improves fuel consumption and reduces emission of contaminants as well as the noise level. An effective means of achieving this is to use a heat reservoir, which can store warm coolant for several days.

~Flame glow plug. *See below*, ~Flame starting system

~Flame starting system. The flame starting system is preferred for larger diesel engines. It works with additional fuel and heats the intake air through combustion. The most important component of the flame starting aid is the sheathed-element glow plug (**Fig. S61**), which measures the fuel. It must be ensured that enough oxygen remains for combustion in the combustion chamber.

~Fuel preheating. Electrical fuel preheating is mainly done for diesel fuels in the winter to avoid paraffin deposits in the fuel filter and clogging of the fuel filter.

~Glow plug. In diesel engines, glow plugs are used as starting aids (**Fig. S62**). During the starting phase, the glow plug is heated in a short time by electrical heating to a temperature of about 850°C. It must be located in the region of the combustion chamber, in which an ignitable mixture is present.

After the engine starts, the glow plug is still electrically heated to improve the start-up process and to

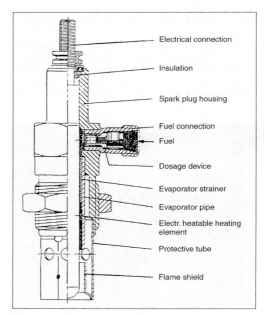

Electrical connection

Insulation

Spark plug housing

Fuel connection

Fuel

Dosage device

Evaporator strainer

Evaporator pipe

Electr. heatable heating element

Protective tube

Flame shield

Fig. S61 Sheathed element glow plug (Source: Mollenhauer)

S

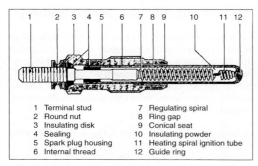

1 Terminal stud	7 Regulating spiral
2 Round nut	8 Ring gap
3 Insulating disk	9 Conical seat
4 Sealing	10 Insulating powder
5 Spark plug housing	11 Heating spiral ignition tube
6 Internal thread	12 Guide ring

Fig. S62 Glow plug (Source: Bosch)

reduce blue-smoke emissions as well as combustion noise (start-up delay). Hydrocarbon and carbon monoxide emissions are reduced as well. In driving operation, the postheating time is a maximum of 180 s. Additional aids are intake air preheating and preheating of the coolant.

~Intake air preheating. Intake air preheating takes place, for example, using the exhaust-gas manifold, through which the intake air is routed. Electric heating cartridges and heating flanges (**Fig. S63**) also can be applied effectively.

~Mixture preheating. The mixture preheating is often used in gasoline engines with carburetors. In the case of downdraft carburetors, an electrically heated hotspot is located in the intake manifold underneath the carburetor, which heats up the air-fuel mixture for better mixture formation during the warm-up phase.

~Start pilot. In some cases at lower intake temperatures, the starting procedure in larger diesel engines is performed using the start pilot injection. An especially ignitable special fuel is injected into the intake air, which enables cold starting under extreme conditions.

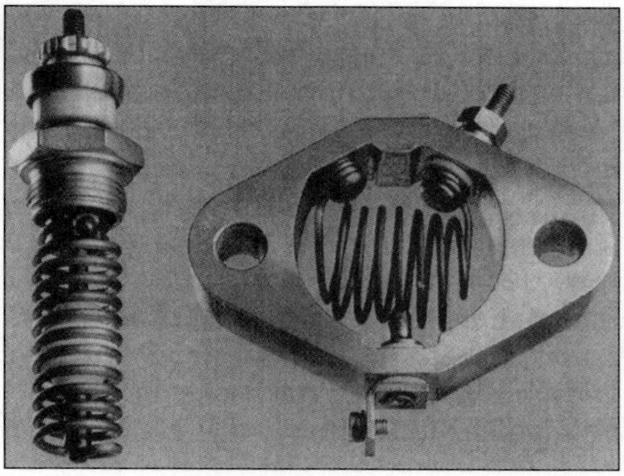

Fig. S63 Heating cartridge and heating flange (Source: Mollenhauer)

~Starting fuel injection. In conventional gasoline engines, an effective measure for improving cold starts is the enrichment of the air-fuel mixture through additional fuel. In engines with carburetors, this takes place using the starter carburetor, and in injection engines through increasing the injected fuel volume. A disadvantage is the increased fuel consumption during the heating-up phase.

Starting aid ignition system →Ignition system, diesel engine/preheat system ~Cold starting aid

Starting evaluation criteria →Ignition system, diesel engine/preheat system ~Cold start

Starting fuel injection →Starting aid

Starting performance →Catalytic converter; →Starting aid; →Starting phase

Starting phase (*also*, →Starting aid). The starting phase of an engine is characterized by the time between the first revolution and the running-up to a stable idle speed. During the first revolutions, the injection and possibly the ignition are synchronized. An increase of the volume of fuel injected and an adaptation of the ignition timing possibly takes place. Depending on the starting temperature, additional starting aids might be necessary.

Starting system →Carburetor

Start-stop function →Starter ~Starter-generators ~~Operating modes of the starter-generator

Start-up enrichment →Carburetor ~Starting systems ~~Requirements

Stationary diesel engines →Piston ring ~Ring set ~~Commercial vehicle diesel engines ~~Marine and stationary diesel engines

Stationary engine operating point. The stationary (steady-state) engine operating point is characterized by a constant speed and a constant load at which the engine operates for a certain amount of time. This state defines the steady-state operating behavior. In contrast, short-term changes of load and speed are present in transient behavior.

Stationary engines (radial engines) →Engine; →Engine concepts ~Reciprocating piston engines ~~Single-shaft engines ~~~Radial engines

Stationary high-voltage distribution →Ignition system, gasoline engine ~High-voltage distribution

Stationary noise measurement →Engine acoustics

Statistical energy analysis (SEA) →Engine acoustics

Steam engine →Engine ~Alternative engines

Steam jet impermeability →Electronic/mechanical engine and transmission control ~Requirements for mechanical and housing concepts

Steel →Piston ring ~Materials

Steel connecting rod →Connecting rod ~Materials ~~Forged steel

Steel crankshaft →Crankshaft ~Materials

Steel oil control ring →Piston ring ~Versions ~~Oil scraper rings

Steel ring →Piston ring ~Materials ~~Steel

Stirling engine →Engine ~Alternative engines

Stirling process →Cycle; →Engine ~Alternative engines

Stoichiometric air-fuel ratio →Air-fuel mixture; →Air-fuel ratio

Stopper →Sealing systems ~Cylinder head gaskets

Storage capability →Particles (Particulates) ~Particulate filter system ~~Particulate filter

Strainer →Lubrication

Stratified charge →Mixture formation ~Mixture formation, gasoline engine

Stratified charge method →Combustion process, gasoline engine; →Injection valves ~Gasoline engine ~~Direct injection ~~~Installation positions

Strength, crankcase →Crankcase ~Crankcase design

Strength computation →Calculation processes ~Application areas ~~Stability

Stroke law →Balancing of masses

Stroke switchover →Variable valve control ~Systems with camshaft ~~Variability between cam and valve ~~~Discontinuously variable systems

Stroke terminating systems →Variable valve control ~Systems with camshaft ~~Variability between cam and valve ~~~Continuously variable systems

Structural dynamics →Calculation processes ~Application areas

Structural optimization →Calculation processes ~Application areas ~~Stability

Structure-borne noise →Engine acoustics

Structure-borne sound decoupling →Sealing systems ~Elastomer sealing systems

SU carburetor →Carburetor ~Constant pressure carburetor

Subdivided combustion chamber →Combustion chamber

Substrate →Electronic/mechanical engine and transmission control ~Electronic components

Suction jet pump →Injection system (components) ~Gasoline engine ~~Fuel feed unit

Suction system →Air intake system; →Intake system

Sulfate formation →Catalytic converter ~Diesel engine catalytic converter

Sulfur →Fuel, diesel engine ~Composition ~~Sulfur content; →Fuel, gasoline engine

Sulfur compounds (*also*, →Fuel, gasoline engine ~Sulfur content). Because of the sulfur in the crude oil, sulfur compounds, especially sulfur dioxide (SO_2), are created during combustion in diesel engines. SO_2 is a colorless gas with a pungent smell which can irritate the mucous membranes. Sulfur dioxide is released into the air and, when mixed with water, forms sulfurous and sulfur acids and acid rain, among other things, damaging vegetation. Another negative effect is catalyst poisoning, especially in case of the NO_x adsorption catalyst. The most effective method to eliminate these problems is to produce sulfur-free fuel.

Sulfur content →Fuel, diesel engine ~Composition ~~Sulfur content; →Fuel, gasoline engine

Sum spectrum →Engine acoustics ~Spectrum

Summer fuel →Fuel, diesel engine ~Properties ~~Cetane value

Super air thermal process →Supercharging ~Miller method

Supercharged engine →Supercharging ~Rotary piston supercharger

Supercharger →Supercharging

Supercharging (*also*, →Air cycling valve; →Fuel consumption ~Variables ~~Engine measures; →Intake

S

system). Supercharging has been used almost exclusively for performance enhancement and to improve torque. The objective is to achieve higher performance for a given displacement or the same performance for a reduced displacement. Additionally, attempts are made to supercharge using small chargers with a larger transmission ratio to reduce fuel consumption.

Supercharging can increase the performance in production vehicles by up to 60%.

The advantages of supercharging are:

- The existing engine series can be utilized to achieve higher performance and higher torques.
- A smaller additional installation space for supercharger and charge air cooler compared to a naturally aspirated engine with the same performance.
- A price advantage in relation to specific power output compared to displacement increase.
- A lower noise level (insulation of exhaust and intake noise) by exhaust gas turbocharging.
- Lower fuel consumption and lower emissions, especially for diesel engines.

Supercharging is almost always used in large engines and truck engines. With a few exceptions, engines with supercharging are also offered for diesel car engines, which otherwise have a lower power output than the equivalent gasoline engine. A larger air mass is supplied to the engine by means of a power-driven machine (fan, supercharger). This results in a larger work output, because the fuel mass per combustion cycle is also increased. The compression ratio must be reduced in gasoline engines to avoid a knocking combustion. However, this reduces the efficiency.

A number of opportunities exist to increase engine performance, which can be seen from the equation for engine power, P_e:

$$P_e = i \cdot n \cdot w_e \cdot V_H,$$

where n = engine speed, w_e = specific work, and V_H = displacement. The factor $i = 0.5$ is used for four-stroke engines and $i = 1$ for two-stroke engines.

- *Engine speed increase* (*n*): This puts a significantly higher load on the components (inertia forces). In addition, the speed increase results in a larger speed range for the auxiliary assemblies (idle to maximum speed). Problems with valve actuators (valve spring design, valve bounce) can also result. Higher speed can at the same time result in a worsening of engine noise. Therefore, speed increases are only used selectively, with the exception of special cases such as engines for sports applications. Increases in speed also result in greater friction losses and, therefore, in increased fuel consumption.
- *Displacement* (V_H): A change of the displacement is possible in two directions: on the one hand, lengthening of the stroke, and on the other hand, an increase in the bore. Stroke lengthening results in an increase of piston speed and an increase in the installation space. Increasing the bore but maintaining the web width results basically in a new engine, be-

cause the cylinder gap increases and this lengthens the engine.
- *Specific work* (*w_e*): Specific work is a value that depends on very many parameters. It depends, among others, on the charge mass in the cylinder, the ignition angle, the air-fuel ratio, the compression ratio, friction, combustion, mixture conditioning, flow losses, and so on.

For the gasoline engine, it can be described as

$$w_e = \eta_e \cdot H_u \cdot \lambda_a \cdot \rho_G / (\lambda \cdot m_{L,St} + 1),$$

and for the diesel engine,

$$w_e = \eta_e \cdot H_u \cdot \lambda_a \cdot \rho_L / (\lambda \cdot m_{L,St}),$$

where

η_e = overall efficiency
H_u = lower thermal value of the fuel
λ_a = volumetric efficiency
ρ_G and ρ_L = mixture or air density
λ_a = air mass ratio, and
$m_{L,St}$ = stoichiometric air mass.

Specific work can be increased by adjusting the mixture or air density if the values η_e, λ_a, and λ are kept constant (H_u, m_L, and St are fuel parameters). This principle is used for supercharging.

Literature: K. Zinner: Aufladung von Verbrennungsmotoren, Grundlagen, Berechnungen, Ausführungen, Springer Publishers, Berlin, 1980. — H. Pucher: Aufladung von Verbrennungsmotoren, Expert Publishers, Sindelfingen, 1985.

~Air cycle valve →Air cycling valve

~Blow-off valve (*also*, →Valve ~Exhaust gas control valves ~~Supercharging pressure control valve). The blow-off valve is used for the adjustment of the supercharging pressure during exhaust gas turbocharging. Two principal types are possible: the corresponding supercharging pressure can be set on the air side through venting of the compressed air or, depending on the desired supercharging pressure, part of the exhaust gas can be directed around the turbine by a bypass (wastegate). Supercharging pressure control on the exhaust gas side is exclusively used today for energy reasons. The overrun control valve can be integrated in the turbine housing or it can be mounted separately. The maximum cross-sectional area of the valve must be dimensioned so that the desired supercharging pressure can be adjusted for large exhaust gas volumes so that the exhaust gas back pressure does not exceed the intake manifold pressure (negative scavenging loop). The blow-off valve is thermally a heavily loaded component and requires a corresponding material selection.

~Charge air cooler (*also*, →Radiator). Compression of the air results in a temperature increase, which in turn results in reduced fill of the engine cylinder. The compressed air can be cooled to increase the density. A charge air cooler is used for this purpose. Air-air radiators are normally used, which means that the su-percharged air is cooled in the charge air cooler by the outside air which flows through the radiator. The tem-perature of the supercharged air after it has been in the charge air cooler is always higher than the outside temperature because the efficiency of the charge air cooler is always less than 100%. Engines with a charge air cooler have a better power output and lower fuel consumption than those without one.

An air-water radiator is an alternative to the radiator type described above. It has an inferior effectiveness if it uses the engine cooling water, but it has advantages with respect to space requirements and positioning. The rea-son for the inferior effectiveness is found in the higher temperature of the cooling water compared to ambient air.

The charge air cooler generates a pressure loss of approximately 100 mbar on the intake air side depend-ing on its size and airflow. The thermal efficiency of charge air coolers is approximately 80%.

~Charging process. The following charging pro-cesses are available:

- *Mechanical supercharging*. This requires a direct connection of the supercharger to the crankshaft.
- *Exhaust gas turbocharging*. The exhaust gas energy is used for the exhaust gas turbocharging to power a turbine. The turbine wheel is mounted on a shaft with a supercharger (compressor) wheel, which com-presses the air up to the supercharging pressure. A heat exchanger, the charge air cooler, is often mounted between the supercharger and engine inlet.
- *Other processes*. A number of other processes or systems exist such as compound operation (combi-nation of mechanical and exhaust gas turbocharg-ing), register charging, multistage supercharging, hyperbaric supercharging (additional fuel and air will be supplied to the turbine with this process), and compound charging.

Exhaust gas turbocharging has now been accepted for diesel and gasoline car engines; so, too, in some cases has mechanical supercharging; however, the blast wave supercharger has not.

The scavenge loop is positive for the supercharged engine, which means that the gas exchange work pro-vided is larger than the work from a naturally aspirated engine. **Fig. S64** shows the basic differences between the scavenge work. Definitions: p_U is the atmospheric pressure, p_A is the exhaust pressure, p_L is the super-charging pressure, V_c is the compression volume, and V_h is the displacement.

The idealized description of the course of the pro-cess of the low pressure loop is anticlockwise for a nat-urally aspirated engine during charge exchanges (the scavenge process), which means the work will be neg-ative; however, the loop is clockwise for the super-charged engine, thereby producing positive work. This, however, is only true as long as a positive scavenging loop exists, which means the pressure in the intake pipe, p_S, is higher than the exhaust gas back pressure, p_A. The condition $p_S < p_A$ exists at a certain engine

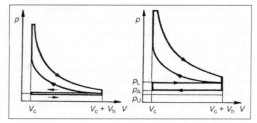

Fig. S64 *p-V* diagram of a supercharged and a non-supercharged engine

speed if the turbine cross-sectional area is too small for supercharged engines with very high mass airflow rate. This has a negative impact, due to the valve over-laps, on the scavenging characteristics (residual gas) and, therefore, on the volumetric efficiency.

~Compound. *See below*, ~Turbo-compound super-charging

~Compressor. Compressors are systems that are used to increase the density of the intake air for the purpose of supercharging. Radial compressors are mostly used for engine installations.

~Compressor air mass flow curve. The compressor air mass flow curve describes the supply characteris-tics of the compressor analogous to the engine air mass flow curve. The mass flow decreases with increasing pressure ratio in the supercharger such as a Roots su-percharger. A radial compressor shows basically the same characteristics; however, the characteristic curves are flatter, which means that, in comparison to the su-percharger, the mass flow decreases more strongly with an increasing pressure ratio.

~Compressor map. *See below*, ~Turbo supercharger

~Comprex. *See below*, ~Pressure wave supercharger

~Consumption concept. The supercharging of gaso-line engines was used in the past mainly for power in-creases. The pressure ratio had to be reduced due to the higher knocking sensitivity of these engines. Super-charged gasoline engines have higher fuel consump-tion than gasoline naturally aspirated engines.

Supercharging can, however, also be used as a fuel consumption-improving concept. The approach uses systems which increase the torque in the lower engine speed range or achieve the maximum torque at low en-gine speeds. This, however, results in a surrender of maximum power at high engine speeds, with a con-comitant increase of the acceleration characteristics (elasticity). Fuel consumption advantages result from a reduction of engine speed or from a shift of operating points into a range that has a better efficiency if the original vehicle performance is reestablished with the help of a higher total gear ratio.

S

Another improvement is achieved if an engine with a small displacement gets supercharged, which results in the same power as that produced by an engine with a larger displacement. Fuel savings are achieved by operating at higher load ratios (smaller throttling). The total potential for fuel consumption reduction is between 8 and 19% based on the US test.

Gasoline engines with direct injection are especially well qualified for supercharging. The knocking tendency decreases due to the fuel vaporization in the combustion chamber (internal cooling). This means that the compression ratio must only be reduced marginally.

The durability and the characteristics of the engine will be improved by the possible engine speed decrease.

Literature: P. Langen, J. Mallog, M. Theissen, R. Zielinski: Aufladung zur Verbrauchsreduzierung, MTZ 54 (1993) 10. — H.-D. Erdmann: Mehrventiltechnik und Aufladung als Verbrauchskonzept, Engineering Academy of Esslingen, 1994, No. 1901 6/64103.

~Dynamic charging. A dynamic charging effect is achieved by utilizing the wave characteristics of exhaust gas or fresh mixture gas columns. The control times are chosen in such a way that a pressure as high as possible is achieved and, therefore, a high mixture density can be attained in the cylinder when the intake closes (*also*, →Supercharging ~Resonance induction).

A different form of dynamic charging is pulse turbocharging. The kinetic energy of the exhaust gas flow is used to transfer an increased energy to the turbine rotor. This makes a fast acceleration of the turbine possible and delivers a higher supercharging pressure.

~Electrically supported supercharging. The voluntary commitment of the automotive industry for a significant reduction of CO_2 emissions to 140 CO_2 g/km

before the year 2008 has especially supported the idea for reduction of engine displacement (down-sizing). This is, however, only acceptable if the driving characteristics stay almost unchanged compared to engines with larger displacements. Supercharging is one measure to compensate for the disadvantages of an engine with small displacement. This assumes, however, that the known torque weakness of the exhaust gas turbocharger at low speeds, especially during the transient operation, can be eliminated. Current measures make this possible only on a limited basis. These are the measures:

- Use of small turbocharger, which has disadvantages at high mass airflow rates or high engine speeds (high exhaust gas back pressure, negative scavenging loop)
- Variable nozzle turbocharger
- Series connection of superchargers with different sizes

All these measures are, however, still connected to the energy supply in the exhaust gas. Another promising possibility to increase the supercharging pressure at low engine speeds is the feeding of external energy, such as electrical energy, into the supercharger system.

Two different arrangements are available, the eBooster charge system and the electrically supported turbocharger.

~~eBooster. For the so-called eBooster, an electronically powered supercharger is connected in series with the ATL system. This type of two-stage compression achieves a higher supercharging pressure ratio for those operating points of the engine that have small exhaust gas mass flow. The electrical power must be provided from the vehicle electrical system. A schematic diagram of an eBooster charging system is shown in **Fig. S65**. Plans are to mass-produce this system in 2006.

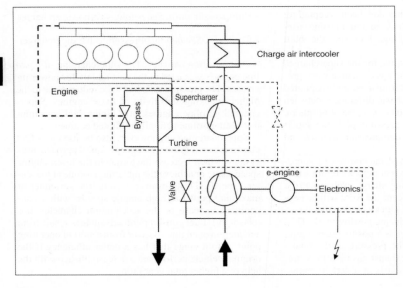

Fig. S65
Schematic description of an eBooster supercharging

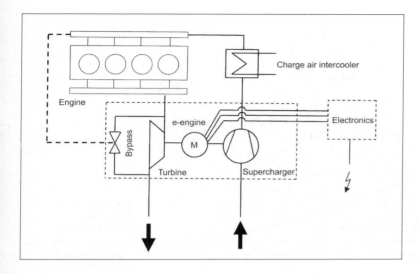

Charge air intercooler

Engine

e-engine

Electronics

Bypass

Turbine

M

Supercharger

Fig. S66
Schematic description of
an electrically supported
turbocharging

~~**Electrically supported turbocharger.** The electrically supported turbocharger has an asynchronous electrical motor, which sits on a shaft between compressor and turbine. Depending on the electrical energy supplied, this results in a significant improvement of the vehicle response; the increase of the absolute supercharging pressure is limited by the compressor selection due to the principles of single-stage compression. **Fig. S66** shows the diagram of an electrically supported turbocharger.

Fig. S67 shows a cross section through an electrically supported turbocharger. No dates for mass production are known.

Literature: H. Zellbeck, J. Friedrich, C. Berger: Die elektrisch unterstützte Abgasturboaufladung als neues Auflade-konzept, MTZ 60 (1999). – P. Hoeker, J.W. Jaisle, S. Münz: Der eBooster, Schlüsselkomponente eines neuen Aufladesystems von BorgWarner Turbo Systems für Pkw, 22nd International Engine Symposium, April 2001. – R. van Basshuysen, F. Schäfer (Eds.): Handbuch Verbrennungsmotor, Wiesbaden, Vieweg Publishers, 2002.

~**Engine air mass flow curve.** The engine air mass flow curve can be constructed by recording the super-

charging pressure ratio (p_{2t}/p_u is the stagnation pressure ratio) over the mass or over the volume flow. It is shown in the performance characteristic map in **Fig. S68**.

The air mass, m_{LM} , flowing through the engine is defined as

$$w_{LM} = \lambda_l \cdot \rho_L \cdot n \cdot i \cdot V_H ,$$

which must be equal to the mass that flows through the compressor; that is,

$$m_{LM} = m_{LV} ,$$

where λ_l = volumetric efficiency, ρ_L = density of the air, n = engine speed, $i = 0.5$ for a four-stroke engine, and V_H = engine displacement.

The four-stroke engine is equivalent to a supercharged engine with reversed behavior during the charge exchange. The reason is that the engine is in the intake phase for $p_{2t} \approx p_s$ (pressure in intake pipe) and that the mass flow increases due to the increasing pressure ratio, p_{2t}/p_{1t} . The ratio p_{2t}/p_{1t} is the pressure ratio across the compressor. The slope of the curves depends on the valve overlap (scavenging losses). The curves become more flat if the valve overlaps increase, which means that the scavenging losses increase. The mass flow must be increased to achieve the same pressure ratio.

~**Engine speed decrease.** The engine speed decrease is characterized by operating conditions at constant engine mean effective pressure and variable engine speed. With respect to the exhaust gas turbocharging, this results in a flatter characteristic line for the supercharger (**Fig. S69**). This condition exists in the vehicle in certain ranges close to the full-load line and reaches its most critical point just before the limitation of the supercharging pressure—for example, through supercharging pressure control. In general, this is where the distance to the pump limit is smallest.

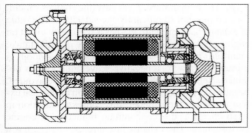

Fig. S67 Section through an electrically supported
turbocharger.

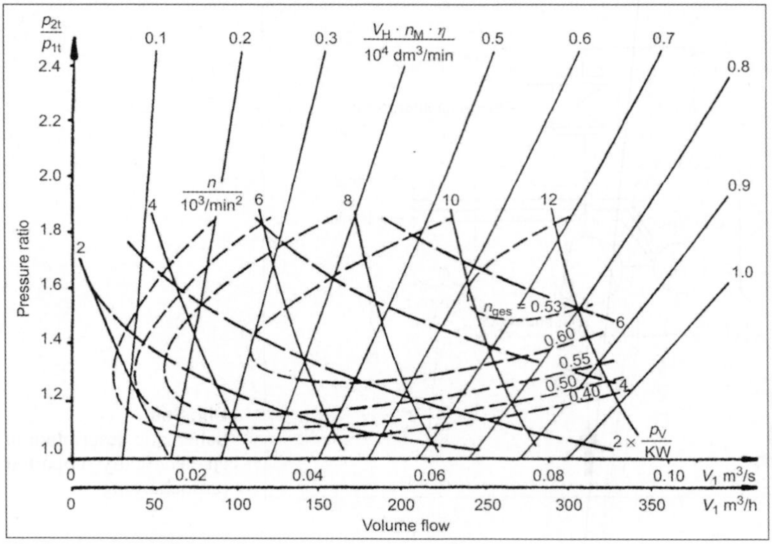

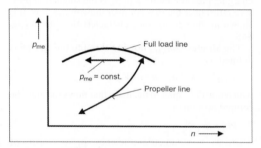

Fig. S68
Performance characteristic map of a rotary-piston supercharger with engine air mass flow curves

Fig. S69 Engine speed decrease

~Exhaust gas turbocharging. This type of supercharging uses the exhaust gas energy (thermal, kinetic, and potential) to power a radial compressor (which compresses the air to the desired supercharging pressure) with the help of a turbine. Ram induction and pulse turbocharging are two different methods of supercharging engines.

The advantages of exhaust gas turbocharging are:

• Known and proven technology
• Flexible positioning on the engine or in the engine compartment
• Compact design
• Attenuation of the exhaust and intake noise
• Lower fuel consumption and reduced pollution emissions for the diesel engine
• Relatively low cost

The disadvantages of exhaust gas turbocharging are:

• Relatively poor torque back-up in the lower engine speed range, especially for diesel engines
• Delayed vehicle response in lower speed ranges, especially for gasoline engines

• Higher fuel consumption for performance improvements in gasoline engines with current designs

The disadvantages can be reduced by using a number of smaller turbochargers in gasoline engines. In this case, the supercharger delivers high supercharging pressures even for low engine speeds and, therefore, for low air mass flow rates. This results in a better vehicle response and higher torque at lower engine speeds by forgoing maximum power; in addition, fuel consumption is reduced. Gasoline engines with direct injection have fuel consumption advantages, because a higher compression ratio can be selected due to internal cooling of the charge as the mixture is generated in the combustion chamber.

Supercharger and turbine are mounted on a single shaft; the supercharger is powered by the turbine, which receives its energy from the exhaust gas (primarily potential energy, due to the higher pressure). Radial turbines are used almost exclusively.

Fig. S70 shows schematically the designations of the air and exhaust gas conditions during exhaust gas turbocharging.

A very important factor is, among other things, the condition of the gas upstream of the turbine—that is, T_3, p_3. The temperature, T_3, in gasoline engines should, for example, not significantly exceed 950°C. The value of exhaust back pressure, p_3, has a significant effect on engine performance. If $p_3 < p_s$, where p_s is the intake manifold pressure, a positive scavenging loop is present; if $p_3 > p_s$, a negative scavenging loop is obtained. Negative scavenging loops should be avoided, because they result in a large proportion of residual exhaust gases in the cylinder; this causes a reduction in fresh charge and a concomitant reduction in engine performance. The exhaust pressure, p_3 or p_{Abg}, has a significant influence on the power, torque, and fuel consumption of the engine.

S

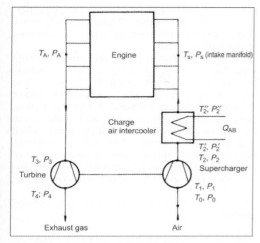

Fig. S70 Status description for the supercharged engine

For heavily loaded powertrains, the bearing housing of the exhaust gas turbocharger in gasoline engines is cooled with liquids (connected to the cooling circuit of the engine) to prevent oil carbonization inside the turbocharger. The rotational speed of the turbocharger shaft is extremely high, and can exceed more than 150,000 rpm depending on the engine operating conditions (e.g., operation at high altitudes). This is the reason why the turbine housing must be made of high-strength, thermostable, and scale-resistant material. Normally Niresist material (high nickel content) is used to provide acceptable safety in case the turbine wheel bursts. Recently, turbine housings have also been designed with high-temperature sheet metals to save weight. In addition, the heat capacity is lower, which enables the exhaust gas treatment system to warm up faster.

Radial compressors are used for exhaust gas turbocharging in cars and trucks. The supercharging pressure reduces disproportionately with reduced speed if compared with a supercharged engine. This is why the development of inlet manifold pressure is slower in an engine with exhaust gas turbocharging than in one with a mechanically coupled supercharger.

There are several alternatives for the design of an engine turbocharger system. A major criterion for car engines is the vehicle response. This means that a high supercharging pressure must be present at low engine speeds and low exhaust gas mass flows—that is, low loads. This requires the selection of a small turbine cross section, which means that a "small" charger is used. If high engine power output is the objective, a turbine with a large cross-sectional area is recommended. This results in bad vehicle response at low speeds.

Another possibility to modify the engine operating characteristics is to choose a variable nozzle turbocharger (VNT); this influences the inflow to the turbine through an adjustment of the stator geometry.

This is the reason why such an approach was until now only technologically possible in diesel engines and not in gasoline engines with their high exhaust gas temperatures. However, these systems are now being developed for gasoline engines. The use of turbine wheels made out of ceramics can further improve the vehicle response.

Future developments will be:

- Reduction of the moment of inertia by the use of ceramic turbine wheels
- Weight reduction and installation space reduction
- Bearing improvement of the turbines or supercharger shafts (roller, air, magnet bearings; therefore eliminating of the oil feed from the engine circuit)

~~Exhaust gas turbocharger. The exhaust gas turbocharger comprises a group consisting of supercharger (compressor) and turbine.

~Fan. Radial fans or radial compressors are exclusively used for exhaust gas turbocharging. This means that supercharging pressures of 3 bar absolute are possible in car applications. Radial compressors are compact and cost-effective. The maximum efficiencies are in the range of 80%.

A number of systems are used for mechanical supercharging. The most important representatives are the G-supercharger, which uses a spiral system, the rotary-piston supercharger, the vane-type supercharger, and the screw-type supercharger (Lisholm supercharger).

Some of these systems are shown schematically in **Fig. S71**.

~Generator mode. The generator mode is an engine operating at constant engine speed. This is required due to high demands on the generator for constant rotary frequency. The generator mode establishes itself in the engine map by load changes at constant engine speeds.

Different gas conditions at the turbine intake, due to load changes of the engine, result in different supercharging pressures. The generator mode with exhaust gas turbocharging is, therefore, a mode alongside the engine air mass flow curve, which belongs to the corresponding generator speed. **Fig. S72** shows the relationship.

~Group efficiency. The total efficiency factor of the charge group is called group efficiency, η_{TL}. It is defined as the product of all the individual efficiencies.

$$\eta_{TL} = \eta_{mV} \cdot \eta_{sV} \cdot \eta_{mT} \cdot \eta_{sT},$$

where η_{mV} and η_{mT} = the mechanical efficiencies of the compressor and turbine, and η_{sV} and η_{sT} = the isentropic efficiencies of compressor and turbine.

~G-supercharger. *See below,* ~Rotary-piston supercharger

~High pressure supercharging. High pressure supercharging is not a clearly defined term. For large engines it can, for example, be achieved by multistage

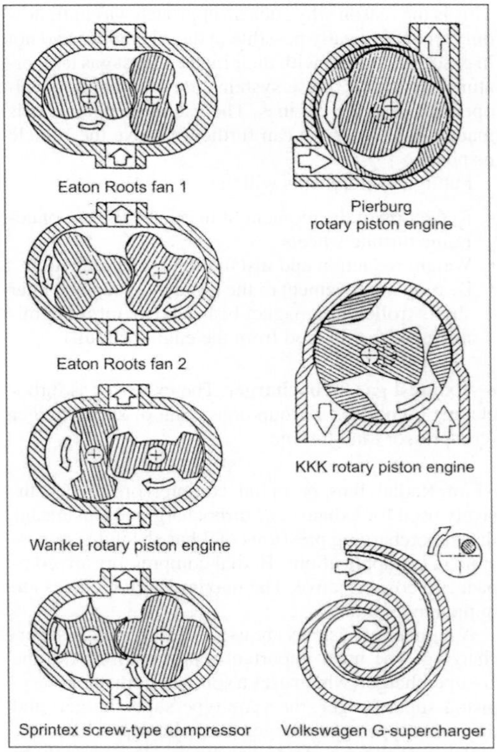

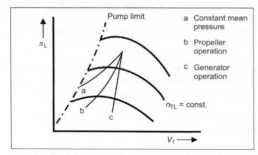

Fig. S71 Mechanical supercharging system

supercharging and by a mean effective pressure of approximately 24 bar or higher. High pressure supercharging for car engines occurs if mean effective pressure of approximately 18 bar or higher are achieved by supercharging. Mean effective pressures of about 25 bar are achieved in diesel car engines with direct injection. Tests with multistage supercharging for car engines are currently being performed.

~**Hyperbaric supercharging.** It is the objective of hyperbaric supercharging to provide additional energy

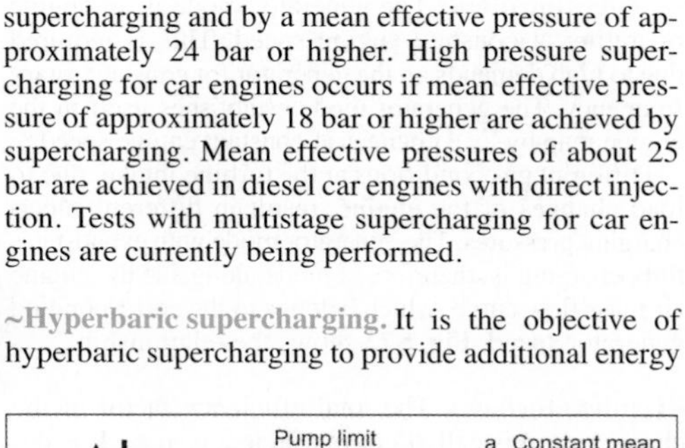

Fig. S72 Engine air mass flow curve and generator mode characteristic curve c

in front of the turbine to make a larger enthalpy difference available to the turbine. This results in faster acceleration of the turbine and, therefore, in an improved response of the engine.

There are two alternatives for achieving hyperbaric supercharging. First, the engine can be operated extremely rich. This results in a high ratio of chemical energy in the exhaust gas, which can be used in front of the turbine through an additional air supply and subsequent secondary combustion. Second, fuel and air can be supplied additionally in front of the turbine, which also results in an increase of the enthalpy difference across the turbine. This alternative, however, is not economical, because additional fuel is required. Therefore, it is only used in special cases, such as in sport engines.

~**Isentropic compressor work** →Isentropic enthalpy gradient

~**Isentropic enthalpy gradient.** The power absorbed by the turbine or the power required at the compressor can be determined by the isentropic enthalpy change across the turbine or across the compressor. The isentropic enthalpy change is equivalent to the isentropic compressor or turbine work. The actual power supplied or absorbed can be calculated if the volume flow, the isentropic efficiency, and the mechanical efficiency are known.

The isentropic enthalpy change at the compressor is calculated as

$$\Delta h_{sV} = R_1 \cdot T_1 \frac{\kappa_1}{\kappa_1 - 1} \cdot \left[\left(\frac{p_2}{p_1} \right)^{\frac{\kappa - 1}{\kappa_1}} - 1 \right].$$

The isentropic enthalpy change at the turbine is calculated as

$$\Delta h_{sT} = R_3 \cdot T_3 \frac{\kappa_3}{\kappa_3 - 1} \cdot \left[1 - \left(\frac{p_2}{p_1} \right)^{\frac{\kappa_3 - 1}{\kappa_3}} \right].$$

The indices 1 and 3 identify air and exhaust gas conditions, respectively; p_2/p_1 is the pressure ratio across the compressor; p_4/p_3 is the pressure ratio across the turbine; κ identifies the ratio of specific heats, c_p/c_v; R represents the gas constant; and T represents the temperature.

~**Isentropic turbine work** →Isentropic enthalpy gradient

~**Lysholm compressor.** *See below*, ~Miller method (super air thermal process)

~**Mechanical supercharging.** In most cases, the crankshaft connects the compressor mechanically to the engine for mechanical supercharging. The characteristic curves of the engine and the compressor determine the operating characteristics of the system.

A fixed compressor speed is achieved for each engine speed. The operating points on the performance characteristic map are the crossing points of both characteristic curves. The supercharging pressure falls with declining speeds, and as a result the density of the media also falls and so does the mean effective pressure of the engine. An important fact for mechanical supercharging is that the power for the supercharger must be delivered by the engine. The effective power output is

$$P_e = P_i \eta_m - P_V,$$

where P_e and P_i represent the effective induced power of the engine, P_V is the power that must be delivered by the engine for the supercharger, and η_m is the mechanical efficiency. A switch-off clutch can be installed to avoid the power loss of the mechanical supercharger under partial load. However, this increases complexity and cost.

The advantages of mechanical supercharging are:

- Fast engine response at all engine speeds
- High supercharging pressure at low engine speeds (compared to exhaust gas turbocharging)

The disadvantages are:

- The absorbed supercharger power must be delivered by the engine
- Large installation space required
- Not difficult to vary its position because of mechanical connection to the engine
- The installation space of the supercharger will become disproportionately large. This is the reason why it is only used in gasoline engines with relatively small displacements (<2 liter), to compensate for the torque weakness of engines with small capacities.
- High cost

~Miller method (super air thermal process). The method developed by Miller is different from conventional supercharging processes due to a modified intake closing process. The objective is to reduce the fresh charge temperature. This is possible in two alternative ways. The intake valve can be closed before the piston arrives at bottom dead center during the intake cycle; or part of the compression stroke is not used for compression (very late closing of the intake valve).

The compression begins in the first case from a lower temperature level, which is at the expense of the conventionally achievable supercharging volume that can be achieved with the same supercharging pressure. In the second case, part of the fresh charge will be returned into the intake system via the open intake valve. However, in both cases the temperature is reduced at the end of compression; the compression ratio is lower than the expansion ratio. Early intake closing was first shown by Miller for a diesel engine. Based on present-day experience, this process is not very effective. A modification of the process with very late intake closing and its applications in a gasoline engine offer many advantages. The disadvantage of a smaller fill will be compensated for by an adequate supercharging ratio. The Lysholm compressor with charge air cooler has proven to be the optimum supercharging device.

The advantages of the process are found in higher thermal efficiency and a lower tendency to knocking, which is possible due to early ignition angles. Lower maximum cylinder pressures and lower exhaust gas temperatures are also results of this process.

The Miller method can also be interpreted as a process that has divided the compression between the engine and the compressor differently.

Literature: K. Zinner: Aufladung von Verbrennungsmotoren, 2nd Edition, Springer Publishers, 1980. — K. Katamura, et al.: Development of Automotive Miller Cycle Gasoline Engine, SAE 945008, 1994. — R.H. Miller: Supercharging and Internal Cooling Cycle for High Output, ASME 69, 1947.

~Multistage supercharging. The desired supercharging pressure for multistage supercharging is achieved across several compressor stages with intermediate cooling if required. Usually a maximum of two stages is used in large engines. These engines achieve mean effective pressures of 25 bar. Increasing vehicle masses and large demands on the driving dynamics lead inevitably also to higher engine performance required in the car. Multistage supercharging offers a large potential for satisfying this requirement in diesel car engines. The expected specific power for supercharged diesel engines with two-stage supercharging is currently approximately 70 kW/L and approximately 25 bar mean effective pressure.

~Pressure wave supercharger (Comprex). Pressure wave supercharging has a special place among supercharging processes. It is a combination of mechanical supercharging and the extraction of energy from the exhaust gas; this is shown schematically in **Fig. S73**. The mechanical element is a cylindrical rotor with longitudinal channels; this is driven by the engine.

The exhaust gas energy (mostly the pressure energy, not the thermal energy) is used for the compression of the inlet air. This energy exchange is performed through pressure waves. The pressure wave super-

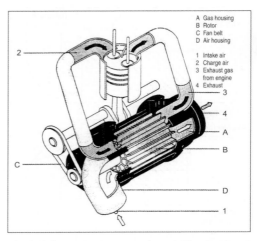

A Gas housing
B Rotor
C Fan belt
D Air housing

1 Intake air
2 Charge air
3 Exhaust gas from engine
4 Exhaust

Fig. S73 Pressure wave supercharger (Comprex)

S

charger is also called a Comprex charger. The process is characterized by the alternative presence of hot exhaust gas and cold fresh air in the channels of the rotor.

The pressure equalization is faster than the gas flow if two gas flows with different pressures are joined. This is especially true if the gas flows through narrow but smooth channels.

The channels, or cells, are arranged axially and are open at both ends. The rotor is driven by a belt drive, and only the bearing friction must be overcome. The housing has channels for the supply of fresh air and to transfer the compressed air to the engine; it also has the gas intake and exhaust channels to and from the engine and to the exhaust. The openings are alternately opened and closed during the rotation of the rotor (**Fig. S74**).

The rotor channels are filled with fresh air at atmospheric pressure at the beginning of the process. The opening of the outlet port is closed.

As the rotor turns, these cells pass in front of the inlet opening of the exhaust gases that come from the engine. These gases cause a pressure wave in the rotor. The air has the same pressure and the same speed as the incoming gas after the pressure wave. The other end of the rotor cell is still closed. The rotor must be rotated further to maintain the flow achieved by the pressure wave. The other end of the cell, namely the outlet vent of the charge air, will be opened. The compressed air (up to 2 bar) arrives at the engine. The flow must be decelerated before the exhaust gas arrives on the fresh air side. The rotor cell opening will be closed for that purpose, and this creates a rarefaction (negative pressure) wave, which decelerates the flow. The closing of the charge air side follows. An air cushion is still present on the air side. This is a high pressure cycle. The low pressure cycle has the task to release the exhaust gas into the exhaust gas pipe. The exhaust vent to the exhaust pipe is opened for this purpose.

The pressure drop between rotor and exhaust pipe creates an expansion wave, which transports the gas into the exhaust. The air side will be opened for filling the rotor cells with fresh air and the process begins anew, as another wave arrives at the exhaust vent.

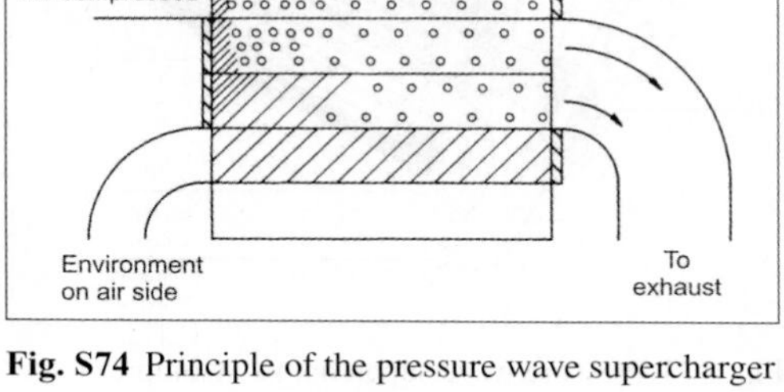

The advantages of the pressure wave supercharger are:

- Good transient characteristics (fast vehicle response).
- High torque at low speed.
- Exhaust gas recirculation is easy to implement.

The disadvantages of the pressure wave supercharger are:

- It is back-pressure sensitive on the intake and exhaust gas side (air filter, soot filter, oxidizing catalyst).
- The system must be bypassed at the start, because otherwise exhaust gas gets into the intake system.
- Large construction volume and high weight.
- Requires drive from the engine (but only to overcome the friction during the rotation of the wheel).
- Only feasible for diesel engines (intake throttling and low exhaust gas temperatures are absent).
- High expenditures to reduce the noise.

There is only one application in mass production so far.

~**Propeller operation.** Propeller operation is defined as

$$p_{me} \propto n_M^2 ,$$

where n_M = engine speed and p_{me} = mean effective pressure.

This operation is important for ship engines because the received propeller torque depends on the square of the propeller speed. This results in a characteristic curve between operation with constant mean effective pressure and generator mode for the exhaust gas turbocharged engine (**Fig. S72**, characteristic curve b).

~**Pulse converter.** Pulse converters convert the kinetic energy of the exhaust gas into potential energy to achieve, for example, a larger pressure ratio through the turbine. Generally, they consist of jets that are connected to the individual exhaust tracts, a venturi tube, and a downstream diffuser for the conversion of kinetic energy into pressure energy. **Fig. S75** shows a diagram of a pulse converter. Combining the exhaust tracts, the geometry, and the flow cross-sectional areas are the main design parameter. This system enables pressure peaks of up to 1.4 bar.

Literature: J. Tielemann: Berechnung der Vorgänge im Multi-entry Pulse Converter, MTZ 51 (1990) 4.

S

Fig. S74 Principle of the pressure wave supercharger

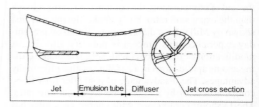

Jet   Emulsion tube   Diffuser   Jet cross section

Fig. S75 Schematic description of a pulse converter

~Pulse supercharging. *See below*, ~Pulse turbo-charging; *also*, →Air cycling valve ~Air cycling valve functions ~~Torque increase ~~~Dynamic charging (pulse charging)

~Pulse turbocharging. The pulse turbocharging uses the kinetic (impulse) part due to the high speed of the exhaust gas in front of the turbine. The volume of the exhaust gas pipe should be as small as possible for this application. The pipe will be filled quickly, when the exhaust valve is opened. A pressure wave develops, which is processed by the turbine. This maintains a larger part of the kinetic energy of the exhaust gas, and it will be used in the turbine. Only cylinders that do not interfere with each other can be combined in multiple-cylinder engines. In this application, the turbine is loaded dynamically.

Literature: F. Pischinger: Verbrennungsmotoren, Volume I, Lecture Reprint, 10/94, Technical University Aachen.

~Radial compressor. *See below*, ~Turbo supercharger

~Ram induction. For the ram induction, the exhaust branches of the individual cylinders will be connected to a collector tank, which is mounted in front of the turbine.

This process uses mainly the pressure energy contained in this exhaust gas plenum chamber (**Fig. S76**).

~Register supercharging. Two or more compressor groups can be connected in parallel for multistage supercharging. During the operation, the total exhaust gas mass will first flow through a turbine at low engine speeds, and the second or more turbines will be switched on for higher mass airflow rates (higher engine speed). The advantages of this design are found in a better vehicle response and higher mean effective pressures. This makes higher supercharging pressures of over 3 bar for cars available. And this results in specific power outputs of over 80 kW/L.

Literature: N.N.: Vectra OPC mit Hochleistungs-Dieselmotor, MTZ 65 1/2004. – B.G. Cooper, I.J. Penny, S. Whelan: A New Diesel System Methodology to Meet the Demands of Low Engine-Out NOx and High Performance for Future Global Markets, VDI Reports 1808, 2003, p. 41. – G. Hack: Der Doppler-Effekt, auto-motor und sport, 4/2004.

~Resonance induction. The gas column in the intake pipe of the engine receives a certain excitation frequency, which is dependent on engine speed and number of cylinders. A supercharging effect takes place, and this matches the resonance frequency of the intake system. This assumes that groups of cylinders with the same firing intervals are connected through short pipes to a chamber, the so-called resonance chamber (**Fig. S77**).

A shift of the maximum torque is possible depending on the length and diameter of the pipes and the volume of the resonance chamber. An increase of the specific work up to 10% is only possible within a limited engine speed range, which means that a torque increase across the total engine speed band is not necessarily possible. The solution is so-called variable geometry (switchover) intake systems, which permit different resonance tube lengths dependent on the engine speed. The design of the resonance system is primarily oriented toward the following criteria:

* Engine displacement
* Resonance speed
* Volume of the resonance chamber
* Length and diameter of the resonance tubes

Literature: H. Seiffert: Die charakteristischen Merkmale der Schwingrohr- und Resonanzaufladung bei Verbrennungsmotoren, XIX Fisita Congress, Melbourne, Australia, SAE 82032, 1982. – H. Duelli, B. Geringer, F. Bauer: Möglichkeiten der Saugrohrentwicklung zur Verbesserung des motorischen Betriebsverhaltens durch Berechnung und Versuch, 6th Vienna Engine Symposium, VDI Progress Reports,

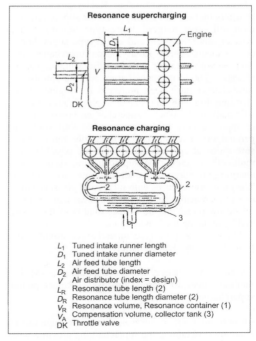

L_1 Tuned intake runner length
D_1 Tuned intake runner diameter
L_2 Air feed tube length
D_2 Air feed tube diameter
V Air distributor (index = design)
L_R Resonance tube length (2)
D_R Resonance tube length diameter (2)
V_R Resonance volume, Resonance container (1)
V_A Compensation volume, collector tank (3)
DK   Throttle valve

Fig. S77 Intake runner and resonance induction

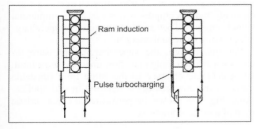

Fig. S76 Ram induction and pulse turbocharging

S

Series 6, No. 173. — H.P. Lenz: Gemischbildung bei Gasolinemotoren, Vienna/New York, Springer Publishers, Volume
6, 1990.

~Resonance tube supercharging. For this process, a
pressure wave will be induced into an intake pipe by
the piston motion. The intake pipe is connected to an
air distributor. Each cylinder has a separate intake pipe
(**Fig. S77**). This pressure wave gets to the intake valve
shortly before it closes and causes an increased fill
based on the higher density. A rarefaction (negative
pressure) wave travels through the intake pipe at the
beginning of the intake phase, and it will be reflected
as pressure wave at the end of the open intake pipe in
the air distributor. Short intake runners result in performance increases at high engine speeds. However, long
intake pipes are required if a high torque at low engine speeds is desired. This often represents a problem, because extremely long (>800 mm) and small
bore intake pipes are required for low engine speeds
and for increased torques, but these cannot be installed in the given installation space. Torque increases of 15–20% are achievable. The volume of the
air distributor has a secondary influence on the torque
characteristics.

~Roots supercharger. See below, ~Rotary-piston
supercharger

~Rotary-piston compressor. See below, ~Rotary-piston
supercharger

~Rotary-piston supercharger. Rotary-piston superchargers include, for example: Roots superchargers,
multicell superchargers, screw-type superchargers,
Wankel superchargers, and spiral-type superchargers (G-superchargers). Rotary-piston superchargers are
devices with rotating positive displacement superchargers. The roots supercharger and the spiral-type
supercharger have been preferred for engine supercharging. **Fig. S68** shows an example of the characteristic map for the Wankel supercharger. Pressure ratios
of approximately 1.8 bar are possible with Roots superchargers, generally two or three lobe rotors; pressure ratios of approximately 5 bar are possible with
multicell or screw-type superchargers. High supercharging pressures and small flow rates are required.
The mass flow reduces with increasing pressure ratios
due to increasing leakage losses. These can be minimized by improving the surface quality of the rotors
and by reducing the leakage gap.

The use of a spiral-type supercharger provides improved delivery characteristics.

~Screw-type supercharger. See above, ~Miller
method (super air thermal process), ~Rotary-piston
supercharger

~Single stage supercharging. The single stage supercharging is obtaining the supercharging pressure with
one supercharger stage.

~Spiral-type supercharger. See above, ~Rotary-pis-
ton supercharger

~Super air thermal process. See above, ~Miller method
(super air thermal process)

~Supercharged engine. See above, ~Rotary-piston
supercharger

~~VTG →Supercharging ~Variable nozzle turbocharger (VNT)

~Supercharger. See above, ~Fan; see below, ~Turbo
supercharger

~Supercharging pressure. The supercharging pressure is the compressor downstream pressure or the upstream engine intake pressure in the intake pipe. This
pressure varies depending on the application. It is approximately 2 to 3 bar for supercharged car engines at
full load.

~Supercharging pressure control (also, →Electronic
open- and closed-loop control). The characteristics of
the supercharging pressure relative to engine speed are
a major variable for influencing the engine and, therefore, the road performance at full load. Supercharging
pressure is related to each engine speed. Control of the
supercharging pressure is required to get the desired
relationship between the changing operating conditions (intake air temperature, altitude, humidity, exhaust gas temperature, exhaust pressure, coolant temperature, oil temperature, etc.) and the pressure in the
manifold. Alternative approaches are interventions on
the air or the exhaust gas side. The boost control valve
is opened if the target charge pressure is reached on the
exhaust side and the exhaust gas flows around the turbine into the exhaust gas pipe. Manipulation of the exhaust gas condition in front of the turbine is possible
through the change of the opening cross section of the
boost control valve. It is possible to manipulate the supercharging pressure on the air side by venting the already supercharged air. This alternative has not been
used because it is inefficient.

~Supercharging pressure ratio. The supercharging
pressure ratio is generally the pressure ratio between
the induction and pressure sides of the compressor.
This is either based on the actual atmospheric pressure
or on a pressure based on standard conditions.

~Turbo supercharger. Turbo superchargers are exclusively radial compressors in car and commercial
vehicle engines. They are used to produce supercharging pressure (potential energy).

For this design, the supercharging pressure increases with decreasing mass flow up to the surge limit.
An acceptable distance to the surge limit (unstable
range of the compressor) is required in real applications. **Fig. S78** shows the performance characteristic
map of a radial compressor.

~Turbocharger. See above, ~Exhaust gas turbocharging

S

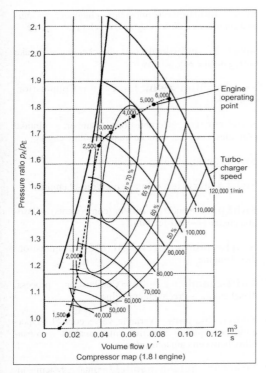

Fig. S78 Radial compressor map

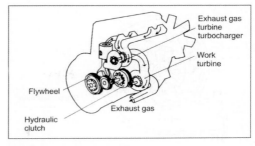

Fig. S79 Turbocompounding

the downstream turbine is transferred directly—for example, by gears—to the engine crankshaft (**Fig. S79**). The vibrations of the system are dampened by a hydraulic clutch in the reduction gear. The engine performance for the diesel engine can be improved by 10% and the fuel consumption by 5% with the help of this additional work turbine. The reason for this gain is found in the improved utilization of the exhaust gas energy if compared to the pure exhaust gas turbocharging. Turbo-compound supercharging has so far only been used in commercial vehicle engines.

The combination of naturally aspirated engines and additional work turbines is the most meaningful combination for the gasoline engine due to the higher exhaust gas availability for the gasoline engine process.

Literature: G. Woschni, F. Bergbauer: Verbesserung von Kraftstoffverbrauch und Betriebsverhalten von Verbrennungsmotoren durch Turbocompounding, MTZ 51 (1990) 3.

~**Turbocharger main equation.** The turbocharger main equation is derived from the performance output of compressor and turbine, and it shows the compressor pressure ratio, π_V, based on parameters from the compressor and the turbine.

$$\pi_V = \left[1 + \frac{\dot{m}_T}{\dot{m}_V} \cdot K_1 \cdot \frac{T_3}{T_1} \cdot \eta_{TL}\left(1 - \frac{p_4}{p_3} \right)^{\frac{k_A - 1}{k_A}} \right]^{3,5}$$

where

$\dot{m}_T$ = mass flow through the turbine,
$\dot{m}_V$ = mass flow through the compressor,
K_1 = constant,
T_3 = temperature in front of the turbine,
T_1 = compressor intake temperature,
η_{TL} = group efficiency factor,

$\dfrac{p_4}{p_3}$ = pressure drop across the turbine, and
κ_A = isentropic exponent (c_p/c_v) of the exhaust gas.

~**Turbo-compound supercharging** (*also*, →Engine concepts ~Composite systems ~~Turbo compound systems). These are supercharged diesel or gasoline engines with exhaust gas turbocharging, charge air cooling, and a downstream work turbine. The power of

~**Twin-turbo.** These are charge systems—which consist, for example, of two turbochargers—that are not connected in series or in a multistage position but in parallel. Application areas include, for example, the supercharging of V-engines, where each bank is powered by its own turbocharger.

~**Vane-type supercharger.** *See above*, ~Fan

~**Variable nozzle turbocharger (VNT).** An adjustment of the turbine intake geometry is required to better adapt the characteristics of the flow engine (turbine of a turbocharger) to the operating characteristics of the engine. This can basically be done in several ways. An adaptation is possible by the design of adjustable guide blades at the circumference of the turbine wheel (**Fig. S80**), and it is also possible by an adjustable spiral housing tongue. This allows the adaptation of the exhaust gas flow depending on the load or the mass throughput. The operating characteristics for smaller exhaust gas mass flows can also be improved, in addition to the vehicle response during load changes. Additionally, mean effective pressure, power, engine fuel consumption, and exhaust gas will be influenced positively.

These systems are currently mainly used in diesel engines because of their low exhaust gas temperature and, therefore, their limited thermal component stresses.

965

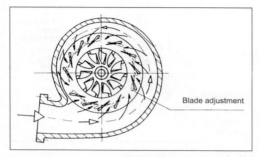

Fig. S80 Exhaust gas turbine with adjustable guide blades

However, these systems are now also developed for gasoline engines.

~**Vehicle response.** Vehicle response is defined by the delay time starting from the actuation of the gas pedal until a torque development can be felt.

~**Wankel supercharger.** *See above*, ~Rotary-piston supercharger

Supporting layer →Sealing systems ~Cylinder head seals

Surface ignition →Glow ignition

Surface performance →Piston

Surface pressure →Camshaft ~Cam width; →Piston ring ~Contact pressure

Surface utilization valves →Valve arrangement ~Number of valves

Surface-type filter →Particles (Particulates) ~Particulate filter system ~~Particulate filter ~~~Operating mode of particulate filters

Surface-volume ratio →Combustion chamber

Surge impedance →Engine acoustics ~Rated noise impedance

Swash plate engines →Engine concepts ~~Engines without crankshaft

Swirl →Charge movement; →Intake port, diesel engine; →Intake port, gasoline engine

Swirl chamber →Combustion chamber; →Combustion process, diesel engine ~Prechamber engine

Swirl channel →Intake port, diesel engine; →Intake port, gasoline engine ~Swirl

Swirl combustion chamber →Combustion chamber

Swirl control →Combustion chamber; →Intake port, diesel engine; →Intake port, gasoline engine

Swirl factor →Charge movement

Swirl generation →Intake port, diesel engine

Swirl port →Injection system (components) ~Gasoline engine ~~Fuel-feed unit

Swirl-chamber diesel engine →Combustion, diesel engine

Swirl-chamber engine →Combustion process, diesel engine ~Prechamber engine

Switchable cam →Variable valve control ~Operation principles, ~Systems with camshaft ~~Variability between cam and valve

Switchable intake manifold →Intake system ~Intake manifold ~~Intake manifold design ~~~Variable intake manifold

Switching bucket tappet →Variable valve control ~Operation principles, ~Systems with camshaft ~~Variability between cam and valve

Switch-on spark →Ignition system, gasoline engine ~Ignition ~~Ignition spark

Synthetic diesel fuel →Fuel, diesel engine

Synthetic fiber mats →Filter ~Intake air filter ~~Filter media

Synthetic gasoline →Fuel, gasoline engine

Synthetic lubricating oils →Oil ~Body ~~Basic oils ~~~Fully synthetic oils, ~~~Semi-synthetic oils

Synthetic oil →Oil ~Body ~~Basic oils ~~~Fully synthetic oils

Systems without camshaft →Variable valve control

S

T

TAN →Oil ~Properties and characteristic values
~~Neutralization capacity

Tandem engines →Engine concepts ~Reciprocating piston engines ~~Single shaft engines

Tangential force (*also*, →Balancing of masses; →Crankshaft drive; →Piston ring). The tangential force curve in a reciprocating piston engine shows the engine torque as a function of the crank angle by using the appropriate crankshaft radius. The tangential force is shown in the torque diagram (**Fig. T1**) at a crankshaft drive.

The tangential force is not constant but changes periodically with the crank angle. This means that the engine torque, M, also changes periodically.

The crankshaft radius, r, and the tangential force, F_T, result in the following torque levels:

$$M = r \cdot F_T.$$

The periodic change of the tangential force, F_T, is a result of the crank angle, φ, and the position of the connecting rod in relationship to the cylinder axis corresponding to the angle, β (**Fig. T1**), given by the following equation:

$$F_T = F \cdot \frac{\sin(\varphi + \beta)}{\cos \beta},$$

where F represents the force that results from the gas and inertia forces. The values F_N, F_P, and F_R in **Fig. T1** represent the normal, the con rod, and the radial force.

The change of the tangential force with crank angle is shown in **Fig. T2** for a single-cylinder engine. It can be seen that there is a large variation in the tangential force. These variations in tangential force result in a degree of nonuniformity in the driving of the engine, which represents itself in engine speed variations. This can be compensated for partly by using a flywheel. These variations decrease with a larger number of cylinders in the engine.

Multicylinder engines produce a resultant tangential force which consists of the sum of the tangential forces of the individual cylinders related to the firing sequence (**Fig. T3**). The resultant tangential force is shown in **Fig. T3** as a solid line.

Tangential port →Intake port, gasoline engine

Tank venting →Emission measurements ~Test type IV

Tank venting emissions →Emission measurements ~Test type IV

Tank venting system →Injection system (components) ~Gasoline engine ~~Fuel tank systems

Taper, injection nozzle hole →Injection valves ~Diesel engine ~~Hole-type nozzle

Taper face Napier ring →Piston ring ~Versions ~~Oil scraper rings

Taper face ring →Piston ring ~Versions ~~Compression rings

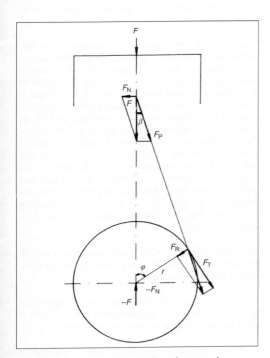

Fig. T1 Forces in a reciprocating piston engine

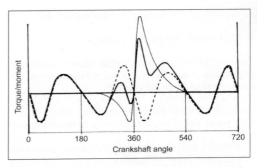

Fig. T2 Variation of the tangential force in a single-cylinder engine

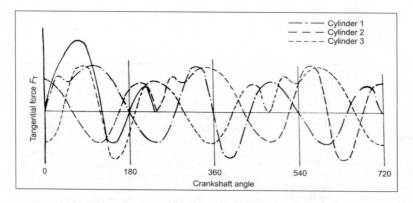

Fig. T3
Variation of the tangential force in a three-cylinder engine (Source: Küttner)

TBN →Oil ~Properties and characteristic values ~~Neutralization capacity

Telemetry →Engine ~Racing engines ~~World Rally Car, or WRC

Temperature, catalytic converter →Catalytic converter ~Light-off temperature

Temperature, charge air →Supercharging ~Charge air cooler

Temperature, combustion chamber →Combustion chamber temperature

Temperature, coolant →Radiator ~Coolant ~~Temperature

Temperature, intake air →Intake air ~Air temperature

Temperature, oil →Oil ~Properties and characteristic values ~~Viscosity ~~~Temperature dependency

Temperature dependency, oil →Oil ~Properties and characteristic values ~~Viscosity

Temperature gradient →Radiator

Temperature resistance →Sealing systems

Temperature sensors →Sensors

Temperature stresses →Electronic/mechanical engine and transmission control ~Requirements for mechanical and housing concepts

Temperature-dependent starting-stop →Injection system, fuel ~Diesel engine ~~Serial injection pump

Tension computation →Calculation processes ~Application areas ~~Stability

Tensioning system →Sealing systems

Test conditions—noise emission →Engine acoustics ~Exterior noise

Test cycle →Emission measurements ~Test type I ~~Perform driving cycle, ~FTP-75

Test cycle for particulate filter →Particles (Particulates) ~Legal limit values

Test fuel →Emission measurements

Test methods, emissions measurement →Emission measurements

Tetraethyl lead →Fuel, gasoline engine ~Antiknock additives

Texaco CCS. The Texaco CCS process (controlled combustion process) is a process for stratified charging of externally ignited engines. A wide injection spray is directed close to the spark plug. The moment of ignition is adapted so that the first mixture cloud is ignited. The further course of the injection is then determined exclusively by the framework conditions (speed) of the fuel injection. In the area of the spark-plug a combustion zone is created, which is supplied with oxygen by the swirling air in the cylinder and with fuel by the injection nozzle. The process includes a very complex combination of injection and ignition timing, load status, injection speed, and air turbulence. Ideally, this creates a quasi-standing flame front. **Fig. T4** shows the principle of the Texaco CCS process.

Thermal bridges →Electronic/mechanical engine and transmission control ~Requirements for mechanical and housing concepts

Thermal class →Electronic/mechanical engine and transmission control ~Requirements for mechanical and housing concepts ~~Installation location

Thermal cold-start valve (also, →Carburetor ~Starting systems ~~Functional systems). The ther-

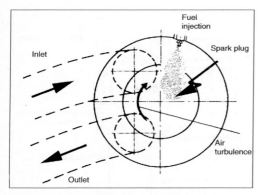

Fig. T4 Principle of the Texaco CCS process (according to Urlaub)

mal cold-start valve was used as an extra valve in addition to the automatic choke for carbureted engines to avoid a speed reduction of the engine or an engine stalling with automatic transmissions when drive is engaged or both. For this purpose, an additional, exactly measured mixture volume, depending on the automatic choke setting, was added.

Thermal conductivity of cylinder and fin material →Crankcase ~Crankcase construction ~~Ribbing

Thermal cracking →Soot generation

Thermal efficiency →Cycle; →Engine efficiency

Thermal expansion, piston →Piston

Thermal heat →Heat release characteristics

Thermal heat exchanger →Coolant

Thermal ignition. Thermal ignition in contrast to external ignition is the spontaneous ignition of the air-fuel mixture—for example, through temperature reached in the combustion chamber after compression.

~Auto-ignition. In contrast to the gasoline engine, the diesel engine works with auto-ignition. Because of the high compression ratio in a diesel engine, $\varepsilon = 17$–21, the induced air is heated up to values of between 800 and 900 K so that the injected fuel can ignite spontaneously.

~Glow ignition →Glow ignition

~Hot bulb ignition. The hot bulb ignition was used for older models of agricultural vehicles. Before startup, the cylinder head was preheated from the outside with a burner to enable a cold start.

Thermal insulation for pintle-type nozzles →Injection valves ~Diesel engine ~~Pintle-type nozzle

Thermal loading →Air cycling valve ~Air cycling valve functions

Thermal loss →Heat flow

Thermal management →Electronic/mechanical engine and transmission control ~Requirements for mechanical and housing concepts; →Fuel consumption ~Variables ~~Engine measures

Thermal range →Spark plug

Thermal reactor →Exhaust gas thermal reactor; →Pollutant aftertreatment

Thermal simulation →Electronic/mechanical engine and transmission control ~Requirements for mechanical and housing concepts

Thermal stress →Engine damage ~Stress

Thermal stresses. Thermal stresses result from different temperatures in a component or from the restriction of the elongation that would normally result from a temperature rise of the component. Thermal stresses inside a component are proportional to the temperature difference—for example, from the inside of the wall to the outside. It is important to know the distribution of thermal stresses in a component when designing it—these can be determined with sufficient accuracy by analyzing the complex geometry of the component using the finite element method (FEM).

Thermal stressing →Piston ring ~Shaping

Thermal-shock calculation →Calculation processes ~Application areas ~~Stability

Thermal-time switch. A cold-start valve was activated during the engine starting procedure to ensure the cold start of engines with older mixture formation systems—for example, L-Jetronic and K-Jetronic systems. This injected additional fuel into the intake manifold. The thermal time switch controls the fuel supply depending on the engine temperature and the time since starting.

Thermodynamic optimization →Fuel consumption ~Variables ~~Engine measures

Thermodynamic processes →Cycle

Thermodynamics →Air intake system; →Combustion; →Cycle

Thermodynamics calculation →Calculation processes ~Application areas

Thermo-siphon cooling →Cooling circuit

Thermostat →Radiator ~Coolant

T

Thermostat valve →Lubrication ~Control and safety components, ~Lubricating systems

Thin-wall casting →Crankcase ~Weight, crankcase ~~Reduction of specific weight of material

Third harmonic analysis →Engine acoustics

Three valves →Valve arrangement ~Number of valves

Three valves and one spark plug →Valve arrangement ~Number of valves ~~Three valves

Three valves and two spark plugs →Valve arrangement ~Number of valves ~~Three valves

Three-bed catalytic converter →Catalytic converter ~Three-way catalytic converter

Three-component bearing →Bearings ~Bearing positions ~~Crankshaft main bearings

Three-dimensional cam →Variable valve control ~Operation principles, ~Systems with camshaft ~~Three-dimensional cams

Three-dimensional flow simulation (3D), transient →Intake system ~Intake manifold ~~Computation process

Three-phase alternator →Alternator

Three-shaft engines →Engine concepts ~Reciprocating piston engines ~~Multiple-shaft engines

Three-stage intake manifold →Intake system ~Intake manifold ~~Intake manifold design

Three-valve engine →Valve arrangement ~Number of valves

Three-way catalytic converter →Catalytic converter

Threshold frequency →Engine acoustics

Threshold speed →Speed

Throttle dashpot. Throttle dashpots are used for gasoline engines with carburetors and fuel injection. They are used to ensure combustion during trailing throttle by delaying the closing of the throttle valve. This increases the intake manifold vacuum at a lower rate. The pressure reduction supports a fast evaporation of the fuel film from the wall, and this prevents the mixture from getting too rich. This used to be a problem, especially for engines with carburetors or central injection, because of the long travel from the fuel injector to the combustion chamber.

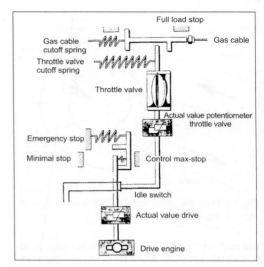

Fig. T5 Diagram of throttle valve actuator (Source: *MTZ*)

Lower hydrocarbon peaks in the exhaust gas are the result of a reduced shut-off speed. In addition, less exhaust gas is sucked back into the intake manifold. This maintains the ignitability of the mixture. The schematic illustration of a throttle dashpot can be seen in **Fig. T5**.

Throttle flange →Air limiter

Throttle valve →Carburetor; →Restrictor

Throttle valve actuator (*also*, →Carburetor ~Equipment). The throttle valve actuator is used for idle speed control directly above the position or the opening of the throttle valve. The bypass idle speed controller, which is normally used, is eliminated. The idle speed is controlled by a control actuator with electric motor. The required positioning of the throttle valve is performed by a bearing controller. The idle speed control system performs an adjustment using an electric motor, and this is superimposed over the mechanical actuation. The correct filling will always be adjusted during idle speed, depending on the engine temperature and the load on the engine from the accessories such as the alternator or the steering booster pump. The control range depends on the engine displacement. **Fig. T6** shows a throttle valve actuator with idle speed controlling. It covers an air mass of about 9–50 kg/h for a 1.8-L engine. Design details can be seen in **Fig. T7**.

Literature: H.D. Erdmann, et al.: Der neue Vierzylindermotor von Audi mit Fünfventiltechnik, Special Issue MTZ 56 (1995) 1 and 2.

Throttle valve control →Air cycling valve ~Air cycling valve functions; →Load ~Load control; →Variable valve control ~Load control

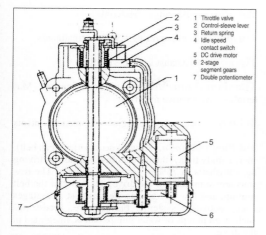

2	1 Throttle valve
3	2 Control-sleeve lever
4	3 Return spring
	4 Idle speed
	contact switch
	5 DC drive motor
	6 2-stage
1	segment gears
	7 Double potentiometer

Fig. T6 Throttle valve actuator (Source: *MTZ*)

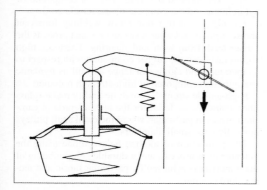

Fig. T7 Throttle dashpot

Throttle valve housing (*also*, →Throttle valve actuator). The throttle valve housing contains the throttle and the spring package that resets the throttle valve, the throttle valve potentiometer, the throttle control linkage, and the idle controller.

It is either a separate component or integrated into the intake manifold. The throttle valve housing is

heated with water in some engines to prevent throttle valve icing.

Throttle valve icing →Fuel, gasoline engine ~Additives

Throttle valve location (*also*, →Restrictor). Engines with quantitative control (gasoline engines with conventional combustion process) have one or more throttle valves. These are primarily positioned in the mixture-forming systems (carburetor or central injection) or in the intake manifold or intake tract. The engine reacts faster to the gas transfer if the throttle valve is positioned close to the engine. This is why one throttle valve is used for each cylinder in sports engines — the throttle valves can be arranged directly at the engine intake in this design version.

Throttle valves are also used for switchover intake systems. They are used here to change the intake manifold lengths and/or to join together several intake manifold arms. A more even torque progression can be achieved through a change of the tuned intake runners.

On turbocharged engines, the throttle valve can be placed before or after the turbocharger (**Fig. T8**). Placing the throttle valve between the supercharger and the engine is the most common arrangement. The limiting pressure ratio of the turbocharger may be exceeded if the throttle valve is suddenly closed. This problem can be solved by a bypass valve, for example, by blowing off or recirculating the charge air.

However, the placing of the throttle valve in the direction of the flow in front of the supercharger has advantages during acceleration. A significantly increased volume must be compressed to the desired pressure in front of the engine intake, because the pressure at the supercharger intake is low. This, however, is done with a better compressor efficiency and with higher turbocharger speeds, and this improves the response of the device. The precondition for such a solution is an oil-tight gasket on the turbocharger shaft, because oil is sucked into the supercharger as a result of the increased vacuum during part load if this condition is not met. This results in higher oil consumption, an increase in the hydrocarbon emissions, and a tendency to knock.

Throttle valves are also used in diesel engines in special cases and for special functions (e.g., for the

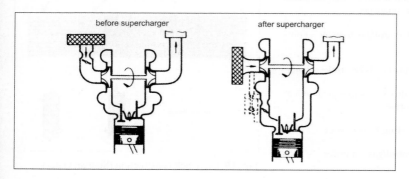

before supercharger   after supercharger

Fig. T8
Throttle valve
arrangement (Source:
Pischinger)

971

generation of a pressure difference between the exhaust gas and intake manifold to bring about exhaust gas recirculation).

Throttle valve potentiometer →Carburetor

Throttle valve switch. The throttle valve switch is generally attached to the throttle valve housing and is actuated by the throttle valve shaft. The throttle valve switch consists of a throttle valve potentiometer with full load and idle speed contact. The idle speed contact closes below a certain opening angle, and the full-load switch closes above a certain opening angle. Functions such as idle speed control, idle speed stabilization, and overrun fuel cutoff are displayed with the corresponding ignition angle depending on transmission of an appropriate signal.

The recognition of the full load creates mixture enrichment and special ignition angles depending on the speed.

The throttle valve switch can also be used as a load sensor, for example, for transmission control (automatic transmission) or for supercharged engines.

Throttle-free load control →Air cycling valve ~Air cycling valve functions; →Variable valve control ~Load control

Throttling →Load ~Load control; →Variable valve control ~Load control

Throttling losses →Charge transfer ~Charge transfer loss

Throttling pintle nozzle →Injection valves ~Diesel engine ~~Pintle-type nozzle

Throw, crankshaft →Crankshaft

Throw arrangement →Crankshaft ~Throw

Thyristor ignition →Ignition system, gasoline engine ~High-voltage generation ~~High-voltage capacitor ignition

Tightness to spray water →Electronic/mechanical engine and transmission control ~Requirements for mechanical and housing concepts

Time cross-section →Control/gas transfer ~Four-stroke engine ~~Timing

Timing →Control/gas transfer ~Four-stroke engine

Timing chain →Chain drive; →Valve gear ~Gear components ~~Camshaft drive

Timing chain whine →Engine acoustics

Timing diagram →Control/gas transfer ~Two-stroke engine

Tin alloys →Bearings ~Materials

Tin-plating →Piston ring ~Surface treatments

Tip noise →Exhaust system ~Muffler

Titanium connecting rod →Connecting rod ~Materials ~~Alternative materials

Toluene →Fuel, gasoline engine ~Aromatics contents

Toothed belt (*also*, →Engine accessories ~Fan belt). Increasingly today, the toothed belt is used for driving the camshafts in 75% of European engines. The reasons are: simplicity of the drive, flexibility of the belt routing, and low friction, as well as cost advantages over alternative systems. Furthermore, accessories such as oil pumps or water pumps can be integrated in the drive.

~Body. The toothed belt (**Fig. T9**) is a compound of nylon fabric, rubber compound, and tensile cord.

The high-strength nylon fabric webbing forms an abrasion-proof and wear-resistant coat and protects the rubber teeth from wear and shearing. There are high requirements on the rubber compound with respect to temperature and aging resistance as well as dynamic strength. It is composed of high-strength polymer.

High-tensile strength at high flexibility makes glass fiber a suitable material for the tensile cords in camshaft drives in particular, where the crankshaft pulleys normally have small diameters.

The production of toothed belts is done using the vulcanization process, and a durable compounding of the materials is achieved by particular fabric and tensile cord coatings.

~Characteristics. **Fig. T10** represents the most important characteristics of toothed belts. The tooth height plus the web height make the total thickness of

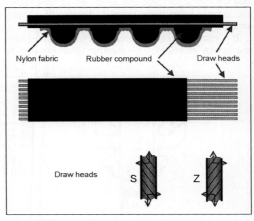

Fig. T9 Drive belt construction (Source: Gates)

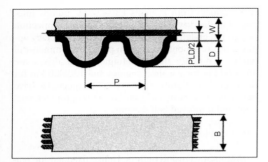

Fig. T10 Toothed belt characteristics (Source: Gates)

the toothed belt. The effective line distance—the distance from the web area to the center of the tensile cords—depends on the design of the toothed belt, the fabric thickness, and the diameters of the tensile cords. The width of the toothed belt is chosen according to the dynamic alternating load, which normally would be between 20 and 28 mm for passenger car engines, but for individual applications it might be up to 32 mm.

The profile of toothed pulley is be determined by its diameter. The effective diameter results from the number of teeth and pitch, and the outside diameter of the toothed pulley is reduced accordingly by the PLD (**Fig. T11**).

~Toothed belt profiles. The first camshaft belts were based on the Power Grip trapezoidal tooth formation. Circular-arc-type profiles (Power Grip HTD, as in "high torque drive") were developed because of increased requirements relating to load transmission, jump-over safety, and noise. With circular sections, in comparison with the trapezoid section, the forces are introduced into the tooth more evenly and peaks in tension are prevented (**Fig. T12**). Current applications exclusively employ circular arc-shaped profiles (HTD).

When the HTD profiles were introduced into the market, it had to be taken into account that some motor vehicle manufacturers continued using the existing trapezoidal toothed pulleys.

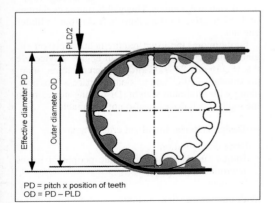

PD = pitch x position of teeth
OD = PD – PLD

Fig. T11 Toothed pulley characteristics (Source: Gates)

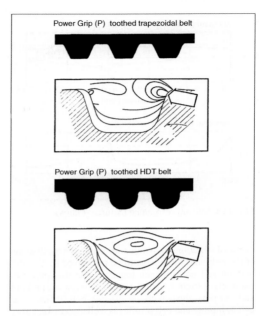

Fig. T12 Evolution of toothed belt profiles (Source: Gates)

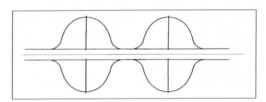

Fig. T13 Double-sided toothed belt (Source: Gates)

The profiles were optimized in terms of tooth root radius, tooth surface form and tooth height (power function profiles) such that they could be used on the existing trapezoidal toothed pulleys. The associated toothed pulleys of type ZA (C or CF tooth) and type B (B or BF tooth) are specified in ISO 9011.

The HTD profiles were developed and patented by Gates. Use of this circular arc-type profile made an essential improvement with respect to noise reduction, space, load transmission, and, as a result, the lifespan of the belt.

HTD profiles can also be used on a double-sided toothed belt (**Fig. T13**) with balancer-shaft drives.

Literature: M. Arnold, M. Farrenkopf, S. McNamara: Zahnriementriebe mit Motorlebensdauer für zukünftige Motoren, MTZ 62 (2001) 2. — R. van Basshuysen, F. Schäfer: Handbuch Verbrennungsmotor, 2nd edition, Vieweg Publishers, 2002.

Toothed belt drive (*also*, →Engine accessories; →Valve gear). The dynamic load on the drive consists of

T

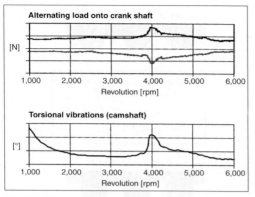

Fig. T14 System resonance (Source: Gates)

torsional vibrations, dynamic forces, and run oscillations of the belt, and these have to be optimized to give the best overall performance together. For this purpose, parameters such as tensioner characteristics, preload and damping, the belt characteristics, belt stiffness and belt damping, and belt profile, as well as camshaft gear wheel moments of inertia have to be coordinated such that the dynamic stress in the system is minimized. **Fig. T14** shows two important characteristics for the dynamics of a toothed belt drive, the alternating load on the crankshaft and the torsional vibrations of the camshaft. The resonance of the system, here at 4000 rpm, is reduced to a minimum through optimum layout of the system, and this has to be checked throughout the operating life of the drive.

Fig. T15 shows examples of typical applications in

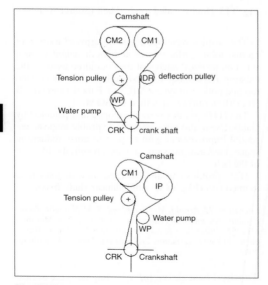

Fig. T15 Application examples (Source: Gates)

two engines. Both systems have water pumps integrated in the drive. Diesel engines have the injection pumps integrated in the primary belt drive in many applications (distributor injection pump or common-rail pump). Currently the operating lives of drives are 160,000 km for gasoline engines and 120,000 km for diesel engines. Future engines can be expected to have belt service lives of 240,000 km with optimized systems and improved belt designs.

~Belt tensioning systems

~~**Automatic tension pulleys.** Unlike fixed-tension pulleys, these offer the advantage of strongly increased dynamic forces in the camshaft drive at simultaneously higher lifespan demands.

An automatic tension pulley compensates for both increased temperature-related tension and belt elongation and maintains a constant high tension. The most widespread tensioner in use is the mechanical, friction-damped compact tensioner. Hydraulic tension pulleys also are employed in some applications that have the very high dynamic forces present in the toothed-belt drive system.

~~**Camshaft drive.** Today, 75% of European engines have toothed belts driving their camshafts. This percentage is related to the advantages of toothed belt drive, such as drive simplicity, flexibility of belt routing, low friction, as well as cost advantages over alternative driving systems. Furthermore, accessories such as oil pumps and water pumps can be integrated into the drive. Circular arc-shaped profiles are exclusively deployed.

The most important requirement for the toothed belt drive system is synchronization of the camshaft to the crankshaft throughout the service life of the engine. This is an important criterion for complying with emission restrictions even after prolonged mileages. Toothed belt elongations below 0.1% of the belt length can be maintained by correct selection of the toothed belt materials, by employment of an automatic tensioning device, as well as through optimized system dynamics. This yields timing deviations of 1–5° relative to the crankshaft on four-cylinder engines.

Furthermore, the usual requirements apply on engine manufacturing relating to an engine service life that is currently 240,000 km, with temperatures of approximately 120°C, and as small a space as possible, along with minimum weight. Disturbing noises from the toothed belt drive are also not acceptable.

~~**Design criteria.** Important criteria are:

• Drive configuration
• Torque curves and the dynamic peripheral forces determined from them
• Belt data

With this information, quantities such as run lengths, wrap angles, and lifespan with regard to various modes of defect can be evaluated. Further components such as deflection pulleys and tension pulleys can be dimensioned with the aid of the dynamic forces.

The following data should be accepted for toothed-belt systems in order to keep to the currently demanded lifespan of 240,000 km:

a. *Recommended minimum wrap angles.*
—Camshaft/injection pump 100°
—Accessory pulley 90°
—Tension roller (smooth or toothed) minimum 30°
better >70°
—Deflection pulley (smooth or toothed) 30°

b. *Periodic gear meshing.* This means that the same belt tooth would always engage in the same pulley gaps, which has to be avoided because otherwise unequal belt wear or belt damage would occur.

c. *Run lengths.* Free run lengths should not be in the range of 75 to 130 mm, so as to avoid resonance noises at idle speed.

d. *Minimum diameter of toothed pulleys and deflection pulleys.*

• Tooth pitch 9.525 mm, 18 teeth (54.57 mm diameter)
• Tooth pitch 8.00 mm, 21 teeth (53.48 mm diameter)
• Pulleys, 52 mm diameter without gear teeth

e. *Tolerances of the toothed pulleys and deflection pulleys.* Of particular importance in this case are concentric running/wobble, outside diameter taper, parallelism of bore and gearing, pitch errors, and surface roughness.

The belt needs to be guided on a pulley through flanged wheels to avoid toothed-belt runoff; this is mostly done on the crankshaft pulley where the crankshaft vibration damper can serve as the front flanged wheel. The rear pulley is attached to the toothed crankshaft pulley. Complex multivalve drives may require further flanged wheels, depending on number of pulleys and deflection pulleys—the flanges should be placed on toothed pulleys and not on deflection pulleys. It is necessary to ensure that exact alignment with the other pulleys is ensured if toothed pulleys with flanged wheels are employed. Toothed pulleys and rollers with one flanged wheel only, or without any, are dimensioned wider than the belt width to ensure reliable belt running. The width of toothed pulleys, as well as geometric dimensioning of the axial guide pulleys, is illustrated in **Fig. T16**.

~~Fixed-tension pulleys. In the past, eccentrically pivoted deflection pulleys were predominantly used (**Fig. T17**). The disadvantages of using fixed-tension pulleys are the temperature-dependent buildup that is conditional on engine expansion and the drop in tension by belt elongation and belt wear throughout the running period. These disadvantages cannot be counterbalanced.

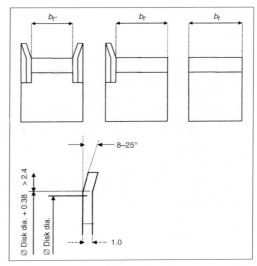

Fig. T16 Pulley width and belt routing (Source: Gates)

Literature: M. Arnold, M. Farrenkopf, St. McNamara: Zahn-riementriebe mit Motorlebensdauer für zukünftige Motoren, 9th Aachen Colloquium on Vehicle and Engine Technology.

Toothed belt profiles →Toothed belt

Toothed belt whining →Engine acoustics

Fig. T17 Belt tensioning systems (Source: Gates)

Toothed pulley →Toothed belt ~Characteristics

Top ring →Piston ring

Torque. Torque represents the "turning effort" that an engine supplies. The torque of an engine should, therefore, be as high as possible and its maximum value should occur at as low an engine speed as possible. This results in a good driving performance from a relatively small displacement engine, which results in smaller friction losses and reduced gas transfer work in a gasoline engine by means of de-throttling. Direct benefits are reduced fuel consumption and reduced emission of pollutants.

Engines may be supercharged to increase torque and, hence, power. The torque, M_d (Nm), is computed as follows for four-cycle engines:

$$M_d = \frac{w_e \cdot V_H}{4 \cdot \pi}$$

or

$$M_d = \frac{P_e}{\omega},$$

where V_H = displacement, w_e = specific work, P_e = brake power output, and ω = angular velocity.

~Diesel engine. Naturally aspirated engines with two valves, which are rarely used in cars anymore, have a maximum specific torque of about 65 Nm/L at engine speeds of 2000 rpm to 2200 rpm. The maximum specific torque with turbocharging is approximately 110 Nm/L for a speed of about 2000 rpm. These values relate to prechamber engines, while diesel engines with direct injection and turbocharging have maximum torques of up to 150 Nm/L with a tendency to increase further; hence, turbocharged engines have higher torques than naturally aspirated gasoline engines. Diesel engines with four-valve technology and turbocharging achieve 170 Nm/L at 1900 rpm.

~Gasoline engine. Naturally aspirated gasoline engines achieve maximum torques of about 100 Nm/L. This is true for engines with intake manifold injection and for engines with direct injection. Large torques at low speeds can only be implemented with additional technologies such as switchover (variable geometry) intake systems, four-valve technology, and/or variable valve timing if they are not to interfere with overall engine performance.

Gasoline engines with turbocharging achieve torques of about 140 Nm/L. These values are achieved at engine speed levels of 1800 to 2000 rpm assuming the right turbocharger is selected. **Fig. T18** compares examples of the variation in torque in a gasoline engine with a mechanical turbocharger with those of a naturally aspirated gasoline engine, and **Fig. T19** compares the torque curves for two-, four-, and five-valve engines. The maximum torque in the latter case is achieved at very high speeds.

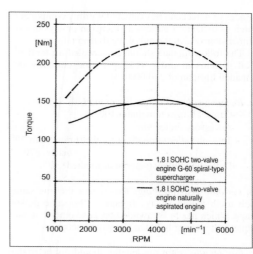

Fig. T18 Torque variation for a turbocharged and a naturally aspired engine

Torque boost →Air cycling valve ~Air cycling valve functions ~~Torque increase; →Intake system ~Intake manifold ~~Intake manifold design

Torque controlled tightening →Engine bolts ~Tightening procedure

Torque increase →Air cycling valve ~Air cycling valve functions; →Intake system ~Intake manifold ~~Intake manifold design

Torque motor →Actuators ~General Purpose Actuator (GPA) ~~Electric motors

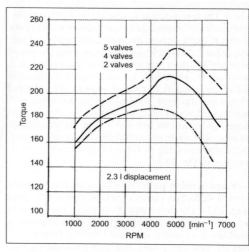

Fig. T19 Torque variation for two-, four-, and five-valve gasoline engines

Torsional vibration →Torsional vibrations

Torsional vibration absorber →Torsional vibrations ~Vibration damper

Torsional vibration damper →Torsional vibrations ~Vibration damper

Torsional vibrations. Torsional vibrations are vibrations of engine shafts that are stimulated by periodically changing torques and are superimposed on the basic rotation of the engine. The strongest stimulation comes from the main harmonic vibration of the gas and inertia forces through the crankshaft drive and the crankshaft.

Rigid body movements are generated close to the idle speed range (degree of nonuniformity), and elastic torsional vibrations are generated in the medium and higher speed ranges.

The torsional resonant frequencies are stimulated and forced into resonance by several harmonic vibrations of the gas and inertia forces over the operating speed range. The torques that stimulate these vibrations result from the kinematic conversion of the forces on the spin axis of the crankshaft. The Campell diagram in **Fig. T20** shows a graphical description of the resonance points in relation to the speed range.

Coupled axial and radial vibrations are generated though the geometry of the crankshaft and the force

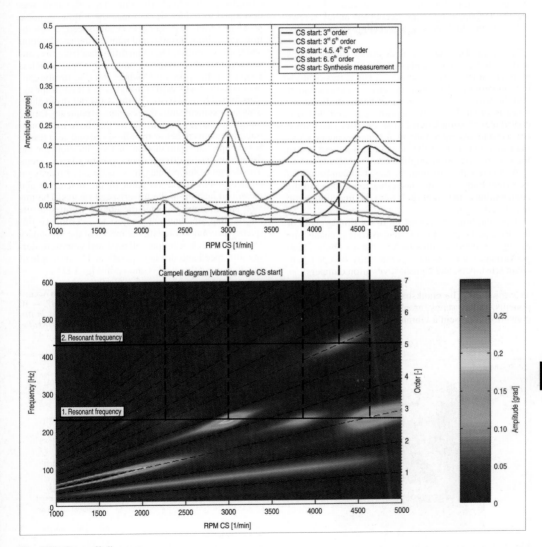

Fig. T20 Campell diagram

applied through the connecting rod; these vibrations can emit airborne sound mainly through the belt pulley and the engine housing. They also stimulate bending vibration in the engine housing through the bearing positions on the crankshaft, and as a result, airborne sound is emitted from the surface of the crankcase. The axial vibrations of the crankshaft also stimulate air sound through the belt pulley.

~**Belt drive.** The belt drive, together with its accessories, is stimulated by the nonuniform engine rotation especially at low speed ranges, and this results in noise and increased belt wear. Satisfactory compromises can be achieved by using adapted belt tensioners as damping elements, or freewheels and decoupling elements at the generator. Vibrations can, however, be completely decoupled by using decoupled pulley screws similar to the dual-mass flywheel (**Fig. T21**).

~**Camshaft.** The camshaft is driven by timing chains or drive belts; gear transmissions are also used in commercial vehicles. The elements create a torsional vibration system, which is stimulated by the crankshaft and the valve gear. The injection pump in diesel engines is often driven by the camshaft, and this also contributes to stimulation of the vibration. To reduce frictional losses, more four-stroke engines are employing roller rockers for valve actuation, and these in turn increase the vibration amplitudes.

As a first approximation, the camshaft can be treated as a rigid body, and the elasticity can be found in the connections to the crankshaft or to the injection pump in a diesel engine.

The vibration stresses set up in the timing chain or the drive belt reduce the lifespan as well as change the timing. Countermeasures include duplex chains, high performance drive belts, hydraulically damped belt, chain tensioners, and camshaft vibration dampers.

~**Crankshaft.** The crankshaft and the attached components (pulley screw, crankshaft drive, flywheel, or converter) represent a vibratory system.

Torsional vibrations have an important effect on the durability of the crankshaft. The resonance frequencies of the torsional vibrations are normally lower than the bending or axial vibrations of the crankshaft, because the crankshaft throws reduce the torsional stiffness significantly, while the bending stiffness is increased by the multiple bearings of the crankshaft. Strong increases in the amplitudes of torsional vibration result in resonance cases, and these can generate noticeable noise and in extreme cases can break the crankshaft. The typical first torsional resonance frequency of an inline car engine with six cylinders is about 200–400 Hz; higher torsional resonance frequencies tend to be of no practical importance.

~**Dual-mass flywheel.** The dual-mass flywheel was developed to reduce or eliminate the following noise problems: transmission rattle at idle speed, acceleration and trailing throttle, and hum during acceleration and trailing throttle. The main difference from the conventional flywheel is the secondary flywheel, which is coupled to the primary flywheel (that is, firmly connected to the crankshaft) through a soft torsion spring. This allows a rigid design of the converter drive plate. This arrangement shifts the resonance frequency for the vibrations that are responsible for the above-mentioned noise problems to below the idle speed and, as a result, it decouples the stimulation of the vibration for the drivetrain during driving operations to a large extent. The design of the primary mass is based primarily on an acceptable speed variation of the belt drive and clean combustion in the lower speed ranges. The torsion spring must transmit the engine torques in the elastic map area and must adjust to a resonance frequency in connection with the inertia of the secondary flywheel, which decouples all torsional vibrations during idle speed and driving operation. The principle of the dual-mass flywheel is shown in **Fig. T22**.

Literature: W. Reik: Das Zweimassenschwungrad, 1st Aachen Colloquium on Vehicle and Engine Technology, 1987. — W. Reik: Torsionsschwingungen und Getriebegeräusche, in Automobilindustrie 1, 1987. — H. Maass, H. Klier: Kräfte,

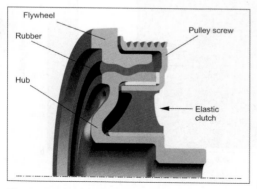

Fig. T21 Decoupled pulley screws

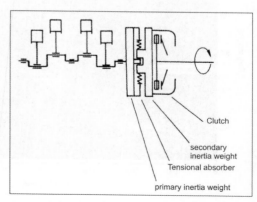

Fig. T22 Principle of the dual-mass flywheel

Momente und deren Ausgleich in der Verbrennungskraftmaschine, Vienna/New York, Springer Publishers, 1981. — K.E. Hafner, H. Mass: Torsionsschwingungen in der Verbrennungskraftmaschine, Vienna/New York, Springer Publishers, 1985.

~Dual-mass torsional vibration damper. *See above,* ~Dual-mass flywheel

~Flywheel. The inertia masses of the flywheel attached to the crankshaft and the transmission are used to reduce the nonuniform rate of engine rotation, especially when operated near idle speed. Emissions legislation, fuel consumption, acoustics, and so on, require an idle speed as low as possible, which can be achieved by electronic engine management, among other things. A large flywheel permits a low idle speed, but it reduces the acceleration characteristics of the engine and increases the torsional load on the crankshaft. Engines with high idle speed—for example, for car racing applications—basically have no flywheel anymore.

The rotating mass on the crankshaft side is important for engines with dual-mass flywheels; the mass of the torque converter is important for engines with automatic transmission. The dual-mass flywheel runs through its resonance frequency on its way to the idle speed and decouples the torsional vibrations (transmission rattle, acceleration hum) from the drivetrain at idle speed and driving operations. Torsional vibration absorbers are often used with conventional flywheels to improve the acoustics in the drivetrain (transmission output), and they are adjusted to the relevant resonances.

Complete decoupling from the drivetrain by the dual-mass flywheel and from the belt drive by the decoupled pulley screw can result in inadmissibly high speed variations at the crankshaft, especially in diesel engines.

The flywheel inertia is designed in such a way that acceptable speed variations are achieved with respect to the impact on the drivetrain and the belt drive. Very high inertias contribute to bad acceleration characteristics, especially in diesel engines, and also increase the stress on the crankshaft. The latter can be reduced by the torsional vibration absorber.

~Multimass torsional vibration damper. *See above,* ~Dual-mass flywheel

~Nonuniformity. The rigid body vibrations of an engine shaft in the region of medium speed are defined by the equation

$$\delta = (n_{max} - n_{min}) / n_{avg},$$

where δ is called nonuniformity.

It is especially noticeable during the acceleration of the engine under load, and it stimulates strong torsional vibrations in the aggregates on the front end through the belt drive and in the drivetrain.

The vibration amplitudes are reduced directly in proportion to an increase of the inertia masses of the crankshaft and especially the flywheel and are increased proportional to the mean effective pressure by a process depending on gas forces.

~Vibration damper. A spring-mass system is connected to the crankshaft to split the resonance frequency into two intrinsic shapes. Its mass is coupled by steel or rubber springs with or without hydraulic damping. The vibration damper is a resonance vibrator and consists of a hub connected on the end of the crankshaft, a flywheel that acts as inertia mass, and a layer of material that transmits the forces between them.

The inertia of the flywheel part of the damper should be at least 10% of the mass of the crankshaft plus the main flywheel. A large damping device (viscous, parallel, or material) is required for the optimum design of the engine and for component durability. The most important basic types of damper are shown in **Fig. T23**.

Rubber torsional vibration absorbers are increasingly being designed with an axial rubber contour to achieve a reduction of the axial sound emission in addition to the reduction in torsional vibrations. Radial rubber contours also reduce bending and tumbling vibrations.

Total air-fuel ratio (*also,* →Air-fuel ratio). The term "total air-fuel ratio" is used for mixed-mode operation with several fuels. The total air-fuel ratio can be defined as follows if, for example, gasoline and methanol are used as mixed fuels:

$$\lambda = \frac{m_{air}}{m_{methanol} m_{stoich,methanol} + m_{gasoline} m_{stoich,gasoline}}$$

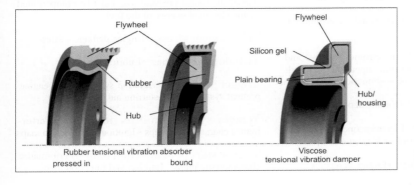

Flywheel

Flywheel

Silicon gel

Rubber

Plain bearing

Hub

Hub/housing

Rubber tensional vibration absorber
pressed in   bound

Viscose
tensional vibration damper

Fig. T23
Basic types

where m_{air} = actual air mass, $m_{methanol}$ = actual methanol mass, $m_{stoich,gasoline}$ = stoichiometric air mass of gasoline, $m_{stoich,methanol}$ = stoichiometric air mass of methanol, and $m_{gasoline}$ = actual gasoline mass.

The total air-fuel ratio in gasoline engines is in a first approximation independent of the particular location in the combustion chamber due to its homogeneous mixture.

The fuel in diesel engines is not homogeneous after the injection, which means that an average total air-fuel ratio must be defined. The value of the air-fuel ratio ranges locally between zero (pure fuel) and infinity (pure air).

Total contamination →Fuel, diesel engine ~Properties

Total fuel contamination →Fuel, diesel engine ~Properties

Total separation rate →Filter ~Intake air filter ~~Filter characteristics

Total system check →Emission measurements ~Test type I

Total-loss lubrication →Two-stroke engine

Toxicology →Oil ~Safety and environmental aspects

Tractor oils →Oil ~Classification, specifications, and quality requirements ~~Special engine oils

Trailing throttle →Carburetor ~Equipment; →Load

Transfer function →Engine acoustics ~Acoustic transfer function, ~Structure-borne sound transfer function

Transient engine operation →Load

Transistor ignition →Ignition system, gasoline engine ~Ignition timing sensor

Transition holes →Carburetor ~Systems

Transition systems →Carburetor ~Systems

Transmission adaptation →Fuel consumption ~Variables

Transmission control →Electronic open- and closed-loop control ~Electronic open- and closed-loop control, gasoline engine ~~Functions

Transmission degree →Engine acoustics

Transmission frame →Electronic/mechanical engine and transmission control ~Electronic components

Transmission grinding →Engine acoustics

Transmission noise →Engine acoustics

Transmission oil cooler →Radiator ~Liquid/air radiator, ~Liquid/liquid radiator, ~Oil cooler

Transmission oil leak-tightness →Electronic/mechanical engine and transmission control ~Requirements for mechanical and housing concepts

Transmission rattle →Engine acoustics

Transmission whine →Engine acoustics ~Transmission noise

Transmission whistle →Engine acoustics ~Transmission noise

Transversely mounted engine →Engine installation ~Transverse installation

Trap (HC) →Pollutant aftertreatment ~Exhaust gas treatment with three-way catalytic converter ~HC absorber

Trap channel →Intake port, gasoline engine ~Design

Trapped pressure. The trapped pressure for naturally aspirated engines is the pressure in the combustion chamber after the suction phase has been completed (intake valves closed) in relation to bottom dead center. It is normally lower than the atmospheric pressure due to the intake resistance or the intake pressure losses from the air filters, the collection volumes, the intake pipes, the throttles and valves, and so on. Currently, work is focused on minimizing the pressure losses for reasons of performance and consumption—for example, by streamlining the design of the intake system and by introducing larger valve cross-sectional areas. Multiple intake valves are the state of the art. Intake pressure losses can be overcompensated for by supercharging. The simplest forms of supercharging are intake runner and resonance induction supercharging. More effective are turbochargers or mechanical superchargers, whose supercharging pressures are in the 2-bar range for car engines.

The intake pressure loss must be minimized as far as possible, because the required work for the compensation of the intake pressure loss must be compensated by the chemical energy of the fuel.

Tribological stress →Engine damage ~Stress

Tribology →Friction; →Lubrication

Trigger glow plug →Ignition system, diesel engine/preheat system ~Cold starting aid

Trigger start main injection map →Performance characteristic maps ~Ignition and injection maps

Trigger start pilot injection map →Performance characteristic maps ~Ignition and injection maps

T

Triple carburetor →Carburetor ~Design types

Triple-layer materials →Bearings ~Bearing materials

Triple-mass fly wheel →Flywheel

Trochoidal pump →Lubrication ~Oil pump

Trochoids →Engine ~Alternative engines ~~Wankel engine

Truck diesel engines →Lubrication ~Lubricating systems; →Piston ring ~Ring set

Trunk piston engine. The reciprocating engine is also called a trunk piston engine as long as it is not a crosshead engine. The forces acting on the piston from the combustion chamber are transferred directly to the crankshaft via the piston pin and the connecting rod. The piston has to straighten itself in the cylinder bore. The connecting rod, which translates the reciprocating piston movement into a rotary movement of the crankshaft, performs a vibrating movement. The part of the connecting rod mass that is straightened with the piston reciprocates, while part of the mass at the crankshaft end rotates—the center of gravity completes an elliptical course.

In contrast to the trunk piston engine, the crosshead engine has an additional straightening mechanism between the piston and the connecting rod. This structural solution can be found in large engines.

Tubular heater →Exhaust system ~Exhaust gas manifold

Tumble →Charge movement

Tuned intake runner →Intake system ~Intake manifold ~~Intake manifold design; →Supercharging

Tuned intake runner system →Intake manifold

Turbine →Supercharging ~Exhaust gas turbocharging

Turbine geometry →Supercharging ~Variable nozzle turbocharger (VNT)

Turbo engine →Supercharging ~Variable nozzle turbocharger

Turbocharger →Supercharging ~Exhaust gas turbocharging

Turbocharger balancing →Balancing of masses ~Balancing

Turbocharger main equation →Supercharging

Turbocharger speed →Supercharging ~Mechanical supercharging

Turbo-compound →Supercharging ~Turbo-compound supercharging

Turbo-compound charging →Supercharging

Turbo-compound systems →Engine concepts ~Composite systems

Turbo-MAP →Sensors ~Pressure sensors

Turbo-supercharger →Supercharging

Turbulence system →Intake system

Turbulent charge motion →Charge movement

Turbulent flow →Charge movement; →Flame quenching

Twin needle nozzle →Injection valves ~Diesel engine ~~Vario nozzle

Twin-cylinder engines →Engine concepts ~Reciprocating piston engines ~~Multiple-shaft engines ~~~Twin-shaft engines ~~~~Dual inline engines (or parallel and twin engines)

Twin-piston engine →Engine concepts ~Reciprocating piston engines ~~Single-shaft engines

Twin-shaft engines →Engine concepts ~Reciprocating piston engines ~~Multiple-shaft engines

Twin-tube shock absorber →Exhaust system

Two valve →Valve arrangement ~Number of valves

Two-barrel carburetor →Carburetor ~Design types

Two-bed catalytic converter →Catalytic converter ~Dual-bed catalytic converter

Two-sided keystone ring →Piston ring ~Versions ~~Compression rings ~~~Keystone rings

Two-stage carburetor →Carburetor ~Design types, ~~Multistage carburetor

Two-stage manifold →Intake system~Intake manifold~~Intake manifold design

Two-stage supercharging →Supercharging ~Multistage supercharging

Two-stroke engine (also, →Control/gas transfer; →Lubrication ~Lubricating systems). A combustion engine is called a two-stroke engine when it operates according to the two-stroke principle (see DIN 1940). The major characteristic of the two-stroke principle is

T

that a complete duty cycle takes place each revolution of the engine (i.e., two piston strokes). Only the compression of the fresh charge (compression period) and the expansion of burned charge (expansion period) on two-stroke engines are considered as individual processes. Removal of the burned charge and introduction of fresh charge into the cylinder (i.e., the cylinder scavenging process) are made simultaneously via the opened charge transfer elements in the crank angle region around the BDC. As opposed to the process in four-stroke engines, induction of fresh mixture and expulsion of exhaust gas in two-stroke engines are achieved by way of the piston in the simplest design, because the operating process of two-stroke engines runs at crankshaft rotational speed. A separate scavenging pump is required to generate a scavenging pressure difference for the scavenging process—this can be produced by the underside of the piston and the crankcase volume in the simplest case (crankcase chamber scavenging pump). The two-stroke principle can be achieved both with the gasoline-engine combustion principle (two-stroke gasoline engine) and the diesel combustion principle (two-stroke diesel engine). Two-stroke engines are produced over a broad range of displacements from 0.16 cm^3 (power approximately 20 W for model aircraft propulsion) up to more than 1.5 m^3 (cylinder power >3000 kW for marine propulsion). Specific power outputs per liter of more than 250 kW/L are reached using (nonsupercharged) two-stroke gasoline engines. The basic reasons why two-stroke gasoline engines are primarily confined to leisure and sports equipment, as well as small two-wheeled vehicles and implements, are lifespan, smooth running, and emission of pollutants, particularly in road vehicles. Two-stroke diesel engines are primarily limited to large engines at the present time. Two-stroke engines, however, have large shares of these market segments.

Development of four-stroke engines with corresponding power output is competing with two-stroke gasoline engines because of more stringent requirements related to the emission of pollutants in low-power engines nowadays. However, the high demands in terms of costs, performance, power weight, design volume, ruggedness, and position-independent low-maintenance operation associated with two-stroke engines make their replacement by four-stroke engines difficult. It is likely that two-stroke gasoline engines will retain their major importance in certain market segments for the foreseeable future at least because of the increased application of technical measures, such as improvement of the main fresh air supply, optimization of the scavenging processes, leaner basic engine tuning, the use of secondary air systems and oxidation catalysts, and the transition to direct gasoline injection. Owing to the availability of direct gasoline injection systems that are ready for production, it is even possible that two-stroke gasoline engines, over the medium term, may gain higher significance in individual partial market segments, such as motorcycle engines with displacements above 50 cm^2. Low power-to-weight ratios, rapid engine heating after cold start, the possibility of reducing the number of cylinders along with the favorable torque characteristics and the option for long geared-down transmission concepts make two-stroke engines attractive, in particular for small and fuel-efficient vehicles (three-liter or two-liter passenger cars). Numerous publications on new concepts for passenger car two-stroke diesel and gasoline engines rekindled interest in two-stroke engines as passenger car powerplants in the 1990s. However, series launches of passenger-car two-stroke engines are not foreseeable in the United States and Europe in the years to come against the background of high consumer and legislative requirements (emission of pollutants, lifespan, smooth running/noise) on these engines, various individual problems not satisfactorily resolved in the sense of production capability, and conceptual shortcomings of individual concepts. On the other hand, the series launch of an Orbital twin-cylinder two-stroke gasoline engine in a passenger car of the Indonesian maker Texmako seems to be imminent. The low-speed, two-stroke diesel engine with uniflow scavenging will keep its dominant position as the propulsion unit for large cargo ships and tankers for a foreseeable time owing to its high level of engineering, high dependability, and favorable fuel consumption.

~Acoustics. Noise emissions from two-stroke engines generally have more high-frequency components than those from four-stroke engines due to the doubling of the firing frequency. High residual gas contents at part load cause cyclic fluctuations in the combustion process and the development of harsh noise particularly in two-stroke gasoline engines with external mixture formation. Thermal distortions in the port-interrupted cylinder liner that need to be coped with by widening of piston running clearance as well as employment of antifriction crankshaft and connecting-rod bearings normally lead to comparatively high mechanical noise levels in conventional two-stroke gasoline engines.

~Asymmetric timing diagram →Control/gas transfer ~Two-stroke engine ~~Timing diagram

~Blue smoke. Blue smoke develops from incomplete combustion of the oil in the exhaust gas of two-stroke engines with lubrication from oil in the mixture. Reduction or avoidance of blue smoke is possible using exactly metered feed of the lubricating-oil (automatic lubrication).

~Bypass flow. See below, ~Charge transfer

~Charge transfer (also, see below, ~Scavenging process). The characteristic feature of the charge transfer on two-stroke engines is the displacement of exhaust gas from the cylinder by the fresh charge delivered by a scavenging pump. There are a number of different scavenging processes from which to choose. As opposed to the four-stroke engine, the control of charge transfer on the two-stroke engine is possible using the

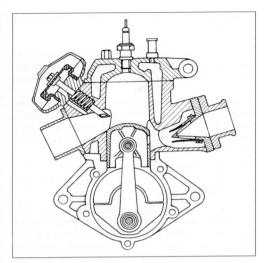

Fig. T24 Sectional representation of a two-stroke gas-oline engine with loop scavenging, crankcase scavenging pump, intake-system reed valves, and constant-depression exhaust control

piston (piston port control) in the simplest case, because its operating cycle takes place at crankshaft rotational speed. Pressure waves excited by the intake and exhaust processes exercise an essential influence on the charge transfer in a two-stroke engine. Depending on the concept selected, modern two-stroke engines are equipped on the fresh-gas side, among others, with rotary intake valves, reed valves, reed-valve bypass controls, resonance chambers, and, at the exhaust side, control valves and barrel valves for reduction of adverse influences of pressure waves and enhancement of cylinder filling. **Fig. T24** shows the sectional representation of a modern two-stroke gasoline engine with loop scavenging, a crankcase scavenging pump, intake-system reed valves, and constant-depression exhaust control. Provided the delivery characteristic of the scavenging pump allows low loss reduction of the delivery rate independent of the engine speed, the charge transfer work can be significantly reduced at part load by reducing the volume of scavenging gas and simultaneously increasing the quantity of residual gas remaining in the cylinder.

~~Bypass flow. Bypass flow is understood to occur when fresh gas gets directly into the exhaust system during charge transfer.

~Charge transfer work. Charge transfer work is the work required to perform the transfer of charge in the engine. It has a negative impact on engine efficiency and should, therefore, be minimized as far as possible. In two-stroke engines this charge transfer work is performed by the crankcase scavenging pump or a separate scavenging fan. The charge transfer work on

crankcase-scavenged two-stroke engines decreases at throttled part-load operation.

~Cooling. While in four-stroke engines 60–80% of the cooling heat has to be dissipated via the cylinder head and approximately 20–40% via the cylinder liner, two-stroke engines with pure piston port control (loop scavenging) show roughly reverse proportions. Because of this and particularly due to the ignition frequency being twice that in a four-stroke engine, the specific thermal load per unit area is 40–60% higher than that of comparable four-strokes. In addition, the asymmetrical thermal load on the cylinder and resultant distortions of the cylinder, which occur mainly in loop-scavenged engines, make heat transfer difficult from the piston via piston rings to the cylinder liner. Limitation of the component temperatures at full load is an essential prerequisite for control of engine wear, satisfactory dependability, and lifespan, as well as free determination of the fuel/air ratios in the map with regard to pollution aspects. The following measures will help, among others, to reduce the thermal stress in components: limitation of the displacement of individual cylinders, limitation of the rated speeds, careful dimensioning of cylinder cooling systems—using liquid cooling preferably—particularly in the region of the scavenging ports and exhaust ports (limiting cylinder distortions), dispensing with the crankcase as a scavenging pump, employing oil spray cooling of the piston in conjunction with a cooling duct piston, and, finally selection of a scavenging process that limits heating up of the piston, if necessary.

~Crankcase injection. *See below,* ~Mixture formation

~Crankcase precompression. A vacuum pressure that sucks fresh charge into the crankcase will develop during the compression process in the cylinder; this will be followed by compression. At the same time, the combustion gases will expand in the cylinder, as long as the scavenging ducts are shut. When the scavenging ducts open, fresh charge will flow into the cylinder at the same time as exhaust gases flow out through the exhaust port. Fresh charge can flow into the exhaust system at the same time (bypass flow) during this process.

~Crankcase scavenging pump. *See below,* ~Scavenging fan

~Crankshaft web control. Crankshaft web control is carried out using the same principle as the rotary intake valve control; in this case, the latter is replaced with the crankshaft web.

~Cross-flow scavenging. *See below,* ~Scavenging process

~Cylinder wall injection. *See below,* ~Mixture formation

~Diesel engine. In addition to the gasoline combustion process, the two-stroke cycle can also be combined with the diesel combustion process. Although two-stroke diesel engines are used to some extent as small stationary and tractor engines as well as truck power-plants, the two-stroke diesel engine is of no real significance in this market segment or as a power unit for passenger cars. The reasons for the insignificance of the two-stroke diesel engine in these segments are founded on the increasing requirements relating to lifespan, lubricating-oil consumption, and emission of exhaust pollutants that could not be met to a satisfactory degree using simply designed two-stroke diesel engines. In addition, the performance advantages of two-stroke diesel engines have dwindled because of the increasing employment of exhaust turbocharging on four-stroke diesel engines. In the field of large engines, in contrast, the low-speed, uniflow-scavenged and high-pressure exhaust-gas turbocharged two-stroke diesel engine has gained dominance in propulsion of large cargo ships and tankers. The advantages of the two-stroke diesel engine, among other things, relate to the drivetrain vibrations on engines with a low number of cylinders, the torque characteristics, power-to-weight ratio, cold-start performance, engine heating after cold start, and untreated NO_x emissions; all these features rekindled the interest in two-stroke diesel engines in the 1990s. AVL, Toyota, Yamaha, and Daihatsu published development projects on the realization of two-stroke diesel engines for passenger-cars in the recent years. **Fig. T25** shows a sectional view of the passenger car two-stroke diesel engine designed by AVL. Choice of the combustion process is strongly determined by predefining a certain scavenging concept on two-stroke diesel engines. It is relatively simple to generate swirl flow in the cylinder of two-stroke diesel engines with uniflow scavenging. On the other hand, chamber combustion processes often were given preference in loop-scavenged two-stroke diesel engines

because of deviating flow conditions in the cylinder around the TDC.

~Direct injection (internal mixture formation). Direct injection into the cylinder takes place before or after closing of all control units (valves/ports).

~Emission of pollutants. Exhaust gases from two-stroke engines generally contain the same pollutants as those from four-stroke engines but with the difference that C4 to C8 hydrocarbons prevail in the two-stroke engine because of the

- Scavenging method (scavenging process),
- Mixture formation (mixture formation process), and
- Lubricating system.

Small air ratios (λ) cause high CO emissions. This primarily occurs in two-stroke gasoline engines in the part load and idling ranges, where misfiring happens with many scavenging systems because of deficiencies in the mixture formation and high residual gas contents in the cylinder (HC emissions). Emissions of oxides of nitrogen from the engines are rather low due to the high quantity of residual gas in the charge, the low effective compression ratio, and the low mean effective pressures.

~Engine. The engine components on two-stroke engines with crankcase scavenging pumps are lubricated using mixture lubrication or total-loss lubrication because of the way that the fresh charge is ducted through the crankcase. The pistons of these engines have only compression piston rings and no oil scraper rings. The main bearings and connecting-rod bearings are fitted with antifriction (ball, roller) bearings, which require comparatively little lubricating oil. The crankshaft is assembled either from component parts or split deep-groove ball bearings or the roller bearings are

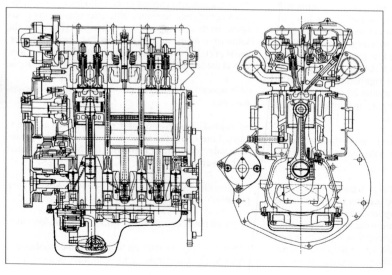

Fig. T25
Longitudinal and sectional view of a passenger car uniflow-scavenged two-stroke diesel engine (AVL)

used for mounting the antifriction bearings. In multi-cylinder engines the crankcases are sealed against each other via shaft sealing rings. The crankshaft webs are often cylindrical in form to reduce the dead-space volume of the crankcase scavenging pump. In V-engines, one thin disk each is arranged on the crankpins between both piston rod stems of a throw (outboard engines, for instance) that seal the two crankcases thus formed against each other by relative movement between outside disk diameter and inside diameter of the crankcase using seal rings. In piston-controlled two-stroke engines with a separately arranged scavenging pump, the intake of fresh gas into the crankcase and the oversupply of lubricating oil onto the cylinder liner has to be prevented by using a sufficiently long piston skirt and an additional piston ring in the region of the bottom edge of the piston skirt. The overall engine heights are greater, particularly in long-stroke two-stroke engines.

~Exhaust gas. The exhaust emissions of two-stroke engines may differ considerably from those of four-stroke engines because of differences in gas transfer, engine layout, and engine lubrication. The normally limited part load scavenging gas volumes result in high residual gas contents in the cylinder (internal exhaust gas recirculation). This leads to comparatively low emissions of oxides of nitrogen from two-stroke engines at part load in particular. In two-stroke gasoline engines with external mixture formation (carburetor, intake manifold injection), part of the fresh gas is already scavenged into the exhaust unburned during the scavenging process, depending on the quality of the scavenging process. These fresh mixture losses, as well as combustion misfires due to the high content of exhaust gas in the cylinder at low part loads, result in high hydrocarbon emissions of conventional two-stroke gasoline engines. The transition to direct injection results in a drastic reduction of the fresh mixture losses and HC emissions from two-stroke gasoline engines. Further causes for high HC emissions from two-stroke gasoline engines derive from comparatively high lubricating-oil consumption (blue smoke).

~~Exhaust odor. Odor emissions from two-stroke engines primarily result from the HC emissions, which are 10 times higher than from a comparable four-stroke engine and are influenced by the following factors:

- Scavenging system (scavenging process)
- Fuel (diesel or gasoline)
- Lubricating systems
- Lubricating oil type (mineral, synthetic)
- Secondary exhaust treatment
- Mixture preparation (mixture generation process)
- Additives in fuel

Depending on these engine conditions, a more or less pronounced offensive smell develops in the exhaust gas; it can be reduced by appropriate measures.

~~Secondary exhaust treatment. Equipment for secondary exhaust treatment such as secondary air injection, oxidation catalysts, and NO_x accumulator-type catalytic converters significantly aid in reducing emission of pollutants of two-stroke gasoline engines. Limited volumes of scavenging air and less space or area speeds in catalytic converters or particle filters on two-stroke engines cause comparatively high exhaust gas temperatures for part loads so that relatively good conditions will basically prevail in these areas of the operating map. Realization of the $\lambda = 1$ control combined with a three-way catalytic converter however, causes difficulties in practice because of more or less distinct scavenging-gas losses.

~Fresh mixture losses. In two-stroke engines, the exhaust gas is "pushed" out of the cylinder by fresh gas flowing in during the scavenging process. Depending on engine load condition and the quality of the scavenging process, and deviating from the ideal borderline case of displacement scavenging, part of the fresh gas passes directly out of the cylinder (bypass scavenging), or it passes out after intermixing with the exhaust gas. These fresh mixture losses are the major cause of high fuel consumption and high HC emissions in two-stroke gasoline engines with external mixture formation (carburetor, intake manifold injection). In two-stroke diesel engines with exhaust turbocharging, the fresh mixture losses (scavenging-air losses) would reduce the exhaust gas temperatures and hence turbine work.

~Friction loss. Friction loss is defined as that power by which the overall power output (at the crankshaft) is less than the indicated power. In two-stroke engines, friction is normally less than in four-stroke engines because there are no friction-afflicted engine parts such as valves, oil pump, and so on, and antifriction (ball, roller) bearings are primarily used in place of plain bearings.

~Fuel consumption. The specific fuel consumption particularly of two-stroke gasoline engines is essentially influenced by the scavenging process applied, the kind of scavenging-air supply, and above all by the selection of the mixture formation system. The positive features of two-stroke engines are the potential relating to improving fuel consumption, the low power-to-weight ratios, low charge transfer work and compression losses at part load, rapid engine heating after cold start, favorable torque characteristics (high torque at low speeds), and, obviously as a matter of principle, high internal exhaust gas recirculation rates at part load. In conventional two-stroke gasoline engines with external mixture formation, combustion misfires at lower part load and loss of fresh mixture at higher engine loads result in comparatively high fuel consumption and hydrocarbon emissions. As series-produced motorcycle and outboard engines demonstrate, the transition to direct injection systems can reduce fuel consumption by up to 40%, cumulatively. This means that, in practice, fuel consumption is below the fuel consumption of comparable four-stroke gasoline engines in some cases. Low-

T

speed, uniflow-scavenged marine diesel engines currently represent the only essential field of application for two-stroke diesel engines. Their overall efficiencies have reached up to 54% corresponding to 156 g/kWh specific fuel consumption, which represents the top value of all thermal powerplants. This was achieved by consequent optimization, particularly pertaining to the scavenging system, stroke-to-bore ratio, material technology, and supercharging technique.

~Gas dynamics. Gas-dynamic effects on two-stroke engines exert a comparatively great influence on charge transfer, primarily because of the open gas transfer (direct contact of fresh gas and exhaust gas during the scavenging process). By optimizing gas transfer (enhancement of engine torque) and avoiding detrimental influences from pressure waves following charge transfer, it is possible to use, for example, blade valves, rotary valves, multichamber intake silencers on the intake side, and, on the exhaust-gas side, exhaust systems with a diffuser and opposing cone. Normally, there is a gas-dynamic conflict of objectives when dimensioning the intake and exhaust systems, because the particular measures only yield optimal conditions over a certain speed range.

~Gasoline direct injection. *See below*, ~Mixture formation

~Gasoline engine. Combustion of the compressed air-fuel mixture on two-stroke gasoline engines is initiated by time-controlled external ignition. Two-stroke gasoline engines have dominant market shares in the field of powerplant for small implements such as chain saws and grass trimmers and also for small motorcycles, snowmobiles, very light aircraft, outboard engines, and jet-ski drives. This dominance can be explained by their low weight, mechanical ruggedness, small design dimensions, low power-to-weight ratios, and low-maintenance operation. These engines are practically all equipped with loop scavenging and a crankcase scavenging pump. Against the background of tightened requirements on pollutant emissions, these engines are increasingly being fitted with lean mixture setting and precatalytic converters, secondary air systems, equipment for charge stratification, or main fresh air supply or direct fuel injection systems. The percentage of two-stroke gasoline engines driving passenger cars has declined to insignificance worldwide because of severe limitations on the pollutant emissions, lifespan, and smooth running. Later concepts of two-stroke engines for driving passenger cars have the objective, among others, of reducing pollutant emissions and fuel consumption by turning to direct fuel injection. **Fig. T26** shows a cross-sectional representation of an 800 cm³ displacement, twin-cylinder, two-stroke gasoline engine manufactured by Orbital, which is intended to be employed as a standard-production application by Texmako, the Indonesian passenger-car manufacturer.

~Head loop scavenging. *See below*, ~Scavenging process ~~Reverse scavenging

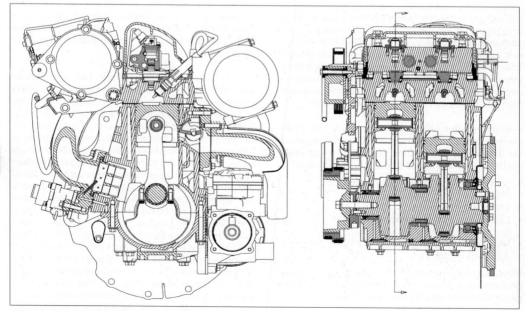

Fig. T26 Cross-sectional representation of a two-cylinder two-stroke gasoline engine for employment in passenger cars (Orbital)

~Intake manifold injection. *See below,* ~Mixture formation

~Lateral scavenging. *See below,* ~Scavenging process ~~Uniflow scavenging

~Lubrication. Two-stroke engines do not allow lubrication of the engine components by oil-sump lubrication because of fresh-gas that is ducted through the crankcase. On refueling, the lubricating oil for these engines is added to the fuel as total-loss lubrication with usual mixing ratios of 1:16, 1:25, 1:50, or 1:100. Later concepts have the supply of lubricating oil controlled by speed-dependent and—if necessary—load-dependent techniques and delivered by a metering pump into the intake port or crankcase or as a partial flow against the cylinder wall (total-loss lubrication/automatic lubrication). High consumption of lubricating oil promotes oil-derived deposits on pistons and combustion chamber walls as well as in the exhaust system, causing bluish opacity to develop (blue smoke) in the smelly exhaust gas. In two-stroke engines without a crankcase scavenge pump, the engine components can be conventionally lubricated through pressure circulation lubrication from the oil sump. The consumption of lubricating oil on these engines is still normally considerably above that of comparable four-stroke engines, particularly because of the openings in the cylinder walls for the scavenging ports and exhaust ports.

~~**Automatic lubrication.** Automatic lubrication is total-loss lubrication. The oil is in a separate vessel and is delivered by a pump into the intake manifold as a metered quantity depending on the speed or load.

~~**Fresh-oil lubrication.** As opposed to oil-sump lubrication, the engine receives fresh motor oil from a reservoir (automatic lubrication) only or in premixed form with gasoline (mixture lubrication).

~~**Lubricating oil.** Lubricating oil for two-stroke engines with mixture lubrication and total-loss lubrication differs significantly from motor oils for four-stroke engines because of the requirements requiring good solubility in fuel, low-residue and low-smoke combustion, low environmental pollution of the emitted oil and its combustion products, easy environmental degradability, and good boundary lubrication capability even if highly diluted with fuel. In addition to semi-synthetic lubricating oils, bio-oils with vegetable oil–based components are increasingly used for their good boundary lubrication capability and biological biodegradability.

~~**Mixture lubrication.** Mixture lubrication is total-loss lubrication used on small two-stroke engines in which the crankcase is used as the scavenging pump. Lubricating oil is admixed to the gasoline (normally in ratios from 1:25 to 1:100) and is delivered into the interior of the engine with the fuel. Lubrication is achieved by a fine oil vapor coating the engine parts and cylinder liners.

~~**Total-loss lubrication.** Two-stroke engines very often operate with total-loss lubrication. The characteristic is that the fresh oil is exclusively used for engine lubrication and cannot be recycled because it is partly burned along with the fuel and partly escapes through the exhaust port without being burned.

~Main fresh air supply. Main fresh air supply represents an approach for decreasing both fuel consumption and emission of pollutants, particularly in small two-stroke gasoline engines with external mixture formation, as used in small implements and two-wheeled vehicles not involving high technology. Fuel scavenging losses during the scavenging process are thereby reduced by prepositioning of air before the fuel-air mixture. A further objective of fresh gas supply is improvement of fresh-gas combustion conditions at lower part load by reducing the exhaust residuals in the cylinder and particularly in the region of the spark plug, as well as through stratified-charge effects.

~Mixture formation. While internal mixture formation is used in two-stroke diesel engines as a matter of principle, both external mixture generation (carburetor, intake manifold injection) and internal mixture formation (direct injection) are employed in two-stroke gasoline engines. The loss of fresh mixture encountered when using external mixture generation is inevitable due to scavenging the cylinder with an air-fuel mixture. For this reason, the transition to internal formation of the mixture (gasoline direct injection) is considered to be obligatory in engines with high demands on fuel mileage and exhaust emissions, such as two-stroke passenger car gasoline engines.

~~**Crankcase injection.** In this case, the fuel is either injected into the crankcase during the compression process or during transfer into the transfer duct (scavenge port injection).

~~**Cylinder wall injection.** In this the injection valve is located in the cylinder wall above one of the scavenging ducts. The fuel spray is directed against the jet of scavenging-air leaving the facing scavenging duct.

That way charge stratification (stratified scavenging process) is realized. The fuel intermixes with the scavenging air during the scavenging phase. Injection normally starts when the piston is close to TDC.

~~**Gasoline direct injection.** In conventional two-stroke engines with external mixture formation (carburetors and also intake manifold injection), a major part (up to 30%) of the fuel-air mixture, depending on the quality of the scavenging process, is passed into the exhaust because of the intermixing of fresh charge and exhaust gas or immediate bypass scavenging during the scavenging process at high engine loads. In addition, the high content of residual gas causes combustion misfires

T

at low loads. These two effects result in high hydrocarbon emissions and high fuel consumption in conventional two-stroke gasoline engines. Transition to gasoline direct injection opens up the opportunity of injecting fuel into the cylinder shortly before or after closing the exhaust components (i.e., ports or valves). The cylinder may also be scavenged more strongly for the reduction of the exhaust gas contents at part load because it is scavenged with pure air in direct injection applications. Engine operation without misfire up to idling is possible, assuming correct dimensioning of scavenging and injection processes, with the objective of positioning of flammable mixture at the spark plug/stratified charge. Fuel injection—with its corresponding advantages and disadvantages—is possible from the cylinder head, cylinder wall, or scavenging duct (semi-direct injection) when gasoline direct injection is used. The first current serial applications of gasoline direct injection on small motorcycle engines and outboard engines apply the principle of air-supported injection (Synerject; **Fig. T27**) or pressure modulation by pressure impulse (Ficht). Experiences from series production of 50 cm³ motorcycles equipped with air-supported gasoline direct injection confirm, among other things, that employment of direct gasoline injection can reduce the fuel consumption of two-stroke gasoline engines with a concurrent decrease in emission of exhaust pollutants below the level of comparable four-stroke gasoline engines.

~~**Intake manifold injection.** In this case, the fuel is either injected in a time-controlled manner into the intake duct of each individual cylinder or into the collecting pipe of several engine cylinders when collective injection is used. The injection needs to take place at a time of high air velocity. High injection pressures are advantageous to obtain fine distribution of the fuel, but the spray should not splash onto the wall of the intake manifold. Intake manifold injection simulates an ideally adjustable carburetor.

~~**Mixture formation process.** Possible mixture formation processes for two-stroke engines are:

- Mixture formation in the carburetor
- Gasoline injection
  - Direct gasoline injection
  - Intake manifold injection
  - Semi-direct injection
  - Cylinder wall injection
  - Crankcase injection
  - Scavenge port injection
- Mixture induction

~~**Scavenge port injection.** In this case, the injection valve is located in one of the scavenging ducts and the fuel is injected into the scavenging duct. Injection would normally start before the scavenging port opens.

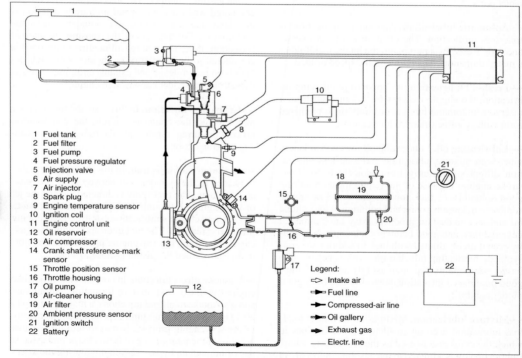

1 Fuel tank
2 Fuel filter
3 Fuel pump
4 Fuel pressure regulator
5 Injection valve
6 Air supply
7 Air injector
8 Spark plug
9 Engine temperature sensor
10 Ignition coil
11 Engine control unit
12 Oil reservoir
13 Air compressor
14 Crank shaft reference-mark sensor
15 Throttle position sensor
16 Throttle housing
17 Oil pump
18 Air-cleaner housing
19 Air filter
20 Ambient pressure sensor
21 Ignition switch
22 Battery

Legend:
⇨ Intake air
➡ Fuel line
➡ Compressed-air line
➡ Oil gallery
➡ Exhaust gas
— Electr. line

Fig. T27 Operating schematic diagram of an air-supported direct injection system for two-stroke gasoline engines (Synerject)

~~Semi-direct injection. This is applied on two-stroke engines with crankcase scavenging. The injection nozzle is located in the transfer duct. Normally, the fuel would be sprayed directly onto the hot piston crown through the open scavenging duct, where it would evaporate and intermix with the scavenging air during the following processes. This layout combines the advantages of scavenge port injection and low-pressure direct injection.

~Mode of operation *See below*, ~Operating process; *also*, →Control/gas transfer ~Two-stroke engine

~Noise. Noise from two-stroke engines is predominantly generated by:

- Pressure waves in the fluid in the intake and exhaust systems
- Pressure rise in the combustion space (structure-borne sound emission by cylinder, housing, exhaust walls, etc.)

Noises can be minimized by optimizing measures on the intake and exhaust systems and on the engine structure. The character of the noise, compared with that from a four-stroke engine, is rather high-pitched due to the fast firing order.

~Operating process. One complete working cycle is passed through each revolution of the crankshaft on two-stroke engines. **Fig. T28** shows the *p-V* diagram of a two-stroke gasoline engine and the working cylinder in schematic representation. The first stroke (bottom dead center, or BDC, to top dead center, or TDC): Inlet of the precompressed fresh gases through scav-

enging ports or intake valves into the cylinder with the aid of the scavenging pump, accompanied by exhaust gas displacement from the cylinder through the exhaust opening(s). Interruption of fresh gas supply from scavenge port closure (Ss) and from exhaust port closure (EC), then compression up to the TDC; immediately before TDC, initiation of combustion by the spark. The second stroke (TDC to BDC): Combustion and expansion of the cylinder charge, from exhaust port opening (EO) advance deflagration, and from scavenge port opening (SO) initiation of further cylinder scavenging. Between 15 and 46% of the piston stroke cannot be utilized for the generation of work because the expansion process is already completed with the outlet ports opening at between 45 and 85° before BDC for execution of the charge-transfer.

~Piston port control (piston edge control). With piston port control, intake of the fresh gases into the cylinder is controlled by the piston (piston edge control), and, depending on the scavenging process, flow of exhaust gas out of the cylinder is controlled by the piston, too. In comparison with valve control, large cross-sectional areas can be opened and closed within short periods of time so that scavenging methods using pure piston port control allow high rated speeds. Compression of the piston rings into the ports has a detrimental effect on the piston rings and port edges due to the mechanical stress. For this reason, the permissible port widths have to be limited and rotation of the piston rings prevented (risk of the ring-end entering the ports), if necessary.

~Predischarge. Predischarge is defined as that difference between the angles of opening of the exhaust ports and scavenging ports. This predischarge causes the cylinder pressure to have decreased sufficiently at the onset of scavenging to let the scavenging flow from the crankcase or scavenge-gas accumulator enter the cylinder.

~Pressure waves. *See above*, ~Gas dynamics

~Reverse scavenging. *See below*, ~Scavenging process

~Rotary intake slide valve. In order to attain high volumetric efficiency across the entire range of service speeds at full load, the first two-stroke gasoline engines were equipped with rotary valves about 100 years ago. Starting from tube and rotary sleeve valves, one design with rotary disk valve prevailed on high-performance two-stroke engines in which an axially freely adjustable, thin circle-segment disk—usually arranged parallel to the crankshaft web—controls the intake of fresh gas into the crankcase. Because the volumetric efficiency can also be enhanced and achieved more evenly by employing blade valves (reed valves) in the intake system, the significance of rotary intake slide valves on two-stroke gasoline engines has decayed in recent years.

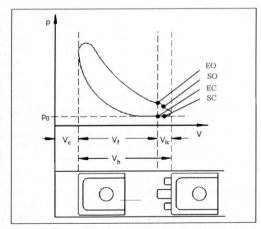

Fig. T28 *p-V* diagram of a piston-controlled two-stroke engine: V_c, compression volume; V_f, cylinder charge volume; V_h, displacement; V_s, charging loss volume through scavenging and exhaust ports; EO, exhaust port opens; EC, exhaust port closes; SO, scavenging ports open; Ss, scavenging ports close.

T

~Scavenge port injection. *See above*, ~Mixture formation

~Scavenging fan. While the pressure drop for charge transfer in four-stroke engines is developed by the motion of the piston and the exhaust and intake processes, two-stroke engines require the necessary scavenging-pressure differential to be generated by a scavenging fan. The crankcase scavenging pump represents the simplest form of scavenging fan for two-stroke engines. It uses the cyclically changing volume of the crankcase and the bottom of the piston for compressing and delivering fresh gas. The intake of fresh gas into the crankcase is controlled via the bottom of the piston or a window in the piston skirt, often in conjunction with blade (reed) valves or rotary valves. The flow of the compressed fresh gases from the crankcase into the cylinder via transfer ducts is also controlled by the piston. Ducting the working fluid through the crankcase presupposes that the main and connecting-rod (particularly the big-end) bearings are fitted with antifriction bearings featuring lower friction losses and requiring less lubricating oil than plain bearings.

The advantages of the widely used crankcase scavenging pump, particularly on small two-stroke engines, are its compact design, modest added costs, steep compressor pressure rise characteristic, and the small power required to drive it, especially at part load. The primary disadvantages are its limited volumetric efficiency and grave constraints relating to engine component lubrication and piston cooling. Moreover, the crankcases need to be sealed from each other in multicylinder engines. Crankshaft-driven superchargers are an alternative to the crankcase scavenge pump and can be either positive-displacement devices (reciprocating-piston compressors or rotary-piston superchargers) or flow devices that can be employed for scavenging or supercharging of two-stroke engines, if necessary. Mechanically driven reciprocating-piston compressors need much more space and require high production expenditure. In rotary-piston superchargers, the supply of fresh fluid or its compression is affected by the displacement effects of rotating elements or pistons. This group of superchargers includes Roots compressors, vane-type superchargers, rotary-piston superchargers, spiral-type superchargers (G-superchargers), and screw-type compressors (**Fig. T29**). Similar to reciprocating-piston compressors, rotary-piston superchargers have steep compressor characteristics, generally at medium compressor efficiencies, in which case the mass flows are approximately proportional to their driven speed. Radial compressors are preferred as flow devices (turbo-superchargers) on small engines in which the flow rate varies approximately linearly with the speed and the pressure varies approximately in proportion to the square of the driven speed. Employment of modern radial compressors could yield high compressor efficiencies yet at compact dimensions. The application of separate compressors driven by the crankshaft, geared up if necessary, offers the advantage that the crankshaft drive of two-stroke engines is

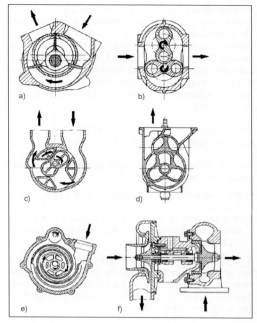

Fig. T29 Overview of various compressor or supercharging design types: (a) vane-type supercharger, (b) roots compressor, (c) rotary-piston supercharger, (d) screw-type compressor, (e) spiral-type supercharger (G-supercharger), (f) turbocharger

supported by plain bearings with pressure circulation lubrication, as on conventional four-stroke engines, and that the pistons can be efficiently cooled using oil-spray cooling and a cooling duct, if appropriate. In addition, volumetric efficiencies can easily be higher than those of the crankcase scavenging pump. The unfavorable features compared with the crankcase scavenging pump are additional assembly expenditure and the higher power required to drive the separately driven compressor. A drive mechanism with variable step-up ratio of the supercharger is additionally required for optimal adaptation of the scavenging-air volume to the particular point on the engine performance map when dispensing with the loss-prone throttle control.

~Scavenging process. The scavenging process on two-stroke engines is defined as the simultaneous removal of burned mixture from the cylinder and the introduction of fresh fluid into the cylinder. The objective is to displace the exhaust gas from the cylinder by the inflow of fresh charge without losing a significant amount of fresh mixture in the process. Many scavenging concepts are available for implementing the charge transfer, including cross-flow scavenging, reverse-flow scavenging, fountain scavenging, loop scavenging, rotary scavenging, reverse scavenging (Schnuerle/MAN), uniflow scavenging with exhaust valves, uniflow

T

scavenging with opposed pistons, and head loop scavenging, as well as various dual-piston scavenging concepts. Except for head loop scavenging, entry of fresh gases into the cylinder is made via piston-controlled ports in the cylinder wall, and outflow of the burned cylinder charge occurs via exhaust ports or exhaust valves. Schnuerle reverse scavenging and uniflow scavenging with exhaust valves have major significance in terms of current practical applications.

With Schnuerle reverse scavenging, fresh gas enters the cylinder via two to six scavenging ducts (transfer ducts) that are normally arranged in mirrored-symmetry to the central axis of the exhaust ports and in opposite direction from the outflowing exhaust gas. The scavenging flows deflect each other and form an ascending flow of fresh gases on the cylinder wall opposite the exhaust port, and this flow reverses its direction near the cylinder head, displacing the exhaust gas from the cylinder. With uniflow scavenging (with exhaust valves), fresh gas enters the cylinder through intake ports arranged over the cylinder periphery, and this displaces the exhaust gas from the cylinder, usually through several exhaust valves that are arranged in the cylinder head and operated at crankshaft rotational

speed. **Fig. T30** shows an overview of various scavenging methods with their respective pros and cons.

~~**Cross-flow scavenging.** Cross-flow scavenging is brought about by deflection of the scavenging flow toward the cylinder head by means of a deflector piston; this avoids direct losses of the fresh mixture through the exhaust duct.

~~**Lateral scavenging** →Two-stroke engine ~Uniflow scavenging (lateral scavenging)

~~**Open gas transfer.** Open gas transfer means that the scavenging duct and exhaust duct are simultaneously open over an extended period of time during the scavenging process and that the exhaust gas is mainly displaced from the cylinder by the fresh gas flowing in.

~~**Reverse scavenging.** With reverse scavenging, the scavenging ducts or valves are arranged such that the scavenging flows are pointing upward and in the direction of the cylinder wall, thus forming a self-stabilizing upward-pointing aggregate scavenging flow. This requires at least two or more scavenging ducts or valves.

Scavenging Concept	Pros	Cons
1. Reverse scavenging 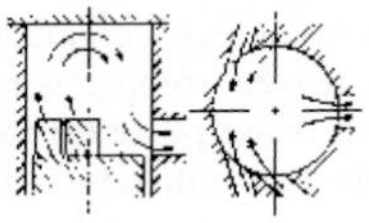	• Compact design dimensions • High revolutions possible • Combustion chamber recess can be arranged in the cylinder head well cooled • Simple design when doing without slide valve	• Asymmetric timing diagram only possible using accessory equipment (slide valve) • Asymmetric thermal piston load • Piston rings particularly endangered by scavenging ports and exhaust ports • Comparatively difficult generation of charge swirl
2. Uniflow scavenging using outlet valves 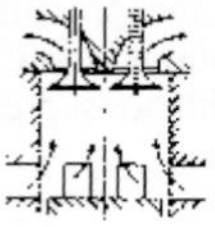	• Good scavenging efficiencies/low air requirement • Simple generation and influence in combustion-chamber swirl possible • Combustion process can be largely adopted from four-stroke engines • Asymmetric timing diagram possible without accessory equipment	• Larger design height compared to 1 • Complex/optimized timing gear required for realizing large effective strokes and low consumptions
3. Uniflow scavenging with counter piston 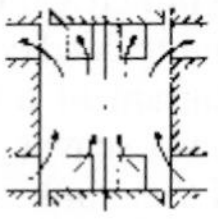	• Minimization of the combustion chamber surfaces heated up during the high-pressure phase • Asymmetric timing diagram solely possible through piston-head edge control • Good scavenging efficiencies/low air requirement	• Large assembly expenditure • Large bottom-to-top height (side-to-side width) • Extreme thermal load of the exhaust port controlling piston • No conventional combustion process applicable due to grave limitations pertaining to fuel-injector mount arrangement
4. Head reverse scavenging 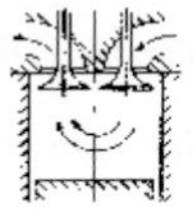	• Engine design very similar to four-stroke engines • No piston rings endangerment through scavenging ports and exhaust ports	• Low scavenging efficiencies/high air requirement • Grave increase of charge change work and consumptions at high revolutions due to limited opening time cross sections • No conventional diesel process applicable

Fig. T30 Comparison of various scavenging concepts

T

With head loop scavenging, this principle is reversed such that the scavenging flow is deflected downward by intake valves. Then the flow is deflected on the piston surface in the direction of the exhaust valves that are also located in the cylinder head. This system, too, has its intake and exhaust control elements simultaneously opened over wide ranges of crank angle, and this constitutes a risk of bypass flow in this case as well.

~~Schnuerle reverse scavenging. *See above*, ~Scavenging process ~~Reverse scavenging

~~Stratified scavenging process. The objective of the stratified scavenging process is to attain stratification between the burned gases, scavenging air and rich air-fuel mixture. The scavenging air jets leaving the main scavenging ducts separate the rich air-fuel mixture coming from the secondary scavenging ducts from the burned gases in the cylinder, thus preventing, to a great extent, the loss of fuel by the direct path into the exhaust tract during the scavenging period (bypass flow).

~~Uniflow scavenging (lateral scavenging). The fresh charge enters the cylinder through ports (very rarely through valves) during the scavenging process; the burned gas escapes through controlled valves or ports as exhaust gas. The main flow direction of travel for the fresh charge and the exhaust gas during the scavenging process is longitudinally along the cylinder axis. The employment of camshaft-actuated valves enables the so-called asymmetrical timings, relative to the dead centers, to be achieved. Therefore, because of its low mixing of fresh charge and exhaust gas, uniflow scavenging is better than the other concepts in terms of scavenging efficiency.

~Schnuerle loop scavenging. *See above*, ~Scavenging process ~~Reverse scavenging

~Semi-direct injection. *See above*, ~Mixture formation

~Smooth running. Generally speaking, the two-stroke engine features smooth engine operation because of the rapid sequence of power cycles and the relatively small degree of nonuniformity in the crankshaft rotational motion.

In two-stroke gasoline engines with external formation of the mixture, however, so-called idle misfiring (four-stroking) will occur on idling because of insufficient scavenging or too high a content of residual gas in the cylinder (no ignitable fresh charge), causing irregular engine operation as a consequence.

~Specific power output. Two-stroke engines ought to develop double the power compared with equally large four-stroke engines because they have twice as many power cycles at the same engine speed. In practice, however, a well-engineered two-stroke engine, depending on the respective scavenging process, outputs significantly less than twice the four-stroke engine power at the same displacement because of incomplete filling of the cylinder and losses during charge transfer (reduced effective piston stroke due to charge transfer).

~Stratified scavenging process. *See above*, ~Scavenging process

~Supercharging. The prerequisite for charging two-stroke engines is that fresh gases are trapped in the cylinder by premature closing of the exhaust ports/valves (asymmetric timing diagram) or turbine back pressure effects, if a turbocharger is being employed. Because exhaust valves or slides cannot be closed abruptly, and because too-early opening of the exhaust components is undesirable, there is a problem with mechanical supercharging of high-speed two-stroke engines compared with corresponding naturally aspirated engines, because a larger quantity of gas has to flow through an exhaust opening-time cross section with a reducing tendency (increased throttling losses and charge transfer work). Supercharging is basically also possible when employed in two-stroke engines with symmetrical timing diagrams. Preconditions for positive scavenging-pressure differential include high scavenging efficiencies, low-loss exhaust gas routing to the turbine (high exhaust gas temperatures), and high overall efficiency of the turbocharger. Engine operation at part load will become possible using the mechanic or electric support of the turbocharger or series connection of an additional mechanically driven compressor in series configuration aside from using pulse turbocharging or employment of compressors with adjustable turbine geometry (guide vane adjustment, sliding casing and double-helix charger). Exhaust turbocharging with charge-air cooling is successfully employed for ship propulsion using modern low-speed diesel engines with uniflow scavenging. Mean effective pressures of up to 20 bar and specific fuel consumptions below 160 g/kWh are achieved by these engines with boost pressure ratios of about 3.5.

~Timing diagram. The timing diagram shows opening and closing times of the valves or ports on the two-stroke engine. It is possible to see the periods over which the ports remain open. In two-stroke engines with conventional piston port control (e.g., reverse scavenging) the open periods are (symmetrical timing diagram) symmetrical about the engine dead centers. **Fig. T31** shows the gas-transfer diagram of a two-stroke gasoline engine with a crankcase scavenging pump and piston-head edge control of the crankcase intake.

~Total-loss lubrication. *See above*, ~Lubrication

~Two-stroke diesel engine. *See above*, ~Diesel engine

~Two-stroke gasoline engine. *See above*, ~Gasoline engine

~Uniflow scavenging. *See above*, ~Scavenging process

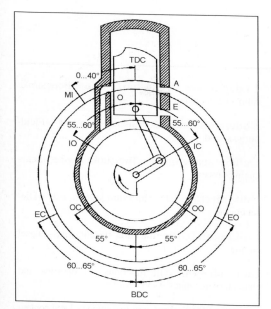

Fig. T31 Gas-transfer diagram of a two-stroke gasoline engine with crankcase scavenging pump and piston-head edge control of the crankcase intake. E, exhaust port; TDC, top dead center; EO, exhaust opens; BDC, bottom dead center; EC, exhaust closes; O, transfer duct; I, intake port; OO, transfer duct opens; IO, intake opens; OC, transfer duct closes; IC, intake closes; MI, moment of ignition.

~Wear. Signs of wear on engines will generally develop on all components that are in contact with each other and that perform relative movement against each other (bearings, piston rings, cylinder wall, etc.). In two-stroke engines with piston-head edge control of the ports, the piston rings and cylinder walls in the port region mainly are affected because of frequently passed-over port edges.

The wear process can, however, be kept low by providing sufficient heat dissipation and by having the engine components designed in a geometrically correct fashion. On the piston, signs of wear such as erosive wear occur primarily on ring grooves and pin bores; there is material fatigue due to the alternating thermal load on the piston crown; and solid-body friction between piston and bushing also causes wear.

Literature: U. Meinig: Standortbestimmung des Zweitaktmotors als Pkw-Antrieb, Parts 1–4, MTZ 62 (2001, July–Nov.) 7/ 8, 9, 10, 11. — B.S. Cumming: Opportunities and Challenges for 2-Stroke Engines, Contribution to 3rd Aachen Colloquium on Vehicle and Engine Technology, Aachen, 15–17.10.1991. — K. Mollenhauer (Ed.): Handbuch Dieselmotoren, Berlin/ Heidelberg/New York, Springer Publishers, 1997. — M. Krämer, J. Abthoff, F. Duvinage, Ch. Enderle, M. Paule, St. Pischinger, J. Willand: Der 2-Takt DE–Dieselmotor mit Common-Rail-Einspritzung als Antrieb für Personenkraftwagen, 18th International Vienna Engine Symposium 24–25 April 1997, VDI Progress Reports Series 12, No. 306, Düsseldorf, VDI Publishers, 1997. — R. Knoll, P. Prenninger, G. Feichtinger: 2-Takt-Prof. List Dieselmotor, der Komfortmotor für zukünftige kleine Pkw-Antriebe 17th International Vienna Engine Symposium 25–26 April 1996, VDI Progress Reports Series 12, No. 267, Düsseldorf, VDI Publishers, 1996. — K. Nomura, N. Nakamura: Development of a New Two-Stroke Engine with Poppet-Valves: Toyota S-2 Engine, in a New Generation of Two-Stroke Engines for the Future? P. Duret (Ed.) and Editions Technip, Paris, 1993, pp. 53–62. — K. Zinner: Aufladung von Verbrennungsmotoren, Grundlagen–Berechnung–Ausführung, 3rd Edition, Berlin/Heidelberg/New York/Tokyo, Springer Publishers, 1985. — G.P. Blair: Design and Simulation of Two-Stroke Engines, Warrendale, PA, SAE Publishers, 1996, ISBN 1-5 6091-685-0. — M. Nuti: Emissions from Two-Stroke Engines, Warrendale, PA, SAE Publishers, 1998. — C. Stan (Ed.): Direkteinspritzsysteme für Otto- und Dieselmotoren, Berlin/Heidelberg, Springer Publishers, 1999. — R. van Basshuysen, F. Schäfer (Eds.): Handbuch Verbrennungsmotor, Vieweg Publishers, 2002. — H.-H. Braess, U. Seiffert (Eds.): Vieweg Handbuch Kraftfahrzeugtechnik. Braunschweig/Wiesbaden, Vieweg Publishers, 2000. — v. Donkelaar, et al.: Moderne Zweitakt-Motorenschmierung, Sindelfingen, Expert Publishers, 1987.

Two-stroke gasoline engine oils →Oil ~classification, specifications, and quality requirements

Two-stroke piston →Piston

Type approval →Emission measurements ~Test type I, ~Test type II, ~Test type III, ~Test type IV

Type formula →Spark plug

U

ULEV →Emission limits

Unburned hydrocarbons →Emission limits; →Emission measurements; →Emissions

Uncontrolled combustion process →Glow ignition; →Knocking

Uncontrolled three-way catalytic converter →Catalytic converter ~Three-way catalytic converter

Underfloor encapsulation →Engine acoustics ~Engine noise

Underfloor engine →Engine concepts ~Reciprocating piston engine ~~Location of cylinders ~~~Horizontal engines (under-floor engines)

Underfloor noise encapsulation →Engine acoustics ~Engine noise

Underhood catalytic converter →Pollutant aftertreatment ~Aftertreatment concept with a three-way catalytic converter

Underhood encapsulation →Engine acoustics ~Engine noise

Underhood installation →Electronic/mechanical engine and transmission control

Underhood noise encapsulation →Engine acoustics ~Engine noise

Undersquare stroke-bore ratio →Bore-stroke ratio ~Undersquare

Underwater sound →Engine acoustics ~Fluid-borne sound

Uniflow scavenging →Two-stroke engine ~Scavenging process

Unit injector (UI) →Injection system, fuel ~Diesel engine

Unit pump (UP) →Injection system, fuel ~Diesel engine

Units, oil →Oil ~Oil functions, ~Properties and characteristic values ~~Viscosity

Unleaded fuel →Fuel, gasoline engine ~Lead content

Updraft carburetor →Carburetor ~Design types

US emissions test →Emission limits

Used oil →Oil ~Safety and environmental aspects

Used oil analysis →Oil ~Oil maintenance

Used oil collection →Oil ~Safety and environmental aspects ~~Used oil

U-type engine →Engine concepts ~Reciprocating piston engines ~~Single-shaft engines ~~~Twin-piston engines

V

V angle →Engine concepts ~Reciprocating piston engines ~~Single-shaft engines ~~~V-engines

Vacuum →Air intake system; →Crankcase

Vacuum advance mechanism →Ignition system, gasoline engine ~Ignition ~~Spark control

Vacuum pump. Vacuum pumps are used to generate a vacuum. While the vacuum available in the intake tract of conventionally driven spark-ignition engines can be used as simple control medium, it is not available for engines that are operated using quality regulation of the mixture. This is the case with diesel engines and compression-ignition engines, and these require employment of vacuum pumps, mostly designed as vane-type pumps. The drive can be from the camshaft, for instance, but electrically driven, demand-controlled pumps will be employed more frequently in the future to avoid permanent drive and the higher fuel consumption associated with that.

Vacuum spark advance →Ignition system, gasoline engine ~High-voltage distribution ~~Ignition distributor (conventional)

Vacuum unit →Ignition system, gasoline engine ~High-voltage distribution ~~Ignition distributor (conventional)

Vacuum-regulated crankcase ventilation →Crankcase ~Crankcase ventilation

Valve (also, →Air cycling valve). Valves are engine components, mainly poppet values, that for the most part shut off or open flow cross sections. In essence, they control the engine gas exchange processes. Other functions, such as control of the supercharging pressure and control of exhaust gas recirculation, however, are also achieved by poppet valves.

~Exhaust gas control valves. There are two essential functions for valves outside the control of the gas exchange process: valves for control of supercharging pressure (supercharging pressure control valve) and valves for control of exhaust gas recirculation (EGR valve).

~~EGR valve (exhaust gas return valve). These valves control for exhaust gas recirculation to the engine intake system (**Fig. V1**), and have the purpose of charge dilution and hence reduction of oxides of nitrogen (NO_x). EGR valves are exposed to temperatures up to approximately 800°C. In this case, 21–4 N(X 53 Cr Mn Ni N 21–9) has proved the most serviceable of the valve materials available, since in practical terms

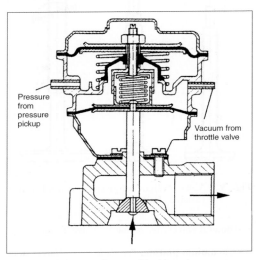

Pressure from pressure pickup

Vacuum from throttle valve

Fig. V1 EGR valve (Source: Pischinger)

the valves are exposed only to thermal stress. In addition to this, there is also corrosive and mechanical stress to a minor degree.

~~**Supercharging pressure control valve.** The supercharging pressure control valve for turbocharging, also known as a blow-off valve, limits the boost pressure of the exhaust gas turbocharger and is exposed to temperatures of approximately 1000°C in the gasoline engine for a short time, while the thermal load in the diesel engine is about 850°C. The selection of appropriate materials must be made on this basis. In general, the material 21–4 N(X 53 Cr Mn Ni N 21–91) is satisfactory for diesel engines, but high-temperature resistant materials such as Nimonic 80 A(Ni Cr 20 Ti A1) are used for gasoline engines. **Fig. V2** represents typical versions.

~Gas transfer valves. These control the gas exchange processes and seal the combustion chamber against the intake and exhaust gas sides. **Fig. V3** shows an example of an installed valve.

The intake valves are less thermally stressed than the exhaust valves and are cooled by being flushed with fresh gas as well as by thermal conduction at their seats. Exhaust valves, however, are exposed to high thermal stresses and chemical corrosion. Both valve types are manufactured from different materials corresponding to their function. It can be assumed that the valves are subjected to approximately 200 million load cycles, sometimes at very high temperatures, during the engine service life. **Fig. V4** shows the most important valve nomenclature.

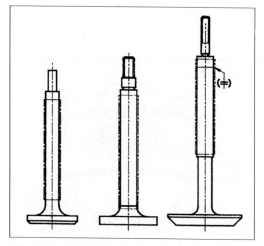

Fig. **V2** Valve designs of overrun control valves (Source: TRW)

~~**Bimetal valves.** *See below,* ~~Valve designs

~~**Chromium plating.** *See below,* ~~Valve stem ~~~Surface treatment

~~**Differential angle.** *See below,* ~~Valve head

~~**Exhaust valve.** The exhaust valve opens the exhaust ports for releasing the burnt gases at the end of the expansion stroke. Exhaust valves are exposed to high temperatures and the corroding influences of the exhaust gas. They are often of hollow design for improving heat dissipation and filled with sodium that transports heat from the valve head to the stem when in liquid state.

~~**Hollow chamfer.** The transition from the valve stem to the valve head is called a hollow chamfer. The design has to consider the bending stress and abrasion through corrosion. The design of the hollow chamfer has a large effect on the inherent rigidity of the valve.

~~**Hollow head valve.** *See below,* ~~Valve designs ~~~Hollow stem valves

~~**Hollow stem valves.** *See below,* ~~Valve designs

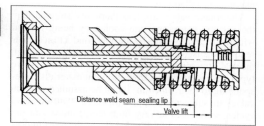

Fig. V3 Hollow stem valve when installed (Source TRW)

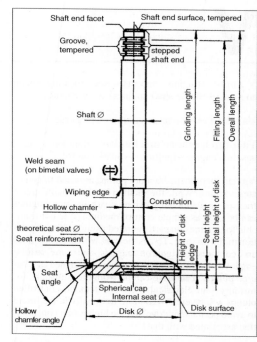

Fig. **V4** Designations on a valve (Source: TRW)

~~**Installation length.** *See below,* ~~Valve stem

~~**Intake valve.** The intake valve opens the intake port to the combustion chamber in the cylinder head. Thereby air or mixture can flow into the combustion chamber. The low temperature of the air or mixture, compared to the exhaust gas, results in the intake valve being stressed less than the exhaust valve.

~~**Monometal valve.** *See below,* ~~Valve designs

~~**Seat angle.** *See below,* ~~Valve seat

~~**Sodium-filled valves.** The hollow space in the valve stem is filled with sodium to approximately 60% of the volume when hollow stem valves are employed for lowering the valve temperature. The melting point of sodium is 97.5°C so that it is in liquid form during engine operation. Heat transfer is made by the sodium from the thermally highly stressed valve head into the stem through the oscillating valve motion.

~~**Special valves.** Motor sport poses the highest demands on valves where it is essential that they withstand extreme stresses temporarily. For instance, very light valve gears—therefore, also, light valves—are required to reach extremely high speeds in motor sports.

In addition to the use of hollow head valves, a further step to weight reduction is selection of the material. For instance, titanium allows weight reduction of about 40% compared with steel, but it does not have a very high thermal stability. Hence, particularly intensive thermal dissipation has to be applied when using titanium as exhaust valve material; this can be achieved by using hollow head valves in connection with high-temperature conducting seat rings.

~~Surface treatment. *See below,* ~~Valve stem

~~Valve designs. In essence, valves are subdivided into three main groups: bimetal valves, hollow stem valves, and monometal valves.

~~~Bimetal valves. Bimetal valves provide an ideal combination of materials that is optimal for stem and head. They usually have a hot-worked head, based on the above-mentioned methods, which is then connected to the stem by friction welding (**Fig. V5**).

The preferred material pairings are X53CrMnNiN219, X50CrMnNiNb219, X60CrMnMoVNbN2110, NiCr20TiAl for the valve head and X45CrSi93 for the valve stem.

The weld seam on the head should be drawn such that it is one-half lift inside the guide with the valve closed or 6 mm above the scraper lip. The length of the cylindrical part must be at least one and a half times the stem diameter for production-engineering reasons. Bimetal valves can also be armored on the seat.

~~~Hollow stem valves. These are employed mostly on the exhaust side. Sometimes, under special circumstances, they are used on the intake side to lower the temperatures, predominantly in the hollow chamfer and head region, and also for weight reduction. Distinctions can be drawn between valves with hollow stems and valves with hollow heads.

Hollow stem valves are those with their stems provided with a hole that is in line with the valve axis and is externally permanently closed. In case of hollow head valves, the valve disk is hollow.

The sodium required for heat transfer moves freely in the hollow space of the valve stem. Part of the heat generated on the hollow chamfer and valve head is conducted through the liquid sodium to the valve guide and released to the cooling circuit (**Fig. V6**).

The hollow space is filled up to approximately 60% with metallic sodium when hollow stem valves are employed for lowering of temperature. Depending on engine speed, the liquid sodium (melting point 97.5°C) has a corresponding "shaker effect" in the hollow valve space, transporting heat from the valve head into the valve stem. The lowering of temperature that can be attained at optimal heat dissipation and least running clearance is 80–150°C.

Hollow valve variants:

- "Pipe on full" design: A hardenable valve-stem end piece (full) is fixed by means of friction welding onto the basic body (pipe) that was drilled from the end of the stem.
- "Close drawn" design: Manufacturing this variant is much more expensive than the previous design. The basic body is also drilled from the end of the stem. Closing of the bore is achieved by inductive heating with subsequent "cutting to size." The valve-stem end piece is attached by means of friction welding. Close-drawn hollow stem valves are predominantly used for high-performance and aircraft engine applications.
- Hollow head valve: This valve represents a further measure for weight reduction and thermal dissipation from the center of the valve head. As opposed to the previously described methods, these valves are drilled and machined from the disk side. The opening is closed by inserting a cover using a special method. These valves, which are expensive to manufacture, are predominantly employed in motor sports (**Fig. V7**).

Hollow stem valves can be manufactured from a 5-mm stem diameter on up. The bore diameters are about 60% of the stem diameter. The bore of the valve has to end approximately 10 mm before the running path of the sealing lip so as not to expose the valve stem seals to excessively high temperatures. Attention has

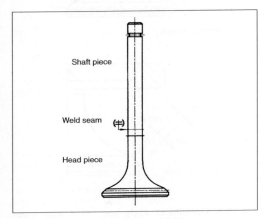

Fig. V5 Bimetal valve (Source: TRW)

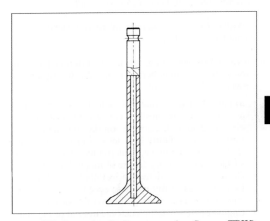

Fig. V6 Hollow valve (hollow stem valve; Source: TRW)

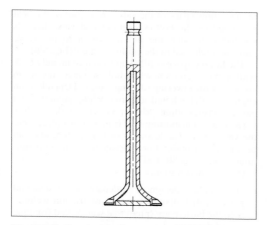

Fig. V7 Hollow head valve (Source: TRW)

to be given to the different clearance between the valve stem and the valve guide compared to solid valves. Valve sticking is reduced by employing slightly conical stems to compensate for the temperature gradient.

Hollow stem valves can be monometal ones; however, it is more common to have bimetal valves with following material combinations: head piece X53CrMnNiN219, X50CrMnNiNb219, and NiCr20TiAl; stem piece X45CrSi93.

~~~Monometal valves.
Monometal valves can be manufactured according to the hot-extrusion method and the upsetting method. The starting point in case of the hot-extrusion method is a rod section having a diameter of about two-thirds of the finished disk diameter and a length that corresponds to the volume of the blank to be manufactured.

When using the upsetting method, a ground rod section having a diameter slightly larger than the valve stem diameter is heated on one end and formed to sort of a "pear" by pushing the rod forward, and this pear is then die-formed to become a valve head.

~~Valve disk.
The underside of the valve head is designated as valve disk.

~~Valve failures.
There are many different types of valve failure. Among them, it is possible to distinguish between failures due to:

- Manufacturing or material defects: Among these are, for example, mechanical processing deficiencies, hardening deficiencies on the stem and in the turned groove, faulty heat and surface treatments, faults in producing the blank, faulty microstructure, surface flaws, wrong degree of material purity.
- Design flaws and defective installation such as too loose or tight valve guides, eccentricity between valve guide and seat ring, faulty valve design.
- Engine operation deficiencies: Thermal or mechanical overload, for example, through hot corrosion,

changes of the microstructure, lack of valve rotation, unbalanced stress on cam follower arms or rocker arms, valve clearance not optimal, shutoff faults.
- Influence of fuel and lubricating oil—for instance, due to sulfur.

~~Valve head.
The valve head performs the sealing function together with the valve seat insert. The basis for the valve design is the theoretical valve seat diameter. The total disk height depends on the respective combustion pressure and the mean component temperature on the valve. This temperature should be determined by means of, for example, an FE (finite element) analysis. Practice shows that values between 7 and 10% of the valve head diameter are standard.

The disk height determines the rigidity of the valve head and depends on the valve seat angle: at 45°, approximately 50% total disk height, and at 30°, about 55–60% total disk height.

Generally the seat angle is 45°, but seat angles of 30 and 20° are also selected to reduce the seat wear. Small seat angles are indispensable in gas engines. A difference of at least 5° is necessary between seat angle and hollow chamfer angle (**Fig. V8**).

A differential angle between valve seat and seat ring helps seal the combustion chamber side of the valve more efficiently through initial linear contact. The valve seat width must be greater than the seat ring bearing width.

Spherical caps on the valve disk surface are provided for purposes of weight reduction, combustion-chamber control, or as a distinguishing feature between the intake and exhaust valve or similar valves.

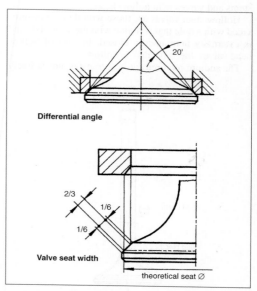

Fig. V8 Differential angle and valve seat width (Source: TRW)

V

An optimal design of the transition from the hollow chamfer to the valve stem can only be found from tests on the engine.

~~**Valve length compensation.** During driving under various operating conditions, considerable elongations will occur compared to the installed condition at room temperature; this is caused by different thermal expansions of the cylinder head and the components for valve control and the valves themselves. Engineering solutions that result in no malfunctions caused by the above-mentioned elongations are called valve length compensation.

~~**Valve mass.** Valve mass represents the oscillating masses that exert the inertia force brought about by the acceleration. This inertia force influences the dimensions of valve gear, camshaft, and valve spring and exerts an influence on the frictional behavior and noise. It is, therefore, necessary to keep the valve mass as low as possible. This can be done via the materials employed and valve design. **Fig.V9** shows a comparison between valve masses.

~~**Valve materials.** Valve materials need to withstand particularly high and multiple operational demands. The major properties required of valve materials are:

* Strength
* Wear resistance
* Corrosion resistance
* Thermal conductivity
* Temperature resistance

The demands made on a valve include sufficient dynamic strength at elevated temperatures, resistance to wear, resistance to high-temperature corrosion and oxidation, as well as resistance to corrosion.

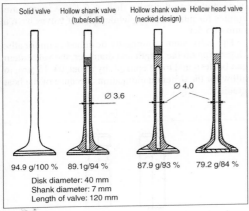

94.9 g/100 %   89.1g/94 %   87.9 g/93 %   79.2 g/84 %

Disk diameter: 40 mm
Shank diameter: 7 mm
Length of valve: 120 mm

Fig. V9 Comparison of valve masses of different valve designs (Source: TRW)

Standard valve materials are:

* Ferritic-martensitic valve steels: X 45 Cr Si 9 3 is used as standard solution for monometallic intake valves, and in case of bimetal valves, it is employed exclusively as material for the stems. X 85 Cr Mo V 18 2 is a higher-alloyed material and employed as material for intake valves, where the load level both in thermal and mechanical regard does not allow using the Cr-Si material.
* Austenitic valve steels: Austenitic CrMn steels have proven themselves as an economical solution in this case. Material X 53 Cr Mn Ni N 21-9 (21-4N) has gained large currency to be considered as the classic material for exhaust valves and also for hollow stem valves.
* Valve materials with high nickel content: If the CrMn steels do not suffice, it is necessary to resort to materials with a high nickel content. This is required where utmost operational safety is of the essence, such as in fracture and corrosion resistance (aircraft engines, racing applications, highly supercharged diesel engines, and heavy-fuel service).

Special materials in the form of powder-metallurgical (PM) valve steels are available. This material has qualities that have positive effects on strength and resistance to hot corrosion. The technological properties of valve steels can be enhanced by specific heat treatments. This can save resorting to premium-quality alloys in many cases. Martensitic valve steels are generally annealed. In case of austenitic steels, both hardness and strength can be enhanced using so-called dispersion hardening.

~~**Valve rotation** →Valve rotating devices

~~**Valve seat.** The sealing face between port and combustion chamber is of conical design and serves as the valve seat. The cone angle of this surface is called the seat angle. Seat angles of 45 or 30° are common. A large seat angle helps the sealing effect but causes higher operating forces on opening as well as higher compressive stress on the valve-seat surface. Smaller seat angles reduce the compressive stress but also the sealing effect of the valve seats.

The seat of the exhaust valve is thermally stressed and also suffers corrosion. For this reason the valves are armored with special alloys. This method is also carried out for intake valves in isolated cases although material-conditioned martensitic hardening would normally be applied in this case.

~~~**Seat hardening.** Seat wear on intake valves made from martensitic materials is reduced by inductive seat hardening. Care should be taken to ensure that seat temperatures do not exceed 550–600°C in order to maintain sufficient hardness under operating conditions.

~~~**Valve seat armoring.** Valve armoring designates the welding application of wear-resistant materials onto locations that are particularly exposed to wear.

The valve seat-faces are usually "armored." The valve stem butt ends are also armored, but more rarely.

Armoring can reduce wear and enhance the sealing efficiency. Methods of valve armoring include:

- The fusion welding processes in which the armoring rod material is applied using an acetylene oxygen flame
- The electric plasma-transferred Arc (PTA) process in which the pulverized armoring material is fluxed on in a plasma arc and applied to the workpiece.

These armoring methods are applied for hollow stem valves, bimetal valves, and sometimes also for monometal valves. Care should be taken to ensure that valve temperatures of 550 to 600°C are not exceeded to keep the hardening decay of the inductively hardened valve seat within acceptable limits.

~~**Valve stem.** The cylindrical part of a valve is designated as the valve stem and is designed to guide the valve, transmit power, and dissipate heat. The stem is reduced at the first turned groove for locking the valve collets and also at the scraper lip or transition to the hollow chamfer. A scraper lip is fitted on the exhaust port side by reducing the stem diameter in order to limit the build-up of oil-derived deposits (**Fig. V10**). The scraper lip is about one-half lift inside the valve guide when the valve is closed. The weld should rest in the valve guide in the event that there is bending in the valve during the closing process—for example, by cylinder-head distortion or a nonaxial fault. That is why the friction welding seam is run at least one-half lift into the valve guide.

Depending on the tribological conditions, it is necessary to finish the surfaces of the valve stems by chromium-plating or nitriding.

~~~**Installation length.** The installation length is the distance from the theoretical valve seat to the groove center on the valve stem.

~~~**Surface treatment.** The following techniques are employed:

- Hard chrome-plating of the valve stem: The production process, material selection, and operating conditions may possibly necessitate standard valves to be chromium-plated on the valve stem running zone. A chrome layer of 3–7 μm covers both valve materials in case of standard bimetal valves. In the truck or large-engine field, stronger chrome layers, up to 25 μm, can be employed in case of high loads or increased wear.
- Polish-grinding: The stem has to be polish-ground, in the case of a chromium-plated surface at any rate, for the removal of adherent chromium buds and to equalize unevenness. The roughness will be a maximum Ra 0.2 (nonchrome-plated maximum Ra 0.4) after the polishing operation; this has a very favorable impact on valve guide wear, allowing minimum guide clearance.
- Nitriding the valves: Both bath- and plasma-nitriding are applied. The nitriding layers, approximately 10–30 μm thick, are very hard on the case (about 1000 HV 0.025) and particularly wear-resistant. Valves subjected to salt-bath nitriding are polish-ground as chromium-plated valves are.

~~~**Valve stem butt end.** The stem butt ends of valves with multiple turned grooves, for the purpose of supporting free valve rotation, are always inductively hardened in the region of the cone-locating piece to avoid wear. For the same reason, small hard-metal plates or hardenable stem material are welded onto the end of the stem in case of very high surface pressures through the valve actuator. Valves with single grooves are rarely hardened but also in this case little hard-metal plates or hardenable materials are welded onto the stem end for wear protection. **Fig. V11** shows types of valve grooves.

The measurement from stem end to groove center must not fall below 2.5 mm. The chamfered edge on the stem end is either a chamfer under 45 or 30° or rounded off to make automatic valve installation easier.

~~~**Valve stem diameter.** The valve stem diameter is in proportion to the valve disk. A 6:1 disk:stem ratio applies for intake valves, while exhaust valves have a ratio of 5.5:1.

The valve stem is normally designed cylindrically. Depending on the length and diameter, the valve stem can be designed to be conical, by about 10–15 μm, to allow for the different elongations because of the heat gradient.

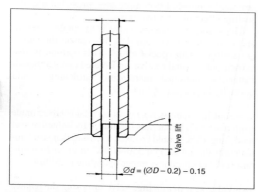

Fig. V10 Valve stem with constriction and scraper lip (Source: TRW)

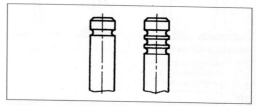

Fig. V11 Groove types in valve stems (Source: TRW)

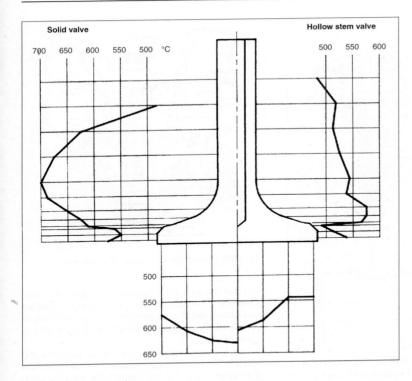

Fig. V12
Typical temperature
profiles for two valve
designs (Source: TRW)

~~**Valve stem seal.** This serves to ensure sufficient stem lubrication and prevent excessive flow of lubricating oil. Spring-loaded elastomer cups are normally employed.

~~**Valve temperature.** The valve materials are dependent on the thermal stress of the valve, among other factors. The anticipated thermal load can be determined by means, for instance, of temperature measuring valves or thermocouples. **Fig. V12** shows typical temperature profiles for two valve designs.

~~**Versions.** *See above,* ~~ Valve designs

~~**Wiping edge.** *See above,* ~~ Valve stem

~Overrun control valve. *See above,* ~Exhaust gas control valves ~~Supercharging pressure control valve

Literature: TRW Thompson GmbH & Co. KG: Handbuch, 7th issue, 1991. — R. Milbach, TRW Thompson GmbH & Co. KG: Ventilschäden und ihre Ursache, 5th issue, 1989.

Valve acceleration (*also,* → Valve gear). The gas transfer valves are accelerated or decelerated on opening and closing. High valve accelerations are required to attain large time-area integrals. The permitted valve acceleration diagram is determined primarily by the rigidity and mass of the moved and power-transmitting parts of the valve control mechanism. The rigidity of all the power-transmitting parts needs to be as high as

possible, and the mass of all moving parts has to be minimized if a high permissible valve acceleration is to be achieved.

Valve actuation (*also,* → Valve gear). Mechanical systems are commonly used for valve actuation. In these the valve is actuated through intermediate components by a cam that is mounted on a camshaft. Systems are being developed now that allow electromagnetic or electrohydraulic actuation of the valve. These designs have the advantage that they do not require a camshaft. **Fig. V13** shows an example of electromechanical valve actuation.

Valve angle → Valve arrangement

Valve armoring → Valve ~ Gas transfer valves ~~ Valve seat

Valve arrangement. The design layout of the valves influences nearly all significant engine parameters such as power, torque, maximum engine speed, combustion, and exhaust gas emissions, and also combustion chamber geometry, engine weight, and maintenance and production costs.

~Lateral valves. As opposed to overhead valves, lateral valves are the economic design variant for valve control with low overall height (**Fig. V14**). The cost advantages result from a low number of components

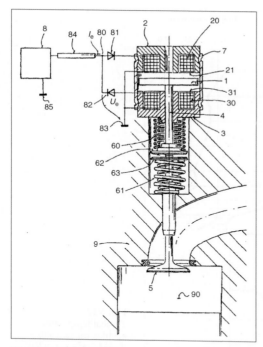

Fig. V13 Valve actuation with an electromechanical actuator (Source: Published German patent application DE19852168)

and a simple cylinder head. The disadvantages of this design are low volumetric efficiency and, thus, low power, slow and incomplete combustion due to unfavorable combustion chamber geometry, and, hence, poor thermal efficiency and a lower level of exhaust-gas quality.

~Laterally arranged valves. Laterally arranged valves do not yield optimal combustion chamber designs (surface-volume ratio) and are, therefore, no longer employed in passenger car engines.

~Number of valves. The number of valves has an impact on nearly all engine parameters. Four-stroke engines on the market have two, four, five, or six valves per cylinder.

One intake valve and one exhaust valve are often employed in cylinder heads when a particularly inexpensive variant has to be manufactured. Except for enhancing the mixture flow rate, two intake valves bring about an extension of the degrees of freedom for generating particular charge movement—for example, through different valve lift patterns of the two valves or through shutting the intake port. Engines with four valves—two exhaust valves and two intake valves—are now the most common engine type manufactured, whether diesel and gasoline.

The objective of considering the number of valves when designing an engine is, for example, a cylinder

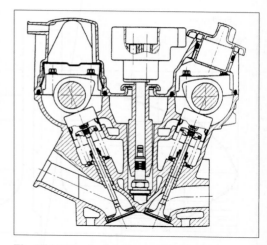

Fig. V14 Lateral valve (Source: Pischinger)

head that can be built as compactly as possible, large aspiratory cross sections for gas exchange, compact combustion chamber in gasoline engines resulting in short flame travel and central location of spark plug or injector nozzle.

Such systems can be described using geometric parameters. This enables optimization of such systems to give, for example, the maximum valve opening cross sections. For gasoline engines these geometric parameters are, for example,

- Valve area ratio, φ, reflecting the total intake area versus total exhaust area ratio
- Valve angle on exhaust and intake sides
- Spark plug inclination angle and diameter
- Displacement of the spark plug center from the cylinder center
- Diameter of the cylinder bore
- Height of the cylinder head
- Distance of the valves from each other, the spark plug, and the cylinder wall

~~Five valves. A cylinder head with five valves has two exhaust valves and three intake valves. When considering the geometric relations, additional special aspects become apparent that contrast with what is observed with the numbers of valves considered above.

The respective valve angle of the "center" valve is different from that of the two outer valves. It must be observed whether all three intake valves are directed toward the camshaft center and have the same length. Conditional on the smaller inclination angle of the center intake valve, its valve disk or its seat ring is no longer positioned on the same level as that of the other two intake valves (**Fig. V15**). Five valves per cylinder are employed rather rarely. Important quantities here are:

- The surface ratio, φ, of the intake valve to the exhaust valve

V

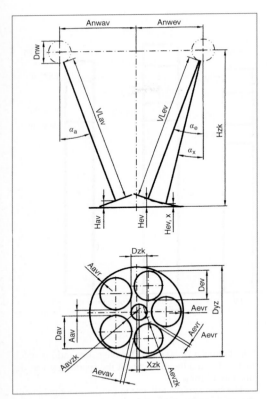

Fig. V15 Schematic representation of the geometric conditions on a cylinder head with five valves

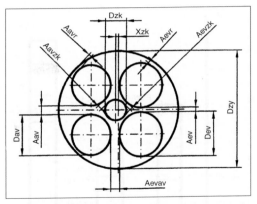

Fig. V16 Schematic representation of the geometric conditions on a cylinder head with four valves

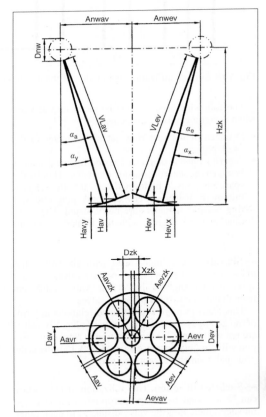

Fig. V17 Schematic representation of the geometric conditions on a cylinder head with six valves

- Valve angle of exhaust, α_a, and intake, α_e
- Height of the cylinder head, H_{ch}
- Diameter of the camshaft base radius, D_{nw}
- Diameter of the cylinder bore, D_{cy}
- Diameter of the spark plug, D_{sp}
- Distance between the seat ring outside diameter of the intake valves, A_{iv}, and the exhaust valves, A_{ev}
- Distance between the intake-side seat ring outside diameter and cylinder bore, A_{ivr}
- Distance between the intake-side seat ring outside diameter and spark plug, A_{isp}
- Distance between the intake-side seat ring outside diameter and exhaust-side seat ring outside diameter, A_{eav}
- Distance between the exhaust-side seat ring outside diameter and cylinder bore, A_{evr}
- Distance between the exhaust-side seat ring outside diameter and spark plug, A_{evsp}

This parameter set allows, for example, determination of size and coordinates of the individual valves, coordinates of the spark plug, valve length, gradient of the center intake valve, and distances of the camshafts.

~~Four valves. The engine with four valves has two intake and two exhaust valves (**Fig. V16**). The influ-

ence of the spark plug on valve size decreases from four valves. Rather, the distance of the valves from one another is of higher importance. The spark plug is given a more or less central position.

V

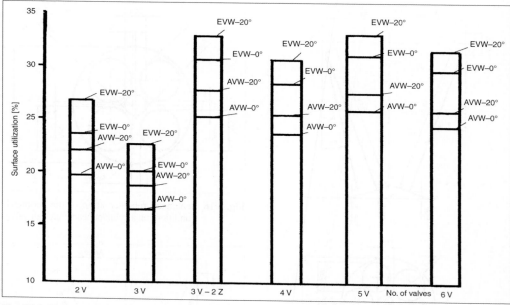

Fig. V18 Surface utilization as function of the number of valves

The following aspects must be taken into account when designing a four-valve engine:

- Distance of the seat ring outside diameter to the spark plug
- Seat ring outside diameter considering the distances between the seat ring outside diameters. Hereby the spark plug has to be located such that the selected minimum distances between spark plug and seat ring outside diameters are not violated. The spark plug may be aligned with the intake or exhaust valve.

~~Six valves. A cylinder head with six valves has three intake valves and three exhaust valves. The respective valve angle of the "center" valve is different from that of the two neighboring valves.

This leads to more sophisticated relations in terms of the geometric position of the spark plug. Six valves are not usual, because the exhaust and refill cross section will possibly decrease again. **Fig. V17** reflects the principal geometrical representation.

~~Surface utilization. The utilization of the surface can be assessed by using appropriate geometric parameters, including the valve distances to each other, to the spark plug, and to the cylinder wall as well as the given spark plug diameter. Surface utilization is thereby defined as the quotient of overall intake and exhaust area divided by the cylinder bore area.

Fig. V18 shows, as an example, the surface utilization for the intake valve and exhaust valve angles of 0 and 20° in percent. The relations are presented for a

bore diameter of 80 mm. The valve area ratio is specified as 1.2 and the diameter of the spark plug was defined as 14 mm.

The best surface utilization for the marginal conditions selected is obtained for an engine with five valves and a valve angle of 0°—that is, the aspiration cross sections will reach their maximum. The surface utilization moves toward the cylinder head with three valves and two spark plugs at rising valve angle. As the example of a cylinder head having six valves shows, the maximum possible aspiration cross section does not necessarily increase with a growing number of valves.

Fig. V19 gives an overview of surface utilization against the diameter of the cylinder bore. The surface ratio, φ, is 1.2 and the valve angles are 20°.

In relation to the aspiration cross sections, it is not a good idea to employ six valves in the range of the cylinder bores for automobile engines. A similar rule applies for a cylinder head with five valves for bore diameters greater than approximately 90 mm.

It should be mentioned that an arrangement with three valves and two spark plugs makes for highly efficient utilization of the aspiration cross section. The exact and individual consideration of web widths, which were not varied for reasons of clarity, at this point may, however, shift the amount slightly but will not change the principle.

Surface utilization with four valves will drop below an approximate bore diameter of 100 mm. This is because, in this case, the distances between seat ring outside diameters are not decisive for the valve area, but rather for the selected distance of the outer seat ring di-

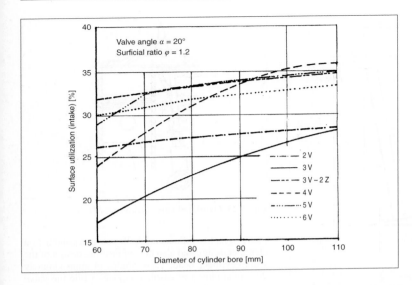

Fig. V19
Surface utilization as
function of the
cylinder-bore diameter

ameters to the spark plug. This applies similarly to the
cylinder head with five valves from a bore diameter of
about 70 mm onward.

~~Three valves. A cylinder head with three valves
has two intake valves and one exhaust valve. There are
two variants: one with one spark plug and the second
with two spark plugs. The exhaust valve is larger than
either intake valve, but the overall intake valve cross
section is larger than that of the exhaust.

~~~Three valves and one spark plug. The spark plug
takes a nearly central position close to the cylinder axis
and about central from the three valves (**Fig. V20**).

~~~Three valves and two spark plugs. The objective in the case of a cylinder head with three valves and
two spark plugs can be, for instance, to achieve:

- Maximum areas for the intake and exhaust valves,
  minimum combustion distances from both spark
  plugs (dual ignition)
- Different ignition points for both spark plugs

The principal geometric conditions can be taken from
Fig. V21.

~~Two valves. The cylinder head with two valves
often has only one camshaft. The valves are normally
positioned more or less vertically in the cylinder head

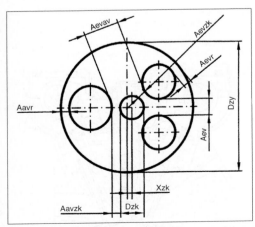

Fig. V20 Schematic representation of the geometric
conditions for a cylinder head with three
valves and one spark plug

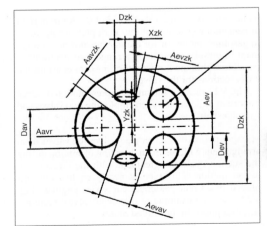

Fig. V21 Schematic representation of the geometric
conditions on a cylinder head with three
valves and two spark plugs

V

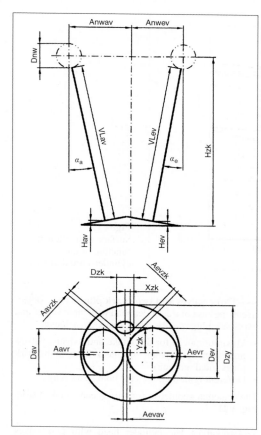

Fig. V22 Schematic representation of the geometric arrangement on a cylinder head with two valves

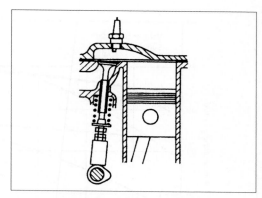

Fig. V23 Overhead valves (Source: *MTZ*)

plane—that is, with a valve angle of 0°. An important dimensioning criterion for the spark plug arrangement in gasoline engines or for the injector nozzle arrangement in diesel engines is to arrange it as centrally as possible within the limitations of maximum aspiration cross sections for the valves.

Fig. V22 shows a schematic representation of some important dimensions when employing two valves and two camshafts.

~Suspended valves. Overhead valves are valves in an arrangement with the main direction of motion during the suction cycle in agreement with the direction of piston motion. This type is nearly exclusively employed in current passenger-car series engines. **Fig. V23** shows an example of overhead valves: there are still exceptions in the United States.

~Valve angle (relative to cylinder axis). The valve angle is termed as that angle that encloses valve axis and cylinder axis. The valve angle is an important geo-

metric quantity influencing both port and combustion chamber design and also the engineering design of the whole cylinder head. Valve angles of approximately 20° are common in gasoline engines while the valve angles on direct injection diesel engines are usually between 0 and about 10°. Older engine designs show an angle of around 0° also on gasoline engines, for the exhaust valve at least. **Figs V24 and V25** show valve angles in modern gasoline and diesel engines.

Valve bonnet →Valve cover

Valve bonnet modules →Sealing systems ~Modules

Valve chrome-plating →Valve ~Gas transfer valves ~~Valve stem ~~~Surface treatment

Valve clearance (*also*, →Valve). Valve clearance is defined as the intended distance between cam heel and cam follower (or other elements in power transmission

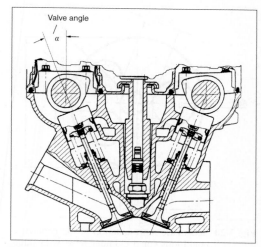

Fig. V24 Valve angles in a gasoline engine

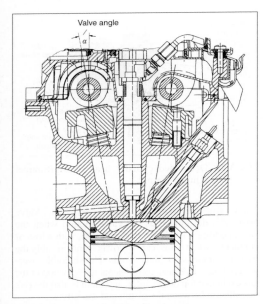

Fig. V25 Valve angles on a diesel engine

between the cam and the valve). The valve clearance compensates for wear and different thermal expansion of the valve gear components, the cylinder head, and the engine block.

~Hydraulic clearance compensation. Hydraulic valve-clearance (tappet clearance) elements are employed for automatically compensating longitudinal expansion and wear. They consist of an oil-filled piston-cylinder unit, a return valve, and an adjuster spring. **Figs V26 and V27** show hydraulic clearance compensation elements.

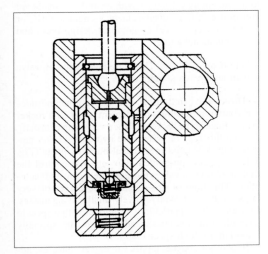

Fig. V26 Hydraulic compensating element

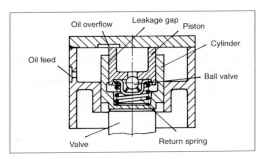

Fig. V27 Hydraulic overhead bucket tappet

Inside a housing there is a piston with integrated return valve; both elements can be moved relative to each other, forming a defined leakage gap on the contact surface, and they are pushed apart by an internal spring.

With the return valve closed, during valve lift a high pressure will build up inside the space confined by the housing and piston. A very small quantity of oil will reach the storage space in the piston through the very narrow leakage gap. With the valve closed and the cam heel snug, the inside spring pushes the hydraulic elements apart until the valve clearance is balanced. This generates a differential pressure opening the return valve and there will be fresh supply of oil required for the compensation process. The advantages of hydraulic compensation are:

- No measurement or adjustment work required during cylinder head installation because the hydraulic element compensates all tolerances
- No maintenance
- Constant timing at all operating points
- Acoustic advantages through low opening and closing ramps and low opening and closing velocities

The oil circulation has to satisfy certain requirements relating to the oil pressure and aeration to make these advantages available and, in particular, compressibility has to be avoided. Moreover, it is necessary that the cam heel be manufactured with narrow form tolerances.

~Mechanical valve clearance adjustment. Valve clearance (tappet clearance) setting is made with screws, adjusting shims, or overhead bucket tappets. The required setting accuracy has to be taken into account when dimensioning the ramps on valve opening and closing. Wear-related enlargement of valve clearance can be counteracted by resetting the play. Changes of clearance due to thermal expansion (cold/warm engine) cannot be compensated. These conditions lead to significant variation of clearance, requiring steep ramps with high opening and closing velocities; the consequence is a change of valve timing with all its negative repercussions.

The advantages of mechanical valve clearance adjustment against comparable valve gear components with hydraulic clearance compensation are:

- Higher stiffness
- Reduction of frictional losses
- Cost advantages

Fig. V28 shows mechanical options of clearance adjustment.

Valve clearance compensation →Valve clearance ~Hydraulic clearance compensation

Valve closure velocity →Variable valve control

Valve cone →Valve keeper

Valve control →Control/gas transfer ~Four-stroke-engine; →Valve gear; →Variable valve control

Valve cover. Valve covers cover and seal the valve gear against the upper side of the cylinder head. In general, there is an elastomer seal between the valve cover and cylinder head; this is frequently inserted into a component groove to avoiding overloading. This type of seal allows sealing of strongly deforming components at approximately 20 to 30% compression degree. In addition to safe sealing against oil, the valve cover is an acoustically relevant component. Therefore, careful design of this component (e.g., ribbing) is important, and it is also beneficial to obtain acoustical decoupling between cylinder head and valve cover. The requirements imposed on such acoustically decoupled systems are:

- Decoupling of structure-borne noise
- Safe component screw-fitting
- Sealing
- Preassembly of the component parts

Fig. V29 shows such a decoupled system for a cylinder head cover.

Valve deactivation →Variable valve control

Valve designs →Valve ~Gas transfer valves

Valve differential angle →Valve ~Gas transfer valves ~~Valve head

Valve disk →Valve ~Gas transfer valves

Valve failures →Valve ~Gas transfer valves

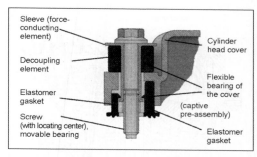

Fig. V29 Decoupled cylinder head cover (Source: Klinger Elring)

Valve floating (bounce) (*also*, →Valve). Valve floating is designated as the valve motion when the valve no longer follows the cam profile, with a loss of contact between cams, cam followers, and possibly the transfer elements at high speeds. Considerable overloads of the valve gear components will develop at the gas exchange top dead center, and there is also the risk of collision between the piston and the valves.

Valve gear (*also*, →Mass balancing mechanism). Poppet valves are nearly exclusively used on combustion engines for control of charge exchange. The mechanism for actuating the valves is called the valve gear (**Fig. V30**). The valve gear has the task of opening and closing the valves in synchronism with the crankshaft rotation, and this requires the employment of transmission elements (valve gear components) from the crankshaft to the valves. In the case of high-speed and high-power engines, this presupposes a system with low elasticity. Engines with overhead camshafts (single overhead camshaft: OHC; double overhead camshaft: DOHC), which can be considered to be the technological state of the art, offer these features in automobile engines. Engines with central camshafts are still in use in large-displacement V-engines. OHC or DOHC concepts excel by compact design types and economic manufacture.

~Actuation of valves. There are two types of valve actuation: direct and indirect.

~~Direct valve actuation. The direct driveline does not contain a transmission element, such as a rocker arm or finger-type rocker between valve and camshaft. Among this type of system are hydraulic bucket tappets (**Fig. V31**) and mechanical bucket tappets (**Fig. V32**); and column-guided elements that control the motion of several valves through direct camshaft actuation. This kind of drive enables the design of very "rigid" valve gears with relatively small moving masses. Low elasticity is particularly important with respect to strength at high engine speeds, and this is especially important for high-speed engines.

In addition, bucket tappet valve gears have the advantage of low design height for the cylinder head.

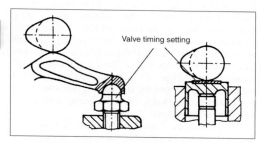

Fig. V28 Mechanical valve clearance compensation

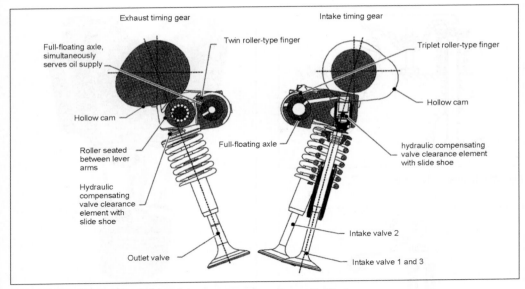

Fig. V30 Valve gear of an Audi five-valve engine (Source: Braess/Seiffert)

Bucket tappets can be found in many different applications, including two- or four-valve gasoline and diesel engines.

~~Indirect valve actuation. The indirect valve gear has intermediate transmission elements between camshaft and valve—for example, in form of finger-type rockers or rocker arms. **Fig. V33** shows a schematic diagram.

~Design types. On conventional combustion engines, valve actuation is made by the camshaft that rotates at half crankshaft speed on four-stroke engines. Valve gear can be classified by:

- The number of valves: A range from two to five valves is sufficient in large-series production.
- The number of camshafts: one or two camshafts.
- The position of camshafts: overhead camshafts (**Fig. V34**) or central camshafts (**Fig. V35**).

- Systems without camshaft: for example, electromechanical valve actuation is at the development stage, and here valve actuation is made electromagnetically (**Fig. V36**).

~Elastic valve gear. Simplified treatment of the valve gear as a solid valve train is not accurate enough in many cases. Therefore, the elasticity of the individual elements has to be taken into account for more detailed tests. This leads to a single-mass replacement system for flexible valve actuation analysis. There can be considerable differences in the behavior compared to that of the solid valve gear, as **Fig. V37** shows.

~Fully variable valve gears →Variable valve control

~Gear components. In this context, all components that are necessary to ensure functioning of the valves

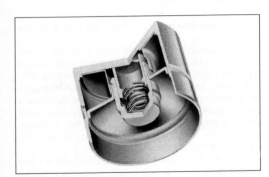

Fig. V31 Hydraulic bucket tappet (Source: INA)

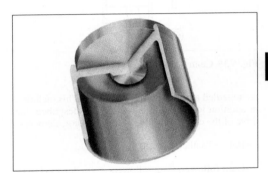

Fig. V32 Mechanical bucket tappet (Source: INA)

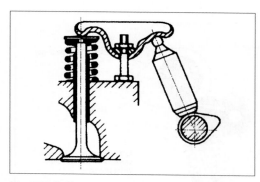

Fig. V33 Indirect valve gear

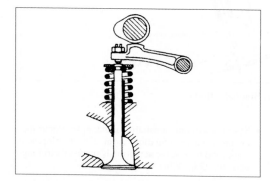

Fig. V34 Overhead camshaft

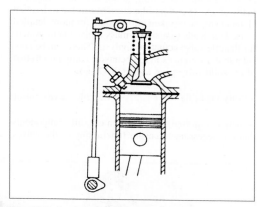

Fig. V35 Central camshaft

are regarded as valve gear components. This includes, in addition to the components already described, the drive of the camshafts and the transmission elements.

~~Belt →Toothed belt

~~Cam follower. Cam followers are understood to be all elements that transform the contour defined by the

Fig. V36 Electromechanical valve actuation

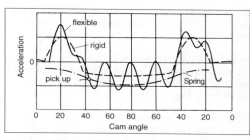

Fig. V37 Valve gear acceleration (Source: Pischinger)

cam on the camshaft into the lifting motion of a gas transfer valve. Cam followers can be, for example, rocker arms or cam followers, bucket tappets, or roller tappets.

~~Camshaft →Camshaft

~~Camshaft bearing →Bearings ~Bearing positions

~~Camshaft drive. The drive of the camshaft is normally performed using chains or a toothed belt—gears are sometimes used for specialty applications. **Fig. V38** shows, for example, the drive in a W-type engine using a chain: transmission using double chains on a central intermediate shaft is made from a double chain sprocket attached to the crankshaft. Further transmission is made from there using a simplex chain to the right or left cylinder bank in each case.

Fig. V39 shows a drive with toothed belt. The toothed belt, starting from the crankshaft, wraps the drive for the auxiliary aggregates and the camshaft gear wheels.

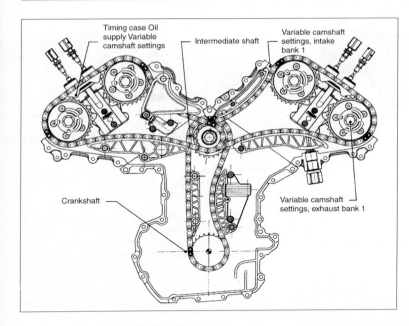

Fig. V38
Camshaft drive by chain
(Source: *MTZ*)

~~Finger-type rocker. *See below,* ~Indirect valve gear
~~Rocker arm valve gear

~~Follower arm. *See below,* ~Indirect valve gear
~~Rocker arm valve gear

~~Pushrod. There are many pushrod designs, and these are adapted to their respective applications. In addition to the bucket tappet (the direct driving element), pot tappets, mushroom tappets, flat tappets, and roller tappets serve as transfer elements to indirectly drive the valves. In addition to these rigid push rods, there are

push rods with hydraulic valve clearance compensation. The advantages of the latter are better acoustics and maintenance-free design, but disadvantageous are lower rigidity and great mass. The preferred application of roller tappets is in diesel engines, due to higher load.

~~~Bucket tappet. *See above,* ~Actuation of valves
~~Direct valve actuation

~~Rocker arm. *See below,* ~Indirect valve gear
~~Rocker arm valve train

~~Tappet rod. These may be manufactured either from solid material or from steel pipes. In the case of solid material, the tappet or the rocker-arm rests are formed by upsetting the ends; pressure pieces are put on in the case of steel pipe.

~~Timing chain →Chain drive

~~Twin roller chain →Chain drive ~Multiple chain

~~Valve guide →Valve guide

~~Valve spring →Valve spring

~~Valve-stem seal →Valve stem seal

~Indirect valve gear. This group of valve gears includes rocker arms or finger-type rockers; the latter are also termed cam followers.

~~Cam follower. *See below,* ~~Rocker arm valve gear

~~Cam follower drive. *See below,* ~~Rocker arm valve gear

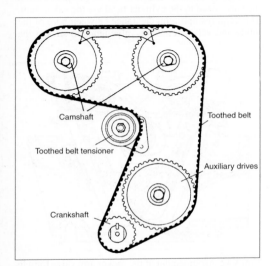

Fig. V39 Camshaft drive with toothed belt

V

Fig. V40 Roller-type finger with hydraulic supporting
element

~~**Rocker arm valve gear.** The rocker arm or cam
follower is characterized by having bearings at one
end while the other rests on the valve. Engagement
with the camshaft takes place either in the middle of
the lever or on the valve-side end. A roller-type finger
is employed in many cases to reduce friction, particularly in the lower speed range (**Fig. V40**). A contact
line is reached between the cam follower and camshaft
via a roller pivoted in the finger. This leads to a significant reduction, up to 30%, of the valve gear frictional
torque compared to a cam sliding on the cam follower.
The disadvantage, however, is the reduced damping of
the camshaft torsional vibrations, which has an effect
on the drive elements—that is, the chain or belt.

The advantages over the rocker arm valve train are
smaller forces and automatic valve clearance compensation without increasing the elasticity of the system.

Finger-type rockers can be economically manufactured from sheet metal or by precision casting methods; the latter allow large creative scope for constructional development (stiffness, inertia). However, the cost
advantages of a cam follower manufactured from sheet
metal are so big that precision-cast cam followers are
employed in special circumstances only. On the whole,
roller-type fingers are less stiff than bucket tappets.

~~**Rocker arm valve train. Fig. V41** shows examples of rocker arm valve trains; on the left, a forged one
and on the right, one made from metal sheet. A typical
characteristic is that their bearings are in the middle of
the rocker arm. The cam of the camshaft acts on one
side of the rocker arm, either directly or via suitable push
rods, and on the other side of the rocker arm pushes on
the valve. The high supporting forces in the pivot point
place severe requirements on the bearings. The point
of application of force should preferably be aligned
with the valve axis to keep the shearing forces loading
the valve guide as small as possible. High stiffness,
low mass, favorable sliding conditions, and minimiza-

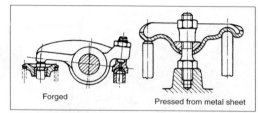

Forged   Pressed from metal sheet

Fig. V41 Rocker arm valve train (Source: Pischinger)

tion of rocker arm manufacturing cost are the main
consideration in design. **Fig. V42** shows a rocker arm
valve train on a four-valve engine.

~Solid valve gear. The solid valve gear model is a
method for calculating forces by or onto the valve gear
components. Excepting the valve spring, all the valve
gear components are thereby assumed to be rigid—
that is, without elasticity. This allows for checking the
frictional locking between the valve and cam. All involved masses—that is, valve spring force as well as
moments of inertia—are included in this calculation,
and the quantities involved have to have their values
reduced to the cams, for instance.

Valve gear friction →Friction

Valve guide. The valve guide has to guide the valve
such that it is perfectly positioned in the seal seat of the
valve seat insert. This guiding function needs lubrication that is delivered through the clearance between
valve stem and valve guide; it is also necessary to
maintain the oil consumption as low as possible. Combinations of materials featuring a certain inherent lubrication behavior are advantageous. Furthermore, the
valve guide ensures that heat can be dissipated from
the valve head to the cylinder head via the valve stem.
This requires optimum clearance between bore of the
guide and valve stem. The valve tends to stick if the

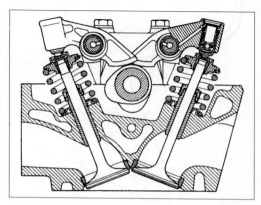

Fig. V42 Arrangement of the rocker arm valve train in
the cylinder head (Source: Pischinger)

V

clearance is too narrow, but excessive clearance would obstruct heat removal. The aim should be to obtain the minimum possible valve guide clearance. Also, the end of the valve guide should not project into the exhaust-gas stream, otherwise there is the risk of valve guide expansion and entry of combustion residues into the guide bore. Proper functioning of the valve requires maintaining the offset between the centers of the valve guide and seat ring within certain limits. Excessive offset creates, in the main, a strong deflection of the valve disk relative to the stem. This excessive load can cause early failure, and other consequences can be leakage, poor heat transfer, and high oil consumption.

~Center offset. *See below,* ~Design

~Design. Valve guides have cylindrical geometries with their ends differently formed depending on the detailed design.

The outside diameter of the valve guide has to be adjusted to the cylinder head bore. Interferences from 0.02 to 0.05 mm are used on cast iron cylinder heads, and from 0.04 to 0.08 mm for aluminum cylinder heads are usual for the press fit in the cylinder head [3]. The minimum wall thickness for powder-metal valve guides is 1.8 mm; generally, the larger the wall thickness, the shorter the valve guide. The inside diameter of a nonfitted valve guide is usually unmachined; it is finished in the cylinder head simultaneous with seat ring machining. This minimizes the possible offset of centers between valve guide and seat ring. Values of offset for a new engine are in the region of 0.02 to 0.03 mm [2].

Use of the maximum possible installation length is basically advantageous on valve guides for keeping valve tipping tendency as low as possible; the length should be at least 40% of the valve length [2].

~Installation length. *See above,* ~Design

~Materials. The essential materials or material groups employed for valve guides are:

- Powder metal materials. These can be ferrous or nonferrous materials. Alloying constituents of Cu, P, and Sn are usual for the ferrous materials, because these improve the dimensional accuracy, thermal conductivity, and mechanical properties. Oil is embedded in the pores of the sintered material to produce better lubricating behavior.
- Nonferrous materials. Materials on a copper base are preferred in this case. These materials play, however, no role in large-series production.
- Nonferrous metals. These are preferably used as wrought material on copper base (Cu-Zn compounds).
- Cast iron/cast steel. Cast-iron valve guides on a ferrous basis are employed particularly in commercial vehicle engines.

An important feature of all materials used for valve guides is resistance to wear. This resistance exhibits itself particularly on the valve guide ends, and the port

side shows higher wear than the cam side because of the higher thermal load. Thermal conductivity is important for reduction of the valve temperature, particularly for exhaust valves. Thermal conductivities of ferrous materials are between 6.2 and 7.1 W/mK and for nonferrous metals are in the region of at least 8 W/mK.

A high coefficient of thermal expansion for the material employed is decisive for the strength of the bond between the cylinder head and valve guide, but it also influences the valve clearance. Usual values of thermal expansion are between 9 and 22×10^{-6} depending on the material employed. Nonporous materials are advantageous due to their high-percentage contact area, brought about by their low density.

~Operational stress. The load on the valve guide comes from forces that are introduced into the valve guide via the valve stem from tilting of the valve. The forces consist of [1]:

- Friction force, F_F, on the valve-seat surface
- Shearing force, F_S caused by the valve spring
- Normal force, F_N, which also acts on the valve-seat surface
- Gas force, F_G, on the valve disk

The bending moments resulting from these forces have to be counteracted by forces at both ends of the valve guide. **Fig. V43** shows the forces on the valve that exert the operating stress on the valve guide.

The oscillating valve motion leads to pressure build-up inside the valve guide at both ends. There will be a solid contact shortly after reversal of the valve motion, which will subsequently reverse from static friction back to sliding friction. The operating stresses inside the valve guide are primarily caused by:

- Valve gear: Compared with overhead bucket tappets, rocker arms exert up to five times higher side

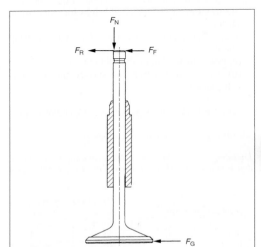

Fig. V43 Forces on the valve (Source: Bleistahl)

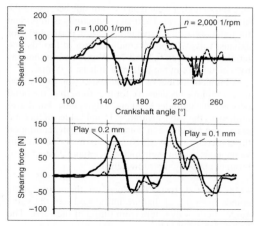

Fig. V44 Shearing forces at different speeds [1] (engine operating, valve clearance 0.1 mm, valve guide clearance 45 μm, oil temperature 50°C, rocker-arm type valve gear)

forces. **Fig. V44**, top, shows the typical shear force pattern of a rocker arm valve train on a valve guide.

- Valve clearance: An increase of the valve clearance by, for example, 0.1 mm could cause an increase of the shearing force by 22% [1]. **Fig. V44**, bottom, shows the conditions on a rocker arm at different valve clearances, a speed of 1000 rpm, a valve guide clearance of 45 μm, and an oil temperature of 60°C.
- Valve guide clearance: The ideal aim is to have the minimum possible valve guide clearance for safe guidance and exact positioning of valve seat and seat ring. This will result in better heat transfer and decreases the risk of valve canting. **Fig. V45** summarizes some recommended values for valve guide clearances.
- Valve: The main impacts on valve guide wearing quality are:
(a) heat transfer from valve stem to guide
(b) valve stem material and surface finish (valve, valve stem)

~Valve guide clearance. *See above,* ~Operational stress

~Wall thickness. *See above,* ~Design

Literature: [1] M. Meinecke: Öltransportmechanismen an den Ventilen von 4-Takt-Dieselmotoren, FVV, Final Report

Shank Diameter in mm	Valve Guide Play in μm	
	Intake	Exhaust
6–7	10–35	25–50
8–9	20–30	35–60
10–12	40–70	55–85

Fig. V45 Guide values for valve guide clearances [2]

Project No. 556, Institute for Frictive and Mechanical Engineering, Technical University of Clausthal, 1994. — [2] A. Linke, F. Ludwig: Handbuch TRW Motorenteile, TRW Motorkomponenten GmbH, 7th issue, 1991. — [3] N. Funabashi, et. al.: US-Japan PM Valve Guide History and Technology, Proceedings of the International Symposium on Valve Train Systems Design and Materials, ASM, 1998.

Valve hardening →Valve ~Gas transfer valves ~~Valve seat ~~~Seat hardening

Valve head →Valve ~Gas transfer valves

Valve installation length →Valve ~Gas transfer valves ~~Valve stem ~~Installation length

Valve keeper. The task of valve keepers is to connect valve spring plate and valve such that the valve spring will always hold the valve at its required position. Drop-forged valve keepers for valve stem diameters up to 12.7 mm are the state of the art, and material grades C 10 or SAE 1010 are used for these.

The valve keepers can be subdivided according to their function into:

- Clamping connection by which frictional locking between the valve, valve keeper, and spring seat is obtained
- Nonclamping connection that allows free valve rotation

Clamping connection. Clamping valve keepers (collets) transfer the force through frictional engagement. This requires a gap between the valve keeper halves. Valve keepers with cone angles of 14, 15, and 10° are used. Valve keepers with smaller cone angles produce tighter clamping and are, therefore, particularly suitable for engines with very high speeds. Carburized (480–610 HV 1) or nitrided (≥400 HV 1) valve keepers are recommended for highly stressed clamped joints. **Fig. V46** shows an example for the installed position of a clamping valve keeper.

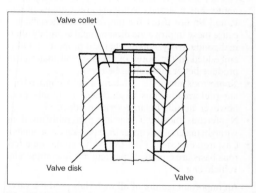

Fig. V46 Installation principle for valve keepers with clamping connection (Source: TRW)

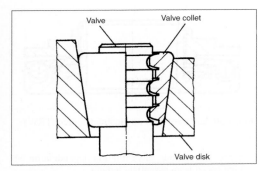

Fig. V47 Installation principle for valve keepers with nonclamping connection (Source: TRW)

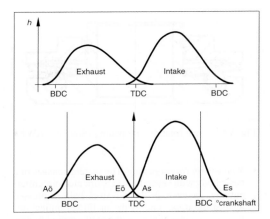

Fig. V48 Valve lift curve

Nonclamping connections. Connection is made using valve keepers (collets) having a 14°15' cone angle. When installed, the valve keeper halves support each other on the inner attachment faces. This renders a clearance between the valve keepers and the valve stem, thus making valve rotation in the spring seat possible. Thereby the resonance-borne excitation for valve rotation, the eccentric action of the rocker arm on the valve stem end, and the pulse caused by overhead bucket tappet rotation are supportive.

In the case of nonclamping connections, the forces are transferred through three to four beads. That is why case hardening of the valve keepers is necessary. **Fig. V47** shows an example for the installed position of a nonclamping valve keeper.

~Production. Valve keepers (collets) are drop-forged from shaped strip steel. Multigroove valve keepers are carburized and ground on their joint faces. Other designs can be nonhardened, carburized, or nitrided, according to choice. Production-conditioned and depending on design, the exterior jacket can be shaped concavely around the center region by up to 0.06 mm— a convex exterior jacket is not permissible.

The cone length of the spring seat has to be long enough so that the valve keepers will not project from any side when in firmly installed condition. The cone-shaped shell must on no account have a convex shape and should serve as the reference face for the geometric tolerances of the valve spring plate.

Valve length compensation →Valve ~Gas transfer valves

Valve lift →Control/gas transfer ~Four-stroke engine ~~Variable valve timing ~~~Valve-lift curve; →Valve lift curve; →Variable valve control ~Variation parameters

Valve lift curve. The representation of the valve lift against the camshaft or crankshaft rotation (usually in degrees) is called valve lift curve (**Fig. V48**). It is possible to integrate and differentiate valve lift curves

for simple analysis. The integral of the lift curve over the angle of rotation produces the so-called valve area integral. The first derivative of the lift curve with respect to angle of rotation is valve velocity; the second derivative of the lift curve with respect to angle of rotation is valve acceleration.

Valve lubrication →Lubrication ~Oil demand

Valve mass →Valve ~Gas transfer valves ~~Valve mass

Valve materials →Valve ~Gas transfer valves

Valve opening duration →Control/gas transfer ~Four-stroke engine ~~Open period

Valve opening time →Control/gas transfer ~Four-stroke engine

Valve overlap →Control/gas transfer ~Four-stroke engine ~~Variable valve timing; →Variable valve control ~Variation parameters

Valve push rod →Valve gear ~Gear components ~~Pushrod

Valve rotating devices. Regular valve rotation is of decisive importance for proper functioning of the valve. This equalizes the valve head temperatures and avoids leaks through distortion. In addition, it tends to prevent deposits on the valve seat. Forced rotating devices are employed only in cases in which natural valve rotation is insufficient—for example, in large engines or at low speeds.

At higher speeds the torque generated through compression and relief of the valve springs is sufficient to cause the valves to rotate. Valve rotation is assisted by so-called nonclamping valve keepers.

~Function. Valve rotating devices work on two principles:

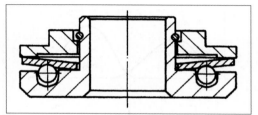

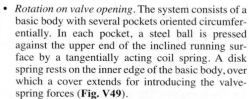

Fig. V49 Valve rotation on opening (Source: TRW)

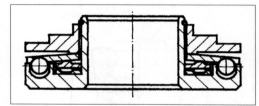

Fig. V50 Valve rotation on closing (Source: TRW)

- *Rotation on valve opening.* The system consists of a basic body with several pockets oriented circumferentially. In each pocket, a steel ball is pressed against the upper end of the inclined running surface by a tangentially acting coil spring. A disk spring rests on the inner edge of the basic body, over which a cover extends for introducing the valve-spring forces (**Fig. V49**).

  The disk spring is flattened by the increasing valve-spring force when the valve opens. In so doing, it forces the balls in the pockets of the basic body to roll on their inclined running surfaces, which is itself rolling on the balls. The disk-spring pressure against the inner edge of the basic body decreases because of support by the balls so that this will cause sliding. The cover and disk spring, however, are connected by frictional forces so as to introduce no torsion. The relative rotation between the disk spring/cover and basic body is transferred to the valve through the cover, valve spring, spring seat, and valve keepers. When the valve closes, the disk spring and the balls are relieved and are pushed back to their initial position without rolling by the tangential springs.

  It has to be remembered that the spring ends of a coil spring will twist in the opposite direction when compressed and return to their initial position when relieved. This rotational effect will continue, and the balls in the basic body pockets are inserted such that the effects from rotation device and valve-spring twist will be additive when the valve opens, but the valve-spring return rotation is solely effective on valve closing. The difference between the two rotations results in the actual valve rotation angle per lift.

- *Rotation on valve closing.* This principle is preferred as an upper installation because its function there is less impaired by soiling (**Fig. V50**). The function of this valve rotation device is the opposite of the valve rotating mode at valve opening.

  Both types can generally be used both as lower and upper designs, and the lower configuration is preferred in high-speed engines because it does not increase the inertia forces of the valve gear.

  In the case of the upper configuration, the rotation device replaces the spring seat. It is employed in low-speed engines or in cases in which the lower configuration cannot be accommodated for lack of space. Continuous valve rotation dependent on engine speed is an essential element.

~Roto cap. Valve rotation device in which the valve rotation occurs on valve opening.

~Rotomat. Valve rotation device in which the valve rotation occurs on valve closing.

Valve rotation →Valve rotating devices

Valve seal →Valve stem seal

Valve seat →Valve ~Gas transfer valves

Valve seat angle →Valve ~Gas transfer valves ~~Valve stem

Valve seat hardening →Valve ~Gas transfer valves ~~Valve seat ~~~Seat hardening

Valve shutoff. It may be desirable to shut off cylinders to reduce fuel consumption and minimize the emission of pollutants. The basic idea of this downsizing is to enhance the specific work of individual cylinders in the part-load range in order to run these cylinders at an operating point with lower specific fuel consumption. Therefore, cylinders are shut off as compensation at certain driving situations—for instance, during city journeys. Eight-cylinder and 12-cylinder engines particularly lend themselves to the cylinder shutoff concept. These engines allow shutting off of half the cylinders at low load and speed demand. The cylinders to be shut off can be identified from the engine firing order, based on the requirement of maintaining an even firing order when in shut-off service. Eight-cylinder engines lend themselves to shutting off two cylinders from each bank, while one whole bank can be disabled on a 12-cylinder engine.

There are disadvantages if the cylinder shutoff is solely achieved by interrupting the fuel feed. These include increased flow losses, cooling of the shut-off cylinders, and possible problems with the lambda control. An assessment of techniques ends in showing that cylinder shutoff using valve disablement is the only reasonable solution.

The possibility of changing the charge movement (valve gear, valve stoppage) represents a further advantage of valve shutoff.

V

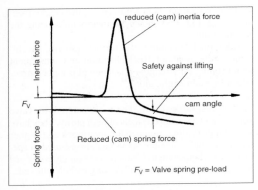

Fig. V51 Masses and spring forces on valve gear

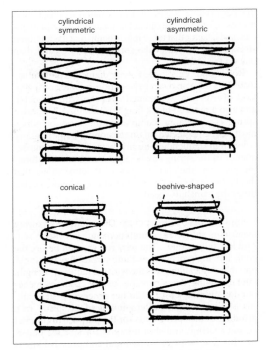

Fig. V52 Types of designs for valve springs

Valve spring. Valve springs are used to close the intake and exhaust valves. Their task is to keep the stationary valve closed and the accelerating valve in contact with the cam or rocker arm. **Fig. V51** illustrates the masses and spring forces on the valve gear. In this case the spring force has to be high enough to have sufficient reserves to maintain the friction locking in the valve gear: acceleration, valve gear elasticity, as well as possible overspeeding have to be considered, too.

~Beehive spring. *See below,* ~Design types

~Conical spring. *See below,* ~Design types

~Cylindrical spring. *See below,* ~Design types

~Cylindrically asymmetric spring. *See below,* ~Design types

~Design types. Valve springs for the intake and exhaust are almost exclusively coil springs with round or oval cross section. Established design types of manufactured valve springs are (**Fig. V52**):

- Cylindrical coil springs as single, double, or triple springs
- Cylindrical asymmetric coil spring
- Conical spring
- Beehive spring

Asymmetrically wound springs have less dynamic spring mass. The tighter pitches are directed toward the cylinder head for the purpose of corresponding progression: quality-assuring measures prevent faulty installation. The spring mass of a conical valve spring is less than that of a cylindrical spring. In addition, the block height is slightly smaller so that a smaller spring seat can be used on the valve. The disadvantage is less progression compared to a cylindrical spring.

The "beehive spring," made up of a cylindrical and a conical part that abuts the spring seat, is employed if the integrated valve stem seal prevents a conical design of the whole spring. Also, this design type has

advantages pertaining to dynamic masses. Moreover, the required progression is adjustable over the cylindrical part of the spring.

To minimize space and weight, valve springs are dimensioned for high maximum and fluctuating loads, which demands high-strength, pure materials to achieve a high number of stress cycles. The so-called pneumatic spring (air spring) valves have established themselves in racing engines because the performance of metal valve springs is limited by material characteristics. In the case of air springs, a confined gas volume serves as the elastic medium.

~External valve spring. When using two valve springs per valve, it is possible to distinguish between the external and internal valve springs. The result of using two valve springs is a compact design. A further advantage is that the engine can continue operating at low speed with a broken spring.

~Internal valve spring. *See above,* ~External valve spring

~Load on the valve spring. The valve spring, usually a cylindrical coil spring, is subjected to bending and torsion under load. The torsional load is decisive for fatigue strength dimensioning—that is, the spring loading is limited by the permissible torsional load. The permissible spring force is, among others, a function of

V

the permissible torsional load (material, production process), spring wire diameter, mean spring coil diameter, and a factor that takes into account that the torsional stress on the inside of the spring wire is greater than that on the outside.

Round and "egg-shaped" wire diameters are primarily used. Using an oval cross section, a more consistent load distribution is obtained over the wire diameter, and also a reduced block height is obtained. The valve spring is a dynamic component the mass of which must not be neglected on dimensioning it, particularly when high frequencies are involved. The resonant frequency of the spring can be estimated from the vibrating mass and the spring stiffness. Practice shows that the resonant frequency of the spring should be ten times the camshaft speed.

~Manufacture, valve spring. The initial valve spring material is subjected to high demands in order to ensure low failure rates. The wire rod is pared before the cold-drawing process to exclude surface flaws as far as possible. Annealing gives the wire the required strength; after annealing, the wire is checked for cracks by means of eddy-current sensors. Further processing steps are: stress relief annealing after winding of the spring, face grinding of the spring ends for the purpose of parallel spring seating, and shot peening for surface densification and build-up of residual compressive stress in the subsurface zone. The residual compressive stresses are superimposed on the tensile stresses that

occur in operation, hence preventing the cracks from spreading. **Fig. V53** shows the factors influencing the fatigue limit.

Additional annealing can further enhance the fatigue limits by about 10%. Nitriding and subsequent shot peening for strength enhancement is a good practice for certain applications.

~Materials, valve spring. The usual materials for valve springs are CrSi-alloyed steels of high purity. They have fewer high-melting, nonmetallic inclusions and higher tensile strength. Titanium alloys can also be used in special cases. High-strength wires are increasingly employed as well, wires which are CrSiV- or CrSiNiV alloyed.

~Pneumatic valve spring (air spring). Conventional valve springs represent a mechanical system having a speed limit of approximately 14,000 rpm if they are to comply with the geometrical limitations. Pneumatic valve springs in conjunction with light valves (titanium) are the solution for many motor sport applications if higher power is to be reached through an increase in engine speed. **Fig. V54** depicts the pneumatic valve spring principle

~Resonant frequency, spring. See above, ~Load on the valve spring

~Spring characteristic. The valve spring characteristic is defined as the spring force plotted against the

Product Form / Production Step		Factors of Fatigue Strength				
		Degree of Purity	Surface	Characteristic Mechanical Values	Structure	Residual Stress
Liquid steel	Smelting and refining	•		•		
Slab/block	Casting	•				
Billet	Hot-roll	•	•			
Wire rod	Hot-roll	•	•		•	
	Peeling		•			
	Patenting			•	•	
Valve spring wire	Cold-drawing		•	•		
	Final oil hardening and tempering			•	•	
Valve spring	Winding		•			•
	Stress relief annealing			•	•	•
	Spring end face grinding					
	Shot peening		•	(•)		•
	Warm pre-setting					•

Fig. V53 Factors influencing the fatigue limit of valve springs [1]

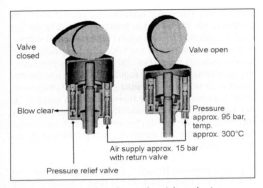

Fig. V54 Pneumatic valve spring (air spring)

Literature: [1] T. Muhr: Zur Konstruktion von Ventilfedern in hochbeanspruchten Verbrennungsmotoren, Diss., RWTH Aachen University, 1992.

Valve stem →Valve ~Gas transfer valves ~~Valve stem

Valve stem seal (*also*, →Sealing; →Valve ~Gas transfer valves). Valve stem seals are in general spring-loaded elastomer cups. Sufficient oil supply is necessary for building up a hydrodynamic lubricant film in the valve stem and guide contact zone, so valve stem seals have to run defined quantities of oil through the stem sealing on the one hand, but they also need to prevent increased passage lubricating oil and, thus, increased oil consumption. The typical characteristic values for the oil flow are 0.07–1 cm³/100 h. The oil flow rate depends, among other things, on the port-side pressure conditions—for instance, on supercharged engines. **Fig. V56** shows the arrangement of a valve stem seal.

Valve surface treatment →Valve ~Gas transfer valves ~~Valve stem

Valve temperature →Valve ~Gas transfer valves

Valve throat →Valve ~Gas transfer valves ~~Hollow chamfer

Valve timing →Control/gas transfer ~Four-stroke engine ~~Timing

Valve with external drive →Air cycling valve ~Design

Valves (sodium filled) →Valve ~Gas transfer valves ~~Valve designs ~~Hollow stem valves

spring deflection. The characteristic of the load deflection curve is dependent on the design and dimensioning of the spring. Valve spring characteristics are dimensioned progressively rising at high spring rate for good dynamic properties.

~Spring loading. *See above*, ~Load on the valve spring

Valve spring plate. The valve spring plate is an element that couples the valve spring with the valve via the valve keepers. It serves at the same time as the supporting surface or support for the valve spring. They are usually produced using an extrusion method. Certain manufacturing tolerances need to be complied with in the manufacturing process because spring seats are normally not reworked. **Fig. V55** is an example of a valve spring plate.

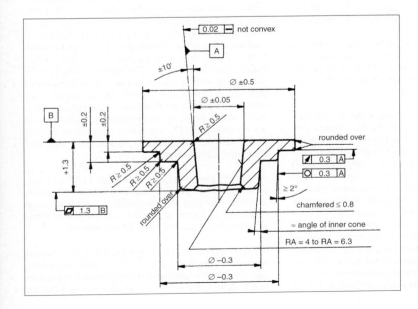

Fig. V55
Valve spring plate
(Source: TRW)

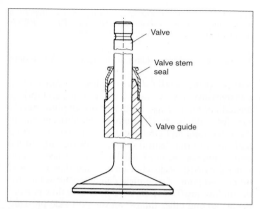

Fig. V56 Arrangement of the valve stem seal

Vane-type supercharger →Supercharging ~Fan

Vapor bubbles →Fuel, gasoline engine ~Vapor lock

Vapor lock →Fuel, diesel engine; →Fuel, gasoline engine ~Vapor lock

Vapor pressure →Fuel, gasoline engine

Vaporization, fuel (*also*, →Fuel, gasoline engine ~Vapor lock). Fuels for conventional engines are liquid and consist of a mixture of hydrocarbons. Burning the fuel in an engine requires conditioning and intermixing with air. The fuel is transformed from the liquid to the gaseous phase by evaporation and can be mixed more or less homogeneously with air. This conditioning either takes place in the engine combustion space (diesel engine, direct injection gasoline engine) or outside the combustion space (intake manifold injection).

~**Boiling characteristic.** The boiling characteristic or volatility is determined from the boiling curve and vapor pressure. It is one of the most important criteria for assessing gasoline fuels that change to the vapor state between 30 and 205–210°C.

~**Boiling curve.** When performing a boiling analysis according to DIN EN ISO 3405, the fuel sample employed is evaporated with variable power heating at a specified temperature rise of 1°C/min and then condensed. The resultant boiling curve is very informative for the application of the fuel in the engine. A well-balanced boiling characteristic is an essential precondition for operating any motor vehicle with a gasoline engine under all the conditions encountered.

Such light—that is, low-boiling point—components are decisive for fast starting of a cold engine, good response, and low exhaust emissions during the warm-up phase. Too high a proportion of these, however, can cause vapor lock and increased evaporation losses in summer. Cold, damp weather can also lead to throttle

valve icing. Too many high-boiling components can condense on the cylinder walls, particularly during cold operation, thus diluting the oil film and oil supply. Too few components in the middle boiling range lead to poor running properties and possibly to jerking during acceleration. The fuel requirements are just the opposite after stopping the hot engine and restarting soon after. Components of the fuel system can become so hot under unfavorable conditions that an excessively large portion of fuel evaporates, and this can cause vapor lock in the fuel pump or vapor bubbles in the injection lines.

~**Boiling process.** Because fuels are a mixture of many hydrocarbons, they do not have a unique boiling point but rather a boiling range. Diesel fuel starts to evaporate at approximately 180°C and stops evaporating at about 380°C. This behavior is not so important compared with gasoline, because the mixture preparation takes place right inside the combustion space. However, with diesel fuel an inordinately high proportion of high boiling point components, aromatics in particular—and a high final boiling point—would produce larger droplets in the injection spray. The ignition delay caused by this will influence the course of combustion negatively, and this would result in an increase in noise levels and the tendency to produce soot. On the other hand, a certain degree of high volatility would be a positive factor in terms of cold-start performance, because an excessively high portion of more volatile components causes evaporation immediately at the injection nozzle, which would disturb the distribution of fuel in the combustion space.

~**Evaporative heat.** Evaporation of fuel causes cooling. High latent heat, therefore, causes strong cooling of the air-fuel mixture, which brings about better internal cooling and better filling of the combustion space and, thus, higher performance. The pronounced increase in the quantity of air-fuel mixture after fuel evaporation enables higher mean effective pressures—for instance, with methanol than with gasoline—and offers higher thermodynamic efficiency. The higher boiling point in comparison with the start of boiling and the lower vapor pressure in connection with the strong cooling that occurs with high latent heat fuels require special measures such as intake system preheating, particularly at low temperatures. The theoretical temperature drop of the mixture without preheating is 120°C for methanol compared with gasoline.

~**Reformulated fuels** (*also*, →Fuel, diesel engine ~Composition; →Fuel, gasoline engine). This is understood to mean changing the composition and/or physical characteristic values of the fuel with the goal of minimizing exhaust gas and evaporative emissions.

~**Vapor pressure.** The equilibrium temperature-dependent pressure in a closed vessel with evaporating fuel is termed the (saturated) vapor pressure. In connection with the other volatility criteria, it has an influence on cold and hot-start, the cold-running performance of an engine, and also the evaporation losses from a vehicle. In essence, it is determined by the lightest components, such as butane, present at the initial boiling point. The current standard is DIN EN 12 (RVP = Reid vapor pressure), which embodied the determination according to Reid until 1999. The testing temperature is 37.8°C (100°F) at a vapor-liquid ratio of 4:1. The wet Reid method is generally considered to be sufficiently accurate.

Different volatility grades, depending on ambient temperature, have been specified in the requirements standard. For instance, eight grades were specified for particular evaporated fuel quantities against vapor pressure, which correlates with hot start and hot running behavior. Because fuel is exposed to higher temperatures than previously, particularly before and in the injector nozzles in modern fuel-injected engines, an additional measurement method (measuring in the range −40 to 100°C) was developed. The steam-liquid ratio is 3:2 with this dry method based on the test apparatus according to Grabner. It is universally accepted that excessively low vapor pressure—that is, slowly evaporating fuel—will result in insufficient starting and cold-running performance, while excessively high vapor pressure would mean problems with the hot-start performance and hot-running behavior.

Vapor pressure determination has also changed with the introduction of DIN EN 228 on 2/01/2000. The Reid method was replaced by the DVPE (dry vapor pressure equivalent) according to DIN EN 13016-1. The DVPE is calculated from the ASVP (air saturated vapor pressure), determined (e.g., in the Grabner apparatus) with a vapor-liquid ratio of 4:1.

~**Vaporization enthalpy.** The difference in enthalpy between saturated vapor and boiling liquid at the same pressure and temperature is termed the vaporization enthalpy or latent heat of vaporization.

Vaporization, oil (*also*, →Vaporization, fuel). Evaporation losses strongly depend on viscosity and the type of base liquid used. In earlier times, when mineral oil raffinates were exclusively used as basic oil, it was generally accepted that the lighter the oil, the higher the evaporation loss. The basic liquids in the current high-performance engine oils with the "new technology" (such as special raffinates, hydrocracked oils, synthetic hydrocarbons, and esters) show unequal reductions of evaporation losses at the same viscosity. These are indispensable for the longer periods in the engine between oil changes.

Vaporization enthalpy →Vaporization, fuel

Variability between cam and valve →Variable valve control ~Systems with camshaft

Variable camshaft adjustment →Control/gas transfer; →Variable valve control ~Systems with camshaft ~~Camshaft phasing systems

Variable combustion chamber →Displacement

Variable compression ratio →Compression ~Variable compression; →Piston

Variable displacement →Displacement

Variable drive of ancillary components →Engine accessories ~Variable engine

Variable (geometry) intake manifold →Intake system ~Intake manifold ~~Intake manifold design

Variable nozzle turbocharger →Supercharging

Variable orifice nozzle →Injection valves ~Diesel engine ~~Vario nozzle

Variable parameters for valve control →Variable valve control

Variable register nozzle →Injection valves ~Diesel engine ~~Vario nozzle

Variable speed governor →Injection system (components) ~Diesel engine ~~Mechanical control ~~~Control maps

Variable swirl →Charge movement

Variable timing →Control/gas transfer ~Four-stroke engine

Variable valve control. The term variable valve control is used when the valve opening cross section of the intake valves or exhaust valves (also designated as valve lift function) is varied in size and varied with time and/or position relative to the engine crank angle position. Characteristic variable parameters are valve lift, opening time, and the steepness of the opening/closing curves, as well as the combinations obtainable from these (**Fig. V57**). Some publications designate phasing, valve lift, and opening time as primary characteristics of variable lift functions while the steepness of the opening/closing edges are classified as secondary characteristics. Another classification concerns the variability of valve controls, distinguishing between phase adjusters and partly and fully variable valve controls. A further classification differentiates discontinuously adjusting systems (e.g., stroke switch-over, valve switch-off) and continuously adjusting systems (phase adjuster, continuously adjustable valve opening cross sections). Internal combustion engines are normally operated using valve lift functions that represent a compromise for the charge exchange in several respects. That is why variable valve timing is increasingly employed in modern spark-ignition engines, because it offers potential for optimization, particularly for spark-ignition engines but also for diesel engines.

~**Braking operation.** *See below,* ~Engine brake

V

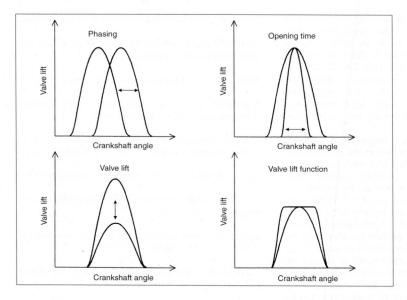

Fig. V57
Variation options for
variable valve timings

~Cam phase setter. *See below*, ~Systems with camshaft ~~Camshaft phasing systems

~Camshaft phasing. *See below*, ~Systems with
camshaft

~Charge movement. Particular charge movement during the suction process is used for generating the mechanical mixture generation, for promoting mixing
processes before and after combustion, and for generation of swirl and tumble flow in the cylinder. Charge
movement stabilizes combustion, enhances residual
gas compatibility, reduces NO_x formation, expands the
lean operating limit, and reduces soot generation depending on the engine and combustion principle. Variable valve-gear systems are suitable for generating
charge movements that can either change the valve
opening cross section (systems for stroke switch-
over), the valve switch-off, or the continuous change
of the intake stroke progression) or that enable particular pressure differences through freely selectable
opening times on the intake valve time (e.g., electromagnetic valve-gear systems).

~Charge-exchange. In addition to the primary tasks
of controlling the charge exchange, variable valve control mechanisms fulfill additional functions that improve the properties of engines, particularly within the
wide operating range for vehicle use. In this, the suction phase has greater significance than the exhaust
process. The tasks of variable valve timing are:

• Control of the internal exhaust gas recirculation
• Improvement of maximum cylinder filling over the
  entire speed range
• Charge movement generation
• Charge volume control
• Avoidance of charge-exchange losses

~Charge-exchange losses. The reduction of charge-
exchange losses is an important task of the variable
valve control mechanism on spark-ignition engines.
These losses are generated through expansion losses
on the initial opening of the exhaust valves against
overpressure in the combustion chamber, through flow
losses on exhausting the exhaust gas, and through
losses as the intake air fills the cylinder. In the case of
engines with throttle control, the charge-exchange
losses increase with falling load because of intake
air throttling. About 40% of the indicated work performed needs to be used for charge exchange in an
operating range close to idling (**Fig. V58**). Further
fill losses develop at full load as a consequence of
fixed intake closure control times. Late closing at low
speeds causes fresh air to be pushed back into the intake port; early closing at high speeds blocks the intake into the combustion chamber before the flow has
finished.

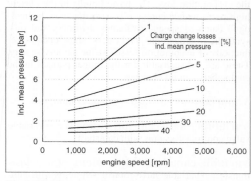

Fig. V58 Charge-exchange losses at throttle control

~Combustion chamber feedback. *See below*, ~Residual gas control

~Cylinder shutoff. Cylinder shutoff is an effective measure for reducing fuel consumption, particularly on large-displacement engines with six or more cylinders. The cylinders that are not shut off are operated at a higher load when cylinders are shut off at low part-load operation. This enhances the combustion efficiency, and the charge-exchange losses are reduced at the same time. Cylinder shutoff can, in principle, be achieved by interrupting fuel feed and through non-throttled, nonignited operation of the shut-off cylinders. Disadvantages such as increased flow losses, cooling of the shut-off cylinders, and confused lambda control signals can be ignored for series production. Cylinder shutoff is only applied in a promising way in combination with a complete stoppage of all valves. The systems that can be used for these purposes correspond to those used for valve stoppage.

~Delayed intake valve closure. *See below*, ~Load control

~Delayed intake valve opening. *See below*, ~Load control

~Early intake valve closure. *See below*, ~Load control

~Effect of fully variable valve control. **Fig. V59** shows the effects and influences of variable valve controls. Charge exchange, the flow condition, and cylinder charge are directly influenced by the valve opening patterns and act in turn on the engine-related target values of fuel consumption, exhaust gas emissions, engine smoothness, torque, and power. In addition, charge-exchange losses and valve train friction directly influence fuel consumption, as does the cylinder charge, which directly influences the power and torque produced through the volumetric efficiency or energy value in the mixture.

~~**Engine smoothness.** Engine smoothness is particularly relevant for spark-ignition engines in an operating range close to idling. Reduction of the valve overlap avoids exhaust gas blowback into the intake tract and results in low amounts of residual gas. Engine smoothness is decisively improved and the reduced cyclical fluctuations could possibly be used to bring about a reduction in the idle speed. Enhanced charge movements at low fills support combustion and improve engine smoothness.

~~**Exhaust gas emissions.** Internal exhaust gas recirculation using variable valve overlap exercises considerable influence on exhaust gas emissions. As a rule, low amounts of residual gas at low loads contribute to stabilizing combustion, thereby decreasing the HC emissions. High amounts of residual gas in the mean-load range result in a decrease of process peak temperatures and thereby in a reduction of the NO_x emissions. Particular charge movements in a gasoline engine would normally lead to reduced HC emissions, especially when operating with high amounts of residual gas, during lean operation, or with throttle-free load control. In diesel engines, the charge movement initiated at inflow intensifies the mixing processes during combustion, thus decisively contributing to the degradation of soot during main combustion and afterburning phases.

~~**Fuel consumption.** A low-loss charge exchange process—for example, by throttle-free load control—yields a reduction in fuel consumption, especially at part load. A precondition is that the variable valve control mechanisms do not offset these advantages through increased friction. Mixture conditioning, mixing processes (e.g., in diesel-engine combustion), residual gas compatibility, and lean operating capability can be improved by generating particular charge movements, which yield further fuel consumption improvements in most cases. Optimization of the amount of residual gas by variable valve overlap cross-sectional areas supports

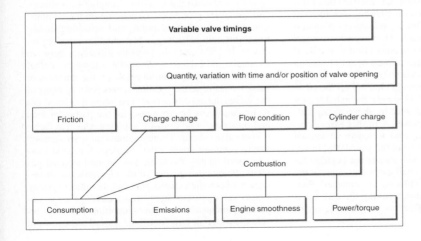

Fig. V59
Effects and influences of variable valve controls

low fuel consumption through stable combustion, reduction of the charge-exchange losses through dethrottling, and reduction of the wall-heat losses by lower peak combustion temperatures. Cylinder shutoff leads very effectively to reduction of fuel consumption through the improved higher load efficiencies and reduced charge-exchange losses.

~~**Torque/power.** High torque and high power presuppose high cylinder filling. Decisive from the view of variable valve control mechanisms are flow-optimized valve opening cross sections and exact intake valve closure at the time when the flow of the fresh mixture has come to a standstill. This point in time will shift from the bottom dead center in direction retarded with increasing speed because of the inertia of the air movement. In addition, variable timing allows the dynamic recharging effects because of intake runner and resonance induction. An important prerequisite for translating cylinder filling into torque include sufficient antiknock properties in the charge, which in turn requires maximum expulsion of residual gas exhaust. High charge movements, however, normally impede high specific torques and powers because of raised resistances to fluid flow and an increased knocking tendency.

~**Electric systems.** *See below,* ~Systems without camshaft

~**Electrohydraulic systems.** *See below,* ~Systems without camshaft

~**Electromagnetic systems.** *See below,* ~Systems without camshaft

~**Electromechanical systems.** *See below,* ~Systems without camshaft

~**Engine brake.** First suggestions to use the piston engine simultaneously as the engine brake by changing the valve timing accordingly go back as far as to the beginning of the twentieth century. A patent specification by Sauerer in 1904 (patent no. 158915) describes a braking method by which the pressure-volume diagram of a four-stroke engine is split into two suction and compression periods each. The exhaust valve and intake valve are alternately opened close to top dead center without fuel feed and closed again at around bottom dead center. In this way, very effective braking powers can be achieved through blowing off the compressed air at engine TDC. However, the exhaust cam needs to be displaced to advance by about 160 to 180° CA to reach full braking power. A further early braking method from Panhard and Levassor is known through patent no. 229880 from the year 1908. In addition to the exhaust cam with its angular position remaining unchanged, two small cams can depress the exhaust valve by axially displacing the camshaft. Panhard and Levassor suggested blocking the carburetor on the intake side. **Fig. V60** represents the corresponding timing diagram.

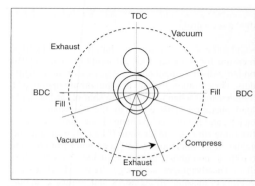

Fig. V60 Timing diagram of Panhard-Levassor (Source: *der Motorwagen*, issue 27, 1921)

Engine brakes are common on utility vehicles today. In addition to systems utilizing the compression work of the compression cycle for braking (decompression brakes) by blowing off in the region of ignition TDC, there is also the so-called exhaust brake. Back pressure is built up by closing a shut-off element in the exhaust against which the pistons have to push out the exhaust. A commonly used decompression brake is the Jakobs brake. It is preferred for engines with pump-jet injection systems. The pump lift is used for lesser exhaust valve opening or additional decompression valve at the ignition TDC by attaching a hydraulic unit to the cylinder head.

~**Exhaust.** *See below,* ~Variation parameters

~**Exhaust gas recirculation, internal.** Exhaust-gas blowback from the exhaust port into the combustion chamber or into the intake system during the valve overlap period is designated as internal exhaust gas recirculation (EGR). Internal exhaust gas recirculation largely determines the proportion of exhaust residuals in the fresh mixture of the cylinder-charge, which influences engine-related properties such as fuel consumption, exhaust gas emissions, engine smoothness, and antiknock properties. Depending on the combustion process, operating point, and operating conditions, different exhaust gas recirculation rates are required. In addition to the pressure difference between the intake and exhaust systems, the volume of internal exhaust gas recirculation depends on the duration of valve overlap and the opening cross-sectional areas of the valves during this period. The internal exhaust gas recirculation can be optimized for the particular load and operating conditions by applying selective variation of valve overlap. Throttle-controlled spark-ignition engines require small valve overlaps for good low-idle-speed quality to enable low internal exhaust gas recirculation rates in spite of the high pressure difference between the exhaust port and the intake system. Larger valve overlaps in the mid-load range are an important prerequisite for higher proportion of exhaust residuals, which in turn contribute toward reductions in fuel consumption and NO_x emissions, in particular. Many automotive engines have variable valve timing for variation of the valve overlap as standard-production equipment. The most widespread systems are those in which the intake and/or exhaust camshaft can be rotated relative to the crankshaft, either discretely in two stages or continuously between two end positions. In the case of spark-ignition engines, the internal exhaust gas recirculation achieved with these phase adapters shows important advantages compared to external exhaust gas recirculation:

- No additional connections and devices between the exhaust port and intake system are necessary to transport and dose of the recirculated exhaust gas. The transmission lines and contamination of the dosing device are avoided with cooled exhaust gas.
- Exact exhaust gas recirculation rate dosing during transient operation.
- The individual cylinders are equally treated on camshaft phase shifting. The risk of uneven distribution of the recirculated exhaust gas volume is minimized.
- Fuel evaporation is supported by the higher temperatures of internal exhaust gas recirculation. This has an advantageous effect on the engine mixture generation, through throttle free load control in particular. HC emissions with internal exhaust gas recirculation are less than with external exhaust gas recirculation.

~Exhaust port feedback. *See below*, ~Residual gas control

~Exhaust residual amount. The residual gas amount is defined as the ratio of the quantity of exhaust gas contained in the fresh mixture after the end of charge exchange to the total cylinder charge. The residual exhaust gas is composed of the quantity of exhaust gas remaining in the compression volume and the quantity of exhaust gas flowing from the exhaust port into the intake system. Decisive for the residual-gas amount is the time-based valve overlap cross-sectional area and the pressure difference present between intake and exhaust gas systems that exists then. The amount of residual gas is given particular attention in spark-ignition engines with throttle control. The drop in the intake manifold pressure with decreasing load in engines with throttle control produces a progressive rise of the amount of residual gas down to the idling point; this can lead to impairment of engine smoothness. Therefore, dimensioning of the valve overlap on spark-ignition engines has to be made taking into account tolerable levels of irregular idling. **Fig. V61** shows the general dependence of the residual gas volume on engine load for throttle-controlled spark-ignition engines with fixed valve timing. In comparison to that, a process-optimized curve of the residual gas amount for spark-ignited engines is also shown. A specific reduction of the residual gas amount in an operating range close to idling enhances ignition and combustion duration, thus providing for stable combustion conditions. Residual gas compatibility in the combustion process improves as the cylinder filling increases, whereby

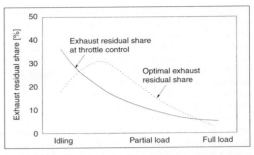

Fig. V61 Residual gas volume inside spark-ignition engine

higher amounts of residual gas effectively decrease the process peak temperatures, thus reducing wall-heat losses and oxides of nitrogen emissions. Additional advantages in fuel consumption derive from de-throttling as a result of higher charge volumes (fresh mixture + residual gas). The maximum torque demands the minimum amount of residual gas, on the one hand to fill the cylinder with the maximum fresh charge of air-fuel mixture, and on the other to achieve the best possible antiknock properties.

~Exhaust valve control time. *See below*, ~Variation parameters

~Fully variable valve control. Fully variable valve control enables continuous variation of valve opening cross section. Basically, three methods are appropriate:

- Variable maximum valve lift
- Variable closing against maximum valve lift
- Variable opening/closing times independent from each other at maximum lift

One application of fully variable valve controls is low-loss load control of spark-ignition engines. Appropriate systems, used primarily for the intake valves, are being developed for these purposes. In addition, the fully variable valve controls have to effectively support spark-ignition engine-type mixture conditioning. On the one hand, this is mechanically achieved by means of high rates of charge volume inflow through smaller valve opening cross sections. Furthermore, using variable valve overlap achieves thermal mixture conditioning and optimal control of the ideal cylinder charge composition of fuel, air, and residual gas. Currently, fully variable valve control is primarily applied in spark-ignition engine concepts (throttle valve replacement) rather than in production engines. Such approaches can be applied in diesel engines in the future, because intensification of charge movement in parts of the performance characteristic map using small valve lifts (possibly in combination with intake valve masking) improves, for example, mixture conditioning and, thus, emission behavior. It may be possible to do without conventional swirl channels, thus improving the charging behavior. Further need for fully variable valve controls results from future unconventional control concepts—for instance, for starting and warm-up, braking operation, and new combustion processes such as CAI and HCCI. With fully variable valve control deployed, there is potential both on the intake and exhaust side to support HCCI by achieving mixture homogenization through small intake-sided valve gaps controlling self-ignition by means of control of the exhaust-side residual gas content.

Fully variable valve controls can generally be designed based on mechanically variable valve control mechanisms with camshafts (in particular systems of classes with variability between cams and valves or DOHC systems) or on hydraulic systems with camshafts or electromechanical valve actuation.

~History. Efforts to improve engine properties through intervention in valve control have a history that goes back to the beginning of the twentieth century. As early as in 1901, Daimler engines designed by Maybach already had intake valves with variable lifts and opening periods; these, in combination with spark timing, controlled the speed and power. A patent specification from Louis Renault on mechanically variable control of upright valves was registered by him in 1902 (**Fig. V62**). A patent specification from Haltenberger from 1918 describes a phase adjuster using an axially movable, straight-toothed/helical-toothed sleeve between the camshaft and drive gears (**Fig. V63**). The subject of variable valve timing then returned to the agenda at the end of the 1960s and beginning of the 1970s. With expansion of the useful speed range of automotive engines to over 6000 rpm, it became more and more difficult to find a satisfying compromise between idling performance and performance at nominal

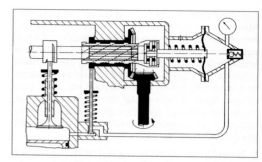

Fig. V63 System for phase shifting by Haltenberger, 1918

power. The first serious tests for resolving the problem of optimally matching control times and opening cross sections to the respective speeds go back to a completely new valve control system by a Fiat engineer, G. Torazza, at the beginning of the 1970s, which was protected by patent in the United States in 1972 (**Fig. V64**). Since that time, a flood of patent registrations has taken place with a peak of 150 to 325 patent registrations annually so far in the 1990s. The development of variable valve control mechanisms is enhanced by the revolutionary progress of electronics in engine technology and the potentialities of precise control and timing associated with them. Alfa-Romeo made the first introduction of a variable valve control mechanism in series production in 1983—a phase adjuster for intake valves found its first employment in a two-liter four-cylinder engine. Since the early 1990s, variable

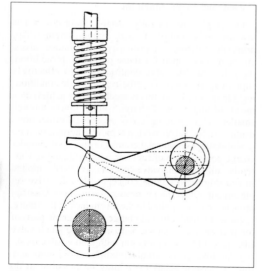

Fig. V62 Variable valve control mechanism by Louis Renault, 1902

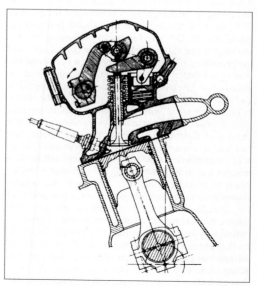

Fig. V64 Variable valve control mechanism by Torazza (Source: Hütten)

valve control mechanisms have been employed progressively in newly developed engines step-by-step.

The following milestones for the introduction of valve gear variability into series production need to be mentioned:

- 1980: electromechanical system for valve and cylinder shutoff on a two-valve, V-8 engine (Cadillac)
- 1983: Two-point phase adjuster for the intake valves of a two-liter, four-cylinder engine (Alfa-Romeo)
- 1983: mechanical valve shutoff using oil pressure on a 400-cc, four-cylinder motorcycle engine as predecessor of the VTEC system (Honda)
- 1989: Mechanical stroke switch-over using oil pressure (VTEC) on a 1.6-liter, four-cylinder engine (Honda)
- 1991: Intake camshaft adjuster "VarioCam" via hydraulically actuated chain tensioner (Porsche)
- 1991: Eccentric cam-follower arm bearing adjuster on a 400-cc, four-cylinder motorcycle engine (Suzuki)
- 1993: Continuous intake-camshaft phasing VANOS with separate oil pump on a three-liter, six-cylinder, inline engine (BMW)
- 1994: 2.3-liter engine using the Miller method in the Mazda Millenia
- 1997: Uneven drive of the intake camshaft with hydraulic control on a 1.8-liter, four-cylinder engine (Rover/MG)
- 1999: Valve-lift switch-over via switching bucket tappets in combination with a camshaft phasing device (VarioCam Plus) on a 3.6-liter boxer engine (Porsche)
- 2001: Mechanically variable intake stroke progressions (Valvetronic) in combination with dual camshaft phasing device (Double-VANOS) on a 1.8-liter, four-cylinder engine (BMW)
- 2003: Mechanically variable intake stroke progressions (Valvetronic) and dual camshaft phasing device (Double-VANOS) in combination with direct gasoline injection on a six-liter, 12-cylinder engine (BMW)
- Probably 2004: Debut of a V-8 engine with cylinder shutoff (displacement on demand) in various vehicles (General Motors)

~Hydraulic systems. *See below*, ~Systems with camshaft, ~Systems without camshaft

~Influencing variables. *See above*, ~Effect of fully variable valve control

~Intake. *See below*, ~Variation parameters

~Intake port feedback. *See below*, ~Residual gas control

~Intake valve control time. *See below*, ~Variation parameters

~Load control. To calculate internal combustion engine performance, we use

$$P_e = i \cdot n \cdot p_{me} \cdot V_h \cdot z,$$

where P_e = effective power output, i = number of cycles, n = speed, p_{me} = effective mean pressure, V_h = volume of cylinder, and z = number of cylinders

Expansion of this power equation leads to the following connection:

$$P_e = \frac{p_{ES} H_u}{R_i T_{ES}} \cdot \frac{\frac{(\varepsilon + 1)}{\varepsilon} \{V_h = f(\alpha_{ES})\}}{(1 + \lambda L_{St}(1 + \alpha))} \cdot z \cdot i \cdot n \cdot \eta_e$$

The following factors must be considered when applying this equation to evaluate the vehicle performance. Assuming a fixed ratio between the engine speed and drive speed of the road wheels, speed cannot be used as an independent parameter for power control. Therefore, it is common to speak of load control for automotive engines. The calorific value, H_u, stoichiometric air demand, L_{st}, and gas constant, R_i, represent fuel-dependent quantities that also cannot be used as independent parameters for load control. The air ratio, λ, and, to a limited degree, the exhaust-gas portion, δ, also reflect quantities that are used for quality regulation in diesel engines or in engines with direct injection and stratified charge for load or power control. The quality regulation standard for spark-ignition engines requires specific control of the mass of fresh mixture. This regulation depends on the parameters for pressure, p_{ES}, temperature, T_{ES}, and cylinder volume displaced after intake valve closing, V_h = f (ES). The time-weighted average working volume of the engine represents a further effective variable for load control as a result of displacement, V_h, number of cylinders, z, and number of cycles, i. The utilization of efficiency, η_e, for load control contradicts the objective of optimum process control and, therefore, can be neglected. **Fig. V65** shows the various options of load control in principle.

Quality control via variable cross-sectional intake opening times represents a low-loss throttle control alternative, which is of interest. Typical load-control methods using variable intake stroke functions, shown in **Fig. V66**, are:

- Early intake valve closure
- Delayed inlet closure
- Delayed inlet opening

However, quality control via exhaust gas recirculation is narrowly confined because of its repercussion on the combustion process. This procedure is only suitable in combination with additional measures of load control.

Procedures with a variable number of cylinders or cycles promise significant potentials for fuel savings, particularly in the case of multicylinder engines (six or more cylinders). A prerequisite is the presence of valve-gear systems that allow complete shut-down of the flow control devices and, thereby, gas-tight closing of the cylinders. In addition to reduction of the charge-exchange losses with cylinder shutoff, cylinder charge enhancement also contributes to efficiency enhancement of the nonshut-off cylinders. The qualifying consideration when this method of load control is used is the impair-

V

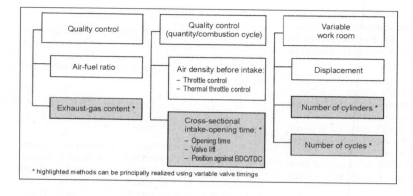

Fig. V65
Load control options
on internal combustion
engines

ment of passenger comfort associated with the longer ignition intervals. In the current state of the art, this method is only suitable in combination with other load control measures.

~~Delayed inlet (valve) closure. In this case, for load control (also termed return motion control), the intake valve stays open beyond the BDC, and the excess charge volume is pushed back into the intake port during the subsequent upward motion of the piston. This results in part of the charge volume passing through the intake valve twice and, in comparison with the "early intake valve closure" method, this approach causes enhanced flow losses, particularly at very low loads. These losses are not, however, directly available for mixture conditioning. Moreover, the intake valve closing time may be later than the optimum ignition point. The delayed ignition necessitated by this leads to efficiency impairments. Additional larger cyclical combustion fluctuations due to different mixture qual-

ities result in reduced efficiency with the "delayed inlet (valve) closure" load control method compared to the early intake valve closure method.

~~Delayed inlet (valve) opening. In this load control method, the intake valve is opened well after TDC and after closing of the exhaust valve, with more or less high vacuums in the combustion chamber. The charge volume flowing into the cylinder is the result of the pressure difference between intake port and cylinder at the time of intake valve opening and is affected by the subsequent valve opening cross-sectional area. Low fills would either need very small valve area integrals (the integration of the instantaneous cross-sectional area with crank angle) or, with delayed inlet closure, very late intake closing times so that the excess charge volume can be pushed back. The high rates of inflow at delayed inlet opening provide good mixture conditioning, high turbulence, and rapid combustion but also contribute to elevated flow losses. This procedure is

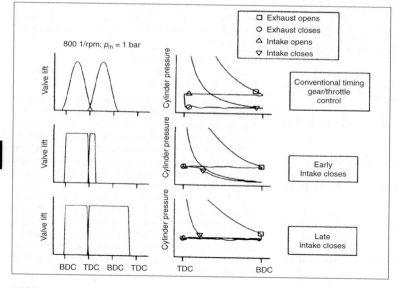

Fig. V66
Load control method
(Source: Pischinger/
Salber)

particularly suitable for the cold-start and warm-up phases of operation. The internal exhaust gas recirculation rate can be controlled via the exhaust closure time.

~~Early intake valve closure. In this load control method, the intake valve is closed early, during the downstroke at a time when the required amount of fresh mixture is in the cylinder. The cross section applied during the short intake valve opening period determines the amount of remaining charge-exchange losses. The minimum, however, is not the sole decisive factor in total efficiency at the best possible rate. The reduction of the valve lift associated with shortening the intake valve opening time certainly produces slightly higher flow losses but has overcompensating advantages for the following combustion process. On the one hand, mixture conditioning is effectively supported by the enhanced rates of inflow, which initially would be impeded in throttle-free load control without the induction manifold vacuum. In addition, the internal flow in the cylinder can be specifically generated for the operating point, and this is particularly advantageous because of the very long time between intake valve closure and the ignition point.

~Mechanical systems. *See below,* ~Systems with camshaft, ~Systems without camshaft

~Miller method. In 1947, US citizen Ralph Miller developed the Miller method for four-stroke piston engines. This principle was first applied for low-speed diesel engines and gas engines before Mazda employed it in a V-6 spark-ignition engine with mechanical supercharging for the first time.

The Miller method is characterized by a changed compression stroke with delayed closing of the intake valve to a time when the piston has already returned to upward motion. Part of the already induced fresh charge escapes through the open intake valve into the intake port and the compression process commences later than normal. A supercharger compensates for the charge loss by compressing the intake air into the combustion chamber. The effective compression ratio is reduced through the late closing of the intake valve, and the temperature before ignition and during combustion is kept lower. As a result, the knocking tendency is reduced in spark-ignition engines and higher supercharging pressures can be achieved and, thus, higher specific power output. A further characteristic of the Miller method is an enlarged expansion ratio compared to the effective compression ratio (**Fig. V67**).

~Mixture generation. Achievement of the fuel consumption advantages of throttle-free load control on spark-ignition engines presupposes good mixture generation. In throttle-control systems, other mechanisms of action for conditioning the air-fuel mixture are required instead of the induction-manifold vacuum. Among them are thermal and mechanical mixture generation. Thermal mixture generation can be supported by hot, internally recirculated exhaust gas to some de-

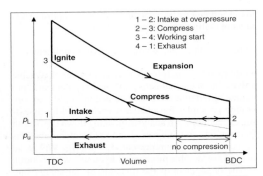

Fig. V67 Pressure-volume diagram of the Miller method

gree; however, this has detrimental effects on the rate of combustion in the operating range close to idling. Significantly more effective is mechanical mixture generation with the objective of generating the maximum contact surface for mass transfer between fuel and oxygen. Generally speaking, there are two suitable measures:

- Fuel carburetion through the mixture generator
- Generation of high flow velocities of the air-fuel mixture

High flow velocities in the open cross sections of the intake valves can be very effectively controlled, specifically with the aid of variable valve gear systems. Suitable systems for these purposes are those that change the open cross section depending on load and speed (systems for valve stroke switch-over, valve switch-off, or continuous change of the intake stroke progression), or that can achieve specific pressure differences between the intake port and combustion chamber via freely selectable opening times (e.g., solenoid valve-gear systems) at the intake-opens time. **Fig. V68** shows an example of the variation of flow energies for valve control employing continuously variable intake stroke functions (VVH system) generated in the valve gap. The remaining charge-exchange

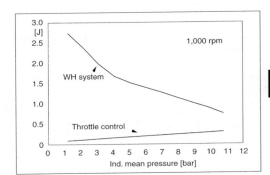

Fig. V68 Flow energy at the valve gap at variable intake-stroke control

losses amount to 10–20% of the throttle control losses in an operating range close to idling.

~Opening time. See below, ~Variation parameters

~Operation principles. The large number of variable valve control mechanisms published in patent and specialized literature are classified according to their operating principles for a better overview and for assessing expenditure. An analysis breaking down the overall function into suitable partial and rudimentary functions is recommended as represented, with **Fig. V69** as an example of variable mechanical-hydraulic valve control.

The breakdown of the transmission of energy into the valve lift allows general classification of variable valve control mechanisms:

- Mechanical
- Hydraulic
- Electrical
- Pneumatic
- Combinations of the above systems

A further option for classification is based on the concept of the "camshaft" and distinguishes between systems with and without camshaft actuation. Solutions without a camshaft can be directly classified into mechanically, hydraulically, electrically, or pneumatically actuated systems depending on the kind of energy transfer. The number of solutions based on camshaft-actuated function principles enables further division of this group depending on where the variability occurs. **Fig. V70** shows a systematic breakdown of variable valve control mechanisms according to the criteria mentioned.

According to the state of the art, development of variable valve control mechanisms focuses on the following systems:

- Systems with camshaft:
  —Camshaft adjustment systems (phase adjusters)
  —Switchable lift characteristics
  —Variable opening time (irregular cam drive)
  —Continuous change of lift connected with change of opening time
- Systems without camshaft, particularly with electrical (electromechanical) actuation
- Variable opening and closing

~Phase adjuster. See below, ~Systems with camshaft
~~Camshaft adjustment systems

~Phasing. See below, ~Variation parameters

~Power control. Power control for the implementation of the driver command is generally required for short-haul motor vehicle operation. The task is reduced on the engine to load control, in the case of fixed ratio of the engine speed to vehicle speed defined by the wheels.

~Requirements. The desired valve lifts are derived from the different operating conditions, the engine-related target values, and the requirements with respect to the technical processes, which in turn determine the principle by which the valve controls are varied (**Fig. V71**).

The engine-related target values for the different operating conditions of the engine in the vehicle, such as idling, part load, full load, cold start, warm-up, or transient operation should be met at the best possible rate. The emission limits—similar to those for consumption, torque, and power—represent uncompromising target values defined by legislators in the same way that customers expect smooth running qualities from the engines as a matter of course. The engine developer bases process engineering specifications on these target values, for example:

- Increase of maximum cylinder filling
- Variable residual gas control according to the various operating conditions
- Low-loss charge volume control
- Generation of charge movements in the cylinder
- Individual cylinder cutoffs

The number of demands with respect to the technical processes determines the demands on the variation of the valve lift with the parameters:

- Phasing of intake/exhaust
- Valve-lift-pattern fill factor/steepness of the opening/closing edges
- Partial stroke
- Continuously variable lift and variable opening time
- Freely variable opening/closing times
- Valve switch-off /cylinder shutoff

Together with the analysis of function and cost, the operating principle of variable valve control is derived from these requirements.

~Residual gas control. Internal residual gas control is made in the region of the charge exchange TDC

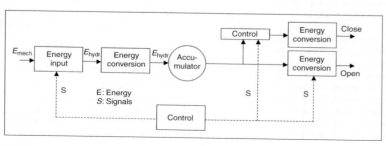

Fig. V69
Function analysis of a hydraulic variable valve control mechanism
(Source: Wellman)

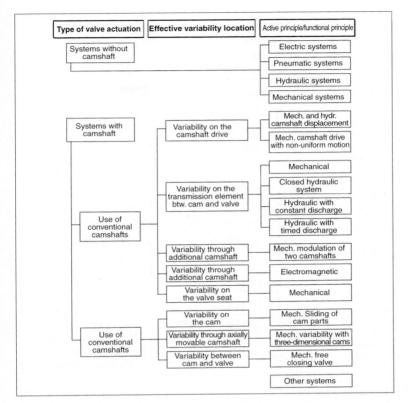

Fig. V70
Systematic breakdown of
variable valve control
mechanisms (Source:
Hannibal)

through the exhaust valve closure and intake valve
opening times. Three methods of residual gas control
via internal exhaust gas recirculation variable valve
timing are normally distinguished (**Fig. V72**):

- Intake port feedback
- Exhaust port feedback

- Combustion chamber feedback

~~**Combustion chamber feedback.** With combustion chamber feedback, the exhaust valve closes before the charge exchange TDC and before the intake valve opens. This causes part of the exhaust gas to

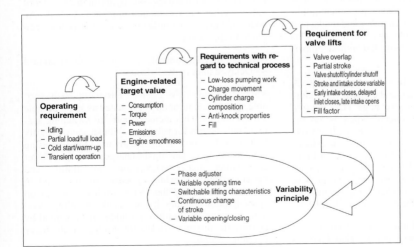

Fig. V71
Demands on variable
valve controls

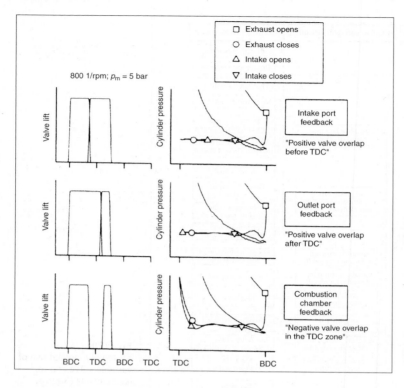

800 1/rpm; p_m = 5 bar

☐ Exhaust opens
○ Exhaust closes
△ Intake opens
▽ Intake closes

Intake port feedback

"Positive valve overlap before TDC"

Outlet port feedback

"Positive valve overlap after TDC"

Combustion chamber feedback

"Negative valve overlap in the TDC zone"

BDC TDC BDC TDC TDC BDC

Fig. V72
Residual gas control method (Source: Pischinger/Salber)

remain in the combustion chamber and be compressed until TDC or until the intake valve opens. The time of intake opening may be determined by various criteria. Premature opening at overpressure in the combustion chamber leads to a pulsed return flow of exhaust gas into the intake port and can be specifically utilized for thorough mixing of exhaust gas and fresh mixture and for removal of the wall-film. This will, however, cause additional charge-exchange losses. The intake valve is correctly opened to minimize the charge-exchange losses when there is an approximate pressure equilibrium between the combustion chamber and intake port and after the piston downstroke has commenced. This kind of residual gas control attains the highest temperature level in the cylinder.

~~Exhaust port feedback. In exhaust port feedback systems, intake valve opening and exhaust valve closing times follow the charge exchange TDC. During this action, the intake valve generally opens before the exhaust valve closure—this is called a positive valve overlap. The piston sucks exhaust gas from the exhaust port back into the cylinder during its downward motion. The quantity of recirculated exhaust gas is primarily determined by the exhaust valve closing time. Use of the thermal energy in the exhaust gas for mixture conditioning is delayed in the combustion chamber.

~~Intake port feedback. Intake port feedback is characterized by a negative valve overlap (exhaust

valve closure) before the charge exchange TDC. The piston presses exhaust gas into the intake port through the intake cross section during the last phase of the exhaust cycle. The exhaust gas intermixes there with the fresh charge and is re-induced in the following part of the charge process. Pushing back hot exhaust gas into the intake port effectively supports thermal mixture conditioning, particularly in throttle-free load control. The quantity of exhaust gas overflow into the intake port can be controlled specifically through the order of valve overlap and its position in relation to TDC. An additional intake manifold pressure control can be employed supportively.

~Return motion control. *See above*, ~Load control ~~Delayed inlet (valve) closure

~Semi-variable valve control. Discontinuously variable systems are normally designated as semi-variable valve controls. This includes both lift-switching systems and systems for valve and cylinder shutoff. Partly variable valve controls are usually used for supporting combustion by enhancing the charge movement (faster combustion, higher exhaust-gas compatibility, and better capability of achieving leaner mixtures) and, in addition, decreasing the charge-exchange losses on spark-ignition engines by partial de-throttling. Systems for cylinder shutoff offer additional potential by shifting the driven cylinders toward higher efficiency operating points.

~Stroke switch-over. *See above*, ~Mixture generation; s*ee below*, ~Systems with camshaft ~~Variability between cam and valve ~~Discontinuously variable systems

~Switchable cams. *See below*, ~Systems with camshaft ~~Variability between cam and valve

~Switching bucket tappets. *See above*, ~Operation principles; *also*, →Variable valve control

~Systems with camshaft. Depending on the operating principle and kind of variability, variable valve gear systems with camshafts can be classified as:

- Camshaft phasing systems
- Three-dimensional cam systems
- Systems with irregular cam drive
- Variations between cam and valve
  —Discontinuously variable systems (stroke switch-over)
  —Continuously variable systems
  (a) Stroke-deactivating systems
  (b) Lift-adding systems

~~**Camshaft phasing systems** (*also*, →Camshaft ~Adjustment). Camshaft adjustment systems, also known as phase adapters, are the most widespread variable valve control mechanisms in standard production for automotive engines. The camshafts are rotated against each other and against the crankshaft, depending on load and speed during operation. The cam profiles themselves—and, thus, valve lift and opening time—remain unchanged. Camshaft adjustment systems are employed for variation of the valve overlap. The main functional properties of camshaft adjustment systems are:

- Improvement of low-idle-speed quality
- Control of the internal exhaust gas recirculation for fuel consumption and emissions reduction
- Optimization of the maximum torque curve across a wide speed range

Engines with separate intake and exhaust camshafts are naturally suitable for the employment of camshaft phasing systems. Camshaft phasing systems have been in series production since the early 1980s (1983, Alfa-Romeo). They first were developed as two-position actuators with two end positions—advanced and retarded—which also allowed intermediate positions. Today, continuously adjusting systems increasingly are used. The following systems have established preeminence among known operating principles:

- Axial-piston displacement
- Sliding-vane displacement
- Belt or chain displacement

With axial-piston adjusters (**Fig. V73**), a helical-toothed and axially-sliding hub makes a rotating connection between the driven camshaft timing gear wheel and the camshaft. Double helical toothing al-

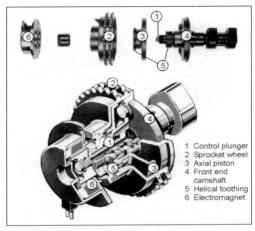

1 Control plunger
2 Sprocket wheel
3 Axial piston
4 Front end camshaft
5 Helical toothing
6 Electromagnet

Fig. V73 Camshaft phasing system in axial piston design (Mercedes-Benz)

lows large rotation angles (up to 60°CA) at small axial path movements. The set piston is normally placed into the desired position and held by oil pressure that is controlled via solenoids. In some cases, additional oil pumps are used to cope with the required forces and the short setting times desired and to achieve independence from engine oil pressure. These pumps raise the engine oil pressure to a higher level, allowing a flexible adjustment of the camshafts independent of engine speed and operating temperature. Following initial introduction by Alfa-Romeo in 1983, companies such as Mercedes-Benz (**Fig. V73**), Nissan, Ford, and BMW followed up with further embodiments of the axial piston design. The "VANOS" system (BMW) first employed as a two-position actuator in 1991 has also been used as a continuous phase adjuster since 1993. This enabled adjustment ranges of 60°CA for the intake valves and 46°CA for the exhaust valves to be achieved in the M3 engine of model year 2001.

Camshaft adjusters using the sliding-vane principle offer an economic alternative to the systems with helical toothing. Sliding-vane adjusters (**Fig. V74**) consist of two main modules: the internal part, firmly connected to the camshaft as a rotor, and the external part that is either chain-driven or driven by the crankshaft via a toothed belt. The connection between the two parts is achieved with an oil chamber that is subdivided by rotor vanes into two working chambers, to the left and right of the vanes. The working chambers are charged, via a proportioning valve, with oil pressure depending on engine speed, load, and engine temperature. This allows stable setting of any angular camshaft position relative to the crankshaft between the specified end positions. The oil supply is achieved from the engine oil system.

A further option for phase shifting is offered by a controllable toothed belt or chain tensioner by which the free length of the continuously variable transmis-

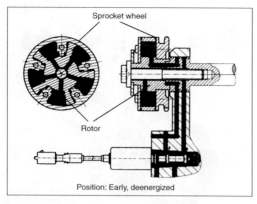

Fig. **V74** Camshaft phasing system in sliding-vane design (Hydraulic Ring)

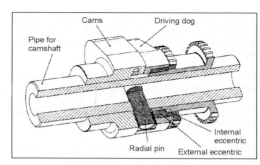

Fig. **V76** Variable valve timing with irregular cam drive (VAST system)

sion is varied. A system of this kind is the hydraulic ring chain-adjusting system (**Fig. V75**) that Porsche (VarioCam) introduced into series production as early as in 1991. It was then used by Audi and the VW group as well following the introduction of the five-valve technology. Characteristic of this system is the direct drive of the exhaust camshaft from the crankshaft via a toothed belt or chain as well as a transverse drive between exhaust camshaft and intake camshaft via chain. A hydraulically driven chain tensioner is integrated into this transverse drive, which takes over the adjustment of the intake camshaft against the exhaust camshaft within the space between both camshafts. The adjustment ranges of implemented chain-adjusting systems are in the region of 15–30°CA.

~~**Irregular cam drive.** Systems with irregular cam drive, also termed oscillating cams, are a special type of variable camshaft drive. These allow adjustment of opening time, nip, or phasing. The main effects of these systems are:

- Improvement of engine smoothness by use of small valve overlaps when idling
- Increase of torque in the low- and medium-speed range through matched intake closing times

- Enhancement of specific power at high speeds through large intake lift functions

Known representatives of this class are the systems from Elrod/Nelson (Harmonic Drive), and Mitchell/Mechadyne or Korostenski (VAST; **Fig. V76**). With this principle, a nonaligned drive shaft is put into a hollow camshaft, driving it with an adjustable eccentricity. The camshaft drive is made uneven through changes in eccentricity, and the opening time of the valve lift function is also changed. In 1997, a comparable system with a continuously variable opening time on the intake valve was put into series production in an MG/Rover roadster with a 1.8-L engine.

~~**Three-dimensional cams.** Variable valve timing with phase adjuster and three-dimensional cams allows superimposed valve lift variation as well as change of phasing. The Titolo system is considered to be the classic representative of this class. The solution shown in **Fig. V77** requires an adapter between the overhead bucket-tappet assembly and cams because of the conical and asymmetrical cams so as to ensure linear contact at all lifts. Phase shifting is made through helical toothing between camshaft and drive gear. A flyweight governor serves to control the oil pressure for displacing

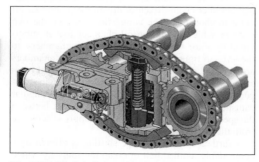

Fig. **V75** Chain adjuster (Hydraulic Ring)

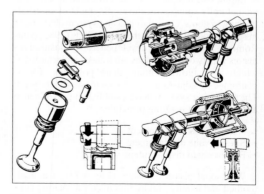

Fig. **V77** Variable valve timing with conical asymmetrical cams (Fiat, Titolo)

the camshaft. This system was tested on a Ferrari V-8 engine for torque and performance improvement.

~~Variability between cam and valve. As a rule, systems with variabilities between cams and valves enable displacement of opening time and valve lift and allow for setting of more or less tailored valve lift diagrams for varying load and speed ranges. It is necessary to distinguish between systems with discrete (stepped) changeover of valve lift diagrams and systems attaining continuous change of opening time and valve lift. Another breakdown distinguishes between solutions with mechanical-hydraulic variation and purely hydraulic variation between cam and valve.

~~~Continuously variable systems. A further step toward enhancing variability of valve-gear systems is systems with continuously variable valve lift functions. These systems already meet most of the requirements for an improved engine process, including:

- Low-loss load control of internal combustion engines
- Control of the internal exhaust gas recirculation
- Full load-speed optimization over the whole speed range
- Generation of charge movement in the cylinder

The mechanical systems for continuously variable valve timing are subdivided into lift-adding systems and lift-terminating systems.

~~~~Lift-adding systems. Lift-adding systems act physically on the valves as an adding gear via a pick-off element. A typical representative of this type is Meta's VVH system (**Fig. V78**). The system is actuated by counterrotating camshafts running at the same speed. Opening/closing processes are distributed to both camshafts. The first camshaft, directly driven by the crankshaft, checks the opening behavior with its ascending slope in the direction of rotation, while the second camshaft determines the closing process with its descending slope. The pick-off element is designed as a sled with roller pick-off. For symmetry reasons, the two outer rollers are in mesh with two identical cams, and an inner roller senses the central cams of the opening shaft. The sled runs on a cam follower arm and transfers the sum of the cam-lobe lifts of both camshafts to the intake valves. As with conventional timing gears, the play between the valve stem end and the cam follower arm can be compensated for by a hydraulic compensating element. The valve lift diagrams can be continuously changed from zero lift to maximum lift by changing the relative phasings of the camshafts to each other. **Fig. V79** shows a family of the resultant valve lift functions. As opposed to most stroke terminating systems, the maximum stroke and intake closure time on lift-adding systems is shifted toward the advanced position with decreasing stroke, and at the same time, the ratio of valve-lift height against opening time decreases significantly less than on lift-terminating systems.

The closing camshaft drive and rotation is obtained

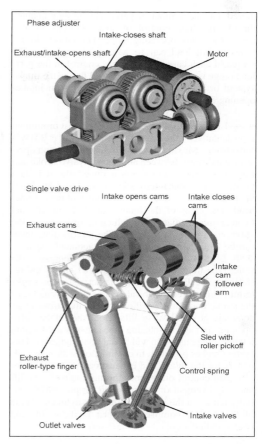

Fig. V78 Lift-adding system for continuously variable intake-stroke control (Meta VVH)

through a four-wheel coupling gear that can be shifted using an electric motor in response to the driver commands. **Fig. V78** shows the VVH system in its application for the intake valves of a four-valve engine. The exhaust opening and intake-opening cams are arranged on a common shaft to minimize the space re-

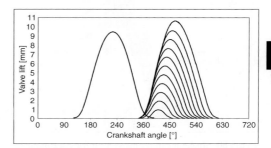

Fig. V79 Continuously variable intake valve lift diagrams for a lift-adding system (Meta VVH)

quired and the number of parts. This system enables throttle-free load control of spark-ignition engines across the entire operating range. The intake-closure times at full load can be adjusted to optimize the torque in each case. Residual gas control in the part-load range is possible using a control flap in the intake system or a phase adjuster for the common intake-opening/exhaust camshaft.

~~~~**Lift-terminating systems.** The most prominent representative of the lift-terminating system is BMW's Valvetronic. Maximum valve lifts, depending on operating conditions, between almost zero valve stroke and full valve stroke can be continuously adjusted via a position-adjustable intermediate lever. **Fig. V80** shows the intake valve gear as a sectional view of the cylinder head. The cams of the intake camshaft initially act upon an intermediate lever, which in turn actuates the valve via a roller-type finger. The intermediate lever can be adjusted via an electrically actuated eccentric shaft so that the intermediate lever depresses the roller-type finger downward. **Fig. V81** shows the resulting valve lift diagrams. Characteristic for stroke terminating systems is the almost negligible change of the stroke maximum position and the flat stroke characteristic at small valve lifts. Both the intake valve opening start and the closing time can be varied within a larger range using an additional phase adjuster. This system makes possible, in conjunction with phase adjusters for the intake and exhaust camshafts, throttle-free load control on engines through intake valve lift and improvement of the full-load torque. BMW initially launched the system in the four-cylinder engine of the 3-series compact and then in the eight- and twelve-cylinder engines of the 7-series.

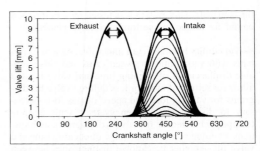

Fig. V81 Continuously variable intake valve lift characteristics of a stroke terminating system (BMW Valvetronic)

Further representatives of mechanical valve stroke-terminating systems being developed now are those from Ina-Eco Valve, Mahle-FEV-MVVT, Nissan VEL, and Delphi VVA, among others.

Fiat is developing a timing gear with electrohydraulic control elements for achieving continuously variable intake valve lift (UNI-AIR system; **Fig. V82**). The intake cam is pressed against a piston-driven oil pad, which in turn actuates the intake valve via a push-rod. Valve lift and open period can be continuously changed by regulating the volume of the oil pad via a solenoid valve and pressure accumulator. Major development tasks for this system include a valve brake to ensure exact valve closing at acoustically permissible valve closure velocities across all engine operating range, and achieving independence from the oil viscosity. This system is also suitable both for throttle-free load control and optimization of cylinder filling at full load.

~~~**Discontinuously variable systems.** The main effects of discontinuously variable systems for stroke switch-over, or switch-over between different cam contours, are:

- Reduction of fuel consumption through reduced charge-exchange losses at partial lift
- Improvement of engine smoothness at idle through small valve overlap as well as enhanced intake flow rates
- Improvement of residual gas compatibility or lean operating capability through enhanced charge movement
- Torque enhancement in the low-end torque region through simultaneous closing of the intake valves
- Power increase at high speeds through extended lift functions

Mechanical systems for changeover between different cam contours have been used in series production of passenger car engines since the end of the 1980s. Honda, building on preceding valve shut-off solutions on motorcycle engines, came to the market with its VTEC (variable valve timing and lift, electronic control) in 1989. A four-valve DOHC engine was fitted with three cams per valve pair that are in contact with

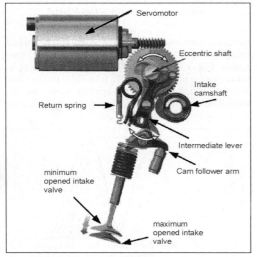

Fig. V80 Stroke terminating system for continuously variable intake valve stroke progression (BMW Valvetronic)

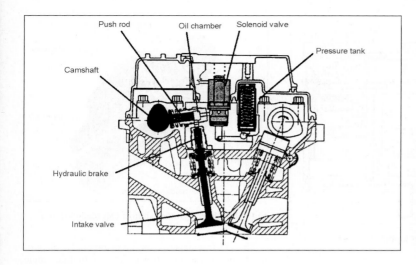

Fig. V82
Electrohydraulic systems
for variable intake stroke
progressions (Fiat)

the associated cam follower arms (**Fig. V83**). The outer cams actuate the valves directly via the outer cam follower arms at part-load operation and low-to-medium speeds. At high speed, the spring-loaded center cam follower arm synchronizes, "idling" with the steeper center cam. The outer cam follower arms are engaged with the center cam follower arm via a hydraulically actuated bolt. The valves now follow the center steeper and more filling cams. This system not only changes the intake valve and exhaust valve timings, but also changes the valve lift. At low speeds,

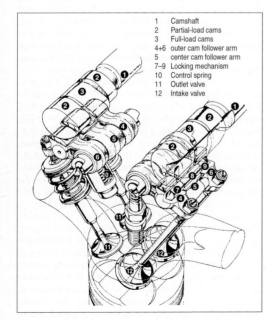

1	Camshaft
2	Partial-load cams
3	Full-load cams
4+6	outer cam follower arm
5	center cam follower arm
7–9	Locking mechanism
10	Control spring
11	Outlet valve
12	Intake valve

Fig. V83 Stroke switch-over system (Honda VTEC)

shorter control times and a reduced valve lift enhance the rate of inflow and also influence the filling and low-end torque at the same time. The larger opening cross sections at high speeds improve the engine aspiration capability and contribute to high specific powers. Honda introduced the VTEC-E system, in additon to the VTEC system, in 1991. This system has a similar working principle, although it shuts off one intake valve nearly completely except for a residual lift of 0.65 mm. This generates intense charge movement in the cylinder, which is specifically utilized via lean operation or EGR for fuel saving purposes. Mitsubishi introduced a similar system for valve-lift switch-over or valve shutoff in 1992 (MIVEC system).

Also in series application are switching bucket tappets with switchable cam contours as well as lever-actuated timing gears. In 1999, Porsche was the first motor vehicle manufacturer to introduce such a system under the name of VarioCam Plus in combination with a mechanical camshaft phasing device. The switching bucket tappet (**Fig. V84**), consists of an internal and external housing that can be connected to a small, hydraulically actuated set piston against the return force of a spring. The result is that an internal partial-lift cam (3 mm valve lift) or two symmetrically arranged outer full-lift cams (10 mm valve lift) are alternatively contacted. The IAV (Ingenieurgesellschaft Auto und Verkehr) solution was to have locking balls couple the two tappet parts instead of the set pistons.

Systems with transfer elements that switch between cams and valve are primarily developed and deployed for enhancing specific engine performance, with concurrent improvement of the part-load behavior of the engines. A further application of switchable transfer elements between cams and valve is to switch off individual valves up to shutoff of whole cylinders. Cylinder shutoff systems are comparable with each other in principle. When in switched-off mode, however, the

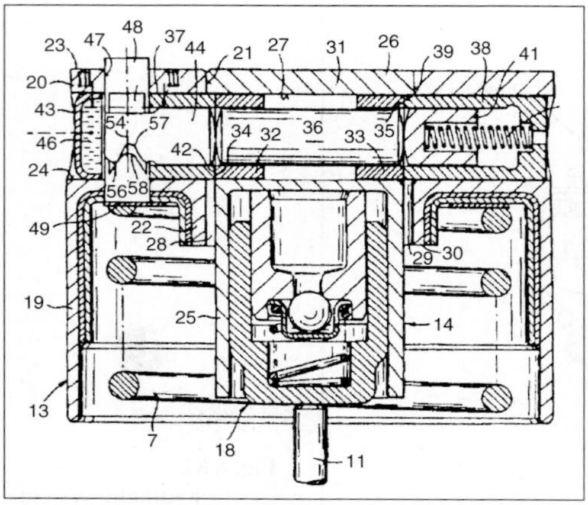

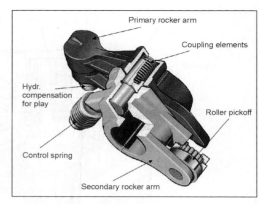

Fig. **V84** Switching bucket tappet (Porsche)

Fig. **V85** Valve/cylinder shutoff (Mercedes-Benz)

part of the cam-actuated pickoff element carries out an idle stroke without actuating the valve. Both the intake valves and the exhaust valves of the cylinders to be shut off are usually immobilized in case of cylinder shutoff. This prevents unnecessary pumping losses, cooling of the combustion chamber walls of the shut-off cylinders, and prevention of disturbance of the lambda control of the cylinders that are not shut off. In low-level part-load operation, it is possible to achieve fuel economy improvements of up to 20% and, in the NEFZ (new European driving cycle), fuel savings of 6–8%, depending on the vehicle. After GM launched a short-lived eight-cylinder Cadillac in 1980, it was only in recent years that cylinder shutoff has been reintroduced as a serious subject of engine development. Mitsubishi launched a 1.6-L four-cylinder engine with cylinder shutoff on the Japanese market and Mercedes-Benz followed as the first European motor vehicle manufacturer with its eight-cylinder engine in the new S-class in 1998 (**Fig. V85**). It is possible that there will be wider introduction of cylinder shutoff in the coming years in combination with further efforts in the direction of fuel economy, particularly for automotive engines with six cylinders. This will apply especially to the American market.

~Systems without camshaft. Systems without a camshaft offer the largest potential for independent, freely selectable opening and closing times. The main features of these systems are:

- Throttle-free load control on spark-ignition engines
- Free controllability of the internal exhaust gas recirculation
- Enhancement of maximum torque over the entire speed range
- Unconventional control concepts for start, warmup, low part load, coast load/trailing throttle, new combustion methods (e.g., controlled autoignition, CAI; homogeneous charge compression ignition, HCCI)

The electromagnetic—or more precisely, electromechanical—valve control using a "vibrating single mass device" based on the Pischinger/Kreuter patent of 1980 represents one of the best-known solutions among the different concepts for variable valve control without a camshaft (mechanical, hydraulic, electrical, pneumatic). The starting point for this system is a center position where an anchor is clamped between two springs independent of load and corresponding to an upper and lower electromagnet at equal distances (**Fig. V86**).

The anchor is in contact with the valve, which is also in a semi-open position at this time. The two electromagnets define the valve end positions in the open and closed positions. The system is put into its start position from where the solenoid-controlled valves are opened and closed independently of each other by alternately exciting the solenoids with the inherent frequency of the vibrating single mass device. However, this system is not suitable for diesel engines, because the combustion-chamber gap (bumping clearance) between cylinder head and piston is too small. The duration for one opening or closing process is about 3–4 ms, regardless of engine speed. This makes for very steep opening and closing curves, particularly at low speeds. At engine speeds between 6000 and 7000 rpm, the lift characteristics level off with increasing speed, approaching the conventional lift function of a camshaft-actuated valve. The motion process can be monitored by adaptive control, ensuring a safe landing of the valves without anchor swing-back and low landing speed (soft landing). This system offers the greatest degree of freedom pertaining to valve opening and closing times, allowing all options of throttle-free load control as well as various concepts of internal residual-gas control. The maximum valve lift remains unchanged without any additional measures. Supporting effects for mixture preparation with throttle-free load control are based on thermal energy with appropriate exhaust gas recirculation strategies or are specifically generated by high rates of inflow at retarded intake valve opening. Rising charge-exchange losses, in-

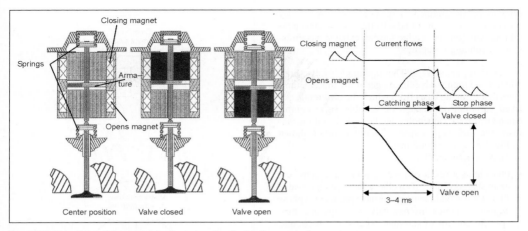

Fig. V86 Principle of the variable electromechanical valve control

creasing intake noise, as well as raised requirements on control accuracy have to be observed, however. This system allows for stoppage, on a particular cycle, of individual valves or complete cylinders without requiring additional measures. Developments concerning electromechanical valve actuation have been achieved for the most part by international motor vehicle manufacturers. This technology becomes interesting in combination with other automotive developments, such as the 42-volt vehicle electrical system or the crankshaft starter-generator, by means of which generator efficiencies of more than about 80% are achieved. A definite date for the start of production of electromechanical valve actuation has not been set yet.

~**Three-dimensional cam.** *See above*, ~Operation principles, ~Systems with camshaft

~**Throttle control.** *See above*, ~Exhaust residual amount

~**Valve closure velocity.** The acceptable valve closure velocity is among the most significant demands on variable valve control. A maximum valve closure velocity of 0.5 m/s should not be exceeded, according to the state of development of modern timing gears. This represents an important challenge for the development of new systems, particularly in view of the way the masses to be moved grow as the valve gear becomes increasingly complex. End position brakes for checking the valve closing motion are often used on hydraulic systems. The valve closure velocities of electromechanical valve-gear systems designed as spring/mass oscillators are nearly independent of engine speed. These systems are also the subject of intense further development into possibilities for control of valve closure velocity at high levels of operational safety. Intelligent control of the magnetic current feed is being undertaken (soft landing).

~**Valve lift pattern.** Variable valve lift patterns can be subdivided into primary and secondary characteristics. Primary characteristics are start of lift (opening time), end of lift (closing time), the resulting opening period, and the maximum lift. Among the secondary characteristics is steepness of the opening/closing edges.

~**Valve overlap.** *See below*, ~Variation parameters

~**Valve stoppage.** Both specific charge movements in the combustion chamber of multiple valve engines can be generated and individual cylinders can be deactivated (switched off) during part-load operation using valve stoppage. In four-valve engines, a swirl flow is generated by stoppage of one intake valve per cylinder, which is normally more effective than the method involving port switch-off or a swirl channel. This measure is suitable both for gasoline and diesel engines. In spark-ignition engines with homogeneous mixtures, generation of the mixture by valve stoppage promotes more stable combustion, thereby improving exhaust gas compatibility and improving the lean operating capability. When the stratification process is used, valve stoppage can be specifically utilized for air-guided combustion. During the main combustion, swirl flow effectively supports mixture-controlled diffusion combustion in diesel engines and promotes soot oxidation during afterburning.

Mechanical valve control with variations between cam and valve are suitable for valve stoppage and, in particular, systems with locking devices generally allow lift switch-over. Well-known solutions on the market are systems from Honda (VTEC-E) and Mitsubishi (MIVEC). Generally all directly actuated valve-gear systems—without camshaft—are suitable for valve stoppage.

Particularly high demands are made on the locking mechanism in switchable solutions for valve stoppage. The switch-over process has to function dependably and reproducibly under all relevant operating condi-

V

tions so that the engine management can allow exactly for abrupt changes in cylinder filling and the detrimental repercussions on engine torque, exhaust gas emission, or running properties.

~**Variation parameters.** The opening/closing times and the maximum valve lift are generally those parameters that form the basis from which the differently derived variation parameters of variable valve control can be deduced. Depending on the system and mechanism, the various principles of variable timing shown in **Fig. V87** can be combined.

These are, in detail:

- Phasing as a result of relative rotation, against the crankshaft position, of an otherwise unchanged valve lift pattern. Continuously adjustable phasing via phase adjusters enables different valve overlaps and, thus, specific control of internal exhaust gas recirculation. In addition, optimization of intake closure depending on speed is achieved by rotating the intake camshaft at full load.
- The open period as the difference between opening and closing time. Systems with variable drives are particularly suitable. Also, this system mainly serves for variation of the valve overlap and speed-dependent adaptation of intake closure at full load.
- The discontinuously variable lift functions at which different lift functions are sensed or valves are immobilized altogether. These systems can be used for partial de-throttling of spark-ignition engines within limits, for generation of enhanced charge movements, for variation of the valve overlap, or for improvement of the maximum cylinder filling within an extended speed range. The most consistent application of these switchable systems is cylinder shutoff for transposing the nonswitched-off cylinders to specific load points.
- The continuously variable valve lift function can best be utilized for throttle-free load control on spark-ignition engines in combination with a correspondingly changeable intake closure time. At the same time, the optimum intake-closure times can be set for any engine speed at full load. A continuous change of the maximum valve stroke without a corresponding change of the opening or closing times can at best be utilized for charge movement generation within a limited range. This principle is not suitable as pure filling control due to the strong increase in flow losses at the now maximum lifts.
- Opening/closing times that are almost independent of each other can be achieved with directly actuated timing gears such as electromechanical valve controls. These systems open and close at a constant rate over time so that the steepness of the slopes depends on engine speed. The high variability of these systems allows optimization of nearly all process-relevant quantities and procedures. This includes the different methods of load and residual gas control as well as optimization of full-load timings related to maximum fill and minimum amounts of resid-

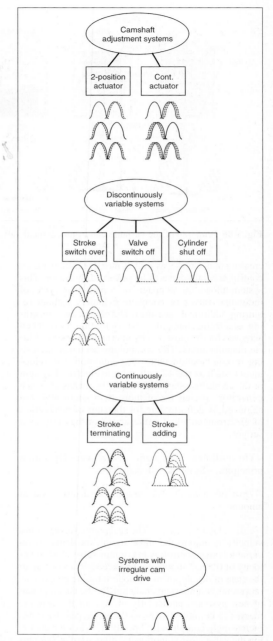

Fig. V87 Principles of variable valve control mechanisms

ual gas. Specific charge movements can be generated through unconventional control methods (delayed inlet opens) or valve stoppage. Complete cylinder shutoffs are also feasible. Changes in valve stroke, generally speaking, require additional measures.

V

Literature: N. Wellmann: Untersuchung des Einflusses variabler Hubfunktionen des Einlassventils, Dissertation, RWTH Aachen University, 1989. — W. Hannibal: Vergleichende Untersuchung verschiedener variabler Ventilsteuerungen für Serien-Ottomotoren, Dissertation, University of Stuttgart, 1993. — R. Stone, E. Kwan: Variable Valve Actuation Mechanism and the Potential for their Application, SAE 890673. — H. Hütten: Schnelle Motoren, Motorbuch Publishers, 1994. — S. Pischinger, W. Salber: Möglichkeiten zur Verbesserung des Kaltstart-, Warmlauf- und Instationärverhaltens mittels variabler Ventilsteuerzeiten, Progress Reports VDI, Series 12, No. 376, p. 191 ff. — C. Wirsum: Motorbremsen, Untersuchungen über die Bremswirkung des Antriebsmotors und Konstruktionsvorschläge, The Best from "Der Motorwagen 1902–1922," Steiger Publishers, 1988, pp. 600–605. — A. Titolo: The Fiat Variable Valve Timing. Internationaler Congress, "Der Fahrzeug-Ottomotor unter neuen europäischen Randbedingungen," Aachen, October 1985. — W. Hannibal: Anforderungen, Funktionsprinzipien und technische Bewertung variabler Ventilsteuerungen, Conference, "Variable Ventilsteuerungen bei 4-Takt-Motoren," House of Engineering, 18 July 1997. — L. Bernard, A. Ferrari, C. Vafidis, F. Vattaneo: Developments in Fiat's Electro-Hydraulic Variable Valve Actuation System, Conference, "Variable Ventilsteuerung," House of Engineering, 28/29 March 2000. — K.-H. Oehling: Anforderungen und Potenziale moderner Ventiltriebskonzepte, Conference, "Variable Ventilsteuerungen bei 4-Takt-Motoren," House of Engineering, 18 July 1997. — M. Lechner: Möglichkeiten und Grenzen vollvariabler Ventilsteuerungen, Conference, "Variable Ventilsteuerungen bei 4-Takt-Motoren," House of Engineering, 18 July 1997. — R. van Basshuysen, F. Schäfer: Handbuch Verbrennungsmotor, Vieweg Publishers, 2002.

Variable valve drive →Fuel consumption ~Variables ~~Engine measures

Variables, fuel consumption →Fuel consumption ~Variables

Variables, particle generation →Particles (Particulates) ~Formation

Variables, variable valve control →Variable valve control ~Effect of fully variable valve control

Vario nozzle →Injection valves ~Diesel engine

V-belt →Engine accessories; →Toothed belt

V-belt drive →Engine accessories

Vegetable oil →Fuel, diesel engine ~Alternative fuels

Vegetable oil methylester →Fuel, diesel engine ~Alternative fuels

Vehicle conditioning →Emission measurements ~FTP-75 ~~Vehicle preparation

Vehicle electrical system. The vehicle electrical system consists of the components for energy storage in the form of a battery, or the current generator, and the consumers. All of these components are connected by a network of electrically conductive elements (cables). The type of wiring has an effect on voltage and battery charge status. Vehicle electrical systems of the future may possibly consist of a two-battery system; one battery functioning as a starter battery and the other responsible for power supply to the consumers. The latter battery need only deliver significantly lower current levels. The engine control system separates the two systems, which prevents a voltage drop in the vehicle electrical system during the start phase.

Future energy supply systems may also have a 42-V vehicle electrical system in addition to a 12-V system. This system uses so-called starter-generators. They combine the starter and the generator in one machine.

Vehicle engine →Engine

Vehicle features (consumers) →Fuel consumption ~Variables

Vehicle interior noise →Air intake system ~Acoustics ~~Legislation

Vehicle mass →Fuel consumption ~Variables

Vehicle preparation →Emission measurements ~FTP-75

Vehicle resistance →Emission measurements ~FTP-75 ~~Dynamometer adjustment; →Fuel consumption ~Variables

Vehicle response →Air cycling valve ~Air cycling valve functions ~~Torque increase; →Supercharging

Velocity →Engine acoustics ~Sound particle velocity

V-engine →Engine concepts ~Reciprocating piston engines ~~Single-shaft engines ~~~V-engines

V-engine, 60° →Balancing of masses ~Inertia forces ~~V-engine

V-engine, 90° →Balancing of masses ~Inertia forces ~~V-engine

V-engines →Balancing of masses ~Inertia forces; →Engine concepts ~Reciprocating piston engines ~~Single-shaft engines

Venturi →Carburetor

Venturi tube →Carburetor ~Systems

Vertical drive shaft →Camshaft ~Drive

Vertical engines →Engine; →Engine concepts

Vertical shaft →Engine acoustics

Vertical valve →Valve gear

Vertical valves →Valve arrangement; →Valve gear

Vibration analysis of operating mode →Engine acoustics

Vibration damper →Torsional vibrations

Vibration damper, crankshaft →Torsional vibrations ~Vibration damper

Vibration insulation →Engine acoustics

Vibration stresses →Electronic/mechanical engine and transmission control ~Requirements for mechanical and housing concepts

Virtual engine development (*also*, →Calculation processes ~Application areas; →Simulation process; →Simultaneous engineering). The shortness of technology cycles, the necessity of cost reduction, the constant requirements for high quality, more flexible adaptation to respective basic conditions, and many more constraints have made virtual development of engines an indispensable approach. Virtual engine development would normally be applied in the concept phase, when the engine can be digitally described using—for instance, CAD/CAE tools and simulation. In a more general sense, virtual engine development represents all the development activities that optimize components, systems, and modules not in combination with hardware. Optimization takes the form of geometric optimization and computational dimensioning. The data thus created form the basis for basic evaluation of the engine, for example, with regard to cost, function, and weight.

Elements of virtual engine development are, for instance, conclusiveness of various CAD module coactions, dimensioning of components for charge transfer, kinematics and dynamics of the valve gear, engine cooling, components, and service strength.

Components optimized using virtual CAD are, for example, crankcase, cylinder head gasket, cylinder head, and cylinder-head bolting. Virtual CAE components in this connection are finite element models of the water jacket, structure, and combustion system. The virtual CAE assembly resulting from this then would be a finite element base model: the basic engine.

The employment of virtual technologies in engine development, however, goes far beyond these possibilities. The digital description of the engine, which should be as comprehensive as possible, also contains descriptions such as:

- Physical properties (stiffness, description of the flow processes, etc.)
- Geometry (space, structures, tolerances, installed position, etc.)
- Technical documentation (finished part drawings, test reports, etc.)

as well as elements such as

- Information on aspects of production and environment

- Administrative information (scheduling, bills of materials, etc.) and can be linked to the digital description of the production process in a digitally based factory

The future goal of this approach will be to largely lift the test and quality requirements on the first prototypes to as high as possible a level and to shorten development times by means of virtual support of many functions.

Literature: Der virtuelle Produktentstehungsprozess im Automobil- und Motorenbau, 5th Automotive Engineering Conference, Wiesbaden, 2001.

Viscosity (*also*, →Fuel, diesel engine ~Properties; →Oil ~Properties and characteristic values). Liquid viscosity is a measure of the resistance with which the liquid opposes flow. Therefore, it is important to know the viscosity both with respect to the pumping capability of the liquid and how the liquid responds when squeezed through jets (fuels, heating oils) and between surfaces (e.g., lubricating oils, hydraulic oils). The viscosity of engine oil depends strongly on temperature, for instance.

The absolute unit used for defining viscosity is the poise (P), which is measured in the centimeter-gram-second system as dyne-second per square centimeter. In a simpler way, viscosity is determined in stokes (St) in routine tests that do not allow for liquid density.

Viscosity classes →Oil ~Properties and characteristic values ~~Viscosity

Viscosity increase →Oil ~Power during operation ~~Aging of oil

Viscosity index (VI) →Oil ~Properties and characteristic values ~~Viscosity

Viscosity index improvers (VII) →Oil ~Body ~~Additives

Volatility →Fuel, gasoline engine

Voltage →Ignition system, gasoline engine ~High-voltage generation

Voltage regulator →Electronic/mechanical engine and transmission control ~Electronic components

Voltage requirement →Ignition system, gasoline engine ~Ignition

Voltage supply →Electronic/mechanical engine and transmission control ~Electronic components ~~Voltage regulator

Volume →Engine acoustics ~Psychoacoustic parameters

Volume control screw →Carburetor

Volume control system →Carburetor ~Systems

Volumetric efficiency (*also*, →Cylinder charge; →Delivery degree). The power output of a combustion engine depends on, among other things, the cylinder charge. The cylinder charge is defined through the volumetric efficiency and the delivery degree. The volumetric efficiency, λ_a, is a measure of the fresh charge trapped in the engine cylinder and can be calculated as

$$\lambda_a = \frac{m_g}{m_{th}} = \frac{m_g}{V_H \cdot \rho_{th}},$$

where m_g = trapped mass in the engine cylinder; m_{th} = theoretical mass that could be contained in the engine cylinder at ambient conditions in naturally-aspirated engines or at the pressure in the intake manifold for turbocharged engines, just after either the compressor or the charge air intercooler; and ρ_{th} = charge density.

For the trapped mass, m_g, the following applies for gasoline engines: $m_g = m_L + m_{Kr}$; and for diesel engines, $m_g = m_L$, where: m_L = trapped mass of air in cylinder and m_{Kr} = trapped mass of fuel in cylinder.

If, for a gasoline engine, ρ_{th} is set equal to ρ_G, the density of the ambient air, or that just upstream of the intake valve, then the trapped mass is

$$m_G = V_G \cdot \rho_G,$$

where V_G is the trapped volume—that is, the volume of mixture per combustion cycle.

For a gasoline engine with a cylinder volume, V_h, this gives the following for the volumetric efficiency:

$$\lambda_a = \frac{V_G}{V_h}.$$

For a diesel engine with V_L as volume of air trapped at ambient conditions, this means

$$\lambda_a = \frac{V_L}{V_h}.$$

Volumetric flow control valve →Injection system (components) ~Diesel engine ~~Control valves

VP15, VP37, distributor pump (Bosch), edge-controlled →Injection system, fuel ~Diesel engine ~~Distributor pumps (DP)

VP29, VP30 distributor pump (Bosch), cam timing controlled →Injection system, fuel ~Diesel engine ~~Distributor pumps (DP)

VP 44 radial piston distributor pump →Injection system, fuel ~Diesel engine ~~Distributor pumps (DP)~~~Radial piston distributor injection pump

VR engines →Engine concepts ~Reciprocating piston engines ~~Single-shaft engines

VW PCI process →Combustion process, gasoline engine

W

Wall distributed injection →Combustion process, diesel engine ~Direct injection ~~Single-spray process, wall-applied; →Combustion process, gasoline engine ~Internal mixture formation

Wall film (*also*, →Mixture formation ~Mixture formation, gasoline engine). The fuel deposits on the intake manifold walls of gasoline engines or on combustion chamber walls of both gasoline and diesel engines are termed wall films.

~Diesel engine. In the case of diesel engines, the wall film is of particular interest in engines with direct injection. The wall film is deliberately produced by applying the injection spray onto the spherical bowl in the piston when a single fuel spray is employed. The air swirl removes the fuel film applied to the wall by, for example, evaporating the film, thus leading to "soft" combustion. This method has not been used on passenger car engines because of disadvantages related to engine power output, fuel consumption, and hydrocarbon emissions.

With the multiple-spray method, the injected fuel should not hit the wall—this keeps the emissions of pollutants low and avoids oil dilution. Therefore, it is necessary to find the correct coordination between injection pressure, hole diameter of the injection nozzles, and the number of injection orifices.

~Gasoline engine. In gasoline engines, particularly those with carbureted mixture preparation, fuel adsorption on the intake manifold wall, as a film, leads to great problems. The risk is highest during warm-up, because the fuel components with a high boiling point cannot yet evaporate or disintegrate into fine droplets. This does not have a negative impact on steady-state operation. However, when the throttle is opened rapidly, the wall film does not reach the cylinder as quickly as the fuel portion that has evaporated into the airstream, thus producing short-term lean mixtures which in turn lead to engine misfires. To remedy this, a carburetor accelerator pump that is actuated by the throttle injects extra fuel into the throat zone of the carburetor. This has a disadvantageous impact on emission of pollutants due to over-enrichment and on the fuel consumption.

Fuel films applied to the walls in the combustion space of gasoline engines are harmful, too. They increase HC emissions, wash off the oil film, and can lead to oil dilution. Wall-applied films on engines with direct injection can additionally lead to soot generation.

Wall film formation →Air-fuel mixture

Wall surface →Flame quenching

Wall temperature →Combustion chamber internal wall temperature

Wall-controlled combustion process →Injection valves ~Gasoline engine ~~Direct injection ~~~Installation positions

Wall-directed injection →Combustion process, diesel engine ~Direct injection ~~Single-spray process, wall-applied

Wankel engine →Engine ~Alternative engines

Wankel supercharger →Supercharging ~Rotary-piston supercharger

Warm-up (*also*, →Carburetor ~Starting systems ~~Requirements). Engine warm-up is understood to be the phase of engine operation from starting to when it reaches the operating temperature. Increased fuel feed (also known as warm-up enrichment) is required to make up for the higher frictional losses during this phase and to compensate for the fuel that condenses on the intake manifold walls in carbureted engines. These effects cause increased fuel consumption during warm-up. Also, increased emissions result, in particular in the form of unburned hydrocarbons, during this phase. The goal of engine development is to keep the time from cold start to reaching operating temperature as short as possible.

Warm-up control →Electronic open- and closed-loop control ~Electronic open- and closed-loop control, gasoline engine ~~Functions

Warm-up enrichment →Warm-up

Warm-up phase →Warm-up

Wash coat →Catalytic converter ~Coating

Waste disposal →Oil ~Safety and environmental aspects ~~Used oil

Waste heat. Combustion engines utilize the chemical energy of fuels, which they convert into mechanical energy. This cannot be done without losses. The losses are primarily heat losses in the form of "waste heat" in the exhaust gas, in the cooling water, or in the oil cooler and heat conduction into the engine parts, which in turn is transferred to the air by convection and heat radiation. The waste heat from the engine can be used in many ways: for example, for the charging of a heat accumulator system, for the heating of intake air or the mixture during the warm-up phase, for exhaust gas turbocharging, and for a faster catalytic converter startup. The waste heat of the cooling water is mainly used for passenger compartment heating in winter. Newer systems use the engine waste heat in heat accumulator systems—for example, to heat the engine faster after a cold start. This accelerates the starting performance of the catalytic converter (reduced HC emissions during the starting phase), and it reduces the losses due to higher friction in the engine during the warm-up phase, which results in better efficiency.

Waste heat utilization →Exhaust gas "heat"; →Heat accumulator; →Waste heat

Watchdog function →Electronic/mechanical engine and transmission control ~Electronic components ~~Voltage regulator

Water →Coolant; →Oil ~Power during operation ~~Contamination by foreign matter; →Radiator

Water chamber. In water-cooled engines, the water chamber is understood to be areas in the block and head that are surrounded by water for cooling reasons. These areas have to be dimensioned such that heat is dissipated optimally from the respective component zones. The goal is to attain an even thermal loading of the component.

Water circuit →Radiator ~Coolant circuit

Water content →Fuel, diesel engine ~Composition; →Fuel, gasoline engine

Water cooler →Radiator ~Liquid/liquid radiator

Water Ecology Act →Oil ~Safety and environmental aspects ~~Used oil

Water heat exchanger →Radiator ~Liquid/liquid radiator

Water injection. Water injection can be applied for decreasing peak temperatures during combustion. This has positive repercussions in the case of oxides of nitrogen emissions and soot generation. Emissions, compared with the standard engine, can be halved at water quantities between 25 and 40%.

Literature: R. Palus, C. Simon: Einfluss der geschichteten Wassereinspritzung auf das Abgas- und Verbrauchsverhalten eines Dieselmotors mit Direkteinspritzung, MTZ (2004) 1, p. 49.

Water jacket →Water chamber

Water lubrication →Cooling, engine

Water pump →Coolant pump

Water separator →Filter ~Intake air filter ~~Air filter elements ~~~Water separation; →Injection system (components) ~Diesel engine ~~Fuel filter

Water-cooled crankcase →Crankcase ~Crankcase construction ~~Water-cooled

Water-fuel emulsion →Fuel, diesel engine ~Emulsions

Water-heated intake manifold. Particularly in engines with carburetors, the water-heated intake manifold served to minimize condensation and the depositing of fuel at the intake manifold wall. The negative outcomes without water heating were deterioration of running characteristics, poorer emission behavior, and less high fuel consumption.

Wear, bearings →Bearings ~Operational damage

Wear, camshaft →Camshaft ~Wear

Wear protection →Oil ~Oil functions

Wear reducers →Oil ~Body ~~Additives

Wear resistance →Bearings ~Materials ~~Qualities

Web →Piston ~Fire land

Web width →Sealing systems ~Cylinder head gaskets

Weight, crankcase →Crankcase

Weight, engine →Engine weight

Weight, F1 engine →Engine ~Racing engines ~~Formula 1

Weight, piston →Piston

Weight reduction by structural optimization →Crankcase ~Weight, crankcase

Weight-to-power ratio (*also*, →Power output ~Specific power). The weight-to-power ratio, m_p, is defined as the ratio of the engine mass, m_M, to the effective power output, P_e. It represents the reciprocal value of the specific power output:

$$m_p = \frac{m_M}{P_e}.$$

Wet cylinder liners →Crankcase ~Crankcase design ~~Cylinder ~~~Insert technology

White smoke. White smoke can occur after cold starting of diesel engines when the liquid fuel wets the cold walls of the combustion chamber. The high concentration of unburned HCs is visible as white smoke in the exhaust gas; this develops through evaporation of fuel from the combustion chamber walls. The white smoke disappears with the increasing temperature of the combustion chamber walls.

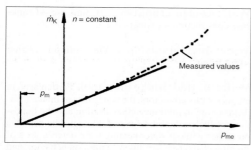

Fig. W1 Willians line

Willians line (*also*, →Friction ~Friction measuring techniques). The Willians line is an approximate method for determining the mechanical losses in an engine (**Fig. W1**). The fuel consumption and mean effective pressure are determined at constant speed on the engine test bench.

Plotting the measured values of fuel consumption against mean effective pressure and drawing a tangent to the fuel consumption curve at $p_{me} = 0$ will result in the tangent intersecting the abscissa at a negative value of p_{me}. The value from the coordinate origin to the point of intersection represents the mean friction pressure. This method is accurate only for diesel engines and has sufficient accuracy only in the lower load range because the losses due to throttling also will be allowed for.

Winding →Piston ring ~Shaping

Window piston →Piston ~Design

Window test bench →Engine acoustics ~Noise reduction ~~Measurement of noise reduction

Windowing technique →Engine acoustics

Winter fuel →Fuel, diesel engine ~Properties ~~Low-temperature behavior

Work. The work provided by an engine is often specified as the specific work (kJ/m³), although the torque (Nm) and the mean effective pressure (bar) are also commonly used as an indication of the work output. The work is a result of converting the chemical energy in the fuel into mechanical energy, and the net work takes the losses into consideration.

Work of expulsion →Charge transfer ~Charge transfer loss, ~Charge transfer work

Work per liter. Work per liter is a specific variable (specific work). It describes the ratio of work output to displacement. A possible and often-used dimension is kJ/dm³. This parameter is independent of the displacement, thereby permitting comparisons between engines with different displacements. A physically equivalent variable is the mean effective pressure with the dimensions of bar.

Work process →Balancing of masses; →Combustion process

Working medium. The working medium in a combustion engine is, for example, a gas or a gas-steam mixture. Depending on the time of observation and the combustion process used, the combustion chamber is filled with air (during compression in the diesel engine or during direct injection in the gasoline engine), a more or less homogeneous air-fuel mixture (gasoline engine with intake manifold injection), or an inhomogeneous air-fuel mixture (in the diesel engine and gasoline engine with direct injection from the time of the fuel injection). The working medium exits the engine as exhaust gas, and the exhaust gas composition is typical for combustion engines. The emissions can be minimized by influencing the combustion process.

World rally car (WRC) →Engine ~Racing engines

W-type engines →Engine concepts ~Reciprocating piston engines ~~Single-shaft engines

X

X-engines →Engine concepts ~Reciprocating piston engines ~~Single-shaft engines

Y

Yield-point controlled tightening process →Engine bolts ~Tightening procedure

Z

Zeldovich mechanism. The Zeldovich mechanism is the set of chemical equations describing the basic formation of NO. In a truncated form,

$$N_2 + O \Leftrightarrow NO + N$$

$$O_2 + N \Leftrightarrow NO + O$$

Zenith-Stromberg CD carburetor →Carburetor ~Constant pressure carburetor

Zeolites →Catalytic converter

Zero emission (*also*, →Emission limits). Zero emission means that there is no emission of pollutants when operating a motor vehicle (zero-emission vehicle; ZEV). This is, however, currently feasible only in motor vehicles equipped with electric drives. When using fossil fuels, it has to be considered at the same time that emissions are generated at other locations—for example, at the powerplant producing electrical energy. The overall efficiency of energy transformation is normally higher at these locations, and the production of emissions can be moved out of conurbations. Hydrogen obtained from renewable energies—for example, solar or wind energy—represents an alternative that will produce zero emissions of CO_2.

Zero load →Load

ZEV (Zero-emission vehicle) →Emission limits; →Zero emission